STUDY GUIDE TO ACCOMPANY
CAREY: ORGANIC CHEMISTRY

FRANCIS A. CAREY
Department of Chemistry
University of Virginia

ROBERT C. ATKINS
Department of Chemistry
James Madison University

McGRAW-HILL BOOK COMPANY

New York St. Louis San Francisco Auckland Bogotá
Hamburg London Madrid Mexico Milan
Montreal New Delhi Panama Paris São Paulo
Singapore Sydney Tokyo Toronto

STUDY GUIDE TO ACCOMPANY CAREY: ORGANIC CHEMISTRY

567890 SEMSEM 9210

ISBN 0-07-009833-6

This book was set in Times Roman. The editors were Karen S. Misler, Randi B. Kashan, and David A. Damstra; the designer was Rafael Hernandez; the production supervisor was Leroy A. Young.
Semline, Inc., was printer and binder.

CONTENTS

PREFACE

It is our hope that in writing this *Study Guide* we will be able to assist the student in making the study of organic chemistry a more meaningful and worthwhile experience. To achieve maximum effectiveness a study guide should be more than just an answer book. What we have written was designed with that goal in mind.

The study guide contains detailed solutions to all the problems in the text. Learning how to solve a problem is, in our view, more important than merely knowing the correct answer. To that end we have included solutions sufficiently detailed to provide the student with the steps leading to the solution of each problem.

In addition, each chapter includes summaries headed "Important Terms and Concepts" and/or "Important Reactions." These are intended to provide brief overviews of the major points and topics presented in each chapter of the text. At the conclusion of each chapter is a "Self-Test," designed to test the student's mastery of the material.

The completion of this guide was made possible with the help of numerous people who contributed their time and talent. Special thanks go to Professors Marye Anne Fox, University of Texas at Austin; and Raymond C Fort, University of Maine for reading and commenting on the entire manuscript. We also thank Professors William H Bunnelle, University of Missouri; and Roy G Garvey, North Dakota State University for their assistance in reading portions of the work. We also wish to acknowledge the assistance and understanding of Randi Kashan of McGraw-Hill. Lastly, we thank our wives and families for their understanding of the long hours invested in this work.

Francis A. Carey
Robert C. Atkins

TO THE STUDENT

Before beginning the study of organic chemistry, a few words about "how to do it" are in order. You've probably heard that organic chemistry is difficult; there's no denying that. It need not be overwhelming, though, when approached with the right frame of mind and with sustained effort.

First of all you should realize that organic chemistry tends to "build" on itself. That is, once you have learned a reaction or concept, you will find it being used again and again later on. In this way it is quite different from general chemistry, which tends to be much more compartmentalized. In organic chemistry you will continually find previously learned material cropping up and being used to explain and to help you understand new topics. Often, for example, you will see the preparation of one class of compounds using reactions of other classes of compounds studied earlier in the year.

How to keep track of everything? It might be possible to memorize every bit of information presented to you, but you would still lack a fundamental understanding of the subject. It is far better to *generalize* as much as possible.

You will find that the early chapters of the text will emphasize concepts of *reaction theory*. These will be used, as the various classes of organic molecules are presented, to describe *mechanisms* of organic reactions. A relatively few fundamental mechanisms suffice to describe almost every reaction you will encounter. Once learned and understood, these mechanisms provide a valuable means of categorizing the reactions of organic molecules.

There will still be numerous facts to learn in the course of the year, however. For example, chemical reagents necessary to carry out specific reactions must be learned. You might find a study aid known as *flash cards* to be helpful. These take many forms, but one idea is to use 3 × 5 cards. As an example of how the cards might be used, consider the reduction of alkenes (compounds with carbon-carbon double bonds) to alkanes (compounds containing only carbon-carbon single bonds). The front of the card might look like this:

$$\text{Alkenes} \xrightarrow{?} \text{alkanes}$$

On the reverse of the card would be found the reagents necessary for this reaction:

$$H_2\text{, Pt or Pd catalyst}$$

The card can actually be studied in two ways. You may ask yourself: What reagents will convert alkenes into alkanes? or, using the back of the card: What chemical reaction is carried out with hydrogen and a platinum or palladium catalyst? This is by no means the only way to use the cards—be creative! Just making up the cards will help you to study.

While study aids such as flash cards will prove helpful, there is only one way to truly master the subject matter in organic chemistry—*work out problems*! The more you work, the more you will learn. Almost certainly the grade you will receive will be a reflection of your ability to solve problems. Don't just think over the problems either; write them out as if you were handing them in as a problem set to be graded. Also, be careful of how you use the study guide. The solutions contained in this book have been intended to provide explanations to help you understand the problem. Be sure to write out *your* solution to the problem first, and only then look it up to see if what you have done is correct.

One concern frequently expressed by students is that they feel they understand the material but don't do as well as expected on tests. One way to overcome this is to "test" yourself. Each chapter in the study guide has a self-test at the end. Work the problems in these tests *without* looking up how to solve them in the text. You'll find it is much harder this way, but it is also a closer approximation to what will be expected of you while taking a test in class.

Success in organic chemistry depends on skills in analytical reasoning. Many of the problems you will be asked to solve require you to proceed through a series of logical steps to the correct answer. Most of the concepts of organic chemistry are fairly simple; stringing them together in a coherent fashion is where the challenge lies. By doing exercises conscientiously you should see a significant increase in your overall reasoning ability. Enhancement of your analytical powers is just one fringe benefit enjoyed by those students who attack the course rather than simply attend it.

Gaining a mastery of organic chemistry is hard work. It is hoped that the hints and suggestions outlined here will be helpful to you and that you will find your efforts rewarded with a knowledge and understanding of an important area of science.

STUDY GUIDE TO ACCOMPANY
CAREY: ORGANIC CHEMISTRY

CHAPTER

1

CHEMICAL BONDING: THE LEWIS APPROACH

IMPORTANT TERMS AND CONCEPTS

Atoms and Electrons (Sec. 1.1) The electronic organization (configuration) of all atoms may be specified by describing the *occupied orbitals* on the atom. An *orbital* is the region of space in which there is a high probability of finding an electron.

By using *quantum numbers*, each orbital may be completely described. The first shell contains only the 1*s* orbital; the second shell contains the 2*s* orbital and three 2*p* orbitals. The *s* orbitals are spherical; the *p* orbitals are orthogonal to each other, and are described as being "dumbbell-shaped". They are designated $2p_x$, $2p_y$, and $2p_z$, respectively.

According to the *Pauli exclusion principle* no two electrons may have the same set of quantum numbers. Since the first three quantum numbers describe the orbital and the fourth quantum number, known as *spin*, may take only one of two possible values ($+\frac{1}{2}$ or $-\frac{1}{2}$), a maximum of two electrons may exist in the same orbital.

Hund's rule states that when electrons enter orbitals of equal energy (for example the three 2*p* orbitals), they do so with parallel spin in separate orbitals.

Ionization Energy (Sec. 1.2) The energy required to completely remove an electron from the highest-energy occupied orbital of an atom. Ionization is an *endothermic* process, and the *ionization energy* describes the enthalpy change for the reaction

$$M(g) \longrightarrow M^+(g) + e^-$$

The general trend is for ionization energy to increase going from left to right across the periodic table.

Electron Affinity (Sec. 1.2) Electron affinity is the energy *liberated* when an electron is added to a neutral atom, that is, the energy change for the process

$$X(g) + e^- \longrightarrow X^-(g)$$

Ionic Bonding (Sec. 1.3) Ionic bonding is bonding that tends to occur between a metal and a nonmetal, that is, a species of low ionization energy and one of high electron affinity. The transfer of one or more electrons from the metal to the nonmetal gives rise to *ions* of opposite charge. An *ionic* or *electrostatic bond* results from the coulombic attraction of the oppositely charged ions for each other.

Covalent Bonding (Sec. 1.4) When nonmetals bond together, the forces which hold the atoms together result from *shared electron pair bonds* or *covalent bonds.* A discrete unit or collection of atoms held together by covalent bonds is known as a *molecule.*

The most popular method for representing the structure of a molecule is with a *Lewis structure,* or "electron dot" formula. Each valence electron in an atom is represented by a dot, and the dots are paired to form covalent bonds in the molecule. Not all the valence electrons on every element are *shared,* that is, part of a covalent bond. They may also exist as *unshared pairs* in an atom of the molecule. Note in the following examples that each pair of dots may be replaced by a line, representing a *single bond* between two atoms.

$$\mathrm{H\!:\!\ddot{\underset{\cdot\cdot}{O}}\!:} \atop \mathrm{H} \quad \text{or} \quad \mathrm{H{-}\ddot{O}\!:} \atop \mathrm{|} \atop \mathrm{H} \qquad\qquad \mathrm{H\!:\!\ddot{\underset{\cdot\cdot}{N}}\!:\!H} \atop \mathrm{H} \quad \text{or} \quad \mathrm{H{-}\ddot{N}{-}H} \atop \mathrm{|} \atop \mathrm{H}$$

For elements in the first row of the periodic table, there must be eight electrons in the outer shell of the element (dots in a Lewis formula), that is, the *octet* of the atom must be satisfied.

Polar Bonds (Sec. 1.5) *Electronegativity* is a measure of the tendency of an atom to draw electron density away from another atom to which it is bonded and toward itself. When two atoms of different electronegativity are bonded together, the more electronegative atom will attract electrons, while the less electronegative (or more *electropositive*) atom will donate electrons. The resulting unequal distribution of electron charge, or *dipole,* gives rise to a covalent bond that is *polarized*; such a bond is referred to as a *polar bond.*

Formal Charge (Sec. 1.6) The charge value assigned to an atom in a covalent molecule is the *formal charge* present on that atom. It may be computed as follows:

$$\text{Formal charge} = (\text{group number}) - (\text{electron count})$$

where electron count is the sum of half the number of shared electrons plus the number of unshared electrons.

Certain commonly encountered examples are worth noting:

1. Tetravalent N is positive, e.g., $\overset{+}{N}H_4$, $\overset{+}{N}(CH_3)_4$
2. Trivalent O is positive, e.g., $H{-}\overset{\cdot\cdot +}{\underset{|\atop H}{O}}{-}H$, $CH_3{-}\overset{\cdot\cdot +}{\underset{|\atop CH_3}{O}}{-}CH_3$
3. Univalent O is negative, e.g., $H{-}\overset{\cdot\cdot -}{\underset{\cdot\cdot}{O}}\!:$, $CH_3{-}\overset{\cdot\cdot -}{\underset{\cdot\cdot}{O}}\!:$

Note that in all the above cases the octet of each atom other than H has been satisfied.

Multiple Bonds (Sec. 1.7) More than one pair of electrons may be shared between two atoms in a molecule. A *double bond* results when two electron pairs (four electrons) are shared; a *triple bond* results from sharing three electron pairs (six electrons). Remember that the octet of a second row element must not be exceeded when writing Lewis structures. This can become especially important when writing formulas which involve multiple bonds. Some examples include:

$$\mathrm{H_2C{=}\ddot{O}\!:} \qquad \mathrm{H_2C{=}CH_2} \qquad \mathrm{H{-}C{\equiv}C{-}H} \qquad \mathrm{H{-}C{\equiv}N\!:}$$

Resonance (Sec. 1.8) For certain chemical substances, the bonding picture depicted by a single Lewis structure does not fully describe the electronic structure of the substance. For example, the three C—O bonds in the carbonate ion, CO_3^{2-}, are equivalent. Yet the Lewis structure shown below suggests two oxygen atoms connected by single bonds and a third oxygen atom attached by a double bond:

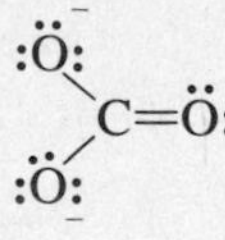

In order to completely describe the bonding of carbonate two additional Lewis structures are necessary:

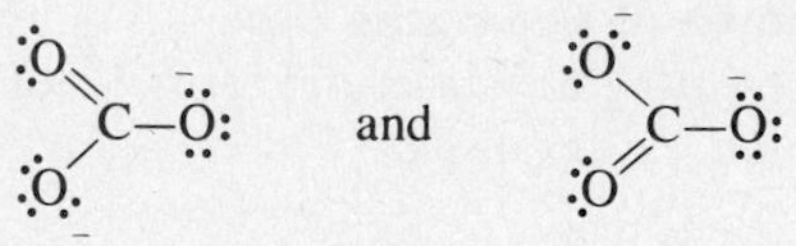

and

The three structures shown are *resonance contributors* to the "true" bonding picture of carbonate ion. They are *not* in equilibrium with each other, but all contribute simultaneously to the electronic structure of the substance. The true bonding picture could be thought of as a *resonance hybrid* to which each of the individual *resonance structures* contributes. A double-headed arrow is used to indicate resonance, as shown below:

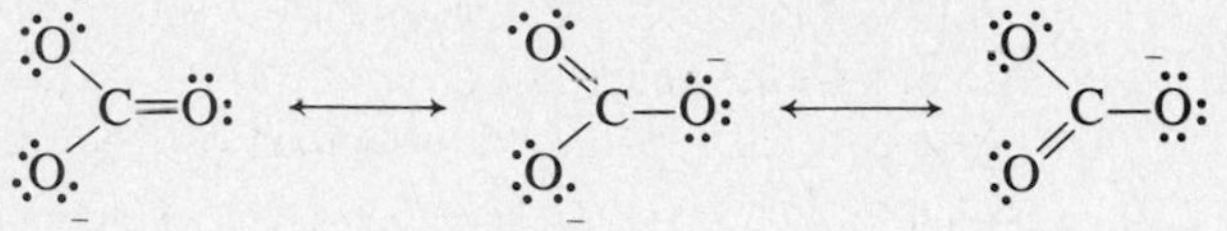

The smearing of electron density as depicted by the preceding resonance structures is an example of *electron delocalization.* An important rule to remember in writing resonance structures is that atoms may not change position—only electrons may differ in their location.

VSEPR Theory (Sec. 1.9) The *valence-shell electron pair repulsion* (*VSEPR*) *theory* provides a method of explaining molecular shapes. The fundamental idea is that electron pairs, both bonded and nonbonded, will orient themselves as far away from each other as possible around the central atom in a molecule. Therefore, an atom having two electron pairs would exist in a linear shape; one with three would be trigonal planar; and one with four tetrahedral. Note that the electrons in multiple bonds are treated together. The following examples illustrate the three basic shapes.

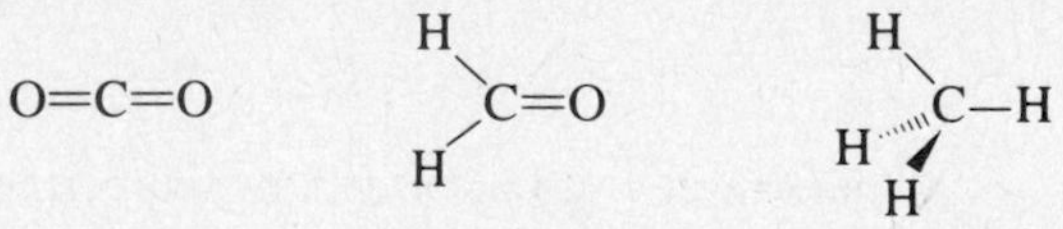

O=C=O

Linear

Trigonal planar

Tetrahedral

An additional consideration of VSEPR is that nonbonded electron pairs exert a greater repulsion than bonded electron pairs. This provides an explanation for the differences in bond angle when one considers methane, ammonia, and water:

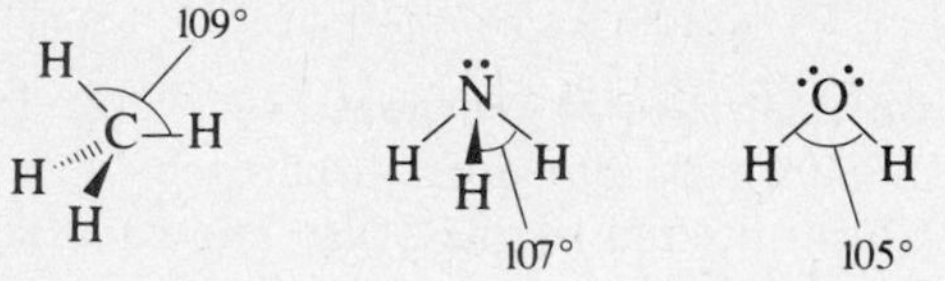

109°

107°

105°

Molecular Dipole Moments (Sec. 1.10) Only for a diatomic molecule (such as HF) can it be stated with certainty that a polar bond will give rise to a polar molecule. *Molecular polarity* is shape-dependent, that is, the dipoles of the individual polar bonds may cancel the effect of each other, depending on the shape of the molecule. For example, CO_2, which is linear, is a nonpolar molecule, although each carbon-oxygen bond is polar. On the other hand, formaldehyde, H_2CO, is polar; there is no cancellation of the dipole of the CO bond.

$$\begin{array}{c} \longleftarrow\!\!\!+ \\ O{=}C{=}O \\ +\!\!\!\longrightarrow \end{array} \qquad\qquad \begin{array}{l} H \quad +\!\!\!\longrightarrow \\ \quad\;\; \diagdown C{=}O \\ H \diagup \end{array}$$

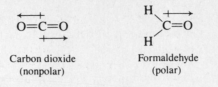

Carbon dioxide (nonpolar) Formaldehyde (polar)

Structural Formulas (Sec. 1.11) The Lewis notation is the most widely used method for representing the structural formulas of organic molecules. *Condensed structural formulas* may be used. For example:

$$(CH_3)_2CHCH_2CH_3 \equiv \begin{array}{ccccccccc} & & H & & H & & H & & H \\ & & | & & | & & | & & | \\ H & - & C & - & C & - & C & - & C & - & H \\ & & | & & | & & | & & | \\ & & H & & | & & H & & H \\ & & H & - & C & - & H & & \\ & & & & | & & & & \\ & & & & H & & & & \end{array}$$

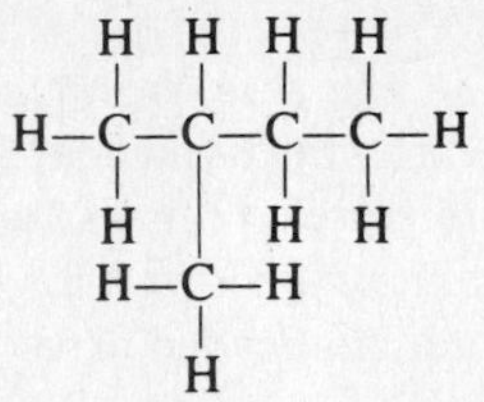

$$(C_2H_5)_2O \equiv \begin{array}{ccccccccccccc} & & H & & H & & & & H & & H & & \\ & & | & & | & & & & | & & | & & \\ H & - & C & - & C & - & O & - & C & - & C & - & H \\ & & | & & | & & & & | & & | & & \\ & & H & & H & & & & H & & H & & \end{array}$$

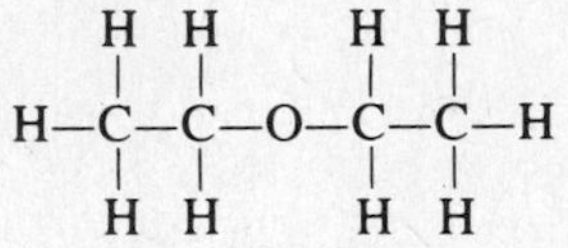

"Stick" formulas are often used to represent complex structures in which the carbons and hydrogens are omitted:

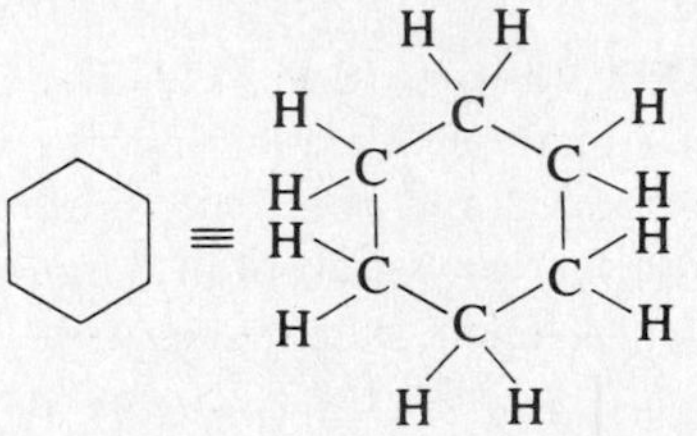

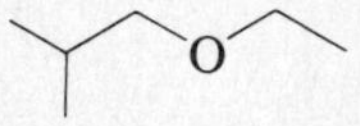

$$\text{(cyclohexane ring)} \equiv \begin{array}{ccccc} & & H\;\;H & & \\ & H & C & H & \\ H & C & & C & H \\ H & C & & C & H \\ & H & C & H & \\ & & H\;\;H & & \end{array}$$

(stick formula with O) $\equiv (CH_3)_2CHCH_2OCH_2CH_3$

Some simple rules are worth remembering:

Hydrogen:	Forms only one bond
Carbon:	Always has four bonds in a stable molecule
Oxygen:	Has two bonds in a neutral molecule and two unshared electron pairs
Nitrogen:	Has three bonds in a neutral molecule and one unshared electron pair

Constitutional Isomers (Sec. 1.12) Two or more substances which have the same molecular formula but differ from each other in some way are known as *isomers* of each other. Constitutional isomers, also known as *structural isomers,* differ in their atomic connections. For example each of the following structures has the formula C_3H_8O, yet they represent different substances:

$$CH_3{-}CH_2{-}CH_2{-}O{-}H \qquad CH_3{-}O{-}CH_2{-}CH_3 \qquad \begin{array}{c} CH_3{-}CH{-}CH_3 \\ | \\ O{-}H \end{array}$$

SOLUTIONS TO TEXT PROBLEMS

1.1 Electron configurations of elements are derived by applying the following principles:

(*a*) The number of electrons in a neutral atom is equal to its atomic number Z.
(*b*) The maximum number of electrons in any orbital is two.
(*c*) Electrons are added to orbitals in order of increasing energy, filling the 1*s* orbital before any electrons occupy the 2*s* level. The 2*s* orbital is filled before any of the 2*p* orbitals, and the 3*s* orbital is filled before any of the 3*p* orbitals.
(*d*) All the 2*p* orbitals are of equal energy, and each is singly occupied before any is doubly occupied. The same holds for the 3*p* orbitals.

With this as background, the electron configuration of the third-row elements is derived as follows:

Na ($Z = 11$)	$1s^22s^22p^63s^1$
Mg ($Z = 12$)	$1s^22s^22p^63s^2$
Al ($Z = 13$)	$1s^22s^22p^63s^23p_x^1$
Si ($Z = 14$)	$1s^22s^22p^63s^23p_x^1\,3p_y^1$
P ($Z = 15$)	$1s^22s^22p^63s^23p_x^1\,3p_y^1\,3p_z^1$
S ($Z = 16$)	$1s^22s^22p^63s^23p_x^2\,3p_y^1\,3p_z^1$
Cl ($Z = 17$)	$1s^22s^22p^63s^23p_x^2\,3p_y^2\,3p_z^1$
Ar ($Z = 18$)	$1s^22s^22p^63s^23p_x^2\,3p_y^2\,3p_z^2$

1.2 A positively charged ion is formed when an electron is removed from a neutral atom. The equation representing the ionization of carbon and the electron configurations of the neutral atom and the ion are:

$$\underset{1s^22s^22p_x^12p_y^1}{C} \longrightarrow \underset{1s^22s^22p_x^1}{C^+} + e^-$$

The ionization energy of carbon is 261 kcal/mol, over twice that of lithium.

1.3 The electron configurations of the designated positive ions are:

	Ion	Z	Number of electrons in ion	Electron configuration of ion
(*a*)	Na^+	11	10	$1s^22s^22p^6$
(*b*)	He^+	2	1	$1s^1$
(*c*)	H^-	1	2	$1s^2$
(*d*)	O^-	8	9	$1s^22s^22p_x^2\,2p_y^2\,2p_z^1$
(*e*)	Cl^-	17	18	$1s^22s^22p^63s^23p^6$
(*f*)	Ca^{2+}	20	18	$1s^22s^22p^63s^23p^6$

Those with a rare gas configuration are Na^+, H^-, Cl^-, and Ca^{2+}.

1.4 The ionization of sodium, indeed the ionization of any atom, is an endothermic process. An amount of energy equal to its ionization energy is required to dislodge an electron. Capture of an electron by a halogen atom gives off energy equal to its electron affinity; thus, chlorine captures an electron in an exothermic reaction.

1.5 (*b*) Nitrogen is a group V element and so contributes five valence electrons, while three hydrogen atoms contribute one each.

Combine :Ṅ· (with one dot below) and three H· to write a Lewis structure for ammonia

```
   H
   ..
  :N:H
   ..
   H
```

In the Lewis structure for ammonia, nitrogen has eight electrons in its valence shell; two are unshared and six are grouped in three pairs.

(*c*) There are 26 valence electrons to be accounted for in NF_3; each fluorine contributes 7 and nitrogen contributes 5.

Combine :N· and three :F· to write a Lewis structure for NF_3 :F: :N:F: :F:

Both nitrogen and fluorine are surrounded by eight electrons in NF_3.

(*d*) Phosphorus, like nitrogen, is in group V of the periodic table and so contributes five valence electrons. Each chlorine contributes seven.

Combine :P· and three :Cl· to write a Lewis structure for PCl_3 :Cl: :P:Cl: :Cl:

The octet rule is satisfied for both phosphorus and chlorine.

(*e*) There is only one way to combine the atoms in CH_3Cl so as to produce a valid Lewis structure. Carbon can form four bonds while hydrogen and chlorine form one each in their stable compounds.

Combine ·C· with :Cl· and three H· to write the Lewis structure H H:C:Cl: H

(*f*) In order to satisfy the Lewis rules, C_2H_6 must have a carbon-carbon bond.

Combine two ·C· and six H· to write the Lewis structure of ethane H H H:C:C:H H H

There are a total of 14 valence electrons distributed as shown. Each carbon is surrounded by eight electrons.

1.6 The direction of a bond dipole is governed by the electronegativity of the atoms involved. Among the halogens the order of electronegativity is F > Cl > Br > I. Therefore fluorine attracts electrons away from chlorine in FCl and chlorine attracts electrons away from iodine in ICl.

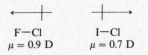

F—Cl $\mu = 0.9$ D   I—Cl $\mu = 0.7$ D

1.7 (*b*) The electron count of carbon is five; there are two electrons in an unshared pair, and three electrons are counted as carbon's share of the three covalent bonds to hydrogen.

H:C:H H — two electrons "owned" by carbon; one of the electrons in each C—H bond "belongs" to carbon

An electron count of five is one more than that for a neutral carbon atom. The formal charge on carbon is −1, as is the net charge on this species.

(*c*) This species has one less electron than that of the preceding problem. None of the atoms bears a formal charge. The species is neutral.

H—Ċ—H (with H below)   Electron count of carbon = $1 + \frac{1}{2}(6) = 4$

unshared electron   electrons shared in covalent bonds

(*d*) The formal charge of carbon in this species is +1. Its only electrons are those in its three covalent bonds to hydrogen, so its electron count is three. This corresponds to one less electron than in a neutral carbon atom, giving it a unit positive charge.

(*e*) In this species the electron count of carbon is four or, exactly as in (*c*), that of a neutral carbon atom. Its formal charge is zero and the species is neutral.

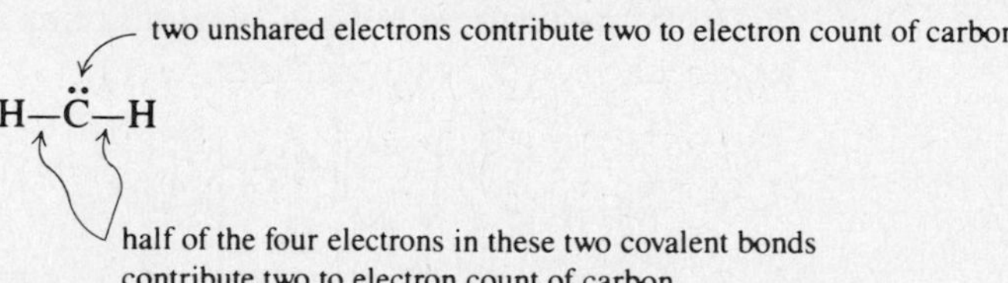

1.8 (*b*) Combining its constituent atoms and the associated electrons gives a structure that contains an unpaired electron on carbon and one on oxygen.

·Ö: H· ·Ċ· ·H Combine to write ·Ö: | H—Ċ—H

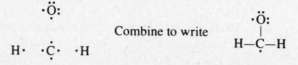

An additional carbon-oxygen bond may be written by pairing the two odd electrons.

·Ö: | H—Ċ—H Pair odd electrons to write Ö: ‖ H—C—H

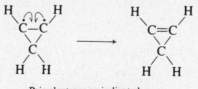

This is the Lewis structure for formaldehyde.

(*c*) A central carbon atom is flanked by two oxygens in carbon dioxide. Since the molecule has a total of 16 valence electrons, the octets of all three atoms may be satisfied by a structure that has two carbon-oxygen double bonds.

Combine ·Ċ· and two ·Ö· to write :Ö—Ċ—Ö:

Pair the remaining electrons :Ö—Ċ—Ö: to write :Ö=C=Ö:

(*d*) Since we are told that the molecule contains a three-membered ring, there is only one way to assemble the atomic framework. Pairing the odd electrons gives a carbon-carbon double bond.

H(C—C)H over C(H)(H) ⟶ H(C=C)H over C(H)(H)

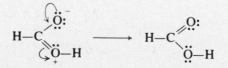

Pair electrons as indicated so as to give a double bond

(*e*) The order of bond connections is given as NCCCH and the atoms bring a total of 18 valence electrons to the molecule. The octet rule can be satisfied by a structure in which each atom (except hydrogen) participates in a triple bond.

:N≡C—C≡C—H

1.9 (*b*) The dipolar Lewis structure given can be transformed to one which has no charge separation by moving electron pairs as shown:

H—C(—Ö:⁻)(=Ö⁺—H) ⟶ H—C(=Ö:)(—Ö—H)

(*c*) Move electrons toward the positive charge. Sharing the lone pair gives an additional covalent bond and avoids the separation of opposite charges.

$$^{+}CH_2{-}\ddot{C}H_2^{-} \longrightarrow CH_2{=}CH_2$$

(*d*) Octets of electrons at all the carbon atoms can be produced by moving the electrons toward the site of positive charge.

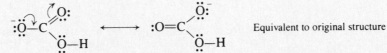

$$^{+}CH_2{-}CH{=}CH{-}\ddot{C}H_2^{-} \longrightarrow CH_2{=}CH{-}CH{=}CH_2$$

(*e*) As in the previous example, move the electron pairs toward the carbon atom that has only six.

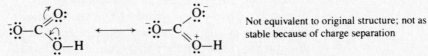

$$^{+}CH_2{-}CH{=}CH{-}\ddot{O}:^{-} \longrightarrow CH_2{=}CH{-}CH{=}\ddot{O}:$$

1.10 (*b*) Move electrons from the negatively charged oxygen, as shown by the curved arrows.

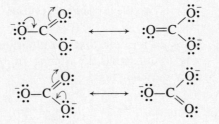

$$^{-}:\ddot{O}{-}C(=\ddot{O}:){-}\ddot{O}{-}H \longleftrightarrow :\ddot{O}{=}C(\ddot{O}:^{-}){-}\ddot{O}{-}H$$

Equivalent to original structure

The ion shown is bicarbonate ion. The resonance interaction shown is more important than an alternative one involving delocalization of lone pair electrons in the OH group.

$$^{-}:\ddot{O}{-}C(=\ddot{O}:){-}\ddot{O}{-}H \longleftrightarrow ^{-}:\ddot{O}{-}C(\ddot{O}:^{-}){=}\overset{+}{O}{-}H$$

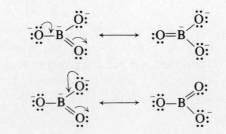

Not equivalent to original structure; not as stable because of charge separation

(*c*) All three oxygens are equivalent in carbonate ion. Either negatively charged oxygen can serve as the donor atom.

$$^{-}:\ddot{O}{-}C(=\ddot{O}:){-}\ddot{O}:^{-} \longleftrightarrow :\ddot{O}{=}C(\ddot{O}:^{-}){-}\ddot{O}:^{-}$$

$$^{-}:\ddot{O}{-}C(=\ddot{O}:){-}\ddot{O}:^{-} \longleftrightarrow ^{-}:\ddot{O}{-}C(\ddot{O}:^{-}){=}\ddot{O}:$$

(*d*) Resonance in borate ion is exactly analogous to that in carbonate.

$$^{-}:\ddot{O}{-}B^{-}(\ddot{O}:^{-}){=}\ddot{O}: \longleftrightarrow :\ddot{O}{=}B^{-}(\ddot{O}:^{-}){-}\ddot{O}:^{-}$$

and

$$^{-}:\ddot{O}{-}B^{-}(\ddot{O}:^{-}){=}\ddot{O}: \longleftrightarrow ^{-}:\ddot{O}{-}B^{-}(=\ddot{O}:){-}\ddot{O}:^{-}$$

1.11 (*b*) Boron is surrounded by four bonded pairs of electrons and has a tetrahedral arrangement of those bonds.

$$BF_4^{-}$$

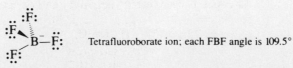

Tetrafluoroborate ion; each FBF angle is 109.5°

(*c*) There are three bonded pairs and one unshared electron pair around oxygen in hydronium ion. They are distributed in an approximately tetrahedral fashion. The geometry of the ion as defined by its atoms is trigonal pyramidal.

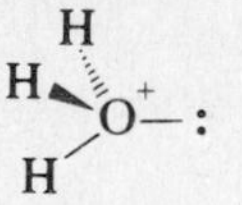

H_3O^+ has the same arrangement of electrons as $:NH_3$

(*d*) Carbon in CH_3Cl has four shared electron pairs. The molecule is tetrahedral.

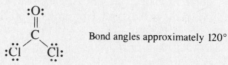

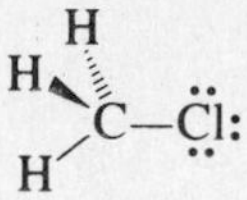

(*e*) Carbon is bonded to three groups in $COCl_2$; it is connected to oxygen by a double bond. For the purposes of the VSEPR method, a double bond is considered to be like a single bond but slightly more repulsive. The maximum separation of the electrons is achieved in a trigonal planar geometry.

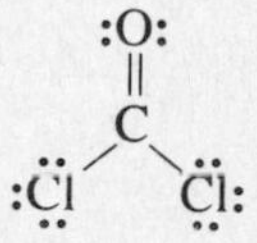

Bond angles approximately 120°

(*f*) Carbon has two substituents in HCN. Electrons in the triple bond to nitrogen are separated most from those in the C—H bond when the molecule is linear.

H—C≡N Bond angle is 180°

1.12 (*b*) Water is a bent molecule, so the individual O—H bond moments do not cancel. Water has a dipole moment.

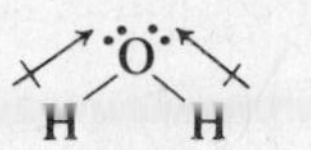

Individual OH bond moments in water

Direction of net dipole moment

(*c*) Methane, CH_4, is perfectly tetrahedral, so the individual (small) C—H bond moments cancel. Methane has no dipole moment.

(*d*) Methyl chloride has a dipole moment.

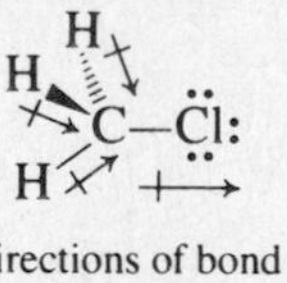

Directions of bond moments in CH_3Cl

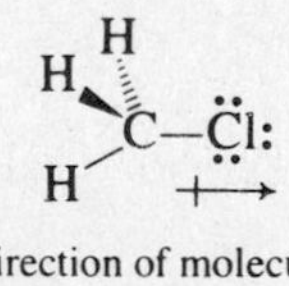

Direction of molecular dipole moment

(*e*) Oxygen is more electronegative than carbon and attracts electrons from it. The formaldehyde molecule has a dipole moment.

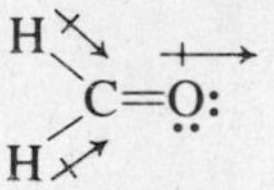

Direction of bond moments in formaldehyde

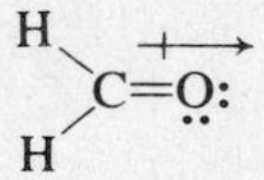

Direction of molecular dipole moment

(*f*) Nitrogen is more electronegative than carbon. Hydrogen cyanide has a dipole moment.

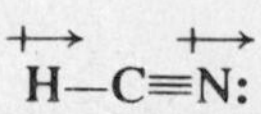

H—C≡N:

Direction of bond moments in HCN

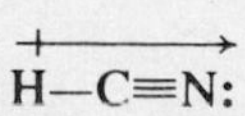

H—C≡N:

Direction of molecular dipole moment

1.13 (*b*) Each carbon has four bonds and so has no unshared electron pairs. Oxygen has two bonds and so has two unshared electron pairs. The octet rule is satisfied for both carbon and oxygen.

```
     H   H
     |   |
   H—C———C—H
      \ ‥ /
        O
        ‥
```

(*c*) A neutral nitrogen that has three bonds bears one unshared electron pair.

```
       H   H
       |   |
     H—C———C—H
   H  /     \  H
    C   ‥   C
   H   \ N /   H
         |
         H
```

(*d*) The total number of valence electrons in this molecule is 24: each carbon contributes 4, each hydrogen 1, and the boron atom 3. The structural formula given contains 12 covalent bonds and accounts for all the electrons. There are no unshared electron pairs.

```
          H
          |
        H—C—H
   H      |      H
    \     B     /
     C  /   \  C
   H   \     /   H
        H   H
```

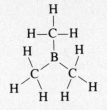

1.14 (*b*) Two CH_2Cl groups are linked together in this molecule.

```
   H  H
   |  |
 H—C——C—H      can be rewritten as ClCH2CH2Cl
   |  |
   Cl Cl
```

(*c*) Three CH_3 groups are bonded to a central carbon atom, on which the fourth substituent is hydrogen.

```
   H   H   H                                 H
   |   |   |                                 |
 H—C———C———C—H      can be rewritten as  CH3—C—CH3
   |   |   |                                 |
   H   |   H                                CH3
     H—C—H
       |
       H
```

or even more simply as $(CH_3)_3CH$.

(*d*) There are two CH_3CH_2 groups bonded to oxygen in this compound.

```
   H  H     H  H
   |  |     |  |
 H—C——C—O—C——C—H      can be rewritten as    CH3CH2OCH2CH3
   |  |     |  |
   H  H     H  H
```

or more simply as $(CH_3CH_2)_2O$.

1.15 (*b*) This compound has a four-carbon chain to which are appended two other carbons.

```
                            CH3 H
                             |  |
 [skeletal]  is equivalent to  CH3—C——C—CH3   which may be rewritten as   (CH3)2CHCH(CH3)2
                             |  |
                             H  CH3
```

(*c*) The carbon skeleton is the same as that of the compound in (*b*), but one of the terminal carbons bears an OH group in place of one of its hydrogens.

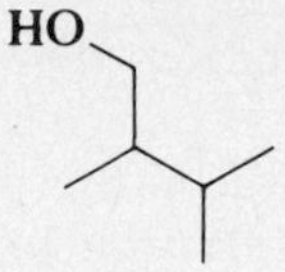

HO (bond-line structure) is equivalent to

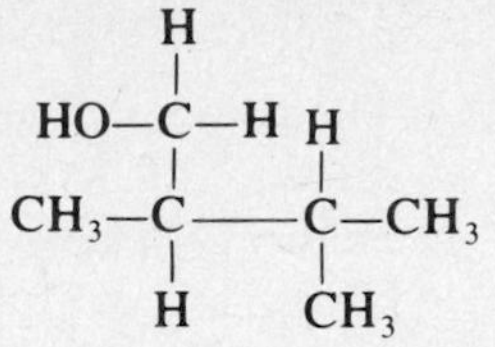

$$\begin{array}{ccc} & H & \\ & | & \\ HO{-} & C{-}H & H \\ & | & | \\ CH_3{-} & C{-}{-}{-} & C{-}CH_3 \\ & | & | \\ & H & CH_3 \end{array}$$

which may be rewritten as

$$\begin{array}{l} CH_3CHCH(CH_3)_2 \\ \quad\;\; | \\ \quad CH_2OH \end{array}$$

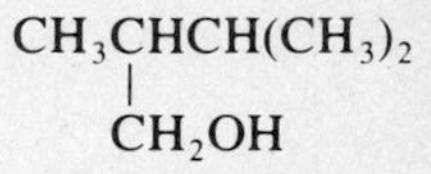

(*d*) The compound is a six-membered ring that bears a $-C(CH_3)_3$ substituent.

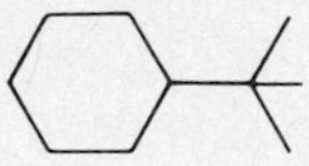

(bond-line structure) is equivalent to

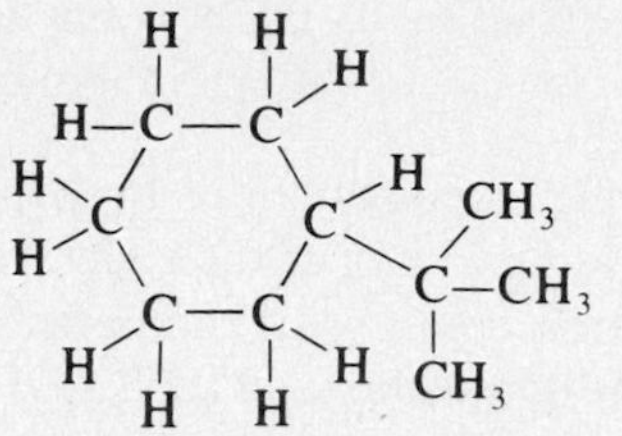

$$\begin{array}{ccccc} & H & H & & \\ & | & | \diagup H & & \\ H{-} & C{-} & C & & \\ H\diagdown C & & C \diagup H & CH_3 & \\ H\diagup & & \diagdown & C{-}CH_3 & \\ & C{-} & C & | & \\ H\diagup & | & | \diagdown H & CH_3 & \\ & H & H & & \end{array}$$

which may be rewritten as

(cyclohexane ring)$-C(CH_3)_3$

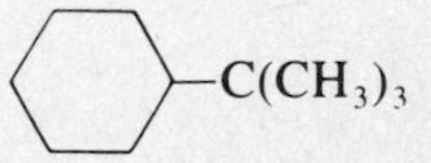

1.16 (*b*) C_4H_{10}

$$CH_3CH_2CH_2CH_3 \quad \text{and} \quad \begin{array}{l} CH_3CHCH_3 \\ \quad\;\;\; | \\ \quad\;\; CH_3 \end{array}$$

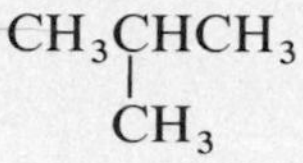

(*c*) C_5H_{12}

$$CH_3CH_2CH_2CH_2CH_3, \quad \begin{array}{l} CH_3CHCH_2CH_3 \\ \quad\;\;\; | \\ \quad\;\; CH_3 \end{array} \quad \text{and} \quad \begin{array}{c} CH_3 \\ | \\ CH_3{-}C{-}CH_3 \\ | \\ CH_3 \end{array}$$

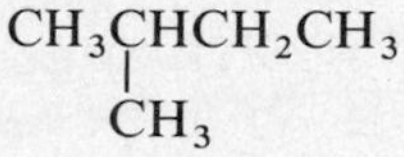

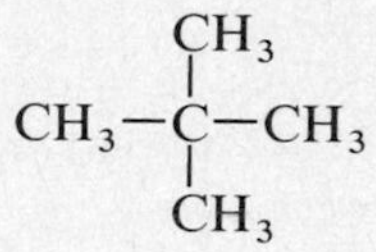

(*d*) $C_2H_4Cl_2$

$$CH_3CHCl_2 \quad \text{and} \quad ClCH_2CH_2Cl$$

(*e*) C_4H_9Br

$$CH_3CH_2CH_2CH_2Br, \quad \begin{array}{l} CH_3CHCH_2CH_3, \\ \quad\;\;\; | \\ \quad\;\;\, Br \end{array} \quad \begin{array}{l} CH_3CHCH_2Br \\ \quad\;\;\; | \\ \quad\;\; CH_3 \end{array} \quad \text{and} \quad \begin{array}{c} CH_3 \\ | \\ CH_3{-}C{-}Br \\ | \\ CH_3 \end{array}$$

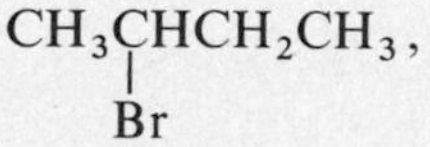

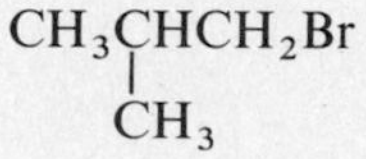

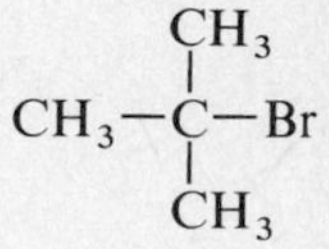

(*f*) C_3H_8O

$$CH_3CH_2CH_2OH, \quad \begin{array}{l} CH_3CHCH_3 \\ \quad\;\;\; | \\ \quad\;\; OH \end{array} \quad \text{and} \quad CH_3OCH_2CH_3$$

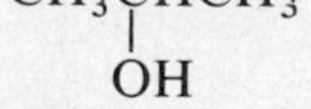

1.17 All these species are characterized by the formula :X≡Y:, and each atom has an electron count of five.

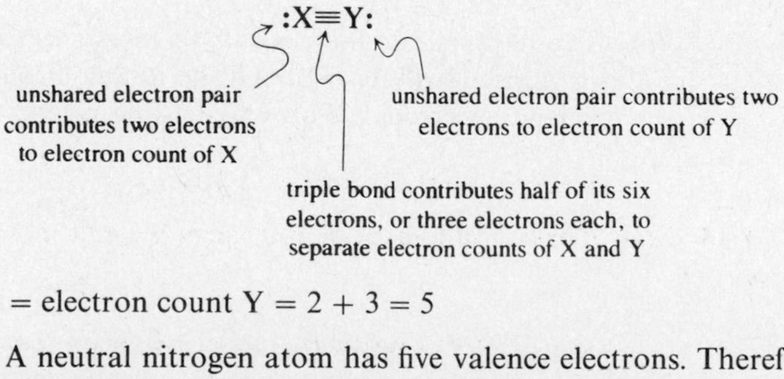

:X≡Y:

unshared electron pair contributes two electrons to electron count of X

unshared electron pair contributes two electrons to electron count of Y

triple bond contributes half of its six electrons, or three electrons each, to separate electron counts of X and Y

Electron count X = electron count Y = 2 + 3 = 5

(*a*) :N≡N: A neutral nitrogen atom has five valence electrons. Therefore, each atom is electrically neutral in molecular nitrogen.

(*b*) :C≡N: Nitrogen, as before, is electrically neutral. A neutral carbon has four valence electrons, so carbon in this species, with an electron count of five, has a unit negative charge. The species is cyanide anion; its net charge is −1.

(*c*) :C≡C: There are two negatively charged carbon atoms in this species. It is a dianion; its net charge is −2.

(*d*) :N≡O: Here again is a species with a neutral nitrogen atom. Oxygen, with an electron count of five, has one less electron in its valence shell than a neutral oxygen atom. Oxygen has a formal charge of +1; the net charge is +1.

(*e*) :C≡O: Carbon has a formal charge of −1; oxygen has a formal charge of +1. Carbon monoxide is a neutral molecule.

1.18 All these species are of the type :Ÿ=X=Ÿ:. Atom X has an electron count of four, corresponding to half of the eight shared electrons in its four covalent bonds. Each atom Y has an electron count of six; comprising it are four unshared electrons plus half of the four electrons in the double bond of each Y to X.

(*a*) :Ö=C=Ö: Oxygen, with an electron count of six, and carbon, with an electron count of four, both correspond to the respective neutral atoms in the number of electrons they "own." Carbon dioxide is a neutral molecule, and neither carbon nor oxygen has a formal charge in this Lewis structure.

(*b*) :N̈=N=N̈: The two terminal nitrogens each have an electron count (six) that is one more than a neutral atom and thus each has a formal charge of −1. The central N has an electron count (four) that is one less than a neutral nitrogen; it has a formal charge of +1. The net charge on the species is (−1 + 1 − 1), or −1.

(*c*) :Ö=N=Ö: As in (*b*), the central nitrogen has a formal charge of +1. As in (*a*), each oxygen is electrically neutral. The net charge is +1.

1.19 (*a*) The problem specifies that ionic bonding is present and that the anion is tetrahedral, so $Na^+(BH_4)^-$ is a reasonable candidate for the species in question. The tetrahedral anion $(BH_4)^-$ has four electron pairs around boron, and boron has a formal charge of −1.

Na^+ H—$\bar{B}$(H)(H)(H) Sodium borohydride

(*b*) Aluminum, like boron, is a group III element, and the ion $(AlH_4)^-$ is tetrahedral. Lithium is the cation.

Li^+ H—$\bar{Al}$(H)(H)(H) Lithium aluminum hydride

(*c*) The tetrahedral anion is sulfate, $(SO_4)^{2-}$. Sulfur contributes six valence electrons, each oxygen contributes six, and the double negative charge accounts for two more. The 32 valence electrons are arranged as indicated.

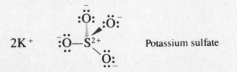

$2K^+$ $:\ddot{O}:^-$ / $:\ddot{O}:^-$ / $^-:\ddot{O}$—S^{2+} / $\ddot{O}:^-$ Potassium sulfate

Sulfur and each oxygen have complete octets. The formal charge of sulfur is +2, and the formal charge of each oxygen is −1. The net charge on sulfate is −2.

(*d*) The tetrahedral anion is phosphate, $(PO_4)^{3-}$. Phosphorus contributes five valence electrons, each oxygen contributes six, and the triple negative charge contributes three more. The 32 valence electrons are arranged as shown.

$3Na^+$ $\quad$ $^-$:Ö—P(:Ö:$^-$)(:Ö:$^-$)(:Ö:$^-$), P$^+$ $\quad$ Sodium phosphate

Phosphorus and oxygen have complete octets. The formal charge of phosphorus is +1, and the formal charge of each oxygen is −1. The net charge on phosphate is −3.

1.20 (*a*) Species A, B, and C have the same molecular formula, the same atomic positions, and the same number of electrons. They differ only in the arrangement of their electrons. Therefore, they are resonance forms of a single compound.

$H_2\bar{C}$:—$\overset{+}{N}$≡N:	H_2C=$\overset{+}{N}$=N̈:$^-$	$H_2\overset{+}{C}$—N̈=N̈:$^-$
A	B	C

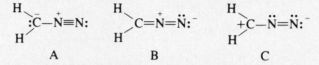

(*b*) Structure A has a formal charge of −1 on carbon.
(*c*) Structure C has a formal charge of +1 on carbon.
(*d*) Structures A and B have formal charges of +1 on the internal nitrogen.
(*e*) Structures B and C have a formal charge of −1 on the terminal nitrogen.
(*f*) All resonance forms of a particular species must have the same net charge. In this case, the net charge on A, B, and C is zero.
(*g*) Both A and B have the same number of covalent bonds, but the negative charge is on a more electronegative atom in B (nitrogen) than it is in A (carbon). Structure B is more stable than A.
(*h*) Structure B is more stable than C. Structure B has one more covalent bond, all of its atoms have octets of electrons, and it has a lesser degree of charge separation than C.

1.21 (*a*) These two structures are resonance forms of azide anion since they have the same atomic positions and the same number of electrons.

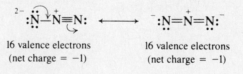

$^{2-}$:N̈—$\overset{+}{N}$≡N: ⟷ $^-$:N̈=$\overset{+}{N}$=N̈:$^-$

16 valence electrons (net charge = −1) $\quad$ 16 valence electrons (net charge = −1)

(*b*) The two structures have different numbers of electrons and therefore they are not resonance forms.

$^{2-}$:N̈—$\overset{+}{N}$≡N: $\quad$:N̈—$\overset{2+}{N}$=N̈:$^-$

16 valence electrons (net charge −1) $\quad$ 14 valence electrons (net charge +1)

(*c*) These two structures have different numbers of electrons; they are not resonance forms.

$^{2-}$:N̈—$\overset{+}{N}$≡N: $\quad$ $^{2-}$:N̈—N̈$^-$—N̈:$^{2-}$

16 valence electrons (net charge = −1) $\quad$ 20 valence electrons (net charge = −5)

1.22 Structure F has 10 electrons surrounding nitrogen, but nitrogen is limited to 8 electrons. Structure F is incorrect.

CH_2=N(—CH_3)=Ö: $\quad$ Not a valid Lewis structure

1.23 (*a*) In this resonance form, the carbon atom of carbon dioxide has only six electrons surrounding it. Move electrons as indicated by the curved arrow so as to generate a more stable Lewis structure in which all the atoms have complete octets.

$$:\ddot{O}=\overset{+}{C}-\ddot{\underset{..}{O}}:^- \longleftrightarrow :\ddot{O}=C=\ddot{O}:$$

(*b*) By moving electron pairs toward oxygen, the negative charge may be placed on a more electronegative atom. Oxygen is more electronegative than nitrogen.

$$:\ddot{O}=C=\ddot{N}:^- \longleftrightarrow {}^-:\ddot{\underset{..}{O}}-C\equiv N:$$

(*c*) One of the oxygen atoms in this structure of ozone has only six electrons surrounding it. A more stable Lewis structure, derived by moving electrons as shown, has complete octets at all three oxygens. Notice also that there is less charge separation in the more stable structure.

$$:\overset{+}{\underset{..}{O}}-\ddot{\underset{..}{O}}-\ddot{\underset{..}{O}}:^- \longleftrightarrow :\ddot{O}=\overset{+}{\underset{..}{O}}-\ddot{\underset{..}{O}}:^-$$

(*d*) By moving electrons toward positive charge and away from negative charge, a structure with complete octets around all second-row elements, as well as no charge separation, is generated.

$$H_2\overset{+}{C}-\ddot{C}-\ddot{\underset{..}{O}}:^- \longleftrightarrow H_2C=C=\ddot{O}:$$

(*e*) Negative charge can be placed on the most electronegative atom (oxygen) in this molecule by moving electrons as indicated.

$$H_2\overset{-}{\ddot{C}}-C(H)=\ddot{O}: \longleftrightarrow H_2C=C(H)-\ddot{\underset{..}{O}}:^-$$

(*f*) This exercise is similar to (*e*). Electron reorganization allows the negative charge to be placed on nitrogen.

$$H_2\overset{-}{\ddot{C}}-C\equiv N: \longleftrightarrow H_2C=C=\ddot{N}:^-$$

(*g*) Octets of electrons are present around both carbon and oxygen if an oxygen unshared pair is moved toward the positively charged carbon to give an additional covalent bond.

$$H-\overset{+}{C}=\ddot{O}: \longleftrightarrow H-C\equiv\overset{+}{O}:$$

(*h*) This exercise is similar to (*g*); move electrons from oxygen to carbon so as to produce an additional bond and satisfy the octet rule for each atom.

$$H_2\overset{+}{C}-\ddot{\underset{..}{O}}H \longleftrightarrow H_2C=\overset{+}{\underset{..}{O}}H$$

(*i*) By moving electrons from the site of negative charge toward the positive charge, a structure that has no charge separation is generated.

$$H_2\overset{-}{\ddot{C}}-\ddot{N}=\overset{+}{N}H_2 \longleftrightarrow H_2C=\ddot{N}-\ddot{N}H_2$$

1.24 Oxygen is surrounded by a complete octet of electrons in each structure but has a different "electron count" in each one because the proportion of shared to unshared pairs is different.

(*a*) $CH_3\ddot{\underset{\cdot\cdot}{O}}:$

Electron count $= 6 + \frac{2}{2} = 7$; formal charge $= -1$

(*b*) $CH_3\ddot{\underset{\cdot\cdot}{O}}CH_3$

Electron count $= 4 + \frac{4}{2} = 6$; formal charge $= 0$

(*c*) $CH_3\ddot{O}(CH_3)CH_3$

Electron count $= 2 + \frac{6}{2} = 5$; formal charge $= +1$

1.25 (*a*) All three carbons must be bonded together, and each one has four bonds. Therefore the molecular formula C_3H_8 uniquely corresponds to:

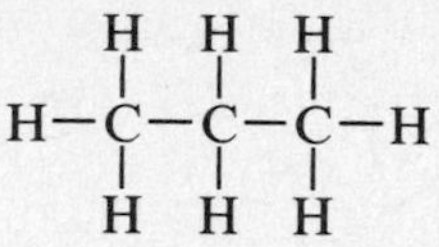

H–C(H)(H)–C(H)(H)–C(H)(H)–H ($CH_3CH_2CH_3$)

(*b*) With two fewer hydrogen atoms than the preceding compound, either C_3H_6 must contain a carbon-carbon double bond or its carbons must be arranged in a ring; thus the structures below are constitutional isomers.

$CH_2{=}CHCH_3$ and cyclo-($CH_2{-}CH_2{-}CH_2$)

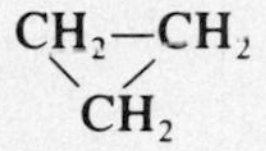

(*c*) The molecular formula C_3H_4 is satisfied by the structures

$CH_2{=}C{=}CH_2$ $HC{\equiv}CCH_3$ and cyclo-($HC{=}CH{-}CH_2$)

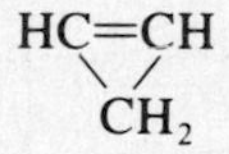

1.26 (*a*) The only atomic arrangements of C_3H_6O that contain only single bonds must have a ring as part of their structure.

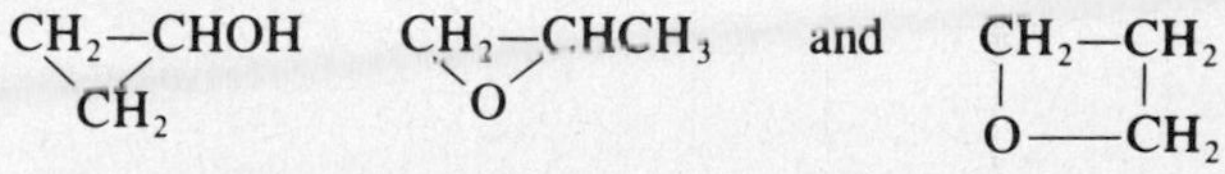

cyclo-($CH_2{-}CHOH{-}CH_2$) cyclo-($CH_2{-}CHCH_3{-}O$) and cyclo-($CH_2{-}CH_2{-}CH_2{-}O$)

(*b*) Structures corresponding to C_3H_6O are possible in noncyclic compounds if they contain a carbon-carbon or carbon-oxygen double bond.

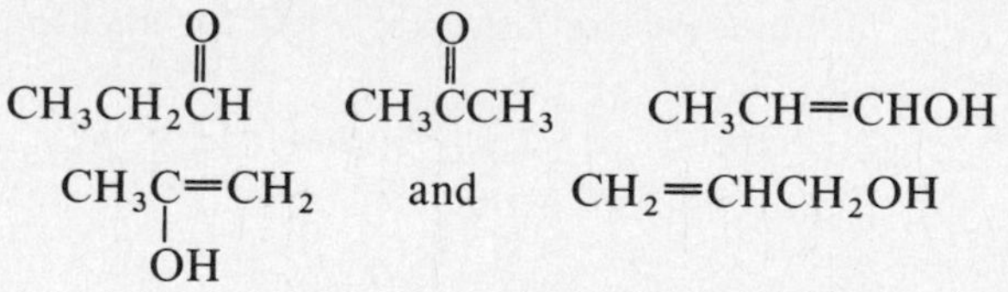

$CH_3CH_2CH{=}O$ $CH_3C({=}O)CH_3$ $CH_3CH{=}CHOH$

$CH_3C(OH){=}CH_2$ and $CH_2{=}CHCH_2OH$

1.27 (*a*) Sodium chloride is ionic; it has a unit positive charge and a unit negative charge separated from each other. Hydrogen chloride has a polarized bond but is a covalent compound. Sodium chloride has a larger dipole moment. The measured values are as shown.

$Na^+ Cl^-$ (μ 9.4 D) is more polar than H–Cl (μ 1.1 D)

(*b*) Fluorine is more electronegative than chlorine, so its bond to hydrogen is more polar, as the measured dipole moments indicate.

$\overset{+\!\!\longrightarrow}{H{-}F}$ (μ 1.7 D) is more polar than $\overset{+\!\!\longrightarrow}{H{-}Cl}$ (μ 1.1 D) (*c*) Boron trifluoride is planar. Its individual B—F bond dipoles cancel. It has no dipole moment.

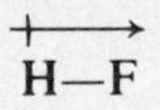

H—F is more polar than BF_3

μ 1.7 D μ 0 D

(*d*) A carbon-chlorine bond is strongly polar; carbon-hydrogen and carbon-carbon bonds are only weakly polar.

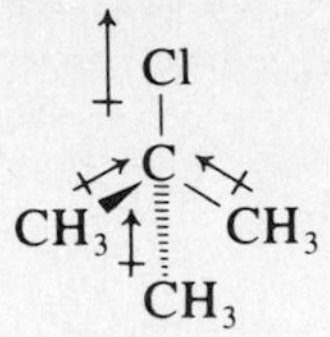

$(CH_3)_3CCl$ is more polar than $(CH_3)_3CH$

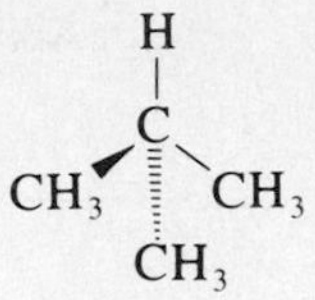

μ 2.1 D μ 0.1 D

(*e*) A carbon-fluorine bond in CCl_3F opposes the polarizing effect of the chlorines. The carbon-hydrogen bond in $CHCl_3$ reinforces it. Therefore $CHCl_3$ has a larger dipole moment.

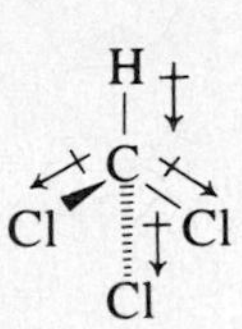

$HCCl_3$ is more polar than $FCCl_3$

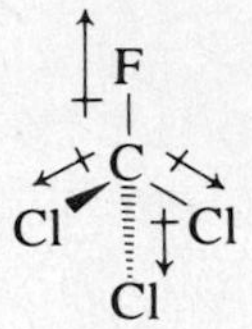

μ 1.0 D μ 0.5 D

(*f*) Oxygen is more electronegative than nitrogen; its bonds to carbon and hydrogen are more polar than the corresponding bonds formed by nitrogen.

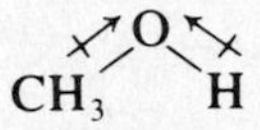

CH_3OH is more polar than CH_3NH_2

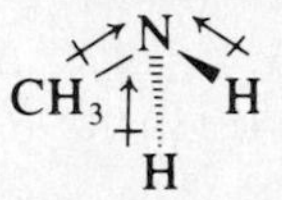

μ 1.7 D μ 1.3 D

(*g*) The Lewis structure for CH_3NO_2 has a formal charge of +1 on nitrogen, making it more electron-attracting than the uncharged nitrogen of CH_3NH_2.

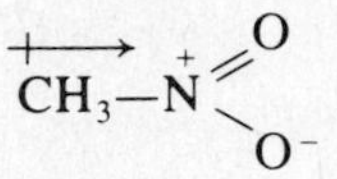

$CH_3—\overset{+}{N}(=O)O^-$ is more polar than $CH_3—\ddot{N}H_2$

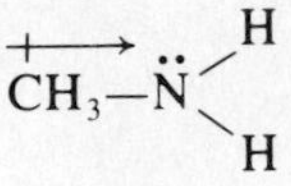

μ 3.1 D μ 1.3 D

1.28 (*a*) There are four electron pairs around carbon in $:\bar{C}H_3$; they are arranged in a tetrahedral fashion. The species is trigonal pyramidal.

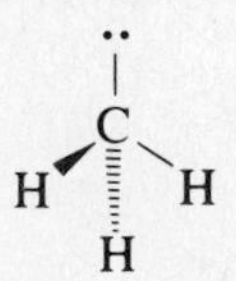

(*b*) Only three electron pairs are present in $\overset{+}{C}H_3$, so it is trigonal planar.

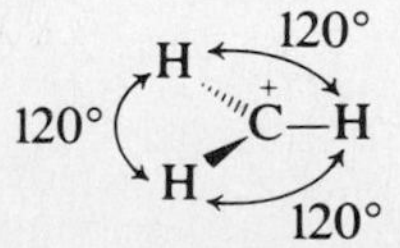

120° 120° 120°

(*c*) As in (*b*), there are three electron pairs. When these electron pairs are arranged in a plane, the atoms in $:CH_2$ are not colinear. The atoms of this species are arranged in a bent structure according to VSEPR considerations.

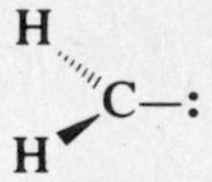

H
C—:
H

The HCH angle is slightly less than 120° because interactions involving the unshared pair are more repulsive than those of the bonded pairs

1.29 The structures, written in a form so as to indicate hydrogen substituents and unshared electrons, are as given below. Remember, a neutral carbon has four bonds, a neutral nitrogen has three bonds plus one unshared pair, and a neutral oxygen has two bonds plus two unshared pairs. Halogen substituents have one bond and three unshared pairs.

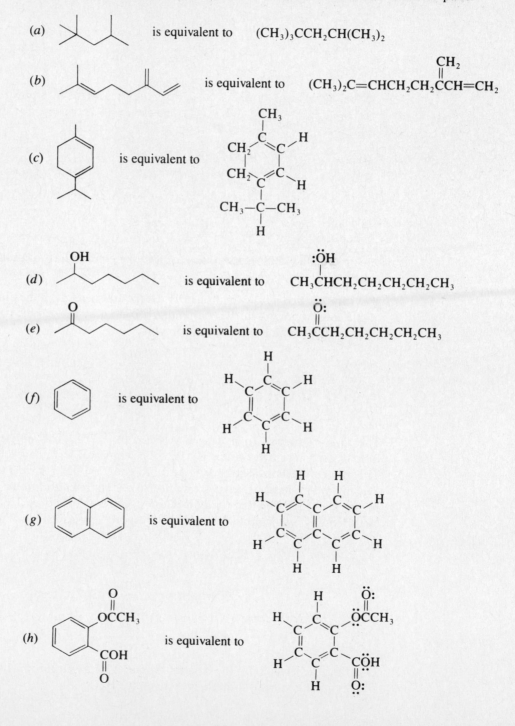

(*a*) 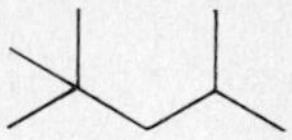 is equivalent to $(CH_3)_3CCH_2CH(CH_3)_2$

(*b*) 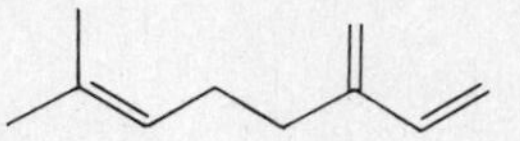 is equivalent to $(CH_3)_2C{=}CHCH_2CH_2\overset{\overset{\displaystyle CH_2}{\|}}{C}CH{=}CH_2$

(*c*) 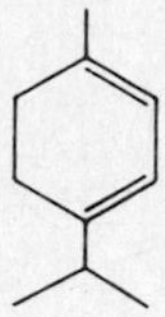 is equivalent to

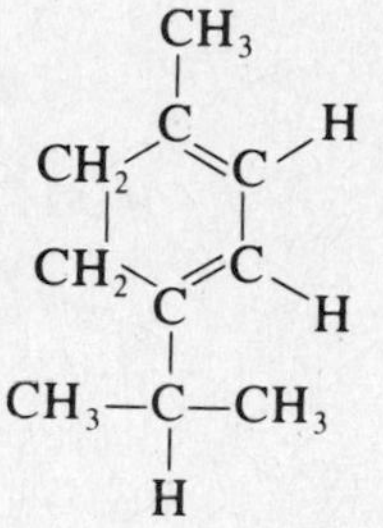

CH₃ … H / CH₂ C C / CH₂ C C H / CH₃—C—CH₃ / H

(*d*) OH is equivalent to $CH_3\overset{\overset{\displaystyle :\ddot{O}H}{|}}{C}HCH_2CH_2CH_2CH_2CH_3$

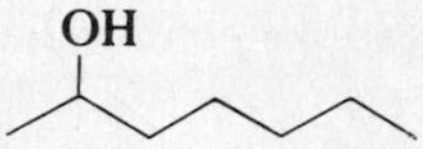

(*e*) O is equivalent to $CH_3\overset{\overset{\displaystyle \ddot{O}:}{\|}}{C}CH_2CH_2CH_2CH_2CH_3$

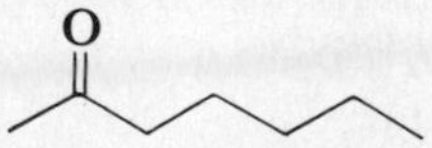

(*f*)  is equivalent to

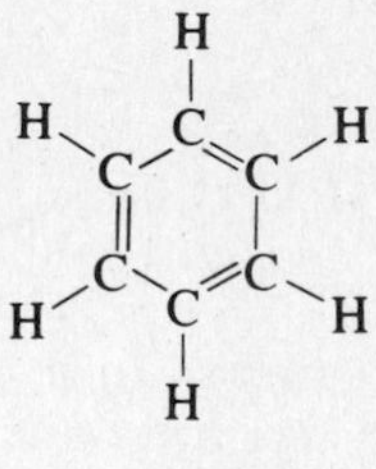

H H C H C C C C H C H H

(*g*) 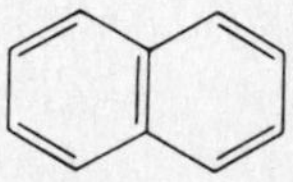 is equivalent to

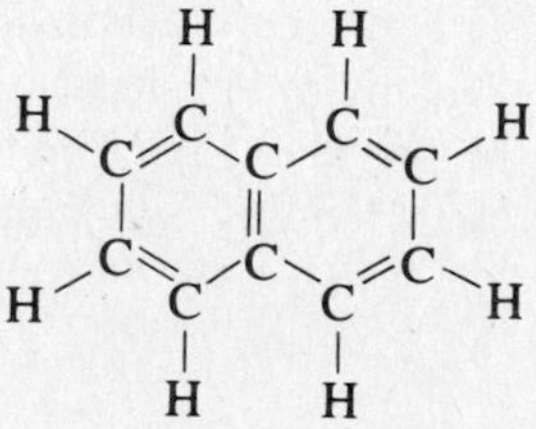

H H H C C C H C C C C C H C H H

(*h*) $\overset{\overset{\displaystyle O}{\|}}{O}CCH_3$ COH $\|$ O is equivalent to

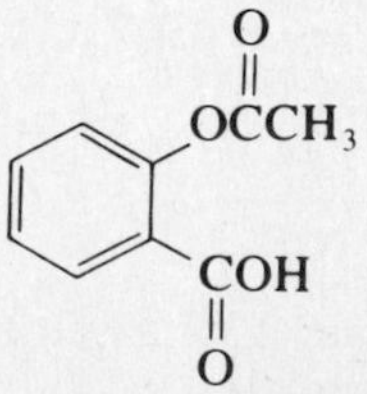

H $\ddot{O}:$ H C $\ddot{O}CCH_3$ C C C C H C C $\ddot{C}OH$ H $\ddot{O}:$

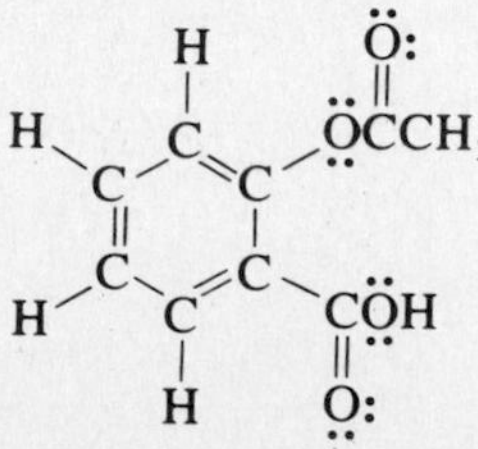

(*i*) 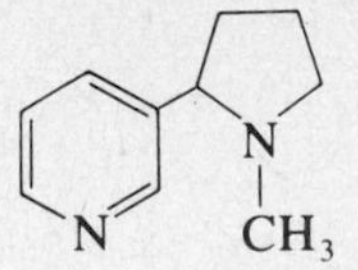 is equivalent to

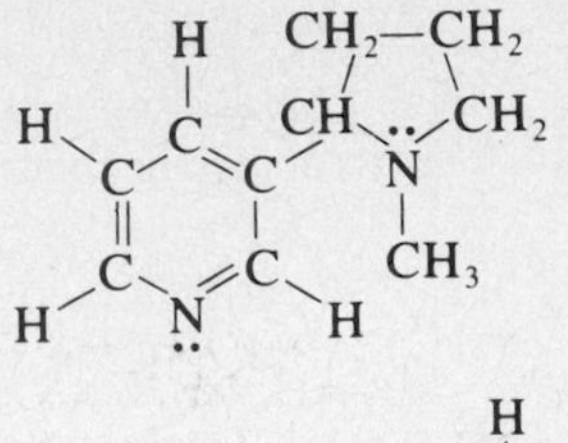

(*j*) 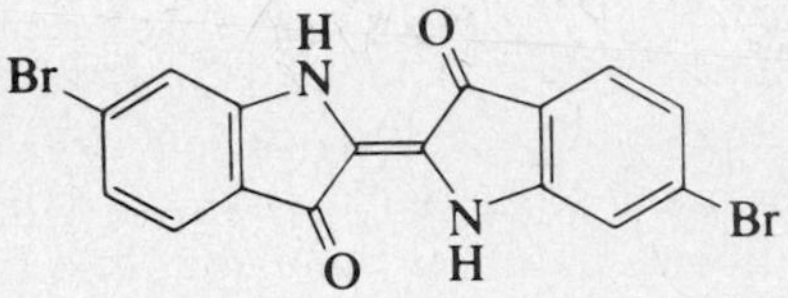 is equivalent to

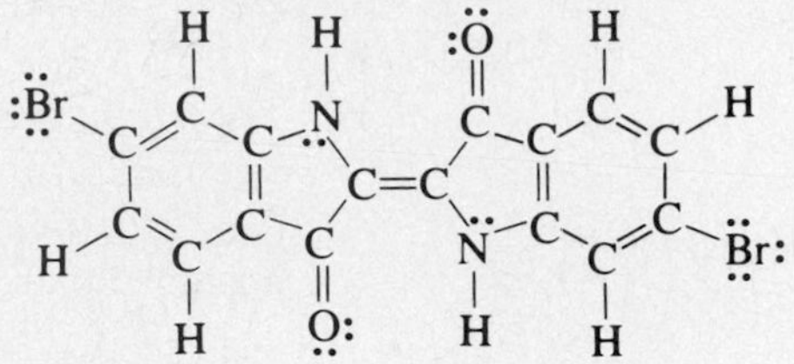

(*k*) 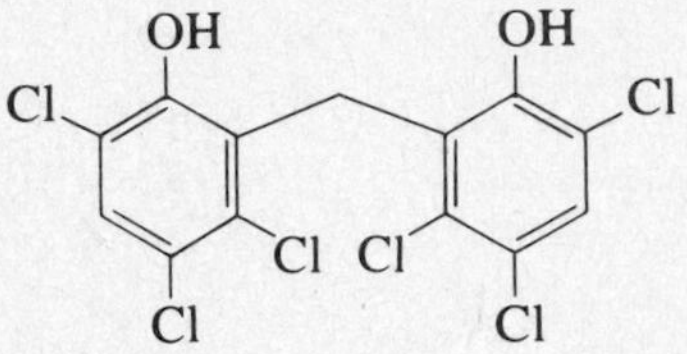 is equivalent to

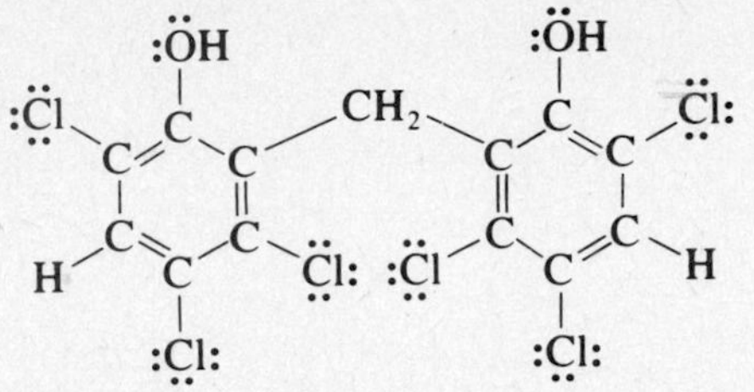

1.30 (*a*) C_8H_{18} (*b*) $C_{10}H_{16}$
(*c*) $C_{10}H_{16}$ (*d*) $C_7H_{16}O$
(*e*) $C_7H_{14}O$ (*f*) C_6H_6
(*g*) $C_{10}H_8$ (*h*) $C_9H_8O_4$
(*i*) $C_{10}H_{14}N_2$ (*j*) $C_{16}H_8Br_2N_2O_2$
(*k*) $C_{13}H_6Cl_6O_2$

Isomers are different compounds that have the same molecular formula; only (*b*) and (*c*) are isomers.

1.31 (*a*) The sum of carbon and hydrogen is nearly 100 percent, indicating that these are the only elements present in the compound. Dividing by the atomic weight of each element gives the relative number of atoms present.

$$\text{Carbon:} \quad \frac{92.61}{12} = 7.72 \text{ atoms C}$$

$$\text{Hydrogen:} \quad \frac{7.01}{1} = 7.01 \text{ atoms H}$$

The ratio of carbon to hydrogen is 7.72/7.01 = 1.10. The simplest formula with this ratio of carbon to hydrogen is $C_{11}H_{10}$. The empirical formula is $C_{11}H_{10}$.

(*b*) The sum of carbon and hydrogen, 56.28 + 6.07 = 62.35 percent, is far short of 100 percent, which indicates that the molecule contains an additional element, assumed to be oxygen. Therefore, the percentage of oxygen is taken to be 37.65 percent. This gives, after dividing by the atomic weights, the respective proportions of atoms:

$$\text{Carbon:} \quad \frac{56.28}{12} = 4.69 \text{ atoms C}$$

$$\text{Hydrogen:} \quad \frac{6.07}{1} = 6.07 \text{ atoms H}$$

$$\text{Oxygen:} \quad \frac{37.65}{16} = 2.35 \text{ atoms O}$$

Dividing each result by the smallest value, 2.35, gives the C/H/O ratio 1.99 : 2.58 : 1. The simplest empirical formula is therefore $C_4H_5O_2$.

(*c*) The sum of carbon and hydrogen, 50.79 + 6.52 = 57.31 percent, indicates an oxygen content of 42.69 percent. The relative numbers of atoms then are:

$$\text{Carbon:} \quad \frac{50.79}{12} = 4.23 \text{ atoms C}$$

$$\text{Hydrogen:} \quad \frac{6.52}{1} = 6.52 \text{ atoms H}$$

$$\text{Oxygen:} \quad \frac{42.69}{16} = 2.67 \text{ atoms O}$$

Dividing each of these by 2.67 gives a C/H/O ratio of 1.58 : 2.44 : 1. The simplest empirical formulas with this ratio of elements are $C_6H_{10}O_4$ and $C_8H_{12}O_5$.

		× 4	× 5
Carbon:	1.58	6.32	7.90
Hydrogen:	2.44	9.76	12.20
Oxygen:	1	4	5

The correspondence is better for $C_8H_{12}O_5$, which is the correct empirical formula.

(*d*) The sum of carbon, hydrogen, and nitrogen is 70.11 percent. The oxygen content is therefore 29.89 percent.

$$\text{Carbon:} \quad \frac{55.65}{12} = 4.64; \quad \frac{4.64}{0.46} = 10.1$$

$$\text{Hydrogen:} \quad \frac{7.99}{1} = 7.99; \quad \frac{7.99}{0.46} = 17.4$$

$$\text{Nitrogen:} \quad \frac{6.47}{14} = 0.46; \quad \frac{0.46}{0.46} = 1.0$$

$$\text{Oxygen:} \quad \frac{29.89}{16} = 1.87; \quad \frac{1.87}{0.46} = 4.1$$

The simplest empirical formula is $C_{10}H_{17}NO_4$.

(*e*) The sum of C, H, N, and Br is 83.89 percent. The oxygen content is therefore 16.11 percent.

$$\text{Carbon:} \quad \frac{45.94}{12} = 3.83; \quad \frac{3.83}{0.35} = 10.9$$

$$\text{Hydrogen:} \quad \frac{4.99}{1} = 4.99; \quad \frac{4.99}{0.35} = 14.3$$

$$\text{Nitrogen:} \quad \frac{5.13}{14} = 0.37; \quad \frac{0.37}{0.35} = 1.06$$

$$\text{Bromine:} \quad \frac{27.83}{80} = 0.35; \quad \frac{0.35}{0.35} = 1.0$$

$$\text{Oxygen:} \quad \frac{16.11}{16} = 1.01; \quad \frac{1.01}{0.35} = 2.89$$

The simplest empirical formula is $C_{11}H_{14}BrNO_3$.

1.32 (*a*) The oxygen content of an organic compound is not normally determined experimentally but is estimated by difference. Since the sum of the amounts of carbon and hydrogen is 46.6 percent, the amount of oxygen is taken as 100 − 46.6 = 53.4 percent.

We now convert the mass ratios to ratios of atoms by dividing the percentages by the atomic weights.

A 100-g sample of glucose contains:

$$39.8 \text{ g carbon} \equiv \frac{39.8}{12} = 3.32 \text{ g atoms C}$$

$$6.8 \text{ g hydrogen} \equiv \frac{6.8}{1} = 6.8 \text{ g atoms H}$$

$$53.4 \text{ g oxygen} \equiv \frac{53.4}{16} = 3.34 \text{ g atoms O}$$

These amounts are converted to whole number ratios by dividing by the smallest number.

$$\text{Carbon:} \quad \frac{3.32}{3.32} = 1.0$$

$$\text{Hydrogen:} \quad \frac{6.8}{3.32} = 2.0$$

$$\text{Oxygen:} \quad \frac{3.34}{3.32} = 1.0$$

Therefore, the empirical formula is CH_2O. The mass of a CH_2O unit is 30. Since the molecular weight is 175 ± 10, there must be six CH_2O units per molecule. Therefore, the molecular formula of glucose is $C_6H_{12}O_6$.

(*b*) For 2, 4-D, the sum of carbon, hydrogen, and chlorine is 43.6 + 2.8 + 32.0 = 78.4 per cent. Therefore the compound contains 21.6 per cent oxygen.

Carbon:	43.6/12 = 3.63	3.63/0.90 ≡ 4.0
Hydrogen:	2.8/1 = 2.8	2.8/0.90 ≡ 3.1
Chlorine:	32.0/35.5 = 0.90	0.90/0.90 ≡ 1.0
Oxygen:	21.6/16 = 1.35	1.35/0.90 ≡ 1.5

The empirical formula is therefore $C_8H_6Cl_2O_3$. Since the molecular weight of this unit is 221, it also corresponds to the molecular formula.

SELF-TEST

PART A

A-1. Write the electronic configuration of the stable ion of sulfur.

A-2. Nicotine is a physiologically active compound found in tobacco. Combustion analysis showed nicotine to be 74.0% C, 8.69% H, and 17.3% N. The molecular weight was found to be 162. What is the molecular formula of nicotine?

A-3. Determine the formal charge on nitrogen in the following species:

$$:\ddot{N}{=}C{=}\ddot{S}:$$

A-4. Write a correct Lewis structure for methylamine, CH_5N.

A-5. What is the molecular formula of the structure shown below?

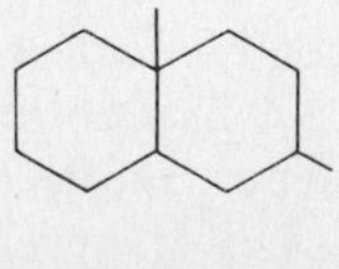

A-6. Account for the fact that all bonds in SO_3 are the same by drawing the appropriate Lewis structure(s).

A-7. Cortisone has been found to contain 69.96% carbon and 7.83% hydrogen by weight. Experimental measurements have shown the molecular weight to be about 360. What is the molecular formula?

A-8. What is the correct molecular formula for the substance shown below?

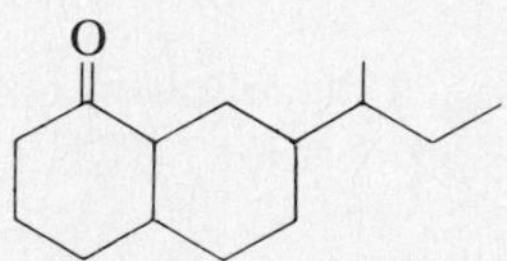

A-9. The cyanate ion contains three atoms arranged in the order OCN. Write a Lewis structure for this species and assign a formal charge to each atom. What is the net charge of the ion?

A-10. Deduce the shape of NCl_3 using VSEPR theory and draw a three-dimensional representation of the molecule.

PART B

B-1. Which of the following is most likely to have ionic bonds?
(*a*) HCl (*b*) Na_2O (*c*) N_2O (*d*) NCl_3

B-2. Which of the following has the highest electron affinity?
(*a*) Li (*b*) B (*c*) F (*d*) C

B-3. Which of the following is *not* an isomer of compound I?

I	II	III	IV
$CH_2{-}CH{-}CH_3$ (with O bridging CH_2 and CH)	$CH_3{-}CH_2{-}\overset{\displaystyle O}{\overset{\|}{C}}{-}H$	$CH_3{-}\overset{\displaystyle O}{\overset{\|}{C}}{-}CH_3$	$CH_3CH{=}CH$ (with OH on the last CH)

(*a*) II (*b*) IV
(*c*) II and III (*d*) None of these (all are isomers)

B-4. In which of the following bonds is oxygen the positive end of the dipole?
(*a*) O—F (*b*) O—N (*c*) O—S (*d*) O—H

B-5. Ionic bonds are likely to form under which of the following sets of conditions?
(*a*) One atom has a low ionization energy and the other has a high electron affinity.
(*b*) One atom has a low ionization energy and the other has a low electron affinity.
(*c*) Both atoms are small.
(*d*) Each atom has a low ionization energy.

B-6. Which of the following is not an electronic configuration for an atom in its ground state?
(*a*) $1s^2 2s^2 2p_x^2\, 2p_y^1\, 2p_z^1$ (*b*) $1s^2 2s^2 2p_x^2\, 2p_y^2\, 2p_z^0$
(*c*) $1s^2 2s^2 2p_x^2\, 2p_y^2\, 2p_z^1$ (*d*) $1s^2 2s^2 2p_x^2\, 2p_y^2\, 2p_z^2$

B-7. The formal charge on the phosphorus atom in $(CH_3)_4P$ is
(*a*) 0 (*b*) -1 (*c*) $+1$ (*d*) $+2$

The following two questions refer to the hypothetical compounds shown below.

I	II	III	IV
A—B—A	$A{=}\ddot{B}{-}A$	A=B=A	$A{-}\ddot{B}{-}A$ (with A bonded below B)

B-8. Which substance(s) is (are) linear?
(*a*) I only (*b*) I and III (*c*) I and II (*d*) III only

B-9. Assuming A is more electronegative than B, which substance(s) is (are) polar?
(*a*) I and III (*b*) II only (*c*) IV only (*d*) II and IV

B-10. The total number of *unshared pairs* of electrons in the molecule 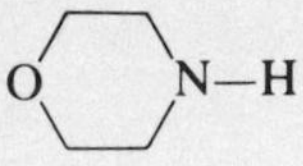 O⟨ring⟩N—H is
(*a*) 0 (*b*) 1 (*c*) 2 (*d*) 3

B-11. Which of the following contains a triple bond?
(*a*) SO_2 (*b*) HCN (*c*) C_2H_4 (*d*) NH_3

CHAPTER

2

ALKANES

IMPORTANT TERMS AND CONCEPTS

Molecular Orbitals (Secs. 2.1 to 2.3) In the previous chapter the concept of sharing electrons between atoms to form covalent bonds was presented. A more detailed description of these bonds is provided by viewing the orbitals on individual atoms as overlapping and interacting with each other. These regions of interaction are known as *molecular orbitals* (MO's). The method of describing bonding in molecules is known as the *linear combination of atomic orbitals-molecular orbital* (LCAO-MO) approach.

Constructive interference, or *in-phase* overlap of two individual atomic orbitals gives rise to what is called a *bonding molecular orbital.* An electron in a bonding MO is of lower energy than one in the atomic orbitals from which the MO is formed. The result is an *attractive force* between the two atoms.

The subtractive interaction between two atomic orbitals gives rise to an *antibonding molecular orbital.* An electron in an antibonding MO is of higher energy than one in the atomic orbitals from which the antibonding MO arises, resulting in a *repulsion* of the two atoms. When the number of electrons in bonding MO's exceeds the number in antibonding MO's, the net force between the two atoms is attractive, resulting in formation of a stable molecule.

The bonding MO's formed by the overlap of atomic orbitals have the region of highest electron density concentrated between the two nuclei. The electron distribution is symmetric around the internuclear axis. A molecular orbital having this characteristic is known as a σ (*sigma*) *orbital.* Two nuclei bonded by electrons in a σ orbital are said to be connected by a σ *bond.*

Bonding in Methane and Ethane (Secs. 2.4 to 2.6) Two methods may be used to describe the bonding in a molecule such as methane, CH_4. As shown in text Figure 2.7, the atomic orbitals of the carbon and hydrogen atoms may be viewed as forming a set of molecular orbitals which encompass the entire molecule.

A second approach, known as the *localized molecular orbital model,* is widely used for describing the bonding in complex molecules. It departs significantly from the previous MO approach in that the bonds, and therefore the molecular orbitals which form them, are viewed as being localized between pairs of atoms. The molecule is then pictured as a collection of two-center bonds.

The ground-state electronic configuration of carbon ($1s^2$, $2s^2$, $2p^2$) is poorly suited for covalent bonding and for explaining the observed facts regarding carbon compounds. This configuration only provides two orbital vacancies, and these are in *p* orbitals situated perpendicular to each other. The four carbon-hydrogen bonds of methane are equivalent and are directed toward the corners of a tetrahedron (a *tetrahedral* arrangement).

These problems may be resolved by viewing the ground-state atomic orbitals as mixing together by a process known as *orbital hybridization* to form a new set of orbitals. Each of these new *hybrid orbitals* has characteristics of the ground-state atomic orbitals from which they are derived.

In the case of methane, mixing the 2*s* orbital of the carbon atom with the three 2*p* orbitals yields four new sp^3-*hybrid* orbitals. Each C—H σ bond is formed from the overlap of an sp^3 hybrid orbital with a hydrogen 1*s* orbital, as shown in text Figure 2.10. These orbitals are arranged in a tetrahedral fashion around the carbon atom.

Bonding in ethane, C_2H_6, may be explained in a similar manner. Noteworthy is the fact that hybrid orbitals from each of two carbons may overlap to form an sp^3-sp^3 σ bond. The carbon-carbon single bond framework of larger organic molecules is constructed of orbital overlaps similar to that in ethane.

Alkane Nomenclature (Secs. 2.7 to 2.11) Alkanes are members of an important class of organic compounds known as *hydrocarbons*, which contain only carbon and hydrogen. In alkanes the carbon framework consists of only single bonds (hydrogen can, of course, never have more than a single bond). The general formula for alkanes is C_nH_{2n+2}. For example, the following all represent formulas of alkanes:

$$CH_4, \quad C_5H_{12}, \quad C_{20}H_{42}$$

For C_4H_{10} and higher alkanes it is possible to have *constitutional isomers*, that is, more than one compound having the same formula. In order to distinguish between isomers a method of *systematic nomenclature* has been developed by the International Union of Pure and Applied Chemistry (IUPAC).

According to the IUPAC nomenclature rules, alkanes are named as derivatives of unbranched parent alkanes. Although often referred to as the *straight-chain* isomer, it should be remembered that carbon forms bonds in a tetrahedral framework, and the chain of carbon atoms is anything but "straight." The names of the more common unbranched parent alkanes are listed in text Table 2.1.

Some branched alkanes are often given "common" names. Examples include isobutane and isopentane, shown below.

$$(CH_3)_2CH\text{—}CH_3 \qquad (CH_3)_2CH\text{—}CH_2\text{—}CH_3$$

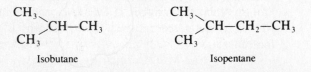

Isobutane   Isopentane

Branched hydrocarbons such as these provide examples of compounds that contain hydrocarbon *substituent* groups. These groups are named according to the number of carbons they contain. The stem of the name is taken from one of the alkane names in Table 2.1 and the suffix *-ane* is replaced by *-yl*. For example:

CH_3—	Methyl group
CH_3CH_2—	Ethyl group
$CH_3CH_2CH_2CH_2CH_2$—	Pentyl group

Note that each of these alkyl groups has one less hydrogen atom than the parent hydrocarbon from which its name is derived.

There are an enormous number of isomers possible for larger alkanes (more than 300,000 $C_{20}H_{42}$ isomers, for example); therefore unique common names for them all would be impossible. The IUPAC rules provide a means of identifying, with a unique name, each and every possible alkane isomer. In fact, as you will see, all organic compounds may be named by following the appropriate IUPAC rules.

Step 1: Identify the longest *continuous* carbon chain. The stem of the compound name is based on the name of the alkane having this number of carbon atoms (from text Table 2.1).

Step 2: Number the longest chain from the end that allows the lowest numbers to be assigned to the carbon atoms bearing alkyl group substituents.

Step 3: Identify the alkyl substituents by name and specify the carbon atoms to which they are attached by number. If a substituent appears more than once, the appropriate prefix di-, tri-, tetra-, etc. is used.

Examples of compounds named according to these rules are the following:

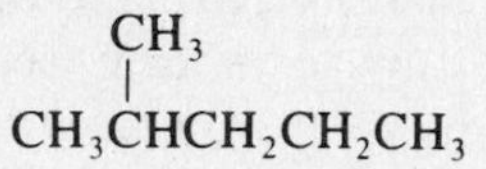

$$\begin{array}{c} CH_3 \\ | \\ CH_3CHCH_2CH_2CH_3 \end{array} \qquad \text{2-Methylpentane}$$

$$(CH_3)_3CCH_2CH(CH_3)_2 \equiv \begin{array}{c} \quad CH_3 \qquad\quad CH_3 \\ \quad | \qquad\qquad\quad | \\ CH_3{-}C{-}CH_2{-}C{-}CH_3 \\ \quad | \qquad\qquad\quad | \\ \quad CH_3 \qquad\qquad H \end{array}$$

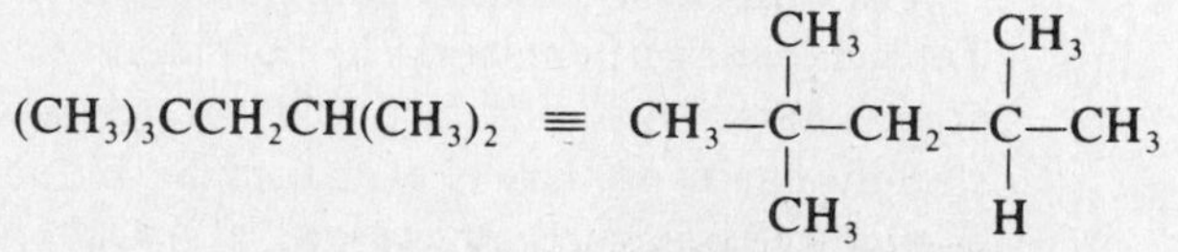

2,2,4-Trimethylpentane

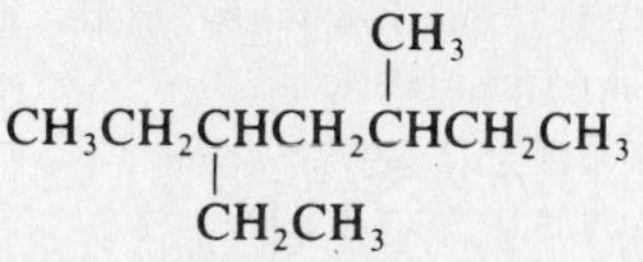

$$\begin{array}{c} \qquad\qquad CH_3 \\ \qquad\qquad | \\ CH_3CH_2CHCH_2CHCH_2CH_3 \\ | \qquad\qquad \\ CH_2CH_3 \qquad\qquad \end{array} \qquad \text{3-Ethyl-5-methylheptane}$$

Cycloalkane Nomenclature (Sec. 2.12) Alkanes whose carbon backbone forms a ring are commonly encountered in organic chemistry. Such cyclic alkanes, or *cycloalkanes* have the general formula C_nH_{2n}. The stem names (from text Table 2.1) are preceded by "cyclo." Therefore cyclopentane C_5H_{10} would be represented as:

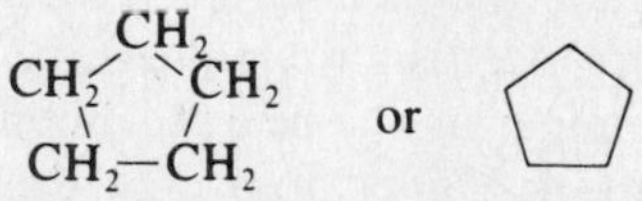

$$\begin{array}{c} CH_2 \\ CH_2 \diagup \quad \diagdown CH_2 \\ \diagdown \qquad \diagup \\ CH_2{-}CH_2 \end{array} \quad \text{or} \quad \text{(pentagon)}$$

Cyclopentane

Substituents are named and numbered as described previously. As the following examples demonstrate, a number is not necessary when the ring contains only one substituent. When there are two or more, the first group in the IUPAC name is assigned location 1 on the ring, and the numbering around the ring is done to keep the other numbers as low as possible.

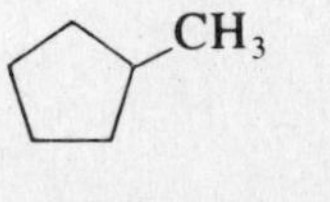

Methylcyclopentane (ring with CH_3) | 1,2-Dimethylcyclopentane (ring with CH_3, CH_3) | 1-Ethyl-3-methylcyclopentane (ring with CH_2CH_3, CH_3)

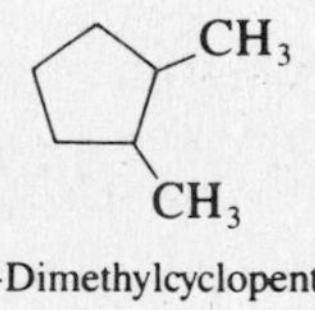

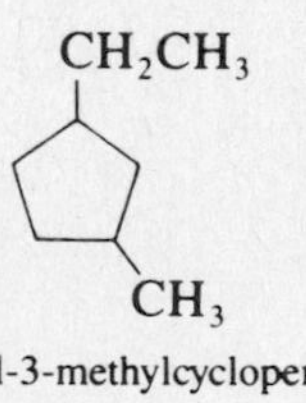

Physical Properties (Sec. 2.14) Consideration of the relationship between the structure of a substance and its boiling point leads to a discussion of *intermolecular forces.* These forces may be classified in three categories:

1. Hydrogen bonding
2. Dipole-dipole attractions
3. Attractive van der Waals forces

Dipole-dipole attractions exist when one polar molecule interacts with another, that is, when the negative end of one molecule attracts the positive end of another. *Hydrogen bonding*, a specific case of dipole-dipole attraction, occurs when hydrogen is attached to a small electronegative element, such as O, N, or F. In the case of water the hydrogen bonding forces are particularly strong, having about 5 percent of the strength of a covalent bond. These attractive forces tend to keep the substance in the liquid state until a temperature much higher than would otherwise be expected is reached. That is, in order for the substance to reach its boiling point, all the hydrogen bonds must be broken. As an example, hydrogen sulfide (in which hydrogen bonding is very weak) may be compared with water; the boiling point of H_2S is $-62°C$, whereas that of water $+100°C$!

Van der Waals attractions constitute the predominant intermolecular force in nonpolar substances. Also known as *induced dipole-induced dipole* attractions, these forces arise from temporary distortion of the electron distribution of molecules in close proximity to each other. The temporary dipoles which result give rise to a weak attraction between the molecules.

The observation that boiling points increase with increasing molecular size, as seen in text Table 2.2, may be explained by considering the attractive van der Waals forces present in the liquid state. Also, the fact that branched alkanes have lower boiling points than their straight-chain isomers is explained by the more compact shape, and hence smaller surface area, of the branched molecules.

Combustion (Sec. 2.15) Reaction with oxygen is known as *combustion*. The general equation for the combustion of an alkane is:

$$C_nH_{2n+2} + \left(\frac{3n+1}{2}\right)O_2 \longrightarrow nCO_2 + (n+1)H_2O + \text{heat}$$

The heat released in the combustion reaction of a substance is known as its *heat of combustion*. By convention the heat of combustion is given by $-\Delta H^0$, where

$$\Delta H^0 = H^0_{\text{products}} - H^0_{\text{reactants}}$$

and H^0 represents the *enthalpy*, or heat content, of a substance in its standard state. Heats of combustion are listed as positive numbers, as combustion reactions are always *exothermic*. Therefore, ΔH^0 is always a negative number.

By comparing the heat of combustion of one isomer of a substance with another, information may be gained regarding the relative energies of the compounds. That is, an isomer which is more *stable* will give off less heat on combustion than one which is of higher energy and therefore less stable.

SOLUTIONS TO TEXT PROBLEMS

2.1 The orbital energy diagram of H_2^+ is the same as that of H_2 except that it contains just one electron instead of two.

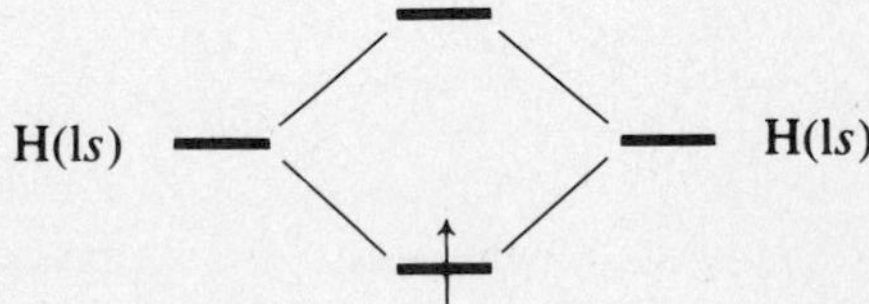

An electron in the bonding orbital of H_2^+ is of lower energy than when it occupies the 1*s* orbital of a hydrogen atom. The system H_2^+ is more stable than a separate hydrogen atom and a proton. The reaction:

$$H_2^+ \longrightarrow H\cdot + H^+$$

is endothermic. Energy must be put into the system to break the bond in H_2^+. (The bond dissociation energy of H_2^+ has been estimated to be 61 kcal/mol.)

2.2 (*b*) Each hydrogen contributes one 1*s* orbital while nitrogen contributes a 2*s* and three 2*p* orbitals. Therefore, NH_3 has seven molecular orbitals. The eight valence electrons of ammonia fill the four orbitals that are of lowest energy.

(*c*) Two fluorine atoms each contribute four orbitals (2*s*, $2p_x$, $2p_y$, $2p_z$), giving a total of eight molecular orbitals for F_2. There are a total of 14 valence electrons (seven from each fluorine atom), so seven of the eight molecular orbitals are occupied.

(*d*) Carbon and oxygen each contribute four atomic orbitals, one 2*s* and three 2*p* orbitals. There are eight molecular orbitals in CO; carbon has four valence electrons and oxygen has six. These 10 valence electrons occupy the five lowest-energy molecular orbitals.

(*e*) There are 12 molecular orbitals in CO_2. Carbon contributes a 2*s* orbital and three 2*p* orbitals; each oxygen does likewise. The number of valence electrons is 16, 4 from carbon and 6 from each oxygen. These 16 electrons occupy the eight molecular orbitals that are of lowest energy.

(*f*) Ethane has 14 molecular orbitals. Two carbons contribute four each, for a total of eight, and each of six hydrogens contributes one. Since there are 14 valence electrons, 7 of the 14 molecular orbitals are occupied.

2.3 Each of the carbons in propane is bonded to four other atoms. All the carbon atoms in propane are sp^3 hybridized.

```
    H  H  H
    |  |  |
 H—C—C—C—H
    |  |  |
    H  H  H
```

All 10 of the bonds in propane are σ bonds. The eight C—H σ bonds arise from C(sp^3)-H(1*s*) overlap. The two C—C σ bonds arise from C(sp^3)-C(sp^3) overlap.

2.4 The structures of the C_6H_{14} isomers are presented in the text. Recall that carbon-skeleton formulas omit individual hydrogen substituents and that carbon atoms are understood to be present at the end of each line and at each bend in the chain.

C_6H_{14} *isomer*	*Carbon skeleton*
$CH_3CH_2CH_2CH_2CH_2CH_3$	
$CH_3CHCH_2CH_2CH_3$ with CH_3 on C-2	or
$CH_3CCH_2CH_3$ with two CH_3 on C-2	
$CH_3CH—CHCH_3$ with CH_3 on C-2 and C-3	
$CH_3CH_2CHCH_2CH_3$ with CH_3 on C-3	

2.5 (*b*) Octacosane is not listed in Table 2.1, but its structure can be deduced from its systematic name. The suffix *-cosane* pertains to alkanes that contain 20 to 29 carbons in their longest continuous chain. The prefix octa- means "eight." Therefore, octacosane is the unbranched alkane having 28 carbon atoms. It is $CH_3(CH_2)_{26}CH_3$.

2.6 The ending -hexadecane reveals that the longest continuous carbon chain has 16 carbon atoms.

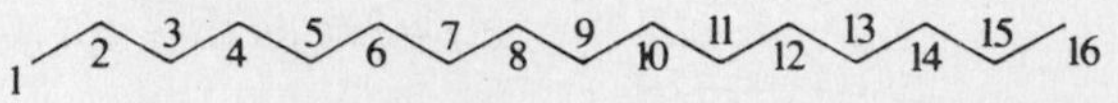

1 2 3 4 5 6 7 8 9 10 11 12 13 14 15 16

There are four methyl groups ("tetramethyl") and they are located at carbons 2, 6, 10, and 14.

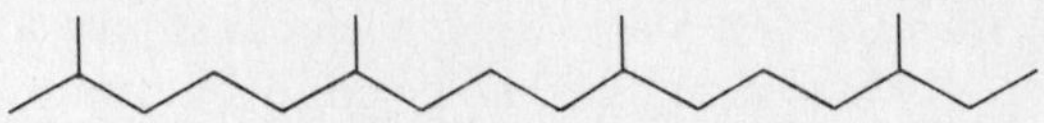

2,6,10,14-Tetramethylhexadecane
(phytane)

2.7 (*b*) The structures of the C_5H_{12} isomers and their common names are given in the text. The systematic name of the unbranched isomer is pentane (Table 2.1).

$CH_3CH_2CH_2CH_2CH_3$ IUPAC name: **pentane**
common name: *n*-pentane

The isomer $(CH_3)_2CHCH_2CH_3$ has four carbons in the longest continuous chain and so is named as a derivative of butane. Since it has a methyl group at C-2, it is 2-methylbutane.

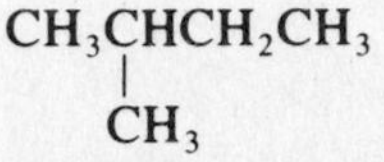

$$\begin{array}{l} CH_3CHCH_2CH_3 \\ \quad\;\; | \\ \quad\; CH_3 \end{array}$$

methyl group at C-2

IUPAC name: **2-methylbutane**
Common name: isopentane

The isomer $(CH_3)_4C$ has three carbons in its longest continuous chain and so is named as a derivative of propane. There are two methyl groups at C-2, so it is a 2,2-dimethyl derivative of propane.

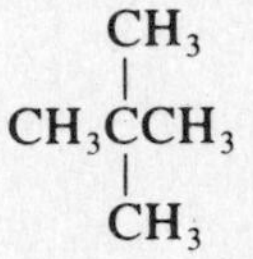

$$\begin{array}{c} CH_3 \\ | \\ CH_3CCH_3 \\ | \\ CH_3 \end{array}$$

IUPAC name: **2,2-dimethylpropane**
Common name: neopentane

(*c*) First write out the structure in more detail and identify the longest continuous carbon chain.

$$\begin{array}{ccccccccc} & & CH_3 & & & & H & & \\ & & | & & & & | & & \\ CH_3 & - & C & - & CH_2 & - & C & - & CH_3 \\ & & | & & & & | & & \\ & & CH_3 & & & & CH_3 & & \end{array}$$

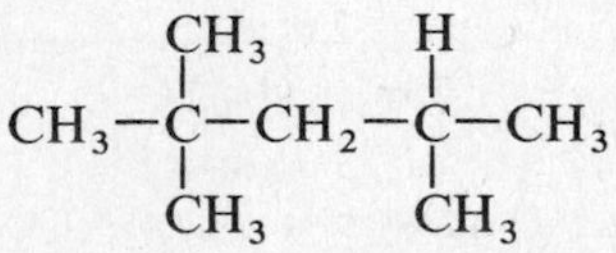

There are five carbon atoms in the longest chain and so the compound is named as a derivative of pentane. This five-carbon chain has three methyl substituents attached to it, making it a trimethyl derivative of pentane. We number the chain in the direction that gives the lowest numbers to the substituents at the first point of difference.

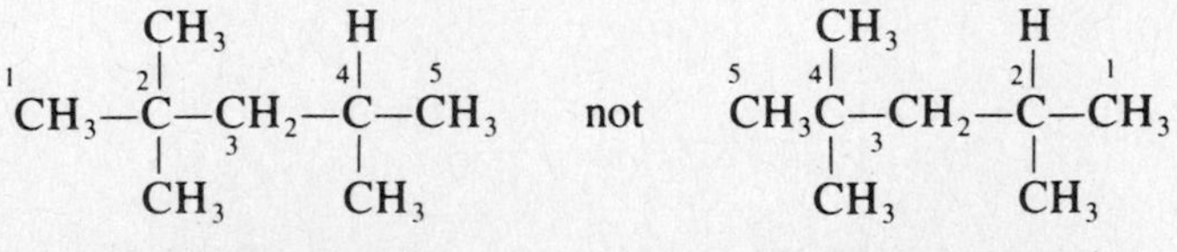

$$\begin{array}{ccccccccc} & & CH_3 & & & & H & & \\ {}^1 & & {}^2| & & & & {}^4| & & {}^5 \\ CH_3 & - & C & - & CH_2 & - & C & - & CH_3 \\ & & | & & {}^3 & & | & & \\ & & CH_3 & & & & CH_3 & & \end{array} \quad \text{not} \quad \begin{array}{ccccccc} & CH_3 & & & & H & \\ {}^5 & {}^4| & & & & {}^2| & {}^1 \\ CH_3 & C & - & CH_2 & - & C & -CH_3 \\ & | & {}^3 & & & | & \\ & CH_3 & & & & CH_3 & \end{array}$$

2,2,4-Trimethylpentane   2,4,4-Trimethylpentane

(*d*) The longest continuous chain in $(CH_3)_3CC(CH_3)_3$ contains four carbon atoms.

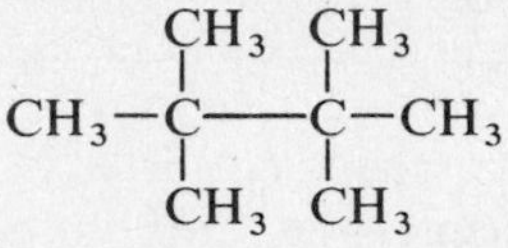

$$\begin{array}{ccccccc} & & CH_3 & & CH_3 & & \\ & & | & & | & & \\ CH_3 & - & C & — & C & - & CH_3 \\ & & | & & | & & \\ & & CH_3 & & CH_3 & & \end{array}$$

The compound is named as a tetramethyl derivative of butane; it is 2,2,3,3-tetramethylbutane.

2.8 There are three C_5H_{11} alkyl groups derived from pentane. One is primary while two are secondary.

$CH_3CH_2CH_2CH_2CH_2—$ Primary alkyl group

$CH_3CH_2CH_2\underset{|}{C}HCH_3$ Secondary alkyl group

$CH_3CH_2\underset{|}{C}HCH_2CH_3$ Secondary alkyl group

There are four alkyl groups derived from $(CH_3)_2CHCH_2CH_3$. Two are primary, one is secondary, and one is tertiary.

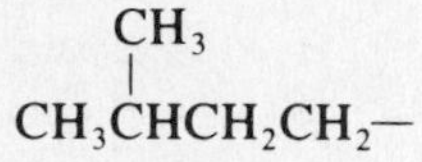

$CH_3\overset{CH_3}{\overset{|}{C}}HCH_2CH_2—$ Primary alkyl group

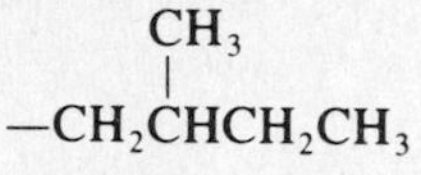

$—CH_2\overset{CH_3}{\overset{|}{C}}HCH_2CH_3$ Primary alkyl group

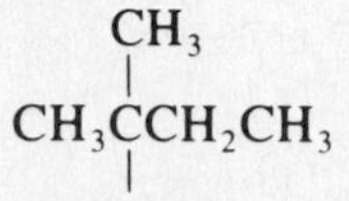

$CH_3\overset{CH_3}{\overset{|}{\underset{|}{C}}}CH_2CH_3$ Tertiary alkyl group

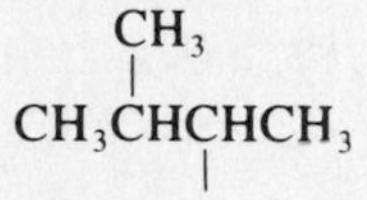

$CH_3\overset{CH_3}{\overset{|}{C}}H\underset{|}{C}HCH_3$ Secondary alkyl group

Only the neopentyl group $(CH_3)_3CCH_2$- (primary) is derivable from neopentane.

2.9 (*b*) The compound is named as a derivative of hexane because it has six carbons in its longest continuous chain.

$$\overset{6}{C}H_3\overset{5}{C}H_2\overset{4}{C}HCH_2\overset{3}{}\overset{2}{C}H\overset{1}{C}H_3$$
$$\text{with } CH_3CH_2 \text{ on C-4 and } CH_3 \text{ on C-2}$$

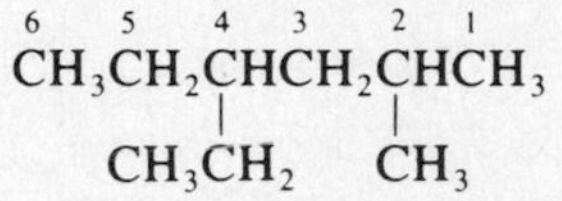

The chain is numbered so as to give the lowest number to the substituent that appears closest to the end of the chain. In this case it is numbered so that the substituents are located at C-2 and C-4 rather than at C-3 and C-5. In alphabetical order the groups are ethyl and methyl; they are listed in alphabetical order in the name. The compound is 4-ethyl-2-methylhexane.

(*c*) The longest continuous chain is shown in the structure; it contains 10 carbon atoms. The structure also shows the numbering scheme that gives the lowest numbers to the substituent at the first point of difference.

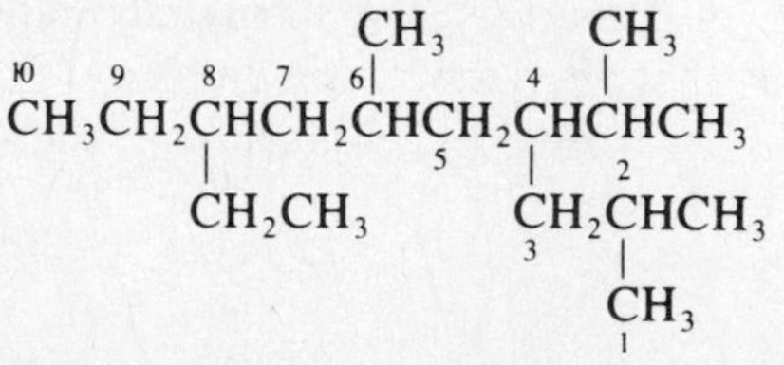

$$\overset{10}{C}H_3\overset{9}{C}H_2\overset{8}{C}H\overset{7}{C}H_2\overset{6}{C}H\overset{5}{C}H_2\overset{4}{C}H\overset{}{C}HCH_3$$
$$\text{(CH}_3\text{ on C-6; CH}_3\text{ on C-4 branch carbon; CH}_2CH_3\text{ on C-8; } \underset{3}{C}H_2\underset{2}{C}H\underset{1}{C}H_3\text{ branch)}$$

In alphabetical order, the substituents are ethyl (at C-8), isopropyl (at C-4), and two methyl groups (at C-2 and C-6). The alkane is 8-ethyl-4-isopropyl-2,6-dimethyldecane.

2.10 (*b*) There are 10 carbon atoms in the ring in this cycloalkane. It is cyclodecane.

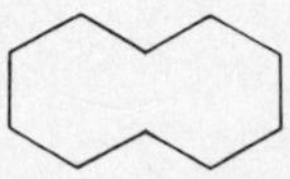

(*c*) In alphabetical order, an isopropyl group and two methyl groups are substituents on a cyclodecane ring. The numbering pattern is chosen so as to give the lowest numbers to the substituents at the first point of difference between them. The correct name is 4-isopropyl-1,1-dimethylcyclodecane.

(CH$_3$)$_2$CH— [cyclodecane ring numbered 1–10 with two CH$_3$ groups at C-1]

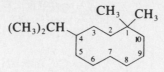

(*d*) In this compound there are two cyclopropyl groups attached to the same terminal carbon of a four-carbon chain. The compound is 1,1-dicyclopropylbutane.

▷—CHCH$_2$CH$_2$CH$_3$ (second cyclopropyl on the CH)

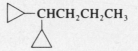

(*e*) There are four carbon atoms in the longest continuous chain in this compound. Cyclopropyl substituents are present at C-1 and C-2. The IUPAC name is 1,2-dicyclopropylbutane.

▷—CH$_2$CHCH$_2$CH$_3$ (second cyclopropyl on the CH)

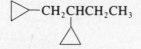

(*f*) When two cycloalkyl groups are attached by a single bond, the compound is named as a cycloalkyl-substituted cycloalkane. This compound is cyclohexylcyclohexane.

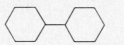

2.11 Hydrogen fluoride (bp 19°C) has a higher boiling point than neon (bp −246°C) because of association due to hydrogen bonding.

$$\overset{\delta+}{H}-\overset{\delta-}{\ddot{\underset{..}{F}}}:\cdots\overset{\delta+}{H}-\overset{\delta-}{\ddot{\underset{..}{F}}}:\cdots\overset{\delta+}{H}-\overset{\delta-}{\ddot{\underset{..}{F}}}:\cdots\overset{\delta+}{H}-\overset{\delta-}{\ddot{\underset{..}{F}}}:$$

2.12 (*b*) While there is no entry in Table 2.3 for tetradecane ($C_{14}H_{30}$), the heats of combustion of dodecane ($C_{12}H_{26}$) and hexadecane ($C_{16}H_{34}$) are given. The heat of combustion of an alkane increases by 156 kcal/mol for each additional CH_2 group. Therefore, the heat of combustion of tetradecane can be calculated from that of dodecane:

$$\text{Heat of combustion of tetradecane} = \begin{array}{l}\text{heat of combustion}\\ \text{of dodecane}\end{array} + 2(156)\text{ kcal/mol}$$

$$= 1933\text{ kcal/mol} + 312\text{ kcal/mol}$$

$$= 2245\text{ kcal/mol}$$

Alternatively, the calculation may be based on the heat of combustion of hexadecane:

$$\text{Heat of combustion of tetradecane} = \begin{array}{l}\text{heat of combustion}\\ \text{of hexadecane}\end{array} - 2(156)\text{ kcal/mol}$$

$$= 2558\text{ kcal/mol} - 312\text{ kcal/mol}$$

$$= 2246\text{ kcal/mol}$$

(*c*) Icosane (Table 2.1) is $C_{20}H_{42}$. It has four more methylene (CH_2) groups than hexadecane, the last entry in Table 2.3. Therefore, its calculated heat of combustion is 4×156 kcal/mol higher.

$$\text{Heat of combustion of icosane} = \begin{array}{l}\text{heat of combustion}\\ \text{of hexadecane}\end{array} + 4(156)\text{ kcal/mol}$$

$$= 2558\text{ kcal/mol} + 624\text{ kcal/mol}$$

$$= 3182\text{ kcal/mol}$$

2.13 Two factors that influence the heats of combustion of alkanes are: (1) the number of carbon atoms; and (2) the extent of chain branching. Pentane, isopentane, and neopentane are all C_5H_{12}; hexane is C_6H_{14}. Hexane has the largest heat of combustion. Branching leads to a lower heat of combustion; neopentane is the most branched and has the lowest heat of combustion.

Hexane	$CH_3(CH_2)_4CH_3$	Heat of combustion 995.0 kcal/mol
Pentane	$CH_3CH_2CH_2CH_2CH_3$	Heat of combustion 845.3 kcal/mol
Isopentane	$(CH_3)_2CHCH_2CH_3$	Heat of combustion 843.4 kcal/mol
Neopentane	$(CH_3)_4C$	Heat of combustion 839.9 kcal/mol

2.14 It is best to approach problems of this type systematically. Since the problem requires all the isomers of C_7H_{16} to be written, begin with the unbranched isomer heptane.

$CH_3CH_2CH_2CH_2CH_2CH_2CH_3$ Heptane

Two isomers have six carbons in their longest continuous chain. One bears a methyl substituent at C-2, the other a methyl substituent at C-3.

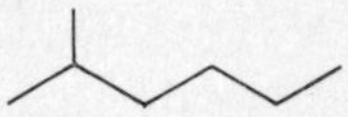

$(CH_3)_2CHCH_2CH_2CH_2CH_3$ 2-Methylhexane

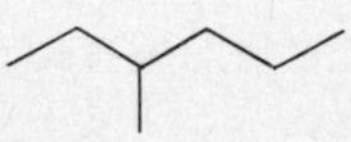

$CH_3CH_2\underset{\displaystyle CH_3}{\underset{|}{C}}HCH_2CH_2CH_3$ 3-Methylhexane

Now consider all the isomers that have two methyl groups as substituents on a five-carbon continuous chain.

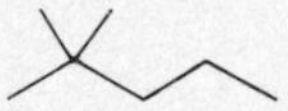

$(CH_3)_3CCH_2CH_2CH_3$ 2,2,-Dimethylpentane

$(CH_3CH_2)_2C(CH_3)_2$ 3,3-Dimethylpentane

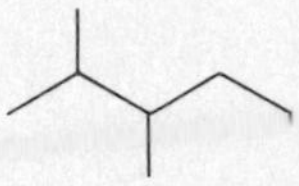

$(CH_3)_2CH\underset{\displaystyle CH_3}{\underset{|}{C}}HCH_2CH_3$ 2,3-Dimethylpentane

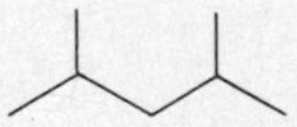

$(CH_3)_2CHCH_2CH(CH_3)_2$ 2,4-Dimethylpentane

There is one isomer characterized by an ethyl substituent on a five-carbon chain.

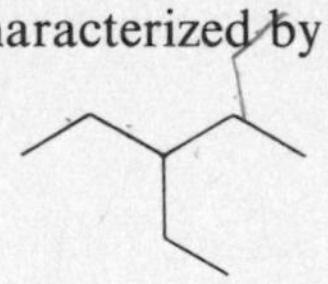

$(CH_3CH_2)_3CH$ 3-Ethylpentane

The remaining isomer has three methyl substituents attached to a four-carbon chain.

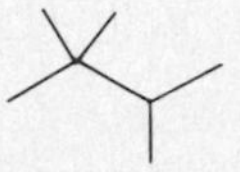

$(CH_3)_3CCH(CH_3)_2$ 2,2,3-Trimethylbutane

2.15 (*a*) The ending -hexane tells us that the longest continuous chain in 3-ethylhexane has six carbons.

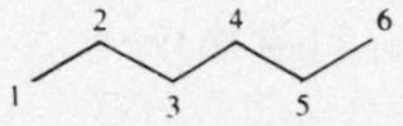

This six-carbon chain bears an ethyl group (CH_3CH_2) at the third carbon.

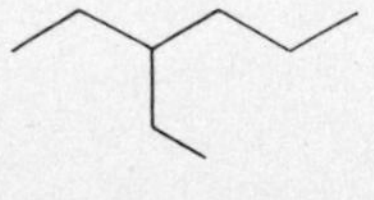

or $(CH_3CH_2)_2CHCH_2CH_2CH_3$

(*b*) The longest continuous chain contains nine carbon atoms. Begin the problem by writing and numbering the carbon skeleton of nonane.

Now add two methyl groups (one to C-2 and the other to C-3) and an isopropyl group (to C-6) to give a structural formula for 2,3-dimethyl-6-isopropylnonane.

or $CH_3\overset{CH_3}{C}H\underset{CH_3}{C}HCH_2CH_2\overset{CH(CH_3)_2}{C}HCH_2CH_2CH_3$

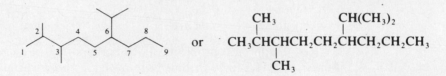

(*c*) To the carbon skeleton of heptane (seven carbons) add a *tert*-butyl group to C-4 and a methyl group to C-3 to give 4-*tert*-butyl-3-methylheptane.

or $CH_3CH_2\underset{CH_3}{C}H\overset{C(CH_3)_3}{C}HCH_2CH_2CH_3$

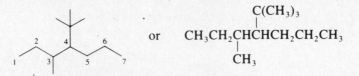

(*d*) An isobutyl group is $-CH_2CH(CH_3)_2$. The structure of 4-isobutyl-1,1-dimethylcyclohexane is as shown.

or $(CH_3)_2$-cyclohexane-$CH_2CH(CH_3)_2$

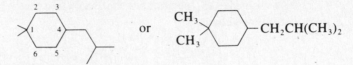

(*e*) A *sec*-butyl group is $CH_3\underset{|}{C}HCH_2CH_3$. *sec*-Butylcycloheptane has a *sec*-butyl group on a seven-membered ring.

or $CH_3CHCH_2CH_3$

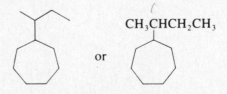

(*f*) Dicyclopropylmethane has two cyclopropyl groups as substituents on the same carbon.

$-CH_2-$

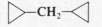

(*g*) A cyclobutyl group is a substituent on a five-membered ring in cyclobutylcyclopentane.

2.16 Isomers are different compounds that have the same molecular formula. In all these problems the safest approach is to write a structural formula, then count the number of carbons and hydrogens.

(*a*) Among this group of compounds, only butane and isobutane have the same molecular formula; only these two are isomers.

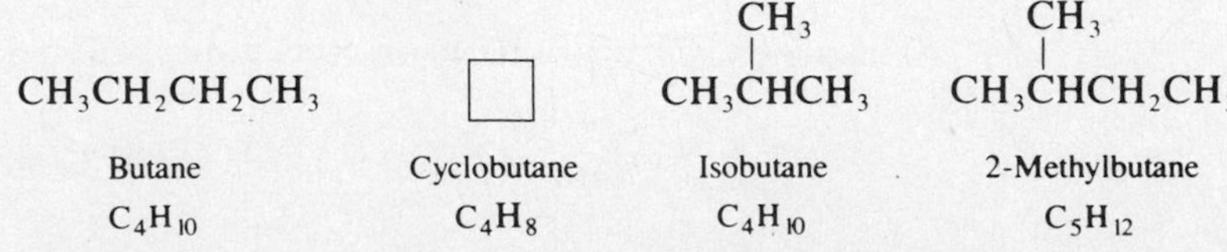

Butane	Cyclobutane	Isobutane	2-Methylbutane
C_4H_{10}	C_4H_8	C_4H_{10}	C_5H_{12}

(*b*) The two compounds that are isomers, i.e., those that have the same molecular formula, are 2,2-dimethylpentane and 2,2,3-trimethylbutane.

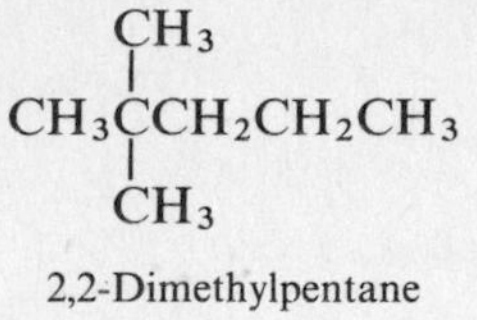

$$\begin{array}{c} CH_3 \\ | \\ CH_3CCH_2CH_2CH_3 \\ | \\ CH_3 \end{array} \qquad \begin{array}{c} CH_3\ \ CH_3 \\ |\quad\ \ | \\ CH_3C{-}{-}CHCH_3 \\ | \\ CH_3 \end{array}$$

2,2-Dimethylpentane C_7H_{16} — 2,2,3-Trimethylbutane C_7H_{16}

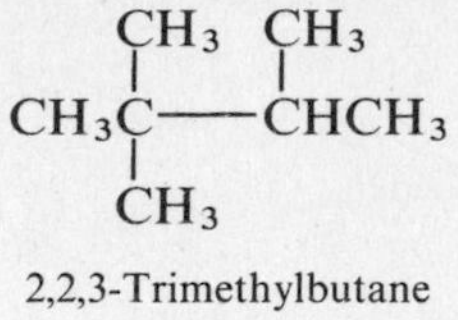

Cyclopentane and neopentane are not isomers of the above compounds nor are they isomers of each other.

$$\begin{array}{c} CH_3 \\ | \\ CH_3CCH_3 \\ | \\ CH_3 \end{array}$$

Cyclopentane C_5H_{10} — Neopentane C_5H_{12}

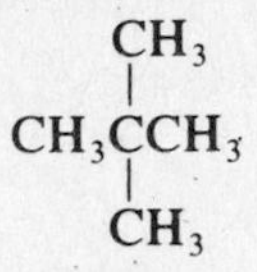

(*c*) The compounds that are isomers are cyclohexane, methylcyclopentane, and 1,1,2-trimethylcyclopropane.

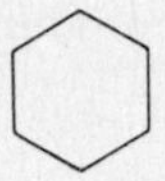

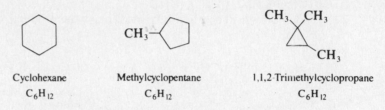

Cyclohexane C_6H_{12} — Methylcyclopentane (CH_3–) C_6H_{12} — 1,1,2-Trimethylcyclopropane (CH_3, CH_3, CH_3) C_6H_{12}

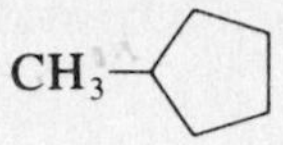

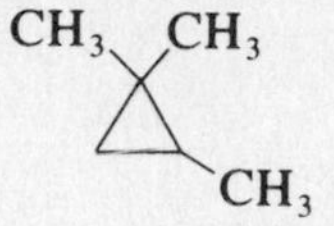

Hexane, $CH_3CH_2CH_2CH_2CH_2CH_3$, has the molecular formula C_6H_{14}; it is not an isomer of the others.

(*d*) The three that are isomers all have the molecular formula C_5H_{10}.

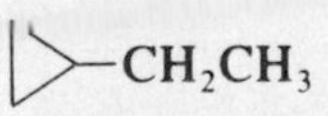

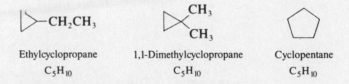

Ethylcyclopropane ($-CH_2CH_3$) C_5H_{10} — 1,1-Dimethylcyclopropane (CH_3, CH_3) C_5H_{10} — Cyclopentane C_5H_{10}

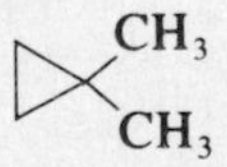

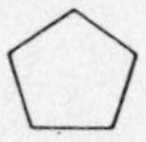

1-Cyclopropylpropane is not an isomer of the others. Its molecular formula is C_6H_{12}.

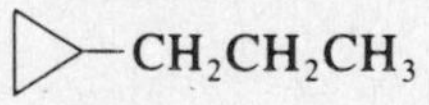

$$\triangleright{-}CH_2CH_2CH_3$$

(*e*) Only 4-methyltetradecane and pentadecane are isomers. Both have the molecular formula $C_{15}H_{32}$.

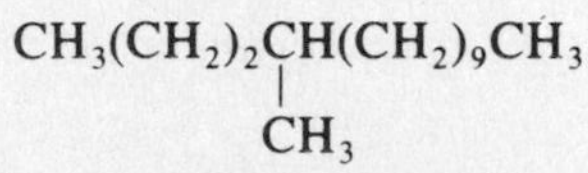

$$\begin{array}{c} CH_3(CH_2)_2CH(CH_2)_9CH_3 \\ | \\ CH_3 \end{array} \qquad CH_3(CH_2)_{13}CH_3$$

4-Methyltetradecane $C_{15}H_{32}$ — Pentadecane $C_{15}H_{32}$

$$\begin{array}{c} CH_3\ \ CH_3 \\ |\qquad | \\ CH_3CHCHCHCH(CH_2)_4CH_3 \\ |\qquad | \\ CH_3\ \ CH_3 \end{array} \qquad \begin{array}{c} CH_3CH_2CH_2CH(CH_2)_5CH_3 \\ | \\ \diamond \end{array}$$

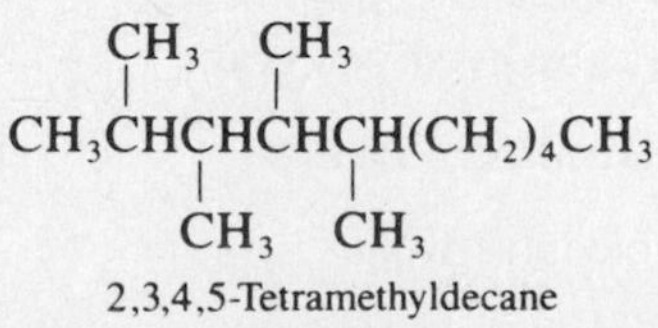

2,3,4,5-Tetramethyldecane $C_{14}H_{30}$ — 4-Cyclobutyldecane $C_{14}H_{28}$

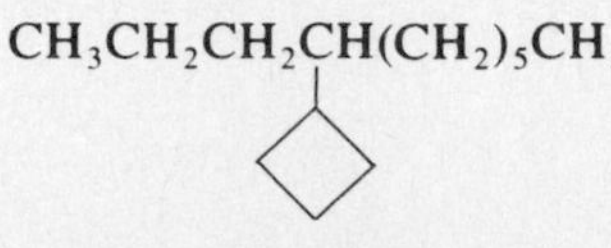

2.17 Alkanes are characterized by the molecular formula C_nH_{2n+2}. The value of n can be calculated on the basis of the fact that the molecular weight is 240, and the atomic weights of carbon and hydrogen are 12 and 1, respectively.

$$12n + 1(2n + 2) = 240$$
$$14n = 238$$
$$n = 17$$

The molecular formula of the alkane is $C_{17}H_{36}$. Since the problem specifies that the carbon chain is unbranched, the hydrocarbon is heptadecane, $CH_3(CH_2)_{15}CH_3$.

2.18 Since it is an alkane, the sex attractant of the tiger moth has a molecular formula of C_nH_{2n+2}. The number of carbons and hydrogens may be calculated from its molecular weight.

$$12n + 1(2n + 2) = 254$$
$$14n = 252$$
$$n = 18$$

The molecular formula of the alkane is $C_{18}H_{38}$. In the problem it is stated that the sex attractant is a 2-methyl-branched alkane. It is therefore 2-methylheptadecane, $(CH_3)_2CHCH_2(CH_2)_{13}CH_3$.

2.19 When any hydrocarbon is burned in air, the products of combustion are carbon dioxide and water.

(*a*) $CH_3(CH_2)_8CH_3 + \frac{31}{2}O_2 \longrightarrow 10\ CO_2 + 11\ H_2O$

Decane Oxygen Carbon dioxide Water

(*b*) [Cyclodecane] $+ 15\ O_2 \longrightarrow 10\ CO_2 + 10\ H_2O$

Cyclodecane Oxygen Carbon dioxide Water

(*c*) [Methylcyclononane]$-CH_3 + 15\ O_2 \longrightarrow 10\ CO_2 + 10\ H_2O$

Methylcyclononane Oxygen Carbon dioxide Water

(*d*) [Cyclopentylcyclopentane] $+ \frac{29}{2}O_2 \longrightarrow 10\ CO_2 + 9\ H_2O$

Cyclopentylcyclopentane Oxygen Carbon dioxide Water

2.20 The heats of combustion in Table 2.3 are given in kilocalories per mole. To determine the quantity of heat evolved per unit mass of material, divide the heat of combustion by the molecular weight.

Methane	Heat of combustion	= 212.8 kcal/mol
	Molecular weight	= 16 g/mol
	Heat evolved per gram	= 13.2 kcal/g
Butane	Heat of combustion	= 687.4 kcal/mol
	Molecular weight	= 58 g/mol
	Heat evolved per gram	= 11.8 kcal/g

When equal masses of methane and butane are compared, methane evolves more heat when it is burned.

Equal volumes of gases contain an equal number of moles, so when equal volumes of methane and butane are compared, the one with the greater heat of combustion in kilocalories per mole gives off more heat. Butane evolves more heat when it is burned than does an equal volume of methane.

2.21 When comparing heats of combustion of alkanes, two factors are of importance.

1. The heats of combustion of alkanes increase with increasing molecular weight.
2. An unbranched alkane has a greater heat of combustion than a branched isomer of equal molecular weight.

(*a*) In the group hexane, heptane, and octane, three unbranched alkanes are being compared. Octane (C_8H_{18}) has the greatest molecular weight and has the greatest heat of combustion. Hexane (C_6H_{14}) has the lowest molecular weight and the lowest heat of combustion. The measured values in this group are given below:

Hexane	Heat of combustion 995.0 kcal/mol
Heptane	Heat of combustion 1151.3 kcal/mol
Octane	Heat of combustion 1307.5 kcal/mol

(*b*) Isobutane has a lower molecular weight than either pentane or isopentane and so is the member of the group with the lowest heat of combustion. Isopentane is a 2-methyl-branched isomer of pentane and so has a lower heat of combustion. Pentane has the highest heat of combustion among these compounds.

Isobutane	$(CH_3)_3CH$	Heat of combustion 685.4 kcal/mol
Isopentane	$(CH_3)_2CHCH_2CH_3$	Heat of combustion 843.4 kcal/mol
Pentane	$CH_3CH_2CH_2CH_2CH_3$	Heat of combustion 845.3 kcal/mol

(*c*) Isopentane and neopentane each have a lower molecular weight than 2-methylpentane, which therefore has the greatest heat of combustion. Neopentane is more highly branched than isopentane; neopentane has the lowest heat of combustion.

Neopentane	$(CH_3)_4C$	Heat of combustion 839.9 kcal/mol
Isopentane	$(CH_3)_2CHCH_2CH_3$	Heat of combustion 843.4 kcal/mol
2-Methylpentane	$(CH_3)_2CHCH_2CH_2CH_3$	Heat of combustion 993.6 kcal/mol

(*d*) Chain branching has a small effect on heat of combustion; molecular weight has a much larger effect. The highest-molecular-weight alkane in this group is 3,3-dimethylpentane; it has the greatest heat of combustion. Pentane, the lowest molecular weight alkane, has the smallest heat of combustion.

Pentane	$CH_3CH_2CH_2CH_2CH_3$	Heat of combustion 845.3 kcal/mol
3-Methylpentane	$(CH_3CH_2)_2CHCH_3$	Heat of combustion 994.1 kcal/mol
3,3-Dimethylpentane	$(CH_3CH_2)_2C(CH_3)_2$	Heat of combustion 1148.3 kcal/mol

(*e*) In this series the heat of combustion increases with increasing molecular weight. Ethylcyclopentane has the lowest heat of combustion; ethylcycloheptane has the greatest.

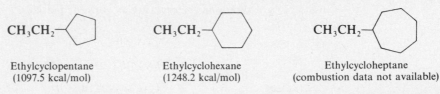

CH_3CH_2–(cyclopentyl)
Ethylcyclopentane
(1097.5 kcal/mol)

CH_3CH_2–(cyclohexyl)
Ethylcyclohexane
(1248.2 kcal/mol)

CH_3CH_2–(cycloheptyl)
Ethylcycloheptane
(combustion data not available)

2.22 The problem specifies that the second-row element is sp^3 hybridized in each of the compounds. Therefore, any unshared electron pairs occupy sp^3 hybridized orbitals and bonded pairs are located in σ bonds.

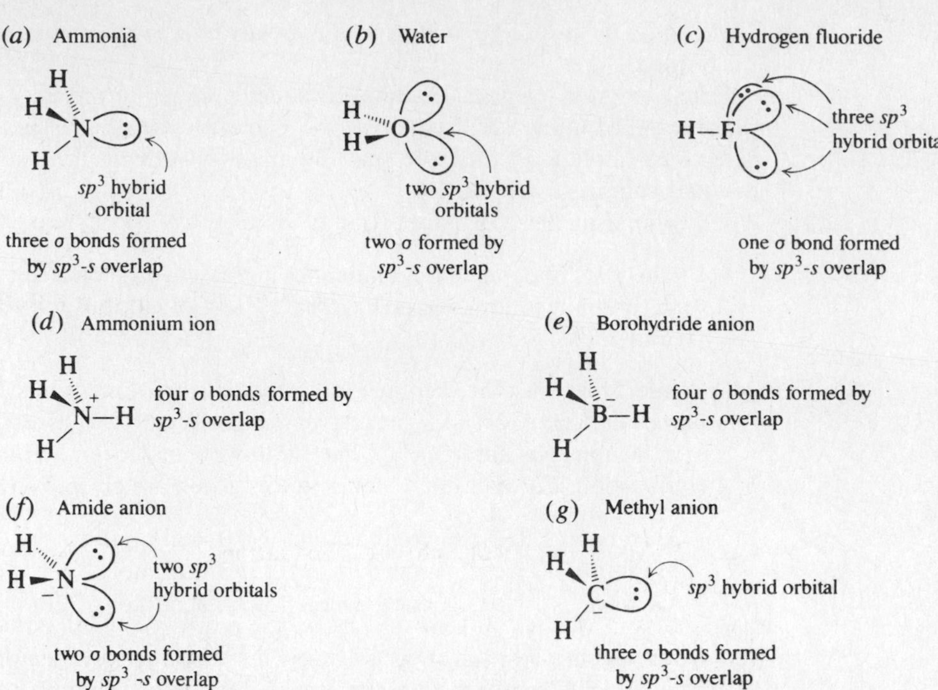

2.23 A bonding interaction exists when two orbitals overlap "in phase" with each other, that is, when the algebraic signs of their wave functions are the same in the region of overlap. The orbital shown below is a bonding orbital. It involves overlap of an *s* orbital with the lobe of a *p* orbital of the same sign.

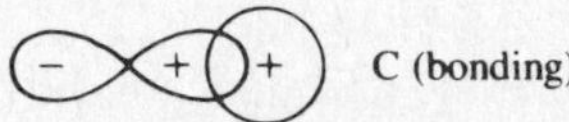

On the other hand, the overlap of an *s* orbital with the lobe of a *p* orbital of opposite sign is antibonding.

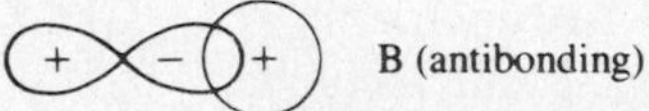

Overlap in the manner shown below is nonbonding. Both the positive lobe and the negative lobe of the *p* orbital overlap with the spherically symmetric *s* orbital. The bonding overlap between the *s* orbital and one lobe of the *p* orbital is exactly canceled by an antibonding interaction between the *s* orbital and the lobe of opposite sign.

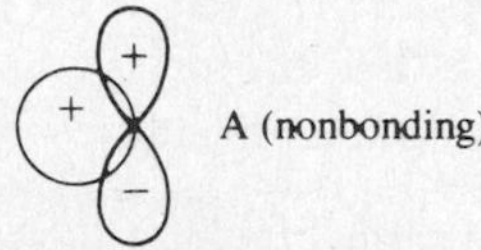

2.24 The hydrogens that are added so as to give a tetrahedral orientation of the C—H bonds of methane are shown in the drawing as darkened circles.

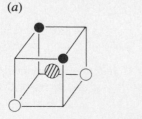

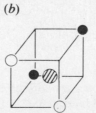

SELF-TEST

PART A

A-1. Write the structure and give the common name of each of the four-carbon alkyl groups.

A-2. (*a*) Write a structural formula for 3-isopropyl-2,4-dimethylpentane.
(*b*) How many methyl groups are there in this compound? How many isopropyl groups?

A-3. Give the systematic (i.e., IUPAC) name for the substance shown below.

A-4. The compound in the previous question contains _____ primary carbon(s), _____ secondary carbon(s), and _____ tertiary carbons.

A-5. Write a balanced chemical equation for the complete combustion of 2,3-dimethylpentane.

A-6. Write structural formulas and give the names of all the constitutional isomers of C_5H_{10} that contain a ring.

A-7. Draw the structure indicated by the name 2,3-dimethyl-3-propylpentane. Is that name the correct one for this substance? If not, write the correct name.

A-8. Draw the structure of the octane isomer which has the highest boiling point.

PART B

B-1. Choose the response which best describes the compounds shown below:

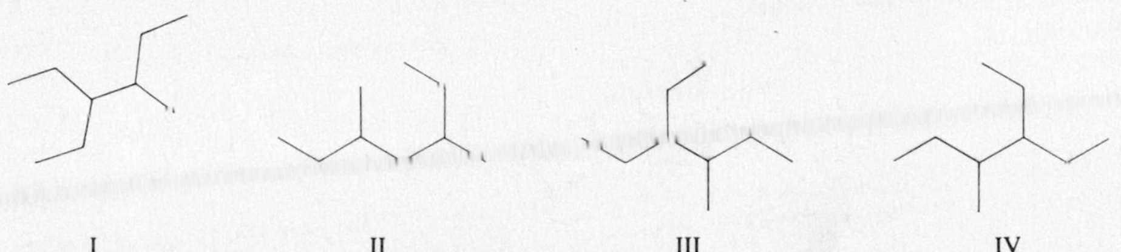

I II III IV

(*a*) I, III, and IV represent the same compound.
(*b*) I and III are isomers of II and IV.
(*c*) I and IV are isomers of II and III.
(*d*) All the structures represent the same compound.

B-2. The molecular orbital electronic configuration of Be_2 is $(\sigma_{1s})^2\ (\sigma^*_{1s})^2\ (\sigma_{2s})^2\ (\sigma^*_{2s})^2$. One would predict that Be_2 would be held together by
(*a*) One bond (*b*) Two bonds (*c*) Four bonds (*d*) No bonds

B-3. Which of the following is a correct name according to the IUPAC rules?
(*a*) 2-Methylcyclohexane
(*b*) 3,4-Dimethylpentane
(*c*) 2-Ethyl-2-methylpentane
(*d*) 3-Ethyl-2-methylpentane

B-4. Following are the structures of four isomers of hexane. Which of the names given will correctly identify an additional isomer?

$CH_3CH_2CH_2CH_2CH_2CH_3$ $(CH_3)_3CCH_2CH_3$
$(CH_3)_2CHCH_2CH_2CH_3$ $(CH_3)_2CHCH(CH_3)CH_3$

(*a*) 2-Methylpentane (*b*) 2,3-Dimethylbutane
(*c*) 2-Ethylbutane (*d*) 3-Methylpentane

B-5. Which of the following structures is termed an "isopentyl" group?

(*a*) $CH_3CH_2CH_2CH_2CH_2—$ (*b*) $(CH_3)_2CHCH_2CH_2—$
(*c*) $CH_3CH_2\underset{\displaystyle CH_2CH_3}{\underset{|}{C}}H—$ (*d*) $(CH_3)_3CCH_2—$

B-6. Rank the following substances in decreasing order of heats of combustion (largest → smallest)

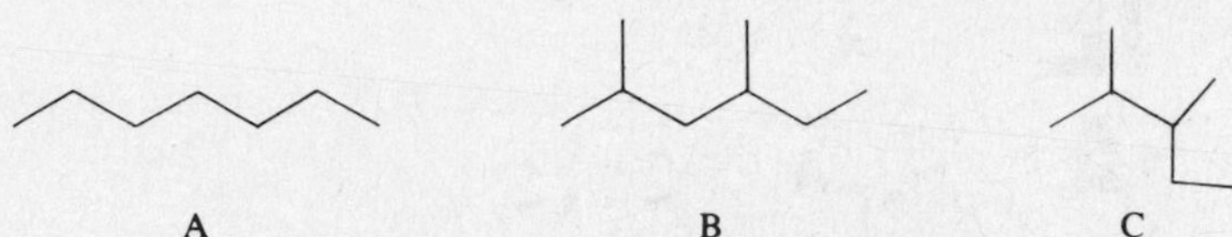

(*a*) B > A > C (*b*) B > C > A
(*c*) C > A > B (*d*) C > B > A

B-7. What is the total number of σ bonds present in the molecule shown?

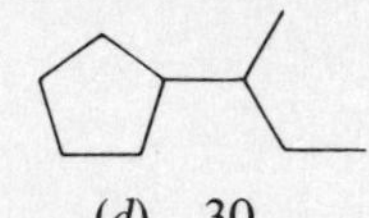

(*a*) 18 (*b*) 26 (*c*) 27 (*d*) 30

B-8. Which of the following substances is *not* an isomer of 3-ethyl-2-methylpentane?

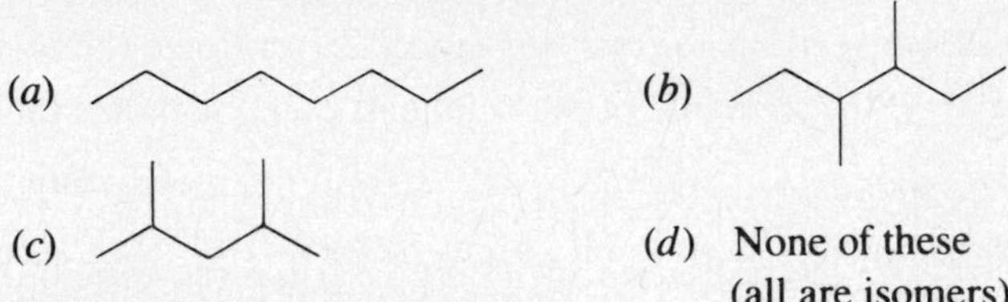

(*d*) None of these (all are isomers)

B-9. Which of the following isomers has the weakest attractive van der Waals forces in the liquid state?

(*a*) $(CH_3)_3CCH_2CH_3$ (*b*) $CH_3CH_2CH_2CH(CH_3)_2$
(*c*) $CH_3CH_2CH_2CH_2CH_2CH_3$ (*d*) The attractive van der Waals forces of all isomers are the same.

CHAPTER

3

CONFORMATIONS OF ALKANES AND CYCLOALKANES

IMPORTANT TERMS AND CONCEPTS

Conformations (Sec. 3.1) Structures that differ by rotation around single bonds are said to have different conformations. The study of this subject is known as *conformational analysis.* As shown in text Figure 3.1 for the molecule ethane, three types of representations are used to represent conformations; they are *wedge and dash* formulas, *sawhorse* drawings, and *Newman projections.*

Ethane Conformations (Secs. 3.1, 3.2) Of the many possible ethane conformations, those having minimum and maximum energy, respectively, are known as *staggered* and *eclipsed.* The angle between adjacent C—H bonds, known as the *torsion angle*, is 60° in the staggered conformation and 0° in the eclipsed conformation.

A *potential energy diagram*, as shown in text Figure 3.3, may be used to depict the energy changes as a molecule of ethane undergoes carbon-carbon bond rotation. As may be seen from the diagram, the staggered conformation is *more stable* than the eclipsed by 2.9 kcal/mol. This difference is due to *torsional strain* in the eclipsed form resulting from repulsion of bonds on adjacent atoms. The difference in energy between the staggered and the eclipsed conformations is termed the *rotational energy barrier* for ethane.

Conformational Analysis of Butane and Higher Alkanes (Secs. 3.3, 3.4) Butane, when viewed looking down the central bond (between C-2 and C-3), has two distinct staggered conformations known as *gauche* and *anti.* In the former the torsion angle between the methyl groups is 60°; in the latter it is 180°. The two eclipsed conformations differ as to whether a methyl group is eclipsed with another methyl or with a hydrogen. The relative energies of the butane conformations may be seen in text Figure 3.4. Note that both the gauche and anti conformations are energy minima, with anti being more stable by 0.8 kcal/mol. This energy difference may be accounted for by *van der Waals strain* between the methyl groups, that is, their electrons repel one another.

Higher alkanes are most stable in the all-anti conformation, which is the normal conformation adopted when alkanes crystallize. More efficient packing is possible than would occur with other conformations.

Cycloalkanes (Sec. 3.6) Deviation of the C—C—C bond angles from the ideal tetrahedral value in cyclic hydrocarbons gives rise to destabilization resulting from *angle strain.* This phenomenon is most significant in cyclopropane, which is the only cycloalkane in which all the carbon atoms lie in a plane.

Cyclohexane Conformations (Secs. 3.7 to 3.9) Two of the nonplanar conformations of cyclohexane are named the *chair* and the *boat.* While both conformations have bond angles close to tetrahedral, the chair conformation is more stable than the boat because, in addition to being free of angle strain, it is free of torsional strain as well. A third conformation, the *twist* or *skew boat,* is less stable than the chair but more stable than the boat.

The carbon-hydrogen bonds in cyclohexane may be classified as being of one of two types, *axial* and *equatorial.* As shown in text Figure 3.9, the axial C—H bonds point alternately straight up and down from the ring. The equatorial bonds on the other hand, alternate slightly up and down around the "equator" of the ring.

By a process of *conformational inversion,* also known as *ring flipping,* the axial and equatorial positions on the ring may be interchanged, that is, a group which is axial becomes equatorial, and vice versa. To draw the ring-flipping process, carbon atoms pointing "up" in the chair cyclohexane are moved "down"; those pointing "down" are moved "up."

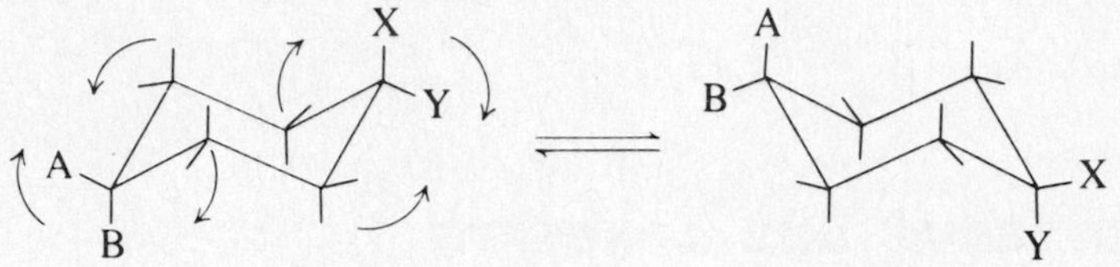

Substituted Cyclohexanes (Secs. 3.10, 3.11) In a monosubstituted cyclohexane, in which one hydrogen is replaced by a substituent such as a methyl group, the two chair conformations are no longer equivalent. In one the methyl substituent is axial, in the other it is equatorial.

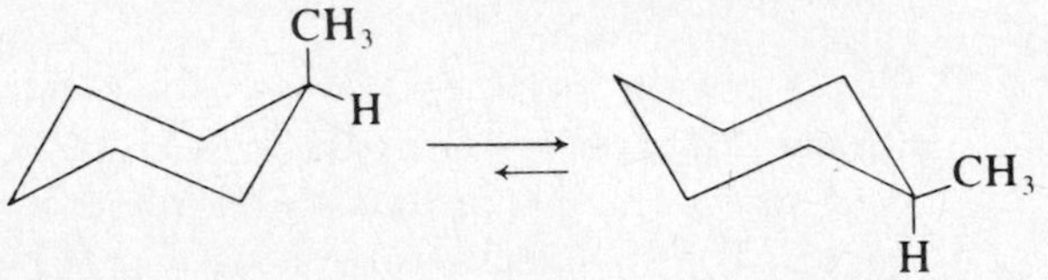

Axial methyl — Equatorial methyl

The equatorial orientation of the methyl group is more stable and is the predominant form present (about 95 percent). Fewer van der Waals repulsions exist when the methyl group is equatorial, which results in that conformation being more stable.

In general, *any* alkyl substituent is more stable in the equatorial orientation. As the group becomes bulkier, the preference for this orientation increases, being particularly pronounced for a *tert*-butyl group. At equilibrium the amount of axial *tert*-butylcyclohexane is too small to measure.

In disubstituted cyclohexanes the axial and equatorial orientations of each alkyl group must be determined in order to evaluate the relative stability of the two ring conformers, remembering that groups are more stable in the equatorial position. Groups on the same side of the ring are said to be *cis*; those on opposite sides are *trans* to each other. The possible orientations of alkyl disubstituted cyclohexanes (where *a* denotes axial and *e* denotes equatorial) are:

1,2-cis	*a, e* (larger group *e*)
1,2-trans	*e, e* favored over *a, a*
1,3-cis	*e, e* favored over *a, a*
1,3-trans	*a, e* (larger group *e*)
1,4-cis	*a, e* (larger group *e*)
1,4-trans	*e, e* favored over *a, a*

Polycyclic Ring Systems (Sec. 3.15) When two or more atoms of a cycloalkane are common to more than one ring, the compounds are called *polycyclic*; they may be *bicyclic*, *tricyclic*, etc. When a single carbon atom is common to two rings, the substance is called a *spirocyclic* substance.

Bicylopentane Spiropentane

The IUPAC name for a bicyclic substance is assigned by identifying the atoms common to more than one ring and naming the number of atoms in each "bridge" connecting these carbons. For example,

Bicyclo[4.3.0]nonane Bicyclo[2.2.2]octane

where the asterisks denote bridgehead carbons.

Among the most important bicyclic hydrocarbons are *cis*- and *trans*-decalin. These ring systems are frequently found in natural substances such as steroids.

cis-Decalin *trans*-Decalin

SOLUTIONS TO TEXT PROBLEMS

3.1 (*b*) The sawhorse formula contains four carbon atoms in an unbranched chain. The compound is butane, $CH_3CH_2CH_2CH_3$.

(*c*) Rewrite the structure so as to better show its constitution. The compound is $CH_3CH_2CH(CH_3)_2$; it is 2-methylbutane (isopentane).

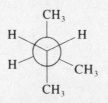

(*d*) In this structure, we are sighting down the C-3—C-4 bond of a six-carbon chain. It is $CH_3CH_2CH_2\underset{|}{C}HCH_2CH_3$ (with CH_3 on the branching carbon), or 3-methylhexane.

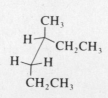

3.2 All the staggered conformations of propane are equivalent to each other and all its eclipsed conformations are equivalent to each other. The energy diagram resembles that of ethane in that it is a symmetrical one.

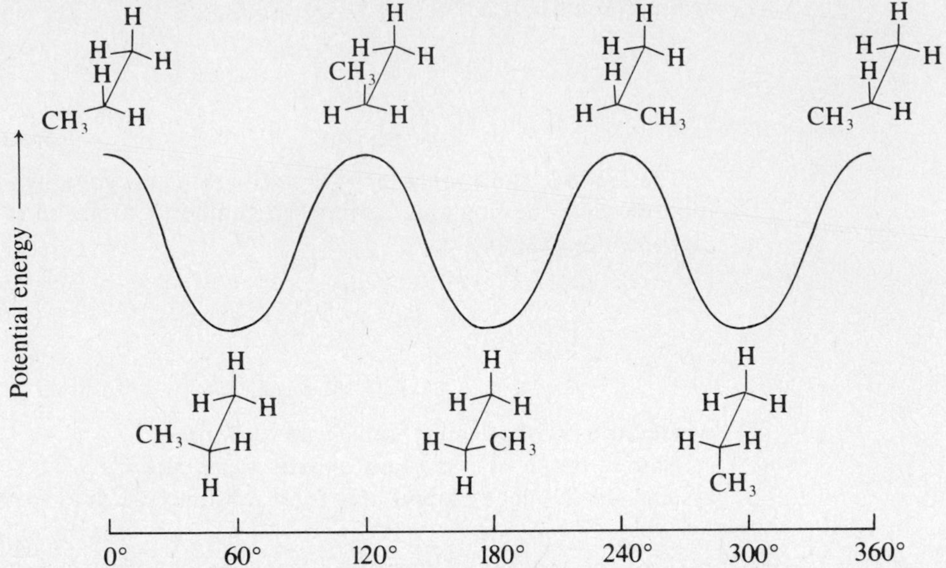

The activation energy for internal rotation in propane is expected to be somewhat higher than that in ethane because its eclipsed conformation incorporates an unfavorable van der Waals interaction between the methyl group and an eclipsed hydrogen. This interaction is, however, less repulsive than the corresponding methyl-methyl interaction of butane, which makes the activation energy for internal rotation less for propane than for butane.

3.3 The formula for the ring angles of a regular polygon is given in the problem as:

$$\frac{(n-2)}{(n)}(180°)$$

where n is the number of sides.
For cyclododecane, $n = 12$ and:

$$\frac{(n-2)}{(n)}(180°) = \frac{10}{12}(180°) = 150°$$

3.4 (*b*) In order to be gauche, substituents X and A must be related by a 60° torsion angle. Therefore, if A is axial as specified in the problem, X must be equatorial.

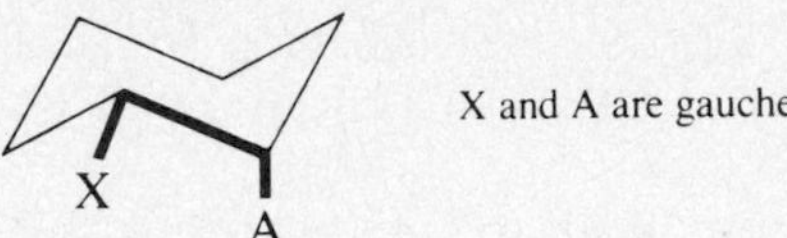

(*c*) In order for substituent X at C-1 to be anti to C-3, it must be equatorial.

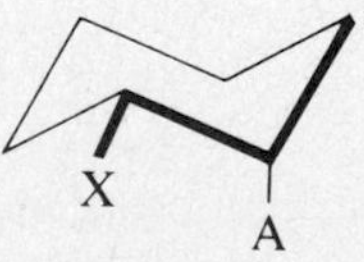

(*d*) When X is axial at C-1, it is gauche to C-3.

X

3

A

3.5 (*b*) A methyl group is axial when it is "up" at C-1 but is equatorial when it is up at C-4 according to the numbering scheme given in the problem. Since substituents are more stable when they occupy equatorial rather than axial sites, a methyl group that is up at C-1 is less stable than one which is up at C-4.

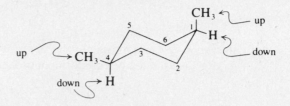

(*c*) An equatorial substituent at C-3 is "down."

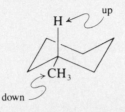

3.6 A *tert*-butyl group is much larger than a methyl group and has a greater preference for the equatorial position. The most stable conformation of 1-*tert*-butyl-1-methylcyclohexane has an axial methyl group and an equatorial *tert*-butyl group.

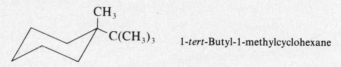

1-*tert*-Butyl-1-methylcyclohexane

3.7 When comparing two stereoisomeric cyclohexane derivatives, the more stable stereoisomer is the one with the greater number of its substituents in equatorial orientations. Rewrite the structures as chair conformations in order to see which substituents are axial and which are equatorial.

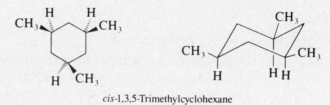

cis-1,3,5-Trimethylcyclohexane

All methyl groups are equatorial in *cis*-1,3,5-trimethylcyclohexane. It is more stable than *trans*-1,3,5-trimethylcyclohexane (shown below), which has one axial methyl group in its most stable conformation.

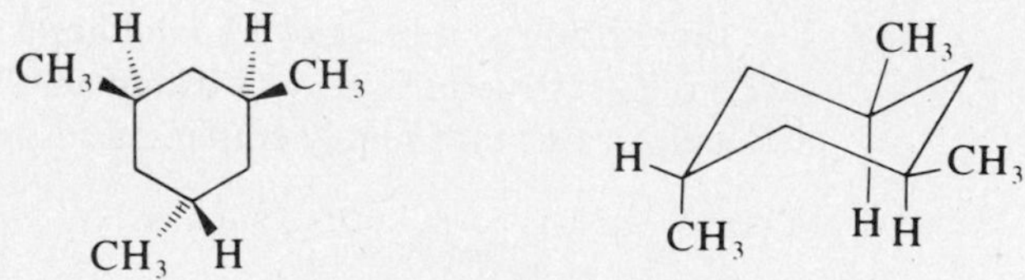

trans-1,3,5-Trimethylcyclohexane

3.8 In each of these problems, a *tert*-butyl group is the larger substituent and will be equatorial in the most stable conformation. Draw a chair conformation of cyclohexane, add an equatorial *tert*-butyl group, then add the remaining substituent so as to give the required *cis* or *trans* relationship to the *tert*-butyl group.

(*b*) In *cis*-1-*tert*-butyl-3-methylcyclohexane both substituents are up; therefore the C-3 methyl group is equatorial.

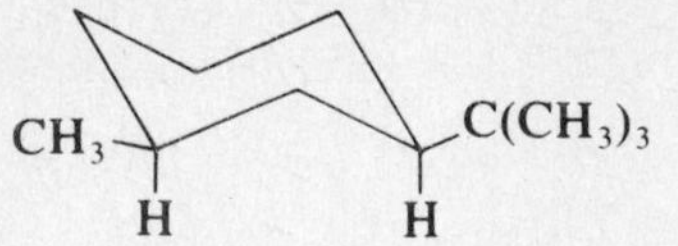

CH_3 H H $C(CH_3)_3$

(*c*) In *trans*-1-*tert*-butyl-4-methylcyclohexane the *tert*-butyl group is up; therefore the C-4 methyl group must be equatorial in order to be down.

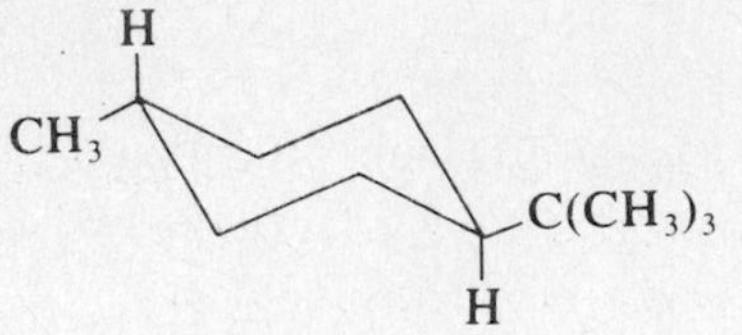

H CH_3 $C(CH_3)_3$ H

(*d*) In *cis*-1-*tert*-butyl-4-methylcyclohexane both groups are up; therefore the C-4 methyl group is axial.

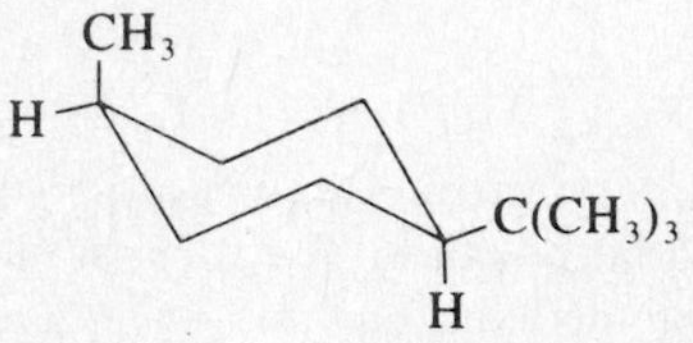

CH_3 H $C(CH_3)_3$ H

3.9 The trans stereoisomer has a lower heat of combustion than *cis*-1,2-dimethylcyclopropane, which indicates that it is more stable. The two methyl groups are close together in the cis isomer and it is the van der Waals repulsion between them that increases the energy content of this stereoisomer.

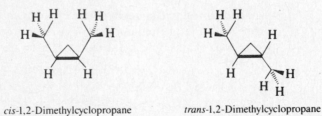

cis-1,2-Dimethylcyclopropane

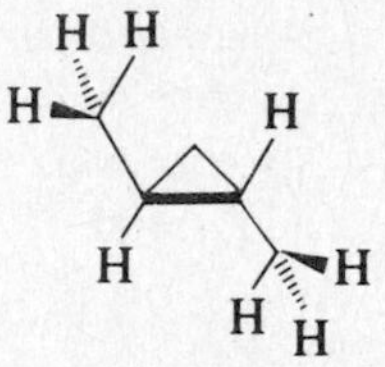

trans-1,2-Dimethylcyclopropane

3.10 (*b*) This bicyclic compound contains nine carbon atoms. The two carbons that are common to both rings are spanned by a five-carbon bridge and a two-carbon bridge. The 0 in the name bicyclo[5.2.0]nonane tells us that the third bridge has no atoms in it—the carbons that are common to both rings are directly attached to each other.

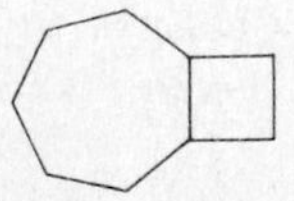

Bicyclo[5.2.0]nonane

(*c*) The three bridges in bicyclo[3.1.1]heptane contain three carbons, one carbon, and one carbon. The structure can be written in a form which shows the actual shape of the molecule or one that simply emphasizes its constitution.

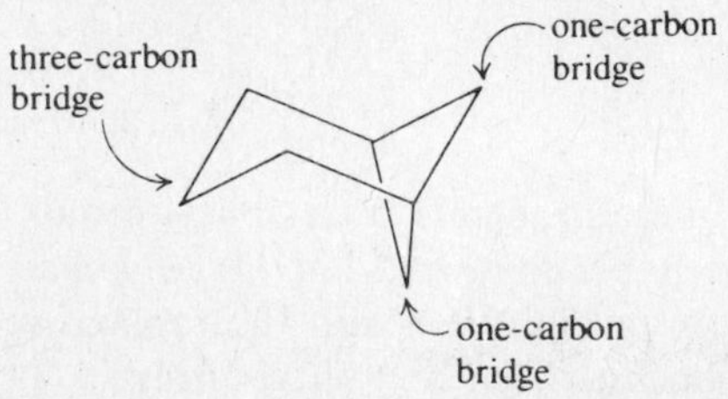

three-carbon bridge

one-carbon bridge

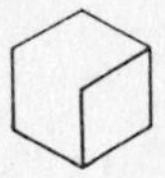

one-carbon bridge

(*d*) Bicyclo[3.3.0]octane has two five-membered rings that share a common side.

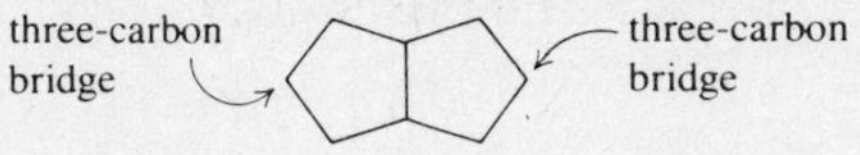

three-carbon bridge

three-carbon bridge

3.11 (*a*) The four atoms of hydrogen peroxide can be coplanar, as shown in the anti and the eclipsed conformations.

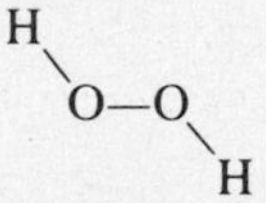

Anti

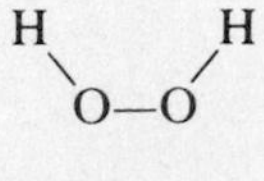

Eclipsed

(*b*) Recall that a neutral nitrogen atom has three covalent bonds and an unshared electron pair. The three bonds are arranged in a trigonal pyramidal manner around each nitrogen in hydrazine (H_2NNH_2).

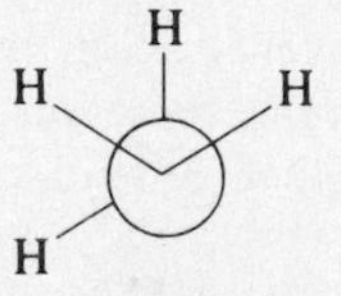

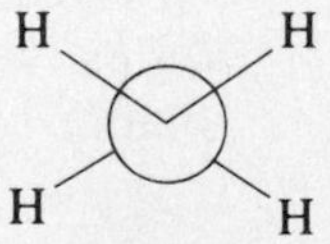

3.12 The most stable conformations of alkanes are the staggered ones. There are two different staggered conformations of 2-methylbutane.

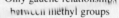

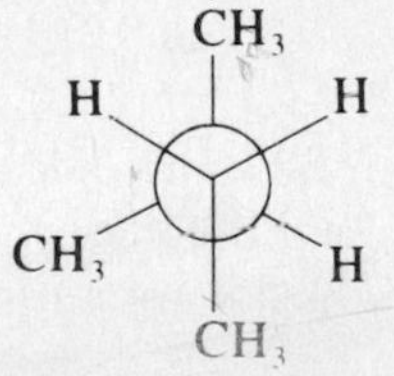

Most stable conformation; two of the methyl groups are anti

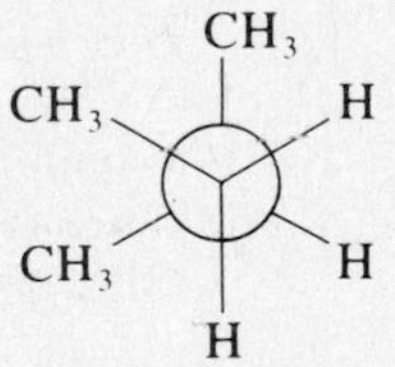

Only gauche relationships between methyl groups

3.13 The structure given is not the most stable conformation because the bonds of the methyl group are eclipsed with those of the ring carbon to which it is attached. The most stable conformation has the bonds of the methyl group and its attached carbon in a staggered relationship.

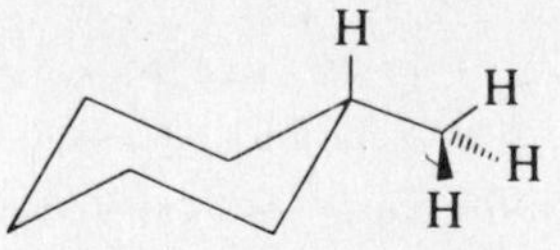

Bonds of methyl group eclipsed with those of attached carbon

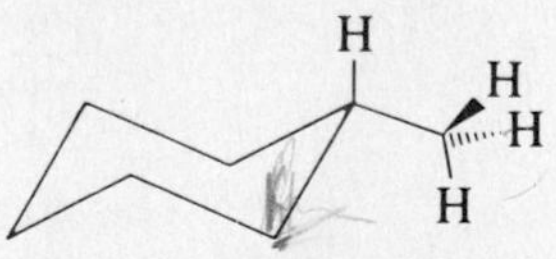

Bonds of methyl group staggered with those of attached carbon

3.14 Structure A has its methyl group eclipsed with the ring bonds and is less stable than B. The methyl group in structure B has its bonds and those of its attached ring carbon in a staggered relationship.

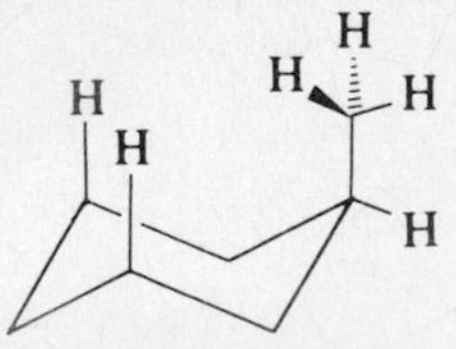

A (less stable)

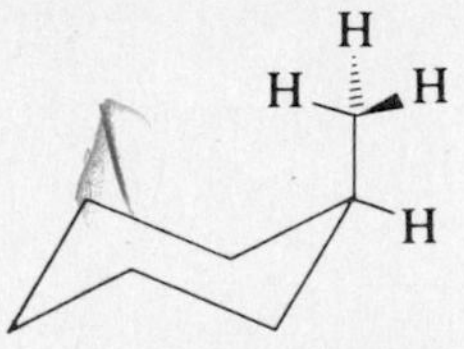

B (more stable)

Further, two of the hydrogens of the methyl group of A are uncomfortably close to two axial hydrogens of the ring.

3.15 (*a*) By rewriting the structures in a form that shows the order of their atomic connections, it is apparent that the two structures are constitutional isomers.

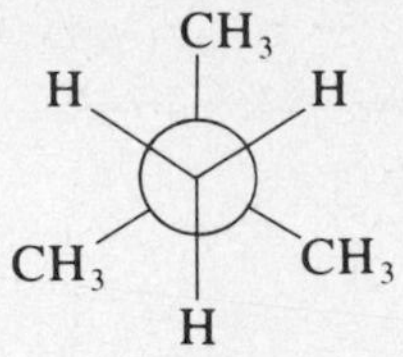

is equivalent to $CH_3C(CH_3)_3$ (neopentane)

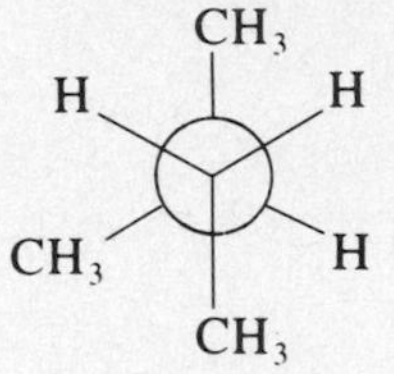

is equivalent to $CH_3CH_2CH(CH_3)_2$ (isopentane)

(*b*) Both structural formulas represent hexane. They are different conformations related by rotation about the C-2–C-3 bond and the C-3–C-4 bond.

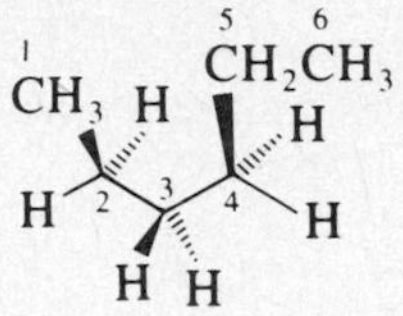

rotate about C-2–C-3 bond

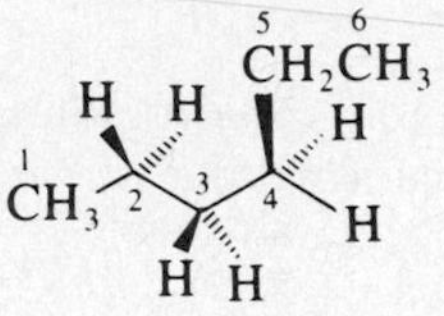

rotate about C-3–C-4 bond

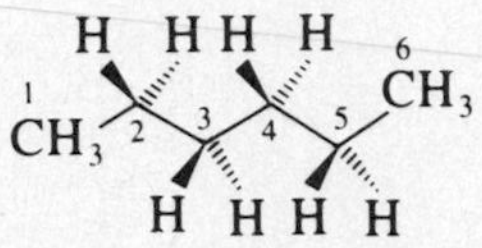

(*c*) The two compounds have the same constitution; both are $(CH_3)_2CHCH(CH_3)_2$. The Newman projections represent different staggered conformations of the same molecule: in one the hydrogen substituents are anti to each other, whereas in the other they are gauche.

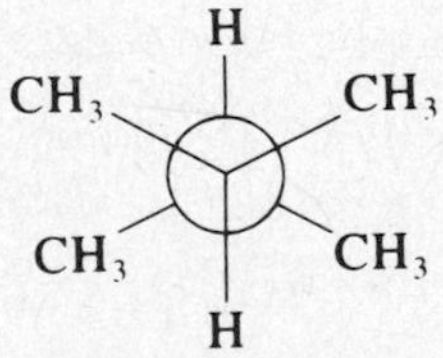

and

are different conformations of 2,3-dimethylbutane

Hydrogens at C-2 and C-3 are anti

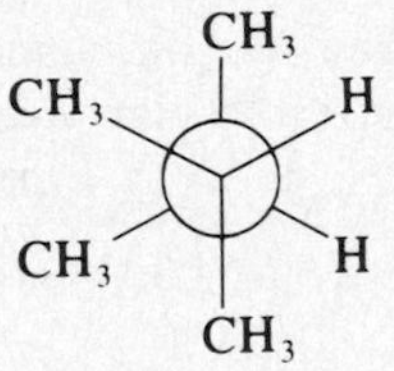

Hydrogens at C-2 and C-3 are gauche

(*d*) The methyl and the ethyl groups are cis in the first structure but trans in the second. The two compounds are stereoisomers; they have the same constitution but differ in the arrangement of their atoms in space.

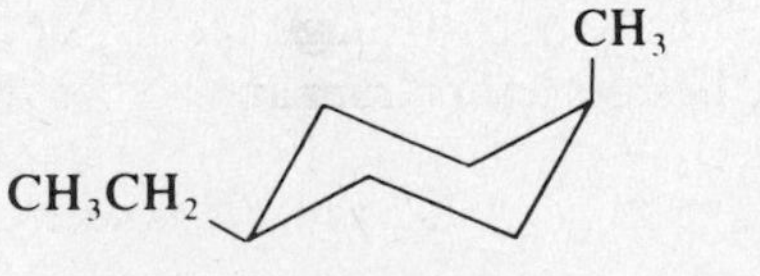

cis-1-Ethyl-4-methylcyclohexane (both alkyl groups are up)

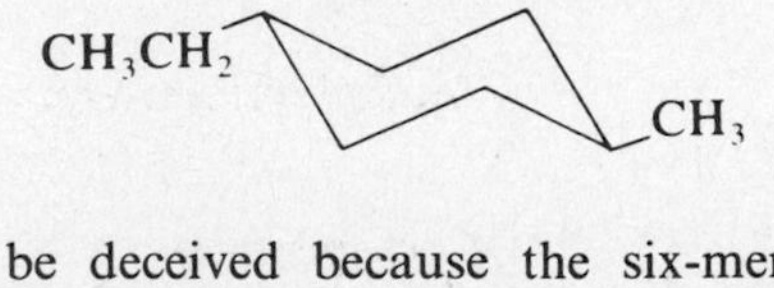

trans-1-Ethyl-4-methylcyclohexane (ethyl group is down; methyl group is up)

Do not be deceived because the six-membered rings look like ring-flipped forms. Remember, chair-chair interconversion converts all the equatorial bonds to axial and vice versa. Here the ethyl group is equatorial in both structures.

(*e*) The two structures have the same constitution but differ in the arrangement of their atoms in space; they are stereoisomers. They are not different conformations of the same molecule because they are not related by rotation about C—C bonds. In the first structure the methyl group is trans to the darkened bonds in the structure shown below, whereas in the second it is cis to these bonds.

CH_3 H — methyl is trans to these bonds

H CH_3 — methyl is cis to these bonds

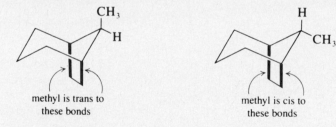

3.16 There are five C_5H_8 constitutional isomers that do not contain double or triple bonds.

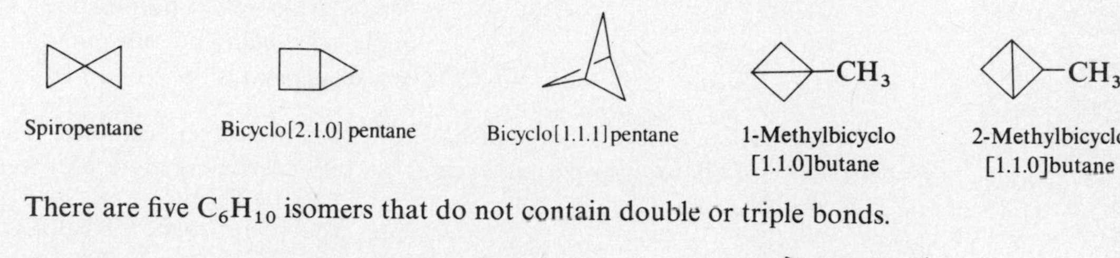

3.17 There are five C_6H_{10} isomers that do not contain double or triple bonds.

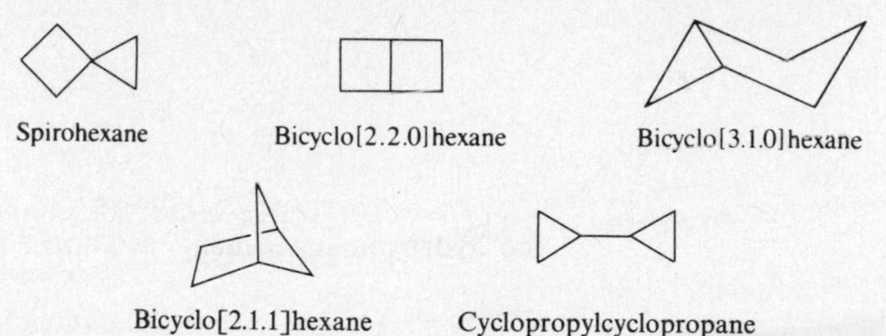

3.18 (*a*) The heat of combustion, in kilocalories per mole, is highest for the hydrocarbon of highest molecular weight. Thus, cyclopropane, even though it is more strained than cyclobutane or cyclopentane, has the lowest heat of combustion.

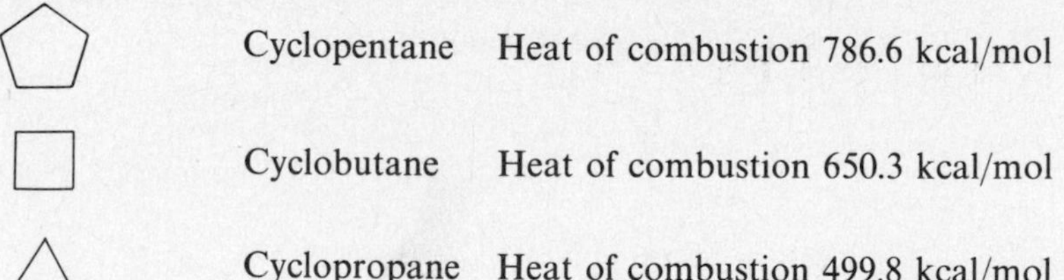

(*b*) All the compounds in the group ethylcyclopropane, methylcyclobutane, and cyclopentane are isomers, since they all have the molecular formula C_5H_{10}. Ethylcyclopropane has the most angle strain and so has the highest heat of combustion. Cyclopentane has the least strain and so has the lowest heat of combustion.

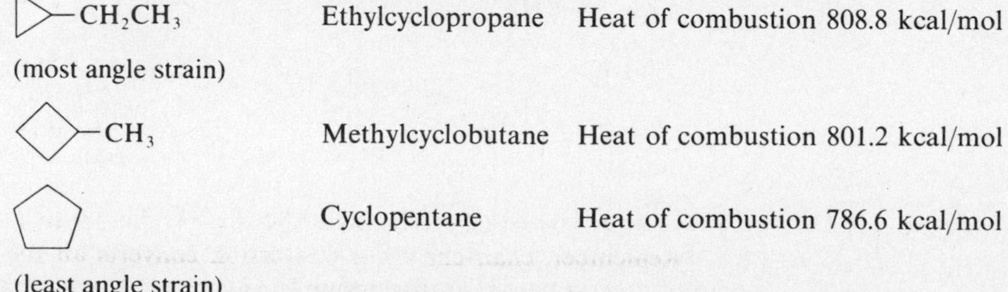

(*c*) All these compounds have the molecular formula C_7H_{14}. They are isomers, so the one with the most strain will have the highest heat of combustion.

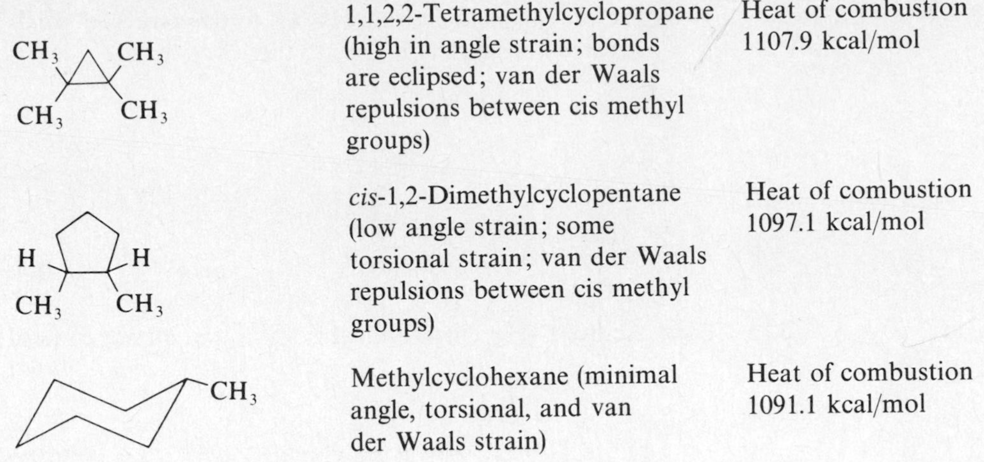

1,1,2,2-Tetramethylcyclopropane (high in angle strain; bonds are eclipsed; van der Waals repulsions between cis methyl groups) — Heat of combustion 1107.9 kcal/mol

cis-1,2-Dimethylcyclopentane (low angle strain; some torsional strain; van der Waals repulsions between cis methyl groups) — Heat of combustion 1097.1 kcal/mol

Methylcyclohexane (minimal angle, torsional, and van der Waals strain) — Heat of combustion 1091.1 kcal/mol

(*d*) These hydrocarbons all have different molecular formulas. Their heats of combustion (in kilocalories per mole) decrease with decreasing molecular weight.

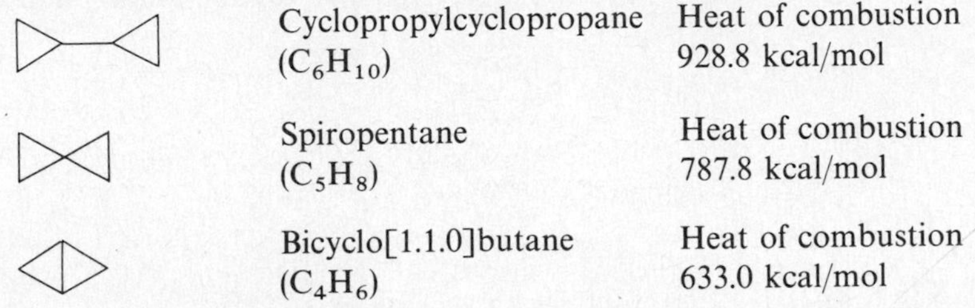

Cyclopropylcyclopropane (C_6H_{10}) — Heat of combustion 928.8 kcal/mol

Spiropentane (C_5H_8) — Heat of combustion 787.8 kcal/mol

Bicyclo[1.1.0]butane (C_4H_6) — Heat of combustion 633.0 kcal/mol

(*e*) Bicyclo[3.3.0]octane and bicyclo[5.1.0]octane are isomers and their heats of combustion can be compared on the basis of their relative strains. The three-membered ring in bicyclo[5.1.0]octane imparts a significant amount of angle strain to this isomer, making it less stable than bicyclo[3.3.0]octane. The third hydrocarbon, bicyclo[4.3.0]nonane, has a higher molecular weight than either of the others and has the largest heat of combustion.

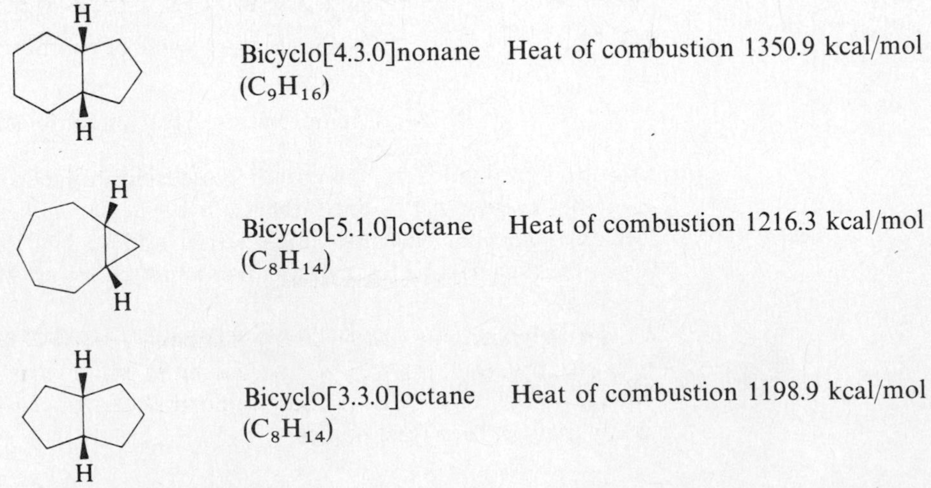

Bicyclo[4.3.0]nonane (C_9H_{16}) — Heat of combustion 1350.9 kcal/mol

Bicyclo[5.1.0]octane (C_8H_{14}) — Heat of combustion 1216.3 kcal/mol

Bicyclo[3.3.0]octane (C_8H_{14}) — Heat of combustion 1198.9 kcal/mol

3.19 All the staggered conformations about the C-2–C-3 bond of 2,2-dimethylpropane are equivalent to each other and of equal energy; they represent potential energy minima. All the eclipsed conformations are equivalent and represent potential energy maxima.

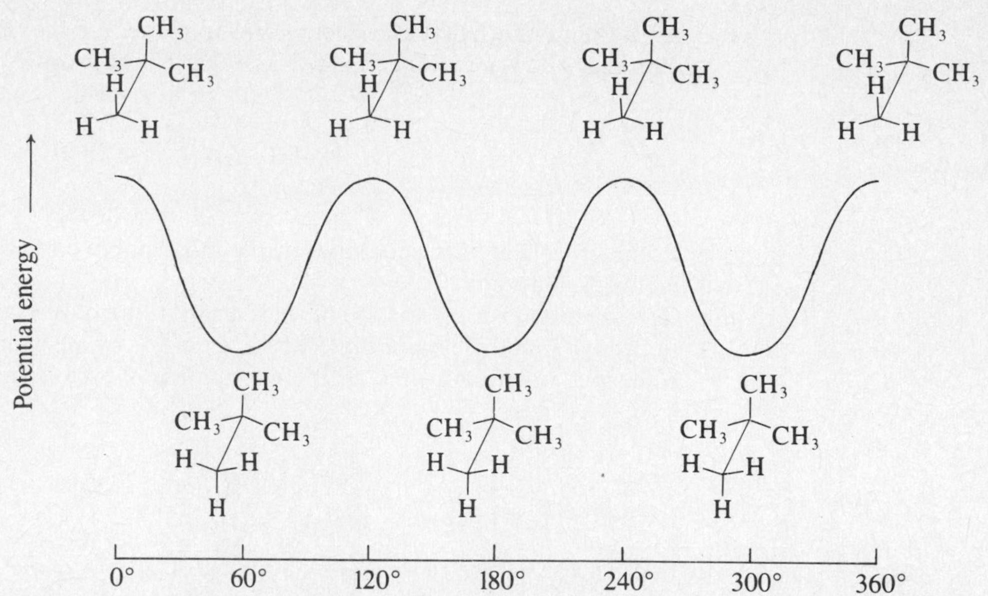

The shape of the potential energy profile for internal rotation in 2,2-dimethylpropane more closely resembles that of ethane than that of butane.

3.20 (*a*) The structural formula corresponding to 2,2,5,5-tetramethylhexane is $(CH_3)_3CCH_2CH_2C(CH_3)_3$. The substituents at C-3 are two hydrogens and a *tert*-butyl group. The substituents at C-4 are the same as those at C-3. The most stable conformation has the large *tert*-butyl groups anti to each other.

$C(CH_3)_3$ H H H H $C(CH_3)_3$

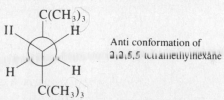

Anti conformation of 2,2,5,5-tetramethylhexane

(*b*) The zigzag conformation of 2,2,5,5-tetramethylhexane is an alternative way of expressing the same conformation implied in the Newman projection of (*a*). It is more complete, however, in that it also shows the spatial arrangement of the atoms in the *tert*-butyl substituents.

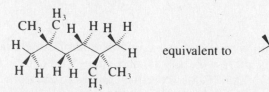

equivalent to

2,2,5,5-Tetramethylhexane

(*c*) An isopropyl group is more conformationally demanding than a methyl group and so will occupy an equatorial site in the most stable conformation of *cis*-1-isopropyl-3-methylcyclohexane. Draw a chair conformation of cyclohexane and place an isopropyl group in an equatorial position.

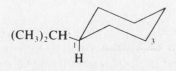

$(CH_3)_2CH$ 1 3 H

Notice that the equatorial methyl group is up on the carbon atom to which it is attached. Add a methyl group to C-3 so that it is also up.

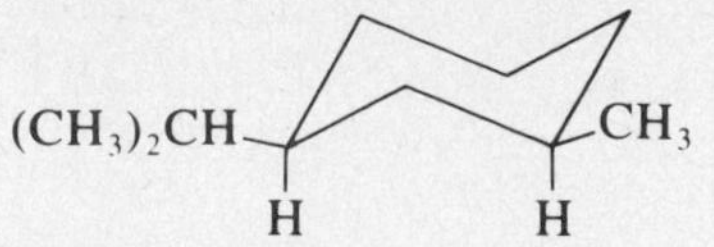

Both substituents are equatorial in the most stable conformation of *cis*-1-isopropyl-3-methylcyclohexane.

(*d*) One substituent is up and the other is down in the most stable conformation of *trans*-1-isopropyl-3-methylcyclohexane. Begin as in (*c*) by placing an isopropyl group in an equatorial orientation on a chair conformation of cyclohexane.

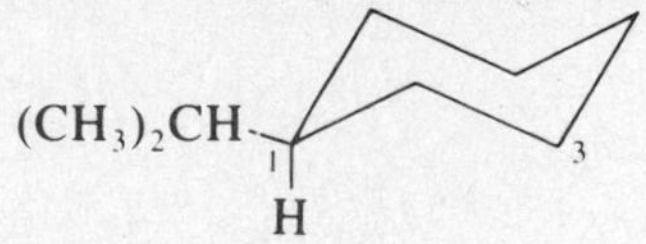

In order to be trans to the C-1 isopropyl group, the C-3 methyl group must be down.

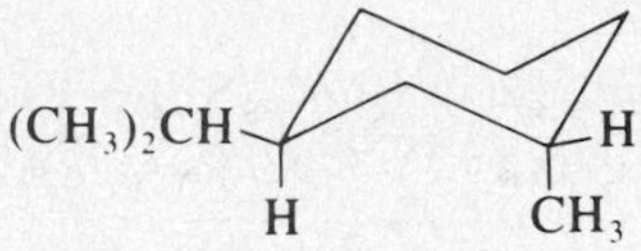

The isopropyl group is thus equatorial and the methyl group axial in the most stable conformation.

(*e*) In order to be cis to each other, one substituent must be axial and the other equatorial when they are located at positions 1 and 4 on a cyclohexane ring.

Place the larger substituent (the *tert*-butyl group) at the equatorial site and the smaller substituent (the ethyl group) at the axial one.

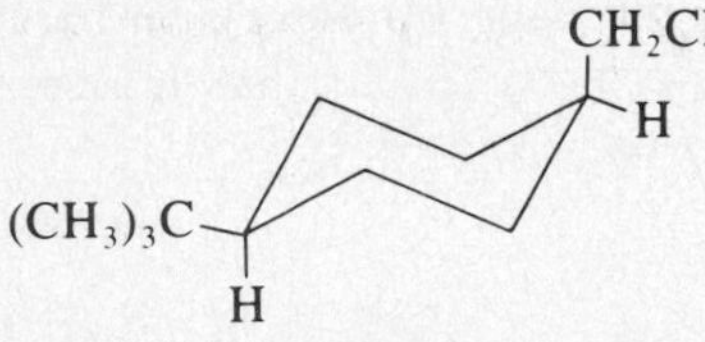

(*f*) First write a chair conformation of cyclohexane, then add two methyl groups at C-1 and draw in the axial and equatorial bonds at C-3 and C-4. Next, add methyl groups to C-3 and C-4 so that they are cis to each other. There are two different ways that this can be accomplished—either the C-3 and C-4 methyl groups are both up or they are both down.

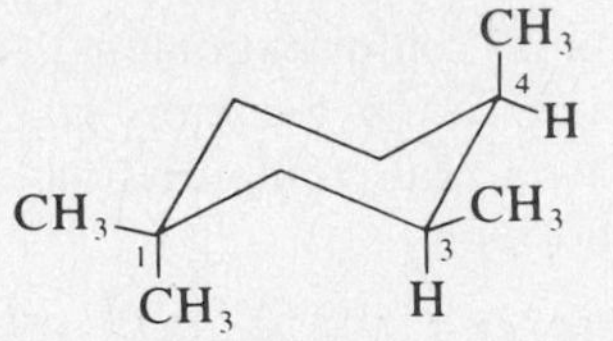

More stable chair conformation: C-3 methyl group is equatorial; axial C-1 methyl group is not engaged in van der Waals repulsion with C-3 methyl

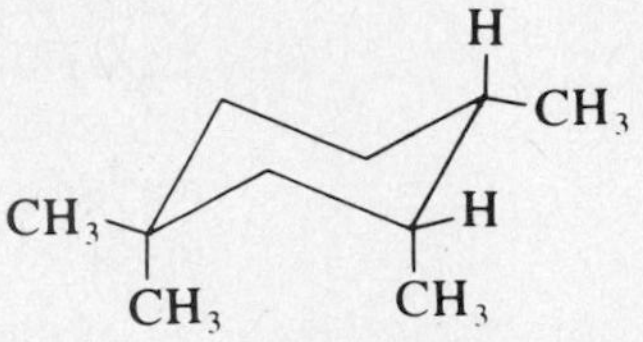

Less stable chair conformation: C-3 methyl group is axial; axial C-1 and C-3 methyl groups engaged in strong van der Waals repulsion

(*g*) Translate the projection formula to a chair conformation.

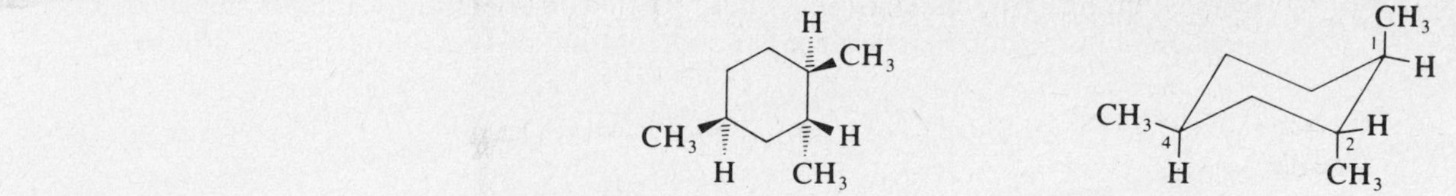

Check to see if this is the most stable conformation by writing its ring-flipped form.

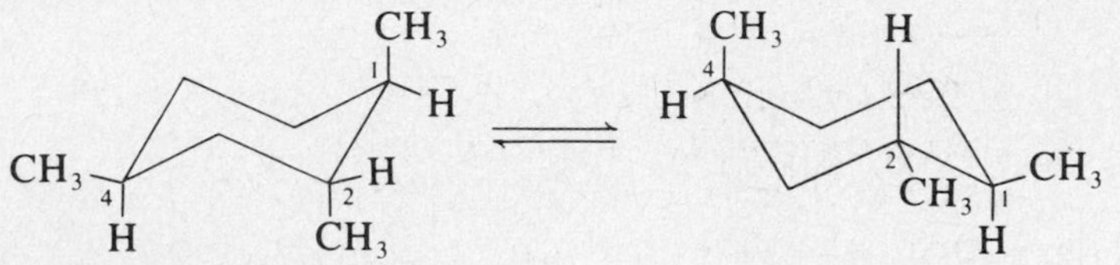

Less stable conformation: two axial methyl groups

More stable conformation: one axial methyl group

The ring-flipped form, with two equatorial methyl groups and one axial methyl group, is more stable than the originally drawn conformation, with two axial and one equatorial methyl groups.

3.21 Begin by writing each of the compounds in its most stable conformation. Compare them by examining their conformations for sources of strain, particularly van der Waals strain arising from groups located too close together in space.

(*a*) Its axial methyl group makes the cis stereoisomer of 1-isopropyl-2-methylcyclohexane less stable than the trans.

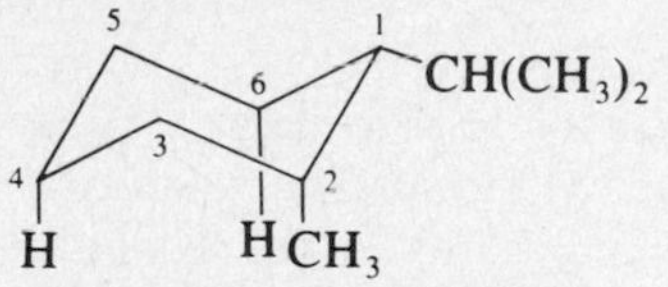

cis-1-Isopropyl-2-methylcyclohexane (less stable stereoisomer)

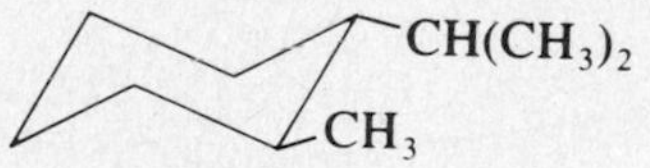

trans-1-Isopropyl-2-methylcyclohexane (more stable stereoisomer)

The axial methyl group in the cis stereoisomer is involved in an unfavorable van der Waals repulsion with the C-4 and C-6 axial hydrogens indicated in the drawing.

(*b*) Both groups are equatorial in the cis stereoisomer of 1-isopropyl-3-methylcyclohexane; cis is more stable than trans in 1,3-disubstituted cyclohexanes.

cis-1-Isopropyl-3-methylcyclohexane (more stable stereoisomer; both groups are equatorial)

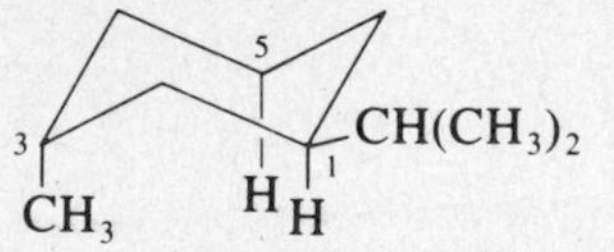

trans-1-Isopropyl-3-methylcyclohexane (less stable stereoisomer; methyl group is axial and involved in van der Waals repulsion with axial hydrogens at C-1 and C-5)

(*c*) The more stable stereoisomer of 1,4-disubstituted cyclohexanes is the trans; both alkyl groups are equatorial in *trans*-1-isopropyl-4-methylcyclohexane.

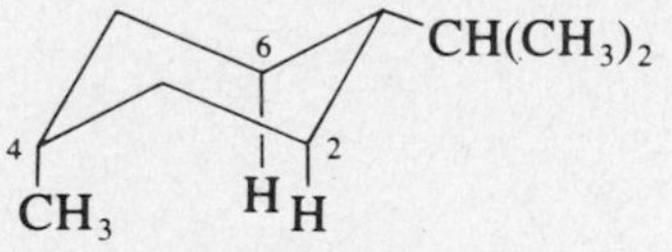

cis-1-Isopropyl-4-methylcyclohexane (less stable stereoisomer; methyl group is axial and involved in van der Waals repulsion with axial hydrogens at C-2 and C-6)

trans-1-Isopropyl-4-methylcyclohexane (more stable stereoisomer; both groups are equatorial)

(*d*) The first stereoisomer of 1,2,4-trimethylcyclohexane is the more stable one. All its methyl groups are equatorial in its most stable conformation. The most stable conformation of the second stereoisomer has one axial and two equatorial methyl groups.

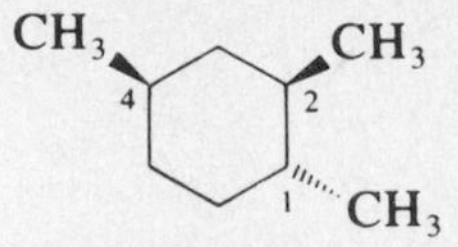

More stable stereoisomer

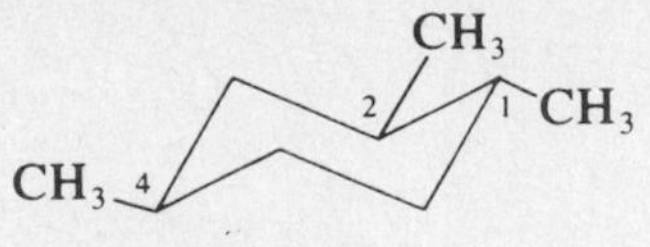

All methyl groups equatorial in most stable conformation

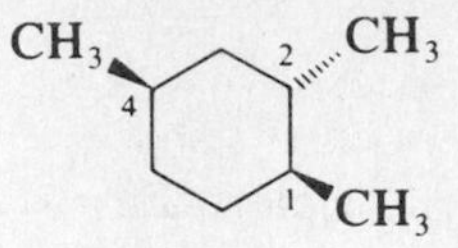

Less stable stereoisomer

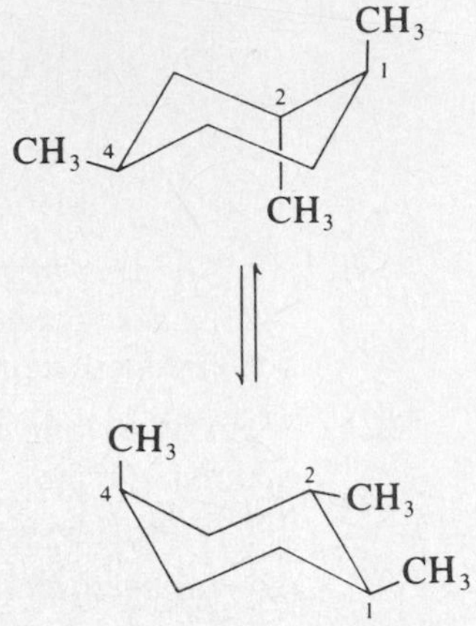

One axial methyl group in most stable conformation

(*e*) The first stereoisomer of 1,2,4-trimethylcyclohexane is the more stable one here, as it was in (*d*). All its methyl groups are equatorial, while one of the methyl groups is axial in the most stable conformation of the second stereoisomer.

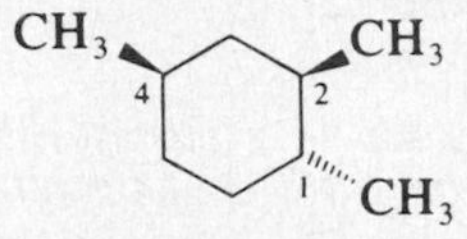

(More stable stereoisomer)

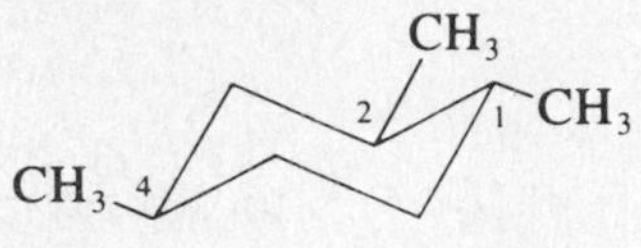

(All methyl groups equatorial in most stable conformation)

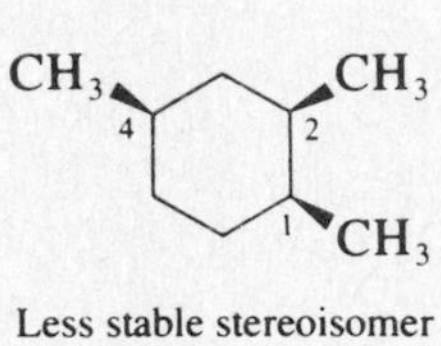

Less stable stereoisomer

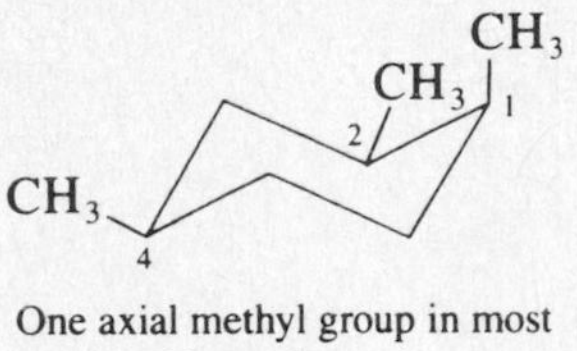

One axial methyl group in most stable conformation

(*f*) Each stereoisomer of 2,3-dimethylbicyclo[3.2.1]octane has one axial and one equatorial methyl group. The first one, however, has a close contact between its axial methyl group and both methylene groups of the two-carbon bridge. The second stereoisomer has a van der Waals repulsion with only one axial methylene group; it is more stable.

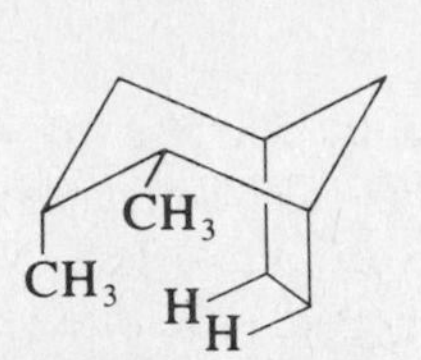

Less stable stereoisomer (more van der Waals strain)

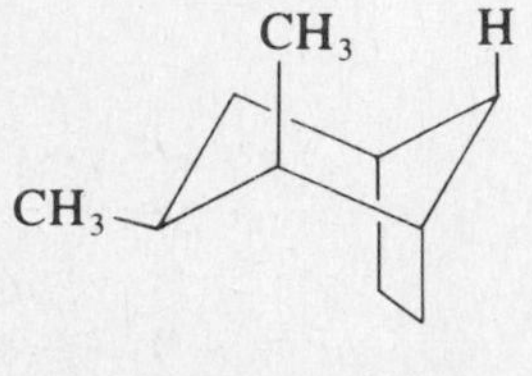

More stable stereoisomer (less van der Waals strain)

3.22 First write structural formulas showing the relative stereochemistries and the preferred conformations of the two stereoisomers of 1,1,3,5-tetramethylcyclohexane.

CH_3 CH_3 CH_3 CH_3 written in its most stable conformation as CH_3 CH_3 CH_3 CH_3

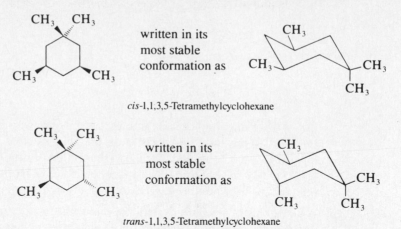

cis-1,1,3,5-Tetramethylcyclohexane

CH_3 CH_3 CH_3 CH_3 written in its most stable conformation as CH_3 CH_3 CH_3 CH_3

trans-1,1,3,5-Tetramethylcyclohexane

The cis stereoisomer is more stable than the trans. It exists in a conformation with only one axial methyl group, while the trans stereoisomer has two axial methyl groups in close contact with one another. The trans stereoisomer is destabilized by van der Waals strain.

3.23 Both the structures have approximately the same degree of angle strain and of torsional strain. Structure D has more van der Waals strain than C because two pairs of hydrogens (shown below) approach each other at distances that are rather close.

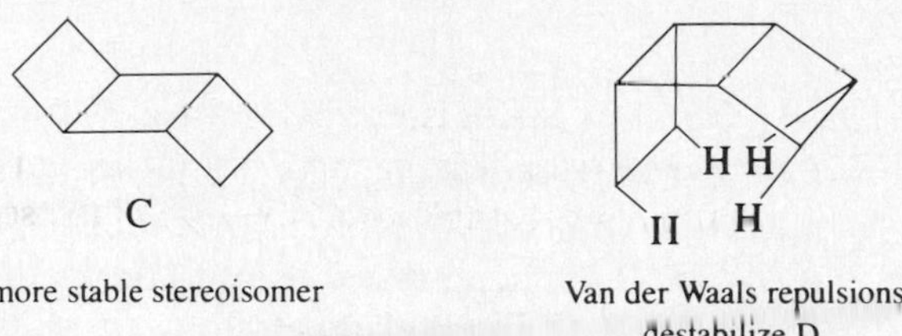

more stable stereoisomer

Van der Waals repulsions destabilize D

3.24 Conformational representations of the two different forms of glucose are drawn in the usual way. An oxygen atom is present in the six-membered ring, and we are told in the problem that the ring exists in a chair conformation.

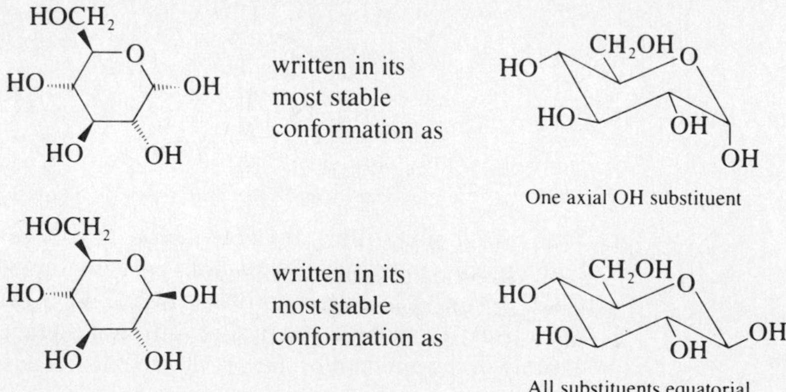

One axial OH substituent

All substituents equatorial

The two structures are not interconvertible by ring flipping; therefore they are not different conformations of the same molecule. Remember, ring flipping transforms all axial substituents to equatorial ones and vice versa. The two structures differ with respect to only one substituent; they are stereoisomers of each other.

3.25 This problem is primarily an exercise in correctly locating equatorial and axial positions in cyclohexane rings that are joined together into a steroid skeleton. Parts (*a*) through (*e*) are concerned with positions 1, 4, 7, 11, and 12 in that order. The diagram below shows the orientation of axial and equatorial bonds at each of those positions.

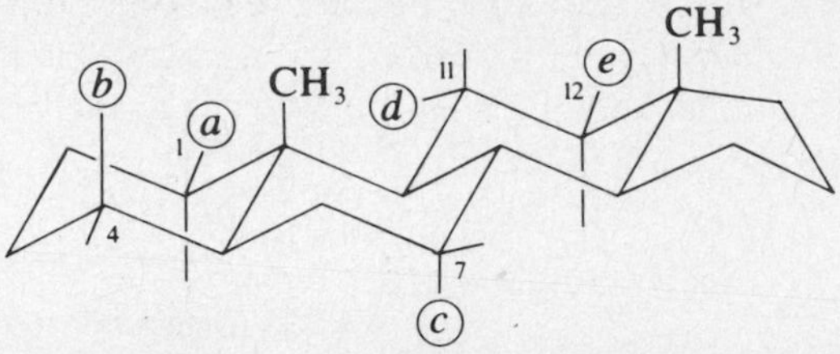

Both methyl groups are up

(*a*) At C-1 the bond that is cis to the methyl groups is equatorial (up).
(*b*) At C-4 the bond that is cis to the methyl groups is axial (up).
(*c*) At C-7 the bond that is trans to the methyl groups is axial (down).
(*d*) At C-11 the bond that is trans to the methyl groups is equatorial (down).
(*e*) At C-12 the bond that is cis to the methyl groups is equatorial (up).

3.26 Analyze this problem in exactly the same way as the preceding one by locating the axial and equatorial bonds at each position. It will be seen that the only differences are those at C-1 and C-4.

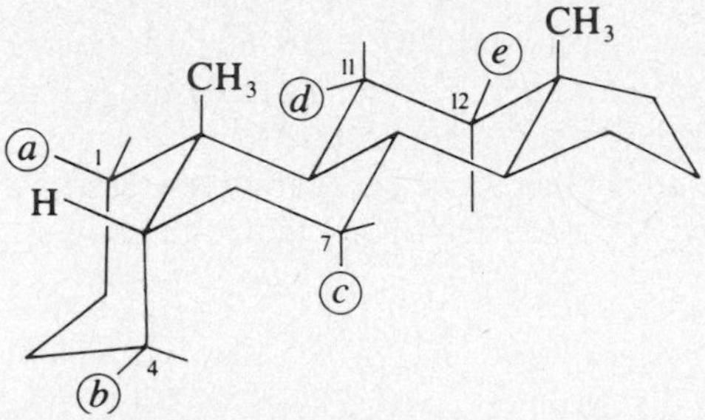

Both methyl groups are up

(*a*) At C-1 the bond that is cis to the methyl groups is axial (up).
(*b*) At C-4 the bond that is cis to the methyl groups is equatorial (up).
(*c*) At C-7 the bond that is trans to the methyl groups is axial (down).
(*d*) At C-11 the bond that is trans to the methyl groups is equatorial (down).
(*e*) At C-12 the bond that is cis to the methyl groups is equatorial (up).

3.27 (*a*) The torsion angle between chlorine substituents is 60° in the gauche conformation and 180° in the anti conformation of $ClCH_2CH_2Cl$.

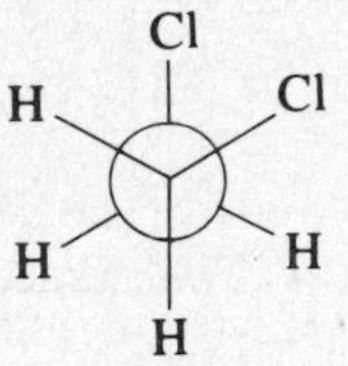

Gauche
(can have a dipole moment)

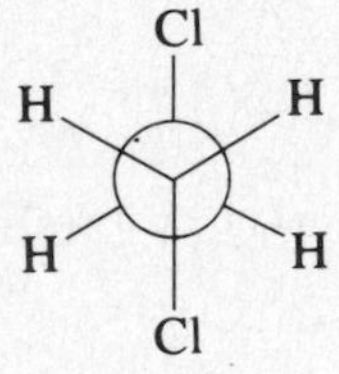

Anti
(cannot have a dipole moment)

(*b*) The anti conformation of $ClCH_2CH_2Cl$ has a center of symmetry; all its individual bond dipole moments cancel and this conformation has no dipole moment. Since $ClCH_2CH_2Cl$ has a dipole moment of 1.12 D, it can exist entirely in the gauche conformation or it can be a mixture of anti and gauche conformations, but it cannot exist entirely in the anti conformation. Statement 1 is false.

3.28 (*a*) The planar and nonplanar conformations of *trans*-1,3-dibromocyclobutane are as shown.

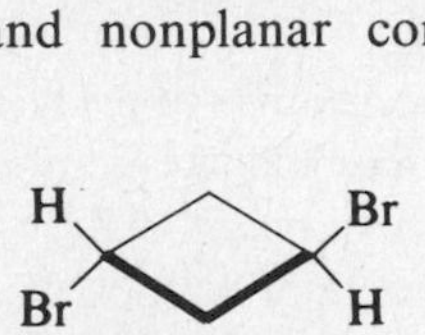

Four-membered ring planar
(cannot have a dipole moment)

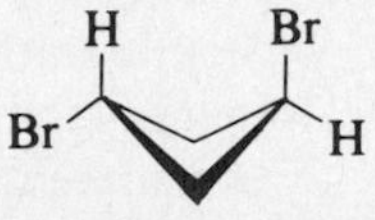

Four-membered ring nonplanar
(can have a dipole moment)

(*b*) The planar conformation of *trans*-1,3-dibromocyclobutane has a center of symmetry and cannot have a dipole moment. The nonplanar conformation can have a dipole moment. Since *trans*-1,3-dibromocyclobutane has a dipole moment of 1.10 D, it may exist entirely in the nonplanar conformation or it may exist as a mixture of the planar and nonplanar forms, but it cannot exist entirely in the planar conformation. Statement 1 is false.

SELF-TEST

PART A

A-1. Draw the gauche conformation of $CH_3CH_2CH_2Cl$ using both a Newman and a sawhorse projection.

A-2. Considering the C_5H_{10} isomers which contain a ring:

(*a*) Which isomer contains both primary and secondary hydrogens but no tertiary hydrogens?

(*b*) Which isomer has the smallest heat of combustion?

(*c*) Which isomer contains a single methyl group?

A-3. Write the structure of the most stable conformation of the less stable stereoisomer of 1-isopropyl-3-methylcyclohexane.

A-4. Draw a Newman projection of the least stable conformation of butane.

A-5. Draw the most stable conformation of the following substance:

Br
CH_3
$C(CH_3)_3$

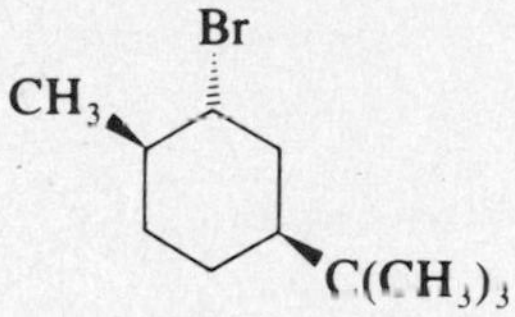

Which substituents are axial and which equatorial?

A-6. Two isomers of a cyclic substance have heats of combustion of 1248.3 kcal/mol (isomer A) and 1246.8 kcal/mol (isomer B). Which isomer is more stable and which more strained?

A-7. Draw the structure of *cis*-bicyclo[4.3.0]nonane.

PART B

B-1. Which of the listed terms best describes the relationship between the methyl groups in the chair conformation of the substance shown?

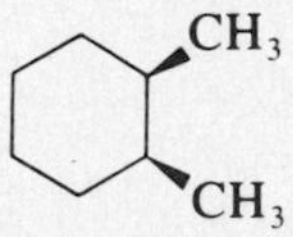

CH_3
CH_3

(*a*) eclipsed (*b*) trans (*c*) anti (*d*) gauche

B-2. Rank the following substances in order of decreasing heat of combustion (largest → smallest)

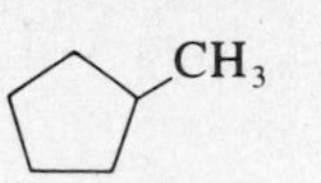

CH_3 A

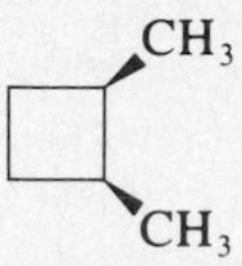

CH_3, CH_3 B

C

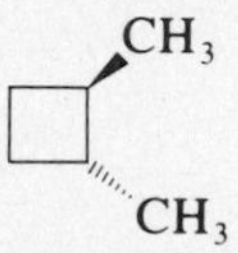

CH_3, CH_3 D

(*a*) A > B > D > C (*b*) B > D > A > C
(*c*) C > D > B > A (*d*) A > C > B > D

B-3. Which of the following statements best describes the most stable conformation of trans-1,3-dimethylcyclohexane?
(*a*) Both methyl groups are axial.
(*b*) Both methyl groups are equatorial.
(*c*) One methyl group is axial, the other equatorial.
(*d*) The molecule is severely strained and cannot exist.

B-4. Compare the stability of the following two compounds:
I *cis*-1-Bromo-3-methylcyclohexane
II *trans*-1-Bromo-3-methylcyclohexane
(*a*) I is more stable
(*b*) II is more stable
(*c*) I and II are of equal stability
(*d*) No comparison can be made

B-5. What, if anything, can be said about the magnitude of the equilibrium constant *K* for the following process?

H, $CH(CH_3)_2$, H, CH_3 ⇌ H, CH_3, H, $CH(CH_3)_2$

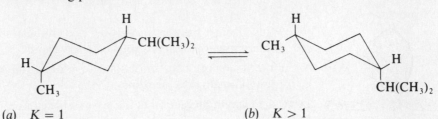

(*a*) $K = 1$ (*b*) $K > 1$
(*c*) $K < 1$ (*d*) No estimate of *K* can be made

B-6. The two structures shown below are ______ each other.

(*a*) identical with (*b*) stereoisomers of
(*c*) constitutional isomers of (*d*) different conformations of

B-7. Which one of the following energy versus rotation curves best describes propane?

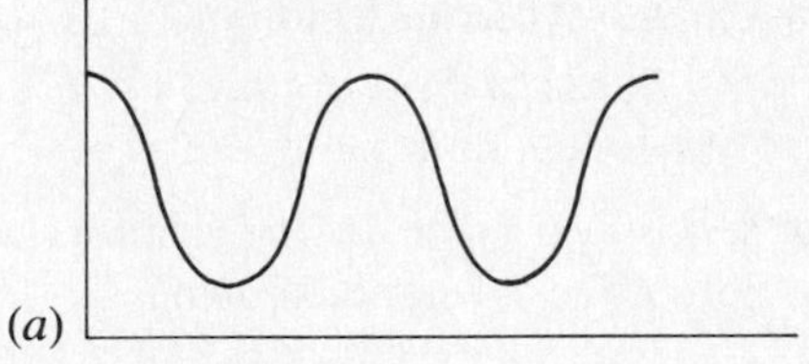
(*a*)

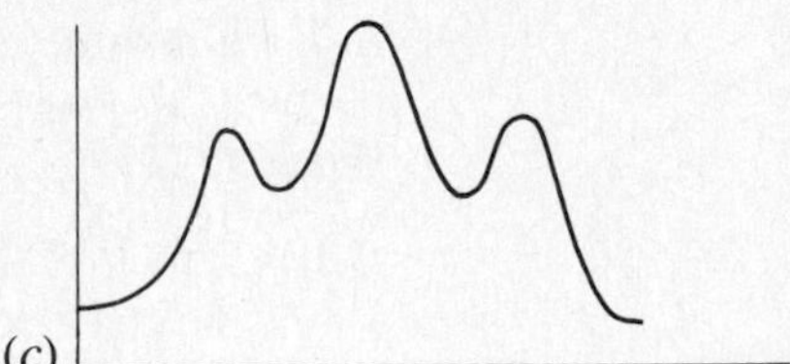
(*c*)

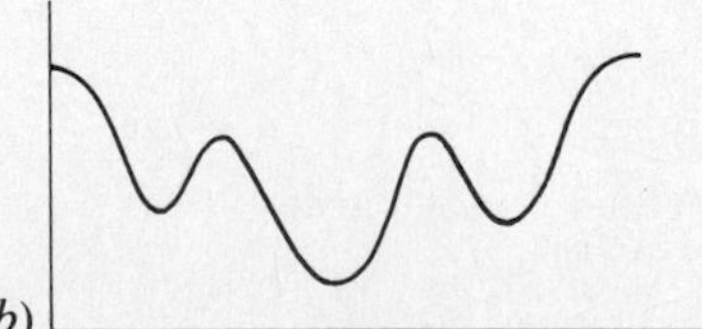
(*b*)

(*d*) both (*b*) and (*c*)

B-8. Which of the following statements is *not* true concerning the conformational interconversion of *trans*-1,2-diethylcyclohexane?
(*a*) An axial group will be changed into the equatorial position.
(*b*) The energy of repulsions present in the molecule will be changed.
(*c*) Formation of the cis substance will result.
(*d*) One conformational form will be greatly favored over the other.

CHAPTER 4

INTRODUCTION TO FUNCTIONAL GROUPS. ALCOHOLS AND ALKYL HALIDES

IMPORTANT TERMS AND CONCEPTS

Nomenclature (Secs. 4.1, 4.2) Alcohols are a class of organic compounds characterized by the presence of the *hydroxyl* (—OH) *functional group*. The —OH group of an alcohol is bonded to an sp^3 hybridized carbon and may be thought of as replacing one hydrogen of a parent alkane. Similarly, alkyl halides result from replacement of one or more hydrogens of an alkane with a halogen atom. The actual synthesis of alkyl halides, however, can begin with an alcohol or an alkane and appropriate reagents.

Alcohol and alkyl halide nomenclature is straightforward, using the general rules established in Chapter 2 for alkanes. The steps are:

1. Find the longest chain containing the functional group (—OH or halide).
2. Number the chain, keeping the —OH number as low as possible; halides are numbered as substituents with the same ranking as alkyl groups.
3. Alkyl substituents are named and numbered as usual.
4. For an alcohol, replace the *-e* ending of the alkane parent name with *-ol*, as shown in the following examples.

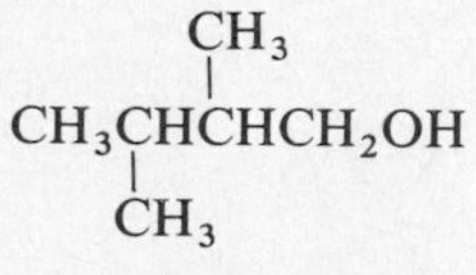

$$\begin{array}{c} CH_3 \\ | \\ CH_3CHCHCH_2OH \\ | \\ CH_3 \end{array}$$

2,3-Dimethyl-1-butanol

OH, CH_2CH_3 (cyclohexane ring)

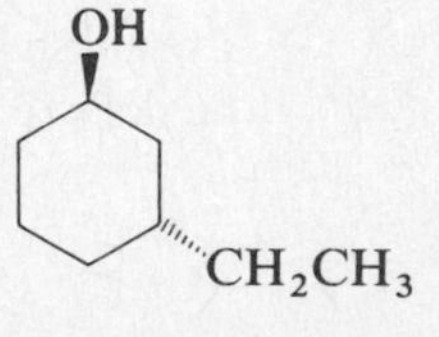

trans-3-Ethylcyclohexanol

5. For an alkyl halide, precede the stem name with the appropriate *halo-* prefix, as shown below.

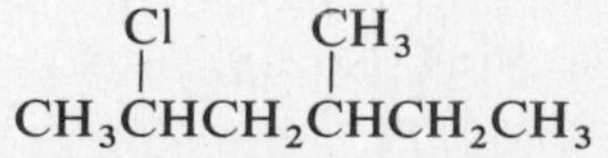

$$\begin{array}{c} Cl \quad\quad CH_3 \\ | \quad\quad\quad | \\ CH_3CHCH_2CHCH_2CH_3 \end{array}$$

2-Chloro-4-methylhexane

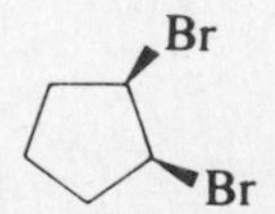

Br, Br (cyclopentane ring)

cis-1,2-Dibromocyclopentane

Alcohols and alkyl halides may be classified as primary (1°), secondary (2°), or tertiary (3°) according to the substitution of the carbon bearing the functional group. Note that -X is often used to represent the halogen in alkyl halides.

$$\underset{\text{Primary}}{R{-}\overset{\displaystyle H}{\underset{\displaystyle H}{C}}{-}OH(\text{-}X)} \qquad \underset{\text{Secondary}}{R{-}\overset{\displaystyle H}{\underset{\displaystyle R}{C}}{-}OH(\text{-}X)} \qquad \underset{\text{Tertiary}}{R{-}\overset{\displaystyle R}{\underset{\displaystyle R}{C}}{-}OH(\text{-}X)}$$

Keep in mind that it is the carbon that bears the functional group that determines whether the group is described as being primary, secondary, or tertiary. Therefore, 2-methyl-1-propanol, $(CH_3)_2CHCH_2OH$, is a primary alcohol.

Physical Properties (Sec. 4.4) The bent C—O—H unit of alcohols and the polar nature of both the C—O and the O—H bonds result in alcohols being polar molecules. The dipole moments of most alcohols are similar to that of water. Likewise, alkyl halides are polar compounds owing to the polar nature of the carbon–halogen bond.

Hydrogen bonding is an important intermolecular attractive force in alcohols and provides an explanation for boiling points that are unusually high when compared with other molecules of similar molecular size.

Acid-Base Properties of Alcohols (Secs. 4.5, 4.6) Alcohols may act as both *Brönsted bases* and *Brönsted acids,* that is, they may act both as *proton acceptors* (*Brönsted bases*) and as *proton donors* (*Brönsted acids*).

Acceptance of a proton in an acid-base reaction yields the *conjugate acid* of the alcohol:

$$R{-}OH + HA \rightleftharpoons R{-}\overset{+}{O}H_2 + A^-$$

Donation of a proton results in formation of the *conjugate base* of the alcohol, known as an *alkoxide ion.*

$$R{-}OH + OH^- \rightleftharpoons R{-}O^- + H_2O$$

The strength of an acid in solution may be determined from the *equilibrium constant K* for the process:

$$HA + H_2O \rightleftharpoons H_3O^+ + A^-$$

$$K = \frac{[H_3O^+][A^-]}{[HA][H_2O]}$$

Presuming the concentration of water to be a constant, the *acid dissociation constant* K_a may be defined as:

$$K[H_2O] = K_a = \frac{[H_3O^+][A^-]}{[HA]}$$

A frequently used way to express acid strength is pK_a, defined as:

$$pK_a = -\log(K_a)$$

As a result, numbers that are not exponentials and that increase with *decreasing* acid strength are used. For example

$$K_a = 10^{-16} \qquad pK_a = 16 \qquad \text{(stronger acid)}$$

$$K_a = 10^{-25} \qquad pK_a = 25 \qquad \text{(weaker acid)}$$

An additional means of describing acid and base strength is to note that *the stronger the acid, the weaker the conjugate base and vice versa.*

Carbocation Intermediates (Secs. 4.10 to 4.13) The *substitution reaction* whereby alcohols react with hydrogen halides to form alkyl halides proceeds through formation of a *carbocation* intermediate. The mechanism of this reaction is outlined in the reaction summary presented later in this chapter.

A carbocation is a *trivalent* intermediate in which the central carbon bears a formal charge of +1. Both the VSEPR and the orbital hybridization bonding model predict the carbon to be trigonal planar. According to the latter theory, the positively charged carbon adopts an sp^2 hybridization state, as shown in text Figure 4.6.

Substituted carbocations may be classified as *primary* (1°), *secondary* (2°), or *tertiary* (3°), according to the number of substituents other than hydrogen attached to the positively charged carbon.

H | R—C+ | H	R | R—C+ | H	R | R—C+ | R
Primary carbocation	Secondary carbocation	Tertiary carbocation

Alkyl substituents attached to the positively charged carbon stabilize carbocations. The result is the following *carbocation stability series*:

$$\text{Tertiary} > \text{Secondary} > \text{Primary} > CH_3^+$$

Most stable ... Least stable

Stabilization of carbocations by alkyl groups is the result of two effects: the *inductive effect*, in which the alkyl groups are electron-releasing, and *delocalization* of electrons directly into the vacant $2p$ orbital of the carbocation by the filled σ orbitals of the alkyl groups.

Carbocations, having a deficiency of electron density in the carbon valence shell (the octet is not satisfied), are said to be highly *electrophilic*, that is, "electron-seeking" or "electron-loving." An electron-deficient species such as a carbocation is able to react readily with an electron-rich species known as a *nucleophile* (meaning "nucleus-loving"). Such a *nucleophilic substitution reaction* is one example of an important class of organic reactions.

Reaction Mechanisms (Secs. 4.14, 4.15) The overall rate of a multistep process is determined by the rate of the *slow*, or *rate-determining, step* of the process. In the reaction of alcohols with hydrogen halides, formation of the carbocation intermediate is the rate-determining step of the reaction.

The energy changes which take place as a reaction proceeds to form products may be described by using a *reaction coordinate diagram* (or *potential energy diagram*) as shown in text Figures 4.10 and 4.12 through 4.15.

The rate constant for each step in a chemical process is related to the magnitude of the *activation energy*, E_{act}, of the step. A reaction step proceeds faster (has a larger value of k) when the activation energy is smaller. The rate-determining step of a multistep process therefore has the largest activation energy.

When a carbocation is formed, as in the reaction of alcohols with hydrogen halides, the stability of the *transition state* is determined by the stability of the carbocation. The more stable the carbocation, the lower the activation energy of the reaction step leading to its formation. Therefore a *reactivity series* for alcohols may be written that parallels the series for carbocation stability:

$$R_3COH > R_2CHOH > RCH_2OH > CH_3OH$$

Tertiary, most reactive	Secondary	Primary	Least reactive

Free Radicals (Secs. 4.18, 4.19) An alkyl group containing a trivalent carbon that possesses an unpaired electron is known as an *alkyl free radical.* Alkyl groups stabilize free radicals in a manner parallel to that described for carbocations, resulting in a similar stability series:

$$\underset{\text{Most stable}}{R_3C\cdot} > R_2CH\cdot > RCH_2\cdot > \underset{\text{Least stable}}{CH_3\cdot}$$

Relative stabilities of free radicals may also be described by using *bond dissociation energies* (BDE's). The BDE is the energy required for the *homolytic* dissociation of a bond, that is, the BDE equals the enthalpy change (ΔH^0) for the process

$$X{-}Y \longrightarrow X\cdot + Y\cdot$$

As shown in text Figure 4.17, formation of a secondary free radical requires less energy than formation of a primary free radical.

Halogenation of Methane and Ethane (Secs. 4.20 to 4.23) Alkanes undergo *substitution reactions* with halogens to give *alkyl halides.* For example, the chlorination of ethane is shown below.

Bond dissociation energies (described above) may be used to determine if a particular halogenation reaction is *exothermic* or *endothermic.* For a reaction to be exothermic, the energy released in bond formation must be greater than the energy absorbed in bond rupture, that is, the enthalpy change ΔH^0 must be negative. As an example, BDE data from text Table 4.4 may be used to calculate the enthalpy change for chlorination of ethane:

$$\underset{98\ \text{kcal/mol}}{CH_3CH_2{-}H} + \underset{58\ \text{kcal/mol}}{Cl{-}Cl} \longrightarrow \underset{81\ \text{kcal/mol}}{CH_3CH_2{-}Cl} + \underset{103\ \text{kcal/mol}}{H{-}Cl}$$

Bond breaking: 156 kcal/mol — Bond formation: 184 kcal/mol

$$\Delta H^0 = \text{BDE (bonds broken)} - \text{BDE (bonds formed)}$$

$$\Delta H^0 = (156 - 184)\ \text{kcal/mol} = -28\ \text{kcal/mol}$$

To correlate the enthalpy change of a reaction with whether or not the process is *spontaneous* the *free energy change* must be considered. The free energy change is related to the enthalpy change by the expression

$$\Delta G^0 = \Delta H^0 - T\Delta S^0$$

where ΔS^0 is the *entropy* change of the process.

The change in free energy also allows the equilibrium constant K for a process to be calculated according to the relationship

$$\Delta G^0 = -RT \ln K$$

In free-radical halogenations ΔS^0 is usually small, and so the enthalpy change is a good approximation of the change in free energy and therefore can be used to calculate a reasonably close value of the equilibrium constant. For an exothermic reaction, in which ΔH^0 is negative, K will be greater than unity.

The *reaction mechanism* provides a look at the details, on a molecular level, of how reactions proceed; in the case of alkane halogenations, the pathway is a *free-radical chain reaction.* The steps, illustrated for the chlorination of ethane, are:

1. *Initiation*, in which a halogen molecule dissociates to form free radicals

$$Cl_2 \xrightarrow[\text{light}]{\text{heat or}} 2\,Cl\cdot$$

2. *Propagation*, in which the halogen radical reacts with the alkane to form new free radicals.

$$Cl\cdot + CH_3CH_3 \longrightarrow CH_3\dot{C}H_2 + H{-}Cl$$
$$CH_3\dot{C}H_2 + Cl_2 \longrightarrow CH_3CH_2{-}Cl + Cl\cdot$$

3. *Termination*, in which free radicals are consumed by reacting with each other.

$$2\,Cl\cdot \longrightarrow Cl_2$$
$$2\,CH_3\dot{C}H_2 \longrightarrow CH_3CH_2CH_2CH_3$$
$$CH_3\dot{C}H_2 + Cl\cdot \longrightarrow CH_3CH_2Cl$$

Halogenation of Higher Alkanes (Secs. 4.23, 4.24) Alkane chlorination has a low selectivity for the replacement of tertiary rather than secondary or primary hydrogens, although a greater selectivity of tertiary substitution is observed for bromination of alkanes. The predominant product formed in halogenation of alkanes is that arising from formation of the *most stable free radical*, according to the stability series noted earlier.

IMPORTANT REACTIONS

Reaction of Alcohols with Metals (Sec. 4.7)

General: $2R{-}OH + 2M \longrightarrow 2R{-}O^- + 2M^+ + H_2$

Example: $2(CH_3)_2CHOH + 2K \longrightarrow 2(CH_3)_2CHO^-K^+ + H_2$

PREPARATION OF ALKYL HALIDES

Reaction of Alcohols with Hydrogen Halides (Secs. 4.9, 4.10)

General: $R{-}OH + H{-}X \longrightarrow R{-}X + H_2O$

Examples:

$$(CH_3)_2\overset{\overset{\large OH}{|}}{C}CH_2CH_3 \xrightarrow{HBr} (CH_3)_2\overset{\overset{\large Br}{|}}{C}CH_2CH_3 + H_2O$$

1-methylcyclohexanol (cyclohexane ring bearing CH_3 and OH on the same carbon) $\xrightarrow{HI}$ 1-iodo-1-methylcyclohexane (ring bearing CH_3 and I)

Reaction mechanism:

1. $ROH + HX \overset{fast}{\rightleftharpoons} R{-}\overset{+}{O}H_2 + X^-$

2. $R{-}\overset{+}{O}H_2 \xrightarrow{slow} R^+ + H_2O$

3. $R^+ + X^- \xrightarrow{fast} R{-}X$

Unhindered (primary and some secondary) react by:

$$X^- + R{-}\overset{+}{O}H_2 \longrightarrow RX + H_2O$$

Reaction of Alcohols with Thionyl Chloride (Sec. 4.17)

General: $R{-}OH + SOCl_2 \xrightarrow{pyridine} R{-}Cl$

Example: $CH_3\underset{\underset{\large OH}{|}}{C}HCH(CH_3)_2 + SOCl_2 \xrightarrow{pyridine} CH_3\underset{\underset{\large Cl}{|}}{C}HCH(CH_3)_2$

Reaction of Alcohols with Phosphorus Tribromide (Sec. 4.17)

General: $R{-}OH + PBr_3 \longrightarrow R{-}Br$

Example:

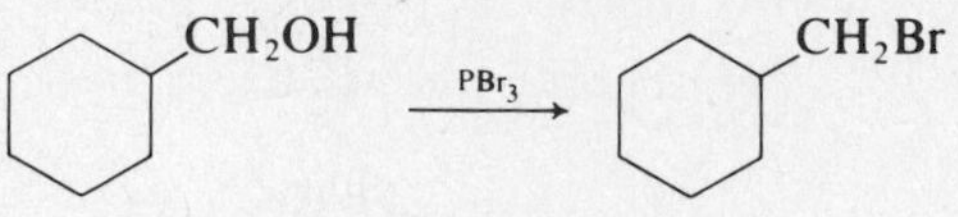

$$\text{cyclohexyl-}CH_2OH \xrightarrow{PBr_3} \text{cyclohexyl-}CH_2Br$$

Free-Radical Halogenation of Alkanes (Secs. 4.20 to 4.24)

Examples:

$$CH_3{-}\underset{}{\overset{CH_3}{\overset{|}{C}H}}{-}CH_3 + Br_2 \xrightarrow[\text{light}]{\text{heat or}} CH_3{-}\underset{\underset{Br}{|}}{\overset{CH_3}{\overset{|}{C}}}{-}CH_3$$

(major)

$$CH_3CH_2CH_2CH_3 + Cl_2 \xrightarrow[\text{light}]{\text{heat or}} CH_3CH_2CH_2CH_2Cl + CH_3CH_2\underset{\underset{Cl}{|}}{C}HCH_3$$

SOLUTIONS TO TEXT PROBLEMS

4.1 There are four C_4H_9 alkyl groups, so there are four isomeric C_4H_9Cl alkyl chlorides. Their structures and common names are:

$CH_3CH_2CH_2CH_2Cl$	*n*-Butyl chloride
$CH_3\underset{\underset{Cl}{\mid}}{C}HCH_2CH_3$	*sec*-Butyl chloride
$CH_3\underset{\underset{CH_3}{\mid}}{C}HCH_2Cl$	Isobutyl chloride
$CH_3\underset{\underset{Cl}{\mid}}{\overset{\overset{CH_3}{\mid}}{C}}CH_3$	*tert*-Butyl chloride

4.2 Alcohols containing four carbons may have their hydroxyl functional group attached to a butyl, *sec*-butyl, isobutyl, or *tert*-butyl group. Their structures and systematic IUPAC names are:

$CH_3CH_2CH_2CH_2OH$	1-Butanol
$CH_3\underset{\underset{OH}{\mid}}{C}HCH_2CH_3$	2-Butanol
$CH_3\underset{\underset{CH_3}{\mid}}{C}HCH_2OH$	2-Methyl-1-propanol
$CH_3\underset{\underset{OH}{\mid}}{\overset{\overset{CH_3}{\mid}}{C}}CH_3$	2-Methyl-2-propanol

4.3 The structures of the C_4H_9Cl isomers were given in the solution to Problem 4.1. Their systematic names are:

$CH_3CH_2CH_2CH_2Cl$	1-Chlorobutane
$CH_3CH(Cl)CH_2CH_3$	2-Chlorobutane
$CH_3CH(CH_3)CH_2Cl$	1-Chloro-2-methylpropane
$CH_3C(CH_3)(Cl)CH_3$	2-Chloro-2-methylpropane

4.4 Alcohols are classified as primary, secondary, or tertiary according to the number of carbon substituents attached to the carbon that bears the hydroxyl group.

$CH_3CH_2CH_2C(H)(H)$—OH	Primary alcohol (one alkyl group bonded to $—CH_2OH$)
$CH_3—C(H)(OH)—CH_2CH_3$	Secondary alcohol (two alkyl groups bonded to $>CHOH$)
$(CH_3)_2CH—C(H)(H)—OH$	Primary alcohol (one alkyl group bonded to $—CH_2OH$)
$CH_3—C(CH_3)(CH_3)—OH$	Tertiary alcohol (three alkyl groups bonded to $\geq COH$)

4.5 Dipole moment is the product of charge and distance. Although the electron distribution in the carbon-chlorine bond is more polarized than that in the carbon-bromine bond, this effect is counterbalanced by the longer carbon-bromine bond distance.

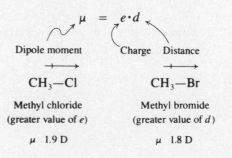

$$\mu = e \cdot d$$

Dipole moment   Charge   Distance

$CH_3—Cl$	$CH_3—Br$
Methyl chloride (greater value of e)	Methyl bromide (greater value of d)
μ 1.9 D	μ 1.8 D

4.6 Sulfuric acid is $HOS(=O)_2OH$. When it transfers a proton to a base, the anion $HOSO_3^-$ (hydrogen sulfate) is formed. In this case, the base is methanol. It accepts a proton from sulfuric acid to give the corresponding oxonium ion.

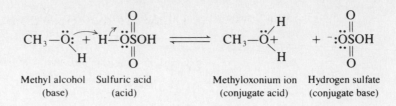

$$CH_3-\ddot{O}(H): + H-\ddot{O}SO_2OH \rightleftharpoons CH_3-\ddot{O}H_2^+ + {}^-:\ddot{O}SO_2OH$$

Methyl alcohol (base) | Sulfuric acid (acid) | Methyloxonium ion (conjugate acid) | Hydrogen sulfate (conjugate base)

4.7 (*b*) As described in the sample solution, $K_a = 10^{-pK_a}$; therefore K_a for vitamin C is given by the expression:

$$K_a = 10^{-4.17}$$

$$K_a = 10^{-4} \times 10^{-0.17} = 10^{-5} \times 10^{+0.83}$$

$$K_a = 6.7 \times 10^{-5}$$

(*c*) Similarly, $K_a = 1.8 \times 10^{-4}$ for formic acid (pK_a 3.75).
(*d*) $K_a = 6.5 \times 10^{-2}$ for oxalic acid (pK_a 1.19).

In ranking the acids in order of decreasing acidity, remember that the larger the equilibrium constant K_a, the stronger the acid and the lower the pK_a value, the stronger the acid.

Acid	K_a	pK_a
Oxalic (strongest)	6.5×10^{-2}	1.19
Aspirin	3.3×10^{-4}	3.48
Formic acid	1.8×10^{-4}	3.75
Vitamin C (weakest)	6.7×10^{-5}	4.17

4.8 The reaction that takes place is:

$$H_2\ddot{N}:^- + H-\ddot{O}CH_3 \rightleftharpoons H_3N: + {}^-:\ddot{O}CH_3$$

Amide ion (stronger base) | Methanol (stronger acid) $K_a = 10^{-16}$ | Ammonia (weaker acid) $K_a = 10^{-36}$ | Methoxide ion (weaker base)

The position of equilibrium lies to the right, to favor formation of the weaker acid and the weaker base.

4.9 (*b*) Hydrogen chloride converts tertiary alcohols to tertiary alkyl chlorides.

$$(CH_3CH_2)_3COH + HCl \longrightarrow (CH_3CH_2)_3CCl + H_2O$$

3-Ethyl-3-pentanol | Hydrogen chloride | 3-Chloro-3-ethylpentane | Water

(*c*) 1-Tetradecanol is a primary alcohol having an unbranched 14-carbon chain. Hydrogen bromide reacts with primary alcohols to give the corresponding primary alkyl bromide.

$$CH_3(CH_2)_{12}CH_2OH + HBr \longrightarrow CH_3(CH_2)_{12}CH_2Br + H_2O$$

1-Tetradecanol | Hydrogen bromide | 1-Bromotetradecane | Water

4.10 There are two $C_4H_9^+$ carbocations that have unbranched carbon chains:

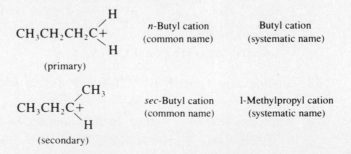

$$CH_3CH_2CH_2CH_2^+$$

(primary)

n-Butyl cation (common name) | Butyl cation (systematic name)

$$CH_3CH_2\overset{+}{C}H(CH_3)$$

(secondary)

sec-Butyl cation (common name) | 1-Methylpropyl cation (systematic name)

Two $C_4H_9^+$ carbocations have branched carbon chains:

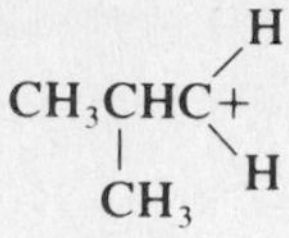

$$CH_3\underset{\underset{CH_3}{|}}{C}HC^+H_2$$
(primary)

Isobutyl cation (common name) — 2-Methylpropyl cation (systematic name)

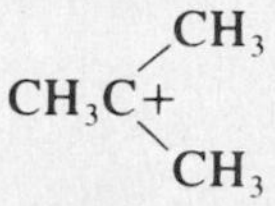

$$CH_3C^+(CH_3)_2$$
(tertiary)

tert-Butyl cation (common name) — 1,1-Dimethylethyl cation (systematic name)

4.11 The order of carbocation stability is tertiary > secondary > primary. There is only one $C_5H_{11}^+$ carbocation that is tertiary, so that is the most stable one.

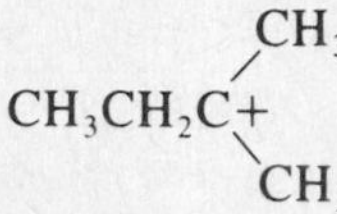

$$CH_3CH_2C^+(CH_3)_2$$

tert-Pentyl cation (common name) — 1,1-Dimethylpropyl cation (systematic name)

4.12 1-Butanol is a primary alcohol; 2-butanol is a secondary alcohol. A carbocation intermediate is possible in the reaction of 2-butanol with hydrogen bromide but not in the corresponding reaction of 1-butanol.

The mechanism of the reaction of 1-butanol with hydrogen bromide proceeds by displacement of water by bromide from the protonated form of the alcohol.

Protonation of the alcohol

$$CH_3CH_2CH_2CH_2\ddot{O}H + H{-}\ddot{B}r: \longrightarrow CH_3CH_2CH_2CH_2\ddot{O}^+H_2 + :\ddot{B}r:^-$$

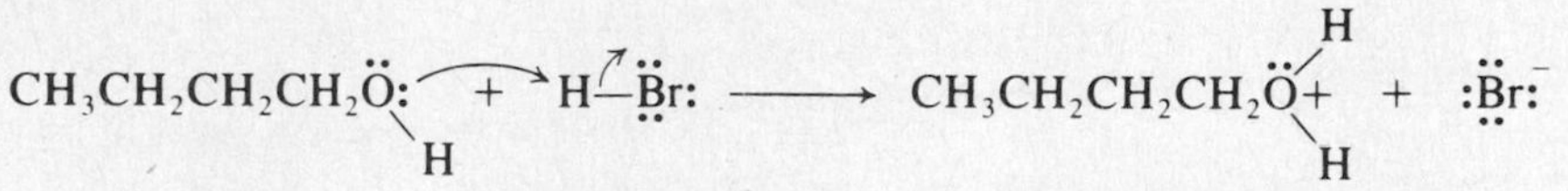

1-Butanol — Hydrogen bromide — Butyloxonium ion — Bromide

Displacement of water by bromide

$$:\ddot{B}r:^- + CH_3CH_2CH_2CH_2{-}\ddot{O}^+H_2 \longrightarrow CH_3CH_2CH_2CH_2Br + H\ddot{O}H$$

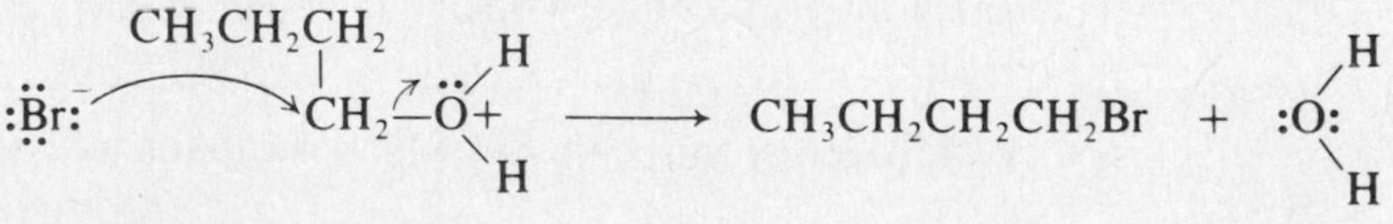

Bromide ion — Butyloxonium ion — 1-Bromobutane — Water

The reaction of 2-butanol with hydrogen bromide involves a carbocation intermediate.

Protonation of the alcohol

$$CH_3CH_2CH(\ddot{O}H)CH_3 + H{-}\ddot{B}r: \longrightarrow CH_3CH_2CH(\ddot{O}^+H_2)CH_3 + :\ddot{B}r:^-$$

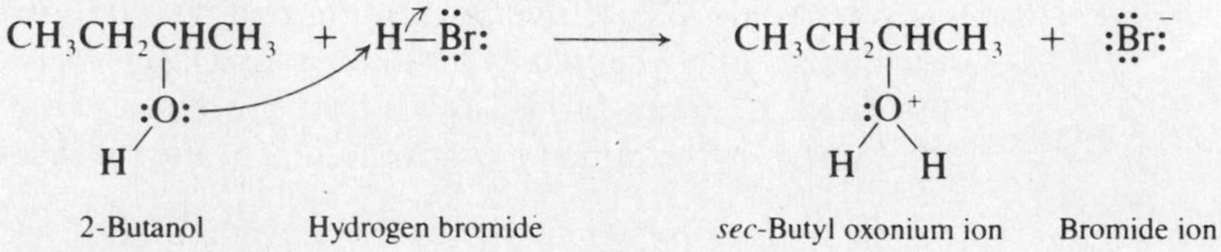

2-Butanol — Hydrogen bromide — *sec*-Butyl oxonium ion — Bromide ion

Dissociation of the oxonium ion

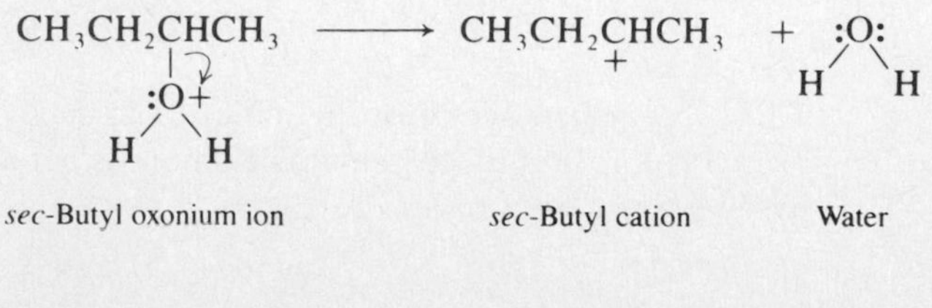

$$CH_3CH_2CH(\ddot{O}^+H_2)CH_3 \longrightarrow CH_3CH_2\overset{+}{C}HCH_3 + H\ddot{O}H$$

sec-Butyl oxonium ion — *sec*-Butyl cation — Water

Capture of *sec*-butyl cation by bromide

$$:\ddot{\underset{..}{Br}}:^{-} + \underset{+}{\overset{CH_3CH_2}{\overset{|}{C}HCH_3}} \longrightarrow \underset{\underset{Br}{|}}{CH_3CH_2CHCH_3}$$

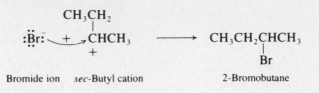

Bromide ion *sec*-Butyl cation 2-Bromobutane

4.13 (*b*) Writing the equations for carbon-carbon bond cleavage in propane and in 2-methylpropane, we see that a primary ethyl radical is produced by a cleavage of propane while a secondary isopropyl radical is produced by cleavage of 2-methylpropane.

$$CH_3CH_2—CH_3 \longrightarrow CH_3\dot{C}H_2 + \cdot CH_3$$

Propane Ethyl radical Methyl radical

$$\overset{CH_3}{\overset{|}{CH_3CHCH_3}} \longrightarrow CH_3\dot{C}HCH_3 + \cdot CH_3$$

2-Methylpropane Isopropyl radical Methyl radical

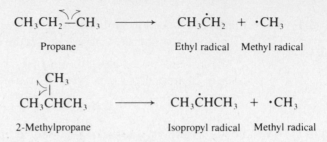

A secondary radical is more stable than a primary one, so carbon-carbon bond cleavage of 2-methylpropane requires less energy than carbon-carbon bond cleavage of propane.

(*c*) Carbon-carbon bond cleavage of 2,2-dimethylpropane gives a tertiary radical.

$$CH_3\overset{CH_3}{\underset{CH_3}{C}}CH_3 \longrightarrow CH_3—\overset{CH_3}{\underset{CH_3}{C\cdot}} + \cdot CH_3$$

2,2-Dimethylpropane *tert*-Butyl radical Methyl radical

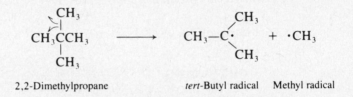

As noted in part (*b*), a secondary radical is produced on carbon-carbon bond cleavage of 2-methylpropane. Therefore we expect a lower carbon-carbon bond dissociation energy for 2,2-dimethylpropane than for 2-methylpropane, as a tertiary radical is more stable than a secondary one.

4.14 Writing the structural formula for ethyl chloride reveals that there are two nonequivalent sets of hydrogen atoms, either of which is capable of being replaced by chlorine.

$$CH_3CH_2Cl \xrightarrow[\text{light or heat}]{Cl_2} CH_3CHCl_2 + ClCH_2CH_2Cl$$

Ethyl chloride 1,1-Dichloroethane 1,2-Dichloroethane

The two dichlorides are 1,1-dichloroethane and 1,2-dichloroethane.

4.15 Propane has six primary versus two secondary hydrogens. In the chlorination of propane the relative proportions of hydrogen atom removal are given by the product of the statistical distribution and the relative rate per hydrogen. Given that a secondary hydrogen is abstracted 3.9 times as fast as a primary one, we write the expression for the amount of chlorination at the primary relative to that at the secondary position as:

$$\frac{\text{Number of primary hydrogens} \times \text{rate of abstraction of a primary hydrogen}}{\text{Number of secondary hydrogens} \times \text{rate of abstraction of a secondary hydrogen}}$$

$$= \frac{6 \times 1}{2 \times 3.9} = \frac{0.77}{1.00}$$

Thus, the percentage of propyl chloride formed is 0.77/1.77, or 43 percent, and that of isopropyl chloride is 57 percent. (The amounts actually observed are propyl 45 percent, isopropyl 55 percent.)

4.16 (*a*) Cyclobutanol has a hydroxyl group attached to a four-membered ring.

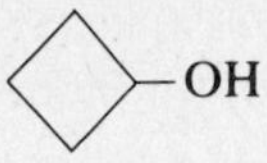

Cyclobutanol—OH   Cyclobutanol

(*b*) *sec*-Butyl alcohol is the common name for 2-butanol.

$$CH_3\underset{\displaystyle OH}{\underset{|}{C}}HCH_2CH_3 \qquad \textit{sec}\text{-Butyl alcohol}$$

(*c*) The hydroxyl group is at C-3 of an unbranched seven-carbon chain in 3-heptanol.

$$CH_3CH_2\underset{\displaystyle OH}{\underset{|}{C}}HCH_2CH_2CH_2CH_3 \qquad \text{3-Heptanol}$$

(*d*) 1-Pentadecanol has 15 carbon atoms in an unbranched chain. The hydroxyl group is attached at the end of the chain.

$$CH_3(CH_2)_{13}CH_2OH \qquad \text{1-Pentadecanol}$$

(*e*) A chlorine at C-2 is on the opposite side of the ring from the C-1 hydroxyl group in *trans*-2-chlorocyclopentanol. Note that it is not necessary to assign a number to the carbon that bears the hydroxyl group; naming the compound as a derivative of cyclopentanol automatically requires the hydroxyl group to be located at C-1.

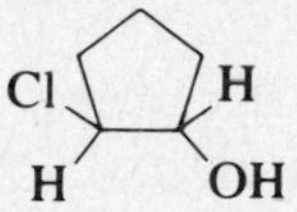

Cl, H, H, OH   *trans*-2-Chlorocyclopentanol

(*f*) The longest continuous chain has six carbon atoms in 4-methyl-2-hexanol. The hydroxyl group is at C-2, the methyl is at C-4.

$$CH_3\underset{\displaystyle OH}{\underset{|}{C}}HCH_2\underset{\displaystyle CH_3}{\underset{|}{C}}HCH_2CH_3 \qquad \text{4-Methyl-2-hexanol}$$

(*g*) This compound is an alcohol in which the longest continuous chain that incorporates the hydroxyl function has eight carbons. It bears chlorine substituents at C-2 and C-6 and methyl and hydroxyl groups at C-4.

$$CH_3\underset{\displaystyle Cl}{\underset{|}{C}}HCH_2\overset{\displaystyle CH_3}{\overset{|}{\underset{\displaystyle OH}{\underset{|}{C}}}}CH_2\underset{\displaystyle Cl}{\underset{|}{C}}HCH_2CH_3 \qquad \text{2,6-Dichloro-4-methyl-4-octanol}$$

(*h*) The hydroxyl group is at C-1 in *trans*-4-*tert*-butylcyclohexanol; the *tert*-butyl group is at C-4. The structure of the compound can be represented as shown below at the left; the structure at the right depicts it in its most stable conformation.

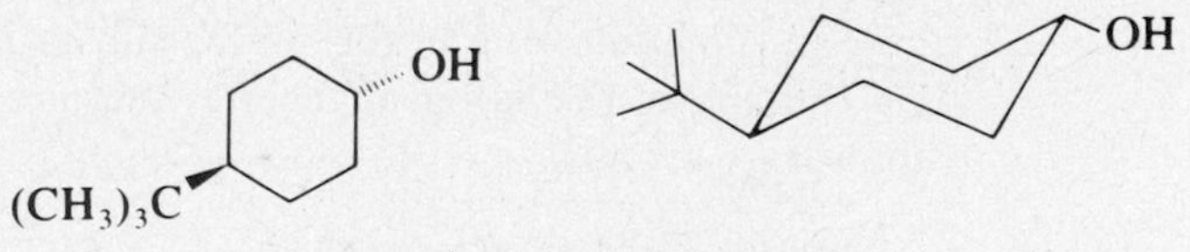

OH $(CH_3)_3C$ OH

trans-4-*tert*-Butylcyclohexanol

(*i*) Each carbon bears a substituent in this compound; it is named as a derivative of 2-propanol, with the halogen substituents cited in alphabetical order.

$$BrCH_2\underset{\displaystyle OH}{\underset{|}{C}}HCH_2I \qquad \text{1-Bromo-3-iodo-2-propanol}$$

(*j*) The cyclopropyl group is on the same carbon as the hydroxyl group in 1-cyclopropylethanol.

$$\text{(cyclopropyl)}-\underset{\displaystyle CH_3}{\underset{|}{CHOH}} \qquad \text{1-Cyclopropylethanol}$$

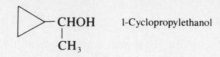

(*k*) The cyclopropyl group and the hydroxyl group are on adjacent carbons in 2-cyclopropylethanol.

$$\text{(cyclopropyl)}-CH_2CH_2OH \qquad \text{2-Cyclopropylethanol}$$

(*l*) Chlorotrifluoromethane has a single carbon, which bears a chlorine and three fluorine substituents.

$$F-\overset{\displaystyle F}{\overset{|}{\underset{\displaystyle Cl}{\underset{|}{C}}}}-F \qquad \text{Chlorotrifluoromethane}$$

4.17 (*a*) This compound has an unbranched nine-carbon chain and so is named as a functionally substituted derivative of nonane. It bears an iodo substituent at C-1; it is 1-iodononane.

$$CH_3(CH_2)_8I \qquad \text{1-Iodononane}$$

(*b*) This compound has a five-carbon chain that bears a methyl substituent and a bromine. The numbering scheme that gives the lower number to the substituent closest to the end of the chain is chosen. Therefore, bromine is at C-1 and methyl is a substituent at C-4.

$$CH_3\underset{\displaystyle CH_3}{\underset{|}{C}}HCH_2CH_2CH_2Br \qquad \text{1-Bromo-4-methylpentane}$$

(*c*) This compound has the same carbon skeleton as the compound in the previous problem but bears a hydroxyl group in place of the bromine and so is named as a derivative of 1-pentanol.

$$CH_3\underset{\displaystyle CH_3}{\underset{|}{C}}HCH_2CH_2CH_2OH \qquad \text{4-Methyl-1-pentanol}$$

(*d*) This molecule is a derivative of ethane and bears three chlorines and one bromine. The name 2-bromo-1,1,1-trichloroethane gives a lower number at the first point of difference than 1-bromo-2,2,2-trichloroethane and is the correct IUPAC name.

$$Cl_3CCH_2Br \qquad \text{2-Bromo-1,1,1-trichloroethane}$$

(*e*) This compound is a constitutional isomer of the preceding one. Regardless of which carbon the numbering begins at, the substitution pattern is 1,1,2,2. Therefore, alphabetical ranking of the halogens dictates the direction of numbering. Begin with the carbon that bears bromine.

$$Cl_2CH\underset{\displaystyle Cl}{\underset{|}{C}}HBr \qquad \text{1-Bromo-1,2,2-trichloroethane}$$

(*f*) This is a trifluoro derivative of ethanol. The direction of numbering is dictated by the hydroxyl group, which is at C-1 in ethanol.

$$CF_3CH_2OH \qquad \text{2,2,2-Trifluoroethanol}$$

(*g*) Italicized prefixes are not considered when ranking substituents alphabetically. Therefore, *tert*-butyl precedes fluoro. Both substituents are on the same side of the ring, so they are cis.

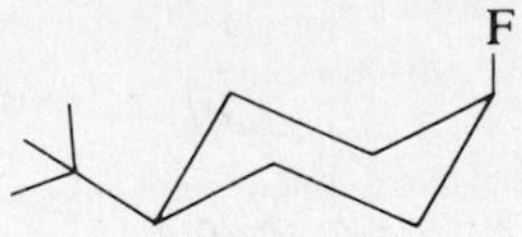

cis-1-*tert*-Butyl-4-fluorocyclohexane

(*h*) Here the compound is named as a derivative of cyclohexanol, so numbering begins at the carbon that bears the hydroxyl group.

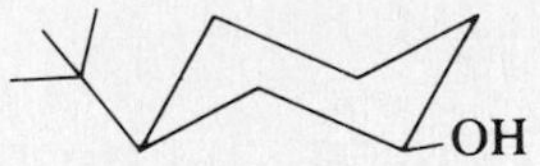

cis-3-*tert*-Butylcyclohexanol

(*i*) This alcohol has its hydroxyl group attached to C-2 of a three-carbon continuous chain; it is named as a derivative of 2-propanol.

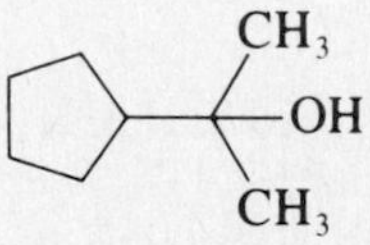

2-Cyclopentyl-2-propanol

(*j*) The longest continuous chain that contains the hydroxyl group has six carbons. Number in the direction that assigns the lower number to the hydroxyl group, placing the hydroxyl group at C-2 and the two methyl groups at C-5.

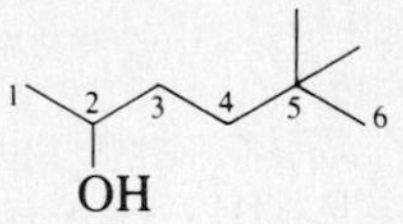

5,5-Dimethyl-2-hexanol

(*k*) The six carbons that form the longest continuous chain have substituents at C-2, C-3, and C-5 when numbering proceeds in the direction that gives the lowest locants to substituents at the first point of difference. The substituents are cited in alphabetical order.

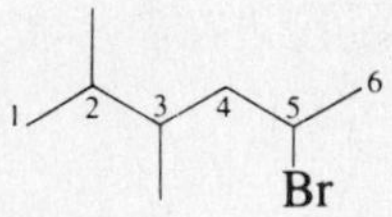

5-Bromo-2,3-dimethylhexane

Had numbering begun in the opposite direction, the locants would be 2,4,5 rather than 2,3,5.

(*l*) Now hydroxyl controls the numbering because the compound is named as an alcohol.

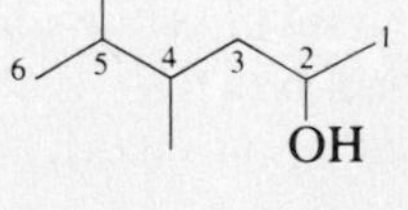

4,5-Dimethyl-2-hexanol

4.18 (*a*) A single carbon bears two chlorines and two fluorines; Cl_2CF_2 is dichlorodifluoromethane. The halogens are cited in alphabetical order; no numerical locants are needed.

(*b*) This compound is named as a derivative of ethane.

$ClCF_2CF_2Cl$ 1,2-Dichloro-1,1,2,2-tetrafluoroethane

(*c*) F_2CBrCl is bromochlorodifluoromethane.

(*d*) The correct name for $F_3CCHBrCl$ is 2-bromo-2-chloro-1,1,1-trifluoroethane.

(*e*) A cyclobutane ring has eight fluorine substituents. Since all the possible positions bear fluorines, no numerical locants are necessary.

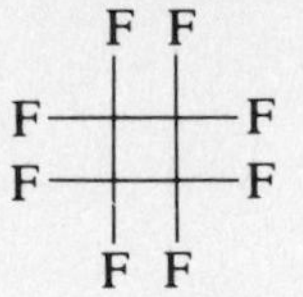

Octafluorocyclobutane

4.19 Primary alcohols are alcohols in which the hydroxyl group is attached to a carbon atom which has one alkyl substituent and two hydrogens. There are four primary alcohols which have the molecular formula $C_5H_{12}O$.

$CH_3CH_2CH_2CH_2CH_2OH$ — 1-Pentanol

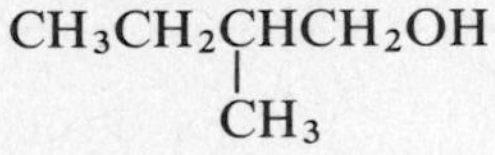

$CH_3CH_2CH(CH_3)CH_2OH$ — 2-Methyl-1-butanol

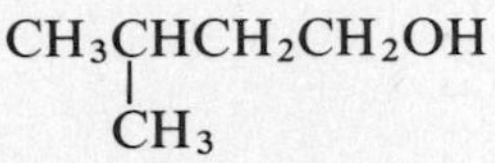

$CH_3CH(CH_3)CH_2CH_2OH$ — 3-Methyl-1-butanol

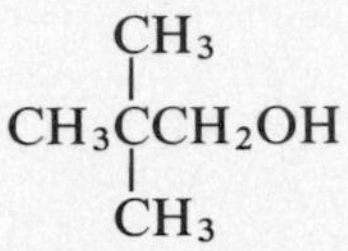

$CH_3C(CH_3)_2CH_2OH$ — 2,2-Dimethyl-1-propanol

Secondary alcohols are alcohols in which the hydroxyl group is attached to a carbon atom which has two alkyl substituents and one hydrogen. There are three secondary alcohols of molecular formula $C_5H_{12}O$.

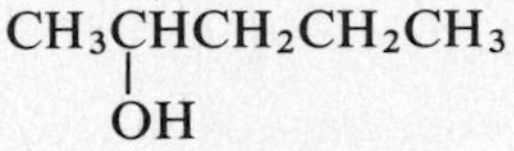

$CH_3CH(OH)CH_2CH_2CH_3$ — 2-Pentanol

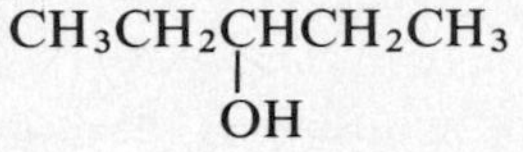

$CH_3CH_2CH(OH)CH_2CH_3$ — 3-Pentanol

$CH_3CH(CH_3)CH(OH)CH_3$ — 3-Methyl-2-butanol

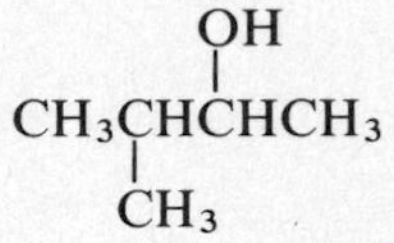

Only 2-methyl-2-butanol is a tertiary alcohol (three alkyl substituents on the hydroxyl-bearing carbon).

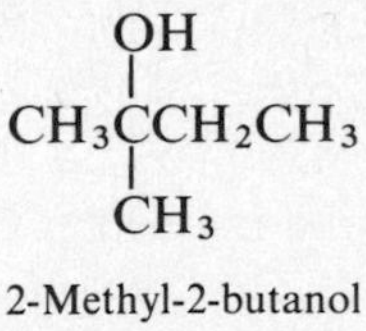

$CH_3C(OH)(CH_3)CH_2CH_3$

2-Methyl-2-butanol

4.20 The first methylcyclohexanol to be considered is 1-methylcyclohexanol. The preferred chair conformation will have the larger methyl group in an equatorial orientation while the smaller hydroxyl group will be axial.

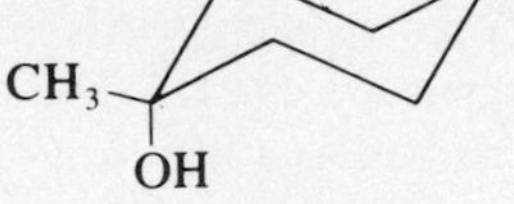

CH_3, OH

Most stable conformation of 1-methylcyclohexanol

In the other isomers methyl and hydroxyl will be in a 1,2, 1,3, or 1,4 relationship and can be cis or trans in each. We can write the preferred conformation by recognizing that the methyl group will always be equatorial and the hydroxyl either equatorial or axial.

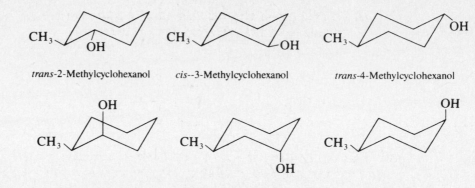

trans-2-Methylcyclohexanol *cis*--3-Methylcyclohexanol *trans*-4-Methylcyclohexanol

cis-2-Methylcyclohexanol *trans*-3-Methylcyclohexanol *cis*-4-Methylcyclohexanol

4.21 The assumption is incorrect for the 3-methylcyclohexanols. *cis*-3-Methylcyclohexanol is more stable than *trans*-3-methylcyclohexanol because the methyl group and the hydroxyl group are both equatorial in the cis while one substituent must be axial in the trans.

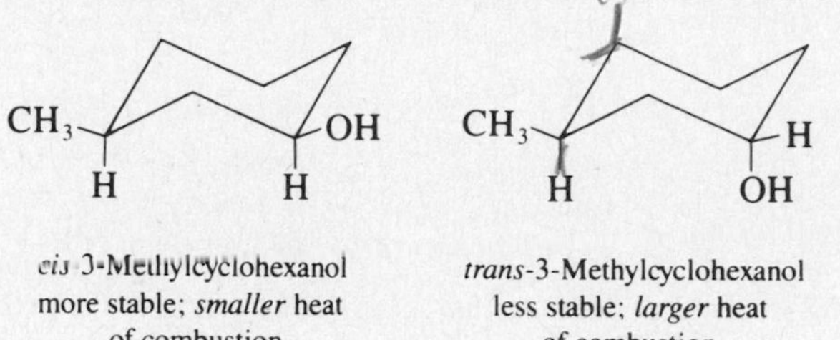

cis-3-Methylcyclohexanol more stable; *smaller* heat of combustion

trans-3-Methylcyclohexanol less stable; *larger* heat of combustion

4.22 (*a*) The most stable conformation will be the one with all the substituents equatorial.

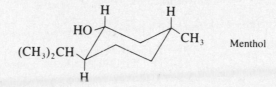

Menthol

The hydroxyl group is down. It is trans to the isopropyl group (which is up) and cis to the methyl group (which is down).

(*b*) Instead of all three substituents being equatorial, one or two of them may be axial. Since neomenthol is the second most stable stereoisomer, we choose the structure with *one* axial substituent. Further, we choose the structure with the smallest substituent (the hydroxyl group) as the axial one. Therefore, neomenthol is as shown below:

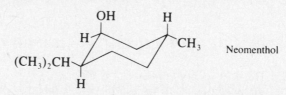

Neomenthol

4.23 In all these reactions the negatively charged atom abstracts a proton from an acid.

(*a*)

$$HI + HO^- \rightleftharpoons I^- + H_2O$$

Hydrogen iodide: acid (stronger acid, $K_a \sim 10^{10}$); Hydroxide ion: base; Iodide ion: conjugate base; Water: conjugate acid (weaker acid, $K_a \sim 10^{-16}$)

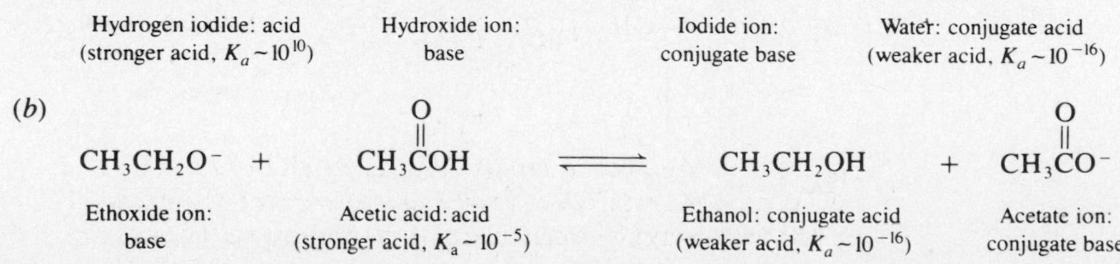

(*b*)

$$CH_3CH_2O^- + CH_3\overset{\overset{O}{\|}}{C}OH \rightleftharpoons CH_3CH_2OH + CH_3\overset{\overset{O}{\|}}{C}O^-$$

Ethoxide ion: base; Acetic acid: acid (stronger acid, $K_a \sim 10^{-5}$); Ethanol: conjugate acid (weaker acid, $K_a \sim 10^{-16}$); Acetate ion: conjugate base

(c) $HF + H_2N^- \rightleftharpoons F^- + H_3N$

Hydrogen fluoride: acid (stronger acid, $K_a \sim 10^{-4}$); Amide ion: base; Fluoride ion: conjugate base; Ammonia: conjugate acid (weaker acid, $K_a \sim 10^{-36}$)

(d) $CH_3\overset{O}{\overset{\|}{C}}O^- + HCl \rightleftharpoons CH_3\overset{O}{\overset{\|}{C}}OH + Cl^-$

Acetate ion: base; Hydrogen chloride: acid (stronger acid, $K_a \sim 10^7$); Acetic acid: conjugate acid (weaker acid, $K_a \sim 10^{-5}$); Chloride ion: conjugate base

(e) $(CH_3)_3CO^- + H_2O \rightleftharpoons (CH_3)_3COH + HO^-$

tert-Butoxide ion: base; Water: acid (stronger acid, $K_a \sim 10^{-16}$); *tert*-Butyl alcohol: conjugate acid (weaker acid, $K_a \sim 10^{-18}$); Hydroxide ion: conjugate base

(f) $(CH_3)_2CHOH + H_2N^- \rightleftharpoons (CH_3)_2CHO^- + H_3N$

Isopropyl alcohol: acid (stronger acid, $K_a \sim 10^{-17}$); Amide ion: base; Isopropoxide ion: conjugate base; Ammonia: conjugate acid (weaker acid, $K_a \sim 10^{-36}$)

(g) $F^- + H_2SO_4 \rightleftharpoons HF + HSO_4^-$

Fluoride ion: base; Sulfuric acid: acid (stronger acid, $K_a \sim 10^5$); Hydrogen fluoride: conjugate acid (weaker acid, $K_a \sim 10^{-4}$); Hydrogen sulfate ion: conjugate base

4.24 This problem illustrates the reactions of a primary alcohol with the reagents described in the chapter.

(a) $2\,CH_3CH_2CH_2CH_2OH + 2\,Li \longrightarrow 2\,CH_3CH_2CH_2CH_2O^-Li^+ + H_2$

1-Butanol → Lithium butoxide

(b) $2\,CH_3CH_2CH_2CH_2OH + 2\,Na \longrightarrow 2\,CH_3CH_2CH_2CH_2O^-Na^+ + H_2$

Sodium butoxide

(c) $2\,CH_3CH_2CH_2CH_2OH + 2\,K \longrightarrow 2\,CH_3CH_2CH_2CH_2O^-K^+ + H_2$

Potassium butoxide

(d) $CH_3CH_2CH_2CH_2OH + NaNH_2 \longrightarrow CH_3CH_2CH_2CH_2O^-Na^+ + NH_3$

Sodium butoxide

(e) $CH_3CH_2CH_2CH_2OH \xrightarrow[\text{heat}]{HBr} CH_3CH_2CH_2CH_2Br$

1-Bromobutane

(f) $CH_3CH_2CH_2CH_2OH \xrightarrow[\text{heat}]{NaBr,\ H_2SO_4} CH_3CH_2CH_2CH_2Br$

1-Bromobutane

(g) $CH_3CH_2CH_2CH_2OH \xrightarrow{PBr_3} CH_3CH_2CH_2CH_2Br$

1-Bromobutane

(h) $CH_3CH_2CH_2CH_2OH \xrightarrow{SOCl_2} CH_3CH_2CH_2CH_2Cl$

1-Chlorobutane

4.25 Similar reactions occur with secondary alcohols.

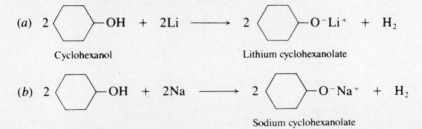

(a) 2 (cyclohexyl)–OH + 2Li ⟶ 2 (cyclohexyl)–O^-Li^+ + H_2

Cyclohexanol → Lithium cyclohexanolate

(b) 2 (cyclohexyl)–OH + 2Na ⟶ 2 (cyclohexyl)–O^-Na^+ + H_2

Sodium cyclohexanolate

(c) 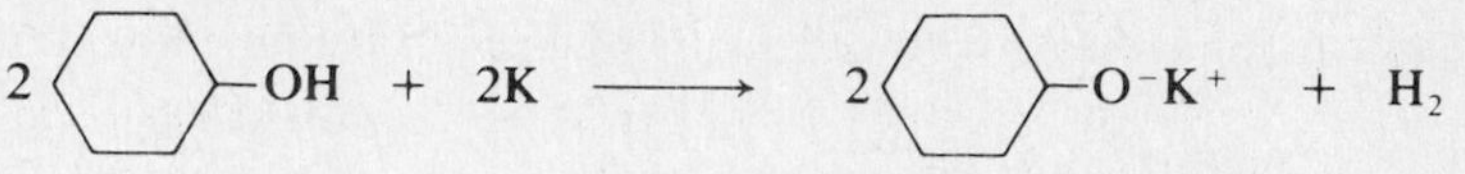 2 (cyclohexyl)–OH + 2K ⟶ 2 (cyclohexyl)–O^-K^+ + H_2

Potassium cyclohexanolate

(d) 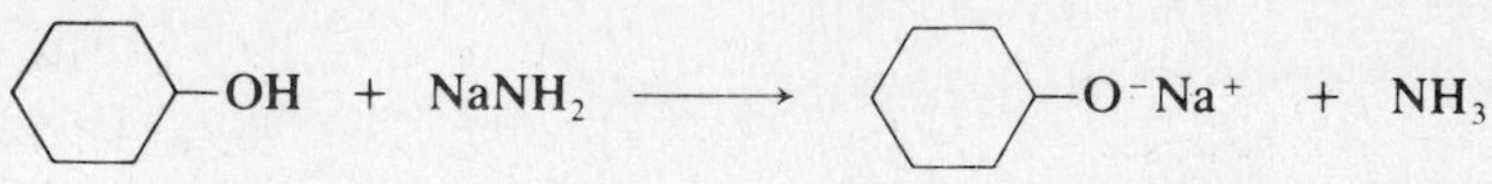 (cyclohexyl)–OH + $NaNH_2$ ⟶ (cyclohexyl)–O^-Na^+ + NH_3

Sodium cyclohexanolate

(e) 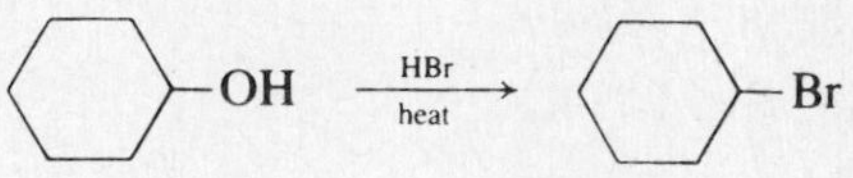 (cyclohexyl)–OH $\xrightarrow[\text{heat}]{\text{HBr}}$ (cyclohexyl)–Br

Bromocyclohexane

(f) 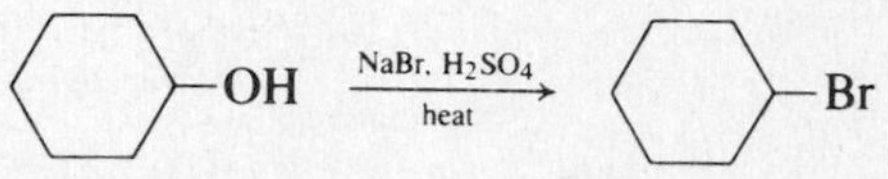 (cyclohexyl)–OH $\xrightarrow[\text{heat}]{\text{NaBr, } H_2SO_4}$ (cyclohexyl)–Br

Bromocyclohexane

(g) 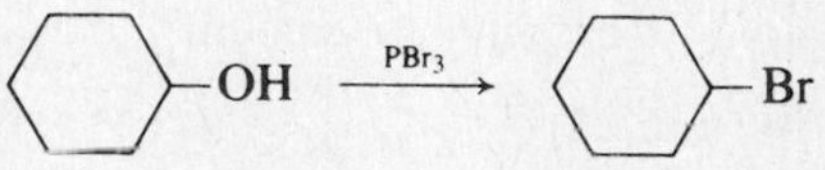 (cyclohexyl)–OH $\xrightarrow{PBr_3}$ (cyclohexyl)–Br

Bromocyclohexane

(h) 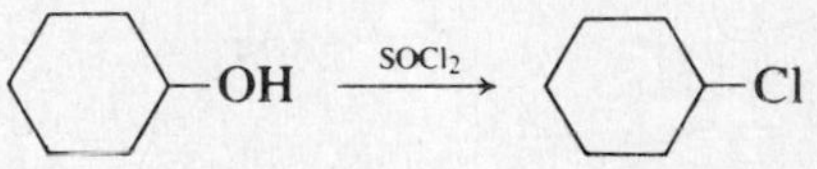 (cyclohexyl)–OH $\xrightarrow{SOCl_2}$ (cyclohexyl)–Cl

Chlorocyclohexane

4.26 (a) This reaction was used to convert the primary alcohol to the corresponding bromide in 60 percent yield.

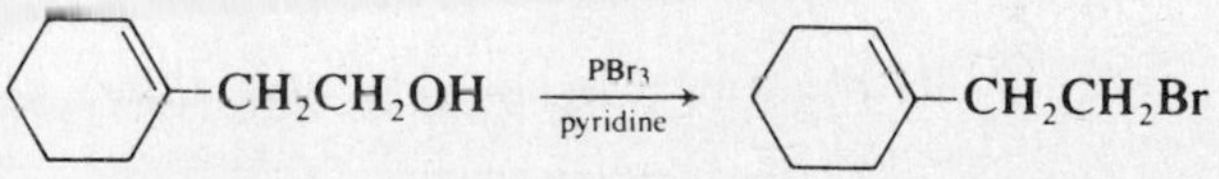

(cyclohexenyl)–CH_2CH_2OH $\xrightarrow[\text{pyridine}]{PBr_3}$ (cyclohexenyl)–CH_2CH_2Br

(b) Thionyl chloride treatment of this secondary alcohol gave the chloro derivative in 59 percent yield.

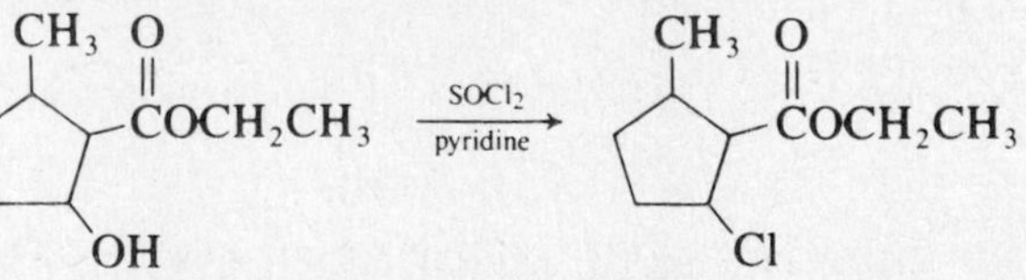

(cyclopentane with CH_3, $\overset{O}{\overset{\|}{C}}OCH_2CH_3$, OH) $\xrightarrow[\text{pyridine}]{SOCl_2}$ (cyclopentane with CH_3, $\overset{O}{\overset{\|}{C}}OCH_2CH_3$, Cl)

(c) The starting material is a tertiary alcohol and reacted readily with hydrogen chloride to form the corresponding chloride in 67 percent yield.

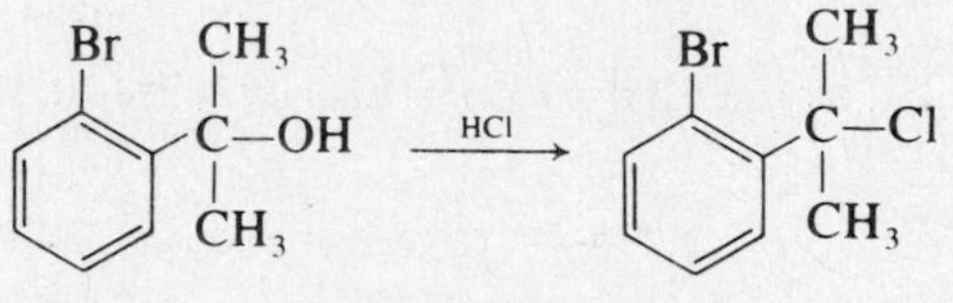

(o-bromophenyl)–$C(CH_3)_2$–OH $\xrightarrow{HCl}$ (o-bromophenyl)–$C(CH_3)_2$–Cl

(d) Both primary alcohol functional groups were converted to primary bromides; the yield was 88 percent.

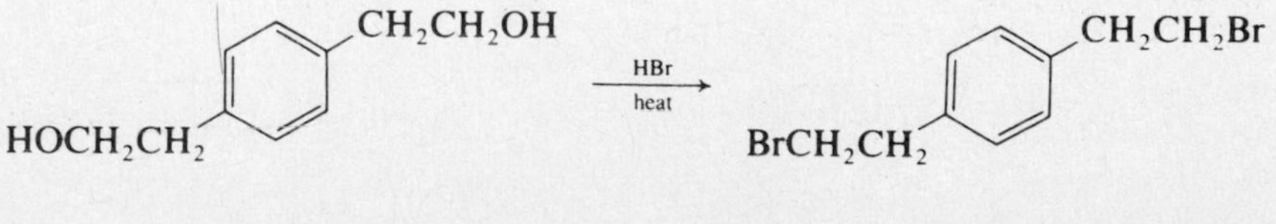

$HOCH_2CH_2$–C_6H_4–CH_2CH_2OH $\xrightarrow[\text{heat}]{\text{HBr}}$ $BrCH_2CH_2$–C_6H_4–CH_2CH_2Br

(*e*) Alcohols react with potassium to give potassium alkoxides, usually in quantitative yield.

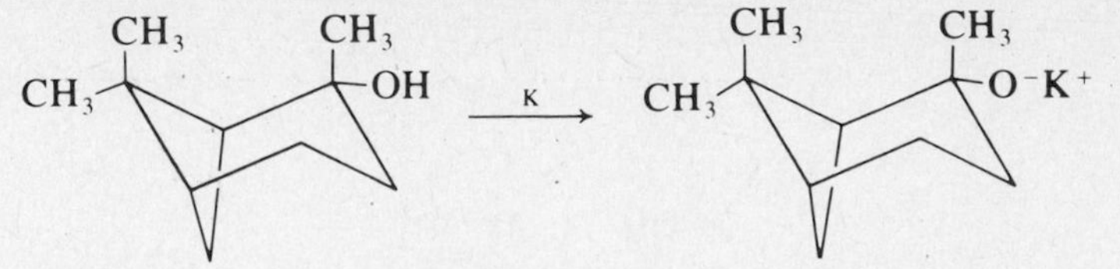

(*f*) The substance that is chlorinated is adamantane. All adamantane's methylene hydrogens are equivalent to each other, and all its methine hydrogens are equivalent. Thus, only two monochlorides are possible.

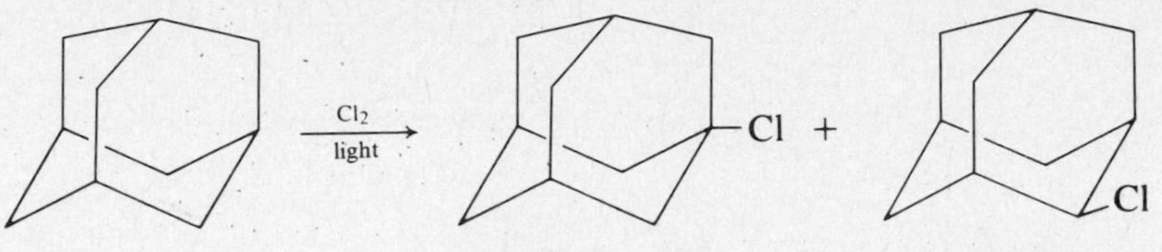

Adamantane $\xrightarrow[\text{light}]{Cl_2}$ 1-Chloroadamantane + 2-Chloroadamantane

(*g*) Bromination is very selective for substitution of tertiary hydrogens. The reactant, 2,3-dimethylbutane, has two equivalent tertiary hydrogens, both of which are replaced in the reaction with two moles of bromine to give an 89 percent yield of the dibromide.

$$(CH_3)_2CHCH(CH_3)_2 \xrightarrow[h\nu,\ \text{heat}]{2\ Br_2} (CH_3)_2C(Br)C(Br)(CH_3)_2$$

2,3-Dimethylbutane → 2,3-Dibromo-2,3-dimethylbutane

4.27 The order of reactivity of alcohols with hydrogen halides is tertiary > secondary > primary > methyl.

$$ROH + HBr \longrightarrow RBr + H_2O$$

Reactivity of Alcohols with Hydrogen Bromide

Part	*More reactive alcohol*	*Less reactive alcohol*
(*a*)	$CH_3CH(OH)CH_2CH_3$ 2-Butanol: secondary	$CH_3CH_2CH_2CH_2OH$ 1-Butanol: primary
(*b*)	$CH_3CH(OH)CH_2CH_3$ 2-Butanol: secondary	$CH_3CH_2CH(CH_3)CH_2OH$ 2-Methyl-1-butanol: primary
(*c*)	$(CH_3)_2C(OH)CH_2CH_3$ 2-Methyl-2-butanol: tertiary	$CH_3CH(OH)CH_2CH_3$ 2-Butanol: secondary
(*d*)	$CH_3CH(OH)CH_2CH_3$ 2-Butanol	$CH_3CH(CH_3)CH_2CH_3$ 2-Methylbutane: not an alcohol; does not react with HBr

Reactivity of Alcohols with Hydrogen Bromide (*continued*)

Part	*More reactive alcohol*	*Less reactive alcohol*
(*e*)	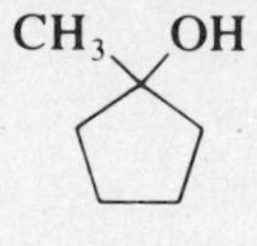 1-Methylcyclopentanol: tertiary	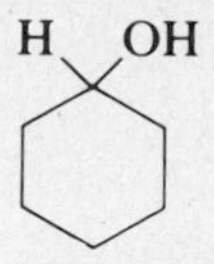 Cyclohexanol: secondary
(*f*)	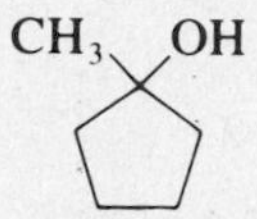 1-Methylcyclopentanol: tertiary	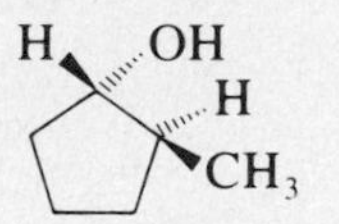 *trans*-2-Methylcyclopentanol: secondary
(*g*)	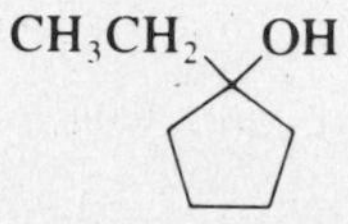 1-Ethylcyclopentanol: tertiary	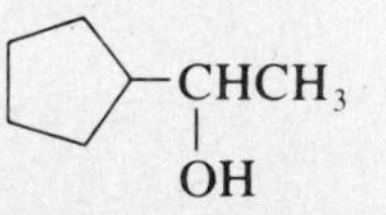 1-Cyclopentylethanol: secondary

4.28 (*a*) The ideal bond angles at the positively charged carbon in carbocations are 120° (sp^2 hybridization). Cyclopropyl cation, because of its geometry, has a C—$\overset{+}{C}$—C bond angle that is much smaller than 120°. Cyclopropyl cation is relatively unstable because angle strain at its positively charged carbon precludes sp^2 hybridization.

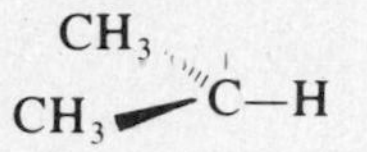

Isopropyl cation (more stable)  — Cyclopropyl cation (less stable)

(*b*) Carbocations are stabilized by electron release from substituents. Fluorine is very electronegative and a CF_3 group is far less able to donate electrons to a carbocation than is a CH_3 group. Indeed, the effect of a CF_3 group is to withdraw rather than to donate electrons.

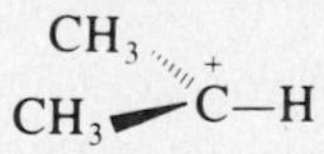

Isopropyl cation (more stable) 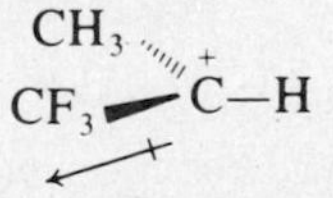 — 1-(Trifluoromethyl)ethyl cation (less stable)

4.29 Free radicals and carbocations are both stabilized by substituents that donate electrons to a carbon atom with an unfilled $2p$ orbital. A carbocation is much more sensitive to electron release from alkyl substituents than is a free radical because it has one less electron than a free radical and is positively charged.

4.30 The acid dissociation constants of ethanol and 2,2,2-trifluoroethanol are represented by the equations shown:

$$\underset{\text{Ethanol}}{CH_3CH_2OH} \rightleftharpoons \underset{\text{Proton}}{H^+} + \underset{\text{Ethoxide ion}}{CH_3CH_2O^-}$$

$$\underset{\text{2,2,2-Trifluoroethanol}}{CF_3CH_2OH} \rightleftharpoons \underset{\text{Proton}}{H^+} + \underset{\text{2,2,2-Trifluoroethoxide ion}}{CF_3CH_2O^-}$$

Structural features that stabilize the products cause the position of equilibrium to favor products relative to reactants. Fluorine substituents attract electrons from the negatively charged oxygen of 2,2,2-trifluoroethoxide ion, disperse the charge, and stabilize the anion. The position of equilibrium lies farther to the right for ionization of 2,2,2-trifluoroethanol than for ethanol because 2,2,2-trifluoroethoxide ion is more stabilized than is ethoxide ion. 2,2,2-Trifluoroethanol is the stronger acid; its measured K_a is 4×10^{-12} (pK_a 11.4). The acid dissociation constant of ethanol is 10^{-16} (pK_a 16).

4.31 Examine the equations to ascertain which bonds are made and which are broken. Then use the bond dissociation energies in Table 4.4 to calculate ΔH^0 for each reaction.

(*a*) $(CH_3)_2CH{-}OH + H{-}F \longrightarrow (CH_3)_2CH{-}F + H{-}OH$

92 kcal/mol 136 kcal/mol 105 kcal/mol 119 kcal/mol
Bond breaking: 228 kcal/mol Bond making: 224 kcal/mol

$$\Delta H^0 = \text{energy cost of breaking bonds} - \text{energy given off in making bonds}$$

$$\Delta H^0 = 228 \text{ kcal/mol} - 224 \text{ kcal/mol}$$

$$\Delta H^0 = +4 \text{ kcal/mol}$$

The reaction of isopropyl alcohol with hydrogen fluoride is endothermic.

(*b*) $(CH_3)_2CH{-}OH + H{-}Cl \longrightarrow (CH_3)_2CH{-}Cl + H{-}OH$

92 kcal/mol 103 kcal/mol 81 kcal/mol 119 kcal/mol
Bond breaking: 195 kcal/mol Bond making: 200 kcal/mol

$$\Delta H^0 = \text{energy cost of breaking bonds} - \text{energy given off in making bonds}$$

$$\Delta H^0 = 195 \text{ kcal/mol} - 200 \text{ kcal/mol}$$

$$\Delta H^0 = -5 \text{ kcal/mol}$$

The reaction of isopropyl alcohol with hydrogen chloride is exothermic.

(*c*) $CH_3\underset{\displaystyle H}{\underset{|}{C}}HCH_3 + H{-}Cl \longrightarrow (CH_3)_2CH{-}Cl + H{-}H$

94.5 kcal/mol 103 kcal/mol 81 kcal/mol 104 kcal/mol
Bond breaking: 197.5 kcal/mol Bond making: 185 kcal/mol

$$\Delta H^0 = \text{energy cost of breaking bonds} - \text{energy given off in making bonds}$$

$$\Delta H^0 = 197.5 \text{ kcal/mol} - 185 \text{ kcal/mol}$$

$$\Delta H^0 = +12.5 \text{ kcal/mol}$$

The reaction of propane with hydrogen chloride is endothermic.

4.32 In the statement of the problem you are told that the starting material is neopentane, that the reaction is one of fluorination, meaning that F_2 is a reactant, and that the product is $(CF_3)_4C$. You need to complete the equation by realizing that HF is also formed in the fluorination of alkanes. Therefore, the balanced equation is:

$$(CH_3)_4C + 12\,F_2 \longrightarrow (CF_3)_4C + 12\,HF$$

4.33 The reaction is free-radical chlorination, and substitution occurs at all possible positions that bear a replaceable hydrogen. Write the structure of the starting material and identify the nonequivalent hydrogens.

$$Cl{-}\overset{\displaystyle F}{\underset{\displaystyle F}{C}}{-}\overset{\displaystyle Cl}{\underset{\displaystyle H}{C}}{-}CH_3$$

1,2-Dichloro-1,1-difluoropropane

The problem states that one of the products is 1,2,3-trichloro-1,1-difluoropropane. This compound arises by substitution of one of the methyl hydrogens by chlorine. We are told that the other product is an isomer of 1,2,3-trichloro-1,1-difluoropropane; therefore, it must be formed by replacement of the hydrogen at C-2.

$$\mathrm{Cl{-}C(F)_2{-}C(Cl)(H){-}CH_2Cl} \qquad \mathrm{Cl{-}C(F)_2{-}C(Cl)_2{-}CH_3}$$

1,2,3-Trichloro-1,1-difluoropropane 1,2,2-Trichloro-1,1-difluoropropane

4.34 Free-radical chlorination leads to substitution at each distinct position that bears a hydrogen substituent. This problem essentially requires you to recognize structures that possess various numbers of nonequivalent hydrogens.

(*a*) Neopentane is the C_5H_{12} isomer that gives a single monochloride, since all the hydrogens in neopentane are equivalent.

$$\mathrm{CH_3C(CH_3)_2CH_3} \xrightarrow[\text{light}]{\mathrm{Cl_2}} \mathrm{CH_3C(CH_3)_2CH_2Cl}$$

Neopentane Neopentyl chloride

(*b*) The C_5H_{12} isomer that has three nonequivalent sets of hydrogens is pentane. It yields three isomeric monochlorides on free-radical chlorination.

$$\mathrm{CH_3CH_2CH_2CH_2CH_3} \xrightarrow{\mathrm{Cl_2}} \begin{cases} \mathrm{ClCH_2CH_2CH_2CH_2CH_3} & \text{1-Chloropentane} \\ \mathrm{CH_3CH(Cl)CH_2CH_2CH_3} & \text{2-Chloropentane} \\ \mathrm{CH_3CH_2CH(Cl)CH_2CH_3} & \text{3-Chloropentane} \end{cases}$$

Pentane

(*c*) Isopentane forms four different monochlorides.

$$\mathrm{CH_3CH(CH_3)CH_2CH_3} \xrightarrow{\mathrm{Cl_2}} \begin{cases} \mathrm{ClCH_2CH(CH_3)CH_2CH_3} & \text{1-Chloro-2-methylbutane} \\ \mathrm{(CH_3)_2C(Cl)CH_2CH_3} & \text{2-Chloro-2-methylbutane} \\ \mathrm{(CH_3)_2CHCH(Cl)CH_3} & \text{2-Chloro-3-methylbutane} \\ \mathrm{(CH_3)_2CHCH_2CH_2Cl} & \text{1-Chloro-3-methylbutane} \end{cases}$$

Isopentane

(*d*) In order that only two dichlorides be formed, the starting alkane must have a structure that is rather symmetrical; that is, one in which most (or all) the hydrogens are equivalent. Neopentane satisfies this requirement.

$$\mathrm{CH_3C(CH_3)_2CH_3} \xrightarrow[\text{light}]{\mathrm{2Cl_2}} \mathrm{CH_3C(CH_3)_2CHCl_2} + \mathrm{ClCH_2C(CH_3)_2CH_2Cl}$$

Neopentane 1,1-Dichloro-2,2-dimethylpropane 1,3-Dichloro-2,2-dimethylpropane

4.35 (*a*) Heptane has five methylene groups, which on chlorination together contribute 85 percent of the total monochlorinated product.

$$\underset{}{CH_3(CH_2)_5CH_3} \longrightarrow \underset{15\%}{CH_3(CH_2)_5CH_2Cl} + \underset{85\%}{(\text{2-chloro} + \text{3-chloro} + \text{4-chloro})}$$

Since the problem specifies that attack at each methylene group is equally probable, the five methylene groups each give rise to 85/5, or 17 percent of the monochloride product.

Since C-2 and C-6 of heptane are equivalent to each other, we calculate that 2-chloroheptane will constitute 34 percent of the monochloride fraction. Similarly, C-3 and C-5 are equivalent, so there should be 34 percent of 3-chloroheptane. The remainder, 17 percent, is 4-chloroheptane.

These predictions are very close to the observed proportions.

	Calculated, %	*Observed, %*
2-Chloro	34	35
3-Chloro	34	34
4-Chloro	17	16

(*b*) There are a total of 20 methylene hydrogens in dodecane, $CH_3(CH_2)_{10}CH_3$. The 19 percent 2-chlorododecane that is formed arises by substitution of any of the four equivalent methylene hydrogens at C-2 and C-11. Therefore, the total amount of substitution of methylene hydrogens must be:

$$\frac{20}{4} \times 19\% = 95\%$$

The remaining 5 percent corresponds to substitution of methyl hydrogens at C-1 and C-12. The proportion of 1-chlorododecane in the monochloride fraction is 5 percent.

4.36 A molecular weight of 72 corresponds to a molecular formula of C_5H_{12}. Since only one monochloride is formed, all the hydrogens must be equivalent, which indicates that the compound is neopentane.

$$CH_3C(CH_3)_2CH_3 \xrightarrow[\text{light}]{Cl_2} CH_3C(CH_3)_2CH_2Cl \xrightarrow[\text{light}]{Cl_2} ClCH_2C(CH_3)_2CH_2Cl + CH_3C(CH_3)_2CHCl_2$$

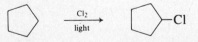

4.37 The only molecular formula possible for a hydrocarbon that has a molecular weight of 70 is C_5H_{10}. Since only one monochloride is formed, all the hydrogens in this compound must be equivalent. The hydrocarbon is cyclopentane.

$$\text{Cyclopentane} \xrightarrow[\text{light}]{Cl_2} \text{Cyclopentyl chloride}$$

Cyclopentane Cyclopentyl chloride

4.38 There are two isomers having the molecular formula C_3H_7Cl, 1-chloropropane and 2-chloropropane.

$$CH_3CH_2CH_2Cl \xrightarrow[\text{light}]{Cl_2} ClCH_2CH_2CH_2Cl + CH_3CHClCH_2Cl + CH_3CH_2CHCl_2$$

1-Chloropropane

$$CH_3CHClCH_3 \xrightarrow[\text{light}]{Cl_2} CH_3CHClCH_2Cl + CH_3CCl_2CH_3$$

2-Chloropropane

Therefore A is 2-chloropropane and B is 1-chloropropane.

4.39 (*a*) The monobromination of 2,2-dimethylbutane can give two different primary bromides, C and D

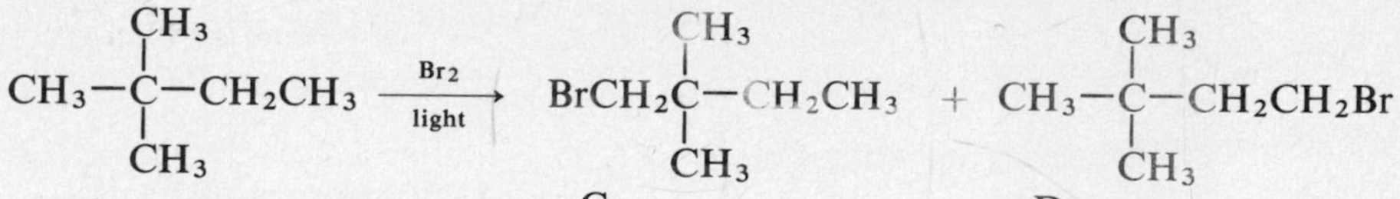

$$\underset{\text{2,2-Dimethylbutane}}{CH_3-\overset{CH_3}{\underset{CH_3}{\overset{|}{\underset{|}{C}}}}-CH_2CH_3} \xrightarrow[\text{light}]{Br_2} \underset{\text{1-Bromo-2,2-dimethylbutane}}{\underset{C}{BrCH_2\overset{CH_3}{\underset{CH_3}{\overset{|}{\underset{|}{C}}}}-CH_2CH_3}} + \underset{\text{1-Bromo-3,3-dimethylbutane}}{\underset{D}{CH_3-\overset{CH_3}{\underset{CH_3}{\overset{|}{\underset{|}{C}}}}-CH_2CH_2Br}}$$

and a secondary bromide, E

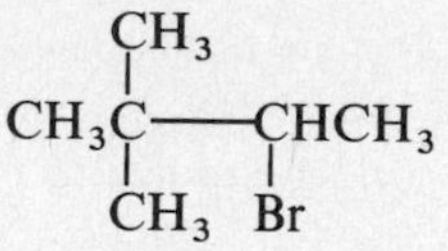

$$\underset{E}{CH_3\overset{CH_3}{\underset{CH_3}{\overset{|}{\underset{|}{C}}}}-\underset{Br}{\underset{|}{C}}HCH_3} \quad \text{3-Bromo-2,2-dimethylbutane}$$

(*b*) If we let x equal the rate of abstraction of a primary hydrogen, then $82x$ is the rate of abstraction of a secondary hydrogen and c, d, and e, the relative amounts of products C, D, and E, respectively, are given by:

$$c = 9x$$
$$d = 3x$$
$$e = 2(82)x$$

Normalizing to 100 percent we get

$$9x + 3x + 164x = 100$$

$$x = \frac{100}{176} = 0.57$$

and

$$c = 9(0.57) = 5.1\% \text{ 1-bromo-2,2-dimethylbutane}$$
$$d = 3(0.57) = 1.7\% \text{ 1-bromo-3,3-dimethylbutane}$$
$$e = 93.2\% \text{ 2-bromo-3,3-dimethylbutane}$$

4.40 (*a*) Acid-catalyzed hydrogen-deuterium exchange takes place by a pair of Brönsted acid base reactions.

$$\underset{\text{Base}}{R-\ddot{\underset{..}{O}}-H} + \underset{\text{Acid}}{D_3O^+} \rightleftharpoons \underset{\text{Conjugate acid}}{R-\overset{D}{\overset{|}{\underset{..}{O^+}}}-H} + \underset{\text{Conjugate base}}{D_2O}$$

$$\underset{\text{Acid}}{R-\overset{D}{\overset{|}{\underset{..}{O^+}}}-H} + \underset{\text{Base}}{D_2O} \rightleftharpoons \underset{\text{Conjugate base}}{R-\ddot{\underset{..}{O}}-D} + \underset{\text{Conjugate acid}}{D-\overset{H}{\overset{|}{\underset{..}{O^+}}}-D}$$

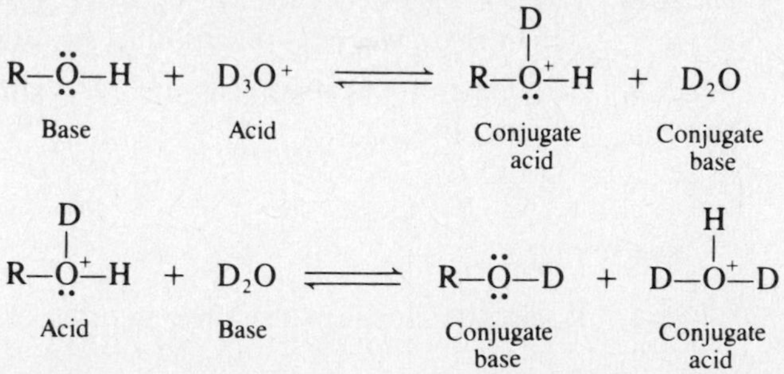

(*b*) Base-catalyzed hydrogen-deuterium exchange occurs by a different pair of Brönsted acid-base equilibria.

$$\underset{\text{Acid}}{R-\ddot{\underset{..}{O}}-H} + \underset{\text{Base}}{DO^-} \rightleftharpoons \underset{\text{Conjugate base}}{R-\ddot{\underset{..}{O}}:^-} + \underset{\text{Conjugate acid}}{DOH}$$

$$\underset{\text{Base}}{R-\ddot{\underset{..}{O}}:^-} + \underset{\text{Acid}}{D_2O} \rightleftharpoons \underset{\text{Conjugate acid}}{R-\ddot{\underset{..}{O}}-D} + \underset{\text{Conjugate base}}{DO^-}$$

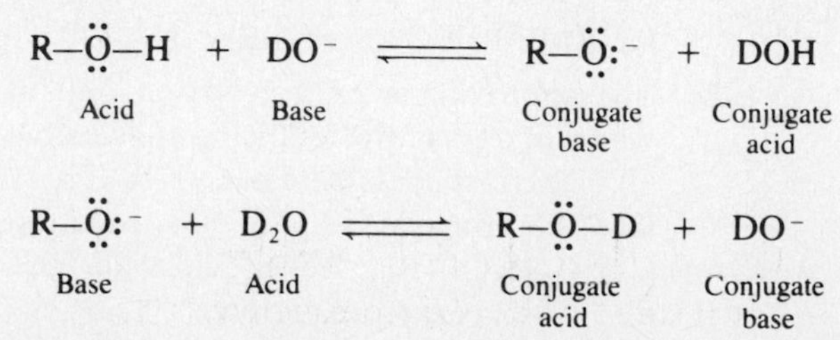

SELF-TEST

PART A

A-1. Give the correct IUPAC systematic name for the following compounds:

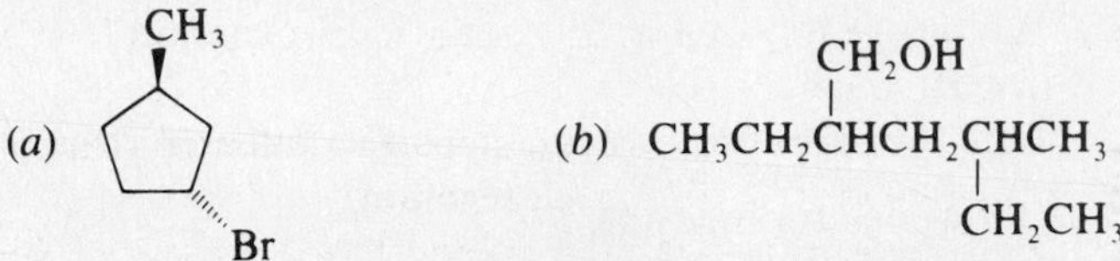

A-2. Draw the structures of the following substances:
(*a*) 2-Chloro-1-iodo-2-methylheptane
(*b*) *cis*-3-Isopropylcyclohexanol

A-3. What are the structures of the conjugate acid and the conjugate base of CH_3OH?

A-4. The acid ionization constant of acetic acid is 1.8×10^{-5}. Calculate the pK_a of this substance.

A-5. Supply the missing component for each of the following reactions:

(*a*) $(CH_3)_2CHCH_2OH \xrightarrow{SOCl_2} ?$

(*b*) $? \xrightarrow{HBr} CH_3CH_2\overset{\displaystyle Br}{\overset{|}{C}}(CH_3)_2$

(*c*) cyclohexyl–OH $\xrightarrow{?}$ cyclohexyl–$O^- Na^+$

A-6. Write the products of the acid-base reaction below, and identify the stronger acid and base and the conjugate of each. Will the equilibrium lie to the left ($K < 1$) or to the right ($K > 1$)? (The pK_a values are given in Problem B-2.)

$$CH_3CH_2O^- + NH_3 \rightleftharpoons$$

A-7. (*a*) How many different free radicals can possibly be produced in the reaction between chlorine atoms and 2,4-dimethylpentane? (*b*) Write their structures. (*c*) Which is the most stable? Which is the least stable?

A-8. Write a balanced chemical equation for the reaction of chlorine with the pentane isomer that gives only one product on monochlorination.

A-9. Write the propagation steps for the light-initiated reaction of bromine with methylcyclohexane.

PART B

B-1. Rank the following substances in order of increasing boiling point (lowest → highest):

$CH_3CH_2CH_2CH_2OH$	$(CH_3)_2CHOCH_3$	$(CH_3)_3COH$
A	B	C

(*a*) A < B < C (*b*) B < A < C (*c*) B < C < A (*d*) C < B < A

B-2. The approximate pK_a of NH_3 is 36; that of C_2H_5OH is 16. From this information which (if any) of the following conclusions is correct?
(*a*) The conjugate base of C_2H_5OH is stronger.
(*b*) The conjugate base of NH_3 is stronger.
(*c*) C_2H_5OH is a weaker acid than NH_3.
(*d*) None of these is correct.

B-3. When 2-propanol is treated with metallic sodium:
(*a*) Sodium isopropoxide is formed
(*b*) The sodium is oxidized
(*c*) The acidic hydrogen is reduced
(*d*) All these occur

B-4. The activation energy (E_{act}) of a given reaction is unrelated to which of the following parameters?
(*a*) The rate of the slowest step of a multistep reaction
(*b*) The rate of the overall reaction
(*c*) The heat absorbed or given off by the reaction (ΔH)
(*d*) The stability of the transition state

B-5. What is the decreasing stability order (most stable → least stable) of the following carbocations?

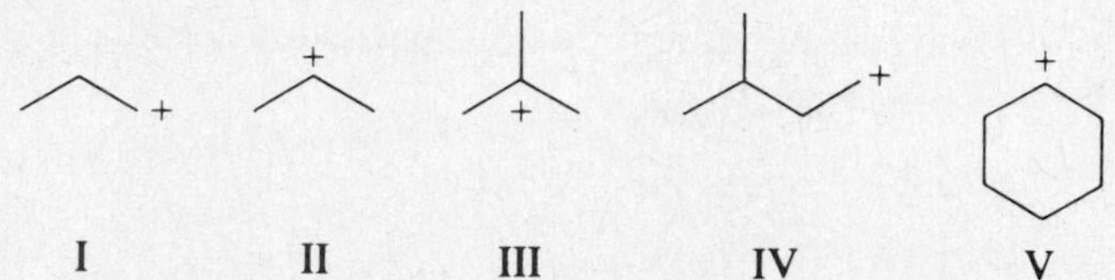

(*a*) III > II > I > IV > V
(*b*) I ≈ IV > II ≈ V > III
(*c*) III > II ≈ V > I ≈ IV
(*d*) III > I ≈ IV > II ≈ V

B-6. Which of the following is least able to serve as a nucleophile in a chemical reaction?
(*a*) Br^- (*b*) OH^- (*c*) NH_3 (*d*) CH_3^+

B-7. Thiols are alcohol analogs in which the oxygen has been replaced by sulfur (e.g., CH_3SH). Given the fact that the S—H bond is less polar than the O—H bond, which of the following statements comparing thiols and alcohols is correct?
(*a*) Hydrogen bonding forces are weaker in thiols.
(*b*) Hydrogen bonding forces are stronger in thiols.
(*c*) Hydrogen bonding forces would be the same.
(*d*) No comparison can be made without additional information.

B-8. Rank the *transition states* which occur during the following reaction steps in order of increasing stability (least → most stable):

(A) $CH_3-\overset{+}{O}H_2 \longrightarrow CH_3^+ + H_2O$

(B) $(CH_3)_3C-\overset{+}{O}H_2 \longrightarrow (CH_3)_3C^+ + H_2O$

(C) $(CH_3)_2CH-\overset{+}{O}H_2 \longrightarrow (CH_3)_2CH^+ + H_2O$

(*a*) A < B < C (*b*) B < C < A (*c*) A < C < B (*d*) B < A < C

B-9. Using the data from appendix B Table B-1, calculate the heat of reaction ΔH^0 for the following

$$CH_3CH_2\cdot + HBr \longrightarrow CH_3CH_3 + Br\cdot$$

(*a*) +16.5 kcal/mol (*b*) +10.5 kcal/mol
(*c*) −16.5 kcal/mol (*d*) −10.5 kcal/mol

For the next three questions, consider the following free-radical reaction:

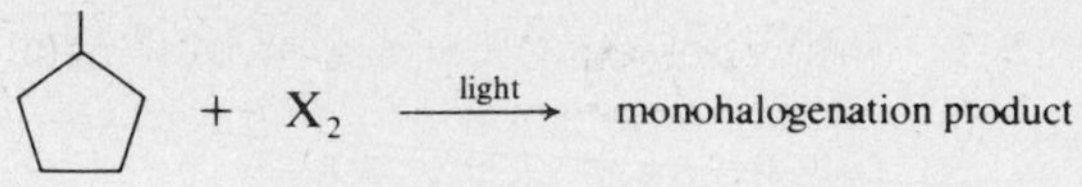

$$\text{(methylcyclopentane)} + X_2 \xrightarrow{\text{light}} \text{monohalogenation product}$$

B-10. Light is involved in which of the following reaction steps?
(*a*) Initiation only
(*b*) Propagation only
(*c*) Termination only
(*d*) Initiation and propagation

B-11. Which of the following statements about the reaction is *not* true?
(*a*) Halogen atoms are consumed in the first propagation step.
(*b*) Halogen atoms are regenerated in the second propagation step.
(*c*) Hydrogen atoms are produced in the first propagation step.
(*d*) Chain termination occurs when two radicals react with each other.

B-12. How many monohalogenation products are possible? (Do not consider stereoisomers.)
(*a*) 2 (*b*) 3 (*c*) 4 (*d*) 5

CHAPTER 5

ALKENES. STRUCTURE AND STABILITY

IMPORTANT TERMS AND CONCEPTS

Alkene Nomenclature (Sec. 5.1) Alkenes are hydrocarbons which contain a carbon-carbon double bond, which is the *functional group* that gives rise to their chemical and physical properties. The double bond is located in the hydrocarbon chain by numbering in the direction that gives the two double-bonded carbons the lower numbers. In the name the number of the first of these carbons precedes the name stem, the ending of which is *-ene* (replacing the *-ane* of the parent alkane).

$CH_2{=}CHCH_2CH_2CH_3$ — 1-Pentene

$CH_3CH_2CH_2CH{=}CHCH_3$ — 2-Hexene

Branched alkenes are named by using the longest chain *containing the double bond* as the parent name and identifying the substituents as usual.

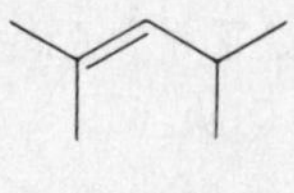

2,4-Dimethyl-2-pentene

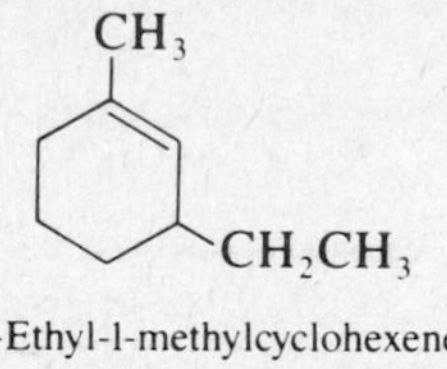

CH_3, CH_2CH_3

3-Ethyl-1-methylcyclohexene

Structure and Bonding (Sec. 5.3) The carbon-carbon double bond of an alkene consists of a σ (sigma) component and a π (pi) component. The carbons of the double bond are sp^2 hybridized. The three hybrid orbitals are utilized to form sigma bonds to three atoms, all of which lie in the same plane as the carbon atom. The remaining p orbitals of the double-bonded carbons are parallel to each other, giving rise to a π *orbital.* This arrangement results in the four atoms attached to the double bond being in one plane. Ethylene is a planar molecule:

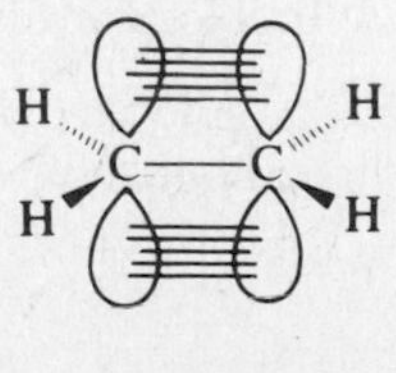

H H C C H H

Stereoisomerism (Sec. 5.5) The π overlap of an alkene double bond inhibits the rotation of one carbon atom relative to the other. The rotation about a carbon-carbon bond is imperceptibly slow; in fact, it is presumed not to occur. Thus *stereoisomers* may result which are not interconvertible. For example, there are two stereoisomers of 2-butene, known as *cis*-2-butene and *trans*-2-butene:

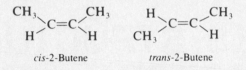

$$\underset{\textit{cis}\text{-2-Butene}}{\overset{CH_3\quad\quad CH_3}{\underset{H\quad\quad\quad H}{C{=}C}}} \qquad \underset{\textit{trans}\text{-2-Butene}}{\overset{H\quad\quad\quad CH_3}{\underset{CH_3\quad\quad H}{C{=}C}}}$$

E-Z System of Nomenclature (Sec. 5.6) Stereoisomeric alkenes may be named by establishing *priorities*, or rankings, of the pair of groups attached to each end of the double bond. The priority ranking is based on the *atomic number* of the groups. If the higher-ranking groups are on the *same* side of the double bond, the alkene is named Z; if they are on opposite sides of the double bond, it is named E. When the atoms attached directly to the double bond are the same, priority rankings are determined outward from the double bond until the first difference is found. For example,

$$\underset{(E)\text{-2-Chloro-2-pentene}}{\overset{CH_3\quad\quad CH_2CH_3}{\underset{Cl\quad\quad\quad H}{C{=}C}}}$$

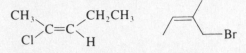

(*E*)-2-Chloro-2-pentene (*Z*)-1-Bromo-2-methyl-2-butene

Alkene Stabilities (Secs. 5.7 to 5.10) Alkyl substitution tends to *stabilize* an alkene. When stereoisomeric alkenes are compared, the trans isomer is generally more stable than the cis (cyclic alkenes would be a notable exception). The following stability series may be written in order of decreasing stability:

$$R_2C{=}CR_2 > R_2C{=}CHR > R_2C{=}CH_2 \sim RCH{=}CHR\ \text{(trans)}$$
$$> RCH{=}CHR\ \text{(cis)} > RCH{=}CH_2 > CH_2{=}CH_2$$

An explanation for the increased stability associated with substituted alkenes is the electron-releasing power of an alkyl group when compared with that of a hydrogen substituent on a double bond.

One method for measuring the stability of alkenes is to determine the heat energy released when the substance undergoes *hydrogenation*. These *heats of hydrogenation*, which decrease with increasing stabilization, may be used to compare the stabilities of both isomeric and nonisomeric alkenes.

Cyclic Alkenes (Secs. 5.11, 5.12) Cycloalkenes having fewer than 12 carbons in the ring are more stable in the cis configuration; in fact, a trans double bond is not stable in a ring smaller than 8 carbons.

Catalytic hydrogenation of alkenes proceeds by a syn addition of hydrogen to the double bond. As a result, hydrogenation of substituted cycloalkenes leads to one stereoisomer predominating over another.

Molecular Formulas (Sec. 5.13) Information regarding the presence or absence of double bonds and rings in a molecule may be gained by consideration of the molecular formula. The *sum of double bonds and rings* (SODAR), is calculated by the equation shown below, which takes into account the difference in H content between the molecule being considered and the saturated acyclic alkane having the same number of carbon atoms.

$$\text{SODAR} = \tfrac{1}{2}(\text{molecular formula of sat'd alkane} - \text{molecular formula of compound})$$

For example, a molecule with the formula C_8H_{12} has a SODAR value of 3; that is, there are a total of three double bonds or rings present.

$$\text{SODAR} = \tfrac{1}{2}(C_8H_{18} - C_8H_{12}) = 3$$

Frequently double bonds and rings may be distinguished from each other on the basis of hydrogenation data. Each double bond will react with one mole of H_2, thus providing a means of determining the number of rings present.

When an oxygen atom is present in the molecule, it is omitted from the SODAR calculation; a halogen is treated as if it were a hydrogen. For example

$$C_5H_9Cl \qquad \text{SODAR} = \tfrac{1}{2}(C_5H_{12} - C_5H_{10}) = 1$$

$$C_3H_8O \qquad \text{SODAR} = \tfrac{1}{2}(C_3H_8 - C_3H_8) = 0$$

IMPORTANT REACTION

Catalytic Hydrogenation (Secs. 5.9, 5.10, 5.12)

General:

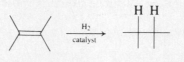

$$\text{C}=\text{C} \xrightarrow[\text{catalyst}]{H_2} \text{H}-\text{C}-\text{C}-\text{H}$$

Example:

$$\text{1,2-dimethylcyclohexene} \xrightarrow[\text{Pt}]{H_2} \textit{cis}\text{-1,2-dimethylcyclohexane}$$

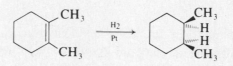

SOLUTIONS TO TEXT PROBLEMS

5.1 (*b*) Writing the structure in more detail, we see that the longest continuous chain contains four carbon atoms.

$$\overset{4}{C}H_3-\overset{3}{C}(CH_3)_2-\overset{2}{C}H=\overset{1}{C}H_2$$

The double bond is located at the end of the chain, so the alkene is named as a derivative of 1-butene. Two methyl groups are substituents at C-3. The correct IUPAC name is 3,3-dimethyl-1-butene.

(*c*) Expanding the structural formula reveals the molecule to be a methyl-substituted derivative of hexene.

$$\overset{1}{C}H_3-\overset{2}{C}(CH_3)=\overset{3}{C}H\overset{4}{C}H_2\overset{5}{C}H_2\overset{6}{C}H_3$$

The chain is numbered so as to give the lowest number to the doubly bonded carbons, which makes the alkene a derivative of 2-hexene (not 4-hexene). Its IUPAC name is 2-methyl-2-hexene.

(*d*) The IUPAC rules require that the chain be numbered in the direction that gives the lower numbers to the doubly bonded carbons rather than the direction that gives the lower numbers to alkyl substituents.

$$\overset{1}{C}H_3-\overset{2}{C}(CH_3)=\overset{3}{C}H\overset{4}{C}(CH_3)_2\overset{5}{C}H_3$$

2,4,4-Trimethyl-2-pentene
(not 2,2,4-trimethyl-3-pentene)

(*e*) First identify the longest continuous chain that includes the double bond, then number it in the direction that gives the lowest numbers to the doubly bonded carbons.

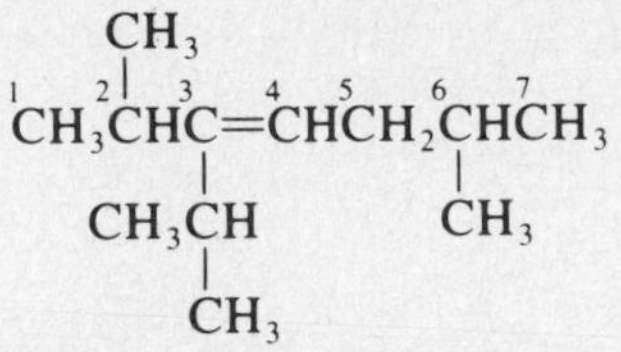

$$\overset{1}{C}H_3\overset{2}{C}H(CH_3)\overset{3}{C}(CH(CH_3)CH_3)=\overset{4}{C}H\overset{5}{C}H_2\overset{6}{C}H(CH_3)\overset{7}{C}H_3$$

This compound is named as a derivative of 3-heptene. It bears methyl substituents at C-2 and C-6, along with an isopropyl group at C-3. The substituents are listed in alphabetical order in the IUPAC name. This alkene is 3-isopropyl-2,6-dimethyl-3-heptene.

5.2 There are three sets of nonequivalent positions on a cyclopentene ring, identified as *a*, *b*, and *c* on the cyclopentene structure shown below.

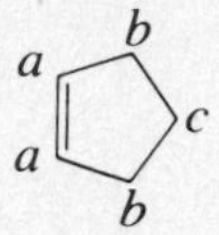

Thus, there are three different monochloro-substituted derivatives of cyclopentene. The carbons that bear the double bond are numbered C-1 and C-2 in each isomer, and the other positions are numbered in sequence in the direction that gives the chlorine-bearing carbon its lower locant.

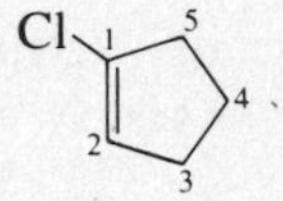

1-Chlorocyclopentene 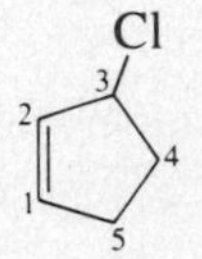 3-Chlorocyclopentene 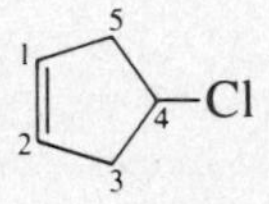 4-Chlorocyclopentene

5.3 Carbon-carbon bond distances decrease in the order sp^3-sp^3 > sp^3-sp^2 > sp^2-sp^2. The experimentally determined values for cyclobutene, shown below, agree completely with this generalization. The C-3—C-4 bond (sp^3-sp^3) is the longest, while the C-1—C-2 bond (sp^2-sp^2) is the shortest.

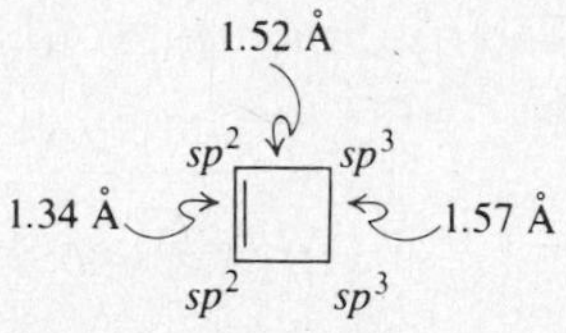

1.52 Å; 1.34 Å; 1.57 Å; sp^2, sp^3, sp^2, sp^3

5.4 Consider first the C_5H_{10} alkenes that have an unbranched carbon chain.

1-Pentene 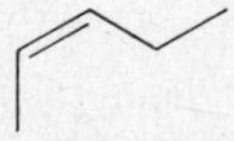 *cis*-2-Pentene 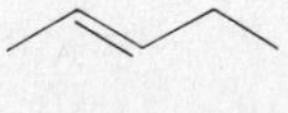 *trans*-2-Pentene

There are three additional isomers. These have a four-carbon chain with a methyl substituent.

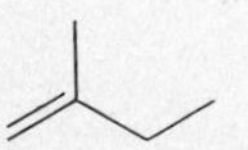

2-Methyl-1-butene 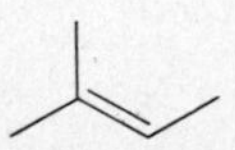 2-Methyl-2-butene 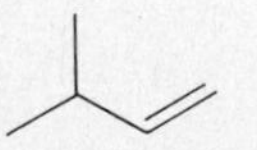 3-Methyl-1-butene

5.5 When the chlorine atoms are cis, the individual C—Cl bond dipoles reinforce each other, but when the chlorine atoms are trans, the individual bond dipoles cancel; *trans*-1,2-dichloroethene has no dipole moment.

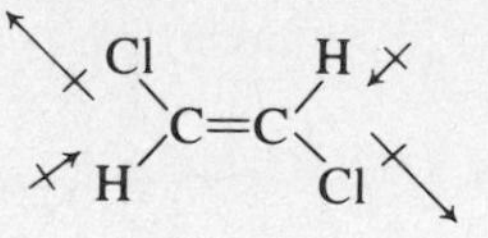

trans-1,2-Dichloroethene
μ 0

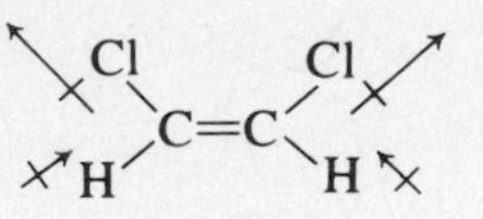

cis-1,2-Dichloroethene
μ 1.8 D

5.6 First, identify the constitution of 9-tricosene. Referring back to Table 2.1 in Section 2.8 of the text, we see that tricosane is the unbranched alkane containing 23 carbon atoms. 9-Tricosene, therefore, contains an unbranched chain of 23 carbons with a double bond between C-9 and C-10. Since the problem specifies that the pheromone has the cis configuration, the first 8 carbons and the last 13 must be on the same side of the C-9—C-10 double bond.

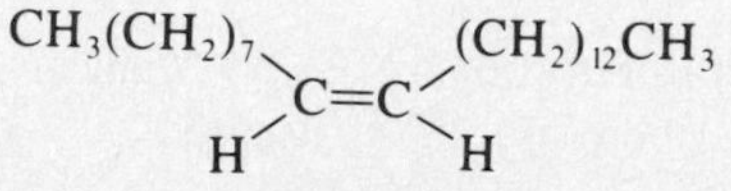

$CH_3(CH_2)_7$ $(CH_2)_{12}CH_3$
C=C
H H

cis-9-Tricosene

5.7 (*b*) One of the carbons bears a methyl group and a hydrogen; methyl is of higher rank than hydrogen. The other carbon bears the groups $-CH_2CH_2F$ and $-CH_2CH_2CH_2CH_3$. At the first point of difference between these two, fluorine is of higher atomic number than carbon, so $-CH_2CH_2F$ is of higher precedence.

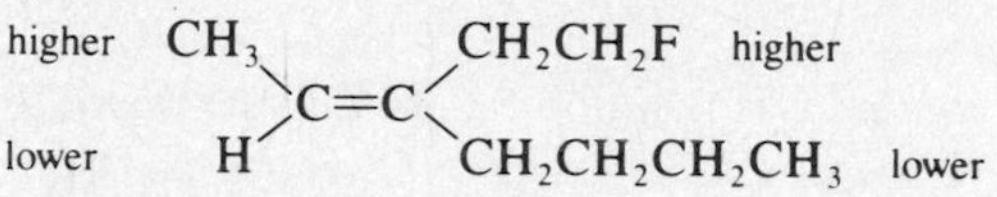

higher CH_3 CH_2CH_2F higher
C=C
lower H $CH_2CH_2CH_2CH_3$ lower

Higher-ranked substituents are on the same side of the double bond; the alkene has the *Z* configuration.

(*c*) One of the carbons bears a methyl group and a hydrogen; as we have seen, methyl is of higher priority. The other carbon bears $-CH_2CH_2OH$ and $-C(CH_3)_3$. Let us analyze these two groups so that we may determine their order of precedence.

$-CH_2CH_2OH$	$-C(CH_3)_3$
$-C(C,H,H)$	$-C(C,C,C)$
Lower priority	Higher priority

We examine the atoms one by one at the point of attachment before proceeding down the chain. Therefore, $-C(CH_3)_3$ outranks $-CH_2CH_2OH$.

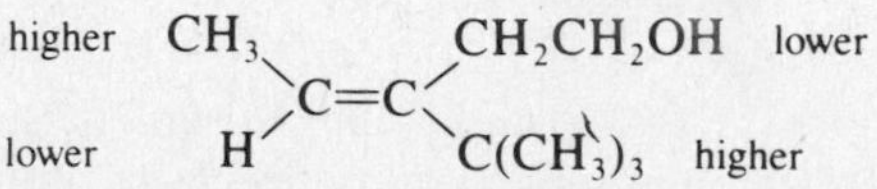

higher CH_3 CH_2CH_2OH lower
C=C
lower H $C(CH_3)_3$ higher

Higher-ranked substituents are on opposite sides; the configuration of the alkene is *E*.

(*d*) The cyclopropyl ring is attached to the double bond by a carbon that bears the substituents (C,C,H) and is therefore of higher precedence than an ethyl group —C(C,H,H).

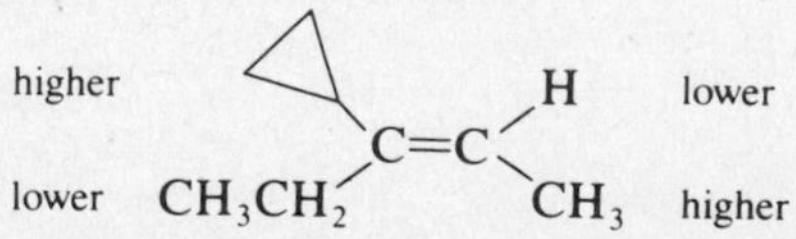

higher (cyclopropyl) H lower
C=C
lower CH_3CH_2 CH_3 higher

Higher-ranked substituents are on opposite sides; the configuration of the alkene is *E*.

5.8 In this problem we are given the constitution of the molecule and told that the double bond configuration is *E*. We have to translate that information into a structure that shows stereochemistry. Identify the substituents on the doubly bonded carbons and rank those on each carbon in order of decreasing precedence. The substituents on one of the doubly bonded carbons of the ant alarm pheromone are:

$CH_3CH_2CH(CH_3)-$ (Higher priority) and $H-$ (Lower priority)

At the other carbon the substituents are:

$-C(=O)CH_2CH_3$ (Higher priority, $-C(O,O,C)$) and $-CH_3$ (Lower priority, $-C(H,H,H)$)

The configuration is (*E*), so higher-ranked substituents are on opposite sides of the double bond. Therefore, the structure is:

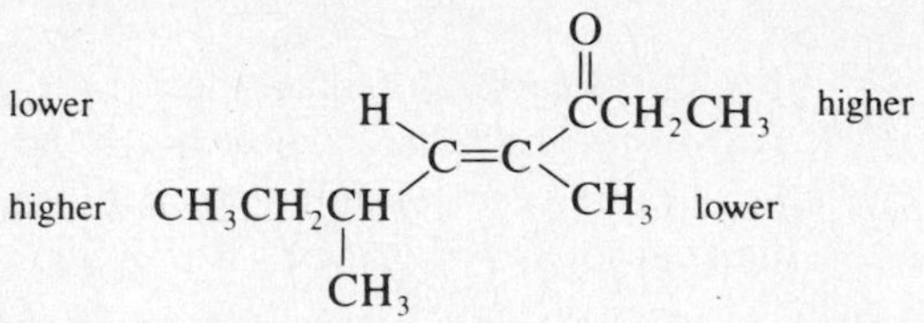

lower H and higher $CH_3CH_2CH(CH_3)$ on one carbon; $C=C$; higher $C(=O)CH_2CH_3$ and lower CH_3 on the other carbon.

5.9 By describing a double bond as *trisubstituted*, we mean that three of the four atoms attached to the doubly bonded carbons are carbon.

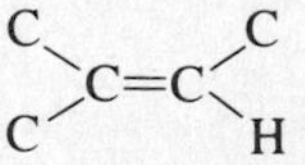

$(C)(C)C=C(C)(H)$ Trisubstituted double bond

Thus, the positions of five carbon atoms are fixed. The sixth carbon may be bonded to any of the three sp^3 hybridized carbons. Therefore, there are three isomeric C_6H_{12} alkenes that have trisubstituted double bonds, namely:

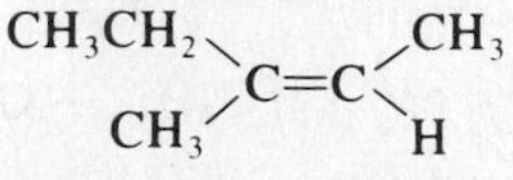

$(CH_3CH_2)(CH_3)C=C(CH_3)(H)$ — (*Z*)-3-Methyl-2-pentene

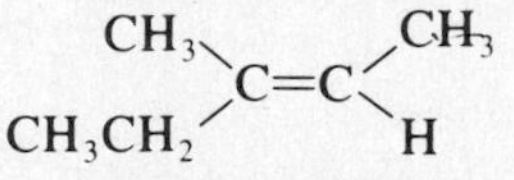

$(CH_3)(CH_3CH_2)C=C(CH_3)(H)$ — (*E*)-3-Methyl-2-pentene

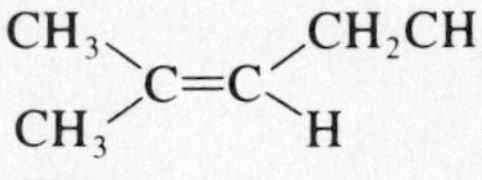

$(CH_3)(CH_3)C=C(CH_2CH_3)(H)$ — 2-Methyl-2-pentene

5.10 Apply the two general rules for alkene stability to rank these compounds. First, more highly substituted double bonds are more stable than less substituted ones. Second, when two double bonds are similarly constituted, the trans stereoisomer is more stable than the cis. Therefore, the predicted order of decreasing stability is:

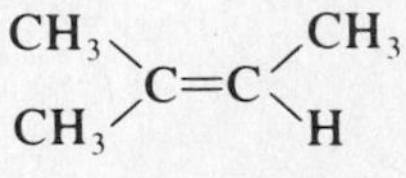

$(CH_3)(CH_3)C=C(CH_3)(H)$ — 2-Methyl-2-butene (trisubstituted): most stable

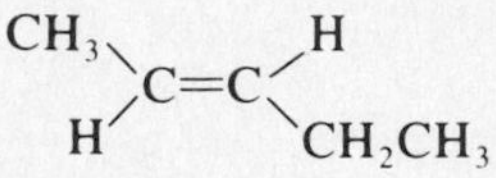

$(CH_3)(H)C=C(H)(CH_2CH_3)$ — (*E*)-2-Pentene (disubstituted)

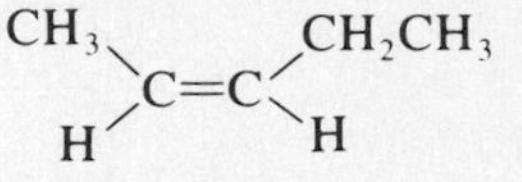

$(CH_3)(H)C=C(CH_2CH_3)(H)$ — (*Z*)-2-Pentene (disubstituted)

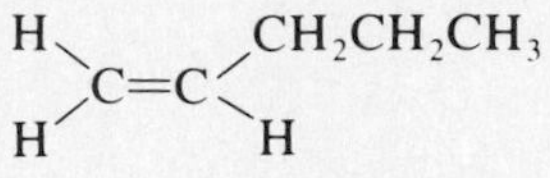

$(H)(H)C=C(CH_2CH_2CH_3)(H)$ — 1-Pentene (monosubstituted): least stable

5.11 Catalytic hydrogenation converts an alkene to an alkane having the same carbon skeleton. Since 2-methylbutane is the product of hydrogenation, all three alkenes must have a four-carbon chain with a one-carbon branch. Therefore, the three alkenes are:

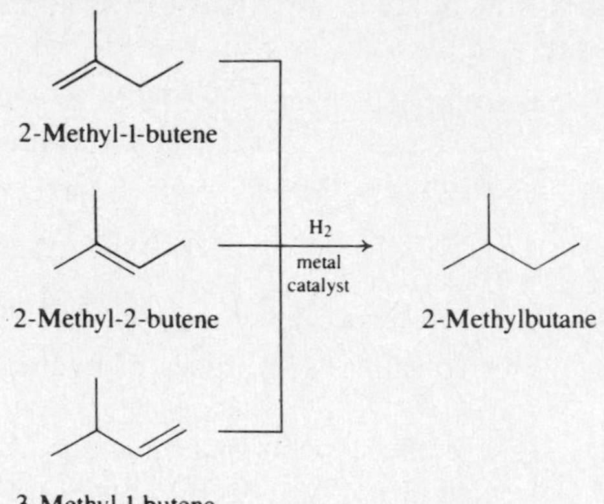

5.12 Double bonds are placed on the parent figure to give the structures shown.

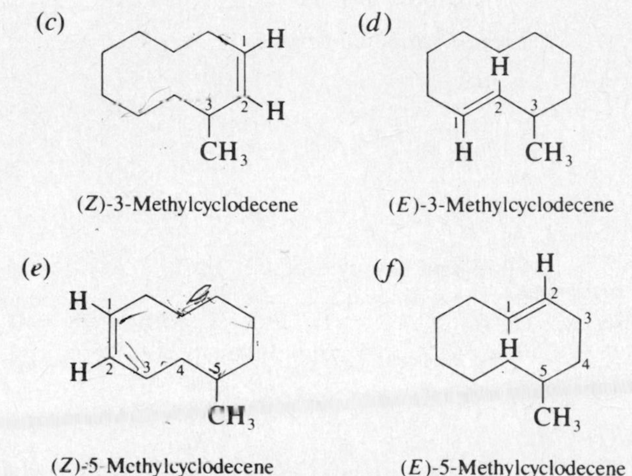

5.13 (*b*) The sum of double bonds and rings (SODAR) is given by the formula:

$$\text{SODAR} = \tfrac{1}{2}(C_nH_{2n+2} - C_nH_x)$$

The compound given contains eight carbons (C_8H_8). Therefore:

$$\text{SODAR} = \tfrac{1}{2}[C_8H_{18} - C_8H_8]$$

$$\text{SODAR} = 5$$

The problem specifies that it consumes two moles of hydrogen, so the compound contains two double bonds (or one triple bond). Since the SODAR is equal to 5, there must be three rings.

(*c*) Chlorine substituents are equivalent to hydrogens when calculating the SODAR. Therefore, consider $C_8H_8Cl_2$ as equivalent to C_8H_{10}. Thus, the SODAR of this compound is four.

$$\text{SODAR} = \tfrac{1}{2}[C_8H_{18} - C_8H_{10}]$$

$$\text{SODAR} = 4$$

If the compound consumes two moles of hydrogen on catalytic hydrogenation, it must therefore contain two rings.

(*d*) Oxygen atoms are ignored when calculating the SODAR. Thus, C_8H_8O is treated as if it were C_8H_8.

$$\text{SODAR} = \tfrac{1}{2}[C_8H_{18} - C_8H_8]$$

$$\text{SODAR} = 5$$

Since the problem specifies that two moles of hydrogen are consumed on catalytic hydrogenation, this compound contains three rings.

(*e*) Ignoring the oxygen atoms in $C_8H_{10}O_2$, we treat this compound as if it were C_8H_{10}.

$$\text{SODAR} = \tfrac{1}{2}[C_8H_{18} - C_8H_{10}]$$

$$\text{SODAR} = 4$$

If it consumes two moles of hydrogen on catalytic hydrogenation, there must be two rings.

(*f*) Ignore the oxygen and treat the chlorine as if it were hydrogen. Thus, C_8H_9ClO is treated as if it were C_8H_{10}. Its SODAR is 4, and it contains two rings.

5.14 A compound with the molecular formula C_5H_{10} must have one ring or one double bond. The ring may be three-, four-, or five-membered.

Three-membered rings:

CH₃, CH₃ on same ring carbon — 1,1-Dimethylcyclopropane

H, H / CH₃, CH₃ — *cis*-1,2-Dimethylcyclopropane

H, CH₃ / CH₃, H — *trans*-1,2-Dimethylcyclopropane

Four- and five-membered rings:

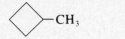

—CH₃ — Methylcyclobutane

Cyclopentane

The alkenes may have an *unbranched* carbon chain:

$CH_2{=}CHCH_2CH_2CH_3$ — 1-Pentene

$(CH_3)(H)C{=}C(CH_2CH_3)(H)$ — *cis*- or (*Z*)-2-Pentene

$(CH_3)(H)C{=}C(H)(CH_2CH_3)$ — *trans*- or (*E*)-2-Pentene

or they may have a *branched* carbon chain:

$CH_2{=}C(CH_3)CH_2CH_3$ — 2-Methyl-1-butene

$(CH_3)_2C{=}CHCH_3$ — 2-Methyl-2-butene

$(CH_3)_2CHCH{=}CH_2$ — 3-Methyl-1-butene

5.15 (*a*) 1-Heptene is $CH_2{=}CH(CH_2)_4CH_3$

(*b*) 3-Ethyl-1-pentene is $CH_2{=}CHCH(CH_2CH_3)_2$

(*c*) 3-Isopropyl-2-methyl-2-hexene is $(CH_3)_2C{=}C(CH(CH_3)_2)CH_2CH_2CH_3$

(*d*) *cis*-3-Octene is $(CH_3CH_2)(H)C{=}C(CH_2CH_2CH_2CH_3)(H)$

(*e*) *trans*-2-Hexene is $(CH_3)(H)C{=}C(H)(CH_2CH_2CH_3)$

(*f*) (*Z*)-3-Methyl-2-hexene is

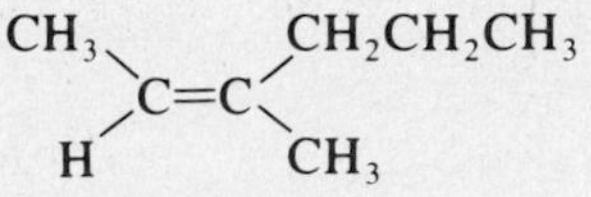

$$\begin{array}{ccc} CH_3 & & CH_2CH_2CH_3 \\ & C{=}C & \\ H & & CH_3 \end{array}$$

(*g*) (*E*)-3-Chloro-2-hexene is

$$\begin{array}{ccc} CH_3 & & CH_2CH_2CH_3 \\ & C{=}C & \\ H & & Cl \end{array}$$

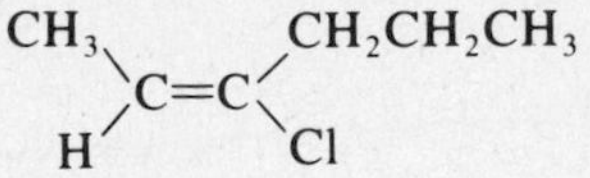

(*h*) 1-Bromocyclohexene is —Br

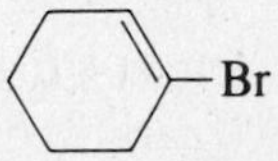

(*i*) 1,3-Dibromocyclohexene is Br Br

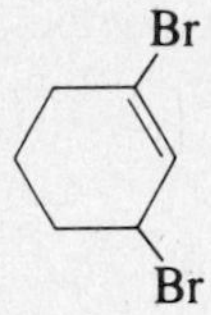

(*j*) 1,6-Dibromocyclohexene is Br Br

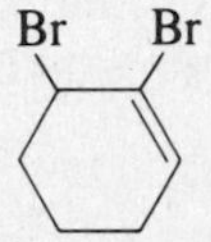

(*k*) Vinylcycloheptane is H

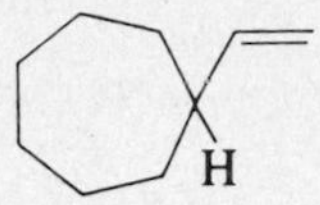

(*l*) 1,1-Diallylcyclopropane is

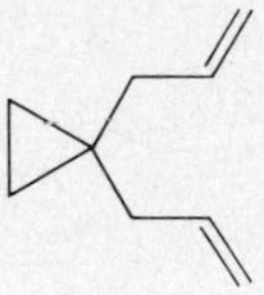

(*m*) *trans*-1-Isopropenyl-3-methylcyclohexane is CH_3 CH_2 CH_3

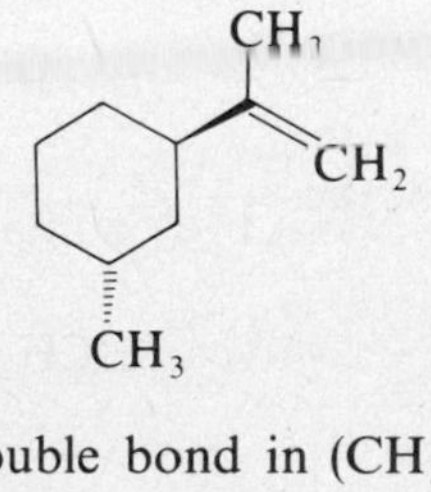

5.16 (*a*) The longest chain that includes the double bond in $(CH_3CH_2)_2C{=}CHCH_3$ contains five carbon atoms, so the parent alkene is a pentene. The numbering scheme which gives the double bond the lowest number is

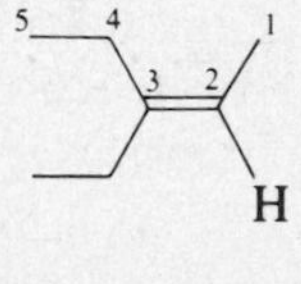

5 4 3 2 1 H

so the compound is 3-ethyl-2-pentene.

(*b*) Write out the structure in detail and identify the longest continuous chain that includes the double bond.

$$\begin{array}{ccc} \overset{1}{C}H_3\overset{2}{C}H_2 & & \overset{5}{C}H_2\overset{6}{C}H_3 \\ & \overset{3}{C}{=}\overset{4}{C} & \\ CH_3CH_2 & & CH_2CH_3 \end{array}$$

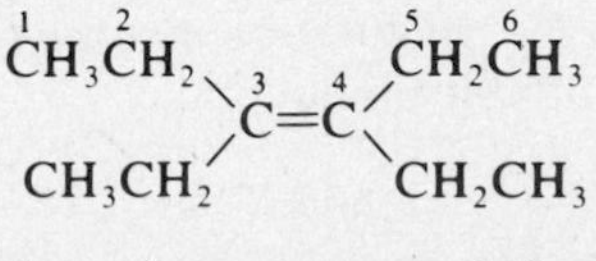

The longest chain contains six carbon atoms and the double bond is between C-3 and C-4. The compound is named as a derivative of 3-hexene. There are ethyl substituents at C-3 and C-4. The complete name is 3,4-diethyl-3-hexene.

(c) Write out the structure completely.

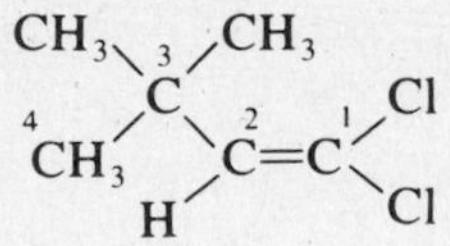

$$\begin{array}{c} CH_3 \quad CH_3 \\ \;\; \overset{3}{C} \quad\quad Cl \\ {}^4CH_3 \quad \overset{2}{C}=\overset{1}{C} \\ H \quad\quad\quad Cl \end{array}$$

The longest carbon chain contains four carbons. Number the chain so as to give the lowest numbers to the doubly bonded carbons and list the substituents in alphabetical order. This compound is 1,1-dichloro-3,3-dimethyl-1-butene.

(d) The longest chain has five carbon atoms, the double bond is at C-1, and there are two methyl substituents. The compound is 4,4-dimethyl-1-pentene.

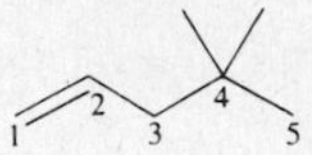

(e) We number this trimethylcyclobutene derivative so as to provide the lowest number for the substituent at the first point of difference. Therefore, we number

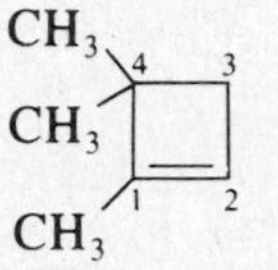

rather than

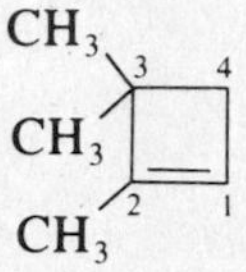

The correct IUPAC name is 1,4,4-trimethylcyclobutene, not 2,3,3-trimethylcyclobutene.

(f) The cyclohexane ring has a 1,2-cis arrangement of vinyl substituents. The compound is *cis*-1,2-divinylcyclohexane.

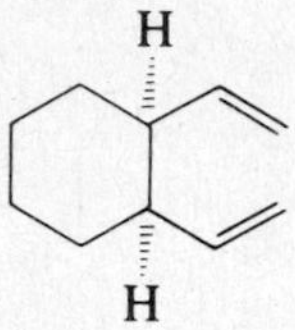

(g) Name this compound as a derivative of cyclohexene. It is 1,2-divinylcyclohexene.

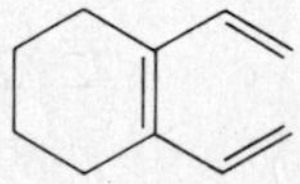

5.17 (a) Go to the end of the name because this tells you how many carbon atoms are present. In the hydrocarbon name 2,6,10,14-tetramethyl-2-pentadecene, the suffix -2-pentadecene reveals that the longest continuous chain has 15 carbon atoms and that there is a double bond between C-2 and C-3. The rest of the name provides the information that there are four methyl groups and that they are located at C-2, C-6, C-10, and C-14.

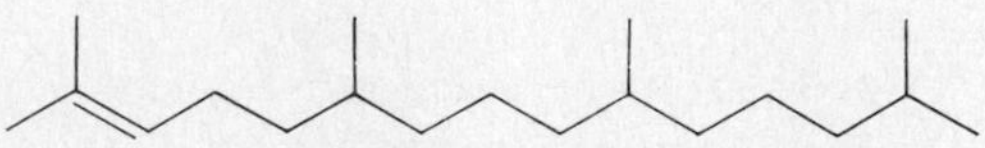

2,6,10,14-Tetramethyl-2-pentadecene

(b) The *E* configuration means that the higher-priority substituents are on opposite sides of the double bond.

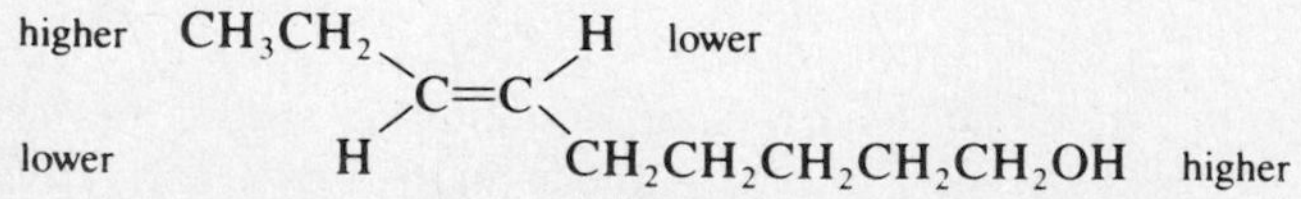

$$\begin{array}{llll} \text{higher} & CH_3CH_2 & H & \text{lower} \\ & \quad C=C & & \\ \text{lower} & H & CH_2CH_2CH_2CH_2CH_2OH & \text{higher} \end{array}$$

(*E*)-6-Nonen-1-ol

(*c*) Geraniol has two double bonds, but only one of them, the one between C-2 and C-3, is capable of stereochemical variation. Of the substituents at C-2 CH_2OH is of higher priority than H. At C-3 CH_2CH_2 outranks CH_3. Higher-priority substituents are on opposite sides of the double bond in the (*E*) isomer; hence geraniol has the structure shown.

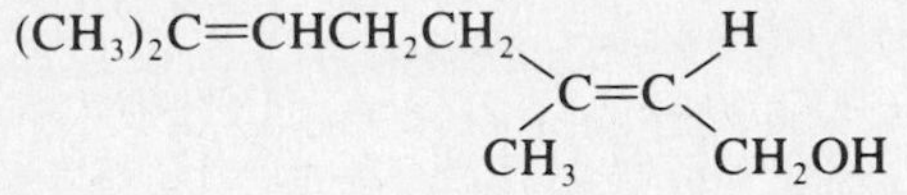

$$\begin{array}{c} (CH_3)_2C{=}CHCH_2CH_2 \quad\quad H \\ \diagdown\!C{=}C\!\diagup \\ CH_3 \quad\quad CH_2OH \end{array}$$

Geraniol

(*d*) Since nerol is a stereoisomer of geraniol, it has the same constitution and differs from geraniol only in having the *Z* configuration of the double bond.

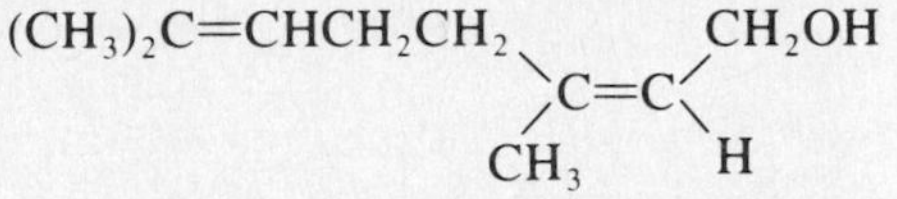

$$\begin{array}{c} (CH_3)_2C{=}CHCH_2CH_2 \quad\quad CH_2OH \\ \diagdown\!C{=}C\!\diagup \\ CH_3 \quad\quad H \end{array}$$

Nerol

(*e*) Beginning at the 6,7 double bond, we see that the propyl group is of higher priority than the methyl group at C-7. Since the 6,7 double bond is *E*, the propyl group must be on the opposite side of the higher-priority substituent at C-6, where the CH_2 fragment has a higher priority than hydrogen. Therefore, we write for the stereochemistry of the 6,7 double bond:

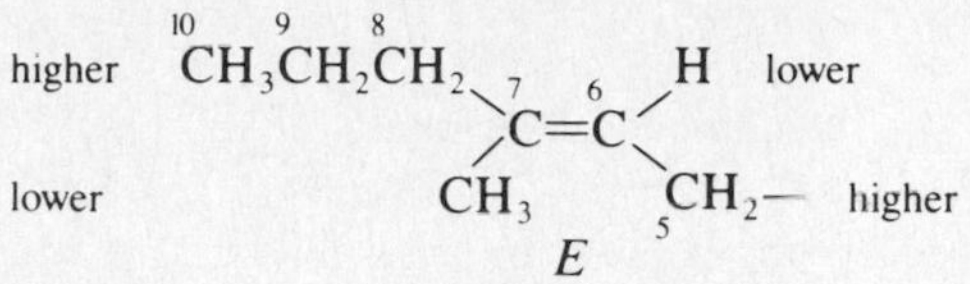

$$\begin{array}{lcl} \text{higher} & \overset{10}{C}H_3\overset{9}{C}H_2\overset{8}{C}H_2 \quad\quad H & \text{lower} \\ & \overset{7}{C}{=}\overset{6}{C} & \\ \text{lower} & CH_3 \quad\quad \underset{5}{C}H_2{-} & \text{higher} \\ & E & \end{array}$$

At C-2 CH_2OH is of higher priority than H, and at C-3 $CH_2CH_2C{-}$ is of higher priority than CH_2CH_3. The double-bond configuration at C-2 is *Z*. Therefore

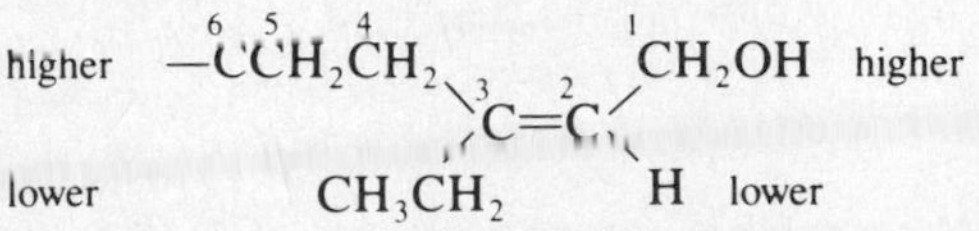

$$\begin{array}{lcl} \text{higher} & {-}\overset{6}{C}\overset{5}{C}H_2\overset{4}{C}H_2 \quad\quad \overset{1}{C}H_2OH & \text{higher} \\ & \overset{3}{C}{=}\overset{2}{C} & \\ \text{lower} & CH_3CH_2 \quad\quad H & \text{lower} \end{array}$$

Combining the two partial structures, we obtain for the full structure of the codling moth's sex pheromone

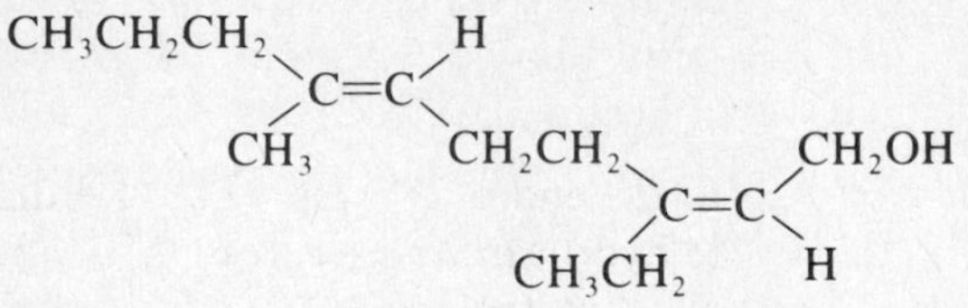

$$\begin{array}{c} CH_3CH_2CH_2 \quad\quad H \\ \diagdown\!C{=}C\!\diagup \\ CH_3 \quad\quad CH_2CH_2 \quad\quad CH_2OH \\ \diagdown\!C{=}C\!\diagup \\ CH_3CH_2 \quad\quad H \end{array}$$

The compound is (2*Z*,6*E*)-3-ethyl-7-methyl-2,6-decadien-1-ol.

(*f*) The sex pheromone of the honeybee is (*E*)-9-oxo-2-decenoic acid, with the structure

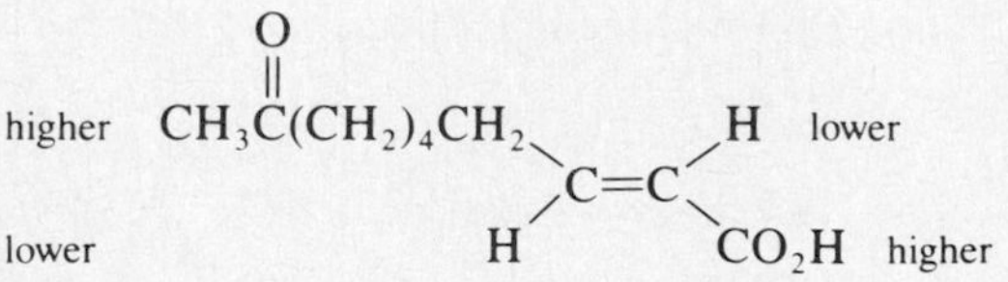

$$\begin{array}{lcl} \text{higher} & CH_3\overset{O}{\overset{\|}{C}}(CH_2)_4CH_2 \quad\quad H & \text{lower} \\ & C{=}C & \\ \text{lower} & H \quad\quad CO_2H & \text{higher} \end{array}$$

(*g*) Looking first at the 2,3 double bond of the cecropia moth's growth hormone

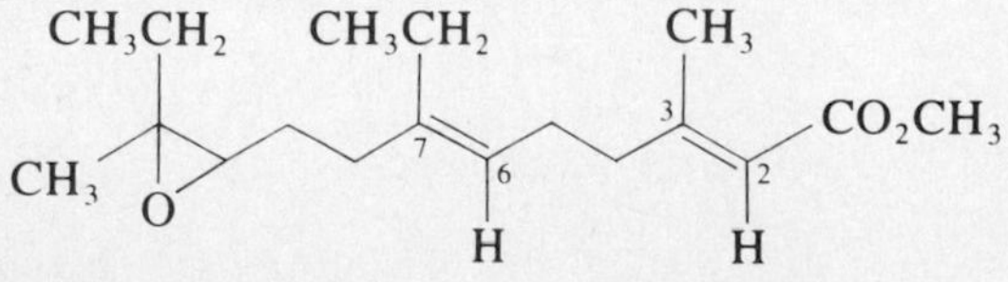

$$\begin{array}{c} CH_3CH_2 \quad CH_3CH_2 \quad\quad CH_3 \\ CH_3\text{–(epoxide, O)}\text{–}CH_2CH_2\text{–}\overset{7}{C}{=}\overset{6}{C}H\text{–}CH_2CH_2\text{–}\overset{3}{C}{=}\overset{2}{C}H\text{–}CO_2CH_3 \end{array}$$

we find that its configuration is *E*, since the higher-priority substituents are on opposite sides of the double bond.

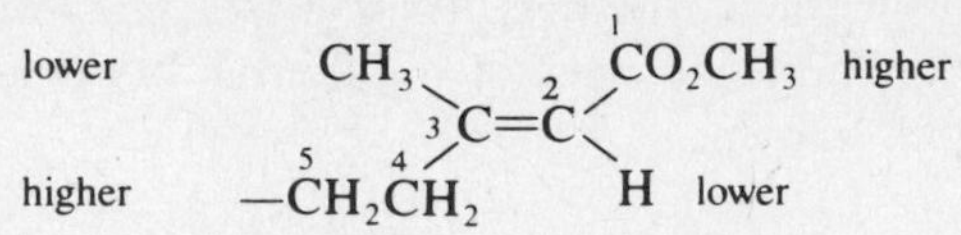

lower CH_3 ... CO_2CH_3 higher (C-1)

$\overset{3}{C}=\overset{2}{C}$

higher $-\overset{5}{C}H_2\overset{4}{C}H_2$ ... H lower

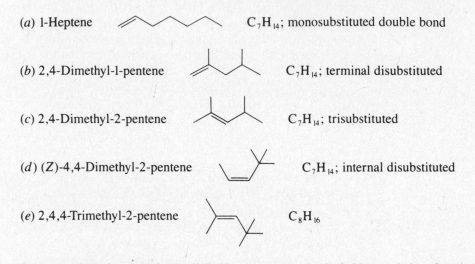

The configuration of the 6,7 double bond is also *E*.

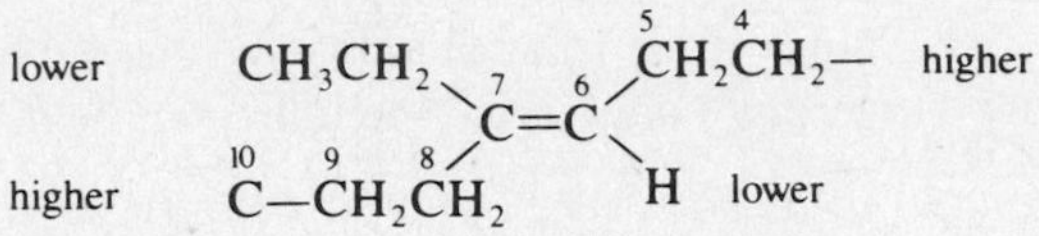

lower CH_3CH_2 ... $\overset{5}{C}H_2\overset{4}{C}H_2-$ higher

$\overset{7}{C}=\overset{6}{C}$

higher $\overset{10}{C}-\overset{9}{C}H_2\overset{8}{C}H_2$ ... H lower

5.18 We need to write the structures out in more detail in order to evaluate their structural relationships in terms of the position and substitution of their double bonds.

(*a*) 1-Heptene — C_7H_{14}; monosubstituted double bond

(*b*) 2,4-Dimethyl-1-pentene — C_7H_{14}; terminal disubstituted

(*c*) 2,4-Dimethyl-2-pentene — C_7H_{14}; trisubstituted

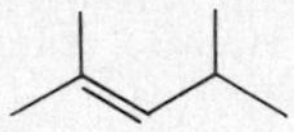

(*d*) (*Z*)-4,4-Dimethyl-2-pentene — C_7H_{14}; internal disubstituted

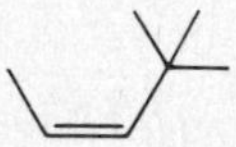

(*e*) 2,4,4-Trimethyl-2-pentene — C_8H_{16}

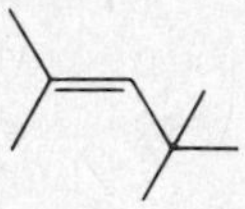

Compounds *a* through *d* are all C_7H_{14}; compound *e* is C_8H_{16} and therefore has the highest heat of combustion value, namely, 1264.9 kcal/mol. Compound *a* has a monosubstituted double bond and so will be the least stabilized of the C_7H_{14} isomers; its heat of combustion is 1113.4 kcal/mol. Compound *d* is a crowded disubstituted double bond, *b* is disubstituted, and *c* is trisubstituted. Therefore the heats of combustion of *d*, *b*, and *c* are 1111.4, 1108.6, and 1107.1 kcal/mol, respectively.

5.19 We need first to write out the structures in more detail to evaluate the substitution patterns at the double bonds.

(*a*) 1-Pentene — Monosubstituted

(*b*) (*E*)-4,4-Dimethyl-2-pentene — trans-Disubstituted

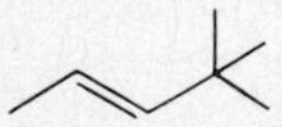

(*c*) (*Z*)-4-Methyl-2-pentene — cis-Disubstituted

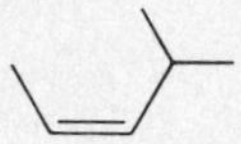

(*d*) (*Z*)-2,2,5,5-Tetramethyl-3-hexene — Two *tert*-butyl groups cis

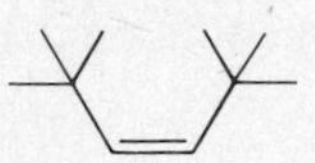

(*e*) 2,4-Dimethyl-2-pentene — Trisubstituted

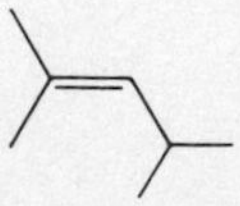

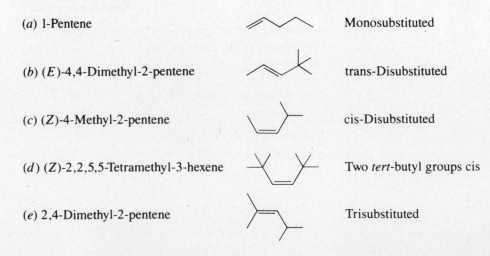

Compound *d*, having two *tert*-butyl groups cis should have the least stable (highest-energy) double bond. The remaining alkenes are arranged in order of increasing stability (decreasing heats of hydrogenation) according to the degree of substitution of the double bond: monosubstituted; cis-disubstituted; trans-disubstituted; trisubstituted. Therefore the heats of hydrogenation are:

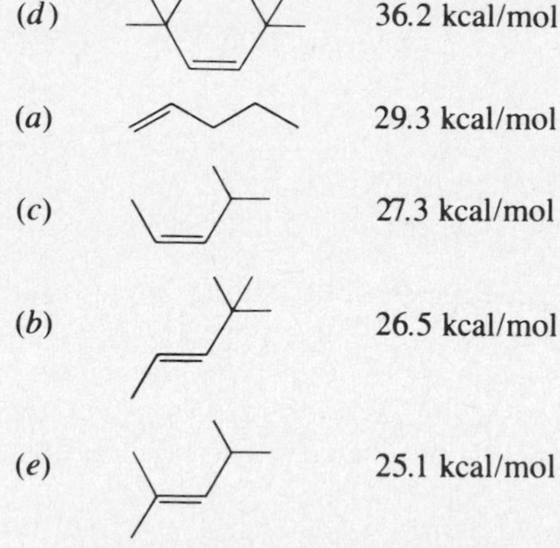

5.20 (*a*) 1-Methylcyclohexene is more stable; it contains a *trisubstituted* double bond while 3-methylcyclohexene has only a disubstituted double bond.

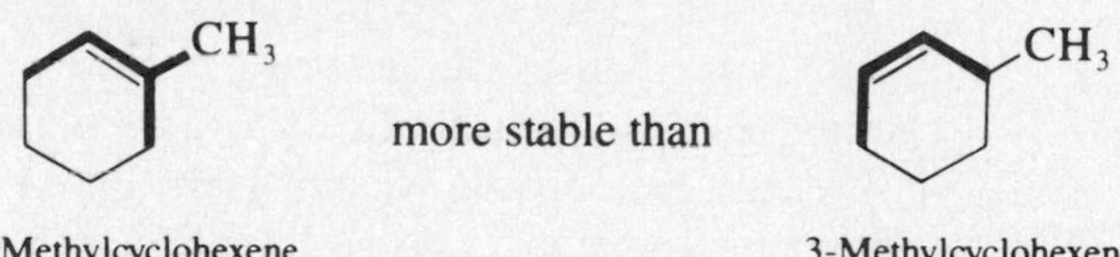

1-Methylcyclohexene 3-Methylcyclohexene

(*b*) Both isopropenyl and allyl are three-carbon alkenyl groups: isopropenyl is $CH_2{=}\underset{|}{C}CH_3$, allyl is $CH_2{=}CHCH_2{-}$.

$CH_2{=}C(CH_3)-$cyclopentyl cyclopentyl$-CH_2CH{=}CH_2$

Isopropenylcyclopentane has a disubstituted double bond and so is predicted to be more stable than allylcyclopentane, in which the double bond is monosubstituted.

(*c*) A double bond in a six-membered ring is less strained than a double bond in a four-membered ring; therefore bicyclo[4.2.0]oct-3-ene is more stable.

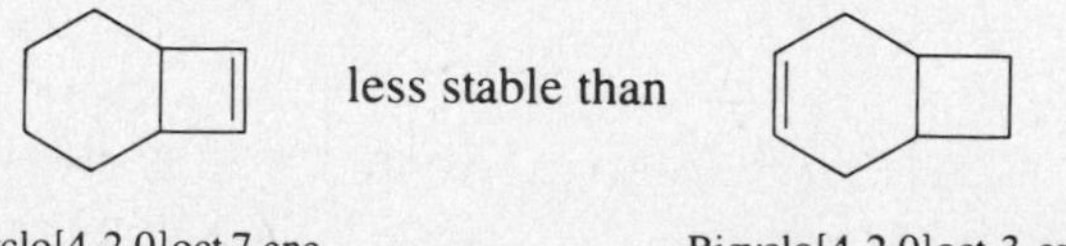

Bicyclo[4.2.0]oct-7-ene Bicyclo[4.2.0]oct-3-ene

(*d*) Cis double bonds are more stable than trans double bonds when the ring is smaller than 11-membered. (*Z*)-Cyclononene has a cis double bond in a 9-membered ring, and is thus more stable than (*E*)-cyclononene.

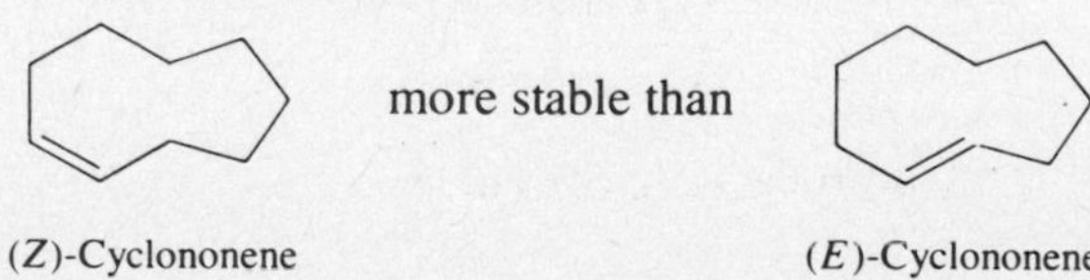

(*Z*)-Cyclononene (*E*)-Cyclononene

(*e*) Trans double bonds are more stable than cis when the ring is large. Here the rings are 18-membered, so (*E*)-cyclooctadecene is more stable than (*Z*)-cyclooctadecene.

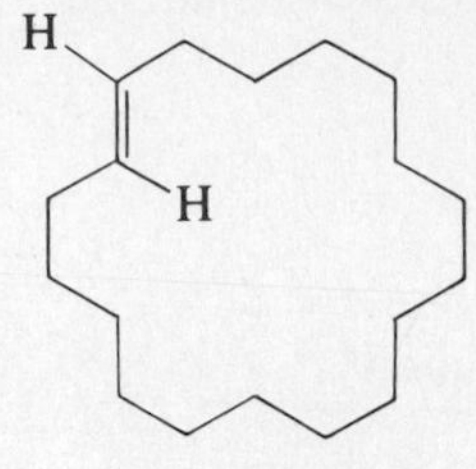

H H more stable than H H

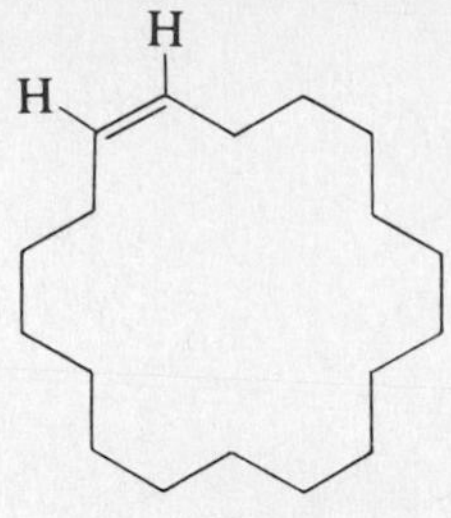

(*E*)-Cyclooctadecene (*Z*)-Cyclooctadecene

5.21 (*a*) Carbon atoms that are involved in double bonds are sp^2 hybridized, with ideal bond angles of 120°. Incorporating an sp^2 hybridized carbon into a three-membered ring leads to more angle strain than incorporation of an sp^3 hybridized carbon. 1-Methylcyclopropene has *two* sp^2 hybridized carbons in a three-membered ring and so has substantially more angle strain than methylenecyclopropane.

$-CH_3$ $=CH_2$

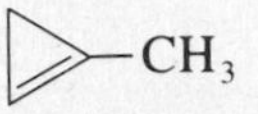

1-Methylcyclopropene 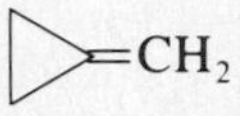 Methylenecyclopropane

The higher degree of substitution at the double bond in 1-methylcyclopropene is not sufficient to offset the increased angle strain, so 1-methylcyclopropene is less stable, and thus has a higher heat of combustion, than methylenecyclopropane.

(*b*) 3-Methylcyclopropene has a disubstituted double bond and two sp^2 hybridized carbons in its three-membered ring. It is the least stable of the isomers.

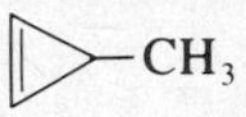

$-CH_3$

3-Methylcyclopropene

5.22 The normal stability order, in which a trisubstituted double bond is more stable than a disubstituted one, is opposed in this instance by an unfavorable van der Waals repulsion between a *tert*-butyl substituent and a cis methyl group in 2,4,4-trimethyl-2-pentene.

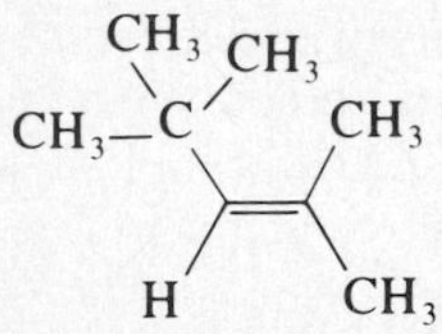

CH_3 CH_3 CH_3-C CH_3 H CH_3

2,4,4-Trimethyl-2-pentene
(destabilized by *tert*-butyl–methyl repulsion)

5.23 In all parts of this exercise we deduce the carbon skeleton on the basis of the alkane formed on hydrogenation of an alkene, then determine what carbon atoms may be connected by a double bond in that skeleton. Problems of this type are best done by using carbon skeleton formulas.

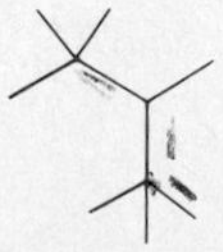

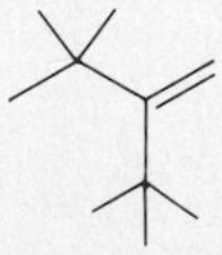

(*a*) Product is 2,2,3,4,4-pentamethylpentane The only possible alkene precursor is

(*b*) Product is 2,3-dimethylbutane

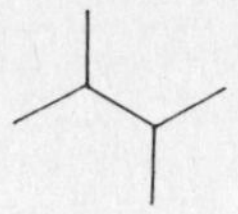

May be formed by hydrogenation of

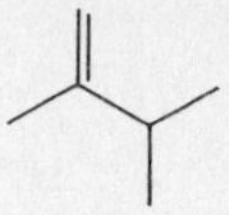

or

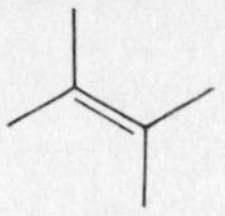

(*c*) Product is methylcyclobutane

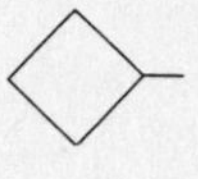

May be formed by hydrogenation of

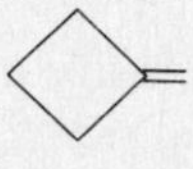

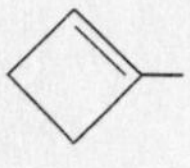

or

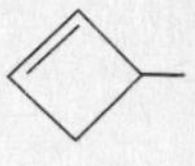

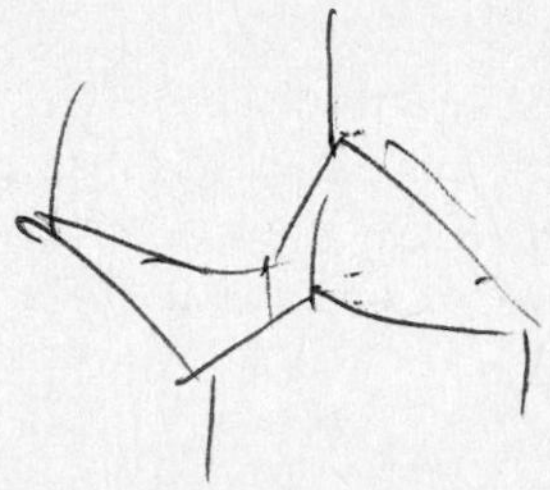

(*d*) Product is *cis*-1,4-dimethylcyclohexane

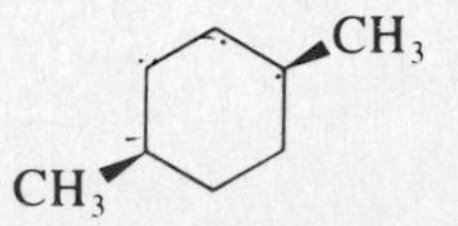

The alkene that gives only the cis isomer on hydrogenation is

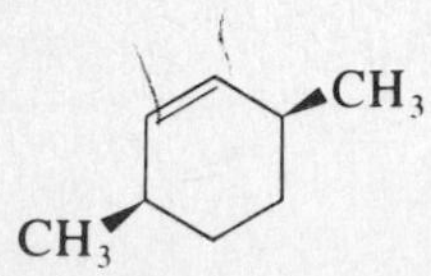

CH_3 CH_3 CH_3 CH_3

5.24 The methyl group in compound B shields one face of the double bond from the catalyst surface, so that hydrogen can be transferred only to the bottom face of the double bond. The methyl group in compound A does not interfere with hydrogen transfer to the double bond.

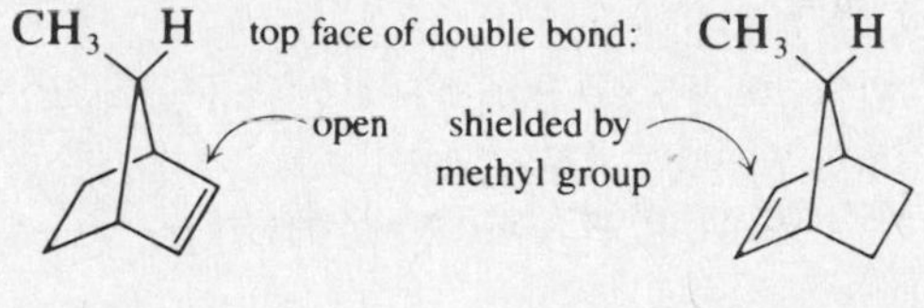

CH_3 H top face of double bond: open — shielded by methyl group CH_3 H

Compound A Compound B

Thus hydrogenation of A is faster than that of B because B contains a more sterically hindered double bond.

5.25 Hydrogen can add to the double bond of 1,4-dimethylcyclopentene either from the same side as the C-4 methyl group or from the opposite side. The two possible products are *cis*- and *trans*-1,3-dimethylcyclopentane.

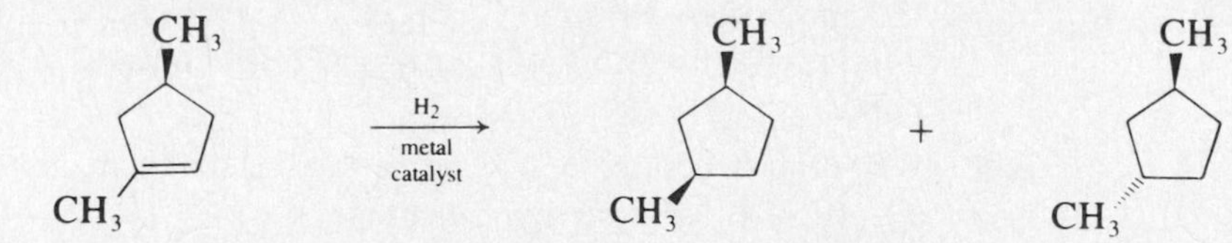

$$\text{1,4-Dimethylcyclopentene} \xrightarrow[\text{metal catalyst}]{H_2} \textit{cis}\text{-1,3,Dimethylcyclopentane} + \textit{trans}\text{-1,3-dimethylcyclopentane}$$

Hydrogen transfer occurs to the less hindered face of the double bond, that is, trans to the C-4 methyl group. Thus, the major product is *cis*-1,3-dimethylcyclopentane.

5.26 Hydrogen can add to either the top face or the bottom face of the double bond. Syn addition to the double bond requires that the methyl groups in the product be cis.

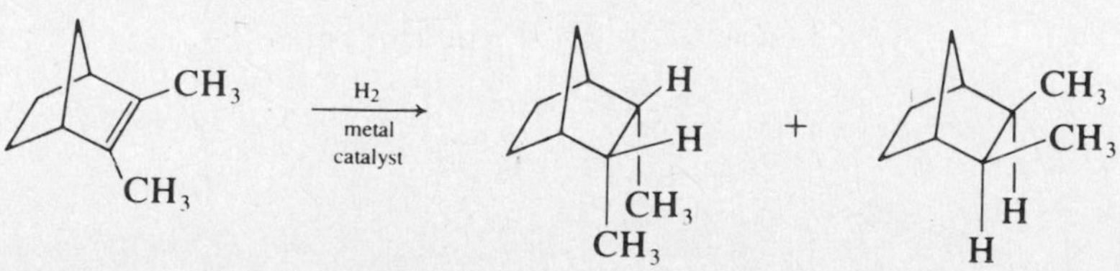

CH_3 CH_3 $\xrightarrow[\text{metal catalyst}]{H_2}$ H H CH_3 CH_3 + CH_3 CH_3 H H

5.27 3-Carene can in theory undergo hydrogenation to give either *cis*-carane or *trans*-carane.

$\xrightarrow[\text{Pt/C ethanol}]{H_2\text{ (100 atm)}}$ or

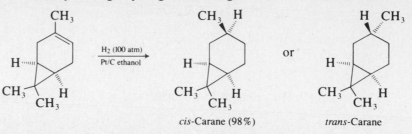

cis-Carane (98%) *trans*-Carane

The exclusive product is *cis*-carane since it corresponds to transfer of hydrogen from the less hindered side.

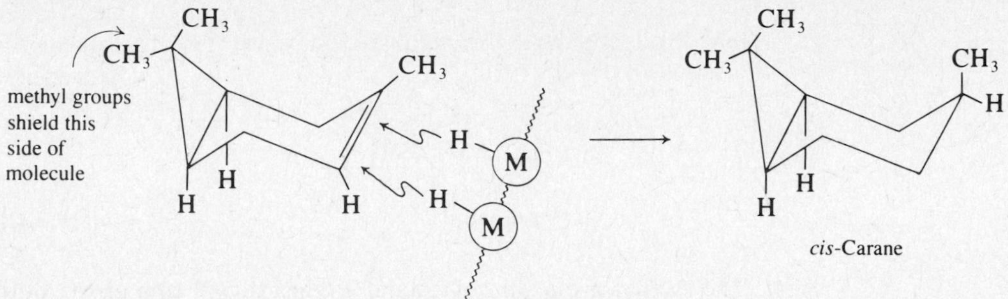

SELF-TEST

PART A

A-1. Write the correct IUPAC name for each of the following.

(*a*) $(CH_3)_3CCH{=}CHCH_3$ (*b*)

A-2. Each of the following is an incorrect name for an alkene. Write the structure and give the correct name for each.

(*a*) 2-Ethyl-3-methyl-2-butene (*b*) 2,5-Dimethylcyclohexene

A-3. (*a*) Write the structures of all the pentene (C_5H_{10}) isomers.

(*b*) Which isomer gives off the most heat on hydrogenation?

(*c*) Which isomer gives off the least heat on hydrogenation?

A-4. The physiologically active constituent of marijuana, tetrahydrocannabinol (THC), has the molecular formula $C_{21}H_{30}O_2$. Exhaustive catalytic hydrogenation results in uptake of four moles of hydrogen. THC contains _____ rings and _____ double bonds.

A-5. Write the structure, clearly indicating the stereochemistry, of the following:

(*a*) (*Z*)-4-Ethyl-3-methyl-3-heptene

(*b*) (*E*)-1,2-Dichloro-3-methyl-2-hexene

A-6. How many different alkenes will yield 2,3-dimethylpentane on catalytic hydrogenation? Draw their structures and name them.

PART B

B-1. Which (if any) of the following has the word "pentene" in the correct IUPAC name?

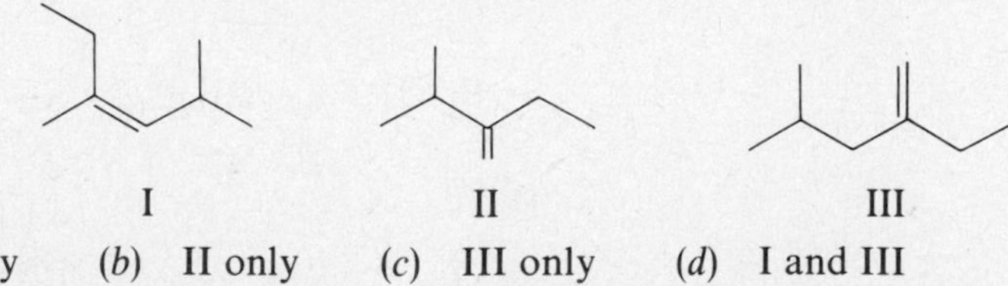

(*a*) I only (*b*) II only (*c*) III only (*d*) I and III

B-2. Rank the following substituent groups in order of decreasing priority according to the Cahn-Ingold-Prelog system.

$$-CH(CH_3)_2 \qquad -CH_2Br \qquad -CH_2CH_2Br$$
A B C

(*a*) B > C > A (*b*) A > C > B (*c*) C > A > B (*d*) B > A > C

B-3. The heats of combustion for the four C_6H_{12} isomers shown are (not necessarily in order): 955.3, 953.6, 950.6, and 949.7 (all in kilocalories per mole). Which of these values is most likely the heat of combustion of isomer I?

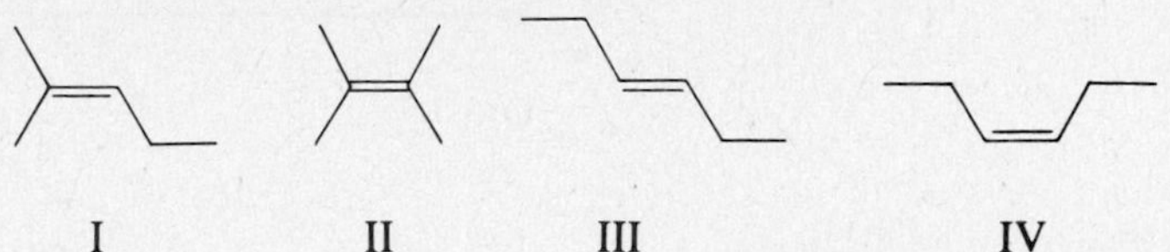

(*a*) 955.3 kcal/mol (*b*) 953.6 kcal/mol
(*c*) 950.6 kcal/mol (*d*) 949.7 kcal/mol

B-4. Referring to the structures in the previous question, what can be said about isomers III and IV?
(*a*) III is more stable by 1.7 kcal/mol.
(*b*) IV is more stable by 1.7 kcal/mol.
(*c*) III is more stable by 3.0 kcal/mol.
(*d*) III is more stable by 0.9 kcal/mol.

B-5. The best structure for (*Z*)-1-chloro-3-methyl-3-hexene is

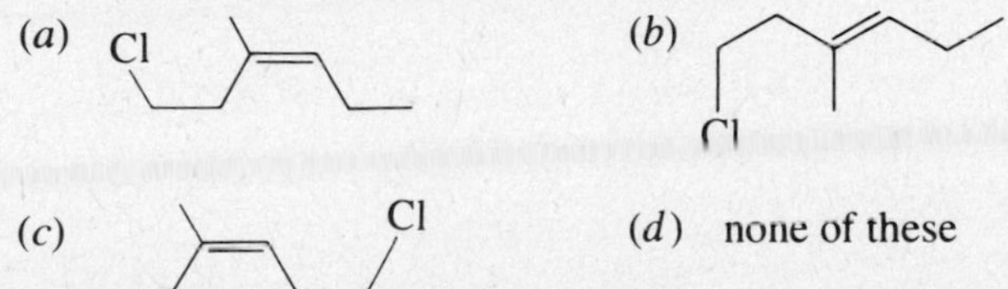

(*d*) none of these

B-6. The heat of hydrogenation of substance I is 28.5 kcal/mol. Which of the following values most likely represents the heat of hydrogenation of II?

$$\begin{matrix} CH_3CH_2 \\ CH_3 \end{matrix}\!\!>\!C{=}CH_2 \qquad \begin{matrix} CH_3 \\ CH_3 \end{matrix}\!\!>\!C{=}CHCH_3$$
I II

(*a*) 30.3 kcal/mol (*b*) 26.7 kcal/mol
(*c*) 29.5 kcal/mol (*d*) 28.5 kcal/mol

B-7. For which, if any, of the following do cis/-trans stereoisomers exist?

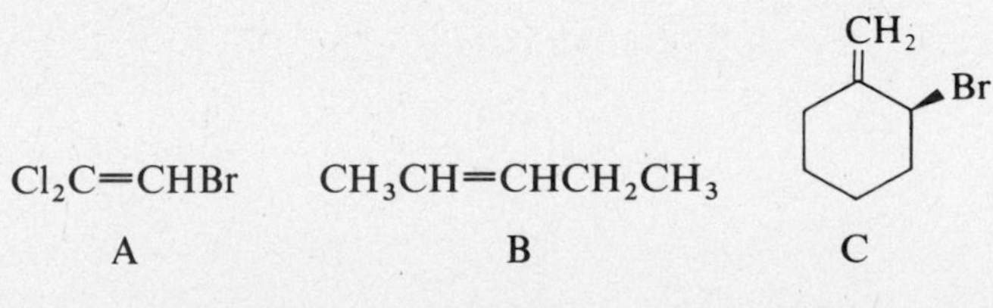

(*a*) C only (*b*) B only (*c*) A and B (*d*) B and C

B-8. A compound having a molecular formula of $C_{20}H_{36}$ is inert to catalytic hydrogenation. Which of the following statements is true?

(*a*) The substance has at least one double bond and two rings.
(*b*) The substance is acyclic (i.e., no rings are present).
(*c*) The substance has two rings.
(*d*) The substance has three rings.

B-9. A compound having the formula $C_{10}H_{10}Cl_2O$ has a SODAR value of ______. Three moles of hydrogen are consumed on catalytic hydrogenation. The substance has ______ double bonds and ______ rings.

	SODAR	*Double bonds*	*Rings*
(*a*)	6	3	3
(*b*)	6	6	0
(*c*)	5	3	2
(*d*)	5	2	3

CHAPTER 6

PREPARATION OF ALKENES. ELIMINATION REACTIONS

IMPORTANT TERMS AND CONCEPTS

Elimination Reactions (Sec. 6.1) Elimination reactions that lead to alkenes are known as *β-elimination reactions*, since the reactions proceed by loss of atoms from *adjacent* carbons on the reactant. The reactions discussed in this chapter, *dehydration of alcohols* and *dehydrohalogenation of alkyl halides*, are examples of *β* elimination reactions.

Zaitsev's Rule (Secs. 6.3, 6.6) Alcohol dehydration is a *regioselective* process; that is, although more than one regioisomer can be formed from a single substrate, one is observed to predominate. *Zaitsev's rule* states that the major alkene product formed by an elimination reaction will correspond to *proton loss from the β carbon having the fewest hydrogen substituents*. The rule applies to both dehydration and dehydrohalogenation reactions. In other words, elimination of alcohols and alkyl halides yields the *most highly substituted* alkene as the major product. As discussed in the previous chapter, the major product is the *most stable alkene*. These reactions are also *stereoselective*—when stereoisomeric products are possible, the more stable stereoisomer predominates. For example, dehydration of 3-pentanol yields a 3 : 1 ratio of *E* (or trans) to *Z* (or cis) isomer of 2-pentene.

$$CH_3CH_2\underset{\displaystyle OH}{\underset{|}{C}}HCH_2CH_3 \xrightarrow[\text{heat}]{H_2SO_4} \underset{\text{minor product}}{\underset{(Z)}{\begin{matrix} CH_3 & & CH_2CH_3 \\ & C{=}C & \\ H & & H \end{matrix}}} + \underset{\text{major product}}{\underset{(E)}{\begin{matrix} CH_3 & & H \\ & C{=}C & \\ H & & CH_2CH_3 \end{matrix}}}$$

E2 Mechanism (Secs. 6.7, 6.8) The dehydrohalogenation of alkyl halides by strong base is considered to be a *concerted* reaction. That is, the entire reaction proceeds in a single step. Such a *β*-elimination reaction is termed an *E2* (*elimination bimolecular*) *reaction*. The rate of an E2 reaction is described by *second-order kinetics*, the general rate equation being

$$\text{Rate} = k[\text{alkyl halide}][\text{base}]$$

The *order of reactivity* of alkyl halides toward E2 elimination is tertiary > secondary > primary. The rate of elimination also depends on the nature of the leaving group and *decreases in the order* I > Br > Cl > F.

E2 reactions occur by an anti elimination, that is, the β hydrogen and the halide leaving group must be in an anti periplanar arrangement for elimination to proceed.

$$B:^- + H\text{–C–C–}X \longrightarrow \text{C=C} + X^- + BH$$

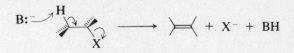

E1 Mechanism for Dehydrohalogenation (Sec. 6.10) An alternative mechanism for elimination takes place when ionization of the halide leaving group occurs *prior* to loss of the β hydrogen. This *unimolecular ionization* of the substrate gives rise to a carbocation intermediate, in a manner similar to that described for the dehydration of alcohols. Such a reaction exhibits *first-order kinetics* and is known as an *E1 (elimination unimolecular) reaction.* Such a mechanism is important only with tertiary and some secondary halides and when there is a low concentration of a weak base. In general however, conditions under which E2 elimination predominates are preferred for the preparation of alkenes from alkyl halides.

IMPORTANT REACTIONS

Dehydration of Alcohols (Secs. 6.2, 6.3)

General:

$$\text{H–C–C–OH} \xrightarrow{H^+} \text{C=C} + H_2O$$

Examples:

$$(CH_3)_2C(OH)CH_2CH_3 \xrightarrow[\text{heat}]{H_2SO_4} \underset{\text{(major)}}{(CH_3)_2C{=}CHCH_3} + CH_2{=}C(CH_3)CH_2CH_3$$

$$\text{2-methylcyclohexanol} \xrightarrow[\text{heat}]{H_3PO_4} \underset{\text{(major)}}{\text{1-methylcyclohexene}} + \text{3-methylcyclohexene}$$

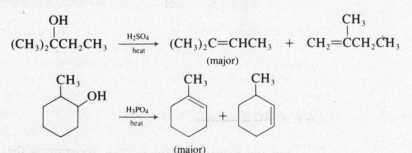

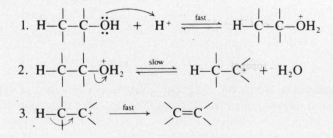

Dehydration Mechanism (Sec. 6.4)

1. $$\text{H–C–C–}\ddot{\text{O}}\text{H} + H^+ \overset{\text{fast}}{\rightleftharpoons} \text{H–C–C–}\overset{+}{\text{O}}H_2$$

2. $$\text{H–C–C–}\overset{+}{\text{O}}H_2 \overset{\text{slow}}{\rightleftharpoons} \text{H–C–C}^+ + H_2O$$

3. $$\text{H–C–C}^+ \xrightarrow{\text{fast}} \text{C=C}$$

Primary and some secondary alcohols undergo dehydration by a mechanism that avoids carbocation formation.

$$B: + \text{H–C–C–}\overset{+}{\text{O}}H_2 \longrightarrow \overset{+}{B}H + \text{C=C} + H_2O$$

Rearrangements (Sec. 6.5)

$$\text{(1,2-dimethylcyclohexanol: OH, H, CH}_3\text{, CH}_3) \xrightarrow[-H_2O]{H^+} \text{(carbocation: H, CH}_3\text{, CH}_3) \longrightarrow \text{(rearranged carbocation: H, CH}_3\text{, CH}_3) \xrightarrow{-H^+} \text{1,2-dimethylcyclohexene (CH}_3\text{, CH}_3)$$

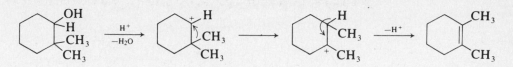

Dehydrohalogenation of Alkyl Halides (Sec. 6.6)

General:

$$\mathrm{H{-}\overset{|}{\underset{|}{C}}{-}\overset{|}{\underset{|}{C}}{-}X + B^- \longrightarrow \;\rangle C{=}C\langle\; + HB + X^-}$$

Examples:

$$\text{1-chloro-1-methylcyclohexane (CH}_3\text{, Cl)} \xrightarrow[CH_3OH]{CH_3O^-} \text{1-methylcyclohexene (CH}_3)$$

$$\mathrm{(CH_3)_2\overset{Br}{\overset{|}{C}}{-}CH(CH_3)_2} \xrightarrow[\substack{EtOH \\ (Et=CH_3CH_2-)}]{KOEt} \mathrm{(CH_3)_2C{=}C(CH_3)_2}$$

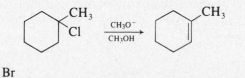

Mechanism of Dehydrohalogenation (Secs. 6.7, 6.8)

$$\mathrm{R\ddot{O}:^-} \curvearrowright \mathrm{H{-}C(A)(B){-}C(A)(B){-}X} \xrightarrow{E2} \mathrm{(A)(B)C{=}C(A)(B)}$$

$$\text{chlorocyclohexane (H, Cl equatorial)} \rightleftharpoons \text{chlorocyclohexane (Cl axial; } \mathrm{R\ddot{O}:^-} \text{ attacks axial H)} \longrightarrow \text{cyclohexene}$$

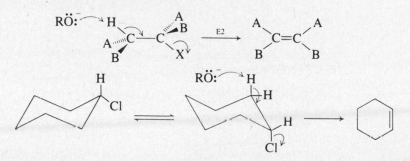

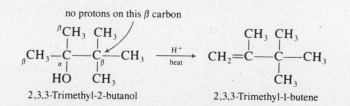

SOLUTIONS TO TEXT PROBLEMS

6.1 Write out the structure of the alcohol, recognizing that the alkene is formed by loss of a hydrogen and a hydroxyl group from adjacent carbons.

(*b*), (*c*) Both 1-propanol and 2-propanol give propene on acid-catalyzed dehydration.

$$\underset{\text{1-Propanol}}{\mathrm{CH_3\overset{\beta}{C}H_2\overset{\alpha}{C}H_2OH}} \xrightarrow[\text{heat}]{H^+} \underset{\text{Propene}}{\mathrm{CH_3CH{=}CH_2}} \xleftarrow[\text{heat}]{H^+} \underset{\text{2-Propanol}}{\mathrm{\overset{\beta}{C}H_3\overset{\alpha}{C}H(OH)\overset{\beta}{C}H_3}}$$

(*d*) Carbon-3 has no hydrogen substituents in 2,3,3-trimethyl-2-butanol. Elimination can involve only the hydroxyl group at C-2 and a hydrogen at C-1.

no protons on this β carbon

$$\underset{\text{2,3,3-Trimethyl-2-butanol}}{\mathrm{{}_{\beta}CH_3{-}\overset{{}^{\beta}CH_3}{\underset{HO}{\overset{|}{\underset{|\,\alpha}{C}}}}{-}\overset{CH_3}{\underset{CH_3}{\overset{|}{\underset{|\,\beta}{C}}}}{-}CH_3}} \xrightarrow[\text{heat}]{H^+} \underset{\text{2,3,3-Trimethyl-1-butene}}{\mathrm{CH_2{=}\overset{CH_3}{\overset{|}{C}}{-}\overset{CH_3}{\underset{CH_3}{\overset{|}{\underset{|}{C}}}}{-}CH_3}}$$

6.2 (*b*) Elimination can involve loss of a hydrogen from the methyl group or from C-2 of the ring in 1-methylcyclohexanol.

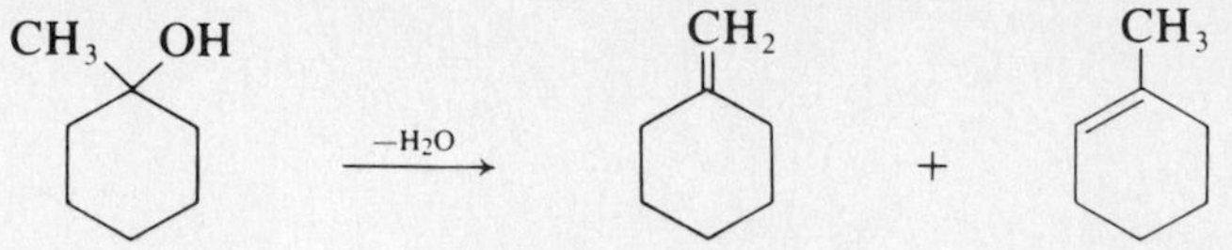

CH_3 OH → ($-H_2O$) CH_2 + CH_3

1-Methylcyclohexanol — Methylenecyclohexane (a disubstituted alkene; minor product) — 1-Methylcyclohexene (a trisubstituted alkene; major product)

According to the Zaitsev rule, the major alkene is the one corresponding to loss of a hydrogen from the alkyl group that has the fewer number of hydrogens. Thus hydrogen is removed from the methylene group rather than from the methyl group, and 1-methylcyclohexene is formed in greater amounts than methylenecyclohexane.

(*c*) The two alkenes are as shown in the equation.

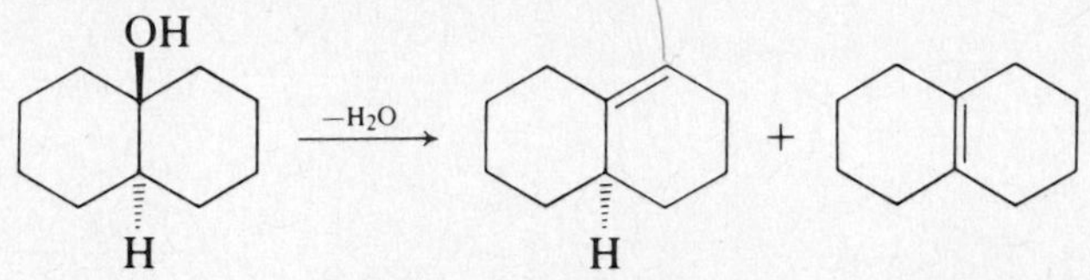

OH, H → ($-H_2O$) H +

Compound has a trisubstituted double bond — Compound has a tetrasubstituted double bond; more stable

The more highly substituted alkene is formed in greater amounts, according to the Zaitsev rule.

6.3 2-Pentanol can undergo dehydration in two different directions, giving either 1-pentene or 2-pentene. 2-Pentene is formed as a mixture of the cis and trans isomers.

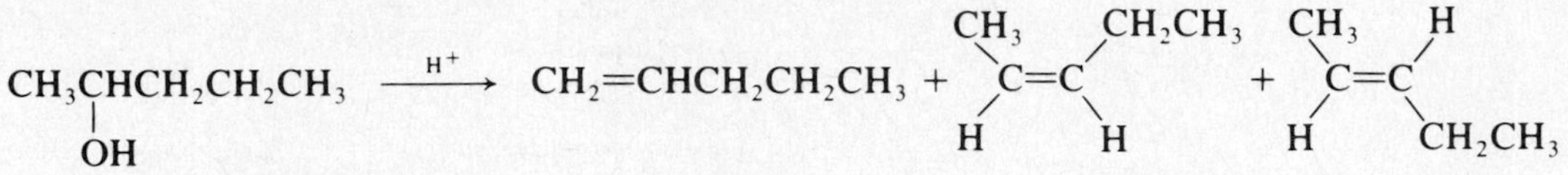

$$CH_3\underset{\displaystyle OH}{\underset{|}{C}}HCH_2CH_2CH_3 \xrightarrow{H^+} CH_2{=}CHCH_2CH_2CH_3 + \underset{H\qquad H}{\overset{CH_3\qquad CH_2CH_3}{C{=}C}} + \underset{H\qquad CH_2CH_3}{\overset{CH_3\qquad H}{C{=}C}}$$

2-Pentanol — 1-Pentene — *cis*-2-Pentene — *trans*-2-Pentene

6.4 (*b*) The site of positive charge in the carbocation is the carbon atom that bears the hydroxyl group in the starting alcohol.

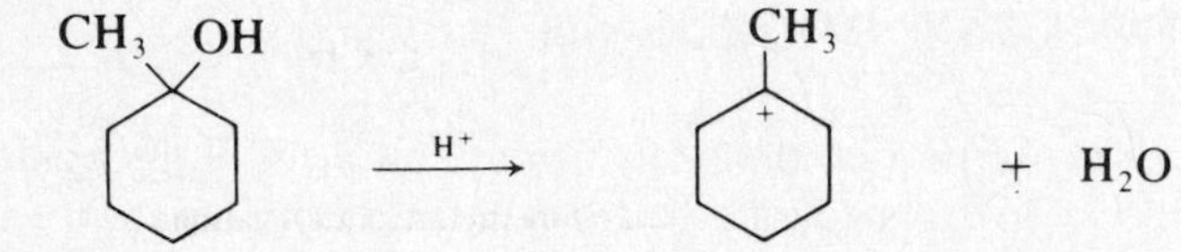

CH_3 OH → (H^+) CH_3 (+) + H_2O

1-Methylcyclohexanol — 1-Methylcyclohexyl cation

The curved arrow notation for proton abstraction by hydrogen sulfate anion from the methyl group is demonstrated in the following equation:

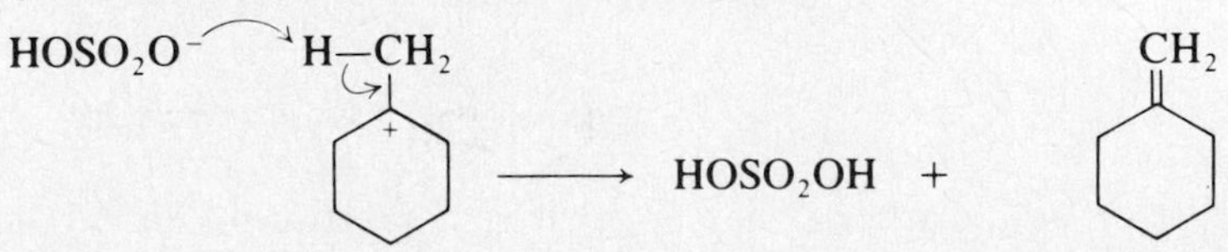

$HOSO_2O^-$ H—CH_2 (+) → $HOSO_2OH$ + CH_2

Methylenecyclohexane

Abstraction of a proton from the ring gives 1-methylcyclohexene.

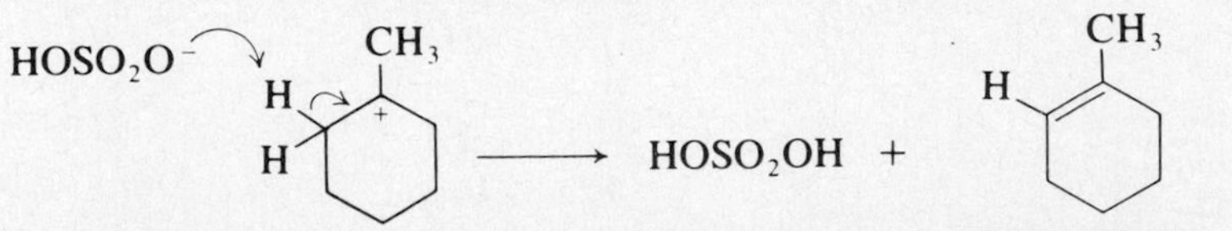

$HOSO_2O^-$ H H CH_3 (+) → $HOSO_2OH$ + H CH_3

1-Methylcyclohexene

(*c*) Loss of the hydroxyl group under conditions of acid catalysis yields a tertiary carbocation.

OH H $\xrightarrow[-H_2O]{H_2SO_4}$ H

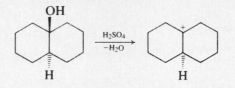

Hydrogen sulfate ion may abstract a proton from an adjacent methylene group to give a trisubstituted alkene.

$HOSO_2O^-$ H H H H → H + $HOSO_2OH$

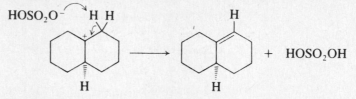

Abstraction of the methine proton affords a tetrasubstituted alkene.

H $^-OSO_2OH$ → + $HOSO_2OH$

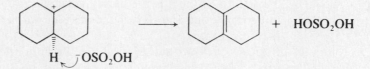

6.5 In writing mechanisms for acid-catalyzed dehydration of alcohols, begin with formation of the carbocation intermediate.

CH_3 CH_3 H OH $\xrightarrow[-H_2O]{H^+}$ CH_3 CH_3 H

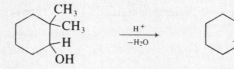

2,2-Dimethylcyclohexanol 2,2-Dimethylcyclohexyl cation

This secondary carbocation can rearrange to a more stable tertiary carbocation by a methyl group shift.

CH_3 CH_3 H → CH_3 CH_3 H

2,2-Dimethylcyclohexyl cation (secondary) 1,2-Dimethylcyclohexyl cation (tertiary)

Loss of a proton from the 1,2-dimethylcyclohexyl cation intermediate yields 1,2-dimethylcyclohexene.

CH_3 H CH_3 $\xrightarrow{-H^+}$ CH_3 CH_3

1,2-Dimethylcyclohexyl cation 1,2-Dimethylcyclohexene

6.6 (*b*) All the hydrogens of *tert*-butyl chloride are equivalent. Loss of any of these hydrogens along with the chlorine substituent yields 2-methylpropene as the only alkene.

$CH_3CCl(CH_3)_2$ → $(CH_3)_2C{=}CH_2$

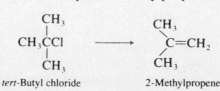

tert-Butyl chloride 2-Methylpropene

(*c*) All the β hydrogens of 3-bromo-3-ethylpentane are equivalent, so β elimination can give only 3-ethyl-2-pentene.

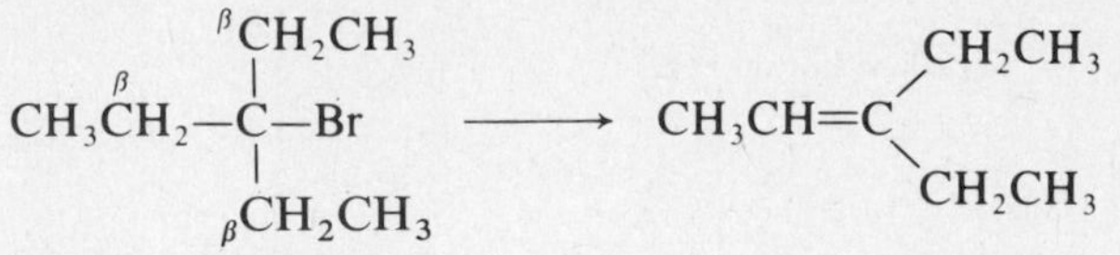

$$CH_3\overset{\beta}{C}H_2-\underset{\underset{_\beta CH_2CH_3}{|}}{\overset{\overset{^\beta CH_2CH_3}{|}}{C}}-Br \longrightarrow CH_3CH=C(CH_2CH_3)_2$$

3-Bromo-3-ethylpentane → 3-Ethyl-2-pentene

(*d*) There are two possible modes of β elimination from 2-bromo-3-methylbutane. Elimination in one direction provides 3-methyl-1-butene; elimination in the other gives 2-methyl-2-butene.

$$\overset{\beta}{C}H_3\underset{\underset{Br}{|}}{C}H\overset{\beta}{C}H(CH_3)_2 \longrightarrow CH_2=CHCH(CH_3)_2 + CH_3CH=C(CH_3)_2$$

2-Bromo-3-methylbutane → 3-Methyl-1-butene (monosubstituted) + 2-Methyl-2-butene (trisubstituted)

The major product is the more highly substituted alkene, 2-methyl-2-butene. It is the more stable one and corresponds to removal of a proton from the carbon that has the fewer hydrogens.

(*e*) Regioselectivity is not an issue here because 3-methyl-1-butene is the only alkene that can be formed by β elimination from 1-bromo-3-methylbutane.

$$BrCH_2\overset{\beta}{C}H_2CH(CH_3)_2 \longrightarrow CH_2=CHCH(CH_3)_2$$

1-Bromo-3-methylbutane → 3-Methyl-1-butene

(*f*) Two alkenes may be formed here. The more highly substituted one is 1-methylcyclohexene, and this is predicted to be the major product in accordance with Zaitsev's rule.

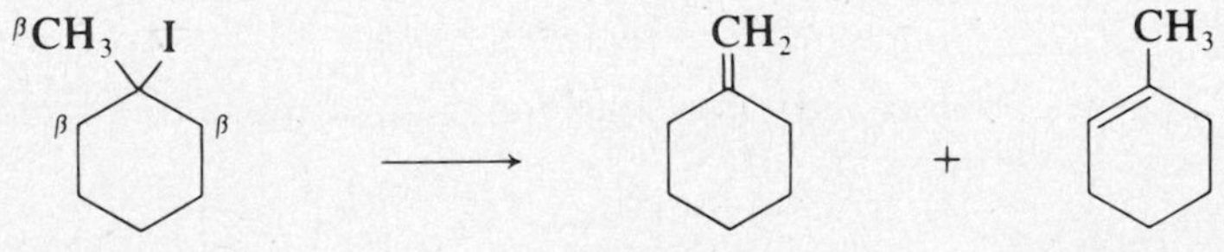

1-Iodo-1-methylcyclohexane ($^\beta CH_3$, I; ring carbons β) → Methylenecyclohexane (CH_2) (disubstituted) + 1-Methylcyclohexene (CH_3) (trisubstituted; major product)

6.7 (*b*) The dehydrohalogenation of 2-bromohexane is similar to that of 2-bromobutane described in (*a*) of this problem. Elimination can occur to give 1-hexene, *cis*-2-hexene, or *trans*-2-hexene.

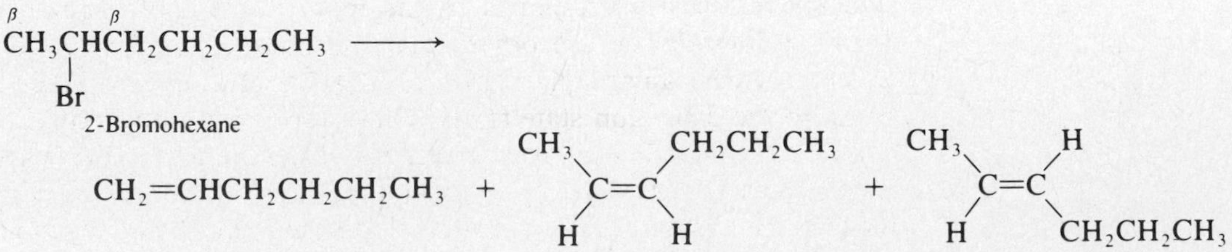

$$\overset{\beta}{C}H_3\underset{\underset{Br}{|}}{C}H\overset{\beta}{C}H_2CH_2CH_2CH_3 \longrightarrow$$

2-Bromohexane

$$CH_2=CHCH_2CH_2CH_2CH_3 + \underset{H}{\overset{CH_3}{}}C=C\underset{H}{\overset{CH_2CH_2CH_3}{}} + \underset{H}{\overset{CH_3}{}}C=C\underset{CH_2CH_2CH_3}{\overset{H}{}}$$

1-Hexene + *cis*-2-Hexene + *trans*-2-Hexene

The experimental results of an elimination reaction of 2-bromohexane using sodium methoxide as the base in methanol at 99°C are in accord with what we have learned of the regioselectivity and stereoselectivity of these reactions. The major product observed was *trans*-2-hexene (54 percent), with *cis*-2-hexene (18 percent) and 1-hexene (28 percent) as the minor components of the reaction mixture.

(*c*) In 2-iodo-4-methylpentane β elimination can involve a proton at C-1 or at C-3. The alkenes that are capable of being formed are 4-methyl-1-pentene, *trans*-4-methyl-2-pentene, and *cis*-4-methyl-2-pentene.

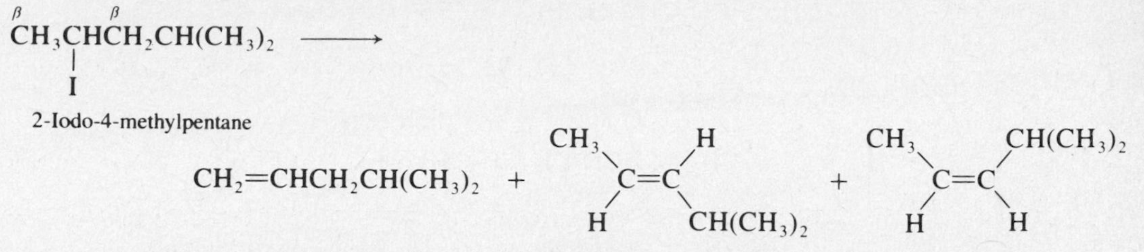

$$\overset{\beta}{C}H_3CH\overset{\beta}{C}H_2CH(CH_3)_2 \text{ (I on C-2)} \longrightarrow$$

2-Iodo-4-methylpentane

$$CH_2{=}CHCH_2CH(CH_3)_2 + (CH_3)HC{=}CH(CH(CH_3)_2) + (CH_3)HC{=}CH(CH(CH_3)_2)$$

4-Methyl-1-pentene; *trans*-4-Methyl-2-pentene; *cis*-4-Methyl-2-pentene

When this reaction was performed with use of potassium *tert*-butoxide in dimethyl sulfoxide, *trans*-4-methyl-2-pentene was formed as the major product (60 percent).

(*d*) Elimination can proceed in only one direction in 3-iodo-2,2-dimethylpentane. A mixture of *cis*- and *trans*-4,4-dimethyl-2-pentene is formed.

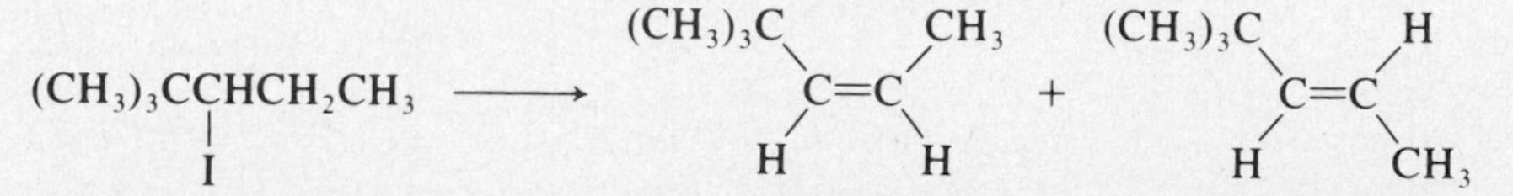

$$(CH_3)_3CCHCH_2CH_3 \text{ (I on C-3)} \longrightarrow (CH_3)_3CHC{=}CHCH_3 + (CH_3)_3CHC{=}CHCH_3$$

3-Iodo-2,2-dimethylpentane; *cis*-4,4-Dimethyl-2-pentene; *trans*-4,4-Dimethyl-2-pentene

6.8 The two starting materials are stereoisomers of each other, so it is reasonable to begin by examining each one in more stereochemical detail. First, write the most stable conformation of each isomer, keeping in mind that isopropyl is the bulkiest substituent of the three and has the greatest preference for an equatorial orientation.

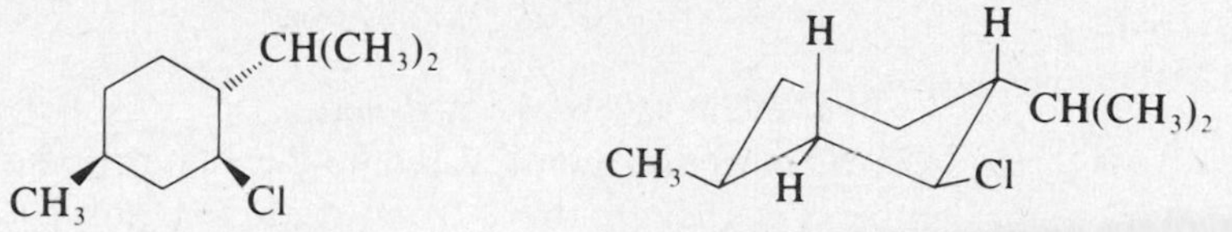

Menthyl chloride

Most stable conformation of menthyl chloride: none of the three β protons are anti to chlorine

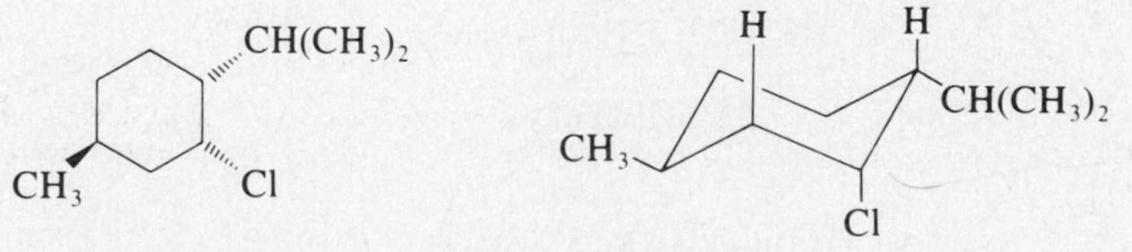

Neomenthyl chloride

Most stable conformation of neomenthyl chloride; each β carbon has a proton that is anti to chlorine

The anti periplanar relationship of halide and proton can be achieved only when the chlorine is axial; this corresponds to the most stable conformation of neomenthyl chloride. Menthyl chloride, on the other hand, must undergo appreciable distortion of its ring in order to achieve an anti periplanar Cl—C—C—H geometry. Strain increases substantially in going to the transition state for E2 elimination in menthyl chloride but not in neomenthyl chloride. Neomenthyl chloride undergoes E2 elimination at the faster rate.

6.9 In all parts of this exercise, write the structure of the alkyl halide in sufficient detail to identify the carbon that bears the halogen and the β carbon atoms that bear at least one hydrogen substituent. These are the carbons that become doubly bonded in the alkene product.

(*a*) 1-Bromohexane can give only 1-hexene under conditions of E2 elimination.

$$BrCH_2\overset{\beta}{C}H_2CH_2CH_2CH_2CH_3 \xrightarrow[E2]{base} CH_2{=}CHCH_2CH_2CH_2CH_3$$

1-Bromohexane; 1-Hexene (only alkene)

(*b*) 2-Bromohexane can give both 1-hexene and 2-hexene on dehydrobromination. The 2-hexene fraction is a mixture of cis and trans stereoisomers.

$$\underset{\text{2-Bromohexane}}{\overset{\beta}{C}H_3\overset{\beta}{C}H(Br)CH_2CH_2CH_3} \xrightarrow[\text{E2}]{\text{base}}$$

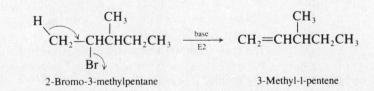

$$\underset{\text{1-Hexene}}{CH_2{=}CHCH_2CH_2CH_2CH_3} + \underset{\textit{cis}\text{-2-Hexene}}{(CH_3)(H)C{=}C(H)(CH_2CH_2CH_3)} + \underset{\textit{trans}\text{-2-Hexene}}{(CH_3)(H)C{=}C(H)(CH_2CH_2CH_3)}$$

(*c*) Both a cis-trans pair of 2-hexenes and a cis-trans pair of 3-hexenes are capable of being formed from 3-bromohexane.

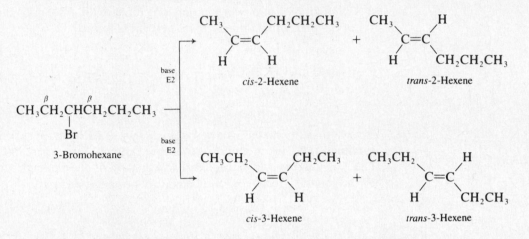

$$\underset{\text{3-Bromohexane}}{CH_3\overset{\beta}{C}H_2CH(Br)\overset{\beta}{C}H_2CH_2CH_3} \xrightarrow[\text{E2}]{\text{base}} \underset{\textit{cis}\text{-2-Hexene}}{(CH_3)(H)C{=}C(H)(CH_2CH_2CH_3)} + \underset{\textit{trans}\text{-2-Hexene}}{(CH_3)(H)C{=}C(H)(CH_2CH_2CH_3)}$$

$$\underset{\text{3-Bromohexane}}{CH_3CH_2CH(Br)CH_2CH_2CH_3} \xrightarrow[\text{E2}]{\text{base}} \underset{\textit{cis}\text{-3-Hexene}}{(CH_3CH_2)(H)C{=}C(H)(CH_2CH_3)} + \underset{\textit{trans}\text{-3-Hexene}}{(CH_3CH_2)(H)C{=}C(H)(CH_2CH_3)}$$

(*d*) Dehydrobromination of 2-bromo-2-methylpentane can involve one of the hydrogens of either a methyl group (C-1) or a methylene group (C-3).

$$\underset{\text{2-Bromo-2-methylpentane}}{CH_3C(CH_3)(Br)CH_2CH_2CH_3} \xrightarrow[\text{E2}]{\text{base}} \underset{\text{2-Methyl-1-pentene}}{CH_2{=}C(CH_3)CH_2CH_2CH_3} + \underset{\text{2-Methyl-2-pentene}}{(CH_3)_2C{=}CHCH_2CH_3}$$

Neither alkene is capable of existing in stereoisomeric forms, so these two are the only products of E2 elimination.

(*e*) 2-Bromo-3-methylpentane can undergo dehydrohalogenation by loss of a proton from either C-1 or C-3. Loss of a proton from C-1 gives 3-methyl-1-pentene.

$$\underset{\text{2-Bromo-3-methylpentane}}{H{-}CH_2{-}CH(Br)CH(CH_3)CH_2CH_3} \xrightarrow[\text{E2}]{\text{base}} \underset{\text{3-Methyl-1-pentene}}{CH_2{=}CHCH(CH_3)CH_2CH_3}$$

Loss of a proton from C-3 gives a mixture of (*E*)- and (*Z*)-3-methyl-2-pentene.

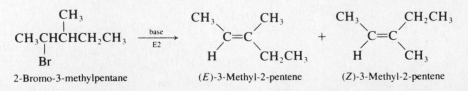

$$\underset{\text{2-Bromo-3-methylpentane}}{CH_3CH(Br)CH(CH_3)CH_2CH_3} \xrightarrow[\text{E2}]{\text{base}} \underset{(E)\text{-3-Methyl-2-pentene}}{(CH_3)(H)C{=}C(CH_3)(CH_2CH_3)} + \underset{(Z)\text{-3-Methyl-2-pentene}}{(CH_3)(H)C{=}C(CH_2CH_3)(CH_3)}$$

(*f*) Three alkenes are possible from 3-bromo-2-methylpentane. Loss of the C-2 proton gives 2-methyl-2-pentene.

$$CH_3-\underset{CH_3}{\overset{H}{C}}-\underset{Br}{C}HCH_2CH_3 \xrightarrow{E2} (CH_3)_2C{=}CHCH_2CH_3$$

3-Bromo-2-methylpentane → 2-Methyl-2-pentene

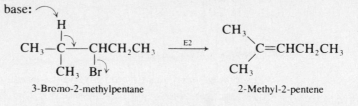

Abstraction of a proton from C-4 can yield either (*E*)- or (*Z*)-4-methyl-2-pentene.

$$(CH_3)_2CH\overset{H}{C}H-\underset{Br}{C}HCH_3 \xrightarrow{E2} (E)\text{-}(CH_3)_2CH(H)C{=}C(H)CH_3 + (Z)\text{-}(CH_3)_2CH(H)C{=}C(H)CH_3$$

3-Bromo-2-methylpentane → (*E*)-4-Methyl-2-pentene + (*Z*)-4-Methyl-2-pentene

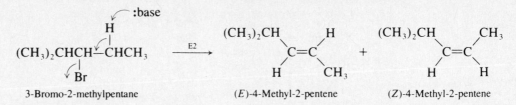

(*g*) Proton abstraction from the C-3 methyl group of 3-bromo-3-methylpentane yields 2-ethyl-1-butene.

$$CH_3CH_2\underset{Br}{\overset{CH_2-H}{C}}CH_2CH_3 \xrightarrow{E2} CH_2{=}C(CH_2CH_3)CH_2CH_3$$

3-Bromo-3-methylpentane → 2-Ethyl-1-butene

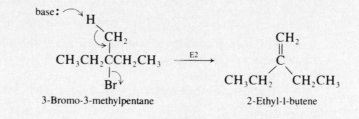

Stereoisomeric 3-methyl-2-pentenes are formed by proton abstraction from C-2.

$$CH_3-\overset{H}{C}H-\underset{Br}{\overset{CH_3}{C}}CH_2CH_3 \xrightarrow{E2} (E)\text{-}CH_3(H)C{=}C(CH_3)CH_2CH_3 + (Z)\text{-}CH_3(H)C{=}C(CH_2CH_3)CH_3$$

3-Bromo-3-methylpentane → (*E*)-3-Methyl-2-pentene + (*Z*)-3-Methyl-2-pentene

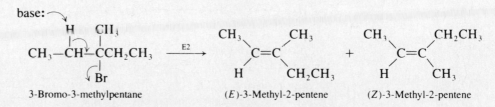

(*h*) Only 3,3-dimethyl-1-butene may be formed under conditions of E2 elimination from 2-bromo-3,3-dimethylbutane.

$$CH_3-\underset{CH_3}{\overset{CH_3}{C}}-\underset{Br}{C}H-CH_2H \xrightarrow{E2} CH_3-\underset{CH_3}{\overset{CH_3}{C}}-CH{=}CH_2$$

2-Bromo-3,3-dimethylbutane → 3,3-Dimethyl-1-butene

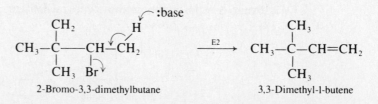

6.10 (*a*) The reaction that takes place with 1-bromo-3,3-dimethylbutane is an E2 elimination involving loss of the bromide at C-1 and abstraction of the proton at C-2 by the strong base potassium *tert*-butoxide, yielding a single alkene.

$$(CH_3)_3CO^- + (CH_3)_3C\overset{H}{C}H-CH_2Br \xrightarrow{E2} (CH_3)_3CCH{=}CH_2$$

1-Bromo-3,3-dimethylbutane → 3,3-Dimethyl-1-butene

(*b*) Two alkenes are capable of being formed in this β elimination reaction.

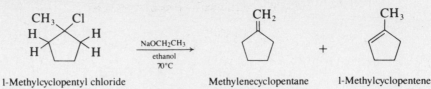

$$\text{1-Methylcyclopentyl chloride} \xrightarrow[\text{ethanol, 70°C}]{NaOCH_2CH_3} \text{Methylenecyclopentane} + \text{1-Methylcyclopentene}$$

The more highly substituted alkene is 1-methylcyclopentene; it is the major product of this reaction. According to Zaitsev's rule, the major alkene is formed by proton removal from the β carbon that has the fewest hydrogen substituents.

(*c*) Acid-catalyzed dehydration of 3-methyl-3-pentanol can lead either to 2-ethyl-1-butene or to a mixture of (*E*)- and (*Z*)-3-methyl-2-pentene.

$$CH_3CH_2C(CH_3)(OH)CH_2CH_3 \xrightarrow[80°C]{H_2SO_4}$$

3-Methyl-3-pentanol

$$(CH_3CH_2)_2C{=}CH_2 + (E)\text{-}CH_3CH{=}C(CH_3)CH_2CH_3 + (Z)\text{-}CH_3CH{=}C(CH_3)CH_2CH_3$$

2-Ethyl-1-butene; (*E*)-3-Methyl-2-pentene; (*Z*)-3-Methyl-2-pentene

The major product is a mixture of the trisubstituted alkenes, (*E*)- and (*Z*)-3-methyl-2-pentene. Of these two stereoisomers the *E* isomer is slightly more stable and is expected to predominate.

(*d*) Acid-catalyzed dehydration of 2,3-dimethyl-2-butanol can proceed in either of two directions.

$$CH_3{-}C(CH_3)(OH){-}CH(CH_3)CH_3 \xrightarrow[120°C,\ -H_2O]{H_3PO_4} CH_2{=}C(CH_3)CH(CH_3)_2 + (CH_3)_2C{=}C(CH_3)_2$$

2,3-Dimethyl-2-butanol; 2,3-Dimethyl-1-butene (disubstituted); 2,3-Dimethyl-2-butene (tetrasubstituted)

The major alkene is the one with the more highly substituted double bond, 2,3-dimethyl-2-butene. Its formation corresponds to Zaitsev's rule in that a proton is lost from the β carbon that has the fewest hydrogen substituents.

(*e*) Only a single alkene is capable of being formed on E2 elimination from this alkyl iodide. Stereoisomeric alkenes are not possible and because all the β hydrogens are equivalent, regioisomers cannot be formed either.

$$CH_3CH(CH_3)CH(I)CH(CH_3)CH_3 \xrightarrow[\text{ethanol, 70°C}]{NaOCH_2CH_3} (CH_3)_2C{=}CHCH(CH_3)_2$$

3-Iodo-2,4-dimethylpentane; 2,4-Dimethyl-2-pentene

(*f*) Despite the structural similarity of this alcohol to the alkyl halide in the preceding part of this problem, its dehydration is more complicated. The initially formed carbocation is secondary and can rearrange to a more stable tertiary carbocation by a hydride shift.

$$CH_3CH(CH_3)CH(OH)CH(CH_3)CH_3 \xrightarrow[-H_2O]{H^+} CH_3C(H)(CH_3){-}\overset{+}{C}HCH(CH_3)CH_3 \longrightarrow CH_3\overset{+}{C}(CH_3)CH_2CH(CH_3)CH_3$$

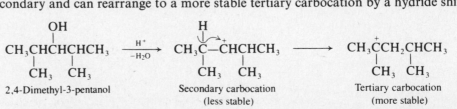

2,4-Dimethyl-3-pentanol; Secondary carbocation (less stable); Tertiary carbocation (more stable)

The tertiary carbocation, once formed, can give either 2,4-dimethyl-1-pentene or 2,4-dimethyl-2-pentene by loss of a proton.

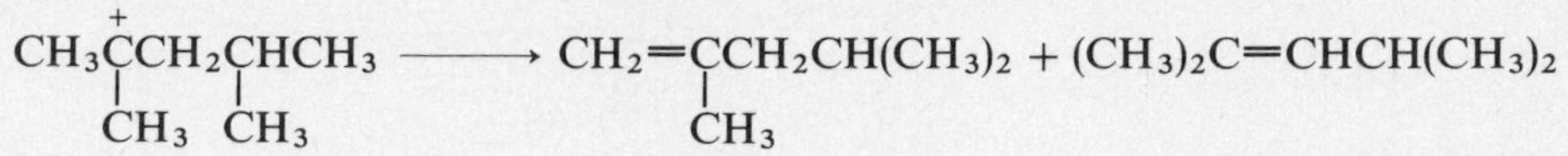

$$\underset{\substack{| \quad\quad\; | \\ CH_3 \;\; CH_3}}{CH_3\overset{+}{C}CH_2CHCH_3} \longrightarrow \underset{\substack{| \\ CH_3}}{CH_2{=}CCH_2CH(CH_3)_2} + (CH_3)_2C{=}CHCH(CH_3)_2$$

2,4-Dimethyl-1-pentene (disubstituted) 2,4-Dimethyl-2-pentene (trisubstituted)

The proton is lost from the methylene group in preference to the methyl group. The major alkene is the more highly substituted one, 2,4-dimethyl-2-pentene.

6.11 In all parts of this problem you need to reason backward from an alkene to a bromide of molecular formula $C_7H_{13}Br$ that gives only the desired alkene under E2 elimination conditions. Recall that the carbon-carbon double bond is formed by loss of a proton from one of the doubly bonded carbons and a bromide from the other.

(*a*) Cycloheptene is the only alkene formed by an E2 elimination reaction of cycloheptyl bromide.

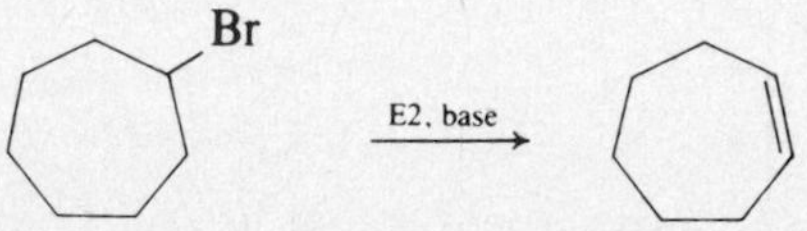

Br; E2. base

Cycloheptyl bromide Cycloheptene

(*b*) (Bromomethyl)cyclohexane is the correct answer. It gives methylenecyclohexane as the *only* alkene under E2 conditions.

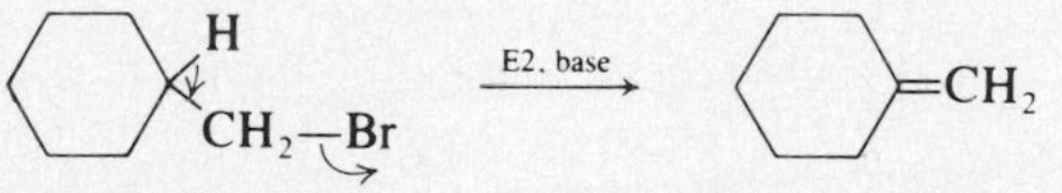

H; CH_2—Br; E2. base; $=CH_2$

(Bromomethyl)cyclohexane Methylenecyclohexane

1-Bromo-1-methylcyclohexane is not correct. It gives a mixture of 1-methylcyclohexene and methylenecyclohexane on elimination.

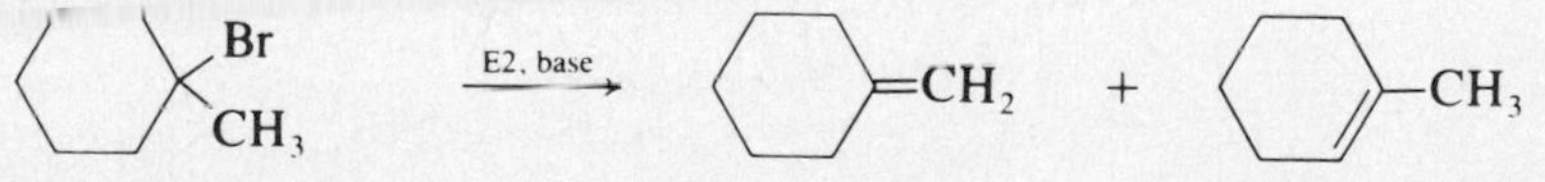

Br; CH_3; E2. base; $=CH_2$ + —CH_3

1-Bromo-1-methylcyclohexane Methylenecyclohexane 1-Methylcyclohexene

(*c*) In order for 4-methylcyclohexene to be the only alkene, the starting alkyl bromide must be 1-bromo-4-methylcyclohexane. Either the cis or trans isomer may be used, although the cis will react more readily, as the more stable conformation (equatorial methyl) has an axial bromine.

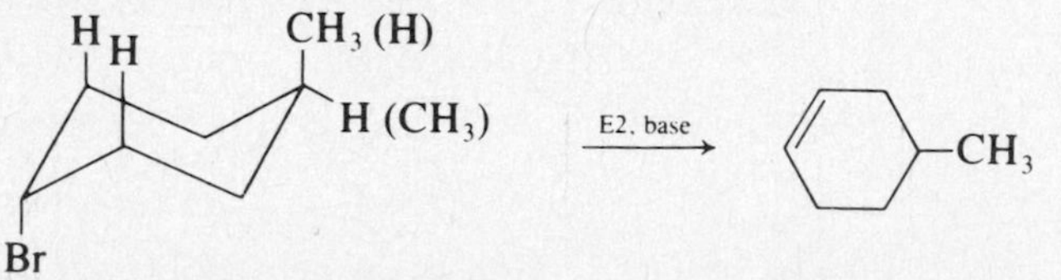

H H; CH_3 (H); H (CH_3); Br; E2. base; —CH_3

cis- or *trans*-1-Bromo-4-methylcyclohexane 4-Methylcyclohexene

1-Bromo-3-methylcyclohexane is incorrect; its dehydrobromination yields a mixture of 3-methylcyclohexene and 4-methylcyclohexene.

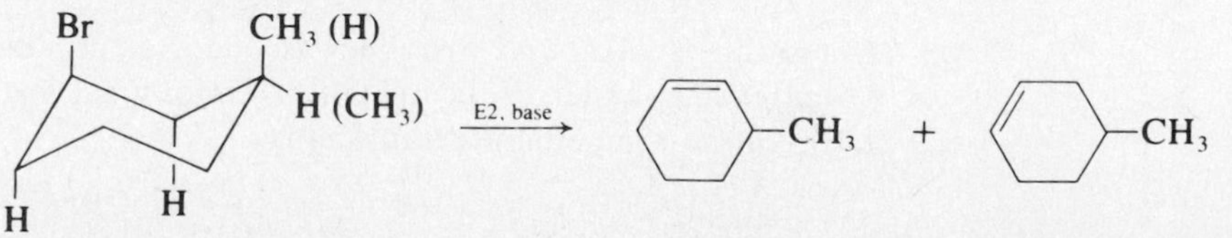

Br; CH_3 (H); H (CH_3); H; H; E2. base; —CH_3 + —CH_3

1-Bromo-3-methylcyclohexene 3-Methylcyclohexene 4-Methylcyclohexene

(*d*) The bromine must be at C-2 in the starting alkyl bromide.

2-Bromo-1,1-dimethylcyclopentane $\xrightarrow{\text{E2, base}}$ 3,3-Dimethylcyclopentene

If the bromine substituent were at C-3, a mixture of 3,3-dimethyl and 4,4-dimethylcyclopentene would be formed.

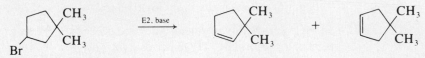

3-Bromo-1,1-dimethylcyclopentane $\xrightarrow{\text{E2, base}}$ 3,3-Dimethylcyclopentene + 4,4-Dimethylcyclopentene

(*e*) The alkyl bromide must be primary in order for the desired alkene to be the only product of E2 elimination.

$C_5H_9-CH_2CH_2Br \xrightarrow{\text{E2, base}} C_5H_9-CH{=}CH_2$

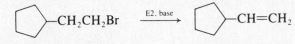

2-Cyclopentylethyl bromide — Vinylcyclopentane

If 1-cyclopentylethyl bromide were used, a mixture of regioisomeric alkenes would be formed, with the desired vinylcyclopentane being the minor component of the mixture.

(*f*) Either *cis*- or *trans*-1-bromo-3-isopropylcyclobutane would be appropriate here.

cis- or *trans*-1-Bromo-3-isopropylcyclobutane $\xrightarrow{\text{E2, base}}$ 3-Isopropylcyclobutene

(*g*) The desired alkene is the exclusive product formed on E2 elimination from 1-bromo-1-*tert*-butylcyclopropane.

1-Bromo-1-*tert*-butylcyclopropane $\xrightarrow{\text{E2, base}}$ 1-*tert*-Butylcyclopropene

6.12 (*a*) Both 1-bromopropane and 2-bromopropane yield propene as the exclusive product of E2 elimination.

$CH_3CH_2CH_2Br$ or $CH_3CHBrCH_3 \xrightarrow{\text{E2, base}} CH_3CH{=}CH_2$

1-Bromopropane — 2-Bromopropane — Propene

(*b*) Isobutene is formed on dehydrobromination of either *tert*-butyl bromide or isobutyl bromide.

$(CH_3)_3CBr$ or $(CH_3)_2CHCH_2Br \xrightarrow{\text{E2, base}} (CH_3)_2C{=}CH_2$

tert-Butyl bromide — Isobutyl bromide — Isobutene

(*c*) A tetrabromoalkane is required as the starting material in order to form a tribromoalkene under E2 elimination conditions. Either 1,1,2,2-tetrabromoethane or 1,1,1,2-tetrabromoethane is satisfactory.

$Br_2CHCHBr_2$ or $BrCH_2CBr_3 \xrightarrow{\text{E2, base}} BrCH{=}CBr_2$

1,1,2,2-Tetrabromoethane — 1,1,1,2-Tetrabromoethane — 1,1,2-Tribromoethene (*d*) The bromine substituent may be at either C-2 or C-3.

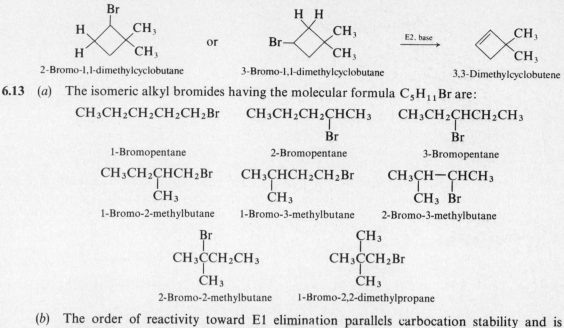

2-Bromo-1,1-dimethylcyclobutane or 3-Bromo-1,1-dimethylcyclobutane $\xrightarrow{\text{E2, base}}$ 3,3-Dimethylcyclobutene

6.13 (*a*) The isomeric alkyl bromides having the molecular formula $C_5H_{11}Br$ are:

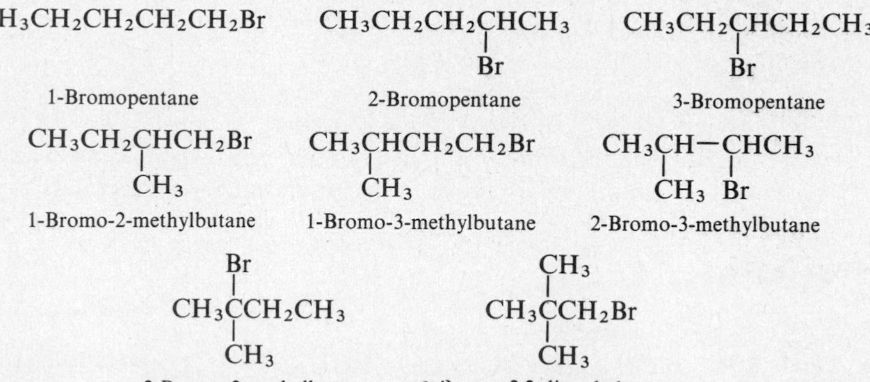

(*b*) The order of reactivity toward E1 elimination parallels carbocation stability and is tertiary > secondary > primary. The tertiary bromide 2-bromo-2-methylbutane will undergo E1 elimination at the fastest rate.

(*c*) The order of reactivity toward E2 elimination reflects the stability of the developing double bond and the weakness of the carbon-halogen bond; it is tertiary > secondary > primary. Here also 2-bromo-2-methylbutane will react fastest.

(*d*) Neopentyl bromide (1-bromo-2,2-dimethylpropane) has no hydrogens on the β carbon and so cannot form an alkene directly by an E2 process.

$$(CH_3)_3C\text{—}CH_2\text{—}Br \xrightarrow{\text{E1}} CH_3\overset{+}{C}CH_2CH_3\,(CH_3) \xrightarrow{-H^+} CH_3C{=}CHCH_3\,(CH_3)$$

The only available pathway is E1 with rearrangement.

(*e*) Only the primary bromides shown can give both regiospecific elimination and no possibility of cis-trans isomers.

$$CH_3CH_2CH_2CH_2CH_2Br \xrightarrow{\text{base, E2}} CH_3CH_2CH_2CH{=}CH_2$$

1-Bromopentane → 1-pentene

$$CH_3CH_2CH(CH_3)CH_2Br \xrightarrow{\text{base, E2}} CH_3CH_2C(CH_3){=}CH_2$$

1-Bromo-2-methylbutane → 2-Methyl-1-butene

$$CH_3CH(CH_3)CH_2CH_2Br \xrightarrow{\text{base, E2}} CH_3CH(CH_3)CH{=}CH_2$$

1-Bromo-3-methylbutane → 3-Methyl-1-butene

(*f*) Elimination in 3-bromopentane will give (*E*)- and (*Z*)-2-pentene only.

$$CH_3CH_2CH(Br)CH_2CH_3 \xrightarrow{\text{base, E2}} (E)\text{-}CH_3CH{=}CHCH_2CH_3 + (Z)\text{-}CH_3CH{=}CHCH_2CH_3$$

3-Bromopentane → (*E*)-2-Pentene + (*Z*)-2-Pentene

(*g*) Three alkenes can be formed from 2-bromopentane.

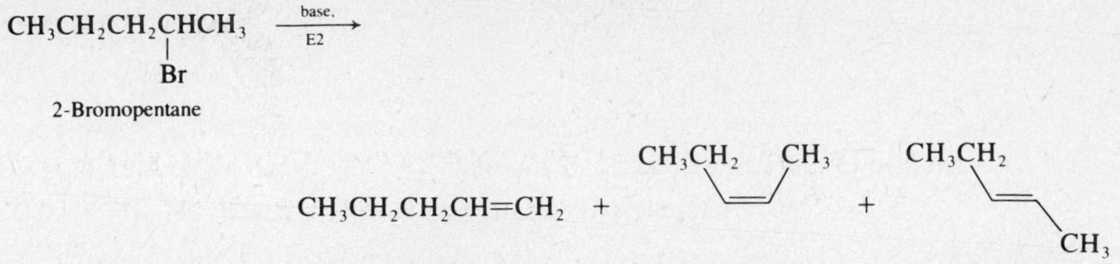

$$CH_3CH_2CH_2\underset{\displaystyle Br}{\underset{|}{C}}HCH_3 \xrightarrow[E2]{\text{base,}}$$

2-Bromopentane

$$CH_3CH_2CH_2CH{=}CH_2 \quad + \quad (Z)\text{-}CH_3CH_2CH{=}CHCH_3 \quad + \quad (E)\text{-}CH_3CH_2CH{=}CHCH_3$$

1-Pentene (*Z*)-2-Pentene (*E*)-2-Pentene

6.14 The reaction conditions are those of E2 elimination. The substrate dependence of the elimination rate decreases in the orders tertiary > secondary > primary and I > Br > Cl > F.

(*a*)

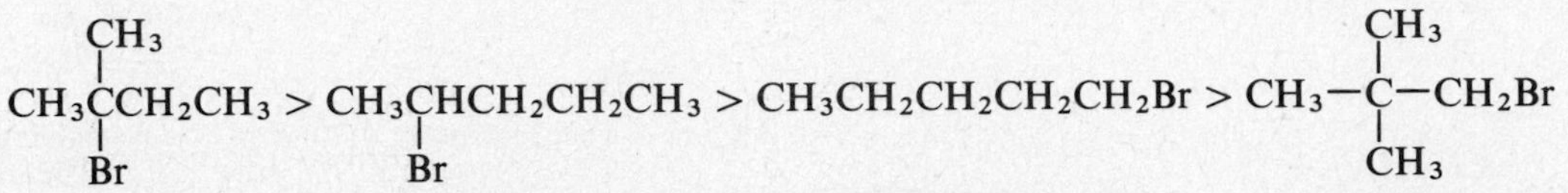

$$CH_3\overset{\displaystyle CH_3}{\underset{\displaystyle Br}{C}}CH_2CH_3 > CH_3\underset{\displaystyle Br}{C}HCH_2CH_2CH_3 > CH_3CH_2CH_2CH_2CH_2Br > CH_3{-}\overset{\displaystyle CH_3}{\underset{\displaystyle CH_3}{C}}{-}CH_2Br$$

2-Bromo-2-methylbutane (3°); 2-Bromopentane (2°); 1-Bromopentane (1°); 1-Bromo-2,2-dimethylpropane (cannot undergo E2 elimination)

(*b*)

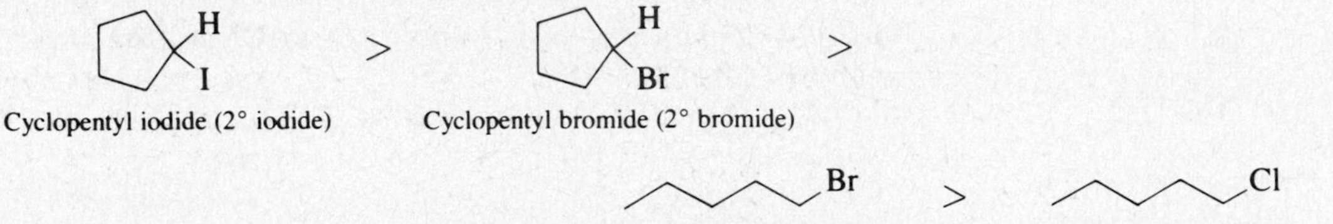

Cyclopentyl iodide (2° iodide) > Cyclopentyl bromide (2° bromide) > 1-Bromopentane (1° bromide) > 1-Chloropentane (1° chloride)

(*c*)

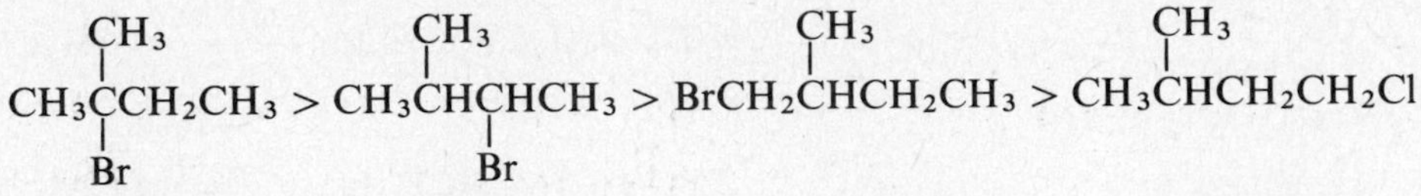

$$CH_3\overset{\displaystyle CH_3}{\underset{\displaystyle Br}{C}}CH_2CH_3 > CH_3\overset{\displaystyle CH_3}{C}H\underset{\displaystyle Br}{C}HCH_3 > BrCH_2\overset{\displaystyle CH_3}{C}HCH_2CH_3 > CH_3\overset{\displaystyle CH_3}{C}HCH_2CH_2Cl$$

2-Bromo-2-methylbutane (3° bromide); 2-Bromo-3-methylbutane (2° bromide); 1-Bromo-2-methylbutane (1° bromide); 1-Chloro-3-methylbutane (1° chloride)

(*d*)

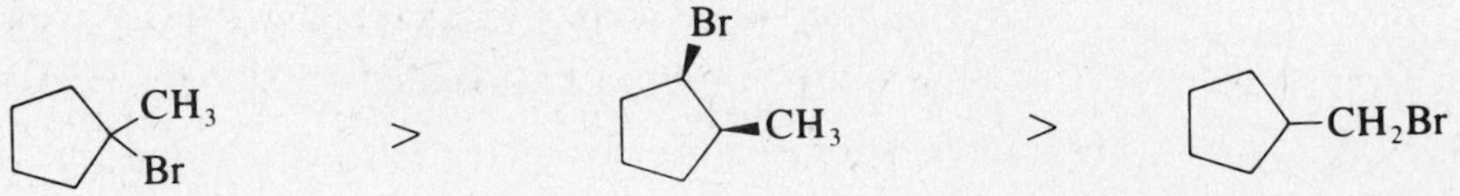

1-Methylcyclopentyl bromide (3°) > *cis*-2-Methylcyclopentyl bromide (2°) > (Bromomethyl)cyclopentane (1°)

(*e*) In an E2 reaction the preferred spatial relationship between the leaving group and the proton which is lost is anti. In a six-membered ring the halogen must be able to adopt an axial orientation.

Br CH$_3$ > CH$_3$ Br >

1-Methylcyclohexyl bromide (3°) *cis*-4-Methylcyclohexyl bromide (2° axial bromide)

CH$_3$ Br > CH$_3$ F

trans-4-Methylcyclohexyl bromide (2° equatorial bromide) *trans*-4-Methylcyclohexyl fluoride (2° equatorial fluoride)

6.15 (*a*) The isomeric $C_5H_{12}O$ alcohols are:

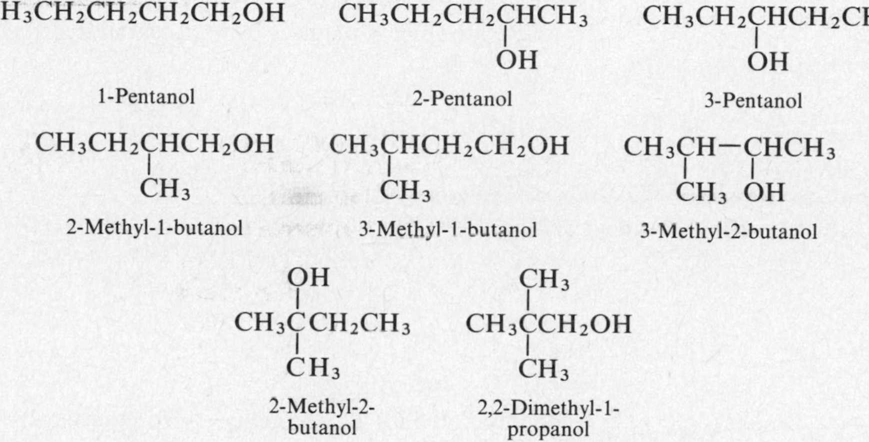

(*b*) The order of reactivity in alcohol dehydration is tertiary > secondary > primary. The only tertiary alcohol in the group is 2-methyl-2-butanol (*tert*-amyl alcohol). It will dehydrate fastest.

(*c*) The most stable C_5H_{11} carbocation is the tertiary carbocation.

$$CH_3\overset{+}{C}(CH_3)CH_2CH_3$$ 1,1-Dimethylpropyl cation

(*d*) A proton may be lost from C-1 or C-3:

$$CH_3\overset{+}{C}(CH_3)CH_2CH_3 \longrightarrow CH_2{=}C(CH_3)CH_2CH_3 + CH_3C(CH_3){=}CHCH_3$$

1,1-Dimethylpropyl cation; 2-Methyl-1-butene (minor alkene); 2-Methyl-2-butene (major alkene)

(*e*) In order for the 1,1-dimethylpropyl cation to be formed by a process involving a hydride shift, the starting alcohol must have the same carbon skeleton as the 1,1-dimethylpropyl cation.

$CH_3\overset{+}{C}(CH_3)CH_2CH_3$ has same carbon skeleton as

$$HOCH_2C(H)(CH_3)CH_2CH_3 \quad \text{and} \quad CH_3C(H)(CH_3){-}CH(OH)CH_3$$

$$HOCH_2\overset{H}{\underset{CH_3}{C}}CH_2CH_3 \xrightarrow{H^+} H\overset{+}{O}(H){-}CH_2{-}\overset{H}{\underset{CH_3}{C}}CH_2CH_3 \longrightarrow CH_3{-}\overset{+}{\underset{CH_3}{C}}CH_2CH_3$$

2-Methyl-1-butanol

$$CH_3\overset{H}{\underset{CH_3}{C}}{-}\underset{OH}{CH}CH_3 \xrightarrow[-H_2O]{H^+} CH_3{-}\overset{H}{\underset{CH_3}{C}}{-}\overset{+}{C}HCH_3 \longrightarrow CH_3{-}\overset{+}{\underset{CH_3}{C}}{-}CH_2CH_3$$

3-Methyl-2-butanol

While the same carbon skeleton is necessary, it alone is not sufficient; the alcohol must also have its hydroxyl group on the carbon atom adjacent to the carbon which bears the migrating hydrogen. Thus, 3-methyl-1-butanol cannot form a tertiary carbocation by a hydride shift. It requires two sequential hydride shifts.

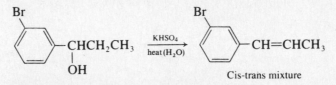

$$CH_3\underset{CH_3}{C}HCH_2CH_2OH \xrightarrow{H^+} CH_3\underset{CH_3}{C}H\overset{H}{C}H{-}CH_2{-}\overset{+}{O}H_2 \longrightarrow CH_3\overset{H}{\underset{CH_3}{C}}{-}\overset{+}{C}HCH_3 \longrightarrow CH_3\overset{+}{\underset{CH_3}{C}}CH_2CH_3$$

3-Methyl-1-butanol

(*f*) Neopentyl alcohol (2,2-dimethyl-1-propanol) can yield a tertiary carbocation by a process involving a methyl shift.

$$CH_3\overset{CH_3}{\underset{CH_3}{C}}{-}CH_2OH \xrightarrow{H^+} CH_3\overset{CH_3}{\underset{CH_3}{C}}{-}CH_2{-}\overset{+}{O}H_2 \longrightarrow CH_3\overset{+}{\underset{CH_3}{C}}{-}CH_2CH_3$$

2,2-Dimethyl-1-propanol

6.16 (*a*) Heating an alcohol in the presence of an acid catalyst ($KHSO_4$) leads to dehydration with formation of an alkene. In this alcohol elimination can occur in only one direction to give a mixture of cis and trans alkenes.

$$m\text{-}BrC_6H_4{-}\underset{OH}{C}HCH_2CH_3 \xrightarrow[\text{heat}(H_2O)]{KHSO_4} m\text{-}BrC_6H_4{-}CH{=}CHCH_3$$

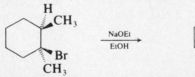

Cis-trans mixture

(*b*) Alkyl halides undergo E2 elimination on being heated with potassium *tert*-butoxide.

$$ICH_2CH(OCH_2CH_3)_2 \xrightarrow[(CH_3)_3COH,\ \text{heat}]{KOC(CH_3)_3} CH_2{=}C(OCH_2CH_3)_2$$

(*c*) The exclusive product of this reaction is 1,2-dimethylcyclohexene.

$$\text{1-Bromo-}trans\text{-1,2-dimethylcyclohexane} \xrightarrow[EtOH]{NaOEt} \text{1,2-dimethylcyclohexene}$$

1-Bromo-*trans*-1,2-dimethylcyclohexane

1,2-Dimethylcyclohexene (100%)

(*d*) Elimination can occur only in one direction to give the alkene shown.

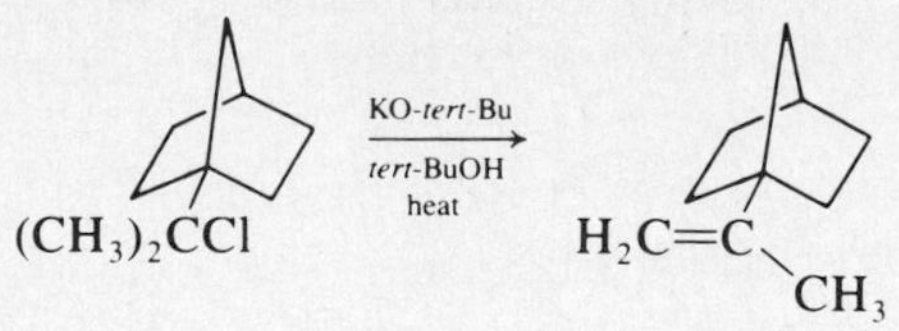

$(CH_3)_2CCl$ —(KO-*tert*-Bu, *tert*-BuOH, heat)→ $H_2C{=}C$ CH_3

(*e*) The reaction is a conventional one of alcohol dehydration and proceeds as written in 76 to 78 percent yield.

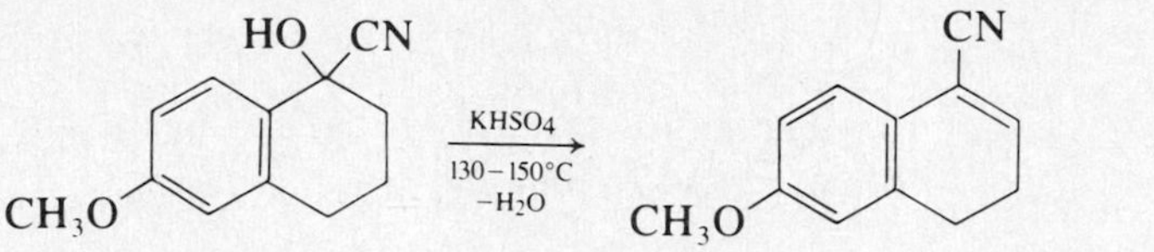

HO CN, CH_3O —($KHSO_4$, 130–150°C, $-H_2O$)→ CN, CH_3O

(*f*) Dehydration of citric acid occurs, giving aconitic acid.

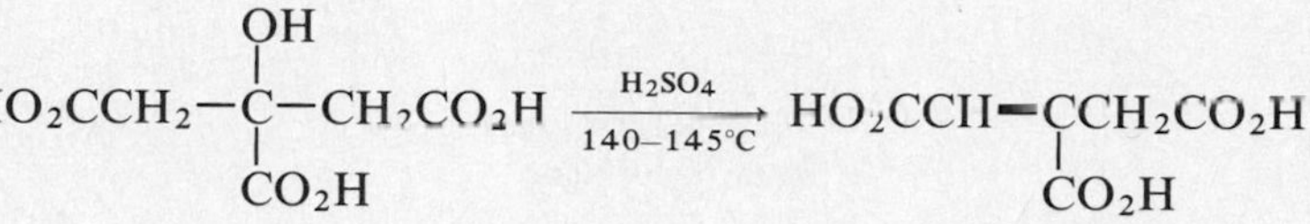

$$HO_2CCH_2{-}\overset{OH}{\underset{CO_2H}{C}}{-}CH_2CO_2H \xrightarrow[140\text{–}145°C]{H_2SO_4} HO_2CCH{=}\underset{CO_2H}{C}CH_2CO_2H$$

Citric acid — Aconitic acid

(*g*) Sequential double dehydrohalogenation gives the diene.

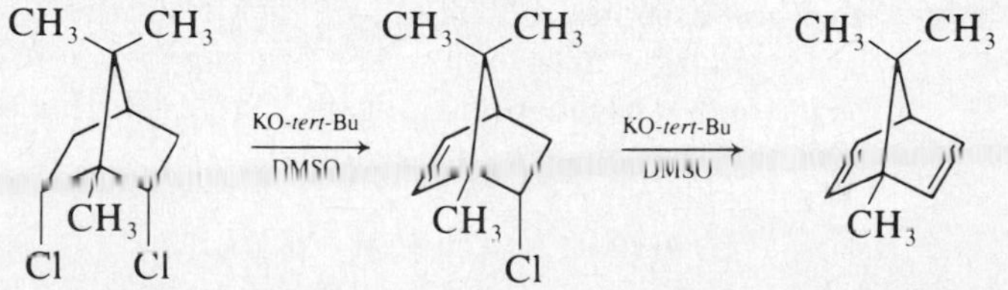

CH_3 CH_3, CH_3, Cl, Cl —(KO-*tert*-Bu, DMSO)→ CH_3 CH_3, CH_3, Cl —(KO-*tert*-Bu, DMSO)→ CH_3 CH_3, CH_3

Bornylene (83%)

(*h*) In spite of the complexity of the starting material, this example has been reported in the chemical literature and elimination proceeds in the usual way.

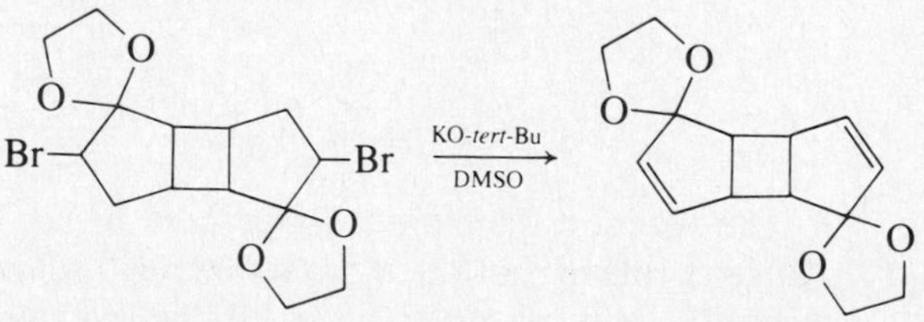

Br, Br —(KO-*tert*-Bu, DMSO)→ (84%)

(*i*) Again, we have a fairly complicated substrate, but notice that it is well disposed toward E2 elimination of the axial bromide.

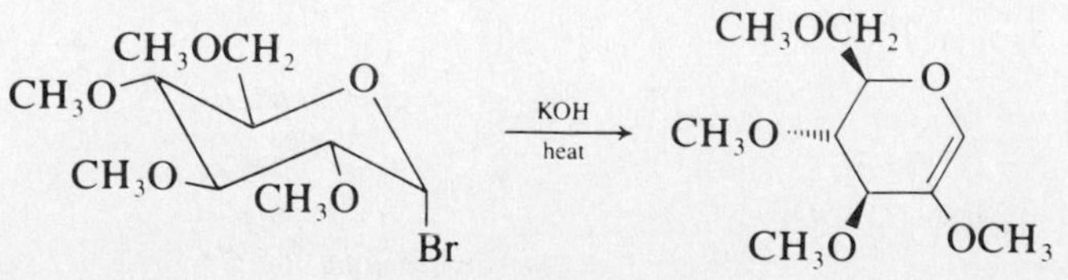

CH_3OCH_2, CH_3O, CH_3O, CH_3O, Br —(KOH, heat)→ CH_3OCH_2, CH_3O, CH_3O, OCH_3

(*j*) In the most stable conformation of this compound, chlorine occupies an axial site, so it is ideally situated to undergo an E2 elimination reaction by way of an anti arrangement in the transition state.

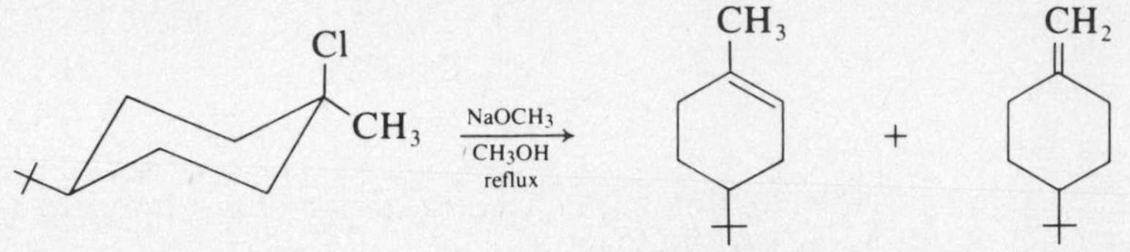

Cl, CH_3 —($NaOCH_3$, CH_3OH, reflux)→ CH_3 + CH_2

4-*tert*-Butyl-1-methyl cyclohexene (95%) | 4-*tert*-Butyl-(methylene) cyclohexane (5%)

The minor product is the less highly substituted isomer, in which the double bond is exocyclic to the ring.

6.17 Let us examine the experimental data more closely.

$$(CH_3)_3CCH_2C(Br)(CH_3)_2 \xrightarrow[\text{EtOH, 70°C}]{\text{KOEt}} (CH_3)_3CCH_2C(CH_3){=}CH_2 + (CH_3)_3CCH{=}C(CH_3)_2$$

2-Bromo-2,4,4-trimethylpentane | 2,4,4-Trimethyl-1-pentene, 86% | 2,4,4-Trimethyl-2-pentene, 14%

The less substituted alkene is formed in greater amount than the more substituted alkene. Thus, the regioselectivity does not correspond to the generalization that dehydrohalogenation tends to produce the more highly substituted alkene, but remember that this generalization assumes that the more highly substituted alkene is also the more stable one. The hydrogenation data tell us that this happens not to be true in this particular case; here the heats of hydrogenation indicate that 2,4,4-trimethyl-1-pentene is more stable than 2,4,4-trimethyl-2-pentene by 1.2 kcal/mol.

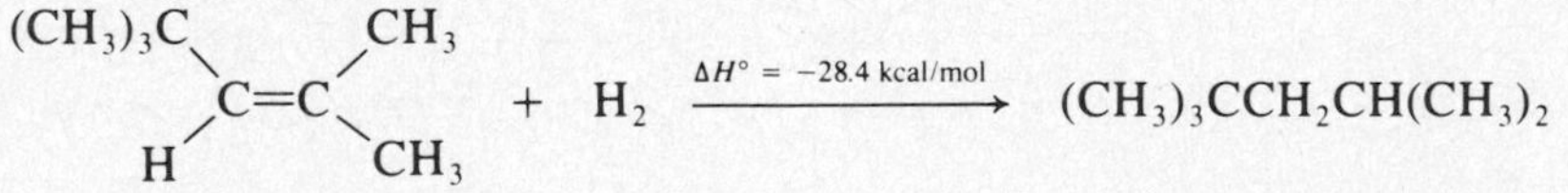

$$(CH_3)_3C(H)C{=}C(CH_3)_2 + H_2 \xrightarrow{\Delta H° = -28.4 \text{ kcal/mol}} (CH_3)_3CCH_2CH(CH_3)_2$$

2,4,4-Trimethyl-2-pentene | 2,2,4-Trimethylpentane

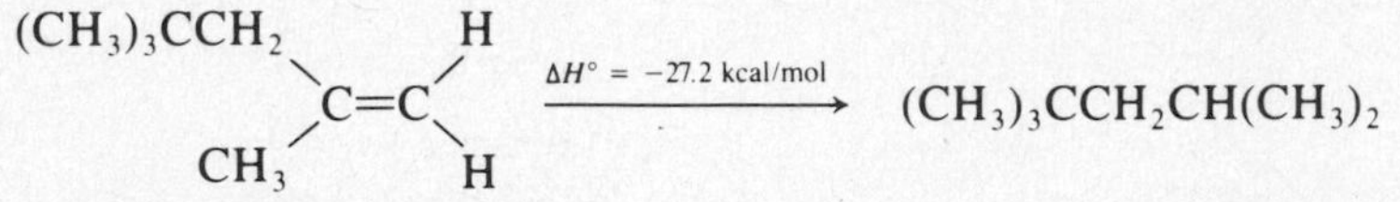

$$(CH_3)_3CCH_2(CH_3)C{=}CH_2 \xrightarrow{\Delta H° = -27.2 \text{ kcal/mol}} (CH_3)_3CCH_2CH(CH_3)_2$$

2,4,4-Trimethyl-1-pentene | 2,2,4-Trimethylpentane

The reaction therefore does proceed in accordance with the more modern statement of Zaitsev's rule in that the more stable regioisomer predominates. The reason the more substituted alkene turns out to be less stable is a steric effect arising from a van der Waals repulsion between a methyl group and the *tert*-butyl group in 2,4,4-trimethyl-2-pentene.

6.18 Begin by writing chemical equations for the processes specified in the problem. First consider rearrangement by way of a hydride shift:

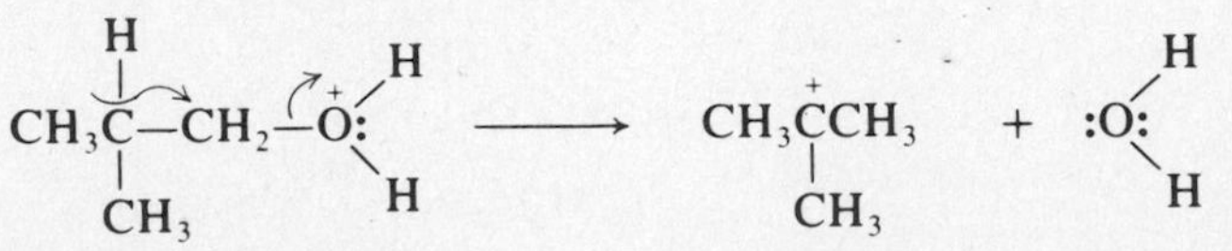

$$CH_3C(H)(CH_3){-}CH_2{-}\overset{+}{O}H_2 \longrightarrow CH_3\overset{+}{C}(CH_3)CH_3 + H_2\ddot{O}:$$

Isobutyloxonium ion | *tert*-Butyl cation | Water

Rearrangement by way of a methyl group shift is as follows:

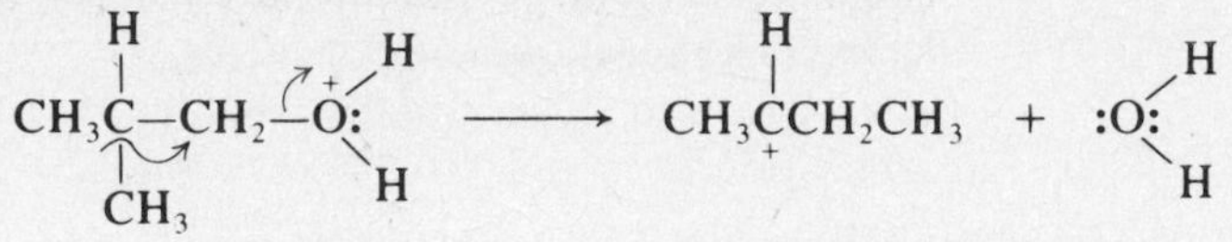

$$CH_3\underset{\underset{CH_3}{|}}{\overset{\overset{H}{|}}{C}}-CH_2-\overset{+}{O}H_2 \longrightarrow CH_3\overset{\overset{H}{|}}{\underset{+}{C}}CH_2CH_3 + H_2\ddot{O}:$$

Isobutyloxonium ion — *sec*-Butyl cation — Water

A hydride shift gives a tertiary carbocation; a methyl migration gives a secondary carbocation. It is reasonable to expect that rearrangement will occur so as to produce the more stable of these two carbocations because the activated complex has carbocation character at the carbon that bears the migrating group. We predict that rearrangement proceeds by a hydride shift rather than a methyl shift as the group that remains behind in this process stabilizes the carbocation better.

6.19 Rearrangement proceeds by migration of a hydrogen or an alkyl group from the carbon atom adjacent to the positively charged carbon.

(*a*) A propyl cation is primary and rearranges to an isopropyl cation, which is secondary, by migration of a hydrogen with its pair of electrons.

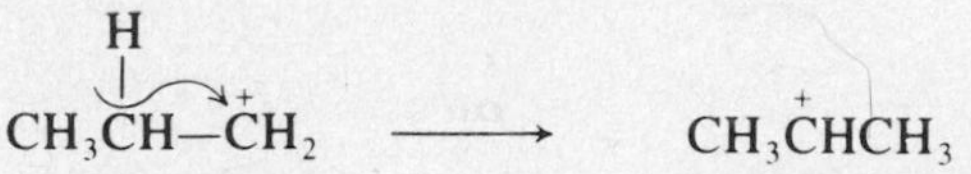

$$CH_3\overset{\overset{H}{|}}{C}H-\overset{+}{C}H_2 \longrightarrow CH_3\overset{+}{C}HCH_3$$

1-Propyl cation (primary, less stable) — Isopropyl cation (secondary, more stable)

(*b*) A hydride shift transforms the secondary carbocation to a tertiary one.

$$CH_3\underset{\underset{CH_3}{|}}{\overset{\overset{H}{|}}{C}}-\overset{+}{C}HCH_3 \longrightarrow CH_3\underset{\underset{CH_3}{|}}{\overset{+}{C}}CH_2CH_3$$

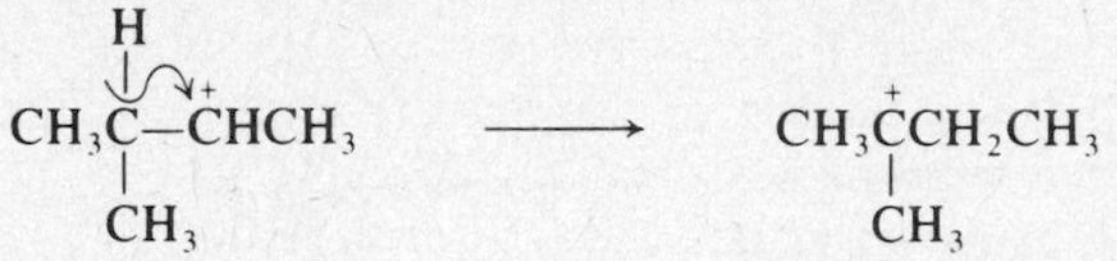

1,2-Dimethylpropyl cation (secondary, less stable) — 1,1-Dimethylpropyl cation (tertiary, more stable)

This hydride shift occurs in preference to methyl migration, which would produce the same secondary carbocation. (Verify this by writing appropriate structural formulas.)

(*c*) Migration of a methyl group converts this secondary carbocation to a tertiary one.

$$CH_3\underset{\underset{CH_3}{|}}{\overset{\overset{CH_3}{|}}{C}}-\overset{+}{C}HCH_3 \longrightarrow CH_3\underset{\underset{CH_3}{|}}{\overset{+}{C}}-\overset{\overset{CH_3}{|}}{C}HCH_3$$

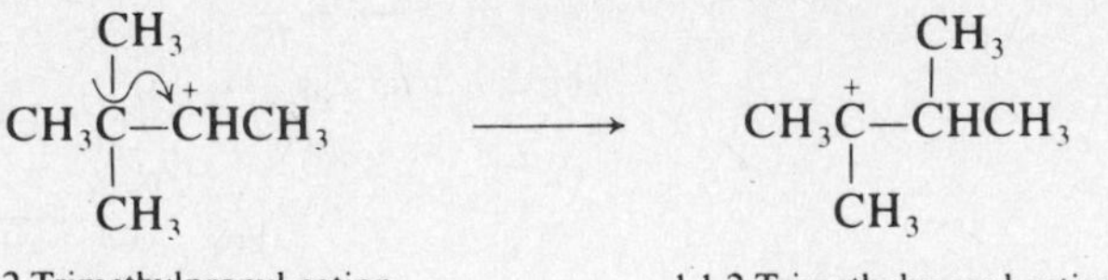

1,2,2-Trimethylpropyl cation (secondary, less stable) — 1,1,2-Trimethylpropyl cation (tertiary, more stable)

(*d*) The group that shifts in this case is the entire ethyl group.

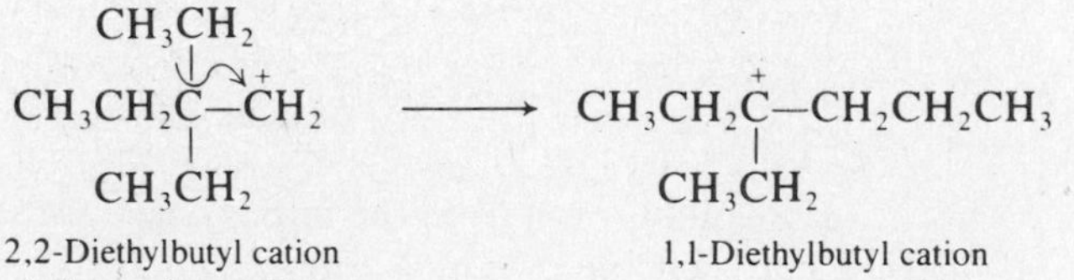

$$CH_3CH_2\underset{\underset{CH_3CH_2}{|}}{\overset{\overset{CH_3CH_2}{|}}{C}}-\overset{+}{C}H_2 \longrightarrow CH_3CH_2\underset{\underset{CH_3CH_2}{|}}{\overset{+}{C}}-CH_2CH_2CH_3$$

2,2-Diethylbutyl cation (primary, less stable) — 1,1-Diethylbutyl cation (tertiary, more stable)

(*e*) Migration of a hydride from the ring carbon that bears the methyl group produces a tertiary carbocation.

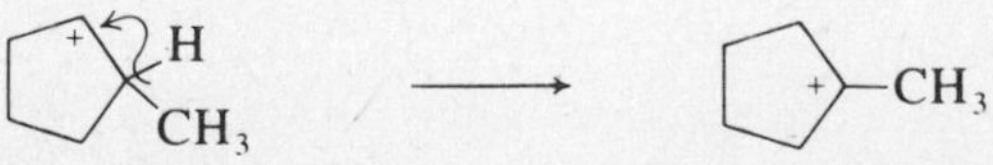

2-Methylcyclopentyl cation (secondary, less stable) $\longrightarrow$ 1-Methylcyclopentyl cation (tertiary, more stable)

6.20 (*a*) Note that the starting material is an alcohol and that it is treated with an acid. The product is an alkene but its carbon skeleton is different from that of the starting alcohol. The reaction is one of alcohol dehydration accompanied by rearrangement at the carbocation stage. Begin by writing the step in which the alcohol is converted to a carbocation.

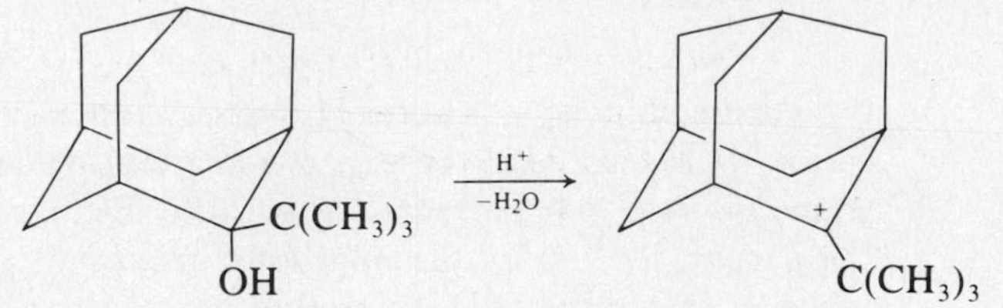

The carbocation is tertiary and relatively stable. However, migration of a methyl group from the *tert*-butyl substituent converts it to an isomeric carbocation, which is also tertiary.

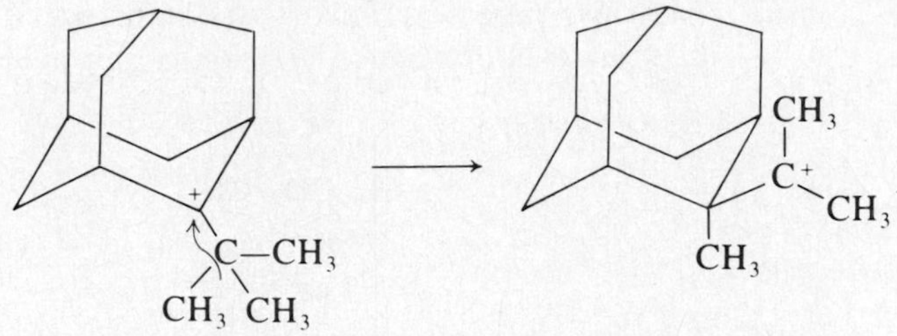

Loss of a proton from this carbocation gives the observed product.

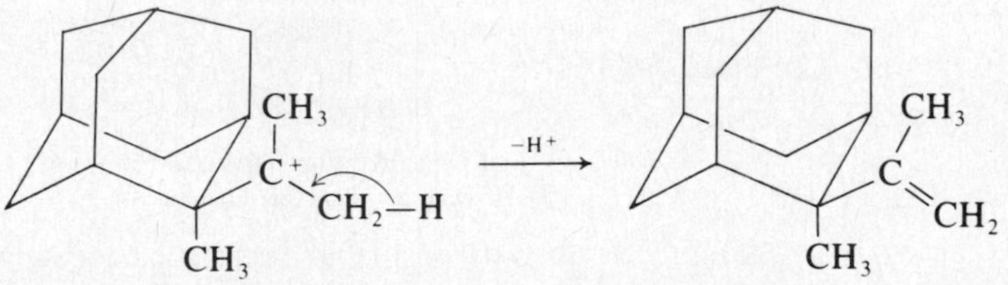

(*b*) Here also we have an alcohol dehydration reaction accompanied by rearrangement. The first formed carbocation is secondary.

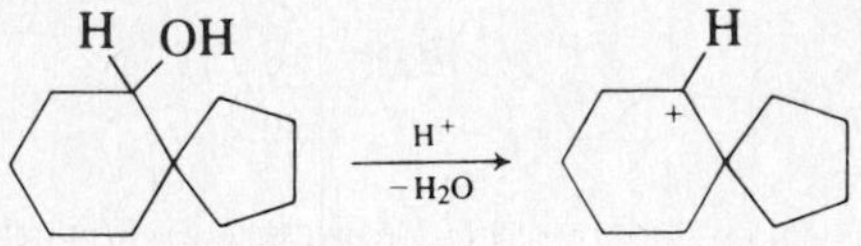

This cation can rearrange to a tertiary carbocation by an alkyl group shift.

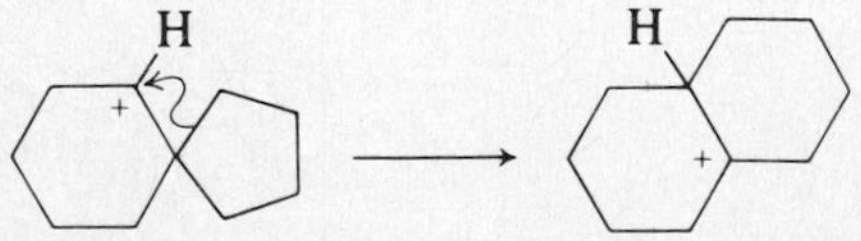

Loss of a proton from the tertiary carbocation gives the observed alkene.

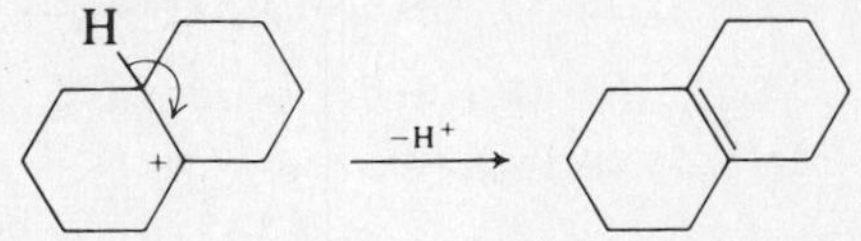

(*c*) The reaction begins as a normal alcohol dehydration in which the hydroxyl group is protonated by the acid catalyst and then loses water from the oxonium ion to give a carbocation.

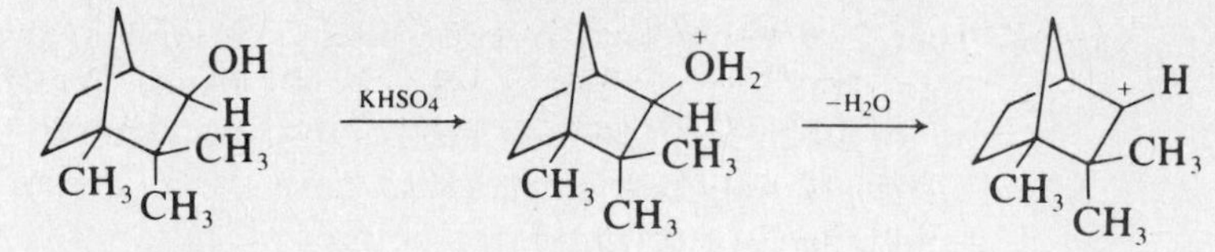

4-Methylcamphenilol

We see that the final product, 1-methylsantene, has a rearranged carbon skeleton corresponding to a methyl shift, so we consider the rearrangement of the initially formed secondary carbocation to a tertiary ion.

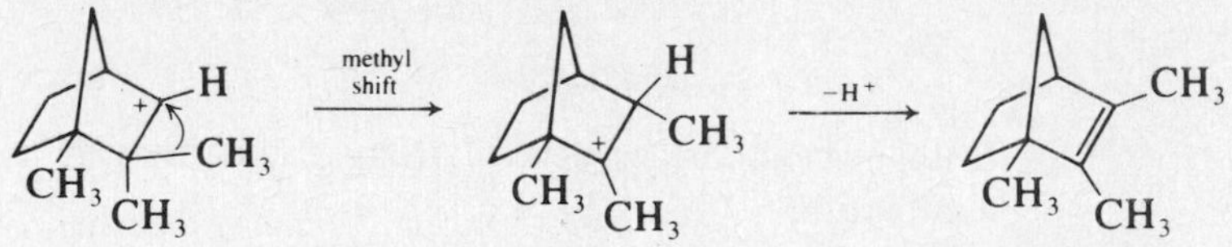

1-Methylsantene

Deprotonation of the tertiary carbocation yields 1-methylsantene.

6.21 The secondary carbocation can, as we have seen, rearrange by a methyl shift (Problem 6.5). It can also rearrange by migration of one of the ring bonds.

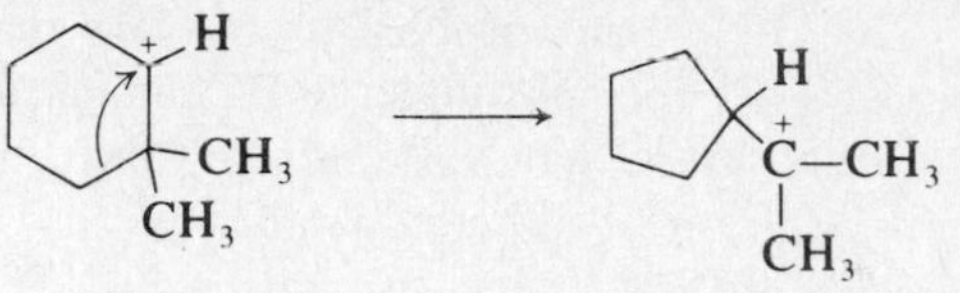

Secondary carbocation — Tertiary carbocation

The tertiary carbocation formed by this rearrangement can lose a proton to give the observed by-product.

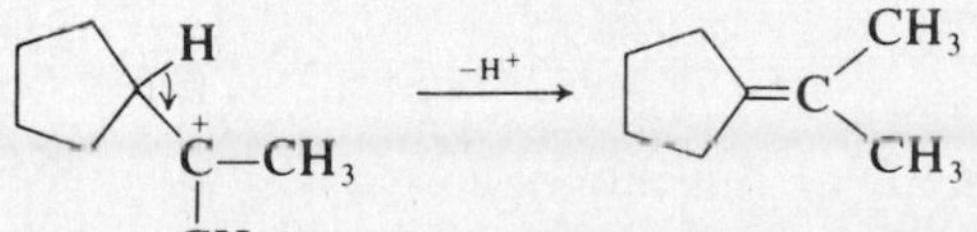

Isopropylidenecyclopentane

6.22 The carbon skeleton is revealed by the hydrogenation experiment. Compounds B and C must have the same carbon skeleton as 3-ethylpentane.

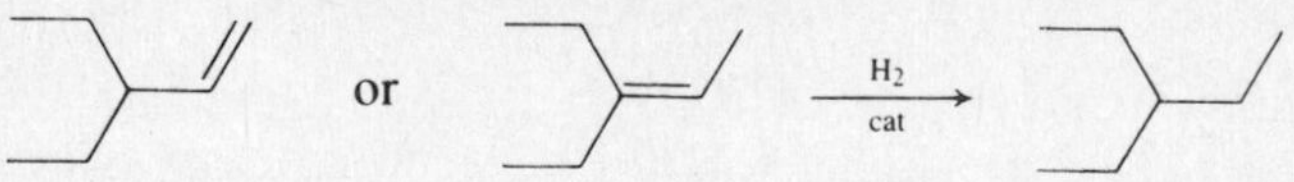

There are three alkyl bromides having this carbon skeleton, namely 1-bromo-3-ethylpentane, 2-bromo-3-ethylpentane, and 3-bromo-3-ethylpentane. Of these three only 2-bromo-3-ethylpentane will give two alkenes on dehydrobromination.

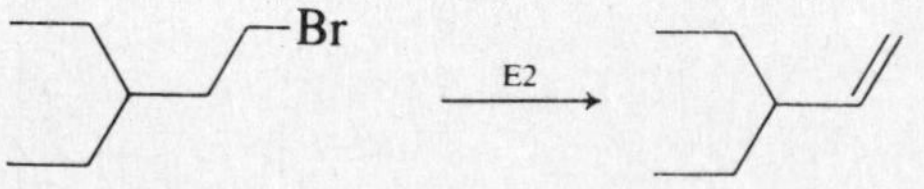

1-Bromo-3-ethylpentane $\xrightarrow{E2}$ (only product)

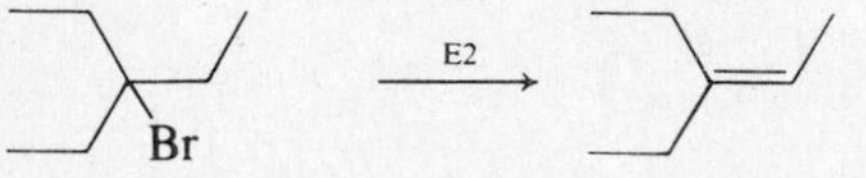

3-Bromo-3-ethylpentane $\xrightarrow{E2}$ (only product)

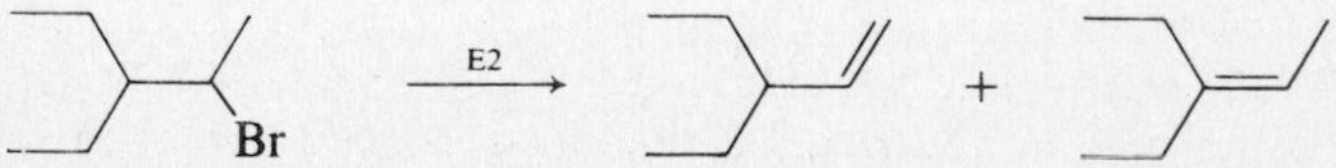

2-Bromo-3-ethylpentane $\xrightarrow{E2}$ 3-Ethyl-1-pentene + 3-Ethyl-2-pentene

Therefore, compound A must be 2-bromo-3-ethylpentane. Dehydrobromination of A will follow the Zaitsev rule, so the major alkene (compound B) is 3-ethyl-2-pentene and the minor alkene (compound C) is 3-ethyl-1-pentene.

6.23 There are only two alkanes that have the molecular formula C_4H_{10}; they are butane and isobutane (2-methylpropane), both of which give two monochlorides on free-radical chlorination. However, dehydrochlorination of one of the monochlorides derived from butane yields a mixture of alkenes.

$$\underset{\text{2-Chlorobutane}}{CH_3\underset{\substack{|\\ Cl}}{C}HCH_2CH_3} \xrightarrow[\text{dimethyl sulfoxide}]{KOC(CH_3)_3} \underset{\text{1-Butene}}{CH_2{=}CHCH_2CH_3} + \underset{\text{2-Butene (cis + trans)}}{CH_3CH{=}CHCH_3}$$

Both monochlorides derived from isobutane yield only isobutene under conditions of E2 elimination.

$$\underset{\textit{tert}\text{-Butyl chloride}}{(CH_3)_3CCl} \quad \text{or} \quad \underset{\text{Isobutyl chloride}}{(CH_3)_2CHCH_2Cl} \xrightarrow[\text{dimethyl sulfoxide}]{KOC(CH_3)_3} \underset{\text{Isobutene}}{(CH_3)_2C{=}CH_2}$$

Therefore, compound D is isobutane, the two alkyl chlorides are *tert*-butyl chloride and isobutyl chloride, and alkene E is isobutene (2-methylpropene).

6.24 The key to this problem is the fact that one of the alkyl chlorides of molecular formula $C_6H_{13}Cl$ does not undergo E2 elimination. Therefore, it must have a structure in which the carbon atom that is β to the chlorine bears no hydrogen substituents. This $C_6H_{13}Cl$ isomer is 1-chloro-2,2-dimethylbutane.

$$CH_3CH_2\overset{\substack{CH_3\\ |}}{\underset{\substack{|\\ CH_3}}{C}}CH_2Cl$$

1-Chloro-2,2-dimethylbutane
(cannot form an alkene)

Identifying this monochloride derivative gives us the carbon skeleton. The starting alkane (compound F) must be 2,2-dimethylbutane. Its free-radical halogenation gives three different monochlorides.

$$\underset{\substack{\text{2,2-Dimethylbutane}\\ \text{(compound F)}}}{CH_3CH_2\overset{\substack{CH_3\\ |}}{\underset{\substack{|\\ CH_3}}{C}}CH_3} \xrightarrow[\text{light}]{Cl_2} \underset{\substack{\text{1-Chloro-2,2-}\\ \text{dimethylbutane}}}{CH_3CH_2\overset{\substack{CH_3\\ |}}{\underset{\substack{|\\ CH_3}}{C}}CH_2Cl} + \underset{\substack{\text{3-Chloro-2,2-}\\ \text{dimethylbutane}}}{CH_3\underset{\substack{|\\ Cl}}{C}H{-}\overset{\substack{CH_3\\ |}}{\underset{\substack{|\\ CH_3}}{C}}CH_3} + \underset{\substack{\text{1-Chloro-3,3-}\\ \text{dimethylbutane}}}{ClCH_2CH_2\overset{\substack{CH_3\\ |}}{\underset{\substack{|\\ CH_3}}{C}}CH_3}$$

Both 3-chloro-2,2-dimethylbutane and 1-chloro-3,3-dimethylbutane give only 3,3-dimethyl-1-butene on E2 elimination.

$$\underset{\substack{\text{3-Chloro-2,2-}\\ \text{dimethylbutane}}}{CH_3\underset{\substack{|\\ Cl}}{C}H{-}\overset{\substack{CH_3\\ |}}{\underset{\substack{|\\ CH_3}}{C}}CH_3} \quad \text{or} \quad \underset{\substack{\text{1-Chloro-3,3-}\\ \text{dimethylbutane}}}{ClCH_2CH_2\overset{\substack{CH_3\\ |}}{\underset{\substack{|\\ CH_3}}{C}}CH_3} \xrightarrow[(CH_3)_3COH]{KOC(CH_3)_3} \underset{\substack{\text{3,3-Dimethyl-1-butene}\\ \text{(alkene G)}}}{CH_2{=}CH\overset{\substack{CH_3\\ |}}{\underset{\substack{|\\ CH_3}}{C}}CH_3}$$

6.25 The information that compound I gives 2,4-dimethylpentane on catalytic hydrogenation establishes its carbon skeleton.

$$\text{Compound I} \xrightarrow[\text{catalyst}]{H_2} \text{2,4-Dimethylpentane}$$

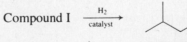

2,4-Dimethylpentane

Compound I is an alkene derived from compound H—an alkyl bromide of molecular formula $C_7H_{15}Br$. We are told that compound H is not a primary alkyl bromide. Therefore, compound H can be only:

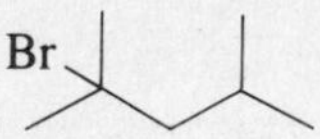

Br or Br

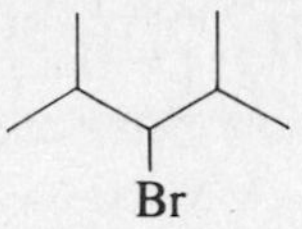

Since compound H gives a single alkene on being treated with sodium ethoxide in ethanol, it can only be 3-bromo-2,4-dimethylpentane, and compound I must be 2,4-dimethyl-2-pentene.

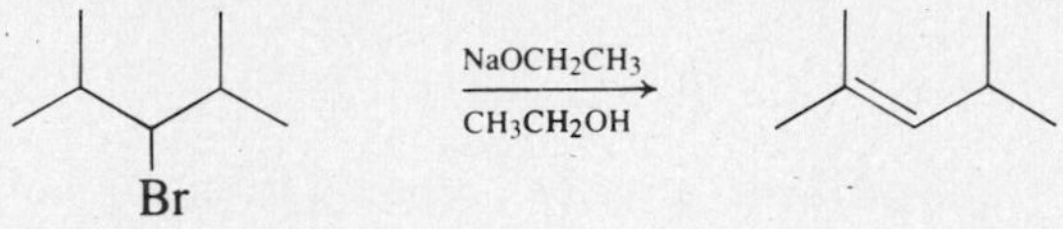

$\xrightarrow[CH_3CH_2OH]{NaOCH_2CH_3}$

3-Bromo-2,4-dimethylpentane (compound H)

2,4-Dimethyl-2-pentene (compound I)

6.26 Alkene L must have the same carbon skeleton as its hydrogenation product, 2,3,3,4-tetramethylpentane.

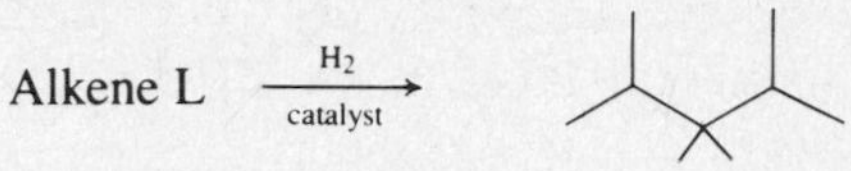

Alkene L $\xrightarrow[catalyst]{H_2}$

2,3,3,4-Tetramethylpentane

Therefore alkene L can only be 2,3,3,4-tetramethyl-1-pentene. The two alkyl bromides, compounds J and K, that give this alkene on dehydrobromination have their bromine substituents at C-1 and C-2, respectively.

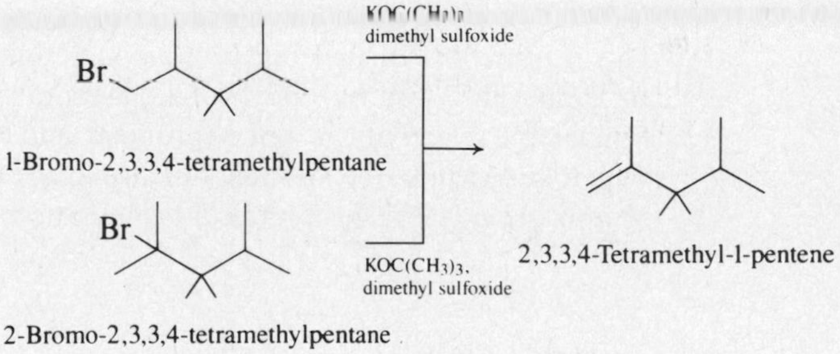

$KOC(CH_3)_3$, dimethyl sulfoxide

1-Bromo-2,3,3,4-tetramethylpentane

2-Bromo-2,3,3,4-tetramethylpentane

$KOC(CH_3)_3$, dimethyl sulfoxide

2,3,3,4-Tetramethyl-1-pentene

6.27 The only alcohol (compound M) that can undergo acid-catalyzed dehydration to alkene N without rearrangement is the one shown in the equation.

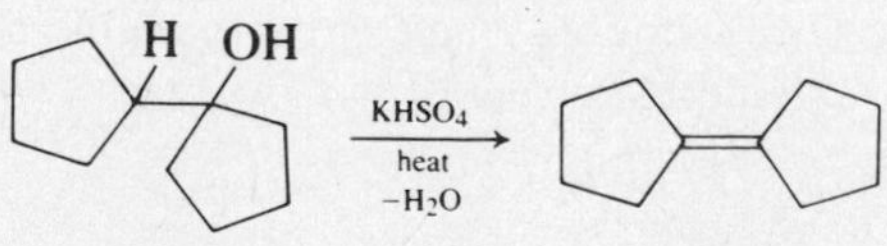

H OH

$\xrightarrow[heat, -H_2O]{KHSO_4}$

Alcohol M

Alkene N

Dehydration of alcohol M also yields an isomeric alkene under these conditions.

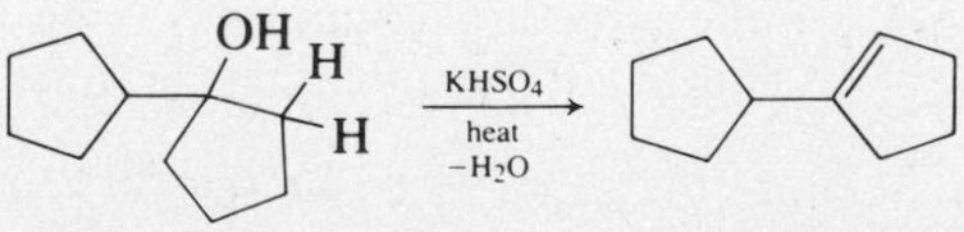

OH H H

$\xrightarrow[heat, -H_2O]{KHSO_4}$

Alcohol M

Alkene O

6.28 (*a*) Up to this point the only method that you have learned for preparing alkanes is hydrogenation of alkenes. Therefore, convert 2-methyl-2-hexanol to a mixture of alkenes by acid-catalyzed dehydration, then hydrogenate.

$$(CH_3)_2\underset{\displaystyle OH}{\underset{|}{C}}CH_2CH_2CH_2CH_3 \xrightarrow[\text{heat}]{H_2SO_4} (CH_3)_2C{=}CHCH_2CH_2CH_3 + CH_2{=}\underset{\displaystyle CH_3}{\underset{|}{C}}CH_2CH_2CH_2CH_3$$

2-Methyl-2-hexanol

$$\downarrow H_2, Pt$$

$$(CH_3)_2CHCH_2CH_2CH_2CH_3$$

2-Methylhexane

(*b*) Alkenes may be prepared from alcohols by dehydration and from alkyl halides by dehydrohalogenation. The starting material is an alkane, so it needs to be converted to an alcohol or to an alkyl halide in order to prepare the desired alkene. You have not yet learned how to prepare alcohols but have seen in Chapter 4 that alkanes may be converted to alkyl halides by free-radical halogenation. Free-radical bromination is highly selective for substitution of tertiary hydrogens.

Methylcyclopentane (H, CH_3) $\xrightarrow[\text{light, heat}]{Br_2}$ 1-Bromo-1-methylcyclopentane (Br, CH_3)

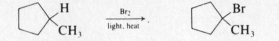

Dehydrohalogenation yields the desired 1-methylcyclopentene.

1-Bromo-1-methylcyclopentane (Br, CH_3) $\xrightarrow[\text{CH}_3\text{CH}_2\text{OH, heat}]{NaOCH_2CH_3}$ 1-Methylcyclopentene ($-CH_3$)

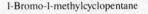

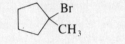

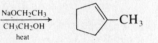

(*c*) This synthesis is carried out in a manner similar to the previous one—halogenation of an alkane followed by dehydrohalogenation. Since all the protons of cyclodecane are equivalent, selectivity is not a problem and free-radical chlorination is an acceptable way to introduce the desired functional group, although polychlorination may be a complicating factor. Free-radical bromination is also effective.

Cyclodecane $\xrightarrow[\text{light}]{Cl_2}$ Cyclodecyl chloride (Cl) $\xrightarrow[\text{CH}_3\text{CH}_2\text{OH, heat}]{NaOCH_2CH_3}$ (*Z*)-Cyclodecene

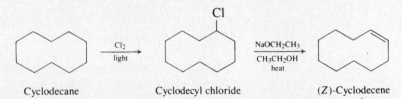

A cis double bond is more stable than a trans in a 10-membered ring and is formed stereoselectively.

SELF-TEST

PART A

A-1. Write structural formulas for the reactant or product(s) omitted from each of the following. If more than one product is formed, indicate the major one.

(*a*) $(CH_3)_2\underset{\displaystyle OH}{\underset{|}{C}}CH_2CH_2CH_3 \xrightarrow[\text{heat}]{H_2SO_4} ?$

(*b*) Cl CH_3 (1-chloro-1-methylcyclohexane) $\xrightarrow[CH_3OH]{NaOCH_3}$?

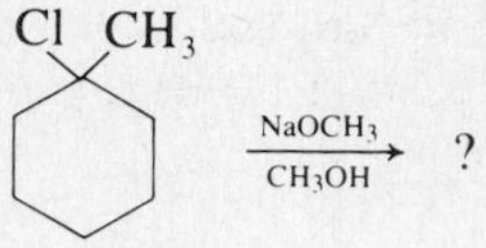

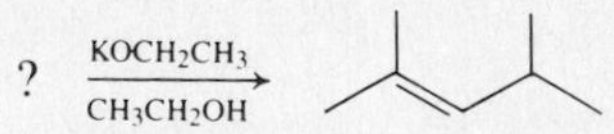

(*c*) ? $\xrightarrow[CH_3CH_2OH]{KOCH_2CH_3}$ [alkene] (only alkene formed)

(*d*) $(CH_3)_2\underset{|}{C}OH$ $\xrightarrow{H_3PO_4}$?
$C(CH_3)_3$

A-2. Write the structure of the $C_6H_{13}Br$ isomer that is not capable of undergoing E2 elimination.

A-3. Outline a mechanism detailing the steps in the following reaction.

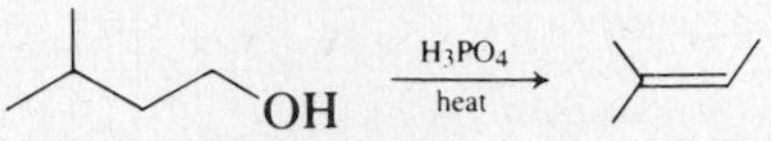

[alcohol] OH $\xrightarrow[heat]{H_3PO_4}$ [alkene] + other alkenes

A-4. Using perspective drawings (of chair cyclohexanes), explain formation of the major E2 product from the following reaction:

CH_3, Br (cyclohexane) $\xrightarrow{E2}$?

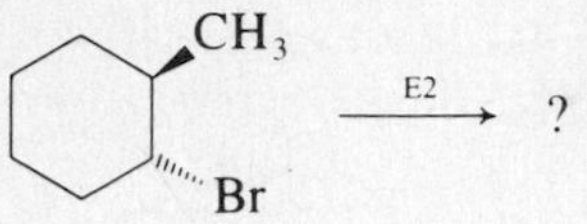

A-5. Compare the relative rate of reaction of *cis*- and *trans*-3-chloroisopropylcyclohexane with sodium methoxide in methanol.

A-6. Provide a synthesis of 1-methylcyclohexene starting with methylcyclohexane.

A-7. Compound A, on reaction with bromine in the presence of light, gave as the major product compound B ($C_9H_{19}Br$). Reaction of B with sodium ethoxide in ethanol gave 3-ethyl-4,4-dimethyl-2-pentene as the only alkene. Identify compounds A and B.

PART B

B-1. Rank the following alcohols in order of decreasing reactivity (fastest → slowest) toward dehydration with 85% H_3PO_4:

A: $(CH_3)_2CHCH_2CH_2OH$

B: $(CH_3)_2\overset{OH}{C}CH_2CH_3$

C: $(CH_3)_2CH\overset{OH}{C}HCH_3$

(*a*) B > C > A (*b*) A > C > B (*c*) B > A > C (*d*) A > B > C

B-2. Rank the following in order of increasing reactivity (least → most) toward E2 elimination.

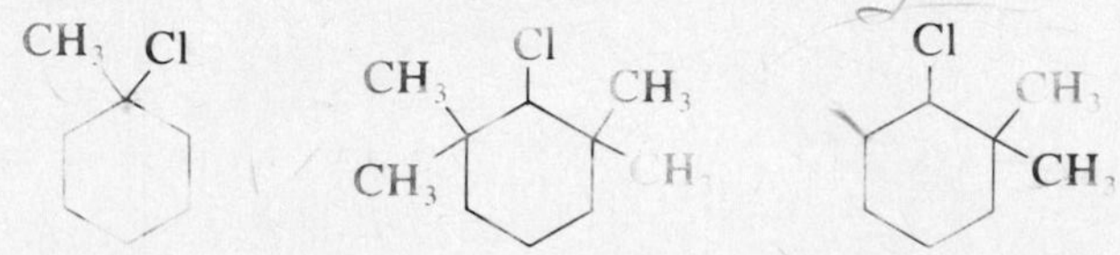

A: CH_3 Cl (1-chloro-1-methylcyclohexane)

B: Cl, CH_3, CH_3, CH_3, CH_3 (cyclohexane)

C: Cl, CH_3, CH_3 (cyclohexane)

(*a*) A < C < B (*b*) B < C < A (*c*) B < A < C (*d*) C < B < A

B-3. The reaction

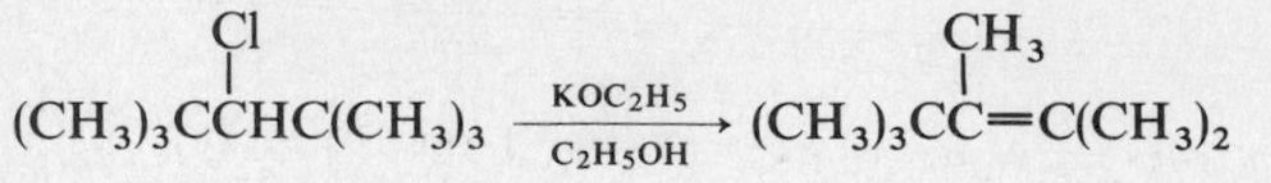

$$(CH_3)_3C\overset{Cl}{\overset{|}{C}}HC(CH_3)_3 \xrightarrow[C_2H_5OH]{KOC_2H_5} (CH_3)_3\overset{CH_3}{\overset{|}{C}}C{=}C(CH_3)_2$$

is best described as:

(*a*) A poor synthetic route
(*b*) A good synthetic route
(*c*) Of variable quality, depending on the temperature
(*d*) Of variable quality, depending on the reaction time

B-4. Consider the following reaction:

OH $\xrightarrow[heat]{H_3PO_4}$ +

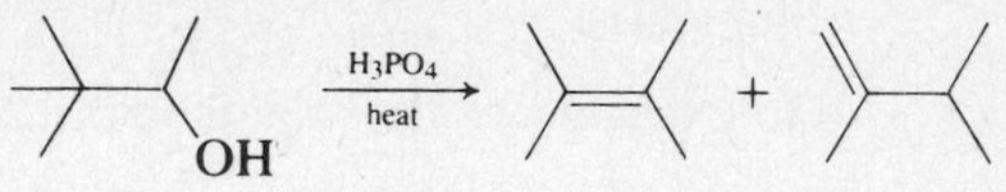

Which response contains all the correct statements about this process and no incorrect ones?

1. Dehydration
2. E2 mechanism
3. Carbon skeleton migration
4. Most stable carbocation forms
5. Single step reaction

(*a*) 1, 3 (*b*) 1, 2, 3 (*c*) 1, 2, 5 (*d*) 1, 3, 4

B-5. Select the formula or statement representing the major product(s) of the following reaction:

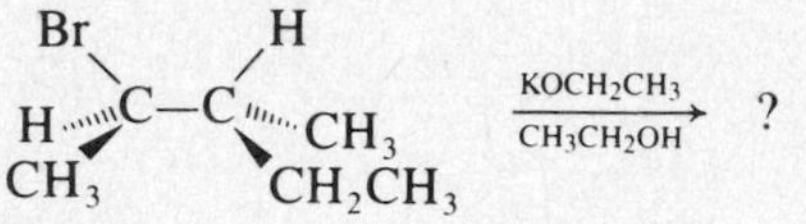

Br, H, CH$_3$ on C — C with H, CH$_3$, CH$_2$CH$_3$ $\xrightarrow[CH_3CH_2OH]{KOCH_2CH_3}$?

(*a*) H, CH$_3$ / C=C / CH$_3$, CH$_2$CH$_3$

(*b*) H, CH$_2$CH$_3$ / C=C / CH$_3$, CH$_3$

(*c*) $CH_2{=}CH\overset{CH_3}{\overset{|}{C}}HCH_2CH_3$

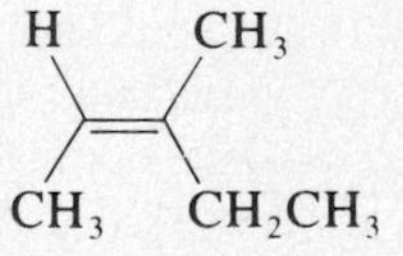

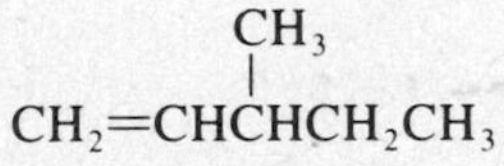

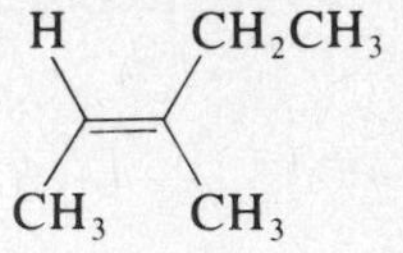

(*d*) Both (*a*) and (*b*) form in approximately equal amounts

CHAPTER 7

REACTIONS OF ALKENES. ELECTROPHILIC ADDITION REACTIONS

IMPORTANT TERMS AND CONCEPTS

Electrophilic addition reactions constitute perhaps the most varied class of alkene reactions. In general, an ionic species E^+Y^- or a polarizable molecule $\overset{\delta+}{E}-\overset{\delta-}{Y}$ adds to the double bond.

$$\overset{\delta+}{E}-\overset{\delta-}{Y} + \;>C=C<\; \longrightarrow -\overset{E}{\overset{|}{C}}-\overset{Y}{\overset{|}{C}}-$$

Markovnikov's rule (Secs. 7.2, 7.3) The addition of a hydrogen halide H—X to an alkene is a regioselective reaction, that is, addition to an unsymmetrical alkene results in predominance of one regioisomer. *Markovnikov's rule* states that the hydrogen atom adds to the carbon having the greater number of hydrogen substituents.

$$H-X + R_2C=CH_2 \longrightarrow R_2\overset{X}{\overset{|}{C}}-\overset{H}{\overset{|}{C}}H_2$$

Consideration of the mechanism of these reactions provides a theoretical explanation of the rule. The more stable carbocation forms as an intermediate in the process, which results in the more substituted halide product being formed.

$$R_2C=CH_2 \xrightarrow{H-X} \begin{cases} R_2CH-\overset{+}{C}H_2 \text{ (1° carbocation) (not formed)} \\ R_2\overset{+}{C}-CH_3 \text{ (3° carbocation)} \xrightarrow{X^-} R_2\overset{X}{\overset{|}{C}}-CH_3 \end{cases}$$

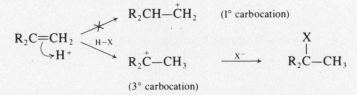

As discussed in Chapter 4, the transition state leading to formation of a tertiary carbocation is more stable than that leading to a secondary carbocation, which in turn is more stable than that leading to a primary. An *alkene reactivity series* may therefore be written.

$$R_2C=CH_2 > RCH=CH_2 > CH_2=CH_2$$

which reflects the order of stability of the carbocations formed in the rate-determining step.

Le Chatelier's Principle (Sec. 7.6) The statement that a *system at equilibrium adjusts so as to minimize any stress applied to it* is known as *Le Chatelier's principle.* An example is provided by acid-catalyzed hydration of an alkene and the acid-catalyzed dehydration of an alcohol, reactions which are the reverse of one another.

$$R_2C{=}CH_2 + H_2O \overset{H^+}{\rightleftharpoons} R_2C(OH){-}CH_3$$

Addition of water shifts the equilibrium to the right, favoring alcohol formation. Removal of water (or keeping the concentration of water low) shifts the equilibrium to the left, favoring the formation of alkene.

The *principle of microscopic reversibility* is also demonstrated by the equilibrium shown above. Any intermediates or transition states which appear on the lowest-energy pathway of the forward process must also appear, in the reverse order, on the lowest-energy pathway of the reverse process.

IMPORTANT REACTIONS

Addition of Hydrogen Halides (Secs. 7.1 to 7.3)

General:

$$\rangle C{=}C\langle \; + \; H{-}X \longrightarrow {-}\overset{X}{\underset{|}{\overset{|}{C}}}{-}\overset{H}{\underset{|}{\overset{|}{C}}}{-}$$

(X = Cl, Br, or I)

Examples:

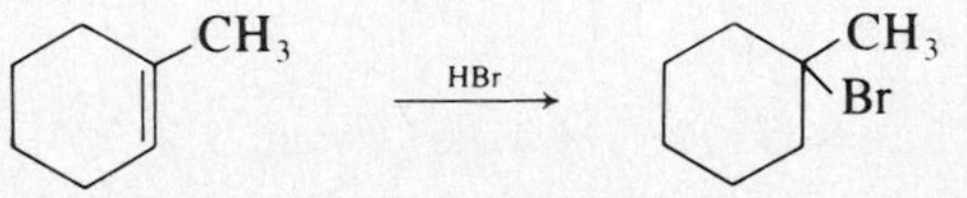

$$\text{1-methylcyclohexene} \xrightarrow{HBr} \text{1-bromo-1-methylcyclohexane } (CH_3, Br)$$

$$(CH_3)_2C{=}CHCH_3 \xrightarrow{HCl} (CH_3)_2C(Cl)CH_2CH_3$$

Mechanism:

$$R_2C{=}CH_2 \overset{H{-}X}{\rightleftharpoons} R_2\overset{+}{C}{-}CH_3 + X^-$$

$$R_2\overset{+}{C}{-}CH_3 \xrightarrow{X^-} R_2C(X){-}CH_3$$

Free-Radical Addition of Hydrogen Bromide (Sec. 7.4)

General:

$$\rangle C{=}C\langle \; + \; H{-}Br \xrightarrow{ROOR} {-}\overset{H}{\underset{|}{\overset{|}{C}}}{-}\overset{Br}{\underset{|}{\overset{|}{C}}}{-}$$

Example:

$$(CH_3)_2C{=}CHCH_3 + HBr \xrightarrow{ROOR} (CH_3)_2CHCH(Br)CH_3$$

Mechanism:

Initiation: $$\mathrm{ROOR \longrightarrow 2\,RO\cdot}$$

$$\mathrm{RO\cdot + H{-}Br \longrightarrow ROH + Br\cdot}$$

Propagation: $$\mathrm{R_2C{=}CH_2 + Br\cdot \longrightarrow R_2\dot{C}{-}CH_2Br}$$

$$\mathrm{R_2\dot{C}{-}CH_2Br + H{-}Br \longrightarrow R_2CH{-}CH_2Br + Br\cdot}$$

Addition of Sulfuric Acid; Alcohol Formation (Sec. 7.5)

General:

$$\mathrm{>C{=}C< + H_2SO_4 \longrightarrow {-}\underset{|}{\overset{H}{\overset{|}{C}}}{-}\underset{|}{\overset{OSO_3H}{\overset{|}{C}}}{-} \xrightarrow[heat]{H_2O} {-}\underset{|}{\overset{H}{\overset{|}{C}}}{-}\underset{|}{\overset{OH}{\overset{|}{C}}}{-}}$$

Example:

$$\mathrm{CH_3CH{=}CH_2 + H_2SO_4 \longrightarrow CH_3\overset{OSO_3H}{\overset{|}{C}}HCH_3}$$

$$\mathrm{CH_3\overset{OSO_3H}{\overset{|}{C}}HCH_3 \xrightarrow[heat]{H_2O} CH_3\overset{OH}{\overset{|}{C}}HCH_3}$$

Acid-Catalyzed Hydration (Sec. 7.6)

General:

$$\mathrm{>C{=}C< \xrightarrow{H_3O^+} {-}\underset{|}{\overset{H}{\overset{|}{C}}}{-}\underset{|}{\overset{OH}{\overset{|}{C}}}{-}}$$

Examples:

$$\mathrm{(CH_3)_2C{=}CH_2 \xrightarrow{H_3O^+} (CH_3)_3C{-}OH}$$

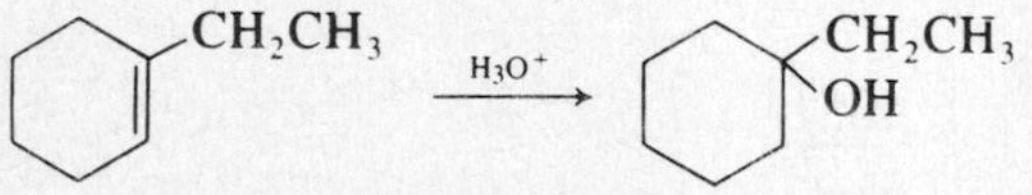

$$\text{1-ethylcyclohexene (ring}{-}\mathrm{CH_2CH_3}) \xrightarrow{H_3O^+} \text{cyclohexane ring bearing } \mathrm{CH_2CH_3} \text{ and } \mathrm{OH}$$

Synthesis of Alcohols by Oxymercuration-Demercuration (Sec. 7.7)

General:

$$\mathrm{>C{=}C< \xrightarrow[THF-H_2O]{Hg(O_2CCH_3)_2} \xrightarrow[HO^-]{NaBH_4} {-}\underset{|}{\overset{H}{\overset{|}{C}}}{-}\underset{|}{\overset{OH}{\overset{|}{C}}}{-}}$$

Example:

$$\mathrm{(CH_3)_2CHCH{=}CH_2 \xrightarrow[2.\ NaBH_4,\ HO^-]{1.\ Hg(OAc)_2,\ THF\text{-}H_2O} (CH_3)_2CH\overset{OH}{\overset{|}{C}}HCH_3}$$

Synthesis of Alcohols by Hydroboration-Oxidation (Secs. 7.8 to 7.10)

General:

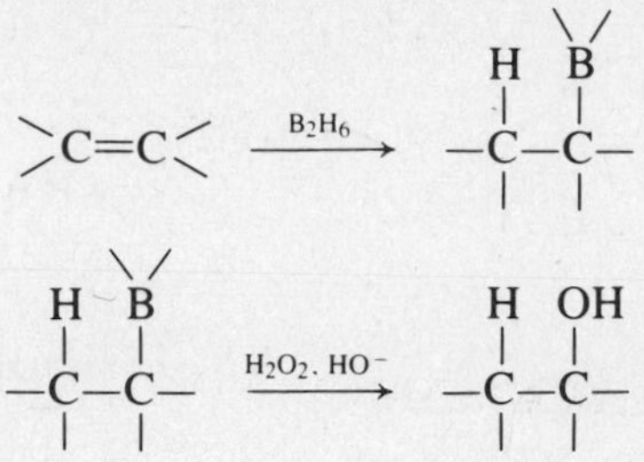

$$\rangle C{=}C\langle \xrightarrow{B_2H_6} -\overset{H}{\underset{|}{C}}-\overset{B\lessgtr}{\underset{|}{C}}-$$

$$-\overset{H}{\underset{|}{C}}-\overset{B\lessgtr}{\underset{|}{C}}- \xrightarrow{H_2O_2,\ HO^-} -\overset{H}{\underset{|}{C}}-\overset{OH}{\underset{|}{C}}-$$

Examples:

$$\text{1-methylcyclohexene} \xrightarrow[2.\ H_2O_2,\ HO^-]{1.\ B_2H_6} \text{trans-2-methylcyclohexanol } (CH_3,\ H,\ OH)$$

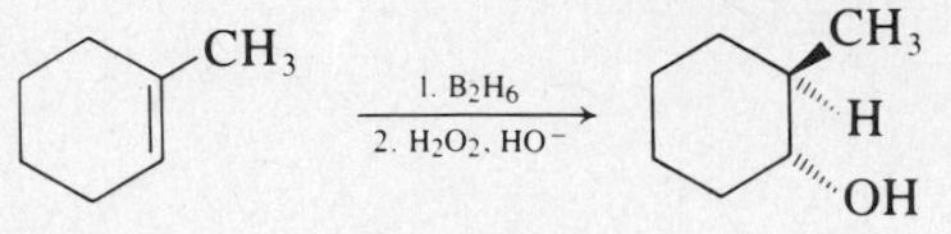

$$(CH_3)_2C{=}CH_2 \xrightarrow[2.\ H_2O_2,\ HO^-]{1.\ B_2H_6} (CH_3)_2CH{-}CH_2OH$$

Addition of Halogens (Secs. 7.11 to 7.13)

General:

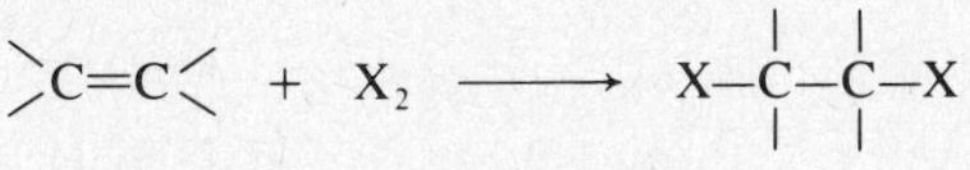

$$\rangle C{=}C\langle + X_2 \longrightarrow X-\underset{|}{\overset{|}{C}}-\underset{|}{\overset{|}{C}}-X$$

(X = Cl, Br, or I)

Examples:

$$\text{cyclohexene} + Br_2 \longrightarrow \text{trans-1,2-dibromocyclohexane}$$

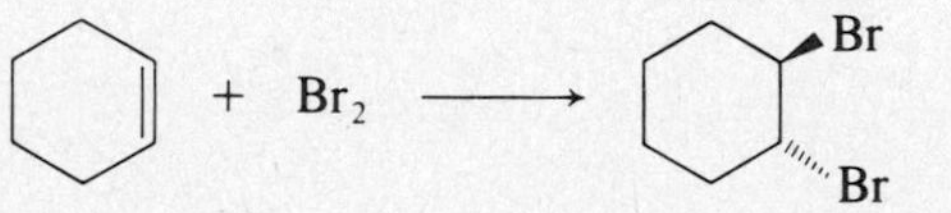

$$CH_3CH_2CH{=}CH_2 \xrightarrow{Cl_2} CH_3CH_2\overset{Cl}{\overset{|}{C}}HCH_2Cl$$

Mechanism:

$$\rangle C{=}C\langle \xrightarrow{Br_2} \text{bromonium ion } (\overset{+}{Br} \text{ bridging } C{-}C)$$

$$\text{bromonium ion} + \overset{-}{Br} \longrightarrow Br{-}C{-}C{-}Br \text{ (anti addition)}$$

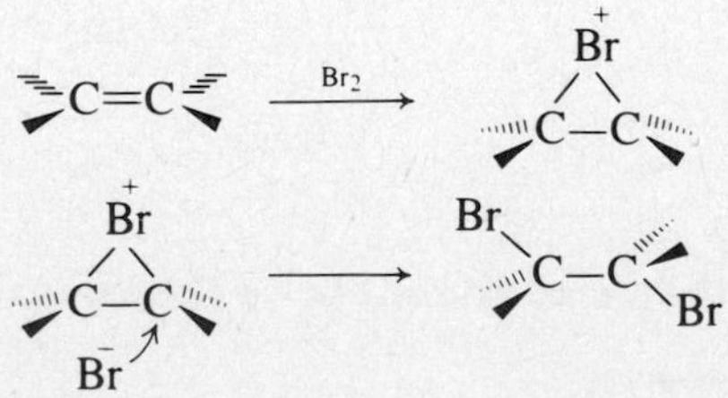

Addition of Hypohalous Acids (Sec. 7.14)

General:

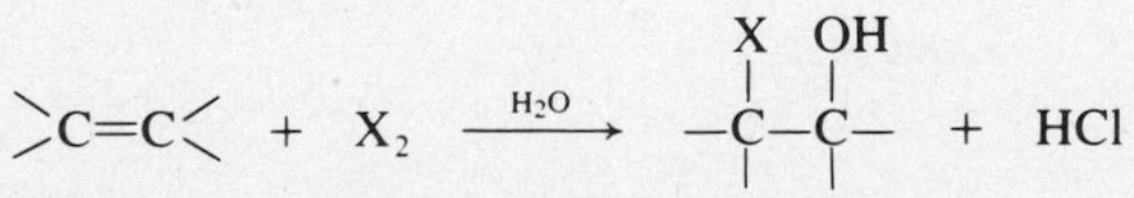

$$\rangle C{=}C\langle + X_2 \xrightarrow{H_2O} -\overset{X}{\underset{|}{C}}-\overset{OH}{\underset{|}{C}}- + HCl$$

(X = Cl, Br)

Examples:

$$(CH_3)_2C{=}CH_2 + Cl_2 \xrightarrow{H_2O} (CH_3)_2\overset{OH}{\overset{|}{C}}{-}CH_2Cl + HCl$$

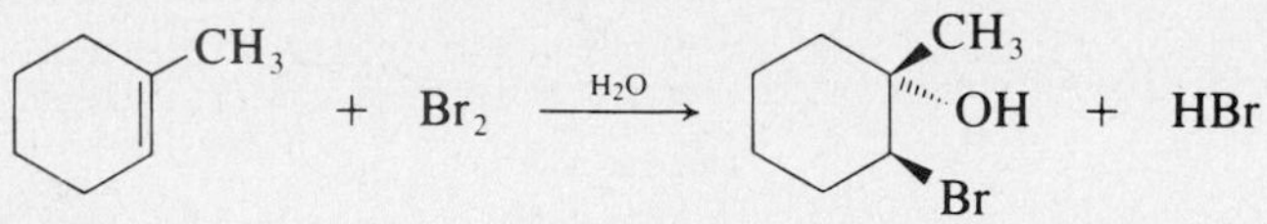

$$\text{1-methylcyclohexene} + Br_2 \xrightarrow{H_2O} \text{2-bromo-1-methylcyclohexanol (CH}_3\text{, OH, Br)} + HBr$$

Epoxidation (Sec. 7.15)

General:

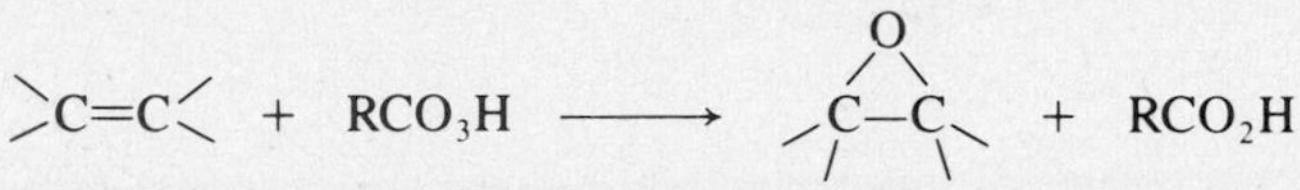

$$>C{=}C< + RCO_3H \longrightarrow >\underset{O}{C{-}C}< + RCO_2H$$

Examples:

$$\text{cyclohexene} + CH_3CO_3H \longrightarrow \text{cyclohexene oxide} + CH_3CO_2H$$

$$\underset{H}{\overset{CH_3}{>}}C{=}C\underset{H}{\overset{CH_3}{<}} + CH_3CO_3H \longrightarrow CH_3\underset{H}{C}{-}\overset{O}{}\underset{H}{C}CH_3 + CH_3CO_2H$$

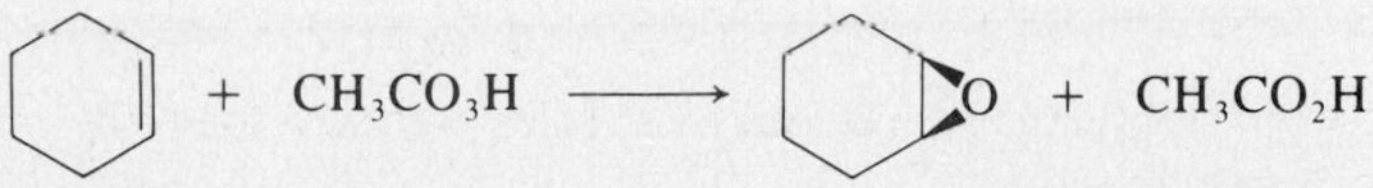

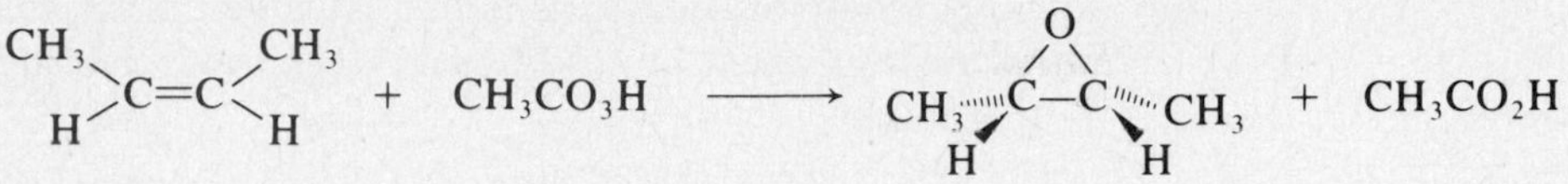

Ozonolysis (Sec. 7.16)

General:

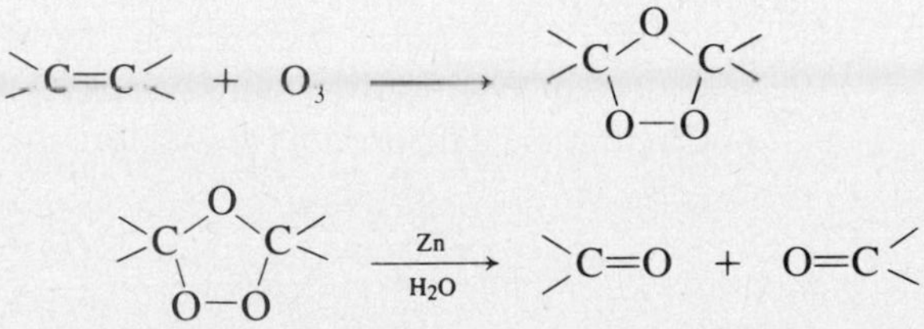

$$>C{=}C< + O_3 \longrightarrow >C\overset{O}{}C< \ (\text{O–O})$$

$$>C\overset{O}{}C< \ (\text{O–O}) \xrightarrow[H_2O]{Zn} >C{=}O + O{=}C<$$

Examples:

$$(CH_3)_2C{=}CHCH_3 \xrightarrow[2.\ Zn,\ H_2O]{1.\ O_3} (CH_3)_2C{=}O + CH_3\overset{O}{\overset{\|}{C}}H$$

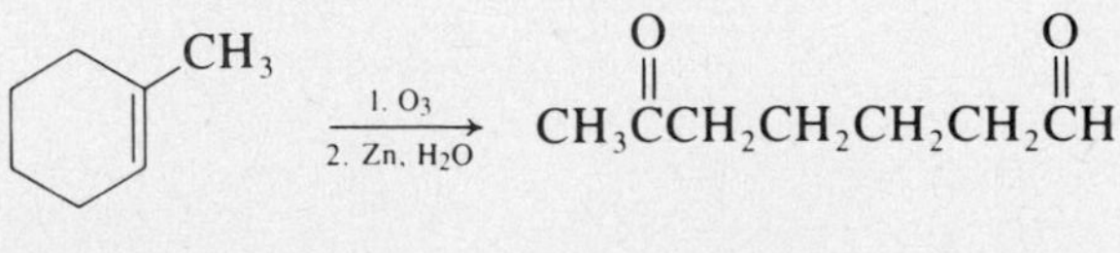

$$\text{1-methylcyclohexene} \xrightarrow[2.\ Zn,\ H_2O]{1.\ O_3} CH_3\overset{O}{\overset{\|}{C}}CH_2CH_2CH_2CH_2\overset{O}{\overset{\|}{C}}H$$

Permanganate Cleavage (Sec. 7.17)

General:

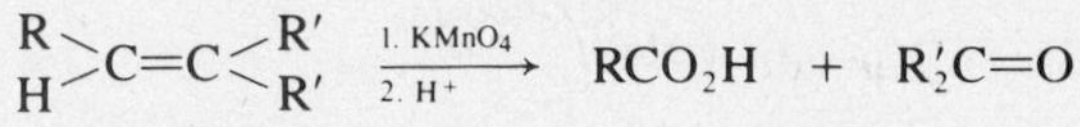

$$\underset{H}{\overset{R}{>}}C{=}C\underset{R'}{\overset{R'}{<}} \xrightarrow[2.\ H^+]{1.\ KMnO_4} RCO_2H + R'_2C{=}O$$

Example:

$$(CH_3)_2C{=}CHCH_2CH_3 \xrightarrow[2.\ H^+]{1.\ KMnO_4} (CH_3)_2C{=}O + CH_3CH_2CO_2H$$

SOLUTIONS TO TEXT PROBLEMS

7.1 (*b*) Begin by writing out the structure of the starting alkene. Identify the doubly bonded carbon that has the greater number of hydrogen substituents; this is the one to which the proton of hydrogen chloride adds. Chloride adds to the carbon atom of the double bond that has the fewer hydrogen substituents.

chlorine adds to this carbon → $(CH_3)(CH_3CH_2)C{=}CH_2$ ← hydrogen adds to this carbon

2-Methyl-1-butene

$\xrightarrow{HCl}$ $CH_3CH_2C(CH_3)(Cl)CH_3$

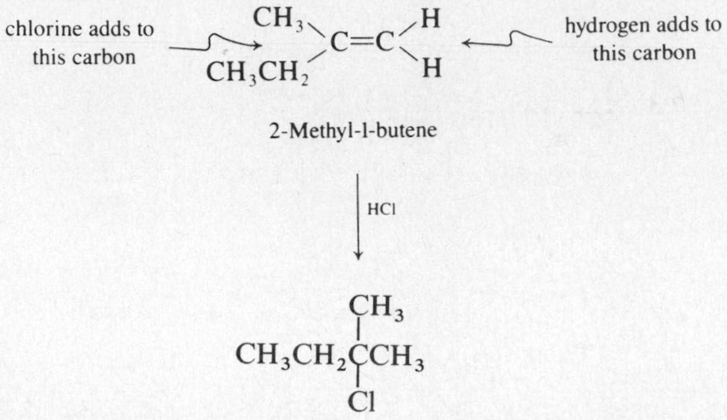

By applying Markovnikov's rule, we see that the major product is 2-chloro-2-methylbutane.

(*c*) Regioselectivity of addition is not an issue here because the two carbons of the double bond are equivalent in *cis*-2-butene. Hydrogen chloride adds to *cis*-2-butene to give 2-chlorobutane.

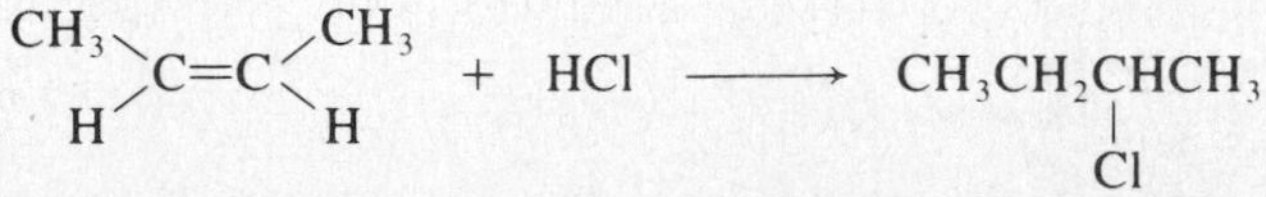

$$CH_3CH{=}CHCH_3 + HCl \longrightarrow CH_3CH_2CH(Cl)CH_3$$

cis-2-Butene — Hydrogen chloride — 2-Chlorobutane

(*d*) One end of the double bond has no hydrogen substituents while the other end has one. In accordance with Markovnikov's rule, the proton adds to the carbon that already has one hydrogen. The product is 1-chloro-1-ethylcyclohexane.

$$CH_3CH{=}C_6H_{10} + HCl \longrightarrow (CH_3CH_2)(Cl)C_6H_{10}$$

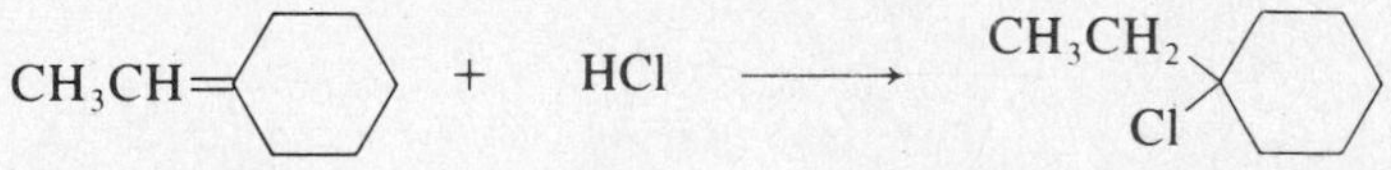

Ethylidenecyclohexane — Hydrogen chloride — 1-Chloro-1-ethylcyclohexane

7.2 (*b*) A proton is transferred to the terminal carbon atom of 2-methyl-1-butene so as to produce a tertiary carbocation.

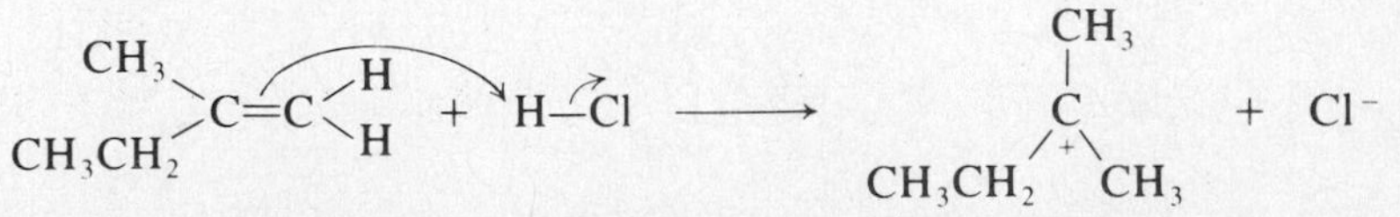

$$(CH_3)(CH_3CH_2)C{=}CH_2 + H{-}Cl \longrightarrow (CH_3)_2\overset{+}{C}CH_2CH_3 + Cl^-$$

2-Methyl-1-butene — Hydrogen chloride — *tert*-Pentyl cation — Chloride

This is the carbocation that leads to the observed product, 2-chloro-2-methylbutane.

(*c*) A secondary carbocation is an intermediate in the reaction of *cis*-2-butene with hydrogen chloride.

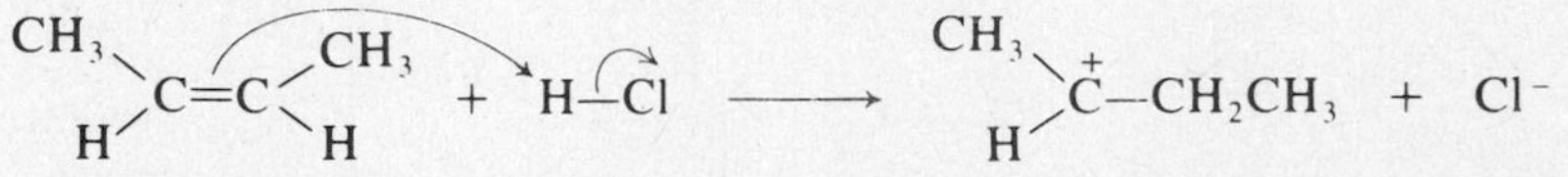

$$CH_3CH{=}CHCH_3 + H{-}Cl \longrightarrow CH_3\overset{+}{C}H{-}CH_2CH_3 + Cl^-$$

cis-2-Butene — Hydrogen chloride — *sec*-Butyl cation — Chloride

Capture of this carbocation by chloride gives *sec*-butyl chloride (2-chlorobutane).

(*d*) A tertiary carbocation is formed by proton transfer from hydrogen chloride to the indicated alkene.

$$CH_3CH{=}C_6H_{10} + H{-}Cl \longrightarrow CH_3CH_2{-}\overset{+}{C}_6H_{10} + Cl^-$$

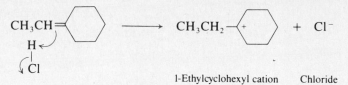

1-Ethylcyclohexyl cation — Chloride

7.3 The structure of allyl bromide is $CH_2{=}CHCH_2Br$. Its reaction with hydrogen bromide in accordance with Markovnikov's rule proceeds by addition of a proton to the doubly bonded carbon that has the greater number of hydrogens attached to it.

Addition according to Markovnikov's rule:

$$\underset{\text{Allyl bromide}}{CH_2{=}CHCH_2Br} + \underset{\text{Hydrogen bromide}}{HBr} \longrightarrow \underset{\text{1,2-Dibromopropane}}{CH_3\underset{\substack{|\\Br}}{C}HCH_2Br}$$

Addition of hydrogen bromide contrary to Markovnikov's rule leads to 1,3-dibromopropane.

Addition contrary to Markovnikov's rule:

$$\underset{\text{Allyl bromide}}{CH_2{=}CHCH_2Br} + \underset{\text{Hydrogen bromide}}{HBr} \longrightarrow \underset{\text{1,3-Dibromopropane}}{BrCH_2CH_2CH_2Br}$$

7.4 (*b*) Hydrogen bromide adds to 2-methyl-1-butene in accordance with Markovnikov's rule when peroxides are absent. The product is 2-bromo-2-methylbutane.

$$\underset{\text{2-Methyl-1-butene}}{(CH_3)(CH_3CH_2)C{=}CH_2} + \underset{\text{Hydrogen bromide}}{HBr} \longrightarrow \underset{\text{2-Bromo-2-methylbutane}}{CH_3CH_2\overset{\substack{CH_3\\|}}{\underset{\substack{|\\Br}}{C}}CH_3}$$

The opposite orientation is observed when peroxides are present. The product is 1-bromo-2-methylbutane.

$$\underset{\text{2-Methyl-1-butene}}{(CH_3)(CH_3CH_2)C{=}CH_2} + \underset{\text{Hydrogen bromide}}{HBr} \xrightarrow{\text{peroxides}} \underset{\text{1-Bromo-2-methylbutane}}{CH_3CH_2\overset{\substack{CH_3\\|}}{\underset{\substack{|\\H}}{C}}CH_2Br}$$

(*c*) Both ends of the double bond in *cis*-2-butene are equivalently substituted, so the same product (2-bromobutane) is formed by hydrogen bromide addition regardless of whether the reaction is carried out in the presence of peroxides or in their absence.

$$\underset{\textit{cis}\text{-2-Butene}}{(CH_3)HC{=}CH(CH_3)} + \underset{\text{Hydrogen bromide}}{HBr} \longrightarrow \underset{\text{2-Bromobutane}}{CH_3CH_2\underset{\substack{|\\Br}}{C}HCH_3}$$

(*d*) A tertiary bromide is formed on addition of hydrogen bromide to ethylidenecyclohexane in the absence of peroxides.

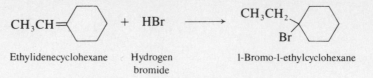

$CH_3CH{=}C_6H_{10} + HBr \longrightarrow$ 1-Bromo-1-ethylcyclohexane (CH_3CH_2 and Br on the same ring carbon)

Ethylidenecyclohexane   Hydrogen bromide   1-Bromo-1-ethylcyclohexane

The regioselectivity of addition is reversed in the presence of peroxides, and the product is 1-bromo-1-cyclohexylethane.

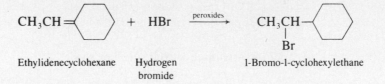

$CH_3CH{=}C_6H_{10} + HBr \xrightarrow{\text{peroxides}} CH_3CH(Br){-}C_6H_{11}$

Ethylidenecyclohexane   Hydrogen bromide   1-Bromo-1-cyclohexylethane

7.5 The presence of hydroxide ion in the second step is incompatible with the medium in which the reaction is carried out. The reaction as shown in step 1

$$1.\ (CH_3)_2C{=}CH_2 + H_3O^+ \longrightarrow (CH_3)_3C^+ + H_2O$$

is performed in acidic solution. There are, for all practical purposes, no hydroxide ions in aqueous acid, the strongest base present being water itself. It is quite important to pay attention to the species that are actually present in the reaction medium whenever you formulate a reaction mechanism.

7.6 Electrophilic addition of hydrogen chloride to 2-methylpropene as outlined in the mechanism of Section 7.3 proceeds through a carbocation intermediate. This mechanism is the reverse of the E1 elimination. The E2 mechanism is concerted—it does not involve an intermediate.

7.7 The two alcohols are 2-pentanol and 3-pentanol.

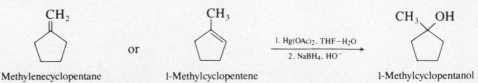

$$\underset{trans\text{-2-Pentene}}{CH_3CH{=}CHCH_2CH_3} \xrightarrow[\text{2. NaBH}_4,\ HO^-]{\text{1. Hg(OAc)}_2,\ \text{THF}-H_2O} \underset{\text{2-Pentanol}}{CH_3\underset{OH}{CH}CH_2CH_2CH_3} + \underset{\text{3-Pentanol}}{CH_3CH_2\underset{OH}{CH}CH_2CH_3}$$

The substituents at either end of the double bond are so similar that there is little regioselective differentiation between them.

7.8 (*b*) The tertiary alcohol 1-methylcyclopentanol may be prepared either from methylenecyclopentane or from 1-methylcyclopentene.

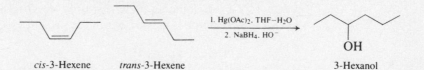

Methylenecyclopentane or 1-Methylcyclopentene $\xrightarrow[\text{2. NaBH}_4,\ HO^-]{\text{1. Hg(OAc)}_2,\ \text{THF}-H_2O}$ 1-Methylcyclopentanol

(*c*) Either the cis or the trans isomer of 3-hexene may be used as the starting alkene in order to prepare 3-hexanol by oxymercuration-demercuration.

cis-3-Hexene   *trans*-3-Hexene $\xrightarrow[\text{2. NaBH}_4,\ HO^-]{\text{1. Hg(OAc)}_2,\ \text{THF}-H_2O}$ 3-Hexanol (OH)

2-Hexene (either *cis*- or *trans*-$CH_3CH{=}CHCH_2CH_2CH_3$) is not a satisfactory answer because there is no reason to believe that addition will be regioselective in the direction that gives the desired 3-hexanol. A mixture of 2- and 3-hexanol will be produced.

7.9 (*b*) The carbon-carbon double bond is symmetrically substituted in *cis*-2-butene, so the regioselectivity of hydroboration-oxidation is not an issue. Hydration of the double bond gives 2-butanol.

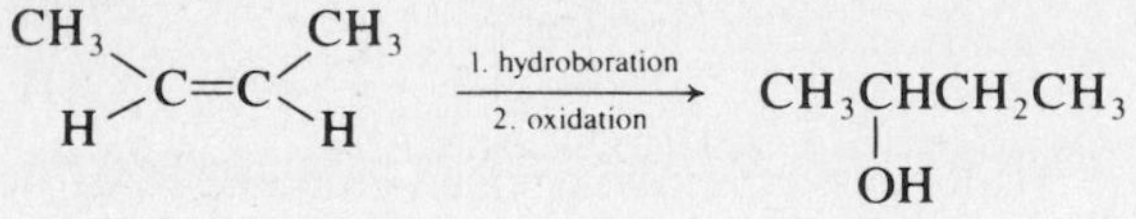

$$\underset{cis\text{-2-Butene}}{(CH_3)HC{=}CH(CH_3)} \xrightarrow[\text{2. oxidation}]{\text{1. hydroboration}} \underset{\text{2-Butanol}}{CH_3\underset{|}{\overset{}{C}}HCH_2CH_3 \; (OH)}$$

(*c*) Hydroboration-oxidation of alkenes is a method that leads to anti-Markovnikov hydration of the double bond.

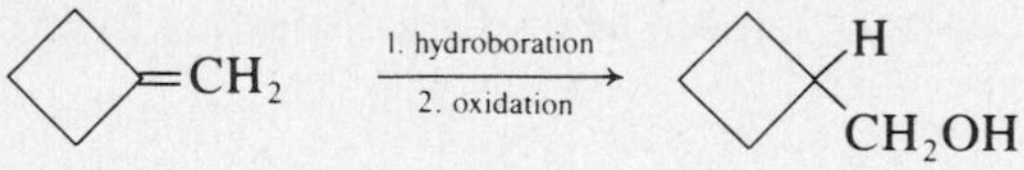

Methylenecyclobutane ($=CH_2$) $\xrightarrow[\text{2. oxidation}]{\text{1. hydroboration}}$ Cyclobutylmethanol (H, CH_2OH)

(*d*) Hydroboration-oxidation of cyclopentene gives cyclopentanol.

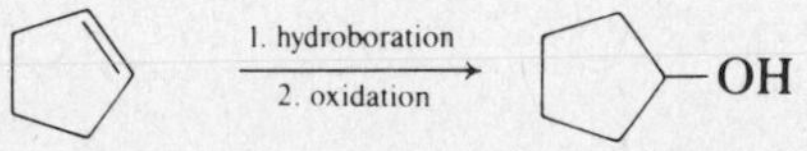

Cyclopentene $\xrightarrow[\text{2. oxidation}]{\text{1. hydroboration}}$ Cyclopentanol (—OH)

(*e*) When alkenes are converted to alcohols by hydroboration-oxidation, the hydroxyl group is introduced at the less substituted carbon of the double bond.

$$\underset{\text{3-Ethyl-2-pentene}}{CH_3CH{=}C(CH_2CH_3)_2} \xrightarrow[\text{2. oxidation}]{\text{1. hydroboration}} \underset{\text{3-Ethyl-2-pentanol}}{CH_3\underset{\displaystyle OH}{\underset{|}{C}}HCH(CH_2CH_3)_2}$$

(*f*) The less substituted carbon of the double bond in 3-ethyl-1-pentene is at the end of the chain. It is this carbon that bears the hydroxyl group in the product of hydroboration-oxidation.

$$\underset{\text{3-Ethyl-1-pentene}}{CH_2{=}CHCH(CH_2CH_3)_2} \xrightarrow[\text{2. oxidation}]{\text{1. hydroboration}} \underset{\text{3-Ethyl-1-pentanol}}{HOCH_2CH_2CH(CH_2CH_3)_2}$$

7.10 Bromine adds anti to the double bond of 1-bromocyclohexene to give 1,1,2-tribromocyclohexane. The radioactive bromines (^{82}Br) are vicinal and trans to each other.

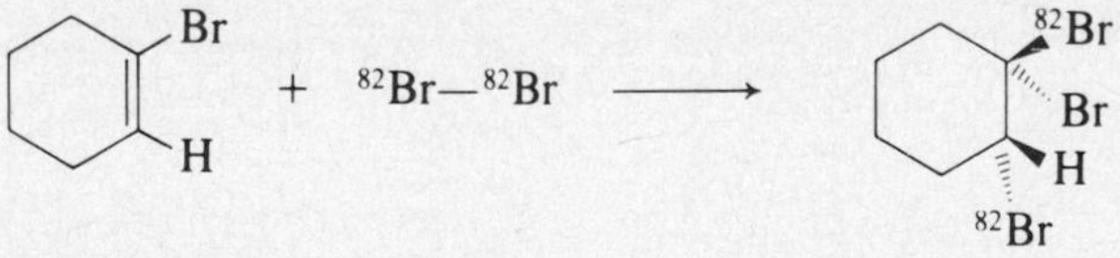

1-Bromocyclohexene (Br, H) + $^{82}Br{-}^{82}Br$ (Bromine) $\longrightarrow$ 1,1,2-Tribromocyclohexane (^{82}Br, Br, H, ^{82}Br)

7.11 Alkyl substituents on the double bond increase the reactivity of the alkene toward addition of bromine.

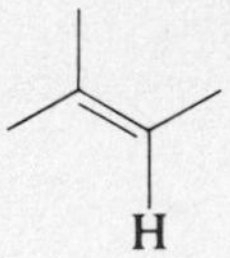

2-Methyl-2-butene (trisubstituted double bond; most reactive)

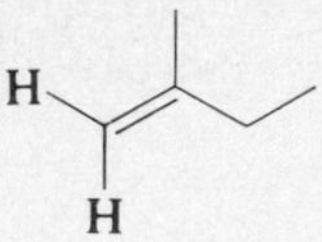

2-Methyl-1-butene (disubstituted double bond)

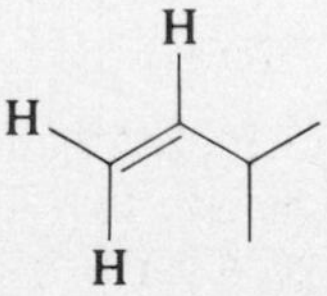

3-Methyl-1-butene (monosubstituted double bond; least reactive)

7.12 The structure of disparlure is as shown.

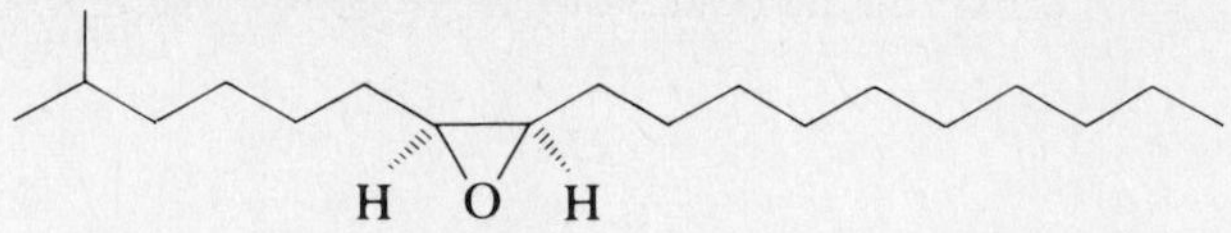

H O H

Its longest continuous chain contains 18 carbon atoms, so it is named as an epoxy derivative of octadecane. Number the chain in the direction that gives the lowest number to the carbons that bear oxygen. Thus, disparlure is *cis*-2-methyl-7,8-epoxyoctadecane.

7.13 Disparlure can be prepared by epoxidation of the corresponding alkene. Epoxidation is stereospecific; cis alkenes yield cis epoxides. Therefore, *cis*-2-methyl-7-octadecene is the alkene chosen to prepare disparlure by epoxidation.

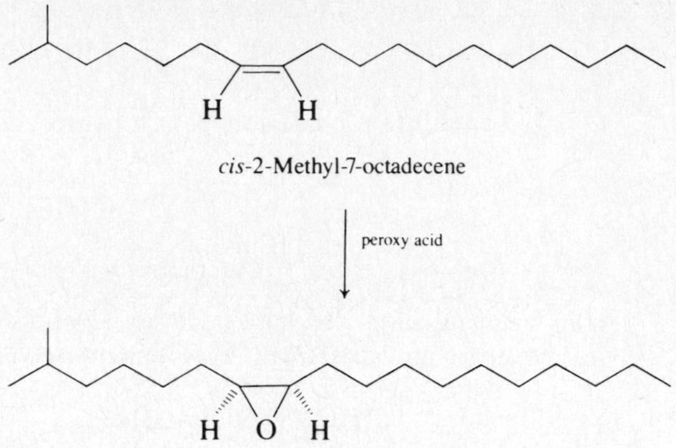

H H

cis-2-Methyl-7-octadecene

↓ peroxy acid

H O H

Disparlure

7.14 The products of ozonolysis are formaldehyde and 4,4-dimethyl-2-pentanone.

$$(H)_2C{=}O \qquad O{=}C(CH_3)CH_2C(CH_3)_3$$

Formaldehyde 4,4-Dimethyl-2-pentanone

The two carbons that were doubly bonded to each other in the alkene become the carbons that are doubly bonded to oxygen in the products of ozonolysis. Therefore, mentally remove the oxygens and connect these two carbons by a double bond to reveal the structure of the starting alkene.

$$(H)_2C{=}C(CH_3)CH_2C(CH_3)_3$$

2,4,4-Trimethyl-1-pentene

7.15 Each of the doubly bonded carbons of the alkene becomes the carbon of a C=O group in the product of permanganate oxidation. A hydrogen substituent on a double bond is replaced by a hydroxyl group (—CH=C becomes —C(=O)OH).

Bicyclo[2.2.1]hept-2-ene therefore becomes *cis*-1,3-cyclopentanedicarboxylic acid.

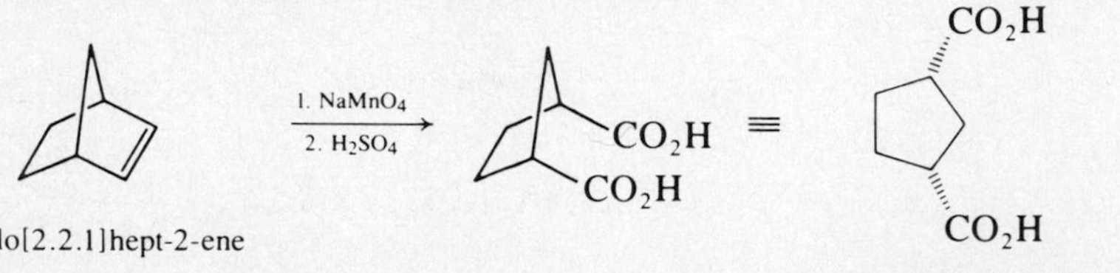

Bicyclo[2.2.1]hept-2-ene $\xrightarrow[2.\ H_2SO_4]{1.\ NaMnO_4}$ CO_2H, CO_2H ≡ CO_2H, CO_2H

cis-1,3-Cyclopentanedicarboxylic acid

7.16 All that is necessary to convert 2,4,4-trimethyl-1-pentene and 2,4,4-trimethyl-2-pentene to 2,2,4-trimethylpentane is catalytic hydrogenation of the double bond.

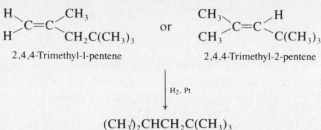

$$\underset{\text{2,4,4-Trimethyl-1-pentene}}{(H)_2C{=}C(CH_3)CH_2C(CH_3)_3} \quad \text{or} \quad \underset{\text{2,4,4-Trimethyl-2-pentene}}{(CH_3)_2C{=}CHC(CH_3)_3}$$

$$\downarrow H_2, Pt$$

$$\underset{\text{2,2,4-Trimethylpentane}}{(CH_3)_2CHCH_2C(CH_3)_3}$$

7.17 This problem illustrates the reactions of alkenes with various reagents and requires application of the Markovnikov rule in the addition of unsymmetrical electrophiles.

(*a*) Markovnikov addition of hydrogen chloride to 1-pentene will give 2-chloropentane.

$$\underset{\text{1-Pentene}}{CH_2{=}CHCH_2CH_2CH_3} + HCl \longrightarrow \underset{\text{2-Chloropentane}}{CH_3\underset{\displaystyle Cl}{\underset{|}{C}}HCH_2CH_2CH_3}$$

(*b*) Ionic addition of hydrogen bromide will give 2-bromopentane.

$$CH_2{=}CHCH_2CH_2CH_3 + HBr \longrightarrow \underset{\text{2-Bromopentane}}{CH_3\underset{\displaystyle Br}{\underset{|}{C}}HCH_2CH_2CH_3}$$

(*c*) The presence of peroxides in the reaction medium will cause free-radical addition of hydrogen bromide, and anti-Markovnikov regioselectivity will be observed.

$$CH_2{=}CHCH_2CH_2CH_3 + HBr \xrightarrow{\text{peroxides}} \underset{\text{1-Bromopentane}}{BrCH_2CH_2CH_2CH_2CH_3}$$

(*d*) Hydrogen iodide will add according to Markovnikov's rule.

$$CH_2{=}CHCH_2CH_2CH_3 + HI \longrightarrow \underset{\text{2-Iodopentane}}{CH_3\underset{\displaystyle I}{\underset{|}{C}}HCH_2CH_2CH_3}$$

(*e*) Dilute sulfuric acid will cause hydration of the double bond with regioselectivity in accord with Markovnikov's rule.

$$CH_2{=}CHCH_2CH_2CH_3 + H_2O \xrightarrow{H_2SO_4} \underset{\text{2-Pentanol}}{CH_3\underset{\displaystyle OH}{\underset{|}{C}}HCH_2CH_2CH_3}$$

(*f*) Hydroboration-oxidation of an alkene brings about anti-Markovnikov hydration of the double bond; 1-pentanol will be the product.

$$CH_2{=}CHCH_2CH_2CH_3 \xrightarrow[\text{2. } H_2O_2,\ HO^-]{\text{1. } B_2H_6} \underset{\text{1-Pentanol}}{HOCH_2CH_2CH_2CH_2CH_3}$$

(*g*) Oxymercuration-demercuration is a method for hydration of an alkene according to Markovnikov's rule.

$$CH_2{=}CHCH_2CH_2CH_3 \xrightarrow[\text{2. } NaBH_4,\ HO^-]{\text{1. } Hg(OAc)_2,\ THF\text{-}H_2O} \underset{\text{2-Pentanol}}{CH_3\underset{\displaystyle OH}{\underset{|}{C}}HCH_2CH_2CH_3}$$

(*h*) Bromine adds across the double bond to give the vicinal dibromide.

$$CH_2{=}CHCH_2CH_2CH_3 + Br_2 \xrightarrow{CCl_4} BrCH_2\underset{\displaystyle Br}{\underset{|}{C}}HCH_2CH_2CH_3$$

1,2-Dibromopentane

(*i*) Vicinal bromohydrins are formed when bromine in water adds to alkenes. Markovnikov orientation is observed if one considers attack by bromonium ion to be followed by interception of the cation by water as a nucleophile.

$$CH_2{=}CHCH_2CH_2CH_3 + Br_2 \xrightarrow{H_2O} BrCH_2\underset{\displaystyle OH}{\underset{|}{C}}HCH_2CH_2CH_3$$

1-Bromo-2-pentanol

(*j*) Epoxidation of the alkene occurs on treatment with peroxy acids.

$$CH_2{=}CHCH_2CH_2CH_3 + CH_3CO_2OH \longrightarrow \underset{\diagdown O \diagup}{CH_2{-}CH}CH_2CH_2CH_3 + CH_3CO_2H$$

1,2-Epoxypentane   Acetic acid

(*k*) Ozone reacts with alkenes to give ozonides.

$$CH_2{=}CHCH_2CH_2CH_3 + O_3 \xrightarrow{\text{acetic acid}} \overset{O}{CH_2\qquad CHCH_2CH_2CH_3}\atop{O{-}O}$$

3-Propyl-1,2,4-trioxolane

(*l*) When the ozonide in (*k*) is hydrolyzed in the presence of zinc, formaldehyde and butanal are formed.

$$\overset{O}{CH_2\qquad CHCH_2CH_2CH_3}\atop{O{-}O} \xrightarrow[Zn]{H_2O} H\overset{O}{\overset{\|}{C}}H + \overset{O}{\overset{\diagdown}{}}CCH_2CH_2CH_3 \atop {H \diagup}$$

Formaldehyde   Butanal

(*m*) The reaction that occurs when alkenes are cleaved by potassium permanganate can be represented in general terms by the equation.

$$\begin{matrix} R & & R' \\ & C{=}C & \\ H & & R'' \end{matrix} \xrightarrow[2.\ H^+]{1.\ KMnO_4} \begin{matrix} R & \\ & C{=}O \\ HO & \end{matrix} + \begin{matrix} & R' \\ O{=}C & \\ & R'' \end{matrix}$$

Thus, cleavage of 1-pentene proceeds as follows:

$$\begin{matrix} H & & CH_2CH_2CH_3 \\ & C{=}C & \\ H & & H \end{matrix} \xrightarrow[2.\ H^+]{1.\ KMnO_4} \begin{matrix} HO & \\ & C{=}O \\ HO & \end{matrix} + \begin{matrix} & CH_2CH_2CH_3 \\ O{=}C & \\ & OH \end{matrix}$$

1-Pentene   Carbonic acid   Butanoic acid

Carbonic acid, however, is unstable and decomposes to carbon dioxide and water.

$$\begin{matrix} HO & \\ & C{=}O \\ HO & \end{matrix} \longrightarrow H_2O + CO_2$$

Carbonic acid   Water   Carbon dioxide

Thus, oxidation of 1-pentene by potassium permanganate gives butanoic acid, carbon dioxide, and water.

7.18 When we compare the reactions of 2-methyl-2-butene with the analogous reactions of 1-pentene, we find that the reactions proceed in a similar manner.

(*a*) $(CH_3)_2C{=}CHCH_3 + HCl \longrightarrow (CH_3)_2\underset{\displaystyle Cl}{\underset{|}{C}}CH_2CH_3$

2-Methyl-2-butene — 2-Chloro-2-methylbutane

(*b*) $(CH_3)_2C{=}CHCH_3 + HBr \longrightarrow (CH_3)_2\underset{\displaystyle Br}{\underset{|}{C}}CH_2CH_3$

2-Bromo-2-methylbutane

(*c*) $(CH_3)_2C{=}CHCH_3 + HBr \xrightarrow{\text{peroxides}} (CH_3)_2CH\underset{\displaystyle Br}{\underset{|}{C}}HCH_3$

2-Bromo-3-methylbutane

(*d*) $(CH_3)_2C{=}CHCH_3 + HI \longrightarrow (CH_3)_2\underset{\displaystyle I}{\underset{|}{C}}CH_2CH_3$

2-Iodo-2-methylbutane

(*e*) $(CH_3)_2C{=}CHCH_3 + H_2O \xrightarrow{H_2SO_4} (CH_3)_2\underset{\displaystyle OH}{\underset{|}{C}}CH_2CH_3$

2-Methyl-2-butanol

(*f*) $(CH_3)_2C{=}CHCH_3 \xrightarrow[\text{2. } H_2O_2,\ HO^-]{\text{1. } B_2H_6} (CH_3)_2CH\underset{\displaystyle OH}{\underset{|}{C}}HCH_3$

3-Methyl-2-butanol

(*g*) $(CH_3)_2C{=}CHCH_3 \xrightarrow[\text{2. } NaBH_4,\ HO^-]{\text{1. } Hg(OAc)_2,\ THF-H_2O} (CH_3)_2\underset{\displaystyle OH}{\underset{|}{C}}CH_2CH_3$

2-Methyl-2-butanol

(*h*) $(CH_3)_2C{=}CHCH_3 + Br_2 \xrightarrow{CCl_4} (CH_3)_2\overset{\displaystyle Br}{\overset{|}{C}}\underset{\displaystyle Br}{\underset{|}{C}}HCH_3$

2,3-Dibromo-2-methylbutane

(*i*) $(CH_3)_2C{=}CHCH_3 + Br_2 \xrightarrow{H_2O} (CH_3)_2\overset{\displaystyle Br}{\overset{|}{C}}\underset{\displaystyle OH}{\underset{|}{C}}HCH_3$

3-Bromo-2-methyl-2-butanol

(*j*) $(CH_3)_2C{=}CHCH_3 + CH_3CO_2OH \longrightarrow (CH_3)_2C{-}CHCH_3 \text{ (bridged by O)} + CH_3CO_2H$

2,2,3-Trimethyloxirane

(*k*) $(CH_3)_2C{=}CHCH_3 + O_3 \longrightarrow$ [ring: $(CH_3)_2C$ and $C(H)(CH_3)$ joined by –O– and –O–O–]

3,3,5-Trimethyl-1,2,4-trioxolane

(*l*) [3,3,5-trimethyl-1,2,4-trioxolane] $\xrightarrow[Zn]{H_2O} CH_3\overset{\displaystyle O}{\overset{\|}{C}}CH_3 + H{-}\overset{\displaystyle O}{\overset{\|}{C}}CH_3$

Acetone — Acetaldehyde

(*m*) The products of oxidation of 2-methyl-2-butene by potassium permanganate are acetic acid and acetone.

$$(CH_3)(H)C{=}C(CH_3)_2 \xrightarrow[2.\ H^+]{1.\ KMnO_4} CH_3C({=}O)OH + O{=}C(CH_3)_2$$

2-Methyl-2-butene — Acetic acid — Acetone

7.19 Cycloalkenes undergo the same kinds of reactions as do noncyclic ones.

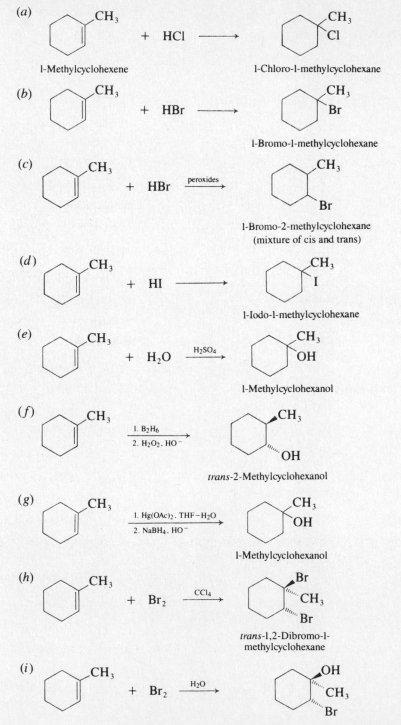

(*a*) 1-Methylcyclohexene + HCl ⟶ 1-Chloro-1-methylcyclohexane

(*b*) 1-Methylcyclohexene + HBr ⟶ 1-Bromo-1-methylcyclohexane

(*c*) 1-Methylcyclohexene + HBr $\xrightarrow{\text{peroxides}}$ 1-Bromo-2-methylcyclohexane (mixture of cis and trans)

(*d*) 1-Methylcyclohexene + HI ⟶ 1-Iodo-1-methylcyclohexane

(*e*) 1-Methylcyclohexene + H_2O $\xrightarrow{H_2SO_4}$ 1-Methylcyclohexanol

(*f*) 1-Methylcyclohexene $\xrightarrow[2.\ H_2O_2,\ HO^-]{1.\ B_2H_6}$ *trans*-2-Methylcyclohexanol

(*g*) 1-Methylcyclohexene $\xrightarrow[2.\ NaBH_4,\ HO^-]{1.\ Hg(OAc)_2,\ THF-H_2O}$ 1-Methylcyclohexanol

(*h*) 1-Methylcyclohexene + Br_2 $\xrightarrow{CCl_4}$ *trans*-1,2-Dibromo-1-methylcyclohexane

(*i*) 1-Methylcyclohexene + Br_2 $\xrightarrow{H_2O}$ *trans*-2-Bromo-1-methylcyclohexanol (*j*) 1-Methylcyclohexene + CH_3CO_2OH ⟶ 1,2-Epoxy-1-methylcyclohexane + CH_3CO_2H

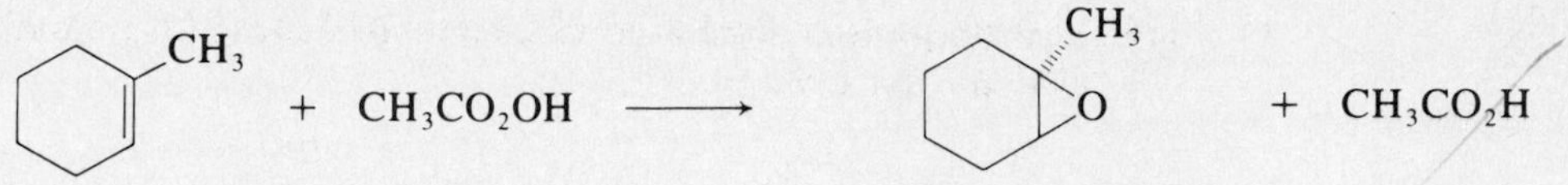

1,2-Epoxy-l-methylcyclohexane

(*k*) 1-Methylcyclohexene + O_3 ⟶

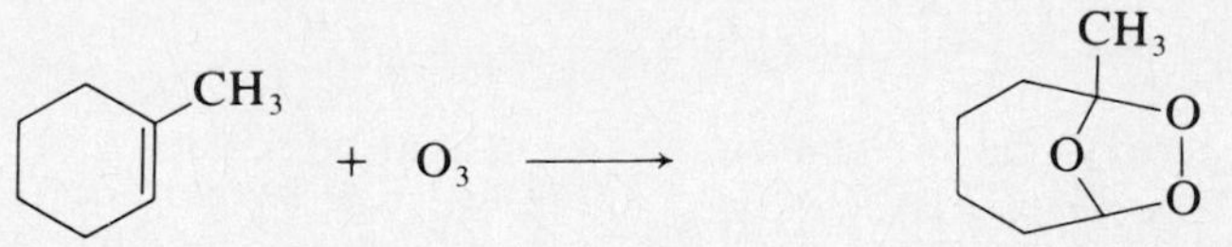

1-Methyl-7,8,9-trioxabicyclo[4.2.1]nonane

(*l*) Ozonide $\xrightarrow[\text{Zn}]{H_2O}$ 6-Oxoheptanal $\equiv$ $CH_3\overset{O}{\overset{\|}{C}}CH_2CH_2CH_2CH_2\overset{O}{\overset{\|}{C}}H$

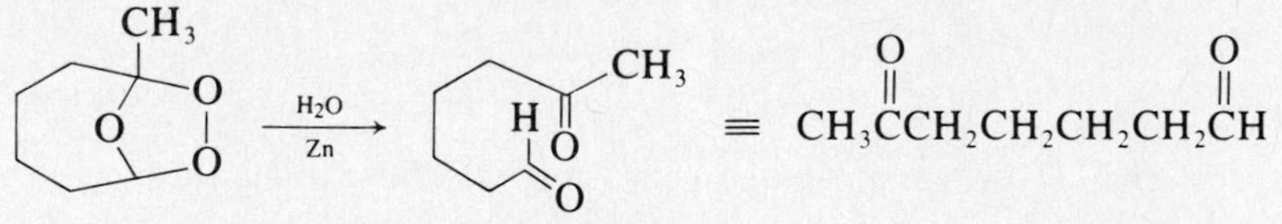

6-Oxoheptanal

(*m*) Permanganate cleaves the double bond to give a noncyclic compound containing all seven carbon atoms of 1-methylcyclohexene.

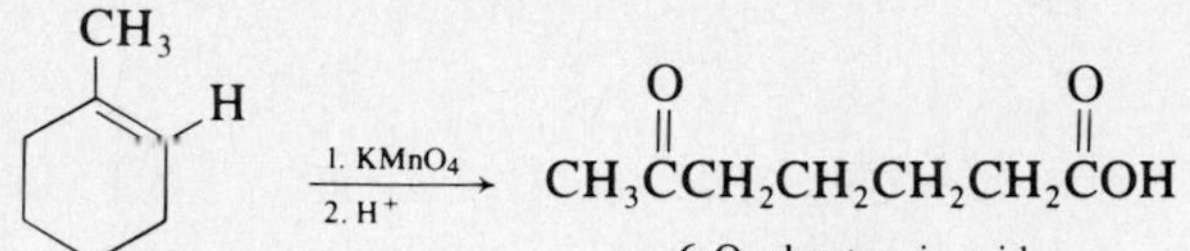

$$\text{1-Methylcyclohexene} \xrightarrow[2.\ H^+]{1.\ KMnO_4} CH_3\overset{O}{\overset{\|}{C}}CH_2CH_2CH_2CH_2\overset{O}{\overset{\|}{C}}OH$$

6-Oxoheptanoic acid

7.20 (*a*) The desired transformation is the conversion of an alkene to a vicinal dibromide.

$$CH_3CH{=}C(CH_2CH_3)_2 \xrightarrow[CCl_4]{Br_2} CH_3\underset{Br}{\underset{|}{C}}H\underset{Br}{\underset{|}{C}}(CH_2CH_3)_2$$

3 Ethyl-2-pentene 2,3-Dibromo-3-ethylpentane

(*b*) Markovnikov addition of hydrogen chloride is indicated.

$$CH_3CH{=}C(CH_2CH_3)_2 \xrightarrow{HCl} CH_3CH_2\underset{Cl}{\underset{|}{C}}(CH_2CH_3)_2$$

3-Chloro-3-ethylpentane

(*c*) Free-radical addition of hydrogen bromide will produce the required anti-Markovnikov orientation.

$$CH_3CH{=}C(CH_2CH_3)_2 \xrightarrow[\text{peroxides}]{HBr} CH_3\underset{Br}{\underset{|}{C}}HCH(CH_2CH_3)_2$$

2-Bromo-3-ethylpentane

(*d*) Acid-catalyzed hydration will occur in accordance with Markovnikov's rule to yield the desired tertiary alcohol.

$$CH_3CH{=}C(CH_2CH_3)_2 \xrightarrow[H_2SO_4]{H_2O} CH_3CH_2\underset{OH}{\underset{|}{C}}(CH_2CH_3)_2$$

3-Ethyl-3-pentanol

(*e*) A superior procedure for the Markovnikov hydration of alkenes is oxymercuration-demercuration.

$$CH_3CH{=}C(CH_2CH_3)_2 \xrightarrow[2.\ NaBH_4,\ HO^-]{1.\ Hg(OAc)_2,\ THF\text{-}H_2O} CH_3CH_2\underset{OH}{\underset{|}{C}}(CH_2CH_3)_2$$

3-Ethyl-3-pentanol

(*f*) Hydroboration-oxidation results in hydration of alkenes with a regioselectivity contrary to that of Markovnikov's rule.

$$CH_3CH{=}C(CH_2CH_3)_2 \xrightarrow[\text{2. } H_2O_2,\ HO^-]{\text{1. } B_2H_6} CH_3\underset{\displaystyle OH}{\underset{|}{C}}HCH(CH_2CH_3)_2$$

3-Ethyl-2-pentanol

(*g*) A peroxy acid will convert an alkene to an epoxide.

$$CH_3CH{=}C(CH_2CH_3)_2 \xrightarrow{CH_3CO_2OH}$$ (epoxide: CH_3, H on one carbon; CH_2CH_3, CH_2CH_3 on the other; ring O)

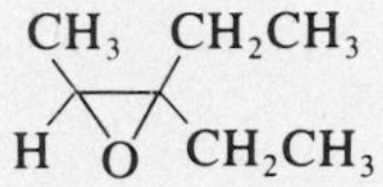

2,2-Diethyl-3-methyloxirane

(*h*) Hydrogenation of alkenes converts them to alkanes.

$$CH_3CH{=}C(CH_2CH_3)_2 \xrightarrow[Pt]{H_2} CH_3CH_2CH(CH_2CH_3)_2$$

3-Ethylpentane

7.21 (*a*) There are four primary alcohols having the molecular formula $C_5H_{12}O$. They are:

$CH_3CH_2CH_2CH_2CH_2OH$ — 1-Pentanol

$CH_3CH_2\overset{\displaystyle CH}{\overset{|}{C}}HCH_2OH$ — 2-Methyl-1-butanol

$(CH_3)_2CHCH_2CH_2OH$ — 3-Methyl-1-butanol

$(CH_3)_3CCH_2OH$ — 2,2-Dimethyl-1-propanol (neopentyl alcohol)

Neopentyl alcohol cannot be prepared by hydration of an alkene because no alkene can have the same carbon skeleton as neopentyl alcohol.

(*b*) Hydroboration-oxidation of alkenes is the method of choice for converting terminal alkenes to primary alcohols.

$$CH_3CH_2CH_2CH{=}CH_2 \xrightarrow[\text{2. } H_2O_2,\ HO^-]{\text{1. } B_2H_6} CH_3CH_2CH_2CH_2CH_2OH$$

1-Pentene → 1-Pentanol

$$CH_3CH_2\underset{\displaystyle CH_3}{\underset{|}{C}}{=}CH_2 \xrightarrow[\text{2. } H_2O_2,\ HO^-]{\text{1. } B_2H_6} CH_3CH_2\underset{\displaystyle CH_3}{\underset{|}{C}}HCH_2OH$$

2-Methyl-1-butene → 2-Methyl-1-butanol

$$(CH_3)_2CHCH{=}CH_2 \xrightarrow[\text{2. } H_2O_2,\ HO^-]{\text{1. } B_2H_6} (CH_3)_2CHCH_2CH_2OH$$

3-Methyl-1-butene → 3-Methyl-1-butanol

(*c*) The three secondary alcohols are:

$CH_3\underset{\displaystyle OH}{\underset{|}{C}}HCH_2CH_2CH_3$ — 2-Pentanol

$CH_3CH_2\underset{\displaystyle OH}{\underset{|}{C}}HCH_2CH_3$ — 3-Pentanol

$CH_3\underset{\displaystyle OH}{\underset{|}{C}}HCH(CH_3)_2$ — 3-Methyl-2-butanol

2-Pentanol can be prepared from 1-pentene by Markovnikov hydration of the double bond.

$$CH_2{=}CHCH_2CH_2CH_3 \xrightarrow[\text{2. } NaBH_4,\ HO^-]{\text{1. } Hg(OAc)_2,\ THF\text{-}H_2O} CH_3\underset{\displaystyle OH}{\underset{|}{C}}HCH_2CH_2CH_3$$

1-Pentene → 2-Pentanol

3-Methyl-2-butanol can be prepared by Markovnikov hydration of 3-methyl-1-butene

$$\underset{\text{3-Methyl-1-butene}}{CH_2{=}CHCH(CH_3)_2} \xrightarrow[\text{2. NaBH}_4\text{, HO}^-]{\text{1. Hg(OAc)}_2\text{, THF-H}_2\text{O}} \underset{\text{3-Methyl-2-butanol}}{CH_3\underset{\displaystyle OH}{\underset{|}{C}}HCH(CH_3)_2}$$

or by anti-Markovnikov hydration of 2-methyl-2-butene.

$$\underset{\text{2-Methyl-2-butene}}{CH_3CH{=}C(CH_3)_2} \xrightarrow[\text{2. H}_2\text{O}_2\text{, HO}^-]{\text{1. B}_2\text{H}_6} \underset{\text{3-Methyl-2-butanol}}{CH_3\underset{\displaystyle OH}{\underset{|}{C}}HCH(CH_3)_2}$$

3-Pentanol is formally derived by hydration of 2-pentene. However, there is no method which allows 3-pentanol to be made by hydration of 2-pentene without 2-pentanol being formed as well in comparable amounts.

$$\underset{\text{2-Pentene}}{CH_3CH{=}CHCH_2CH_3} \longrightarrow \underset{\text{2-Pentanol}}{CH_3\underset{\displaystyle OH}{\underset{|}{C}}HCH_2CH_2CH_3} + \underset{\text{3-Pentanol}}{CH_3CH_2\underset{\displaystyle OH}{\underset{|}{C}}HCH_2CH_3}$$

Therefore, 3-pentanol cannot be efficiently prepared from an alkene.

(*d*) The only tertiary alcohol is 2-methyl-2-butanol. It can be made by Markovnikov hydration of 2-methyl-1-butene or of 2-methyl-2-butene.

$$\underset{\text{2-Methyl-1-butene}}{CH_2{=}\underset{\displaystyle CH_3}{\underset{|}{C}}CH_2CH_3} \xrightarrow[\text{2. NaBH}_4\text{, HO}^-]{\text{1. Hg(OAc)}_2\text{, THF-H}_2\text{O}} \underset{\text{2-Methyl-2-butanol}}{(CH_3)_2\underset{\displaystyle OH}{\underset{|}{C}}CH_2CH_3}$$

$$\underset{\text{2-Methyl-2-butene}}{(CH_3)_2C{=}CHCH_3} \xrightarrow[\text{2. NaBH}_4\text{, HO}^-]{\text{1. Hg(OAc)}_2\text{, THF-H}_2\text{O}} \underset{\text{2-Methyl-2-butanol}}{(CH_3)_2\underset{\displaystyle OH}{\underset{|}{C}}CH_2CH_3}$$

7.22 (*a*) Because the double bond is symmetrically substituted, the same addition product is formed under either ionic or free-radical conditions. Peroxides are absent, so addition takes place by an ionic mechanism to give 3-bromohexane. (It does not matter whether the starting material is *cis*- or *trans*-3-hexene; both give the same product.)

$$\underset{\text{3-Hexene}}{CH_3CH_2CH{=}CHCH_2CH_3} + \underset{\text{Hydrogen bromide}}{HBr} \xrightarrow{\text{no peroxides}} \underset{\substack{\text{3-Bromohexane} \\ \text{(observed yield 76\%)}}}{CH_3CH_2CH_2\underset{\displaystyle Br}{\underset{|}{C}}HCH_2CH_3}$$

(*b*) The combination of potassium iodide and phosphoric acid reacts with alkenes in the same way that HI does.

$$\underset{\text{2,3-Dimethyl-2-butene}}{(CH_3)_2C{=}C(CH_3)_2} \xrightarrow[H_3PO_4]{KI} \underset{\substack{\text{2-Iodo-2,3-dimethylbutane} \\ \text{(observed yield 91\%)}}}{(CH_3)_2CH\underset{\displaystyle I}{\underset{|}{C}}(CH_3)_2}$$

(*c*) As noted above, potassium iodide and phosphoric acid react with alkenes to transfer a proton and iodide to the double bond. The regioselectivity of addition corresponds to Markovnikov's rule.

$$\underset{\text{1-Hexene}}{CH_2{=}CHCH_2CH_2CH_2CH_3} \xrightarrow[H_3PO_4]{KI} \underset{\substack{\text{2-Iodohexane} \\ \text{(observed yield 93\%)}}}{CH_3\underset{\displaystyle I}{\underset{|}{C}}HCH_2CH_2CH_2CH_3}$$

(*d*) In the presence of peroxides hydrogen bromide adds with a regioselectivity opposite to that predicted by Markovnikov's rule. The product is the corresponding primary bromide.

$$(CH_3)_2CHCH_2CH_2CH_2CH{=}CH_2 \xrightarrow[\text{peroxides}]{\text{HBr}} (CH_3)_2CHCH_2CH_2CH_2CH_2CH_2Br$$

6-Methyl-1-heptene — 1-Bromo-6-methylheptane (observed yield 92%)

(*e*) Oxymercuration-demercuration of alkenes gives alcohols that correspond to addition of the elements of water to the double bond according to Markovnikov's rule. Rearrangements do not occur.

$$(CH_3)_2C{=}CHC(CH_3)_3 \xrightarrow[\text{2. NaBH}_4\text{, HO}^-]{\text{1. Hg(OAc)}_2\text{, THF-H}_2\text{O}} (CH_3)_2\underset{\substack{|\\ OH}}{C}CH_2C(CH_3)_3$$

2,4,4-Trimethyl-2-pentene — 2,4,4-Trimethyl-2-pentanol (observed yield 86%)

(*f*) Hydroboration-oxidation of alkenes leads to hydration of the double bond with a regioselectivity contrary to Markovnikov's rule and without rearrangement of the carbon skeleton.

$$CH_2{=}C[C(CH_3)_3]_2 \xrightarrow[\text{2. H}_2\text{O}_2\text{, HO}^-]{\text{1. B}_2\text{H}_6} HOCH_2\overset{\substack{C(CH_3)_3\\ |}}{C}HC(CH_3)_3$$

2-*tert*-Butyl-3,3-dimethyl-1-butene — 2-*tert*-Butyl-3,3-dimethyl-1-butanol (observed yield 65%)

(*g*) Hydroboration-oxidation of alkenes leads to syn hydration of double bonds.

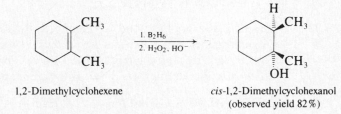

$$\text{1,2-dimethylcyclohexene} \xrightarrow[\text{2. H}_2\text{O}_2\text{, HO}^-]{\text{1. B}_2\text{H}_6} \text{cyclohexane with H, CH}_3 \text{ at C-2 and CH}_3\text{, OH at C-1}$$

1,2-Dimethylcyclohexene — *cis*-1,2-Dimethylcyclohexanol (observed yield 82%)

(*h*) Bromine adds across the double bond of alkenes to give vicinal dibromides.

$$CH_2{=}\underset{\substack{|\\ CH_3}}{C}CH_2CH_2CH_3 + Br_2 \xrightarrow{CHCl_3} BrCH_2\overset{\substack{Br\\ |}}{\underset{\substack{|\\ CH_3}}{C}}CH_2CH_2CH_3$$

2-Methyl-1-pentene — 1,2-Dibromo-2-methylpentane (observed yield 60%)

(*i*) In aqueous solution bromine reacts with alkenes to give bromohydrins. Bromine is the electrophile in this reaction and adds to the carbon that has the greater number of hydrogen substituents.

$$(CH_3)_2C{=}CHCH_3 + Br_2 \xrightarrow{H_2O} (CH_3)_2\underset{\substack{|\\ OH}}{C}{-}\overset{\substack{Br\\ |}}{C}HCH_3$$

2-Methyl-2-butene — Bromine — 3-Bromo-2-methyl-2-butanol (observed yield 77%)

(*j*) Compounds of the type $\mathrm{R\overset{O}{\overset{\|}{C}}OOH}$ are peroxy acids and react with alkenes to give epoxides.

$$\underset{\text{2,3-Dimethyl-2-butene}}{(CH_3)_2C{=}C(CH_3)_2} + \underset{\text{Peroxyacetic acid}}{CH_3\overset{O}{\overset{\|}{C}}OOH} \longrightarrow \underset{\substack{\text{2,2,3,3-Tetramethyloxirane}\\ \text{(observed yield 70–80\%)}}}{(CH_3)_2C\overset{O}{—}C(CH_3)_2} + \underset{\text{Acetic acid}}{CH_3\overset{O}{\overset{\|}{C}}OH}$$

(*k*) The double bond is cleaved by ozonolysis. Each of the doubly bonded carbons becomes doubly bonded to oxygen in the product.

$$\text{bicyclic alkene} \xrightarrow[\text{2. } H_2O]{\text{1. } O_3} \text{Cyclodecan-1,6-dione}$$

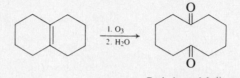

Cyclodecan-1,6-dione
(observed yield 45%)

7.23 (*a*) There is no direct, one-step transformation that moves a hydroxyl group from one carbon to another, so it is not possible to convert 2-propanol to 1-propanol in a single reaction. Analyze the problem by reasoning backward. 1-Propanol is a primary alcohol. What reactions do we have available for the preparation of primary alcohols? One way is by the hydroboration-oxidation of terminal alkenes.

$$\underset{\text{Propene}}{CH_3CH{=}CH_2} \xrightarrow{\text{hydroboration-oxidation}} \underset{\text{1-Propanol}}{CH_3CH_2CH_2OH}$$

The problem now becomes the preparation of propene from 2-propanol. The simplest way is by acid-catalyzed dehydration.

$$\underset{\text{2-Propanol}}{CH_3\underset{\substack{|\\OH}}{C}HCH_3} \xrightarrow{H^+,\ \text{heat}} \underset{\text{Propene}}{CH_3CH{=}CH_2}$$

After analyzing the problem in terms of overall strategy, present the synthesis in detail showing the reagents required in each step. Thus, the answer is:

$$\underset{\text{2-Propanol}}{CH_3\underset{\substack{|\\OH}}{C}HCH_3} \xrightarrow[\text{heat}]{H_2SO_4} \underset{\text{Propene}}{CH_3CH{=}CH_2} \xrightarrow[\text{2. } H_2O_2,\ HO^-]{\text{1. } B_2H_6} \underset{\text{1-Propanol}}{CH_3CH_2CH_2OH}$$

(*b*) We analyze this synthetic exercise in a manner similar to the preceding one. There is no direct way to move a bromine from C-2 in 2-bromopropane to C-1 in 1-bromopropane. However, we can prepare 1-bromopropane from propene by free-radical addition of hydrogen bromide in the presence of peroxides.

$$\underset{\text{Propene}}{CH_3CH{=}CH_2} + \underset{\text{Hydrogen bromide}}{HBr} \xrightarrow{\text{peroxides}} \underset{\text{1-Bromopropane}}{CH_3CH_2CH_2Br}$$

We prepare propene from 2-bromopropane by dehydrohalogenation.

$$\underset{\text{2-Bromopropane}}{CH_3\underset{\substack{|\\Br}}{C}HCH_3} \xrightarrow{E2} \underset{\text{Propene}}{CH_3CH{=}CH_2}$$

Sodium ethoxide in ethanol is a suitable base-solvent system for this conversion. Sodium methoxide in methanol or potassium *tert*-butoxide in *tert*-butyl alcohol could also be used, as could potassium hydroxide in ethanol.

Combining these two transformations gives the complete synthesis.

$$\underset{\text{2-Bromopropane}}{\underset{\displaystyle\text{Br}}{CH_3\overset{|}{C}HCH_3}} \xrightarrow[CH_3CH_2OH,\ \text{heat}]{NaOCH_2CH_3} \underset{\text{Propene}}{CH_3CH{=}CH_2} \xrightarrow[\text{peroxides}]{HBr} \underset{\text{1-Bromopropane}}{CH_3CH_2CH_2Br}$$

(*c*) Planning your strategy in a forward direction can lead to problems when the conversion of 2-bromopropane to 1,2-dibromopropane is considered. There is a temptation to try to simply add the second bromine by free-radical halogenation.

$$\underset{\text{2-Bromopropane}}{\underset{\displaystyle\text{Br}}{CH_3\overset{|}{C}HCH_3}} \xrightarrow{Br_2,\ \text{light and heat}} \underset{\text{1,2-Dibromopropane}}{\underset{\displaystyle\text{Br}}{CH_3\overset{|}{C}HCH_2Br}}$$

This is incorrect! There is no reason to believe that the second bromine will be introduced exclusively at C-1. In fact, the selectivity rules for bromination tell us that 2,2-dibromopropane is the expected major product.

The best approach is to reason backward. 1,2-Dibromopropane is a vicinal dibromide, and we prepare vicinal dibromides by adding elemental bromine to alkenes.

$$\underset{\text{Propene}}{CH_3CH{=}CH_2} + \underset{\text{Bromine}}{Br_2} \longrightarrow \underset{\text{1,2-Dibromopropane}}{\underset{\displaystyle\text{Br}}{CH_3\overset{|}{C}HCH_2Br}}$$

As described in the preceding exercise, we prepare propene from 2-bromopropane by E2 elimination. Therefore, the correct synthesis is:

$$\underset{\text{2-Bromopropane}}{\underset{\displaystyle\text{Br}}{CH_3\overset{|}{C}HCH_3}} \xrightarrow[CH_3CH_2OH]{NaOCH_2CH_3} \underset{\text{Propene}}{CH_3CH{=}CH_2} \xrightarrow{Br_2} \underset{\text{1,2-Dibromopropane}}{\underset{\displaystyle\text{Br}}{CH_3\overset{|}{C}HCH_2Br}}$$

(*d*) Do not attempt to reason forward and convert 2-propanol to 1-bromo-2-propanol by free-radical bromination. Reason backward! The desired compound is a vicinal bromohydrin, and vicinal bromohydrins are prepared by adding bromine to alkenes in aqueous solution. The correct solution is:

$$\underset{\text{2-Propanol}}{\underset{\displaystyle\text{OH}}{CH_3\overset{|}{C}HCH_3}} \xrightarrow[\text{heat}]{H_2SO_4} \underset{\text{Propene}}{CH_3CH{=}CH_2} \xrightarrow[H_2O]{Br_2} \underset{\text{1-Bromo-2-propanol}}{\underset{\displaystyle\text{OH}}{CH_3\overset{|}{C}HCH_2Br}}$$

(*e*) Here we have another problem where reasoning forward can lead to trouble. If we try to conserve the oxygen of 2-propanol so that it becomes the oxygen of 1,2-epoxypropane, we need a reaction in which this oxygen becomes bonded to C-1.

$$\underset{\text{2-Propanol}}{\underset{\displaystyle\text{OH}}{CH_3\overset{|}{C}HCH_3}} \longrightarrow \underset{\text{1,2-Epoxypropane}}{\underset{\diagdown\,O\,\diagup}{CH_3CH{-}CH_2}}$$

No synthetic method for such a single-step transformation exists!

By reasoning backward, recalling that epoxides are made from alkenes by reaction with peroxy acids, we develop a proper synthesis.

$$\underset{\text{2-Propanol}}{CH_3\underset{\substack{|\\OH}}{C}HCH_3} \xrightarrow[\text{heat}]{H_2SO_4} \underset{\text{Propene}}{CH_3CH{=}CH_2} \xrightarrow{CH_3\overset{O}{\overset{\|}{C}}OOH} \underset{\text{1,2-Epoxypropane}}{CH_3CH{-}CH_2 \text{ (with O bridging)}}$$

(*f*) *tert*-Butyl alcohol and isobutyl alcohol have the same carbon skeleton; all that is required is to move the hydroxyl group from C-1 to C-2. As pointed out in (*a*) of this problem, we cannot do that directly but we can do it in two efficient steps through a synthesis that involves hydration of an alkene.

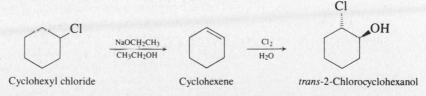

$$\underset{\text{Isobutyl alcohol}}{(CH_3)_2CHCH_2OH} \xrightarrow[\text{heat}]{H_2SO_4} \underset{\text{Isobutene}}{(CH_3)_2C{=}CH_2} \xrightarrow[\text{2. NaBH}_4\text{, HO}^-]{\text{1. Hg(OAc)}_2\text{, THF-H}_2\text{O}} \underset{tert\text{-Butyl alcohol}}{(CH_3)_3COH}$$

Note the use of the oxymercuration-demercuration sequence to cause hydration of the alkene with the desired regioselectivity. (Acid-catalyzed hydration could also be used in this step.)

(*g*) The strategy of this exercise is similar to that of the preceding one. Convert the starting material to an alkene by an elimination reaction, followed by electrophilic addition to the double bond.

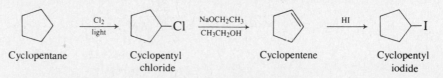

$$\underset{\text{Isobutyl iodide}}{(CH_3)_2CHCH_2I} \xrightarrow[(CH_3)_3COH\text{, heat}]{KOC(CH_3)_3} \underset{\text{Isobutene}}{(CH_3)_2C{=}CH_2} \xrightarrow[\text{KI, }H_3PO_4]{\text{HI or}} \underset{tert\text{-Butyl iodide}}{(CH_3)_3CI}$$

(*h*) This problem is similar to the one in (*d*) in that it requires the preparation of a halohydrin from an alkyl halide. The strategy is the same. Convert the alkyl halide to an alkene, then form the halohydrin by treatment with the appropriate halogen in aqueous solution.

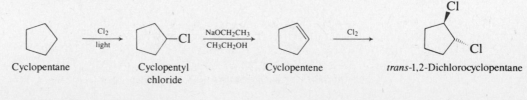

$$\text{Cyclohexyl chloride} \xrightarrow[CH_3CH_2OH]{NaOCH_2CH_3} \text{Cyclohexene} \xrightarrow[H_2O]{Cl_2} trans\text{-2-Chlorocyclohexanol}$$

(*i*) Halogenation of an alkane is required here. Iodination of alkanes, however, is not a feasible reaction because it is endothermic. We can make alkyl iodides from alcohols or from alkenes by treatment with HI (or with KI and phosphoric acid). A reasonable synthesis using reactions that have been presented to this point proceeds as shown:

$$\text{Cyclopentane} \xrightarrow[\text{light}]{Cl_2} \text{Cyclopentyl chloride} \xrightarrow[CH_3CH_2OH]{NaOCH_2CH_3} \text{Cyclopentene} \xrightarrow{HI} \text{Cyclopentyl iodide}$$

(*j*) Dichlorination of cyclopentane under free-radical conditions is not a realistic approach to the introduction of two chlorines in a trans-1,2 relationship without contamination by isomeric dichlorides. Vicinal dichlorides are prepared by electrophilic addition of chlorine to alkenes. The stereochemistry of addition is anti.

$$\text{Cyclopentane} \xrightarrow[\text{light}]{Cl_2} \text{Cyclopentyl chloride} \xrightarrow[CH_3CH_2OH]{NaOCH_2CH_3} \text{Cyclopentene} \xrightarrow{Cl_2} trans\text{-1,2-Dichlorocyclopentane}$$

(*k*) The desired compound contains all five carbon atoms of cyclopentane but is not cyclic. Two dicarboxylic acid functions are present. We know that cleavage of carbon-carbon double bonds by potassium permanganate leads to two carbonyl groups, which suggests the synthesis shown in the following equation.

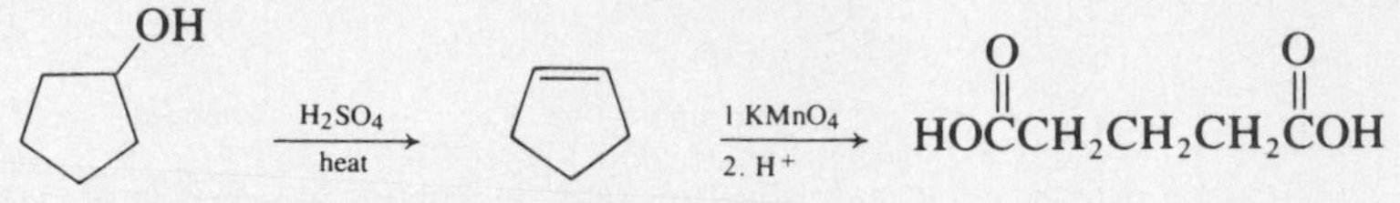

$$\text{Cyclopentanol} \xrightarrow[\text{heat}]{H_2SO_4} \text{Cyclopentene} \xrightarrow[\text{2. } H^+]{\text{1. } KMnO_4} HO\overset{O}{\overset{\|}{C}}CH_2CH_2CH_2\overset{O}{\overset{\|}{C}}OH \;\; \text{(Pentanedioic acid)}$$

Cyclopentanol Cyclopentene Pentanedioic acid

7.24 The two products formed by addition of hydrogen bromide to 1,2-dimethylcyclohexene cannot be regioisomers. Stereoisomers are possible, however.

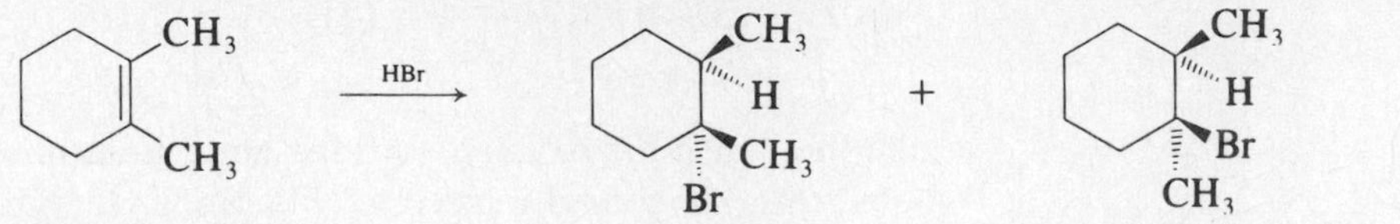

$$\text{1,2-Dimethylcyclohexene} \xrightarrow{HBr} \textit{cis}\text{-1,2-Dimethylcyclohexyl bromide} + \textit{trans}\text{-1,2-Dimethylcyclohexyl bromide}$$

1,2-Dimethylcyclohexene *cis*-1,2-Dimethylcyclohexyl bromide *trans*-1,2-Dimethylcyclohexyl bromide

The same two products are formed from 1,6-dimethylcyclohexene because addition of hydrogen bromide follows Markovnikov's rule in the absence of peroxides.

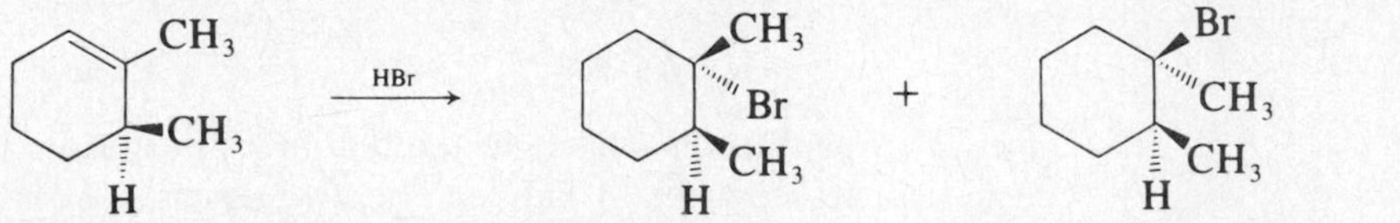

$$\text{1,6-Dimethylcyclohexene} \xrightarrow{HBr} \textit{cis}\text{-1,2-Dimethylcyclohexyl bromide} + \textit{trans}\text{-1,2-Dimethylcyclohexyl bromide}$$

1.6-Dimethylcyclohexene *cis*-1,2-Dimethylcyclohexyl bromide *trans*-1,2-Dimethylcyclohexyl bromide

7.25 The reaction is one of electrophilic addition of hydrogen chloride to the double bond. Two regioisomeric modes of addition are indicated below:

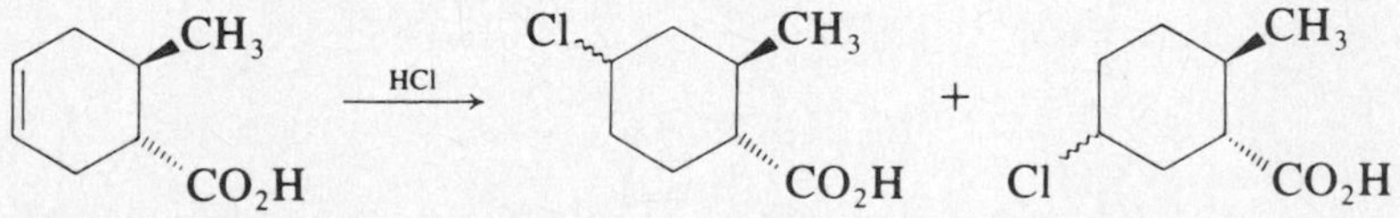

$$\text{A} \xrightarrow{HCl} \text{(Cl, } CH_3\text{, } CO_2H\text{ cyclohexane)} + \text{(Cl, } CH_3\text{, } CO_2H\text{ cyclohexane)}$$

A

Each of these regioisomeric modes can lead to stereoisomers with the chlorine either cis or trans to the methyl group. The four products are:

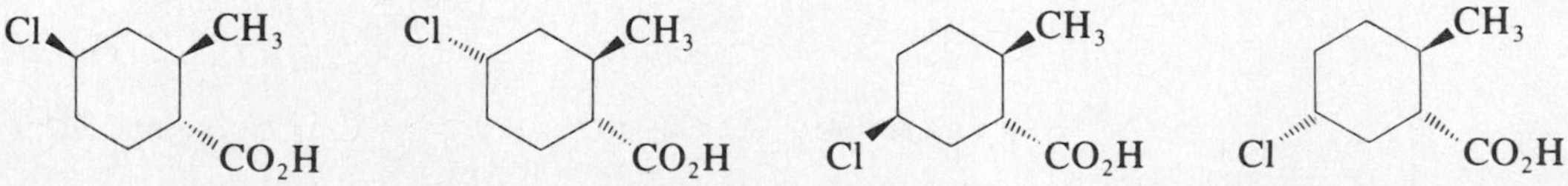

[Structures: four chloro-methylcyclohexanecarboxylic acid stereoisomers bearing Cl, CH_3, CO_2H]

7.26 The regiochemistry of oxymercuration-demercuration follows Markovnikov's rule faithfully. Therefore the products must be the two stereoisomeric tertiary alcohols; one has the hydroxyl group cis to the 3- and 4-methyl groups, the other has the hydroxyl group trans.

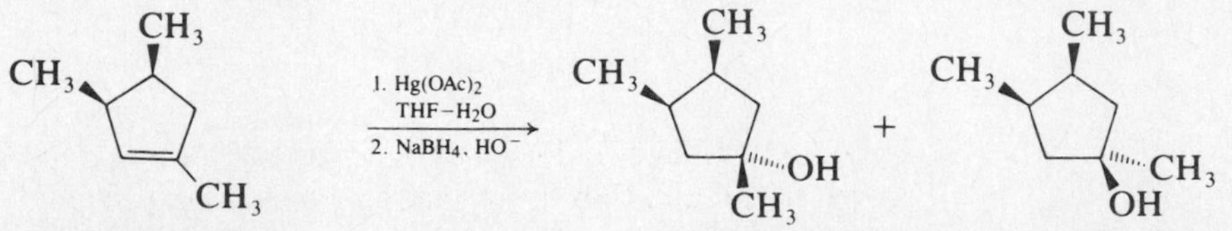

$$\textit{cis}\text{-1,3,4-Trimethylcyclopentene} \xrightarrow[\text{2. } NaBH_4\text{, } HO^-]{\text{1. } Hg(OAc)_2\text{, } THF-H_2O} \text{(OH cis)} + \text{(OH trans)}$$

cis-1,3,4-Trimethylcyclopentene

7.27 Two different trans dibromides arising by anti addition of bromine to the double bond are possible.

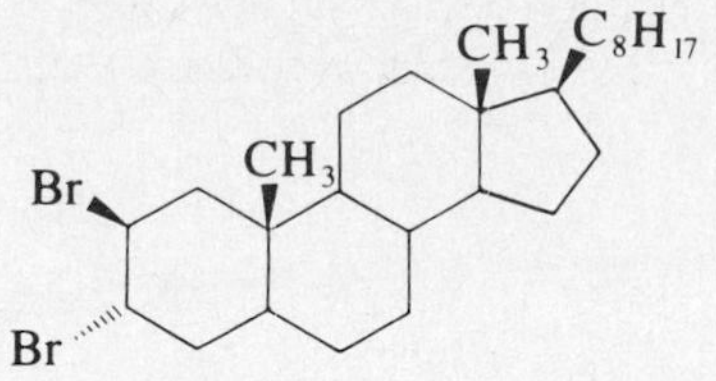

CH_3 C_8H_{17} CH_3 Br Br and CH_3 C_8H_{17} CH_3 Br Br

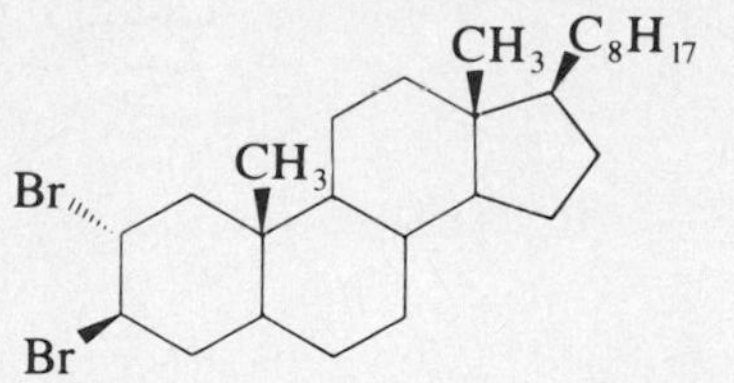

7.28 The carbon skeletons of B, C, and D are identified by the observation that isopentane is the product of hydrogenation of each of them. Therefore, the alkenes must be isomeric methylbutenes.

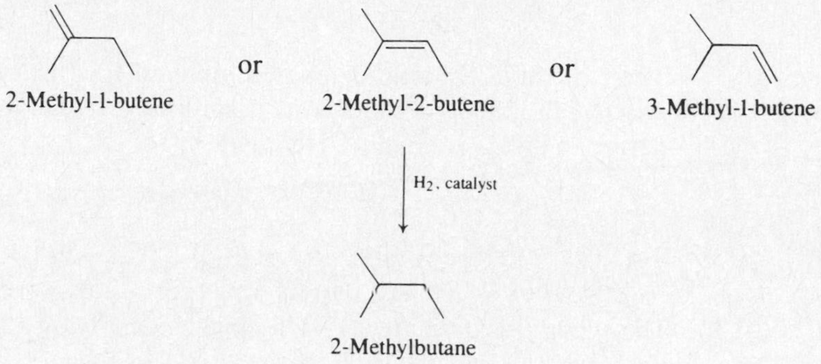

2-Methyl-1-butene or 2-Methyl-2-butene or 3-Methyl-1-butene

H_2, catalyst

2-Methylbutane

Only 2-methyl-1-butene and 2-methyl-2-butene can give a tertiary alcohol on oxymercuration-demercuration.

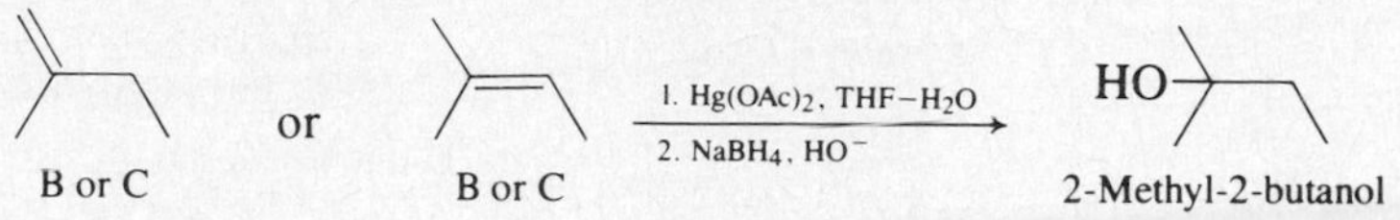

B or C or B or C $\xrightarrow[\text{2. NaBH}_4\text{, HO}^-]{\text{1. Hg(OAc)}_2\text{, THF–H}_2\text{O}}$ HO 2-Methyl-2-butanol

Only 2-methyl-1-butene and 3-methyl-1-butene can give primary alcohols on hydroboration-oxidation.

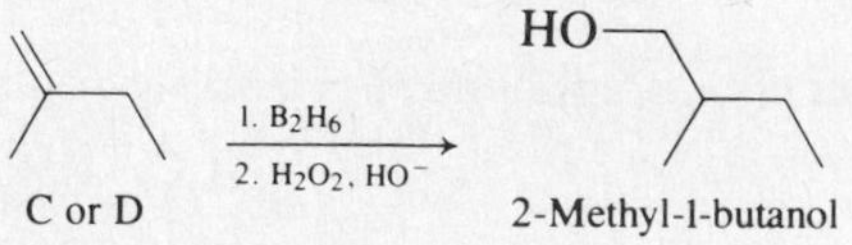

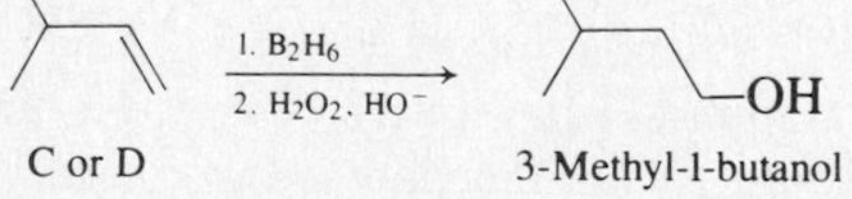

C or D $\xrightarrow[\text{2. H}_2\text{O}_2\text{, HO}^-]{\text{1. B}_2\text{H}_6}$ HO 2-Methyl-1-butanol

C or D $\xrightarrow[\text{2. H}_2\text{O}_2\text{, HO}^-]{\text{1. B}_2\text{H}_6}$ OH 3-Methyl-1-butanol

Therefore, C must be 2-methyl-1-butene; B must be 2-methyl-2-butene; D must be 3-methyl-1-butene.

7.29 Electrophilic addition of hydrogen iodide should occur in accordance with Markovnikov's rule.

$$CH_2{=}CHC(CH_3)_3 \underset{\text{KOH, nPrOH}}{\overset{\text{HI}}{\rightleftharpoons}} CH_3\underset{\substack{|\\ I}}{C}HC(CH_3)_3$$

3,3-Dimethyl-1-butene 2-Iodo-3,3-dimethylbutane

Treatment of 2-iodo-3,3-dimethylbutane with alcoholic potassium hydroxide should bring about E2 elimination to regenerate the starting alkene. Hence, compound E is 2-iodo-3,3-dimethylbutane.

The carbocation intermediate formed in the addition of hydrogen iodide to the alkene is one which can rearrange by a methyl group migration.

$$CH_2{=}CHC(CH_3)_3 + H^+ \longrightarrow CH_3{-}\overset{+}{C}H{-}C(CH_3)_2CH_3 \longrightarrow CH_3{-}CH(CH_3){-}\overset{+}{C}(CH_3)CH_3$$

$$\downarrow I^- \qquad\qquad \downarrow I^-$$

$$CH_3CH(I){-}C(CH_3)_3 \qquad (CH_3)_2CH{-}C(I)(CH_3)_2$$

Compound E   Compound F (2-iodo-2,3-dimethylbutane)

Therefore, a likely candidate for compound F is the one with a rearranged carbon skeleton, 2-iodo-2,3-dimethylbutane. This is confirmed by the fact that compound F undergoes elimination to give 2,3-dimethyl-2-butene.

$$\underset{F}{(CH_3)_2CH{-}C(I)(CH_3)_2} \xrightarrow{E2} \underset{\text{2,3-Dimethyl-2-butene}}{(CH_3)_2C{=}C(CH_3)_2}$$

7.30 The ozonolysis data are useful in quickly identifying alkenes G and H.

$$\text{Compound G} \longrightarrow HCH(=O) + (CH_3)_3C\overset{O}{\overset{\|}{C}}C(CH_3)_3$$

Therefore, compound G is 2-*tert*-butyl-3,3-dimethyl-1-butene.

$$(CH_3)_3C\overset{CH_2}{\overset{\|}{C}}C(CH_3)_3 \quad \text{Compound G}$$

$$\text{Compound H} \longrightarrow HCH(=O) + CH_3\overset{O}{\overset{\|}{C}}{-}C(CH_3)_2{-}C(CH_3)_3$$

Therefore, compound H is 2,3,3,4,4-pentamethyl-1-pentene.

$$CH_3\overset{H_2C}{\overset{\|}{C}}{-}C(CH_3)_2{-}C(CH_3)_3 \quad \text{Compound H}$$

Compound H has a carbon skeleton different from the alcohol which produced it by dehydration. We are therefore led to consider a carbocation rearrangement.

$$(CH_3)_3C{-}C(CH_3)(OH){-}C(CH_3)_3 \xrightarrow{H^+} CH_3C(CH_3)_2{-}\overset{+}{C}(CH_3)C(CH_3)_3 \longrightarrow \underset{\text{Compound G}}{(CH_3)_3C{-}\overset{CH_2}{\overset{\|}{C}}C(CH_3)_3}$$

$$\downarrow \text{methyl migration}$$

$$CH_3\overset{+}{C}(CH_3){-}C(CH_3)_2{-}C(CH_3)_3 \longrightarrow \underset{\text{Compound H}}{CH_3\overset{H_2C}{\overset{\|}{C}}{-}C(CH_3)_2{-}C(CH_3)_3}$$

7.31 The important clue to deducing the structures of I and J is the ozonolysis product K. Remembering that the two carbonyl carbons of K must have been joined by a double bond in the precursor J, we write

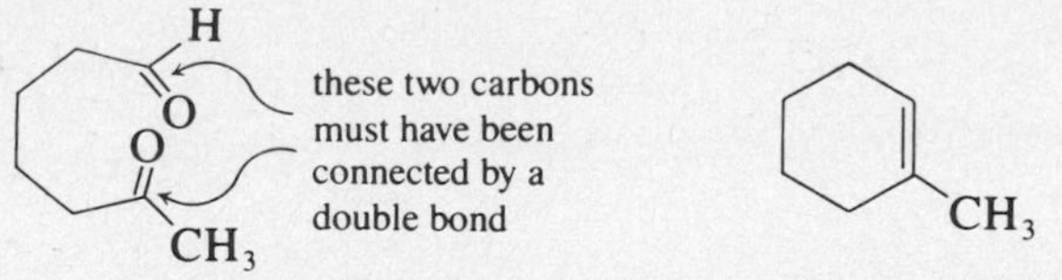

these two carbons must have been connected by a double bond

Compound K Compound J

The tertiary bromide which gives compound J on dehydrobromination is 1-methylcyclohexyl bromide.

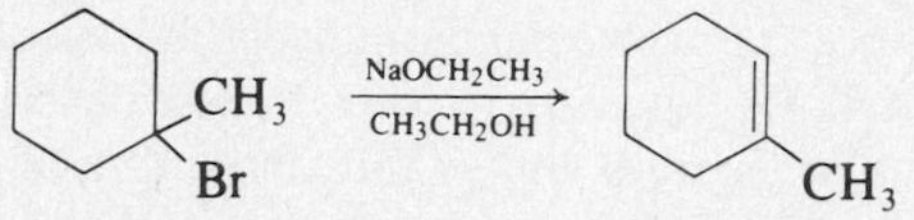

$$\xrightarrow[CH_3CH_2OH]{NaOCH_2CH_3}$$

Compound I Compound J

When tertiary halides are treated with base, they undergo E2 elimination. The regioselectivity of elimination of tertiary halides follows the Zaitsev rule.

7.32 Since santene and 1,3-diacetylcyclopentane (compound L) contain the same number of carbon atoms, the two carbonyl carbons of the diketone must have been connected by a double bond in santene. Therefore, the structure of santene must be

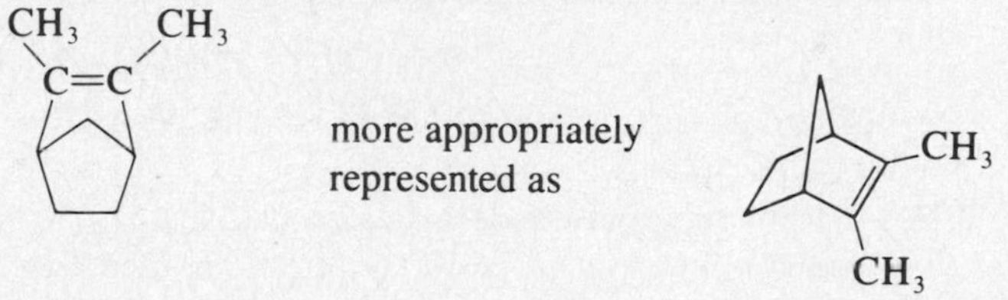

more appropriately represented as

7.33 (*a*) Compound M contains 9 of the 10 carbons and 14 of the 16 hydrogens of sabinene. Ozonolysis has led to the separation of one carbon and two hydrogens from the rest of the molecule. The carbon and the two hydrogens must have been lost as formaldehyde, $H_2C{=}O$. This H_2C unit was originally doubly bonded to the carbonyl carbon of compound M. Therefore, sabinene must have the structure shown in the equation representing its ozonolysis.

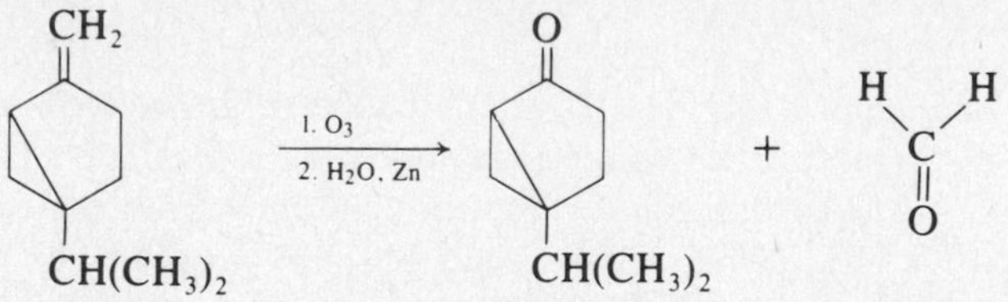

$$\xrightarrow[2.\ H_2O,\ Zn]{1.\ O_3}$$

Sabinene Compound M Formaldehyde

(*b*) Compound N contains all 10 of the carbons and all 16 of the hydrogens of Δ^3-carene. The two carbonyl carbons of compound N must have been linked by a double bond in Δ^3-carene.

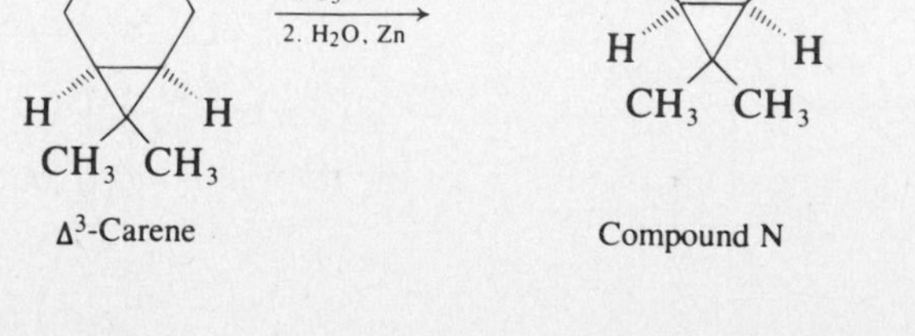

$$\xrightarrow[2.\ H_2O,\ Zn]{1.\ O_3}$$

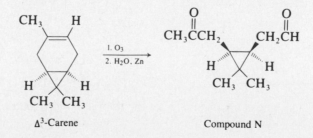

Δ^3-Carene Compound N

(*c*) Compound O contains all the carbon atoms of α-pinene. The two carbonyl carbons of compound O must have been doubly bonded in α-pinene, and α-pinene must be bicyclic.

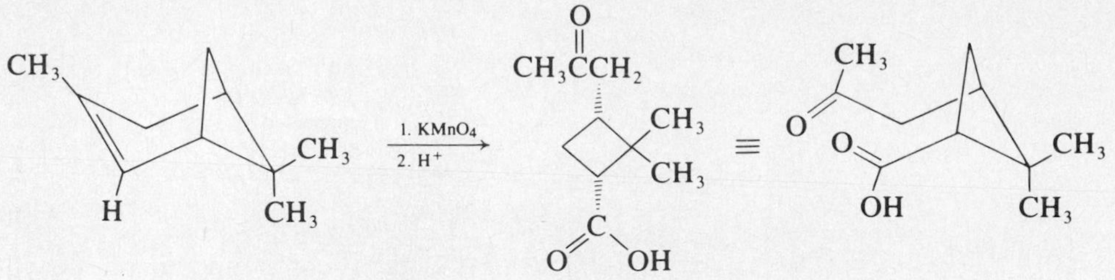

CH_3 H CH_3 CH_3 $\xrightarrow[2.\ H^+]{1.\ KMnO_4}$ $CH_3\overset{O}{\overset{\|}{C}}CH_2$ CH_3 CH_3 C O OH ≡ CH_3 O O OH CH_3 CH_3

α-Pinene Compound O

7.34 Since the molecular formula of the sex attractant of the female housefly is $C_{23}H_{46}$, it has a SODAR equal to 1 (see text Section 5.13). It takes up one mole of hydrogen on catalytic hydrogenation and so must have one double bond and no rings. The position of the double bond is revealed by the ozonolysis data.

$$C_{23}H_{46} \xrightarrow[2.\ H_2O,\ Zn]{1.\ O_3} CH_3(CH_2)_7\overset{O}{\overset{\|}{C}}H + CH_3(CH_2)_{12}\overset{O}{\overset{\|}{C}}H$$

An unbranched 9-carbon unit and an unbranched 14-carbon unit comprise the carbon skeleton, and these two units must be connected by a double bond. Therefore, the housefly sex attractant has the constitution:

$$CH_3(CH_2)_7CH{=}CH(CH_2)_{12}CH_3 \quad \text{9-Tricosene}$$

The data cited in the problem do not permit the stereochemistry of this natural product to be determined.

7.35 The hydrogenation data tell us that $C_{19}H_{38}$ contains one double bond and has the same carbon skeleton as 2,6,10,14-tetramethylpentadecane. We locate the double bond at C-2 on the basis of the fact that acetone, $(CH_3)_2C{=}O$, is obtained on ozonolysis. The structures of the natural product and the aldehyde produced on its ozonolysis are shown below.

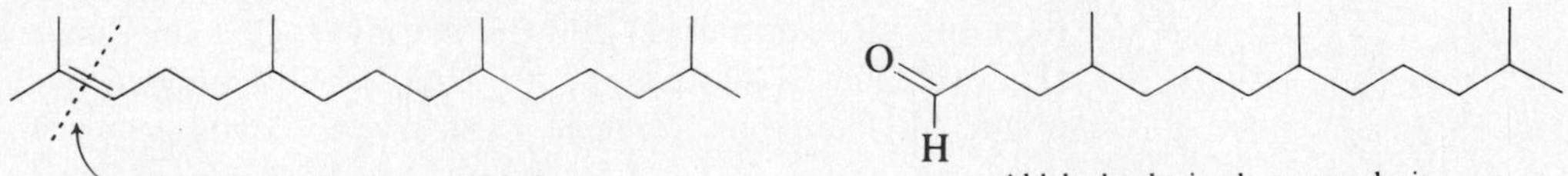

Ozonolysis cleaves molecule here

O H

Aldehyde obtained on ozonolysis

7.36 Since $H\overset{O}{\overset{\|}{C}}CH_2\overset{O}{\overset{\|}{C}}H$ is one of the products of its ozonolysis, the sex attractant of the arctiid moth must contain the unit $=CHCH_2CH=$. This unit must be bonded to an unbranched 12-carbon unit at one end and an unbranched six-carbon unit at the other in order to give $CH_3(CH_2)_{10}CH{=}O$ and $CH_3(CH_2)_4CH{=}O$ on ozonolysis.

$$CH_3(CH_2)_{10}CH{=}CHCH_2CH{=}CH(CH_2)_4CH_3$$

Sex attractant of arctiid moth
(dotted lines show positions of cleavage on ozonolysis

$\downarrow$ 1. O_3 2. H_2O, Zn

$$CH_3(CH_2)_{10}\overset{O}{\overset{\|}{C}}H + H\overset{O}{\overset{\|}{C}}CH_2\overset{O}{\overset{\|}{C}}H + H\overset{O}{\overset{\|}{C}}(CH_2)_4CH_3$$

The stereochemistry of the double bonds cannot be determined on the basis of the available information.

7.37 Compound P contains all 15 carbon atoms of cedrene. To deduce the structure of cedrene connect the carbonyl carbons by a double bond and replace the —OH group by —H. Permanganate oxidation of cedrene proceeds as shown in the equation.

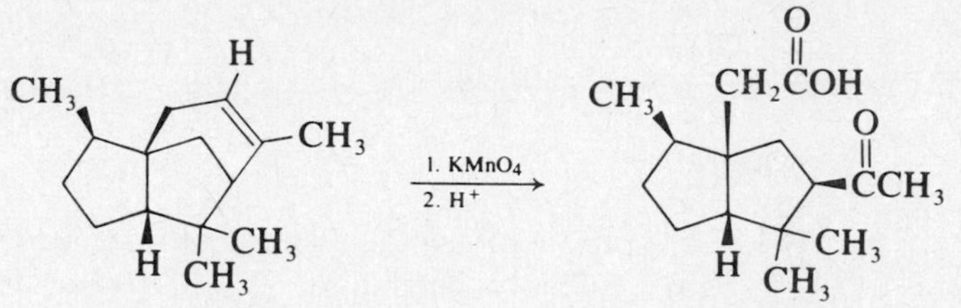

Cedrene $\xrightarrow[2.\ H^+]{1.\ KMnO_4}$ Compound P

SELF-TEST

PART A

A-1. Write structural formulas for the reactant, reagent, or product omitted from each of the following.

(*a*) $(CH_3)_2C{=}CHCH_3 \xrightarrow{H_2SO_4(\text{dilute})} ?$

(*b*) (cyclohexylidene)$=CH_2 \xrightarrow{?}$ (cyclohexyl)$-CH_2Br$

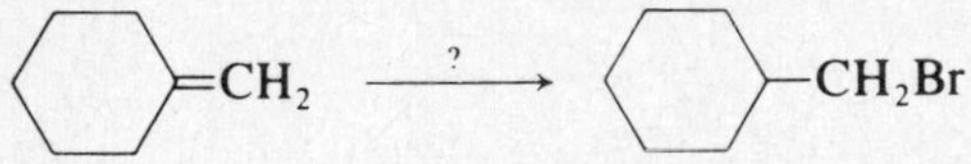

(*c*) $? \xrightarrow[2.\ H_2O,\ Zn]{1.\ O_3}$ 2-oxocyclohexyl–$CH_2CH_2CH_2\overset{\displaystyle O}{\overset{\|}{C}}H$

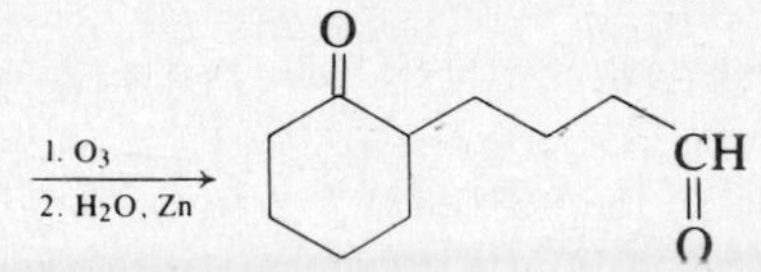

(*d*) $(CH_3)_3CCH{=}CH_2 \xrightarrow{?} (CH_3)_3C\overset{\displaystyle OH}{\overset{|}{C}}HCH_3$

(*e*) $CH_3CH_2\overset{\displaystyle CH_3}{\overset{|}{C}}{=}CH_2 \xrightarrow[H_2O]{Br_2} ?$

A-2. Provide a sequence of reaction steps to carry out the following conversations. Write the structure of each intermediate product.

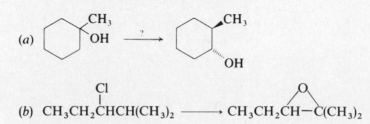

(*a*) 1-methylcyclohexanol (CH_3, OH) $\xrightarrow{?}$ trans-2-methylcyclohexanol (CH_3, OH)

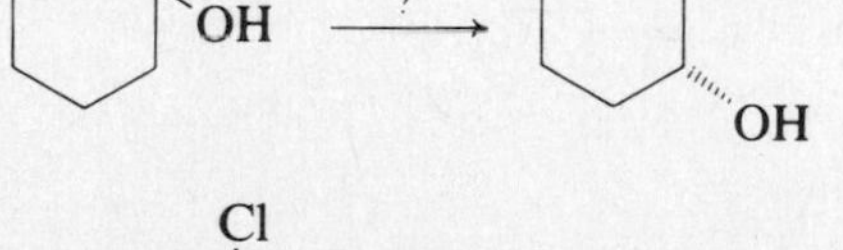

(*b*) $CH_3CH_2\overset{\displaystyle Cl}{\overset{|}{C}}HCH(CH_3)_2 \longrightarrow CH_3CH_2CH{-}C(CH_3)_2$ (epoxide, O bridging)

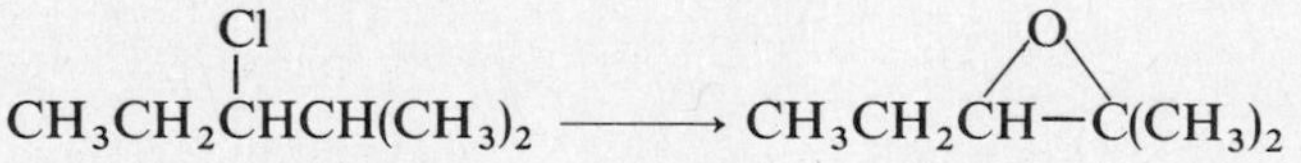

A-3. A hydrocarbon A (C_6H_{12}) undergoes reaction with HBr to yield B ($C_6H_{13}Br$). Treatment of B with sodium ethoxide in ethanol yields C, an isomer of A. Reaction of C with ozone followed by treatment with water and zinc gives acetone, $(CH_3)_2C{=}O$, as the only organic product. Provide structures for A, B, and C, and outline the reaction pathway.

A-4. Provide a detailed mechanism describing the reaction of 1-butene with HBr in the presence of peroxides.

A-5. Chlorine reacts with an alkene to give the 2,3-dichlorobutane isomer whose structure is shown. What is the structure and name of the alkene? Outline a mechanism for the reaction.

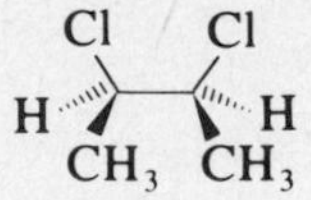

Cl Cl
H H
CH_3 CH_3

PART B

B-1. The product from the reaction of 1-pentene with Cl_2 in H_2O is named:
(*a*) 1-Chloro-2-pentanol (*b*) 2-Chloro-2-pentanol
(*c*) 1-Chloro-1-pentanol (*d*) 2-Chloro-1-pentanol

B-2. In the reaction of a reagent such as HBr with an alkene, the first step of the reaction is the ______ to the alkene.
(*a*) Fast addition of an electrophile
(*b*) Fast addition of a nucleophile
(*c*) Slow addition of an electrophile
(*d*) Slow addition of a nucleophile

B-3. The reaction sequence indicated below

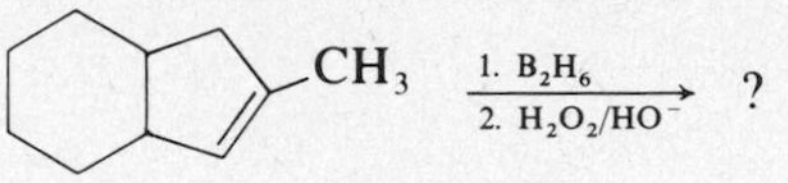

CH_3 —— 1. B_2H_6 / 2. H_2O_2/HO^- → ?

gives as the major product:

(*a*) CH_3 OH
(*b*) CH_3 OH
(*c*) CH_3 OH
(*d*) CH_3 OH

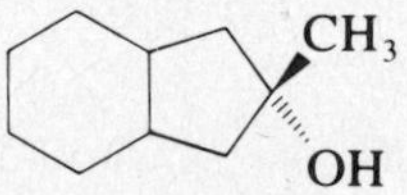

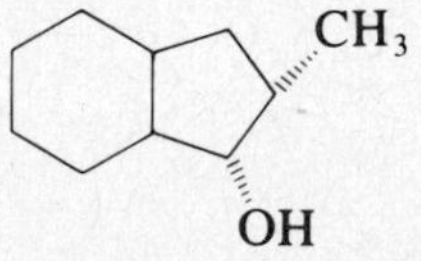

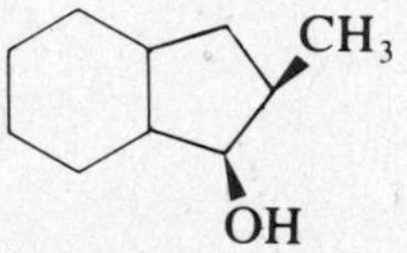

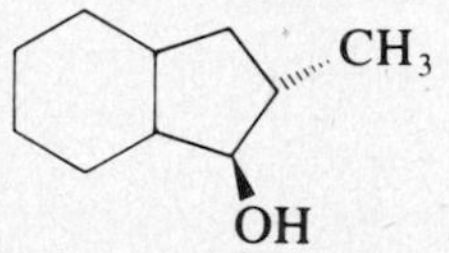

B-4. The trifluoromethyl group ($-CF_3$) is electron withdrawing and has a tendency to destabilize an adjacent positive charge. The reaction

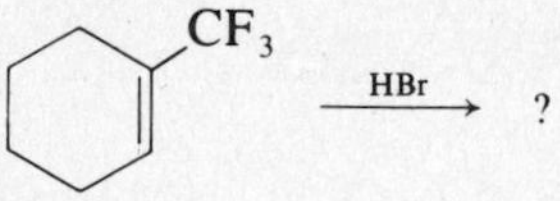

CF_3 —— HBr → ?

gives as the major product:

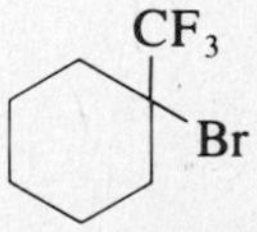

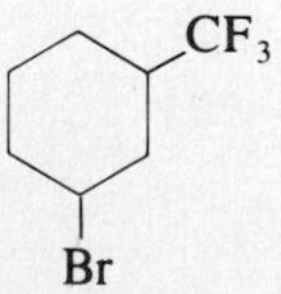

(*a*) CF_3 Br
(*b*) CF_3 Br
(*c*) CF_3 Br
(*d*) None of these

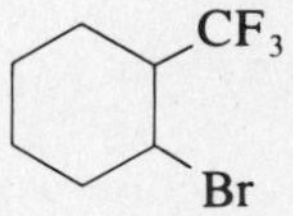

B-5. Markovnikov's rule "works" because:
(*a*) The most stable transition state is the one leading to the more substituted carbocation.
(*b*) The nucleophile adds during the second step of the ionic reaction.
(*c*) The electrophile adds to the less substituted end of the double bond.
(*d*) All of these are true.

B-6. Treatment of 2-methyl-2-butene with HBr in the presence of peroxide yields

(*a*) A primary alkyl bromide
(*b*) A secondary alkyl bromide
(*c*) A tertiary alkyl bromide
(*d*) A vicinal dibromide

B-7. The strongest evidence for the formation of a bridged bromonium ion as an intermediate in the addition of Br_2 to an alkene is:

(*a*) Markovnikov's rule
(*b*) Zaitsev's rule
(*c*) The regioselectivity of the reaction
(*d*) The stereospecificity of the reaction

B-8. The reaction

$$(CH_3)_2C{=}CH_2 + Br\cdot \longrightarrow (CH_3)_2\dot{C}{-}CH_2Br$$

is an example of a(n) _____ step in a radical chain reaction.

(*a*) Initiation (*b*) Propagation
(*c*) Termination (*d*) Heterolytic cleavage

CHAPTER

8

STEREOCHEMISTRY

IMPORTANT TERMS AND CONCEPTS

Molecular Chirality (Secs. 8.1 to 8.3) Any molecule which is *not superposable* on its mirror image is said to be *chiral*; a molecule which is superposable on its mirror image is *achiral*. The two nonsuperposable mirror images are *stereoisomers* of each other, that is, they differ in the arrangement of their atoms in space. Molecules that are nonsuperposable mirror images of each other are said to be *enantiomers* of each other. Stereoisomers that are not mirror images are *diastereomers* of each other; for example, (*E*)- and (*Z*)-2-butene are diastereomers.

Recognizing the presence or absence of *elements of symmetry* is important in determining if a molecule is chiral. Any structure which contains a *plane of symmetry* will be *achiral*, and likewise any structure which contains a *center of symmetry* will also be *achiral*. Note that only one of these symmetry elements need be present and that the plane is usually easier to recognize when evaluating a structural formula.

The most common chiral organic molecules possess a carbon atom having four different groups attached, which is known as the *chiral center* or the *chiral carbon atom*. For example, carbon-2 of 2-chlorobutane is the chiral center in this molecule.

H, C—Cl, CH_3, CH_2CH_3 H, Cl—C, CH_3, CH_2CH_3

Optical Activity (Sec. 8.4) An *optically active* substance is one which *rotates the plane of polarized light* as measured in an instrument known as a *polarimeter*. A substance which does not rotate the plane of polarized light is said to be *optically inactive*.

In order for a substance to exhibit optical activity, *one enantiomer of a chiral substance* must be present in excess over the other. The *observed rotation* is denoted by α. An equal mixture of the two enantiomers of a chiral substance is known as a *racemic mixture* and is *optically inactive*. The percent enantiomeric excess or *optical purity* of a substance is defined as

$$\text{Optical purity} = \text{percent of one enantiomer} - \text{percent of other enantiomer}$$

The two enantiomers of a chiral substance differ *only* in the direction of rotation of polarized light. That is, if one enantiomer has an optical rotation of $+\alpha$, that of the other will be $-\alpha$. The positive and negative rotations are termed *dextrorotatory* and *levorotatory*, respectively. Frequently the abbreviations *d* and *l* are used, as well as (+) and (−). A racemic mixture of a substance is often termed (*d*, *l*) or (±).

In order to catalog the optical activity of a substance, the *specific rotation*, [α], has been defined as follows.

$$[\alpha] = \frac{100\alpha}{c \cdot l}$$

where α = observed rotation
c = concentration of solution, g/100 mL
l = length of polarimeter tube, decimeters

Absolute and Relative Configuration (Secs. 8.5, 8.6) The precise arrangement of substituents at a chiral center is its *absolute configuration.* The *relative configuration* of a chiral center indicates the spatial relationship when one compound (or chiral center) is compared with another. It is important to note that there is no definite relationship between sign of rotation and absolute configuration for an optically active substance.

The absolute configuration of a chiral center is best specified by using the *Cahn-Ingold-Prelog* or *R-S* system. This method was first encountered in Chapter 5 for describing *E* and *Z* alkenes. The rules are as follows:

1. Identify the substituents attached to the chiral atom and rank them in order of decreasing precedence.
2. Orient the molecule so that the *lowest*-ranking substituent is pointing away from you. Note this will often be hydrogen.
3. The three highest ranking substituents will appear, when viewed as in step 2, pointing out as if they were spokes of a wheel. If the order of decreasing precedence is *clockwise*, the chiral center is *R*. If the order is *counterclockwise* (anticlockwise), the configuration is *S*.

In the following example, the numbers represent group rankings according to the Cahn-Ingold-Prelog rules, where 1 indicates the highest-ranked group and 4 the lowest.

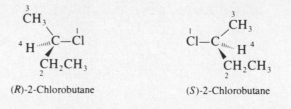

3 CH_3; 1 Cl; 4 H; 2 CH_2CH_3 — C

(*R*)-2-Chlorobutane

1 Cl; 3 CH_3; 4 H; 2 CH_2CH_3 — C

(*S*)-2-Chlorobutane

From the example, an important relationship may be seen: *the enantiomer of a molecule having the R configuration is S, and vice versa.*

Stereochemistry of Chemical Reactions (Sec. 8.8) Many chemical reactions lead to formation of a chiral product from an achiral starting material. The chiral product is however, *optically inactive*, being a *racemic mixture*, that is, equal amounts of the two enantiomers have been produced. As a general rule, optically inactive products result when optically inactive substrates react with optically inactive reagents. This rule holds whether the reactants and reagents are achiral or racemic mixtures of a chiral substance.

Optically active reactants may or may not lead to formation of optically active products, depending on the nature of the reaction. When an achiral intermediate forms in the reaction, optically inactive products (a *racemic mixture*) will result. Specific examples of reactions involving chiral reactants will be discussed in the next chapter.

Molecules with Two or More Chiral Centers (Secs. 8.9, 8.10) The maximum number of stereoisomers a compound may possess is 2^n, where n equals the number of chiral centers. Each chiral center may be R or S. Note however, that the number of distinct stereoisomers may be *less* than the maximum. If, for example, a plane of symmetry is present in the molecule, the mirror images of such a stereoisomer *will* be superposable on and identical to each other. These molecules, which are achiral, may contain two or more chiral carbon atoms; they are known as *meso forms*. The achiral meso form is a *diastereomer* with respect to each of the chiral stereoisomers, as it is a stereoisomer that is not an enantiomer.

2,3-Dibromobutane, shown below, provides an example.

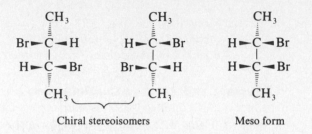

CH₃	CH₃	CH₃
Br—C—H	H—C—Br	H—C—Br
H—C—Br	Br—C—H	H—C—Br
CH₃	CH₃	CH₃

Chiral stereoisomers (first two); Meso form (third)

Reactions Which Form Diastereomers (Sec. 8.11) The addition reaction of halogens with alkenes is an example of a *stereospecific reaction*, that is, stereoisomeric reactants yield stereoisomeric products. The reaction of Br_2 with the isomeric 2-butenes yields products that are diastereomers of each other. As shown in text Figure 8.10, the Z, or cis, isomer reacts to give a racemic mixture of enantiomers; the E, or trans, isomer yields the meso form. Epoxidation of alkenes provides another example of a stereospecific reaction which yields diastereomeric products.

Resolution (Sec. 8.12) The separation of a racemic mixture into its enantiomeric components is known as *resolution*. The racemic substance is allowed to react with an optically active chiral reagent, frequently a naturally occurring acid or base. The product is a mixture of *diastereomers*, which may have physical properties (such as solubility) sufficiently different to allow their separation. The appropriate chemical steps then allow the original substance to be recovered in pure enantiomeric form.

Chiral Atoms Other than Carbon (Sec. 8.14) The rules employed for identifying chiral carbon atoms apply to other atoms as well. In the case of nitrogen and phosphorus, pyramidal inversion must be slow in order for individual enantiomers to exist. The absolute configuration of any chiral center may be specified with the Cahn-Ingold-Prelog rules, with the provision that an unshared electron pair is considered to be the lowest-ranking substituent.

SOLUTIONS TO TEXT PROBLEMS

8.1 (*b*) None of the carbons in 3-bromopentane has four different groups attached to it, so we would expect 3-bromopentane to be *achiral*. We can show this by testing for superposability of mirror image forms.

First, represent the two mirror image forms of 3-bromopentane.

Reference structure: $(CH_3CH_2)(CH_3CH_2)C(H)Br$ — and — Mirror image: $H{-}C(CH_2CH_3)(CH_2CH_3)Br$

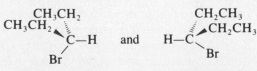

Reference structure   Mirror image

Turn the mirror image form so that it is oriented in the same fashion as the reference structure.

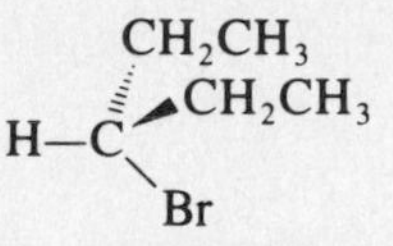

Mirror image — is equivalent to — Reoriented mirror image

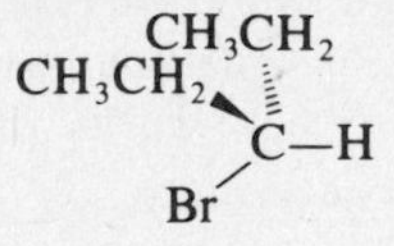

Mirror image representations of 3-bromopentane are superposable, so the molecule is achiral.

(*c*) Carbon-2 of 1-bromo-2-methylbutane has four different groups attached to it, so the molecule should be chiral.

$$BrCH_2-\underset{|}{C}HCH_2CH_3 \quad \text{(with } CH_3 \text{ on C-2)}$$

1-Bromo-2-methylbutane

The two mirror image forms of 1-bromo-2-methylbutane are:

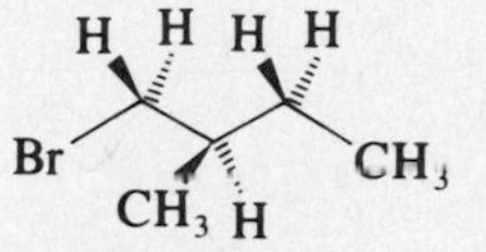

Reference structure — and — Mirror image

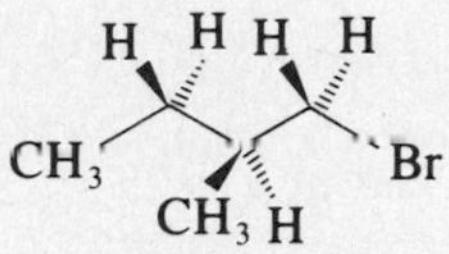

Reorienting the structure on the right, we get:

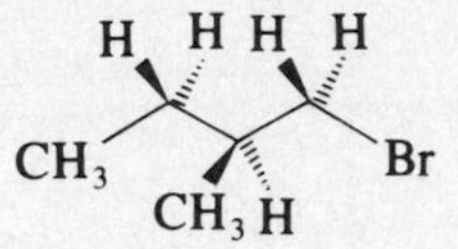

Mirror image — is equivalent to — Reoriented mirror image

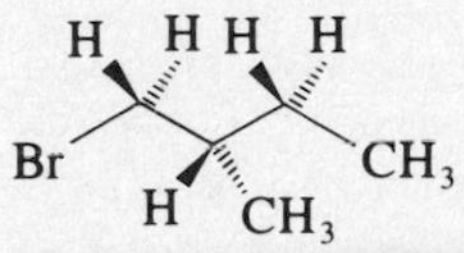

When this representation is compared with the original reference structure, we see:

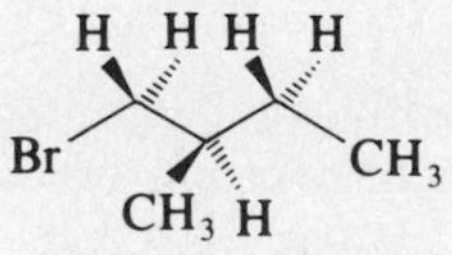

Reference structure 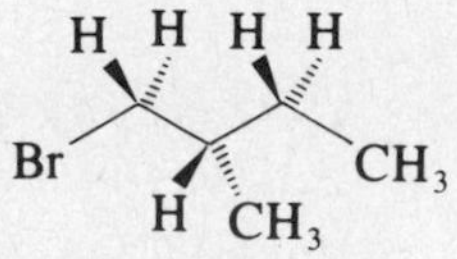 — Reoriented mirror image

The two are not superposable; therefore 1-bromo-2-methylbutane is chiral.

(*d*) The two mirror image representations of 2-bromo-2-methylbutane are:

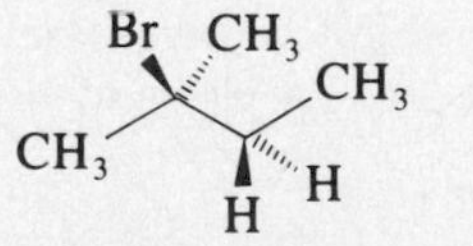

Reference structure — and — Mirror image

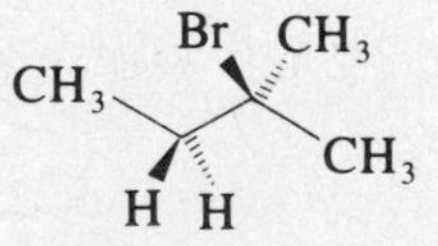

Reorienting the structure on the right, we get:

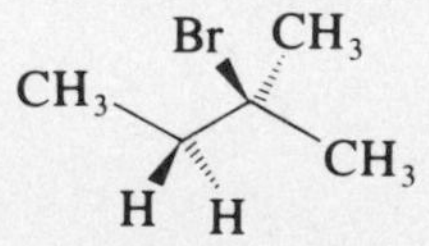

Mirror image 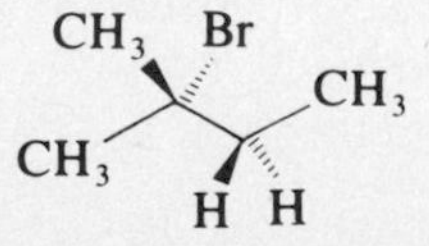 — Reoriented mirror image

On comparing the reoriented mirror image with the original reference structure,

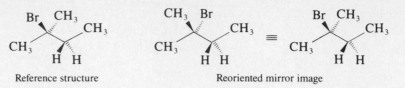

Reference structure

Reoriented mirror image

we see that the two represent only different conformations of the same substance, as rotation around the C-2–C-3 bond permits them to be superposed. Therefore 2-bromo-2-methylbutane is achiral.

8.2 (*b*) There are *two* planes of symmetry in (*Z*)-1,2-dichloroethene, of which one is the plane of the molecule, and the second plane of symmetry bisects the carbon-carbon bond. There is no center of symmetry. The molecule is achiral.

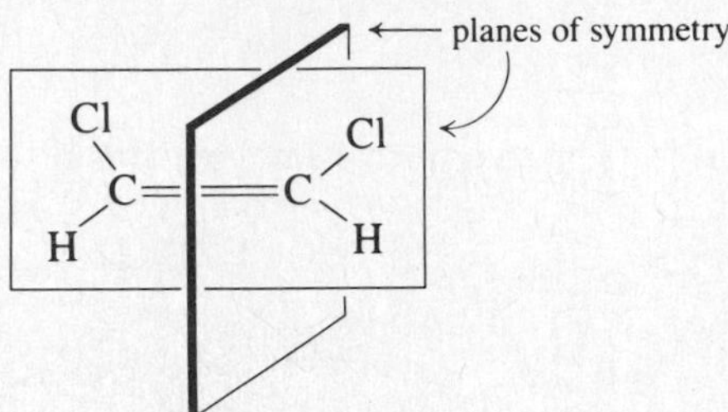

(*c*) There is a plane of symmetry in *cis*-1,2-dichlorocyclopropane which bisects the C-1–C-2 bond and passes through C-3. The molecule is achiral.

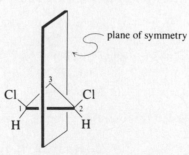

(*d*) *trans*-1,2-Dichlorocyclopropane has neither a plane of symmetry nor a center of symmetry; the only symmetry element present is an axis of symmetry. The molecule is chiral.

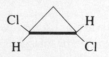

8.3 (*c*) The C-2 atom is a chiral center in 1-bromo-2-methylbutane, as it has four different substituents: H, CH_3, CH_3CH_2, and $BrCH_2$.

$$BrCH_2-\underset{CH_3}{\overset{H}{C}}-CH_2CH_3$$

(*d*) There are no chiral centers in 2-bromo-2-methylbutane.

$$CH_3-\underset{CH_3}{\overset{Br}{C}}-CH_2CH_3$$

8.4 (*c*) Carbon-2 is a chiral center in 1,1,2-trimethylcyclobutane.

CH_3 CH_3 H CH_3

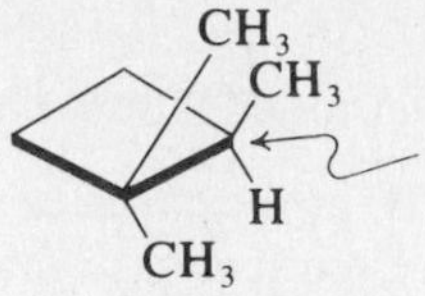

A chiral center; the four substituents to which it is directly bonded [H, CH_3, CH_2, and $C(CH_3)_2$] are all different from one another

(*d*) 1,1,3-Trimethylcyclobutane has no chiral centers.

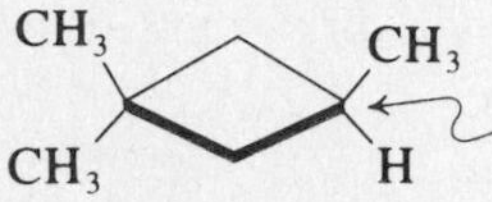

CH_3 CH_3 CH_3 H

Not a chiral center; two of its substituents are the same

8.5 The equation relating specific rotation [α] to observed rotation α is

$$[\alpha] = \frac{100\alpha}{cl}$$

The concentration *c* is expressed in grams per 100 mL and the length *l* of the polarimeter tube in decimeters. Since the problem specifies the concentration as 0.3 g per 15 mL and the path length as 10 cm, the specific rotation [α] is:

$$[\alpha] = \frac{100(-0.78^\circ)}{100\left(\frac{0.3\text{ g}}{15\text{ mL}}\right)\left(\frac{10\text{ cm}}{10\text{ cm/dm}}\right)}$$

$$[\alpha] = -39^\circ$$

8.6 From the previous problem, the specific rotation of natural cholesterol is $[\alpha] = -39^\circ$. The mixture of natural (−)-cholesterol and synthetic (+)-cholesterol specified in this problem has a specific rotation [α] of -13°. It is 33.3 percent optically pure.

$$\text{Optical purity} = \%\ (-)\text{-cholesterol} - \%\ (+)\text{-cholesterol}$$

$$33.3\% = \%\ (-)\text{-cholesterol} - [100 - \%\ (-)\text{-cholesterol}]$$

$$133.3\% = 2\ [\%\ (-)\text{-cholesterol}]$$

$$66.7\% = \%\ (-)\text{-cholesterol}$$

The mixture is two-thirds natural (−)-cholesterol and one-third synthetic (+)-cholesterol.

8.7 (*b*) The solution to this problem is exactly analogous to the sample solution given in the text to (*a*).

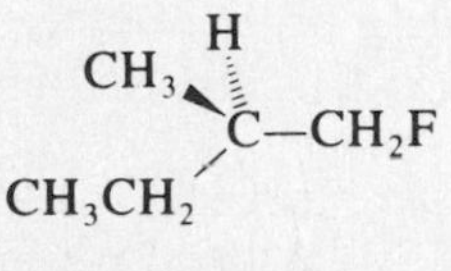

H CH_3 C—CH_2F CH_3CH_2 (+)-1-Fluoro-2-methylbutane

Order of precedence: $CH_2F > CH_3CH_2 > CH_3 > H$

The lowest-ranked substituent (H) at the chiral center points away from the reader so the molecule is oriented properly as drawn. The three higher-ranked substituents trace a clockwise path from CH_2F to CH_2CH_3 to CH_3.

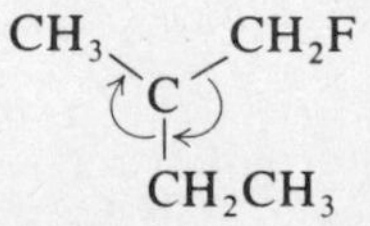

CH_3 CH_2F C CH_2CH_3

The absolute configuration is *R*; the compound is (*R*)-1-fluoro-2-methylbutane.

(c) The highest-ranked substituent at the chiral center of 1-bromo-2-methylbutane is CH_2Br, and the lowest-ranked substituent is H. Of the remaining two, ethyl outranks methyl.

$$\text{Order of precedence: } CH_2Br > CH_2CH_3 > CH_3 > H$$

The lowest-ranking substituent (H) at the chiral center is directed toward the reader and therefore the molecule needs to be reoriented so that H points in the opposite direction.

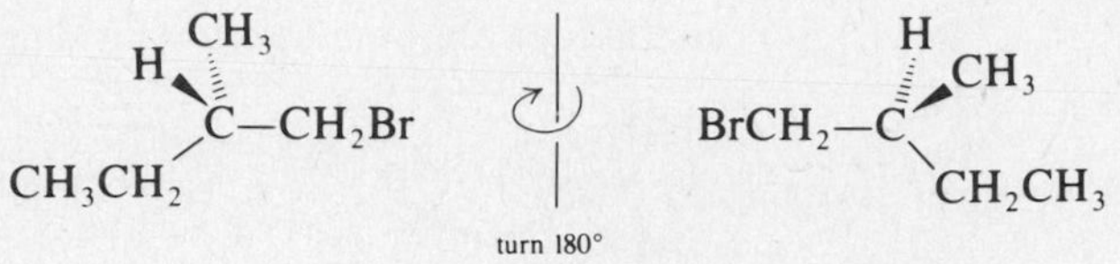

turn 180°

(+)-1-Bromo-2-methylbutane

The three highest-ranking substituents trace a counterclockwise path when the lowest-ranked substituent is held away from the reader.

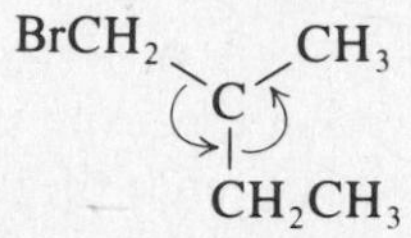

The absolute configuration is *S*, and thus the compound is (*S*)-1-bromo-2-methylbutane.

(d) The highest-ranked substituent at the chiral center of 3-buten-2-ol is the hydroxyl group, and the lowest-ranked substituent is H. Of the remaining two, vinyl outranks methyl.

$$\text{Order of precedence: } HO > CH_2{=}CH > CH_3 > H$$

The lowest-ranking substituent (H) at the chiral center is directed away from the reader. We see that the order of decreasing precedence appears in counterclockwise manner.

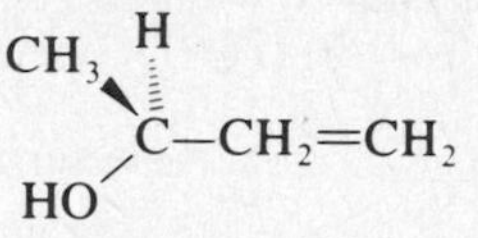

(+)-3-Buten-2-ol

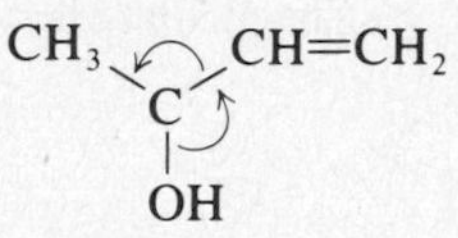

The absolute configuration is *S*, and the compound is (*S*)-3-buten-2-ol.

8.8 (b) The chiral center is the one which bears the methyl group. Its substituents are:

$$-CF_2CH_2 > -CH_2CF_2 > CH_3 > H$$

Highest priority — Lowest priority

When the lowest-priority substituent points away from the reader, the remaining three must appear in descending order of precedence in a counterclockwise fashion in the *S* enantiomer. Therefore, (*S*)-1,1-difluoro-2-methylcyclopropane is

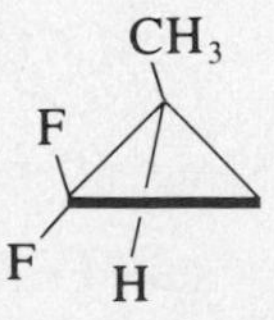

8.9 The reaction which leads to 1,4-dichloro-2-methylbutane from (*S*)-(+)-1-chloro-2-methylbutane is

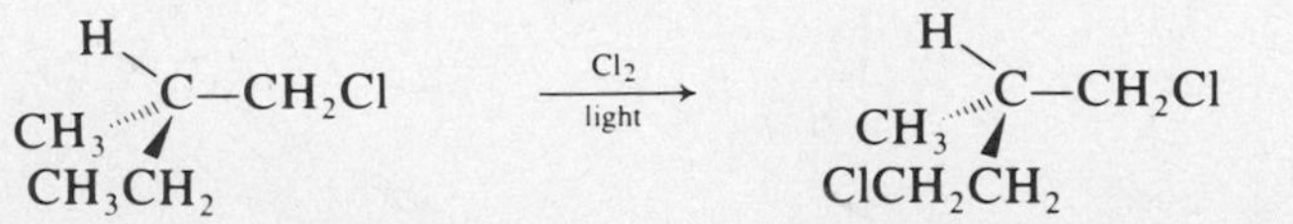

$$\xrightarrow[\text{light}]{Cl_2}$$

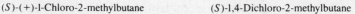

(*S*)-(+)-1-Chloro-2-methylbutane — (*S*)-1,4-Dichloro-2-methylbutane

The reaction does not involve any of the bonds to the chiral center, so the arrangement of the various groups in the product must be the same as in the starting material. Its configuration is determined to be *S* by the Cahn-Ingold-Prelog system based on the order of substituent priorities:

$$\underset{\text{Highest priority}}{CH_2Cl} > CH_2CH_2Cl > CH_3 > \underset{\text{Lowest priority}}{H}$$

Both the starting material and the product are chiral. Since the starting material was a single enantiomer and no bonds were made or broken at its chiral center during the reaction the product is optically active.

8.10 There are four stereoisomeric forms of 3-amino-2-butanol. They are:

(2*R*, 3*R*) and its enantiomer (2*S*, 3*S*)
(2*R*, 3*S*) and its enantiomer (2*S*, 3*R*)

In the text we are told that the (2*R*, 3*R*) stereoisomer is a liquid. Its enantiomer (2*S*, 3*S*) has the same physical properties and so must also be a liquid. The text notes that the (2*R*, 3*S*) stereoisomer is a solid (mp 49°C). Therefore, its enantiomer (2*S*, 3*R*) must be the other stereoisomer that is a crystalline solid.

8.11 The anti conformation of *meso*-2,3-butanediol has a center of symmetry, namely, the midpoint of the C-2–C-3 bond.

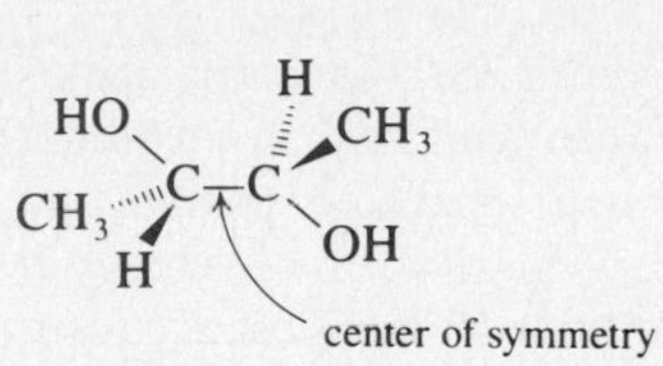

8.12 There is a plane of symmetry in the cis stereoisomer of 1,3-dimethylcyclohexane, so it is an achiral substance—it is a meso form.

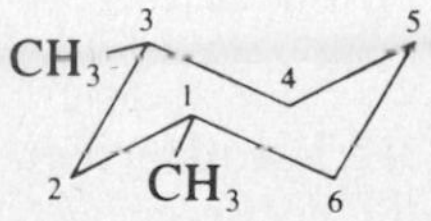

Plane of symmetry passes through C-2 and C-5 and bisects the ring

The trans stereoisomer is chiral. It is not a meso form.

8.13 A molecule with three chiral centers has 2^3, or 8, stereoisomers. The eight combinations of *R* and *S* chiral centers are:

	Chiral center 1 2 3		*Chiral center* 1 2 3
Isomer 1	*R R R*	Isomer 5	*S S S*
Isomer 2	*R R S*	Isomer 6	*S S R*
Isomer 3	*R S R*	Isomer 7	*S R S*
Isomer 4	*S R R*	Isomer 8	*R S S*

8.14 2-Ketohexoses have three chiral centers. They are marked with asterisks in the structural formula.

$$\mathrm{HOCH_2\overset{\overset{O}{\|}}{C}\underset{OH}{\overset{*}{C}H}\underset{OH}{\overset{*}{C}H}\underset{OH}{\overset{*}{C}H}CH_2OH}$$

No meso forms are possible, so there are a total of 2^3, or 8, stereoisomeric 2-ketohexoses.

8.15 Syn addition of bromine to (*Z*)-2-butene would lead to the meso dibromide.

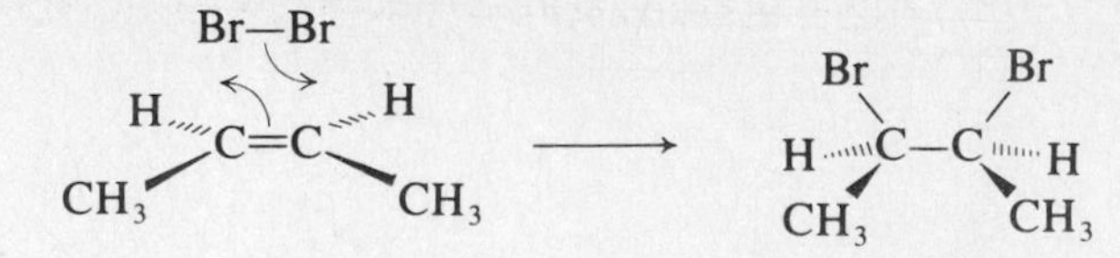

(*Z*)-2-Butene → *meso*-2,3-Dibromobutane

This is not the pathway that is observed, of course. Addition of bromine to an alkene is an anti addition process.

8.16 The tartaric acids incorporate two equivalently substituted chiral centers. (+)-Tartaric acid, as noted in the text, is the 2*R*, 3*R* stereoisomer. There will be two additional stereoisomers, the enantiomeric (−)-tartaric acid (2*S*, 3*S*) and an optically inactive meso form.

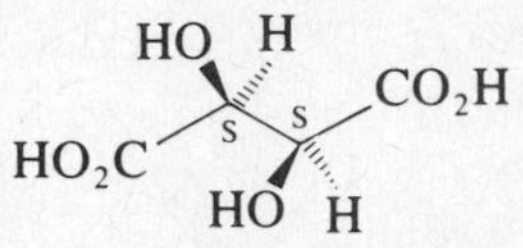

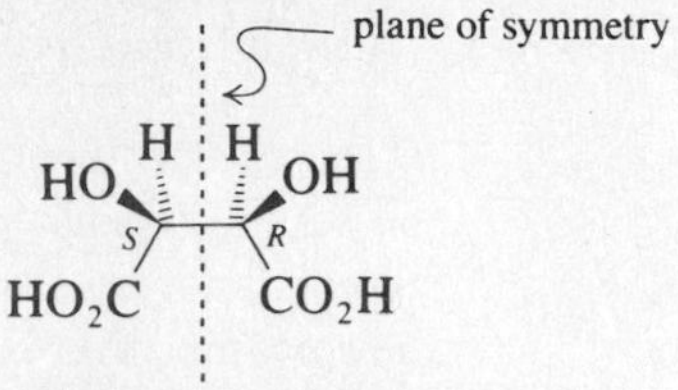

plane of symmetry

(2*S*, 3*S*)-Tartaric acid, mp 170°C, $[\alpha]_D$ −12°

meso-Tartaric acid, mp 140°C, optically inactive

Interestingly, (−)-tartaric acid is itself a natural product present in the leaves of bauhinia, a Central African bush, from which it can be extracted with hot water.

8.17 No. Pasteur separated an optically inactive racemic mixture into two optically active enantiomers. A meso form is achiral, is identical to its mirror image, and is incapable of being separated into optically active forms.

8.18 The more soluble salt must have the opposite configuration at the chiral center of 1-phenylethylamine, i.e., the *S* configuration. The malic acid used in the resolution is a single enantiomer, *S*. Therefore the more soluble salt in this particular case is (*S*)-1-phenylethylammonium (*S*)-malate.

8.19 (*a*) Carbon-2 is a chiral center in 3-chloro-1,2-propanediol. Carbon-2 has two equivalent substituents in 2-chloro-1,3-propanediol, where it is not a chiral center.

$ClCH_2\overset{*}{C}H(OH)CH_2OH$ $HOCH_2CH(Cl)CH_2OH$

3-Chloro-1,2-propanediol, chiral   2-Chloro-1,3-propanediol, achiral

(*b*) The primary bromide is achiral; the secondary bromide contains a chiral center and is chiral.

$CH_3CH{=}CHCH_2Br$ $CH_3\overset{*}{C}H(Br)CH{=}CH_2$

Achiral   Chiral

(*c*) The left-hand structure is achiral, as it has a plane of symmetry and contains no chiral centers. The other structure is chiral; it contains two nonequivalent chiral centers.

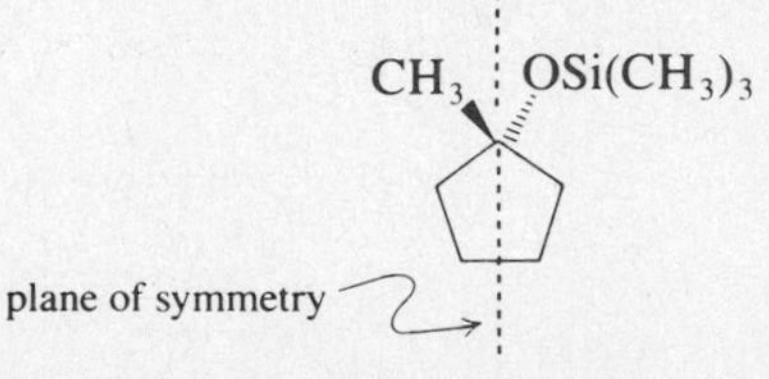

CH_3 $OSi(CH_3)_3$ CH_3 $OSi(CH_3)_3$

plane of symmetry

Achiral 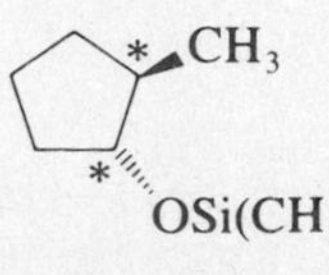   Chiral

(*d*) 1,1,2-Trimethylcyclopropane is chiral; C-2 is a chiral center. *trans*-1,2,3-Trimethylcyclopropane is achiral; it has a plane of symmetry.

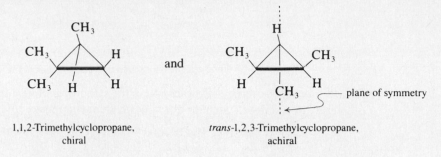

and

plane of symmetry

1,1,2-Trimethylcyclopropane, chiral

trans-1,2,3-Trimethylcyclopropane, achiral

(*e*) The compound on the left is chiral; it has two different chiral centers. The compound on the right is achiral; it has a plane of symmetry.

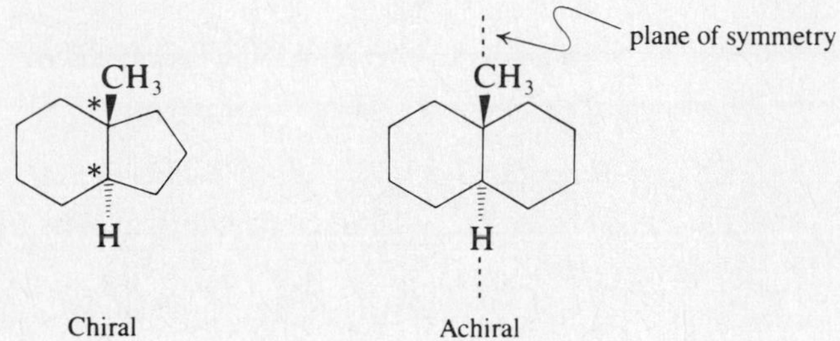

Chiral Achiral

(*f*) The structure on the left is chiral; the other is an achiral meso form.

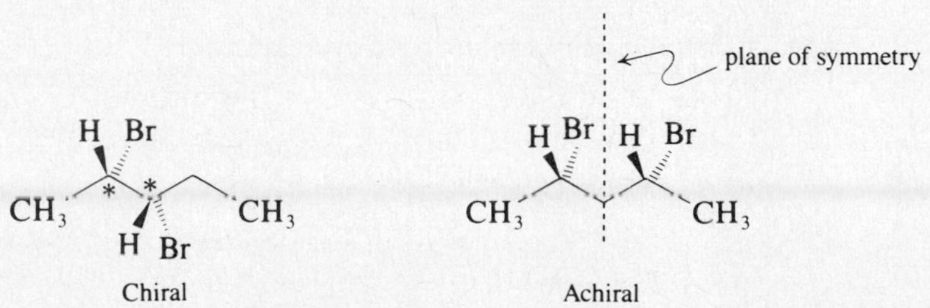

Chiral Achiral

(*g*) The first structure is achiral; it has a plane of symmetry.

Plane of symmetry passes through C-1, C-4, and C-7

The second structure cannot be superposed on its mirror image; it is chiral.

Reference structure Mirror image Reoriented mirror image

(*h*) The first structure is chiral; it is not superposable on its mirror image.

Reference structure Mirror image Reoriented mirror image

The second structure is achiral; it has a plane of symmetry.

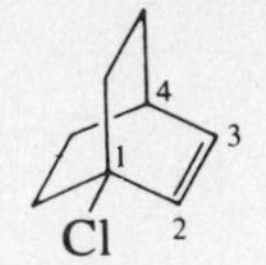

Cl 2 1 4 3 is equivalent to 3 2 4 1 Cl

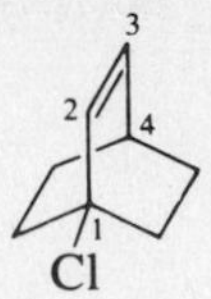

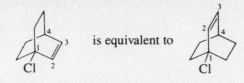

Achiral; plane of symmetry passes through C-1, C-2, C-3, and C-4

8.20 In an earlier exercise (Problem 4.19) the structures of all the isomeric $C_5H_{12}O$ alcohols were presented. Those which do not contain an asymmetric center and which are *achiral* are:

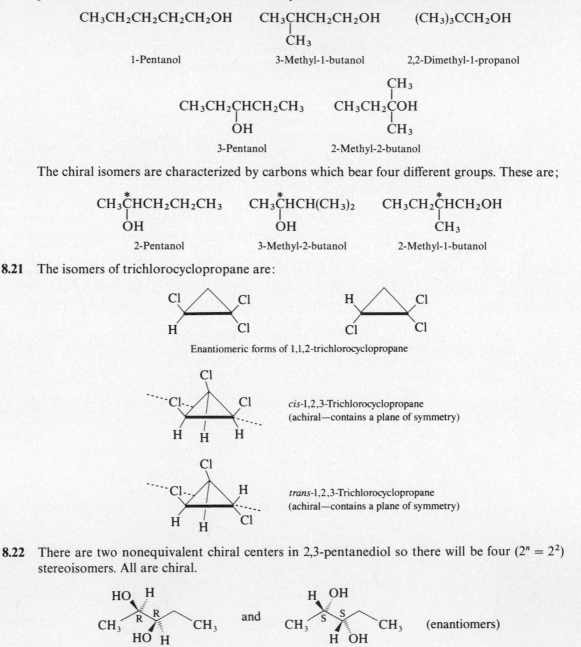

$CH_3CH_2CH_2CH_2CH_2OH$ — 1-Pentanol

$CH_3CHCH_2CH_2OH$ (with CH_3 on C-3) — 3-Methyl-1-butanol

$(CH_3)_3CCH_2OH$ — 2,2-Dimethyl-1-propanol

$CH_3CH_2CHCH_2CH_3$ (with OH on C-3) — 3-Pentanol

CH_3CH_2COH (with two CH_3 on C-2) — 2-Methyl-2-butanol

The chiral isomers are characterized by carbons which bear four different groups. These are;

$CH_3\overset{*}{C}HCH_2CH_2CH_3$ (with OH) — 2-Pentanol

$CH_3\overset{*}{C}HCH(CH_3)_2$ (with OH) — 3-Methyl-2-butanol

$CH_3CH_2\overset{*}{C}HCH_2OH$ (with CH_3) — 2-Methyl-1-butanol

8.21 The isomers of trichlorocyclopropane are:

Enantiomeric forms of 1,1,2-trichlorocyclopropane

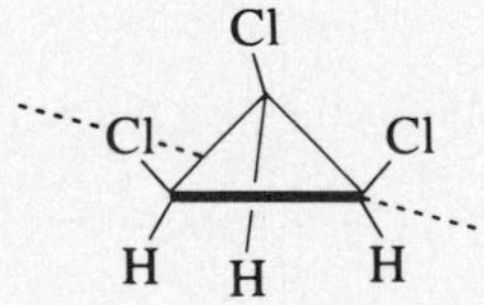

cis-1,2,3-Trichlorocyclopropane (achiral—contains a plane of symmetry)

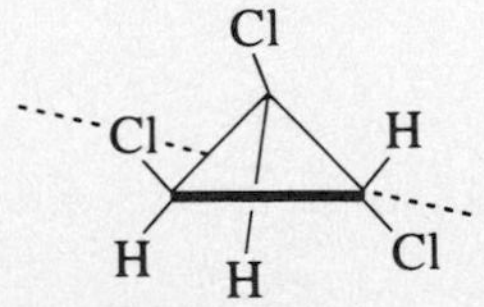

trans-1,2,3-Trichlorocyclopropane (achiral—contains a plane of symmetry)

8.22 There are two nonequivalent chiral centers in 2,3-pentanediol so there will be four ($2^n = 2^2$) stereoisomers. All are chiral.

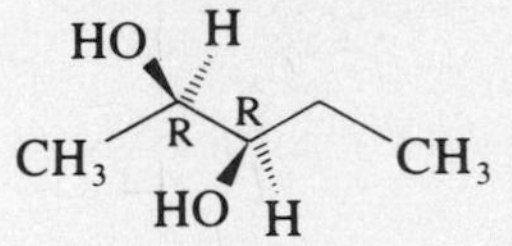

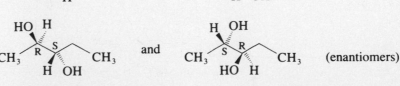

(2R,3R) and (2S,3S) (enantiomers)

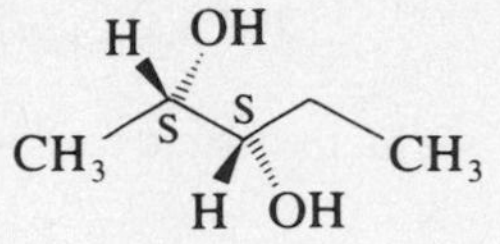

(2R,3S) and (2S,3R) (enantiomers)

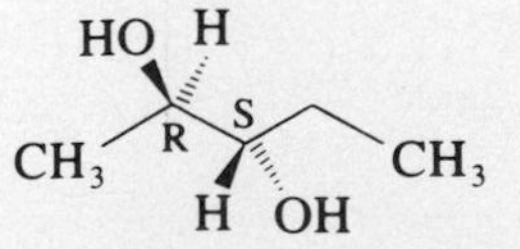

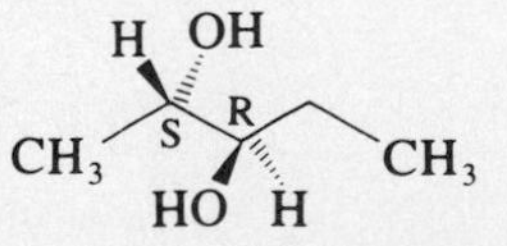

The two chiral centers in 2,4-pentanediol are equivalently substituted, so we need to be aware of the fact that there will be meso forms.

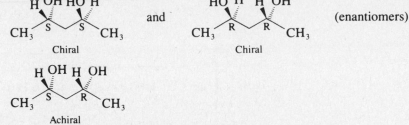

H OH HO H / CH_3 S S CH_3 — Chiral — and — HO H H OH / CH_3 R R CH_3 — Chiral (enantiomers)

H OH H OH / CH_3 S R CH_3

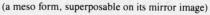

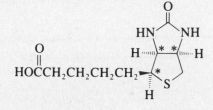

Achiral
(a meso form, superposable on its mirror image)

8.23 (*a*) There are three chiral centers in biotin. Naturally occurring (+)-biotin is the stereoisomer shown.

O; HN NH; H * * H; $HOCCH_2CH_2CH_2CH_2$ * S; H

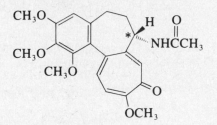

(*b*) Colchicine has only one chiral center.

CH_3O; CH_3O; CH_3O; H; * $NHCCH_3$ (O); O; OCH_3

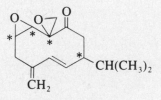

(*c*) There are four chiral centers in periplanone B.

O O O; * * *; * $CH(CH_3)_2$; CH_2

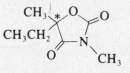

(*d*) Only one chiral center is present in paramethadione.

CH_3 * O O; CH_3CH_2; N; O CH_3

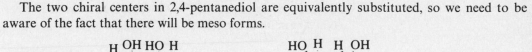

(*e*) Calciferol has six chiral centers.

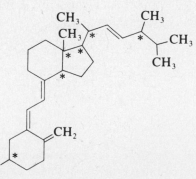

CH_3 CH_3; CH_3 * * CH_3; * *; CH_3; *; CH_2; *; HO

(*f*) There are six chiral centers in *S*-adenosylmethionine. Five of the chiral atoms are carbon; the sixth is the sulfur atom.

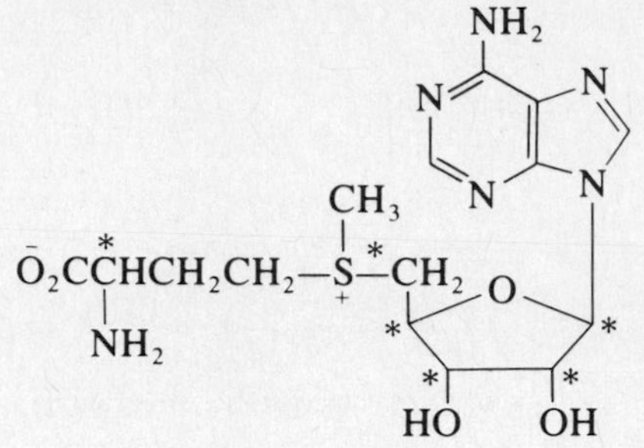

$$\bar{O}_2C\overset{*}{C}HCH_2CH_2-\overset{+}{\overset{*}{S}}(CH_3)-CH_2-\text{ribose}(\overset{*}{C}\times 4)-\text{adenine}$$

8.24 (*a*) (−)-2-Octanol has the *R* configuration at C-2. The order of substituent preference is:

$$HO > CH_2CH_2 > CH_3 > H$$

The molecule is oriented so that the lowest-ranking substituent is directed away from the reader and the order of decreasing precedence is clockwise.

(*b*) In order of decreasing sequence rule preference, the four substituents at the chiral center of monosodium L-glutamate are:

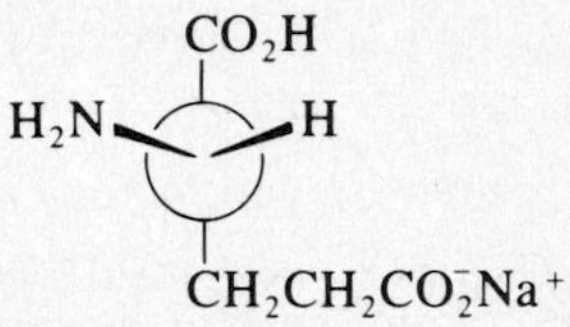

$$NH_2 > CO_2H > CH_2 > H$$

When the molecule is oriented so that the lowest-ranking substituent (hydrogen) is directed away from the reader the other three substituents are arranged as shown.

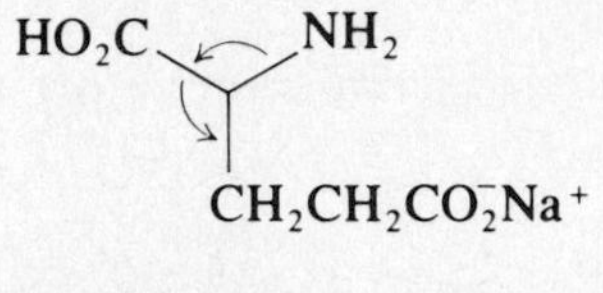

The order of decreasing ranking is counterclockwise; the absolute configuration is *S*.

(*c*) (+)-2-Phenylbutanoic acid is already shown with its lowest-ranking substituent (H) pointing away from the reader.

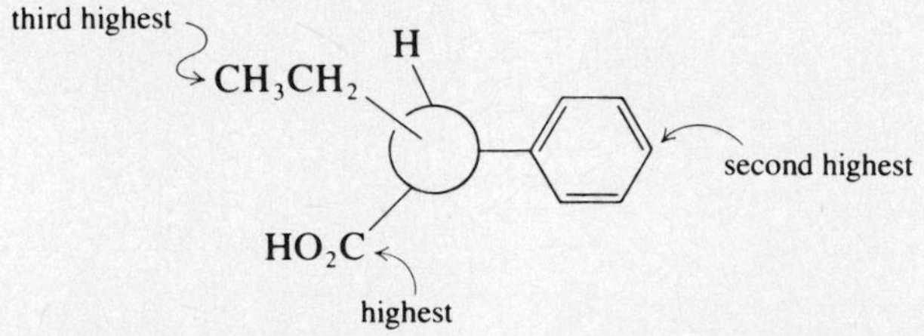

third highest

second highest

highest

The order of decreasing ranking of the three highest-priority substituents appears counterclockwise in this orientation. The absolute configuration is *S*.

(*d*) The NH_3^+ group is the highest-ranking substituent in 5-hydroxytryptophan.

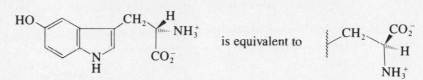

$$NH_3^+ > CO_2^- > CH_2 > H$$

Highest ranking — Lowest ranking

HO–(indole)–CH_2–C(H)(NH_3^+)(CO_2^-) is equivalent to CH_2–C(CO_2^-)(H)(NH_3^+)

The absolute configuration is *S*.

8.25 (*a*) The two compounds are constitutional isomers. Their IUPAC names clearly reflect this difference.

CH_3CHCH_2Br (with OH on C-2) and CH_3CHCH_2OH (with Br on C-2)

1-Bromo-2-propanol — 2-Bromo-1-propanol

(*b*) The two structures have the same constitution. Test them for superposability. To do this we need to place them in comparable orientations.

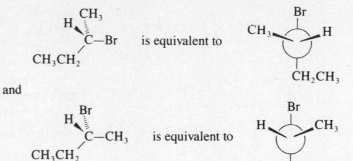

CH_3, H, C–Br, CH_3CH_2 is equivalent to Br, CH_3, H, CH_2CH_3

and

Br, H, C–CH_3, CH_3CH_2 is equivalent to Br, H, CH_3, CH_2CH_3

The two are nonsuperposable mirror images of each other. They are enantiomers.

To check this conclusion, work out the absolute configuration of each using the Cahn-Ingold-Prelog system.

CH_3, H, C–Br, CH_3CH_2 — Br, H, C–CH_3, CH_3CH_2

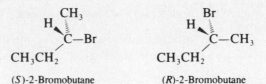

(*S*)-2-Bromobutane — (*R*)-2-Bromobutane

(*c*) Again, place the structures in comparable orientations and examine them for superposability.

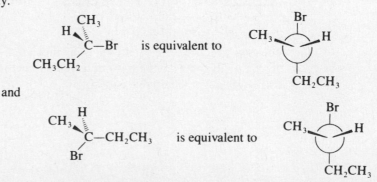

CH_3, H, C–Br, CH_3CH_2 is equivalent to Br, CH_3, H, CH_2CH_3

and

H, CH_3, C–CH_2CH_3, Br is equivalent to Br, CH_3, H, CH_2CH_3

The two structures represent the same compound, as they are superposable. (As a check, notice that both represent the *S* configuration.)

(*d*) If we reorient the first structure

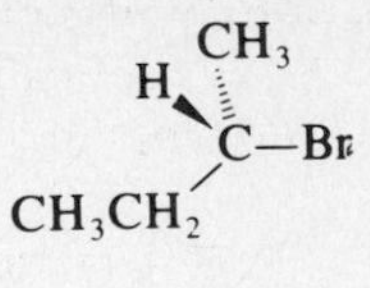

CH_3, H, C—Br, CH_3CH_2 becomes CH_2CH_3, CH_3, Br, H

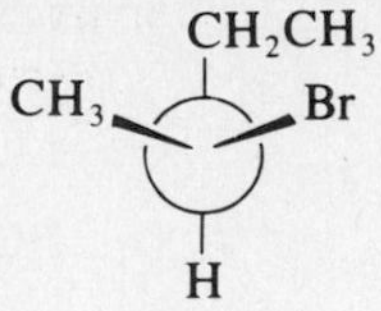

which is the enantiomer of

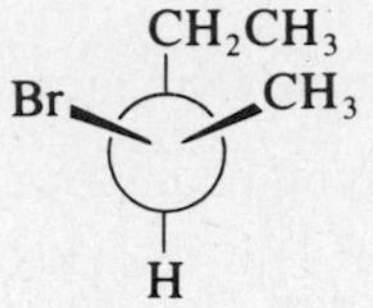

CH_2CH_3, Br, CH_3, H

As a check, the first structure is seen to have the *S* configuration, while the second has the *R* configuration.

(*e*) As drawn, the two structures are mirror images of each other; however, they represent an achiral molecule. The two structures are superposable mirror images and are not stereoisomers but identical.

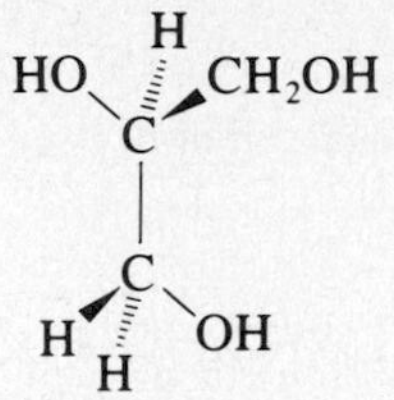

HO, H, CH_2OH, C, C, H, H, OH and $HOCH_2$, H, OH, C, C, HO, H, H are both $HOCH_2CHCH_2OH$ (OH)

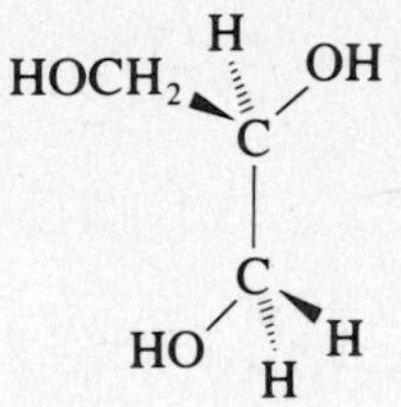

(*f*) The two structures—one cis, the other trans—are stereoisomers which are not mirror images; they are diastereomers.

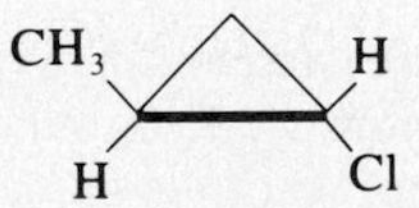

CH_3, H, H, Cl

trans-1-Chloro-2-methylcyclopropane

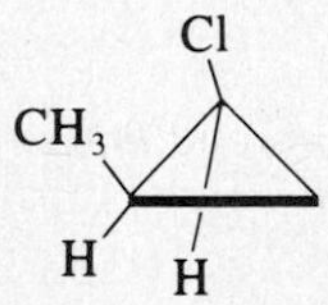

Cl, CH_3, H, H

cis-1-Chloro-2-methylcyclopropane

(*g*) Both structures are *cis*-1-chloro-2-methylcyclopropanes. They are related as an object and its nonsuperposable mirror image; they are enantiomers.

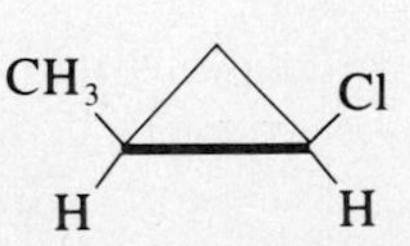

CH_3, Cl, H, H and Cl, CH_3, H, H which is equivalent to Cl, CH_3, H, H

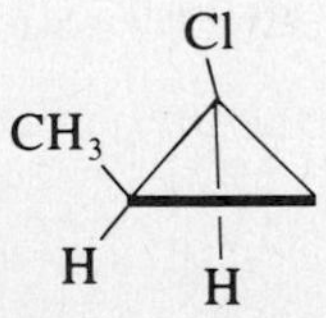

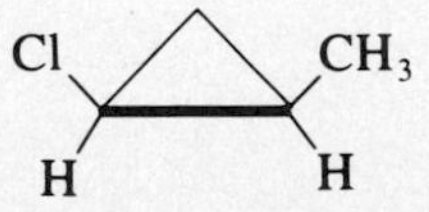

(*h*) The two structures are enantiomers, as they are nonsuperposable mirror images. Checking their absolute configurations reveals one to be *R*, the other *S*. Both have the *E* configuration at the double bond.

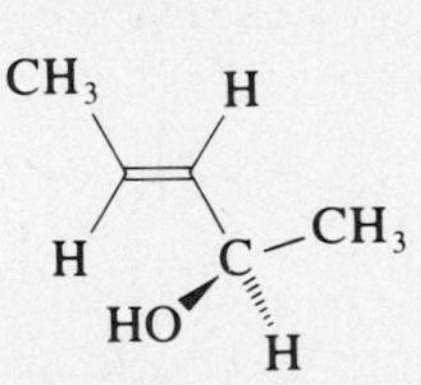

CH_3, H, H, C, CH_3, HO, H

(2*R*, 3*E*)-3-Penten-2-ol

CH_3, H, H, C, CH_3, H, OH

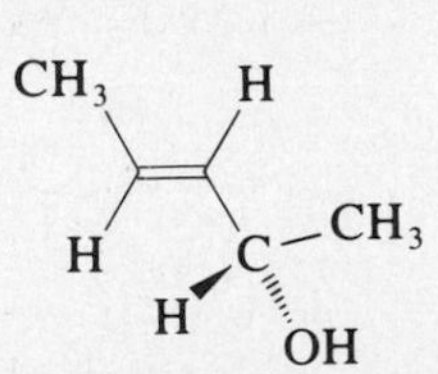

(2*S*, 3*E*)-3-Penten-2-ol

(*i*) These two structures are identical; both have the *E* configuration at the double bond and the *R* configuration at the chiral center.

Alternatively, we can show their superposability by rotating the second structure 180° about an axis passing through the doubly bonded carbons.

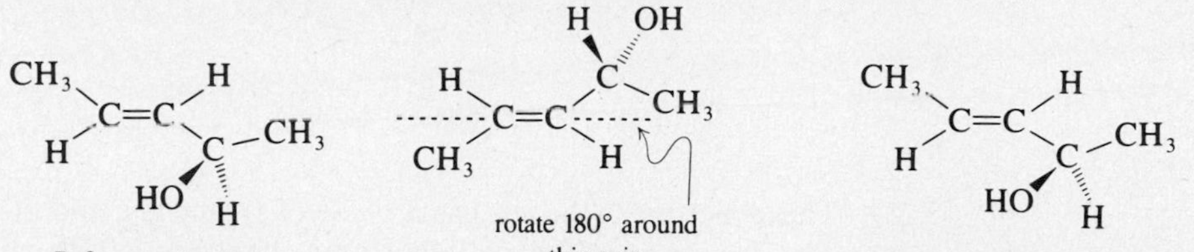

Reference structure — rotate 180° around this axis — Identical to reference structure

(*j*) One structure has a cis double bond, the other a trans double bond; therefore, the two are diastereomers.

(2*R*, 3*E*)-3-Penten-2-ol — (2*S*, 3*Z*)-3-Penten-2-ol

(*k*) Here it will be helpful to reorient the second structure so that it may be more readily compared with the first.

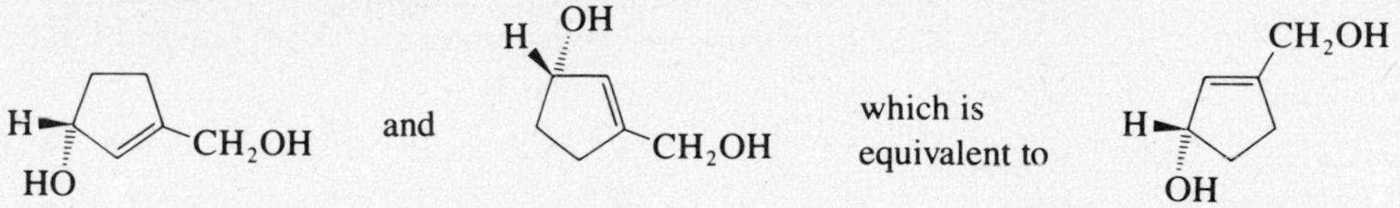

Reference structure — and — which is equivalent to — Enantiomer of reference structure

The two compounds are enantiomers.

Examining their absolute configurations confirms the enantiomeric nature of the two compounds.

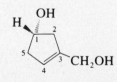

R — *S*

(*l*) These two compounds differ in their atomic connections; they are constitutional isomers.

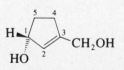

3-Hydroxymethyl-2-cyclopenten-1-ol — 3-Hydroxymethyl-3-cyclopenten-1-ol

(*m*) In order to better compare these two structures, place them both in the same format.

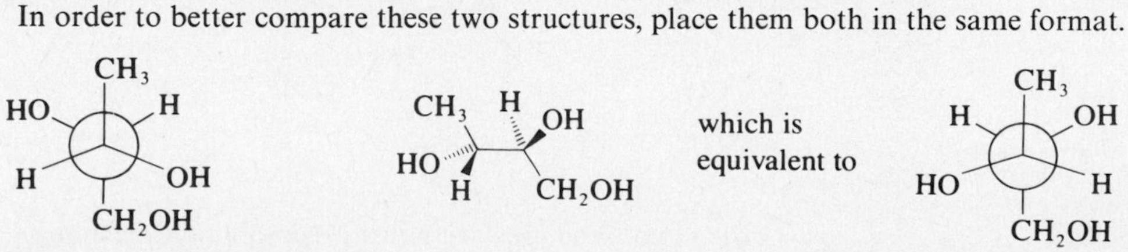

which is equivalent to

The two are enantiomers.

(*n*) Since *cis*-1,3-dimethylcyclopentane has a plane of symmetry, it is achiral and cannot have an enantiomer. The two structures are identical.

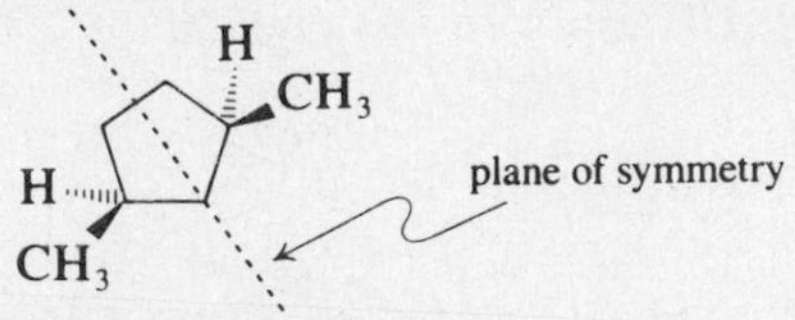

H CH$_3$ H CH$_3$ plane of symmetry

(*o*) The structure at the left has a plane of symmetry, and all three of its substituents are cis. The structure at the right is chiral, with one substituent trans to the other two. The compounds are stereoisomers but not enantiomers; therefore they are diastereomers.

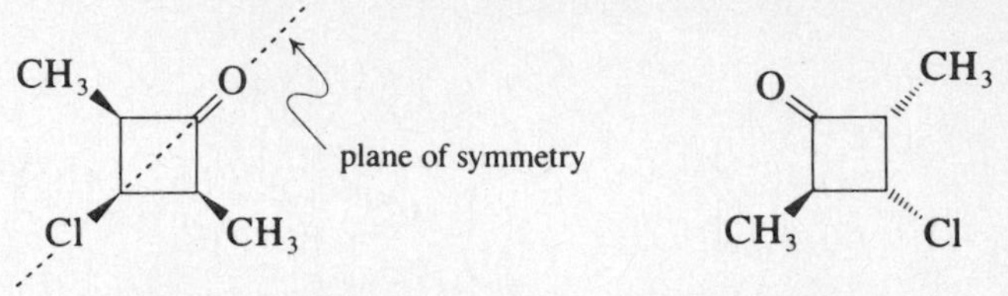

CH$_3$ O plane of symmetry Cl CH$_3$ O CH$_3$ CH$_3$ Cl

Achiral Chiral

(*p*) These structures are diastereomers, i.e., stereoisomers which are not mirror images. They have the same configuration at C-3 but opposite configurations at C-2.

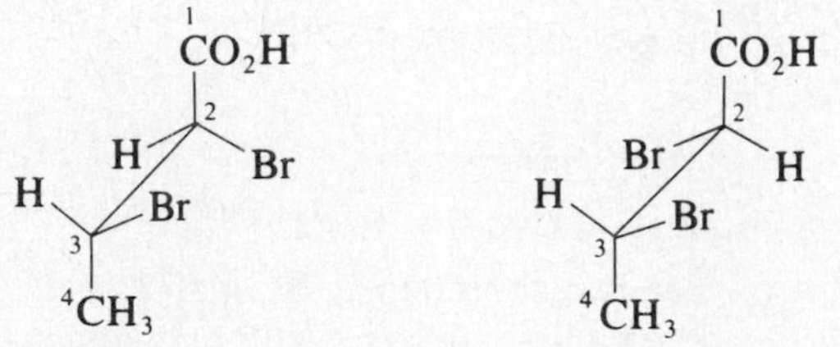

2*R*, 3*R* 2*S*, 3*R*

(*q*) Reorientation is required so that we may better compare these two structures.

The reorientation operation consists of a 180° rotation about an axis passing through the midpoint of the C-2–C-3 bond.

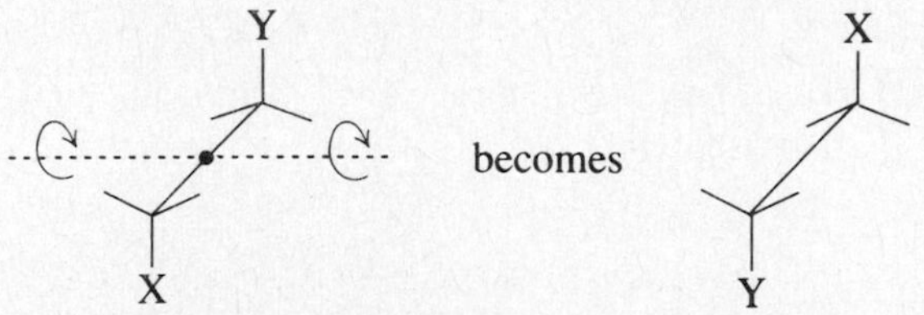

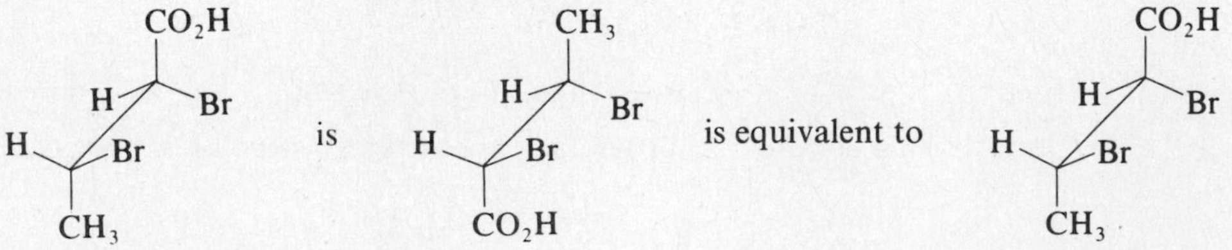

becomes

is is equivalent to

Reference structure Identical to reference structure

(*r*) These two structures are nonsuperposable mirror images of a molecule with two nonequivalent chiral centers; they are enantiomers.

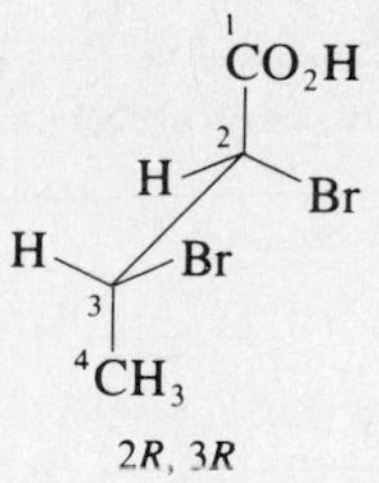

2R, 3R

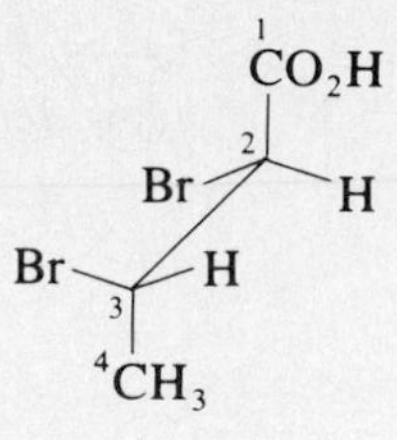

2S, 3S

(s) The two structures are stereoisomers which are not enantiomers; they are diastereomers.

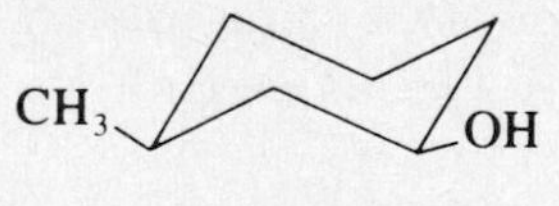

cis-3-Methylcyclohexanol

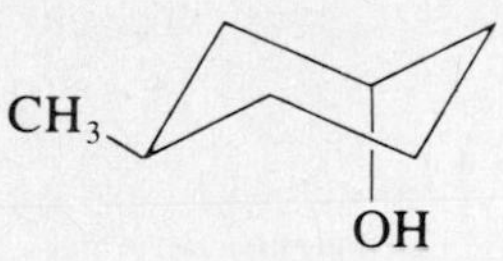

trans-3-Methylcyclohexanol

(t) These two structures, *cis*- and *trans*-4-*tert*-butylcyclohexyl iodide, are diastereomers.

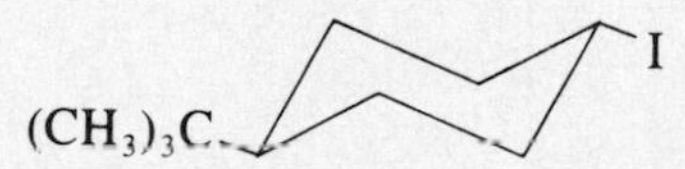

Trans

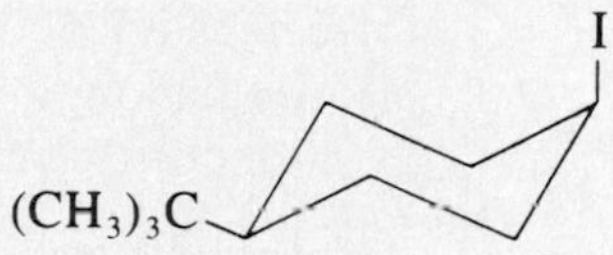

Cis

(u) The two structures are nonsuperposable mirror images; they are enantiomers.

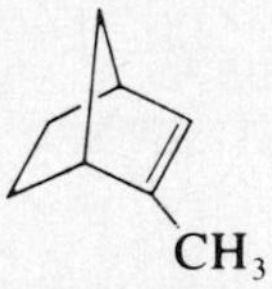

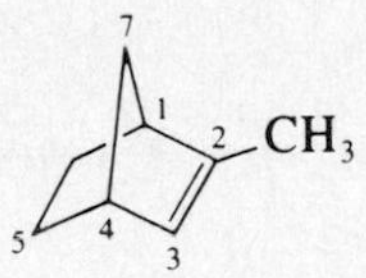

is equivalent to

Reference structure

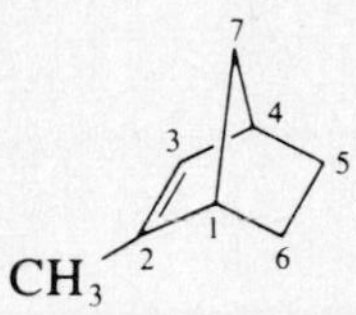

Enantiomer of reference structure

(v) The two structures are identical.

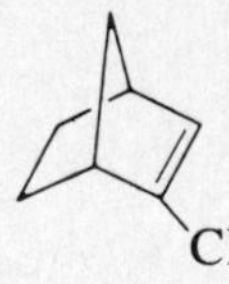

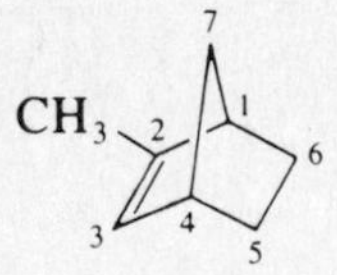

is equivalent to

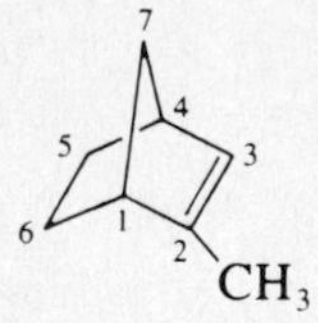

Reference structure

Identical to reference structure

(w) As represented, the two structures are mirror images of each other, but because the molecule is achiral (it has a plane of symmetry), the two must be superposable. They represent the identical compound.

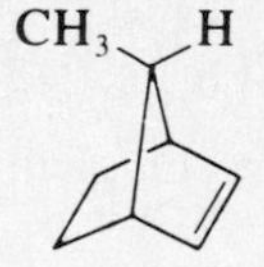

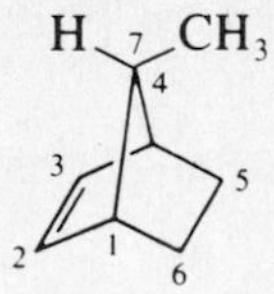

is equivalent to

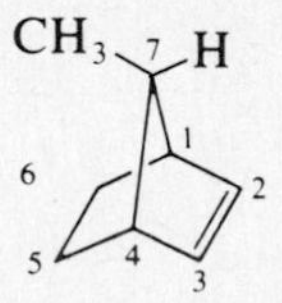

Reference structure

Identical to reference structure

The plane of symmetry passes through C-7 and bisects the C-2–C-3 bond and the C-5–C-6 bond.

(*x*) The structures are stereoisomers but not enantiomers; they are diastereomers. (Both are achiral and so cannot have enantiomers.)

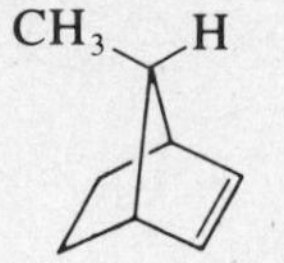

CH_3 H and H CH_3 are stereoisomers but not mirror images

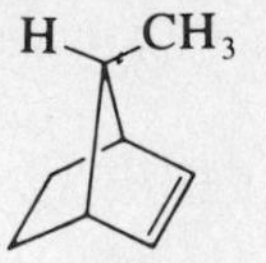

Achiral Achiral

8.26 In order to write a stereochemically accurate representation of ectocarpene, it is best to begin with the configuration of the chiral center, which we are told is *S*.

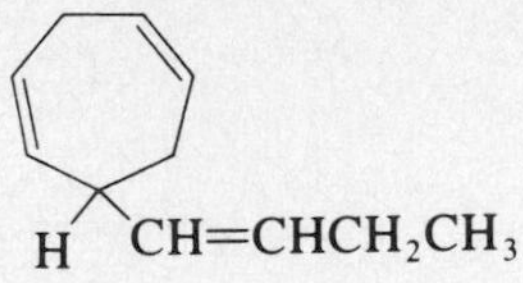

H $CH{=}CHCH_2CH_3$

Clearly, hydrogen is the lowest-ranking substituent at the chiral center; among the other three substituents, two are part of the ring and the third is the four-carbon side chain. The priority rankings of these groups are determined by systematically working along the chain.

The substituents

$-CH{=}CHCH_2CH{=}CHCH_2$ (Ring) $-CH{=}CHCH_2CH_3$ (Side chain) $-CH_2CH{=}CHCH_2CH{=}CH$ (Ring)

are considered as if they were

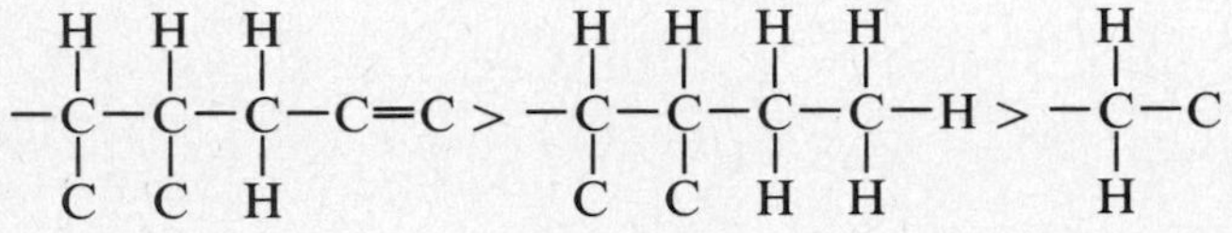

$$-\underset{C}{\overset{H}{C}}-\underset{C}{\overset{H}{C}}-\underset{H}{\overset{H}{C}}-C{=}C > -\underset{C}{\overset{H}{C}}-\underset{C}{\overset{H}{C}}-\underset{H}{\overset{H}{C}}-\underset{H}{\overset{H}{C}}-H > -\underset{H}{\overset{H}{C}}-C$$

(Ring) (Side chain) (Ring)

Orienting the molecule with the hydrogen away from the reader

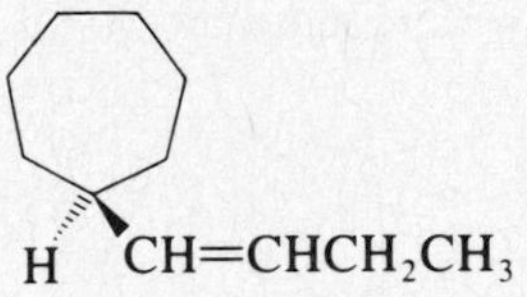

H $CH{=}CHCH_2CH_3$

we place the double bonds in the ring so that the order of decreasing sequence rule precedence is counterclockwise:

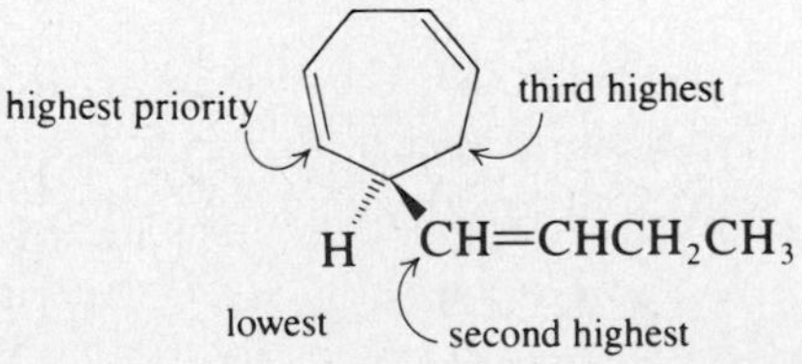

highest priority third highest

H $CH{=}CHCH_2CH_3$

lowest second highest

Finally, since all the double bonds are cis, the complete structure becomes:

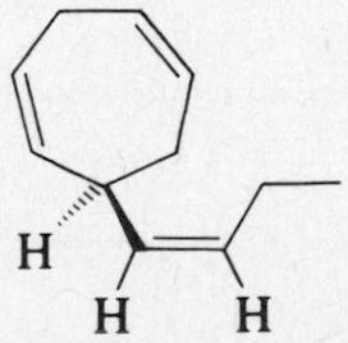

H

H H

8.27 (*a*) Multifidene has two chiral centers and three double bonds. Neither the ring double bond nor the double bond of the vinyl substituent can give rise to stereoisomers, but there is the possibility of *E* and *Z* isomers of the butenyl side chain. Therefore there are eight (2^3) possible stereoisomers. We can rationalize them as:

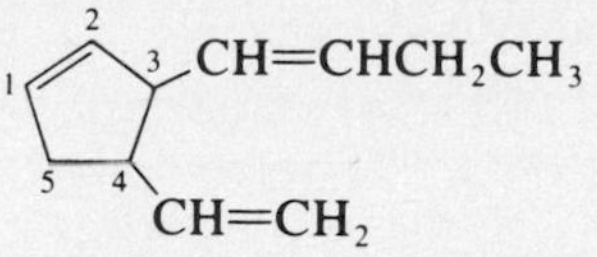

$CH{=}CHCH_2CH_3$ (at C-3), $CH{=}CH_2$ (at C-4) on a cyclopentene ring numbered 1–5

Stereoisomer	C-3	C-4	Butenyl double bond	
1	*R*	*R*	*E*	enantiomers
2	*S*	*S*	*E*	
3	*R*	*R*	*Z*	enantiomers
4	*S*	*S*	*Z*	
5	*R*	*S*	*E*	enantiomers
6	*S*	*R*	*E*	
7	*R*	*S*	*Z*	enantiomers
8	*S*	*R*	*Z*	

(*b*) Given the information that the alkenyl substituents are cis to each other, the number of stereoisomers is reduced by half. Therefore, four stereoisomers are possible.

(*c*) Knowing that the butenyl group has a *Z* double bond reduces the number of possibilities by half. Two stereoisomers are possible.

(*d*) The two stereoisomers are:

and

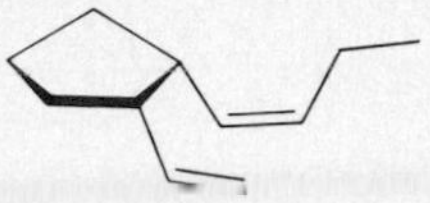

(*e*) These two stereoisomers are enantiomers. They are nonsuperposable mirror images.

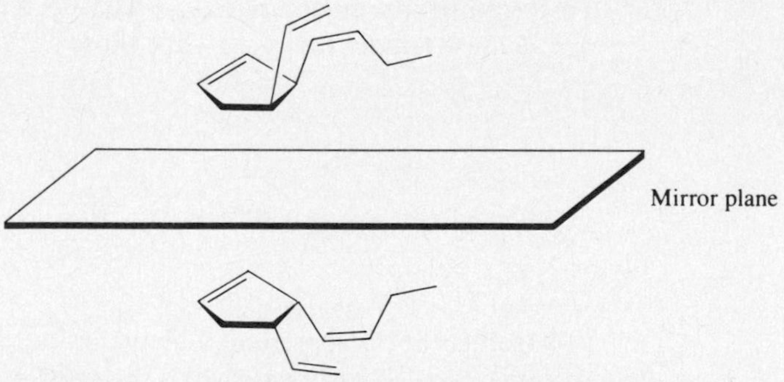

Mirror plane

Naturally occurring (+)-multifidene has the 3*S*, 4*S* configuration.

8.28 In a substance with more than one chiral center, each center is independently specified as *R* or *S*. Streptimidone has two chiral centers and two double bonds. Only the internal double bond is capable of stereoisomerism.

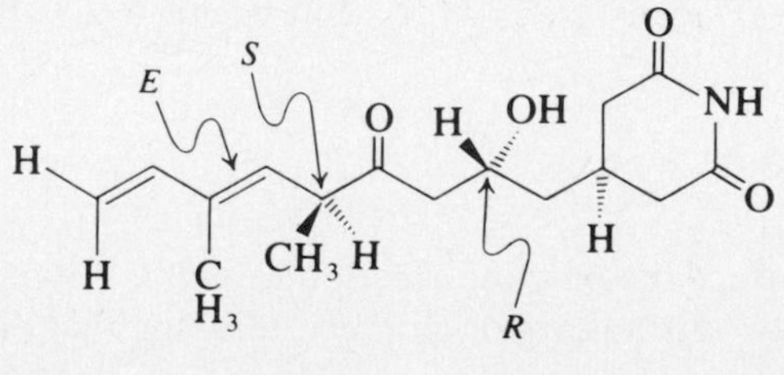

E S O H OH NH O H CH_3 H H C H_3 H R

The three stereochemical variables give rise to eight (2^3) stereoisomers, of which one is streptimidone and a second is the enantiomer of streptimidone. The remaining six stereoisomers are diastereomers of streptimidone.

8.29 (*a*) The first step is to set out the constitution of menthol, which we are told is 2-isopropyl-5-methylcyclohexanol.

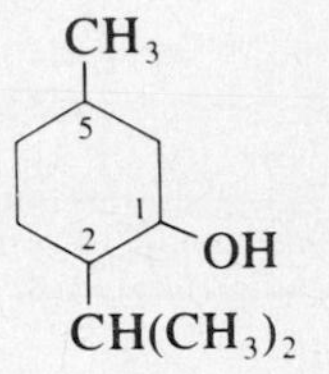

2-Isopropyl-5-methylcyclohexanol

Since the configuration at C-1 is *R* in (−)-menthol, the hydroxyl group must be "up" in our drawing.

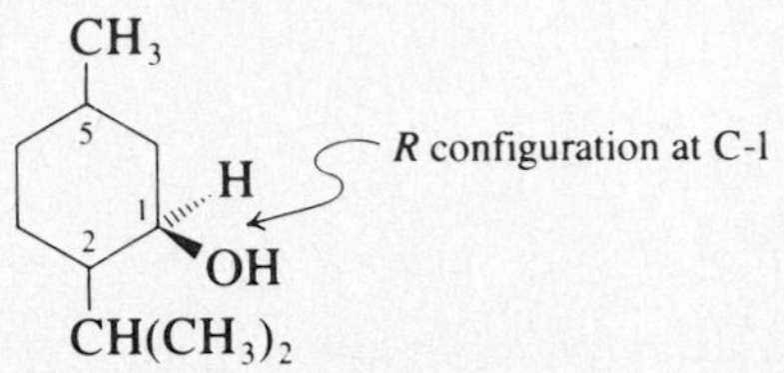

R configuration at C-1

Because menthol is the most stable stereoisomer of this constitution, all three of its substituents must be equatorial. Therefore, we draw the chair form of the preceding structure, which has the hydroxyl group equatorial and up, placing isopropyl and methyl groups so as to preserve the *R* configuration at C-1.

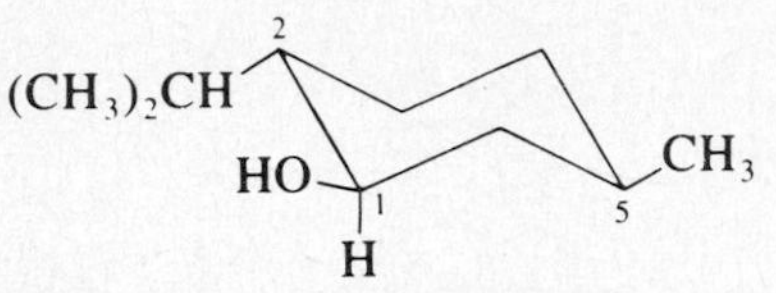

(−)-Menthol

(*b*) To transform the structure of (−)-menthol to that of (+)-isomenthol, the configuration at C-5 must remain the same while those at C-1 and C-2 are inverted.

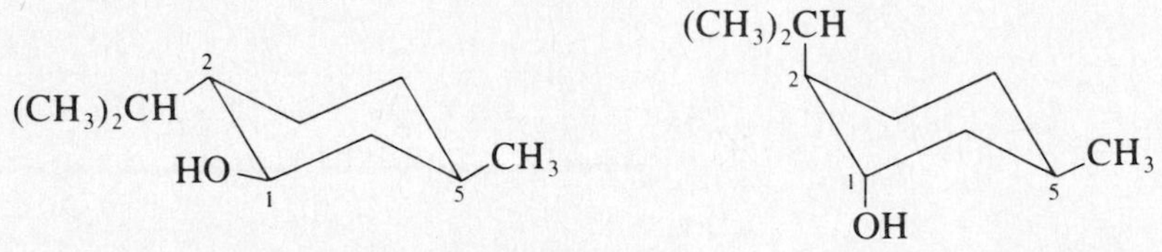

(+)-Isomenthol is represented above in its correct configuration, but the conformation with two axial substituents is not the most stable one. The ring-flipped form will be the preferred conformation of (+)-isomenthol.

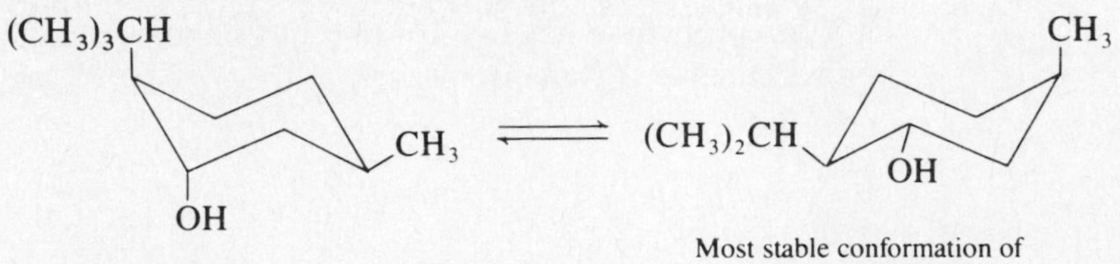

Most stable conformation of (+)-isomenthol

8.30 Since the only information available about the compound is its optical activity, examine the two structures for chirality, recalling that only chiral substances can be optically active.

The structure with the six-membered ring has a plane of symmetry passing through C-1 and C-4. It is achiral and cannot be optically active.

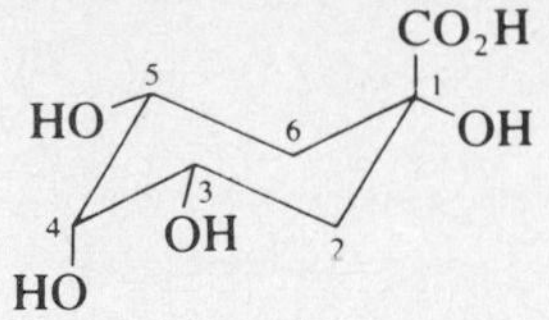

Achiral; $[\alpha]_D$ 0°

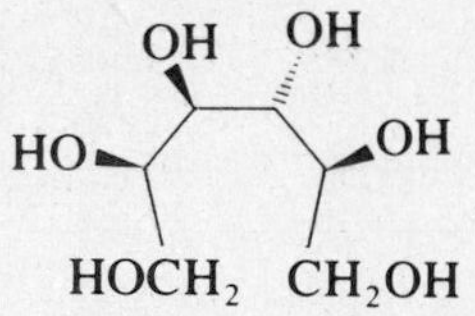

Chiral; can be optically active

The open-chain structure has neither a plane of symmetry nor a center of symmetry; it is not superposable on its mirror image and so is chiral. It can be optically active and is more likely to be the correct choice.

8.31 (*a*) The equation which relates specific rotation $[\alpha]_D$ to observed rotation α is

$$[\alpha]_D = \frac{100\alpha}{cl}$$

where c is concentration in grams per 100 mL and l is path length in decimeters.

$$[\alpha]_D = \frac{100(-5.20°)}{\left(\frac{2.0\text{ g}}{100\text{ mL}}\right)(2\text{ dm})}$$

$$[\alpha]_D = -130°$$

(*b*) The optical purity of the resulting solution is 10/15, or 66.7 percent, since 10 g of optically pure fructose has been mixed with 5 g of racemic fructose. Therefore, the specific rotation will be two-thirds (10/15) of the specific rotation of optically pure fructose

$$[\alpha]_D = \tfrac{2}{3}(-130°) = -87°$$

8.32 (*a*) The reaction of 1-butene with hydrogen iodide is one of electrophilic addition. It follows Markovnikov's rule and yields a racemic mixture of (*R*)- and (*S*)-2-iodobutane.

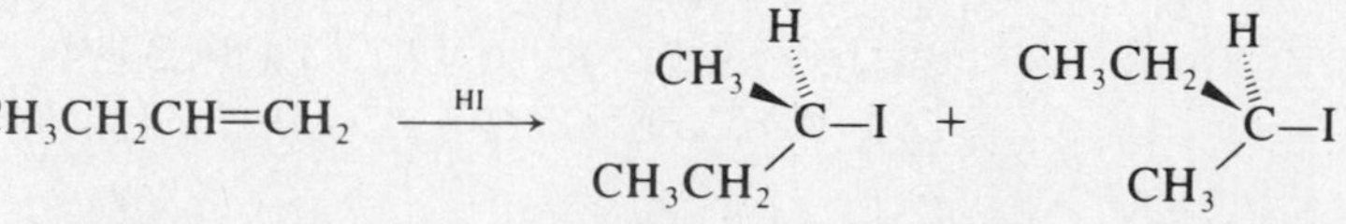

$CH_3CH_2CH{=}CH_2 \xrightarrow{HI}$ (*R*)-2-Iodobutane + (*S*)-2-Iodobutane

1-Butene (*R*)-2-Iodobutane (*S*)-2-Iodobutane

(*b*) This is a free-radical addition of hydrogen bromide, leading to the formation of equal amounts of (*R*)- and (*S*)-2-bromobutane.

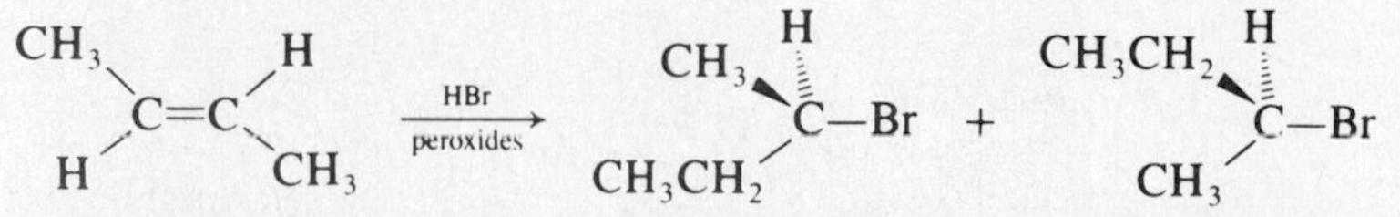

(*E*)-2-Butene $\xrightarrow[\text{peroxides}]{HBr}$ (*R*)-2-Bromobutane + (*S*)-2-Bromobutane

(*c*) Bromine adds anti to carbon-carbon double bonds to give vicinal dibromides.

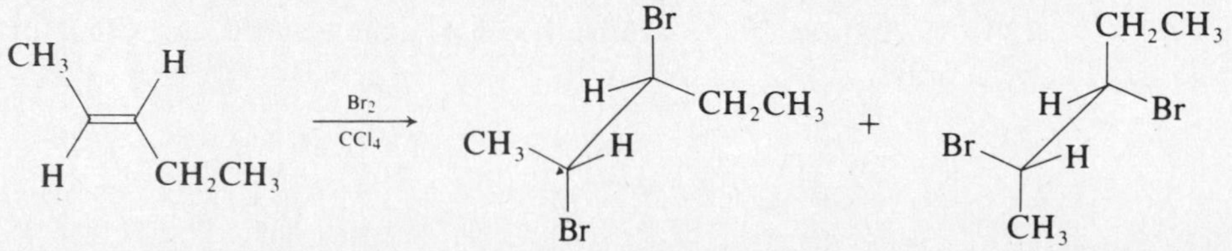

(*E*)-2-Pentene $\xrightarrow[CCl_4]{Br_2}$ (2*R*, 3*S*)-2,3-Dibromopentane + (2*S*, 3*R*)-2,3-Dibromopentane

The two stereoisomers are enantiomeric and are formed in equal amounts.

(*d*) Two enantiomers are formed in equal amounts in this reaction, involving electrophilic addition of bromine to (*Z*)-2-pentene. These two are diastereomeric with those formed in (*c*).

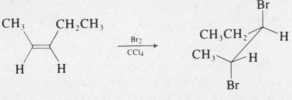

(*Z*)-2-Pentene $\xrightarrow[CCl_4]{Br_2}$ (2*R*, 3*R*)-2,3-Dibromopentane + (2*S*, 3*S*)-2,3-Dibromopentane

(*e*) Epoxidation of 1-butene yields a racemic epoxide mixture.

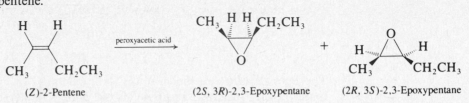

$CH_3CH_2CH{=}CH_2$ $\xrightarrow{\text{peroxyacetic acid}}$ (*S*)-2-Ethyloxirane + (*R*)-2-Ethyloxirane

(*f*) Two enantiomeric epoxides are formed in equal amounts on epoxidation of (*Z*)-2-pentene.

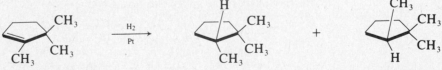

(*Z*)-2-Pentene $\xrightarrow{\text{peroxyacetic acid}}$ (2*S*, 3*R*)-2,3-Epoxypentane + (2*R*, 3*S*)-2,3-Epoxypentane

The reaction is a stereospecific syn addition. The cis alkyl groups in the starting alkene remain cis in the product epoxide.

(*g*) The starting material is achiral so even though a chiral product is formed, it is a racemic mixture of enantiomers and is optically inactive.

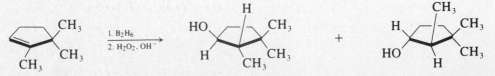

1,5,5-Trimethylcyclopentene $\xrightarrow[Pt]{H_2}$ (*R*)-1,1,2-Trimethylcyclopentane + (*S*)-1,1,2-Trimethylcyclopentane

(*h*) Recall that hydroboration-oxidation leads to anti-Markovnikov hydration of the double bond.

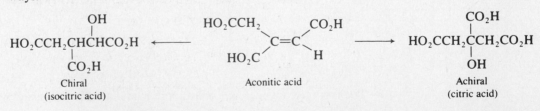

1,5,5-Trimethylcyclopentene $\xrightarrow[2.\ H_2O_2,\ OH^-]{1.\ B_2H_6}$ (1*S*, 2*S*)-2,3,3-Trimethylcyclopentanol + (1*R*, 2*R*)-2,3,3-Trimethylcyclopentanol

The product has two chiral centers. It is formed as a racemic mixture of enantiomers.

8.33 Hydration of the double bond of aconitic acid can occur in two regiochemically distinct ways.

$HO_2CCH_2CH(CO_2H)CH(OH)CO_2H$ ⟵ Aconitic acid [$(HO_2CCH_2)(HO_2C)C{=}C(CO_2H)H$] ⟶ $HO_2CCH_2C(OH)(CO_2H)CH_2CO_2H$

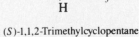

Chiral (isocitric acid) — Aconitic acid — Achiral (citric acid)

One of the hydration products lacks a chiral carbon. It must be citric acid, the achiral, optically inactive isomer. The other one has two different chiral centers and must be isocitric acid, the optically active isomer.

8.34 The molecular formula (C_6H_{10}) and the fact that A contains a five-membered ring tells us that it must also contain a double bond. The molecular formula (C_6H_{10}) differs from that of the corresponding alkane (C_6H_{14}) by four hydrogens, and since each double bond or ring causes a decrease of two hydrogens, the sum of rings and double bonds in A must be two. Therefore, there are the following possibilities for A.

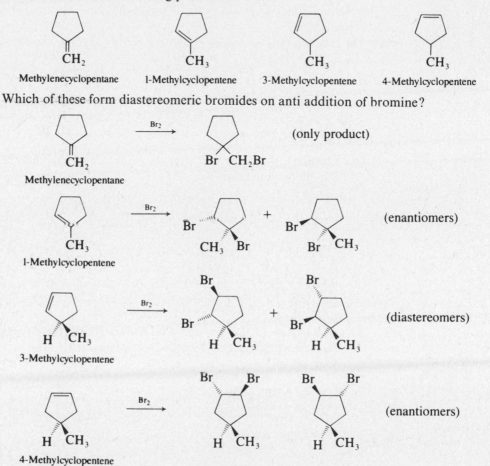

CH_2 Methylenecyclopentane — CH_3 1-Methylcyclopentene — CH_3 3-Methylcyclopentene — CH_3 4-Methylcyclopentene

Which of these form diastereomeric bromides on anti addition of bromine?

Methylenecyclopentane (CH_2) $\xrightarrow{Br_2}$ Br CH_2Br (only product)

1-Methylcyclopentene (CH_3) $\xrightarrow{Br_2}$ Br, CH_3 Br + Br, Br CH_3 (enantiomers)

3-Methylcyclopentene (H CH_3) $\xrightarrow{Br_2}$ Br, Br, H CH_3 + Br, Br, H CH_3 (diastereomers)

4-Methylcyclopentene (H CH_3) $\xrightarrow{Br_2}$ Br Br, H CH_3 + Br Br, H CH_3 (enantiomers)

Therefore, A must be 3-methylcyclopentene. (However, on the basis of the information available, we cannot tell whether it is the *R* or the *S* enantiomer or is racemic.)

8.35 Dehydration of this tertiary alcohol can yield 2,3-dimethyl-1-pentene or 2,3-dimethyl-2-pentene. Only the terminal alkene in this case is chiral.

$$CH_3CH_2\overset{*}{C}(CH_3)HC(OH)(CH_3)_2 \longrightarrow CH_3CH_2\overset{*}{C}H(CH_3)C(=CH_2)CH_3 + CH_3CH_2C(CH_3){=}C(CH_3)_2$$

2,3-Dimethyl-2-pentanol (chiral, optically pure) — 2,3-Dimethyl-1-pentene (chiral, optically pure) — 2,3-Dimethyl-2-pentene (achiral, optically inactive)

$\downarrow$ H_2, Pt (each alkene)

$$CH_3CH_2\overset{*}{C}H(CH_3)CH(CH_3)_2 \qquad CH_3CH_2\overset{*}{C}H(CH_3)CH(CH_3)_2$$

2,3-Dimethylpentane (chiral, optically pure) — 2,3-Dimethylpentane (chiral, optically inactive)

The 2,3-dimethyl-1-pentene formed in the dehydration reaction must be optically pure because it arises from optically pure alcohol by a reaction which does not involve any of the bonds to the chiral center. When optically pure 2,3-dimethyl-1-pentene is hydrogenated, it must yield optically pure 2,3-dimethylpentane—again, no bonds to the chiral center are involved in this step.

The 2,3-dimethyl-2-pentene formed in the dehydration reaction is achiral and must yield racemic 2,3-dimethylpentene on hydrogenation.

Since the alkane is 50 percent optically pure, the alkene fraction must have contained equal amounts of optically pure 2,3-dimethyl-1-pentene and its achiral isomer 2,3-dimethyl-2-pentene.

8.36 The reaction of (*R*)-3-buten-2-ol with a peroxy acid may be represented as:

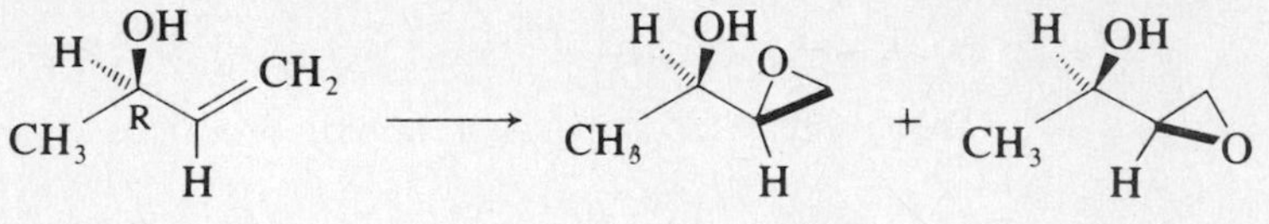

(*R*)-3-Buten-2-ol → Minor stereoisomer + [second epoxide]

(*a*) Oxygen may be transferred to the front face or back face of the double bond. Therefore, the major stereoisomer is:

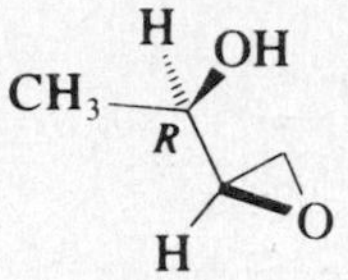

(*b*) The two epoxides have the same configuration (*R*) at their secondary alcohol carbons but opposite configurations at the chiral center of the oxirane ring; they are diastereomers.

(*c*) In addition to the two epoxides shown above, the corresponding enantiomers will be formed when racemic 3-buten-2-ol is epoxidized. The relative amounts of each of the four will be:

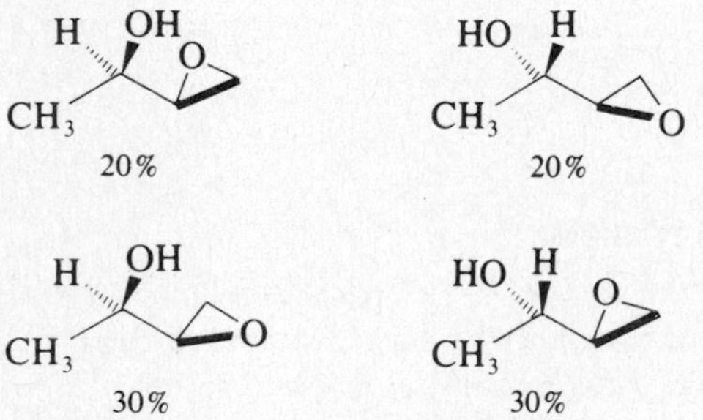

20% 20% Enantiomeric forms of minor diastereomer, totaling 40%

30% 30% Enantiomeric forms of major diastereomer, totaling 60%

SELF-TEST

PART A

A-1. For each of the following pairs of drawings, identify the molecules as chiral or achiral and tell whether each pair represents molecules that are enantiomers, diastereomers, or identical.

(*a*) ① and ②

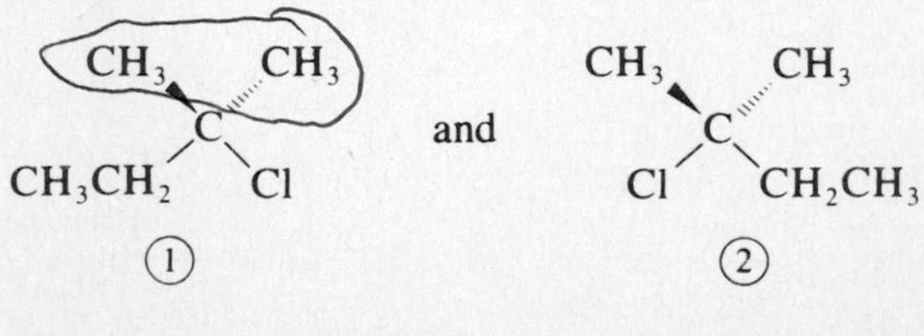

(*b*) 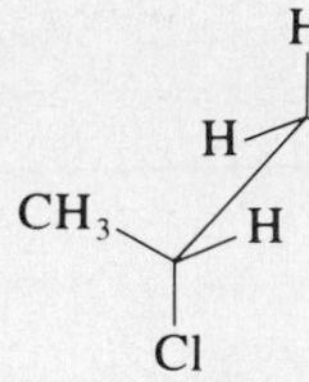 and 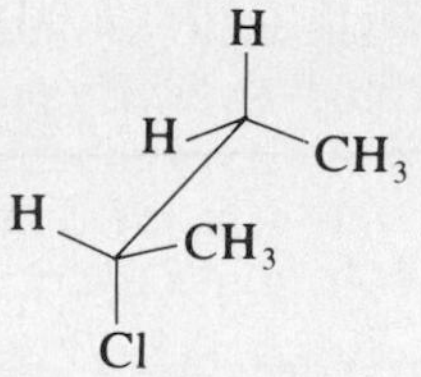

③ ④

(*c*) 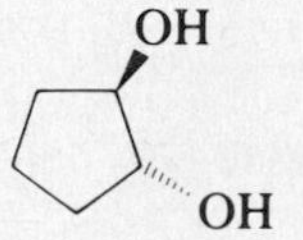 and 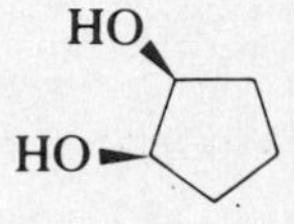

⑤ ⑥

(*d*) 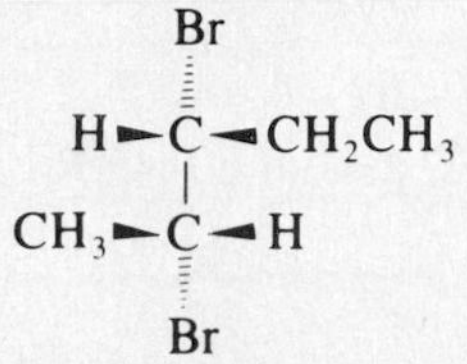 and 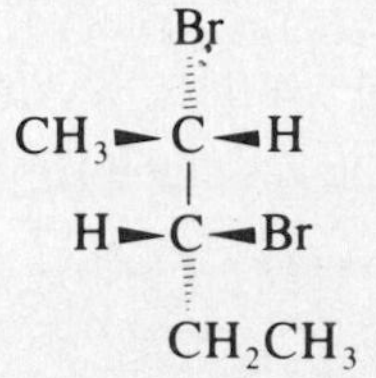

⑦ ⑧

A-2. Specify the configuration of each chiral carbon in the preceding problem, using the Cahn-Ingold-Prelog *R-S* system.

A-3. For each of the following predict the maximum number of stereoisomers possible. For which of these will meso forms be possible?

(*a*) 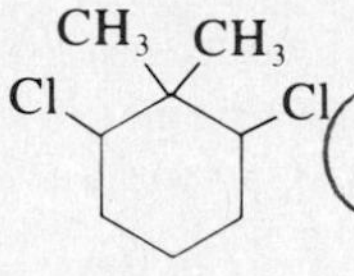 (*b*) $CH_3\overset{Br}{\overset{|}{C}}H\overset{Br}{\overset{|}{C}}H\overset{Cl}{\overset{|}{C}}HCH_3$

A-4. Using the skeleton shown as a guide, draw a perspective view of (2*R*, 3*R*)-3-chloro-2-butanol.

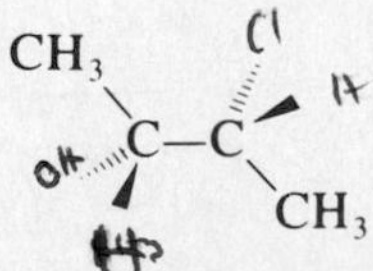

A-5. The specific rotation of pure (−)-cholesterol is −39°. What is the specific rotation of a sample of cholesterol containing 10 percent (+)-cholesterol as the only contaminant?

A-6. Write the organic product(s) expected from each of the following reactions. Show each stereoisomer if more than one forms.
(*a*) 1,5,5-Trimethylcyclopentene and hydrogen bromide
(*b*) (*E*)-2-Butene and chlorine (Cl_2)

PART B

B-1. The structure of (*S*)-2-fluorobutane is best represented by

(*a*) $CH_3\overset{F}{\overset{|}{C}}HCH_2CH_3$

(*c*) 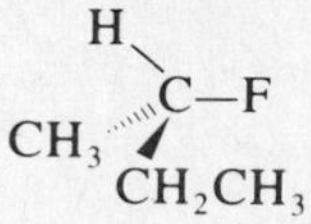

(*b*) 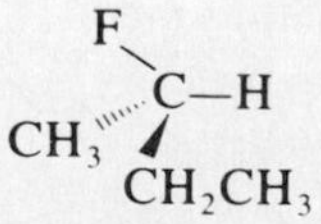

(*d*) 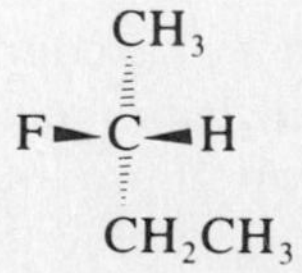

B-2. The major product(s) from the reaction of Br_2 with (*Z*)-3-hexene is (are)
(*a*) Optically pure
(*b*) A racemic mixture of enantiomers
(*c*) The meso form
(*d*) Both the racemic mixture and the meso form

B-3. Which of the following compounds are meso?

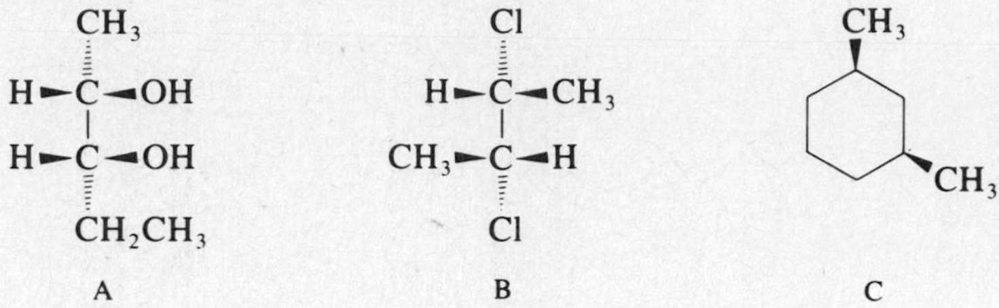

(*a*) A only (*b*) C only (*c*) A and B (*d*) B and C

B-4. The 2,3-dichloropentane whose structure is shown is

CH_3
H—C—Cl
Cl—C—H
CH_2CH_3

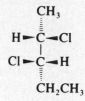

(*a*) 2*R*, 3*R* (*b*) 2*R*, 3*S* (*c*) 2*S*, 3*R* (*d*) 2*S*, 3*S*

B-5. The separation of enantiomers into their pure forms is termed
(*a*) Racemization (*b*) Resolution
(*c*) Isomerization (*d*) Equilibration

B-6. Order the groups below in order of *R*-*S* ranking (1 is highest):

$-CH(CH_3)_2$ A $-CH_2CH_2Br$ B $-CH_2Br$ C $-C(CH_3)_3$ D

	1	2	3	4
(*a*)	C	B	D	A
(*b*)	A	D	B	C
(*c*)	C	D	A	B
(*d*)	C	D	B	A

B-7. A meso compound:
(*a*) Is an achiral molecule which contains chiral atoms
(*b*) Contains a plane of symmetry or a center of symmetry
(*c*) Is optically inactive
(*d*) Is characterized by all of the above

B-8. The reaction sequence

H, CH_3, CH_2CH_3 —C—CH=CH_2 $\xrightarrow[\text{2. NaBH}_4\text{, HO}^-]{\text{1. Hg(OAc)}_2\text{, THF–H}_2\text{O}}$

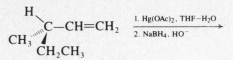

will yield:
(*a*) A pair of products which are enantiomers
(*b*) A single product which is optically active
(*c*) A pair of products which are diastereomers
(*d*) A pair of products one of which is meso

B-9. Which of the following structures are equivalent (identical)?

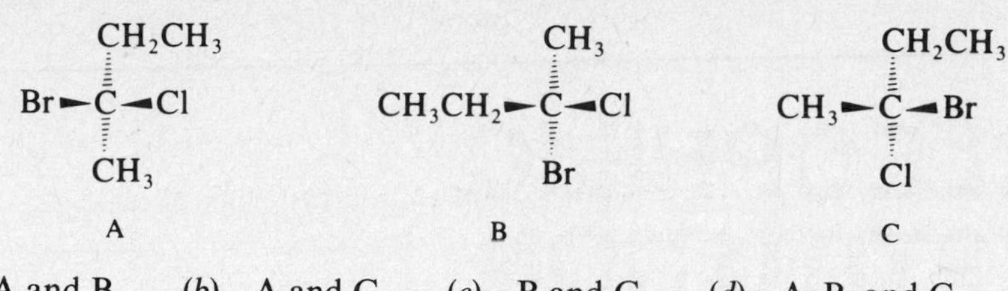

(*a*) A and B (*b*) A and C (*c*) B and C (*d*) A, B, and C

NUCLEOPHILIC SUBSTITUTION REACTIONS

IMPORTANT TERMS AND CONCEPTS

Nucleophilic Substitution (Secs. 9.1 to 9.3) A nucleophilic substitution reaction may be described in the following general terms:

$$\underset{\text{Substrate}}{R{-}X} + \underset{\text{Nucleophile}}{[Y{:}]^-} \longrightarrow \underset{\text{Product}}{R{-}Y} + \underset{\text{Leaving group}}{[X{:}]^-}$$

The *substrate* is normally an alkyl halide in which the halogen is bonded to an sp^3 hybridized carbon. The *nucleophile* is a Lewis base, that is, a species containing an unshared pair of electrons. The *leaving group* must be a weaker base than the nucleophile in order for the substitution reaction to be effective, that is, to shift the equilibrium to the right.

$$\underset{\text{Substrate}}{R{-}X} + \underset{\text{Stronger base}}{Y^-} \rightleftharpoons \underset{\text{Product}}{R{-}Y} + \underset{\text{Weaker base}}{X^-}$$

Bimolecular Nucleophilic Substitution (Secs. 9.4 to 9.7) One class of nucleophilic substitution reactions exhibits *second-order kinetics* and is known as a *bimolecular nucleophilic substitution*, or S_N2, reaction. The reaction, of which a simple example is

$$CH_3Br + OH^- \longrightarrow CH_3OH + Br^-$$

is a single-step process in which *both* the substrate and the nucleophile contribute to the activated complex.

An S_N2 reaction proceeds with *inversion of configuration* at the substrate carbon. That is, backside displacement of the leaving group occurs upon attack by the nucleophile.

$$\text{Nuc:}^- + \text{(b)(c)(a)C{-}L} \longrightarrow \text{Nuc{-}C(b)(c)(a)} + \text{:L}^-$$

The reactivity of substrates toward bimolecular substitution is controlled by *steric effects*;

bulky substrates are less likely to undergo S_N2 reaction. The reactivity order may be summarized as follows:

$$\underset{\text{Most reactive}}{CH_3X} > \underset{1^\circ}{RCH_2X} > \underset{2^\circ}{R_2CHX} > \underset{\substack{3^\circ \\ \text{Least reactive}}}{R_3CX}$$

Increased substitution on the carbon adjacent to that undergoing reaction will also decrease the reactivity. For example, $(CH_3)_2CHCH_2X$ reacts more slowly than $CH_3CH_2CH_2X$.

Nucleophiles (Sec. 9.8) Nucleophiles are Lewis bases and are often anions. Neutral nucleophiles frequently participate in *solvolysis* reactions, that is, the nucleophilic species is the solvent. Examples include water (in which case the reaction is known as *hydrolysis*) and alcohols.

Nucleophilicity is a measure of how reactive a nucleophile is in displacing a leaving group from a substrate. A negatively charged nucleophile is more reactive than its neutral conjugate acid. For example, an alkoxide RO^- is more nucleophilic than the corresponding alcohol ROH. In comparing elements of the same group on the periodic chart in protic solvents, we see that larger anions further down on the chart are less solvated and are better nucleophiles. Therefore I^- is the most nucleophilic halide, and RS^- is more nucleophilic than RO^- in a protic solvent. Protic solvents include water, alcohols, and carboxylic acids.

Unimolecular Nucleophilic Substitution (Secs. 9.9 to 9.12) The class of substitution reactions which exhibit *first-order kinetics*, that is, which have a *unimolecular* rate-determining step, are known as S_N1 reactions. The mechanism of these reactions involves two steps, the first of which is a unimolecular ionization resulting in formation of a carbocation. This step is the *slow*, rate-determining step of the reaction. In the second step the carbocation is rapidly captured by the nucleophile.

$$R_3C{-}L \xrightarrow{\text{slow}} R_3C^+L^- \xrightarrow[\text{fast}]{\text{Nuc:}} R_3C{-}Nuc$$

Reactivity in S_N1 substitution is governed by *electronic effects* and parallels ease of carbocation formation by the substrate.

$$\underset{\substack{3^\circ \\ \text{Most reactive}}}{R_3CX} > \underset{2^\circ}{R_2CHX} > \underset{1^\circ}{RCH_2X} > \underset{\text{Least reactive}}{CH_3X}$$

Although the intermediate carbocation is planar (the charged carbon is sp^2 hybridized), only partial racemization occurs upon S_N1 reaction of an optically active substrate. As shown in text Figure 9.8, a carbocation–halide ion pair forms, allowing the leaving group to shield one side of the carbocation from attack by the nucleophile.

Rearrangement of the carbon skeleton may also accompany an S_N1 reaction. As discussed in Chapter 6, carbocations may rearrange (by migration of a hydrogen or an alkyl group) to form a more stable cation, which then undergoes further reaction. For example, hydrolysis of 2-bromo-3-methylbutane yields mainly 2-methyl-2-butanol.

$$CH_3\underset{\substack{|\\Br}}{CH}\overset{\substack{CH_3\\|}}{C}HCH_3 \xrightarrow{H_2O} CH_3\underset{\substack{|\\OH}}{\overset{\substack{CH_3\\|}}{C}}CH_2CH_3$$

Substitution versus Elimination (Sec. 9.14) Two factors are important in determining whether substitution or elimination will predominate in a particular chemical system; these are the *structure of the substrate* and the *basicity of the anion*. Primary substrates will tend to favor S_N2 substitution; increasing steric hindrance will cause more E2 elimination to occur.

The principal reaction of *secondary and tertiary alkyl halides* with alkoxide bases is *elimination.* Nucleophiles that are *weaker bases* than alkoxides, for example CN^-, will favor substitution with secondary as well as primary substrates. The following table summarizes these observations.

	Nucleophile	
Substrate	*Strong base*[a]	*Weak base*
Primary	S_N2	S_N2
Secondary	E2	S_N2, S_N1, or E1
Tertiary	E2	E2, E1, or S_N1

[a] K_a of conjugate acid less than 10^{-12}

Sulfonate Esters (Sec. 9.15) Reaction of an alcohol with a sulfonyl chloride yields a sulfonate ester derivative. The *p*-toluenesulfonates or tosylates are frequently employed as substrates in nucleophilic substitution reactions.

$$ROH + CH_3\text{—}C_6H_4\text{—}SO_2Cl \xrightarrow{\text{pyridine}} CH_3\text{—}C_6H_4\text{—}SO_2OR \quad (ROTs)$$

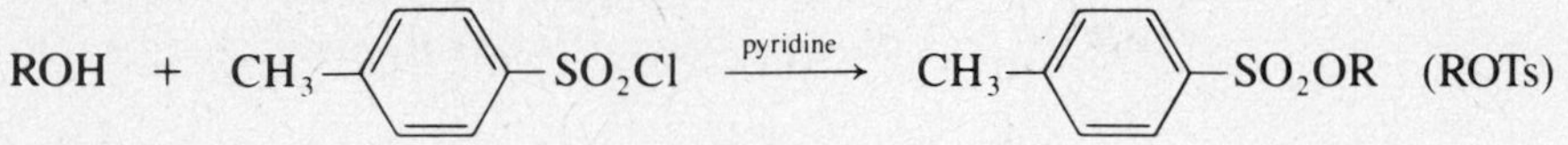

$$ROTs + Y\!:^- \longrightarrow R\text{—}Y + OTs^-$$

IMPORTANT REACTIONS

S_N2 SUBSTITUTION REACTIONS

Examples:

$$(CH_3)_2CHCH_2Br \xrightarrow{CN^-} (CH_3)_2CHCH_2CN$$

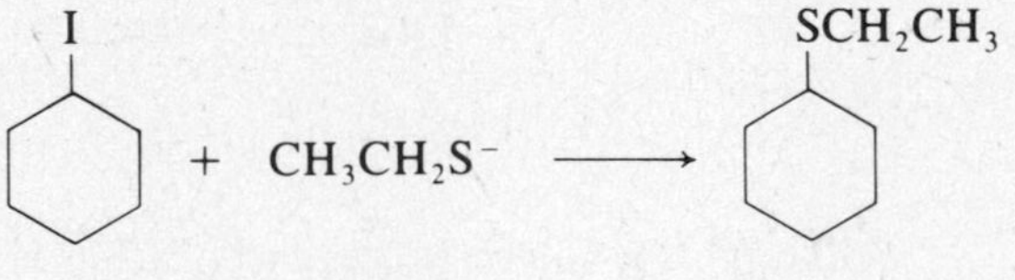

$$\text{cyclohexyl–I} + CH_3CH_2S^- \longrightarrow \text{cyclohexyl–}SCH_2CH_3$$

$$(CH_3)_3CO^- + CH_3Br \longrightarrow (CH_3)_3COCH_3$$

$$(R)\text{-}CH_3\overset{Br}{\overset{|}{C}}HCH_2CH_3 \xrightarrow{I^-} (S)\text{-}CH_3\overset{I}{\overset{|}{C}}HCH_2CH_3$$

S_N1 SUBSTITUTION REACTIONS

Examples:

$$(CH_3)_2\overset{Br}{\overset{|}{C}}CH_2CH_3 \xrightarrow{C_2H_5OH} (CH_3)_2\overset{OCH_2CH_3}{\overset{|}{C}}CH_2CH_3 + \text{alkenes}$$

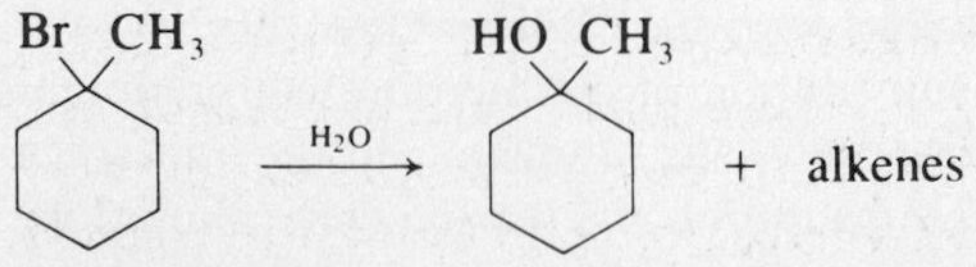

$$\text{1-bromo-1-methylcyclohexane (Br, CH}_3\text{)} \xrightarrow{H_2O} \text{1-methylcyclohexanol (HO, CH}_3\text{)} + \text{alkenes}$$

SOLUTIONS TO TEXT PROBLEMS

9.1 Identify the nucleophilic anion in each salt. The nucleophilic anion replaces bromine as a substituent on carbon.

(*b*) Potassium *tert*-butoxide serves as a source of the nucleophilic anion $(CH_3)_3CO^-$.

$$(CH_3)_3CO^- + CH_3Br \longrightarrow (CH_3)_3COCH_3 + Br^-$$

tert-Butoxide ion (nucleophile) | Methyl bromide | *tert*-Butyl methyl ether (product) | Bromide ion

(*c*) Lithium azide is a source of the azide ion

$$:\overset{-}{\ddot{N}}=\overset{+}{N}=\overset{-}{\ddot{N}}:$$

It reacts with methyl bromide to give methyl azide.

$$:\overset{-}{\ddot{N}}=\overset{+}{N}=\overset{-}{\ddot{N}}: + CH_3Br \longrightarrow CH_3\ddot{N}=\overset{+}{N}=\overset{-}{\ddot{N}}: + Br^-$$

Azide ion (nucleophile) | Methyl bromide | Methyl azide (product) | Bromide ion

(*d*) The nucleophilic anion in KCN is cyanide ($:\bar{C}{\equiv}N:$). The carbon atom is negatively charged and is normally the site of nucleophilic reactivity.

$$:N{\equiv}C:^- + CH_3Br \longrightarrow CH_3C{\equiv}N: + Br^-$$

Cyanide ion (nucleophile) | Methyl bromide | Methyl cyanide (product) | Bromide ion

(*e*) The anion in sodium hydrogen sulfide (NaSH) is ^-SH.

$$HS^- + CH_3Br \longrightarrow CH_3SH + Br^-$$

Hydrogen sulfide ion | Methyl bromide | Methanethiol | Bromide ion

9.2 Write out the structure of the starting material. Notice that it contains a primary bromide and a primary chloride. Bromide is a better leaving group than chloride and is the one that is displaced faster by the nucleophilic cyanide ion.

$$ClCH_2CH_2CH_2Br \xrightarrow[\text{ethanol-water}]{\text{NaCN}} ClCH_2CH_2CH_2C{\equiv}N$$

1-Bromo-3-chloropropane | 3-Chloropropyl cyanide

9.3 No, the two-step sequence is not consistent with the observed behavior for the hydrolysis of methyl bromide. The rate-determining step in the two-step sequence shown is the first step, ionization of methyl bromide to give methyl cation.

1. $CH_3Br \xrightarrow{\text{slow}} CH_3^+ + Br^-$
2. $CH_3^+ + HO^- \xrightarrow{\text{fast}} CH_3OH$

In such a sequence the nucleophile would not participate in the reaction until after the rate-determining step is past, and the reaction rate would depend only on the concentration of methyl bromide and be independent of the concentration of hydroxide ion.

$$\text{Rate} = k[CH_3Br]$$

The indicated kinetic behavior is first-order, not second-order, as is actually observed for methyl bromide hydrolysis.

9.4 The example given in the text illustrates inversion of configuration in the S_N2 hydrolysis of (*S*)-(+)-2-bromooctane, which yields (*R*)-(−)-2-octanol. The hydrolysis of (*R*)-(−)-2-bromooctane exactly mirrors that of its enantiomer and yields (*S*)-(+)-2-octanol.

Hydrolysis of racemic 2-bromooctane gives racemic 2-octanol. Remember, optically inactive reactants must yield optically inactive products.

9.5 Sodium iodide in acetone is a reagent that converts alkyl chlorides and bromides into alkyl iodides by an S_N2 mechanism. Pick the alkyl halide that is most reactive toward S_N2 displacement.

(*b*) 1-Bromopentane is a primary alkyl halide and so is more reactive than 3-bromopentane, which is secondary.

$BrCH_2CH_2CH_2CH_2CH_3$ — 1-Bromopentane (primary; more reactive in S_N2)

$CH_3CH_2\underset{\displaystyle Br}{\underset{|}{C}}HCH_2CH_3$ — 3-Bromopentane (secondary; less reactive in S_N2)

Crowding increases at the transition state for S_N2 reactions. The less crowded alkyl halide reacts faster.

(*c*) Both halides are secondary, but fluoride is quite a poor leaving group in nucleophilic substitution reactions. Alkyl chlorides are more reactive than alkyl fluorides.

$CH_3\underset{\displaystyle Cl}{\underset{|}{C}}HCH_2CH_2CH_3$ — 2-Chloropentane (more reactive)

$CH_3\underset{\displaystyle F}{\underset{|}{C}}HCH_2CH_2CH_3$ — 2-Fluoropentane (less reactive)

(*d*) A secondary alkyl bromide reacts faster under S_N2 conditions than a tertiary one.

$CH_3\underset{\displaystyle Br}{\underset{|}{C}}HCH_2CH_2\underset{\displaystyle CH_3}{\underset{|}{C}}HCH_3$ — 2-Bromo-5-methylhexane (secondary; more reactive in S_N2)

$CH_3\overset{\displaystyle CH_3}{\overset{|}{\underset{\displaystyle Br}{\underset{|}{C}}}}CH_2CH_2CH_2CH_3$ — 2-Bromo-2-methylhexane (tertiary; less reactive in S_N2)

(*e*) The number of carbons does not matter as much as the degree of substitution at the reaction site. The primary alkyl bromide is more reactive than the secondary.

$BrCH_2(CH_2)_8CH_3$ — 1-Bromodecane (primary; more reactive in S_N2)

$CH_3\underset{\displaystyle Br}{\underset{|}{C}}HCH_3$ — 2-Bromopropane (secondary; less reactive in S_N2)

9.6 The reactivity of an alkyl halide in an S_N1 reaction is dictated by the ease with which it ionizes to form a carbocation. Tertiary alkyl halides are the most reactive, methyl halides the least reactive.

(*b*) Cyclopentyl iodide ionizes to form a secondary carbocation while the carbocation from 1-methylcyclopentyl iodide is tertiary. The tertiary halide is more reactive.

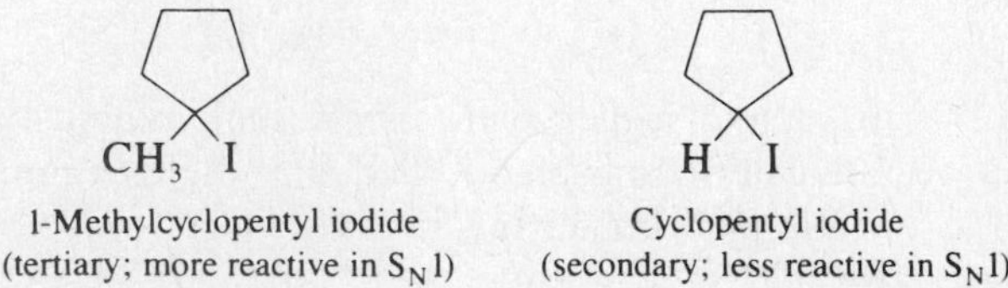

1-Methylcyclopentyl iodide (tertiary; more reactive in S_N1)   Cyclopentyl iodide (secondary; less reactive in S_N1)

(*c*) Cyclopentyl bromide ionizes to a secondary carbocation. Neopentyl bromide is a primary alkyl halide and is therefore less reactive.

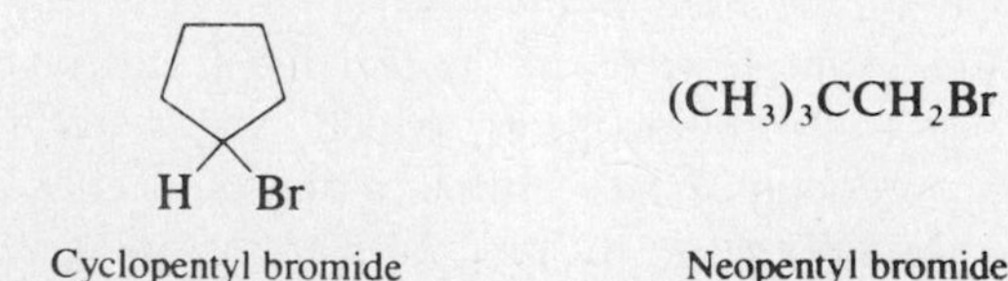

$(CH_3)_3CCH_2Br$

Cyclopentyl bromide (secondary; more reactive in S_N1)   Neopentyl bromide (primary; less reactive in S_N1)

(*d*) Iodide is a better leaving group than chloride in both S_N1 and S_N2 reactions.

$(CH_3)_3CI$ — *tert*-Butyl iodide (more reactive)

$(CH_3)_3CCl$ — *tert*-Butyl chloride (less reactive)

9.7 The alkyl halide is tertiary and so undergoes hydrolysis by an S_N1 mechanism. The carbocation can be captured by water at either face. A mixture of the axial and the equatorial alcohols is formed.

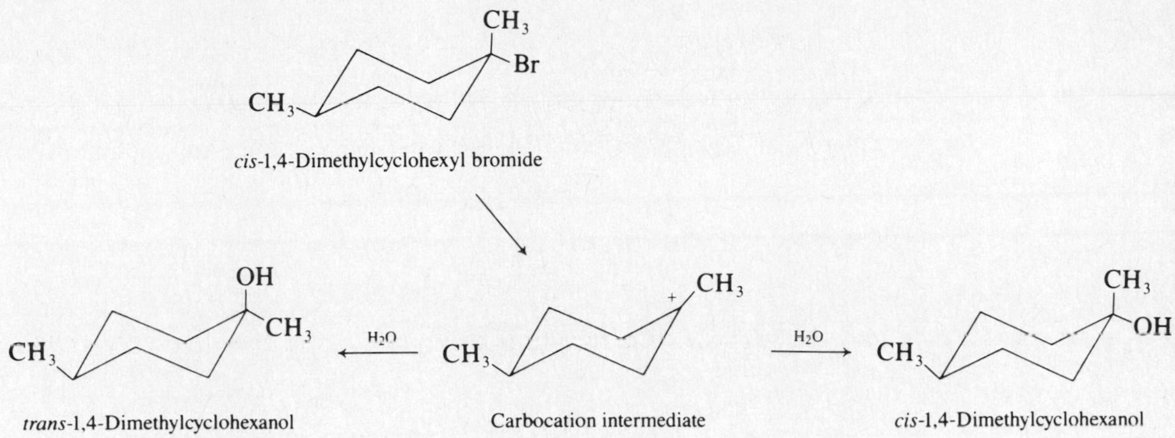

cis-1,4-Dimethylcyclohexyl bromide

trans-1,4-Dimethylcyclohexanol $\xleftarrow{H_2O}$ Carbocation intermediate $\xrightarrow{H_2O}$ *cis*-1,4-Dimethylcyclohexanol

The same two substitution products are formed from *trans*-1,4-dimethylcyclohexyl bromide because it undergoes hydrolysis via the same carbocation intermediate.

9.8 Write chemical equations illustrating each rearrangement process.

Hydride shift

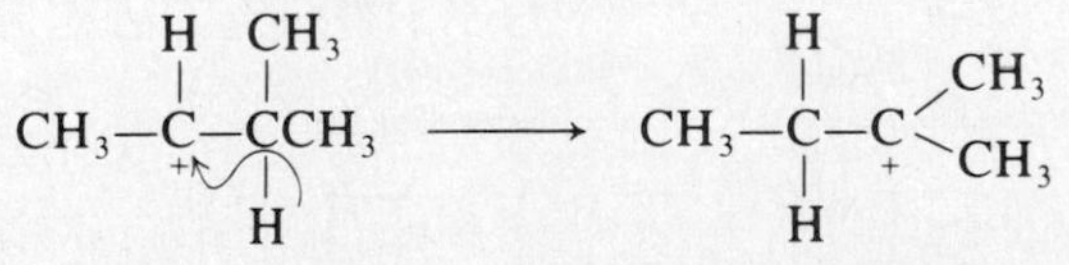

$$CH_3\overset{+}{C}H-C(H)(CH_3)CH_3 \longrightarrow CH_3CH_2-\overset{+}{C}(CH_3)_2$$

Tertiary carbocation

Methyl shift

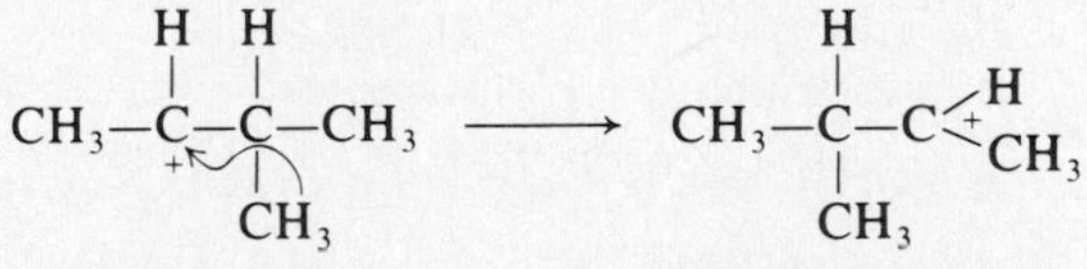

$$CH_3\overset{+}{C}H-CH(CH_3)-CH_3 \longrightarrow CH_3-CH(CH_3)-\overset{+}{C}H-CH_3$$

Secondary carbocation

Rearrangement by a hydride shift is observed because it converts a secondary carbocation to a more stable tertiary one. A methyl shift gives a secondary carbocation—the same carbocation as the one that existed prior to rearrangement.

9.9 (*b*) Ethyl bromide is a primary alkyl halide and reacts with the potassium salt of cyclohexanol by substitution.

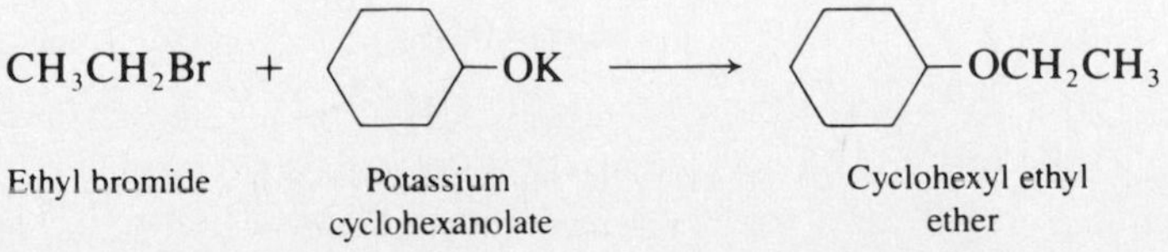

$$CH_3CH_2Br + C_6H_{11}OK \longrightarrow C_6H_{11}OCH_2CH_3$$

Ethyl bromide; Potassium cyclohexanolate; Cyclohexyl ethyl ether (*c*) No strong base is present in this reaction; the nucleophile is methanol itself, not methoxide. It reacts with *sec*-butyl bromide by substitution, not elimination.

$$\underset{\text{sec-Butyl bromide}}{CH_3\underset{\substack{|\\Br}}{C}HCH_2CH_3} \xrightarrow{CH_3OH} \underset{\text{sec-Butyl methyl ether}}{CH_3\underset{\substack{|\\OCH_3}}{C}HCH_2CH_3}$$

(*d*) Secondary alkyl halides react with alkoxide bases by E2 elimination.

$$\underset{\text{sec-Butyl bromide}}{CH_3\underset{\substack{|\\Br}}{C}HCH_2CH_3} \xrightarrow{NaOCH_3} \underset{\substack{\text{2-Butene}\\\text{(major product; mixture}\\\text{of cis and trans)}}}{CH_3CH{=}CHCH_3} + \underset{\text{1-Butene}}{CH_2{=}CHCH_2CH_3}$$

9.10 Alkyl *p*-toluenesulfonates are prepared from alcohols and *p*-toluenesulfonyl chloride.

$$\underset{\text{1-Octadecanol}}{CH_3(CH_2)_{16}CH_2OH} + \underset{\text{p-Toluenesulfonyl chloride}}{CH_3\text{—}C_6H_4\text{—}\overset{O}{\underset{O}{\overset{\|}{\underset{\|}{S}}}}Cl} \xrightarrow{\text{pyridine}} \underset{\text{Octadecyl p-toluenesulfonate}}{CH_3(CH_2)_{16}CH_2O\overset{O}{\underset{O}{\overset{\|}{\underset{\|}{S}}}}\text{—}C_6H_4\text{—}CH_3} + \underset{\text{Hydrogen chloride}}{HCl}$$

9.11 (*b*)

$$\underset{\text{Iodide ion}}{I^-} + \underset{\text{Octadecyl p-toluenesulfonate}}{CH_3(CH_2)_{16}CH_2OTs} \longrightarrow \underset{\text{Octadecyl iodide}}{CH_3(CH_2)_{16}CH_2I} + \underset{\text{p-Toluenesulfonate anion}}{TsO^-}$$

(*c*)

$$\underset{\text{Cyanide ion}}{\bar{C}{\equiv}N} + \underset{\text{Octadecyl p-toluenesulfonate}}{CH_3(CH_2)_{16}CH_2OTs} \longrightarrow \underset{\text{Octadecyl cyanide}}{CH_3(CH_2)_{16}CH_2C{\equiv}N} + \underset{\text{p-Toluenesulfonate anion}}{TsO^-}$$

(*d*)

$$\underset{\text{Hydrogen sulfide ion}}{HS^-} + \underset{\text{Octadecyl p-toluenesulfonate}}{CH_3(CH_2)_{16}CH_2OTs} \longrightarrow \underset{\text{1-Octadecanethiol}}{CH_3(CH_2)_{16}CH_2SH} + \underset{\text{p-Toluenesulfonate anion}}{TsO^-}$$

(*e*)

$$\underset{\text{Butanethiolate ion}}{CH_3CH_2CH_2CH_2S^-} + \underset{\text{Octadecyl p-toluenesulfonate}}{CH_3(CH_2)_{16}CH_2OTs} \longrightarrow$$

$$\underset{\text{Butyl octadecyl thioether}}{CH_3(CH_2)_{16}CH_2SCH_2CH_2CH_2CH_3} + \underset{\text{p-Toluenesulfonate anion}}{T_sO^-}$$

9.12 The hydrolysis of (*S*)-(+)-1-methylheptyl *p*-toluenesulfonate proceeds with inversion of configuration, giving the *R* enantiomer of 2-octanol.

$$\underset{\substack{(S)\text{-(+)-1-Methylheptyl}\\ p\text{-toluenesulfonate}}}{CH_3(CH_2)_5,\ CH_3,\ H\text{—}C\text{—}OTs} \xrightarrow{H_2O} \underset{(R)\text{-(−)-2-Octanol}}{HO\text{—}C(H)(CH_2)_5CH_3(CH_3)}$$

In Section 9.15 of the text we are told that optically pure (*S*)-(+)-1-methylheptyl *p*-toluenesulfonate is prepared from optically pure (*S*)-(+)-2-octanol having a specific rotation $[\alpha]_D^{25}$ + 9.9°. The conversion of an alcohol to a *p*-toluenesulfonate proceeds with complete *retention* of configuration. Therefore hydrolysis of this *p*-toluenesulfonate with *inversion* of configuration yields optically pure (*R*)-(−)-2-octanol of $[\alpha]_D^{25}$ − 9.9°.

9.13 All the reactions are S_N2 displacements on a primary alkyl halide, so they should proceed cleanly.

(*a*) $CH_3CH_2CH_2Br \xrightarrow[\text{acetone}]{\text{NaI}} CH_3CH_2CH_2I$

1-Bromopropane → 1-Iodopropane

(*b*) $CH_3CH_2CH_2Br \xrightarrow[\text{acetic acid}]{CH_3\overset{O}{\overset{\|}{C}}ONa} CH_3CH_2CH_2O\overset{O}{\overset{\|}{C}}CH_3$

Propyl acetate

(*c*) $CH_3CH_2CH_2Br \xrightarrow[\text{ethanol}]{NaOCH_2CH_3} CH_3CH_2CH_2OCH_2CH_3$

Ethyl propyl ether

(*d*) $CH_3CH_2CH_2Br \xrightarrow[\text{DMSO}]{\text{NaCN}} CH_3CH_2CH_2CN$

1-Cyanopropane

(*e*) $CH_3CH_2CH_2Br \xrightarrow[\text{ethanol-water}]{NaN_3} CH_3CH_2CH_2N_3$

1-Azidopropane

(*f*) $CH_3CH_2CH_2Br \xrightarrow[\text{ethanol}]{\text{NaSH}} CH_3CH_2CH_2SH$

1-Propanethiol

(*g*) $CH_3CH_2CH_2Br \xrightarrow[\text{ethanol}]{NaSCH_3} CH_3CH_2CH_2SCH_3$

Methyl propyl thioether

9.14 In the corresponding reactions of 2-bromopropane, the possibility of elimination is more pronounced.

(*a*) Iodide is weakly basic and very nucleophilic, so substitution is the principal reaction.

$$CH_3\underset{\underset{Br}{|}}{C}HCH_3 \xrightarrow[\text{acetone}]{\text{NaI}} CH_3\underset{\underset{I}{|}}{C}HCH_3$$

2-Bromopropane → 2-Iodopropane

(*b*) Sodium acetate is weakly basic (K_a for acetic acid is 1.8×10^{-5}) so the product should be isopropyl acetate.

$$CH_3\underset{\underset{Br}{|}}{C}HCH_3 \xrightarrow[\text{acetic acid}]{CH_3\overset{O}{\overset{\|}{C}}ONa} CH_3\underset{\underset{O\underset{\underset{O}{\|}}{C}CH_3}{|}}{C}HCH_3$$

Isopropyl acetate

(*c*) Bases as strong as or stronger than hydroxide react with secondary alkyl halides by E2 elimination. Ethoxide is comparable with hydroxide in basicity.

$$CH_3\underset{\underset{Br}{|}}{C}HCH_3 \xrightarrow{NaOCH_2CH_3} CH_3CH{=}CH_2$$

Propene

(By way of confirmation, it is known that the elimination substitution ratio is 87 : 13 for the reaction of isopropyl bromide with KOH in ethanol-water.)

(*d*) Sodium cyanide is a weak enough base and a good enough nucleophile that substitution is favored with most secondary alkyl halides.

$$\underset{\displaystyle\text{Br}}{\text{CH}_3\text{CHCH}_3} \xrightarrow[\text{DMSO}]{\text{NaCN}} \underset{\displaystyle\text{CN}}{\text{CH}_3\text{CHCH}_3}$$

2-Cyanopropane

(*e*) Sodium azide is an even weaker base than cyanide and a better nucleophile. Substitution occurs.

$$\underset{\displaystyle\text{Br}}{\text{CH}_3\text{CHCH}_3} \xrightarrow[\text{ethanol-water}]{\text{NaN}_3} \underset{\displaystyle\text{N}_3}{\text{CH}_3\text{CHCH}_3}$$

2-Azidopropane

(*f*) Sodium hydrogen sulfide is a relatively weak base and reacts with secondary alkyl halides by substitution.

$$\underset{\displaystyle\text{Br}}{\text{CH}_3\text{CHCH}_3} \xrightarrow{\text{NaSH}} \underset{\displaystyle\text{SH}}{\text{CH}_3\text{CHCH}_3}$$

2-Propanethiol

(*g*) Substitution occurs with sodium methanethiolate.

$$\underset{\displaystyle\text{Br}}{\text{CH}_2\text{CHCH}_3} \xrightarrow{\text{NaSCH}_3} \underset{\displaystyle\text{SCH}_3}{\text{CH}_3\text{CHCH}_3}$$

Isopropyl methyl thioether

9.15 The alkyl halide is tertiary and reacts with anionic nucleophiles by elimination.

$$\text{CH}_3\overset{\displaystyle\text{Br}}{\underset{\displaystyle\text{CH}_3}{\text{C}}}\text{CH}_3 \xrightarrow{\text{anionic nucleophile}} \text{CH}_2{=}\text{C(CH}_3)_2$$

2-Bromo-2-methylpropane (*tert*-butyl bromide) → 2-Methylpropene (isobutene)

9.16 (*a*) The substrate is a primary alkyl bromide and reacts with sodium iodide in acetone to give the corresponding iodide.

$$\text{BrCH}_2\overset{\displaystyle\text{O}}{\overset{\|}{\text{C}}}\text{OCH}_2\text{CH}_3 \xrightarrow[\text{acetone}]{\text{NaI}} \text{ICH}_2\overset{\displaystyle\text{O}}{\overset{\|}{\text{C}}}\text{OCH}_2\text{CH}_3$$

Ethyl bromoacetate → Ethyl iodoacetate (89%)

(*b*) Primary alkyl chlorides react readily with sodium iodide in acetone to yield the corresponding iodides.

$$O_2N\text{-}C_6H_4\text{-}CH_2Cl \xrightarrow[\text{acetone}]{\text{NaI}} O_2N\text{-}C_6H_4\text{-}CH_2I$$

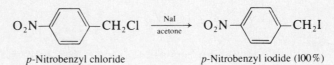

p-Nitrobenzyl chloride → *p*-Nitrobenzyl iodide (100%)

(*c*) An analogous reaction occurs with sodium acetate to yield an acetate ester.

$$O_2N\text{-}C_6H_4\text{-}CH_2Cl \xrightarrow[\text{acetic acid}]{\text{CH}_3\overset{\text{O}}{\overset{\|}{\text{C}}}\text{ONa}} O_2N\text{-}C_6H_4\text{-}CH_2O\overset{\displaystyle\text{O}}{\overset{\|}{\text{C}}}CH_3$$

p-Nitrobenzyl acetate (78 – 82%)

(*d*) The only leaving group in the substrate is bromide. Neither of the carbon-oxygen bonds is susceptible to cleavage by nucleophilic attack.

$$CH_3CH_2OCH_2CH_2Br \xrightarrow[\text{ethanol-water}]{\text{NaCN}} CH_3CH_2OCH_2CH_2CN$$

2-Bromoethyl ethyl ether → 2-Cyanoethyl ethyl ether (52–58%)

(*e*) Hydrolysis of the primary chloride yields the corresponding alcohol.

$$NC-C_6H_4-CH_2Cl \xrightarrow{H_2O,\ HO^-} NC-C_6H_4-CH_2OH$$

p-Cyanobenzyl chloride → *p*-Cyanobenzyl alcohol (85%)

(*f*) The substrate is a primary chloride.

$$ClCH_2\overset{O}{\overset{\|}{C}}OC(CH_3)_3 \xrightarrow[\text{acetone-water}]{NaN_3} N_3CH_2\overset{O}{\overset{\|}{C}}OC(CH_3)_3$$

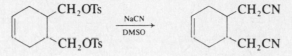

tert-Butyl chloroacetate → *tert*-Butyl azidoacetate (92%)

(*g*) Both primary toluenesulfonate leaving groups are replaced by cyanide.

4,5-Bis(toluenesulfonyloxymethyl)cyclohexene (CH_2OTs, CH_2OTs) $\xrightarrow[\text{DMSO}]{\text{NaCN}}$ (CH_2CN, CH_2CN)

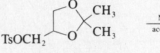

4,5-Bis-(cyanomethyl)cyclohexene

(*h*) Primary alkyl tosylates yield iodides on treatment with sodium iodide in acetone.

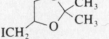

$TsOCH_2$–(2,2-dimethyl-1,3-dioxolane, CH_3, CH_3) $\xrightarrow[\text{acetone}]{\text{NaI}}$ ICH_2–(2,2-dimethyl-1,3-dioxolane, CH_3, CH_3)

(2,2-Dimethyl-1,3-dioxolan-4-yl)-methyl *p*-toiuenesulfonate → 2,2-Dimethyl-4-(iodomethyl)-1,3-dioxolane (60%)

(*i*) The nucleophilic sulfur displaces chloride from the $-CH_2Cl$ group.

$$C_6H_5-CH_2Cl + C_6H_5-SNa \longrightarrow C_6H_5-CH_2S-C_6H_5$$

Benzyl chloride + Sodium benzenethiolate → Benzyl phenyl thioether (60%)

(*j*), (*k*) Nucleophilic substitution takes place with inversion of configuration. The trans chloride yields a cis substitution product.

Cl–cyclohexane–$C(CH_3)_3$ + C_6H_5–SNa ⟶ C_6H_5S–cyclohexane–$C(CH_3)_3$

trans-4-*tert*-Butylcyclohexyl chloride + Sodium benzenethiolate → *cis*-4-*tert*-Butylcyclohexyl phenyl thioether

The cis chloride yields a trans substitution product.

Cl–cyclohexane–$C(CH_3)_3$ + NaS–C_6H_5 ⟶ C_6H_5S–cyclohexane–$C(CH_3)_3$

cis-4-*tert*-Butylcyclohexyl chloride + Sodium benzenethiolate → *trans*-4-*tert*-Butylcyclohexyl phenyl thioether

(*l*) Cyanide displaces chloride from the $-CH_2Cl$ group.

2-(Chloromethyl)furan (furan-CH_2Cl) $\xrightarrow{NaCN}$ furan-$CH_2C{\equiv}N$

2-(Chloromethyl)furan — 2-(Cyanomethyl)furan (90%)

(*m*) Sulfur displaces bromide from ethyl bromide.

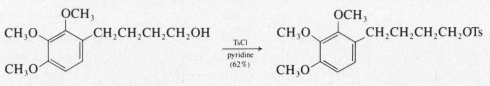

furan-CH_2SNa + CH_3CH_2Br $\longrightarrow$ furan-$CH_2SCH_2CH_3$

Sodium (2-furyl)-methanethiolate — Ethyl bromide — Ethyl (2-furyl)methyl thioether (80%)

(*n*) The first reaction is one in which a substituted alcohol is converted to a *p*-toluene-sulfonate ester. This is followed by an S_N2 displacement with lithium iodide.

2,3,4-$(CH_3O)_3C_6H_2$-$CH_2CH_2CH_2CH_2OH$ $\xrightarrow[\text{pyridine (62\%)}]{TsCl}$ 2,3,4-$(CH_3O)_3C_6H_2$-$CH_2CH_2CH_2CH_2OTs$

4-(2,3,4-Trimethoxyphenyl)-1-butanol

$\downarrow$ LiI, acetone (88%)

2,3,4-$(CH_3O)_3C_6H_2$-$CH_2CH_2CH_2CH_2I$

4-(2,3,4-Trimethoxyphenyl)-1-iodobutane

9.17 The isomers of C_4H_9Cl are:

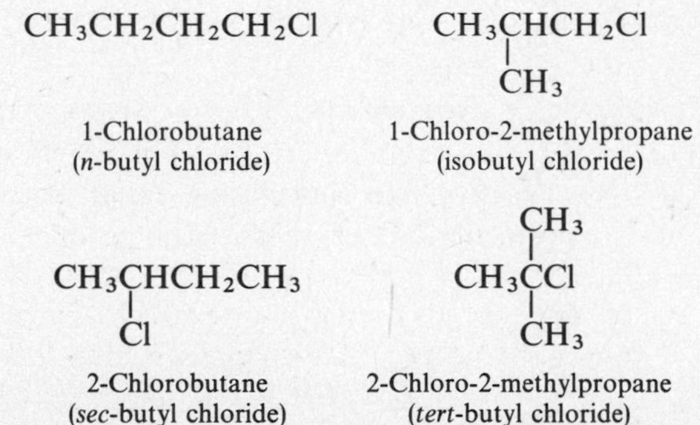

The reaction conditions (sodium iodide in acetone) are typical for an S_N2 process.

The order of S_N2 reactivity is primary > secondary > tertiary, and branching of the chain close to the site of substitution hinders reaction. The unbranched primary halide *n*-butyl chloride wili be the most reactive and the tertiary halide *tert*-butyl chloride the least. Therefore the order of reactivity will be: 1-chlorobutane > 1-chloro-2-methylpropane > 2-chlorobutane > 2-chloro-2-methylpropane.

9.18 (*a*) Iodide is a better leaving group than bromide, so 1-iodobutane should undergo S_N2 attack by cyanide faster than 1-bromobutane.

(*b*) The reaction conditions are typical for an S_N2 process. The methyl branch in 1-chloro-2-methylbutane sterically hinders attack at C-1. The unbranched isomer, 1-chloropentane, reacts faster.

$$\begin{array}{c} CH_3 \\ | \\ CH_3CH_2CHCH_2Cl \end{array} \qquad CH_3CH_2CH_2CH_2CH_2Cl$$

1-Chloro-2-methylbutane is more sterically hindered, therefore less reactive

1-Chloropentane is less sterically hindered, therefore more reactive

(c) Hexyl chloride is a primary alkyl halide and cyclohexyl chloride is secondary. Azide ion is a good nucleophile so the S_N2 reactivity rules apply; primary is more reactive than secondary.

$$CH_3CH_2CH_2CH_2CH_2CH_2Cl$$

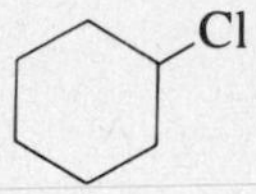

Hexyl chloride is primary, therefore more reactive in S_N2

Cyclohexyl chloride is secondary, therefore less reactive in S_N2

(d) Neopentyl bromide is too hindered to react with the weakly nucleophilic ethanol by an S_N2 reaction, and since it is a primary alkyl halide, it is less reactive in S_N1 reactions, although some reaction may occur with simultaneous methyl migration. *tert*-Butyl bromide will react with ethanol by an S_N1 mechanism at a reasonable rate owing to formation of a tertiary carbocation.

$$(CH_3)_3CBr \qquad (CH_3)_3CCH_2Br$$

tert-Butyl bromide; more reactive in S_N1 solvolysis

Neopentyl bromide; relatively unreactive in nucleophilic substitution reactions

(e) Solvolysis of alkyl halides in aqueous formic acid is faster for those which form carbocations readily. The S_N1 reactivity order applies here: secondary > primary.

$$\begin{array}{c} CH_3CHCH_2CH_3 \\ | \\ Br \end{array} \qquad (CH_3)_2CHCH_2Br$$

sec-Butyl bromide; secondary, therefore more reactive in S_N1

Isobutyl bromide; primary, therefore less reactive in S_N1

(f) 1-Chlorobutane is a primary alkyl halide and so should react by a direct displacement or S_N2 mechanism. Sodium methoxide is more basic than sodium acetate and is a better nucleophile. Reaction will occur faster with sodium methoxide than with sodium acetate.

(g) Azide ion is a very good nucleophile, whereas *p*-toluenesulfonate is a very good leaving group but a very poor nucleophile. In an S_N2 reaction with 1-chlorobutane, sodium azide will react faster than sodium *p*-toluenesulfonate.

9.19 (a) The starting material incorporates both a primary chloride and a secondary chloride. The nucleophile (iodide) attacks the less hindered primary position.

$$\begin{array}{c} ClCH_2CH_2CHCH_2CH_3 \\ | \\ Cl \end{array} \xrightarrow[\text{acetone}]{\text{NaI}} \begin{array}{c} ICH_2CH_2CHCH_2CH_3 \\ | \\ Cl \end{array}$$

1,3-Dichloropentane

3-Chloro-1-iodopentane ($C_5H_{10}ClI$)

(b) Nucleophilic substitution of the first bromide by sulfur occurs in the usual way.

$$^-SCH_2CH_2S^- + \begin{array}{c} CH_2{-}Br \\ | \\ CH_2{-}Br \end{array} \longrightarrow {}^-SCH_2CH_2SCH_2CH_2Br$$

The product of this step cyclizes by way of an intramolecular nucleophilic substitution.

$$^{-}S\begin{matrix}CH_2-CH_2\\CH_2-CH_2\\|\\Br\end{matrix}S \longrightarrow \text{1,4-Dithiane } (C_4H_8S_2)$$

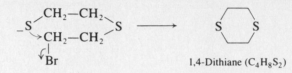

(*c*) The nucleophile is a dianion (S^{2-}). Two nucleophilic substitution reactions take place; the second of the two leads to intramolecular cyclization.

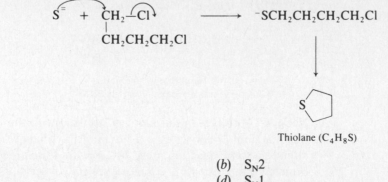

$$S^{=} + CH_2{-}Cl \text{ (}CH_2CH_2CH_2Cl\text{)} \longrightarrow {}^{-}SCH_2CH_2CH_2CH_2Cl \longrightarrow \text{Thiolane } (C_4H_8S)$$

9.20 (*a*) S_N2 (*b*) S_N2
(*c*) E2 (*d*) S_N1
(*e*) E2 (*f*) S_N2, E2
(*g*) S_N2, E2 (*h*) S_N1, E1
(*i*) S_N1, E1 (*j*) S_N1, S_N2, E1, E2

9.21 (*a*) Alkyl fluorides cannot be prepared directly from alcohols, but other alkyl halides can be converted to alkyl fluorides by nucleophilic substitution. Therefore, first convert ethanol to ethyl bromide (or chloride or iodide).

$$\underset{\text{Ethanol}}{CH_3CH_2OH} + \underset{\text{Hydrogen bromide}}{HBr} \xrightarrow{\text{heat}} \underset{\text{Ethyl bromide}}{CH_3CH_2Br} + \underset{\text{Water}}{H_2O}$$

Then use fluoride ion to displace bromide.

$$\underset{\text{Ethyl bromide}}{CH_3CH_2Br} \xrightarrow{KF} \underset{\text{Ethyl fluoride}}{CH_3CH_2F}$$

Alternatively, ethanol could be converted to its *p*-toluenesulfonate ester and then treated with potassium fluoride. This reaction has been reported in the chemical literature and leads to ethyl fluoride in 86 percent yield.

(*b*) Cyclopentyl cyanide can be prepared from a cyclopentyl halide by a nucleophilic substitution reaction. The first task, therefore, is to convert cyclopentane to a cyclopentyl halide.

$$\underset{\text{Cyclopentane}}{C_5H_{10}} + \underset{\text{Chlorine}}{Cl_2} \xrightarrow{\text{light or heat}} \underset{\text{Cyclopentyl chloride}}{C_5H_9Cl} + \underset{\text{Hydrogen chloride}}{HCl}$$

$$\underset{\text{Cyclopentyl chloride}}{C_5H_9Cl} + \underset{\text{Sodium cyanide}}{NaCN} \xrightarrow[\text{or DMSO}]{\text{ethanol-water}} \underset{\text{Cyclopentyl cyanide}}{C_5H_9CN} + \underset{\text{Sodium chloride}}{NaCl}$$

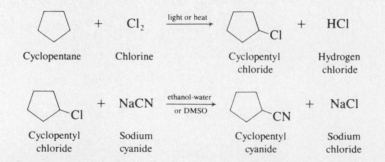

An analogous sequence involving cyclopentyl bromide could be used.

(c) Cyclopentene can serve as a precursor to a cyclopentyl halide.

Cyclopentene + HBr ⟶ Cyclopentyl bromide

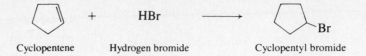

(Cyclopentene, Hydrogen bromide, Cyclopentyl bromide)

Once cyclopentyl bromide has been prepared, it is converted to cyclopentyl cyanide by nucleophilic substitution, as shown in (b).

(d) Dehydration of cyclopentanol yields cyclopentene, which can then be converted to cyclopentyl cyanide, as shown in (c).

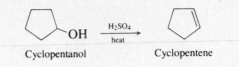

$$\text{Cyclopentanol (C}_5\text{H}_9\text{OH)} \xrightarrow[\text{heat}]{H_2SO_4} \text{Cyclopentene}$$

(e) Two cyano groups are required here, both of which must be introduced in nucleophilic substitution reactions. The substrate in the key reaction is $BrCH_2CH_2Br$.

$$BrCH_2CH_2Br + 2\,NaCN \xrightarrow[\text{DMSO}]{\text{ethanol-water or}} N{\equiv}CCH_2CH_2C{\equiv}N$$

1,2-Dibromoethane; Sodium cyanide; 1,2-Dicyanoethane (succinonitrile)

1,2-Dibromoethane is prepared from ethylene (ethene). Therefore, the overall synthesis from ethyl alcohol is formulated as shown:

$$CH_3CH_2OH \xrightarrow[\text{heat}]{H_2SO_4} CH_2{=}CH_2 \xrightarrow{Br_2} BrCH_2CH_2Br \xrightarrow{NaCN} NCCH_2CH_2CN$$

Ethyl alcohol; Ethylene; 1,2-dibromoethane; 1,2-Dicyanoethane

(f) In this synthesis a primary alkyl chloride must be converted to a primary alkyl iodide. This is precisely the kind of transformation for which sodium iodide in acetone is used.

$$(CH_3)_2CHCH_2Cl + NaI \xrightarrow{\text{acetone}} (CH_3)_2CHCH_2I + NaCl$$

Isobutyl chloride; Sodium iodide; Isobutyl iodide; Sodium chloride

(g) First convert *tert*-butyl chloride into an isobutyl halide.

$$(CH_3)_3CCl \xrightarrow{NaOCH_3} (CH_3)_2C{=}CH_2 \xrightarrow[\text{peroxides}]{HBr} (CH_3)_2CHCH_2Br$$

tert-Butyl chloride; Isobutene; Isobutyl bromide

Treating isobutyl bromide with sodium iodide in acetone converts it to isobutyl iodide.

$$(CH_3)_2CHCH_2Br \xrightarrow[\text{acetone}]{NaI} (CH_3)_2CHCH_2I$$

Isobutyl bromide; Isobutyl iodide

A second approach is by way of isobutyl alcohol.

$$(CH_3)_3CCl \xrightarrow{NaOCH_3} (CH_3)_2C{=}CH_2 \xrightarrow[\text{2. } H_2O_2,\ HO^-]{\text{1. } B_2H_6} (CH_3)_2CHCH_2OH$$

tert-Butyl chloride; Isobutene; Isobutyl alcohol

Isobutyl alcohol is then treated directly with hydrogen iodide (or KI, H_3PO_4) or converted to its *p*-toluenesulfonate ester, which reacts with sodium iodide in acetone in a manner analogous to that of isobutyl bromide.

(*h*) First introduce a leaving group into the molecule by converting isopropyl alcohol to an isopropyl halide. Then convert the resulting isopropyl halide to isopropyl azide by a nucleophilic substitution reaction.

$$\underset{\text{Isopropyl alcohol}}{CH_3\underset{\displaystyle OH}{\underset{|}{C}}HCH_3} \xrightarrow{HBr} \underset{\text{Isopropyl bromide}}{CH_3\underset{\displaystyle Br}{\underset{|}{C}}HCH_3} \xrightarrow{NaN_3} \underset{\text{Isopropyl azide}}{CH_3\underset{\displaystyle N_3}{\underset{|}{C}}HCH_3}$$

(*i*) In this synthesis 1-propanol must be first converted to an isopropyl halide.

$$\underset{\text{1-Propanol}}{CH_3CH_2CH_2OH} \xrightarrow[\text{heat}]{H_2SO_4} \underset{\text{Propene}}{CH_3CH{=}CH_2} \xrightarrow{HI} \underset{\text{Isopropyl iodide}}{CH_3\underset{\displaystyle I}{\underset{|}{C}}HCH_3}$$

Once an isopropyl halide has been obtained, it can be treated with sodium azide in a nucleophilic substitution reaction.

$$\underset{\text{Isopropyl iodide}}{CH_3\underset{\displaystyle I}{\underset{|}{C}}HCH_3} + \underset{\text{Sodium azide}}{NaN_3} \longrightarrow \underset{\text{Isopropyl azide}}{CH_3\underset{\displaystyle N_3}{\underset{|}{C}}HCH_3} + \underset{\text{Sodium iodide}}{NaI}$$

(*j*) First write out the structure of the starting material and of the product so as to determine their relationship in three dimensions.

$$\underset{(R)\text{-}sec\text{-Butyl alcohol}}{(CH_3)(CH_3CH_2)\overset{H}{C}\text{—}OH} \qquad \underset{(S)\text{-}sec\text{-Butyl azide}}{N_3\text{—}\overset{H}{C}(CH_3)(CH_2CH_3)}$$

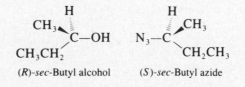

The hydroxyl group must be replaced by azide with inversion of configuration. First, however, a leaving group must be introduced and it must be introduced in such a way that the configuration at the chiral center is not altered. The best way to do this is to convert (*R*)-*sec*-butyl alcohol to its corresponding *p*-toluenesulfonate ester.

$$\underset{(R)\text{-}sec\text{-Butyl alcohol}}{(CH_3)(CH_3CH_2)\overset{H}{C}\text{—}OH} \xrightarrow[\text{pyridine}]{p\text{-toluenesulfonyl chloride}} \underset{(R)\text{-}sec\text{-Butyl } p\text{-toluenesulfonate}}{(CH_3)(CH_3CH_2)\overset{H}{C}\text{—}OSO_2\text{—}C_6H_4\text{—}CH_3}$$

Next, convert the *p*-toluenesulfonate to the desired azide by an S_N2 reaction.

$$\underset{(R)\text{-}sec\text{-Butyl } p\text{-toluenesulfonate}}{(CH_3)(CH_3CH_2)\overset{H}{C}\text{—}OSO_2\text{—}C_6H_4\text{—}CH_3} \xrightarrow{NaN_3} \underset{(S)\text{-}sec\text{-Butyl azide}}{N_3\text{—}\overset{H}{C}(CH_3)(CH_2CH_3)}$$

(*k*) This problem is carried out in exactly the same way as the preceding one, except the nucleophile in the second step is HS^-.

$$\underset{(R)\text{-}sec\text{-Butyl alcohol}}{(CH_3)(CH_3CH_2)\overset{H}{C}\text{—}OH} \xrightarrow[\text{pyridine}]{p\text{-toluenesulfonyl chloride}} \underset{(R)\text{-}sec\text{-Butyl } p\text{-toluenesulfonate}}{(CH_3)(CH_3CH_2)\overset{H}{C}\text{—}OSO_2\text{—}C_6H_4\text{—}CH_3} \xrightarrow{NaSH} \underset{(S)\text{-2-Butanethiol}}{HS\text{—}\overset{H}{C}(CH_3)(CH_2CH_3)}$$

9.22 (*a*) The two possible combinations of alkyl bromide and alkoxide ion which might yield *tert*-butyl methyl ether are:

1. $CH_3Br + (CH_3)_3CO^- \xrightarrow{\text{fast}} (CH_3)_3COCH_3$

Methyl bromide — *tert*-Butoxide ion — *tert*-Butyl methyl ether

2. $CH_3O^- + (CH_3)_3CBr \xrightarrow{\text{slow}} (CH_3)_3COCH_3$

Methoxide ion — *tert*-Butyl bromide — *tert*-Butyl methyl ether

We choose the first approach because it is an S_N2 reaction on the unhindered substrate, methyl bromide. The second approach requires an S_N2 reaction on a hindered tertiary alkyl halide, a very poor choice. Indeed, we would expect that the reaction of methoxide ion with *tert*-butyl bromide would not give any ether at all but would proceed entirely by E2 elimination.

$$CH_3O^- + (CH_3)_3CBr \xrightarrow{\text{fast}} CH_3OH + CH_2{=}C(CH_3)_2$$

Methanol — 2-Methylpropene

(*b*) Again, the better alternative is to choose the less hindered alkyl halide to permit substitution to predominate over elimination.

Cyclopentyl–OK + CH_3Br ⟶ Cyclopentyl–OCH_3

Potassium cyclopentoxide — Bromomethane — Cyclopentyl methyl ether

An attempt to prepare this compound by the reaction

Cyclopentyl–Cl + CH_3ONa ⟶ Cyclopentyl–OCH_3

Chlorocyclopentane — Sodium methoxide — Cyclopentyl methyl ether

gave cyclopentyl methyl ether in only 24 percent yield. Cyclopentene was isolated in 31 percent yield.

(*c*) Neopentyl halides are too sterically hindered to be good candidates for this synthesis. The only practical method is:

$$(CH_3)_3CCH_2OK + CH_3CH_2Br \longrightarrow (CH_3)_3CCH_2OCH_2CH_3$$

Potassium neopentoxide — Bromoethane — Ethyl neopentyl ether

(*d*) The bond to oxygen does not involve a chiral center, so the configuration at the chiral center remains unchanged during the conversion of the alkyl bromide to the ether.

$$(R)\text{-}CH_3CH_2\underset{\displaystyle CH_3}{\underset{|}{C}}HCH_2Br + (CH_3)_2CHOK \longrightarrow (R)\text{-}CH_3CH_2\underset{\displaystyle CH_3}{\underset{|}{C}}HCH_2OCH(CH_3)_2$$

(*R*)-1-Bromo-2-methylbutane — Potassium isopropoxide — (*R*)-1-Isopropoxy-2-methylbutane

9.23 (*a*) In order to convert *trans*-2-methylcyclopentanol to *cis*-2-methylcyclopentyl acetate the hydroxyl group must be replaced by acetate with inversion of configuration. Hydroxide is a poor leaving group and so must first be converted to a good leaving group. The best choice is *p*-toluenesulfonate because this can be prepared by a reaction which alters none of the bonds to the chiral center.

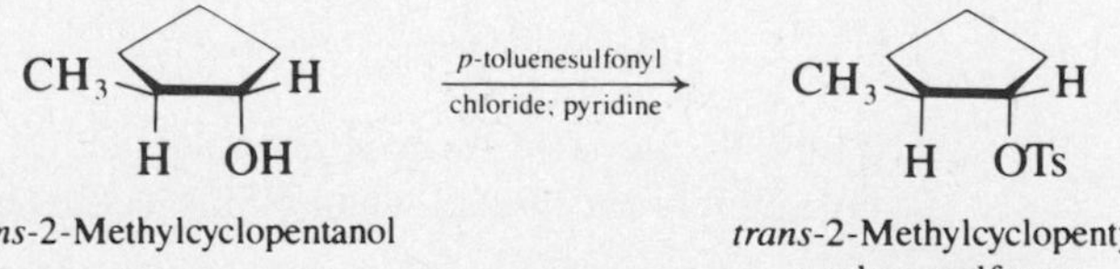

trans-2-Methylcyclopentanol — *trans*-2-Methylcyclopentyl *p*-toluenesulfonate

Treatment of the *p*-toluenesulfonate with potassium acetate in acetic acid will proceed with inversion of configuration to give the desired product.

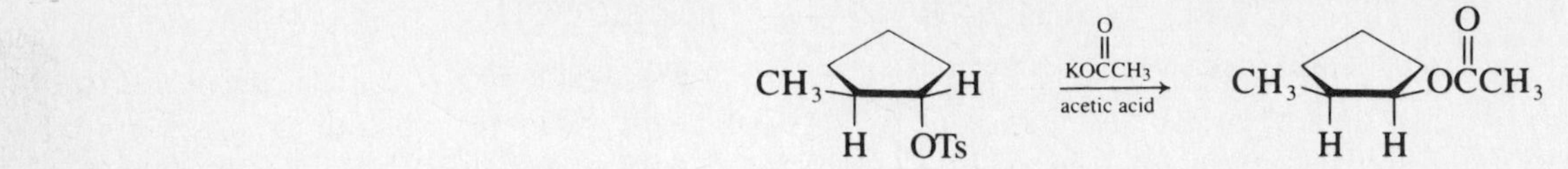

trans-2-Methylcyclopentyl p-toluenesulfonate $\xrightarrow[\text{acetic acid}]{KOCCH_3}$ cis-2-Methylcyclopentyl acetate

trans-2-Methylcyclopentyl *p*-toluenesulfonate — *cis*-2-Methylcyclopentyl acetate

(*b*) In order to decide on the best sequence of reactions, we must begin by writing structural formulas to determine what kinds of transformations are required.

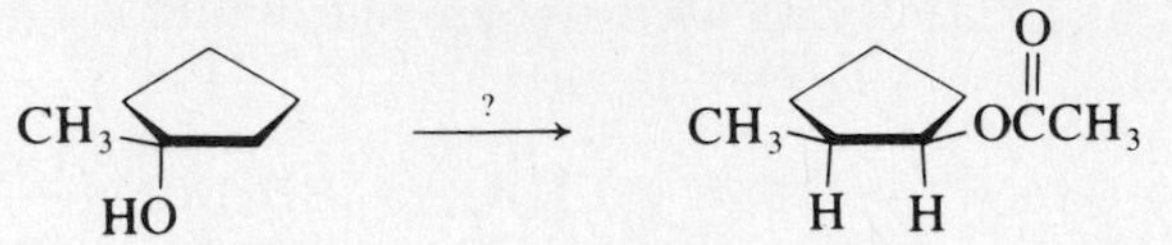

1-Methylcyclopentanol $\xrightarrow{?}$ *cis*-2-Methylcyclopentyl acetate

We already know from (*a*) of this problem how to convert *trans*-2-methylcyclopentanol to *cis*-2-methylcyclopentyl acetate, so all that is really necessary is to design a synthesis of *trans*-2-methylcyclopentanol. Therefore:

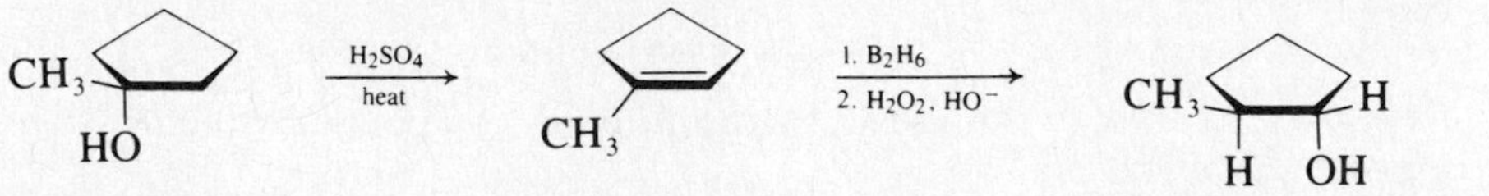

1-Methylcyclopentanol $\xrightarrow[\text{heat}]{H_2SO_4}$ 1-Methylcyclopentene $\xrightarrow[2.\ H_2O_2,\ HO^-]{1.\ B_2H_6}$ *trans*-2-Methylcyclopentanol

Hydroboration-oxidation converts 1-methylcyclopentene to the desired alcohol by anti-Markovnikov syn hydration of the double bond. The resulting alcohol is then converted to its *p*-toluenesulfonate ester and treated with acetate ion as in (*a*) to give *cis*-2-methylcyclopentyl acetate.

9.24 (*a*) If each act of substitution occurred with retention of configuration, there would be no observable racemization; $k_{rac} = 0$.

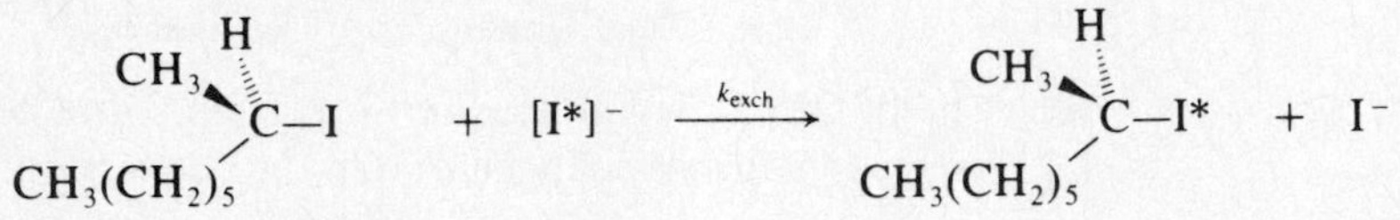

$CH_3(CH_2)_5CH(CH_3)\text{—I} + [I^*]^- \xrightarrow{k_{exch}} CH_3(CH_2)_5CH(CH_3)\text{—I}^* + I^-$

(*R*)-(−)-2-Iodooctane (* indicates radioactive label) — [(*R*)-(−)-2-Iodooctane]*

Therefore $k_{rac}/k_{exch} = 0$.

(*b*) If each act of exchange proceeds with inversion of configuration, (*R*)-(−)-2-iodooctane will be transformed to radioactively labeled (*S*)-(+)-2-iodooctane.

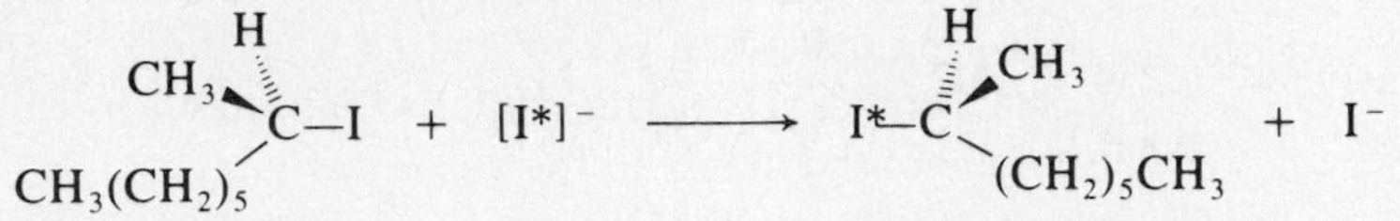

$CH_3(CH_2)_5CH(CH_3)\text{—I} + [I^*]^- \longrightarrow I^*\text{—}CH(CH_3)(CH_2)_5CH_3 + I^-$

(*R*)-(−)-2-Iodooctane — [(*S*)-(+)-2-Iodooctane]*

Thus, if one starts with 100 molecules of (*R*)-(−)-2-iodooctane, the compound will be completely racemized when 50 molecules have become radioactive. Therefore

$$k_{rac}/k_{exch} = 2$$

(*c*) If radioactivity is incorporated in a stereorandom fashion, then 2-iodooctane will be 50 percent racemized when 50 percent of it has reacted. Therefore,

$$k_{rac}/k_{exch} = 1$$

In fact, Hughes found that the rate of racemization was twice the rate of incorporation of radioactive iodide. This experiment provided strong evidence for the belief that bimolecular nucleophilic substitution proceeds stereospecifically with inversion of configuration.

9.25 The substrate is a tertiary alkyl bromide and can undergo S_N1 substitution and E1 elimination under these reaction conditions. Elimination in either of two directions to give regioisomeric alkenes can also occur.

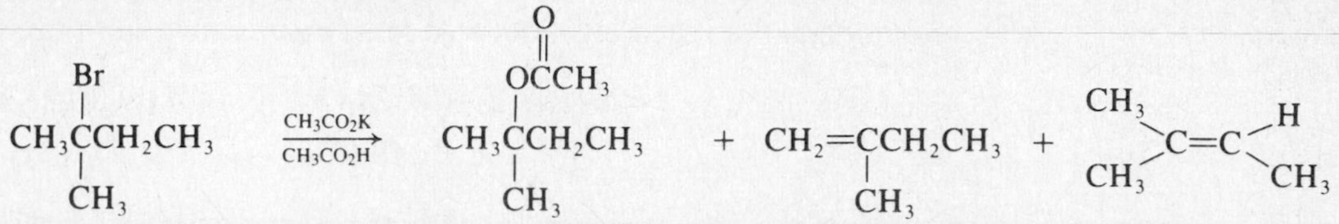

$$CH_3\underset{\underset{CH_3}{|}}{\overset{\overset{Br}{|}}{C}}CH_2CH_3 \xrightarrow[CH_3CO_2H]{CH_3CO_2K} CH_3\underset{\underset{CH_3}{|}}{\overset{\overset{O\overset{O}{\overset{\|}{C}}CH_3}{|}}{C}}CH_2CH_3 + CH_2{=}\underset{\underset{CH_3}{|}}{C}CH_2CH_3 + (CH_3)_2C{=}C(H)CH_3$$

2-Bromo-2-methylbutane; 1,1-Dimethylpropyl acetate; 2-Methyl-1-butene; 2-Methyl-2-butene

9.26 (*a*) Tertiary alkyl halides undergo nucleophilic substitution only by way of carbocations: S_N1 is the most likely mechanism for solvolysis of the 2-halo-2-methylbutanes.

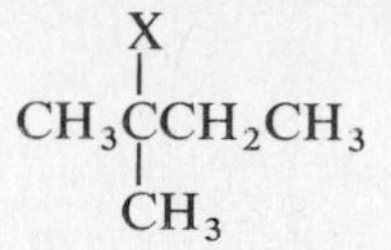

$$CH_3\underset{\underset{CH_3}{|}}{\overset{\overset{X}{|}}{C}}CH_2CH_3$$

2-Halo-2-methylbutanes are tertiary alkyl halides

(*b*) Tertiary alkyl halides can undergo either E1 or E2 elimination. Since no alkoxide base is present, solvolytic elimination most likely occurs by an E1 mechanism.

(*c*), (*d*) Iodides react faster than bromides in substitution and elimination reactions irrespective of whether the mechanism is E1, E2, S_N1, or S_N2.

(*e*) Solvolysis in aqueous ethanol can give rise to an alcohol or an ether as product, depending on whether the carbocation is captured by water or ethanol.

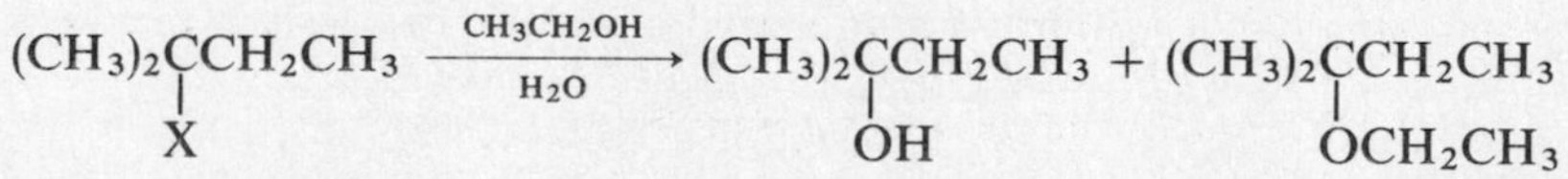

$$(CH_3)_2\underset{\underset{X}{|}}{C}CH_2CH_3 \xrightarrow[H_2O]{CH_3CH_2OH} (CH_3)_2\underset{\underset{OH}{|}}{C}CH_2CH_3 + (CH_3)_2\underset{\underset{OCH_2CH_3}{|}}{C}CH_2CH_3$$

2-Methyl-2-pentanol; Ethyl *tert*-pentyl ether

(*f*) Elimination can yield either of two isomeric alkenes.

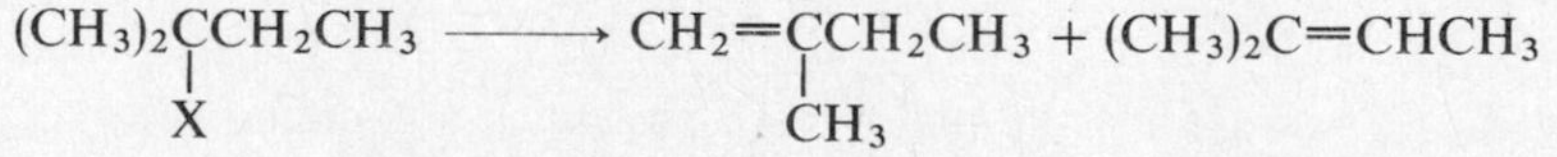

$$(CH_3)_2\underset{\underset{X}{|}}{C}CH_2CH_3 \longrightarrow CH_2{=}\underset{\underset{CH_3}{|}}{C}CH_2CH_3 + (CH_3)_2C{=}CHCH_3$$

2-Methyl-1-butene; 2-Methyl-2-butene

Zaitsev's rule predicts that 2-methyl-2-butene should be the major alkene.

(*g*) The product distribution is determined by what happens to the carbocation intermediate. If the carbocation is free of its leaving group, its fate will be the same no matter whether the leaving group is bromide or iodide.

9.27 Solvolysis of 1,2-dimethylpropyl *p*-toluenesulfonate in acetic acid is expected to give one substitution product and two alkenes.

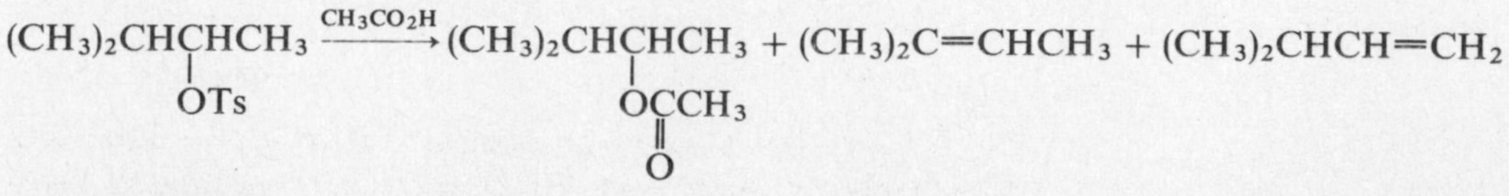

$$(CH_3)_2CH\underset{\underset{OTs}{|}}{C}HCH_3 \xrightarrow{CH_3CO_2H} (CH_3)_2CH\underset{\underset{O\underset{\underset{O}{\|}}{C}CH_3}{|}}{C}HCH_3 + (CH_3)_2C{=}CHCH_3 + (CH_3)_2CHCH{=}CH_2$$

1,2-Dimethylpropyl *p*-toluenesulfonate; 1,2-Dimethylpropyl acetate; 2-Methyl-2-butene; 3-Methyl-1-butene

Since five products are formed, we are led to consider the possibility of carbocation rearrangements in S_N1 and E1 solvolysis.

$$\underset{\text{1,2-Dimethylpropyl cation}}{(CH_3)_2CH\overset{+}{C}HCH_3} \longrightarrow \underset{\text{1,1-Dimethylpropyl cation}}{(CH_3)_2\overset{+}{C}CH_2CH_3} \longrightarrow$$

$$\underset{\text{1,1-Dimethylpropyl acetate}}{(CH_3)_2C(OCCH_3)CH_2CH_3 \text{ (with C=O)}} + \underset{\text{2-Methyl-2-butene}}{(CH_3)_2C{=}CHCH_3} + \underset{\text{2-Methyl-1-butene}}{CH_2{=}C(CH_3)CH_2CH_3}$$

Since 2-methyl-2-butene is a product common to both carbocation intermediates, a total of five different products are accounted for. There are two substitution products:

$$\underset{\text{1,2-Dimethylpropyl acetate}}{(CH_3)_2CHCH(OCCH_3)CH_3 \text{ (O=C)}} \qquad \underset{\text{1,1-Dimethylpropyl acetate}}{(CH_3)_2C(OCCH_3)CH_2CH_3 \text{ (O=C)}}$$

and three elimination products:

$$\underset{\text{2-Methyl-2-butene}}{(CH_3)_2C{=}CHCH_3} \qquad \underset{\text{3-Methyl-1-butene}}{(CH_3)_2CHCH{=}CH_2} \qquad \underset{\text{2-Methyl-1-butene}}{CH_2{=}C(CH_3)CH_2CH_3}$$

9.28 Solution A contains both acetate ion and methanol as nucleophiles. Acetate is more nucleophilic than methanol so the major observed reaction is:

$$\underset{\text{Methyl iodide}}{CH_3I} + \underset{\text{Acetate}}{CH_3\overset{O}{\overset{\|}{C}}O^-} \xrightarrow{\text{methanol}} \underset{\text{Methyl acetate}}{CH_3O\overset{O}{\overset{\|}{C}}CH_3}$$

Solution B prepared by adding potassium methoxide to acetic acid rapidly undergoes an acid-base transfer process:

$$\underset{\substack{\text{Methoxide}\\\text{(stronger base)}}}{CH_3O^-} + \underset{\substack{\text{Acetic acid}\\\text{(stronger acid)}}}{CH_3\overset{O}{\overset{\|}{C}}OH} \longrightarrow \underset{\substack{\text{Methanol}\\\text{(weaker acid)}}}{CH_3OH} + \underset{\substack{\text{Acetate}\\\text{(weaker base)}}}{CH_3\overset{O}{\overset{\|}{C}}O^-}$$

Thus the major base present is not methoxide but acetate. Therefore methyl iodide reacts with acetate anion in solution B to give methyl acetate.

9.29 Alkyl chlorides arise by the reaction sequence:

$$\underset{\substack{\text{Primary}\\\text{alcohol}}}{RCH_2OH} + \underset{\substack{p\text{-Toluenesulfonyl}\\\text{chloride}}}{CH_3{-}C_6H_4{-}SO_2Cl} + \underset{\text{Pyridine}}{C_5H_5N} \longrightarrow$$

$$\underset{\text{Primary alkyl }p\text{-toluenesulfonate}}{RCH_2OSO_2{-}C_6H_4{-}CH_3} + \underset{\substack{\text{Pyridinium}\\\text{chloride}}}{C_5H_5\overset{+}{N}H \quad Cl^-}$$

$$RCH_2OSO_2{-}C_6H_4{-}CH_3 + Cl^- \longrightarrow RCH_2Cl + CH_3{-}C_6H_4{-}SO_3^-$$

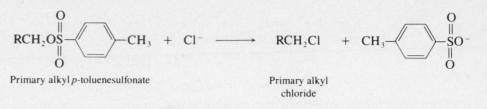

Primary alkyl *p*-toluenesulfonate — Primary alkyl chloride

The reaction proceeds to form the alkyl *p*-toluenesulfonate as expected, but the chloride anion formed in this step subsequently acts as a nucleophile and displaces *p*-toluenesulfonate from RCH_2OTs.

9.30 Iodide ion is both a better nucleophile than cyanide and a better leaving group than bromide. Therefore, the two reactions shown are together faster than the reaction of cyclopentyl bromide with sodium cyanide alone.

$$C_5H_9Br \xrightarrow{NaI} C_5H_9I \xrightarrow{NaCN} C_5H_9CN$$

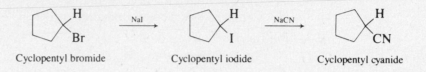

Cyclopentyl bromide — Cyclopentyl iodide — Cyclopentyl cyanide

SELF-TEST

PART A

A-1. Write the correct structure of the reactant, reagent, or product omitted from each of the following. Clearly indicate stereochemistry where it is important.

(*a*) $CH_3CH_2CH_2CH_2Br \xrightarrow[CH_3CH_2OH]{CH_3CH_2ONa} ?$

(*b*) $? \xrightarrow{NaCN}$ (cyclopentane bearing CN (hashed) and CH_3 (wedged) on adjacent carbons)

(*c*) 1-Chloro-3-methylbutane + sodium azide $\longrightarrow$?

(*d*) Chlorocyclohexane $\xrightarrow[CH_3OH]{CH_3ONa}$? (major)

A-2. Choose the best pair of reactants to form the following product by an S_N2 reaction.

$$(CH_3)_2CHSCH_2CH_2CH_3$$

A-3. Outline the chemical steps necessary to synthesize (*R*)-2-iodobutane from (*S*)-2-butanol as the starting material.

A-4. Hydrolysis of 2-chloro-3,3-dimethylbutane yields 2,3-dimethyl-2-butanol as the major product. Explain this observation, using chemical structures to outline the mechanism of the reaction.

A-5. Identify the class of reaction (e.g., E2, etc.) and write the kinetic equation for the solvolysis of *tert*-butyl bromide in methanol.

A-6. Provide a brief explanation why the halogen exchange reaction shown is an acceptable synthetic method.

$$CH_3CH(Br)CH_3 + NaI \xrightarrow{acetone} CH_3CH(I)CH_3 + NaBr$$

PART B

B-1. The bimolecular substitution reaction

$$CH_3Br + OH^- \longrightarrow CH_3OH + Br^-$$

is represented by the kinetic equation:
(*a*) Rate = $k[CH_3Br]^2$
(*b*) Rate = $k[CH_3Br][OH^-]$
(*c*) Rate = $k[CH_3Br] + k[OH^-]$
(*d*) Rate = $k/[CH_3Br][OH^-]$

B-2. The solvolysis reactions of optically active halides generally proceed with
(*a*) Complete racemization
(*b*) Partial racemization
(*c*) Complete retention of configuration
(*d*) Complete inversion of configuration

B-3. For the reaction

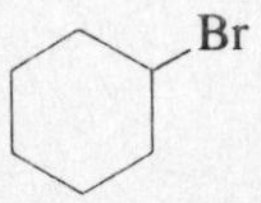

$$\text{(cyclohexyl bromide, } C_6H_{11}Br) + CH_3CH_2O^-Na^+ \longrightarrow$$

the major product is formed by
(*a*) An S_N1 reaction
(*b*) An S_N2 reaction
(*c*) An E1 reaction
(*d*) An E2 reaction

B-4. Which of the following statements pertaining to an S_N2 reaction are true?
1. The rate of reaction is independent of the concentration of the nucleophile.
2. The nucleophile attacks carbon on the side of the molecule opposite the group being displaced.
3. The reaction proceeds with simultaneous bond formation and bond rupture.
4. Partial racemization of an optically active substrate results.

(*a*) 1, 4 (*b*) 1, 3, 4 (*c*) 2, 3 (*d*) All

B-5. Ethanol reacts with *tert*-butyl bromide according to the equation shown below. If the concentration of ethanol is doubled, by what factor will the rate of the reaction change?

$$(CH_3)_3CBr + CH_3CH_2OH \longrightarrow (CH_3)_3COCH_2CH_3$$

(*a*) Remain the same
(*b*) Increase by a factor of 2
(*c*) Increase by a factor of 4
(*d*) Decrease by a factor of 2

B-6. Which of the following phrases are *not* correctly associated with S_N1 reactions?
1. Rearrangements are possible
2. Rate affected by solvent polarity
3. Strength of nucleophile important in determining rate
4. Reactivity series tertiary > secondary > primary
5. Proceed with complete inversion of configuration

(*a*) 3, 5 (*b*) 5 only (*c*) 2, 3, 5 (*d*) 3 only

B-7. Rank the following in order of decreasing rate of solvolysis with aqueous ethanol (fastest → slowest):

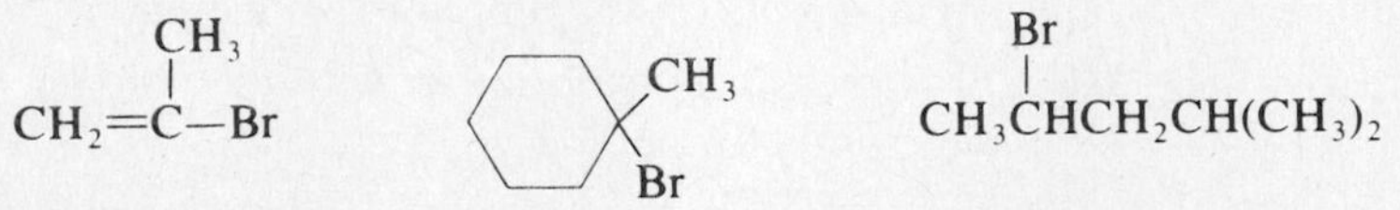

A: $CH_2{=}C(CH_3){-}Br$

B: 1-bromo-1-methylcyclohexane (ring carbon bearing CH_3 and Br)

C: $CH_3CH(Br)CH_2CH(CH_3)_2$

(*a*) B > A > C (*b*) A > B > C (*c*) B > C > A (*d*) A > C > B

B-8. Rank the following species in order of decreasing nucleophilicity in a polar protic solvent (most → least nucleophilic)

$$\underset{A}{CH_3CH_2CH_2O^-} \qquad \underset{B}{CH_3CH_2CH_2S^-} \qquad \underset{C}{CH_3CH_2\overset{\overset{O}{\|}}{C}O^-}$$

(*a*) C > A > B (*b*) B > C > A (*c*) A > C > B (*d*) B > A > C

CHAPTER

10

ALKYNES

IMPORTANT TERMS AND CONCEPTS

Nomenclature (Sec. 10.2) Alkynes are hydrocarbons having a carbon-carbon triple bond and are characterized by the general formula C_nH_{2n-2}. The usual IUPAC rules are followed in naming alkynes, the *-ane* suffix of the parent alkane being replaced with *-yne*. For example

$$CH_3-C\equiv C-\underset{}{\overset{\displaystyle CH_3}{\overset{|}{C}}}HCH_2CH_3$$

4-Methyl-2-hexyne

As noted below, the C—C≡C—C unit of an alkyne is linear; therefore stereoisomerism around the triple bond is not possible.

Structure and Bonding (Sec. 10.4) The triple-bond carbons of an alkyne are *sp* hybridized. The hybrid orbitals on each carbon atom are oriented 180° from each other; that is, they are in a linear arrangement. The triple bond itself consists of one σ component from overlap of *sp* hybrid orbitals and two π components at right angles to each other from overlap of the unhybridized *p* orbitals on each carbon atom. The resulting bonding arrangement is depicted in text Figure 10.2.

Acidity of Acetylenic C—H Bonds (Sec. 10.6) Terminal alkynes, i.e., those having the triple bond between C-1 and C-2 of the carbon chain, are about 10^{19} times more acidic than terminal alkenes. A typical pK_a for a terminal alkyne is about 26. The greater *s* character of an *sp* orbital (50 percent) compared with an sp^2 orbital (33 percent) explains the greater stability of the acetylide anion. Acetylide anions may be generated by reaction of a terminal acetylene with a suitably strong base, as shown below.

$$\underset{\text{Stronger acid}}{CH_3CH_2C\equiv CH} + \underset{\text{Stronger base}}{NaNH_2} \longrightarrow \underset{\text{Weaker base}}{CH_3CH_2C\equiv C{:}^-Na^+} + \underset{\text{Weaker acid}}{NH_3}$$

IMPORTANT REACTIONS

PREPARATION OF ALKYNES

Alkylation of Acetylide Anions (Sec. 10.7)

$$HC{\equiv}C^-Na^+ + CH_3CH_2CH_2Br \longrightarrow CH_3CH_2CH_2C{\equiv}CH$$

$$CH_3CH_2C{\equiv}C^-Na^+ + C_6H_5CH_2Br \longrightarrow C_6H_5CH_2C{\equiv}CCH_2CH_3$$

Elimination Reactions (Sec. 10.8)

$$(CH_3)_2CHCH_2CHCl_2 \xrightarrow{NaNH_2} (CH_3)_2CHC{\equiv}CH$$

$$CH_3CH_2CH(Br)CH_2Br \xrightarrow{NaNH_2} CH_3CH_2C{\equiv}CH$$

REACTIONS OF ALKYNES

Hydrogenation (Sec. 10.10)

$$R{-}C{\equiv}C{-}R \xrightarrow[Pt]{H_2} RCH_2CH_2R$$

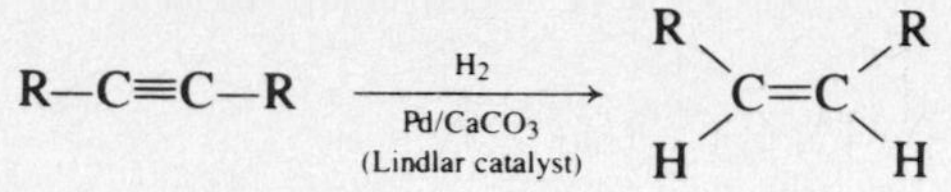

$$R{-}C{\equiv}C{-}R \xrightarrow[\substack{Pd/CaCO_3\\(\text{Lindlar catalyst})}]{H_2} (R)(H)C{=}C(R)(H) \quad \text{(cis isomer only)}$$

Metal-Ammonia Reduction (Sec. 10.11)

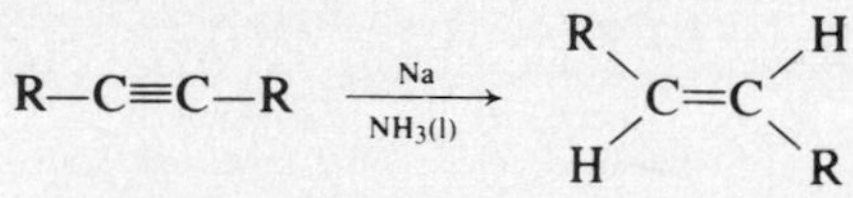

$$R{-}C{\equiv}C{-}R \xrightarrow[NH_3(l)]{Na} (R)(H)C{=}C(H)(R) \quad \text{(trans isomer only)}$$

Hydrogen Halide Addition (Sec. 10.12)

$$CH_3CH_2C{\equiv}CH \xrightarrow[\text{excess}]{HBr} CH_3CH_2CBr_2CH_3$$

(formed via $CH_3CH_2C(Br){=}CH_2$)

Hydration (Sec. 10.13)

$$R{-}C{\equiv}CH \xrightarrow[Hg^{2+}]{H_2O,\ H^+} R{-}\overset{O}{\overset{\|}{C}}{-}CH_3$$

Example:

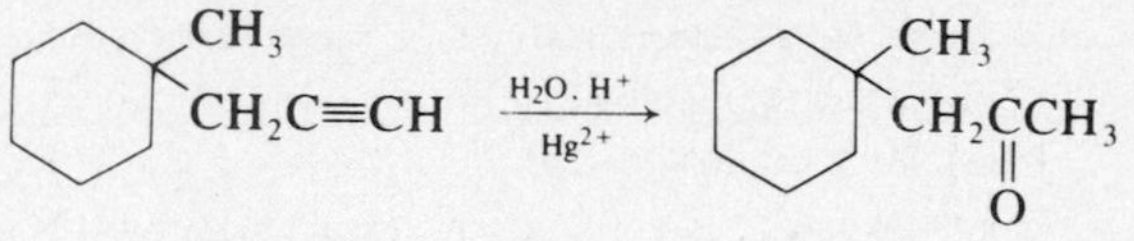

$$\text{1-methyl-1-}(CH_2C{\equiv}CH)\text{cyclohexane} \xrightarrow[Hg^{2+}]{H_2O,\ H^+} \text{1-methyl-1-}(CH_2\overset{O}{\overset{\|}{C}}CH_3)\text{cyclohexane}$$

Halogen Addition (Sec. 10.14)

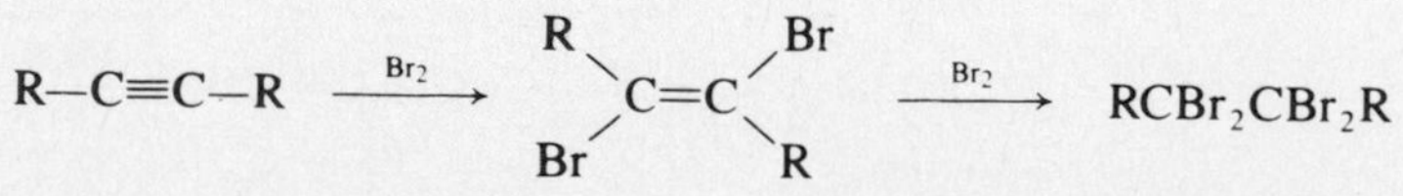

$$R{-}C{\equiv}C{-}R \xrightarrow{Br_2} (R)(Br)C{=}C(Br)(R) \xrightarrow{Br_2} RCBr_2CBr_2R$$

Oxidation (Sec. 10.15)

$$RC{\equiv}CR' \xrightarrow[\text{1. } O_3 \text{ 2. } H_2O]{KMnO_4 \text{ or}} RCO_2H + R'CO_2H$$

SOLUTIONS TO TEXT PROBLEMS

10.1 A triple bond may connect C-1 and C-2 or C-2 and C-3 in an unbranched chain of five carbons.

$$\underset{\text{1-Pentyne}}{CH_3CH_2CH_2C{\equiv}CH} \qquad \underset{\text{2-Pentyne}}{CH_3CH_2C{\equiv}CCH_3}$$

One of the C_5H_8 isomers has a branched carbon chain.

$$\underset{\substack{|\\CH_3}}{CH_3CHC{\equiv}CH} \quad \text{3-Methyl-1-butyne}$$

10.2 The methyl group in 1-butyne is attached to an sp^3 hybridized carbon, so the $CH_3{-}C$ bond in 1-butyne is longer than in 2-butyne, where the methyl carbon is bonded to an sp hybridized carbon.

$$\overset{sp^3\quad sp^3}{CH_3{-}CH_2C{\equiv}CH} \qquad \overset{sp^3\quad sp}{CH_3{-}C{\equiv}C{-}CH_3}$$

longer bond — shorter bond

10.3 (*b*) A proton is transferred from acetylene to ethyl anion.

$$HC{\equiv}C{-}H + :\bar{C}H_2CH_3 \underset{}{\overset{K>>1}{\rightleftharpoons}} HC{\equiv}\bar{C}: + CH_3CH_3$$

Acetylene (stronger acid)	Ethyl anion (stronger base)	Acetylide ion (weaker base)	Ethane (weaker acid)
K_a 10^{-26} (pK_a 26)			$K_a \sim 10^{-62}$ ($pK_a \sim 62$)

The position of equilibrium lies to the right. Ethyl anion is a very powerful base and deprotonates acetylene quantitatively.

(*c*) Amide ion is not a strong enough base to remove a proton from ethylene. The equilibrium lies to the left.

$$CH_2{=}CH{-}H + :\ddot{\bar{N}}H_2 \overset{K<<1}{\rightleftharpoons} CH_2{=}\ddot{\bar{C}}H + :NH_3$$

Ethylene (weaker acid)	Amide ion (weaker base)	Vinyl anion (stronger base)	Ammonia (stronger acid)
$K_a \sim 10^{-45}$ ($pK_a \sim 45$)			K_a 10^{-36} (pK_a 36)

(*d*) Alcohols are stronger acids than ammonia; the position of equilibrium lies to the right.

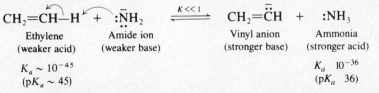

$$CH_3C{\equiv}CCH_2\ddot{O}{-}H + :\ddot{\bar{N}}H_2 \overset{K>>1}{\rightleftharpoons} {}^-CH_3C{\equiv}CCH_2\ddot{O}:^- + :NH_3$$

2-Butyn-1-ol (stronger acid)	Amide ion (stronger base)	2-Butyn-1-oxide anion (weaker base)	Ammonia (weaker acid)
$K_a \sim 10^{-16}$ to 10^{-20} $pK_a \sim 16-20$			K_a 10^{-36} pK_a 36

10.4 (*b*) The desired alkyne has a methyl group and a butyl group attached to a $-C{\equiv}C-$ unit. Therefore, two alkylations of acetylene are required—one with a methyl halide, the other with a butyl halide.

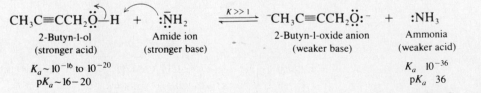

$$\underset{\text{Acetylene}}{HC{\equiv}CH} \xrightarrow[\text{2. } CH_3Br]{\text{1. } NaNH_2,\ NH_3} \underset{\text{Propyne}}{CH_3C{\equiv}CH} \xrightarrow[\text{2. } CH_3CH_2CH_2CH_2Br]{\text{1. } NaNH_2,\ NH_3} \underset{\text{2-Heptyne}}{CH_3C{\equiv}CCH_2CH_2CH_2CH_3}$$

It does not matter whether the methyl group or the butyl group is introduced first; the order of steps shown in the above synthetic scheme may be inverted.

(*c*) An ethyl group and a propyl group need to be introduced as substituents on a —C≡C— unit. As in (*b*), it does not matter which of the two is introduced first.

$$\underset{\text{Acetylene}}{HC{\equiv}CH} \xrightarrow[\text{2. }CH_3CH_2CH_2Br]{\text{1. }NaNH_2,\ NH_3} \underset{\text{1-Pentyne}}{CH_3CH_2CH_2C{\equiv}CH} \xrightarrow[\text{2. }CH_3CH_2Br]{\text{1. }NaNH_2,\ NH_3} \underset{\text{3-Heptyne}}{CH_3CH_2CH_2C{\equiv}CCH_2CH_3}$$

10.5 Both 1-pentyne and 2-pentyne can be prepared by alkylating acetylene. All the alkylation steps involve nucleophilic substitution of a methyl or primary alkyl halide.

$$\underset{\text{Acetylene}}{HC{\equiv}CH} \xrightarrow[\text{2. }CH_3CH_2CH_2Br]{\text{1. }NaNH_2,\ NH_3} \underset{\text{1-Pentyne}}{CH_3CH_2CH_2C{\equiv}CH}$$

$$\underset{\text{Acetylene}}{HC{\equiv}CH} \xrightarrow[\text{2. }CH_3CH_2Br]{\text{1. }NaNH_2,\ NH_3} \underset{\text{1-Butyne}}{CH_3CH_2C{\equiv}CH} \xrightarrow[\text{2. }CH_3Br]{\text{1. }NaNH_2,\ NH_3} \underset{\text{2-Pentyne}}{CH_3CH_2C{\equiv}CCH_3}$$

The third isomer, 3-methyl-1-butyne, cannot be prepared by alkylation of acetylene because it requires a secondary alkyl halide as the alkylating agent. The reaction that takes place elimination, not substitution.

$$\underset{\text{Acetylide ion}}{HC{\equiv}\bar{C}:} + \underset{\text{Isopropyl bromide}}{CH_3\underset{\displaystyle Br}{\underset{|}{C}}HCH_3} \xrightarrow{\text{E2}} \underset{\text{Acetylene}}{HC{\equiv}CH} + \underset{\text{Propene}}{CH_2{=}CHCH_3}$$

10.6 Each of the dibromides shown yields 3,3-dimethyl-1-butyne when subjected to double dehydrohalogenation with strong base.

$$\underset{\text{2,2-Dibromo-3,3-dimethylbutane}}{(CH_3)_3C\overset{\displaystyle Br}{\overset{|}{\underset{\displaystyle Br}{\underset{|}{C}}}}CH_3} \quad \text{or} \quad \underset{\text{1,1-Dibromo-3,3-dimethylbutane}}{(CH_3)_3CCH_2CHBr_2}$$

$$\text{or} \quad \underset{\text{1,2-Dibromo-3,3-dimethylbutane}}{(CH_3)_3C\underset{\displaystyle Br}{\underset{|}{C}}HCH_2Br} \xrightarrow[\text{2. }H_2O]{\text{1. }3NaNH_2} \underset{\text{3,3-Dimethyl-1-butyne}}{(CH_3)_3CC{\equiv}CH}$$

10.7 (*b*) The first task is to convert 1-propanol to propene.

$$\underset{\text{1-Propanol}}{CH_3CH_2CH_2OH} \xrightarrow[\text{heat}]{H_2SO_4} \underset{\text{Propene}}{CH_3CH{=}CH_2}$$

Once propene is available, it is converted to 1,2-dibromopropane and then to propyne as described in the sample solution for (*a*).

(*c*) Treat isopropyl bromide with a base to effect dehydrohalogenation.

$$\underset{\text{Isopropyl bromide}}{(CH_3)_2CHBr} \xrightarrow{NaOCH_2CH_3} \underset{\text{Propene}}{CH_3CH{=}CH_2}$$

Next, convert propene to propyne as in (*a*) and (*b*).

(*d*) The starting material contains only two carbon atoms, so an alkylation step is needed at some point. Propyne arises by alkylation of acetylene, so the last step in the synthesis is:

$$\underset{\text{Acetylene}}{HC{\equiv}CH} \xrightarrow[\text{2. } CH_3Br]{\text{1. } NaNH_2,\ NH_3} \underset{\text{Propyne}}{CH_3C{\equiv}CH}$$

The designated starting material, 1,1-dichloroethane, is a geminal dihalide and can be used to prepare acetylene by a double dehydrohalogenation.

$$\underset{\text{1,1-Dichloroethane}}{CH_3CHCl_2} \xrightarrow[\text{2. } H_2O]{\text{1. } NaNH_2,\ NH_3} \underset{\text{Acetylene}}{HC{\equiv}CH}$$

(*e*) The first task is to convert ethyl alcohol to acetylene. Once acetylene is prepared it can be alkylated with a methyl halide.

$$\underset{\text{Ethyl alcohol}}{CH_3CH_2OH} \xrightarrow[\text{heat}]{H_2SO_4} \underset{\text{Ethylene}}{CH_2{=}CH_2} \xrightarrow{Br_2} \underset{\text{1,2-Dibromoethane}}{BrCH_2CH_2Br} \xrightarrow{NaNH_2} \underset{\text{Acetylene}}{HC{\equiv}CH}$$

$$\underset{\text{Acetylene}}{HC{\equiv}CH} \xrightarrow[\text{2. } CH_3Br]{\text{1. } NaNH_2,\ NH_3} \underset{\text{Propyne}}{CH_3C{\equiv}CH}$$

10.8 Hydrogenation over Lindlar palladium converts an alkyne to a cis alkene. Therefore, oleic acid has the structure indicated in the equation.

$$\underset{\text{Stearolic acid}}{CH_3(CH_2)_7C{\equiv}C(CH_2)_7CO_2H} \xrightarrow[\text{Lindlar Pd}]{H_2} \underset{\text{Oleic acid}}{\overset{CH_3(CH_2)_7\quad\quad(CH_2)_7CO_2H}{\underset{H\quad\quad\quad H}{C{=}C}}}$$

Hydrogenation of alkynes over platinum leads to alkanes.

$$\underset{\text{Stearolic acid}}{CH_3(CH_2)_7C{\equiv}C(CH_2)_7CO_2H} \xrightarrow[Pt]{H_2} \underset{\text{Stearic acid}}{CH_3(CH_2)_{16}CO_2H}$$

10.9 Alkynes are converted to trans alkenes on reduction with sodium in liquid ammonia.

$$\underset{\text{Stearolic acid}}{CH_3(CH_2)_7C{\equiv}C(CH_2)_7CO_2H} \xrightarrow[\text{2. } H_3O^+]{\text{1. Na, } NH_3} \underset{\text{Elaidic acid}}{\overset{CH_3(CH_2)_7\quad\quad H}{\underset{H\quad\quad (CH_2)_7CO_2H}{C{=}C}}}$$

10.10 The proper double-bond stereochemistry may be achieved by using 2-heptyne as a starting material in the final step. Lithium-ammonia reduction of 2-heptyne gives the trans alkene; hydrogenation over Lindlar palladium gives the cis isomer. Therefore, the first task is the alkylation of propyne to 2-heptyne.

$$\underset{\text{Propyne}}{CH_3C{\equiv}CH} \xrightarrow[\text{2. } CH_3CH_2CH_2CH_2Br]{\text{1. } NaNH_2,\ NH_3} \underset{\text{2-Heptyne}}{CH_3C{\equiv}CCH_2CH_2CH_2CH_3}$$

$$\xrightarrow{\text{Li, } NH_3} \underset{(E)\text{-2-Heptene}}{\overset{CH_3\quad\quad H}{\underset{H\quad\quad CH_2CH_2CH_2CH_3}{C{=}C}}}$$

$$\xrightarrow[\text{Lindlar Pd}]{H_2} \underset{(Z)\text{-2-Heptene}}{\overset{CH_3\quad\quad CH_2CH_2CH_2CH_3}{\underset{H\quad\quad H}{C{=}C}}}$$

10.11 First convert ethyl bromide to acetylene; then add hydrogen bromide across the carbon–carbon triple bond.

$$\underset{\text{Ethyl bromide}}{CH_3CH_2Br} \xrightarrow[\text{DMSO, heat}]{KOC(CH_3)_3} \underset{\text{Ethylene}}{CH_2{=}CH_2} \xrightarrow{Br_2} \underset{\text{1,2-Dibromoethane}}{BrCH_2CH_2Br} \xrightarrow[\text{2. } H_2O]{\text{1. } NaNH_2}$$

$$HC{\equiv}CH \xrightarrow{HBr} \underset{\text{Vinyl bromide}}{CH_2{=}CHBr}$$

Alternatively, treatment of 1,2-dibromoethane with a base such as KOH or $NaOCH_2CH_3$ brings about a single dehydrohalogenation to give vinyl bromide.

$$\underset{\text{1,2-Dibromoethane}}{BrCH_2CH_2Br} \xrightarrow{NaOCH_2CH_3} \underset{\text{Vinyl bromide}}{CH_2{=}CHBr}$$

10.12 (*b*) Addition of hydrogen chloride to vinyl chloride gives the geminal dichloride 1,1-dichloroethane.

$$\underset{\text{Vinyl chloride}}{CH_2{=}CHCl} \xrightarrow{HCl} \underset{\text{1,1-Dichloroethane}}{CH_3CHCl_2}$$

(*c*) Since 1,1-dichloroethane can be prepared by adding two moles of hydrogen chloride to acetylene, first convert vinyl bromide to acetylene by dehydrohalogenation.

$$\underset{\text{Vinyl bromide}}{CH_2{=}CHBr} \xrightarrow[\text{2. } H_2O]{\text{1. } NaNH_2,\ NH_3} \underset{\text{Acetylene}}{HC{\equiv}CH} \xrightarrow{2HCl} \underset{\text{1,1-Dichloroethane}}{CH_3CHCl_2}$$

(*d*) As in (*c*), first convert the designated starting material to acetylene.

$$\underset{\text{1,1-Dibromoethane}}{CH_3CHBr_2} \xrightarrow[\text{2. } H_2O]{\text{1. } NaNH_2,\ NH_3} \underset{\text{Acetylene}}{HC{\equiv}CH} \xrightarrow{2HCl} \underset{\text{1,1-Dichloroethane}}{CH_3CHCl_2}$$

10.13 The enol arises by addition of water across the triple bond.

$$\underset{\text{2-Butyne}}{CH_3C{\equiv}CCH_3} \xrightarrow[H_2SO_4]{H_2O,\ Hg^{2+}} \underset{\text{2-Buten-2-ol (enol form)}}{\left[CH_3\overset{}{\underset{OH}{C}}{=}CHCH_3\right]} \longrightarrow \underset{\text{2-Butanone}}{CH_3\overset{O}{\overset{\|}{C}}CH_2CH_3}$$

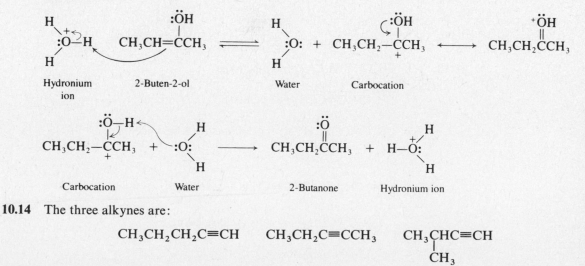

The mechanism described in Figure 10.6 is adapted to the case of 2-butyne hydration as shown:

$$\underset{\text{Hydronium ion}}{H_2\overset{+}{O}{-}H} + \underset{\text{2-Buten-2-ol}}{CH_3CH{=}\overset{:\ddot{O}H}{\overset{|}{C}}CH_3} \rightleftharpoons \underset{\text{Water}}{H_2\ddot{O}:} + \underset{\text{Carbocation}}{CH_3CH_2{-}\overset{:\ddot{O}H}{\overset{|}{\underset{+}{C}}}CH_3} \longleftrightarrow CH_3CH_2\overset{^{+}\ddot{O}H}{\overset{\|}{C}}CH_3$$

$$\underset{\text{Carbocation}}{CH_3CH_2{-}\overset{:\ddot{O}{-}H}{\overset{|}{\underset{+}{C}}}CH_3} + \underset{\text{Water}}{H_2\ddot{O}:} \longrightarrow \underset{\text{2-Butanone}}{CH_3CH_2\overset{:\ddot{O}}{\overset{\|}{C}}CH_3} + \underset{\text{Hydronium ion}}{H{-}\overset{+}{\ddot{O}}H_2}$$

10.14 The three alkynes are:

$$CH_3CH_2CH_2C{\equiv}CH \qquad CH_3CH_2C{\equiv}CCH_3 \qquad CH_3\underset{CH_3}{\underset{|}{C}}HC{\equiv}CH$$

Electron-releasing alkyl substituents increase the rate of electrophilic addition to alkynes. The disubstituted alkyne $CH_3CH_2C{\equiv}CCH_3$ is the most reactive of the three isomers.

10.15 Each of the carbons that are part of $-CO_2H$ groups was once part of a $-C{\equiv}C-$ unit. The two fragments $CH_3(CH_2)_4CO_2H$ and $HO_2CCH_2CH_2CO_2H$ account for only 10 of the original 16 carbons. The full complement of carbons can be accommodated by assuming that two molecules of $CH_3(CH_2)_4CO_2H$ are formed, along with one molecule of $HO_2CCH_2CH_2CO_2H$. The starting alkyne is therefore deduced from the ozonolysis data to be as shown:

$$CH_3(CH_2)_4C{\equiv}CCH_2CH_2C{\equiv}C(CH_2)_4CH_3$$

$$\swarrow \qquad \downarrow \qquad \searrow$$

$$CH_3(CH_2)_4CO_2H \qquad HO_2CCH_2CH_2CO_2H \qquad HO_2C(CH_2)_4CH_3$$

10.16 The unknown compound must have a terminal triple bond ($RC{\equiv}CH$) because these are the only compounds that give a red precipitate on reaction with cuprous chloride in aqueous ammonia. The hydrogenation data reveal that it must have the same carbon skeleton as 2-methylpentane.

$$\underset{\text{4-Methyl-1-pentyne}}{CH_3\underset{|}{\underset{CH_3}{C}}HCH_2C{\equiv}CH} + 2H_2 \xrightarrow{\text{metal catalyst}} \underset{\text{2-Methylpentane}}{CH_3\underset{CH_3}{\underset{|}{C}}HCH_2CH_2CH_3}$$

The unknown compound must be 4-methyl-1-pentyne.

10.17 There are three isomers that have unbranched carbon chains:

$$\underset{\text{1-Hexyne}}{CH_3CH_2CH_2CH_2C{\equiv}CH} \qquad \underset{\text{2-Hexyne}}{CH_3CH_2CH_2C{\equiv}CCH_3} \qquad \underset{\text{3-Hexyne}}{CH_3CH_2C{\equiv}CCH_2CH_3}$$

Next consider all the alkynes with a single methyl branch.

$$\underset{\text{4-Methyl-1-pentyne}}{CH_3\underset{CH_3}{\underset{|}{C}}HCH_2C{\equiv}CH} \qquad \underset{\text{3-Methyl-1-pentyne}}{CH_3CH_2\underset{CH_3}{\underset{|}{C}}HC{\equiv}CH} \qquad \underset{\text{4-Methyl-2-pentyne}}{CH_3\underset{CH_3}{\underset{|}{C}}HC{\equiv}CCH_3}$$

There is one isomer with two methyl branches. None is possible with an ethyl branch.

$$\underset{\text{3,3-Dimethyl-1-butyne}}{CH_3\overset{CH_3}{\underset{CH_3}{C}}C{\equiv}CH}$$

10.18 (*a*) $\overset{5}{C}H_3\overset{4}{C}H_2\overset{3}{C}H_2\overset{2}{C}{\equiv}\overset{1}{C}H$ is 1-pentyne

(*b*) $\overset{5}{C}H_3\overset{4}{C}H_2\overset{3}{C}{\equiv}\overset{2}{C}\overset{1}{C}H_3$ is 2-pentyne

(*c*) $\overset{1}{C}H_3\overset{2}{C}{\equiv}\overset{3}{C}\overset{4}{\underset{CH_3}{C}}H\overset{5}{\underset{CH_3}{C}}H\overset{6}{C}H_3$ is 4,5-dimethyl-2-hexyne

(*d*) $\triangleright\!-\overset{5}{C}H_2\overset{4}{C}H_2\overset{3}{C}H_2\overset{2}{C}{\equiv}\overset{1}{C}H$ is 5-cyclopropyl-1-pentyne

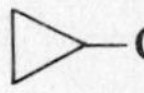

(*e*) $\overset{13}{C}H_2\overset{1}{C}{\equiv}\overset{2}{C}\overset{3}{C}H_2$ (in a thirteen-membered ring) is cyclotridecyne

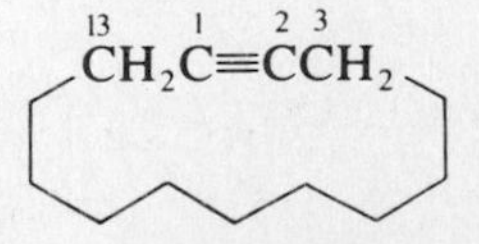

(*f*)
```
        4 5 6 7 8 9
CH3CH2CH2CH2CHCH2CH2CH2CH2CH3   is 4-butyl-2-nonyne
            |
            C≡CCH3
            3  2 1
```
(Parent chain must contain the triple bond)

(*g*)
```
   CH3  CH3
    |    |
CH3CC≡CCCH3   is 2,2,5,5-tetramethyl-3-hexyne
    |    |
   CH3  CH3
```

(*h*)
```
   CH3      CH3
    |        |
CH3C—CH—C—CH3   is 3-tert-butyl-4,4-dimethyl-1-pentyne
    |   |    |
  H3C   C   CH3
        |||
        C
        H
```

10.19 (*a*) 1-Octyne is $HC{\equiv}CCH_2CH_2CH_2CH_2CH_2CH_3$

(*b*) 2-Octyne is $CH_3C{\equiv}CCH_2CH_2CH_2CH_2CH_3$

(*c*) 3-Octyne is $CH_3CH_2C{\equiv}CCH_2CH_2CH_2CH_3$

(*d*) 4-Octyne is $CH_3CH_2CH_2C{\equiv}CCH_2CH_2CH_3$

(*e*) 2,5-Dimethyl-3-hexyne is
```
CH3CHC≡CCHCH3
    |    |
   CH3  CH3
```

(*f*) 4-Ethyl-1-hexyne is
```
CH3CH2CHCH2C≡CH
      |
     CH2CH3
```

(*g*) Ethynylcyclohexane is (cyclohexane ring)—C≡CH

(*h*) 3-Ethyl-3-methyl-1-pentyne is
```
       CH3
        |
CH3CH2CC≡CH
        |
       CH2CH3
```

10.20 Ethynylcyclohexane has the molecular formula C_8H_{12}. All the other compounds are C_8H_{14}.

10.21 The formation of a red precipitate with ammoniacal cuprous chloride requires that there be a terminal triple bond. The alkynes which will give a red precipitate are 1-octyne, 4-ethyl-1-hexyne, ethynylcyclohexane, and 3-ethyl-3-methyl-1-pentyne.

$$RC{\equiv}CH + Cu^+ \longrightarrow \underset{\text{Red precipitate}}{RC{\equiv}CCu} + H^+$$

10.22 The carbon skeleton of the unknown acetylenic amino acid must be the same as that of homoleucine. The structure of homoleucine is such that there is only one possible location for a carbon–carbon triple bond in an acetylenic precursor.

```
                 O                                 O
                 ||                                ||
HC≡CCHCH2CHCOH      --H2, Pt-->   CH3CH2CHCH2CHCOH
     |      |                            |     |
    CH3    NH2                          CH3   NH2
  C7H11NO2                             Homoleucine
```

10.23 (*a*) $\underset{\text{1,1-Dichlorohexane}}{CH_3CH_2CH_2CH_2CH_2CHCl_2} \xrightarrow[\text{2. } H_2O]{\text{1. } NaNH_2,\ NH_3} \underset{\text{1-Hexyne}}{CH_3CH_2CH_2CH_2C{\equiv}CH}$

(*b*) $CH_3CH_2CH_2CH_2CH{=}CH_2 \xrightarrow[CCl_4]{Br_2} CH_3CH_2CH_2CH_2CH(Br)CH_2Br$

1-Hexene → 1,2-Dibromohexane

$\xrightarrow[2.\ H_2O]{1.\ NaNH_2,\ NH_3} CH_3CH_2CH_2CH_2C{\equiv}CH$

1-Hexyne

(*c*) $HC{\equiv}CH \xrightarrow[NH_3]{NaNH_2} HC{\equiv}C^-Na^+ \xrightarrow{CH_3CH_2CH_2CH_2Br} CH_3CH_2CH_2CH_2C{\equiv}CH$

Acetylene → 1-Hexyne

(*d*) $CH_3CH_2CH_2CH_2CH_2CH_2I \xrightarrow[DMSO]{KOC(CH_3)_3} CH_3CH_2CH_2CH_2CH{=}CH_2$

1-Iodohexane → 1-Hexene

1-Hexene is then converted to 1-hexyne as in (*b*).

(*e*) $(E)\text{-}CH_3CH_2CH_2CH_2CH{=}CHBr \xrightarrow[NH_3]{NaNH_2} CH_3CH_2CH_2CH_2C{\equiv}CH$

(*E*)-1-Bromohexene → 1-Hexyne

10.24 (*a*) Working backwards from the final product, it can be seen that preparation of 1-butyne will allow the desired carbon skeleton to be constructed.

$$CH_3CH_2C{\equiv}CCH_2CH_3 \quad \text{prepared from} \quad CH_3CH_2C{\equiv}C{:}^- + BrCH_2CH_3$$

3-Hexyne

The desired intermediate 1-butyne is available by halogenation followed by dehydrohalogenation of 1-butene.

$$CH_3CH_2CH{=}CH_2 + Br_2 \longrightarrow CH_3CH_2CH(Br)CH_2Br \xrightarrow[2.\ H_2O]{1.\ NaNH_2,\ NH_3} CH_3CH_2C{\equiv}CH$$

1-Butene → 1-Butyne

Reaction of the anion of 1-butyne with ethyl bromide completes the synthesis.

$$CH_3CH_2C{\equiv}CH \xrightarrow[NH_3]{NaNH_2} CH_3CH_2C{\equiv}C{:}^-Na^+ \xrightarrow{CH_3CH_2Br} CH_3CH_2C{\equiv}CCH_2CH_3$$

1-Butyne → 3-Hexyne

(*b*) Dehydrohalogenation of 1,1-dichlorobutane yields 1-butyne. The synthesis is completed as in (*a*).

$$CH_3CH_2CH_2CHCl_2 \xrightarrow[2.\ H_2O]{1.\ NaNH_2,\ NH_3} CH_3CH_2C{\equiv}CH$$

1,1-Dichlorobutane → 1-Butyne

(*c*) Once again dehydrohalogenation with strong base yields 1-butyne, allowing the synthesis to be completed as in (*a*).

$$CH_3CH_2CH{=}CHCl \xrightarrow[2.\ H_2O]{1.\ NaNH_2,\ NH_3} CH_3CH_2C{\equiv}CH$$

1-Chloro-1-butene → 1-Butyne

(*d*) $HC{\equiv}CH \xrightarrow[NH_3]{NaNH_2} HC{\equiv}C^-Na^+ \xrightarrow{CH_3CH_2Br} HC{\equiv}CCH_2CH_3$

Acetylene → 1-Butyne

1-Butyne is converted to 3-hexyne as in (*a*).

10.25 A single dehydrobromination step occurs in the conversion of 1,2-dibromodecane to $C_{10}H_{19}Br$. Bromine may be lost from C-1 to give 2-bromo-1-decene.

$$\underset{\text{1,2-Dibromodecane}}{BrCH_2\underset{\substack{|\\Br}}{C}H(CH_2)_7CH_3} \xrightarrow[\text{ethanol-water}]{KOH} \underset{\text{2-Bromo-1-decene}}{CH_2{=}\underset{\substack{|\\Br}}{C}(CH_2)_7CH_3}$$

Loss of bromine from C-2 gives (*E*)- and (*Z*)-1-bromo-1-decene.

$$\underset{\text{1,2-Dibromodecane}}{BrCH_2\underset{\substack{|\\Br}}{C}H(CH_2)_7CH_3} \xrightarrow[\text{ethanol-water}]{KOH} \underset{(E)\text{-1-Bromo-1-decene}}{(H)(Br)C{=}C((CH_2)_7CH_3)(H)} + \underset{(Z)\text{-1-Bromo-1-decene}}{(Br)(H)C{=}C((CH_2)_7CH_3)(H)}$$

10.26 (*a*) $$\underset{\text{1-Hexyne}}{CH_3CH_2CH_2CH_2C{\equiv}CH} + 2H_2 \xrightarrow{Pt} \underset{\text{Hexane}}{CH_3CH_2CH_2CH_2CH_2CH_3}$$

(*b*) $$\underset{\text{1-Hexyne}}{CH_3CH_2CH_2CH_2C{\equiv}CH} + H_2 \xrightarrow{\text{Lindlar Pd}} \underset{\text{1-Hexene}}{CH_3CH_2CH_2CH_2CH{=}CH_2}$$

(*c*) $$\underset{\text{1-Hexyne}}{CH_3CH_2CH_2CH_2C{\equiv}CH} \xrightarrow[NH_3]{Li} \underset{\text{1-Hexene}}{CH_3CH_2CH_2CH_2CH{=}CH_2}$$

(*d*) $$\underset{\text{1-Hexyne}}{CH_3CH_2CH_2CH_2C{\equiv}CH} \xrightarrow[NH_3]{NaNH_2} \underset{\text{Sodium 1-hexynide}}{CH_3CH_2CH_2CH_2C{\equiv}C^-Na^+}$$

(*e*) $$\underset{\text{Sodium 1-hexynide}}{CH_3CH_2CH_2CH_2C{\equiv}C^-Na^+} + \underset{\text{1-Bromobutane}}{CH_3CH_2CH_2CH_2Br} \longrightarrow \underset{\text{5-Decyne}}{CH_3CH_2CH_2CH_2C{\equiv}CCH_2CH_2CH_2CH_3}$$

(*f*) $$\underset{\text{Sodium 1-hexynide}}{CH_3CH_2CH_2CH_2C{\equiv}C^-Na^+} + \underset{\textit{tert}\text{-Butyl bromide}}{(CH_3)_3CBr} \longrightarrow \underset{\text{1-Hexyne}}{CH_3CH_2CH_2CH_2C{\equiv}CH} + \underset{\text{2-Methylpropene (isobutene)}}{(CH_3)_2C{=}CH_2}$$

(*g*) $$\underset{\text{1-Hexyne}}{CH_3CH_2CH_2CH_2C{\equiv}CH} \xrightarrow[\text{(1 mol)}]{HCl} \underset{\text{2-Chloro-1-hexene}}{CH_3CH_2CH_2CH_2\underset{\substack{|\\Cl}}{C}{=}CH_2}$$

(*h*) $$\underset{\text{1-Hexyne}}{CH_3CH_2CH_2CH_2C{\equiv}CH} \xrightarrow[\text{(2 mol)}]{HCl} \underset{\text{2,2-Dichlorohexane}}{CH_3CH_2CH_2CH_2\overset{\substack{Cl\\|}}{\underset{\substack{|\\Cl}}{C}}CH_3}$$

(*i*) $$\underset{\text{1-Hexyne}}{CH_3CH_2CH_2CH_2C{\equiv}CH} \xrightarrow[\text{(1 mol)}]{Cl_2} \underset{(E)\text{-1,2-Dichloro-1-hexene}}{(CH_3CH_2CH_2CH_2)(Cl)C{=}C(Cl)(H)}$$

(*j*) $CH_3CH_2CH_2CH_2C{\equiv}CH \xrightarrow[\text{(2 mol)}]{Cl_2} CH_3CH_2CH_2CH_2CCl_2CHCl_2$

1-Hexyne → 1,1,2,2-Tetrachlorohexane

(*k*) $CH_3CH_2CH_2CH_2C{\equiv}CH \xrightarrow[HgSO_4]{H_2O,\ H_2SO_4} CH_3CH_2CH_2CH_2\overset{O}{\overset{\|}{C}}CH_3$

1-Hexyne → 2-Hexanone

(*l*) $CH_3CH_2CH_2CH_2C{\equiv}CH \xrightarrow[NH_3,\ H_2O]{CuCl} CH_3CH_2CH_2CH_2C{\equiv}CCu$

1-Hexyne → Cuprous 1-hexynide

(*m*) $CH_3CH_2CH_2CH_2C{\equiv}CH \xrightarrow[NH_3,\ H_2O]{AgNO_3} CH_3CH_2CH_2CH_2C{\equiv}CAg$

1-Hexyne → Silver 1-hexynide

(*n*) $CH_3CH_2CH_2CH_2C{\equiv}CH \xrightarrow[2.\ H_2O]{1.\ O_3} CH_3CH_2CH_2CH_2\overset{O}{\overset{\|}{C}}OH + HO\overset{O}{\overset{\|}{C}}H$

1-Hexyne → Pentanoic acid + Formic acid

10.27 (*a*) $CH_3CH_2C{\equiv}CCH_2CH_3 + 2H_2 \xrightarrow{Pt} CH_3CH_2CH_2CH_2CH_2CH_3$

3-Hexyne → Hexane

(*b*) $CH_3CH_2C{\equiv}CCH_2CH_3 + H_2 \xrightarrow[\text{Pb(OAc)}_2,\ \text{quinoline (Lindlar palladium)}]{Pd/CaCO_3}$ (CH₃CH₂ and CH₂CH₃ cis; H and H cis) $CH_3CH_2CH{=}CHCH_2CH_3$

3-Hexyne → (*Z*)-3-Hexene

(*c*) $CH_3CH_2C{\equiv}CCH_2CH_3 \xrightarrow[NH_3]{Li} CH_3CH_2CH{=}CHCH_2CH_3$ (CH₃CH₂ and CH₂CH₃ trans)

3-Hexyne → (*E*)-3-Hexene

(*d*) $CH_3CH_2C{\equiv}CCH_2CH_3 \xrightarrow[\text{(1 mol)}]{HCl} CH_3CH_2C(Cl){=}CHCH_2CH_3$ (CH₃CH₂ and H cis; Cl and CH₂CH₃ cis)

3-Hexyne → (*Z*)-3-Chloro-3-hexene

(*e*) $CH_3CH_2C{\equiv}CCH_2CH_3 \xrightarrow[\text{(2 mol)}]{HCl} CH_3CH_2CCl_2CH_2CH_2CH_3$

3-Hexyne → 3,3-Dichlorohexane

(*f*) $CH_3CH_2C{\equiv}CCH_2CH_3 \xrightarrow[\text{(1 mol)}]{Cl_2} CH_3CH_2C(Cl){=}C(Cl)CH_2CH_3$ (CH₃CH₂ and Cl cis; Cl and CH₂CH₃ cis)

3-Hexyne → (*E*)-3,4-Dichloro-3-hexene

(*g*) $CH_3CH_2C{\equiv}CCH_2CH_3 \xrightarrow[\text{(2 mol)}]{Cl_2} CH_3CH_2CCl_2{-}CCl_2CH_2CH_3$

3-Hexyne → 3,3,4,4-Tetrachlorohexane

(*h*) $CH_3CH_2C{\equiv}CCH_2CH_3 \xrightarrow[HgSO_4]{H_2O,\ H_2SO_4} CH_3CH_2\overset{\overset{O}{\|}}{C}CH_2CH_2CH_3$

3-Hexyne → 3-Hexanone

(*i*) $CH_3CH_2C{\equiv}CCH_2CH_3 \xrightarrow[NH_3]{CuCl}$ no reaction (requires $RC{\equiv}CH$)

3-Hexyne

(*j*) $CH_3CH_2C{\equiv}CCH_2CH_3 \xrightarrow[NH_3]{AgNO_3}$ no reaction (requires $RC{\equiv}CH$)

3-Hexyne

(*k*) $CH_3CH_2C{\equiv}CCH_2CH_3 \xrightarrow[2.\ H_2O]{1.\ O_3} 2\,CH_3CH_2\overset{\overset{O}{\|}}{C}OH$

3-Hexyne → Propanoic acid

10.28 The two carbons of the triple bond are similarly but not identically substituted in $CH_3C{\equiv}CCH_2CH_2CH_2CH_3$. Two regioisomeric enols are formed, each of which gives a different ketone.

$CH_3C{\equiv}CCH_2CH_2CH_2CH_3 \xrightarrow[HgSO_4]{H_2O,\ H_2SO_4}$

2-Heptyne

$CH_3\overset{\overset{OH}{|}}{C}{=}CHCH_2CH_2CH_2CH_3 + CH_3CH{=}\overset{\overset{OH}{|}}{C}CH_2CH_2CH_2CH_3$

2-Hepten-2-ol + 2-Hepten-3-ol

↓ ↓

$CH_3\overset{\overset{O}{\|}}{C}CH_2CH_2CH_2CH_2CH_3$ and $CH_3CH_2\overset{\overset{O}{\|}}{C}CH_2CH_2CH_2CH_3$

2-Heptanone and 3-Heptanone

10.29 The alkane formed by hydrogenation of (*S*)-3-methyl-1-pentyne is achiral; it cannot be optically active.

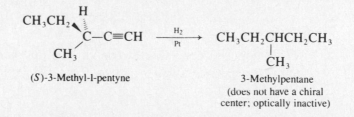

$(CH_3CH_2)(CH_3)(H)C{-}C{\equiv}CH \xrightarrow[Pt]{H_2} CH_3CH_2\underset{\underset{CH_3}{|}}{C}HCH_2CH_3$

(*S*)-3-Methyl-1-pentyne → 3-Methylpentane (does not have a chiral center; optically inactive)

The product of hydrogenation of (*S*)-4-methyl-1-hexyne is optically active because a chiral center is present in the starting material and is carried through to the product.

$(CH_3CH_2)(CH_3)(H)C{-}CH_2C{\equiv}CH \xrightarrow[Pt]{H_2} (CH_3CH_2)(CH_3)(H)CCH_2CH_2CH_3$

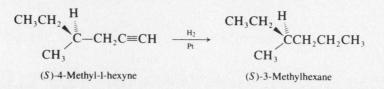

(*S*)-4-Methyl-1-hexyne → (*S*)-3-Methylhexane

Both (*S*)-3-methyl-1-pentyne and (*S*)-4-methyl-1-hexyne yield optically active products when their triple bonds are reduced to double bonds.

10.30 (*a*) The dihaloalkane contains both a primary alkyl chloride and a primary alkyl iodide functional group. Iodide is a better leaving group than chloride and is the one replaced by acetylide.

$$\underset{\text{Sodium acetylide}}{NaC{\equiv}CH} + \underset{\text{1-Chloro-6-iodohexane}}{ClCH_2CH_2CH_2CH_2CH_2CH_2I} \longrightarrow \underset{\text{8-Chloro-1-octyne}}{ClCH_2CH_2CH_2CH_2CH_2CH_2C{\equiv}CH}$$

(*b*) Both vicinal dibromide functions are converted to alkyne units on treatment with excess sodium amide.

$$\underset{\text{1,2,5,6-Tetrabromohexane}}{BrCH_2\underset{\displaystyle Br}{\underset{|}{C}}HCH_2CH_2\underset{\displaystyle Br}{\underset{|}{C}}HCH_2Br} \xrightarrow[\text{2. } H_2O]{\text{1. excess } NaNH_2,\ NH_3} \underset{\text{1,5-Hexadiyne}}{HC{\equiv}CCH_2CH_2C{\equiv}CH}$$

(*c*) The starting material is a geminal dichloride. Potassium *tert*-butoxide in dimethyl sulfoxide is a sufficiently strong base to convert it to an alkyne.

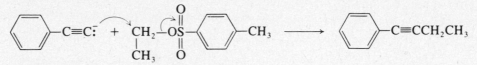

$$\underset{\text{1,1-Dichloro-1-cyclopropylethane}}{\triangleright\!-\overset{\displaystyle Cl}{\overset{|}{\underset{\displaystyle Cl}{\underset{|}{C}}}}CH_3} \xrightarrow[\text{heat}]{KOC(CH_3)_3,\ \text{DMSO}} \underset{\text{Ethynylcyclopropane}}{\triangleright\!-C{\equiv}CH}$$

(*d*) Alkyl *p*-toluenesulfonates react similarly to alkyl halides in nucleophilic substitution reactions. The alkynide nucleophile displaces the *p*-toluenesulfonate leaving group from ethyl *p*-toluenesulfonate.

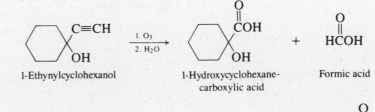

$$\underset{\text{Phenylacetylide ion}}{C_6H_5-C{\equiv}\ddot{C}{:}^-} + \underset{\text{Ethyl } p\text{-toluenesulfonate}}{\underset{\displaystyle CH_3}{\underset{|}{C}}H_2-O\overset{\displaystyle O}{\overset{\|}{\underset{\displaystyle O}{\underset{\|}{S}}}}-C_6H_4-CH_3} \longrightarrow \underset{\text{1-Phenyl-1-butyne}}{C_6H_5-C{\equiv}CCH_2CH_3}$$

(*e*) Both carbons of a $-C{\equiv}C-$ unit are converted to carboxyl groups ($-CO_2H$) on ozonolysis.

$$\underset{\text{Cyclodecyne}}{\text{(cyclodecyne)}} \xrightarrow[\text{2. } H_2O]{\text{1. } O_3} \underset{\text{Decanedioic acid}}{HO\overset{\displaystyle O}{\overset{\|}{C}}(CH_2)_8\overset{\displaystyle O}{\overset{\|}{C}}OH}$$

(*f*) Ozonolysis cleaves the carbon–carbon triple bond.

$$\underset{\text{1-Ethynylcyclohexanol}}{\text{(cyclohexane with } C{\equiv}CH \text{ and } OH\text{)}} \xrightarrow[\text{2. } H_2O]{\text{1. } O_3} \underset{\text{1-Hydroxycyclohexanecarboxylic acid}}{\text{(cyclohexane with } \overset{\displaystyle O}{\overset{\|}{C}}OH \text{ and } OH\text{)}} + \underset{\text{Formic acid}}{H\overset{\displaystyle O}{\overset{\|}{C}}OH}$$

(*g*) Hydration of a terminal carbon–carbon triple bond converts it to a $-\overset{\displaystyle O}{\overset{\|}{C}}CH_3$ group.

$$\underset{\text{3,5-Dimethyl-1-hexyn-3-ol}}{CH_3\underset{\displaystyle CH_3}{\underset{|}{C}}HCH_2\overset{\displaystyle OH}{\overset{|}{\underset{\displaystyle CH_3}{\underset{|}{C}}}}C{\equiv}CH} \xrightarrow[HgO]{H_2O,\ H_2SO_4} \underset{\text{3-Hydroxy-3,5-dimethyl-2-hexanone}}{CH_3\underset{\displaystyle CH_3}{\underset{|}{C}}HCH_2\overset{\displaystyle HO}{\overset{|}{\underset{\displaystyle CH_3}{\underset{|}{C}}}}-\overset{\displaystyle O}{\overset{\|}{C}}CH_3}$$

(*h*) Sodium-in-ammonia reduction of an alkyne yields a trans alkene. The stereochemistry of a double bond that is already present in the molecule is, of course, not altered during the process.

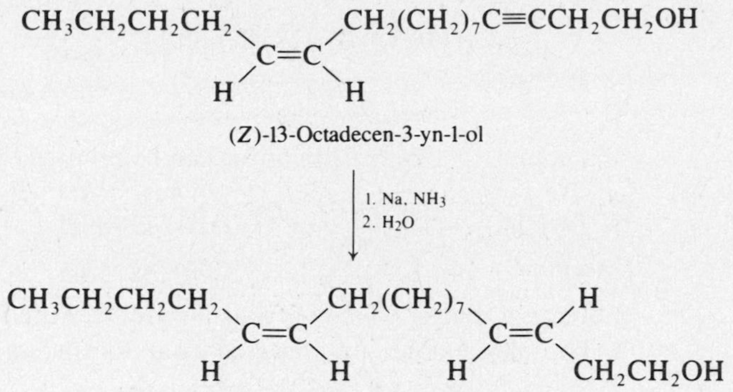

$$\underset{H}{\overset{CH_3CH_2CH_2CH_2}{\diagdown}}C{=}C\underset{H}{\overset{CH_2(CH_2)_7C{\equiv}CCH_2CH_2OH}{\diagup}}$$

(*Z*)-13-Octadecen-3-yn-1-ol

$$\downarrow \text{1. Na, NH}_3\text{; 2. H}_2\text{O}$$

$$\underset{H}{\overset{CH_3CH_2CH_2CH_2}{\diagdown}}C{=}C\underset{H}{\overset{CH_2(CH_2)_7}{\diagup}}\;\; \underset{H}{}C{=}C\underset{CH_2CH_2OH}{\overset{H}{}}$$

(3*E*, 13*Z*)-3,13-Octadecadien-1-ol

(*i*) The primary chloride leaving group is displaced by the alkynide nucleophile.

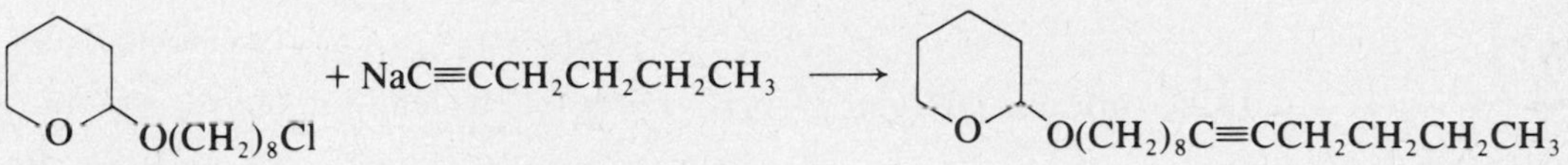

$$\text{(tetrahydropyranyl)}{-}O(CH_2)_8Cl + NaC{\equiv}CCH_2CH_2CH_2CH_3 \longrightarrow \text{(tetrahydropyranyl)}{-}O(CH_2)_8C{\equiv}CCH_2CH_2CH_2CH_3$$

8-Chlorooctyl tetrahydropyranyl ether — Sodium 1-hexynide — 9-Tetradecyn-1-yl tetrahydropyranyl ether

(*j*) Hydrogenation of the triple bond over the Lindlar catalyst converts it to a cis alkene.

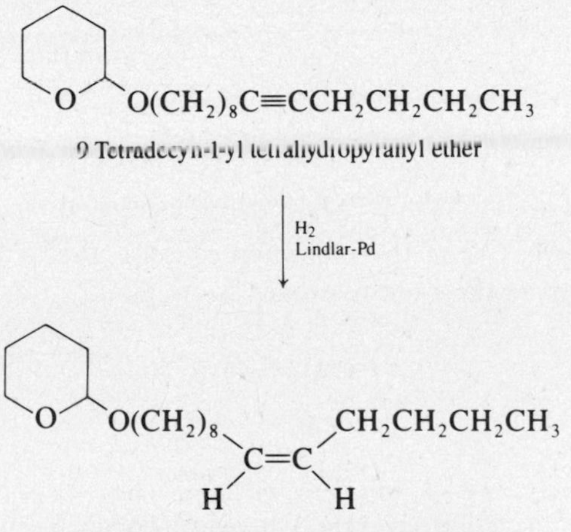

$$\text{(tetrahydropyranyl)}{-}O(CH_2)_8C{\equiv}CCH_2CH_2CH_2CH_3$$

9-Tetradecyn-1-yl tetrahydropyranyl ether

$$\downarrow H_2,\ \text{Lindlar-Pd}$$

$$\text{(tetrahydropyranyl)}{-}\underset{H}{\overset{O(CH_2)_8}{\diagdown}}C{=}C\underset{H}{\overset{CH_2CH_2CH_2CH_3}{\diagup}}$$

(*Z*)-9-Tetradecen-1-yl tetrahydropyranyl ether

10.31 Ketones such as 2-heptanone may be readily prepared by hydration of terminal acetylenes. Thus, if we had 1-heptyne, it could be converted to 2-heptanone.

$$HC{\equiv}C(CH_2)_4CH_3 \xrightarrow[H_2SO_4,\ HgSO_4]{H_2O} CH_3\overset{\overset{\displaystyle O}{\|}}{C}(CH_2)_4CH_3$$

1-Heptyne — 2-Heptanone

Acetylene, as we have seen in earlier problems, can be converted to 1-heptyne by alkylation.

$$HC{\equiv}CH \xrightarrow[NH_3]{NaNH_2} HC{\equiv}C^-Na^+$$

$$HC{\equiv}C^-Na^+ + CH_3CH_2CH_2CH_2CH_2Br \longrightarrow HC{\equiv}C(CH_2)_4CH_3$$

10.32 Apply the technique of reasoning backward to gain a clue as to how to attack this synthetic problem. A reasonable final step is the formation of the *Z* double bond by semi-hydrogenation of an alkyne over Lindlar palladium.

$$\underset{\text{9-Tricosyne}}{CH_3(CH_2)_7C{\equiv}C(CH_2)_{12}CH_3} \xrightarrow[\text{Lindlar-Pd}]{H_2} \underset{(Z)\text{-9-Tricosene}}{\overset{CH_3(CH_2)_7\qquad(CH_2)_{12}CH_3}{\underset{H\qquad\qquad H}{C{=}C}}}$$

The necessary alkyne 9-tricosyne can be prepared by a double alkylation of acetylene.

$$\underset{\text{Acetylene}}{HC{\equiv}CH} \xrightarrow[\text{2. } CH_3(CH_2)_7Br]{\text{1. } NaNH_2,\ NH_3} \underset{\text{1-Decyne}}{CH_3(CH_2)_7C{\equiv}CH} \xrightarrow[\text{2. } CH_3(CH_2)_{12}Br]{\text{1. } NaNH_2,\ NH_3} \underset{\text{9-Tricosyne}}{CH_3(CH_2)_7C{\equiv}C(CH_2)_{12}CH_3}$$

It does not matter which alkyl group is introduced first.

The alkyl halides are prepared from the corresponding alcohols.

$$\underset{\text{1-Octanol}}{CH_3(CH_2)_7OH} \xrightarrow[\text{or } PBr_3]{HBr} \underset{\text{1-Bromooctane}}{CH_3(CH_2)_7Br}$$

$$\underset{\text{1-Tridecanol}}{CH_3(CH_2)_{12}OH} \xrightarrow[\text{or } PBr_3]{HBr} \underset{\text{1-Bromotridecane}}{CH_3(CH_2)_{12}Br}$$

10.33 (*a*) 2,2-Dibromopropane is prepared by addition of hydrogen bromide to propyne.

$$\underset{\text{Propyne}}{CH_3C{\equiv}CH} + \underset{\text{Hydrogen bromide}}{2HBr} \longrightarrow \underset{\text{2,2-Dibromopropane}}{CH_3\overset{Br}{\underset{Br}{C}}CH_3}$$

The designated starting material, 1,1-dibromopropane, is converted to propyne by a double dehydrohalogenation.

$$\underset{\text{1,1-Dibromopropane}}{CH_3CH_2CHBr_2} \xrightarrow[\text{2. } H_2O]{\text{1. } NaNH_2,\ NH_3} \underset{\text{Propyne}}{CH_3C{\equiv}CH}$$

(*b*) As in the preceding exercise, first convert the designated starting material to propyne, then add hydrogen bromide.

$$\underset{\text{1,2-Dibromopropane}}{CH_3\underset{Br}{C}HCH_2Br} \xrightarrow[\text{2. } H_2O]{\text{1. } NaNH_2,\ NH_3} \underset{\text{Propyne}}{CH_3C{\equiv}CH} \xrightarrow{2HBr} \underset{\text{2,2-Dibromopropane}}{CH_3\overset{Br}{\underset{Br}{C}}CH_3}$$

(*c*) First convert 1-bromopropene to propyne, then add one mole of hydrogen chloride across the triple bond. Addition occurs in accordance with Markovnikov's rule.

$$\underset{\text{1-Bromopropene}}{BrCH{=}CHCH_3} \xrightarrow[\text{2. } H_2O]{\text{1. } NaNH_2,\ NH_3} \underset{\text{Propyne}}{HC{\equiv}CCH_3} \xrightarrow{HCl} \underset{\text{2-Chloropropene}}{CH_2{=}\underset{Cl}{C}CH_3}$$

(*d*) Instead of trying to introduce two additional chlorines into 1,2-dichloropropane by free-radical substitution (a mixture of products would result), convert the vicinal dichloride to propyne, then add two moles of Cl_2.

$$\underset{\text{1,2-Dichloropropane}}{CH_3\underset{Cl}{C}HCH_2Cl} \xrightarrow[\text{2. } H_2O]{\text{1. } NaNH_2} \underset{\text{Propyne}}{CH_3C{\equiv}CH} \xrightarrow{2Cl_2} \underset{\text{1,1,2,2-Tetrachloropropane}}{CH_3\overset{Cl}{\underset{Cl}{C}}CHCl_2}$$

(*e*) The required carbon skeleton can be constructed by alkylating acetylene with ethyl bromide.

$$\underset{\text{Acetylene}}{HC{\equiv}CH} \xrightarrow[NH_3]{NaNH_2} \underset{\text{Sodium acetylide}}{HC{\equiv}\bar{C}{:}Na^+} \xrightarrow{CH_3CH_2Br} \underset{\text{1-Butyne}}{HC{\equiv}CCH_2CH_3}$$

Addition of hydrogen iodide to 1-butyne gives 2,2-diiodobutane.

$$\underset{\text{1-Butyne}}{HC{\equiv}CCH_2CH_3} + \underset{\text{Hydrogen iodide}}{2HI} \longrightarrow \underset{\text{2,2-Diiodobutane}}{CH_3CI_2CH_2CH_3}$$

(*f*) The six-carbon chain is available by alkylation of acetylene with 1-bromobutane.

$$\underset{\text{Acetylene}}{HC{\equiv}CH} \xrightarrow[\text{2. } CH_3CH_2CH_2CH_2Br]{\text{1. } NaNH_2,\ NH_3} \underset{\text{1-Hexyne}}{HC{\equiv}CCH_2CH_2CH_2CH_3}$$

The alkylating agent, 1-bromobutane, is prepared from 1-butene by free-radical (anti-Markovnikov) addition of hydrogen bromide.

$$\underset{\text{1-Butene}}{CH_3CH_2CH{=}CH_2} + \underset{\text{Hydrogen bromide}}{HBr} \xrightarrow{\text{peroxides}} \underset{\text{1-Bromobutane}}{CH_3CH_2CH_2CH_2Br}$$

Once 1-hexyne is prepared, it can be converted to 1-hexene by semihydrogenation or by sodium-ammonia reduction.

$$\underset{\text{1-Hexyne}}{CH_3CH_2CH_2CH_2C{\equiv}CH} \xrightarrow[\text{or Na, } NH_3]{H_2,\ \text{Lindlar-Pd}} \underset{\text{1-Hexene}}{CH_3CH_2CH_2CH_2CH{=}CH_2}$$

(*g*) Dialkylation of acetylene with 1-bromobutane gives the necessary 10-carbon chain.

$$\underset{\text{Acetylene}}{HC{\equiv}CH} \xrightarrow[\text{2. } CH_3CH_2CH_2CH_2Br]{\text{1. } NaNH_2,\ NH_3} \underset{\text{1-Hexyne}}{CH_3CH_2CH_2CH_2C{\equiv}CH}$$

$$\xrightarrow[\text{2. } CH_3CH_2CH_2CH_2Br]{\text{1. } NaNH_2,\ NH_3} \underset{\text{5-Decyne}}{CH_3CH_2CH_2CH_2C{\equiv}CCH_2CH_2CH_2CH_3}$$

Hydrogenation of 5-decyne yields decane.

$$\underset{\text{5-Decyne}}{CH_3(CH_2)_3C{\equiv}C(CH_2)_3CH_3} \xrightarrow[Pt]{H_2} \underset{\text{Decane}}{CH_3(CH_2)_3CH_2CH_2(CH_2)_3CH_3}$$

(*h*) A standard method for converting alkenes to alkynes is to add Br_2, then carry out a double dehydrohalogenation.

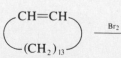

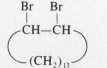

$$\underset{\text{Cyclopentadecene}}{\overbrace{CH{=}CH}\ \ (CH_2)_{13}} \xrightarrow{Br_2} \underset{\text{1,2-Dibromocyclopentadecane}}{\overbrace{CHBr{-}CHBr}\ \ (CH_2)_{13}} \xrightarrow[NH_3]{NaNH_2} \underset{\text{Cyclopentadecyne}}{\overbrace{C{\equiv}C}\ \ (CH_2)_{13}}$$

(*i*) Alkylation of the triple bond gives the required carbon skeleton.

$$\text{1-Ethynylcyclohexene (cyclohexenyl–C}\equiv\text{CH)} \xrightarrow[\text{2. } CH_3Br]{\text{1. } NaNH_2,\ NH_3} \text{1-Propynyl-1-cyclohexene (cyclohexenyl–C}\equiv\text{CCH}_3\text{)}$$

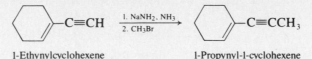

1-Ethynylcyclohexene 1-Propynyl-1-cyclohexene

Semihydrogenation over the Lindlar catalyst converts the carbon–carbon triple bond to a cis double bond.

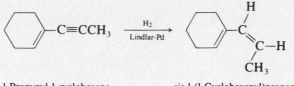

$$\text{cyclohexenyl–C}\equiv\text{CCH}_3 \xrightarrow[\text{Lindlar-Pd}]{H_2} \text{cyclohexenyl–C(H)=C(H)–CH}_3$$

1-Propynyl-1-cyclohexene *cis*-1-(1-Cyclohexenyl)propene

10.34 Attack this problem by first planning a synthesis of 4-methyl-2-pentyne from any starting material in a single step. Two different alkyne alkylations suggest themselves.

$$CH_3C{\equiv}CCH(CH_3)_2 \begin{cases} (a) \text{ from } CH_3C{\equiv}\ddot{C}{:}^- \text{ and } BrCH(CH_3)_2 \\ (b) \text{ from } CH_3I \text{ and } {:}\bar{C}{\equiv}CCH(CH_3)_2 \end{cases}$$

4-Methyl-2-pentyne

Isopropyl bromide is a secondary alkyl halide and cannot be used to alkylate $CH_3C{\equiv}\bar{C}{:}$ according to (*a*). Therefore, a reasonable last step is the alkylation of $(CH_3)_2CHC{\equiv}CH$ via reaction of its anion with methyl iodide.

The next question that arises from this analysis is the origin of $(CH_3)_2CHC{\equiv}CH$. One of the available starting materials is 1,1-dichloro-3-methylbutane. It can be converted to $(CH_3)_2CHC{\equiv}CH$ by a double dehydrohalogenation. Therefore, the complete synthesis is:

$$(CH_3)_2CHCH_2CHCl_2 \xrightarrow[\text{2. } H_2O]{\text{1. } NaNH_2,\ NH_3} (CH_3)_2CHC{\equiv}CH \xrightarrow[\text{2. } CH_3I]{\text{1. } NaNH_2} (CH_3)_2CHC{\equiv}CCH_3$$

1,1-Dichloro-3-methylbutane 3-Methyl-1-butyne 4-Methyl-2-pentyne

10.35 The reaction that produces compound A is reasonably straightforward. Compound A is 14-bromo-1-tetradecyne.

$$NaC{\equiv}CH + Br(CH_2)_{12}Br \longrightarrow Br(CH_2)_{12}C{\equiv}CH$$

Sodium acetylide 1,12-Dibromododecane Compound A ($C_{14}H_{25}Br$)

Treatment of compound A with sodium amide converts it to compound B. Compound B on ozonolysis gives a diacid that retains all the carbon atoms of B. Therefore, compound B must be a cyclic alkyne, formed by an intramolecular alkylation.

$$Br(CH_2)_{12}C{\equiv}CH \xrightarrow{NaNH_2} \text{cyclo-}[C{\equiv}C(CH_2)_{12}] \xrightarrow[\text{2. } H_2O]{\text{1. } O_3} HO\overset{O}{\overset{\|}{C}}(CH_2)_{12}\overset{O}{\overset{\|}{C}}OH$$

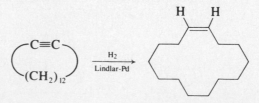

Compound A Compound B

Compound B is cyclotetradecyne.

Hydrogenation of compound B over Lindlar palladium yields *cis*-cyclotetradecene (compound C).

$$\text{cyclo-}[C{\equiv}C(CH_2)_{12}] \xrightarrow[\text{Lindlar-Pd}]{H_2} \text{cis-cyclotetradecene (HC=CH in ring)}$$

Compound C ($C_{14}H_{26}$)

Hydrogenation over platinum gives cyclotetradecane (compound D).

$$\text{cyclo-}[-C{\equiv}C-(CH_2)_{12}-] \xrightarrow[\text{Pt}]{H_2} \text{cyclotetradecane}$$

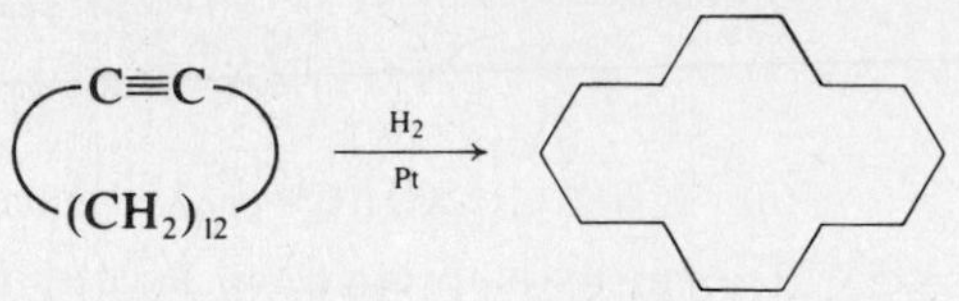

Compound D ($C_{14}H_{28}$)

Sodium–ammonia reduction of compound B yields *trans*-cyclotetradecene.

$$\text{cyclo-}[-C{\equiv}C-(CH_2)_{12}-] \xrightarrow{\text{Na, NH}_3} \textit{trans}\text{-cyclotetradecene}$$

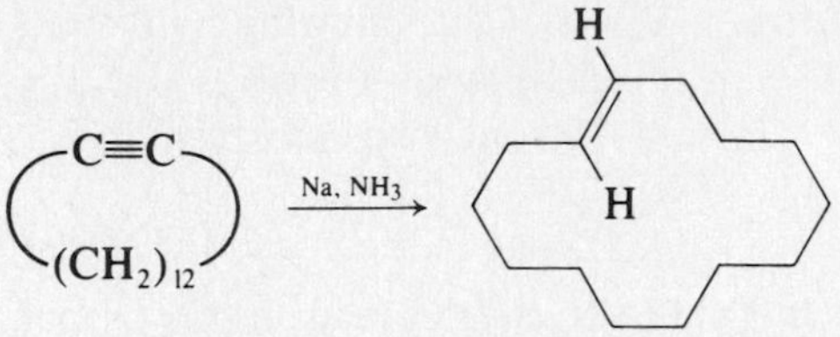

Compound E ($C_{14}H_{26}$)

The cis and trans isomers of cyclotetradecene are both converted to $HO_2C(CH_2)_{12}CO_2H$ on oxidation with potassium permanganate, whereas cyclotetradecane does not react with potassium permanganate.

SELF-TEST

PART A

A-1. Provide the correct IUPAC names for the following:

(*a*) $CH_3C{\equiv}CCHCH(CH_3)_2$ (with CH_3 on the CH carbon)

(*b*) $CH_3CH_2CH_2CHCHC{\equiv}CH$ (with CH_2CH_3 on the first CH and $CH_2CH_2CH_3$ on the second CH)

A-2. Give the structure of the reactant, reagent, or product omitted from each of the following reactions.

(*a*) $CH_3CH_2CH_2C{\equiv}CH \xrightarrow{\text{HCl (1 mol)}} ?$

(*b*) $CH_3CH_2CH_2C{\equiv}CH \xrightarrow{?} CH_3CH_2CH_2\overset{\displaystyle O}{\overset{\|}{C}}CH_3$

(*c*) $CH_3C{\equiv}CCH_3 \xrightarrow[\text{Lindlar Pd}]{H_2} ?$

(*d*) $? \xrightarrow[\text{2. } CH_3CH_2Br]{\text{1. } NaNH_2} (CH_3)_2CHC{\equiv}CCH_2CH_3$

A-3. Only one of the two reactions shown below is effective in the synthesis of 4-methyl-2-hexyne. Which one is that, and why is the other not effective?

(I) $CH_3CH_2\overset{\displaystyle Br}{\overset{|}{C}}HCH_3 + CH_3C{\equiv}C^-Na^+ \longrightarrow$

(II) $CH_3CH_2\overset{\displaystyle CH_3}{\overset{|}{C}}HC{\equiv}C^-Na^+ + CH_3I \longrightarrow$

A-4. Outline a series of steps, using any necessary organic and inorganic reagents, for the preparation of:

(*a*) 3-Hexyne from 1-butene

(*b*) $CH_3\overset{\overset{\displaystyle O}{\|}}{C}CH_2CH(CH_3)_2$ from acetylene

A-5. Treatment of propyne in successive steps with sodium amide, 1-bromobutane, and sodium in liquid ammonia yields as the final product ______.

PART B

B-1. Which of the following statements is *not* true concerning pK_a?

(*a*) The larger the pK_a value, the weaker the acid.
(*b*) Strong acids have small pK_a values.
(*c*) $pK_a = \log K_a$
(*d*) None of these—all the statements are correct.

B-2. The best set of reagents for preparing (*Z*)-2-pentene from 2-pentyne is:

(*a*) H_2 + Lindlar catalyst ($Pd/CaCO_3$, $Pb(OAc)_2$, quinoline)
(*b*) Na in liquid NH_3
(*c*) excess H_2 + Pt catalyst
(*d*) 1. Cl_2; 2. $NaNH_2$

B-3. Referring to the following equilibrium (R = alkyl group)

$$RCH_2CH_3 + RC{\equiv}C\!:^- \rightleftharpoons RCH_2\ddot{C}H_2^- + RC{\equiv}C{-}H$$

(*a*) $K << 1$; the equilibrium would lie to the left.
(*b*) $K >> 1$; the equilibrium would lie to the right.
(*c*) $K = 1$; equal amounts of all species would be present.
(*d*) Not enough information is given; the structure of R must be known.

B-4. Which of the following is an effective way to prepare 1-pentyne?

(*a*) 1-Pentene $\xrightarrow[\text{2. } NaNH_2\text{, heat}]{\text{1. } Cl_2}$

(*b*) Acetylene $\xrightarrow[\text{2. } CH_3CH_2CH_2Br]{\text{1. } NaNH_2}$

(*c*) 1,1-Dichloropentane $\xrightarrow[\text{2. } H_2O]{\text{1. } NaNH_2\text{, } NH_3}$

(*d*) All of these are effective.

B-5. Which alkyne yields butanoic acid ($CH_3CH_2CH_2CO_2H$) as the only organic product upon treatment with ozone followed by hydrolysis?

(*a*) 1-Butyne (*b*) 4-Octyne (*c*) 1-Pentyne (*d*) 2-Hexyne

B-6. Which of the following produces a significant amount of acetylide ion on reaction with acetylene?

(*a*) Conjugate base of CH_3OH (pK_a 16)
(*b*) Conjugate base of H_2 (pK_a 35)
(*c*) Conjugate base of H_2O (pK_a 16)
(*d*) Both (*a*) and (*c*).

CHAPTER 11

CONJUGATION IN ALKADIENES AND ALLYLIC SYSTEMS

IMPORTANT TERMS AND CONCEPTS

Allylic Carbocations (Sec. 11.2) A carbocation in which the charge is on a carbon atom adjacent to a double bond is said to be *allylic*. Such intermediates are stabilized by delocalization of the electrons of the double bond.

$$\rangle C{=}\overset{|}{C}{-}\overset{+}{C}\langle \longleftrightarrow \rangle\overset{+}{C}{-}\overset{|}{C}{=}C\langle$$

Allylic carbocations are more stable than simple alkyl carbocations.

Allylic Free Radicals (Sec. 11.3) An intermediate in which a carbon bearing an unpaired electron is adjacent to a double bond is known as an allylic free radical. Allylic radicals are stabilized by delocalization of the unpaired electron and π electrons over the three-carbon framework.

$$\rangle C{=}\overset{|}{C}{-}\dot{C}\langle \longleftrightarrow \rangle\dot{C}{-}\overset{|}{C}{=}C\langle$$

Classes of Dienes (Sec. 11.5) Three classes of dienes may be described according to the relationship between the double bonds. They are:

Isolated:	One or more sp^3 carbon atoms separate the double bonds.
Conjugated:	The double bonds are joined by one single bond, forming a continuous chain of orbitals.
Cumulated:	The double bonds share a common atom (these dienes are also known as *allenes*).

Diene Stabilities (Secs. 11.7, 11.8) As shown by heats of hydrogenation, conjugated dienes are about 3 to 4 kcal/mol more stable than isolated dienes. This stabilization may be explained as arising from *delocalization* of the π electrons. The stabilization is also known as *conjugation energy*. The *s*-trans conformation of a conjugated diene is more stable than the *s*-cis.

Direct and Conjugate Addition (Sec. 11.10) Two modes of reaction are possible when a conjugated diene undergoes an addition reaction. Addition across one of the double bonds is known as *direct*, or *1,2 addition*; addition to the ends of the diene system is known as *conjugate*, or *1,4 addition.*

$$\text{H—X} + \text{>C=C—C=C<} \longrightarrow \underset{\substack{\text{Direct}\\\text{(1,2 addition)}}}{\text{—C(H)—C(X)—C=C<}} + \underset{\substack{\text{Conjugate}\\\text{(1,4 addition)}}}{\text{—C(H)—C=C—C(X)—}}$$

At low temperature the electrophilic addition reaction of a conjugated diene is subject to *kinetic control*, also known as *rate control*. The proportion of products from the reaction is governed by their relative rates of formation, and once formed, the products do not interconvert or equilibrate with one another.

At room temperature or above, the reaction is subject to *thermodynamic control*, also known as *equilibrium control*, which means that the product distribution is governed by the relative stabilities of the products. The products are in equilibrium with each other and so the most stable one predominates regardless of which product is formed fastest.

IMPORTANT REACTIONS

Allylic Halogenation (Sec. 11.4)

General:

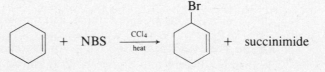

$$\text{>C=C—C(H)—} \xrightarrow[\text{heat}]{N\text{-bromosuccinimide (NBS)}} \text{>C=C—C(Br)—}$$

Example:

$$\text{cyclohexene} + \text{NBS} \xrightarrow[\text{heat}]{CCl_4} \text{3-bromocyclohexene} + \text{succinimide}$$

Preparation of Dienes (Sec. 11.6)

Example:

$$CH_2{=}CH\underset{\displaystyle Br}{C}HCH_2CH_3 \xrightarrow[\text{heat}]{\text{KOH}} CH_2{=}CHCH{=}CHCH_3$$

Hydrogen Halide Addition to Conjugated Dienes (Sec. 11.10)

Example:

$$CH_2{=}CH{-}CH{=}CH_2 + HCl \longrightarrow \underset{\text{1,2 Addition}}{CH_2{=}CH{-}\underset{\displaystyle Cl}{C}HCH_3} + \underset{\text{1,4 Addition}}{\underset{\displaystyle Cl}{C}H_2{-}CH{=}CHCH_3}$$

Halogen Addition to Conjugated Dienes (Sec. 11.11)

Example:

$$CH_2{=}CH{-}CH{=}CH_2 + Cl_2 \longrightarrow \underset{\text{1,2 Addition}}{CH_2{=}CH{-}\underset{\displaystyle Cl}{C}HCH_2Cl} + \underset{\text{1,4 Addition}}{\underset{\displaystyle Cl}{C}H_2{-}CH{=}CH{-}\underset{\displaystyle Cl}{C}H_2}$$

Diels-Alder Reaction (Sec. 11.12)

General:

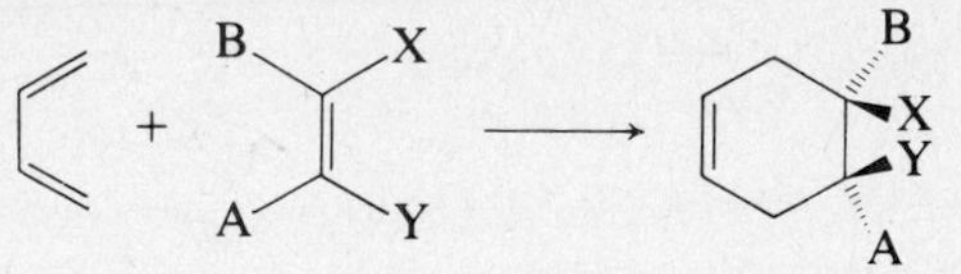

B X A Y

Example:

H O O H O O H O O H O O

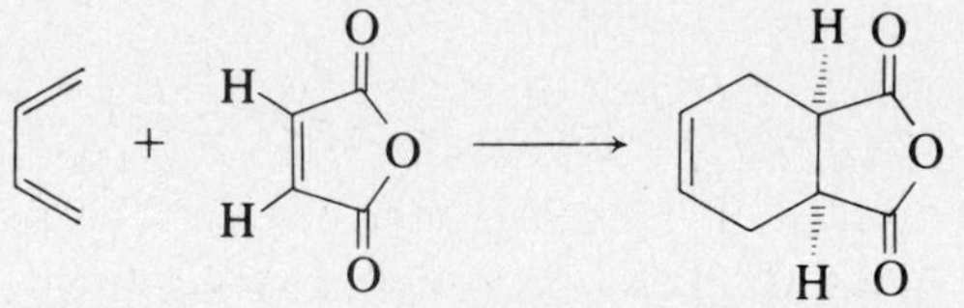

SOLUTIONS TO TEXT PROBLEMS

11.1 As noted in the sample solution to (*a*), a pair of electrons is moved from the double bond toward the positively charged carbon.

(*b*) $CH_2{=}C{-}\overset{+}{C}H_2$ (with CH_3 on C-2) $\longleftrightarrow$ $\overset{+}{C}H_2{-}C{=}CH_2$ (with CH_3 on C-2)

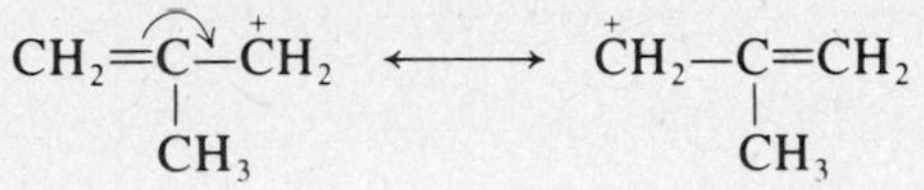

(*c*) $=C(CH_3)_2$ $\longleftrightarrow$ $-\overset{+}{C}(CH_3)_2$

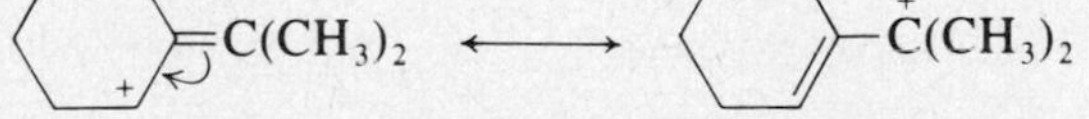

11.2 In order for two isomeric halides to yield the same carbocation on ionization, they must have the same carbon skeleton. They may have their leaving group at a different location, but the carbocations must become equivalent by allylic resonance.

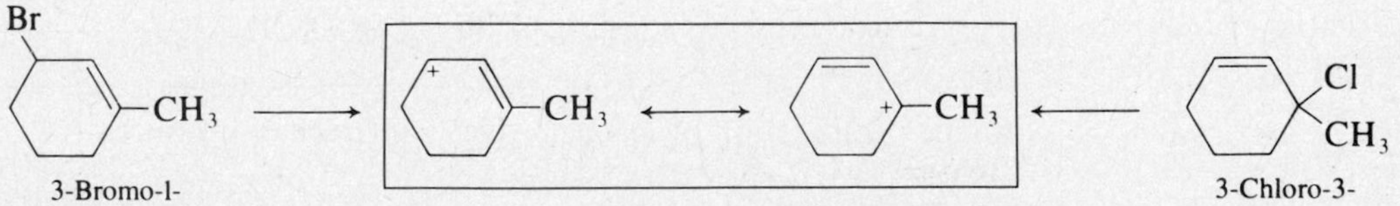

Br $-CH_3$ $\longrightarrow$ $-CH_3$ $\longleftrightarrow$ $-CH_3$ $\longleftarrow$ Cl CH_3

3-Bromo-1-methylcyclohexene

3-Chloro-3-methylcyclohexene

Br $-CH_3$ $\longrightarrow$ $-CH_3$ not an allylic carbocation

4-Bromo-1-methylcyclohexene

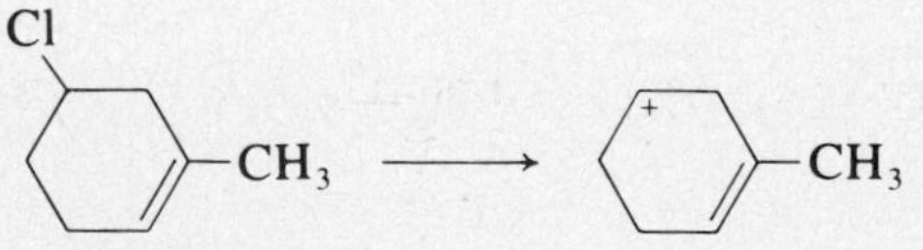

Cl $-CH_3$ $\longrightarrow$ $-CH_3$ not an allylic carbocation

5-Chloro-1-methylcyclohexene

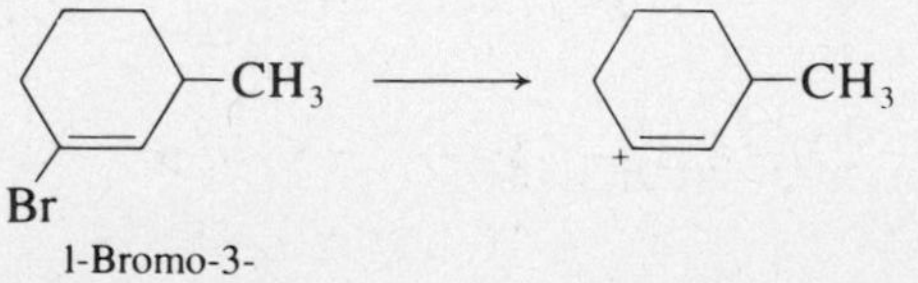

$-CH_3$ $\longrightarrow$ $-CH_3$ not an allylic carbocation

Br

1-Bromo-3-methylcyclohexene

11.3 The allylic hydrogens are the ones shown in the structural formulas.

(*b*) 1-Methylcyclohexene

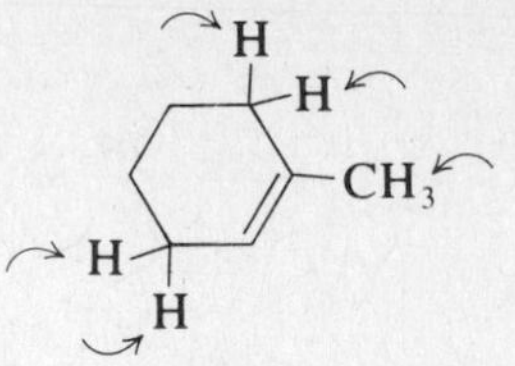

(*c*) 2,3,3-Trimethyl-1-butene

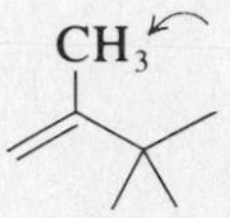

(*d*) 1-Octene

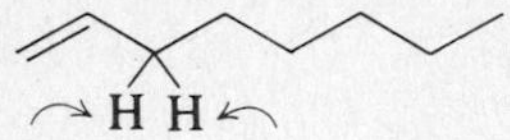

11.4 Write both resonance forms of the allylic radicals produced by hydrogen atom abstraction from the alkene.

$$(CH_3)_3CC(=CH_2)CH_3 \longrightarrow (CH_3)_3CC(=CH_2)\dot{C}H_2 \longleftrightarrow (CH_3)_3CC(\dot{C}H_2)=CH_2$$

2,3,3-Trimethyl-1-butene

Both resonance forms are equivalent, so 2,3,3-trimethyl-1-butene gives a single bromide on treatment with *N*-bromosuccinimide (NBS).

$$(CH_3)_3CC(CH_3)=CH_2 \xrightarrow{NBS} (CH_3)_3CC(CH_2Br)=CH_2$$

2,3,3-Trimethyl-1-butene → 2-(Bromomethyl)-3,3-dimethyl-1-butene

Hydrogen atom abstraction from 1-octene gives a radical in which the unpaired electron is delocalized between two nonequivalent positions.

$$CH_2{=}CHCH_2(CH_2)_4CH_3 \longrightarrow$$

1-Octene

$$CH_2{=}CH\dot{C}H(CH_2)_4CH_3 \longleftrightarrow \dot{C}H_2CH{=}CH(CH_2)_4CH_3$$

Allylic bromination of 1-octene gives a mixture of products.

$$CH_2{=}CHCH_2(CH_2)_4CH_3 \xrightarrow{NBS} CH_2{=}CHCH(Br)(CH_2)_4CH_3 + BrCH_2CH{=}CH(CH_2)_4CH_3$$

1-Octene; 3-Bromo-1-octene; 1-Bromo-2-octene (cis and trans)

11.5 (*b*) All the double bonds in humulene are isolated from each other.

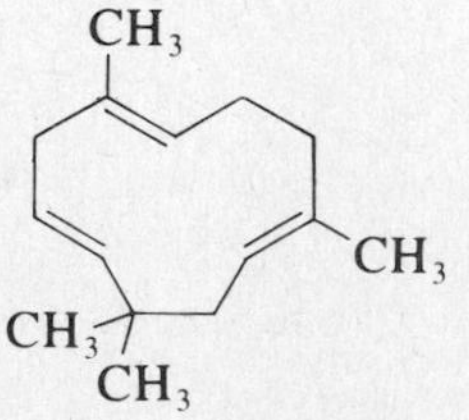

Humulene

(*c*) The C-1 and C-3 double bonds of cembrene are conjugated to each other.

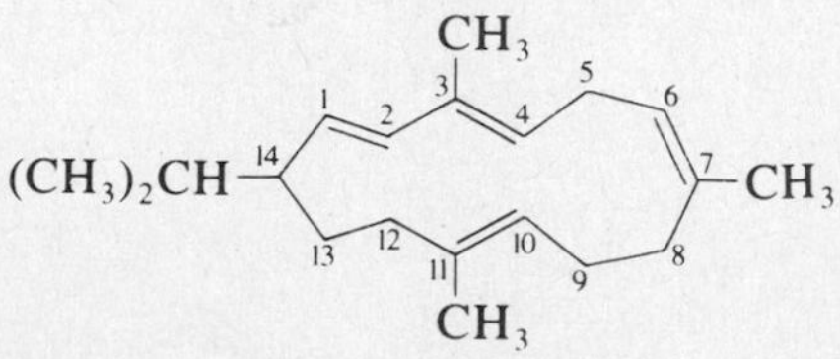

Cembrene

The double bonds at C-6 and C-10 are isolated from each other and from the conjugated diene system.

(*d*) The sex attractant of the dried-bean beetle has a cumulated diene system involving C-4, C-5, and C-6. This allenic system is conjugated with the C-2 double bond.

$$CH_3(CH_2)_6CH_2\overset{6}{C}H=\overset{5}{C}=\overset{4}{C}H\overset{3}{C}H=\overset{2}{C}H\overset{1}{C}O_2CH_3$$

11.6 Since all three double bonds are conjugated in *cis*-alloocimene, it is more stable and has a smaller heat of combustion than myrcene.

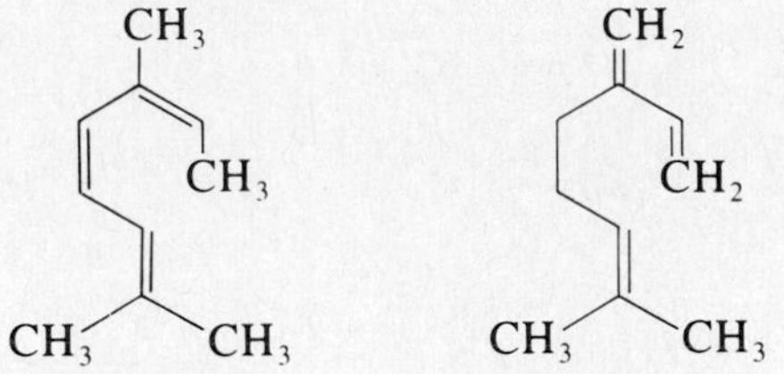

cis-Alloocimene (heat of combustion 1481.3 kcal/mol)

Myrcene (heat of combustion 1490.5 kcal/mol)

In addition to the extended conjugation of *cis*-alloocimene, two of its double bonds are trisubstituted and one is disubstituted, while in myrcene one is trisubstituted, one disubstituted, and one is monosubstituted.

11.7 The two double bonds of 2-methyl-1,3-butadiene are not equivalent, so two different products of direct addition are possible, along with one conjugate addition product.

$$CH_2{=}C(CH_3)CH{=}CH_2 \xrightarrow{Br_2} BrCH_2C(Br)(CH_3)CH{=}CH_2 + CH_2{=}C(CH_3)CH(Br)CH_2Br + BrCH_2C(CH_3){=}CHCH_2Br$$

2-Methyl-1,3-butadiene; 3,4-Dibromo-3-methyl-1-butene (direct addition); 3,4-Dibromo-2-methyl-1-butene (direct addition); 1,4-Dibromo-2-methyl-2-butene (conjugate addition)

11.8 The molecular formula of the product, $C_{10}H_9ClO_2$, is that of a 1 : 1 Diels-Alder adduct between 2-chloro-1,3-butadiene and benzoquinone.

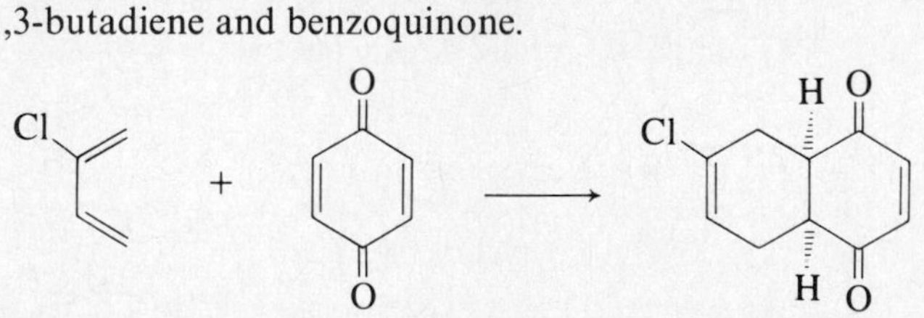

2-Chloro-1,3-butadiene + Benzoquinone ⟶ $C_{10}H_9ClO_2$

11.9 "Unravel" the Diels-Alder adduct as described in the sample solution to (*a*).

(*b*) Diels-Alder adduct is prepared from Diene + Dienophile (cyano groups are cis)

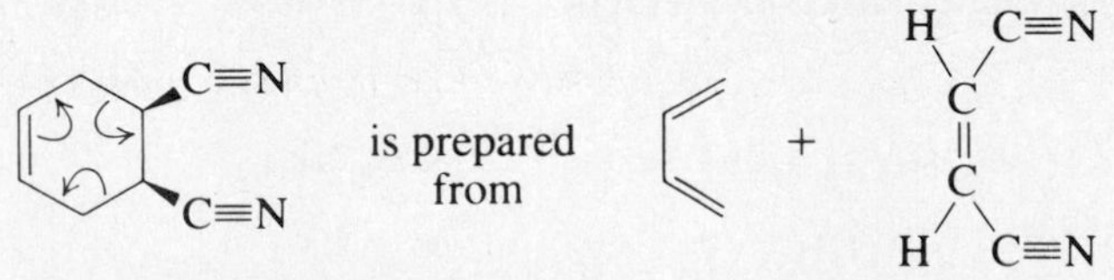

(c) 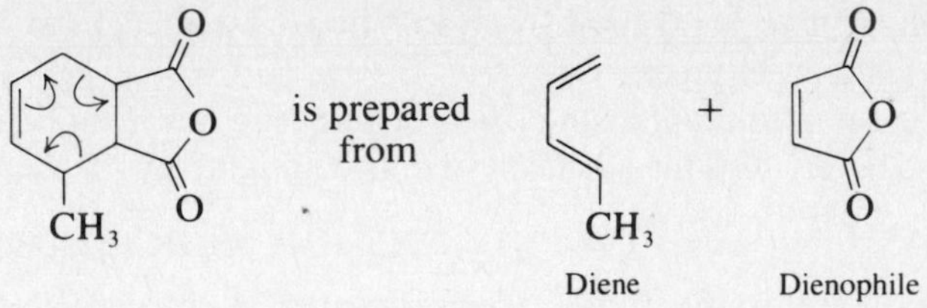 [structure, with CH_3] is prepared from [diene, CH_3] + [dienophile]

Diene Dienophile

(d) 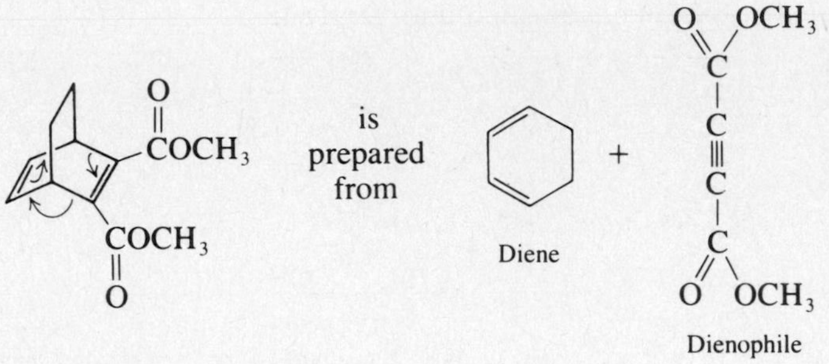 [structure with $\overset{O}{\|}COCH_3$ and $COCH_3$ groups] is prepared from [cyclohexadiene] + $CH_3OC(=O)C{\equiv}CC(=O)OCH_3$

Diene Dienophile

11.10 Dienes and trienes are named according to the IUPAC convention by replacing the *-ane* ending of the alkane with *-adiene* or *-atriene* and locating the positions of the double bonds by number. The stereoisomers are identified as *E* or *Z* according to the rules established in Chapter 5.

(*a*) 3,4-Octadiene $CH_3CH_2CH{=}C{=}CHCH_2CH_2CH_3$

(*b*) (*E*, *E*)-3,5-Octadiene [structure: CH_3, H, H, H, H, CH_2CH_3]

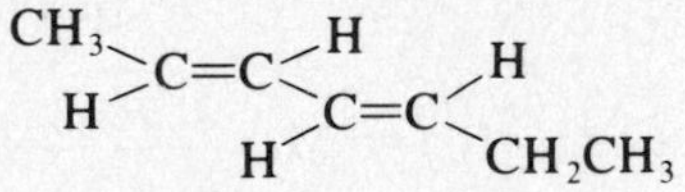

(*c*) (*Z*, *Z*)-1,3-Cyclooctadiene

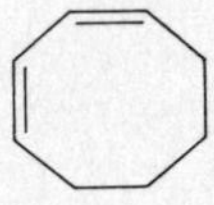

(*d*) (*Z*, *Z*)-1,4-Cyclooctadiene

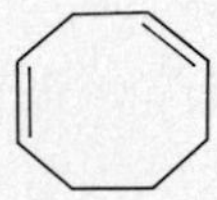

(*e*) (*Z*, *E*)-1,5-Cyclooctadiene

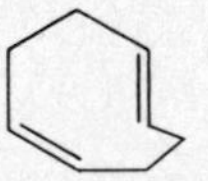

(*f*) (2*E*, 4*Z*, 6*E*)-2,4,6-Octatriene [structure: CH_3, H H, CH_3, H, H, H H]

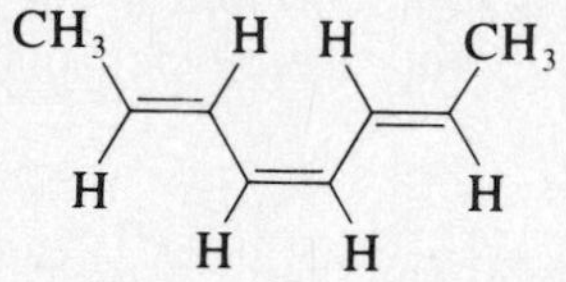

(*g*) 5-Allyl-1,3-cyclopentadiene [structure: H, $CH_2CH{=}CH_2$]

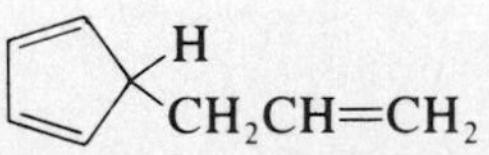

(*h*) *trans*-1,2-Divinylcyclopropane [structure: $CH_2{=}CH$, H, H, $CH{=}CH_2$]

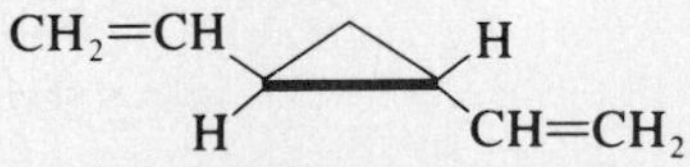

(*i*) 2,4-Dimethyl-1,3-pentadiene $CH_2{=}C(CH_3)CH{=}C(CH_3)CH_3$

11.11 (*a*) $CH_2{=}CH(CH_2)_5CH{=}CH_2$ 1,8-Nonadiene

(*b*) $(CH_3)_2C{=}C(CH_3)C(CH_3){=}C(CH_3)_2$ 2,3,4,5-Tetramethyl-2,4-hexadiene (*c*) $CH_2{=}CH{-}CH(CH{=}CH_2){-}CH{=}CH_2$ 3-Vinyl-1,4-pentadiene

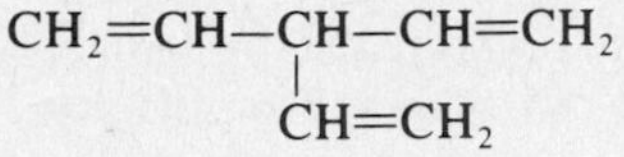

(*d*) 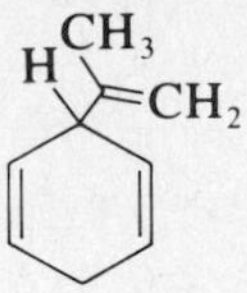 3-Isopropenyl-1,4-cyclohexadiene

(*e*) 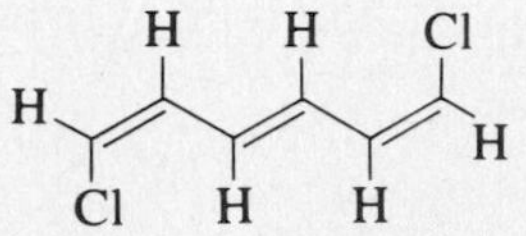 (1*Z*, 3*E*, 5*Z*)-1,6-Dichloro-1,3,5-hexatriene

(*f*) $CH_2{=}C{=}CHCH{=}CHCH_3$ 1,2,4-Hexatriene

(*g*) 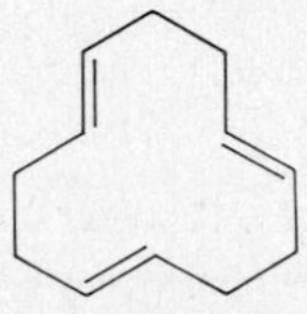 (1*E*, 5*E*, 9*E*)-1,5,9-Cyclododecatriene

(*h*) $(CH_3)(CH_3CH_2)C{=}C(CH{=}CH_2)(CH_2CH_3)$ (*E*)-3-Ethyl-4-methyl-1,3-hexadiene

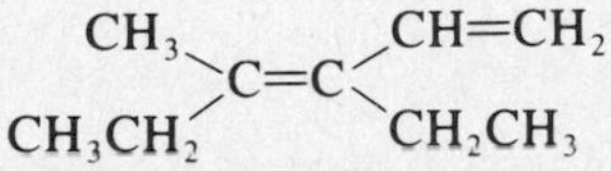

11.12 (*a*) Since the product is 2,3-dimethylbutane we know that the carbon skeleton of the starting material must be

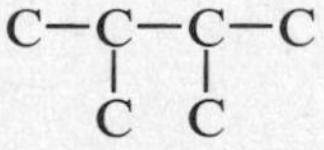

C–C(C)–C(C)–C

Since 2,3-dimethylbutane is C_6H_{14} and the starting material is C_6H_{10}, *two* molecules of H_2 must have been taken up and the starting material must have two double bonds. The starting material can only be 2,3-dimethyl-1,3-butadiene.

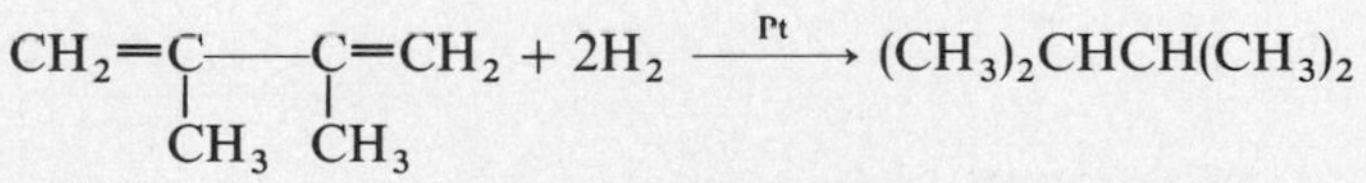

$$CH_2{=}C(CH_3){-}C(CH_3){=}CH_2 + 2H_2 \xrightarrow{Pt} (CH_3)_2CHCH(CH_3)_2$$

(*b*) Write the carbon skeleton corresponding to 2,2,6,6-tetramethylheptane.

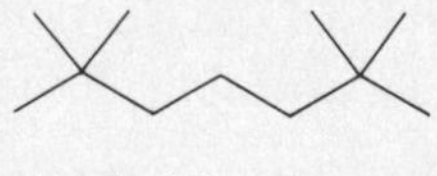

Compounds of molecular formula $C_{11}H_{20}$ have a SODAR of 2. The only compounds with the proper carbon skeleton with a SODAR of 2 are the alkyne and the allene shown.

$(CH_3)_3CC{\equiv}CCH_2C(CH_3)_3$ — 2,2,6,6-Tetramethyl-3-heptyne

$(CH_3)_3CCH{=}C{=}CHC(CH_3)_3$ — 2,2,6,6-Tetramethyl-3,4-heptadiene

11.13 The important piece of information which allows us to complete the structure properly is that the ant repellent is an *allenic* substance. The allenic unit cannot be incorporated into the ring because the three carbons must be colinear. Therefore, the only possible constitution is

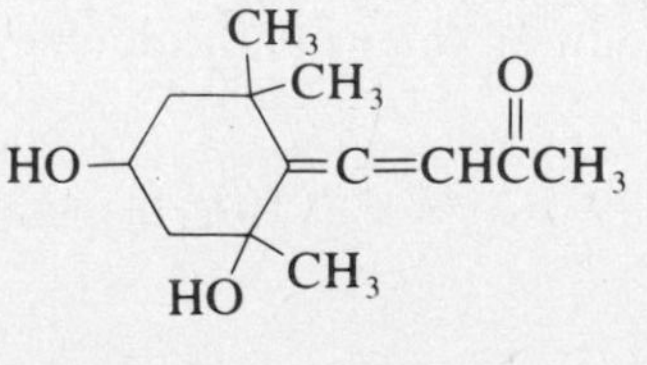

HO–(cyclohexane ring bearing CH_3, CH_3; HO, CH_3)$=C{=}CHCCH_3$ (with C=O)

11.14 (*a*) The desired allylic alcohol can be prepared by hydrolysis of an allylic halide. Cyclopentene can be converted to an allylic bromide by free-radical bromination with *N*-bromosuccinimide (NBS).

Cyclopentene $\xrightarrow[\text{heat, peroxides}]{\text{NBS}}$ 3-Bromocyclopentene $\xrightarrow[Na_2CO_3]{H_2O}$ 2-Cyclopenten-1-ol

(*b*) Reaction of the allylic bromide from (*a*) with sodium iodide in acetone converts it to the corresponding iodide.

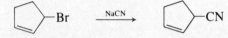

3-Bromocyclopentene $\xrightarrow[\text{acetone}]{\text{NaI}}$ 3-Iodocyclopentene

(*c*) Nucleophilic substitution by cyanide converts the allylic bromide to 3-cyanocyclopentene.

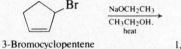

3-Bromocyclopentene $\xrightarrow{\text{NaCN}}$ 3-Cyanocyclopentene

(*d*) Reaction of the allylic bromide with a strong base will yield cyclopentadiene by an E2 elimination.

3-Bromocyclopentene $\xrightarrow[CH_3CH_2OH,\ \text{heat}]{NaOCH_2CH_3}$ 1,3-Cyclopentadiene

(*e*) Cyclopentadiene formed in (*d*) is needed in order to form the required Diels-Alder adduct.

1,3-Cyclopentadiene $\xrightarrow{CH_3OCC{\equiv}CCOCH_3}$ (with C=O on each carbonyl carbon) Dimethyl bicyclo[2.2.1]heptadiene-2,3-dicarboxylate

11.15 The starting material in all cases is 2,3-dimethyl-1,3-butadiene.

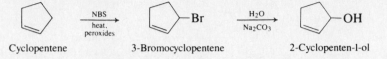

$$CH_2{=}C(CH_3){-}C(CH_3){=}CH_2$$

2,3-Dimethyl-1,3-butadiene

(*a*) Hydrogenation of both double bonds will occur to yield 2,3-dimethylbutane.

$$CH_2{=}C(CH_3){-}C(CH_3){=}CH_2 \xrightarrow[Pt]{H_2} (CH_3)_2CHCH(CH_3)_2$$

(*b*) Direct addition of one mole of hydrogen chloride will give the product of Markovnikov addition to one of the double bonds, 3-chloro-2,3-dimethyl-1-butene.

$$CH_2{=}C(CH_3){-}C(CH_3){=}CH_2 \xrightarrow{HCl} (CH_3)_2CCl{-}C(CH_3){=}CH_2$$

(*c*) Conjugate addition will lead to double-bond migration and produce 1-chloro-2,3-dimethyl-2-butene.

$$CH_2{=}C(CH_3){-}C(CH_3){=}CH_2 \xrightarrow{HCl} (CH_3)_2C{=}C(CH_3)CH_2Cl$$

(*d*) The direct addition product is 3,4-dibromo-2,3-dimethyl-1-butene.

$$CH_2{=}C(CH_3){-}C(CH_3){=}CH_2 \xrightarrow{Br_2} BrCH_2C(CH_3)(Br){-}C(CH_3){=}CH_2$$

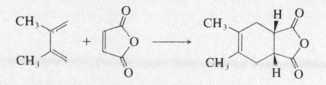

(*e*) The conjugate addition product will be 1,4-dibromo-2,3-dimethyl-2-butene.

$$CH_2{=}C(CH_3){-}C(CH_3){=}CH_2 \xrightarrow{Br_2} BrCH_2C(CH_3){=}C(CH_3)CH_2Br$$

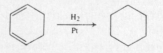

(*f*) Bromination of both double bonds will lead to 1,2,3,4-tetrabromo-2,3-dimethylbutane irrespective of whether the first addition step occurs by direct or conjugate addition.

$$CH_2{=}C(CH_3){-}C(CH_3){=}CH_2 \xrightarrow{2Br_2} BrCH_2C(CH_3)(Br){-}C(CH_3)(Br)CH_2Br$$

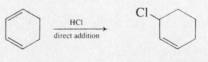

(*g*) The reaction of a diene with maleic anhydride is a Diels-Alder reaction.

[Structure: 2,3-dimethyl-1,3-butadiene + maleic anhydride → Diels-Alder adduct (CH₃, CH₃ on cyclohexene ring; H, H at ring fusion; anhydride O, O, O)]

11.16 The starting material in all cases is 1,3-cyclohexadiene.

[Structure: 1,3-cyclohexadiene]

(*a*) Cyclohexane will be the product of hydrogenation of 1,3-cyclohexadiene:

$$\text{1,3-cyclohexadiene} \xrightarrow[Pt]{H_2} \text{cyclohexane}$$

(*b*) Direct addition will occur according to Markovnikov's rule to give 3-chlorocyclohexene

$$\text{1,3-cyclohexadiene} \xrightarrow[\text{direct addition}]{HCl} \text{3-chlorocyclohexene}$$

3-Chlorocyclohexene

(*c*) The product of conjugate addition is 3-chlorocyclohexene also. Direct addition and conjugate addition of hydrogen chloride to 1,3-cyclohexadiene give the same product.

$$\text{1,3-cyclohexadiene} \xrightarrow[\text{conjugate addition}]{HCl} \text{3-chlorocyclohexene}$$

3-Chlorocyclohexene

(*d*) Bromine can add directly to one of the double bonds to give 3,4-dibromocyclohexene:

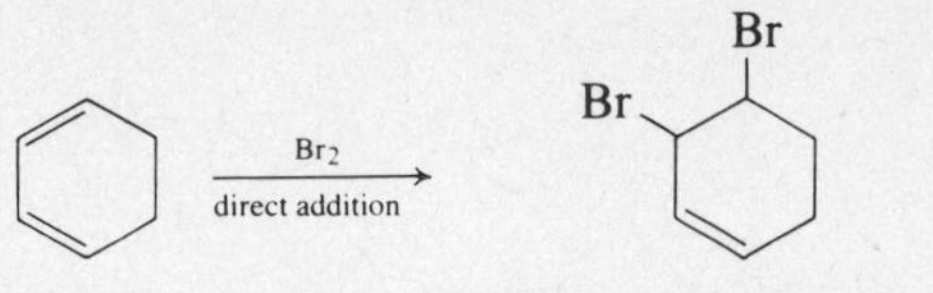

$$\text{1,3-Cyclohexadiene} \xrightarrow[\text{direct addition}]{Br_2} \text{3,4-Dibromocyclohexene}$$

3,4-Dibromocyclohexene

(*e*) Conjugate addition of bromine will give 3,6-dibromocyclohexene:

$$\text{1,3-Cyclohexadiene} \xrightarrow[\text{conjugate addition}]{Br_2} \text{3,6-Dibromocyclohexene}$$

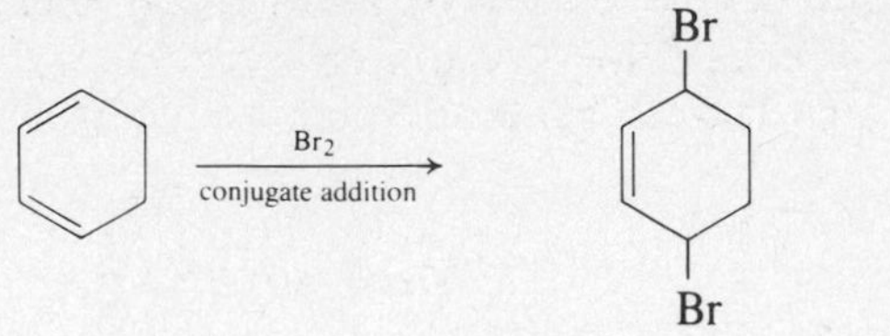

3,6-Dibromocyclohexene

(*f*) Addition of two moles of bromine will yield 1,2,3,4-tetrabromocyclohexane.

$$\text{1,3-Cyclohexadiene} \xrightarrow{2Br_2} \text{1,2,3,4-Tetrabromocyclohexane}$$

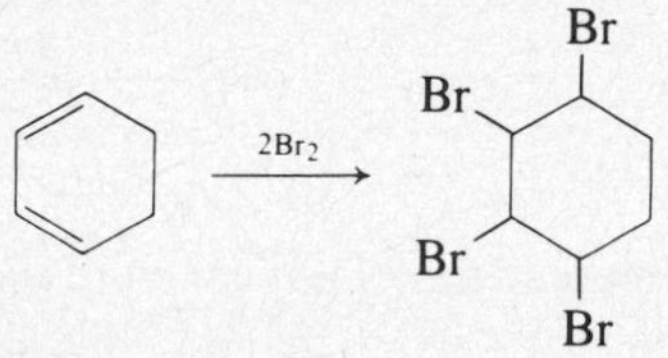

(*g*) The constitution of the Diels-Alder adduct of 1,3-cyclohexadiene and maleic anhydride will have a bicyclo[2.2.2]octyl carbon skeleton.

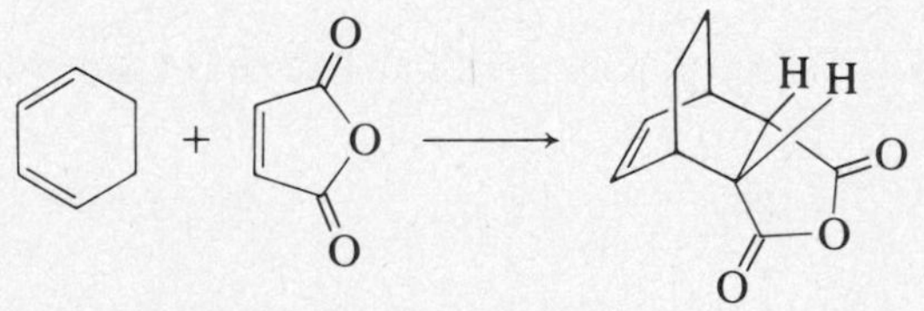

$$\text{1,3-Cyclohexadiene} + \text{maleic anhydride} \longrightarrow \text{Diels-Alder adduct}$$

11.17 (*a*) Protonation of 2-methyl-1,3-butadiene will occur at C-1 to give the delocalized ion shown. The positive charge in this intermediate is borne predominantly by the tertiary carbon, so chloride attack takes place preferentially at the tertiary carbon under conditions of kinetic control.

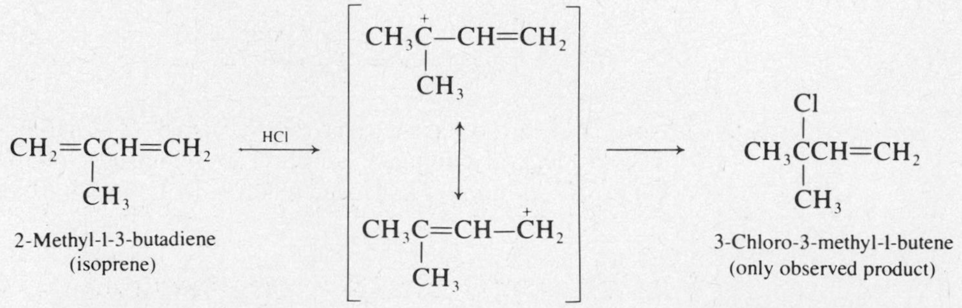

$$CH_2{=}C(CH_3)CH{=}CH_2 \xrightarrow{HCl} \left[CH_3\overset{+}{C}(CH_3){-}CH{=}CH_2 \longleftrightarrow CH_3C(CH_3){=}CH{-}\overset{+}{C}H_2 \right] \longrightarrow CH_3CCl(CH_3)CH{=}CH_2$$

2-Methyl-1-3-butadiene (isoprene)

3-Chloro-3-methyl-1-butene (only observed product)

(*b*) Under conditions of thermodynamic control the tertiary halide will be converted to the more stable primary halide with an internal double bond.

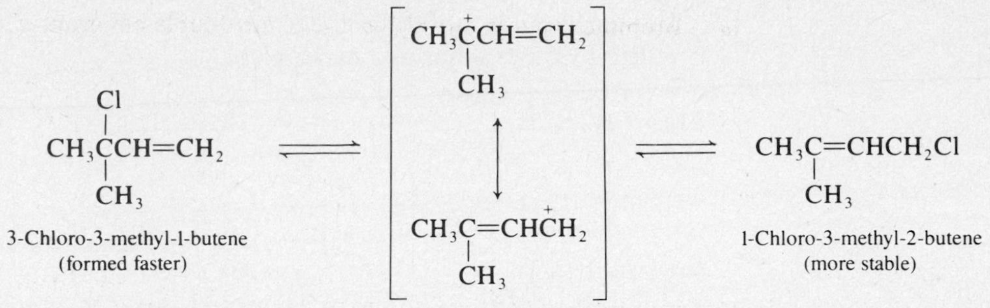

$$\underset{\substack{\text{3-Chloro-3-methyl-1-butene}\\ \text{(formed faster)}}}{CH_3\overset{Cl}{\underset{CH_3}{C}}CH{=}CH_2} \rightleftharpoons \left[CH_3\overset{+}{\underset{CH_3}{C}}CH{=}CH_2 \longleftrightarrow CH_3\underset{CH_3}{C}{=}CH\overset{+}{C}H_2 \right] \rightleftharpoons \underset{\substack{\text{1-Chloro-3-methyl-2-butene}\\ \text{(more stable)}}}{CH_3\underset{CH_3}{C}{=}CHCH_2Cl}$$

11.18 Compound A must be a conjugated diene because the product of bromine addition corresponds to that of conjugate addition.

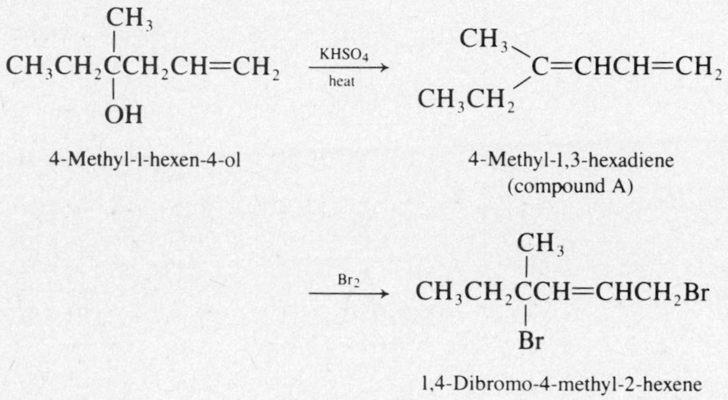

$$\underset{\text{4-Methyl-1-hexen-4-ol}}{CH_3CH_2\overset{CH_3}{\underset{OH}{C}}CH_2CH{=}CH_2} \xrightarrow[\text{heat}]{KHSO_4} \underset{\substack{\text{4-Methyl-1,3-hexadiene}\\ \text{(compound A)}}}{(CH_3)(CH_3CH_2)C{=}CHCH{=}CH_2}$$

$$\xrightarrow{Br_2} \underset{\text{1,4-Dibromo-4-methyl-2-hexene}}{CH_3CH_2\overset{CH_3}{\underset{Br}{C}}CH{=}CHCH_2Br}$$

11.19 Compound C must arise by way of a Diels-Alder reaction between compound B and dimethyl acetylenedicarboxylate. Therefore, compound B must have a conjugated diene system.

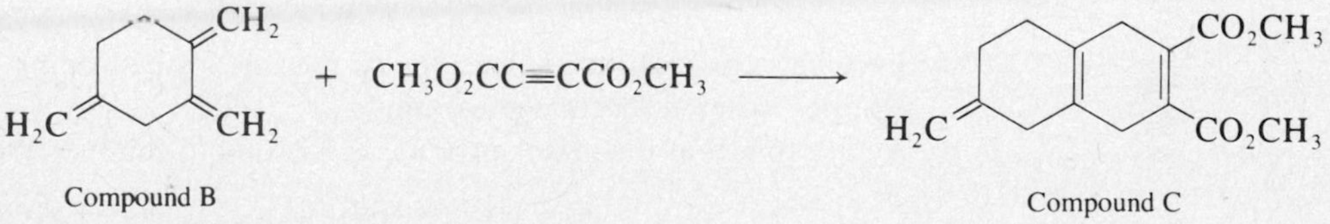

Compound B (1,3,5-tris(methylene)cyclohexane, with $H_2C{=}$ and two $=CH_2$ groups) + $CH_3O_2CC{\equiv}CCO_2CH_3$ ⟶ Compound C (bicyclic diene bearing $H_2C{=}$ and two CO_2CH_3 groups)

Compound B　　　　Compound C

11.20 (*a*) Solvolysis of $(CH_3)_2C{=}CHCH_2Cl$ in ethanol proceeds by an S_N1 mechanism and involves a carbocation intermediate.

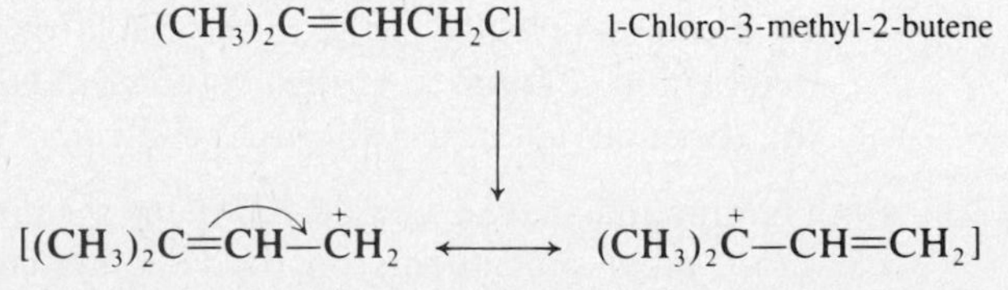

$$\underset{\text{1-Chloro-3-methyl-2-butene}}{(CH_3)_2C{=}CHCH_2Cl}$$

$$\downarrow$$

$$[(CH_3)_2C{=}CH{-}\overset{+}{C}H_2 \longleftrightarrow (CH_3)_2\overset{+}{C}{-}CH{=}CH_2]$$

This carbocation has some of the character of a tertiary carbocation. It is more stable and is therefore formed faster than allyl cation, $CH_2{=}CH\overset{+}{C}H_2$.

(*b*) An allylic carbocation is formed from the alcohol in the presence of an acid catalyst.

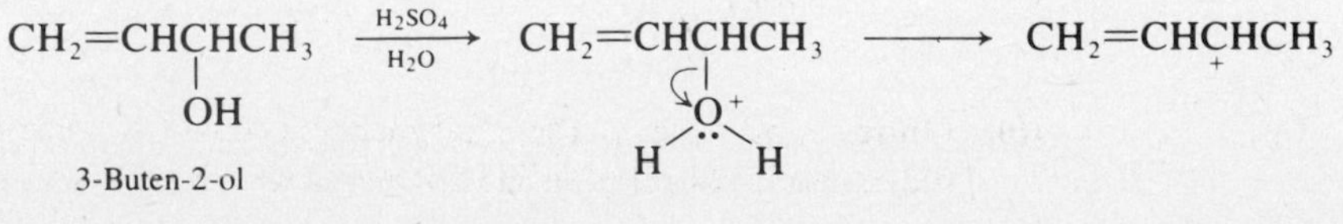

$$\underset{\text{3-Buten-2-ol}}{CH_2{=}CH\underset{OH}{C}HCH_3} \xrightarrow[H_2O]{H_2SO_4} CH_2{=}CH\underset{\overset{+}{O}H_2}{C}HCH_3 \longrightarrow CH_2{=}CH\overset{+}{C}HCH_3$$

This carbocation is a delocalized one and can be captured at either end of the allylic system by water acting as a nucleophile.

$$\boxed{\begin{array}{c} CH_2{=}CH\overset{+}{C}HCH_3 \\ \updownarrow \\ \overset{+}{C}H_2CH{=}CHCH_3 \end{array}} \xrightarrow{H_2O} \begin{cases} CH_2{=}CHCH(\overset{+}{O}H_2)CH_3 \longrightarrow CH_2{=}CHCH(OH)CH_3 \quad \text{3-Buten-2-ol} \\ H_2\overset{+}{O}{-}CH_2CH{=}CHCH_3 \longrightarrow HOCH_2CH{=}CHCH_3 \quad \text{2-Buten-1-ol} \end{cases}$$

(*c*) Hydrogen bromide converts the alcohol to an allylic carbocation. Bromide ion captures this carbocation at either end of the delocalized allylic system.

$$\underset{\text{2-Buten-1-ol}}{CH_3CH{=}CHCH_2OH} \xrightarrow{HBr} CH_3CH{=}CHCH_2{-}\overset{+}{O}H_2 \longrightarrow CH_3CH{=}CH\overset{+}{C}H_2$$

$$\boxed{\begin{array}{c} CH_3CH{=}CH\overset{+}{C}H_2 \\ \updownarrow \\ CH_3\overset{+}{C}HCH{=}CH_2 \end{array}} \xrightarrow{Br^-} \begin{cases} CH_3CH{=}CHCH_2Br \quad \text{1-Bromo-2-butene} \\ CH_3CH(Br)CH{=}CH_2 \quad \text{3-Bromo-1-butene} \end{cases}$$

(*d*) The same delocalized carbocation is formed from 3-buten-2-ol as from 2-buten-1-ol.

$$\underset{\text{3-Buten-2-ol}}{CH_3CH(OH)CH{=}CH_2} \xrightarrow{HBr} CH_3\overset{+}{C}HCH{=}CH_2 \longleftrightarrow CH_3CH{=}CH\overset{+}{C}H_2$$

Since this carbocation is the same as the one formed in (*c*), it gives the same mixture of products when it reacts with bromide.

(*e*) We are told that the major product is 1-bromo-2-butene, not 3-bromo-1-butene.

$$\underset{\substack{\text{1-Bromo-2-butene} \\ \text{(major)}}}{BrCH_2CH{=}CHCH_3} \qquad \underset{\substack{\text{3-Bromo-1-butene} \\ \text{(minor)}}}{CH_2{=}CHCH(Br)CH_3}$$

The major product is the more stable one. It is a primary rather than a secondary halide and contains a more substituted double bond. Therefore the reaction is governed by thermodynamic (equilibrium) control.

11.21 Both compounds have a conjugated diene system and so are even on that score. α-Terpinene has more alkyl substituents on its diene system than does α-phellandrene and so will be more stable and have a lower heat of hydrogenation.

[Structures: α-phellandrene and α-terpinene, each a ring bearing CH_3 and $CH(CH_3)_2$ with H substituents on the diene system]

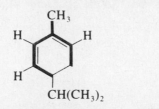

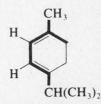

α-Phellandrene: three hydrogen substituents on diene system; measured heat of hydrogenation 52.9 kcal/mol

α-Terpinene: two hydrogen substituents on diene system; measured heat of hydrogenation 50.2 kcal/mol

11.22 (*a*) The two equilibria are:

For (*E*)-1,3-pentadiene:

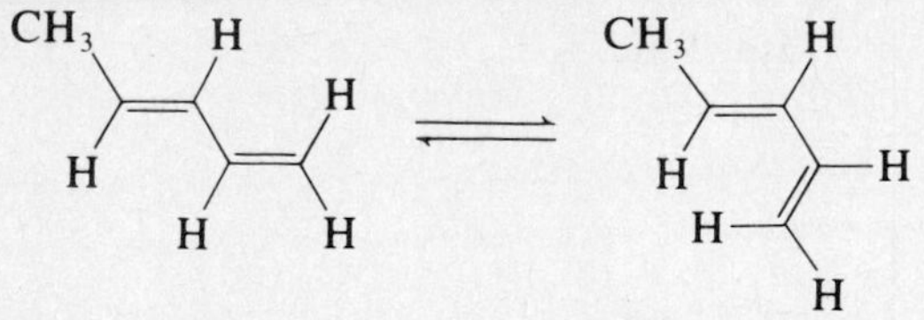

s-trans ⇌ s-cis

For (*Z*)-1,3-pentadiene:

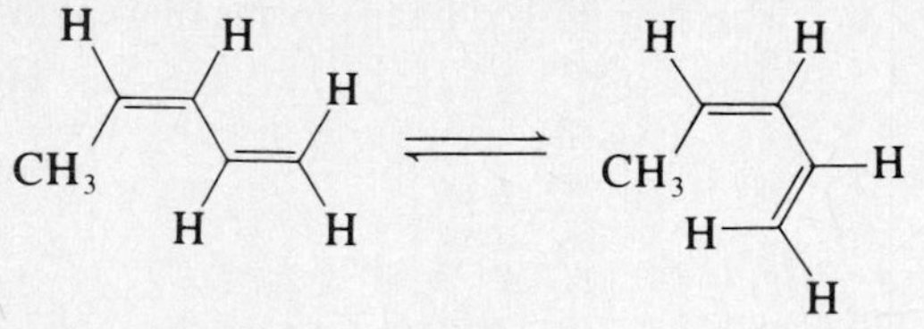

s-trans ⇌ s-cis

(*b*) The s-cis conformation of (*Z*)-1,3-pentadiene is destabilized by a van der Waals repulsion involving the methyl group.

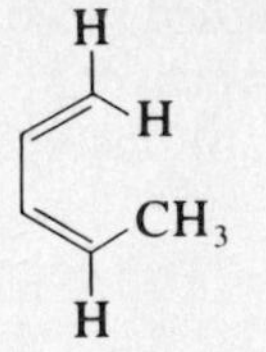

Methyl-hydrogen repulsion

s-Cis conformation of (*Z*)-1,3-pentadiene

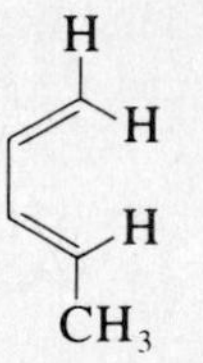

Hydrogen-hydrogen repulsion

s-Cis conformation of (*E*)-1,3-pentadiene

The equilibrium favors the s-trans conformation of (*Z*)-1,3-pentadiene more than it does that of the *E* isomer because the s-cis conformation of the *Z* isomer has more van der Waals strain.

11.23 Compare the mirror image forms of each compound for superposability.

(*a*) 2-Methyl-2,3-pentadiene

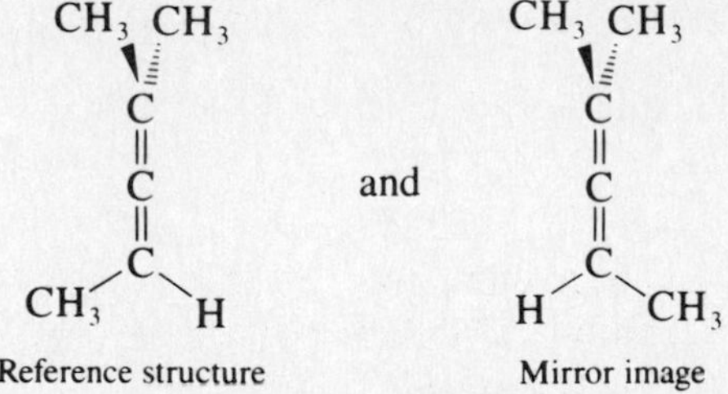

Reference structure and Mirror image

Rotation of the mirror image 180° around an axis passing through the three carbons of the C=C=C unit demonstrates that the reference structure and its mirror image are superposable.

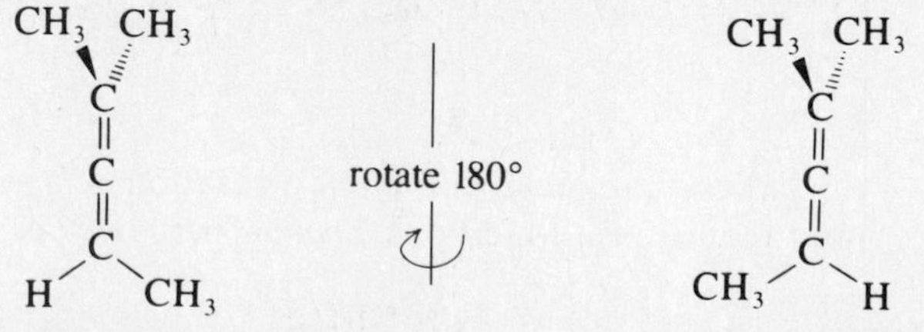

Mirror image — rotate 180° — Reoriented mirror image

2-Methyl-2,3-pentadiene is an achiral allene.

(*b*) The stereochemical situation here is the same as in (*a*). In (*a*) substituent R was a methyl group; in (*b*) it is an ethyl group. Changing the nature of group R does not affect the superposability of the reference structure and its mirror image. 2-Methyl-2,3-hexadiene is achiral.

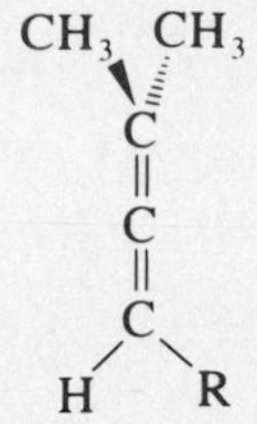

CH_3 CH_3 C=C=C H R

R=CH_3 2-Methyl-2,3-pentadiene

R=CH_2CH_2 2-Methyl-2,3-hexadiene

(*c*) The two mirror image forms of 4-methyl-2,3-hexadiene are as shown:

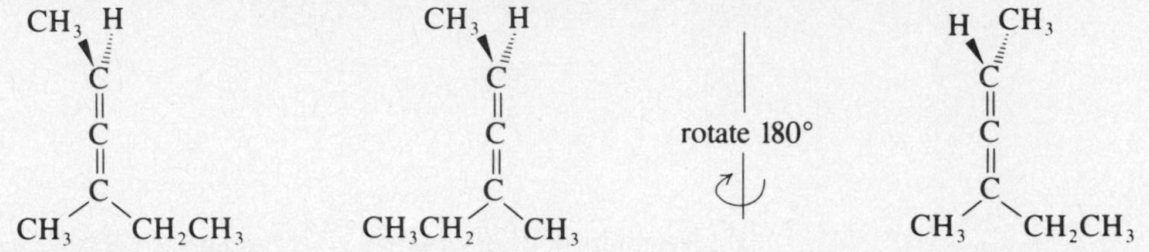

Reference structure — Mirror image — rotate 180° — Reoriented mirror image

The two structures cannot be superposed. 4-Methyl-2,3-hexadiene is chiral. Rotation of either representation 180° around an axis that passes through the three carbons of the C=C=C unit leads to superposition of the groups at the "bottom" carbon but not at the "top."

(*d*) 2,4-Dimethyl 2,3-pentadiene is achiral. Its two mirror image forms are superposable.

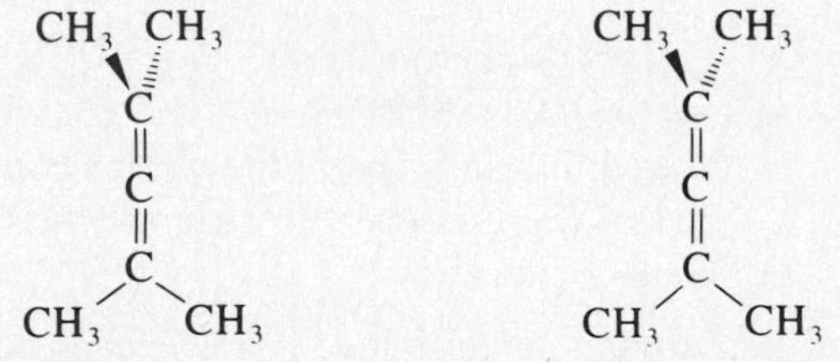

Reference structure — Mirror image

The molecule has two planes of symmetry defined by the three carbons of each CH_3CCH_3 unit.

11.24 Reaction (*a*) is an electrophilic addition of bromine to an alkene; the appropriate reagent is *bromine in carbon tetrachloride.*

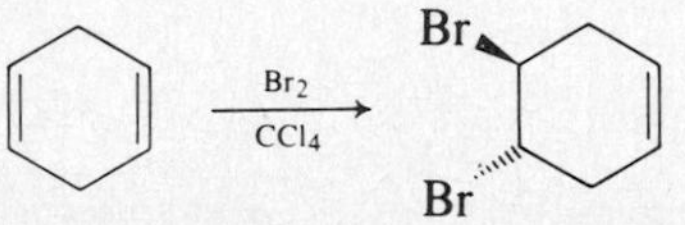

$\xrightarrow[CCl_4]{Br_2}$ (74%)

Reaction (*b*) is an epoxidation of an alkene, for which almost any peroxy acid could be used. *Peroxybenzoic acid* was actually used.

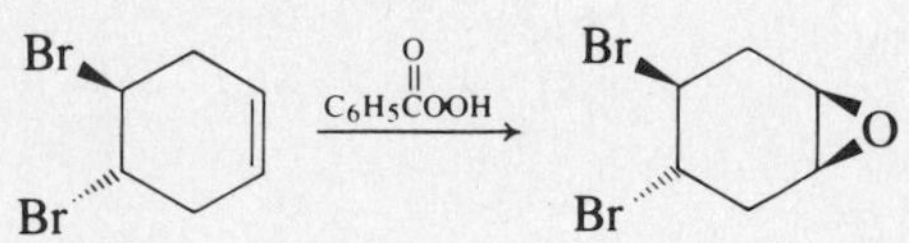

$\xrightarrow{C_6H_5COOH}$ (69%)

Reaction (*c*) is an elimination reaction of a vicinal dibromide to give a conjugated diene and requires E2 conditions. *Sodium methoxide in methanol* was used.

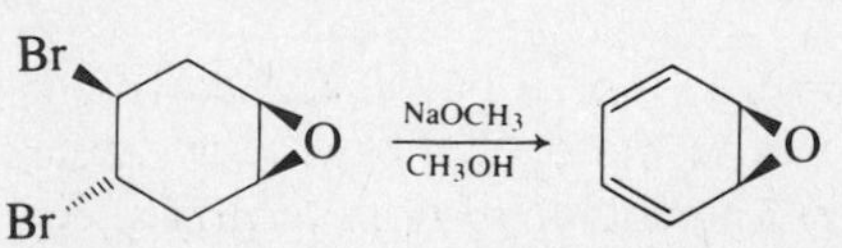

$\xrightarrow[CH_3OH]{NaOCH_3}$ (80%)

Reaction (*d*) is a Diels-Alder reaction in which the dienophile is *maleic anhydride*. The dienophile adds from the side opposite that of the epoxide ring.

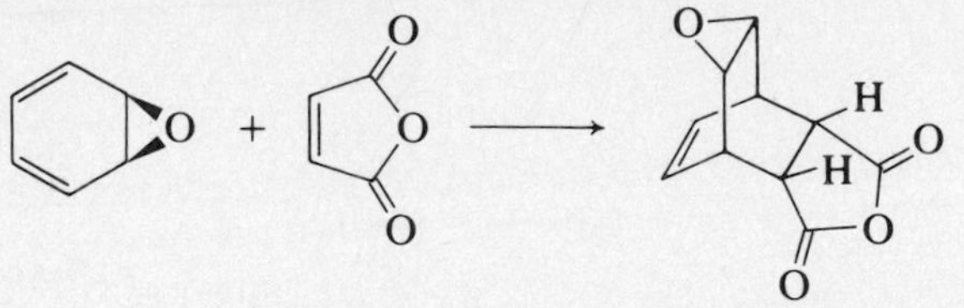

(100%)

11.25 To predict the constitution of the Diels-Alder adducts, we can ignore the substituents and simply remember that the fundamental process is

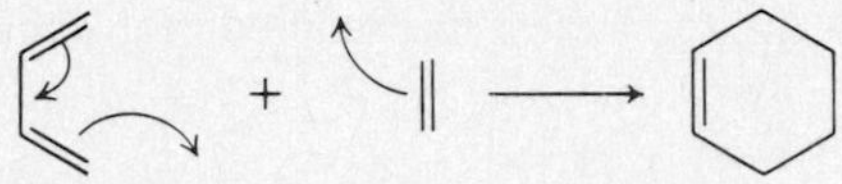

(*a*) 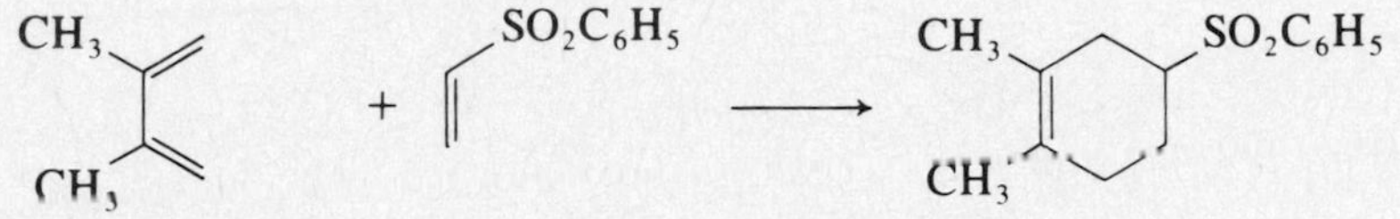 2,3-Dimethyl-1,3-butadiene

(*b*)

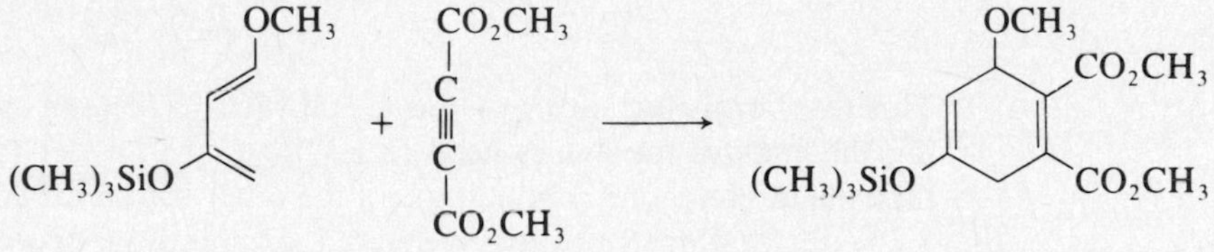

(*c*)

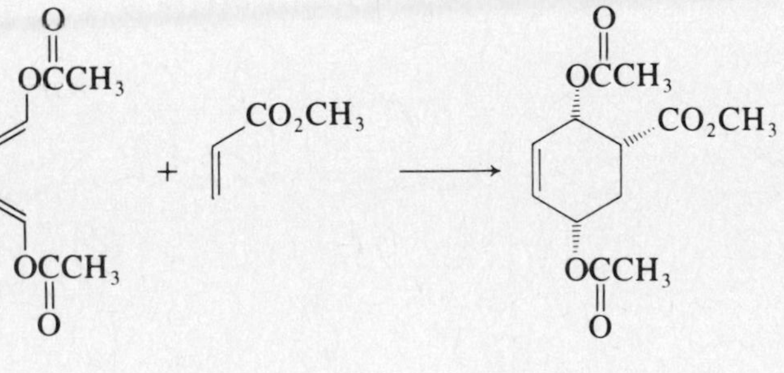

(*d*)

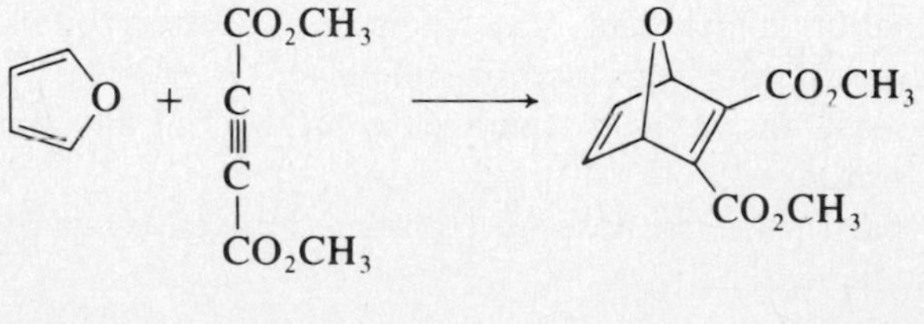

(*e*)

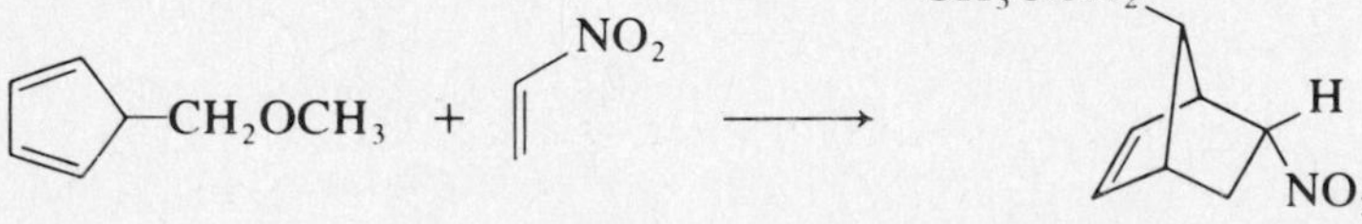

11.26 The carbon skeleton of dicyclopentadiene must be the same as that of its hydrogenation product and dicyclopentadiene must contain two double bonds since two moles of hydrogen are consumed in its hydrogenation ($C_{10}H_{12} \rightarrow C_{10}H_{16}$).

The molecular formula of dicyclopentadiene ($C_{10}H_{12}$) is twice that of 1,3-cyclopentadiene (C_5H_6), and its carbon skeleton suggests that 1,3-cyclopentadiene is undergoing a Diels-Alder reaction with itself. Therefore:

C_5H_6 + C_5H_6 ⟶ Dicyclopentadiene $C_{10}H_{12}$ $\xrightarrow[\text{Pt}]{2H_2}$ $C_{10}H_{16}$

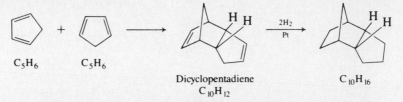

One molecule of 1,3-cyclopentadiene acts as the diene and the other acts as the dienophile in this Diels-Alder reaction.

11.27 (*a*) Analyze the reaction of two butadiene molecules by the Woodward-Hoffmann rules by examining the symmetry properties of the highest occupied molecular orbital (HOMO) of one diene and the lowest unoccupied molecular orbital (LUMO) of the other.

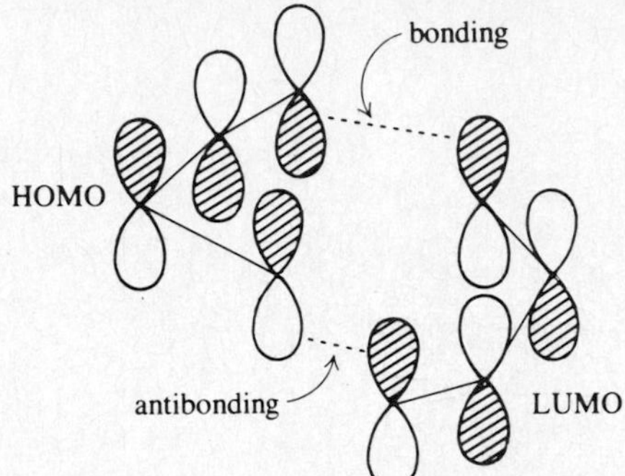

This reaction is forbidden by the Woodward-Hoffmann rules. Both interactions involving the ends of the dienes need to be bonding in order for concerted cycloaddition to take place. Here, one is bonding and the other is antibonding.

(*b*) Since allyl cation is positively charged, examine the process in which electrons "flow" from the HOMO of ethylene to the LUMO of allyl cation.

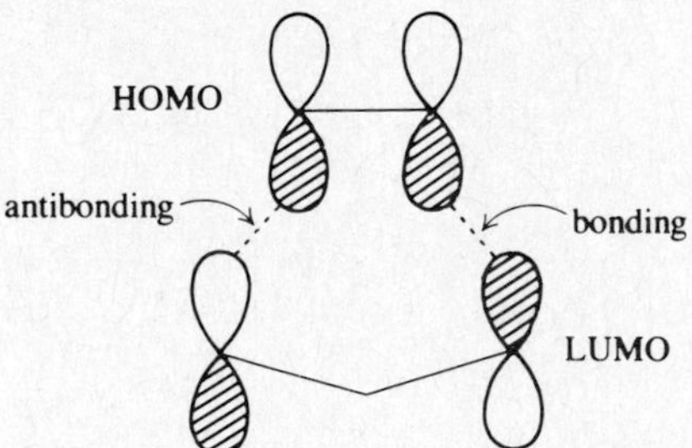

This reaction is forbidden. The symmetries of the orbitals are such that one interaction is bonding and the other is antibonding.

The same answer is obtained if the HOMO of allyl cation and the LUMO of ethylene are examined.

(*c*) In this part of the exercise we consider the LUMO of allyl cation and the HOMO of 1,3-butadiene.

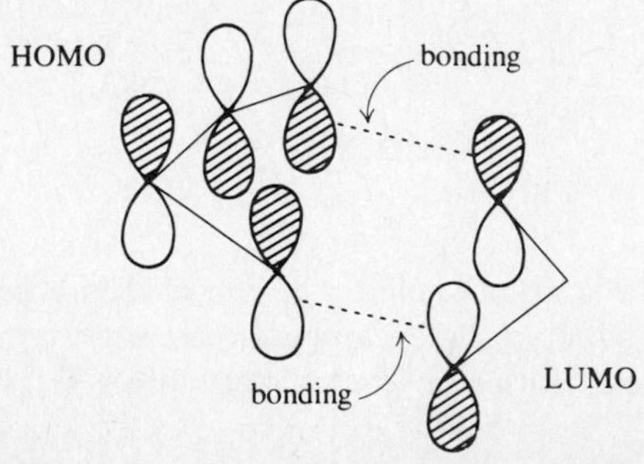

This reaction is allowed by the Woodward-Hoffmann rules. Both interactions are bonding.

The same prediction would be arrived at if the HOMO of allyl cation and LUMO of 1,3-butadiene were the orbitals considered.

11.28 Since oxygen has two unpaired electrons, it can abstract a hydrogen atom from the allylic position of cyclohexene to give a free-radical intermediate.

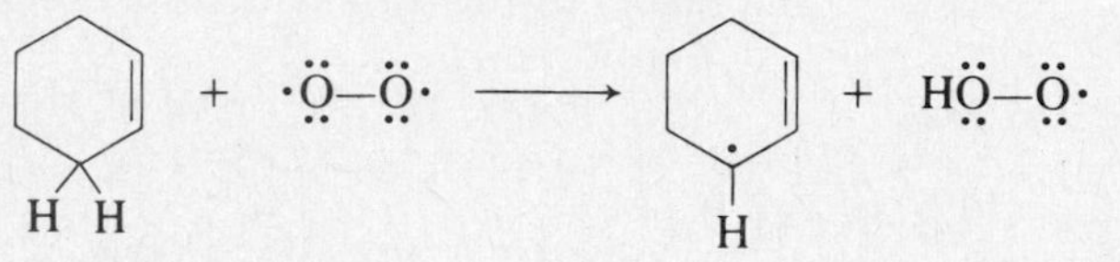

H H + ·Ö—Ö· ⟶ H + HÖ—Ö·

The cyclohexenyl radical is resonance-stabilized. It reacts further via the following two propagation steps:

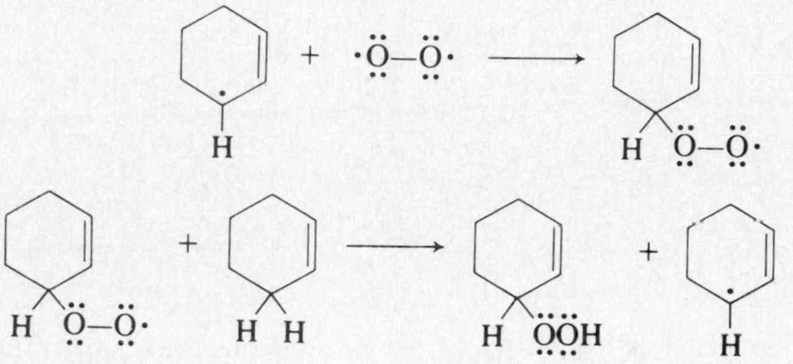

H + ·Ö—Ö· ⟶ H Ö—Ö·

H Ö—Ö· + H H ⟶ H ÖÖH + H

SELF-TEST

PART A

A-1. Give the structures of all the isomeric alkadienes of molecular formula C_5H_8, ignoring stereoisomers. Indicate which are conjugated and which are allenes.

A-2. Provide the IUPAC name for each of the conjugated dienes of the previous problem, *including stereoisomers.*

A-3. Hydrolysis of 3-bromo-3-methylcyclohexene yields two isomeric alcohols. Draw their structures and the structure of the intermediate that leads to their formation.

A-4. Give the chemical structure of the reactant, reagent, or product omitted from each of the following:

(*a*) $CH_3CH{=}CHCH{=}CHCH_3 \xrightarrow{Br_2}$? (two products)

(*b*) ? $\xrightarrow{\text{Diels-Alder}}$ [O, O, O]

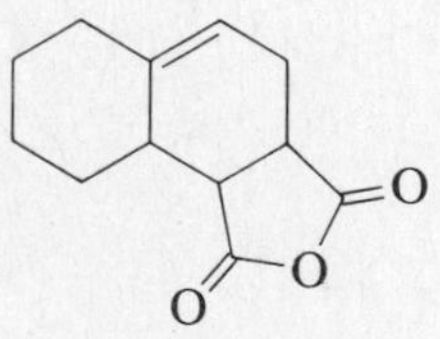

(*c*) ⟶ Br + Br

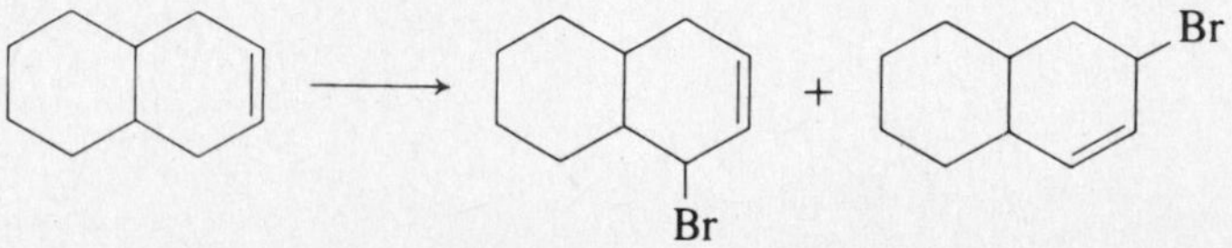

A-5. One of the isomeric conjugated dienes having the formula C_6H_8 is not able to react with a dienophile in a Diels-Alder reaction. Draw the structure of this compound.

PART B

B-1. 2,3-Pentadiene, $CH_3CH{=}C{=}CHCH_3$, is
(*a*) A planar substance
(*b*) An allene
(*c*) A conjugated diene
(*d*) A substance capable of cis-trans isomerism

B-2. Rank the following carbocations in order of increasing stability (least → most)

$$\underset{\text{A}}{CH_3\overset{+}{C}HCH_3} \qquad \underset{\text{B}}{CH_3\overset{+}{C}H{-}CH{=}CHCH_3} \qquad \underset{\text{C}}{(CH_3)_3C\overset{+}{C}H_2}$$

(*a*) A < B < C (*b*) B < C < A (*c*) C < A < B (*d*) B < A < C

B-3. Hydrogenation of cyclohexene releases 28.6 kcal/mol of heat. Which of the following most likely represents the observed heat of hydrogenation of 1,3-cyclohexadiene?
(*a*) 55.4 kcal/mol (*b*) 57.2 kcal/mol
(*c*) 59.0 kcal/mol (*d*) 28.6 kcal/mol

B-4. Which of the following is *not* a proper resonance form of 1,3-cyclohexadiene?

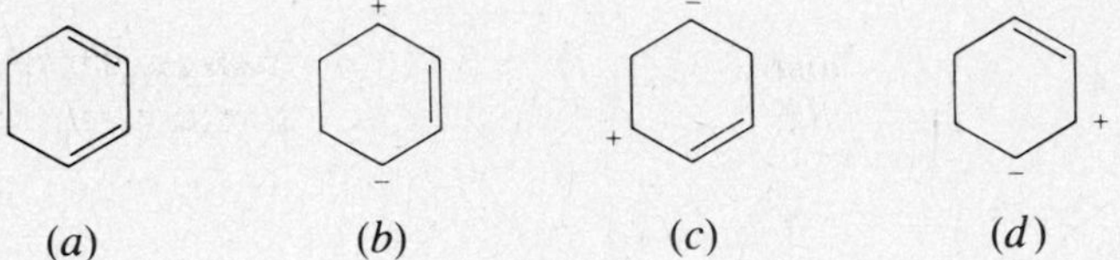

(*a*) (*b*) (*c*) (*d*)

B-5. For the following reactions the major products are shown.

$$CH_2{=}CH{-}CH{=}CH_2 \xrightarrow{\text{HCl}} CH_2{=}CH\underset{\text{Cl}}{\underset{|}{C}}HCH_3 \xrightarrow{+25°C} \underset{\text{Cl}}{\underset{|}{C}}H_2CH{=}CHCH_3$$

These provide an example of $\underset{1}{\_\_\_\_\_}$ control at low temperature and $\underset{2}{\_\_\_\_\_}$ control at higher temperature.

	1	2
(*a*)	kinetic	thermodynamic
(*b*)	thermodynamic	kinetic
(*c*)	kinetic	kinetic
(*d*)	thermodynamic	thermodynamic

B-6. Which of the C—H bonds shown below would have the smallest bond dissociation energy?

(*a*) $CH_3\overset{H\,\leftarrow}{\overset{|}{C}}HCH_3$ (*b*) $CH_3CH{=}\overset{H\,\leftarrow}{\overset{|}{C}}H$

(*c*) $CH_3CH_2\overset{H\,\leftarrow}{\overset{|}{C}}H_2$ (*d*) $CH_2{=}CH{-}\overset{H\,\leftarrow}{\overset{|}{C}}HCH_3$

CHAPTER

12

ARENES AND AROMATICITY

IMPORTANT TERMS AND CONCEPTS

Structure and Bonding of Benzene (Secs. 12.3 to 12.8) Benzene, as the parent *aromatic hydrocarbon* or *arene*, is best represented as a resonance hybrid of the two Kekulé structures

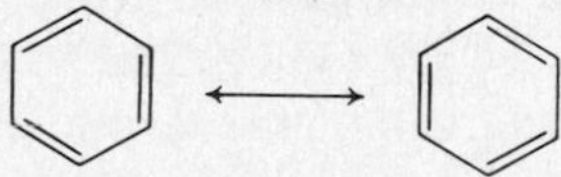

Benzene is planar and all six carbon-carbon bonds are the same length. Each carbon is sp^2 hybridized; the unhybridized p orbitals are all parallel, and their overlap generates a continuous π system encompassing all the carbon atoms of the ring.

Experiments have shown the heat of hydrogenation of benzene to be 36 kcal/mol *less* than that expected for a hypothetical 1,3,5-cyclohexatriene. This *stabilization energy* is also known as the *resonance energy* or *delocalization energy* of benzene.

Hückel's Rule (Sec. 12.10) Hückel's rule states that only those molecules which are planar, monocyclic, conjugated polyenes having $(4n + 2)$ π electrons (where n is an integer) will be aromatic. That is, molecules containing 2, 6, 10, etc. π electrons in a planar cyclic array will exhibit aromatic stabilization. Such molecules are termed *annulenes*; benzene therefore is [6]-annulene.

Aromatic Ions (Sec. 12.11) The preceding statement of Hückel's rule applies to planar ions as well as to neutral molecules. Therefore the cycloheptatrienyl cation is aromatic, having six π electrons distributed over a planar cyclic array of seven p orbitals. Likewise, the cyclopentadienide anion contains six π electrons in a five-p-orbital framework. In both cases aromaticity imparts special stability to the ions, making their formation more favorable than would be anticipated otherwise.

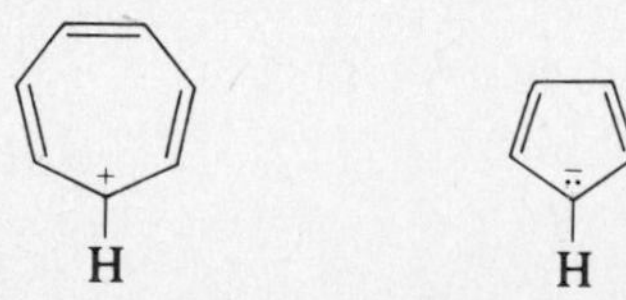

Cycloheptatrienyl cation Cyclopentadienide anion

Aromatic Heterocyclic Compounds (Sec. 12.12) A *heterocycle* is a molecule having an atom other than carbon as part of the ring framework. These heteroatoms may contribute an unshared electron pair toward satisfying Hückel's rule. For example, pyrrole and furan are aromatic.

Pyrrole Furan

Nomenclature of Benzene Derivatives (Sec. 12.15) Disubstituted benzene derivatives are named by use of the prefixes *ortho-* (*o*-), *meta-* (*m*-), and *para-* (*p*-), which refer to 1,2 disubstitution, 1,3 disubstitution, and 1,4 disubstitution, respectively.

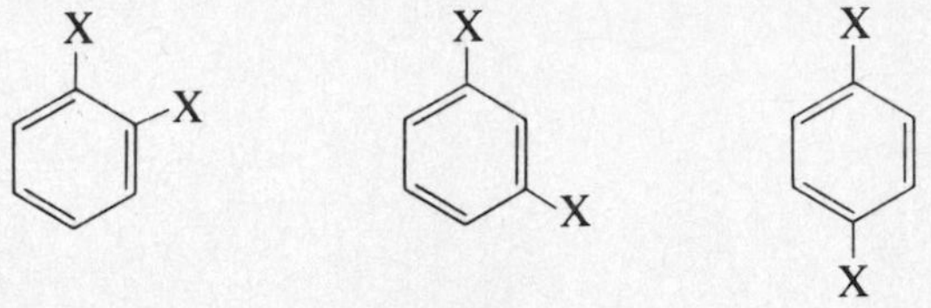

Ortho Meta Para

When three or more substituents are present on the ring, the substituents are numbered to specify their position; the *o*-, *m*-, and *p*- prefixes are not used.

Benzylic Conjugation (Sec. 12.16) Benzylic cations and radicals are highly stabilized species in which conjugation with the benzene ring results in delocalization of the positive charge in the cation and the unpaired electron in the radical.

Benzyl cation

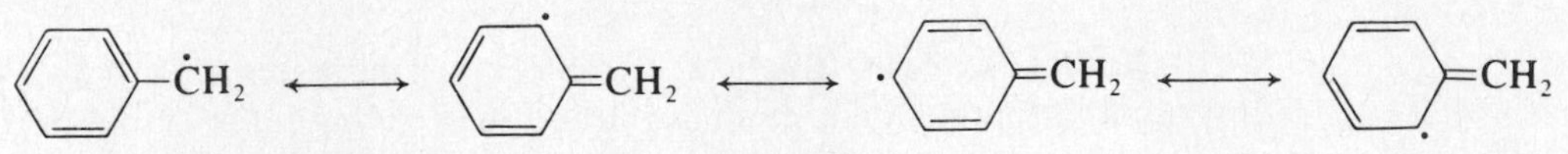

Benzyl radical

IMPORTANT REACTIONS

Free-Radical Halogenation (Sec. 12.17)

General:

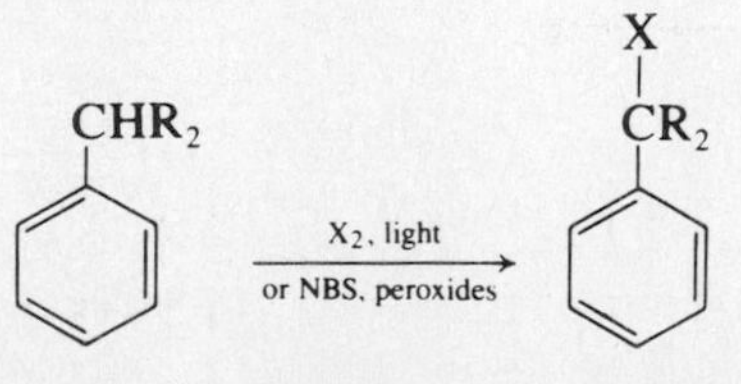

$$C_6H_5CHR_2 \xrightarrow[\text{or NBS, peroxides}]{X_2,\ \text{light}} C_6H_5CR_2X$$

Examples:

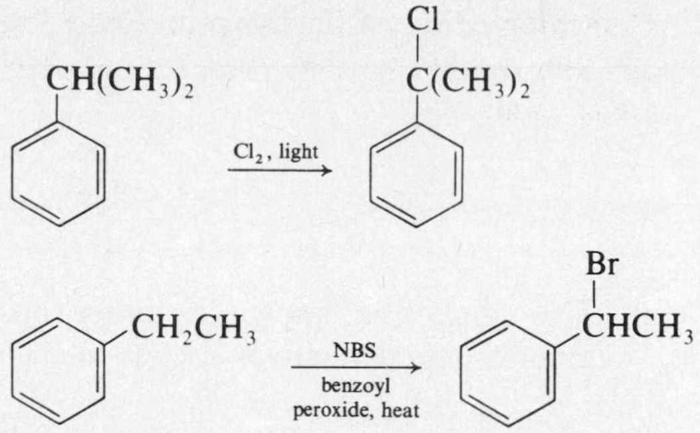

$C_6H_5CH(CH_3)_2 \xrightarrow{Cl_2,\ light} C_6H_5C(Cl)(CH_3)_2$

$C_6H_5CH_2CH_3 \xrightarrow[\text{benzoyl peroxide, heat}]{NBS} C_6H_5CHBrCH_3$

Alkylbenzene Oxidation (Sec. 12.18)

General:

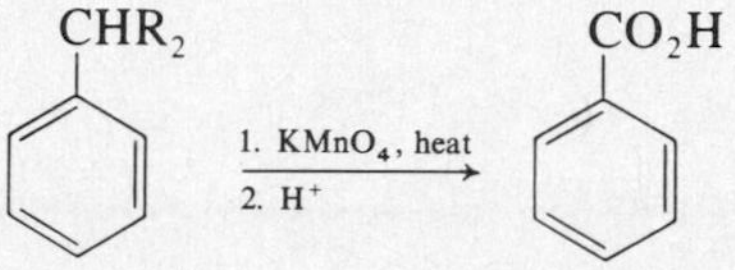

$C_6H_5CHR_2 \xrightarrow[2.\ H^+]{1.\ KMnO_4,\ heat} C_6H_5CO_2H$

Example:

$m\text{-}CH_3C_6H_4CH_2CH_3 \xrightarrow[2.\ H^-]{1.\ KMnO_4,\ heat} m\text{-}C_6H_4(CO_2H)_2$

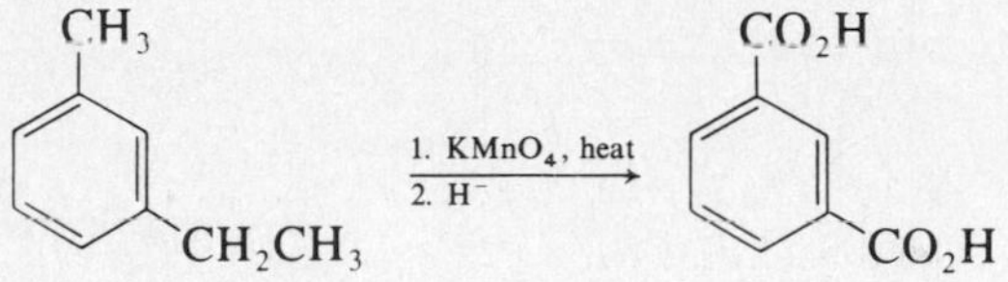

Alkenylbenzenes (Secs. 12.20, 12.21)

Preparation:

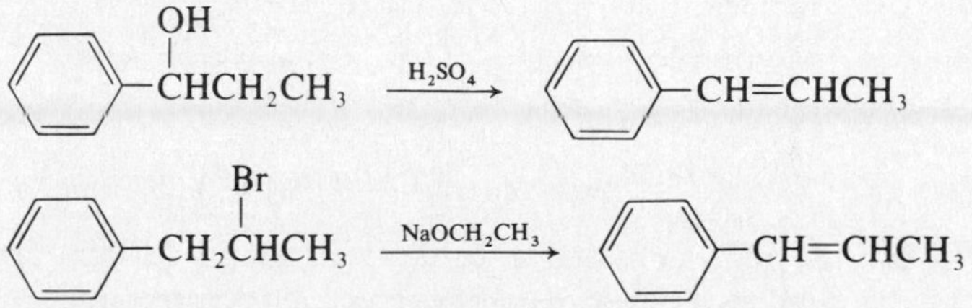

$C_6H_5CH(OH)CH_2CH_3 \xrightarrow{H_2SO_4} C_6H_5CH{=}CHCH_3$

$C_6H_5CH_2CHBrCH_3 \xrightarrow{NaOCH_2CH_3} C_6H_5CH{=}CHCH_3$

Reactions:

$C_6H_5CH{=}CHCH_3 \xrightarrow{HBr} C_6H_5CHBrCH_2CH_3$

$C_6H_5CH{=}CHCH_3 \xrightarrow[2.\ NaBH_4,\ HO^-]{1.\ Hg(OAc)_2,\ THF{-}H_2O} C_6H_5CH(OH)CH_2CH_3$

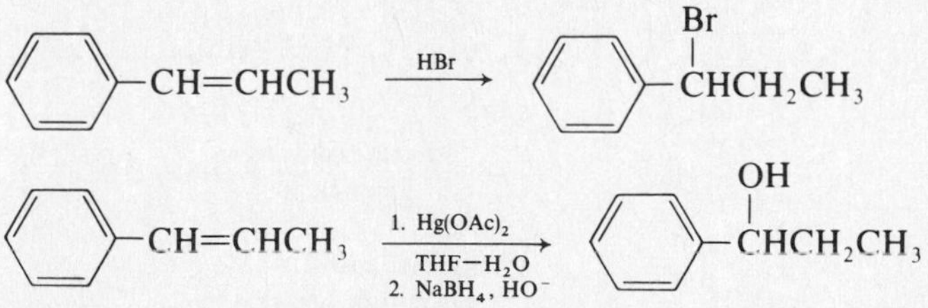

SOLUTIONS TO TEXT PROBLEMS

12.1 (*b*) This hydrocarbon is known as *prismane*; it has the shape of a prism. All its carbons are equivalent and all its hydrogens are equivalent. Only one monosubstitution product is possible.

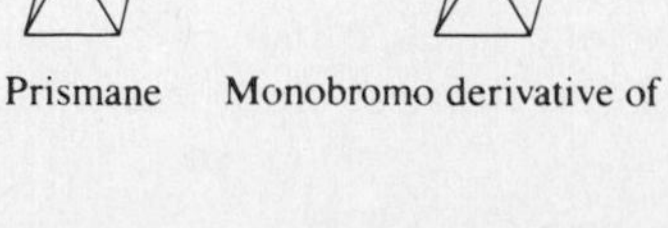

Prismane   Monobromo derivative of prismane

(c) All the hydrogens of this compound are equivalent. If substitution by bromine could be achieved, only a single monobromo derivative would be obtained.

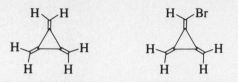

(d) This compound, known as *fulvene*, has three different pairs of vinyl hydrogens. Three monobromo derivatives are possible.

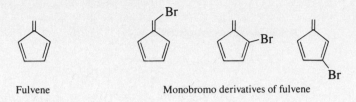

Fulvene Monobromo derivatives of fulvene

12.2 Toluene is $C_6H_5CH_3$; it has a methyl group attached to a benzene ring.

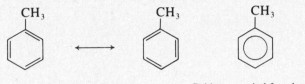

Kekulé forms of toluene Robinson symbol for toluene

Benzoic acid has a $-CO_2H$ substituent on the benzene ring.

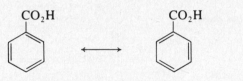

Kekulé forms of benzoic acid Robinson symbol for benzoic acid

12.3 Styrene contains a benzene ring and will be appreciably stabilized by resonance, which makes it lower in energy than cyclooctatetraene.

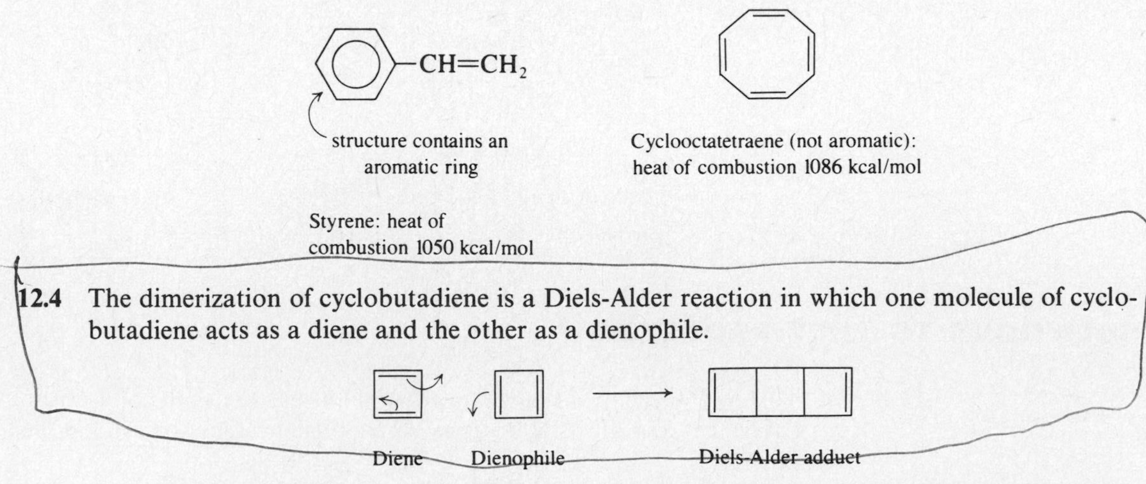

12.4 The dimerization of cyclobutadiene is a Diels-Alder reaction in which one molecule of cyclobutadiene acts as a diene and the other as a dienophile.

12.5 (b) Since 12 2p orbitals contribute to the cyclic conjugated system of [12]-annulene, there will be 12 π molecular orbitals. These MO's are arranged so that one is of highest energy, one is of lowest energy, and the remaining 10 are found in pairs between the highest- and lowest-energy orbitals. There are 12 π electrons so the lowest five orbitals are each doubly occupied, while each of the next two orbitals—orbitals of equal energy—is singly occupied.

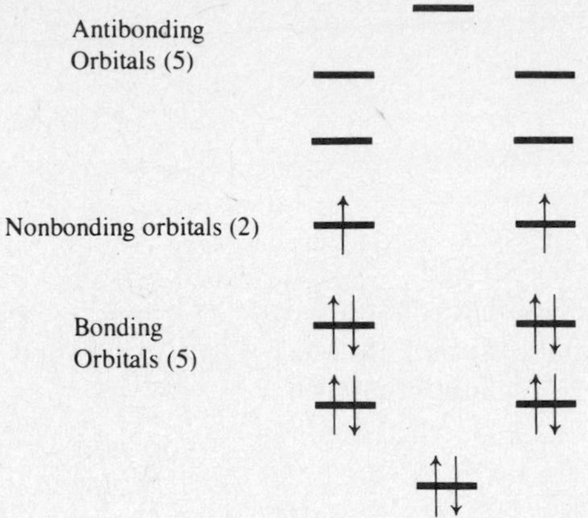

12.6 (*b*) Cyclononatetraenide anion has 10 π electrons; it is aromatic. The 10 π electrons are most easily seen by writing a Lewis structure for the anion: there are 2 π electrons for each of four double bonds, and the negatively charged carbon contributes 2.

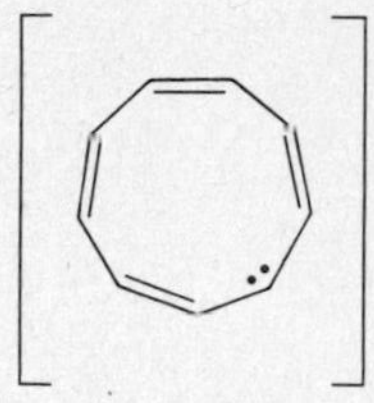

12.7 (*b*) There are eight hydrogen atoms in naphthalene, divided into two sets of four equivalent hydrogens. Those at C-1, C-4, C-5, and C-8 are equivalent, as are those at C-2, C-3, C-6, and C-7. Thus, only two monobromo derivatives of naphthalene are possible.

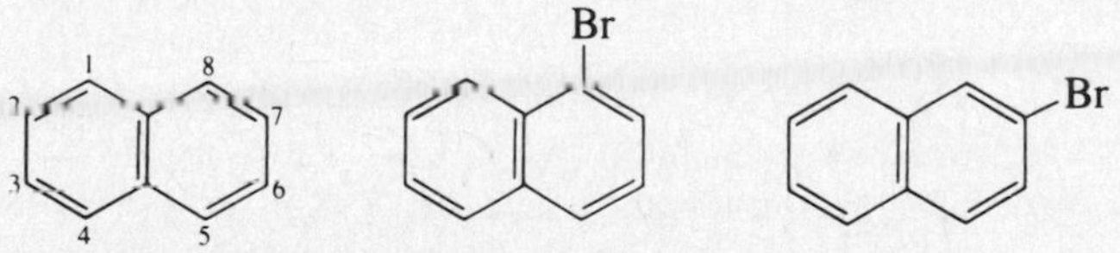

(*c*) Phenanthrene can form a total of five monobromo derivatives

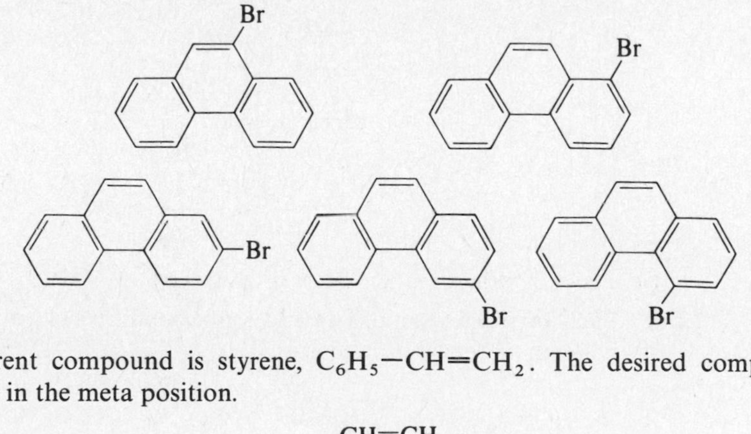

12.8 (*b*) The parent compound is styrene, $C_6H_5{-}CH{=}CH_2$. The desired compound has a chlorine in the meta position.

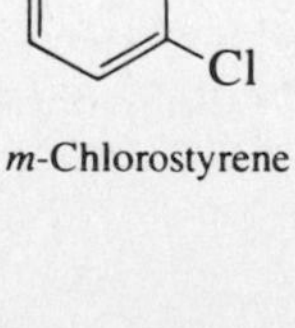

m-Chlorostyrene

(c) The parent compound is aniline, $C_6H_5NH_2$. p-Nitroaniline is therefore:

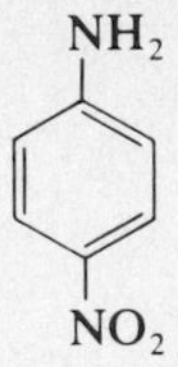

NH_2

NO_2

p-Nitroaniline

12.9 The dihydronaphthalene in which the double bond is conjugated with the aromatic ring is more stable; thus 1,2-dihydronaphthalene has a lower heat of hydrogenation than 1,4-dihydronaphthalene.

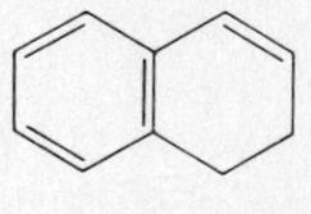

1,2-Dihydronaphthalene (heat of hydrogenation 24.1 kcal/mol)

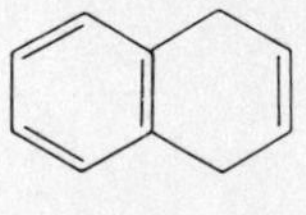

1,4-Dihydronaphthalene (heat of hydrogenation 27.1 kcal/mol)

12.10 (b) Only the benzylic hydrogen is replaced by bromine in the reaction of 4-methyl-3-nitroanisole with *N*-bromosuccinimide.

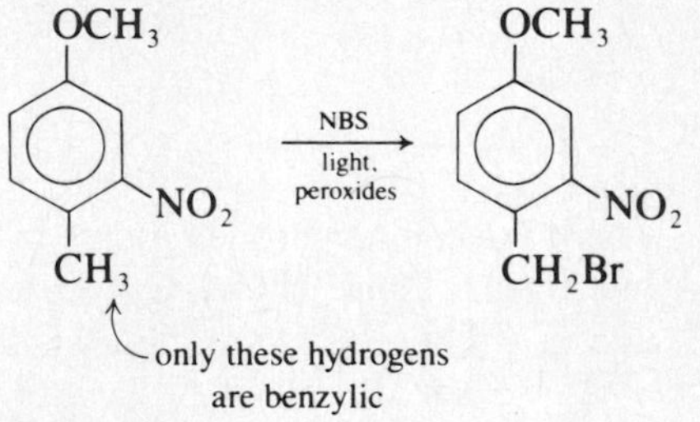

OCH_3 NO_2 CH_3 → (NBS, light, peroxides) → OCH_3 NO_2 CH_2Br

only these hydrogens are benzylic

12.11 The molecular formula of the product is $C_{12}H_{14}O_4$. Since it contains four oxygens, the product must have two $-CO_2H$ groups. None of the hydrogens of a *tert*-butyl substituent on a benzene ring is benzylic, so this group is inert to oxidation. Only the methyl groups of 4-*tert*-butyl-1,2-dimethylbenzene are susceptible to oxidation; therefore the product is 4-*tert*-butylbenzene-1,2-dicarboxylic acid.

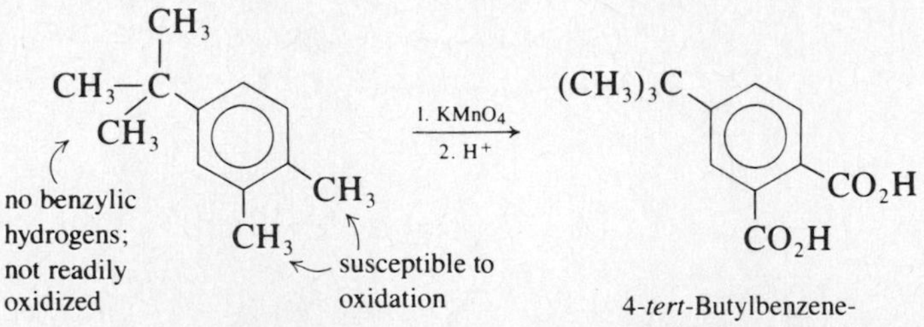

CH_3, CH_3-C, CH_3 — no benzylic hydrogens; not readily oxidized

CH_3, CH_3 — susceptible to oxidation

1. $KMnO_4$
2. H^+

$(CH_3)_3C$, CO_2H, CO_2H

4-*tert*-Butylbenzene-1,2-dicarboxylic acid

12.12 Each of these reactions involves nucleophilic substitution of the S_N2 type at the benzylic position of benzyl bromide.

(b) $(CH_3)_3CO^-$ CH_2-Br ⟶ $CH_2OC(CH_3)_3$

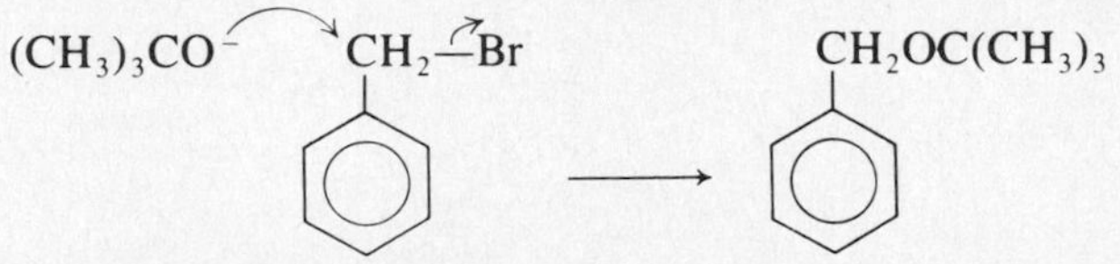

tert-Butoxide ion — Benzyl bromide — Benzyl *tert*-butyl ether

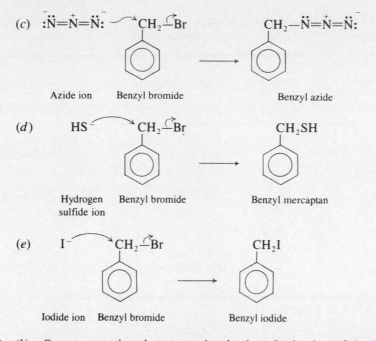

(c) $:\ddot{\bar{N}}=\overset{+}{N}=\ddot{\bar{N}}:$ + $C_6H_5CH_2-Br$ ⟶ $C_6H_5CH_2-\ddot{\bar{N}}-\overset{+}{N}=\ddot{N}:$

Azide ion Benzyl bromide Benzyl azide

(d) HS^- + $C_6H_5CH_2-Br$ ⟶ $C_6H_5CH_2SH$

Hydrogen sulfide ion Benzyl bromide Benzyl mercaptan

(e) I^- + $C_6H_5CH_2-Br$ ⟶ $C_6H_5CH_2I$

Iodide ion Benzyl bromide Benzyl iodide

12.13 (b) Oxymercuration-demercuration leads to hydration of double bonds in accordance with Markovnikov's rule, as described in Section 7.7. The product is the tertiary alcohol.

$$C_6H_5C(CH_3)=CH_2 \xrightarrow[2.\ NaBH_4]{1.\ Hg(OAc)_2,\ THF-H_2O} C_6H_5C(OH)(CH_3)-CH_3$$

2-Phenylpropene 2-Phenyl-2-propanol (100%)

(c) The regioselectivity of alcohol formation is reversed on hydroboration-oxidation.

$$C_6H_5C(CH_3)=CH_3 \xrightarrow[2.\ H_2O_2,\ HO^-]{1.\ B_2H_6} C_6H_5CH(CH_3)CH_2OH$$

2-Phenylpropene 2-Phenyl-1-propanol (92%)

(d) Bromine adds to alkenes in aqueous solution to give bromohydrins. A water molecule acts as a nucleophile, attacking the bromonium ion at the carbon that can bear most of the positive charge, which in this case is the benzylic carbon.

$$C_6H_5CH=CH_2 \xrightarrow[H_2O]{Br_2} C_6H_5CH(OH)CH_2Br$$

Styrene 2-Bromo-1-phenylethanol (82%)

(e) Peroxy acids convert alkenes to epoxides. Peroxybenzoic acid is an oxygen-transfer agent, just like other peroxy acids such as peroxyacetic acid.

$$C_6H_5CH=CH_2 + C_6H_5\overset{O}{\overset{\|}{C}}OOH \longrightarrow C_6H_5\underset{\backslash O/}{CH-CH_2} + C_6H_5\overset{O}{\overset{\|}{C}}OH$$

Styrene Peroxybenzoic acid 2-Phenyloxirane (69–75%) Benzoic acid

12.14 Since the problem requires that the benzene ring be monosubstituted, all that need to be examined are the various isomeric forms of the C_4H_9 substituent.

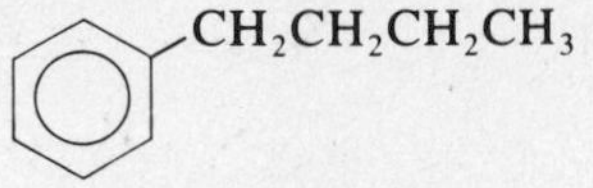

$C_6H_5CH_2CH_2CH_2CH_3$

Butylbenzene
(1-phenylbutane)

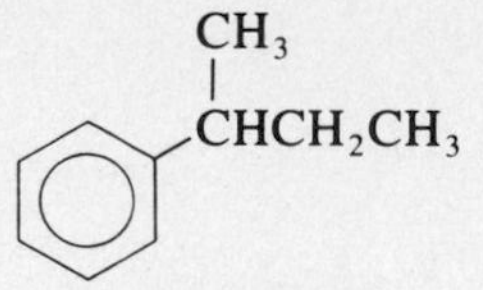

$C_6H_5CH(CH_3)CH_2CH_3$

sec-Butylbenzene
(2-phenylbutane)

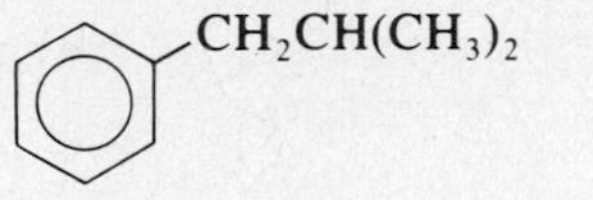

$C_6H_5CH_2CH(CH_3)_2$

Isobutylbenzene
(2-methyl-1-phenylpropane)

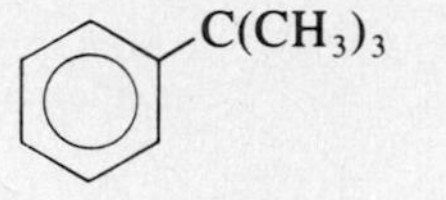

$C_6H_5C(CH_3)_3$

tert-Butylbenzene
(2-methyl-2-phenylpropane)

These are the four constitutional isomers. *sec*-Butylbenzene is chiral and so exists in enantiomeric *R* and *S* forms.

12.15 (*a*) A neopentyl group is $-CH_2C(CH_3)_3$, so neopentylbenzene is

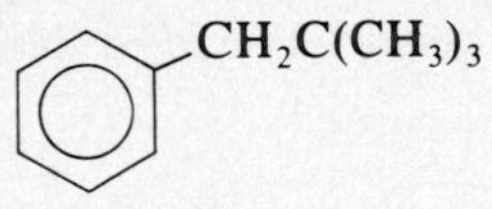

$C_6H_5CH_2C(CH_3)_3$

(*b*) An allyl substituent is $-CH_2CH{=}CH_2$.

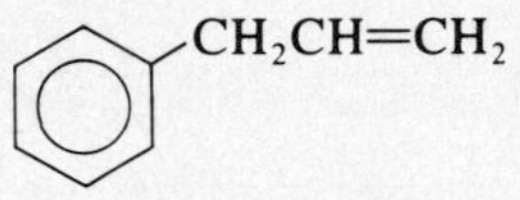

$C_6H_5CH_2CH{=}CH_2$

Allylbenzene

(*c*) A phenyl group replaces one of the hydrogens of acetylene in phenylacetylene.

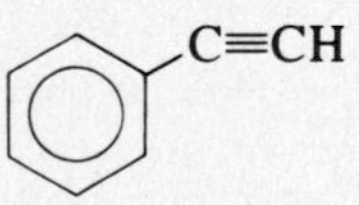

$C_6H_5C{\equiv}CH$ Phenylacetylene
(phenylethyne)

(*d*) The constitution of 1-phenyl-1-butene is $C_6H_5CH{=}CHCH_2CH_3$. The *E* stereoisomer is

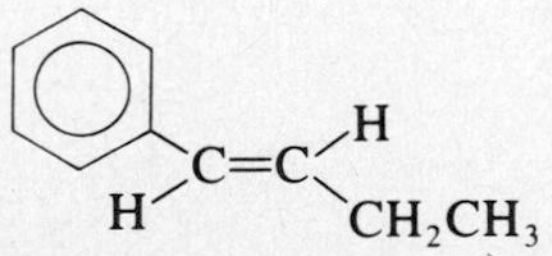

$C_6H_5(H)C{=}C(H)CH_2CH_3$ (*E*)-1-Phenyl-1-butene

The two higher-priority substituents, phenyl and ethyl, are on opposite sides of the double bond.

(*e*) The constitution of 2-phenyl-2-butene is $CH_3C(C_6H_5){=}CHCH_3$. The *Z* stereoisomer is

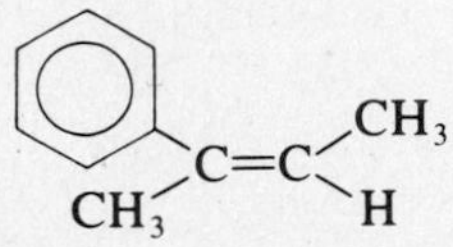

$C_6H_5(CH_3)C{=}C(CH_3)H$ (*Z*)-2-Phenyl-2-butene

The two higher-priority substituents, phenyl and methyl, are on the same side of the double bond.

(*f*) 1-Phenylethanol is chiral and has the constitution $CH_3\underset{\displaystyle OH}{\underset{|}{C}}HC_6H_5$. Among the substituents attached to the chiral center, the order of decreasing sequence rule priority is

$$HO > C_6H_5 > CH_3 > H$$

In the *R* enantiomer the three highest-priority substituents must appear in a clockwise sense in proceeding from higher priority to next lower priority when the lowest-priority substituent is directed away from the reader.

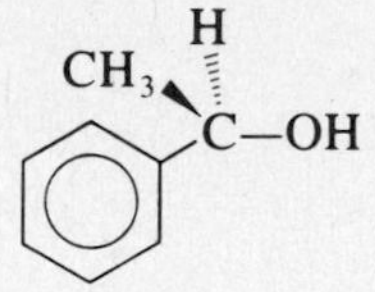

(*R*)-1-Phenylethanol

(*g*) In *p*-chlorophenol the benzene ring bears a chlorine and a hydroxyl substituent in a 1,4 substitution pattern.

p-Chlorophenol

(*h*) Benzenecarboxylic acid is an alternative IUPAC name for benzoic acid.

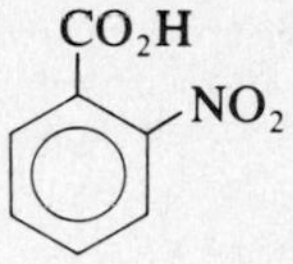

2-Nitrobenzenecarboxylic acid

(*i*) Two isopropyl groups are in a 1,4-relationship in *p*-diisopropylbenzene.

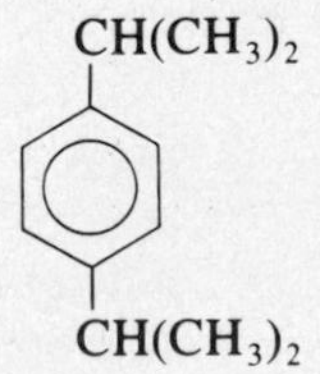

p-Diisopropylbenzene

(*j*) Aniline is $C_6H_5NH_2$. Therefore

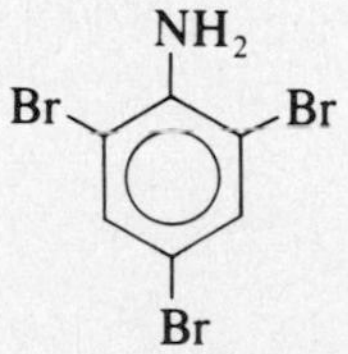

2,4,6-Tribromoaniline

(*k*) Acetophenone is acetylbenzene, $C_6H_5\overset{O}{\overset{\|}{C}}CH_3$. Therefore

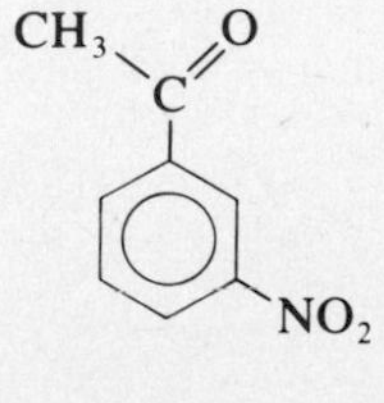

m-Nitroacetophenone

(*l*) Benzenesulfonic acid is $C_6H_5SO_3H$. Therefore

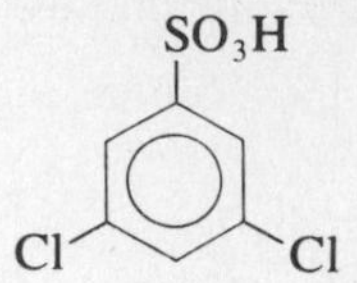

3,5-Dichlorobenzenesulfonic acid

12.16 (*a*) There are three isomeric nitrotoluenes, as the nitro group can be ortho, meta, or para to the methyl group.

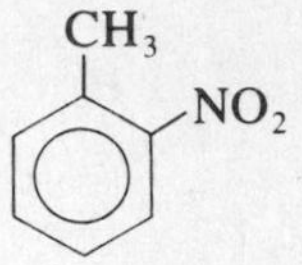

o-Nitrotoluene (2-nitrotoluene) 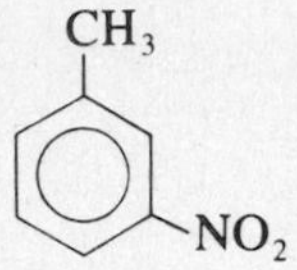 *m*-Nitrotoluene (3-nitrotoluene) 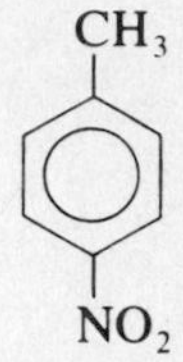 *p*-Nitrotoluene (4-nitrotoluene)

(*b*) Benzoic acid is $C_6H_5CO_2H$. In the isomeric dichlorobenzoic acids, two of the ring hydrogens of benzoic acid have been replaced by chlorines. The isomeric dichlorobenzoic acids are:

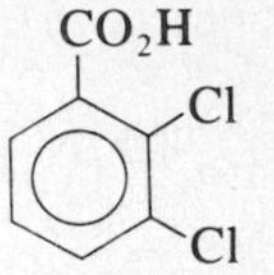

2,3-Dichlorobenzoic acid 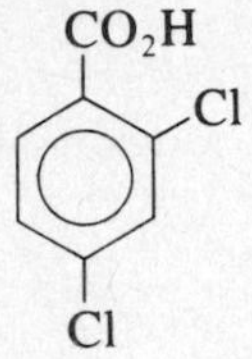 2,4-Dichlorobenzoic acid 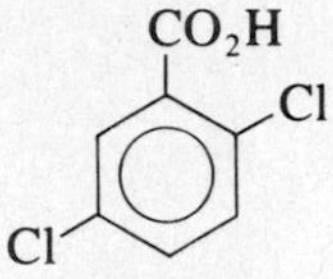 2,5-Dichlorobenzoic acid

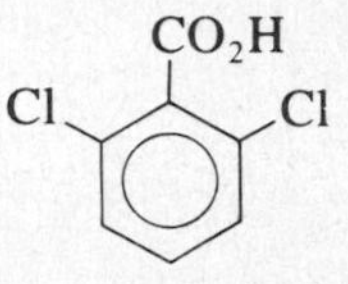

2,6-Dichlorobenzoic acid 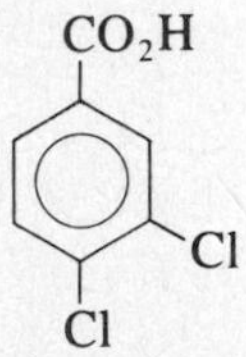 3,4-Dichlorobenzoic acid 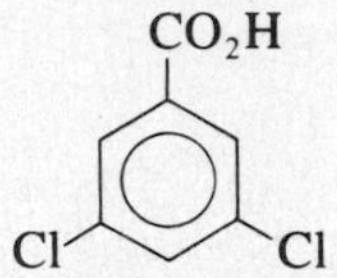 3,5-Dichlorobenzoic acid

The prefixes *o*-, *m*-, and *p*- may not be used in trisubstituted arenes; numerical prefixes are used. Note also that "benzenecarboxylic" may be used in place of "benzoic."

(*c*) In the various tribromophenols we are dealing with tetrasubstitution on a benzene ring. Again, *o*-, *m*-, and *p*- are not valid prefixes. The hydroxyl group is assigned position 1 because the base name is phenol (hydroxybenzene).

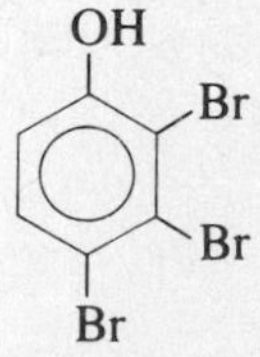

2,3,4-Tribromophenol 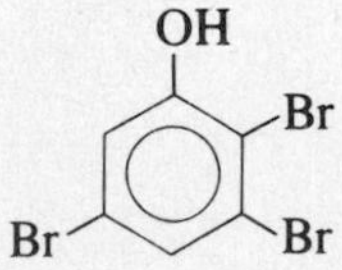 2,3,5-Tribromophenol 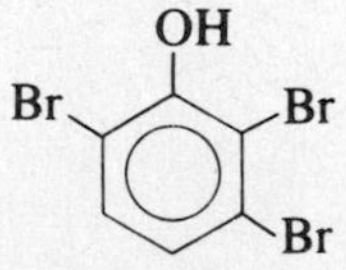 2,3,6-Tribromophenol

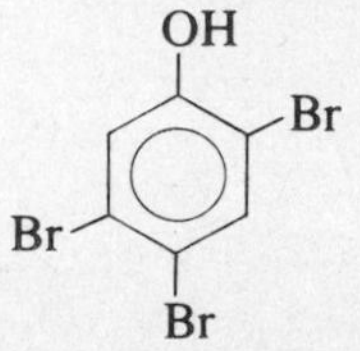

2,4,5-Tribromophenol 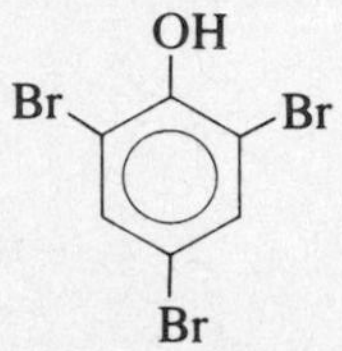 2,4,6-Tribromophenol 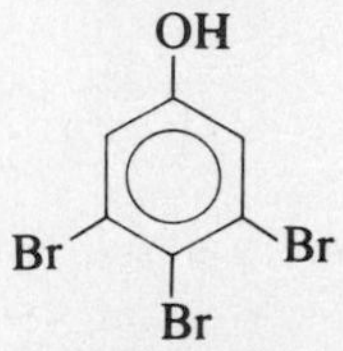 3,4,5-Tribromophenol

(*d*) There are only three tetrafluorobenzenes. The two hydrogen substituents may be ortho, meta, or para with respect to each other.

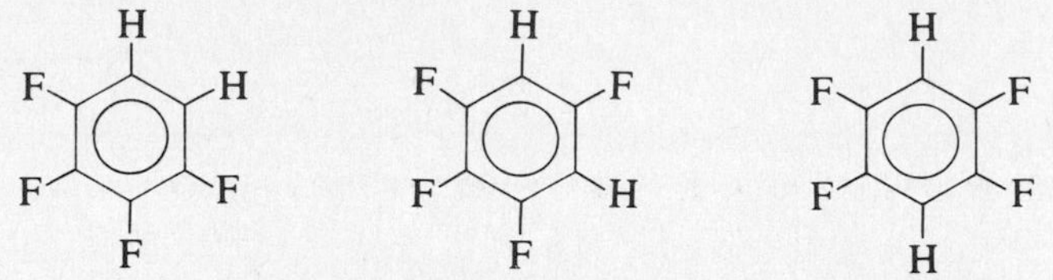

1,2,3,4-Tetrafluorobenzene 1,2,3,5-Tetrafluorobenzene 1,2,4,5-Tetrafluorobenzene

(*e*) There are only two naphthalenesulfonic acids.

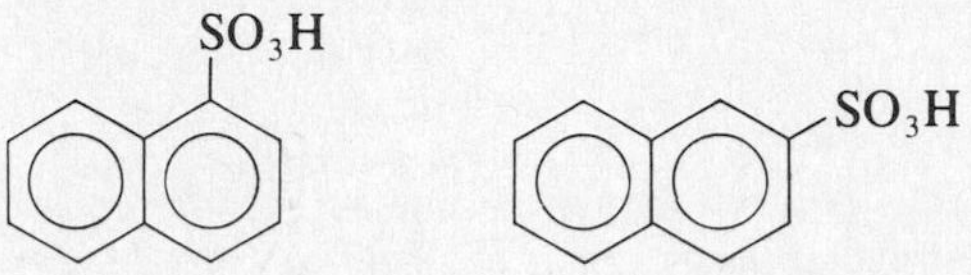

Naphthalene-1-sulfonic acid Naphthalene-2-sulfonic acid

12.17 There are three isomeric trimethylbenzenes.

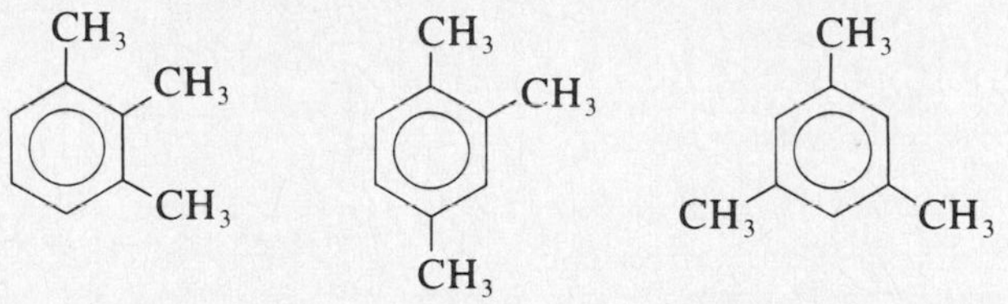

1,2,3-Trimethylbenzene 1,2,4-Trimethylbenzene 1,3,5-Trimethylbenzene

Their relative stabilities are determined by steric effects. Mesitylene (the 1,3,5-trisubstituted isomer) is the most stable because none of its methyl groups are ortho to any other methyl group. Ortho substituents on a benzene ring can, depending on their size, exert repulsive van der Waals interactions upon each other, much as can cis substituents on a carbon–carbon double bond. Because the carbon–carbon bond length in benzene is somewhat longer than in an alkene, these effects are smaller in magnitude, however. The 1,2,4-substitution pattern has one methyl-methyl repulsion between ortho substituents. The least stable isomer is the 1,2,3-trimethyl derivative because it is the most crowded. The energy differences between isomers are relatively small, heats of combustion being 1242.4, 1241.6, and 1241.2 kcal/mol for the 1,2,3-, 1,2,4-, and 1,3,5- isomers, respectively.

12.18 *p*-Dichlorobenzene has a center of symmetry. Each of its individual bond moments is balanced by an identical bond dipole oriented opposite to it. *p*-Dichlorobenzene has no dipole moment.

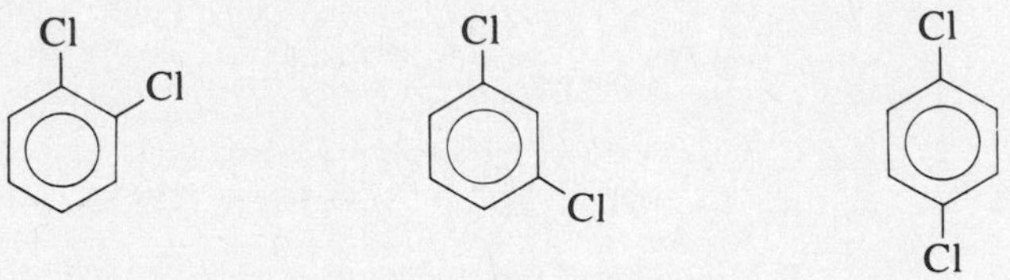

o-Dichlorobenzene $\mu = 2.27$ D *m*-Dichlorobenzene $\mu = 1.48$ D *p*-Dichlorobenzene $\mu = 0$ D

12.19 The shortest carbon-carbon bond in styrene is the double bond of the vinyl substituent; its length is much the same as the double bond length of any other alkene. The carbon-carbon bond lengths of the ring are intermediate between single and double bond lengths. The longest carbon-carbon bond is the sp^2 to sp^2 single bond connecting the vinyl group to the benzene ring.

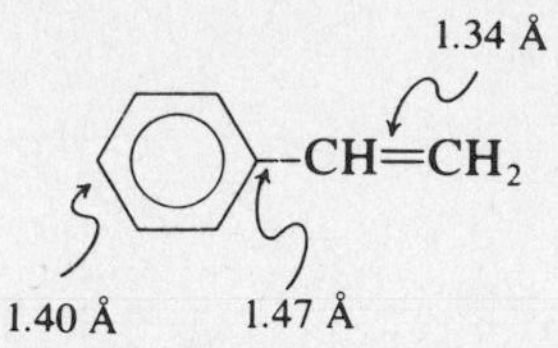

1.34 Å

$-CH{=}CH_2$

1.40 Å 1.47 Å

12.20 (*a*) The better dipolar resonance structure is A because it has an aromatic cyclopentadienide anion bonded to an aromatic cyclopropenyl cation. In structure B neither ring is aromatic.

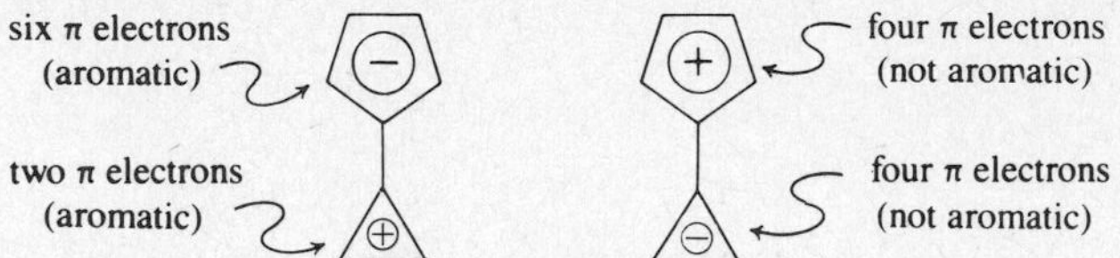

(*b*) Structure D can be stabilized by resonance involving the dipolar form.

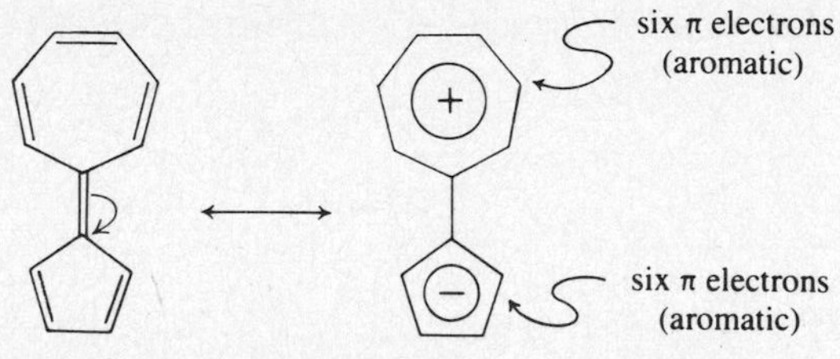

six π electrons
(aromatic)

six π electrons
(aromatic)

D

Comparable stabilization is not possible in C because neither a cyclopropenyl system nor a cycloheptatrienyl system is aromatic in its anionic form. Both are aromatic as cations.

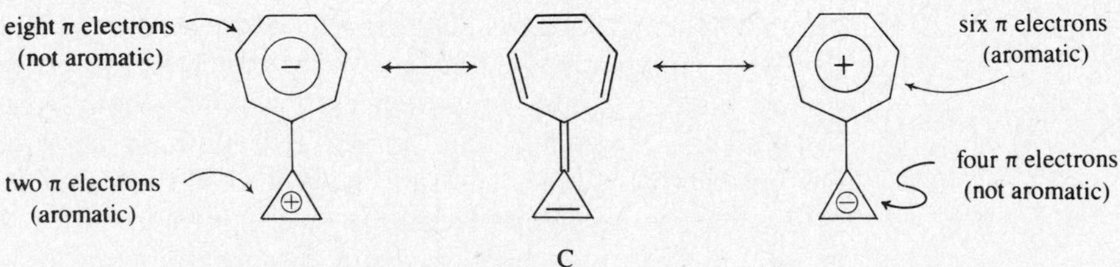

12.21 (*a*) In the structure shown for naphthalene, one ring but not the other corresponds to a Kekulé form of benzene. We say that one ring is *benzenoid* and the other is not.

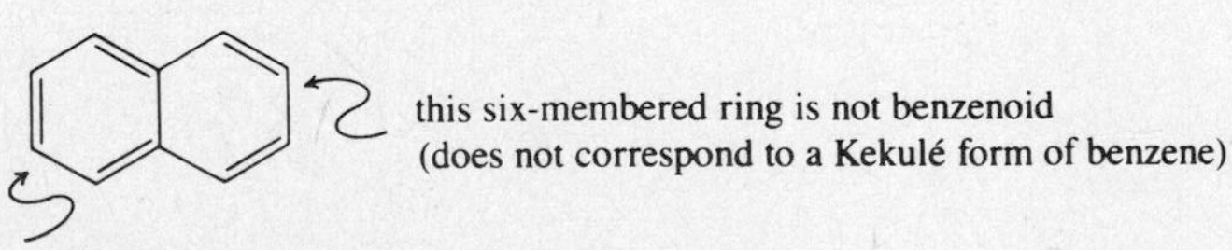

this six-membered ring is not benzenoid
(does not correspond to a Kekulé form of benzene)

this six-membered ring is benzenoid
(corresponds to a Kekulé form of benzene)

By rewriting the benzenoid ring in its alternative Kekulé form, *both* rings become benzenoid.

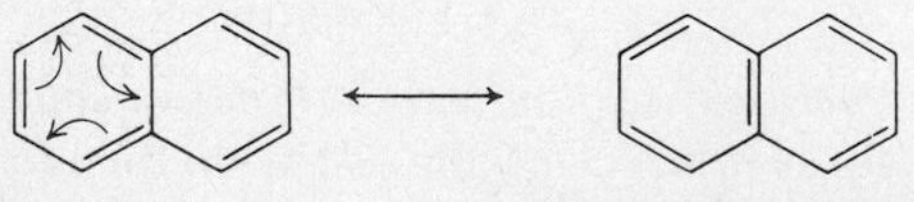

Both rings are benzenoid

(*b*) Here a cyclobutadiene ring is fused to benzene. By writing the alternative resonance form of cyclobutadiene, the six-membered ring becomes benzenoid.

(*c*) The structure portrayed for phenanthrene contains two terminal benzenoid rings and a nonbenzenoid central ring. All three rings may be represented in benzenoid forms by converting one of the terminal six-membered rings to its alternative Kekulé form as shown:

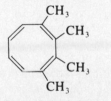

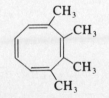

Central ring not benzenoid All three rings benzenoid

(*d*) Neither of the six-membered rings is benzenoid in the structure shown. By writing the cyclooctatetraene portion of the molecule in its alternative representation, the two six-membered rings become benzenoid.

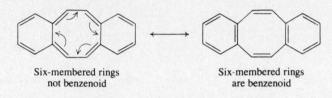

Six-membered rings not benzenoid Six-membered rings are benzenoid

12.22 Cyclooctatetraene is not aromatic. 1,2,3,4-Tetramethylcyclooctatetraene and 1,2,3,8-tetramethylcyclooctatetraene are different compounds.

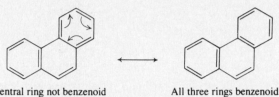

1,2,3,4-Tetramethylcyclooctatetraene 1,2,3,8-Tetramethylcyclooctatetraene

Leo A. Paquette at the Ohio State University synthesized each of these compounds independently of the other and showed them to be stable enough to be stored separately without interconversion.

12.23 (*a*) Cycloundecapentaene is *not aromatic*. Its π system is not conjugated; it is interrupted by an sp^3 hybridized carbon.

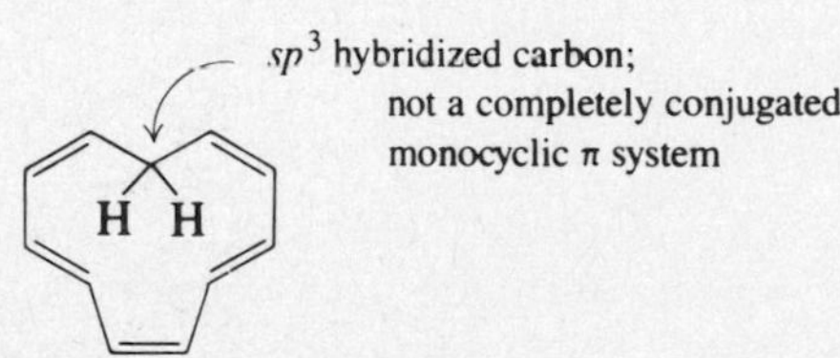

(*b*) Cycloundecapentaenyl radical is *not aromatic*. Its π system is completely conjugated and monocyclic but contains 11 π electrons—a number not equal to $(4n + 2)$ where n is an integer.

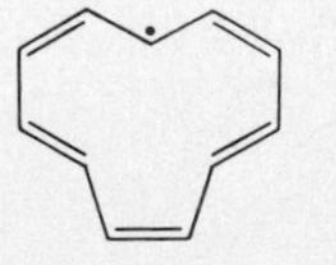

There are 11 electrons in the conjugated π system. The five double bonds contribute 10 π electrons; the odd electron of the radical is the eleventh.

(*c*) Cycloundecapentaenyl cation is *aromatic*. It includes a completely conjugated π system which contains 10 π electrons (10 equals $4n + 2$ where $n = 2$).

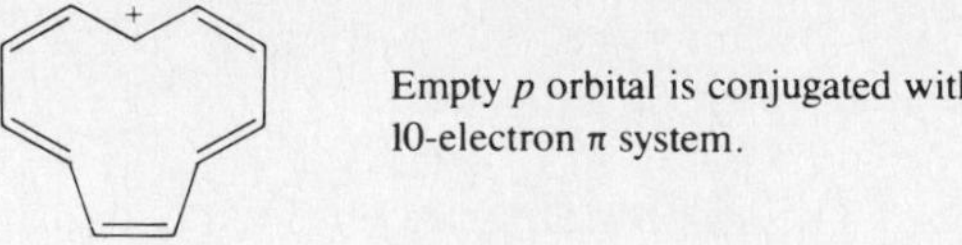

(*d*) Cycloundecapentadienide anion is *not aromatic*. It contains 12 π electrons and is thus a $4n$, not a $(4n + 2)$, system.

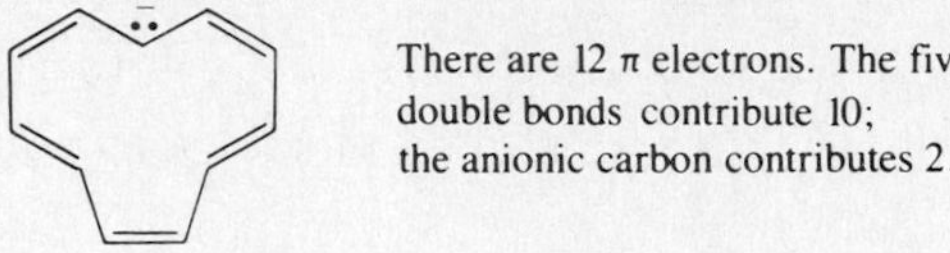

There are 12 π electrons. The five
double bonds contribute 10;
the anionic carbon contributes 2.

12.24 (*a*) Cyclooctatetraene dianion has *two* more electrons than does cyclooctatetraene. The orbital diagram in Figure 12.7 is used and the additional two electrons are placed in the lowest-energy unfilled orbitals. They are added one by one to the half-filled nonbonding orbitals of the neutral species.

Using the arrangement of orbitals from Figure 12.7:

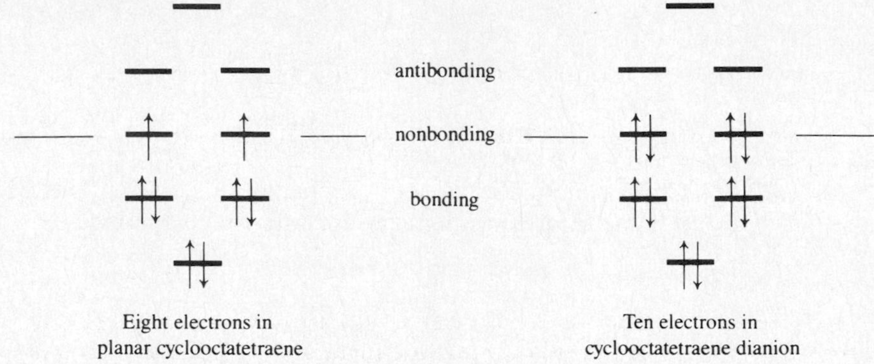

(*b*) Since cyclooctatetraene dianion contains two more electrons than cyclooctatetraene, the conversion that must be accomplished is

$$+ \; 2\,e^- \longrightarrow$$

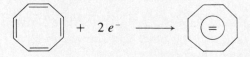

A strong *reducing* agent must be used. Chemically, this can be accomplished by using a metal characterized by a low ionization potential, such as potassium.

$$+ \; 2\,K \longrightarrow \quad + \; 2\,K^+$$

12.25 (*a*) This molecule, called oxepin, is *not aromatic*. There are three double bonds each of which contributes two π electrons and an oxygen atom which contributes two π electrons to the conjugated system, giving a total of eight π electrons. Only one of the two unshared pairs on oxygen can contribute to the π system; the other unshared pair is in an sp^2 hybridized orbital and cannot interact with it.

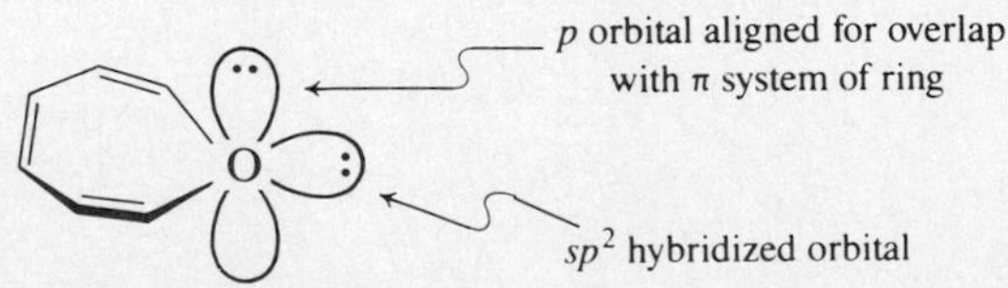

(*b*) This compound, called azonine, has 10 electrons in a completely conjugated planar monocyclic π system and therefore satisfies Hückel's rule for $(4n + 2)$ π electrons where $n = 2$. There are eight π electrons from the conjugated tetraene and two electrons contributed by the nitrogen unshared pair.

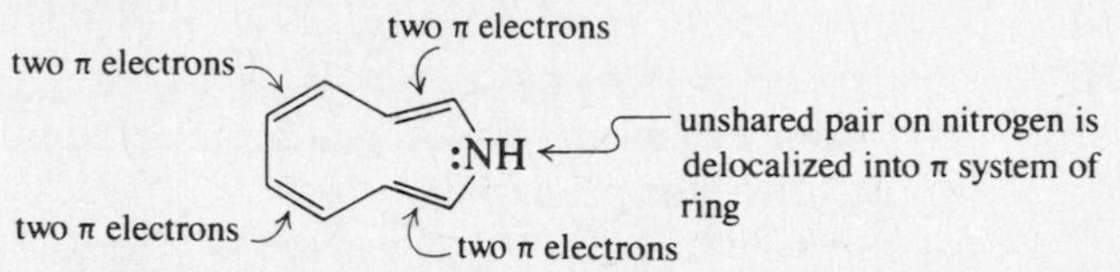

two π electrons
two π electrons
:NH
unshared pair on nitrogen is delocalized into π system of ring
two π electrons
two π electrons

(*c*) Borazole, sometimes called inorganic benzene, is *aromatic*. Six π electrons are contributed by the unshared pairs of the three nitrogen atoms. Each boron contributes a *p* orbital to maintain the conjugated system but no electrons.

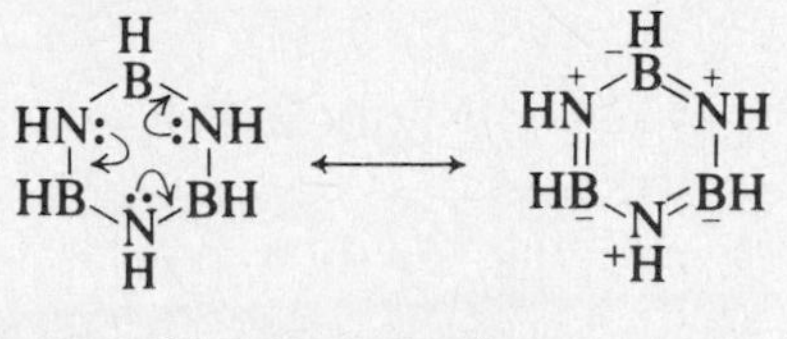

(*d*) Dioxane has eight π electrons and is *not aromatic.*

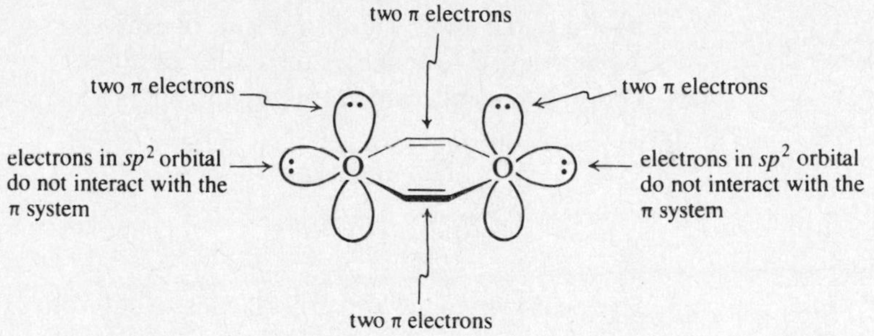

two π electrons
two π electrons
two π electrons
electrons in sp^2 orbital do not interact with the π system
electrons in sp^2 orbital do not interact with the π system
two π electrons

12.26 Styrene reacts as a typical alkene toward most reagents, with the side-chain double bond serving as the functional group. Reactions involving the benzene ring, which will be discussed in the next chapter, do not take place under these conditions. The activation energy for such reactions is higher, as the aromaticity of the benzene ring is disrupted during the initial step.

(*a*) Bromine adds to the side-chain double bond.

$$C_6H_5{-}CH{=}CH_2 \xrightarrow{Br_2} C_6H_5{-}CHBrCH_2Br$$

Styrene → 1,2-Dibromo-1-phenylethane

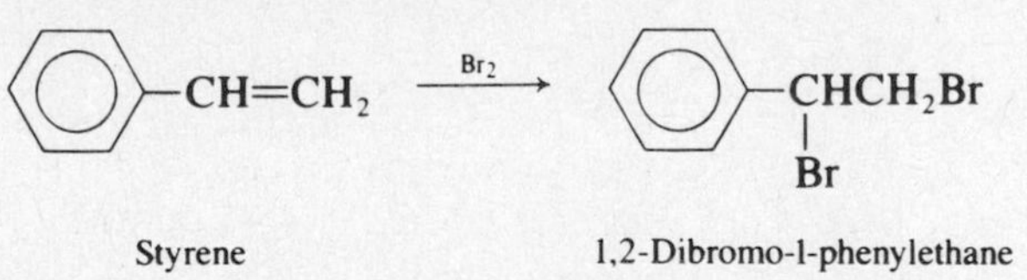

(*b*) The regiochemistry of addition of hydrogen chloride to styrene is controlled by the protonation step, which leads to the formation of the more stable carbocation.

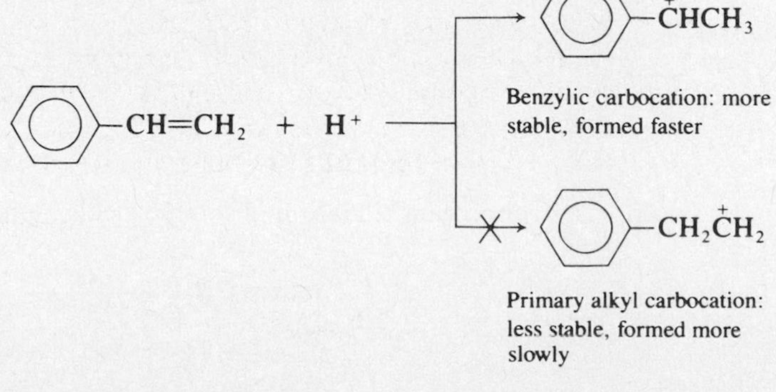

$$C_6H_5{-}CH{=}CH_2 + H^+ \longrightarrow C_6H_5{-}\overset{+}{C}HCH_3$$

Benzylic carbocation: more stable, formed faster

$$C_6H_5{-}CH{=}CH_2 + H^+ \not\longrightarrow C_6H_5{-}CH_2\overset{+}{C}H_2$$

Primary alkyl carbocation: less stable, formed more slowly

Therefore

$$C_6H_5\text{-}CH{=}CH_2 \xrightarrow{HCl} C_6H_5\text{-}CH(Cl)CH_3$$

Styrene 1-Chloro-1-phenylethane

(*c*) The mechanism of hydrogen bromide addition to alkenes under free-radical conditions involves the propagation steps

$$C_6H_5\text{-}CH{=}CH_2 + Br\cdot \longrightarrow C_6H_5\text{-}\dot{C}HCH_2Br$$

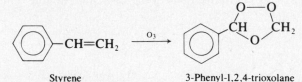

Styrene Benzylic radical

$$C_6H_5\text{-}\dot{C}HCH_2Br + HBr \longrightarrow C_6H_5\text{-}CH_2CH_2Br + Br\cdot$$

2-Phenylethyl bromide

The regioselectivity of the first of these propagation steps is determined by the stability of the alkyl radical produced. A bromine atom will add to the double bond so as to produce the more stable radical, in this case a benzylic radical, which then leads to 2-phenylethyl bromide as the major product.

(*d*) Oxidizing agents convert alkyl- and alkenylbenzenes to benzoic acid.

$$C_6H_5\text{-}CH{=}CH_2 \xrightarrow[2.\ H^+]{1.\ KMnO_4} C_6H_5\text{-}CO_2H + CO_2 + H_2O$$

Styrene Benzoic acid

(*e*) Hypochlorous acid converts alkenes to chlorohydrins.

$$C_6H_5\text{-}CH{=}CH_2 \xrightarrow[H_2O]{Cl_2} C_6H_5\text{-}CH(OH)CH_2Cl$$

Styrene 2-Chloro-1-phenylethanol

(*f*) Ozone reacts with alkenes to yield ozonides.

$$C_6H_5\text{-}CH{=}CH_2 \xrightarrow{O_3} C_6H_5\text{-}\underset{\text{(O–O, O ring)}}{CH}\quad CH_2$$

Styrene 3-Phenyl-1,2,4-trioxolane

(*g*) Hydrolysis of ozonides leads to the formation of carbonyl compounds.

$$\text{3-Phenyl-1,2,4-trioxolane} \xrightarrow[Zn]{H_2O} C_6H_5\text{-}CH{=}O + HCH{=}O$$

3-Phenyl-1,2,4-trioxolane Benzaldehyde Formaldehyde

(*h*) Hydroboration-oxidation of styrene yields 2-phenylethanol.

$$C_6H_5\text{-}CH{=}CH_2 \xrightarrow[2.\ H_2O_2,\ HO^-]{1.\ B_2H_6} C_6H_5\text{-}CH_2CH_2OH$$

Styrene 2-Phenylethanol

(*i*) Epoxides are formed when peroxy acids react with alkenes.

C_6H_5—CH=CH_2 + *m*-$ClC_6H_4CO_2OH$ ⟶ C_6H_5—CH—CH_2 (O bridge)

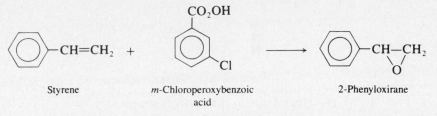

Styrene | *m*-Chloroperoxybenzoic acid | 2-Phenyloxirane

(*j*) The side-chain double bond of styrene is hydrogenated faster than the aromatic ring.

C_6H_5—CH=CH_2 $\xrightarrow[Pt]{H_2}$ C_6H_5—CH_2CH_3

Styrene | Ethylbenzene

(*k*) Eventually, the ring is hydrogenated as well.

C_6H_5—CH=CH_2 $\xrightarrow[Pt]{4H_2}$ C_6H_{11}—CH_2CH_3

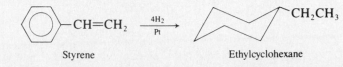

Styrene | Ethylcyclohexane

12.27 All the specific reactions in this problem have been reported in the chemical literature with results as indicated.

(*a*) Hydroboration-oxidation of alkenes leads to syn anti-Markovnikov hydration of the double bond.

C_6H_5 (1-phenylcyclobutene) $\xrightarrow[2.\ H_2O_2,\ HO^-]{1.\ B_2H_6}$ C_6H_5, H, H, OH (cyclobutanol)

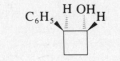

1-Phenylcyclobutene | *trans*-2-Phenyl-1-cyclobutanol (82%)

(*b*) The compound contains a substituted benzene ring and an alkene-like double bond. When hydrogenation of this compound was carried out under the usual conditions, the alkene-like double bond was hydrogenated cleanly.

CH_2CH_3 (indene) + H_2 $\xrightarrow{Pt}$ CH_2CH_3 (indan)

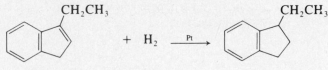

1-Ethylindene | 1-Ethylindan (80%)

(*c*) Free-radical chlorination will lead to substitution of benzylic hydrogens. The starting material contains four benzylic hydrogens, all of which may eventually be replaced.

$(C_6H_5)_2$C(H)—C_6H_4—CH_3 $\xrightarrow[CCl_4,\ light]{Cl_2}$ $(C_6H_5)_2$C(Cl)—C_6H_4—CCl_3

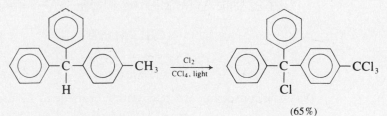

(65%)

(*d*) Epoxidation of alkenes is stereospecific.

C_6H_5(H)C=C(H)C_6H_5 $\xrightarrow[acetic\ acid]{CH_3CO_2OH}$ C_6H_5(H)C—C(H)C_6H_5 (O bridge)

(*E*)-1,2-Diphenylethene | *trans*-1,2-Diphenyloxirane (78–83%)

(*e*) The reaction is one of acid-catalyzed alcohol dehydration.

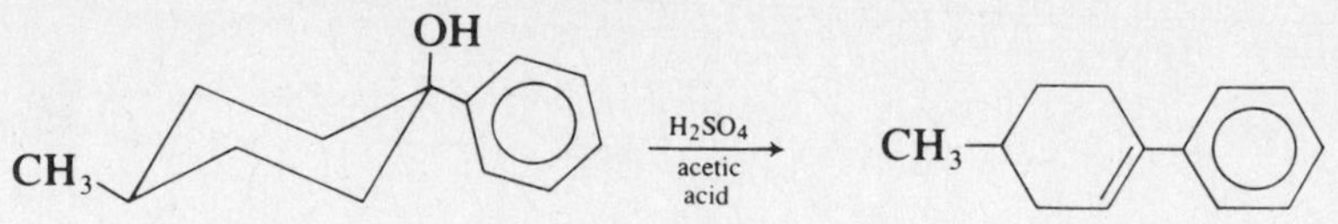

$\xrightarrow[\text{acetic acid}]{H_2SO_4}$

cis-4-Methyl-1-phenylcyclohexanol — 4-Methyl-1-phenylcyclohexene (81%)

(*f*) This reaction illustrates identical reactivity at two equivalent sites in a molecule. Both alcohol functions are tertiary and benzylic and undergo acid-catalyzed dehydration readily.

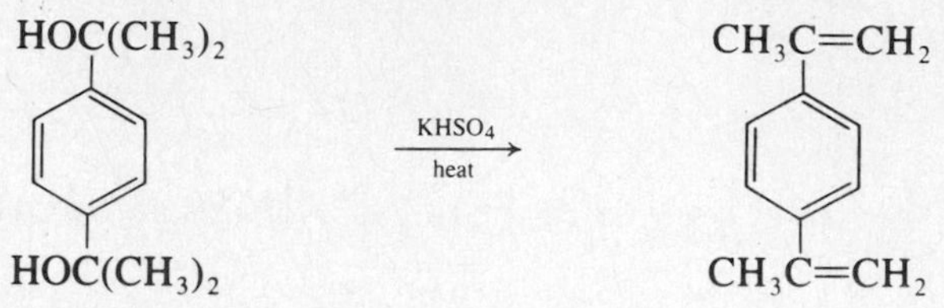

$HOC(CH_3)_2$ ... $HOC(CH_3)_2$ $\xrightarrow[\text{heat}]{KHSO_4}$ $CH_3C{=}CH_2$ ... $CH_3C{=}CH_2$

1,4-Di-(2-hydroxy-2-propyl)benzene — 1,4-Di-(2-propenyl)benzene (68%)

(*g*) The compound shown is DDT (standing for the nonsystematic name "dichlorodiphenyltrichloroethane"). It undergoes β elimination to form an alkene.

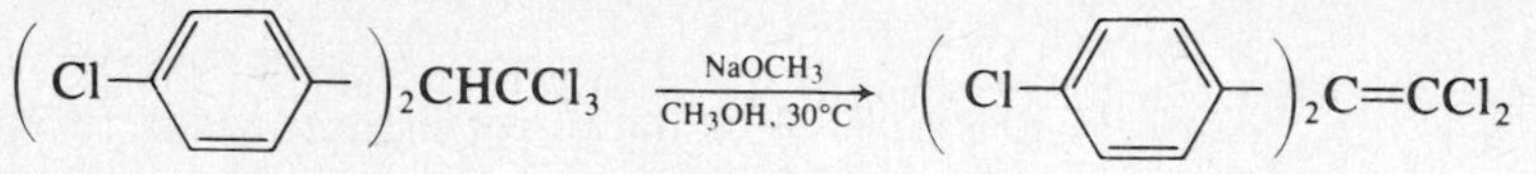

$(Cl\text{-}C_6H_4\text{-})_2CHCCl_3 \xrightarrow[CH_3OH,\ 30°C]{NaOCH_3} (Cl\text{-}C_6H_4\text{-})_2C{=}CCl_2$

(100%)

(*h*) Alkyl side chains on naphthalene undergo reactions analogous to those of alkyl groups on benzene.

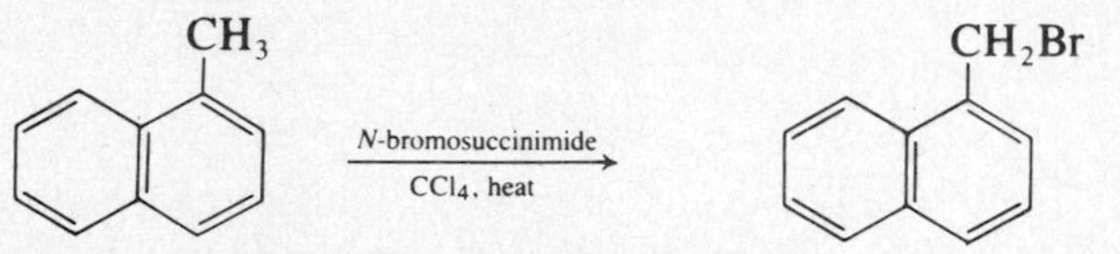

CH_3 $\xrightarrow[CCl_4,\ \text{heat}]{N\text{-bromosuccinimide}}$ CH_2Br

1-Methylnaphthalene — 1-(Bromomethyl)naphthalene (46%)

(*i*) Potassium carbonate is a weak base. Hydrolysis of the primary benzylic halide converts it to an alcohol.

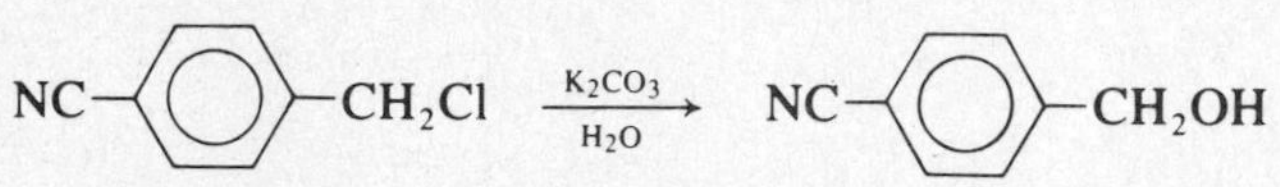

$NC\text{-}C_6H_4\text{-}CH_2Cl \xrightarrow[H_2O]{K_2CO_3} NC\text{-}C_6H_4\text{-}CH_2OH$

p-Cyanobenzyl chloride — *p*-Cyanobenzyl alcohol (85%)

12.28 (*a*) The conversion of ethylbenzene to 1-phenylethyl bromide is a benzylic bromination. It can be achieved by using either bromine or *N*-bromosuccinimide.

$C_6H_5CH_2CH_3 \xrightarrow[\text{or NBS, heat}]{Br_2,\ \text{light}} C_6H_5CH(Br)CH_3$

Ethylbenzene — 1-Phenylethyl bromide

(*b*) The conversion of 1-phenylethyl bromide to 1,2-dibromo-1-phenylethane

$C_6H_5CH(Br)CH_3 \longrightarrow C_6H_5CH(Br)CH_2Br$

cannot be achieved cleanly in a single step. What must be done is to reason backward from the target molecule. Ask yourself: How can I make 1,2-dibromo-1-phenylethane in one step from anything? Vicinal dibromides are customarily prepared by addition of bromine to alkenes. This suggests that 1,2-dibromo-1-phenylethane can be prepared by the reaction

$$C_6H_5CH{=}CH_2 + Br_2 \longrightarrow C_6H_5\underset{\displaystyle Br}{\underset{|}{C}}HCH_2Br$$

Styrene — 1,2-Dibromo-1-phenylethane

The necessary alkene, styrene, is available by dehydrohalogenation of the given starting material, 1-phenylethyl bromide.

$$C_6H_5\underset{\displaystyle Br}{\underset{|}{C}}HCH_3 \xrightarrow[CH_3CH_2OH]{NaOCH_2CH_3} C_6H_5CH{=}CH_2$$

1-Phenylethyl bromide — Styrene

Thus, by reasoning backward from the target molecule, the synthetic scheme becomes apparent.

$$C_6H_5\underset{\displaystyle Br}{\underset{|}{C}}HCH_3 \xrightarrow[CH_3CH_2OH]{NaOCH_2CH_3} C_6H_5CH{=}CH_2 \xrightarrow{Br_2} C_6H_5\underset{\displaystyle Br}{\underset{|}{C}}HCH_2Br$$

1-Phenylethyl bromide — Styrene — 1,2-Dibromo-1-phenylethane

(*c*) The conversion of styrene to phenylacetylene cannot be carried out in a single step. However, as was pointed out in Chapter 10, a standard sequence for converting alkenes to alkynes consists of bromine addition followed by a double dehydrohalogenation in strong base.

$$C_6H_5CH{=}CH_2 \xrightarrow{Br_2} C_6H_5\underset{\displaystyle Br}{\underset{|}{C}}HCH_2Br \xrightarrow[NH_3]{NaNH_2} C_6H_5C{\equiv}CH$$

Styrene — 1,2-Dibromo-1-phenylethane — Phenylacetylene

(*d*) The conversion of phenylacetylene to *n*-butylbenzene requires both a hydrogenation step and a carbon–carbon bond formation step. The acetylene function is essential for carbon–carbon bond formation by alkylation. Therefore, the correct sequence is:

$$C_6H_5C{\equiv}CH \xrightarrow[NH_3]{NaNH_2} C_6H_5C{\equiv}C^-Na^+$$

Phenylacetylene

$$C_6H_5C{\equiv}C^-Na^+ + CH_3CH_2Br \longrightarrow C_6H_5C{\equiv}CCH_2CH_3$$

$$C_6H_5C{\equiv}CCH_2CH_3 \xrightarrow[Pt]{H_2} C_6H_5CH_2CH_2CH_2CH_3$$

n-Butylbenzene

(*e*) There is no synthetic method for moving a hydroxyl group from one carbon to another in a single step.

$$C_6H_5CH_2CH_2OH \longrightarrow C_6H_5\underset{\displaystyle OH}{\underset{|}{C}}HCH_3$$

2-Phenylethanol — 1-Phenylethanol

However, the target molecule, 1-phenylethanol, can be prepared by hydration of styrene in accordance with Markovnikov's rule—say, by oxymercuration-demercuration.

$$C_6H_5CH{=}CH_2 \xrightarrow[\text{2. } NaBH_4]{\text{1. } Hg(OAc)_2,\ THF\text{-}H_2O} C_6H_5\underset{\displaystyle OH}{\underset{|}{C}}HCH_3$$

Styrene — 1-Phenylethanol

All that remains to complete the synthesis is to convert the starting material, 2-phenylethanol, to styrene. The best method is to first convert the primary alcohol to the corresponding bromide, then carry out an elimination of the bromide under E2 conditions.

$$\underset{\text{2-Phenylethanol}}{C_6H_5CH_2CH_2OH} \xrightarrow{PBr_3} \underset{\text{2-Phenylethyl bromide}}{C_6H_5CH_2CH_2Br} \xrightarrow[(CH_3)_3COH]{KOC(CH_3)_3} \underset{\text{Styrene}}{C_6H_5CH{=}CH_2}$$

This route is preferred over acid-catalyzed dehydration because styrene tends to polymerize readily in the presence of acids.

(*f*) The transformation corresponds to alkylation of acetylene, so the alcohol must first be converted to a species with a good leaving group such as its halide derivative.

$$\underset{\text{2-Phenylethanol}}{C_6H_5CH_2CH_2OH} \xrightarrow{PBr_3} \underset{\text{2-Phenylethyl bromide}}{C_6H_5CH_2CH_2Br}$$

$$\underset{\text{2-Phenylethyl bromide}}{C_6H_5CH_2CH_2Br} + \underset{\text{Sodium acetylide}}{NaC{\equiv}CH} \longrightarrow \underset{\text{4-Phenyl-1-butyne}}{C_6H_5CH_2CH_2C{\equiv}CH}$$

(*g*) The target compound is a bromohydrin. Bromohydrins are formed by addition of hypobromous acid to alkenes.

$$\underset{\text{2-Phenylethyl bromide}}{C_6H_5CH_2CH_2Br} \xrightarrow[(CH_3)_3COH]{KOC(CH_3)_3} \underset{\text{Styrene}}{C_6H_5CH{=}CH_2} \xrightarrow[H_2O]{Br_2} \underset{\text{2-Bromo-1-phenylethanol}}{C_6H_5\underset{|}{C}HCH_2Br \atop \quad OH}$$

12.29 The stability of free-radicals is reflected in their ease of formation. Toluene, which forms a benzyl radical, reacts with bromine 64,000 times faster than does ethane, which forms a primary alkyl radical. Ethylbenzene, which forms a secondary benzylic radical, reacts 1 million times faster than ethane.

$$\underset{\text{Toluene}}{C_6H_5{-}CH_3} + Br\cdot \longrightarrow \underset{\text{Primary benzylic radical}}{C_6H_5{-}\dot{C}H_2} + HBr$$

$$\underset{\text{Ethylbenzene (most reactive)}}{C_6H_5{-}CH_2CH_3} + Br\cdot \longrightarrow \underset{\text{Secondary benzylic radical}}{C_6H_5{-}\dot{C}HCH_3} + HBr$$

$$\underset{\text{Ethane (least reactive)}}{CH_3CH_3} + Br\cdot \longrightarrow \underset{\text{Primary radical}}{CH_3\dot{C}H_2} + HBr$$

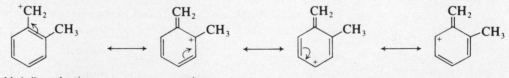

12.30 A good way to develop alternative resonance structures for carbocations is to move electron pairs toward sites of positive charge.

[Resonance structures: $^+CH_2$ / CH_3 ring ⟷ CH_2 / CH_3 ring (+) ⟷ CH_2 / CH_3 ring (+) ⟷ CH_2 / CH_3 ring (+)]

o-Methylbenzyl cation — tertiary carbocation

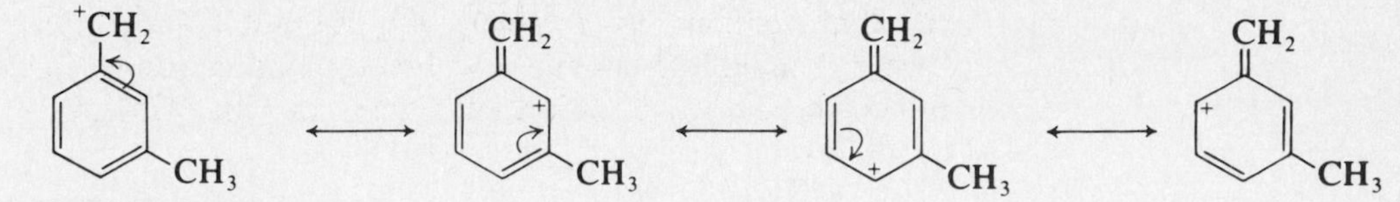

m-Methylbenzyl cation

Only one of the Lewis structures shown is a tertiary carbocation. *o*-Methylbenzyl cation has tertiary carbocation character; *m*-methylbenzyl cation does not.

12.31 Remembering from earlier chapters that tertiary alcohols are converted to carbocations on treatment with acids:

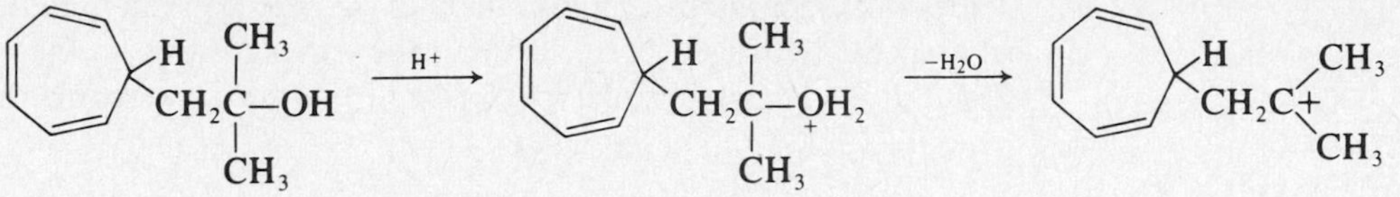

The tertiary carbocation can form the more stable and aromatic tropylium cation by fragmentation, yielding 2-methylpropene.

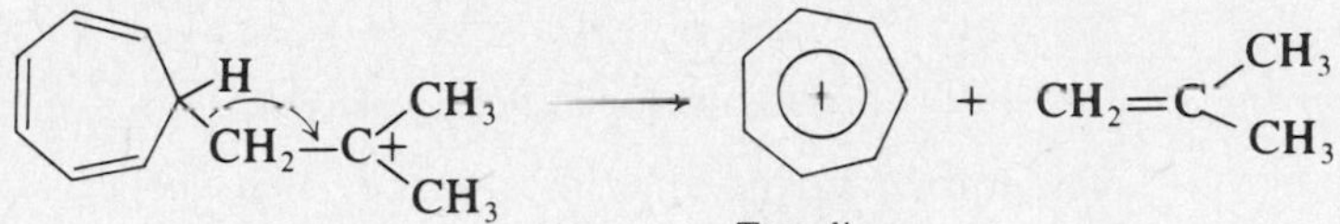

Tropylium cation (aromatic)

12.32 2,3,5-Trimethoxybenzoic acid has the structure shown. The three methoxy groups occupy the same positions in this oxidation product that they did in compound E. The carboxylic acid function must have arisen by oxidation of the $-CH_2CH{=}C(CH_3)_2$ side chain. Therefore

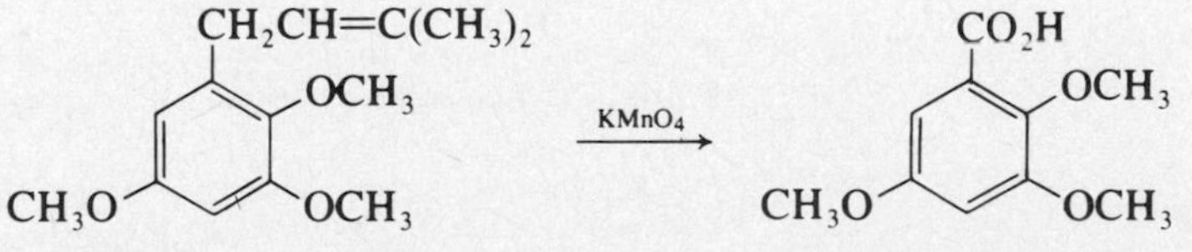

Compound E ($C_{14}H_{20}O_3$) → ($KMnO_4$) → 2,3,5-Trimethoxybenzoic acid

12.33 Hydroboration-oxidation leads to stereospecific syn addition of the elements of water across a carbon–carbon double bond. The regiochemistry of addition is opposite to that predicted by Markovnikov's rule.

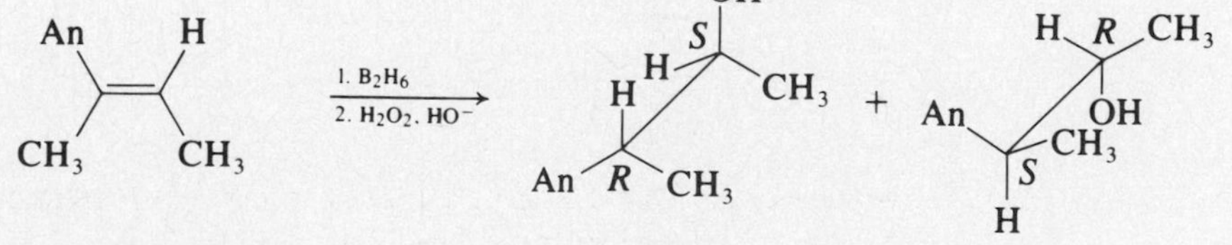

(*E*)-2-(*p*-Anisyl)-2-butene → (1. B_2H_6; 2. H_2O_2, HO^-) → 2*S*,3*R* + 2*R*,3*S*

An = $CH_3O-C_6H_4-$

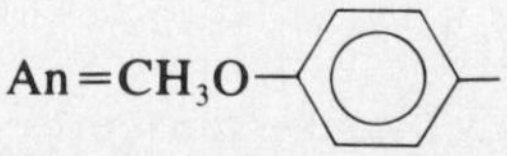

Alcohol F is a racemic mixture of the 2*S*, 3*R* and 2*R*, 3*S* enantiomers of 3-(*p*-anisyl)-2-butanol.

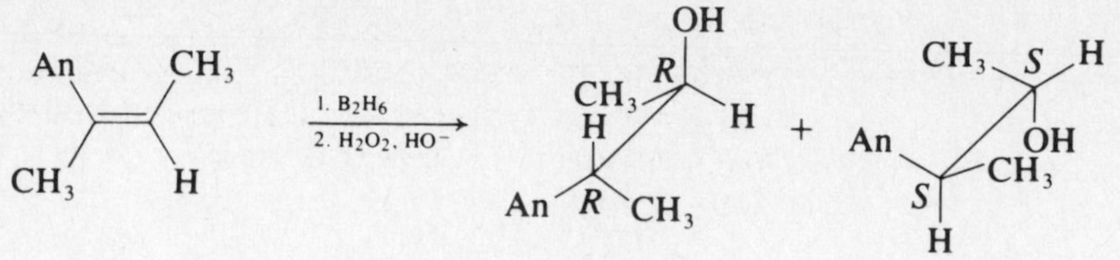

1. B_2H_6
2. H_2O_2, HO^-

(Z)-2-(p-Anisyl)-2-butene → (2R,3R) + (2S,3S)

Alcohol G is a racemic mixture of the 2R, 3R and 2S, 3S enantiomers of 3-(p-anisyl)-2-butanol. Alcohols F and G are stereoisomers which are not enantiomers; they are diastereomers.

12.34 Dehydrohalogenation of alkyl halides is stereospecific, requiring an anti arrangement between the hydrogen being lost and the leaving group in the activated complex. Therefore (Z)-1,2-diphenylpropene must be formed from the diastereomer shown below.

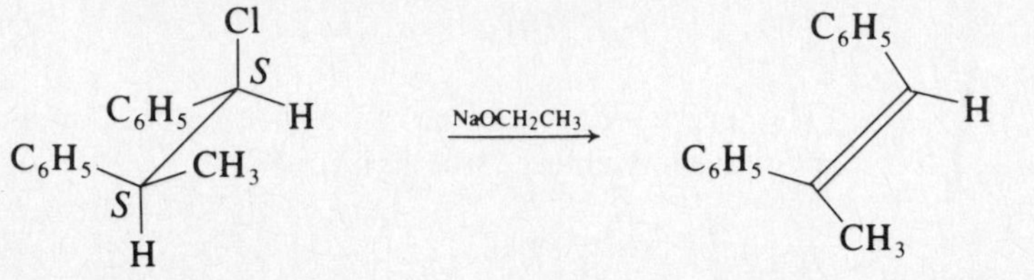

(1S, 2S)-1-Chloro-1,2-diphenylpropane —$NaOCH_2CH_3$→ (Z)-1,2-Diphenylpropene (90%)

The mirror image chloride, 1R, 2R, will also give the Z alkene. In fact, the reaction was carried out on a racemic mixture of the 1R, 2R and 1S, 2S stereoisomers.

The E isomer is formed from either the 1R, 2S or the 1S, 2R chloride (or from a racemic mixture of the two).

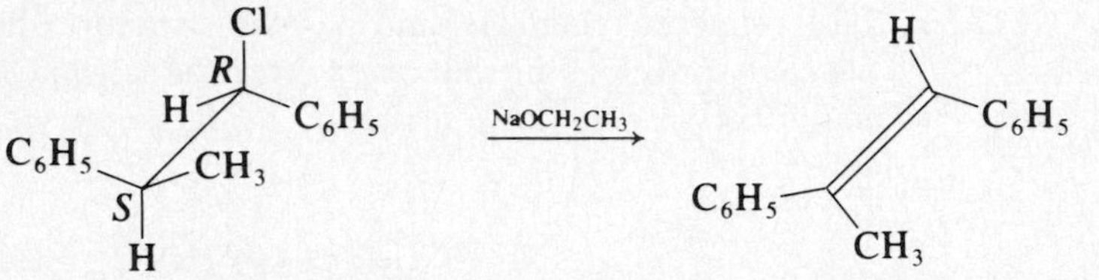

(1R,2S)-1-Chloro-1,2-diphenylpropane —$NaOCH_2CH_3$→ (E)-1,2-Diphenylpropene (87%)

SELF-TEST

PART A

A-1. Give the IUPAC name of each of the following:

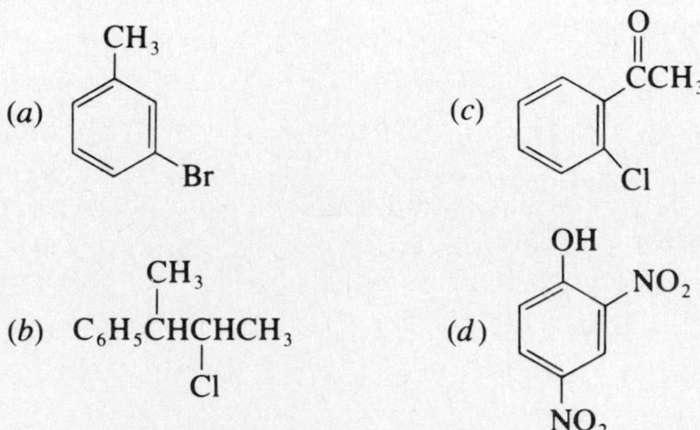

A-2. Draw the structure corresponding to each of the following:
(a) 3-Chlorobenzenecarboxylic acid
(b) p-Nitroanisole
(c) 2,4-Dimethylaniline

A-3. Write a positive (+) or negative (−) charge at the appropriate position so that each of the following structures contains the proper number of π electrons to permit it to be considered an aromatic ion. For purposes of this problem ignore strain effects that might destabilize the molecule.

(*a*) (*b*)

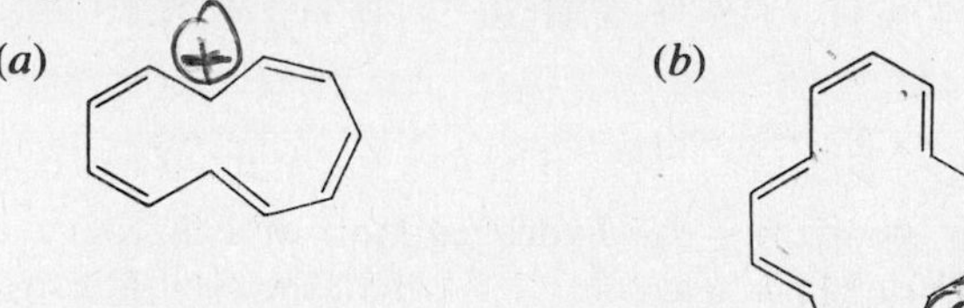

A-4. How many π electrons are counted toward satisfying Hückel's rule in the following substance? Is it aromatic?

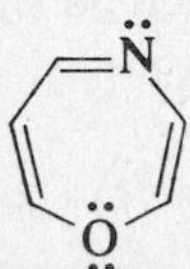

A-5. Azulene, shown below, is highly polar. Draw a dipolar resonance structure to explain this fact.

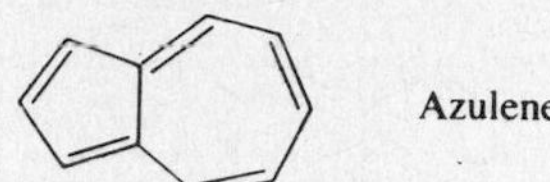

A-6. Give the reactant, reagent, or product omitted from each of the following.

(*c*)

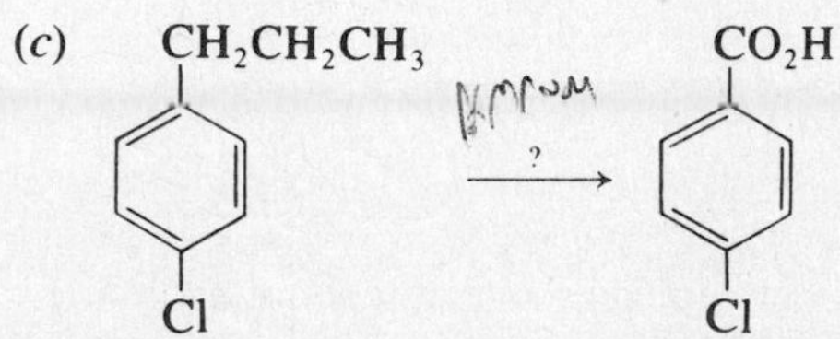

A-7. Provide two methods for the synthesis of 1-bromo-1-phenylpropane from an aromatic hydrocarbon.

PART B

B-1. The number of possible dichloronitrobenzene isomers is
(*a*) 3 (*b*) 4 (*c*) 6 (*d*) 8

B-2. Which of the following statements is correct concerning the class of reactions to be expected for benzene and cyclooctatetraene?
(*a*) Both substances would undergo addition reactions.
(*b*) Both substances would undergo substitution reactions.
(*c*) Benzene would undergo substitution; cyclooctatetraene would undergo addition.
(*d*) Benzene would undergo addition; cyclooctatetraene would undergo substitution.

B-3. Which, if any, of the following structures represents an aromatic species?

(*a*) (*b*) (*c*)

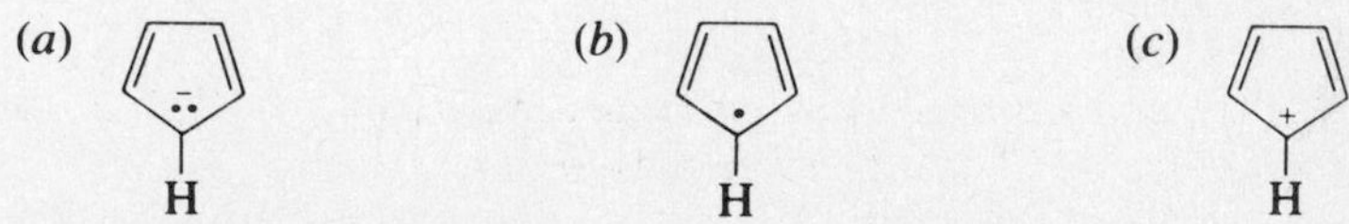

(*d*) None of these is aromatic

B-4. Rank the following compounds in order of increasing rate of solvolysis in aqueous acetone (slowest → fastest).

$$\underset{\text{A}}{(CH_3)_2CHCH_2CH_2Br} \qquad \underset{\text{B}}{(CH_3)_2CH\overset{\overset{\text{Br}}{|}}{C}HCH_3} \qquad \underset{\text{C}}{C_6H_5\overset{\overset{\text{Br}}{|}}{C}HCH(CH_3)_2}$$

(*a*) A < B < C (*b*) B < A < C (*c*) C < B < A (*d*) A < C < B

B-5. When comparing the hydrogenation of benzene with that of a hypothetical 1,3,5-cyclohexatriene, benzene ______ than the cyclohexatriene.

(*a*) Absorbs 36 kcal/mol more heat
(*b*) Absorbs 36 kcal/mol less heat
(*c*) Gives off 36 kcal/mol more heat
(*d*) Gives off 36 kcal/mol less heat

B-6. The reaction

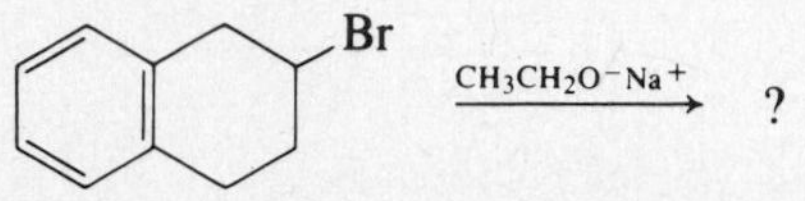

Br $\xrightarrow{CH_3CH_2O^-Na^+}$?

gives as the major elimination product

(*a*) (*b*)

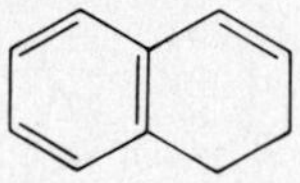

(*c*) *Equal* amounts of (*a*) and (*b*) (*d*) Neither (*a*) nor (*b*)

B-7. The molecule whose structure is shown below

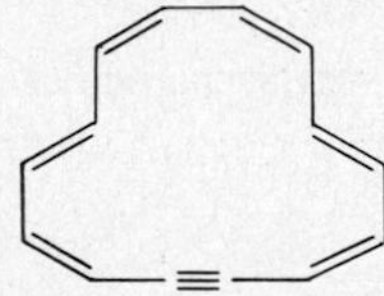

(*a*) Does not obey Hückel's rule and is not aromatic
(*b*) Obeys Hückel's rule and is aromatic
(*c*) Forms an aromatic cation by loss of one electron
(*d*) Forms an aromatic dianion by gaining two electrons

B-8. State which of the listed compounds is *not* an effective substrate for the process

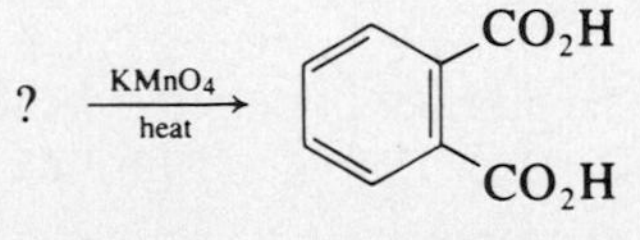

? $\xrightarrow[\text{heat}]{KMnO_4}$ CO_2H, CO_2H

(*a*) CH_2CH_3, $C(CH_3)_3$ (*b*)

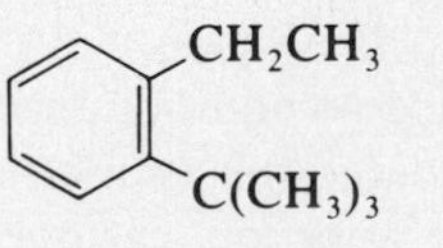

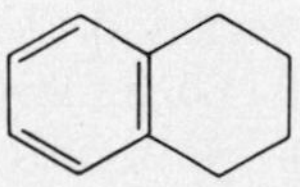

(*c*) 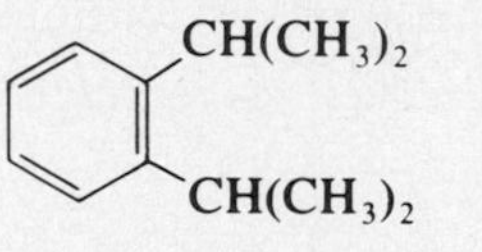 $CH(CH_3)_2$, $CH(CH_3)_2$ (*d*) CH_3, CH_3

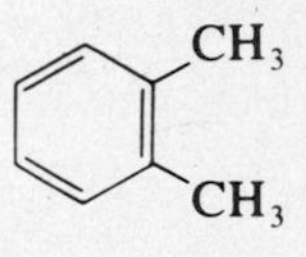

CHAPTER

13

REACTIONS OF ARENES. ELECTROPHILIC AROMATIC SUBSTITUTION REACTIONS

IMPORTANT TERMS AND CONCEPTS

General Mechanism of Electrophilic Aromatic Substitution (Sec. 13.2) Electrophilic aromatic substitution is the most widely encountered class of reaction involving the aromatic ring. A general mechanism may be used to describe the various specific reactions: an *electrophile*, i.e., an electron-deficient, often positively charged species, adds to the aromatic ring; the intermediate *cyclohexadienyl cation* thereby formed is resonance-stabilized; and loss of a proton from it generates the aromatic product.

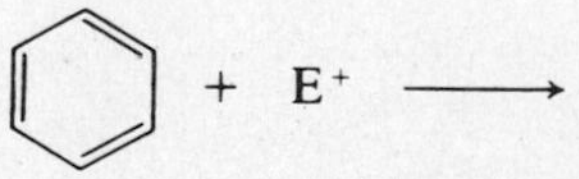

$$C_6H_6 + E^+ \longrightarrow$$

(E = electrophilic species)

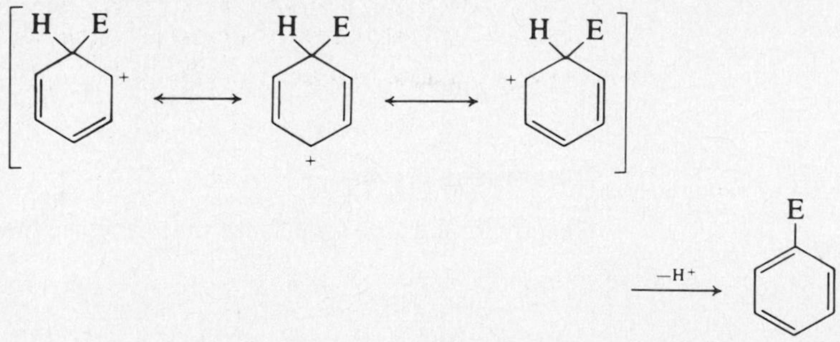

$$\left[C_6H_6E^+ \longleftrightarrow C_6H_6E^+ \longleftrightarrow C_6H_6E^+ \right] \xrightarrow{-H^+} C_6H_5E$$

(E = NO_2, SO_3H, halogen, R, $\overset{O}{\overset{\|}{C}}R$)

Specific examples of electrophilic aromatic substitution reactions are tabulated on the following page.

Rate and Orientation Effects (Secs. 13.9 to 13.14) Once a substituent is attached to an aromatic ring, its nature affects the rate of reaction (relative to benzene) and the orientation (ortho, meta, or para relative to the group attached to the ring) of further substitution.

Reaction (text section)	Typical reagents	Electrophile	Typical product
Nitration (13.1, 13.3)	HNO_3/H_2SO_4	NO_2^+	$C_6H_5{-}NO_2$
Sulfonation (13.1, 13.4)	H_2SO_4/heat (SO_3/H_2SO_4)	SO_3	$C_6H_5{-}SO_3H$
Halogenation (13.1, 13.5)	$Cl_2/FeCl_3$ $Br_2/FeBr_3$	"Cl^+"[a] "Br^+"[b]	$C_6H_5{-}Cl$ $C_6H_5{-}Br$
Friedel-Crafts:			
Alkylation (13.1, 13.6)	$R{-}Cl/AlCl_3$	R^+	$C_6H_5{-}R$
($R{-}OH$, alkenes/H_2SO_4) *Note:* Rearrangements of R are possible.			
Acylation (13.1, 13.7)	$RCCl/AlCl_3$ (RC(=O)Cl) ($RCOCR/AlCl_3$) (RC(=O)OC(=O)R)	$R{-}C{\equiv}O^+$	$C_6H_5{-}CR$ (C(=O)R)

(*a*) Electrophile is $Cl_2/FeCl_3$ complex.
(*b*) Electrophile is $Br_2/FeBr_3$ complex.

Activating groups

Activating groups are those that increase the rate of substitution. Examples include:

$$-OH,\ -OR,\ -O\overset{O}{\overset{\|}{C}}R,\ -NH_2,\ -NH\overset{O}{\overset{\|}{C}}CH_3,\ -R\ (\text{alkyl}),\ C_6H_5$$

All these groups are *ortho, para directors.* That is, further substitution of the ring will occur in the positions ortho and para to the directing group.

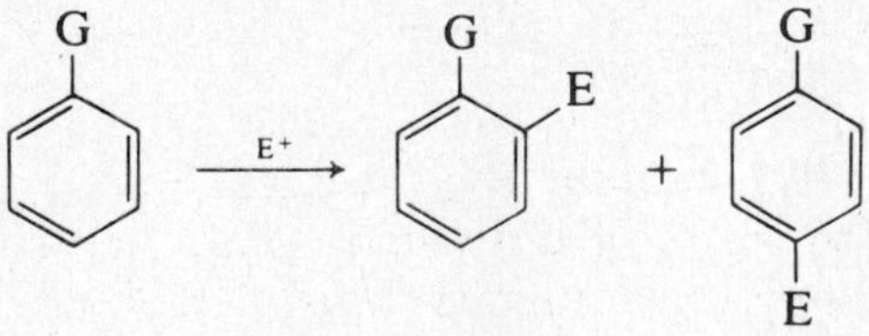

G–C6H5 —E+→ (ortho G/E) + (para G/E)

Ortho Para

Deactivating groups

Deactivating groups are those that decrease the rate of substitution. Examples include:

$$-CF_3,\ -\overset{O}{\overset{\|}{C}}R,\ (R = \text{H, alkyl, aryl, OH, OR}'),\ -C{\equiv}N,\ -NO_2,\ -SO_3H,\ -\overset{+}{N}R_3$$

These groups are meta directors, directing further substitution on the ring to the position meta to that occupied by the directing group.

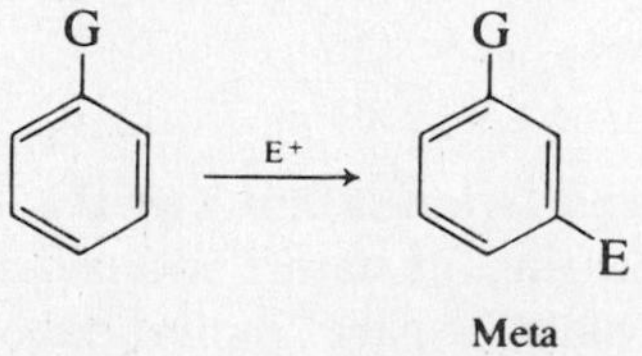

G–C6H5 —E+→ (meta G/E)

Meta

The *halogens* (F, Cl, Br, I) are an important exception. They are *slightly deactivating* ortho-para directors.

Remember: It is the group *already* attached to the ring (*not* the one being added) which determines rate and orientation effects. For example, nitration of anisole ($C_6H_5OCH_3$) proceeds *faster* than nitration of benzene and gives a mixture of the ortho and para products.

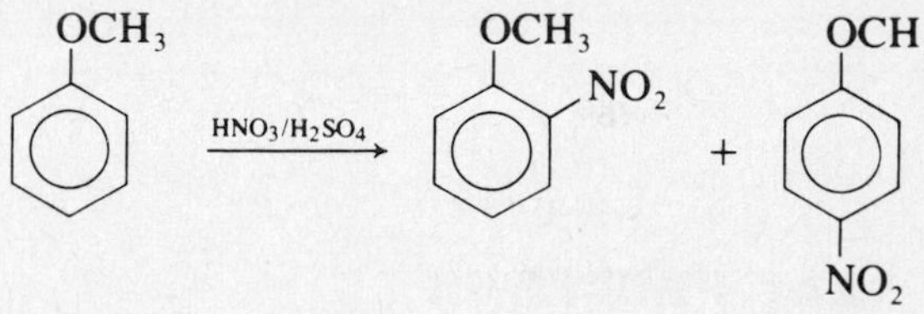

$$C_6H_5OCH_3 \xrightarrow{HNO_3/H_2SO_4} o\text{-}NO_2C_6H_4OCH_3 + p\text{-}NO_2C_6H_4OCH_3$$

Multiple Substitution (Sec. 13.16) When two or more groups are attached to the benzene ring, orientation is controlled by the *activating groups*. For example,

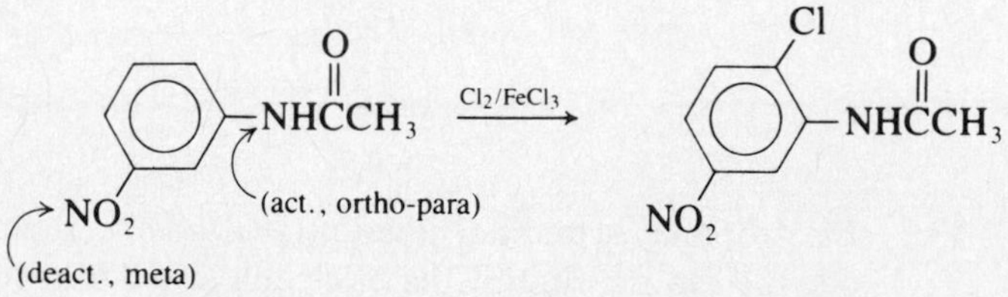

$$m\text{-}NO_2C_6H_4NHCCH_3(=O) \xrightarrow{Cl_2/FeCl_3} \text{Cl-substituted } NO_2C_6H_3NHCCH_3(=O)$$

NO_2 (deact., meta); $NHCCH_3$ (act., ortho-para)

Naphthalene Substitution (Sec. 13.17) Electrophilic aromatic substitution of naphthalene occurs faster at the 1 (or α) position. For example

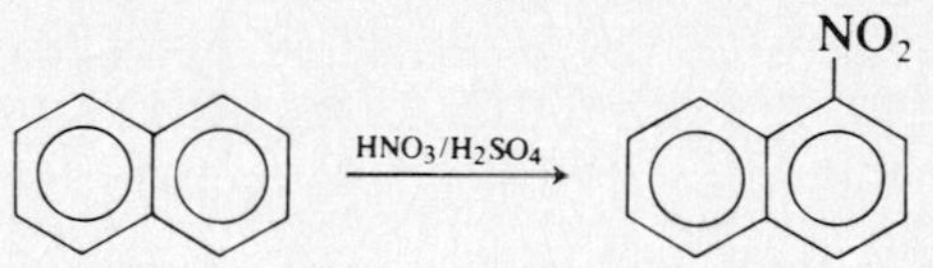

$$C_{10}H_8 \xrightarrow{HNO_3/H_2SO_4} 1\text{-}NO_2C_{10}H_7$$

IMPORTANT REACTIONS

Alkylbenzene Synthesis (Secs. 13.7, 13.8) Friedel–Crafts *acylation* followed by *reduction* of the ketone provides a method of synthesis that avoids carbocation rearrangement (which may accompany Friedel-Crafts alkylations).

Friedel–Crafts acylation:

$$C_6H_6 + CH_3CH_2\overset{O}{\overset{\|}{C}}Cl \xrightarrow{AlCl_3} C_6H_5\overset{O}{\overset{\|}{C}}CH_2CH_3 + HCl$$

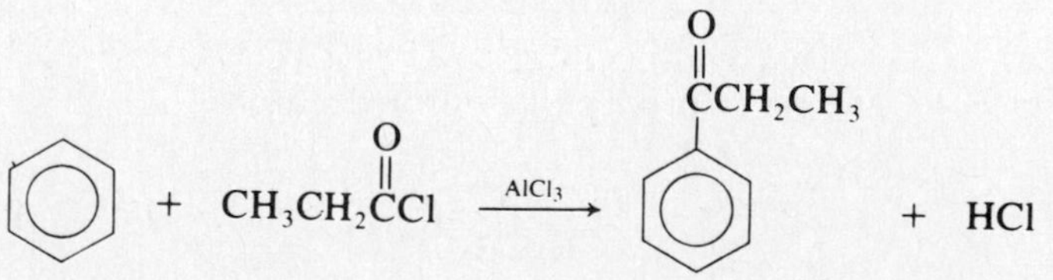

Clemmensen reduction:

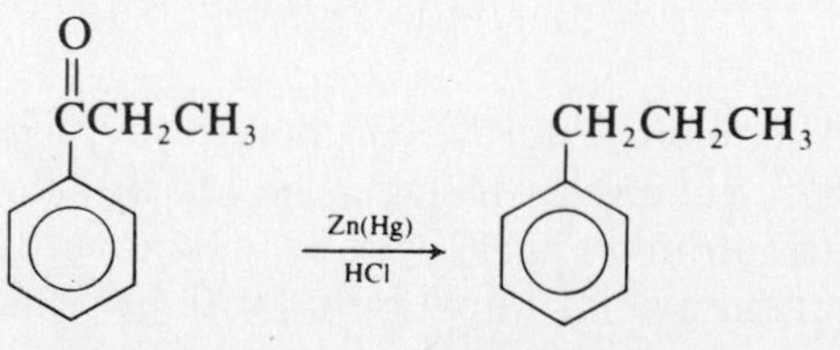

$$C_6H_5\overset{O}{\overset{\|}{C}}CH_2CH_3 \xrightarrow[HCl]{Zn(Hg)} C_6H_5CH_2CH_2CH_3$$

Wolff–Kishner reduction:

$$C_6H_5CCH_2CH_3\ (\text{C=O}) \xrightarrow[\text{heat}]{H_2NNH_2,\ KOH} C_6H_5CH_2CH_2CH_3$$

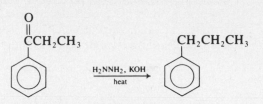

SOLUTIONS TO TEXT PROBLEMS

13.1 To show electron delocalization in a cyclohexadienyl cation intermediate, generate alternative resonance structures by moving pairs of π electrons toward sites of positive charge.

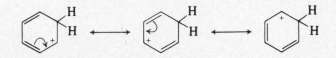

13.2 Electrophilic aromatic substitution leads to replacement of one of the hydrogens directly attached to the ring by the electrophile. All four of the ring hydrogens of *p*-xylene are equivalent, so it does not matter which one is replaced by the nitro group.

$$\text{p-Xylene} \xrightarrow{^+NO_2} \text{1,4-Dimethyl-2-nitrobenzene}$$

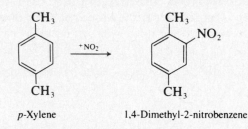

p-Xylene 1,4-Dimethyl-2-nitrobenzene

13.3 The aromatic ring of 1,2,4,5-tetramethylbenzene has two equivalent hydrogen substituents. Sulfonation of the ring leads to replacement of one of them by $-SO_3H$.

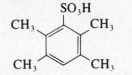

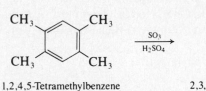

$$\text{1,2,4,5-Tetramethylbenzene} \xrightarrow[H_2SO_4]{SO_3} \text{2,3,5,6-Tetramethylbenzenesulfonic acid}$$

1,2,4,5-Tetramethylbenzene 2,3,5,6-Tetramethylbenzenesulfonic acid

13.4 The major product is isopropylbenzene.

$$C_6H_6 + CH_3CH_2CH_2Cl \xrightarrow{AlCl_3} C_6H_5CH_2CH_2CH_3 + C_6H_5CH(CH_3)_2$$

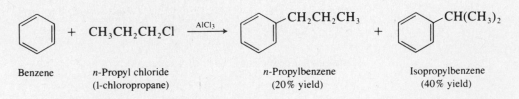

Benzene; *n*-Propyl chloride (1-chloropropane); *n*-Propylbenzene (20% yield); Isopropylbenzene (40% yield)

Aluminum chloride coordinates with *n*-propyl chloride to give a weak Lewis acid–Lewis base complex, which can be attacked by benzene to yield *n*-propylbenzene or can undergo an intramolecular hydride shift to produce isopropyl cation. Isopropylbenzene arises by reaction of isopropyl cation with benzene.

13.5 Isopropylbenzene arises from the reaction of isopropyl cation with benzene. Isopropyl cation is formed by protonation of propene with hydrogen fluoride.

$$CH_3CH{=}CH_2 + H{-}F \longrightarrow CH_3\overset{+}{C}HCH_3 + F^-$$

Propene, Hydrogen fluoride, Isopropyl cation, Fluoride ion

$$CH_3\overset{+}{C}HCH_3 + C_6H_6 \longrightarrow \text{cyclohexadienyl cation (H, }CH(CH_3)_2) \longrightarrow C_6H_5{-}CH(CH_3)_2 + H^+$$

Isopropyl cation, Benzene, Cyclohexadienyl cation intermediate, Isopropylbenzene

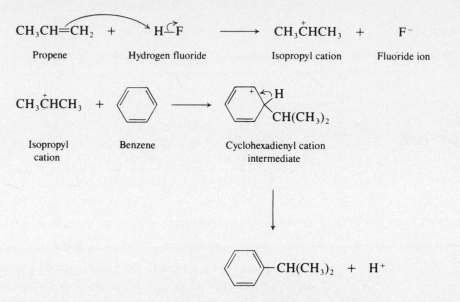

13.6 (*b*) Neopentyl systems exhibit a pronounced tendency to undergo cationic rearrangements by shift of a methyl group. A Friedel–Crafts alkylation of benzene using neopentyl chloride would not be a satisfactory method to prepare neopentylbenzene; the best way to prepare this compound is by Friedel–Crafts acylation followed by Clemmensen reduction.

$$(CH_3)_3C\overset{O}{\overset{\|}{C}}Cl + C_6H_6 \xrightarrow{AlCl_3} (CH_3)_3C\overset{O}{\overset{\|}{C}}{-}C_6H_5 \xrightarrow{Zn(Hg),\ HCl} (CH_3)_3CCH_2{-}C_6H_5$$

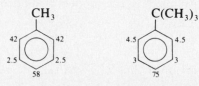

2,2-Dimethylpropanoyl chloride, Benzene, 2,2-Dimethylpropanoylbenzene, Neopentylbenzene

13.7 (*b*) Partial rate factors for nitration of toluene and *tert*-butylbenzene, relative to a single position of benzene are as shown:

	Toluene (CH_3)	*tert*-Butylbenzene ($C(CH_3)_3$)
ortho	42, 42	4.5, 4.5
meta	2.5, 2.5	3, 3
para	58	75

The sum of these partial rate factors is 147 for toluene, 90 for *tert*-butylbenzene. Toluene is 147/90, or 1.7 times, more reactive than *tert*-butylbenzene.

(*c*) The product distribution for nitration of *tert*-butylbenzene is determined from the partial rate factors.

$$\text{Ortho:} \quad \frac{2\,(4.5)}{90} = 10\%$$

$$\text{Meta:} \quad \frac{2\,(3)}{90} = 6.7\%$$

$$\text{Para:} \quad \frac{75}{90} = 83.3\%$$

13.8 (*b*) Attack by bromine at the position meta to the amino group gives a cyclohexadienyl cation intermediate in which delocalization of the nitrogen lone pair cannot participate in dispersal of the positive charge.

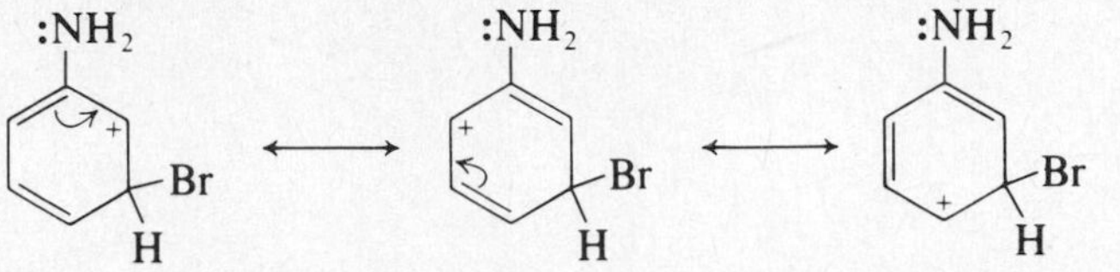

$:NH_2$ ... Br, H ⟷ $:NH_2$ ... Br, H ⟷ $:NH_2$ ... Br, H

(*c*) Attack at the position para to the amino group yields a cyclohexadienyl cation intermediate that is stabilized by delocalization of the electron pair of the amino group.

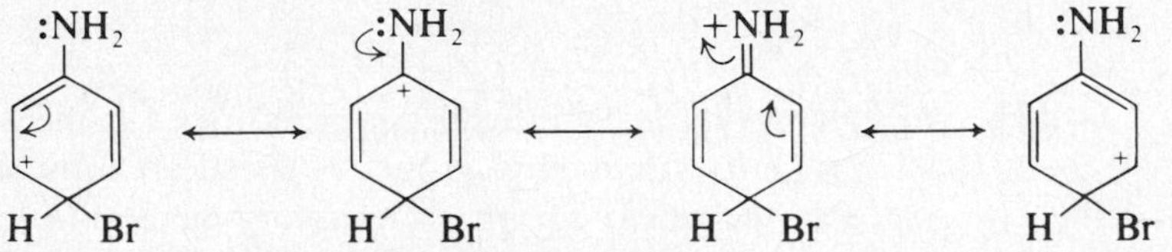

$:NH_2$ ... H Br ⟷ $:NH_2$ ... H Br ⟷ $\overset{+}{N}H_2$ ... H Br ⟷ $:NH_2$ ... H Br

13.9 Electrophilic aromatic substitution in biphenyl is best understood by considering one ring as the functional group and the other as a substituent. An aryl substituent is ortho, para-directing. Nitration of biphenyl gives a mixture of *o*-nitrobiphenyl and *p*-nitrobiphenyl.

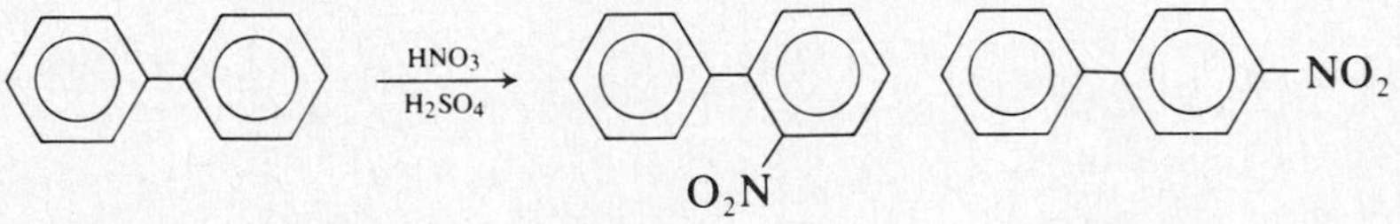

Biphenyl $\xrightarrow[H_2SO_4]{HNO_3}$ *o*-Nitrobiphenyl (37%) (O_2N) + *p*-Nitrobiphenyl (63%) ($-NO_2$)

13.10 (*b*) The carbonyl group attached directly to the ring is a signal that the substituent is a meta-directing group. Nitration of methyl benzoate yields methyl *m*-nitrobenzoate.

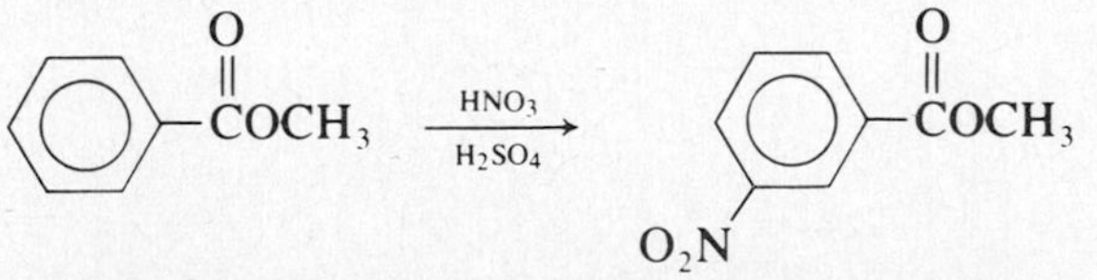

Methyl benzoate ($C_6H_5COCH_3$, with C=O) $\xrightarrow[H_2SO_4]{HNO_3}$ Methyl *m*-nitrobenzoate (O_2N, $COCH_3$) (isolated in 81–85% yield)

(*c*) The acyl group in propiophenone is meta-directing— the carbonyl is attached directly to the ring. The product is *m*-nitropropiophenone.

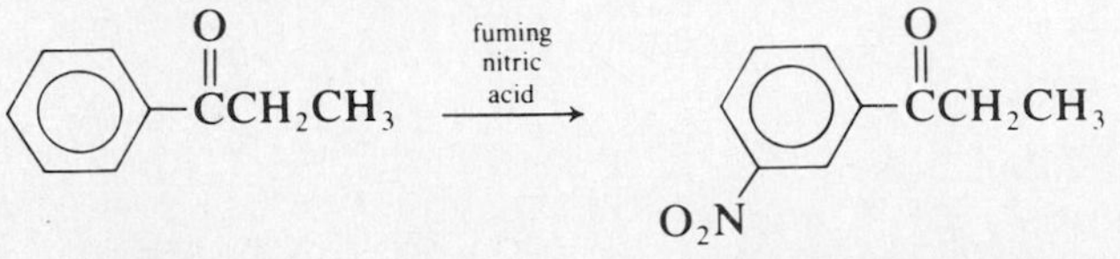

Propiophenone ($C_6H_5CCH_2CH_3$, with C=O) $\xrightarrow{\text{fuming nitric acid}}$ *m*-Nitropropiophenone (O_2N, CCH_2CH_3) (isolated in 60% yield)

13.11 Writing the structures out in more detail reveals that the substituent $-\overset{+}{N}(CH_3)_3$ lacks the unshared electron pair of $-N(CH_3)_2$.

$-\overset{+}{N}(CH_3)_3$ $\quad\quad$ $-\ddot{N}(CH_3)_2$ $\quad\quad$ $-\overset{+}{N}(=O)O^-$

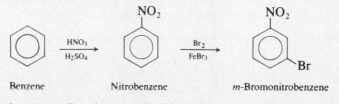

This unshared pair is responsible for the powerful activating effect of an $-N(CH_3)_2$ group. On the other hand, the nitrogen in $-\overset{+}{N}(CH_3)_3$ is positively charged and in that respect resembles the nitrogen of a nitro group. On these bases, we expect the substituent $-\overset{+}{N}(CH_3)_3$ to be deactivating and meta-directing.

13.12 The product that is obtained when benzene is subjected to bromination and nitration depends on the order in which the reactions are carried out. A nitro group is meta-directing, so if it is introduced prior to the bromination step, *m*-bromonitrobenzene is obtained.

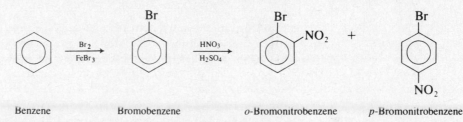

Benzene $\xrightarrow[H_2SO_4]{HNO_3}$ Nitrobenzene $\xrightarrow[FeBr_3]{Br_2}$ *m*-Bromonitrobenzene

Bromine is an ortho, para–directing group. If it is introduced first, nitration of the resulting bromobenzene yields a mixture of *o*-bromonitrobenzene and *p*-bromonitrobenzene.

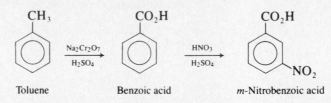

Benzene $\xrightarrow[FeBr_3]{Br_2}$ Bromobenzene $\xrightarrow[H_2SO_4]{HNO_3}$ *o*-Bromonitrobenzene + *p*-Bromonitrobenzene

13.13 A straightforward approach to the synthesis of *m*-nitrobenzoic acid involves preparation of benzoic acid by oxidation of toluene, followed by nitration. The carboxyl group of benzoic acid is meta-directing. Nitration of toluene prior to oxidation would lead to a mixture of ortho and para products.

Toluene $\xrightarrow[H_2SO_4]{Na_2Cr_2O_7}$ Benzoic acid $\xrightarrow[H_2SO_4]{HNO_3}$ *m*-Nitrobenzoic acid

13.14 (*b*) Halogen substituents are ortho, para–directing, and the disposition in *m*-dichlorobenzene is such that their effects reinforce each other. The major product is 2,4-dichloro-1-nitrobenzene. Substitution at the position between the two chlorines is slow because it is a sterically hindered position.

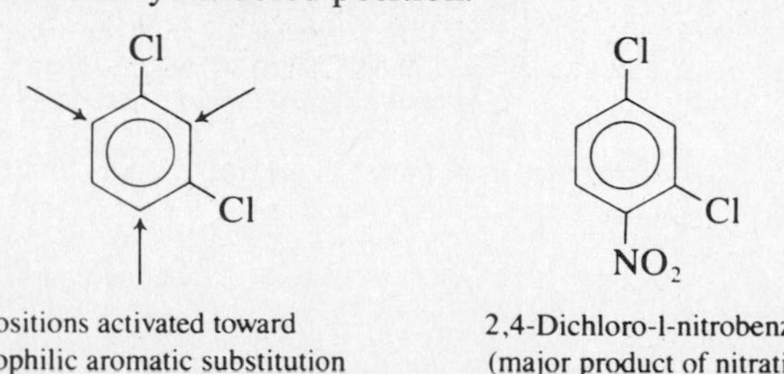

Positions activated toward electrophilic aromatic substitution in *m*-dichlorobenzene

2,4-Dichloro-1-nitrobenzene (major product of nitration)

(c) Nitro groups are meta-directing. Both nitro groups of *m*-dinitrobenzene direct an incoming substituent to the same position in an electrophilic aromatic substitution reaction. Nitration of *m*-nitrobenzene yields 1,3,5-trinitrobenzene.

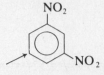

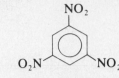

Both nitro groups of *m*-dinitrobenzene direct electrophile to same position

1,3,5-Trinitrobenzene (principal product of nitration of *m*-dinitrobenzene)

(d) A methoxy group is ortho, para–directing and a carbonyl group is meta-directing. The open positions of the ring that are activated by the methoxy group in *p*-methoxyacetophenone are also those that are meta to the carbonyl, and so the directing effects of the two substituents reinforce one another. Nitration of *p*-methoxyacetophenone yields 4-methoxy-3-nitroacetophenone.

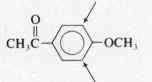

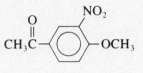

Positions ortho to the methoxy group are meta to the carbonyl

4-Methoxy-3-nitroacetophenone

(e) The methoxy group of *p*-methylanisole activates the positions that are ortho to it; the methyl activates those ortho to itself. Methoxy is a more powerful activating substituent than methyl, so nitration occurs ortho to the methoxy group.

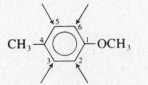

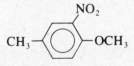

Methyl activates C-3 and C-5; methoxy activates C-2 and C-6

4-Methyl-2-nitroanisole (principal product of nitration)

(f) All the substituents in 2,6-dibromoanisole are ortho, para–directing, and their effects are felt at different positions. However, the methoxy group is a far more powerful activating substituent than bromine, so it controls the regioselectivity of nitration.

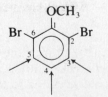

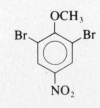

Methoxy directs toward C-4; bromines direct toward C-3 and C-5

2,6-Dibromo-4-nitroanisole (principal product of nitration)

13.15 (a) Nitration of benzene is the archetypical electrophilic aromatic substitution reaction.

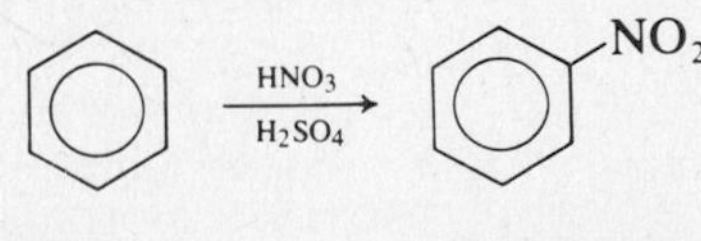

Benzene Nitrobenzene

(*b*) Nitrobenzene is much less reactive than benzene toward electrophilic aromatic substitution. The nitro group on the ring is a meta director.

NO_2 $\xrightarrow[H_2SO_4]{HNO_3}$ NO_2, NO_2

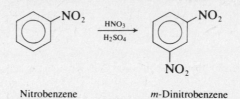

Nitrobenzene *m*-Dinitrobenzene

(*c*) Toluene is more reactive than benzene in electrophilic aromatic substitution. A methyl substituent is an ortho, para director.

CH_3 $\xrightarrow[FeBr_3]{Br_2}$ CH_3, Br + CH_3, Br

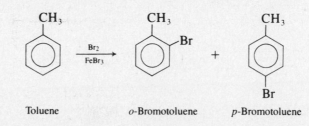

Toluene *o*-Bromotoluene *p*-Bromotoluene

(*d*) Trifluoromethyl is deactivating and meta-directing.

CF_3 $\xrightarrow[FeBr_3]{Br_2}$ CF_3, Br

(Trifluoromethyl)benzene *m*-Bromo(trifluoromethyl)benzene

(*e*) Anisole is ortho, para–directing, strongly activated toward electrophilic aromatic substitution, and readily sulfonated in sulfuric acid.

OCH_3 $\xrightarrow{H_2SO_4}$ OCH_3, SO_3H + OCH_3, SO_3H

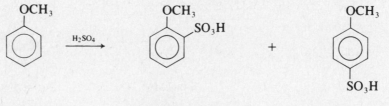

Anisole *o*-Methoxybenzenesulfonic acid *p*-Methoxybenzenesulfonic acid

Sulfur trioxide, of course, could be added to the sulfuric acid and would facilitate reaction. The para isomer is the predominant product.

(*f*) Acetanilide is quite similar to anisole in its behavior toward electrophilic aromatic substitution.

$HNCCH_3$ (O) $\xrightarrow{H_2SO_4}$ $HNCCH_3$ (O), SO_3H + $HNCCH_3$ (O), SO_3H

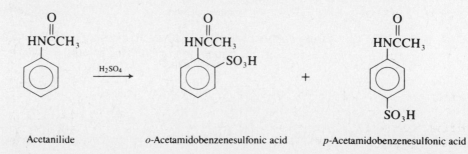

Acetanilide *o*-Acetamidobenzenesulfonic acid *p*-Acetamidobenzenesulfonic acid

(*g*) Bromobenzene is less reactive than benzene. A bromine substituent is ortho, para–directing.

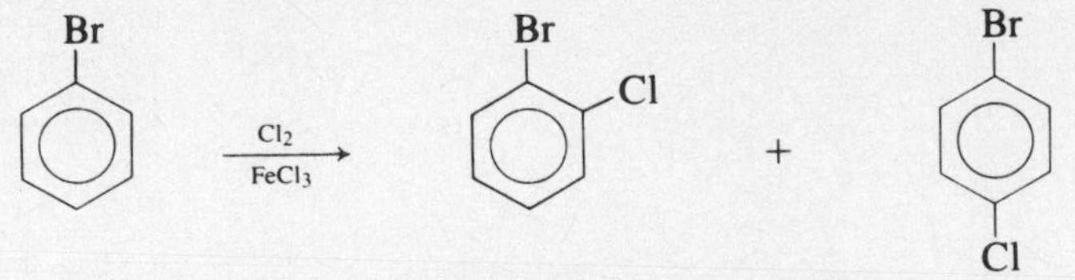

Br, Cl, Br, Cl, Br, $\xrightarrow[FeCl_3]{Cl_2}$, +, Cl

Bromobenzene o-Bromochlorobenzene p-Bromochlorobenzene

(*h*) Anisole is a reactive substrate toward Friedel–Crafts alkylation and yields a mixture of *o*- and *p*-benzylated products when treated with benzyl chloride and aluminum chloride.

OCH_3, CH_2Cl, +, $\xrightarrow{AlCl_3}$, OCH_3, $CH_2C_6H_5$, +, OCH_3, $CH_2C_6H_5$

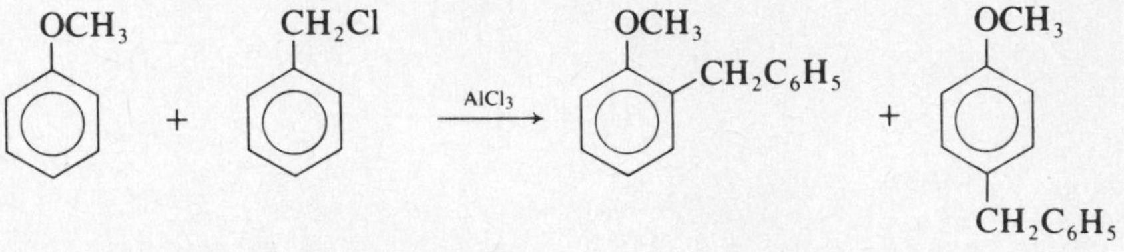

Anisole Benzyl chloride o-Benzylanisole p-Benzylanisole

(*i*) Benzene will undergo acylation with benzoyl chloride and aluminum chloride.

O, CCl, +, $\xrightarrow{AlCl_3}$, O, C

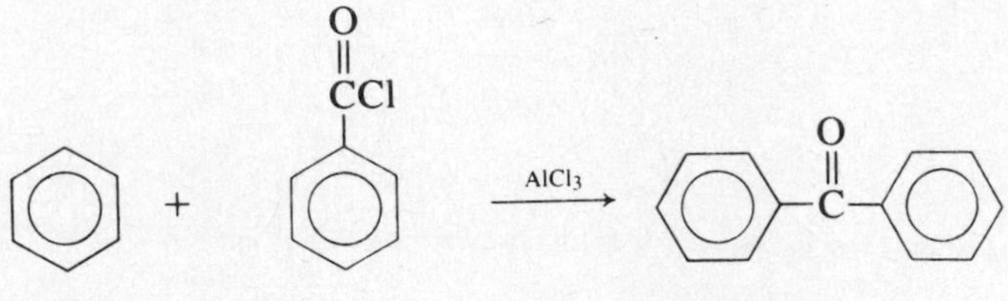

Benzene Benzoyl chloride Benzophenone

(*j*) A benzoyl substituent is meta-directing and deactivating.

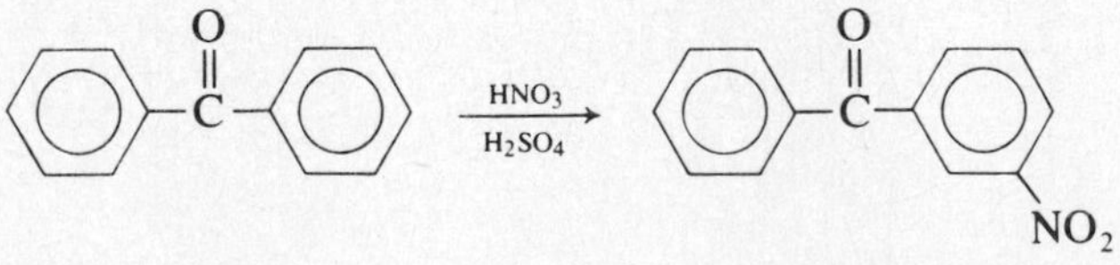

O, C, $\xrightarrow[H_2SO_4]{HNO_3}$, O, C, NO_2

Benzophenone m-Nitrobenzophenone

(*k*) Clemmensen reduction conditions involve treating a ketone with zinc amalgam and concentrated hydrochloric acid.

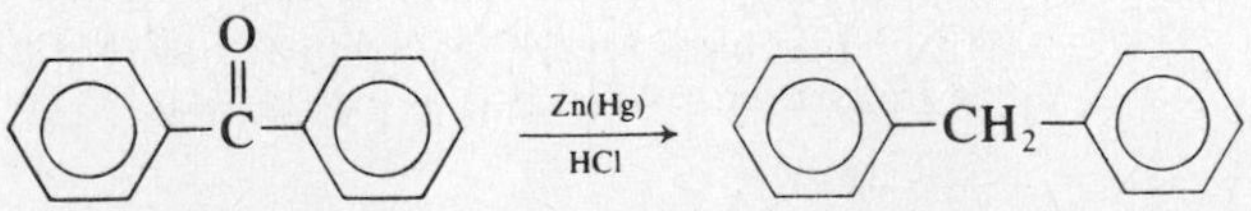

O, C, $\xrightarrow[HCl]{Zn(Hg)}$, CH_2

Benzophenone Diphenylmethane

(*l*) Wolff–Kishner reduction utilizes hydrazine, a base, and a high-boiling alcohol solvent to reduce ketone functions to methylene groups.

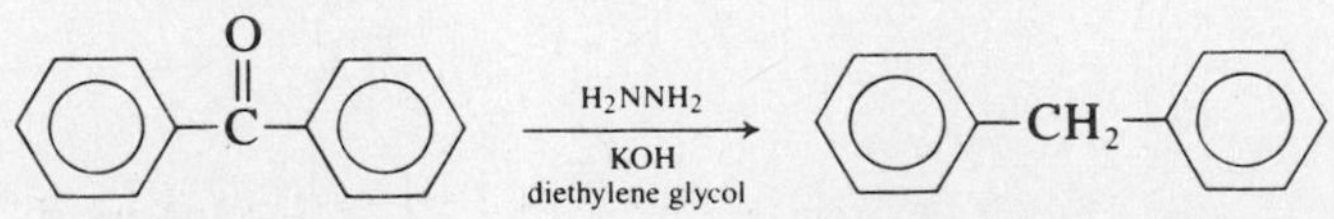

O, C, $\xrightarrow[\text{KOH, diethylene glycol}]{H_2NNH_2}$, CH_2

Benzophenone Diphenylmethane

13.16 (*a*) There are three principal resonance forms of the cyclohexadienyl cation intermediate formed by attack of bromine on *p*-xylene.

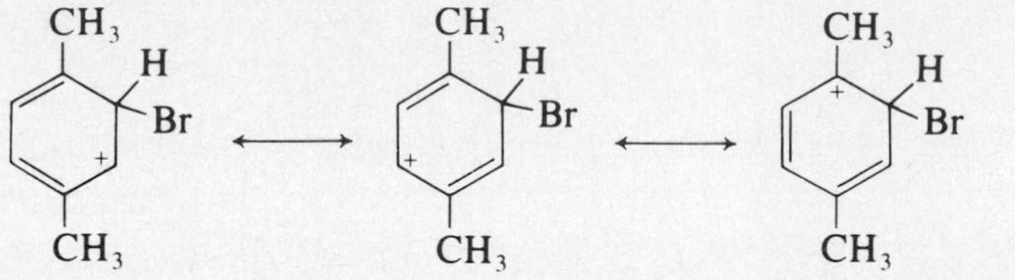

Any one of these resonance forms is a satisfactory answer to the question. Because of its tertiary carbocation character this carbocation is more stable than the corresponding intermediate formed from benzene.

(*b*) Chlorination of *m*-xylene will give predominantly 4-chloro-1,3-dimethylbenzene.

Cl_2 via

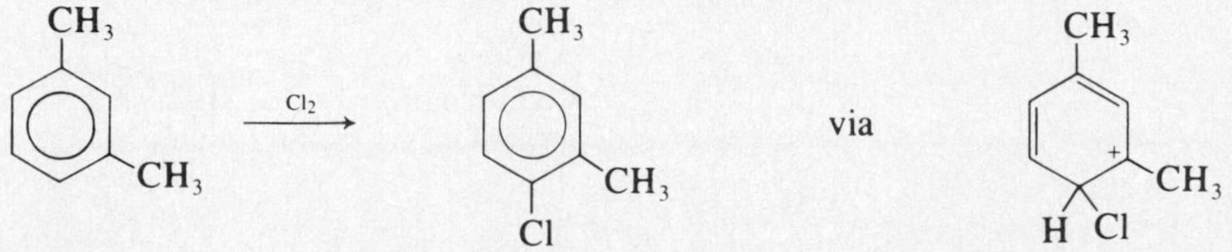

m-Xylene 4-Chloro-1,3-dimethylbenzene More stable cyclohexadienyl cation

The intermediate shown (or any of its resonance forms) is more stable for steric reasons than

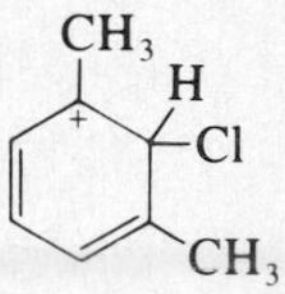

Less stable cyclohexadienyl cation

The cyclohexadienyl cation intermediate leading to 4-chloro-1,3-dimethylbenzene is more stable and is formed faster than the intermediate leading to chlorobenzene because of its tertiary carbocation character.

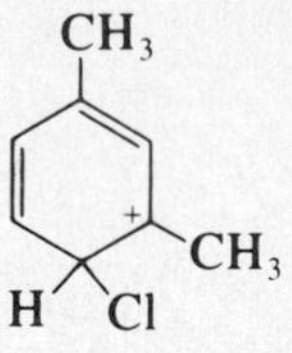

more stable than

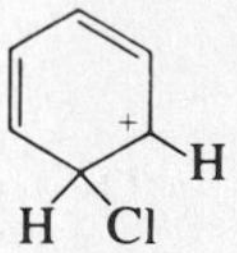

(*c*) The most stable carbocation intermediate formed during nitration of acetophenone is the one corresponding to meta attack.

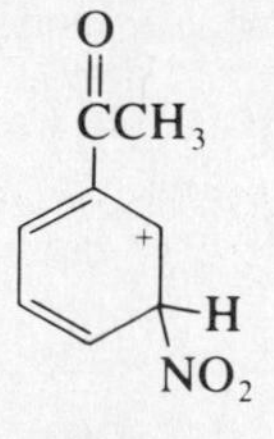

more stable than 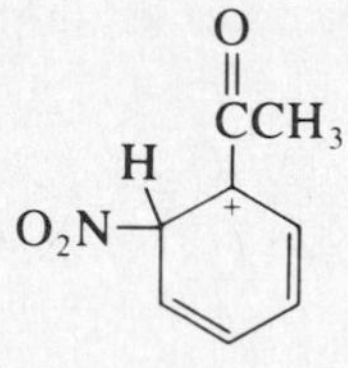 or

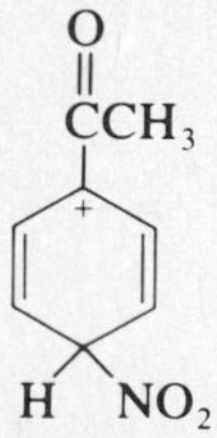

An acetyl group is electron-withdrawing and destabilizes a carbocation to which it is attached. The most stable carbocation intermediate in the nitration of acetophenone is

less stable and is formed more slowly than is the corresponding carbocation formed during nitration of benzene.

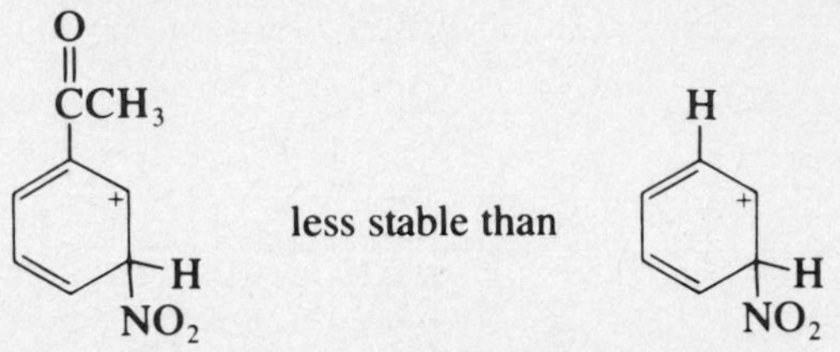

less stable than

(*d*) The methoxy group in anisole is strongly activating and ortho, para–directing. For steric reasons and because of inductive electron withdrawal by oxygen the intermediate leading to para substitution is the most stable.

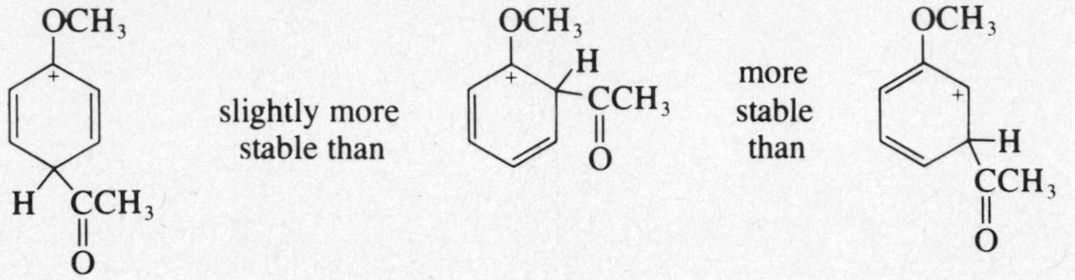

slightly more stable than ... more stable than

Of the various resonance forms for the most stable intermediate, the most stable one has eight electrons around each atom.

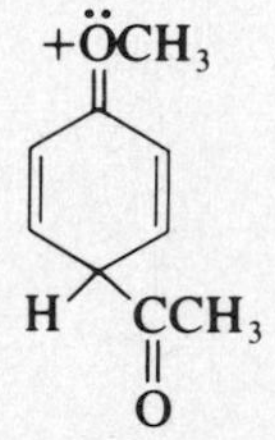

This intermediate is much more stable than the corresponding one from benzene.

(*e*) An isopropyl group is an activating substituent and is ortho, para–directing. Attack at the ortho position is sterically hindered. The most stable intermediate is

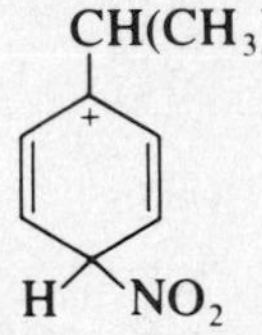

or any of its resonance forms. Because of its tertiary carbocation character this cation is more stable than the corresponding cyclohexadienyl cation intermediate from benzene.

(*f*) A nitro substituent is deactivating and meta-directing. The most stable cyclohexadienyl cation formed in the bromination of nitrobenzene is:

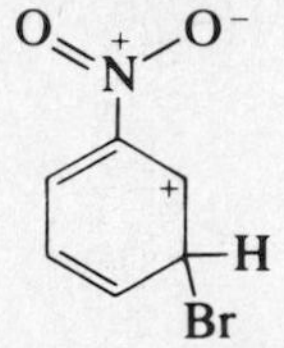

This ion is less stable than the cyclohexadienyl cation formed during bromination of benzene.

13.17 (*a*) Toluene is more reactive than chlorobenzene in electrophilic aromatic substitution reactions because a methyl substituent is activating while a halogen substituent is deactivating. Both are ortho, para–directing, however. Nitration of toluene is faster than nitration of chlorobenzene.

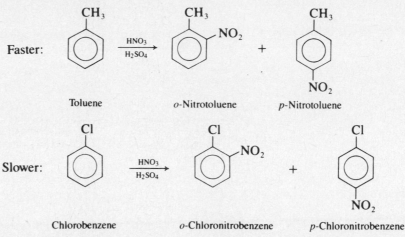

Faster: Toluene $\xrightarrow[H_2SO_4]{HNO_3}$ *o*-Nitrotoluene + *p*-Nitrotoluene

Slower: Chlorobenzene $\xrightarrow[H_2SO_4]{HNO_3}$ *o*-Chloronitrobenzene + *p*-Chloronitrobenzene

(*b*) A fluorine substituent is not nearly as strongly deactivating as is a trifluoromethyl group. The reaction that takes place is Friedel–Crafts alkylation of fluorobenzene.

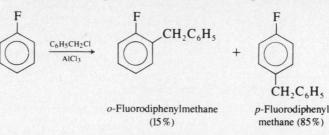

$C_6H_5F \xrightarrow[AlCl_3]{C_6H_5CH_2Cl}$ *o*-Fluorodiphenylmethane (15%) + *p*-Fluorodiphenyl methane (85%)

Strongly deactivated aromatic compounds do not undergo Friedel–Crafts reactions.

$C_6H_5CF_3 + C_6H_5CH_2Cl \xrightarrow{AlCl_3}$ no reaction

(*c*) A carbonyl group directly bonded to a benzene ring strongly *deactivates* it toward electrophilic aromatic substitution. Methyl benzoate is much less reactive than benzene.

$C_6H_5C(=O)OCH_3 \longleftrightarrow C_6H_5\overset{+}{C}(O^-)OCH_3$

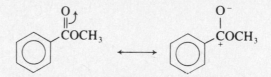

An oxygen substituent directly attached to the ring strongly *activates* it toward electrophilic aromatic substitution. Phenyl acetate is much more reactive than benzene or methyl benzoate.

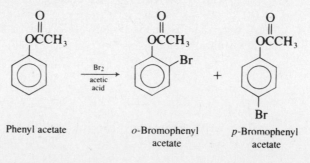

Phenyl acetate ($C_6H_5OCCH_3$, with C=O) $\xrightarrow[\text{acetic acid}]{Br_2}$ *o*-Bromophenyl acetate + *p*-Bromophenyl acetate

Bromination of methyl benzoate requires more vigorous conditions; catalysis by ferric bromide is required for bromination of deactivated aromatic rings.

(*d*) Acetanilide is strongly activated toward electrophilic aromatic substitution and reacts faster than nitrobenzene, which is strongly deactivated.

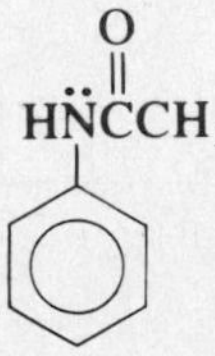

Acetanilide (lone pair on nitrogen can stabilize cyclohexadienyl cation intermediate)

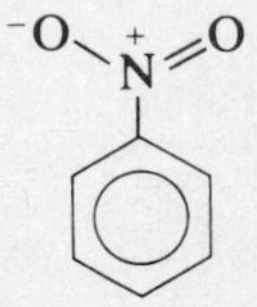

Nitrobenzene (nitrogen is positively charged and is electron-withdrawing)

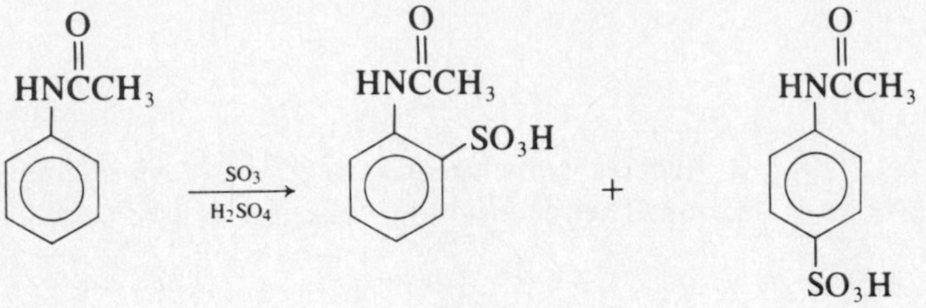

$\xrightarrow[H_2SO_4]{SO_3}$

Acetanilide

o-Acetamidobenzenesulfonic acid

+

p-Acetamidobenzenesulfonic acid

(*e*) Both substrates are of the type

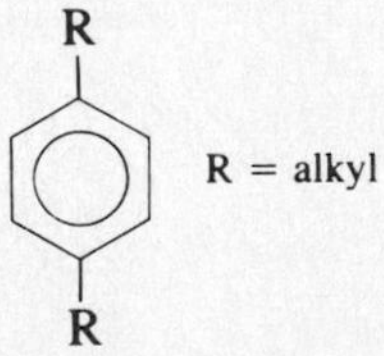

R = alkyl

and are activated toward Friedel–Crafts acylation. Since electronic effects are comparable, we look to differences in steric factors and conclude that reaction will be faster for $R = CH_3$ than for $R = (CH_3)_3C—$.

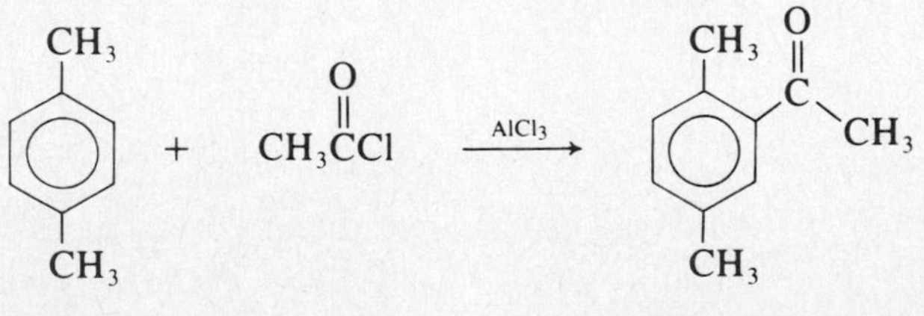

p-Xylene + Acetyl chloride (CH_3CCl) $\xrightarrow{AlCl_3}$ 2,5-Dimethylacetophenone

(*f*) A phenyl substituent is activating and ortho, para-directing. Biphenyl will undergo chlorination readily.

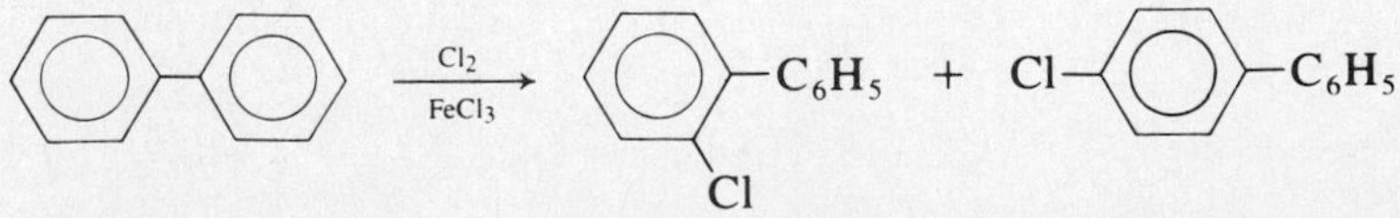

Biphenyl $\xrightarrow[FeCl_3]{Cl_2}$ *o*-Chlorobiphenyl + *p*-Chlorobiphenyl

Each benzene ring of benzophenone is deactivated by the carbonyl group.

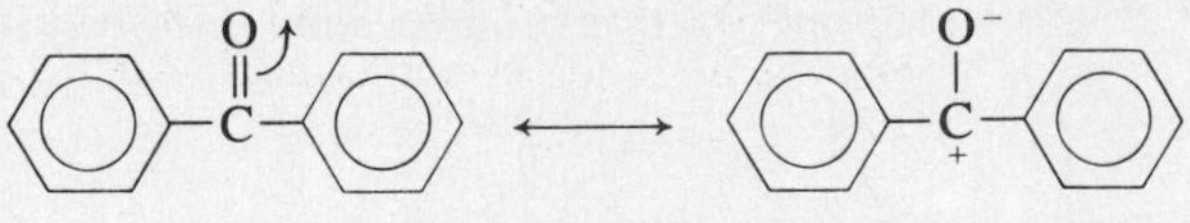

Benzophenone

Benzophenone is much less reactive than biphenyl in electrophilic aromatic substitution reactions.

13.18 Reactivity toward electrophilic aromatic substitution increases with increasing number of electron-releasing substituents. Benzene, with no methyl substituents, is the least reactive, followed by toluene, with one methyl group. 1,3,5-Trimethylbenzene, with three methyl substituents, is the most reactive.

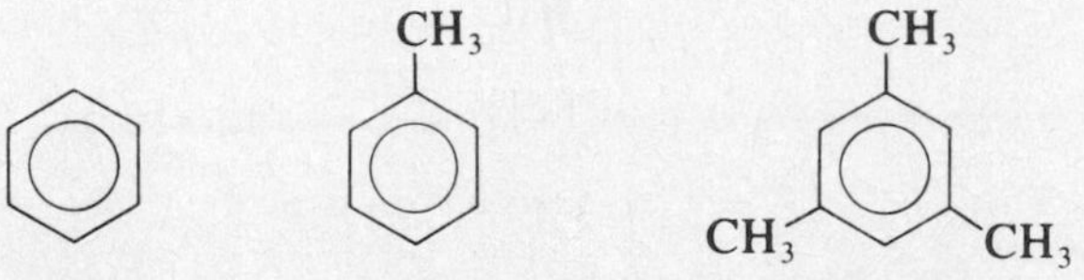

	Benzene	Toluene	1,3,5-Trimethylbenzene
Relative reactivity:	1	60	2×10^7

o-Xylene and *m*-xylene are intermediate in reactivity between toluene and 1,3,5-trimethylbenzene. Of the two, *m*-xylene is more reactive than *o*-xylene because the activating effects of the two methyl groups reinforce each other.

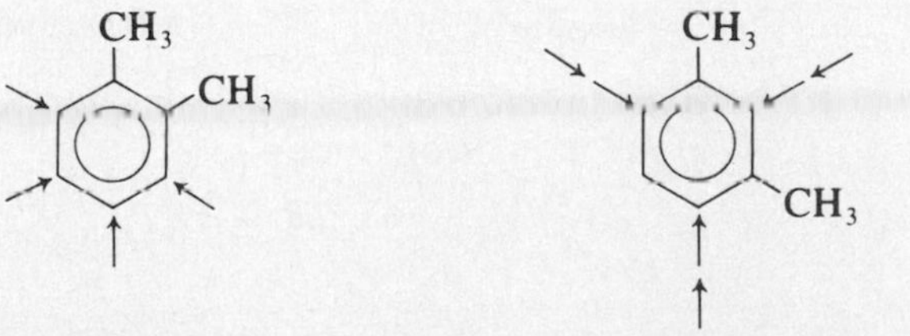

	o-Xylene (all positions somewhat activated)	*m*-Xylene (activating effects reinforce each other)
Relative reactivity:	5×10^2	5×10^4

13.19 (*a*) Chlorine is ortho, para-directing, carboxyl is meta-directing. The positions that are ortho to the chlorine are meta to the carboxyl, so both substituents direct an incoming electrophile to the same position. Introduction of the second nitro group at the remaining position that is ortho to the chlorine puts it meta to the carboxyl and meta to the first nitro group.

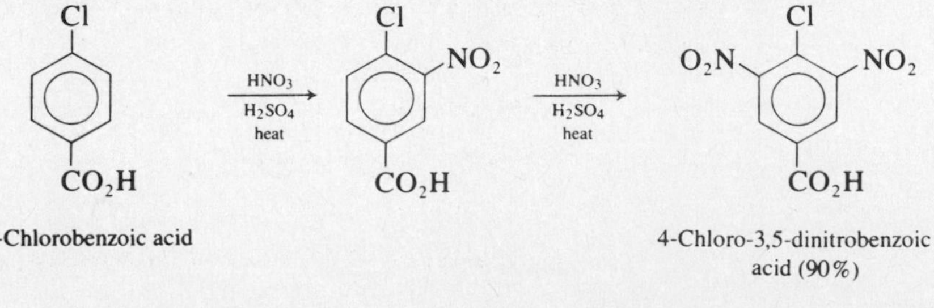

$$\text{Cl-C}_6\text{H}_4\text{-CO}_2\text{H} \xrightarrow[\text{H}_2\text{SO}_4,\ \text{heat}]{\text{HNO}_3} \text{Cl(NO}_2\text{)C}_6\text{H}_3\text{CO}_2\text{H} \xrightarrow[\text{H}_2\text{SO}_4,\ \text{heat}]{\text{HNO}_3} \text{Cl(O}_2\text{N)(NO}_2\text{)C}_6\text{H}_2\text{CO}_2\text{H}$$

p-Chlorobenzoic acid → 4-Chloro-3,5-dinitrobenzoic acid (90%)

(*b*) An amino group is one of the strongest activating substituents. The para and both ortho positions are readily substituted in aniline. When aniline is treated with excess bromine, 2,4,6-tribromoaniline is formed in quantitative yield.

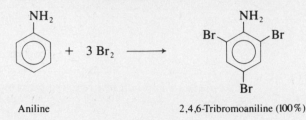

Aniline + 3 Br_2 ⟶ 2,4,6-Tribromoaniline (100%)

(*c*) The positions ortho and para to the amino group in *o*-aminoacetophenone are the ones most activated toward electrophilic aromatic substitution.

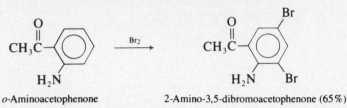

o-Aminoacetophenone $\xrightarrow{Br_2}$ 2-Amino-3,5-dibromoacetophenone (65%)

(*d*) The carboxyl group in benzoic acid is meta-directing, so nitration gives *m*-nitrobenzoic acid. The second nitration step introduces a nitro group meta to both the carboxyl group and the first nitro group.

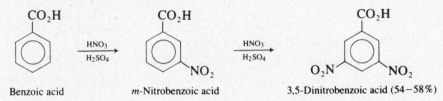

Benzoic acid $\xrightarrow[H_2SO_4]{HNO_3}$ *m*-Nitrobenzoic acid $\xrightarrow[H_2SO_4]{HNO_3}$ 3,5-Dinitrobenzoic acid (54–58%)

(*e*) Both bromine substituents are introduced ortho to the strongly activating hydroxyl group in *p*-nitrophenol.

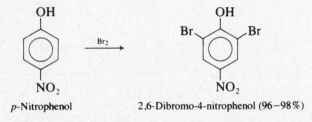

p-Nitrophenol $\xrightarrow{Br_2}$ 2,6-Dibromo-4-nitrophenol (96–98%)

(*f*) Friedel–Crafts alkylation occurs when biphenyl is treated with *tert*-butyl chloride and ferric chloride (a Lewis acid catalyst); the product of monosubstitution is *p-tert*-butylbiphenyl. All the positions of the ring that bears the *tert*-butyl group are sterically hindered, so the second alkylation step introduces a *tert*-butyl group at the para position of the second ring.

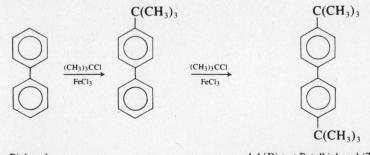

Biphenyl $\xrightarrow[FeCl_3]{(CH_3)_3CCl}$ *p-tert*-butylbiphenyl $\xrightarrow[FeCl_3]{(CH_3)_3CCl}$ 4,4′-Di-*tert*-Butylbiphenyl (70%)

(*g*) Disulfonation of phenol occurs at positions ortho and para to the hydroxyl group. The ortho, para product predominates over the ortho, ortho one.

OH; $\xrightarrow{H_2SO_4}$ OH, SO_3H, SO_3H

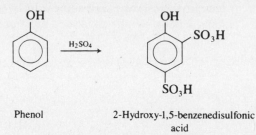

Phenol — 2-Hydroxy-1,5-benzenedisulfonic acid

13.20 When carrying out each of the following syntheses, evaluate how the structure of the product differs from that of benzene or toluene, that is, determine which groups have been substituted on the benzene ring or altered in some way. The sequence of reaction steps when multiple substitution is desired is important; recall that some groups direct ortho, para and others meta.

(*a*) Isopropylbenzene may be prepared by a Friedel–Crafts alkylation of benzene with isopropyl chloride (or bromide, or iodide).

$$C_6H_6 + (CH_3)_2CHCl \xrightarrow{AlCl_3} C_6H_5{-}CH(CH_3)_2$$

Benzene — Isopropyl chloride — Isopropylbenzene

It would not be appropriate to use propyl chloride and trust that a rearrangement reaction would lead to isopropylbenzene, because a mixture of propylbenzene and isopropylbenzene would be obtained.

Isopropylbenzene may also be prepared by alkylation of benzene with propene in the presence of a proton donor. Liquid hydrogen fluoride is often used in reactions of this type.

$$C_6H_6 + CH_3CH{=}CH_2 \xrightarrow{HF} C_6H_5{-}CH(CH_3)_2$$

Benzene — Propene — Isopropylbenzene

(*b*) As the isopropyl and sulfonic acid groups are para to each other, the first group introduced on the ring must be the ortho, para director, i.e., the isopropyl group. Therefore we may use the product of (*a*), isopropylbenzene, in this synthesis. An isopropyl group is a fairly bulky ortho, para director, so sulfonation of isopropylbenzene gives mainly *p*-isopropylbenzenesulfonic acid.

$$(CH_3)_2CH{-}C_6H_5 \xrightarrow[H_2SO_4]{SO_3} (CH_3)_2CH{-}C_6H_4{-}SO_3H$$

Isopropylbenzene — *p*-Isopropylbenzenesulfonic acid

A sulfonic acid group is meta-directing so the order of steps must be alkylation followed by sulfonation rather than the reverse.

(*c*) Free-radical halogenation of isopropylbenzene occurs with high regioselectivity at the benzylic position. *N*-Bromosuccinimide (NBS) is a good reagent to use for benzylic bromination reactions.

$$C_6H_5{-}CH(CH_3)_2 \xrightarrow[\text{or NBS}]{Br_2,\ \text{light}} C_6H_5{-}C(Br)(CH_3){-}CH_3$$

Isopropylbenzene — 2-Bromo-2-phenylpropane

(*d*) Toluene is an obvious starting material for the preparation of 4-*tert*-butyl-2-nitrotoluene. Two possibilities both involving nitration and alkylation of toluene, present themselves; the problem to be addressed is in what order to carry out the two steps. Friedel–Crafts alkylation must precede nitration.

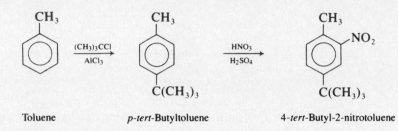

Toluene $\xrightarrow[AlCl_3]{(CH_3)_3CCl}$ *p-tert*-Butyltoluene $\xrightarrow[H_2SO_4]{HNO_3}$ 4-*tert*-Butyl-2-nitrotoluene

Introduction of the nitro group as the first step is not a satisfactory approach since Friedel–Crafts reactions cannot be carried out on nitro-substituted aromatic compounds.

(*e*) Two electrophilic aromatic substitution reactions need to be performed, chlorination and Friedel–Crafts acylation. The order in which the reactions are carried out is important; chlorine is an ortho, para director, and the acetyl group is a meta director. Since the groups are meta in the desired compound, introduce the acetyl group first.

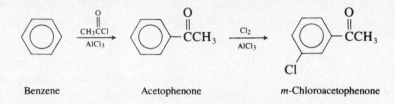

Benzene $\xrightarrow[AlCl_3]{CH_3CCl}$ Acetophenone $\xrightarrow[AlCl_3]{Cl_2}$ *m*-Chloroacetophenone

(*f*) Reverse the order of steps in (*e*) above to prepare *p*-chloroacetophenone.

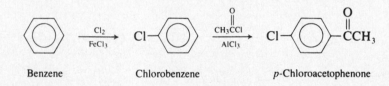

Benzene $\xrightarrow[FeCl_3]{Cl_2}$ Chlorobenzene $\xrightarrow[AlCl_3]{CH_3CCl}$ *p*-Chloroacetophenone

Friedel–Crafts reactions can be carried out on halobenzenes but not on arenes that are more strongly deactivated.

(*g*) Here again the problem involves two successive electrophilic aromatic substitution reactions, in this case using toluene as the initial substrate. The proper sequence is Friedel–Crafts acylation first, followed by bromination of the ring.

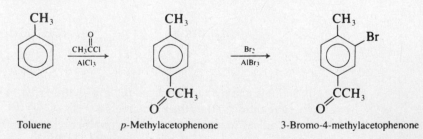

Toluene $\xrightarrow[AlCl_3]{CH_3CCl}$ *p*-Methylacetophenone $\xrightarrow[AlBr_3]{Br_2}$ 3-Bromo-4-methylacetophenone

If the sequence of steps had been reversed, with halogenation preceding acylation, the first intermediate would be *o*-bromotoluene, Friedel–Crafts acylation of which would give a complex mixture of products because both groups are ortho, para–directing. On the other hand, the orienting effects of the two groups in *p*-methylacetophenone reinforce each other, so its bromination is highly regioselective and in the desired direction.

(*h*) Recalling that alkyl groups attached to the benzene ring by CH_2 may be prepared by reduction of the appropriate ketone, we may reduce 3-bromo-4-methylacetophenone, as prepared in the preceding exercise, by the Clemmensen or Wolff–Kishner procedure to give 2-bromo-4-ethyltoluene.

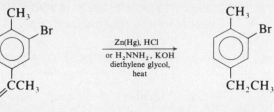

3-Bromo-4-methylacetophenone $\xrightarrow[\text{or } H_2NNH_2\text{, KOH, diethylene glycol, heat}]{\text{Zn(Hg), HCl}}$ 2-Bromo-4-ethyltoluene

(*i*) This is a relatively straightforward synthetic problem. Bromine is an ortho, para–directing substituent, nitro is meta-directing. Nitrate first, then brominate to give 1-bromo-3-nitrobenzene.

Benzene $\xrightarrow[H_2SO_4]{HNO_3}$ Nitrobenzene $\xrightarrow[AlBr_3]{Br_2}$ 1-Bromo-3-nitrobenzene

(*j*) Take advantage of the ortho, para–directing properties of bromine to prepare 1-bromo-2,4-dinitrobenzene. Brominate first, then nitrate under conditions that lead to disubstitution. The nitro groups are introduced at positions ortho and para to the bromine and meta to each other.

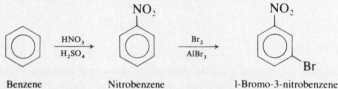

Benzene $\xrightarrow{Br_2,\ FeBr_3}$ Bromobenzene $\xrightarrow[H_2SO_4\text{, heat}]{HNO_3}$ 1-Bromo-2,4-dinitrobenzene

(*k*) While bromo and nitro substituents are readily introduced by electrophilic aromatic substitution, the only methods we have available so far to prepare carboxylic acids is by oxidation of alkyl side chains. Thus, use toluene as a starting material, planning to convert the methyl group to a carboxyl group by oxidation. Nitrate next; nitro and carboxyl are both meta-directing groups, so the bromination in the last step occurs with the proper regioselectivity.

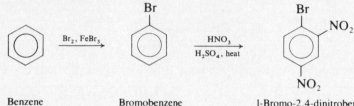

Toluene $\xrightarrow[H_2SO_4]{K_2Cr_2O_7}$ Benzoic acid $\xrightarrow[H_2SO_4]{HNO_3}$ 3-Nitrobenzoic acid $\xrightarrow[AlBr_3]{Br_2}$ 3-Bromo-5-nitrobenzoic acid

If bromination is performed prior to nitration, the bromine substituent will direct an incoming electrophile to positions ortho and para to itself, giving the wrong orientation of substituents in the product.

(*l*) Again toluene is a suitable starting material, with its methyl group serving as the source of the carboxyl substituent. The orientation of the substituents in the final product requires that the methyl group be retained until the final step.

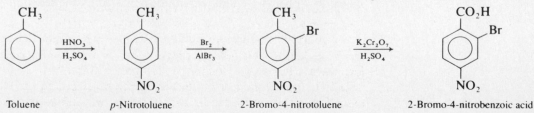

Toluene $\xrightarrow[H_2SO_4]{HNO_3}$ *p*-Nitrotoluene $\xrightarrow[AlBr_3]{Br_2}$ 2-Bromo-4-nitrotoluene $\xrightarrow[H_2SO_4]{K_2Cr_2O_7}$ 2-Bromo-4-nitrobenzoic acid

Nitration must precede bromination, as in the previous part, in order to prevent formation of an undesired mixture of isomers.

(*m*) Friedel–Crafts alkylation of benzene with benzyl chloride (or benzyl bromide) is a satisfactory route to diphenylmethane.

$$C_6H_6 + C_6H_5CH_2Cl \xrightarrow{AlCl_3} C_6H_5CH_2C_6H_5$$

Benzene, Benzyl chloride, Diphenylmethane

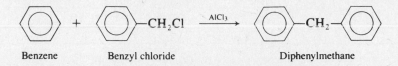

Benzyl chloride is prepared by free-radical chlorination of toluene.

$$C_6H_5CH_3 \xrightarrow[\text{light or heat}]{Cl_2} C_6H_5CH_2Cl$$

Toluene, Benzyl chloride

Alternatively, benzene could have been subjected to Friedel–Crafts acylation with benzoyl chloride to give benzophenone. Clemmensen or Wolff–Kishner reduction of benzophenone would then furnish diphenylmethane.

(*n*) 1-Phenyloctane cannot be prepared efficiently by direct alkylation of benzene because of the probability that rearrangement will occur. Indeed, a mixture of 1-phenyloctane and 2-phenyloctane is formed under the usual Friedel–Crafts conditions, along with 3-phenyloctane.

$$\underset{\text{Benzene}}{C_6H_6} + \underset{\text{1-Bromooctane}}{CH_3(CH_2)_6CH_2Br} \xrightarrow{AlBr_3}$$

$$\underset{\text{1-Phenyloctane (40\%)}}{C_6H_5CH_2(CH_2)_6CH_3} + \underset{\text{2-Phenyloctane (30\%)}}{C_6H_5CH(CH_3)(CH_2)_5CH_3} + \underset{\text{3-Phenyloctane (30\%)}}{C_6H_5CH(CH_2CH_3)(CH_2)_4CH_3}$$

A method which permits the synthesis of 1-phenyloctane free of isomeric compounds is acylation followed by reduction.

$$\underset{\text{Benzene}}{C_6H_6} + \underset{\text{Octanoyl chloride}}{CH_3(CH_2)_6\overset{O}{\overset{\|}{C}}Cl} \xrightarrow{AlCl_3} \underset{\text{Octanoylbenzene}}{C_6H_5\overset{O}{\overset{\|}{C}}(CH_2)_6CH_3}$$

$$\xrightarrow{Zn(Hg),\ HCl} \underset{\text{1-Phenyloctane}}{C_6H_5CH_2(CH_2)_6CH_3}$$

Alternatively, Wolff–Kishner conditions (hydrazine, potassium hydroxide, diethylene glycol) could be used in the reduction step.

(*o*) Direct alkenylation of benzene under Friedel–Crafts reaction conditions does not take place so 1-phenyl-1-octene cannot be prepared by the reaction

$$\underset{\text{Benzene}}{C_6H_6} + \underset{\text{1-Chloro-1-octene}}{ClCH{=}CH(CH_2)_5CH_3} \xrightarrow{AlCl_3} \underset{\text{1-Phenyl-1-octene}}{C_6H_5CH{=}CH(CH_2)_5CH_3}$$

No! Reaction only effective with *alkyl* halides, not 1-haloalkenes.

However, having already prepared 1-phenyloctane in (*n*), we can functionalize the benzylic position by bromination and then carry out a dehydrohalogenation to obtain the target compound.

$$C_6H_5CH_2(CH_2)_6CH_3 \xrightarrow[\text{or NBS}]{Br_2} C_6H_5\underset{\displaystyle Br}{\underset{|}{C}}H(CH_2)_6CH_3 \xrightarrow[CH_3OH]{KOCH_3} C_6H_5CH{=}CH(CH_2)_5CH_3$$

1-Phenyloctane — 1-Phenyl-1-bromooctane — 1-Phenyl-1-octene

(*p*) 1-Phenyl-1-octyne cannot be prepared in one step from benzene; 1-haloalkynes are not suitable reactants for a Friedel–Crafts process. In Chapter 10, however, we learned that alkynes may be prepared from the corresponding alkene:

$$RC{\equiv}CR \quad \text{obtained from} \quad \underset{\displaystyle Br}{\underset{|}{RCH}}{-}\underset{\displaystyle Br}{\underset{|}{CHR}} \quad \text{obtained from} \quad RCH{=}CHR$$

Using the alkene prepared in (*o*)

$$C_6H_5CH{=}CH(CH_2)_5CH_3 \xrightarrow{Br_2} C_6H_5\underset{\displaystyle Br}{\underset{|}{C}}H\underset{\displaystyle Br}{\underset{|}{C}}H(CH_2)_5CH_3 \xrightarrow[NH_3]{NaNH_2} C_6H_5C{\equiv}C(CH_2)_5CH_3$$

1-Phenyl-1-octene — 1,2-Dibromo-1-phenyloctane — 1-Phenyl-1-octyne

13.21 (*a*) Methoxy is an ortho, para–directing substituent. All that is required to prepare *p*-methoxybenzenesulfonic acid is to sulfonate anisole.

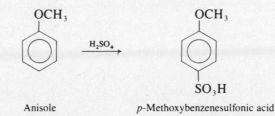

Anisole (OCH_3 on benzene) $\xrightarrow{H_2SO_4}$ *p*-Methoxybenzenesulfonic acid (OCH_3, SO_3H para)

(*b*) In reactions involving disubstitution of anisole, the better strategy is to introduce the para substituent first. The methoxy group is ortho, para–directing but para substitution is preferred.

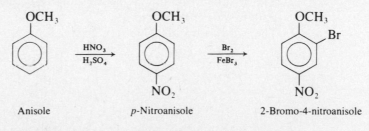

Anisole (OCH_3) $\xrightarrow[H_2SO_4]{HNO_3}$ *p*-Nitroanisole (OCH_3, NO_2) $\xrightarrow[FeBr_3]{Br_2}$ 2-Bromo-4-nitroanisole (OCH_3, Br, NO_2)

(*c*) Reversing the order of the steps yields 4-bromo-2-nitroanisole.

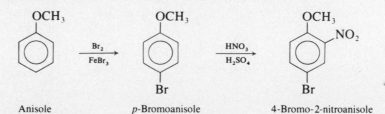

Anisole (OCH_3) $\xrightarrow[FeBr_3]{Br_2}$ *p*-Bromoanisole (OCH_3, Br) $\xrightarrow[H_2SO_4]{HNO_3}$ 4-Bromo-2-nitroanisole (OCH_3, NO_2, Br)

(*d*) Direct introduction of a vinyl substituent onto an aromatic ring is not a feasible reaction. *p*-Methoxystyrene must be prepared in an indirect way by adding an ethyl side chain, then taking advantage of the reactivity of the benzylic position by bromination (e.g., with *N*-bromosuccinimide) and dehydrohalogenation.

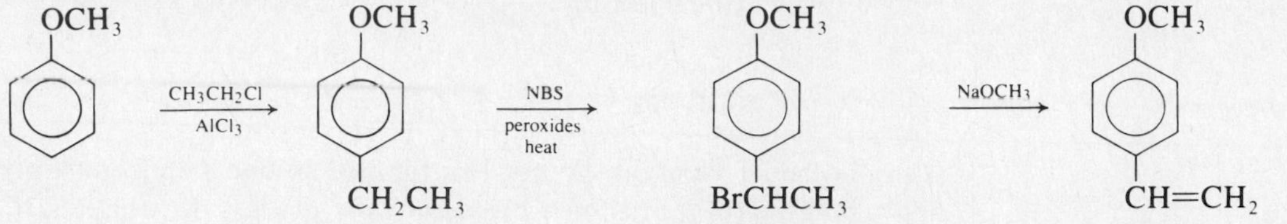

OCH_3 OCH_3 OCH_3 OCH_3

$\xrightarrow[AlCl_3]{CH_3CH_2Cl}$ $\xrightarrow[\text{peroxides, heat}]{NBS}$ $\xrightarrow{NaOCH_3}$

CH_2CH_3 $BrCHCH_3$ $CH{=}CH_2$

Anisole *p*-Ethylanisole 1-(*p*-Methoxyphenyl)ethyl bromide *p*-Methoxystyrene

13.22 (*a*) Methyl is an ortho, para–directing substituent and toluene yields mainly *o*-nitrotoluene and *p*-nitrotoluene on mononitration. Some *m*-nitrotoluene is also formed.

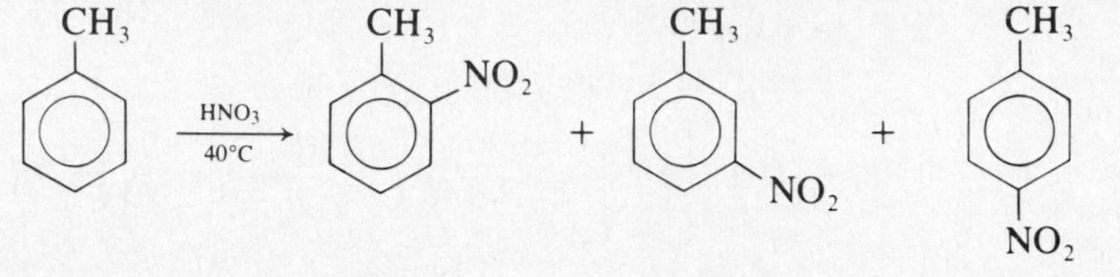

CH_3 $\xrightarrow[40°C]{HNO_3}$ CH_3, NO_2 + CH_3, NO_2 + CH_3, NO_2

Toluene *o*-Nitrotoluene *m*-Nitrotoluene *p*-Nitrotoluene

(*b*) There are six isomeric dinitrotoluenes. They are:

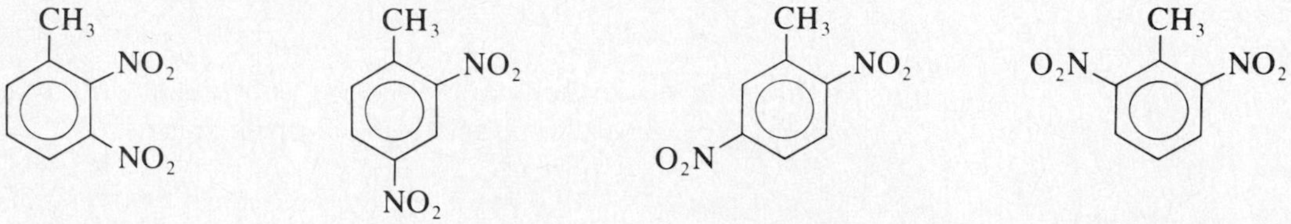

2,3-Dinitrotoluene 2,4-Dinitrotoluene 2,5-Dinitrotoluene 2,6-Dinitrotoluene

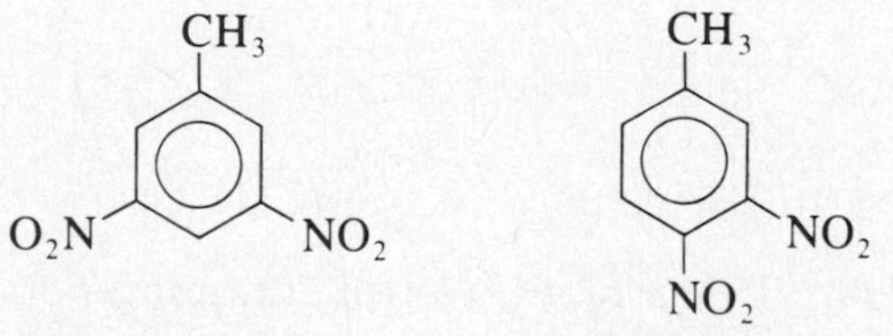

3,5-Dinitrotoluene 3,4-Dinitrotoluene

The least likely product is 3,5-dinitrotoluene since neither of its nitro groups is ortho or para to the methyl group.

(*c*) There are six trinitrotoluene isomers.

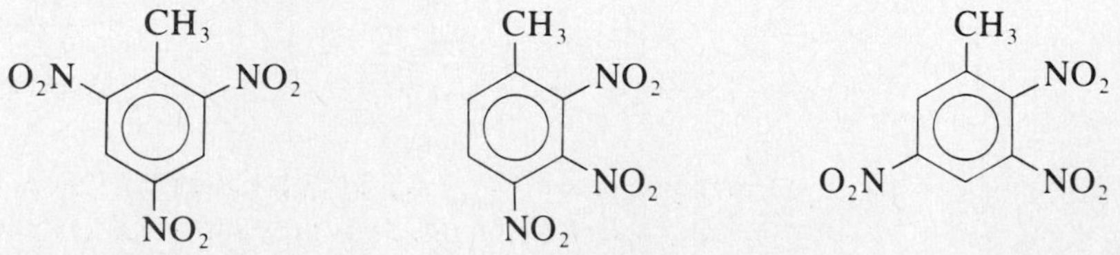

2,4,6-Trinitrotoluene 2,3,4-Trinitrotoluene 2,3,5-Trinitrotoluene

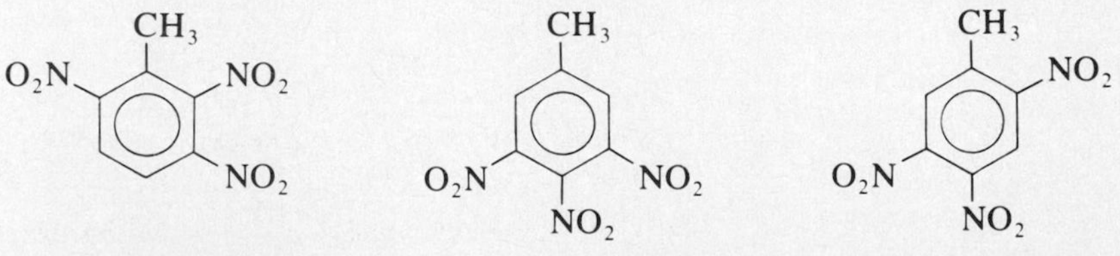

2,3,6-Trinitrotoluene 3,4,5-Trinitrotoluene 2,4,5-Trinitrotoluene

The most likely major product is 2,4,6-trinitrotoluene since all the positions activated by the methyl group are substituted. This is, in fact, the compound commonly known as TNT.

13.23 (*a*) From *o*-xylene:

$$o\text{-Xylene} + \underset{\text{Acetyl chloride}}{CH_3\overset{O}{\overset{\|}{C}}Cl} \xrightarrow{AlCl_3} \text{3,4-Dimethylacetophenone (94\%)}$$

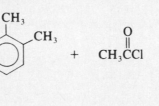

o-Xylene — Acetyl chloride — 3,4-Dimethylacetophenone (94%)

(*b*) From *m*-xylene:

$$m\text{-Xylene} + CH_3\overset{O}{\overset{\|}{C}}Cl \xrightarrow{AlCl_3} \text{2,4-Dimethylacetophenone (86\%)}$$

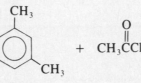

m-Xylene — 2,4-Dimethylacetophenone (86%)

(*c*) From *p*-xylene:

$$p\text{-Xylene} + CH_3\overset{O}{\overset{\|}{C}}Cl \xrightarrow{AlCl_3} \text{2,5-Dimethylacetophenone (99\%)}$$

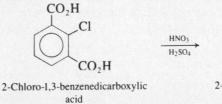

p-Xylene — 2,5-Dimethylacetophenone (99%)

13.24 (*a*) Nitration of the ring takes place para to the ortho, para–directing chlorine substituent; this position is also meta to the meta-directing carboxyl groups.

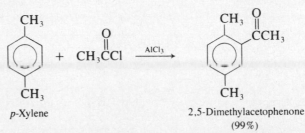

$$\text{2-Chloro-1,3-benzenedicarboxylic acid} \xrightarrow[H_2SO_4]{HNO_3} \text{2-Chloro-5-nitro-1,3-benzenedicarboxylic acid (86\%)}$$

2-Chloro-1,3-benzenedicarboxylic acid — 2-Chloro-5-nitro-1,3-benzenedicarboxylic acid (86%)

(*b*) Bromination of the ring occurs at the only available position activated by the amino group, a powerful activating substituent and an ortho, para director. This position is meta to the meta-directing trifluoromethyl group and to the meta-directing nitro group.

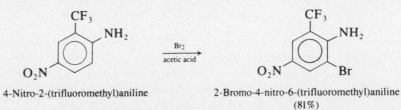

$$\text{4-Nitro-2-(trifluoromethyl)aniline} \xrightarrow[\text{acetic acid}]{Br_2} \text{2-Bromo-4-nitro-6-(trifluoromethyl)aniline (81\%)}$$

4-Nitro-2-(trifluoromethyl)aniline — 2-Bromo-4-nitro-6-(trifluoromethyl)aniline (81%)

(*c*) This may be approached as a problem in which there are two aromatic rings. One of them bears two activating substituents and so is more reactive than the other, which bears only one activating substituent. Of the two activating substituents (—OH and C_6H_5—), the hydroxyl substituent is the more powerful and controls the regioselectivity of substitution.

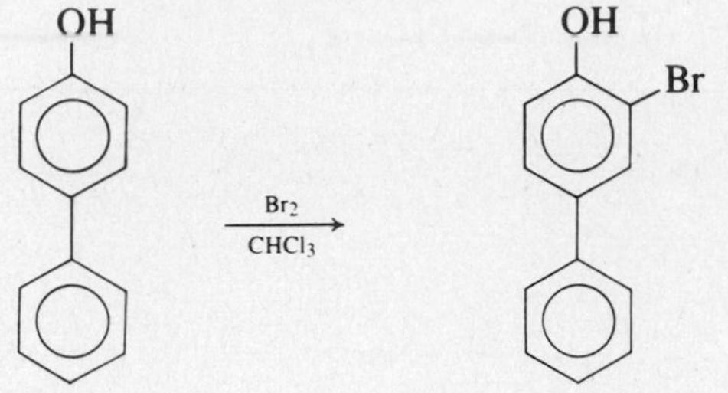

Br_2, $CHCl_3$

p-Phenylphenol → 2-Bromo-4-phenylphenol (90%)

(*d*) Both substituents are activating, nitration occurring readily even in the absence of sulfuric acid; both are ortho, para–directing and comparable in activating power. Therefore, the position at which substitution takes place is as shown below:

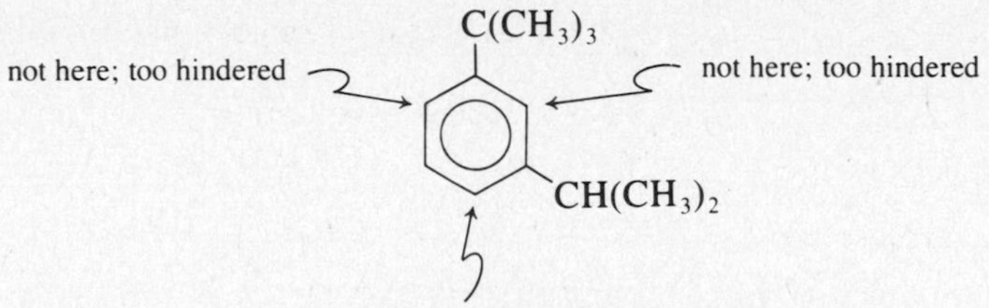

$C(CH_3)_3$

not here; too hindered

not here; too hindered

$CH(CH_3)_2$

ortho to isopropyl
para to *tert*-butyl

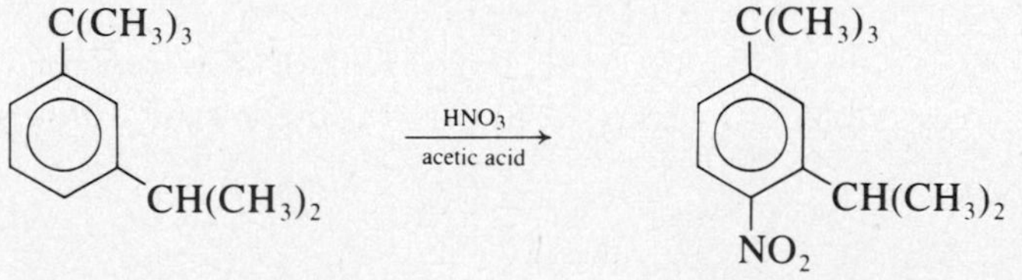

HNO_3, acetic acid

1-Isopropyl-3-*tert*-butylbenzene → 2-Isopropyl-1-nitro-4-*tert*-butylbenzene (78%)

(*e*) Succinic anhydride functions as an acylating agent in Friedel–Crafts reactions. The aromatic substrate is a naphthalene derivative. Naphthalene undergoes electrophilic aromatic substitution faster at C-1 than at any other position. Therefore

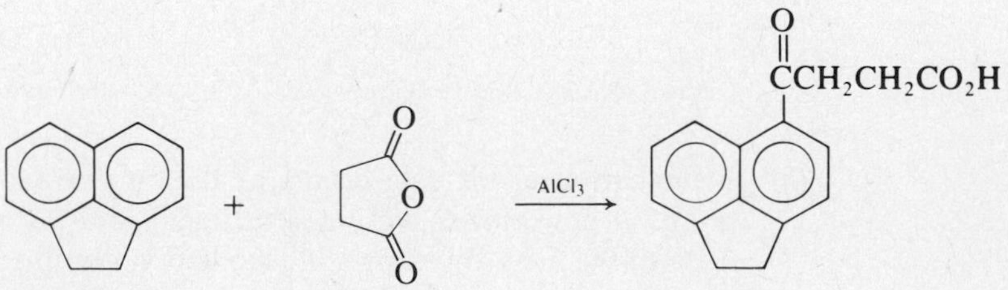

$\overset{O}{\overset{\|}{C}}CH_2CH_2CO_2H$

Acenaphthene + Succinic anhydride $\xrightarrow{AlCl_3}$ (formed in 81% yield)

Notice also that substitution has occurred para to the alkyl substituent (the five-membered ring).

(*f*) The reaction that occurs with arenes and acid anhydrides in the presence of aluminum chloride is Friedel–Crafts acylation. The methoxy group is the more powerful activating substituent, so acylation occurs para to it.

$$o\text{-}FC_6H_4OCH_3 + CH_3\overset{O}{\overset{\|}{C}}O\overset{O}{\overset{\|}{C}}CH_3 \xrightarrow{AlCl_3} CH_3\overset{O}{\overset{\|}{C}}\text{-}C_6H_3(F)\text{-}OCH_3 + CH_3CO_2H$$

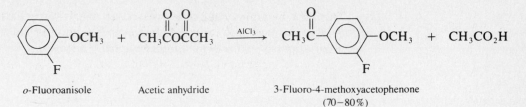

o-Fluoroanisole — Acetic anhydride — 3-Fluoro-4-methoxyacetophenone (70–80%)

(*g*) The isopropyl group is ortho, para–directing and the nitro group is meta-directing. In this case their orientation effects reinforce each other. Electrophilic aromatic substitution takes place ortho to isopropyl and meta to nitro.

$$4\text{-}(CH_3)_2CH\text{-}C_6H_4\text{-}NO_2 \xrightarrow[H_2SO_4]{HNO_3} 2,4\text{-}(NO_2)_2C_6H_3CH(CH_3)_2$$

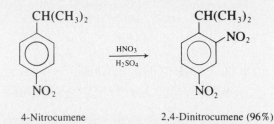

4-Nitrocumene — 2,4-Dinitrocumene (96%)

(*h*) In the presence of an acid catalyst (H_2SO_4) isobutene is converted to *tert*-butyl cation, which then attacks the aromatic ring ortho to the strongly activating methoxy group.

$$(CH_3)_2C{=}CH_2 + H^+ \longrightarrow (CH_3)_3C^+$$

$$4\text{-}CH_3C_6H_4OCH_3 + (CH_3)_3C^+ \longrightarrow 2\text{-}(CH_3)_3C\text{-}4\text{-}CH_3C_6H_3OCH_3$$

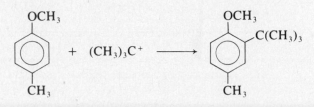

In this particular example, 2-*tert*-butyl-4-methylanisole was isolated in 98 percent yield.

(*i*) There are two things to consider in this problem: (1) In which ring does bromination occur? and (2) What is the orientation of substitution in that ring? All the substituents are activating groups, so substitution will take place in the ring which bears the greater number of substituents. Orientation is governed by the most powerful activating substituent, the hydroxyl group. Both positions ortho to the hydroxyl group are already substituted, so bromination takes place para to it. The product shown was isolated from the bromination reaction in 100 percent yield.

$$C_6H_5CH_2\text{-}C_6H_2(CH_3)(OH)(CH_3) \xrightarrow{Br_2} C_6H_5CH_2\text{-}C_6H(CH_3)(OH)(CH_3)(Br)$$

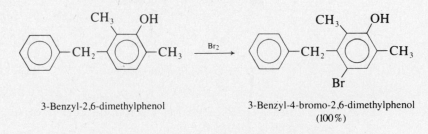

3-Benzyl-2,6-dimethylphenol — 3-Benzyl-4-bromo-2,6-dimethylphenol (100%)

(*j*) Wolff–Kishner reduction converts benzophenone to diphenylmethane.

$$C_6H_5\overset{O}{\overset{\|}{C}}C_6H_5 \xrightarrow[\text{triethylene glycol, heat}]{H_2NNH_2,\ KOH} C_6H_5CH_2C_6H_5$$

Benzophenone — Diphenylmethane (83%)

(*k*) Fluorine is an ortho, para-directing substituent. It undergoes Friedel–Crafts alkylation on being treated with benzyl chloride and aluminum chloride to give a mixture of *o*-fluorodiphenylmethane and *p*-fluorodiphenylmethane.

Fluorobenzene + $C_6H_5CH_2Cl$ $\xrightarrow{AlCl_3}$ *o*-Fluorodiphenylmethane (15%) + *p*-Fluorodiphenylmethane (85%)

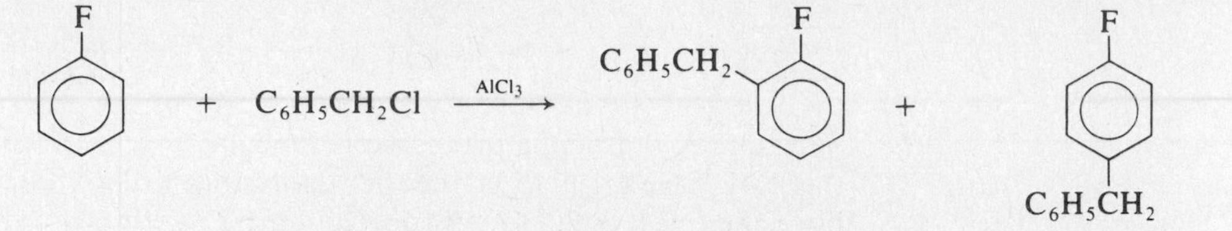

Fluorobenzene | Benzyl chloride | *o*-Fluorodiphenylmethane (15%) | *p*-Fluorodiphenylmethane (85%)

(*l*) The $-NH\overset{O}{\overset{\|}{C}}CH_3$ substituent is a more powerful activator than the ethyl group. It directs Friedel–Crafts acylation primarily to the position para to itself.

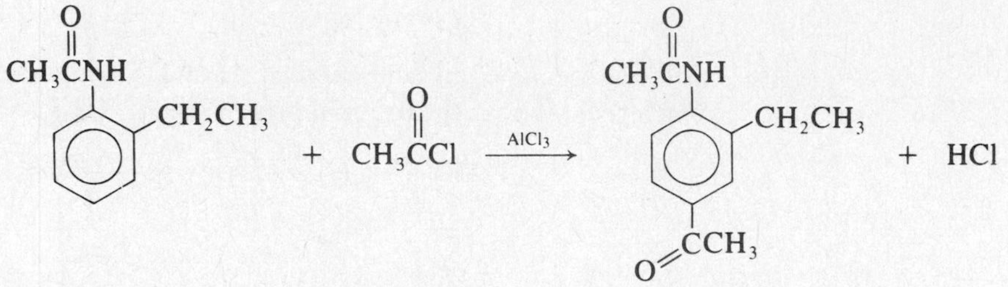

o-Ethylacetanilide + $CH_3\overset{O}{\overset{\|}{C}}Cl$ $\xrightarrow{AlCl_3}$ 4-Acetamido-3-ethylacetophenone (57%) + HCl

o-Ethylacetanilide | Acetyl chloride | 4-Acetamido-3-ethylacetophenone (57%)

(*m*) Protonation of 1-octene yields a secondary carbocation, which attacks benzene.

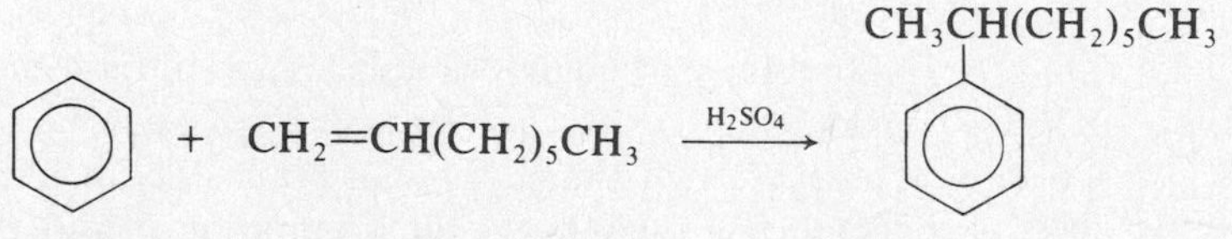

Benzene + $CH_2{=}CH(CH_2)_5CH_3$ $\xrightarrow{H_2SO_4}$ $C_6H_5CH(CH_3)(CH_2)_5CH_3$

Benzene | 1-Octene | 2-Phenyloctane (84%)

(*n*) Clemmensen reduction converts the carbonyl group to a CH_2 unit.

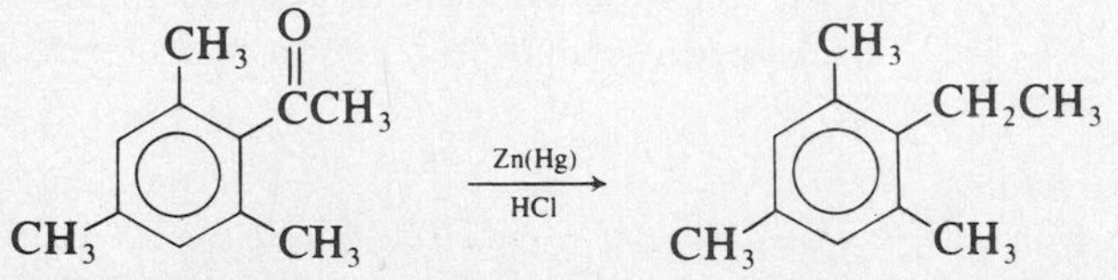

2,4,6-Trimethylacetophenone $\xrightarrow[HCl]{Zn(Hg)}$ 2-Ethyl-1,3,5-trimethylbenzene (74%)

13.25 In a Friedel–Crafts acylation reaction an acyl chloride or acid anhydride reacts with an arene to yield an aryl ketone.

$$ArH + R\overset{O}{\overset{\|}{C}}Cl \xrightarrow{AlCl_3} Ar\overset{O}{\overset{\|}{C}}R$$

The ketone carbonyl is bonded directly to the ring. Therefore, in each of these problems you should identify the bond between the aromatic ring and the carbonyl group and realize that it arises as shown above.

(*a*) The compound is derived from benzene and $C_6H_5CH_2\overset{O}{\overset{\|}{C}}Cl$. The observed yield in this reaction is 82 to 83 percent.

$$C_6H_5\overset{O}{\overset{\|}{C}}{-}CH_2{-}C_6H_5 \quad \text{arises from} \quad C_6H_5{-}H \text{ and } Cl\overset{O}{\overset{\|}{C}}CH_2{-}C_6H_5$$

(*b*) The presence of the $Ar\overset{O}{\overset{\|}{C}}CH_2CH_2CO_2H$ unit suggests an acylation reaction using succinic anhydride.

$$2,5\text{-}(CH_3)_2C_6H_3\overset{O}{\overset{\|}{C}}CH_2CH_2CO_2H \quad \text{arises from} \quad CH_3{-}C_6H_4{-}CH_3 \quad \text{and} \quad \text{succinic anhydride}$$

In practice, this reaction has been carried out in 55 percent yield.

(*c*) Two routes seem possible here but only one actually works. The only effective combination is:

$$O_2N{-}C_6H_4{-}\overset{O}{\overset{\|}{C}}Cl + C_6H_6 \xrightarrow{AlCl_3} O_2N{-}C_6H_4{-}\overset{O}{\overset{\|}{C}}{-}C_6H_5$$

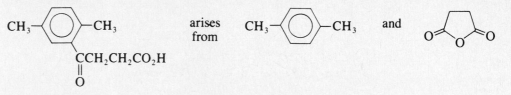

p-Nitrobenzoyl chloride — Benzene — *p*-Nitrobenzophenone (87%)

The alternative combination

$$O_2N{-}C_6H_5 + Cl\overset{O}{\overset{\|}{C}}{-}C_6H_5 \xrightarrow{AlCl_3} \text{no reaction}$$

fails because it requires a Friedel–Crafts reaction on a strongly deactivated aromatic ring (nitrobenzene).

(*d*) Here also two routes seem possible but only one is successful in practice. The valid synthetic route is:

$$3,5\text{-}(CH_3)_2C_6H_3{-}\overset{O}{\overset{\|}{C}}Cl + C_6H_6 \xrightarrow{AlCl_3} 3,5\text{-}(CH_3)_2C_6H_3{-}\overset{O}{\overset{\|}{C}}{-}C_6H_5$$

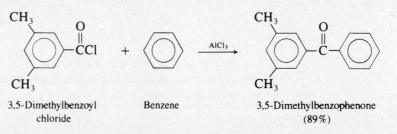

3,5-Dimethylbenzoyl chloride — Benzene — 3,5-Dimethylbenzophenone (89%)

The alternative combination will not give 3,5-dimethylbenzophenone because of the ortho, para–directing properties of the methyl substituents in *m*-xylene. The product will be 2,4-dimethylbenzophenone.

$$m\text{-}(CH_3)_2C_6H_4 + Cl\overset{O}{\overset{\|}{C}}{-}C_6H_5 \xrightarrow{AlCl_3} 2,4\text{-}(CH_3)_2C_6H_3{-}\overset{O}{\overset{\|}{C}}{-}C_6H_5$$

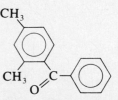

m-Xylene — Benzoyl chloride — 2,4-Dimethylbenzophenone

(*e*) The combination shown below is not effective because it involves a Friedel–Crafts reaction on a deactivated aromatic.

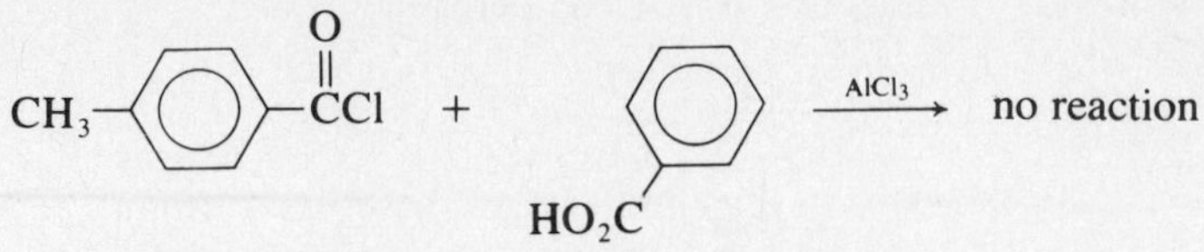

CH_3–C₆H₄–CCl(=O) + C₆H₅–HO_2C $\xrightarrow{AlCl_3}$ no reaction

p-Methylbenzoyl chloride   Benzoic acid

Therefore, the combination below utilizing toluene seems appropriate:

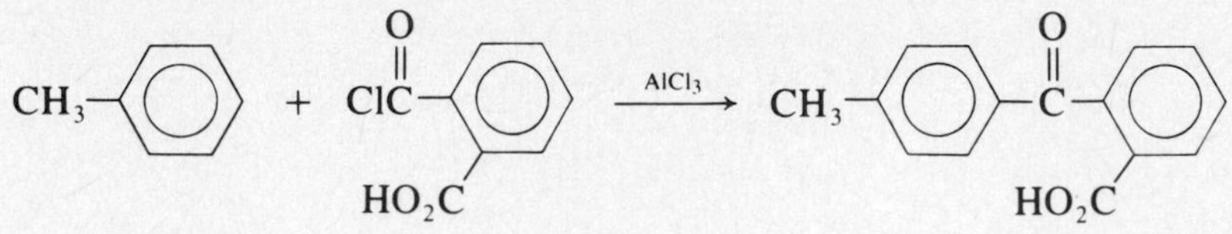

CH_3–C₆H₅ + ClC(=O)–C₆H₄–HO_2C $\xrightarrow{AlCl_3}$ CH_3–C₆H₄–C(=O)–C₆H₄–HO_2C

The actual sequence used phthalic anhydride in a reaction analogous to that seen earlier which employs succinic anhydride.

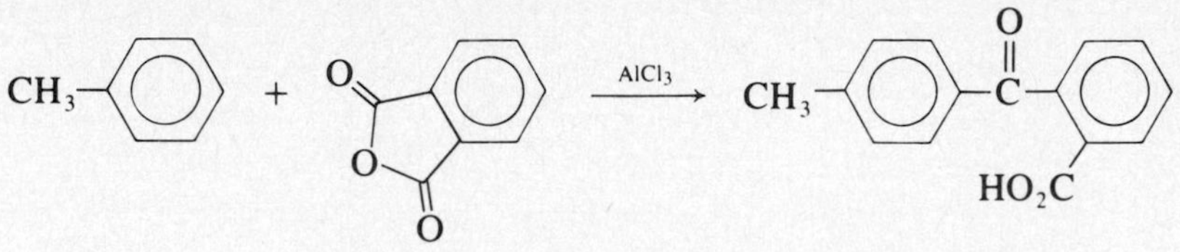

CH_3–C₆H₅ + Phthalic anhydride $\xrightarrow{AlCl_3}$ CH_3–C₆H₄–C(=O)–C₆H₄–HO_2C

Toluene   Phthalic anhydride   *o*-(4-Methylbenzoyl)benzoic acid (96%)

13.26 (*a*) The problem to be confronted here is that two meta-directing groups are para to each other in the product. However, by recognizing that the carboxylic acid function can be prepared by oxidation of the isopropyl group

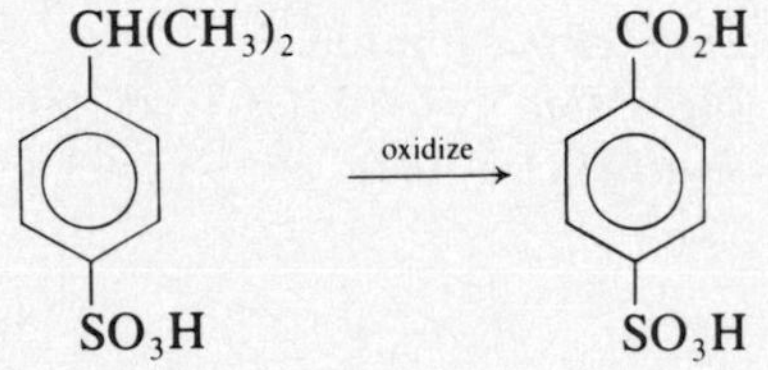

$CH(CH_3)_2$–C₆H₄–SO_3H $\xrightarrow{\text{oxidize}}$ CO_2H–C₆H₄–SO_3H

we have a reasonable last step in the synthesis. Happily, the key intermediate has its sulfonic acid group para to the ortho, para-directing isopropyl group, which suggests the following approach:

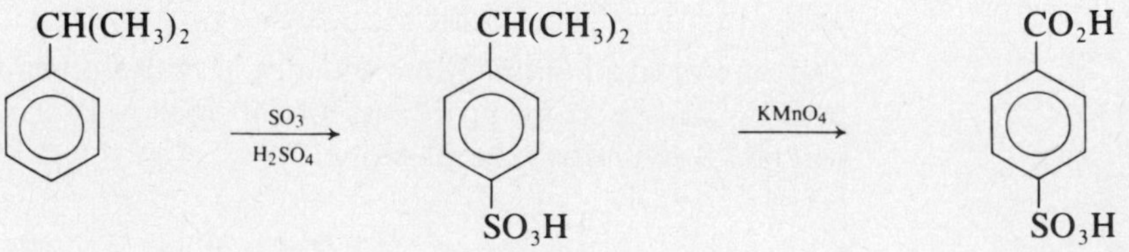

$CH(CH_3)_2$–C₆H₅ $\xrightarrow[H_2SO_4]{SO_3}$ $CH(CH_3)_2$–C₆H₄–SO_3H $\xrightarrow{KMnO_4}$ CO_2H–C₆H₄–SO_3H

Isopropylbenzene   *p*-Isopropylbenzenesulfonic acid   *p*-Carboxylbenzenesulfonic acid

(*b*) In this problem two methyl groups must be oxidized to carboxylic acid functions and a *tert*-butyl group must be introduced, most likely by a Friedel–Crafts reaction. Since Friedel–Crafts alkylations cannot be performed on deactivated aromatic rings, oxidation must *follow*, not precede, alkylation. Therefore, the reaction sequence shown seems appropriate.

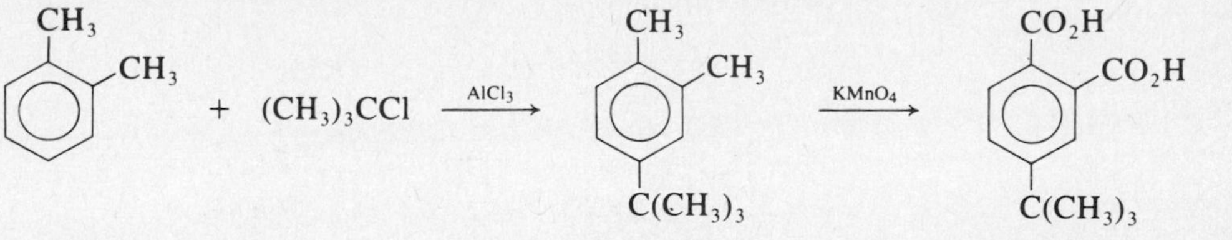

CH_3 CH_3 + $(CH_3)_3CCl$ $\xrightarrow{AlCl_3}$ CH_3 CH_3 $C(CH_3)_3$ $\xrightarrow{KMnO_4}$ CO_2H CO_2H $C(CH_3)_3$

o-Xylene — 4-*tert*-Butyl-1,2-dimethylbenzene — 4-*tert*-Butylbenzene-1,2-dicarboxylic acid

In practice, zinc chloride was used as the Lewis acid to catalyze the Friedel–Crafts reaction (64 percent yield). Oxidation of the methyl groups occurs preferentially because the *tert*-butyl group has no benzylic hydrogens. The yield, though, is rather poor—only 35 percent.

(*c*) This problem requires performing two Friedel–Crafts reactions—one an alkylation, the other an acylation. Remember, an acyl substituent is strongly deactivating and prevents a subsequent Friedel–Crafts reaction from being carried out; therefore the alkylation should be performed first. Notice also that considering the relative reactivities of the two sites leads to the same conclusion. All the aromatic ring positions of indan are activated because they are either ortho or para to an alkyl group.

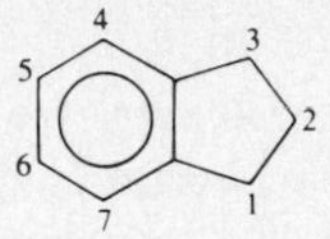

4 3 5 2 6 1 7

The C-5 and C-6 positions are less sterically hindered than are C-4 and C-7; thus the first substituent to be introduced must go on C-5 or C-6. Therefore *tert*-butyl must be introduced first.

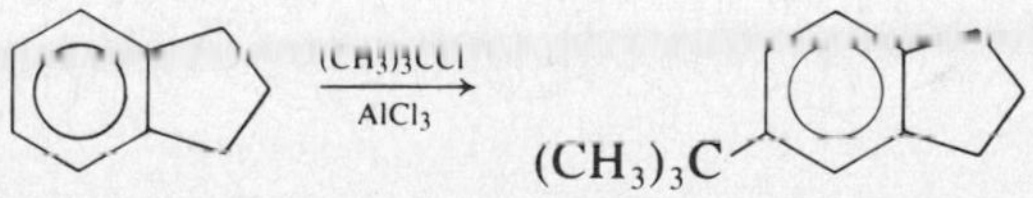

$\xrightarrow[AlCl_3]{(CH_3)_3CCl}$ $(CH_3)_3C$

Acetylation is now more sterically hindered at C-5 and C-7 by the *tert*-butyl group than it is at C-4 by the CH_2 of the five-membered ring.

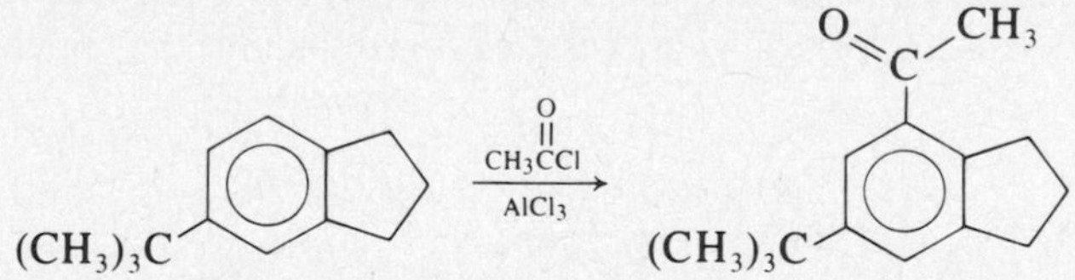

$(CH_3)_3C$ $\xrightarrow[AlCl_3]{CH_3\overset{O}{\overset{\|}{C}}Cl}$ O=C—CH_3 $(CH_3)_3C$

The target molecule, 4-acetyl-6-*tert*-butylindan, was synthesized in just this way, the only difference being that the *tert*-butyl group was introduced by using *tert*-butyl alcohol and sulfuric acid as the source of *tert*-butyl cation.

(*d*) *m*-Dimethoxybenzene is a strongly activated aromatic and so will undergo electrophilic aromatic substitution readily. The ring position between the two methoxyl groups is sterically hindered and less reactive than the other positions activated by virtue of being ortho or para to a methoxyl group.

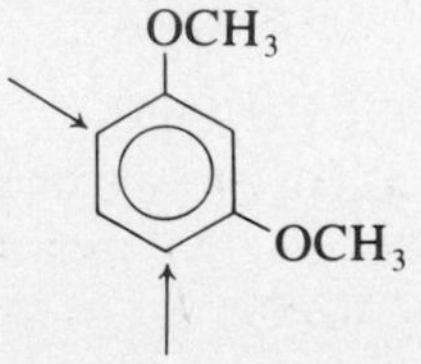

OCH_3 OCH_3

Arrows indicate equivalent ring positions strongly activated by methoxyl groups.

Because Friedel–Crafts reactions may not be performed on deactivated aromatic rings, the *tert*-butyl group must be introduced before the nitro group. Therefore, the correct sequence is:

OCH$_3$, OCH$_3$ (benzene) → OCH$_3$, OCH$_3$, C(CH$_3$)$_3$ (benzene) → O_2N, OCH$_3$, OCH$_3$, C(CH$_3$)$_3$ (benzene)

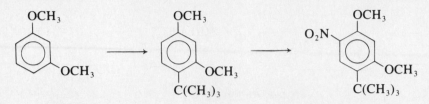

This is essentially the procedure actually followed. Alkylation was effected, however, not with *tert*-butyl chloride and aluminum chloride but with 2-methylpropene and phosphoric acid.

$$(CH_3)_2C{=}CH_2 \xrightarrow{H_3PO_4}$$

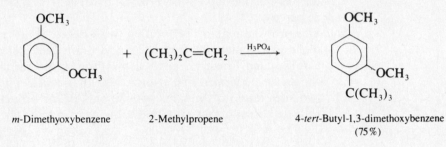

m-Dimethyoxybenzene + 2-Methylpropene → 4-*tert*-Butyl-1,3-dimethoxybenzene (75%)

Nitration was carried out in the usual way. The orientation of nitration is controlled by the more powerfully activating methoxyl groups rather than by the weakly activating *tert*-butyl.

$$\xrightarrow[H_2SO_4]{HNO_3}$$

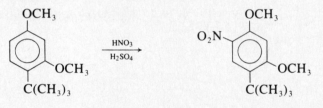

4-*tert*-Butyl-6-nitro-1,3-dimethoxybenzene

13.27 The first step is a Friedel–Crafts acylation reaction. The use of a cyclic anhydride introduces both the acyl and carboxyl groups into the molecule.

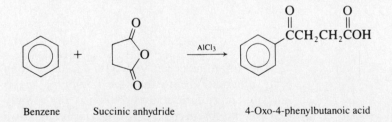

Benzene + Succinic anhydride $\xrightarrow{AlCl_3}$ $C_6H_5\overset{O}{\overset{\|}{C}}CH_2CH_2\overset{O}{\overset{\|}{C}}OH$

4-Oxo-4-phenylbutanoic acid

The second step is a reduction of the ketone carbonyl to a methylene group. A Clemmensen reduction is normally used for this step.

$$C_6H_5\overset{O}{\overset{\|}{C}}CH_2CH_2\overset{O}{\overset{\|}{C}}OH \xrightarrow[HCl]{Zn(Hg)} C_6H_5CH_2CH_2CH_2\overset{O}{\overset{\|}{C}}OH$$

4-Oxo-4-phenylbutanoic acid → 4-Phenylbutanoic acid

The cyclization phase of the process is an intramolecular Friedel–Crafts acylation reaction. It requires conversion of the carboxylic acid to the acyl chloride— thionyl chloride is a suitable reagent— then treatment with aluminum chloride.

$C_6H_5CH_2CH_2CH_2COOH$ $\xrightarrow{SOCl_2}$ $C_6H_5CH_2CH_2CH_2COCl$ $\xrightarrow{AlCl_3}$ 1-tetralone

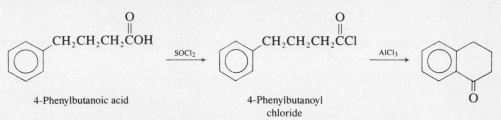

4-Phenylbutanoic acid — 4-Phenylbutanoyl chloride

13.28 Intramolecular Friedel–Crafts acylation reactions which produce five-membered or six-membered rings occur readily. Cyclization must take place at the position ortho to the reacting side chain.

$C_6H_5CH_2CH_2COCl$ $\xrightarrow{AlCl_3}$ 1-indanone and $C_6H_5CH_2CH_2CH_2COCl$ $\xrightarrow{AlCl_3}$ 1-tetralone

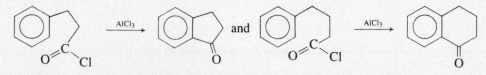

(*a*) A five-membered cyclic ketone is formed here:

$(CH_3)_3C-C_6H_4-C(CH_3)_2CH_2COCl$ $\xrightarrow{AlCl_3}$ 3,3-dimethyl-6-tert-butyl-1-indanone (46%)

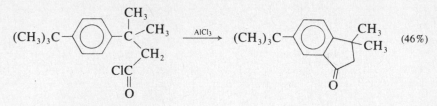

(*b*) This intramolecular Friedel–Crafts acylation takes place to form a six-membered cyclic ketone in excellent yield.

2-phenylcyclohexyl-CH_2CCl (O) $\xrightarrow{AlCl_3}$ tricyclic ketone (93%)

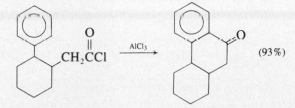

(*c*) In this case two aromatic rings are available for attack by the acylium ion. The more reactive ring is the one which bears the two activating methoxyl groups, and cyclization occurs on it.

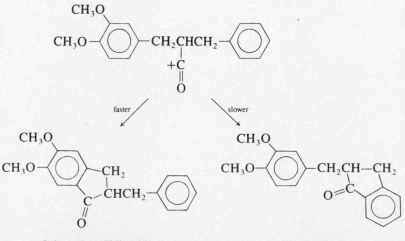

$(CH_3O)_2C_6H_3-CH_2CHCH_2-C_6H_5$ with $+C{=}O$ on CH

faster → $(CH_3O)_2$-indanone with $CHCH_2-C_6H_5$: Only product (78% yield)

slower → $(CH_3O)_2C_6H_3-CH_2CH-CH_2$ cyclized onto phenyl ring via $O{=}C$: (not observed)

13.29 (*a*) To determine the total rate of chlorination of biphenyl relative to that of benzene, we add up the *partial rate factors* for all the positions in each substrate and compare them.

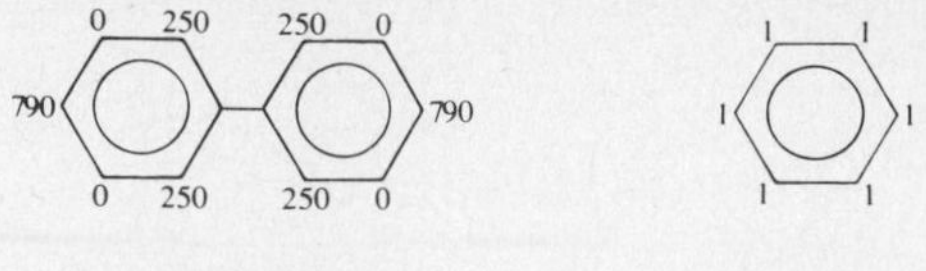

Biphenyl (sum = 2580) Benzene (sum = 6)

$$\text{Relative rate of chlorination: } \frac{\text{Biphenyl}}{\text{Benzene}} = \frac{2580}{6} = \frac{430}{1}$$

(*b*) The relative rate of attack at the para compared with the ortho positions is given by the ratio of their partial rate factors.

$$\frac{\text{Para}}{\text{Ortho}} = \frac{1580}{1000}$$

Therefore, 15.8 g of *p*-chlorobiphenyl is formed for every 10 g of *o*-chlorobiphenyl.

13.30 The isomerization is triggered by protonation of the aromatic ring, an electrophilic attack by HCl catalyzed by $AlCl_3$.

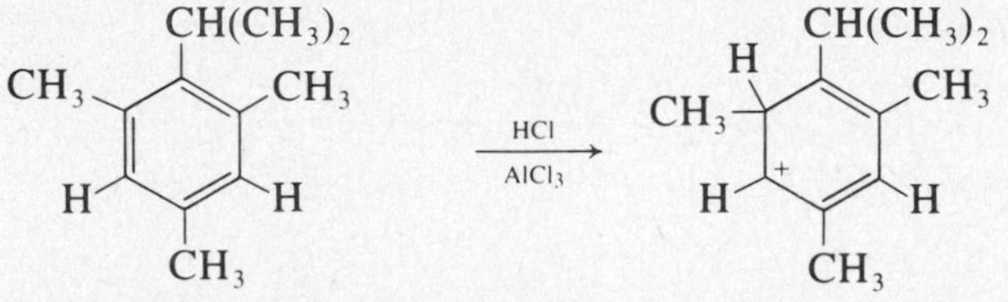

2-Isopropyl-1,3,5-trimethylbenzene

The carbocation then rearranges by a methyl shift, and the rearranged cyclohexadienyl cation loses a proton to form the isomeric product.

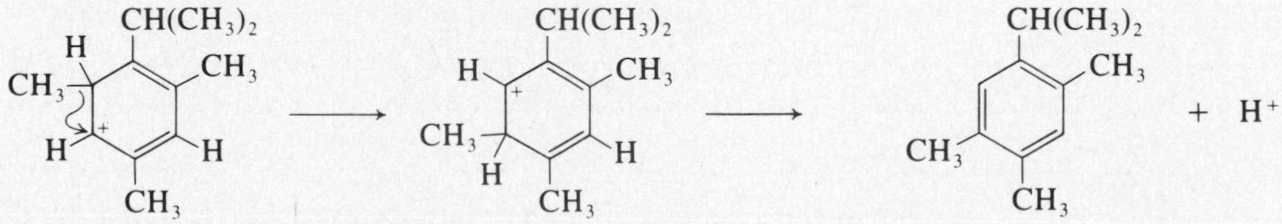

1-Isopropyl-2,4,5-trimethylbenzene

The driving force for rearrangement is relief of the adverse steric interaction between the isopropyl group and one of its adjacent methyl groups. Isomerization is catalytic with respect to H^+, as protonation generates the necessary carbocation intermediate and rearomatization occurs by loss of a proton.

13.31 The relation of Compound A to the starting material is

$$C_6H_5(CH_2)_5\overset{O}{\overset{\|}{C}}Cl \longrightarrow A + HCl$$
$$(C_{12}H_{15}ClO) \qquad (C_{12}H_{14}O)$$

The starting acyl chloride has lost the elements of HCl in the formation of A. Because A forms benzene-1,2-dicarboxylic acid on oxidation, it must have two carbon substituents ortho to each other.

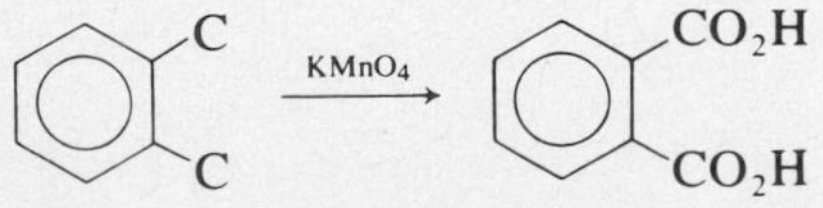

These facts suggest the following process

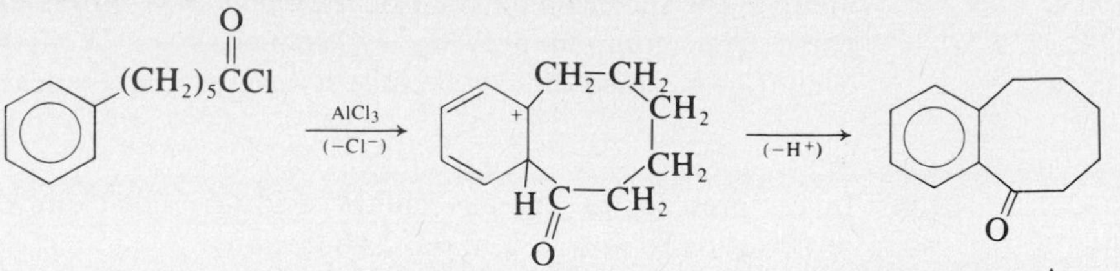

$$C_6H_5(CH_2)_5CCl(=O) \xrightarrow[(-Cl^-)]{AlCl_3} \text{cyclohexadienyl cation} \xrightarrow{(-H^+)} \text{A}$$

A

The reaction leading to A is an intramolecular Friedel–Crafts acylation. Since cyclization to form an eight-membered ring is difficult, it must be carried out in dilute solution to minimize competition with intermolecular acylation.

13.32 While hexamethylbenzene has no positions available at which ordinary electrophilic aromatic *substitution* might occur, electrophilic *attack* on the ring can still take place to form a cyclohexadienyl cation.

$$\text{hexamethylbenzene} \xrightarrow[AlCl_3]{CH_3Cl} \text{heptamethylcyclohexadienyl cation}$$

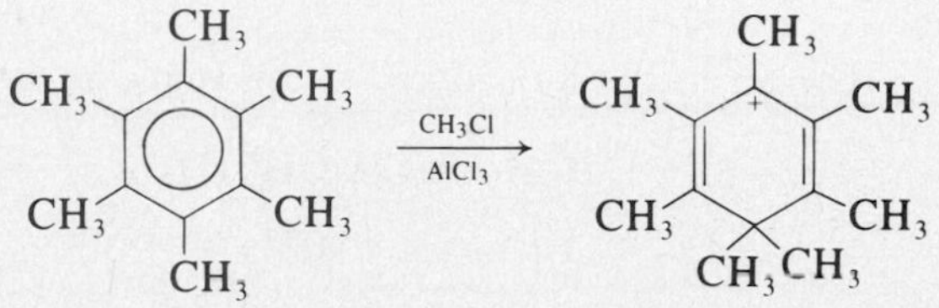

Compound B is the tetrachloroaluminate ($AlCl_4^-$) salt of the carbocation shown. It undergoes deprotonation on being treated with aqueous sodium bicarbonate.

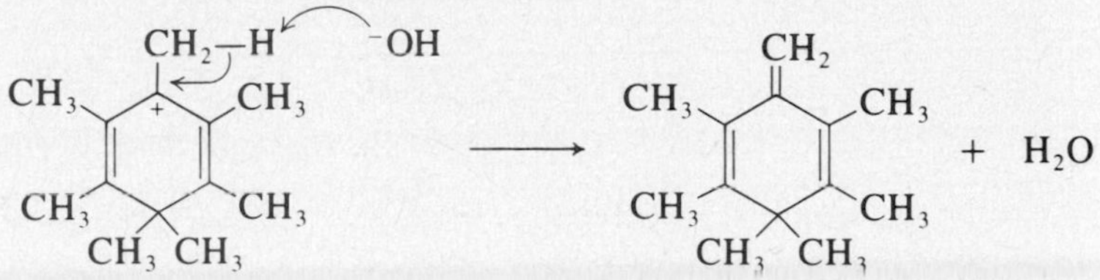

$$\text{cation} + {}^-OH \longrightarrow \text{C} + H_2O$$

C

13.33 By examining the structure of the target molecule, compound F, we see that the bond indicated below joins two fragments which are related to the given starting materials D and E.

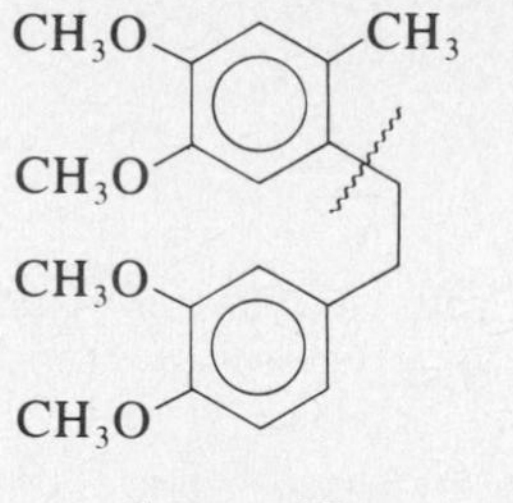

Compound F

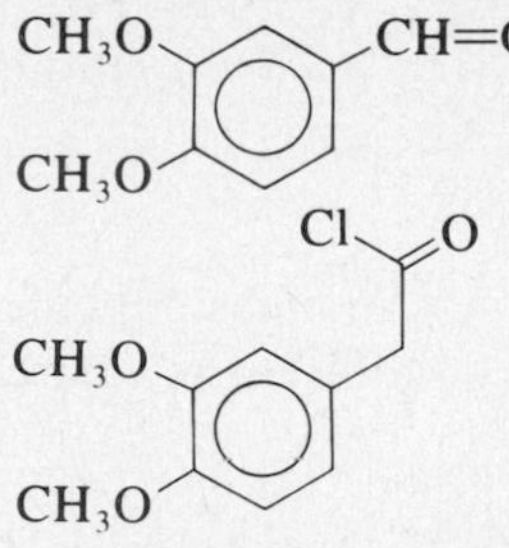

D

E

The bond connecting the two fragments can be made by a Friedel–Crafts acylation-reduction sequence using the acid chloride E.

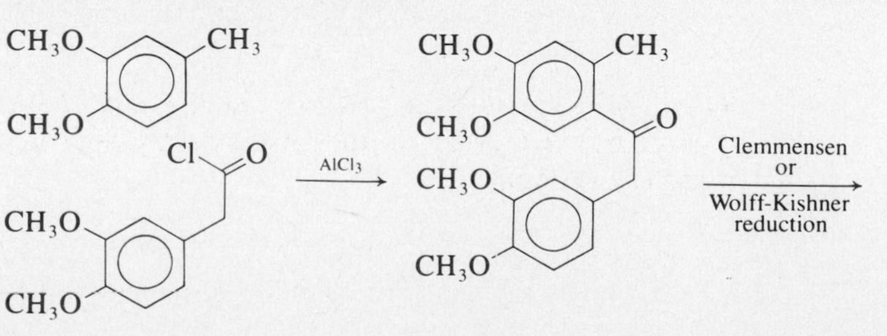

$$\xrightarrow{AlCl_3} \xrightarrow[\text{Wolff-Kishner reduction}]{\text{Clemmensen or}} \text{F}$$

The orientation is right; attack is para to one of the methoxy groups and ortho to the methyl. The substrate for the Friedel–Crafts acylation reaction, 3,4-dimethoxytoluene, is prepared from compound D by a Clemmensen or Wolff–Kishner reduction. Compound D cannot be acylated directly because it bears a strongly deactivating $-\overset{\displaystyle O}{\overset{\|}{C}}H$ substituent.

13.34 In the presence of aqueous sulfuric acid, the side-chain double bond of styrene undergoes protonation to form a benzylic carbocation:

$$\underset{\text{Styrene}}{C_6H_5CH{=}CH_2} + H^+ \longrightarrow \underset{\text{1-Phenylethyl cation}}{C_6H_5\overset{+}{C}HCH_3}$$

This carbocation then reacts with a molecule of styrene in the manner we have seen earlier (Chapter 7) for alkene dimerization.

$$C_6H_5\overset{+}{C}HCH_3 + C_6H_5CH{=}CH_2 \longrightarrow C_6H_5\overset{+}{C}HCH_2\underset{\displaystyle CH_3}{\underset{|}{C}}HC_6H_5$$

The carbocation produced in this step can lose a proton to form 1,3-diphenyl-1-butene

$$C_6H_5\overset{+}{C}HCH_2\underset{\displaystyle CH_3}{\underset{|}{C}}HC_6H_5 \longrightarrow \underset{\text{1,3-Diphenyl-1-butene}}{C_6H_5CH{=}CH\underset{\displaystyle CH_3}{\underset{|}{C}}HC_6H_5} + H^+$$

or it can undergo a cyclization reaction in what amounts to an intramolecular Friedel–Crafts alkylation

CH_3, CH, CH_2, ^+CH, C_6H_5 → CH_3, C_6H_5 + H^+

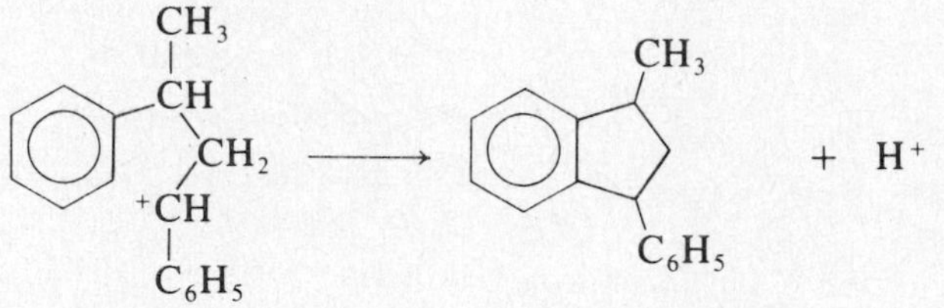

1-Methyl-3-phenylindan

13.35 The alcohol is tertiary and benzylic. In the presence of sulfuric acid a carbocation is formed.

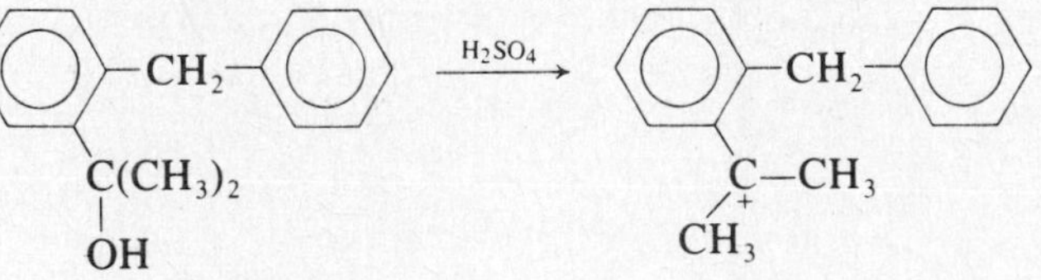

CH_2, $C(CH_3)_2$, OH $\xrightarrow{H_2SO_4}$ CH_2, $\overset{+}{C}-CH_3$, CH_3

An intramolecular Friedel–Crafts alkylation reaction follows, in which the carbocation attacks the adjacent aromatic ring.

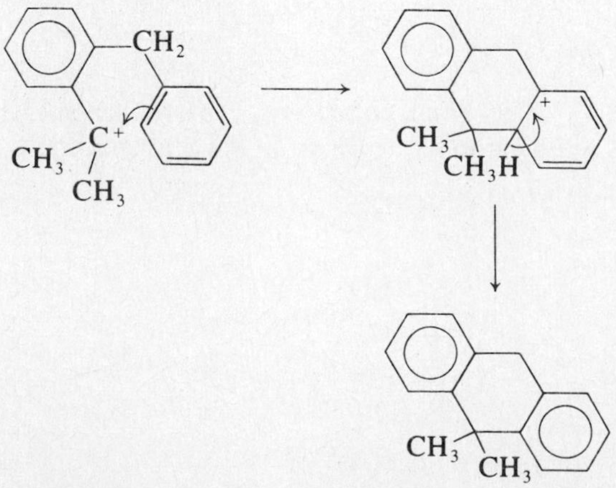

CH_2, C^+, CH_3, CH_3 → CH_3, CH_3H, + → CH_3, CH_3

SELF-TEST

PART A

A-1. Write the three most stable resonance contributors for the cyclohexadienyl cation found in the ortho bromination of toluene.

A-2. Give the major product(s) for each of the following reactions. Indicate if the reaction proceeds faster or slower than the corresponding reaction of benzene.

(*a*) $C_6H_5NO_2 \xrightarrow[H_2SO_4]{HNO_3}$?

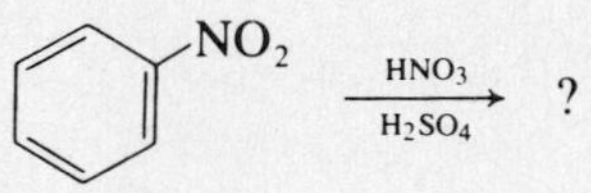

(*b*) $C_6H_5CH_2CH_3 \xrightarrow[FeBr_3]{Br_2}$?

(*c*) $C_6H_5Cl \xrightarrow[H_2SO_4]{SO_3}$?

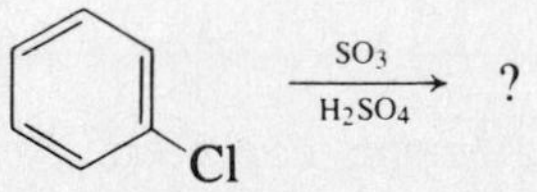

A-3. Write the formula of the electrophilic species present in each reaction of the preceding problem.

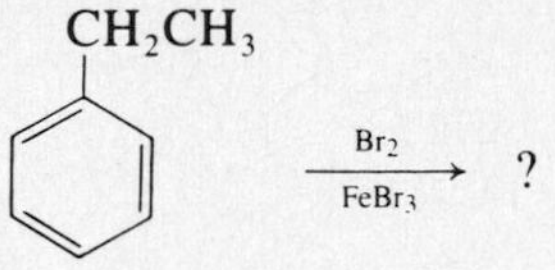

A-4. Provide the reactant, reagent, or product omitted from each of the following:

(*a*) $p\text{-}CH_3C_6H_4NO_2 + Br_2 \xrightarrow{Fe}$?

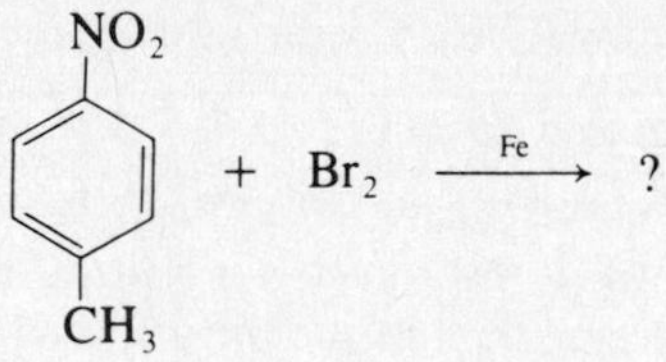

(*b*) ? $\xrightarrow[HCl]{Zn(Hg)}$ $p\text{-}(CH_3)_3C\text{—}C_6H_4\text{—}CH_2CH(CH_3)_2$

(*c*) $C_6H_5CH_2CH_2CH_2CHClCH_3 \xrightarrow{AlCl_3}$?

(*d*) $C_6H_5OCH_3 \xrightarrow{?}$ $o\text{-}CH_3OC_6H_4\overset{O}{\overset{\|}{C}}C_6H_5$ + $p\text{-}CH_3OC_6H_4\text{—}C(\text{=}O)\text{—}C_6H_5$

(*e*) $m\text{-}CH_3OC_6H_4CN \xrightarrow[FeCl_3]{Cl_2}$?

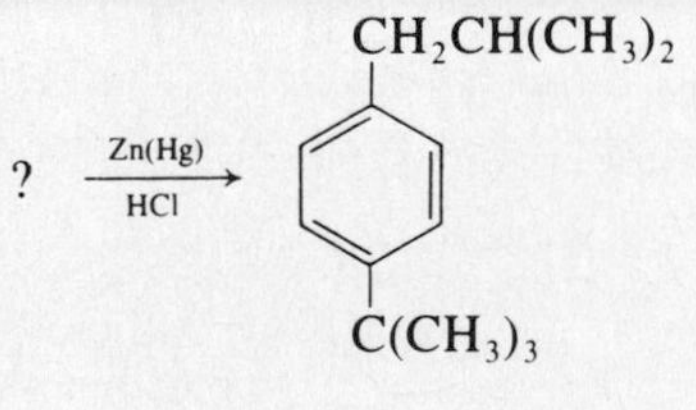

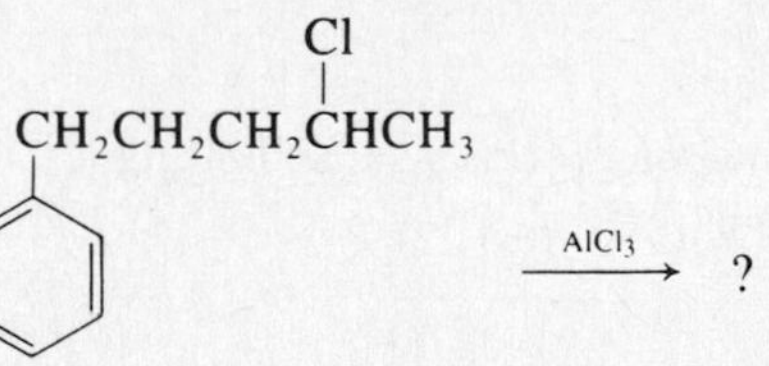

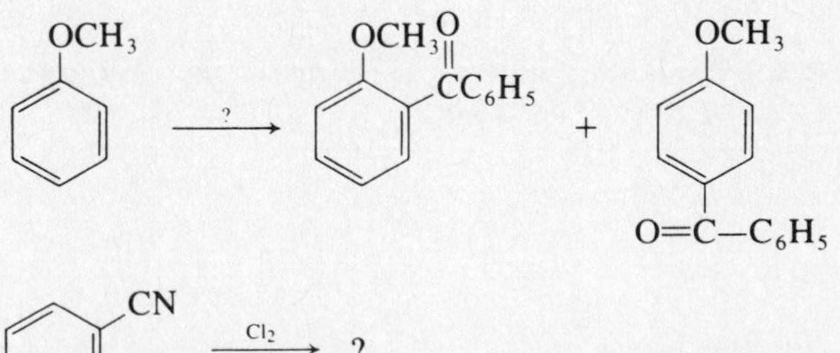

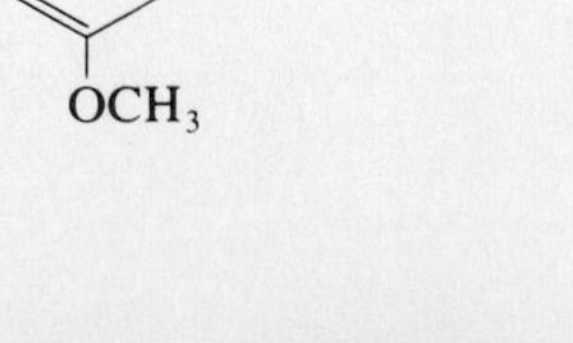

A-5. Provide the necessary reagents for each of the following transformations. More than one step may be necessary.

(*a*) C_6H_5—$CH(CH_3)_2$ $\xrightarrow{?}$ HO_3S—C_6H_4—CO_2H

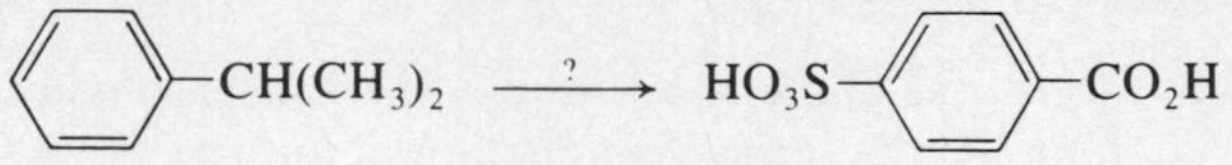

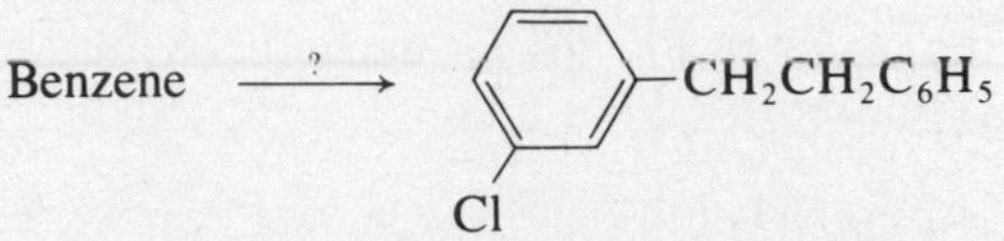

(*b*) Benzene $\xrightarrow{?}$ (3-chlorophenyl)—$CH_2CH_2C_6H_5$ (Cl)

(*c*) C_6H_5—CH(=O) $\xrightarrow{?}$ CH_3C(=O)—C_6H_4—CH_3

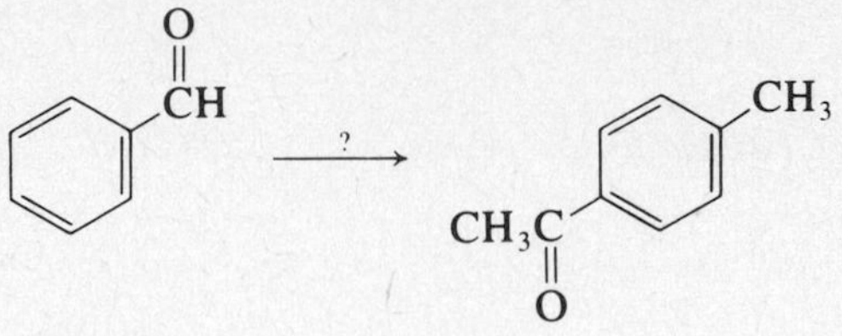

PART B

B-1. Consider the following statements concerning the effect of a trifluoromethyl group, $-CF_3$, on an electrophilic aromatic substitution.

1. The CF_3 group will activate the ring.
2. The CF_3 group will deactivate the ring.
3. The CF_3 group will be a meta director.
4. The CF_3 group will be an ortho, para director.

Which of the above are correct?

(*a*) 1, 3 (*b*) 1, 4 (*c*) 2, 3 (*d*) 2, 4

B-2. Which of the following resonance structures is not a contributor to the cyclohexadienyl cation intermediate in the nitration of benzene?

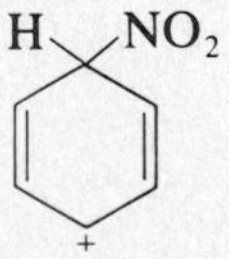

(*a*) H, NO_2, + (*c*) H, NO_2, +

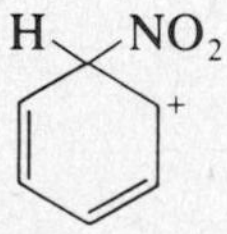

(*b*) NO_2, H, H, + (*d*) None of these (all are contributors)

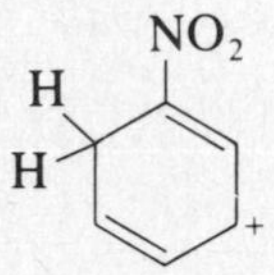

B-3. All the following groups are activating ortho, para directors when attached to a benzene ring except

(*a*) $-OCH_3$ (*b*) $-NHCCH_3$ (C=O)

(*c*) $-Cl$ (*d*) $-N(CH_3)_2$

B-4. Rank the following in terms of increasing reactivity toward nitration with HNO_3, H_2SO_4 (least → most)

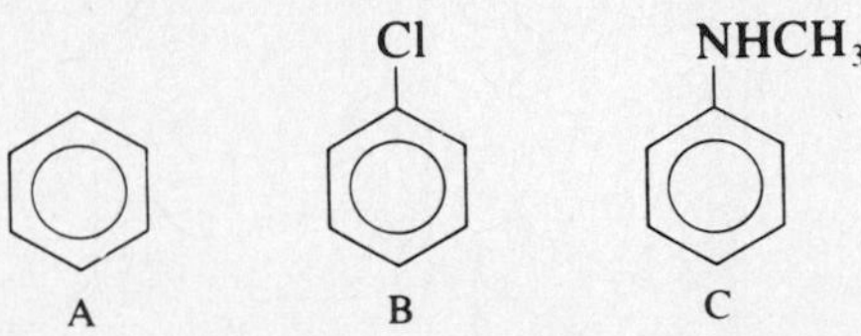

(*a*) A < B < C (*b*) B < A < C

(*c*) C < A < B (*d*) C < B < A

B-5. For the reaction

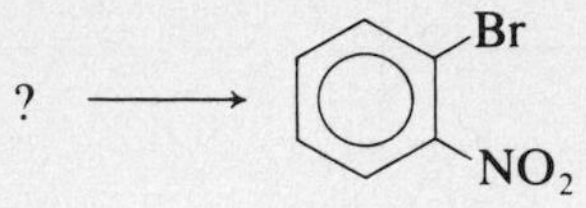

? ⟶ (o-bromonitrobenzene: Br, NO_2)

the best reactants are:

(*a*) $C_6H_5Br + HNO_3, H_2SO_4$ (*b*) $C_6H_5NO_2 + Br_2, FeBr_3$
(*c*) $C_6H_5Br + H_2SO_4$, heat (*d*) $C_6H_5NO_2 + HBr$

B-6. For the reaction

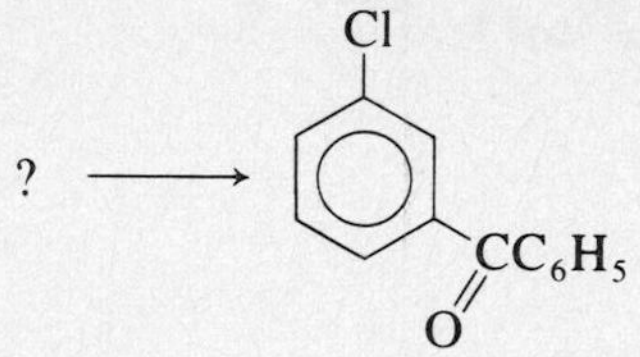

? ⟶ (m-chlorobenzophenone: Cl, $\overset{\displaystyle O}{\|}{C}C_6H_5$)

the best reactants are

(*a*) $C_6H_5Cl + C_6H_5\overset{O}{\overset{\|}{C}}Cl, AlCl_3$ (*b*) $C_6H_5\overset{O}{\overset{\|}{C}}C_6H_5 + Cl_2, FeCl_3$
(*c*) $C_6H_5CH_2C_6H_5$ followed by oxidation with $KMnO_4$
(*d*) None of these yields the desired product

B-7. The reaction

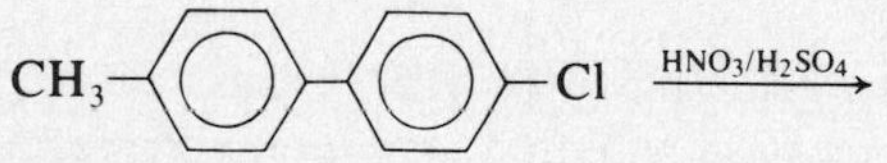

$CH_3-C_6H_4-C_6H_4-Cl \xrightarrow{HNO_3/H_2SO_4}$

gives as the major product:

(*a*) $CH_3-C_6H_4-C_6H_3(NO_2)-Cl$ (NO_2 ortho to Cl) (*c*) $CH_3-C_6H_3(NO_2)-C_6H_4-Cl$ (NO_2 ortho to CH_3)

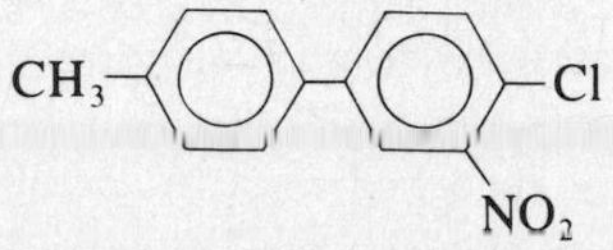

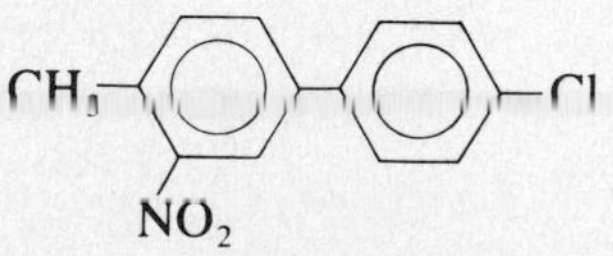

(*b*) $CH_3-C_6H_4-C_6H_3(NO_2)-Cl$ (NO_2 ortho to the biphenyl bond) (*d*) $CH_3-C_6H_3(NO_2)-C_6H_4-Cl$ (NO_2 ortho to the biphenyl bond)

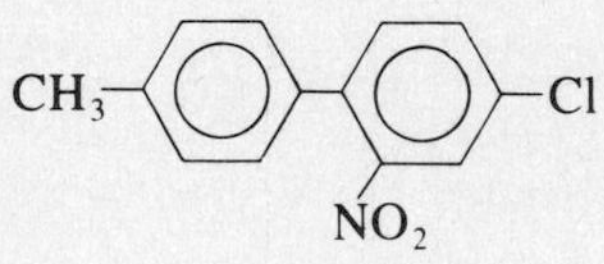

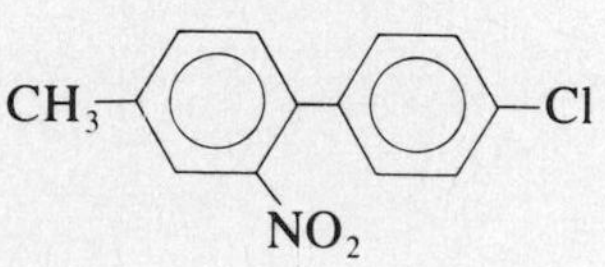

CHAPTER

14

SPECTROSCOPY

IMPORTANT TERMS AND CONCEPTS

Electromagnetic Radiation (Secs. 14.1, 14.2) All the important spectroscopic techniques are based on absorption of energy by a molecule. This energy is derived from electromagnetic radiation in nuclear magnetic resonance (nmr) spectroscopy, infrared (ir) spectroscopy, and ultraviolet-visible (uv-vis) spectroscopy. Important relationships for electromagnetic radiation are that *frequency is inversely proportional to wavelength* and that *energy is proportional to frequency.*

Electromagnetic radiation is absorbed by a molecule when the energy of the photons is equal to the difference in energy between two *quantized* energy states. The energy states for each type of spectroscopy are:

nmr:	nuclear spin states
ir:	vibrational energy states
uv-vis:	electronic energy states

Proton Nuclear Magnetic Resonance (^{1}H Nmr) Spectroscopy (Secs. 14.3 to 14.6) In the presence of an external magnetic field, a proton may absorb energy in the radiofrequency range to "flip" from one nuclear spin state to the other. When the proton is bound in a molecule, different chemical environments will cause a difference in the precise frequency at which absorption occurs. This difference is known as the *chemical shift.*

Chemical shifts are measured relative to tetramethylsilane (TMS), as a standard and are expressed in parts per million (ppm).

$$\text{Chemical shift } (\delta) = \frac{\text{position of signal of interest (Hz)} - \text{position of TMS signal (Hz)}}{\text{spectrometer frequency (MHz)}}$$

Interpreting Proton Nmr Spectra (Secs. 14.7 to 14.11) Three important aspects of an nmr spectrum must be considered when analyzing an unknown. They are:

1. *The number of signals:* A separate nmr signal is observed for each of the protons in a substance that have nonequivalent chemical environments. All protons which are equivalent to each other have the same chemical shift.

2. *The intensity of the signals as measured by the area under each peak:* The area under each signal, determined electronically by a process known as *integration*, is a measure of the *relative* number of protons giving rise to the signal. For example, ethyl bromide and diethyl ether both give two nmr signals having area ratios of 2 (methylene hydrogens) to 3 (methyl hydrogens).

$$\underset{\text{Ethyl bromide}}{CH_3CH_2Br} \qquad \underset{\text{Diethyl ether}}{CH_3CH_2OCH_2CH_3}$$

3. *The multiplicity, or splitting, of each signal:* The number of peaks into which a signal is split, the *multiplicity*, is equal to one more than the number of hydrogen atoms vicinal to the one being observed. Common patterns are:

Singlet:	no splitting;	no vicinal H
Doublet:	two peaks;	one vicinal H
Triplet:	three peaks;	two vicinal H
Quartet:	four peaks;	three vicinal H

Only those hydrogens that have different chemical shifts split each other. Splitting is usually too small to be detectable if there are more than three single bonds separating the protons.

The following examples illustrate how the nmr spectrum of a compound is interpreted in terms of the three categories presented above

2-Butanone: $\underset{a}{CH_3}-\overset{O}{\overset{\|}{C}}-\underset{b}{CH_2}-\underset{c}{CH_3}$

signals: 3
peak areas: a, 3H; b, 2H; c, 3H
multiplicity: a, singlet; b, quartet, c, triplet

1,2-Dimethoxyethane: $\underset{a}{CH_3}-O-\underset{b}{CH_2}-\underset{b}{CH_2}-O-\underset{a}{CH_3}$

signals: 2
peak areas: a, 6H; b, 4H
multiplicity: a, singlet; b, singlet

Carbon-13 Nuclear Magnetic Resonance Spectroscopy (Secs. 14.14, 14.15) By using instruments known as *Fourier transform* (*FT*) *nmr spectrometers*, the carbon skeleton of a molecule can be deduced by recording the nmr spectrum of the ^{13}C isotope present at natural abundance. When *broad-band decoupling* is used, all the peaks in a ^{13}C nmr spectrum appear as singlets. Chemical shifts are measured relative to TMS, as with ^1H nmr.

When *off-resonance decoupling* methods are used, the number of hydrogens attached to each carbon in a molecule may be determined. A methine carbon (CH) appears as a doublet, methylene (CH_2) as a triplet, and methyl (CH_3) as a quartet.

One important distinction of FT nmr is that signal intensities are distorted and peak areas are not as easily obtained as in a ^1H nmr spectrum. However, the chemical shift range of ^{13}C nmr is so broad that each nonequivalent carbon in a molecule usually gives rise to a separate signal, and these signals overlap to a much smaller extent than is typical for a ^1H nmr spectrum.

Infrared Spectroscopy (Sec. 14.16) The primary usefulness of ir spectroscopy is in identifying the presence of certain functional groups within molecules. An ir spectrum is usually recorded in *wave numbers*, with units of cm^{-1}. Their region ranges from 4000 cm^{-1} (high-energy end) to 625 cm^{-1} (low-energy end).

Absorptions characteristic of particular functional groups are usually found in the region 4000 to 1600 cm^{-1}. In the region 1300 to 625 cm^{-1}, known as the *fingerprint region*, variations of peak patterns are observed for different compounds even if the same functional groups are present.

Ultraviolet-Visible Spectroscopy (Sec. 14.17) Ultraviolet-visible spectroscopy probes the electron distribution in a molecule and is particularly useful when conjugated π electron systems are present. Absorption positions are generally expressed as wavelengths, measured in nanometers (1 nm $= 10^{-9}$ m). The visible region corresponds to 800 to 400 nm and the ultraviolet region to 400 to 200 nm.

Mass Spectrometry (Sec. 14.18) Mass spectrometry examines what happens to a molecule when it is bombarded with high-energy electrons. Ionization of a molecule by *electron impact* gives rise to a positively charged species, known as the *molecular ion*, having the same mass as the neutral molecule. The molecular ion then undergoes *fragmentation* to give various species of lower mass.

A mass spectrometer allows all the ions formed by a particular molecule to be separated according to the mass to charge ratio, m/z, of each ion. The *mass spectrum* obtained is characteristic of the particular compound.

SOLUTIONS TO TEXT PROBLEMS

14.1 The same equation is employed to calculate the chemical shifts as was used in the sample solution to (*a*).

(*b*) Iodoform (CHI_3), 322 Hz at 60 MHz

$$\delta = \frac{322 \text{ Hz}}{60 \times 10^6 \text{ Hz}} \times 10^6 = 5.37 \text{ ppm}$$

(*c*) Methyl chloride (CH_3Cl), 184 Hz at 60 MHz

$$\delta = \frac{184 \text{ Hz}}{60 \times 10^6 \text{ Hz}} \times 10^6 = 3.07 \text{ ppm}$$

14.2 (*b*) There are four nonequivalent sets of protons bonded to carbon in 1-butanol as well as a fifth distinct type of proton, the one bonded to oxygen. There should be five signals in the ^1H nmr spectrum of 1-butanol.

$CH_3CH_2CH_2CH_2OH$ (↑ ↑ ↑ ↑ ↑) Five different proton environments in 1-butanol; five signals

(*c*) Apply the "proton replacement" test to butane.

$CH_3CH_2CH_2CH_3$ — Butane

$BrCH_2CH_2CH_2CH_3$ — 1-Bromobutane

$CH_3CHCH_2CH_3$ (Br on C-2) — 2-Bromobutane

$CH_3CH_2CHCH_3$ (Br on C-3) — 2-Bromobutane

$CH_3CH_2CH_2CH_2Br$ — 1-Bromobutane

There are *two* different types of protons in butane; it will exhibit *two* signals in its ^1H nmr spectrum.

(*d*) Like butane, 1,4-dibromobutane has two different types of protons. This can be illustrated by using the chlorine atom as a test group.

$$BrCH_2CH_2CH_2CH_2Br \quad BrCH(Cl)CH_2CH_2CH_2Br \quad BrCH_2CH(Cl)CH_2CH_2Br$$

1,4-Dibromobutane 1,4-Dibromo-1-chlorobutane 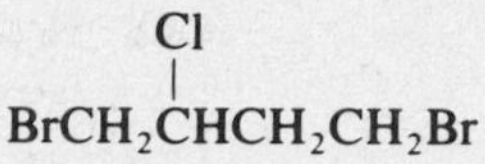 1,4-Dibromo-2-chlorobutane

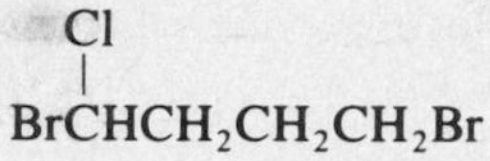

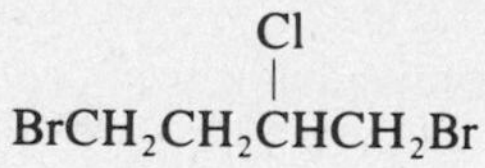

$$BrCH_2CH_2CH(Cl)CH_2Br \quad BrCH_2CH_2CH_2CH(Cl)Br$$

1,4-Dibromo-2-chlorobutane 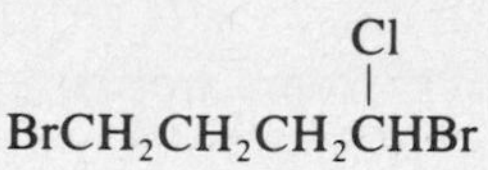 1,4-dibromo-1-chlorobutane

The ^1H nmr spectrum of 1,4-dibromobutane is expected to consist of two signals.

(*e*) All the carbons in 2,2-dibromobutane are different from each other, so protons attached to one carbon are not equivalent to the protons attached to any of the other carbons. This compound should have *three* signals in its ^1H nmr spectrum.

$$\underset{\uparrow}{CH_3}C(Br)_2\underset{\uparrow}{CH_2}\underset{\uparrow}{CH_3}$$

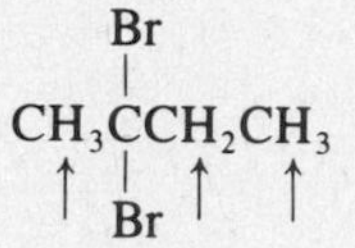

There are three nonequivalent sets of protons in 2,2-dibromobutane

(*f*) All the protons in 2,2,3,3-tetrabromobutane are equivalent. Its ^1H nmr spectrum will consist of one signal.

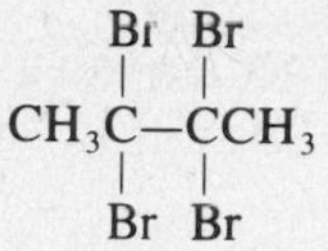

$$CH_3C(Br)_2{-}C(Br)_2CH_3$$

2,2,3,3-Tetrabromobutane

(*g*) There are *four* nonequivalent sets of protons in 1,1,4-tribromobutane. It will exhibit four signals in its ^1H nmr spectrum.

$$Br\underset{\uparrow}{C}(Br)(H)\underset{\uparrow}{CH_2}\underset{\uparrow}{CH_2}\underset{\uparrow}{CH_2}Br$$

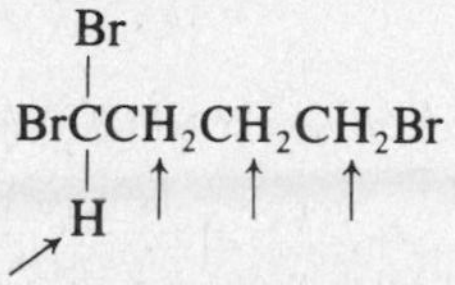

1,1,4-Tribromobutane

(*h*) The seven protons of 1,1,1-tribromobutane belong to three nonequivalent sets, and hence the ^1H nmr spectrum will consist of three signals.

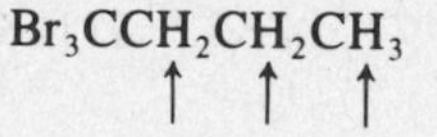

$$Br_3C\underset{\uparrow}{CH_2}\underset{\uparrow}{CH_2}\underset{\uparrow}{CH_3}$$

1,1,1-Tribromobutane

14.3 (*b*) Apply the replacement test to each of the protons of 1,1-dibromoethene.

$$Br_2C{=}CH_2 \qquad Br_2C{=}C(Cl)H \qquad Br_2C{=}C(H)Cl$$

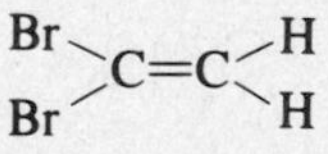

1,1-Dibromoethene 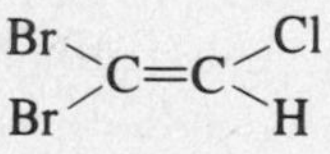 1,1-Dibromo-2-chloroethene 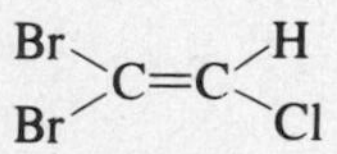 1,1-Dibromo-2-chloroethene

Replacement of one proton by a test group (Cl) gives exactly the same compound as replacement of the other. The two protons of 1,1-dibromoethene are equivalent, and there is only one signal in the ^1H nmr spectrum of this compound.

(*c*) The replacement test reveals that both protons of *cis*-1,2-dibromoethene are equivalent.

$$Br(H)C{=}C(Br)H \qquad Br(Cl)C{=}C(Br)H \qquad Br(H)C{=}C(Br)Cl$$

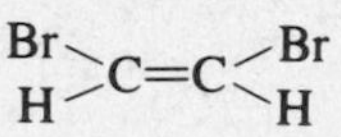

cis-1,2-Dibromoethene 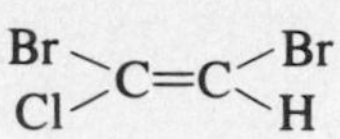 (*Z*)-1,2-Dibromo-2-chloroethene 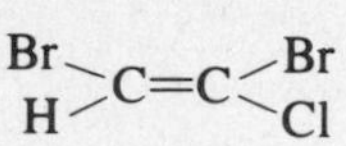 (*Z*)-1,2-Dibromo-2-chloroethene

Because both protons are equivalent, the ^1H nmr spectrum of *cis*-1,2-dibromoethene consists of a single signal.

(*d*) Both protons of *trans*-1,2-dibromoethene are equivalent; each is cis to a bromine substituent.

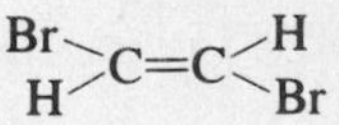

$$\begin{matrix}Br\\H\end{matrix}\!>\!C{=}C\!<\!\begin{matrix}H\\Br\end{matrix}$$ *trans*-1,2-Dibromoethene (single signal in the ^{1}H nmr spectrum)

(*e*) There are *four* nonequivalent sets of protons in allyl bromide.

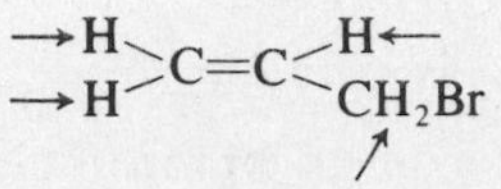

$$\begin{matrix}\rightarrow H\\\rightarrow H\end{matrix}\!>\!C{=}C\!<\!\begin{matrix}H\leftarrow\\CH_2Br\\\nearrow\end{matrix}$$ Allyl bromide (four signals in the ^{1}H nmr spectrum)

(*f*) The protons of a single methyl group are equivalent to each other, but all the methyl groups of 2-methyl-2-butene are nonequivalent. The vinyl proton is unique.

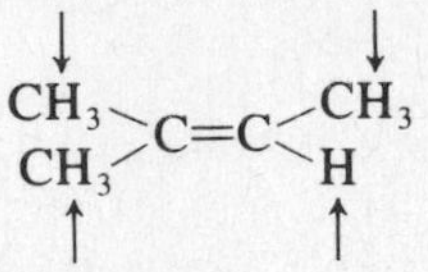

$$\begin{matrix}\downarrow\\CH_3\\CH_3\\\uparrow\end{matrix}\!>\!C{=}C\!<\!\begin{matrix}\downarrow\\CH_3\\H\\\uparrow\end{matrix}$$ 2-Methyl-2-butene (four signals in the ^{1}H nmr spectrum)

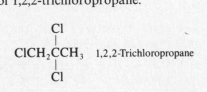

14.4 (*b*) The three methyl protons of 1,1,1-trichloroethane (Cl_3CCH_3) are equivalent. They have the same chemical shift and do not split each other. The ^{1}H nmr spectrum of Cl_3CCH_3 consists of a single sharp peak.

(*c*) Separate signals will be seen for the methylene protons and for the methine proton of 1,1,2-trichloroethane.

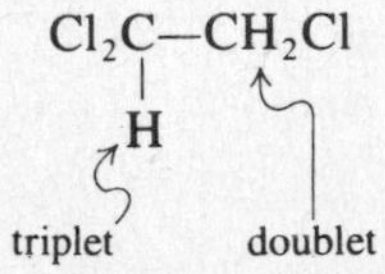

$$Cl_2\underset{\substack{|\\H}}{C}{-}CH_2Cl$$ 1,1,2-Trichloroethane

H: triplet; CH_2: doublet

The methine proton splits the signal for the methylene protons into a doublet. The two methylene protons split the methine proton's signal into a triplet.

(*d*) Examine the structure of 1,2,2-trichloropropane.

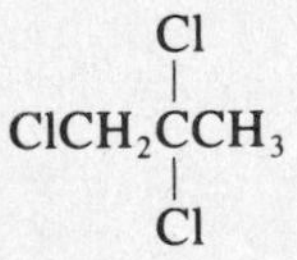

$$ClCH_2\overset{\substack{Cl\\|}}{\underset{\substack{|\\Cl}}{C}}CH_3$$ 1,2,2-Trichloropropane

The ^{1}H nmr spectrum exhibits a signal for the two equivalent methylene protons and one for the three equivalent methyl protons. Both these signals are sharp singlets. The protons of the methyl group and the methylene group are separated by more than three bonds and do not split each other.

(*e*) The methine proton of 1,1,1,2-tetrachloropropane splits the signal of the methyl protons into a doublet; it is split into a quartet by the three methyl protons.

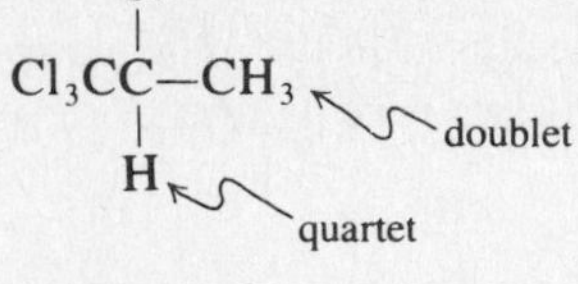

$$Cl_3C\overset{\substack{Cl\\|}}{\underset{\substack{|\\H}}{C}}{-}CH_3$$ 1,1,1,2-Tetrachloropropane

CH_3: doublet; H: quartet

14.5 (*b*) The ethyl group appears as a triplet-quartet pattern and the methyl group as a singlet.

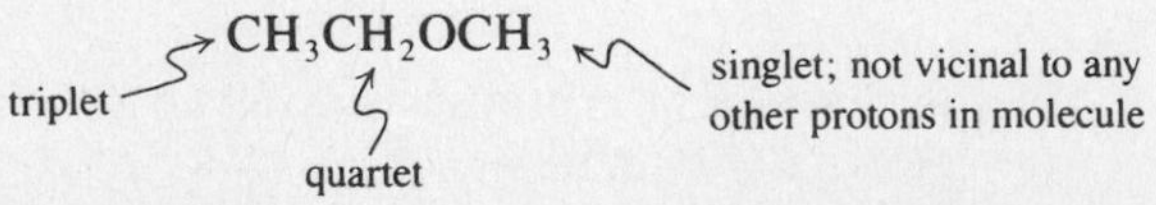

$$CH_3CH_2OCH_3$$

CH_3 (of ethyl): triplet; CH_2: quartet; OCH_3: singlet; not vicinal to any other protons in molecule (*c*) The two ethyl groups of diethyl ether are equivalent to each other. The two methyl groups appear as a single triplet and the two methylene groups as a single quartet.

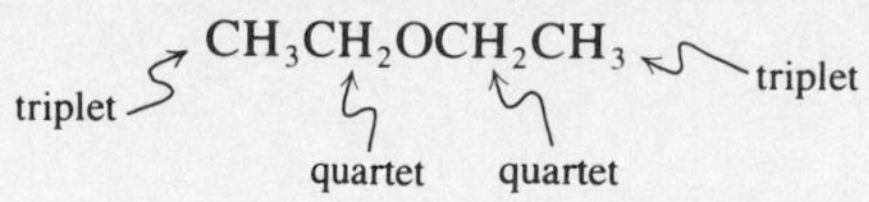

$CH_3CH_2OCH_2CH_3$

triplet, quartet, quartet, triplet

(*d*) The two ethyl groups of *p*-diethylbenzene are equivalent to each other and give rise to a single triplet-quartet pattern.

$CH_3CH_2-C_6H_4-CH_2CH_3$ (four aromatic H)

Three signals: CH_3 triplet; CH_2 quartet; aromatic H singlet

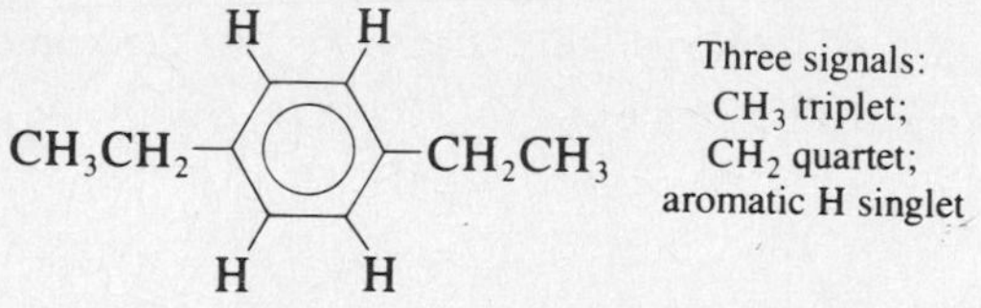

All four protons of the aromatic ring are equivalent, have the same chemical shift, and split neither each other nor any of the signals of the ethyl group.

(*e*) There are four nonequivalent sets of protons in this compound.

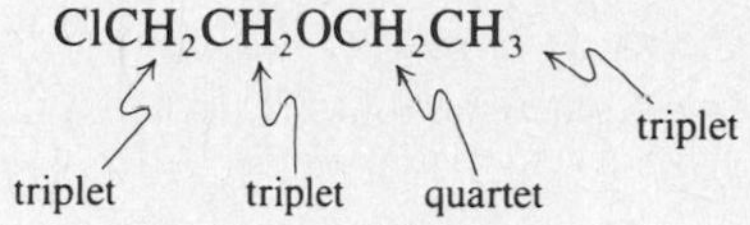

$ClCH_2CH_2OCH_2CH_3$

triplet, triplet, quartet, triplet

Vicinal protons in the $ClCH_2CH_2O$ group split one another, as do those in the CH_3CH_2O group.

(*f*) The two methine protons are nonequivalent and each splits the signal of the other into a doublet. The two ethyl groups are equivalent and give a single triplet-quartet pattern.

$Cl_2CH-CH(OCH_2CH_3)_2$

one triplet for CH_3 groups, plus one quartet for CH_2 groups

doublet, doublet

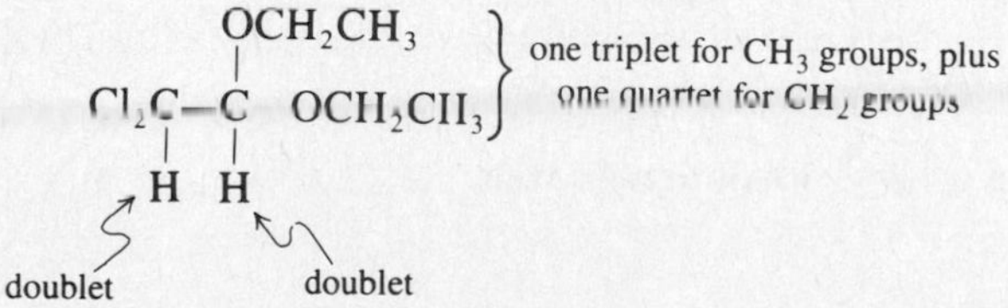

14.6 The ^{1}H nmr spectrum of Figure 14.19 is that of $Cl_2CHCH(OCH_2CH_3)_2$. All three compounds contain two equivalent ethyl groups, but only $Cl_2CHCH(OCH_2CH_3)_2$ has its remaining two protons in different environments. The two methine protons split each other's signal into a doublet.

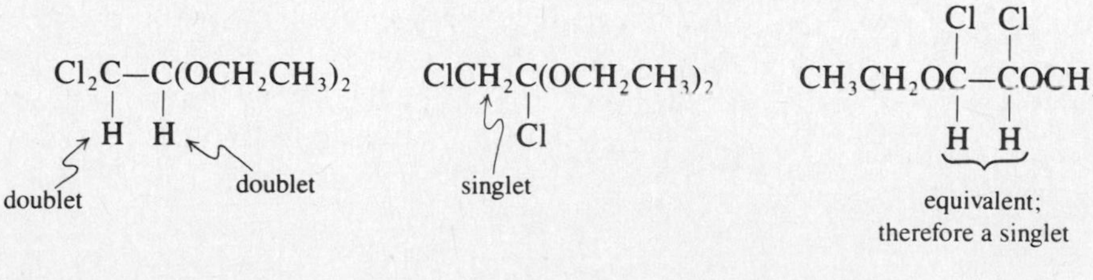

$Cl_2CH-CH(OCH_2CH_3)_2$ — doublet, doublet

$ClCH_2CCl(OCH_2CH_3)_2$ — singlet

$CH_3CH_2OCHCl-CHClOCH_2CH_3$ — H H equivalent; therefore a singlet

14.7 (*b*) The two methyl carbons of the isopropyl group are equivalent.

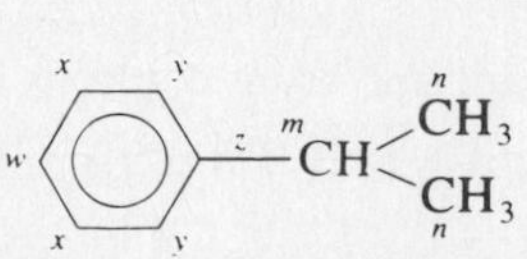

$C_6H_5-CH(CH_3)_2$ (ring carbons w, x, y, z; CH m; CH_3 n)

There are four different types of carbons in the aromatic ring and two different types in the isopropyl group. The ^{13}C nmr spectrum of isopropylbenzene contains *six* signals.

(*c*) The methyl substituent at C-2 is different from those at C-1 and C-3.

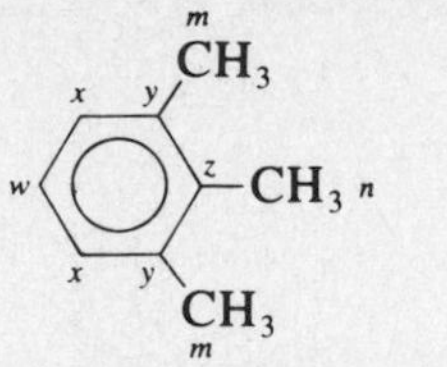

The four nonequivalent ring carbons and the two different types of methyl carbons give rise to a ^{13}C nmr spectrum that contains *six* signals.

(*d*) The three methyl carbons of 1,2,4-trimethylbenzene are different from one another.

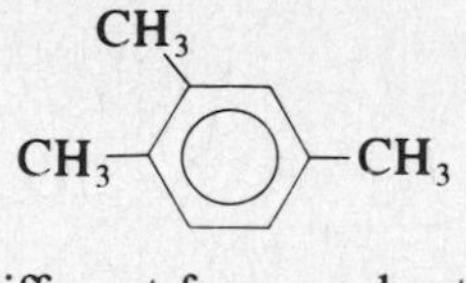

Also, all the ring carbons are different from each other. The nine different carbons give rise to *nine* separate signals.

(*e*) All the methyl carbons of 1,3,5-trimethylbenzene are equivalent.

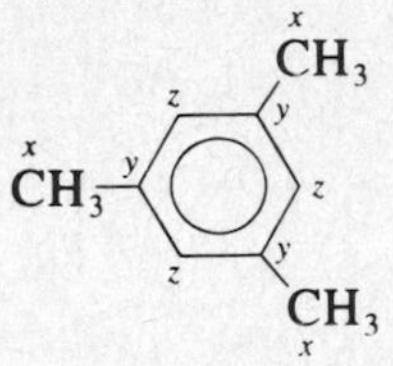

Because of its high symmetry 1,3,5-trimethylbenzene has only *three* signals in its ^{13}C nmr spectrum.

14.8 The splitting rules for ^{13}C nmr are summarized in the sample solution to (*a*) given in the text. In all the solutions presented below, multiplicities are cited as *s* (singlet), *d* (doublet), *t* (triplet), *q* (quartet).

(*b*) Isopropylbenzene

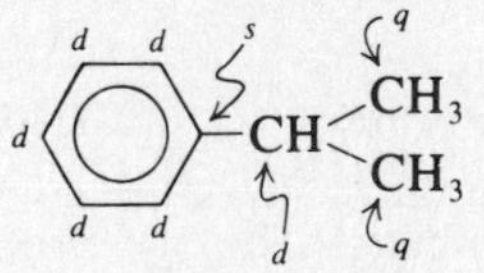

(*d*) 1,2,4-Trimethylbenzene

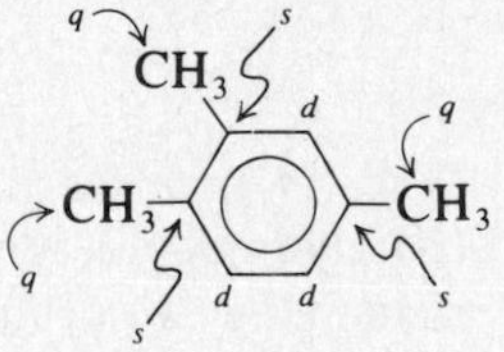

(*c*) 1,2,3-Trimethylbenzene

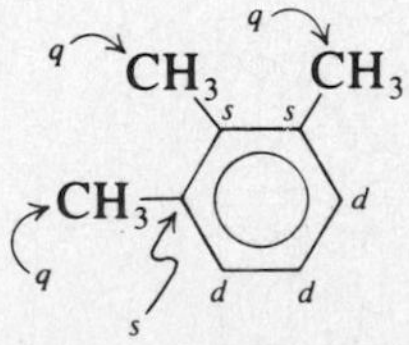

(*e*) 1,3,5-Trimethylbenzene

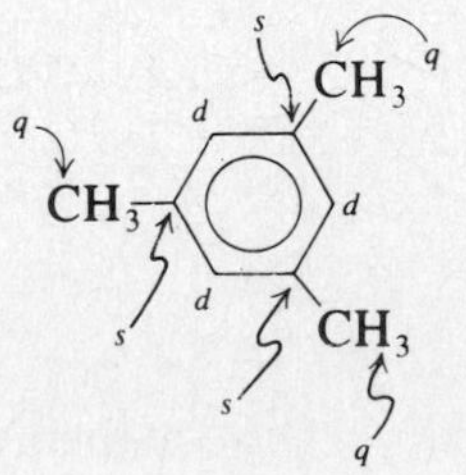

14.9 The ^{13}C nmr spectrum contains three different sp^3 hybridized carbons and six different aromatic protons. This spectrum is in accord only with 1,2,4-trimethylbenzene.

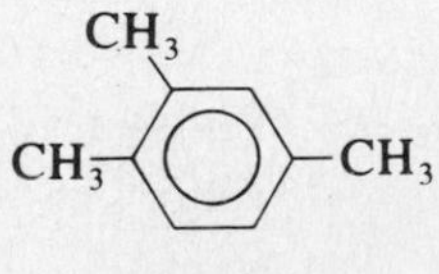

14.10 The infrared spectrum of Figure 14.30 is void of absorption in the 1600 to 1800 cm^{-1} region, so the unknown compound cannot contain a carbonyl (C=O) group. Therefore, it cannot be acetophenone or benzoic acid.

There is a broad, intense absorption at 3300 cm^{-1} attributable to a hydroxyl group. While both phenol and benzyl alcohol are possibilities, the peaks at 2800 to 2900 cm^{-1} reveal the presence of hydrogen bonded to sp^3 hybridized carbon. All carbons are sp^2 hybridized in phenol. The infrared spectrum is that of *benzyl alcohol*.

14.11 (*b*) The distribution of molecular-ion peaks in *o*-dichlorobenzene is identical to that in the para isomer. As the sample solution to (*a*) in the text describes, peaks at *m/z* 146, 148, and 150 are present for the molecular ion.

(*c*) The two isotopes of bromine are ^{79}Br and ^{81}Br. When both bromines of *p*-bromobenzene are ^{79}Br, the molecular ion appears at *m/z* 234. When one is ^{79}Br and the other is ^{81}Br, *m/z* for the molecular ion is 236. When both bromines are ^{81}Br, *m/z* for the molecular ion is 238.

(*d*) The combinations of ^{35}Cl, ^{37}Cl, ^{79}Br, and ^{81}Br in *p*-bromochlorobenzene and the values of *m/z* for the corresponding molecular ion are as shown.

(^{35}Cl, ^{79}Br) $m/z = 190$
(^{37}Cl, ^{79}Br) or (^{35}Cl, ^{81}Br) $m/z = 192$
(^{37}Cl, ^{81}Br) $m/z = 194$

14.12 The base peak in the mass spectrum of alkylbenzenes corresponds to carbon-carbon bond cleavage at the benzylic carbon.

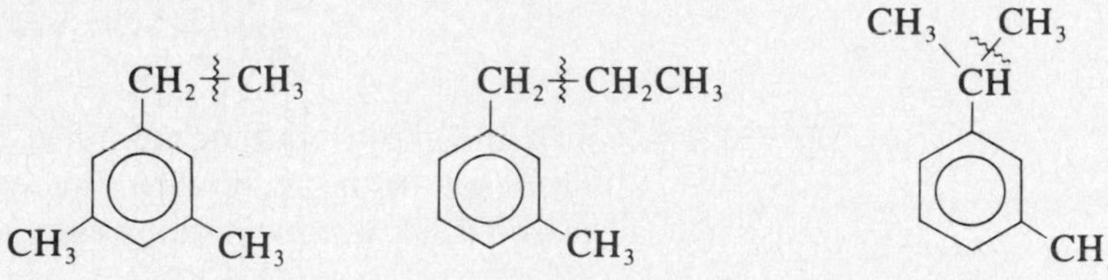

CH_2⫶CH_3 (3,5-dimethyl substituted ring) — Base peak: $C_9H_{11}^+$, *m/z* 119

CH_2⫶CH_2CH_3 (3-methyl substituted ring) — Base peak: $C_8H_9^+$, *m/z* 105

$(CH_3)_2CH$ (3-methyl substituted ring, cleavage of CH–CH_3) — Base peak: $C_9H_{11}^+$, *m/z* 119

14.13 Since each compound exhibits only a single peak in its ^{1}H nmr spectrum, all the hydrogen substituents are equivalent in each one. Structures are assigned on the basis of their molecular formulas and chemical shifts.

(*a*) This compound has the molecular formula C_8H_{18} and so must be an alkane. The 18 hydrogens are contributed by six equivalent methyl groups.

$(CH_3)_3CC(CH_3)_3$ 2,2,3,3-Tetramethylbutane ($\delta = 0.9$ ppm)

(*b*) A hydrocarbon with the molecular formula C_5H_{10} has a SODAR of 1 and so is either a cycloalkane or an alkene. Since all 10 hydrogens are equivalent, this compound must be cyclopentane.

Cyclopentane ($\delta = 1.5$ ppm)

(*c*) The chemical shift of the eight equivalent hydrogens in C_8H_8 is $\delta = 5.8$ ppm, which is consistent with protons attached to a carbon-carbon double bond.

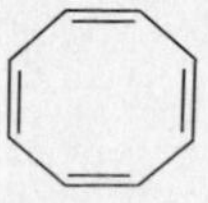

1,3,5,7-Cyclooctatetraene ($\delta = 5.8$ ppm)

(*d*) The compound C_4H_9Br has no rings or double bonds. The nine hydrogens belong to three equivalent methyl groups.

$(CH_3)_3CBr$ *tert*-Butyl bromide ($\delta = 1.8$ ppm)

(*e*) This dichloride has no rings or double bonds. The four equivalent hydrogens are present as two $-CH_2Cl$ groups.

$ClCH_2CH_2Cl$ 1,2-Dichloroethane ($\delta = 3.7$ ppm)

(*f*) All three hydrogens in $C_2H_3Cl_3$ must be part of the same methyl group in order to be equivalent.

CH_3CCl_3 1,1,1-Trichloroethane ($\delta = 2.7$ ppm)

(*g*) This compound has no rings or double bonds. In order to have eight equivalent hydrogens it must have four equivalent methylene groups.

$$\begin{array}{c} CH_2Cl \\ | \\ ClCH_2CCH_2Cl \\ | \\ CH_2Cl \end{array}$$

1,3-Dichloro-2,2-di(chloromethyl)propane (δ=3.7 ppm)

14.14 (*a*) A five-proton signal at $\delta = 7.1$ ppm indicates a monosubstituted aromatic ring. With a SODAR of 4, C_8H_{10} contains this monosubstituted aromatic ring and no other rings or multiple bonds. The triplet-quartet pattern at high field suggests an ethyl group.

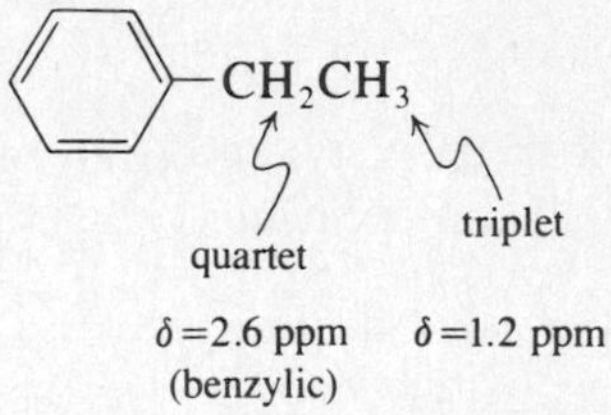

$C_6H_5-CH_2CH_3$ Ethylbenzene

quartet δ=2.6 ppm (benzylic) — triplet δ=1.2 ppm

(*b*) The SODAR of 4 and the five-proton multiplet at $\delta = 7.0$ to 7.5 ppm are accommodated by a monosubstituted aromatic ring. The remaining four carbons and nine hydrogens are most reasonably a *tert*-butyl group since all nine hydrogens are equivalent.

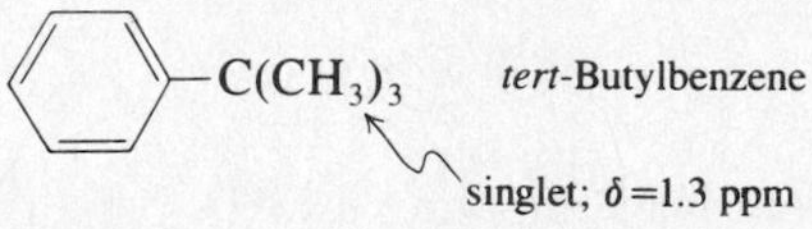

$C_6H_5-C(CH_3)_3$ *tert*-Butylbenzene

singlet; δ=1.3 ppm

(*c*) Its molecular formula requires that C_6H_{14} be an alkane. The doublet-heptet pattern is consistent with an isopropyl group, and the total number of protons requires that two of these groups be present.

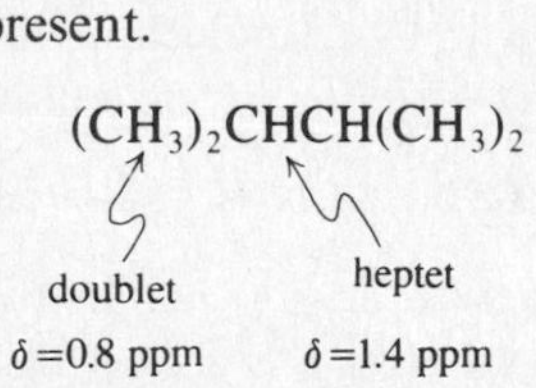

$(CH_3)_2CHCH(CH_3)_2$ 2,3-Dimethylbutane

doublet δ=0.8 ppm — heptet δ=1.4 ppm

(*d*) The molecular formula C_6H_{12} requires the presence of one double bond or ring. A peak at $\delta = 5.1$ ppm is consistent with $-C{=}CH$, so the compound is a noncyclic alkene. The vinyl proton gives a triplet signal, so the group $C{=}CHCH_2$ is present. The ^1H nmr spectrum shows the presence of the structural units:

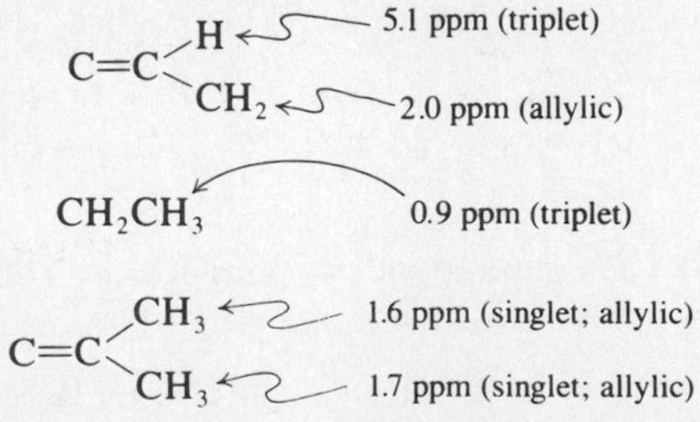

$C{=}C(H)(CH_2)$: H — 5.1 ppm (triplet); CH_2 — 2.0 ppm (allylic)

CH_2CH_3: CH_3 — 0.9 ppm (triplet)

$C{=}C(CH_3)(CH_3)$: CH_3 — 1.6 ppm (singlet; allylic); CH_3 — 1.7 ppm (singlet; allylic)

Putting all these fragments together yields a unique structure:

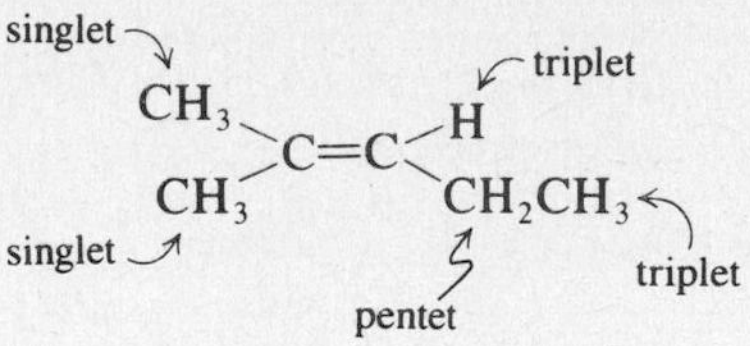

$$\begin{matrix} \text{singlet} \rightarrow CH_3 \\ \text{singlet} \rightarrow CH_3 \end{matrix} \!\!> C{=}C <\!\! \begin{matrix} H \leftarrow \text{triplet} \\ CH_2CH_3 \leftarrow \text{triplet} \end{matrix}$$

(CH_2: pentet) 2-Methyl-2-pentene

(*e*) The compound $C_4H_6Cl_4$ contains no double bonds or rings. There are no high-field peaks ($\delta = 0.5$ to 1.5 ppm), so there are no methyl groups. Therefore at least one chlorine substituent must be at each end of the chain. The most likely structure has the four chlorines divided into two groups of two.

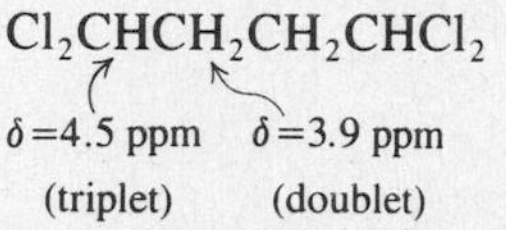

$$Cl_2CHCH_2CH_2CHCl_2$$ 1,1,4,4-Tetrachlorobutane

$\delta = 4.5$ ppm (triplet) $\delta = 3.9$ ppm (doublet)

(*f*) The molecular formula $C_4H_6Cl_2$ indicates the presence of one double bond or ring. A signal at $\delta = 5.7$ ppm is consistent with a proton attached to a doubly bonded carbon. The following structural units are present:

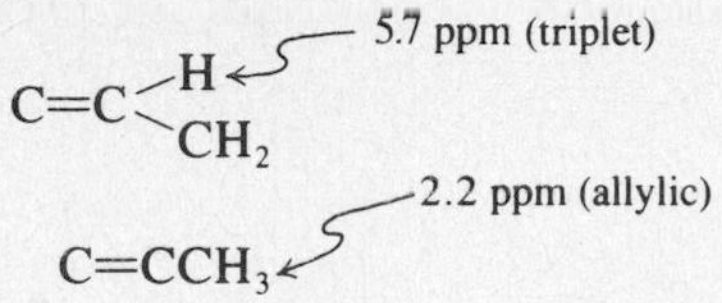

$$C{=}C <\!\! \begin{matrix} H \\ CH_2 \end{matrix}$$ (H: 5.7 ppm (triplet))

$$C{=}CCH_3$$ (CH_3: 2.2 ppm (allylic))

In order for the methyl group to appear as a singlet and the methylene group to appear as a doublet, the chlorine substituents must be distributed as shown:

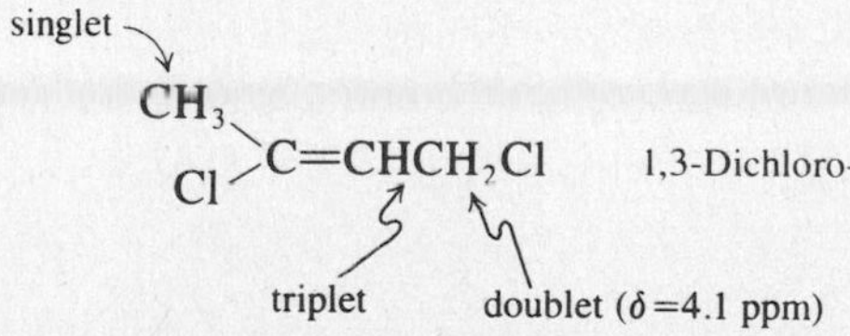

$$\begin{matrix} CH_3 \\ Cl \end{matrix} \!\!> C{=}CHCH_2Cl$$ 1,3-Dichloro-2-butene

(CH_3: singlet; CH: triplet; CH_2: doublet ($\delta = 4.1$ ppm))

The stereochemistry of the double bond (*E* or *Z*) is not revealed by the ^1H nmr spectrum.

(*g*) A molecular formula of C_3H_7ClO is consistent with the absence of rings and multiple bonds. None of the signals is equivalent to three protons, so no methyl groups are present. There are three methylene groups, all of which are different from each other. Therefore, the compound is:

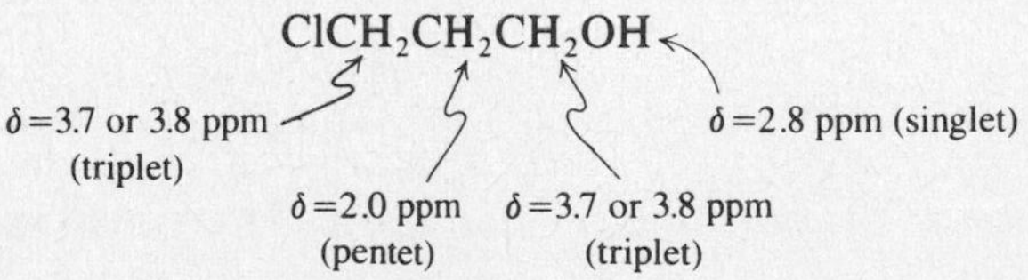

$$ClCH_2CH_2CH_2OH$$

$ClCH_2$: $\delta = 3.7$ or 3.8 ppm (triplet); CH_2: $\delta = 2.0$ ppm (pentet); CH_2O: $\delta = 3.7$ or 3.8 ppm (triplet); OH: $\delta = 2.8$ ppm (singlet)

(*h*) The compound has a molecular formula of $C_{14}H_{14}$ and a SODAR of 8. With a 10-proton signal at $\delta = 7.1$ ppm, a logical conclusion is that there are two monosubstituted benzene rings. The other four protons belong to two equivalent methylene groups.

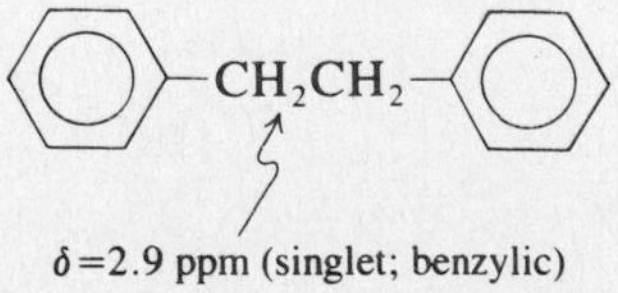

$$C_6H_5{-}CH_2CH_2{-}C_6H_5$$ 1,2-Diphenylethane

$\delta = 2.9$ ppm (singlet; benzylic)

14.15 The compounds of molecular formula C_4H_9Cl are the isomeric chlorides: butyl, isobutyl, *sec*-butyl, and *tert*-butyl chloride.

(*a*) All nine methyl protons of *tert*-butyl chloride $(CH_3)_3CCl$ are equivalent; its ^1H nmr spectrum has only one peak.

(*b*) A doublet at $\delta = 3.4$ ppm indicates a $-CH_2Cl$ group attached to a carbon that bears a single proton.

$$(CH_3)_2CHCH_2Cl$$

$\delta = 3.4$ ppm (doublet)

Isobutyl chloride

(*c*) A triplet at $\delta = 3.5$ ppm means that there is a methylene group attached to the carbon that bears the chlorine.

$$CH_3CH_2CH_2CH_2Cl$$

$\delta = 3.5$ ppm (triplet)

Butyl chloride

(*d*) There are two nonequivalent methyl groups in this compound.

$\delta = 1.5$ ppm (doublet) → $CH_3CHCH_2CH_3$ (with Cl on C-2) ← $\delta = 1.0$ ppm (triplet)

sec-Butyl chloride

14.16 Compounds with the molecular formula C_3H_5Br have either one ring or one double bond.

(*a*) The two small peaks at $\delta = 5.3$ and 5.5 ppm have chemical shifts consistent with each peak being due to a vinyl proton (C=CH). The remaining three protons belong to an allylic methyl group ($\delta = 2.3$ ppm).

The compound cannot be $CH_3CH{=}CHBr$ because the methyl signal would be split into a doublet. Isomer A can only be

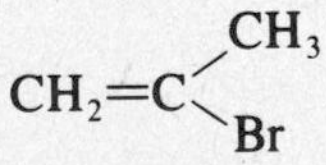

$$CH_2{=}C(CH_3)Br$$

2-Bromo-1-propene

(*b*) Two of the carbons of isomer B have chemical shifts characteristic of sp^2 hybridized carbon. One of these bears two protons ($\delta = 118.8$, triplet); the other bears one proton ($\delta = 134.2$ ppm, doublet). The remaining carbon is sp^3 hybridized and bears two hydrogens. Isomer B is allyl bromide.

$$CH_2{=}CHCH_2Br$$

$\delta = 118.8$ ppm (triplet) $\delta = 134.2$ ppm (doublet) $\delta = 32.6$ ppm (triplet)

Allyl bromide

(*c*) All the carbons are sp^3 hybridized in this isomer. Two of the carbons belong to equivalent methylene groups, and the other bears only one hydrogen. Isomer C is cyclopropyl bromide.

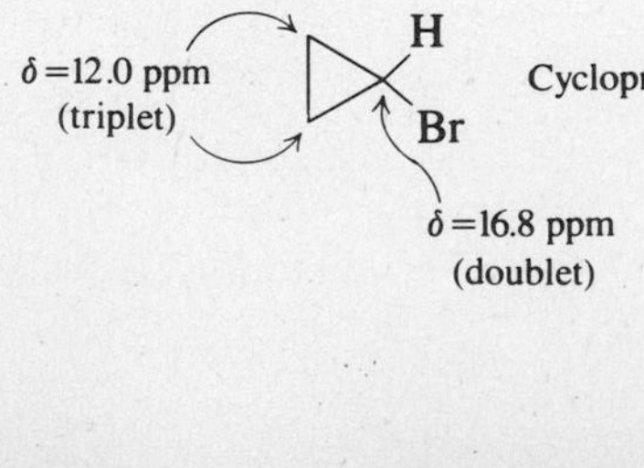

$\delta = 12.0$ ppm (triplet) — H, Br — $\delta = 16.8$ ppm (doublet)

Cyclopropyl bromide

14.17 All these compounds have the molecular formula $C_4H_{10}O$. They have neither multiple bonds nor rings.

(*a*) There are two equivalent methyl groups because there is a quartet at $\delta = 18.9$ ppm corresponding to two carbons. There is one carbon that bears a single hydrogen. The least shielded carbon, presumably the one bonded to oxygen, has two hydrogen substituents because its signal is a triplet.

Putting all the information together reveals this compound to be isobutyl alcohol.

$(CH_3)_2CHCH_2OH$

$\delta = 18.9$ ppm (quartet) $\delta = 30.8$ ppm (doublet) $\delta = 69.4$ ppm (triplet)

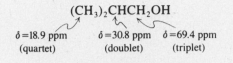

Isobutyl alcohol

(*b*) There are four distinct peaks in this compound, so none of the four carbons is equivalent to any of the others. The signal for the least shielded carbon is a doublet, so the oxygen is attached to a secondary carbon. Only one carbon appears at low field; the compound is an alcohol, not an ether. Therefore:

$\delta = 69.2$ ppm (doublet)

$CH_3CHCH_2CH_3$ (with OH on C-2)

$\delta = 22.7$ ppm (quartet) $\delta = 32.0$ ppm (triplet) $\delta = 10.0$ ppm (quartet)

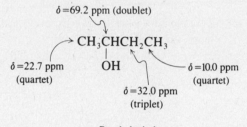

sec-Butyl alcohol

(*c*) Signals for three equivalent methyl carbons indicate that this isomer is *tert*-butyl alcohol. This assignment is reinforced by the observation that the least shielded carbon has no hydrogens attached to it.

$(CH_3)_3COH$

$\delta = 31.2$ ppm (quartet) $\delta = 68.9$ ppm (singlet)

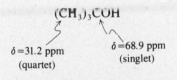

tert-Butyl alcohol

14.18 The molecular formula of C_6H_{14} for each of these isomers requires that all of them be alkanes.

(*a*) This compound contains only methyl and methine carbons.

$(CH_3)_2CHCH(CH_3)_2$

$\delta = 19.1$ ppm (quartet) $\delta = 33.9$ ppm (doublet)

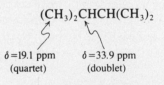

2,3-Dimethylbutane

(*b*) This isomer has no methine carbons, and there are two different kinds of methylene groups.

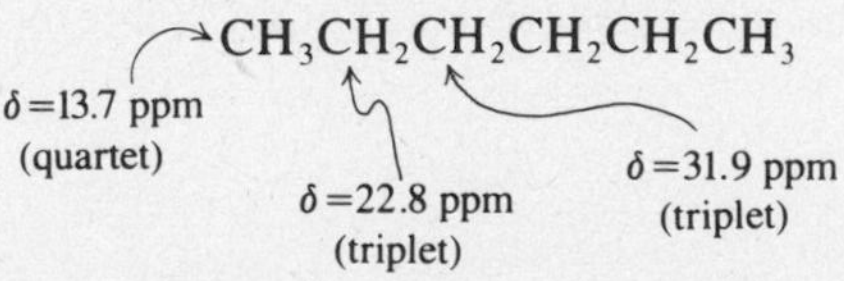

Hexane

(c) Methyl, methylene, and methine carbons are all present in this isomer. There are two different kinds of methyl groups.

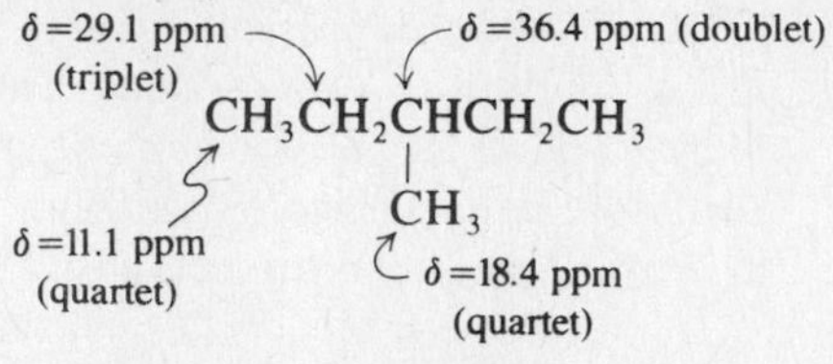

3-Methylpentane

(d) This isomer contains a quaternary carbon in addition to a methylene group and two different kinds of methyl groups.

$$CH_3-\overset{\displaystyle CH_3}{\underset{\displaystyle CH_3}{\overset{|}{\underset{|}{C}}}}-CH_2CH_3$$

2,2-Dimethylbutane

δ=28.7 ppm (quartet); δ=30.2 ppm (singlet); δ=36.5 ppm (triplet); δ=8.5 ppm (quartet)

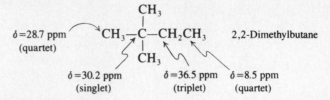

(e) This isomer contains two different kinds of methyl groups, two different kinds of methylenes, and a methine carbon.

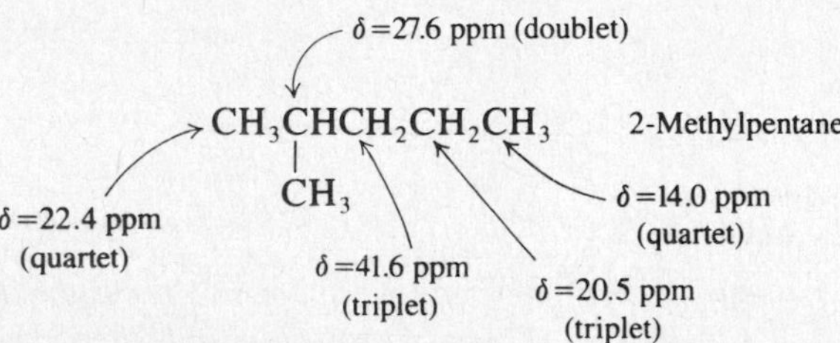

14.19 The SODAR of compound D (C_4H_6) is 2. It can have two double bonds, two rings, one ring and one double bond, or one triple bond.

The chemical shift data indicate that two carbons are sp^3 hybridized and two are sp^2. The most reasonable structure that is consistent with ^{13}C nmr data is cyclobutene.

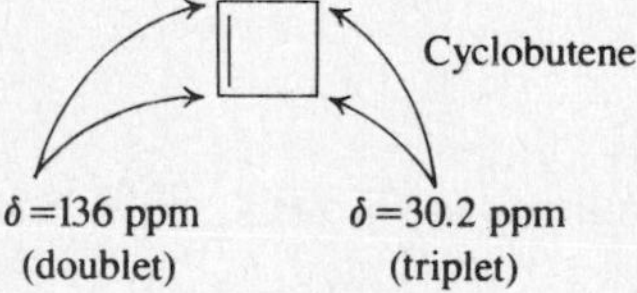

Compound D cannot be 1- or 2-methylcyclopropene. Neither of the carbon signals is a quartet, so compound D cannot contain a methyl group.

14.20 Each of the carbons in compound E gives its ^{13}C nmr signal at relatively low field; it is likely that each one bears an electron-withdrawing substituent. Compound E is:

$$ClCH_2\underset{\displaystyle OH}{\underset{|}{C}}HCH_2OH$$

3-Chloro-1,2-propanediol

δ=46.8 ppm (triplet); δ=72.0 ppm (doublet); δ=63.5 ppm (triplet)

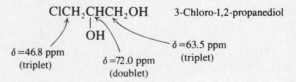

The regioisomeric compound 2-chloro-1,3-propanediol

$$(HOCH_2\underset{\displaystyle Cl}{\underset{|}{C}}HCH_2OH)$$

cannot be correct. The C-1 and C-3 positions are equivalent; the ^{13}C nmr spectrum of this compound exhibits only two peaks, not three.

14.21 (*a*) All the hydrogens are equivalent in *p*-dichlorobenzene, therefore it has the simplest ^1H nmr spectrum of the three compounds chlorobenzene, *o*-dichlorobenzene, and *p*-dichlorobenzene.

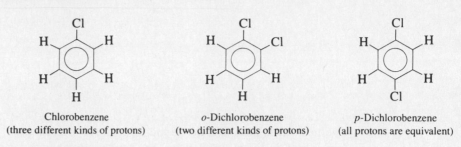

Chlorobenzene (three different kinds of protons) | *o*-Dichlorobenzene (two different kinds of protons) | *p*-Dichlorobenzene (all protons are equivalent)

(*b*) through (*d*) In addition to giving the simplest ^1H nmr spectrum, *p*-dichlorobenzene gives the simplest ^{13}C nmr spectrum. It has two peaks in its ^{13}C nmr spectrum, chlorobenzene has four, and *o*-dichlorobenzene has three.

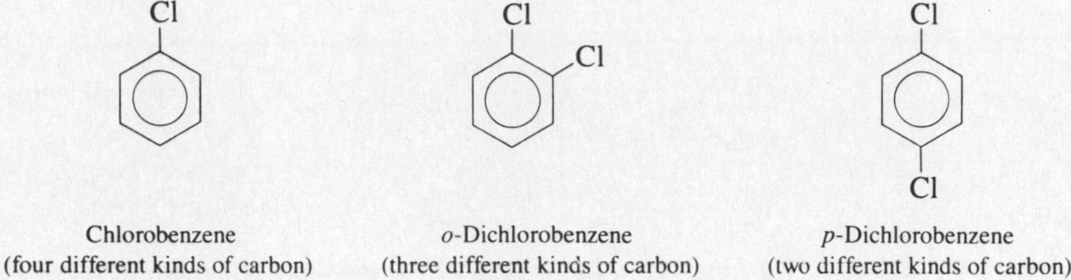

Chlorobenzene (four different kinds of carbon) | *o*-Dichlorobenzene (three different kinds of carbon) | *p*-Dichlorobenzene (two different kinds of carbon)

14.22 Compounds F and G ($C_{10}H_{14}$) have SODARs of 4. Both have peaks in the $\delta = 7$ ppm region of their ^1H nmr spectra, so the SODAR of 4 can be accommodated by a benzene ring.

Compound F has only two aromatic protons and the ring is therefore tetrasubstituted. All four substituents are methyl groups, and all the methyl groups are equivalent.

$\delta = 2.2$ ppm

$\delta = 6.9$ ppm

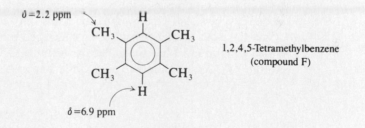

1,2,4,5-Tetramethylbenzene (compound F)

In Compound G both aromatic protons are equivalent and there are four methyl groups divided into two pairs.

$\delta = 6.8$ ppm

$\delta = 2.2$ and 2.3 ppm

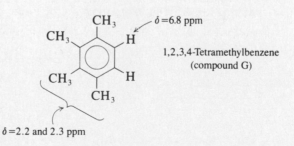

1,2,3,4-Tetramethylbenzene (compound G)

14.23 Since Compound H has a five-proton signal at $\delta = 7.2$ ppm and a SODAR of 4, we conclude that six of its eight carbons belong to a monosubstituted aromatic ring. The infrared spectrum exhibits absorption at 3300 cm^{-1}, indicating the presence of a hydroxyl group. Compound H is an alcohol. A three-proton doublet at $\delta = 1.4$ ppm, along with a one-proton quartet $\delta = 4.7$ ppm, signals the presence of a CH_3CH unit.

Compound H is 1-phenylethanol.

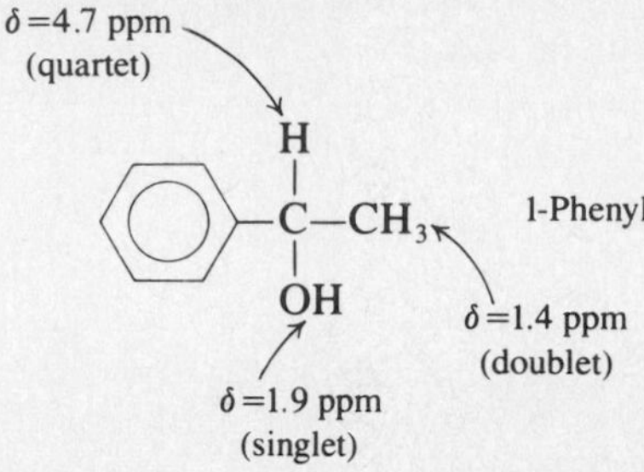

δ=4.7 ppm (quartet)

H | C₆H₅—C—CH₃ | OH

1-Phenylethanol (compound H)

δ=1.4 ppm (doublet)

δ=1.9 ppm (singlet)

14.24 The peak at highest m/z in the mass spectrum of compound I is $m/z = 134$; this is likely to correspond to the molecular ion. Among the possible molecular formulas, $C_{10}H_{14}$ correlates best with the information from the ^1H nmr spectrum. What is evident is that there is a signal due to aromatic protons, as well as a triplet-quartet pattern of an ethyl group. A molecular formula of $C_{10}H_{14}$ suggests a benzene ring that bears two ethyl groups. Since the signal for the aryl protons is so sharp, they are probably equivalent. Compound I is *p*-diethylbenzene.

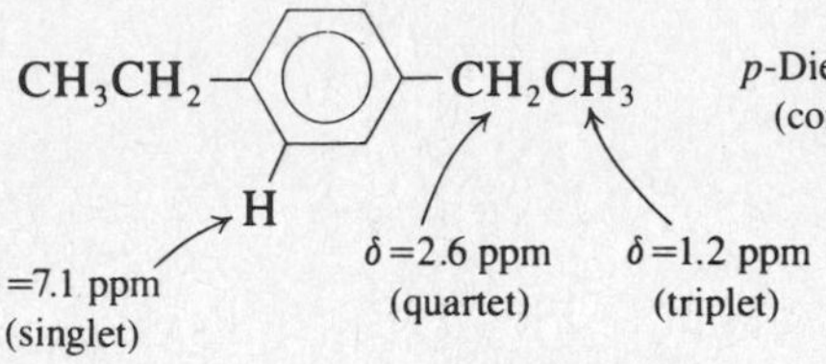

CH_3CH_2—C₆H₄—CH_2CH_3

p-Diethylbenzene (compound I)

δ=7.1 ppm (singlet) — H

δ=2.6 ppm (quartet)

δ=1.2 ppm (triplet)

14.25 The most prominent peak in the infrared spectrum of compound J is at 1725 cm^{-1}, a region characteristic of C=O stretching vibrations.

The ^1H nmr spectrum shows only two sets of signals, a triplet at $\delta = 1.1$ ppm and a quartet at $\delta = 2.4$ ppm. Compound J contains a CH_3CH_2 group as the only source of its protons.

There are three peaks in its ^{13}C nmr spectrum, one of which is at very low field. The signal at $\delta = 211.4$ ppm is in the region characteristic of carbons of C=O groups.

If one assumes that Compound J contains only carbon, hydrogen, and one oxygen atom and that the peak at highest m/z in its mass spectrum (m/z 86) corresponds to the molecular ion, then Compound J has the molecular formula $C_5H_{10}O$.

All the information points to compound J having the structure shown.

$$CH_3CH_2\overset{\overset{\displaystyle O}{\|}}{C}CH_2CH_3$$

3-Pentanone

14.26 [18]-Annulene has *two* different kinds of protons; the 12 protons on the outside periphery of the ring are different from the 6 on the inside.

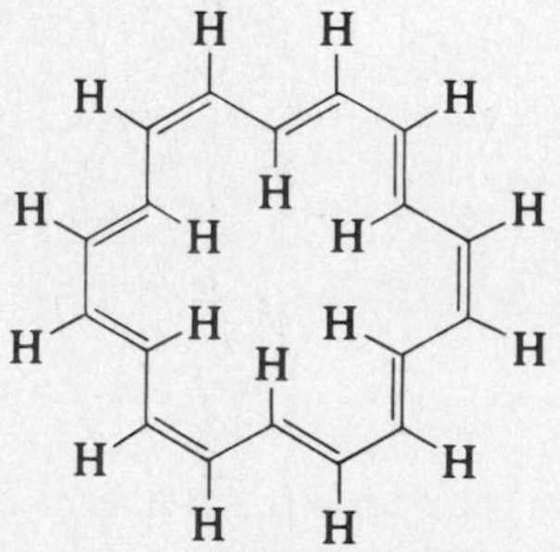

These different environments explain why the ^1H nmr spectrum contains two peaks in a 2 : 1 ratio. The less intense signal, that for the interior protons, is more shielded than the signal for the outside protons. The reason for this has to do with the magnetic field induced by the circulating π electrons of this aromatic ring, which reinforces the applied field in the region of the outside protons but opposes it in the interior of the ring.

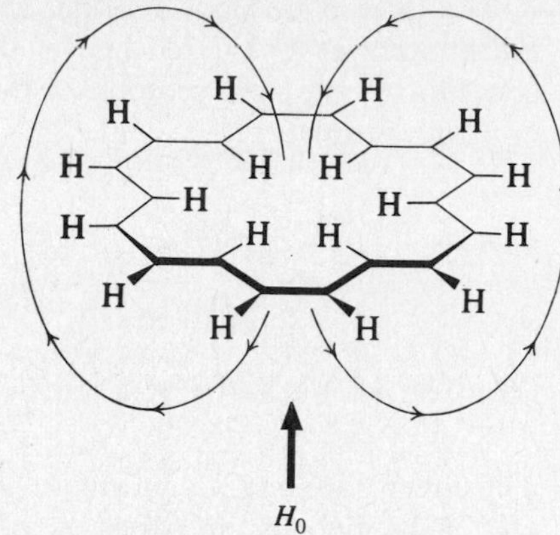

Protons inside the ring are shielded by the induced field to a significant extent— so much so that their signal appears at $\delta = -1.9$ ppm.

14.27 (*a*) The nuclear spin of ^{19}F is $\pm\frac{1}{2}$, i.e., the same as that of a proton. The splitting rules for ^{19}F–^{1}H couplings are the same as those for ^{1}H–^{1}H. Thus the single fluorine atom of CH_3F splits the signal for the protons of the methyl group into a *doublet*.

(*b*) The set of three equivalent protons of CH_3F splits the signal for fluorine into a *quartet*.

(*c*) The proton signal in CH_3F is a doublet centered at $\delta = 4.3$ ppm. The separation between the two halves of this doublet is 45 Hz, which is equivalent to 0.75 ppm at 60 MHz (60 Hz = 1 ppm). Thus one line of the doublet appears at $\delta = (4.3 + 0.375)$ ppm and the other at $\delta = (4.3 - 0.375)$ ppm.

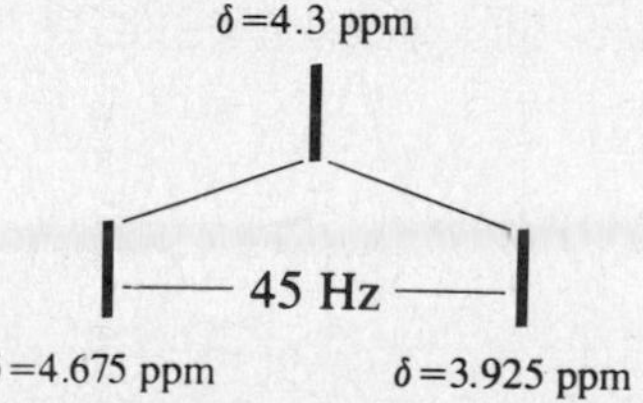

(*d*) The splitting is the same (45 Hz) whether the spectrum is run at 60 MHz or at 100 MHz. However, at 100 MHz, 45 Hz is equivalent to 0.45 ppm. Therefore:

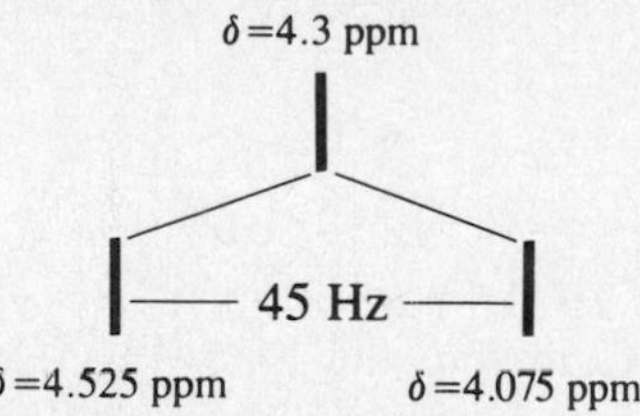

14.28 (*a*) Both hydrogens are equivalent in 1,1,1,2-tetrafluoroethane, so there is one signal. This signal is split into a doublet by the fluorine at C-2 and each line of the doublet is split into a quartet by the three fluorines at C-1. The proton signal is split into a doublet of quartets, or a total of eight peaks.

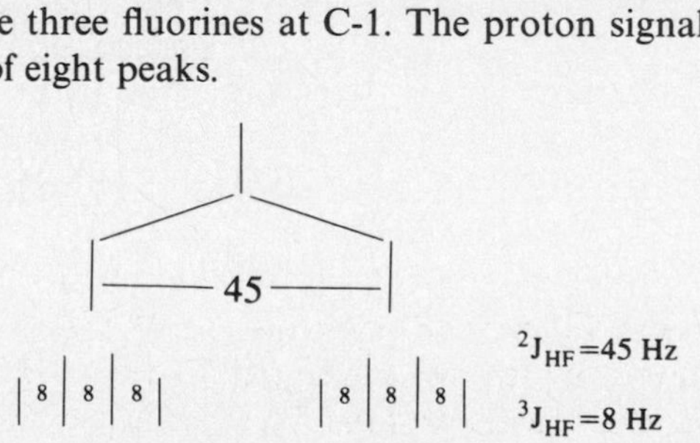

45

8 8 8 8 8 8

$^2J_{HF}$=45 Hz

$^3J_{HF}$=8 Hz

(*b*) There are two different kinds of fluorine in 1,1,1,2-tetrafluoroethane, so there are two distinct signals.

The three equivalent fluorines at C-1 are split into a doublet by vicinal coupling to the fluorine at C-2. Each line of this doublet is further split into a triplet by vicinal coupling to the two protons at C-2. There are six peaks for the fluorines at C-1, a doublet of triplets.

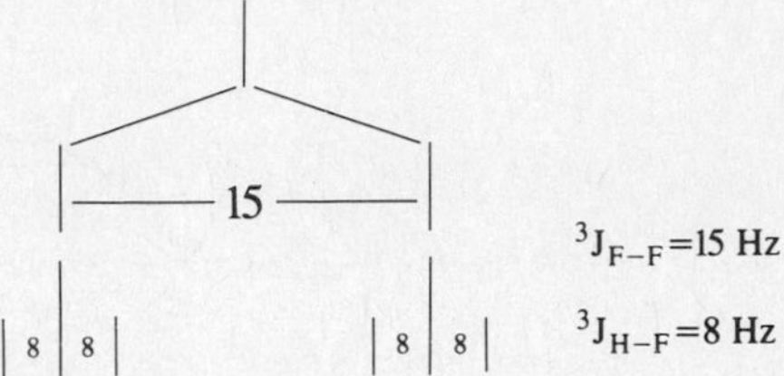

The signal for the C-2 fluorine is split into a triplet by the two geminal hydrogens at C-2. Each line of the triplet is split into a quartet by vicinal coupling to the three equivalent fluorines at C-1. The signal for the C-2 fluorine appears as a 12-line pattern (a triplet of quartets).

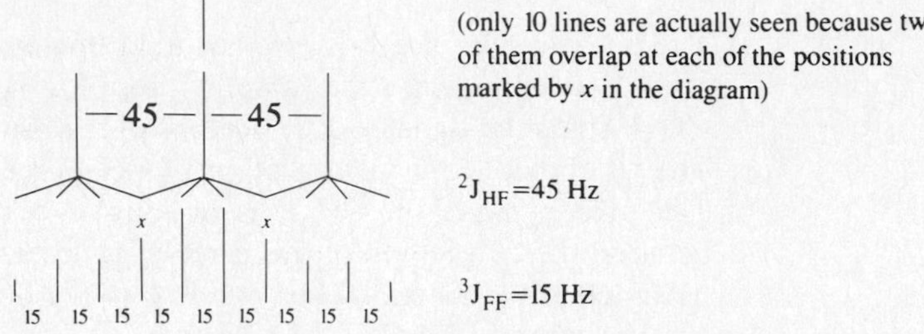

14.29 Since ^{31}P has a spin of $\pm\frac{1}{2}$, it is capable of splitting the ^1H nmr signal of protons in the same molecule. The problem stipulates that the methyl protons are coupled through three bonds to phosphorus in trimethyl phosphite.

$$CH_3O\text{—}P\begin{matrix}\diagup OCH_3\\ \diagdown OCH_3\end{matrix}$$

(*a*) The reciprocity of splitting requires that the protons split the ^{31}P signal of phosphorus. There are 9 equivalent protons, so the ^{31}P signal is split into *10* peaks.

(*b*) Each peak in the ^{31}P multiplet is separated from the next by a value equal to the ^1H–^{31}P coupling constant of 12 Hz. There are nine such intervals in a 10-line multiplet, so the separation is 108 Hz between the highest- and lowest-field peaks in the multiplet.

SELF-TEST

PART A

A-1. Complete the following table relating to ^1H nmr spectra by supplying the missing data for entries 1 through 4.

	Spectrometer frequency	*Chemical shift ppm*	*Hz*
(*a*)	60 MHz	____ 1	366
(*b*)	220 MHz	4.35	____ 2
(*c*)	____ 3	3.50	210
(*d*)	100 MHz	____ 4 of	TMS

A-2. Indicate the number of signals to be expected and the multiplicity of each in the ^{1}H nmr spectrum of each of the following substances:

(*a*) $BrCH_2CH_2CH_2Br$

(*b*) $CH_3CH_2\underset{Cl}{\overset{Cl}{C}}CH_2CH_3$

(*c*) $CH_3OCH_2\overset{O}{\overset{\|}{C}}OCH_3$

A-3. Two isomeric compounds having the molecular formula $C_6H_{12}O_2$ both gave ^{1}H nmr spectra consisting of only two singlets. Given the chemical shifts and integrations shown, identify both compounds.

Compound A $\delta = 1.45$ ppm (9H) Compound B: $\delta = 1.20$ ppm (9H)
$\delta = 1.95$ ppm (3H) $\delta = 3.70$ ppm (3H)

A-4. Identify each of the following compounds based on the ir and ^{1}H nmr information provided:

(*a*) $C_{10}H_{12}O$: ir: 1710 cm^{-1}
nmr: $\delta = 1.0$ ppm (triplet, 3H)
$\delta = 2.4$ ppm (quartet, 2H)
$\delta = 3.6$ ppm (singlct, 2H)
$\delta = 7.2$ ppm (singlet, 5H)

(*b*) $C_6H_{14}O_2$: ir: 3400 cm^{-1}
nmr: $\delta = 1.2$ ppm (singlet, 12H)
$\delta = 2.0$ ppm (broad singlet, 2H)

(*c*) $C_{10}H_{16}O_6$: ir: 1740 cm ^{1}
nmr: $\delta = 1.3$ ppm (triplet, 9H)
$\delta = 4.2$ ppm (quartet, 6H)
$\delta = 4.4$ ppm (singlet, 1H)

(*d*) C_4H_7NO: ir: 2240 cm^{-1}
3400 cm^{-1} (broad)
nmr: $\delta = 1.65$ ppm (singlet, 6H)
$\delta = 3.7$ ppm (singlet, 1H)

A-5. Considering the ^{13}C nmr spectrum of the substance shown, give (*a*) the number of signals expected and (*b*) the number of peaks each of these signals would be split into in the off-resonance decoupling mode.

$Cl-C_6H_4-\overset{O}{\overset{\|}{C}}CH_2CH_3$

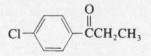

PART B

B-1. The ^{1}H nmr spectrum of acetone consists of a singlet with a chemical shift of 2.07 ppm. What was the spectrometer frequency of the instrument used if this chemical shift equaled 186 Hz?

(*a*) 60 MHz (*b*) 90 MHz (*c*) 100 MHz
(*d*) Need more information to determine

B-2. How many signals are expected in the ^{1}H nmr spectrum of $CH_3CH_2OCH_2OCH_2CH_3$?

(*a*) 12 (*b*) 5 (*c*) 4 (*d*) 3

B-3. The relationship between magnetic field intensity and the energy difference between nuclear spin states is:

(a) They are independent of each other.
(b) They are directly proportional.
(c) They are inversely proportional.
(d) The relationship varies from molecule to molecule.

B-4. An infrared spectrum exhibits a broad band in the 3000 to 3500 cm^{-1} region and a strong peak at 1710 cm^{-1}. Which of the following substances best fits the data?

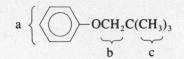

(a) $C_6H_5CH_2CH_2OH$ (b) $C_6H_5CH_2\overset{O}{\overset{\|}{C}}OH$

(c) $C_6H_5CH_2\overset{O}{\overset{\|}{C}}CH_3$ (d) $C_6H_5CH_2\overset{O}{\overset{\|}{C}}OCH_3$

B-5. Considering the ^1H nmr spectrum of the following substance, which set of protons appears furthest downfield relative to TMS?

a {C_6H_5}$-OCH_2C(CH_3)_3$ (b = OCH_2, c = $C(CH_3)_3$)

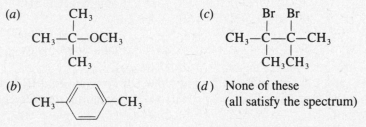

(a) a (b) b (c) c (d) Need more information to determine

B-6. Which of the following substances does *not* satisfy a proton nmr spectrum consisting of only two peaks?

(a) $CH_3-C(CH_3)_2-OCH_3$

(b) $CH_3-C_6H_4-CH_3$ (para)

(c) $CH_3-C(Br)(CH_3)-C(Br)(CH_3)-CH_3$

(d) None of these (all satisfy the spectrum)

B-7. The multiplicity of the "a" protons in the ^1H nmr spectrum of the following substance is:

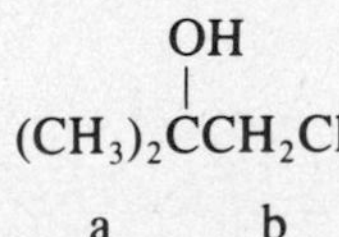

$(CH_3)_2C(OH)CH_2Cl$ (a = $(CH_3)_2$, b = CH_2)

(a) Singlet (b) Doublet (c) Triplet (d) Quartet

B-8. How many signals are expected in the ^{13}C nmr spectrum of the following substance?

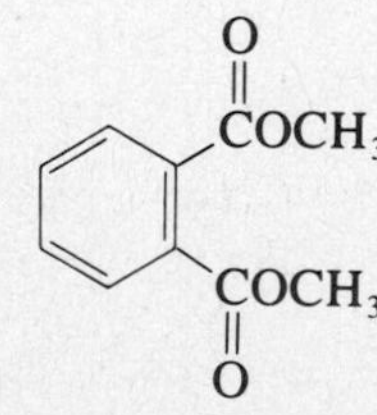

(a) 5 (b) 6 (c) 8 (d) 10

CHAPTER

15

ORGANOMETALLIC COMPOUNDS

IMPORTANT TERMS AND CONCEPTS

Bonding and Nomenclature (Secs. 15.1, 15.2) Organic compounds that possess carbon-metal bonds are classified as *organometallic compounds.* Recalling the Pauling scale of electronegativities (from Chapter 1), it may be seen that carbon is *more* electronegative than the metallic elements. Therefore an organometallic compound contains a carbon atom that is *negatively polarized.*

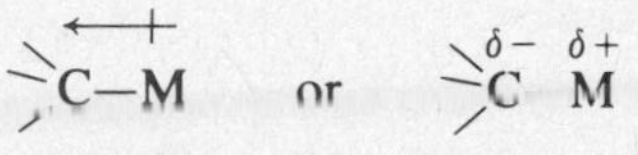

$$\overset{\longleftarrow\!\!+}{\gt C{-}M} \quad \text{or} \quad \gt\overset{\delta-}{C}\;\overset{\delta+}{M}$$

(M = a metallic element)

Covalently bonded organometallic compounds are, as a result of the bonding shown above, said to have *carbanionic character.* It is this property which is responsible for the synthetic utility of organometallic reagents.

Organometallic compounds are named as substituted derivatives of metals, as shown in the following examples.

$(CH_3CH_2)_2Cd$ — Diethylcadmium

C_6H_5MgBr — Phenylmagnesium bromide

Planning a Synthesis; Retrosynthetic Analysis (Sec. 15.9) The technique of reasoning backward from a target molecule to suitable starting materials in planning a synthesis is known as *retrosynthetic analysis.* Organometallic reagents have as their primary synthetic utility the ability to form new carbon-carbon bonds. Breaking (or disconnecting) carbon-carbon bonds of the target molecule will allow potential precursors to be identified.

The synthesis of 1-phenyl-1-propanol provides an example of the technique. An open arrow is used to represent a retrosynthetic step.

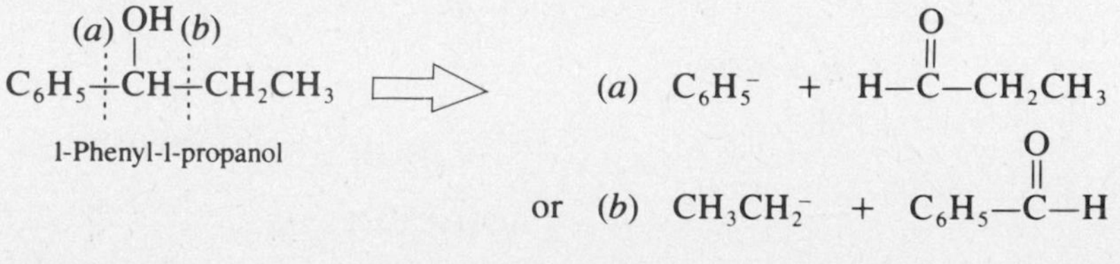

$$C_6H_5 \overset{(a)}{\vdots} \underset{|}{\overset{OH}{CH}} \overset{(b)}{\vdots} CH_2CH_3 \Longrightarrow (a)\ C_6H_5^- + H{-}\overset{O}{\overset{\|}{C}}{-}CH_2CH_3$$

1-Phenyl-1-propanol

$$\text{or } (b)\ CH_3CH_2^- + C_6H_5{-}\overset{O}{\overset{\|}{C}}{-}H$$

Synthetic routes:

(*a*) $C_6H_5Br + Mg \longrightarrow C_6H_5MgBr$

$$C_6H_5MgBr \xrightarrow[2.\ H_3O^+]{1.\ HC(=O)CH_2CH_3} C_6H_5CH(OH)CH_2CH_3$$

or (*b*) $CH_3CH_2Br + Mg \longrightarrow CH_3CH_2MgBr$

$$CH_3CH_2MgBr \xrightarrow[2.\ H_3O^+]{1.\ C_6H_5C(=O)H} CH_3CH_2CH(OH)C_6H_5$$

IMPORTANT REACTIONS

ORGANOLITHIUM COMPOUNDS

Preparation (Sec. 15.3)

General:

$$RX + 2\,Li \longrightarrow RLi + LiX$$

(X = halogen)

Example:

$$(CH_3)_2CHBr + 2\,Li \longrightarrow (CH_3)_2CHLi + LiBr$$

Synthesis of Alcohols (Sec. 15.7)

General:

$$RLi + \rangle C{=}O \longrightarrow R{-}\overset{|}{\underset{|}{C}}{-}O^-Li^+ \xrightarrow{H_3O^+} R{-}\overset{|}{\underset{|}{C}}{-}OH$$

Examples:

$$(CH_3)_2CHLi \xrightarrow[2.\ H_3O^+]{1.\ CH_3C(=O)H} CH_3CH(OH)CH(CH_3)_2$$

$$C_6H_5Li \xrightarrow[2.\ H_3O^+]{1.\ C_6H_5C(=O)CH_3} (C_6H_5)_2C(OH)CH_3$$

ORGANOMAGNESIUM COMPOUNDS (GRIGNARD REAGENTS)

Preparation (Sec. 15.4)

General:

$$RX + Mg \longrightarrow RMgX$$

(X = halogen)

Examples:

$$C_6H_5Br + Mg \xrightarrow{\text{diethyl ether}} C_6H_5MgBr$$

$$(CH_3)_3CCl + Mg \xrightarrow{\text{diethyl ether}} (CH_3)_3CMgCl$$

Synthesis of Alcohols (Sec. 15.6)

General:

$$\mathrm{RMgX} + \;\rangle\mathrm{C{=}O} \longrightarrow \mathrm{R{-}\overset{|}{\underset{|}{C}}{-}O^{-+}MgX}$$

$$\mathrm{R{-}\overset{|}{\underset{|}{C}}{-}OMgX} \xrightarrow{H_3O^+} \mathrm{R{-}\overset{|}{\underset{|}{C}}{-}OH} + \mathrm{HOMgX}$$

Examples:

$$\mathrm{C_6H_5MgBr} \xrightarrow[2.\ H_3O^+]{1.\ CH_3\overset{O}{\overset{\|}{C}}H} \mathrm{C_6H_5\overset{OH}{\overset{|}{C}}HCH_3}$$

$$\mathrm{(CH_3)_3CMgCl} \xrightarrow[2.\ H_3O^+]{1.\ H_2C{=}O} \mathrm{(CH_3)_3CCH_2OH}$$

Acid-Base Reactions (Sec. 15.5)

General:

$$\mathrm{R{-}M + R'OH \longrightarrow RH + R'O^-M^+}$$

Examples:

$$\mathrm{C_6H_5{-}Li} + \mathrm{D_2O} \longrightarrow \mathrm{C_6H_5{-}D} + \mathrm{LiOD}$$

$$\mathrm{(CH_3)_3CMgBr + CH_3CH_2OH \longrightarrow (CH_3)_3CH + BrMgOCH_2CH_3}$$

Reaction with Esters; Synthesis of Tertiary Alcohols (Sec. 15.10)

General:

$$2\,\mathrm{RMgX} \xrightarrow[2.\ H_3O^+]{1.\ R'\overset{O}{\overset{\|}{C}}OR''} \mathrm{R{-}\overset{OH}{\overset{|}{\underset{R}{\underset{|}{C}}}}{-}R'} + \mathrm{R''OH}$$

Example:

$$2\,\mathrm{CH_3CH_2MgBr} \xrightarrow[2.\ H_3O^+]{1.\ C_6H_5CO_2CH_3} \mathrm{C_6H_5\overset{OH}{\overset{|}{C}}(CH_2CH_3)_2} + \mathrm{CH_3OH}$$

Organocopper Reagents (Sec. 15.11)

Synthesis of hydrocarbons:

1. $\mathrm{2\,CH_3CH_2Li + CuBr \longrightarrow (CH_3CH_2)_2CuLi + LiBr}$
2. $\mathrm{(CH_3CH_2)_2CuLi + C_6H_5CH_2I \longrightarrow C_6H_5CH_2CH_2CH_3 + CH_3CH_2Cu + LiI}$

Organozinc Intermediates (Sec. 15.12)

$$\mathrm{(CH_3)_2C\!\begin{smallmatrix}\diagup CH_2Br\\ \diagdown CH_2Br\end{smallmatrix}} + \mathrm{Zn} \xrightarrow{C_2H_5OH} \mathrm{(CH_3)_2C\!\begin{smallmatrix}\diagup CH_2\\ \diagdown CH_2\end{smallmatrix}} + \mathrm{ZnBr_2}$$

$$\text{cyclohexene} + \mathrm{CH_2I_2} \xrightarrow{Zn(Cu)} \text{bicyclo[4.1.0]heptane}$$

SOLUTIONS TO TEXT PROBLEMS

15.1 (*b*) Magnesium bears a cyclohexyl substituent and a chlorine. Chlorine is named in its anionic form. The compound is cyclohexylmagnesium chloride.

(*c*) Cadmium bears two ethyl groups. This organometallic is diethylcadmium.

(*d*) This substance is iodomethylzinc iodide.

15.2 (*b*) The alkyl bromide precursor to *sec*-butyllithium must be *sec*-butyl bromide.

$$CH_3\underset{\displaystyle Br}{\underset{|}{C}}HCH_2CH_3 \;+\; 2Li \longrightarrow CH_3\underset{\displaystyle Li}{\underset{|}{C}}HCH_2CH_3 \;+\; LiBr$$

2-Bromobutane (*sec*-butyl bromide) — 1-Methylpropyllithium (*sec*-butyllithium)

(*c*) Benzylsodium is an organometallic in which the alkyl group is $C_6H_5CH_2$. Benzyl bromide is the correct precursor.

$$C_6H_5CH_2Br \;+\; 2Na \longrightarrow C_6H_5CH_2Na \;+\; NaBr$$

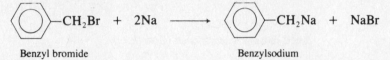

Benzyl bromide — Benzylsodium

(*d*) The compound *m*-fluorophenylpotassium is derived from the aryl halide *m*-bromofluorobenzene.

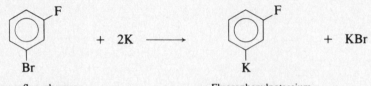

$$m\text{-}FC_6H_4Br \;+\; 2K \longrightarrow m\text{-}FC_6H_4K \;+\; KBr$$

m-Bromofluorobenzene — *m*-Fluorophenylpotassium

Since *m*-fluorophenylpotassium is ionic, an alternative way of representing its structure is

$$[m\text{-}FC_6H_4:]^-\;K^+$$

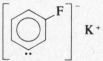

15.3 (*b*) Allyl chloride is converted to allylmagnesium chloride on reaction with magnesium.

$$CH_2{=}CHCH_2Cl \xrightarrow[\text{diethyl ether}]{\text{Mg}} CH_2{=}CHCH_2MgCl$$

Allyl chloride — Allylmagnesium chloride

(*c*) The carbon-iodine bond of iodocyclobutane is replaced by a carbon-magnesium bond in the Grignard reagent.

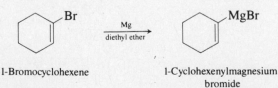

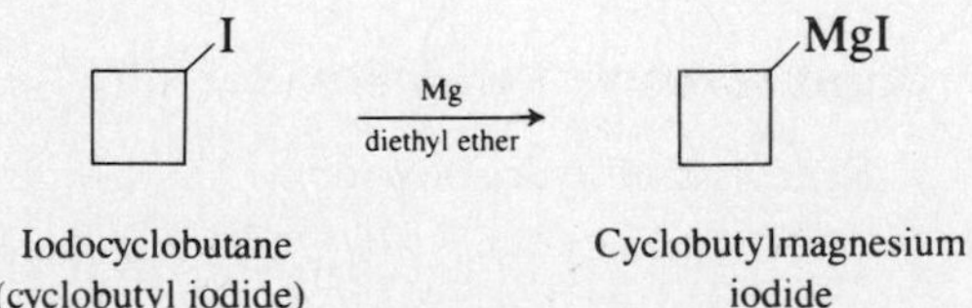

Iodocyclobutane (cyclobutyl iodide) — Cyclobutylmagnesium iodide

(*d*) Bromine is attached to sp^2 hybridized carbon in 1-bromocyclohexene. The product of its reaction with magnesium has a carbon-magnesium bond in place of the carbon-bromine bond.

$$\text{1-Bromocyclohexene} \xrightarrow[\text{diethyl ether}]{\text{Mg}} \text{1-Cyclohexenylmagnesium bromide}$$

1-Bromocyclohexene — 1-Cyclohexenylmagnesium bromide

15.4 (*b*) 1-Hexanol will protonate butyllithium since its hydroxyl group is a proton donor only slightly weaker than water. This proton transfer reaction could be used to prepare lithium hexanolate.

$$\underset{\substack{\text{1-Hexanol}\\(n\text{-hexyl alcohol})}}{CH_3CH_2CH_2CH_2CH_2CH_2OH} + \underset{\text{Butyllithium}}{CH_3CH_2CH_2CH_2Li} \longrightarrow \underset{\text{Butane}}{CH_3CH_2CH_2CH_3} + \underset{\text{Lithium hexanolate}}{CH_3CH_2CH_2CH_2CH_2CH_2OLi}$$

(*c*) The proton donor here is benzenethiol.

$$\underset{\text{Benzenethiol}}{C_6H_5SH} + \underset{\text{Butyllithium}}{CH_3CH_2CH_2CH_2Li} \longrightarrow \underset{\text{Butane}}{CH_3CH_2CH_2CH_3} + \underset{\substack{\text{Lithium}\\\text{benzenethiolate}}}{C_6H_5SLi}$$

15.5 (*b*) Propylmagnesium bromide reacts with benzaldehyde by addition to the carbonyl group:

$$CH_3CH_2CH_2-MgBr + C_6H_5CH{=}O \xrightarrow{\text{diethyl ether}} C_6H_5\overset{CH_3CH_2CH_2}{\underset{H}{C}}-OMgBr \xrightarrow{H_3O^+} \underset{\text{1-Phenyl-1-butanol}}{C_6H_5\underset{OH}{C}HCH_2CH_2CH_3}$$

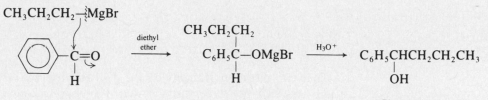

(*c*) Tertiary alcohols result from the reaction of Grignard reagents and ketones:

$$CH_3CH_2CH_2MgBr + \text{cyclohexanone} \xrightarrow[\text{2. } H_3O^+]{\text{1. diethyl ether}} \text{1-}(CH_2CH_2CH_3)\text{-1-}OH\text{ cyclohexane}$$

The product is 1-propylcyclohexanol.

(*d*) The starting material is a ketone and so reacts with a Grignard reagent to give a tertiary alcohol.

$$\underset{\text{Propylmagnesium bromide + 2-butanone}}{CH_3CH_2CH_2-MgBr + (CH_3)(CH_3CH_2)C{=}O} \xrightarrow{\text{diethyl ether}} CH_3-\overset{CH_3CH_2CH_2}{\underset{CH_3CH_2}{C}}-OMgBr \xrightarrow{H_3O^+} \underset{\text{3-Methyl-3-hexanol}}{CH_3CH_2CH_2\overset{CH_3}{\underset{CH_2CH_3}{C}}OH}$$

15.6 (*b*) The target alcohol is secondary and so can be prepared by the reaction of a Grignard reagent with an aldehyde. The two retrosynthetic transformations are:

$$C_6H_5\underset{OH}{C}HCH_2CH_3 \Longrightarrow C_6H_5^- \quad H\underset{O}{\overset{\|}{C}}CH_2CH_3$$

and

$$C_6H_5\underset{OH}{C}HCH_2CH_3 \Longrightarrow C_6H_5\overset{O}{\overset{\|}{C}}H \quad \bar{C}H_2CH_3$$

Therefore, two plausible syntheses are:

$$C_6H_5MgBr + CH_3CH_2\overset{O}{\overset{\|}{C}}H \xrightarrow[2.\ H_3O^+]{1.\ \text{diethyl ether}} C_6H_5\underset{OH}{\underset{|}{C}}HCH_2CH_3$$

Phenylmagnesium bromide · Propanal (propionaldehyde) · 1-Phenyl-1-propanol

$$\text{and}\quad CH_3CH_2MgBr + C_6H_5\overset{O}{\overset{\|}{C}}H \xrightarrow[2.\ H_3O^+]{1.\ \text{diethyl ether}} C_6H_5\underset{OH}{\underset{|}{C}}HCH_2CH_3$$

Ethylmagnesium bromide · Benzaldehyde · 1-Phenyl-1-propanol

(c) The target alcohol is tertiary and so is prepared by addition of a Grignard reagent to a ketone.

$$CH_3^- \;\; C_6H_5\overset{O}{\overset{\|}{C}}{-}CH_3 \Longleftarrow CH_3{-}\overset{OH}{\overset{|}{\underset{C_6H_5}{\underset{|}{C}}}}{-}CH_3 \Longrightarrow CH_3\overset{O}{\overset{\|}{C}}CH_3 \;\; C_6H_5^-$$

Since two of the substituents on the hydroxyl-bearing carbon are methyl groups, there are only *two*, not three, distinct ways of preparing 2-phenyl-2-propanol by a Grignard reaction.

$$CH_3MgI + C_6H_5\overset{O}{\overset{\|}{C}}CH_3 \xrightarrow[2.\ H_3O^+]{1.\ \text{diethyl ether}} C_6H_5\overset{OH}{\overset{|}{\underset{CH_3}{\underset{|}{C}}}}CH_3$$

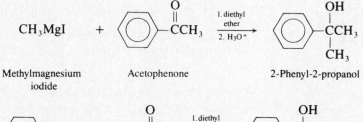

Methylmagnesium iodide · Acetophenone · 2-Phenyl-2-propanol

$$C_6H_5MgBr + CH_3\overset{O}{\overset{\|}{C}}CH_3 \xrightarrow[2.\ H_3O^+]{1.\ \text{diethyl ether}} C_6H_5{-}\overset{OH}{\overset{|}{\underset{CH_3}{\underset{|}{C}}}}{-}CH_3$$

Phenylmagnesium bromide · Acetone · 2-Phenyl-2-propanol

15.7 (b) The preparation of 6-methyl-6-undecanol has been described in the chemical literature. It was achieved in 75 percent yield by the reaction of pentylmagnesium bromide with ethyl acetate.

$$2\,CH_3CH_2CH_2CH_2CH_2MgBr + CH_3\overset{O}{\overset{\|}{C}}OCH_2CH_3 \xrightarrow[2.\ H_3O^+]{1.\ \text{diethyl ether}} CH_3\overset{OH}{\overset{|}{C}}(CH_2CH_2CH_2CH_2CH_3)_2$$

Pentylmagnesium bromide · Ethyl acetate · 6-Methyl-6-undecanol

(c) The two phenyl substituents arise by addition of a phenyl Grignard reagent to an ester of cyclopropanecarboxylic acid.

$$2\,C_6H_5MgBr + C_3H_5\overset{O}{\overset{\|}{C}}OCH_3 \xrightarrow[2.\ H_3O^+]{1.\ \text{diethyl ether}} (C_6H_5)_2\underset{OH}{\underset{|}{C}}{-}C_3H_5$$

Phenylmagnesium bromide · Methyl cyclopropanecarboxylate · 1-Cyclopropyl-1,1-diphenylmethanol (*d*) All the substituents bonded to the $-\overset{|}{\underset{|}{C}}OH$ group in this compound are phenyl groups.

Therefore use an ester of benzoic acid and phenylmagnesium bromide.

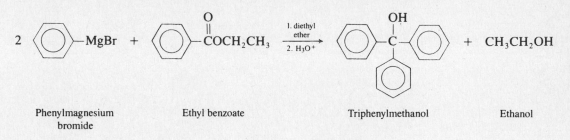

$$2\ C_6H_5MgBr + C_6H_5\overset{O}{\overset{\|}{C}}OCH_2CH_3 \xrightarrow[2.\ H_3O^+]{1.\ \text{diethyl ether}} (C_6H_5)_3COH + CH_3CH_2OH$$

Phenylmagnesium bromide | Ethyl benzoate | Triphenylmethanol | Ethanol

15.8 (*b*) Of the three methyl groups of 1,3,3-trimethylcyclopentene, only the one connected to the double bond can be attached by way of an organocuprate reagent. Attachment of either of the other methyls would involve a tertiary carbon, a process that does not occur very efficiently.

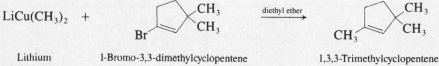

$$LiCu(CH_3)_2 + \text{1-Bromo-3,3-dimethylcyclopentene} \xrightarrow{\text{diethyl ether}} \text{1,3,3-Trimethylcyclopentene}$$

Lithium dimethylcuprate | 1-Bromo-3,3-dimethylcyclopentene | 1,3,3-Trimethylcyclopentene

15.9 Either the C-1–C-2 bond or C-2–C-3 bond of methylcyclopropane can be formed in the cyclization reaction.

Mentally disconnect C-1–C-2 bond:

$$\text{Methylcyclopropane} \Rightarrow CH_3CHBrCH_2CH_2Br$$

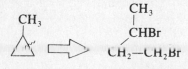

1,3-Dibromobutane

Mentally disconnect C-2–C-3 bond:

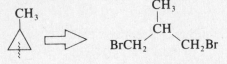

$$\text{Methylcyclopropane} \Rightarrow BrCH_2CH(CH_3)CH_2Br$$

1,3-Dibromo-2-methylpropane

Both 1,3-dibromobutane and 1,3-dibromo-2-methylpropane are suitable starting materials for the preparation of methylcyclopropane by zinc-promoted cyclization.

15.10 (*b*) Methylenecyclobutane is the appropriate precursor to the spirohexane shown.

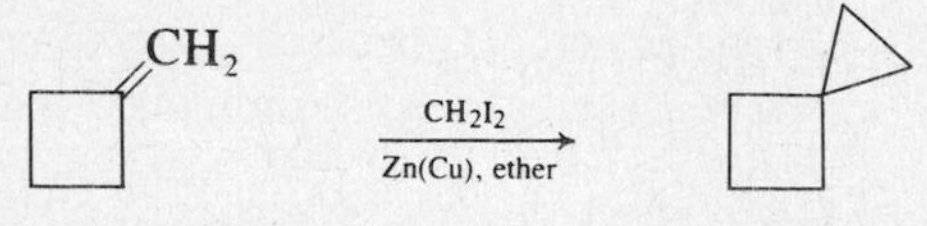

Methylenecyclobutane | Spiro[3.2]hexane (22%)

(*c*) Cyclobutene reacts with the Simmons-Smith reagent to give bicyclo[2.1.0]pentane.

$$\text{Cyclobutene} \xrightarrow[\text{Zn(Cu), ether}]{CH_2I_2} \text{Bicyclo[2.1.0]pentane}$$

Cyclobutene | Bicyclo[2.1.0]pentane (47%)

(*d*) The appropriate reaction is

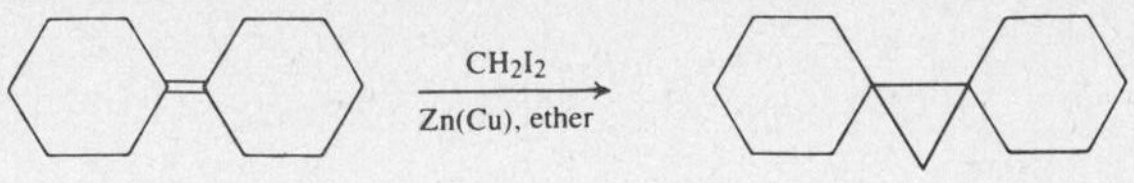

$$\text{Bicyclohexylidene} \xrightarrow[\text{Zn(Cu), ether}]{CH_2I_2} \text{Dispiro[5.1.5.0]tridecane (87\%)}$$

Bicyclohexylidene   Dispiro[5.1.5.0]tridecane (87%)

15.11 (*a*) Cyclopentyllithium is

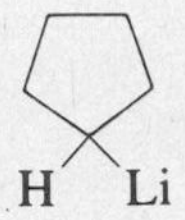

H Li

It has a carbon-lithium bond. It satisfies the requirement for classification as an organometallic compound.

(*b*) Ethoxymagnesium chloride does not have a carbon-metal bond. It is not an organometallic compound.

$$CH_3CH_2OMgCl \quad \text{or} \quad CH_3CH_2O^-Mg^{2+}Cl^-$$

(*c*) 2-Phenylethylmagnesium iodide is an example of a Grignard reagent. It is an organometallic compound.

$$C_6H_5{-}CH_2CH_2MgI$$

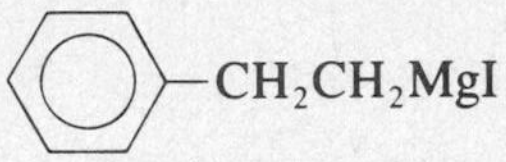

(*d*) Lithium divinylcuprate has two vinyl groups bonded to copper. It is an organometallic.

$$Li^+(CH_2{=}CH{-}\bar{C}u{-}CH{=}CH_2)$$

(*e*) Mercuric acetate is not an organometallic compound. Two acetate groups are ionically bonded to mercury through oxygen, not carbon.

$$Hg^{2+}(\bar{O}\overset{O}{\overset{\|}{C}}CH_3)_2$$

(*f*) Benzylpotassium is represented as

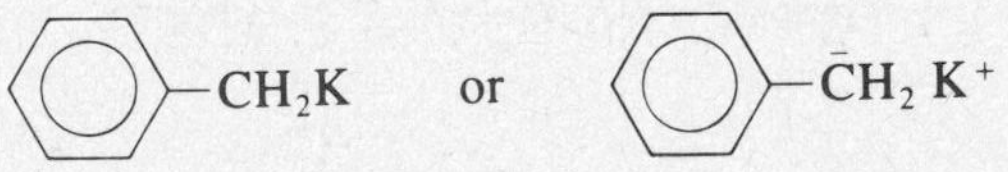

$$C_6H_5{-}CH_2K \quad \text{or} \quad C_6H_5{-}\bar{C}H_2\ K^+$$

It has a carbon-potassium bond and thus is an organometallic compound.

(*g*) Sodium *p*-toluenesulfonate has its sodium atom ionically bonded to oxygen. It is not an organometallic compound.

$$CH_3{-}C_6H_4{-}\overset{O}{\overset{\|}{\underset{\underset{O}{\|}}{S}}}O^-Na^+$$

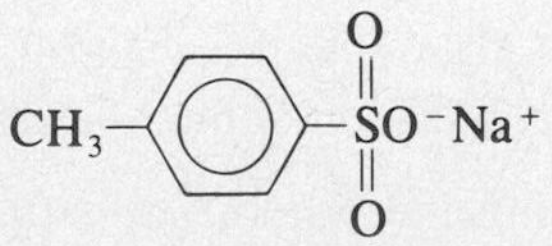

15.12 (*a*) Grignard reagents such as pentylmagnesium iodide are prepared by reaction of magnesium with the corresponding alkyl halide.

$$CH_3CH_2CH_2CH_2CH_2I + Mg \xrightarrow{\text{diethyl ether}} CH_3CH_2CH_2CH_2CH_2MgI$$

1-Iodopentane   Pentylmagnesium iodide

(*b*) Trialkylboranes are most conveniently prepared by addition of diborane to alkenes.

$$6\,CH_3CH_2CH_2CH{=}CH_2 + B_2H_6 \longrightarrow 2(CH_3CH_2CH_2CH_2CH_2)_3B$$

1-Pentene   Tripentylborane

(c) Alkyllithiums are formed by reaction of lithium with an alkyl halide.

$$\underset{\substack{\text{1-Halopentane}\\ (X = Cl,\ Br,\ or\ I)}}{CH_3CH_2CH_2CH_2CH_2X} + 2\,Li \longrightarrow \underset{\text{Pentyllithium}}{CH_3CH_2CH_2CH_2CH_2Li} + LiX$$

(d) Lithium dialkylcuprates arise by the reaction of an alkyllithium with a cuprous salt.

$$\underset{\substack{\text{Pentyllithium}\\ \text{(from } c)}}{2\,CH_3CH_2CH_2CH_2CH_2Li} + \underset{(X = Cl,\ Br,\ or\ I)}{CuX} \longrightarrow \underset{\text{Lithium dipentylcuprate}}{LiCu(CH_2CH_2CH_2CH_2CH_3)_2} + LiX$$

(e) Dialkylcadmium reagents are normally prepared by the reaction of Grignard reagents with cadmium chloride.

$$\underset{\substack{\text{Pentylmagnesium iodide}\\ \text{[from } (a)]}}{2\,CH_3CH_2CH_2CH_2CH_2MgI} + CdCl_2 \xrightarrow{\text{diethyl ether}} \underset{\text{Dipentylcadmium}}{(CH_3CH_2CH_2CH_2CH_2)_2Cd} + 2\,MgClI$$

15.13 (a) In comparing a carbon-lithium bond and a carbon-boron bond, the more polar bond will involve the more electropositive atom. Metallic character and electropositivity increase in going down the periodic table but decrease in going across a row. Lithium is more electropositive than boron, therefore

$$CH_3\overset{\delta-}{C}H_2{-}\overset{\delta+}{Li} \quad \text{has a more polar carbon-metal bond than} \quad CH_3\overset{\delta-}{C}H_2{-}\overset{\delta+}{B}(CH_2CH_3)_2$$

Table 1.2 (Section 1.5) lists electronegativity values for a number of elements according to the Pauling scale. Lithium's electronegativity is 1.0, which makes it more electropositive than boron with an electronegativity of 2.0.

(b) Beryllium and magnesium are both group II elements. Metallic character and electropositivity increase in going down the periodic table, so magnesium is more electropositive than beryllium and should have a more polar bond to carbon.

$$\overset{\delta-}{C}H_3{-}\overset{\delta+}{Mg}{-}\overset{\delta-}{C}H_3 \quad \text{has more polar carbon-metal bonds than} \quad \overset{\delta-}{C}H_3{-}\overset{\delta+}{Be}{-}\overset{\delta-}{C}H_3$$

According to Table 1.2, the Pauling electronegativity of magnesium is 1.2 while that of beryllium is 1.5—thus magnesium is in fact more electropositive than beryllium.

(c) Magnesium is more electropositive than silicon, since both are third-row elements and electropositivity and metallic character decrease in going across a row. The electronegativity of magnesium is 1.2 on the Pauling scale while that of silicon is 1.8.

$$\overset{\delta-}{C}H_3{-}\overset{\delta+}{Mg}CH_3 \quad \text{has more polar carbon-metal bonds than} \quad \overset{\delta-}{C}H_3{-}\overset{\delta+}{Si}(CH_3)_3$$

(d) Sodium and aluminum are both third-row elements. Sodium is more electropositive than aluminum, their respective Pauling electronegativities from Table 1.2 being 0.9 and 1.5, respectively. The carbon-sodium bond is more polar than the carbon-aluminum bond—indeed, methylsodium is ionic.

$$\overset{-}{C}H_3Na^+ \quad \text{has a more polar carbon-metal bond than} \quad \overset{\delta-}{C}H_3\overset{\delta+}{Al}H_2$$

15.14 (a) $\underset{\text{1-Bromopropane}}{CH_3CH_2CH_2Br} + 2\,Li \xrightarrow{\text{diethyl ether}} \underset{\text{Propyllithium}}{CH_3CH_2CH_2Li} + LiBr$

(*b*) $CH_3CH_2CH_2Br + Mg \xrightarrow{\text{diethyl ether}} CH_3CH_2CH_2MgBr$

1-Bromopropane → Propylmagnesium bromide

(*c*) $CH_3CH(I)CH_3 + 2\,Li \xrightarrow{\text{diethyl ether}} CH_3CH(Li)CH_3 + LiI$

2-Iodopropane → Isopropyllithium

(*d*) $CH_3CH(I)CH_3 + Mg \xrightarrow{\text{diethyl ether}} CH_3CH(MgI)CH_3$

2-Iodopropane → Isopropylmagnesium iodide

(*e*) $CH_3CH(I)CH_3 \xrightarrow[\text{acetic acid}]{\text{Zn}} CH_3CH_2CH_3$

2-Iodopropane → Propane

(*f*) $2\,CH_3CH_2CH_2Li + CuI \longrightarrow (CH_3CH_2CH_2)_2CuLi$

Propyllithium → Lithium dipropylcuprate

(*g*) $(CH_3CH_2CH_2)_2CuLi + CH_3CH_2CH_2CH_2Br \longrightarrow CH_3CH_2CH_2CH_2CH_2CH_2CH_3$

Lithium dipropylcuprate + 1-Bromobutane → Heptane

(*h*) $(CH_3CH_2CH_2)_2CuLi + C_6H_5I \longrightarrow C_6H_5CH_2CH_2CH_3$

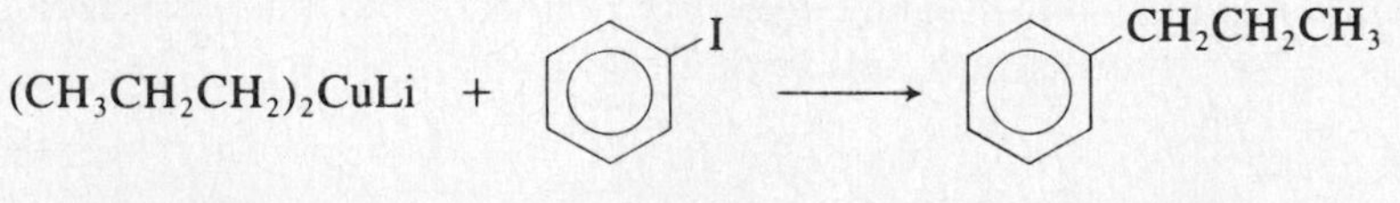

Lithium dipropylcuprate + Iodobenzene → Propylbenzene

(*i*) $CH_3CH_2CH_2MgBr \xrightarrow[\text{DCl}]{D_2O} CH_3CH_2CH_2D$

Propylmagnesium bromide → 1-Deuteriopropane

(*j*) $CH_3CH(MgI)CH_3 \xrightarrow[\text{DCl}]{D_2O} CH_3CH(D)CH_3$

Isopropylmagnesium iodide → 2-Deuteriopropane

(*k*) $CH_3CH_2CH_2Li + HCH{=}O \xrightarrow[\text{2. } H_3O^+]{\text{1. diethyl ether}} CH_3CH_2CH_2CH_2OH$

Propyllithium → 1-Butanol

(*l*) $CH_3CH_2CH_2MgBr + C_6H_5CH{=}O \xrightarrow[\text{2. } H_3O^+]{\text{1. diethyl ether}} C_6H_5CH(OH)CH_2CH_2CH_3$

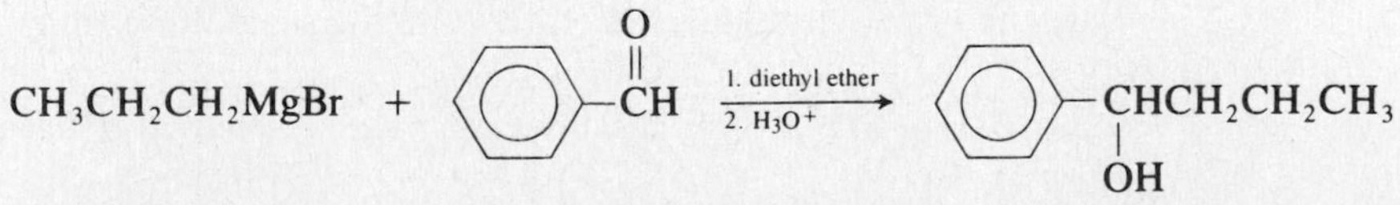

Propylmagnesium bromide + Benzaldehyde → 1-Phenyl-1-butanol

(*m*) $CH_3CH(Li)CH_3$ + cycloheptanone $\xrightarrow[\text{2. } H_3O^+]{\text{1. diethyl ether}}$ 1-$(CH_3)_2CH$, 1-OH cycloheptane

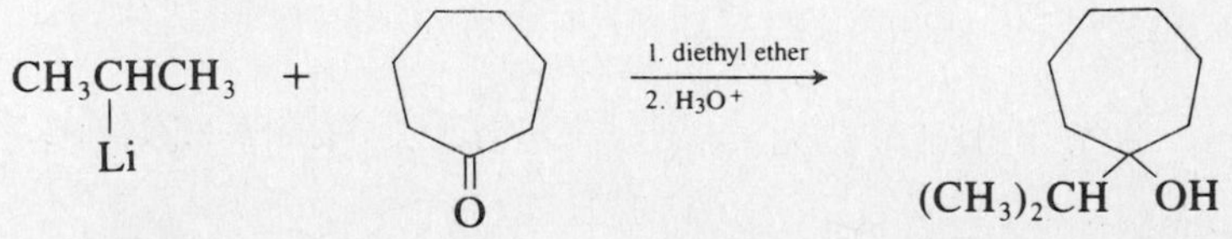

Isopropyllithium + Cycloheptanone → 1-(Isopropyl)cycloheptanol

(*n*) $CH_3CH(MgI)CH_3$ + $CH_3\overset{O}{\overset{\|}{C}}CH_2CH_3$ $\xrightarrow[2.\ H_3O^+]{1.\ \text{diethyl ether}}$ $CH_3CH(CH_3){-}C(OH)(CH_3)CH_2CH_3$

Isopropylmagnesium iodide — 2-Butanone — 2,3-Dimethyl-3-pentanol

(*o*) $CH_3CH_2CH_2MgBr$ + $C_6H_5\overset{O}{\overset{\|}{C}}OCH_3$ $\xrightarrow[2.\ H_3O^+]{1.\ \text{diethyl ether}}$ $C_6H_5\overset{OH}{\overset{|}{C}}(CH_2CH_2CH_3)_2$

Propylmagnesium bromide — Methyl benzoate — 4-Phenyl-4-heptanol

(*p*) $CH_2{=}CH(CH_2)_5CH_3$ $\xrightarrow[\text{Zn(Cu)},\ \text{diethyl ether}]{CH_2I_2}$ $CH_2{-}CH(CH_2)_5CH_3$ (ring closed by CH_2)

1-Octene — 1-Cyclopropylhexane

(*q*) $(CH_3)(H)C{=}C(H)((CH_2)_6CH_3)$ $\xrightarrow[\text{Zn(Cu)},\ \text{diethyl ether}]{CH_2I_2}$ cyclopropane bearing CH_3, H and H, $(CH_2)_6CH_3$ (trans)

(*E*)-2-Decene — *trans*-1-Heptyl-2-methylcyclopropane

(*r*) $(CH_3CH_2)(H)C{=}C((CH_2)_5CH_3)(H)$ $\xrightarrow[\text{Zn(Cu)},\ \text{diethyl ether}]{CH_2I_2}$ cyclopropane bearing CH_3CH_2, H and $(CH_2)_5CH_3$, H (cis)

(*Z*)-3-Decene — *cis*-1-Ethyl-2-hexylcyclopropane

(*s*) $ICH_2CH(I)CH_3$ $\xrightarrow[\text{ethanol}]{Zn}$ $CH_2{=}CHCH_3$

1,2-Diiodopropane — Propene

(*t*) $ICH_2CH_2CH_2I$ $\xrightarrow[\text{ethanol}]{Zn}$ $CH_2{-}CH_2$ (ring closed by CH_2)

1,3-Diiodopropane — Cyclopropane

15.15 In the solutions to 15.15*a* through *e* shown, the Grignard reagent butylmagnesium bromide is used. In each case the use of butyllithium would be equally satisfactory.

(*a*) 1-Pentanol is a primary alcohol having one more carbon atom than 1-bromobutane. This suggests the reaction of a Grignard reagent with formaldehyde.

$$CH_3CH_2CH_2CH_2Br \xrightarrow[\text{diethyl ether}]{Mg} CH_3CH_2CH_2CH_2MgBr \xrightarrow[2.\ H_3O^+]{1.\ H\overset{O}{\overset{\|}{C}}H} CH_3CH_2CH_2CH_2CH_2OH$$

1-bromobutane — Butylmagnesium bromide — 1-Pentanol

(*b*) 2-Hexanol is a secondary alcohol having two more carbon atoms than 1-bromobutane. It may be prepared by reaction of ethanal with butylmagnesium bromide.

$$CH_3CH_2CH_2CH_2Br \xrightarrow[\text{diethyl ether}]{Mg} CH_3CH_2CH_2CH_2MgBr \xrightarrow[2.\ H_3O^+]{1.\ CH_3\overset{O}{\overset{\|}{C}}H} CH_3CH_2CH_2CH_2\underset{OH}{\underset{|}{C}H}CH_3$$

1-Bromobutane — Butylmagnesium bromide — 2-Hexanol

(*c*) 1-Phenyl-1-pentanol is a secondary alcohol. This suggests that it can be prepared from butylmagnesium bromide and an aldehyde; benzaldehyde is the appropriate one.

$$CH_3CH_2CH_2CH_2MgBr + C_6H_5CH{=}O \xrightarrow[2.\ H_3O^+]{1.\ ether} C_6H_5CH(OH)CH_2CH_2CH_2CH_3$$

Butylmagnesium bromide — Benzaldehyde — 1-Phenyl-1-pentanol

(*d*) The target molecule 3-methyl-3-heptanol has the structure

$$CH_3CH_2CH_2CH_2{-}C(CH_3)(OH){-}CH_2CH_3$$

By retrosynthetically disconnecting the butyl group from the carbon that bears the hydroxyl substituent, we see that the appropriate starting ketone is 2-butanone.

$$CH_3CH_2CH_2CH_2{-}C(CH_3)(OH){-}CH_2CH_3 \Longrightarrow CH_3CH_2CH_2CH_2^- \quad O{=}C(CH_3)CH_2CH_3$$

2-Butanone

Therefore

$$CH_3CH_2CH_2CH_2MgBr + CH_3\overset{O}{\overset{\|}{C}}CH_2CH_3 \xrightarrow[2.\ H_3O^+]{1.\ diethyl\ ether} CH_3CH_2CH_2CH_2C(CH_3)(OH)CH_2CH_3$$

Butylmagnesium bromide — 2-Butanone — 3-Methyl-3-heptanol

(*e*) 1-Butylcyclobutanol is a tertiary alcohol. The appropriate ketone is cyclobutanone.

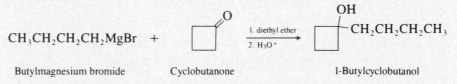

$$CH_3CH_2CH_2CH_2MgBr + \text{cyclobutanone} \xrightarrow[2.\ H_3O^+]{1.\ diethyl\ ether} \text{1-(OH)-1-}(CH_2CH_2CH_2CH_3)\text{cyclobutane}$$

Butylmagnesium bromide — Cyclobutanone — 1-Butylcyclobutanol

15.16 (*a*) Conversion of bromobenzene to benzyl alcohol requires formation of the corresponding Grignard reagent and its reaction with formaldehyde.

$$C_6H_5Br \xrightarrow[\text{diethyl ether}]{Mg} C_6H_5MgBr \xrightarrow[2.\ H_3O^+]{1.\ H\overset{O}{\overset{\|}{C}}H} C_6H_5CH_2OH$$

Bromobenzene — Phenylmagnesium bromide — Benzyl alcohol

(*b*) The product is a secondary alcohol and is formed by reaction of phenylmagnesium bromide with hexanal.

$$C_6H_5MgBr + CH_3CH_2CH_2CH_2CH_2\overset{O}{\overset{\|}{C}}H \longrightarrow C_6H_5CH(OH)CH_2CH_2CH_2CH_2CH_3$$

Phenylmagnesium bromide — Hexanal — 1-Phenyl-1-hexanol

(*c*) Since the desired product is a secondary alcohol, phenylmagnesium bromide should be treated with an aldehyde. The appropriate aldehyde is benzaldehyde.

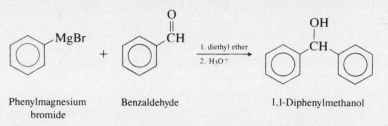

$$C_6H_5MgBr + C_6H_5CH{=}O \xrightarrow[\text{2. } H_3O^+]{\text{1. diethyl ether}} C_6H_5CH(OH)C_6H_5$$

Phenylmagnesium bromide　　Benzaldehyde　　1,1-Diphenylmethanol

(*d*) The target molecule is a tertiary alcohol, which requires that phenylmagnesium bromide react with a ketone. By mentally disconnecting the phenyl group from the carbon that bears the hydroxyl, we see that the appropriate ketone is 4-heptanone.

$$CH_3CH_2CH_2C(OH)(C_6H_5)CH_2CH_2CH_3 \Longrightarrow CH_3CH_2CH_2\overset{O}{\overset{\|}{C}}CH_2CH_2CH_3 + C_6H_6$$

4-Phenyl-4-heptanol　　4-Heptanone

The synthesis is therefore:

$$C_6H_5MgBr + CH_3CH_2CH_2\overset{O}{\overset{\|}{C}}CH_2CH_2CH_3 \xrightarrow[\text{2. } H_3O^+]{\text{1. diethyl ether}} CH_3CH_2CH_2C(OH)(C_6H_5)CH_2CH_2CH_3$$

Phenylmagnesium bromide　　4-Heptanone　　4-Phenyl-4-heptanol

(*e*) Reaction of phenylmagnesium bromide with cyclooctanone will give the desired tertiary alcohol.

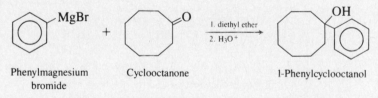

$$C_6H_5MgBr + \text{cyclooctanone} \xrightarrow[\text{2. } H_3O^+]{\text{1. diethyl ether}} \text{1-phenylcyclooctanol}$$

Phenylmagnesium bromide　　Cyclooctanone　　1-Phenylcyclooctanol

15.17 In these problems the principles of retrosynthetic analysis are applied. The alkyl groups attached to the carbon that bears the hydroxyl group are mentally disconnected to reveal the Grignard reagent and carbonyl compound.

(*a*)

$$CH_3CH_2\underset{OH}{\underset{|}{C}}HCH_2CH(CH_3)_2$$

5-Methyl-3-hexanol

$$\left[CH_3CH_2\underset{O}{\underset{\|}{C}}H + XMgCH_2CH(CH_3)_2\right] \qquad \left[CH_3CH_2MgX + H\underset{O}{\underset{\|}{C}}CH_2CH(CH_3)_2\right]$$

Propanal　　Isobutylmagnesium halide　　Ethylmagnesium halide　　3-Methylbutanal

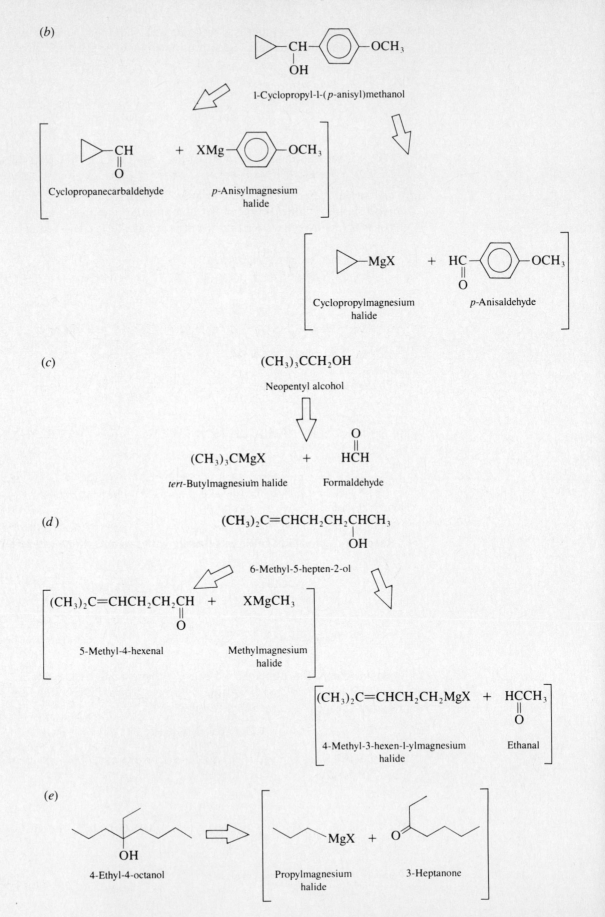

(*b*)

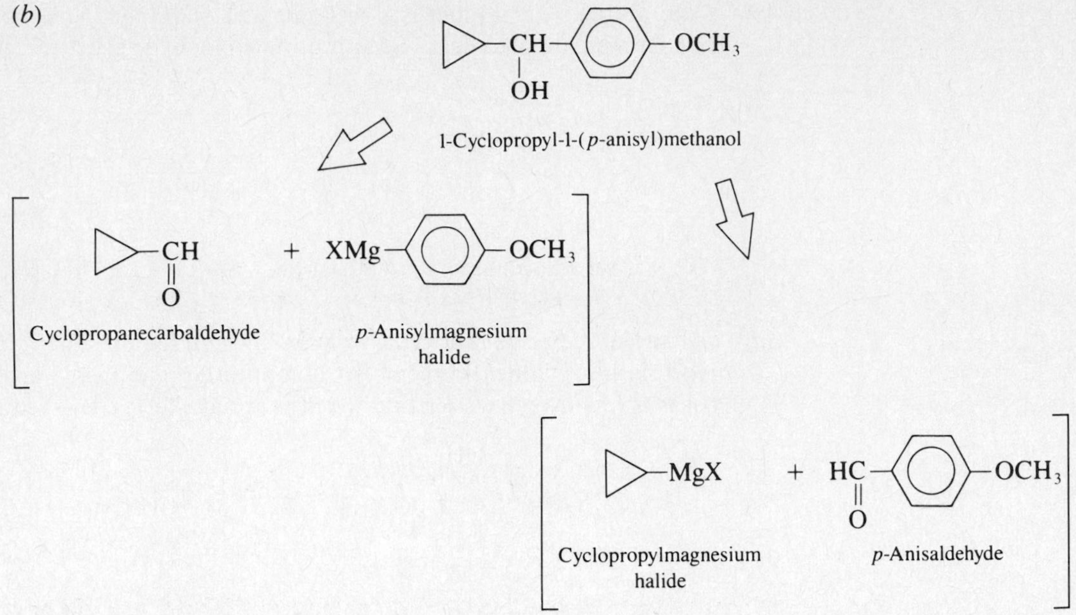

1-Cyclopropyl-1-(*p*-anisyl)methanol

[Cyclopropanecarbaldehyde + *p*-Anisylmagnesium halide]

[Cyclopropylmagnesium halide + *p*-Anisaldehyde]

(*c*)

$(CH_3)_3CCH_2OH$

Neopentyl alcohol

$(CH_3)_3CMgX$ + HCH (C=O)

tert-Butylmagnesium halide Formaldehyde

(*d*)

$(CH_3)_2C{=}CHCH_2CH_2CHCH_3$ (OH)

6-Methyl-5-hepten-2-ol

[$(CH_3)_2C{=}CHCH_2CH_2CH$ (C=O) + $XMgCH_3$]

5-Methyl-4-hexenal Methylmagnesium halide

[$(CH_3)_2C{=}CHCH_2CH_2MgX$ + $HCCH_3$ (C=O)]

4-Methyl-3-hexen-1-ylmagnesium halide Ethanal

(*e*)

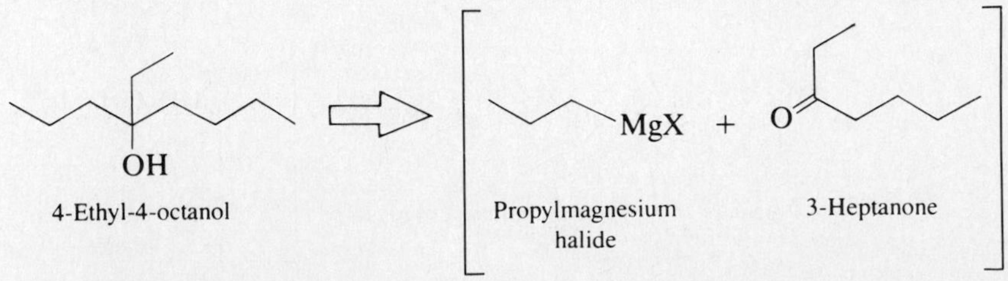

4-Ethyl-4-octanol (OH)

[MgX + O]

Propylmagnesium halide 3-Heptanone

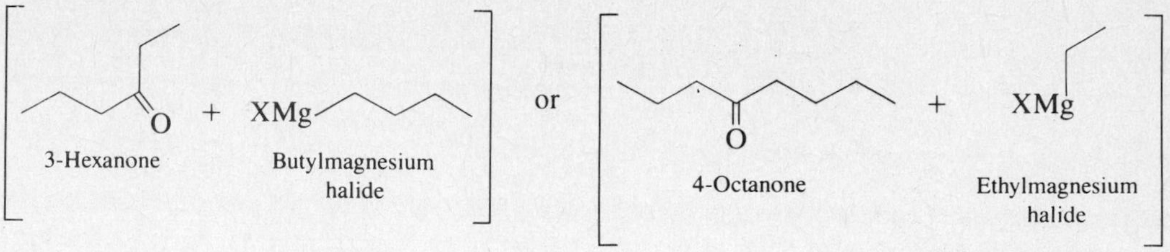

[3-Hexanone + Butylmagnesium halide] or [4-Octanone + Ethylmagnesium halide]

3-Hexanone + XMg–(butyl)

Butylmagnesium halide

4-Octanone + XMg–(ethyl)

Ethylmagnesium halide

15.18 (*a*) Meparfynol is a tertiary alcohol and so can be prepared by addition of a carbanionic species to a ketone. Use the same reasoning which applies to the synthesis of alcohols from Grignard reagents. On mentally disconnecting one of the bonds to the carbon bearing the hydroxyl group

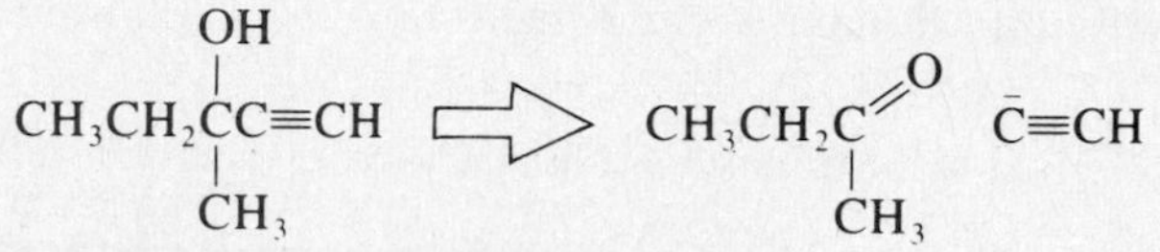

$$CH_3CH_2C(OH)(CH_3)C{\equiv}CH \Longrightarrow CH_3CH_2C(=O)CH_3 \quad \bar{C}{\equiv}CH$$

we see that the addition of acetylide ion to 2-butanone will provide the target molecule.

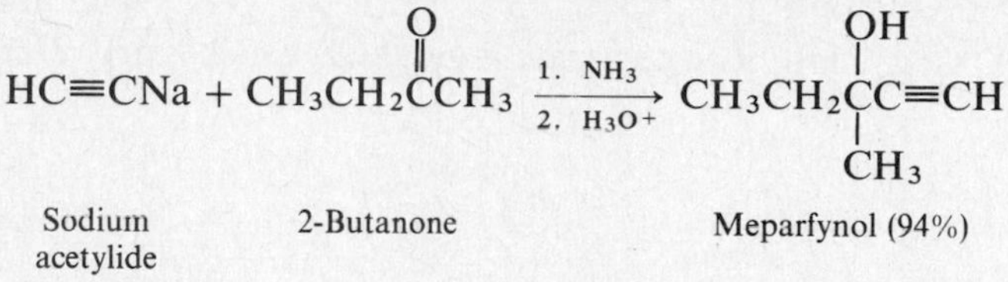

$$\underset{\text{Sodium acetylide}}{HC{\equiv}CNa} + \underset{\text{2-Butanone}}{CH_3CH_2\overset{O}{\overset{\|}{C}}CH_3} \xrightarrow[2.\ H_3O^+]{1.\ NH_3} \underset{\text{Meparfynol (94\%)}}{CH_3CH_2C(OH)(CH_3)C{\equiv}CH}$$

The alternative, reaction of a Grignard reagent with an alkynyl ketone, is not acceptable in this case. The acidic terminal alkyne C—H would protonate, and hence destroy, the anionic Grignard reagent.

(*b*) Diphepanol is a tertiary alcohol and so may be prepared by reaction of a Grignard or organolithium reagent with a ketone. Retrosynthetically, two possibilities seem reasonable.

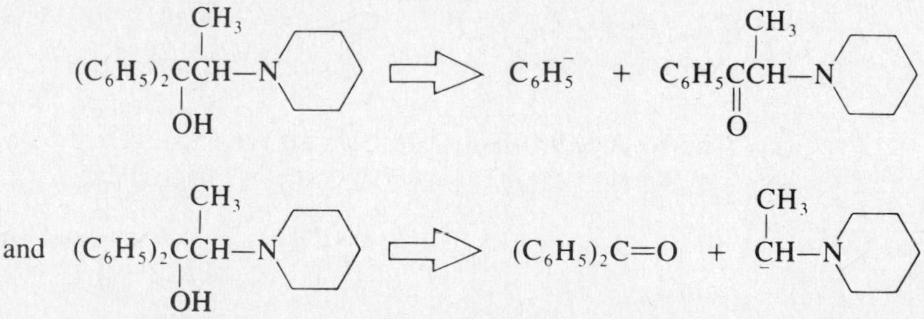

$$(C_6H_5)_2C(OH)CH(CH_3)\text{—N(piperidine)} \Longrightarrow C_6H_5^- + C_6H_5C(=O)CH(CH_3)\text{—N(piperidine)}$$

and

$$(C_6H_5)_2C(OH)CH(CH_3)\text{—N(piperidine)} \Longrightarrow (C_6H_5)_2C{=}O + \bar{C}H(CH_3)\text{—N(piperidine)}$$

In principle either strategy is acceptable; in practice the one involving phenylmagnesium bromide is used.

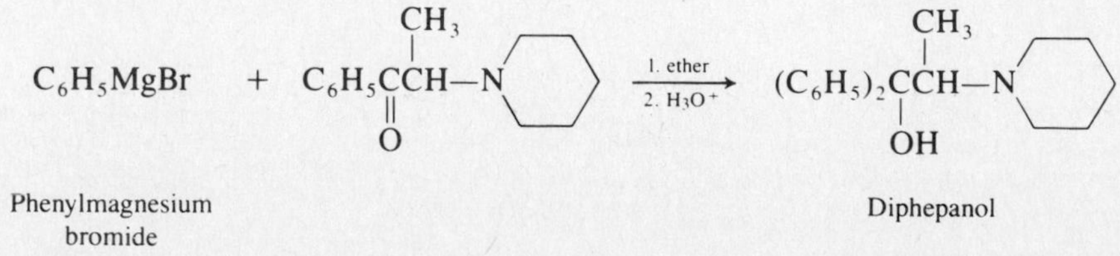

$$\underset{\text{Phenylmagnesium bromide}}{C_6H_5MgBr} + C_6H_5C(=O)CH(CH_3)\text{—N(piperidine)} \xrightarrow[2.\ H_3O^+]{1.\ ether} \underset{\text{Diphepanol}}{(C_6H_5)_2C(OH)CH(CH_3)\text{—N(piperidine)}}$$

(*c*) A reasonable last step in the synthesis of mestranol is the addition of sodium acetylide to the ketone shown.

CH_3 O ... CH_3O — 1. NaC≡CH, NH_3; 2. H_3O^+ → CH_3 OH, C≡CH ... CH_3O

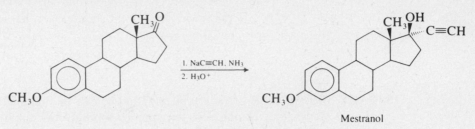

Mestranol

Acetylide anion adds to the carbonyl from the less sterically hindered side. The methyl group shields the top face of the carbonyl and so acetylide adds from the bottom.

15.19 (*a*) Sodium acetylide adds to ketones to give tertiary alcohols.

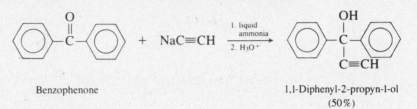

Benzophenone + NaC≡CH — 1. liquid ammonia; 2. H_3O^+ → 1,1-Diphenyl-2-propyn-1-ol (50%)

(*b*) The substrate is a ketone, which reacts with ethyllithium to yield a tertiary alcohol.

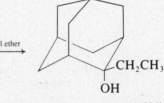

2-Adamantanone + CH_3CH_2Li — 1. diethyl ether; 2. H_3O^+ → 2-Ethyl-2-adamantanol (83%)

(*c*) The first step is conversion of bromocyclopentene to the corresponding Grignard reagent, which then reacts with formaldehyde to give a primary alcohol.

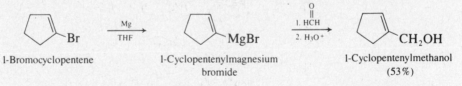

1-Bromocyclopentene — Mg, THF → 1-Cyclopentenylmagnesium bromide — 1. HCH (O); 2. H_3O^+ → 1-Cyclopentenylmethanol (53%)

(*d*) The reaction is one in which an alkene is converted to a cyclopropane through use of the Simmons-Smith reagent, iodomethylzinc iodide.

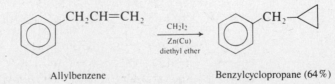

Allylbenzene ($C_6H_5CH_2CH{=}CH_2$) — CH_2I_2, Zn(Cu), diethyl ether → Benzylcyclopropane (64%)

(*e*) Methylene transfer using the Simmons-Smith reagent is stereospecific. The trans arrangement of substituents in the alkene is carried over to the cyclopropane product.

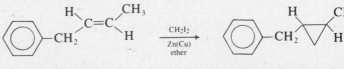

(*E*)-1-Phenyl-2-butene — CH_2I_2, Zn(Cu), ether → *trans*-1-Benzyl-2-methylcyclopropane (50%)

(*f*) Lithium dimethylcuprate transfers a methyl group, which substitutes for iodine on the iodoalkene. Even halogens on sp^2 hybridized carbon are reactive in substitution reactions with lithium dialkylcuprates.

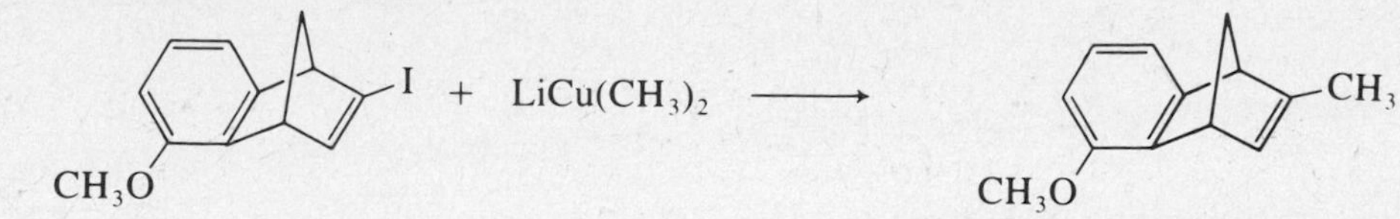

CH_3O ... $-I$ + $LiCu(CH_3)_2$ ⟶ CH_3O ... $-CH_3$

2-Iodo-8-methoxybenzonorbornadiene

8-Methoxy-2-methylbenzonorbornadiene (73%)

(*g*) The product of the reaction is the corresponding alkane, formed by reduction of the $-CH_2I$ function to $-CH_3$. The reaction does not affect any of the bonds to the chiral center. By working through the stereochemistry, we see that the absolute configuration of the product is *R*.

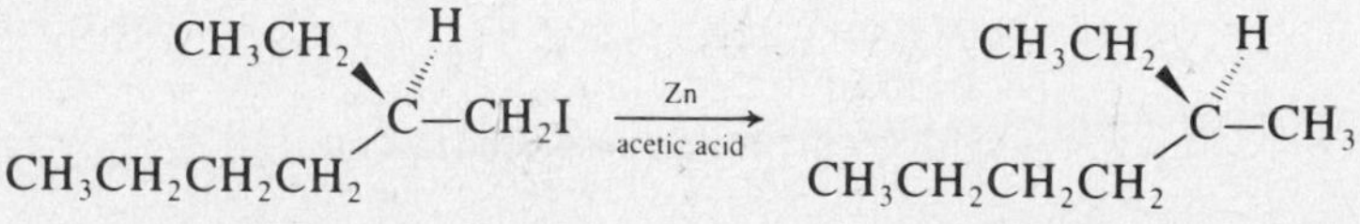

CH_3CH_2, H, $C-CH_2I$, $CH_3CH_2CH_2CH_2$ $\xrightarrow[\text{acetic acid}]{\text{Zn}}$ CH_3CH_2, H, $C-CH_3$, $CH_3CH_2CH_2CH_2$

(*R*)-2-Ethyl-1-iodohexane

(*R*)-2-Methylheptane

(*h*) Zinc reacts with vicinal dibromides by dehalogenation. The product is an alkene.

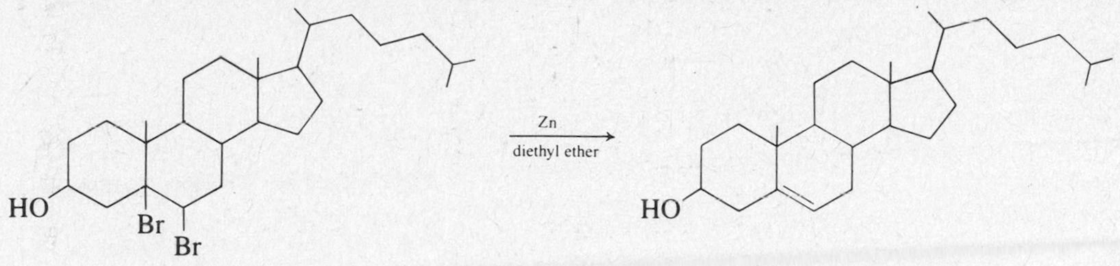

HO, Br, Br $\xrightarrow[\text{diethyl ether}]{\text{Zn}}$ HO

5,6-Dibromocholestanol

Cholesterol (78–80%)

(*i*) Cyclopropanes result when 1,3-dihalides are treated with zinc.

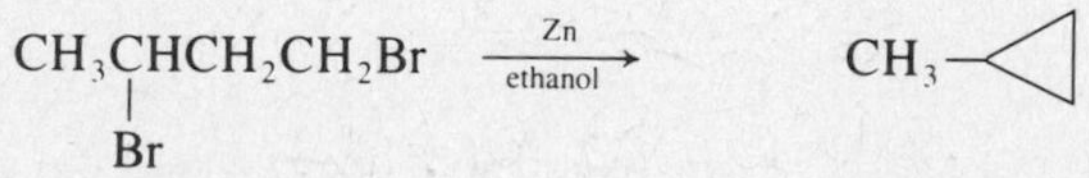

$CH_3CHCH_2CH_2Br$ (with Br on C-3) $\xrightarrow[\text{ethanol}]{\text{Zn}}$ CH_3–cyclopropane

1,3-Dibromobutane

Methylcyclopropane (70–88%)

15.20 Phenylmagnesium bromide reacts with 4-*tert*-butylcyclohexanone as shown:

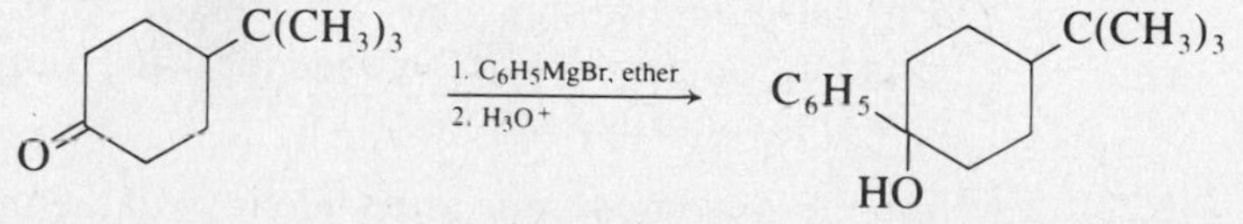

O, $C(CH_3)_3$ $\xrightarrow[\text{2. } H_3O^+]{\text{1. } C_6H_5MgBr\text{, ether}}$ C_6H_5, HO, $C(CH_3)_3$

4-*tert*-Butylcyclohexanone

4-*tert*-butyl-1-phenylcyclohexenol

The phenyl substituent can be introduced either cis or trans to the *tert*-butyl group. Therefore the two alcohols are stereoisomers (diastereomers).

Dehydration of either alcohol yields 4-*tert*-butyl-1-phenylcyclohexene.

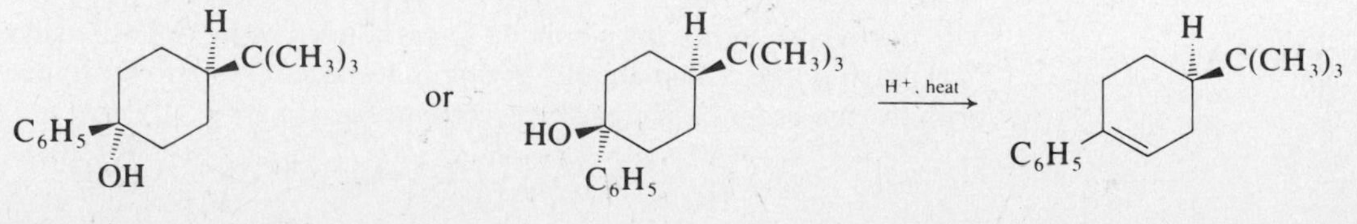

H, $C(CH_3)_3$, C_6H_5, OH or H, $C(CH_3)_3$, HO, C_6H_5 $\xrightarrow{H^+\text{, heat}}$ H, $C(CH_3)_3$, C_6H_5

15.21 (*a*) By working through the sequence of reactions that occur when ethyl formate reacts with a Grignard reagent, we can see that this combination leads to *secondary alcohols.*

$$\underset{\text{Grignard reagent}}{RMgX} + \underset{\text{Ethyl formate}}{H\overset{O}{\overset{\|}{C}}OCH_2CH_3} \longrightarrow \underset{\text{Aldehyde}}{R\overset{O}{\overset{\|}{C}}H} + CH_3CH_2OMgX$$

$$\xrightarrow{\text{1. RMgX, diethyl ether; 2. } H_3O^+} \underset{\text{Secondary alcohol}}{\underset{\underset{OH}{|}}{RCHR}}$$

This is simply because the substituent on the carbonyl carbon of the ester, in this case a hydrogen, is carried through and becomes a substituent on the hydroxyl-bearing carbon of the alcohol.

(*b*) Diethyl carbonate has the potential to react with *three* moles of a Grignard reagent.

$$\underset{\text{Grignard reagent}}{RMgX} + \underset{\text{Diethyl carbonate}}{CH_3CH_2O\overset{O}{\overset{\|}{C}}OCH_2CH_3} \longrightarrow \underset{\text{Ester}}{R\overset{O}{\overset{\|}{C}}OCH_2CH_3} + CH_3CH_2OMgX$$

$$\xrightarrow{\text{RMgX}} \underset{\text{Ketone}}{R\overset{O}{\overset{\|}{C}}R} + CH_3CH_2OMgX$$

$$\xrightarrow{\text{RMgX, then usual workup}} \underset{\text{Tertiary alcohol}}{R\overset{\overset{R}{|}}{\underset{\underset{OH}{|}}{C}}R}$$

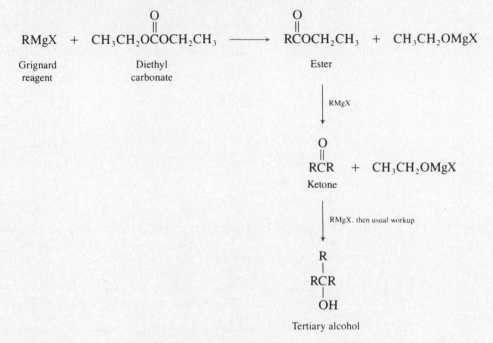

The tertiary alcohols that are formed by the reaction of diethyl carbonate with Grignard reagents have three identical R groups attached to the carbon that bears the hydroxyl substituent.

15.22 If we use the 2-bromobutane given, along with the information that the reaction occurs with net inversion of configuration, the stereochemical course of the reaction may be written as:

$$(CH_3CH_2)(CH_3)(Br)C{-}H \xrightarrow{LiCu(C_6H_5)_2} (CH_3)(CH_3CH_2)(C_6H_5)C{-}H$$

The phenyl group becomes bonded to carbon from the opposite side of the leaving group.

Applying the Cahn-Ingold-Prelog notational system described in Section 8.6 to the product, the order of decreasing precedence is

$$C_6H_5 > CH_3CH_2 > CH_3 > H$$

Orienting the molecule so that the lowest-priority substituent (H) is away from us, we see that the order of decreasing precedence is clockwise.

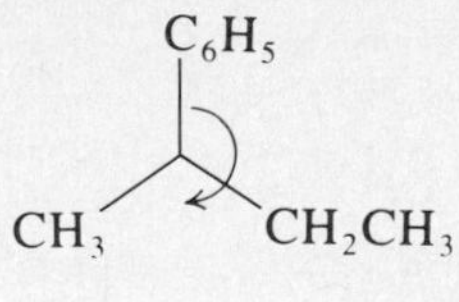

C_6H_5

CH_3 CH_2CH_3

The absolute configuration is *R*.

15.23 The substrates are secondary alkyl *p*-toluenesulfonates, so elimination is expected to compete with substitution. Compound B is formed in both reactions and has the molecular formula of 4-*tert*-butylcyclohexene. Since the two *p*-toluenesulfonates are diastereomers, it is likely that compounds A and C, especially since they have the same molecular formula, are also diastereomers. Assuming that the substitution reactions proceed with inversion of configuration, we conclude that the products are as shown.

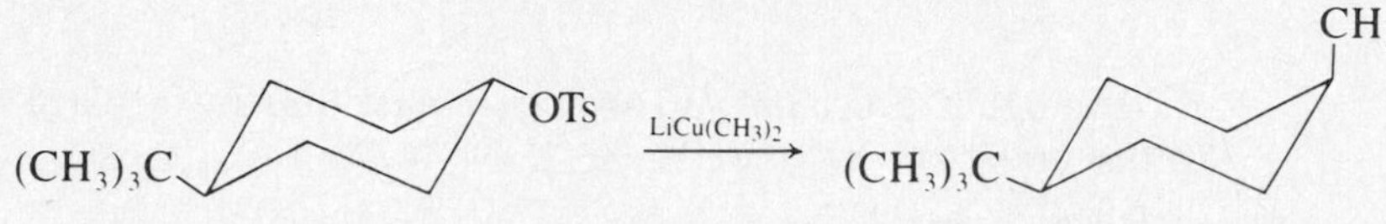

$(CH_3)_3C$–cyclohexyl–OTs $\xrightarrow{LiCu(CH_3)_2}$ $(CH_3)_3C$–cyclohexyl–CH_3 + $(CH_3)_3C$–cyclohexenyl

trans-4-*tert*-Butylcyclohexyl *p*-toluenesulfonate

cis-4-*tert*-Butyl-1-methylcyclohexane (compound A, $C_{11}H_{22}$)

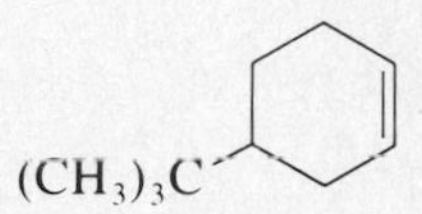

4-*tert*-Butylcyclohexene (compound B, $C_{10}H_{18}$)

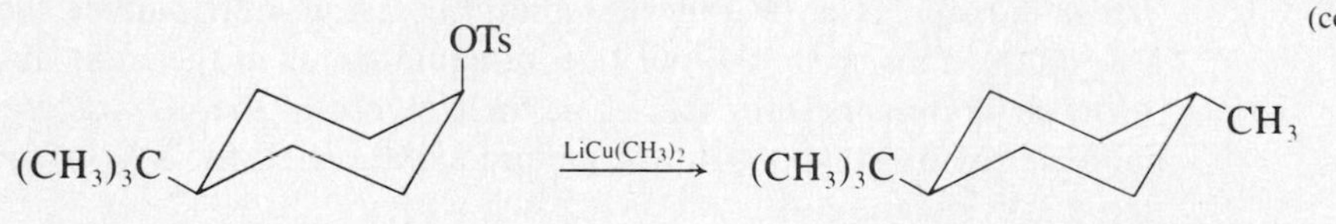

$(CH_3)_3C$–cyclohexyl–OTs $\xrightarrow{LiCu(CH_3)_2}$ $(CH_3)_3C$–cyclohexyl–CH_3 + $(CH_3)_3C$–cyclohexenyl

cis-4-*tert*-Butylcyclohexyl *p*-toluenesulfonate

trans-4-*tert*-Butyl-1-methylcyclohexane (compound C, $C_{11}H_{22}$)

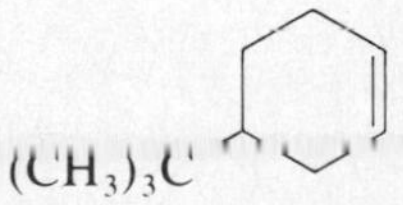

Compound B

Inversion of configuration is borne out by the fact given in the problem that compound C is more stable than compound A. Both substituents are equatorial in C; the methyl group is axial in A.

15.24 (*a*) The desired 1-deuteriobutane can be obtained by reaction of D_2O with *n*-butyllithium or *n*-butylmagnesium bromide.

$$CH_3CH_2CH_2CH_2Li + D_2O \rightarrow CH_3CH_2CH_2CH_2D$$

Butyllithium, Deuterium oxide, 1-Deuteriobutane

or

$$CH_3CH_2CH_2CH_2MgBr + D_2O \rightarrow CH_3CH_2CH_2CH_2D$$

Butylmagnesium bromide

Preparation of the organometallics requires an alkyl bromide, which is synthesized from the corresponding alcohol.

$$CH_3CH_2CH_2CH_2OH \xrightarrow[\text{or HBr}]{PBr_3} CH_3CH_2CH_2CH_2Br$$

1-Butanol (*n*-butyl alcohol) → 1-Bromobutane (*n*-butyl bromide)

$$CH_3CH_2CH_2CH_2Li \xleftarrow[\text{ether}]{Li} CH_3CH_2CH_2CH_2Br \xrightarrow[\text{ether}]{Mg} CH_3CH_2CH_2CH_2MgBr$$

Butyllithium, Butyl bromide, Butylmagnesium bromide

(*b*) In a sequence identical to that of (*a*) in design but using 2-butanol as the starting material, 2-deuteriobutane may be prepared.

$$\underset{\text{2-Butanol}}{\underset{\displaystyle\text{OH}}{CH_3\underset{|}{C}HCH_2CH_3}} \xrightarrow{PBr_3} \underset{\text{2-Bromobutane}}{\underset{\displaystyle\text{Br}}{CH_3\underset{|}{C}HCH_2CH_3}} \xrightarrow[\text{ether}]{Mg} \underset{\textit{sec}\text{-Butylmagnesium bromide}}{\underset{\displaystyle\text{MgBr}}{CH_3\underset{|}{C}HCH_2CH_3}}$$

$$\xrightarrow{D_2O} \underset{\text{2-Deuteriobutane}}{\underset{\displaystyle\text{D}}{CH_3\underset{|}{C}HCH_2CH_3}}$$

An analogous procedure involving *sec*-butyllithium in place of the Grignard reagent can be used.

15.25 A Grignard reagent exists in solution as an equilibrium mixture of an alkylmagnesium halide and a dialkylmagnesium.

$$2\,RMgX \rightleftharpoons R_2Mg + MgX_2$$

When dioxane is added to the solution, magnesium halides, being insoluble in dioxane, precipitate. This shifts the position of equilibrium to the right, increasing the concentration of the dialkylmagnesium species at the expense of the alkylmagnesium halide. In effect, the halide is removed from solution by precipitation, so the alkylmagnesium halide is converted to a dialkylmagnesium.

15.26 All the protons in benzene are equivalent. In diphenylmethane and in triphenylmethane, protons are attached to either the sp^2 hybridized carbons of the ring or to the sp^3 hybridized carbon between the rings. The large difference in acidity between diphenylmethane and benzene suggests that it is not a ring proton that is lost on ionization in diphenylmethane but rather a proton from the methylene group.

$$C_6H_5-CH_2-C_6H_5 \rightleftharpoons C_6H_5-\overset{H}{\underset{\cdot\cdot}{C}}{}^{-}-C_6H_5 + H^+$$

Diphenylmethane

The anion produced is stabilized by resonance. It is a *benzylic* anion.

$$C_6H_5-\overset{H}{\underset{\cdot\cdot}{C}}{}^{-}-C_6H_5 \longleftrightarrow {}^{-}\!:C_6H_5{=}\overset{H}{C}-C_6H_5 \quad \text{etc.}$$

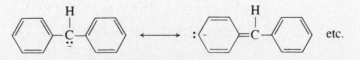

Both rings are involved in delocalizing the negative charge. The anion from triphenylmethane is stabilized by resonance involving all three rings.

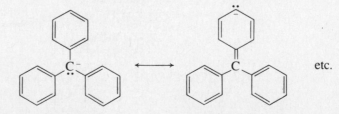

$$(C_6H_5)_3\ddot{C}^{-} \longleftrightarrow (C_6H_5)_2C{=}C_6H_5{:}^{-} \quad \text{etc.}$$

Delocalization of the negative charge by resonance is not possible in the anion of benzene. The pair of unshared electrons in phenyl anion is in an sp^2 hybrid orbital that does not interact with the π system.

not delocalized into π system

15.27 The meso form of 2,3-dibromobutane is the stereoisomer shown; syn elimination, that is, loss of both bromine atoms from the same side will yield (*Z*)-2-butene.

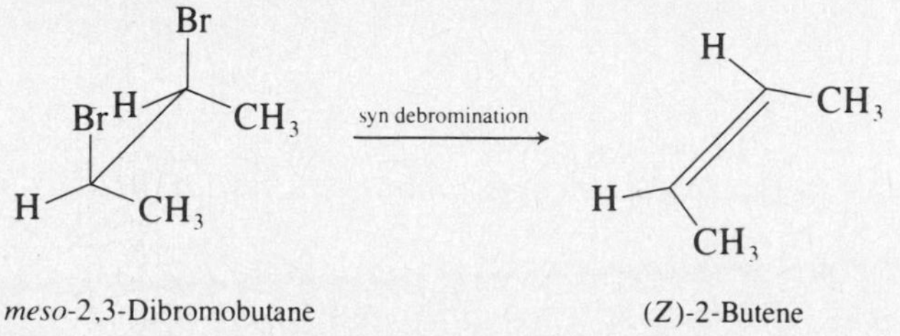

syn debromination

meso-2,3-Dibromobutane (*Z*)-2-Butene

This is at variance with the experimental observation that *meso*-2,3-dibromobutane gives (*E*)-2-butenc. Therefore, a syn elimination may be ruled out. Now let us look at the anti conformation of the meso form.

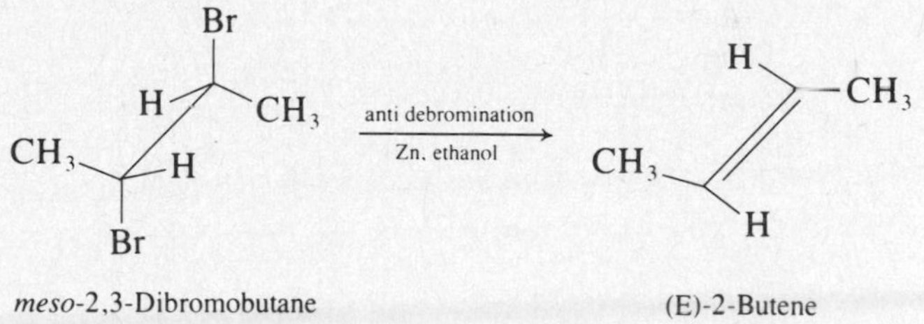

anti debromination
Zn, ethanol

meso-2,3-Dibromobutane (E)-2-Butene

Loss of bromines from opposite sides, that is, anti elimination, leads to the observed product, (*E*)-2-butene.

Similarly, syn debromination of the chiral diastereomer gives the wrong stereoisomeric form of the product.

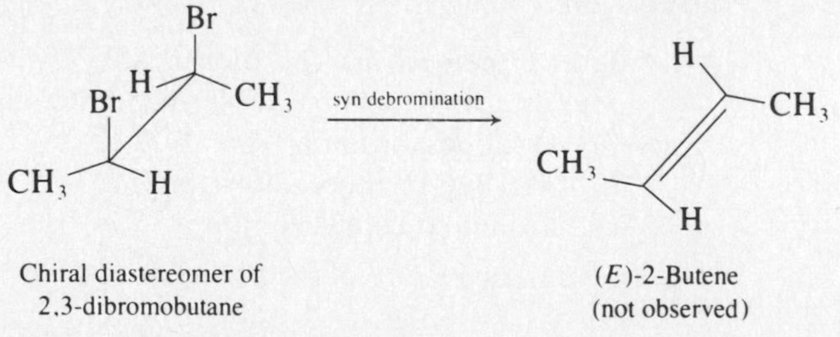

syn debromination

Chiral diastereomer of 2,3-dibromobutane

(*E*)-2-Butene (not observed)

Anti debromination of this diastereomer, however, is consistent with the observed stereochemistry.

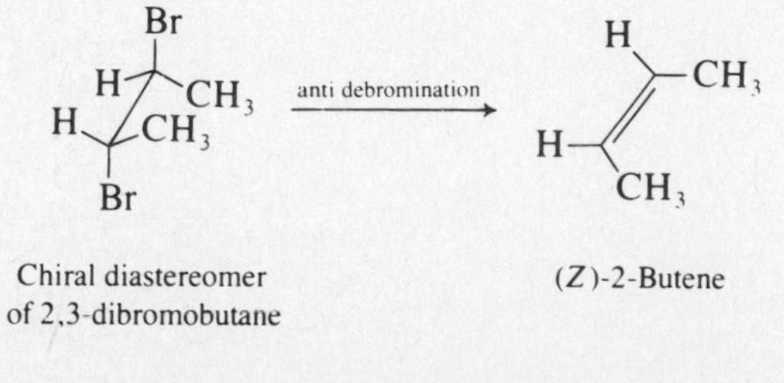

anti debromination

Chiral diastereomer of 2,3-dibromobutane

(*Z*)-2-Butene

15.28 (*a*) Compound D, as indicated by its molecular formula, is a dibromide. The selectivity of free-radical bromination is high and favors substitution of tertiary hydrogens. The starting alkane has two tertiary hydrogens that are capable of being replaced by bromine atoms, so Compound D is most reasonably the dibromide shown.

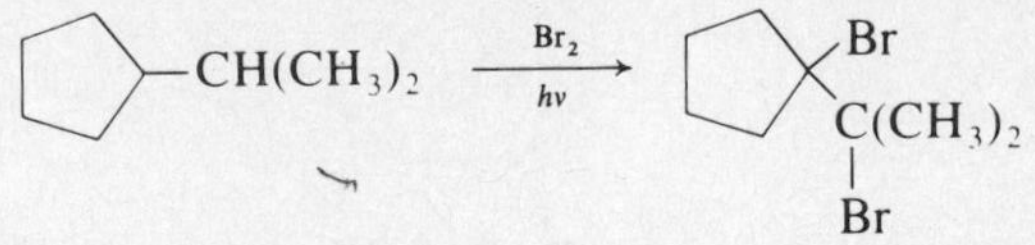

$$\text{C}_5\text{H}_9\text{—CH(CH}_3)_2 \xrightarrow[h\nu]{\text{Br}_2} \text{C}_5\text{H}_8(\text{Br})\text{—C(CH}_3)_2\text{Br} \quad \text{(Compound D)}$$

Compound D is a vicinal dibromide. Vicinal dibromides undergo dehalogenation on being treated with zinc to yield alkenes.

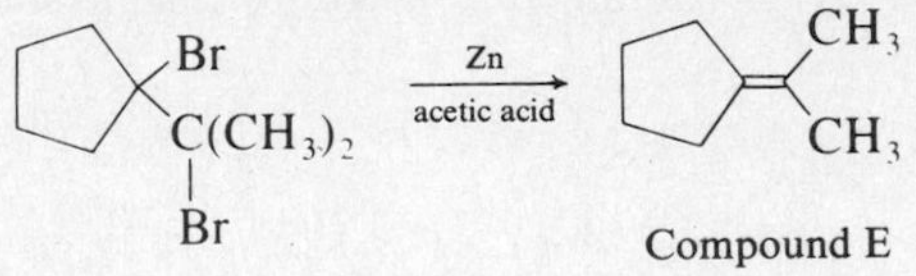

$$\text{C}_5\text{H}_8(\text{Br})\text{—C(CH}_3)_2\text{Br} \xrightarrow[\text{acetic acid}]{\text{Zn}} \text{C}_5\text{H}_8\text{=C(CH}_3)_2$$

Compound E

(*b*) *N*-Bromosuccinimide reacts with the starting material by benzylic bromination.

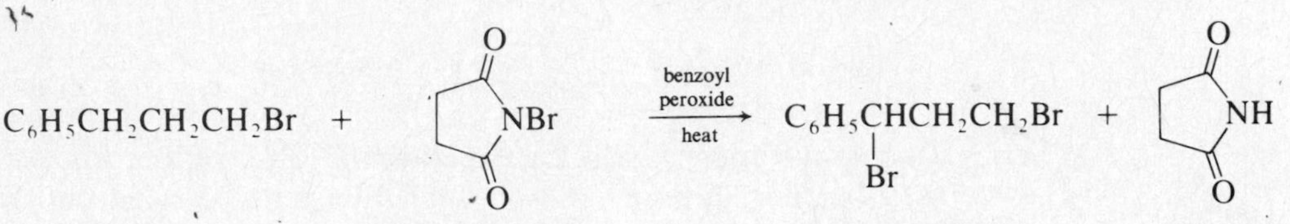

$$C_6H_5CH_2CH_2CH_2Br + \text{NBS} \xrightarrow[\text{heat}]{\text{benzoyl peroxide}} C_6H_5CHBrCH_2CH_2Br + \text{succinimide}$$

1-Bromo-3-phenylpropane; *N*-Bromosuccinimide; 1,3-Dibromo-1-phenylpropane (Compound F); Succinimide

Treatment with zinc converts the 1,3-dibromide to a cyclopropane derivative.

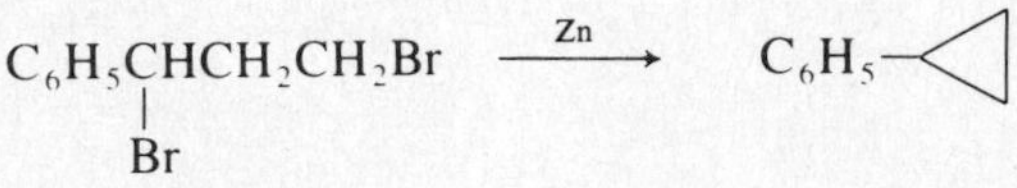

$$C_6H_5CHBrCH_2CH_2Br \xrightarrow{\text{Zn}} C_6H_5\text{—}C_3H_5$$

1,3-Dibromo-1-phenylpropane; Cyclopropylbenzene (Compound G)

SELF-TEST

PART A

A-1. Give a method for the preparation of each of the following organometallic compounds, using appropriate starting materials:

(*a*) Cyclohexyllithium
(*b*) *tert*-Butylmagnesium bromide
(*c*) Lithium dibenzylcuprate

A-2. Give the structure of the product obtained by each of the following reaction schemes:

(*a*) $CH_3CO_2CH_2CH_3 \xrightarrow[2.\ H_3O^+]{1.\ 2C_6H_5MgBr} ?$

(*b*) $(CH_3)_2CHCH_2Li \xrightarrow{D_2O} ?$

(*c*) $CH_3\text{—}C_6H_4\text{—}Br \xrightarrow[3.\ H_3O^+]{1.\ Mg;\ 2.\ H_2C{=}O} ?$ (*p*-bromotoluene)

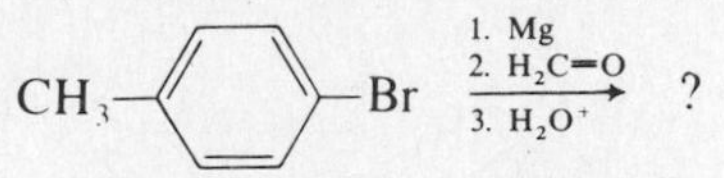

(*d*) Cyclohexanone $\xrightarrow[2.\ H_3O^+]{1.\ CH_3CH_2Li} ?$

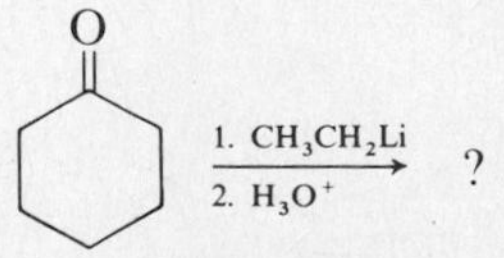

A-3. Give the structure of the organometallic reagent necessary to carry out each of the following:

(*a*) 1-iodo-2-methylcyclohexene $\xrightarrow{?}$ 1-methyl-2-propylcyclohexene (CH_3, $CH_2CH_2CH_3$)

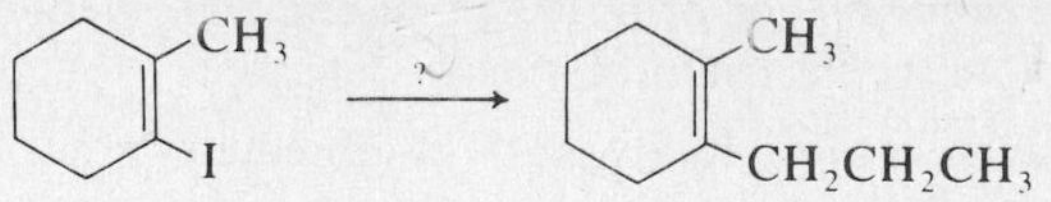

(*b*) $C_6H_5CH_2CO_2CH_3 \xrightarrow[2.\ H_3O^+]{1.\ ?} HC(OH)(CH(CH_3)_2)CH_2C_6H_5$

(*c*) 1,2-dimethylcyclohexene $\longrightarrow$ 1,6-dimethylbicyclo[4.1.0]heptane (CH_3, CH_3)

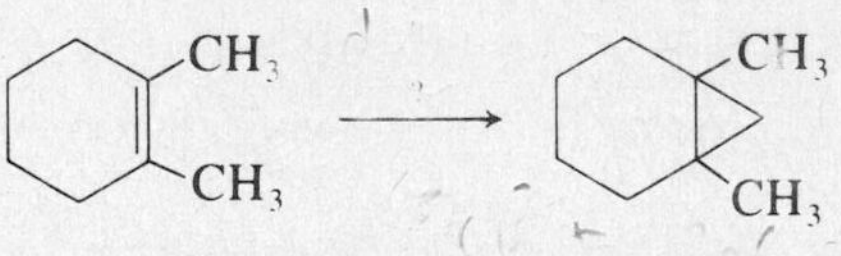

A-4. Several common organic solvents are shown below. Which ones would be suitable for use in the preparation of a Grignard reagent? For those which are not suitable, give a brief reason why.

A: $CH_3CH_2CH_2CH_2OCH_2CH_2CH_2CH_3$ B: $CH_3OCH_2CH_2OCH_3$ C: $HOCH_2CH_2OH$

D: $CH_3\overset{O}{\overset{\|}{C}}OCH_2CH_3$ E: tetrahydrofuran (five-membered ring containing O) F: $CH_3\overset{O}{\overset{\|}{C}}OH$

A-5. Show by a series of chemical equations how you could prepare octane from 1-butanol as the source of all its carbon atoms.

A-6. Synthesis of the following alcohol is possible by three schemes utilizing Grignard reagents. Give the reagents necessary to carry out each of them.

$$(CH_3)_2CH\overset{OH}{\overset{|}{C}}(CH_3)_2$$

A-7. Using ethylbenzene and any other necessary organic or inorganic reagents, outline a synthesis of 3-phenyl-2-butanol.

PART B

B-1. Which (if any) of the following would *not* be classified as an organometallic substance?
(*a*) Triethylaluminum
(*b*) Ethylmagnesium iodide
(*c*) Potassium *tert*-butoxide
(*d*) None of these (all are organometallics)

B-2. Rank the following species in order of increasing polarity of the carbon-metal bond (least → most polar):

A: CH_3CH_2MgCl B: CH_3CH_2Na C: $(CH_3CH_2)_3Al$

(*a*) C < A < B (*b*) B < A < C (*c*) A < C < B (*d*) B < C < A

B-3. The following reaction scheme

$$HOCH_2CH_2CH_2Br \xrightarrow[\substack{2.\ C_6H_5CH=O \\ 3.\ H_3O^+}]{1.\ Mg} HOCH_2CH_2CH_2\overset{OH}{\overset{|}{C}}HC_6H_5$$

(*a*) Is a good synthetic route
(*b*) Is a poor synthetic route
(*c*) May be good or poor, depending on solvent used
(*d*) Needs more information to determine

B-4. Arrange the following intermediates in order of decreasing basicity (strongest → weakest):

$$\underset{\text{A}}{CH_2{=}CHNa} \qquad \underset{\text{B}}{CH_3CH_2Na} \qquad \underset{\text{C}}{CH_3CH_2ONa} \qquad \underset{\text{D}}{HC{\equiv}CNa}$$

(*a*) B > A > D > C (*b*) D > A > B > C
(*c*) C > D > A > B (*d*) C > B > D > A

B-5. Give the major product of the following reaction:

$$(E)\text{-2-pentene} \xrightarrow{CH_2I_2,\ Zn(Cu)} ?$$

(*a*) *cis*-1-Ethyl-2-methylcyclopropane
(*b*) *trans*-1-Ethyl-2-methylcyclopropane
(*c*) 1-Ethyl-1-methylcyclopropane
(*d*) An equimolar mixture of (*a*) and (*b*)

B-6. Which of the following reagents would be effective for the following reaction sequence?

$$C_6H_5C{\equiv}CH \xrightarrow[\text{3. } H_3O^+]{\text{1. ?}\quad \text{2. } H_2C{=}O} C_6H_5C{\equiv}CCH_2OH$$

(*a*) Sodium ethoxide (*b*) Mercuric acetate
(*c*) Butyllithium (*d*) Potassium hydroxide

CHAPTER 16

ALCOHOLS

IMPORTANT TERMS AND CONCEPTS

Oxygen Containing Functional Groups (Sec. 16.2) Several classes of organic compounds contain one or more oxygen atoms. Arranged in order of increasing oxidation states, these groups are:

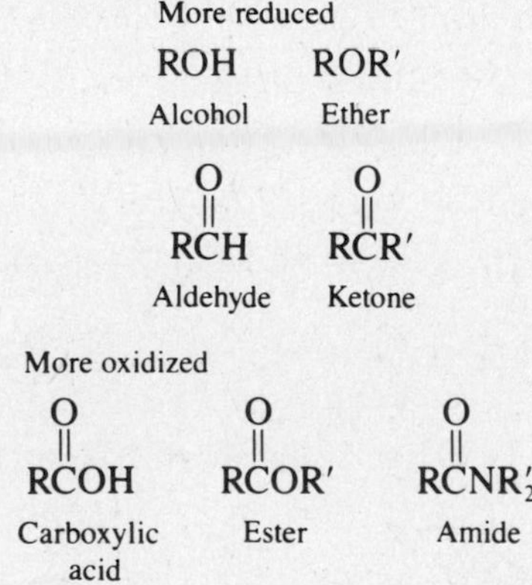

Keeping in mind the order of progression of these groups will aid in determining what types of reaction conditions are needed for their interconversion. For example, an alcohol must be oxidized to yield an aldehyde or ketone. Likewise, formation of an alcohol from an ester must involve a reduction pathway.

$$\text{Alcohol} \xrightarrow{\text{oxidizing agent}} \text{aldehyde or ketone}$$

$$\text{Ester} \xrightarrow{\text{reducing agent}} \text{alcohol}$$

Preparation of Alcohols (Secs. 16.3 to 16.7) In previous chapters several methods for the preparation of alcohols have been discussed. These include:

1. Hydration of alkenes (Chapter 7)
   (*a*) Acid-catalyzed hydration
   (*b*) Oxymercuration-demercuration
   (*c*) Hydroboration

2. Hydrolysis of alkyl halides (Chapter 9)
3. Grignard (and organolithium) reactions with carbonyl derivatives (Chapter 15)

A summary of these methods is given in text Table 16.1. The preparation of alcohols frequently involves the reduction of carbonyl derivatives. The general transformations are summarized below.

Reduction reactions:

$$\text{Aldehydes} \longrightarrow \text{primary alcohols}$$
$$\text{Ketones} \longrightarrow \text{secondary alcohols}$$
$$\left.\begin{array}{l}\text{Carboxylic acids}\\ \text{and esters}\end{array}\right\} \longrightarrow \text{primary alcohols}$$

Methods for carrying out these transformations include *catalytic hydrogenation* and reduction with the *metal hydrides* lithium aluminum hydride ($LiAlH_4$) and sodium borohydride ($NaBH_4$). A summary of the functional groups reduced by these reagents follows.

Reagent	*Functional groups reduced*
H_2, catalyst	Aldehydes, ketones, esters
$LiAlH_4$	Aldehydes, ketones, acids, esters
$NaBH_4$	Aldehydes, ketones (esters reduced slowly)

Specific examples of these reactions are given in the reaction summary which follows this section.

Alcohols may also be prepared by epoxide ring opening reactions and the *hydroxylation* of alkenes.

Reactions of Alcohols (Secs. 16.8 to 16.14) General classes of alcohol reactions may be described which involve the breaking of bonds. The three general possibilities are:

1. O—H bond breaking results in formation of alkoxides:

$$-\overset{|}{\underset{|}{C}}-O\wr H \longrightarrow -\overset{|}{\underset{|}{C}}-O^-$$

2. C—O bond breaking results in substitution reactions:

$$-\overset{|}{\underset{|}{C}}\wr O-H \longrightarrow -\overset{|}{\underset{|}{C}}-X$$

3. O—H and adjacent C—H bonds are broken in oxidation reactions (1° and 2°):

$$-\overset{H}{\underset{|}{C}}-O\wr H \longrightarrow \;\rangle C{=}O$$

Examples of alcohol reactions are present in the following section.

IMPORTANT REACTIONS

PREPARATION OF ALCOHOLS

Reduction of Aldehydes and Ketones (Sec. 16.3)

General:

$$R-\overset{O}{\overset{\|}{C}}-H \xrightarrow{[H]} RCH_2OH$$

$$R-\overset{O}{\overset{\|}{C}}-R' \xrightarrow{[H]} R\overset{OH}{\overset{|}{C}}HR'$$

Examples:

Catalytic Hydrogenation:

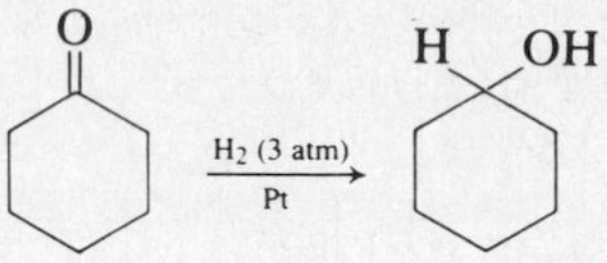

$$\text{cyclohexanone} \xrightarrow[\text{Pt}]{H_2\ (3\ \text{atm})} \text{cyclohexanol}$$

Metal Hydrides:

$$CH_3CH_2\overset{O}{\overset{\|}{C}}H \xrightarrow[C_2H_5OH]{NaBH_4} CH_3CH_2CH_2OH$$

$$C_6H_5\overset{O}{\overset{\|}{C}}CH_3 \xrightarrow[2.\ H_2O]{1.\ LiAlH_4} C_6H_5\overset{OH}{\overset{|}{C}}HCH_3$$

Reduction of Carboxylic Acids (Sec. 16.4)

General:

$$R\overset{O}{\overset{\|}{C}}OH \xrightarrow[2.\ H_2O]{1.\ LiAlH_4} RCH_2OH$$

Example:

$$(CH_3)_3CCO_2H \xrightarrow[2.\ H_2O]{1.\ LiAlH_4} (CH_3)_3CCH_2OH$$

Reduction of Esters (Sec. 16.5)

General:

$$R\overset{O}{\overset{\|}{C}}OR' \xrightarrow{[H]} RCH_2OH + R'OH$$

Examples:

$$(CH_3)_2CHCO_2CH_3 \xrightarrow[\text{Cu chromite}]{H_2\ (200\ \text{atm})} (CH_3)_2CHCH_2OH + CH_3OH$$

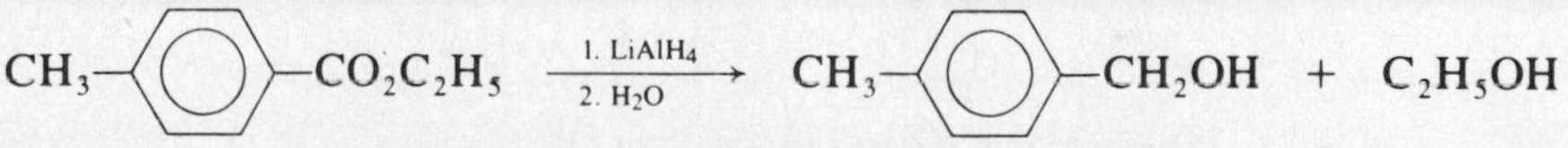

$$CH_3-C_6H_4-CO_2C_2H_5 \xrightarrow[2.\ H_2O]{1.\ LiAlH_4} CH_3-C_6H_4-CH_2OH + C_2H_5OH$$

Preparation of Alcohols from Epoxides (Sec. 16.6)

General:

$$RMgX \xrightarrow[2.\ H_3O^+]{1.\ CH_2-CH_2\ (\text{epoxide, }O)} RCH_2CH_2OH$$

Example:

$$C_6H_5MgBr \xrightarrow[2.\ H_3O^+]{1.\ \text{ethylene oxide}} C_6H_5CH_2CH_2OH$$

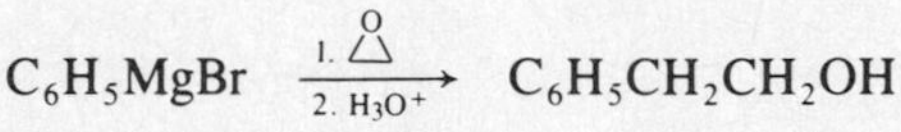

Preparation of Diols (Sec. 16.7)

General:

$$R_2C{=}CR_2' \xrightarrow{[O]} R_2\overset{OH}{\overset{|}{C}}-\overset{OH}{\overset{|}{C}}R_2'$$

Examples:

$$CH_3CH_2CH{=}CH_2 \xrightarrow[\text{cold}]{KMnO_4,\ H_2O} CH_3CH_2\overset{OH}{\overset{|}{C}}HCH_2OH$$

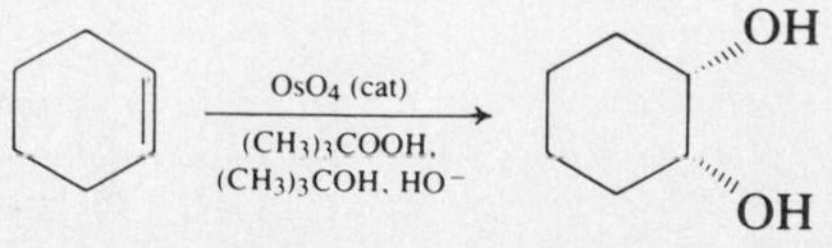

$$\text{cyclohexene} \xrightarrow[(CH_3)_3COOH,\ (CH_3)_3COH,\ HO^-]{OsO_4\ (\text{cat})} \text{cis-1,2-cyclohexanediol}$$

REACTIONS OF ALCOHOLS

Formation of Ethers (Sec. 16.9)

General:

$$2\ ROH \xrightarrow[\text{heat}]{H^+} ROR + H_2O \quad \text{(R primary)}$$

Example:

$$2\ CH_3CH_2CH_2CH_2OH \xrightarrow[\text{heat}]{H^+} CH_3CH_2CH_2CH_2OCH_2CH_2CH_2CH_3$$

Esterification (Sec. 16.10)

General:

$$ROH + R'\overset{O}{\overset{\|}{C}}OH \overset{H^+}{\rightleftharpoons} R'\overset{O}{\overset{\|}{C}}OR + H_2O$$

(Alcohol or acid in excess; or water removed as formed)

$$ROH + R'\overset{O}{\overset{\|}{C}}{-}X \xrightarrow{\text{pyridine}} R'\overset{O}{\overset{\|}{C}}OR$$

(Acyl chloride or anhydride)

Examples:

$$CH_3CH_2OH + C_6H_5CH_2CO_2H \xrightarrow{H^+} C_6H_5CH_2CO_2CH_2CH_3 + H_2O$$

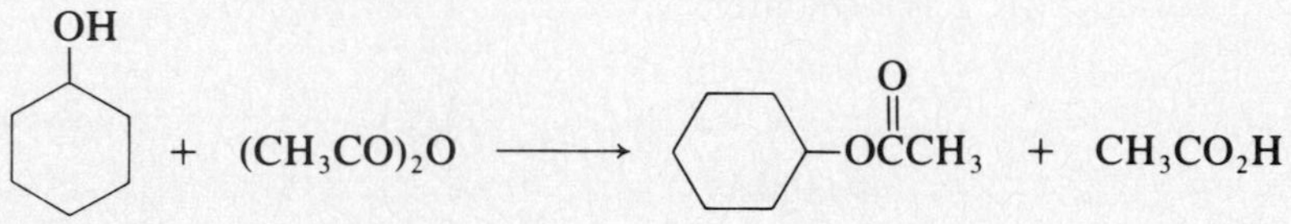

$$\text{cyclohexanol } (C_6H_{11}OH) + (CH_3CO)_2O \longrightarrow C_6H_{11}{-}O\overset{O}{\overset{\|}{C}}CH_3 + CH_3CO_2H$$

$$(CH_3)_2CHOH + C_6H_5\overset{O}{\overset{\|}{C}}Cl \xrightarrow{\text{pyridine}} C_6H_5\overset{O}{\overset{\|}{C}}OCH(CH_3)_2$$

Oxidation Reactions (Sec. 16.12)

General:

$$RCH_2OH \xrightarrow{[O]} R\overset{O}{\overset{\|}{C}}H \xrightarrow{[O]} RCO_2H$$

$$R_2CHOH \xrightarrow{[O]} R\overset{O}{\overset{\|}{C}}R$$

Examples:

$$C_6H_5CH_2OH \xrightarrow[H^+,\ H_2O]{K_2Cr_2O_7} C_6H_5CO_2H$$

$$(CH_3)_2CHCH_2OH \xrightarrow{(C_5H_5N)_2CrO_3} (CH_3)_2CH\overset{O}{\overset{\|}{C}}H$$

Oxidative Cleavage of Vicinal Diols (Sec. 16.14)

General:

$$R{-}\underset{OH}{\overset{R}{C}}{-}\underset{OH}{\overset{R'}{C}}{-}R' \xrightarrow{HIO_4} R_2C{=}O + R'_2C{=}O$$

Examples:

$$(CH_3)_2\overset{\displaystyle OH}{\overset{|}{C}}\text{—}CH_2OH \xrightarrow{HIO_4} (CH_3)_2C\text{=}O + H_2C\text{=}O$$

$$\text{cyclopentane-1,2-diol} \xrightarrow{HIO_4} H\overset{O}{\overset{\|}{C}}CH_2CH_2CH_2\overset{O}{\overset{\|}{C}}H$$

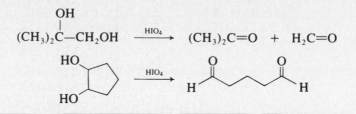

SOLUTIONS TO TEXT PROBLEMS

16.1 (*b*) When carbon monoxide is converted to carbon dioxide, it is oxidized. Compound B must serve as a source of oxygen to oxidize CO to CO_2 and itself be reduced to H_2. Water is a reasonable candidate substance for compound B.

$$\underset{\substack{\text{Carbon monoxide}\\\text{reducing agent}}}{CO} + \underset{\substack{\text{Water}\\\text{oxidizing agent}\\\text{(compound B)}}}{H_2O} \longrightarrow \underset{\text{Hydrogen}}{H_2} + \underset{\text{Carbon dioxide}}{CO_2}$$

(*c*) The relationship between CO and $H\overset{O}{\overset{\|}{C}}OH$ is that they differ by the elements of water, which suggests the equation:

$$\underset{\text{Carbon monoxide}}{CO} + \underset{\substack{\text{Water}\\\text{(compound C)}}}{H_2O} \longrightarrow \underset{\substack{\text{Methanoic acid}\\\text{(formic acid)}}}{H\overset{O}{\overset{\|}{C}}OH}$$

Reactions of this type, in which the elements of water are *added* to the substrate, involve no change in oxidation state at carbon and are not oxidation-reduction processes.

Carbon monoxide and methanoic (formic) acid are in equivalent oxidation states.

16.2 The two primary alcohols 1-butanol and 2-methyl-1-propanol can be prepared by hydrogenation of the corresponding aldehydes:

$$\underset{\text{Butanal}}{CH_3CH_2CH_2\overset{O}{\overset{\|}{C}}H} \xrightarrow{H_2,\ Ni} \underset{\text{1-Butanol}}{CH_3CH_2CH_2CH_2OH}$$

$$\underset{\substack{\text{2-Methylpropanal}\\\text{(isobutyraldehyde)}}}{(CH_3)_2CH\overset{O}{\overset{\|}{C}}H} \xrightarrow{H_2,\ Ni} \underset{\substack{\text{2-Methyl-1-propanol}\\\text{(isobutyl alcohol)}}}{(CH_3)_2CHCH_2OH}$$

The secondary alcohol 2-butanol arises by hydrogenation of a ketone.

$$\underset{\text{2-Butanone}}{CH_3\overset{O}{\overset{\|}{C}}CH_2CH_3} \xrightarrow{H_2,\ Ni} \underset{\substack{\text{2-Butanol}\\(\textit{sec}\text{-butyl alcohol})}}{CH_3\underset{\underset{\displaystyle OH}{|}}{C}HCH_2CH_3}$$

Tertiary alcohols such as 2-methyl-2-propanol $(CH_3)_3COH$ (*tert*-butyl alcohol) cannot be prepared by hydrogenation of a carbonyl compound.

16.3 (*b*) A deuterium atom is transferred from $NaBD_4$ to the carbonyl group of acetone.

$$CH_3\overset{\overset{D-\bar{B}D_3}{\curvearrowright}}{C}(CH_3){=}O \longrightarrow CH_3\overset{D}{\underset{CH_3}{C}}-O\bar{B}D_3 \xrightarrow{3\,(CH_3)_2C=O} \left(CH_3\overset{D}{\underset{CH_3}{C}}O\right)_4\bar{B}$$

On reaction with CH_3OD, deuterium is transferred from the alcohol to the oxygen of $[(CH_3)_2CDO]_4\bar{B}$.

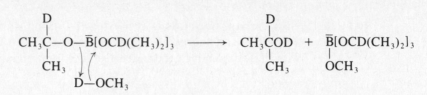

$$CH_3\overset{D}{\underset{CH_3}{C}}-O-\bar{B}[OCD(CH_3)_2]_3 \;\; (D-OCH_3) \longrightarrow CH_3\overset{D}{\underset{CH_3}{C}}OD + \underset{OCH_3}{\bar{B}}[OCD(CH_3)_2]_3$$

Overall:

$$\underset{\text{Acetone}}{(CH_3)_2C{=}O} \xrightarrow[CH_3OD]{NaBD_4} \underset{\text{2-Propanol-2-}d\text{-}O\text{-}d}{(CH_3)_2\overset{D}{C}OD}$$

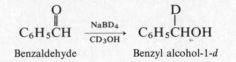

(*c*) In this case $NaBD_4$ serves as a deuterium donor to carbon while CD_3OH is a proton (not deuterium) donor to oxygen.

$$\underset{\text{Benzaldehyde}}{C_6H_5\overset{O}{\overset{\|}{C}}H} \xrightarrow[CD_3OH]{NaBD_4} \underset{\text{Benzyl alcohol-1-}d}{C_6H_5\overset{D}{C}HOH}$$

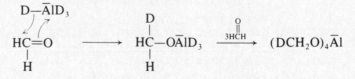

(*d*) Lithium aluminum deuteride is a deuterium donor to the carbonyl carbon of formaldehyde.

$$\underset{H}{HC}{=}O \;(D-\bar{A}lD_3) \longrightarrow \underset{H}{H\overset{D}{C}}-O\bar{A}lD_3 \xrightarrow{3H\overset{O}{\overset{\|}{C}}H} (DCH_2O)_4\bar{A}l$$

On hydrolysis with D_2O, the oxygen-aluminum bond is cleaved and DCH_2OD is formed.

$$\bar{A}l(OCH_2D)_4 \xrightarrow{D_2O} \underset{\text{Methanol-1-}d\text{-}O\text{-}d}{DCH_2OD}$$

16.4 It is the bond between the carbonyl group and the ester oxygen that is cleaved on hydrogenolysis. The carbonyl group becomes the $-CH_2OH$ group of a primary alcohol. Thus, reduction of isopropyl phenylacetate proceeds as follows:

$$\underset{\text{Isopropyl phenylacetate}}{C_6H_5CH_2\overset{O}{\overset{\|}{C}}OCH(CH_3)_2} \xrightarrow{H_2,\ \text{copper chromite}} \underset{\text{2-Phenylethanol}}{C_6H_5CH_2CH_2OH} + \underset{\text{Isopropyl alcohol}}{(CH_3)_2CHOH}$$

Isopropyl alcohol is one of the reduction products, but 2-phenylethanol, not benzyl alcohol, is the other alcohol.

Reduction of isopropyl benzoate does produce equimolar amounts of isopropyl alcohol and benzyl alcohol.

$$\underset{\text{Isopropyl benzoate}}{C_6H_5\overset{O}{\overset{\|}{C}}OCH(CH_3)_2} \xrightarrow{H_2,\ \text{copper chromite}} \underset{\text{Benzyl alcohol}}{C_6H_5CH_2OH} + \underset{\text{Isopropyl alcohol}}{(CH_3)_2CHOH}$$

Benzyl alcohol is one of the products of reduction of benzyl isobutyrate, but isobutyl alcohol, not isopropyl alcohol, is the other.

$$\underset{\text{Benzyl isobutyrate}}{(CH_3)_2CH\overset{O}{\overset{\|}{C}}OCH_2C_6H_5} \xrightarrow{H_2,\ \text{copper chromite}} \underset{\text{Isobutyl alcohol}}{(CH_3)_2CHCH_2OH} + \underset{\text{Benzyl alcohol}}{C_6H_5CH_2OH}$$

16.5 Catalytic hydrogenolysis could not be used to prepare oleyl alcohol from butyl oleate because the carbon-carbon double bond would be hydrogenated under these conditions.

$$\underset{\text{Butyl oleate}}{CH_3(CH_2)_7CH{=}CH(CH_2)_7\overset{O}{\overset{\|}{C}}OCH_2CH_2CH_2CH_3} \xrightarrow[\text{copper chromite}]{H_2,\ \text{pressure}}$$

$$\underset{\text{1-Octadecanol}}{CH_3(CH_2)_7CH_2CH_2(CH_2)_7CH_2OH} + \underset{\text{1-Butanol}}{CH_3CH_2CH_2CH_2OH}$$

Lithium aluminum hydride, however, does not reduce isolated carbon-carbon double bonds and could be used to convert butyl oleate to oleyl alcohol.

16.6 (*b*) Reaction with ethylene oxide results in the addition of a $-CH_2CH_2OH$ unit to the Grignard reagent. Cyclohexylmagnesium bromide (or chloride) is the appropriate reagent.

$$\underset{\text{Cyclohexylmagnesium bromide}}{C_6H_{11}MgBr} + \underset{\text{Ethylene oxide}}{\underset{\diagdown O \diagup}{CH_2{-}CH_2}} \xrightarrow[2.\ H_3O^+]{1.\ \text{diethyl ether}} \underset{\text{2-Cyclohexylethanol}}{C_6H_{11}CH_2CH_2OH}$$

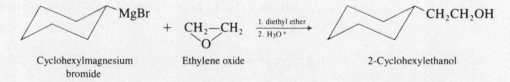

The reaction carried out as shown gave 2-cyclohexylethanol in 50 percent yield.

(*c*) When 1-nonanol is prepared by reaction of a Grignard reagent with ethylene oxide, the oxygen, C-1, and C-2 are all derived from the epoxide.

$$CH_3(CH_2)_5CH_2 \wr CH_2{-}CH_2OH \Longrightarrow CH_3(CH_2)_5CH_2^- + \underset{\diagdown O \diagup}{CH_2{-}CH_2}$$

Therefore we choose the Grignard reagent from 1-bromoheptane.

$$\underset{\text{1-Bromoheptane}}{CH_3(CH_2)_5CH_2Br} \xrightarrow[\text{diethyl ether}]{Mg} \underset{\text{Heptylmagnesium bromide}}{CH_3(CH_2)_5CH_2MgBr}$$

$$\underset{\text{Heptylmagnesium bromide}}{CH_3(CH_2)_5CH_2MgBr} + \underset{\diagdown O \diagup}{CH_2{-}CH_2} \xrightarrow[2.\ H_3O^+]{1.\ \text{diethyl ether}} \underset{\text{1-Nonanol}}{CH_3(CH_2)_5CH_2CH_2CH_2OH}$$

The synthesis of 1-nonanol from 1-bromoheptane has been carried out this way in 55 percent yield.

(*d*) Again, the $-CH_2CH_2OH$ unit is derived from ethylene oxide.

CH_2CH_2OH (1-naphthyl) ⇨ (1-naphthyl anion) + CH_2-CH_2 (O)

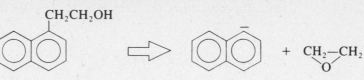

The synthesis shown, based on the Grignard reagent of 1-bromonaphthalene, has been carried out in 76 percent yield.

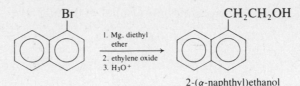

1-Bromonaphthalene (Br) $\xrightarrow[\text{2. ethylene oxide; 3. } H_3O^+]{\text{1. Mg, diethyl ether}}$ CH_2CH_2OH (1-naphthyl)

2-(α-naphthyl)ethanol

16.7 Only the hydroxyl groups on C-1 and C-4 can be involved since only these two can lead to a five-membered cyclic ether.

$$HOCH_2\underset{\displaystyle OH}{\underset{|}{C}}HCH_2CH_2OH \xrightarrow[\text{heat}]{H^+} \text{(HO-substituted oxolane)} + H_2O$$

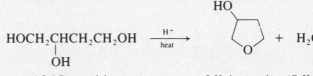

1,2,4-Butanetriol — 3-Hydroxyoxolane ($C_4H_8O_2$)

Any other combination of hydroxyl groups would lead to a strained three-membered or four-membered ring and is unfavorable under conditions of acid catalysis.

16.8 Alcohols react with carboxylic acids in the presence of an acid catalyst to give an ester and water.

(*b*) $C_6H_5CH_2OH + (CH_3)_2CHCH_2\overset{O}{\overset{\|}{C}}OH \xrightarrow{H^+} (CH_3)_2CHCH_2\overset{O}{\overset{\|}{C}}OCH_2C_6H_5 + H_2O$

Benzyl alcohol — 3-Methylbutanoic acid — Benzyl 3-methylbutanoate — Water

(*c*) The relationship of the molecular formula of the ester ($C_{10}H_{10}O_4$) to that of the starting dicarboxylic acid ($C_8H_6O_4$) indicates that the diacid reacted with two moles of methanol to form a diester.

$$2\,CH_3OH + HO\overset{O}{\overset{\|}{C}}-C_6H_4-\overset{O}{\overset{\|}{C}}OH \xrightarrow{H^+} CH_3O\overset{O}{\overset{\|}{C}}-C_6H_4-\overset{O}{\overset{\|}{C}}OCH_3$$

Methanol — 1,4-Benzenedicarboxylic acid — Dimethyl 1,4-benzenedicarboxylate

16.9 While neither *cis*- nor *trans*-4-*tert*-butylcyclohexanol is a chiral molecule, the stereochemical course of their reactions with acetic anhydride becomes evident when the relative stereochemistry of the ester function is examined for each case. The cis alcohol yields the cis acetate.

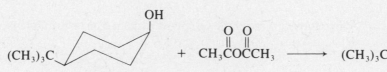

$(CH_3)_3C$–(cyclohexane)–OH + $CH_3\overset{O}{\overset{\|}{C}}O\overset{O}{\overset{\|}{C}}CH_3$ ⟶ $(CH_3)_3C$–(cyclohexane)–$O\overset{O}{\overset{\|}{C}}CH_3$

cis-4-*tert*-Butylcyclohexanol — Acetic anhydride — *cis*-4-*tert*-Butylcyclohexyl acetate

The trans alcohol yields the trans acetate.

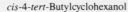

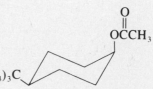

$(CH_3)_3C$–(cyclohexane)–OH + $CH_3\overset{O}{\overset{\|}{C}}O\overset{O}{\overset{\|}{C}}CH_3$ ⟶ $(CH_3)_3C$–(cyclohexane)–$O\overset{O}{\overset{\|}{C}}CH_3$

trans-4-*tert*-Butylcyclohexanol — Acetic anhydride — *trans*-4-Butylcyclohexyl acetate

16.10 Glycerol has three hydroxyl groups, each of which is converted to a nitrate ester function on reaction with nitric acid. Nitroglycerin has the structure shown.

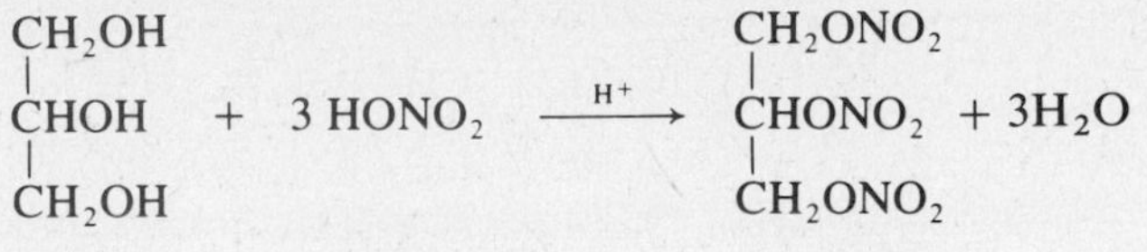

$$\begin{array}{l} CH_2OH \\ | \\ CHOH \\ | \\ CH_2OH \end{array} + 3\,HONO_2 \xrightarrow{H^+} \begin{array}{l} CH_2ONO_2 \\ | \\ CHONO_2 \\ | \\ CH_2ONO_2 \end{array} + 3H_2O$$

Glycerol   Nitric acid   Nitroglycerin

16.11 (*b*) The substrate is a secondary alcohol and so gives a ketone on oxidation with sodium dichromate. 2-Octanone has been prepared in 92 to 96 percent yield under these reaction conditions.

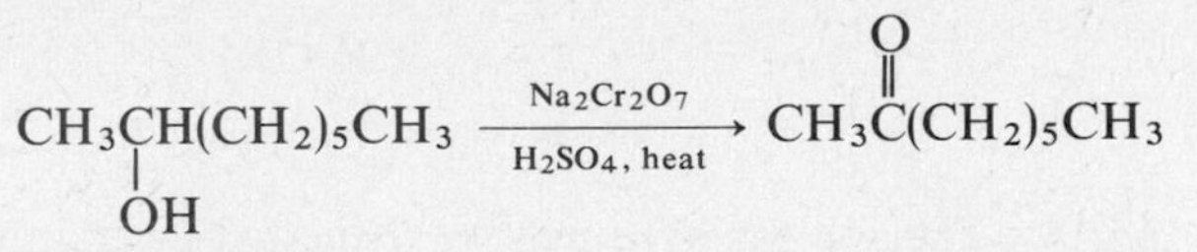

$$CH_3\underset{\displaystyle OH}{\underset{|}{C}}H(CH_2)_5CH_3 \xrightarrow[H_2SO_4,\ heat]{Na_2Cr_2O_7} CH_3\overset{\displaystyle O}{\overset{\|}{C}}(CH_2)_5CH_3$$

2-Octanol   2-Octanone

(*c*) The alcohol is primary, so oxidation can produce either an aldehyde or a carboxylic acid depending on the reaction conditions. Here the oxidation is carried out under anhydrous conditions with Collins' reagent. The product, the corresponding aldehyde, has been obtained in 70 to 84 percent yield.

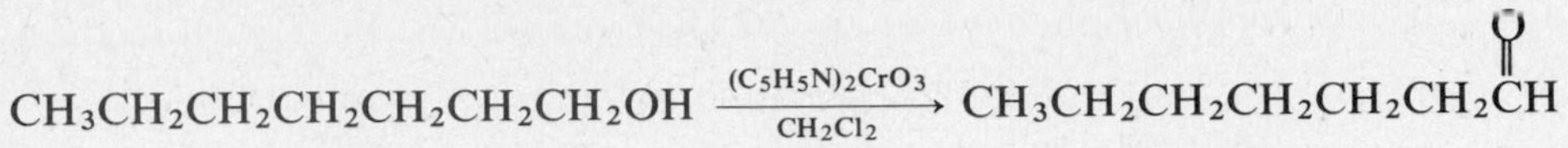

$$CH_3CH_2CH_2CH_2CH_2CH_2CH_2OH \xrightarrow[CH_2Cl_2]{(C_5H_5N)_2CrO_3} CH_3CH_2CH_2CH_2CH_2CH_2\overset{\displaystyle O}{\overset{\|}{C}}H$$

1-Heptanol   Heptanal

(*d*) The alcohol is a cyclic secondary alcohol and yields a cyclic ketone on oxidation. This particular alcohol (menthol) yields the corresponding ketone (menthone) in 83 to 85 percent yield.

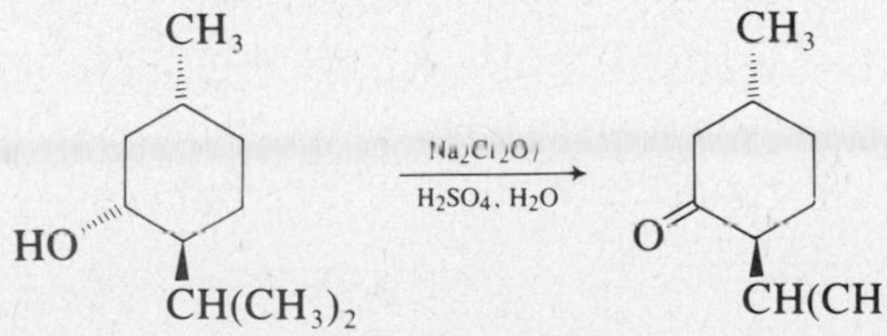

CH$_3$ ... HO ... CH(CH$_3$)$_2$ $\xrightarrow[H_2SO_4,\ H_2O]{Na_2Cr_2O_7}$ CH$_3$ ... O ... CH(CH$_3$)$_2$

16.12 (*b*) Biological oxidation of CH_3CD_2OH leads to loss of one of the C-1 deuterium atoms to NAD^+. The dihydropyridine ring of the reduced form of the coenzyme will bear a single deuterium substituent.

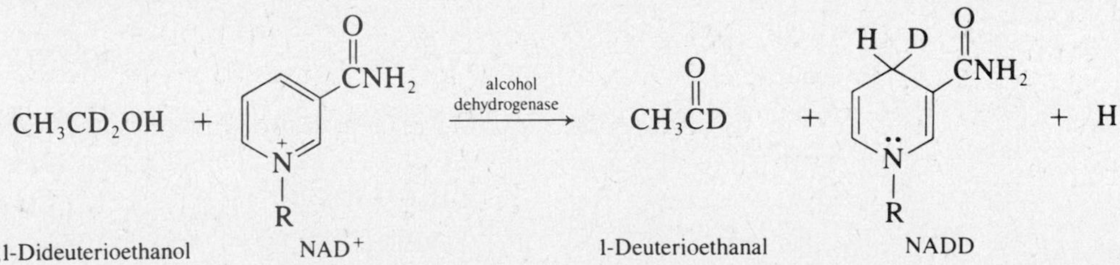

CH_3CD_2OH + NAD$^+$ (pyridinium ring with $\overset{O}{\overset{\|}{C}}NH_2$, $\overset{+}{N}$–R) $\xrightarrow{\text{alcohol dehydrogenase}}$ $CH_3\overset{O}{\overset{\|}{C}}D$ + (dihydropyridine ring with H, D, $\overset{O}{\overset{\|}{C}}NH_2$, $\ddot{N}$–R) + H^+

1,1-Dideuterioethanol   NAD$^+$ 1-Deuterioethanal   NADD

(*c*) The deuterium atom of CH_3CH_2OD is lost as D^+. The reduced form of the coenzyme contains no deuterium.

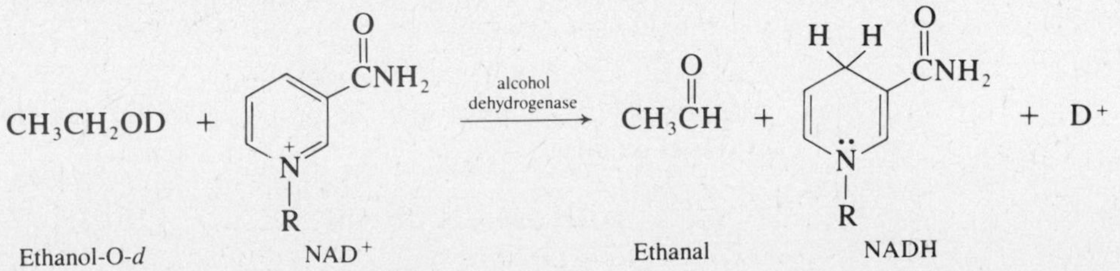

CH_3CH_2OD + NAD$^+$ (pyridinium ring with $\overset{O}{\overset{\|}{C}}NH_2$, $\overset{+}{N}$–R) $\xrightarrow{\text{alcohol dehydrogenase}}$ $CH_3\overset{O}{\overset{\|}{C}}H$ + (dihydropyridine ring with H, H, $\overset{O}{\overset{\|}{C}}NH_2$, $\ddot{N}$–R) + D^+

Ethanol-O-*d*   NAD$^+$ Ethanal   NADH

16.13 (*b*) Oxidation of the carbon-oxygen bonds to carbonyl groups accompanies their cleavage.

$$\underset{\text{1-Phenyl-5-methyl-2,3-hexanediol}}{(CH_3)_2CHCH_2\underset{\substack{|\\HO}}{CH}-\underset{\substack{|\\OH}}{C}HCH_2C_6H_5} \xrightarrow{HIO_4} \underset{\text{3-Methylbutanal}}{(CH_3)_2CHCH_2\overset{\substack{O\\||}}{C}H} + \underset{\text{2-Phenylethanal}}{H\overset{\substack{O\\||}}{C}CH_2C_6H_5}$$

(*c*) The CH_2OH group is cleaved from the ring as formaldehyde to leave cyclopentanone.

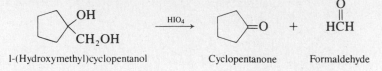

$$\underset{\text{1-(Hydroxymethyl)cyclopentanol}}{\text{cyclopentane-1-OH-1-}CH_2OH} \xrightarrow{HIO_4} \underset{\text{Cyclopentanone}}{\text{cyclopentane}{=}O} + \underset{\text{Formaldehyde}}{H\overset{\substack{O\\||}}{C}H}$$

16.14 The molecular weight of 2-methyl-2-butanol is 88. A peak in its mass spectrum at m/z 70 corresponds to loss of water from the molecular ion. The peaks at m/z 73 and m/z 59 represent stable cations corresponding to the cleavages shown in the equation:

$$CH_3-\underset{\substack{|\\CH_3}}{\overset{\substack{+\cdot\ddot{O}H\\|}}{C}}-CH_2CH_3$$

$$\cdot CH_3 + \underset{m/z\ 73}{CH_3\overset{\substack{+\ddot{O}H\\||}}{C}CH_2CH_3} \qquad\qquad \underset{m/z\ 59}{CH_3\overset{\substack{+\ddot{O}H\\||}}{C}CH_3} + \cdot CH_2CH_3$$

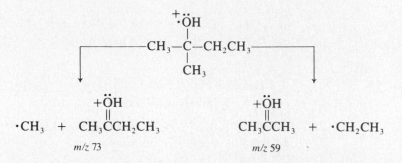

16.15 (*a*) The appropriate alkene for the preparation of 1-butanol by a hydroboration-oxidation sequence is 1-butene. Remember, hydroboration-oxidation leads to hydration of alkenes with a regioselectivity opposite to that seen in acid-catalyzed hydration and in oxymercuration-demercuration.

$$\underset{\text{1-Butene}}{CH_3CH_2CH{=}CH_2} \xrightarrow[\text{2. }H_2O_2,\ HO^-]{\text{1. }B_2H_6} \underset{\text{1-Butanol}}{CH_3CH_2CH_2CH_2OH}$$

(*b*) 1-Butanol can be prepared by reaction of a Grignard reagent with formaldehyde.

$$CH_3CH_2CH_2CH_2OH \Longrightarrow CH_3CH_2\bar{C}H_2 + H\overset{\substack{O\\||}}{C}H$$

An appropriate Grignard reagent is propylmagnesium bromide.

$$\underset{\text{1-Bromopropane}}{CH_3CH_2CH_2Br} \xrightarrow[\text{diethyl ether}]{Mg} \underset{\text{Propylmagnesium bromide}}{CH_3CH_2CH_2MgBr}$$

$$CH_3CH_2CH_2MgBr + H\overset{\substack{O\\||}}{C}H \xrightarrow[\text{2. }H_3O^+]{\text{1. diethyl ether}} \underset{\text{1-Butanol}}{CH_3CH_2CH_2CH_2OH}$$

(*c*) Alternatively, 1-butanol may be prepared by the reaction of a Grignard reagent with ethylene oxide.

$$CH_3CH_2CH_2CH_2OH \Longrightarrow CH_3\bar{C}H_2 + \underset{\diagdown O \diagup}{CH_2-CH_2}$$

In this case, ethylmagnesium bromide would be used.

$$\underset{\text{Ethyl bromide}}{CH_3CH_2Br} \xrightarrow[\text{diethyl ether}]{\text{Mg}} \underset{\text{Ethylmagnesium bromide}}{CH_3CH_2MgBr}$$

$$CH_3CH_2MgBr + \underset{\text{Ethylene oxide}}{CH_2{-}CH_2 \text{ (O bridge)}} \xrightarrow[\text{2. } H_3O^+]{\text{1. diethyl ether}} \underset{\text{1-Butanol}}{CH_3CH_2CH_2CH_2OH}$$

(*d*) Primary alcohols may be prepared by reduction of the carboxylic acid having the same number of carbons. Among the reagents we have discussed, the only one that is effective in the reduction of carboxylic acids is lithium aluminum hydride. The four-carbon carboxylic acid butanoic acid is the proper substrate.

$$\underset{\text{Butanoic acid}}{CH_3CH_2CH_2\overset{O}{\overset{\|}{C}}OH} \xrightarrow[\text{2. } H_3O^+]{\text{1. } LiAlH_4\text{, diethyl ether}} \underset{\text{1-Butanol}}{CH_3CH_2CH_2CH_2OH}$$

(*e*) Hydrogenolysis of esters is normally performed by using a copper chromite catalyst at elevated temperatures and pressures. The correct methyl ester is methyl butanoate.

$$\underset{\text{Methyl butanoate}}{CH_3CH_2CH_2\overset{O}{\overset{\|}{C}}OCH_3} \xrightarrow[\text{copper chromite}]{H_2} \underset{\text{1-Butanol}}{CH_3CH_2CH_2CH_2OH} + \underset{\text{Methanol}}{CH_3OH}$$

(*f*) 1-Butanol, along with ethanol, is formed when ethyl butanoate is reduced with sodium in ethanol.

$$\underset{\text{Ethyl butanoate}}{CH_3CH_2CH_2\overset{O}{\overset{\|}{C}}OCH_2CH_3} \xrightarrow[\text{ethanol}]{\text{Na}} \underset{\text{1-Butanol}}{CH_3CH_2CH_2CH_2OH} + \underset{\text{Ethanol}}{CH_3CH_2OH}$$

(*g*), (*h*) Reduction of an aldehyde or ketone does not change the carbon skeleton.

(*g*) Since 1-butanol is primary alcohol having four carbons, butanal must be the aldehyde which is hydrogenated. Suitable catalysts are nickel, palladium, platinum, and ruthenium.

$$\underset{\text{Butanal}}{CH_3CH_2CH_2\overset{O}{\overset{\|}{C}}H} \xrightarrow{H_2\text{, Pt}} \underset{\text{1-Butanol}}{CH_3CH_2CH_2CH_2OH}$$

(*h*) Sodium borohydride reduces aldehydes and ketones efficiently. It does not reduce carboxylic acids and its reaction with esters is too slow to be of synthetic value.

$$\underset{\text{Butanal}}{CH_3CH_2CH_2\overset{O}{\overset{\|}{C}}H} \xrightarrow[\text{water, ethanol, or methanol}]{NaBH_4} \underset{\text{1-Butanol}}{CH_3CH_2CH_2CH_2OH}$$

16.16 (*a*) Both (*Z*)- and (*E*)-2-butene yield 2-butanol on hydroboration-oxidation.

$$\underset{(Z)\text{- or }(E)\text{-2-Butene}}{CH_3CH{=}CHCH_3} \xrightarrow[\text{2. } H_2O_2\text{, } HO^-]{\text{1. } B_2H_6} \underset{\text{2-Butanol}}{CH_3\underset{OH}{\underset{|}{C}}HCH_2CH_3}$$

Since the alkene is symmetrically substituted, regioselectivity is not an issue.

(*b*) Oxymercuration-demercuration of (*Z*)- and (*E*)-2-butene yields 2-butanol. Here also, regioselectivity is not a consideration.

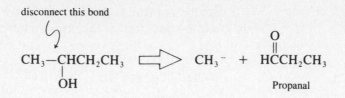

$$\underset{\text{(Z)- or (E)-2-Butene}}{CH_3CH{=}CHCH_3} \xrightarrow[\text{2. NaBH}_4\text{, HO}^-]{\text{1. Hg(O}_2\text{CCH}_3)_2\text{, THF-H}_2\text{O}} \underset{\text{2-Butanol}}{CH_3\underset{\substack{|\\ OH}}{C}HCH_2CH_3}$$

1-Butene could also be used, as the hydration follows Markovnikov's rule.

(*c*) Disconnection of one of the bonds to the carbon that bears the hydroxyl group reveals a feasible route using a Grignard reagent and propanal.

disconnect this bond

$$CH_3{-}\underset{\substack{|\\ OH}}{C}HCH_2CH_3 \Longrightarrow CH_3^- + \underset{\text{Propanal}}{H\overset{\substack{O\\ \|}}{C}CH_2CH_3}$$

The synthetic sequence is:

$$\underset{\text{Methyl bromide}}{CH_3Br} \xrightarrow[\text{diethyl ether}]{\text{Mg}} \underset{\text{Methylmagnesium bromide}}{CH_3MgBr} \xrightarrow[\text{2. H}_3\text{O}^+]{\text{1. CH}_3\text{CH}_2\overset{\text{O}}{\overset{\|}{\text{C}}}\text{H}} \underset{\text{2-Butanol}}{CH_3\underset{\substack{|\\ OH}}{C}HCH_2CH_3}$$

(*d*) There is another disconnection related to a synthetic route using a Grignard reagent and acetaldehyde.

$$CH_3\underset{\substack{|\\ OH}}{C}H{-}CH_2CH_3 \Longrightarrow \underset{\text{Acetaldehyde}}{CH_3\overset{\substack{O\\ \|}}{C}H} + CH_3CH_2^-$$

disconnect this bond

$$\underset{\text{Ethyl bromide}}{CH_3CH_2Br} \xrightarrow[\text{diethyl ether}]{\text{Mg}} \underset{\text{Ethylmagnesium bromide}}{CH_3CH_2MgBr} \xrightarrow[\text{2. H}_3\text{O}^+]{\text{1. CH}_3\overset{\text{O}}{\overset{\|}{\text{C}}}\text{H}} \underset{\text{2-Butanol}}{CH_3\underset{\substack{|\\ OH}}{C}H_2CHCH_3}$$

(*e–h*) Since 2-butanol is a secondary alcohol, it can be prepared by reduction of a ketone having the same carbon skeleton, in this case 2-butanone. All four reducing agents indicated in the equations are satisfactory.

$$\underset{\text{2-Butanone}}{CH_3\overset{\substack{O\\ \|}}{C}CH_2CH_3} \xrightarrow[\text{(or Pt, Ni, etc.)}]{\text{H}_2\text{, Pd}} \underset{\text{2-Butanol}}{CH_3\underset{\substack{|\\ OH}}{C}HCH_2CH_3}$$

$$\underset{\text{2-Butanone}}{CH_3\overset{\substack{O\\ \|}}{C}CH_2CH_2} \xrightarrow[\text{ethanol}]{\text{Na}} \underset{\text{2-Butanol}}{CH_3\underset{\substack{|\\ OH}}{C}HCH_2CH_3}$$

$$\underset{\text{2-Butanone}}{CH_3\overset{O}{\overset{\|}{C}}CH_2CH_3} \xrightarrow[CH_3OH]{NaBH_4} \underset{\text{2-Butanol}}{CH_3\underset{OH}{\underset{|}{C}}HCH_2CH_3}$$

$$\underset{\text{2-Butanone}}{CH_3\overset{O}{\overset{\|}{C}}CH_2CH_3} \xrightarrow[2.\ H_2O]{1.\ LiAlH_4} \underset{\text{2-Butanol}}{CH_3\underset{OH}{\underset{|}{C}}HCH_2CH_3}$$

16.17 (*a*) The desired compound is a tertiary alcohol. The alkene utilized must have the same carbon skeleton as the product. Oxymercuration-demercuration of 2-methylpropene is appropriate.

$$\underset{\text{2-Methylpropene}}{(CH_3)_2C{=}CH_2} \xrightarrow[2.\ NaBH_4,\ HO^-]{1.\ Hg(O_2CCH_3)_2,\ THF\text{-}H_2O} \underset{\text{2-Methyl-2-propanol}}{(CH_3)_3COH}$$

(*b*) All the carbon-carbon disconnections are equivalent.

$$CH_3-\overset{CH_3}{\overset{|}{\underset{CH_3}{\underset{|}{C}}}}-OH \Longrightarrow CH_3^- + \underset{\text{Acetone}}{CH_3\overset{O}{\overset{\|}{C}}CH_3}$$

The synthesis via a Grignard reagent and acetone is:

$$\underset{\text{Methyl bromide}}{CH_3Br} \xrightarrow[\text{diethyl ether}]{Mg} \underset{\text{Methylmagnesium bromide}}{CH_3MgBr} \xrightarrow[2.\ H_3O^+]{1.\ CH_3\overset{O}{\overset{\|}{C}}CH_3} \underset{\text{2 Methyl 2 propanol}}{(CH_3)_3COH}$$

(*c*) An alternative route to 2-methyl-2-propanol is addition of a Grignard reagent to an ester.

$$CH_3-\overset{CH_3}{\overset{|}{\underset{CH_3}{\underset{|}{C}}}}-OH \Longrightarrow 2CH_3^- + \underset{\text{Acetate ester}}{CH_3\overset{O}{\overset{\|}{C}}OR}$$

Any acetate ester can be used here. Let us use ethyl acetate.

$$\underset{\text{Methyl bromide}}{CH_3Br} \xrightarrow[\text{diethyl ether}]{Mg} \underset{\text{Methylmagnesium bromide}}{CH_3MgBr} \xrightarrow[2.\ H_3O^+]{1.\ CH_3\overset{O}{\overset{\|}{C}}OCH_2CH_3} \underset{\text{2-Methyl-2-propanol}}{(CH_3)_3COH}$$

16.18 In order to identify which alcohols might arise by sodium borohydride reduction of carbonyl compounds, write equations for the reduction of all possible aldehydes and ketones that have the molecular formula $C_5H_{10}O$.

$$\underset{\text{Pentanal}}{CH_3CH_2CH_2CH_2\overset{O}{\overset{\|}{C}}H} \xrightarrow[CH_3OH]{NaBH_4} \underset{\text{1-Pentanol}}{CH_3CH_2CH_2CH_2CH_2OH}$$

$$CH_3CH_2CH_2\overset{O}{\overset{\|}{C}}CH_3 \xrightarrow[CH_3OH]{NaBH_4} CH_3CH_2CH_2\underset{OH}{\underset{|}{C}H}CH_3$$

2-Pentanone → 2-Pentanol

$$CH_3CH_2\overset{O}{\overset{\|}{C}}CH_2CH_3 \xrightarrow[CH_3OH]{NaBH_4} CH_3CH_2\underset{OH}{\underset{|}{C}H}CH_2CH_3$$

3-Pentanone → 3-Pentanol

$$CH_3CH_2\underset{CH_3}{\underset{|}{C}H}\overset{O}{\overset{\|}{C}}H \xrightarrow[CH_3OH]{NaBH_4} CH_3CH_2\underset{CH_3}{\underset{|}{C}H}CH_2OH$$

2-Methylbutanal → 2-Methyl-1-butanol

$$(CH_3)_2CHCH_2\overset{O}{\overset{\|}{C}}H \xrightarrow[CH_3OH]{NaBH_4} (CH_3)_2CHCH_2CH_2OH$$

3-Methylbutanal → 3-Methyl-1-butanol

$$(CH_3)_2CH\overset{O}{\overset{\|}{C}}CH_3 \xrightarrow[CH_3OH]{NaBH_4} (CH_3)_2CH\underset{OH}{\underset{|}{C}H}CH_3$$

3-Methyl-2-butanone → 3-Methyl-2-butanol

$$(CH_3)_3C\overset{O}{\overset{\|}{C}}H \xrightarrow[CH_3OH]{NaBH_4} (CH_3)_3CCH_2OH$$

2,2-Dimethylpropanal → 2,2-Dimethyl-1-propanol

These are all the *primary* and *secondary alcohols* of molecular formula $C_5H_{12}O$. The tertiary alcohol $(CH_3)_2\underset{OH}{\underset{|}{C}}CH_2CH_3$ (2-methyl-2-butanol) cannot be prepared by reduction of a carbonyl compound.

16.19 (*a*) The suggested synthesis

$$CH_3CH_2CH_2CH_3 \xrightarrow[\text{light or heat}]{Br_2} CH_3CH_2CH_2CH_2Br \xrightarrow{KOH} CH_3CH_2CH_2CH_2OH$$

Butane → 1-Bromobutane → 1-Butanol

is a poor one because bromination of butane yields a mixture of 1-bromobutane and 2-bromobutane, 2-bromobutane being the major product.

$$CH_3CH_2CH_2CH_3 \xrightarrow[\text{light or heat}]{Br_2} CH_3CH_2CH_2CH_2Br + CH_3\underset{Br}{\underset{|}{C}H}CH_2CH_3$$

Butane → 1-Bromobutane (minor product) + 2-Bromobutane (major product)

(*b*) The suggested synthesis

$$(CH_3)_3CH \xrightarrow[\text{light or heat}]{Br_2} (CH_3)_3CBr \xrightarrow{KOH} (CH_3)_3COH$$

2-Methylpropane → 2-Bromo-2-methylpropane → 2-Methyl-2-propanol

will fail because the reaction of 2-bromo-2-methylpropane with potassium hydroxide will proceed by elimination rather than by substitution. The first step in the process, selective bromination of 2-methylpropane to 2-bromo-2-methylpropane, is satisfactory because bromination is selective for substitution of tertiary hydrogens in the presence of secondary and primary ones.

(*c*) Benzyl alcohol, unlike 1-butanol and 2-methyl-2-propanol, can be prepared effectively by this method.

$$C_6H_5CH_3 \xrightarrow[\text{light or heat}]{Br_2} C_6H_5CH_2Br \xrightarrow{KOH} C_6H_5CH_2OH$$

Toluene — Benzyl bromide — Benzyl alcohol

Bromination of toluene is selective for the benzylic position. Benzyl bromide cannot undergo elimination, so nucleophilic substitution of bromide by hydroxide will work well.

16.20 Glucose contains five hydroxyl groups and an aldehyde functional group. Its hydrogenation will not affect the hydroxyl groups but will reduce the aldehyde to a primary alcohol.

$$\text{Glucose} \xrightarrow[\text{Ni, 140°C}]{H_2\text{ (120 atm)}} \text{Sorbitol}$$

Glucose — Sorbitol

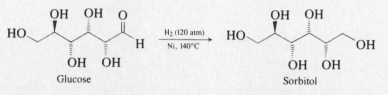

16.21 (*a*) 1-Phenylethanol is a secondary alcohol and so can be prepared by the reaction of a Grignard reagent with an aldehyde. One combination is phenylmagnesium bromide and acetaldehyde.

$$C_6H_5CH(OH)CH_3 \Longrightarrow C_6H_5MgBr + HC(=O)CH_3$$

1-Phenylethanol — Phenylmagnesium bromide — Acetaldehyde

Grignard reagents, phenylmagnesium bromide in this case, are always prepared by reaction of magnesium metal with the corresponding halide. Starting with bromobenzene, a suitable synthesis is described by the sequence:

$$C_6H_5Br \xrightarrow[\text{diethyl ether}]{Mg} C_6H_5MgBr \xrightarrow[2.\ H_3O^+]{1.\ CH_3CHO} C_6H_5CH(OH)CH_3$$

Bromobenzene — Phenylmagnesium bromide — 1-Phenylethanol

(*b*) An alternative disconnection of 1-phenylethanol reveals a second route using benzaldehyde and a methyl Grignard reagent.

$$C_6H_5CH(OH)CH_3 \Longrightarrow C_6H_5C(=O)H + CH_3MgI$$

1-Phenylethanol — Benzaldehyde — Methylmagnesium iodide

Equations representing this approach are:

$$CH_3I \xrightarrow[\text{diethyl ether}]{Mg} CH_3MgI \xrightarrow[2.\ H_3O^+]{1.\ C_6H_5CHO} C_6H_5CH(OH)CH_3$$

Iodomethane — Methylmagnesium iodide — 1-Phenylethanol

(*c*) Aldehydes are, in general, obtainable by oxidation of the corresponding primary alcohol. By recognizing that benzaldehyde can be obtained by Collins' oxidation of benzyl alcohol, we write:

$$\underset{\text{Benzyl alcohol}}{C_6H_5CH_2OH} \xrightarrow[CH_2Cl_2]{(C_5H_5N)_2CrO_3} \underset{\text{Benzaldehyde}}{C_6H_5\overset{O}{\overset{\|}{C}}H} \xrightarrow[\text{2. } H_3O^+]{\text{1. } CH_3MgI,\ \text{diethyl ether}} \underset{\text{1-Phenylethanol}}{C_6H_5\underset{OH}{\underset{|}{C}}HCH_3}$$

(*d*) Styrene, $C_6H_5CH{=}CH_2$, and the desired product 1-phenylethanol have the same carbon skeleton. The conversion therefore requires hydration of the double bond with a regioselectivity that follows Markovnikov's rule. This is best done by oxymercuration-demercuration.

$$\underset{\text{Styrene}}{C_6H_5CH{=}CH_2} \xrightarrow[\text{2. } NaBH_4,\ HO^-]{\text{1. } Hg(O_2CCH_3)_2,\ THF{-}H_2O} \underset{\text{1-Phenylethanol}}{C_6H_5\underset{OH}{\underset{|}{C}}HCH_3}$$

(*e*) The conversion of acetophenone to 1-phenylethanol is a reduction.

$$\underset{\text{Acetophenone}}{C_6H_5\overset{O}{\overset{\|}{C}}CH_3} \xrightarrow{\text{reducing agent}} \underset{\text{1-Phenylethanol}}{C_6H_5\underset{OH}{\underset{|}{C}}HCH_3}$$

Any of a number of reducing agents could be used. These include:

1. $NaBH_4$, CH_3OH
2. $LiAlH_4$ in diethyl ether, then H_2O
3. H_2 and a Pt, Pd, Ni, or Ru catalyst
4. Sodium in ethanol

(*f*) Benzene can be employed as the ultimate starting material in a synthesis of 1-phenylethanol. Friedel-Crafts acetylation of benzene gives acetophenone, which can then be reduced as in (*e*) above.

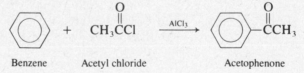

$$\underset{\text{Benzene}}{C_6H_6} + \underset{\text{Acetyl chloride}}{CH_3\overset{O}{\overset{\|}{C}}Cl} \xrightarrow{AlCl_3} \underset{\text{Acetophenone}}{C_6H_5\overset{O}{\overset{\|}{C}}CH_3}$$

Acetic anhydride ($CH_3\overset{O}{\overset{\|}{C}}O\overset{O}{\overset{\|}{C}}CH_3$) can be used in place of acetyl chloride.

16.22 2-Phenylethanol is an ingredient in many perfumes, to which it imparts a roselike fragrance. Numerous methods have been employed for its synthesis.

(*a*) As a primary alcohol having two more carbon atoms than bromobenzene, it can be formed by reaction of a Grignard reagent, phenylmagnesium bromide, with ethylene oxide.

$$C_6H_5CH_2CH_2OH \Longrightarrow C_6H_5MgBr + \overset{O}{CH_2{-}CH_2}$$

The desired reaction sequence is therefore:

$$\underset{\text{Bromobenzene}}{C_6H_5Br} \xrightarrow[\text{diethyl ether}]{Mg} \underset{\text{Phenylmagnesium bromide}}{C_6H_5MgBr} \xrightarrow[\text{2. } H_3O^+]{\text{1. } CH_2{-}CH_2\text{ (O)}} \underset{\text{2-Phenylethanol}}{C_6H_5CH_2CH_2OH}$$

(*b*) Hydration of styrene with a regioselectivity contrary to that of Markovnikov's rule is required. This is accomplished readily by hydroboration-oxidation.

$$\underset{\text{Styrene}}{C_6H_5CH{=}CH_2} \xrightarrow[\text{2. } H_2O_2,\ HO^-]{\text{1. } B_2H_6,\ \text{diglyme}} \underset{\text{2-Phenylethanol}}{C_6H_5CH_2CH_2OH}$$

(*c*) Phenylacetylene can serve as a precursor to 2-phenylethanol if it is first converted to styrene.

$$\underset{\text{Phenylacetylene}}{C_6H_5C{\equiv}CH} \xrightarrow[H_2,\ Pd/CaCO_3,\ Pb(OAc)_2,\ \text{quinoline (Lindlar cat.)}]{\text{Li, } NH_3 \text{ or}} \underset{\text{Styrene}}{C_6H_5CH{=}CH_2}$$

Hydration of styrene to 2-phenylethanol as in (*b*) completes the synthesis. Alternatively, hydroboration-oxidation of phenylacetylene to yield 2-phenylethanal followed by reduction [see (*d*)] also yields 2-phenylethanol.

(*d*) Reduction of aldehydes yields primary alcohols.

$$\underset{\text{2-Phenylethanal}}{C_6H_5CH_2\overset{\overset{O}{\|}}{C}H} \xrightarrow{\text{reducing agent}} \underset{\text{2-Phenylethanol}}{C_6H_5CH_2CH_2OH}$$

Among the reducing agents that could be (and have been) used are:

1. $NaBH_4$, CH_3OH
2. $LiAlH_4$ in diethyl ether, then H_2O
3. H_2 and a Pt, Pd, Ni, or Ru catalyst

Sodium in alcohol is not suitable for reduction of aldehydes because of competing side reactions.

(*e*) Esters are readily reduced to primary alcohols by lithium aluminum hydride.

$$\underset{\text{Ethyl 2-phenylethanoate}}{C_6H_5CH_2\overset{\overset{O}{\|}}{C}OCH_2CH_3} \xrightarrow[\text{2. } H_2O]{\text{1. } LiAlH_4,\ \text{diethyl ether}} \underset{\text{2-Phenylethanol}}{C_6H_5CH_2CH_2OH}$$

Sodium-in-alcohol reduction also works well here, and catalytic hydrogenation has been used for reduction of ethyl phenylethanoate to 2-phenylethanol.

(*f*) The only reagent that is suitable for the direct reduction of carboxylic acids to primary alcohols is lithium aluminum hydride.

$$\underset{\text{2-Phenylethanoic acid}}{C_6H_5CH_2\overset{\overset{O}{\|}}{C}OH} \xrightarrow[\text{2. } H_2O]{\text{1. } LiAlH_4,\ \text{diethyl ether}} \underset{\text{2-Phenylethanol}}{C_6H_5CH_2CH_2OH}$$

Alternatively, the carboxylic acid could be esterified with ethanol and the resulting ethyl phenylethanoate reduced by any of the methods of (*e*).

$$\underset{\text{2-Phenylethanoic acid}}{C_6H_5CH_2\overset{\overset{O}{\|}}{C}OH} + \underset{\text{Ethanol}}{CH_3CH_2OH} \xrightarrow{H^+} \underset{\text{Ethyl 2-phenylethanoate}}{C_6H_5CH_2\overset{\overset{O}{\|}}{C}OCH_2CH_3} \xrightarrow{\text{reduce as in } (e)} \underset{\text{2-Phenylethanol}}{C_6H_5CH_2CH_2OH}$$

16.23 (*a*) The target molecule 2-hexanol may be mentally disconnected as shown to a four-carbon unit and a two-carbon unit.

$$CH_3\underset{\underset{OH}{|}}{C}H{-}CH_2CH_2CH_2CH_3 \Longrightarrow CH_3\overset{\overset{O}{\|}}{C}H + {}^-CH_2CH_2CH_2CH_3$$

The alternative disconnection to CH_3^- and $HCCH_2CH_2CH_2CH_3$ (with C=O on the first carbon) reveals a plausible approach to 2-hexanol but is inconsistent with the requirement of the problem that limits starting materials to four carbons or less. The five-carbon aldehyde would have to be prepared first, making for a lengthy overall synthetic scheme.

An appropriate synthesis based on alcohols as starting materials is:

$$\underset{\text{Ethanol}}{CH_3CH_2OH} \xrightarrow[CH_2Cl_2]{(C_5H_5N)_2CrO_3} \underset{\text{Ethanal (acetaldehyde)}}{CH_3\overset{O}{\overset{\|}{C}}H}$$

$$\underset{\text{1-Butanol}}{CH_3CH_2CH_2CH_2OH} \xrightarrow{PBr_3} \underset{\text{1-Bromobutane}}{CH_3CH_2CH_2CH_2Br} \xrightarrow[\text{diethyl ether}]{Mg} \underset{\text{Butylmagnesium bromide}}{CH_3CH_2CH_2CH_2MgBr}$$

$$\underset{\text{Butylmagnesium bromide}}{CH_3CH_2CH_2CH_2MgBr} + \underset{\text{Ethanal}}{CH_3\overset{O}{\overset{\|}{C}}H} \xrightarrow[2.\ H_3O^+]{1.\ \text{diethyl ether}} \underset{\text{2-Hexanol}}{CH_3\underset{OH}{\underset{|}{C}}HCH_2CH_2CH_2CH_3}$$

(*b*) Oxidation of 2-hexanol from (*a*) yields 2-hexanone.

$$\underset{\text{2-Hexanol}}{CH_3\underset{OH}{\underset{|}{C}}HCH_2CH_2CH_2CH_3} \xrightarrow[H_2SO_4,\ \text{water}]{Na_2Cr_2O_7} \underset{\text{2-Hexanone}}{CH_3\overset{O}{\overset{\|}{C}}CH_2CH_2CH_2CH_3}$$

Collins' reagent can also be used for this transformation.

(*c*) In order to obtain hexanoic acid from alcohols having four carbons or less, a two-carbon chain extension must be carried out. This suggests reaction of a Grignard reagent with ethylene oxide. The resulting alcohol, 1-hexanol, must then be oxidized to yield the desired carboxylic acid. The retrosynthetic path for this approach is as follows:

$$CH_3(CH_2)_4CO_2H \Longrightarrow CH_3(CH_2)_4CH_2OH \Longrightarrow CH_3CH_2CH_2CH_2MgBr + \overset{O}{CH_2{-}CH_2}$$

The reaction sequence therefore becomes:

$$\underset{\text{1-Bromobutane}}{CH_3CH_2CH_2CH_2Br} \xrightarrow[\text{diethyl ether}]{Mg} \underset{\text{Butylmagnesium bromide}}{CH_3CH_2CH_2CH_2MgBr} \xrightarrow[2.\ H_3O^+]{1.\ \overset{O}{CH_2{-}CH_2}} \underset{\text{1-Hexanol}}{CH_3CH_2CH_2CH_2CH_2CH_2OH}$$

$$\underset{\text{1-Hexanol}}{CH_3(CH_2)_4CH_2OH} \xrightarrow[2.\ H^+]{1.\ KMnO_4,\ HO^-} \underset{\text{Hexanoic acid}}{CH_3(CH_2)_4CO_2H}$$

1-Bromobutane can be prepared by reaction of 1-butanol with PBr_3 or with HBr.

$$\underset{\text{1-Butanol}}{CH_3CH_2CH_2CH_2OH} \xrightarrow[\text{or HBr}]{PBr_3} \underset{\text{1-Bromobutane}}{CH_3CH_2CH_2CH_2Br}$$

Given the constraints of the problem, we prepare ethylene oxide by the sequence:

$$\underset{\text{Ethanol}}{CH_3CH_2OH} \xrightarrow[\text{heat}]{H_2SO_4} \underset{\text{Ethylene}}{CH_2{=}CH_2} \xrightarrow{CH_3\overset{O}{\overset{\|}{C}}OOH} \underset{\diagdown O \diagup}{CH_2{-}CH_2}$$

(*d*) Fischer esterification of hexanoic acid with ethanol produces ethyl hexanoate.

$$\underset{\text{Hexanoic acid}}{CH_3(CH_2)_4CO_2H} + \underset{\text{Ethanol}}{CH_3CH_2OH} \xrightarrow{H^+} \underset{\text{Ethyl hexanoate}}{CH_3(CH_2)_4\overset{O}{\overset{\|}{C}}OCH_2CH_3}$$

(*e*) Vicinal diols are normally prepared by hydroxylation of alkenes with osmium tetroxide and *tert*-butyl hydroperoxide (or with cold potassium permanganate).

$$\underset{\text{2-Methylpropene}}{(CH_3)_2C{=}CH_2} \xrightarrow[\substack{(CH_3)_3COOH,\ HO^- \\ (CH_3)_3COH}]{OsO_4} \underset{\text{2-Methyl-1,2-propanediol}}{(CH_3)_2\underset{\underset{HO}{|}}{C}CH_2OH}$$

The required alkene is available by dehydration of 2-methyl-2-propanol.

$$\underset{\substack{\text{2-Methyl-2-propanol} \\ (\textit{tert}\text{-butyl alcohol})}}{(CH_3)_3COH} \xrightarrow[\text{heat}]{H_3PO_4} \underset{\text{2-Methylpropene}}{(CH_3)_2C{=}CH_2}$$

(*f*) The desired aldehyde can be prepared by oxidation of the corresponding primary alcohol.

$$\underset{\substack{\text{2,2-Dimethyl-1-propanol} \\ \text{(neopentyl alcohol)}}}{(CH_3)_3CCH_2OH} \xrightarrow[CH_2Cl_2]{(C_5H_5N)_2CrO_3} \underset{\text{2,2-Dimethylpropanal}}{(CH_3)_3\overset{O}{\overset{\|}{C}}H}$$

Neopentyl alcohol is available by reaction of a *tert*-butyl Grignard reagent with formaldehyde, as shown by the disconnection

$$(CH_3)_3CCH_2OH \Longrightarrow (CH_3)_3CMgCl + H_2C{=}O$$

$$\underset{\text{Methanol}}{CH_3OH} \xrightarrow[CH_2Cl_2]{(C_5H_5N)_2CrO_3} \underset{\text{Formaldehyde}}{H\overset{O}{\overset{\|}{C}}H}$$

$$\underset{\substack{\text{2-Methyl-2-propanol} \\ (\textit{tert}\text{-butyl} \\ \text{alcohol})}}{(CH_3)_3COH} \xrightarrow{HCl} \underset{\substack{\text{2-Chloro-2-methylpropane} \\ (\textit{tert}\text{-butyl chloride})}}{(CH_3)_3CCl} \xrightarrow[\text{Diethyl ether}]{Mg} \underset{\substack{\text{1,1-Dimethylethylmagnesium} \\ \text{chloride } (\textit{tert}\text{-butylmagnesium} \\ \text{chloride})}}{(CH_3)_3CMgCl}$$

$$\underset{\substack{\text{1,1-Dimethylethylmagnesium} \\ \text{chloride } (\textit{tert}\text{-butylmagnesium} \\ \text{chloride})}}{(CH_3)_3CMgCl} \xrightarrow[2.\ H_3O^+]{1.\ CH_2O,\ \text{diethyl ether}} \underset{\substack{\text{2,2-Dimethyl-1-propanol} \\ \text{(neopentyl alcohol)}}}{(CH_3)_3CCH_2OH}$$

(*g*) The simplest route to this primary chloride from benzene is through the corresponding alcohol. The first step is the two-carbon chain extension used in Problem 16.22*a*.

$$C_6H_6 \xrightarrow[FeBr_3]{Br_2} C_6H_5Br \xrightarrow[\text{2. } CH_2\text{—}CH_2 \text{ (O bridged); 3. } H_3O^+]{\text{1. Mg, diethyl ether}} C_6H_5CH_2CH_2OH$$

Benzene — Bromobenzene — 2-Phenylethanol

$$C_6H_5CH_2CH_2OH \xrightarrow{SOCl_2} C_6H_5CH_2CH_2Cl$$

2-Phenylethanol — 1-Chloro-2-phenylethane (2-phenylethyl chloride)

(*h*) A Friedel-Crafts acylation is the best approach to the target ketone.

$$C_6H_6 + (CH_3)_2CH\overset{O}{\overset{\|}{C}}Cl \xrightarrow{AlCl_3} C_6H_5\overset{O}{\overset{\|}{C}}CH(CH_3)_2$$

Benzene — 2-Methylpropanoyl chloride — 2-Methyl-1-phenyl-1-propanone

Since carboxylic acid chlorides are prepared from the corresponding acids, we write:

$$(CH_3)_2CHCH_2OH \xrightarrow[H_2SO_4,\ heat]{K_2Cr_2O_7} (CH_3)_2CH\overset{O}{\overset{\|}{C}}OH \xrightarrow{SOCl_2} (CH_3)_2CH\overset{O}{\overset{\|}{C}}Cl$$

2-Methyl-1-propanol — 2-Methylpropanoic acid — 2-Methylpropanoyl chloride

(*i*) Wolff-Kishner or Clemmensen reduction of the ketone just prepared in (*h*) affords isobutylbenzene.

$$C_6H_5\overset{O}{\overset{\|}{C}}CH(CH_3)_2 \xrightarrow[\text{or Zn (Hg), HCl}]{H_2NNH_2,\ HO^-,\ \text{triethylene glycol, heat}} C_6H_5CH_2CH(CH_3)_2$$

2-Methyl-1-phenyl-1-propanone — Isobutylbenzene

A less direct approach requires three steps:

$$C_6H_5\overset{O}{\overset{\|}{C}}HCH(CH_3)_2 \xrightarrow[CH_3OH]{NaBH_4} C_6H_5\underset{OH}{\underset{|}{C}}CH(CH_3)_2$$

2-Methyl-1-phenyl-1-propanone — 2-Methyl-1-phenyl-1-propanol

$$\downarrow H_2SO_4,\ heat$$

$$C_6H_5CH_2CH(CH_3)_2 \xleftarrow[Pt]{H_2} C_6H_5CH{=}C(CH_3)_2$$

Isobutylbenzene — 2-Methyl-1-phenylpropene

16.24 (*a*), (*b*) Primary alcohols react in two different ways on being heated with acid catalysts: they can condense to form dialkyl ethers or undergo dehydration to yield alkenes. Ether formation is favored at lower temperature and alkene formation is favored at higher temperature.

$$\underset{\text{1-Propanol}}{2CH_3CH_2CH_2OH} \xrightarrow[140^\circ C]{H_2SO_4} \underset{\text{Dipropyl ether}}{CH_3CH_2CH_2OCH_2CH_2CH_3} + \underset{\text{Water}}{H_2O}$$

$$\underset{\text{1-Propanol}}{CH_3CH_2CH_2OH} \xrightarrow[200^\circ C]{H_2SO_4} \underset{\text{Propene}}{CH_3CH{=}CH_2} + \underset{\text{Water}}{H_2O}$$

(*c*) Collins' reagent oxidizes primary alcohols to aldehydes.

$$\underset{\text{1-Propanol}}{CH_3CH_2CH_2OH} \xrightarrow[CH_2Cl_2]{(C_5H_5N)_2CrO_3} \underset{\text{Propanal}}{CH_3CH_2\overset{\overset{O}{\|}}{C}H}$$

(*d*) Potassium permanganate oxidizes primary alcohols to carboxylic acids.

$$\underset{\text{1-Propanol}}{CH_3CH_2CH_2OH} \xrightarrow[2.\ H^+]{1.\ KMnO_4,\ HO^-} \underset{\text{Propanoic acid}}{CH_3CH_2\overset{\overset{O}{\|}}{C}OH}$$

(*e*) Potassium dichromate in sulfuric acid oxidizes primary alcohols to carboxylic acids.

$$\underset{\text{1-Propanol}}{CH_3CH_2CH_2OH} \xrightarrow[H_2SO_4,\ heat]{K_2Cr_2O_7,} \underset{\text{Propanoic acid}}{CH_3CH_2\overset{\overset{O}{\|}}{C}OH}$$

(*f*) Alcohols react with active metals such as sodium to give alkoxides.

$$\underset{\text{1-Propanol}}{2\,CH_3CH_2CH_2OH} + \underset{\text{Sodium}}{2\,Na} \longrightarrow \underset{\text{Sodium 1-propanolate}}{2\,CH_3CH_2CH_2ONa} + \underset{\text{Hydrogen}}{H_2}$$

(*g*) Amide ion, a strong base, abstracts a proton from 1-propanol to form ammonia and 1-propanolate ion. This is an acid-base reaction.

$$\underset{\text{1-Propanol}}{CH_3CH_2CH_2OH} + \underset{\text{Sodium amide}}{NaNH_2} \longrightarrow \underset{\text{Sodium 1-propanolate}}{CH_3CH_2CH_2ONa} + \underset{\text{Ammonia}}{NH_3}$$

(*h*) Acetate ion is a weaker base than 1-propanolate. The equilibrium

$$\underset{\text{1-Propanol}}{CH_3CH_2CH_2OH} + \underset{\text{Acetate}}{CH_3\overset{\overset{O}{\|}}{C}O^-} \rightleftharpoons \underset{\text{1-Propanolate}}{CH_3CH_2CH_2O^-} + \underset{\text{Acetic acid}}{CH_3\overset{\overset{O}{\|}}{C}OH}$$

strongly favors starting materials. 1-Propanol does not react with sodium acetate in any significant way.

(*i*) With acetic acid and in the presence of an acid catalyst, 1-propanol is converted to its acetate ester.

$$\underset{\text{1-Propanol}}{CH_3CH_2CH_2OH} + \underset{\text{Acetic acid}}{CH_3\overset{\overset{O}{\|}}{C}OH} \overset{HCl}{\rightleftharpoons} \underset{\text{Propyl acetate}}{CH_3\overset{\overset{O}{\|}}{C}OCH_2CH_2CH_3} + \underset{\text{Water}}{H_2O}$$

This is an equilibrium process that slightly favors products.

(*j*) Alcohols react with *p*-toluenesulfonyl chloride to give *p*-toluenesulfonate esters.

$$\underset{\text{1-Propanol}}{CH_3CH_2CH_2OH} + \underset{p\text{-Toluenesulfonyl chloride}}{CH_3-C_6H_4-SO_2Cl} \xrightarrow{\text{pyridine}} \underset{\text{Propyl } p\text{-toluenesulfonate}}{CH_3CH_2CH_2O\overset{O}{\underset{O}{\overset{\|}{\underset{\|}{S}}}}-C_6H_4-CH_3} + HCl$$

(*k*) Acyl chlorides convert alcohols to esters.

$$\underset{\text{1-Propanol}}{CH_3CH_2CH_2OH} + \underset{p\text{-Methoxybenzoyl chloride}}{CH_3O-C_6H_4-\overset{O}{\overset{\|}{C}}Cl} \xrightarrow{\text{pyridine}} \underset{\text{Propyl } p\text{-methoxybenzoate}}{CH_3CH_2CH_2O\overset{O}{\overset{\|}{C}}-C_6H_4-OCH_3} + HCl$$

(*l*) The reagent is benzoic anhydride. Carboxylic acid anhydrides react with alcohols to give esters.

$$\underset{\text{1-Propanol}}{CH_3CH_2CH_2OH} + \underset{\text{Benzoic anhydride}}{C_6H_5\overset{O}{\overset{\|}{C}}O\overset{O}{\overset{\|}{C}}C_6H_5} \xrightarrow{\text{pyridine}} \underset{\text{Propyl benzoate}}{CH_3CH_2CH_2O\overset{O}{\overset{\|}{C}}C_6H_5} + \underset{\text{Benzoic acid}}{C_6H_5\overset{O}{\overset{\|}{C}}OH}$$

(*m*) The reagent is succinic anhydride, a cyclic anhydride. Esterification occurs, but in this case the resulting ester and carboxylic acid functions remain part of the same molecule.

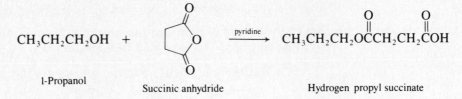

$$\underset{\text{1-Propanol}}{CH_3CH_2CH_2OH} + \underset{\text{Succinic anhydride}}{\text{(cyclic } C_4H_4O_3)} \xrightarrow{\text{pyridine}} \underset{\text{Hydrogen propyl succinate}}{CH_3CH_2CH_2O\overset{O}{\overset{\|}{C}}CH_2CH_2\overset{O}{\overset{\|}{C}}OH}$$

16.25 (*a*), (*b*) The reaction of 2-propanol with sulfuric acid at elevated temperatures yields propene and diisopropyl ether.

$$\underset{\text{2-Propanol}}{CH_3\underset{OH}{\underset{|}{C}}HCH_3} \xrightarrow[\text{heat}]{H_2SO_4} \underset{\text{Propene}}{CH_3CH{=}CH_2} + \underset{\text{Diisopropyl ether}}{(CH_3)_2CHOCH(CH_3)_2}$$

Propene is the major product regardless of whether the temperature is 140°C (*a*) or 200°C (*b*). The fraction of diisopropyl ether formed is greater at the lower temperature.

(*c*) The secondary alcohol 2-propanol is oxidized to the ketone acetone by Collins' reagent.

$$\underset{\text{2-Propanol}}{CH_3\underset{OH}{\underset{|}{C}}HCH_3} \xrightarrow[CH_2Cl_2]{(C_5H_5N)_2CrO_3} \underset{\text{Acetone}}{CH_3\overset{O}{\overset{\|}{C}}CH_3}$$

(*d*) Potassium permanganate oxidizes secondary alcohols to ketones, although overoxidation may occur. Chromium reagents [(*c*) and (*e*)] are more commonly used.

$$\underset{\text{2-Propanol}}{CH_3\underset{\underset{OH}{|}}{C}HCH_3} \xrightarrow[2.\ H^+]{1.\ KMnO_4,\ HO^-} \underset{\text{Acetone}}{CH_3\overset{\overset{O}{\|}}{C}CH_3}$$

(*e*) Chromium(VI) reagents oxidize 2-propanol to acetone.

$$\underset{\text{2-Propanol}}{CH_3\underset{\underset{OH}{|}}{C}HCH_3} \xrightarrow[H_2SO_4,\ H_2O]{K_2Cr_2O_7,} \underset{\text{Acetone}}{CH_3\overset{\overset{O}{\|}}{C}CH_3}$$

(*f*) Sodium reacts with alcohols to yield the corresponding sodium alkoxide.

$$\underset{\text{2-Propanol}}{(CH_3)_2CHOH} + \underset{\text{Sodium}}{2\,Na} \longrightarrow \underset{\text{Sodium 2-propanolate}}{(CH_3)_2CHONa} + \underset{\text{Hydrogen}}{H_2}$$

(*g*) The reaction of 2-propanol with sodium amide is an acid-base reaction.

$$\underset{\substack{\text{2-Propanol}\\ \text{(stronger acid)}}}{(CH_3)_2CHOH} + \underset{\substack{\text{Sodium amide}\\ \text{(stronger base)}}}{NaNH_2} \longrightarrow \underset{\substack{\text{Sodium 2-propanolate}\\ \text{(weaker base)}}}{(CH_3)_2CHONa} + \underset{\substack{\text{Ammonia}\\ \text{(weaker acid)}}}{NH_3}$$

(*h*) No net reaction is observed here. An equilibrium is established but it overwhelmingly favors starting materials.

$$\underset{\substack{\text{2-Propanol}\\ \text{(weaker acid)}}}{(CH_3)_2CHOH} + \underset{\substack{\text{Sodium acetate}\\ \text{(weaker base)}}}{NaO\overset{\overset{O}{\|}}{C}CH_3} \rightleftharpoons \underset{\substack{\text{Sodium 2-propanolate}\\ \text{(stronger base)}}}{(CH_3)_2CHONa} + \underset{\substack{\text{Acetic acid}\\ \text{(stronger acid)}}}{CH_3\overset{\overset{O}{\|}}{C}OH}$$

(*i*) Esterification occurs when 2-propanol is allowed to react with acetic acid in the presence of an acid catalyst.

$$\underset{\text{2-Propanol}}{(CH_3)_2CHOH} + \underset{\text{Acetic acid}}{CH_3\overset{\overset{O}{\|}}{C}OH} \xrightleftharpoons{HCl} \underset{\text{Isopropyl acetate}}{CH_3\overset{\overset{O}{\|}}{C}OCH(CH_3)_2} + H_2O$$

An equilibrium is established.

(*j*) Secondary alcohols react with *p*-toluenesulfonyl chloride to yield secondary alkyl *p*-toluenesulfonate esters.

$$\underset{\text{2-Propanol}}{(CH_3)_2CHOH} + \underset{p\text{-Toluenesulfonyl chloride}}{CH_3-C_6H_4-SO_2Cl} \xrightarrow{\text{pyridine}} \underset{\text{Isopropyl }p\text{-toluenesulfonate}}{(CH_3)_2CHO\overset{\overset{O}{\|}}{\underset{\underset{O}{\|}}{S}}-C_6H_4-CH_3} + HCl$$

(*k*) Ester formation is the characteristic reaction between an alcohol and an acyl chloride.

$$\underset{\text{2-Propanol}}{(CH_3)_2CHOH} + \underset{p\text{-Methoxybenzoyl chloride}}{CH_3O-C_6H_4-\overset{\overset{O}{\|}}{C}Cl} \xrightarrow{\text{pyridine}} \underset{\text{Isopropyl }p\text{-methoxybenzoate}}{(CH_3)_2CHO\overset{\overset{O}{\|}}{C}-C_6H_4-OCH_3} + HCl$$

(*l*) An acid anhydride converts an alcohol to an ester.

$$(CH_3)_2CHOH + C_6H_5\overset{O}{\overset{\|}{C}}O\overset{O}{\overset{\|}{C}}C_6H_5 \xrightarrow{\text{pyridine}} (CH_3)_2CHO\overset{O}{\overset{\|}{C}}C_6H_5 + C_6H_5\overset{O}{\overset{\|}{C}}OH$$

2-Propanol — Benzoic anhydride — Isopropyl benzoate — Benzoic acid

(*m*) One of the carbonyl groups of succinic anhydride is converted to an ester group, the other to a carboxylic acid group.

$$(CH_3)_2CHOH + \text{succinic anhydride} \xrightarrow{\text{pyridine}} (CH_3)_2CHO\overset{O}{\overset{\|}{C}}CH_2CH_2\overset{O}{\overset{\|}{C}}OH$$

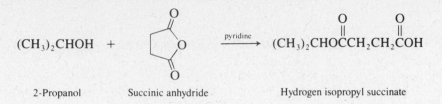

2-Propanol — Succinic anhydride — Hydrogen isopropyl succinate

16.26 (*a*) On being heated in the presence of sulfuric acid, tertiary alcohols undergo elimination.

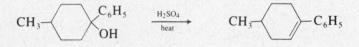

$$\text{4-Methyl-1-phenylcyclohexanol} \xrightarrow[\text{heat}]{H_2SO_4} \text{4-Methyl-1-phenylcyclohexene (81\%)}$$

4-Methyl-1-phenylcyclohexanol — 4-Methyl-1-phenylcyclohexene (81%)

(*b*) The substrate is a primary alcohol and is oxidized by Cr(VI) in the form of Collins' reagent to an aldehyde. The double bonds are unaffected.

$$CH_2{=}CHCH{=}CHCH_2CH_2CH_2OH \xrightarrow[CH_2Cl_2]{(C_5H_5N)_2CrO_3} CH_2{=}CHCH{=}CHCH_2CH_2\overset{O}{\overset{\|}{C}}H$$

4,6-Heptadien-1-ol — 4,6-Heptadienal (87%)

(*c*) The combination of reagents specified converts alkenes to vicinal diols.

$$(CH_3)_2C{=}C(CH_3)_2 \xrightarrow[(CH_3)_3COH,\ HO^-]{(CH_3)_3COOH,\ OsO_4(cat)} (CH_3)_2\underset{HO}{\underset{|}{C}}-\underset{OH}{\underset{|}{C}}(CH_3)_2$$

2,3-Dimethyl-2-butene — 2,3-Dimethyl-2,3-butanediol (72%)

(*d*) Hydroboration-oxidation of the double bond takes place with a regioselectivity that is opposite to Markovnikov's rule. The elements of water are added in a stereospecific syn fashion.

$$\text{1-Phenylcyclobutene} \xrightarrow[2.\ H_2O_2,\ HO^-]{1.\ B_2H_6,\ \text{diglyme}} \textit{trans}\text{-2-Phenylcyclobutanol}$$

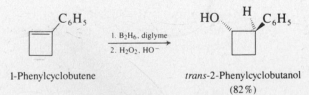

1-Phenylcyclobutene — *trans*-2-Phenylcyclobutanol (82%)

(*e*) The starting material has both ether and ketone functional groups. Sodium-in-alcohol reduction of the ketone carbonyl does not affect the ether function.

$$\text{Xanthone} \xrightarrow[\text{ethanol}]{Na} \text{Xanthydrol}$$

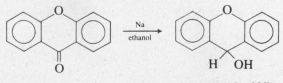

Xanthone — Xanthydrol (91–95%)

(*f*) Chromic acid oxidizes the secondary alcohol to the corresponding ketone but does not affect the triple bond.

$$CH_3\underset{\displaystyle OH}{\underset{|}{C}}HC{\equiv}C(CH_2)_3CH_3 \xrightarrow[\text{acetone}]{H_2CrO_4 \atop H_2SO_4,\ H_2O} CH_3\overset{\displaystyle O}{\overset{\|}{C}}C{\equiv}C(CH_2)_3CH_3$$

3-Octyn-2-ol — 3-Octyn-2-one (80%)

(*g*) The starting material is oleic acid. Hydroxylation of the double bond proceeds with syn stereospecificity.

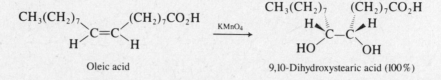

$$\underset{\text{Oleic acid}}{CH_3(CH_2)_7(H)C{=}C(H)(CH_2)_7CO_2H} \xrightarrow{KMnO_4} \underset{\text{9,10-Dihydroxystearic acid (100\%)}}{CH_3(CH_2)_7(H)(HO)C{-}C(H)(OH)(CH_2)_7CO_2H}$$

(*h*) Lithium aluminum hydride reduces carbonyl groups efficiently but does not normally react with double bonds.

$$CH_3\overset{\displaystyle O}{\overset{\|}{C}}CH_2CH{=}CHCH_2\overset{\displaystyle O}{\overset{\|}{C}}CH_3 \xrightarrow[2.\ H_2O]{1.\ LiAlH_4,\ \text{diethyl ether}} CH_3\underset{\displaystyle OH}{\underset{|}{C}}HCH_2CH{=}CHCH_2\underset{\displaystyle OH}{\underset{|}{C}}HCH_3$$

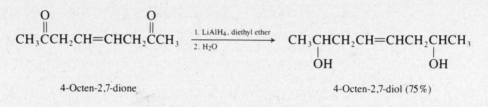

4-Octen-2,7-dione — 4-Octen-2,7-diol (75%)

(*i*) Alcohols react with acyl chlorides to yield esters. The O—H bond is broken in this reaction; the C—O bond of the alcohol remains intact on ester formation.

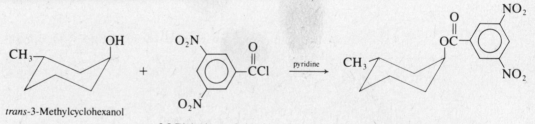

trans-3-Methylcyclohexanol + 3,5-Dinitrobenzoyl chloride $\xrightarrow{\text{pyridine}}$ *trans*-3-Methylcyclohexyl 3,5-dinitrobenzoate (74%)

(*j*) Carboxylic acid anhydrides react with alcohols to give esters. Here too, the spatial orientation of the C—O bond remains intact.

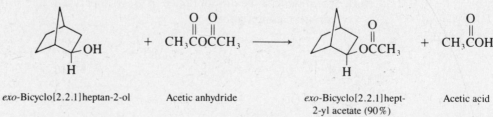

exo-Bicyclo[2.2.1]heptan-2-ol + $CH_3\overset{O}{\overset{\|}{C}}O\overset{O}{\overset{\|}{C}}CH_3$ (Acetic anhydride) $\longrightarrow$ *exo*-Bicyclo[2.2.1]hept-2-yl acetate (90%) + $CH_3\overset{O}{\overset{\|}{C}}OH$ (Acetic acid)

(*k*) In spite of its complexity, the molecule shown reacts in a straightforward way. Its functional groups include a ketone carbonyl, a double bond, and a vicinal diol. The only one of those capable of undergoing a reaction with periodic acid is the vicinal diol, which undergoes the oxidative carbon-carbon bond cleavage reaction that is characteristic of this reagent.

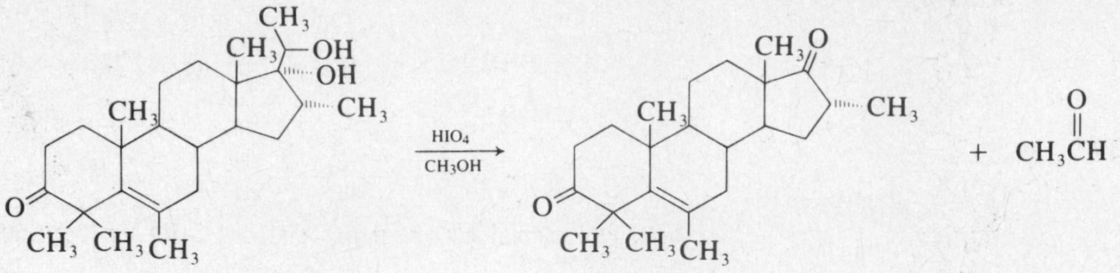

$\xrightarrow[\text{CH}_3\text{OH}]{\text{HIO}_4}$ + CH_3CHO

(90%)

(*l*) The substrate is a carboxylic acid and undergoes Fischer esterification with methanol.

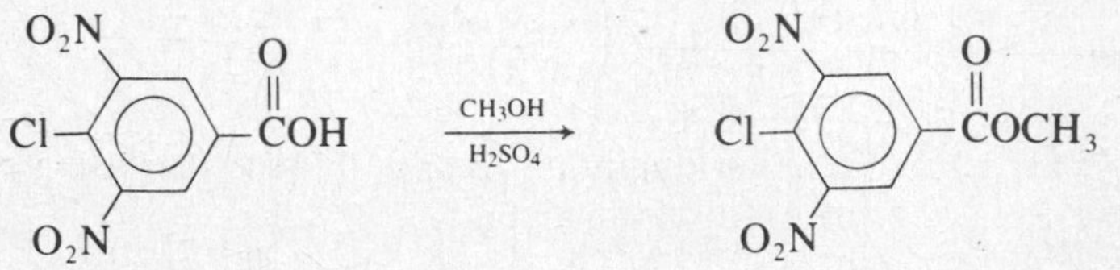

$\xrightarrow[\text{H}_2\text{SO}_4]{\text{CH}_3\text{OH}}$

4-Chloro-3,5-dinitrobenzoic acid Methyl 4-chloro-3,5-dinitrobenzoate (96%)

(*m*) Lithium aluminum hydride reduces the ester to a primary alcohol but does not affect the carbon-carbon double bond. The primary alcohol has been isolated in this reaction in 77 percent yield. The other product is ethyl alcohol.

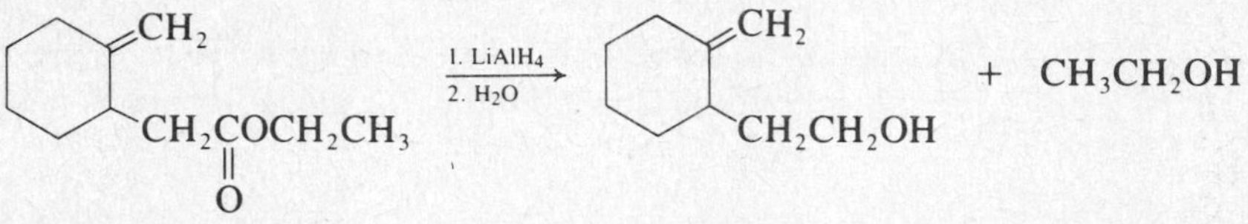

$CH_2COCH_2CH_3$ $\xrightarrow[\text{2. H}_2\text{O}]{\text{1. LiAlH}_4}$ CH_2CH_2OH + CH_3CH_2OH

(*n*) Both ester functions are cleaved by reduction with lithium aluminum hydride. The product is a diol.

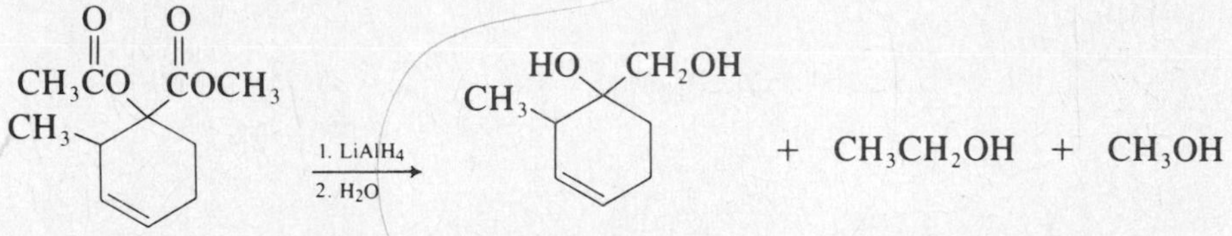

$\xrightarrow[\text{2. H}_2\text{O}]{\text{1. LiAlH}_4}$ + CH_3CH_2OH + CH_3OH

(96%)

(*o*) Treatment of the diol obtained in (*n*) with periodic acid brings about its cleavage to two carbonyl compounds.

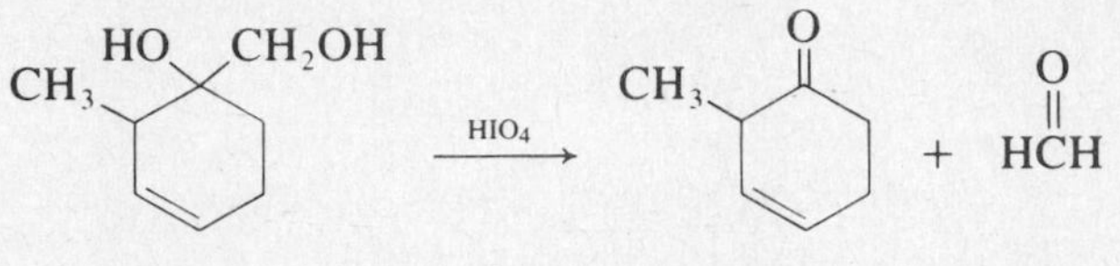

$\xrightarrow{\text{HIO}_4}$ + $HCHO$

(74%)

16.27 Hydroxylation of alkenes with potassium permanganate leads to introduction of vicinal hydroxyl groups at the same face of a carbon-carbon double bond. It is a syn addition.

Applying this reasoning to the question of the hydroxylation of maleic acid, we see that the meso stereoisomer of tartaric acid is formed.

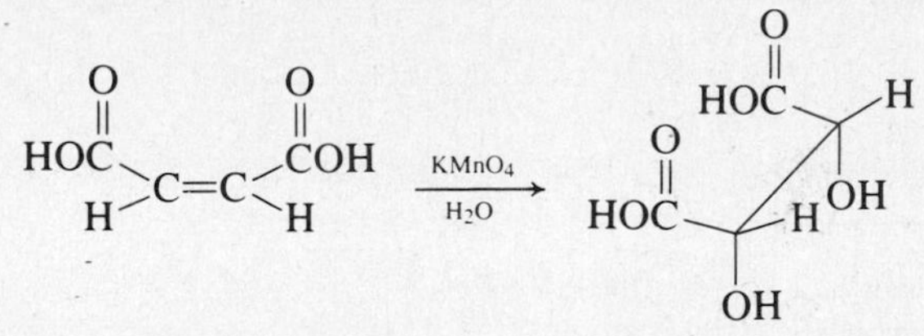

Maleic acid $\xrightarrow[H_2O]{KMnO_4}$ *meso*-Tartaric acid

Fumaric acid yields a racemic mixture of the enantiomeric chiral isomers of tartaric acid.

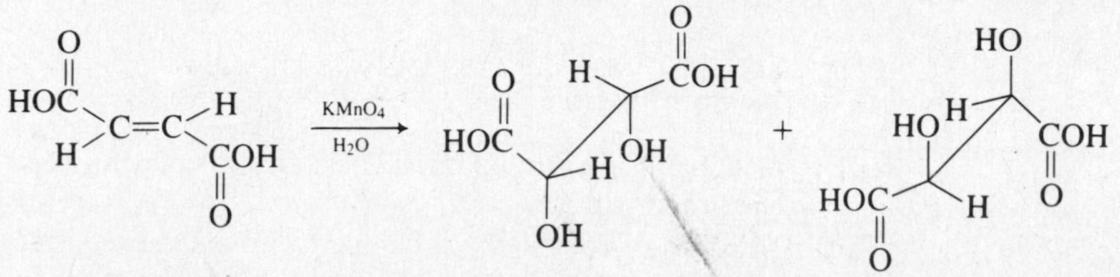

Fumaric acid $\xrightarrow[H_2O]{KMnO_4}$ Racemic tartaric acid

16.28 Hydroxylation under these conditions proceeds by stereospecific syn addition of hydroxyl groups to the alkene. The diol from (*Z*)-4-octene has a plane of symmetry in its eclipsed conformation; it is the meso stereoisomer of 4,5-octanediol.

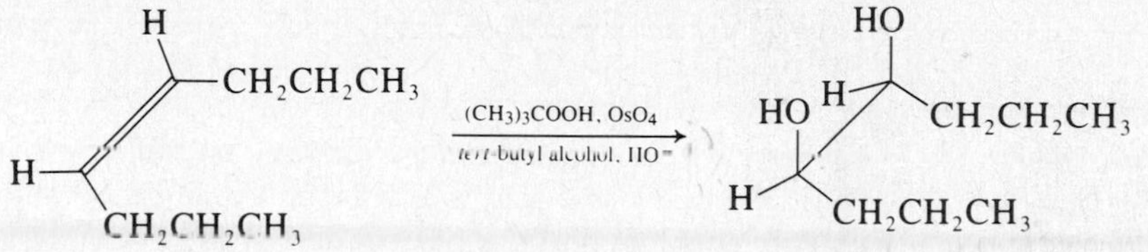

(*Z*)-4-Octene $\xrightarrow[\text{tert-butyl alcohol, } HO^-]{(CH_3)_3COOH,\ OsO_4}$ *meso*-4,5-Octanediol

Hydroxylation of the *E* alkene yields the chiral diastereomer as a racemic mixture of enantiomers.

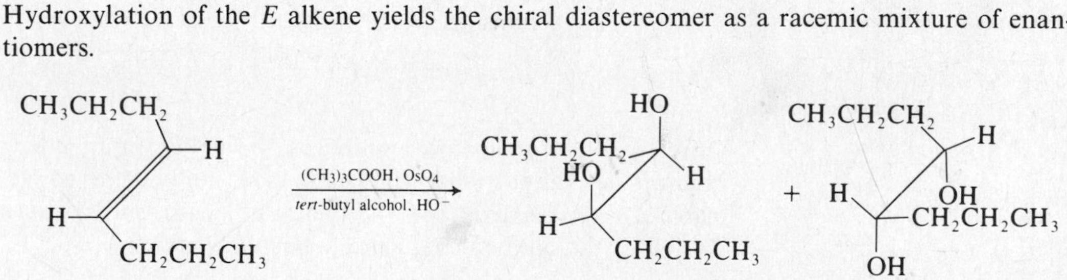

(*E*)-4-Octene $\xrightarrow[\text{tert-butyl alcohol, } HO^-]{(CH_3)_3COOH,\ OsO_4}$ Racemic 4,5-octanediol

16.29 (*a*) The target bromide can be made if the corresponding alcohol is available.

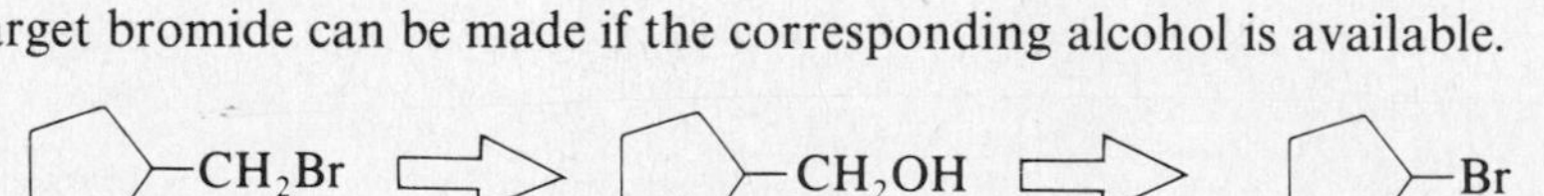

Cyclopentyl–CH_2Br ⇨ Cyclopentyl–CH_2OH ⇨ Cyclopentyl–Br

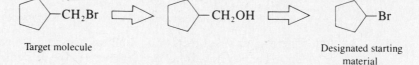

Target molecule; Designated starting material

When the problem is viewed from the retrosynthetic perspective above, a reasonable synthetic sequence becomes apparent.

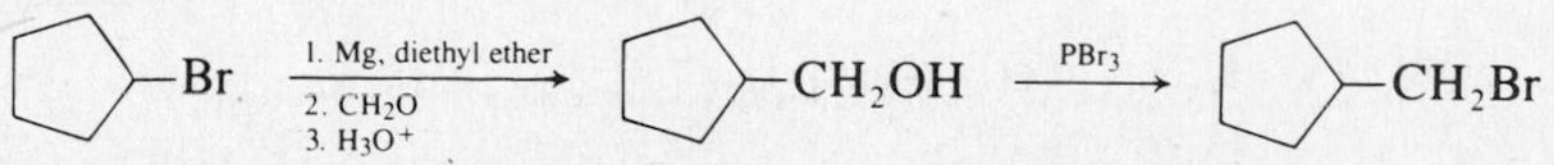

Cyclopentyl–Br $\xrightarrow{\text{1. Mg, diethyl ether; 2. } CH_2O\text{; 3. } H_3O^+}$ Cyclopentyl–CH_2OH $\xrightarrow{PBr_3}$ Cyclopentyl–CH_2Br

(*b*) The desired product, an alkene, may be prepared from an alcohol, which is in turn available from reaction of the designated starting material with a Grignard reagent.

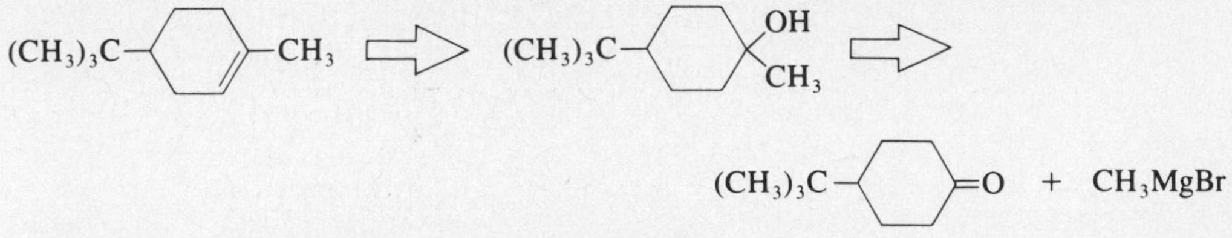

$(CH_3)_3C$–⟨cyclohexene⟩–CH_3 ⟹ $(CH_3)_3C$–⟨cyclohexane⟩(OH)(CH_3) ⟹

$(CH_3)_3C$–⟨cyclohexane⟩=O + CH_3MgBr

The correct reaction sequence therefore becomes:

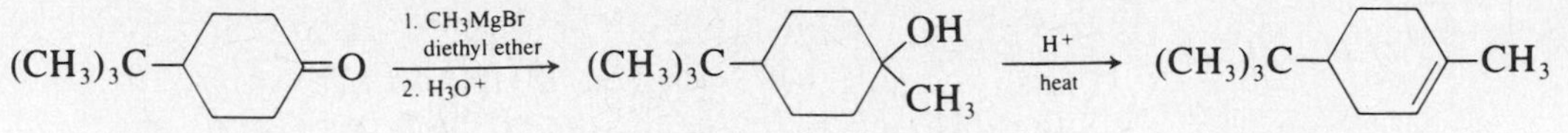

$(CH_3)_3C$–⟨cyclohexane⟩=O $\xrightarrow[2.\ H_3O^+]{1.\ CH_3MgBr,\ diethyl\ ether}$ $(CH_3)_3C$–⟨cyclohexane⟩(OH)(CH_3) $\xrightarrow[heat]{H^+}$ $(CH_3)_3C$–⟨cyclohexene⟩–CH_3

4-*tert*-Butylcyclohexanone | 4-*tert*-Butyl-1-methylcyclohexanol | 4-*tert*-Butyl-1-methylcyclohexene

As actually performed, methyllithium was used in place of the Grignard reagent. Either one is acceptable.

(*c*) The task of converting a ketone to an alkene requires first the reduction of the ketone to an alcohol and then dehydration. In practice the two-step transformation has been carried out in 54 percent yield by treating the ketone with sodium borohydride and then heating the resulting alcohol with *p*-toluenesulfonic acid.

(ketone) $\xrightarrow[CH_3OH]{NaBH_4}$ (alcohol, OH) $\xrightarrow[heat]{H^+}$ (alkene)

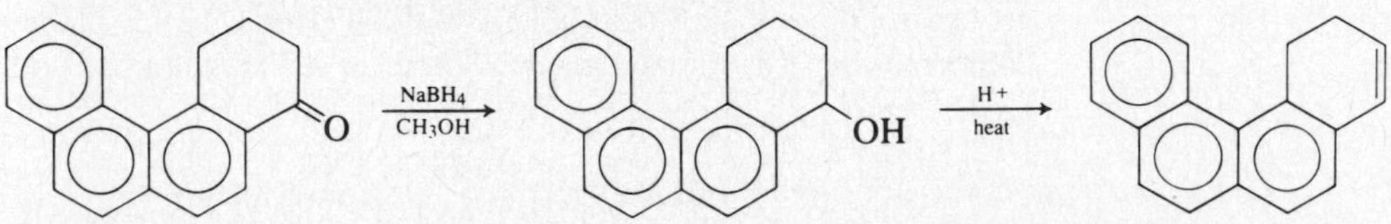

Of course, sodium borohydride may be replaced by other suitable reducing agents and *p*-toluenesulfonic acid is not the only acid that could be used in the dehydration step.

(*d*) This problem and the next one illustrate the value of reasoning backward. The desired product, cyclohexanol, can be prepared cleanly from cyclohexanone.

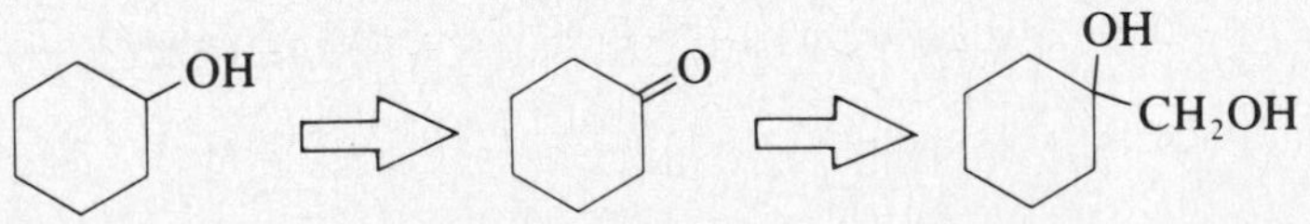

(cyclohexanol, OH) ⟹ (cyclohexanone, O) ⟹ (1-hydroxycyclohexyl: OH, CH_2OH)

Once cyclohexanone is recognized to be a key intermediate, the synthetic pathway becomes apparent—what is needed is a method to convert the indicated starting material to cyclohexanone. The reagent ideally suited to this task is periodic acid. Therefore, the synthetic sequence to be followed is:

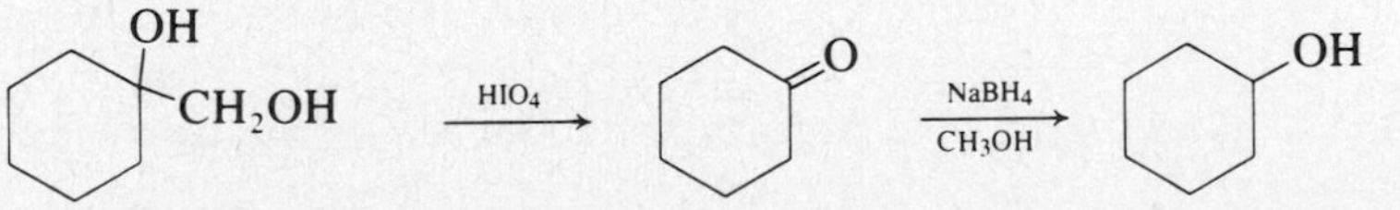

(OH, CH_2OH) $\xrightarrow{HIO_4}$ (cyclohexanone, O) $\xrightarrow[CH_3OH]{NaBH_4}$ (cyclohexanol, OH)

(1-Hydroxycyclohexyl)methanol | Cyclohexanone | Cyclohexanol

(*e*) There is no direct method that allows a second hydroxyl group to be introduced at C-2 of 1-phenylcyclohexanol in a single step. We recognize the product as a vicinal diol and recall that such compounds are available by hydroxylation of alkenes.

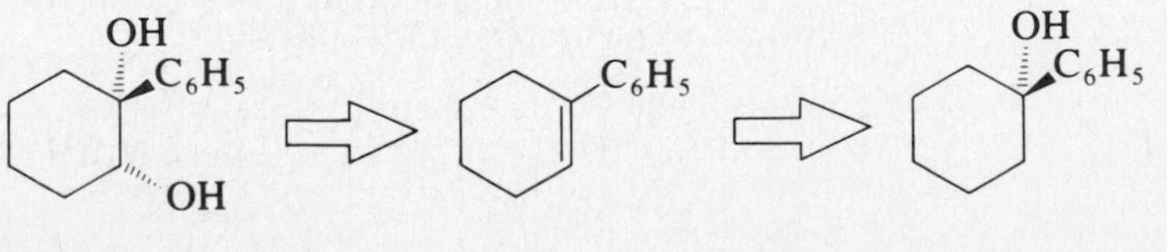

(OH, C_6H_5, OH) ⟹ (C_6H_5 cyclohexene) ⟹ (OH, C_6H_5)

This tells us that we must first dehydrate the tertiary alcohol, then hydroxylate the resulting alkene.

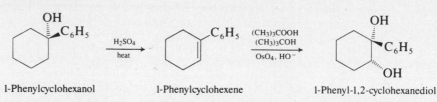

$$\text{1-Phenylcyclohexanol} \xrightarrow[\text{heat}]{H_2SO_4} \text{1-Phenylcyclohexene} \xrightarrow[OsO_4,\ HO^-]{(CH_3)_3COOH,\ (CH_3)_3COH} \text{1-Phenyl-1,2-cyclohexanediol}$$

The syn stereoselectivity of the hydroxylation step ensures that the product will have its hydroxyl groups cis as the problem requires.

16.30 Since the target molecule is an eight-carbon secondary alcohol and the problem restricts our choices of starting materials to alcohols of five carbons or less, we are led to consider building up the carbon chain by a Grignard reaction.

$$\underset{\text{4-Methyl-3-heptanol}}{CH_3CH_2\underset{OH}{CH}\text{—}\overset{CH_3}{CH}CH_2CH_2CH_3} \Longrightarrow CH_3CH_2\overset{O}{\overset{\|}{C}}H + {}^{-}\overset{CH_3}{CH}CH_2CH_2CH_3$$

The disconnection shown leads to a three-carbon aldehyde and a five-carbon Grignard reagent. Starting with the corresponding alcohols, the following synthetic scheme seems reasonable.

First propanal is prepared.

$$\underset{\text{1-Propanol}}{CH_3CH_2CH_2OH} \xrightarrow[\text{pyridine}]{(C_5H_5N)_2CrO_3} \underset{\text{Propanal}}{CH_3CH_2\overset{O}{\overset{\|}{C}}H}$$

After converting 2-pentanol to its bromo derivative, a solution of the Grignard reagent is prepared.

$$\underset{\text{2-Pentanol}}{CH_3\underset{OH}{CH}CH_2CH_2CH_3} \xrightarrow{PBr_3} \underset{\text{2-Bromopentane}}{CH_3\underset{Br}{CH}CH_2CH_2CH_3} \xrightarrow[\text{diethyl ether}]{Mg} \underset{\text{1-Methylbutylmagnesium bromide}}{CH_3\underset{MgBr}{CH}CH_2CH_2CH_3}$$

Reaction of the Grignard reagent with the aldehyde yields the desired 4-methyl-3-heptanol.

$$\underset{\text{1-Methylbutylmagnesium bromide}}{CH_3\underset{MgBr}{CH}CH_2CH_2CH_3} + \underset{\text{Propanal}}{CH_3CH_2\overset{O}{\overset{\|}{C}}H} \xrightarrow[2.\ H_3O^+]{1.\ \text{diethyl ether}} \underset{\text{4-Methyl-3-heptanol}}{CH_3\underset{HOCHCH_2CH_3}{CH}CH_2CH_2CH_3}$$

16.31 Our target molecule is void of functionality and so requires us to focus attention on the carbon skeleton. Notice that it can be considered to arise from three ethyl groups.

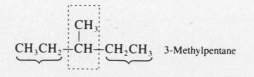

$$\underbrace{CH_3CH_2}\text{—}\overset{CH_3}{CH}\text{—}\underbrace{CH_2CH_3} \quad \text{3-Methylpentane}$$

Considering the problem retrosynthetically, we can see that a key intermediate having the carbon skeleton of the desired product is 3-methyl-3-pentanol. This becomes apparent from the fact that alkanes may be prepared from alkenes, which in turn are available from alcohols. The desired alcohol may be prepared from reaction of an acetate ester with a Grignard reagent, ethylmagnesium bromide.

$$CH_3CH_2\underset{}{\overset{CH_3}{\overset{|}{C}}}HCH_2CH_3 \Longrightarrow CH_3CH_2\underset{\underset{OH}{|}}{\overset{\overset{CH_3}{|}}{C}}CH_2CH_3 \Longrightarrow CH_3\overset{\overset{O}{\|}}{C}OR + 2CH_3CH_2MgBr$$

The carbon skeleton can be assembled in one step by the reaction of ethylmagnesium bromide and ethyl acetate.

$$2CH_3CH_2MgBr + CH_3\overset{\overset{O}{\|}}{C}OCH_2CH_3 \xrightarrow[2.\ H_3O^+]{1.\ \text{diethyl ether}} CH_3\underset{\underset{CH_2CH_3}{|}}{\overset{\overset{OH}{|}}{C}}CH_2CH_3$$

Ethylmagnesium bromide — Ethyl acetate — 3-Methyl-3-pentanol

The resulting tertiary alcohol is converted to the desired hydrocarbon by acid-catalyzed dehydration and catalytic hydrogenation of the resulting mixture of alkenes.

$$CH_3\underset{\underset{CH_2CH_3}{|}}{\overset{\overset{OH}{|}}{C}}CH_2CH_3 \xrightarrow{H^+} CH_3\underset{\underset{CH_2CH_3}{|}}{C}{=}CHCH_3 + CH_2{=}C(CH_2CH_3)_2$$

3-Methyl-3-pentanol — 3-Methyl-2-pentene (cis + trans) — 2-Ethyl-1-butene

$$\xrightarrow{H_2,\ Ni} CH_3CH(CH_2CH_3)_2$$

3-Methylpentane

Since the problem requires that ethanol be the ultimate starting material, we need to show the preparation of the ethylmagnesium bromide and ethyl acetate used in constructing the carbon skeleton.

$$CH_3CH_2OH \xrightarrow{PBr_3} CH_3CH_2Br \xrightarrow[\text{diethyl ether}]{Mg} CH_3CH_2MgBr$$

Ethanol — Ethyl bromide — Ethylmagnesium bromide

$$CH_3CH_2OH \xrightarrow[H_2SO_4,\ \text{heat}]{K_2Cr_2O_7} CH_3\overset{\overset{O}{\|}}{C}OH$$

Ethanol — Acetic acid

$$CH_3\overset{\overset{O}{\|}}{C}OH + CH_3CH_2OH \xrightarrow{H^+} CH_3\overset{\overset{O}{\|}}{C}OCH_2CH_3$$

Acetic acid — Ethyl acetate

16.32 Retrosynthetically, we can see that the cis carbon-carbon double bond is available by hydrogenation of the corresponding alkyne over the Lindlar catalyst.

$$CH_3CH_2CH{=}CHCH_2CH_2OH \Longrightarrow CH_3CH_2C{\equiv}CCH_2CH_2OH$$

The $-CH_2CH_2OH$ unit can be appended to an alkynide anion by reaction with ethylene oxide.

$$CH_3CH_2C{\equiv}CCH_2CH_2OH \Longrightarrow CH_3CH_2C{\equiv}C{:}^- + \underset{\diagdown O \diagup}{CH_2{-}CH_2}$$

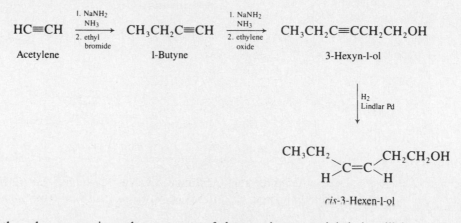

The alkynide anion is derived from 1-butyne by alkylation of acetylene. This analysis suggests the following synthetic sequence.

$$\underset{\text{Acetylene}}{HC{\equiv}CH} \xrightarrow[\text{2. ethyl bromide}]{\text{1. NaNH}_2\text{, NH}_3} \underset{\text{1-Butyne}}{CH_3CH_2C{\equiv}CH} \xrightarrow[\text{2. ethylene oxide}]{\text{1. NaNH}_2\text{, NH}_3} \underset{\text{3-Hexyn-1-ol}}{CH_3CH_2C{\equiv}CCH_2CH_2OH}$$

$$\xrightarrow{H_2,\ \text{Lindlar Pd}} \underset{\textit{cis}\text{-3-Hexen-1-ol}}{(CH_3CH_2)(H)C{=}C(H)(CH_2CH_2OH)}$$

16.33 Even though we are given the structure of the starting material, it is still better to reason backward from the target molecule rather than forward from the starting material.

The desired product contains a cyano ($-CN$) group. The only method we have seen so far for introducing such a function into a molecule is by nucleophilic substitution. Therefore, the last step in the synthesis must be:

$$m\text{-}CH_3OC_6H_4CH_2X + CN^- \longrightarrow m\text{-}CH_3OC_6H_4CH_2CN + X^-$$

This step should work very well since the substrate is a primary benzylic halide, cannot undergo elimination, and is very reactive in S_N2 reactions.

The primary benzylic halide can be prepared from the corresponding alcohol by any of a number of methods.

$$m\text{-}CH_3OC_6H_4CH_2OH \longrightarrow m\text{-}CH_3OC_6H_4CH_2X$$

Suitable reagents include HBr, PBr_3, or $SOCl_2$.

There remains now only to prepare the primary alcohol from the given starting aldehyde, which is accomplished by reduction.

$$m\text{-}CH_3OC_6H_4\overset{O}{\overset{\|}{C}}H \longrightarrow m\text{-}CH_3OC_6H_4CH_2OH$$

Reduction can be achieved by catalytic hydrogenation, with lithium aluminum hydride, or with sodium borohydride.

The actual sequence of reactions as carried out is as shown:

m-Methoxybenzaldehyde $\xrightarrow[\text{(100\% yield)}]{H_2, PtO_2 \text{ ethanol}}$ m-Methoxybenzyl alcohol

m-Methoxybenzyl alcohol $\xrightarrow[\text{(98\% yield)}]{\text{HBr, benzene}}$ m-Methoxybenzyl bromide

m-Methoxybenzyl bromide $\xrightarrow[\text{(87\% yield)}]{\text{NaCN ethanol, water}}$ m-Methoxybenzyl cyanide

(Structures: m-$CH_3OC_6H_4CHO$; m-$CH_3OC_6H_4CH_2OH$; m-$CH_3OC_6H_4CH_2Br$; m-$CH_3OC_6H_4CH_2CN$)

m-Methoxybenzaldehyde *m*-Methoxybenzyl alcohol

m-Methoxybenzyl cyanide *m*-Methoxybenzyl bromide

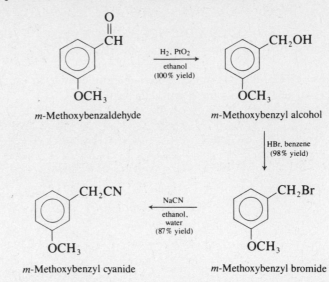

Another three-step synthesis, which is reasonable but does not involve an alcohol as an intermediate, is:

m-Methoxybenzaldehyde $\xrightarrow{\text{Clemmensen or Wolff-Kishner reduction}}$ m-Methoxytoluene

m-Methoxytoluene $\xrightarrow{N\text{-bromosuccinimide, } h\nu}$ m-Methoxybenzyl bromide

m-Methoxybenzyl bromide $\xrightarrow{CN^-}$ m-Methoxybenzyl cyanide

(Structures: m-$CH_3OC_6H_4CHO$; m-$CH_3OC_6H_4CH_3$; m-$CH_3OC_6H_4CH_2Br$; m-$CH_3OC_6H_4CH_2CN$)

m-Methoxybenzaldehyde *m*-Methoxytoluene

m-Methoxybenzyl cyanide *m*-Methoxybenzyl bromide

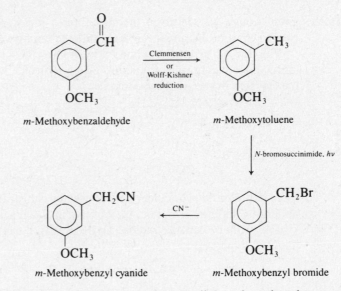

16.34 (*a*) Addition of hydrogen chloride to cyclopentadiene takes place by way of the most stable carbocation. In this case it is an allylic carbocation.

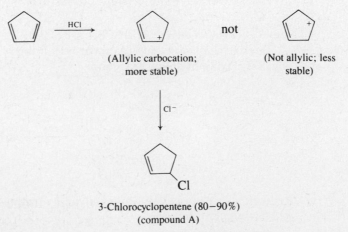

Cyclopentadiene $\xrightarrow{HCl}$ (Allylic carbocation; more stable) not (Not allylic; less stable)

(Allylic carbocation; more stable) $\xrightarrow{Cl^-}$ 3-Chlorocyclopentene (80–90%) (compound A)

Cl

3-Chlorocyclopentene (80–90%)
(compound A)

Hydrolysis of 3-chlorocyclopentene gives the corresponding alcohol. Sodium bicarbonate in water is a weakly basic solvolytic medium.

$$\text{Compound A (3-chlorocyclopentene, Cl)} \xrightarrow[H_2O]{NaHCO_3} \text{2-Cyclopenten-1-ol (88\%) (compound B) (OH)}$$

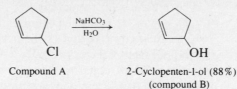

Oxidation of compound B (a secondary alcohol) gives the ketone 2-cyclopenten-1-one.

$$\text{Compound B (OH)} \xrightarrow[H_2SO_4,\ H_2O]{Na_2Cr_2O_7} \text{2-Cyclopenten-1-one (60–68\%) (compound C) (O)}$$

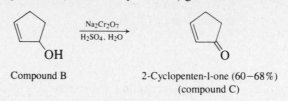

(*b*) Thionyl chloride converts alcohols to alkyl chlorides.

$$\underset{\text{5-Hexen-2-ol}}{CH_2{=}CHCH_2CH_2\underset{|}{\overset{}{C}}HCH_3 \ (OH)} \xrightarrow[pyridine]{SOCl_2} \underset{\text{5-Chloro-1-hexene (compound D)}}{CH_2{=}CHCH_2CH_2CHCH_3 \ (Cl)}$$

Ozonation, followed by reduction of the ozonide, cleaves the carbon-carbon double bond.

$$\underset{\text{(Compound D)}}{CH_2{=}CHCH_2CH_2CHCH_3 \ (Cl)} \xrightarrow[\text{2. reductive workup}]{\text{1. } O_3} \underset{\text{4-Chloropentanal (compound E)}}{H\overset{O}{\overset{\|}{C}}CH_2CH_2CHCH_3 \ (Cl)} + \underset{\text{Formaldehyde}}{H\overset{O}{\overset{\|}{C}}H}$$

Reduction of compound E yields the corresponding alcohol.

$$\underset{\text{4-Chloropentanal}}{H\overset{O}{\overset{\|}{C}}CH_2CH_2CHCH_3 \ (Cl)} \xrightarrow{NaBH_4} \underset{\text{4-Chloro-1-pentanol (compound F)}}{HOCH_2CH_2CH_2CHCH_3 \ (Cl)}$$

16.35 The difference between the two ethers is that 1-*O*-benzylglycerol contains a vicinal diol function while 2-*O*-benzylglycerol does not. Periodic acid will react with 1-*O*-benzylglycerol but not with 2-*O*-benzylglycerol.

$$\underset{\text{1-}O\text{-Benzylglycerol}}{C_6H_5CH_2OCH_2CHCH_2OH \ (OH)} \xrightarrow{HIO_4} \underset{\text{2-Benzyloxyacetaldehyde}}{C_6H_5CH_2OCH_2\overset{O}{\overset{\|}{C}}H} + \underset{\text{Formaldehyde}}{H\overset{O}{\overset{\|}{C}}H}$$

$$\underset{\text{2-}O\text{-Benzylglycerol}}{HOCH_2CHCH_2OH \ (OCH_2C_6H_5)} \xrightarrow{HIO_4} \text{no reaction}$$

16.36 (*a*) An intense, broad absorption at about 3400 cm^{-1} in its infrared spectrum indicates that compound G is an alcohol, and the presence of peaks in the C—O stretching region

(1025 to 1200 cm^{-1}) supports this conclusion. A signal at $\delta = 3.5$ ppm in the ^1H nmr spectrum, equivalent to one proton, can be assigned to the OH proton of an alcohol.

The presence of aromatic rings is indicated by the low hydrogen-to-carbon ratio ($C_{14}H_{14}O$) of its molecular formula and by the presence of weak absorbances in the 1670 to 2000 cm^{-1} region of the ir spectrum. The difference of the molecular formula ($C_{14}H_{14}O$) from C_nH_{2n+2} ($C_{14}H_{30}$) tells us that the sum of double bonds plus rings is eight. A SODAR equal to 8 can be accommodated by two benzene rings. That these two benzene rings are monosubstituted is shown by the presence of 10 aromatic protons in the nmr spectrum at $\delta = 7$ to 7.5 ppm and by a peak in the ir spectrum at 700 cm^{-1}.

At this point we have identified three structural units accounting for the oxygen, 12 carbons, and 11 hydrogens.

C_6H_5- $\quad$ C_6H_5- $\quad$ and $\quad$ OH

The three remaining hydrogens are part of a methyl group. These methyl protons appear as a singlet in the nmr at $\delta = 1.9$ ppm. The methyl group must be attached to a carbon that bears no hydrogens.

$CH_3-\overset{|}{\underset{|}{C}}-$

Putting these structural units together, we arrive at 1,1-diphenylethanol as the correct structure of compound G.

$C_6H_5-\overset{CH_3}{\underset{OH}{C}}-C_6H_5$

1,1-Diphenylethanol
(compound G)

(*b*) Compound H is an alcohol. It exhibits an O—H stretching vibration in its ir spectrum at 3350 cm^{-1} and C—O stretching between 1050 and 1100 cm^{-1}. The ^1H nmr spectrum contains a one-proton singlet at $\delta = 2.7$ ppm, assignable to the hydroxyl proton.

An aromatic ring is indicated by the molecular formula which requires a SODAR of 4 and by a strong band at about 800 cm^{-1}, characteristic of a para-substituted benzene ring. Para substitution is suggested as well by the four-proton AB system at $\delta = 6.8$ to 7.5 ppm in the nmr spectrum.

The hydrogen on the carbon that bears the hydroxyl group appears as a triplet centered at $\delta = 4.4$ ppm, which tells us that it is attached to a methylene group. Since its chemical shift is shifted to lower field than normal, it is probably benzylic. A reasonable conclusion points to the partial structure

$-C_6H_4-\overset{H}{\underset{OH}{C}}-CH_2-$

$\delta = 4.4$ ppm (H); $\delta = 1.6$ ppm (CH_2); $\delta = 2.7$ ppm (OH)

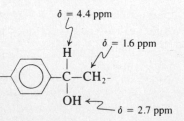

There is also a methyl triplet at $\delta = 0.8$ ppm in the nmr, which must arise from a methyl group attached to the methylene unit. There remains only to add the bromine atom to the para position of the ring to complete the structure.

$Br-C_6H_4-\underset{OH}{CH}CH_2CH_3$

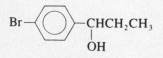

1-(*p*-Bromophenyl)-1-propanol
(compound H)

16.37 The ratio of carbon to hydrogen in the molecular formula is C_nH_{2n+2} ($C_8H_{18}O_2$), so compound I has no double bonds or rings; both oxygen atoms must be attached to carbon by single bonds. Hydroxyl stretching in the ir spectrum (3200 cm^{-1}) and a two-proton singlet at $\delta = 2.0$ ppm in the ^1H nmr spectrum suggest a diol. Compound I cannot be a vicinal diol, however, since it does not react with periodic acid.

The nmr spectrum is rather simple; all peaks are singlets and multiples of 2 with respect to their integrated areas. The 12-proton singlet at $\delta = 1.2$ ppm must correspond to four equivalent methyl groups and the four-proton singlet to two equivalent methylene groups. No nonequivalent protons can be vicinal, since no splitting is observed.

Compound I must be 2,5-dimethyl-1,2-hexanediol.

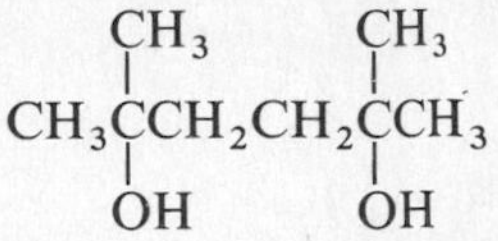

$$CH_3\overset{\overset{CH_3}{|}}{\underset{\underset{OH}{|}}{C}}CH_2CH_2\overset{\overset{CH_3}{|}}{\underset{\underset{OH}{|}}{C}}CH_3$$

16.38 (*a*) This compound has only two different types of carbons. One type of carbon comes at low field and since its ^{13}C peak is a singlet, has no hydrogen substituents. It is most likely a carbon bonded to oxygen and three other carbons. The other peak is a quartet, ascribable to methyl groups. The spectrum is consistent with this compound being *tert*-butyl alcohol.

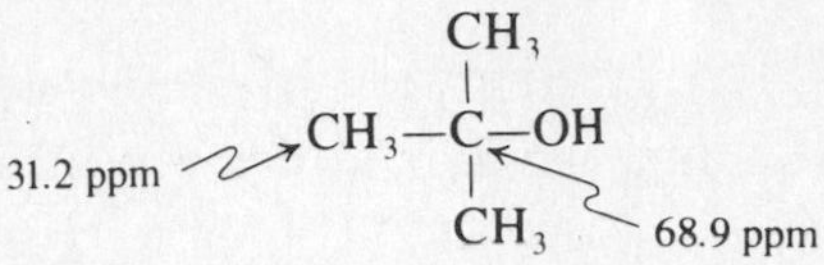

$$CH_3-\overset{\overset{CH_3}{|}}{\underset{\underset{CH_3}{|}}{C}}-OH$$

31.2 ppm → CH_3; 68.9 ppm → C

(*b*) There are four different types of carbons in this compound. The only $C_4H_{10}O$ isomers that have four nonequivalent carbons are $CH_3CH_2CH_2CH_2OH$, $CH_3\underset{\underset{OH}{|}}{C}HCH_2CH_3$, and $CH_3OCH_2CH_2CH_3$. The lowest field signal, the one at 69.2 ppm from the carbon that bears the oxygen substituent, is a doublet and therefore has only one hydrogen attached to it. The compound is therefore *sec*-butyl alcohol.

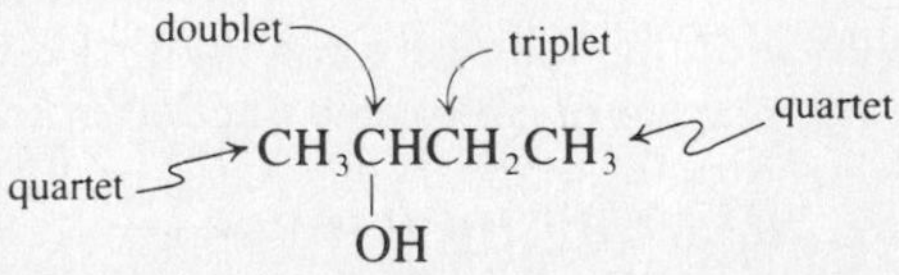

$$CH_3\underset{\underset{OH}{|}}{C}HCH_2CH_3$$

quartet → CH_3; doublet → CH; triplet → CH_2; quartet → CH_3

(*c*) This compound has two equivalent methyl groups as indicated by the quartet of area 2 at 18.9 ppm. Its lowest-field carbon is a triplet, so the group $-CH_2O$ must be present. The compound must be isobutyl alcohol.

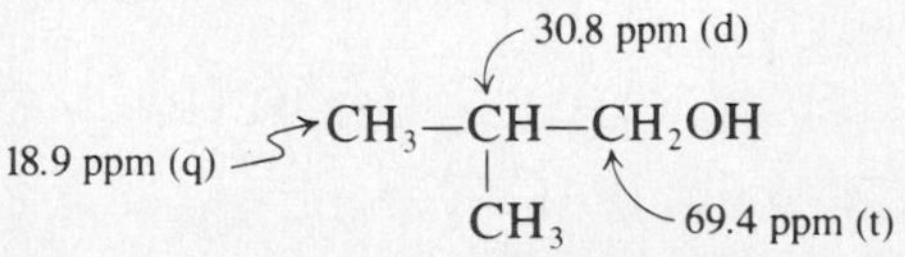

$$CH_3-\underset{\underset{CH_3}{|}}{C}H-CH_2OH$$

18.9 ppm (q) → CH_3; 30.8 ppm (d) → CH; 69.4 ppm (t) → CH_2OH

16.39 Compound J has only three carbons, none of which can be methyl carbons because the ^{13}C nmr spectrum contains no quartets. Two of the carbon signals are triplets and so arise from CH_2 groups; the other is a doublet and so corresponds to a CH group. The only structure consistent with the observed data is that of 3-chloro-1,2-propanediol.

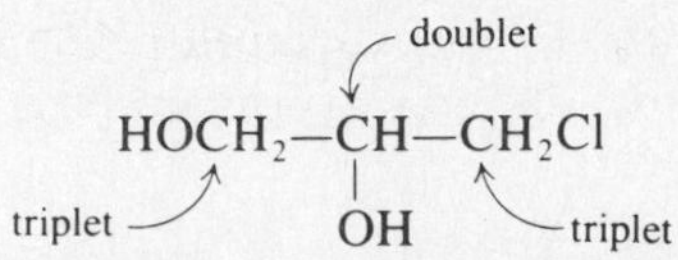

$$HOCH_2-\underset{\underset{OH}{|}}{C}H-CH_2Cl$$

triplet → $HOCH_2$; doublet → CH; triplet → CH_2Cl

The structure $HOCH_2\underset{\underset{Cl}{|}}{C}HCH_2OH$ cannot be correct. It would exhibit only two peaks in its ^{13}C nmr spectrum because the two terminal carbons are equivalent to each other.

16.40 The observation of a peak at m/z 31 in the mass spectrum of compound K suggests the presence of a primary alcohol. This fragment is most likely $CH_2{=}\overset{+}{O}H$. Based on this fact and the appearance of four different carbons in the ^{13}C nmr spectrum, compound K is 2-ethyl-1-butanol.

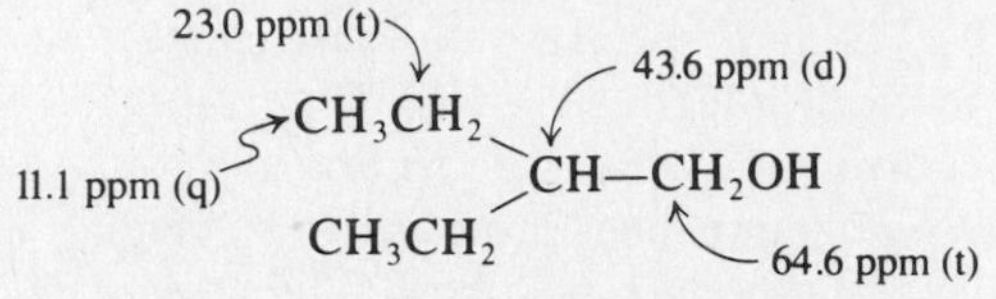

$$(CH_3CH_2)_2CH{-}CH_2OH$$

23.0 ppm (t); 43.6 ppm (d); 11.1 ppm (q); 64.6 ppm (t)

SELF-TEST

PART A

A-1. For each of the following reactions give the structure of the missing reactant or reagent.

(*a*) ? $\xrightarrow[\text{2. } H_2O]{\text{1. } LiAlH_4}$ cyclohexane bearing H and OH on one carbon

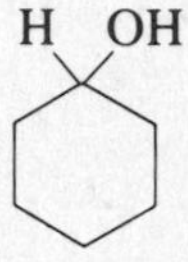

(*b*) ? + $2CH_3CH_2MgBr$ $\xrightarrow[\text{2. } H_3O^+]{\text{1. diethyl ether}}$ $C_6H_5C(OH)(CH_2CH_3)_2$ + CH_3CH_2OH

(*c*) $C_6H_5CH_2C(CH_3){=}CH_2$ $\xrightarrow{?}$ $C_6H_5CH_2CH(CH_3)CH_2OH$

(*d*) 1-methylcyclohexene (CH_3) $\xrightarrow{?}$ cyclohexane with CH_3, OH, OH

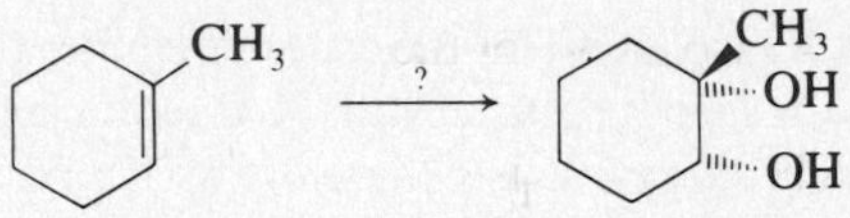

A-2. For the following reactions of 2-phenylethanol, $C_6H_5CH_2CH_2OH$, give the correct reagent or product(s) omitted from the equation.

(*a*) $C_6H_5CH_2CH_2OH$ $\xrightarrow[CH_2Cl_2]{(C_5H_5N)_2CrO_3}$?

(*b*) $C_6H_5CH_2CH_2OH$ $\xrightarrow{?}$ $CH_3CO_2CH_2CH_2{-}C_6H_5$

(*c*) $C_6H_5CH_2CH_2OH$ (2 moles) $\xrightarrow[\text{heat}]{H^+}$ H_2O + ?

(*d*) $C_6H_5CH_2CH_2OH$ $\xrightarrow{?}$ $C_6H_5CH_2CO_2H$

A-3. Outline two synthetic schemes for the preparation of 3-methyl-1-butanol using different Grignard reagents.

A-4. Give the structure of the reactant, reagent, or product omitted from each of the following. Show stereochemistry where important.

(*a*) cyclohexane with OH, H on one carbon and OH, CH_3 on the adjacent carbon $\xrightarrow{HIO_4}$?

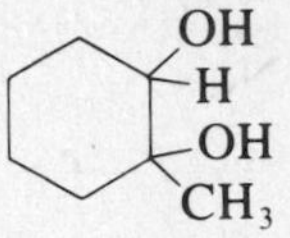

(*b*) ? (a diol) $\xrightarrow[\text{heat}]{H^+}$ tetrahydropyran ring with O and CH_3

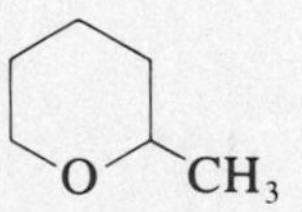

(*c*) ? $\xrightarrow[\text{cold}]{KMnO_4,\ H_2O}$ 2,3-Butanediol (chiral diastereomer)

A-5. Provide structures for compounds A to C in the following reaction scheme:

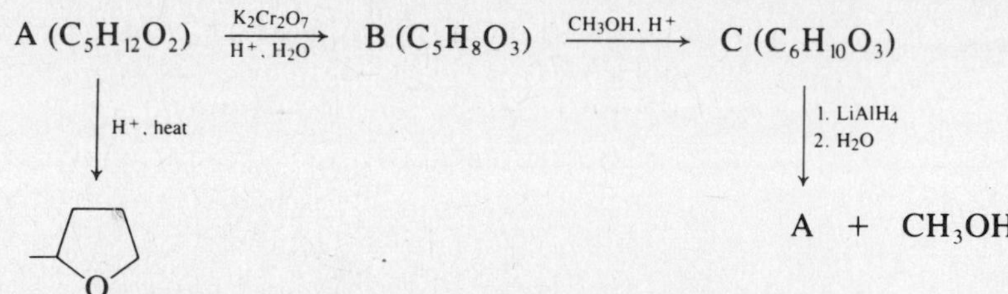

A-6. Using any necessary organic or inorganic reagents, outline a scheme for the preparation of ethyl 3-phenylpropanoate (D) from toluene.

$$C_6H_5CH_3 \xrightarrow{?} \underset{D}{C_6H_5CH_2CH_2CO_2C_2H_5}$$

PART B

B-1. The following chemical change (R = alkyl group) may be best described as

(*a*) An oxidation (*b*) A reduction
(*c*) An elimination (*d*) An addition

B-2. Which of the following would yield a secondary alcohol after the indicated reaction, followed by aqueous hydrolysis if necessary?
(*a*) $LiAlH_4$ + a ketone (*b*) CH_3CH_2MgBr + an aldehyde
(*c*) 2-Butene + mercuric acetate, then $NaBH_4$ (*d*) All of these

B-3. What is the major product of the following reaction?

O
$\xrightarrow[CH_3OH]{NaBH_4}$?
CO_2H

(*a*) OH, CH_2OH (*c*) OH, CO_2H

(*b*) O, CH_2OH (*d*) OH, CO_2CH_3

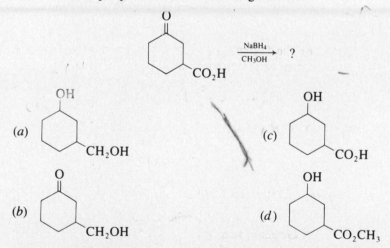

B-4. Which of the esters shown, after reduction with $LiAlH_4$ and aqueous workup, will yield two molecules of only a single alcohol?
(*a*) $CH_3CH_2CO_2CH_2CH_3$ (*b*) $C_6H_5CO_2C_6H_5$
(*c*) $C_6H_5CO_2CH_2C_6H_5$ (*d*) None of these

B-5. For the reaction shown below, select the statement which best describes the situation.

$$RCH_2OH + (C_5H_5N)_2CrO_3 \longrightarrow$$

(*a*) The alcohol is oxidized to an acid, and the Cr(VI) is reduced.
(*b*) The alcohol is oxidized to an aldehyde, and the Cr(VI) is reduced.
(*c*) The alcohol is reduced to an aldehyde, and the Cr(III) is oxidized.
(*d*) The alcohol is oxidized to a ketone, and the Cr(VI) is reduced.

B-6. What is the product from the following esterification?

$$C_6H_5CH_2CO_2H + CH_3CH_2{-}^{18}OH \xrightarrow[\text{heat}]{H^+} ?$$

(*a*) $C_6H_5CH_2\overset{^{18}O}{\overset{\|}{C}}OCH_2CH_3$

(*c*) $C_6H_5CH_2\overset{^{18}O}{\overset{\|}{C}}{-}^{18}OCH_2CH_3$

(*b*) $C_6H_5CH_2\overset{O}{\overset{\|}{C}}{-}^{18}OCH_2CH_3$

(*d*) $CH_3CH_2\overset{^{18}O}{\overset{\|}{C}}OCH_2C_6H_5$

B-7. The following substance acts as a coenzyme in which of the following biological reactions?

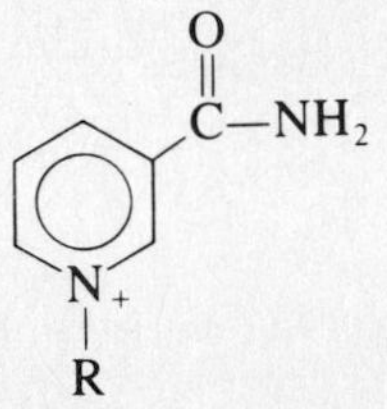

(R = adenine dinucleotide)

(*a*) Alcohol oxidation
(*b*) Ketone reduction
(*c*) Epoxide formation
(*d*) None of these

CHAPTER 17

ETHERS AND EPOXIDES

IMPORTANT TERMS AND CONCEPTS

Nomenclature (Sec. 17.1) Ethers are described as *symmetrical* or *unsymmetrical* depending on whether the two groups bonded to oxygen are the same or different. For example

$CH_3CH_2CH_2OCH_2CH_2CH_3$ Dipropyl ether (symmetrical)

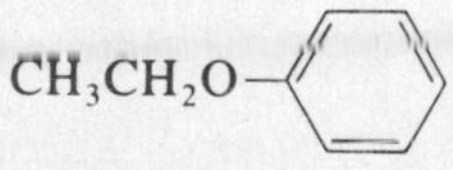

$CH_3CH_2O-C_6H_5$ Ethyl phenyl ether (unsymmetrical)

Epoxides are cyclic ethers whose oxygen atom is part of a three-membered ring.

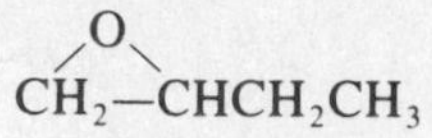

$$\overset{O}{CH_2-CHCH_2CH_3}$$

1,2-Epoxybutane

Structure, Bonding, and Physical Properties (Secs. 17.2, 17.3) The alkyl (or aryl) groups bonded to an ether oxygen are arranged in an approximately tetrahedral fashion. The polar nature of the carbon-oxygen bond and the nonlinear arrangement of groups causes most ethers to be polar molecules.

The boiling points of ethers are similar to those of alkanes of the same chain length and are substantially lower than those of comparably constituted alcohols. Ethers, although polar, are not able to form hydrogen-bonded aggregates as do alcohols. Ethers are, however, able to serve as donors of unshared electron pairs. Therefore their solubility in water tends to parallel that of alcohols.

Crown Ethers (Sec. 17.4) Another property of ethers explained by the presence of unshared electron pairs on oxygen is the ability to coordinate metal ions.

$$R\ddot{\underset{\cdot\cdot}{O}}R + M^+ \rightleftharpoons R\overset{+}{\ddot{O}}R\ (\text{with } O-M)$$

A series of *macrocyclic polyethers* has been prepared. These *crown ethers* are capable of complexing metal ions, with resulting drastic alteration of the solubility and chemical reactivity of inorganic salts in nonpolar media. An example, 12-crown-4, is shown below.

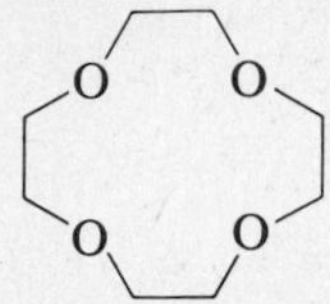

IMPORTANT REACTIONS

Preparation of Ethers (Secs. 17.6 to 17.8)

Williamson synthesis

General:

$$RO^- + \underset{(1^\circ\text{ best})}{R'X} \longrightarrow ROR' + X^-$$

Example:

$$(CH_3)_2CHO^- + CH_3CH_2Br \longrightarrow (CH_3)_2CHOCH_2CH_3$$

Solvomercuration-demercuration

General:

$$>C=C< \; + \; ROH \xrightarrow[\text{2. NaBH}_4\text{, HO}^-]{\text{1. Hg(OAc)}_2} H-\overset{|}{\underset{|}{C}}-\overset{|}{\underset{|}{C}}-OR$$

Examples:

$$CH_3CH=CH_2 \xrightarrow[\text{2. NaBH}_4\text{, HO}^-]{\text{1. Hg(OAc)}_2\text{, CH}_3\text{OH}} (CH_3)_2CHOCH_3$$

$$C_6H_5CH=CH_2 \xrightarrow[\text{2. NaBH}_4\text{, HO}^-]{\text{1. Hg(OAc)}_2\text{, C}_2\text{H}_5\text{OH}} C_6H_5\overset{\overset{OCH_2CH_3}{|}}{C}HCH_3$$

Reactions of Ethers (Secs. 17.9, 17.10)

Acid-catalyzed cleavage

General:

$$ROR' + 2\,HX \xrightarrow{\text{heat}} RX + R'X + H_2O$$

Example:

$$C_6H_5CH_2OCH_2CH_3 \xrightarrow[\text{heat}]{\text{HI}} C_6H_5CH_2I + CH_3CH_2I + H_2O$$

Preparation of Epoxides (Secs. 17.11, 17.12)

From alkenes by epoxidation (Chap. 7)

General:

$$R_2C=CR_2 \; + \; R'\overset{O}{\overset{\|}{C}}OOH \longrightarrow R_2\overset{O}{\widehat{C-C}}R_2 \; + \; R'CO_2H$$

Example:

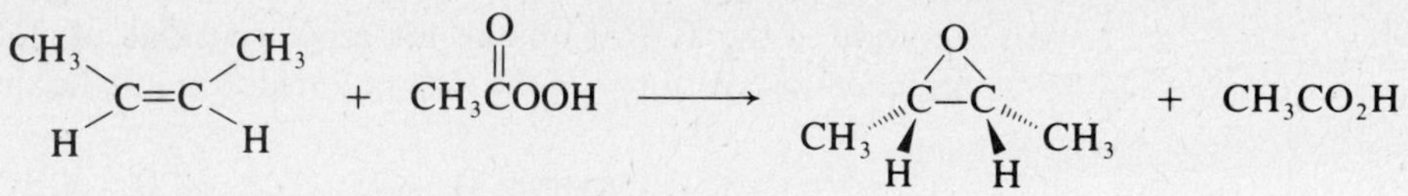

$$\underset{H}{\overset{CH_3}{}}\!\!\!>C{=}C<\!\!\!\underset{H}{\overset{CH_3}{}} + CH_3\overset{O}{\overset{\|}{C}}OOH \longrightarrow CH_3\underset{H}{}\!\overset{O}{\widehat{C{-}C}}\!\underset{H}{}CH_3 + CH_3CO_2H$$

From vicinal halohydrins

General:

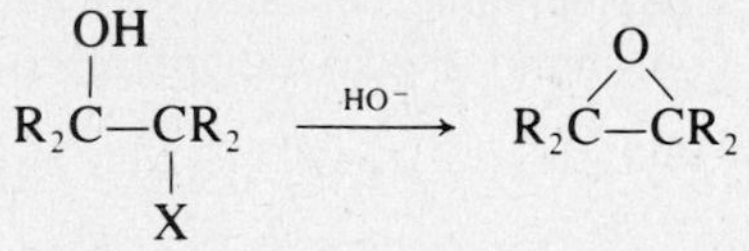

$$R_2\overset{OH}{\overset{|}{C}}{-}\underset{X}{\underset{|}{C}}R_2 \xrightarrow{HO^-} R_2\overset{O}{\widehat{C{-}C}}R_2$$

Example:

$$CH_3CH_2\overset{OH}{\overset{|}{C}}HCH_2Br \xrightarrow{HO^-} CH_3CH_2\overset{O}{\widehat{CH{-}CH_2}}$$

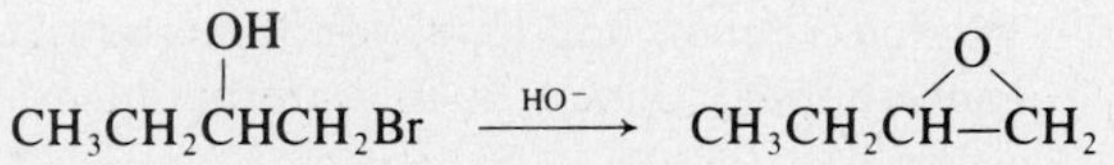

Reactions of Epoxides (Secs. 17.13 to 17.15)

Nucleophilic ring opening

General:

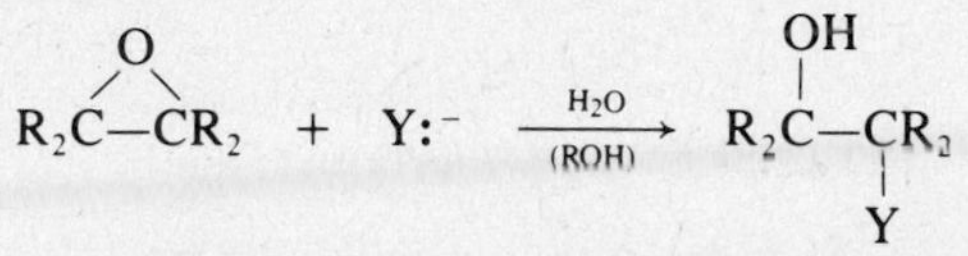

$$R_2\overset{O}{\widehat{C{-}C}}R_2 + Y{:}^- \xrightarrow[(ROH)]{H_2O} R_2\overset{OH}{\overset{|}{C}}{-}\underset{Y}{\underset{|}{C}}R_2$$

Examples:

$$C_6H_5MgBr + \xrightarrow[2.\ H_3O^+]{1.\ \overset{O}{\widehat{CH_2{-}CH_2}}} C_6H_5CH_2CH_2OH$$

$$\text{cyclohexene oxide} + CH_3CH_2O^- \xrightarrow{CH_3CH_2OH} \text{trans-2-ethoxycyclohexanol } (OCH_2CH_3,\ OH)$$

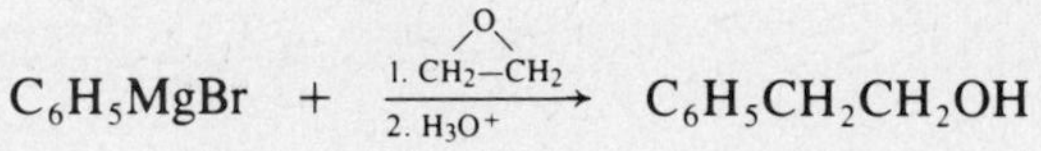

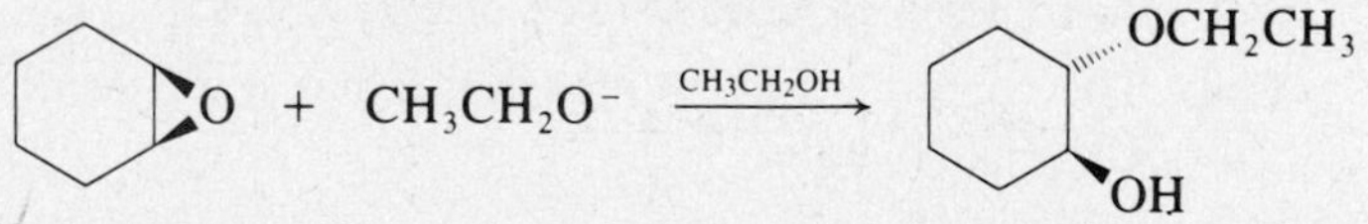

Acid catalyzed ring opening

General:

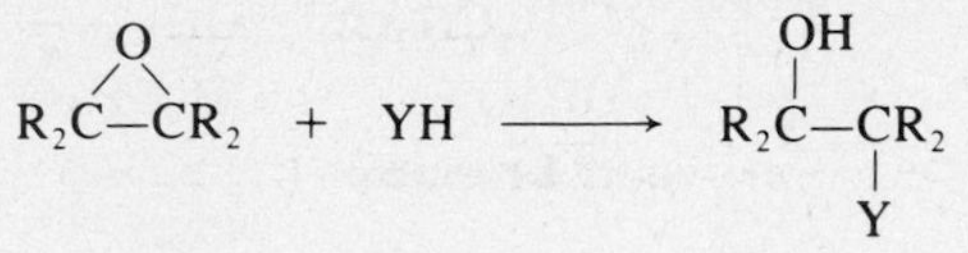

$$R_2\overset{O}{\widehat{C{-}C}}R_2 + YH \longrightarrow R_2\overset{OH}{\overset{|}{C}}{-}\underset{Y}{\underset{|}{C}}R_2$$

Example:

$$C_6H_5\overset{O}{\widehat{CH{-}CH_2}} + HBr \longrightarrow C_6H_5\overset{Br}{\overset{|}{C}}HCH_2OH$$

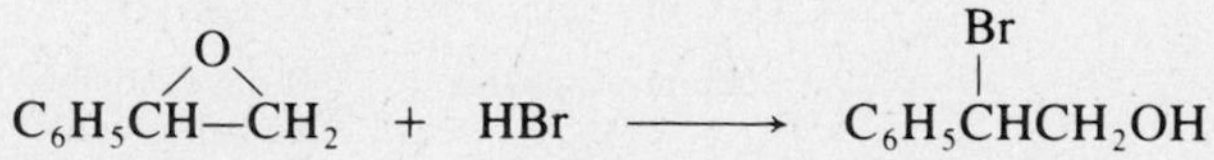

SOLUTIONS TO TEXT PROBLEMS

17.1 (*b*) Oxirane is the IUPAC name for ethylene oxide. A chloromethyl group ($ClCH_2—$) is attached to position 2 of the ring in 2-(chloromethyl)oxirane.

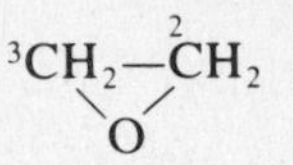

$$\underset{\text{Oxirane}}{\overset{3}{C}H_2—\overset{2}{C}H_2 \text{ (ring via O)}} \qquad \underset{\text{2-(chloromethyl)oxirane}}{CH_2—CHCH_2Cl \text{ (ring via O)}}$$

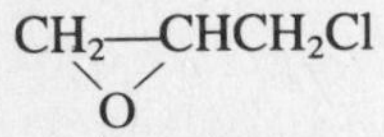

This compound is more commonly known by its trivial name, epichlorohydrin.

(*c*) Epoxides may be named by adding the prefix epoxy to the IUPAC name of a parent compound, specifying both atoms to which the oxygen is attached by number.

$$\underset{\text{1-Butene}}{CH_2{=}CHCH_2CH_3} \qquad \underset{\text{3,4-Epoxy-1-butene}}{CH_2{=}CHCH—CH_2 \text{ (ring via O)}}$$

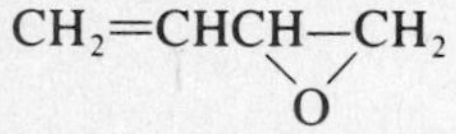

17.2 1,2-Epoxybutane and tetrahydrofuran both have the molecular formula C_4H_8O, i.e., they are constitutional isomers, so it is appropriate to compare their heats of combustion directly.

$$C_4H_8O + \frac{11}{2}O_2 \longrightarrow 4CO_2 + 4H_2O$$

The strained three-membered ring of 1,2-epoxybutane causes it to have more internal energy than tetrahydrofuran and its combustion is more exothermic.

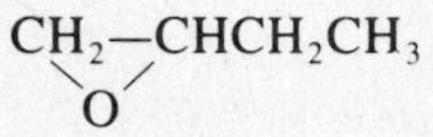

$CH_2—CHCH_2CH_3$ (ring via O)
1,2-Epoxybutane;
heat of combustion 609.1 kcal/mol

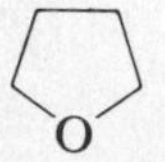

Tetrahydrofuran;
heat of combustion 597.8 kcal/mol

17.3 An ether can function only as a proton acceptor in a hydrogen bond, while an alcohol can be either a proton acceptor or a donor. Therefore the only hydrogen bond possible between an ether and an alcohol is the one shown.

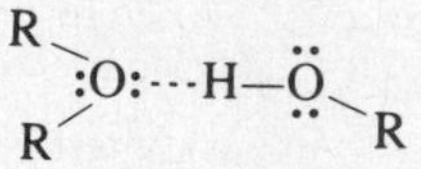

$$R_2\ddot{O}{:}\cdots H—\ddot{O}—R$$

17.4 The compound is 1,4-dioxane; it has a six-membered ring and two oxygens separated by $CH_2—CH_2$ units.

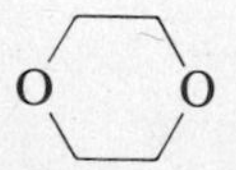

1,4-dioxane ("6-crown-2")

17.5 (*b*) Methyl propyl ether can be prepared by the reaction of sodium methoxide with a propyl halide:

$$\underset{\text{Sodium methoxide}}{CH_3ONa} + \underset{\substack{\text{1-Bromopropane}\\(n\text{-propyl bromide})}}{CH_3CH_2CH_2Br} \longrightarrow \underset{\substack{\text{Methyl propyl}\\\text{ether}}}{CH_3OCH_2CH_2CH_3} + \underset{\substack{\text{Sodium}\\\text{bromide}}}{NaBr}$$

It can also be made by the reaction of sodium propoxide with a methyl halide.

$$\underset{\text{Sodium propoxide}}{CH_3CH_2CH_2ONa} + \underset{\substack{\text{Bromomethane}\\\text{(methyl bromide)}}}{CH_3Br} \longrightarrow \underset{\substack{\text{Methyl propyl}\\\text{ether}}}{CH_3OCH_2CH_2CH_3} + \underset{\substack{\text{Sodium}\\\text{bromide}}}{NaBr}$$

(*c*) The two routes to benzyl ethyl ether are:

$$C_6H_5CH_2ONa + CH_3CH_2Br \longrightarrow C_6H_5CH_2OCH_2CH_3 + NaBr$$

Sodium benzyloxide | Bromoethane (ethyl bromide) | Benzyl ethyl ether | Sodium bromide

$$C_6H_5CH_2Br + CH_3CH_2ONa \longrightarrow C_6H_5CH_2OCH_2CH_3 + NaBr$$

Benzyl bromide | Sodium ethoxide | Benzyl ethyl ether | Sodium bromide

17.6 (*b*) A primary carbon and a secondary carbon are attached to the ether oxygen. The secondary carbon can only be derived from the alkoxide, as secondary alkyl halides cannot be used in the preparation of ethers by the Williamson method. The only effective route uses an allyl halide and sodium isopropoxide.

$$(CH_3)_2CHONa + CH_2{=}CHCH_2Br \longrightarrow CH_2{=}CHCH_2OCH(CH_3)_2 + NaBr$$

Sodium isopropoxide | Allyl bromide | Allyl isopropyl ether | Sodium bromide

Elimination will be the major reaction of an isopropyl halide with an alkoxide base.

(*c*) Here the ether is a mixed primary-tertiary one. The best combination is the one that utilizes the primary alkyl halide.

$$(CH_3)_3COK + C_6H_5CH_2Br \longrightarrow (CH_3)_3COCH_2C_6H_5 + NaBr$$

Potassium *tert*-butoxide | Benzyl bromide | Benzyl *tert*-butyl ether | Sodium bromide

The reaction between $(CH_3)_3CBr$ and $C_6H_5CH_2O^-$ will be one of elimination, not substitution.

17.7 (*b*) The alkene must be cyclohexene and the solvent for the solvomercuration must be methanol.

$$\text{Cyclohexene} \xrightarrow[\text{2. NaBH}_4\text{, HO}^-]{\text{1. Hg(OAc)}_2\text{, CH}_3\text{OH}} \text{Cyclohexyl methyl ether } (C_6H_{11}OCH_3)$$

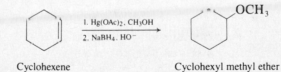

Cyclohexene | Cyclohexyl methyl ether

This reaction has been carried out in the manner shown to give the desired product in 100 percent yield.

(*c*) In principle solvomercuration of ethylene could be employed as a synthesis of the desired ethyl 1-methylpentyl ether.

$$CH_2{=}CH_2 \xrightarrow[\text{2. NaBH}_4\text{, HO}^-]{\text{1. Hg(OAc)}_2\text{, CH}_3\text{CH(OH)CH}_2\text{CH}_2\text{CH}_2\text{CH}_3} CH_3CH(OCH_2CH_3)CH_2CH_2CH_2CH_3$$

Ethylene | Ethyl 1-methylpentyl ether

Since ethylene is a gas, however, it is more convenient to prepare alkyl ethyl ethers by the alternative procedure using ethanol as the solvent in a solvomercuration reaction.

$$CH_2{=}CHCH_2CH_2CH_2CH_3 \xrightarrow[\text{2. NaBH}_4\text{, HO}^-]{\text{1. Hg(OAc)}_2\text{, CH}_3\text{CH}_2\text{OH}} CH_3CH(OCH_2CH_3)CH_2CH_2CH_2CH_3$$

1-Hexene | Ethyl 1-methylpentyl ether (98%)

The orientation of addition corresponds to that of Markovnikov's rule. Notice that 2-hexene ($CH_3CH{=}CHCH_2CH_2CH_3$) would not be suitable as a starting material because it would likely lead to formation of two regioisomeric ethyl ethers.

(*d*) There are two solvomercuration routes to methyl 1,1-dimethylpropyl ether. One involves a terminal alkene:

$$\underset{\text{2-Methyl-1-butene}}{CH_2{=}C(CH_3)CH_2CH_3} \xrightarrow[\text{2. NaBH}_4\text{, HO}^-]{\text{1. Hg(OAc)}_2\text{, CH}_3\text{OH}} \underset{\text{Methyl 1,1-dimethylpropyl ether (100\%)}}{(CH_3)_2C(OCH_3)CH_2CH_3}$$

The other uses 2-methyl-2-butene:

$$\underset{\text{2-Methyl-2-butene}}{(CH_3)_2C{=}CHCH_3} \xrightarrow[\text{2. NaBH}_4\text{, HO}^-]{\text{1. Hg(OAc)}_2\text{, CH}_3\text{OH}} \underset{\text{Methyl 1,1-dimethylpropyl ether}}{(CH_3)_2C(OCH_3)CH_2CH_3}$$

17.8 (*b*)

$$\underset{\text{Dipentyl ether}}{CH_3CH_2CH_2CH_2CH_2OCH_2CH_2CH_2CH_2CH_3} + \underset{\text{Oxygen}}{15O_2} \longrightarrow \underset{\text{Carbon dioxide}}{10CO_2} + \underset{\text{Water}}{11H_2O}$$

(*c*)

$$\underset{\text{Tetrahydrofuran}}{\text{(cyclic } C_4H_8O)} + \underset{\text{Oxygen}}{\frac{11}{2}O_2} \longrightarrow \underset{\text{Carbon dioxide}}{4CO_2} + \underset{\text{Water}}{4H_2O}$$

17.9 (*b*) If benzyl bromide is the only organic product from reaction of a dialkyl ether with hydrogen bromide, then both alkyl groups attached to oxygen must be benzyl.

$$\underset{\text{Dibenzyl ether}}{C_6H_5CH_2OCH_2C_6H_5} \xrightarrow[\text{heat}]{\text{HBr}} \underset{\text{Benzyl bromide}}{2\,C_6H_5CH_2Br} + \underset{\text{Water}}{H_2O}$$

(*c*) Since *one mole of a dihalide*, rather than two moles of a monohalide, is produced per mole of ether, the ether must be cyclic.

$$\underset{\text{Tetrahydropyran (oxane)}}{\text{(cyclic } C_5H_{10}O)} \xrightarrow[\text{heat}]{\text{2HBr}} \underset{\text{1,5-Dibromopentane}}{BrCH_2CH_2CH_2CH_2CH_2Br} + \underset{\text{Water}}{H_2O}$$

17.10 Dissociation of the oxonium ion is reasonable since a tertiary carbocation results.

$$\underset{\text{Di-}tert\text{-butyloxonium ion}}{(CH_3)_3C{-}\overset{+}{\ddot{O}}H{-}C(CH_3)_3} \longrightarrow \underset{tert\text{-Butyl alcohol}}{(CH_3)_3C\ddot{O}H} + \underset{tert\text{-Butyl cation}}{(CH_3)_3C^+}$$

Capture of *tert*-butyl cation by chloride then occurs rapidly.

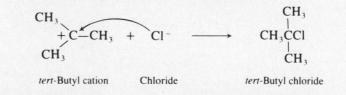

$$\underset{tert\text{-Butyl cation}}{(CH_3)_3C^+} + \underset{\text{Chloride}}{Cl^-} \longrightarrow \underset{tert\text{-Butyl chloride}}{(CH_3)_3CCl}$$

The *tert*-butyl alcohol generated in the dissociation step goes on to form *tert*-butyl chloride by way of a carbocation intermediate, as described in Chapter 4.

17.11 The cis epoxide is achiral. It is a meso form containing a plane of symmetry. The trans isomer is chiral; its two mirror image representations are not superposable.

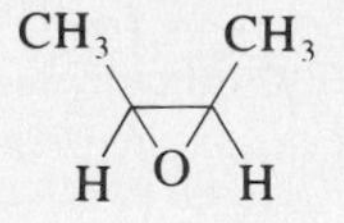

CH_3 CH_3 / H O H

cis-2,3-Epoxybutane (plane of symmetry passes through oxygen and midpoint of carbon–carbon bond)

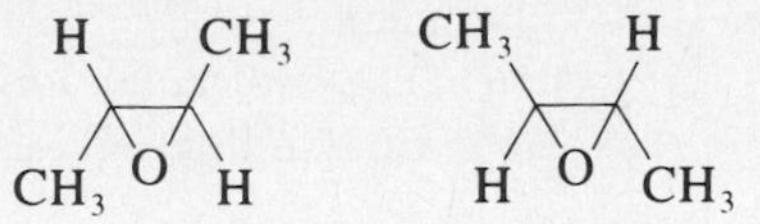

H CH_3 / CH_3 O H CH_3 H / H O CH_3

Nonsuperposable mirror image forms of *trans*-2,3-epoxybutane

Neither the cis nor trans epoxide is optically active when formed from the alkene. The cis epoxide is achiral; it cannot be optically active. The trans epoxide is capable of optical activity but is formed as a racemic mixture since achiral starting materials, the alkene and hypobromous acid, are used.

17.12 (*b*) Azide ion $[N{=}N{=}N]^-$ is a good nucleophile, reacting readily with ethylene oxide to yield 2-azidoethanol.

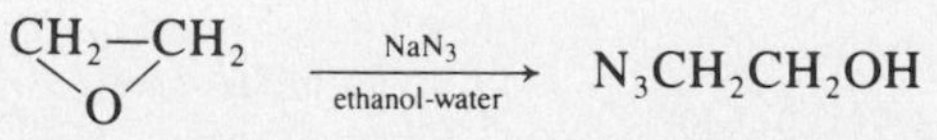

$$\underset{\text{Ethylene oxide}}{CH_2{-}CH_2 \text{ (O bridge)}} \xrightarrow[\text{ethanol-water}]{NaN_3} \underset{\text{2-Azidoethanol}}{N_3CH_2CH_2OH}$$

(*c*) Ethylene oxide is hydrolyzed to ethylene glycol in the presence of aqueous base.

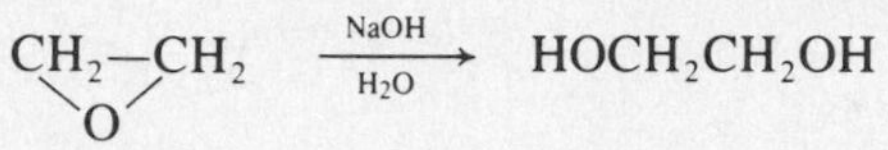

$$\underset{\text{Ethylene oxide}}{CH_2{-}CH_2 \text{ (O bridge)}} \xrightarrow[H_2O]{NaOH} \underset{\text{Ethylene glycol}}{HOCH_2CH_2OH}$$

(*d*) Phenyllithium reacts with ethylene oxide in a manner similar to that of a Grignard reagent.

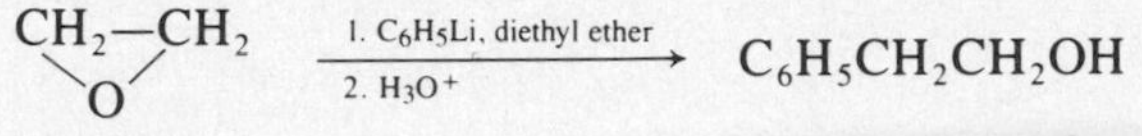

$$\underset{\text{Ethylene oxide}}{CH_2{-}CH_2 \text{ (O bridge)}} \xrightarrow[\text{2. } H_3O^+]{\text{1. } C_6H_5Li\text{, diethyl ether}} \underset{\text{2-Phenylethanol}}{C_6H_5CH_2CH_2OH}$$

(*e*) The nucleophilic species here is the acetylenic anion $CH_3CH_2C{\equiv}C^-$, which attacks a carbon atom of ethylene oxide to give 3-hexyn-1-ol.

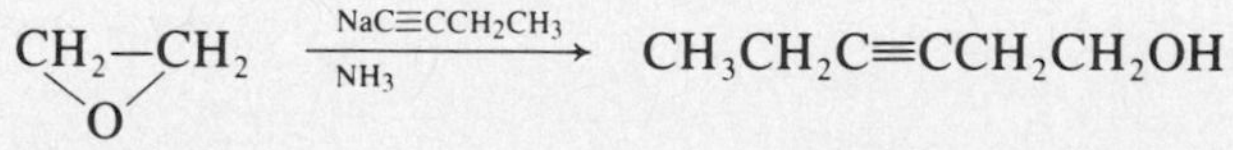

$$\underset{\text{Ethylene oxide}}{CH_2{-}CH_2 \text{ (O bridge)}} \xrightarrow[NH_3]{NaC{\equiv}CCH_2CH_3} \underset{\text{3-Hexyn-1-ol (48\%)}}{CH_3CH_2C{\equiv}CCH_2CH_2OH}$$

17.13 Nucleophilic attack at C-2 of the starting epoxide will be faster than attack at C-1 because C-1 is more sterically hindered. Compound A, corresponding to attack at C-1, is not as likely as compound B. Compound B not only arises by methoxide ion attack at C-2 but also satisfies the stereochemical requirement that epoxide ring opening take place with inversion of configuration at the site of substitution. Compound B is correct. Compound C, while it is formed by methoxide substitution at the less crowded carbon of the epoxide, is wrong stereochemically. It requires substitution with retention of configuration, a process contrary to the normal mode of epoxide ring opening.

17.14 Acid-catalyzed nucleophilic ring opening proceeds by attack of methanol at the more substituted carbon of the protonated epoxide. Inversion of configuration is observed at the site of attack. The correct product is compound A.

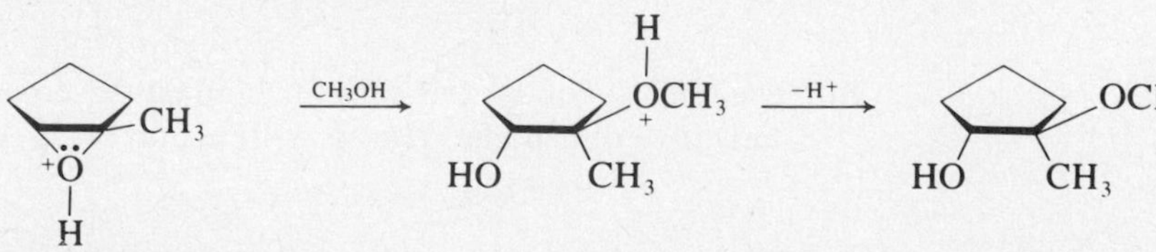

CH_3 / $^+\ddot{O}$ / H → (CH_3OH) → H / $\overset{+}{O}CH_3$ / HO CH_3 → ($-H^+$) → OCH_3 / HO CH_3

Protonated form of 1-methyl-1,2-epoxycyclopentane

Compound A

17.15 Dammarenediol can be formed from the same carbocation intermediate that leads to dammaradienol. Capture of the carbocation by water acting as a nucleophile yields dammarenediol.

17.16 The molecular ion from *sec*-butyl ethyl ether can also fragment by cleavage of a carbon-carbon bond in its ethyl group to give an oxygen-stabilized cation of m/z 87.

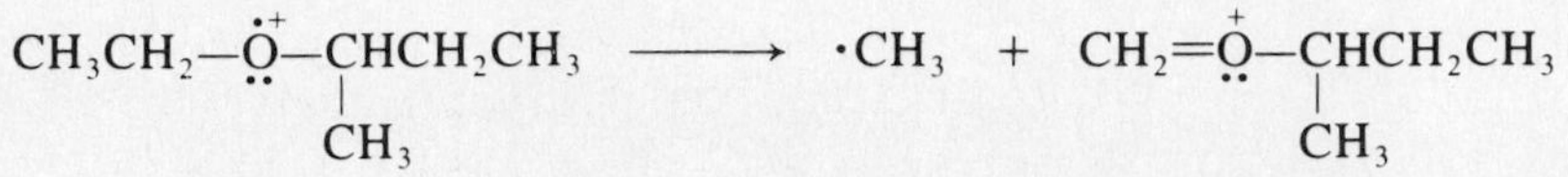

$$\mathrm{CH_3CH_2{-}\overset{\bullet+}{\underset{\cdot\cdot}{O}}{-}\underset{\displaystyle CH_3}{\underset{|}{C}}HCH_2CH_3 \longrightarrow \cdot CH_3 + CH_2{=}\overset{+}{\underset{\cdot\cdot}{O}}{-}\underset{\displaystyle CH_3}{\underset{|}{C}}HCH_2CH_3}$$

m/z 87

17.17 All the constitutionally isomeric ethers of molecular formula $C_5H_{12}O$ belong to one of two general groups, $CH_3OC_4H_9$ and $CH_3CH_2OC_3H_7$. Thus, we have

$CH_3OCH_2CH_2CH_2CH_3$ — Butyl methyl ether

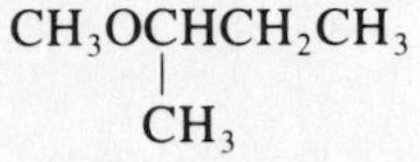

$CH_3OCH(CH_3)CH_2CH_3$ — *sec*-Butyl methyl ether

$CH_3OCH_2CH(CH_3)_2$ — Isobutyl methyl ether

$CH_3OC(CH_3)_3$ — *tert*-Butyl methyl ether

$CH_3CH_2OCH_2CH_2CH_3$ — Ethyl propyl ether

and

$CH_3CH_2OCH(CH_3)_2$ — Ethyl isopropyl ether

These ethers could also have been named as "alkoxyalkanes." Thus, *sec*-butyl methyl ether would become 2-methoxybutane.

17.18

Isoflurane: $CF_3{-}CHCl{-}O{-}CHF_2$

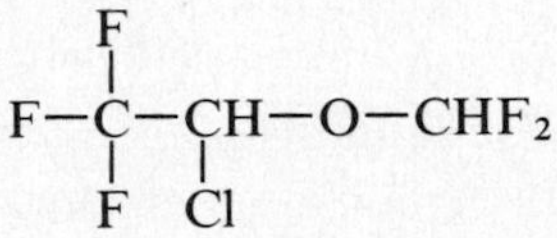

1-Chloro-2,2,2-trifluoroethyl difluoromethyl ether

Enflurane: $Cl{-}CHF{-}CF_2{-}O{-}CHF_2$

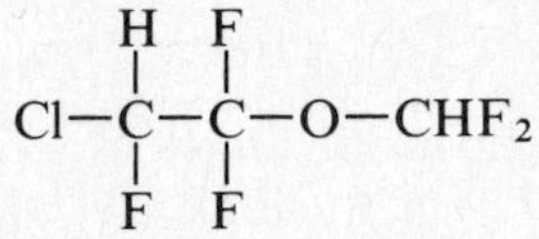

2-Chloro-1,1,2-trifluoroethyl difluoromethyl ether

17.19 (*a*) The parent compound is cyclopropane. It has a three-membered epoxide function, and thus a reasonable name is epoxycyclopropane. Numbers locating positions of attachment (as in "1,2-epoxycyclopropane") are not necessary because no other structures (1,3 or 2,3) are possible here.

Epoxycyclopropane

(*b*) The longest continuous carbon chain has seven carbons so the compound is named as a derivative of heptane. The epoxy function bridges C-2 and C-4. Therefore

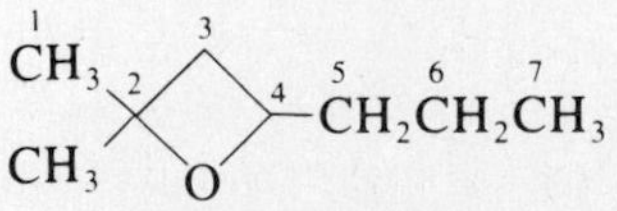

[structure: oxetane ring with $(CH_3)_2$ at C-2, C-3, and $CH_2CH_2CH_3$ (C-5, C-6, C-7) at C-4; numbered 1–7] is 2-methyl-2,4-epoxyheptane. An alternative name is 2,2-dimethyl-4-propyloxetane.

(*c*) The oxygen atom bridges the C-1 and C-4 atoms of a cyclohexane ring.

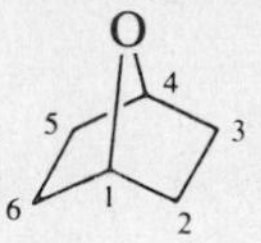

1,4-Epoxycyclohexane

Alternatively this compound may be named 7-oxabicyclo[2.2.1]heptane.

(*d*) There are eight carbon atoms continuously linked and bridged by an oxygen. We name the compound as an epoxy derivative of cyclooctane.

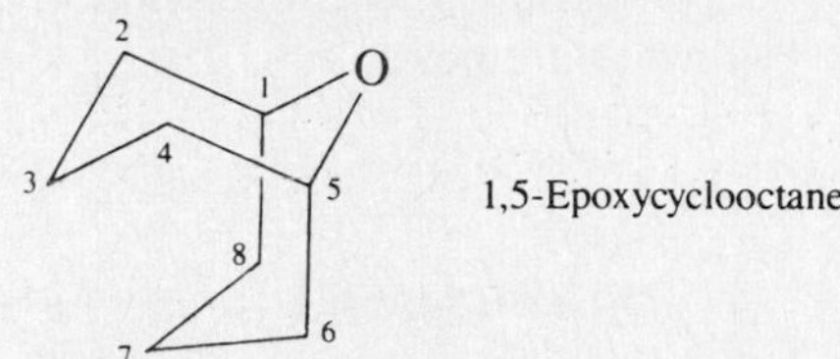

Again, an alternative name may be used: 9-oxabicyclo[3.3.1]nonane.

17.20 (*a*) There are three methyl-substituted oxanes.

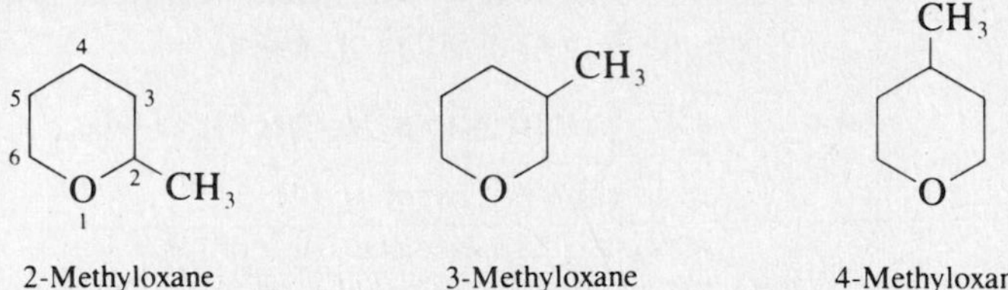

The 2- and 3-methyl derivatives are chiral and exist as *R* and *S* enantiomers. The 4-methyl derivative is achiral.

(*b*) 1,3-Dioxane has two oxygens in the six-membered ring separated by a CH_2 group.

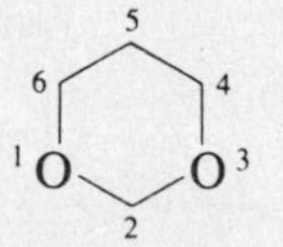

1,3-Dioxane

In 1,4-dioxane the two oxygens are on opposite sides of the ring.

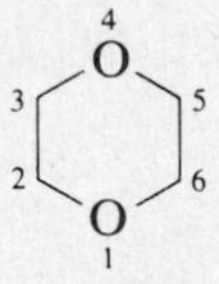

1,4-Dioxane

1,3,5-Trioxane has the structure

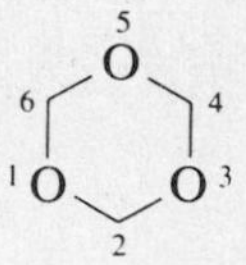

1,3,5-Trioxane

(*c*) There is no C—O—C unit in 1,2-dioxane so it is not classified as an ether.

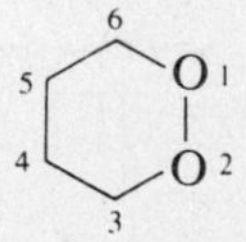

1,2-Dioxane

It is a cyclic peroxide since its ring includes an O—O bond.

17.21 Intramolecular hydrogen bonding between the hydroxyl group and the ring oxygens is possible when the hydroxyl group is axial but not when it is equatorial.

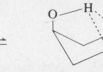

Less stable conformation; no intramolecular hydrogen bonding ⇌ More stable conformation; stabilized by hydrogen bonding

17.22 (*a*) Consideration of the formation of 2-ethoxypentane retrosynthetically reveals that two pathways are possible.

$$CH_3CH(OCH_2CH_3)CH_2CH_2CH_3 \Rightarrow CH_3CH(O^-)CH_2CH_2CH_3 + CH_3CH_2X \text{ or } CH_3CH(X)CH_2CH_2CH_3 + CH_3CH_2O^-$$

The first pathway will yield the desired ether from the reaction of sodium 2-pentanolate with ethyl bromide.

$$\underset{\text{Sodium 2-pentanolate}}{CH_3CH(ONa)CH_2CH_2CH_3} + \underset{\text{Ethyl bromide}}{CH_3CH_2Br} \longrightarrow \underset{\text{2-Ethoxypentane}}{CH_3CH(OCH_2CH_3)CH_2CH_2CH_3} + NaBr$$

This reaction is expected to work well since it involves nucleophilic substitution on a primary alkyl halide. Conversely, the alternative reaction

$$\underset{\text{2-Bromopentane}}{CH_3CH(Br)CH_2CH_2CH_3} + \underset{\text{Sodium ethoxide}}{CH_3CH_2ONa} \longrightarrow \underset{\text{2-Ethoxypentane}}{CH_3CH(OCH_2CH_3)CH_2CH_2CH_3} + NaBr$$

will fail because elimination rather than substitution is the reaction exhibited by secondary alkyl halides with alkoxide bases.

(*b*) For solvomercuration-demercuration we choose an alkene and an alcohol.

$$\underset{\text{1-Pentene}}{CH_2{=}CHCH_2CH_2CH_3} \xrightarrow[\text{2. NaBH}_4\text{, HO}^-]{\text{1. Hg(O}_2\text{CCH}_3)_2\text{, CH}_3\text{CH}_2\text{OH}} \underset{\text{2-Ethoxypentane}}{CH_3CH(OCH_2CH_3)CH_2CH_2CH_3}$$

Had we used 2-pentene instead of 1-pentene, a mixture of 2-ethoxypentane and 3-ethoxypentane would have been formed.

$$\underset{\text{2-Pentene}}{CH_3CH{=}CHCH_2CH_3} \longrightarrow \underset{\text{2-Ethoxypentane}}{CH_3CH(OCH_2CH_3)CH_2CH_2CH_3} + \underset{\text{3-Ethoxypentane}}{CH_3CH_2CH(OCH_2CH_3)CH_2CH_3}$$

Alternatively, ethene could be used as the alkene component along with 2-pentanol as the solvent.

$$\underset{\text{Ethene}}{CH_2{=}CH_2} \xrightarrow[\text{2. NaBH}_4\text{, HO}^-]{\text{1. Hg(O}_2\text{CCH}_3)_2\text{, 2-pentanol}} \underset{\text{2-Ethoxypentane}}{CH_3CH(OCH_2CH_3)CH_2CH_2CH_3}$$

17.23 The desired ether can be seen to arise by Markovnikov addition of the elements of methanol across the double bond of 2-methylpropene.

$$\underset{\text{2-Methylpropene}}{CH_3\underset{\displaystyle CH_3}{\underset{|}{C}}{=}CH_2} + \underset{\text{Methanol}}{CH_3OH} \xrightarrow{H^+} \underset{\textit{tert}\text{-Butyl methyl ether}}{CH_3\overset{\displaystyle CH_3}{\overset{|}{\underset{\displaystyle CH_3}{\underset{|}{C}}}}OCH_3}$$

This mode of addition can be accomplished by solvomercuration-demercuration.

$$\underset{\text{2-Methylpropene}}{(CH_3)_2C{=}CH_2} \xrightarrow[\text{2. NaBH}_4\text{, HO}^-]{\text{1. Hg(O}_2\text{CCH}_3)_2\text{, CH}_3\text{OH}} \underset{\textit{tert}\text{-Butyl methyl ether}}{(CH_3)_3COCH_3}$$

The commercial synthesis is a direct one, involving acid-catalyzed addition of methanol to 2-methylpropene.

$$\underset{\text{2-Methylpropene}}{CH_3\underset{\displaystyle CH_3}{\underset{|}{C}}{=}CH_2} \xrightarrow{H^+} CH_3\overset{\displaystyle CH_3}{\overset{|}{\underset{\displaystyle CH_3}{\underset{|}{C}}}}{+} \xrightarrow{CH_3OH} CH_3\overset{\displaystyle CH_3}{\overset{|}{\underset{\displaystyle H_3C}{\underset{|}{C}}}}{-}\underset{\displaystyle H}{\underset{|}{\overset{+}{O}}}CH_3 \xrightarrow{-H^+} \underset{\textit{tert}\text{-Butyl methyl ether}}{CH_3\overset{\displaystyle CH_3}{\overset{|}{\underset{\displaystyle CH_3}{\underset{|}{C}}}}OCH_3}$$

17.24 The ethers that are to be prepared are:

$$\underset{\text{Methyl propyl ether}}{CH_3OCH_2CH_2CH_3} \qquad \underset{\text{Isopropyl methyl ether}}{CH_3OCH(CH_3)_2} \qquad \text{and} \qquad \underset{\text{Diethyl ether}}{CH_3CH_2OCH_2CH_3}$$

First examine the preparation of each ether by the Williamson method. Methyl propyl ether can be prepared in two ways.

$$\underset{\text{Sodium methoxide}}{CH_3ONa} + \underset{\text{Propyl bromide}}{CH_3CH_2CH_2Br} \longrightarrow \underset{\text{Methyl propyl ether}}{CH_3OCH_2CH_2CH_3}$$

$$\underset{\text{Methyl bromide}}{CH_3Br} + \underset{\text{Sodium propoxide}}{CH_3CH_2CH_2ONa} \longrightarrow \underset{\text{Methyl propyl ether}}{CH_3OCH_2CH_2CH_3}$$

Either combination is satisfactory. The necessary reagents are prepared as shown.

$$\underset{\text{Methanol}}{CH_3OH} \xrightarrow{Na} \underset{\text{Sodium methoxide}}{CH_3ONa}$$

$$\underset{\text{1-Propanol}}{CH_3CH_2CH_2OH} \xrightarrow[\text{(or HBr)}]{PBr_3} \underset{\text{Propyl bromide}}{CH_3CH_2CH_2Br}$$

$$\underset{\text{Methanol}}{CH_3OH} \xrightarrow[\text{(or HBr)}]{PBr_3} \underset{\text{Methyl bromide}}{CH_3Br}$$

$$\underset{\text{1-Propanol}}{CH_3CH_2CH_2OH} \xrightarrow{Na} \underset{\text{Sodium propoxide}}{CH_3CH_2CH_2ONa}$$

Isopropyl methyl ether is best prepared by the reaction,

$$\underset{\text{Methyl bromide}}{CH_3Br} + \underset{\text{Sodium isopropoxide}}{(CH_3)_2CHONa} \longrightarrow \underset{\text{Isopropyl methyl ether}}{CH_3OCH(CH_3)_2}$$

The reaction of sodium methoxide with isopropyl bromide will proceed mainly by elimination. Methyl bromide is prepared as shown previously; sodium isopropoxide is prepared by adding sodium to isopropyl alcohol.

Diethyl ether may be prepared as outlined:

$$\underset{\text{Ethanol}}{CH_3CH_2OH} \xrightarrow{Na} \underset{\text{Sodium ethoxide}}{CH_3CH_2ONa}$$

$$\underset{\text{Ethanol}}{CH_3CH_2OH} \xrightarrow[\text{(or HBr)}]{PBr_3} \underset{\text{Ethyl bromide}}{CH_3CH_2Br}$$

$$\underset{\text{Sodium ethoxide}}{CH_3CH_2ONa} + \underset{\text{Ethyl bromide}}{CH_3CH_2Br} \longrightarrow \underset{\text{Diethyl ether}}{CH_3CH_2OCH_2CH_3} + \underset{\text{Sodium bromide}}{NaBr}$$

17.25 (*a*) This reaction represents a typical Williamson ether synthesis wherein an alkoxide reacts with a primary alkyl halide.

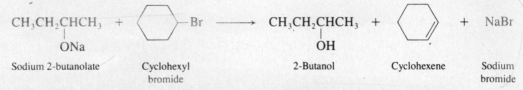

$$\underset{\text{Sodium 2-butanolate}}{CH_3CH_2CH(ONa)CH_3} + \underset{\text{Ethyl bromide}}{CH_3CH_2Br} \longrightarrow \underset{\substack{\textit{sec}\text{-Butyl ethyl ether}\\ \text{(2-ethoxybutane)}}}{CH_3CH_2CH(OCH_2CH_3)CH_3} + \underset{\text{Sodium bromide}}{NaBr}$$

(*b*) Secondary alkyl halides react with alkoxide bases by E2 elimination as the major pathway. The Williamson ether synthesis is not a useful reaction with secondary alkyl halides.

$$\underset{\text{Sodium 2-butanolate}}{CH_3CH_2CH(ONa)CH_3} + \underset{\text{Cyclohexyl bromide}}{C_6H_{11}Br} \longrightarrow \underset{\text{2-Butanol}}{CH_3CH_2CH(OH)CH_3} + \underset{\text{Cyclohexene}}{C_6H_{10}} + \underset{\text{Sodium bromide}}{NaBr}$$

(*c*) The starting material is a primary allylic halide, a structural class known to be reactive in nucleophilic substitution reactions of the S_N2 type.

$$\underset{\text{1-Chloro-2-butene}}{CH_3CH{=}CHCH_2Cl} \xrightarrow{KOC(CH_3)_3} \underset{\text{2-Butenyl \textit{tert}-butyl ether (61\%)}}{CH_3CH{=}CHCH_2OC(CH_3)_3}$$

The reaction produces an ether and is an example of the Williamson ether synthesis.

(*d*) The potassium alkoxide acts as a nucleophile toward iodoethane to yield an ethyl ether.

$$(CH_3CH_2)(CH_3)(C_6H_5)C{-}O^-K^+ + CH_3CH_2I \longrightarrow \underset{(R)\text{-2-Ethoxy-2-phenylbutane}}{(CH_3CH_2)(CH_3)(C_6H_5)C{-}OCH_2CH_3}$$

The ether product has the same absolute configuration as the starting alkoxide because no bonds to the chiral center are made or broken in the reaction.

(*e*) Vicinal halohydrins are converted to epoxides on being treated with base.

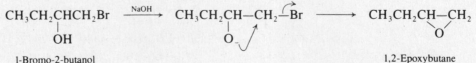

$$\underset{\text{1-Bromo-2-butanol}}{CH_3CH_2CH(OH)CH_2Br} \xrightarrow{NaOH} CH_3CH_2CH(O^-){-}CH_2{-}Br \longrightarrow \underset{\text{1,2-Epoxybutane}}{CH_3CH_2CH{-}CH_2 \text{ (bridged by O)}}$$

(*f*) The reaction conditions are appropriate for ether formation by solvomercuration-demercuration of an alkene.

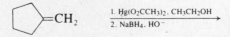

$$\underset{\text{Methylenecyclopentane}}{C_5H_8{=}CH_2} \xrightarrow[\text{2. NaBH}_4\text{, HO}^-]{\text{1. Hg(O}_2\text{CCH}_3)_2\text{, CH}_3\text{CH}_2\text{OH}} \underset{\text{1-Ethoxy-1-methylcyclopentane}}{C_5H_8(CH_3)(OCH_2CH_3)}$$

Addition of the elements of ethanol proceeds in accordance with Markovnikov's rule.

(*g*) The reactants, an alkene plus a peroxy acid, are customary ones for epoxide preparation. The reaction is a stereospecific syn addition of oxygen to the double bond.

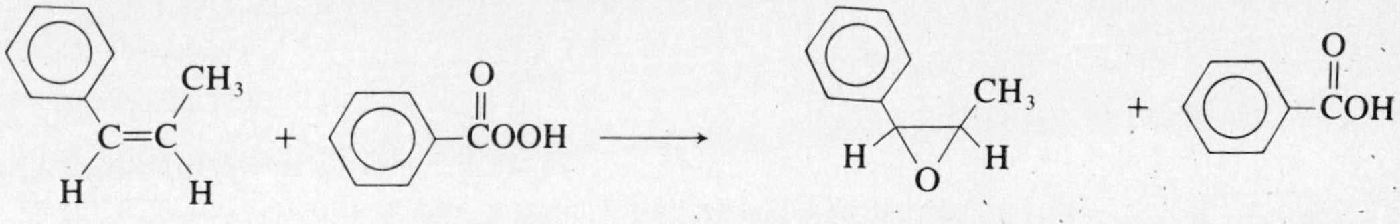

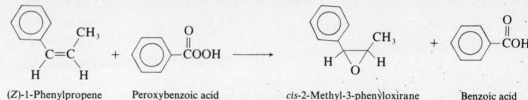

(*Z*)-1-Phenylpropene + Peroxybenzoic acid ⟶ *cis*-2-Methyl-3-phenyloxirane + Benzoic acid

(*h*) Azide ion is a good nucleophile and attacks the epoxide function. Substitution occurs at carbon with inversion of configuration. The product is *trans*-2-azidocyclohexanol.

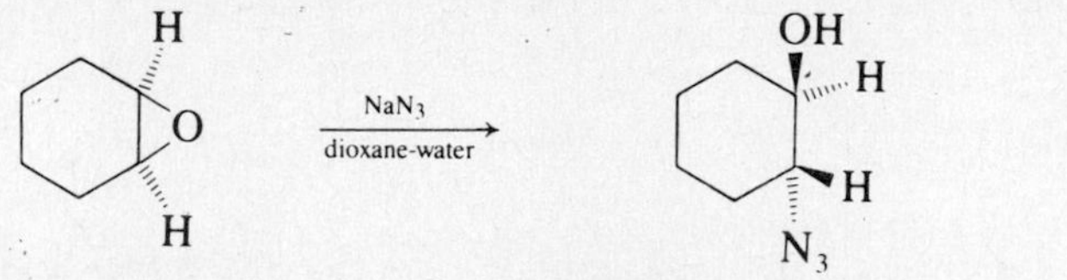

1,2-Epoxycyclohexane $\xrightarrow[\text{dioxane-water}]{NaN_3}$ *trans*-2-Azidocyclohexanol (61%)

(*i*) Ammonia is a nucleophile capable of reacting with epoxides. It attacks the less hindered carbon of the epoxide function.

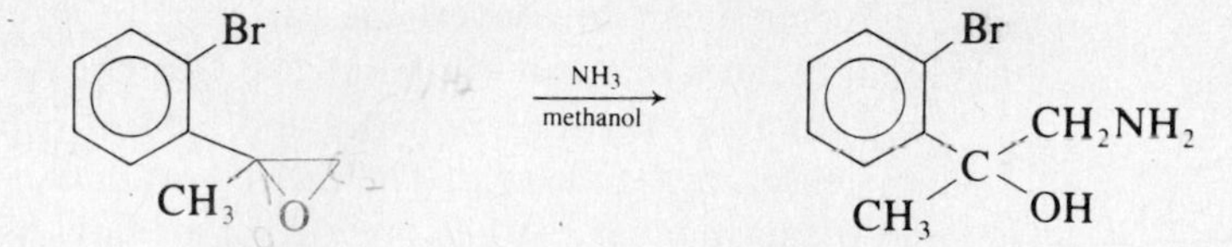

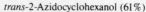

2-(*o*-Bromophenyl)-2-methyloxirane $\xrightarrow[\text{methanol}]{NH_3}$ 1-Amino-2-(*o*-bromophenyl)-2-propanol

Aryl halides do not react with nucleophiles under these conditions, so the bromine substituent on the ring is unaffected.

(*j*) Methoxide ion attacks the less substituted carbon of the epoxide ring with inversion of configuration.

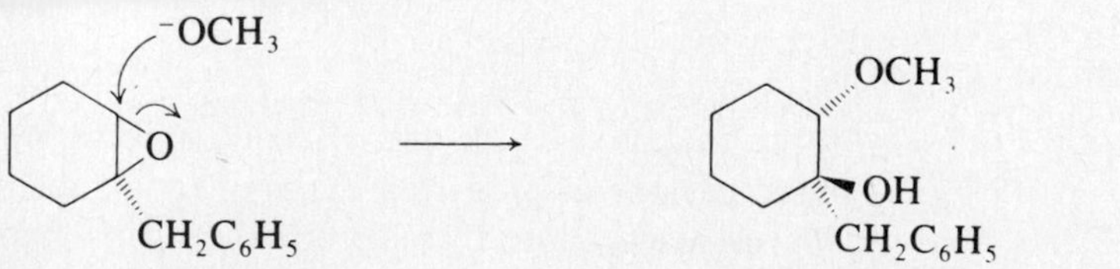

1-Benzyl-1,2-epoxycyclohexane ⟶ 1-Benzyl-*trans*-2-methoxycyclohexanol (98%)

17.26 The proper approach to this problem is to first write the equations given in full stereochemical detail.

(*a*)

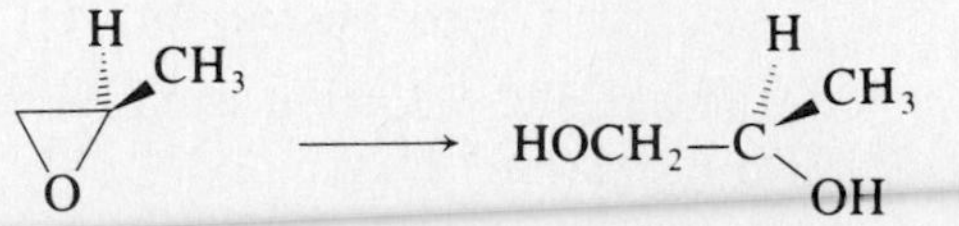

⟶ (*R*)-1,2-Propanediol

It now becomes clear that the arrangement of groups around the chiral center remains unchanged in going from starting materials to products. Therefore, choose conditions such that the nucleophile attacks the CH_2 group of the epoxide rather than the chiral center. Base-catalyzed hydrolysis is required; aqueous sodium hydroxide is appropriate.

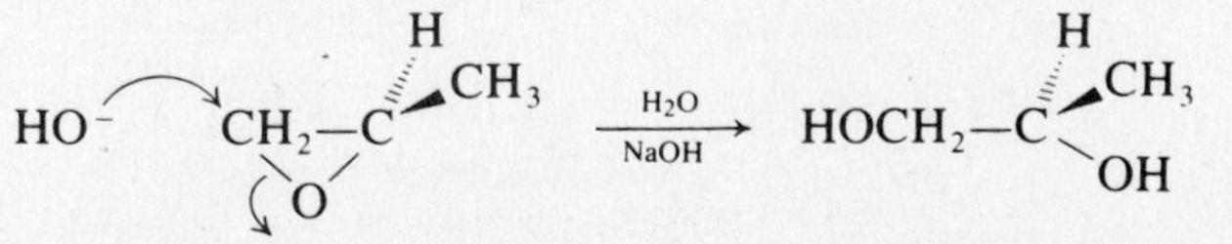

$HO^- \; CH_2{-}C(H)(CH_3)$ (epoxide) $\xrightarrow[NaOH]{H_2O}$ $HOCH_2{-}C(H)(CH_3)OH$

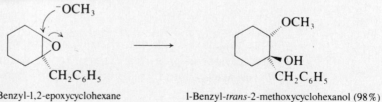

The nucleophile (hydroxide ion) attacks the less hindered carbon of the epoxide ring.

(*b*) [epoxide of propene with H and CH_3 on chiral carbon] $\longrightarrow$ $HOCH_2—C(OH)(H)(CH_3)$

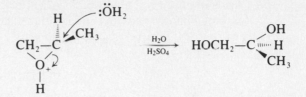

(*S*)-1,2-Propanediol

Inversion of configuration at the chiral center is required. Therefore the nucleophile must attack the chiral center, and acid-catalyzed hydrolysis should be chosen. Dilute sulfuric acid would be satisfactory.

$$\text{:}\ddot{O}H_2 \text{ attacking } CH_2—C(H)(CH_3) \text{ (protonated epoxide, } O^+—H) \xrightarrow[H_2SO_4]{H_2O} HOCH_2—C(OH)(H)(CH_3)$$

The nucleophile (a water molecule) attacks that carbon atom of the ring which can better support a positive charge. Carbocation character develops at the transition state and is better supported by the carbon atom that is more highly substituted.

17.27 (*a*) There is a temptation to try to do this transformation in a single step by using a reducing agent to convert the carbonyl to a methylene group. No reagent is available that reduces esters in this way! The Clemmensen and Wolff-Kishner reduction methods are suitable only for aldehydes and ketones. The best way to approach this problem is by reasoning backwards. The desired product is an ether. Ethers can be prepared by the Williamson ether synthesis involving an alkyl halide and an alkoxide ion.

$$C_6H_5—CH_2OCH_3 \Rightarrow C_6H_5—CH_2X + CH_3O^-$$

or

$$C_6H_5—CH_2OCH_3 \Rightarrow C_6H_5—CH_2O^- + CH_3X$$

Both the alkyl halide and alkoxide ion are prepared from alcohols. The problem then becomes one of preparing the appropriate alcohol (or alcohols) from the starting ester. This is readily done by reduction by any of a number of methods.

$$C_6H_5—\overset{O}{\overset{\|}{C}}OCH_3 \xrightarrow[2.\ H_2O]{1.\ LiAlH_4} C_6H_5—CH_2OH + CH_3OH$$

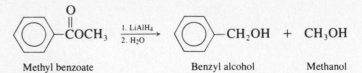

Methyl benzoate — Benzyl alcohol — Methanol

(Alternatively, reduction could be carried out with sodium in alcohol or by catalytic hydrogenation.) Then

$$CH_3OH \xrightarrow{Na} CH_3ONa$$

Methanol — Sodium methoxide

$$C_6H_5—CH_2OH \xrightarrow{HBr} C_6H_5—CH_2Br$$

Benzyl alcohol — Benzyl bromide

and

$$C_6H_5—CH_2Br + NaOCH_3 \longrightarrow C_6H_5—CH_2OCH_3 + NaBr$$

Benzyl bromide — Sodium methoxide — Benzyl methyl ether

The following sequence is also appropriate once methanol and benzyl alcohol are obtained by reduction of methyl benzoate.

$$C_6H_5CH_2OH \xrightarrow{Na} C_6H_5CH_2ONa$$

Benzyl alcohol — Sodium benzyloxide

$$CH_3OH \xrightarrow{PBr_3} CH_3Br$$

Methanol — Bromomethane

and

$$C_6H_5CH_2ONa + CH_3Br \longrightarrow C_6H_5CH_2OCH_3 + NaBr$$

Sodium benzyloxide — Bromomethane — Benzyl methyl ether

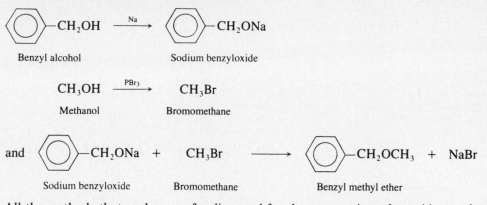

(*b*) All the methods that we have so far discussed for the preparation of epoxides are based on alkenes as starting materials. This leads us to consider the partial retrosynthesis shown.

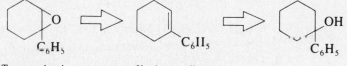

Target molecule (1-phenyl-1,2-epoxycyclohexane, C_6H_5) ⇒ Key intermediate (1-phenylcyclohexene, C_6H_5) ⇒ 1-phenylcyclohexanol (OH, C_6H_5)

The key intermediate, 1-phenylcyclohexene, is both a proper precursor to the desired epoxide and readily available from the given starting materials. A reasonable synthesis is:

$$\text{1-phenylcyclohexanol (OH, } C_6H_5) \xrightarrow[\text{heat}]{H^+} \text{1-phenylcyclohexene } (C_6H_5) \xrightarrow{CH_3COOH\text{ (O, peroxy)}} \text{epoxide (O, } C_6H_5)$$

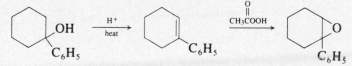

Preparation of the required tertiary alcohol, 1-phenylcyclohexanol, completes the synthesis.

$$\text{Cyclohexanol} \xrightarrow[H_2SO_4]{K_2Cr_2O_7} \text{Cyclohexanone}$$

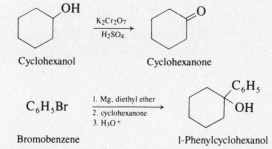

Cyclohexanol — Cyclohexanone

$$C_6H_5Br \xrightarrow[\text{2. cyclohexanone; 3. } H_3O^+]{\text{1. Mg, diethyl ether}} \text{1-phenylcyclohexanol } (C_6H_5, OH)$$

Bromobenzene — 1-Phenylcyclohexanol

(*c*) The necessary carbon skeleton can be assembled through the reaction of a Grignard reagent with 1,2-epoxypropane.

$$C_6H_5CH_2\underset{Br}{C}HCH_3 \Longrightarrow C_6H_5CH_2\underset{OH}{C}HCH_3 \Longrightarrow C_6H_5^- + CH_2\!\overset{O}{-}\!CHCH_3$$

The reaction sequence is therefore:

$$C_6H_5MgBr + \underset{O}{CH_2-CHCH_3} \longrightarrow C_6H_5CH_2\underset{OH}{C}HCH_3$$

Phenylmagnesium bromide (from bromobenzene and magnesium) — 1,2-Epoxypropane — 1-Phenyl-2-propanol

The alcohol obtained in this step need only be treated with phosphorus tribromide to obtain the desired product.

$$\underset{\text{1-Phenyl-2-propanol}}{C_6H_5CH_2\underset{|}{\overset{}{C}}HCH_3 \atop \quad OH} \xrightarrow{PBr_3} \underset{\text{2-Bromo-1-phenylpropane}}{C_6H_5CH_2\underset{|}{C}HCH_3 \atop \quad Br}$$

The epoxide required in the first step, 1,2-epoxypropane, is prepared as follows from isopropyl alcohol:

$$\underset{\substack{\text{2-Propanol}\\\text{(isopropyl}\\\text{alcohol)}}}{CH_3\underset{|}{C}HCH_3 \atop OH} \xrightarrow[\text{heat}]{H_2SO_4} \underset{\text{Propene}}{CH_3CH{=}CH_2} \xrightarrow{CH_3\overset{O}{\overset{\|}{C}}OOH} \underset{\text{1,2-Epoxypropane}}{CH_3CH{-}CH_2 \atop \diagdown O \diagup}$$

(*d*) Since the target molecule is an ether, it ultimately derives from two alcohols.

$$C_6H_5CH_2CH_2CH_2OCH_2CH_3 \Longrightarrow C_6H_5CH_2CH_2CH_2OH + CH_3CH_2OH$$

Our first task is to assemble 3-phenyl-1-propanol from the designated starting material benzyl alcohol. This requires formation of a primary alcohol with the original carbon chain extended by two carbons. The standard method for this transformation involves reaction of a Grignard reagent with ethylene oxide.

$$\underset{\text{Benzyl alcohol}}{C_6H_5CH_2OH} \xrightarrow{PBr_3} \underset{\text{Benzyl bromide}}{C_6H_5CH_2Br} \xrightarrow[\substack{2.\ CH_2{-}CH_2\ (\diagdown O \diagup)\\3.\ H_3O^+}]{1.\ \text{Mg, diethyl ether}} \underset{\text{3-Phenyl-1-propanol}}{C_6H_5CH_2CH_2CH_2OH}$$

Once 3-phenyl-1-propanol has been prepared, its conversion to the corresponding ethyl ether can be accomplished in either of two ways.

$$\underset{\text{3-Phenyl-1-propanol}}{C_6H_5CH_2CH_2CH_2OH} \xrightarrow{PBr_3} \underset{\text{1-Bromo-3-phenylpropane}}{C_6H_5CH_2CH_2CH_2Br}$$

$$\downarrow \text{NaOCH}_2\text{CH}_3\text{, ethanol}$$

$$\underset{\substack{\text{Ethyl 3-phenyl-1-propyl ether}\\\text{(1-ethoxy-3-phenyl propane)}}}{C_6H_5CH_2CH_2CH_2OCH_2CH_3}$$

Alternatively

$$\underset{\text{3-Phenyl-1-propanol}}{C_6H_5CH_2CH_2CH_2OH} \xrightarrow[2.\ CH_3CH_2Br]{1.\ Na} \underset{\text{Ethyl 3-phenyl-1-propyl ether}}{C_6H_5CH_2CH_2CH_2OCH_2CH_3}$$

The reagents in each step are prepared from ethanol.

$$\underset{\text{Ethanol}}{CH_3CH_2OH} \xrightarrow{Na} \underset{\text{Sodium ethoxide}}{CH_3CH_2ONa}$$

$$\underset{\text{Ethanol}}{CH_3CH_2OH} \xrightarrow{PBr_3} \underset{\text{Ethyl bromide}}{CH_3CH_2Br}$$

(e) The target epoxide can be prepared in a single step from the corresponding alkene.

$$\text{Bicyclo[2.2.2]-oct-2-ene} \xrightarrow{CH_3COOOH} \text{2,3-Epoxybicyclo[2.2.2]octane}$$

Bicyclo[2.2.2]-oct-2-ene 2,3-Epoxybicyclo[2.2.2]octane

Disconnections show that this alkene is available through a Diels-Alder reaction.

$$\text{Bicyclo[2.2.2]-oct-2-ene} \Rightarrow \text{1,3-cyclohexadiene} + CH_2{=}CH_2$$

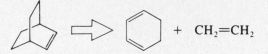

The reaction of 1,3-cyclohexadiene with ethylene gives the desired substance.

$$\text{1,3-Cyclohexadiene} + CH_2{=}CH_2 \longrightarrow \text{Bicyclo[2.2.2]-oct-2-ene}$$

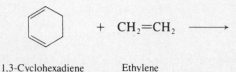

1,3-Cyclohexadiene Ethylene Bicyclo[2.2.2]-oct-2-ene

1,3-Cyclohexadiene is one of the given starting materials. Ethylene is prepared from ethanol.

$$CH_3CH_2OH \xrightarrow[\text{heat}]{H_2SO_4} CH_2{=}CH_2$$

Ethanol Ethylene (ethene)

17.28 (a) A reasonable mechanism is one that parallels the usual one for acid-catalyzed ether formation from alcohols, modified to accommodate these particular starting materials and products. Begin with protonation of one of the oxygen atoms of ethylene glycol.

$$HOCH_2CH_2OH + H_2SO_4 \rightleftharpoons HOCH_2CH_2{-}\overset{+}{\ddot{O}}(H){-}H + HSO_4^-$$

Ethylene glycol

The protonated alcohol then reacts in the usual way with another molecule of alcohol to give an ether. (This ether is known as diethylene glycol.)

$$HOCH_2CH_2{-}\overset{+}{\ddot{O}}(H){-}H + HOCH_2CH_2\ddot{O}H \xrightarrow{-H_2O} HOCH_2CH_2{-}\overset{+}{\ddot{O}}(H){-}CH_2CH_2OH \xrightarrow{-H^+} HOCH_2CH_2OCH_2CH_2OH$$

Diethylene glycol

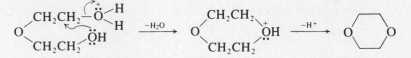

Diethylene glycol then undergoes intramolecular ether formation to yield 1,4-dioxane.

$$HOCH_2CH_2OCH_2CH_2OH + H_2SO_4 \rightleftharpoons HOCH_2CH_2OCH_2CH_2{-}\overset{+}{\ddot{O}}(H){-}H + HSO_4^-$$

$$O(CH_2CH_2{-}\overset{+}{\ddot{O}}H_2)(CH_2CH_2\ddot{O}H) \xrightarrow{-H_2O} O(CH_2CH_2)_2\overset{+}{\ddot{O}}H \xrightarrow{-H^+} \text{1,4-dioxane}$$

(b) The substrate is a primary alkyl halide and reacts with sodium hydroxide by nucleophilic substitution.

$$ClCH_2CH_2OCH_2CH_2Cl + HO^- \longrightarrow HOCH_2CH_2OCH_2CH_2Cl + Cl^-$$

Bis(2-chloroethyl) ether

The product of nucleophilic substitution of one of the chloroethyl groups now has an alcohol function and a primary chloride built into the same molecule. It contains the requisite functionality to undergo an intramolecular Williamson reaction.

$$HOCH_2CH_2OCH_2CH_2Cl + HO^- \rightleftharpoons {}^-OCH_2CH_2OCH_2CH_2Cl + H_2O$$

$$\text{(}^-OCH_2CH_2OCH_2CH_2\text{—Cl, cyclic)} \longrightarrow \text{1,4-Dioxane} + Cl^-$$

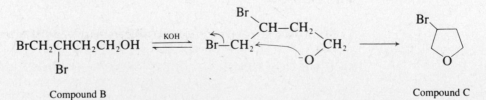

1,4-Dioxane

17.29 (*a*) The first step is a standard Grignard synthesis of a primary alcohol using formaldehyde. Compound A can only be 3-buten-1-ol.

$$\underset{\text{Allyl bromide}}{CH_2{=}CHCH_2Br} \xrightarrow[\text{2. } CH_2{=}O \quad \text{3. } H_3O^+]{\text{1. Mg}} \underset{\text{3-Buten-1-ol (compound A)}}{CH_2{=}CHCH_2CH_2OH}$$

Addition of bromine to the carbon-carbon double bond of 3-buten-1-ol takes place readily to yield the vicinal dibromide.

$$\underset{\text{3-Buten-1-ol}}{CH_2{=}CHCH_2CH_2OH} \xrightarrow{Br_2} \underset{\text{3,4-Dibromo-1-butanol (compound B)}}{BrCH_2\underset{\displaystyle Br}{\underset{|}{C}}HCH_2CH_2OH}$$

When compound B is treated with potassium hydroxide, it loses the elements of HBr to give compound C. Since further treatment of compound C with potassium hydroxide converts it to D by a second dehydrobromination, a reasonable candidate for C is 3-bromotetrahydrofuran.

$$\underset{\text{3,4-Dibromo-1-butanol (compound B)}}{BrCH_2\underset{\displaystyle Br}{\underset{|}{C}}HCH_2CH_2OH} \xrightarrow[25°C]{KOH} \text{3-Bromotetrahydrofuran (compound C)} \xrightarrow[\text{heat}]{KOH} \text{Compound D}$$

3,4-Dibromo-1-butanol (compound B) — 3-Bromotetrahydrofuran (compound C) — Compound D

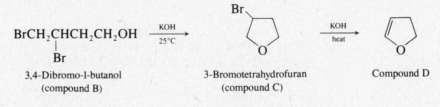

Ring closure occurs by an intramolecular Williamson reaction.

$$\underset{\text{Compound B}}{BrCH_2\underset{\displaystyle Br}{\underset{|}{C}}HCH_2CH_2OH} \overset{KOH}{\rightleftharpoons} Br\text{—}CH_2\text{—}CH(Br)\text{—}CH_2\text{—}CH_2\text{—}O^- \longrightarrow \text{Compound C}$$

Compound B — Compound C

Dehydrohalogenation of compound C converts it to the final product D.

The alternative series of events, in which double bond formation precedes ring closure, is unlikely because it requires nucleophilic attack by the alkoxide on a vinyl bromide.

$$BrCH_2\underset{\displaystyle Br}{\underset{|}{C}}HCH_2CH_2OH \xrightarrow{KOH} BrCH{=}CHCH_2CH_2OH \overset{KOH}{\rightleftharpoons} BrCH{=}CHCH_2CH_2O^-$$

(cyclization of this intermediate does not occur)

(*b*) Lithium aluminum hydride reduction of the starting ketone can yield either a cis or trans bromohydrin.

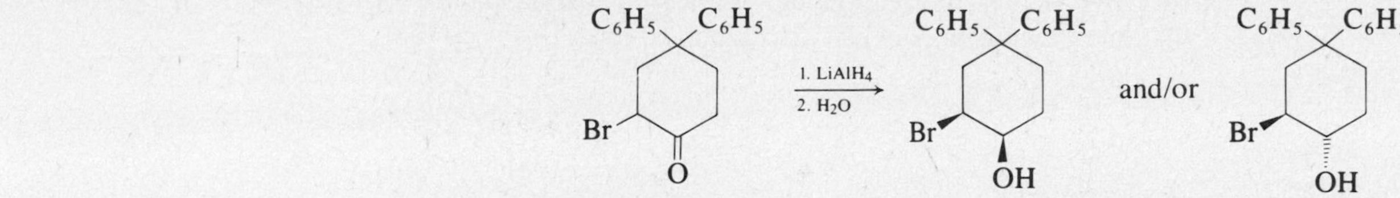

C_6H_5 C_6H_5 ... Br ... O → (1. $LiAlH_4$ 2. H_2O) C_6H_5 C_6H_5 ... Br ... OH and/or C_6H_5 C_6H_5 ... Br ... OH

2-Bromo-4,4-diphenylcyclohexanone

At this stage in our consideration of the problem we cannot choose which of these two diastereomers is the correct structure of compound E. We note, however, that compound F is formed by loss of a hydrogen and a bromine from compound E on treatment with potassium hydroxide. A working hypothesis is that compound F is an epoxide. Conversion of a bromohydrin to an epoxide requires a trans relationship between the hydroxyl group and bromine. Therefore, compound E is most likely the trans bromohydrin.

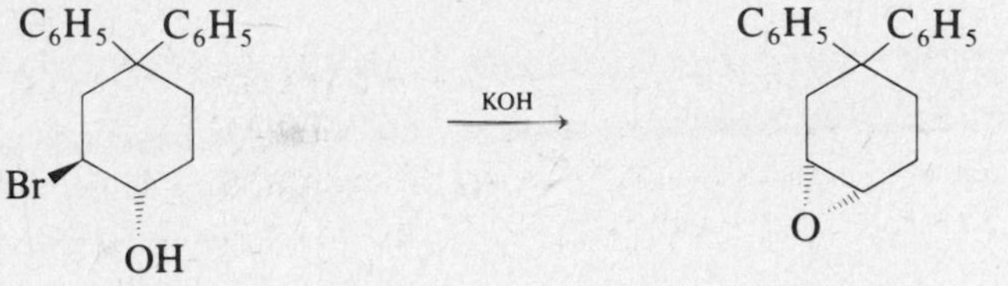

C_6H_5 C_6H_5 ... Br ... OH → (KOH) C_6H_5 C_6H_5 ... O

trans-2-Bromo-4,4-diphenylcyclohexanol (compound E) 4,4-Diphenyl-1,2-epoxycyclohexane (compound F)

On treatment with HBr, compound F can undergo two distinct modes of ring opening. Both are stereospecific, yielding trans bromohydrins.

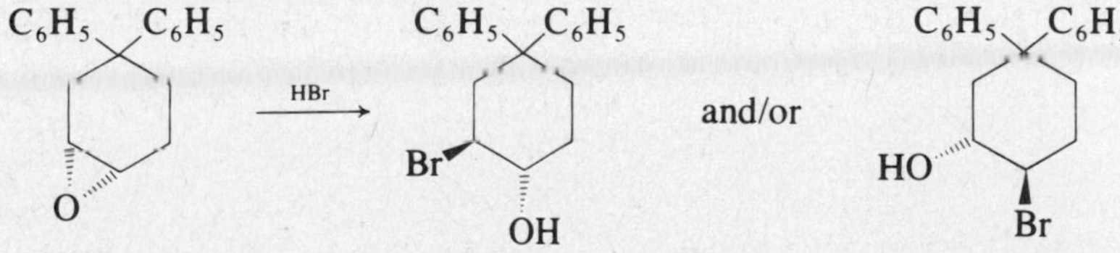

C_6H_5 C_6H_5 ... O → (HBr) C_6H_5 C_6H_5 ... Br ... OH and/or C_6H_5 C_6H_5 ... HO ... Br

Compound F Compound E *trans*-2-Bromo-5,5-diphenylcyclohexanol

Since oxidation of compound G yields a ketone (compound H) having a constitution that is different from the original ketone, compound G must be *trans*-2-bromo-5,5-diphenylcyclohexanol.

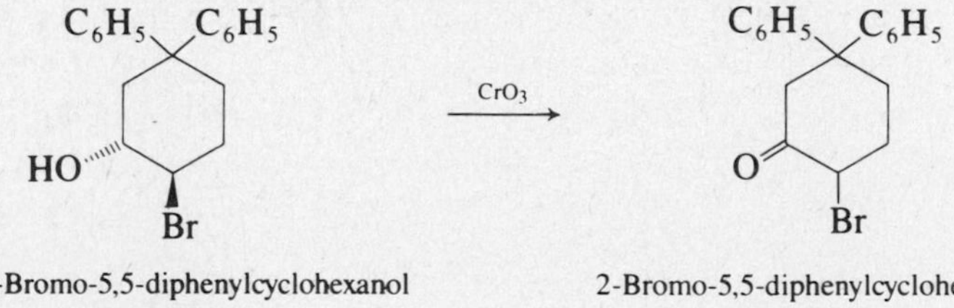

C_6H_5 C_6H_5 ... HO ... Br → (CrO_3) C_6H_5 C_6H_5 ... O ... Br

trans-2-Bromo-5,5-diphenylcyclohexanol (compound G) 2-Bromo-5,5-diphenylcyclohexanone (compound H)

(*c*) Since it gives an epoxide on treatment with a peroxy acid, compound I must be an alkene; more specifically, it is 1,2-dimethylcyclopentene.

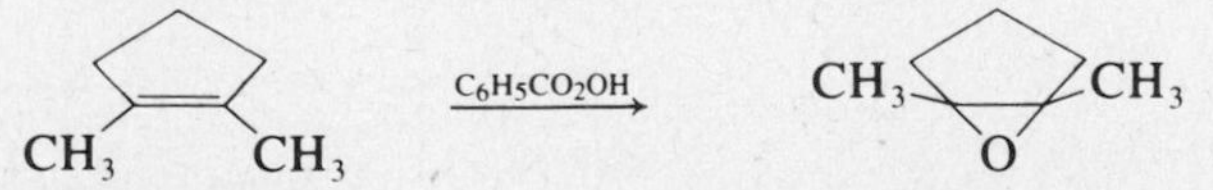

CH_3 CH_3 → ($C_6H_5CO_2OH$) CH_3 O CH_3

1,2-Dimethylcyclopentene (compound I) 1,2-Dimethyl-1,2-epoxycyclopentane (compound K)

Compounds J and L have the same molecular formula, $C_7H_{14}O_2$, but J is a liquid while L is a crystalline solid. Their molecular formulas correspond to the addition of two OH groups to compound I. Potassium permanganate brings about syn hydroxylation of an alkene; therefore compound J must be the cis diol.

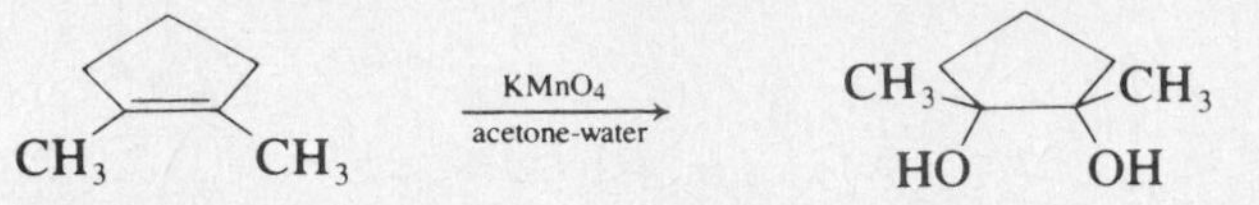

1,2-Dimethylcyclopentene (compound I) $\xrightarrow[\text{acetone-water}]{KMnO_4}$ *cis*-1,2-Dimethylcyclopentane-1,2-diol (compound J)

Acid-catalyzed hydrolysis of an epoxide yields a trans diol (compound L).

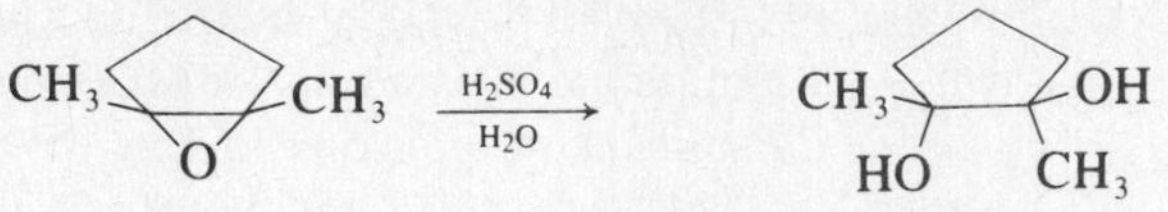

$\xrightarrow[H_2O]{H_2SO_4}$ *trans*-1,2-Dimethylcyclopentane-1,2-diol (compound L)

17.30 Since cineole contains no double or triple bonds, it must be bicyclic based on its molecular formula.

$$\begin{array}{rcl} C_{10}H_{22} & & \text{molecular formula of corresponding alkane} \\ -C_{10}H_{18}O & & \text{molecular formula of cineole} \\ \hline 4\,H & \equiv & \text{two double bonds and/or rings (SODAR = 2)} \end{array}$$

When cineole reacts with hydrogen chloride, one of the rings is broken and water is formed.

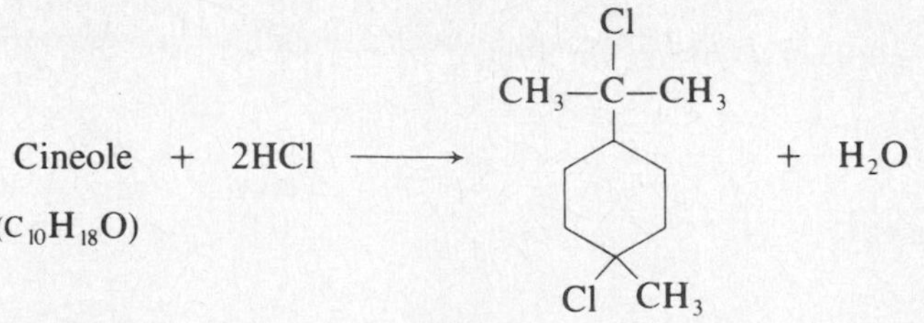

$$\underset{(C_{10}H_{18}O)}{\text{Cineole}} + 2HCl \longrightarrow \underset{(C_{10}H_{18}Cl_2)}{\text{1-chloro-4-(1-chloro-1-methylethyl)-1-methylcyclohexane}} + H_2O$$

The reaction that takes place is hydrogen halide–promoted ether cleavage. In such a reaction with excess hydrogen halide, the C—O—C unit is cleaved and two carbon-halogen bonds are formed. This suggests that cineole is a cyclic ether because the product contains both newly formed carbon-halogen bonds. A reasonable structure consistent with these facts is:

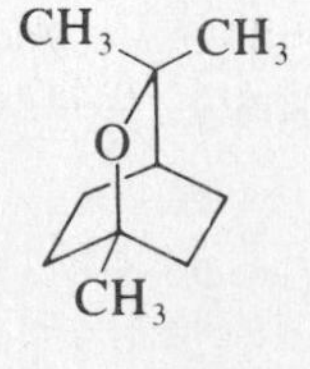

Cineole

17.31 (*a*) Since all the peaks in the ^1H nmr spectrum of this ether are singlets, none of the protons can be vicinal to any other nonequivalent proton. The only $C_5H_{12}O$ ether that satisfies this requirement is *tert*-butyl methyl ether.

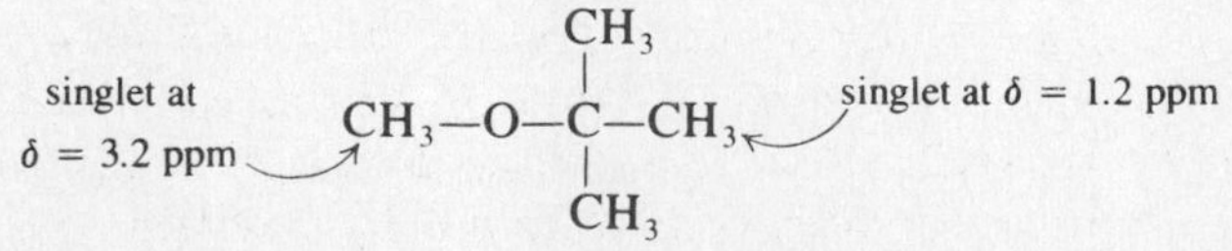

$$CH_3—O—C(CH_3)_3$$

singlet at $\delta = 3.2$ ppm (CH_3O); singlet at $\delta = 1.2$ ppm ($C(CH_3)_3$)

(*b*) A doublet-heptet pattern is characteristic of an isopropyl group. There are two isomeric $C_5H_{12}O$ ethers that contain an isopropyl group, ethyl isopropyl ether and isobutyl methyl ether.

$(CH_3)_2CHOCH_2CH_3$ Ethyl isopropyl ether

$(CH_3)_2CHCH_2OCH_3$ Isobutyl methyl ether

The methine proton in isobutyl methyl ether will be split into more than a heptet, however, because in addition to being split by two methyl groups, it is coupled to the two protons in the methylene group. Thus, isobutyl methyl ether does not have the correct splitting pattern to be the answer. The correct answer is ethyl isopropyl ether.

(*c*) The low-field signals are due to the protons on the carbon atoms of the C—O—C linkage. Since one is a doublet, it must be vicinal to only one other proton. Therefore, we can specify the partial structure:

C—O—C(H)—C(H)(C)—C

low-field doublet → H

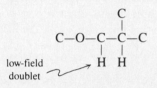

This partial structure contains all the carbon atoms in the molecule. Fill in the remaining valences with hydrogen atoms to reveal isobutyl methyl ether as the correct choice.

$CH_3—O—CH_2—CH(CH_3)_2$

low-field singlet; low-field doublet

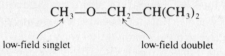

(*d*) Here again, signals at low field arise from protons on the carbons of the C—O—C unit. One of these signals is a quartet and so corresponds to a proton on a carbon bearing a methyl group.

$CH_3—C(H)—O—C$

quartet → H

The other carbon of the C—O—C unit has a hydrogen substituent that is split into a triplet. Therefore, this hydrogen must be attached to a carbon that bears a methylene group.

$CH_3—C(H)—O—C(H)—CH_2—$

quartet → H; H ← triplet

These data permit us to complete the structure by adding an additional carbon and the requisite number of hydrogens in such a way that the protons attached to the carbons of the ether linkage are not split further. The correct structure is ethyl propyl ether.

$CH_3CH_2OCH_2CH_2CH_3$

quartet; triplet

17.32 A good way to address this problem is to consider the dibromide derived by treatment of compound M with hydrogen bromide. The presence of an nmr signal equivalent to four protons in the aromatic region at $\delta = 7.3$ ppm indicates that this dibromide contains a disubstituted aromatic ring. The four remaining protons appear as a sharp singlet at $\delta = 4.7$ ppm and are most reasonably contained in two equivalent methylene groups of the type $ArCH_2Br$. Since the dibromide contains all the carbons and hydrogens of the starting material and is derived from it by treatment with hydrogen bromide, it is likely that compound M is a cyclic ether in which a CH_2OCH_2 unit spans two of the carbons of a benzene ring. This can only occur when the positions involved are ortho to each other. Therefore

Compound M (δ = 5.1 ppm) + 2 HBr ⟶ o-$C_6H_4(CH_2Br)_2$ (δ = 4.7 ppm) + H_2O

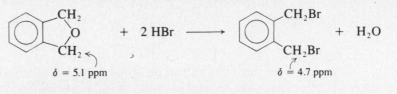

Compound M

17.33 The molecular formula of compound N ($C_{12}H_{18}O_2$) lacks eight hydrogens as compared with an alkane having the same number of carbon atoms; this indicates a total of four double bonds plus rings. A para-substituted aromatic ring is indicated by oxidation of compound N to benzene-1,4-dicarboxylic acid. The benzene ring contributes the equivalent of three double bonds plus one ring, so both of the side chains are saturated and neither contains a ring. At this point we know the location of 8 of the 12 carbon atoms of compound N.

C—C_6H_4—C

Turning now to the ^{1}H nmr spectrum, we see the triplet-quartet pattern characteristic of an ethyl group. Since there are six methyl protons and four methylene protons, as determined by integration of the nmr spectrum, there must be two equivalent ethyl groups in the molecule. Further, because the chemical shift of the methylene protons is at δ = 3.4 ppm, the ethyl group is bonded to oxygen, i.e., there are two $CH_3CH_2O—$ groups in compound N. The two ethoxy groups and the partial structure deduced previously account for all the carbon atoms in compound N. A reasonable candidate structure is:

$CH_3CH_2OCH_2$—C_6H_4—$CH_2OCH_2CH_3$

quartet at δ = 3.4 ppm; triplet at δ = 1.2 ppm; singlet at δ = 4.4 ppm; ring hydrogens are singlets at δ = 7.2 ppm

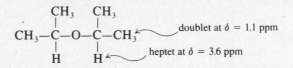

This structure satisfies all the requirements of the ^{1}H nmr spectrum. All the aromatic ring protons are equivalent and so appear as a sharp singlet, and both benzylic methylene groups are equivalent and appear as a sharp singlet at δ = 4.4 ppm. Their low chemical shift results from their benzylic nature and the fact that they are bonded to oxygen.

17.34 (*a*) Compound O ($C_6H_{14}O$) has the same number of hydrogen atoms as a six-carbon alkane and so contains neither double bonds nor rings. There are no O—H absorbances in the 3300 cm^{-1} region of its infrared spectrum, so it is not an alcohol. Completely saturated substances containing oxygen can only be alcohols or ethers; therefore, compound O must be an ether. This conclusion is supported by the observation of a strong C—O stretching vibration at 1100 cm^{-1}.

The ^{1}H nmr spectrum of compound O shows a doublet-heptet pattern typical of an isopropyl group as its only signals. Compound O is diisopropyl ether.

$(CH_3)_2CH—O—CH(CH_3)_2$

doublet at δ = 1.1 ppm (CH_3); heptet at δ = 3.6 ppm (H)

(*b*) The infrared spectrum of compound P shows C—H absorption above 3000 cm^{-1}, indicating an aromatic ring. A strong band at 700 cm^{-1} points to monosubstitution on that ring. Since there are no O—H peaks in the infrared spectrum, the C—O stretch at 1100 cm^{-1} tells us that P is an ether, not an alcohol.

The ^1H nmr spectrum is quite simple, consisting of a 4-proton singlet at $\delta = 4.5$ ppm and a 10-proton singlet at $\delta = 7.3$ ppm. The low-field signal is assigned to two monosubstituted phenyl groups. The signal at $\delta = 4.5$ ppm arises from two CH_2O units. Compound P is dibenzyl ether.

$$C_6H_5CH_2OCH_2C_6H_5$$

(*c*) The infrared spectrum of compound Q ($C_9H_{10}O$) is devoid of O—H bands but has an absorption at 1100 cm^{-1} characteristic of an ether. Also present are both aliphatic and aromatic C—H stretching bands. A strong band at 740 cm^{-1} indicates ortho disubstitution on a benzene ring.

The molecular formula of compound Q is consistent with a total of five double bonds plus rings. One benzene ring contributes three double bonds and one ring, leaving one other double bond or ring to be accounted for.

In addition to a peak for four aromatic protons at $\delta = 7.0$ ppm, the ^1H nmr spectrum of compound Q has a sharp two-proton singlet at $\delta = 4.7$ ppm. The downfield nature of this signal and its lack of splitting suggest that it corresponds to a methylene group of the $ArCH_2O$ type. The two remaining triplets at $\delta = 3.9$ ppm and $\delta = 2.7$ ppm are methylene groups connected to each other (CH_2CH_2). The structure most consistent with all these data is:

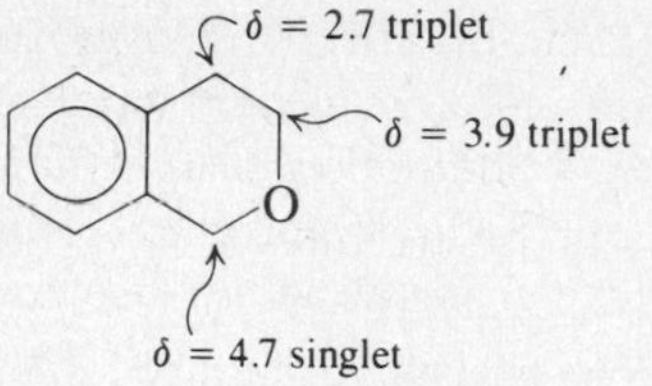

$\delta = 2.7$ triplet

$\delta = 3.9$ triplet

O

$\delta = 4.7$ singlet

SELF-TEST

PART A

A-1. Write the structures of all the isomeric ethers of molecular formula $C_4H_{10}O$ and give the correct name for each.

A-2. Give the structure of the product obtained from each of the following reactions. Show stereochemistry where it is important.

(*a*) CH_3CH_2—C(H)(OH)—CH_3 $\xrightarrow[\text{2. }CH_3I]{\text{1. Na}}$?

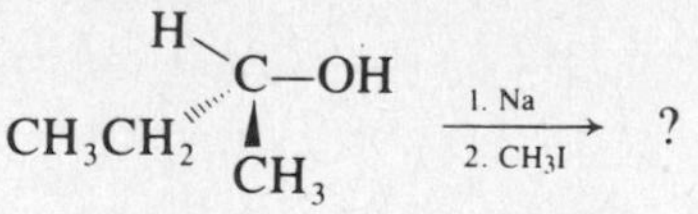

(*b*) (*Z*)-2-butene $\xrightarrow[\text{2. }H_3O^+]{\text{1. }CH_3CO_3H}$?

(*c*) $C_6H_5\overset{O}{CH-CH_2}$ $\xrightarrow{HI}$?

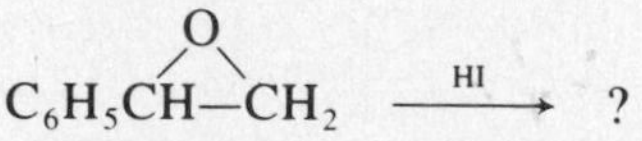

(*d*) [dimethylcyclohexene oxide: CH_3, CH_3, O] $\xrightarrow[CH_3CH_2SH]{CH_3CH_2SNa}$?

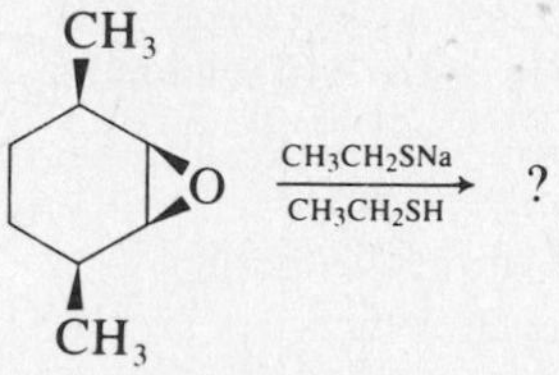

A-3. Outline schemes for the preparation of cyclohexyl ethyl ether using (*a*) the Williamson method and (*b*) oxymercuration-demercuration.

A-4. Outline a synthesis of 2-ethoxyethanol, $CH_3CH_2OCH_2CH_2OH$, using ethanol as the source of all the carbon atoms.

A-5. Provide structures for compounds A and B in the following reaction scheme:

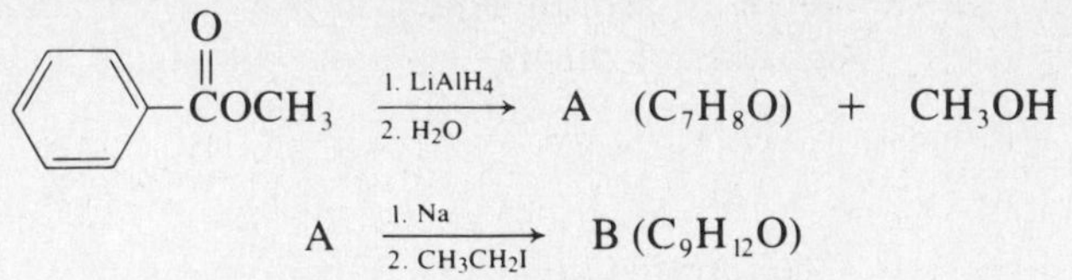

$$C_6H_5COCH_3 \xrightarrow[2.\ H_2O]{1.\ LiAlH_4} A\ (C_7H_8O) + CH_3OH$$

$$A \xrightarrow[2.\ CH_3CH_2I]{1.\ Na} B\ (C_9H_{12}O)$$

PART B

B-1. Consider the following two schemes for preparing isopropyl methyl ether, $(CH_3)_2CHOCH_3$:

$$(CH_3)_2CHO^-Na^+ + CH_3I \longrightarrow \text{(scheme I)}$$

$$(CH_3)_2CHI + CH_3O^-Na^+ \longrightarrow \text{(scheme II)}$$

Which gives the maximum yield of the desired product?
(*a*) Scheme I is preferred.
(*b*) Scheme II is preferred.
(*c*) Both schemes work equally well.
(*d*) Neither scheme is effective.

B-2. The most effective pair of reagents to prepare cyclohexyl ethyl ether is:
(*a*) Sodium cyclohexanolate + ethyl bromide (bromoethane)
(*b*) Sodium cyclohexanolate + ethanol
(*c*) Sodium ethoxide + bromocyclohexane
(*d*) Cyclohexanol + bromoethane

B-3. The *best* choice of reactant(s) for the following conversion is

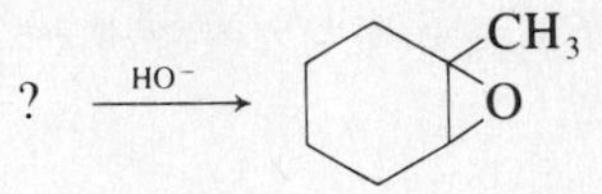

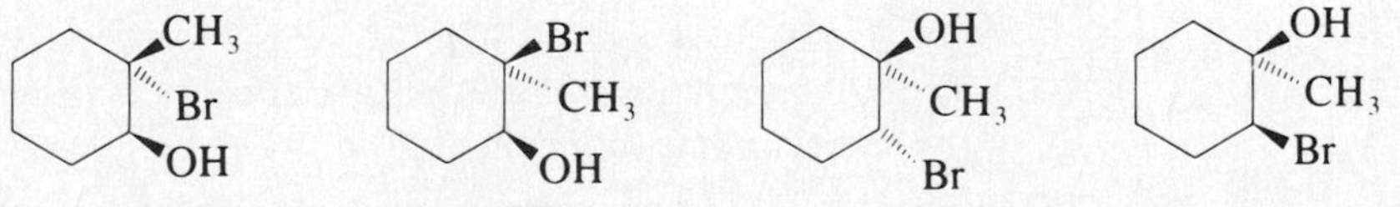

$$? \xrightarrow{HO^-} \text{1-methylcyclohexene oxide } (CH_3, O)$$

A: CH_3, Br, OH B: Br, CH_3, OH C: OH, CH_3, Br D: OH, CH_3, Br

(*a*) A and C (*b*) C only (*c*) C and D (*d*) B only

B-4. For which of the following ethers would the ^1H nmr spectrum consist of only two singlets?

(*a*) $CH_3OC(CH_3)_2CH_3$ (*c*) 2,2-dimethyloxirane (O, CH_3, CH_3)

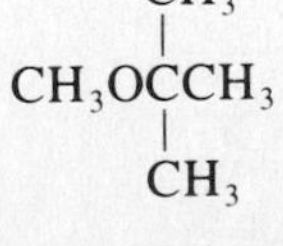

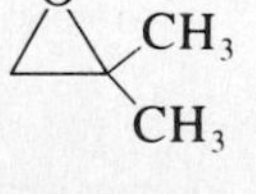

(*b*) $CH_3OCH_2CH_2OCH_3$ (*d*) All of these

B-5. Heating a particular ether with HBr yielded a single organic product. Which of the following conclusions may be reached?
(*a*) The reactant was a methyl ether.
(*b*) The reactant was a symmetrical ether.
(*c*) The reactant was a cyclic ether.
(*d*) Both (*b*) and (*c*) are correct.

CHAPTER

18

ALDEHYDES AND KETONES. NUCLEOPHILIC ADDITION TO THE CARBONYL GROUP

IMPORTANT TERMS AND CONCEPTS

Nomenclature (Sec. 18.1) In naming both aldehydes and ketones, the base name is derived from the longest chain of carbon atoms containing the carbonyl group. In aldehydes the carbonyl carbon is always C-1 and the *-e* ending of the alkane base name is replaced by *-al.* For example

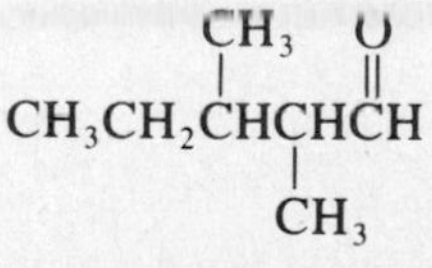

$$CH_3CH_2\underset{CH_3}{\overset{CH_3}{CH}}\underset{}{CH}\overset{O}{\overset{\|}{C}}H$$

2,3-Dimethylpentanal

$$(CH_3)_2CHCH_2CH_2\underset{CH_2CH_3}{CH}\overset{O}{\overset{\|}{C}}H$$

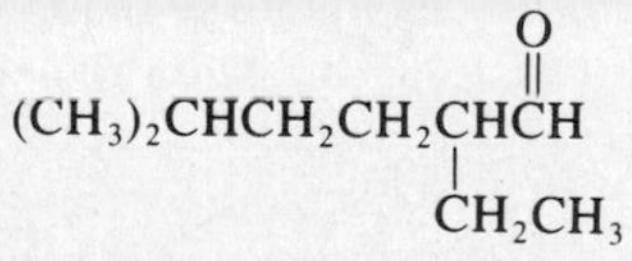

2-Ethyl-5-methylhexanal

Ketones are named so that the carbonyl position is the lowest possible. In addition, the *-e* ending of the alkane base name is replaced by *-one.* For example

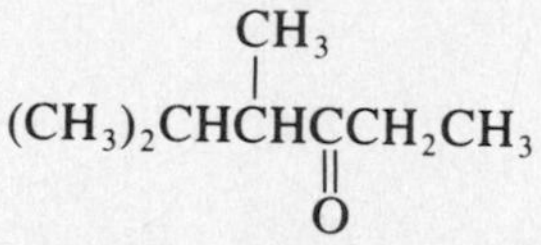

$$(CH_3)_2CH\overset{CH_3}{CH}\underset{O}{\underset{\|}{C}}CH_2CH_3$$

4,5-Dimethyl-3-hexanone

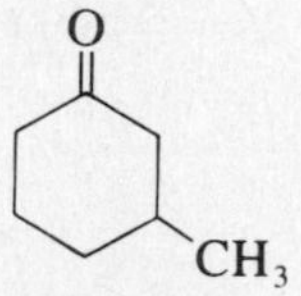

3-Methylcyclohexanone

The Carbonyl Group (Sec. 18.2) Understanding the structure and bonding of the carbonyl group is fundamental to being able to explain the chemical reactivity of aldehydes and ketones. The carbon utilizes sp^2 hybridized orbitals to bond to the oxygen and two other groups. The unhybridized p orbital on carbon forms a π bond to oxygen through overlap with an oxygen $2p$ orbital.

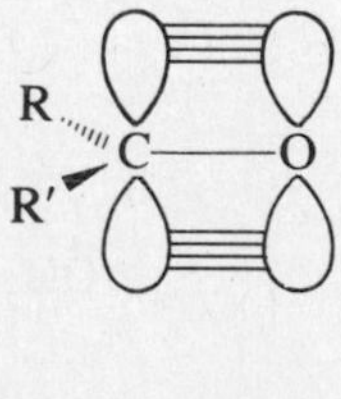

The polarity of the carbonyl group is explained by the greater electronegativity of oxygen as compared with carbon. The polarization of the carbonyl group may be represented as

$$\overset{\delta+}{\gt C}=\overset{\delta-}{O} \quad \text{or} \quad \overset{\longmapsto}{\gt C=O}$$

The orientation of this dipole explains why *nucleophilic reagents* react with the carbon and *electrophilic reagents* react with the oxygen of the carbonyl group.

Preparation of Aldehydes and Ketones (Secs. 18.5 to 18.7) Several methods have been discussed in previous chapters and are summarized in text Table 18.2. These include:

Ozonolysis of alkenes	Chapter 7
Hydration of alkynes	Chapter 10
Friedel-Crafts acylation	Chapter 13
Oxidation of alcohols	Chapter 16

Methods presented in this chapter for the preparation of aldehydes and ketones are outlined in the reaction summary which follows this section.

Reactions of Aldehydes and Ketones (Secs. 18.8 to 18.19) The principal types of reactivity exhibited by aldehydes and ketones may be divided into two classes: (1) nucleophilic addition to the carbonyl group; and (2) electrophilic substitution at the α carbon by way of enol and enolate intermediates. This latter class of reactions will be discussed in the next chapter.

Nucleophilic addition reactions that are typical of aldehydes and ketones include:

Acetal formation, Secs. 18.10, 18.11
Cyanohydrin formation, Sec. 18.12
Reactions with amines, Secs. 18.13, 18.14
Reactions with ammonia derivatives, Sec. 18.15
The Wittig reaction, Secs. 18.16, 18.17

Examples of these reactions are given in the section which follows.

The hydration reaction (Sec. 18.9), while not a synthetically useful process, provides an example through which a general statement may be made about the mechanisms of nucleophilic addition reactions to carbonyls.

In an acidic medium protonation of the carbonyl oxygen is the *first* step of the reaction. The protonated species thus formed is more susceptible to attack by a nucleophile, resulting in *acid catalysis* of the reaction. This increased reactivity may be understood by considering the resonance contributions to the cationic species.

Acid-catalyzed hydration:

$$R_2C{=}O + H_3O^+ \rightleftharpoons R_2C{=}\overset{+}{O}H + H_2O$$

$$R_2C{=}\overset{+}{O}H + H_2O \rightleftharpoons R_2\underset{{}^+OH_2}{\underset{|}{C}}{-}OH \xrightleftharpoons{H_2O} R_2\underset{OH}{\underset{|}{C}}{-}OH + H_3O^+$$

In neutral or basic medium, where protonation of the carbonyl oxygen cannot occur, addition of the nucleophile is the first step of the reaction. This most often occurs when a strong, anionic nucleophile is present.

Base-catalyzed hydration:

$$R_2C{=}O + HO^- \rightleftharpoons R_2\underset{OH}{\underset{|}{C}}{-}O^-$$

$$R_2\underset{OH}{\underset{|}{C}}{-}O^- + H_2O \rightleftharpoons R_2\underset{OH}{\underset{|}{C}}{-}OH + HO^-$$

IMPORTANT REACTIONS

Preparation of Aldehydes (Sec. 18.6)

General:

$$RCO_2H \xrightarrow{SOCl_2} R\overset{O}{\overset{\|}{C}}Cl \xrightarrow[\text{Pd/BaSO}_4,\ \text{heat}]{H_2} R\overset{O}{\overset{\|}{C}}H$$

Example:

$$C_6H_5CH_2CO_2H \xrightarrow[2.\ H_2,\ Pd/BaSO_4,\ heat]{1.\ SOCl_2} C_6H_5CH_2\overset{O}{\overset{\|}{C}}H$$

Preparation of Ketones (Sec. 18.7)

Using dialkyl cuprates:

General:

$$LiCuR_2 + R'\overset{O}{\overset{\|}{C}}Cl \longrightarrow R'\overset{O}{\overset{\|}{C}}R + RCu + LiCl$$

Example:

$$LiCu(CH_2CH_3)_2 + C_6H_5\overset{O}{\overset{\|}{C}}Cl \longrightarrow C_6H_5\overset{O}{\overset{\|}{C}}CH_2CH_3 + CH_3CH_2Cu + LiCl$$

Using organocadmium reagents:

General:

$$2\,RMgX + CdCl_2 \longrightarrow R_2Cd + 2\,MgXCl$$

$$R_2Cd + R'\overset{O}{\overset{\|}{C}}Cl \longrightarrow R'\overset{O}{\overset{\|}{C}}R + RCdCl$$

Example:

$$2\,CH_3MgBr + CdCl_2 \longrightarrow (CH_3)_2Cd + 2\,MgBrCl$$

$$(CH_3)_2Cd + CH_3CH_2\overset{O}{\overset{\|}{C}}Cl \longrightarrow CH_3CH_2\overset{O}{\overset{\|}{C}}CH_3 + CH_3CdCl$$

Reactions of Aldehydes and Ketones: Acetal Formation (Secs. 18.10, 18.11)

General:

$$R\overset{O}{\overset{\|}{C}}R + 2R'OH \overset{H^+}{\rightleftharpoons} R{-}\underset{OR'}{\overset{OR'}{C}}{-}R + H_2O$$

Mechanism:

$$R_2C{=}O \overset{H^+}{\rightleftharpoons} R_2C{=}\overset{+}{O}H \overset{R'OH}{\rightleftharpoons} R_2\underset{H\overset{+}{O}R'}{C}{-}OH \overset{-H^+}{\rightleftharpoons} R_2\underset{OR'}{C}{-}OH$$

$$R_2\underset{OR'}{C}{-}OH \overset{H^+}{\rightleftharpoons} R_2\underset{OR'}{C}{-}\overset{+}{O}H_2 \overset{-H_2O}{\rightleftharpoons} R_2\underset{OR'}{C^+} \overset{R'OH}{\rightleftharpoons} R_2\underset{OR'}{C}{-}\underset{H}{\overset{+}{O}}R' \overset{-H^+}{\rightleftharpoons} R_2\underset{OR'}{C}{-}OR'$$

Examples:

$$C_6H_5\overset{\overset{O}{\|}}{C}H + 2\,C_2H_5OH \xrightleftharpoons{H^+} C_6H_5CH(OC_2H_5)_2 + H_2O$$

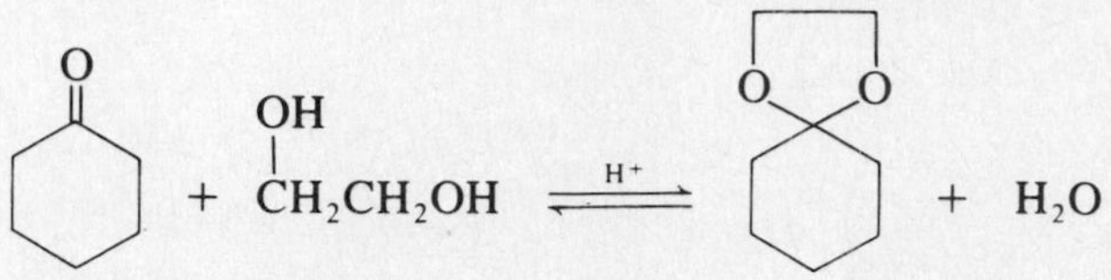

$$\text{cyclohexanone} + \underset{\displaystyle CH_2CH_2OH}{\overset{OH}{|}} \xrightleftharpoons{H^+} \text{(cyclic ketal: 1,4-dioxaspiro[4.5]decane)} + H_2O$$

Cyanohydrin Formation (Sec. 18.12)

General:

$$R\overset{\overset{O}{\|}}{C}R' + HC{\equiv}N \xrightarrow{CN^-} R-\overset{\overset{OH}{|}}{\underset{\underset{C\equiv N}{|}}{C}}-R'$$

Example:

$$C_6H_5\overset{\overset{O}{\|}}{C}H + HCN \xrightarrow{CN^-} C_6H_5\overset{\overset{OH}{|}}{\underset{\underset{CN}{|}}{C}}H$$

Reaction with Primary Amines (Sec. 18.13)

General:

$$R_2C{=}O + R'NH_2 \longrightarrow R_2C{=}NR' + H_2O$$

Mechanism:

Nucleophilic addition stage

$$R_2C{=}O + R'NH_2 \rightleftharpoons R_2\overset{\overset{OH}{|}}{C}-NHR'$$

Elimination stage

$$R_2\overset{\overset{OH}{|}}{C}-NHR' \rightleftharpoons R_2C{=}NR' + H_2O$$

Examples:

$$C_6H_5\overset{\overset{O}{\|}}{C}CH_3 + CH_3CH_2NH_2 \longrightarrow C_6H_5\overset{\overset{N-CH_2CH_3}{\|}}{C}CH_3$$

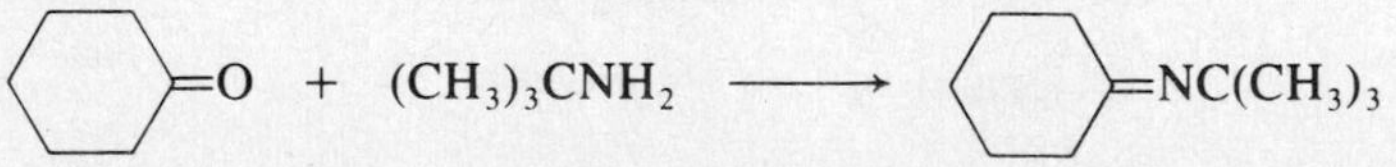

$$\text{cyclohexanone}{=}O + (CH_3)_3CNH_2 \longrightarrow \text{cyclohexylidene}{=}NC(CH_3)_3$$

Reaction with Secondary Amines; Formation of Enamines (Sec. 18.14)

General:

$$RCH_2\overset{\overset{O}{\|}}{C}R' + R''_2NH \longrightarrow RCH{=}\overset{\overset{NR''_2}{|}}{C}R' + H_2O$$

Example:

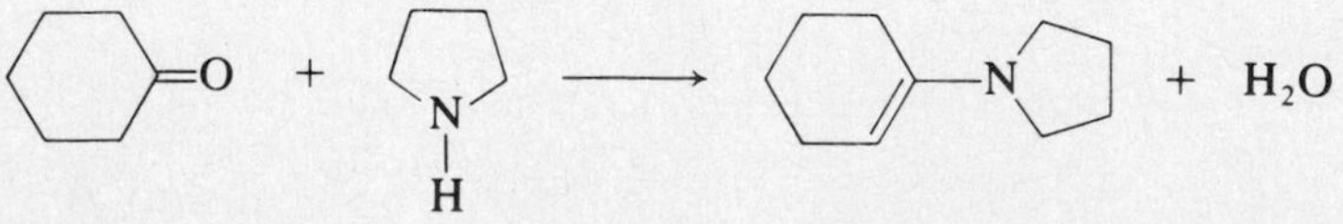

$$\text{cyclohexanone}{=}O + \text{pyrrolidine (N–H)} \longrightarrow \text{1-(1-cyclohexenyl)pyrrolidine} + H_2O$$

Reaction with Ammonia Derivatives (Sec. 18.15)

General:

$$R_2C{=}O + Z{-}NH_2 \longrightarrow R_2C{=}N{-}Z + H_2O$$

Examples:

Oximes (Z=OH)

$$CH_3CH_2\overset{\overset{O}{\|}}{C}H + NH_2OH \longrightarrow CH_3CH_2CH{=}NOH + H_2O$$

Hydrazones (Z = NHR′)

$$\text{cyclohexanone}{=}O + C_6H_5NHNH_2 \longrightarrow \text{cyclohexylidene}{=}NNHC_6H_5 + H_2O$$

The Wittig Reaction (Secs. 18.16, 18.17)

General:

$$R_2C{=}O + (C_6H_5)_3\overset{+}{P}{-}\overset{-}{\ddot{C}}R'_2 \longrightarrow R_2C{=}CR'_2 + (C_6H_5)_3\overset{+}{P}{-}\overset{-}{O}$$

Mechanism:

$$R_2C{=}O + R'_2\overset{-}{\ddot{C}}{-}\overset{+}{P}(C_6H_5)_3 \longrightarrow R{-}\underset{O^-}{\overset{R}{C}}{-}\underset{\overset{+}{P}(C_6H_5)_3}{\overset{R'}{C}}{-}R' \longrightarrow R_2C{=}CR'_2 + (C_6H_5)_3\overset{+}{P}{-}\overset{-}{O}$$

Examples:

$$\text{cyclohexanone} + (C_6H_5)_3\overset{+}{P}{-}\overset{-}{\ddot{C}}HCH_3 \longrightarrow \text{cyclohexylidene}{=}CHCH_3 + (C_6H_5)_3\overset{+}{P}{-}\overset{-}{O}$$

$$C_6H_5\overset{\overset{O}{\|}}{C}H + (C_6H_5)_3\overset{+}{P}{-}\overset{-}{\ddot{C}}\text{(cyclohexylidene)} \longrightarrow C_6H_5CH{=}\text{cyclohexylidene} + (C_6H_5)_3\overset{+}{P}{-}\overset{-}{O}$$

Oxidation of Aldehydes (Sec. 18.19)

General:

$$R\overset{\overset{O}{\|}}{C}H \xrightarrow{[O]} RCO_2H$$

Example:

$$CH_3CH_2CH_2\overset{\overset{O}{\|}}{C}H \xrightarrow[H_2SO_4,\ H_2O]{Na_2Cr_2O_7} CH_3CH_2CH_2CO_2H$$

SOLUTIONS TO TEXT PROBLEMS

18.1 (*b*) The longest continuous chain in glutaraldehyde has five carbons and terminates in aldehyde functions at both ends. Pentanedial is an acceptable IUPAC name for this compound.

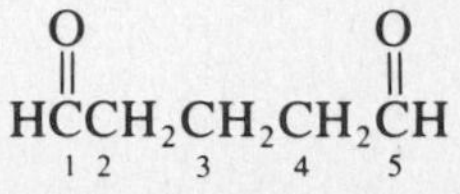

$$\underset{1\ 2}{\overset{\text{O}}{\overset{\|}{\text{HC}}}}\underset{3}{\text{CH}_2}\underset{4}{\text{CH}_2}\text{CH}_2\underset{5}{\overset{\text{O}}{\overset{\|}{\text{CH}}}}$$ Pentanedial (glutaraldehyde)

(*c*) The three-carbon parent chain has a double bond between C-2 and C-3 and a phenyl substituent at C-3.

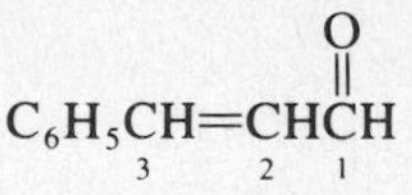

$$\text{C}_6\text{H}_5\underset{3}{\text{CH}}{=}\underset{2}{\text{CH}}\underset{1}{\overset{\text{O}}{\overset{\|}{\text{CH}}}}$$ 3-Phenyl-2-propenal (cinnamaldehyde)

(*d*) Vanillin can be named as a derivative of benzaldehyde. Remember to cite the remaining substituents in alphabetical order.

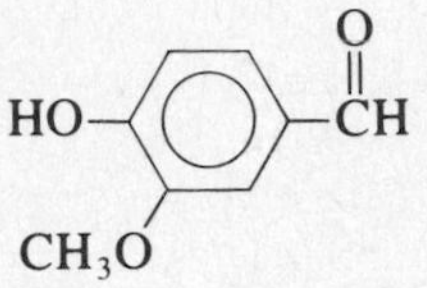

$$\text{HO}-\text{C}_6\text{H}_3(\text{OCH}_3)-\overset{\text{O}}{\overset{\|}{\text{CH}}}$$ 4-Hydroxy-3-methoxybenzaldehyde (vanillin)

18.2 (*b*) The structure that corresponds to benzyl *tert*-butyl ketone is

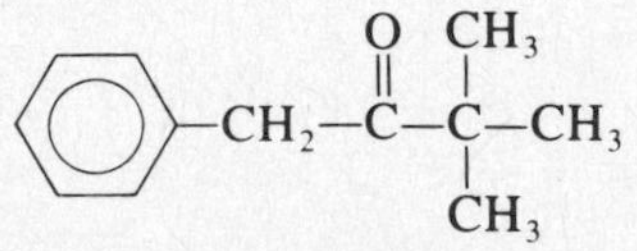

$$\text{C}_6\text{H}_5-\text{CH}_2-\overset{\text{O}}{\overset{\|}{\text{C}}}-\underset{\text{CH}_3}{\underset{|}{\overset{\text{CH}_3}{\overset{|}{\text{C}}}}}-\text{CH}_3$$

The longest carbon chain contains four carbon atoms; the compound is therefore named as a derivative of butane. An acceptable IUPAC name is 3,3-dimethyl-1-phenyl-2-butanone.

(*c*) First write the structure from the name given. Ethyl isopropyl ketone has an ethyl group and an isopropyl group bonded to a carbonyl group.

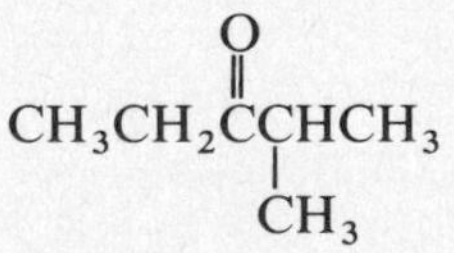

$$\text{CH}_3\text{CH}_2\overset{\text{O}}{\overset{\|}{\text{C}}}\underset{\text{CH}_3}{\underset{|}{\text{CH}}}\text{CH}_3$$

Ethyl isopropyl ketone may be alternatively named 2-methyl-3-pentanone. Its longest continuous chain has five carbons. The carbonyl carbon is C-3 irrespective of the direction in which the chain is numbered, so we choose the one that gives the lower number to the position that bears the methyl group.

(*d*) Methyl neopentyl ketone has a methyl group and a neopentyl group bonded to a carbonyl group.

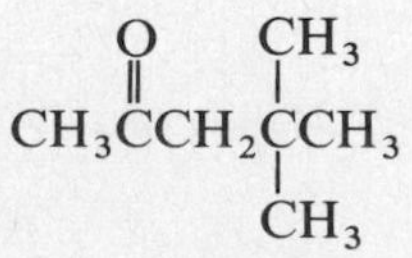

$$\text{CH}_3\overset{\text{O}}{\overset{\|}{\text{C}}}\text{CH}_2\underset{\text{CH}_3}{\underset{|}{\overset{\text{CH}_3}{\overset{|}{\text{C}}}}}\text{CH}_3$$

The longest continuous chain has five carbons and the carbonyl carbon is C-2. Thus, methyl neopentyl ketone may also be named 4,4-dimethyl-2-pentanone.

(*e*) The structure corresponding to allyl methyl ketone is

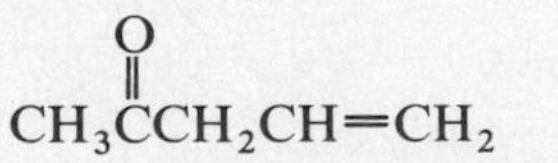

$$\text{CH}_3\overset{\text{O}}{\overset{\|}{\text{C}}}\text{CH}_2\text{CH}{=}\text{CH}_2$$

Since the carbonyl group is given the lowest possible number in the chain, an appropriate name is 4-penten-2-one.

18.3 No. Lithium aluminum hydride is the only reagent we have discussed that is capable of reducing carboxylic acids (Sec. 16.4).

18.4 In addition to lithium aluminum hydride, sodium in alcohol (the Bouveault-Blanc reduction) and hydrogenation over copper chromite are satisfactory methods for reducing esters to primary alcohols (Sec. 16.5).

18.5 The target molecule, 2-butanone, may in theory be prepared two ways, as shown by the following disconnections.

$$CH_3 \vdots \overset{O}{\overset{\|}{C}}CH_2CH_3 \Longrightarrow LiCu(CH_3)_2 + Cl\overset{O}{\overset{\|}{C}}CH_2CH_3$$

or

$$CH_3\overset{O}{\overset{\|}{C}} \vdots CH_2CH_3 \Longrightarrow LiCu(CH_2CH_3)_2 + CH_3\overset{O}{\overset{\|}{C}}Cl$$

The statement of the problem requires use of ethanol as the source of all carbon atoms. Therefore the second of these approaches is the desired one. The reaction is:

$$\underset{\text{Lithium diethylcuprate}}{LiCu(CH_2CH_3)_2} + \underset{\text{Acetyl chloride}}{CH_3\overset{O}{\overset{\|}{C}}Cl} \xrightarrow[-78°C]{\text{diethyl ether}} \underset{\text{2-Butanone}}{CH_3\overset{O}{\overset{\|}{C}}CH_2CH_3} + CH_3CH_2Cu + LiCl$$

The necessary reagents are prepared from ethyl alcohol by standard methods.

$$\underset{\text{Ethyl alcohol}}{CH_3CH_2OH} \xrightarrow[\text{or } PBr_3]{HBr} \underset{\text{Ethyl bromide}}{CH_3CH_2Br} \xrightarrow{Li} \underset{\text{Ethyllithium}}{CH_3CH_2Li} \xrightarrow{CuI} \underset{\text{Lithium diethylcuprate}}{LiCu(CH_2CH_3)_2}$$

$$\underset{\text{Ethyl alcohol}}{CH_3CH_2OH} \xrightarrow[\text{or } K_2Cr_2O_7,\ H_2SO_4]{KMnO_4} \underset{\text{Acetic acid}}{CH_3\overset{O}{\overset{\|}{C}}OH} \xrightarrow{SOCl_2} \underset{\text{Acetyl chloride}}{CH_3\overset{O}{\overset{\|}{C}}Cl}$$

18.6 Acetic acid can be converted to 2-butanone by the sequence:

$$\underset{\text{Acetic acid}}{CH_3\overset{O}{\overset{\|}{C}}OH} \xrightarrow{SOCl_2} \underset{\text{Acetyl chloride}}{CH_3\overset{O}{\overset{\|}{C}}Cl} \xrightarrow{(CH_3CH_2)_2Cd} \underset{\text{2-Butanone}}{CH_3\overset{O}{\overset{\|}{C}}CH_2CH_3}$$

The necessary reagent, diethylcadmium, is prepared from acetic acid according to the following series of steps:

$$\underset{\text{Acetic acid}}{CH_3\overset{O}{\overset{\|}{C}}OH} \xrightarrow[2.\ H_2O]{1.\ LiAlH_4} \underset{\text{Ethanol}}{CH_3CH_2OH} \xrightarrow[\text{or } PBr_3]{HBr} \underset{\text{Ethyl bromide}}{CH_3CH_2Br}$$

$$\downarrow \text{Mg, diethyl ether}$$

$$\underset{\text{Diethylcadmium}}{(CH_3CH_2)_2Cd} \xleftarrow{CdCl_2} \underset{\text{Ethylmagnesium bromide}}{CH_3CH_2MgBr}$$

18.7 (*a*) The hemiacetal intermediate corresponds to addition of methanol to the carbonyl group.

$$\underset{\text{Acetaldehyde}}{CH_3\overset{O}{\overset{\|}{C}}H} + \underset{\text{Methanol}}{CH_3OH} \overset{H^+}{\rightleftharpoons} \underset{\text{Hemiacetal intermediate}}{CH_3\overset{OCH_3}{\overset{|}{C}}HOH}$$

(*b*) This hemiacetal is converted to a carbocation by protonation followed by loss of a water molecule from the protonated form.

$$CH_3\overset{\overset{\displaystyle OCH_3}{|}}{C}HOH \underset{}{\overset{H^+}{\rightleftharpoons}} CH_3\overset{\overset{\displaystyle OCH_3}{|}}{C}H\text{—}\overset{+}{\ddot{O}}H_2 \overset{-H_2O}{\rightleftharpoons} CH_3\overset{+}{C}HOCH_3$$

Hemiacetal intermediate — Carbocation intermediate

The two principal forms of this carbocation are:

$$CH_3\text{—}\overset{+}{C}(H)\text{—}\ddot{O}CH_3 \longleftrightarrow CH_3\text{—}C(H)\text{=}\overset{+}{\ddot{O}}CH_3$$

(*c*) The dimethyl acetal of acetaldehyde is $CH_3CH(OCH_3)_2$.

18.8 (*a*) Ethanol adds to benzaldehyde to form a hemiacetal.

$$C_6H_5\overset{\overset{\displaystyle O}{\|}}{C}H + CH_3CH_2OH \overset{H^+}{\rightleftharpoons} C_6H_5\overset{\overset{\displaystyle OCH_2CH_3}{|}}{C}HOH$$

Benzaldehyde — Ethanol — Hemiacetal intermediate

(*b*) Protonation of the hemiacetal triggers its conversion to a carbocation intermediate.

$$C_6H_5\overset{\overset{\displaystyle OCH_2CH_3}{|}}{C}HOH \overset{H^+}{\rightleftharpoons} C_6H_5\overset{+}{C}HOCH_2CH_3 + H_2O$$

Hemiacetal intermediate — Carbocation intermediate

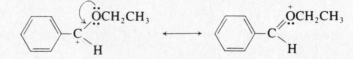

The two principal resonance forms of this carbocation are:

$$C_6H_5\text{—}\overset{+}{C}(H)\text{—}\ddot{O}CH_2CH_3 \longleftrightarrow C_6H_5\text{—}C(H)\text{=}\overset{+}{\ddot{O}}CH_2CH_3$$

Other resonance forms involving the aromatic ring are also reasonable, for example:

$$C_6H_5\text{—}\overset{+}{C}(H)\text{—}\ddot{O}CH_2CH_3 \longleftrightarrow (\overset{+}{C_6H_5})\text{=}C(H)\text{—}\ddot{O}CH_2CH_3$$

(*c*) The diethyl acetal of benzaldehyde is $C_6H_5CH(OCH_2CH_3)_2$.

18.9 (*b*) 1,3-Propanediol forms acetals that contain a six-membered 1,3-dioxane ring.

$$C_6H_5\overset{\overset{\displaystyle O}{\|}}{C}H + HOCH_2CH_2CH_2OH \xrightarrow{H^+} \text{2-phenyl-1,3-dioxane (ring O–C–O, C-2 bears } C_6H_5 \text{ and H)}$$

Benzaldehyde — 1,3-Propanediol — 2-Phenyl-1,3-dioxane

(*c*) The cyclic acetal derived from isobutyl methyl ketone and ethylene glycol bears an isobutyl group and a methyl group at C-2 of a 1,3-dioxolane ring.

$$(CH_3)_2CHCH_2\overset{\overset{\displaystyle O}{\|}}{C}CH_3 + HOCH_2CH_2OH \xrightarrow{H^+} \text{1,3-dioxolane ring, C-2 bears } (CH_3)_2CHCH_2 \text{ and } CH_3$$

Isobutyl methyl ketone — Ethylene glycol — 2-Isobutyl-2-methyl-1,3-dioxolane (*d*) Since the starting diol is 2,2-dimethyl-1,3-propanediol, the cyclic acetal is six-membered and bears two methyl substituents at C-5 in addition to isobutyl and methyl groups at C-2.

$$(CH_3)_2CHCH_2\overset{O}{\overset{\|}{C}}CH_3 + HOCH_2\underset{CH_3}{\overset{CH_3}{C}}CH_2OH \xrightarrow{H^+} \text{2-Isobutyl-2,5,5-trimethyl-1,3-dioxane}$$

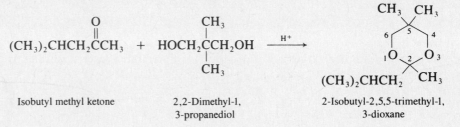

Isobutyl methyl ketone; 2,2-Dimethyl-1,3-propanediol; 2-Isobutyl-2,5,5-trimethyl-1,3-dioxane

18.10 (*b*) Cyanohydrins of ketones have a more favorable equilibrium constant for dissociation than do cyanohydrins of aldehydes. Crowding is relieved to a greater extent when a ketone cyanohydrin dissociates, and a more stable carbonyl group is formed. The measured dissociation constants are:

$$C_6H_5\overset{OH}{\overset{|}{C}}HCN \rightleftharpoons C_6H_5\overset{O}{\overset{\|}{C}}H + HCN \qquad K = 4.7 \times 10^{-3}$$

Benzaldehyde cyanohydrin; Benzaldehyde

$$C_6H_5\underset{CH_3}{\overset{OH}{C}}CN \rightleftharpoons C_6H_5\overset{O}{\overset{\|}{C}}CH_3 + HCN \qquad K = 1.3$$

Acetophenone cyanohydrin; Acetophenone

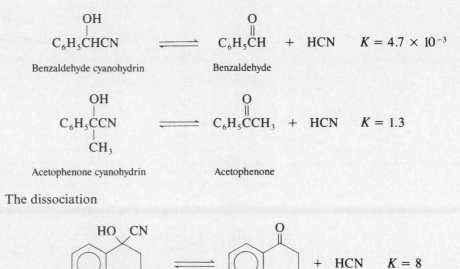

(*c*) The dissociation

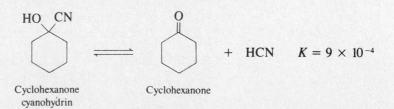

$$\alpha\text{-Tetralone cyanohydrin} \rightleftharpoons \alpha\text{-Tetralone} + HCN \qquad K = 8$$

has a more favorable dissociation constant than

$$\text{Cyclohexanone cyanohydrin} \rightleftharpoons \text{Cyclohexanone} + HCN \qquad K = 9 \times 10^{-4}$$

because conjugation of the carbonyl group with the aromatic ring stabilizes α-tetralone.

18.11 (*b*) Nucleophilic addition of butylamine to benzaldehyde gives the carbinolamine.

$$C_6H_5\overset{O}{\overset{\|}{C}}H + CH_3CH_2CH_2CH_2NH_2 \longrightarrow C_6H_5\overset{OH}{\overset{|}{C}}H{-}\underset{H}{N}CH_2CH_2CH_2CH_3$$

Benzaldehyde; Butylamine; Carbinolamine intermediate

Dehydration of the carbinolamine produces the imine.

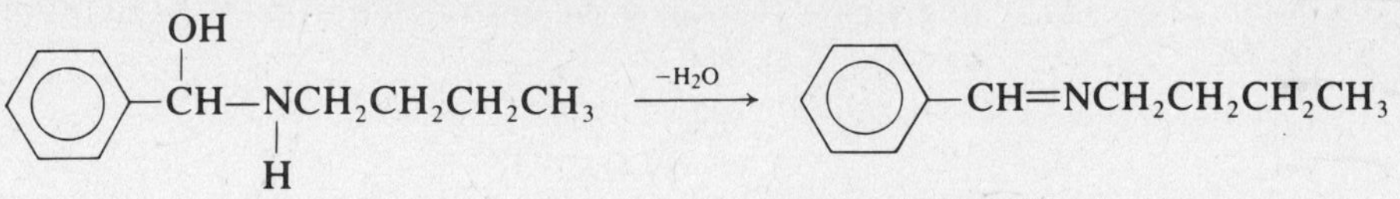

$$C_6H_5CH(OH)NHCH_2CH_2CH_2CH_3 \xrightarrow{-H_2O} C_6H_5CH{=}NCH_2CH_2CH_2CH_3$$

N-Benzylidenebutylamine

(c) Cyclohexanone and *tert*-butylamine react according to the equation

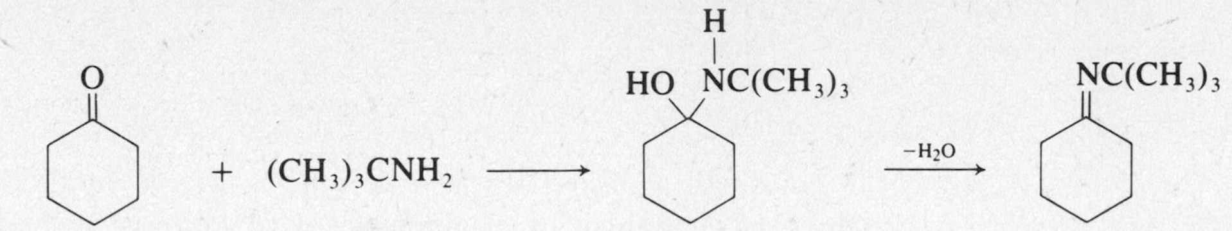

$$\text{Cyclohexanone} + (CH_3)_3CNH_2 \longrightarrow \text{1-hydroxy-1-}[NHC(CH_3)_3]\text{cyclohexane} \xrightarrow{-H_2O} \text{cyclohexylidene}{=}NC(CH_3)_3$$

Cyclohexanone — *tert*-Butylamine — Carbinolamine intermediate — N-Cyclohexylidene-*tert*-butylamine

(d)

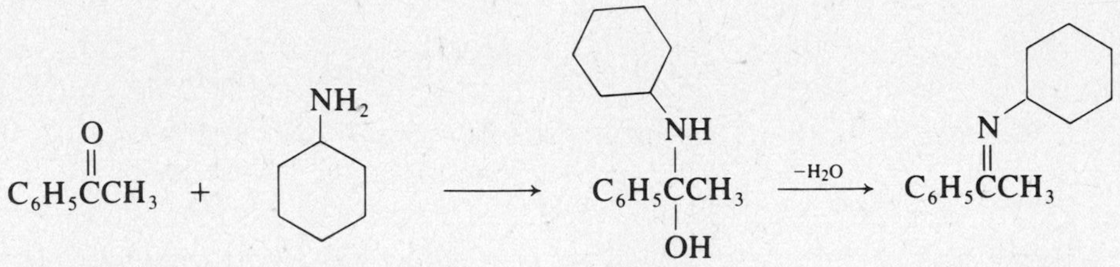

$$C_6H_5CCH_3(=O) + \text{cyclohexyl}NH_2 \longrightarrow C_6H_5C(OH)(NH\text{-cyclohexyl})CH_3 \xrightarrow{-H_2O} C_6H_5C(=N\text{-cyclohexyl})CH_3$$

Acetophenone — Cyclohexylamine — Carbinolamine intermediate — N-(1-Phenylethylidene)-cyclohexylamine

18.12 (b) Pyrrolidine, a secondary amine, adds to 3-pentanone to give a carbinolamine.

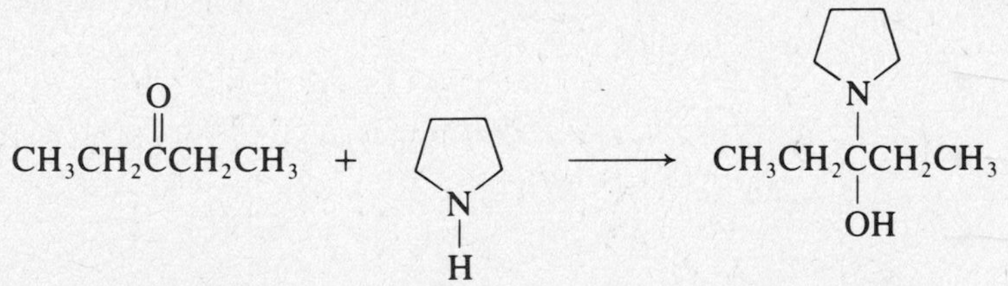

$$CH_3CH_2CCH_2CH_3(=O) + \text{pyrrolidine (N–H)} \longrightarrow CH_3CH_2C(OH)(N\text{-pyrrolidinyl})CH_2CH_3$$

3-Pentanone — Pyrrolidine — Carbinolamine intermediate

Dehydration produces the enamine.

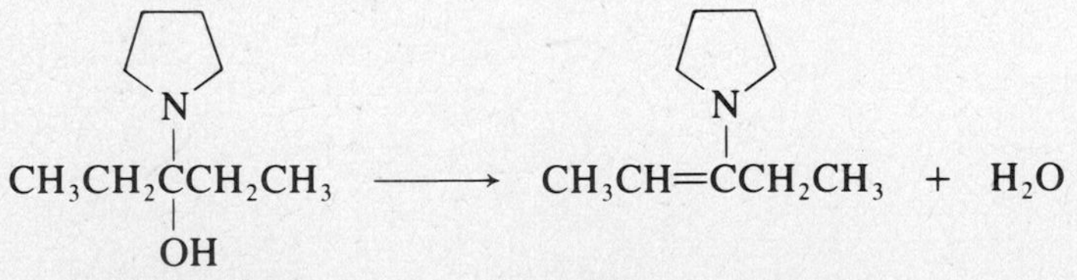

$$CH_3CH_2C(OH)(N\text{-pyrrolidinyl})CH_2CH_3 \longrightarrow CH_3CH{=}C(N\text{-pyrrolidinyl})CH_2CH_3 + H_2O$$

Carbinolamine intermediate — 3-Pyrrolidino-2-pentene

(c)

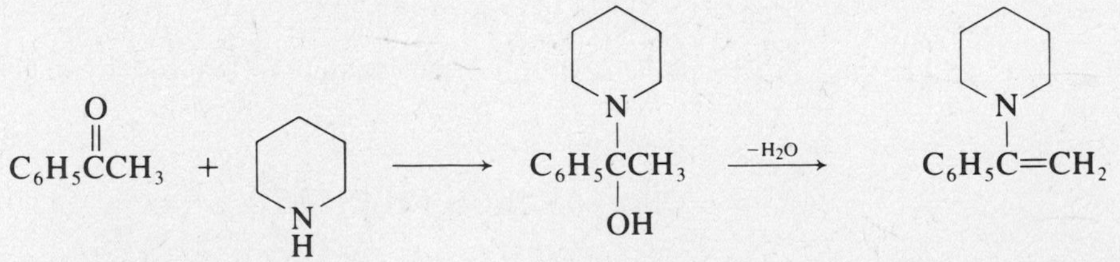

$$C_6H_5CCH_3(=O) + \text{piperidine (N–H)} \longrightarrow C_6H_5C(OH)(N\text{-piperidinyl})CH_3 \xrightarrow{-H_2O} C_6H_5C(N\text{-piperidinyl}){=}CH_2$$

Acetophenone — Piperidine — Carbinolamine intermediate — 1-Piperidino-1-phenylethene

18.13 (*b*) The product is a mixture of the *E* and *Z* isomers of 1-phenylpropene.

$$C_6H_5CH{=}O + (C_6H_5)_3\overset{+}{P}\!-\!\overset{-}{\ddot{C}}HCH_3 \longrightarrow$$

Benzaldehyde; Ethylidenetriphenylphosphorane

$$C_6H_5CH{=}CHCH_3 + (C_6H_5)_3\overset{+}{P}\!-\!O^-$$

1-Phenylpropene (98%, a mixture of *E* and *Z* stereoisomers); Triphenylphosphine oxide

The major alkene is the *Z* (cis) stereoisomer. The reasons for this stereoselectivity are not fully understood.

(*c*) Formaldehyde is often used as the carbonyl component in Wittig reactions.

$$\overset{O}{\overset{\|}{HCH}} + (C_6H_5)_3\overset{+}{P}\!-\!\overset{-}{\ddot{C}}HC_6H_5 \longrightarrow CH_2{=}CHC_6H_5 + (C_6H_5)_3\overset{+}{P}\!-\!O^-$$

Formaldehyde; Benzylidenetriphenylphosphorane; Styrene (75%); Triphenylphosphine oxide

(*d*) Here we see an example of the Wittig reaction applied to diene synthesis by use of an ylide containing a carbon-carbon double bond.

$$CH_3CH_2CH_2\overset{O}{\overset{\|}{C}}H + (C_6H_5)_3\overset{+}{P}\!-\!\overset{-}{\ddot{C}}HCH{=}CH_2 \longrightarrow$$

Butanal; Allylidenetriphenylphosphorane

$$CH_3CH_2CH_2CH{=}CHCH{=}CH_2 + (C_6H_5)_3\overset{+}{P}\!-\!O^-$$

1,3-Heptadiene (52%); Triphenylphosphine oxide

(*e*) Methylene transfer from methylenetriphenylphosphorane is one of the most commonly used Wittig reactions.

$$C_6H_{11}\overset{O}{\overset{\|}{C}}CH_3 + (C_6H_5)_3\overset{+}{P}\!-\!\overset{-}{\ddot{C}}H_2 \longrightarrow$$

Cyclohexyl methyl ketone; Methylenetriphenylphosphorane

$$C_6H_{11}\overset{CH_2}{\overset{\|}{C}}CH_3 + (C_6H_5)_3\overset{+}{P}\!-\!O^-$$

2-Cyclohexylpropene (66%); Triphenylphosphine oxide

18.14 (*b*) There are two Wittig reaction routes that lead to 1-pentene. One is represented retrosynthetically by the disconnection

$$CH_3CH_2CH_2CH{=}CH_2 \Longrightarrow CH_3CH_2CH_2\overset{O}{\overset{\|}{C}}H + (C_6H_5)_3\overset{+}{P}\!-\!\overset{-}{\ddot{C}}H_2$$

1-Pentene; Butanal; Methylenetriphenylphosphorane

The other route is

$$CH_3CH_2CH_2CH{=}CH_2 \Longrightarrow CH_3CH_2CH_2\overset{-}{\ddot{C}}H\!-\!\overset{+}{P}(C_6H_5)_3 + \overset{O}{\overset{\|}{HCH}}$$

1-Pentene; Butylidenetriphenylphosphorane; Formaldehyde

(*c*) The two retrosyntheses are:

$$C_6H_5CH_2CH{=}C(CH_2CH_3)_2 \Longrightarrow C_6H_5CH_2\overset{\overset{\large O}{\|}}{C}H + (C_6H_5)_3\overset{+}{P}{-}\overset{..}{\bar{C}}(CH_2CH_3)_2$$

3-Ethyl-1-phenyl-2-pentene — 2-Phenylethanal — (1-Ethylpropylidene)triphenylphosphorane

and

$$C_6H_5CH_2CH{=}C(CH_2CH_3)_2 \Longrightarrow$$

3-Ethyl-1-phenyl-2-pentene

$$C_6H_5CH_2\overset{..}{\bar{C}}H{-}\overset{+}{P}(C_6H_5)_3 + CH_3CH_2\overset{\overset{\large O}{\|}}{C}CH_2CH_3$$

2-Phenylethylidenetriphenylphosphorane — 3-Pentanone

(*d*) Cyclopentanone is the carbonyl component in one of the Wittig routes.

Isopropylidenecyclopentane $\Longrightarrow$ Cyclopentanone ($=$O) $+$ $(C_6H_5)_3\overset{+}{P}{-}\overset{..}{\bar{C}}(CH_3)_2$

Isopropylidenecyclopentane — Cyclopentanone — Isopropylidenetriphenylphosphorane

Acetone is the carbonyl component in the other.

Isopropylidenecyclopentane $\Longrightarrow$ cyclopentylidene$\overset{..}{\bar{C}}{-}\overset{+}{P}(C_6H_5)_3$ $+$ $CH_3\overset{\overset{\large O}{\|}}{C}CH_3$

Isopropylidenecyclopentane — Cyclopentylidenetriphenylphosphorane — Acetone

18.15 (*a*) First consider all the isomeric aldehydes of molecular formula $C_5H_{10}O$.

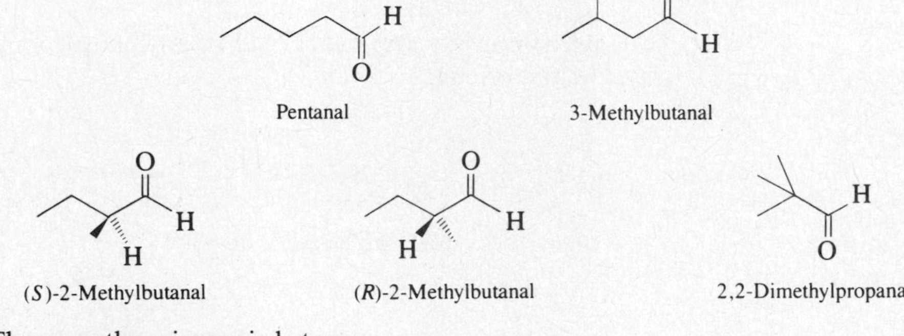

Pentanal — 3-Methylbutanal

(*S*)-2-Methylbutanal — (*R*)-2-Methylbutanal — 2,2-Dimethylpropanal

There are three isomeric ketones.

2-Pentanone — 3-Pentanone — 3-Methyl-2-butanone

(*b*) Reduction of an aldehyde to a primary alcohol does not introduce a chiral center into the molecule. Therefore the only aldehydes that yield chiral alcohols on reduction are those that already contain a chiral center.

(*S*)-2-Methylbutanal $\xrightarrow{NaBH_4}$ (*S*)-2-Methyl-1-butanol

(*S*)-2-Methylbutanal — (*S*)-2-Methyl-1-butanol

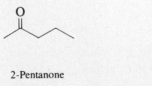

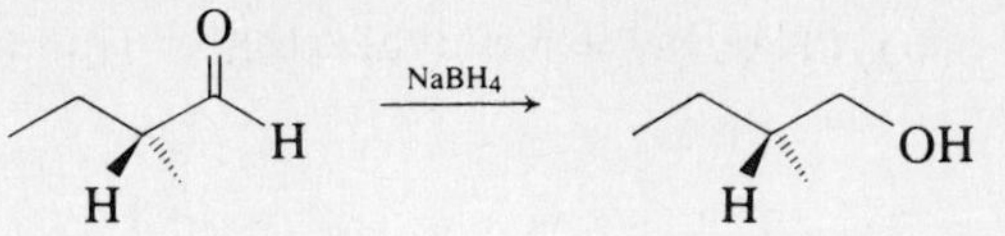

NaBH$_4$

(R)-2-Methylbutanal → (R)-2-Methyl-1-butanol

Among the ketones, 2-pentanone and 3-methyl-2-butanone are reduced to chiral alcohols.

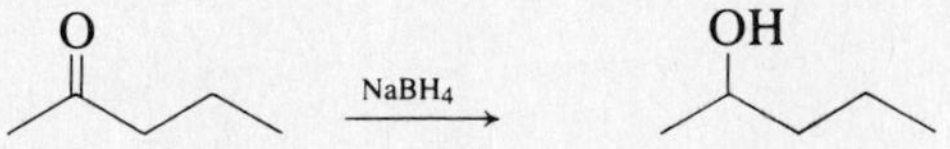

2-Pentanone →(NaBH$_4$) 2-Pentanol (chiral but racemic)

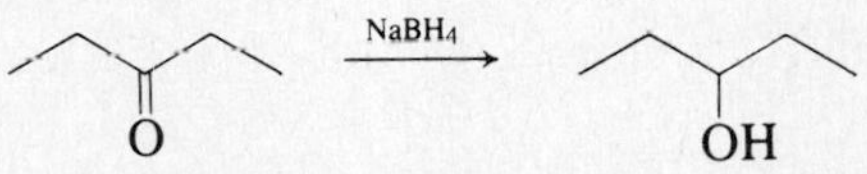

3-Pentanone →(NaBH$_4$) 3-Pentanol (achiral)

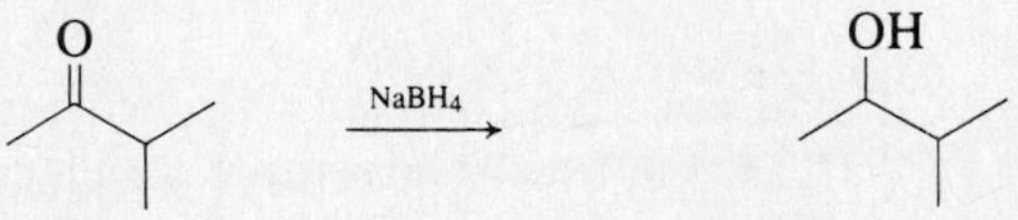

3-Methyl-2-butanone →(NaBH$_4$) 3-Methyl-2-butanol (chiral but racemic)

(c) All the aldehydes yield chiral alcohols on reaction with methylmagnesium iodide. Thus

$$C_4H_9\overset{O}{\overset{\|}{C}}H \xrightarrow[2.\ H_3O^+]{1.\ CH_3MgI} C_4H_9\underset{OH}{\overset{H}{C}}CH_3$$

A chiral center is introduced in each case. None of the ketones yields chiral alcohols

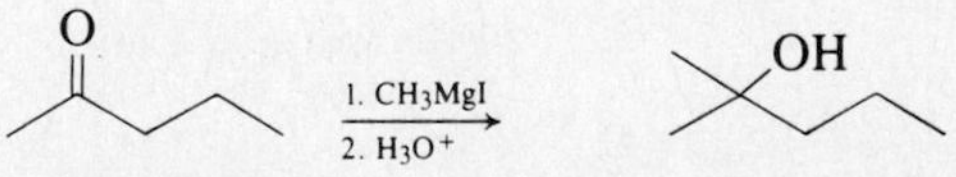

2-Pentanone →(1. CH$_3$MgI; 2. H$_3$O$^+$) 2-Methyl-2-pentanol (achiral)

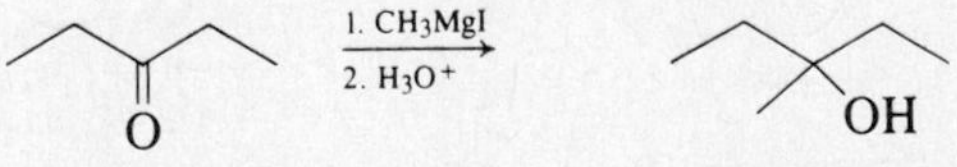

3-Pentanone →(1. CH$_3$MgI; 2. H$_3$O$^+$) 3-Methyl-3-pentanol (achiral)

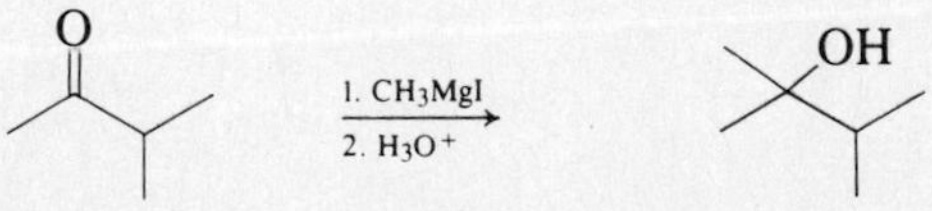

3-Methyl-2-butanone →(1. CH$_3$MgI; 2. H$_3$O$^+$) 2,3-Dimethyl-2-butanol (achiral)

18.16 (a) Chloral is the trichloro derivative of acetaldehyde.

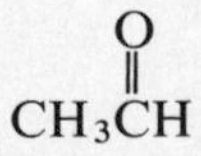

$$CH_3\overset{O}{\overset{\|}{C}}H \qquad Cl-\underset{Cl}{\overset{Cl}{C}}-\overset{O}{\overset{\|}{C}}H$$

Acetaldehyde 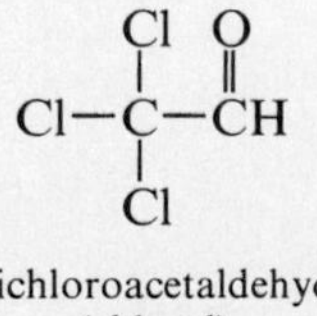 — Trichloroacetaldehyde (chloral)

(*b*) Pivaldehyde has two methyl groups attached to C-2 of propanal.

$$CH_3CH_2\overset{O}{\overset{\|}{C}}H \qquad CH_3\underset{CH_3}{\overset{CH_3}{C}}-\overset{O}{\overset{\|}{C}}H$$

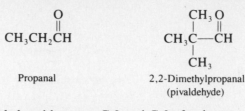

Propanal — 2,2-Dimethylpropanal (pivaldehyde)

(*c*) Acrolein has a double bond between C-2 and C-3 of a three-carbon aldehyde.

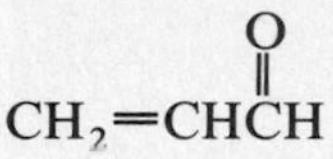

$$CH_2{=}CH\overset{O}{\overset{\|}{C}}H$$ 2-Propenal (acrolein)

(*d*) Crotonaldehyde has a trans double bond between C-2 and C-3 of a four-carbon aldehyde.

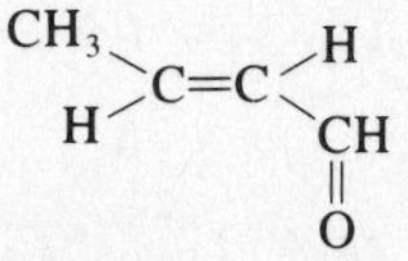

(*E*)-2-Butenal (crotonaldehyde)

(*e*) Citral has two double bonds, one between C-2 and C-3 and the other between C-6 and C-7. The one at C-2 has the *E* configuration. There are methyl substituents at C-3 and C-7.

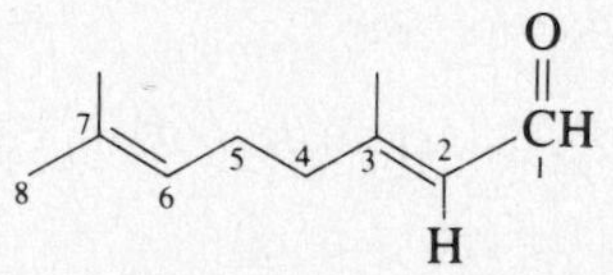

(*E*)-3,7-Dimethyl-2,6-octadienal (citral)

(*f*) The structure of pinacolone is more easily revealed in its systematic name.

$$\equiv \quad CH_3\overset{O}{\overset{\|}{C}}C(CH_3)_3$$

3,3-Dimethyl-2-butanone (pinacolone)

(*g*) Deoxybenzoin (benzyl phenyl ketone) has a benzyl group and a phenyl group attached to a carbonyl group.

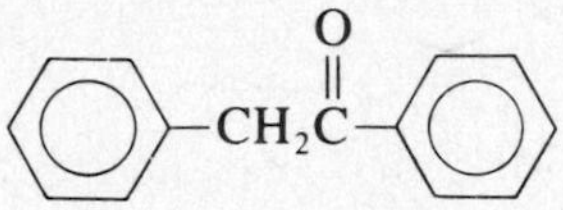

$$C_6H_5CH_2\overset{O}{\overset{\|}{C}}C_6H_5$$ Benzyl phenyl ketone

(*h*) Diacetone alcohol is

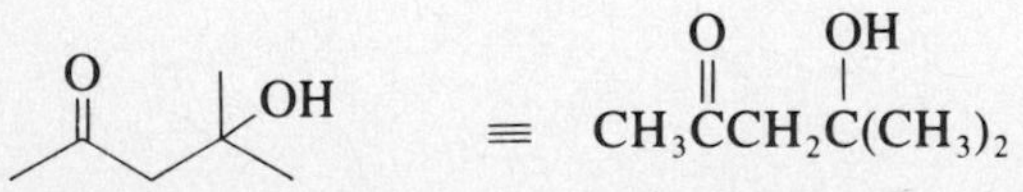

$$\equiv \quad CH_3\overset{O}{\overset{\|}{C}}CH_2\overset{OH}{\overset{|}{C}}(CH_3)_2$$

4-Hydroxy-4-methyl-2-pentanone

(*i*) The systematic name for mesityl oxide tells us that the correct structure is as shown.

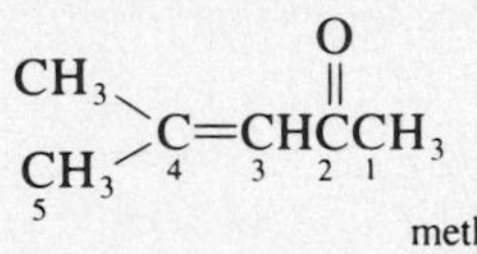

$$(CH_3)_2C{=}CH\overset{O}{\overset{\|}{C}}CH_3$$ 4-Methyl-3-penten-2-one

(carbons numbered 5, 4, 3, 2, 1)

4-Methyl: methyl substituent at C-4; -3-penten-: five carbons in main chain with double bond at C-3; -2-one: C-2 is a ketone carbonyl

(*j*) The parent ketone is 2-cyclohexenone.

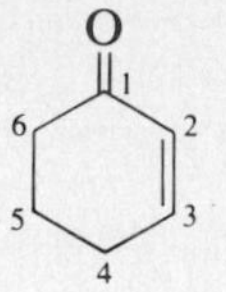

2-Cyclohexenone

Carvone has an isopropenyl group at C-5 and a methyl group at C-2.

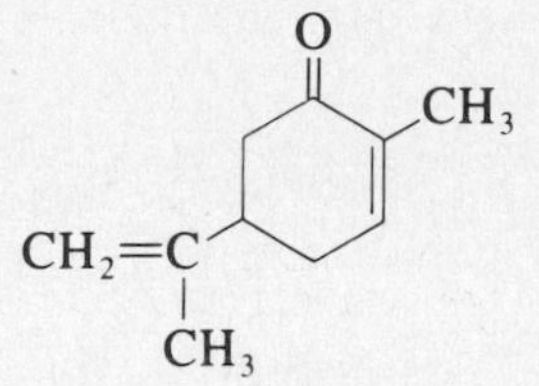

5-Isopropenyl-2-methyl-2-cyclohexenone (carvone)

(*k*) Biacetyl is 2,3-butanedione. It has a four-carbon chain that incorporates ketone carbonyls at C-2 and C-3.

$$CH_3\overset{O}{\overset{\|}{C}}\overset{O}{\overset{\|}{C}}CH_3$$ 2,3-Butanedione (biacetyl)

(*l*) Dimedone has two carbonyl groups in a 1,3 relationship on a six-membered ring. There are two methyl substituents at C-5.

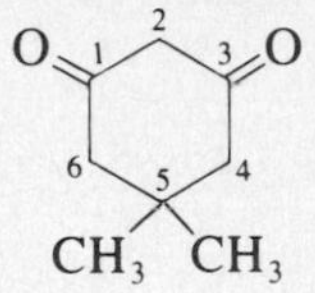

5,5-Dimethyl-1,3-cyclohexanedione (dimedone)

(*m*) Dypnone is

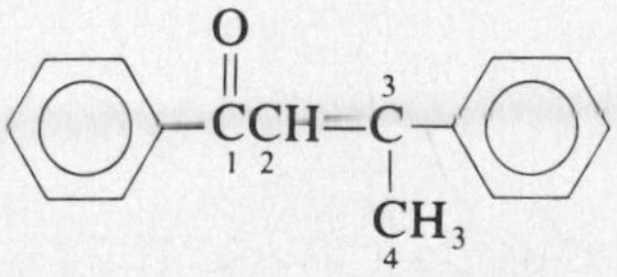

$$C_6H_5\overset{O}{\overset{\|}{C}}CH{=}\underset{CH_3}{\underset{|}{C}}C_6H_5$$ 1,3-Diphenyl-2-buten-1-one (dypnone)

18.17 (*a*) Lithium aluminum hydride reduces aldehydes to primary alcohols.

$$CH_3CH_2\overset{O}{\overset{\|}{C}}H \xrightarrow[2.\ H_2O]{1.\ LiAlH_4} CH_3CH_2CH_2OH$$

Propanal → 1-Propanol

(*b*) Sodium borohydride reduces aldehydes to primary alcohols.

$$CH_3CH_2\overset{O}{\overset{\|}{C}}H \xrightarrow[CH_3OH]{NaBH_4} CH_3CH_2CH_2OH$$

Propanal → 1-Propanol

(*c*) Aldehydes can be reduced to primary alcohols by catalytic hydrogenation.

$$CH_3CH_2\overset{O}{\overset{\|}{C}}H \xrightarrow[Ni]{H_2} CH_3CH_2CH_2OH$$

Propanal → 1-Propanol

(*d*) Aldehydes react with Grignard reagents to form secondary alcohols.

$$CH_3CH_2\overset{O}{\overset{\|}{C}}H \xrightarrow[2.\ H_3O^+]{1.\ CH_3MgI,\ \text{diethyl ether}} CH_3CH_2\overset{OH}{\overset{|}{C}}HCH_3$$

Propanal → 2-Butanol

(*e*) Sodium acetylide adds to the carbonyl group of propanal to give an acetylenic alcohol.

$$\underset{\text{Propanal}}{CH_3CH_2\overset{\overset{O}{\|}}{C}H} \xrightarrow[\text{2. } H_3O^+]{\text{1. } HC\equiv CNa,\ \text{liquid ammonia}} \underset{\text{1-Pentyn-3-ol}}{CH_3CH_2\overset{\overset{OH}{|}}{C}HC\equiv CH}$$

(*f*) Alkyl- or aryllithium reagents react with aldehydes in much the same way that Grignard reagents do.

$$\underset{\text{Propanal}}{CH_3CH_2\overset{\overset{O}{\|}}{C}H} \xrightarrow[\text{2. } H_3O^+]{\text{1. } C_6H_5Li,\ \text{diethyl ether}} \underset{\text{1-Phenyl-1-propanol}}{CH_3CH_2\underset{\underset{OH}{|}}{C}HC_6H_5}$$

(*g*) Aldehydes are converted to acetals on reaction with alcohols in the presence of an acid catalyst.

$$\underset{\text{Propanal}}{CH_3CH_2\overset{\overset{O}{\|}}{C}H} + \underset{\text{Methanol}}{2CH_3OH} \xrightarrow{\text{HCl}} \underset{\text{Propanal dimethyl acetal}}{CH_3CH_2CH(OCH_3)_2}$$

(*h*) Cyclic acetal formation occurs when aldehydes react with ethylene glycol.

$$\underset{\text{Propanal}}{CH_3CH_2\overset{\overset{O}{\|}}{C}H} + \underset{\text{Ethylene glycol}}{HOCH_2CH_2OH} \xrightarrow[\text{benzene}]{p\text{-toluenesulfonic acid}} \underset{\text{2-Ethyl-1,3-dioxolane}}{\text{(cyclic acetal: O–CH}_2\text{–CH}_2\text{–O ring, C bearing H and } CH_2CH_3)}$$

(*i*) Aldehydes react with primary amines to yield imines.

$$\underset{\text{Propanal}}{CH_3CH_2\overset{\overset{O}{\|}}{C}H} + \underset{\text{Aniline}}{C_6H_5NH_2} \xrightarrow{-H_2O} \underset{N\text{-Propylideneaniline}}{CH_3CH_2CH{=}NC_6H_5}$$

(*j*) Secondary amines combine with aldehydes to yield enamines.

$$\underset{\text{Propanal}}{CH_3CH_2\overset{\overset{O}{\|}}{C}H} + \underset{\text{Dimethylamine}}{(CH_3)_2NH} \longrightarrow \underset{\text{1-(Dimethylamino)propene}}{CH_3CH{=}\overset{\overset{N(CH_3)_2}{|}}{C}H}$$

(*k*) Oximes are formed on reaction of hydroxylamine with aldehydes.

$$\underset{\text{Propanal}}{CH_3CH_2\overset{\overset{O}{\|}}{C}H} \xrightarrow{H_2NOH} \underset{\text{Propanal oxime}}{CH_3CH_2CH{=}NOH}$$

(*l*) Hydrazine reacts with aldehydes to form hydrazones.

$$\underset{\text{Propanal}}{CH_3CH_2\overset{\overset{O}{\|}}{C}H} \xrightarrow{H_2NNH_2} \underset{\text{Propanal hydrazone}}{CH_3CH_2CH{=}NNH_2}$$

(*m*) Hydrazone formation is the first step in the Wolff-Kishner reduction (Sec. 13.8).

$$\underset{\text{Propanal hydrazone}}{CH_3CH_2CH{=}NNH_2} \xrightarrow[\text{triethylene glycol, heat}]{\text{NaOH}} \underset{\text{Propane}}{CH_3CH_2CH_3} + N_2$$

(*n*) Reaction of an aldehyde with *p*-nitrophenylhydrazine is analogous to that with hydrazine.

$$\underset{\text{Propanal}}{CH_3CH_2\overset{\overset{\displaystyle O}{\|}}{C}H} + \underset{p\text{-Nitrophenylhydrazine}}{O_2N-C_6H_4-NHNH_2} \longrightarrow \underset{\substack{\text{Propanal}\\ p\text{-nitrophenylhydrazone}}}{CH_3CH_2CH{=}NNH-C_6H_4-NO_2} + H_2O$$

(*o*) Semicarbazide converts aldehydes to the corresponding semicarbazone.

$$\underset{\text{Propanal}}{CH_3CH_2\overset{\overset{\displaystyle O}{\|}}{C}H} + \underset{\text{Semicarbazide}}{H_2NNH\overset{\overset{\displaystyle O}{\|}}{C}NH_2} \longrightarrow \underset{\text{Propanal semicarbazone}}{CH_3CH_2CH{=}NNH\overset{\overset{\displaystyle O}{\|}}{C}NH_2} + H_2O$$

(*p*) Phosphorus ylides convert aldehydes to alkenes by a Wittig reaction.

$$\underset{\text{Propanal}}{CH_3CH_2\overset{\overset{\displaystyle O}{\|}}{C}H} + \underset{\text{Ethylidenetriphenylphosphorane}}{(C_6H_5)_3\overset{+}{P}-\overset{-}{\ddot{C}}HCH_3} \longrightarrow \underset{\text{2-Pentene}}{CH_3CH_2CH{=}CHCH_3} + \underset{\text{Triphenylphosphine oxide}}{(C_6H_5)_3\overset{+}{P}-\overset{-}{O}}$$

(*q*) Acidification of solutions of sodium cyanide generates HCN, which reacts with aldehydes to form cyanohydrins.

$$\underset{\text{Propanal}}{CH_3CH_2\overset{\overset{\displaystyle O}{\|}}{C}H} + \underset{\text{Hydrogen cyanide}}{HCN} \longrightarrow \underset{\text{1-Cyano-1-propanol}}{CH_3CH_2\overset{\overset{\displaystyle OH}{|}}{C}HCN}$$

(*r*) Silver oxide oxidizes aldehydes to carboxylic acids with formation of metallic silver.

$$\underset{\text{Propanal}}{CH_3CH_2\overset{\overset{\displaystyle O}{\|}}{C}H} \xrightarrow{Ag_2O} \underset{\text{Propanoic acid}}{CH_3CH_2CO_2H}$$

(*s*) Chromic acid oxidizes aldehydes to carboxylic acids.

$$\underset{\text{Propanal}}{CH_3CH_2\overset{\overset{\displaystyle O}{\|}}{C}H} \xrightarrow{H_2CrO_4} \underset{\text{Propanoic acid}}{CH_3CH_2CO_2H}$$

(*t*) Potassium permanganate is quite a strong oxidizing agent. It oxidizes aldehydes to carboxylic acids.

$$\underset{\text{Propanal}}{CH_3CH_2\overset{\overset{\displaystyle O}{\|}}{C}H} \xrightarrow[2.\ H^+]{1.\ KMnO_4} \underset{\text{Propanoic acid}}{CH_3CH_2CO_2H}$$

18.18 (*a*) Lithium aluminum hydride reduces ketones to secondary alcohols.

$$\text{Cyclopentanone} \xrightarrow[\text{2. } H_2O]{\text{1. } LiAlH_4} \text{Cyclopentanol}$$

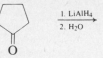

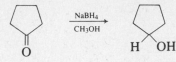

Cyclopentanone — Cyclopentanol (H, OH)

(*b*) Sodium borohydride converts ketones to secondary alcohols.

$$\text{Cyclopentanone} \xrightarrow[CH_3OH]{NaBH_4} \text{Cyclopentanol}$$

(*c*) Catalytic hydrogenation of ketones yields secondary alcohols.

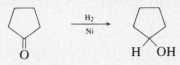

$$\text{Cyclopentanone} \xrightarrow[Ni]{H_2} \text{Cyclopentanol}$$

(*d*) Grignard reagents react with ketones to form tertiary alcohols.

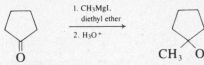

$$\text{Cyclopentanone} \xrightarrow[\text{2. } H_3O^+]{\text{1. } CH_3MgI,\ \text{diethyl ether}} \text{1-Methylcyclopentanol } (CH_3,\ OH)$$

(*e*) Addition of sodium acetylide to cyclopentanone yields a tertiary acetylenic alcohol.

$$\text{Cyclopentanone} \xrightarrow[\text{2. } H_3O^+]{\text{1. } HC{\equiv}CNa,\ \text{liquid ammonia}} \text{1-Ethynylcylopentanol } (HC{\equiv}C,\ OH)$$

(*f*) Phenyllithium adds to the carbonyl group of cyclopentanone to yield 1-phenylcyclopentanol.

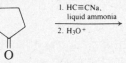

$$\text{Cyclopentanone} \xrightarrow[\text{2. } H_3O^+]{\text{1. } C_6H_5Li,\ \text{diethyl ether}} \text{1-Phenylcyclopentanol } (C_6H_5,\ OH)$$

(*g*) The equilibrium constant for acetal formation from ketones is generally unfavorable.

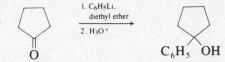

$$\text{Cyclopentanone} + 2CH_3OH \text{ (Methanol)} \xrightarrow{HCl} \text{Cyclopentanone dimethyl acetal } (CH_3O,\ OCH_3) \qquad K < 1$$

(*h*) Cyclic acetal formation is favored even for ketones.

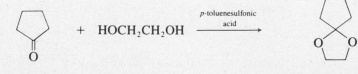

$$\text{Cyclopentanone} + HOCH_2CH_2OH \text{ (Ethylene glycol)} \xrightarrow{p\text{-toluenesulfonic acid}} \text{1,4-Dioxaspiro[4.4]nonane}$$

(*i*) Ketones react with primary amines to form imines.

Cyclopentanone + $C_6H_5NH_2$ ⟶ (cyclopentylidene =NC_6H_5)

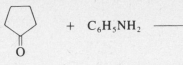

Cyclopentanone Aniline *N*-Cyclopentylideneaniline

(*j*) Dimethylamine reacts with cyclopentanone to yield an enamine.

Cyclopentanone + $(CH_3)_2NH$ ⟶ (cyclopentene–$N(CH_3)_2$)

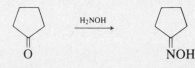

Cyclopentanone Dimethylamine 1-(Dimethylamino)cyclopentene

(*k*) An oxime is formed when cyclopentanone is treated with hydroxylamine.

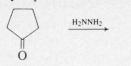

Cyclopentanone $\xrightarrow{H_2NOH}$ (=NOH)

Cyclopentanone Cyclopentanone oxime

(*l*) Hydrazine reacts with cyclopentanone to form a hydrazone.

Cyclopentanone $\xrightarrow{H_2NNH_2}$ (=NNH_2)

Cyclopentanone Cyclopentanone hydrazone

(*m*) Heating a hydrazone in base with a high-boiling alcohol as solvent converts it to an alkane.

(=NNH_2) $\xrightarrow[\text{triethylene glycol, heat}]{NaOH}$ Cyclopentane + N_2

Cyclopentanone hydrazone Cyclopentane

(*n*) A *p*-nitrophenylhydrazone is formed.

Cyclopentanone + $O_2N-C_6H_4-NHNH_2$ ⟶ (=$NNH-C_6H_4-NO_2$)

Cyclopentanone *p*-Nitrophenylhydrazine Cyclopentanone *p*-nitrophenylhydrazone

(*o*) Cyclopentanone is converted to a semicarbazone on reaction with semicarbazide.

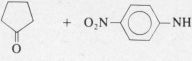

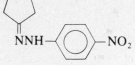

Cyclopentanone + $H_2NNHCNH_2$ (C=O) ⟶ (=$NNHCNH_2$, C=O) + H_2O

Cyclopentanone Semicarbazide Cyclopentanone semicarbazone

(*p*) A Wittig reaction takes place, forming ethylidenecyclopentane.

Cyclopentanone + $(C_6H_5)_3\overset{+}{P}-\overset{-}{\ddot{C}}HCH_3$ ⟶ (=$CHCH_3$) + $(C_6H_5)_3\overset{+}{P}-\overset{-}{O}$

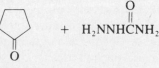

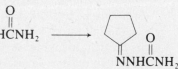

Cyclopentanone Ethylidenetriphenylphosphorane Ethylidenecyclopentane Triphenylphosphine oxide

(q) Cyanohydrin formation takes place.

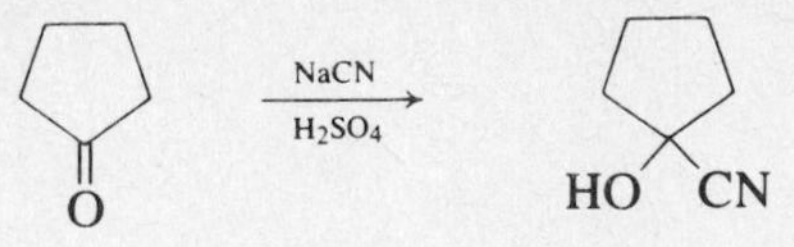

Cyclopentanone $\xrightarrow[H_2SO_4]{NaCN}$ 1-Cyanocyclopentanol

(r) No reaction occurs between silver oxide and cyclopentanone.

(s), (t) Cyclopentanone is not oxidized readily with either chromic acid or potassium permanganate. On heating with either of these reagents, oxidative degradation to a variety of products takes place.

18.19 (a) The first step in analyzing this problem is to write the structure of the starting ketone in stereochemical detail.

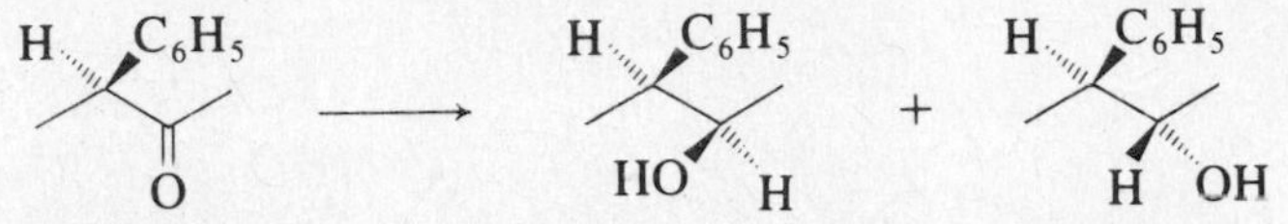

(S)-3-Phenyl-2-butanone ⟶ (2R, 3S)-3-Phenyl-2-butanol + (2S, 3S)-3-Phenyl-2-butanol

Reduction of the ketone introduces a new chiral center, which may have either the *R* or *S* configuration; the configuration of the original chiral center is unaffected. In practice it is observed that the 2*R*,3*S* diastereomer is formed in greater amounts than the 2*S*,3*S* (ratio 2.5 : 1 for $LiAlH_4$ reduction).

(b) Reduction of the ketone can yield either *cis*- or *trans*-4-*tert*-butylcyclohexanol.

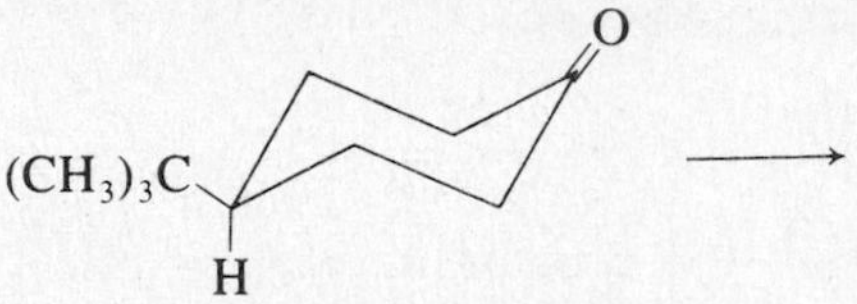

4-*tert*-Butylcyclohexanone ⟶

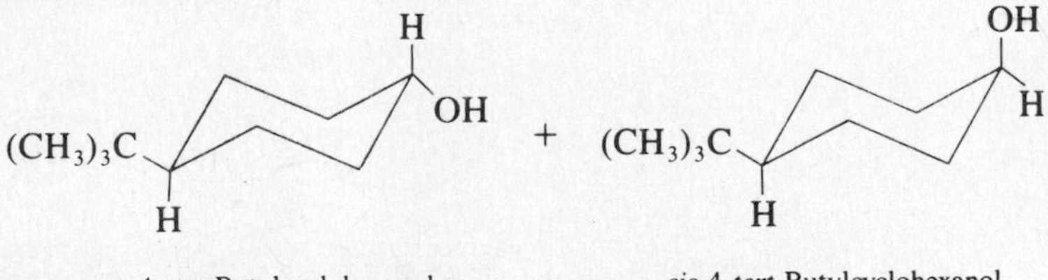

trans-4-*tert*-Butylcyclohexanol + *cis*-4-*tert*-Butylcyclohexanol

It has been observed that the major product obtained on reduction with either lithium aluminum hydride or sodium borohydride is the trans alcohol (trans/cis ~ 9 : 1). Although the equatorial alcohol is more stable, the main reason for this stereoselectivity is thought to involve torsional strain in the transition state leading to the axial product.

(c) The two diastereomeric alcohols are *cis*- and *trans*-1,2-cyclobutanediol.

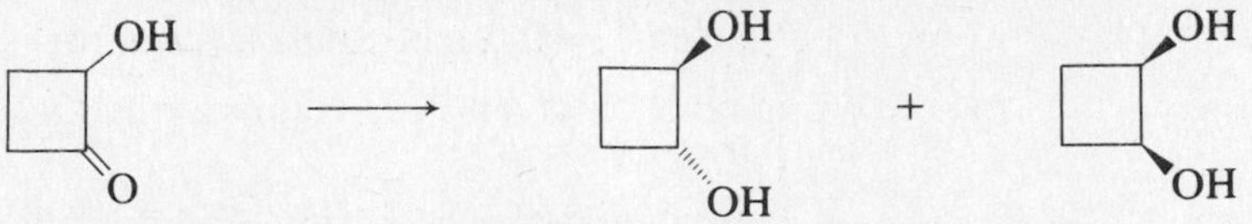

2-Hydroxycyclobutanone ⟶ *trans*-1,2-Cyclobutanediol + *cis*-1,2-Cyclobutanediol

Both products are formed in approximately equal amounts on reduction of 2-hydroxycyclobutanone with lithium aluminum hydride.

(d) The two reduction products are the exo and endo alcohols.

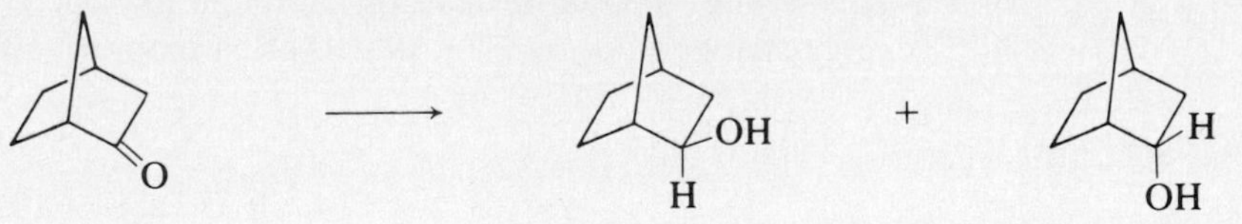

Bicyclo[2.2.1]heptan-2-one → *exo*-Bicyclo[2.2.1]heptan-2-ol + *endo*-Bicyclo[2.2.1]heptan-2-ol

The major product is observed to be the endo alcohol (endo/exo 9 : 1) for reduction with $NaBH_4$ or $LiAlH_4$. The stereoselectivity observed in this reaction is due to decreased steric hindrance to attack of the hydride reagent from the exo face of the molecule, giving rise to the endo alcohol.

(e) The hydroxyl group may be on the same side as the double bond or on the opposite side.

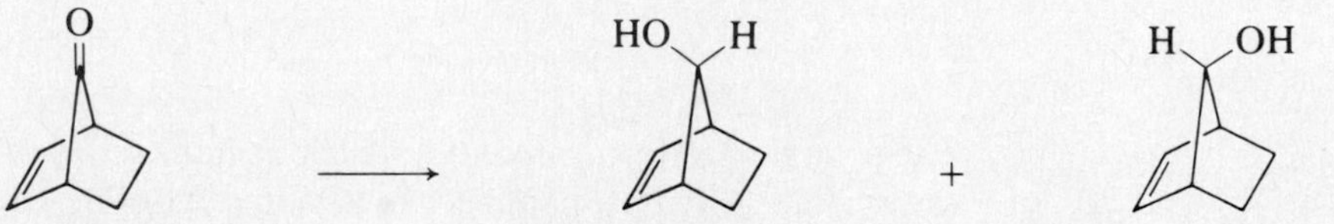

Bicyclo[2.2.1]hept-2-en-7-one → *syn*-Bicyclo[2.2.1]hept-2-en-7-ol + *anti*-Bicyclo[2.2.1]hept-2-en-7-ol

The anti alcohol is observed to be formed in greater amounts (85 : 15) on reduction of the ketone with $LiAlH_4$. Steric factors governing attack of the hydride reagent again explain the major product observed.

18.20 (a) Aldehydes undergo nucleophilic addition faster than ketones; benzaldehyde is reduced by sodium borohydride more rapidly than is acetophenone. The measured relative rates are:

$$k_{rel} = C_6H_5\overset{O}{\overset{\|}{C}}H / C_6H_5\overset{O}{\overset{\|}{C}}CH_3 = 440$$

(b) The carbonyl group of benzaldehyde is stabilized through conjugation with the aromatic ring.

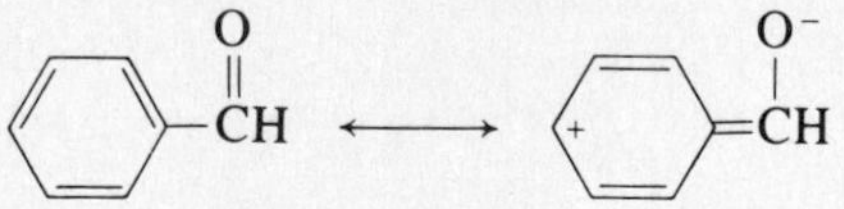

$$C_6H_5\overset{O}{\overset{\|}{C}}H \longleftrightarrow {}^{+}C_6H_5{=}\overset{O^-}{\overset{|}{C}}H$$

This reduces the electrophilicity of the aldehyde group and makes it less reactive than acetaldehyde toward nucleophilic addition of semicarbazide.

$$k_{rel} = CH_3\overset{O}{\overset{\|}{C}}H / C_6H_5\overset{O}{\overset{\|}{C}}H = 180$$

(c) The presence of an electronegative substituent on the α carbon atom causes a dramatic increase in K_{hydr}. Trichloroacetaldehyde (chloral) is almost completely converted to its gem diol (chloral hydrate) in aqueous solution.

$$Cl_3C\overset{O}{\overset{\|}{C}}H + H_2O \longrightarrow Cl_3C\overset{OH}{\underset{OH}{C}}H$$

Trichloroacetaldehyde (chloral) → 1-Hydroxy-2,2,2-trichloroethanol (choral hydrate)

Electron-withdrawing groups—and a Cl_3C-substituent is strongly electron-withdrawing—destabilize carbonyl groups to which they are attached and make the energy change favoring the products of nucleophilic addition more favorable.

$$K_{rel} = Cl_3\overset{O}{\overset{\|}{C}}H/CH_3\overset{O}{\overset{\|}{C}}H \cong 20{,}000$$

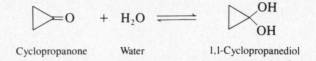

(*d*) The hybridization at carbon changes from sp^2 to sp^3 on addition of water to a carbonyl group. Since the preferred bond angles for sp^2 and sp^3 hybridized carbon are 120° and 109.5°, respectively, there is more angle strain in small-ring ketones than in their hydrates. The carbonyl group of cyclopropanone is more strained than the corresponding hydrate. Some angle strain is relieved in the process:

$$\text{Cyclopropanone} + H_2O \rightleftharpoons \text{1,1-Cyclopropanediol}$$

Cyclopropanone Water 1,1-Cyclopropanediol

Less angle strain is present in cyclopentanone, so there is little driving force for its hydration. Cyclopropanone has a more favorable equilibrium constant for hydration; indeed, it exists mainly as its hydrate in aqueous solution.

(*e*) Recall that the equilibrium constants for nucleophilic addition to carbonyl groups are governed by a combination of electronic effects and steric effects. Electronically there is little difference between acetone and 3,3-dimethyl-2-butanone, but sterically there is a significant difference. The cyanohydrin products are more crowded than the starting ketones, so the bulkier the alkyl groups that are attached to the carbonyl, the more strained and less stable will be the cyanohydrin.

$$\begin{matrix}R\\CH_3\end{matrix}\!\!>C{=}O + HCN \rightleftharpoons CH_3\underset{R}{\underset{|}{\overset{OH}{\overset{|}{C}}}}CN$$

Ketone Hydrogen cyanide Cyanohydrin [less strained for R = CH_3 than for R = $C(CH_3)_3$]

$$K_{rel} = CH_3\overset{O}{\overset{\|}{C}}CH_3/CH_3\overset{O}{\overset{\|}{C}}C(CH_3)_3 = 40$$

(*f*) Steric effects influence the rate of nucleophilic addition to these two ketones. Carbon is on its way from tricoordinate to tetracoordinate at the transition state. Alkyl groups are forced closer together in the activated complex than they are in the ketone.

$$\underset{sp^2}{\begin{matrix}R\\CH_3\end{matrix}\!\!>C{=}O} \xrightarrow{NaBH_4} \left[\begin{matrix}R\\CH_3\!-\!C\overset{\delta-}{\cdots}O\\H\cdots\overset{\delta-}{B}H_3\end{matrix}\right] \longrightarrow \underset{sp^3}{\begin{matrix}R\\CH_3\!-\!C\!-\!\overset{-}{O}BH_3\\H\end{matrix}}$$

Activated complex

The activated complex is of lower energy when R is smaller. Acetone (R is methyl) is reduced faster than 3,3-dimethyl-2-butanone (R is *tert*-butyl).

$$k_{rel} = CH_3\overset{O}{\overset{\|}{C}}CH_3/CH_3\overset{O}{\overset{\|}{C}}C(CH_3)_3 = 12$$

(*g*) In this problem we examine the rate of hydrolysis of acetals to the corresponding ketone or aldehyde. The rate-determining step is carbocation formation.

$$R_2C(OCH_2CH_3)\overset{+}{O}(H)CH_2CH_3 \longrightarrow R_2\overset{+}{C}{-}OCH_2CH_3 + CH_3CH_2OH$$

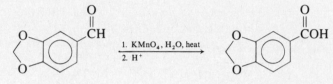

Hybridization at carbon changes from sp^3 to sp^2; crowding at this carbon is relieved as the carbocation is formed. The more crowded acetal ($R = CH_3$) forms a carbocation faster than the less crowded one ($R = H$). Another factor of even greater importance is the extent of stabilization of the carbocation intermediate; the more stable carbocation ($R = CH_3$) is formed faster than the less stable one ($R = H$).

$$k_{rel} = \frac{(CH_3)_2C(OCH_2CH_3)_2}{CH_2(OCH_2CH_3)_2} = 1.8 \times 10^7$$

18.21 (*a*) Oxidation of aldehydes with potassium permanganate converts them to carboxylic acids.

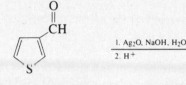

Piperonal $\xrightarrow[2.\ H^+]{1.\ KMnO_4,\ H_2O,\ heat}$ Piperonylic acid (90–96%)

(*b*) Silver oxide is a mild oxidant, which converts aldehydes to carboxylic acids.

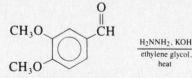

Thiophene-3-carbaldehyde $\xrightarrow[2.\ H^+]{1.\ Ag_2O,\ NaOH,\ H_2O}$ Thiophene-3-carboxylic acid (95%)

(*c*) Heating an aldehyde with hydrazine and base in a high-boiling alcohol solvent converts the aldehyde function to a methyl group by a Wolff-Kishner reaction.

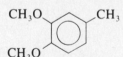

3,4-Dimethoxybenzaldehyde $\xrightarrow[\text{ethylene glycol, heat}]{H_2NNH_2,\ KOH}$ 3,4-Dimethoxytoluene (79%)

(*d*) The ketone carbonyl of the starting material reacts with hydroxylamine to form an oxime.

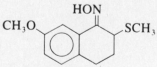

7-Methoxy-2-thiomethyl-1-tetralone $\xrightarrow[\text{ethanol}]{H_2NOH}$ 7-Methoxy-2-thiomethyl-1-tetralone oxime (64%)

(*e*) Acyl chlorides yield ketones on treatment with dialkylcadmium reagents.

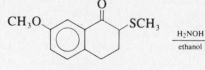

$$CH_3(CH_2)_{16}\overset{O}{\overset{\|}{C}}Cl + (C_6H_5CH_2CH_2)_2Cd \longrightarrow CH_3(CH_2)_{16}\overset{O}{\overset{\|}{C}}CH_2CH_2C_6H_5$$

Octadecanoyl chloride + *Bis*-(2-phenylethyl)cadmium ⟶ 1-Phenylicosan-3-one (66%)

(*f*) The reaction is a Wolff-Kishner reduction of a ketone.

$\xrightarrow[\text{diethylene glycol, heat}]{H_2NNH_2,\ KOH}$

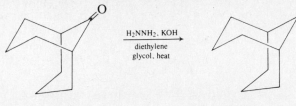

Bicyclo[3.3.1]nonan-2-one → Bicyclo[3.3.1]nonane (54%)

(*g*) Hydrogen cyanide adds to carbonyl groups to form cyanohydrins.

$$C_6H_5\overset{O}{\overset{\|}{C}}CH_3 \xrightarrow[\text{2. HCl}]{\text{1. NaCN, ether-water}} C_6H_5\underset{CH_3}{\overset{OH}{C}}CN$$

Acetophenone → 1-Cyano-1-phenylethanol

(*h*) The reagent is a secondary amine known as morpholine. Secondary amines react with ketones to give enamines.

$$C_6H_5\overset{O}{\overset{\|}{C}}CH_3 + HN\text{(morpholine ring)}O \xrightarrow{H^+} C_6H_5\overset{OH}{C}CH_3(\text{N-morpholino}) \xrightarrow{H^+(-H_2O)} C_6H_5C{=}CH_2(\text{N-morpholino})$$

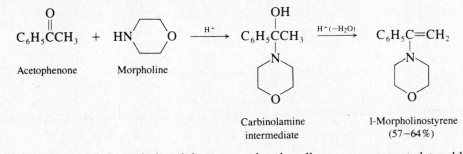

Acetophenone; Morpholine; Carbinolamine intermediate; 1-Morpholinostyrene (57–64%)

(*i*) The oxo reaction is an industrial process whereby alkenes are converted to aldehydes when heated with carbon monoxide and hydrogen under pressure in the presence of a cobalt octacarbonyl catalyst.

$$\text{Cyclohexene} + CO + H_2 \xrightarrow[\text{benzene, 120°C, pressure}]{Co_2(CO)_8} \text{Cyclohexane-}\overset{O}{\overset{\|}{C}}H$$

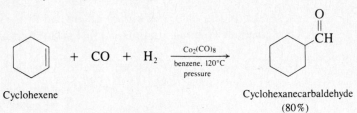

Cyclohexene → Cyclohexanecarbaldehyde (80%)

(*j*) Acid-catalyzed hydrolysis of acetals yields ketones.

$$(CH_3O)_2\text{-cyclopropene} \xrightarrow[H_2SO_4]{H_2O} \text{cyclopropenone} + 2CH_3OH$$

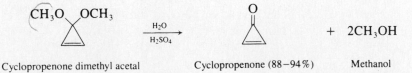

Cyclopropenone dimethyl acetal; Cyclopropenone (88–94%); Methanol

In this case the ketone is cyclopropenone, an interesting small molecule characterized by a surprising level of stability and believed to have a significant amount of aromatic character. A dipolar resonance form is an oxy-substituted cyclopropenyl cation.

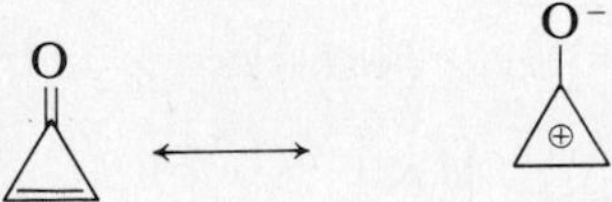

(two π electrons in three-membered ring)

(*k*) The reagent is semicarbazide, which reacts with aldehydes to yield derivatives known as semicarbazones. The carbon-oxygen double bond of the carbonyl group is replaced by a carbon-nitrogen double bond of a semicarbazone.

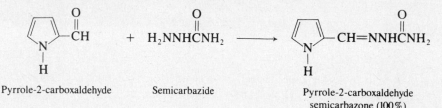

$$\text{Pyrrole-2-carboxaldehyde} + H_2NNHCONH_2 \longrightarrow \text{Pyrrole-2-CH=NNHCONH}_2$$

Pyrrole-2-carboxaldehyde | Semicarbazide | Pyrrole-2-carboxaldehyde semicarbazone (100%)

18.22 Of the three carbonyl groups in the starting triketone, two are stabilized by conjugation with an aromatic ring.

$$C_6H_5COCOCOC_6H_5$$

stabilized | stabilized

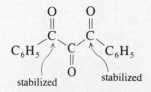

The central carbonyl group is destabilized because it bears two powerful electron-withdrawing groups. It is the one that forms the stable hydrate.

$$C_6H_5COC(OH)_2COC_6H_5$$

(Structure of hydrate)

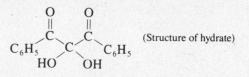

18.23 Hydration of formaldehyde by $H_2{}^{17}O$ produces a *gem*-diol in which the labeled and unlabeled hydroxyl groups are equivalent. When this *gem*-diol reverts to formaldehyde, loss of either of the hydroxyl groups is equally likely except for a small isotope effect favoring loss of the common isotope, which leads to eventual replacement of the mass 16 isotope of oxygen by ^{17}O.

$$^{16}O{=}CH_2 + H_2{}^{17}O \rightleftharpoons (H^{17}O)(H^{16}O)CH_2 \rightleftharpoons {}^{17}O{=}CH_2 + H_2{}^{16}O$$

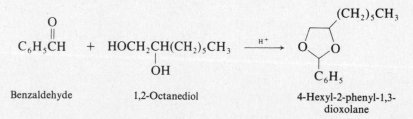

This reaction has been monitored by ^{17}O nmr spectroscopy; ^{17}O gives an nmr signal but ^{16}O does not.

18.24 First write out the chemical equation for the reaction that takes place. Vicinal diols (1,2-diols) react with aldehydes to give cyclic acetals.

$$C_6H_5CHO + HOCH_2CH(OH)(CH_2)_5CH_3 \xrightarrow{H^+} \text{4-hexyl-2-phenyl-1,3-dioxolane}$$

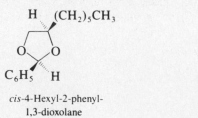

Benzaldehyde | 1,2-Octanediol | 4-Hexyl-2-phenyl-1,3-dioxolane

Notice that the phenyl and hexyl substituents may be either cis or trans to each other. The two products are the cis and trans stereoisomers.

cis-4-Hexyl-2-phenyl-1,3-dioxolane | *trans*-4-Hexyl-2-phenyl-1,3-dioxolane

18.25 Cyclic hemiacetals are formed by intramolecular nucleophilic addition of a hydroxyl group to a carbonyl.

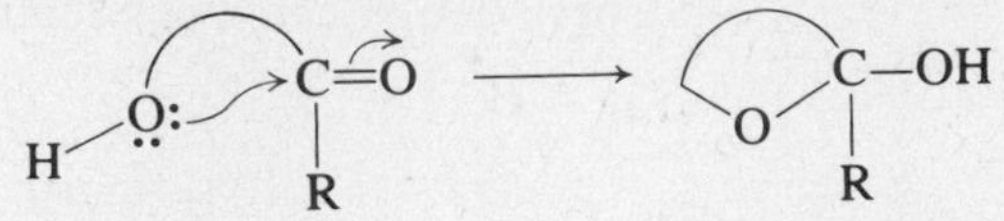

Cyclic hemiacetal

The ring oxygen is derived from the hydroxyl group; the carbonyl oxygen becomes the hydroxyl oxygen of the hemiacetal.

(*a*) This compound is the cyclic hemiacetal of 5-hydroxypentanal.

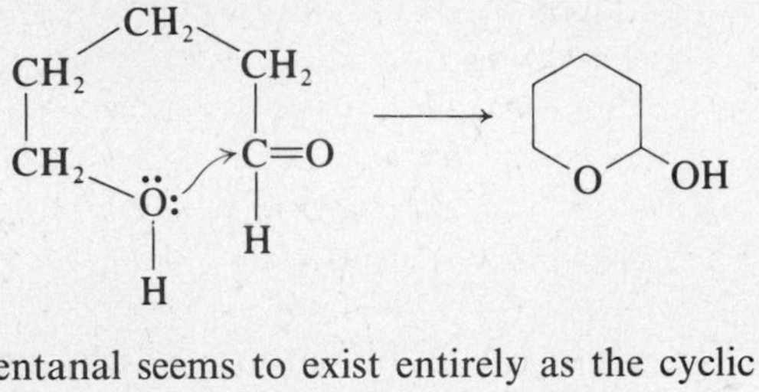

Indeed, 5-hydroxypentanal seems to exist entirely as the cyclic hemiacetal. Its infrared spectrum is devoid of absorption in the carbonyl region.

(*b*) The carbon connected to two oxygens is the one that is derived from the carbonyl group. Using retrosynthetic symbolism, disconnect the ring oxygen from this carbon.

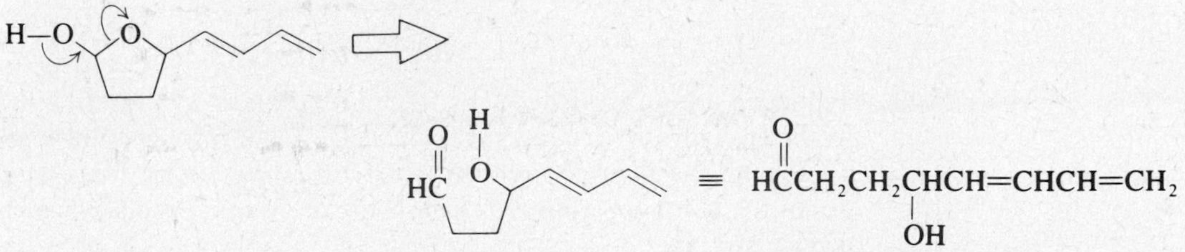

$$\equiv HCCH_2CH_2CHCH{=}CHCH{=}CH_2$$

4-Hydroxy-5,7-octadienal

The next four compounds are all cyclic acetals. The original carbonyl group is identifiable as the one that bears two oxygen substituents, which originate as hydroxyl oxygens of a diol.

(*c*)

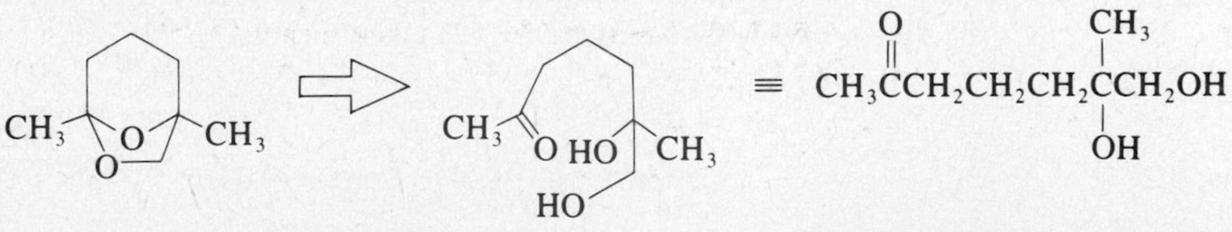

Frontalin

$$\equiv CH_3CCH_2CH_2CH_2CCH_2OH$$

6-Methyl-6,7-dihydroxy-2-heptanone

(*d*)

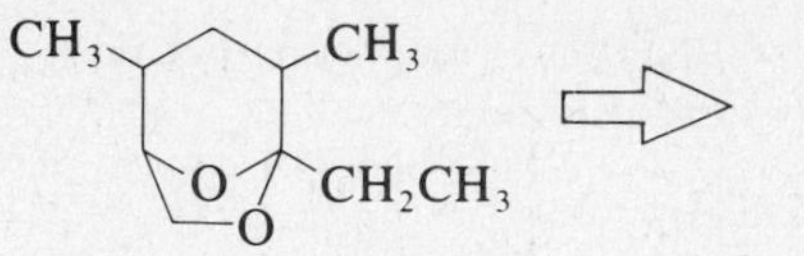

Multistriatin

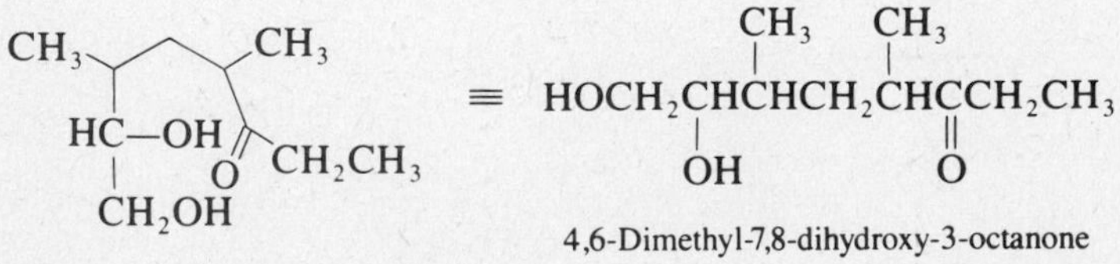

$$\equiv HOCH_2CHCHCH_2CHCCH_2CH_3$$

4,6-Dimethyl-7,8-dihydroxy-3-octanone

(e)

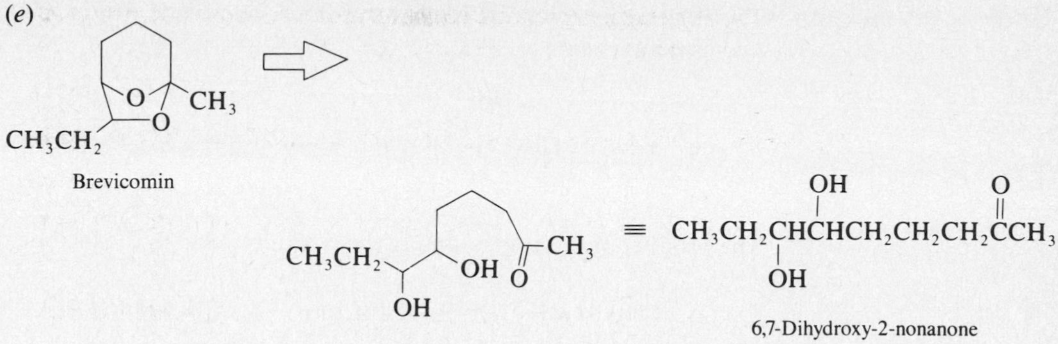

Brevicomin

$$CH_3CH_2CHCHCH_2CH_2CH_2\overset{O}{\overset{\|}{C}}CH_3$$ (with OH on C-6 and C-7)

6,7-Dihydroxy-2-nonanone

(f)

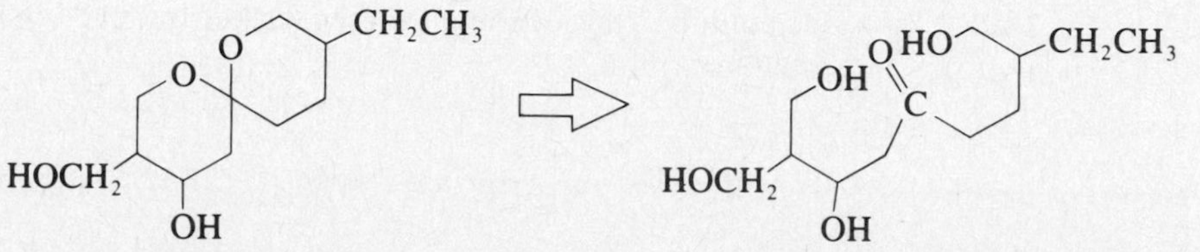

Taloromycin A

2,8-Di(hydroxymethyl)-1,3-dihydroxy-5-decanone

18.26 (a) The Z stereoisomer of $CH_3CH{=}NCH_3$ has its higher-priority substituents on the same side of the double bond.

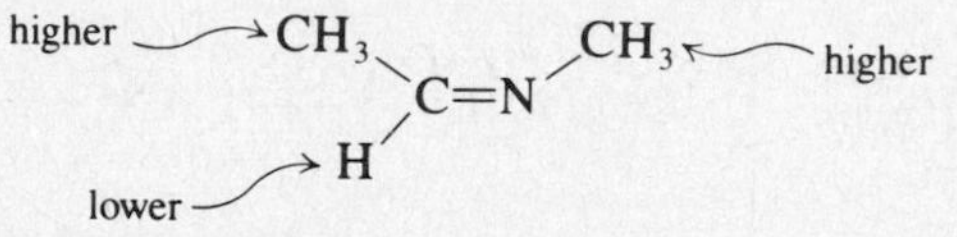

higher → CH_3, CH_3 ← higher; lower → H

$$\text{(CH}_3\text{)(H)C=N(CH}_3\text{)}$$

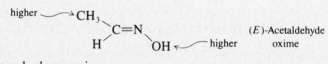

(Z)-N-Ethylidenemethylamine

The lone pair of nitrogen need not be shown. It is considered a "phantom ligand," lower in priority ranking than any other.

(b) Higher-priority groups are on opposite sides of the carbon-nitrogen double bond in the E oxime of acetaldehyde.

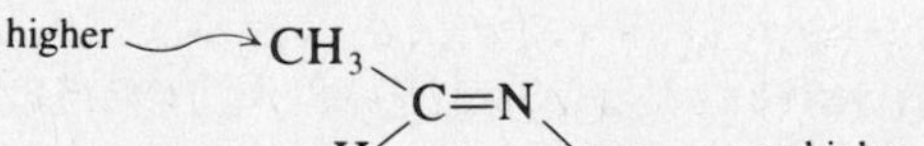

higher → CH_3; OH ← higher

$$\text{(CH}_3\text{)(H)C=N–OH}$$

(E)-Acetaldehyde oxime

(c) (Z)-2-Butanone hydrazone is:

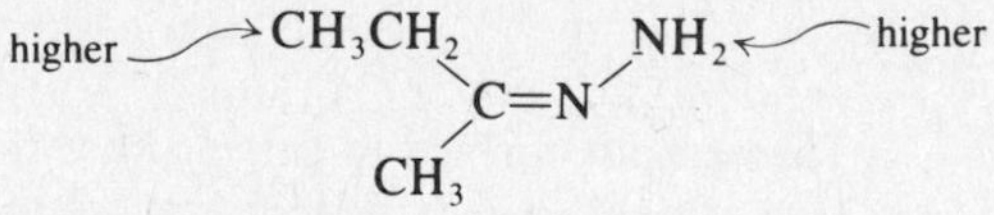

higher → CH_3CH_2; NH_2 ← higher

$$\text{(CH}_3\text{CH}_2\text{)(CH}_3\text{)C=N–NH}_2$$

(d) (E)-Acetophenone semicarbazone is:

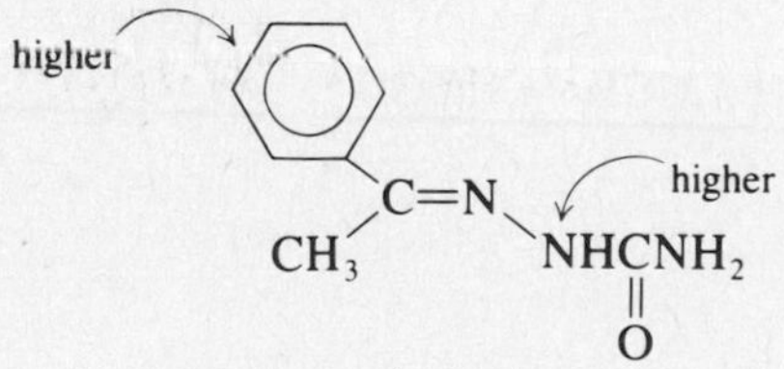

higher → C_6H_5; higher → $NHCNH_2$

$$\text{(C}_6\text{H}_5\text{)(CH}_3\text{)C=N–NH}\overset{O}{\overset{\|}{C}}\text{NH}_2$$

18.27 By analogy to other processes in which derivatives of ammonia react with aldehydes, the first step is nucleophilic addition to the carbonyl group.

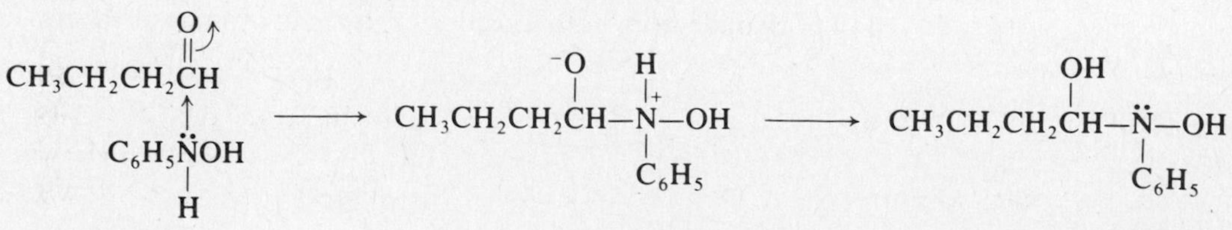

$$CH_3CH_2CH_2\overset{O}{\overset{\|}{C}}H + C_6H_5\ddot{N}(H)OH \longrightarrow CH_3CH_2CH_2\overset{^-O}{\overset{|}{C}}H\text{–}\overset{H}{\overset{|+}{N}}(C_6H_5)\text{–OH} \longrightarrow CH_3CH_2CH_2\overset{OH}{\overset{|}{C}}H\text{–}\ddot{N}(C_6H_5)\text{–OH}$$

The intermediate formed in this step loses hydroxide ion in a step that is assisted by the nitrogen lone pair.

$$CH_3CH_2CH_2CH(OH)\text{—}\ddot{N}(C_6H_5)\text{—}OH \longrightarrow CH_3CH_2CH_2CH{=}\overset{+}{N}(C_6H_5)\text{—}OH + {}^-OH$$

The product of this step is simply the conjugate acid of the nitrone.

$$CH_3CH_2CH_2CH{=}\overset{+}{N}(C_6H_5)\text{—}O\text{—}H \longrightarrow CH_3CH_2CH_2CH{=}\overset{+}{N}(O^-)C_6H_5 + H^+$$

18.28 (*a*) Nucleophilic ring opening of the epoxide occurs by attack of methoxide at the less hindered carbon.

$$(CH_3)_3C(Cl)\overset{O}{\widehat{C\text{—}CH_2}} + {}^-OCH_3 \longrightarrow (CH_3)_3C\text{—}C(O^-)(Cl)\text{—}CH_2OCH_3$$

The anion formed in this step loses a chloride ion to form the carbon-oxygen double bond of the product.

$$(CH_3)_3C\text{—}C(O^-)(Cl)\text{—}CH_2OCH_3 \longrightarrow (CH_3)_3C\text{—}\overset{O}{\overset{\|}{C}}\text{—}CH_2OCH_3 + Cl^-$$

(*b*) Nucleophilic addition of methoxide ion to the aldehyde carbonyl generates an oxyanion, which can close to an epoxide by an intramolecular nucleophilic substitution reaction.

$$(CH_3)_3CCH(Cl)\overset{O}{\overset{\|}{C}}H + {}^-OCH_3 \longrightarrow (CH_3)_3CCH(Cl)\text{—}CH(O^-)OCH_3 \longrightarrow (CH_3)_3C\overset{O}{\widehat{CH\text{—}CH}}OCH_3 + Cl^-$$

The epoxide formed in this process then undergoes nucleophilic ring opening on attack by a second methoxide ion.

$$(CH_3)_3C\overset{O}{\widehat{CH\text{—}CH}}OCH_3 + {}^-OCH_3 \longrightarrow (CH_3)_3CCH(O^-)\text{—}CH(OCH_3)OCH_3 \longrightarrow (CH_3)_3CCH(OH)CH(OCH_3)OCH_3$$

18.29 First draw accurate three-dimensional representations of the cyanohydrins formed by addition of cyanide from both directions. Attack of cyanide at the top face is shown by:

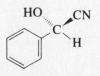

$$C_6H_5\text{—}C(OH)(CN)H$$

The order of decreasing sequence rule precedence is HO > CN > C_6H_5 > H. This is the *R* enantiomer, so we need go no further. Addition of cyanide to the top face yields the *R* enantiomer. Addition to the bottom face yields the *S* enantiomer.

18.30 (*a*) The target molecule is the diethyl acetal of acetaldehyde.

$$CH_3CH(OCH_2CH_3)_2 \Longrightarrow CH_3\overset{O}{\overset{\|}{C}}H,\ CH_3CH_2OH$$

Acetaldehyde diethyl acetal

Acetaldehyde may be prepared by oxidation of ethanol.

$$\underset{\text{Ethanol}}{CH_3CH_2OH} \xrightarrow[CH_2Cl_2]{(C_5H_5N)_2CrO_3} \underset{\text{Acetaldehyde}}{CH_3\overset{O}{\overset{\|}{C}}H}$$

Reaction with ethanol in the presence of hydrogen chloride yields the desired acetal.

$$\underset{\text{Acetaldehyde}}{CH_3\overset{O}{\overset{\|}{C}}H} + \underset{\text{Ethanol}}{2\,CH_3CH_2OH} \xrightarrow{HCl} \underset{\text{Acetaldehyde diethyl acetal}}{CH_3CH(OCH_2CH_3)_2}$$

(*b*) In this case the target molecule is a cyclic acetal of acetaldehyde.

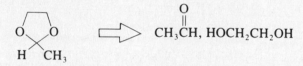

$$\text{2-Methyl-1,3-dioxolane} \Longrightarrow CH_3\overset{O}{\overset{\|}{C}}H,\ HOCH_2CH_2OH$$

2-Methyl-1,3-dioxolane

Acetaldehyde has been prepared in (*a*). Recalling that glycols are available from the hydroxylation of alkenes, ethylene glycol may be prepared by the sequence

$$\underset{\text{Ethanol}}{CH_3CH_2OH} \xrightarrow[\text{heat}]{H_2SO_4} \underset{\text{Ethylene}}{CH_2{=}CH_2} \xrightarrow[(CH_3)_3COH,\ HO^-]{OsO_4,\ (CH_3)_3COOH} \underset{\text{1,2-Ethanediol}}{HOCH_2CH_2OH}$$

Hydrolysis of ethylene oxide is also reasonable.

$$\underset{\text{Ethylene}}{CH_2{=}CH_2} \xrightarrow{CH_3CO_2OH} \underset{\text{Ethylene oxide}}{CH_2{-}CH_2\ (\text{bridged by O})} \xrightarrow[HO^-]{H_2O} \underset{\text{1,2-Ethanediol}}{HOCH_2CH_2OH}$$

Reaction of acetaldehyde with 1,2-ethanediol yields the cyclic acetal.

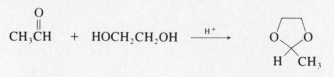

$$\underset{\text{Acetaldehyde}}{CH_3\overset{O}{\overset{\|}{C}}H} + \underset{\text{1,2-Ethanediol}}{HOCH_2CH_2OH} \xrightarrow{H^+} \text{2-Methyl-1,3-dioxolane}$$

2-Methyl-1,3-dioxolane

(*c*) The target molecule is, in this case, the cyclic acetal of ethylene glycol and formaldehyde.

$$\text{1,3-Dioxolane} \Longrightarrow H\overset{O}{\overset{\|}{C}}H,\ HOCH_2CH_2OH$$

1,3-Dioxolane

The preparation of ethylene glycol was described in (*b*). One method of preparing formaldehyde is ozonolysis.

$$\underset{\text{Ethanol}}{CH_3CH_2OH} \xrightarrow[\text{heat}]{H_2SO_4} \underset{\text{Ethylene}}{CH_2{=}CH_2} \xrightarrow[\text{2. } H_2O,\ Zn]{\text{1. } O_3} \underset{\text{Formaldehyde}}{2HCH(=O)}$$

Another method is periodate cleavage of ethylene glycol.

$$\underset{\text{1,2-Ethanediol}}{HOCH_2CH_2OH} \xrightarrow{HIO_4} \underset{\text{Formaldehyde}}{2\,HCH(=O)}$$

Cyclic acetal formation is then carried out in the usual way.

$$\underset{\text{Formaldehyde}}{HCH(=O)} + \underset{\text{1,2-Ethanediol}}{HOCH_2CH_2OH} \xrightarrow{H^+} \text{1,3-Dioxolane}$$

(*d*) Acetylenic alcohols are best prepared from carbonyl compounds and acetylide anions.

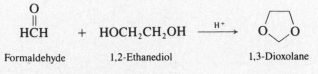

$$\underset{\text{3-Butyn-2-ol}}{CH_3CH(OH)C{\equiv}CH} \Rightarrow CH_3CH(=O),\ ^-C{\equiv}CH$$

Acetaldehyde is available as in (*a*). Alkynes such as acetylene are available from the corresponding alkene by bromination followed by double dehydrobromination. Using ethylene, prepared in (*b*), the sequence becomes:

$$\underset{\text{Ethylene}}{CH_2{=}CH_2} \xrightarrow{Br_2} \underset{\text{1,2-Dibromoethane}}{BrCH_2CH_2Br} \xrightarrow[NH_3]{NaNH_2} \underset{\text{Acetylene}}{HC{\equiv}CH}$$

Then

$$\underset{\text{Acetylene}}{HC{\equiv}CH} \xrightarrow{NaNH_2} \underset{\text{Sodium acetylide}}{HC{\equiv}CNa} \xrightarrow[\text{2. } H_3O^+]{\text{1. } CH_3CH(=O)} \underset{\text{3-Butyn-2-ol}}{HC{\equiv}CCH(OH)CH_3}$$

(*e*) The target aldehyde may be prepared from the corresponding alcohol.

$$\underset{\text{3-Butynal}}{HC(=O)CH_2C{\equiv}CH} \Rightarrow \underset{\text{3-Butyn-1-ol}}{HOCH_2CH_2C{\equiv}CH}$$

The best route to this alcohol is through reaction of an acetylide ion with ethylene oxide.

$$\underset{\substack{\text{Sodium acetylide}\\ \text{[prepared in } (d)]}}{HC{\equiv}CNa} + \underset{\substack{\text{Ethylene oxide}\\ \text{[prepared in } (b)]}}{CH_2{-}CH_2\ (\text{-O-})} \xrightarrow[\text{2. } H_3O^+]{\text{1. diethyl ether}} \underset{\text{3-Butyn-1-ol}}{HC{\equiv}CCH_2CH_2OH}$$

Oxidation with Collins' reagent is appropriate for the final step.

$$\underset{\text{3-Butyn-1-ol}}{HC{\equiv}CCH_2CH_2OH} \xrightarrow[CH_2Cl_2]{(C_5H_5N)_2CrO_3} \underset{\text{3-Butynal}}{HC{\equiv}CCH_2\overset{O}{\overset{\|}{C}}H}$$

18.31 (*a*) Friedel-Crafts acylation of benzene with benzoyl chloride is a direct route to benzophenone.

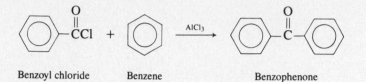

$$\underset{\text{Benzoyl chloride}}{C_6H_5\overset{O}{\overset{\|}{C}}Cl} + \underset{\text{Benzene}}{C_6H_6} \xrightarrow{AlCl_3} \underset{\text{Benzophenone}}{C_6H_5\overset{O}{\overset{\|}{C}}C_6H_5}$$

(*b*) Acyl chlorides may be converted to ketones using either organocadmium or organocuprate reagents.

$$\underset{\text{Bromobenzene}}{C_6H_5Br} \xrightarrow[\text{diethyl ether}]{Mg} \underset{\text{Phenylmagnesium bromide}}{C_6H_5MgBr} \xrightarrow{CdCl_2} \underset{\text{Diphenylcadmium}}{(C_6H_5)_2Cd}$$

$$\xrightarrow{C_6H_5\overset{O}{\overset{\|}{C}}Cl} \underset{\text{Benzophenone}}{C_6H_5\overset{O}{\overset{\|}{C}}C_6H_5}$$

$$\underset{\text{Bromobenzene}}{C_6H_5Br} \xrightarrow[\text{diethyl ether}]{Li} \underset{\text{Phenyllithium}}{C_6H_5Li} \xrightarrow{CuBr} \underset{\text{Lithium diphenylcuprate}}{(C_6H_5)_2CuLi}$$

$$\xrightarrow{C_6H_5\overset{O}{\overset{\|}{C}}Cl} \underset{\text{Benzophenone}}{C_6H_5\overset{O}{\overset{\|}{C}}C_6H_5}$$

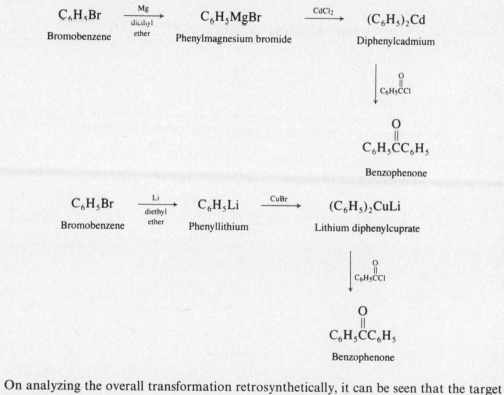

(*c*) On analyzing the overall transformation retrosynthetically, it can be seen that the target molecule may be prepared by a Grignard synthesis followed by oxidation of the alcohol formed.

$$C_6H_5\overset{O}{\overset{\|}{C}}C_6H_5 \Longrightarrow C_6H_5\overset{OH}{\overset{|}{C}}HC_6H_5 \Longrightarrow C_6H_5\overset{O}{\overset{\|}{C}}H + C_6H_5MgBr$$

In the desired synthesis, benzyl alcohol must first be oxidized to benzaldehyde.

$$\underset{\text{Benzyl alcohol}}{C_6H_5CH_2OH} \xrightarrow[CH_2Cl_2]{(C_5H_5N)_2CrO_3} \underset{\text{Benzaldehyde}}{C_6H_5\overset{O}{\overset{\|}{C}}H}$$

Reaction of benzaldehyde with the Grignard reagent of bromobenzene followed by oxidation of the resulting secondary alcohol gives benzophenone.

$$\underset{\text{Benzaldehyde}}{C_6H_5\overset{O}{\overset{\|}{C}}H} + \underset{\text{Phenylmagnesium bromide}}{C_6H_5MgBr} \xrightarrow[\text{2. } H_3O^+]{\text{1. diethyl ether}} \underset{\text{Diphenylmethanol}}{C_6H_5\underset{OH}{\underset{|}{C}}HC_6H_5} \xrightarrow{(C_5H_5N)_2CrO_3,\ CH_2Cl_2} \underset{\text{Benzophenone}}{C_6H_5\overset{O}{\overset{\|}{C}}C_6H_5}$$

(*d*) Hydrolysis of diphenylmethyl bromide yields the corresponding alcohol, which can be oxidized to benzophenone as in (*c*).

$$\underset{\text{Diphenylmethyl bromide}}{C_6H_5\underset{Br}{\underset{|}{C}}HC_6H_5} \xrightarrow{H_2O} \underset{\text{Diphenylmethanol}}{C_6H_5\underset{OH}{\underset{|}{C}}HC_6H_5} \xrightarrow{\text{oxidize}} \underset{\text{Benzophenone}}{C_6H_5\overset{O}{\overset{\|}{C}}C_6H_5}$$

(*e*) The starting material is the dimethyl acetal of benzophenone. All that is required is acid-catalyzed hydrolysis.

$$\underset{\text{Dimethoxydiphenylmethane}}{C_6H_5\overset{OCH_3}{\overset{|}{\underset{OCH_3}{\underset{|}{C}}}}C_6H_5} + \underset{\text{Water}}{2H_2O} \xrightarrow{H^+} \underset{\text{Benzophenone}}{C_6H_5\overset{O}{\overset{\|}{C}}C_6H_5} + \underset{\text{Methanol}}{2CH_3OH}$$

(*f*) Oxidative cleavage of the alkene yields benzophenone. Either ozonolysis or permanganate oxidation may be used.

$$\underset{\text{1,1,2,2-Tetraphenylethene}}{(C_6H_5)_2C{=}C(C_6H_5)_2} \xrightarrow[\text{or } KMnO_4,\text{ heat}]{O_3,\text{ then } H_2O} \underset{\text{Benzophenone}}{2(C_6H_5)_2C{=}O}$$

18.32 The two alcohols given as starting materials contain all the carbon atoms of the desired product.

$$CH_3(CH_2)_8CH{=}CHCH_2CH{=}CHCH_2CH \,\vdots\!{=}\, CHCH{=}CH_2 \Longrightarrow$$

$$\underset{\text{3,6-Hexadecadien-1-ol}}{CH_3(CH_2)_8CH{=}CHCH_2CH{=}CHCH_2CH_2OH} \quad \text{and} \quad \underset{\text{Allyl alcohol}}{HOCH_2CH{=}CH_2}$$

What is needed is to attach the two groups together so that the two primary alcohol carbons become doubly bonded to each other. This can be accomplished by using a Wittig reaction as the key step.

$$\underset{\text{3,6-Hexadecadien-1-ol}}{CH_3(CH_2)_8CH{=}CHCH_2CH{=}CHCH_2CH_2OH} \xrightarrow[CH_2Cl_2]{(C_5H_5N)_2CrO_3} \underset{\text{3,6-Hexadecadienal}}{CH_3(CH_2)_8CH{=}CHCH_2CH{=}CHCH_2\overset{O}{\overset{\|}{C}}H}$$

$$CH_2{=}CHCH_2OH \xrightarrow{PBr_3} CH_2{=}CHCH_2Br \xrightarrow{(C_6H_5)_3P} (C_6H_5)_3\overset{+}{P}CH_2CH{=}CH_2\ Br^-$$

Allyl alcohol — Allyl bromide — Allyltriphenylphosphonium bromide

$$\xrightarrow{CH_3CH_2CH_2CH_2Li,\ THF} (C_6H_5)_3\overset{+}{P}{-}\overset{-}{\ddot{C}}HCH{=}CH_2$$

Allylidenetriphenylphosphorane

$$CH_3(CH_2)_8CH{=}CHCH_2CH{=}CHCH_2\overset{O}{\overset{\|}{C}}H + (C_6H_5)_3\overset{+}{P}{-}\overset{\cdot\cdot}{\overset{-}{C}}HCH{=}CH_2 \longrightarrow$$

3,6-Hexadecadienal — Allylidenetriphenylphosphorane

$$CH_3(CH_2)_8CH{=}CHCH_2CH{=}CHCH_2CH{=}CHCH{=}CH_2$$

1,3,6,9-Nonadecatetraene

Alternatively, allyl alcohol could be oxidized to $CH_2{=}CHCHO$ for subsequent reaction with the ylide derived from $CH_3(CH_2)_8CH{=}CHCH_2CH{=}CHCH_2CH_2OH$ via its bromide and triphenylphosphonium salt.

18.33 Since compound B corresponds to a hydrogenation product and its infrared spectrum shows that it is a ketone, its most reasonable constitution is

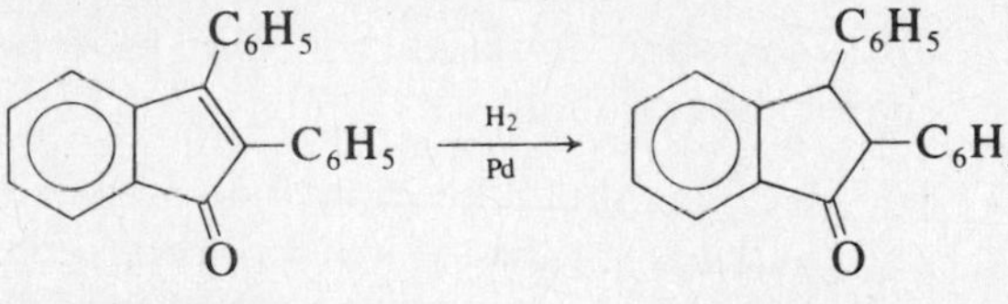

$C_{21}H_{14}O$ (2,3-diphenylindenone) $\xrightarrow[Pd]{H_2}$ $C_{21}H_{16}O$ (compound B)

Compound A presumably arises by way of an intramolecular Friedel-Crafts acylation.

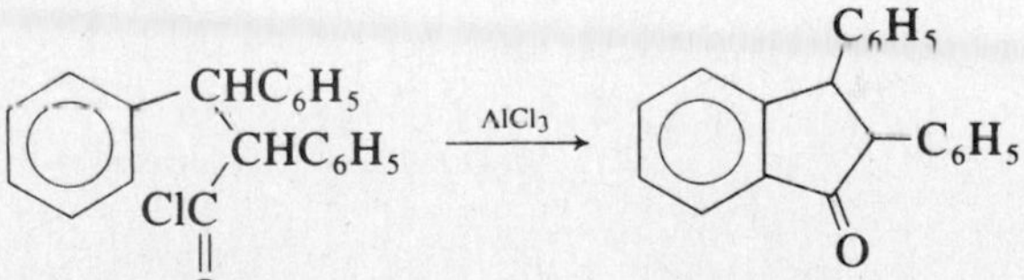

$C_{21}H_{17}OCl$ ($C_6H_5CHC_6H_5$–CHC_6H_5–$ClC{=}O$) $\xrightarrow{AlCl_3}$ $C_{21}H_{16}O$ (compound A)

Compounds A and B have the same constitution but are isomers; thus they must be stereoisomers. Compound B, arising via hydrogenation, is most likely to be the cis isomer.

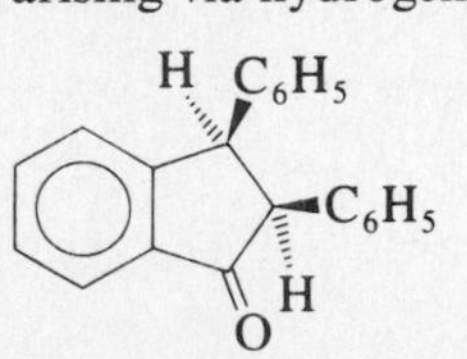

Compound B: syn hydrogenation yields cis stereoisomer

The acylium ion leading to compound A can cyclize to give either a trans or cis arrangement of phenyl groups. It yields the more stable trans arrangement because that is the one that has the largest separation of bulky groups in the transition state.

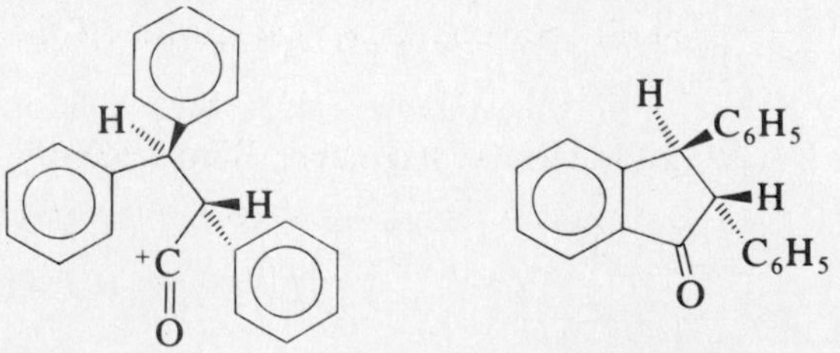

Compound A

18.34 The expected course of the reaction would be hydrolysis of the acetal to the corresponding aldehyde.

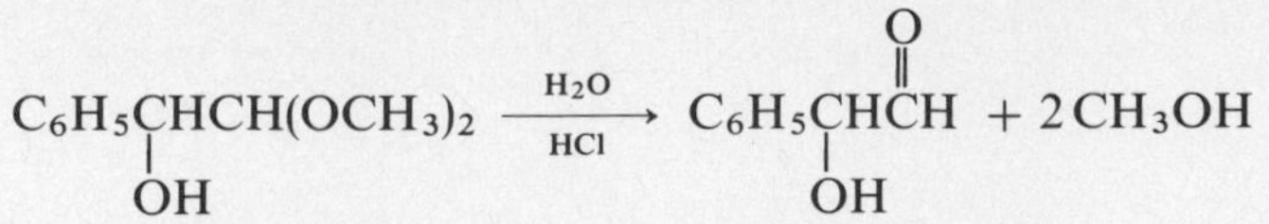

$$\underset{\text{Compound C (mandelaldehyde dimethyl acetal)}}{C_6H_5\underset{|}{\overset{}{C}}HCH(OCH_3)_2 \atop \quad OH} \xrightarrow[HCl]{H_2O} \underset{\text{Mandelaldehyde}}{C_6H_5CH\overset{O}{\overset{\|}{C}}H} + \underset{\text{Methanol}}{2CH_3OH}$$

Compound C (mandelaldehyde dimethyl acetal) — Mandelaldehyde — Methanol

The molecular formula of the observed product (compound D, $C_{16}H_{16}O_4$) is exactly twice that of mandelaldehyde. This suggests that it might be a dimer of mandelaldehyde resulting from hemiacetal formation between the hydroxyl group of one mandelaldehyde molecule and the carbonyl group of another.

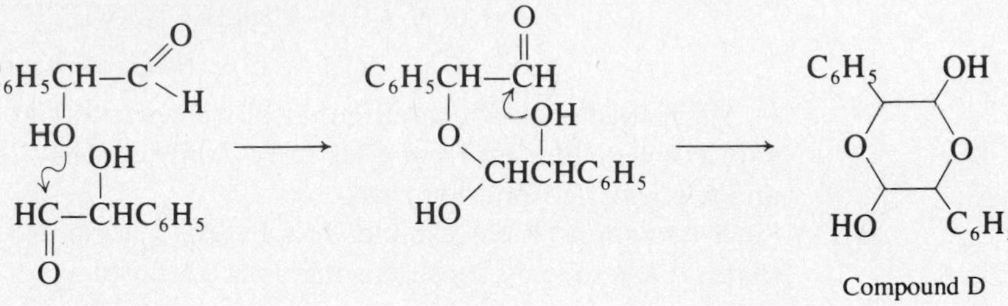

Compound D

Since compound D lacks carbonyl absorption in its infrared spectrum, the cyclic structure shown above is indicated.

18.35 (*a*) Recalling that alkanes may be prepared by hydrogenation of the appropriate alkene, a synthesis of the desired product becomes apparent. What is needed is to convert $-C{=}O$ into $-C{=}CH_2$; a Wittig reaction is appropriate.

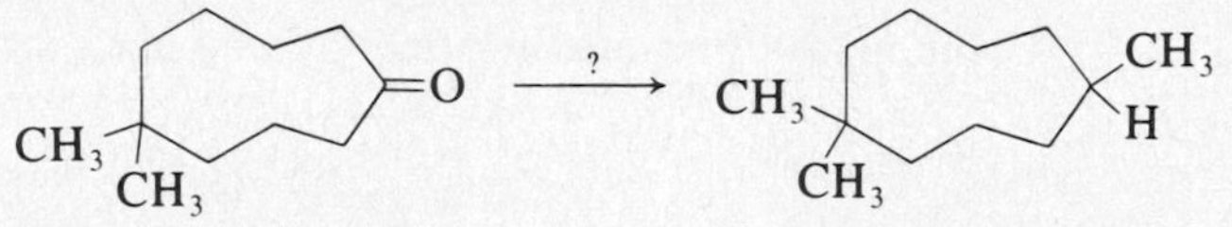

5,5-Dimethylcyclononanone $\xrightarrow{?}$ 1,1,5-Trimethylcyclononane

The two-step procedure that was followed used a Wittig reaction to form the carbon-carbon bond, then catalytic hydrogenation of the resulting alkene.

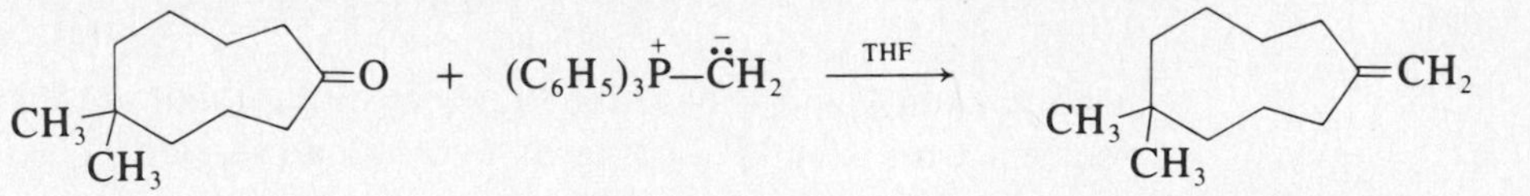

5,5-Dimethylcyclononanone + $(C_6H_5)_3\overset{+}{P}-\overset{-}{\ddot{C}}H_2$ $\xrightarrow{THF}$ 5,5-Dimethyl-1-methylenecyclononane (59%)

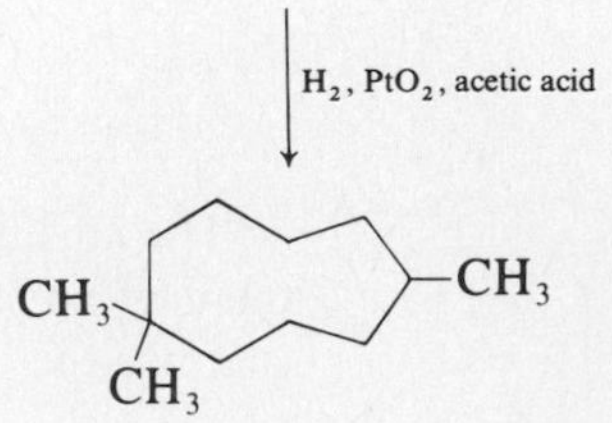

$\xrightarrow{H_2,\ PtO_2,\ \text{acetic acid}}$ 1,1,5-Trimethylcyclononane (73%)

(*b*) In putting together the carbon skeleton of the target molecule, a methyl group has to be added to the original carbonyl carbon.

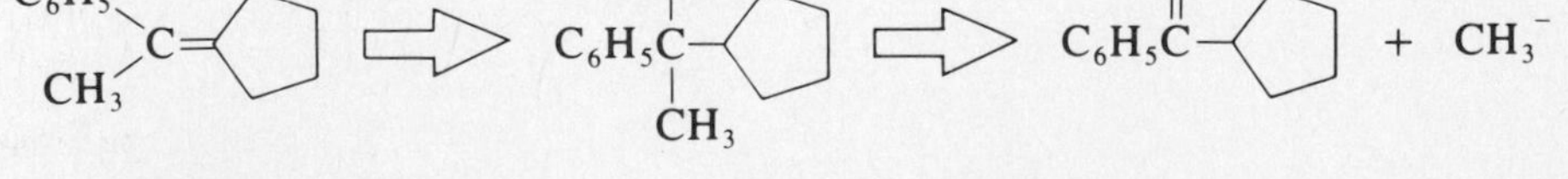

$$(C_6H_5)(CH_3)C{=}C_5H_8 \Rightarrow C_6H_5\overset{OH}{\underset{CH_3}{C}}-C_5H_9 \Rightarrow C_6H_5\overset{O}{\overset{\|}{C}}-C_5H_9 + CH_3^-$$

The logical way to do this is by way of a Grignard reagent.

$$C_6H_5C(=O)\text{-cyclopentyl} + CH_3MgI \xrightarrow[2.\ H_3O^+]{1.\ \text{diethyl ether}} C_6H_5C(OH)(CH_3)\text{-cyclopentyl}$$

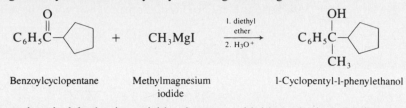

Benzoylcyclopentane — Methylmagnesium iodide — 1-Cyclopentyl-1-phenylethanol

Acid-catalyzed dehydration yields the more highly substituted alkene, the desired product, in accordance with the Zaitsev rule.

$$C_6H_5C(OH)(CH_3)\text{-cyclopentyl} \xrightarrow[\text{heat}]{H_3PO_4} (C_6H_5)(CH_3)C{=}\text{cyclopentylidene}$$

1-Cyclopentyl-1-phenylethanol — (1-Phenylethylidene)cyclopentane

(*c*) Analyzing the transformation retrosynthetically, keeping in mind the starting materials stated in the problem, we see that the carbon skeleton may be constructed in a straightforward manner.

$$o\text{-}CH_3C_6H_4C(=CH_2)(CH_2)_3CH{=}CH_2 \Rightarrow o\text{-}CH_3C_6H_4C(=O)(CH_2)_3CH{=}CH_2 + (C_6H_5)_3\overset{+}{P}\text{—}\ddot{C}H_2^{-}$$

$$o\text{-}CH_3C_6H_4CH(OH)(CH_2)_3CH{=}CH_2 \Rightarrow o\text{-}CH_3C_6H_4MgBr + HC(=O)(CH_2)_3CH{=}CH_2$$

Proceeding with the synthesis in the forward direction, reaction between the Grignard reagent of *o*-bromotoluene and 5-hexenal produces most of the desired carbon skeleton.

$$o\text{-}CH_3C_6H_4MgBr + CH_2{=}CHCH_2CH_2CH_2CH{=}O \xrightarrow[2.\ H_3O^+]{1.\ \text{diethyl ether}} o\text{-}CH_3C_6H_4CH(OH)(CH_2)_3CH{=}CH_2$$

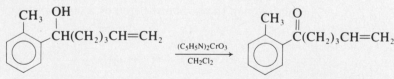

o-Methylphenylmagnesium bromide — 5-Hexenal — 1-(*o*-Methylphenyl)-5-hexen-1-ol

Oxidation of the resulting alcohol to the ketone followed by a Wittig reaction leads to the final product.

$$o\text{-}CH_3C_6H_4CH(OH)(CH_2)_3CH{=}CH_2 \xrightarrow[CH_2Cl_2]{(C_5H_5N)_2CrO_3} o\text{-}CH_3C_6H_4C(=O)(CH_2)_3CH{=}CH_2$$

1-(*o*-Methylphenyl)-5-hexen-1-ol — 1-(*o*-Methylphenyl)-5-hexen-1-one

$$\xrightarrow{(C_6H_5)_3\overset{+}{P}\text{—}\overset{-}{C}H_2} o\text{-}CH_3C_6H_4C(=CH_2)(CH_2)_3CH{=}CH_2$$

2-(*o*-Methylphenyl)-1,6-heptadiene

Acid-catalyzed dehydration of the corresponding tertiary alcohol would not be suitable because the major elimination product would have the more highly substituted double bond.

$$\text{2-}(o\text{-Methylphenyl})\text{C}(CH_3)(OH)(CH_2)_3CH{=}CH_2 \xrightarrow[\text{heat}]{H^+} \text{2-}(o\text{-Methylphenyl})\text{C}(CH_3){=}CHCH_2CH_2CH{=}CH_2$$

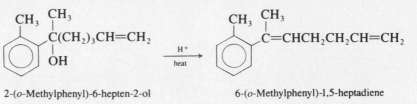

2-(*o*-Methylphenyl)-6-hepten-2-ol → 6-(*o*-Methylphenyl)-1,5-heptadiene

(*d*) The most efficient synthesis is two steps. First extend the chain by two carbon atoms.

$$CH_3(CH_2)_{16}\overset{O}{\overset{\|}{C}}Cl + (CH_3CH_2)_2Cd \longrightarrow CH_3(CH_2)_{16}\overset{O}{\overset{\|}{C}}CH_2CH_3$$

Octadecanoyl chloride, Diethyl cadmium → 3-Icosanone (62%)

(Lithium diethylcuprate could be used for the same transformation.) Wolff-Kishner reduction of the ketone affords the desired icosane.

$$CH_3(CH_2)_{16}\overset{O}{\overset{\|}{C}}CH_2CH_3 \xrightarrow[\text{base, heat in high-boiling alcohol}]{H_2NNH_2} CH_3(CH_2)_{16}CH_2CH_2CH_3$$

3-Icosanone → Icosane (73%)

(*e*) Remember that terminal acetylenes can serve as sources of methyl ketones by hydration.

$$CH_3\overset{O}{\overset{\|}{C}}CH_2CH_2\overset{O}{\overset{\|}{C}}(CH_2)_5CH_3 \Longrightarrow HC{\equiv}CCH_2CH_2\overset{O}{\overset{\|}{C}}(CH_2)_5CH_3$$

This gives us a clue as to how to proceed, since the acetylenic ketone may be prepared from the starting acetylenic alcohol.

$$HC{\equiv}CCH_2CH_2\overset{O}{\overset{\|}{C}}(CH_2)_5CH_3 \Longrightarrow HC{\equiv}CCH_2CH_2\overset{O}{\overset{\|}{C}}H + CH_3(CH_2)_5^-$$

$$\Downarrow$$

$$HC{\equiv}CCH_2CH_2CH_2OH$$

The first synthetic step is oxidation of the primary alcohol to the aldehyde and construction of the carbon skeleton by a Grignard reaction.

$$HC{\equiv}CCH_2CH_2CH_2OH \xrightarrow[CH_2Cl_2]{(C_5H_5N)_2CrO_3} HC{\equiv}CCH_2CH_2\overset{O}{\overset{\|}{C}}H$$

4-Pentyn-1-ol → 4-Pentynal

$$\downarrow \text{1. } CH_3(CH_2)_5MgBr \quad \text{2. } H_3O^+$$

$$HC{\equiv}CCH_2CH_2\underset{OH}{\underset{|}{C}}H(CH_2)_5CH_3$$

1-Undecyn-5-ol

Oxidation of the secondary alcohol to a ketone and hydration of the terminal triple bond complete the synthesis.

$$HC{\equiv}CCH_2CH_2CH(OH)(CH_2)_5CH_3 \xrightarrow[CH_2Cl_2]{(C_5H_5N)_2CrO_3} HC{\equiv}CCH_2CH_2C(=O)(CH_2)_5CH_3$$

1-Undecyn-5-ol → 1-Undecyn-5-one

$$\xrightarrow{H_2O,\ H_2SO_4,\ HgSO_4} CH_3C(=O)CH_2CH_2C(=O)(CH_2)_5CH_3$$

2,5-Undecanedione

(*f*) The desired product is a benzylic ether. In order to prepare it, the aldehyde must first be reduced to the corresponding primary alcohol. Sodium borohydride was used in the preparation described in the literature, but lithium aluminum hydride or catalytic hydrogenation would also be possible. Once the alcohol is prepared, it can be converted to its alkoxide ion and this alkoxide ion treated with methyl iodide.

$$\text{(3-Cl, 2-CH}_3\text{)C}_6\text{H}_3\text{CHO} \xrightarrow[H_2O]{NaBH_4} \text{(3-Cl, 2-CH}_3\text{)C}_6\text{H}_3\text{CH}_2\text{OH} \xrightarrow[2.\ CH_3I]{1.\ NaH} \text{(3-Cl, 2-CH}_3\text{)C}_6\text{H}_3\text{CH}_2\text{OCH}_3$$

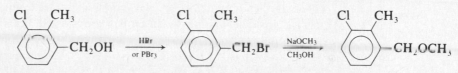

Alternatively, the alcohol could be treated with hydrogen bromide or with phosphorus tribromide to give the benzylic bromide and the bromide then allowed to react with sodium methoxide.

$$\text{(3-Cl, 2-CH}_3\text{)C}_6\text{H}_3\text{CH}_2\text{OH} \xrightarrow[\text{or } PBr_3]{HBr} \text{(3-Cl, 2-CH}_3\text{)C}_6\text{H}_3\text{CH}_2\text{Br} \xrightarrow[CH_3OH]{NaOCH_3} \text{(3-Cl, 2-CH}_3\text{)C}_6\text{H}_3\text{CH}_2\text{OCH}_3$$

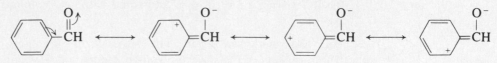

18.36 We need to assess the extent of resonance donation to the carbonyl group by the π electrons of the aromatic ring. Such resonance for the case of benzaldehyde may be written as:

$$C_6H_5CH{=}O \longleftrightarrow {}^{+}C_6H_5{=}CH{-}O^{-} \,(\text{ortho}) \longleftrightarrow {}^{+}C_6H_5{=}CH{-}O^{-} \,(\text{ortho}) \longleftrightarrow {}^{+}C_6H_5{=}CH{-}O^{-} \,(\text{para})$$

Electron-releasing groups such as methoxy at positions ortho and para to the aldehyde function increase the "single bond character" of the aldehyde by stabilizing the dipolar resonance forms and increasing their contribution to the overall electron distribution in the molecule. Electron-withdrawing groups such as nitro decrease this single bond character. The aldehyde with the lowest carbonyl stretching frequency is 2,4,6-trimethoxybenzaldehyde; the one with the highest is 2,4,6-trinitrobenzaldehyde. The measured values are:

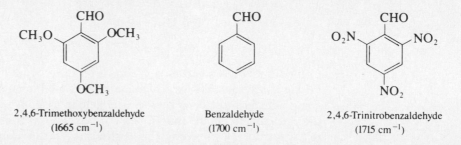

2,4,6-Trimethoxybenzaldehyde	Benzaldehyde	2,4,6-Trinitrobenzaldehyde
($1665\ cm^{-1}$)	($1700\ cm^{-1}$)	($1715\ cm^{-1}$)

18.37 A signal in the ^1H nmr spectrum at $\delta = 9.8$ ppm tells us that compound E is an aldehyde rather than a ketone. This aldehyde signal appears as a triplet, so C-2 must bear two hydrogens. The only C_4H_8O aldehyde that satisfies this requirement is butanal. We can assign all the signals as follows:

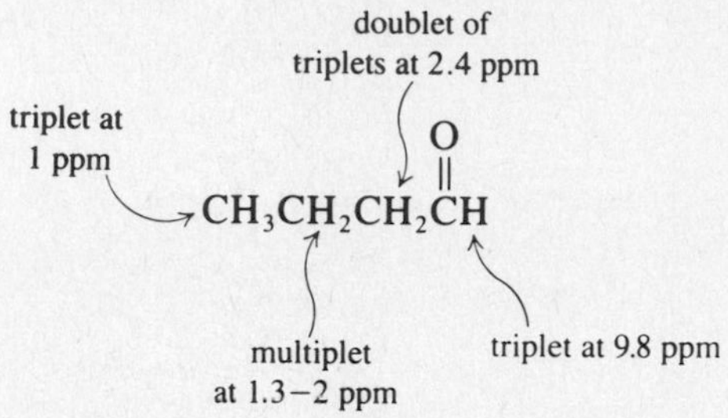

$$\overset{\text{triplet at 1 ppm}}{CH_3}\underset{\text{multiplet at 1.3–2 ppm}}{CH_2}\overset{\text{doublet of triplets at 2.4 ppm}}{CH_2}\underset{\text{triplet at 9.8 ppm}}{\overset{O}{\overset{\|}{C}}H}$$

18.38 (*a*) A carbonyl group is evident in the strong infrared absorption at 1710 cm^{-1}. Since all the ^1H nmr signals are singlets, there are no nonequivalent hydrogens in a vicinal or "three bond" relationship. The three-proton signal at $\delta = 2.1$ ppm and the two-proton signal at $\delta = 2.4$ ppm can be understood as arising from a $CH_2\overset{O}{\overset{\|}{C}}CH_3$ unit. The intense nine proton singlet at $\delta = 1.0$ ppm is due to the three equivalent methyl groups of a $(CH_3)_3C$ unit. The compound is methyl neopentyl ketone.

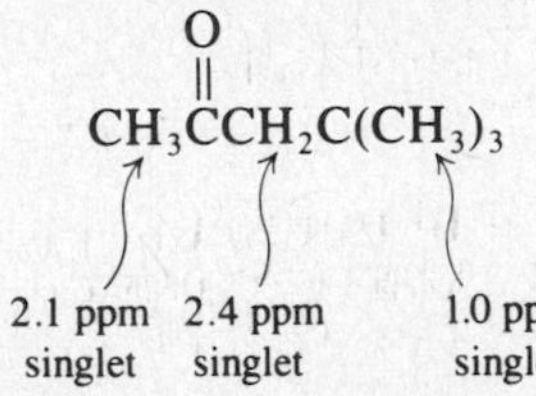

$$\underset{\substack{2.1\text{ ppm}\\ \text{singlet}}}{CH_3}\overset{O}{\overset{\|}{C}}\underset{\substack{2.4\text{ ppm}\\ \text{singlet}}}{CH_2}C\underset{\substack{1.0\text{ ppm}\\ \text{singlet}}}{(CH_3)_3} \qquad \text{Compound F}$$

(*b*) Often the compounds with the simplest infrared and nuclear magnetic resonance spectra are the ones whose structures are most difficult to determine. The compound has the molecular formula $C_6H_{10}O_2$, which tells us that it has a total of two double bonds and rings ($C_6H_{14} - C_6H_{10} = 4\,H$). A strong carbonyl band in the infrared at 1710 cm^{-1} tells us that at least one of these elements of unsaturation is a ketone function.

The ^1H nmr spectrum has only two peaks, both singlets, at $\delta = 2.2$ and 2.7 ppm. Their intensity ratio (6 : 4) is consistent with two equivalent methyl groups and two equivalent methylene groups. The chemical shifts are appropriate for $CH_3\overset{O}{\overset{\|}{C}}$ and $CH_2\overset{O}{\overset{\|}{C}}$. The simplicity of the spectrum can be understood if we are dealing with a symmetrical diketone.

The correct structure is:

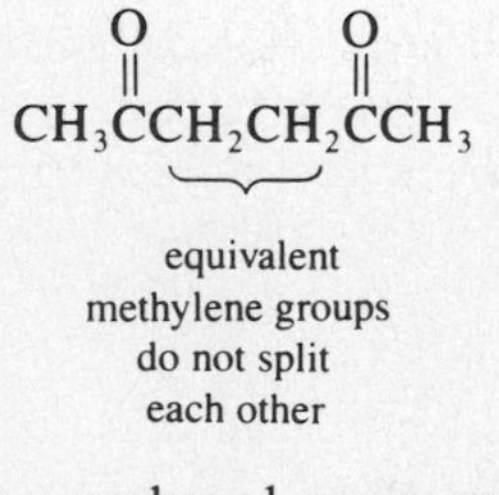

$$CH_3\overset{O}{\overset{\|}{C}}\underbrace{CH_2CH_2}_{\substack{\text{equivalent}\\ \text{methylene groups}\\ \text{do not split}\\ \text{each other}}}\overset{O}{\overset{\|}{C}}CH_3 \qquad \text{2,5-Hexanedione (compound G)}$$

(*c*) Compound H has a carbonyl group, as revealed by an intense peak at 1710 cm^{-1} in its infrared spectrum. The lack of an aldehyde signal in its ^1H nmr spectrum indicates that compound H must be a ketone. The ketone carbonyl contributes the only element of unsaturation to compound H, based on its molecular formula of $C_6H_{12}O_3$. The infrared spectrum shows strong C—O stretching in the range 1050 to 1120 cm^{-1}, so it is reasonable to assume that the two remaining oxygens are part of ether linkages. (Alcohols are ruled out because there are no hydroxyl peaks in the infrared spectrum.)

A sharp singlet at $\delta = 2.2$ ppm in the ^{1}H nmr spectrum is consistent with a methyl ketone. Two of compound H's six carbons are accounted for by $CH_3\overset{O}{\overset{\|}{C}}$. A six-proton singlet at $\delta = 3.4$ ppm is reasonably assigned to two methoxyl groups. A doublet at $\delta = 2.6$ ppm and a triplet at $\delta = 4.7$ ppm are the only spin-coupled signals in the spectrum. They suggest a $-CH_2-\overset{|}{\underset{|}{C}}-H$ unit and lead to the conclusion that compound H is the acetal shown below.

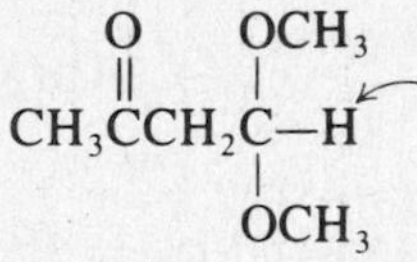

$$CH_3\overset{O}{\overset{\|}{C}}CH_2\overset{OCH_3}{\overset{|}{\underset{OCH_3}{\underset{|}{C}}}}-H$$

low-field (δ = 4.7 ppm) triplet; methine carbon bears two electron-withdrawing oxygen substituents

18.39 (*a*) A monosubstituted aromatic ring is indicated by a five-proton signal in the ^{1}H nmr spectrum of compound I at $\delta = 7.2$ ppm and by strong infrared absorption at 700 and 735 cm^{-1}. A strong carbonyl stretching band is seen at 1710 cm^{-1}. The nmr spectrum reveals a three-proton singlet at $\delta = 2.0$ ppm, typical of a methyl ketone. The remaining signal is a two-proton singlet at $\delta = 3.7$ ppm for a methylene group. The most reasonable structure for this compound is

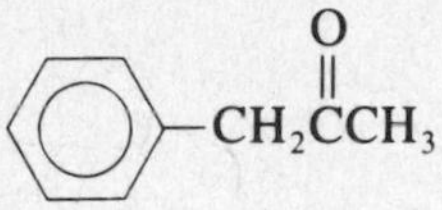

$$C_6H_5-CH_2\overset{O}{\overset{\|}{C}}CH_3$$

Benzyl methyl ketone (compound I)

The shift to lower field of the methylene signal is a result of deshielding by both the aromatic ring and the carbonyl group.

(*b*) A number of factors indicate an aromatic ring in compound J. The molecular formula C_9H_9ClO has a low hydrogen/carbon ratio consistent with a total of five double bonds and rings; an aromatic ring can account for four of these elements of unsaturation. A weak C—H absorption at 3010 cm^{-1} in the infrared spectrum suggests aromatic C—H groups. Similarly the ^{1}H nmr spectrum shows an AB quartet in the aromatic region at $\delta = 7.3$ to 8 ppm. This pattern is typical of para substitution, as is the presence of a moderately intense peak in the infrared at 840 cm^{-1}.

The characteristic triplet-quartet pattern of an ethyl group is evident in the nmr spectrum, with the methylene quartet appearing at $\delta = 3$ ppm. The position of this signal indicates that the methylene group is attached to some electron-withdrawing group; that this electron-withdrawing group is a ketone carbonyl is revealed by an intense infrared absorption at 1685 cm^{-1}. The double bond of the carbonyl group is the fifth element of unsaturation.

When all the information—a para-disubstituted benzene, an ethyl group bonded to a carbonyl, and a chlorine substituent—is put together, it points to *p*-chloropropiophenone as the correct structure.

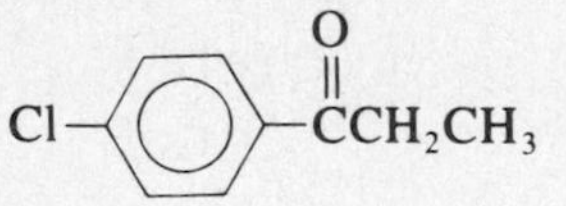

$$Cl-C_6H_4-\overset{O}{\overset{\|}{C}}CH_2CH_3$$

p-Chloropropiophenone (compound J)

Notice that isomeric structures such as

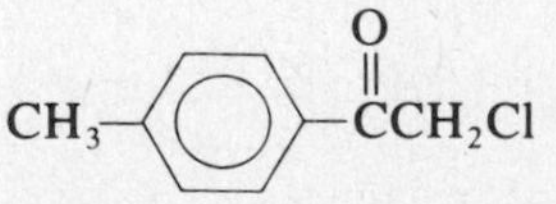

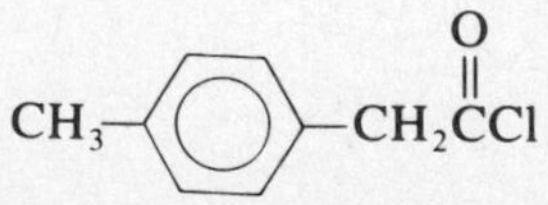

$$CH_3-C_6H_4-\overset{O}{\overset{\|}{C}}CH_2Cl \qquad ClCH_2-C_6H_4-\overset{O}{\overset{\|}{C}}CH_3 \qquad CH_3-C_6H_4-CH_2\overset{O}{\overset{\|}{C}}Cl$$

are ruled out on the basis of nmr splitting patterns and chemical shifts. In each of these incorrectly identified isomers the methyl and methylene signals would appear as singlets.

(c) The most prominent peak in the infrared spectrum of compound K is a carbonyl absorption at 1695 cm^{-1}. A monosubstituted aromatic ring is indicated by strong absorptions at 685 and 710 cm^{-1} and the five-proton ^1H nmr pattern between 7.2 and 8.2 ppm.

A doublet-quartet pattern in the nmr spectrum indicates a $CH_3{-}CH$ group. The chemical shift of the quartet ($\delta = 5.2$ ppm) is at relatively low field and is accommodated by a methine group deshielded by both a bromine and a carbonyl group. The compound is:

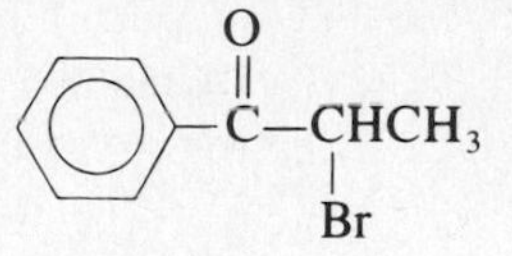

$$C_6H_5{-}\overset{O}{\overset{\|}{C}}{-}\underset{Br}{\underset{|}{C}}HCH_3$$

1-Bromoethyl phenyl ketone (compound K)

18.40 With a molecular formula of $C_7H_{14}O$, compound L has a SODAR of 1. Since we are told that it is a ketone, it has no rings or double bonds other than the one belonging to its C=O group. The peak at 211 ppm in the ^{13}C nmr spectrum corresponds to the carbonyl carbon. There are only three signals in the spectrum, so there are only three types of carbons other than the carbonyl carbon. This suggests that the symmetrical ketone 4-heptanone is compound L.

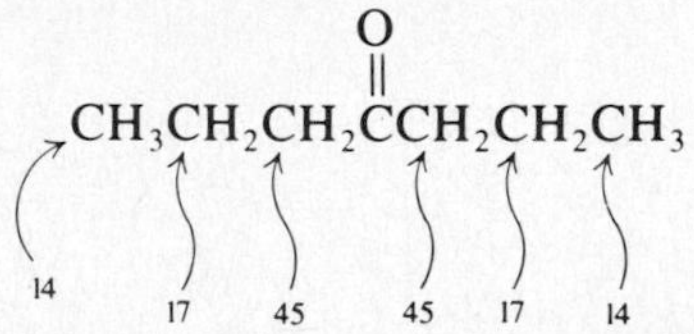

$$CH_3CH_2CH_2\overset{O}{\overset{\|}{C}}CH_2CH_2CH_3$$

14 17 45 45 17 14 (all shifts in ppm)

18.41 Compounds M and N are isomers and have a SODAR of 5. Signals in the region 125 to 140 ppm in their ^{13}C nmr spectra suggest an aromatic ring, and a peak at 200 ppm indicates a carbonyl group. An aromatic ring contributes one ring and three double bonds and a carbonyl group contributes one double bond, so the SODAR of 5 is satisfied by a benzene ring and a carbonyl group. The carbonyl group is attached directly to the benzene ring, as evidenced by the presence of a peak at m/z 105 in the mass spectra of compounds M and N.

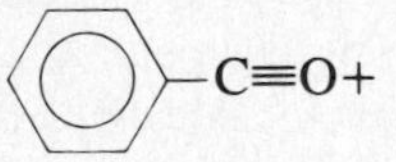

$$C_6H_5{-}C{\equiv}O^+ \quad m/z\ 105$$

Each ^{13}C nmr spectrum shows four aromatic signals, so the rings are monosubstituted.

Compound M has three unique carbons apart from its benzoyl group and so must be phenyl propyl ketone. Compound N has only two signals other than those of the benzoyl group and so must be isopropyl phenyl ketone.

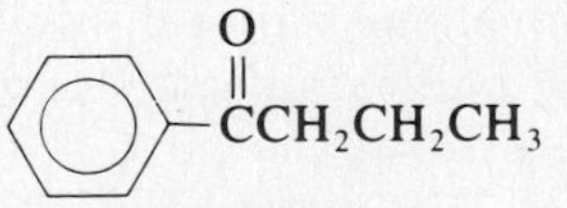

$$C_6H_5{-}\overset{O}{\overset{\|}{C}}CH_2CH_2CH_3 \qquad C_6H_5{-}\overset{O}{\overset{\|}{C}}CH(CH_3)_2$$

Compound M 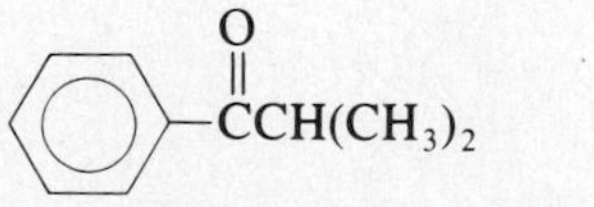 Compound N

SELF-TEST

PART A

A-1. Give the correct IUPAC name for each of the following:

(a) 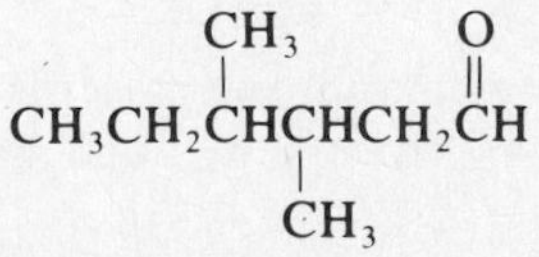 $CH_3CH_2\underset{CH_3}{\underset{|}{\overset{CH_3}{\overset{|}{C}}}}HCHCH_2\overset{O}{\overset{\|}{C}}H$

(*b*) $(CH_3)_3C\overset{O}{\overset{\|}{C}}CH_2CH(CH_3)_2$

(*c*) O Br CH_3

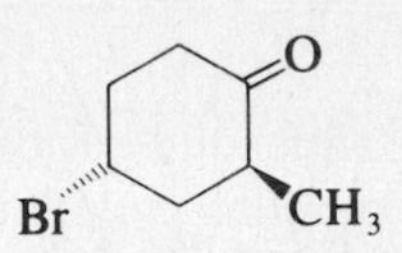

A-2. Write the structural formulas for:
(*a*) (*E*)-3-Hexen-2-one
(*b*) 3-Cyclopropyl-2,4-pentanedione
(*c*) 3-Ethyl-4-phenylpentanal

A-3. For each of the following reactions supply the structure of the missing reactant, reagent, or product:

(*a*) cyclohexanone (=O) + HCN $\xrightarrow{CN^-}$?

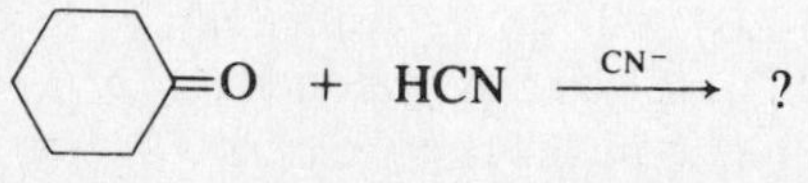

(*b*) $C_6H_5\overset{O}{\overset{\|}{C}}H$ + ? ⟶ $C_6H_5CH{=}NOH$

(*c*) $(CH_3)_2CHCH$ O O $\xrightarrow{H_2O,\ H^+}$? (two products)

(*d*) cyclopentanone (=O) + ? ⟶ cyclopentylidene =$CHCH_2CH_3$

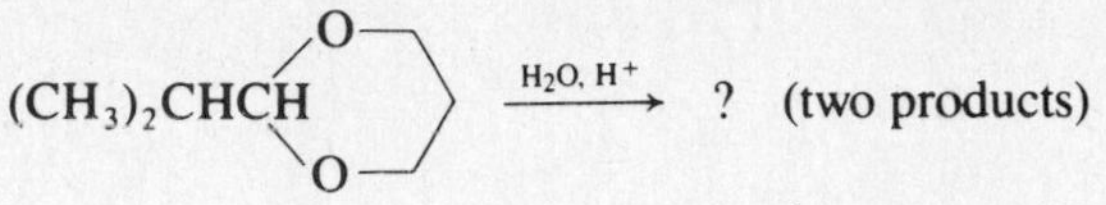

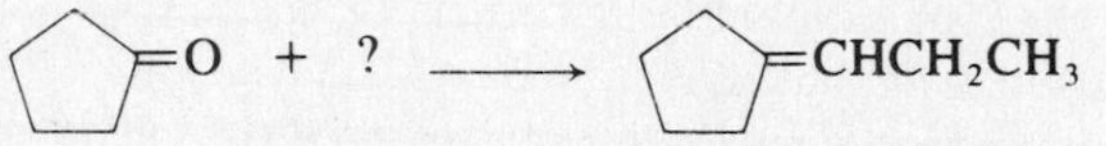

(*e*) $C_6H_5CH_2\overset{O}{\overset{\|}{C}}Cl$ $\xrightarrow{?}$ $C_6H_5CH_2\overset{O}{\overset{\|}{C}}H$

(*f*) 2-methylcyclohexanone (=O, CH_3) + $C_6H_5NHNH_2$ ⟶ ?

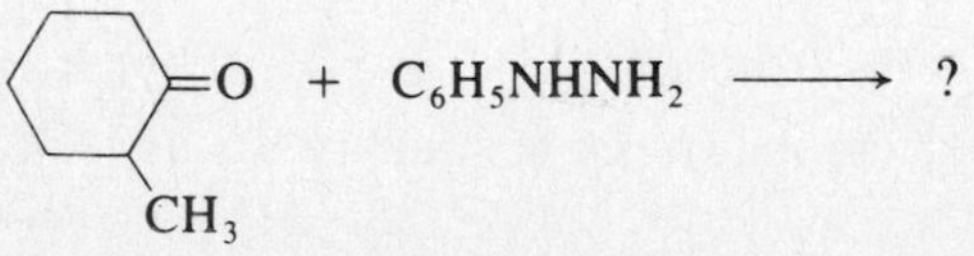

A-4. Write the structures of the products, compounds A and B, of the reaction schemes shown.

(*a*) $(C_6H_5)_3P$ + $(CH_3)_2CHBr$ $\xrightarrow[2.\ CH_3\overset{O}{\overset{\|}{C}}CH_2CH_3]{1.\ C_4H_9Li}$ Compound A

(*b*) CH_3CH_2Br $\xrightarrow[2.\ CdCl_2]{1.\ Mg,\ dry\ ether}$ Compound B
3. $CH_3CH_2\overset{O}{\overset{\|}{C}}Cl$

A-5. Outline reaction schemes to carry out each of the following interconversions, using any necessary organic or inorganic reagents.

(*a*) $(CH_3)_2C{=}O$ to $(CH_3)_2\overset{O}{C{-}}CHCH_3$ (epoxide)

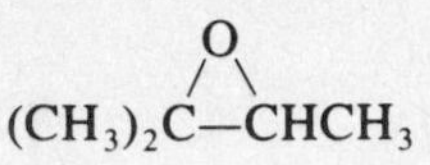

(*b*) 4-hydroxy-2-methylcyclohexanone (O, CH_3, OH) to 4-hydroxy-2,4-dimethylcyclohexanone (O, CH_3, HO CH_3)

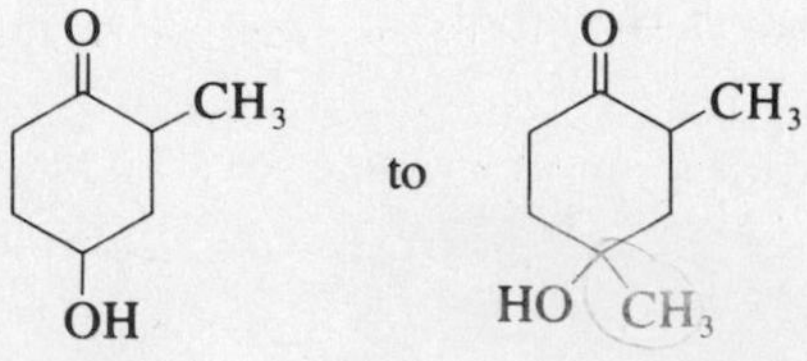

PART B

B-1. When a nucleophile encounters a ketone, the site of attack is:

(*a*) The carbon of the carbonyl
(*b*) The oxygen of the carbonyl
(*c*) Both the carbon and oxygen, with equal probability
(*d*) No attack occurs—ketones do not react with nucleophiles

B-2. Which of the following reaction schemes is (are) effective in the preparation of 3-methyl-2-butanone?

$$CH_3\overset{O}{\overset{\|}{C}}CH(CH_3)_2$$

1. $(CH_3)_2CH\overset{O}{\overset{\|}{C}}Cl + LiCu(CH_3)_2$

2. $(CH_3)_2CH\overset{O}{\overset{\|}{C}}Cl + CH_3MgI$; then H_3O^+

3. $(CH_3)_2CH\overset{O}{\overset{\|}{C}}Cl + (CH_3)_2Cd$

4. $CH_3\overset{O}{\overset{\|}{C}}Cl + (CH_3)_2CHCl + AlCl_3$

(*a*) 1, 2, and 3 (*b*) 1 and 3
(*c*) 1 only (*d*) 1, 3, and 4

B-3. Rank the following in order of increasing tendency to form a stable hydrate (least → most likely):

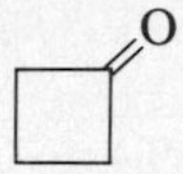

A: cyclobutanone B: cyclopropanone C: $(CH_3)_3C\overset{O}{\overset{\|}{C}}C(CH_3)_3$

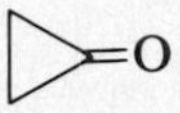

(*a*) A < B < C (*b*) C < A < B (*c*) B < A < C (*d*) B < C < A

B-4. The structure shown below would be best classified as a(n)

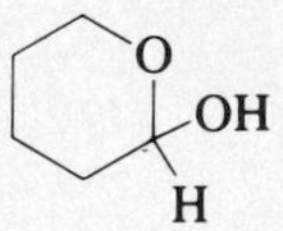

(tetrahydropyran ring with OH and H on the carbon adjacent to O)

(*a*) Acetal (*b*) Hemiacetal
(*c*) Hydrate (*d*) Cyanohydrin

B-5. Which of the following pairs of reactants is most effective in forming an enamine?

(*a*) $CH_3CH_2\overset{O}{\overset{\|}{C}}H + [(CH_3)_2CH]_2NH$

(*b*) $(CH_3)_3C\overset{O}{\overset{\|}{C}}H + (CH_3)_2NH$

(*c*) cyclohexanone $+ (CH_3)_3CNH_2$

(*d*) None of these forms an enamine.

CHAPTER

19

ENOLS, ENOLATES, AND ENAMINES

IMPORTANT TERMS AND CONCEPTS

Enolization (Secs. 19.4, 19.5) Enols are related to aldehydes or ketones by a proton-transfer equilibrium known as *keto-enol tautomerism.*

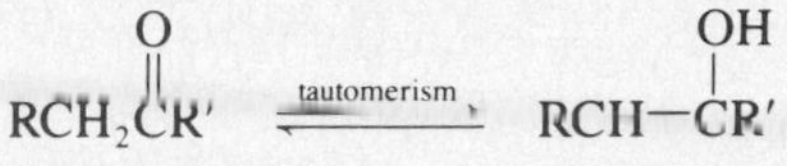

$$\underset{\text{Keto form}}{RCH_2\overset{O}{\overset{\|}{C}}R'} \xrightleftharpoons{\text{tautomerism}} \underset{\text{Enol form}}{RCH{-}\overset{OH}{\overset{|}{C}}R'}$$

Only a small amount of enol is present in a simple aldehyde or ketone. In 1,3-dicarbonyl compounds, such as 2,4-pentanedione, the enol isomer is of comparable stability with the keto form and may predominate in the equilibrium.

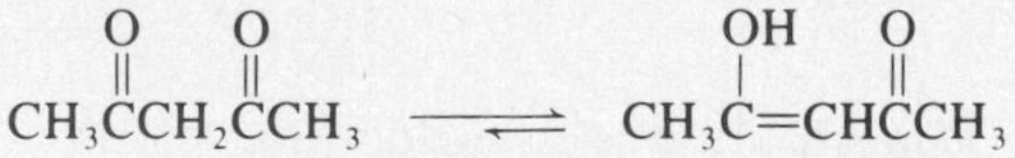

$$\underset{\text{Keto form (20\%)}}{CH_3\overset{O}{\overset{\|}{C}}CH_2\overset{O}{\overset{\|}{C}}CH_3} \rightleftharpoons \underset{\text{Enol form (80\%)}}{CH_3\overset{OH}{\overset{|}{C}}{=}CH\overset{O}{\overset{\|}{C}}CH_3}$$

Enolate Anions (Sec. 19.6) The conjugate base of an aldehyde or ketone is known as an *enolate.* The K_a's are in the range of those for alcohols and water, from 10^{-16} to 10^{-20}.

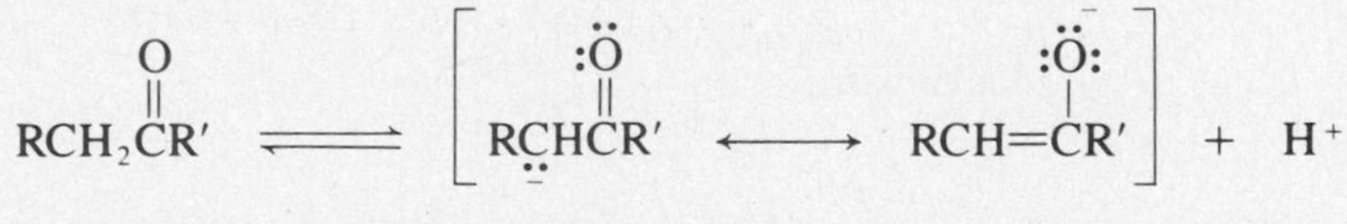

$$RCH_2\overset{O}{\overset{\|}{C}}R' \rightleftharpoons \left[R\underset{\cdot\cdot}{\overset{-}{C}}H\overset{:\ddot{O}}{\overset{\|}{C}}R' \longleftrightarrow RCH{=}\overset{:\ddot{O}:^-}{\overset{|}{C}}R' \right] + H^+$$

β-Diketones are even more acidic, owing to delocalization of the negative charge over the five-atom framework, as shown below:

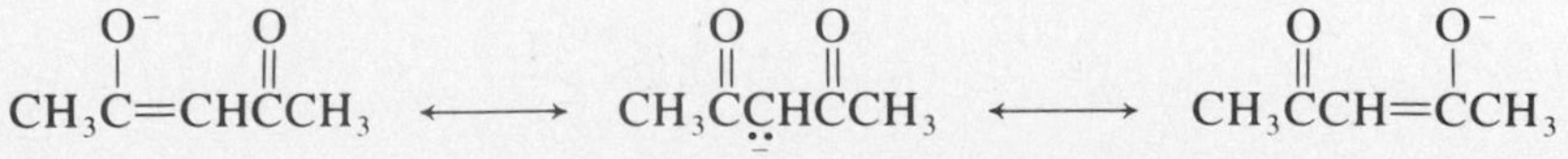

$$CH_3\overset{O^-}{\overset{|}{C}}{=}CH\overset{O}{\overset{\|}{C}}CH_3 \longleftrightarrow CH_3\overset{O}{\overset{\|}{C}}\underset{\cdot\cdot}{\overset{-}{C}}H\overset{O}{\overset{\|}{C}}CH_3 \longleftrightarrow CH_3\overset{O}{\overset{\|}{C}}CH{=}\overset{O^-}{\overset{|}{C}}CH_3$$

Conjugation in α,β-Unsaturated Aldehydes and Ketones (Secs. 19.13, 19.14) The electron delocalization resulting from overlap of the carbon-carbon double bond and the carbonyl group in an α,β-unsaturated aldehyde or ketone results in unique chemical properties for these compounds.

$$-\overset{\beta}{C}=\overset{\alpha}{C}-\underset{\underset{O}{\|}}{C}- \longleftrightarrow -\overset{+}{C}-C=\underset{\underset{O^-}{|}}{C}-$$

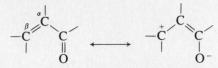

Electrophilic reagents such as bromine or peroxy acids react more slowly with the double bond of an α,β-unsaturated aldehyde or ketone than with simple alkenes. The β carbon is electrophilic and susceptible to attack by nucleophiles. Such an addition is termed *conjugate addition* or 1,4-*addition.*

$$>C=C<^{\overset{O}{\|}}_{CR} + HY \xrightarrow{Y^-} -\underset{Y}{\overset{|}{C}}-\underset{H}{\overset{|}{C}}-\overset{O}{\overset{\|}{C}}R$$

IMPORTANT REACTIONS

The Haloform Reaction (Sec. 19.7)

General:

$$R\overset{O}{\overset{\|}{C}}CH_3 \xrightarrow[HO^-]{X_2} R\overset{O}{\overset{\|}{C}}O^- + CHX_3 \quad (X = Cl, Br, I)$$

Mechanism:

$$R\overset{O}{\overset{\|}{C}}CH_3 \overset{HO^-}{\rightleftharpoons} R\overset{O^-}{\overset{|}{C}}=CH_2 \xrightarrow{X_2} R\overset{O}{\overset{\|}{C}}CH_2X \xrightarrow{HO^-,\,X_2} R\overset{O}{\overset{\|}{C}}CHX_2 \xrightarrow[X_2]{HO^-} R\overset{O}{\overset{\|}{C}}CX_3$$

$$R\overset{O}{\overset{\|}{C}}CX_3 \overset{HO^-}{\rightleftharpoons} R\underset{OH}{\overset{O^-}{\overset{|}{\underset{|}{C}}}}-CX_3 \longrightarrow R\overset{O}{\overset{\|}{C}}OH + :\bar{C}X_3 \longrightarrow RCO_2^- + HCX_3$$

Example:

$$C_6H_5\overset{O}{\overset{\|}{C}}CH_3 \xrightarrow{HO^-,\,I_2} C_6H_5CO_2^- + HCI_3$$

The Aldol Condensation (Sec. 19.9)

General:

$$2\,RCH_2\overset{O}{\overset{\|}{C}}H + HO^- \rightleftharpoons RCH_2\overset{OH}{\overset{|}{C}}H-\underset{R}{\underset{|}{C}}H\overset{O}{\overset{\|}{C}}H \xrightarrow{heat} RCH_2CH=\underset{R}{\underset{|}{C}}\overset{O}{\overset{\|}{C}}H + H_2O$$

Mechanism:

$$RCH_2\overset{O}{\overset{\|}{C}}H + HO^- \rightleftharpoons H_2O + RCH=\overset{O^-}{\overset{|}{C}}H + H\underset{O}{\underset{\|}{C}}CH_2R \rightleftharpoons$$

$$\begin{matrix} RCH-\overset{O}{\overset{\|}{C}}H \\ | \\ HCCH_2R \\ | \\ O^- \end{matrix} \overset{H_2O}{\rightleftharpoons} RCH_2\overset{OH}{\overset{|}{C}}H\underset{R}{\underset{|}{C}}H\overset{O}{\overset{\|}{C}}H$$

Example:

$$2\,CH_3CH_2\overset{O}{\overset{\|}{C}}H + HO^- \rightleftharpoons CH_3CH_2\overset{OH}{\overset{|}{C}}H\underset{CH_3}{\underset{|}{C}}H\overset{O}{\overset{\|}{C}}H \xrightarrow{heat} CH_3CH_2CH{=}\underset{CH_3}{\underset{|}{\overset{O}{\overset{\|}{C}}}}\!\!\!\!\!CH$$

Mixed (Crossed) Aldol Condensation (Sec. 19.11)

General:

$$RCH_2\overset{O}{\overset{\|}{C}}H + R'\overset{O}{\overset{\|}{C}}H \underset{}{\overset{HO^-}{\rightleftharpoons}} R'\overset{OH}{\overset{|}{C}}H\underset{R}{\underset{|}{C}}H\overset{O}{\overset{\|}{C}}H \xrightarrow{heat} R'CH{=}\underset{R}{\underset{|}{C}}\overset{O}{\overset{\|}{C}}H$$

Example:

$$C_6H_5\overset{O}{\overset{\|}{C}}H + CH_3CH_2\overset{O}{\overset{\|}{C}}H \xrightarrow[heat]{HO^-} C_6H_5CH{=}\underset{CH_3}{\underset{|}{C}}\overset{O}{\overset{\|}{C}}H$$

Conjugate Additions (Secs. 19.14, 19.15)

Michael addition:

1,3-cyclohexanedione + methyl vinyl ketone $\xrightarrow{NaOC_2H_5}$ 2-(3-oxobutyl)-1,3-cyclohexanedione

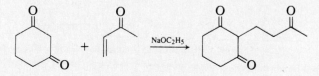

Organocopper reagents:

$LiCu(CH_2CH_3)_2$ + 2-cyclohexenone $\xrightarrow[2.\ H_2O]{1.\ ether}$ 3-ethylcyclohexanone (CH_2CH_3)

Alkylation of Enolate Anions (Sec. 19.16)

General:

$$R_2CH\overset{O}{\overset{\|}{C}}R' \xrightarrow{base} R_2C{=}\overset{O^-}{\overset{|}{C}}R' \xrightarrow{R''X} R_2\underset{R''}{\underset{|}{C}}\overset{O}{\overset{\|}{C}}R'$$

Example:

1,3-cyclohexanedione + $C_6H_5CH_2Br$ $\xrightarrow{KOH}$ 2-($CH_2C_6H_5$)-1,3-cyclohexanedione

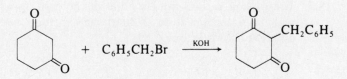

Alkylation of Enamines (Sec. 19.17)

General:

$$R\overset{NR'_2}{\overset{|}{C}}{=}CR_2 + R''CH_2X \longrightarrow R\overset{+NR'_2}{\overset{\|}{C}}{-}\underset{CH_2R''}{\underset{|}{C}}R_2 \longrightarrow R\overset{O}{\overset{\|}{C}}CR_2CH_2R'' + NHR'_2$$

Example:

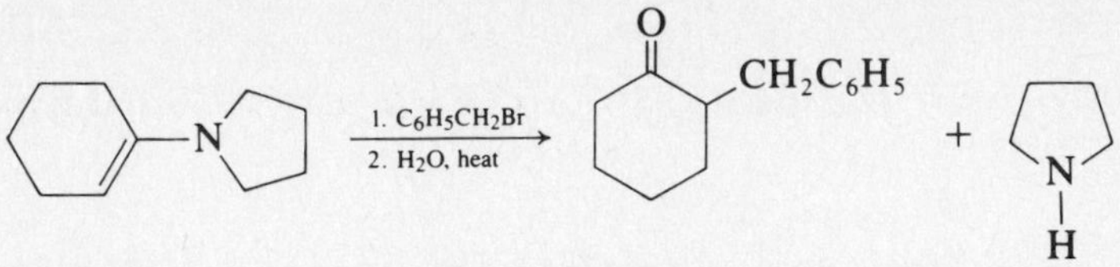

1-(1-Cyclohexenyl)pyrrolidine $\xrightarrow[\text{2. } H_2O\text{, heat}]{\text{1. } C_6H_5CH_2Br}$ 2-($CH_2C_6H_5$)cyclohexanone + pyrrolidine (N–H)

SOLUTIONS TO TEXT PROBLEMS

19.1 (*b*) There are no α hydrogen atoms in 2,2-dimethylpropanal, as the α carbon atom bears three methyl groups.

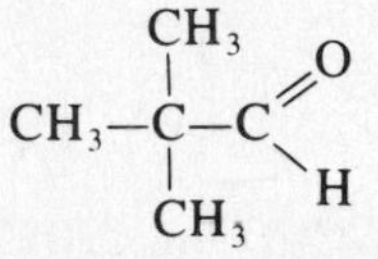

$$CH_3-C(CH_3)_2-CH{=}O \qquad \text{2,2-Dimethylpropanal}$$

(*c*) All three protons of the methyl group as well as the two benzylic protons are α hydrogens.

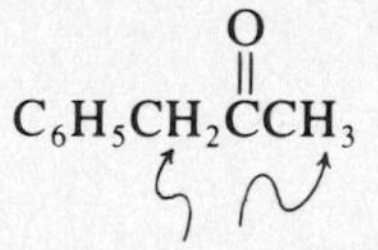

$$C_6H_5CH_2\overset{O}{\overset{\|}{C}}CH_3 \qquad \text{Benzyl methyl ketone}$$

five α hydrogens

(*d*) Cyclohexanone has four equivalent α hydrogens.

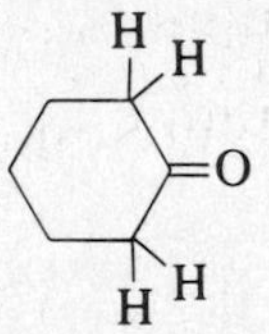

Cyclohexanone (the hydrogens indicated are the α hydrogens)

19.2 As shown in the general equation and the examples, halogen substitution is specific for the α carbon atom. The ketone 2-butanone has two nonequivalent α carbons, so substitution is possible at both positions. Both 1-chloro-2-butanone and 3-chloro-2-butanone are formed in the reaction.

$$\underset{\text{2-Butanone}}{CH_3\overset{O}{\overset{\|}{C}}CH_2CH_3} + \underset{\text{Chlorine}}{Cl_2} \xrightarrow{H^+} \underset{\text{1-Chloro-2-butanone}}{ClCH_2\overset{O}{\overset{\|}{C}}CH_2CH_3} + \underset{\text{3-Chloro-2-butanone}}{CH_3\overset{O}{\overset{\|}{C}}\underset{Cl}{\underset{|}{C}}HCH_3}$$

19.3 The carbon-carbon double bond of the enol always involves the original carbonyl carbon and the α carbon atom. 2-Butanone can form two different enols, each of which yields a different chloro ketone.

$$\underset{\text{2-Butanone}}{CH_3\overset{O}{\overset{\|}{C}}CH_2CH_3} \underset{}{\overset{\text{slow}}{\rightleftharpoons}} \underset{\text{1-Buten-2-ol (enol)}}{CH_2{=}\overset{OH}{\overset{|}{C}}CH_2CH_3} \xrightarrow[\text{fast}]{Cl_2} \underset{\text{1-Chloro-2-butanone}}{ClCH_2\overset{O}{\overset{\|}{C}}CH_2CH_3}$$

$$\underset{\text{2-Butanone}}{CH_3\overset{O}{\overset{\|}{C}}CH_2CH_3} \overset{\text{slow}}{\rightleftharpoons} \underset{\text{2-Buten-2-ol (enol)}}{CH_3\overset{OH}{\overset{|}{C}}{=}CHCH_3} \xrightarrow[\text{fast}]{Cl_2} \underset{\text{3-Chloro-2-butanone}}{CH_3\overset{O}{\overset{\|}{C}}\underset{Cl}{\underset{|}{C}}HCH_3}$$

19.4 Chlorine attacks the enol at its carbon-carbon double bond.

$$CH_2{=}C(\ddot{O}H)CH_2CH_3 + Cl{-}Cl \longrightarrow ClCH_2{-}C(={}^{+}\ddot{O}H)CH_2CH_3 + Cl^-$$

$$CH_3C(\ddot{O}H){=}CHCH_3 + Cl{-}Cl \longrightarrow CH_3C(={}^{+}\ddot{O}H){-}CH(Cl)CH_3 + Cl^-$$

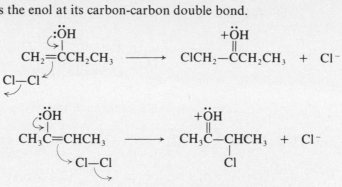

19.5 (*b*) Acetophenone can enolize only in the direction of the methyl group.

$$C_6H_5CCH_3\ (C{=}O) \rightleftharpoons C_6H_5C(OH){=}CH_2$$

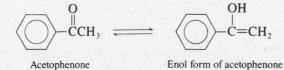

Acetophenone — Enol form of acetophenone

(*c*) The enol form of cyclohexanone is shown.

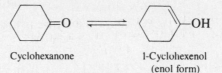

Cyclohexanone ⇌ 1-Cyclohexenol (enol form)

(*d*) Enolization of 2-methylcyclohexanone can take place in two different directions.

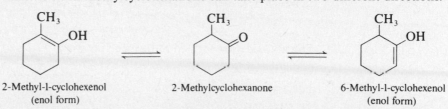

2-Methyl-1-cyclohexenol (enol form) ⇌ 2-Methylcyclohexanone ⇌ 6-Methyl-1-cyclohexenol (enol form)

19.6 (*b*) The C-3 methylene group is the one that is involved in the enolization process giving rise to the two most stable enols. Either the C-2 or C-4 carbonyl group can also be involved, so two isomeric enols are possible.

$$CH_3C(O)CH{=}C(HO)CH_2CH_3 \rightleftharpoons CH_3C(O)CH_2C(O)CH_2CH_3 \rightleftharpoons CH_3C(OH){=}CHC(O)CH_2CH_3$$

4-Hydroxy-3-hexen-2-one — 2,4-Hexanedione — 5-Hydroxy-4-hexen-3-one

(*c*) Enolization of the central methylene group can involve either of the two carbonyl groups.

$$C_6H_5C(O)CH{=}C(HO)CH_3 \rightleftharpoons C_6H_5C(O)CH_2C(O)CH_3 \rightleftharpoons C_6H_5C(OH){=}CHC(O)CH_3$$

3-Hydroxy-1-phenyl-2-buten-1-one — 1-Phenyl-1,3-butanedione — 4-Hydroxy-4-phenyl-3-buten-2-one

19.7 (*b*) Proton abstraction from the methylene group of 1-phenyl-1,3-butanedione yields a carbanion.

$$C_6H_5C(O)CH_2C(O)CH_3 + HO^- \longrightarrow C_6H_5C(O)\ddot{C}HC(O)CH_3 + H_2O$$

The three principal resonance forms of this anion are:

$$C_6H_5C(O)CH{=}C(O^-)CH_3 \longleftrightarrow C_6H_5C(O)\ddot{C}HC(O)CH_3 \longleftrightarrow C_6H_5C(O^-){=}CHC(O)CH_3$$

(*c*) Deprotonation at C-2 of this β-dicarbonyl compound yields the carbanion shown.

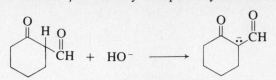

The three most stable resonance forms of the anion are:

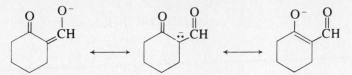

19.8 (*b*) Approaching this problem in the same way as that of (*a*), write the structure of the enolate ion from 2-methylbutanal.

$$CH_3CH_2CH(CH_3)CH{=}O + HO^- \rightleftharpoons CH_3CH_2\ddot{C}^-(CH_3)CH{=}O \longleftrightarrow CH_3CH_2C(CH_3){=}CHO^-$$

2-Methylbutanal — Enolate of 2-methylbutanal

This enolate adds to the carbonyl group of the aldehyde.

$$CH_3CH_2CH(CH_3)CH{=}O + {}^-{:}C(CH_3)(CH_2CH_3)HC{=}O \longrightarrow CH_3CH_2CH(CH_3)CH(O^-){-}C(CH_3)(CH_2CH_3)HC{=}O$$

2-Methylbutanal — Enolate of 2-methylbutanal

The alkoxide ion formed in this step accepts a proton from solvent to yield the product of aldol addition.

$$CH_3CH_2CH(CH_3)CH(O^-){-}C(CH_3)(CH_2CH_3)HC{=}O + H_2O \longrightarrow CH_3CH_2CH(CH_3)CH(OH){-}C(CH_3)(CH_2CH_3)HC{=}O + HO^-$$

2-Ethyl-2,4-dimethyl-3-hydroxyhexanal

(*c*) The aldol addition product of 3-methylbutanal can be identified through the same mechanistic approach.

$$(CH_3)_2CHCH_2CH{=}O + HO^- \rightleftharpoons (CH_3)_2CH\ddot{C}^-HCH{=}O + H_2O$$

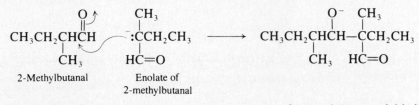

3-Methylbutanal — Enolate of 3-methylbutanal

$$(CH_3)_2CHCH_2CH{=}O + {:}\bar{C}HCH(CH_3)_2(HC{=}O) \longrightarrow (CH_3)_2CHCH_2CH(O^-){-}CH(HC{=}O)CH(CH_3)_2$$

3-Methylbutanal — Enolate of 3-methylbutanal

$$\xrightarrow{H_2O} (CH_3)_2CHCH_2CH(OH){-}CH(HC{=}O)CH(CH_3)_2 + OH^-$$

3-Hydroxy-2-isopropyl-5-methylhexanal

19.9 Dehydration of the aldol addition product involves loss of a proton from the α carbon atom and hydroxide from the β carbon atom.

$$R_2C(OH){-}CH(H)CHO + {}^-OH \longrightarrow R_2C{=}CHCH{=}O + H_2O + HO^-$$

(*b*) The product of aldol addition of 2-methylbutanal has no α hydrogens. It cannot dehydrate to an aldol condensation product.

$$2\ CH_3CH_2CH(CH_3)CH{=}O \overset{HO^-}{\rightleftharpoons} CH_3CH_2CH(CH_3)CH(OH){-}C(CH_3)(HC{=}O)CH_2CH_3$$

2-Methylbutanal — (no protons on α carbon atom)

(*c*) Aldol condensation is possible with 3-methylbutanal.

$$2\ (CH_3)_2CHCH_2CH{=}O \overset{HO^-}{\rightleftharpoons} (CH_3)_2CHCH_2CH(OH)CH(HC{=}O)CH(CH_3)_2 \xrightarrow{-H_2O} (CH_3)_2CHCH_2CH{=}C(HC{=}O)CH(CH_3)_2$$

3-Methylbutanal — 2-Isopropyl-5-methyl-2-hexenal

19.10 Aldol addition involves reaction of the ketone enolate of one cyclohexanone molecule with the carbonyl group of another.

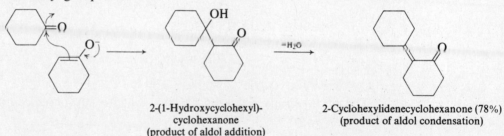

(cyclohexanone + cyclohexanone enolate) ⟶ 2-(1-Hydroxycyclohexyl)-cyclohexanone (product of aldol addition) $\xrightarrow{-H_2O}$ 2-Cyclohexylidenecyclohexanone (78%) (product of aldol condensation)

19.11 (*b*) The only enolate that can be formed from *tert*-butyl methyl ketone arises by proton abstraction from the methyl group.

$$(CH_3)_3CCCH_3(=O) + HO^- \rightleftharpoons (CH_3)_3CC(=O)\ddot{C}H_2^-$$

tert-Butyl methyl ketone — Enolate of *tert*-butyl methyl ketone

This enolate adds to the carbonyl group of benzaldehyde to give the mixed aldol addition product, which then dehydrates under the reaction conditions.

$$C_6H_5CH{=}O + {:}\bar{C}H_2C(=O)C(CH_3)_3 \longrightarrow C_6H_5CH(O^-)CH_2C(=O)C(CH_3)_3$$

Benzaldehyde — Enolate of *tert*-butyl methyl ketone

$$\xrightarrow{H_2O} C_6H_5CH(OH)CH_2C(=O)C(CH_3)_3 \xrightarrow{-H_2O} C_6H_5CH{=}CHC(=O)C(CH_3)_3$$

Product of mixed aldol addition — 4,4-Dimethyl-1-phenyl-1-penten-3-one (product of mixed aldol condensation)

(c) The enolate of cyclohexanone adds to benzaldehyde. Dehydration of the mixed aldol addition product takes place under the reaction conditions to give the mixed aldol condensation product called benzylidenecyclohexanone.

$$\text{Cyclohexanone} + C_6H_5\overset{\displaystyle O}{\overset{\|}{C}}H \xrightarrow{HO^-} \text{Benzylidenecyclohexanone } (2\text{-}(=CHC_6H_5)\text{cyclohexanone})$$

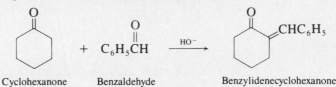

Cyclohexanone Benzaldehyde Benzylidenecyclohexanone

19.12 Mesityl oxide is an α,β-unsaturated ketone. Traces of acids or bases can catalyze its isomerization so that some of the less stable β,γ-unsaturated isomer is present.

$$(CH_3)_2C{=}CH\overset{\displaystyle O}{\overset{\|}{C}}CH_3 \rightleftharpoons H_2C{=}C(CH_3){-}CH_2\overset{\displaystyle O}{\overset{\|}{C}}CH_3$$

Mesityl oxide; 4-methyl-3-penten-2-one (more stable) 4-Methyl-4-penten-2-one (less stable)

19.13 It is the enolate of dibenzyl ketone that adds to methyl vinyl ketone in the conjugate addition step.

$$C_6H_5CH_2\overset{\displaystyle O}{\overset{\|}{C}}CH_2C_6H_5 + CH_2{=}CH\overset{\displaystyle O}{\overset{\|}{C}}CH_3 \xrightarrow[CH_3OH]{NaOCH_3} C_6H_5CH_2\overset{\displaystyle O}{\overset{\|}{C}}CHC_6H_5(CH_2CH_2\overset{\displaystyle O}{\overset{\|}{C}}CH_3)$$

Dibenzyl ketone Methyl vinyl ketone 1,3-Diphenyl-2,6-heptanedione

via:

$$C_6H_5CH_2\overset{\displaystyle O}{\overset{\|}{C}}\ddot{C}^-HC_6H_5 + CH_2{=}CH\overset{\displaystyle O}{\overset{\|}{C}}CH_3 \longrightarrow C_6H_5CH_2\overset{\displaystyle O}{\overset{\|}{C}}CHC_6H_5{-}CH_2{-}\ddot{C}^-H\overset{\displaystyle O}{\overset{\|}{C}}CH_3$$

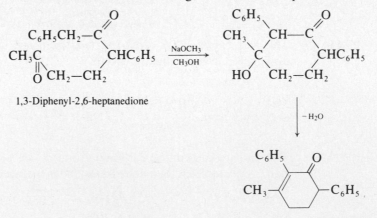

The intramolecular aldol condensation that gives the observed product is:

$$\text{1,3-Diphenyl-2,6-heptanedione} \xrightarrow[CH_3OH]{NaOCH_3} \text{cyclic aldol: } C(CH_3)(OH) \text{ ring with } CH(C_6H_5){-}C(=O){-}CHC_6H_5{-}CH_2{-}CH_2 \xrightarrow{-H_2O} \text{3-Methyl-2,6-diphenyl-2-cyclohexen-1-one}$$

1,3-Diphenyl-2,6-heptanedione

3-Methyl-2,6-diphenyl-2-cyclohexen-l-one

19.14 A second solution to the synthesis of 4-methyl-2-octanone by conjugate addition of a lithium dialkylcuprate reagent to an α,β-unsaturated ketone is revealed by the disconnection shown:

$$CH_3CH_2CH_2CH_2\underset{\displaystyle CH_3}{\underset{|}{C}}HCH_2\overset{\displaystyle O}{\overset{\|}{C}}CH_3 \Longrightarrow CH_3CH_2CH_2CH_2CH{=}CH\overset{\displaystyle O}{\overset{\|}{C}}CH_3 \quad {}^-CH_3$$

disconnect this bond

According to this disconnection, the methyl group is derived from lithium dimethylcuprate.

$$\mathrm{CH_3CH_2CH_2CH_2CH{=}CH\overset{O}{\overset{\|}{C}}CH_3} + \mathrm{LiCu(CH_3)_2} \longrightarrow \mathrm{CH_3CH_2CH_2CH_2\underset{CH_3}{\underset{|}{C}}HCH_2\overset{O}{\overset{\|}{C}}CH_3}$$

3-Octen-2-one — Lithium dimethylcuprate — 4-Methyl-2-octanone

19.15 (*b*) Alkylation of the pyrrolidine enamine of cyclohexanone with 1-bromobutane (or 1-iodobutane), followed by hydrolysis, will yield the desired compound.

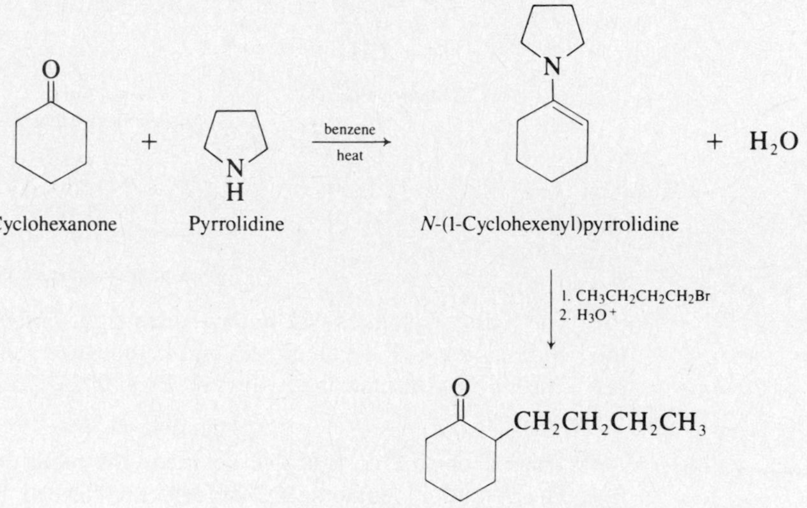

Cyclohexanone + Pyrrolidine $\xrightarrow[\text{heat}]{\text{benzene}}$ N-(1-Cyclohexenyl)pyrrolidine + H_2O

1. $CH_3CH_2CH_2CH_2Br$
2. H_3O^+

2-Butylcyclohexanone ($CH_2CH_2CH_2CH_3$)

(*c*) Examine this problem by the retrosynthetic method. Disconnect a bond to the α carbon atom.

$$\mathrm{CH_3CH_2\overset{O}{\overset{\|}{C}}\underset{CH_3}{\underset{|}{C}}H \wr CH_2C_6H_5} \Longrightarrow \mathrm{CH_3CH_2\overset{O}{\overset{\|}{C}}\underset{CH_3}{\underset{|}{C}}H^-} \quad \mathrm{\overset{+}{C}H_2C_6H_5}$$

2-Methyl-1-phenyl-3-pentanone

This disconnection reveals the synthetic plan to be followed. A benzyl halide is needed as the alkylating agent and an enamine of 3-pentanone is the nucleophile.

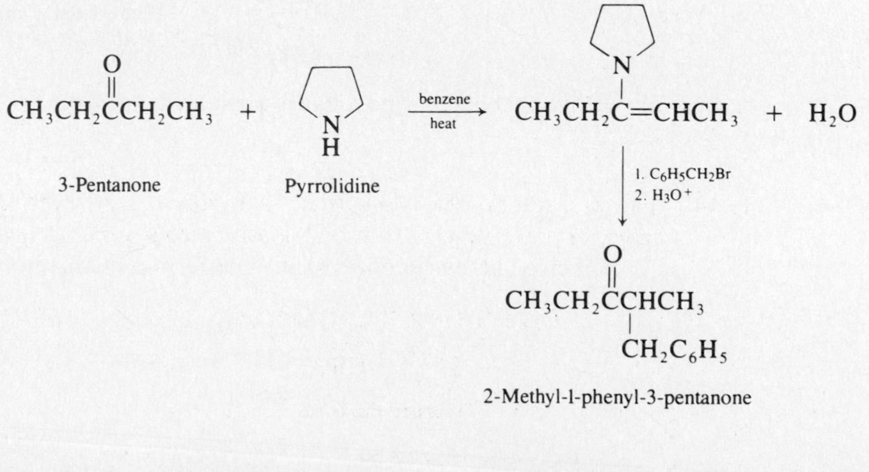

$$\mathrm{CH_3CH_2\overset{O}{\overset{\|}{C}}CH_2CH_3} + \text{Pyrrolidine} \xrightarrow[\text{heat}]{\text{benzene}} \mathrm{CH_3CH_2\overset{N(pyrrolidinyl)}{\overset{|}{C}}{=}CHCH_3} + \mathrm{H_2O}$$

3-Pentanone — Pyrrolidine

1. $C_6H_5CH_2Br$
2. H_3O^+

$$\mathrm{CH_3CH_2\overset{O}{\overset{\|}{C}}\underset{CH_2C_6H_5}{\underset{|}{C}}HCH_3}$$

2-Methyl-1-phenyl-3-pentanone

19.16 (*a*) In addition to the double bond of the carbonyl group, there must be a double bond or ring elsewhere in the molecule in order to satisfy the molecular formula C_4H_6O. There are a total of seven isomers. They are:

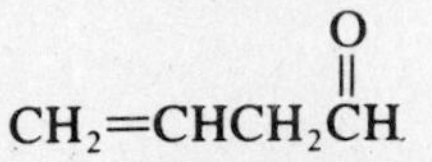

$CH_2{=}CHCH_2\overset{O}{\overset{\|}{C}}H$ — 3-Butenal

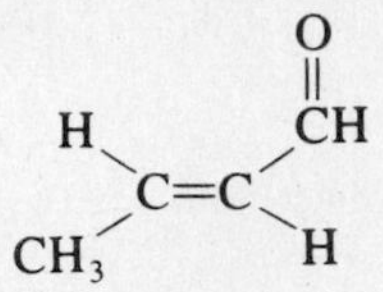

(E)-2-Butenal: H and CH_3 on one carbon, $\overset{O}{\overset{\|}{C}}H$ and H on the other (H/CH on same side top) — (*E*)-2-Butenal

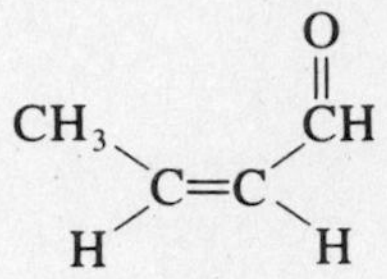

CH_3, H on one carbon; $\overset{O}{\overset{\|}{C}}H$, H on the other — (*Z*)-2-Butenal

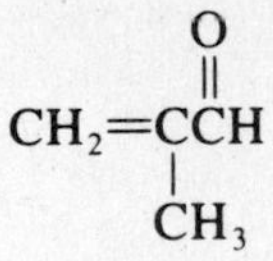

$CH_2{=}\underset{CH_3}{\underset{|}{C}}\overset{O}{\overset{\|}{C}}H$ — 2-Methylpropenal

$CH_2{=}CH\overset{O}{\overset{\|}{C}}CH_3$ — 3-Buten-2-one (methyl vinyl ketone)

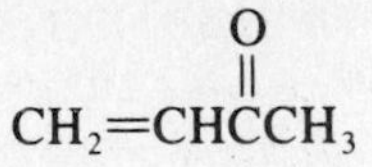

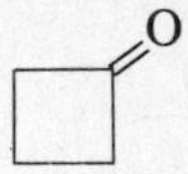

Cyclobutanone

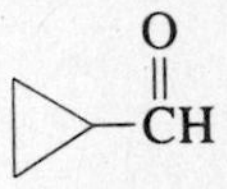

Cyclopropyl$-\overset{O}{\overset{\|}{C}}H$ — Cyclopropanecarbaldehyde

(*b*) The *E* and *Z* isomers of 2-butenal are stereoisomers.
(*c*) None of the C_4H_6O aldehydes and ketones are chiral.
(*d*) The α,β-unsaturated aldehydes are (*E*)- and (*Z*)-$CH_3CH{=}CHCHO$; and $CH_2{=}\underset{CH_3}{\underset{|}{C}}CHO$.
There is one α,β-unsaturated ketone in the group: $CH_2{=}CH\overset{O}{\overset{\|}{C}}CH_3$.
(*e*) The *E* and *Z* isomers of 2-butenal are formed by the aldol condensation of acetaldehyde.
(*f*) The isomer that has four equivalent α hydrogens is cyclobutanone.

19.17 (*a*) 2-Methylpropanal has the greater enol content (since the other compound contains no enol at all).

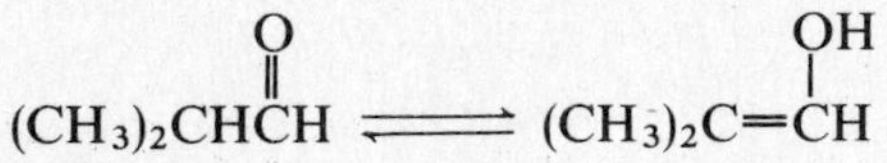

$$\underset{\text{2-Methylpropanal}}{(CH_3)_2CH\overset{O}{\overset{\|}{C}}H} \rightleftharpoons \underset{\text{Enol form}}{(CH_3)_2C{=}\overset{OH}{\overset{|}{C}}H}$$

Although the enol content of 2-methylpropanal is quite small, the compound is nevertheless capable of enolization, whereas the other compound, 2,2-dimethylpropanal, cannot enolize—it has no α hydrogens.

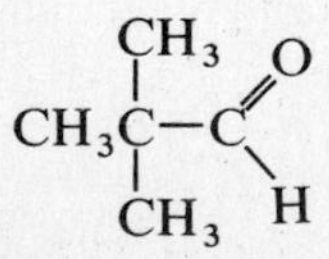

$$CH_3\underset{CH_3}{\overset{CH_3}{C}}-\overset{O}{\overset{\|}{C}}H$$ (Enolization is impossible)

(*b*) Benzophenone has no α hydrogens; it cannot form an enol.

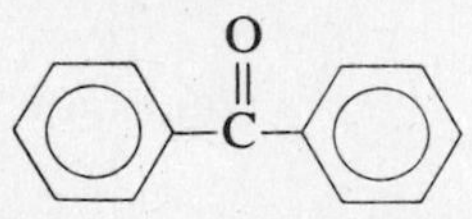

$$C_6H_5\overset{O}{\overset{\|}{C}}C_6H_5$$ Enolization is impossible)

Dibenzyl ketone enolizes slightly to form a small amount of enol.

$$\underset{\text{Dibenzyl ketone}}{C_6H_5CH_2\overset{O}{\overset{\|}{C}}CH_2C_6H_5} \rightleftharpoons \underset{\text{Enol form}}{C_6H_5CH{=}\overset{OH}{\overset{|}{C}}CH_2C_6H_5}$$

(*c*) Here we are comparing a simple ketone, dibenzyl ketone, with a β-diketone. The β-diketone enolizes to a much greater extent than the simple ketone because its enol form is stabilized by conjugation of the double bond with the remaining carbonyl group and by intramolecular hydrogen bonding.

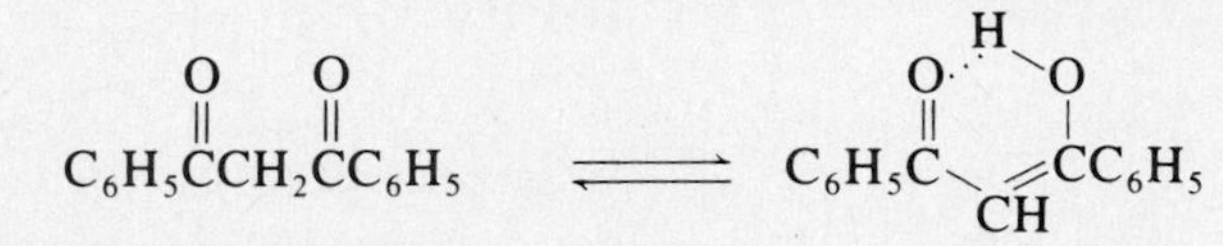

$$C_6H_5\overset{O}{\overset{\|}{C}}CH_2\overset{O}{\overset{\|}{C}}C_6H_5 \rightleftharpoons C_6H_5C(=O\cdots H-O)C{=}CH\,CC_6H_5$$

1,3-Diphenyl-1,3-propanedione — Enol form

(*d*) The enol content of cyclohexanone is quite small, whereas the enol form of 2,4-cyclohexadienone is phenol and therefore enolization is essentially complete.

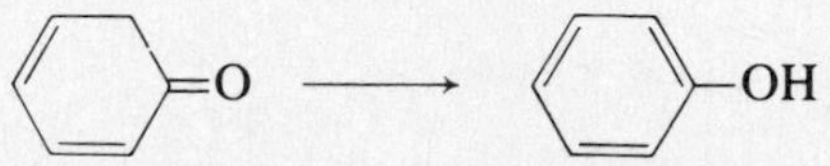

Keto form → Enol form (much more stable)

(*e*) A small amount of enol is in equilibrium with cyclopentanone.

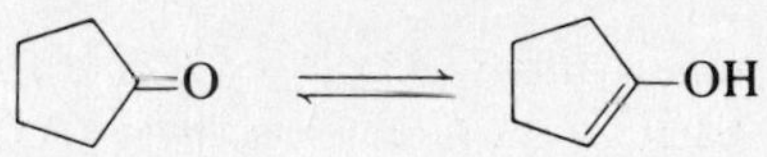

Cyclopentanone ⇌ Enol form

Cyclopentadienone does not form a stable enol. Enolization would lead to a highly strained allene-type compound.

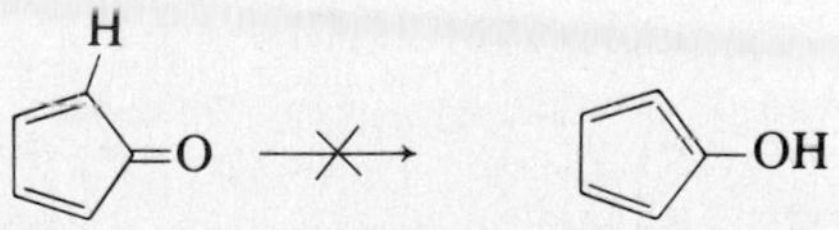

(Not stable; highly strained)

(*f*) The β-diketone is more extensively enolized.

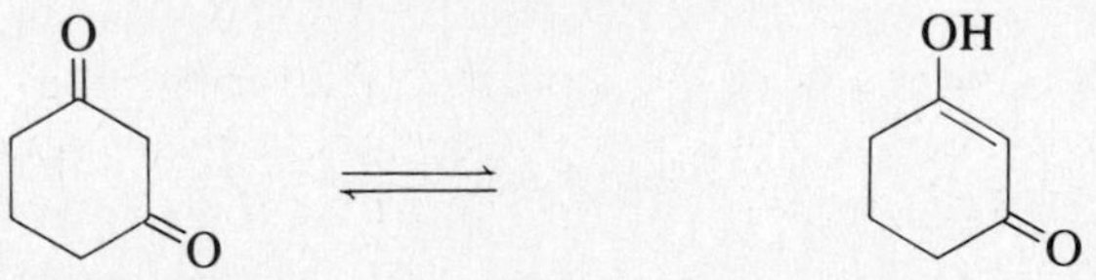

1,3-Cyclohexanedione ⇌ Enol form (double bond conjugated with carbonyl group)

The double bond of the enol form of 1,4-cyclohexanedione is not conjugated with the carbonyl group. Its enol content is expected to be similar to that of cyclohexanone.

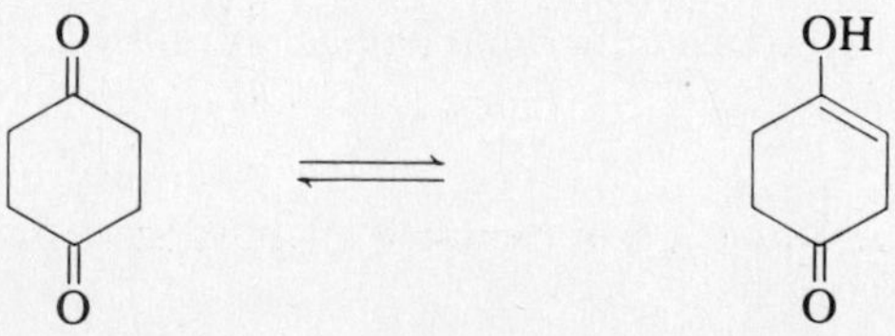

1,4-Cyclohexanedione ⇌ Enol form (not particularly stable; double bond and carbonyl group not conjugated)

19.18 (*a*) Chlorination of 3-phenylpropanal under conditions of acid catalysis occurs via the enol form and yields the α-chloro derivative.

$$C_6H_5CH_2CH_2\overset{O}{\overset{\|}{C}}H + Cl_2 \xrightarrow{\text{acetic acid}} C_6H_5CH_2\underset{Cl}{\underset{|}{C}}H\overset{O}{\overset{\|}{C}}H + HCl$$

3-Phenylpropanal → 2-Chloro-3-phenylpropanal

(*b*) Aldehydes undergo aldol addition on treatment with base.

$$2\,C_6H_5CH_2CH_2\overset{O}{\overset{\|}{C}}H \xrightarrow[\text{ethanol, }10^\circ C]{NaOH} C_6H_5CH_2CH_2\underset{OH}{\underset{|}{C}}H\overset{HC=O}{\overset{|}{C}}HCH_2C_6H_5$$

3-Phenylpropanal → 2-Benzyl-3-hydroxy-5-phenylpentanal

(*c*) Dehydration of the aldol addition product occurs when the reaction is carried out at elevated temperature.

$$2\,C_6H_5CH_2CH_2\overset{O}{\overset{\|}{C}}H \xrightarrow[\text{ethanol, }70^\circ C]{NaOH} C_6H_5CH_2CH_2CH{=}\overset{HC=O}{\overset{|}{C}}CH_2C_6H_5$$

3-Phenylpropanal → 2-Benzyl-5-phenyl-2-pentenal

(*d*) Sodium borohydride reduces the aldehyde function to the corresponding primary alcohol.

$$C_6H_5CH_2CH_2CH{=}\overset{HC=O}{\overset{|}{C}}CH_2C_6H_5 \xrightarrow[\text{ethanol}]{NaBH_4} C_6H_5CH_2CH_2CH{=}\overset{CH_2OH}{\overset{|}{C}}CH_2C_6H_5$$

2-Benzyl-5-phenyl-2-pentenal → 2-Benzyl-5-phenyl-2-penten-1-ol

(*e*) A characteristic reaction of α,β-unsaturated carbonyl compounds is their tendency to undergo conjugate addition on treatment with weakly basic nucleophiles.

$$C_6H_5CH_2CH_2CH{=}\overset{HC=O}{\overset{|}{C}}CH_2C_6H_5 \xrightarrow[H^+]{NaCN} C_6H_5CH_2CH_2\underset{CN}{\underset{|}{C}}H\overset{HC=O}{\overset{|}{C}}HCH_2C_6H_5$$

2-Benzyl-5-phenyl-2-pentenal → 2-Benzyl-3-cyano-5-phenylpentanal

(*f*) Ylides react with α,β-unsaturated carbonyl compounds at the carbonyl group.

$$C_6H_5CH_2CH_2CH{=}\overset{HC=O}{\overset{|}{C}}CH_2C_6H_5 \xrightarrow{(C_6H_5)_3\overset{+}{P}-\ddot{C}H_2^-} C_6H_5CH_2CH_2CH{=}\overset{CH=CH_2}{\overset{|}{C}}CH_2C_6H_5$$

2-Benzyl-5-phenyl-2-pentenal → 3-Benzyl-6-phenyl-1,3-hexadiene

19.19 (*a*) Ketones undergo α halogenation by way of their enol form.

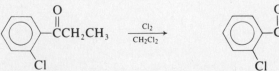

$$o\text{-}ClC_6H_4\overset{O}{\overset{\|}{C}}CH_2CH_3 \xrightarrow[CH_2Cl_2]{Cl_2} o\text{-}ClC_6H_4\overset{O}{\overset{\|}{C}}\underset{Cl}{\underset{|}{C}}HCH_3$$

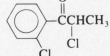

o-Chloropropiophenone (*o*-chlorophenyl ethyl ketone) → 1-Chloroethyl *o*-chlorophenyl ketone (94%)

(*b*) Methyl ketones undergo haloform cleavage on treatment with chlorine, bromine, or iodine in base.

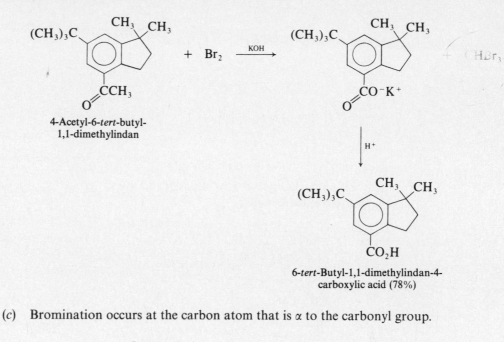

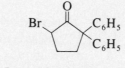

4-Acetyl-6-*tert*-butyl-1,1-dimethylindan + Br_2 $\xrightarrow{KOH}$ potassium carboxylate (CO^-K^+) + $CHBr_3$ $\xrightarrow{H^+}$ 6-*tert*-Butyl-1,1-dimethylindan-4-carboxylic acid (78%)

(*c*) Bromination occurs at the carbon atom that is α to the carbonyl group.

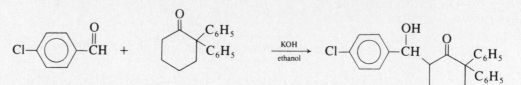

2,2-Diphenylcyclopentanone $\xrightarrow[\text{diethyl ether}]{Br_2}$ 2-Bromo-5,5-diphenylcyclopentanone (76%)

(*d*) The reaction is a mixed aldol condensation. The enolate of 2,2-diphenylcyclohexanone reacts with *p*-chlorobenzaldehyde. Elimination of the aldol addition product occurs readily to yield the α,β-unsaturated ketone as the isolated product.

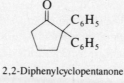

p-Chlorobenzaldehyde + 2,2-Diphenylcyclohexanone $\xrightarrow[\text{ethanol}]{KOH}$ aldol addition product (Not isolated) $\xrightarrow{-H_2O}$ 2-(*p*-Chlorobenzylidene)-6,6-diphenylcyclohexanone (84%)

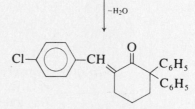

(*e*) The aldehyde given as the starting material may not be familiar to you; it is called furfural and is based on a furan unit as an aromatic ring. Furfural cannot form an enolate. It reacts with the enolate of acetone in a manner much as benzaldehyde would.

$$\text{Furfural (furyl–CHO)} + CH_3\overset{O}{\overset{\|}{C}}CH_3 \xrightarrow[\text{water}]{\text{NaOH}} \text{furyl–}\underset{OH}{\underset{|}{C}H}CH_2\overset{O}{\overset{\|}{C}}CH_3 \text{ (Not isolated)}$$

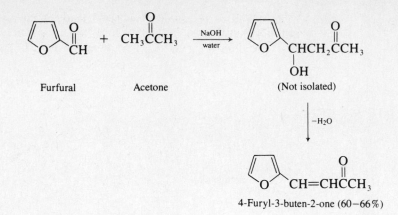

Furfural — Acetone — (Not isolated)

$$\xrightarrow{-H_2O} \text{furyl–}CH{=}CH\overset{O}{\overset{\|}{C}}CH_3$$

4-Furyl-3-buten-2-one (60–66%)

(*f*) Lithium dialkylcuprates transfer an alkyl group to the β carbon atom of α,β-unsaturated ketones.

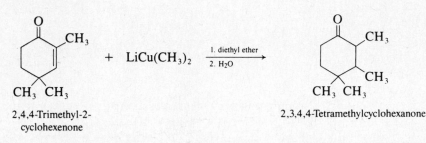

$$\text{2,4,4-Trimethyl-2-cyclohexenone} + LiCu(CH_3)_2 \xrightarrow[\text{2. } H_2O]{\text{1. diethyl ether}} \text{2,3,4,4-Tetramethylcyclohexanone}$$

2,4,4-Trimethyl-2-cyclohexenone — 2,3,4,4-Tetramethylcyclohexanone

A mixture of stereoisomers was obtained in 67 percent yield in this reaction.

(*g*) There are two nonequivalent α carbon atoms in the starting ketone. While enolate formation is possible at either position, only reaction at the methylene carbon leads to an intermediate that can undergo dehydration.

$$\text{1-decalone} \xrightarrow[C_6H_5CHO]{HO^-} \text{2-(}\underset{}{\overset{OH}{\overset{|}{C}H}}C_6H_5\text{)-1-decalone} \xrightarrow{-H_2O} \text{2-(}{=}CHC_6H_5\text{)-1-decalone}$$

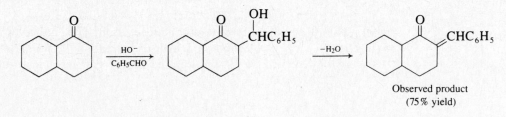

Observed product (75% yield)

Reaction at the other α position gives an intermediate that cannot dehydrate.

$$\text{1-decalone} \underset{}{\overset{HO^-,\ C_6H_5CHO}{\rightleftharpoons}} \text{8a-(}C_6H_5\overset{OH}{\overset{|}{C}H}\text{)-1-decalone}$$

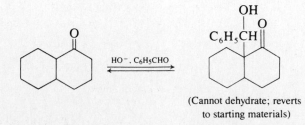

(Cannot dehydrate; reverts to starting materials)

19.20 (*a*) Conversion of 3-pentanone to 2-bromo-3-pentanone is best accomplished by acid-catalyzed bromination via the enol. Bromine in acetic acid is the customary reagent for this transformation.

$$CH_3CH_2\overset{O}{\overset{\|}{C}}CH_2CH_3 \xrightarrow[\text{acetic acid}]{Br_2} CH_3\underset{Br}{\underset{|}{C}H}\overset{O}{\overset{\|}{C}}CH_2CH_3$$

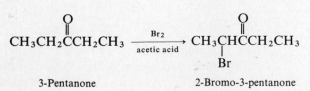

3-Pentanone — 2-Bromo-3-pentanone

Base-promoted halogenation is less useful because it is difficult to limit the reaction to monohalogenation. Also, 2-bromo-3-pentanone, once formed, tends to undergo nucleophilic substitution under the basic conditions of its formation.

$$CH_3CH_2\overset{O}{\overset{\|}{C}}CH_2CH_3 \xrightarrow[HO^-]{Br_2} CH_3\underset{Br}{\underset{|}{C}H}\overset{O}{\overset{\|}{C}}CH_2CH_3$$

$$\xrightarrow[HO^-]{Br_2} CH_3\underset{Br}{\underset{|}{\overset{Br}{\overset{|}{C}}}}-\overset{O}{\overset{\|}{C}}CH_2CH_3$$

$$\xrightarrow{HO^-} CH_3\underset{OH}{\underset{|}{C}H}\overset{O}{\overset{\|}{C}}CH_2CH_3$$

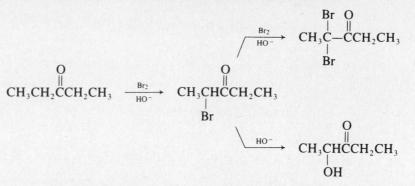

(*b*) Once 2-bromo-3-pentanone has been prepared, its dehydrohalogenation by base converts it to the desired α,β-unsaturated ketone 1-penten-3-one.

$$CH_3\underset{Br}{\underset{|}{C}H}\overset{O}{\overset{\|}{C}}CH_2CH_3 \xrightarrow{KOC(CH_3)_3} CH_2{=}CH\overset{O}{\overset{\|}{C}}CH_2CH_3$$

2-Bromo-3-pentanone — 1-Penten-3-one

Potassium *tert*-butoxide is a good base for bringing about elimination reactions of secondary alkyl halides; suitable solvents include *tert*-butyl alcohol and dimethyl sulfoxide.

(*c*) Reduction of the carbonyl group of 1-penten-3-one converts it to the desired alcohol.

$$CH_2{=}CH\overset{O}{\overset{\|}{C}}CH_2CH_3 \xrightarrow[2.\ H_2O]{1.\ LiAlH_4,\ \text{diethyl ether}} CH_2{=}CH\underset{OH}{\underset{|}{C}H}CH_2CH_3$$

1-Penten-3 one — 1-Penten-3-ol

Catalytic hydrogenation would not be suitable for this reaction because reduction of the double bond would accompany carbonyl reduction.

(*d*) Conversion of 3-pentanone to 3-hexanone requires addition of a methyl group to the β carbon atom.

$$CH_3CH_2CH_2\overset{O}{\overset{\|}{C}}CH_2CH_3 \Longrightarrow CH_3^- + CH_2{=}CH\overset{O}{\overset{\|}{C}}CH_2CH_3$$

The best way to add an alkyl group to the β carbon of a ketone is via conjugate addition of a dialkylcuprate reagent to an α,β-unsaturated ketone.

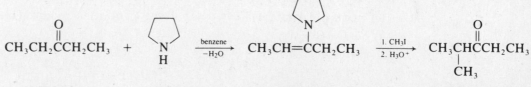

$$CH_2{=}CH\overset{O}{\overset{\|}{C}}CH_2CH_3 \xrightarrow[2.\ H_2O]{1.\ LiCu(CH_3)_2} CH_3CH_2CH_2\overset{O}{\overset{\|}{C}}CH_2CH_3$$

1-Penten-3-one [prepared as described in (*b*)] — 3-Hexanone

(*e*) In this problem a methyl group must be added to the α carbon atom of 3-pentanone. α-Alkylation of ketones can be achieved by way of the corresponding enamine.

$$CH_3CH_2\overset{O}{\overset{\|}{C}}CH_2CH_3 + \text{pyrrolidine (N–H)} \xrightarrow[-H_2O]{\text{benzene}} CH_3CH{=}\underset{|}{\overset{N(\text{pyrrolidinyl})}{C}}CH_2CH_3 \xrightarrow[2.\ H_3O^+]{1.\ CH_3I} CH_3\underset{CH_3}{\underset{|}{C}H}\overset{O}{\overset{\|}{C}}CH_2CH_3$$

3-Pentanone — Pyrrolidine — Pyrrolidine enamine of 3-pentanone — 2-Methyl-3-pentanone (*f*) The compound to be prepared is the mixed aldol condensation product of 3-pentanone and benzaldehyde.

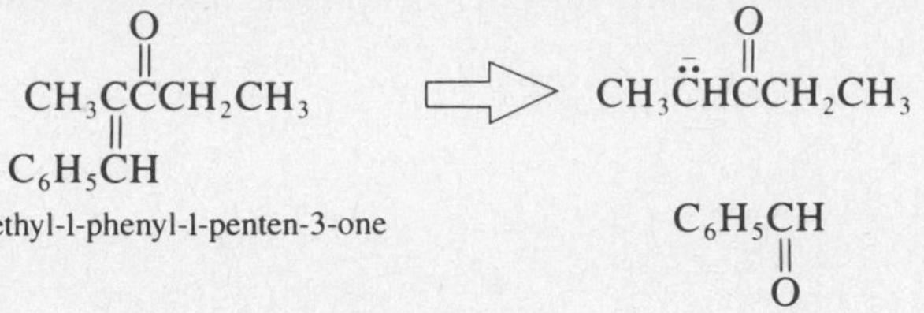

$$\underset{\underset{C_6H_5CH}{\|}}{CH_3\overset{\overset{O}{\|}}{C}CCH_2CH_3} \Longrightarrow CH_3\overset{-}{\ddot{C}}H\overset{\overset{O}{\|}}{C}CH_2CH_3 \qquad C_6H_5\underset{\underset{O}{\|}}{CH}$$

2-Methyl-1-phenyl-1-penten-3-one

The desired reaction sequence is:

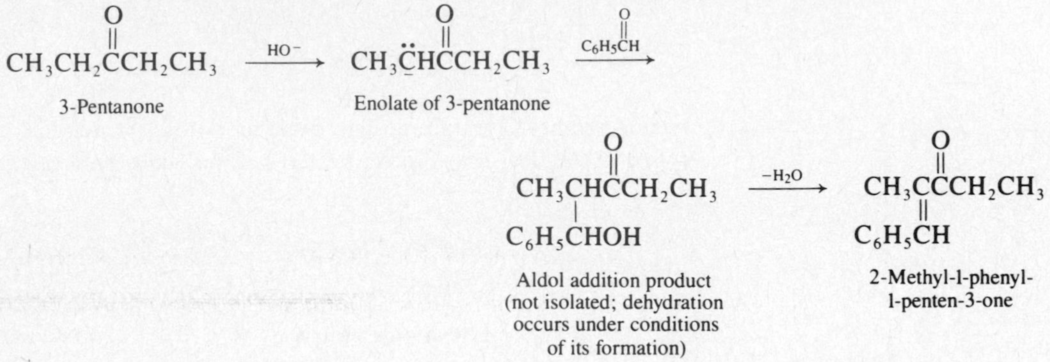

$$CH_3CH_2\overset{\overset{O}{\|}}{C}CH_2CH_3 \xrightarrow{HO^-} CH_3\overset{-}{\ddot{C}}H\overset{\overset{O}{\|}}{C}CH_2CH_3 \xrightarrow{C_6H_5\overset{O}{\overset{\|}{C}}H}$$

3-Pentanone — Enolate of 3-pentanone

$$\underset{\underset{C_6H_5CHOH}{|}}{CH_3CH\overset{\overset{O}{\|}}{C}CH_2CH_3} \xrightarrow{-H_2O} \underset{\underset{C_6H_5CH}{\|}}{CH_3C\overset{\overset{O}{\|}}{C}CH_2CH_3}$$

Aldol addition product (not isolated; dehydration occurs under conditions of its formation) — 2-Methyl-1-phenyl-1-penten-3-one

19.21 All these problems begin in the same way, with exchange of all the α protons for deuterium (Sec. 19.8).

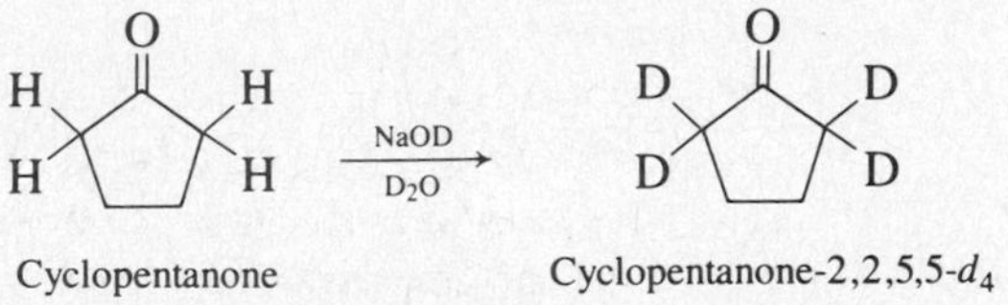

$$\text{Cyclopentanone} \xrightarrow[D_2O]{NaOD} \text{Cyclopentanone-2,2,5,5-}d_4$$

Once the tetradeuterated cyclopentanone has been prepared, functional group transformations are employed to convert it to the desired products.

(*a*) Reduction of the carbonyl group can be achieved by using any of the customary reagents.

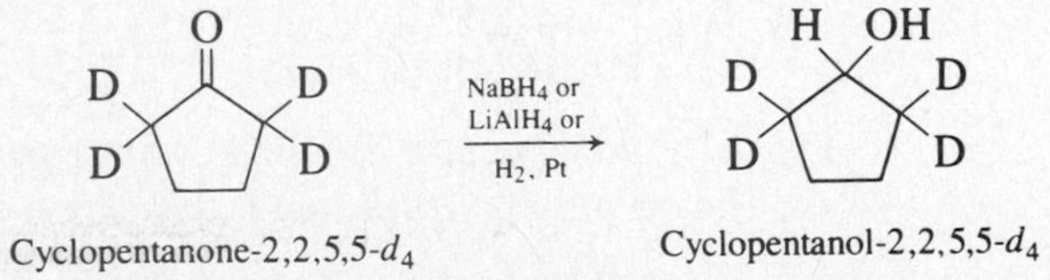

$$\text{Cyclopentanone-2,2,5,5-}d_4 \xrightarrow[H_2,\ Pt]{NaBH_4\ \text{or}\ LiAlH_4\ \text{or}} \text{Cyclopentanol-2,2,5,5-}d_4$$

(*b*) Acid-catalyzed dehydration of the alcohol prepared in (*a*) yields the desired alkene.

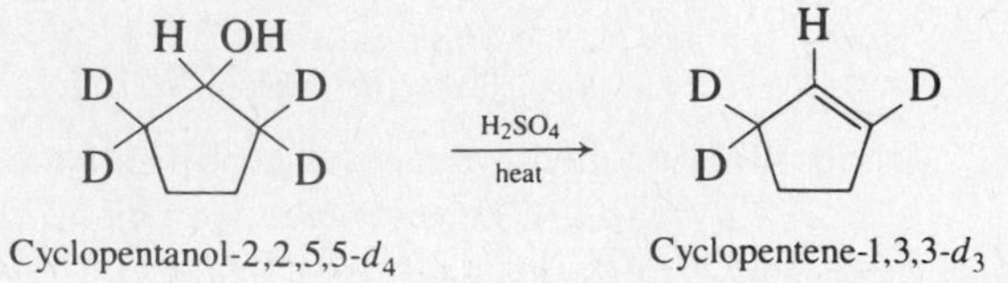

$$\text{Cyclopentanol-2,2,5,5-}d_4 \xrightarrow[\text{heat}]{H_2SO_4} \text{Cyclopentene-1,3,3-}d_3$$

(*c*) Catalytic hydrogenation of the alkene in (*b*) yields cyclopentane-1,1,3-d_3.

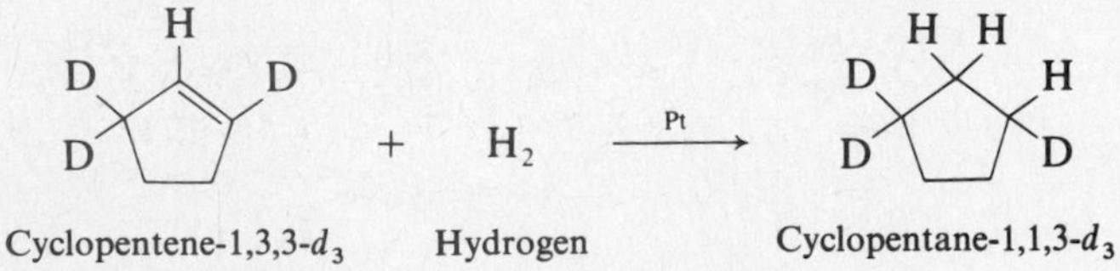

$$\text{Cyclopentene-1,3,3-}d_3 + \underset{\text{Hydrogen}}{H_2} \xrightarrow{Pt} \text{Cyclopentane-1,1,3-}d_3$$

(*d*) Carbonyl reduction of the tetradeuterated ketone under Wolff-Kishner conditions furnishes the desired product.

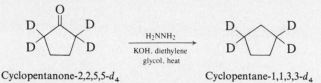

$$\text{Cyclopentanone-2,2,5,5-}d_4 \xrightarrow[\text{KOH, diethylene glycol, heat}]{H_2NNH_2} \text{Cyclopentane-1,1,3,3-}d_4$$

Alternatively, Clemmensen reduction conditions (Zn, HCl) might be used.

19.22 (*a*) The oxo process converts alkenes to aldehydes having one more carbon atom by reaction with carbon monoxide and hydrogen in the presence of a cobalt octacarbonyl catalyst.

$$\underset{\text{Propene}}{CH_3CH{=}CH_2} + \underset{\text{Carbon monoxide}}{CO} + \underset{\text{Hydrogen}}{H_2} \xrightarrow{Co_2(CO)_8} \underset{\text{Butanal}}{CH_3CH_2CH_2\overset{O}{\overset{\|}{C}}H}$$

(*b*) Aldol condensation of acetaldehyde to 2-butenal (crotonaldehyde), followed by catalytic hydrogenation of the carbon-carbon double bond, gives butanal.

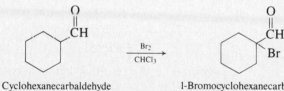

$$\underset{\text{Acetaldehyde}}{2\,CH_3\overset{O}{\overset{\|}{C}}H} \xrightarrow[\text{heat}]{\text{NaOH}} \underset{\text{2-Butenal}}{CH_3CH{=}CH\overset{O}{\overset{\|}{C}}H} \xrightarrow[\text{Ni}]{H_2} \underset{\text{Butanal}}{CH_3CH_2CH_2\overset{O}{\overset{\|}{C}}H}$$

19.23 (*a*) The first conversion is the α halogenation of an aldehyde. As described in Sec. 19.2, this particular conversion has been achieved in 80 percent yield simply by treatment with bromine in chloroform.

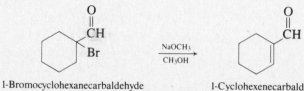

$$\text{Cyclohexanecarbaldehyde} \xrightarrow[CHCl_3]{Br_2} \text{1-Bromocyclohexanecarbaldehyde}$$

Dehydrohalogenation of this compound can be accomplished under E2 conditions by treatment with base. Sodium methoxide in methanol would be appropriate, for example, although almost any alkoxide could be employed to dehydrohalogenate this tertiary bromide.

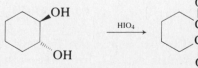

$$\text{1-Bromocyclohexanecarbaldehyde} \xrightarrow[CH_3OH]{NaOCH_3} \text{1-Cyclohexenecarbaldehyde}$$

As the reaction was actually carried out, the bromide was heated with the weak base *N,N*-diethylaniline to effect dehydrobromination in 71 percent yield.

(*b*) Cleavage of vicinal diols to carbonyl compounds can be achieved by using periodic acid (HIO_4) (Sec. 16.14).

$$\textit{trans}\text{-1,2-Cyclohexanediol} \xrightarrow{HIO_4} \text{1,6-Hexanedial}$$

The conversion of this dialdehyde to cyclopentene-1-carbaldehyde is an intramolecular aldol condensation and is achieved by treatment with potassium hydroxide.

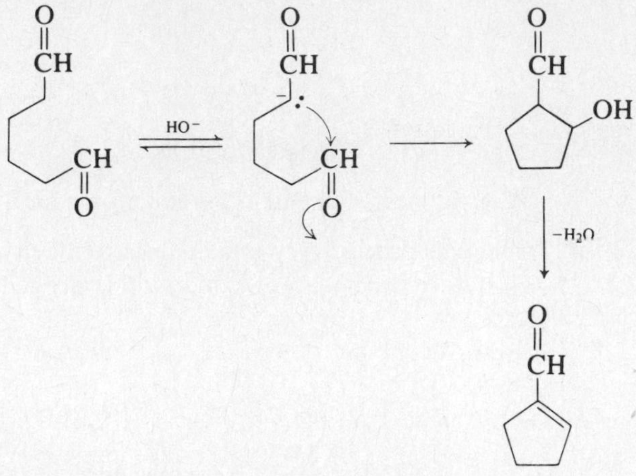

Cyclopentene-1-carbaldehyde

As the reaction was actually carried out, cyclopentene-1-carbaldehyde was obtained in 58 percent yield from *trans*-1,2-cyclohexanediol by this method.

(*c*) The first transformation requires an oxidative cleavage of a carbon-carbon double bond. Ozonolysis followed by hydrolysis in the presence of zinc is indicated.

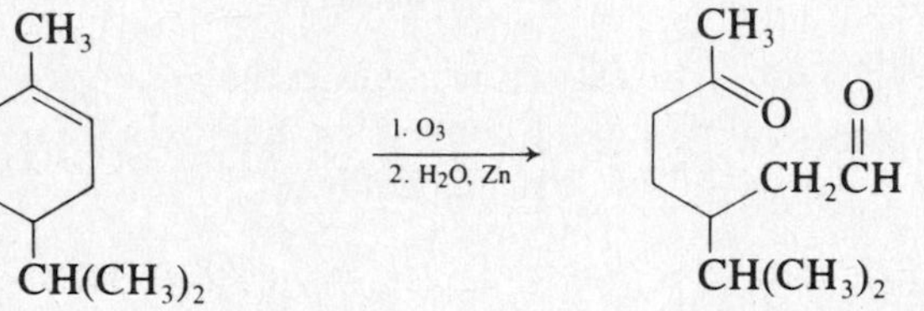

4-Isopropyl-1-methylcyclohexene (*p*-menthene) → (1. O_3; 2. H_2O, Zn) → 3-Isopropyl-6-oxoheptanal

Cyclization of the resulting ketoaldehyde is an intramolecular aldol condensation. Base is required.

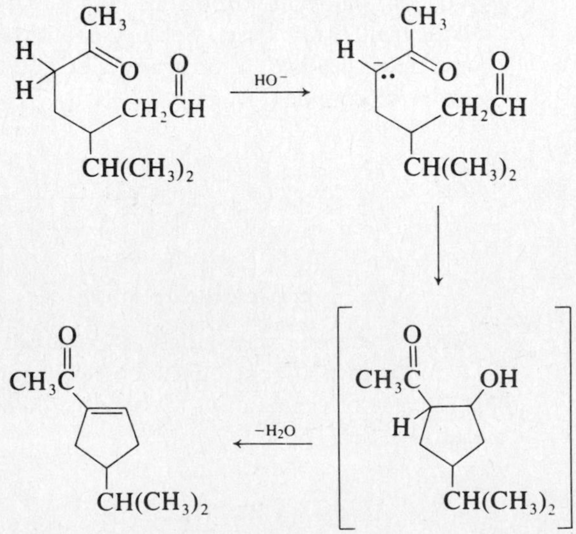

(*d*) The first step in this synthesis is the hydration of the alkene function to an alcohol. Notice that this hydration must take place with a regioselectivity opposite to that of Markovnikov's rule and therefore requires a hydroboration-oxidation sequence.

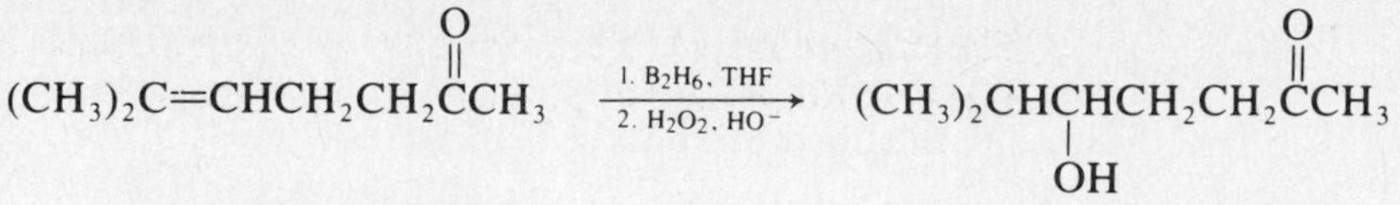

6-Methyl-5-hepten-2-one 5-Hydroxy-6-methyl-2-heptanone

Conversion of the secondary alcohol function to a carbonyl group can be achieved with any of a number of oxidizing agents.

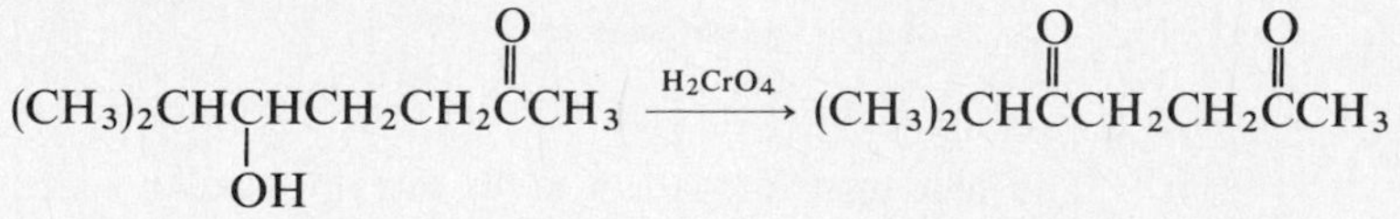

5-Hydroxy-6-methyl-2-heptanone 6-Methyl-2,5-heptanedione

Cyclization of the dione to the final product is a base-catalyzed intramolecular aldol condensation and was accomplished in 71 percent yield by treatment of the dione with a 2 percent solution of sodium hydroxide in aqueous ethanol.

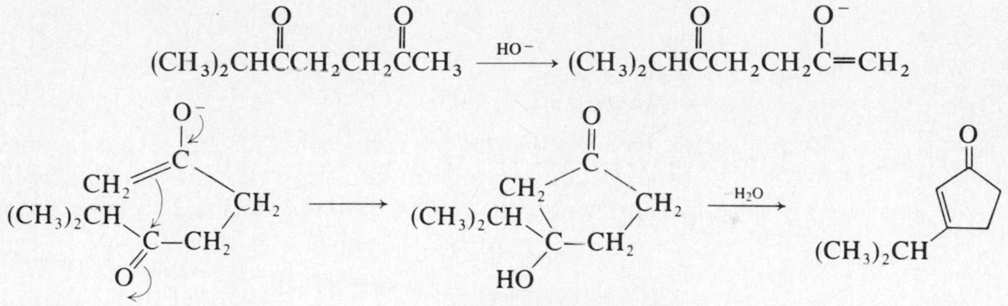

19.24 The first transformation is the aldol condensation of acetone.

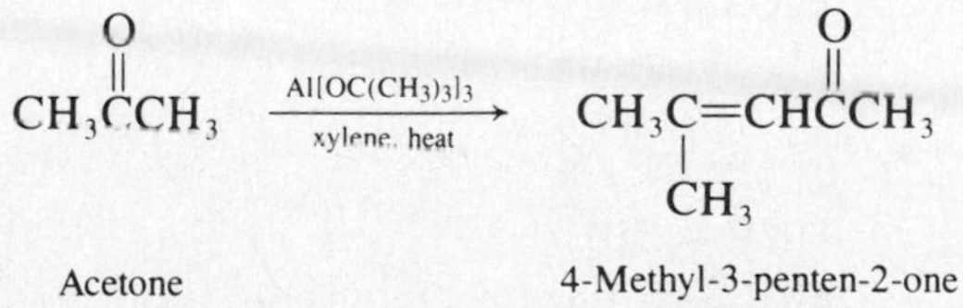

Acetone 4-Methyl-3-penten-2-one (mesityl oxide)

Adding a phenyl substituent to the β carbon atom requires a conjugate addition reaction. An organocuprate reagent is satisfactory here.

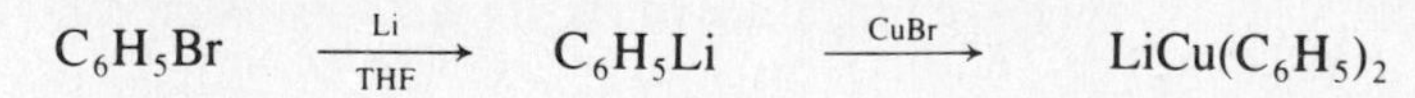

Bromobenzene Phenyllithium Lithium diphenylcuprate

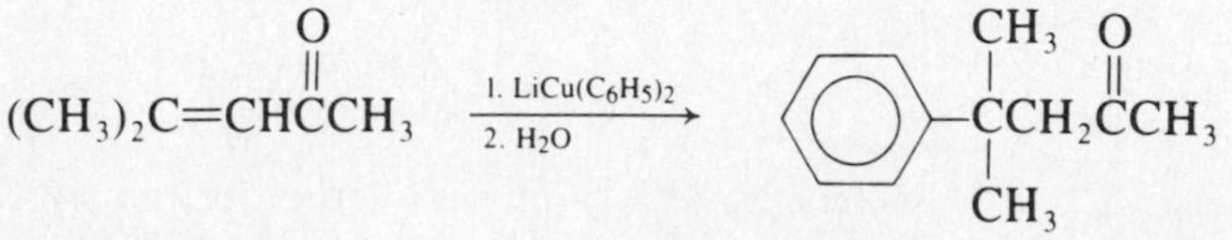

Mesityl oxide 4-Methyl-4-phenyl-2-pentanone

In the next step the methyl ketone must be converted to a carboxylic acid by cleaving the methyl group attached to the carbonyl. Degradation by the haloform reaction is the best way to achieve this. As the reaction was actually carried out, chlorine was used, but bromine or iodine could also have been used.

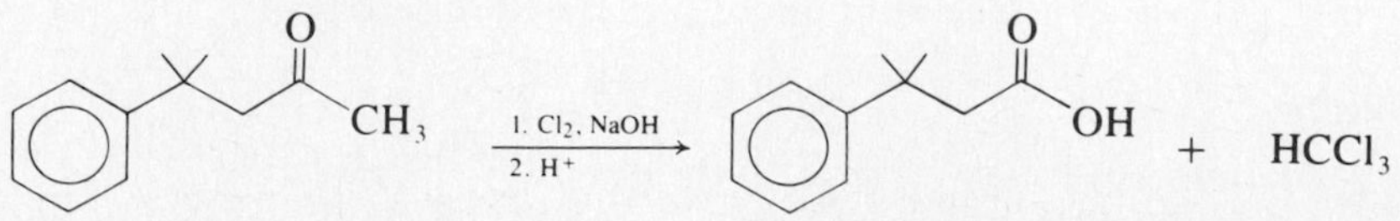

4-Methyl-4-phenyl-2-pentanone 3-Methyl-3-phenylbutanoic acid (61%) Chloroform

The *tert*-butyl group is attached to the aromatic ring by a Friedel-Crafts alkylation reaction. The side chain is an alkyl group and is ortho, para–directing, and activating. Steric effects cause para substitution to be favored.

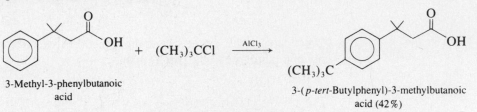

3-Methyl-3-phenylbutanoic acid + $(CH_3)_3CCl$ $\xrightarrow{AlCl_3}$ 3-(*p*-*tert*-Butylphenyl)-3-methylbutanoic acid (42%)

The last transformation is an intramolecular Friedel-Crafts acylation. First the carboxylic acid must be converted to the corresponding acyl chloride with thionyl chloride; then cyclization is effected in the presence of aluminum chloride.

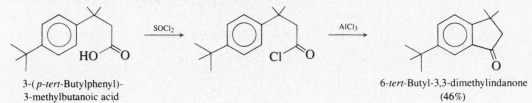

3-(*p*-*tert*-Butylphenyl)-3-methylbutanoic acid $\xrightarrow{SOCl_2}$ (acyl chloride) $\xrightarrow{AlCl_3}$ 6-*tert*-Butyl-3,3-dimethylindanone (46%)

19.25 (*a*) By realizing that the primary alcohol function of the target molecule can be introduced by reduction of an aldehyde, it can be seen that the required carbon skeleton is the same as that of the aldol addition product of 2-methylpropanal.

$$(CH_3)_2CHCH(OH)C(CH_3)_2CH_2OH \Longrightarrow (CH_3)_2CHCH(OH)C(CH_3)_2CH{=}O \Longrightarrow 2\,(CH_3)_2CHCH{=}O$$

The synthetic sequence is:

$$(CH_3)_2CHCH{=}O \xrightarrow[\text{ethanol}]{\text{NaOH}} (CH_3)_2CHCH(OH)C(CH_3)_2CH{=}O \xrightarrow{NaBH_4} (CH_3)_2CHCH(OH)C(CH_3)_2CH_2OH$$

2-Methylpropanal (isobutyraldehyde) — 3-Hydroxy-2,2,4-trimethylpentanal — 2,2,4-Trimethyl-1,3-pentanediol

The starting aldehyde is prepared by oxidation of 2-methyl-1-propanol.

$$(CH_3)_2CHCH_2OH \xrightarrow[CH_2Cl_2]{(C_5H_5N)_2CrO_3} (CH_3)_2CHCH{=}O$$

2-Methyl-1-propanol (isobutyl alcohol) — 2-Methylpropanal (isobutyraldehyde)

(*b*) Retrosynthetic analysis of the desired product shows that the carbon skeleton can be constructed by a mixed aldol condensation between benzaldehyde and propanal.

$$C_6H_5CH{=}C(CH_3)CH_2OH \Longrightarrow C_6H_5CH{=}C(CH_3)CH{=}O \Longrightarrow C_6H_5CH{=}O + CH_3CH_2CH{=}O$$

The reaction scheme therefore becomes:

$$C_6H_5CH{=}O + CH_3CH_2CH{=}O \xrightarrow{HO^-} C_6H_5CH{=}C(CH_3)CH{=}O$$

Benzaldehyde — Propanal — 2-Methyl-3-phenyl-2-propenal

Reduction of the aldehyde to the corresponding primary alcohol gives the desired compound.

$$C_6H_5CH{=}C(CH_3)CHO \xrightarrow[\text{or } NaBH_4,\ CH_3OH]{LiAlH_4,\ \text{then } H_2O} C_6H_5CH{=}C(CH_3)CH_2OH$$

2-Methyl-3-phenyl-2-propenal → 2-Methyl-3-phenyl-2-propen-1-ol

The starting materials for the mixed aldol condensation, benzaldehyde and propanal, are prepared by oxidation of benzyl alcohol and 1-propanol, respectively.

$$C_6H_5CH_2OH \xrightarrow[CH_2Cl_2]{(C_5H_5N)_2CrO_3} C_6H_5CHO$$

Benzyl alcohol → Benzaldehyde

$$CH_3CH_2CH_2OH \xrightarrow[CH_2Cl_2]{(C_5H_5N)_2CrO_3} CH_3CH_2CHO$$

1-Propanol → Propanal

(*c*) The cyclohexene ring in this case can be assembled by a Diels-Alder reaction.

trans-4-Benzoyl-5-(4-methylphenyl)cyclohexene ⇒ 1,3-butadiene + trans-$CH_3C_6H_4CH{=}CHCC_6H_5$ (C=O)

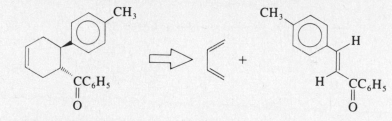

1,3-Butadiene is one of the given starting materials; the α,β-unsaturated ketone is the mixed aldol condensation product of 4-methylbenzaldehyde and acetophenone.

$$CH_3\text{–}C_6H_4\text{–}CH{=}CHCC_6H_5\ (C{=}O) \Rightarrow CH_3\text{–}C_6H_4\text{–}CHO + CH_3C(O)\text{–}C_6H_5$$

The complete synthetic sequence is:

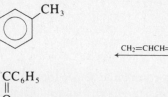

$$CH_3\text{–}C_6H_4\text{–}CH_2OH \xrightarrow[CH_2Cl_2]{(C_5H_5N)_2CrO_3} CH_3\text{–}C_6H_4\text{–}CHO$$

4-Methylbenzyl alcohol → 4-Methylbenzaldehyde

$$CH_3\text{–}C_6H_4\text{–}CHO + CH_3C(O)\text{–}C_6H_5 \xrightarrow{NaOH,\ \text{ethanol}} CH_3\text{–}C_6H_4\text{–}CH{=}CHC(O)\text{–}C_6H_5$$

4-Methylbenzaldehyde + Acetophenone

$$CH_3\text{–}C_6H_4\text{–}CH{=}CHC(O)\text{–}C_6H_5 \xrightarrow{CH_2{=}CHCH{=}CH_2} \text{4-Benzoyl-5-(4-methylphenyl)cyclohexene}$$

4-Benzoyl-5-(4-methylphenyl)cyclohexene

α,β-Unsaturated ketones are good dienophiles in Diels-Alder reactions.

19.26 It is the carbon atom flanked by two carbonyl groups that is involved in the enolization of terreic acid.

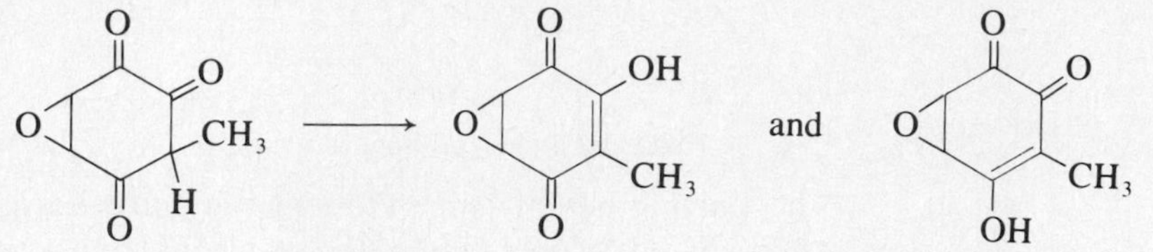

Terreic acid → Enol A and Enol B

Of these two structures, enol A, with its double bond conjugated to two carbonyl groups, is more stable than enol B, in which the double bond is conjugated to only one carbonyl.

19.27 (*a*) At first glance this transformation looks to be an internal oxidation-reduction reaction. An aldehyde function is reduced to a primary alcohol, while a secondary alcohol is oxidized to a ketone.

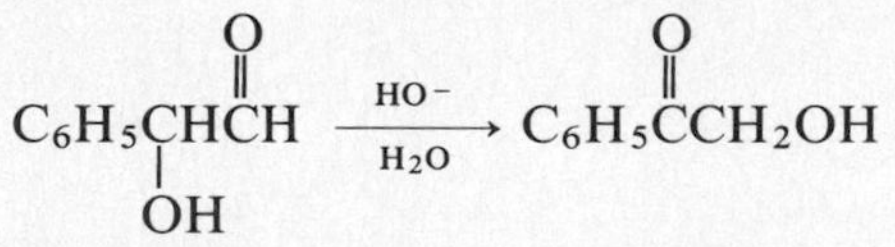

$$C_6H_5\underset{\displaystyle OH}{\underset{|}{C}}H\overset{\displaystyle O}{\overset{\|}{C}}H \xrightarrow[H_2O]{HO^-} C_6H_5\overset{\displaystyle O}{\overset{\|}{C}}CH_2OH$$

Mandelaldehyde (compound A) — Benzoylmethanol (compound B)

Once one realizes that enolization can occur, however, a simpler explanation, involving only proton transfer reactions, emerges.

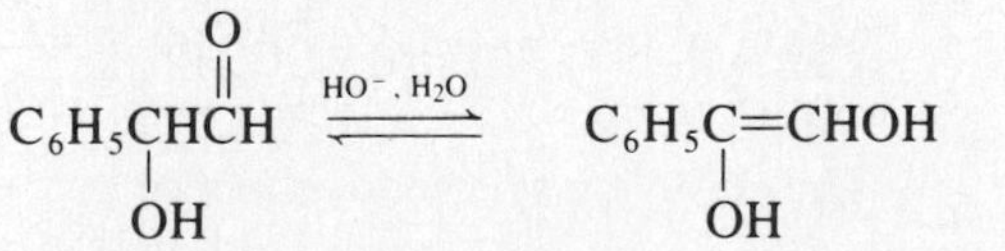

$$C_6H_5\underset{\displaystyle OH}{\underset{|}{C}}H\overset{\displaystyle O}{\overset{\|}{C}}H \overset{HO^-,\ H_2O}{\rightleftharpoons} C_6H_5\underset{\displaystyle OH}{\underset{|}{C}}{=}CHOH$$

Compound A — Enol form of compound A

The enol form of compound A is an enediol; it is at the same time the enol form of compound B. The enediol can revert to compound A or to compound B.

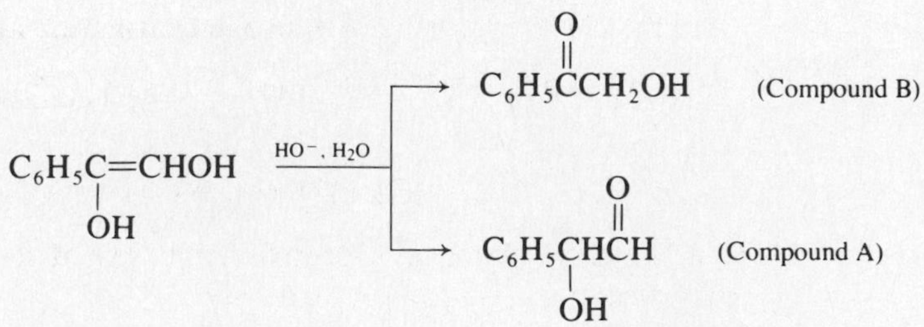

$$C_6H_5\underset{\displaystyle OH}{\underset{|}{C}}{=}CHOH \xrightarrow{HO^-,\ H_2O} C_6H_5\overset{\displaystyle O}{\overset{\|}{C}}CH_2OH \quad \text{(Compound B)}$$

$$C_6H_5\underset{\displaystyle OH}{\underset{|}{C}}{=}CHOH \xrightarrow{HO^-,\ H_2O} C_6H_5\underset{\displaystyle OH}{\underset{|}{C}}H\overset{\displaystyle O}{\overset{\|}{C}}H \quad \text{(Compound A)}$$

At equilibrium compound B predominates because it is more stable than A. A ketone carbonyl is more stabilized than an aldehyde, and the carbonyl in B is conjugated with the benzene ring.

(*b*) The isolated product is the double hemiacetal formed between two molecules of compound A.

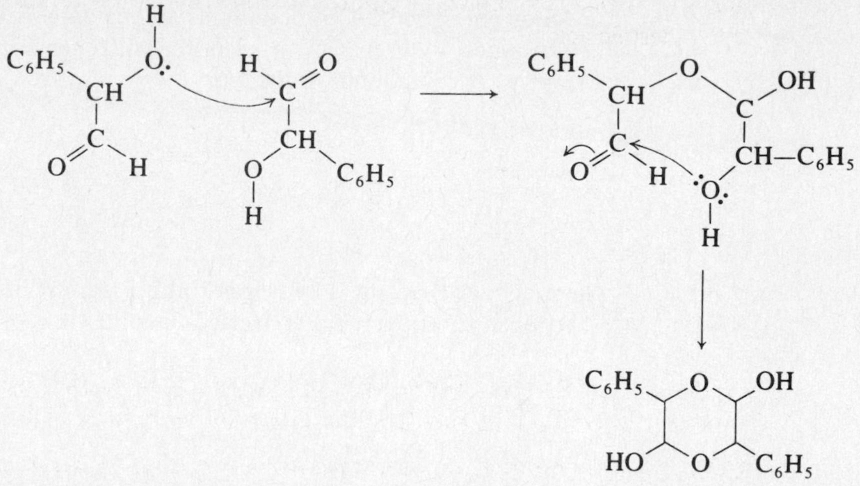

Compound C

19.28 (*a*) The only chiral center in piperitone is adjacent to a carbonyl group. Base-catalyzed enolization causes this carbon to lose its stereochemical integrity.

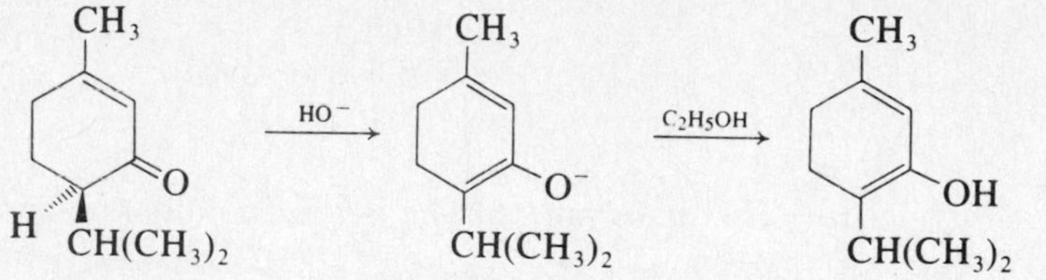

(−)-Piperitone $\xrightarrow{HO^-}$ Enolate of piperitone $\xrightarrow{C_2H_5OH}$ Enol of piperitone

Both the enolate and enol of piperitone are achiral and can revert only to a racemic mixture of piperitones.

(*b*) The enol formed from menthone can revert to either menthone or isomenthone.

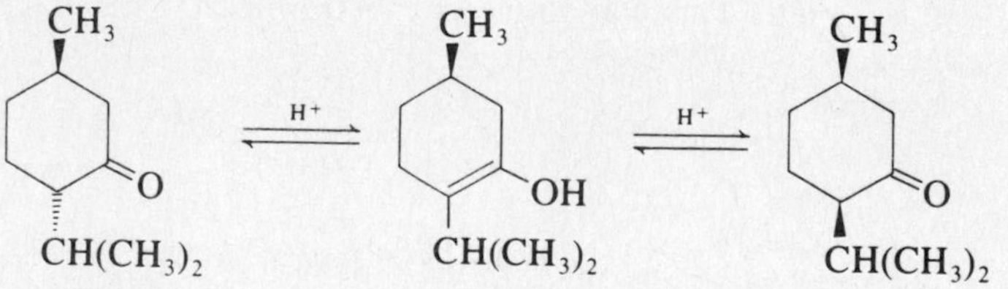

Menthone $\overset{H^+}{\rightleftharpoons}$ Enol form $\overset{H^+}{\rightleftharpoons}$ Isomenthone

Only the stereochemistry at the α carbon atom is affected by enolization. The other chiral center in menthone (the one bearing the methyl group) is not affected.

19.29 In all parts of this problem the bonding change that takes place is described by the general equation:

$$HX{-}N{=}Z \rightleftharpoons X{=}N{-}ZH$$

The more stable form, and therefore the one that is favored in the equilibrium, has the greater sum of individual bond energies.

(*a*) The compound given is nitrosoethane. Nitrosoalkanes are less stable than their oxime isomers formed by proton transfer.

$$CH_3CH(H){-}N{=}O \rightleftharpoons CH_3CH{=}N{-}OH$$

Nitrosoethane (less stable) — Acetaldehyde oxime (more stable)

(*b*) You may recognize this compound as an enamine. It is slightly different, however, from the enamines we discussed earlier in that nitrogen bears a hydrogen substituent. Stable enamines are compounds of the type

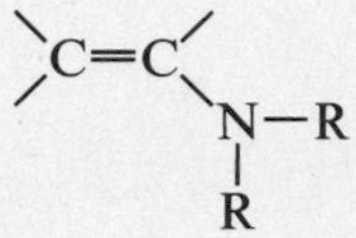

$$\text{>C=C<}\text{N(R)}-\text{R}$$

where neither R group is hydrogen; both R's must be alkyl or aryl. Enamines that bear a hydrogen substituent are converted to imines in a proton transfer equilibrium.

$$(CH_3)_2C{=}CH{-}N(H)CH_3 \rightleftharpoons (CH_3)_2CH{-}CH{=}NCH_3$$

Enamine (less stable) Imine (more stable)

(*c*) The compound given is known as a nitronic acid; its more stable tautomeric form is a nitroalkane.

$$CH_3CH{=}\overset{+}{N}(O^-)(OH) \rightleftharpoons CH_3CH_2{-}\overset{+}{N}(O^-){=}O$$

Nitronic acid Nitroalkane

(*d*) The six-membered ring is benzenoid in the tautomeric form derived from the compound given.

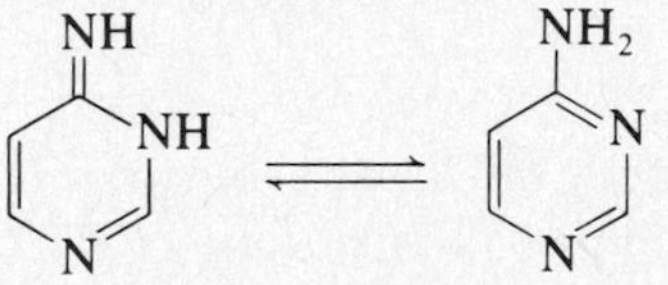

NH ... NH ... N ⇌ NH$_2$ ... N ... N

(*e*) This compound is called isourea. Urea has a carbon-oxygen double bond and is more stable.

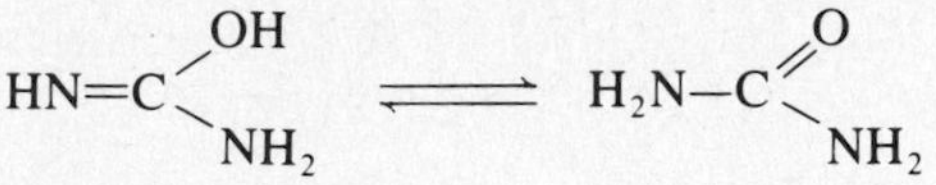

$$HN{=}C(OH)(NH_2) \rightleftharpoons H_2N{-}C({=}O)(NH_2)$$

Isourea (less stable) Urea (more stable)

19.30 (*a*) In the presence of base an enolate anion is formed from the starting ketone.

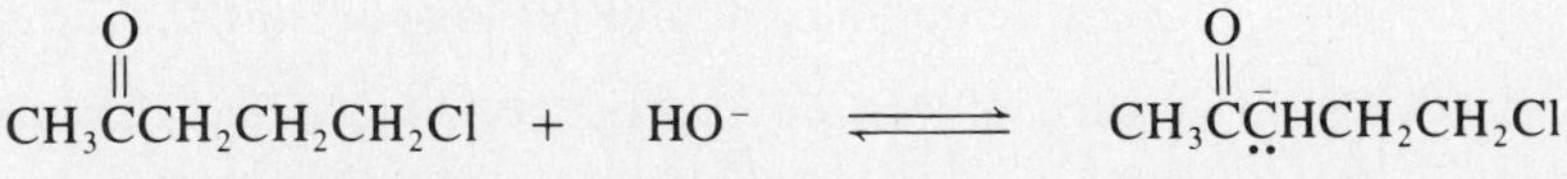

$$CH_3\overset{O}{\overset{\|}{C}}CH_2CH_2CH_2Cl + HO^- \rightleftharpoons CH_3\overset{O}{\overset{\|}{C}}\ddot{\underset{-}{C}}HCH_2CH_2Cl$$

5-Chloro-2-pentanone Hydroxide ion Enolate of 5-chloro-2-pentanone

The enolate has a nucleophilic carbon (the negatively charged α carbon) and a site at which nucleophilic substitution can occur (the $-CH_2Cl$ group). Intramolecular nucleophilic substitution yields the observed product.

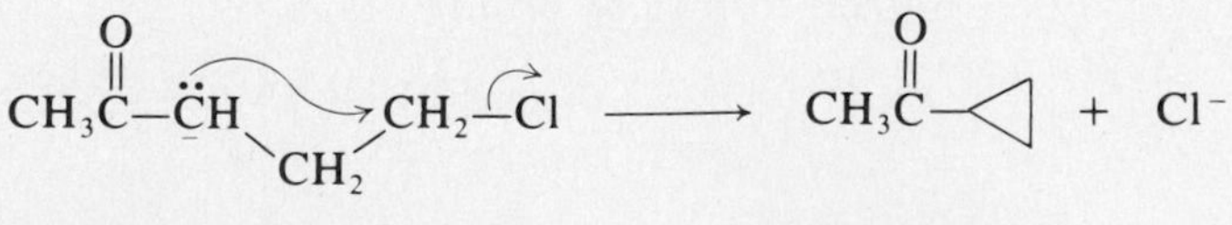

$$CH_3\overset{O}{\overset{\|}{C}}{-}\ddot{C}H(CH_2)CH_2{-}Cl \longrightarrow CH_3\overset{O}{\overset{\|}{C}}{-}\triangleleft + Cl^-$$

Enolate of 5-chloro-2-pentanone Cyclopropyl methyl ketone

While alkylation of a ketone with a separate alkyl halide molecule is usually difficult, *intramolecular* alkylation reactions can be carried out effectively.

(*b*) Here again we have an intramolecular alkylation of a ketone. The enolate formed by proton abstraction from the α carbon atom carries out a nucleophilic attack on the carbon that bears the leaving group.

$$\text{2-}(CH_2CH_2CH_2CH_2Br)\text{cyclohexanone} \xrightarrow{KOC(CH_3)_3} \text{enolate } (O^-)\text{ with } CH_2CH_2CH_2CH_2Br$$

$$\text{enolate } O^- \text{ attacks } CH_2\text{–Br} \longrightarrow \text{spiro[4.5]decan-6-one}$$

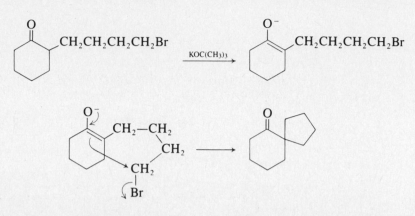

(*c*) The starting material, known as *citral*, is converted to the two products by a reversal of an aldol condensation. The first step is conjugate addition of hydroxide.

$$(CH_3)_2C{=}CHCH_2CH_2\overset{CH_3}{\overset{|}{C}}{=}CH{-}\overset{O}{\overset{\|}{C}}H \;\xrightarrow{\;^-OH\;}\; (CH_3)_2C{=}CHCH_2CH_2\overset{CH_3}{\overset{|}{C}}(OH){-}CH{=}\overset{O^-}{\overset{|}{C}}H$$

$$\xrightarrow{H_2O} (CH_3)_2C{=}CHCH_2CH_2\overset{CH_3}{\overset{|}{C}}(OH){-}CH_2\overset{O}{\overset{\|}{C}}H$$

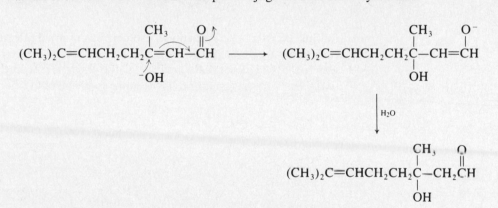

The product of this conjugate addition is a β-hydroxy ketone. It undergoes base-catalyzed cleavage to the observed products.

$$(CH_3)_2C{=}CHCH_2CH_2\overset{CH_3}{\overset{|}{C}}(OH){-}CH_2\overset{O}{\overset{\|}{C}}H \;\underset{}{\overset{HO^-}{\rightleftharpoons}}\; (CH_3)_2C{=}CHCH_2CH_2\overset{CH_3}{\overset{|}{C}}(O^-){-}CH_2{-}\overset{O}{\overset{\|}{C}}H$$

$$\longrightarrow (CH_3)_2C{=}CHCH_2CH_2\overset{O}{\overset{\|}{C}}CH_3 + CH_2{=}\overset{O^-}{\overset{|}{C}}H$$

$$CH_2{=}\overset{O^-}{\overset{|}{C}}H \xrightarrow{H_2O} CH_3\overset{O}{\overset{\|}{C}}H$$

(*d*) The product is formed by an intramolecular aldol condensation reaction.

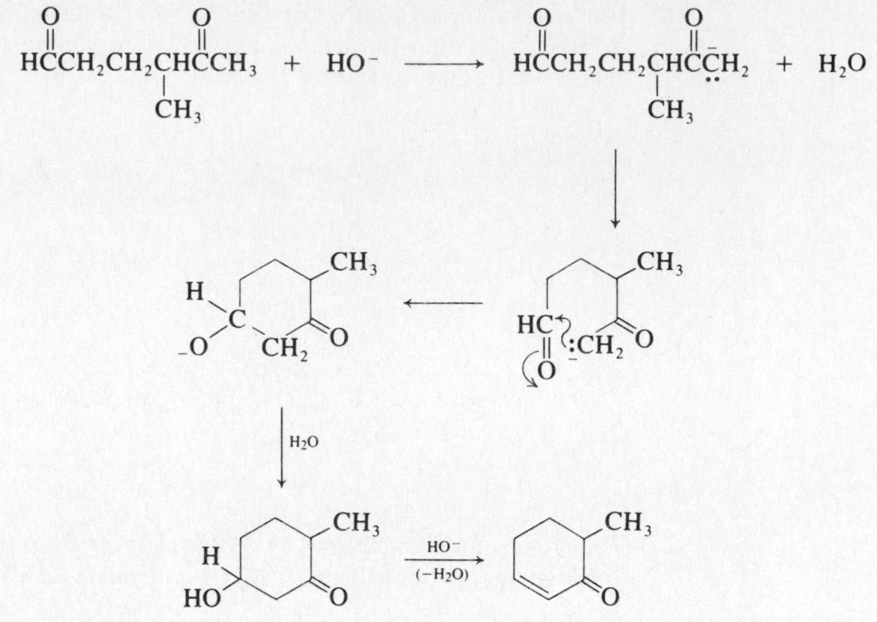

$$\overset{O}{\overset{\|}{H C}}CH_2CH_2\underset{CH_3}{\underset{|}{C}}H\overset{O}{\overset{\|}{C}}CH_3 + HO^- \longrightarrow \overset{O}{\overset{\|}{H C}}CH_2CH_2\underset{CH_3}{\underset{|}{C}}H\overset{O}{\overset{\|}{C}}\ddot{C}H_2^- + H_2O$$

CH₃ H C ⁻O CH₂ O ⟵ CH₃ HC :CH₂⁻ O O

H₂O

$$\text{(3-hydroxy-2-methylcyclohexanone: H, HO, CH}_3\text{, O)} \xrightarrow[(-H_2O)]{HO^-} \text{(2-methylcyclohex-2-enone: CH}_3\text{, O)}$$

(*e*) In this problem stereochemical isomerization involving a proton attached to the α carbon atom of a ketone takes place. Enolization of the ketone yields an intermediate in which the stereochemical integrity of the α carbon is lost. Reversion to ketone eventually leads to the formation of the more stable stereoisomer at equilibrium.

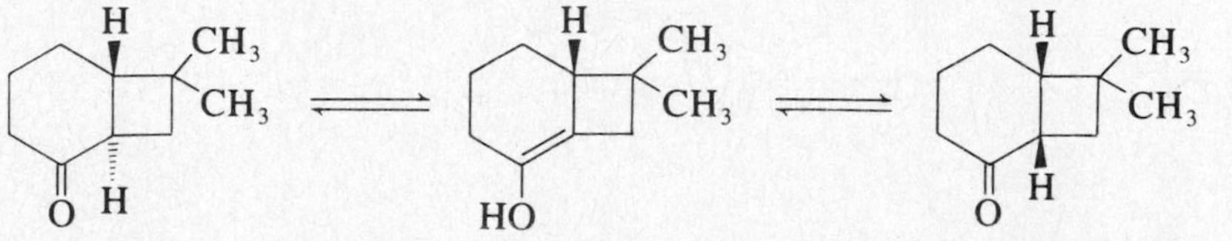

H CH₃ CH₃ O H ⇌ H CH₃ CH₃ HO ⇌ H CH₃ CH₃ O H

Less stable ketone; starting material — Enol — More stable ketone; preferred at equilibrium

The rate of enolization is increased by heating, or by base catalysis. The cis ring fusion in the product is more stable than the trans because there are not enough atoms in the six-membered ring to span trans-1,2 positions in the four-membered ring without excessive strain.

(*f*) The first step is base-catalyzed iodination at the α carbon atom via an enolate anion intermediate.

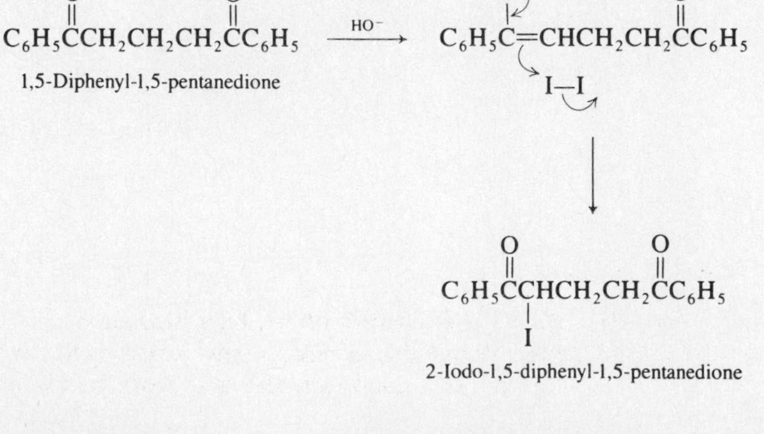

$$C_6H_5\overset{O}{\overset{\|}{C}}CH_2CH_2CH_2\overset{O}{\overset{\|}{C}}C_6H_5 \xrightarrow{HO^-} C_6H_5\overset{O^-}{\overset{|}{C}}{=}CHCH_2CH_2\overset{O}{\overset{\|}{C}}C_6H_5 \quad I{-}I$$

1,5-Diphenyl-1,5-pentanedione

$$\downarrow$$

$$C_6H_5\overset{O}{\overset{\|}{C}}\underset{I}{\underset{|}{C}}HCH_2CH_2\overset{O}{\overset{\|}{C}}C_6H_5$$

2-Iodo-1,5-diphenyl-1,5-pentanedione

When the iodo derivative is converted to an enolate by proton abstraction from C-4, cyclization by way of an intramolecular S_N2 reaction ensues to give the observed product.

$$C_6H_5\overset{O}{\overset{\|}{C}}\underset{I}{\underset{|}{C}}HCH_2CH_2\overset{O}{\overset{\|}{C}}C_6H_5 \xrightarrow{HO^-} C_6H_5\underset{O}{\underset{\|}{C}}\overset{I}{C}H{-}CH_2\,\ddot{C}^-H\underset{O}{\underset{\|}{C}}C_6H_5 \xrightarrow{-I^-} trans\text{-}C_6H_5\overset{O}{\overset{\|}{C}}{-}C_3H_4{-}\overset{O}{\overset{\|}{C}}C_6H_5$$

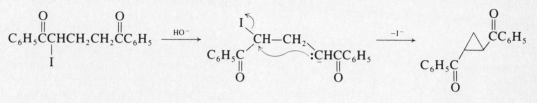

2-Iodo-1,5-diphenyl-1,5-pentanedione

trans-1,2-Dibenzoylcyclopropane

(*g*) Working backwards from the product, we can see that the transformation involves two aldol condensations, one intermolecular and the other intramolecular.

$$\text{2,3,4,5-tetraphenylcyclopentadienone} \Longrightarrow C_6H_5C(=C(C_6H_5)\overset{O}{\overset{\|}{C}}C_6H_5){-}\overset{O}{\overset{\|}{C}}{-}\ddot{C}^-HC_6H_5 \Longrightarrow C_6H_5\ddot{C}^-H\overset{O}{\overset{\|}{C}}CH_2C_6H_5 \;+\; C_6H_5\overset{O}{\overset{\|}{C}}\overset{O}{\overset{\|}{C}}C_6H_5$$

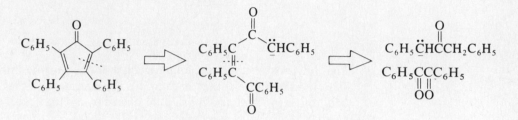

The first reaction is a mixed aldol condensation between the enolate of dibenzyl ketone and one of the carbonyl groups of the dione.

$$\underset{\text{Dibenzyl ketone}}{C_6H_5CH_2\overset{O}{\overset{\|}{C}}CH_2C_6H_5} + \underset{\text{Benzil}}{C_6H_5\overset{O}{\overset{\|}{C}}\overset{O}{\overset{\|}{C}}C_6H_5} \longrightarrow \begin{array}{l} C_6H_5C\overset{O}{\overset{\|}{C}}CH_2C_6H_5 \\ \quad\| \\ C_6H_5C\underset{O}{\underset{\|}{C}}C_6H_5 \end{array}$$

This is followed by an intramolecular aldol condensation.

$$\begin{array}{l} C_6H_5C{-}\overset{O}{\overset{\|}{C}}{-}\ddot{C}^-HC_6H_5 \\ \quad\| \\ C_6H_5C{-}\underset{O}{\underset{\|}{C}}C_6H_5 \end{array} \longrightarrow \text{2,3,4,5-tetraphenyl-4-hydroxycyclopent-2-enone} \longrightarrow \text{2,3,4,5-tetraphenylcyclopentadienone}$$

2,3,4,5-Tetraphenylcyclopentadienone

(*h*) This is a fairly difficult problem since it is not obvious at the outset which of the two possible enolates of benzyl ethyl ketone is the one that undergoes conjugate addition to the enone. A good idea here is to work backwards from the final product—in effect, do a retrosynthetic analysis. The first step is to recognize that the enone arises by dehydration of a β-hydroxy ketone.

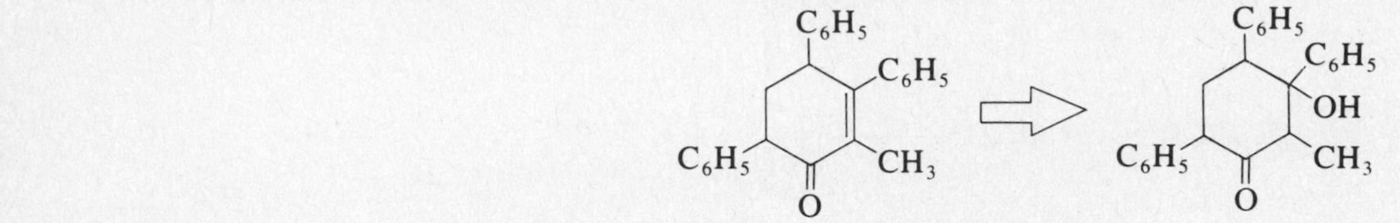

Now, mentally disconnect the bond between the α carbon atom and the carbon that bears the hydroxyl group to reveal the intermediate that undergoes intramolecular aldol condensation.

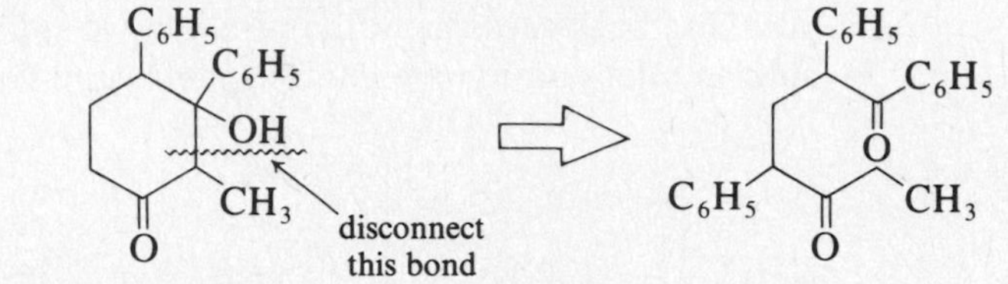

disconnect this bond

The β-hydroxy ketone is the intermediate formed in the intramolecular aldol addition step, and the diketone that leads to it is the intermediate that is formed in the conjugate addition step. The relationship of the starting materials to the intermediates and product is now more evident.

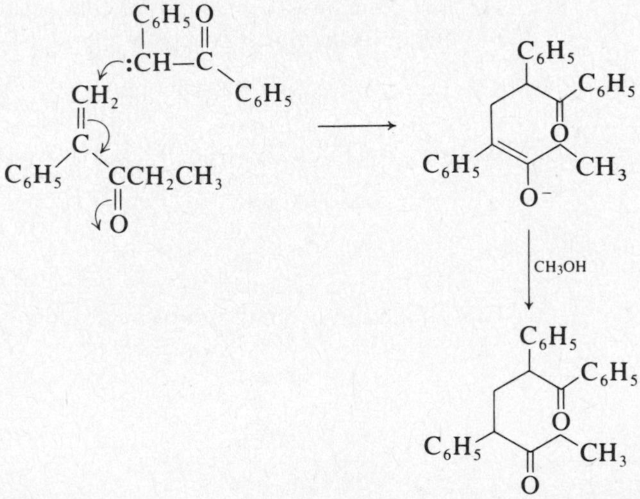

CH_3OH

Intermediate formed in conjugate addition step

19.31 (*a*) The reduced C=O stretching frequency of α,β-unsaturated ketones is consistent with an enhanced degree of single bond character as compared with simple dialkyl ketones.

$$\mathrm{R\overset{O}{\overset{\|}{C}}R'} \longleftrightarrow \mathrm{R\overset{O^-}{\overset{|}{\underset{+}{C}}}R'}$$

$$\mathrm{R_2C{=}CH{-}\overset{O}{\overset{\|}{C}}R'} \longleftrightarrow \mathrm{R_2C{=}CH{-}\overset{O^-}{\overset{|}{\underset{+}{C}}}R'} \longleftrightarrow \mathrm{R_2\overset{+}{C}{-}CH{=}\overset{O^-}{\overset{|}{C}}R'}$$

Resonance is more important in α,β-unsaturated ketones. Conjugation of the carbonyl group with the carbon-carbon double bond increases opportunities for electron delocalization.

(*b*) Even more single bond character is indicated in the carbonyl group of cyclopropenone than in that of typical α,β-unsaturated ketones. The dipolar resonance form contributes substantially to the electron distribution because of the aromatic character of the three-membered ring.

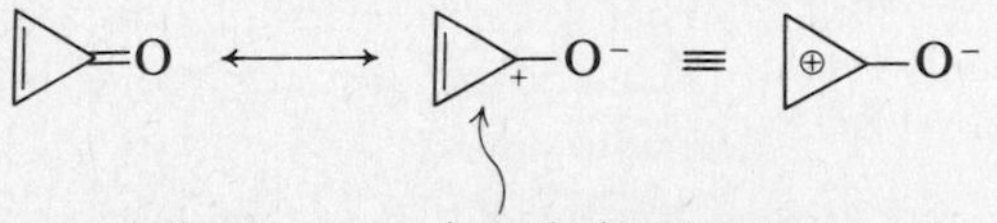

equivalent to an oxyanion-substituted cyclopropenyl cation

(*c*) The dipolar resonance form is a more important contributor to the electron distribution in diphenylcyclopropenone than in benzophenone.

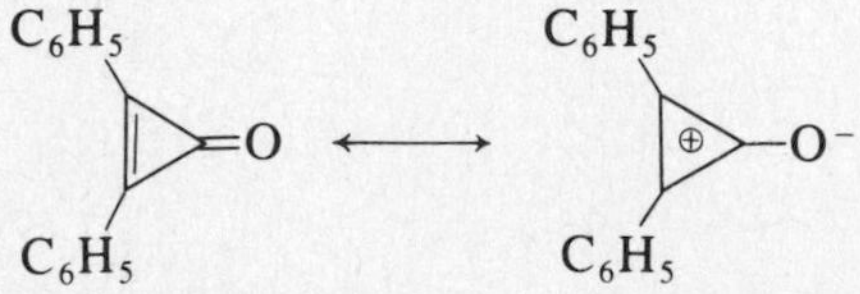

is more pronounced than

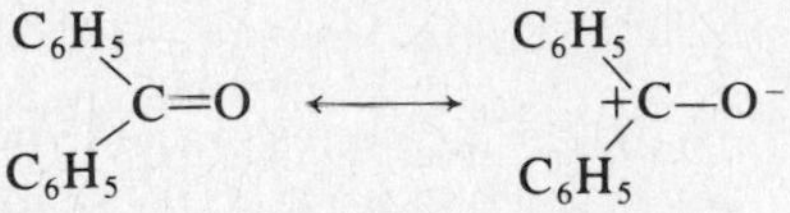

$$(C_6H_5)_2C{=}O \longleftrightarrow (C_6H_5)_2\overset{+}{C}{-}O^-$$

The dipolar resonance form of diphenylcyclopropenone has aromatic character. Its stability leads to increased charge separation and a larger dipole moment.

(*d*) Decreased electron density at the β carbon atom of an α,β-unsaturated ketone is responsible for its decreased shielding. The decreased electron density arises from the polarization of its π electrons as represented by a significant contribution of the dipolar resonance form.

$$CH_2{=}CH{-}\overset{O}{\overset{\|}{C}}R \longleftrightarrow \overset{+}{C}H_2{-}CH{=}\overset{O^-}{\overset{|}{C}}R$$

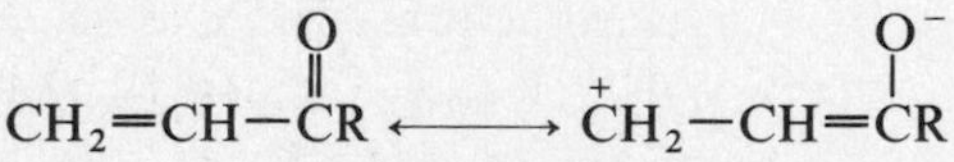

19.32 Bromination can occur at either of the two α carbon atoms.

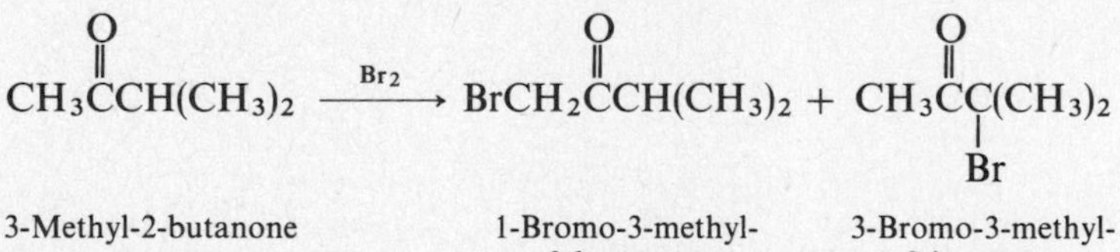

$$CH_3\overset{O}{\overset{\|}{C}}CH(CH_3)_2 \xrightarrow{Br_2} BrCH_2\overset{O}{\overset{\|}{C}}CH(CH_3)_2 + CH_3\overset{O}{\overset{\|}{C}}\underset{Br}{\underset{|}{C}}(CH_3)_2$$

3-Methyl-2-butanone; 1-Bromo-3-methyl-2-butanone; 3-Bromo-3-methyl-2-butanone

The ^1H nmr spectrum of the major product, compound D, is consistent with the structure of 1-bromo-3-methyl-2-butanone. The minor product E is identified as 3-bromo-3-methyl-2-butanone on the basis of its nmr spectrum.

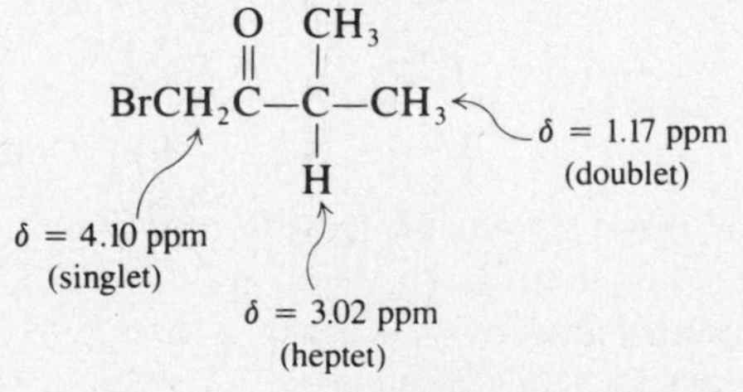

Compound D: $BrCH_2\overset{O}{\overset{\|}{C}}{-}\underset{H}{\underset{|}{\overset{CH_3}{\overset{|}{C}}}}{-}CH_3$

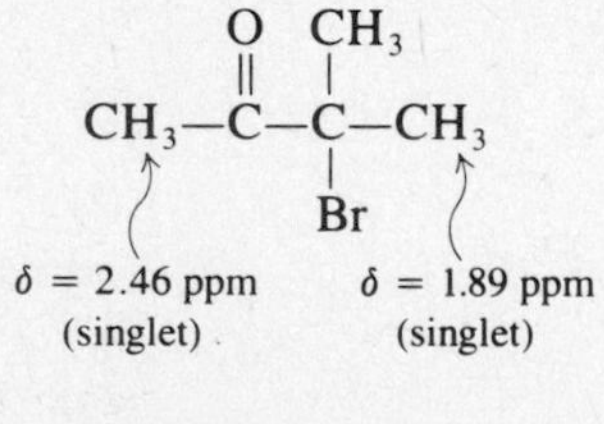

$\delta = 4.10$ ppm (singlet) [$BrCH_2$]; $\delta = 3.02$ ppm (heptet) [H]; $\delta = 1.17$ ppm (doublet) [CH_3]

Compound E: $CH_3{-}\overset{O}{\overset{\|}{C}}{-}\underset{Br}{\underset{|}{\overset{CH_3}{\overset{|}{C}}}}{-}CH_3$

$\delta = 2.46$ ppm (singlet) [$CH_3C{=}O$]; $\delta = 1.89$ ppm (singlet) [$C(CH_3)_2$]

Compound D   Compound E

SELF-TEST

PART A

A-1. Write the correct structure(s) for each of the following:

(*a*) The two enol forms of 2-butanone

(*b*) The enolate ion derived from reaction of 1,3-cyclohexanedione with sodium methoxide

A-2. Give the correct structures for compounds A through D in the following reaction schemes:

(*a*) $2\ C_6H_5CH_2\overset{O}{\overset{\|}{C}}H \xrightarrow[\substack{2.\ \text{heat} \\ (-H_2O)}]{1.\ HO^-} A$

(*b*) $CH_3CH_2CH{=}CH\overset{O}{\overset{\|}{C}}CH_2CH_3 + LiCu(CH_2CH_3)_2 \xrightarrow[2.\ H_2O]{1.\ \text{ether}} B$

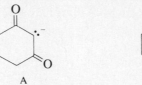

(*c*) $C_6H_5\overset{O}{\overset{\|}{C}}CH_3$ + pyrrolidine (N–H) $\xrightarrow[(-H_2O)]{H^+} C \xrightarrow[2.\ H_2O]{1.\ CH_2{=}CHCH_2Br} D$

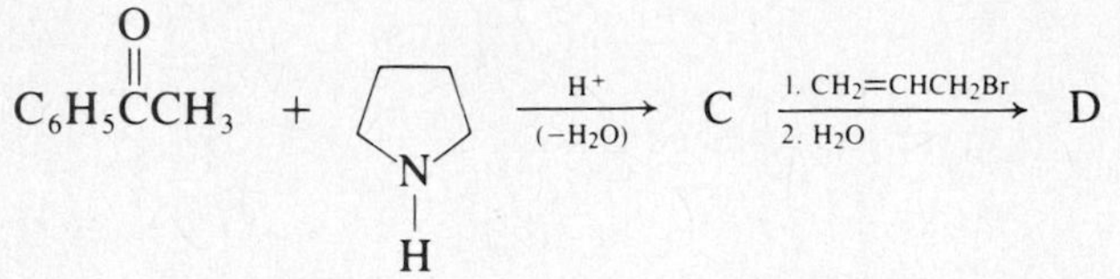

A-3. Write the structures of all the possible aldol addition products which may be obtained by reaction of a mixture of propanal and 2-methylpropanal with base.

$CH_3CH_2\overset{O}{\overset{\|}{C}}H$ — Propanal

$CH_3\overset{CH_3}{\overset{|}{C}}H\underset{O}{\underset{\|}{C}}H$ — 2-Methylpropanal

A-4. Using any necessary organic or inorganic reagents, outline a synthesis of 1,3-butanediol from ethanol as the only source of carbons.

A-5. Outline a series of reaction steps which will allow the preparation of F from 1,3-cyclopentanedione, E.

E (1,3-cyclopentanedione) $\xrightarrow{?}$ F (2-ethyl-2-($CH_2CH_2\overset{O}{\overset{\|}{C}}CH_3$)-1,3-cyclopentanedione)

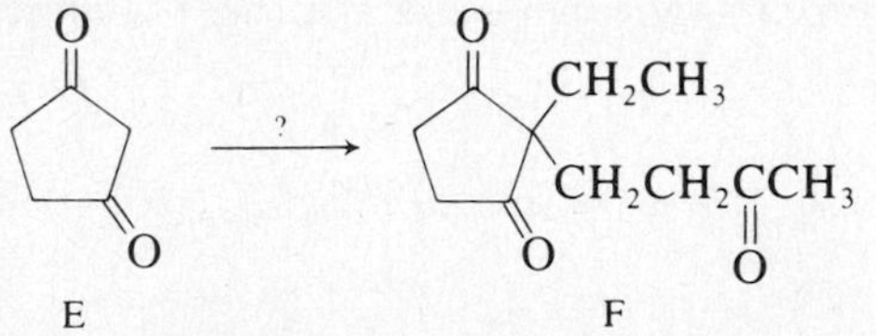

E F

PART B

B-1. When enolate A is compared with enolate B

A (enolate of 1,3-cyclohexanedione, carbanion at C-2) B (enolate of cyclohexanone, carbanion at C-2)

A B

which of the following statements is true?

(*a*) A is more stable than B.

(*b*) B is more stable than A.

(*c*) A and B have the same stability.

(*d*) No comparison of stability can be made.

B-2. Which one of the following molecules is most likely to contain deuterium ($^2H = D$) after reaction with NaOD in D_2O?

(*a*) $C_6H_5\overset{O}{\overset{\|}{C}}H$ (*b*) $C_6H_5CH_2\overset{O}{\overset{\|}{C}}H$

(*c*) $C_6H_5\overset{O}{\overset{\|}{C}}C(CH_3)_3$ (*d*) $(CH_3)_3C\overset{O}{\overset{\|}{C}}C(CH_3)_3$

B-3. Which of the following RX compounds is (are) the best alkylating agent(s) in the alkylation reaction shown?

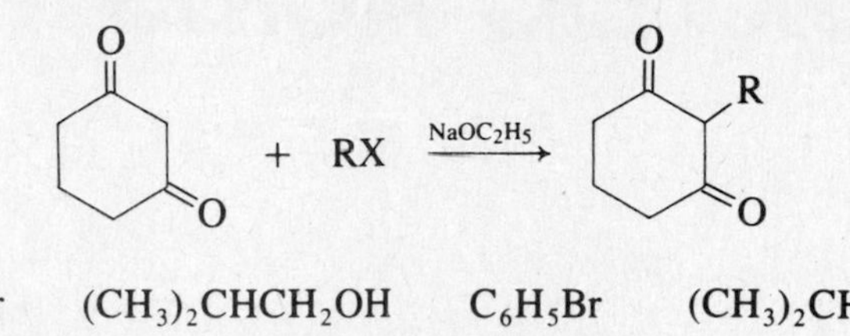

$(CH_3)_3CBr$	$(CH_3)_2CHCH_2OH$	C_6H_5Br	$(CH_3)_2CHCH_2Br$
I	II	III	IV

(*a*) I and IV (*b*) IV only
(*c*) II and IV (*d*) I, II, and IV

B-4. Which of the following pairs of aldehydes gives a single product in a crossed aldol condensation?

(*a*) $C_6H_5CH_2\overset{O}{\overset{\|}{C}}H + C_6H_5\overset{O}{\overset{\|}{C}}H$ (*c*) $C_6H_5\overset{O}{\overset{\|}{C}}H + H_2C{=}O$

(*b*) $C_6H_5\overset{O}{\overset{\|}{C}}H + (CH_3)_3C\overset{O}{\overset{\|}{C}}H$ (*d*) $CH_3\overset{O}{\overset{\|}{C}}H + (CH_3)_2CH\overset{O}{\overset{\|}{C}}H$

B-5. Which of the following compounds, after treatment with Br_2 in aqueous NaOH, forms $HCBr_3$?

(*a*) $C_6H_5\overset{CH_3}{\overset{|}{C}}H{-}\overset{O}{\overset{\|}{C}}H$ (*c*) $(CH_3)_3C\overset{O}{\overset{\|}{C}}H$

(*b*) $C_6H_5CH_2\overset{O}{\overset{\|}{C}}CH_3$ (*d*)

B-6. Which of the following forms an enol to the greatest extent?

(*a*) $CH_3CH_2\overset{O}{\overset{\|}{C}}H$ (*c*) $CH_3\overset{O}{\overset{\|}{C}}CH_2\overset{O}{\overset{\|}{C}}H$

(*b*) $CH_3\overset{O}{\overset{\|}{C}}CH_2CH_2\overset{O}{\overset{\|}{C}}CH_3$ (*d*) $CH_3\overset{O}{\overset{\|}{C}}\overset{O}{\overset{\|}{C}}CH_2CH_3$

CHAPTER 20

CARBOXYLIC ACIDS

IMPORTANT TERMS AND CONCEPTS

Nomenclature (Sec. 20.1) Systematic names are derived by locating the longest carbon atom chain containing the *carboxyl group* and replacing the *-e* ending of the corresponding alkane with *-oic acid.* The locants used to identify substituents are always numbered by beginning with the carboxyl carbon. For example

$C_6H_5CH_2CH_2CO_2H$

3-Phenylpropanoic acid

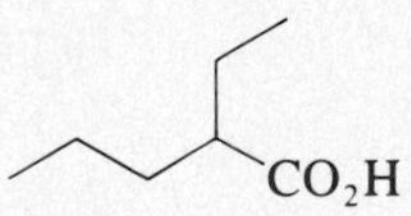

CO_2H

2-Ethylpentanoic acid

As noted in text Table 20.1, certain carboxylic acids are referred to by accepted "common" names. Noteworthy among these are formic and acetic acids (methanoic and ethanoic acid by the IUPAC rules).

HCO_2H Formic acid CH_3CO_2H Acetic acid

Structure, Bonding, and Physical Properties (Secs. 20.2, 20.3) The carbon of the carboxyl group is sp^2 hybridized, and it and the three atoms bonded to it lie in the same plane. As a result of strong hydrogen bonding attractions, carboxylic acids exist as cyclic dimers, which often persist in the gas phase. The melting and boiling points of carboxylic acids are much higher than would be expected on the basis of molecular weight.

Acid-Base Properties (Secs. 20.4 to 20.7) Carboxylic acids are the most acidic class of organic compounds containing only C, H, and O atoms. They are, however, considered weak acids, having ionization constants of about 10^{-5}.

$$\underset{\text{Carboxylic acid}}{R{-}CO_2H} \rightleftharpoons H^+ + \underset{\text{Carboxylate ion}}{R{-}CO_2^-} \qquad K_a \sim 10^{-5}$$

The acidity of carboxylic acids may be explained by the resonance stabilization of the carboxylate ion. The negative charge is delocalized and shared equally by both oxygens.

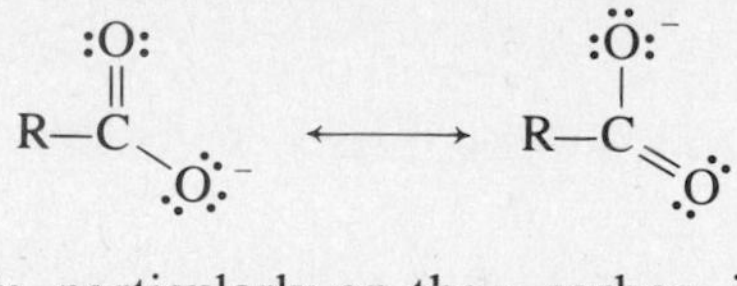

$$\text{R—C(=O)—O}^- \longleftrightarrow \text{R—C(—O}^-\text{)=O}$$

Electronegative substituents, particularly on the α carbon, increase the acidity of carboxylic acids. For example, the pK_a of chloroacetic acid 2.9, whereas that of acetic acid is 4.7.

$ClCH_2CO_2H$ (Chloroacetic acid) is more acidic than CH_3CO_2H (Acetic acid)

Benzoic acid is slightly more acidic than acetic acid, owing to the attachment of the carboxyl group to an sp^2 carbon atom. Substituents on the benzene ring affect the acidity of benzoic acid derivatives, the net effect arising as a result of the combination of inductive, field, resonance, and solvation effects.

Esterification (Secs. 20.14, 20.15) A mechanistic model for many of the reactions of carboxylic acid derivatives is provided by the acid-catalyzed esterification of carboxylic acids. A key aspect of the mechanism shown below is the formation of a *tetrahedral intermediate*, in which the carboxyl carbon has undergone a change from sp^2 hybridization to sp^3.

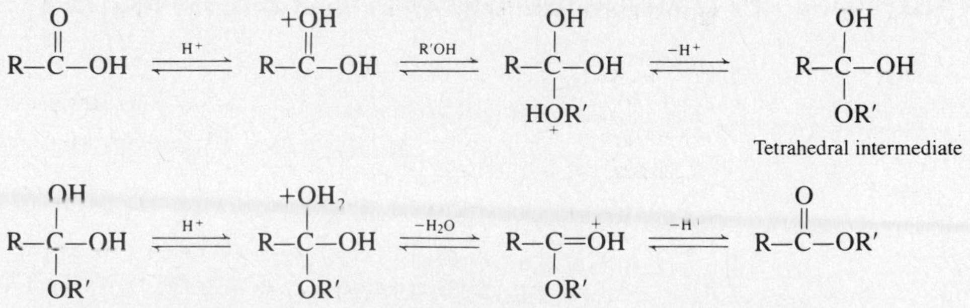

$$\text{R—C(=O)—OH} \underset{}{\overset{H^+}{\rightleftharpoons}} \text{R—C(=}\overset{+}{\text{O}}\text{H)—OH} \overset{R'OH}{\rightleftharpoons} \text{R—C(OH)(OH)—}\overset{+}{\text{HOR}'} \overset{-H^+}{\rightleftharpoons} \text{R—C(OH)(OH)—OR}'$$

Tetrahedral intermediate

$$\text{R—C(OH)(OH)—OR}' \overset{H^+}{\rightleftharpoons} \text{R—C(}\overset{+}{\text{O}}\text{H}_2\text{)(OH)—OR}' \overset{-H_2O}{\rightleftharpoons} \text{R—C(=}\overset{+}{\text{O}}\text{H)—OR}' \overset{-H^+}{\rightleftharpoons} \text{R—C(=O)—OR}'$$

Hydroxy acids may undergo an intramolecular esterification to form cyclic esters known as *lactones*. For example, 4-hydroxybutanoic acid cyclizes to form a γ-*lactone*.

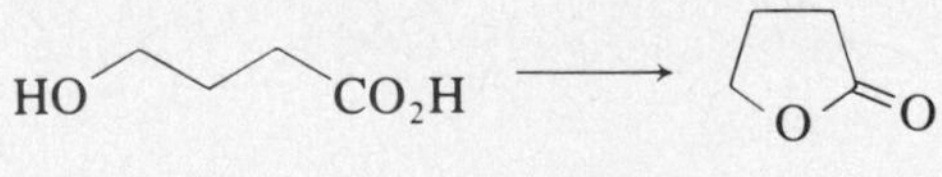

$$\text{HO—CH}_2\text{CH}_2\text{CH}_2\text{—CO}_2\text{H} \longrightarrow \gamma\text{-butyrolactone}$$

Spectroscopic Properties (Sec. 20.18) Carboxylic acids have characteristic hydroxyl absorptions in the 3500 to 2500 cm^{-1} region of the infrared spectrum. In addition, the carbonyl of the carboxyl group appears as a strong band near 1700 cm^{-1}. The hydroxyl proton of a carboxylic acid is very deshielded relative to most other protons in the nmr spectrum and appears about 10 to 12 ppm downfield from tetramethylsilane (TMS).

IMPORTANT REACTIONS

SYNTHESIS OF CARBOXYLIC ACIDS

Carboxylation of Grignard Reagents (Sec. 20.11)

General:

$$RMgX + CO_2 \longrightarrow RCO_2MgX \xrightarrow{H^+, H_2O} RCO_2H$$

Example:

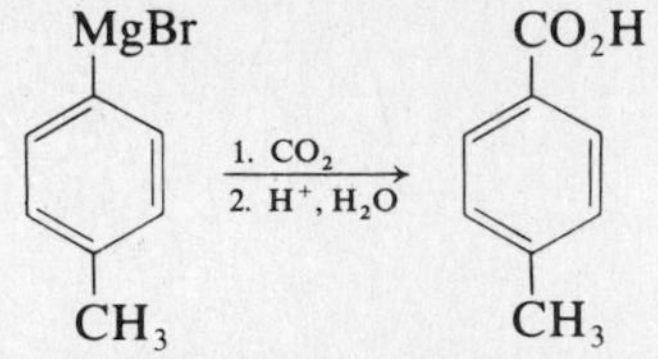

$$p\text{-}CH_3C_6H_4MgBr \xrightarrow[2.\ H^+,\ H_2O]{1.\ CO_2} p\text{-}CH_3C_6H_4CO_2H$$

Preparation and Hydrolysis of Nitriles (Sec. 20.12)

General:

$$\underset{(1^\circ \text{ halides best})}{R{-}X} + CN^- \longrightarrow R{-}CN \xrightarrow{H^+,\ H_2O} R{-}CO_2H$$

Example:

$$(CH_3)_2CHCH_2CH_2Br + CN^- \longrightarrow (CH_3)_2CHCH_2CH_2CN \xrightarrow{H^+,\ H_2O} (CH_3)_2CHCH_2CH_2CO_2H$$

REACTIONS OF CARBOXYLIC ACIDS

α-Halogenation; the Hell-Volhard-Zelinsky Reaction (Sec. 20.16)

General:

$$RCH_2CO_2H \xrightarrow{Br_2,\ PCl_3} RCH(Br)CO_2H$$

Example:

$$CH_3{-}C_6H_4{-}CH_2CO_2H \xrightarrow{Br_2,\ PCl_3} CH_3{-}C_6H_4{-}CH(Br)CO_2H$$

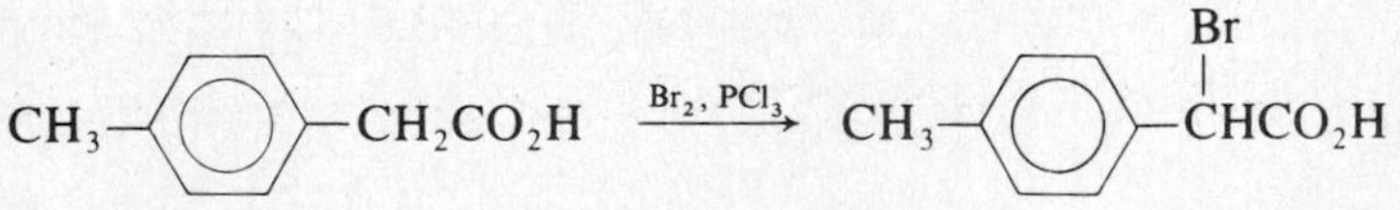

Decarboxylation (Sec. 20.17)

Malonic acid derivatives:

General:

$$HO_2CC(R)(R')CO_2H \xrightarrow{\text{heat}} RCH(R')CO_2H + CO_2$$

Example:

$$CH_3CH_2C(CH_3)(CO_2H)_2 \xrightarrow{\text{heat}} CH_3CH_2CH(CH_3)CO_2H$$

β-Keto acids:

General:

$$RC(=O)CH(R')CO_2H \xrightarrow{\text{heat}} RC(=O)CH_2R' + CO_2$$

Example:

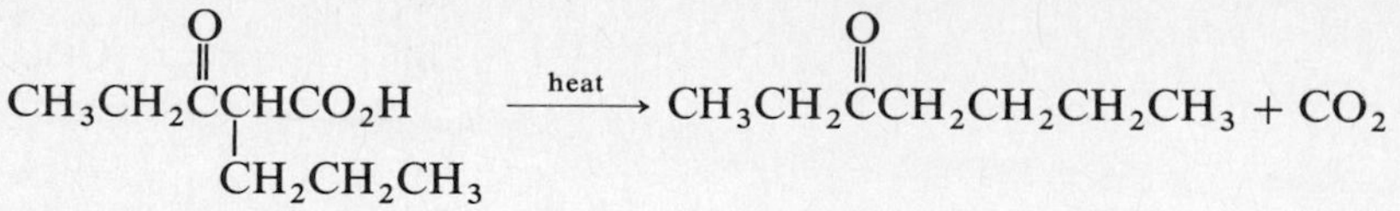

$$CH_3CH_2\overset{O}{\overset{\|}{C}}\underset{CH_2CH_2CH_3}{\underset{|}{C}H}CO_2H \xrightarrow{\text{heat}} CH_3CH_2\overset{O}{\overset{\|}{C}}CH_2CH_2CH_2CH_3 + CO_2$$

SOLUTIONS TO TEXT PROBLEMS

20.1 (*b*) The longest continuous chain in pivalic acid contains three carbons, and therefore it is named as a derivative of propanoic acid. It is *2,2-dimethylpropanoic acid.*

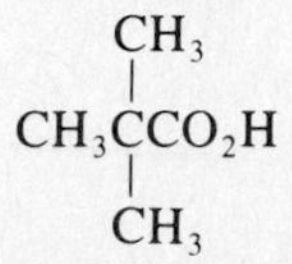

$$CH_3\underset{CH_3}{\underset{|}{\overset{CH_3}{\overset{|}{C}}}}CO_2H$$ 2,2-Dimethylpropanoic acid (pivalic acid)

(*c*) The four carbon atoms of crotonic acid form a continuous chain. Since there is a double bond between C-2 and C-3, it is one of the stereoisomers of 2-butenoic acid. The stereochemistry of the double bond is *E*.

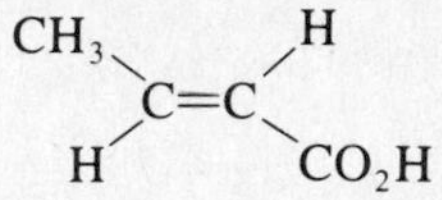

$$\begin{matrix} CH_3 & & H \\ & C{=}C & \\ H & & CO_2H \end{matrix}$$ (*E*)-2-Butenoic acid (crotonic acid)

(*d*) Oxalic acid is a dicarboxylic acid that contains two carbons. It is *ethanedioic acid.*

$$HO_2CCO_2H$$ Ethanedioic acid (oxalic acid)

(*e*) The four carbons of maleic acid form a continuous chain with a double bond between C-2 and C-3; it is the *Z* stereoisomer of *butenedioic acid*. No number is necessary to locate the position of the double bond because the structure permits it to be only at the position shown.

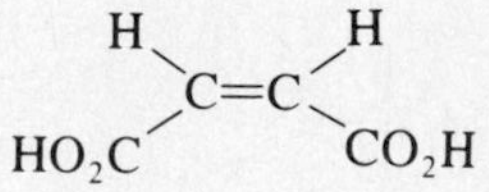

$$\begin{matrix} H & & H \\ & C{=}C & \\ HO_2C & & CO_2H \end{matrix}$$ (*Z*)-Butenedioic acid (maleic acid)

(*f*) The systematic name for $C_6H_5CO_2H$ is benzenecarboxylic acid. Since there is a methyl group at the para position, the compound shown is *p-methylbenzenecarboxylic acid* or *4-methylbenzenecarboxylic acid.*

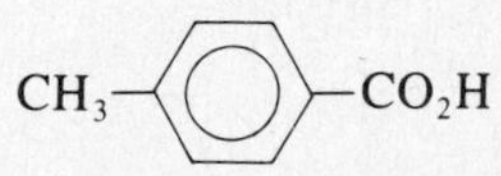

$$CH_3{-}C_6H_4{-}CO_2H$$ *p*-Methylbenzenecarboxylic acid or 4-methylbenzenecarboxylic acid (*p*-toluic acid)

20.2 Ionization of peroxyacetic acid yields an anion that cannot be stabilized by resonance in the same way that acetate can.

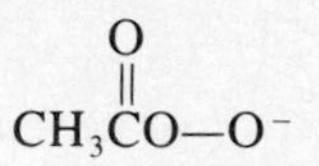

$$CH_3\overset{O}{\overset{\|}{C}}O{-}O^-$$ Delocalization of negative charge into carbonyl group is not possible in peroxyacetate ion

20.3 (*b*) The acid-base reaction between acetic acid and *tert*-butoxide ion is represented by the equation:

$$\underset{\substack{\text{Acetic acid}\\ \text{(stronger acid)}}}{CH_3CO_2H} + \underset{\substack{\textit{tert}\text{-Butoxide}\\ \text{(stronger base)}}}{(CH_3)_3CO^-} \rightleftharpoons \underset{\substack{\text{Acetate ion}\\ \text{(weaker base)}}}{CH_3CO_2^-} + \underset{\substack{\textit{tert}\text{-Butyl alcohol}\\ \text{(weaker acid)}}}{(CH_3)_3COH}$$

Alcohols are weaker acids than carboxylic acids, so the equilibrium lies to the right.

(*c*) Bromide ion is the conjugate base of hydrogen bromide, a strong acid.

$$\underset{\text{Acetic acid (weaker acid)}}{CH_3CO_2H} + \underset{\text{Bromide ion (weaker base)}}{Br^-} \rightleftharpoons \underset{\text{Acetate ion (stronger base)}}{CH_3CO_2^-} + \underset{\text{Hydrogen bromide (stronger acid)}}{HBr}$$

In this case, the position of equilibrium favors the starting materials because acetic acid is a weaker acid than hydrogen bromide.

(*d*) Acetylide ion is a rather strong base and acetylene, with a K_a of 10^{-25}, is a much weaker acid than acetic acid. The position of equilibrium favors the formation of products.

$$\underset{\text{Acetic acid (stronger acid)}}{CH_3CO_2H} + \underset{\text{Acetylide ion (stronger base)}}{HC{\equiv}\bar{C}:} \rightleftharpoons \underset{\text{Acetate ion (weaker base)}}{CH_3CO_2^-} + \underset{\text{Acetylene (weaker acid)}}{HC{\equiv}CH}$$

(*e*) Nitrate ion is a very weak base; it is the conjugate base of the strong acid nitric acid. The position of equilibrium lies to the left.

$$\underset{\text{Acetic acid (weaker acid)}}{CH_3CO_2H} + \underset{\text{Nitrate ion (weaker base)}}{NO_3^-} \rightleftharpoons \underset{\text{Acetate ion (stronger base)}}{CH_3CO_2^-} + \underset{\text{Nitric acid (stronger acid)}}{HNO_3}$$

(*f*) Amide ion is a very strong base. The position of equilibrium lies to the right.

$$\underset{\text{Acetic acid (stronger acid)}}{CH_3CO_2H} + \underset{\text{Amide ion (stronger base)}}{H_2N^-} \rightleftharpoons \underset{\text{Acetate ion (weaker base)}}{CH_3CO_2^-} + \underset{\text{Ammonia (weaker acid)}}{NH_3}$$

20.4 (*b*) Propanoic acid is similar to acetic acid in its acidity. A hydroxyl group at C-2 is electron-withdrawing and stabilizes the carboxylate ion of lactic acid by a combination of inductive and field effects.

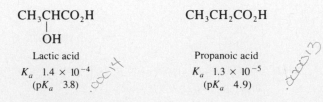

$CH_3CH(OH)-C(=O)O^-$ — Hydroxyl group stabilizes negative charge by attracting electrons

Lactic acid is more acidic than propanoic acid. The measured ionization constants are:

$CH_3CH(OH)CO_2H$	$CH_3CH_2CO_2H$
Lactic acid	Propanoic acid
K_a 1.4×10^{-4}	K_a 1.3×10^{-5}
(pK_a 3.8)	(pK_a 4.9)

(*c*) Changing the hybridization of C-2 from sp^3 to sp^2 increases its *s* character and makes it more electron-withdrawing; electron-withdrawing groups increase acidity. Acrylic acid, $CH_2{=}CHCO_2H$, is expected to be a stronger acid than 2-methylpropanoic acid.

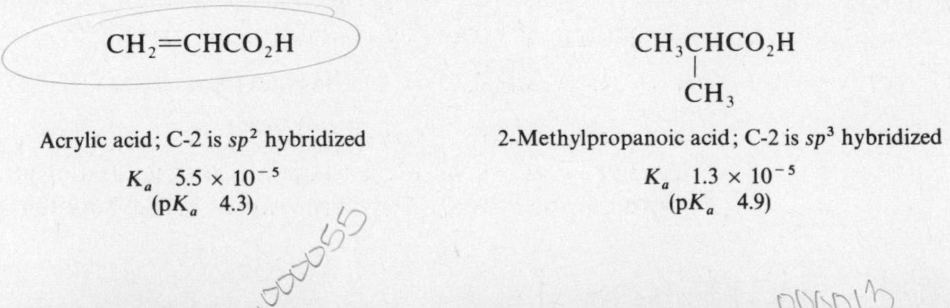

(*d*) The hybridization at C-2 is the same in both compounds, but a carbonyl group is more strongly withdrawing than a carbon-carbon double bond. Pyruvic acid is a stronger acid than acrylic acid.

$$\underset{\text{Pyruvic acid}}{CH_3\overset{O}{\overset{\|}{C}}CO_2H} \qquad \underset{\text{Acrylic acid}}{CH_2{=}CHCO_2H}$$

Pyruvic acid: K_a 5.1 × 10^{-4} (pK_a 3.3)

Acrylic acid: K_a 5.5 × 10^{-5} (pK_a 4.3)

(*e*) Viewing the two compounds as substituted derivatives of acetic acid, RCH_2CO_2H, we judge $CH_3\overset{O}{\overset{\|}{\underset{\underset{O}{\|}}{S}}}$ to be strongly electron-withdrawing and acid-strengthening, whereas an ethyl group has only a small effect.

$$CH_3\overset{O}{\overset{\|}{\underset{\underset{O}{\|}}{S}}}CH_2CO_2H \qquad CH_3CH_2CH_2CO_2H$$

Methanesulfonylacetic acid: K_a 4.3 × 10^{-3} (pK_a 2.4)

Butanoic acid: K_a 1.5 × 10^{-5} (pK_a 4.8)

20.5 The compound can only be a carboxylic acid; no other class containing only carbon, hydrogen, and oxygen can be more acidic. A reasonable choice is $HC{\equiv}CCO_2H$; C-2 is *sp* hybridized and therefore rather electron-withdrawing and acid-strengthening. This is borne out by its measured ionization constant K_a, which is 1.4 × 10^{-2} (pK_a 1.8).

20.6 For carbonic acid, the "true K_1" is given by

$$\text{True } K_1 = \frac{[H^+][HCO_3^-]}{[H_2CO_3]}$$

since

$$4.3 \times 10^{-7} = \frac{[H^+][HCO_3^-]}{[CO_2]}$$

then

$$[H^+][HCO_3^-] = (4.3 \times 10^{-7})[CO_2]$$

and

$$\text{True } K_1 = \frac{(4.3 \times 10^{-7})[CO_2]}{[H_2CO_3]}$$

$$\text{True } K_1 = \frac{(4.3 \times 10^{-7})(99.7)}{(0.3)} = 1.4 \times 10^{-4}$$

Thus, when corrected for the small degree to which carbon dioxide is hydrated, it can be seen that carbonic acid is actually a stronger acid than acetic acid. Carboxylic acids dissolve in sodium bicarbonate solution because the equilibrium leading to carbon dioxide formation is favorable, not because carboxylic acids are stronger acids than carbonic acid.

20.7 (*b*) 2-Chloroethanol has been converted to 3-hydroxypropanoic acid by way of the corresponding nitrile.

$$\underset{\text{2-Chloroethanol}}{HOCH_2CH_2Cl} \xrightarrow[H_2O]{NaCN} \underset{\text{2-Cyanoethanol}}{HOCH_2CH_2CN} \xrightarrow{\text{hydrolysis}} \underset{\text{3-Hydroxypropanoic acid}}{HOCH_2CH_2CO_2H}$$

The presence of the hydroxyl group in 2-chloroethanol precludes the preparation of a Grignard reagent from this material, so any attempt at the preparation of 3-hydroxypropanoic acid via the Grignard reagent of 2-chloroethanol is certain to fail.

(*c*) Grignard reagents can be prepared from tertiary halides and react in the expected manner with carbon dioxide. The procedure shown is entirely satisfactory.

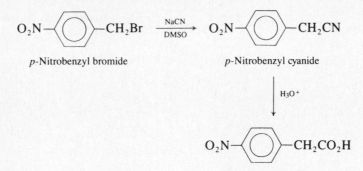

$$\underset{\textit{tert}\text{-Butyl chloride}}{(CH_3)_3CCl} \xrightarrow[\text{diethyl ether}]{Mg} \underset{\textit{tert}\text{-Butylmagnesium chloride}}{(CH_3)_3CMgCl} \xrightarrow[2.\ H_3O^+]{1.\ CO_2} \underset{\substack{\text{2,2-Dimethylpropanoic acid} \\ (61-70\%)}}{(CH_3)_3CCO_2H}$$

Preparation by way of the nitrile will not be feasible. Rather than reacting with sodium cyanide by substitution, *tert*-butyl chloride will undergo elimination exclusively. The S_N2 reaction with cyanide ion is limited to primary and secondary alkyl halides.

(*d*) The sequence shown below is satisfactory.

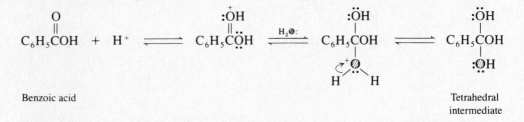

$$\underset{p\text{-Nitrobenzyl bromide}}{O_2N\text{—}C_6H_4\text{—}CH_2Br} \xrightarrow[DMSO]{NaCN} \underset{p\text{-Nitrobenzyl cyanide}}{O_2N\text{—}C_6H_4\text{—}CH_2CN}$$

$$\xrightarrow{H_3O^+} \underset{p\text{-Nitrophenylacetic acid}}{O_2N\text{—}C_6H_4\text{—}CH_2CO_2H}$$

Nitro groups are incompatible with Grignard reagents, so the alternate route via carboxylation of the Grignard reagent is not suitable.

20.8 Incorporation of ^{18}O into benzoic acid proceeds by a mechanism analogous to that of esterification. The nucleophile that adds to the protonated form of benzoic acid is ^{18}O-enriched water (the ^{18}O atom is represented by the shaded letter O in the following equations).

$$\underset{\text{Benzoic acid}}{C_6H_5\overset{O}{\overset{\|}{C}}OH} + H^+ \rightleftharpoons C_6H_5\overset{\overset{+}{:\ddot{O}H}}{\overset{\|}{C}}\ddot{O}H \overset{H_2\text{●}:}{\rightleftharpoons} C_6H_5\overset{:\ddot{O}H}{\overset{|}{C}}OH\ (\text{with } \overset{+}{\text{●}}H_2) \rightleftharpoons \underset{\text{Tetrahedral intermediate}}{C_6H_5\overset{:\ddot{O}H}{\overset{|}{C}}OH\ (\text{with } :\text{●}H)}$$

The three hydroxyl groups of the tetrahedral intermediate are equivalent except that one of them is labeled with ^{18}O. Any one of these three hydroxyl groups may be lost in the dehydration step; when the hydroxyl group that is lost is unlabeled, an ^{18}O label is retained in the benzoic acid.

$$\underset{\text{Tetrahedral intermediate}}{C_6H_5\overset{OH}{\overset{|}{C}}OH\ (\text{with }\text{●}H)} + H^+ \rightleftharpoons C_6H_5C(=\overset{+}{O}H)(\text{●}H) + H_2O$$

$$\xrightarrow{-H^+} \underset{^{18}\text{O-Enriched benzoic acid}}{C_6H_5C(=O)\text{●}H + C_6H_5C(OH)=\text{●}}$$

20.9 (*b*) The 16-membered ring of 15-pentadecanolide is formed from 15-hydroxypentadecanoic acid.

disconnect this bond

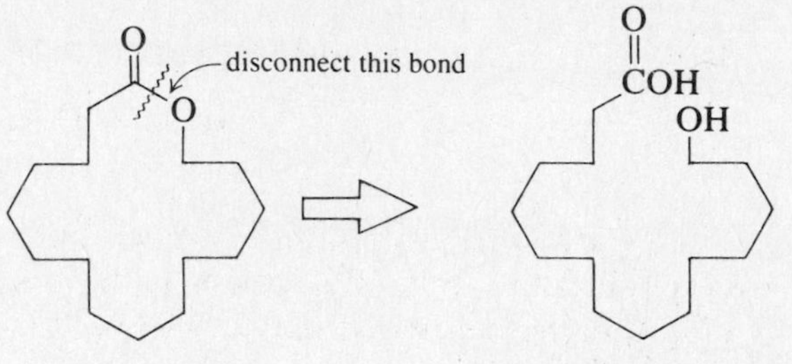

15-Pentadecanolide ⇒ 15-Hydroxypentadecanoic acid

(*c*) Vernolepin has two lactone rings, which can be related to two hydroxy acid combinations.

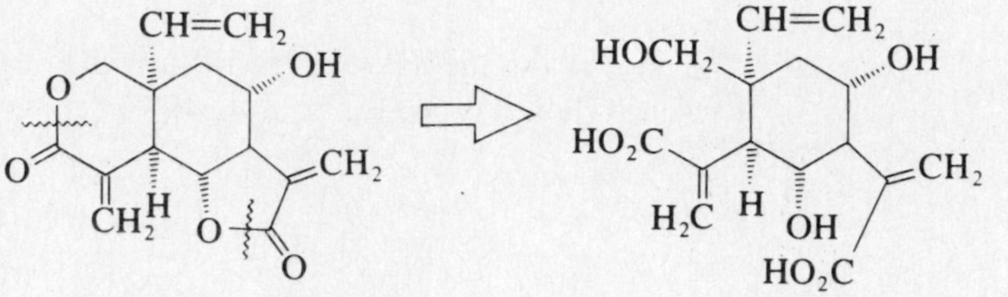

Be sure to keep the relative stereochemistry unchanged. Remember, the carbon-oxygen bond of an alcohol remains intact when the alcohol reacts with a carboxylic acid to give an ester.

20.10 Alkyl chlorides and bromides undergo nucleophilic substitution when treated with sodium iodide in acetone (Sec. 9.2). A reasonable approach is to brominate octadecanoic acid at its α carbon atom, then replace the bromine substituent with iodine by nucleophilic substitution.

$$CH_3(CH_2)_{15}CH_2CO_2H \xrightarrow{Br_2,\ PCl_3} CH_3(CH_2)_{15}\underset{\displaystyle Br}{\underset{|}{C}}HCO_2H \xrightarrow[acetone]{NaI} CH_3(CH_2)_{15}\underset{\displaystyle I}{\underset{|}{C}}HCO_2H$$

Octadecanoic acid — 2-Bromooctadecanoic acid — 2-Iodooctadecanoic acid

The same procedure could be used by starting with chlorination of octadecanoic acid.

20.11 (*b*) The substrate is recognized as a derivative of malonic acid. It undergoes efficient thermal decarboxylation in the manner shown.

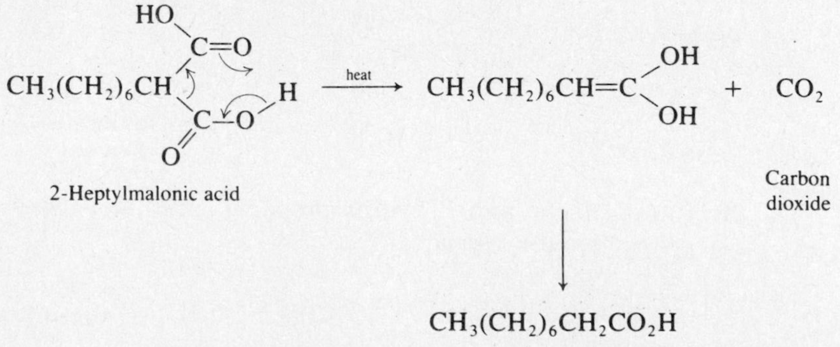

$$CH_3(CH_2)_6CH(CO_2H)_2 \xrightarrow{heat} CH_3(CH_2)_6CH{=}C(OH)_2 + CO_2$$

2-Heptylmalonic acid — Carbon dioxide

$$\downarrow$$

$$CH_3(CH_2)_6CH_2CO_2H$$

Nonanoic acid

(*c*) Decarboxylation of this compound gives cyclohexanecarboxylic acid.

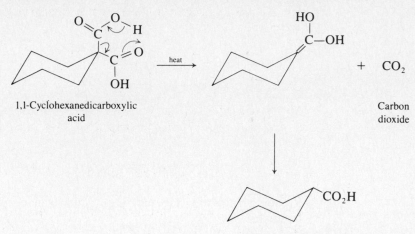

1,1-Cyclohexanedicarboxylic acid $\xrightarrow{\text{heat}}$ (enol intermediate) + CO_2 (Carbon dioxide) $\longrightarrow$ Cyclohexanecarboxylic acid ($C_6H_{11}CO_2H$)

(*d*) The phenyl and methyl substituents attached to C-2 of malonic acid play no role in the decarboxylation process.

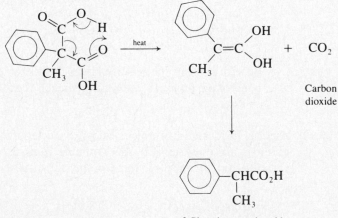

$C_6H_5C(CH_3)(CO_2H)_2 \xrightarrow{\text{heat}} C_6H_5(CH_3)C{=}C(OH)_2 + CO_2$ (Carbon dioxide)

$\longrightarrow C_6H_5CHCO_2H$ with CH_3 on the CH

2-Phenylpropanoic acid

20.12 (*b*) The thermal decarboxylation of β-keto acids resembles that of substituted malonic acids. The structure of 2,2-dimethylacetoacetic acid and the equation representing its decarboxylation were given in the text. The overall process involves the bonding changes shown.

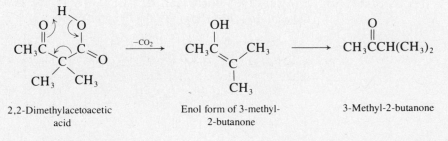

$CH_3COC(CH_3)_2CO_2H \xrightarrow{-CO_2} CH_3C(OH){=}C(CH_3)CH_3 \longrightarrow CH_3\overset{O}{\overset{\|}{C}}CH(CH_3)_2$

2,2-Dimethylacetoacetic acid; Enol form of 3-methyl-2-butanone; 3-Methyl-2-butanone

20.13 (*a*) Lactic acid (2-hydroxypropanoic acid) is a three-carbon carboxylic acid that bears a hydroxyl group at C-2.

$\overset{3}{C}H_3\overset{2}{C}H(OH)\overset{1}{C}O_2H$ 2-Hydroxypropanoic acid

(*b*) The parent name *ethanoic acid* tells us that the chain that includes the carboxylic acid function contains only two carbons. A hydroxyl group and a phenyl substituent are present at C-2.

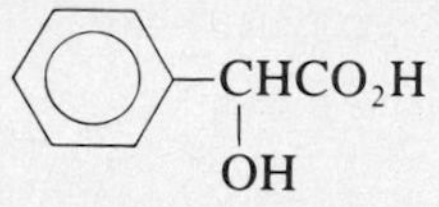

$$C_6H_5\text{-}\underset{\displaystyle OH}{\underset{|}{C}}HCO_2H$$ 2-Hydroxy-2-phenylethanoic acid (mandelic acid)

(*c*) The parent hydrocarbon is *tetradecane*, a linear chain of 14 carbons. The terminal methyl group is transformed to a carboxyl function in tetradecanoic acid.

$$CH_3(CH_2)_{12}\overset{\displaystyle O}{\overset{\|}{C}}OH$$ Tetradecanoic acid (myristic acid)

(*d*) Undecane is a linear hydrocarbon with 11 carbon atoms, undecanoic acid is the corresponding carboxylic acid, and *undecenoic acid* is an 11-carbon carboxylic acid that contains a double bond. Since the carbon chain is numbered beginning with the carboxyl group, 10-undecenoic acid has its double bond at the opposite end of the chain from the carboxyl group.

$$CH_2{=}CH(CH_2)_8CO_2H$$ 10-Undecenoic acid (undecylenic acid)

(*e*) Mevalonic acid has a five-carbon chain with hydroxyl groups at C-3 and C-5, along with a methyl group at C-3.

$$HOCH_2CH_2\overset{\displaystyle CH_3}{\overset{|}{\underset{\displaystyle OH}{\underset{|}{C}}}}CH_2CO_2H$$ 3,5-Dihydroxy-3-methylpentanoic acid (mevalonic acid)

(*f*) The constitution represented by the systematic name 2-methyl-2-butenoic acid gives rise to two stereoisomers.

$$CH_3CH{=}\underset{\displaystyle CH_3}{\underset{|}{C}}CO_2H$$ 2-Methyl-2-butenoic acid

Tiglic acid is the *E* isomer, and the *Z* isomer is known as angelic acid. The higher-priority substituents, methyl and carboxyl, are placed on opposite sides of the double bond in tiglic acid and on the same side in angelic acid.

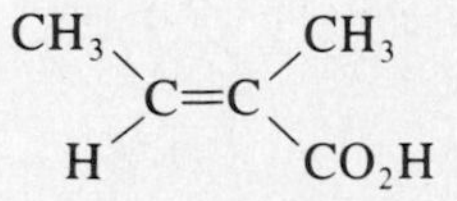

$$(CH_3)(H)C{=}C(CH_3)(CO_2H)$$
(*E*)-2-Methyl-2-butenoic acid (tiglic acid)

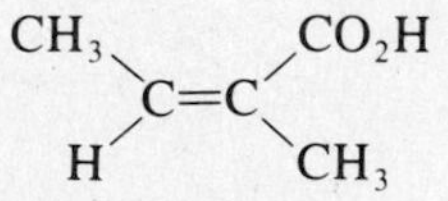

$$(CH_3)(H)C{=}C(CO_2H)(CH_3)$$
(*Z*)-2-Methyl-2-butenoic acid (angelic acid)

(*g*) Butanedioic acid is a four-carbon chain in which both terminal carbons are carboxylic acid groups. Malic acid has a hydroxyl group at C-2.

$$HO_2C\underset{\displaystyle OH}{\underset{|}{C}}HCH_2CO_2H$$ 2-Hydroxybutanedioic acid (malic acid)

(*h*) Each of the carbon atoms of propane bears a carboxyl group as a substituent in 1,2,3-propanetricarboxylic acid. In citric acid C-2 also bears a hydroxyl group.

$$HO_2CCH_2\overset{\displaystyle CO_2H}{\overset{|}{\underset{\displaystyle OH}{\underset{|}{C}}}}CH_2CO_2H$$ 2-Hydroxy-1,2,3-propanetricarboxylic acid (citric acid)

(*i*) There is an aryl substituent at C-2 of propanoic acid in ibuprofen. This aryl substituent is a benzene ring bearing an isobutyl group at the para position.

CH_3CHCO_2H (on benzene ring, para to $CH_2CH(CH_3)_2$)

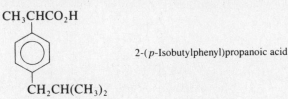

2-(*p*-Isobutylphenyl)propanoic acid

(*j*) Benzenecarboxylic acid is the systematic name for benzoic acid. *Salicylic acid* is a derivative of benzoic acid bearing a hydroxyl group at the position ortho to the carboxyl.

OH / CO_2H (ortho on benzene ring)

o-Hydroxybenzenecarboxylic acid (salicylic acid)

20.14 (*a*) The carboxylic acid contains a linear chain of eight carbon atoms. The parent alkane is *octane*, so the systematic IUPAC name of $CH_3(CH_2)_6CO_2H$ is *octanoic acid*.

(*b*) The compound shown is the potassium salt of octanoic acid. It is *potassium octanoate*.

(*c*) The presence of a double bond in $CH_2{=}CH(CH_2)_5CO_2H$ is indicated by the ending *-enoic acid*. Numbering of the chain begins with the carboxylic acid, so the double bond is between C-7 and C-8. The compound is *7-octenoic acid*.

(*d*) Stereochemistry is systematically described by the *E-Z* notation. Here, the double bond between C-6 and C-7 in octenoic acid has the *Z* configuration; the higher-priority substituents are on the same side.

$CH_3(H)C{=}C(H)(CH_2)_4CO_2H$ (CH_3 and $(CH_2)_4CO_2H$ on the same side)

(*Z*)-6-Octenoic acid

(*e*) A dicarboxylic acid is named as a *dioic acid*. The carboxyl functions are the terminal carbons of an eight-carbon chain; $HO_2C(CH_2)_6CO_2H$ is *octanedioic acid*. It is not necessary to identify the carboxylic acid locations by number, since they can only be at the ends of the chain when the *-dioic acid* name is used.

(*f*) Pick the longest continuous chain that includes both carboxyl groups and name the compound as a *-dioic acid*. This chain contains only three carbons and bears a pentyl group as a substituent at C-2. It is not necessary to specify the position of the pentyl group because it can only be attached to C-2.

$CH_3(CH_2)_4{-}\overset{2}{C}H\overset{1}{C}O_2H$ with ${}_3CO_2H$ on C-2

Pentylpropanedioic acid

Malonic acid is an acceptable synonym for propanedioic acid; this compound may also be named *pentylmalonic acid*.

(*g*) A carboxylic acid function is attached as a substituent on a seven-membered ring. The compound is *cycloheptanecarboxylic acid*.

CO_2H (on cycloheptane ring)

(*h*) The aromatic ring is named as a substituent attached to the eight-carbon carboxylic acid. Numbering of the chain begins with the carboxyl group.

$C_6H_5{-}CH(CH_2CH_3)(CH_2)_4CO_2H$

6-Phenyloctanoic acid

20.15 (*a*) The order of decreasing acidity is:

		K_a	pK_a
Acetic acid	CH_3CO_2H	1.8×10^{-5}	4.7
Ethanol	CH_3CH_2OH	10^{-16}	16
Ethane	CH_3CH_3	$\sim 10^{-46}$	~46

(*b*) Here again, the carboxylic acid is the strongest acid and the hydrocarbon the weakest:

		K_a	pK_a
Benzoic acid	$C_6H_5CO_2H$	6.7×10^{-5}	4.2
Benzyl alcohol	$C_6H_5CH_2OH$	10^{-16}–10^{-18}	16–18
Benzene	C_6H_6	$\sim 10^{-43}$	~43

(*c*) Propanedioic acid is a stronger acid than propanoic acid, since the electron-withdrawing effect of one carbonyl group enhances the ionization of the other. Propanedial is a 1,3-dicarbonyl compound that yields a stabilized enolate; it is more acidic than 1,3-propanediol.

		K_a	pK_a
Propanedioic acid	$HO_2CCH_2CO_2H$	1.4×10^{-3}	2.9
Propanoic acid	$CH_3CH_2CO_2H$	1.3×10^{-5}	4.9
Propanedial	$OHCCH_2CHO$	$\sim 10^{-9}$	~9
1,3-Propanediol	$HOCH_2CH_2CH_2OH$	$\sim 10^{-16}$	~16

(*d*) Trifluoromethanesulfonic acid is by far the strongest acid in the group. It is structurally related to sulfuric acid, but its three fluorine substituents make it much stronger. Fluorine substituents increase the acidity of carboxylic acids and alcohols relative to their nonfluorinated analogs, but not enough to make fluorinated alcohols as acidic as carboxylic acids.

		K_a	pK_a
Trifluoromethanesulfonic acid	CF_3SO_2OH	Very strong acid	
Trifluoroacetic acid	CF_3CO_2H	5.9×10^{-1}	0.2
Acetic acid	CH_3CO_2H	1.8×10^{-5}	4.7
2,2,2-Trifluoroethanol	CF_3CH_2OH	4.2×10^{-13}	12.4
Ethanol	CH_3CH_2OH	$\sim 10^{-16}$	~16

(*e*) The order of decreasing acidity is carboxylic acid > β-diketone > ketone > hydrocarbon.

		K_a	pK_a
Cyclopentanecarboxylic acid	(cyclopentane)–CO_2H	1×10^{-5}	5.0
2,4-Pentanedione	$CH_3\overset{O}{\overset{\|}{C}}CH_2\overset{O}{\overset{\|}{C}}CH_3$	10^{-9}	9
Cyclopentanone	(cyclopentane)=O	10^{-20}	20
Cyclopentene	(cyclopentene)	10^{-45}	45

20.16 (*a*) A trifluoromethyl group is strongly electron-withdrawing and acid-strengthening. Its ability to attract electrons from the carboxylate ion decreases as its distance down the chain increases. 3,3,3-Trifluoropropanoic acid is a stronger acid than 4,4,4-trifluorobutanoic acid.

$CF_3CH_2CO_2H$	$CF_3CH_2CH_2CO_2H$
3,3,3-Trifluoropropanoic acid K_a 9.6 × 10^{-14} (pK_a 3.0)	4,4,4-Trifluorobutanoic acid K_a 6.9 × 10^{-5} (pK_a 4.2)

(*b*) The carbon that bears the carboxyl group in 2-butynoic acid is *sp* hybridized and is, therefore, more electron-withdrawing than the sp^3 hybridized α carbon of butanoic acid. Therefore, the anion of 2-butynoic acid is stabilized better than the anion of butanoic acid and 2-butynoic acid is a stronger acid.

$CH_3C{\equiv}CCO_2H$	$CH_3CH_2CH_2CO_2H$
2-Butynoic acid K_a 2.5 × 10^{-3} (pK_a 2.6)	Butanoic acid K_a 1.5 × 10^{-5} (pK_a 4.8)

(*c*) Cyclohexanecarboxylic acid is a typical aliphatic carboxylic acid and is expected to be similar to acetic acid in acidity. The greater electronegativity of the sp^2 hybridized carbon attached to the carboxyl group in benzoic acid stabilizes benzoate anion better than the corresponding sp^3 hybridized carbon stabilizes cyclohexanecarboxylate. Benzoic acid is a stronger acid.

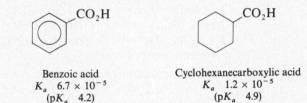

CO_2H CO_2H

Benzoic acid K_a 6.7 × 10^{-5} (pK_a 4.2)	Cyclohexanecarboxylic acid K_a 1.2 × 10^{-5} (pK_a 4.9)

(*d*) Its five fluorine substituents make the pentafluorophenyl group more electron-withdrawing than an unsubstituted phenyl group. Thus, pentafluorobenzoic acid is a stronger acid than benzoic acid.

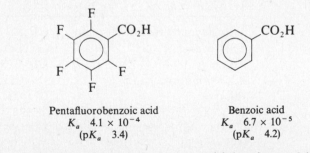

F F CO_2H F F F CO_2H

Pentafluorobenzoic acid K_a 4.1 × 10^{-4} (pK_a 3.4)	Benzoic acid K_a 6.7 × 10^{-5} (pK_a 4.2)

(*e*) The pentafluorophenyl substituent is electron-withdrawing and increases the acidity of a carboxyl group to which it is attached. Its electron-withdrawing effect decreases with distance. Pentafluorobenzoic acid is a stronger acid than *p*-(pentafluorophenyl)benzoic acid.

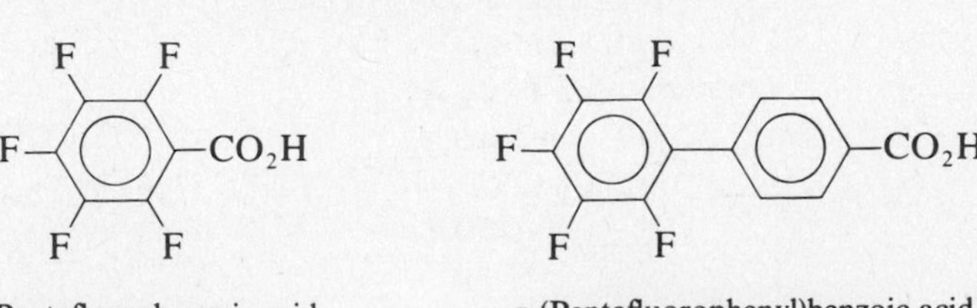

Pentafluorobenzoic acid K_a 4.1 × 10^{-4} (pK_a 3.4)	*p*-(Pentafluorophenyl)benzoic acid (K_a not measured in water, comparable with benzoic acid in acidity)

(*f*) The oxygen of the ring exercises an acidifying effect on the carboxyl group. This effect is largest when the oxygen is attached directly to the carbon that bears the carboxyl group. Furan-2-carboxylic acid is thus a stronger acid than furan-3-carboxylic acid.

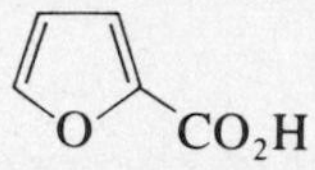

Furan-2-carboxylic acid	Furan-3-carboxylic acid
K_a 6.9×10^{-4}	K_a 1.1×10^{-4}
(pK_a 3.2)	(pK_a 3.9)

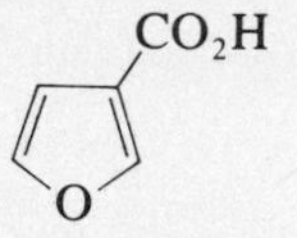

(*g*) Furan-2-carboxylic acid has an oxygen attached to the carbon that bears the carboxyl group, while pyrrole-2-carboxylic acid has a nitrogen in that position. Oxygen is more electronegative than nitrogen and so stabilizes the carboxylate anion better. Furan-2-carboxylic acid is a stronger acid than pyrrole-2-carboxylic acid.

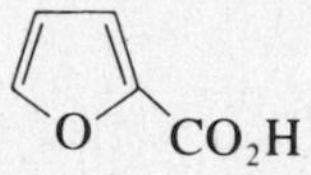

Furan-2-carboxylic acid	Pyrrole-2-carboxylic acid
K_a 6.9×10^{-4}	K_a 3.5×10^{-5}
(pK_a 3.2)	(pK_a 4.4)

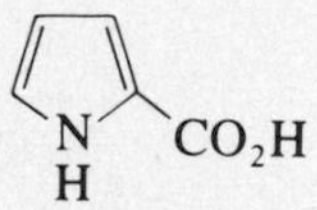

20.17 (*a*) The conversion of 1-butanol to butanoic acid is simply the oxidation of a primary alcohol to a carboxylic acid. Suitable oxidizing agents include potassium permanganate and chromic acid.

$$\underset{\text{1-Butanol}}{CH_3CH_2CH_2CH_2OH} \xrightarrow[2.\ H^+]{1.\ KMnO_4} \underset{\text{Butanoic acid}}{CH_3CH_2CH_2CO_2H}$$

(*b*) Aldehydes may be oxidized to carboxylic acids by any of the oxidizing agents that convert primary alcohols to carboxylic acids or by silver oxide, a mild oxidizing agent that oxidizes aldehydes to carboxylic acids.

$$\underset{\text{Butanal}}{CH_3CH_2CH_2\overset{O}{\overset{\|}{C}}H} \xrightarrow[H_2SO_4]{K_2Cr_2O_7} \underset{\text{Butanoic acid}}{CH_3CH_2CH_2\overset{O}{\overset{\|}{C}}OH}$$

(*c*) The starting material has the same number of carbon atoms as does butanoic acid, so all that is required is a series of functional group transformations. Carboxylic acids may be obtained by oxidation of the corresponding primary alcohol. The alcohol is available from the designated starting material, 1-butene.

$$CH_3CH_2CH_2CO_2H \Longrightarrow CH_3CH_2CH_2CH_2OH \Longrightarrow CH_3CH_2CH{=}CH_2$$

Hydroboration-oxidation of 1-butene yields 1-butanol, which can then be oxidized to butanoic acid as in (*a*).

$$\underset{\text{1-Butene}}{CH_3CH_2CH{=}CH_2} \xrightarrow[2.\ H_2O_2,\ HO^-]{1.\ B_2H_6} \underset{\text{1-Butanol}}{CH_3CH_2CH_2CH_2OH} \xrightarrow[2.\ H^+]{1.\ KMnO_4} \underset{\text{Butanoic acid}}{CH_3CH_2CH_2CO_2H}$$

(*d*) Converting 1-propanol to butanoic acid requires the carbon chain to be extended by one atom. Both methods for achieving this conversion, carboxylation of a Grignard reagent and formation and hydrolysis of a nitrile, begin with alkyl halides. Alkyl halides in turn are prepared from alcohols.

$$CH_3CH_2CH_2CO_2H \Rightarrow CH_3CH_2CH_2MgBr \text{ or } CH_3CH_2CH_2CN$$

$$\Downarrow$$

$$CH_3CH_2CH_2OH \Leftarrow CH_3CH_2CH_2Br$$

Either of the two following procedures is satisfactory.

$$\underset{\text{1-Propanol}}{CH_3CH_2CH_2OH} \xrightarrow[\text{or HBr}]{PBr_3} \underset{\text{1-Bromopropane}}{CH_3CH_2CH_2Br} \xrightarrow[\text{diethyl ether}]{Mg} CH_3CH_2CH_2MgBr \xrightarrow{1.\ CO_2;\ 2.\ H_3O^+} \underset{\text{Butanoic acid}}{CH_3CH_2CH_2CO_2H}$$

$$\underset{\text{1-Propanol}}{CH_3CH_2CH_2OH} \xrightarrow[\text{or HBr}]{PBr_3} \underset{\text{1-Bromopropane}}{CH_3CH_2CH_2Br} \xrightarrow[DMSO]{KCN} \underset{\text{1-Cyanopropane}}{CH_3CH_2CH_2CN} \xrightarrow{H_2O,\ HCl} \underset{\text{Butanoic acid}}{CH_3CH_2CH_2CO_2H}$$

(*e*) Dehydration of 2-propanol to propene followed by free-radical addition of hydrogen bromide affords 1-bromopropane.

$$\underset{\text{2-Propanol}}{CH_3\underset{OH}{\underset{|}{C}}HCH_3} \xrightarrow[\text{heat}]{H_2SO_4} \underset{\text{Propene}}{CH_3CH{=}CH_2} \xrightarrow[\text{peroxides}]{HBr} \underset{\text{1-Bromopropane}}{CH_3CH_2CH_2Br}$$

Once 1-bromopropane has been prepared it is converted to butanoic acid as in (*d*).

(*f*) Oxidative cleavage of 4-octene by potassium permanganate gives two molecules of butanoic acid.

$$\underset{\text{4-Octene}}{CH_3CH_2CH_2CH{=}CHCH_2CH_2CH_3} \xrightarrow[2.\ H^+]{1.\ KMnO_4} \underset{\text{Butanoic acid}}{2CH_3CH_2CH_2CO_2H}$$

(*g*) The carbon skeleton of butanoic acid may be assembled by an aldol condensation of acetaldehyde.

$$CH_3CH_2CH_2CO_2H \Rightarrow CH_3CH{=}CH\overset{O}{\overset{\|}{C}}H \Rightarrow 2\ CH_3\overset{O}{\overset{\|}{C}}H$$

$$\underset{\text{Acetaldehyde}}{2\,CH_3\overset{O}{\overset{\|}{C}}H} \xrightarrow[\text{heat}]{\text{KOH, ethanol}} \underset{\text{2-Butenal}}{CH_3CH{=}CH\overset{O}{\overset{\|}{C}}H}$$

Oxidation of the aldehyde followed by hydrogenation of the double bond yields butanoic acid.

$$\underset{\text{2-Butenal}}{CH_3CH{=}CH\overset{O}{\overset{\|}{C}}H} \xrightarrow{Ag_2O} \underset{\text{2-Butenoic acid}}{CH_3CH{=}CHCO_2H} \xrightarrow[Pt]{H_2} \underset{\text{Butanoic acid}}{CH_3CH_2CH_2CO_2H}$$

(*h*) Ethylmalonic acid belongs to the class of substituted malonic acids that undergo ready thermal decarboxylation. Decarboxylation yields butanoic acid.

$$CH_3CH_2\underset{\underset{CO_2H}{|}}{C}HCO_2H \xrightarrow{\text{heat}} CH_3CH_2CH_2CO_2H + CO_2$$

Ethylmalonic acid (ethylpropanedioic acid) — Butanoic acid — Carbon dioxide

20.18 (*a*) In principle the Friedel-Crafts alkylation of benzene by methyl chloride could be used to prepare ^{14}C-labeled toluene. Once prepared, toluene could be oxidized to benzoic acid.

$$C_6H_6 + \overset{*}{C}H_3Cl \xrightarrow{AlCl_3} C_6H_5\overset{*}{C}H_3$$

Benzene — Methyl chloride — Toluene

$$C_6H_5\overset{*}{C}H_3 \xrightarrow[\text{or } K_2Cr_2O_7,\ H_2SO_4]{1.\ KMnO_4\ 2.\ H^+} C_6H_5\overset{*}{C}O_2H$$

Benzoic acid

In practice monomethylation of benzene by methyl chloride is difficult to achieve. It would be necessary to perform the alkylation with excess benzene and then separate the toluene from higher-boiling dimethyl, trimethyl, etc. derivatives by distillation.

(*b*) Formaldehyde can serve as a one-carbon source if it is attacked by the Grignard reagent derived from bromobenzene.

$$C_6H_6 \xrightarrow[FeBr_3]{Br_2} C_6H_5Br \xrightarrow[\text{diethyl ether}]{Mg} C_6H_5MgBr$$

Benzene — Bromobenzene — Phenylmagnesium bromide

$$C_6H_5MgBr \xrightarrow{1.\ H\overset{*}{C}HO\ \ 2.\ H_3O^+} C_6H_5\overset{*}{C}H_2OH$$

Benzyl alcohol

This sequence yields ^{14}C-labeled benzyl alcohol, which can be oxidized to ^{14}C-labeled benzoic acid.

$$C_6H_5\overset{*}{C}H_2OH \xrightarrow[\text{or } K_2Cr_2O_7,\ H_2SO_4]{1.\ KMnO_4\ 2.\ H^+} C_6H_5\overset{*}{C}O_2H$$

Benzyl alcohol — Benzoic acid

(*c*) Here our ^{14}C source is acetic acid labeled at its carbonyl carbon. The labeled carbon can be attached to the aromatic ring by a Friedel-Crafts acylation reaction.

$$CH_3\overset{*}{C}(=O)OH \xrightarrow{SOCl_2} CH_3\overset{*}{C}(=O)Cl \xrightarrow[AlCl_3]{\text{benzene}} C_6H_5\overset{*}{C}(=O)CH_3$$

Acetic acid — Acetyl chloride — Acetophenone

Degradation of acetophenone to benzoic acid can be accomplished by way of the haloform reaction.

$$C_6H_5\overset{O}{\overset{\|}{C}}\overset{*}{}CH_3 \xrightarrow[2.\ H^+]{1.\ Br_2,\ HO^-} C_6H_5\overset{*}{C}O_2H + CHBr_3$$

Acetophenone — Benzoic acid — Bromoform

(*d*) A direct route to ^{14}C-labeled benzoic acid utilizes a Grignard synthesis employing ^{14}C-labeled carbon dioxide.

$$C_6H_6 \xrightarrow[FeBr_3]{Br_2} C_6H_5Br \xrightarrow{Mg,\ diethyl\ ether} C_6H_5MgBr$$

Benzene — Bromobenzene — Phenylmagnesium bromide

$$C_6H_5MgBr \xrightarrow[2.\ H_3O^+]{1.\ \overset{*}{C}O_2} C_6H_5\overset{*}{C}O_2H$$

Benzoic acid

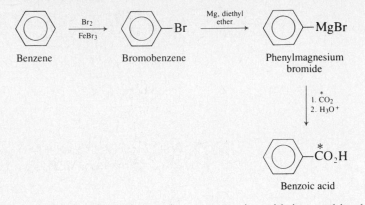

20.19 (*a*) An acid-base reaction takes place when pentanoic acid is combined with sodium hydroxide.

$$CH_3CH_2CH_2CH_2CO_2H + NaOH \longrightarrow CH_3CH_2CH_2CH_2CO_2Na + H_2O$$

Pentanoic acid — Sodium hydroxide — Sodium pentanoate — Water

(*b*) Carboxylic acids react with sodium bicarbonate to give carbonic acid, which dissociates to carbon dioxide and water, so the actual reaction that takes place is:

$$CH_3CH_2CH_2CH_2CO_2H + NaHCO_3 \longrightarrow$$

Pentanoic acid — Sodium bicarbonate

$$CH_3CH_2CH_2CH_2CO_2Na + CO_2 + H_2O$$

Sodium pentanoate — Carbon dioxide — Water

(*c*) Thionyl chloride is a reagent that converts carboxylic acids to the corresponding acyl chlorides.

$$CH_3CH_2CH_2CH_2CO_2H + SOCl_2 \longrightarrow CH_3CH_2CH_2CH_2\overset{O}{\overset{\|}{C}}Cl + SO_2 + HCl$$

Pentanoic acid — Thionyl chloride — Pentanoyl chloride — Sulfur dioxide — Hydrogen chloride

(*d*) Phosphorus tribromide is used to convert carboxylic acids to their acyl bromides.

$$3CH_3CH_2CH_2CH_2CO_2H + PBr_3 \longrightarrow 3CH_3CH_2CH_2CH_2\overset{O}{\overset{\|}{C}}Br + H_3PO_3$$

Pentanoic acid — Phosphorus tribromide — Pentanoyl bromide — Phosphorus acid

(*e*) Carboxylic acids react with alcohols in the presence of acid catalysts to give esters.

$$CH_3CH_2CH_2CH_2CO_2H + C_6H_5CH_2OH \overset{H_2SO_4}{\rightleftharpoons} CH_3CH_2CH_2CH_2\overset{O}{\overset{\|}{C}}OCH_2C_6H_5 + H_2O$$

Pentanoic acid — Benzyl alcohol — Benzyl pentanoate — Water

(*f*) Chlorine is introduced at the α carbon atom of a carboxylic acid. The reaction is catalyzed by a small amount of phosphorus or a phosphorus trihalide and is called the Hell-Volhard-Zelinsky reaction.

$$\underset{\text{Pentanoic acid}}{CH_3CH_2CH_2CH_2CO_2H} + \underset{\text{Chlorine}}{Cl_2} \xrightarrow[\text{(catalyst)}]{PBr_3} \underset{\text{2-Chloropentanoic acid}}{CH_3CH_2CH_2\underset{\displaystyle Cl}{\underset{|}{C}}HCO_2H} + \underset{\text{Hydrogen chloride}}{HCl}$$

The α-halo substituent is derived from the halogen used, not from the phosphorus trihalide.

(*g*) In this case, bromine is introduced at the α carbon.

$$\underset{\text{Pentanoic acid}}{CH_3CH_2CH_2CH_2CO_2H} + Br_2 \xrightarrow[\text{(catalyst)}]{PCl_3} \underset{\text{2-Bromopentanoic acid}}{CH_3CH_2CH_2\underset{\displaystyle Br}{\underset{|}{C}}HCO_2H} + \underset{\text{Hydrogen bromide}}{HBr}$$

(*h*) α-Halo carboxylic acids are reactive substrates in nucleophilic substitution. Iodide acts as a nucleophile to displace bromide from 2-bromopentanoic acid.

$$\underset{\text{2-Bromopentanoic acid}}{CH_3CH_2CH_2\underset{\displaystyle Br}{\underset{|}{C}}HCO_2H} + \underset{\text{Sodium iodide}}{NaI} \xrightarrow{\text{acetone}} \underset{\text{2-Iodopentanoic acid}}{CH_3CH_2CH_2\underset{\displaystyle I}{\underset{|}{C}}HCO_2H} + \underset{\text{Sodium bromide}}{NaBr}$$

(*i*) Aqueous ammonia converts α-halo acids to α-amino acids.

$$\underset{\text{2-Bromopentanoic acid}}{CH_3CH_2CH_2\underset{\displaystyle Br}{\underset{|}{C}}HCO_2H} + \underset{\text{Ammonia}}{2\,NH_3} \longrightarrow \underset{\text{2-Aminopentanoic acid}}{CH_3CH_2CH_2\underset{\displaystyle NH_2}{\underset{|}{C}}HCO_2H} + \underset{\text{Ammonium bromide}}{NH_4Br}$$

(*j*) Lithium aluminum hydride is a powerful reducing agent and reduces carboxylic acids to primary alcohols.

$$\underset{\text{Pentanoic acid}}{CH_3CH_2CH_2CH_2CO_2H} \xrightarrow[\text{2. } H_2O]{\text{1. } LiAlH_4} \underset{\text{1-Pentanol}}{CH_3CH_2CH_2CH_2CH_2OH}$$

(*k*) Phenylmagnesium bromide acts as a base to abstract the carboxylic acid proton.

$$\underset{\text{Pentanoic acid}}{CH_3CH_2CH_2CH_2CO_2H} + \underset{\text{Phenylmagnesium bromide}}{C_6H_5MgBr} \longrightarrow \underset{\text{Bromomagnesium pentanoate}}{CH_3CH_2CH_2CH_2CO_2MgBr} + \underset{\text{Benzene}}{C_6H_6}$$

Grignard reagents are not compatible with carboxylic acids; proton transfer converts the Grignard reagent to the corresponding hydrocarbon.

20.20 (*a*) Conversion of butanoic acid to 1-butanol is a reduction and requires lithium aluminum hydride as the reducing agent.

$$\underset{\text{Butanoic acid}}{CH_3CH_2CH_2\overset{\displaystyle O}{\overset{\|}{C}}OH} \xrightarrow[\text{2. } H_2O]{\text{1. } LiAlH_4} \underset{\text{1-Butanol}}{CH_3CH_2CH_2CH_2OH}$$

(*b*) Carboxylic acids are converted to their corresponding acyl chlorides with thionyl chloride or with phosphorus pentachloride.

$$\underset{\text{Butanoic acid}}{CH_3CH_2CH_2\overset{O}{\overset{\|}{C}}OH} \xrightarrow[\text{or } PCl_5]{SOCl_2} \underset{\text{Butanoyl chloride}}{CH_3CH_2CH_2\overset{O}{\overset{\|}{C}}Cl}$$

(*c*) Carboxylic acids cannot be reduced directly to aldehydes. Among the two-step procedures are the following:

$$\underset{\text{Butanoic acid}}{CH_3CH_2CH_2\overset{O}{\overset{\|}{C}}OH} \xrightarrow[2.\ H_2O]{1.\ LiAlH_4} \underset{\text{1-Butanol}}{CH_3CH_2CH_2CH_2OH} \xrightarrow[CH_2Cl_2]{(C_5H_5N)_2CrO_3} \underset{\text{Butanal}}{CH_3CH_2CH_2\overset{O}{\overset{\|}{C}}H}$$

$$\underset{\text{Butanoic acid}}{CH_3CH_2CH_2\overset{O}{\overset{\|}{C}}OH} \xrightarrow[\text{or } PCl_5]{SOCl_2} \underset{\text{Butanoyl chloride}}{CH_3CH_2CH_2\overset{O}{\overset{\|}{C}}Cl} \xrightarrow[Pd/BaSO_4]{H_2} \underset{\text{Butanal}}{CH_3CH_2CH_2\overset{O}{\overset{\|}{C}}H}$$

(*d*) Remember that alkyl halides are usually prepared from alcohols. Therefore, 1-butanol is needed in order to prepare 1-chlorobutane.

$$\underset{\text{1-Butanol}}{CH_3CH_2CH_2CH_2OH} \xrightarrow{SOCl_2} \underset{\text{1-Chlorobutane}}{CH_3CH_2CH_2CH_2Cl}$$

Conversion of butanoic acid to 1-butanol requires reduction with lithium aluminum hydride as shown in (*a*).

(*e*) Aromatic ketones are frequently prepared by Friedel-Crafts acylation of the appropriate acyl chloride and benzene. Butanoyl chloride, prepared in (*b*), can be used to acylate benzene in a Friedel-Crafts reaction.

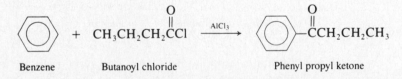

$$\underset{\text{Benzene}}{C_6H_6} + \underset{\text{Butanoyl chloride}}{CH_3CH_2CH_2\overset{O}{\overset{\|}{C}}Cl} \xrightarrow{AlCl_3} \underset{\text{Phenyl propyl ketone}}{C_6H_5{-}\overset{O}{\overset{\|}{C}}CH_2CH_2CH_3}$$

(*f*) There are several ways to prepare 4-octanone using compounds derived from butanoic acid. These may be seen by using disconnections in a retrosynthetic analysis.

$$CH_3CH_2CH_2\overset{O}{\overset{\|}{C}}CH_2CH_2CH_2CH_3 \Rightarrow CH_3CH_2CH_2\overset{O}{\overset{\|}{C}}Cl + (CH_3CH_2CH_2CH_2)_2M \quad (M = Cd \text{ or } CuLi)$$

$$CH_3CH_2CH_2\overset{O}{\overset{\|}{C}}CH_2CH_2CH_2CH_3 \Rightarrow CH_3CH_2CH_2\overset{O}{\overset{\|}{C}}H + CH_3CH_2CH_2CH_2MgBr$$

The reaction schemes which may be used are:

$$\underset{\text{1-Chlorobutane}}{CH_3CH_2CH_2CH_2Cl} \xrightarrow[\text{diethyl ether}]{Mg} \underset{\text{Butylmagnesium chloride}}{CH_3CH_2CH_2CH_2MgCl} \xrightarrow{CdCl_2} \underset{\text{Dibutylcadmium}}{(CH_3CH_2CH_2CH_2)_2Cd}$$

$$\underset{\text{Dibutylcadmium}}{(CH_3CH_2CH_2CH_2)_2Cd} + \underset{\text{Butanoyl chloride}}{CH_3CH_2CH_2\overset{O}{\overset{\|}{C}}Cl} \longrightarrow \underset{\text{4-Octanone}}{CH_3CH_2CH_2\overset{O}{\overset{\|}{C}}CH_2CH_2CH_2CH_3}$$

A similar reaction sequence employs lithium dibutylcuprate.

$$\underset{\text{1-Chlorobutane}}{CH_3CH_2CH_2CH_2Cl} \xrightarrow{Li} \underset{\text{Butyllithium}}{CH_3CH_2CH_2CH_2Li} \xrightarrow{CuBr} \underset{\text{Lithium dibutylcuprate}}{(CH_3CH_2CH_2CH_2)_2CuLi}$$

$$\underset{\text{Lithium dibutylcuprate}}{(CH_3CH_2CH_2CH_2)_2CuLi} + \underset{\text{Butanoyl chloride}}{CH_3CH_2CH_2\overset{\overset{O}{\|}}{C}Cl} \longrightarrow \underset{\text{4-Octanone}}{CH_3CH_2CH_2\overset{\overset{O}{\|}}{C}CH_2CH_2CH_2CH_3}$$

Alternatively, 4-octanol could be prepared and then oxidized to 4-octanone.

$$\underset{\text{Butylmagnesium chloride}}{CH_3CH_2CH_2CH_2MgCl} + \underset{\text{Butanal}}{CH_3CH_2CH_2\overset{\overset{O}{\|}}{C}H} \xrightarrow[\text{2. } H_3O^+]{\text{1. diethyl ether}} \underset{\text{4-Octanol}}{CH_3CH_2CH_2\underset{\underset{OH}{|}}{C}HCH_2CH_2CH_2CH_3}$$

$$\xrightarrow{H_2CrO_4} \underset{\text{4-Octanone}}{CH_3CH_2CH_2\overset{\overset{O}{\|}}{C}CH_2CH_2CH_2CH_3}$$

(*g*) Carboxylic acids are halogenated at their α carbon atom by the Hell-Volhard-Zelinsky reaction.

$$\underset{\text{Butanoic acid}}{CH_3CH_2CH_2CO_2H} \xrightarrow[Br_2]{P} \underset{\text{2-Bromobutanoic acid}}{CH_3CH_2\underset{\underset{Br}{|}}{C}HCO_2H}$$

A catalytic amount of PCl_3 may be used in place of phosphorus in the reaction.

(*h*) Base-promoted dehydrohalogenation of 2-bromobutanoic acid gives 2-butenoic acid.

$$\underset{\text{2-Bromobutanoic acid}}{CH_3CH_2\underset{\underset{Br}{|}}{C}HCO_2H} \xrightarrow[\text{DMSO; 2. } H^+]{\text{1. } KOC(CH_3)_3} \underset{\text{2-Butenoic acid}}{CH_3CH{=}CHCO_2H}$$

20.21 (*a*) The compound to be prepared is *glycine*, an α-amino acid. The amino functional group can be introduced by a nucleophilic substitution reaction on an α-halo acid, which is available by way of the Hell-Volhard-Zelinsky reaction.

$$\underset{\text{Acetic acid}}{CH_3CO_2H} \xrightarrow[P]{Cl_2} \underset{\text{Chloroacetic acid}}{ClCH_2CO_2H} \xrightarrow{NH_3,\ H_2O} \underset{\text{Aminoacetic acid (glycine)}}{H_2NCH_2CO_2H}$$

(*b*) Phenoxyacetic acid is used as a fungicide. It can be prepared by a nucleophilic substitution using sodium phenoxide and chloroacetic acid.

$$\underset{\text{Chloroacetic acid}}{ClCH_2CO_2H} \xrightarrow[\text{2. } H^+]{\text{1. } C_6H_5ONa} \underset{\text{Phenoxyacetic acid}}{C_6H_5OCH_2CO_2H} + NaCl$$

(*c*) Cyanide ion is a good nucleophile and will displace chloride from chloroacetic acid.

$$\underset{\text{Chloroacetic acid}}{ClCH_2CO_2H} \xrightarrow[Na_2CO_3,\ H_2O]{NaCN} \underset{\text{Sodium cyanoacetate}}{N{\equiv}CCH_2CO_2Na} \xrightarrow{H^+} \underset{\text{Cyanoacetic acid}}{N{\equiv}CCH_2CO_2H}$$

(*d*) Cyanoacetic acid, prepared as above, serves as a convenient precursor to malonic acid. Hydrolysis of the nitrile substituent converts it to a carboxyl group.

$$\underset{\text{Cyanoacetic acid}}{N{\equiv}CCH_2CO_2H} \xrightarrow[\text{heat}]{H_2O,\ H^+} \underset{\text{Malonic acid}}{HO_2CCH_2CO_2H}$$

(*e*) Iodoacetic acid is not prepared directly from acetic acid but is derived by nucleophilic substitution of iodide in bromoacetic or chloroacetic acid.

$$\underset{\text{Chloroacetic acid}}{ClCH_2CO_2H} \xrightarrow[\text{acetone}]{NaI} \underset{\text{Iodoacetic acid}}{ICH_2CO_2H}$$

(*f*) Two transformations need to be accomplished, α bromination and esterification. The correct sequence is bromination followed by esterification.

$$\underset{\text{Acetic acid}}{CH_3CO_2H} \xrightarrow[P]{Br_2} \underset{\text{Bromoacetic acid}}{BrCH_2CO_2H} \xrightarrow[H^+]{CH_3CH_2OH} \underset{\text{Ethyl bromoacetate}}{BrCH_2CO_2CH_2CH_3}$$

Reversing the order of steps is not appropriate. It must be the carboxylic acid that is subjected to halogenation, as the Hell-Volhard-Zelinsky reaction is a reaction of carboxylic acids, not esters.

(*g*) The compound shown is an ylide. It can be prepared from ethyl bromoacetate as shown.

$$\underset{\text{Ethyl bromoacetate}}{BrCH_2CO_2CH_2CH_3} + \underset{\text{Triphenylphosphine}}{(C_6H_5)_3P} \longrightarrow$$

$$\underset{Br^-}{(C_6H_5)_3\overset{+}{P}CH_2CO_2CH_2CH_3} \xrightarrow{NaOCH_2CH_3} \underset{\text{Ylide}}{(C_6H_5)_3\overset{+}{P}{-}\overset{-}{\ddot{C}}HCO_2CH_2CH_3}$$

The first step is a nucleophilic substitution of bromide by triphenylphosphine. Treatment of the derived triphenylphosphonium salt with base removes the relatively acidic α proton, forming the ylide. (For a review of ylide formation, refer back to Sec. 18.16.)

(*h*) Reaction of the ylide formed in (*g*) with benzaldehyde gives the desired alkene by a Wittig reaction.

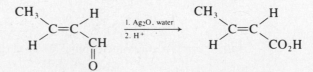

$$\underset{\text{Ylide from }(g)}{(C_6H_5)_3\overset{+}{P}{-}\overset{-}{\ddot{C}}HCO_2CH_2CH_3} + \underset{\text{Benzaldehyde}}{C_6H_5\overset{O}{\overset{\|}{C}}H} \longrightarrow$$

$$\underset{\text{Ethyl cinnamate}}{C_6H_5CH{=}CHCO_2CH_2CH_3} + \underset{\text{Triphenylphosphine oxide}}{(C_6H_5)_3\overset{+}{P}{-}\overset{-}{O}}$$

20.22 (*a*) Silver oxide is a weak oxidizing agent; it oxidizes aldehydes to the corresponding carboxylic acids.

$$\underset{\substack{(E)\text{-2-Butenal}\\(\textit{trans}\text{-Crotonaldehyde})}}{(CH_3)HC{=}CH(CH{=}O)} \xrightarrow[2.\ H^+]{1.\ Ag_2O,\ \text{water}} \underset{\substack{(E)\text{-2-Butenoic acid}\\(\textit{trans}\text{-Crotonic acid})}}{(CH_3)HC{=}CH(CO_2H)}$$

(*b*) Oxidation of the acetyl side chain occurs on treatment of the indicated starting material with potassium permanganate. The methoxy and fluoro substituents are not affected.

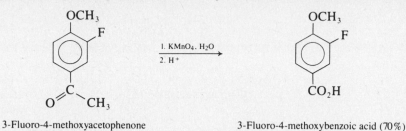

$$\text{3-Fluoro-4-methoxyacetophenone} \xrightarrow[\text{2. } H^+]{\text{1. } KMnO_4,\ H_2O} \text{3-Fluoro-4-methoxybenzoic acid (70\%)}$$

(*c*) Carboxylic acids are converted to ethyl esters when they are allowed to stand in ethanol in the presence of an acid catalyst.

$$(CH_3)(H)C{=}C(CH_3)CO_2H + CH_3CH_2OH \xrightarrow{H_2SO_4} (CH_3)(H)C{=}C(CH_3)COCH_2CH_3 + H_2O$$

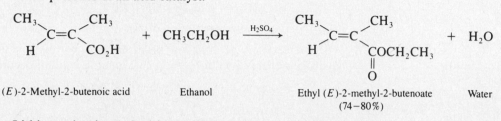

(*E*)-2-Methyl-2-butenoic acid; Ethanol; Ethyl (*E*)-2-methyl-2-butenoate (74–80%); Water

(*d*) Lithium aluminum hydride, $LiAlH_4$, reduces carboxylic acids to primary alcohols. When $LiAlD_4$ is used, deuterium is transferred to the carbonyl carbon.

$$\triangleright\text{—COH (C=O)} \xrightarrow[\text{2. } H_2O]{\text{1. } LiAlD_4\text{, diethyl ether}} \triangleright\text{—CD}_2\text{OH}$$

Cyclopropanecarboxylic acid; 1-Cyclopropyl-1,1-dideuteriomethanol (75%)

Notice that deuterium is bonded only to carbon. The hydroxyl proton is derived from water, not from the reducing agent.

(*e*) In the presence of a catalytic amount of phosphorus, bromine reacts with carboxylic acids to yield the corresponding α-bromo derivative.

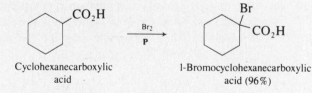

$$\text{Cyclohexanecarboxylic acid} \xrightarrow[\text{P}]{Br_2} \text{1-Bromocyclohexanecarboxylic acid (96\%)}$$

(*f*) Alkyl fluorides are not readily converted to Grignard reagents, so it is the bromine substituent that is attacked by magnesium.

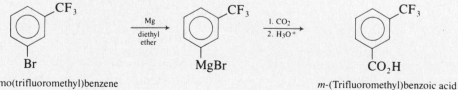

$$m\text{-Bromo(trifluoromethyl)benzene} \xrightarrow[\text{diethyl ether}]{Mg} m\text{-}CF_3C_6H_4MgBr \xrightarrow[\text{2. } H_3O^+]{\text{1. } CO_2} m\text{-(Trifluoromethyl)benzoic acid}$$

(*g*) Cyano substituents are hydrolyzed to carboxyl groups in the presence of acid catalysts.

$$m\text{-}ClC_6H_4CH_2CN \xrightarrow[H_2SO_4\text{, heat}]{H_2O\text{, acetic acid}} m\text{-}ClC_6H_4CH_2CO_2H$$

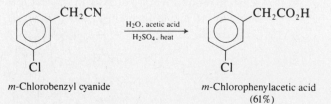

m-Chlorobenzyl cyanide; *m*-Chlorophenylacetic acid (61%)

(*h*) The starting material is a methyl ketone and is cleaved to chloroform and a carboxylic acid under conditions of the haloform reaction.

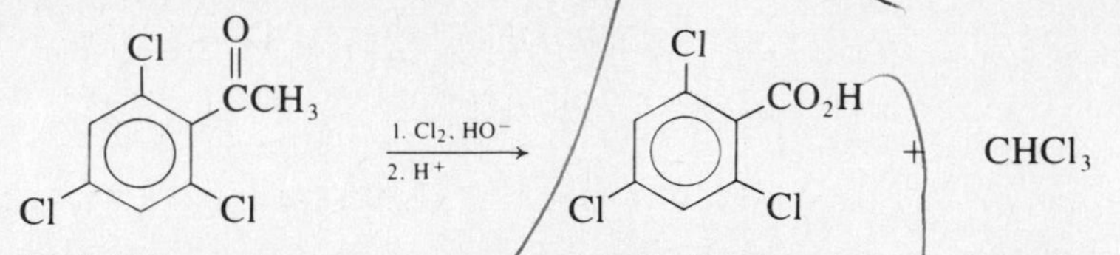

2,4,6-Trichloroacetophenone $\xrightarrow[2.\ H^+]{1.\ Cl_2,\ HO^-}$ 2,4,6-Trichlorobenzoic acid (100%) + $CHCl_3$ Chloroform

(*i*) The carboxylic acid is converted to its methyl ester under these conditions.

$$CH_3C{\equiv}CCO_2H + CH_3OH \xrightarrow{H_2SO_4} CH_3C{\equiv}C\overset{O}{\overset{\|}{C}}OCH_3 + H_2O$$

2-Butynoic acid, Methanol, Methyl 2-butynoate, Water

(*j*) The carboxylic acid function plays no part in this reaction; free-radical addition of hydrogen bromide to the carbon-carbon double bond occurs.

$$CH_2{=}CH(CH_2)_8CO_2H \xrightarrow[\text{benzoyl peroxide}]{HBr} BrCH_2CH_2(CH_2)_8CO_2H$$

10-Undecenoic acid, 11-Bromoundecanoic acid (66–70%)

Recall that hydrogen bromide adds to alkenes in the presence of peroxides with a regioselectivity opposite to that of Markovnikov's rule.

(*k*) Oxidative cleavage of the carbon-carbon double bond takes place.

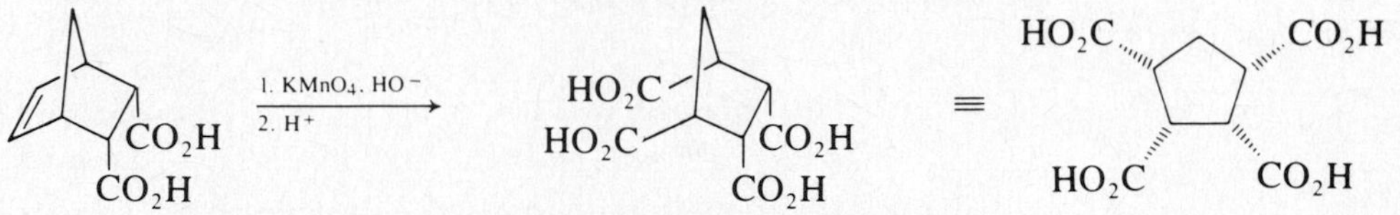

(bicyclic diacid) $\xrightarrow[2.\ H^+]{1.\ KMnO_4,\ HO^-}$ 1,2,3,4-Cyclopentanetetracarboxylic acid

Notice that the four carboxylic acid groups are cis in the product.

20.23 (*a*) The desired product and the starting material have the same carbon skeleton, so all that is required is a series of functional group transformations. Recall that, as seen in Problem 20.17, a carboxylic acid may be prepared by oxidation of the corresponding primary alcohol. The needed alcohol is available from the appropriate alkene.

$$(CH_3)_3COH \xrightarrow[heat]{H^+} (CH_3)_2C{=}CH_2 \xrightarrow[2.\ H_2O_2,\ HO^-]{1.\ B_2H_6} (CH_3)_2CHCH_2OH$$

tert-Butyl alcohol; 2-Methylpropene (isobutene); 2-Methyl-1-propanol (isobutyl alcohol)

$$\xrightarrow[2.\ H^+]{1.\ KMnO_4} (CH_3)_2CHCO_2H$$

2-Methylpropanoic acid

(*b*) The target molecule contains one more carbon than the starting material, so a carbon-carbon bond-forming step is indicated. Two approaches are reasonable; one proceeds by way of nitrile formation and hydrolysis, the other by carboxylation of a Grignard reagent. In either case the key intermediate is isobutyl bromide.

$$(CH_3)_2CHCH_2CO_2H \Longrightarrow (CH_3)_2CHCH_2Br$$

3-Methylbutanoic acid; 1-Bromo-2-methylpropane (isobutyl bromide)

Isobutyl bromide may be prepared by free-radical addition of hydrogen bromide to isobutene.

$$(CH_3)_3COH \xrightarrow[\text{heat}]{H^+} (CH_3)_2C{=}CH_2 \xrightarrow[\text{peroxides}]{HBr} (CH_3)_2CHCH_2Br$$

tert-Butyl alcohol — 2-Methylpropene (isobutene) — 1-Bromo-2-methylpropane (isobutyl bromide)

Another route to isobutyl bromide utilizes the isobutyl alcohol prepared in (*a*).

$$(CH_3)_2CHCH_2OH \xrightarrow{PBr_3} (CH_3)_2CHCH_2Br$$

2-Methyl-1-propanol (isobutyl alcohol) — 1-Bromo-2-methylpropane (isobutyl bromide)

Conversion of isobutyl bromide to the desired acid is then carried out as shown below.

$$(CH_3)_2CHCH_2Br \xrightarrow{KCN} (CH_3)_2CHCH_2CN \xrightarrow[\text{heat}]{H_2O,\ H^+} (CH_3)_2CHCH_2CO_2H$$

$$(CH_3)_2CHCH_2Br \xrightarrow[\text{diethyl ether}]{Mg} (CH_3)_2CHCH_2MgBr \xrightarrow[2.\ H_3O^+]{1.\ CO_2} (CH_3)_2CHCH_2CO_2H$$

Isobutyl bromide — 3-Methylbutanoic acid

(*c*) This synthesis requires extending a carbon chain by two carbon atoms. One way to form dicarboxylic acids is by hydrolysis of dinitriles.

$$HO_2C(CH_2)_5CO_2H \Longrightarrow NC(CH_2)_5CN \Longrightarrow Br(CH_2)_5Br$$

This suggests the following series of steps.

$$HO_2C(CH_2)_3CO_2H \xrightarrow[2.\ H_2O]{1.\ LiAlH_4} HOCH_2(CH_2)_3CH_2OH$$

Pentanedioic acid — 1,5-Pentanediol

$$HOCH_2(CH_2)_3CH_2OH \xrightarrow{HBr \text{ or } PBr_3} BrCH_2(CH_2)_3CH_2Br \xrightarrow{KCN} NCCH_2(CH_2)_3CH_2CN$$

1,5-Dibromopentane — 1,5-Dicyanopentane

$$NCCH_2(CH_2)_3CH_2CN \xrightarrow[\text{heat}]{H_2O,\ H^+} HO_2CCH_2(CH_2)_3CH_2CO_2H \equiv HO_2C(CH_2)_5CO_2H$$

Heptanedioic acid

(*d*) The desired alcohol cannot be prepared directly from the nitrile. It is available, however, by lithium aluminum hydride reduction of the carboxylic acid obtained by hydrolysis of the nitrile.

$$CH_3CH(C_6H_5)CH_2CN \xrightarrow[H^+]{H_2O} CH_3CH(C_6H_5)CH_2\overset{O}{\overset{\|}{C}}OH \xrightarrow[2.\ H_2O]{1.\ LiAlH_4} CH_3CH(C_6H_5)CH_2CH_2OH$$

1-Cyano-2-phenylpropane — 3-Phenylbutanoic acid — 3-Phenyl-1-butanol

(*e*) In spite of the structural similarity between the starting material and the desired product, a one-step transformation cannot be achieved.

Cyclopentyl bromide $\not\longrightarrow$ 1-Bromocyclopentanecarboxylic acid

Instead, recall that α-bromo acids are prepared from carboxylic acids by the Hell-Volhard-Zelinsky reaction.

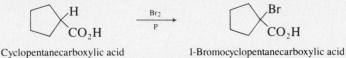

$$\text{Cyclopentanecarboxylic acid} \xrightarrow[\text{P}]{\text{Br}_2} \text{1-Bromocyclopentanecarboxylic acid}$$

Cyclopentanecarboxylic acid 1-Bromocyclopentanecarboxylic acid

The problem now simplifies to one of preparing cyclopentanecarboxylic acid from cyclopentyl bromide. Two routes are possible.

$$\text{C}_5\text{H}_9\text{—Br} \xrightarrow{\text{KCN}} \text{C}_5\text{H}_9\text{—CN} \xrightarrow[\text{H}^+]{\text{H}_2\text{O}} \text{C}_5\text{H}_9\text{—CO}_2\text{H}$$

Cyclopentyl bromide Cyclopentyl cyanide Cyclopentanecarboxylic acid

$$\text{C}_5\text{H}_9\text{—Br} \xrightarrow[\text{diethyl ether}]{\text{Mg}} \text{C}_5\text{H}_9\text{—MgBr} \xrightarrow[\text{2. H}_3\text{O}^+]{\text{1. CO}_2} \text{C}_5\text{H}_9\text{—CO}_2\text{H}$$

Cyclopentyl bromide Cyclopentylmagnesium bromide Cyclopentanecarboxylic acid

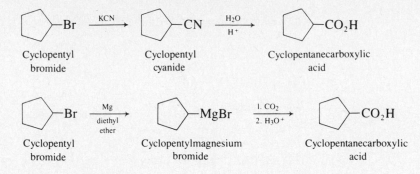

The Grignard route is better; it is a "one-pot" transformation. Converting the secondary bromide to a nitrile will be accompanied by elimination and the procedure requires two separate operations.

(*f*) In this case the halogen substituent is present at the β carbon rather than the α carbon atom of the carboxylic acid. The starting material, a β-chloro unsaturated acid, can lead to the desired carbon skeleton by a Diels-Alder reaction.

$$\text{trans-2-chlorocyclohexanecarboxylic acid} \Longrightarrow \text{trans-2-chloro-4-cyclohexenecarboxylic acid} \Longrightarrow \text{CH}_2\text{=CH—CH=CH}_2 + \text{ClHC=CHCO}_2\text{H}$$

$$\text{CH}_2\text{=CH—CH=CH}_2 + \text{ClHC=CHCO}_2\text{H} \longrightarrow \text{trans-2-chloro-4-cyclohexenecarboxylic acid}$$

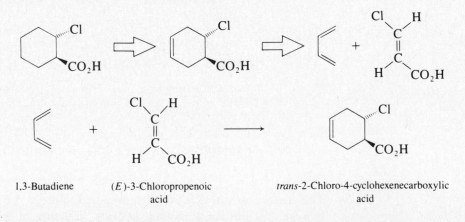

1,3-Butadiene (*E*)-3-Chloropropenoic acid *trans*-2-Chloro-4-cyclohexenecarboxylic acid

The required trans stereochemistry is a consequence of the stereospecificity of the Diels-Alder reaction.

Hydrogenation of the double bond of the Diels-Alder adduct gives the required product.

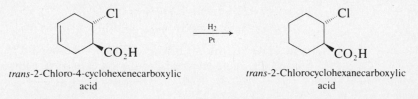

$$\text{trans-2-chloro-4-cyclohexenecarboxylic acid} \xrightarrow[\text{Pt}]{\text{H}_2} \text{trans-2-chlorocyclohexanecarboxylic acid}$$

trans-2-Chloro-4-cyclohexenecarboxylic acid *trans*-2-Chlorocyclohexanecarboxylic acid

(*g*) The conversion of α-pinene to pinic acid was carried out by Adolf von Baeyer in the nineteenth century as part of work aimed at determining the structure of α-pinene. Two different oxidation reactions are involved, the first of which is a cleavage of the carbon-carbon double bond by potassium permanganate.

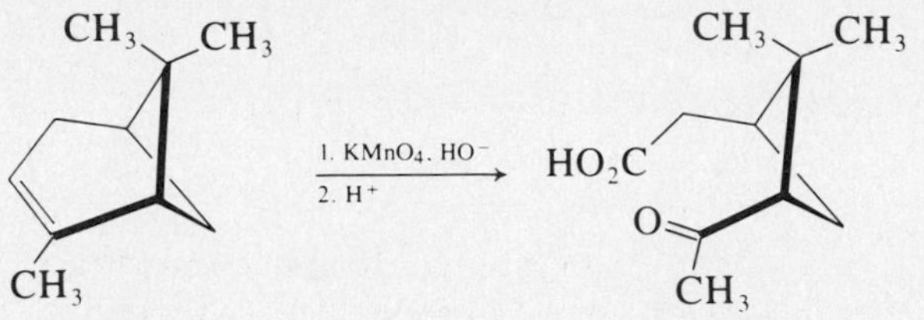

CH_3 CH_3 — 1. $KMnO_4$, HO^-; 2. H^+ → HO_2C, CH_3, CH_3, O, CH_3

α-Pinene　　cis-Pinonic acid

The product of this reaction is known as *cis*-pinonic acid. Cleavage of its methyl ketone function by the haloform reaction yields the desired product, pinic acid.

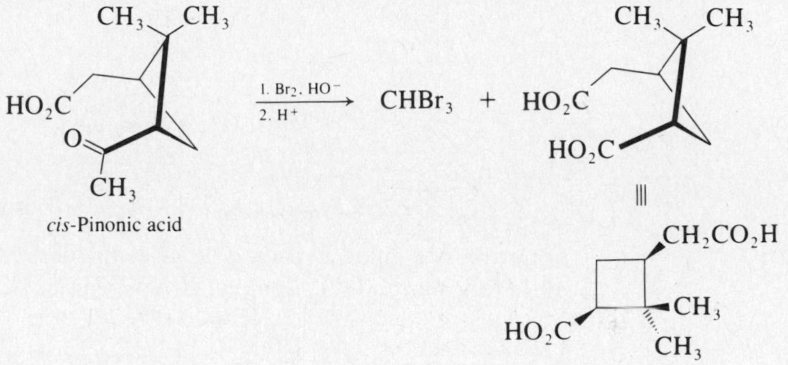

CH_3 CH_3 HO_2C O CH_3 — 1. Br_2, HO^-; 2. H^+ → $CHBr_3$ + HO_2C, HO_2C, CH_3, CH_3 ≡ CH_2CO_2H, CH_3, HO_2C, CH_3

cis-Pinonic acid　　Pinic acid

(*h*) The target molecule is related to the starting material by the retrosynthesis

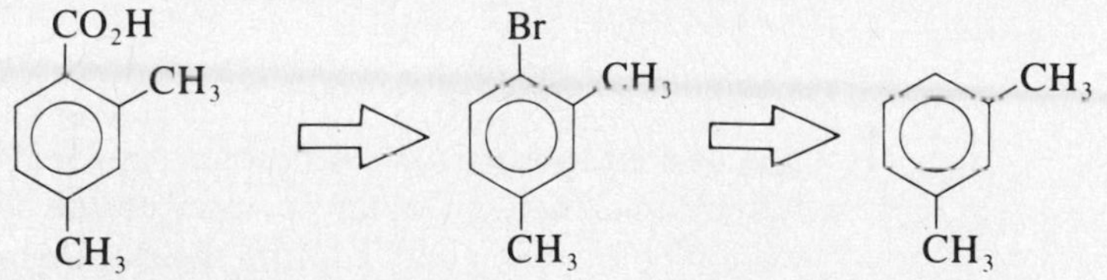

CO_2H, CH_3, CH_3 ⇨ Br, CH_3, CH_3 ⇨ CH_3, CH_3

2,4-Dimethylbenzoic acid　　*m*-Xylene

The necessary bromine substituent can be introduced by electrophilic substitution in the activated aromatic ring of *m*-xylene.

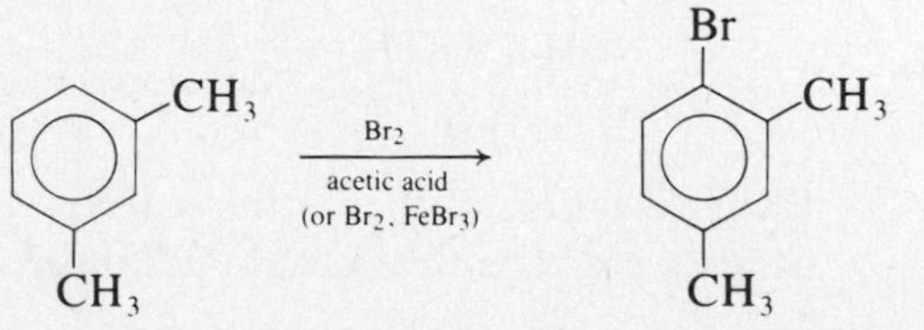

CH_3, CH_3 — Br_2, acetic acid (or Br_2, $FeBr_3$) → Br, CH_3, CH_3

m-Xylene　　1-Bromo-2,4-dimethylbenzene

The aryl bromide cannot be converted to a carboxylic acid by way of the corresponding cyano compound because aryl bromides are not reactive toward nucleophilic substitution. The Grignard route is necessary.

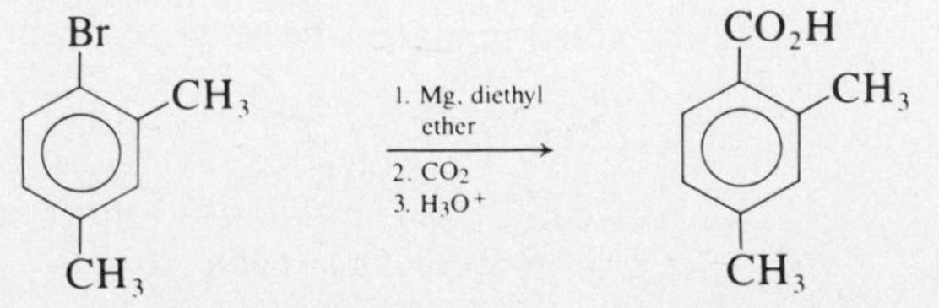

Br, CH_3, CH_3 — 1. Mg, diethyl ether; 2. CO_2; 3. H_3O^+ → CO_2H, CH_3, CH_3

1-Bromo-2,4-dimethylbenzene　　2,4-Dimethylbenzoic acid

(*i*) The relationship of the target molecule to the starting material

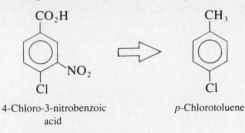

4-Chloro-3-nitrobenzoic acid ⇨ *p*-Chlorotoluene

requires that there be two synthetic operations, oxidation of the methyl group and nitration of the ring. The orientation of the nitro group requires that nitration of the starting material follow oxidation of the methyl group.

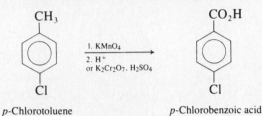

$$p\text{-Chlorotoluene} \xrightarrow[\text{2. } H^+ \text{ or } K_2Cr_2O_7,\ H_2SO_4]{\text{1. } KMnO_4} p\text{-Chlorobenzoic acid}$$

Nitration of *p*-chlorobenzoic acid gives the desired product because the directing effects of the chlorine and the carboxyl groups reinforce each other.

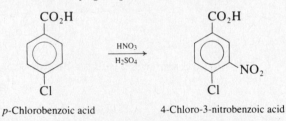

$$p\text{-Chlorobenzoic acid} \xrightarrow[H_2SO_4]{HNO_3} \text{4-Chloro-3-nitrobenzoic acid}$$

(*j*) The desired synthetic route becomes apparent when it is recognized that the *Z* alkene stereoisomer may be obtained from an alkyne which, in turn, is available by carboxylation of the anion derived from the starting material.

$$(CH_3)(H)C{=}C(H)(CO_2H) \Rightarrow CH_3C{\equiv}CCO_2H \Rightarrow CH_3C{\equiv}\bar{C}\!: + CO_2$$

The desired reaction sequence is:

$$\underset{\text{Propyne}}{CH_3C{\equiv}CH} \xrightarrow[NH_3]{NaNH_2} \underset{\text{Propynylsodium}}{CH_3C{\equiv}CNa} \xrightarrow[\text{2. } H_3O^+]{\text{1. } CO_2} \underset{\text{2-Butynoic acid}}{CH_3C{\equiv}CCO_2H}$$

Hydrogenation of the carbon-carbon triple bond of 2-butynoic acid over the Lindlar catalyst converts this compound to the *Z* isomer of 2-butenoic acid.

$$\underset{\text{2-Butynoic acid}}{CH_3C{\equiv}CCO_2H} \xrightarrow[\text{lead acetate (Lindlar Pd)}]{H_2,\ Pd/CaCO_3} \underset{(Z)\text{-2-Butenoic acid}}{(CH_3)(H)C{=}C(H)(CO_2H)}$$

(*k*) The desired product has one less carbon atom than the starting material. Degradation of the starting material can be accomplished by the haloform reaction.

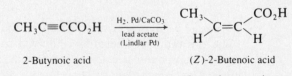

$$(CH_3)_2C{=}CHCCH_3 \text{ (C=O)} \xrightarrow[\text{2. } H^+]{\text{1. } Cl_2,\ HO^-} (CH_3)_2C{=}CHCOH \text{ (C=O)} + CHCl_3$$

4-Methyl-3-penten-2-one (aldol condensation product of acetone) — 3-Methyl-2-butenoic acid (49–53%) — Chloroform

As reported in the literature, the reaction was carried out with chlorine, although bromine or iodine would have been a suitable choice too.

Now that the carbon skeleton of the product has been prepared, all that is required to complete the synthesis is to reduce the carboxylic acid to the primary alcohol. Lithium aluminum hydride is the reagent of choice here.

$$\underset{\text{3-Methyl-2-butenoic acid}}{(CH_3)_2C{=}CH\overset{\overset{O}{\|}}{C}OH} \xrightarrow[2.\ H_2O]{1.\ LiAlH_4,\ \text{diethyl ether}} \underset{\text{3-Methyl-2-buten-1-ol (60\%)}}{(CH_3)_2C{=}CHCH_2OH}$$

20.24 (*a*) The most stable conformation of formic acid is the one that has both hydrogens anti.

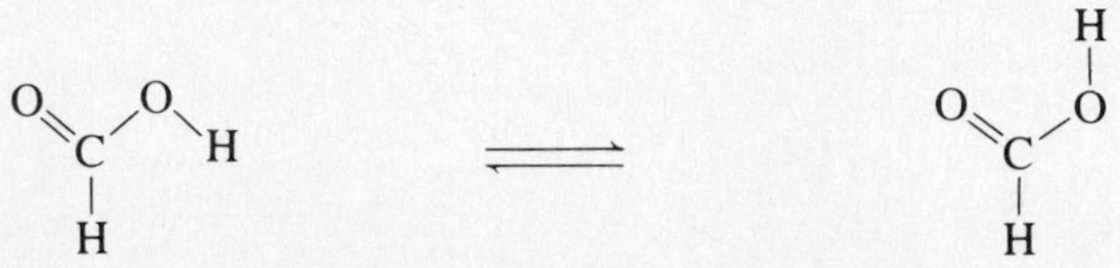

Syn: less stable conformation of formic acid ⇌ Anti: more stable conformation of formic acid

A plausible explanation is that the syn conformation is destabilized by lone pair repulsions.

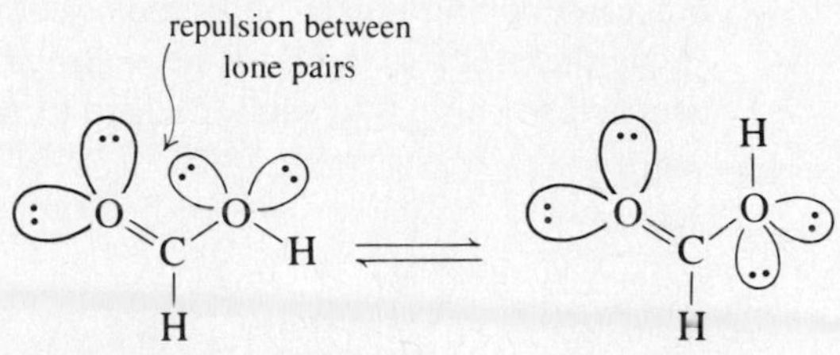

repulsion between lone pairs

Syn ⇌ Anti

(*b*) A dipole moment of zero can mean that the molecule has a center of symmetry. One structure that satisfies this requirement is characterized by intramolecular hydrogen bonding between the two carboxyl groups and an anti relationship between the two carbonyls.

intramolecular hydrogen bond

intramolecular hydrogen bond

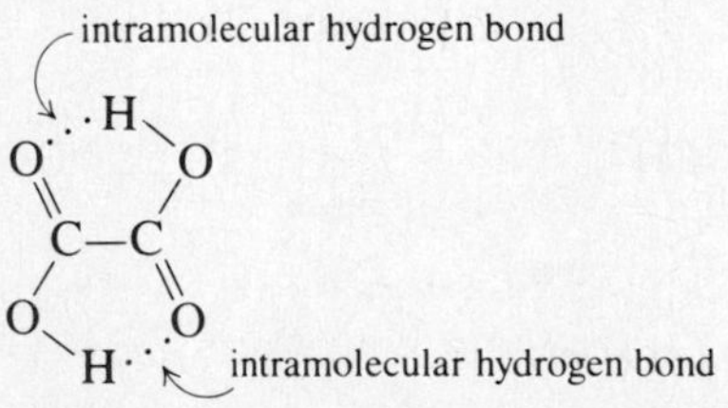

Another possibility is the structure shown below. It also has a center of symmetry and an anti relationship between the two carbonyls.

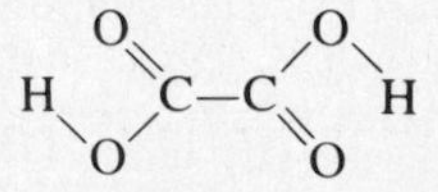

Other centrosymmetric structures can be drawn; these have the two hydrogen atoms out of the plane of the carboxyl groups, however, and are less likely to occur in view of the known planarity of carboxyl groups. Structures in which the carbonyl groups are syn to each other do not have a center of symmetry.

(c) The anion formed on dissociation of o-hydroxybenzoic acid can be stabilized by an intramolecular hydrogen bond.

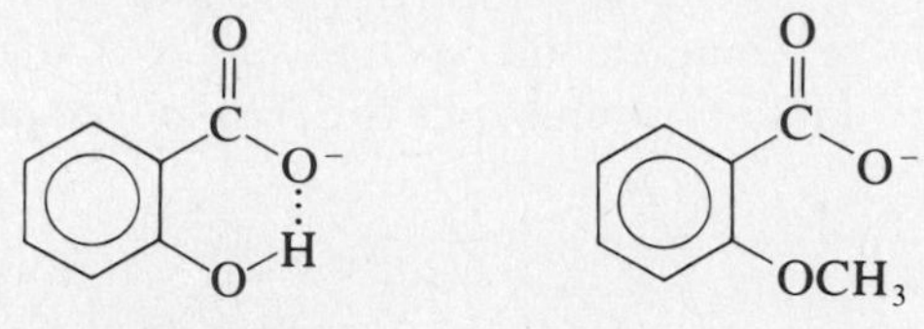

o-Hydroxybenzoate ion (stabilized by hydrogen bonding)

o-Methoxybenzoate ion (hydrogen bonding is not possible)

(d) Lactone formation is possible only when the hydroxyl and carboxyl groups are cis.

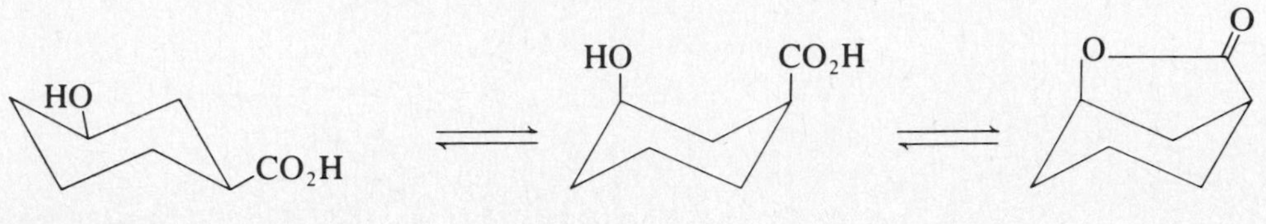

cis-3-Hydroxycyclohexanecarboxylic acid

Lactone

While the most stable conformation of *cis*-3-hydroxycyclohexanecarboxylic acid has both substituents equatorial and is unable to close to a lactone, the diaxial orientation is accessible and is capable of lactone formation.

Neither conformation of *trans*-3-hydroxycyclohexanecarboxylic acid has the substituents close enough to each other to form an unstrained lactone.

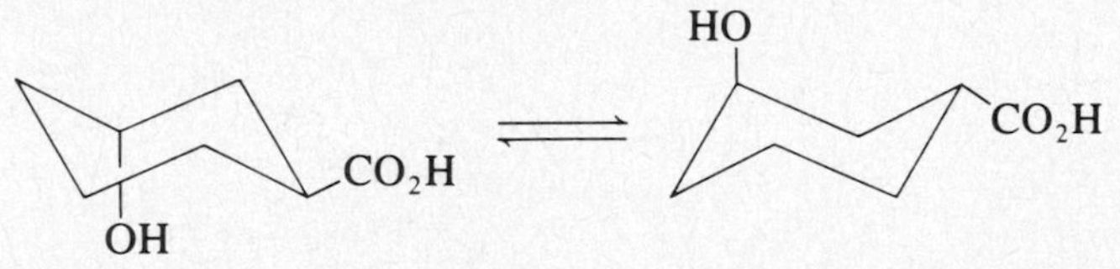

trans-3-Hydroxycyclohexanecarboxylic acid: lactone formation impossible

(e) Ascorbic acid is relatively acidic because ionization of its enolic hydroxyl at C-3 gives an anion that is stabilized by resonance in much the same way as a carboxylate ion; the negative charge is shared by two oxygens.

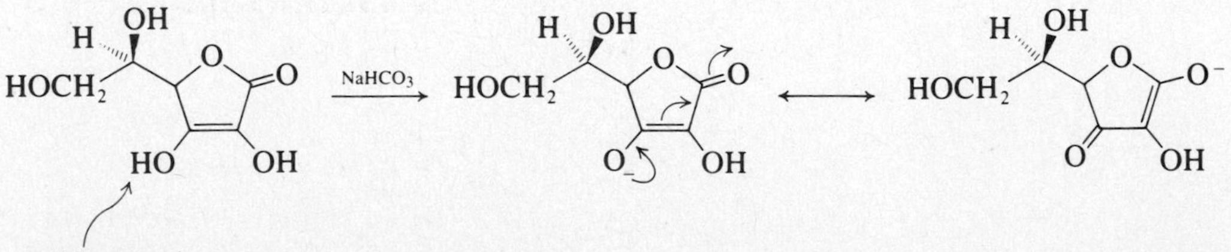

$\xrightarrow{NaHCO_3}$

acidic proton in ascorbic acid

20.25 Dicarboxylic acids in which both carboxyl groups are attached to the same carbon undergo ready thermal decarboxylation to produce the enol form of an acid.

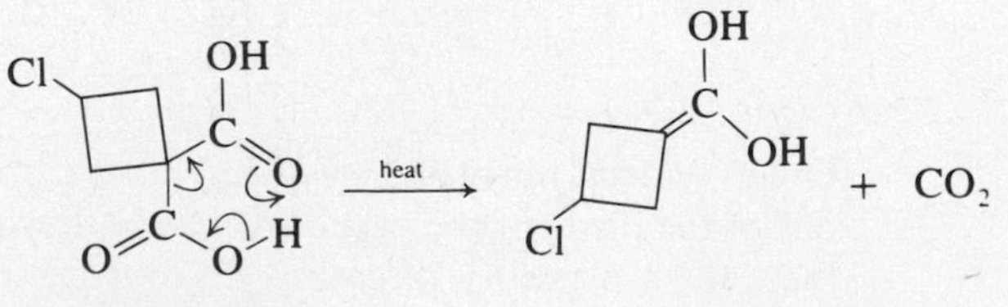

$\xrightarrow{heat}$ + CO_2

Compound A

This enol yields a mixture of *cis*- and *trans*-3-chlorocyclobutanecarboxylic acid. The two products are stereoisomers.

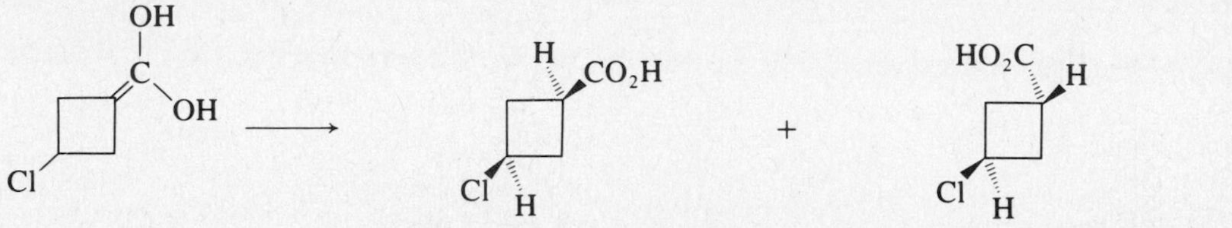

cis-3-Chlorocyclobutanecarboxylic acid *trans*-3-Chlorocyclobutanecarboxylic acid

20.26 Examination of the molecular formula $C_{14}H_{26}O_2$ reveals that the compound contains two elements of unsaturation (SODAR = 2). Since we are told that the compound is a carboxylic acid, one of these elements of unsaturation must be a carbon-oxygen double bond. The other must be a carbon-carbon double bond since the compound undergoes cleavage on treatment with potassium permanganate. Cleavage of the compound serves to locate the position of the double bond.

$$CH_3(CH_2)_7CH{=}CH(CH_2)_3CO_2H \xrightarrow[2.\ H^+]{1.\ KMnO_4} CH_3(CH_2)_7CO_2H + HO_2C(CH_2)_3CO_2H$$

cleavage by $KMnO_4$ occurs here

Nonanoic acid

Pentanedioic acid

The starting acid must be 5-tetradecenoic acid. The stereochemistry of the double bond is not revealed by these experiments.

20.27 Hydrogenation of the starting material is expected to result in reduction of the ketone carbonyl while leaving the carboxyl group unaffected. Since the isolated product lacks a carboxyl group, however, that group must react in some way. The most reasonable reaction is intramolecular esterification to form a γ-lactone.

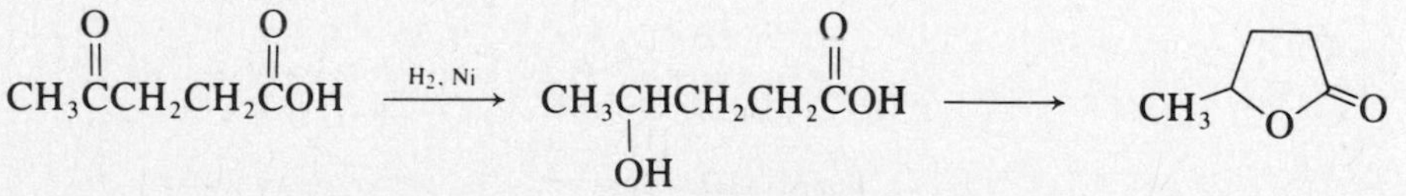

$$CH_3\overset{O}{\overset{\|}{C}}CH_2CH_2\overset{O}{\overset{\|}{C}}OH \xrightarrow{H_2,\ Ni} CH_3\underset{OH}{\underset{|}{C}}HCH_2CH_2\overset{O}{\overset{\|}{C}}OH \longrightarrow$$

Levulinic acid

4-Hydroxypentanoic acid (not isolated)

4-Pentanolide (compound B, $C_5H_8O_2$)

20.28 Compound C is a cyclic acetal and undergoes hydrolysis in aqueous acid to produce acetaldehyde, along with a dihydroxy carboxylic acid.

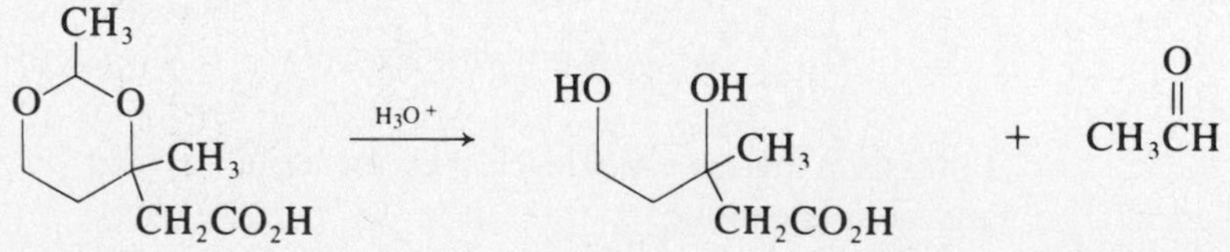

$\xrightarrow{H_3O^+}$

Compound C

3,5-Dihydroxy-3-methylpentanoic acid

Acetaldehyde

The dihydroxy acid that is formed in this step cyclizes to the δ-lactone.

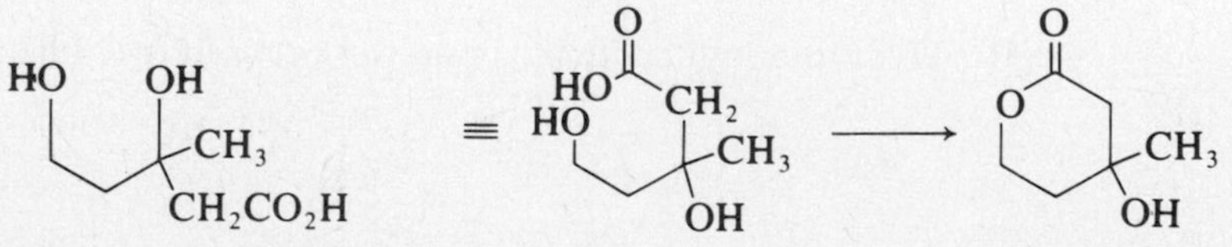

3,5-Dihydroxy-3-methylpentanoic acid

Mevalonolactone

20.29 Compound D is a δ-lactone. To determine its precursor disconnect the ester linkage to a hydroxy acid.

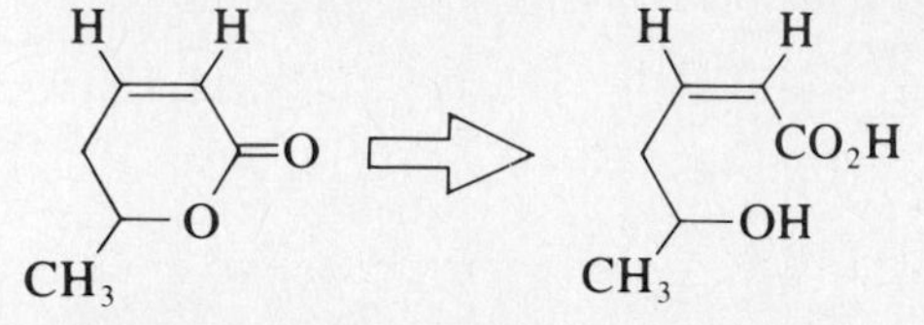

Compound D ⟹ cis-HC=CH(CO$_2$H) hydroxy acid (CH_3CH(OH)CH$_2$CH=CHCO$_2$H)

The precursor has the same carbon skeleton as the designated starting material. All that is necessary is to hydrogenate the double bond of the alkynoic acid to the cis alkene. This can be done by using either the Lindlar catalyst (lead-poisoned palladium-on-calcium carbonate) or palladium on barium sulfate. Cyclization of the hydroxy acid to the lactone is spontaneous.

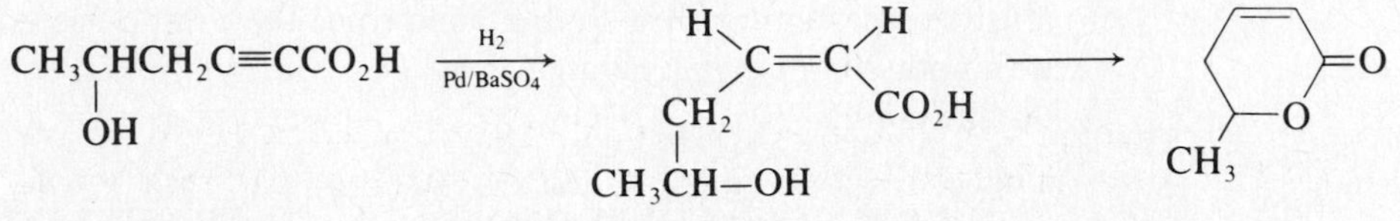

$$CH_3CH(OH)CH_2C{\equiv}CCO_2H \xrightarrow[Pd/BaSO_4]{H_2} CH_3CH(OH)CH_2CH{=}CHCO_2H \longrightarrow \text{Compound D}$$

5-Hydroxy-2-hexynoic acid — (Not isolated) — Compound D

20.30 The reaction of a carboxylic acid with an alcohol in the presence of an acid catalyst yields an ester. Compound E is the ethyl ester of hexadecanoic acid.

$$CH_3(CH_2)_{14}\overset{O}{\overset{\|}{C}}OH + CH_3CH_2OH \xrightarrow{H^+} CH_3(CH_2)_{14}\overset{O}{\overset{\|}{C}}OCH_2CH_3$$

Hexadecanoic acid — Ethanol — Ethyl hexadecanoate (compound E, $C_{18}H_{36}O_2$)

Grignard reagents react with esters to yield tertiary alcohols.

$$CH_3(CH_2)_{14}\overset{O}{\overset{\|}{C}}OCH_2CH_3 + 2C_6H_5MgBr \xrightarrow[2.\ H_3O^+]{1.\ \text{diethyl ether}} CH_3(CH_2)_{14}\overset{OH}{\overset{|}{C}}(C_6H_5)_2$$

Ethyl hexadecanoate — Phenylmagnesium bromide — 1,1-Diphenyl-1-hexadecanol (compound F, $C_{28}H_{42}O$)

Heating compound F, a tertiary benzylic alcohol, in the presence of acid causes it to undergo dehydration.

$$CH_3(CH_2)_{13}CH_2\overset{OH}{\overset{|}{C}}(C_6H_5)_2 \xrightarrow[heat]{H^+} CH_3(CH_2)_{13}CH{=}C(C_6H_5)_2$$

1,1-Diphenyl-1-hexadecanol — 1,1-Diphenyl-1-hexadecene (compound G, $C_{28}H_{40}$)

Potassium permanganate cleaves the double bond in compound G to pentadecanoic acid and benzophenone.

$$CH_3(CH_2)_{13}CH{=}C(C_6H_5)_2 \xrightarrow[2.\ H^+]{1.\ KMnO_4} CH_3(CH_2)_{13}\overset{O}{\overset{\|}{C}}OH + (C_6H_5)_2C{=}O$$

1,1-Diphenyl-1-hexadecene — Pentadecanoic acid — Benzophenone (compound H, $C_{13}H_{10}O$)

20.31 Hydration of the double bond can occur in two different directions.

$$(HO_2C)HC{=}C(CO_2H)(CH_2CO_2H) \xrightarrow{H_2O} HO_2CCH_2\overset{CO_2H}{\underset{OH}{C}}CH_2CO_2H + HO_2C\underset{OH}{C}H\overset{CO_2H}{C}HCH_2CO_2H$$

(*a*) The achiral isomer is citric acid.

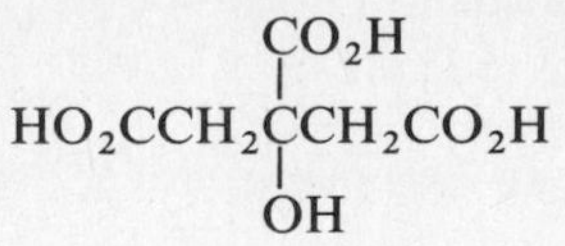

$$\begin{array}{c} CO_2H \\ | \\ HO_2CCH_2CCH_2CO_2H \\ | \\ OH \end{array}$$

Citric acid has no chiral centers

(*b*) The other isomer, isocitric acid, has two chiral centers. Isocitric acid has the constitution

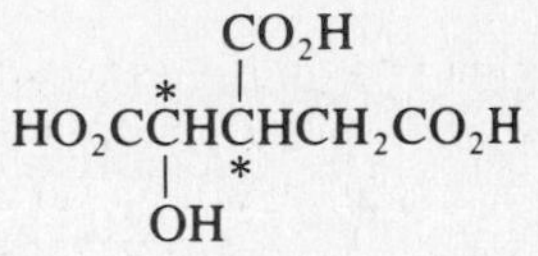

$$HO_2C\overset{*}{C}H(OH)\overset{*}{C}H(CO_2H)CH_2CO_2H$$

Isocitric acid: chiral centers are marked with asterisk

With two chiral centers, there are 2^2, or four stereoisomers represented by this constitution. The one that is actually formed in this enzyme catalyzed reaction is the 2*R*,3*S* isomer.

20.32 Carboxylic acid protons give signals in the range $\delta = 10$ to 12 ppm. A signal in this region suggests the presence of a carboxyl group but tells little about its environment. Thus, in assigning structures to compounds I, J, and K the data that will be most useful are the chemical shifts of the protons other than the carboxyl protons. Compare the three structures:

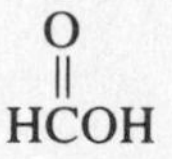

$HCOH$ (C=O)	(H)(HO$_2$C)C=C(H)(CO$_2$H)	$HO_2CCH_2CO_2H$
Formic acid	Maleic acid	Malonic acid

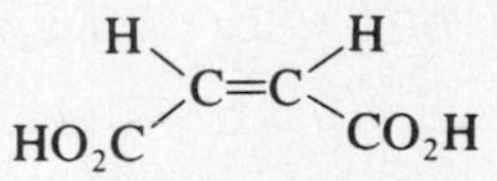

The proton that is diagnostic of structure in formic acid is bonded to a carbonyl group; it is an aldehyde proton. Typical chemical shifts of aldehyde protons are 8 to 10 ppm, and therefore formic acid is compound K.

Compound K $\quad H{-}\overset{O}{\overset{\|}{C}}{-}O{-}H$

$\delta = 8.0$ ppm (H on C); $\delta = 11.4$ ppm (OH)

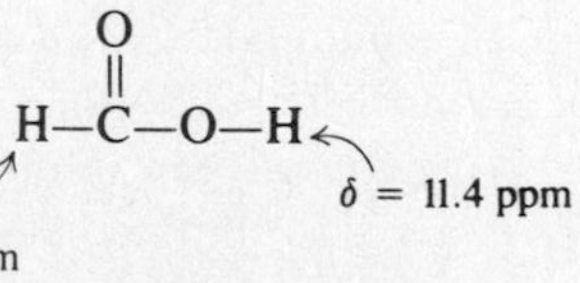

The critical signals in maleic acid are those of the vinyl protons, which normally are found in the range $\delta = 5$ to 7 ppm. Maleic acid is compound J.

Compound J $\quad$ (H)(HO$_2$C)C=C(H)(CO$_2$H)

$\delta = 6.3$ ppm (vinyl H); $\delta = 12.4$ ppm (CO_2H)

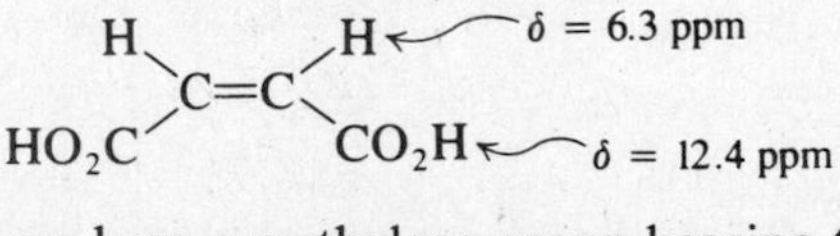

Compound I is malonic acid. Here we have a methylene group bearing two carbonyl substituents. These methylene protons are more shielded than the aldehyde proton of formic acid or the vinyl protons of maleic acid.

Compound I $\quad HO_2CCH_2CO_2H$

$\delta = 3.2$ ppm (CH_2); $\delta = 12.1$ ppm (CO_2H)

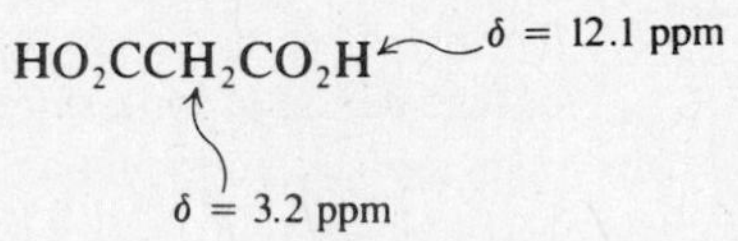

20.33 (*a*) A carboxylic acid is indicated by the presence of a strong, broad absorption in the infrared in the range 2800 to 3400 cm^{-1} and a prominent carbonyl band at 1710 cm^{-1}. A carboxylic acid proton is responsible for the one-proton signal at $\delta = 11.8$ ppm in the nuclear magnetic resonance spectrum of this compound. The six-proton singlet at $\delta = 1.3$ ppm reveals the presence of two equivalent methyl groups. The chlorine substituent is bonded to a methylene group, and the peak corresponding to $-CH_2Cl$ appears at $\delta = 3.6$ ppm. The structure consistent with these data is 3-chloro-2,2-dimethylpropanoic acid.

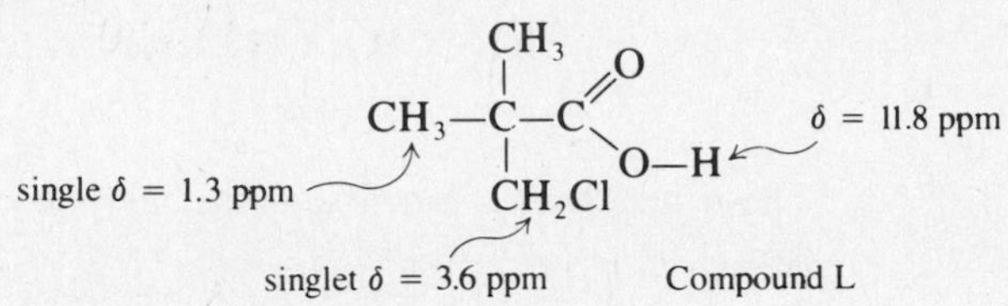

$$\begin{array}{c} CH_3 \\ | \\ CH_3-C-C(=O)-O-H \\ | \\ CH_2Cl \end{array}$$

single δ = 1.3 ppm (the CH_3 groups); singlet δ = 3.6 ppm (CH_2Cl); δ = 11.8 ppm (O—H)

Compound L

(*b*) The infrared spectrum of compound M shows peaks characteristic of a carboxylic acid at 2800 to 3400 cm^{-1} and at 1720 cm^{-1}. This fact, along with its molecular formula ($C_3H_5ClO_2$), permits only two structures:

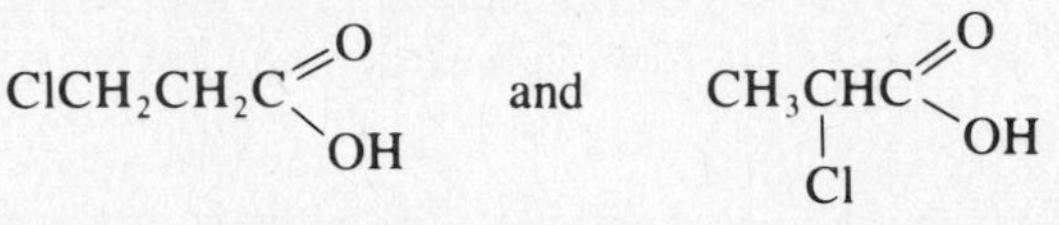

$$ClCH_2CH_2C(=O)OH \quad \text{and} \quad CH_3CHClC(=O)OH$$

3-Chloropropanoic acid and 2-Chloropropanoic acid

Compound M is determined to be 3-chloropropanoic acid on the basis of its ^{1}H nmr spectrum, which shows two triplets at $\delta = 2.9$ and $\delta = 3.8$ ppm.

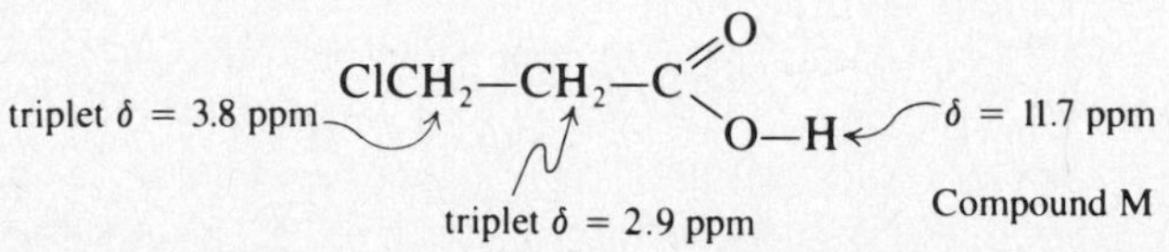

$$ClCH_2-CH_2-C(=O)-O-H$$

triplet δ = 3.8 ppm ($ClCH_2$); triplet δ = 2.9 ppm (CH_2); δ = 11.7 ppm (O—H)

Compound M

Compound M cannot be 2-chloropropanoic acid because that compound's ^{1}H nmr spectrum would show a three-proton doublet for the methyl group and a one-proton quartet for the methine proton.

20.34 (*a*) Compound N, with a molecular formula of $C_5H_8O_2$, has two elements of unsaturation (SODAR = 2). Since we are told that it is a carboxylic acid, one of these must be the carbon-oxygen double bond. With a peak in its ^{1}H nmr spectrum in the vinyl region ($\delta = 5.7$ ppm), a reasonable conclusion is that the second element of unsaturation in compound N is a carbon-carbon double bond. The two peaks at $\delta = 2.0$ and 2.2 ppm are nonequivalent methyl groups and their chemical shift is consistent with their being allylic methyls. Enough information is now available to assign compound N as 3-methyl-2-butenoic acid.

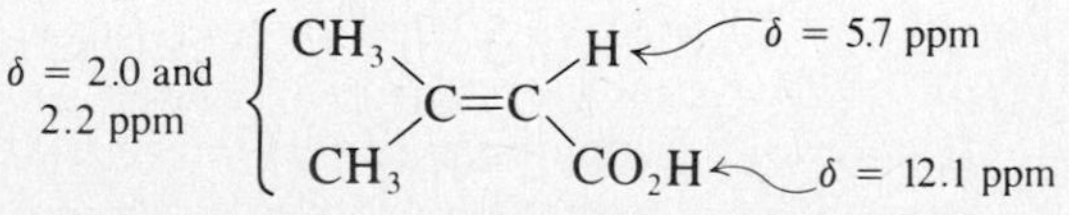

$$(CH_3)_2C=CH-CO_2H$$

δ = 2.0 and 2.2 ppm (the two CH_3 groups); δ = 5.7 ppm (H); δ = 12.1 ppm (CO_2H)

Compound N

The slight splitting of the methyl signals and broadening of the vinyl proton signal are a consequence of long-range coupling.

(*b*) The molecular formula of compound O ($C_{15}H_{14}O_2$) reveals nine elements of unsaturation (SODAR = 9). This is consistent with two monosubstituted aromatic rings (there are 10 aryl hydrogens in the nmr spectrum) and a carboxylic acid group. The two aromatic rings and the carboxyl group account for 13 of the 15 carbon atoms of the molecule. The doublet-triplet pattern in the nmr spectrum suggests the grouping

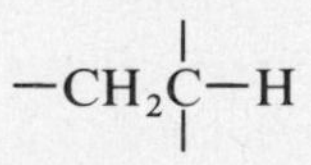

$$-CH_2\overset{|}{\underset{|}{C}}-H$$

The structure most consistent with the data is 3,3-diphenylpropanoic acid.

$$\delta = 10.6\text{ ppm} \rightarrow HO_2CCH_2{-}\underset{\underset{C_6H_5}{|}}{\overset{\overset{C_6H_5}{|}}{C}}{-}H \leftarrow \text{triplet } \delta = 4.6\text{ ppm}$$

(arrow to CH_2: doublet $\delta = 3.0$ ppm)

Compound O

The alternative formulation

$$C_6H_5CH_2{-}\underset{\underset{CO_2H}{|}}{\overset{\overset{C_6H_5}{|}}{C}}{-}H$$

is not consistent with the given information that compound O is achiral; C-2 is a chiral center in 2,3-diphenylpropanoic acid.

SELF-TEST

PART A

A-1. Provide an acceptable IUPAC name for each of the following:

(*a*) $C_6H_5\overset{\overset{CH_3}{|}}{C}H\underset{\underset{CH_3}{|}}{C}HCH_2CH_2CO_2H$

(*b*) cyclohexane ring–CO_2H

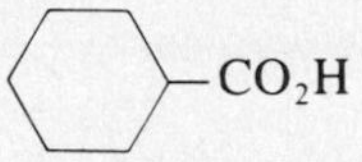

(*c*) $CH_3\overset{\overset{Br}{|}}{C}H\underset{\underset{CH_2CH_3}{|}}{C}HCO_2H$

A-2. Each of the following compounds may be converted into 4-phenylbutanoic acid by one or more reaction steps. Give the reagents and conditions necessary to carry out these conversions.

$C_6H_5CH_2CH_2CH(CO_2H)_2$ $\quad$ $C_6H_5CH_2CH_2CH_2\overset{\overset{O}{\|}}{C}CH_3$ $\quad$ $C_6H_5CH_2CH_2CH_2Br$ (two methods)

A-3. The species whose structure is shown is an intermediate in an esterification reaction. Write the complete, balanced equation for this process.

$$C_6H_5CH_2\underset{\underset{OH}{|}}{\overset{\overset{OH}{|}}{C}}OCH_2CH_3$$

A-4. Give the correct structures for compounds A through D in the following reactions.

(*a*) $(CH_3)_2CHCH_2CH_2CO_2H \xrightarrow{Br_2,\ P} A \xrightarrow[2.\ H^+]{1.\ NaCN} B$

(*b*) $C_6H_5\overset{\overset{CH_3}{|}}{C}HCH_2CH_2Br \xrightarrow[2.\ KMnO_4\quad 3.\ H^+]{1.\ KOC(CH_3)_3} C + HCO_2H$

(*c*) $D \xrightarrow{heat} C_6H_5\overset{\overset{O}{\|}}{C}CH(CH_3)_2 + CO_2$

A-5. Identify the carboxylic acid ($C_4H_7BrO_2$) having the ^{1}H nmr spectrum consisting of:

$\delta = 1.1$ ppm, 3H (triplet)
$\delta = 2.0$ ppm, 2H (pentet)
$\delta = 4.2$ ppm, 1H (triplet)
$\delta = 12.1$ ppm, 1H (singlet)

PART B

B-1. Which of the following is a correct IUPAC name for the compound shown?

$$(CH_3CH_2)_2\overset{\overset{\displaystyle CO_2H}{|}}{C}CH_2CH(CH_2CH_3)_2$$

(*a*) 1,1,3-Triethylhexanoic acid
(*b*) 2,2,4-Triethylhexanoic acid
(*c*) 3,5-Diethyl-3-heptylcarboxylic acid
(*d*) 3,5,5-Triethyl-6-hexanoic acid

B-2. Rank the following substances in order of decreasing acid strength (strongest → weakest):

$CH_3CH_2CH_2CO_2H$ (A) $\quad CH_3CH{=}CHCO_2H$ (B) $\quad CH_3CH_2CH_2CH_2OH$ (C)

$CH_3C{\equiv}CCO_2H$ (D)

(*a*) D > B > A > C (*b*) A > B > D > C
(*c*) C > A > B > D (*d*) B > D > A > C

B-3. Carboxylic acids exist in the vapor state as:

(*a*) Monomeric species (*b*) Cyclic esters
(*c*) Cyclic dimers (*d*) Cyclic trimers

B-4. Which of the following compounds will undergo decarboxylation on heating?

I: cyclohexane-1,3-dione; II: 2-(CO_2CH_3)cyclohexanone; III: 2-oxocyclohexane-CO_2H; IV: cyclohexane-1,2-di(CO_2H)

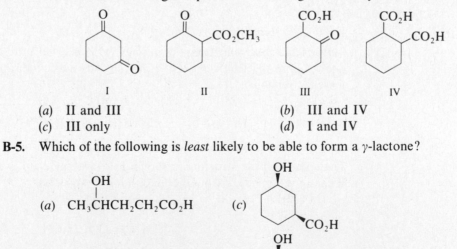

(*a*) II and III (*b*) III and IV
(*c*) III only (*d*) I and IV

B-5. Which of the following is *least* likely to be able to form a γ-lactone?

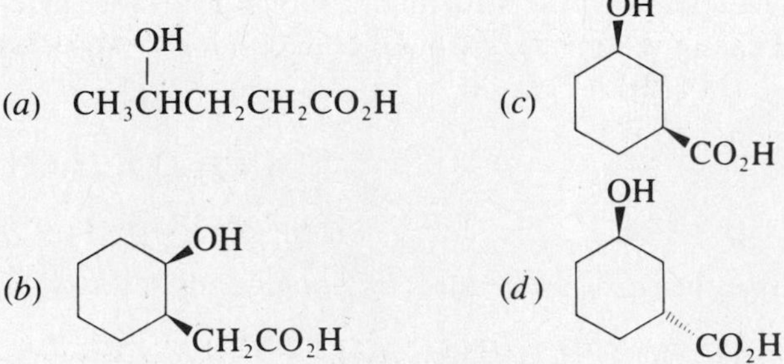

(*a*) $CH_3\overset{\overset{\displaystyle OH}{|}}{C}HCH_2CH_2CO_2H$ (*b*) (*c*) (*d*)

B-6. Rank the following in order of increasing acidity (least acidic → most acidic):

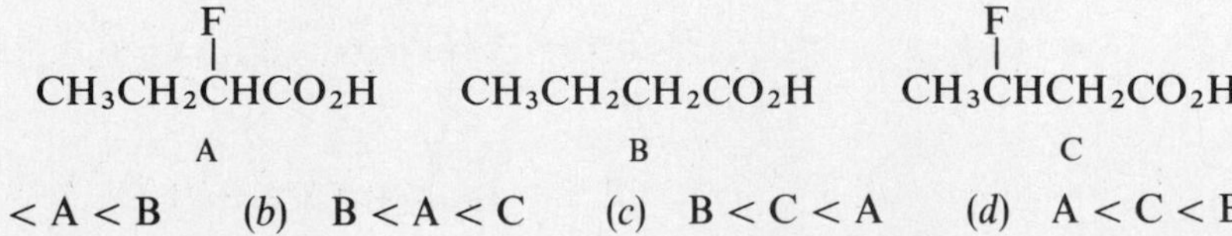

$CH_3CH_2\overset{\overset{\displaystyle F}{|}}{C}HCO_2H$ (A) $\quad CH_3CH_2CH_2CO_2H$ (B) $\quad CH_3\overset{\overset{\displaystyle F}{|}}{C}HCH_2CO_2H$ (C)

(*a*) C < A < B (*b*) B < A < C (*c*) B < C < A (*d*) A < C < B

ACYL TRANSFER REACTIONS

IMPORTANT TERMS AND CONCEPTS

Nomenclature of Carboxylic Acid Derivatives (Sec. 21.1) Important derivatives of carboxylic acids include:

1. Acyl chlorides: $\mathrm{R{-}\overset{\overset{\displaystyle O}{\|}}{C}{-}Cl}$

2. Carboxylic acid anhydrides: $\mathrm{R{-}\overset{\overset{\displaystyle O}{\|}}{C}{-}O{-}\overset{\overset{\displaystyle O}{\|}}{C}{-}R'}$

3. Carboxylic acid esters: $\mathrm{R{-}\overset{\overset{\displaystyle O}{\|}}{C}{-}O{-}R'}$

4. Carboxamides: $\mathrm{R{-}\overset{\overset{\displaystyle O}{\|}}{C}{-}NH_2}$, $\mathrm{R{-}\overset{\overset{\displaystyle O}{\|}}{C}{-}NHR'}$, $\mathrm{R{-}\overset{\overset{\displaystyle O}{\|}}{C}{-}NR'_2}$

The examples shown below illustrate the manner in which systematic names are applied to the various derivatives.

$\mathrm{CH_3CH_2CH_2\overset{\overset{\displaystyle O}{\|}}{C}Cl}$ Butanoyl chloride

$\mathrm{CH_3CH_2\overset{\overset{\displaystyle O}{\|}}{C}O\overset{\overset{\displaystyle O}{\|}}{C}CH_2CH_3}$ Propanoic anhydride

$\mathrm{C_6H_5\overset{\overset{\displaystyle O}{\|}}{C}OCH_2CH_3}$ Ethyl benzoate

$\mathrm{CH_3CH_2\overset{\overset{\displaystyle O}{\|}}{C}NHCH_3}$ *N*-Methylpropanamide

Certain cyclic derivatives are given special names:

Lactones (cyclic esters):

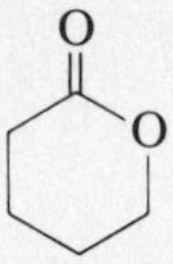

Lactams (cyclic amides, Sec. 21.14):

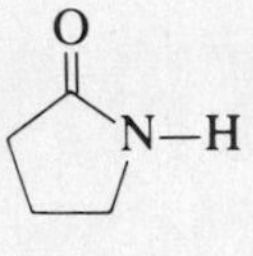

Imides (Sec. 21.15):

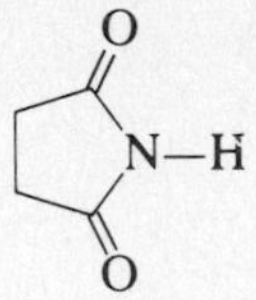

Structure and Reactivity (Sec. 21.2) All the acyl derivatives discussed in this chapter possess one common feature, namely the atom attached to the acyl group has an unshared pair of electrons that is capable of interacting with the carbonyl π system. This may be illustrated with resonance structures.

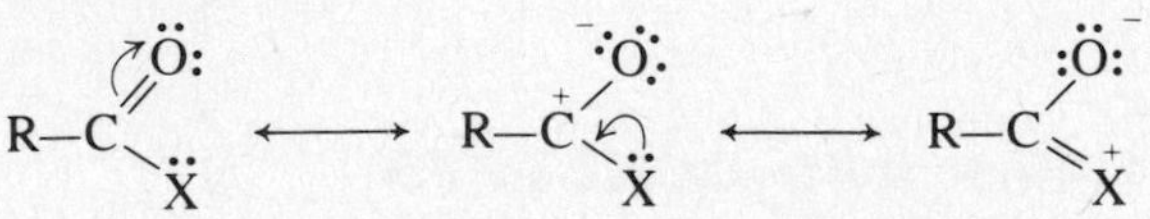

$$\mathrm{R{-}C({=}\ddot{O}{:}){-}\ddot{X}} \longleftrightarrow \mathrm{R{-}\overset{+}{C}({-}\ddot{O}{:}^{-}){-}\ddot{X}} \longleftrightarrow \mathrm{R{-}C({-}\ddot{O}{:}^{-}){=}\overset{+}{X}}$$

The extent of resonance stabilization varies among the derivatives and is a contributing factor in determining the relative reactivity of the groups. The general trend is:

$$\underset{\text{Most reactive}}{\mathrm{R\overset{O}{\overset{\|}{C}}Cl}} > \mathrm{R\overset{O}{\overset{\|}{C}}O\overset{O}{\overset{\|}{C}}R'} > \mathrm{R\overset{O}{\overset{\|}{C}}OR'} > \underset{\text{Least reactive}}{\mathrm{R\overset{O}{\overset{\|}{C}}NR'_2}}$$

Greater resonance stabilization results in lower susceptibility to attack by nucleophiles at the carbonyl carbon.

General Mechanism

The acyl transfer reactions discussed in this chapter all involve reaction of a nucleophilic species with the carbonyl group of an acyl derivative. General mechanisms are shown below.

Acidic medium:

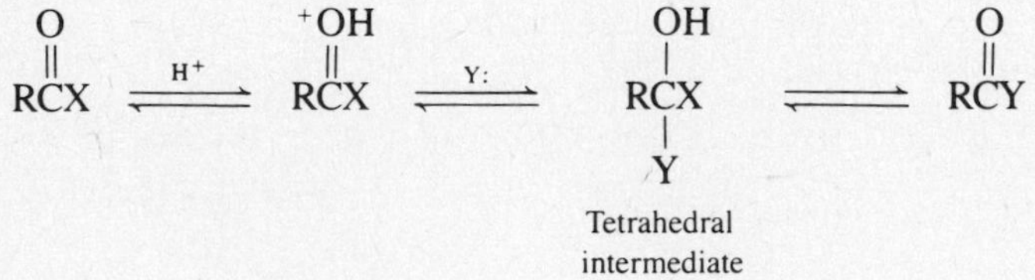

$$\mathrm{R\overset{O}{\overset{\|}{C}}X} \overset{H^+}{\rightleftharpoons} \mathrm{R\overset{{}^+OH}{\overset{\|}{C}}X} \overset{Y:}{\rightleftharpoons} \underset{\text{Tetrahedral intermediate}}{\mathrm{R\overset{OH}{\overset{|}{\underset{Y}{\underset{|}{C}}}}X}} \rightleftharpoons \mathrm{R\overset{O}{\overset{\|}{C}}Y}$$

Basic medium:

$$\mathrm{R\overset{O}{\overset{\|}{C}}X} \overset{Y^-}{\rightleftharpoons} \mathrm{R\overset{O^-}{\overset{|}{\underset{Y}{\underset{|}{C}}}}X} \overset{H^+}{\rightleftharpoons} \underset{\text{Tetrahedral intermediate}}{\mathrm{R\overset{OH}{\overset{|}{\underset{Y}{\underset{|}{C}}}}X}} \overset{-H^+,-X^-}{\rightleftharpoons} \mathrm{R\overset{O}{\overset{\|}{C}}Y}$$

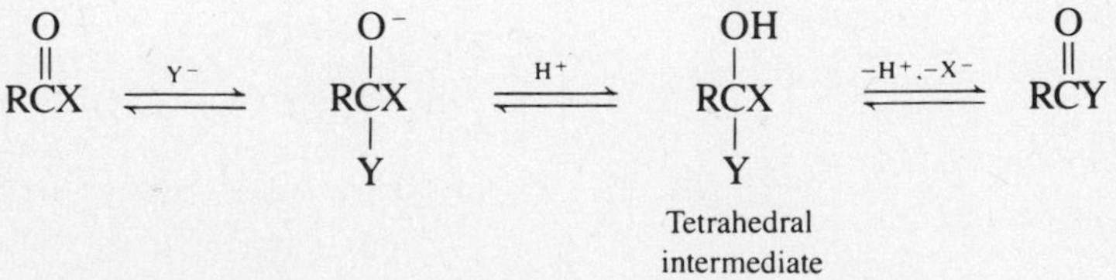

IMPORTANT REACTIONS

ACYL CHLORIDES

Preparation

$$\overset{O}{\overset{\|}{RC}}OH + SOCl_2 \longrightarrow \overset{O}{\overset{\|}{RC}}Cl + SO_2 + HCl$$

Reactions (Sec. 21.3)

$$ArH + \overset{O}{\overset{\|}{RC}}Cl \xrightarrow{AlCl_3} Ar\overset{O}{\overset{\|}{C}}R \quad \text{(Friedel-Crafts acylation, Sec. 13.7)}$$

$$\overset{O}{\overset{\|}{RC}}Cl + H_2 \xrightarrow{Pd/BaSO_4} R\overset{O}{\overset{\|}{C}}H \quad \text{(Rosenmund reduction, Sec. 18.6)}$$

$$\overset{O}{\overset{\|}{RC}}Cl \xrightarrow{LiCuR_2' \text{ (or } R_2'Cd)} R\overset{O}{\overset{\|}{C}}R' \quad \text{(Sec. 18.7)}$$

CARBOXYLIC ACID ANHYDRIDES

Preparation (Sec. 21.4)

$$\overset{O}{\overset{\|}{RC}}Cl + R'CO_2H \xrightarrow{pyridine} \overset{O}{\overset{\|}{RC}}O\overset{O}{\overset{\|}{C}}R'$$

Reactions (Sec. 21.5)

$$ArH + \overset{O}{\overset{\|}{RC}}O\overset{O}{\overset{\|}{C}}R \xrightarrow{AlCl_3} Ar\overset{O}{\overset{\|}{C}}R + \overset{O}{\overset{\|}{RC}}OH \quad \text{(Sec. 13.7)}$$

$$\overset{O}{\overset{\|}{RC}}O\overset{O}{\overset{\|}{C}}R + R'OH \longrightarrow \overset{O}{\overset{\|}{RC}}OR' + \overset{O}{\overset{\|}{RC}}OH \quad \text{(Sec. 16.10)}$$

Reactions of Esters (Secs. 21.8 to 21.11)

Acid-catalyzed hydrolysis (Sec. 21.9)

General:

$$\overset{O}{\overset{\|}{RC}}OR' + H_2O \underset{}{\overset{H^+}{\rightleftharpoons}} \overset{O}{\overset{\|}{RC}}OH + R'OH$$

Example:

$$C_6H_5\overset{O}{\overset{\|}{C}}OCH(CH_3)_2 + H_2O \overset{H^+}{\rightleftharpoons} C_6H_5\overset{O}{\overset{\|}{C}}OH + (CH_3)_2CHOH$$

Mechanism:

$$\overset{O}{\overset{\|}{RC}}OR' \overset{H^+}{\rightleftharpoons} \overset{^+OH}{\overset{\|}{RC}}OR' \overset{+H_2\ddot{O}:}{\rightleftharpoons} R\underset{_+OH_2}{\overset{OH}{C}}OR' \overset{-H^+}{\rightleftharpoons} R\underset{OH}{\overset{OH}{C}}OR'$$

$$R\underset{OH}{\overset{OH}{C}}OR' \overset{H^+}{\rightleftharpoons} R\underset{HO}{\overset{:OH}{C}}-\overset{+}{\underset{H}{O}}R' \rightleftharpoons R'OH + \overset{^+OH}{\overset{\|}{RC}}OH \overset{-H^+}{\rightleftharpoons} \overset{O}{\overset{\|}{RC}}OH$$

Base-Promoted Hydrolysis (Sec. 21.10)

General:

$$\mathrm{R\overset{O}{\overset{\|}{C}}OR' + HO^- \xrightarrow{H_2O} R\overset{O}{\overset{\|}{C}}O^- + R'OH}$$

Example:

$$\mathrm{(CH_3)_2CHCH_2\overset{O}{\overset{\|}{C}}OCH_3 \xrightarrow[H_2O]{HO^-} (CH_3)_2CHCH_2\overset{O}{\overset{\|}{C}}O^- + CH_3OH}$$

Mechanism:

$$\mathrm{R\overset{O}{\overset{\|}{C}}OR' + HO^- \rightleftharpoons R\underset{OH}{\underset{|}{\overset{O^-}{\overset{|}{C}}}}{-}OR' \rightleftharpoons R\overset{O}{\overset{\|}{C}}OH + R'O^-}$$

$$\mathrm{R'O^- + H_2O \longrightarrow R'OH + HO^-}$$

$$\mathrm{R\overset{O}{\overset{\|}{C}}OH + HO^- \longrightarrow R\overset{O}{\overset{\|}{C}}O^- + H_2O}$$

Reaction with Ammonia and Amines (Sec. 21.11)

$$\mathrm{R\overset{O}{\overset{\|}{C}}OR' + NH_3 \longrightarrow R\overset{O}{\overset{\|}{C}}NH_2 + R'OH}$$

$$\mathrm{R\overset{O}{\overset{\|}{C}}OR' + HNR''_2 \longrightarrow R\overset{O}{\overset{\|}{C}}NR''_2 + R'OH}$$

AMIDES

Preparation (Sec. 21.13)

General:

$$\mathrm{2\,R_2NH + R'\overset{O}{\overset{\|}{C}}Cl \longrightarrow R'\overset{O}{\overset{\|}{C}}NR_2 + R_2\overset{+}{N}H_2Cl^-}$$

$$\mathrm{2\,R_2NH + R'\overset{O}{\overset{\|}{C}}O\overset{O}{\overset{\|}{C}}R' \longrightarrow R'\overset{O}{\overset{\|}{C}}NR_2 + R_2\overset{+}{N}H_2\,{}^-O_2CR'}$$

Examples:

$$\mathrm{2(CH_3CH_2)_2NH + C_6H_5\overset{O}{\overset{\|}{C}}Cl \longrightarrow C_6H_5\overset{O}{\overset{\|}{C}}N(CH_2CH_3)_2}$$

$$\mathrm{2\,CH_3NH_2 + CH_3CH_2\overset{O}{\overset{\|}{C}}O\overset{O}{\overset{\|}{C}}CH_2CH_3 \longrightarrow CH_3CH_2\overset{O}{\overset{\|}{C}}NHCH_3}$$

Hydrolysis (Sec. 21.16)

General:

$$\mathrm{R\overset{O}{\overset{\|}{C}}NR'_2 + H_3O^+ \longrightarrow R\overset{O}{\overset{\|}{C}}OH + R'_2\overset{+}{N}H_2}$$

$$\mathrm{R\overset{O}{\overset{\|}{C}}NR'_2 + HO^- \longrightarrow R\overset{O}{\overset{\|}{C}}O^- + R'_2NH}$$

Example:

$$C_6H_5CH_2\overset{O}{\overset{\|}{C}}NHCH_3 + HO^- \xrightarrow{H_2O} C_6H_5CH_2\overset{O}{\overset{\|}{C}}O^- + CH_3NH_2$$

Hofmann Rearrangement (Sec. 21.17)

General:

$$R\overset{O}{\overset{\|}{C}}NH_2 + Br_2 \xrightarrow{HO^-,\ H_2O} RNH_2 + CO_3^{2-} + 2\,Br^-$$

Example:

$$(CH_3)_3C\overset{O}{\overset{\|}{C}}NH_2 + Br_2 \xrightarrow{HO^-,\ H_2O} (CH_3)_3CNH_2$$

Mechanism:

$$R\overset{O}{\overset{\|}{C}}-\underset{H}{\underset{|}{N}}-H + HO^- \rightleftharpoons R\overset{O^-}{\overset{|}{C}}=NH \xrightarrow{Br_2} R\overset{O}{\overset{\|}{C}}NHBr$$

$$Br-Br$$

$$R\overset{O}{\overset{\|}{C}}-\underset{Br}{\underset{|}{N}}-H + HO^- \rightleftharpoons R-\overset{O^-}{\overset{|}{C}}=N-Br \longrightarrow R-N=C=O + Br^-$$

$$R-N=C=O + H_2O \longrightarrow RNH\overset{O}{\overset{\|}{C}}OH \xrightarrow{HO^-} RNH_2 + CO_3^{2-}$$

NITRILES

Preparation (Sec. 21.18)

$$R\overset{O}{\overset{\|}{C}}NH_2 \xrightarrow[heat]{P_4O_{10}(-H_2O)} RC\equiv N$$

Hydrolysis (Sec. 21.19)

$$RC\equiv N + H_2O \xrightarrow{H^+ \text{ or } HO^-} R\overset{O}{\overset{\|}{C}}OH \text{ (or } R\overset{O}{\overset{\|}{C}}O^-) + \overset{+}{N}H_4 \text{ (or } NH_3)$$

Reaction with Grignard Reagents (Sec. 21.20)

$$RC\equiv N \xrightarrow[2.\ H_2O,\ H^+,\ heat]{1.\ R'MgX\ diethyl\ ether} R\overset{O}{\overset{\|}{C}}R'$$

SOLUTIONS TO TEXT PROBLEMS

21.1 (*b*) Carboxylic acid anhydrides bear two acyl groups on oxygen, as in $R\overset{O}{\overset{\|}{C}}O\overset{O}{\overset{\|}{C}}R$. They are named as derivatives of carboxylic acids.

$$CH_3CH_2\underset{C_6H_5}{\underset{|}{C}}H\overset{O}{\overset{\|}{C}}OH \qquad CH_3CH_2\underset{C_6H_5}{\underset{|}{C}}H\overset{O}{\overset{\|}{C}}O\overset{O}{\overset{\|}{C}}\underset{C_6H_5}{\underset{|}{C}}HCH_2CH_3$$

2-Phenylbutanoic acid — 2-Phenylbutanoic anhydride

(*c*) Butyl 2-phenylbutanoate is the butyl ester of 2-phenylbutanoic acid.

$$\underset{\displaystyle C_6H_5}{CH_3CH_2\underset{|}{C}H}\overset{\displaystyle O}{\overset{\|}{C}}OCH_2CH_2CH_2CH_3 \qquad \text{Butyl 2-phenylbutanoate}$$

(*d*) In 2-phenylbutyl butanoate the 2-phenylbutyl group is an alkyl group bonded to oxygen of the ester. It is not involved in the acyl group of the molecule.

$$CH_3CH_2CH_2\overset{\displaystyle O}{\overset{\|}{C}}OCH_2\underset{\displaystyle C_6H_5}{\underset{|}{C}H}CH_2CH_3 \qquad \text{2-Phenylbutyl butanoate}$$

(*e*) The ending *-amide* reveals this to be a compound of the type $R\overset{O}{\overset{\|}{C}}NH_2$.

$$CH_3CH_2\underset{\displaystyle C_6H_5}{\underset{|}{C}H}\overset{\displaystyle O}{\overset{\|}{C}}NH_2 \qquad \text{2-Phenylbutanamide}$$

(*f*) This compound differs from 2-phenylbutanamide in the preceding problem only in that it bears an ethyl substituent on nitrogen.

$$CH_3CH_2\underset{\displaystyle C_6H_5}{\underset{|}{C}H}\overset{\displaystyle O}{\overset{\|}{C}}NHCH_2CH_3 \qquad \text{N-Ethyl-2-phenylbutanamide}$$

(*g*) The *-nitrile* ending signifies a compound of the type $RC{\equiv}N$ containing the same number of carbons as the alkane RCH_3.

$$CH_3CH_2\underset{\displaystyle C_6H_5}{\underset{|}{C}H}C{\equiv}N \qquad \text{2-Phenylbutanenitrile}$$

21.2 The methyl groups in *N,N*-dimethylformamide are nonequivalent; one is cis to oxygen, the other is trans. The two methyl groups have different chemical shifts.

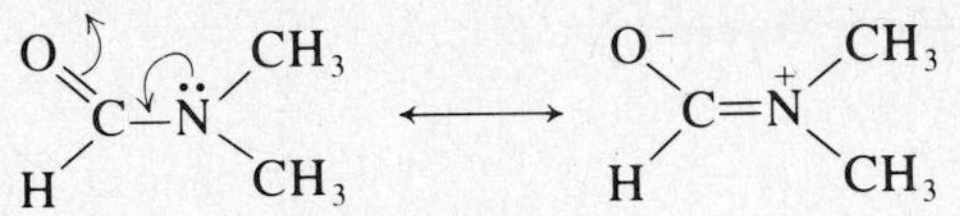

$$(O{=})(H)C{-}\ddot{N}(CH_3)_2 \longleftrightarrow (O^-)(H)C{=}\overset{+}{N}(CH_3)_2$$

Rotation about the carbon-nitrogen bond is required to average the environments of the two methyl groups, but this rotation is relatively slow in amides as the result of the double bond character imparted to the carbon-nitrogen bond, as shown by the resonance structures above.

21.3 (*b*) Benzoyl chloride reacts with benzoic acid to give benzoic anhydride.

$$\underset{\text{Benzoyl chloride}}{C_6H_5\overset{O}{\overset{\|}{C}}Cl} + \underset{\text{Benzoic acid}}{C_6H_5\overset{O}{\overset{\|}{C}}OH} \longrightarrow \underset{\text{Benzoic anhydride}}{C_6H_5\overset{O}{\overset{\|}{C}}O\overset{O}{\overset{\|}{C}}C_6H_5}$$

(*c*) Acyl chlorides react with alcohols to form esters.

$$\underset{\text{Benzoyl chloride}}{C_6H_5\overset{O}{\overset{\|}{C}}Cl} + \underset{\text{Ethanol}}{CH_3CH_2OH} \longrightarrow \underset{\text{Ethyl benzoate}}{C_6H_5\overset{O}{\overset{\|}{C}}OCH_2CH_3}$$

The organic product is the ethyl ester of benzoic acid, ethyl benzoate.

(*d*) Acyl transfer from benzoyl chloride to the nitrogen of methylamine yields the amide *N*-methylbenzamide.

$$\underset{\text{Benzoyl chloride}}{C_6H_5\overset{\overset{\displaystyle O}{\|}}{C}Cl} + \underset{\text{Methylamine}}{CH_3NH_2} \longrightarrow \underset{N\text{-Methylbenzamide}}{C_6H_5\overset{\overset{\displaystyle O}{\|}}{C}NHCH_3}$$

(*e*) In analogy with (*d*), an amide is formed. In this case the product has two methyl groups on nitrogen.

$$\underset{\text{Benzoyl chloride}}{C_6H_5\overset{\overset{\displaystyle O}{\|}}{C}Cl} + \underset{\text{Dimethylamine}}{(CH_3)_2NH} \longrightarrow \underset{N,N\text{-Dimethylbenzamide}}{C_6H_5\overset{\overset{\displaystyle O}{\|}}{C}N(CH_3)_2}$$

21.4 (*b*) Nucleophilic addition of benzoic acid to benzoyl chloride gives the tetrahedral intermediate shown.

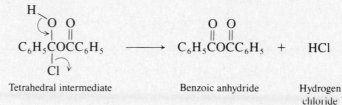

$$\underset{\text{Benzoyl chloride}}{C_6H_5\overset{\overset{\displaystyle O}{\|}}{C}Cl} + \underset{\text{Benzoic acid}}{C_6H_5\overset{\overset{\displaystyle O}{\|}}{C}OH} \longrightarrow \underset{\text{Tetrahedral intermediate}}{C_6H_5\overset{\overset{\displaystyle HO}{|}}{\underset{\underset{\displaystyle Cl}{|}}{C}}-O\overset{\overset{\displaystyle O}{\|}}{C}C_6H_5}$$

Dissociation of the tetrahedral intermediate occurs by loss of chloride and of the proton on the oxygen.

$$\underset{\text{Tetrahedral intermediate}}{C_6H_5\overset{\overset{\displaystyle H-O}{|}}{\underset{\underset{\displaystyle Cl}{|}}{C}}O\overset{\overset{\displaystyle O}{\|}}{C}C_6H_5} \longrightarrow \underset{\text{Benzoic anhydride}}{C_6H_5\overset{\overset{\displaystyle O}{\|}}{C}O\overset{\overset{\displaystyle O}{\|}}{C}C_6H_5} + \underset{\text{Hydrogen chloride}}{HCl}$$

(*c*) Ethanol is the nucleophile that adds to the carbonyl group of benzoyl chloride to form the tetrahedral intermediate.

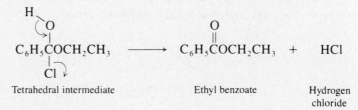

$$\underset{\text{Benzoyl chloride}}{C_6H_5\overset{\overset{\displaystyle O}{\|}}{C}Cl} + \underset{\text{Ethanol}}{CH_3CH_2OH} \longrightarrow \underset{\text{Tetrahedral intermediate}}{C_6H_5\overset{\overset{\displaystyle OH}{|}}{\underset{\underset{\displaystyle Cl}{|}}{C}}OCH_2CH_3}$$

In analogy with (*a*) and (*b*) of this problem, a proton is lost from the hydroxyl group along with chloride to reform the carbon-oxygen double bond.

$$\underset{\text{Tetrahedral intermediate}}{C_6H_5\overset{\overset{\displaystyle H-O}{|}}{\underset{\underset{\displaystyle Cl}{|}}{C}}OCH_2CH_3} \longrightarrow \underset{\text{Ethyl benzoate}}{C_6H_5\overset{\overset{\displaystyle O}{\|}}{C}OCH_2CH_3} + \underset{\text{Hydrogen chloride}}{HCl}$$

(*d*) The tetrahedral intermediate formed from benzoyl chloride and methylamine has a carbon-nitrogen bond.

$$\underset{\text{Benzoyl chloride}}{C_6H_5\overset{\overset{\displaystyle O}{\|}}{C}Cl} + \underset{\text{Methylamine}}{CH_3NH_2} \longrightarrow \underset{\text{Tetrahedral intermediate}}{C_6H_5\overset{\overset{\displaystyle OH}{|}}{\underset{\underset{\displaystyle Cl}{|}}{C}}NHCH_3}$$

Schematically, the dissociation of the tetrahedral intermediate may be shown as:

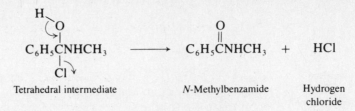

$$\underset{\text{Tetrahedral intermediate}}{C_6H_5\overset{\overset{\text{H}-\text{O}}{|}}{\underset{\underset{\text{Cl}}{|}}{C}}NHCH_3} \longrightarrow \underset{N\text{-Methylbenzamide}}{C_6H_5\overset{\overset{\text{O}}{\|}}{C}NHCH_3} + \underset{\text{Hydrogen chloride}}{HCl}$$

More realistically, it is a second methylamine molecule that abstracts a proton from oxygen.

$$CH_3\ddot{N}H_2 \curvearrowright H-O-\underset{\underset{\text{Cl}}{|}}{C}(C_6H_5)NHCH_3 \longrightarrow \underset{N\text{-Methylbenzamide}}{C_6H_5\overset{\overset{\text{O}}{\|}}{C}NHCH_3} + \underset{\text{Methylammonium chloride}}{CH_3\overset{+}{N}H_3\ Cl^-}$$

(*e*) The intermediates in the reaction of benzoyl chloride with dimethylamine are similar to those in (*d*). The methyl substituents on nitrogen are not directly involved in the reaction.

$$\underset{\text{Benzoyl chloride}}{C_6H_5\overset{\overset{\text{O}}{\|}}{C}Cl} + \underset{\text{Dimethylamine}}{(CH_3)_2NH} \longrightarrow \underset{\text{Tetrahedral intermediate}}{C_6H_5\overset{\overset{\text{OH}}{|}}{\underset{\underset{\text{Cl}}{|}}{C}}N(CH_3)_2}$$

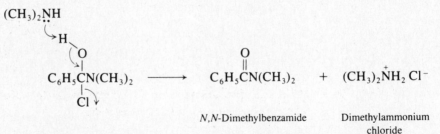

Then

$$(CH_3)_2\ddot{N}H \curvearrowright H-O-\underset{\underset{\text{Cl}}{|}}{C}(C_6H_5)N(CH_3)_2 \longrightarrow \underset{N,N\text{-Dimethylbenzamide}}{C_6H_5\overset{\overset{\text{O}}{\|}}{C}N(CH_3)_2} + \underset{\text{Dimethylammonium chloride}}{(CH_3)_2\overset{+}{N}H_2\ Cl^-}$$

21.5 One equivalent of benzoyl chloride reacts rapidly with water to yield benzoic acid.

$$\underset{\text{Benzoyl chloride}}{C_6H_5\overset{\overset{\text{O}}{\|}}{C}Cl} + \underset{\text{Water}}{H_2O} \longrightarrow \underset{\text{Benzoic acid}}{C_6H_5\overset{\overset{\text{O}}{\|}}{C}OH} + \underset{\text{Hydrogen chloride}}{HCl}$$

The benzoic acid produced in this step reacts with the remaining benzoyl chloride to give benzoic anhydride.

$$\underset{\text{Benzoyl chloride}}{C_6H_5\overset{\overset{\text{O}}{\|}}{C}Cl} + \underset{\text{Benzoic acid}}{C_6H_5\overset{\overset{\text{O}}{\|}}{C}OH} \longrightarrow \underset{\text{Benzoic anhydride}}{C_6H_5\overset{\overset{\text{O}}{\|}}{C}O\overset{\overset{\text{O}}{\|}}{C}C_6H_5} + \underset{\text{Hydrogen chloride}}{HCl}$$

21.6 Acetic anhydride serves as a source of acetyl cation.

$$CH_3\overset{O}{\overset{\|}{C}}O-\overset{O}{\overset{\|}{C}}CH_3 \longrightarrow \ddot{O}=\overset{+}{C}CH_3 \longleftrightarrow :\overset{+}{O}\equiv CCH_3$$

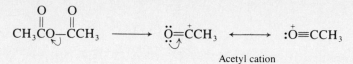

Acetyl cation

21.7 (*b*) Acyl transfer from an acid anhydride to ammonia yields an amide.

$$CH_3\overset{O}{\overset{\|}{C}}O\overset{O}{\overset{\|}{C}}CH_3 + 2NH_3 \longrightarrow CH_3\overset{O}{\overset{\|}{C}}NH_2 + CH_3\overset{O}{\overset{\|}{C}}O^- \overset{+}{N}H_4$$

Acetic anhydride | Ammonia | Acetamide | Ammonium acetate

The organic products are acetamide and ammonium acetate.

(*c*) The reaction of phthalic anhydride with dimethylamine is analogous to that of (*b*) above. The organic products are an amide and the carboxylate salt of an amine.

$$\text{Phthalic anhydride} + 2(CH_3)_2NH \longrightarrow C_6H_4(\overset{O}{\overset{\|}{C}}N(CH_3)_2)(\overset{O}{\overset{\|}{C}}O^-)\ \ H_2\overset{+}{N}(CH_3)_2$$

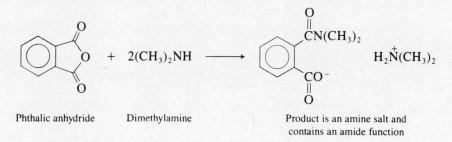

Phthalic anhydride | Dimethylamine | Product is an amine salt and contains an amide function

In this case both the amide function and the ammonium carboxylate salt are incorporated into the same molecule.

(*d*) The disodium salt of phthalic acid is the product of hydrolysis of phthalic acid in excess sodium hydroxide.

$$\text{Phthalic anhydride} + 2NaOH \longrightarrow C_6H_4(\overset{O}{\overset{\|}{C}}O^-\ Na^+)_2 + H_2O$$

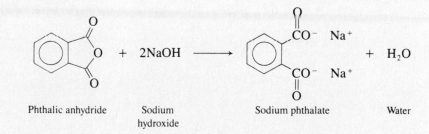

Phthalic anhydride | Sodium hydroxide | Sodium phthalate | Water

21.8 (*b*) The tetrahedral intermediate is formed by nucleophilic addition of ammonia to one of the carbonyl groups of acetic anhydride.

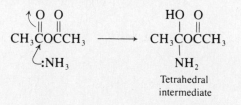

$$CH_3\overset{O}{\overset{\|}{C}}O\overset{O}{\overset{\|}{C}}CH_3 + :NH_3 \longrightarrow CH_3\overset{HO}{\overset{|}{\underset{NH_2}{\underset{|}{C}}}}O\overset{O}{\overset{\|}{C}}CH_3$$

Tetrahedral intermediate

Dissociation of the tetrahedral intermediate occurs by loss of acetate as the leaving group.

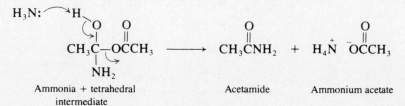

$$H_3N: + CH_3\overset{HO}{\overset{|}{\underset{NH_2}{\underset{|}{C}}}}-O\overset{O}{\overset{\|}{C}}CH_3 \longrightarrow CH_3\overset{O}{\overset{\|}{C}}NH_2 + H_4\overset{+}{N}\ ^-O\overset{O}{\overset{\|}{C}}CH_3$$

Ammonia + tetrahedral intermediate | Acetamide | Ammonium acetate

(*c*) Dimethylamine is the nucleophile; it adds to one of the two equivalent carbonyl groups of phthalic anhydride.

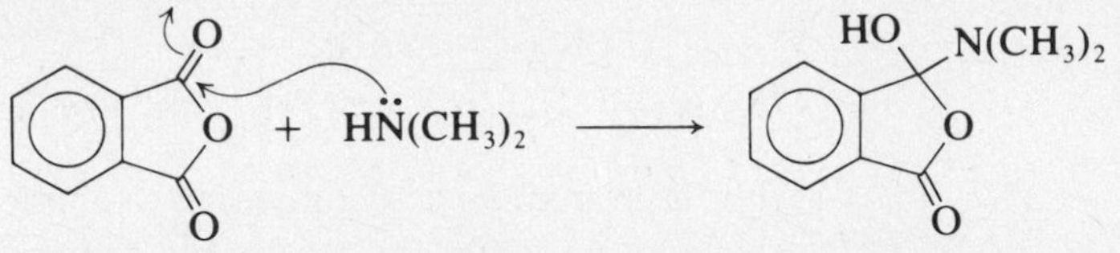

O + HN̈(CH$_3$)$_2$ ⟶ HO N(CH$_3$)$_2$ O O

Phthalic anhydride Dimethylamine Tetrahedral intermediate

A second molecule of dimethylamine abstracts a proton from the tetrahedral intermediate.

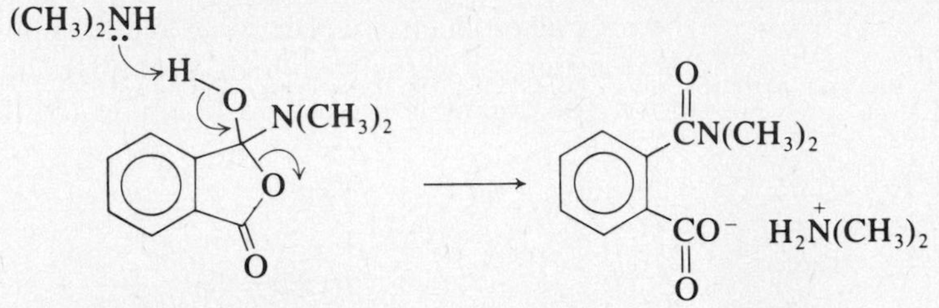

(CH$_3$)$_2$NH H O N(CH$_3$)$_2$ O O ⟶ CN(CH$_3$)$_2$ CO$^-$ H$_2\overset{+}{N}$(CH$_3$)$_2$

Tetrahedral intermediate + second molecule of dimethylamine

Product of reaction

(*d*) Hydroxide acts as a nucleophile to form the tetrahedral intermediate and as a base to facilitate its dissociation.

Hydroxide ion–catalyzed formation of tetrahedral intermediate:

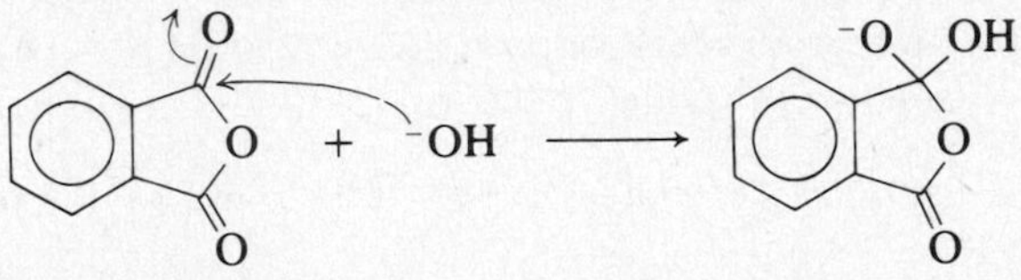

O + $^-$OH ⟶ $^-$O OH O O

Phthalic anhydride

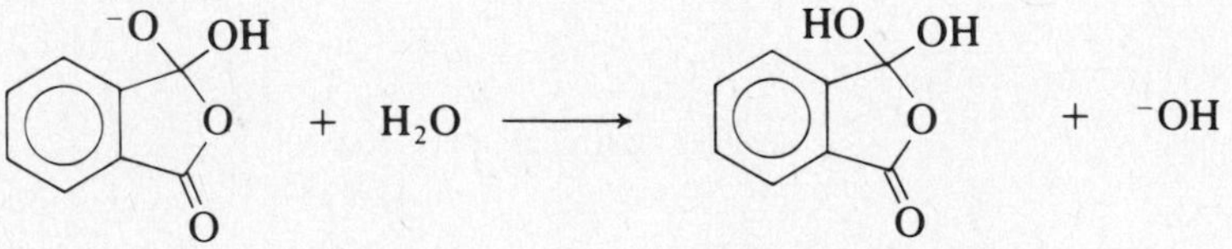

$^-$O OH O O + H$_2$O ⟶ HO OH O O + $^-$OH

Tetrahedral intermediate

Hydroxide ion–promoted dissociation of tetrahedral intermediate:

HO$^-$ H O OH O O ⟶ COH CO$^-$ + H$_2$O

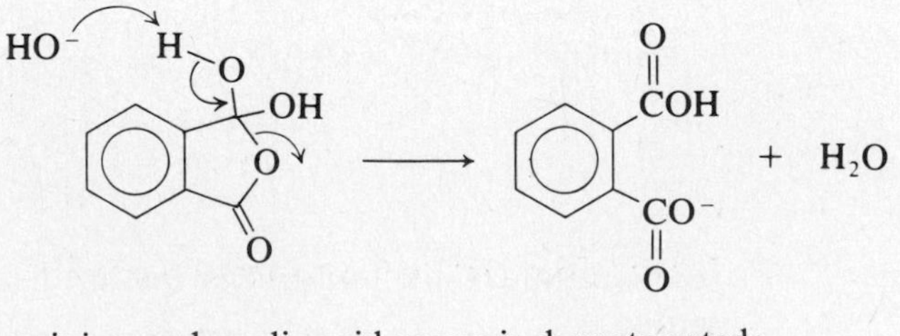

In base the remaining carboxylic acid group is deprotonated:

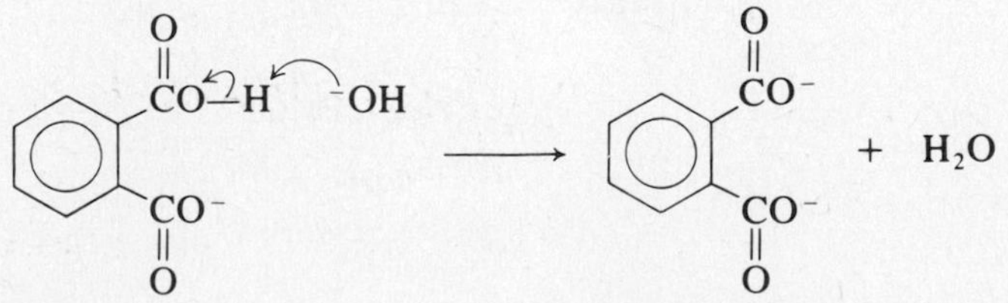

CO–H $^-$OH CO$^-$ ⟶ CO$^-$ CO$^-$ + H$_2$O

21.9 The starting material contains three acetate ester functions. All these ester groups undergo hydrolysis in aqueous sulfuric acid.

$$\underset{\substack{|\\ OCCH_3 \\ \| \\ O}}{CH_3\overset{O}{\overset{\|}{C}}OCH_2CHCH_2CH_2CH_2O\overset{O}{\overset{\|}{C}}CH_3} \xrightarrow[H^+]{H_2O} HOCH_2\underset{\substack{|\\ OH}}{C}HCH_2CH_2CH_2OH + 3CH_3\overset{O}{\overset{\|}{C}}OH$$

1,2,5-Pentanetriol ($C_5H_{12}O_3$) Acetic acid

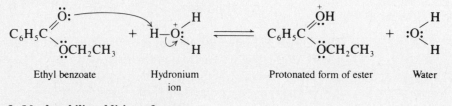

The product is 1,2,5-pentanetriol. Also formed in the hydrolysis of the starting triacetate are three molecules of acetic acid.

21.10

Step 1: Protonation of the carbonyl oxygen

$$C_6H_5C(=\ddot{O}:)\ddot{O}CH_2CH_3 + H-\overset{+}{\ddot{O}}H_2 \rightleftharpoons C_6H_5C(=\overset{+}{O}H)\ddot{O}CH_2CH_3 + :\ddot{O}H_2$$

Ethyl benzoate Hydronium ion Protonated form of ester Water

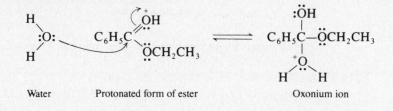

Step 2: Nucleophilic addition of water

$$H_2\ddot{O}: + C_6H_5C(=\overset{+}{O}H)\ddot{O}CH_2CH_3 \rightleftharpoons C_6H_5C(:\ddot{O}H)(\overset{+}{O}H_2)-\ddot{O}CH_2CH_3$$

Water Protonated form of ester Oxonium ion

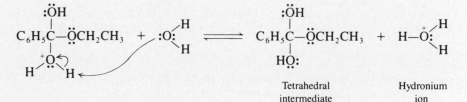

Step 3: Deprotonation of oxonium ion to give neutral form of tetrahedral intermediate

$$C_6H_5C(:\ddot{O}H)(\overset{+}{O}H_2)-\ddot{O}CH_2CH_3 + :\ddot{O}H_2 \rightleftharpoons C_6H_5C(:\ddot{O}H)(H\ddot{O}:)-\ddot{O}CH_2CH_3 + H-\overset{+}{\ddot{O}}H_2$$

Tetrahedral intermediate Hydronium ion

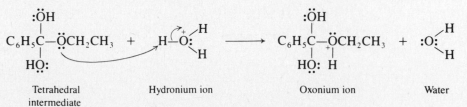

Step 4: Protonation of ethoxy oxygen

$$C_6H_5C(:\ddot{O}H)(H\ddot{O}:)-\ddot{O}CH_2CH_3 + H-\overset{+}{\ddot{O}}H_2 \longrightarrow C_6H_5C(:\ddot{O}H)(H\ddot{O}:)-\overset{+}{\ddot{O}}(H)CH_2CH_3 + :\ddot{O}H_2$$

Tetrahedral intermediate Hydronium ion Oxonium ion Water

Step 5: Dissociation of protonated form of tetrahedral intermediate

This step yields ethyl alcohol and the protonated form of benzoic acid.

$$C_6H_5C(:\ddot{O}H)(:\ddot{O}H)-\overset{+}{\ddot{O}}(H)CH_2CH_3 \rightleftharpoons C_6H_5C(=\overset{+}{O}H):\ddot{O}H + H\ddot{O}CH_2CH_3$$

Oxonium ion Protonated form of benzoic acid Ethyl alcohol

Step 6: Deprotonation of protonated form of benzoic acid

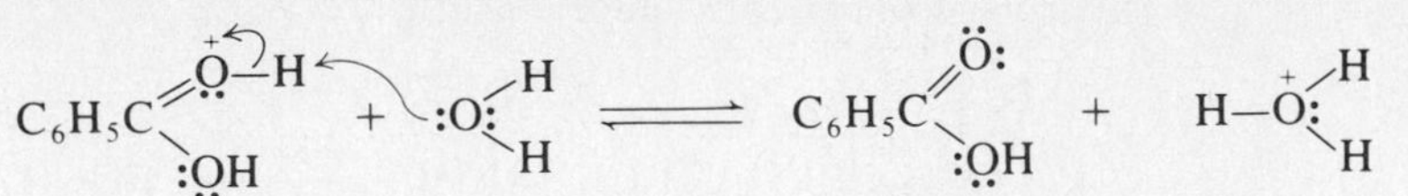

$$C_6H_5C(=\overset{+}{O}H)(\ddot{O}H) + H_2\ddot{O}: \rightleftharpoons C_6H_5C(=\ddot{O}:)(\ddot{O}H) + H_3\overset{+}{O}:$$

Protonated form of benzoic acid — Water — Benzoic acid — Hydronium ion

21.11 To determine which oxygen of γ-butyrolactone becomes labeled with ^{18}O, trace the path of ^{18}O-labeled water (Ø = ^{18}O) as it undergoes nucleophilic addition to the carbonyl group to form the tetrahedral intermediate.

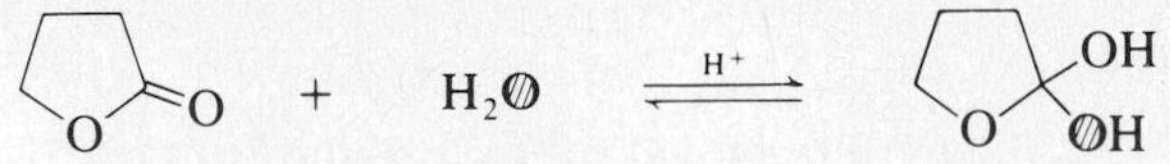

γ-Butyrolactone + H_2Ø (^{18}O-Labeled water) $\overset{H^+}{\rightleftharpoons}$ Tetrahedral intermediate (OH, ØH)

The tetrahedral intermediate can revert to unlabeled γ-butyrolactone by loss of ^{18}O-labeled water. Alternatively it can lose ordinary water to give ^{18}O-labeled lactone.

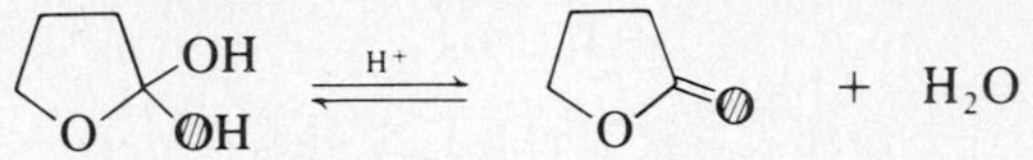

Tetrahedral intermediate $\overset{H^+}{\rightleftharpoons}$ ^{18}O-Labeled γ-butyrolactone + H_2O (Water)

The carbonyl oxygen is the one that is isotopically labeled in the ^{18}O-enriched γ-butyrolactone.

21.12 Based on trimyristin's molecular formula $C_{45}H_{86}O_6$ and on the fact that its hydrolysis gives only glycerol and tetradecanoic acid $CH_3(CH_2)_{12}CO_2H$, it must have the structure shown.

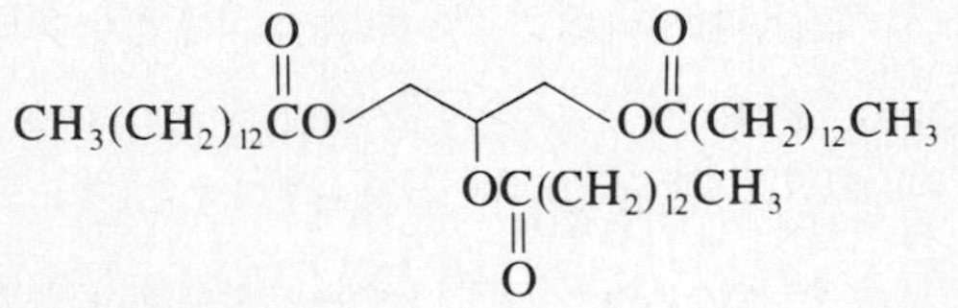

$$CH_3(CH_2)_{12}\overset{O}{\overset{\|}{C}}O-CH_2-CH(O\underset{\|}{\underset{O}{C}}(CH_2)_{12}CH_3)-CH_2-O\overset{O}{\overset{\|}{C}}(CH_2)_{12}CH_3$$

Trimyristin ($C_{45}H_{86}O_6$)

21.13 Since ester hydrolysis in base proceeds by acyl-oxygen cleavage, the ^{18}O label becomes incorporated in acetate ion.

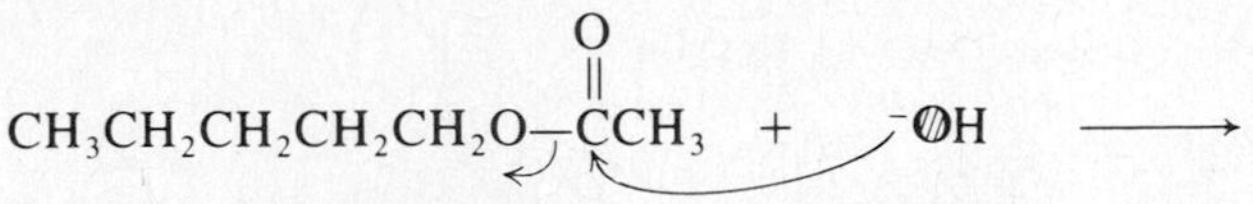

$$CH_3CH_2CH_2CH_2CH_2O-\overset{O}{\overset{\|}{C}}CH_3 + {}^-\text{ØH} \longrightarrow CH_3CH_2CH_2CH_2CH_2OH + {}^-\text{Ø}\overset{O}{\overset{\|}{C}}CH_3$$

Pentyl acetate — Hydroxide ion 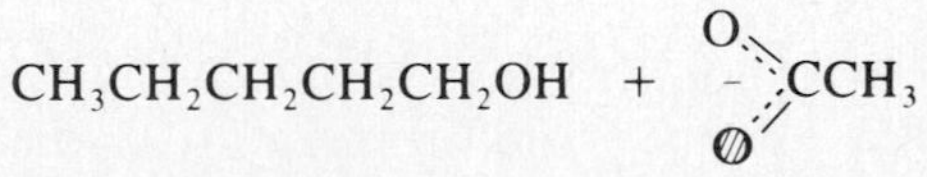 — 1-Pentanol — Acetate ion

21.14

Step 1: Nucleophilic addition of hydroxide ion to the carbonyl group

$$H\ddot{O}:^- + C_6H_5C(=\ddot{O}:)\ddot{O}CH_2CH_3 \rightleftharpoons C_6H_5C(\ddot{O}:^-)(\ddot{O}H)-\ddot{O}CH_2CH_3$$

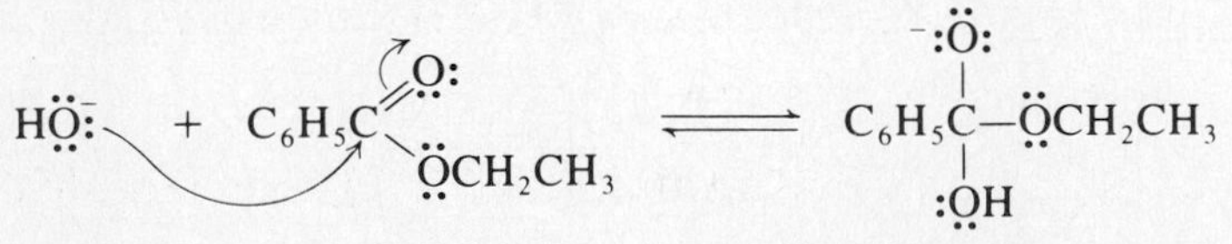

Hydroxide ion — Ethyl benzoate — Anionic form of tetrahedral intermediate

Step 2: Proton transfer from water to give neutral form of tetrahedral intermediate

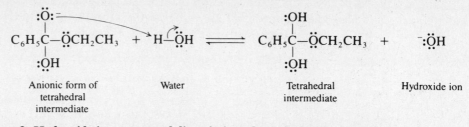

$$C_6H_5\overset{:\ddot{O}:^-}{\underset{:\ddot{O}H}{C}}-\ddot{O}CH_2CH_3 + H-\ddot{O}H \rightleftharpoons C_6H_5\overset{:\ddot{O}H}{\underset{:\ddot{O}H}{C}}-\ddot{O}CH_2CH_3 + {}^-:\ddot{O}H$$

Anionic form of tetrahedral intermediate — Water — Tetrahedral intermediate — Hydroxide ion

Step 3: Hydroxide ion–promoted dissociation of tetrahedral intermediate

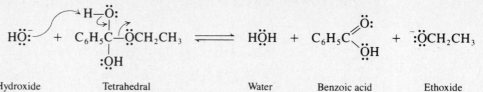

$$H\ddot{O}:^- + C_6H_5\overset{H-\ddot{O}:}{\underset{:\ddot{O}H}{C}}-\ddot{O}CH_2CH_3 \rightleftharpoons H\ddot{O}H + C_6H_5C(=\ddot{O}:)\ddot{O}H + {}^-:\ddot{O}CH_2CH_3$$

Hydroxide ion — Tetrahedral intermediate — Water — Benzoic acid — Ethoxide ion

Step 4: Proton abstraction from benzoic acid

$$C_6H_5C(=\ddot{O}:)\ddot{O}-H + {}^-:\ddot{O}H \longrightarrow C_6H_5C(=\ddot{O}:)\ddot{O}:^- + H\ddot{O}H$$

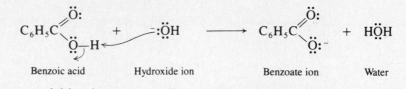

Benzoic acid — Hydroxide ion — Benzoate ion — Water

21.15 The starting material is a lactone, a cyclic ester. The ester function is converted to an amide by nucleophilic acyl substitution.

$$CH_3\ddot{N}H_2 + \text{4-pentanolide} \longrightarrow CH_3NH\overset{O}{\overset{\|}{C}}CH_2CH_2\underset{OH}{\underset{|}{C}}HCH_3$$

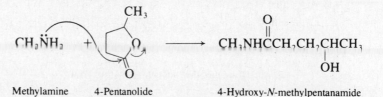

Methylamine — 4-Pentanolide — 4-Hydroxy-*N*-methylpentanamide

21.16 Methanol is the nucleophile that adds to the carbonyl group of the thioester.

$$CH_3\overset{O}{\overset{\|}{C}}SCH_2CH_2OC_6H_5 + CH_3OH \longrightarrow CH_3\overset{OH}{\underset{OCH_3}{C}}-SCH_2CH_2OC_6H_5$$

S-2-Phenoxyethyl ethanethiolate — Methanol — Tetrahedral intermediate

$$\longrightarrow CH_3\overset{O}{\overset{\|}{C}}OCH_3 + HSCH_2CH_2OC_6H_5$$

Methyl acetate — 2-Phenoxyethanethiol

21.17 (*b*) Acetic anhydride is the anhydride that must be used; it transfers an acetyl group to suitable nucleophiles. The nucleophile in this case is methylamine.

$$CH_3\overset{O}{\overset{\|}{C}}O\overset{O}{\overset{\|}{C}}CH_3 + 2CH_3NH_2 \longrightarrow CH_3\overset{O}{\overset{\|}{C}}NHCH_3 + CH_3\overset{O}{\overset{\|}{C}}O^- \ CH_3\overset{+}{N}H_3$$

Acetic anhydride — Methylamine — *N*-Methylacetamide — Methylammonium acetate

(*c*) The acyl group is $\overset{O}{\overset{\|}{HC}}-$. Since the problem specifies that the acyl transfer agent is a methyl ester, methyl formate is one of the starting materials.

$$\underset{\text{Methyl formate}}{\overset{O}{\overset{\|}{HC}}OCH_3} + \underset{\text{Dimethylamine}}{HN(CH_3)_2} \longrightarrow \underset{\text{Dimethyl formamide}}{\overset{O}{\overset{\|}{HC}}N(CH_3)_2} + \underset{\text{Methyl alcohol}}{CH_3OH}$$

21.18 Phthalic anhydride reacts with excess ammonia to give the ammonium salt of a compound known as phthalamic acid.

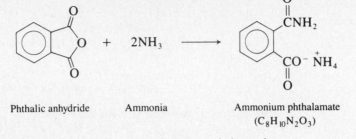

$$\text{Phthalic anhydride} + \underset{\text{Ammonia}}{2NH_3} \longrightarrow \underset{\text{Ammonium phthalamate } (C_8H_{10}N_2O_3)}{C_6H_4(\overset{O}{\overset{\|}{C}}NH_2)(\overset{O}{\overset{\|}{C}}O^-\ \overset{+}{N}H_4)}$$

Phthalimide is formed when ammonium phthalamate is heated.

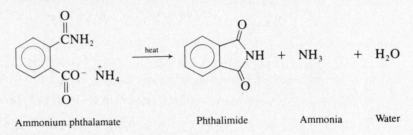

$$\underset{\text{Ammonium phthalamate}}{C_6H_4(\overset{O}{\overset{\|}{C}}NH_2)(\overset{O}{\overset{\|}{C}}O^-\ \overset{+}{N}H_4)} \xrightarrow{\text{heat}} \text{Phthalimide} + \underset{\text{Ammonia}}{NH_3} + \underset{\text{Water}}{H_2O}$$

21.19

Step 1: Protonation of the carbonyl oxygen

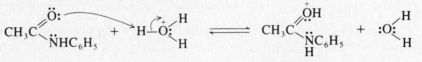

$$\underset{\text{Acetanilide}}{CH_3C(=\ddot{O}:)\ddot{N}HC_6H_5} + \underset{\text{Hydronium ion}}{H-\overset{+}{\ddot{O}}H_2} \rightleftharpoons \underset{\text{Protonated form of amide}}{CH_3C(=\overset{+}{\ddot{O}}H)\ddot{N}HC_6H_5} + \underset{\text{Water}}{:\ddot{O}H_2}$$

Step 2: Nucleophilic addition of water

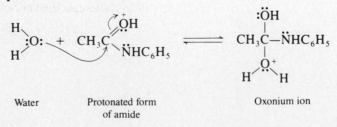

$$\underset{\text{Water}}{H_2\ddot{O}:} + \underset{\text{Protonated form of amide}}{CH_3C(=\overset{+}{\ddot{O}}H)\ddot{N}HC_6H_5} \rightleftharpoons \underset{\text{Oxonium ion}}{CH_3C(:\ddot{O}H)(\overset{+}{\ddot{O}}H_2)-\ddot{N}HC_6H_5}$$

Step 3: Deprotonation of oxonium ion to give neutral form of tetrahedral intermediate

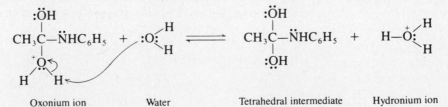

$$\underset{\text{Oxonium ion}}{CH_3C(:\ddot{O}H)(\overset{+}{\ddot{O}}H_2)-\ddot{N}HC_6H_5} + \underset{\text{Water}}{:\ddot{O}H_2} \rightleftharpoons \underset{\text{Tetrahedral intermediate}}{CH_3C(:\ddot{O}H)(:\ddot{O}H)-\ddot{N}HC_6H_5} + \underset{\text{Hydronium ion}}{H-\overset{+}{\ddot{O}}H_2}$$

Step 4: Protonation of amino group of tetrahedral intermediate

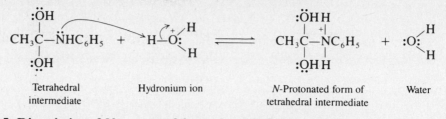

$$CH_3C(OH)_2\ddot{N}HC_6H_5 + H_3O^+ \rightleftharpoons CH_3C(OH)_2\overset{+}{N}H_2C_6H_5 + H_2O$$

Tetrahedral intermediate — Hydronium ion — N-Protonated form of tetrahedral intermediate — Water

Step 5: Dissociation of N-protonated form of tetrahedral intermediate

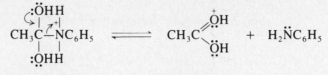

$$CH_3C(OH)_2\overset{+}{N}H_2C_6H_5 \rightleftharpoons CH_3C(=\overset{+}{O}H)OH + H_2\ddot{N}C_6H_5$$

N-Protonated form of tetrahedral intermediate — Protonated form of acetic acid — Aniline

Step 6: Proton transfer processes

$$H_3O^+ + H_2\ddot{N}C_6H_5 \rightleftharpoons H_2O + H_3\overset{+}{N}C_6H_5$$

Hydronium ion — Aniline — Water — Anilinium ion

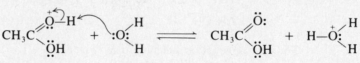

$$CH_3C(=\overset{+}{O}H)OH + H_2O \rightleftharpoons CH_3C(=O)OH + H_3O^+$$

Protonated form of acetic acid — Water — Acetic acid — Hydronium ion

21.20

Step 1: Nucleophilic addition of hydroxide ion to the carbonyl group

$$HO^- + HC(=O)\ddot{N}(CH_3)_2 \longrightarrow HC(O^-)(OH)\ddot{N}(CH_3)_2$$

Hydroxide ion — N,N-Dimethylformamide — Anionic form of tetrahedral intermediate

Step 2: Proton transfer to give neutral form of tetrahedral intermediate

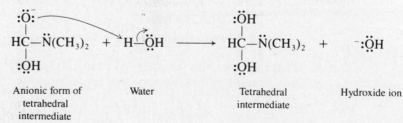

$$HC(O^-)(OH)\ddot{N}(CH_3)_2 + H{-}OH \longrightarrow HC(OH)_2\ddot{N}(CH_3)_2 + {}^-{:}\ddot{O}H$$

Anionic form of tetrahedral intermediate — Water — Tetrahedral intermediate — Hydroxide ion

Step 3: Proton transfer from water to nitrogen of tetrahedral intermediate

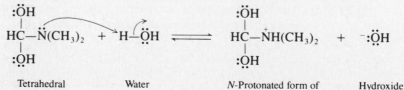

$$HC(OH)_2\ddot{N}(CH_3)_2 + H{-}OH \rightleftharpoons HC(OH)_2\overset{+}{N}H(CH_3)_2 + {}^-{:}\ddot{O}H$$

Tetrahedral intermediate — Water — N-Protonated form of tetrahedral intermediate — Hydroxide ion

Step 4: Dissociation of *N*-protonated form of tetrahedral intermediate

$$HO^- + HC(OH)(OH)\overset{+}{N}H(CH_3)_2 \rightleftharpoons H_2O + HC(=O)OH + HN(CH_3)_2$$

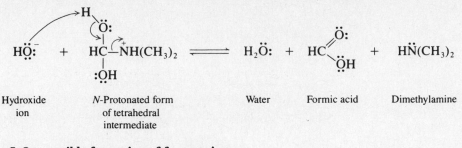

Hydroxide ion | *N*-Protonated form of tetrahedral intermediate | Water | Formic acid | Dimethylamine

Step 5: Irreversible formation of formate ion

$$HC(=O)OH + {}^-OH \longrightarrow HC(=O)O^- + HOH$$

Formic acid | Hydroxide ion | Formate ion | Water

21.21 A synthetic scheme becomes apparent when we recognize that a primary amine may be obtained by Hofmann rearrangement of the primary amide having one more carbon in the skeleton. This amide may, in turn, be prepared from the corresponding carboxylic acid.

$$CH_3CH_2CH_2NH_2 \Longrightarrow CH_3CH_2CH_2\overset{O}{\overset{\|}{C}}NH_2 \Longrightarrow CH_3CH_2CH_2CO_2H$$

The desired reaction scheme is therefore:

$$\underset{\text{Butanoic acid}}{CH_3CH_2CH_2CO_2H} \xrightarrow[2.\ NH_3]{1.\ SOCl_2} \underset{\text{Butanamide}}{CH_3CH_2CH_2\overset{O}{\overset{\|}{C}}NH_2} \xrightarrow[H_2O,\ NaOH]{Br_2} \underset{\text{Propanamine}}{CH_3CH_2CH_2NH_2}$$

21.22 (*a*) Ethanenitrile has the same number of carbon atoms as ethyl alcohol. This suggests a reaction scheme proceeding via an oxime.

$$\underset{\text{Ethyl alcohol}}{CH_3CH_2OH} \longrightarrow \underset{\text{Acetaldoxime}}{CH_3CH{=}NOH} \xrightarrow{P_4O_{10}} \underset{\text{Ethanenitrile}}{CH_3C{\equiv}N}$$

The necessary oxime is prepared from the corresponding aldehyde.

$$\underset{\text{Ethyl alcohol}}{CH_3CH_2OH} \xrightarrow[CH_2Cl_2]{(C_5H_5N)_2CrO_3} \underset{\text{Acetaldehyde}}{CH_3\overset{O}{\overset{\|}{C}}H} \xrightarrow{H_2NOH} \underset{\text{Acetaldoxime}}{CH_3CH{=}NOH}$$

(*b*) Propanenitrile may be prepared from ethyl alcohol by way of a nucleophilic substitution reaction of the corresponding bromide.

$$\underset{\text{Ethyl alcohol}}{CH_3CH_2OH} \xrightarrow[\text{or HBr}]{PBr_3} \underset{\text{Ethyl bromide}}{CH_3CH_2Br} \xrightarrow{NaCN} \underset{\text{Propanenitrile}}{CH_3CH_2CN}$$

21.23 (*a*) The halogen that is attached to the carbonyl group is identified in the name as a separate word following the name of the acyl group.

$$m\text{-}ClC_6H_4\overset{O}{\overset{\|}{C}}Br$$ *m*-Chlorobenzoyl bromide

(*b*) Trifluoroacetic anhydride is the anhydride of trifluoroacetic acid. Notice that it contains six fluorines.

$$CF_3\overset{O}{\overset{\|}{C}}O\overset{O}{\overset{\|}{C}}CF_3 \qquad \text{Trifluoroacetic anhydride}$$

(*c*) This compound is the cyclic anhydride of *cis*-1,2-cyclopropanedicarboxylic acid.

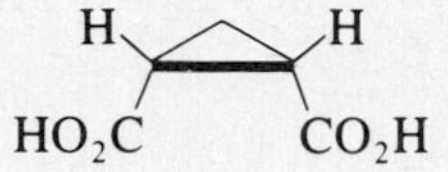

cis-1,2-Cyclopropanedicarboxylic acid

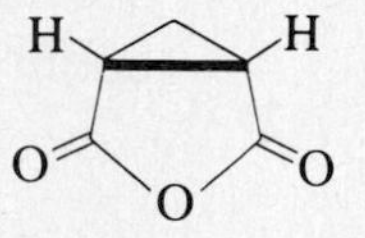

cis-1,2-Cyclopropanedicarboxylic anhydride

(*d*) Ethyl cycloheptanecarboxylate is the ethyl ester of cycloheptanecarboxylic acid.

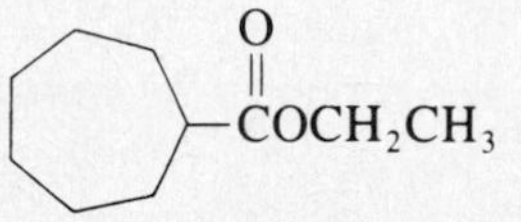

$$\text{C}_7\text{H}_{13}\text{—}\overset{O}{\overset{\|}{C}}OCH_2CH_3 \qquad \text{Ethyl cycloheptanecarboxylate}$$

(*e*) 1-Phenylethyl acetate is the ester of 1-phenylethanol and acetic acid.

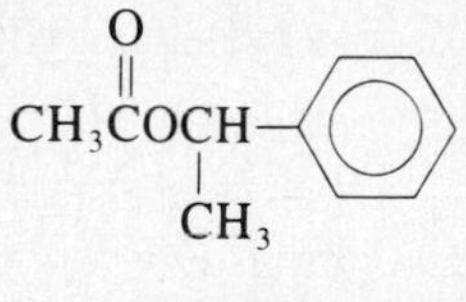

$$CH_3\overset{O}{\overset{\|}{C}}O\underset{CH_3}{\underset{|}{C}}H\text{—}C_6H_5 \qquad \text{1-Phenylethyl acetate}$$

(*f*) 2-Phenylethyl acetate is the ester of 2-phenylethanol and acetic acid.

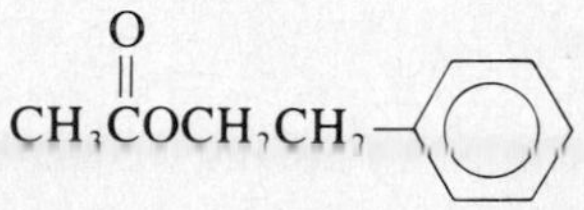

$$CH_3\overset{O}{\overset{\|}{C}}OCH_2CH_2\text{—}C_6H_5 \qquad \text{2-Phenylethyl acetate}$$

(*g*) The parent compound in this case is benzamide. *p*-Ethylbenzamide has an ethyl substituent at the ring position para to the carbonyl group.

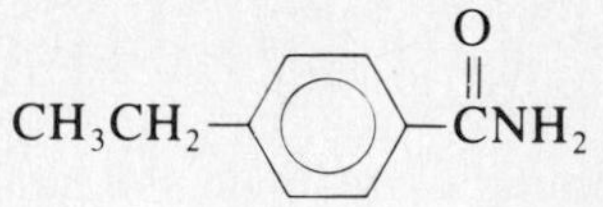

$$CH_3CH_2\text{—}C_6H_4\text{—}\overset{O}{\overset{\|}{C}}NH_2 \qquad p\text{-Ethylbenzamide}$$

(*h*) The parent compound is benzamide. In *N*-ethylbenzamide the ethyl substituent is bonded to nitrogen.

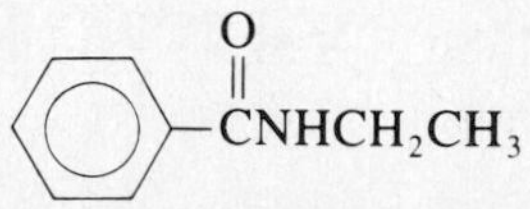

$$C_6H_5\text{—}\overset{O}{\overset{\|}{C}}NHCH_2CH_3 \qquad N\text{-Ethylbenzamide}$$

(*i*) Nitriles are named by adding the suffix -nitrile to the name of the alkane having the same number of carbons. Numbering begins at the nitrile carbon.

$$CH_3CH_2CH_2CH_2\underset{CH_3}{\underset{|}{C}}HC{\equiv}N \qquad \text{2-Methylhexanenitrile}$$

21.24 (*a*) This compound, with a bromine substituent attached to its carbonyl group, is named as an acyl bromide. It is 3-chlorobutanoyl bromide.

$$CH_3\underset{Cl}{\underset{|}{C}}HCH_2\overset{O}{\overset{\|}{C}}Br \qquad \text{3-Chlorobutanoyl bromide}$$

(*b*) The group attached to oxygen, in this case *benzyl*, is identified first in the name of the ester. This compound is the benzyl ester of acetic acid.

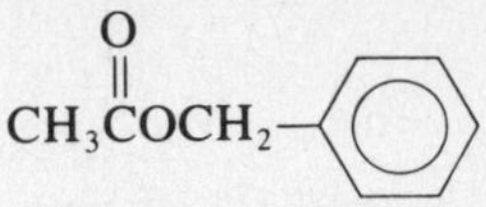

$$CH_3\overset{\overset{O}{\|}}{C}OCH_2-C_6H_5 \quad \text{Benzyl acetate}$$

(*c*) The group attached to oxygen is methyl; this compound is the methyl ester of phenylacetic acid.

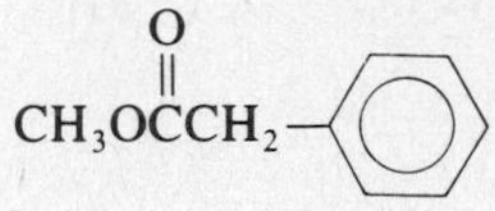

$$CH_3O\overset{\overset{O}{\|}}{C}CH_2-C_6H_5 \quad \text{Methyl phenylacetate}$$

(*d*) This compound contains the functional group $-\overset{\overset{O}{\|}}{C}O\overset{\overset{O}{\|}}{C}-$ and thus is an anhydride of a carboxylic acid. We name the acid, in this case 3-chloropropanoic acid, drop the *acid* part of the name, and replace it by *anhydride*.

$$ClCH_2CH_2\overset{\overset{O}{\|}}{C}O\overset{\overset{O}{\|}}{C}CH_2CH_2Cl \quad \text{3-Chloropropanoic anhydride}$$

(*e*) This compound is a cyclic anhydride, whose parent acid is 3,3-dimethylpentanedioic acid.

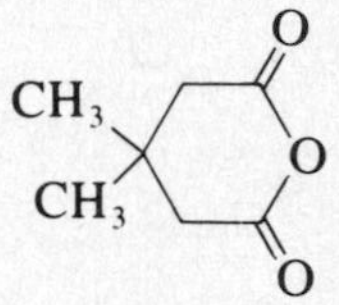

3,3-Dimethylpentanedioic anhydride (3,3-dimethylglutaric anhydride is also acceptable)

(*f*) Nitriles are named by adding *-nitrile* to the name of the alkane having the same number of carbons.

$$CH_3\underset{\underset{CH_3}{|}}{C}HCH_2CH_2C{\equiv}N$$

4-methylpentanenitrile

(*g*) This compound is an amide. We name the corresponding acid, then replace the *-oic acid* suffix by *-amide*.

$$CH_3\underset{\underset{CH_3}{|}}{C}HCH_2CH_2\overset{\overset{O}{\|}}{C}NH_2 \quad \text{4-Methylpentanamide}$$

(*h*) This compound is the *N*-methyl derivative of 4-methylpentanamide.

$$CH_3\underset{\underset{CH_3}{|}}{C}HCH_2CH_2\overset{\overset{O}{\|}}{C}NHCH_3 \quad N\text{-Methyl-4-methylpentanamide}$$

(*i*) The amide nitrogen bears two methyl groups. We designate this as an *N*,*N*-dimethyl amide.

$$CH_3\underset{\underset{CH_3}{|}}{C}HCH_2CH_2\overset{\overset{O}{\|}}{C}N(CH_3)_2 \quad N,N\text{-Dimethyl-4-methylpentanamide}$$

21.25 (*a*) Acetyl chloride acts as an acyl transfer agent to the aromatic ring of bromobenzene. The reaction is a Friedel-Crafts acylation reaction.

$$C_6H_5Br + CH_3\overset{O}{\overset{\|}{C}}Cl \xrightarrow{AlCl_3} o\text{-}BrC_6H_4\overset{O}{\overset{\|}{C}}CH_3 + p\text{-}BrC_6H_4\overset{O}{\overset{\|}{C}}CH_3$$

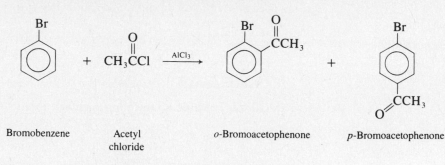

Bromobenzene, Acetyl chloride, *o*-Bromoacetophenone, *p*-Bromoacetophenone

Bromine is an ortho,para–directing substituent.

(*b*) Lithium dialkylcuprates convert acyl chlorides to ketones.

$$C_6H_5\overset{O}{\overset{\|}{C}}Cl + LiCu(CH_3)_2 \longrightarrow C_6H_5\overset{O}{\overset{\|}{C}}CH_3$$

Benzoyl chloride, Lithium dimethylcuprate, Acetophenone

(*c*) Sodium propanoate acts as a nucleophile toward propanoyl chloride. The product is propanoic anhydride.

$$CH_3CH_2\overset{O}{\overset{\|}{C}}O^- + CH_3CH_2\overset{O}{\overset{\|}{C}}{-}Cl \longrightarrow CH_3CH_2\overset{O}{\overset{\|}{C}}O\overset{O}{\overset{\|}{C}}CH_2CH_3$$

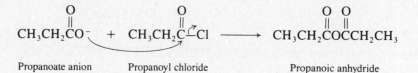

Propanoate anion, Propanoyl chloride, Propanoic anhydride

(*d*) Acyl chlorides convert alcohols to esters.

$$CH_3CH_2CH_2\overset{O}{\overset{\|}{C}}Cl + C_6H_5CH_2OH \longrightarrow CH_3CH_2CH_2\overset{O}{\overset{\|}{C}}OCH_2C_6H_5$$

Butanoyl chloride, Benzyl alcohol, Benzyl butanoate

(*e*) Acyl chlorides react with ammonia to yield amides.

$$Cl{-}C_6H_4{-}\overset{O}{\overset{\|}{C}}Cl + NH_3 \longrightarrow Cl{-}C_6H_4{-}\overset{O}{\overset{\|}{C}}NH_2$$

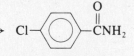

p-Chlorobenzoyl chloride, Ammonia, *p*-Chlorobenzamide

(*f*) The starting material is a cyclic anhydride. Acid anhydrides react with water to yield two carboxylic acid functions; when the anhydride is cyclic, a dicarboxylic acid results.

$$\text{Succinic anhydride} + H_2O \longrightarrow HO\overset{O}{\overset{\|}{C}}CH_2CH_2\overset{O}{\overset{\|}{C}}OH$$

Succinic anhydride, Water, Succinic acid

(*g*) In dilute sodium hydroxide the diacid is converted to its disodium salt.

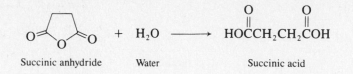

$$\text{Succinic anhydride} + 2NaOH \longrightarrow Na^{+}\,{}^{-}O\overset{O}{\overset{\|}{C}}CH_2CH_2\overset{O}{\overset{\|}{C}}O^{-}\,Na^{+}$$

Succinic anhydride, Sodium hydroxide, Sodium succinate

(*h*) One of the carbonyl groups of the cyclic anhydride is converted to an amide function on reaction with ammonia. The other, the one that would become a carboxylic acid group, is converted to an ammonium carboxylate salt.

$$\text{Succinic anhydride} + 2NH_3 \longrightarrow \overset{+}{N}H_4\,{}^{-}O\overset{O}{\overset{\|}{C}}CH_2CH_2\overset{O}{\overset{\|}{C}}NH_2$$

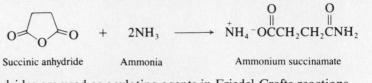

Succinic anhydride — Ammonia — Ammonium succinamate

(*i*) Acid anhydrides are used as acylating agents in Friedel-Crafts reactions.

$$\text{Succinic anhydride} + \text{Benzene} \xrightarrow{AlCl_3} C_6H_5\overset{O}{\overset{\|}{C}}CH_2CH_2\overset{O}{\overset{\|}{C}}OH$$

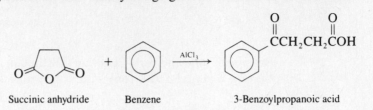

Succinic anhydride — Benzene — 3-Benzoylpropanoic acid

(*j*) The reactant is maleic anhydride; it is a good dienophile in Diels-Alder reactions.

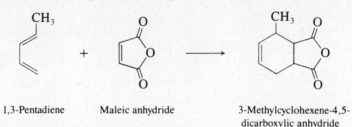

1,3-Pentadiene + Maleic anhydride ⟶ 3-Methylcyclohexene-4,5-dicarboxylic anhydride

(*k*) Acid anhydrides react with alcohols to give an ester and a carboxylic acid.

$$CH_3\overset{O}{\overset{\|}{C}}O\overset{O}{\overset{\|}{C}}CH_3 + CH_3CH_2\underset{OH}{\underset{|}{C}}HCH_2CH_3 \longrightarrow CH_3CH_2\underset{O\underset{O}{\underset{\|}{C}}CH_3}{\underset{|}{C}}HCH_2CH_3 + CH_3CO_2H$$

Acetic anhydride — 3-Pentanol — 3-Pentyl acetate — Acetic acid

(*l*) The starting material is a cyclic ester, a lactone. Esters undergo saponification in aqueous base to give an alcohol and a carboxylate salt.

$$\gamma\text{-Butyrolactone} + NaOH \longrightarrow HOCH_2CH_2CH_2\overset{O}{\overset{\|}{C}}O^{-}Na^{+}$$

γ-Butyrolactone — Sodium hydroxide — Sodium 4-hydroxybutanoate

(*m*) Ammonia reacts with esters to give an amide and an alcohol.

$$\gamma\text{-Butyrolactone} + NH_3 \longrightarrow HOCH_2CH_2CH_2\overset{O}{\overset{\|}{C}}NH_2$$

γ-Butyrolactone — Ammonia — 4-Hydroxybutanamide

(*n*) Lithium aluminum hydride reduces esters to two alcohols; the one derived from the acyl group is a primary alcohol.

$$\gamma\text{-Butyrolactone} \xrightarrow[2.\ H_2O]{1.\ LiAlH_4} HOCH_2CH_2CH_2CH_2OH$$

γ-Butyrolactone — 1,4-Butanediol

(*o*) Grignard reagents react with esters to give tertiary alcohols.

$$\text{γ-Butyrolactone} \xrightarrow[2.\ H_3O^+]{1.\ 2CH_3MgBr} HOCH_2CH_2CH_2C(OH)(CH_3)CH_3$$

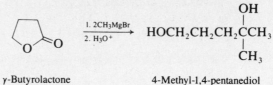

γ-Butyrolactone 4-Methyl-1,4-pentanediol

(*p*) In this reaction methylamine acts as a nucleophile toward the carbonyl group of the ester. The product is an amide.

$$CH_3\ddot{N}H_2 + C_6H_5CH_2\overset{O}{\overset{\|}{C}}-OCH_2CH_3 \longrightarrow C_6H_5CH_2\overset{O}{\overset{\|}{C}}NHCH_3 + CH_3CH_2OH$$

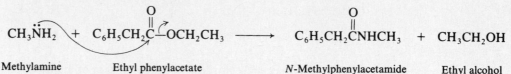

Methylamine Ethyl phenylacetate *N*-Methylphenylacetamide Ethyl alcohol

(*q*) The starting material is a lactam, a cyclic amide. Amides are hydrolyzed in base to amines and carboxylate salts.

$$\text{N-Methylpyrrolidone} + NaOH \longrightarrow CH_3NHCH_2CH_2CH_2\overset{O}{\overset{\|}{C}}O^-Na^+$$

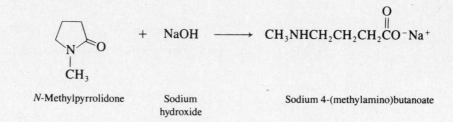

N-Methylpyrrolidone Sodium hydroxide Sodium 4-(methylamino)butanoate

(*r*) In acid solution amides yield carboxylic acids and ammonium salts.

$$\text{N-Methylpyrrolidone} + H_3O^+ \longrightarrow CH_3\overset{+}{N}H_2CH_2CH_2CH_2\overset{O}{\overset{\|}{C}}OH$$

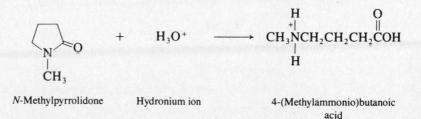

N-Methylpyrrolidone Hydronium ion 4-(Methylammonio)butanoic acid

(*s*) The starting material is a cyclic imide. Both its amide bonds are cleaved by nucleophilic attack by hydroxide ion.

$$\text{N-Methylsuccinimide} + 2NaOH \longrightarrow Na^{+-}O\overset{O}{\overset{\|}{C}}CH_2CH_2\overset{O}{\overset{\|}{C}}O^-Na^+ + CH_3NH_2$$

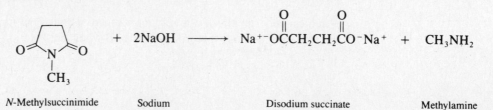

N-Methylsuccinimide Sodium hydroxide Disodium succinate Methylamine

(*t*) In acid the imide undergoes cleavage to give a dicarboxylic acid and the conjugate acid of methylamine.

$$\text{N-Methylsuccinimide} + 2H_2O + HCl \longrightarrow HO\overset{O}{\overset{\|}{C}}CH_2CH_2\overset{O}{\overset{\|}{C}}OH + CH_3\overset{+}{N}H_3Cl^-$$

N-Methylsuccinimide Water Hydrogen chloride Succinic acid Methylammonium chloride (*u*) Acetanilide is hydrolyzed in acid to acetic acid and the conjugate acid of aniline.

$$C_6H_5NHCCH_3 \; (C{=}O) + H_2O + HCl \longrightarrow C_6H_5\overset{+}{N}H_3Cl^- + CH_3COH \; (C{=}O)$$

Acetanilide — Water — Hydrogen chloride — Anilinium chloride — Acetic acid

(*v*) This is another example of amide hydrolysis.

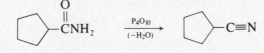

$$C_6H_5CNHCH_3 \; (C{=}O) + H_2O + H_2SO_4 \longrightarrow C_6H_5COH \; (C{=}O) + CH_3\overset{+}{N}H_3 \;\; HSO_4^-$$

N-Methylbenzamide — Water — Sulfuric acid — Benzoic acid — Methylammonium hydrogen sulfate

(*w*) One way to prepare nitriles is by dehydration of amides.

$$\text{cyclopentyl}{-}CNH_2 \; (C{=}O) \xrightarrow[(-H_2O)]{P_4O_{10}} \text{cyclopentyl}{-}C{\equiv}N$$

Cyclopentanecarboxamide — Cyclopentyl cyanide

(*x*) Nitriles are hydrolyzed to carboxylic acids in acidic media.

$$(CH_3)_2CHCH_2C{\equiv}N \xrightarrow[\text{heat}]{HCl,\ H_2O} (CH_3)_2CHCH_2COH \; (C{=}O)$$

3-Methylbutanenitrile — 3-Methylbutanoic acid

(*y*) Nitriles are hydrolyzed in aqueous base to salts of carboxylic acids.

$$CH_3O{-}C_6H_4{-}C{\equiv}N \xrightarrow[\text{heat}]{NaOH,\ H_2O} CH_3O{-}C_6H_4{-}CO^-Na^+ \; (C{=}O) + NH_3$$

p-Methoxybenzonitrile — Sodium *p*-methoxybenzoate — Ammonia

(*z*) Grignard reagents react with nitriles to yield ketones after aqueous acidic workup.

$$CH_3CH_2C{\equiv}N \xrightarrow[2.\ H_3O^+]{1.\ CH_3MgBr} CH_3CH_2CCH_3 \; (C{=}O)$$

Propanenitrile — 2-Butanone

(*aa*) Amides undergo the Hofmann rearrangement on reaction with bromine and base. A methyl carbamate is the product isolated when the reaction is carried out in methanol.

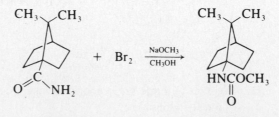

$$\text{7,7-dimethylbicyclo[2.2.1]heptane-1-}C(=O)NH_2 + Br_2 \xrightarrow[CH_3OH]{NaOCH_3} \text{7,7-dimethylbicyclo[2.2.1]heptane-1-}HNCOCH_3 \; (C{=}O)$$

(*bb*) Saponification of the carbamate in (*aa*) gives the corresponding amine.

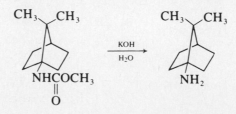

$$\text{7,7-dimethylbicyclo[2.2.1]heptane-1-}NHCOCH_3 \; (C{=}O) \xrightarrow[H_2O]{KOH} \text{7,7-dimethylbicyclo[2.2.1]heptane-1-}NH_2$$

21.26 (*a*) Acetyl chloride is prepared by reaction of acetic acid with thionyl chloride. The first task then is to prepare acetic acid by oxidation of ethanol.

$$\underset{\text{Ethanol}}{CH_3CH_2OH} \xrightarrow[\text{or 1. }KMnO_4\text{ 2. }H^+]{K_2Cr_2O_7,\ H_2SO_4} \underset{\text{Acetic acid}}{CH_3\overset{O}{\overset{\|}{C}}OH} \xrightarrow{SOCl_2} \underset{\text{Acetyl chloride}}{CH_3\overset{O}{\overset{\|}{C}}Cl}$$

(*b*) Acetic acid and acetyl chloride, available from (*a*), can be combined to form acetic anhydride.

$$\underset{\text{Acetic acid}}{CH_3\overset{O}{\overset{\|}{C}}OH} + \underset{\text{Acetyl chloride}}{CH_3\overset{O}{\overset{\|}{C}}Cl} \longrightarrow \underset{\text{Acetic anhydride}}{CH_3\overset{O}{\overset{\|}{C}}O\overset{O}{\overset{\|}{C}}CH_3} + \underset{\text{Hydrogen chloride}}{HCl}$$

$$\text{or}\quad \underset{\text{Acetic acid}}{CH_3\overset{O}{\overset{\|}{C}}OH} \xrightarrow{NaOH} \underset{\text{Sodium acetate}}{CH_3\overset{O}{\overset{\|}{C}}ONa} \xrightarrow{CH_3\overset{O}{\overset{\|}{C}}Cl} \underset{\text{Acetic anhydride}}{CH_3\overset{O}{\overset{\|}{C}}O\overset{O}{\overset{\|}{C}}CH_3}$$

(*c*) Ethanol can be converted to ethyl acetate by reaction with acetic acid, acetyl chloride, or acetic anhydride.

$$\underset{\text{Ethanol}}{CH_3CH_2OH} + \underset{\text{Acetic acid}}{CH_3\overset{O}{\overset{\|}{C}}OH} \xrightarrow{H^+} \underset{\text{Ethyl acetate}}{CH_3\overset{O}{\overset{\|}{C}}OCH_2CH_3} + \underset{\text{Water}}{H_2O}$$

$$\text{or}\quad \underset{\text{Ethanol}}{CH_3CH_2OH} + \underset{\text{Acetyl chloride}}{CH_3\overset{O}{\overset{\|}{C}}Cl} \xrightarrow{\text{pyridine}} \underset{\text{Ethyl acetate}}{CH_3\overset{O}{\overset{\|}{C}}OCH_2CH_3}$$

$$\text{or}\quad \underset{\text{Ethanol}}{CH_3CH_2OH} + \underset{\text{Acetic anhydride}}{CH_3\overset{O}{\overset{\|}{C}}O\overset{O}{\overset{\|}{C}}CH_3} \xrightarrow{\text{pyridine}} \underset{\text{Ethyl acetate}}{CH_3\overset{O}{\overset{\|}{C}}OCH_2CH_3}$$

(*d*) Ethyl bromoacetate is the ethyl ester of bromoacetic acid; thus the first task is to prepare the acid. We use the acetic acid prepared in (*a*), converting it to bromoacetic acid by the Hell-Volhard-Zelinsky reaction.

$$\underset{\text{Acetic acid}}{CH_3CO_2H} \xrightarrow[P]{Br_2} \underset{\text{Bromoacetic acid}}{BrCH_2CO_2H} \xrightarrow[H^+]{CH_3CH_2OH} \underset{\text{Ethyl bromoacetate}}{BrCH_2\overset{O}{\overset{\|}{C}}OCH_2CH_3}$$

Alternatively, bromoacetic acid could be converted to the corresponding acyl chloride, then treated with ethanol. It would be incorrect to try to brominate ethyl acetate; the Hell-Volhard-Zelinsky method requires an acid as starting material, not an ester.

(*e*) The alcohol $BrCH_2CH_2OH$, needed in order to prepare 2-bromoethyl acetate, is prepared from ethanol by way of ethylene.

$$\underset{\text{Ethanol}}{CH_3CH_2OH} \xrightarrow[\text{heat}]{H_2SO_4} \underset{\text{Ethylene}}{CH_2{=}CH_2} \xrightarrow[H_2O]{Br_2} \underset{\text{2-Bromoethanol}}{BrCH_2CH_2OH}$$

$$\text{Then}\quad \underset{\text{2-Bromoethanol}}{BrCH_2CH_2OH} \xrightarrow[\text{or }CH_3\overset{O}{\overset{\|}{C}}Cl]{CH_3\overset{O}{\overset{\|}{C}}O\overset{O}{\overset{\|}{C}}CH_3} \underset{\text{2-Bromoethyl acetate}}{CH_3\overset{O}{\overset{\|}{C}}OCH_2CH_2Br}$$

(*f*) Ethyl cyanoacetate may be prepared from the ethyl bromoacetate obtained in (*d*). The bromide may be displaced by cyanide in a nucleophilic substitution reaction.

$$\underset{\text{Ethyl bromoacetate}}{BrCH_2\overset{O}{\overset{\|}{C}}OCH_2CH_3} \xrightarrow[S_N2]{NaCN} \underset{\text{Ethyl cyanoacetate}}{N{\equiv}CCH_2\overset{O}{\overset{\|}{C}}OCH_2CH_3}$$

(*g*) Reaction of the acetyl chloride prepared in (*a*) or the acetic anhydride from (*b*) with ammonia gives acetamide.

$$\underset{\text{Acetyl chloride}}{CH_3\overset{O}{\overset{\|}{C}}Cl} \quad \text{or} \quad \underset{\text{Acetic anhydride}}{CH_3\overset{O}{\overset{\|}{C}}O\overset{O}{\overset{\|}{C}}CH_3} \xrightarrow{NH_3} \underset{\text{Acetamide}}{CH_3\overset{O}{\overset{\|}{C}}NH_2}$$

(*h*) Methylamine may be prepared from acetamide by a Hofmann rearrangement.

$$\underset{\substack{\text{Acetamide}\\ \text{[prepared as in (g)]}}}{CH_3\overset{O}{\overset{\|}{C}}NH_2} \xrightarrow{Br_2,\ HO^-,\ H_2O} \underset{\text{Methylamine}}{CH_3NH_2}$$

(*i*) The desired hydroxy acid is available from hydrolysis of the corresponding cyanohydrin, which may be prepared by reaction of the appropriate aldehyde with cyanide ion.

$$CH_3\underset{OH}{\underset{|}{C}}H\overset{O}{\overset{\|}{C}}OH \Longrightarrow CH_3\underset{OH}{\underset{|}{C}}HC{\equiv}N \Longrightarrow CH_3\overset{O}{\overset{\|}{C}}H$$

In this synthesis the cyanohydrin is prepared from ethanol by way of acetaldehyde.

$$\underset{\text{Ethanol}}{CH_3CH_2OH} \xrightarrow[CH_2Cl_2]{(C_5H_5N)_2CrO_3} \underset{\text{Acetaldehyde}}{CH_3\overset{O}{\overset{\|}{C}}H} \xrightarrow[H^+]{KCN} \underset{\text{2-Hydroxypropanenitrile}}{CH_3\overset{OH}{\overset{|}{C}}HC{\equiv}N}$$

$$\underset{\text{2-Hydroxypropanenitrile}}{CH_3\underset{OH}{\underset{|}{C}}HC{\equiv}N} \xrightarrow[\substack{\text{or}\\ 1.\ HO^-,\ H_2O,\ \text{heat}\\ 2.\ H^+}]{H_2O,\ H^+,\ \text{heat}} \underset{\text{2-Hydroxypropanoic acid}}{CH_3\underset{OH}{\underset{|}{C}}H\overset{O}{\overset{\|}{C}}OH}$$

21.27 (*a*) Benzoyl chloride is made from benzoic acid. Oxidize toluene to benzoic acid, then treat with thionyl chloride.

$$\underset{\text{Toluene}}{C_6H_5CH_3} \xrightarrow[\substack{\text{or}\\ 1.\ KMnO_4\\ 2.\ H^+}]{K_2Cr_2O_7,\ H_2SO_4} \underset{\text{Benzoic acid}}{C_6H_5\overset{O}{\overset{\|}{C}}OH} \xrightarrow{SOCl_2} \underset{\text{Benzoyl chloride}}{C_6H_5\overset{O}{\overset{\|}{C}}Cl}$$

(*b*) Benzoyl chloride and benzoic acid, both prepared from toluene in (*a*), react with each other to give benzoic anhydride.

$$\underset{\text{Benzoic acid}}{C_6H_5\overset{O}{\overset{\|}{C}}OH} + \underset{\text{Benzoyl chloride}}{C_6H_5\overset{O}{\overset{\|}{C}}Cl} \longrightarrow \underset{\text{Benzoic anhydride}}{C_6H_5\overset{O}{\overset{\|}{C}}O\overset{O}{\overset{\|}{C}}C_6H_5}$$

(*c*) Benzoic acid, benzoyl chloride, and benzoic anhydride have been prepared in (*a*) and (*b*) of this problem. Any of them could be converted to benzyl benzoate on reaction with benzyl alcohol. Thus the synthesis of benzyl benzoate requires the preparation of benzyl alcohol from toluene. This is effected by a nucleophilic substitution reaction of benzyl bromide, in turn prepared by halogenation of toluene.

$$\underset{\text{Toluene}}{C_6H_5CH_3} \xrightarrow[\text{or } Br_2\text{, light}]{N\text{-bromosuccinimide (NBS)}} \underset{\text{Benzyl bromide}}{C_6H_5CH_2Br} \xrightarrow[HO^-]{H_2O} \underset{\text{Benzyl alcohol}}{C_6H_5CH_2OH}$$

Alternatively, recall that primary alcohols may be obtained by reduction of the corresponding carboxylic acid.

$$\underset{\text{Benzoic acid}}{C_6H_5\overset{O}{\overset{\|}{C}}OH} \xrightarrow[2.\ H_2O]{1.\ LiAlH_4} \underset{\text{Benzyl alcohol}}{C_6H_5CH_2OH}$$

$$\text{Then } \underset{\text{Benzoyl chloride}}{C_6H_5\overset{O}{\overset{\|}{C}}Cl} + \underset{\text{Benzyl alcohol}}{C_6H_5CH_2OH} \xrightarrow{\text{pyridine}} \underset{\text{Benzyl benzoate}}{C_6H_5\overset{O}{\overset{\|}{C}}OCH_2C_6H_5}$$

(*d*) Benzamide is prepared by reaction of ammonia with either benzoyl chloride or benzoic anhydride.

$$\underset{\text{Benzoyl chloride}}{C_6H_5\overset{O}{\overset{\|}{C}}Cl} \text{ or } \underset{\text{Benzoic anhydride}}{C_6H_5\overset{O}{\overset{\|}{C}}O\overset{O}{\overset{\|}{C}}C_6H_5} \xrightarrow{NH_3} \underset{\text{Benzamide}}{C_6H_5\overset{O}{\overset{\|}{C}}NH_2}$$

(*e*) Benzonitrile may be prepared by dehydration of benzamide.

$$\underset{\text{Benzamide}}{C_6H_5\overset{O}{\overset{\|}{C}}NH_2} \xrightarrow[\text{heat}]{P_4O_{10}} \underset{\text{Benzonitrile}}{C_6H_5C{\equiv}N}$$

(*f*) Benzyl cyanide is the product of nucleophilic substitution by cyanide ion or benzyl bromide or benzyl chloride. The benzyl halides are prepared by free-radical halogenation of the toluene side chain.

$$\underset{\text{Toluene}}{C_6H_5CH_3} \xrightarrow[\text{light or heat}]{Cl_2} \underset{\text{Benzyl chloride}}{C_6H_5CH_2Cl} \xrightarrow{NaCN} \underset{\text{Benzyl cyanide}}{C_6H_5CH_2C{\equiv}N}$$

$$\text{or } \underset{\text{Toluene}}{C_6H_5CH_3} \xrightarrow[\text{or } Br_2\text{, light}]{NBS} \underset{\text{Benzyl bromide}}{C_6H_5CH_2Br} \xrightarrow{NaCN} \underset{\text{Benzyl cyanide}}{C_6H_5CH_2C{\equiv}N}$$

(*g*) Hydrolysis of benzyl cyanide yields phenylacetic acid.

$$\underset{\text{Benzyl cyanide}}{C_6H_5CH_2C{\equiv}N} \xrightarrow[\text{or } 1.\ NaOH\text{, heat};\ 2.\ H^+]{H_2O\text{, }H^+\text{, heat}} \underset{\text{Phenylacetic acid}}{C_6H_5CH_2\overset{O}{\overset{\|}{C}}OH}$$

Alternatively, the Grignard reagent derived from benzyl bromide may be carboxylated.

$$\underset{\text{Benzyl bromide}}{C_6H_5CH_2Br} \xrightarrow[\text{diethyl ether}]{Mg} \underset{\text{Benzylmagnesium bromide}}{C_6H_5CH_2MgBr} \xrightarrow[2.\ H_2O\text{, }H^+]{1.\ CO_2} \underset{\text{Phenylacetic acid}}{C_6H_5CH_2\overset{O}{\overset{\|}{C}}OH}$$

(*h*) The first goal is to synthesize *p*-nitrobenzoic acid, as this may be readily converted to the desired acid chloride. First convert toluene to *p*-nitrotoluene, then oxidize. Nitration must precede oxidation of the side chain in order to achieve the desired para orientation.

$$C_6H_5-CH_3 \xrightarrow[H_2SO_4]{HNO_3} O_2N-C_6H_4-CH_3 \xrightarrow[\text{or 1. } KMnO_4 \text{ 2. } H^+]{K_2Cr_2O_7,\ H_2SO_4} O_2N-C_6H_4-\overset{O}{\overset{\|}{C}}OH$$

Toluene — *p*-Nitrotoluene (separate from ortho isomer) — *p*-Nitrobenzoic acid

Treatment of *p*-nitrobenzoic acid with thionyl chloride yields *p*-nitrobenzoyl chloride.

$$O_2N-C_6H_4-\overset{O}{\overset{\|}{C}}OH \xrightarrow{SOCl_2} O_2N-C_6H_4-\overset{O}{\overset{\|}{C}}Cl$$

p-Nitrobenzoic acid — *p*-Nitrobenzoyl chloride

(*i*) In order to achieve the correct orientation in *m*-nitrobenzoyl chloride, oxidation of the methyl group must precede nitration.

$$C_6H_5-CH_3 \xrightarrow[\text{or 1. } KMnO_4 \text{ 2. } H^+]{K_2Cr_2O_7,\ H_2SO_4} C_6H_5-CO_2H \xrightarrow[H_2SO_4]{HNO_3} m\text{-}O_2N-C_6H_4-CO_2H$$

Toluene — Benzoic acid — *m*-Nitrobenzoic acid

Once *m*-nitrobenzoic acid has been prepared, it may be converted to the corresponding acyl chloride.

$$m\text{-}O_2N-C_6H_4-\overset{O}{\overset{\|}{C}}OH \xrightarrow{SOCl_2} m\text{-}O_2N-C_6H_4-\overset{O}{\overset{\|}{C}}Cl$$

m-Nitrobenzoic acid — *m*-Nitrobenzoyl chloride

(*j*) A Hofmann rearrangement of benzamide affords aniline.

$$C_6H_5-\overset{O}{\overset{\|}{C}}NH_2 + Br_2 \xrightarrow[H_2O]{HO^-} C_6H_5-NH_2$$

Benzamide [prepared as in (*d*)] — Bromine — Aniline

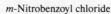

21.28 The problem specifies that $CH_3CH_2\overset{O}{\overset{\|}{C}}\text{Ø}CH_2CH_3$ is to be prepared from ^{18}O-labeled ethyl alcohol (Ø = ^{18}O).

$$CH_3CH_2\overset{O}{\overset{\|}{C}}Cl + CH_3CH_2\text{Ø}H \longrightarrow CH_3CH_2\overset{O}{\overset{\|}{C}}\text{Ø}CH_2CH_3$$

Propionyl chloride — Ethyl alcohol — Ethyl propionate

Thus, we need to prepare ^{18}O-labeled ethyl alcohol from the other designated starting materials, acetaldehyde and ^{18}O-enriched water. First, replace the oxygen of acetaldehyde with ^{18}O by the hydration-dehydration equilibrium in the presence of ^{18}O-enriched water.

$$CH_3\overset{O}{\overset{\|}{C}}H + H_2\text{Ø} \rightleftharpoons CH_3\underset{\text{Ø}H}{\overset{OH}{C}}H \rightleftharpoons CH_3\overset{\text{Ø}}{\overset{\|}{C}}H + H_2O$$

Acetaldehyde — ^{18}O-Enriched water — Hydrate of acetaldehyde — ^{18}O-Enriched acetaldehyde — Water

Once ^{18}O-enriched acetaldehyde has been obtained, it can be reduced to ^{18}O-enriched ethanol.

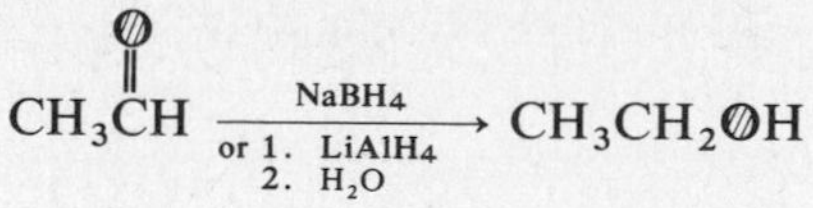

$$CH_3\overset{\overset{\text{Ø}}{\|}}{C}H \xrightarrow[\text{or 1. LiAlH}_4 \quad 2.\ H_2O]{NaBH_4} CH_3CH_2\text{Ø}H$$

21.29 (*a*) The rate-determining step in base-promoted ester hydrolysis is nucleophilic addition of hydroxide to the carbonyl group. The intermediate formed in this step is negatively charged.

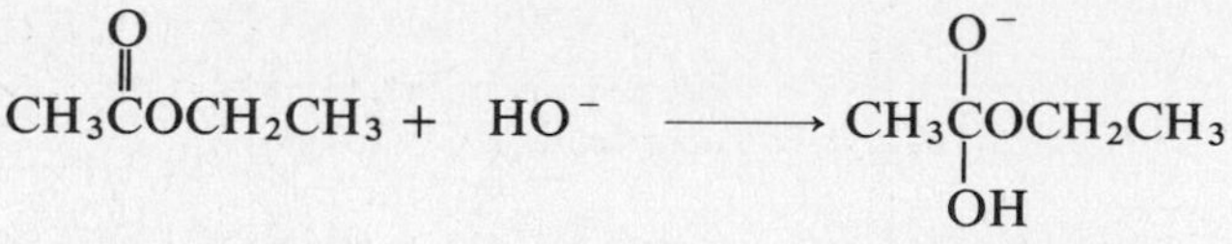

$$\underset{\text{Ethyl acetate}}{CH_3\overset{\overset{O}{\|}}{C}OCH_2CH_3} + \underset{\text{Hydroxide}}{HO^-} \longrightarrow \underset{\text{Rate-determining intermediate}}{CH_3\overset{\overset{O^-}{|}}{\underset{\underset{OH}{|}}{C}}OCH_2CH_3}$$

The electron-withdrawing effect of a CF_3 group stabilizes the intermediate formed in the rate-determining step of ethyl trifluoroacetate saponification.

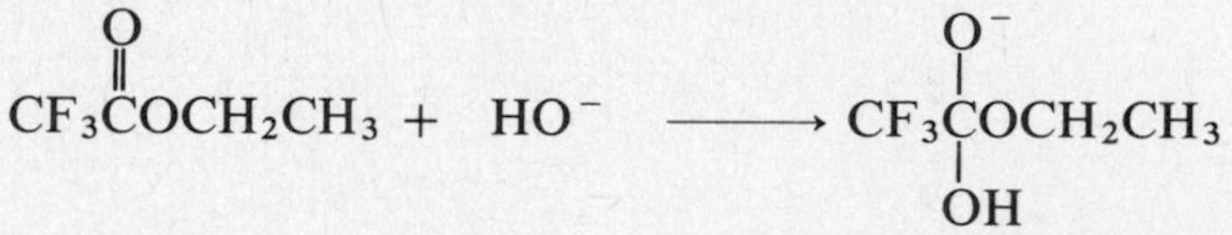

$$\underset{\text{Ethyl trifluoroacetate}}{CF_3\overset{\overset{O}{\|}}{C}OCH_2CH_3} + \underset{\text{Hydroxide}}{HO^-} \longrightarrow \underset{\text{Rate-determining intermediate}}{CF_3\overset{\overset{O^-}{|}}{\underset{\underset{OH}{|}}{C}}OCH_2CH_3}$$

Since the intermediate is more stable, it is formed faster than the one from ethyl acetate.

(*b*) Crowding is increased as the transition state for nucleophilic addition to the carbonyl group is approached. The carbonyl carbon undergoes a change in hybridization from sp^2 to sp^3.

$$\underset{\text{Ethyl trimethylacetate}}{CH_3\overset{\overset{CH_3}{|}}{\underset{\underset{CH_3}{|}}{C}}-\overset{\overset{O}{\|}}{C}OCH_2CH_3} + \underset{\text{Hydroxide ion}}{^-OH} \longrightarrow \underset{\text{Rate-determining intermediate; crowded}}{CH_3-\overset{\overset{CH_3}{|}}{\underset{\underset{CH_3}{|}}{C}}-\overset{\overset{O^-}{|}}{\underset{\underset{OH}{|}}{C}}OCH_2CH_3}$$

The *tert*-butyl group of ethyl trimethylacetate causes more crowding than the methyl group of ethyl acetate; the rate-determining intermediate is less stable and is formed more slowly.

(*c*) We see here another example of a steric effect of a *tert*-butyl group. The intermediate formed when hydroxide adds to the carbonyl group of *tert*-butyl acetate is more crowded and less stable than the corresponding intermediate formed from methyl acetate.

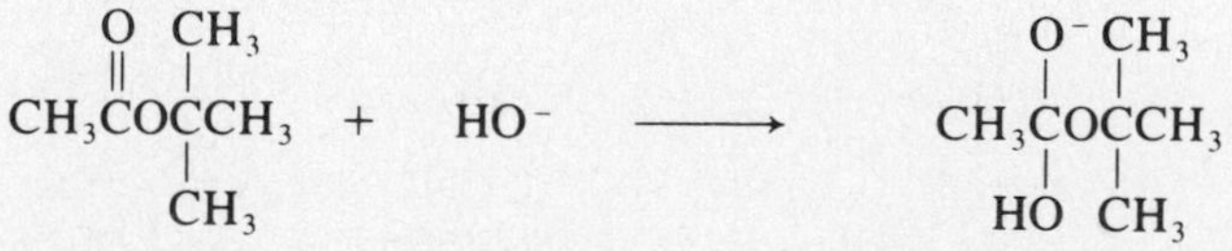

$$\underset{\textit{tert}\text{-Butyl acetate}}{CH_3\overset{\overset{O}{\|}}{C}O\overset{\overset{CH_3}{|}}{\underset{\underset{CH_3}{|}}{C}}CH_3} + \underset{\text{Hydroxide}}{HO^-} \longrightarrow \underset{\text{Rate-determining intermediate; crowded}}{CH_3\overset{\overset{O^-}{|}}{\underset{\underset{HO}{|}}{C}}O\overset{\overset{CH_3}{|}}{\underset{\underset{CH_3}{|}}{C}}CH_3}$$

$$\underset{\text{Methyl acetate}}{CH_3\overset{\overset{O}{\|}}{C}OCH_3} + HO^- \longrightarrow \underset{\text{Rate-determining intermediate; less crowded than intermediate from } \textit{tert}\text{-butyl acetate}}{CH_3\overset{\overset{O^-}{|}}{\underset{\underset{HO}{|}}{C}}OCH_3}$$

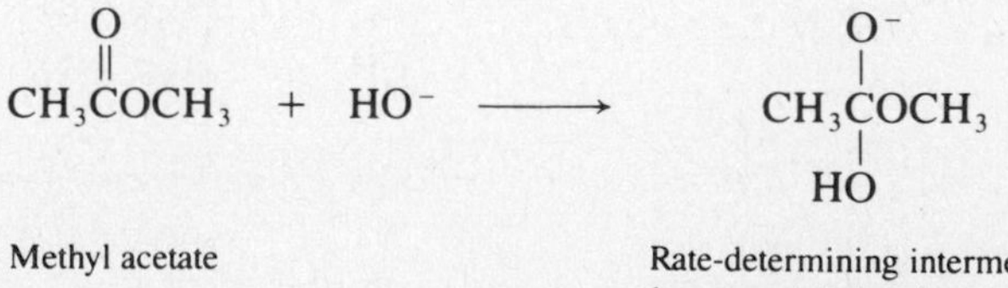

(*d*) Here, as in (*a*) we have an electron-withdrawing substituent increasing the rate of ester saponification. It does so by stabilizing the negatively charged intermediate formed in the rate-determining step.

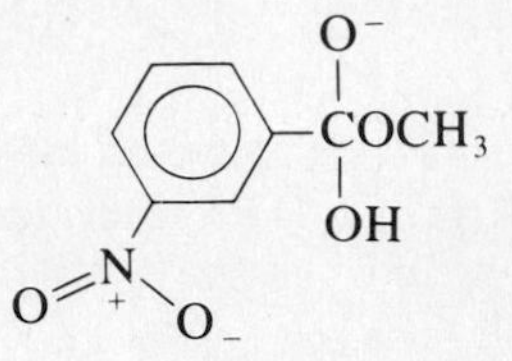

O^- | $-COCH_3$ | OH (with $O=\overset{+}{N}-O^-$ meta substituent) more stable than O^- | $-COCH_3$ | OH

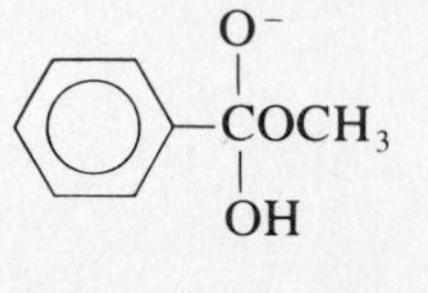

Rate-determining intermediate from methyl *m*-nitrobenzoate

Rate-determining intermediate from methyl benzoate

(*e*) Addition of hydroxide to 4-butanolide introduces torsional strain in the intermediate because of eclipsed bonds. The corresponding intermediate from 5-butanolide is more stable because the bonds are staggered in a six-membered ring.

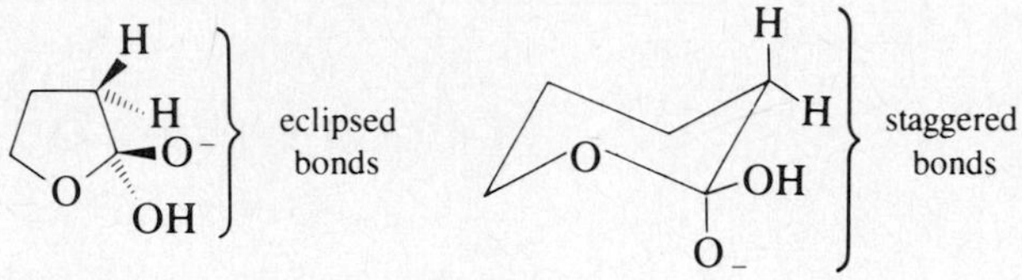

H, H, O^-, O, OH — eclipsed bonds

H, H, O, OH, O^- — staggered bonds

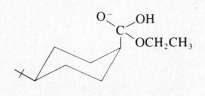

Less stable; formed more slowly

More stable; formed faster

(*f*) Steric crowding increases more when hydroxide adds to the axial carbonyl group.

O^-, OH, C, OCH_2CH_3

O^-, $C-OCH_2CH_3$, OH

Cis diastereomer; greater increase in crowding when carbon changes from sp^2 to sp^3; formed more slowly

Trans diastereomer; smaller increase in crowding when carbon changes from sp^2 to sp^3; formed more rapidly

21.30 Compound A is the *p*-toluenesulfonate ester of *trans*-4-*tert*-butylcyclohexanol. The oxygen atom of the alcohol attacks sulfur of *p*-toluenesulfonyl chloride, so the reaction proceeds with retention of configuration.

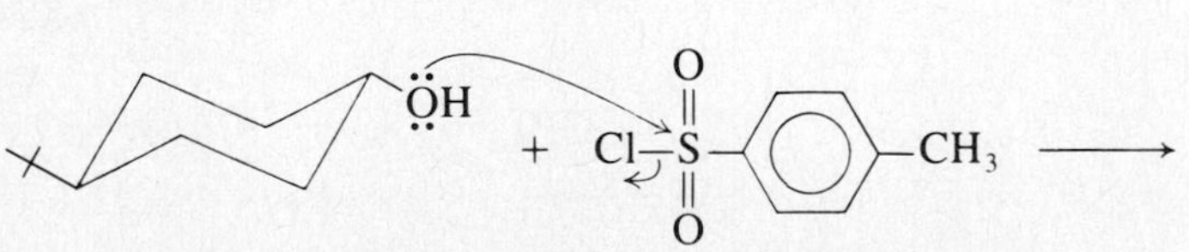

$\ddot{O}H$ + $Cl-\underset{O}{\overset{O}{S}}-C_6H_4-CH_3$ ⟶

trans-4-*tert*-Butylcyclohexanol

p-Toluenesulfonyl chloride

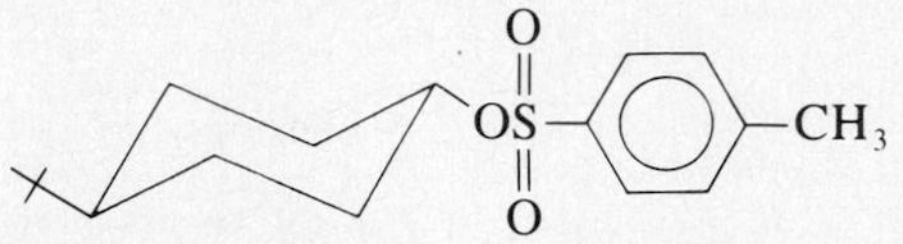

$O\underset{O}{\overset{O}{S}}-C_6H_4-CH_3$

trans-4-*tert*-Butylcyclohexyl *p*-toluenesulfonate (compound A)

The second step is a nucleophilic substitution reaction in which benzoate ion displaces *p*-toluenesulfonate with inversion of configuration.

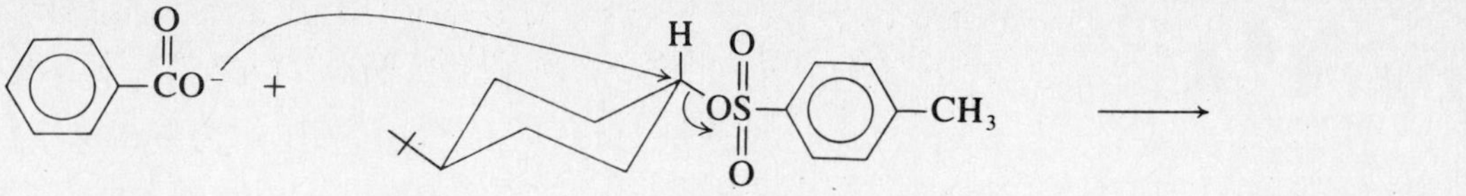

$$C_6H_5CO_2^- + \text{(trans-4-tert-butylcyclohexyl } OSO_2C_6H_4CH_3) \longrightarrow$$

Benzoate ion trans-4-*tert*-Butylcyclohexyl *p*-toluenesulfonate (compound A)

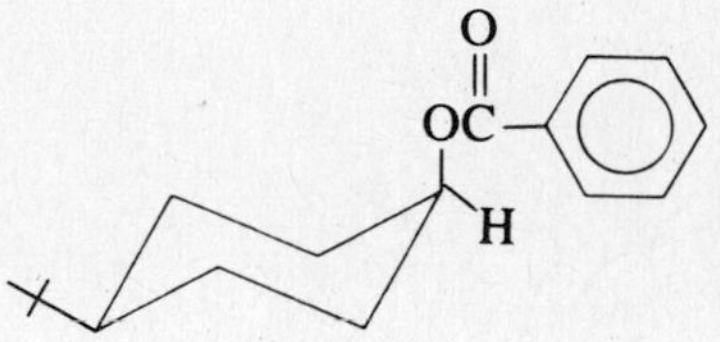

cis-4-*tert*-Butylcyclohexyl benzoate (compound B)

Saponification of *cis*-4-*tert*-butylcyclohexyl benzoate proceeds with acyl-oxygen cleavage to give *cis*-4-*tert*-butylcyclohexanol.

21.31 Step 1 in the ambrettolide synthesis is the formation of a cyclic acetal from acetone and a vicinal diol.

Step 1

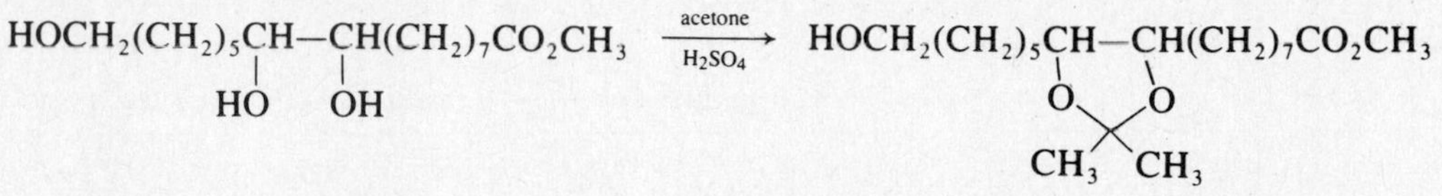

$$HOCH_2(CH_2)_5CH(OH){-}CH(OH)(CH_2)_7CO_2CH_3 \xrightarrow[H_2SO_4]{\text{acetone}} HOCH_2(CH_2)_5CH{-}CH(CH_2)_7CO_2CH_3 \;(\text{O–C(CH}_3)_2\text{–O acetal})$$

Methyl 9,10,16-trihydroxyhexadecanoate Compound C ($C_{20}H_{38}O_5$)

Formation of the cyclic acetal protects two of the hydroxyl groups, leaving the primary hydroxyl available for oxidation to a carboxyl group in step 2.

Step 2

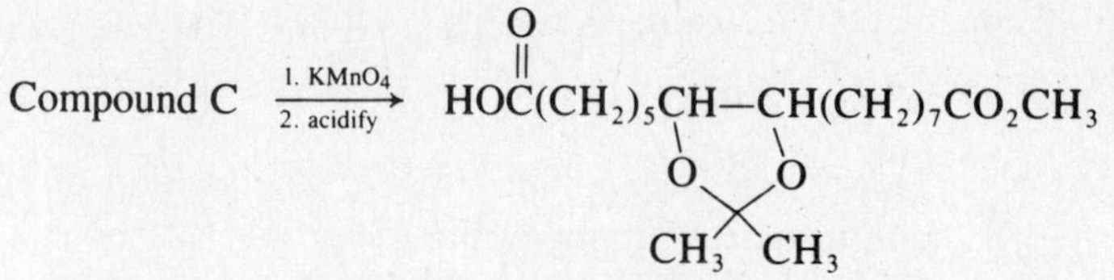

$$\text{Compound C} \xrightarrow[\text{2. acidify}]{\text{1. KMnO}_4} HOC(=O)(CH_2)_5CH{-}CH(CH_2)_7CO_2CH_3 \;(\text{O–C(CH}_3)_2\text{–O acetal})$$

Compound D ($C_{20}H_{36}O_6$)

Sodium-ethanol reduction in step 3 reduces the ester to a primary alcohol. The carboxyl group is not reduced nor is the cyclic acetal affected under these conditions.

Step 3

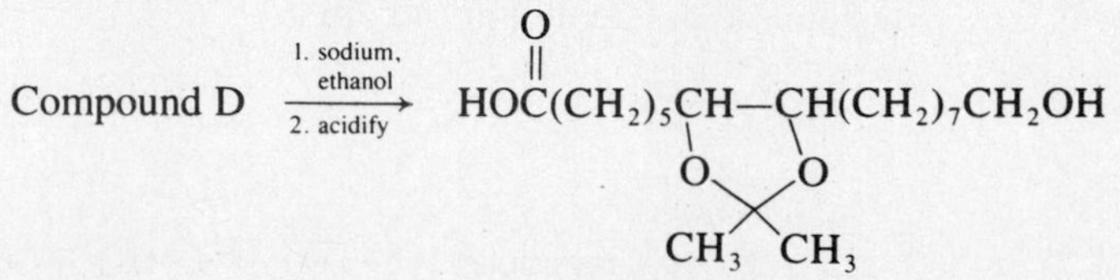

$$\text{Compound D} \xrightarrow[\text{2. acidify}]{\text{1. sodium, ethanol}} HOC(=O)(CH_2)_5CH{-}CH(CH_2)_7CH_2OH \;(\text{O–C(CH}_3)_2\text{–O acetal})$$

Compound E ($C_{19}H_{36}O_5$)

Acid hydrolysis in step 4 removes the acetal protecting group.

Step 4

$$\text{Compound E} \xrightarrow[\text{heat}]{H_2O,\ H^+} \underset{\text{Compound F } (C_{16}H_{32}O_5)}{HO\overset{\overset{\large O}{\|}}{C}(CH_2)_5\underset{\underset{HO}{|}}{C}H-\underset{\underset{OH}{|}}{C}H(CH_2)_7CH_2OH}$$

All three alcohol functions are converted to bromide by reaction with hydrogen bromide in step 5.

Step 5

$$\text{Compound F} \xrightarrow{HBr} \underset{\text{Compound G } (C_{16}H_{29}Br_3O_2)}{HO\overset{\overset{\large O}{\|}}{C}(CH_2)_5\underset{\underset{Br}{|}}{C}H-\underset{\underset{Br}{|}}{C}H(CH_2)_7CH_2Br}$$

Reaction with ethanol in the presence of an acid catalyst converts the carboxylic acid to its ethyl ester in step 6.

Step 6

$$\text{Compound G} \xrightarrow[H_2SO_4]{\text{ethanol}} \underset{\text{Compound H } (C_{18}H_{33}Br_3O_2)}{CH_3CH_2O\overset{\overset{\large O}{\|}}{C}(CH_2)_5\underset{\underset{Br}{|}}{C}H-\underset{\underset{Br}{|}}{C}H(CH_2)_7CH_2Br}$$

Zinc converts vicinal dibromides to alkenes. Of the three bromine substituents in compound H, two of them are vicinal. Step 7 is a dehalogenation reaction.

Step 7

$$\text{Compound H} \xrightarrow[\text{ethanol}]{Zn} \underset{\text{Compound I } (C_{18}H_{33}BrO_2)}{CH_3CH_2O\overset{\overset{\large O}{\|}}{C}(CH_2)_5CH{=}CH(CH_2)_7CH_2Br}$$

Step 8 is a nucleophilic substitution reaction of the S_N2 type. Acetate ion is the nucleophile and displaces bromide from the primary carbon.

Step 8

$$\text{Compound I} \xrightarrow[CH_3CO_2H]{NaO\overset{\overset{O}{\|}}{C}CH_3} \underset{\text{Compound J } (C_{20}H_{36}O_4)}{CH_3CH_2O\overset{\overset{\large O}{\|}}{C}(CH_2)_5CH{=}CH(CH_2)_7CH_2O\overset{\overset{\large O}{\|}}{C}CH_3}$$

Step 9 is ester saponification. It yields a 16-carbon chain having a carboxylic acid function at one end and an alcohol at the other.

Step 9

$$\text{Compound J} \xrightarrow[2.\ H^+]{1.\ KOH,\ \text{ethanol}} \underset{\text{Compound K } (C_{16}H_{30}O_3)}{HO\overset{\overset{\large O}{\|}}{C}(CH_2)_5CH{=}CH(CH_2)_7CH_2OH}$$

Compound K cyclizes to ambrettolide on heating.

21.32 (*a*) This step requires the oxidation of a primary alcohol to an aldehyde. Collins' reagent would be appropriate.

$$\underset{\text{Compound L}}{HOCH_2CH{=}CH(CH_2)_7CO_2CH_3} \xrightarrow[CH_2Cl_2]{(C_5H_5N)_2CrO_3} \underset{\text{Compound M}}{H\overset{O}{\overset{\|}{C}}CH{=}CH(CH_2)_7CO_2CH_3}$$

As reported in the literature, pyridinium dichromate in dichloromethane was used to give the desired aldehyde in 84 percent yield.

(*b*) Conversion of $-\overset{O}{\overset{\|}{C}}H$ to $-CH{=}CH_2$ is a typical case in which a Wittig reaction is appropriate.

$$\underset{\text{Compound M}}{H\overset{O}{\overset{\|}{C}}CH{=}CH(CH_2)_7CO_2CH_3} \xrightarrow{(C_6H_5)_3\overset{+}{P}-\overset{-}{C}H_2} \underset{\substack{\text{Compound N} \\ \text{(observed yield, 53\%)}}}{CH_2{=}CHCH{=}CH(CH_2)_7CO_2CH_3}$$

(*c*) Lithium aluminum hydride was used to reduce the ester to a primary alcohol in 81 percent yield.

$$\underset{\text{Compound N}}{CH_2{=}CHCH{=}CH(CH_2)_7CO_2CH_3} \xrightarrow[2.\ H_2O]{1.\ LiAlH_4} \underset{\text{Compound O}}{CH_2{=}CHCH{=}CH(CH_2)_7CH_2OH}$$

Alternatively, the ester could have been reduced with sodium in alcohol. Catalytic hydrogenation would not be suitable because of the presence of carbon-carbon double bonds in the starting material.

(*d*) The desired sex pheromone is the acetate ester of compound O. Compound O was treated with acetic anhydride to give the acetate ester in 99 percent yield.

$$\underset{\text{Compound O}}{CH_2{=}CHCH{=}CH(CH_2)_7CH_2OH} \xrightarrow[\text{pyridine}]{CH_3\overset{O}{\overset{\|}{C}}O\overset{O}{\overset{\|}{C}}CH_3} \underset{(E)\text{-9,11-Dodecadien-1-yl acetate}}{CH_2{=}CHCH{=}CH(CH_2)_7CH_2O\overset{O}{\overset{\|}{C}}CH_3}$$

Acetyl chloride could have been used in this step instead of acetic anhydride.

21.33 (*a*) Compound P has the molecular formula $C_7H_8O_3$, which corresponds to that of *cis*-1,3-cyclopentanedicarboxylic acid ($C_7H_{10}O_4$) minus one molecule of water. It is the cyclic anhydride of the diacid.

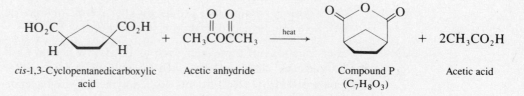

$$\underset{\textit{cis}\text{-1,3-Cyclopentanedicarboxylic acid}}{HO_2C\ (H)\text{-cyclopentane-}(H)\ CO_2H} + \underset{\text{Acetic anhydride}}{CH_3\overset{O}{\overset{\|}{C}}O\overset{O}{\overset{\|}{C}}CH_3} \xrightarrow{\text{heat}} \underset{\substack{\text{Compound P} \\ (C_7H_8O_3)}}{\text{bicyclic anhydride}} + \underset{\text{Acetic acid}}{2CH_3CO_2H}$$

(*b*) Of the two carbonyl groups in the starting material, the ketone carbonyl is more reactive than the ester. (The ester carbonyl is stabilized by electron release from oxygen.)

$$CH_3\overset{O}{\overset{\|}{C}}CH_2CH_2\overset{O}{\overset{\|}{C}}OCH_2CH_3 \xrightarrow{CH_3MgI} CH_3\underset{OMgI}{\underset{|}{\overset{CH_3}{\overset{|}{C}}}}CH_2CH_2\overset{O}{\overset{\|}{C}}OCH_2CH_3$$

Compound Q has the molecular formula $C_6H_{10}O_2$. The initial product forms a cyclic ester, with elimination of ethoxide ion.

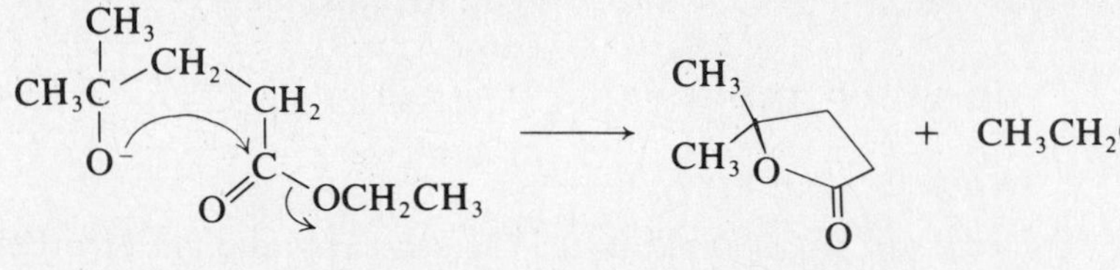

Compound Q

(*c*) Only carboxyl groups that are ortho to each other on a benzene ring are capable of forming a cyclic anhydride.

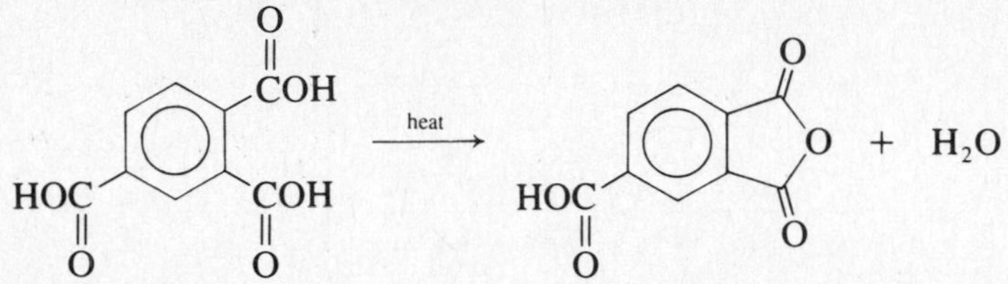

Compound R

21.34 Compound S is an ester but has within it an amine function. Acyl transfer from oxygen to nitrogen converts the ester to a more stable amide.

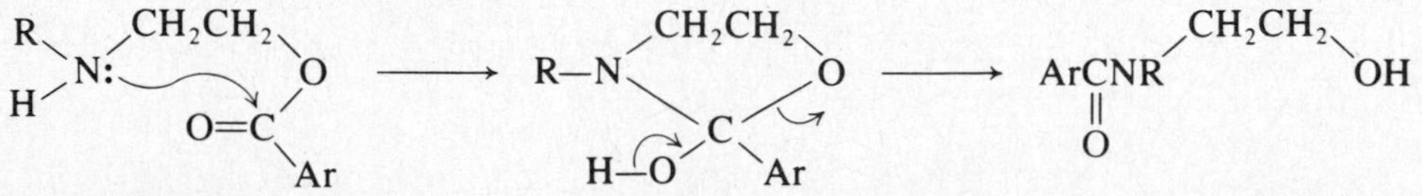

Compound S (Ar = *p*-nitrophenyl) Tetrahedral intermediate Compound T (Ar = *p*-nitrophenyl)

The tetrahedral intermediate is the key intermediate in the reaction.

21.35 (*a*) The rearrangement in this problem is an acyl transfer from nitrogen to oxygen.

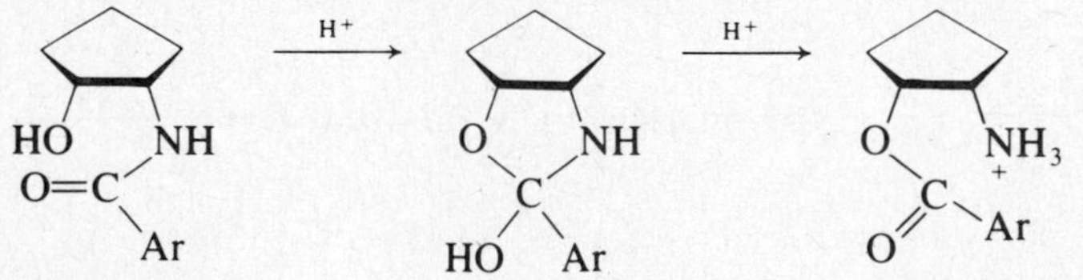

Compound U (Ar = *p*-nitrophenyl) Tetrahedral intermediate Compound V (Ar = *p*-nitrophenyl)

This rearrangement takes place in the indicated direction because it is carried out in acid solution. The amino group is protonated in acid and is no longer nucleophilic.

(*b*) The trans stereoisomer of compound U does not undergo rearrangement because when the oxygen and nitrogen atoms on the five-membered ring are trans, the necessary tetrahedral intermediate would be too strained.

21.36 The ester functions of a polymer such as poly(vinyl acetate) are just like ester functions of simple molecules; they can be cleaved by hydrolysis under either acidic or basic conditions. To prepare poly(vinyl alcohol), therefore, polymerize vinyl acetate to poly(vinyl acetate), then cleave the ester groups by hydrolysis.

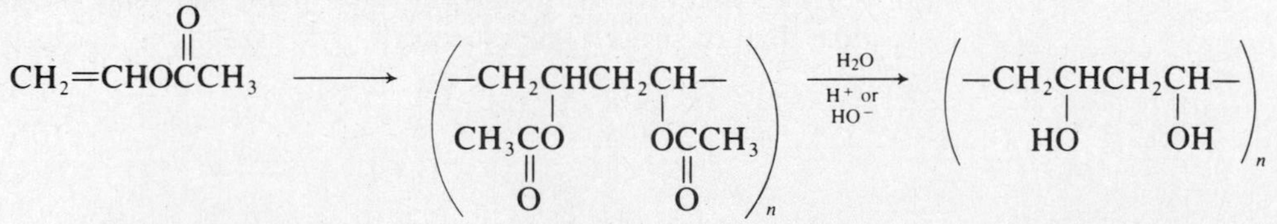

Vinyl acetate Poly(vinyl acetate) Poly(vinyl alcohol)

21.37 First, assume that the peak in the mass spectrum of highest m/z corresponds to the molecular ion. Thus methyl methacrylate is assumed to have a molecular weight of 100. The integrated areas in the ^1H nmr spectrum indicate that there are eight protons in the molecule, which suggests the molecular formula $C_5H_8O_2$. A methyl ester is likely on the basis of the three-proton singlet in the nmr spectrum at $\delta = 3.9$ ppm (OCH_3). Also, there is a prominent peak at m/z 69 in the mass spectrum; this peak corresponds to loss of OCH_3 from the molecular ion.

The ^1H nmr spectrum reveals an allylic methyl group at $\delta = 1.9$ ppm along with two nonequivalent vinyl protons at $\delta = 5.6$ and 6.1 ppm. Two candidate structures present themselves for consideration at this point. The second of these two structures is the correct one.

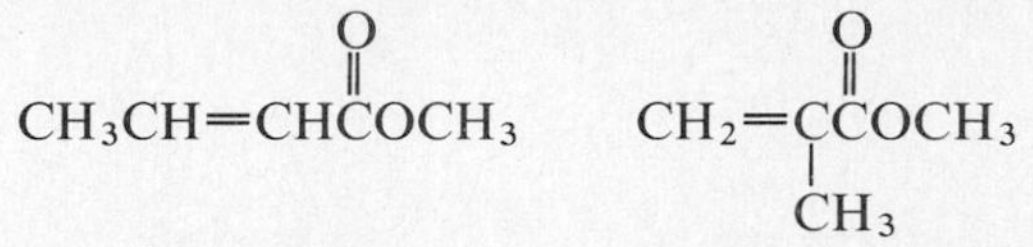

$$CH_3CH{=}CH\overset{\overset{O}{\parallel}}{C}OCH_3 \qquad CH_2{=}\underset{\underset{CH_3}{|}}{C}\overset{\overset{O}{\parallel}}{C}OCH_3$$

Methyl crotonate — Methyl methacrylate

The problem states that methyl methacrylate is made from the cyanohydrin of acetone; therefore it must have the carbon skeleton $C{-}\underset{\underset{C}{|}}{C}{-}C$.

21.38 Compound W ($C_4H_6O_2$) has a SODAR of 2. With two oxygen atoms and a peak in the infrared at 1760 cm^{-1}, it is likely that one of these elements of unsaturation is the carbon-oxygen double bond of an ester. The ^1H nmr spectrum contains a three-proton singlet at $\delta = 2.1$ ppm, which is consistent with a $CH_3\underset{\underset{O}{\parallel}}{C}$ unit. It is likely that compound W is an acetate ester.

The ^{13}C nmr spectrum reveals that the four carbon atoms of the molecule are contained in one each of the structures CH_3, CH_2, and CH, along with the carbonyl carbon. In addition to the two carbons of the acetate group, the remaining two carbons are the CH_2 and CH carbons of a vinyl group, $CH{=}CH_2$. Compound W is vinyl acetate.

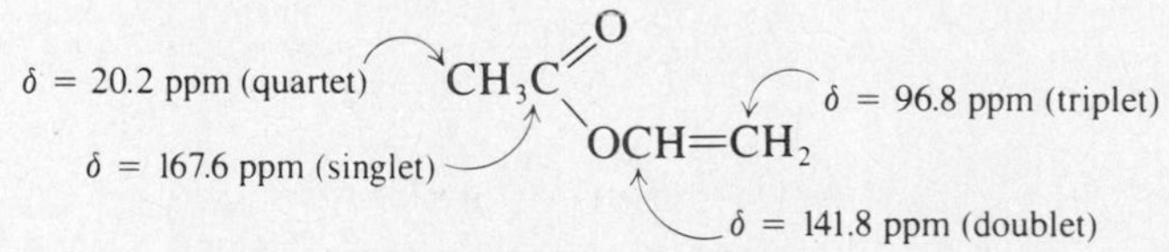

$$CH_3C({=}O)OCH{=}CH_2$$

$\delta = 20.2$ ppm (quartet) (CH_3); $\delta = 167.6$ ppm (singlet) (C=O); $\delta = 96.8$ ppm (triplet) ($=CH_2$); $\delta = 141.8$ ppm (doublet) (OCH=)

Each vinyl proton is coupled to two other vinyl protons; each appears as a doublet of doublets in the ^1H nmr spectrum.

21.39 Compound X contains nitrogen and exhibits a prominent peak in the infrared at 2270 cm^{-1}; it is likely to be a nitrile. Its molecular weight of 69 is consistent with the molecular formula C_4H_7N, and its ^1H nmr spectrum shows the characteristic doublet-heptet pattern of an isopropyl group. Compound X is 2-methylpropanenitrile.

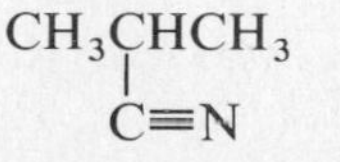

$$CH_3\underset{\underset{C\equiv N}{|}}{C}HCH_3$$

2-Methylpropanenitrile (compound X)

21.40 Compound Y has the characteristic triplet-quartet pattern of an ethyl group in its ^1H nmr spectrum. Since these signals correspond to 10 protons, there must be two equivalent ethyl groups in the molecule. The methylene quartet appears at relatively low field ($\delta = 4.1$ ppm), which is consistent with ethyl groups bonded to oxygen, as in $-OCH_2CH_3$. There is a peak at 1730 cm^{-1}, suggesting that these ethoxy groups reside in ester functions. The molecular formula $C_8H_{14}O_4$ reveals that if two ester groups are present, there can be no rings or double bonds. The remaining four hydrogens are equivalent in the ^1H nmr spectrum, so two equivalent CH_2 groups are present. Compound Y is the diethyl ester of succinic acid.

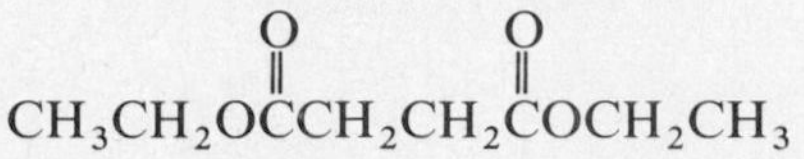

$$CH_3CH_2O\overset{\overset{O}{\parallel}}{C}CH_2CH_2\overset{\overset{O}{\parallel}}{C}OCH_2CH_3$$

Diethyl succinate (compound Y)

21.41 Compound Z contains nitrogen and has a peak in the infrared at 1680 cm^{-1}, suggesting an amide as a possibility. Reinforcing this notion is the appearance of a broad peak, equivalent to two protons, in the ^1H nmr spectrum at $\delta = 6$ to 7 ppm. This is consistent with the $-NH_2$ group of an amide. Since there are only three carbons, the carbon chain cannot be branched and the only point to be determined is the location of the chlorine. The chlorine must be at C-2 because the ^1H nmr shows a three-proton methyl doublet and a one-proton methine quartet. Compound Z is 2-chloropropanamide.

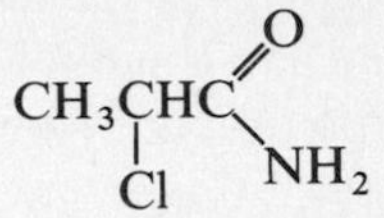

$$CH_3CH(Cl)C(=O)NH_2$$

2-Chloropropanamide (compound Z)

Compound Z is prepared from propanoic acid as shown.

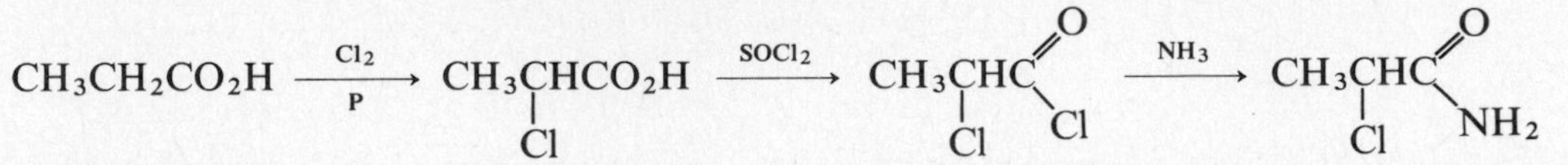

$$\underset{\text{Propanoic acid}}{CH_3CH_2CO_2H} \xrightarrow[P]{Cl_2} \underset{\text{2-Chloropropanoic acid}}{CH_3CH(Cl)CO_2H} \xrightarrow{SOCl_2} \underset{\text{2-Chloropropanoyl chloride}}{CH_3CH(Cl)C(=O)Cl} \xrightarrow{NH_3} \underset{\text{2-Chloropropanamide}}{CH_3CH(Cl)C(=O)NH_2}$$

Propionic acid cannot be converted to its amide and then halogenated at the α carbon because it is the free acid that must be halogenated by the Hell-Volhard-Zelinsky reaction.

SELF-TEST

PART A

A-1. Give a correct name for each of the following acid derivatives.

(*a*) $CH_3CH_2CH_2OC(=O)CH_2CH_2CH_3$

(*b*) $C_6H_5C(=O)NHCH_3$

(*c*) $(CH_3)_2CHCH_2CH_2C(=O)Cl$

A-2. Provide the correct structure of:
(*a*) Benzoic anhydride
(*b*) *N*-(1-Methylpropyl)acetamide
(*c*) Phenyl benzoate

A-3. What is the formula of the reagent(s) needed to carry out each of the following conversions?

(*a*) $C_6H_5CH_2CO_2H \xrightarrow{?} C_6H_5CH_2C(=O)Cl$

(*b*) $(CH_3)_3CC(=O)NH_2 \xrightarrow{?} (CH_3)_3CNH_2$

A-4. Write the structure of the product of each of the following reactions:

(*a*) Cyclohexyl acetate $\xrightarrow[\text{2. } H^+]{\text{1. NaOH, } H_2O}$? (two products)

(*b*) Cyclopentanol + benzoyl chloride $\xrightarrow{\text{pyridine}}$?

(*c*) Phthalic anhydride + $CH_3CH_2OH \xrightarrow{H^+ \text{(cat.)}}$?

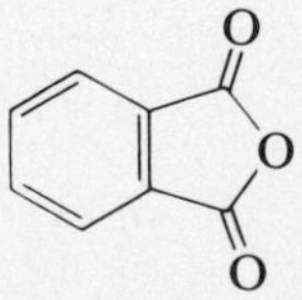

(*d*) Ethyl propanoate + dimethylamine ⟶ ? (two products)

(*e*) $CH_3-C_6H_4-\overset{O}{\overset{\|}{C}}NHCH_3 \xrightarrow[\text{heat}]{H_2O,\ H_2SO_4}$? (two products)

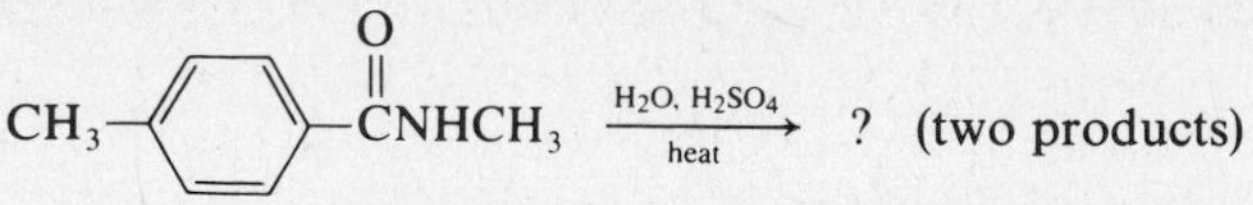

A-5. The following reaction proceeds when the reactant is allowed to stand in pentane. Write the structure of the key intermediate in this process.

$$C_6H_5\overset{O}{\overset{\|}{C}}OCH_2CH_2NHCH_3 \longrightarrow CH_3\underset{C_6H_5C=O}{\underset{|}{N}}CH_2CH_2OH$$

A-6. Give the correct structures, clearly showing stereochemistry, of each compound A through D in the following sequence of reactions:

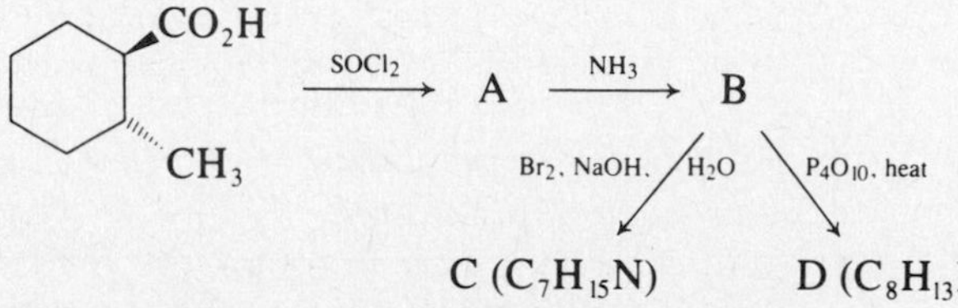

(cyclohexane bearing CO_2H (wedge) and CH_3 (hashed) on adjacent carbons) $\xrightarrow{SOCl_2}$ A $\xrightarrow{NH_3}$ B

B $\xrightarrow{Br_2,\ NaOH,\ H_2O}$ C ($C_7H_{15}N$)

B $\xrightarrow{P_4O_{10},\ heat}$ D ($C_8H_{13}N$)

A-7. Write the structure of the neutral form of the tetrahedral intermediate in the:

(*a*) Acid-catalyzed hydrolysis of methyl acetate
(*b*) Reaction of ammonia with acetic anhydride

A-8. Write the steps necessary to prepare $CH_3-C_6H_4-NH_2$ from $CH_3-C_6H_4-Br$.

PART B

B-1. The most favorable mode of decomposition of the intermediate species shown will yield:

$$C_6H_5-\underset{Cl}{\overset{OH}{C}}-OH$$

(*a*) Benzoic acid and HCl
(*b*) Benzoyl chloride and H_2O
(*c*) Both (*a*) and (*b*) with equal likelihood
(*d*) Neither (*a*) nor (*b*)

B-2. Hydrolysis of propyl butyrate yields:

(*a*) Propanoic acid + butanol
(*b*) Butanoic acid + butanol
(*c*) Butanoic acid + propanol
(*d*) Butanoic acid + propylamine

B-3. Amides are usually ______ alkylamines.

(*a*) more basic than (*b*) less basic than
(*c*) of the same basicity as

B-4. Rank the following in order of increasing reactivity (least → most) toward acid hydrolysis:

$$\underset{A}{CH_3\overset{O}{\overset{\|}{C}}OCH_2CH_3} \qquad \underset{B}{CH_3\overset{O}{\overset{\|}{C}}Cl} \qquad \underset{C}{CH_3\overset{O}{\overset{\|}{C}}NHCH_3}$$

(*a*) A < B < C (*b*) C < A < B
(*c*) A < C < B (*d*) B < A < C

B-5. The structure of *N*-propylacetamide is:

(*a*) $CH_3CH_2\overset{O}{\overset{\|}{C}}NHCH_3$ (*b*) $CH_3\overset{O}{\overset{\|}{C}}N(CH_2CH_2CH_3)_2$

(*c*) $CH_3\overset{O}{\overset{\|}{C}}NHCH_2CH_2CH_3$ (*d*) $CH_3CH{=}NCH_2CH_2CH_3$

B-6. Choose the response which matches the correct functional group classification with the following:

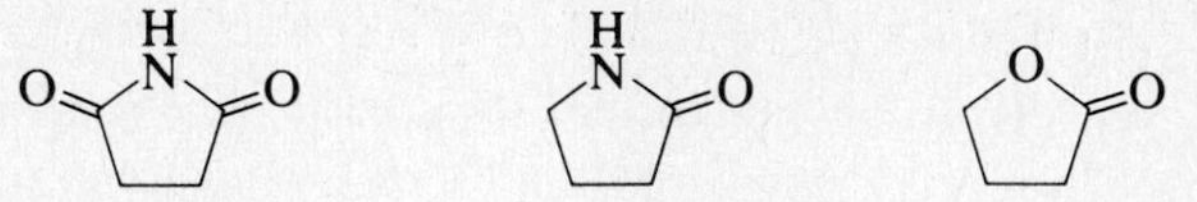

(*a*)	Anhydride	Lactam	Lactone
(*b*)	Lactam	Imide	Lactone
(*c*)	Imide	Lactone	Anhydride
(*d*)	Imide	Lactam	Lactone

ESTER ENOLATES

IMPORTANT REACTIONS

Base-Promoted Ester Condensation. The Claisen Condensation (Sec. 22.1)

General:

$$2\,RCH_2\overset{O}{\overset{\|}{C}}OR' \xrightarrow[2.\ H_3O^+]{1.\ NaOR'} RCH_2\overset{O}{\overset{\|}{C}}\underset{R}{\underset{|}{C}H}\overset{O}{\overset{\|}{C}}OR' + R'OH$$

Example:

$$2CH_3CH_2\overset{O}{\overset{\|}{C}}OCH_2CH_3 \xrightarrow[2.\ H_3O^+]{1.\ NaOCH_2CH_3} CH_3CH_2\overset{O}{\overset{\|}{C}}\underset{CH_3}{\underset{|}{C}H}\overset{O}{\overset{\|}{C}}OCH_2CH_3 + CH_3CH_2OH$$

Mechanism:

$$RCH_2\overset{O}{\overset{\|}{C}}OR' + RO^- \rightleftharpoons R\ddot{C}H\overset{O}{\overset{\|}{C}}OR' \longleftrightarrow RCH{=}\overset{O^-}{\overset{|}{C}}OR'$$

$$R\ddot{C}H\overset{O}{\overset{\|}{C}}OR' + RCH_2\overset{O}{\overset{\|}{C}}OR' \rightleftharpoons RCH_2\overset{^-O}{\overset{|}{\underset{OR'}{\underset{|}{C}}}}-\overset{R}{\overset{|}{C}H}-\overset{O}{\overset{\|}{C}}OR' \rightleftharpoons$$

$$RCH_2\overset{O}{\overset{\|}{C}}\underset{R}{\underset{|}{C}H}\overset{O}{\overset{\|}{C}}OR' + R'O^-$$

$$RCH_2\overset{O}{\overset{\|}{C}}\underset{R}{\underset{|}{C}H}\overset{O}{\overset{\|}{C}}OR' + R'O^- \rightleftharpoons RCH_2\overset{O}{\overset{\|}{C}}\underset{R}{\underset{|}{\ddot{C}^-}}\overset{O}{\overset{\|}{C}}OR' + R'OH$$

Intramolecular Claisen Condensation. The Dieckmann Condensation (Sec. 22.2)

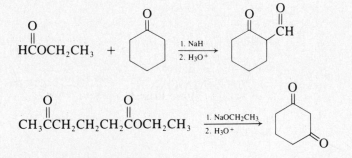

$$R'O\overset{O}{\overset{\|}{C}}(CH_2)_n\overset{O}{\overset{\|}{C}}OR' \xrightarrow[2.\ H_3O^+]{1.\ NaOR'} \text{cyclic } \overset{O}{\overset{\|}{C}}\text{—}CH\text{—}\overset{O}{\overset{\|}{C}}OR' \text{ (ring closed by } (CH_2)_{n-1})$$

Example:

$$CH_3CH_2O\overset{O}{\overset{\|}{C}}(CH_2)_5\overset{O}{\overset{\|}{C}}OCH_2CH_3 \xrightarrow[2.\ H_3O^+]{1.\ NaOCH_2CH_3} \text{2-oxocyclohexane with } \overset{O}{\overset{\|}{C}}OCH_2CH_3$$

Mixed Claisen Condensations (Sec. 22.3)

General:

$$R\overset{O}{\overset{\|}{C}}OCH_2CH_3 + R'CH_2\overset{O}{\overset{\|}{C}}OCH_2CH_3 \xrightarrow[2.\ H_3O^+]{1.\ NaOCH_2CH_3} R\overset{O}{\overset{\|}{C}}\underset{R'}{\underset{|}{C}H}\overset{O}{\overset{\|}{C}}OCH_2CH_3$$

Example:

$$C_6H_5\overset{O}{\overset{\|}{C}}OCH_2CH_3 + CH_3\overset{O}{\overset{\|}{C}}OCH_2CH_3 \xrightarrow[2.\ H_3O^+]{1.\ NaOCH_2CH_3} C_6H_5\overset{O}{\overset{\|}{C}}CH_2\overset{O}{\overset{\|}{C}}OCH_2CH_3$$

Acylation of Ketones (Sec. 22.4)

Examples:

$$H\overset{O}{\overset{\|}{C}}OCH_2CH_3 + \text{cyclohexanone} \xrightarrow[2.\ H_3O^+]{1.\ NaH} \text{2-oxocyclohexane with } \overset{O}{\overset{\|}{C}}H$$

$$CH_3\overset{O}{\overset{\|}{C}}CH_2CH_2CH_2\overset{O}{\overset{\|}{C}}OCH_2CH_3 \xrightarrow[2.\ H_3O^+]{1.\ NaOCH_2CH_3} \text{cyclohexane-1,3-dione}$$

Synthesis of Ketones (Sec. 22.5)

General:

$$R\overset{O}{\overset{\|}{C}}\underset{R'}{\underset{|}{C}H}\overset{O}{\overset{\|}{C}}OCH_2CH_3 \xrightarrow[2.\ H_3O^+]{1.\ HO^-,\ H_2O} R\overset{O}{\overset{\|}{C}}\underset{R'}{\underset{|}{C}H}\overset{O}{\overset{\|}{C}}OH \xrightarrow{\text{heat}} R\overset{O}{\overset{\|}{C}}CH_2R' + CO_2$$

Example:

$$CH_3CH_2\overset{O}{\overset{\|}{C}}\underset{CH_3}{\underset{|}{C}H}\overset{O}{\overset{\|}{C}}OCH_2CH_3 \xrightarrow[2.\ H_3O^+\quad 3.\ \text{heat}]{1.\ HO^-,\ H_2O} CH_3CH_2\overset{O}{\overset{\|}{C}}CH_2CH_3$$

Acetoacetic Ester Synthesis (Sec. 22.6)

General:

$$CH_3\overset{O}{\overset{\|}{C}}CH_2\overset{O}{\overset{\|}{C}}OCH_2CH_3 \xrightarrow[2.\ R-X]{1.\ NaOCH_2CH_3} CH_3\overset{O}{\overset{\|}{C}}\underset{R}{\underset{|}{C}H}\overset{O}{\overset{\|}{C}}OCH_2CH_3 \xrightarrow[2.\ H_3O^+\ 3.\ heat]{1.\ HO^-,\ H_2O} CH_3\overset{O}{\overset{\|}{C}}CH_2R$$

Example:

$$CH_3\overset{O}{\overset{\|}{C}}CH_2\overset{O}{\overset{\|}{C}}OCH_2CH_3 \xrightarrow[2.\ (CH_3)_2CHCH_2Br]{1.\ NaOCH_2CH_3} CH_3\overset{O}{\overset{\|}{C}}\underset{CH_2CH(CH_3)_2}{\underset{|}{C}H}\overset{O}{\overset{\|}{C}}OCH_2CH_3 \xrightarrow{1.\ HO^-,\ H_2O\ 2.\ H_3O^+\ 3.\ heat}$$

$$CH_3\overset{O}{\overset{\|}{C}}CH_2CH_2CH(CH_3)_2$$

Malonic Ester Synthesis (Sec. 22.7)

General:

$$R{-}X + CH_2(CO_2CH_2CH_3)_2 \xrightarrow[CH_3CH_2OH]{NaOCH_2CH_3} RCH(CO_2CH_2CH_3)_2 \xrightarrow[3.\ heat]{1.\ HO^-,\ H_2O\ 2.\ H_3O^+} RCH_2CO_2H$$

Example:

$$CH_2{=}CHCH_2Br + CH_2(CO_2CH_2CH_3)_2 \xrightarrow[2.\ CH_3CH_2OH]{1.\ NaOCH_2CH_3}$$

$$CH_2{=}CHCH_2CH(CO_2CH_2CH_3)_2 \xrightarrow[3.\ heat]{1.\ HO^-,\ H_2O\ 2.\ H_3O^+} CH_2{=}CHCH_2CH_2CO_2H$$

Michael Reaction (Sec. 22.9)

Example:

$$CH_2(CO_2CH_2CH_3)_2 + CH_2{=}CH\overset{O}{\overset{\|}{C}}CH_3 \xrightarrow[CH_3CH_2OH]{KOH} CH_3\overset{O}{\overset{\|}{C}}CH_2CH_2CH(CO_2CH_2CH_3)_2$$

1,4-Addition product

Knoevenagel Condensation (Sec. 22.10)

General:

$$R\overset{O}{\overset{\|}{C}}R' + CH_2(CO_2CH_2CH_3)_2 \xrightarrow{piperidine} \begin{matrix}R\\ \ \ \diagdown \\ \ \ \ \ C{=}C(CO_2CH_2CH_3)_2 + H_2O \\ \ \ \diagup \\ R'\end{matrix}$$

Example:

$$CH_3CH_2\overset{O}{\overset{\|}{C}}H + CH_2(CO_2CH_2CH_3)_2 \xrightarrow{piperidine} CH_3CH_2CH{=}C(CO_2CH_2CH_3)_2$$

Reformatsky Reaction (Sec. 22.11)

General:

$$R\overset{O}{\overset{\|}{C}}R' + R''\underset{Br}{\underset{|}{C}H}\overset{O}{\overset{\|}{C}}OCH_2CH_3 \xrightarrow[2.\ H_3O^+]{1.\ Zn} R\underset{R'}{\underset{|}{\overset{OH}{\overset{|}{C}}}}{-}\underset{R''}{\underset{|}{C}H}\overset{O}{\overset{\|}{C}}OCH_2CH_3$$

Example:

$$\mathrm{CH_3CH_2\overset{O}{\overset{\|}{C}}H} + \mathrm{BrCH_2\overset{O}{\overset{\|}{C}}OCH_2CH_3} \xrightarrow[\text{2. } H_3O^+]{\text{1. Zn}} \mathrm{CH_3CH_2\overset{OH}{\overset{|}{C}}HCH_2\overset{O}{\overset{\|}{C}}OCH_2CH_3}$$

SOLUTIONS TO TEXT PROBLEMS

22.1 Ethyl benzoate cannot undergo the Claisen condensation because it has no protons on its α carbon atom and so cannot form an enolate. Ethyl pentanoate and ethyl phenylacetate can undergo the Claisen condensation.

$$\mathrm{2CH_3CH_2CH_2CH_2\overset{O}{\overset{\|}{C}}OCH_2CH_3} \xrightarrow[\text{2. } H^+]{\text{1. } NaOCH_2CH_3} \mathrm{CH_3CH_2CH_2CH_2\overset{O}{\overset{\|}{C}}\underset{CH_2CH_2CH_3}{\underset{|}{C}}H\overset{O}{\overset{\|}{C}}OCH_2CH_3}$$

Ethyl pentanoate — Ethyl 3-oxo-2-propylheptanoate

$$\mathrm{2C_6H_5CH_2\overset{O}{\overset{\|}{C}}OCH_2CH_3} \xrightarrow[\text{2. } H^+]{\text{1. } NaOCH_2CH_3} \mathrm{C_6H_5CH_2\overset{O}{\overset{\|}{C}}\underset{C_6H_5}{\underset{|}{C}}H\overset{O}{\overset{\|}{C}}OCH_2CH_3}$$

Ethyl phenylacetate — Ethyl 3-oxo-2,4-diphenylbutanoate

22.2 (*b*) The enolate formed by proton abstraction from the α carbon atom of diethyl 4-methylheptanedioate cyclizes to form a six-membered β-keto ester.

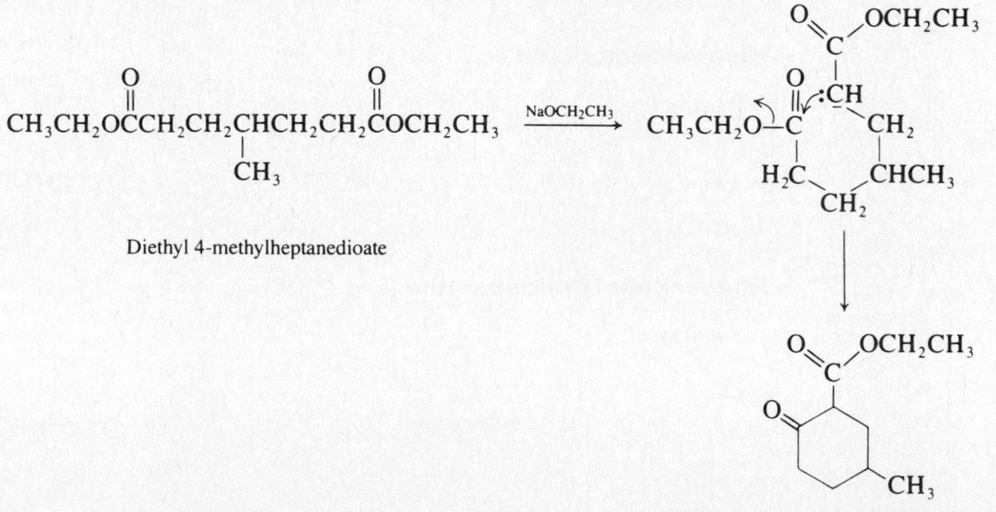

$$\mathrm{CH_3CH_2O\overset{O}{\overset{\|}{C}}CH_2CH_2\underset{CH_3}{\underset{|}{C}}HCH_2CH_2\overset{O}{\overset{\|}{C}}OCH_2CH_3} \xrightarrow{NaOCH_2CH_3}$$

Diethyl 4-methylheptanedioate

↓

Ethyl (5-methyl-2-oxocyclohexane)carboxylate

(*c*) The two α carbons of this diester are not equivalent. Cyclization by attack of the enolate at C-2 gives:

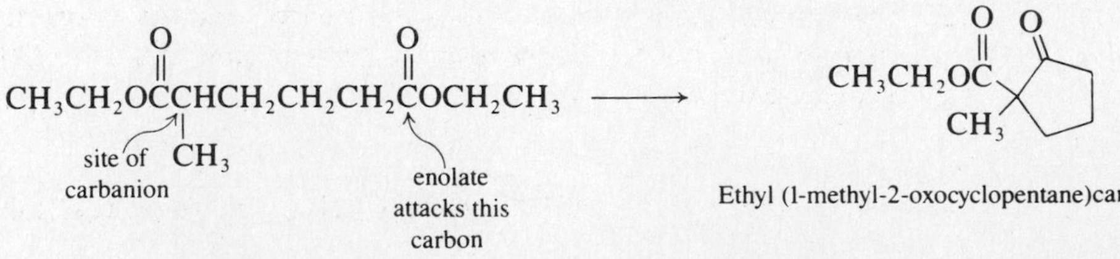

$$\mathrm{CH_3CH_2O\overset{O}{\overset{\|}{C}}\underset{CH_3}{\underset{|}{C}}HCH_2CH_2CH_2\overset{O}{\overset{\|}{C}}OCH_2CH_3} \longrightarrow$$

site of carbanion; enolate attacks this carbon

(product: cyclopentanone with CH_3 and $CH_3CH_2O\overset{O}{\overset{\|}{C}}$ on C-1)

Ethyl (1-methyl-2-oxocyclopentane)carboxylate

This β-keto ester cannot form a stable enolate by deprotonation. It is present in only small amounts at equilibrium. The major product is formed by way of the other enolate.

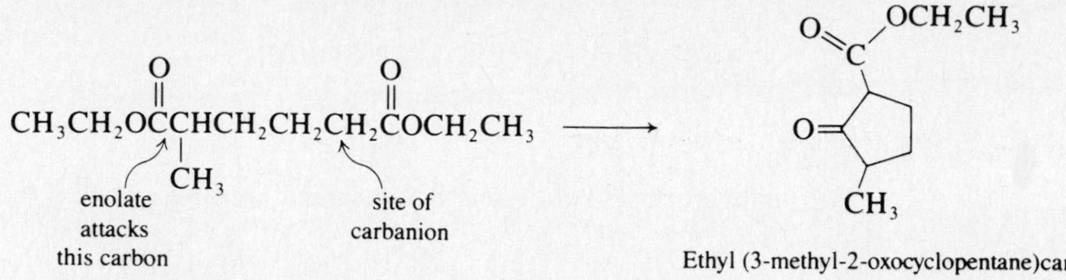

$$CH_3CH_2O\overset{O}{\overset{\|}{C}}\underset{CH_3}{\underset{|}{C}}HCH_2CH_2CH_2\overset{O}{\overset{\|}{C}}OCH_2CH_3 \longrightarrow$$

enolate attacks this carbon; site of carbanion

Ethyl (3-methyl-2-oxocyclopentane)carboxylate

This β-keto ester is converted to a stable enolate on deprotonation, causing the equilibrium to shift in its favor.

22.3 (*b*) Both carbonyl groups of diethyl oxalate are equivalent. The enolate of ethyl phenylacetate attacks one of them.

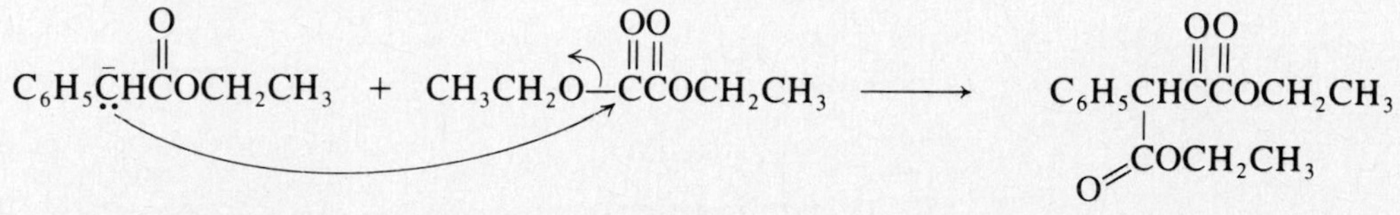

$$C_6H_5\ddot{\bar{C}}H\overset{O}{\overset{\|}{C}}OCH_2CH_3 + CH_3CH_2O-\overset{O}{\overset{\|}{C}}\overset{O}{\overset{\|}{C}}OCH_2CH_3 \longrightarrow C_6H_5\underset{\underset{O}{\|}{COCH_2CH_3}}{\underset{|}{C}}H\overset{O}{\overset{\|}{C}}\overset{O}{\overset{\|}{C}}OCH_2CH_3$$

Diethyl 2-oxo-3-phenylbutanedioate

(*c*) The enolate of ethyl phenylacetate attacks the carbonyl group of ethyl formate.

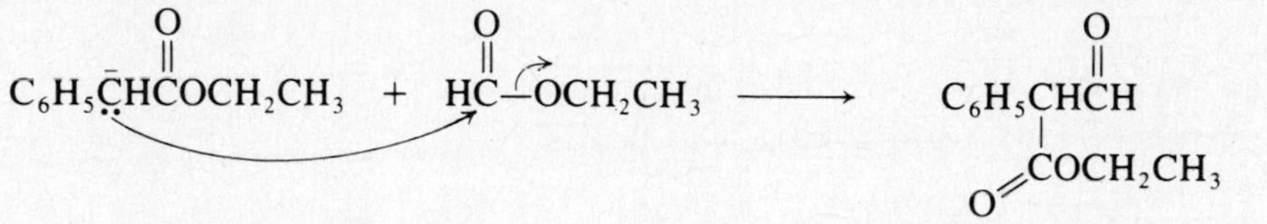

$$C_6H_5\ddot{\bar{C}}H\overset{O}{\overset{\|}{C}}OCH_2CH_3 + H\overset{O}{\overset{\|}{C}}-OCH_2CH_3 \longrightarrow C_6H_5\underset{\underset{O}{\|}{COCH_2CH_3}}{\underset{|}{C}}H\overset{O}{\overset{\|}{C}}H$$

Ethyl 3-oxo-2-phenylpropanoate

22.4 In order for a five-membered ring to be formed, C-5 must be the carbanionic site that attacks the ester carbonyl.

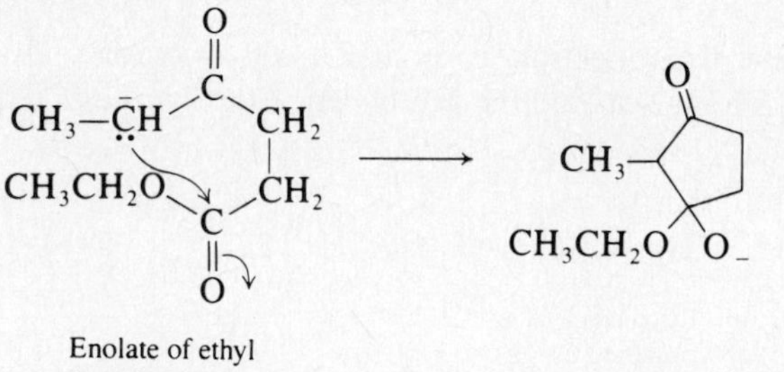

CH₃—C̈H, C(=O), CH₂, CH₂, CH₃CH₂O—C(=O) $\longrightarrow$ CH₃, CH₃CH₂O O⁻

Enolate of ethyl 4-oxohexanoate

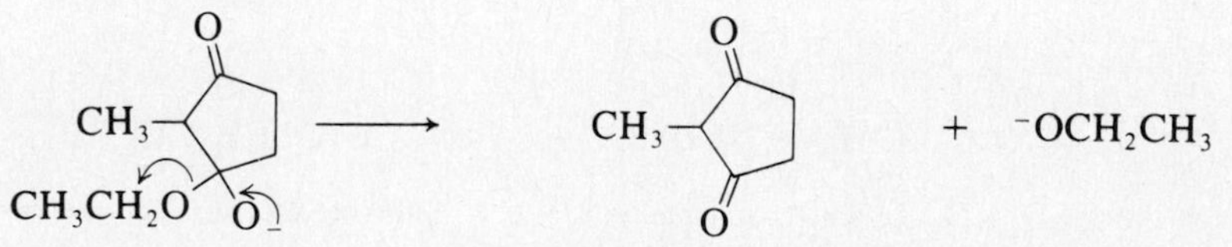

CH₃, CH₃CH₂O O⁻ $\longrightarrow$ CH₃ + $^-OCH_2CH_3$

2-Methyl-1,3-cyclopentanedione

22.5 The desired ketone, cyclopentanone, is derived from the corresponding β-keto ester. This key intermediate is obtained from a Dieckmann condensation of the starting material, diethyl hexanedioate.

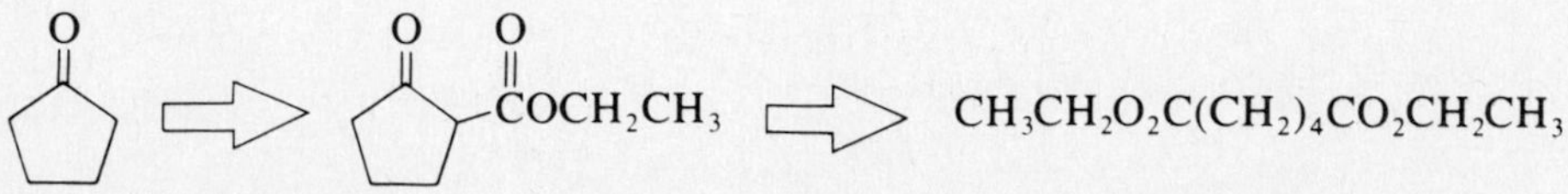

$$\Longrightarrow \overset{O}{\overset{\|}{C}}OCH_2CH_3 \Longrightarrow CH_3CH_2O_2C(CH_2)_4CO_2CH_2CH_3$$

First treat the diester with sodium ethoxide to effect the Dieckmann condensation.

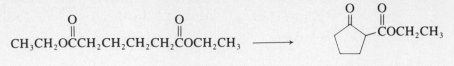

$$CH_3CH_2O\overset{\displaystyle O}{\overset{\|}{C}}CH_2CH_2CH_2CH_2\overset{\displaystyle O}{\overset{\|}{C}}OCH_2CH_3 \longrightarrow \text{(2-oxocyclopentyl)}\overset{\displaystyle O}{\overset{\|}{C}}OCH_2CH_3$$

Ethyl (2-oxocyclopentane)carboxylate

Next convert the β-keto ester to the desired product by saponification and decarboxylation.

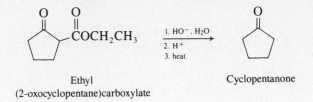

$$\text{Ethyl (2-oxocyclopentane)carboxylate} \xrightarrow[\text{2. } H^+ \quad \text{3. heat}]{\text{1. } HO^-,\ H_2O} \text{Cyclopentanone}$$

Ethyl (2-oxocyclopentane)carboxylate — Cyclopentanone

22.6 (*b*) Write a structural formula for the desired product, then disconnect a bond to the α carbon atom.

$$C_6H_5-CH_2 \vdots CH_2\overset{\displaystyle O}{\overset{\|}{C}}CH_3 \Longrightarrow C_6H_5-CH_2X \quad :\bar{C}H_2\overset{\displaystyle O}{\overset{\|}{C}}CH_3$$

disconnect here — Required alkyl halide — Derived from ethyl acetoacetate

Therefore

$$C_6H_5-CH_2Br + CH_3\overset{\displaystyle O}{\overset{\|}{C}}CH_2\overset{\displaystyle O}{\overset{\|}{C}}OCH_2CH_3 \xrightarrow[\text{2. } HO^-,\ H_2O \quad \text{3. } H^+ \quad \text{4. heat}]{\text{1. } NaOCH_2CH_3} C_6H_5-CH_2CH_2\overset{\displaystyle O}{\overset{\|}{C}}CH_3$$

Benzyl bromide — Ethyl acetoacetate — 4-Phenyl-2-butanone

(*c*) The disconnection approach to retrosynthetic analysis reveals that the preparation of 5-hexen-2-one by the acetoacetic ester synthesis requires an allylic halide.

$$CH_2{=}CHCH_2 \vdots CH_2\overset{\displaystyle O}{\overset{\|}{C}}CH_3 \Longrightarrow CH_2{=}CHCH_2X + :\bar{C}H_2\overset{\displaystyle O}{\overset{\|}{C}}CH_3$$

disconnect here — Required alkyl halide — Derived from ethyl acetoacetate

$$CH_2{=}CHCH_2Br + CH_3\overset{\displaystyle O}{\overset{\|}{C}}CH_2\overset{\displaystyle O}{\overset{\|}{C}}OCH_2CH_3 \xrightarrow[\text{2. } HO^-,\ H_2O \quad \text{3. } H^+ \quad \text{4. heat}]{\text{1. } NaOCH_2CH_3} CH_2{=}CHCH_2CH_2\overset{\displaystyle O}{\overset{\|}{C}}CH_3$$

Allyl bromide — Ethyl acetoacetate — 5-Hexen-2-one

22.7 (*b*) Nonanoic acid has a $CH_3(CH_2)_5CH_2-$ unit attached to the $CH_2\overset{O}{\overset{\|}{C}}OH$ synthon.

$$CH_3(CH_2)_5CH_2 \vdots CH_2\overset{\displaystyle O}{\overset{\|}{C}}OH \Longrightarrow CH_3(CH_2)_5CH_2X + :\bar{C}H_2\overset{\displaystyle O}{\overset{\|}{C}}OH$$

disconnect here — Required alkyl halide — Derived from diethyl malonate

Therefore the anion of diethyl malonate is alkylated with a 1-haloheptane.

$$\underset{\text{1-Bromoheptane}}{CH_3(CH_2)_5CH_2Br} + \underset{\text{Diethyl malonate}}{CH_2(COOCH_2CH_3)_2} \xrightarrow[\text{ethanol}]{NaOCH_2CH_3} \underset{\text{Diethyl heptylmalonate}}{CH_3(CH_2)_5CH_2CH(COOCH_2CH_3)_2}$$

$$\xrightarrow{\text{1. } HO^-, H_2O;\ \text{2. } H^+;\ \text{3. heat}} \underset{\text{Nonanoic acid}}{CH_3(CH_2)_5CH_2CH_2CO_2H}$$

(*c*) Disconnection of the product adjacent to the α carbon reveals the alkyl halide needed to react with the enolate derived from diethyl malonate.

$$CH_3CH_2\underset{CH_3}{\underset{|}{C}}HCH_2 \vdots CH_2\overset{O}{\overset{\|}{C}}OH \Longrightarrow \underset{\text{Required alkyl halide}}{CH_3CH_2\underset{CH_3}{\underset{|}{C}}HCH_2X} + \underset{\text{Derived from diethyl malonate}}{:\bar{C}H_2\overset{O}{\overset{\|}{C}}OH}$$

The necessary alkyl halide in this synthesis is 1-bromo-2-methylbutane.

$$\underset{\text{1-Bromo-2-methylbutane}}{CH_3CH_2\underset{CH_3}{\underset{|}{C}}HCH_2Br} + \underset{\text{Diethyl malonate}}{CH_2(COOCH_2CH_3)_2} \xrightarrow[\text{ethanol}]{NaOCH_2CH_3} \underset{\text{Diethyl 2-(2-methylbutyl)malonate}}{CH_3CH_2\underset{CH_3}{\underset{|}{C}}HCH_2CH(COOCH_2CH_3)_2}$$

$$\xrightarrow{\text{1. } HO^-, H_2O;\ \text{2. } H^+;\ \text{3. heat}} \underset{\text{4-Methylhexanoic acid}}{CH_3CH_2\underset{CH_3}{\underset{|}{C}}HCH_2CH_2\overset{O}{\overset{\|}{C}}OH}$$

(*d*) Once again disconnection reveals the necessary halide which is treated with diethyl malonate.

$$C_6H_5CH_2 \vdots CH_2\overset{O}{\overset{\|}{C}}OH \Longrightarrow \underset{\text{Required halide}}{C_6H_5CH_2X} + \underset{\text{Derived from diethyl malonate}}{:\bar{C}H_2\overset{O}{\overset{\|}{C}}OH}$$

Alkylation of diethyl malonate with benzyl bromide is the first step in the preparation of 3-phenylpropanoic acid.

$$\underset{\text{Benzyl bromide}}{C_6H_5CH_2Br} + \underset{\text{Diethyl malonate}}{CH_2(COOCH_2CH_3)_2} \xrightarrow[\text{ethanol}]{NaOCH_2CH_3} \underset{\text{Diethyl benzylmalonate}}{C_6H_5CH_2CH(COOCH_2CH_3)_2}$$

$$\xrightarrow{\text{1. } HO^-, H_2O;\ \text{2. } H^+;\ \text{3. heat}} \underset{\text{3-Phenylpropanoic acid}}{C_6H_5CH_2CH_2\overset{O}{\overset{\|}{C}}OH}$$

22.8 Retrosynthetic analysis of the formation 3-methyl-2-butanone is carried out in the same way as for other ketones.

$$\underset{\substack{\text{3-Methyl-2-butanone}\\\text{(two disconnections}\\\text{as shown)}}}{CH_3\overset{O}{\overset{\|}{C}}\underset{\underset{CH_3}{|}}{CH}-CH_3} \Longrightarrow \underset{\substack{\text{Derived from}\\\text{ethyl acetoacetate}}}{CH_3\overset{O}{\overset{\|}{C}}\ddot{C}H^-} + 2CH_3X$$

The two alkylation steps are carried out sequentially.

$$\underset{\text{Ethyl acetoacetate}}{CH_3\overset{O}{\overset{\|}{C}}CH_2\overset{O}{\overset{\|}{C}}OCH_2CH_3} \xrightarrow[CH_3Br]{NaOCH_2CH_3} \underset{\text{Ethyl 2-methyl-3-oxobutanoate}}{CH_3\overset{O}{\overset{\|}{C}}\underset{\underset{CH_3}{|}}{CH}\overset{O}{\overset{\|}{C}}OCH_2CH_3} \xrightarrow[CH_3Br]{NaOCH_2CH_3} \underset{\text{Ethyl 2,2-dimethyl-3-oxobutanoate}}{CH_3\overset{O}{\overset{\|}{C}}C(CH_3)_2\overset{O}{\overset{\|}{C}}OCH_2CH_3}$$

$$\xrightarrow{\text{1. } HO^-,\ H_2O\quad \text{2. } H^+\quad \text{3. heat}} \underset{\text{3-Methyl-2-butanone}}{CH_3\overset{O}{\overset{\|}{C}}CH(CH_3)_2}$$

22.9 Alkylation of ethyl acetoacetate with 1,4-dibromobutane gives a product that can cyclize to a five-membered ring.

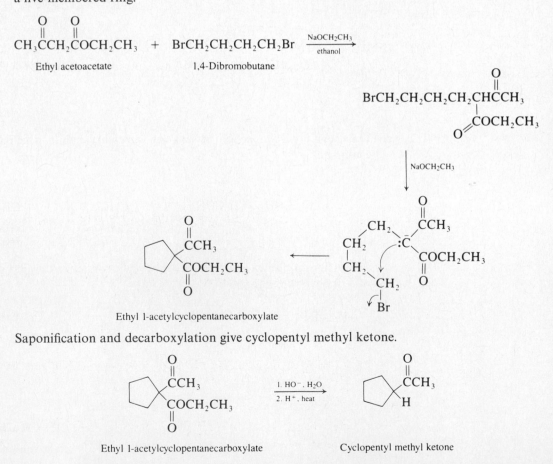

$$\underset{\text{Ethyl acetoacetate}}{CH_3\overset{O}{\overset{\|}{C}}CH_2\overset{O}{\overset{\|}{C}}OCH_2CH_3} + \underset{\text{1,4-Dibromobutane}}{BrCH_2CH_2CH_2CH_2Br} \xrightarrow[\text{ethanol}]{NaOCH_2CH_3} BrCH_2CH_2CH_2CH_2\underset{\underset{O=COCH_2CH_3}{|}}{CH}\overset{O}{\overset{\|}{C}}CH_3$$

$$\xrightarrow{NaOCH_2CH_3} BrCH_2CH_2CH_2CH_2\bar{\ddot{C}}(\overset{O}{\overset{\|}{C}}CH_3)(\overset{O}{\overset{\|}{C}}OCH_2CH_3) \longrightarrow \underset{\text{Ethyl 1-acetylcyclopentanecarboxylate}}{\text{1-(}\overset{O}{\overset{\|}{C}}CH_3\text{)-1-(}\overset{O}{\overset{\|}{C}}OCH_2CH_3\text{)cyclopentane}}$$

Saponification and decarboxylation give cyclopentyl methyl ketone.

$$\underset{\text{Ethyl 1-acetylcyclopentanecarboxylate}}{\text{1-(}\overset{O}{\overset{\|}{C}}CH_3\text{)-1-(}\overset{O}{\overset{\|}{C}}OCH_2CH_3\text{)cyclopentane}} \xrightarrow[\text{2. } H^+,\ \text{heat}]{\text{1. } HO^-,\ H_2O} \underset{\text{Cyclopentyl methyl ketone}}{C_5H_9\overset{O}{\overset{\|}{C}}CH_3}$$

22.10 The last step in the synthesis of pentobarbital is the reaction of the appropriately substituted derivative of diethyl malonate with urea.

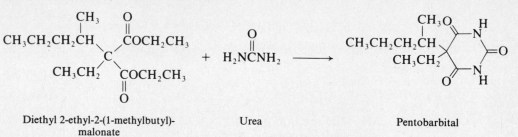

$$CH_3CH_2CH_2CH(CH_3)C(CH_2CH_3)(COOCH_2CH_3)_2 + H_2NCONH_2 \longrightarrow \text{Pentobarbital}$$

Diethyl 2-ethyl-2-(1-methylbutyl)-malonate   Urea   Pentobarbital

The dialkyl derivative of diethyl malonate is made in the usual way. It does not matter whether the ethyl group or the 1-methylbutyl group is introduced first.

$$CH_2(COOCH_2CH_3)_2 \xrightarrow[\text{2. } NaOCH_2CH_3,\ CH_3CH_2Br]{\text{1. } NaOCH_2CH_3,\ CH_3CH_2CH_2CHBrCH_3} CH_3CH_2CH_2CH(CH_3)C(CH_2CH_3)(COOCH_2CH_3)_2$$

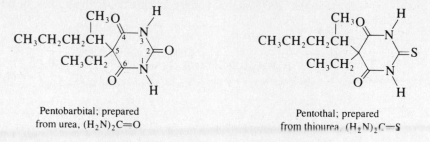

Diethyl malonate   Diethyl 2-ethyl-2-(1-methylbutyl)malonate

22.11 The carbonyl oxygen at C-2 of pentobarbital is replaced by sulfur in Pentothal (thiopental).

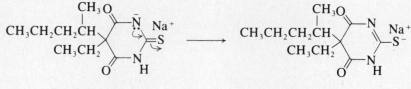

Pentobarbital; prepared from urea, $(H_2N)_2C{=}O$ Pentothal; prepared from thiourea, $(H_2N)_2C{=}S$

The sodium salt of Pentothal is formed by removal of a proton from one of the N—H groups by sodium hydroxide.

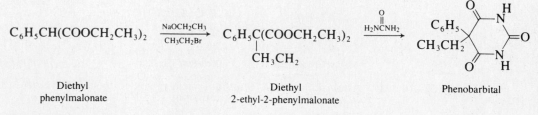

Pentothal sodium

22.12 The synthesis of phenobarbital requires diethyl phenylmalonate as the starting material.

$$C_6H_5CH(COOCH_2CH_3)_2 \xrightarrow[CH_3CH_2Br]{NaOCH_2CH_3} C_6H_5C(CH_2CH_3)(COOCH_2CH_3)_2 \xrightarrow{H_2NCONH_2} \text{Phenobarbital}$$

Diethyl phenylmalonate   Diethyl 2-ethyl-2-phenylmalonate   Phenobarbital

Diethyl phenylmalonate is prepared by a mixed Claisen condensation between ethyl phenylacetate and diethyl carbonate.

$$C_6H_5CH_2COCH_2CH_3 + CH_3CH_2OCOCH_2CH_3 \xrightarrow{NaOCH_2CH_3} C_6H_5CH(COOCH_2CH_3)_2$$

Ethyl phenylacetate   Diethyl carbonate   Diethyl phenylmalonate

22.13 (*b*) The anion of diethyl malonate adds to the β carbon atom of ethyl acrylate.

$$\underset{\text{Diethyl malonate}}{(CH_3CH_2OOC)_2CH_2} + \underset{\text{Ethyl acrylate}}{CH_2{=}CH\overset{O}{\overset{\|}{C}}OCH_2CH_3} \xrightarrow[\text{ethanol}]{NaOCH_2CH_3} \underset{\text{Triethyl 1,1,3-propanetricarboxylate}}{(CH_3CH_2OOC)_2CHCH_2CH_2\overset{O}{\overset{\|}{C}}OCH_2CH_3}$$

(*c*) The reaction of diethyl malonate with ethyl crotonate is analogous to its reaction with ethyl acrylate.

$$\underset{\text{Diethyl malonate}}{(CH_3CH_2OOC)_2CH_2} + \underset{\text{Ethyl crotonate}}{CH_3CH{=}CH\overset{O}{\overset{\|}{C}}OCH_2CH_3} \xrightarrow[\text{ethanol}]{NaOCH_2CH_3} \underset{\text{Triethyl 2-methyl-1,1,3-propanetricarboxylate}}{(CH_3CH_2OOC)_2CH\underset{CH_3}{\underset{|}{C}}HCH_2\overset{O}{\overset{\|}{C}}OCH_2CH_3}$$

(*d*) Michael addition of diethyl malonate to the double bond of diethyl fumarate occurs in the usual way. The stereochemistry of the double bond is not a factor.

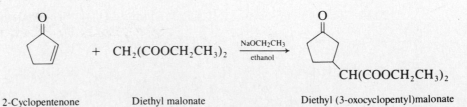

$$\underset{\text{Diethyl malonate}}{(CH_3CH_2OOC)_2CH_2} + \underset{\text{Diethyl fumarate}}{(CH_3CH_2O\overset{O}{\overset{\|}{C}})HC{=}CH(\underset{O}{\underset{\|}{C}}OCH_2CH_3)}$$

$$\xrightarrow[\text{ethanol}]{NaOCH_2CH_3} \underset{\text{Tetraethyl 1,1,2,3-propanetetracarboxylate}}{(CH_3CH_2OOC)_2CH\overset{COOCH_2CH_3}{\overset{|}{C}}HCH_2COOCH_2CH_3}$$

(*e*) The reaction of diethyl malonate with 2-cyclopentenone is an example of a Michael addition to an α,β-unsaturated ketone.

$$\text{2-Cyclopentenone} + \underset{\text{Diethyl malonate}}{CH_2(COOCH_2CH_3)_2} \xrightarrow[\text{ethanol}]{NaOCH_2CH_3} \text{Diethyl (3-oxocyclopentyl)malonate [3-}CH(COOCH_2CH_3)_2\text{ on cyclopentanone]}$$

22.14 (*b*) Analyze 3-methyl-2-butenoic acid retrosynthetically to reveal the source of its structural units.

$$\underset{\text{3-Methyl-2-butenoic acid}}{CH_3\underset{CH_3}{\underset{|}{C}}{=}CH\overset{O}{\overset{\|}{C}}OH} \Longrightarrow \underset{\beta\text{-Hydroxy acid}}{CH_3\overset{OH}{\overset{|}{\underset{CH_3}{\underset{|}{C}}}}{-}CH_2\overset{O}{\overset{\|}{C}}OH} \Longrightarrow CH_3C({=}O)CH_3 + :\bar{C}H_2C({=}O)OH$$

The necessary β-hydroxy acid can be prepared (as its ethyl ester) by a Reformatsky reaction.

$$\underset{\text{Acetone}}{CH_3\overset{O}{\overset{\|}{C}}CH_3} + \underset{\text{Ethyl bromoacetate}}{BrCH_2\overset{O}{\overset{\|}{C}}OCH_2CH_3} \xrightarrow[2.\ H_3O^+]{1.\ Zn} \underset{\text{Ethyl 3-hydroxy-3-methylbutanoate}}{CH_3\overset{OH}{\overset{|}{\underset{CH_3}{\underset{|}{C}}}}{-}CH_2\overset{O}{\overset{\|}{C}}OCH_2CH_3}$$

Dehydration of the β-hydroxy ester followed by ester hydrolysis gives the desired product.

$$\underset{\text{Ethyl 3-hydroxy-3-methylbutanoate}}{(CH_3)_2\overset{OH}{\overset{|}{C}}CH_2\overset{O}{\overset{\|}{C}}OCH_2CH_3} \xrightarrow[\text{heat}]{H_2SO_4} \underset{\text{Ethyl 3-methyl-2-butenoate}}{(CH_3)_2C{=}CH\overset{O}{\overset{\|}{C}}OCH_2CH_3} \xrightarrow[2.\ H^+]{1.\ HO^-,\ H_2O} \underset{\text{3-Methyl-2-butenoic acid}}{(CH_3)_2C{=}CH\overset{O}{\overset{\|}{C}}OH}$$

(*c*) Lithium aluminum hydride reduction of the β-hydroxy ester prepared in the preceding exercise provides the desired diol.

$$\underset{\text{Ethyl 3-hydroxy-3-methylbutanoate (prepared by Reformatsky reaction between acetone and ethyl bromoacetate)}}{(CH_3)_2\underset{OH}{\underset{|}{C}}CH_2\overset{O}{\overset{\|}{C}}OCH_2CH_3} \xrightarrow[2.\ H_2O]{1.\ LiAlH_4} \underset{\text{3-Methyl-1,3-butanediol}}{(CH_3)_2\underset{OH}{\underset{|}{C}}CH_2CH_2OH}$$

(*d*) Disconnection of the product shows that the carbon skeleton of 2,3-dimethyl-1,3-butanediol may be derived from a Reformatsky reaction between acetone and ethyl 2-bromopropanoate.

$$CH_3\overset{OH}{\overset{|}{\underset{CH_3}{\underset{|}{C}}}}{\vdots}\overset{CH_3}{\overset{|}{C}}HCH_2OH \Longrightarrow CH_3\overset{O}{\overset{\|}{C}}CH_3 + Br\underset{CH_3}{\underset{|}{C}}H\overset{O}{\overset{\|}{C}}OCH_2CH_3$$

The β-hydroxy ester resulting from the Reformatsky reaction gives the desired diol upon reduction with lithium aluminum hydride.

$$\underset{\text{Acetone}}{CH_3\overset{O}{\overset{\|}{C}}CH_3} + \underset{\text{Ethyl 2-bromopropanoate}}{CH_3\underset{Br}{\underset{|}{C}}H\overset{O}{\overset{\|}{C}}OCH_2CH_3} \xrightarrow[2.\ H_3O^+]{1.\ Zn} \underset{\text{Ethyl 2-hydroxy-2,3-dimethylbutanoate}}{CH_3\overset{OH}{\overset{|}{\underset{CH_3}{\underset{|}{C}}}}{-}\underset{CH_3}{\underset{|}{C}}H\overset{O}{\overset{\|}{C}}OCH_2CH_3} \xrightarrow[2.\ H_2O]{1.\ LiAlH_4}$$

$$\underset{\text{2,3-Dimethyl-1,3-butanediol}}{CH_3\overset{OH}{\overset{|}{\underset{CH_3}{\underset{|}{C}}}}{-}\underset{CH_3}{\underset{|}{C}}HCH_2OH}$$

22.15 (*b*) The α carbon atom of the ester bears a phenyl substituent and a methyl group. Only the methyl group can be attached to the α carbon by a nucleophilic substitution reaction.

Therefore generate the enolate of methyl phenylacetate with lithium diisopropylamide (LDA) in tetrahydrofuran (THF) and then alkylate with methyl iodide.

$$C_6H_5CH_2CO_2CH_3 \xrightarrow[\text{THF}]{\text{LDA}} C_6H_5CH{=}C(OLi)(OCH_3) \xrightarrow{CH_3I} C_6H_5CH(CH_3)CO_2CH_3$$

Methyl phenylacetate — Enolate of methyl phenylacetate — Methyl 2-phenylpropanoate

(c) The desired product corresponds to an aldol addition product.

$$\text{2-(CH(OH)C}_6\text{H}_5\text{)cyclohexanone} \Rightarrow \text{cyclohexanone enolate (carbanion)} + HCC_6H_5 \;(\text{with C=O})$$

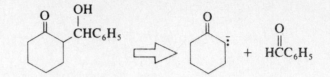

Therefore convert cyclohexanone to its enolate and then treat with benzaldehyde.

$$\text{Cyclohexanone} \xrightarrow[\text{2. } C_6H_5CHO \quad \text{3. } H_2O]{\text{1. LDA, THF}} \text{2-(CH(OH)C}_6\text{H}_5\text{)cyclohexanone}$$

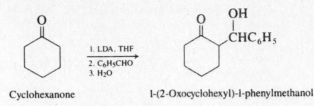

Cyclohexanone — 1-(2-Oxocyclohexyl)-1-phenylmethanol

(d) This product corresponds to the addition of the enolate of *tert*-butyl acetate to cyclohexanone.

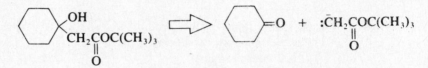

$$\text{1-OH-1-}(CH_2COC(CH_3)_3 \text{ with C=O})\text{cyclohexane} \Rightarrow \text{cyclohexanone (C=O)} + :\bar{C}H_2COC(CH_3)_3 \;(\text{C=O})$$

Generate the enolate of *tert*-butyl acetate with lithium diisopropylamide, then add cyclohexanone.

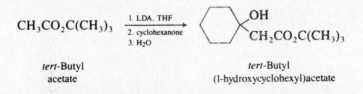

$$CH_3CO_2C(CH_3)_3 \xrightarrow[\text{2. cyclohexanone} \quad \text{3. } H_2O]{\text{1. LDA, THF}} \text{1-OH-1-}(CH_2CO_2C(CH_3)_3)\text{cyclohexane}$$

tert-Butyl acetate — *tert*-Butyl (1-hydroxycyclohexyl)acetate

22.16 In order to undergo a Claisen condensation an ester must have at least two protons on the α carbon:

$$2\,RCH_2\overset{O}{\overset{\|}{C}}OCH_2CH_3 + NaOCH_2CH_3 \longrightarrow \left[RCH_2\overset{O}{\overset{\|}{C}}\ddot{\bar{C}}(R)\overset{O}{\overset{\|}{C}}OCH_2CH_3\right] Na^+ + 2CH_3CH_2OH$$

The equilibrium constant for condensation is unfavorable unless the β-keto ester can be deprotonated to form a stable anion.

(*a*) Among the esters given, ethyl pentanoate and ethyl 3-methylbutanoate undergo the Claisen condensation.

$$\mathrm{CH_3CH_2CH_2CH_2\overset{O}{\overset{\|}{C}}OCH_2CH_3 \xrightarrow[2.\ H^+]{1.\ NaOCH_2CH_3} CH_3CH_2CH_2CH_2\overset{O}{\overset{\|}{C}}\underset{CH_2CH_2CH_3}{\underset{|}{C}}H\overset{O}{\overset{\|}{C}}OCH_2CH_3}$$

Ethyl pentanoate — Ethyl 3-oxo-2-propylheptanoate

$$\mathrm{(CH_3)_2CHCH_2\overset{O}{\overset{\|}{C}}OCH_2CH_3 \xrightarrow[2.\ H^+]{1.\ NaOCH_2CH_3} (CH_3)_2CHCH_2\overset{O}{\overset{\|}{C}}\underset{CH(CH_3)_2}{\underset{|}{C}}H\overset{O}{\overset{\|}{C}}OCH_2CH_3}$$

Ethyl 3-methylbutanoate — Ethyl 2-isopropyl-5-methyl-3-oxohexanoate

(*b*) The Claisen condensation product of ethyl 2-methylbutanoate cannot be deprotonated; the equilibrium constant for its formation is less than 1.

$$\mathrm{CH_3CH_2\underset{CH_3}{\underset{|}{C}}H\overset{O}{\overset{\|}{C}}OCH_2CH_3 \overset{NaOCH_2CH_3}{\rightleftharpoons} CH_3CH_2\underset{CH_3}{\underset{|}{C}}H\overset{O}{\overset{\|}{C}}\underset{CH_2CH_3}{\underset{|}{\overset{CH_3}{\overset{|}{C}}}}COOCH_2CH_3}$$

Ethyl 2-methylbutanoate — No protons on α carbon atom; cannot form stabilized enolate by deprotonation

(*c*) Ethyl 2,2-dimethylpropanoate has no protons on its α carbon; it cannot form the ester enolate required in the first step of the Claisen condensation.

$$\mathrm{CH_3\underset{CH_3}{\underset{|}{\overset{CH_3}{\overset{|}{C}}}}COOCH_2CH_3 + {}^-OCH_2CH_3 \longrightarrow \text{no reaction}}$$

Ethyl 2,2-dimethylpropanoate

22.17 (*a*) The Claisen condensation of ethyl phenylacetate is given by the equation:

$$\mathrm{C_6H_5CH_2\overset{O}{\overset{\|}{C}}OCH_2CH_3 \xrightarrow[2.\ H^+]{1.\ NaOCH_2CH_3} C_6H_5CH_2\overset{O}{\overset{\|}{C}}\underset{C_6H_5}{\underset{|}{C}}H\overset{O}{\overset{\|}{C}}OCH_2CH_3}$$

Ethyl phenylacetate — Ethyl 3-oxo-2,4-diphenylbutanoate

(*b*) Saponification and decarboxylation of this β-keto ester give dibenzyl ketone.

$$\mathrm{C_6H_5CH_2\overset{O}{\overset{\|}{C}}\underset{C_6H_5}{\underset{|}{C}}H\overset{O}{\overset{\|}{C}}OCH_2CH_3 \xrightarrow[2.\ H^+;\ 3.\ heat]{1.\ OH^-,\ H_2O} C_6H_5CH_2\overset{O}{\overset{\|}{C}}CH_2C_6H_5}$$

Ethyl 3-oxo-2,4-diphenylbutanoate — Dibenzyl ketone

(*c*) This process illustrates the alkylation of a β-keto ester with subsequent saponification and decarboxylation.

$$\underset{\text{Ethyl 3-oxo-2,4-diphenylbutanoate}}{C_6H_5CH_2\overset{O}{\overset{\|}{C}}\underset{C_6H_5}{\underset{|}{C}}H\overset{O}{\overset{\|}{C}}OCH_2CH_3} \xrightarrow[CH_2=CHCH_2Br]{NaOCH_2CH_3} C_6H_5CH_2\overset{O}{\overset{\|}{C}}\underset{C_6H_5}{\underset{|}{\overset{CH_2CH=CH_2}{\overset{|}{C}}}}COOCH_2CH_3$$

$$\xrightarrow{\text{1. HO}^-\text{, H}_2\text{O; 2. H}^+\text{; 3. heat}} \underset{\text{1,3-Diphenyl-5-hexen-2-one}}{C_6H_5CH_2\overset{O}{\overset{\|}{C}}\underset{C_6H_5}{\underset{|}{C}}HCH_2CH=CH_2}$$

(*d*) The enolate ion of ethyl phenylacetate attacks the carbonyl carbon of ethyl benzoate.

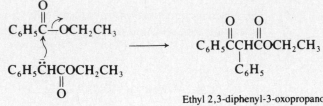

$$C_6H_5\overset{O}{\overset{\|}{C}}-OCH_2CH_3 \quad + \quad C_6H_5\ddot{C}H\underset{O}{\underset{\|}{C}}OCH_2CH_3 \longrightarrow \underset{\text{Ethyl 2,3-diphenyl-3-oxopropanoate}}{C_6H_5\overset{O}{\overset{\|}{C}}\underset{C_6H_5}{\underset{|}{C}}H\overset{O}{\overset{\|}{C}}OCH_2CH_3}$$

(*e*) Saponification and decarboxylation yield benzyl phenyl ketone.

$$\underset{\text{Ethyl 3-oxo-2,3-diphenylpropanoate}}{C_6H_5\overset{O}{\overset{\|}{C}}\underset{C_6H_5}{\underset{|}{C}}H\overset{O}{\overset{\|}{C}}OCH_2CH_3} \xrightarrow[\text{2. H}^+\text{; 3. heat}]{\text{1. HO}^-\text{, H}_2\text{O}} \underset{\text{Benzyl phenyl ketone}}{C_6H_5\overset{O}{\overset{\|}{C}}CH_2C_6H_5}$$

(*f*) This sequence is analogous to that of (*c*).

$$C_6H_5\overset{O}{\overset{\|}{C}}\underset{C_6H_5}{\underset{|}{C}}H\overset{O}{\overset{\|}{C}}OCH_2CH_3 \xrightarrow[CH_2=CHCH_2Br]{NaOCH_2CH_3} C_6H_5\overset{O}{\overset{\|}{C}}\underset{C_6H_5}{\underset{|}{\overset{CH_2CH=CH_2}{\overset{|}{C}}}}COOCH_2CH_3$$

$$\xrightarrow{\text{1. HO}^-\text{, H}_2\text{O; 2. H}^+\text{; 3. heat}} \underset{\text{1,2-Diphenyl-4-penten-1-one}}{C_6H_5\overset{O}{\overset{\|}{C}}\underset{C_6H_5}{\underset{|}{C}}HCH_2CH=CH_2}$$

22.18 (*a*) The Dieckmann condensation is the intramolecular version of the Claisen condensation. It utilizes a diester as starting material.

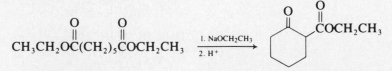

$$\underset{\text{Diethyl heptanedioate}}{CH_3CH_2O\overset{O}{\overset{\|}{C}}(CH_2)_5\overset{O}{\overset{\|}{C}}OCH_2CH_3} \xrightarrow[\text{2. H}^+]{\text{1. NaOCH}_2\text{CH}_3} \text{Ethyl 2-oxocyclohexanecarboxylate (cyclohexanone ring bearing } \overset{O}{\overset{\|}{C}}OCH_2CH_3\text{ at C-2)}$$

Ethyl 2-oxo-cyclohexanecarboxylate

(*b*) Acylation of cyclohexanone with diethyl carbonate yields the same β-keto ester formed in (*a*).

$$\text{Cyclohexanone} + CH_3CH_2OCOCH_2CH_3 \xrightarrow[2.\ H^+]{1.\ NaOCH_2CH_3} \text{Ethyl 2-oxocyclohexanecarboxylate}$$

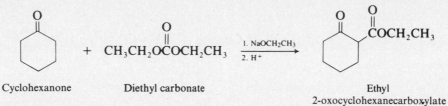

Cyclohexanone Diethyl carbonate Ethyl 2-oxocyclohexanecarboxylate

(*c*) The two most stable enol forms are those that involve the proton on the carbon flanked by the two carbonyl groups.

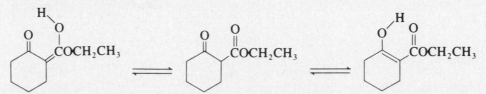

$COCH_2CH_3$ $\rightleftharpoons$ $COCH_2CH_3$ $\rightleftharpoons$ $COCH_2CH_3$

(*d*) Deprotonation of the β-keto ester involves the acidic proton at the carbon flanked by the two carbonyl groups.

$COCH_2CH_3$ $\longleftrightarrow$ $COCH_2CH_3$ $\longleftrightarrow$ $COCH_2CH_3$

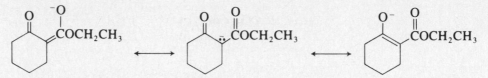

(*e*) The methyl group is introduced by alkylation of the β-keto ester. Saponification and decarboxylation complete the synthesis.

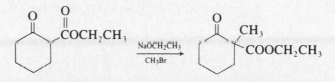

$$\text{Ethyl 2-oxocyclohexanecarboxylate} \xrightarrow[CH_3Br]{NaOCH_2CH_3} \text{Ethyl 1-methyl-2-oxocyclohexanecarboxylate}$$

Ethyl 2-oxocyclohexane-carboxylate Ethyl 1-methyl-2-oxocyclohexane-carboxylate

$$\xrightarrow{1.\ HO^-,\ H_2O;\ 2.\ H^+;\ 3.\ \text{heat}} \text{2-Methylcyclohexanone}$$

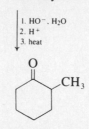

2-Methylcyclohexanone

(*f*) The enolate ion of the β-keto ester [see (*d*)] undergoes Michael addition to the carbon-carbon double bond of acrolein.

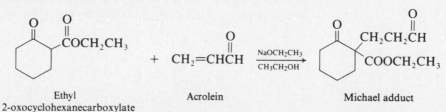

$$\text{Ethyl 2-oxocyclohexanecarboxylate} + CH_2{=}CHCH{=}O \xrightarrow[CH_3CH_2OH]{NaOCH_2CH_3} \text{Michael adduct (}CH_2CH_2CHO,\ COOCH_2CH_3\text{)}$$

Ethyl 2-oxocyclohexanecarboxylate Acrolein Michael adduct

This reaction has been reported in the chemical literature and proceeds in 65 to 75 percent yield.

22.19 (*a*) Ethyl acetoacetate is converted to its enolate ion with sodium ethoxide; this anion then acts as a nucleophile toward 1-bromopentane.

$$\underset{\text{Ethyl acetoacetate}}{CH_3\overset{O}{\overset{\|}{C}}CH_2\overset{O}{\overset{\|}{C}}OCH_2CH_3} + \underset{\text{1-Bromopentane}}{CH_3CH_2CH_2CH_2CH_2Br} \xrightarrow{NaOCH_2CH_3} \underset{\text{Ethyl 2-acetylheptanoate}}{CH_3\overset{O}{\overset{\|}{C}}\underset{CH_2CH_2CH_2CH_2CH_3}{\underset{|}{C}H}\overset{O}{\overset{\|}{C}}OCH_2CH_3}$$

(*b*) Saponification and decarboxylation of the product in (*a*) yield 2-octanone.

$$\underset{\text{Ethyl 2-acetylheptanoate}}{CH_3\overset{O}{\overset{\|}{C}}\underset{(CH_2)_4CH_3}{\underset{|}{C}H}\overset{O}{\overset{\|}{C}}OCH_2CH_3} \xrightarrow[\text{2. } H^+ \text{ 3. heat}]{\text{1. } HO^-,\ H_2O} \underset{\text{2-Octanone}}{CH_3\overset{O}{\overset{\|}{C}}CH_2CH_2CH_2CH_2CH_2CH_3}$$

(*c*) The product derived from the reaction in (*a*) can be alkylated again.

$$\underset{\text{Ethyl 2-acetylheptanoate}}{CH_3\overset{O}{\overset{\|}{C}}\underset{CH_2CH_2CH_2CH_2CH_3}{\underset{|}{C}H}\overset{O}{\overset{\|}{C}}OCH_2CH_3} + CH_3I \xrightarrow{NaOCH_2CH_3} \underset{\text{Ethyl 2-acetyl-2-methylheptanoate}}{CH_3\overset{O}{\overset{\|}{C}}\underset{CH_2CH_2CH_2CH_2CH_3}{\underset{|}{\overset{CH_3}{\overset{|}{C}}}}COOCH_2CH_3}$$

(*d*) The dialkylated derivative of acetoacetic ester formed in (*c*) can be converted to a ketone by saponification and decarboxylation.

$$\underset{\text{Ethyl 2-acetyl-2-methylheptanoate}}{CH_3\overset{O}{\overset{\|}{C}}\underset{CH_2CH_2CH_2CH_2CH_3}{\underset{|}{\overset{CH_3}{\overset{|}{C}}}}COOCH_2CH_3} \xrightarrow[\text{2. } H^+ \text{ 3. heat}]{\text{1. } HO^-,\ H_2O} \underset{\text{3-Methyl-2-octanone}}{CH_3\overset{O}{\overset{\|}{C}}\underset{CH_2CH_2CH_2CH_2CH_3}{\underset{|}{C}H}CH_3}$$

(*e*) The anion of ethyl acetoacetate acts as a nucleophile toward 1-bromo-3-chloropropane. Bromide is a better leaving group than chloride and is displaced preferentially.

$$\underset{\text{Ethyl acetoacetate}}{CH_3\overset{O}{\overset{\|}{C}}CH_2\overset{O}{\overset{\|}{C}}OCH_2CH_3} + \underset{\text{1-Bromo-3-chloropropane}}{BrCH_2CH_2CH_2Cl} \xrightarrow{NaOCH_2CH_3} \underset{\text{Ethyl 2-acetyl-5-chloropentanoate}}{CH_3\overset{O}{\overset{\|}{C}}\underset{CH_2CH_2CH_2Cl}{\underset{|}{C}H}\overset{O}{\overset{\|}{C}}OCH_2CH_3}$$

(*f*) Treatment of the product of (*e*) with sodium ethoxide gives an enolate ion that cyclizes by intramolecular nucleophilic substitution of chloride.

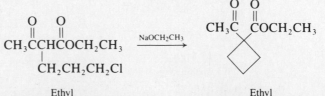

$$\underset{\text{Ethyl 2-acetyl-5-chloropentanoate}}{CH_3\overset{O}{\overset{\|}{C}}\underset{CH_2CH_2CH_2Cl}{\underset{|}{C}H}\overset{O}{\overset{\|}{C}}OCH_2CH_3} \xrightarrow{NaOCH_2CH_3} \underset{\text{Ethyl 1-acetylcyclobutanecarboxylate}}{CH_3\overset{O}{\overset{\|}{C}}\ \overset{O}{\overset{\|}{C}}OCH_2CH_3 \text{ (on cyclobutane C-1)}}$$

(*g*) Cyclobutyl methyl ketone is formed by saponification and decarboxylation of the product in (*f*).

$$\mathrm{CH_3\overset{\overset{O}{\|}}{C}\ \overset{\overset{O}{\|}}{C}OCH_2CH_3\ (\text{on cyclobutane C-1})} \xrightarrow[\text{3. heat}]{\text{1. HO}^-,\ \text{H}_2\text{O};\ \text{2. H}^+} \mathrm{CH_3\overset{\overset{O}{\|}}{C}}\text{–cyclobutyl}$$

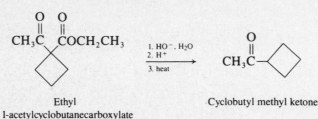

Ethyl 1-acetylcyclobutanecarboxylate — Cyclobutyl methyl ketone

(*h*) Ethyl acetoacetate undergoes Michael addition to phenyl vinyl ketone in the presence of base.

$$\mathrm{CH_3\overset{\overset{O}{\|}}{C}CH_2\overset{\overset{O}{\|}}{C}OCH_2CH_3 + CH_2{=}CH\overset{\overset{O}{\|}}{C}C_6H_5 \xrightarrow[\text{ethanol}]{\text{NaOCH}_2\text{CH}_3} CH_3\overset{\overset{O}{\|}}{C}\underset{\underset{\displaystyle CH_2CH_2\underset{\underset{O}{\|}}{C}C_6H_5}{|}}{C}H\overset{\overset{O}{\|}}{C}OCH_2CH_3}$$

Ethyl acetoacetate — Phenyl vinyl ketone — Ethyl 2-acetyl-5-oxo-5-phenylpentanoate

(*i*) A diketone results from saponification and decarboxylation of the Michael adduct.

$$\mathrm{CH_3\overset{\overset{O}{\|}}{C}\underset{\underset{\displaystyle CH_2CH_2\underset{\underset{O}{\|}}{C}C_6H_5}{|}}{C}H\overset{\overset{O}{\|}}{C}OCH_2CH_3} \xrightarrow[\text{3. H}_2\text{O}]{\text{1. HO}^-,\ \text{H}_2\text{O};\ \text{2. H}^+} \mathrm{CH_3\overset{\overset{O}{\|}}{C}CH_2CH_2CH_2\overset{\overset{O}{\|}}{C}C_6H_5}$$

Ethyl 2-acetyl-5-oxo-5-phenylpentanoate — 1-Phenyl-1,5-hexanedione

(*j*) Ethyl acetoacetate reacts with ketones in the presence of piperidine to give alkenes by way of the Knoevenagel condensation.

$$\mathrm{CH_3\overset{\overset{O}{\|}}{C}CH_2\overset{\overset{O}{\|}}{C}OCH_2CH_3 + CH_3CH_2\overset{\overset{O}{\|}}{C}CH_2CH_3 \xrightarrow{\text{piperidine}} (CH_3CH_2)_2C{=}C\begin{matrix}\diagup \overset{\overset{O}{\|}}{C}CH_3\\ \diagdown \underset{\underset{O}{\|}}{C}OCH_2CH_3\end{matrix}}$$

Ethyl acetoacetate — 3-Pentanone — Ethyl 2-acetyl-3-ethyl-2-pentenoate

22.20 Diethyl malonate reacts with the reagents given in the preceding problem in a manner analogous to that of ethyl acetoacetate.

(*a*)

$$\underset{\text{Diethyl malonate}}{\mathrm{CH_2(COOCH_2CH_3)_2}} + \underset{\text{1-Bromopentane}}{\mathrm{CH_3CH_2CH_2CH_2CH_2Br}} \xrightarrow{\text{NaOCH}_2\text{CH}_3} \mathrm{CH_3CH_2CH_2CH_2CH_2CH(COOCH_2CH_3)_2}$$

Diethyl hexane-1,1-dicarboxylate (diethyl pentylmalonate)

(*b*)

$$\underset{\text{Diethyl 1,1-hexanedicarboxylate}}{\mathrm{CH_3CH_2CH_2CH_2CH_2CH(COOCH_2CH_3)_2}} \xrightarrow[\text{3. heat}]{\text{1. HO}^-,\ \text{H}_2\text{O};\ \text{2. H}^+} \underset{\text{Heptanoic acid}}{\mathrm{CH_3CH_2CH_2CH_2CH_2CH_2\overset{\overset{O}{\|}}{C}OH}}$$

(*c*) $CH_3CH_2CH_2CH_2CH_2CH(COOCH_2CH_3)_2 \xrightarrow[NaOCH_2CH_3]{CH_3I}$

Diethyl 1,1-hexanedicarboxylate

$$CH_3CH_2CH_2CH_2CH_2\underset{\displaystyle CH_3}{\underset{|}{C}}(COOCH_2CH_3)_2$$

Diethyl 2,2-heptanedicarboxylate

(*d*) $CH_3CH_2CH_2CH_2CH_2\underset{\displaystyle CH_3}{\underset{|}{C}}(COOCH_2CH_3)_2 \xrightarrow[\text{2. } H^+ \text{ 3. heat}]{\text{1. } HO^-,\ H_2O} CH_3CH_2CH_2CH_2CH_2\underset{\displaystyle CH_3}{\underset{|}{C}}H\overset{\displaystyle O}{\overset{\|}{C}}OH$

Diethyl 2,2-heptanedicarboxylate → 2-Methylheptanoic acid

(*e*) $CH_2(COOCH_2CH_3)_2 + BrCH_2CH_2CH_2Cl \xrightarrow{NaOCH_2CH_3}$

Diethyl malonate + 1-Bromo-3-chloropropane

$$ClCH_2CH_2CH_2CH(COOCH_2CH_3)_2$$

Diethyl 4-chloro-1,1-butanedicarboxylate

(*f*) $ClCH_2CH_2CH_2CH(COOCH_2CH_3)_2 \xrightarrow{NaOCH_2CH_3}$ cyclobutane-1,1-$(\overset{O}{\overset{\|}{C}}OCH_2CH_3)_2$

Diethyl 4-chloro-1,1-butanedicarboxylate → Diethyl cyclobutane-1,1-dicarboxylate

(*g*) cyclobutane-1,1-$(\overset{O}{\overset{\|}{C}}OCH_2CH_3)_2 \xrightarrow[\text{2. } H^+ \text{ 3. heat}]{\text{1. } HO^-,\ H_2O}$ cyclobutane$-\overset{O}{\overset{\|}{C}}OH$

Diethyl cyclobutane-1,1-dicarboxylate → Cyclobutanecarboxylic acid

(*h*) $CH_2(COOCH_2CH_3)_2 + C_6H_5\overset{\displaystyle O}{\overset{\|}{C}}CH{=}CH_2 \xrightarrow[CH_3CH_2OH]{NaOCH_2CH_3}$

Diethyl malonate + Phenyl vinyl ketone

$$C_6H_5\overset{\displaystyle O}{\overset{\|}{C}}CH_2CH_2CH(COOCH_2CH_3)_2$$

Diethyl 4-oxo-4-phenylbutanedicarboxylate

(*i*) $C_6H_5\overset{\displaystyle O}{\overset{\|}{C}}CH_2CH_2CH(COOCH_2CH_3)_2 \xrightarrow[\text{2. } H^+ \text{ 3. heat}]{\text{1. } HO^-,\ H_2O} C_6H_5\overset{\displaystyle O}{\overset{\|}{C}}CH_2CH_2CH_2\overset{\displaystyle O}{\overset{\|}{C}}OH$

Diethyl 4-oxo-4-phenylbutanedicarboxylate → 5-Oxo-5-phenylpentanoic acid

(*j*) $CH_2(COOCH_2CH_3)_2 + CH_3CH_2\overset{\displaystyle O}{\overset{\|}{C}}CH_2CH_3 \xrightarrow{\text{piperidine}} (CH_3CH_2)_2C{=}C(COOCH_2CH_3)_2$

Diethyl malonate + 3-Pentanone → Diethyl 2-ethyl-1-butene-1,1-dicarboxylate

22.21 (*a*) Both carbonyl groups of diethyl malonate are equivalent, so enolization can occur in either direction.

$$CH_3CH_2O{-}C(OH){=}CH{-}C({=}O){-}OCH_2CH_3 \rightleftharpoons CH_3CH_2O{-}C({=}O){-}CH_2{-}C({=}O){-}OCH_2CH_3 \rightleftharpoons CH_3CH_2O{-}C({=}O){-}CH{=}C(OH){-}OCH_2CH_3$$

Diethyl malonate

(*b*) Ethyl acetoacetate can give three constitutionally isomeric enols.

$$CH_3\overset{O}{\overset{\|}{C}}CH_2\overset{O}{\overset{\|}{C}}OCH_2CH_3$$

Ethyl acetoacetate

$$CH_2{=}C(OH){-}CH_2{-}C({=}O){-}OCH_2CH_3$$

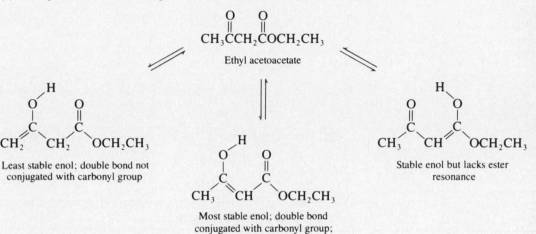

Least stable enol; double bond not conjugated with carbonyl group

$$CH_3{-}C(OH){=}CH{-}C({=}O){-}OCH_2CH_3$$

Most stable enol; double bond conjugated with carbonyl group; ester carbonyl stabilized by resonance

$$CH_3{-}C({=}O){-}CH{=}C(OH){-}OCH_2CH_3$$

Stable enol but lacks ester resonance

(*c*) Bromine reacts with diethyl malonate and ethyl acetoacetate by way of the corresponding enols.

$$CH_2(COOCH_2CH_3)_2 \rightleftharpoons CH_3CH_2O{-}C(OH){=}CH{-}C({=}O){-}OCH_2CH_3 \xrightarrow{Br_2} CH_3CH_2O\overset{O}{\overset{\|}{C}}\underset{Br}{\underset{|}{C}H}\overset{O}{\overset{\|}{C}}OCH_2CH_3$$

Diethyl malonate — Diethyl bromomalonate

$$CH_3\overset{O}{\overset{\|}{C}}CH_2\overset{O}{\overset{\|}{C}}OCH_2CH_3 \rightleftharpoons CH_3\overset{OH}{\overset{|}{C}}{=}CH\overset{O}{\overset{\|}{C}}OCH_2CH_3 \xrightarrow{Br_2} CH_3\overset{O}{\overset{\|}{C}}\underset{Br}{\underset{|}{C}H}\overset{O}{\overset{\|}{C}}OCH_2CH_3$$

Ethyl acetoacetate — Ethyl α-bromoacetoacetate

22.22 (*a*) Recall that Grignard reagents are destroyed by reaction with proton donors. Ethyl acetoacetate is a stronger acid than water; it transfers a proton to a Grignard reagent.

$$CH_3\overset{O}{\overset{\|}{C}}CH_2\overset{O}{\overset{\|}{C}}OCH_2CH_3 + CH_3MgI \longrightarrow CH_4 + CH_3\overset{O}{\overset{\|}{C}}\underset{MgI}{\underset{|}{C}H}\overset{O}{\overset{\|}{C}}OCH_2CH_3$$

Ethyl acetoacetate — Methylmagnesium iodide — Methane — Iodomagnesium salt of ethyl acetoacetate

(*b*) Adding D_2O and DCl to the reaction mixture leads to D^+ transfer to the α carbon atom of ethyl acetoacetate.

$$\underset{\text{Iodomagnesium salt of ethyl acetoacetate}}{CH_3\overset{O}{\overset{\|}{C}}\underset{MgI}{\underset{|}{C}H}\overset{O}{\overset{\|}{C}}OCH_2CH_3} + \underset{\text{Deuterium oxide}}{D_2O} \xrightarrow{DCl} \underset{\text{Ethyl α-deuterioacetoacetate}}{CH_3\overset{O}{\overset{\|}{C}}\underset{D}{\underset{|}{C}H}\overset{O}{\overset{\|}{C}}OCH_2CH_3}$$

22.23 (*a*) Ethyl octanoate undergoes a Claisen condensation to form a β-keto ester on being treated with sodium ethoxide.

$$\underset{\text{Ethyl octanoate}}{CH_3(CH_2)_5CH_2\overset{O}{\overset{\|}{C}}OCH_2CH_3} \xrightarrow[2.\ H^+]{1.\ NaOCH_2CH_3} \underset{\text{Ethyl 2-hexyl-3-oxodecanoate}}{CH_3(CH_2)_5CH_2\overset{O}{\overset{\|}{C}}\underset{(CH_2)_5CH_3}{\underset{|}{C}H}\overset{O}{\overset{\|}{C}}OCH_2CH_3}$$

(*b*) Saponification and decarboxylation of the β-keto ester yield a ketone.

$$\underset{\text{Ethyl 2-hexyl-3-oxodecanoate}}{CH_3(CH_2)_5CH_2\overset{O}{\overset{\|}{C}}\underset{(CH_2)_5CH_3}{\underset{|}{C}H}\overset{O}{\overset{\|}{C}}OCH_2CH_3} \xrightarrow[2.\ H^+\ 3.\ heat]{1.\ NaOH,\ H_2O} \underset{\text{8-Pentadecanone}}{CH_3(CH_2)_5CH_2\overset{O}{\overset{\|}{C}}CH_2(CH_2)_5CH_3}$$

(*c*) On treatment with base, ethyl acetoacetate is converted to its enolate, which reacts as a nucleophile toward 1-bromobutane.

$$\underset{\text{Ethyl acetoacetate}}{CH_3\overset{O}{\overset{\|}{C}}CH_2\overset{O}{\overset{\|}{C}}OCH_2CH_3} + \underset{\text{1-Bromobutane}}{CH_3CH_2CH_2CH_2Br} \xrightarrow{NaOCH_2CH_3} \underset{\text{Ethyl 2-acetylhexanoate}}{CH_3\overset{O}{\overset{\|}{C}}\underset{CH_2CH_2CH_2CH_3}{\underset{|}{C}H}\overset{O}{\overset{\|}{C}}OCH_2CH_3}$$

(*d*) Alkylation of ethyl acetoacetate, followed by saponification and decarboxylation, gives a ketone. The two steps comprise the acetoacetic ester synthesis.

$$\underset{\text{Ethyl 2-acetylhexanoate}}{CH_3\overset{O}{\overset{\|}{C}}\underset{CH_2CH_2CH_2CH_3}{\underset{|}{C}H}\overset{O}{\overset{\|}{C}}OCH_2CH_3} \xrightarrow[2.\ H^+\ 3.\ heat]{1.\ NaOH,\ H_2O} \underset{\text{2-Heptanone}}{CH_3\overset{O}{\overset{\|}{C}}CH_2CH_2CH_2CH_2CH_3}$$

(*e*) An alkylated derivative of ethyl acetoacetate is capable of being alkylated a second time.

$$\underset{\text{Ethyl 2-acetylhexanoate}}{CH_3\overset{O}{\overset{\|}{C}}\underset{CH_2CH_2CH_2CH_3}{\underset{|}{C}H}\overset{O}{\overset{\|}{C}}OCH_2CH_3} + \underset{\text{Iodobutane}}{CH_3CH_2CH_2CH_2I} \xrightarrow[ethanol]{NaOCH_2CH_3} \underset{\text{Ethyl 2-acetyl-2-butylhexanoate}}{CH_3\overset{O}{\overset{\|}{C}}\underset{COOCH_2CH_3}{\underset{|}{C}}(CH_2CH_2CH_2CH_3)_2}$$

(*f*) The dialkylated derivative of acetoacetic ester formed in (*e*) is converted to a ketone by saponification and decarboxylation.

$$\underset{\text{Ethyl 2-acetyl-2-butylhexanoate}}{CH_3\overset{O}{\overset{\|}{C}}\underset{COOCH_2CH_3}{\underset{|}{C}}(CH_2CH_2CH_2CH_3)_2} \xrightarrow[2.\ H^+\ 3.\ heat]{1.\ NaOH} \underset{\text{3-Butyl-2-heptanone}}{CH_3\overset{O}{\overset{\|}{C}}CH(CH_2CH_2CH_2CH_3)_2}$$

(*g*) The enolate of acetophenone attacks the carbonyl group of diethyl carbonate.

$$C_6H_5\overset{O}{\overset{\|}{C}}CH_3 + CH_3CH_2O\overset{O}{\overset{\|}{C}}OCH_2CH_3 \xrightarrow[2.\ H^+]{1.\ NaOCH_2CH_3} C_6H_5\overset{O}{\overset{\|}{C}}CH_2\overset{O}{\overset{\|}{C}}OCH_2CH_3$$

Acetophenone — Diethyl carbonate — 3-Oxo-3-phenylpropanoate

(*h*) Diethyl oxalate acts as an acylating agent toward the enolate of acetone.

$$CH_3\overset{O}{\overset{\|}{C}}CH_3 + CH_3CH_2O\overset{O}{\overset{\|}{C}}\overset{O}{\overset{\|}{C}}OCH_2CH_3 \xrightarrow[2.\ H^+]{1.\ NaOCH_2CH_3} CH_3\overset{O}{\overset{\|}{C}}CH_2\overset{O}{\overset{\|}{C}}\overset{O}{\overset{\|}{C}}OCH_2CH_3$$

Acetone — Diethyl oxalate — Ethyl 2,4-dioxopentanoate (observed yield, 61–66%)

(*i*) The first stage of the malonic ester synthesis is the alkylation of diethyl malonate with an alkyl halide.

$$CH_2(COOCH_2CH_3)_2 + BrCH_2\underset{CH_3}{\underset{|}{C}}HCH_2CH_3 \xrightarrow[ethanol]{NaOCH_2CH_3} CH_3CH_2\underset{CH_3}{\underset{|}{C}}HCH_2CH(COOCH_2CH_3)_2$$

Diethyl malonate — 1-Bromo-2-methylbutane — Diethyl 3-methylpentane-1,1-dicarboxylate

(*j*) Alkylation of diethyl malonate is followed by saponification and decarboxylation to give a carboxylic acid.

$$CH_3CH_2\underset{CH_3}{\underset{|}{C}}HCH_2CH(COOCH_2CH_3)_2 \xrightarrow[2.\ H^+\ \ 3.\ heat]{1.\ NaOH,\ H_2O} CH_3CH_2\underset{CH_3}{\underset{|}{C}}HCH_2CH_2\overset{O}{\overset{\|}{C}}OH$$

Diethyl 3-methylpentane-1,1-dicarboxylate — 4-Methylhexanoic acid (57% yield from 1-bromo-2-methylbutane)

(*k*) The anion of diethyl malonate undergoes Michael addition to 6-methyl-2-cyclohexenone.

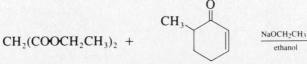

$$CH_2(COOCH_2CH_3)_2 + \text{6-methyl-2-cyclohexenone} \xrightarrow[ethanol]{NaOCH_2CH_3} \text{(cyclohexanone ring with } CH_3 \text{ and } CH(COOCH_2CH_3)_2)$$

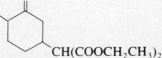

Diethyl malonate — 6-Methyl-2-cyclohexenone — Diethyl 2-(4-methyl-3-oxocyclohexyl)malonate (isolated yield, 50%)

(*l*) Acid hydrolysis converts the diester in (*k*) to a malonic acid derivative, which then undergoes decarboxylation.

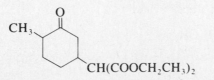

$$\text{(cyclohexanone ring with } CH_3 \text{ and } CH(COOCH_2CH_3)_2) \xrightarrow[heat]{H_2O,\ HCl} \text{(cyclohexanone ring with } CH_3 \text{ and } CH_2COOH)$$

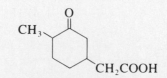

Diethyl 2-(4-methyl-3-oxocyclohexyl)malonate — (4-Methyl-3-oxocyclohexyl)acetic acid (isolated yield, 80%)

(*m*) This reaction is a Knoevenagel condensation.

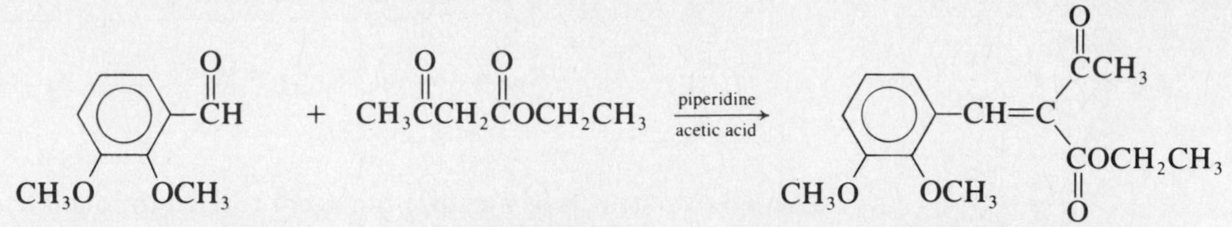

$$\text{2,3-}(CH_3O)_2C_6H_3CH{=}O + CH_3\overset{O}{\overset{\|}{C}}CH_2\overset{O}{\overset{\|}{C}}OCH_2CH_3 \xrightarrow[\text{acetic acid}]{\text{piperidine}} \text{2,3-}(CH_3O)_2C_6H_3CH{=}C(\overset{O}{\overset{\|}{C}}CH_3)\overset{O}{\overset{\|}{C}}OCH_2CH_3$$

2,3-Dimethoxybenzaldehyde Ethyl acetoacetate Ethyl 2-acetyl-3-(2,3-dimethoxyphenyl)propenoate (isolated yield, 64–72%)

(*n*) Lithium diisopropylamide (LDA) is used to convert esters quantitatively to their enolate ions. In this reaction the enolate of *tert*-butyl acetate adds to benzaldehyde.

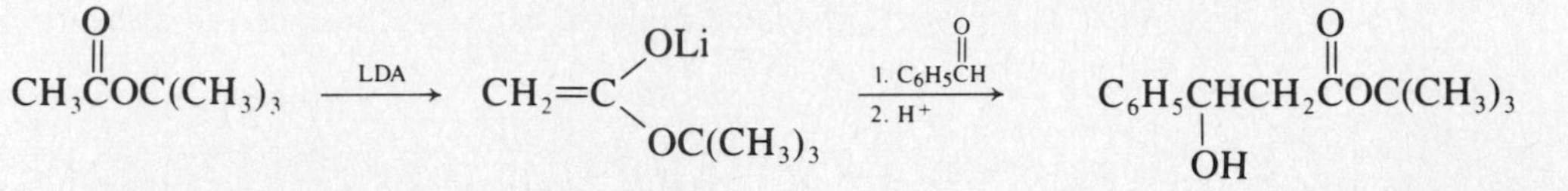

$$CH_3\overset{O}{\overset{\|}{C}}OC(CH_3)_3 \xrightarrow{\text{LDA}} CH_2{=}C(OLi)OC(CH_3)_3 \xrightarrow[\text{2. } H^+]{\text{1. } C_6H_5\overset{O}{\overset{\|}{C}}H} C_6H_5\underset{OH}{\underset{|}{C}}HCH_2\overset{O}{\overset{\|}{C}}OC(CH_3)_3$$

tert-Butyl acetate Lithium enolate of *tert*-butyl acetate *tert*-Butyl 3-hydroxy-3-phenylpropanoate

22.24 (*a*) Both ester functions in this molecule are β to a ketone carbonyl. Hydrolysis is followed by decarboxylation.

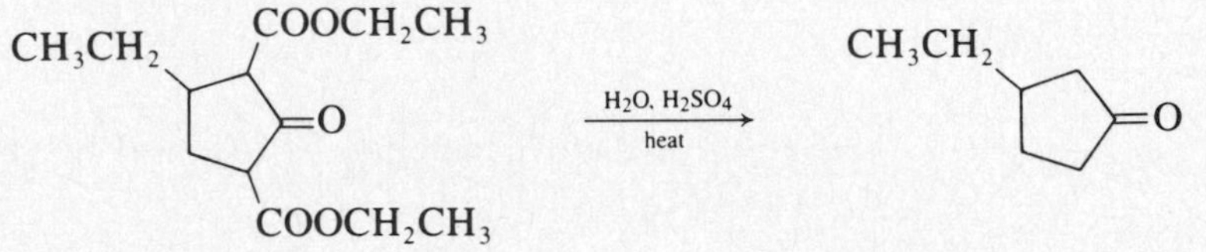

CH₃CH₂, COOCH₂CH₃, =O, COOCH₂CH₃ $\xrightarrow[\text{heat}]{H_2O,\ H_2SO_4}$ CH₃CH₂, =O

Diethyl 3-ethylcyclopentanone-2,5-dicarboxylate 3-Ethylcyclopentanone ($C_7H_{12}O$)

(*b*) Examine each carbon that is α to an ester function to see if it can lead to a five-, six-, or seven-membered cyclic β-keto ester by a Dieckmann cyclization.

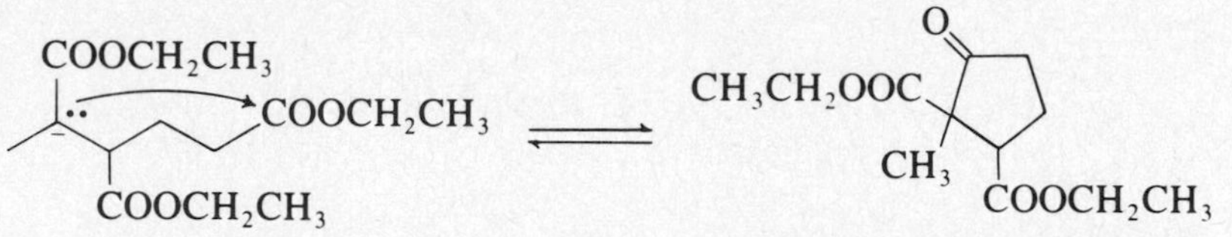

COOCH₂CH₃, COOCH₂CH₃, COOCH₂CH₃ $\rightleftharpoons$ CH₃CH₂OOC, O, CH₃, COOCH₂CH₃

Cyclization to a five-membered ring possible, but β-keto ester cannot be deprotonated to give a stable anion

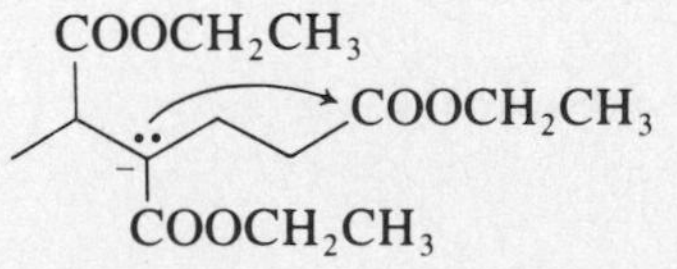

COOCH₂CH₃, COOCH₂CH₃, COOCH₂CH₃

Cyclization not likely; resulting ring is four-membered and highly strained

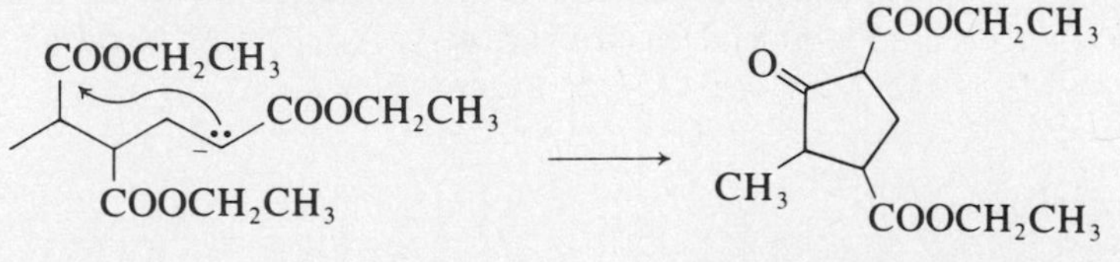

COOCH₂CH₃, COOCH₂CH₃, COOCH₂CH₃ $\longrightarrow$ COOCH₂CH₃, O, CH₃, COOCH₂CH₃

Cyclization gives a five-membered ring; β-keto ester deprotonated under reaction conditions; this is the observed product ($C_{12}H_{18}O_5$)

(*c*) Both ester functions undergo hydrolysis in acid, but decarboxylation occurs only at the carboxyl group that is β to the ketone carbonyl.

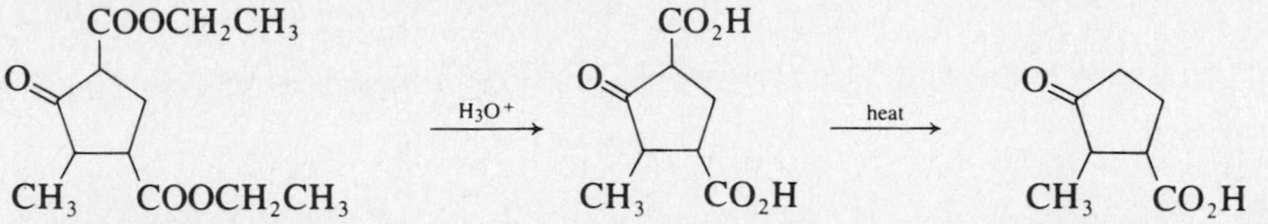

$COOCH_2CH_3$ ... CH_3 $COOCH_2CH_3$ $\xrightarrow{H_3O^+}$ CO_2H ... CH_3 CO_2H $\xrightarrow{heat}$ CH_3 CO_2H

Diethyl 2-methylcyclopentanone-3,5-dicarboxylate

2-Methylcyclopentanone-3-carboxylic acid ($C_7H_{10}O_3$)

(*d*) A Dieckmann condensation occurs, giving a five-membered ring fused to the original three-membered ring.

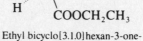

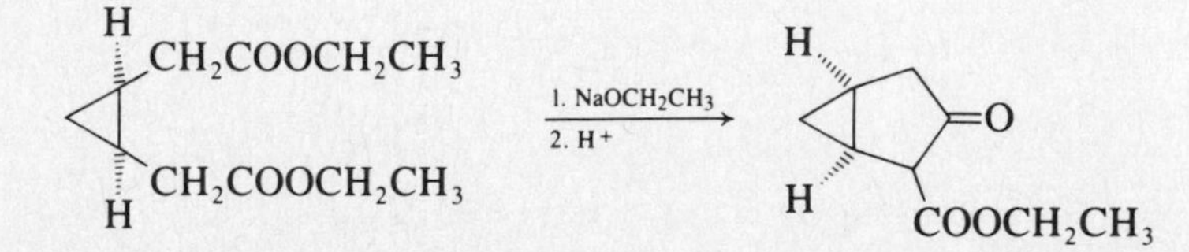

$\xrightarrow{1.\ NaOCH_2CH_3;\ 2.\ H^+}$

Diethyl *cis*-1,2-cyclopropanediacetate

Ethyl bicyclo[3.1.0]hexan-3-one-2-carboxylate ($C_9H_{12}O_3$, 79%)

(*e*) Saponification and decarboxylation convert the β-keto ester to a ketone.

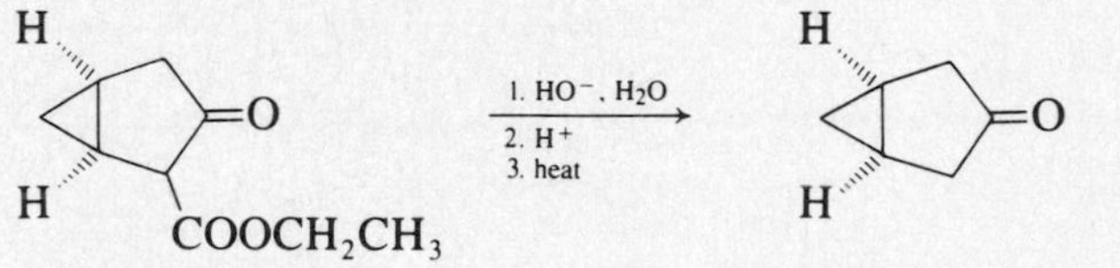

$\xrightarrow{1.\ HO^-,\ H_2O;\ 2.\ H^+;\ 3.\ heat}$

Ethyl bicyclo[3.1.0]hexan-3-one-2-carboxylate

Bicyclo[3.1.0]hexan-3-one (C_6H_8O, 43%)

(*f*) The reactants are functionalized in such a way that two Knoevenagel reactions can take place.

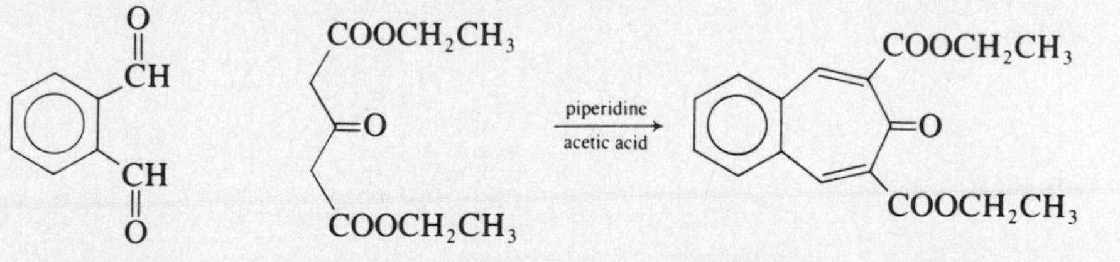

$\xrightarrow{piperidine,\ acetic\ acid}$

Benzene-1,2-dicarbaldehyde

Diethyl 3-oxoglutarate

Diethyl 4,5-benzocycloheptadienone-2,7-dicarboxylate ($C_{17}H_{16}O_5$, 63% yield)

The two Knoevenagel condensations occur sequentially; the first is intermolecular, the second intramolecular.

22.25 (*a*) First write out the structure of 4-phenyl-2-butanone and identify the synthon that is derived from ethyl acetoacetate.

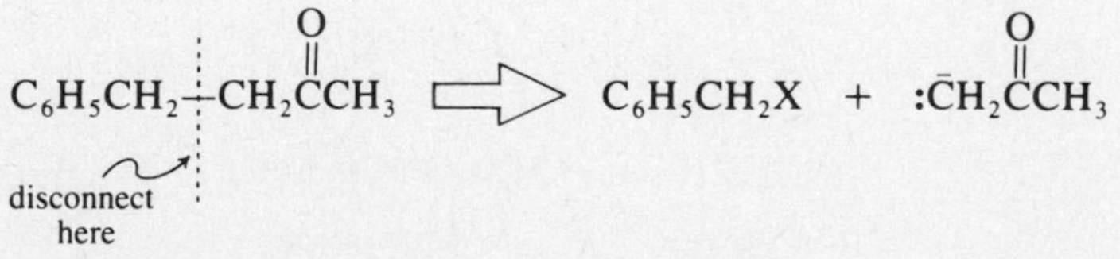

$$C_6H_5CH_2 \vdots CH_2\overset{O}{\overset{\|}{C}}CH_3 \Longrightarrow C_6H_5CH_2X + :\bar{C}H_2\overset{O}{\overset{\|}{C}}CH_3$$

disconnect here

Therefore carry out the acetoacetic ester synthesis using a benzyl halide as the alkylating agent.

$$\underset{\text{Benzyl alcohol}}{C_6H_5CH_2OH} \xrightarrow[\text{or } PBr_3]{HBr} \underset{\text{Benzyl bromide}}{C_6H_5CH_2Br}$$

$$\underset{\text{Ethyl acetoacetate}}{CH_3\overset{O}{\overset{\|}{C}}CH_2\overset{O}{\overset{\|}{C}}OCH_2CH_3} + \underset{\text{Benzyl bromide}}{C_6H_5CH_2Br} \xrightarrow[\text{ethanol}]{NaOCH_2CH_3} \underset{\text{Ethyl 2-benzyl-3-oxobutanoate}}{CH_3\overset{O}{\overset{\|}{C}}\underset{CH_2C_6H_5}{\underset{|}{C}H}\overset{O}{\overset{\|}{C}}OCH_2CH_3}$$

$$\xrightarrow{\text{1. } HO^-,\ H_2O;\ \text{2. } H^+;\ \text{3. heat}} \underset{\text{4-Phenyl-2-butanone}}{CH_3\overset{O}{\overset{\|}{C}}CH_2CH_2C_6H_5}$$

(*b*) Identify the synthon in 3-phenylpropanoic acid that is derived from malonic ester by disconnecting the molecule at its α carbon atom.

$$C_6H_5CH_2 \vdots CH_2\overset{O}{\overset{\|}{C}}OH \Longrightarrow C_6H_5CH_2X + :\bar{C}H_2\overset{O}{\overset{\|}{C}}OH$$

disconnect here

Here as in (*a*), a benzyl halide is the required alkylating agent.

$$\underset{\text{Diethyl malonate}}{CH_2(COOCH_2CH_3)_2} + \underset{\text{Benzyl bromide}}{C_6H_5CH_2Br} \xrightarrow[\text{ethanol}]{NaOCH_2CH_3} \underset{\text{Diethyl benzylmalonate}}{C_6H_5CH_2CH(COOCH_2CH_3)_2}$$

$$\xrightarrow{\text{1. } HO^-,\ H_2O;\ \text{2. } H^+;\ \text{3. heat}} \underset{\text{3-Phenylpropanoic acid}}{C_6H_5CH_2CH_2COOH}$$

(*c*) The target molecule can be prepared by a Knoevenagel condensation of benzaldehyde and diethyl malonate.

$$C_6H_5CH \vdots\!= C(CO_2CH_2CH_3)_2 \Longrightarrow C_6H_5\overset{O}{\overset{\|}{C}}H + CH_2(CO_2CH_2CH_3)_2$$

First benzaldehyde must be prepared by oxidation of benzyl alcohol.

$$\underset{\text{Benzyl alcohol}}{C_6H_5CH_2OH} \xrightarrow{(C_5H_5N)_2CrO_3,\ CH_2Cl_2} \underset{\text{Benzaldehyde}}{C_6H_5\overset{O}{\overset{\|}{C}}H}$$

$$\underset{\text{Benzaldehyde}}{C_6H_5\overset{O}{\overset{\|}{C}}H} + \underset{\text{Diethyl malonate}}{CH_2(COOCH_2CH_3)_2} \xrightarrow[\text{acetic acid}]{\text{piperidine}} \underset{\text{Diethyl benzylidenemalonate}}{C_6H_5CH{=}C(COOCH_2CH_3)_2}$$

(*d*) In this synthesis the desired 1,3-diol function can be derived by reduction of a malonic ester derivative. First propene must be converted to an allyl halide for use as an alkylating agent.

$$CH_2{=}CHCH_2CH(CH_2OH)_2 \Longrightarrow CH_2{=}CHCH_2X + :\bar{C}H(COOCH_2CH_3)_2$$

$$\underset{\text{Propene}}{CH_2{=}CHCH_3} \xrightarrow[\text{heat}]{Cl_2} \underset{\text{Allyl chloride}}{CH_2{=}CHCH_2Cl}$$

$$\underset{\text{Diethyl malonate}}{CH_2(COOCH_2CH_3)_2} + \underset{\text{Allyl chloride}}{CH_2{=}CHCH_2Cl} \xrightarrow[\text{ethanol}]{NaOCH_2CH_3} \underset{\text{Diethyl 2-allylmalonate}}{CH_2{=}CHCH_2CH(COOCH_2CH_3)_2}$$

$$\xrightarrow{1.\ LiAlH_4;\ 2.\ H_2O} \underset{\text{2-Allyl-1,3-propanediol}}{CH_2{=}CHCH_2CH(CH_2OH)_2}$$

(*e*) The desired primary alcohol may be prepared by reduction of the corresponding carboxylic acid, which in turn is available from the malonic ester synthesis using allyl chloride, including saponification and decarboxylation of the diester [prepared in (*d*)].

$$\underset{\text{4-Penten-1-ol}}{CH_2{=}CHCH_2CH_2CH_2OH} \Longrightarrow CH_2{=}CHCH_2CH_2CO_2H \Longrightarrow CH_2{=}CHCH_2CH(CO_2CH_2CH_3)_2$$

The correct sequence of reactions is:

$$\underset{\text{Diethyl 2-allylmalonate [prepared as in (d)]}}{CH_2{=}CHCH_2CH(COOCH_2CH_3)_2} \xrightarrow{1.\ HO^-;\ 2.\ H^+;\ 3.\ heat} \underset{\text{4-Pentenoic acid}}{CH_2{=}CHCH_2CH_2COOH}$$

$$\xrightarrow{1.\ LiAlH_4;\ 2.\ H_2O} \underset{\text{4-Penten-1-ol}}{CH_2{=}CHCH_2CH_2CH_2OH}$$

(*f*) The desired product is an alcohol. It can be prepared by reduction of a ketone, which in turn can be prepared by the acetoacetic ester synthesis.

$$CH_2{=}CHCH_2CH_2\overset{OH}{\overset{|}{C}}HCH_3 \Longrightarrow CH_2{=}CHCH_2CH_2\overset{O}{\overset{\|}{C}}CH_3 \Longrightarrow CH_2{=}CHCH_2X \quad :\bar{C}H_2\overset{O}{\overset{\|}{C}}CH_3$$

Therefore

$$\underset{\text{Ethyl acetoacetate}}{CH_3\overset{O}{\overset{\|}{C}}CH_2\overset{O}{\overset{\|}{C}}OCH_2CH_3} + \underset{\text{Allyl chloride}}{CH_2{=}CHCH_2Cl} \xrightarrow[\text{ethanol}]{NaOCH_2CH_3} CH_3\overset{O}{\overset{\|}{C}}\underset{CH_2CH{=}CH_2}{\underset{|}{C}H}\overset{O}{\overset{\|}{C}}OCH_2CH_3$$

$$\xrightarrow{1.\ HO^-,\ H_2O;\ 2.\ H^+;\ 3.\ heat} CH_3\overset{O}{\overset{\|}{C}}CH_2CH_2CH{=}CH_2 \xrightarrow[CH_3OH]{NaBH_4} \underset{\text{5-Hexen-2-ol}}{CH_3\underset{OH}{\underset{|}{C}H}CH_2CH_2CH{=}CH_2}$$

(*g*) Cyclopropanecarboxylic acid may be prepared by a malonic ester synthesis, as retrosynthetic analysis shows.

$$\triangleright\!-CO_2H \Longrightarrow \triangleright\!<^{CO_2CH_2CH_3}_{CO_2CH_2CH_3} \Longrightarrow \begin{array}{c}CH_2X\\|\\CH_2X\end{array} + CH_2(CO_2CH_2CH_3)_2$$

The desired reaction sequence is:

$$\underset{\text{Diethyl malonate}}{CH_2(COOCH_2CH_3)_2} + \underset{\text{1,2-Dibromoethane}}{BrCH_2CH_2Br} \xrightarrow[\text{ethanol}]{NaOCH_2CH_3} \triangleright\!<^{COOCH_2CH_3}_{COOCH_2CH_3}$$

$$\xrightarrow{\text{1. HO}^-\text{, H}_2\text{O; 2. H}^+\text{; 3. heat}} \triangleright\!-\overset{O}{\overset{\|}{C}}OH$$

Cyclopropanecarboxylic acid

(*h*) Treatment of the diester formed in (*g*) with ammonia gives a diamide.

$$\triangleright\!<^{\overset{O}{\overset{\|}{C}}OCH_2CH_3}_{\underset{O}{\underset{\|}{C}}OCH_2CH_3} \xrightarrow{NH_3} \triangleright\!<^{\overset{O}{\overset{\|}{C}}NH_2}_{\underset{O}{\underset{\|}{C}}NH_2}$$

Diethyl cyclopropane-1,1-dicarboxylate [prepared as in (*g*)] — Cyclopropane-1,1-dicarboxamide

(*i*) We need to extend the carbon chain of the starting material by *four* carbons. One way to accomplish this is by way of a malonic ester synthesis at each end of the chain.

$$\underset{\text{Octanedioic acid}}{HO\overset{O}{\overset{\|}{C}}(CH_2)_6\overset{O}{\overset{\|}{C}}OH} \xrightarrow[\text{2. H}_2\text{O}]{\text{1. LiAlH}_4} HOCH_2(CH_2)_6CH_2OH \xrightarrow[\text{or PBr}_3]{HBr} \underset{\text{1,8-Dibromooctane}}{BrCH_2(CH_2)_6CH_2Br}$$

$$2\ \underset{\text{Diethyl malonate}}{CH_2(COOCH_2CH_3)_2} + \underset{\text{1,8-Dibromooctane}}{Br(CH_2)_8Br} \xrightarrow{NaOCH_2CH_3} (CH_3CH_2OOC)_2CH(CH_2)_8CH(COOCH_2CH_3)_2$$

$$\xrightarrow{\text{1. HO}^-\text{, H}_2\text{O; 2. H}^+\text{; 3. heat}} HO\overset{O}{\overset{\|}{C}}CH_2(CH_2)_8CH_2\overset{O}{\overset{\|}{C}}OH$$

Dodecanoic acid

22.26 The problem states that diphenadione is prepared from 1,1-diphenylacetone and dimethyl 1,2-benzenedicarboxylate. Therefore, disconnect the molecule in a way that reveals the two reactants.

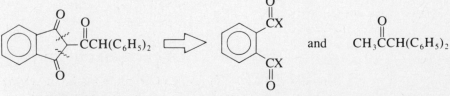

$$\text{Diphenadione (indane-1,3-dione with 2-}\overset{O}{\overset{\|}{C}}CH(C_6H_5)_2\text{)} \Longrightarrow C_6H_4(\overset{O}{\overset{\|}{C}}X)_2 \text{ and } CH_3\overset{O}{\overset{\|}{C}}CH(C_6H_5)_2$$

Diphenadione

Thus all that is required is to treat dimethyl 1,2-benzenedicarboxylate and 1,1-diphenylacetone with base. Two successive acylations of a ketone enolate occur; the first is intermolecular, the second intramolecular.

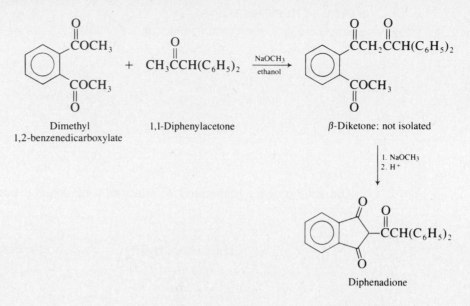

$$\text{Dimethyl 1,2-benzenedicarboxylate } (C_6H_4(\overset{O}{\overset{\|}{C}}OCH_3)_2) + CH_3\overset{O}{\overset{\|}{C}}CH(C_6H_5)_2 \xrightarrow[\text{ethanol}]{NaOCH_3} C_6H_4(\overset{O}{\overset{\|}{C}}CH_2\overset{O}{\overset{\|}{C}}CH(C_6H_5)_2)(\overset{O}{\overset{\|}{C}}OCH_3)$$

Dimethyl 1,2-benzenedicarboxylate | 1,1-Diphenylacetone | β-Diketone: not isolated

1. $NaOCH_3$
2. H^+

$$\text{(indane-1,3-dione)}-\overset{O}{\overset{\|}{C}}CH(C_6H_5)_2$$

Diphenadione

22.27 Esters react with amines to give amides. Each nitrogen of 1,2-diphenylhydrazine reacts with a separate ester function of diethyl butylmalonate.

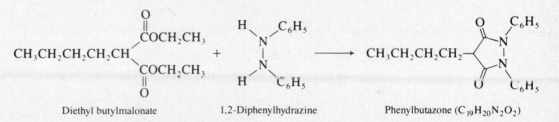

$$CH_3CH_2CH_2CH_2CH(\overset{O}{\overset{\|}{C}}OCH_2CH_3)_2 + C_6H_5NH{-}NHC_6H_5 \longrightarrow CH_3CH_2CH_2CH_2{-}(\text{3,5-dioxo-1,2-diphenylpyrazolidin-4-yl})$$

Diethyl butylmalonate | 1,2-Diphenylhydrazine | Phenylbutazone ($C_{19}H_{20}N_2O_2$)

22.28 The sodium salt of ethyl acetoacetate reacts with chloroacetone in a manner analogous to that of its reaction with alkyl halides.

$$CH_3\overset{O}{\overset{\|}{C}}CH_2\overset{O}{\overset{\|}{C}}OCH_2CH_3 + ClCH_2\overset{O}{\overset{\|}{C}}CH_3 \xrightarrow{NaOCH_2CH_3} CH_3\overset{O}{\overset{\|}{C}}\underset{CH_2\overset{\|}{\underset{O}{C}}CH_3}{\underset{|}{C}H}\overset{O}{\overset{\|}{C}}OCH_2CH_3$$

Ethyl acetoacetate | Chloroacetone | Ethyl 2-acetyl-4-oxopentanoate (compound A, $C_9H_{14}O_4$)

One possible mechanism for the acid-catalyzed conversion of compound A to compound B begins with the protonation of one of the ketone carbonyl groups.

$$CH_3\overset{O}{\overset{\|}{C}}\underset{CH_2\overset{\|}{\underset{O}{C}}CH_3}{\underset{|}{C}H}\overset{O}{\overset{\|}{C}}OCH_2CH_3 \underset{}{\overset{H_2SO_4}{\rightleftharpoons}} CH_3\overset{^+OH}{\overset{\|}{C}}\underset{CH_2\overset{\|}{\underset{O}{C}}CH_3}{\underset{|}{C}H}\overset{O}{\overset{\|}{C}}OCH_2CH_3$$

This is followed by intramolecular nucleophilic attack by the oxygen of the other ketone carbonyl. Subsequent loss of a proton and dehydration complete the process.

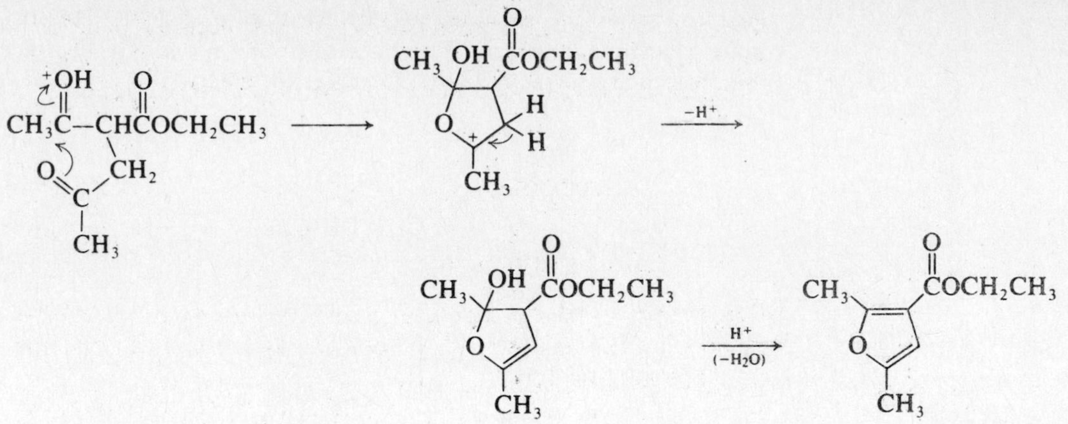

$$CH_3\overset{+OH}{\overset{\|}{C}}-CH\overset{O}{\overset{\|}{C}}OCH_2CH_3 \longrightarrow \xrightarrow{-H^+}$$

$$\xrightarrow[(-H_2O)]{H^+}$$

Compound B

The same product (compound B) is formed if the initial protonation step involves the other ketone function.

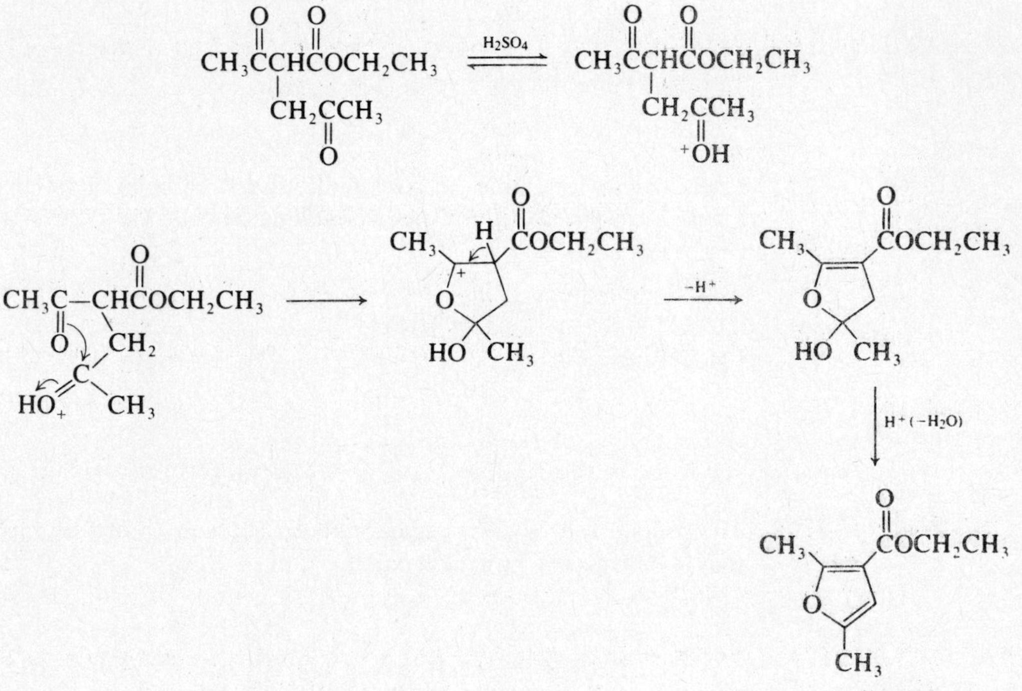

$$CH_3\overset{O}{\overset{\|}{C}}CH\overset{O}{\overset{\|}{C}}OCH_2CH_3 \;(CH_2\overset{O}{\overset{\|}{C}}CH_3) \xrightleftharpoons{H_2SO_4} CH_3\overset{O}{\overset{\|}{C}}CH\overset{O}{\overset{\|}{C}}OCH_2CH_3 \;(CH_2\overset{+OH}{\overset{\|}{C}}CH_3)$$

$$\longrightarrow \xrightarrow{-H^+} \xrightarrow{H^+(-H_2O)}$$

Compound B

We cannot choose between these two very similar pathways; indeed, *both* may be occurring. Other mechanisms are also possible, including some that involve enol forms of compound A.

22.29 Styrene oxide will be attacked by the anion of diethyl malonate at its less hindered ring position.

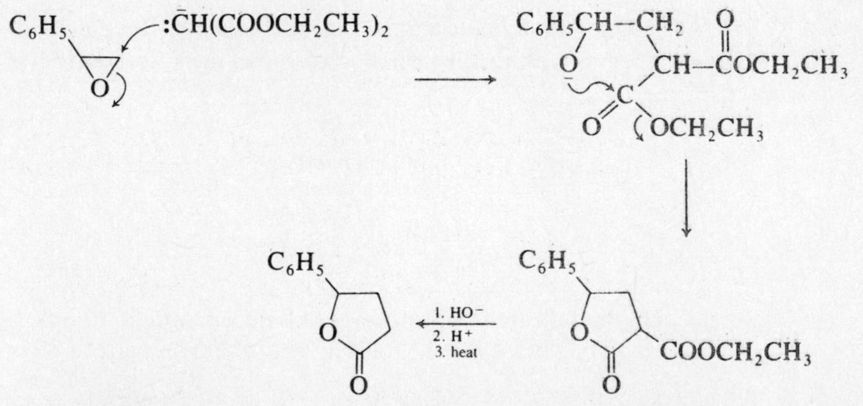

$$C_6H_5\text{-epoxide} + :\bar{C}H(COOCH_2CH_3)_2 \longrightarrow C_6H_5CH(O^-)-CH_2-CH(COOCH_2CH_3)-\overset{O}{\overset{\|}{C}}OCH_2CH_3 \longrightarrow$$

$$\xleftarrow[\text{2. } H^+ \text{ 3. heat}]{\text{1. } HO^-}$$

$COOCH_2CH_3$

The product is 4-phenylbutanolide. It has been prepared in 72 percent yield by this procedure.

22.30 The first task is to convert acetic acid to ethyl chloroacetate.

$$\underset{\text{Acetic acid}}{CH_3\overset{O}{\overset{\|}{C}}OH} \xrightarrow[P]{Cl_2} \underset{\text{Chloroacetic acid}}{ClCH_2\overset{O}{\overset{\|}{C}}OH} \xrightarrow[H^+]{CH_3CH_2OH} \underset{\text{Ethyl chloroacetate}}{ClCH_2\overset{O}{\overset{\|}{C}}OCH_2CH_3}$$

Chlorination must precede esterification because the Hell-Volhard-Zelinsky reaction requires a carboxylic acid, not an ester, as the starting material. The remaining step is a nucleophilic substitution reaction.

$$\underset{\text{Ethyl chloroacetate}}{ClCH_2\overset{O}{\overset{\|}{C}}OCH_2CH_3} \xrightarrow{NaCN} \underset{\text{Ethyl cyanoacetate}}{N{\equiv}CCH_2\overset{O}{\overset{\|}{C}}OCH_2CH_3}$$

22.31 (*a*) The Knoevenagel reaction between ethyl cyanoacetate and benzaldehyde is similar to that of diethyl malonate.

$$\underset{\text{Benzaldehyde}}{C_6H_5\overset{O}{\overset{\|}{C}}H} + \underset{\text{Ethyl cyanoacetate}}{N{\equiv}CCH_2\overset{O}{\overset{\|}{C}}OCH_2CH_3} \longrightarrow \underset{\text{Ethyl 2-cyano-3-phenyl-2-propenoate}}{C_6H_5CH{=}C(\overset{O}{\overset{\|}{C}}OCH_2CH_3)(C{\equiv}N)}$$

(*b*) Michael addition of cyanide ion to the double bond gives a dinitrile.

$$\underset{\text{Ethyl 2-cyano-3-phenyl-2-propenoate}}{C_6H_5CH{=}C(\overset{O}{\overset{\|}{C}}OCH_2CH_3)(C{\equiv}N)} \xrightarrow[H_2O]{NaCN} \underset{\text{Ethyl 2,3-dicyano-3-phenylpropanoate}}{C_6H_5CH(C{\equiv}N)CH(C{\equiv}N)\overset{O}{\overset{\|}{C}}OCH_2CH_3}$$

(*c*) Acid hydrolysis converts both cyano groups and the ester to carboxylic acid functions. The triacid then undergoes decarboxylation.

$$\underset{\text{Ethyl 2,3-dicyano-3-phenylpropanoate}}{C_6H_5CH(C{\equiv}N)CH(C{\equiv}N)\overset{O}{\overset{\|}{C}}OCH_2CH_3} \xrightarrow{H_3O^+} \underset{\text{(Not isolated)}}{\left[C_6H_5CH(COOH)CH(COOH)COOH\right]} \xrightarrow[-CO_2]{heat} \underset{\text{2-Phenylsuccinic acid}}{HO\overset{O}{\overset{\|}{C}}CH(C_6H_5)CH_2\overset{O}{\overset{\|}{C}}OH}$$

The diacid is 2-phenylsuccinic acid. It is converted to a cyclic imide on being heated with methylamine.

$$\underset{\text{2-Phenylsuccinic acid}}{HO_2C{-}CH(C_6H_5){-}CH_2{-}CO_2H} + \underset{\text{Methylamine}}{CH_3NH_2} \xrightarrow{heat} \underset{\text{Phensuximide}}{\text{3-phenyl-N-methylsuccinimide } (C_6H_5\text{ on ring C; } O{=}C{-}N(CH_3){-}C{=}O)}$$

22.32 From the hint given in the problem, it can be seen that synthesis of 2-methyl-2-propyl-1,3-propanediol is required. This diol is obtained by a sequence involving dialkylation of diethyl malonate.

$$(CH_3CH_2CH_2)(CH_3)C(CH_2OCNH_2)_2 \Rightarrow (CH_3CH_2CH_2)(CH_3)C(CH_2OH)_2 \Rightarrow (CH_3CH_2CH_2)(CH_3)C(CO_2CH_2CH_3)_2$$

(in each carbamate group CH_2OCNH_2 the C bears =O)

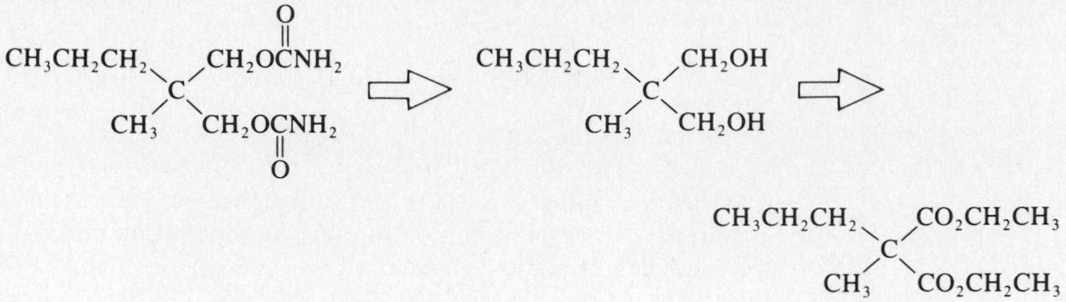

Begin the synthesis by dialkylation of diethyl malonate.

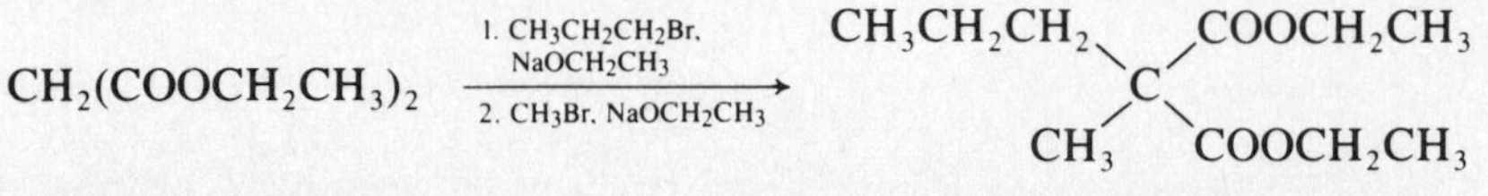

$$\underset{\text{Diethyl malonate}}{CH_2(COOCH_2CH_3)_2} \xrightarrow[\text{2. } CH_3Br,\ NaOCH_2CH_3]{\text{1. } CH_3CH_2CH_2Br,\ NaOCH_2CH_3} \underset{\text{Diethyl 2-methyl-2-propylmalonate}}{(CH_3CH_2CH_2)(CH_3)C(COOCH_2CH_3)_2}$$

Convert the ester functions to primary alcohols by reduction.

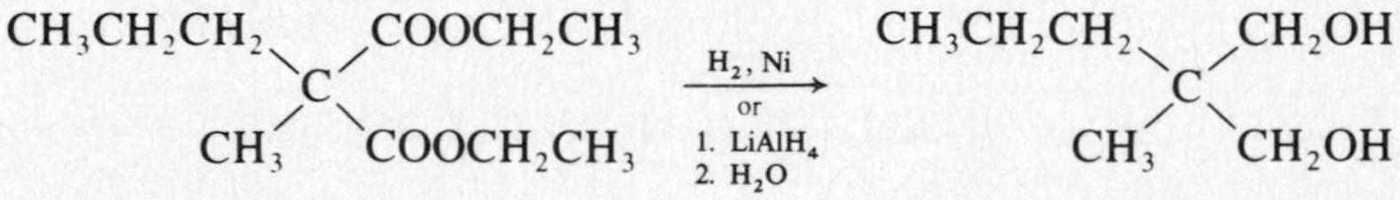

$$\underset{\text{Diethyl 2-methyl-2-propylmalonate}}{(CH_3CH_2CH_2)(CH_3)C(COOCH_2CH_3)_2} \xrightarrow[\text{1. } LiAlH_4 \text{ 2. } H_2O]{H_2,\ Ni \text{ or}} \underset{\text{2-Methyl-2-propyl-1,3-propanediol}}{(CH_3CH_2CH_2)(CH_3)C(CH_2OH)_2}$$

Conversion of the primary alcohol groups to carbamate esters completes the synthesis.

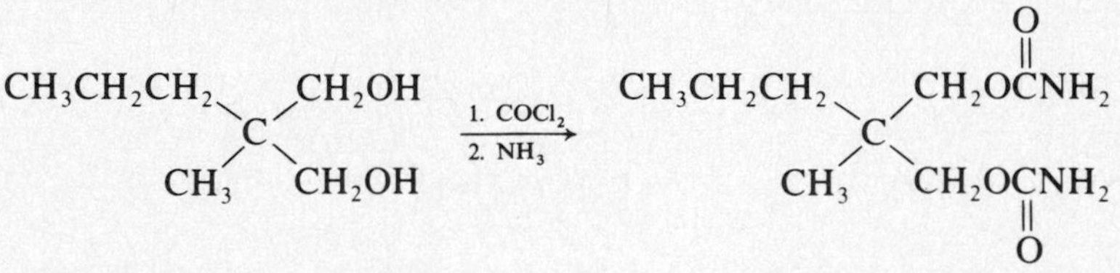

$$\underset{\text{2-Methyl-2-propyl-1,3-propanediol}}{(CH_3CH_2CH_2)(CH_3)C(CH_2OH)_2} \xrightarrow[\text{2. } NH_3]{\text{1. } COCl_2} \underset{\text{Meprobamate}}{(CH_3CH_2CH_2)(CH_3)C(CH_2OCNH_2)_2}$$

(in Meprobamate each CH_2OCNH_2 group has C=O)

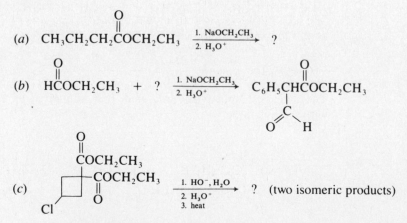

SELF-TEST

PART A

A-1. Give the structure of the reactant, reagent, or product omitted from each of the following:

(*a*) $CH_3CH_2CH_2\overset{O}{\overset{\|}{C}}OCH_2CH_3 \xrightarrow[\text{2. } H_3O^+]{\text{1. } NaOCH_2CH_3} ?$

(*b*) $H\overset{O}{\overset{\|}{C}}OCH_2CH_3 + ? \xrightarrow[\text{2. } H_3O^+]{\text{1. } NaOCH_2CH_3} C_6H_5CH(CHO)\overset{O}{\overset{\|}{C}}OCH_2CH_3$

(the CH bears a $\overset{O}{\|}C{-}H$ (formyl) group)

(*c*) 3-Chlorocyclobutane-1,1-dicarboxylate diethyl ester (ring carbon bearing $\overset{O}{\overset{\|}{C}}OCH_2CH_3$ and $\overset{O}{\overset{\|}{C}}OCH_2CH_3$; Cl on the ring) $\xrightarrow[\text{2. } H_3O^+ \text{ 3. heat}]{\text{1. } HO^-,\ H_2O} ?$ (two isomeric products) (*d*) $CH_3\overset{\overset{O}{\|}}{C}CH_2\overset{\overset{O}{\|}}{C}OCH_2CH_3 \xrightarrow[2.\ C_6H_5CH_2Br]{1.\ NaOCH_2CH_3}$?

(*e*) Product of (*d*) $\xrightarrow{?}$ $C_6H_5CH_2CH_2\overset{\overset{O}{\|}}{C}CH_3$

(*f*) $CH_3\overset{\overset{O}{\|}}{C}\underset{\underset{CO_2H}{|}}{C}HCH_2CO_2H \xrightarrow{\text{heat}} CO_2 + ?$

A-2. Provide the correct structures of compounds A through E in the following reaction sequences:

(*a*) A $\xrightarrow[2.\ H_3O^+]{1.\ NaOCH_2CH_3}$ 2-oxocyclopentane with $\overset{\overset{O}{\|}}{C}OCH_2CH_3$ $\xrightarrow[2.\ CH_3CH_2I]{1.\ NaOCH_2CH_3}$ B $\xrightarrow[2.\ H_3O^+\ \ 3.\ heat]{1.\ HO^-,\ H_2O}$ C

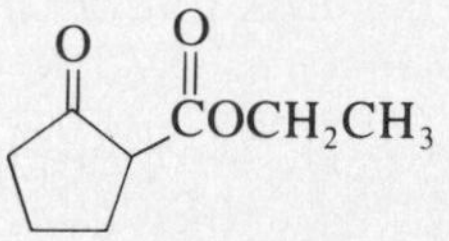

(*b*) $CH_3CH_2CH_2\overset{\overset{O}{\|}}{C}OCH_2CH_3 \xrightarrow[2.\ H_3O^+]{1.\ NaOCH_2CH_3}$ D $\xrightarrow[2.\ H_3O^+\ \ 3.\ heat]{1.\ HO^-,\ H_2O}$ E + CO_2

A-3. Give a series of steps that will enable preparation of each of the following compounds from the starting material(s) given and any other necessary reagents.

(*a*) $CH_3\overset{\overset{O}{\|}}{C}CH_2CH_2\overset{\overset{O}{\|}}{C}OH$ from ethyl acetoacetate

(*b*) $C_6H_5\overset{\overset{O}{\|}}{C}CH_2CH_2CH_2\overset{\overset{O}{\|}}{C}CH_3$ from $C_6H_5\overset{\overset{O}{\|}}{C}CH_3$ and diethyl carbonate

A-4. Write a stepwise mechanism for the reaction of ethyl propanoate with sodium ethoxide in ethanol.

PART B

B-1. Which of the following compounds is the strongest acid?

(*a*) $HCO_2CH_2CH_3$
(*b*) $CH_3CH_2O_2CCH_2CO_2CH_2CH_3$
(*c*) $CH_3CH_2O_2CCH_2CH_2CO_2CH_2CH_3$
(*d*) $CH_3CO_2CH_2CH_3$

B-2. Which of the following will yield a ketone and carbon dioxide following saponification, acidification, and heating?

(*a*) $CH_3CH_2\underset{\underset{\overset{\|}{O}COCH_2CH_3}{|}}{C}HCH_2\overset{\overset{O}{\|}}{C}CH_3$ (branch: $\overset{O}{\|}$COCH$_2$CH$_3$)

(*b*) $CH_3CH_2\underset{|}{C}H\overset{\overset{O}{\|}}{C}OCH_2CH_3$ (branch: $\overset{O}{\|}$COCH$_2$CH$_3$)

(*c*) $CH_3CH_2\underset{|}{C}H\overset{\overset{O}{\|}}{C}CH_3$ (branch: $\overset{O}{\|}$CCH$_2$CH$_3$)

(*d*) $CH_3CH_2\underset{|}{C}H\overset{\overset{O}{\|}}{C}CH_3$ (branch: $\overset{O}{\|}$COCH$_2$CH$_3$)

B-3. Which of the following keto esters is *not* likely to have been prepared by a Claisen condensation?

(*a*) $CH_3CH_2\overset{O}{\overset{\|}{C}}\underset{CH_3}{\underset{|}{C}H}\overset{O}{\overset{\|}{C}}OCH_2CH_3$ (*b*) $C_6H_5\overset{O}{\overset{\|}{C}}\underset{CH_3}{\underset{|}{C}H}\overset{O}{\overset{\|}{C}}OCH_2CH_3$

(*c*) $(CH_3)_2CH\overset{O}{\overset{\|}{C}}\underset{\overset{\|}{O}\!COCH_2CH_3}{\underset{|}{C}}(CH_3)_2$ (*d*) $(CH_3)_2CHCH_2\overset{O}{\overset{\|}{C}}\underset{CH(CH_3)_2}{\underset{|}{C}H}\overset{O}{\overset{\|}{C}}OCH_2CH_3$

B-4. Dieckmann condensation of $CH_3CH_2O\overset{O}{\overset{\|}{C}}(CH_2)_5\overset{O}{\overset{\|}{C}}OCH_2CH_3$ will yield:

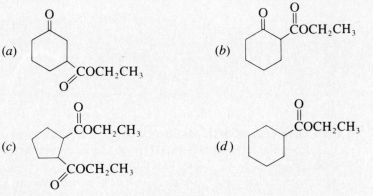

(*a*) O, COCH₂CH₃ → O, $\overset{O}{\overset{\|}{C}}OCH_2CH_3$ (cyclohexanone with ester at 3-position)

(*b*) O, $\overset{O}{\overset{\|}{C}}OCH_2CH_3$ (cyclohexanone with ester at 2-position)

(*c*) $\overset{O}{\overset{\|}{C}}OCH_2CH_3$, $\overset{O}{\overset{\|}{C}}OCH_2CH_3$ (cyclopentane-1,2-diester)

(*d*) $\overset{O}{\overset{\|}{C}}OCH_2CH_3$ (cyclohexane ester)

CHAPTER

23

ALKYLAMINES

IMPORTANT TERMS AND CONCEPTS

Nomenclature (Sec. 23.1) Amines are classified as being either primary, secondary, or tertiary, depending on the number of carbon substituents attached to the nitrogen atom.

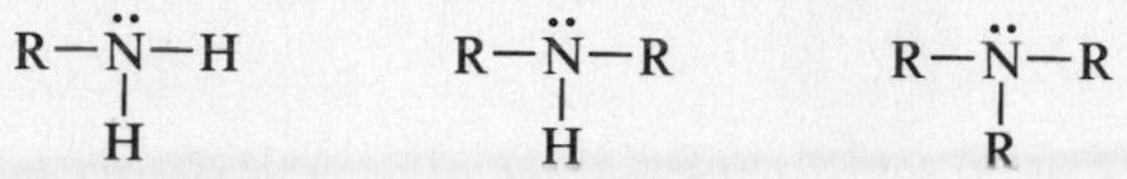

$$\underset{\text{Primary amine}}{R-\ddot{N}(H)-H} \qquad \underset{\text{Secondary amine}}{R-\ddot{N}(H)-R} \qquad \underset{\text{Tertiary amine}}{R-\ddot{N}(R)-R}$$

Primary amines may be named either by adding *-amine* to the alkyl group name or by replacing the *-e* ending of the longest carbon chain with *-amine*. For example

Pentylamine, or 1-pentanamine

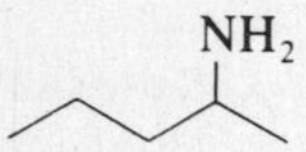

1-Methylbutylamine, or 2-pentanamine

Secondary and tertiary amines are named as *N*-substituted derivatives of primary amines. For example

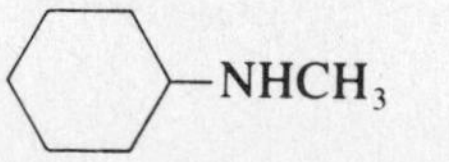

N-Methylcyclohexylamine or, *N*-methylcyclohexanamine

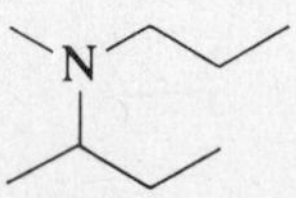

N-Methyl-*N*-propyl-*sec*-butylamine, or *N*-methyl-*N*-propyl-2-butanamine

Structure, Bonding, and Properties (Secs. 23.2, 23.3) Alkylamines have a pyramidal arrangement of the three substituents attached to the nitrogen atom, the fourth valence position being occupied by an unshared pair of electrons. It is this pair of electrons that enables alkylamines to act as bases or nucleophiles. The nitrogen atom is sp^3 hybridized and undergoes rapid pyramidal inversion, as shown below.

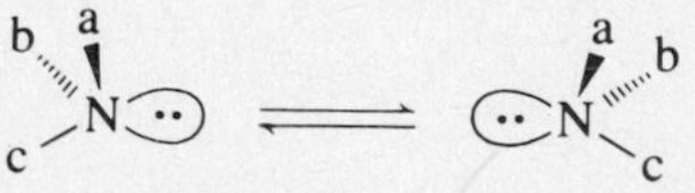

Hydrogen bonding interactions in amines are weaker than those in alcohols; therefore primary and secondary amines have lower boiling points than alcohols of comparable molecular weight. Tertiary amine molecules lack the ability to form hydrogen bonds among themselves, although they may act as hydrogen bond acceptors with molecules of other substances.

Basicity (Sec. 23.4) Amines act as bases by donating their unshared electron pairs to suitable acceptors. They react with acids by proton transfer to form *ammonium salts.*

$$R_3N: + HX \longrightarrow R_3\overset{+}{N}H\ X^-$$

The base strength of an amine is described by the equation

$$R_3N + H_2O \rightleftharpoons R_3\overset{+}{N}H + OH^-$$

The basicity constant K_b is related to the equlibrium constant for this equation by:

$$K_b = K[H_2O] = \frac{[R_3\overset{+}{N}H][OH^-]}{[R_3N]}$$

In aqueous solution the basicity trend of alkylamines may be summarized as:

$$\underset{\text{Least basic}}{NH_3} < RNH_2 \sim R_3N < \underset{\text{Most basic}}{R_2NH}$$

The *gas-phase* basicity trend, however, is somewhat different:

$$\underset{\text{Least basic}}{NH_3} < RNH_2 < R_2NH < \underset{\text{Most basic}}{R_3N}$$

The difference between the trends is explained by the presence of hydrogen bonds between NH protons and water molecules, which cannot be formed in the gas phase.

Hofmann and Cope Eliminations; Mechanistic Details (Secs. 23.14 to 23.18) The Hofmann and Cope eliminations are thermally induced reactions of amine derivatives, which yield alkenes. Hofmann eliminations are elimination reactions of quaternary ammonium hydroxides. Cope eliminations use amine oxides as the starting material. Examples are shown in the next section. The regioselectivity of the Hofmann elimination is such that the *less substituted* of theoretically possible alkenes results from the reaction. The less sterically hindered β hydrogen is removed by the base; therefore the order of preference for deprotonation is:

$$\begin{array}{c}-\overset{+|}{N}-\\ |\\ -\underset{|}{C}-CH_3\end{array} > \begin{array}{c}-\overset{+|}{N}-\\ |\\ -\underset{|}{C}-CH_2-R\end{array} > \begin{array}{c}-\overset{+|}{N}-\\ |\\ -\underset{|}{C}-\underset{\underset{R}{|}}{CH}-R\end{array}$$

The Hofmann elimination is stereospecific. The preferred orientation for elimination is an anti arrangement of the β hydrogen and the nitrogen.

$$(CH_3)_3\overset{+}{N}-C(R_3)(R_4)-C(R_1)(R_2)-H \ \ {}^-OH \xrightarrow{\text{heat}} \begin{array}{c}R_3 \quad\quad R_1\\ C{=}C\\ R_4 \quad\quad R_2\end{array} + :N(CH_3)_3 + H_2O$$

By contrast, the Cope reaction arises from a syn elimination of the β hydrogen and nitrogen.

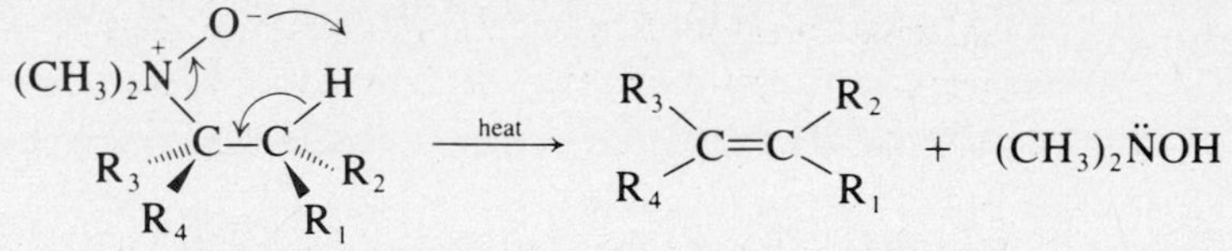

$$(CH_3)_2\overset{+}{N}(O^-)\text{-}C(R_3)(R_4)\text{-}C(H)(R_2)(R_1) \xrightarrow{\text{heat}} R_3R_4C{=}CR_2R_1 + (CH_3)_2\ddot{N}OH$$

Spectroscopic Analysis (Sec. 23.20) The infrared spectra of primary and secondary amines exhibit characteristic N—H stretching vibrations in the range 3000 to 3500 cm^{-1}. Two bands are observed for primary amines (RNH_2), while secondary amines (R_2NH) exhibit only one. The ^1H nmr chemical shifts of N—H protons, like those of hydroxyl protons, are variable and depend on the solvent, concentration, and temperature.

IMPORTANT REACTIONS

PREPARATION OF AMINES

The Gabriel Synthesis (Sec. 23.9)

General:

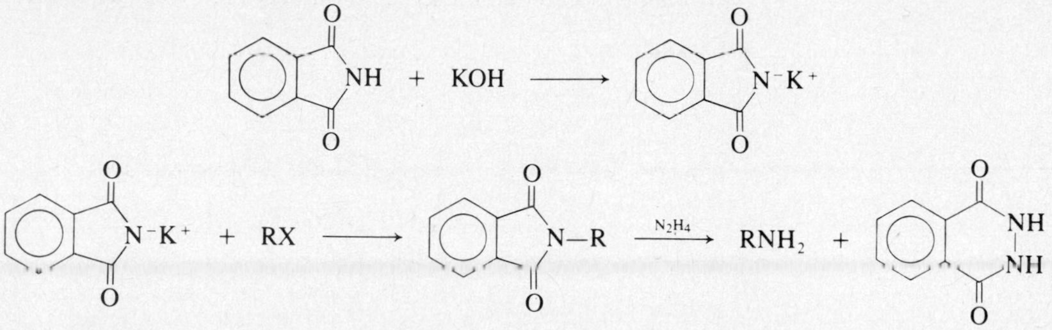

$$\text{Phthalimide (NH)} + KOH \longrightarrow \text{Phthalimide } N^-K^+$$

$$\text{Phthalimide } N^-K^+ + RX \longrightarrow \text{Phthalimide N—R} \xrightarrow{N_2H_4} RNH_2 + \text{phthalhydrazide (NH, NH)}$$

Example:

$$\text{Phthalimide (NH)} \xrightarrow[\text{2. } (CH_3)_2CHCH_2Cl \text{ 3. } N_2H_4]{\text{1. KOH}} (CH_3)_2CHCH_2NH_2$$

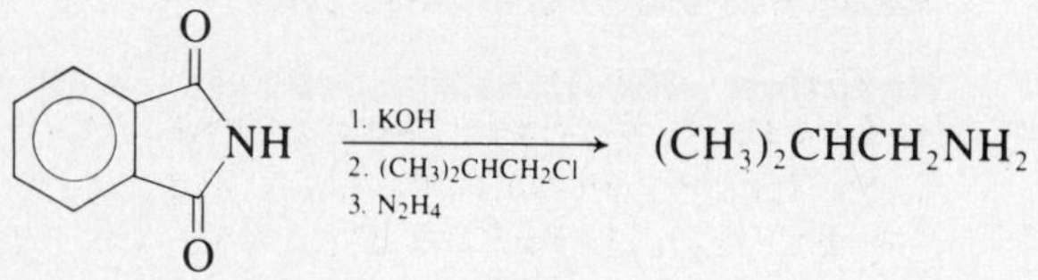

Reduction (Sec. 23.10)

General:

Azides:

$$RN_3 \xrightarrow[\text{2. } H_2O]{\text{1. } LiAlH_4} RNH_2$$

Nitriles:

$$RC{\equiv}N \xrightarrow[\text{2. } H_2O]{\text{1. } LiAlH_4} RCH_2NH_2$$

Oximes:

$$R_2C{=}NOH \xrightarrow[\text{ethanol}]{\text{Na}} R_2CHNH_2$$

Amides:

$$R\overset{O}{\overset{\|}{C}}NR'_2 \xrightarrow[\text{2. } H_2O]{\text{1. } LiAlH_4} RCH_2NR'_2$$

Examples:

$$CH_3CH_2Br \xrightarrow{NaN_3} CH_3CH_2N_3 \xrightarrow[2.\ H_2O]{1.\ LiAlH_4} CH_3CH_2NH_2$$

$$C_6H_5CH_2Cl \xrightarrow{NaCN} C_6H_5CH_2CN \xrightarrow[2.\ H_2O]{1.\ LiAlH_4} C_6H_5CH_2CH_2NH_2$$

$$CH_3\overset{O}{\overset{\|}{C}}CH(CH_3)_2 \xrightarrow{NH_2OH} CH_3\overset{NOH}{\overset{\|}{C}}CH(CH_3)_2 \xrightarrow[ethanol]{Na} CH_3\overset{NH_2}{\overset{|}{C}}HCH(CH_3)_2$$

$$C_6H_5CH_2CO_2H \xrightarrow{SOCl_2} C_6H_5CH_2\overset{O}{\overset{\|}{C}}Cl \xrightarrow{(CH_3CH_2)_2NH} C_6H_5CH_2\overset{O}{\overset{\|}{C}}N(CH_2CH_3)_2 \xrightarrow[2.\ H_2O]{1.\ LiAlH_4} C_6H_5CH_2CH_2N(CH_2CH_3)_2$$

Reductive Amination (Sec. 23.11)

General:

$$R_2C{=}O + R'NH_2 \xrightarrow{[H]} R_2CHNHR'$$

Examples:

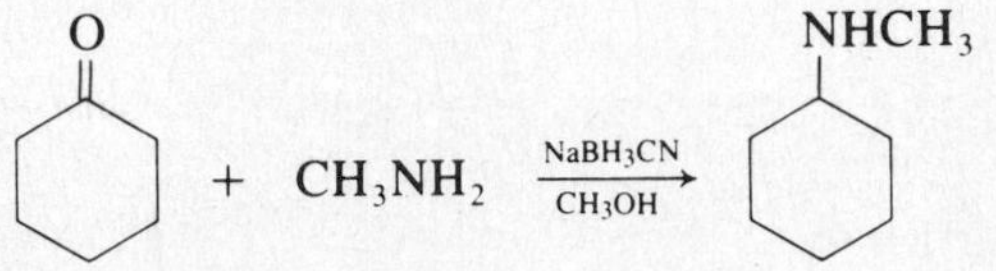

$$\text{cyclohexanone} + CH_3NH_2 \xrightarrow[CH_3OH]{NaBH_3CN} \text{cyclohexyl-}NHCH_3$$

$$C_6H_5CH_2\overset{O}{\overset{\|}{C}}H + NH_3 \xrightarrow{H_2,\ Ni} C_6H_5CH_2CH_2NH_2$$

REACTIONS OF AMINES

Reaction with Alkyl Halides (Sec. 23.13)

General:

$$RNH_2 \xrightarrow{R'X} RNHR' \xrightarrow{R'X} RNR'_2 \xrightarrow{R'X} R\overset{+}{N}R'_3\ X^-$$

Example:

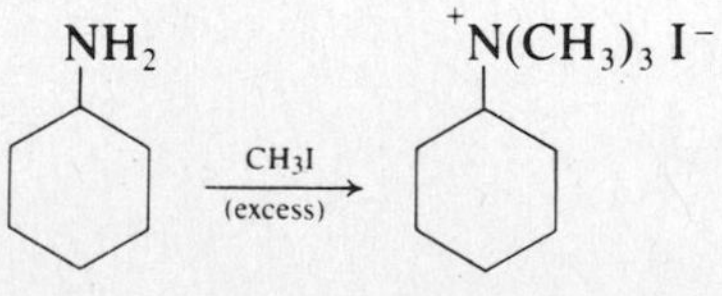

$$\text{cyclohexyl-}NH_2 \xrightarrow[(excess)]{CH_3I} \text{cyclohexyl-}\overset{+}{N}(CH_3)_3\ I^-$$

The Hofmann Elimination (Secs. 23.14 to 23.16)

General:

$$RCH_2CH_2\overset{+}{N}R_3\ I^- \xrightarrow[H_2O]{Ag_2O} RCH_2CH_2\overset{+}{N}R_3\ OH^- \xrightarrow{heat} RCH{=}CH_2 + R'_3N + H_2O$$

Example:

$$(CH_3)_2CHCH_2\overset{+}{N}(CH_3)_2CH_2CH_3\ I^- \xrightarrow[2.\ heat]{1.\ Ag_2O,\ H_2O} (CH_3)_2CHCH_2N(CH_3)_2 + CH_2{=}CH_2$$

The Cope Elimination (Secs. 23.17, 23.18)

General:

$$RCH_2CH_2NR'_2 \xrightarrow{H_2O_2} RCH_2CH_2\overset{O^-}{\overset{|}{\underset{+}{N}}}R'_2 \xrightarrow{\text{heat}} RCH{=}CH_2 + R'_2NOH$$

Example:

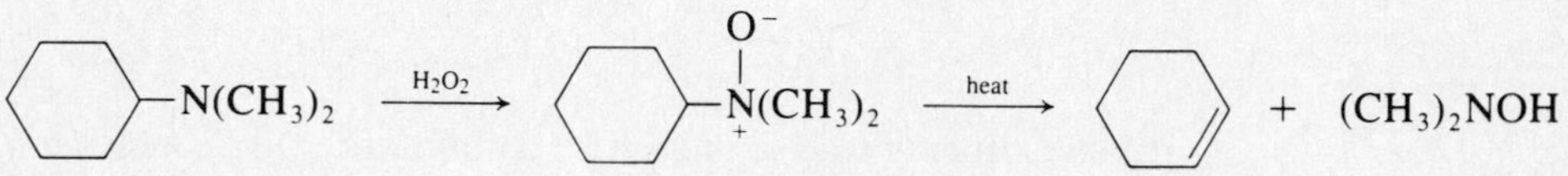

$$\text{C}_6\text{H}_{11}{-}N(CH_3)_2 \xrightarrow{H_2O_2} \text{C}_6\text{H}_{11}{-}\overset{O^-}{\overset{|}{\underset{+}{N}}}(CH_3)_2 \xrightarrow{\text{heat}} \text{cyclohexene} + (CH_3)_2NOH$$

Nitrosation (Sec. 23.19)

General:

$$R_2NH \xrightarrow[H_2O]{NaNO_2,\ HCl} R_2N{-}N{=}O$$

Example:

$$C_6H_5CH_2NHCH_3 \xrightarrow[H_2O]{NaNO_2,\ HCl} C_6H_5CH_2\overset{N=O}{\overset{|}{N}}CH_3$$

SOLUTIONS TO TEXT PROBLEMS

23.1 (*b*) The amino and phenyl groups are both attached to C-1 of an ethyl group.

$$C_6H_5\underset{NH_2}{\underset{|}{C}}HCH_3$$ 1-Phenylethylamine, or 1-phenylethanamine

(*c*) Recall that $(CH_3)_3CCH_2-$ is a neopentyl group. Neopentyl is an acceptable alkyl group name according to IUPAC rules.

$(CH_3)_3CCH_2NH_2$ Neopentylamine, or 2,2-dimethylpropanamine

(*d*) The parent alkyl group is a cyclopropyl ring. It has a methyl group and an amino substituent in a trans relationship.

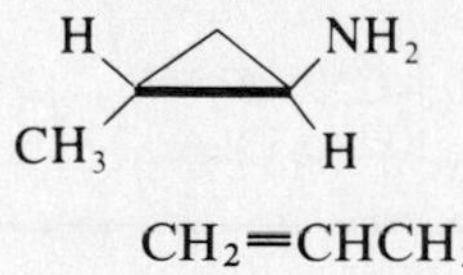

H, NH₂ / CH₃, H on cyclopropane ring: *trans*-2-Methylcyclopropylamine, or *trans*-2-methylcyclopropanamine

(*e*) $CH_2{=}CHCH_2NH_2$ Allylamine, or 2-propen-1-amine

23.2 (*b*) This compound is named as a dimethyl derivative of cycloheptanamine.

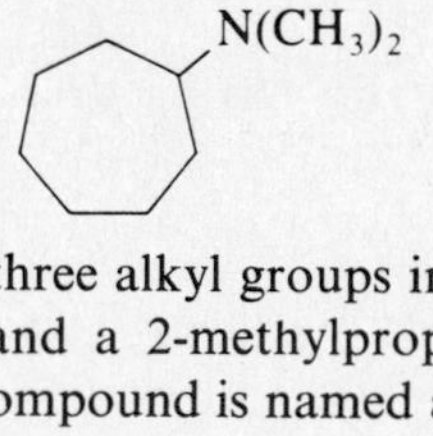

$\text{C}_7\text{H}_{13}{-}N(CH_3)_2$ *N*,*N*-Dimethylcycloheptylamine, or *N*,*N*-dimethylcycloheptanamine

(*c*) Nitrogen bears three alkyl groups in this tertiary alkylamine; they are an ethyl group, a methyl group, and a 2-methylpropyl group. The most complex one is taken as the parent, so this compound is named as a derivative of 2-methyl-1-propanamine

$$(CH_3)_2CHCH_2N(CH_3)(CH_2CH_3)$$

N-Ethyl-*N*,2-dimethyl-1-propanamine

23.3 The four alkyl groups, in order of decreasing sequence rule priority, are:

$$C_6H_5 > C_6H_5CH_2 > CH_2CH_2CH_3 > CH_3$$

Iodine is *not* covalently bonded to nitrogen and so is not included in the list of substituents.

If we orient the molecule with the lowest-priority substituent away from us

$$C_6H_5CH_2(CH_3)\overset{+}{N}{-}CH_2CH_2CH_3(C_6H_5) \qquad C_6H_5CH_2,\ CH_2CH_2CH_3,\ C_6H_5 \text{ on } N$$

the substituents trace a clockwise path in order of decreasing priority. The absolute configuration is *R*.

23.4 (*b*) As was seen in (*a*), a 2-fluoro substituent weakens the basicity of an amine. Successive fluorine substitution reduces the basicity even further. Thus, 2,2-difluoroethylamine is a relatively weak base, but it is a stronger base than its trifluoro analog.

$F_2CHCH_2NH_2$ — 2,2-Difluoroethylamine, stronger base: K_b 1.3×10^{-7} (pK_b 6.9)

$F_3CCH_2NH_2$ — 2,2,2-Trifluoroethylamine, weaker base: K_b 4×10^{-9} (pK_b 8.4)

(*c*) A methoxy group is electron-withdrawing as regards its inductive and field effects. It will reduce the basicity of an amine.

As $CH_3OCH_2CH_2\overset{+}{N}H_3$ is less stabilized than $CH_3CH_2CH_2CH_2\overset{+}{N}H_3$, butylamine is a stronger base than 2-aminoethyl methyl ether.

$CH_3CH_2CH_2CH_2NH_2$ — Butylamine, stronger base: K_b 4.1×10^{-4} (pK_b 3.4)

$CH_3OCH_2CH_2NH_2$ — 2-Aminoethyl methyl ether, weaker base: K_b 2.5×10^{-5} (pK_b 4.6)

23.5 The reaction that leads to allylamine is nucleophilic substitution by ammonia on allyl chloride.

$$\underset{\text{Allyl chloride}}{CH_2{=}CHCH_2Cl} + \underset{\text{Ammonia}}{2NH_3} \longrightarrow \underset{\text{Allylamine}}{CH_2{=}CHCH_2NH_2} + \underset{\text{Ammonium chloride}}{NH_4Cl}$$

Allyl chloride is prepared by free-radical chlorination of propene.

$$\underset{\text{Propene}}{CH_2{=}CHCH_3} + \underset{\text{Chlorine}}{Cl_2} \xrightarrow{400^\circ C} \underset{\text{Allyl chloride}}{CH_2{=}CHCH_2Cl} + \underset{\text{Hydrogen chloride}}{HCl}$$

23.6 (*b*) Isobutylamine is $(CH_3)_2CHCH_2NH_2$. It is a primary amine and can be prepared from a primary alkyl halide by the Gabriel synthesis.

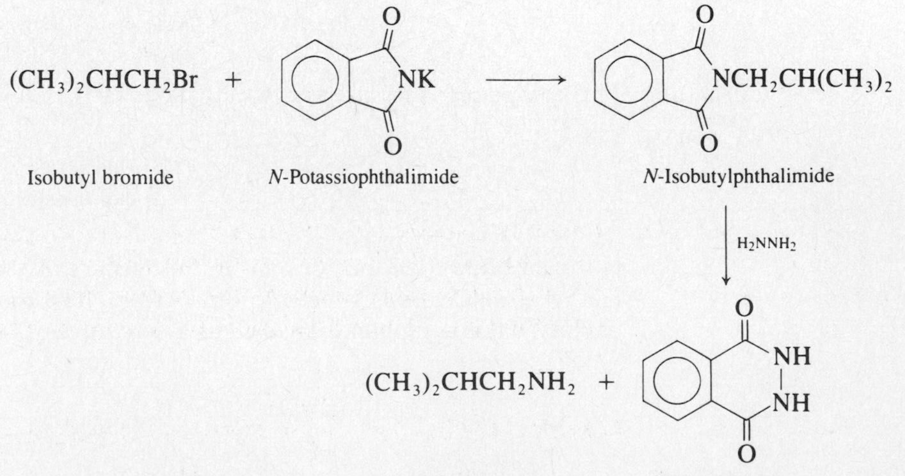

$$\underset{\text{Isobutyl bromide}}{(CH_3)_2CHCH_2Br} + \underset{N\text{-Potassiophthalimide}}{\text{phthalimide-NK}} \longrightarrow \underset{N\text{-Isobutylphthalimide}}{\text{phthalimide-NCH}_2\text{CH(CH}_3)_2}$$

$$\xrightarrow{H_2NNH_2} \underset{\text{Isobutylamine}}{(CH_3)_2CHCH_2NH_2} + \underset{\text{Phthalhydrazide}}{\text{phthalhydrazide (C=O, NH, NH, C=O)}}$$

(*c*) While *tert*-butylamine $(CH_3)_3CNH_2$ is a primary amine, it cannot be prepared by the Gabriel method because it would require an S_N2 reaction on a tertiary alkyl halide in the first step. Elimination occurs instead.

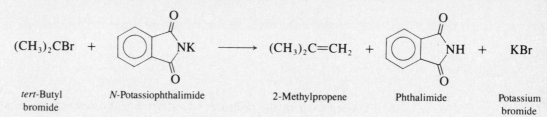

$$(CH_3)_3CBr + \text{N-Potassiophthalimide (NK)} \longrightarrow (CH_3)_2C{=}CH_2 + \text{Phthalimide (NH)} + KBr$$

tert-Butyl bromide; *N*-Potassiophthalimide; 2-Methylpropene; Phthalimide; Potassium bromide

(*d*) The preparation of 2-phenylethylamine by the Gabriel synthesis has been described in the chemical literature.

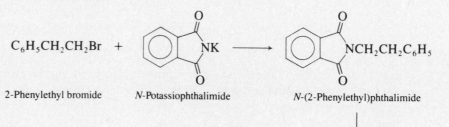

$$C_6H_5CH_2CH_2Br + \text{N-Potassiophthalimide (NK)} \longrightarrow \text{phthalimide-}NCH_2CH_2C_6H_5$$

2-Phenylethyl bromide; *N*-Potassiophthalimide; *N*-(2-Phenylethyl)phthalimide

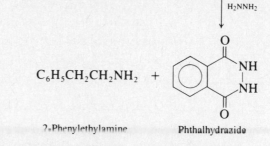

$$\xrightarrow{H_2NNH_2} C_6H_5CH_2CH_2NH_2 + \text{Phthalhydrazide}$$

2-Phenylethylamine; Phthalhydrazide

(*e*) The Gabriel synthesis leads to primary amines; *N*-methylbenzylamine is a secondary amine and cannot be prepared by this method.

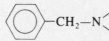

$$C_6H_5-CH_2-N(CH_3)H$$

N-Methylbenzylamine (two carbon substituents on nitrogen; a secondary amine)

(*f*) Aniline cannot be prepared by the Gabriel method. Aryl halides do not undergo nucleophilic substitution under these conditions.

$$C_6H_5Br + \text{N-Potassiophthalimide (NK)} \longrightarrow \text{no reaction}$$

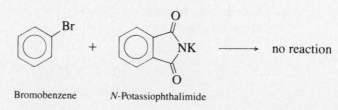

Bromobenzene; *N*-Potassiophthalimide

23.7 (*b*) Nitriles are prepared by nucleophilic substitution reactions of alkyl halides. In this synthesis the final product has the same carbon skeleton as the starting material, and therefore, only substitution reactions are required. 1-Butanol is first converted to 1-bromobutane, then treated with sodium or potassium cyanide to give the desired nitrile.

$$CH_3CH_2CH_2CH_2OH \xrightarrow[\text{or HBr}]{\text{NaBr, } H_2SO_4} CH_3CH_2CH_2CH_2Br \xrightarrow[\text{ethanol-water}]{\text{NaCN}} CH_3CH_2CH_2CH_2CN$$

1-Butanol; 1-Bromobutane; Pentanenitrile

(*c*) On analyzing the desired synthesis retrosynthetically, it can be seen that the desired cyanide is obtained by substitution of the corresponding halide, which in turn is prepared from the carboxylic acid starting material.

$$ArCH_2CN \Longrightarrow ArCH_2Br \Longrightarrow ArCH_2OH \Longrightarrow ArCO_2H$$

$$Ar = F_3C\text{—}C_6H_4\text{—}$$

As reported in the chemical literature, *p*-trifluoromethylbenzoic acid was first reduced to the corresponding alcohol with lithium aluminum hydride.

$$F_3C\text{—}C_6H_4\text{—}CO_2H \xrightarrow[2.\ H_2O]{1.\ LiAlH_4} F_3C\text{—}C_6H_4\text{—}CH_2OH$$

p-(Trifluoromethyl)benzoic acid → *p*-(Trifluoromethyl)benzyl alcohol (96%)

The alcohol was then converted to its bromide with hydrobromic acid. Nucleophilic substitution with sodium cyanide under phase transfer conditions yielded the nitrile.

$$F_3C\text{—}C_6H_4\text{—}CH_2OH \xrightarrow{HBr} F_3C\text{—}C_6H_4\text{—}CH_2Br \xrightarrow[H_2O,\ Bu_4N^+Br^-]{NaCN} F_3C\text{—}C_6H_4\text{—}CH_2CN$$

p-(Trifluoromethyl)benzyl alcohol → *p*-(Trifluoromethyl)benzyl bromide → *p*-(Trifluoromethyl)benzyl cyanide (68%)

(*d*) Oximes are prepared from ketones by reaction with hydroxylamine. The starting material, 2-octanol, must therefore be oxidized to 2-octanone in the first step of the sequence.

$$CH_3(CH_2)_5CH(OH)CH_3 \xrightarrow[H_2SO_4]{Na_2Cr_2O_7} CH_3(CH_2)_5C(=O)CH_3 \xrightarrow{H_2NOH} CH_3(CH_2)_5C(=NOH)CH_3$$

2-Octanol → 2-Octanone (92–96%) → 2-Octanone oxime (90–95%)

(*e*) Amides may be prepared from acyl chlorides and amines. First, convert pentanoic acid to pentanoyl chloride.

$$CH_3CH_2CH_2CH_2CO_2H \xrightarrow{SOCl_2} CH_3CH_2CH_2CH_2C(=O)Cl$$

Pentanoic acid → Pentanoyl chloride

Reaction of pentanoyl chloride with two moles of butylamine yields the desired amide.

$$CH_3CH_2CH_2CH_2C(=O)Cl + 2\,CH_3CH_2CH_2CH_2NH_2 \longrightarrow$$

Pentanoyl chloride + 1-Butanamine

$$CH_3CH_2CH_2CH_2C(=O)NHCH_2CH_2CH_2CH_3 + CH_3CH_2CH_2CH_2\overset{+}{N}H_3\ Cl^-$$

N-Butylpentanamide (81%) + Butylammonium chloride

23.8 (*b*) Dibenzylamine is a secondary amine and can be prepared by reductive amination of benzaldehyde with benzylamine.

$$C_6H_5\overset{O}{\overset{\|}{C}}H + C_6H_5CH_2NH_2 \xrightarrow{H_2,\ Ni} C_6H_5CH_2NHCH_2C_6H_5$$

Benzaldehyde Benzylamine Dibenzylamine

(*c*) *N,N*-Dimethylbenzylamine is a tertiary amine. Its preparation from benzaldehyde requires dimethylamine, a secondary amine.

$$C_6H_5\overset{O}{\overset{\|}{C}}H + (CH_3)_2NH \xrightarrow{H_2,\ Ni} C_6H_5CH_2N(CH_3)_2$$

Benzaldehyde Dimethylamine *N,N*-Dimethylbenzylamine

(*d*) The preparation of *N*-butylpiperidine by reductive amination is described in the text in Section 23.11. An analogous procedure is used to prepare *N*-benzylpiperidine.

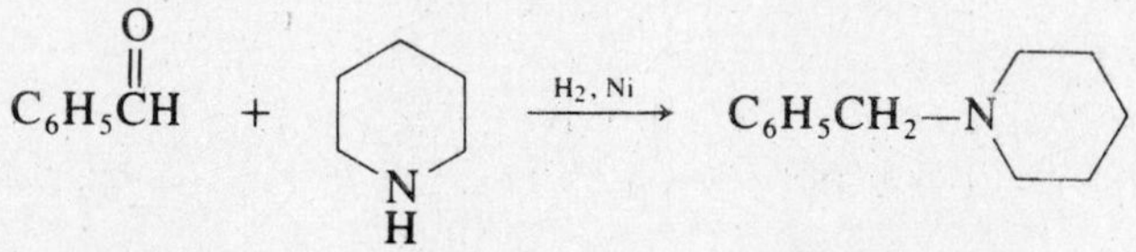

$$C_6H_5\overset{O}{\overset{\|}{C}}H + \text{piperidine (N–H)} \xrightarrow{H_2,\ Ni} C_6H_5CH_2\text{—N(piperidine ring)}$$

Benzaldehyde Piperidine *N*-Benzylpiperidine

23.9 The epoxide ring of ethylene oxide is readily opened by nucleophiles.

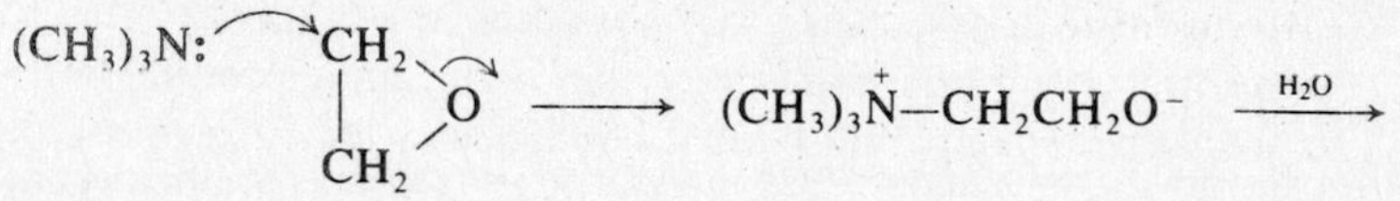

$$(CH_3)_3N: + \text{CH}_2\text{—O—CH}_2 \longrightarrow (CH_3)_3\overset{+}{N}\text{—}CH_2CH_2O^- \xrightarrow{H_2O} (CH_3)_3\overset{+}{N}CH_2CH_2OH \quad HO^-$$

Trimethylamine Ethylene oxide

Choline
[*N,N,N*-trimethyl(2-hydroxyethyl)ammonium hydroxide]

Choline has the structure shown; it is quaternary ammonium hydroxide. It cannot have a nitrogen-oxygen covalent bond as in

$$(CH_3)_3\overset{OH}{\overset{|}{N}}CH_2CH_2OH \quad \text{(not a correct Lewis structure for choline)}$$

because that structure requires nitrogen to have 10 electrons.

23.10 (*b*) There are two points of attachment of the amine nitrogen to the carbon skeleton. Two separate Hofmann elimination sequences are required to expel the nitrogen as trimethylamine.

First Hofmann elimination sequence:

CH_3N(bicyclic amine) $\xrightarrow[\text{2. } Ag_2O,\ H_2O]{\text{1. } CH_3I}$ $(CH_3)_2\overset{+}{N}$(bicyclic) H ^-OH $\xrightarrow[(-H_2O)]{\text{heat}}$ $(CH_3)_2N$:(cycloheptenyl)

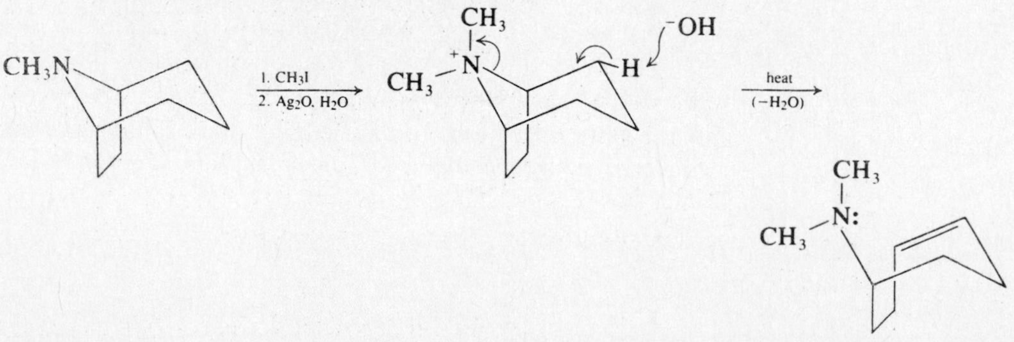

N,N—Dimethyl-4-cycloheptenylamine

Second Hofmann elimination sequence:

$N(CH_3)_2$ $\xrightarrow[\text{2. } Ag_2O,\ H_2O]{\text{1. } CH_3I}$ ^{-}OH, H, $+N(CH_3)_3$ $\xrightarrow[(-H_2O)]{\text{heat}}$ + $(CH_3)_3N:$

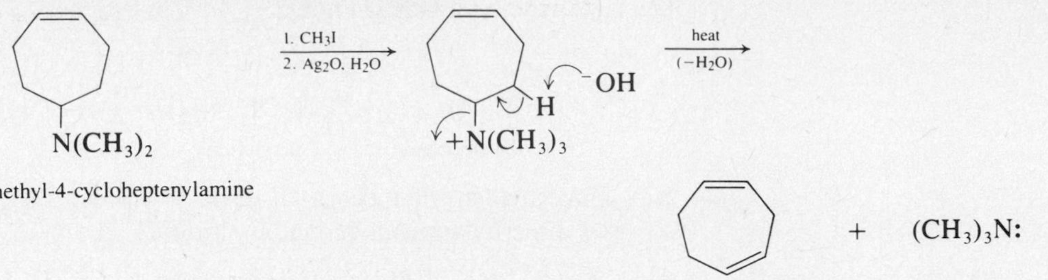

N,N-Dimethyl-4-cycloheptenylamine

1,4-Cycloheptadiene   Trimethylamine

Notice that there are seven contiguous carbons in the starting alkaloid and two points of attachment of nitrogen to this carbon skeleton. When those two carbon-nitrogen bonds are cleaved, a seven-membered cyclic diene results.

(*c*) There are two points of attachment of nitrogen to the remainder of the molecule, so two separate Hofmann eliminations are required.

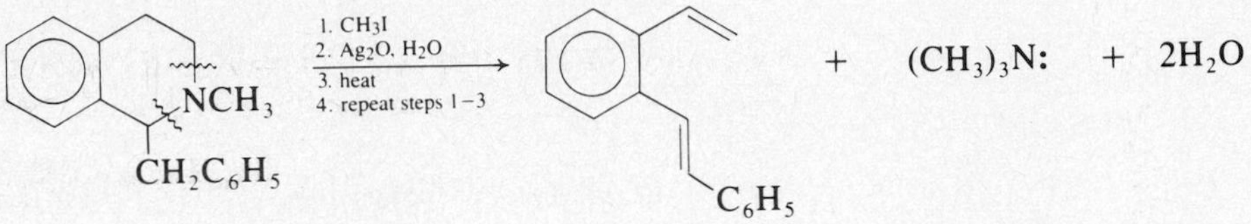

NCH_3, $CH_2C_6H_5$ → (1. CH_3I 2. Ag_2O, H_2O 3. heat 4. repeat steps 1–3) C_6H_5 + $(CH_3)_3N:$ + $2H_2O$

o-Vinyl-*trans*-stilbene   Trimethylamine   Water

As in this example, it is often not necessary to identify the intermediate amino alkene in the first Hofmann elimination since the final product is the same irrespective of the precise order of the individual bond cleavages.

(*d*) Since nitrogen has three points of attachment to the carbon skeleton, three separate Hofmann elimination sequences are required to remove nitrogen as trimethylamine.

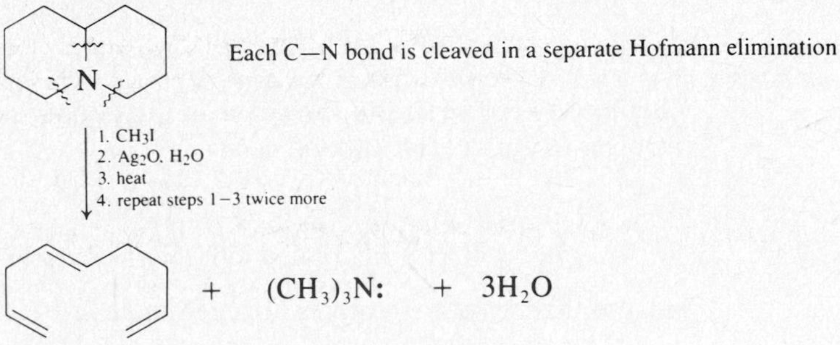

Each C—N bond is cleaved in a separate Hofmann elimination

1. CH_3I
2. Ag_2O, H_2O
3. heat
4. repeat steps 1–3 twice more

+ $(CH_3)_3N:$ + $3H_2O$

1,4,8-Nonatriene   Trimethylamine   Water

23.11 (*b*) First identify the available β hydrogens. Elimination must involve a proton from the carbon atom adjacent to the one that bears the nitrogen.

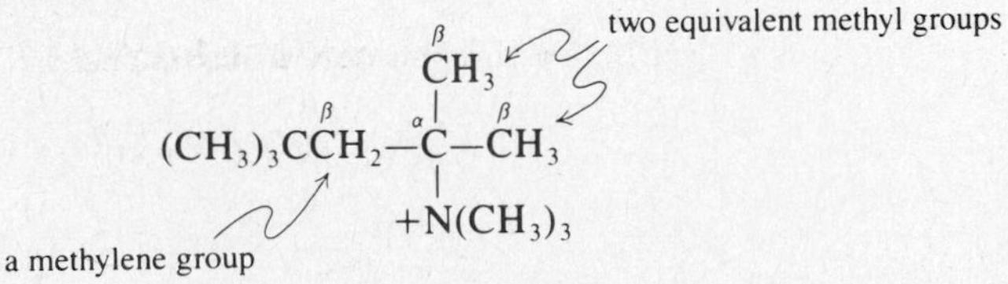

$$(CH_3)_3C\overset{\beta}{C}H_2-\overset{\alpha}{C}(\overset{\beta}{C}H_3)(+N(CH_3)_3)-\overset{\beta}{C}H_3$$

two equivalent methyl groups

a methylene group

It is a proton from one of the methyl groups, rather than one from the more sterically hindered methylene, that is lost on elimination.

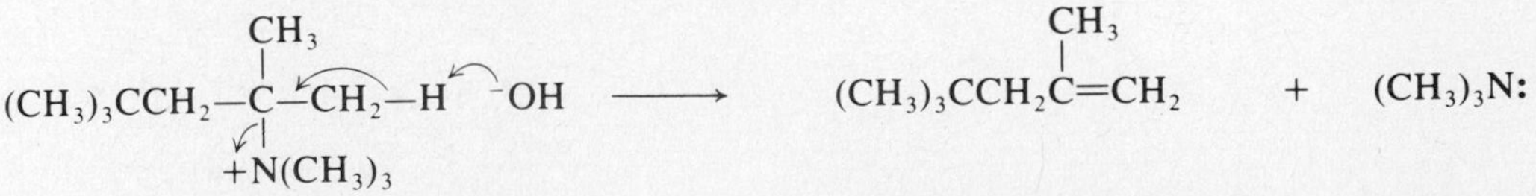

$$(CH_3)_3CCH_2-\underset{+N(CH_3)_3}{\overset{CH_3}{C}}-CH_2-H\ \ ^{-}OH \longrightarrow (CH_3)_3CCH_2\overset{CH_3}{C}=CH_2 + (CH_3)_3N:$$

(1,1,3,3,-Tetramethylbutyl)trimethylammonium hydroxide   2,4,4-Trimethyl-1-pentene (only alkene formed, 70% isolated yield)   Trimethylamine (*c*) The base may abstract a proton from either of two β carbons. Deprotonation of the β methyl carbon yields ethylene.

$$HO^- \; H{-}CH_2{-}CH_2{-}\overset{CH_3}{\underset{CH_3}{\overset{+|}{N}}}CH_2CH_2CH_2CH_3 \xrightarrow[(-H_2O)]{\text{heat}} CH_2{=}CH_2 + (CH_3)_2\ddot{N}CH_2CH_2CH_2CH_3$$

N-Ethyl-*N*,*N*-dimethylbutylammonium hydroxide → Ethylene + *N*,*N*-Dimethylbutylamine

Deprotonation of the β methylene carbon yields 1-butene.

$$CH_3CH_2\overset{CH_3}{\underset{CH_3}{\overset{+|}{N}}}{-}CH_2{-}\underset{H\;\;{}^-OH}{CH}CH_2CH_3 \xrightarrow[(-H_2O)]{\text{heat}} CH_3CH_2\ddot{N}(CH_3)_2 + CH_2{=}CHCH_2CH_3$$

N-Ethyl-*N*,*N*-dimethylbutylammonium hydroxide → *N*,*N*-Dimethylethylamine + 1-Butene

The preferred order of proton removal in Hofmann elimination reactions is β CH_3 > β CH_2 > β CH. Ethylene is the major alkene formed, the observed ratio of ethylene to 1-butene being 98 : 2.

(*d*) In this case the competition is between attack at a β CH_2 group versus attack at a β CH. The methylene group is attacked faster than the methine group.

preferred site of deprotonation; relatively hindered site

$$HO^- \quad CH_3CH_2CH_2{-}\overset{CH_3}{\underset{CH_3}{\overset{+|}{N}}}{-}CH_2CH(CH_3)_2 \xrightarrow[(-H_2O)]{HO^-,\ \text{heat}} CH_3CH{=}CH_2 + (CH_3)_2\ddot{N}CH_2CH(CH_3)_2$$

N-Propyl-*N*,*N*-dimethylisobutylammonium hydroxide → Propene + *N*,*N*-Dimethylisobutylamine

When the elimination reaction was performed, an alkene mixture composed of $CH_3CH{=}CH_2$ (72 percent) and $(CH_3)_2C{=}CH_2$ (28 percent) was isolated in 92 percent yield.

23.12 The principal resonance forms of *N*-nitrosodimethylamine are:

$$(CH_3)_2\ddot{N}{-}\ddot{N}{=}\ddot{O}: \longleftrightarrow (CH_3)_2\overset{+}{N}{=}\ddot{N}{-}\ddot{O}:^-$$

All atoms (except hydrogen) have octets of electrons in each of these structures. Other resonance forms are less stable because they do not have a full complement of electrons around each atom.

23.13 Deamination of *tert*-pentylamine gives products that result from *tert*-pentyl cation. Since neopentylamine gives the same products, it is likely that *tert*-pentyl cation is formed from neopentylamine by way of its diazonium ion. A carbocation rearrangement reaction is indicated.

$$CH_3\overset{CH_3}{\underset{CH_3}{C}}CH_2NH_2 \xrightarrow{HONO} CH_3\overset{CH_3}{\underset{CH_3}{C}}{-}CH_2{-}\overset{+}{N}{\equiv}N \xrightarrow{-N_2} CH_3\overset{CH_3}{\underset{+}{C}}CH_2CH_3$$

Neopentylamine → Neopentyldiazonium ion → *tert*-Pentyl cation

Once formed, *tert*-pentyl cation loses a proton to form an alkene or is captured by water to give an alcohol.

$$\underset{\text{tert-Pentyl cation}}{CH_3\overset{CH_3}{\underset{+}{C}}CH_2CH_3} \xrightarrow{-H^+} \underset{\text{2-Methyl-1-butene}}{CH_2{=}\overset{CH_3}{C}CH_2CH_3} + \underset{\text{2-Methyl-2-butene}}{(CH_3)_2C{=}CHCH_3}$$

$$\xrightarrow{H_2O} \underset{\text{2-Methyl-2-butanol}}{(CH_3)_2\underset{OH}{C}CH_2CH_3}$$

23.14 Amines may be primary, secondary, or tertiary. The $C_4H_{11}N$ primary amines, compounds of the type $C_4H_9NH_2$, and their systematic names are:

$CH_3CH_2CH_2CH_2NH_2$ — Butylamine (1-butanamine)

$(CH_3)_2CHCH_2NH_2$ — Isobutylamine (2-methyl-1-propanamine)

$CH_3\underset{NH_2}{C}HCH_2CH_3$ — *sec*-Butylamine (2-butanamine)

$(CH_3)_3CNH_2$ — *tert*-Butylamine (2-methyl-2-propanamine)

Secondary amines have the general formula R_2NH. Those of molecular formula $C_4H_{11}N$ are:

$(CH_3CH_2)_2NH$ — Diethylamine (*N*-ethylethanamine)

$CH_3\underset{H}{N}CH_2CH_2CH_3$ — *N*-Methylpropylamine (*N*-methyl-1-propanamine)

$CH_3\underset{H}{N}CH(CH_3)_2$ — *N*-Methylisopropylamine (*N*-methyl-2-propanamine)

There is only one tertiary amine (R_3N) of molecular formula $C_4H_{11}N$.

$(CH_3)_2NCH_2CH_3$ — *N,N*-Dimethylethylamine (*N,N*-dimethylethanamine)

23.15 (*a*) Neopentylamine is the IUPAC name for a primary amine that bears a neopentyl group on nitrogen.

$(CH_3)_3CCH_2NH_2$ — Neopentylamine

(*b*) The name 2-ethyl-1-butanamine designates a four-carbon chain terminating in an amino group and bearing an ethyl group at C-2.

$CH_3CH_2\underset{CH_2CH_3}{C}HCH_2NH_2$ — 2-Ethyl-1-butanamine

(*c*) The prefix *N*- in *N*-ethyl-1-butanamine identifies the ethyl group as a substituent on nitrogen in a secondary amine.

$CH_3CH_2CH_2CH_2\underset{H}{N}CH_2CH_3$ — *N*-Ethyl-1-butanamine

(*d*) Dibenzylamine is a secondary amine. It bears two benzyl groups on nitrogen.

$C_6H_5CH_2\underset{H}{N}CH_2C_6H_5$ — Dibenzylamine

(*e*) Tribenzylamine is a tertiary amine.

$(C_6H_5CH_2)_3N$ — Tribenzylamine

(*f*) Tetraethylammonium hydroxide is a quaternary ammonium salt.

$$(CH_3CH_2)_4\overset{+}{N}\ HO^-$$ tetraethylammonium hydroxide

(*g*) This compound is a secondary amine; it bears an allyl substituent on the nitrogen of cyclohexylamine.

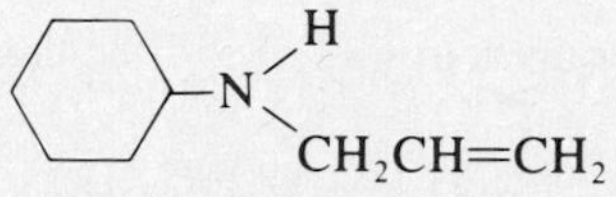

C_6H_{11}–N(H)–$CH_2CH{=}CH_2$ *N*-Allylcyclohexylamine

(*h*) Piperidine is a cyclic secondary amine that contains nitrogen in a six-membered ring. *N*-Allylpiperidine is a tertiary amine.

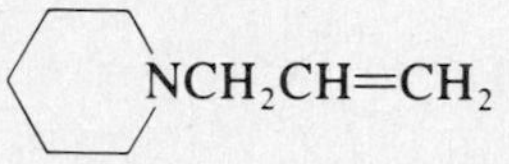

$NCH_2CH{=}CH_2$ (piperidine ring) *N*-Allylpiperidine

(*i*) The compound is the benzyl ester of 2-aminopropanoic acid.

$$CH_3\underset{NH_2}{CH}\overset{O}{\overset{\|}{C}}OCH_2C_6H_5$$ Benzyl 2-aminopropanoate

(*j*) The parent compound is cyclohexanone. The substituent $(CH_3)_2N-$ group is attached to C-4.

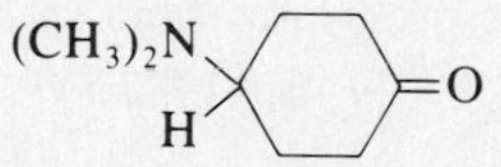

$(CH_3)_2N$(H)–cyclohexane=O 4-(*N*,*N*-Dimethylamino)cyclohexanone

(*k*) The suffix diamine reveals the presence of two amino groups, one at either end of a three-carbon chain that bears two methyl groups at C-2.

$$H_2NCH_2\overset{CH_3}{\underset{CH_3}{C}}CH_2NH_2$$ 2,2-Dimethyl-1,3-propanediamine

23.16 (*a*) An amino group is a substituent at C-4 of butanoic acid in γ-aminobutyric acid.

$$H_2NCH_2CH_2CH_2CO_2H$$ 4-Aminobutanoic acid (γ-aminobutyric acid)

(*b*) Mescaline has a 3,4,5-trimethoxy-substituted benzene ring at C-2 of an ethylamine unit.

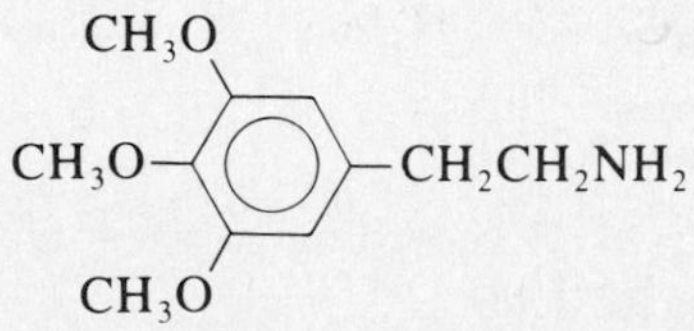

$(CH_3O)_3C_6H_2$–$CH_2CH_2NH_2$ 2-(3,4,5-Trimethoxyphenyl)ethylamine (mescaline)

(*c*) Amino groups are found at C-1 and C-4 of putrescine.

$$H_2NCH_2CH_2CH_2CH_2NH_2$$ 1,4-Butanediamine (putrescine)

(*d*) A phenyl group and an amino group are trans to each other on a three-membered ring in this compound.

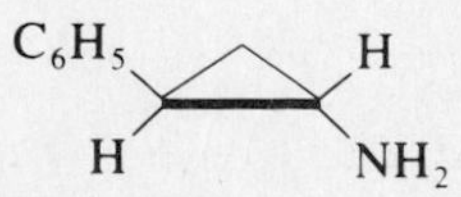

C_6H_5, H / H, NH_2 (cyclopropane) *trans*-2-Phenylcyclopropylamine (tranylcypromine)

(*e*) This compound is a tertiary amine. It bears a benzyl group, a methyl group, and a 2-propynyl group on nitrogen.

$$C_6H_5CH_2-N(CH_3)CH_2C{\equiv}CH$$

N-Benzyl-*N*-methyl-2-propynylamine (pargyline)

(*f*) The amino group is at C-2 of a three-carbon chain that bears a phenyl substituent at its terminus.

$$C_6H_5CH_2CH(NH_2)CH_3$$

1-Phenyl-2-propanamine (amphetamine)

(*g*) This compound is a secondary amine related to the previous one.

$$C_6H_5CH_2CH(N(H)CH_3)CH_3$$

N-Methyl-1-phenyl-2-propanamine (methamphetamine)

(*h*) Phenylephrine is named systematically as an ethanol derivative.

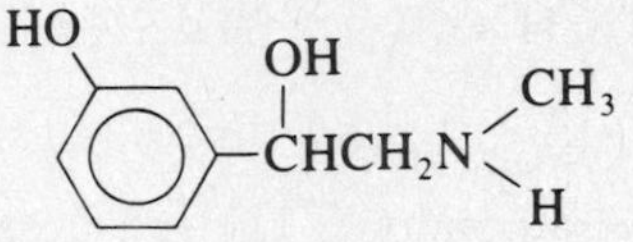

$$m\text{-}HOC_6H_4-CH(OH)CH_2N(H)CH_3$$

1-(*m*-Hydroxyphenyl)-2-(methylamino)ethanol (phenylephrine)

23.17 (*a*) Basicity decreases in proceeding across a row in the periodic table. The increased nuclear charge as one progresses from carbon to nitrogen to oxygen to fluorine causes the electrons to be bound more strongly to the atom and thus less readily shared.

	$H_3\bar{C}:$	>	$H_2\ddot{\bar{N}}:$	>	$H\ddot{O}:^-$	>	$:\ddot{F}:^-$
	Strongest base						Weakest base
K_a of conjugate acid	10^{-60}		10^{-36}		10^{-16}		3.5×10^{-4}

(*b*) The strongest base in this group is amide ion, H_2N^- while the weakest base is water, H_2O. Ammonia is a weaker base than hydroxide ion; the equilibrium lies to the left.

$$\underset{\text{Weaker base}}{NH_3} + \underset{\text{Weaker acid}}{H_2O} \rightleftharpoons \underset{\text{Stronger acid}}{NH_4^+} + \underset{\text{Stronger base}}{OH^-}$$

The correct order is:

$$\underset{\text{Strongest base}}{H_2\ddot{\bar{N}}:} > H\ddot{O}:^- > :NH_3 > \underset{\text{Weakest base}}{H_2\ddot{O}:}$$

(*c*) These anions can be ranked according to their basicity by considering the respective acidities of their conjugate acids.

Base	*Conjugate acid*	*K_a of conjugate acid*
H_2N^-	H_3N	10^{-36}
HO^-	H_2O	10^{-16}
$:\bar{C}{\equiv}N:$	$HC{\equiv}N:$	7.2×10^{-10}
$^-O-\overset{+}{N}(=O)O^-$	$HON(=O)O^-$	2.5×10^{1}

The order of basicities is the opposite of the order of acidities of their conjugate acids.

$$H_2N^- > HO^- > :\bar{C}{\equiv}N: > NO_3^-$$

Strongest base — Weakest base

(*d*) A carbonyl group attached to nitrogen stabilizes its negative charge. The strongest base is the anion that has no carbonyl groups on nitrogen, the weakest base is phthalimide anion, which has two carbonyl groups.

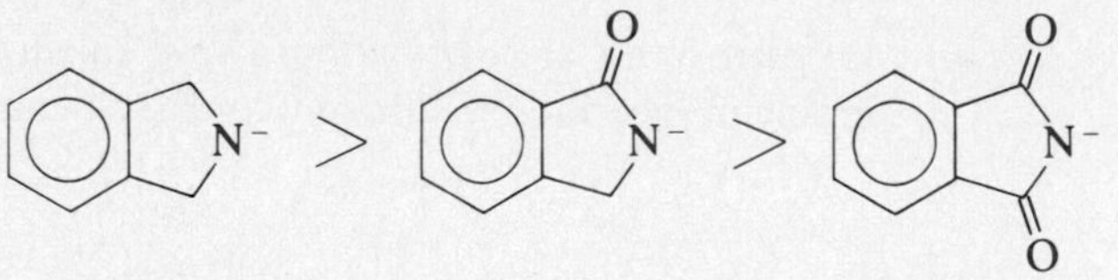

Strongest base — Weakest base

23.18 In all these problems, the amines that are to be compared are of the same type, i.e., both primary, both secondary, or both tertiary. There is no need to consider differences in solvation; only the electronic effects of substituents need to be examined. In each case the important substituent effect is destabilization of the conjugate acid of one of the amines by an electron-withdrawing substituent.

$$RN: + H_2O \rightleftharpoons R\overset{+}{N}{-}H + HO^-$$

The ability of group R to stabilize the ammonium ion decreases as substituents on R become more strongly electron-attracting and the equilibrium constant for protonation of nitrogen decreases. The base-weakening effect of electron-withdrawing substituents arises by a combination of inductive and field effects. Generally, the effect decreases as the number of bonds between the substituent and the amine nitrogen increase.

(*a*) A trichloromethyl group is a powerful electron-withdrawing substituent. Its effect is more pronounced in 2,2,2-trichloroethylamine than in 3,3,3-trichloropropylamine because of the distance factor noted above.

$Cl_3CCH_2CH_2NH_2$	$Cl_3CCH_2NH_2$
3,3,3-Trichloropropylamine	2,2,2-Trichloroethylamine
(stronger base)	(weaker base)
K_b 5×10^{-5}	K_b 3×10^{-9}
(pK_b 4.3)	(pK_b 8.5)

(*b*) Triethylamine is a stronger base than tri(2-chloroethyl)amine. Each chlorine substituent is electronegative and contributes to the base-weakening effect exhibited by electron-attracting groups.

$(CH_3CH_2)_3N$	$(ClCH_2CH_2)_3N$
Triethylamine	Tri(2-chloroethyl)amine
(stronger base)	(weaker base)
K_b 6.3×10^{-4}	K_b 2.5×10^{-10}
(pK_b 3.2)	(pK_b 9.6)

(*c*) As the hybridization of C-2 is changed from sp^3 in $CH_3CH_2CH_2NH_2$ to sp in $HC{\equiv}CCH_2NH_2$, its electron-attracting power increases. Thus, the ion $HC{\equiv}CCH_2\overset{+}{N}H_3$ is destabilized relative to $CH_3CH_2CH_2\overset{+}{N}H_3$ and the equilibrium constant describing its formation is smaller.

$CH_3CH_2CH_2NH_2$	$HC{\equiv}CCH_2NH_2$
Propylamine	2-Propyn-1-amine
(stronger base)	(weaker base)
K_b 5×10^{-4}	K_b 1.6×10^{-6}
(pK_b 3.3)	(pK_b = 5.8)

(*d*) A cyano group is strongly electron-withdrawing and base-weakening.

NCH_2CH_3 (piperidine ring) — NCH_2CN (piperidine ring)

N-Ethylpiperidine (stronger base) K_b 3.2 × 10^{-4} (pK_b 3.5)

N-Cyanomethylpiperidine (weaker base) K_b 3.2 × 10^{-10} (pK_b 9.5)

(*e*) Oxygen is an electron-withdrawing substituent attached to the β carbon atom of an amine in morpholine. Morpholine is a weaker base than piperidine.

NH (ring) — O NH (ring)

Piperidine (stronger base) K_b 4 × 10^{-3} (pK_b 2.4)

Morpholine (weaker base) K_b 2.5 × 10^{-6} (pK_b 5.6)

(*f*) An ester of an α-amino acid is much less basic than an alkylamine because of the strong electron-withdrawing properties of the carbonyl group.

CH_3CHCH_3 with NH_2 — $CH_3CHCO_2CH_2CH_3$ with NH_2

Isopropylamine (stronger base) K_b 4.3 × 10^{-4} (pK_b 3.4)

Ethyl 2-aminopropanoate (weaker base) K_b 6.3 × 10^{-7} (pK_b 6.2)

23.19 (*a*) Looking at the problem retrosynthetically, it can be seen that a variety of procedures are available for preparing ethylamine from ethanol. The methods by which a primary amine may be prepared include:

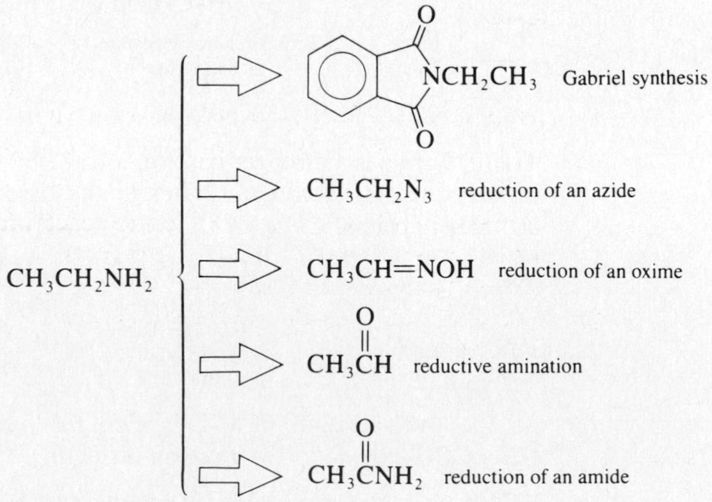

$CH_3CH_2NH_2$ ⇒

- N-ethylphthalimide (NCH_2CH_3 with two C=O on a benzo-fused ring) — Gabriel synthesis
- $CH_3CH_2N_3$ — reduction of an azide
- $CH_3CH{=}NOH$ — reduction of an oxime
- $CH_3\overset{O}{\overset{\|}{C}}H$ — reductive amination
- $CH_3\overset{O}{\overset{\|}{C}}NH_2$ — reduction of an amide

Two of these methods, the Gabriel synthesis and the preparation and reduction of the corresponding azide, begin with ethyl bromide.

$$\underset{\text{Ethanol}}{CH_3CH_2OH} \xrightarrow[\text{or HBr}]{PBr_3} \underset{\text{Ethyl bromide}}{CH_3CH_2Br}$$

$$CH_3CH_2Br + \text{N-Potassiophthalimide } (C_6H_4(CO)_2N^-K^+) \longrightarrow \text{N-Ethylphthalimide } (C_6H_4(CO)_2NCH_2CH_3) \xrightarrow{H_2NNH_2} CH_3CH_2NH_2$$

Ethyl bromide — N-Potassiophthalimide — N-Ethylphthalimide — Ethylamine

$$\underset{\text{Ethyl bromide}}{CH_3CH_2Br} \xrightarrow{NaN_3} \underset{\text{Ethyl azide}}{CH_3CH_2N_3} \xrightarrow[2.\ H_2O]{1.\ LiAlH_4} \underset{\text{Ethylamine}}{CH_3CH_2NH_2}$$

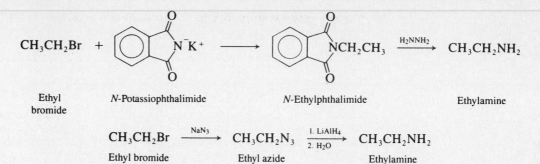

Two others, reduction of an oxime and reductive amination, begin with oxidation of ethanol to acetaldehyde.

$$\underset{\text{Ethanol}}{CH_3CH_2OH} \xrightarrow[CH_2Cl_2]{(C_5H_5N)_2CrO_3} \underset{\text{Acetaldehyde}}{CH_3\overset{O}{\overset{\|}{C}}H}$$

$$\underset{\text{Acetaldehyde}}{CH_3\overset{O}{\overset{\|}{C}}H} \xrightarrow{H_2NOH} \underset{\text{Acetaldehyde oxime}}{CH_3CH{=}NOH} \xrightarrow[\text{ethanol}]{Na} \underset{\text{Ethylamine}}{CH_3CH_2NH_2}$$

$$\underset{\text{Acetaldehyde}}{CH_3\overset{O}{\overset{\|}{C}}H} \xrightarrow[\text{or } NH_4O_2CCH_3,\ NaBH_3CN]{NH_3,\ H_2,\ Ni} \underset{\text{Ethylamine}}{CH_3CH_2NH_2}$$

Another possibility is reduction of acetamide. This requires an initial oxidation of ethanol to acetic acid.

$$\underset{\text{Ethanol}}{CH_3CH_2OH} \xrightarrow[\text{or 1. } KMnO_4\text{, 2. } H^+]{K_2Cr_2O_7,\ H_2SO_4} \underset{\text{Acetic acid}}{CH_3CO_2H} \xrightarrow[2.\ NH_3]{1.\ SOCl_2} \underset{\text{Acetamide}}{CH_3\overset{O}{\overset{\|}{C}}NH_2} \xrightarrow[2.\ H_2O]{1.\ LiAlH_4} \underset{\text{Ethylamine}}{CH_3CH_2NH_2}$$

(*b*) Acylation of ethylamine with acetyl chloride (prepared as above) gives the desired amide.

$$\underset{\text{Acetyl chloride}}{CH_3\overset{O}{\overset{\|}{C}}Cl} + \underset{\text{Ethylamine}}{2\,CH_3CH_2NH_2} \longrightarrow \underset{\text{N-Ethylacetamide}}{CH_3\overset{O}{\overset{\|}{C}}NHCH_2CH_3} + \underset{\text{Ethylammonium chloride}}{CH_3CH_2\overset{+}{N}H_3\ Cl^-}$$

Excess ethylamine can be allowed to react with the hydrogen chloride formed in the acylation reaction. Alternatively, equimolar amounts of acyl chloride and amine can be used in the presence of aqueous hydroxide as the base.

(*c*) Reduction of the *N*-ethylacetamide prepared in the preceding problem yields diethylamine.

$$\underset{\text{N-Ethylacetamide}}{CH_3\overset{O}{\overset{\|}{C}}NHCH_2CH_3} \xrightarrow[2.\ H_2O]{1.\ LiAlH_4} \underset{\text{Diethylamine}}{CH_3CH_2NHCH_2CH_3}$$

Diethylamine can also be prepared by reductive amination of acetaldehyde with ethylamine.

$$\underset{\text{Acetaldehyde}}{CH_3\overset{O}{\overset{\|}{C}}H} + \underset{\text{Ethylamine}}{CH_3CH_2NH_2} \xrightarrow[\text{or } NaBH_3CN]{H_2,\ Ni} \underset{\text{Diethylamine}}{CH_3CH_2NHCH_2CH_3}$$

(*d*) The preparation of *N*,*N*-diethylacetamide is a standard acylation reaction. The necessary acetyl chloride and diethylamine have been prepared in previous parts of this problem.

$$\underset{\text{Acetyl chloride}}{CH_3\overset{O}{\overset{\|}{C}}Cl} + \underset{\text{Diethylamine}}{(CH_3CH_2)_2NH} \xrightarrow{HO^-} \underset{N,N\text{-Diethylacetamide}}{CH_3\overset{O}{\overset{\|}{C}}N(CH_2CH_3)_2}$$

(*e*) Triethylamine arises by reduction of *N*,*N*-diethylacetamide or by reductive amination.

$$\underset{N,N\text{-Diethylacetamide}}{CH_3\overset{O}{\overset{\|}{C}}N(CH_2CH_3)_2} \xrightarrow[2.\ H_2O]{1.\ LiAlH_4} \underset{\text{Triethylamine}}{(CH_3CH_2)_3N}$$

$$\underset{\text{Acetaldehyde}}{CH_3\overset{O}{\overset{\|}{C}}H} + \underset{\text{Diethylamine}}{(CH_3CH_2)_2NH} \xrightarrow[\text{or } NaBH_3CN]{H_2,\ Ni} \underset{\text{Triethylamine}}{(CH_3CH_2)_3N}$$

(*f*) Quaternary ammonium halides are formed by reaction of alkyl halides and tertiary amines.

$$\underset{\text{Ethyl bromide}}{CH_3CH_2Br} + \underset{\text{Triethylamine}}{(CH_3CH_2)_3N} \longrightarrow \underset{\text{Tetraethylammonium bromide}}{(CH_3CH_2)_4\overset{+}{N}\ Br^-}$$

(*g*) Oxidation of tertiary amines yields the corresponding *N*-oxides.

$$\underset{\text{Triethylamine}}{(CH_3CH_2)_3N} \xrightarrow[\text{or } CH_3CO_3H]{H_2O_2} \underset{\text{Triethylamine } N\text{-oxide}}{(CH_3CH_2)_3\overset{+}{N}{-}O^-}$$

(*h*) Heating the amine oxide in (*g*) causes a Cope elimination to occur. In addition to an alkene, Cope elimination results in formation of the desired hydroxylamine.

$$\underset{\text{Triethylamine } N\text{-oxide}}{(CH_3CH_2)_2\overset{+}{N}(O^-)CH_2CH_2{-}H} \xrightarrow{\text{heat}} \underset{N,N\text{-Diethylhydroxylamine}}{(CH_3CH_2)_2NOH} + \underset{\text{Ethylene}}{CH_2{=}CH_2}$$

23.20 (*a*) In this problem a primary alkanamine must be prepared with a carbon chain extended by one carbon. This can be accomplished by way of a nitrile.

$$RCH_2NH_2 \Rightarrow RCN \Rightarrow RBr \Rightarrow ROH$$

$$(R- = CH_3CH_2CH_2CH_2-)$$

The desired reaction sequence is therefore:

$$\underset{\text{1-Butanol}}{CH_3CH_2CH_2CH_2OH} \xrightarrow[\text{or } HBr]{PBr_3} \underset{\text{Butyl bromide}}{CH_3CH_2CH_2CH_2Br} \xrightarrow{NaCN} \underset{\text{Pentanenitrile}}{CH_3CH_2CH_2CH_2CN}$$

$$\xrightarrow[2.\ H_2O]{1.\ LiAlH_4} \underset{\text{1-Pentanamine}}{CH_3CH_2CH_2CH_2CH_2NH_2}$$

(*b*) The carbon chain of *tert*-butyl chloride cannot be extended by a nucleophilic substitution reaction; the S_N2 reaction which would be required on the tertiary halide would not work. Therefore the sequence employed in the previous problem is not effective in this case. The best route is carboxylation of the Grignard reagent and subsequent conversion of the corresponding amide to the desired primary amine product.

$$(CH_3)_3CCH_2NH_2 \Longrightarrow (CH_3)_3C\overset{O}{\overset{\|}{C}}NH_2 \Longrightarrow (CH_3)_3CCO_2H \Longrightarrow (CH_3)_3CCl$$

The reaction sequence to be used is:

$$\underset{tert\text{-Butyl chloride}}{(CH_3)_3CCl} \xrightarrow[\text{2. } CO_2\text{; 3. } H_3O^+]{\text{1. Mg, diethyl ether}} \underset{\text{2,2-Dimethylpropanoic acid}}{(CH_3)_3CCO_2H}$$

Once the carboxylic acid has been obtained, it is converted to the desired amine by reduction of the corresponding amide.

$$\underset{\text{2,2-Dimethylpropanoic acid}}{(CH_3)_3CCO_2H} \xrightarrow[\text{2. } NH_3]{\text{1. } SOCl_2} \underset{\text{2,2-Dimethylpropanamide}}{(CH_3)_3C\overset{O}{\overset{\|}{C}}NH_2} \xrightarrow[\text{2. } H_2O]{\text{1. } LiAlH_4} \underset{\text{Neopentylamine}}{(CH_3)_3CCH_2NH_2}$$

(*c*) Oxidation of cyclohexanol to cyclohexanone gives a suitable substrate for reductive amination.

$$\underset{\text{Cyclohexanol}}{C_6H_{11}{-}OH} \xrightarrow[H_2SO_4]{K_2Cr_2O_7} \underset{\text{Cyclohexanone}}{C_6H_{10}{=}O} \xrightarrow[\text{or } CH_3NH_2,\ NaBH_3CN]{CH_3NH_2,\ H_2,\ Ni} \underset{N\text{-Methylcyclohexylamine}}{C_6H_{11}{-}NHCH_3}$$

(*d*) The desired product is the reduction product of the cyanohydrin of acetone.

$$\underset{\text{2-Cyano-2-propanol}}{CH_3C(OH)(CN)CH_3} \xrightarrow[\text{2. } H_2O]{\text{1. } LiAlH_4} \underset{\text{1-Amino-2-methyl-2-propanol}}{CH_3C(OH)(CH_2NH_2)CH_3}$$

The cyanohydrin is made from acetone in the usual way. Acetone is available by oxidation of isopropyl alcohol.

$$\underset{\text{Isopropyl alcohol}}{CH_3CH(OH)CH_3} \xrightarrow{K_2Cr_2O_7,\ H_2SO_4} \underset{\text{Acetone}}{CH_3\overset{O}{\overset{\|}{C}}CH_3} \xrightarrow[H_2SO_4]{KCN} \underset{\text{2-Cyano-2-propanol}}{CH_3C(OH)(CN)CH_3}$$

(*e*) The target amino alcohol is the product of nucleophilic ring opening of 1,2-epoxypropane by ammonia. Ammonia attacks the less hindered carbon of the epoxide function.

$$\underset{\text{1,2-Epoxypropane}}{CH_3\overset{\frown O \frown}{CH{-}CH_2}} \xrightarrow{NH_3} \underset{\text{1-Amino-2-propanol}}{CH_3CH(OH)CH_2NH_2}$$

The necessary epoxide is formed by epoxidation of propene.

$$\underset{\text{Isopropyl alcohol}}{CH_3CH(OH)CH_3} \xrightarrow[\text{heat}]{H_2SO_4} \underset{\text{Propene}}{CH_3CH{=}CH_2} \xrightarrow{CH_3CO_2OH} \underset{\text{1,2-Epoxypropane}}{CH_3\overset{\frown O \frown}{CH{-}CH_2}}$$

(*f*) The reaction sequence is the same as in the preceding problem except that dimethylamine is used as the nucleophile instead of ammonia.

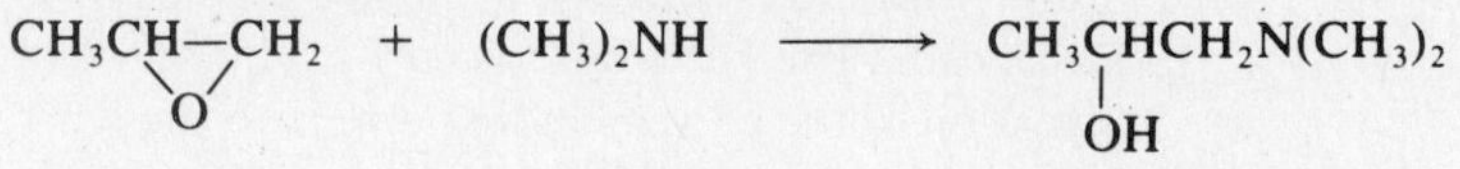

$$CH_3CH\underset{O}{-}CH_2 + (CH_3)_2NH \longrightarrow CH_3\underset{OH}{\underset{|}{C}}HCH_2N(CH_3)_2$$

1,2-Epoxypropane (prepared as in preceding problem) — Dimethylamine — 1-(*N*,*N*-Dimethylamino)-2-propanol

(*g*) The key to performing this synthesis is recognition of the starting material as an acetal of acetophenone. Acetals may be hydrolyzed to carbonyl compounds.

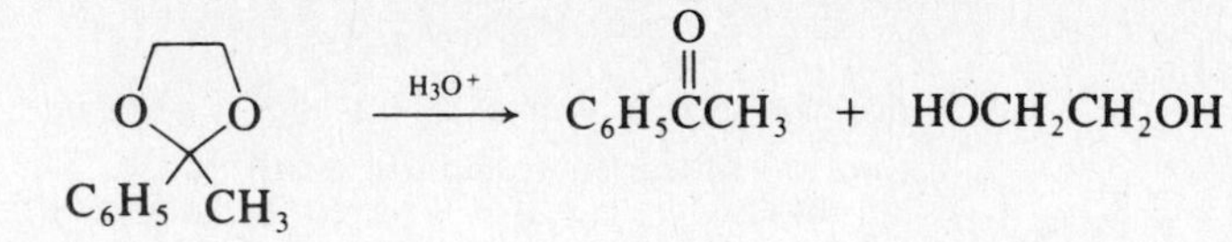

$$\text{2-Methyl-2-phenyl-1,3-dioxolane} \xrightarrow{H_3O^+} C_6H_5\overset{O}{\overset{\|}{C}}CH_3 + HOCH_2CH_2OH$$

2-Methyl-2-phenyl-1,3-dioxolane — Acetophenone — 1,2-Ethanediol

Once acetophenone has been obtained, it may be converted to the required product by reductive amination.

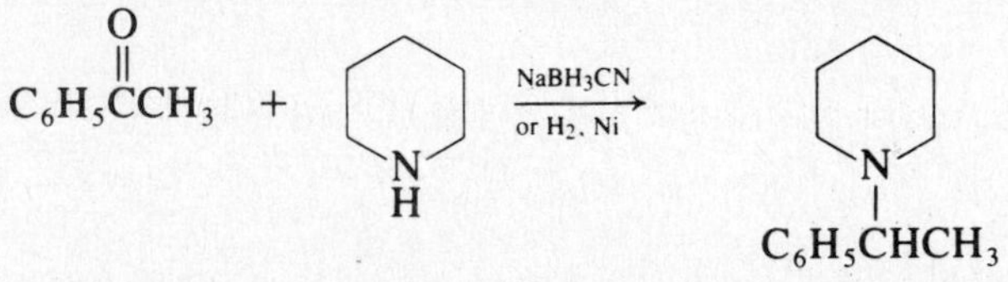

$$C_6H_5\overset{O}{\overset{\|}{C}}CH_3 + \text{piperidine (N–H)} \xrightarrow[\text{or } H_2,\ Ni]{NaBH_3CN} \text{piperidine-}N\text{-}C_6H_5CHCH_3$$

Acetophenone — Piperidine — *N*-(1-Phenylethyl)piperidine

23.21 (*a*) The reaction of alkyl halides with *N*-potassiophthalimide (the first step in the Gabriel synthesis of amines) is a nucleophilic substitution reaction. Alkyl bromides are more reactive than alkyl fluorides, i.e., bromide is a better leaving group than fluoride.

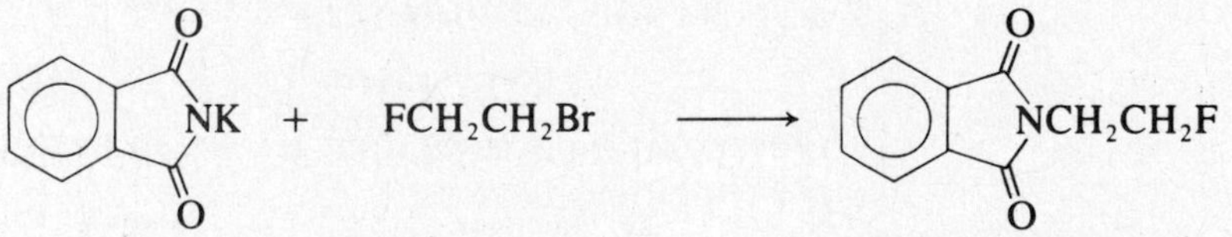

$$\text{Phthalimide-NK} + FCH_2CH_2Br \longrightarrow \text{Phthalimide-}NCH_2CH_2F$$

N-Potassiophthalimide — 2-Bromoethyl fluoride — 2-Phthalimidoethyl fluoride

(*b*) In this example one bromine is attached to a primary and the other to a secondary carbon. Phthalimide anion is a good nucleophile and reacts with alkyl halides by the S_N2 mechanism. It attacks the less hindered primary carbon.

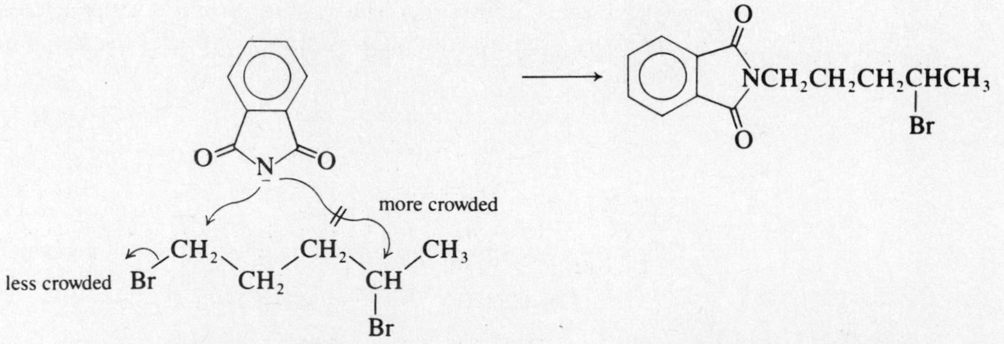

$$\text{Phthalimide anion} + BrCH_2CH_2CH_2\underset{Br}{\underset{|}{C}}HCH_3 \longrightarrow \text{Phthalimide-}NCH_2CH_2CH_2\underset{Br}{\underset{|}{C}}HCH_3$$

less crowded (attack at CH_2Br); more crowded (no attack at CHBr)

1,4-Dibromopentane — *N*-4-Bromopentylphthalimide (only product, 67% yield)

(c) Both bromines are bonded to primary carbons, but branching at the adjacent carbon hinders nucleophilic attack at one of them.

less crowded Br; more crowded Br

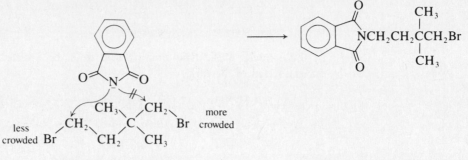

1,4-Dibromo-2,2-dimethylbutane $\longrightarrow$ N-4-Bromo-3,3-dimethylphthalimide (only product, 53% yield)

$$C_6H_5CH_2NCH_2CH_2C(CH_3)_2CH_2Br$$

23.22 (a) Amines are basic and are protonated by hydrogen halides.

$$\underset{\text{Benzylamine}}{C_6H_5CH_2NH_2} + HBr \longrightarrow \underset{\text{Benzylammonium bromide}}{C_6H_5CH_2\overset{+}{N}H_3\ Br^-}$$

(b) Equimolar amounts of benzylamine and sulfuric acid yield benzylammonium hydrogen sulfate as the product.

$$\underset{\text{Benzylamine}}{C_6H_5CH_2NH_2} + \underset{\text{Sulfuric acid}}{HOSO_2OH} \longrightarrow \underset{\text{Benzylammonium hydrogen sulfate}}{C_6H_5CH_2\overset{+}{N}H_3\ \overset{-}{O}SO_2OH}$$

(c) Acetic acid transfers a proton to benzylamine.

$$\underset{\text{Benzylamine}}{C_6H_5CH_2NH_2} + \underset{\text{Acetic acid}}{CH_3\overset{O}{\overset{\|}{C}}OH} \longrightarrow \underset{\text{Benzylammonium acetate}}{C_6H_5CH_2\overset{+}{N}H_3\ \overset{-}{O}\overset{O}{\overset{\|}{C}}CH_3}$$

(d) Acetyl chloride reacts with benzylamine by acyl transfer to form an amide.

$$\underset{\text{Benzylamine}}{2\,C_6H_5CH_2NH_2} + \underset{\text{Acetyl chloride}}{CH_3\overset{O}{\overset{\|}{C}}Cl} \longrightarrow \underset{N\text{-Benzylacetamide}}{CH_3\overset{O}{\overset{\|}{C}}NHCH_2C_6H_5} + \underset{\text{Benzylammonium chloride}}{C_6H_5CH_2\overset{+}{N}H_3\ Cl^-}$$

(e) Acetic anhydride also reacts with benzylamine by acyl transfer.

$$\underset{\text{Benzylamine}}{2\,C_6H_5CH_2NH_2} + \underset{\text{Acetic anhydride}}{CH_3\overset{O}{\overset{\|}{C}}O\overset{O}{\overset{\|}{C}}CH_3} \longrightarrow \underset{N\text{-Benzylacetamide}}{CH_3\overset{O}{\overset{\|}{C}}NHCH_2C_6H_5} + \underset{\text{Benzylammonium acetate}}{C_6H_5CH_2\overset{+}{N}H_3\ \overset{-}{O}\overset{O}{\overset{\|}{C}}CH_3}$$

(f) Primary amines react with ketones to give imines.

$$\underset{\text{Benzylamine}}{C_6H_5CH_2NH_2} + \underset{\text{Acetone}}{CH_3\overset{O}{\overset{\|}{C}}CH_3} \longrightarrow \underset{N\text{-Isopropylidenebenzylamine}}{(CH_3)_2C{=}NHCH_2C_6H_5}$$

(g) These reaction conditions lead to reduction of the imine formed in (f). The reaction is reductive amination.

$$\underset{\text{Benzylamine}}{C_6H_5CH_2NH_2} + \underset{\text{Acetone}}{CH_3\overset{O}{\overset{\|}{C}}CH_3} \xrightarrow{NaBH_3CN} \underset{N\text{-Isopropylbenzylamine}}{(CH_3)_2CHNHCH_2C_6H_5}$$

(*h*) Amines are nucleophilic and bring about the opening of epoxide rings.

$$C_6H_5CH_2NH_2 + \underset{\text{O}}{CH_2{-}CH_2} \longrightarrow C_6H_5CH_2NHCH_2CH_2OH$$

Benzylamine — Ethylene oxide — 2-(*N*-Benzylamino)ethanol

(*i*) In these nucleophilic ring-opening reactions the amine attacks the less sterically hindered carbon of the ring.

$$C_6H_5CH_2NH_2 + \underset{\text{O}}{CH_2{-}CHCH_3} \longrightarrow C_6H_5CH_2NHCH_2\underset{\text{OH}}{CHCH_3}$$

Benzylamine — 1,2-Epoxypropane — 1-(*N*-Benzylamino)-2-propanol

(*j*) With excess methyl iodide amines are converted to quaternary ammonium iodides.

$$C_6H_5CH_2NH_2 + 3\,CH_3I \longrightarrow C_6H_5CH_2\overset{+}{N}(CH_3)_3\ I^-$$

Benzylamine — Methyl iodide — Benzyltrimethylammonium iodide

(*k*) Nitrous acid forms from sodium nitrite in dilute hydrochloric acid. Nitrosation of benzylamine in water gives benzyl alcohol via a diazonium ion intermediate.

$$C_6H_5CH_2NH_2 \xrightarrow[H_2O]{NaNO_2,\ HCl} C_6H_5CH_2\overset{+}{N}{\equiv}N \xrightarrow[H_2O]{-N_2} C_6H_5CH_2OH$$

Benzylamine — Benzyldiazonium ion — Benzyl alcohol

Benzyl chloride will also be formed by attack of chloride on the diazonium ion.

23.23 (*a*) The reaction illustrates the preparation of a secondary amine by reductive amination.

Cyclohexanone + H_2N–cyclohexyl $\xrightarrow{H_2,\ Ni}$ cyclohexyl–NH–cyclohexyl

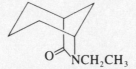

Cyclohexanone — Cyclohexylamine — Dicyclohexylamine (70%)

(*b*) Amides are reduced to amines by lithium aluminum hydride.

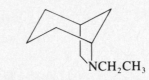

$\xrightarrow[2.\ H_2O]{1.\ LiAlH_4}$

6-Ethyl-6-azabicyclo[3.2.1]octan-7-one ($O{=}$, NCH_2CH_3) — 6-Ethyl-6-azabicyclo[3.2.1]octane (NCH_2CH_3)

(*c*) Treatment of alcohols with *p*-toluenesulfonyl chloride converts them to *p*-toluenesulfonate esters.

$$C_6H_5CH_2CH_2CH_2OH + CH_3{-}C_6H_4{-}SO_2Cl$$

3-Phenyl-1-propanol — *p*-Toluenesulfonyl chloride

↓ pyridine

$$C_6H_5CH_2CH_2CH_2O\overset{O}{\underset{O}{\overset{\|}{\underset{\|}{S}}}}{-}C_6H_4{-}CH_3$$

3-Phenylpropyl *p*-toluenesulfonate

p-Toluenesulfonate is an excellent leaving group in nucleophilic substitution reactions. Dimethylamine is the nucleophile.

$$C_6H_5CH_2CH_2CH_2OSO_2-C_6H_4-CH_3 \;+\; (CH_3)_2NH \longrightarrow C_6H_5CH_2CH_2CH_2N(CH_3)_2$$

3-Phenylpropyl *p*-toluenesulfonate

N,N-Dimethyl-3-phenyl-1-propanamine (86%)

(*d*) The tertiary amine formed in (*c*) is converted to its corresponding amine oxide on oxidation with peroxyacetic acid. Heating the amine oxide causes it to undergo a Cope elimination.

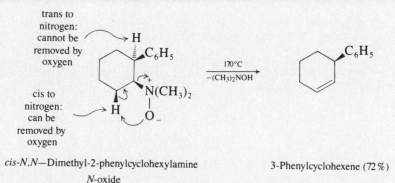

$$C_6H_5CH_2CH_2CH_2N(CH_3)_2 \;+\; CH_3CO_2OH \longrightarrow C_6H_5CH_2CH_2CH_2\overset{O^-}{\overset{|}{N^+}}(CH_3)_2 \;+\; CH_3CO_2H$$

N,N-Dimethyl-3-phenyl-1-propanamine; Peroxyacetic acid; *N,N*-Dimethyl-3-phenyl-1-propanamine *N*-oxide; Acetic acid

$$\xrightarrow{160^\circ C} C_6H_5CH_2CH{=}CH_2 \;+\; (CH_3)_2NOH$$

3-Phenyl-1-propene (91%); *N,N*-Dimethylhydroxylamine

(*e*) Cope elimination reactions require a syn orientation of hydrogen and nitrogen in the activated complex.

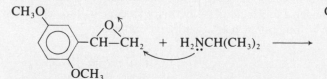

trans to nitrogen: cannot be removed by oxygen

cis to nitrogen: can be removed by oxygen

$$\xrightarrow[-(CH_3)_2NOH]{170^\circ C}$$

cis-N,N—Dimethyl-2-phenylcyclohexylamine *N*-oxide

3-Phenylcyclohexene (72%)

This stereoelectronic requirement leads to a high degree of regioselectivity in the elimination process. The only alkene formed is 3-phenylcyclohexene. Its isomer, 1-phenylcyclohexene, can only be formed by an anti elimination sequence and is not observed.

(*f*) Amines are sufficiently nucleophilic to react with epoxides. Attack occurs at the less substituted carbon of the epoxide.

$$(CH_3O)_2C_6H_3-\underset{\backslash O /}{CH-CH_2} \;+\; H_2NCH(CH_3)_2 \longrightarrow (CH_3O)_2C_6H_3-\overset{OH}{\overset{|}{C}}HCH_2NHCH(CH_3)_2$$

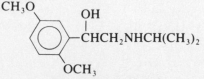

2-(2,5-Dimethoxyphenyl)oxirane; Isopropylamine; 1-(2,5-Dimethoxyphenyl)-2-(isopropylamino)ethanol (67%)

(*g*) α-Halo ketones are reactive substrates in nucleophilic substitution reactions. Dibenzylamine is the nucleophile.

$$(C_6H_5CH_2)_2\ddot{N}H + CH_3\overset{O}{\overset{\|}{C}}CH_2{-}Cl \longrightarrow CH_3\overset{O}{\overset{\|}{C}}CH_2N(CH_2C_6H_5)_2$$

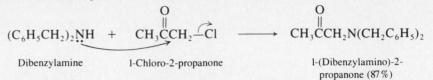

Dibenzylamine 1-Chloro-2-propanone 1-(Dibenzylamino)-2-propanone (87%)

Since the reaction liberates hydrogen chloride, it is carried out in the presence of added base—in this case triethylamine—so as to avoid converting the dibenzylamine to its hydrochloride salt.

(*h*) Quaternary ammonium hydroxides undergo a Hofmann elimination reaction when they are heated. A point to be considered here concerns the regioselectivity of Hofmann eliminations: It is the less hindered β proton that is removed by the base, so that usually the less substituted alkene is formed.

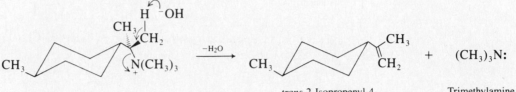

H ⁻OH, CH_3, CH_2, CH_3, $\overset{+}{N}(CH_3)_3$ $\xrightarrow{-H_2O}$ CH_3, CH_3, CH_2 + $(CH_3)_3N$:

trans-2-Isopropenyl-4-methylcyclohexane (98%) Trimethylamine

Elimination to give CH_3–(cyclohexylidene)=$C(CH_3)_2$ does not occur.

(*i*) The combination of sodium nitrite and aqueous acid is a nitrosating agent. Secondary alkanamines react with nitrosating agents to give *N*-nitroso amines as the isolated products.

$$(CH_3)_2CHNHCH(CH_3)_2 \xrightarrow[HCl,\ H_2O]{NaNO_2} (CH_3)_2CH\underset{N=O}{\underset{|}{N}}CH(CH_3)_2$$

Diisopropylamine *N*-Nitrosodiisopropylamine (91%)

23.24 (*a*) 4-Methylpiperidine can participate in intermolecular hydrogen bonding in the liquid phase.

CH_3, H–N:, CH_3, N–H, CH_3, H–N:

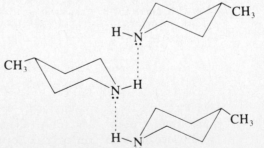

These hydrogen bonds must be broken in order for individual 4-methylpiperidine molecules to escape into the gas phase. *N*-Methylpiperidine lacks a proton bond to nitrogen and so cannot engage in intermolecular hydrogen bonding. Less energy is required to transfer a molecule of *N*-methylpiperidine to the gaseous state, and therefore it has a lower boiling point than 4-methylpiperidine.

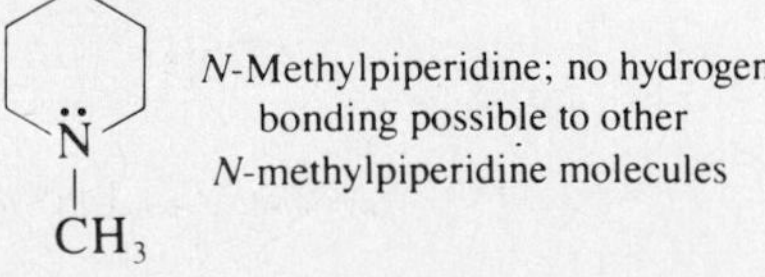

N-Methylpiperidine; no hydrogen bonding possible to other *N*-methylpiperidine molecules

(*b*) The two products are diastereomeric quaternary ammonium chlorides that differ in the configuration at the nitrogen atom.

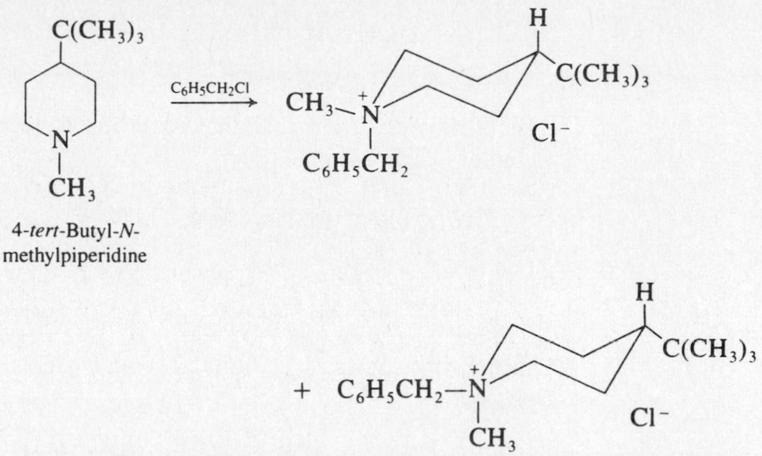

4-*tert*-Butyl-*N*-methylpiperidine $\xrightarrow{C_6H_5CH_2Cl}$ (*trans*/*cis* $C_6H_5CH_2$, CH_3 quaternary ammonium chlorides) Cl^-

(*c*) The activated complex for pyramidal inversion in this compound has a coplanar arrangement of bonds to nitrogen.

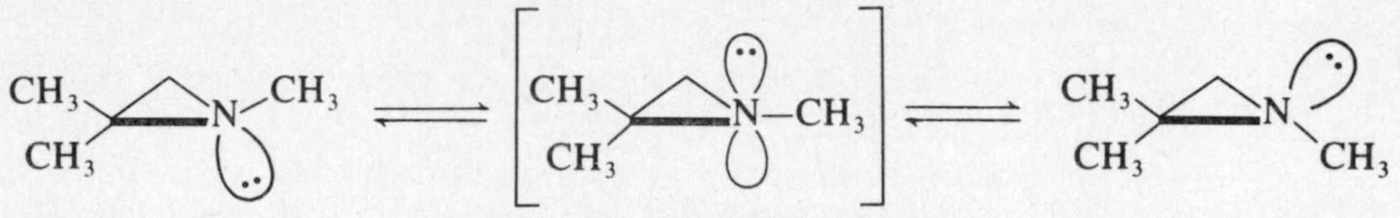

Activated complex

Hybridization of nitrogen changes from sp^3 in the starting state to sp^2 at the transition state. Angle strain increases in going to the transition state because the three-membered ring is less able to accommodate the 120° bond angles associated with sp^2 hybridized nitrogen than the 109.5° bond angles associated with sp^3 hybridization. An increase in angle strain at the transition state translates to a slower rate of reaction.

(*d*) In this reaction two neutral starting materials yield an ionic product.

$$(CH_3CH_2)_3N + CH_3CH_2I \longrightarrow (CH_3CH_2)_4\overset{+}{N}\ I^-$$

Triethylamine Ethyl iodide Tetraethylammonium iodide

The data reveal that the reaction proceeds faster in a polar solvent (acetone) than in a nonpolar one (hexane). The reason for this is that the activated complex is more polar than the starting materials.

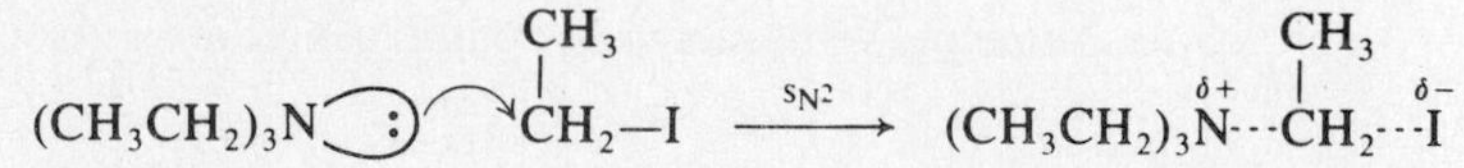

$$(CH_3CH_2)_3N: + CH_3CH_2{-}I \xrightarrow{S_N2} (CH_3CH_2)_3\overset{\delta+}{N}\cdots\underset{\displaystyle CH_3}{CH_2}\cdots\overset{\delta-}{I}$$

Triethylamine Ethyl iodide Representation of activated complex

There is a separation of positive and negative charge in the activated complex. A polar solvent can support this charge separation better than a nonpolar one, lower the energy of the transition state, and cause the reaction to proceed at a faster rate.

(*e*) Tetramethylammonium hydroxide cannot undergo Hofmann elimination. The only reaction that can take place is nucleophilic substitution.

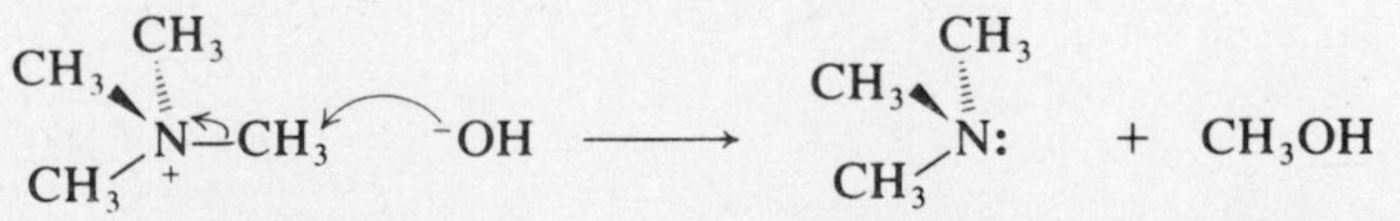

$$(CH_3)_3\overset{+}{N}{-}CH_3\ \ ^-OH \longrightarrow (CH_3)_3N: + CH_3OH$$

Tetramethylammonium hydroxide Trimethylamine Methanol

(*f*) The key intermediate in the reaction of an amine with nitrous acid is the corresponding diazonium ion.

$$CH_3CH_2CH_2NH_2 \xrightarrow[H_2O]{NaNO_2,\ HCl} CH_3CH_2CH_2{-}\overset{+}{N}{\equiv}N:$$

1-Propanamine — Propyldiazonium ion

Loss of nitrogen from this diazonium ion is accompanied by a hydride shift to form a secondary carbocation.

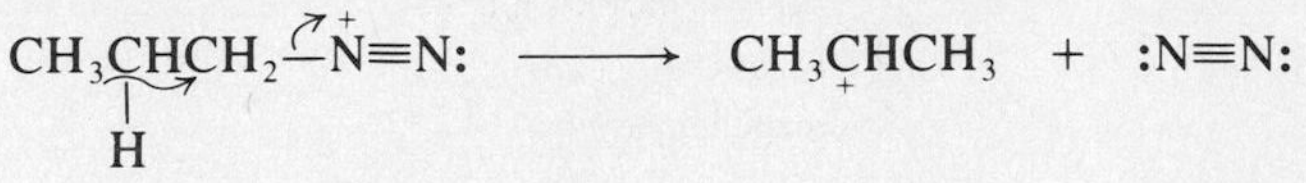

$$CH_3\underset{\displaystyle H}{\underset{|}{C}}HCH_2{-}\overset{+}{N}{\equiv}N: \longrightarrow CH_3\underset{+}{C}HCH_3 + :N{\equiv}N:$$

Propyldiazonium ion — Isopropyl cation — Nitrogen

Capture of isopropyl cation by water yields the major product of the reaction, 2-propanol.

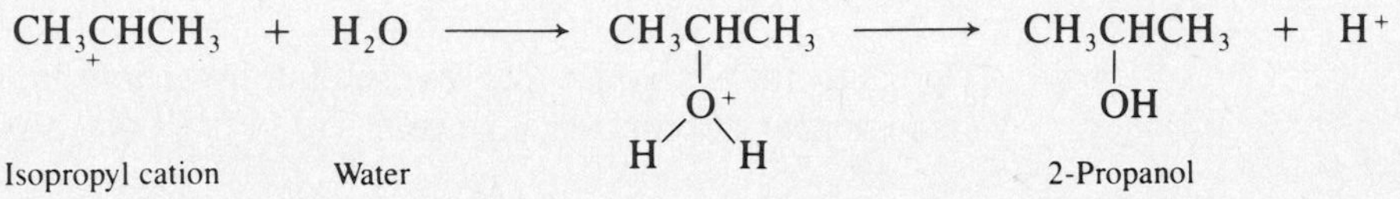

$$CH_3\underset{+}{C}HCH_3 + H_2O \longrightarrow CH_3CHCH_3\ (\text{with } \overset{+}{O}H_2) \longrightarrow CH_3\underset{\displaystyle OH}{\underset{|}{C}}HCH_3 + H^+$$

Isopropyl cation — Water — 2-Propanol

(*g*) The molecular formula of the product indicates that all that has happened has been acceptance by compound A of a proton from perchloric acid. There are two basic sites in the molecule—the amine nitrogen and the carbonyl oxygen. While we would expect the amine group to be more basic, the product does not derive from protonation there because we are told that the product does not contain a carbonyl group. Therefore, consider protonation of the carbonyl oxygen.

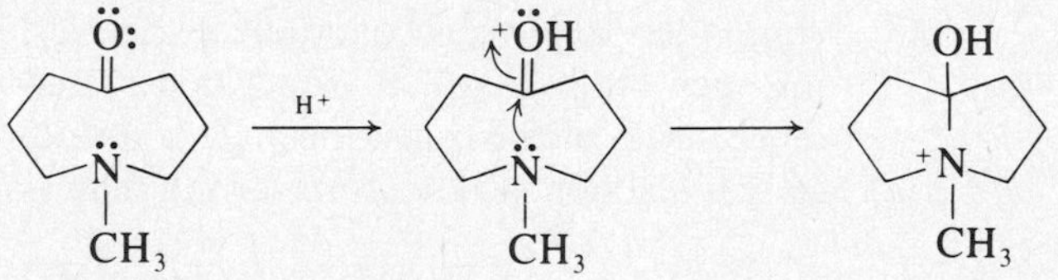

Compound A $\xrightarrow{H^+}$ (protonated carbonyl) $\longrightarrow$ $(C_8H_{16}NO)^+$

The molecule can adopt a conformation in which the amine nitrogen is within bonding distance of the protonated carbonyl. It uses its unshared pair of electrons to form a covalent bond, giving the species shown.

23.25 Alcohols are converted to *p*-toluenesulfonate esters by reaction with *p*-toluenesulfonyl chloride. None of the bonds to the chiral center are affected in this reaction.

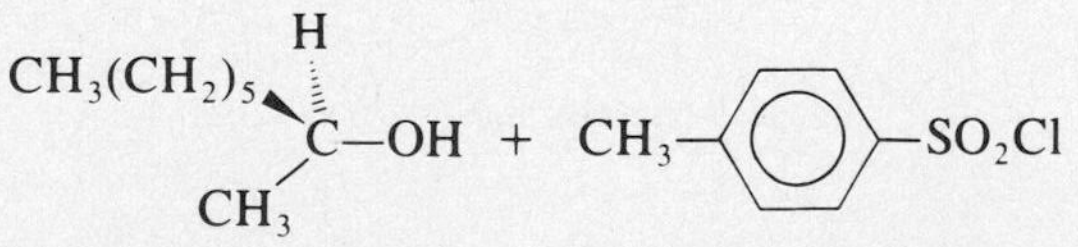

$$CH_3(CH_2)_5C(H)(CH_3){-}OH + CH_3{-}C_6H_4{-}SO_2Cl$$

(*S*)-2-Octanol — *p*-Toluenesulfonyl chloride

↓ pyridine

$$CH_3(CH_2)_5C(H)(CH_3){-}OSO_2{-}C_6H_4{-}CH_3$$

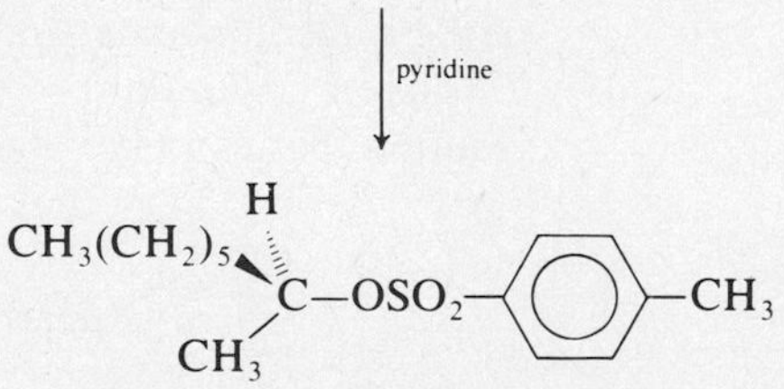

(*S*)-1-Methylheptyl *p*-toluenesulfonate
(compound B)

Displacement of the *p*-toluenesulfonate leaving group by sodium azide proceeds with inversion of configuration.

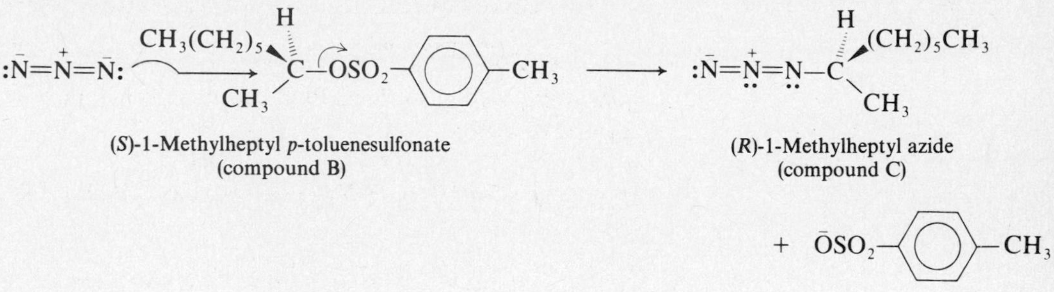

$$:\bar{N}{=}\overset{+}{N}{=}\bar{N}: + CH_3(CH_2)_5(CH_3)(H)C{-}OSO_2{-}C_6H_4{-}CH_3 \longrightarrow :\bar{N}{=}\overset{+}{N}{=}N{-}C(H)(CH_3)(CH_2)_5CH_3$$

(*S*)-1-Methylheptyl *p*-toluenesulfonate (compound B)

(*R*)-1-Methylheptyl azide (compound C)

$$+ \ \bar{O}SO_2{-}C_6H_4{-}CH_3$$

Reduction of the azide yields a primary amine. A nitrogen-nitrogen bond is cleaved; all the bonds to the chiral center remain intact.

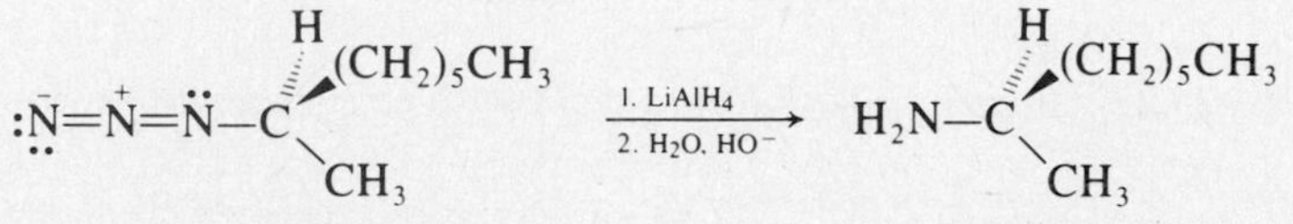

$$:\bar{N}{=}\overset{+}{N}{=}\ddot{N}{-}C(H)(CH_3)(CH_2)_5CH_3 \xrightarrow[\text{2. } H_2O,\ HO^-]{\text{1. } LiAlH_4} H_2N{-}C(H)(CH_3)(CH_2)_5CH_3$$

(*R*)-1-Methylheptyl azide (compound C)

(*R*)-2-Octanamine (compound D)

23.26 Since we are told that compound E has been converted to quinuclidine by treatment with sodium hydroxide, the conversion of compound F to quinuclidine can be achieved if F can be converted to E.

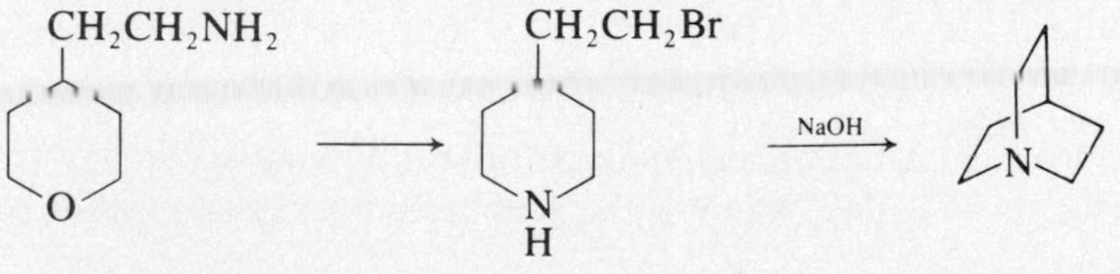

$CH_2CH_2NH_2$ (tetrahydropyran ring, O) $\longrightarrow$ CH_2CH_2Br (piperidine ring, N–H) $\xrightarrow{NaOH}$ quinuclidine (N)

Compound F — Compound E — Quinuclidine

The key intermediate, compound E, can be viewed as the product of an intramolecular nucleophilic substitution reaction. Mentally disconnect one of the bonds to nitrogen:

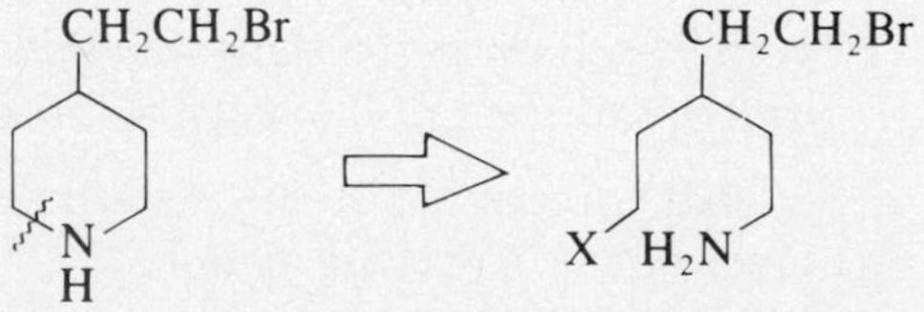

CH_2CH_2Br (piperidine ring, N–H) $\Rightarrow$ CH_2CH_2Br (open chain, X and H_2N)

A compound that has the necessary functionality to undergo the cyclization may be obtained by cleavage of the tetrahydropyran with hydrogen bromide.

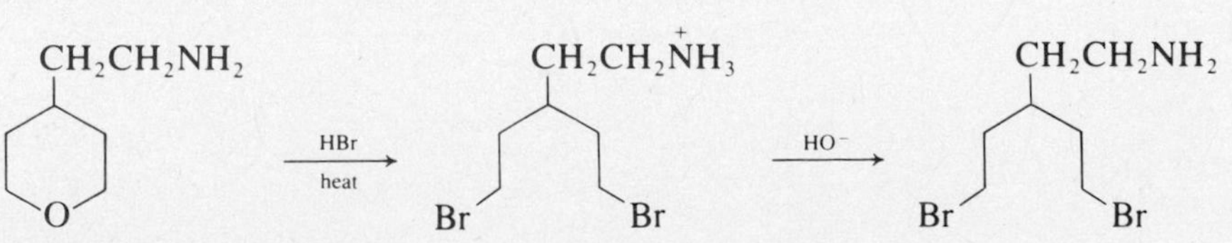

$CH_2CH_2NH_2$ (tetrahydropyran ring, O) $\xrightarrow[\text{heat}]{HBr}$ $CH_2CH_2\overset{+}{N}H_3$ (open chain, Br, Br) $\xrightarrow{HO^-}$ $CH_2CH_2NH_2$ (open chain, Br, Br)

Compound F

Treatment of this amine with base converts it to quinuclidine by way of compound E.

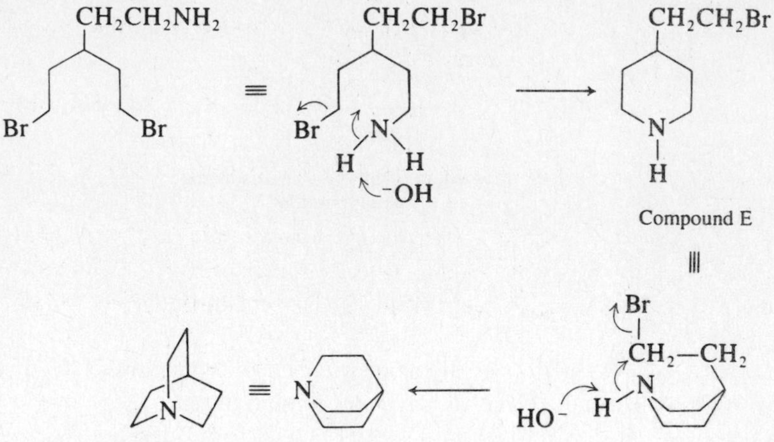

$CH_2CH_2NH_2$ ... Br Br ≡ CH_2CH_2Br ... Br N H H ^-OH ⟶ CH_2CH_2Br ... N H

Compound E

Quinuclidine

23.27 Primary alkylamines react with nitrous acid to form unstable diazonium ion intermediates.

$$(CH_3)_3CCH_2CH_2NH_2 \xrightarrow[HClO_4]{NaNO_2} (CH_3)_3CCH_2CH_2-\overset{+}{N}\equiv N$$

3,3-Dimethyl-1-butanamine — 3,3-Dimethylbutyldiazonium ion

This diazonium ion can undergo solvolysis in water to give the corresponding alcohol.

$$(CH_3)_3CCH_2CH_2-\overset{+}{N}\equiv N \xrightarrow{-N_2} (CH_3)_3CCH_2CH_2-\overset{+}{O}H_2 \xrightarrow{-H^+} (CH_3)_3CCH_2CH_2OH$$

$H_2\ddot{O}:$

3,3-Dimethyl-1-butanol

Alternatively, hydride migration may accompany loss of nitrogen, giving a secondary carbocation, which is then captured by water to yield a secondary alcohol.

$$(CH_3)_3CCH(H)-CH_2-\overset{+}{N}\equiv N \longrightarrow (CH_3)_3C\overset{+}{C}HCH_3 \xrightarrow{H_2O} (CH_3)_3CCH(OH)CH_3$$

3,3-Dimethyl-2-butanol

23.28 Treat this problem in the same way that you would any other one in which a trialkylamine *N*-oxide is subjected to thermolysis. Orient the molecule, or preferably a molecular model, in a way that permits you to see the arrangement of the five atoms ($H-C-C-\overset{+}{N}-O^-$) that are involved in a Cope elimination.

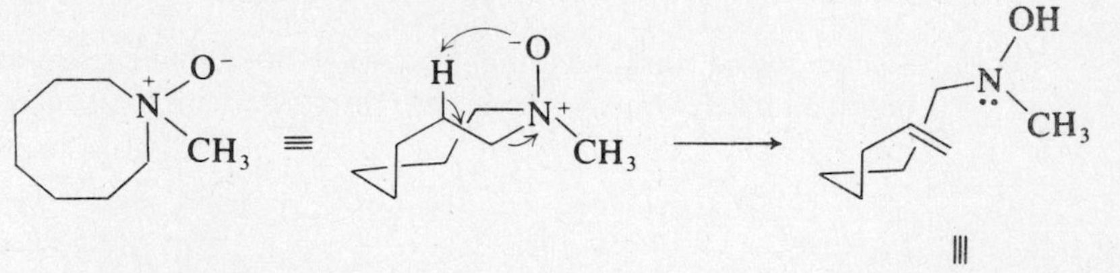

O^- $\overset{+}{N}$ CH_3 ≡ H ^-O N^+ CH_3 ⟶ OH N CH_3

$$CH_2=CH(CH_2)_4CH_2N(OH)CH_3$$

N-(6-Heptenyl)-*N*-methylhydroxylamine

23.29 There are two factors that combine to permit ready conversion of ethyl aminoacetate to ethyl diazoacetate:

1. The rate of loss of nitrogen from the diazonium ion is much slower than that of alkanamines.

$$:N{\equiv}\overset{+}{N}{-}CH_2\overset{\overset{O}{\|}}{C}OCH_2CH_3 \xrightarrow{\text{slow}} :N{\equiv}N: + {}^{+}CH_2\overset{\overset{O}{\|}}{C}OCH_2CH_3$$

   The electron-withdrawing effect of the carbonyl group destabilizes the carbocation and raises the activation energy for its formation.

2. The proton at the α carbon atom of the diazonium salt is relatively acidic and easily removed.

$$:N{\equiv}\overset{+}{N}{-}\underset{\underset{H}{|}}{C}H\overset{\overset{O}{\|}}{C}OCH_2CH_3 \longrightarrow :\overset{-}{\ddot{N}}{=}\overset{+}{N}{=}CH\overset{\overset{O}{\|}}{C}OCH_2CH_3 + H^+$$

Ethyl diazoacetate

   Not only is this proton bonded to the α carbon of an ester, but also the positively charged diazonium group is strongly electron withdrawing and increases the acidity of the proton dramatically.

23.30 (*a*) The overall transformation can be expressed as $RBr \rightarrow RCH_2NH_2$. In many cases this can be carried out via a nitrile, as $RBr \rightarrow RCN \rightarrow RCH_2NH_2$. In this case, however, the substrate is neopentyl bromide, an alkyl halide that reacts very slowly in nucleophilic substitution processes. Carbon-carbon bond formation with neopentyl bromide can be achieved by carboxylation of the corresponding Grignard reagent.

$$\underset{\text{Neopentyl bromide}}{(CH_3)_3CCH_2Br} \xrightarrow[\text{2. } CO_2 \text{ 3. } H_3O^+]{\text{1. Mg}} \underset{\text{3,3-Dimethylbutanoic acid (63\%)}}{(CH_3)_3CCH_2CO_2H}$$

The carboxylic acid can then be converted to the desired amine by reduction of the derived amide.

$$\underset{\text{3,3-Dimethylbutanoic acid}}{(CH_3)_3CCH_2CO_2H} \xrightarrow[\text{2. } NH_3]{\text{1. } SOCl_2} \underset{\text{3,3-Dimethylbutanamide (51\%)}}{(CH_3)_3CCH_2\overset{\overset{O}{\|}}{C}NH_2} \xrightarrow[\text{2. } H_2O]{\text{1. } LiAlH_4} \underset{\text{3,3-Dimethyl-1-butanamine (57\%)}}{(CH_3)_3CCH_2CH_2NH_2}$$

The yields listed in parentheses are those reported in the chemical literature for this synthesis.

(*b*) Consider the starting materials in relation to the desired product:

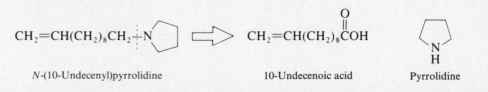

$$\underset{N\text{-(10-Undecenyl)pyrrolidine}}{CH_2{=}CH(CH_2)_8CH_2{-}N\bigcirc} \Longrightarrow \underset{\text{10-Undecenoic acid}}{CH_2{=}CH(CH_2)_8\overset{\overset{O}{\|}}{C}OH} \quad \underset{\text{Pyrrolidine}}{\text{(NH ring)}}$$

The synthetic tasks are to form the necessary carbon-nitrogen bond and to reduce the carbonyl group to a methylene group. This has been accomplished by way of the amide as a key intermediate.

$$CH_2{=}CH(CH_2)_8\overset{\overset{\Large O}{\|}}{C}OH \xrightarrow[\text{2. pyrrolidine}]{\text{1. SOCl}_2} CH_2{=}CH(CH_2)_8\overset{\overset{\Large O}{\|}}{C}{-}N\text{(pyrrolidine ring)}$$

10-Undecenoic acid — N-(10-Undecenoyl)pyrrolidine (75%)

$$\downarrow \text{1. LiAlH}_4,\ \text{2. H}_2\text{O}$$

$$CH_2{=}CH(CH_2)_8CH_2{-}N\text{(pyrrolidine ring)}$$

N-(10-Undecenyl)pyrrolidine (66%)

A second approach utilizes reductive amination following conversion of the starting carboxylic acid to an aldehyde.

$$CH_2{=}CH(CH_2)_8\overset{\overset{\Large O}{\|}}{C}OH \xrightarrow[\text{2. H}_2\text{O}]{\text{1. LiAlH}_4} CH_2{=}CH(CH_2)_8CH_2OH \xrightarrow[\text{CH}_2\text{Cl}_2]{(\text{C}_5\text{H}_5\text{N})_2\text{CrO}_3} CH_2{=}CH(CH_2)_8\overset{\overset{\Large O}{\|}}{C}H$$

10-Undecenoic acid — 10-Undecen-1-ol — 10-Undecenal

The reducing agent in the reductive amination process cannot be hydrogen because that would result in hydrogenation of the double bond. Sodium cyanoborohydride is required.

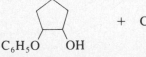

$$CH_2{=}CH(CH_2)_8\overset{\overset{\Large O}{\|}}{C}H + \text{pyrrolidine (N–H)} \xrightarrow{\text{NaBH}_3\text{CN}} CH_2{=}CH(CH_2)_8CH_2{-}N\text{(pyrrolidine ring)}$$

10-Undecenal — Pyrrolidine — N-(10-Undecenyl)pyrrolidine

(*c*) It is stereochemistry that determines the choice of which synthetic method to employ in introducing the amine group. The carbon-nitrogen bond must be formed with inversion of configuration at the alcohol carbon. Conversion of the alcohol to its *p*-toluenesulfonate ester ensures that the leaving group is introduced with exactly the same stereochemistry as the alcohol.

$$\text{cyclopentane}(C_6H_5O)(OH) + CH_3{-}C_6H_4{-}SO_2Cl$$

cis-2-Phenoxycyclopentanol — *p*-Toluenesulfonyl chloride

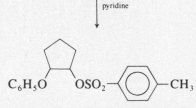

$$\downarrow \text{pyridine}$$

$$\text{cyclopentane}(C_6H_5O)(OSO_2{-}C_6H_4{-}CH_3)$$

cis-2-Phenoxycyclopentyl *p*-toluenesulfonate

Once the leaving group has been introduced with the proper stereochemistry, it can be displaced by a nitrogen nucleophile suitable for subsequent conversion to an amine.

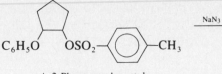

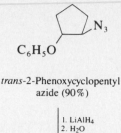

$$\text{cis-2-Phenoxycyclopentyl } p\text{-toluenesulfonate (}C_6H_5O,\ OSO_2C_6H_4CH_3\text{)} \xrightarrow{NaN_3} \text{trans-2-Phenoxycyclopentyl azide (90\%) (}C_6H_5O,\ N_3\text{)}$$

cis-2-Phenoxycyclopentyl p-toluenesulfonate

trans-2-Phenoxycyclopentyl azide (90%)

$$\xrightarrow{1.\ LiAlH_4;\ 2.\ H_2O}$$

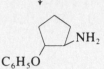

trans-2-Phenoxycyclopentylamine (C_6H_5O, NH_2)

(As actually reported the azide was reduced by hydrogenation over a palladium catalyst, and the amine was isolated as its hydrochloride salt in 66 percent yield.)

(*d*) Recognition that the primary amine is derivable from the corresponding nitrile by reduction

$$C_6H_5CH_2N(CH_3)CH_2CH_2CH_2CH_2NH_2 \Longrightarrow C_6H_5CH_2N(CH_3)CH_2CH_2CH_2C{\equiv}N$$

and that the necessary tertiary amine function can be introduced by a nucleophilic substitution reaction between the two given starting materials suggests the following synthesis:

$$C_6H_5CH_2NH(CH_3) + BrCH_2CH_2CH_2CN \longrightarrow C_6H_5CH_2N(CH_3)CH_2CH_2CH_2CN$$

N-Methylbenzylamine   4-Bromobutanenitrile

$$\xrightarrow{1.\ LiAlH_4;\ 2.\ H_2O} C_6H_5CH_2N(CH_3)CH_2CH_2CH_2CH_2NH_2$$

N-Benzyl-N-methyl-1,4-butanediamine

Alkylation of *N*-methylbenzylamine with 4-bromobutanenitrile has been achieved in 92 percent yield in the presence of potassium carbonate as a weak base to neutralize the hydrogen bromide produced. The nitrile may be reduced with lithium aluminum hydride, as shown in the equation, or by catalytic hydrogenation. Catalytic hydrogenation over platinum gave the desired diamine, isolated as its hydrochloride salt, in 90 percent yield.

(*e*) The only reaction described to this point in the text that leads to *N,N*-disubstituted hydroxylamine derivatives is the Cope elimination. Cope has pointed out that while amine oxide thermolysis may be used in alkene synthesis, it also provides a preparative route to hydroxylamine derivatives from tertiary amines.

$$(CH_3CH_2CH_2CH_2CH_2)_3N \xrightarrow{CH_3CO_2OH} (CH_3CH_2CH_2CH_2CH_2)_3\overset{+}{N}{-}O^-$$

Tripentylamine   Tripentylamine N-oxide (99%)

$$\xrightarrow{160°C} CH_3CH_2CH_2CH{=}CH_2 + (CH_3CH_2CH_2CH_2CH_2)_2NOH$$

1-Pentene (65%)   N,N-Dipentylhydroxylamine (75%)

(*f*) The overall transformation may be viewed retrosynthetically as follows:

$$ArCH_2N(CH_3)_2 \Longrightarrow ArCH_2Br \Longrightarrow ArCH_3$$

$$Ar = NC\text{-}C_6H_4\text{-}$$

The sequence which presents itself begins with benzylic bromination with *N*-bromosuccinimide.

$$\underset{p\text{-Tolunitrile}}{NC\text{-}C_6H_4\text{-}CH_3} \xrightarrow[\text{benzoyl peroxide, } CCl_4 \text{, heat}]{N\text{-bromosuccinimide}} \underset{p\text{-Cyanobenzyl bromide}}{NC\text{-}C_6H_4\text{-}CH_2Br}$$

The reaction shown in the equation has been reported in the chemical literature and gave the benzylic bromide in 60 percent yield.

Treatment of this bromide with dimethylamine gives the desired product. (The isolated yield was 83 percent by this method.)

$$\underset{p\text{-Cyanobenzyl bromide}}{NC\text{-}C_6H_4\text{-}CH_2Br} + \underset{\text{Dimethylamine}}{(CH_3)_2NH} \longrightarrow \underset{N,N\text{-Dimethyl-}p\text{-cyanobenzylamine}}{NC\text{-}C_6H_4\text{-}CH_2N(CH_3)_2}$$

23.31 A standard method for carbon-carbon bond formation using ketones and α-bromo esters is the Reformatsky reaction.

$$\underset{\text{Acetophenone}}{C_6H_5\overset{O}{\overset{\|}{C}}CH_3} + \underset{\text{Ethyl bromoacetate}}{BrCH_2CO_2CH_2CH_3} \xrightarrow[\text{benzene}]{Zn}$$

$$C_6H_5\underset{CH_3}{\overset{OH}{C}}CH_2CO_2CH_2CH_3 \xrightarrow[H^+]{\text{heat}} \underset{\text{Ethyl 3-phenyl-2-butenoate (50\%)}}{C_6H_5\underset{CH_3}{C}{=}CHCO_2CH_2CH_3}$$

Hydrogenation of the double bond is achieved in the usual way.

$$\underset{\text{Ethyl 3-phenyl-2-butenoate}}{C_6H_5\underset{CH_3}{C}{=}CHCO_2CH_2CH_3} \xrightarrow[\text{Pt, ethanol}]{H_2} \underset{\text{Ethyl 3-phenylbutanoate (99\%)}}{C_6H_5\underset{CH_3}{C}HCH_2CO_2CH_2CH_3}$$

Reduction of the ester to the primary alcohol is readily achieved by using lithium aluminum hydride. The primary alcohol is converted to the corresponding bromide with phosphorus tribromide. (Conceivably, HBr could be used instead of PBr_3.)

$$\underset{\text{Ethyl 3-phenylbutanoate}}{C_6H_5\underset{CH_3}{C}HCH_2CO_2CH_2CH_3} \xrightarrow[2.\ H_2O]{1.\ LiAlH_4} \underset{\text{3-Phenyl-1-butanol (98\%)}}{C_6H_5\underset{CH_3}{C}HCH_2CH_2OH} \xrightarrow{PBr_3} \underset{\text{1-Bromo-3-phenylbutane (81\%)}}{C_6H_5\underset{CH_3}{C}HCH_2CH_2Br}$$

Reaction of the bromide with diethylamine gives the desired tertiary amine, which is then transformed to 3-phenyl-1-butene by a standard Cope elimination sequence.

$$\underset{\text{1-Bromo-3-phenylbutane}}{C_6H_5\underset{\displaystyle CH_3}{\underset{|}{C}}HCH_2CH_2Br} + \underset{\substack{\text{Dimethylamine}\\\text{(excess)}}}{HN(CH_3)_2} \longrightarrow \underset{\substack{N,N\text{-Dimethyl-3-phenyl-1-butanamine}\\(98\%)}}{C_6H_5\underset{\displaystyle CH_3}{\underset{|}{C}}HCH_2CH_2N(CH_3)_2}$$

$$\xrightarrow{H_2O_2} C_6H_5\underset{\displaystyle CH_3}{\underset{|}{C}}HCH_2CH_2\overset{\displaystyle O^-}{\overset{|}{\underset{+}{N}}}(CH_3)_2 \xrightarrow{\text{heat}} \underset{\text{3-Phenyl-1-butene}}{C_6H_5\underset{\displaystyle CH_3}{\underset{|}{C}}HCH{=}CH_2} + \underset{N,N\text{-Dimethylhydroxylamine}}{(CH_3)_2NOH}$$

23.32 Weakly basic nucleophiles react with α,β-unsaturated carbonyl compounds by conjugate addition.

$$HY\!: + R_2C{=}CH\overset{\displaystyle O}{\overset{\|}{C}}R' \longrightarrow R_2\underset{\displaystyle Y}{\underset{|}{C}}CH_2\overset{\displaystyle O}{\overset{\|}{C}}R'$$

Ammonia and its derivatives are very prone to react in this way; thus conjugate addition provides a method for the preparation of β-amino carbonyl compounds.

(*a*)

$$\underset{\text{4-Methyl-3-penten-2-one}}{(CH_3)_2C{=}CH\overset{\displaystyle O}{\overset{\|}{C}}CH_3} + \underset{\text{Ammonia}}{NH_3} \longrightarrow \underset{\substack{\text{4-Amino-4-methyl-2-pentanone}\\(63-70\%)}}{(CH_3)_2\underset{\displaystyle NH_2}{\underset{|}{C}}CH_2\overset{\displaystyle O}{\overset{\|}{C}}CH_3}$$

(*b*)

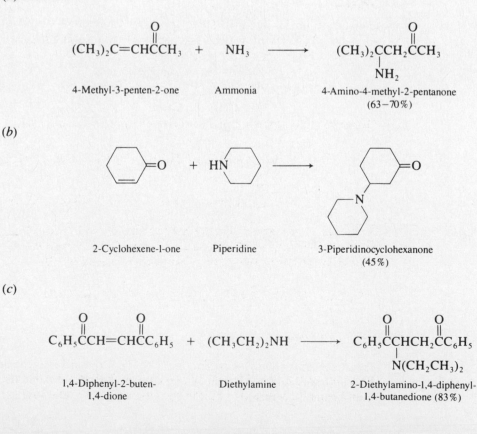

2-Cyclohexene-1-one + Piperidine (HN) ⟶ 3-Piperidinocyclohexanone (45%)

(*c*)

$$\underset{\substack{\text{1,4-Diphenyl-2-buten-}\\\text{1,4-dione}}}{C_6H_5\overset{\displaystyle O}{\overset{\|}{C}}CH{=}CH\overset{\displaystyle O}{\overset{\|}{C}}C_6H_5} + \underset{\text{Diethylamine}}{(CH_3CH_2)_2NH} \longrightarrow \underset{\substack{\text{2-Diethylamino-1,4-diphenyl-}\\\text{1,4-butanedione (83\%)}}}{C_6H_5\overset{\displaystyle O}{\overset{\|}{C}}\underset{\displaystyle N(CH_2CH_3)_2}{\underset{|}{C}}HCH_2\overset{\displaystyle O}{\overset{\|}{C}}C_6H_5}$$

(*d*)

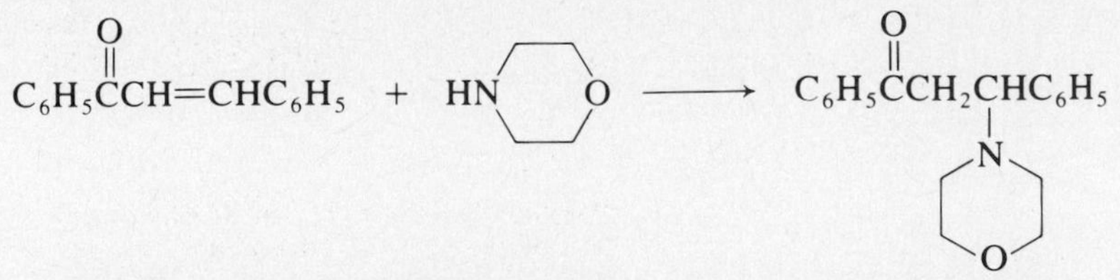

$$C_6H_5\overset{O}{\overset{\|}{C}}CH{=}CHC_6H_5 \ + \ \text{HN}\langle\text{morpholine ring}\rangle\text{O} \longrightarrow C_6H_5\overset{O}{\overset{\|}{C}}CH_2\underset{\text{N(morpholino)}}{\underset{|}{C}H}C_6H_5$$

1,3-Diphenyl-2-propen-1-one — Morpholine — 3-Morpholino-1,3-diphenyl-1-propanone (91%)

(*e*) The conjugate addition reaction that takes place in this case is an intramolecular one and occurs in virtually 100 percent yield.

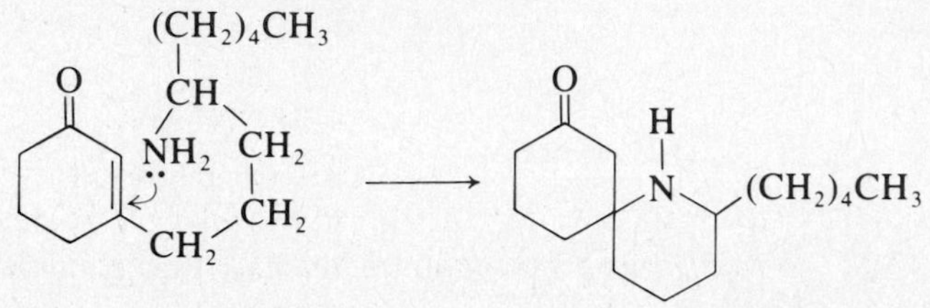

$(CH_2)_4CH_3$, CH, $\ddot{N}H_2$, CH_2, CH_2, CH_2 (cyclohexenone bearing side chain) $\longrightarrow$ spiro compound: cyclohexanone–N(H)–$(CH_2)_4CH_3$ piperidine

23.33 The first step in the synthesis is the conjugate addition of methylamine to ethyl acrylate. Two sequential Michael addition reactions take place.

$$CH_3NH_2 \ + \ CH_2{=}CH\overset{O}{\overset{\|}{C}}OCH_2CH_3 \longrightarrow CH_3NHCH_2CH_2\overset{O}{\overset{\|}{C}}OCH_2CH_3$$

Methylamine — Ethyl acrylate

$$\xrightarrow{CH_2{=}CHCO_2CH_2CH_3} CH_3N(CH_2CH_2CO_2CH_2CH_3)_2$$

Conversion of this intermediate to the desired *N*-methyl-4-piperidone requires a Dieckmann condensation followed by decarboxylation of the resulting β-keto ester.

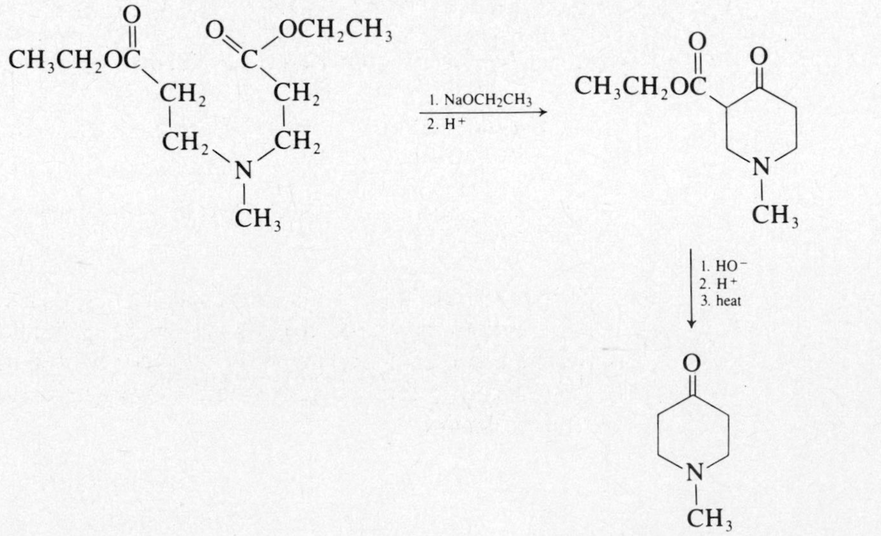

$$CH_3CH_2O\overset{O}{\overset{\|}{C}}CH_2CH_2N(CH_3)CH_2CH_2\overset{O}{\overset{\|}{C}}OCH_2CH_3 \xrightarrow[\text{2. } H^+]{\text{1. } NaOCH_2CH_3} \text{3-}(CH_3CH_2O\overset{O}{\overset{\|}{C}})\text{-1-}CH_3\text{-4-piperidone}$$

$$\xrightarrow{\text{1. } HO^-;\ \text{2. } H^+;\ \text{3. heat}}$$

N-Methyl-4-piperidone

Treatment of *N*-methyl-4-piperidone with the Grignard reagent derived from bromobenzene gives a tertiary alcohol which can be dehydrated to an alkene. Hydrogenation of the alkene completes the synthesis.

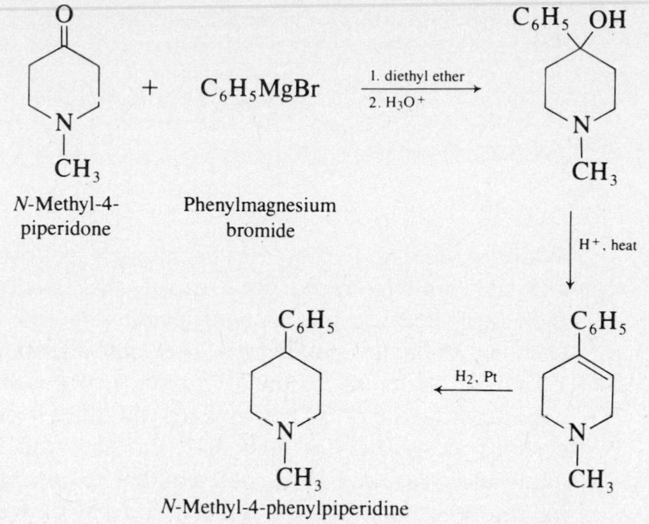

$$\text{N-Methyl-4-piperidone} + C_6H_5MgBr \xrightarrow[\text{2. } H_3O^+]{\text{1. diethyl ether}} \text{4-hydroxy-4-phenyl-N-methylpiperidine} \xrightarrow{H^+,\ \text{heat}} \text{N-methyl-4-phenyl-1,2,3,6-tetrahydropyridine} \xrightarrow{H_2,\ Pt} \text{N-Methyl-4-phenylpiperidine}$$

N-Methyl-4-piperidone; Phenylmagnesium bromide; N-Methyl-4-phenylpiperidine

23.34 The infrared spectrum of compound H exhibits only a single peak at 3330 cm^{-1} in the N—H stretching region. This suggests that it is a secondary amine, since primary amines show two peaks in this region and tertiary amines none. The ^1H nmr spectrum reveals the presence of five aromatic protons at $\delta = 7.3$ ppm, so compound H must be a monosubstituted benzene. The benzene ring accounts for all four sites of unsaturation suggested by the molecular formula $C_8H_{11}N$. The remaining peaks in the nmr spectrum are singlets, and the integral ratio 5 : 2 : 3 : 2 is consistent with the formulation of compound H as N-methylbenzylamine.

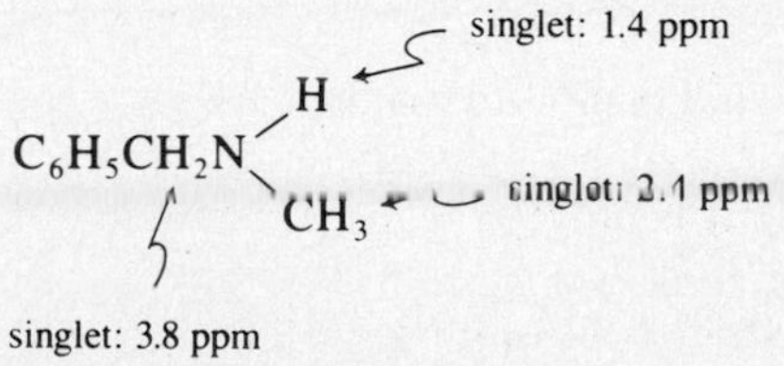

$C_6H_5CH_2N(H)CH_3$

singlet: 1.4 ppm (N—H)
singlet: 2.1 ppm (CH_3)
singlet: 3.8 ppm (CH_2)

23.35 Since there are peaks corresponding to five aromatic protons in the ^1H nmr spectrum of each isomer, compounds I and J (like compound H) each contain a monosubstituted benzene ring. Only four compounds of molecular formula $C_8H_{11}N$ meet this requirement.

$C_6H_5CH_2NHCH_3$	$C_6H_5NHCH_2CH_3$	$C_6H_5CH(NH_2)CH_3$	$C_6H_5CH_2CH_2NH_2$
N-Methylbenzylamine (compound H)	N-Ethylaniline	1-Phenylethylamine	2-Phenylethylamine

Neither ^1H nmr spectrum is consistent with N-ethylaniline, which would exhibit the characteristic triplet-quartet pattern of an ethyl group. While there is a quartet in the spectrum of compound I, it corresponds to only one proton, not the two that an ethyl group requires. The one-proton quartet in compound I arises from an H—C—CH_3 unit. Compound I is 1-phenylethylamine.

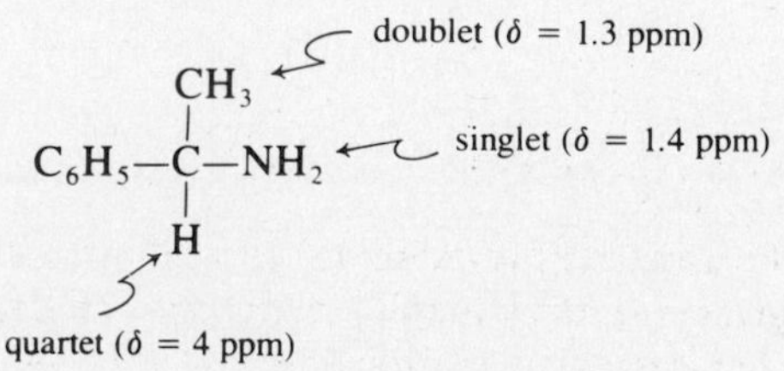

$C_6H_5—C(H)(CH_3)—NH_2$

doublet (δ = 1.3 ppm) (CH_3)
singlet (δ = 1.4 ppm) (NH_2)
quartet (δ = 4 ppm) (H)

Compound J has a ^1H nmr spectrum that fits 2-phenylethylamine.

$$\underbrace{C_6H_5CH_2CH_2}_{\text{overlapping multiplets centered at } \delta = 2.7 \text{ ppm}}\overbrace{NH_2}^{\text{singlet } (\delta = 0.9 \text{ ppm})}$$

While we might expect the signals arising from each set of methylene protons in the CH_2CH_2 unit to appear as a triplet, the actual spectrum is somewhat more complicated than that. Because the nonequivalent methylene groups of the CH_2CH_2 unit have similar chemical shifts, the simple first-order splitting rules do not apply.

23.36 Write the structural formulas for the two possible compounds given in the problem and consider how their ^{13}C nmr spectra will differ from each other. Both will exhibit their methyl carbons as quartets at high field, but they differ in the positions of their methylene and quaternary carbons. A carbon bonded to nitrogen is more shielded than one bonded to oxygen because nitrogen is less electronegative than oxygen.

singlet, lower field → C; triplet, higher field → CH_2

$$(CH_3)_2\underset{\large OH}{\underset{|}{C}}-CH_2NH_2$$

1-Amino-2-methyl-2-propanol

singlet, higher field → C; triplet, lower field → CH_2

$$(CH_3)_2\underset{\large NH_2}{\underset{|}{C}}-CH_2OH$$

2-Amino-2-methyl-1-propanol

In one isomer the lowest-field signal is a singlet, in the other it is a triplet. The spectrum shown in Figure 23.7 shows the lowest-field signal as a triplet. Therefore the compound is 2-amino-2-methyl-1-propanol, $(CH_3)_2\underset{\large NH_2}{\underset{|}{C}}CH_2OH$.

This compound *cannot* be prepared by reaction of ammonia with an epoxide because in basic solution nucleophiles attack epoxides at the less hindered carbon and therefore epoxide ring opening will give 1-amino-2-methyl-2-propanol rather than 2-amino-2-methyl-1-propanol.

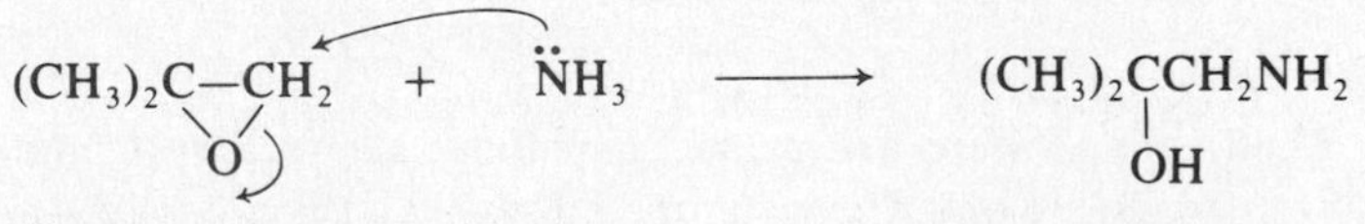

$$\underset{\text{2,2-Dimethyloxirane}}{(CH_3)_2C\overset{O}{—}CH_2} + \underset{\text{Ammonia}}{\ddot{N}H_3} \longrightarrow \underset{\text{1-Amino-2-methyl-2-propanol}}{(CH_3)_2\underset{\large OH}{\underset{|}{C}}CH_2NH_2}$$

SELF-TEST

PART A

A-1. Give an acceptable name for each of the following. Identify each compound as a primary, secondary, or tertiary amine.

(*a*) $CH_3CH_2\overset{\large CH_3}{\overset{|}{\underset{\large NH_2}{\underset{|}{C}}}}CH_3$

(*b*) cyclopentyl$-NHCH_3$

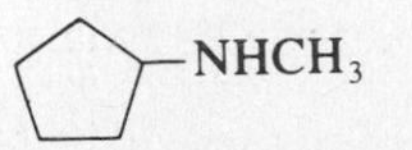

A-2. Provide the correct structure of the reagent omitted from each of the reactions shown below:

(*a*) $C_6H_5CH_2Br \xrightarrow[\text{2. LiAlH}_4 \text{ 3. H}_2\text{O}]{\text{1. ?}} C_6H_5CH_2NH_2$

(*b*) $C_6H_5CH_2Br \xrightarrow[\text{2. LiAlH}_4\text{; 3. H}_2\text{O}]{\text{1. ?}} C_6H_5CH_2CH_2NH_2$

(*c*) $C_6H_5CH_2Br \xrightarrow[\text{2. H}_2\text{NNH}_2]{\text{1. ?}} C_6H_5CH_2NH_2$ + O, NH, NH, O

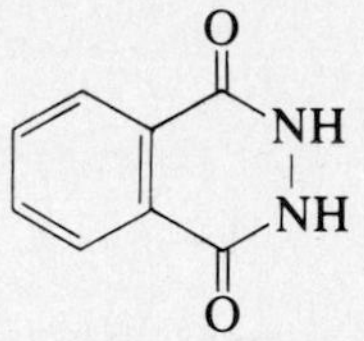

A-3. Provide structures for compounds A through G in the following reaction sequences:

(*a*) $A \xrightarrow{CH_3I} B \xrightarrow[H_2O]{Ag_2O} C \xrightarrow{heat} CH_2{=}CHCH_2CH_2N(CH_3)CH_2CH_3$

(*b*) cyclohexanone + $CH_3CH_2NH_2 \xrightarrow[CH_3OH]{NaBH_3CN} D \xrightarrow[H_2O]{NaNO_2,\ HCl} E$

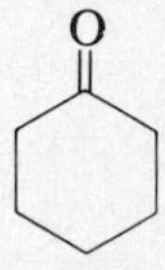

(*c*) (decalin bearing H_3C at ring-fusion carbon and $N(CH_3)_2$ on adjacent carbon) $\xrightarrow{H_2O_2} F \xrightarrow{heat} G + (CH_3)_2NOH$

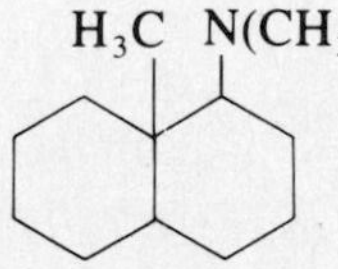

A-4. When compound I is heated, Hofmann elimination occurs; compound II when heated undergoes a Cope elimination. In one of these reactions the deuterium (D) isotope is lost; in the other it is retained. Explain, with chemical structures.

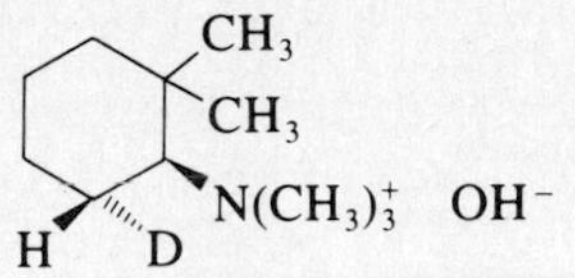

I: cyclohexane with CH_3, CH_3, $N(CH_3)_3^+$ OH^-, H, D

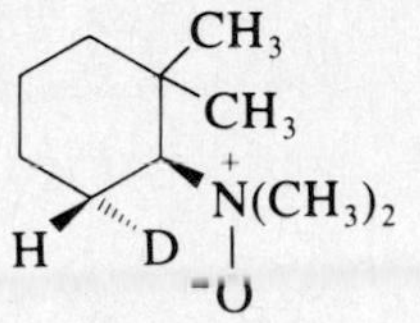

II: cyclohexane with CH_3, CH_3, $\overset{+}{N}(CH_3)_2$–O^-, H, D

PART B

B-1. Rank the following compounds in order of increasing basicity as measured in aqueous solution (least basic → most basic)

$(CH_3CH_2)_2NH$	$(CH_3)_2\overset{+}{N}HCH_2CH_3\ Br^-$	$CH_3CH_2NH_2$	$CF_3CH_2NH_2$
A	B	C	D

(*a*) B < D < A < C (*b*) A < C < D < B
(*c*) B < D < C < A (*d*) B < C < D < A

B-2. The rate of inversion of the sp^3 nitrogen in an alkylamine is:
(*a*) Usually very rapid.
(*b*) Rapid only for primary amines.
(*c*) Usually very slow.
(*d*) Not a factor; the inversion process does not occur.

B-3. Which of the following represents the major isomer of 2-methyl-1-phenyl-1-butene obtained from each of the following reaction schemes?

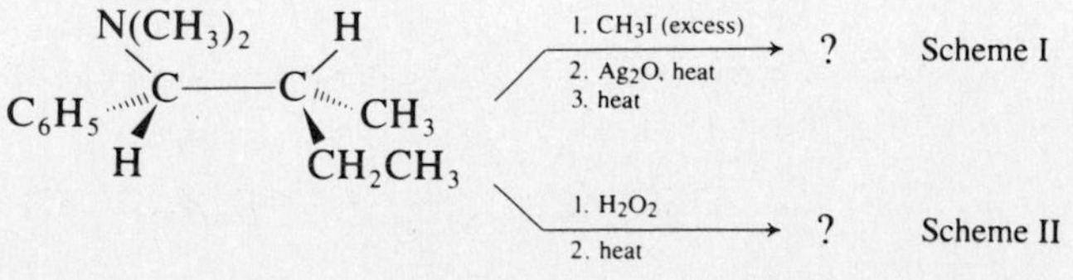

$C_6H_5CH(N(CH_3)_2)$–$CH(CH_3)(CH_2CH_3)$ (stereo structure with H)

$\xrightarrow[\text{2. Ag}_2\text{O, heat; 3. heat}]{\text{1. CH}_3\text{I (excess)}}$? Scheme I

$\xrightarrow[\text{2. heat}]{\text{1. H}_2\text{O}_2}$? Scheme II

	Scheme I	Scheme II
(a)	E isomer	E isomer
(b)	Z isomer	Z isomer
(c)	E isomer	Z isomer
(d)	Z isomer	E isomer

B-4. Which of the following is a secondary amine?
(a) 2-Butanamine
(b) N-Ethyl-2-pentanamine
(c) N-Methylpiperidine
(d) N,N-Dimethylcyclohexylamine

B-5. The reaction

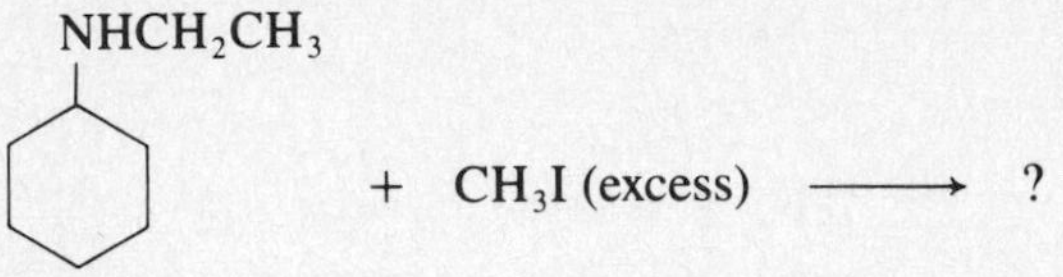

$$\text{C}_6\text{H}_{11}\text{–NHCH}_2\text{CH}_3 + \text{CH}_3\text{I (excess)} \longrightarrow ?$$

gives as final product
(a) A primary amine
(b) A secondary amine
(c) A tertiary amine
(d) A quaternary ammonium salt

B-6. A substance is soluble in dilute aqueous HCl and has a single peak in the region 3200 to 3500 cm^{-1} in its infrared spectrum. Which of the following best fits the data?

(a) C_6H_{11}–$N(CH_3)_2$
(b) C_6H_{11}–NH_2
(c) C_6H_{11}–$NHCH_3$
(d) C_6H_{11}–CO_2H

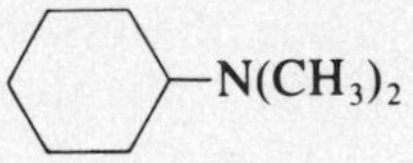

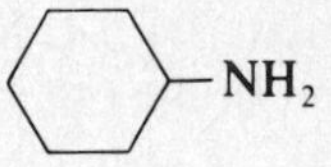

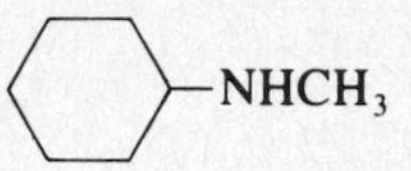

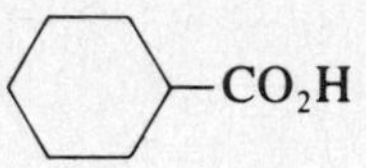

CHAPTER

24

ARYLAMINES

IMPORTANT TERMS AND CONCEPTS

Structure and Bonding (Sec. 24.2) The carbon-nitrogen bond of an arylamine such as aniline is shorter than that of an alkylamine such as cyclohexylamine.

Aniline — Cyclohexylamine

Two factors account for this. First, bonds to sp^2 hybridized carbon atoms are shorter than those to sp^3 hybridized carbon. Second, the carbon-nitrogen bond in an arylamine also has partial double bond character due to resonance delocalization of the lone pair electrons into the aromatic system.

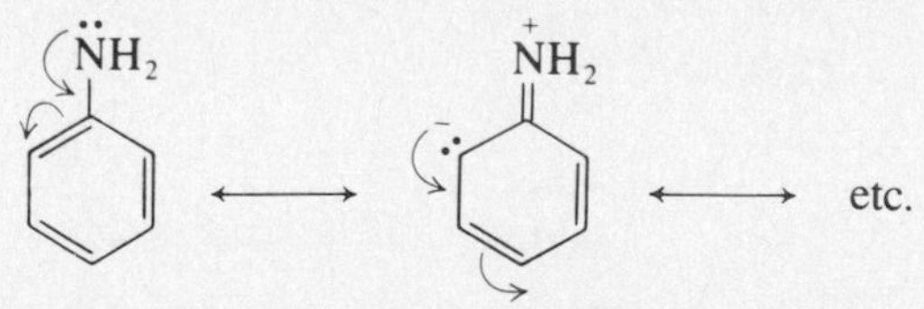

This phenomenon also accounts for the bond angles in aniline being greater than tetrahedral; the nitrogen atom adopts an orbital hybridization that is between sp^3 and sp^2.

Basicity (Sec. 24.4) Arylamines are *less* basic than alkylamines by a factor of about 10^5. The lone pair electrons of an arylamine such as aniline are delocalized through the π system of the aromatic ring; such delocalization is not possible in the corresponding ammonium ion. The resulting stabilization of the amine, compared with the ammonium ion, shifts the acid-base equilibrium to the left, lowering the basicity of an arylamine. Such delocalization is not possible in alkylamines, which lack an adjacent π system.

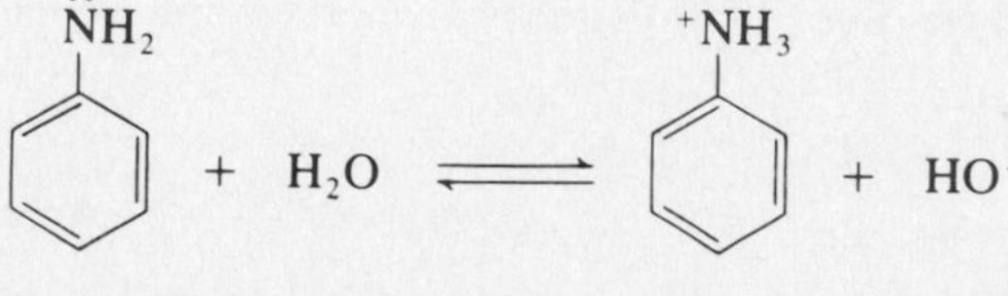

$$C_6H_5\ddot{N}H_2 + H_2O \rightleftharpoons C_6H_5\overset{+}{N}H_3 + HO^- \qquad K_b = 3.8 \times 10^{-10}$$

$$\text{C}_6\text{H}_{11}\ddot{\text{N}}\text{H}_2 + H_2O \rightleftharpoons \text{C}_6\text{H}_{11}\overset{+}{\text{N}}\text{H}_3 + HO^- \qquad K_b = 4.4 \times 10^{-4}$$

Substituents on the aromatic ring affect the basicity of arylamines. In general, electron-donating groups result in an increase in basicity while electron-withdrawing groups tend to decrease it. Electron-withdrawing groups that are conjugated with the amine nitrogen, such as a para nitro group, exert a considerable effect on the basicity of the amine; thus *p*-nitroaniline is almost 4000 times less basic than aniline.

$$p\text{-}O_2N\text{-}C_6H_4\text{-}NH_2 \qquad K_b = 1.0 \times 10^{-13}$$

p-Nitroaniline

IMPORTANT REACTIONS

Preparation of Arylamines (Sec. 24.5)

General:

$$ArH \xrightarrow[H_2SO_4]{HNO_3} ArNO_2 \xrightarrow{[H]} ArNH_2$$

Example:

$$(CH_3)_3C\text{-}C_6H_5 \xrightarrow[H_2SO_4]{HNO_3} (CH_3)_3C\text{-}C_6H_4\text{-}NO_2 \xrightarrow[\text{2. NaOH}]{\text{1. Sn, HCl}} (CH_3)_3C\text{-}C_6H_4\text{-}NH_2$$

REACTIONS OF ARYLAMINES

Electrophilic Aromatic Substitution (Sec. 24.7) The amine group is often protected by acetylation prior to carrying out electrophilic aromatic substitution reactions.

General:

$$ArNH_2 \xrightarrow{CH_3\overset{O}{\overset{\|}{C}}Cl} ArNH\overset{O}{\overset{\|}{C}}CH_3 \xrightarrow{E^+} \text{ortho, para substitution}$$

Example:

$$p\text{-}Br\text{-}C_6H_4\text{-}NH_2 \xrightarrow{CH_3\overset{O}{\overset{\|}{C}}Cl} p\text{-}Br\text{-}C_6H_4\text{-}NH\overset{O}{\overset{\|}{C}}CH_3 \xrightarrow[FeCl_3]{Cl_2} \text{4-Br-2-Cl-}C_6H_3\text{-}NH\overset{O}{\overset{\|}{C}}CH_3$$

Nitrosation (Sec. 24.8) Primary aromatic amines yield aryl diazonium salts on nitrosation.

General:

$$ArNH_2 \xrightarrow[H_2O]{NaNO_2,\ HCl} ArN_2^+\ Cl^-$$

Example:

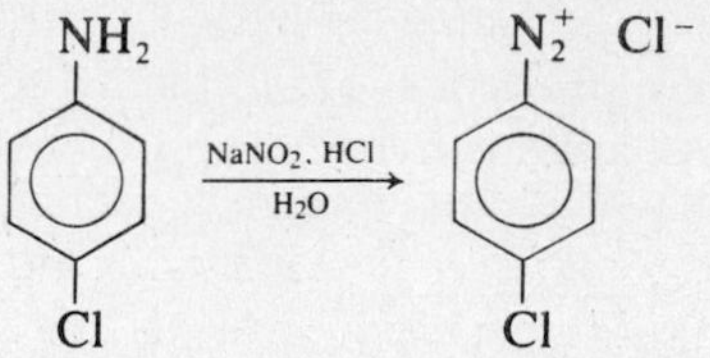

$$p\text{-}ClC_6H_4NH_2 \xrightarrow[H_2O]{NaNO_2,\ HCl} p\text{-}ClC_6H_4N_2^+\ Cl^-$$

Aryl Diazonium Ions (Sec. 24.9) Aryl diazonium salts are useful intermediates in the synthesis of a variety of compounds.

General:

Phenols: $Ar{-}N_2^+ \xrightarrow[heat]{H_2O} Ar{-}OH$

Halides: $Ar{-}N_2^+ \xrightarrow{CuX} Ar{-}X \quad (X = Cl, Br)$

$Ar{-}N_2^+ \xrightarrow{I^-} Ar{-}I$

Cyanides: $Ar{-}N_2^+ \xrightarrow{CuCN} Ar{-}CN$

Deamination: $Ar{-}N_2^+ \xrightarrow{H_3PO_2} Ar{-}H$

Examples:

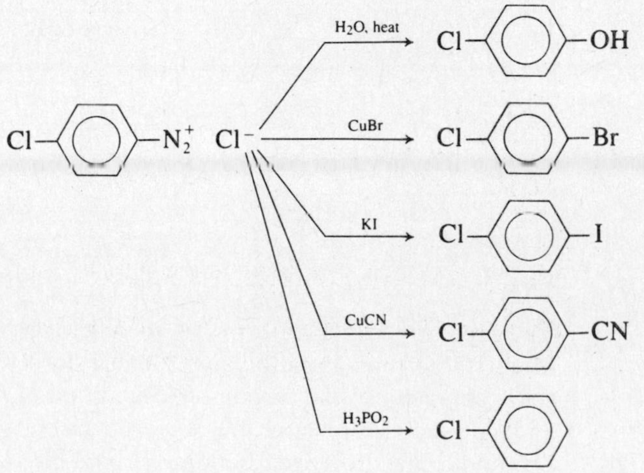

$$Cl{-}C_6H_4{-}N_2^+\ Cl^- \xrightarrow{H_2O,\ heat} Cl{-}C_6H_4{-}OH$$

$$Cl{-}C_6H_4{-}N_2^+\ Cl^- \xrightarrow{CuBr} Cl{-}C_6H_4{-}Br$$

$$Cl{-}C_6H_4{-}N_2^+\ Cl^- \xrightarrow{KI} Cl{-}C_6H_4{-}I$$

$$Cl{-}C_6H_4{-}N_2^+\ Cl^- \xrightarrow{CuCN} Cl{-}C_6H_4{-}CN$$

$$Cl{-}C_6H_4{-}N_2^+\ Cl^- \xrightarrow{H_3PO_2} Cl{-}C_6H_5$$

Azo Coupling (Sec. 24.10)

General:

$$Ar{-}N_2^+ + C_6H_5{-}X \longrightarrow Ar{-}N{=}N{-}C_6H_4{-}X$$

$(X = -OH, -NR_2)$

Example:

$$C_6H_5OH + p\text{-}CH_3C_6H_4N_2^+\ Cl^- \longrightarrow CH_3{-}C_6H_4{-}N{=}N{-}C_6H_4{-}OH$$

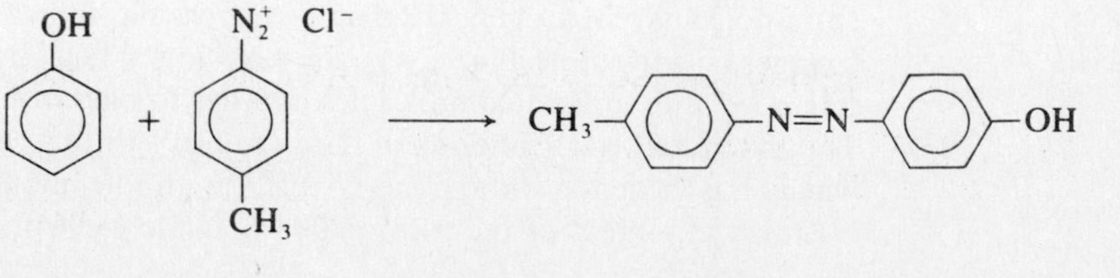

SOLUTIONS TO TEXT PROBLEMS

24.1 (*b*) In alphabetical order, the substituents present on the aniline nucleus are ethyl, isopropyl, and methyl. Their positions are specified as *N*-ethyl, 4-isopropyl, and *N*-methyl.

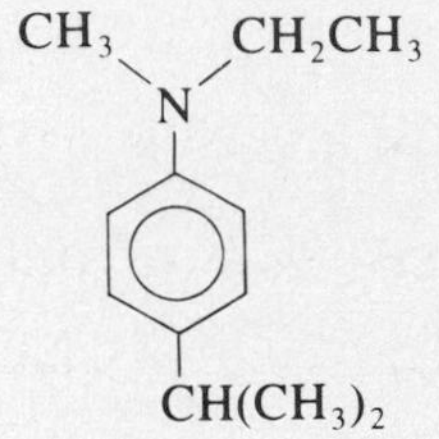

N-Ethyl-4-isopropyl-*N*-methylaniline

(*c*) This compound, rather than being named as a derivative of aniline, is named as a derivative of benzyl alcohol more easily.

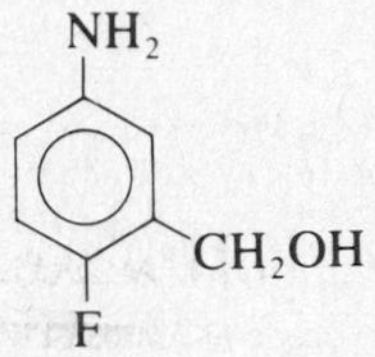

5-Amino-2-fluorobenzyl alcohol

24.2 The amino group and the acetyl group of *p*-aminoacetophenone are directly conjugated to each other.

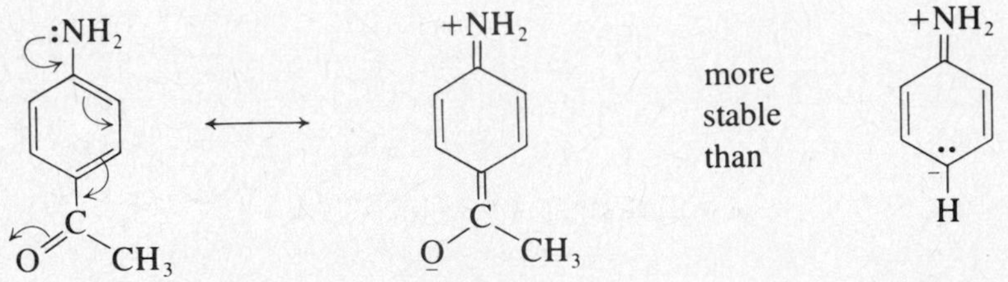

more stable than

Dipolar resonance form of *p*-aminoacetophenone

Dipolar resonance form of aniline

The dipolar resonance form has its negative charge on the electronegative oxygen and is more stable than the dipolar resonance form of aniline itself. Thus, there is more double bond character in the carbon-nitrogen bond of *p*-aminoacetophenone than in that of aniline, and *p*-aminoacetophenone has a shorter carbon-nitrogen bond.

24.3 Intramolecular hydrogen bonding is possible between the amino and carbonyl functions of *o*-aminoacetophenone even in the gas phase.

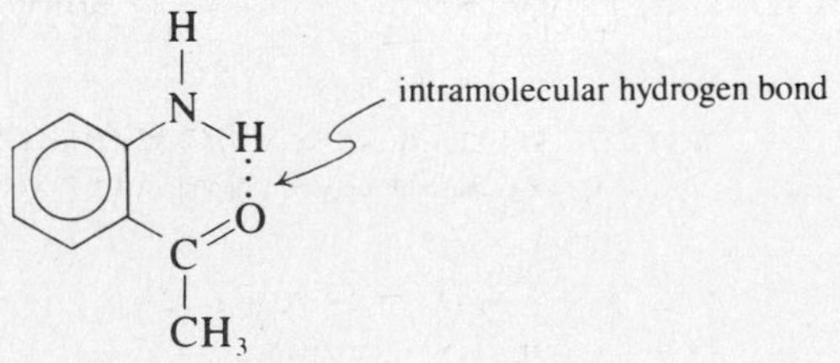

intramolecular hydrogen bond

o-Aminoacetophenone

Intramolecular hydrogen bonding is impossible in *p*-aminoacetophenone, which, however, engages in intermolecular hydrogen bonding. These hydrogen-bonded aggregates must be separated into individual molecules in order to pass from the liquid to the vapor phase.

24.4 The individual contributions of the two nitro groups cancel each other in *p*-dinitrobenzene, which has a center of symmetry and no dipole moment. In *p*-nitroaniline the electron-withdrawing effect of the nitro group and the electron-releasing effect of the amino group reinforce one another, giving the molecule a substantial dipole moment (6.3 D). Direct conjugation of the nitro and amino groups leads to a large separation of charge in *p*-nitroaniline.

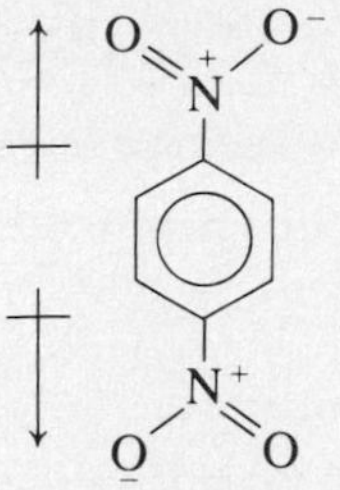

p-Dinitrobenzene, $\mu = 0$

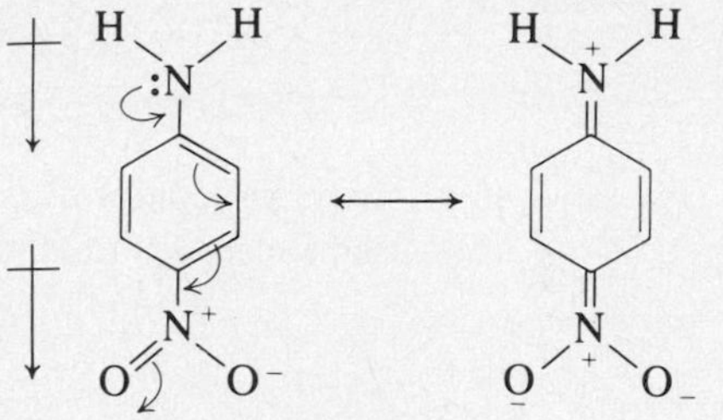

p-Nitroaniline, $\mu = 6.3$ D

24.5 Nitrogen is attached directly to the aromatic ring in tetrahydroquinoline, which is thus an arylamine, and the nitrogen lone pair is delocalized into the π system of the aromatic ring. It is less basic than tetrahydroisoquinoline, in which the nitrogen is insulated from the ring by an sp^3 hybridized carbon.

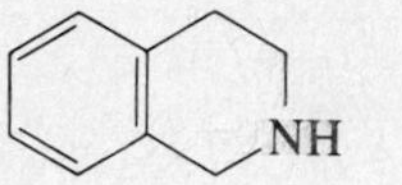

Tetrahydroisoquinoline (an alkylamine): more basic, K_b 2.5×10^{-5} (pK_b 4.6)

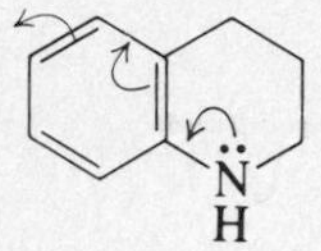

Tetrahydroquinoline (an arylamine): less basic, K_b 1.0×10^{-9} (pK_b 9.0)

24.6 (*b*) There is direct conjugation between the *p*-cyano substituent and the amine lone pair electrons in *p*-cyanoaniline.

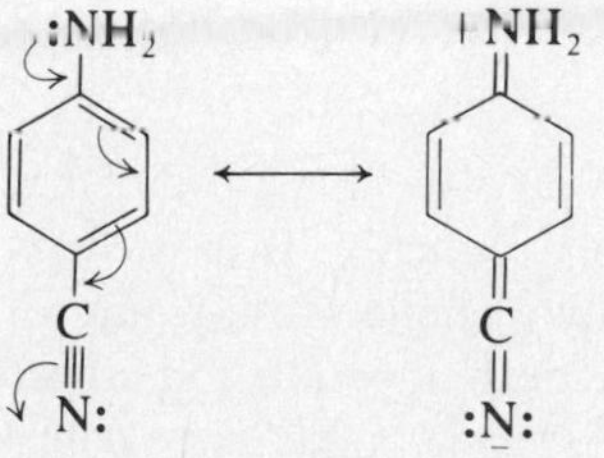

Delocalization of the nitrogen lone pair electrons is a base-weakening effect. Loss of the delocalization upon protonation increases the energy difference between the amine and its conjugate acid relative to an amine in which such conjugation is absent.

(*c*) An acetyl group attached directly to nitrogen as in acetanilide delocalizes the nitrogen lone pair into the carbonyl group. Amides are weaker bases than amines.

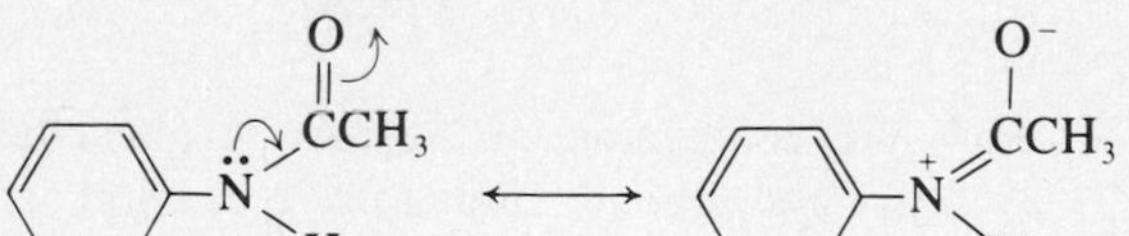

(*d*) An acetyl group in a position para to an amine function is conjugated to it and delocalizes the nitrogen lone pair.

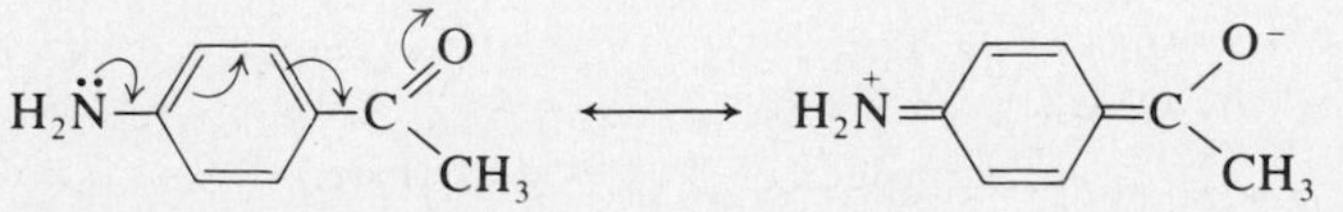

24.7 For each part of this problem, keep in mind that aromatic amines are derived by reduction of the corresponding aromatic nitro compound. Each synthesis should be approached from the standpoint of how best to prepare the necessary nitroaromatic.

$$Ar{-}NH_2 \Longrightarrow Ar{-}NO_2 \Longrightarrow Ar{-}H$$

(Ar = substituted aromatic ring)

(*b*) The para isomer of isopropylaniline may be prepared by a procedure analogous to that used for its ortho isomer in (*a*).

Benzene $\xrightarrow[AlCl_3]{(CH_3)_2CHCl}$ Isopropylbenzene [$CH(CH_3)_2$] $\xrightarrow[H_2SO_4]{HNO_3}$ *o*-Isopropylnitrobenzene [$CH(CH_3)_2$, NO_2] + *p*-Isopropylnitrobenzene [$CH(CH_3)_2$, NO_2]

After separating the ortho, para mixture by distillation, the nitro group of *p*-isopropylnitrobenzene is reduced to yield the desired *p*-isopropylaniline.

[$CH(CH_3)_2$, NO_2] $\xrightarrow[\text{1. Fe, HCl; 2. HO}^-\text{ or 1. Sn, HCl; 2. HO}^-]{H_2,\ Ni;\ or}$ [$CH(CH_3)_2$, NH_2]

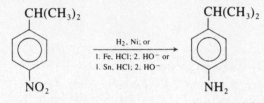

(*c*) The target compound is the reduction product of 1-isopropyl-2,4-dinitrobenzene.

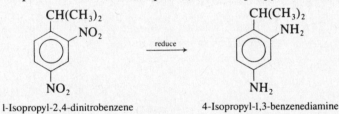

1-Isopropyl-2,4-dinitrobenzene [$CH(CH_3)_2$, NO_2, NO_2] $\xrightarrow{\text{reduce}}$ 4-Isopropyl-1,3-benzenediamine [$CH(CH_3)_2$, NH_2, NH_2]

This reduction is carried out in the same way as reduction of an arene that contains only a single nitro group. In this case hydrogenation over a Raney nickel catalyst gave the desired product in 90 percent yield.

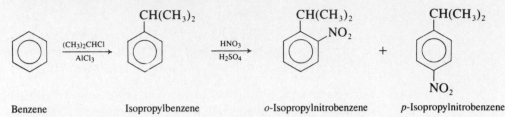

The starting dinitro compound is prepared by nitration of isopropylbenzene.

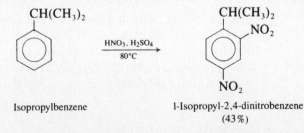

Isopropylbenzene [$CH(CH_3)_2$] $\xrightarrow[80°C]{HNO_3,\ H_2SO_4}$ 1-Isopropyl-2,4-dinitrobenzene (43%) [$CH(CH_3)_2$, NO_2, NO_2]

(*d*) The conversion of *p*-chloronitrobenzene to *p*-chloroaniline was cited as an example in the text to illustrate reduction of aromatic nitro compounds to arylamines. *p*-Chloronitrobenzene is prepared by nitration of chlorobenzene.

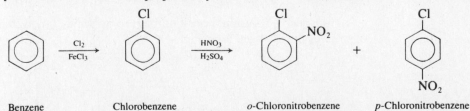

Benzene $\xrightarrow[FeCl_3]{Cl_2}$ Chlorobenzene [Cl] $\xrightarrow[H_2SO_4]{HNO_3}$ *o*-Chloronitrobenzene [Cl, NO_2] + *p*-Chloronitrobenzene [Cl, NO_2]

The para isomer comprises 69 percent of the product in this reaction (30 percent ortho; 1 percent meta). Separation of *p*-chloronitrobenzene and its reduction completes the synthesis.

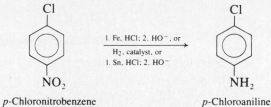

p-Chloronitrobenzene $\xrightarrow[\text{H}_2\text{, catalyst, or 1. Sn, HCl; 2. HO}^-]{\text{1. Fe, HCl; 2. HO}^-\text{, or}}$ *p*-Chloroaniline

Chlorination of nitrobenzene would not be a suitable route to the required intermediate because it would produce mainly *m*-chloronitrobenzene.

(*e*) The synthesis of *m*-aminoacetophenone may be carried out by the scheme shown:

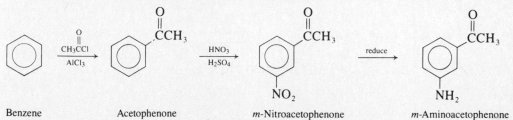

Benzene $\xrightarrow[\text{AlCl}_3]{\text{CH}_3\text{CCl (O)}}$ Acetophenone $\xrightarrow[\text{H}_2\text{SO}_4]{\text{HNO}_3}$ *m*-Nitroacetophenone $\xrightarrow{\text{reduce}}$ *m*-Aminoacetophenone

The acetyl group is attached to the ring by Friedel-Crafts acylation. It is a meta director, and its nitration gives the proper orientation of substituents. The order of the first two steps cannot be reversed because Friedel-Crafts acylation of nitrobenzene is not possible (Section 13.15). Once prepared, *m*-nitroacetophenone can be reduced to *m*-nitroaniline by any of a number of reagents. Indeed, all three reducing combinations described in the text have been employed for this transformation.

	Reducing agent	*Yield,* %
m-Nitroacetophenone	H_2, Pt	94
↓	Fe, HCl	84
m-Aminoacetophenone	Sn, HCl	82

24.8 (*b*) The pattern of substituents in 2,4-dinitroaniline suggests that they can be introduced by dinitration. Since nitration of aniline itself is accompanied by oxidation, the amino group must be protected by conversion to its *N*-acetyl derivative.

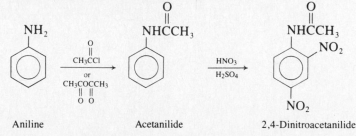

Aniline $\xrightarrow{\text{CH}_3\text{CCl (O) or CH}_3\text{COCCH}_3\text{ (O O)}}$ Acetanilide $\xrightarrow[\text{H}_2\text{SO}_4]{\text{HNO}_3}$ 2,4-Dinitroacetanilide

Hydrolysis of the amide bond in 2,4-dinitroacetanilide furnishes the desired 2,4-dinitroaniline.

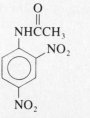

2,4-Dinitroacetanilide $\xrightarrow[\text{1. H}_2\text{O, H}^+\text{; 2. HO}^-]{\text{H}_2\text{O, HO}^-\text{, or}}$ 2,4-Dinitroaniline

(*c*) Retrosynthetically, *p*-aminoacetanilide may be derived from *p*-nitroacetanilide.

$$CH_3\overset{O}{\overset{\|}{C}}NH-C_6H_4-NH_2 \Longrightarrow CH_3\overset{O}{\overset{\|}{C}}NH-C_6H_4-NO_2$$

p-Aminoacetanilide *p*-Nitroacetanilide

This suggests the sequence:

$$H_2N-C_6H_5 \xrightarrow{CH_3\overset{O}{\overset{\|}{C}}O\overset{O}{\overset{\|}{C}}CH_3} CH_3\overset{O}{\overset{\|}{C}}NH-C_6H_5 \xrightarrow[H_2SO_4]{HNO_3} CH_3\overset{O}{\overset{\|}{C}}NH-C_6H_4-NO_2$$

Aniline Acetanilide *p*-Nitroacetanilide (separate from ortho isomer)

reduce | Fe, HCl or Sn, HCl or H_2, Pt

$$CH_3\overset{O}{\overset{\|}{C}}NH-C_6H_4-NH_2$$

p-Aminoacetanilide

24.9 As the example in Section 24.9 of the text illustrates, *m*-bromophenol may be prepared by diazotization of *m*-bromoaniline. The problem simplifies itself, therefore, to the preparation of *m*-bromoaniline. Recognizing that arylamines are ultimately derived from nitroarenes, we derive the retrosynthetic sequence of intermediates:

m-Bromophenol ⟹ *m*-Bromoaniline ⟹ *m*-Bromonitrobenzene ⟹ Nitrobenzene

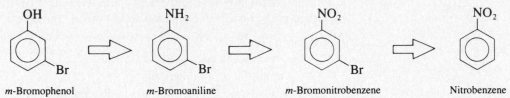

The desired reaction sequence is straightforward, using reactions that have been discussed previously in the text.

$$C_6H_6 \xrightarrow[H_2SO_4]{HNO_3} C_6H_5NO_2 \xrightarrow[Fe]{Br_2} m\text{-}BrC_6H_4NO_2 \xrightarrow[2.\ NaOH]{1.\ Fe,\ HCl} m\text{-}BrC_6H_4NH_2$$

$$\xrightarrow[2.\ H_2O;\ heat]{1.\ NaNO_2,\ H_2SO_4,\ H_2O,\ 0-5°C} m\text{-}BrC_6H_4OH$$

24.10 The key to this problem is to recognize that the iodine substituent in *m*-bromoiodobenzene is derived from an arylamine by diazotization.

m-Bromoiodobenzene ⟹ *m*-Bromoaniline

The preparation of *m*-bromoaniline from benzene has been described in Problem 24.9. All that remains is to write the equation for its conversion to *m*-bromoiodobenzene.

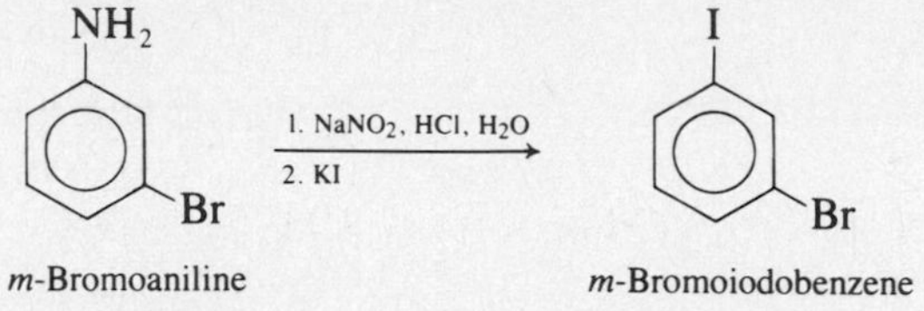

$$m\text{-Bromoaniline} \xrightarrow[2.\ KI]{1.\ NaNO_2,\ HCl,\ H_2O} m\text{-Bromoiodobenzene}$$

m-Bromoaniline — *m*-Bromoiodobenzene

24.11 Since the final step in the preparation of *m*-fluoropropiophenone is shown in the text example immediately preceding this problem, all that is necessary is to describe the preparation of *m*-aminopropiophenone.

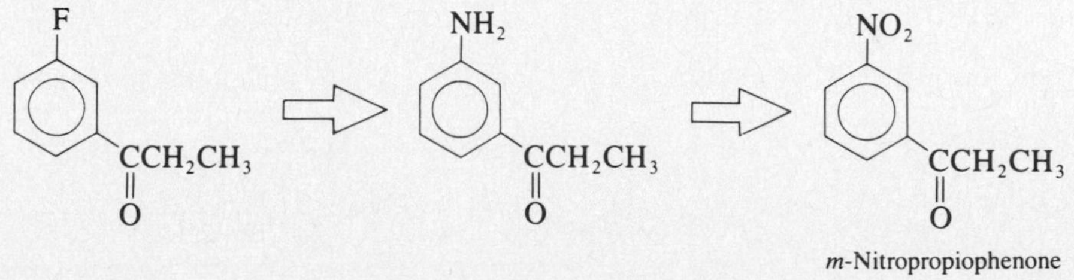

$m\text{-}FC_6H_4CCH_2CH_3$ (C=O) ⇒ $m\text{-}H_2NC_6H_4CCH_2CH_3$ (C=O) ⇒ $m\text{-}O_2NC_6H_4CCH_2CH_3$ (C=O)

m-Nitropropiophenone

Recalling that arylamines are normally prepared by reduction of nitroarenes, we see that *m*-nitropropiophenone is a pivotal synthetic intermediate. It is prepared by nitration of propiophenone, which is analogous to that of acetophenone, shown in Section 13.15. The preparation of propiophenone by Friedel-Crafts acylation of benzene is shown in Section 13.7.

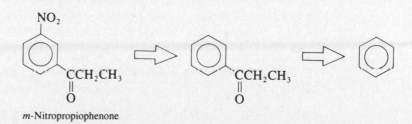

$m\text{-}O_2NC_6H_4CCH_2CH_3$ (C=O) ⇒ $C_6H_5CCH_2CH_3$ (C=O) ⇒ C_6H_6

m-Nitropropiophenone

Reversing the order of introduction of the nitro and propionyl groups is incorrect. It is possible to nitrate propiophenone but not possible to carry out a Friedel-Crafts acylation on nitrobenzene owing to the strong deactivating influence of the nitro group.

24.12 Direct nitration of the prescribed starting material cumene (isopropylbenzene) is not suitable because isopropyl is an ortho, para–directing substituent and will give the target molecule *m*-nitrocumene as only a minor component of the nitration product. However, the conversion of 4-isopropyl-2-nitroaniline to *m*-isopropylnitrobenzene, which was used to illustrate reductive deamination of arylamines, establishes the last step in the synthesis.

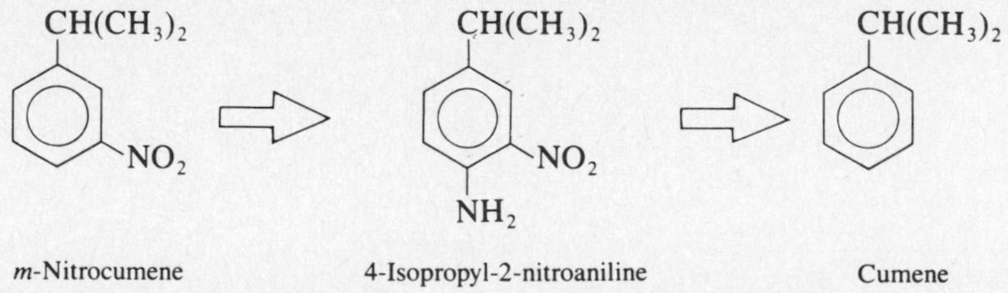

$m\text{-}(CH_3)_2CHC_6H_4NO_2$ ⇒ 4-$(CH_3)_2CH$-2-$NO_2C_6H_3NH_2$ ⇒ $C_6H_5CH(CH_3)_2$

m-Nitrocumene — 4-Isopropyl-2-nitroaniline — Cumene

Our task simplifies itself to the preparation of 4-isopropyl-2-nitroaniline from cumene. The following procedure is a straightforward extension of the reactions and principles developed in this chapter.

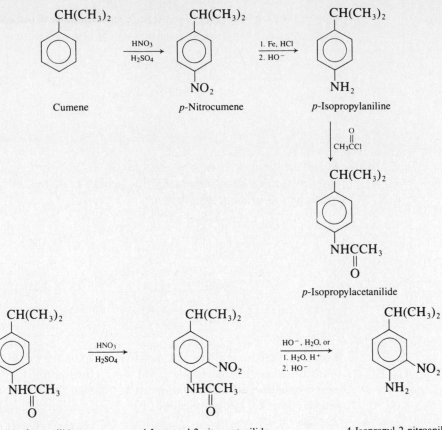

$$\text{Cumene } C_6H_5CH(CH_3)_2 \xrightarrow[H_2SO_4]{HNO_3} p\text{-Nitrocumene} \xrightarrow[2.\ HO^-]{1.\ Fe,\ HCl} p\text{-Isopropylaniline} \xrightarrow{CH_3\overset{O}{\overset{\|}{C}}Cl} p\text{-Isopropylacetanilide}$$

$$p\text{-Isopropylacetanilide} \xrightarrow[H_2SO_4]{HNO_3} \text{4-Isopropyl-2-nitroacetanilide} \xrightarrow[1.\ H_2O,\ H^+;\ 2.\ HO^-]{HO^-,\ H_2O,\ \text{or}} \text{4-Isopropyl-2-nitroaniline}$$

Reductive deamination of 4-isopropyl-2-nitroaniline by diazotization in the presence of ethanol or hypophosphorous acid yields *m*-nitrocumene and completes the synthesis.

24.13 (*b*) The preparation of sulfabenz is modeled after that of other sulfa drugs. Aniline is allowed to react with *p*-acetamidobenzenesulfonyl chloride.

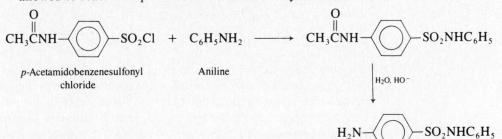

$$CH_3\overset{O}{\overset{\|}{C}}NH\text{—}C_6H_4\text{—}SO_2Cl + C_6H_5NH_2 \longrightarrow CH_3\overset{O}{\overset{\|}{C}}NH\text{—}C_6H_4\text{—}SO_2NHC_6H_5 \xrightarrow{H_2O,\ HO^-} H_2N\text{—}C_6H_4\text{—}SO_2NHC_6H_5$$

p-Acetamidobenzenesulfonyl chloride; Aniline; Sulfabenz

(*c*) Identify the amine that is required in the reaction with *p*-acetamidobenzenesulfonyl chloride as 2-aminopyrimidine by retrosynthetic analysis.

$$H_2N\text{—}C_6H_4\text{—}SO_2\text{—}NH\text{—}(2\text{-pyrimidinyl}) \Longrightarrow CH_3\overset{O}{\overset{\|}{C}}NH\text{—}C_6H_4\text{—}SO_2Cl + H_2N\text{—}(2\text{-pyrimidinyl})$$

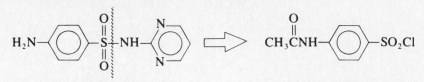

disconnect here in sulfadiazine

p-Acetamidobenzenesulfonyl chloride

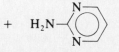

2-Aminopyrimidine

(*d*) The amine that is used to prepare sulfathiazole is the one shown, 2-aminothiazole.

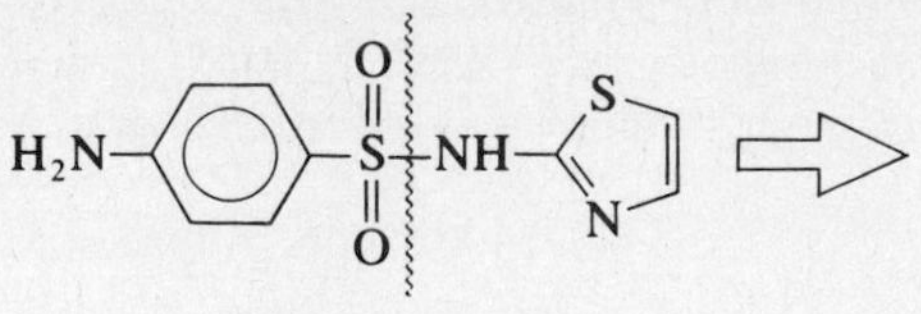

$H_2N-C_6H_4-SO_2-NH-$(2-thiazolyl) ⟹

disconnect here

$CH_3\overset{O}{\overset{\|}{C}}NH-C_6H_4-SO_2Cl$ + H_2N-(2-thiazolyl)

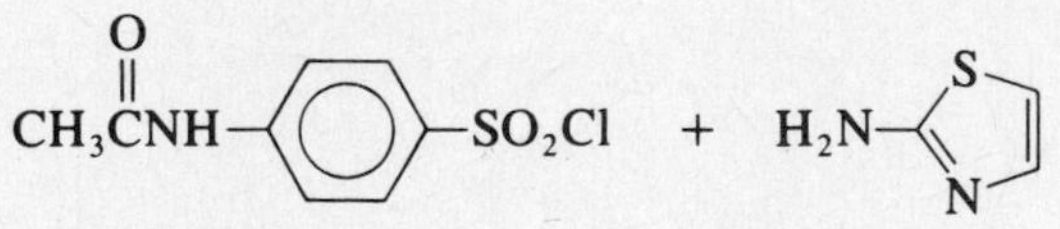

p-Acetamidobenzenesulfonyl chloride   2-Aminothiazole

24.14 (*a*) There are five isomers of C_7H_9N that contain a benzene ring.

$C_6H_5CH_2NH_2$ — Benzylamine

$C_6H_5NHCH_3$ — *N*-Methylaniline

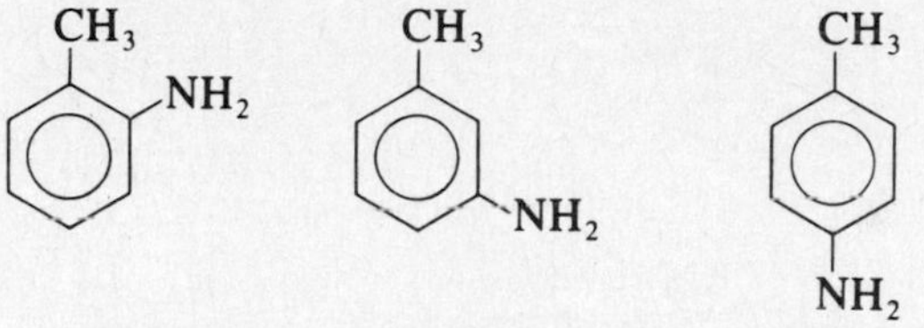

o-Toluidine (*o*-methylaniline)   *m*-Toluidine (*m*-methylaniline)   *p*-Toluidine (*p*-methylaniline)

(*b*) Benzylamine is the strongest base, as its amine group is bonded to an sp^3 hybridized carbon. Benzylamine is a typical alkylamine, with a K_b of 2×10^{-5}. All the other isomers are arylamines, with K_b values in the 10^{-10} range.

(*c*) The formation of *N*-nitrosoamines on reaction with sodium nitrite and hydrochloric acid is a characteristic reaction of secondary amines. The only C_7H_9N isomer in this problem that is a secondary amine is *N*-methylaniline.

$$C_6H_5NHCH_3 \xrightarrow[NaNO_2]{HCl,\ H_2O} C_6H_5N(N{=}O)CH_3$$

N-Methylaniline   *N*-Methyl-*N*-nitrosoaniline

(*d*) Ring nitrosation is a characteristic reaction of tertiary arylamines.

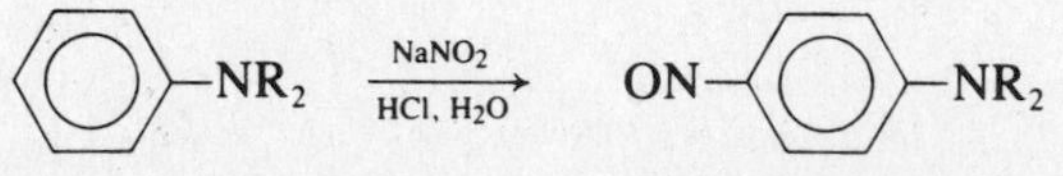

$$C_6H_5-NR_2 \xrightarrow[HCl,\ H_2O]{NaNO_2} ON-C_6H_4-NR_2$$

Tertiary arylamine   *p*-Nitroso-*N*,*N*-dialkylaniline

None of the C_7H_9N isomers in this problem are tertiary amines; hence none will undergo ring nitrosation.

24.15 (*a*) An alkyl substituent on nitrogen is electron-releasing and base-strengthening; thus methylamine is a stronger base than ammonia. An aryl substituent is electron-withdrawing and base-weakening, and so aniline is a weaker base than ammonia.

CH_3NH_2	NH_3	$C_6H_5NH_2$
Methylamine, strongest base:	Ammonia,	Aniline, weakest base:
K_b 4.4×10^{-4}	K_b 1.8×10^{-5}	K_b 3.8×10^{-10}
pK_b 3.4	pK_b 4.7	pK_b 9.4

(*b*) An acetyl group is an electron-withdrawing and base-weakening substituent, especially when bonded directly to nitrogen. Amides are weaker bases than amines, and thus acetanilide is a weaker base than aniline. Alkyl groups are electron-releasing; *N*-methylaniline is a slightly stronger base than aniline.

$C_6H_5NHCH_3$	$C_6H_5NH_2$	$C_6H_5NHC(=O)CH_3$
N-methylaniline, strongest base:	Aniline:	Acetanilide, weakest base:
K_b 8 × 10^{-10}	K_b 3.8 × 10^{-10}	K_b 1 × 10^{-15}
pK_b 9.1	pK_b 9.4	pK_b 15.0

(*c*) Chlorine substituents are slightly electron-withdrawing and methyl groups are slightly electron-releasing. Therefore, 2,4-dimethylaniline is a stronger base than 2,4-dichloroaniline. Nitro groups are strongly electron-withdrawing, their base-weakening effect being especially pronounced when a nitro group is ortho or para to an amino group because the two groups are then directly conjugated.

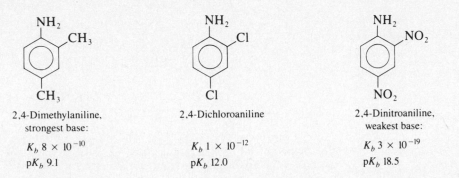

2,4-Dimethylaniline, strongest base:	2,4-Dichloroaniline	2,4-Dinitroaniline, weakest base:
K_b 8 × 10^{-10}	K_b 1 × 10^{-12}	K_b 3 × 10^{-19}
pK_b 9.1	pK_b 12.0	pK_b 18.5

(*d*) Nitro groups are more electron-withdrawing than chlorine and the base-weakening effect of a nitro substituent is greater when it is ortho or para to an amino group than when it is meta to it.

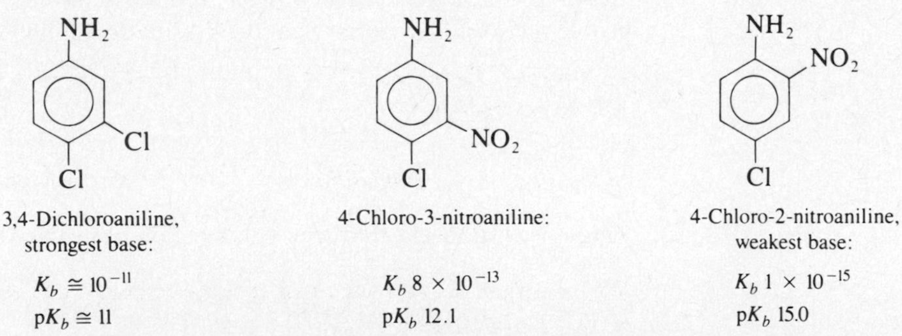

3,4-Dichloroaniline, strongest base:	4-Chloro-3-nitroaniline:	4-Chloro-2-nitroaniline, weakest base:
$K_b \cong 10^{-11}$	K_b 8 × 10^{-13}	K_b 1 × 10^{-15}
p$K_b \cong$ 11	pK_b 12.1	pK_b 15.0

(*e*) According to the principle applied in (*a*) (alkyl groups increase basicity, aryl groups decrease it) the order of decreasing basicity is as shown:

$(CH_3)_2NH$	$C_6H_5NHCH_3$	$(C_6H_5)_2NH$
Dimethylamine, strongest base:	*N*-Methylaniline	Diphenylamine, weakest base:
K_b 5.1 × 10^{-4}	K_b 8 × 10^{-10}	K_b 6 × 10^{-14}
pK_b 3.3	pK_b 9.1	pK_b 13.2

(*f*) Amino groups are strongly electron-releasing, so 1,4-benzenediamine is the strongest base. Methoxy groups are electron-releasing and base-strengthening. Acetyl groups, especially when ortho or para to an amino group, sharply decrease basicity.

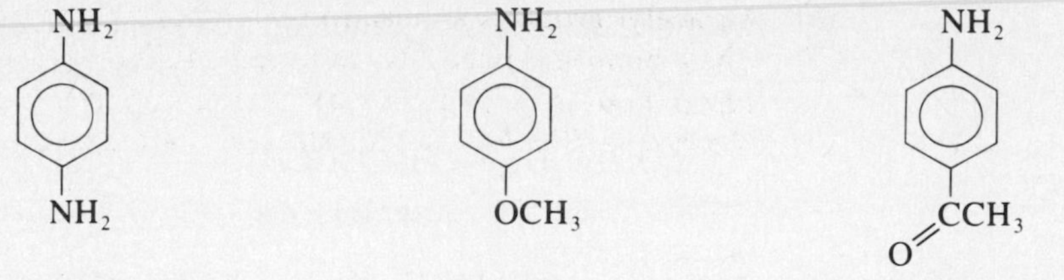

1,4-Benzenediamine, strongest base: K_b 1.6 × 10^{-8}, pK_b 7.8

p-Anisidine: K_b 2.2 × 10^{-9}, pK_b 8.7

p-Aminoacetophenone, weakest base: K_b 5 × 10^{-12}, pK_b 11.3

24.16 The most nucleophilic of the three nitrogen atoms of physostigmine will be the one that reacts with methyl iodide. Nitrogen ⓐ is the most nucleophilic.

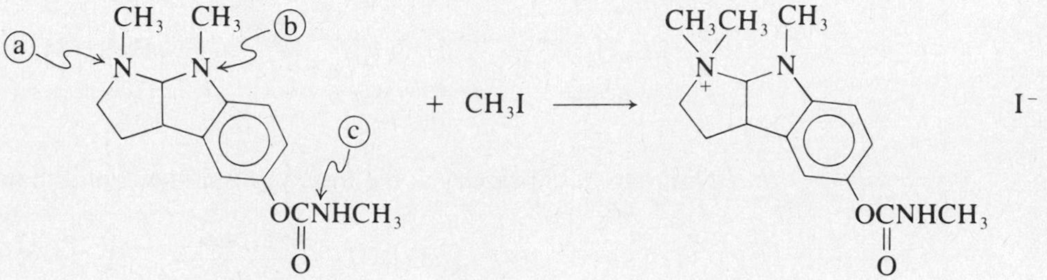

Physostigmine + CH_3I (Methyl iodide) ⟶ "Physostigmine methiodide" I^-

The nitrogen that reacts is the one that is a tertiary alkylamine. Of the other two nitrogens, ⓑ is attached to an aromatic ring and is much less basic and less nucleophilic. The third nitrogen ⓒ is an amide nitrogen; amides are less nucleophilic than amines.

24.17 (*a*) Aniline is a weak base and yields a salt on reaction with hydrogen bromide.

$$C_6H_5NH_2 + HBr \longrightarrow C_6H_5\overset{+}{N}H_3\ Br^-$$

Aniline, Hydrogen bromide, Anilinium bromide

(*b*) Aniline acts as a nucleophile toward methyl iodide. With excess methyl iodide, a quaternary ammonium salt is formed.

$$C_6H_5NH_2 + 3\,CH_3I \longrightarrow C_6H_5\overset{+}{N}(CH_3)_3\ I^-$$

Aniline, Methyl iodide, *N,N,N*-Trimethylanilinium iodide

The reaction is carried out in the presence of a base such as potassium carbonate to neutralize the HI that is formed.

(*c*) Aniline undergoes nucleophilic addition to aldehydes and ketones to form imines.

$$C_6H_5NH_2 + CH_3\overset{O}{\overset{\|}{C}}H \longrightarrow C_6H_5N{=}CHCH_3 + H_2O$$

Aniline, Acetaldehyde, *N*-Phenylacetaldimine, Water

(*d*) When an imine is formed in the presence of hydrogen and a suitable catalyst, reductive amination occurs to give an amine.

$$C_6H_5NH_2 + CH_3\overset{O}{\overset{\|}{C}}H \xrightarrow{H_2,\ Ni} C_6H_5NHCH_2CH_3$$

Aniline, Acetaldehyde, *N*-Ethylaniline

(*e*) Aniline undergoes *N*-acylation on treatment with carboxylic acid anhydrides.

$$2C_6H_5NH_2 + CH_3\overset{\overset{O}{\|}}{C}O\overset{\overset{O}{\|}}{C}CH_3 \longrightarrow C_6H_5NH\overset{\overset{O}{\|}}{C}CH_3 + C_6H_5\overset{+}{N}H_3\ \overset{-}{O}\overset{\overset{O}{\|}}{C}CH_3$$

Aniline — Acetic anhydride — Acetanilide — Anilinium acetate

(*f*) Carboxylic acid chlorides bring about *N*-acylation of arylamines.

$$2C_6H_5NH_2 + C_6H_5\overset{\overset{O}{\|}}{C}Cl \longrightarrow C_6H_5NH\overset{\overset{O}{\|}}{C}C_6H_5 + C_6H_5\overset{+}{N}H_3\ Cl^-$$

Aniline — Benzoyl chloride — Benzanilide — Anilinium chloride

(*g*) Sulfonamides are formed when primary and secondary arylamines react with sulfonyl chlorides.

$$2C_6H_5NH_2 + C_6H_5SO_2Cl \longrightarrow C_6H_5NHSO_2C_6H_5 + C_6H_5\overset{+}{N}H_3\ Cl^-$$

Aniline — Benzenesulfonyl chloride — *N*-Phenylbenzenesulfonamide — Anilinium chloride

(*h*) Nitrosation of primary arylamines yields aryl diazonium salts.

$$C_6H_5NH_2 \xrightarrow[H_2O,\ 0-5^\circ C]{NaNO_2,\ H_2SO_4} C_6H_5\overset{+}{N}\equiv N:\ HSO_4^-$$

Aniline — Benzenediazonium hydrogen sulfate

The replacement reactions that can be achieved by using diazonium salts are illustrated in (*i*) through (*o*). In all cases molecular nitrogen is lost from the ring carbon to which it was attached and is replaced by another substituent.

$C_6H_5\overset{+}{N}\equiv N:\ HSO_4^-$ (Benzenediazonium hydrogen sulfate)

(*i*) $\xrightarrow[heat]{H^+,\ H_2O}$ C_6H_5OH Phenol

(*j*) $\xrightarrow{CuCl}$ C_6H_5Cl Chlorobenzene

(*k*) $\xrightarrow{CuBr}$ C_6H_5Br Bromobenzene

(*l*) $\xrightarrow{CuCN}$ C_6H_5CN Benzonitrile

(*m*) $\xrightarrow{H_3PO_2}$ C_6H_6 Benzene

(*n*) $\xrightarrow{KI}$ C_6H_5I Iodobenzene

(*o*) $\xrightarrow[2.\ heat]{1.\ HBF_4}$ C_6H_5F Fluorobenzene

(*p*) The nitrogens of an aryl diazonium salt are retained on reaction with the electron-rich ring of a phenol. Azo coupling occurs.

$$C_6H_5\overset{+}{N}\equiv N:\ HSO_4^- + C_6H_5OH \longrightarrow C_6H_5N{=}N{-}C_6H_4{-}OH$$

Benzenediazonium hydrogen sulfate — Phenol — *p*-(Azophenyl)phenol (*q*) Azo coupling occurs when aryl diazonium salts react with *N*,*N*-dialkylarylamines.

$$\underset{\text{Benzenediazonium hydrogen sulfate}}{C_6H_5\overset{+}{N}{\equiv}N:\ HSO_4^-} + \underset{N,N\text{-Dimethylaniline}}{C_6H_5N(CH_3)_2} \longrightarrow \underset{p\text{-(Azophenyl)-}N,N\text{-dimethylaniline}}{C_6H_5N{=}N\text{—}C_6H_4\text{—}N(CH_3)_2}$$

24.18 (*a*) Amides are reduced to amines by lithium aluminum hydride.

$$\underset{\text{Acetanilide}}{C_6H_5NH\overset{O}{\overset{\|}{C}}CH_3} \xrightarrow[2.\ H_2O]{1.\ LiAlH_4,\ \text{diethyl ether}} \underset{N\text{-Ethylaniline}}{C_6H_5NHCH_2CH_3}$$

(*b*) Acetanilide is a reactive substrate toward electrophilic aromatic substitution. An acetamido group is ortho, para–directing.

$$\underset{\text{Acetanilide}}{C_6H_5NH\overset{O}{\overset{\|}{C}}CH_3} \xrightarrow[H_2SO_4]{HNO_3} \underset{o\text{-Nitroacetanilide}}{o\text{-}O_2NC_6H_4NH\overset{O}{\overset{\|}{C}}CH_3} + \underset{p\text{-Nitroacetanilide}}{p\text{-}O_2NC_6H_4NH\overset{O}{\overset{\|}{C}}CH_3}$$

(*c*) Sulfonation of the ring occurs.

$$\underset{\text{Acetanilide}}{C_6H_5NH\overset{O}{\overset{\|}{C}}CH_3} \xrightarrow[H_2SO_4]{SO_3} \underset{p\text{-Acetamidobenzenesulfonic acid}}{p\text{-}HO_3SC_6H_4NH\overset{O}{\overset{\|}{C}}CH_3} + \text{ortho isomer}$$

(*d*) Chlorosulfonation of the benzene ring takes place when chlorosulfonic acid is used.

$$\underset{\text{Acetanilide}}{C_6H_5NH\overset{O}{\overset{\|}{C}}CH_3} \xrightarrow{ClSO_3H} \underset{p\text{-Acetamidobenzenesulfonyl chloride}}{p\text{-}ClO_2SC_6H_4NH\overset{O}{\overset{\|}{C}}CH_3} + \text{ortho isomer}$$

(*e*) Arenesulfonyl chlorides react with amines to form sulfonamides.

$$\underset{p\text{-Acetamidobenzenesulfonyl chloride}}{p\text{-}ClO_2SC_6H_4NH\overset{O}{\overset{\|}{C}}CH_3} + \underset{\text{Dimethylamine}}{(CH_3)_2NH} \longrightarrow \underset{p\text{-Acetamido-}N,N\text{-dimethylbenzenesulfonamide}}{p\text{-}(CH_3)_2NO_2SC_6H_4NH\overset{O}{\overset{\|}{C}}CH_3}$$

(*f*) Bromination of the ring takes place.

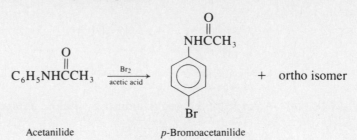

$$\underset{\text{Acetanilide}}{C_6H_5NH\overset{O}{\overset{\|}{C}}CH_3} \xrightarrow[\text{acetic acid}]{Br_2} \underset{p\text{-Bromoacetanilide}}{p\text{-}BrC_6H_4NH\overset{O}{\overset{\|}{C}}CH_3} + \text{ortho isomer}$$

(*g*) Acetanilide undergoes Friedel-Crafts alkylation readily.

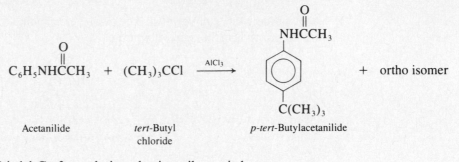

$$\underset{\text{Acetanilide}}{C_6H_5NH\overset{O}{\overset{\|}{C}}CH_3} + \underset{tert\text{-Butyl chloride}}{(CH_3)_3CCl} \xrightarrow{AlCl_3} \underset{p\text{-}tert\text{-Butylacetanilide}}{p\text{-}(CH_3)_3CC_6H_4NH\overset{O}{\overset{\|}{C}}CH_3} + \text{ortho isomer}$$

(*h*) Friedel-Crafts acylation also is easily carried out.

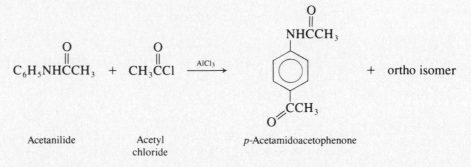

$$\underset{\text{Acetanilide}}{C_6H_5NH\overset{O}{\overset{\|}{C}}CH_3} + \underset{\text{Acetyl chloride}}{CH_3\overset{O}{\overset{\|}{C}}Cl} \xrightarrow{AlCl_3} \underset{p\text{-Acetamidoacetophenone}}{p\text{-}CH_3\overset{O}{\overset{\|}{C}}C_6H_4NH\overset{O}{\overset{\|}{C}}CH_3} + \text{ortho isomer}$$

(*i*) Acetanilide is an amide and can be hydrolyzed when heated with aqueous acid. Under acidic conditions the aniline that is formed exists in its protonated form as the anilinium cation.

$$\underset{\text{Acetanilide}}{C_6H_5NH\overset{O}{\overset{\|}{C}}CH_3} + \underset{\text{Water}}{H_2O} + \underset{\text{Hydrogen chloride}}{HCl} \longrightarrow \underset{\text{Anilinium chloride}}{C_6H_5\overset{+}{N}H_3\ Cl^-} + \underset{\text{Acetic acid}}{HO\overset{O}{\overset{\|}{C}}CH_3}$$

(*j*) Amides are also hydrolyzed in base.

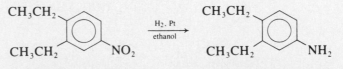

$$\underset{\text{Acetanilide}}{C_6H_5NH\overset{O}{\overset{\|}{C}}CH_3} + \underset{\text{Sodium hydroxide}}{NaOH} \xrightarrow{H_2O} \underset{\text{Aniline}}{C_6H_5NH_2} + \underset{\text{Sodium acetate}}{Na^+\overset{-}{O}\overset{O}{\overset{\|}{C}}CH_3}$$

24.19 (*a*) Catalytic hydrogenation reduces nitro groups to amino groups.

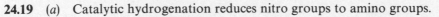

$$\underset{\text{1,2-Diethyl-4-nitrobenzene}}{(CH_3CH_2)_2C_6H_3NO_2} \xrightarrow[\text{ethanol}]{H_2,\ Pt} \underset{\text{3,4-Diethylaniline (93–99\%)}}{(CH_3CH_2)_2C_6H_3NH_2}$$

(*b*) Nitro groups are readily reduced by tin(II) chloride.

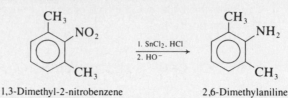

$$\text{1,3-Dimethyl-2-nitrobenzene} \xrightarrow[\text{2. HO}^-]{\text{1. SnCl}_2\text{, HCl}} \text{2,6-Dimethylaniline}$$

This reaction is the first step in a synthesis of the drug *lidocaine*.

(*c*) The amino group of arylamines is nucleophilic and undergoes acylation on reaction with chloroacetyl chloride.

$$\text{2,6-Dimethylaniline} + \underset{\text{Chloroacetyl chloride}}{ClCH_2\overset{O}{\overset{\|}{C}}Cl} \longrightarrow \textit{N}\text{-(Chloroacetyl)-2,6-dimethylaniline}$$

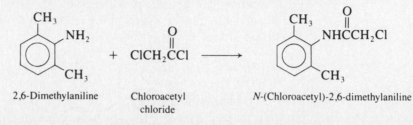

Chloroacetyl chloride is a difunctional compound—it is both an acyl chloride and an alkyl chloride. Acyl chlorides react with nucleophiles faster than do alkyl chlorides, so acylation of the amine nitrogen occurs rather than alkylation.

(*d*) The final step in the synthesis of lidocaine is displacement of the chloride by diethylamine from the α-halo amide formed in (*c*) in a nucleophilic substitution reaction.

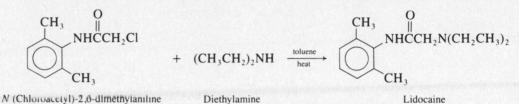

$$\textit{N}\text{-(Chloroacetyl)-2,6-dimethylaniline} + \underset{\text{Diethylamine}}{(CH_3CH_2)_2NH} \xrightarrow[\text{heat}]{\text{toluene}} \text{Lidocaine}$$

The reaction is carried out with excess diethylamine, which acts as a base to neutralize the hydrogen chloride formed.

(*e*) For use as an anesthetic lidocaine is made available as its hydrochloride salt. Of the two nitrogens in lidocaine, the amine nitrogen is more basic than the amide.

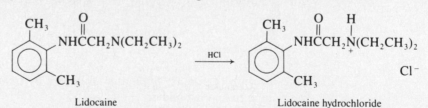

$$\text{Lidocaine} \xrightarrow{\text{HCl}} \text{Lidocaine hydrochloride} \quad (NH\overset{O}{\overset{\|}{C}}CH_2\overset{+}{N}H(CH_2CH_3)_2 \; Cl^-)$$

(*f*) Lithium aluminum hydride reduction of amides is one of the best methods for the preparation of amines, including arylamines.

$$\underset{\textit{N}\text{-Phenylbutanamide}}{C_6H_5NH\overset{O}{\overset{\|}{C}}CH_2CH_2CH_3} \xrightarrow[\text{2. } H_2O]{\text{1. } LiAlH_4} \underset{\textit{N}\text{-Butylaniline (92\%)}}{C_6H_5NHCH_2CH_2CH_2CH_3}$$

(*g*) Arylamines react with aldehydes and ketones in the presence of hydrogen and nickel to give the product of reductive amination.

$$\underset{\text{Aniline}}{C_6H_5NH_2} + \underset{\text{Heptanal}}{CH_3(CH_2)_5\overset{O}{\overset{\|}{C}}H} \xrightarrow{H_2\text{, Ni}} \underset{\textit{N}\text{-Heptylaniline (65\%)}}{C_6H_5NHCH_2(CH_2)_5CH_3}$$

(*h*) Acetanilide is a reactive substrate toward electrophilic aromatic substitution. On reaction with chloroacetyl chloride, it undergoes Friedel-Crafts acylation, primarily at its para position.

$$CH_3\overset{O}{\overset{\|}{C}}NH-C_6H_5 + ClCH_2\overset{O}{\overset{\|}{C}}Cl \xrightarrow{AlCl_3} CH_3\overset{O}{\overset{\|}{C}}NH-C_6H_4-\overset{O}{\overset{\|}{C}}CH_2Cl$$

Acetanilide — Chloroacetyl chloride — *p*-Acetamidophenacyl chloride (79–83%)

Acylation, rather than alkylation, occurs. Acyl chlorides are more reactive than alkyl chlorides toward electrophilic aromatic substitution reactions as a result of the more stable intermediate (acylium ion) formed.

(*i*) Reduction with iron in hydrochloric acid is one of the most common methods for converting nitroarenes to arylamines.

$$Br-C_6H_4-C_6H_4-NO_2 \xrightarrow[2.\ HO^-]{1.\ Fe,\ HCl} Br-C_6H_4-C_6H_4-NH_2$$

4-Bromo-4′-nitrobiphenyl — 4-Amino-4′-bromobiphenyl (94%)

(*j*) Primary arylamines are converted to aryl diazonium salts on treatment with sodium nitrite in aqueous acid. When the aqueous acidic solution containing the diazonium salt is heated, a phenol is formed.

$$Br-C_6H_4-C_6H_4-NH_2 \xrightarrow[H_2O]{NaNO_2,\ H_2SO_4} Br-C_6H_4-C_6H_4-\overset{+}{N}{\equiv}N:$$

4-Amino-4′-bromobiphenyl

$$\xrightarrow[(-N_2)]{H_2O,\ heat} Br-C_6H_4-C_6H_4-OH$$

4-Bromo-4′-hydroxybiphenyl (85%)

(*k*) This problem illustrates the conversion of an arylamine to an aryl chloride by the Sandmeyer reaction.

$$2,6\text{-}(O_2N)_2C_6H_3NH_2 \xrightarrow[H_2O]{NaNO_2,\ H_2SO_4} 2,6\text{-}(O_2N)_2C_6H_3\overset{+}{N}{\equiv}N: \xrightarrow[(-N_2)]{CuCl} 2,6\text{-}(O_2N)_2C_6H_3Cl$$

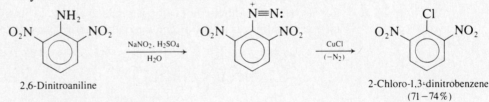

2,6-Dinitroaniline — 2-Chloro-1,3-dinitrobenzene (71–74%)

(*l*) Diazotization of primary arylamines followed by treatment with cuprous bromide converts them to aryl bromides.

$$m\text{-}BrC_6H_4NH_2 \xrightarrow[H_2O]{NaNO_2,\ HBr} m\text{-}BrC_6H_4\overset{+}{N}{\equiv}N: \xrightarrow[(-N_2)]{CuBr} m\text{-}BrC_6H_4Br$$

m-Bromoaniline — *m*-Dibromobenzene (80–87%)

(*m*) Nitriles are formed when aryl diazonium salts react with cuprous cyanide.

$$o\text{-}O_2NC_6H_4NH_2 \xrightarrow[H_2O]{NaNO_2,\ HCl} o\text{-}O_2NC_6H_4\overset{+}{N}{\equiv}N: \xrightarrow[(-N_2)]{CuCN} o\text{-}O_2NC_6H_4CN$$

o-Nitroaniline — *o*-Nitrobenzonitrile (87%)

(*n*) An aryl diazonium salt is converted to an aryl iodide on reaction with potassium iodide.

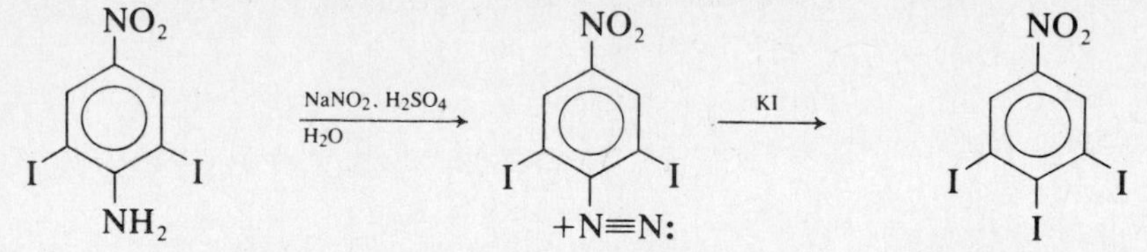

NO_2, I, I, NH_2 $\xrightarrow[H_2O]{NaNO_2,\ H_2SO_4}$ NO_2, I, I, $+N{\equiv}N:$ $\xrightarrow{KI}$ NO_2, I, I, I

2,6-Diiodo-4-nitroaniline

1,2,3-Triiodo-5-nitrobenzene (94–95%)

(*o*) Aryl diazonium fluoroborates are converted to aryl fluorides when heated. Both diazonium salt functions in the starting material undergo this reaction.

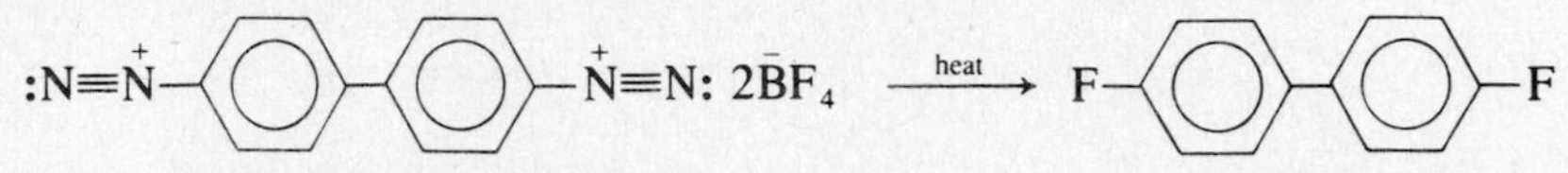

$:N{\equiv}\overset{+}{N}-C_6H_4-C_6H_4-\overset{+}{N}{\equiv}N:\ 2\overset{-}{B}F_4 \xrightarrow{heat} F-C_6H_4-C_6H_4-F$

4,4′-Bis(diazonio)biphenylfluoroborate

4,4′-Difluorobiphenyl (82%)

(*p*) Hypophosphorous acid (H_3PO_2) reduces aryl diazonium salts to arenes.

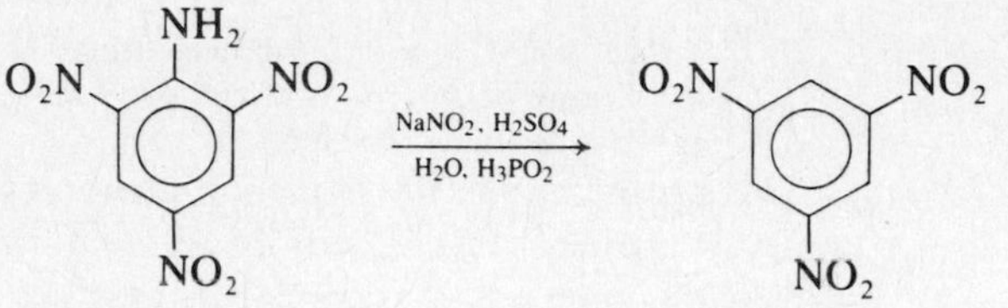

NH_2, O_2N, NO_2, NO_2 $\xrightarrow[H_2O,\ H_3PO_2]{NaNO_2,\ H_2SO_4}$ O_2N, NO_2, NO_2

2,4,6-Trinitroaniline

1,3,5-Trinitrobenzene (60–65%)

(*q*) Ethanol, like hypophosphorous acid, is an effective reagent for the reduction of aryl diazonium salts.

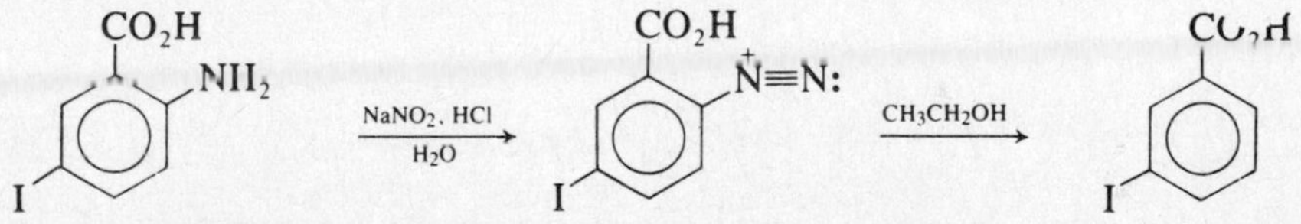

CO_2H, NH_2, I $\xrightarrow[H_2O]{NaNO_2,\ HCl}$ CO_2H, $\overset{+}{N}{\equiv}N:$, I $\xrightarrow{CH_3CH_2OH}$ CO_2H, I

2-Amino-5-iodobenzoic acid

m-Iodobenzoic acid (86–93%)

(*r*) Diazotization of aniline followed by addition of a phenol yields a bright red, diazo-substituted phenol. The diazonium ion acts as an electrophile toward the activated aromatic ring of the phenol.

$$C_6H_5NH_2 \xrightarrow[H_2O]{NaNO_2,\ H_2SO_4} C_6H_5\overset{+}{N}{\equiv}N:\ HSO_4^-$$

Aniline

Benzenediazonium hydrogen sulfate

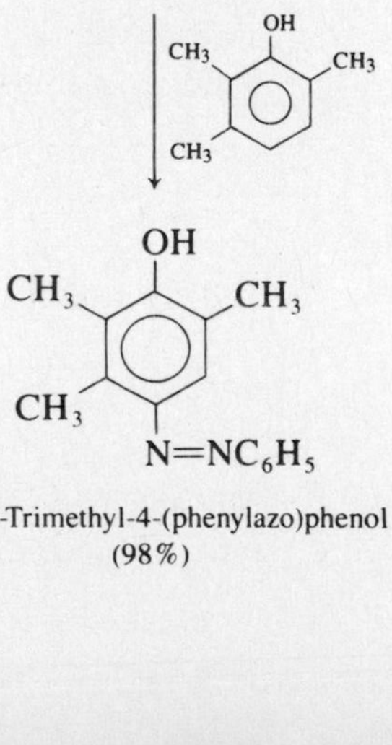

↓ (OH, CH_3, CH_3, CH_3 — 2,3,6-trimethylphenol)

OH, CH_3, CH_3, CH_3, $N{=}NC_6H_5$

2,3,6-Trimethyl-4-(phenylazo)phenol (98%)

(*s*) Nitrosation of *N*,*N*-dialkylarylamines takes place on the ring at the position para to the dialkylamino group.

$(CH_3)_2N$–C₆H₄(3-CH₃) $\xrightarrow[\text{2. HO}^-]{\text{1. NaNO}_2\text{, HCl, H}_2\text{O, 5°C}}$ $(CH_3)_2N$–C₆H₃(3-CH₃)–N=O

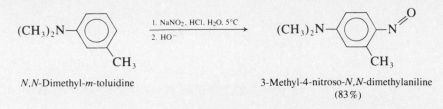

N,*N*-Dimethyl-*m*-toluidine

3-Methyl-4-nitroso-*N*,*N*-dimethylaniline (83%)

(*t*) Amino groups, even those of arylamines, are much more nucleophilic than carboxylic acid functions. The amine is converted to its corresponding sulfonamide.

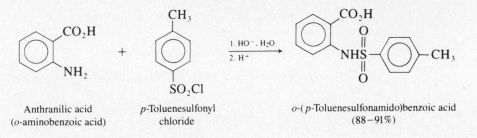

Anthranilic acid (*o*-aminobenzoic acid) + *p*-Toluenesulfonyl chloride $\xrightarrow[\text{2. H}^+]{\text{1. HO}^-\text{, H}_2\text{O}}$ *o*-(*p*-Toluenesulfonamido)benzoic acid (88–91%)

24.20 (*a*) This problem illustrates the application of the Sandmeyer reaction to the preparation of aryl cyanides. Diazotization of *p*-nitroaniline followed by treatment with cuprous cyanide converts it to *p*-nitrobenzonitrile.

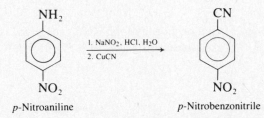

p-Nitroaniline $\xrightarrow[\text{2. CuCN}]{\text{1. NaNO}_2\text{, HCl, H}_2\text{O}}$ *p*-Nitrobenzonitrile

(*b*) An acceptable pathway becomes apparent when it is realized that the amino group in the product is derived from the nitro group of the starting material. Two chlorines are introduced by electrophilic aromatic substitution, the third by a Sandmeyer reaction.

3,4,5-Trichloroaniline ⇒ 3,4,5-trichloronitrobenzene ⇒ 2,6-dichloro-4-nitroaniline ⇒ *p*-nitroaniline

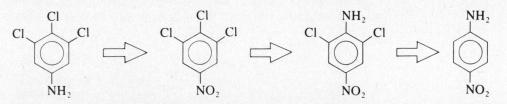

Two of the required chlorine atoms can be introduced by chlorination of the starting material, *p*-nitroaniline.

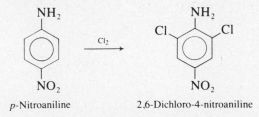

p-Nitroaniline $\xrightarrow{\text{Cl}_2}$ 2,6-Dichloro-4-nitroaniline

The third chlorine can be introduced via the Sandmeyer reaction. Reduction of the nitro group completes the synthesis of 3,4,5-trichloroaniline.

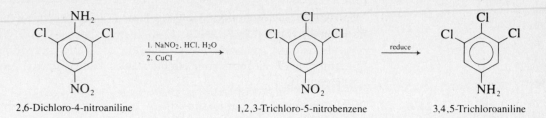

2,6-Dichloro-4-nitroaniline $\xrightarrow[\text{2. CuCl}]{\text{1. NaNO}_2\text{, HCl, H}_2\text{O}}$ 1,2,3-Trichloro-5-nitrobenzene $\xrightarrow{\text{reduce}}$ 3,4,5-Trichloroaniline

The reduction step has been carried out by hydrogenation over Raney nickel in 70 percent yield.

(*c*) The amino group that is present in the starting material has been used to facilitate the introduction of the bromine substituents, then removed by reductive deamination.

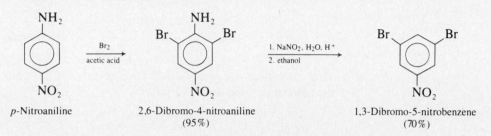

p-Nitroaniline $\xrightarrow[\text{acetic acid}]{Br_2}$ 2,6-Dibromo-4-nitroaniline (95%) $\xrightarrow[\text{2. ethanol}]{\text{1. NaNO}_2\text{, H}_2\text{O, H}^+}$ 1,3-Dibromo-5-nitrobenzene (70%)

Hypophosphorous acid has also been used successfully in the reductive deamination step.

(*d*) Reduction of the nitro group of the 1,3-dibromo-5-nitrobenzene prepared in the preceding part of this problem gives the desired product. The customary reducing agents used for the reduction of nitroarenes would all be suitable.

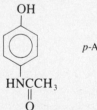

1,3-Dibromo-5-nitrobenzene [prepared from *p*-nitroaniline as in (*c*)] $\xrightarrow{H_2\text{, Ni}}$ 3,5-Dibromoaniline (80%)

(*e*) The synthetic objective is:

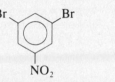

p-Acetamidophenol ($HO-C_6H_4-NHCCH_3$, with C=O)

This compound, known as *acetaminophen* and used as an aspirin substitute to reduce fever and relieve minor pain, may be prepared from *p*-nitroaniline by way of *p*-nitrophenol.

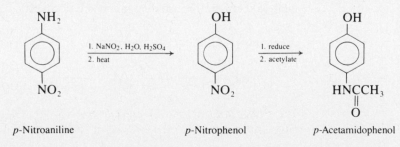

p-Nitroaniline $\xrightarrow[\text{2. heat}]{\text{1. NaNO}_2\text{, H}_2\text{O, H}_2\text{SO}_4}$ *p*-Nitrophenol $\xrightarrow[\text{2. acetylate}]{\text{1. reduce}}$ *p*-Acetamidophenol

Any of the customary reducing agents suitable for converting aryl nitro groups to arylamines (Fe, HCl; Sn, HCl; H_2, Ni) may be used. Acetylation of *p*-aminophenol may be carried out with acetyl chloride or acetic anhydride. The amino group of *p*-aminophenol is more nucleophilic than the hydroxyl group and is acetylated preferentially.

24.21 (*a*) Replacement of an amino substituent by a bromine is readily achieved by the Sandmeyer reaction.

$$\text{o-Anisidine} \xrightarrow[\text{2. CuBr}]{\text{1. NaNO}_2\text{, HBr, H}_2\text{O}} \text{o-Bromoanisole (88–93\%)}$$

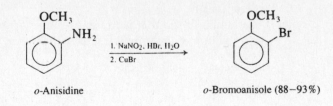

o-Anisidine — *o*-Bromoanisole (88–93%)

(*b*) This conversion demonstrates the replacement of an amino substituent by fluorine via the Schiemann reaction.

$$\text{o-Anisidine} \xrightarrow[\text{2. HBF}_4]{\text{1. NaNO}_2\text{, HCl, H}_2\text{O}} \text{o-}CH_3OC_6H_4\overset{+}{N}{\equiv}N\!:\ \bar{B}F_4 \xrightarrow{\text{heat}} \text{o-Fluoroanisole}$$

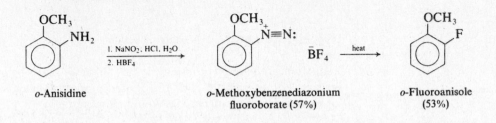

o-Anisidine — *o*-Methoxybenzenediazonium fluoroborate (57%) — *o*-Fluoroanisole (53%)

(*c*) We can use the *o*-fluoroanisole prepared in the preceding exercise to prepare 3-fluoro-4-methoxyacetophenone by Friedel-Crafts acylation.

$$\text{o-Anisidine} \xrightarrow{\text{as in (b)}} \text{o-Fluoroanisole} \xrightarrow[\text{AlCl}_3]{CH_3COCCH_3 \text{ (O, O)}} \text{3-Fluoro-4-methoxyacetophenone}$$

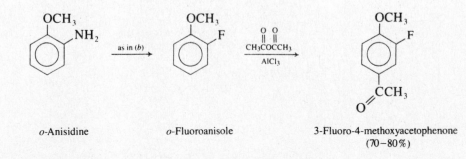

o-Anisidine — *o*-Fluoroanisole — 3-Fluoro-4-methoxyacetophenone (70–80%)

Remember from Section 13.15 that it is the more activating substituent that determines the regioselectivity of electrophilic aromatic substitution when an arene bears two different substituents. Methoxy is a strongly activating substituent; fluorine is slightly deactivating. Friedel-Crafts acylation takes place at the position para to the methoxy group.

(*d*) The *o*-fluoroanisole prepared in (*b*) serves nicely as a precursor to 3-fluoro-4-methoxybenzonitrile via diazonium salt chemistry.

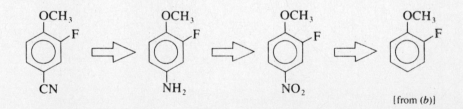

$$\text{3-F-4-}CH_3O\text{-}C_6H_3CN \Rightarrow \text{3-F-4-}CH_3O\text{-}C_6H_3NH_2 \Rightarrow \text{3-F-4-}CH_3O\text{-}C_6H_3NO_2 \Rightarrow \text{o-Fluoroanisole}$$

[from (*b*)]

The desired sequence of reactions to carry out the synthesis is:

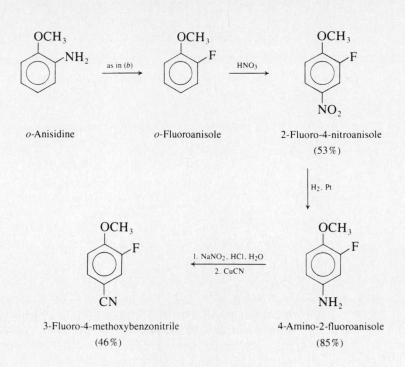

o-Anisidine → (as in (b)) o-Fluoroanisole → (HNO_3) 2-Fluoro-4-nitroanisole (53%) → (H_2, Pt) 4-Amino-2-fluoroanisole (85%) → (1. $NaNO_2$, HCl, H_2O; 2. CuCN) 3-Fluoro-4-methoxybenzonitrile (46%)

Conversion of *o*-fluoroanisole to 4-amino-2-fluoroanisole proceeds in the conventional way by preparation and reduction of a nitro derivative. Once the necessary arylamine is at hand, it is converted to the nitrile by a Sandmeyer reaction.

(*e*) Diazotization followed by hydrolysis of the 4-amino-2-fluoroanisole prepared as an intermediate in (*d*) yields the desired phenol.

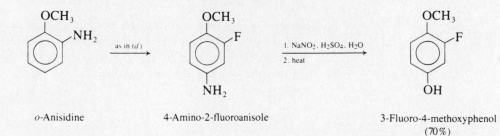

o-Anisidine → (as in (d)) 4-Amino-2-fluoroanisole → (1. $NaNO_2$, H_2SO_4, H_2O; 2. heat) 3-Fluoro-4-methoxyphenol (70%)

24.22 (*a*) The carboxyl group of *p*-aminobenzoic acid can be derived from the methyl group of *p*-toluidine by oxidation. First, however, the nitrogen must be acylated so as to protect the ring from oxidation.

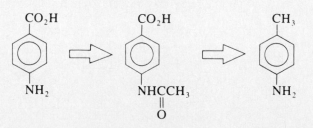

p-Aminobenzoic acid (CO_2H, NH_2) ⇒ CO_2H, $NHCCH_3$ (with C=O) ⇒ *p*-Toluidine (CH_3, NH_2)

The sequence of reactions to be used is:

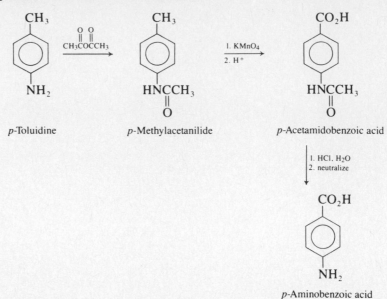

p-Toluidine $\xrightarrow{CH_3COCCH_3 \text{ (anhydride)}}$ p-Methylacetanilide $\xrightarrow{\text{1. } KMnO_4 \text{ 2. } H^+}$ p-Acetamidobenzoic acid $\xrightarrow{\text{1. HCl, } H_2O \text{ 2. neutralize}}$ p-Aminobenzoic acid

(*b*) Attachment of fluoro and propanoyl groups to a benzene ring is required. The fluorine substituent can be introduced by way of the diazonium tetrafluoroborate, the propanoyl group by way of a Friedel-Crafts acylation. Since the fluorine substituent is ortho, para–directing, introducing it first gives the proper orientation of substituents.

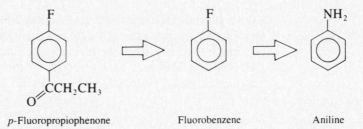

p-Fluoropropiophenone ⇨ Fluorobenzene ⇨ Aniline

Fluorobenzene is prepared from aniline by the Schiemann reaction, shown in Section 24.9. Aniline is, of course, prepared from benzene via nitrobenzene. Friedel-Crafts acylation of fluorobenzene has been carried out with the results shown and gives the required *p*-fluoropropiophenone as the major product.

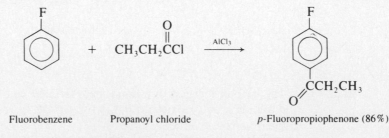

Fluorobenzene + $CH_3CH_2\overset{O}{\overset{\|}{C}}Cl$ (Propanoyl chloride) $\xrightarrow{AlCl_3}$ *p*-Fluoropropiophenone (86%)

(*c*) Our synthetic plan is based on the essential step of forming the fluorine derivative from an amine by way of a diazonium salt.

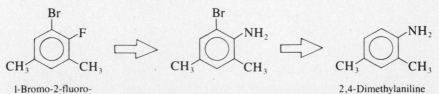

1-Bromo-2-fluoro-3,5-dimethylbenzene ⇨ 2-bromo-4,6-dimethylaniline ⇨ 2,4-Dimethylaniline

The required substituted aniline is derived from *m*-xylene by a standard synthetic sequence.

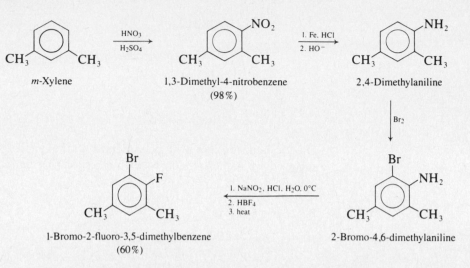

m-Xylene $\xrightarrow[H_2SO_4]{HNO_3}$ 1,3-Dimethyl-4-nitrobenzene (98%) $\xrightarrow[2.\ HO^-]{1.\ Fe,\ HCl}$ 2,4-Dimethylaniline $\xrightarrow{Br_2}$ 2-Bromo-4,6-dimethylaniline $\xrightarrow[2.\ HBF_4;\ 3.\ heat]{1.\ NaNO_2,\ HCl,\ H_2O,\ 0°C}$ 1-Bromo-2-fluoro-3,5-dimethylbenzene (60%)

(*d*) In this problem two nitrogen-containing groups of the starting material are each to be replaced by a halogen substituent. The task is sufficiently straightforward that it may be confronted directly.

Replace amino group by bromine

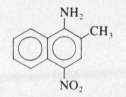

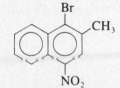

2-Methyl-4-nitro-1-naphthylamine $\xrightarrow[2.\ CuBr]{1.\ NaNO_2,\ HBr,\ H_2O}$ 1-Bromo-2-methyl-4-nitronaphthalene (82%)

Reduce nitro group to amine

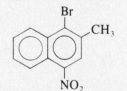

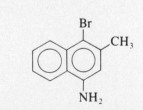

1-Bromo-2-methyl-4-nitronaphthalene $\xrightarrow[2.\ HO^-]{1.\ Fe,\ HCl}$ 4-Bromo-3-methyl-1-naphthylamine

Replace amino group by fluorine

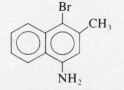

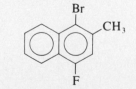

4-Bromo-3-methyl-1-naphthylamine $\xrightarrow[2.\ HBF_4;\ 3.\ heat]{1.\ NaNO_2,\ HCl,\ H_2O,\ 0-5°C}$ 1-Bromo-4-fluoro-2-methylnaphthalene (64%)

(*e*) Bromination of the starting material will introduce the bromine substituent at the correct position, i.e., ortho to the tert-butyl group.

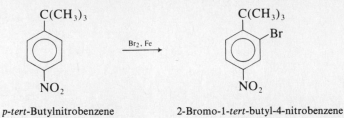

$C(CH_3)_3$ / NO_2 —(Br_2, Fe)→ $C(CH_3)_3$ / Br / NO_2

p-tert-Butylnitrobenzene → 2-Bromo-1-*tert*-butyl-4-nitrobenzene

The desired product will be obtained if the nitro group can be removed. This is achieved by its conversion to the corresponding amine, followed by reductive deamination.

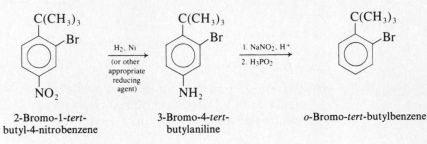

2-Bromo-1-*tert*-butyl-4-nitrobenzene —(H_2, Ni (or other appropriate reducing agent))→ 3-Bromo-4-*tert*-butylaniline —(1. $NaNO_2$, H^+ 2. H_3PO_2)→ *o*-Bromo-*tert*-butylbenzene

(*f*) The proper orientation of the chlorine substituent can only be achieved if it is introduced after the nitro group is reduced.

$C(CH_3)_3$ / Cl ⇨ $C(CH_3)_3$ / Cl / NH_2 ⇨ $C(CH_3)_3$ / NH_2 ⇨ $C(CH_3)_3$ / NO_2

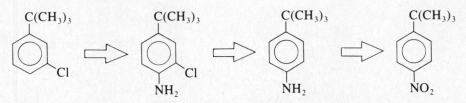

The correct sequence of reactions to carry out this synthesis is shown.

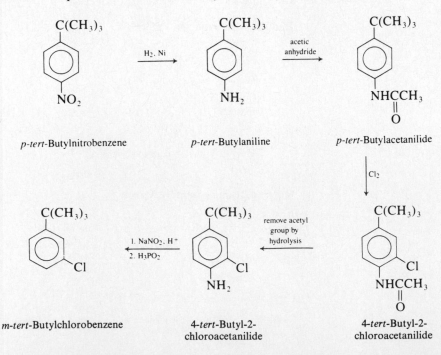

p-tert-Butylnitrobenzene —(H_2, Ni)→ *p-tert*-Butylaniline —(acetic anhydride)→ *p-tert*-Butylacetanilide ($NHCCH_3$, ‖ O)

↓ Cl_2

4-*tert*-Butyl-2-chloroacetanilide ($NHCCH_3$, ‖ O, Cl) —(remove acetyl group by hydrolysis)→ 4-*tert*-Butyl-2-chloroacetanilide (NH_2, Cl) —(1. $NaNO_2$, H^+ 2. H_3PO_2)→ *m-tert*-Butylchlorobenzene (*g*) The orientation of substituents in the target molecule can be achieved by using an amino group to control the regiochemistry of bromination, then removing it by reductive deamination.

CH_3CH_2, Br, CH_3CH_2 ⇨ NH_2, CH_3CH_2, Br, CH_3CH_2 ⇨ NH_2, CH_3CH_2, CH_3CH_2

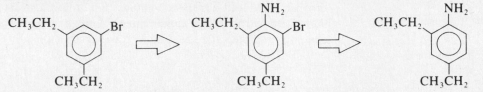

The amino group is introduced in the standard fashion by nitration of an arene followed by reduction.

This analysis leads to the synthesis shown.

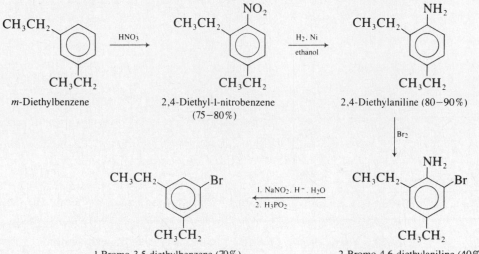

m-Diethylbenzene $\xrightarrow{HNO_3}$ 2,4-Diethyl-1-nitrobenzene (75–80%) $\xrightarrow[\text{ethanol}]{H_2, Ni}$ 2,4-Diethylaniline (80–90%)

2,4-Diethylaniline $\xrightarrow{Br_2}$ 2-Bromo-4,6-diethylaniline (40%) $\xrightarrow[2.\ H_3PO_2]{1.\ NaNO_2,\ H^+,\ H_2O}$ 1-Bromo-3,5-diethylbenzene (70%)

(*h*) In this exercise the two nitrogen substituents are differentiated; one is an amino nitrogen, the other an amide nitrogen. By keeping them differentiated they can be manipulated independently. Remove one amino group completely before deprotecting the other.

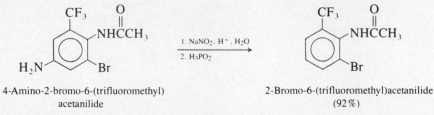

4-Amino-2-bromo-6-(trifluoromethyl) acetanilide $\xrightarrow[2.\ H_3PO_2]{1.\ NaNO_2,\ H^+,\ H_2O}$ 2-Bromo-6-(trifluoromethyl)acetanilide (92%)

Once the acetyl group has been removed by hydrolysis, the molecule is ready for introduction of the iodo substituent by way of a diazonium salt.

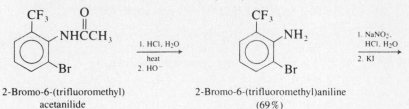

2-Bromo-6-(trifluoromethyl) acetanilide $\xrightarrow[\text{heat; } 2.\ HO^-]{1.\ HCl,\ H_2O}$ 2-Bromo-6-(trifluoromethyl)aniline (69%) $\xrightarrow[2.\ KI]{1.\ NaNO_2,\ HCl,\ H_2O}$ 1-Bromo-2-iodo-3-(trifluoromethyl)benzene (87%)

(*i*) In order to convert the designated starting material to the indicated product, both the nitro group and the ester function must be reduced and a carbon-nitrogen bond must be formed. Converting the starting material to an amide gives the necessary carbon-nitrogen bond and has the advantage that amides can be reduced to amines by lithium aluminum hydride. The amide can be formed intramolecularly by reducing the nitro group to an amine, then heating to cause cyclization.

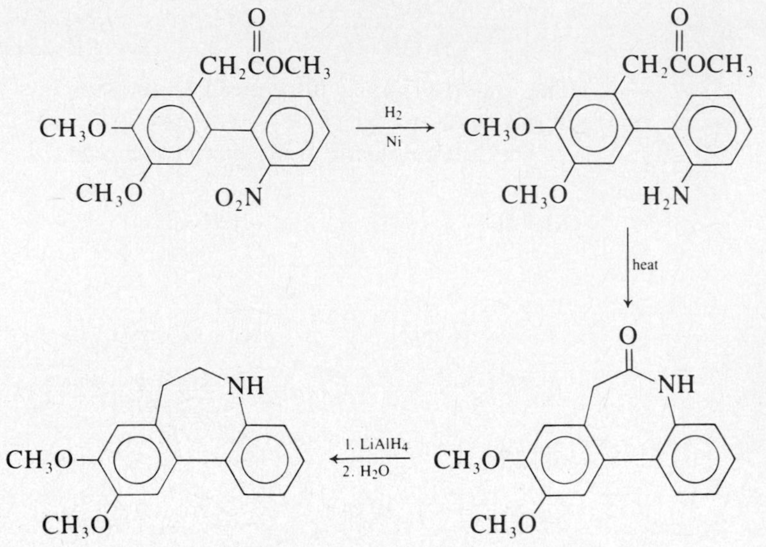

$CH_2\overset{O}{\overset{\|}{C}}OCH_3$, CH_3O, CH_3O, O_2N $\xrightarrow[Ni]{H_2}$ $CH_2\overset{O}{\overset{\|}{C}}OCH_3$, CH_3O, CH_3O, H_2N

$\downarrow$ heat

CH_3O, CH_3O, NH $\xleftarrow[2.\ H_2O]{1.\ LiAlH_4}$ CH_3O, CH_3O, O, NH

The synthesis shown above is the one described in the chemical literature. Other routes are also possible, but the one shown is short and efficient.

24.23 There is no obvious reason why the dimethylamino group in 4-(*N,N*-dimethylamino)-pyridine should be appreciably more basic than it is in *N,N*-dimethylaniline; it is the ring nitrogen of 4-(*N,N*-dimethylamino)pyridine that is more basic. Note that protonation of the ring nitrogen permits delocalization of the dimethylamino lone pair and dispersal of the positive charge.

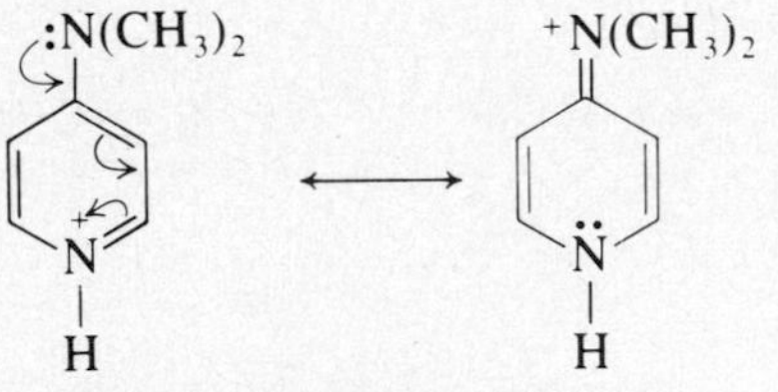

$:N(CH_3)_2$ ... $\overset{+}{N}$—H $\longleftrightarrow$ $^{+}N(CH_3)_2$ ... $\ddot{N}$—H

Most stable protonated form of 4-(*N,N*-dimethylamino)pyridine

24.24 Conversion of a primary amine to a diazonium salt requires that it react in its neutral form. Protonation of an amine destroys its nucleophilicity.

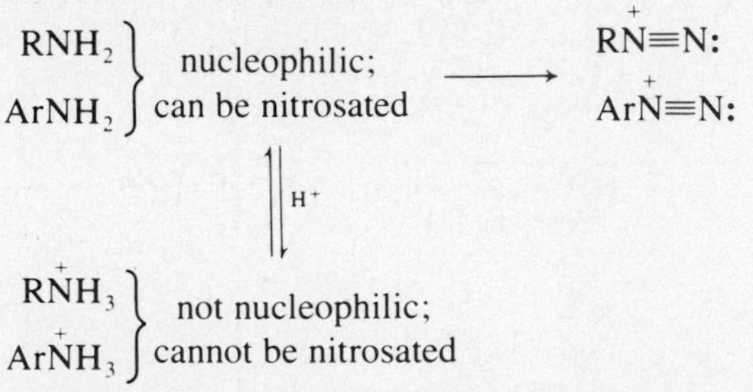

$\left.\begin{matrix} RNH_2 \\ ArNH_2 \end{matrix}\right\}$ nucleophilic; can be nitrosated $\longrightarrow$ $\begin{matrix} R\overset{+}{N}{\equiv}N: \\ Ar\overset{+}{N}{\equiv}N: \end{matrix}$

$\Big\Updownarrow H^+$

$\left.\begin{matrix} R\overset{+}{N}H_3 \\ Ar\overset{+}{N}H_3 \end{matrix}\right\}$ not nucleophilic; cannot be nitrosated

Below pH 3 there is apparently too small a concentration of RNH_2 in equilibrium with $R\overset{+}{N}H_3$ to react with the nitrosating agent, and diazotization does not occur. Arylamines are much less basic than alkylamines, contain a higher concentration of $ArNH_2$ in the presence of $Ar\overset{+}{N}H_3$, and can be nitrosated even at pH 1.

$$H_2N-C_6H_4-CH_2NH_2 + H^+ \rightleftharpoons H_2N-C_6H_4-CH_2\overset{+}{N}H_3$$

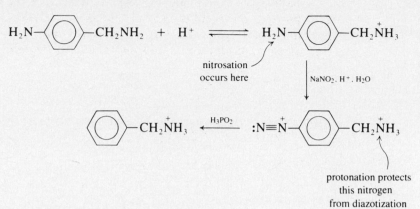

nitrosation occurs here (at H_2N)

$$\xrightarrow{NaNO_2,\ H^+,\ H_2O} :N\equiv\overset{+}{N}-C_6H_4-CH_2\overset{+}{N}H_3 \xrightarrow{H_3PO_2} C_6H_5-CH_2\overset{+}{N}H_3$$

protonation protects this nitrogen from diazotization

SELF-TEST

PART A

A-1. Provide the missing component (reactant, reagent, or product) for each of the following.

(*a*) $CH_3-C_6H_4-NH_2 \xrightarrow[H_2O]{NaNO_2,\ HCl} ?$

(*b*) Product of (*a*) $\xrightarrow{CuBr}$?

(*c*) Product of (*a*) $\xrightarrow{?}$ toluene

(*d*) $CH_3-C_6H_4-NH_2 \xrightarrow{?} CH_3-C_6H_4-NH\overset{O}{\overset{\|}{C}}CH_3$

(*e*) (*m*-ethyl acetanilide: ring with $NH\overset{O}{\overset{\|}{C}}CH_3$ and CH_2CH_3 meta) $\xrightarrow[H_2SO_4]{HNO_3}$?

A-2. Give the series of reaction steps involved in the following synthetic conversions:

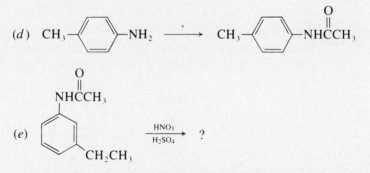

(*a*) $C(CH_3)_3-C_6H_4-CN$ (para) from benzene

(*b*) *m*-Chloroaniline from benzene

(*c*) $C_6H_5N{=}N-C_6H_4-N(CH_3)_2$ from aniline

A-3. *p*-Nitroaniline (A) is less basic than *m*-nitroaniline (B). Using resonance structures, explain the reason for this difference.

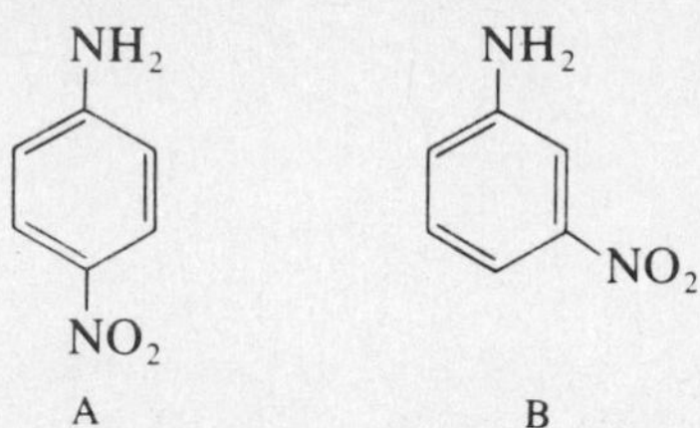

A-4. Identify the strongest and weakest bases among the following:

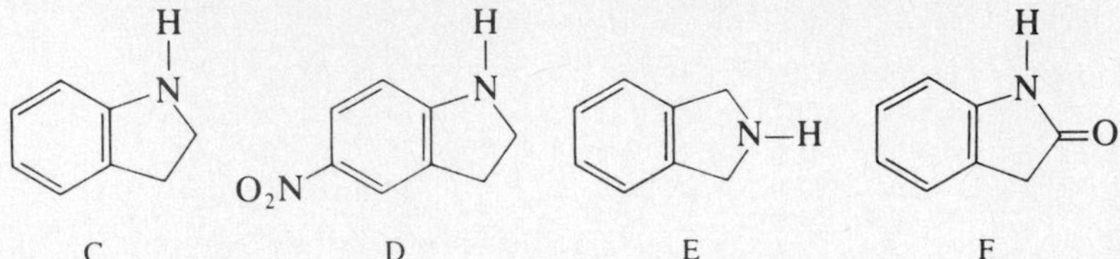

PART B

B-1. Which of the following C_8H_9NO isomers is the weakest base?

(*a*) *o*-Aminoacetophenone
(*b*) *m*-Aminoacetophenone
(*c*) *p*-Aminoacetophenone
(*d*) Acetanilide

B-2. Rank the following compounds in order of increasing basicity (weakest → strongest):

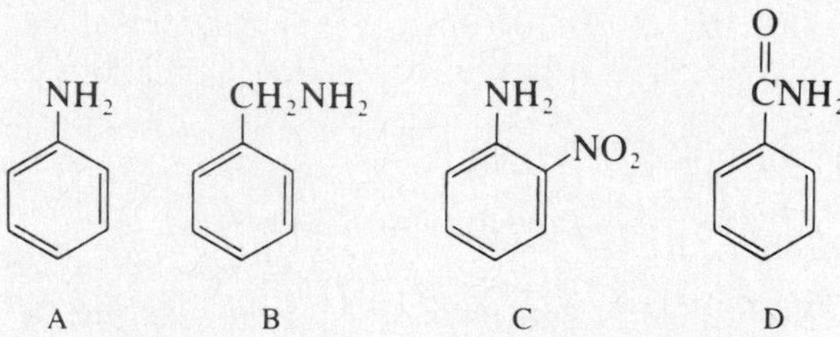

(*a*) D < B < A < C
(*b*) D < A < C < B
(*c*) D < C < A < B
(*d*) B < A < C < D

B-3. Based on comparison of the carbon-nitrogen bond length (of the amino nitrogen) in compounds A and B, which of the following statements is correct?

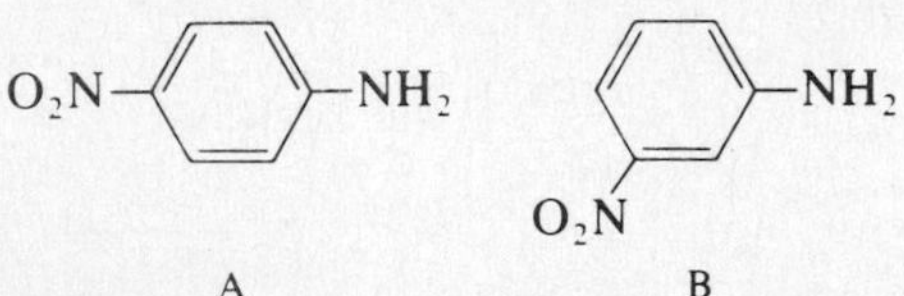

(*a*) The bond in A is shorter.
(*b*) The bond in B is shorter.
(*c*) The bonds in A and B have the same length.
(*d*) No comparison of bond length can be made.

B-4. Which of the following arylamines will *not* form a diazonium salt on reaction with sodium nitrite in hydrochloric acid?

(*a*) *m*-Ethylaniline
(*b*) 4-Chloro-2-nitroaniline
(*c*) *p*-Aminoacetophenone
(*d*) *N*-Ethyl-2-methylaniline

CHAPTER

25

ARYL HALIDES

IMPORTANT TERMS AND CONCEPTS

Nucleophilic Aromatic Substitution Reactions (Secs. 25.4 to 25.7) Under certain circumstances aryl halides are able to undergo nucleophilic aromatic substitution reactions. One of two conditions must be met: the aromatic ring must bear a strongly electron-withdrawing group, such as nitro, or the nucleophile must be a strong base, such as the amide ion.

When the first of these conditions is met, the reaction proceeds by an *addition-elimination mechanism.* As shown below, the intermediate formed on addition of the nucleophile is a resonance-stabilized carbanion. In order to participate in resonance delocalization of the negative charge, the nitro groups must be situated ortho or para to the halide. Elimination of halide ion restores aromaticity to the system, forming the product.

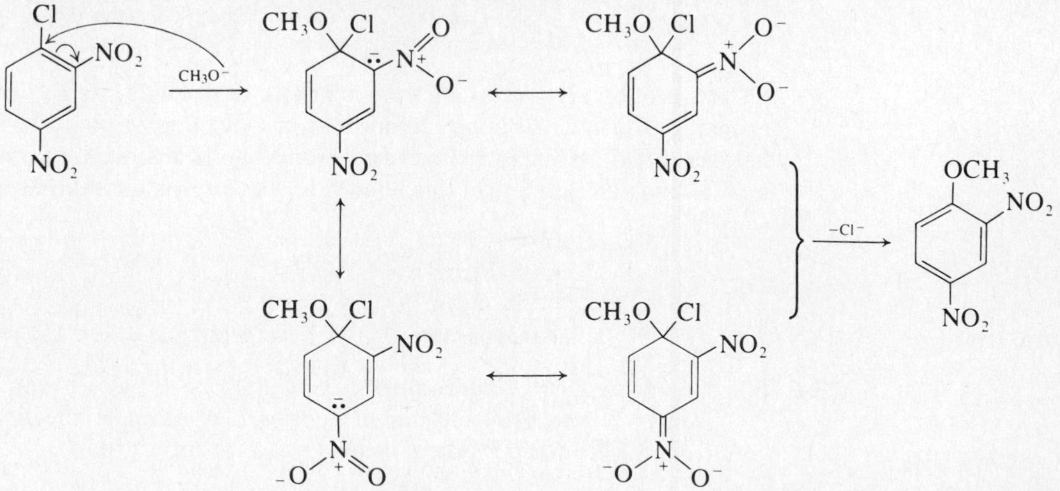

Cl, NO_2, NO_2 $\xrightarrow{CH_3O^-}$ CH_3O Cl, $\ddot{}$ N^+ (=O) O^-, NO_2 $\longleftrightarrow$ CH_3O Cl, $=\overset{+}{N}$ O^- O^-, NO_2

$\updownarrow$

CH_3O Cl, NO_2, $\ddot{}$ $\overset{+}{N}$ ^-O O $\longleftrightarrow$ CH_3O Cl, NO_2, $=\overset{+}{N}$ ^-O O^- $\xrightarrow{-Cl^-}$ OCH_3, NO_2, NO_2

In the presence of a very strong base such as amide ion, aryl halides may undergo a second type of substitution reaction, involving an *elimination-addition mechanism.* The first stage is a base-promoted dehydrohalogenation, forming an intermediate species known as *benzyne*. The second stage involves addition of base (NH_2^-) and protonation of the anion formed.

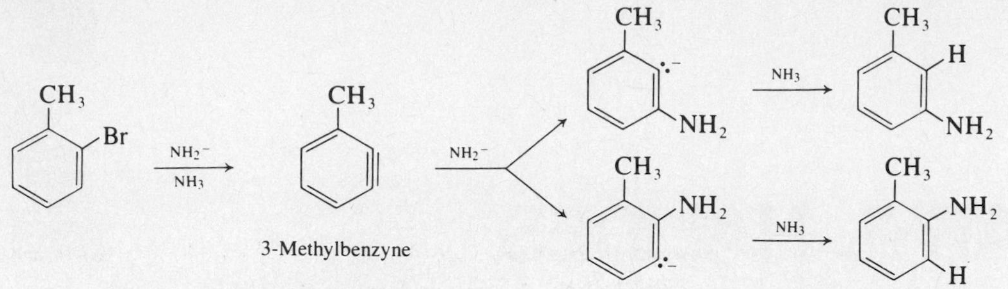

3-Methylbenzyne

Electronegative groups (such as $-OR$ and $-CF_3$) attached to the benzene ring influence the regioselectivity by stabilizing the carbanion formed during the addition phase of the reaction. Alkyl groups have little effect, resulting in a mixture of isomeric substitution products.

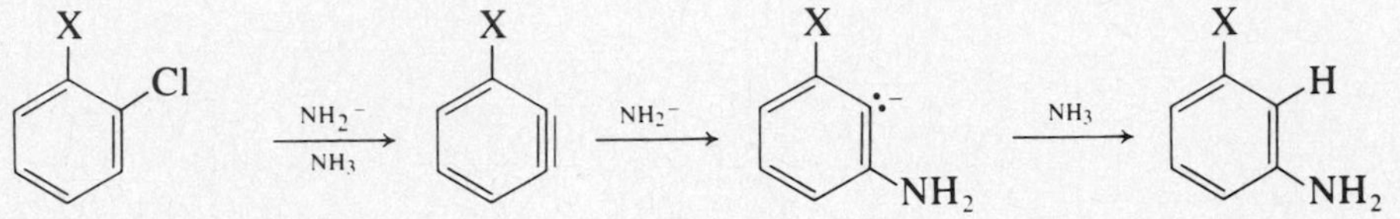

($X = -OR,\ -CF_3$)

(Major product)

SOLUTIONS TO TEXT PROBLEMS

25.1 Among the isomers of C_7H_7Cl are four based on the benzene nucleus, namely, *o*, *m*, and *p*-chlorotoluene and benzyl chloride.

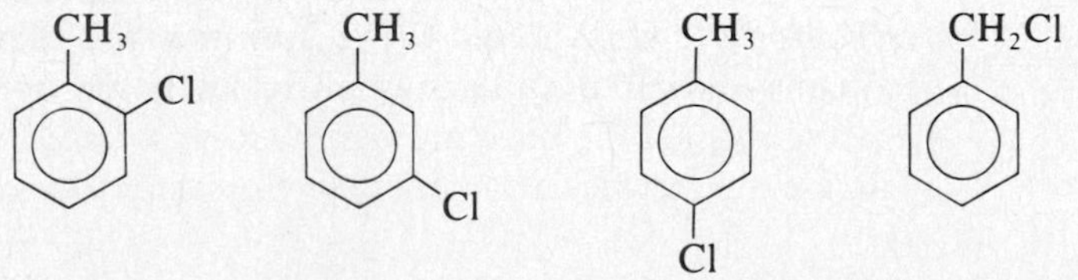

o-Chlorotoluene *m*-Chlorotoluene *p*-Chlorotoluene Benzyl chloride

Of this group only benzyl chloride is not an aryl halide; its halogen is not attached to the aromatic ring but to an sp^3 hybridized carbon. Benzyl chloride has the weakest carbon-halogen bond, its measured carbon-chlorine bond dissociation energy being only 70 kcal/mol. Homolytic cleavage of this bond produces a resonance-stabilized benzyl radical.

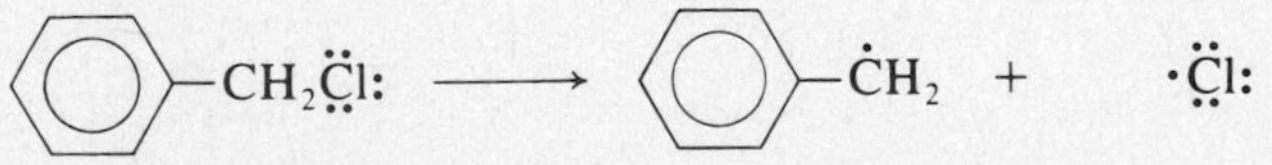

Benzyl chloride Benzyl radical Chlorine atom

25.2 (*b*) Conjugation in *o*-fluoronitrobenzene follows a pattern similar to that of the para isomer. A lone electron pair of fluorine is donated to the ring, while an electron pair from the aromatic π system is donated to the nitro group.

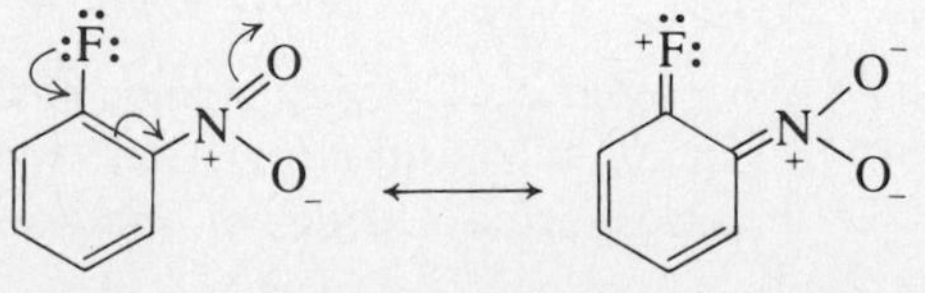

(*c*) Only the nitro group that is para to fluorine can be directly conjugated to it in 4-fluoro-1,2-dinitrobenzene.

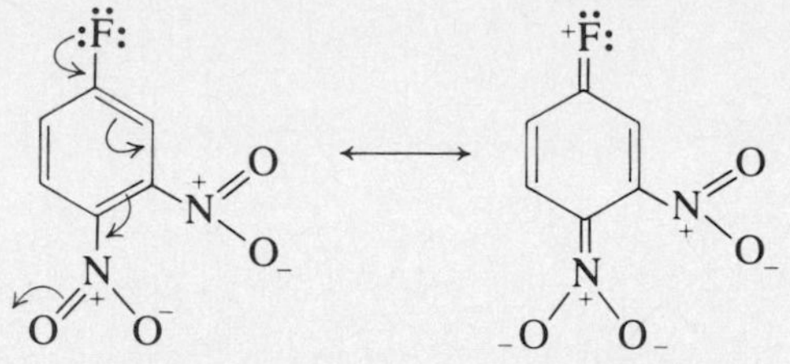

25.3 (*b*) The negatively charged sulfur in $C_6H_5CH_2SNa$ is a good nucleophile, which displaces chloride from 1-chloro-2,4-dinitrobenzene.

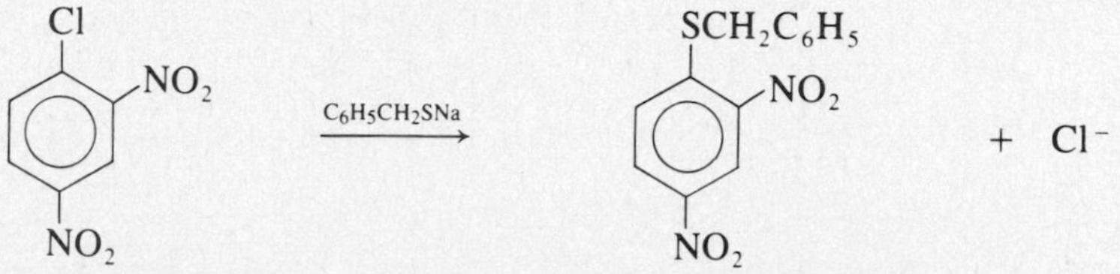

1-Chloro-2,4-dinitrobenzene → ($C_6H_5CH_2SNa$) Benzyl 2,4-dinitrophenyl thioether + Cl^-

(*c*) The nitrogen in methylamine has an unshared electron pair and is nucleophilic; it displaces chloride from 1-chloro-2,4-dinitrobenzene.

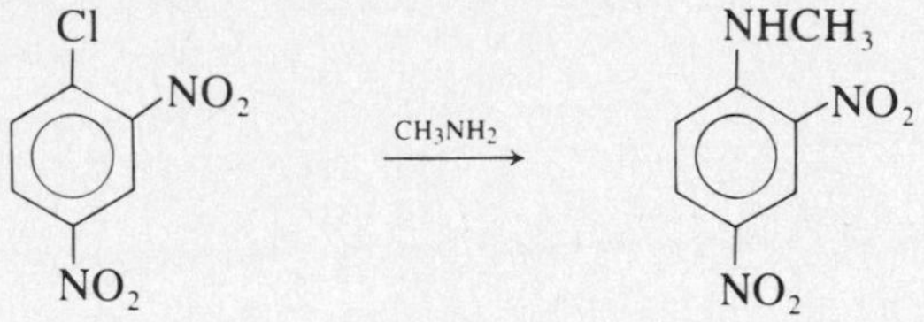

1-Chloro-2,4-dinitrobenzene → (CH_3NH_2) *N*-Methyl-2,4-dinitroaniline

25.4 The positions that are activated toward nucleophilic attack are those that are ortho and para to the nitro group. Among the carbons that bear a bromine leaving group in 1,2,3-tribromo-5-nitrobenzene, only C-2 satisfies this requirement.

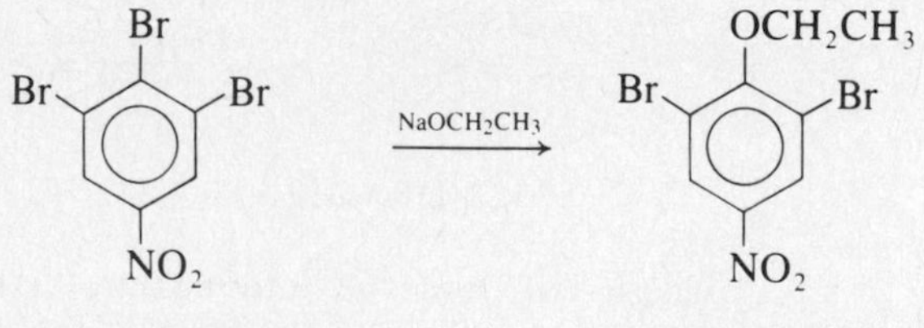

1,2,3-Tribromo-5-nitrobenzene → ($NaOCH_2CH_3$) 1,3-Dibromo-2-ethoxy-5-nitrobenzene

25.5 The presence of an additional electron-withdrawing group in a position para to chlorine makes 1-chloro-2-nitro-4-(trifluoromethyl)benzene more reactive toward nucleophilic aromatic substitution than *o*-chloronitrobenzene. The trifluoromethyl group stabilizes the cyclohexadienyl anion intermediate formed in the rate-determining step by addition of methoxide ion to the aryl halide.

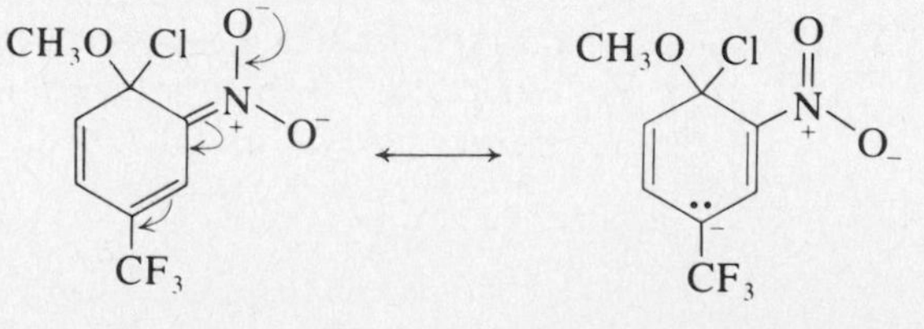

(Electron-withdrawing CF_3 group stabilizes anion)

25.6 The aryl halide is incapable of elimination and so cannot form the benzyne intermediate necessary for substitution by the elimination-addition pathway.

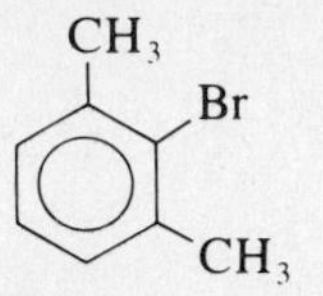

CH_3 Br CH_3

2-Bromo-1,3-dimethylbenzene

(No protons ortho to bromine; elimination is impossible)

25.7 The aryne intermediate from *p*-iodotoluene can undergo addition of hydroxide ion at the position meta to the methyl group or para to it. The two isomeric phenols are *m*- and *p*-cresol.

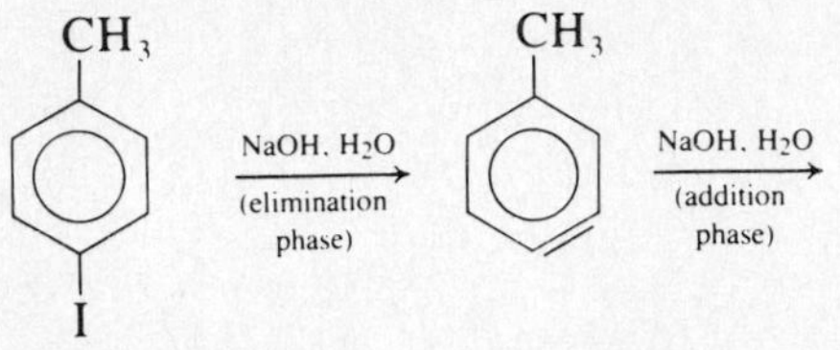

p-Iodotoluene $\xrightarrow[\text{(elimination phase)}]{NaOH,\ H_2O}$ $\xrightarrow[\text{(addition phase)}]{NaOH,\ H_2O}$

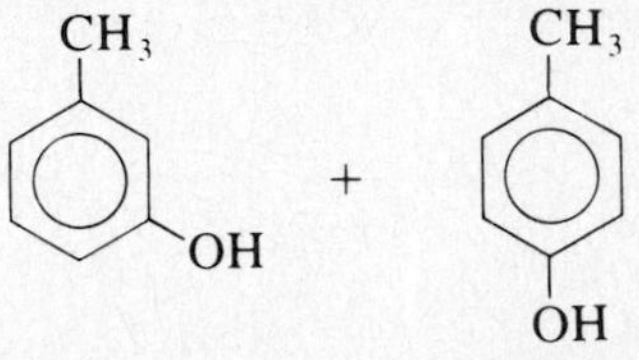

m-Cresol (3-methylphenol) + *p*-Cresol (4-methylphenol)

25.8 (*b*) A mixture of two isomeric benzyne intermediates is possible from *m*-chlorotoluene.

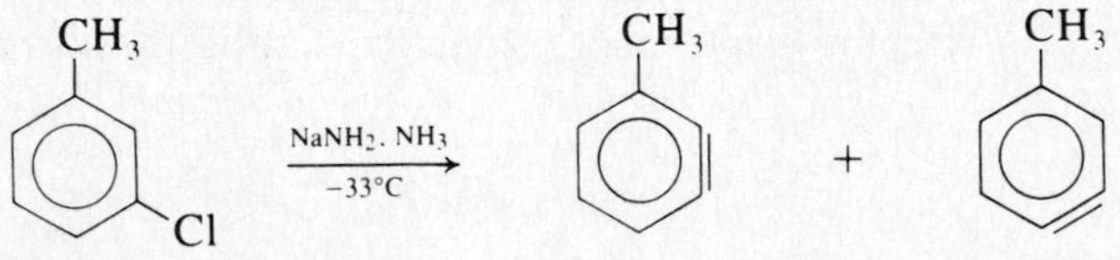

m-Chlorotoluene $\xrightarrow[-33°C]{NaNH_2,\ NH_3}$ 3-Methylbenzyne + 4-Methylbenzyne

These two benzyne intermediates add ammonia to give a total of three isomeric products.

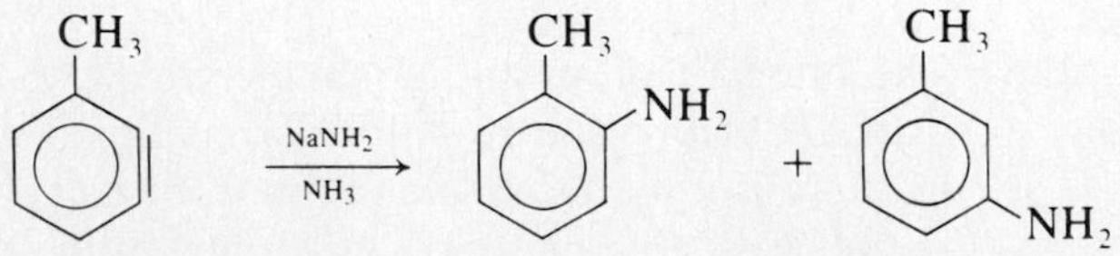

3-Methylbenzyne $\xrightarrow[NH_3]{NaNH_2}$ *o*-Methylaniline + *m*-Methylaniline

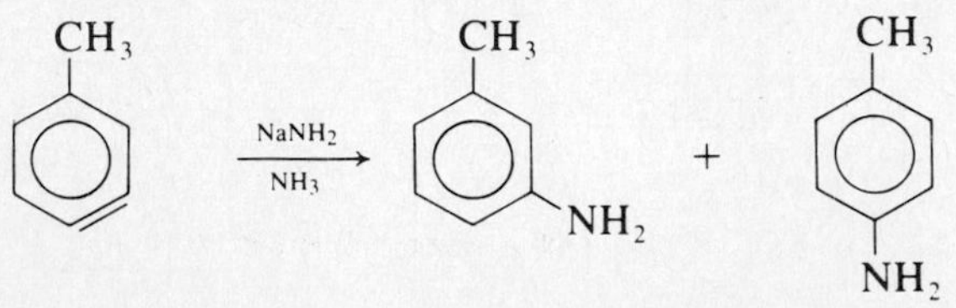

4-Methylbenzyne $\xrightarrow[NH_3]{NaNH_2}$ *m*-Methylaniline + *p*-Methylaniline

(c) Only one benzyne intermediate is possible from p-chlorotoluene, but it can yield two isomeric products in the addition stage of the reaction.

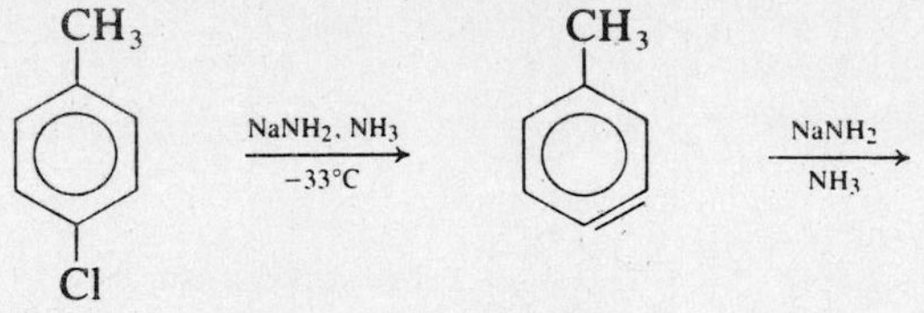

CH_3 CH_3

$\xrightarrow[-33°C]{NaNH_2, NH_3}$ $\xrightarrow[NH_3]{NaNH_2}$

Cl

p-Chlorotoluene 4-Methylbenzyne

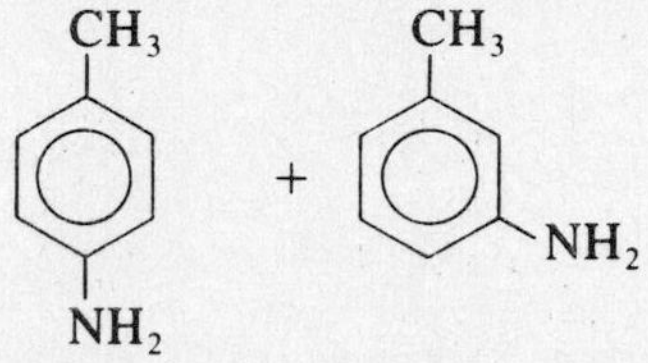

CH_3 CH_3

\+ NH_2

NH_2

p-Methylaniline m-Methylaniline

CH_3

25.9 (a) 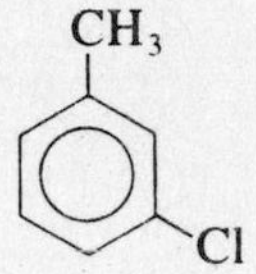 m-Chlorotoluene

Cl

OCH_3

Br Br

(b) 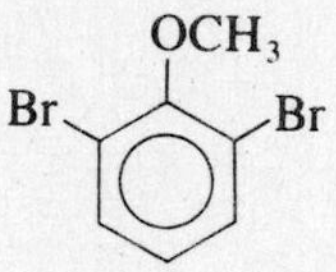 2,6-Dibromoanisole

$CH{=}CH_2$

(c) 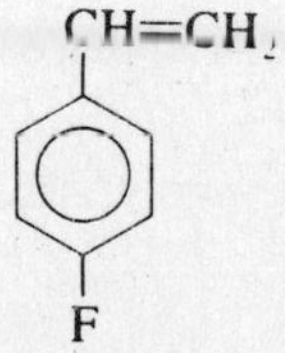 p-Fluorostyrene

F

3 2 2′ 3′

(d) 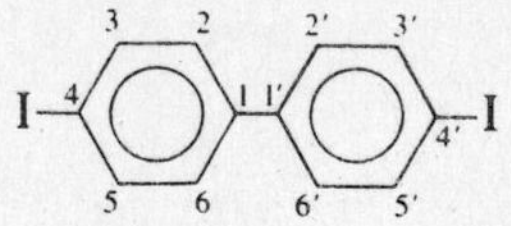 I 4 1 1′ 4′ I 4,4′-Diiodobiphenyl

5 6 6′ 5′

Br

Cl

(e) 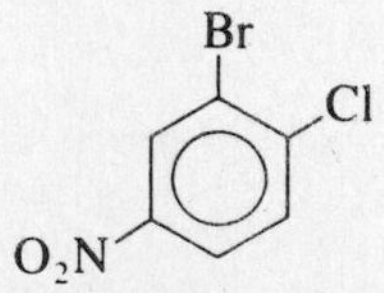 2-Bromo-1-chloro-4-nitrobenzene

O_2N

$ClCHCH_3$

(f)  1-Chloro-1-phenylethane
(*Note:* this compound is not an aryl halide)

CH_2Cl

(g) 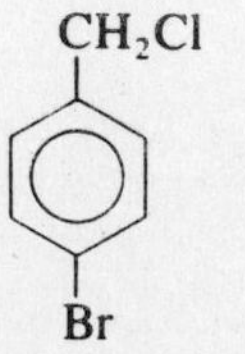 p-Bromobenzyl chloride

Br

(*h*) 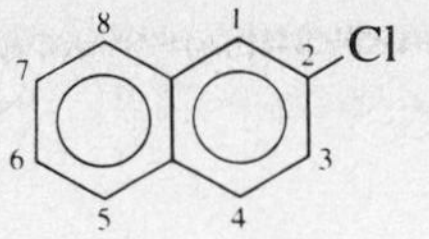 2-Chloronaphthalene

(*i*) 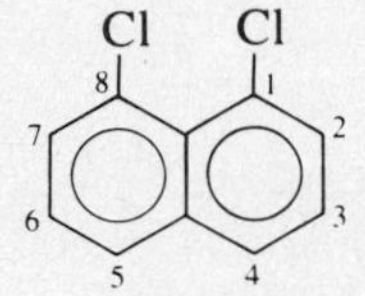 1,8-Dichloronaphthalene

(*j*) 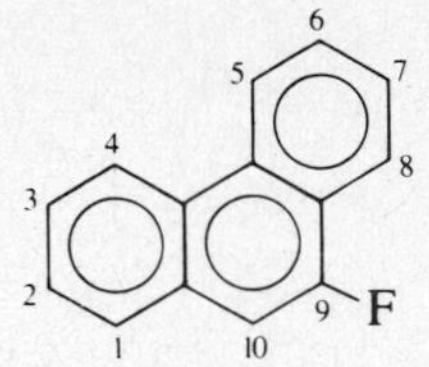 9-Fluorophenanthrene

25.10 (*a*) Chlorine is a weakly deactivating, ortho, para–directing substituent.

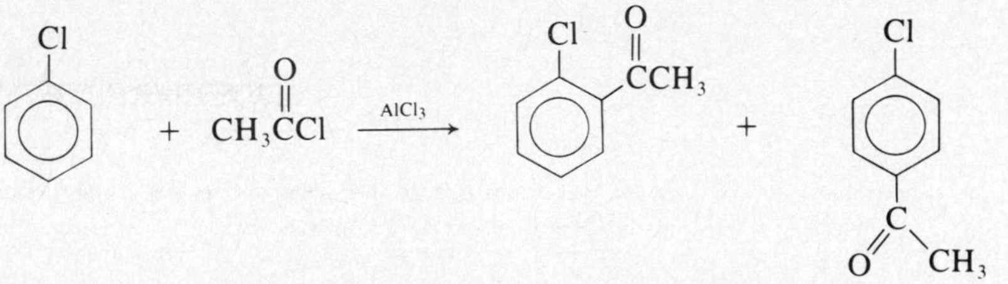

$$C_6H_5Cl + CH_3CCl(=O) \xrightarrow{AlCl_3} o\text{-}ClC_6H_4CCH_3(=O) + p\text{-}ClC_6H_4C(=O)CH_3$$

Chlorobenzene   Acetyl chloride   *o*-Chloroacetophenone   *p*-Chloroacetophenone

(*b*) Bromobenzene reacts with magnesium to give a Grignard reagent.

$$C_6H_5Br + Mg \xrightarrow[\text{ether}]{\text{diethyl}} C_6H_5MgBr$$

Bromobenzene   Phenylmagnesium bromide

(*c*) Protonation of the Grignard reagent in (*b*) converts it to benzene.

$$C_6H_5MgBr \xrightarrow[HCl]{H_2O} C_6H_6$$

Phenylmagnesium bromide   Benzene

(*d*) Aryl halides react with lithium dialkylcuprates in much the same way that alkyl halides do.

$$C_6H_5I + LiCu(CH_3)_2 \xrightarrow[\text{ether}]{\text{diethyl}} C_6H_5CH_3$$

Iodobenzene   Lithium dimethylcuprate   Toluene

(*e*) With a base as strong as sodium amide, nucleophilic aromatic substitution by the elimination-addition mechanism takes place.

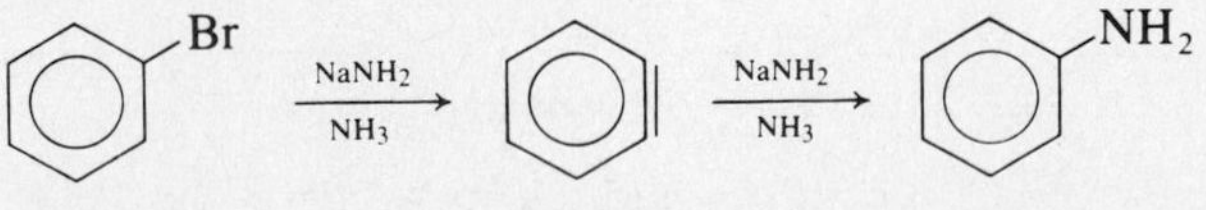

$$C_6H_5Br \xrightarrow[NH_3]{NaNH_2} \text{Benzyne} \xrightarrow[NH_3]{NaNH_2} C_6H_5NH_2$$

Bromobenzene   Benzyne   Aniline

(*f*) The benzyne intermediate from *p*-bromotoluene gives a mixture of *m*- and *p*-methylaniline.

p-Bromotoluene $\xrightarrow[NH_3]{NaNH_2}$ 4-Methylbenzyne $\xrightarrow[NH_3]{NaNH_2}$ m-Methylaniline + p-Methylaniline

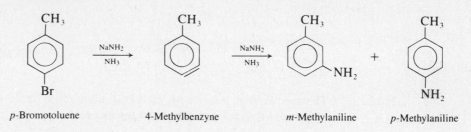

p-Bromotoluene 4-Methylbenzyne *m*-Methylaniline *p*-Methylaniline

(*g*) Nucleophilic aromatic substitution of bromide by ammonia occurs by the addition-elimination mechanism.

1-Bromo-4-nitrobenzene $\xrightarrow{NH_3}$ p-Nitroaniline

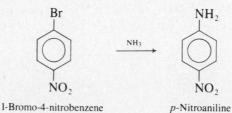

1-Bromo-4-nitrobenzene *p*-Nitroaniline

(*h*) The bromine attached to the benzylic carbon is far more reactive than the one on the ring and is the one replaced by the nucleophile.

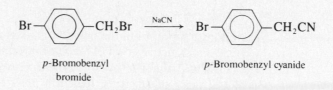

$Br-C_6H_4-CH_2Br \xrightarrow{NaCN} Br-C_6H_4-CH_2CN$

p-Bromobenzyl bromide *p*-Bromobenzyl cyanide

(*i*) The aromatic ring of *N,N*-dimethylaniline is very reactive and is attacked by *p*-chlorobenzenediazonium ion.

$(CH_3)_2N-C_6H_5 + :N\equiv\overset{+}{N}-C_6H_4-Cl \longrightarrow$

N,N-Dimethylaniline *p*-Chlorobenzenediazonium ion

$(CH_3)_2N-C_6H_4-N=N-C_6H_4-Cl$

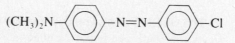

4-(4′-Chlorophenylazo)-*N,N*-dimethylaniline

25.11 (*a*) Since the *tert*-butoxy group replaces fluoride at the position occupied by the leaving group, substitution likely occurs by the addition-elimination mechanism.

o-Fluorotoluene + $(CH_3)_3CO^-$ $\longrightarrow$ [intermediate: CH₃, F, $OC(CH_3)_3$] $\xrightarrow{-F^-}$ tert-Butyl o-methylphenyl ether

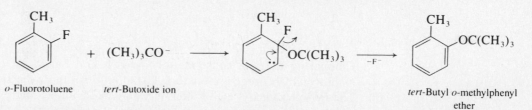

o-Fluorotoluene *tert*-Butoxide ion *tert*-Butyl *o*-methylphenyl ether

(*b*) In nucleophilic aromatic substitution reactions that proceed by the addition-elimination mechanism, aryl fluorides react faster than aryl bromides. Since the aryl bromide is more reactive in this case, it must be reacting by a different mechanism, which is most likely elimination-addition.

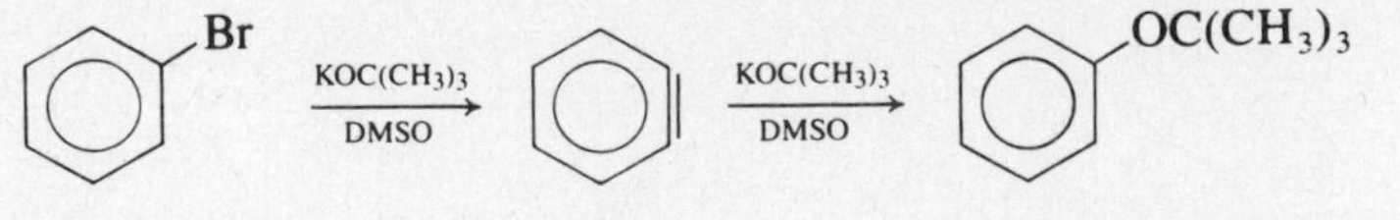
Br
KOC(CH3)3
DMSO
KOC(CH3)3
DMSO
OC(CH3)3

Bromobenzene Benzyne *tert*-Butyl phenyl ether

25.12 (*a*) The species that reacts with carbon dioxide must be 2,6-difluorophenyllithium, formed by proton abstraction from *m*-difluorobenzene.

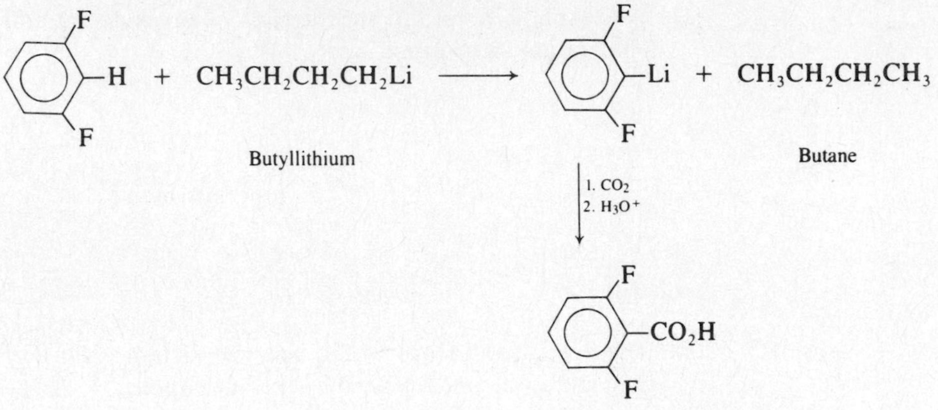
F
H
F
+ CH3CH2CH2CH2Li
Butyllithium
Li
+ CH3CH2CH2CH3
Butane
1. CO2
2. H3O+
CO2H

2,6-Difluorobenzoic acid

Proton abstraction from the position flanked by the fluorine substituents occurs readily because the inductive effect of the fluorines stabilizes the aryl anion.

(*b*) The substrate cannot form a benzyne intermediate by elimination—none of its carbons have hydrogen substituents. Reaction must occur by the addition-elimination mechanism.

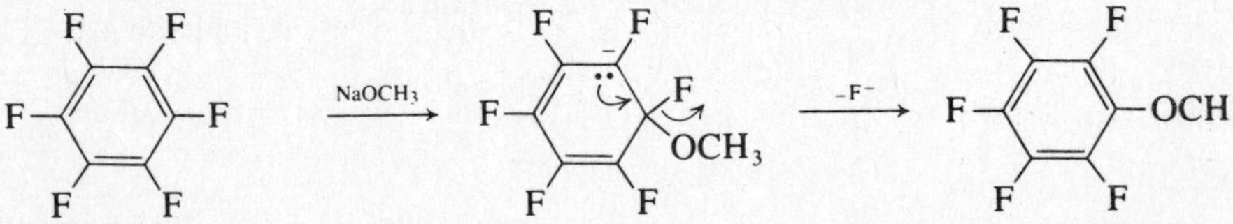
F
NaOCH3
OCH3
-F-

Hexafluorobenzene 2,3,4,5,6-Pentafluoroanisole

The reaction occurs readily because the cyclohexadienyl anion intermediate is stabilized by the combined inductive effect of its six fluorine substituents.

25.13 (*a*) *o*-Chloronitrobenzene is more reactive than chlorobenzene because the cyclohexadienyl anion intermediate is stabilized by the nitro group.

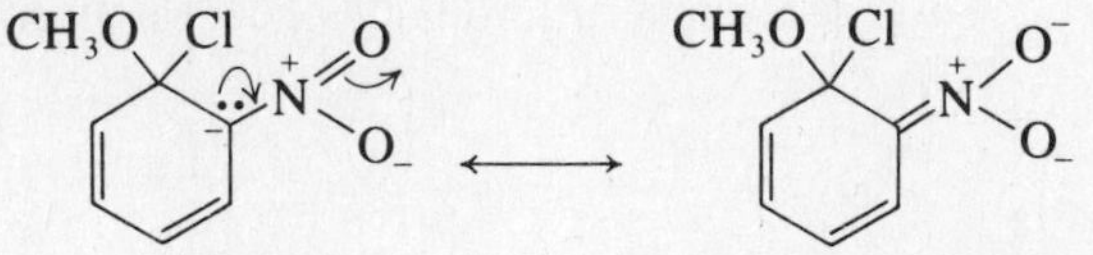
CH3O Cl
N
O
O

Comparing the rate constants for the two aryl halides in this reaction reveals that *o*-chloronitrobenzene is more than 20 billion times more reactive at 50°C.

(*b*) The cyclohexadienyl anion intermediate is more stable, and is formed faster, when the electron-withdrawing nitro group is ortho to chlorine. *o*-Chloronitrobenzene reacts faster than *m*-chloronitrobenzene. The measured difference is a factor of approximately 40,000 at 50°C.

(c) 4-Chloro-3-nitroacetophenone is more reactive because the ring bears two powerful electron-withdrawing groups in positions where they can stabilize the cyclohexadienyl anion intermediate.

CH_3O Cl O N O CH_3 ⟷ CH_3O Cl O N O CH_3 ⟷ CH_3O Cl O N O CH_3

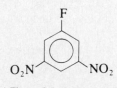

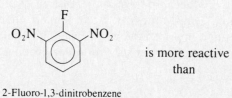

(d) Nitro groups activate aryl halides toward nucleophilic aromatic substitution best when they are ortho or para to the leaving group.

2-Fluoro-1,3-dinitrobenzene is more reactive than 1-Fluoro-3,5-dinitrobenzene

(e) The aryl halide with nitro groups ortho and para to the bromide leaving group is more reactive than the aryl halide with only one nitro group.

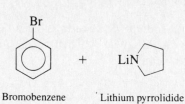

1-Bromo-2,4-dinitrobenzene is more reactive than 1,4-Dibromo-2-nitrobenzene

25.14 (a) The nucleophile is the lithium salt of pyrrolidine, which reacts with bromobenzene by an elimination-addition mechanism.

Bromobenzene + Lithium pyrrolidide ⟶ *N*-Phenylpyrrolidine (observed yield, 84%) + LiBr

(b) The nucleophile in this case is piperidine. The substrate, 1-bromo-2,4-dinitrobenzene, is very reactive in nucleophilic aromatic substitution by the addition-elimination mechanism.

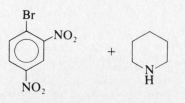

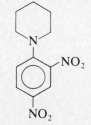

1-Bromo-2,4-dinitrobenzene + Piperidine ⟶ *N*-(2,4-Dinitrophenyl)piperidine

(*c*) Of the two bromine atoms, one is ortho and the other meta to the nitro group. Nitro groups activate positions ortho and para to themselves toward nucleophilic aromatic substitution, so it will be the bromine ortho to the nitro group that is displaced.

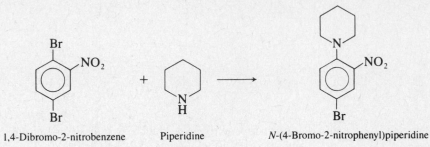

1,4-Dibromo-2-nitrobenzene + Piperidine → *N*-(4-Bromo-2-nitrophenyl)piperidine

25.15 Since isomeric products are formed by reaction of 1- and 2-bromonaphthalene with piperidine at elevated temperatures, it is reasonable to conclude that these reactions do not involve a common intermediate and hence follow an addition-elimination pathway. Piperidine acts as a nucleophile and substitutes for bromine on the same carbon atom from which bromine is lost.

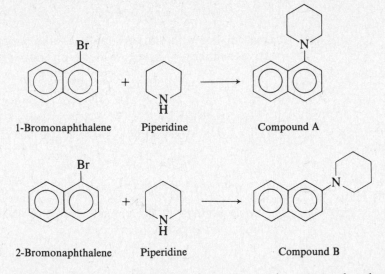

1-Bromonaphthalene + Piperidine → Compound A

2-Bromonaphthalene + Piperidine → Compound B

When the strong base sodium piperidide is used, reaction occurs by the elimination-addition pathway via a "naphthalyne" intermediate. Only one mode of elimination is possible from 1-bromonaphthalene.

1-Bromonaphthalene + sodium piperidide (Na^+) → naphthalyne + piperidine + NaBr

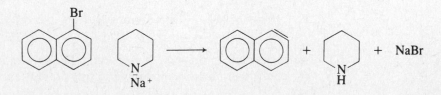

This intermediate can yield both A and B in the addition stage.

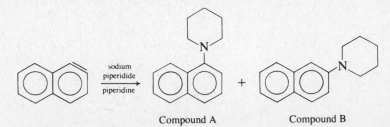

naphthalyne $\xrightarrow[\text{piperidine}]{\text{sodium piperidide}}$ Compound A + Compound B

Two modes of elimination are possible from 2-bromonaphthalene.

elimination stage

addition stage

Compounds A and B

addition stage

Compound B only

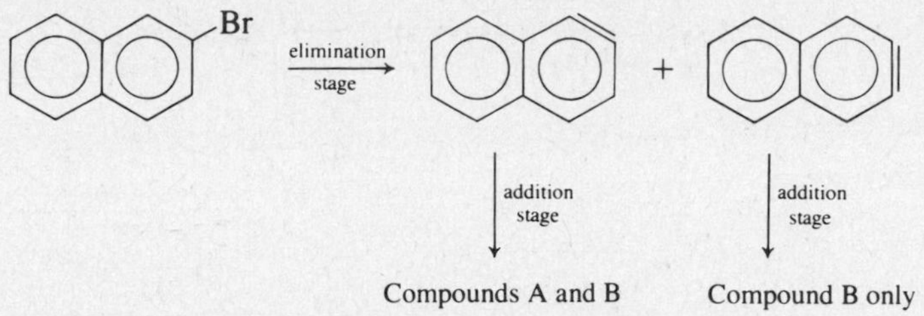

Both naphthalyne intermediates are probably formed from 2-bromonaphthalene since there is no reason to expect elimination to occur only in one direction.

25.16 Reaction of a nitro-substituted aryl halide with a good nucleophile leads to nucleophilic aromatic substitution. Methoxide will displace fluoride from the ring, preferentially at the positions ortho and para to the nitro group.

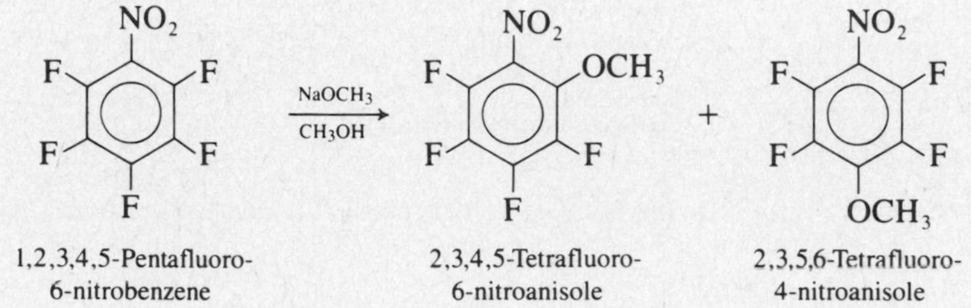

$NaOCH_3$ / CH_3OH

1,2,3,4,5-Pentafluoro-6-nitrobenzene

2,3,4,5-Tetrafluoro-6-nitroanisole

2,3,5,6-Tetrafluoro-4-nitroanisole

25.17 (*a*) Aryl halides undergo nucleophilic aromatic substitution by an elimination-addition mechanism on treatment with strongly basic reagents such as sodium amide. The substrate given in this problem can undergo elimination only in the direction shown.

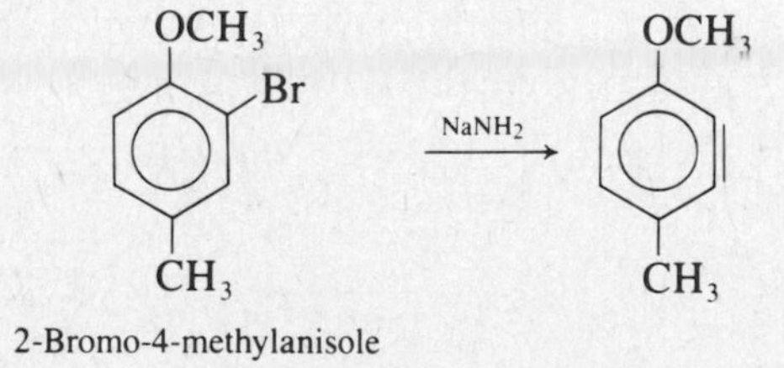

$NaNH_2$

2-Bromo-4-methylanisole

Addition of sodium amide to the aryne intermediate can proceed in two different regioselective senses.

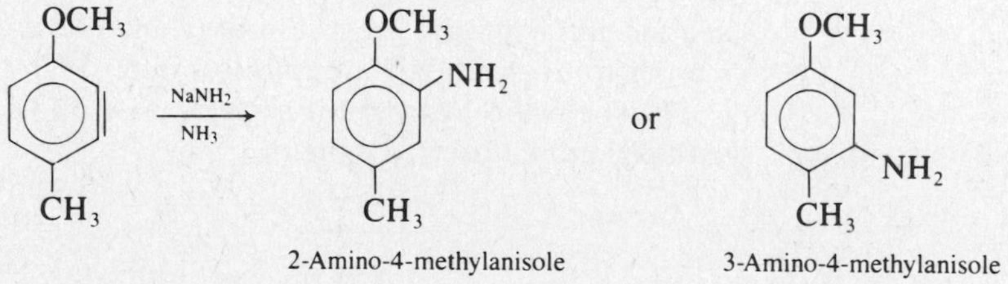

$NaNH_2$ / NH_3

2-Amino-4-methylanisole or 3-Amino-4-methylanisole

A methoxyl group will stabilize a carbanionic site ortho to itself better than a methyl group will, as explained in text Section 25.7.

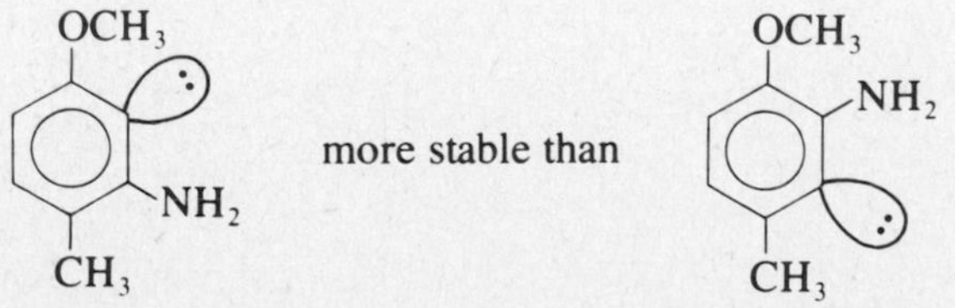

more stable than

Therefore amide ion adds faster to the position meta to the methoxyl group, and 3-amino-4-methylanisole is the product of the reaction. (When the reaction was actually carried out, 3-amino-4-methylanisole was the only product and was isolated in 56 percent yield.)

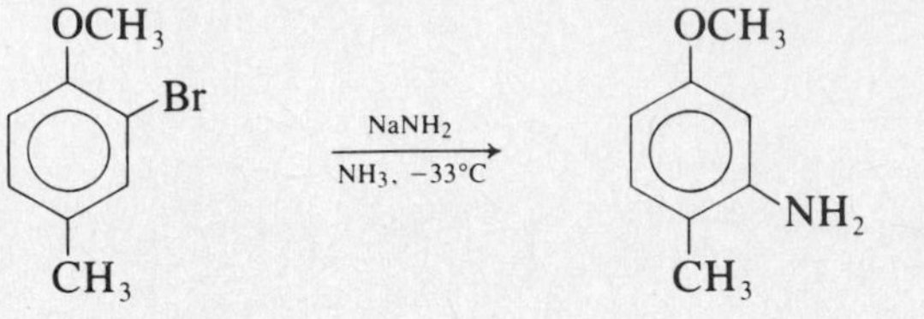

2-Bromo-4-methylanisole $\xrightarrow[NH_3,\ -33°C]{NaNH_2}$ 3-Amino-4-methylanisole

(*b*) Here again only a single benzyne intermediate is formed. Amide adds meta to the methoxy group.

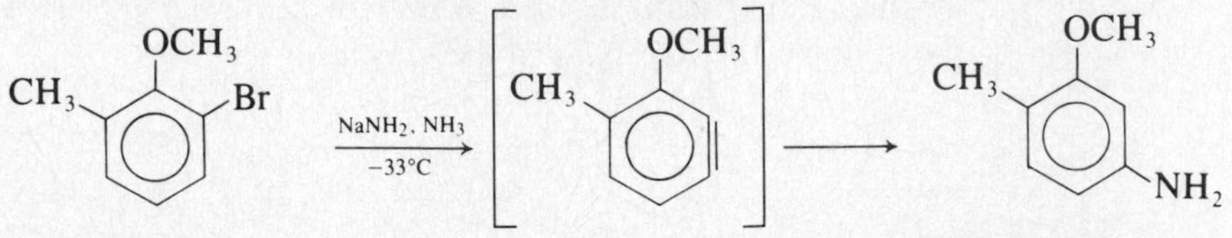

2-Bromo-6-methylanisole $\xrightarrow[-33°C]{NaNH_2,\ NH_3}$ [benzyne] $\longrightarrow$ 5-Amino-2-methylanisole (observed yield 30%)

(*c*) In the elimination stage of this reaction, a benzyne intermediate is formed.

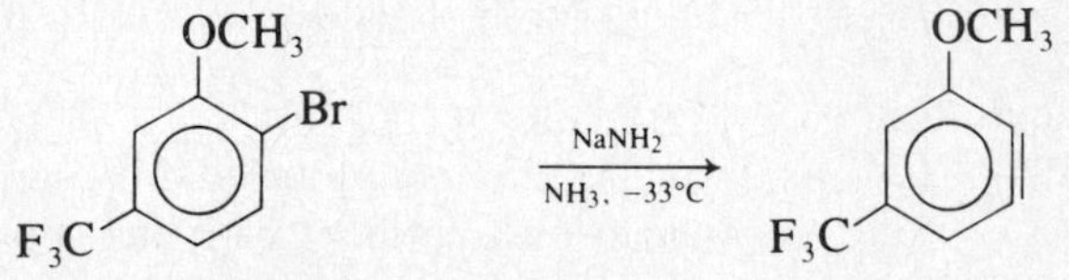

2-Bromo-5-(trifluoromethyl)anisole $\xrightarrow[NH_3,\ -33°C]{NaNH_2}$ [benzyne]

In the addition stage amide adds meta to the methoxy group to produce 3-amino-5-(trifluoromethyl)anisole.

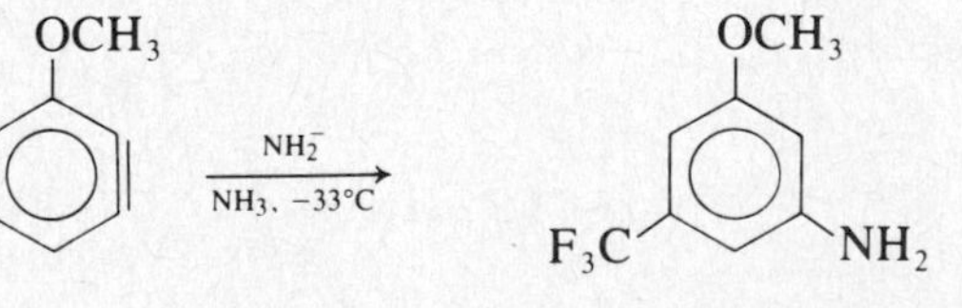

[benzyne] $\xrightarrow[NH_3,\ -33°C]{NH_2^-}$ 3-Amino-5-(trifluoromethyl)anisole (observed yield 71%)

(*d*) In this example two different benzyne intermediates are possible, depending on which chlorine and hydrogen are lost in the elimination stage of the reaction. The proton that is ortho to the methoxyl group is lost faster than the one that is meta to the methoxyl group. The ortho hydrogen is slightly more acidic, as the result of the influence of the methoxyl group, referred to in (*a*).

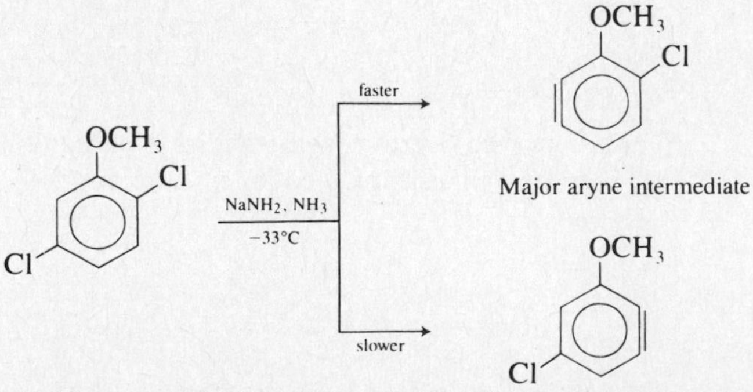

$\xrightarrow[-33°C]{NaNH_2,\ NH_3}$ faster: Major aryne intermediate; slower: Minor aryne intermediate

As we have seen in the three previous examples, arynes of this type tend to undergo addition at the position meta to the methoxyl group.

OCH_3, Cl; $\xrightarrow[-33°C]{NaNH_2, NH_3}$; OCH_3, Cl, H_2N

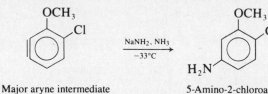

Major aryne intermediate — 5-Amino-2-chloroanisole (observed yield 78%)

(*e*) This reaction is one of nucleophilic aromatic substitution by the addition-elimination mechanism.

Cl, NO_2, CH_3 + $C_6H_5CH_2SK$ $\longrightarrow$ $SCH_2C_6H_5$, NO_2, CH_3

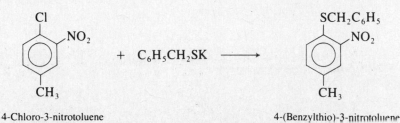

4-Chloro-3-nitrotoluene — 4-(Benzylthio)-3-nitrotoluene

The nucleophile, $C_6H_5CH_2S^-$, displaces chloride directly from the aromatic ring. The product in this case was isolated in 57 percent yield.

(*f*) The nucleophile, hydrazine, will react with 1-chloro-2,4-dinitrobenzene by an addition-elimination mechanism as shown.

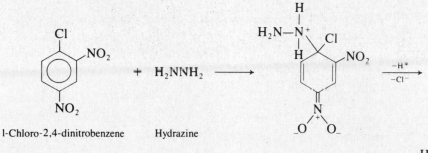

Cl, NO_2, NO_2 + H_2NNH_2 $\longrightarrow$ $H_2N{-}N^+H_2$, Cl, NO_2, N^+, O^-, O^- $\xrightarrow[-Cl^-]{-H^+}$ $H_2N{-}\ddot{N}H$, NO_2, NO_2

1-Chloro-2,4-dinitrobenzene — Hydrazine — 2,4-Dinitrophenylhydrazine

The nitrogen atom of hydrazine has an unshared electron pair and is fairly nucleophilic. The product, 2,4-dinitrophenylhydrazine, is formed in quantitative yield.

(*g*) The problem requires you to track the starting material through two transformations. The first of these is nitration of *m*-dichlorobenzene, an electrophilic aromatic substitution reaction.

Cl, Cl $\xrightarrow[H_2SO_4]{HNO_3}$ Cl, Cl, NO_2

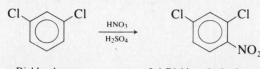

m-Dichlorobenzene — 2,4-Dichloro-1-nitrobenzene

Since the final product of the sequence has four nitrogen atoms ($C_6H_6N_4O_4$), 2,4-dichloro-1-nitrobenzene is an unlikely starting material for the second transformation.

Stepwise nucleophilic aromatic substitution of both chlorines is possible but leads to a compound with the wrong molecular formula ($C_6H_7N_3O_2$).

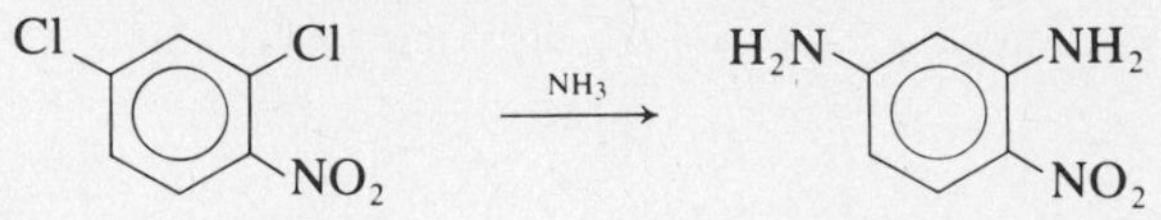

Cl Cl NO$_2$ —NH$_3$→ H$_2$N NH$_2$ NO$_2$

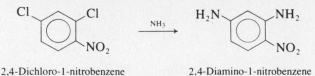

2,4-Dichloro-1-nitrobenzene 2,4-Diamino-1-nitrobenzene

In order to obtain a final product with the correct molecular formula, the original nitration reaction must lead not to a mononitro but to a dinitro derivative. This is reasonable in view of the fact that this reaction is carried out at elevated temperature (120°C).

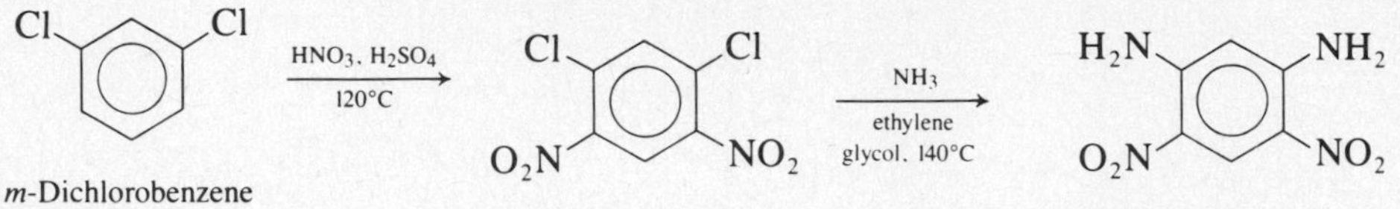

Cl Cl —HNO_3, H_2SO_4, 120°C→ Cl Cl O$_2$N NO$_2$ —NH$_3$, ethylene glycol, 140°C→ H$_2$N NH$_2$ O$_2$N NO$_2$

m-Dichlorobenzene

1,5-Diamino-2,4-dinitrobenzene ($C_6H_6N_4O_4$)

The above two-step sequence has been carried out with product yields of 70 to 71 percent in the first step and 88 to 95 percent in the second step.

(*h*) This problem also involves two transformations, nitration and nucleophilic aromatic substitution. Nitration will take place ortho to chlorine (meta to trifluoromethyl).

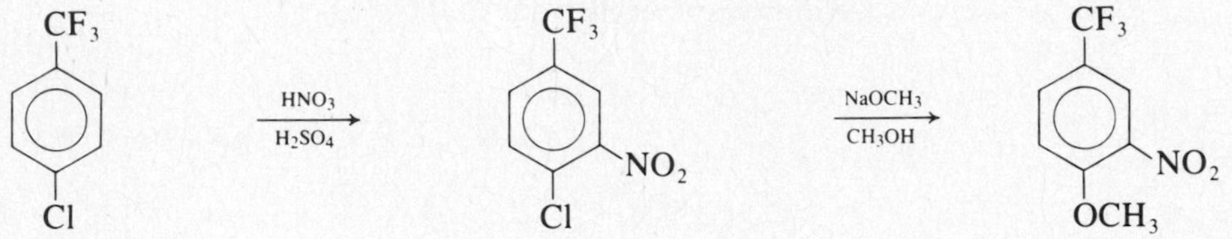

CF$_3$ Cl —HNO_3, H_2SO_4→ CF$_3$ NO$_2$ Cl —$NaOCH_3$, CH_3OH→ CF$_3$ NO$_2$ OCH$_3$

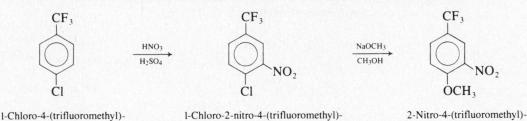

1-Chloro-4-(trifluoromethyl)-benzene 1-Chloro-2-nitro-4-(trifluoromethyl)-benzene 2-Nitro-4-(trifluoromethyl)-anisole

25.18 The reaction of *p*-bromotoluene with aqueous sodium hydroxide at elevated temperature proceeds by way of a benzyne intermediate.

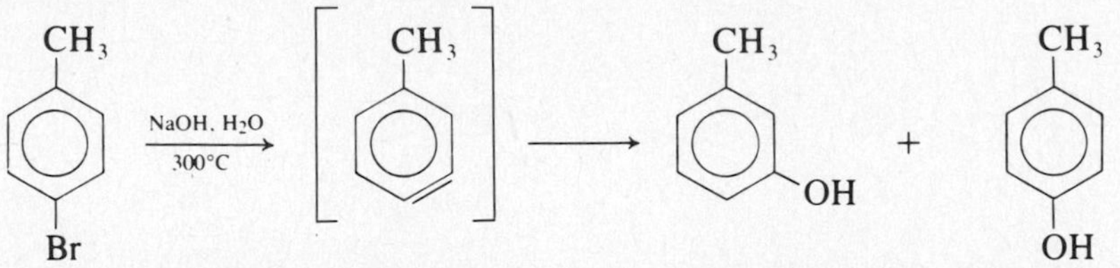

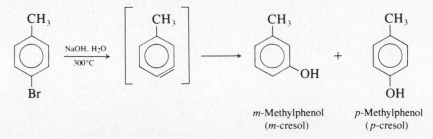

CH$_3$ Br —NaOH, H_2O, 300°C→ [CH$_3$] ⟶ CH$_3$ OH + CH$_3$ OH

m-Methylphenol (*m*-cresol) *p*-Methylphenol (*p*-cresol)

The same benzyne intermediate is formed when *p*-chlorotoluene is the reactant, so the product ratio must be identical regardless of whether the leaving group is bromide or chloride.

25.19 (*a*) Benzyne is formed by loss of nitrogen and carbon dioxide.

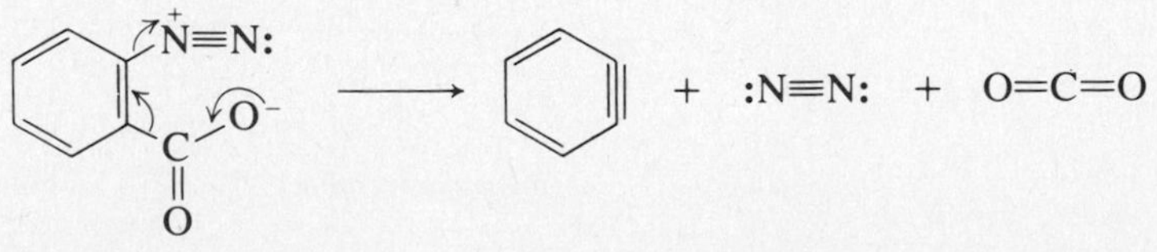

$\overset{+}{N}$≡N: C O O$^-$ ⟶ + :N≡N: + O=C=O

Benzenediazonium-2-carboxylate Benzyne Nitrogen Carbon dioxide (*b*) Organolithium compounds are formed by the reaction of lithium with alkyl and aryl halides. Bromides react much faster than fluorides.

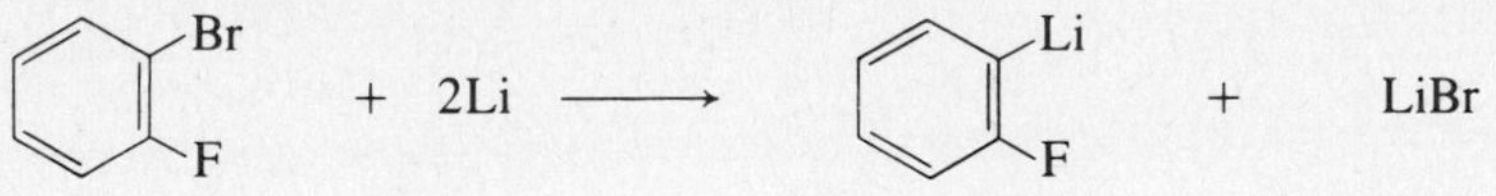

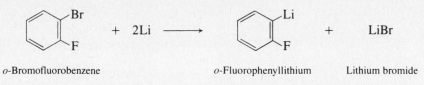

$$o\text{-Bromofluorobenzene} + 2Li \longrightarrow o\text{-Fluorophenyllithium} + LiBr \text{ (Lithium bromide)}$$

This organolithium derivative loses lithium fluoride to form benzyne.

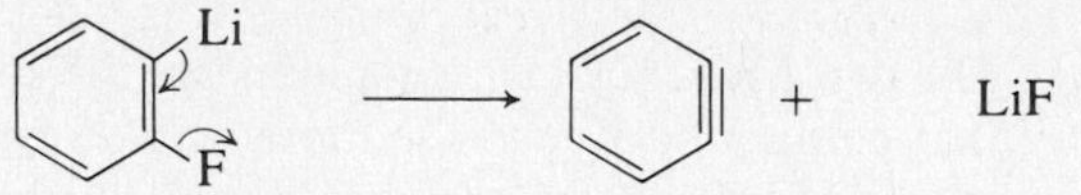

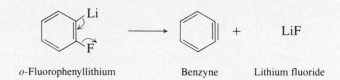

$$o\text{-Fluorophenyllithium} \longrightarrow \text{Benzyne} + LiF \text{ (Lithium fluoride)}$$

25.20 The "triple bond" of benzyne adds to the diene system of furan.

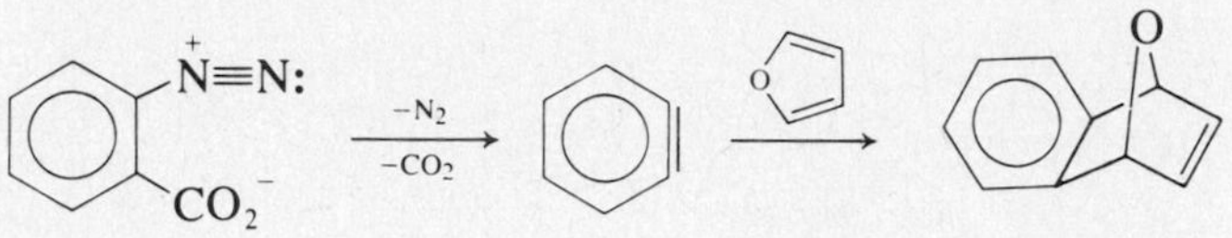

$$\xrightarrow[-CO_2]{-N_2}$$

25.21 (*a*) Ethoxide adds to the aromatic ring to give a cyclohexadienyl anion.

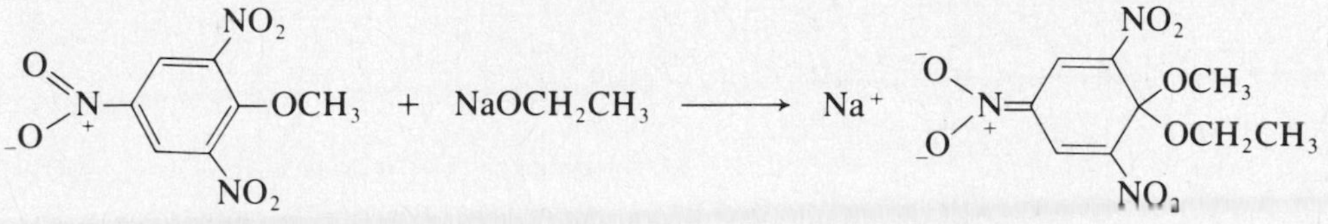

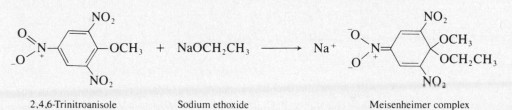

$$\text{2,4,6-Trinitroanisole} + NaOCH_2CH_3 \text{ (Sodium ethoxide)} \longrightarrow Na^+ \text{ Meisenheimer complex}$$

(*b*) The same Meisenheimer complex results when ethyl 2,4,6-trinitrophenyl ether reacts with sodium methoxide.

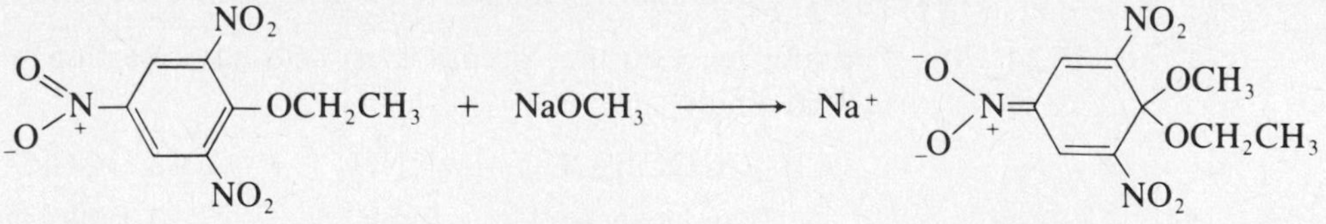

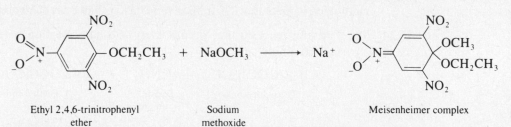

$$\text{Ethyl 2,4,6-trinitrophenyl ether} + NaOCH_3 \text{ (Sodium methoxide)} \longrightarrow Na^+ \text{ Meisenheimer complex}$$

(*c*) The two oxygen substituents on the sp^3 hybridized carbon of the Meisenheimer complex are part of a ring, which most probably arose by an intramolecular cyclization process.

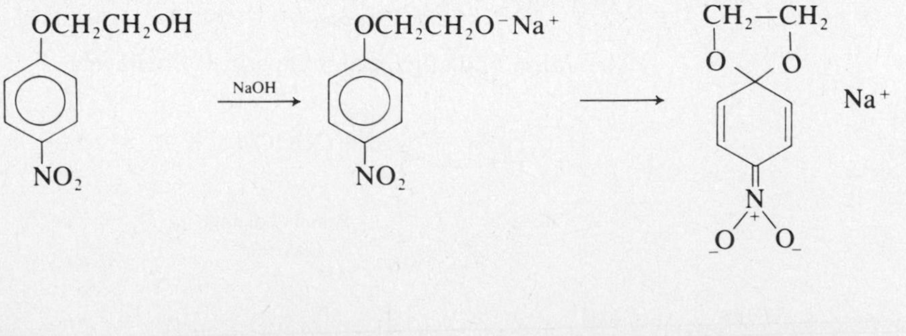

$$OCH_2CH_2OH \xrightarrow{NaOH} OCH_2CH_2O^-Na^+ \longrightarrow Na^+$$

25.22 Methoxide may add to 2,4,6-trinitroanisole either at the ring carbon that bears the methoxyl group or at an unsubstituted ring carbon.

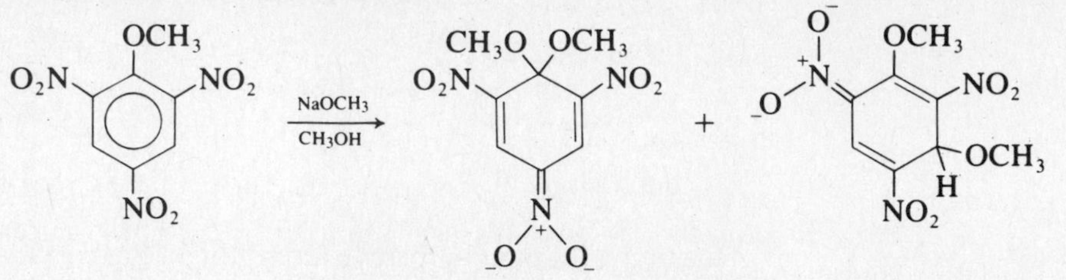

2,4,6-Trinitroanisole $\xrightarrow[CH_3OH]{NaOCH_3}$ A + B

The two Meisenheimer complexes are the sodium salts of the anions shown. It was observed that compound A was the more stable of the two. Compound B was present immediately after adding sodium methoxide to 2,4,6-trinitroanisole but underwent relatively rapid isomerization to A.

25.23 In order to determine whether chlorobenzene or 2-chloropyridine is more reactive in nucleophilic aromatic substitution by the addition-elimination mechanism, compare the relative stability of the intermediates formed in the addition (rate-determining) step.

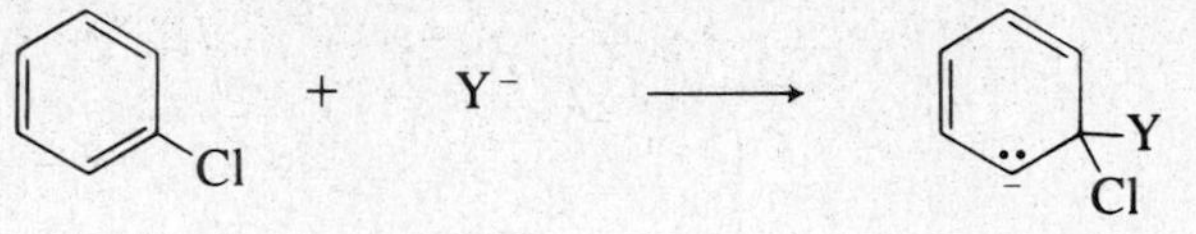

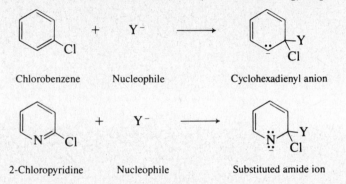

Chlorobenzene + Nucleophile (Y^-) ⟶ Cyclohexadienyl anion

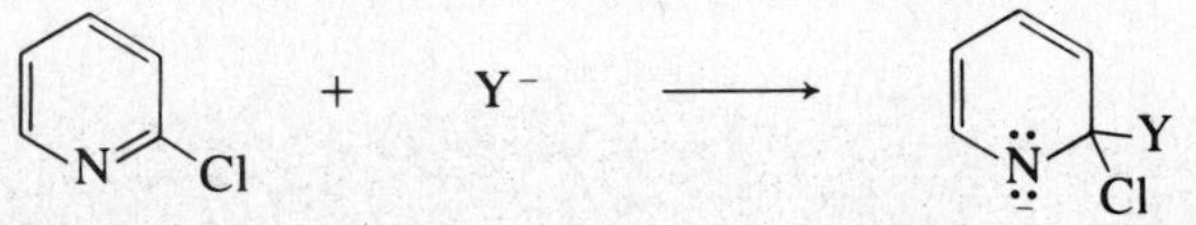

2-Chloropyridine + Nucleophile (Y^-) ⟶ Substituted amide ion

Since nitrogen is more electronegative than carbon, it can support a negative charge better and the anion formed by addition to 2-chloropyridine is more stable than the one formed by addition to chlorobenzene. Therefore, 2-chloropyridine is expected to undergo nucleophilic aromatic substitution by the addition-elimination mechanism faster than does chlorobenzene. This is indeed observed to be the case. At 50°C 2-chloropyridine reacts with sodium methoxide in methanol 230 million times faster than does chlorobenzene.

25.24 (*a*) The first reaction that occurs is an acid-base reaction between diethyl malonate and sodium amide.

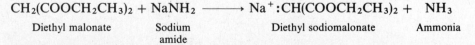

$$CH_2(COOCH_2CH_3)_2 + NaNH_2 \longrightarrow Na^+ :\overset{-}{C}H(COOCH_2CH_3)_2 + NH_3$$

Diethyl malonate; Sodium amide; Diethyl sodiomalonate; Ammonia

A second equivalent of sodium amide converts bromobenzene to benzyne.

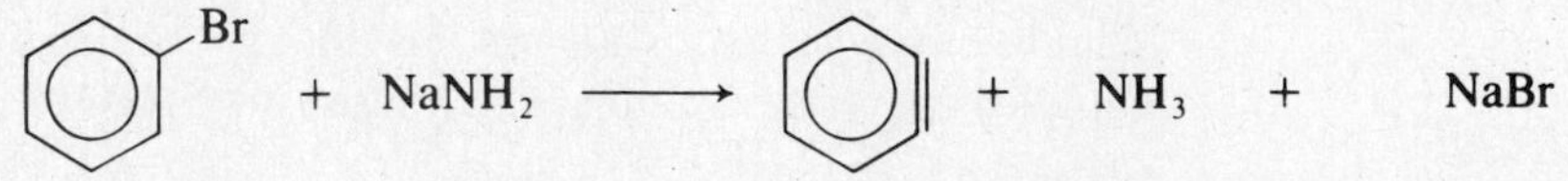

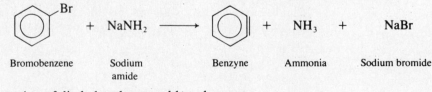

Bromobenzene + $NaNH_2$ (Sodium amide) ⟶ Benzyne + NH_3 (Ammonia) + NaBr (Sodium bromide)

The anion of diethyl malonate adds to benzyne.

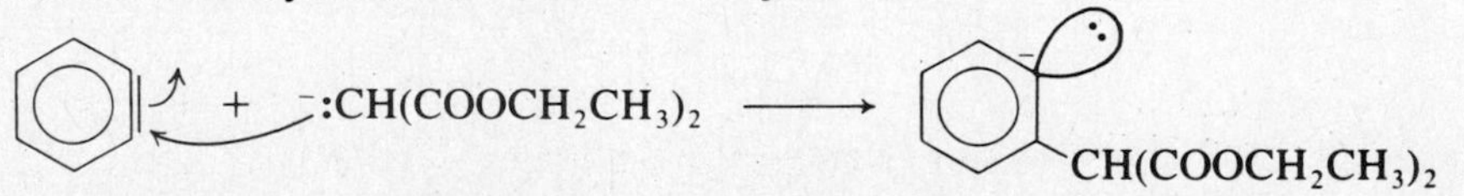

Benzyne + $^-:CH(COOCH_2CH_3)_2$ (Anion of diethyl malonate) ⟶ $C_6H_4^-$–$CH(COOCH_2CH_3)_2$

This anion then abstracts a proton from ammonia to give the observed product.

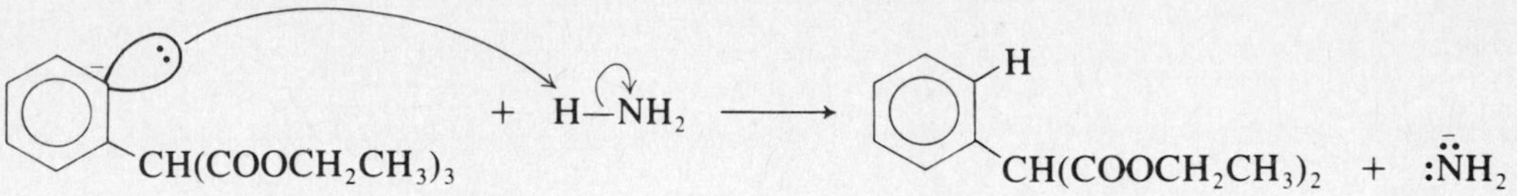

$CH(COOCH_2CH_3)_3$ + $H—NH_2$ ⟶ $CH(COOCH_2CH_3)_2$ + $:\ddot{N}H_2^-$

Ammonia — Diethyl phenylmalonate — Amide anion

(*b*) The ester is deprotonated by the strong base sodium amide, after which the ester enolate undergoes an elimination reaction to form a benzyne intermediate. Cyclization to the final product occurs by intramolecular attack of the ester enolate on the reactive triple bond of the aryne.

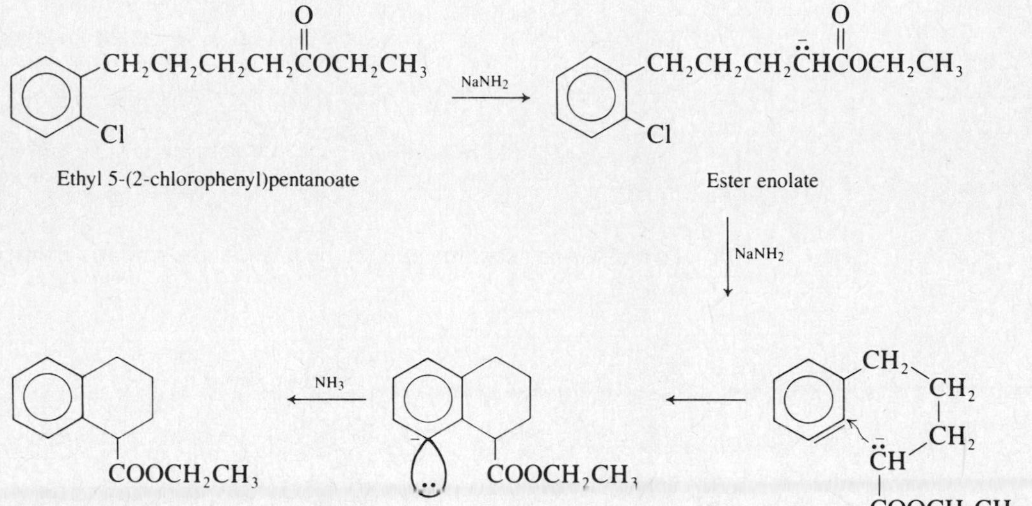

$CH_2CH_2CH_2CH_2COCH_2CH_3$ (Cl) $\xrightarrow{NaNH_2}$ $CH_2CH_2CH_2\ddot{C}HCOCH_2CH_3$ (Cl)

Ethyl 5-(2-chlorophenyl)pentanoate — Ester enolate

NaNH2

Aryne intermediate ⟶ $COOCH_2CH_3$ $\xrightarrow{NH_3}$ $COOCH_2CH_3$

Ethyl 1,2,3,4-tetrahydronaphthalene-1-carboxylate

(*c*) In the presence of very strong bases, aryl halides undergo nucleophilic aromatic substitution by an elimination-addition mechanism. The structure of the product indicates that a nitrogen of the side chain acts as a nucleophile in the addition step.

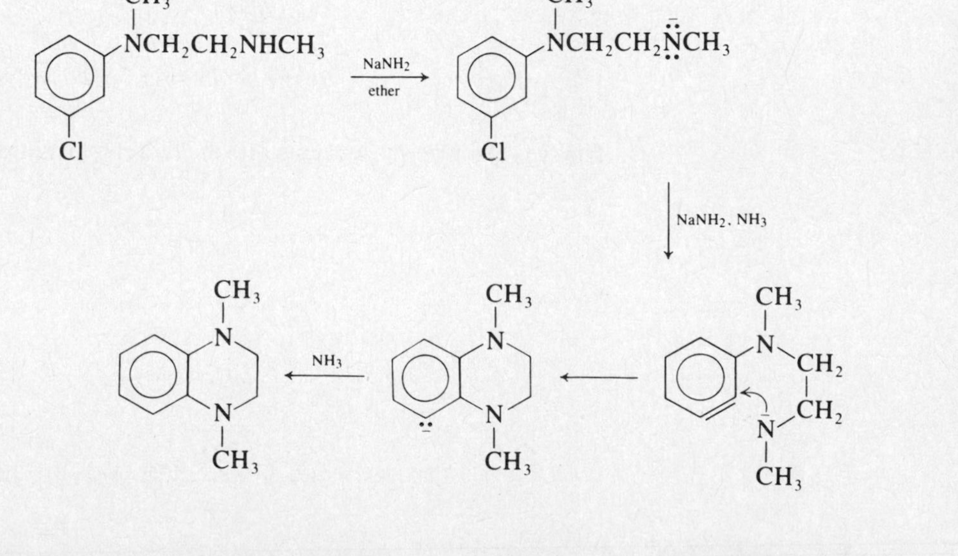

$N(CH_3)CH_2CH_2NHCH_3$ (Cl) $\xrightarrow[\text{ether}]{NaNH_2}$ $N(CH_3)CH_2CH_2\ddot{N}CH_3^-$ (Cl)

NaNH2, NH3

⟶ $\xrightarrow{NH_3}$

25.25 Polychlorinated biphenyls (PCB's) are derived from biphenyl as the base structure. It is numbered as shown.

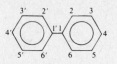

(*a*) There are three monochloro derivatives of biphenyl:

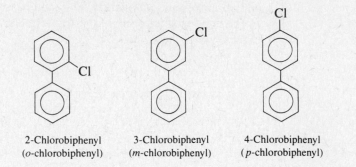

2-Chlorobiphenyl (*o*-chlorobiphenyl) 3-Chlorobiphenyl (*m*-chlorobiphenyl) 4-Chlorobiphenyl (*p*-chlorobiphenyl)

(*b*) Two chlorine substituents may be in the same ring (six isomers):

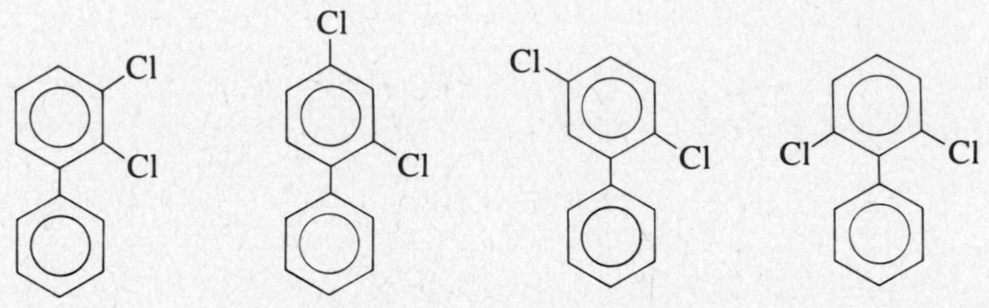

2,3-Dichlorobiphenyl 2,4-Dichlorobiphenyl 2,5-Dichlorobiphenyl 2,6-Dichlorobiphenyl

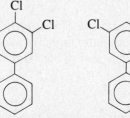

3,4-Dichlorobiphenyl 3,5-Dichlorobiphenyl

The two chlorine substituents may be in different rings (six isomers):

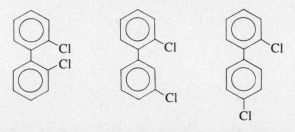

2,2′-Dichlorobiphenyl 2,3′-Dichlorobiphenyl 2,4′-Dichlorobiphenyl

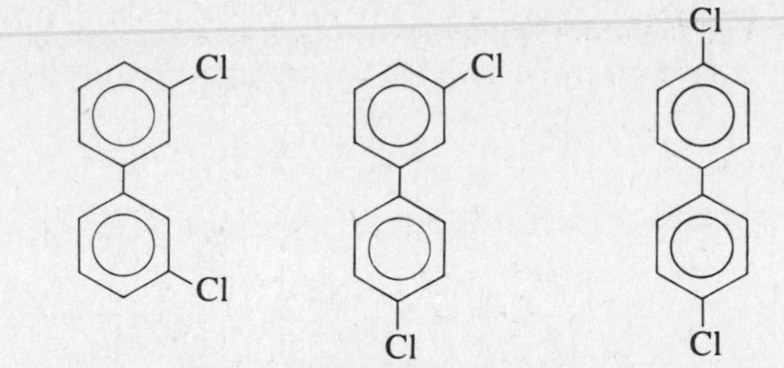

[Structures: 3,3′-Dichlorobiphenyl, 3,4′-Dichlorobiphenyl, 4,4′-Dichlorobiphenyl]

3,3′-Dichlorobiphenyl 3,4′-Dichlorobiphenyl 4,4′-Dichlorobiphenyl

Therefore there are a total of 12 isomeric dichlorobiphenyls.

(*c*) The number of octachlorobiphenyls will be equal to the number of dichlorobiphenyls (12). In both cases we are dealing with a situation in which 8 of the 10 substituents of the biphenyl system are the same and considering how the remaining two may be arranged. In the dichlorobiphenyls described in (*b*), eight substituents are hydrogen and two are chlorine; in the octachlorobiphenyls, eight substituents are chlorine while two are hydrogen.

(*d*) The number of nonachloro isomers (nine chlorine substituents, one hydrogen substituent) must equal the number of monochloro isomers (one chlorine substituent, nine hydrogen substituents). Therefore, there are three nonachloro derivatives of biphenyl.

25.26 The principal isotopes of chlorine are ^{35}Cl and ^{37}Cl. A cluster of five peaks indicates that DDE contains *four* chlorines.

^{35}Cl	^{35}Cl	^{35}Cl	^{35}Cl
^{35}Cl	^{35}Cl	^{35}Cl	^{37}Cl
^{35}Cl	^{35}Cl	^{37}Cl	^{37}Cl
^{35}Cl	^{37}Cl	^{37}Cl	^{37}Cl
^{37}Cl	^{37}Cl	^{37}Cl	^{37}Cl

The peak at m/z 316 therefore corresponds to a compound $C_{14}H_8Cl_4$ in which all four chlorines are ^{35}Cl. The respective molecular formulas indicate that DDE is the dehydrochlorination product of DDT.

$$\underset{\text{DDT}}{C_{14}H_9Cl_5} \xrightarrow{-HCl} \underset{\text{DDE}}{C_{14}H_8Cl_4}$$

The structure of DDT was given in the statement of the problem. This permits the structure of DDE to be assigned.

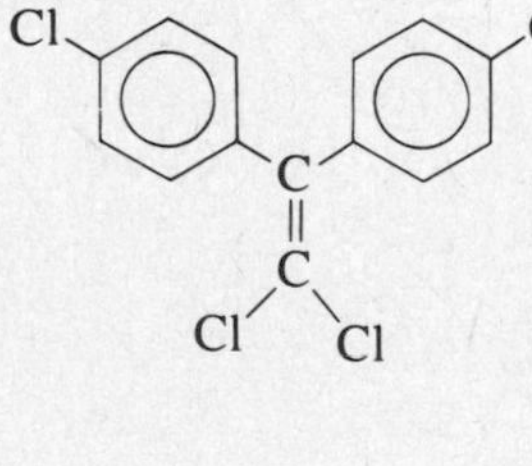

$$(p\text{-}ClC_6H_4)_2C{=}CCl_2$$

DDE (only reasonable dehydrochlorination product of DDT)

SELF-TEST

PART A

A-1. Give the product(s) obtained from each of the following reactions:

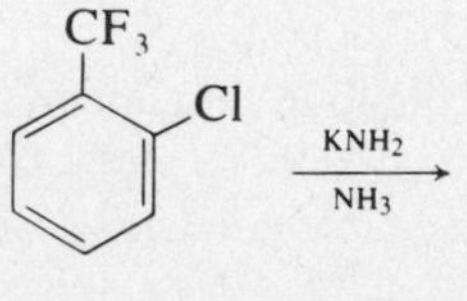

(*a*) 1-chloro-2-(trifluoromethyl)benzene (CF_3, Cl) $\xrightarrow[NH_3]{KNH_2}$? (major) (*b*) 3,4-Dichloronitrobenzene (Cl, Cl, NO_2) $\xrightarrow[CH_3OH]{CH_3O^-}$? (monosubstitution)

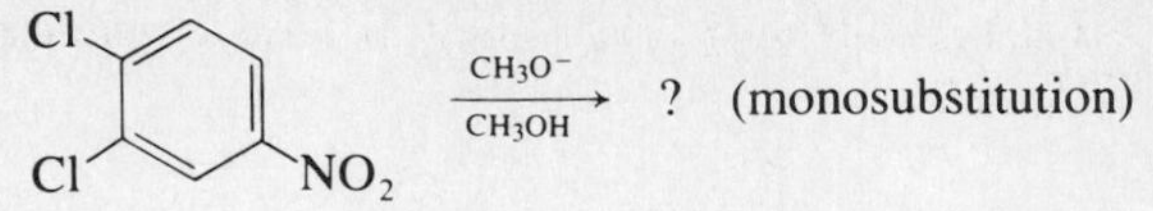

(*c*) 1-*tert*-Butyl-4-iodobenzene ($C(CH_3)_3$, I) $\xrightarrow[NH_3]{NaNH_2}$? (two products)

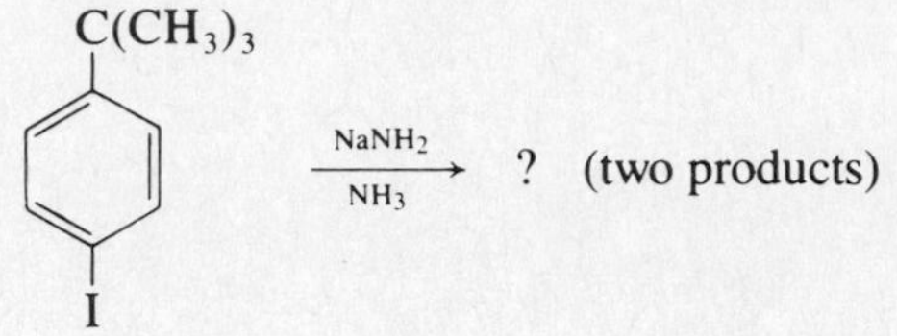

A-2. Suggest synthetic schemes by which chlorobenzene may be converted into:
(*a*) 2,4-Dinitroanisole (1-methoxy-2,4-dinitrobenzene)
(*b*) *p*-Isopropylaniline

A-3. Write a mechanism using resonance structures to show how a nitro group directs ortho, para in a nucleophilic substitution reaction.

PART B

B-1. The reaction

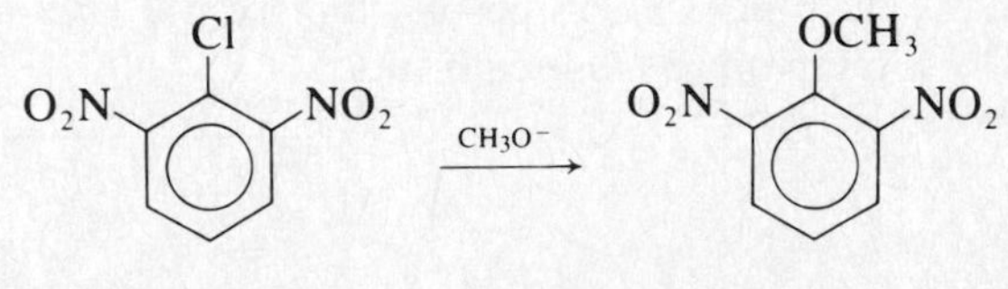

2-Chloro-1,3-dinitrobenzene (Cl, O_2N, NO_2) $\xrightarrow{CH_3O^-}$ 2-methoxy-1,3-dinitrobenzene (OCH_3, O_2N, NO_2)

most likely occurs by which of the following mechanisms?
(*a*) Addition-elimination
(*b*) Elimination-addition
(*c*) Both (*a*) and (*b*)
(*d*) Neither of these

B-2. Rank the following in order of decreasing rate of reaction with ethoxide ion ($CH_3CH_2O^-$) in a nucleophilic aromatic substitution reaction:

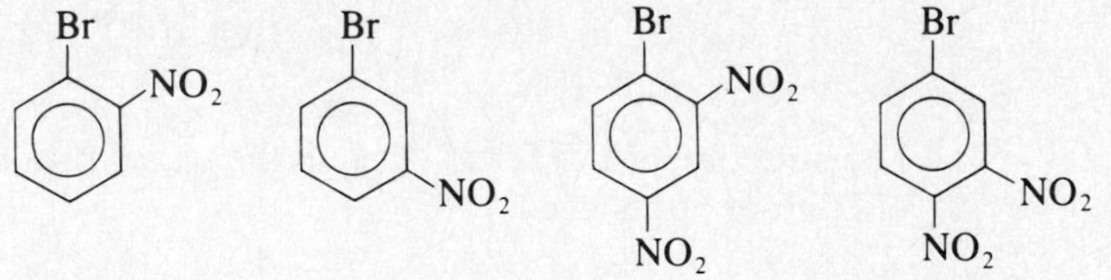

A: Br, NO_2 (ortho)
B: Br, NO_2 (meta)
C: Br, NO_2 (ortho), NO_2 (para)
D: Br, NO_2 (meta), NO_2 (para)

(*a*) C > D > A > B
(*b*) B > A > D > C
(*c*) C > D > B > A
(*d*) D > C > B > A

B-3. The following reaction

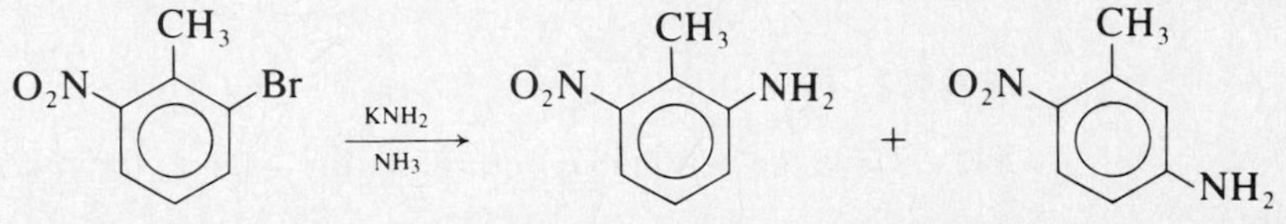

2-Bromo-6-nitrotoluene (CH_3, O_2N, Br) $\xrightarrow[NH_3]{KNH_2}$ 2-methyl-3-nitroaniline (CH_3, O_2N, NH_2) + 3-methyl-4-nitroaniline (CH_3, O_2N, NH_2)

most likely involves which of the following aromatic substitution mechanisms?
(*a*) Addition-elimination
(*b*) Electrophilic substitution
(*c*) Elimination-addition
(*d*) Both (*a*) and (*c*)

B-4. Rank the following aryl halides in order of increasing rate of reaction with CH_3O^- in CH_3OH:

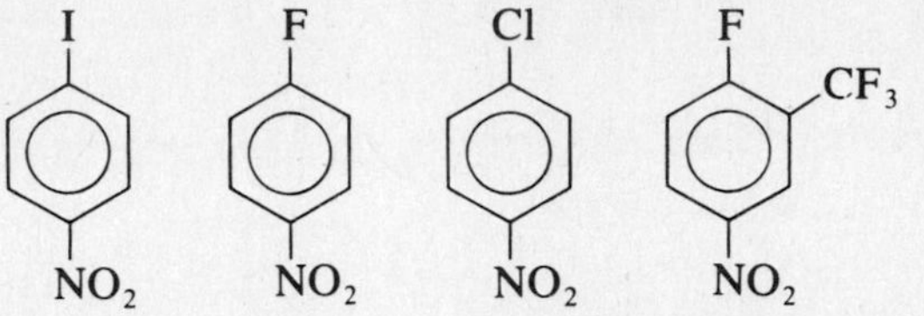

A B C D

(*a*) A < C < D < B (*b*) D < B < C < A
(*c*) B < D < C < A (*d*) A < C < B < D

B-5. Which of the following compounds gives a single benzyne intermediate on reaction with sodium amide?

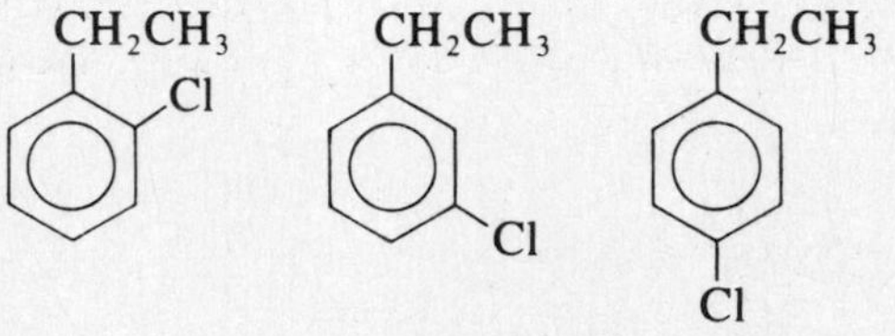

I II III

(*a*) I only (*b*) I and III
(*c*) III only (*d*) I and II

CHAPTER

26

PHENOLS

IMPORTANT TERMS AND CONCEPTS

Structure and Bonding (Sec. 26.2) The carbon-oxygen bond of phenol is somewhat shorter than that of an alcohol for two reasons. The oxygen in phenol is bonded to an sp^2 hybridized carbon of the aromatic ring and the π system of the ring is conjugated with one of the unshared pairs of electrons on the oxygen, which results in partial double bond character.

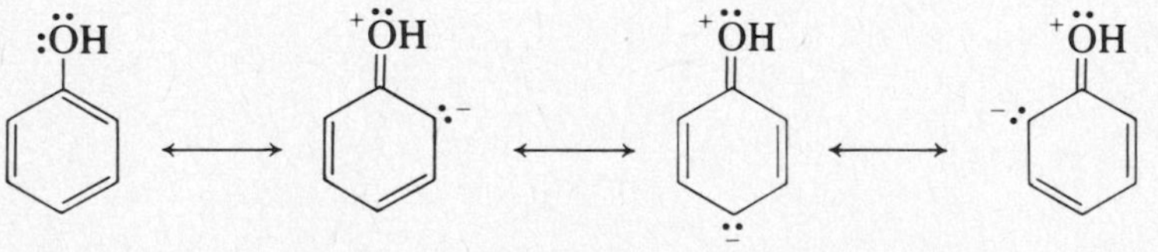

:ÖH ⟷ $^+$ÖH ⟷ $^+$ÖH ⟷ $^+$ÖH

The dipolar resonance forms of phenol also explain the polarity of phenol, which is opposite in direction to that of an alcohol such as methanol.

CH_3—OH — Methanol

—OH — Phenol

Acidity (Secs. 26.4, 26.5) Phenols are approximately 10^6 to 10^{10} times *more* acidic than alcohols. Stabilization of the phenoxide ion and the resulting increased acidity of phenol result from delocalization of the negative charge into the π framework of the aromatic ring.

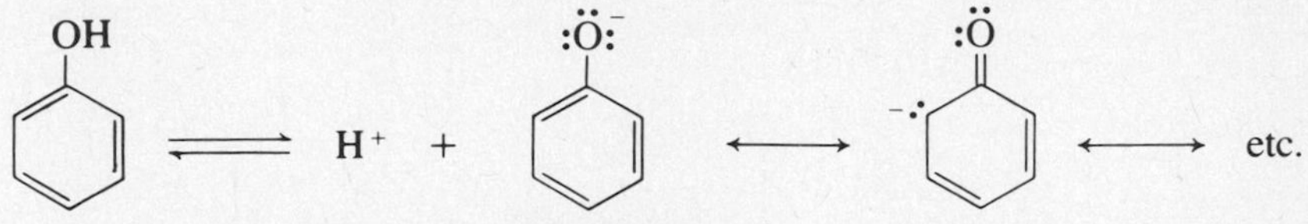

OH ⇌ H^+ + :Ö:$^-$ ⟷ :Ö ⟷ etc.

Phenoxide ion

When strongly electron-withdrawing substituents such as the nitro group are attached to the aromatic ring, substantial increases in acidity of these substituted phenols as against that of phenol itself are noted. For example, *o*- and *p*-nitrophenol are several hundred times more acidic than phenol itself. The largest increase in acidity is noted when the nitro group is directly conjugated with the hydroxyl group, that is, when it is located either ortho or para to the hydroxyl. *m*-Nitrophenol is less acidic than the other nitrophenol isomers; however, it is still more acidic than phenol.

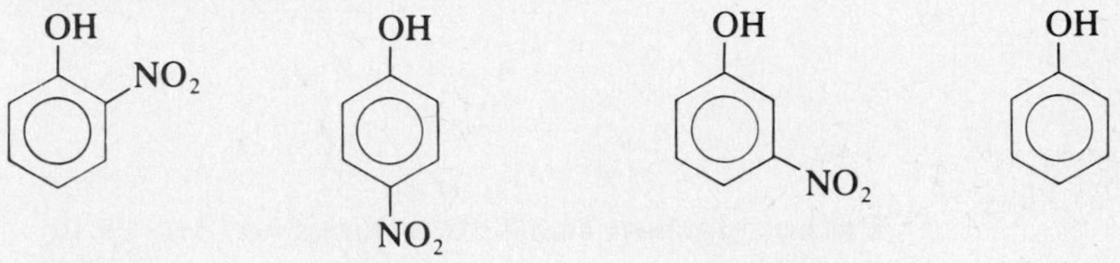

o-Nitrophenol	*p*-Nitrophenol	*m*-Nitrophenol	Phenol
pK_a 7.2	pK_a 7.2	pK_a 8.4	pK_a 10.0

IMPORTANT REACTIONS

Electrophilic Aromatic Substitution (Sec. 26.8)

General:

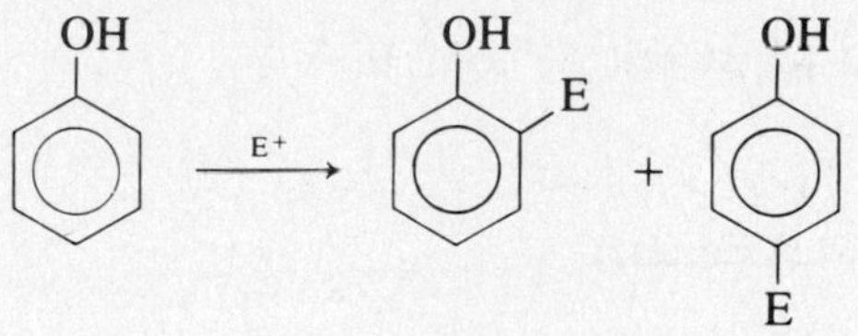

$$C_6H_5OH \xrightarrow{E^+} o\text{-}E\text{-}C_6H_4OH + p\text{-}E\text{-}C_6H_4OH$$

Examples:

$$C_6H_5OH + HNO_3 \xrightarrow[\text{cold}]{H_2O} o\text{-}NO_2C_6H_4OH + p\text{-}NO_2C_6H_4OH$$

$$C_6H_5OH + CH_3\overset{O}{\overset{\|}{C}}Cl \xrightarrow{AlCl_3} o\text{-}HOC_6H_4\overset{O}{\overset{\|}{C}}CH_3 + p\text{-}HOC_6H_4\overset{O}{\overset{\|}{C}}CH_3$$

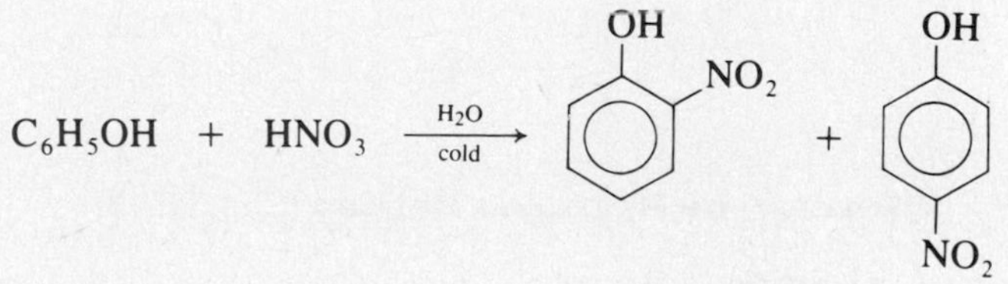

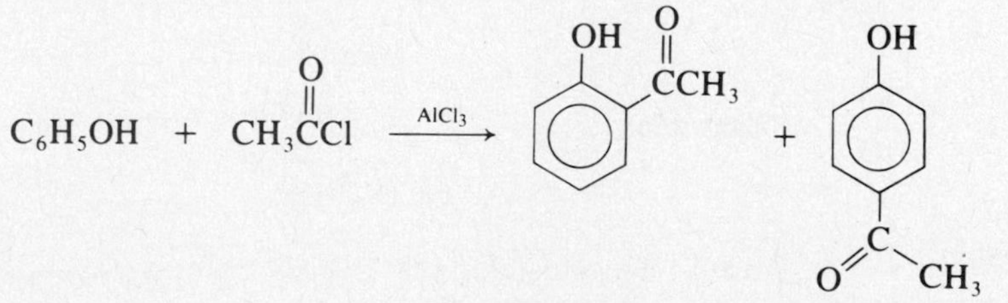

Esterification (Sec. 26.9)

General:

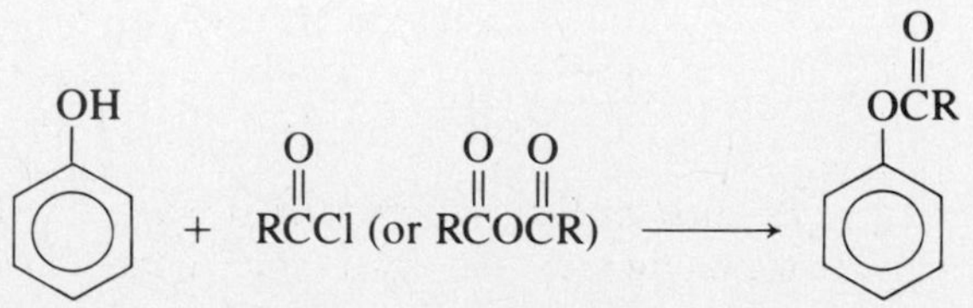

$$C_6H_5OH + R\overset{O}{\overset{\|}{C}}Cl \text{ (or } R\overset{O}{\overset{\|}{C}}O\overset{O}{\overset{\|}{C}}R) \longrightarrow C_6H_5O\overset{O}{\overset{\|}{C}}R$$

Example:

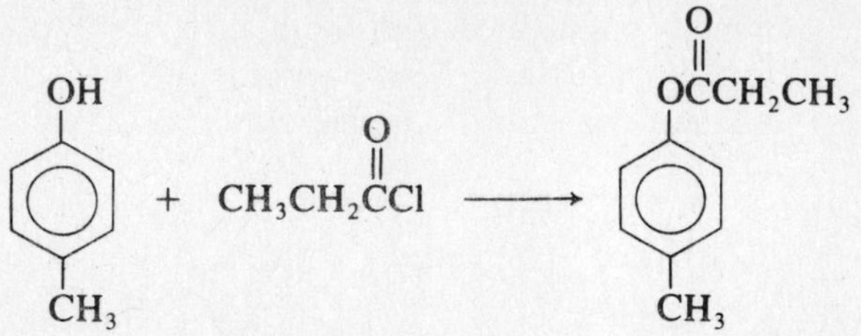

$$p\text{-}CH_3C_6H_4OH + CH_3CH_2\overset{O}{\overset{\|}{C}}Cl \longrightarrow p\text{-}CH_3C_6H_4O\overset{O}{\overset{\|}{C}}CH_2CH_3$$

Carboxylation; The Kolbe Reaction (Sec. 26.10)

Example:

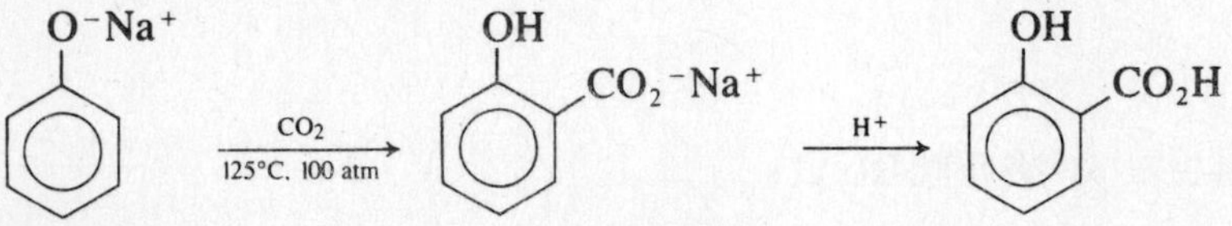

$$C_6H_5O^-Na^+ \xrightarrow[125°C,\ 100\ atm]{CO_2} o\text{-}HOC_6H_4CO_2^-Na^+ \xrightarrow{H^+} o\text{-}HOC_6H_4CO_2H$$

Aryl Ethers

Preparation (Sec. 26.11)

General:

$$ArO^- + RX \longrightarrow ArOR + X^-$$

Example:

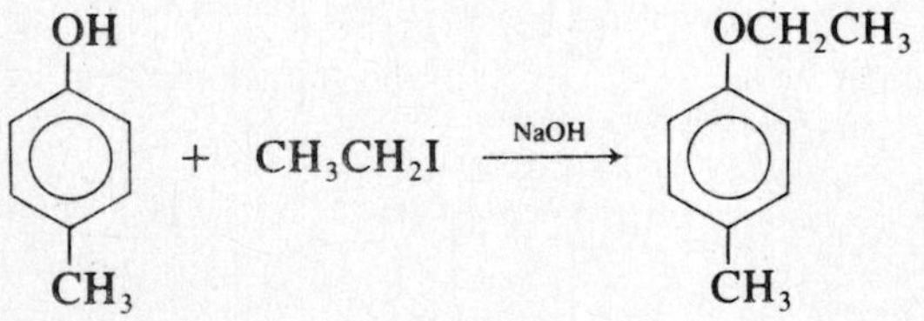

$$p\text{-}CH_3C_6H_4OH + CH_3CH_2I \xrightarrow{NaOH} p\text{-}CH_3C_6H_4OCH_2CH_3$$

Cleavage by Hydrogen Halides (Sec. 26.13)

General:

$$ArOR + HX \longrightarrow ArOH + RX$$

Example:

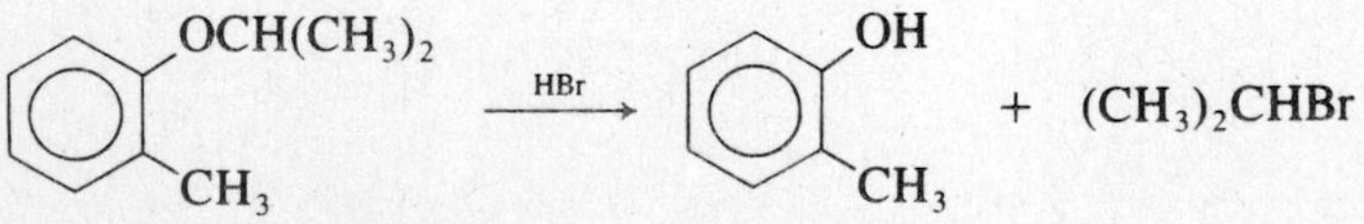

$$o\text{-}CH_3C_6H_4OCH(CH_3)_2 \xrightarrow{HBr} o\text{-}CH_3C_6H_4OH + (CH_3)_2CHBr$$

Oxidation of Phenols; Quinones (Sec. 26.14)

General:

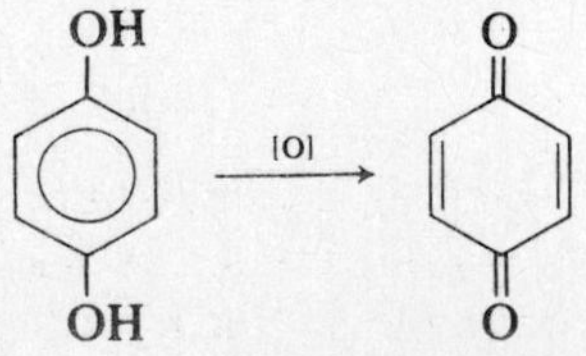

$$p\text{-}HOC_6H_4OH \xrightarrow{[O]} O{=}C_6H_4{=}O$$

Example:

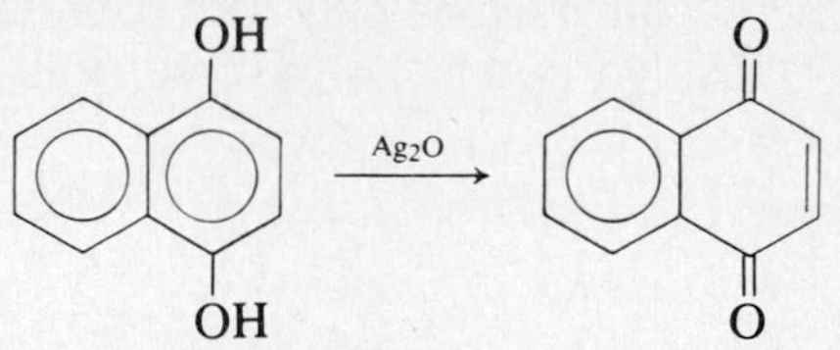

SOLUTIONS TO TEXT PROBLEMS

26.1 (*b*) A benzyl group ($C_6H_5CH_2—$) is ortho to the phenolic hydroxyl group in *o*-benzylphenol.

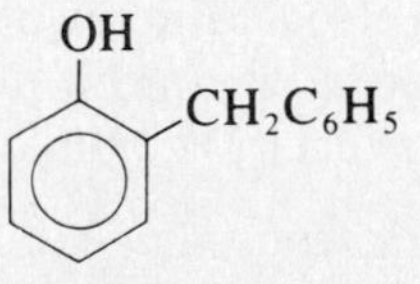

(*c*) Naphthalene is numbered as shown. 3-Nitro-1-naphthol has a hydroxyl group at C-1 and a nitro group at C-3.

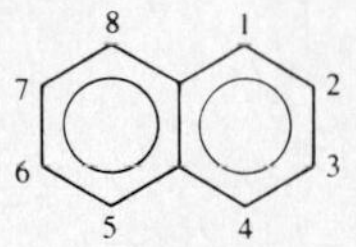

Naphthalene 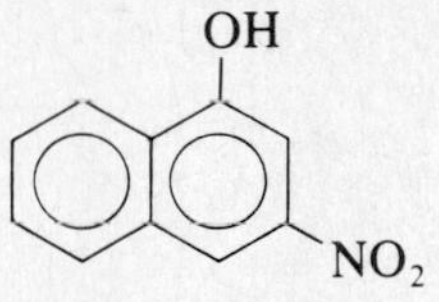 3-Nitro-1-naphthol

(*d*) Resorcinol is 1,3-benzenediol. Therefore 4-chlororesorcinol is

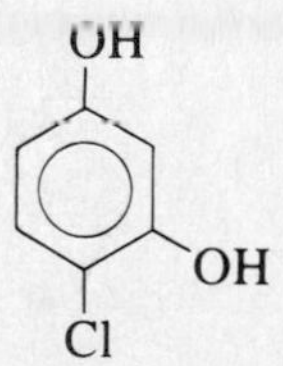

26.2 *p*-Nitrophenol has a larger dipole moment than either phenol or nitrobenzene because the electron-releasing effect of the hydroxyl group and the electron-withdrawing effect of the nitro group reinforce each other. The two groups are directly conjugated; this conjugation can be expressed in resonance terms as shown:

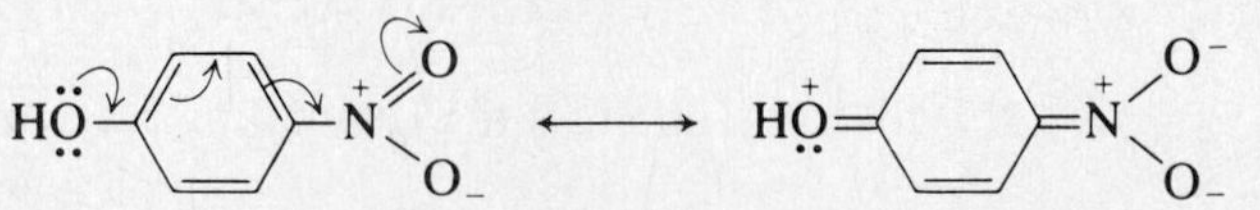

26.3 Intramolecular hydrogen bonding between the hydroxyl group and the ester carbonyl can occur when these groups are ortho to one another.

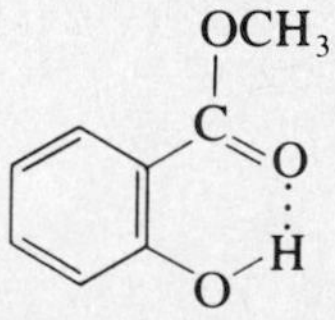

Methyl salicylate

Intramolecular hydrogen bonds form at the expense of intermolecular ones, and intramolecularly hydrogen-bonded species have lower boiling points than intermolecularly hydrogen-bonded ones of the same molecular composition.

26.4 (*b*) A cyano group withdraws electrons from the ring by resonance. A *p*-cyano substituent is conjugated directly with the negatively charged oxygen and stabilizes the anion more than does an *m*-cyano substituent.

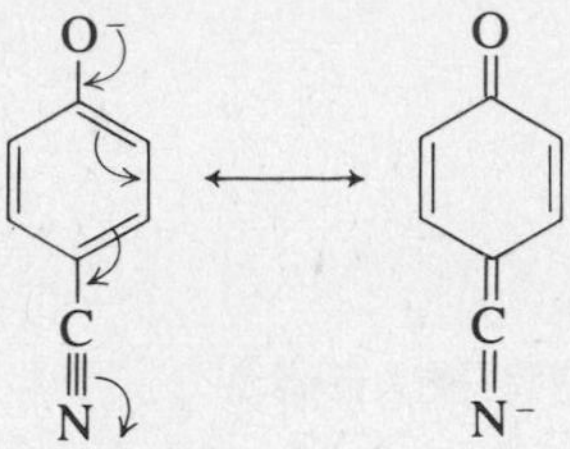

p-Cyanophenol is slightly more acidic than *m*-cyanophenol, the K_a values being 1.0×10^{-8} and 2.8×10^{-9}, respectively.

(*c*) The electron-withdrawing inductive effect of the fluorine substituent will be more pronounced at the ortho position than at the para. *o*-Fluorophenol (K_a 1.9×10^{-9}) is a stronger acid than *p*-fluorophenol (K_a 1.3×10^{-10}).

26.5 The text points out that the reaction proceeds by the addition-elimination mechanism of nucleophilic aromatic substitution.

Under the strongly basic conditions of the reaction, *p*-toluenesulfonic acid is first converted to its anion.

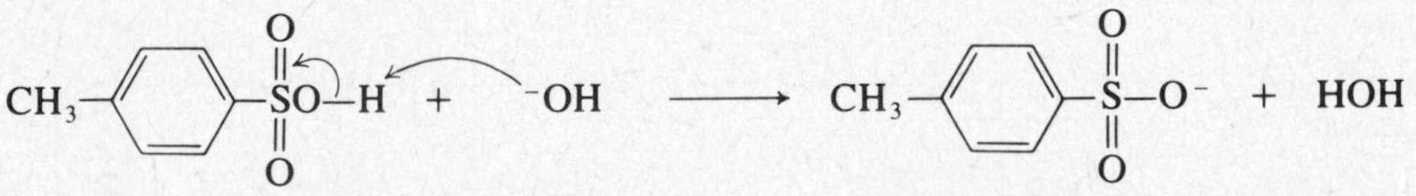

p-Toluenesulfonic acid + Hydroxide ion ⟶ *p*-Toluenesulfonate ion + Water

Nucleophilic addition of hydroxide ion gives a cyclohexadienyl anion intermediate.

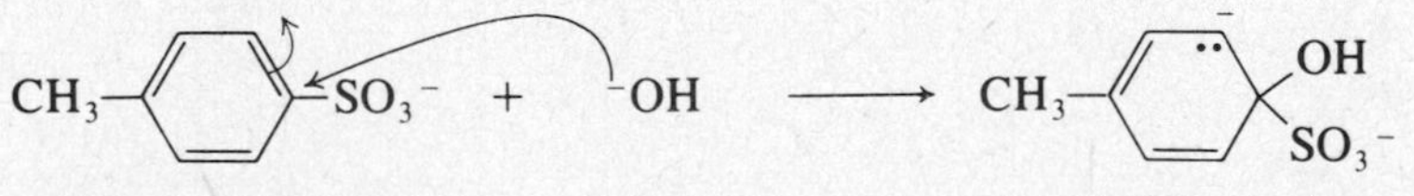

p-Toluenesulfonate ion + Hydroxide ⟶ Cyclohexadienyl anion

Loss of sulfite ion (SO_3^{2-}) gives *p*-cresol.

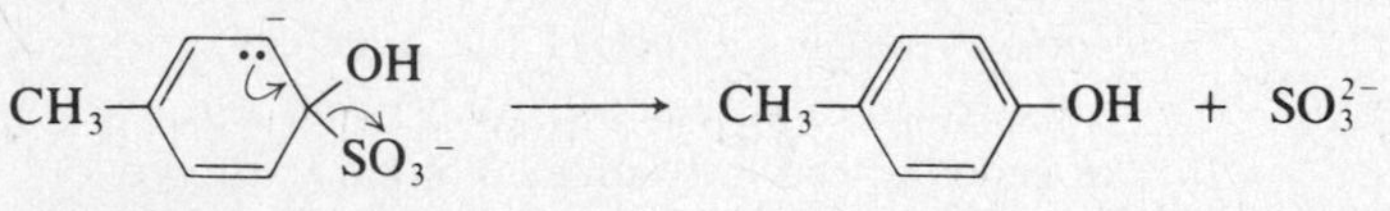

Cyclohexadienyl anion ⟶ *p*-Cresol + SO_3^{2-}

It is also possible that the elimination stage of the reaction proceeds as follows:

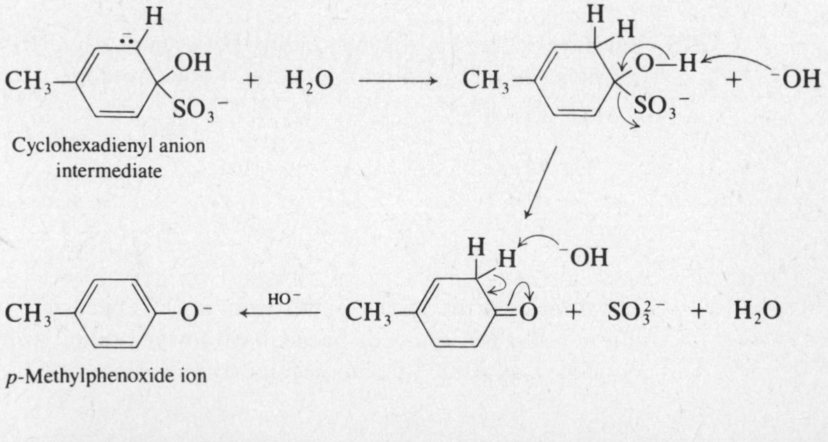

Cyclohexadienyl anion intermediate + H_2O ⟶ ... + ^-OH

... ⟶ ... + SO_3^{2-} + H_2O

p-Methylphenoxide ion

26.6 The text states that the hydrolysis of chlorobenzene in base follows an elimination-addition mechanism.

Chlorobenzene + ^{-}OH ⟶ Benzyne + H_2O + Cl^-

Benzyne + ^{-}OH ⟶ (carbanion)–OH $\xrightarrow{H_2O}$ Phenol

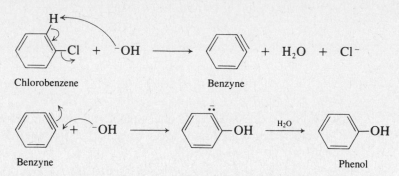

26.7 (*b*) Proton transfer from sulfuric acid to 2-methylpropene gives *tert*-butyl cation. Since the position para to the hydroxyl substituent already bears a bromine, the *tert*-butyl cation attacks the ring at the position ortho to the hydroxyl.

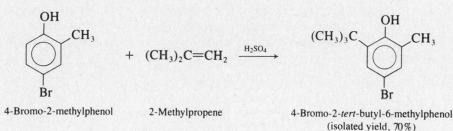

4-Bromo-2-methylphenol + $(CH_3)_2C{=}CH_2$ (2-Methylpropene) $\xrightarrow{H_2SO_4}$ 4-Bromo-2-*tert*-butyl-6-methylphenol (isolated yield, 70%)

(*c*) Acidification of sodium nitrite produces nitrous acid, which nitrosates the strongly activated aromatic ring of phenols.

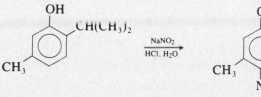

2-Isopropyl-5-methylphenol $\xrightarrow[HCl,\ H_2O]{NaNO_2}$ 2-Isopropyl-5-methyl-4-nitrosophenol (isolated yield, 87%)

(*d*) Friedel-Crafts acylation occurs ortho to the hydroxyl group.

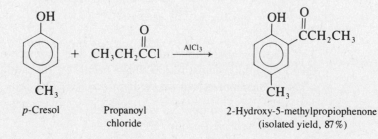

p-Cresol + $CH_3CH_2\overset{O}{\overset{\|}{C}}Cl$ (Propanoyl chloride) $\xrightarrow{AlCl_3}$ 2-Hydroxy-5-methylpropiophenone (isolated yield, 87%)

26.8 (*b*) The hydroxyl group of 2-naphthol is converted to the corresponding acetate ester.

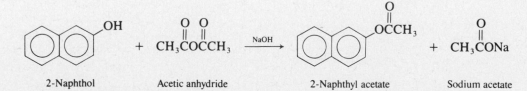

2-Naphthol + $CH_3\overset{O}{\overset{\|}{C}}O\overset{O}{\overset{\|}{C}}CH_3$ (Acetic anhydride) $\xrightarrow{NaOH}$ 2-Naphthyl acetate ($O\overset{O}{\overset{\|}{C}}CH_3$) + $CH_3\overset{O}{\overset{\|}{C}}ONa$ (Sodium acetate)

(*c*) Benzoyl chloride acylates the hydroxyl group of phenol.

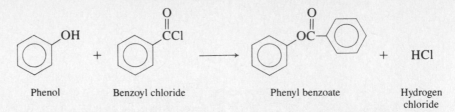

Phenol + Benzoyl chloride (C_6H_5CCl, with C=O) ⟶ Phenyl benzoate ($C_6H_5OC(=O)C_6H_5$) + HCl (Hydrogen chloride)

26.9 Epoxides are sensitive to nucleophilic ring-opening reactions. Phenoxide ion attacks the less hindered carbon atom to yield 1-phenoxy-2-propanol.

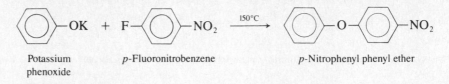

$C_6H_5O^-$ (Phenoxide ion) + $CH_2{-}CHCH_3$ with bridging O (1,2-Epoxypropane) ⟶ $C_6H_5OCH_2CH(OH)CH_3$ (1-Phenoxy-2-propanol)

26.10 The aryl halide must be one that is reactive toward nucleophilic aromatic substitution by the addition-elimination mechanism. *p*-Fluoronitrobenzene is far more reactive than fluorobenzene. The reaction shown yields *p*-nitrophenyl phenyl ether in 92 percent yield.

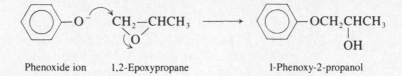

C_6H_5OK (Potassium phenoxide) + $F{-}C_6H_4{-}NO_2$ (*p*-Fluoronitrobenzene) $\xrightarrow{150°C}$ $C_6H_5{-}O{-}C_6H_4{-}NO_2$ (*p*-Nitrophenyl phenyl ether)

26.11 The anion of 2,4,5-trichlorophenol attacks 1,2,4,5-tetrachlorobenzene in the first step of the mechanism below:

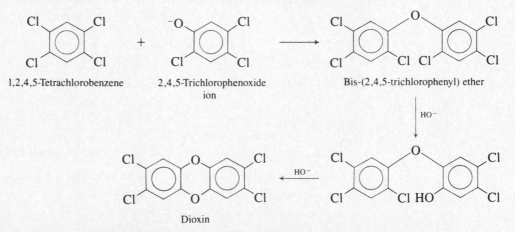

1,2,4,5-Tetrachlorobenzene + 2,4,5-Trichlorophenoxide ion ⟶ Bis-(2,4,5-trichlorophenyl) ether

Bis-(2,4,5-trichlorophenyl) ether $\xrightarrow{HO^-}$ (hydroxy-substituted diaryl ether: Cl, Cl, Cl / HO, Cl, Cl) $\xrightarrow{HO^-}$ Dioxin

The mechanism shown outlines the reaction in terms of reasonable intermediates. Whether the nucleophilic substitution steps occur by the addition-elimination or the elimination-addition mechanism cannot be discerned from the data.

26.12 (*a*) The parent compound is benzaldehyde. Vanillin bears a methoxy group (CH_3O) at C-3 and a hydroxyl group (HO) at C-4.

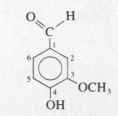

Vanillin (4-hydroxy-3-methoxybenzaldehyde)

(*b*), (*c*) Thymol and carvacrol differ with respect to the position of the hydroxyl group.

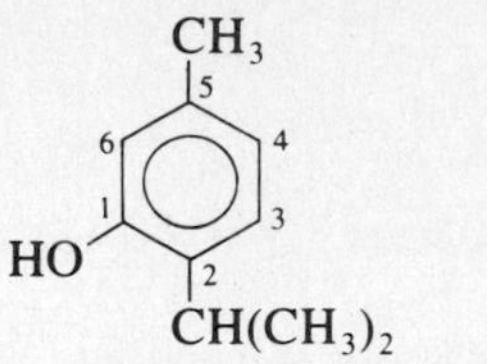

CH3 HO CH(CH3)2 — CH3 HO CH(CH3)2

Thymol (2-isopropyl-5-methylphenol) 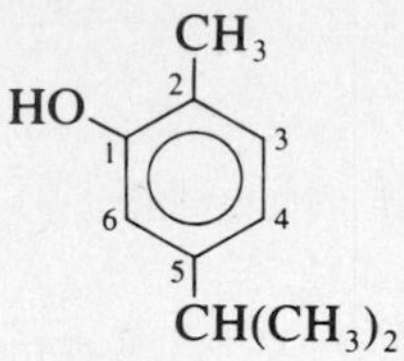 Carvacrol (5-isopropyl-2-methylphenol)

(*d*) An allyl substituent is $-CH_2CH{=}CH_2$.

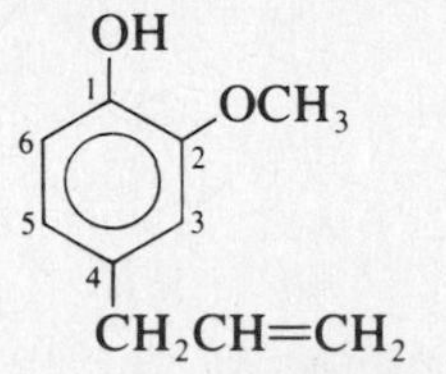

OH OCH3 CH2CH=CH2

Eugenol (4-allyl-2-methoxyphenol)

(*e*) Benzoic acid is $C_6H_5CO_2H$. Gallic acid bears three hydroxyl groups, located at C-3, C-4, and C-5.

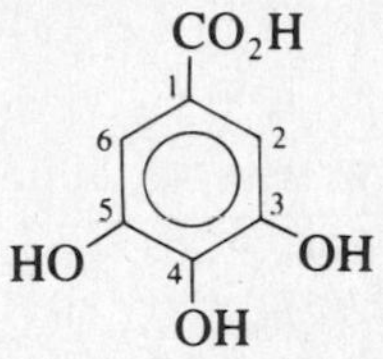

CO2H HO OH OH

Gallic acid (3,4,5-trihydroxybenzoic acid)

(*f*) Benzyl alcohol is $C_6H_5CH_2OH$. Salicyl alcohol bears a hydroxyl group at the ortho position.

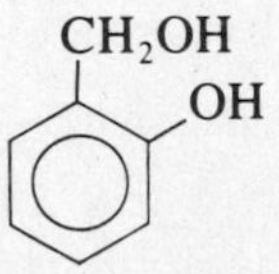

CH2OH OH

Salicyl alcohol (*o*-hydroxybenzyl alcohol)

26.13 (*a*) The compound is named as a derivative of phenol. The substituents (ethyl and nitro) are cited in alphabetical order with numbers assigned in the direction that gives the lowest number at the first point of difference.

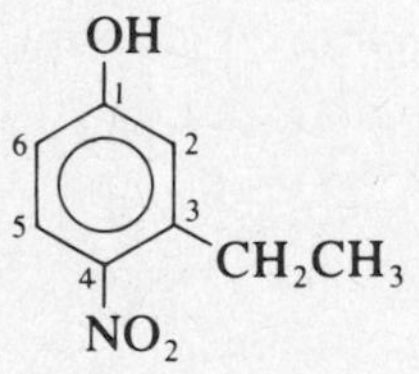

OH CH2CH3 NO2

3-Ethyl-4-nitrophenol

(*b*) An isomer of the above is 4-ethyl-3-nitrophenol.

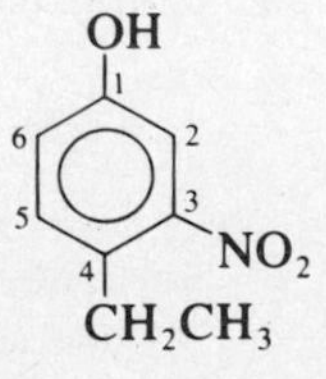

OH NO2 CH2CH3

4-Ethyl-3-nitrophenol

(*c*) The parent compound is phenol. It bears, in alphabetical order, a benzyl group at C-4 and a chlorine at C-2.

4-Benzyl-2-chlorophenol

(*d*) This compound is named as a derivative of anisole, $C_6H_5OCH_3$. Since multiplicative prefixes (di-, tri-, etc.) are not considered when alphabetizing substituents, isopropyl precedes dimethyl.

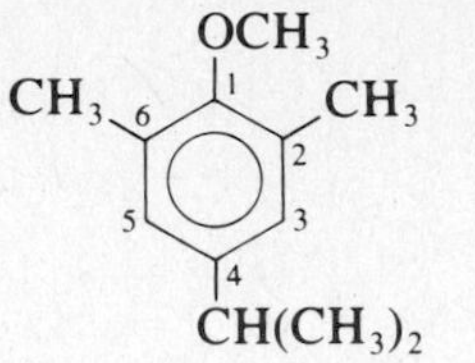

4-Isopropyl-2,6-dimethylanisole

(*e*) The compound is an aryl ester of trichloroacetic acid. The aryl group is 2,5-dichlorophenyl.

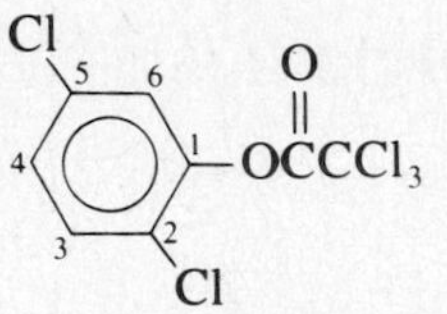

2,5-Dichlorophenyl trichloroacetate

(*f*) Urushiol, with its *o*-dihydroxybenzene nucleus, is a derivative of catechol. The hydrocarbon side chain has 15 carbon atoms and one double bond and so is named pentadecenyl. The double bond is between C-8 and C-9, counting from the point of attachment to the aromatic ring. It has the *Z* or cis configuration.

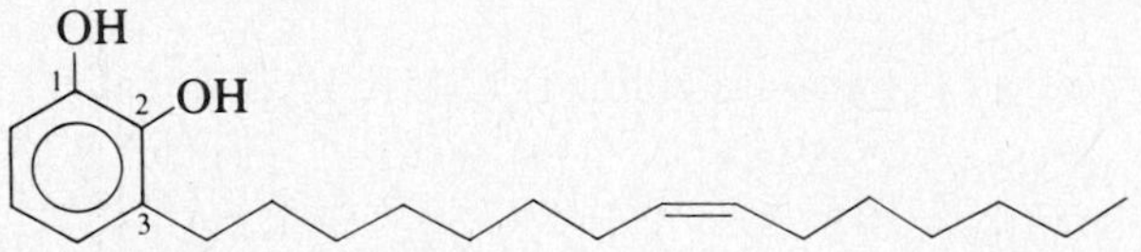

3-[(*Z*)-8-Pentadecenyl]catechol or 3-[(*Z*)-8-pentadecenyl]-1,2-benzenediol

(*g*) A 1,3-dihydroxy benzene is named as a derivative of resorcinol. This compound, a well-known antiseptic, has a hexyl group at C-4.

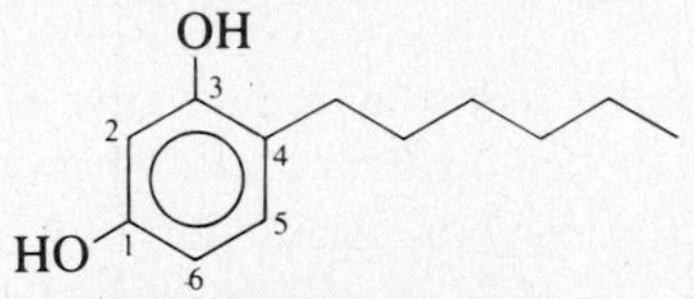

4-Hexylresorcinol or 4-*n*-hexylresorcinol or 4-hexyl-1,3-benzenediol

(*h*) Juglone bears a hydroxyl group at C-5 of a 1,4-naphthoquinone nucleus.

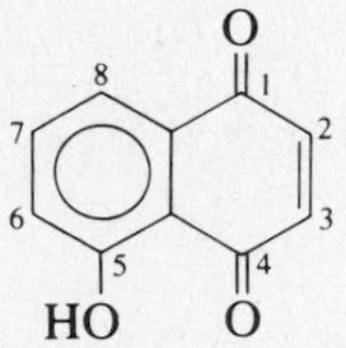

5-Hydroxy-1,4-naphthoquinone

26.14 (*a*) The reaction is an acid-base reaction. Phenol is the acid; sodium hydroxide is the base.

$$C_6H_5OH + NaOH \longrightarrow C_6H_5ONa + H_2O$$

Phenol (stronger acid) + Sodium hydroxide (stronger base) → Sodium phenoxide (weaker base) + Water (weaker acid) (*b*) Sodium phenoxide reacts with ethyl bromide to yield ethyl phenyl ether in a Williamson reaction. Phenoxide ion acts as a nucleophile.

$$\underset{\text{Sodium phenoxide}}{C_6H_5ONa} + \underset{\text{Ethyl bromide}}{CH_3CH_2Br} \longrightarrow \underset{\text{Ethyl phenyl ether}}{C_6H_5OCH_2CH_3} + \underset{\text{Sodium bromide}}{NaBr}$$

(*c*) Dimethyl sulfate is a methylating agent.

$$\underset{\text{Sodium phenoxide}}{C_6H_5ONa} + \underset{\text{Dimethyl sulfate}}{CH_3OS(=O)_2OCH_3} \longrightarrow \underset{\text{Methyl phenyl ether}}{C_6H_5OCH_3} + \underset{\text{Sodium methyl sulfate}}{NaOS(=O)_2OCH_3}$$

The reaction is a nucleophilic substitution (S_N2) at one of the methyl groups of dimethyl sulfate by phenoxide ion acting as the nucleophile.

(*d*) *p*-Toluenesulfonate esters behave much like alkyl iodides in nucleophilic substitution reactions. Phenoxide ion displaces *p*-toluenesulfonate from the primary carbon.

$$\underset{\text{Sodium phenoxide}}{C_6H_5ONa} + \underset{\text{Butyl }p\text{-toluenesulfonate}}{CH_3CH_2CH_2CH_2OS(=O)_2-C_6H_4-CH_3} \longrightarrow$$

$$\underset{\text{Butyl phenyl ether}}{C_6H_5OCH_2CH_2CH_2CH_3} + \underset{\text{Sodium }p\text{-toluenesulfonate}}{NaOS(=O)_2-C_6H_4-CH_3}$$

(*e*) Carboxylic acid anhydrides react with phenoxide anions to yield aryl esters.

$$\underset{\text{Sodium phenoxide}}{C_6H_5ONa} + \underset{\text{Acetic anhydride}}{CH_3\overset{O}{\overset{\|}{C}}O\overset{O}{\overset{\|}{C}}CH_3} \longrightarrow \underset{\text{Phenyl acetate}}{C_6H_5O\overset{O}{\overset{\|}{C}}CH_3} + \underset{\text{Sodium acetate}}{CH_3\overset{O}{\overset{\|}{C}}ONa}$$

(*f*) Acyl chlorides convert phenols to aryl esters.

$$\underset{o\text{-Cresol}}{2\text{-}CH_3C_6H_4OH} + \underset{\text{Benzoyl chloride}}{C_6H_5\overset{O}{\overset{\|}{C}}Cl} \longrightarrow \underset{\text{2-Methylphenyl benzoate}}{2\text{-}CH_3C_6H_4O\overset{O}{\overset{\|}{C}}C_6H_5} + \underset{\text{Hydrogen chloride}}{HCl}$$

(*g*) Phenols react as nucleophiles toward epoxides.

$$\underset{m\text{-Cresol}}{3\text{-}CH_3C_6H_4OH} + \underset{\text{Ethylene oxide}}{CH_2\underset{O}{\diagdown\diagup}CH_2} \longrightarrow \underset{\text{2-(3-Methylphenoxy)ethanol}}{3\text{-}CH_3C_6H_4OCH_2CH_2OH}$$

The reaction as written conforms to the requirements of the problem that a balanced equation be written. Of course, the reaction will be much faster if catalyzed by acid or base, but the catalysts do not enter into the equation representing the overall process.

(*h*) Bromination of the aromatic ring of 2,6-dichlorophenol occurs para to the hydroxy group. The more activating group (—OH) determines the orientation of the product.

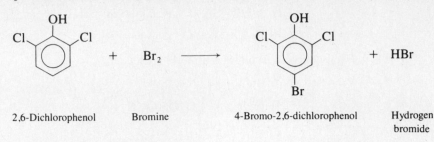

2,6-Dichlorophenol + Bromine (Br_2) → 4-Bromo-2,6-dichlorophenol + Hydrogen bromide (HBr)

(*i*) In aqueous solution bromination occurs at all the open positions that are ortho and para to the hydroxyl group.

p-Cresol + $2Br_2$ $\xrightarrow{H_2O}$ 2,6-Dibromo-4-methylphenol + $2HBr$

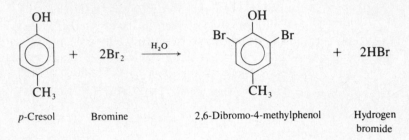

p-Cresol | Bromine | 2,6-Dibromo-4-methylphenol | Hydrogen bromide

(*j*) Hydrogen bromide cleaves ethers to give an alkyl halide and a phenol.

$C_6H_5OCH(CH_3)_2$ + HBr → C_6H_5OH + $(CH_3)_2CHBr$

Isopropyl phenyl ether | Hydrogen bromide | Phenol | Isopropyl bromide

26.15 The three industrial syntheses of phenol listed in Table 26.4 (Section 26.6) of the text all begin with benzene.

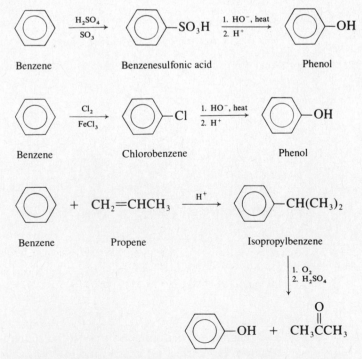

Benzene $\xrightarrow[SO_3]{H_2SO_4}$ $C_6H_5SO_3H$ (Benzenesulfonic acid) $\xrightarrow[2.\ H^+]{1.\ HO^-,\ heat}$ C_6H_5OH (Phenol)

Benzene $\xrightarrow[FeCl_3]{Cl_2}$ C_6H_5Cl (Chlorobenzene) $\xrightarrow[2.\ H^+]{1.\ HO^-,\ heat}$ C_6H_5OH (Phenol)

Benzene + $CH_2{=}CHCH_3$ (Propene) $\xrightarrow{H^+}$ $C_6H_5CH(CH_3)_2$ (Isopropylbenzene)

Isopropylbenzene $\xrightarrow[2.\ H_2SO_4]{1.\ O_2}$ C_6H_5OH (Phenol) + $CH_3\overset{O}{\overset{\|}{C}}CH_3$ (Acetone)

The fourth synthesis utilizes the fact that a phenolic hydroxyl may be introduced by first preparing aniline and then heating the diazonium ion in water.

$$C_6H_5\text{—}OH \Rightarrow C_6H_5\text{—}NH_2 \Rightarrow C_6H_5\text{—}NO_2 \Rightarrow C_6H_6$$

The reaction scheme to be used is:

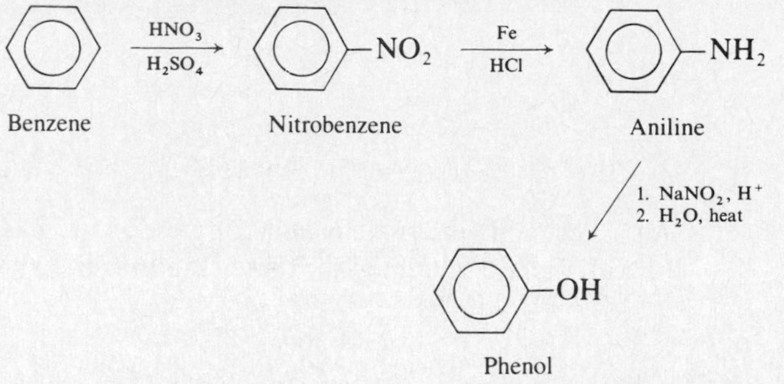

Benzene $\xrightarrow[H_2SO_4]{HNO_3}$ Nitrobenzene $\xrightarrow[HCl]{Fe}$ Aniline $\xrightarrow{1.\ NaNO_2,\ H^+;\ 2.\ H_2O,\ heat}$ Phenol

26.16 (*a*) Strongly electron-withdrawing groups, particularly those such as $-NO_2$, increase the acidity of phenols by resonance stabilization of the resulting phenoxide anion. Electron-releasing substituents such as $-CH_3$ exert a very small acid-weakening effect.

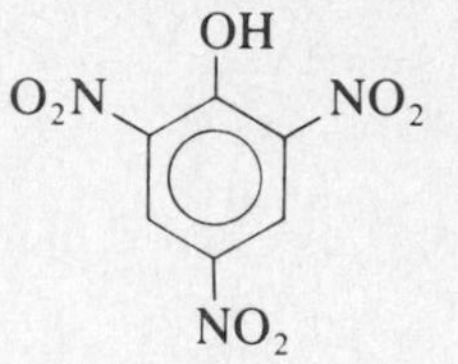

2,4,6-Trinitrophenol, more acidic (K_a 3.8 × 10^{-1}, pK_a 0.42)

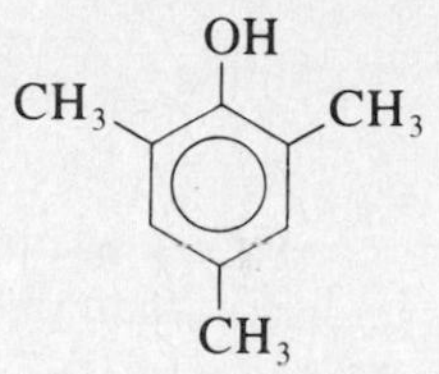

2,4,6-Trimethylphenol, less acidic (K_a 1.3 × 10^{-11}, pK_a 10.88)

Picric acid (2,4,6-trinitrophenol) is a stronger acid by far than 2,4,6-trimethylphenol. All three nitro groups participate in resonance stabilization of the picrate anion.

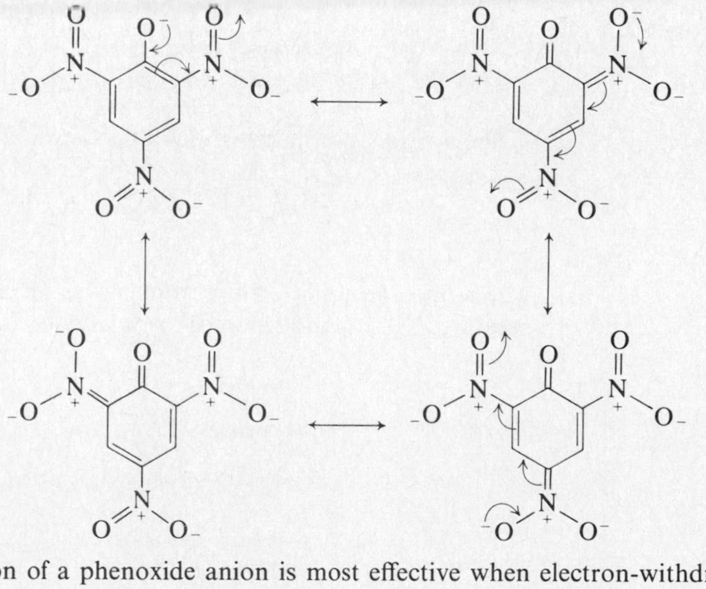

(*b*) Stabilization of a phenoxide anion is most effective when electron-withdrawing groups are present at the ortho and para positions because it is these carbons that bear most of the negative charge in phenoxide anion.

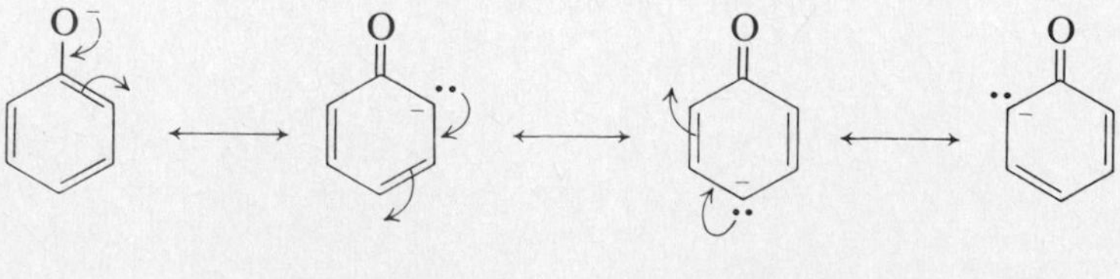

Therefore, 2,6-dichlorophenol is expected to be (and is) a stronger acid than 3,5-dichlorophenol.

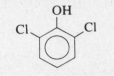

2,6-Dichlorophenol, more acidic
(K_a 1.6 × 10^{-7}, pK_a 6.79)

3,5-Dichlorophenol, less acidic
(K_a 6.5 × 10^{-9}, pK_a 8.19)

(*c*) The same principle is at work here as in (*b*). A nitro group para to the phenol oxygen is directly conjugated to it and stabilizes the anion better than one at the meta position.

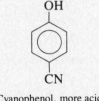

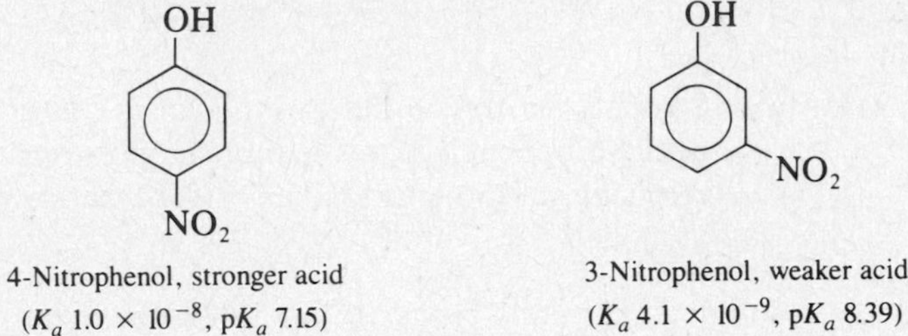

4-Nitrophenol, stronger acid
(K_a 1.0 × 10^{-8}, pK_a 7.15)

3-Nitrophenol, weaker acid
(K_a 4.1 × 10^{-9}, pK_a 8.39)

(*d*) A cyano group is strongly electron-withdrawing, so 4-cyanophenol is a stronger acid than phenol.

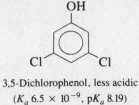

4-Cyanophenol, more acidic
(K_a 1.1 × 10^{-8}, pK_a 7.95)

Phenol, less acidic
(K_a 1 × 10^{-10}, pK_a 10)

There is resonance stabilization of the 4-cyanophenoxide anion:

$$^{-}O-C_6H_4-C{\equiv}N \longleftrightarrow O{=}C_6H_4{=}C{=}N^{-}$$

(*e*) The 5-nitro group in 2,5-dinitrophenol is meta to the hydroxyl group and so does not stabilize the resulting anion as much as does an ortho or para nitro group.

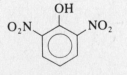

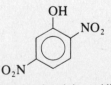

2,6-Dinitrophenol, more acidic
(K_a 2.0 × 10^{-4}, pK_a 3.71)

2,5-Dinitrophenol, less acidic
(K_a 6.0 × 10^{-6}, pK_a 5.22)

26.17 (*a*) The rate-determining step in base-catalyzed ester hydrolysis is formation of the tetrahedral intermediate.

$$ArO\overset{O}{\overset{\|}{C}}CH_3 + HO \underset{}{\overset{slow}{\rightleftharpoons}} ArO\underset{OH}{\overset{O^-}{C}}CH_3$$

Since this intermediate is negatively charged, there will be a small effect favoring its formation when the aryl group bears an electron-withdrawing substituent. Furthermore, this intermediate can either return to starting materials or proceed to products.

$$\text{ArO}\overset{\text{O}^-}{\underset{\text{OH}}{\text{C}}}\text{CH}_3 \rightleftharpoons \text{ArO}^- + \text{CH}_3\overset{\text{O}}{\overset{\|}{\text{C}}}\text{OH} \xrightarrow{\text{HO}^-} \text{ArO}^- + \text{CH}_3\overset{\text{O}}{\overset{\|}{\text{C}}}\text{O}^-$$

The proportion of the tetrahedral intermediate that goes on to products increases as the leaving group ArO^- becomes less basic. This is strongly affected by substituents; electron-withdrawing groups stabilize ArO^-. The prediction is that *m*-nitrophenyl acetate undergoes base-catalyzed hydrolysis faster than phenol. Indeed, this is observed to be the case; *m*-nitrophenyl acetate reacts some 10 times faster than does phenyl acetate at 25°C.

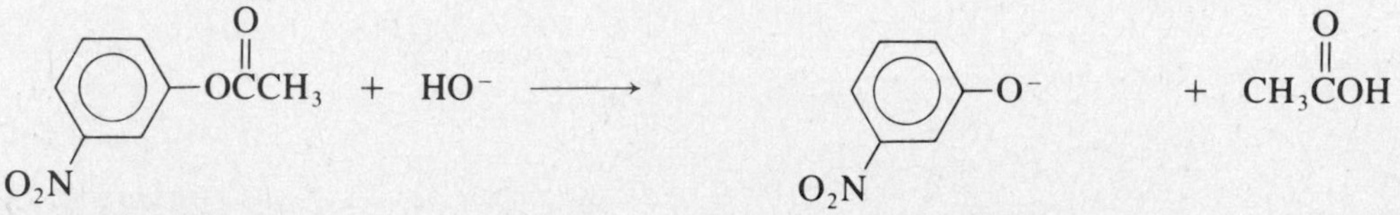

$$m\text{-}O_2N\text{-}C_6H_4\text{-}O\overset{\text{O}}{\overset{\|}{\text{C}}}CH_3 + HO^- \longrightarrow m\text{-}O_2N\text{-}C_6H_4\text{-}O^- + CH_3\overset{\text{O}}{\overset{\|}{\text{C}}}OH$$

m-Nitrophenyl acetate (more reactive)

m-Nitrophenoxide anion (a better leaving group than phenoxide because it is less basic)

(*b*) The same principle applies here as in the preceding question. *p*-Nitrophenyl acetate reacts faster than *m*-nitrophenyl acetate (by about 45 percent) largely because *p*-nitrophenoxide is a better, less basic leaving group than *m*-nitrophenoxide.

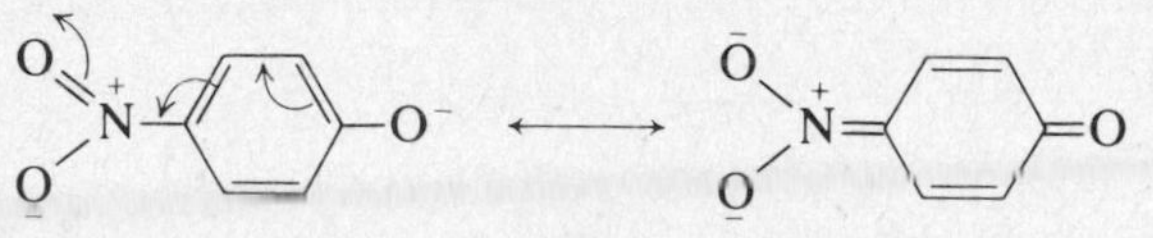

$$O{=}\overset{+}{N}(O^-)\text{-}C_6H_4\text{-}O^- \longleftrightarrow {}^-O\text{-}\overset{+}{N}(O^-){=}C_6H_4{=}O$$

Resonance in *p*-nitrophenoxide is particularly effective because the *p*-nitro group is directly conjugated to the oxyanion; direct conjugation of these groups is absent in *m*-nitrophenoxide.

(*c*) The reaction of ethyl bromide with a phenol is an S_N2 reaction in which the oxygen of the phenol is the nucleophile. The reaction is much faster with sodium phenoxide than with phenol because an anion is more nucleophilic than a corresponding neutral molecule.

Faster reaction:

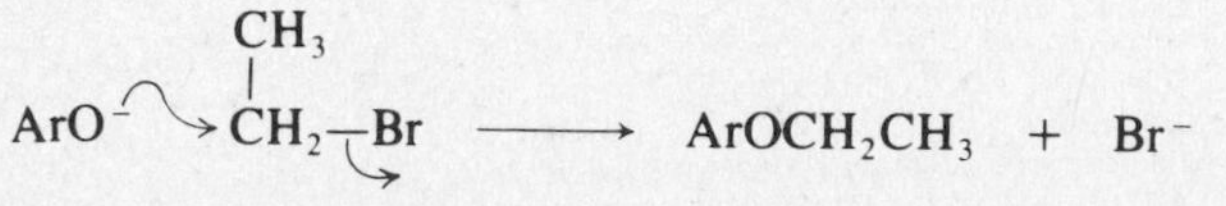

$$\text{ArO}^- + \overset{\text{CH}_3}{|}\!\!\text{CH}_2\text{—Br} \longrightarrow \text{ArOCH}_2\text{CH}_3 + \text{Br}^-$$

Slower reaction:

$$\text{Ar}\ddot{\text{O}}\text{H} \longrightarrow \overset{\text{CH}_3}{|}\!\!\text{CH}_2\text{—Br} \longrightarrow \text{Ar}\overset{+}{\ddot{\text{O}}}(\text{H})\text{CH}_2\text{CH}_3 + \text{Br}^-$$

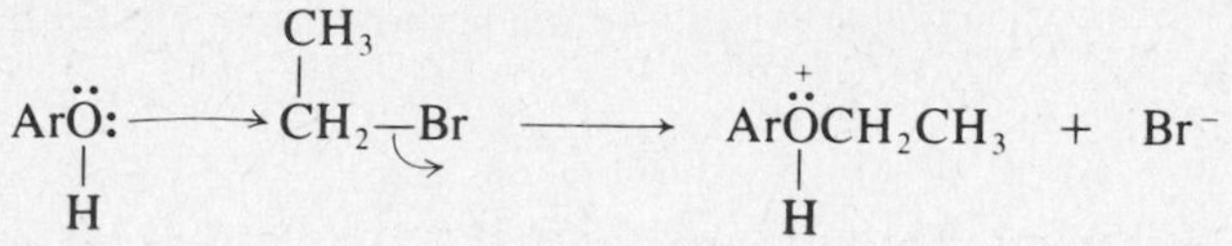

(*d*) The answer here also depends on the nucleophilicity of the attacking species, which is a phenoxide anion in both reactions.

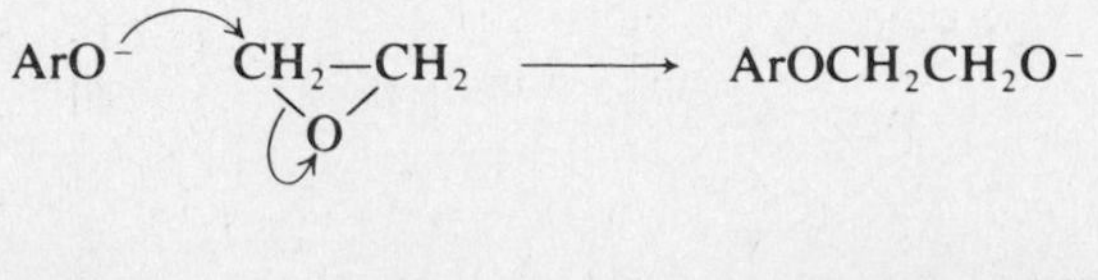

$$\text{ArO}^- + \underset{\text{O}}{\text{CH}_2\text{—CH}_2} \longrightarrow \text{ArOCH}_2\text{CH}_2\text{O}^-$$

The more nucleophilic anion is phenoxide ion because it is more basic than *p*-nitrophenoxide.

C_6H_5—O^-

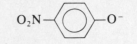

More basic;
better nucleophile

O_2N—C_6H_4—O^-

Better delocalization of negative charge makes this less basic and less nucleophilic

Rate measurements reveal that sodium phenoxide reacts 17 times faster with ethylene oxide (in ethanol at 70°C) than does its *p*-nitro derivative.

(*e*) This reaction is electrophilic aromatic substitution. Since a hydroxy substituent is more activating than an acetate group, phenol undergoes bromination faster than does phenyl acetate.

$$C_6H_5\ddot{O}-\overset{O}{\overset{\|}{C}}CH_3 \longleftrightarrow C_6H_5\overset{+}{O}=\overset{O^-}{\overset{|}{C}}CH_3$$

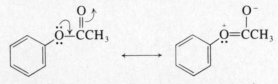

Resonance involving ester group reduces tendency of oxygen to donate electrons to ring

26.18 Nucleophilic aromatic substitution by the elimination-addition mechanism is impossible owing to the absence of any protons that might be abstracted from the substrate. However, the addition-elimination pathway is available.

$$C_6F_6 + HO^- \xrightarrow{\text{slow}} [C_6F_6OH]^- \xrightarrow{\text{fast}} C_6F_5OH + F^-$$

Hexafluorobenzene; Pentafluorophenol

This pathway is favorable because the cyclohexadienyl anion intermediate formed in the rate-determining step is stabilized by the electron-withdrawing inductive effect of its fluorine substituents.

26.19 (*a*) Allyl bromide is a reactive alkylating agent and converts the free hydroxyl group of the aryl compound (a natural product known as *guaiacol*) to its corresponding allyl ether.

$$o\text{-}HOC_6H_4OCH_3 + CH_2{=}CHCH_2Br \xrightarrow[\text{acetone}]{K_2CO_3} o\text{-}CH_2{=}CHCH_2OC_6H_4OCH_3$$

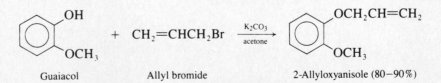

Guaiacol; Allyl bromide; 2-Allyloxyanisole (80–90%)

(*b*) Dimethyl sulfate is a standard reagent for methylation of phenols.

$$2,5\text{-}(HO)_2C_6H_3CCH_3(=O) + (CH_3O)_2SO_2 \xrightarrow[H_2O]{NaOH} 2,5\text{-}(CH_3O)_2C_6H_3CCH_3(=O)$$

2,5-Dihydroxyacetophenone; Dimethyl sulfate; 2,5-Dimethoxyacetophenone (71–74%)

(*c*) Sodium phenoxide acts as a nucleophile in this reaction and is converted to an ether.

$$C_6H_5ONa + ClCH_2CH(OH)CH_2OH \longrightarrow C_6H_5OCH_2CH(OH)CH_2OH$$

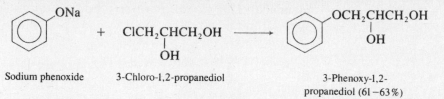

Sodium phenoxide — 3-Chloro-1,2-propanediol — 3-Phenoxy-1,2-propanediol (61–63%)

(*d*) Orientation in nitration is governed by the most activating substituent, in this case the hydroxyl group.

$$\text{Vanillin} \xrightarrow[\text{acetic acid, heat}]{HNO_3} \text{4-Hydroxy-3-methoxy-5-nitrobenzaldehyde (83\%)}$$

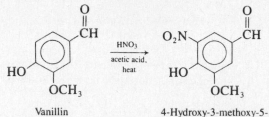

Vanillin — 4-Hydroxy-3-methoxy-5-nitrobenzaldehyde (83%)

(*e*) Hydrogen iodide cleaves aryl ethers to phenols and alkyl iodides.

$$m\text{-}CH_3OC_6H_4CH_2CO_2H + HI \xrightarrow{\text{heat}} m\text{-}HOC_6H_4CH_2CO_2H + CH_3I$$

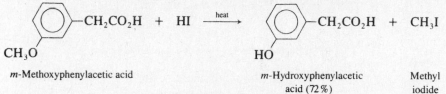

m-Methoxyphenylacetic acid — *m*-Hydroxyphenylacetic acid (72%) — Methyl iodide

(*f*) The hydroxyl group, as the most activating substituent, controls the orientation of electrophilic aromatic substitution. Bromination takes place ortho to the hydroxyl group.

$$\text{2-Ethoxy-4-nitrophenol} + Br_2 \xrightarrow{\text{acetic acid}} \text{2-Bromo-6-ethoxy-4-nitrophenol}$$

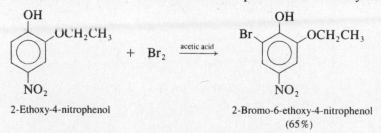

2-Ethoxy-4-nitrophenol — 2-Bromo-6-ethoxy-4-nitrophenol (65%)

(*g*) Oxidation of hydroquinone derivatives (*p*-dihydroxybenzenes) with Cr(VI) reagents is a method for preparing quinones.

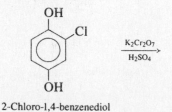

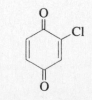

$$\text{2-Chloro-1,4-benzenediol} \xrightarrow[H_2SO_4]{K_2Cr_2O_7} \text{2-Chloro-1,4-benzoquinone}$$

2-Chloro-1,4-benzenediol — 2-Chloro-1,4-benzoquinone (88%)

(*h*) Silver oxide is a mild oxidizing agent, often used for the preparation of quinones from hydroquinones. It will not affect the allyl groups.

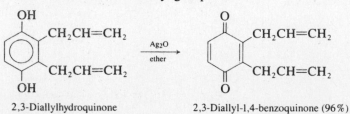

$$\text{2,3-Diallylhydroquinone} \xrightarrow[\text{ether}]{Ag_2O} \text{2,3-Diallyl-1,4-benzoquinone}$$

2,3-Diallylhydroquinone — 2,3-Diallyl-1,4-benzoquinone (96%)

(*i*) Aryl esters undergo a reaction known as the Fries rearrangement on being treated with aluminum chloride, which converts them to acyl phenols. Acylation takes place para to the hydroxyl in this case.

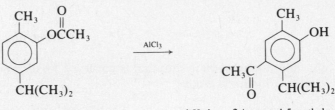

5-Isopropyl-2-methylphenyl acetate $\xrightarrow{AlCl_3}$ 4-Hydroxy-2-isopropyl-5-methylacetophenone (90%)

(*j*) Nucleophilic aromatic substitution takes place to yield a diaryl ether. The nucleophile is the phenoxide ion derived from 2,6-dimethylphenol.

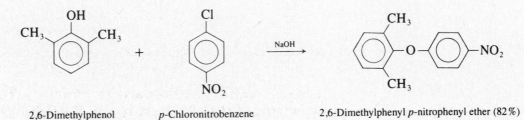

2,6-Dimethylphenol + *p*-Chloronitrobenzene $\xrightarrow{NaOH}$ 2,6-Dimethylphenyl *p*-nitrophenyl ether (82%)

(*k*) Chlorination with excess chlorine occurs at all available positions that are ortho and para to the hydroxyl group.

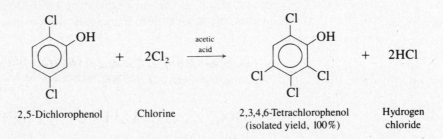

2,5-Dichlorophenol + $2Cl_2$ (Chlorine) $\xrightarrow{\text{acetic acid}}$ 2,3,4,6-Tetrachlorophenol (isolated yield, 100%) + 2HCl (Hydrogen chloride)

(*l*) Amines react with esters to give amides. In the case of a phenyl ester, phenol is the leaving group.

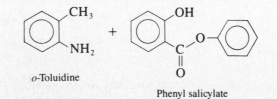

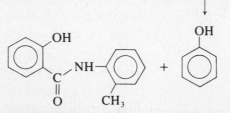

o-Toluidine + Phenyl salicylate $\xrightarrow{\text{heat}}$ *N*-(*o*-Methylphenyl)salicylamide (isolated yield, 73–77%) + Phenol

(*m*) Aryl diazonium salts attack electron-rich aromatic rings, such as those of phenols, to give the products of electrophilic aromatic substitution.

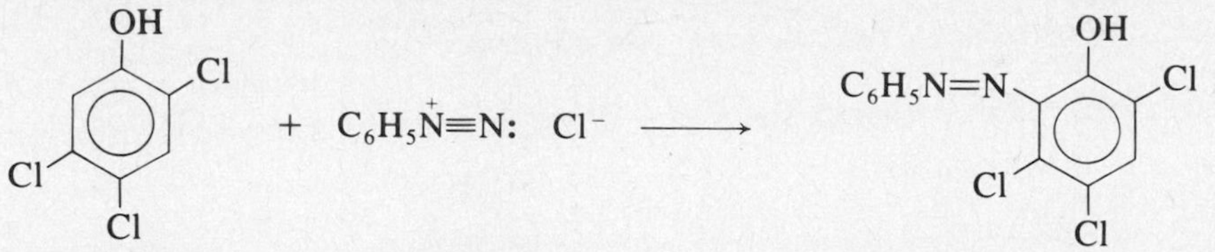

2,4,5-Trichlorophenol + Benzenediazonium chloride ($C_6H_5\overset{+}{N}\equiv N:\ Cl^-$) ⟶ 2-Benzeneazo-3,4,6-trichlorophenol (80%) ($C_6H_5N{=}N$–)

26.20 In the first step *p*-nitrophenol is alkylated on its phenolic oxygen with diethyl sulfate.

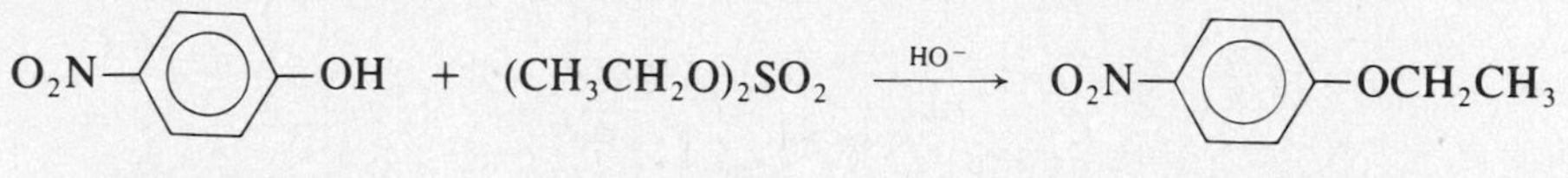

$$O_2N\text{–}C_6H_4\text{–}OH + (CH_3CH_2O)_2SO_2 \xrightarrow{HO^-} O_2N\text{–}C_6H_4\text{–}OCH_2CH_3$$

p-Nitrophenol, Diethyl sulfate, Ethyl *p*-nitrophenyl ether

Reduction of the nitro group gives the corresponding arylamine.

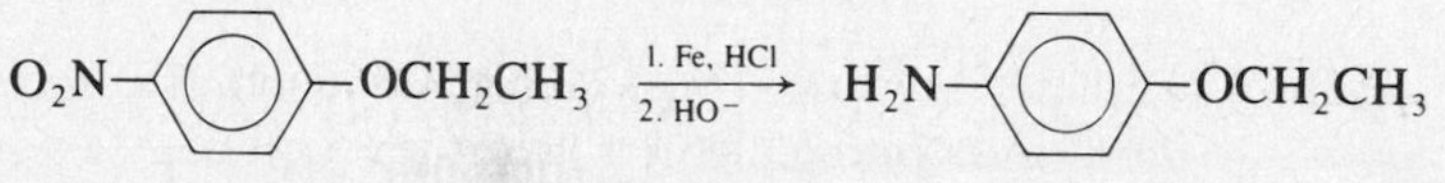

$$O_2N\text{–}C_6H_4\text{–}OCH_2CH_3 \xrightarrow[2.\ HO^-]{1.\ Fe,\ HCl} H_2N\text{–}C_6H_4\text{–}OCH_2CH_3$$

Ethyl *p*-nitrophenyl ether, *p*-Ethoxyaniline

Treatment of *p*-ethoxyaniline with acetic anhydride gives the product of *N*-acylation.

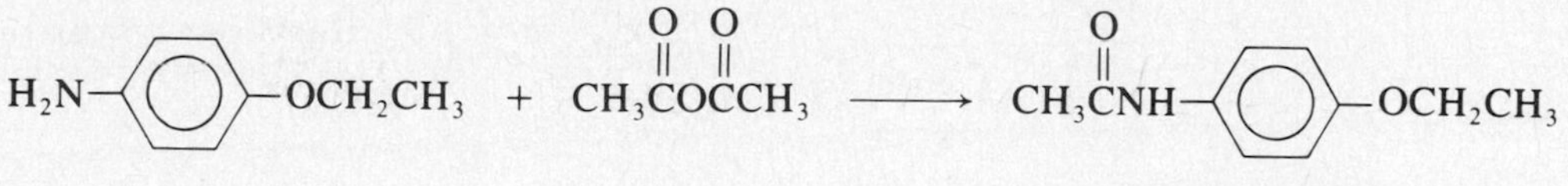

$$H_2N\text{–}C_6H_4\text{–}OCH_2CH_3 + CH_3\overset{O}{\overset{\|}{C}}O\overset{O}{\overset{\|}{C}}CH_3 \longrightarrow CH_3\overset{O}{\overset{\|}{C}}NH\text{–}C_6H_4\text{–}OCH_2CH_3$$

p-Ethoxyaniline, Acetic anhydride, *p*-Ethoxyacetanilide (phenacetin)

26.21 Parts (*a*), (*b*), and (*c*) of this problem comprise the series of steps by which *o*-bromophenol is prepared.

(*a*) Since direct bromination of phenol yields both *o*-bromophenol and *p*-bromophenol, it is essential that the para position be blocked prior to the bromination step. In practice, what is done is to disulfonate phenol, which blocks the para and one of the ortho positions.

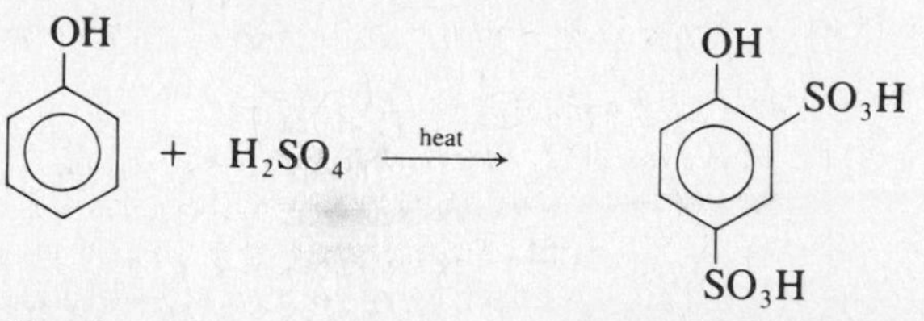

$$\text{Phenol} + H_2SO_4 \xrightarrow{\text{heat}} \text{4-Hydroxy-1,3-benzenedisulfonic acid (compound A)}$$

(OH, SO_3H ortho, SO_3H para)

(*b*) Bromination then can be accomplished cleanly at the open position ortho to the hydroxyl group.

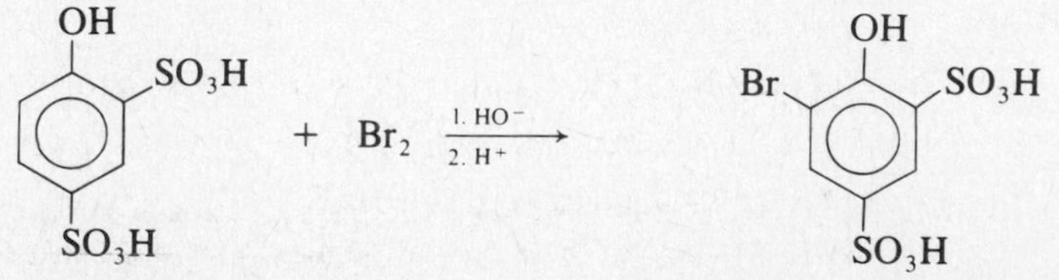

$$\text{Compound A} + Br_2 \xrightarrow[2.\ H^+]{1.\ HO^-} \text{5-Bromo-4-hydroxy-1,3-benzenedisulfonic acid (compound B)}$$

(OH flanked by Br and SO_3H; SO_3H para)

(*c*) After bromination the sulfonic acid groups are removed by acid-catalyzed hydrolysis.

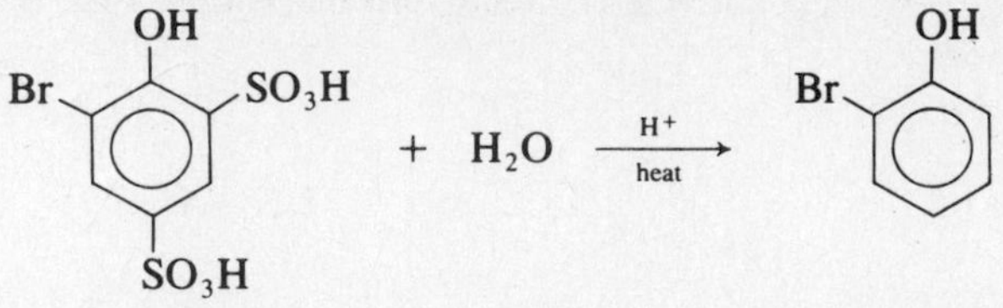

Br, OH, SO_3H, SO_3H + H_2O $\xrightarrow[\text{heat}]{H^+}$ Br, OH

Compound B

o-Bromophenol (compound C)

26.22 Nitration of 3,5-dimethylphenol gives a mixture of the 2-nitro and 4-nitro derivatives.

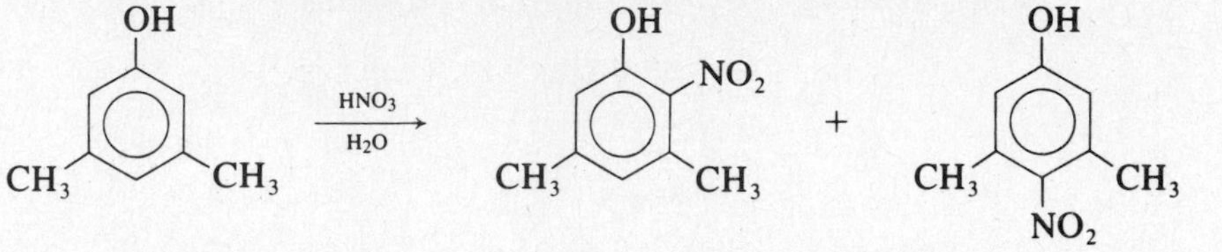

OH, CH_3, CH_3 $\xrightarrow[H_2O]{HNO_3}$ OH, NO_2, CH_3, CH_3 + OH, CH_3, CH_3, NO_2

3,5-Dimethylphenol

3,5-Dimethyl-2-nitrophenol

3,5-Dimethyl-4-nitrophenol

The more volatile compound (compound D), isolated by steam distillation, is the 2-nitro derivative. Intramolecular hydrogen bonding is possible between the nitro group and the hydroxyl group.

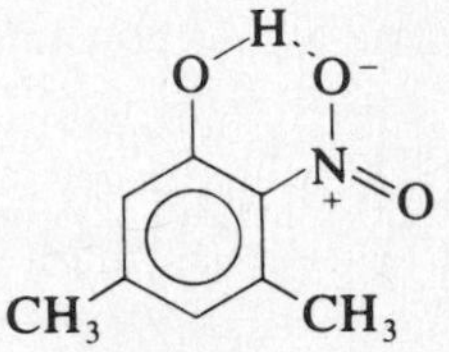

H, O, O^-, $\overset{+}{N}$=O, CH_3, CH_3

Intramolecular hydrogen bonding in 3,5-dimethyl-2-nitrophenol

The 4-nitro derivative participates in intermolecular hydrogen bonds and has a much higher boiling point; it is compound E.

26.23 The relationship between the target molecule and the starting materials tells us that two processes are required, formation of a diaryl ether linkage and nitration of an aromatic ring. The proper order of carrying out these two separate processes is what needs to be considered.

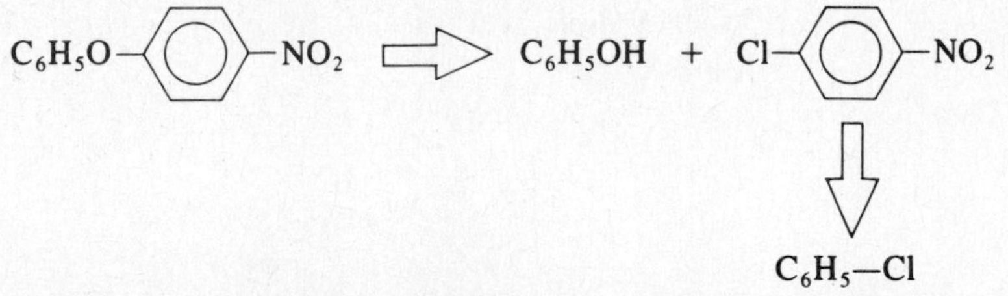

C_6H_5O—C_6H_4—NO_2 $\Rightarrow$ C_6H_5OH + Cl—C_6H_4—NO_2

$\Downarrow$

C_6H_5—Cl

The critical step is ether formation, a step which is feasible for the reactants shown:

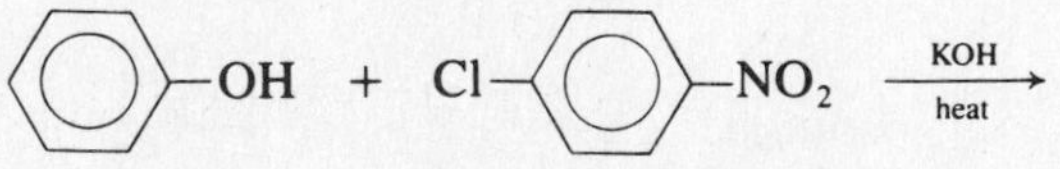

C_6H_5—OH + Cl—C_6H_4—NO_2 $\xrightarrow[\text{heat}]{KOH}$

Phenol

p-Chloronitrobenzene

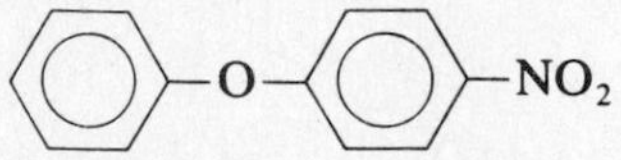

C_6H_5—O—C_6H_4—NO_2

p-Nitrophenyl phenyl ether

The reason this reaction is suitable is that it involves nucleophilic aromatic substitution by the addition-elimination mechanism on a *p*-nitro-substituted aryl halide. Indeed, this reaction has been carried out and gives an 80 to 82 percent yield. Therefore a reasonable synthesis would begin with the preparation of *p*-chloronitrobenzene.

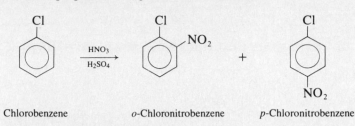

Chlorobenzene $\xrightarrow[H_2SO_4]{HNO_3}$ *o*-Chloronitrobenzene + *p*-Chloronitrobenzene

Separation of the *p*-nitro-substituted aryl halide and reaction with phenoxide ion complete the synthesis.

The following alternative route is less satisfactory:

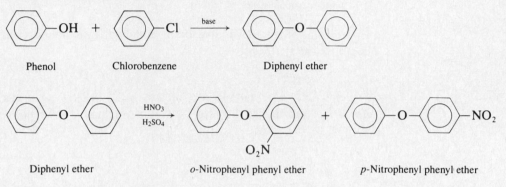

Phenol + Chlorobenzene $\xrightarrow{base}$ Diphenyl ether

Diphenyl ether $\xrightarrow[H_2SO_4]{HNO_3}$ *o*-Nitrophenyl phenyl ether + *p*-Nitrophenyl phenyl ether

The difficulty with this route concerns the preparation of diphenyl ether. Direct reaction of phenoxide with chlorobenzene is very slow because chlorobenzene is a poor substrate for nucleophilic substitution.

A third route is also unsatisfactory because it, too, requires nucleophilic substitution on chlorobenzene.

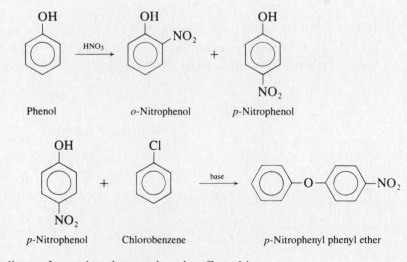

Phenol $\xrightarrow{HNO_3}$ *o*-Nitrophenol + *p*-Nitrophenol

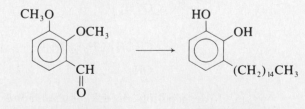

p-Nitrophenol + Chlorobenzene $\xrightarrow{base}$ *p*-Nitrophenyl phenyl ether

26.24 The overall transformation that needs to be effected is:

2,3-Dimethoxybenzaldehyde $\longrightarrow$ 3-Pentadecylcatechol

A reasonable place to begin is with the attachment of the side chain. The aldehyde function allows for chain extension by a Wittig reaction.

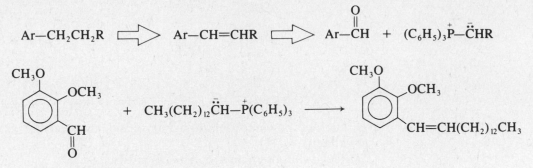

$$\mathrm{Ar{-}CH_2CH_2R} \Longrightarrow \mathrm{Ar{-}CH{=}CHR} \Longrightarrow \mathrm{Ar{-}\overset{O}{\overset{\|}{C}}H} + (\mathrm{C_6H_5})_3\overset{+}{\mathrm{P}}{-}\ddot{\mathrm{C}}\mathrm{HR}$$

$$\text{2,3-Dimethoxybenzaldehyde} + \mathrm{CH_3(CH_2)_{12}\ddot{C}H{-}\overset{+}{P}(C_6H_5)_3} \longrightarrow \text{2,3-}(\mathrm{CH_3O})_2\mathrm{C_6H_3{-}CH{=}CH(CH_2)_{12}CH_3}$$

2,3-Dimethoxybenzaldehyde

Hydrogenation of the double bond and hydrogen halide cleavage of the ether functions complete the synthesis.

$$\text{2,3-}(\mathrm{CH_3O})_2\mathrm{C_6H_3{-}CH{=}CH(CH_2)_{12}CH_3} \xrightarrow[\mathrm{Pt}]{\mathrm{H_2}} \text{2,3-}(\mathrm{CH_3O})_2\mathrm{C_6H_3{-}CH_2CH_2(CH_2)_{12}CH_3} \xrightarrow{\text{HBr, heat}} \text{2,3-}(\mathrm{HO})_2\mathrm{C_6H_3{-}CH_2CH_2(CH_2)_{12}CH_3}$$

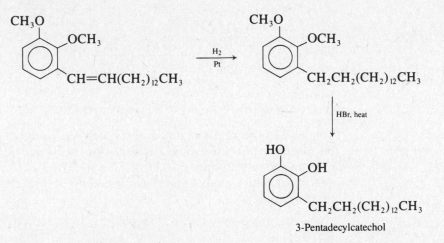

3-Pentadecylcatechol

Other synthetic routes are of course possible. One of the earliest approaches utilized a Grignard reaction to attach the side chain.

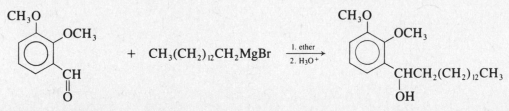

$$\text{2,3-Dimethoxybenzaldehyde} + \mathrm{CH_3(CH_2)_{12}CH_2MgBr} \xrightarrow[\text{2. } \mathrm{H_3O^+}]{\text{1. ether}} \text{2,3-}(\mathrm{CH_3O})_2\mathrm{C_6H_3{-}\underset{OH}{\underset{|}{C}}HCH_2(CH_2)_{12}CH_3}$$

2,3-Dimethoxybenzaldehyde

The resulting secondary alcohol can then be dehydrated to the same alkene intermediate prepared in the preceding synthetic scheme.

$$\text{2,3-}(\mathrm{CH_3O})_2\mathrm{C_6H_3{-}\underset{OH}{\underset{|}{C}}HCH_2(CH_2)_{12}CH_3} \xrightarrow{\mathrm{H^+}} \text{2,3-}(\mathrm{CH_3O})_2\mathrm{C_6H_3{-}CH{=}CH(CH_2)_{12}CH_3}$$

Again, hydrogenation of the double bond and ether cleavage leads to the desired 3-pentadecylcatechol.

26.25 The driving force for this reaction is the stabilization which results from formation of the aromatic ring. A reasonable series of steps is that shown below:

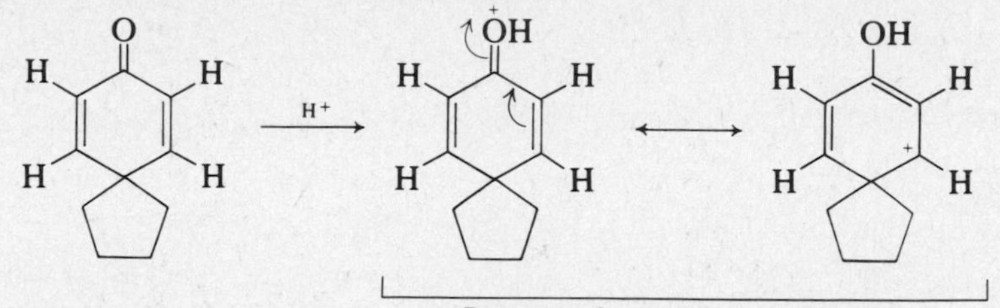

Resonance forms of protonated ketone

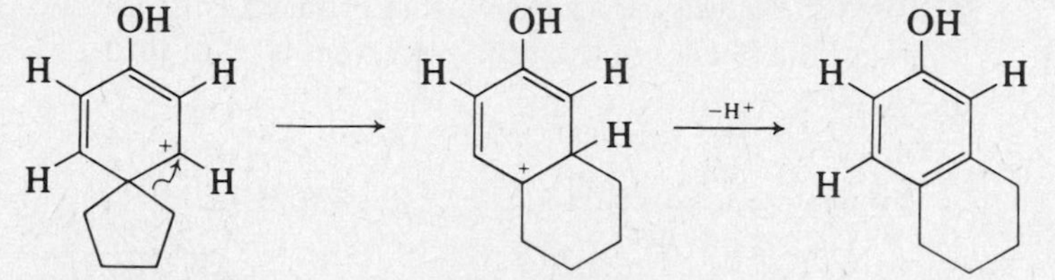

(Protonated ketone can rearrange by alkyl migration)

(Aromatization of this intermediate occurs by loss of a proton)

26.26 Bromination of *p*-hydroxybenzoic acid takes place in the normal fashion at both positions ortho to the hydroxy group.

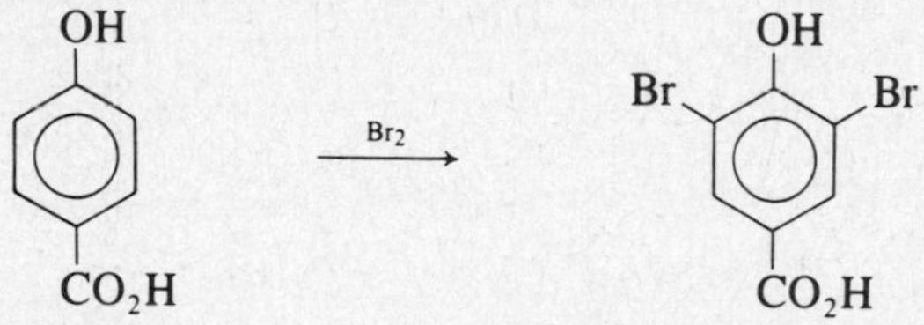

p-Hydroxybenzoic acid → (Br_2) 3,5-Dibromo-4-hydroxybenzoic acid

A third bromination step, this time at the para position, leads to the intermediate shown.

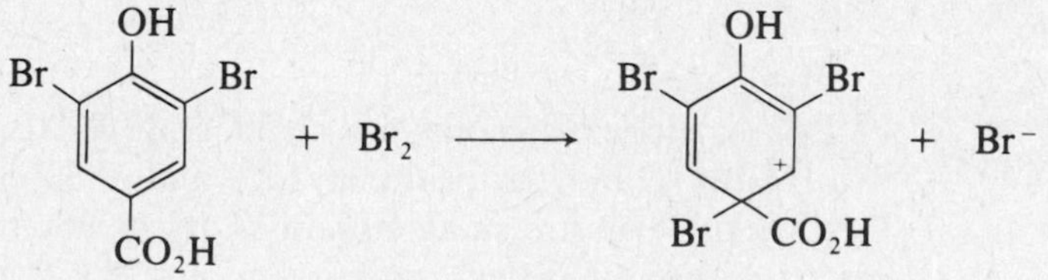

Aromatization of this intermediate occurs by decarboxylation.

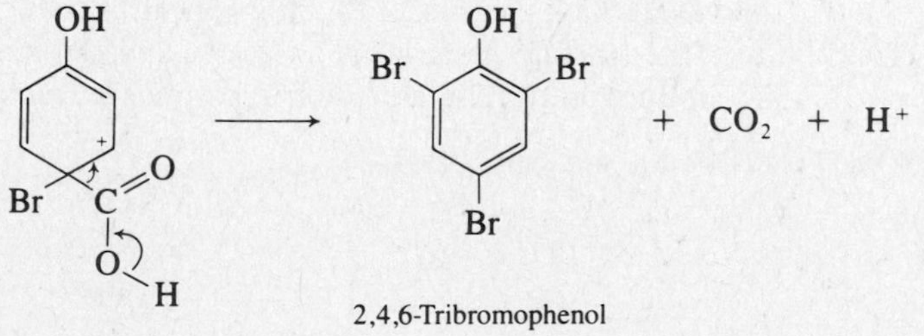

2,4,6-Tribromophenol + CO_2 + H^+

26.27 Electrophilic attack of bromine on 2,4,6-tribromophenol leads to a cationic intermediate.

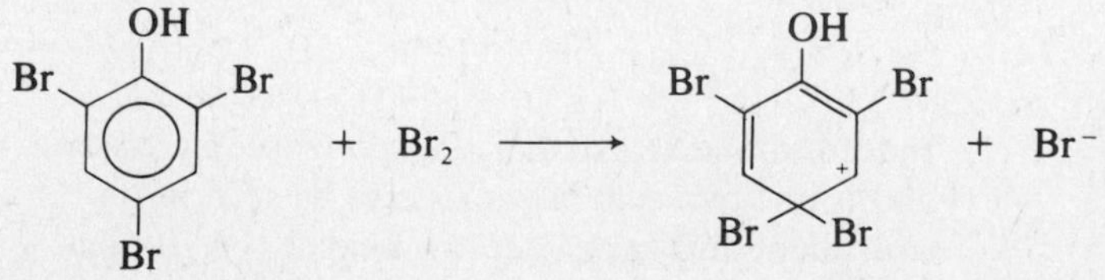

2,4,6-Tribromophenol + Br_2 → + Br^-

Loss of the hydroxyl proton from this intermediate generates the observed product.

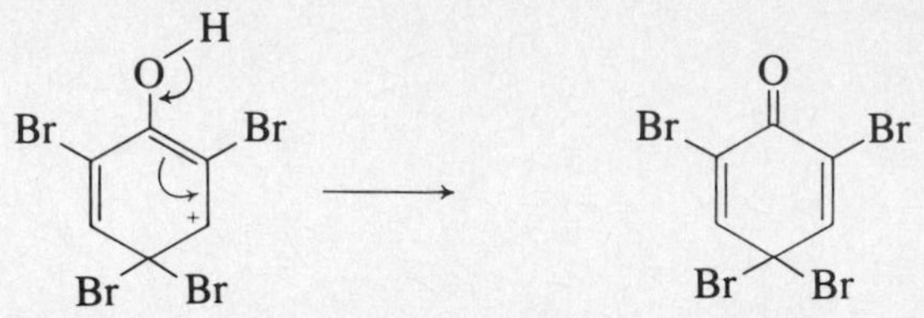

2,4,4,6-Tetrabromocyclohexadienone

26.28 A good way to approach this problem is to assume that bromine attacks the aromatic ring of the phenol in the usual way, that is, para to the hydroxyl group.

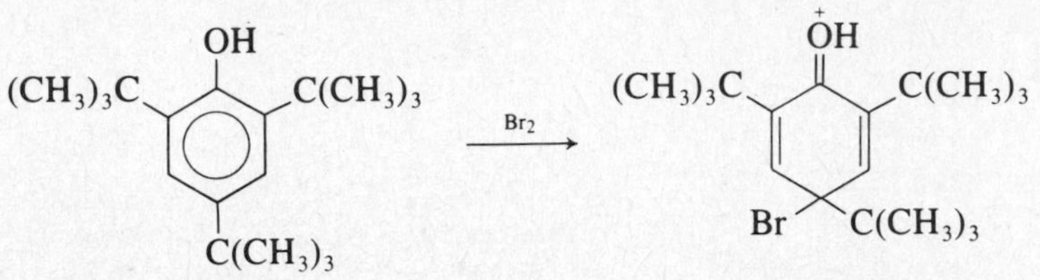

2,4,6-Tri-*tert*-butylphenol

This cation cannot yield the product of electrophilic aromatic substitution by loss of a proton from the ring but can lose a proton from oxygen to give a cyclohexadienone derivative.

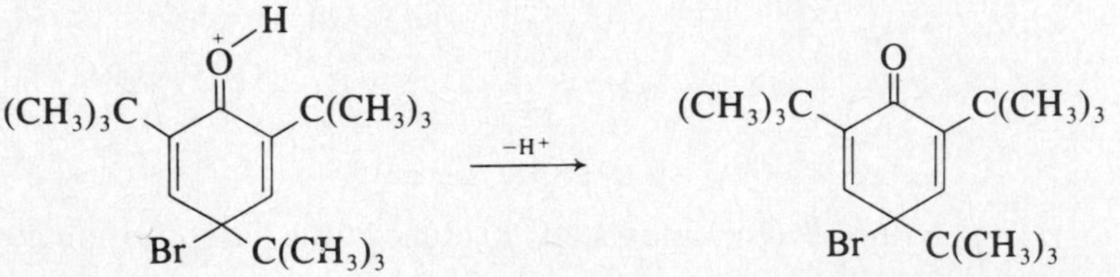

4-Bromo-2,4,6-tri-*tert*-butyl-2,5-cyclohexadienone

This cyclohexadienone is compound F. It has the correct molecular formula ($C_{18}H_{29}BrO$), and the peaks at 1655 and 1630 cm^{-1} in the infrared are consistent with C=O and C=C stretching vibrations. Its symmetry is consistent with the observed ^1H nmr spectrum; two equivalent *tert*-butyl groups at C-2 and C-6 appear as an 18-proton singlet at $\delta = 1.26$ ppm, the other *tert*-butyl group is a 9-proton singlet at $\delta = 1.19$ ppm, and the two equivalent vinyl protons of the ring appear as a singlet at $\delta = 6.90$ ppm.

26.29 Since the starting material is an acetal and the reaction conditions lead to hydrolysis with the production of 1,2-ethanediol, a reasonable reaction course is:

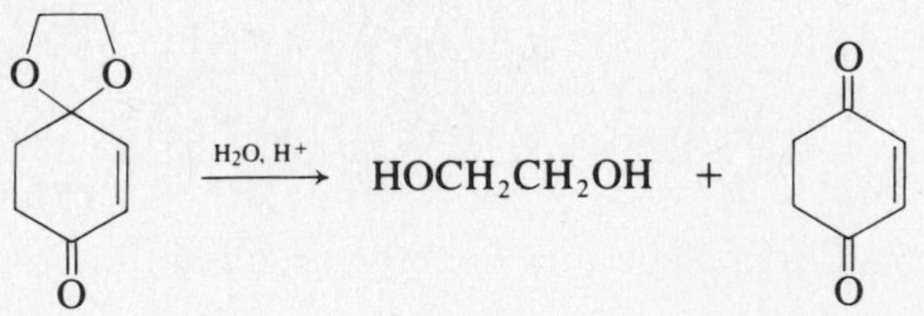

$$\xrightarrow{H_2O,\ H^+} HOCH_2CH_2OH + $$

Compound G 1,2-Ethanediol Compound H

Indeed, dione H satisfies the spectroscopic criteria. Carbonyl bands are seen in the infrared, and it has two sets of protons to be seen in its ^1H nmr spectrum. The two vinyl protons are equivalent and appear at low field, $\delta = 6.70$ ppm; The four methylene protons arc equivalent to each other and are seen at $\delta = 2.88$ ppm.

Compound H is the doubly ketonic tautomeric form of hydroquinone, compound I, to which it isomerizes on standing in water.

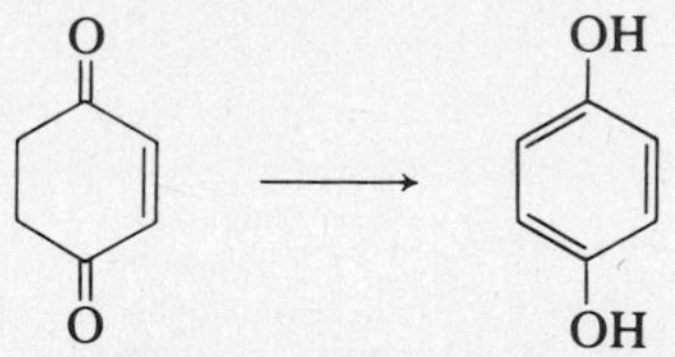

Compound H

Compound I
(hydroquinone)

26.30 The molecular formula of Fomecin A ($C_8H_8O_5$) corresponds to a sum of double bonds and rings (SODAR) equal to 5. Probably an aromatic ring is present, accounting for four of these. Its solubility in dilute sodium hydroxide suggests that Fomecin A is a phenol. A carbonyl group is indicated by formation of a 2,4-dinitrophenylhydrazone derivative and is probably an aldehyde function, since Fomecin A reduces ammoniacal silver nitrate. Thus, the aromatic ring of a phenol and the carbon-oxygen double bond of an aldehyde account for all the elements of unsaturation implied by the molecular formula.

The ^1H nmr data tell us that Fomecin A has a pentasubstituted aromatic ring, since there is only a single aromatic proton in its spectrum at $\delta = 6.53$ ppm. The signal at $\delta = 10.2$ ppm is due to the aldehyde proton. A two-proton singlet at $\delta = 4.58$ ppm is fairly deshielded and is probably a methylene group of the type $ArCH_2O$—.

The location of all the substituents is revealed by the isolation of the tetraacetate J.

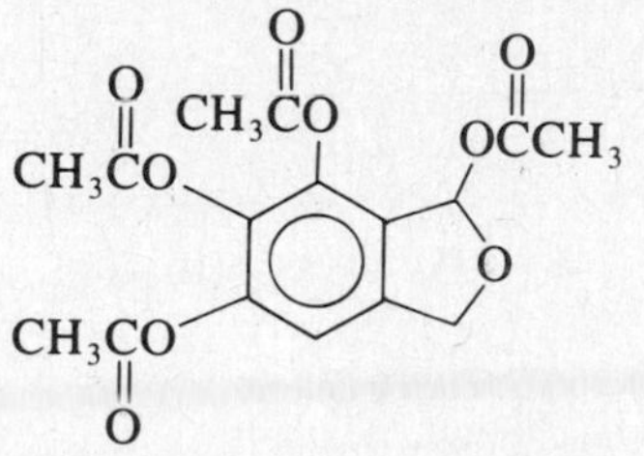

Compound J

Tetraacetate J must have arisen from the corresponding tetrol.

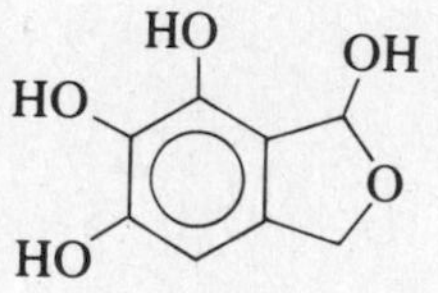

This tetrol, however, does not have an aldehyde group while the nmr evidence tells us that one must be present. Notice that the tetrol shown is a cyclic hemiacetal; its noncyclic form must be Fomecin A.

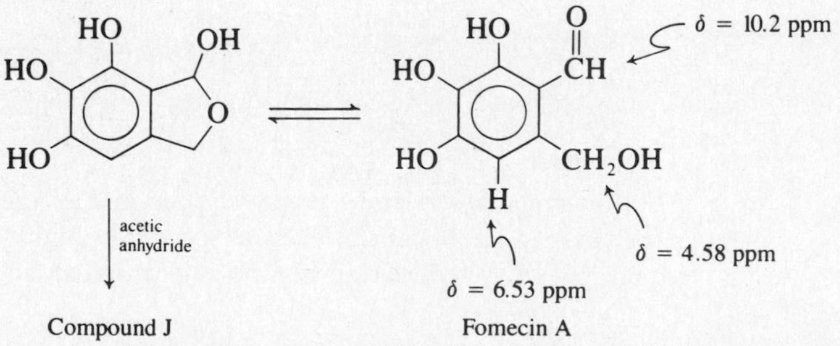

acetic anhydride

$\delta = 10.2$ ppm

$\delta = 4.58$ ppm

$\delta = 6.53$ ppm

Compound J

Fomecin A

26.31 Reduction of 1,4-benzoquinone leads to incorporation of both its double bonds into the aromatic ring of the product. There is a large gain in resonance energy associated with this reduction.

$$\text{1,4-Benzoquinone} + 2e^- + 2H^+ \rightleftharpoons \text{Hydroquinone}$$

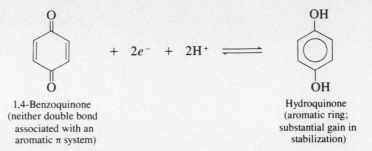

1,4-Benzoquinone (neither double bond associated with an aromatic π system)

Hydroquinone (aromatic ring; substantial gain in stabilization)

One of the double bonds in 1,4-naphthoquinone is not really a double bond at all but is part of an aromatic π system.

$$\text{1,4-Naphthoquinone} + 2e^- + 2H^+ \rightleftharpoons \text{1,4-Naphthalenediol}$$

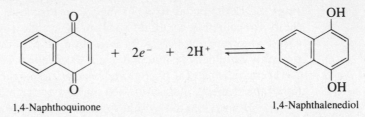

1,4-Naphthoquinone

1,4-Naphthalenediol

The net gain in stabilization energy is less than for 1,4-benzoquinone, so the reaction has a less favorable equilibrium constant. Both double bonds are already part of aromatic π systems in 9,10-anthraquinone, so there is little net stabilization attending its reduction.

$$\text{9,10-Anthraquinone} + 2e^- + 2H^+ \rightleftharpoons \text{9,10-Anthracenediol}$$

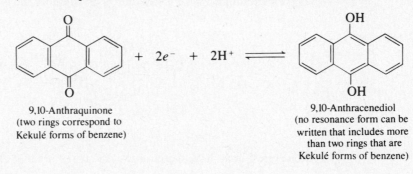

9,10-Anthraquinone (two rings correspond to Kekulé forms of benzene)

9,10-Anthracenediol (no resonance form can be written that includes more than two rings that are Kekulé forms of benzene)

26.32 A reasonable first step is protonation of the hydroxyl oxygen.

$$C_6H_5C(CH_3)_2\ddot{O}{-}\ddot{O}H + H^+ \longrightarrow C_6H_5C(CH_3)_2\ddot{O}{-}\overset{+}{O}H_2$$

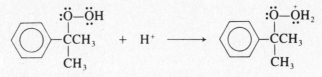

Cumene hydroperoxide

The weak oxygen-oxygen bond can now be cleaved, with loss of water as the leaving group.

$$C_6H_5C(CH_3)_2\ddot{O}{-}\overset{+}{O}H_2 \longrightarrow C_6H_5C(CH_3)_2\ddot{O}{+} + H_2\ddot{O}:$$

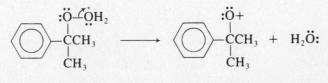

This intermediate bears a positively charged oxygen with only six electrons in its valence shell. Like a carbocation, such a species is highly electrophilic. The electrophilic oxygen attacks the π system of the neighboring aromatic ring to give an unstable intermediate.

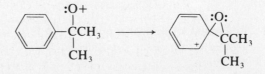

$$C_6H_5C(CH_3)_2\ddot{O}{+} \longrightarrow \text{(cyclic oxonium/carbocation intermediate with } {:}\ddot{O}{:}\text{ bridging ring and } CCH_3 \text{)}$$

Ring opening of this intermediate is assisted by one of the lone pairs of oxygen and restores the aromaticity of the ring.

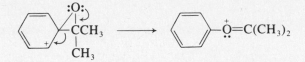

The cation formed by ring opening is captured by a water molecule to yield the hemiacetal product.

$+ H^+$

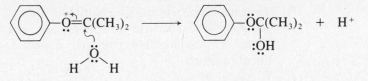

26.33 We are told that the first step is nucleophilic addition of peroxybenzoic acid to propiophenone.

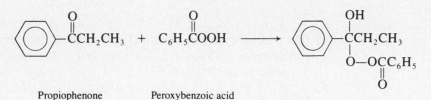

Propiophenone Peroxybenzoic acid

This intermediate has a weak oxygen-oxygen bond. Heterolytic cleavage of benzoate from this intermediate leaves an electrophilic oxygen behind that attacks the neighboring aromatic ring.

$+ C_6H_5CO_2^-$

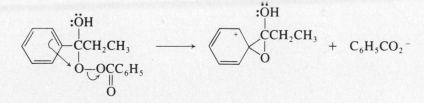

Rearomatization of this intermediate occurs by cleavage of the three-membered ring and is assisted by the electrons of the hydroxyl group.

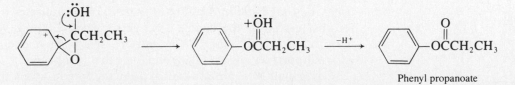

Phenyl propanoate

26.34 (*a*) The molecular formula of compound K ($C_9H_{12}O$) tells us that it has a total of four double bonds and rings (SODAR = 4). Peaks in the $\delta = 6.5$ to 7 ppm region of the ^1H nmr spectrum indicate an aromatic ring, which accounts for six of the nine carbon atoms of K and all its double bonds and rings. Thus any alkyl groups must have only single bonds to all carbons.

The infrared spectrum suggests an alcohol or a phenol, based on a strong absorption at 3300 cm^{-1}. An OH group is also consistent with a broad absorption at $\delta = 5.5$ ppm in the nmr spectrum. A three-proton triplet at high field ($\delta = 0.9$ ppm) must be a CH_3 group attached to a CH_2 unit, which must be different from the CH_2 unit responsible for the two-proton triplet at $\delta = 2.5$ ppm because the triplet nature of the signal at 2.5

ppm means that CH_2 is bonded to another CH_2 and not to CH_3. A propyl group attached to an aromatic ring is a possibility.

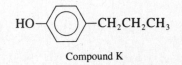

$$Ar{-}CH_2{-}CH_2{-}CH_3$$

$\delta = 2.5$ ppm (benzylic triplet); multiplet at $\delta = 1.5$ ppm; triplet at $\delta = 0.9$ ppm

Integration of the spectrum reveals the aromatic ring to be disubstituted. The characteristic AB quartet with a splitting of about 8 Hz indicates para substitution, permitting the last part of the structure to be assigned. The hydroxyl group can only be located on the aromatic ring and is para to the propyl group. The compound is 4-propylphenol.

$$HO{-}C_6H_4{-}CH_2CH_2CH_3$$

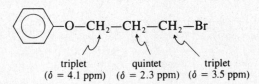

Compound K

(*b*) Compound L has a monosubstituted aromatic ring because there is a five-proton multiplet in the $\delta = 6.5$ to 7.5 ppm region of the ^1H nmr spectrum. The lack of absorption in the infrared spectrum around 3300 cm^{-1} means that no hydroxyl group is present. Since the aromatic ring accounts for all the double bonds and rings of the molecule (SODAR = 4), its oxygen atom can only be part of an ether function.

Two-proton signals in the nmr at $\delta = 3.5$ and 4.1 ppm suggest OCH_2 and CH_2Br. Each one is a triplet, indicating that each CH_2 is itself bonded to a CH_2 group. Compound L is 3-bromopropyl phenyl ether.

$$C_6H_5{-}O{-}CH_2{-}CH_2{-}CH_2{-}Br$$

triplet ($\delta = 4.1$ ppm) quintet ($\delta = 2.3$ ppm) triplet ($\delta = 3.5$ ppm)

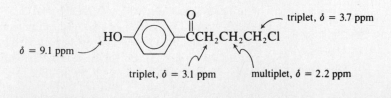

(*c*) Compound M, from its molecular formula, has a total of five double bonds and rings (SODAR = 5). A disubstituted aromatic ring, revealed by two two-proton doublets in the $\delta = 6.5$ to 8 ppm region of the ^1H nmr spectrum, accounts for four of those elements of unsaturation. Thus the remaining atoms must include one double bond or one ring. A strong band in the infrared at 1670 cm^{-1} indicates that this remaining element of unsaturation is the double bond of a carbonyl group. We also see an OH group absorption at 3300 cm^{-1} in the infrared spectrum. A one-proton signal at low field ($\delta = 9.1$ ppm) reveals the OH group to be that of a phenol.

Compound M is a disubstituted aromatic ring in which one of the substituents is a hydroxyl group and the other is a four-carbon chain that bears a chlorine substituent and contains a carbonyl group. The substitution pattern on the ring is para, as indicated by the characteristic AB quartet with a splitting of about 10 Hz. The side chain has three CH_2 groups, as shown by its nmr spectrum.

Compound M must be 3-chloropropyl *p*-hydroxyphenyl ketone.

$$HO{-}C_6H_4{-}\overset{O}{\overset{\|}{C}}CH_2CH_2CH_2Cl$$

$\delta = 9.1$ ppm (OH); triplet, $\delta = 3.1$ ppm; multiplet, $\delta = 2.2$ ppm; triplet, $\delta = 3.7$ ppm

SELF-TEST

PART A

A-1. Which is the stronger acid, *m*-hydroxybenzaldehyde or *p*-hydroxybenzaldehyde? Explain your answer, using resonance structures.

A-2. The cresols are methyl-substituted phenols. Predict the major products to be obtained from the reactions of *o*-, *m*-, and *p*-cresol with dilute nitric acid.

A-3. Give the structure of the product from the reaction of *p*-cresol with propanoyl chloride, $CH_3CH_2\overset{O}{\overset{\|}{C}}Cl$, in the presence of $AlCl_3$. What product is obtained in the absence of $AlCl_3$?

A-4. Provide the structure of the reactant, reagent, or product omitted from each of the following.

(*a*) C_6H_5–$OCH(CH_3)_2$ $\xrightarrow{HBr}$? (two products)

(*b*) ? (two compounds) $\longrightarrow$ $OCH_2CH(CH_3)_2$ benzene ring with CH_3 (meta)

(*c*) O^-Na^+ (benzene ring) $\xrightarrow[2.\ H^+]{1.\ ?}$ OH, CO_2H (benzene ring, ortho)

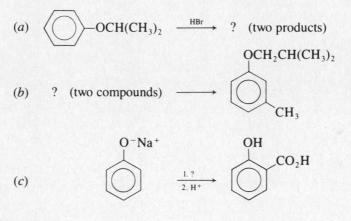

PART B

B-1. Rank the following in order of decreasing acid strength (most acidic → least acidic):

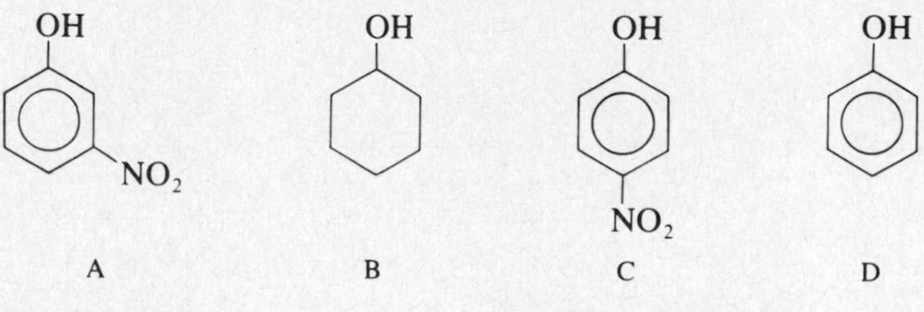

(*a*) B > D > A > C (*b*) C > A > B > D
(*c*) A > C > D > B (*d*) C > A > D > B

B-2. Which of the following choices correctly describes the solubility of benzoic acid (A) and *β*-naphthol (B) in the aqueous solutions shown?

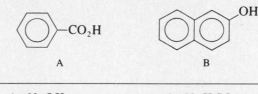

C_6H_5–CO_2H (A)   β-naphthol, OH (B)

	Aq NaOH	*Aq NaHCO$_3$*
(*a*)	A, soluble; B, insoluble	A, insoluble; B, soluble
(*b*)	A, insoluble; B, soluble	A, soluble; B, insoluble
(*c*)	A, soluble; B, soluble	A, soluble; B, insoluble
(*d*)	A, soluble; B, soluble	A, insoluble; B, soluble

B-3. Which of the following phenols has the largest pK_a value (i.e., is least acidic)?

(*a*) Cl–C_6H_4–OH (*b*) CH_3–C_6H_4–OH

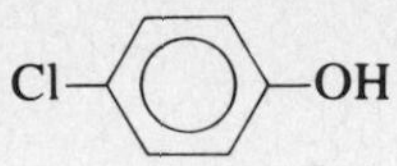

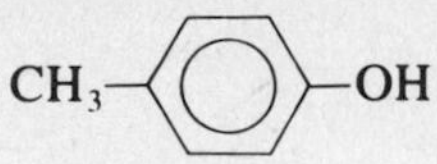

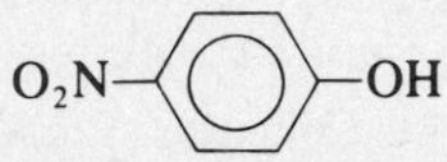

(*c*) O_2N–C_6H_4–OH (*d*) N≡C–C_6H_4–OH

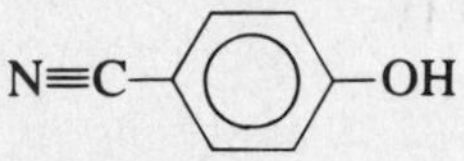

B-4. Which of the following reactions is a more effective method for preparing phenyl propyl ether?

A: $C_6H_5ONa + CH_3CH_2CH_2Br \longrightarrow$

B: $CH_3CH_2CH_2ONa + C_6H_5Br \longrightarrow$

(*a*) Reaction A is more effective.
(*b*) Reaction B is more effective.
(*c*) Both reactions A and B are effective.
(*d*) Neither reaction A nor reaction B is effective.

CHAPTER 27

CARBOHYDRATES

IMPORTANT TERMS AND CONCEPTS

Carbohydrate Classification (Secs. 27.1 to 27.4) The basis for carbohydrate classification is the term *saccharide. Monosaccharides* are simple carbohydrates, that is, those which are not cleaved on hydrolysis to smaller carbohydrates. *Disaccharides* are cleaved to two monosaccharides on hydrolysis. The terms *oligosaccharides* and *polysaccharides* refer to carbohydrates which on hydrolysis yield 3 to 10 and more than 10 monosaccharide units, respectively.

Monosaccharides that are polyhydroxy aldehydes are known as *aldoses*; those that are polyhydroxy ketones are *ketoses*. A carbohydrate having three, four, five, or six carbon atoms is termed a *triose*, *tetrose*, *pentose*, or *hexose*, respectively.

Stereochemical Notation (Sec. 27.2) The stereochemical representation most widely used for carbohydrates is known as a *Fischer projection* formula. As the example below demonstrates, the horizontal line represents bonds projecting above the page, and the vertical line represents bonds projecting into the page.

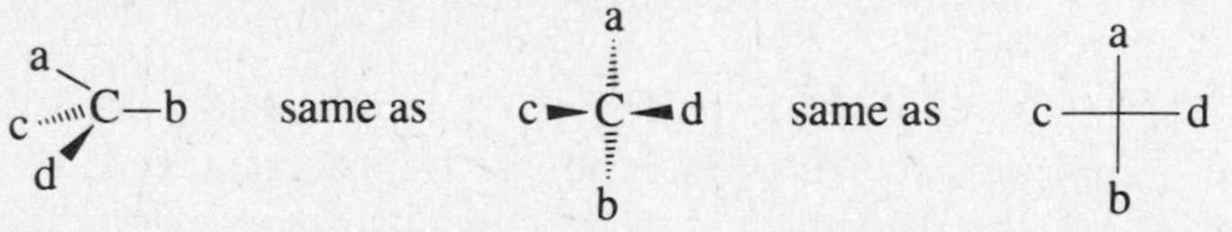

a, b, c, d on C (wedge/dash) same as c ► C ◄ d (a above, b below) same as c—+—d (a above, b below)

Fischer projection

Carbohydrates are identified as being either D or L depending on their relationship to dextrorotatory and levorotatory glyceraldehyde, respectively. That is, a D carbohydrate has the same configuration at its highest numbered chiral center as that of (+)-glyceraldehyde.

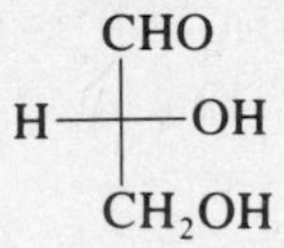

CHO; H—OH; CH_2OH

D-(+)-Glyceraldehyde

CHO; HO—H; HO—H; H—OH; CH_2OH

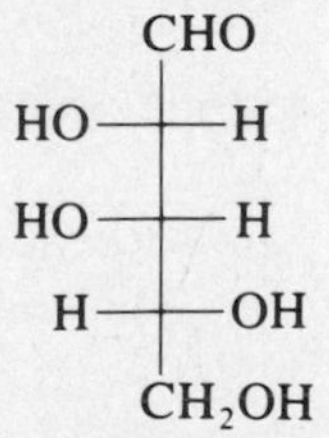

D-Lyxose

Cyclic Forms of Carbohydrates (Secs. 27.6 to 27.8) Aldoses and ketoses exist as *cyclic hemiacetals.* Five-membered ring forms are known as *furanose* forms; the six-membered rings are *pyranose* forms. The ring carbon derived from the carbonyl is called the *anomeric* carbon.

Haworth formulas are often used to represent the cyclic hemiacetal forms of carbohydrates, as shown below. While useful for representing the *configurational* relationships of carbohydrates, it should be recognized that Haworth formulas provide no information as to carbohydrate *conformations.*

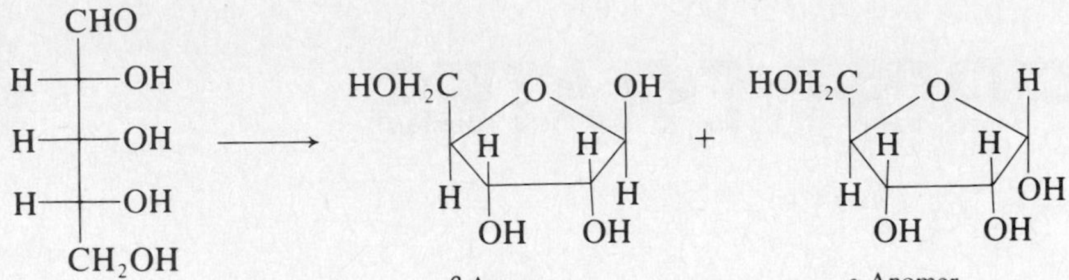

CHO; H—OH; H—OH; H—OH; CH_2OH ⟶ HOH_2C, O, OH, H, H, H, H, OH, OH + HOH_2C, O, H, H, H, H, OH, OH, OH

β Anomer   *α* Anomer

The hydroxyl group on the anomeric carbon may adopt either of two configurations in the cyclic hemiacetal. In the *D* series the hydroxyl is *α* when "down" in the Haworth formula and *β* when "up."

As may be shown by optical rotation measurements, in solution the individual anomers of a carbohydrate rapidly equilibrate with each other by a process known as *mutarotation.*

Glycosides (Secs. 27.13, 27.14) Carbohydrate derivatives formed by the replacement of the anomeric hydroxyl group by some other substituent are known as *glycosides.* Disaccharides are examples of glycosides in which the anomeric hydroxyl has been replaced by a second sugar molecule. Examples include *maltose,* obtained by hydrolysis of starch, and *cellobiose,* obtained by hydrolysis of cellulose.

CARBOHYDRATE REACTIONS

Reduction (Sec. 27.18) Typical procedures for the reduction of carbohydrates employ catalytic hydrogenation or sodium borohydride. The reduction products are known as *alditols.*

Oxidation (Sec. 27.19) Aldoses undergo oxidation of the aldehyde group readily. A solution of copper(II) ion is employed in *Benedict's reagent*; formation of red copper(I) oxide is taken as a positive test. Carbohydrates which give a positive test with Benedict's solution are termed *reducing sugars.* Other reagent solutions used to test for reducing sugars are *Fehling's solution,* a copper(II) tartrate complex, and *Tollens' reagent,* a solution of ammoniacal silver ion.

Carbohydrate derivatives in which the aldehyde group has been oxidized are known as *aldonic acids.* Oxidation with nitric acid leads to formation of *aldaric acids,* in which both the aldehyde and the terminal hydroxyl groups have been oxidized to carboxylic acids.

Osazone Formation (Sec. 27.21) Reaction of aldoses and ketoses with an excess of phenylhydrazine converts two adjacent carbons (the carbonyl and the nearest hydroxyl carbon) to phenylhydrazone units, forming what is known as an *osazone.*

SOLUTIONS TO TEXT PROBLEMS

27.1 (*b*) Redraw the Fischer projection so as to show the orientation of the groups in three dimensions.

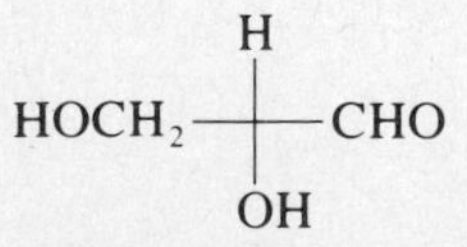

H; $HOCH_2$—CHO; OH   is equivalent to   H; $HOCH_2$►C◄CHO; OH

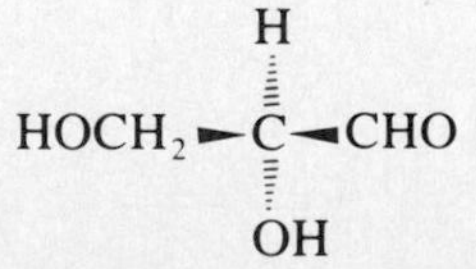

Reorient the three-dimensional representation, putting the aldehyde group at the top and the primary alcohol at the bottom.

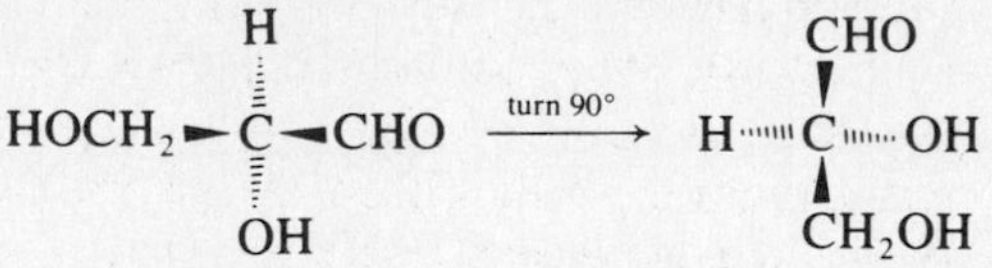

H | $HOCH_2$—C—CHO | OH $\xrightarrow{\text{turn } 90^\circ}$ CHO | H—C—OH | CH_2OH

What results is not equivalent to a proper Fischer projection because the horizontal bonds are directed "back" when they should be "forward." The opposite is true for the vertical bonds. In order to make the drawing correspond to a proper Fischer projection, we need to rotate it 180° around a vertical axis.

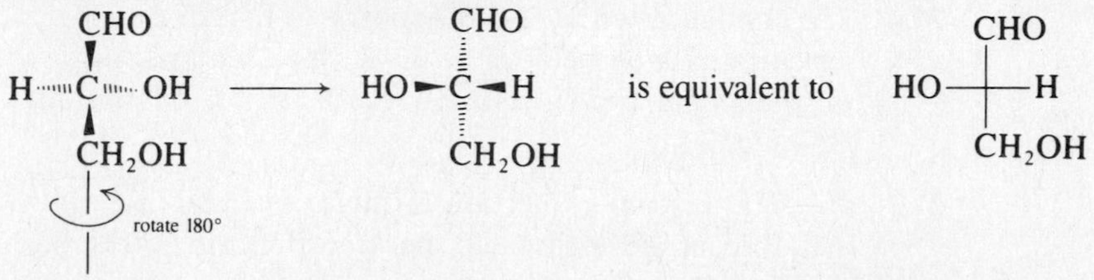

CHO | H—C—OH | CH_2OH (rotate 180°) ⟶ CHO | HO—C—H | CH_2OH is equivalent to CHO | HO—+—H | CH_2OH

Now, having the molecule arranged properly, we see that it is L-glyceraldehyde.

(*c*) Again proceed by converting the Fischer projection into a three-dimensional representation.

CHO | $HOCH_2$—+—H | OH is equivalent to CHO | $HOCH_2$—C—H | OH

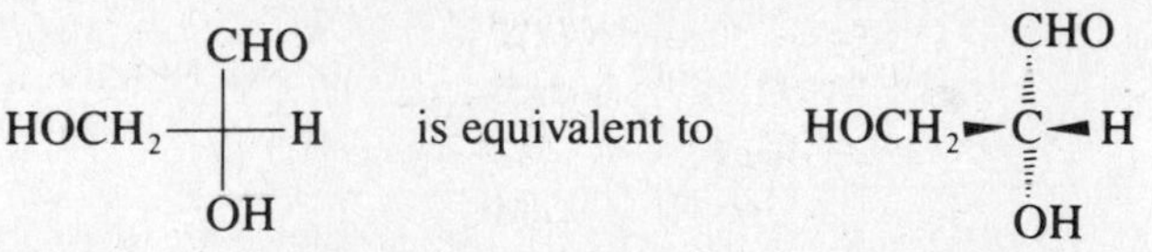

Look at the drawing from a perspective that permits you to see the carbon chain oriented vertically with the aldehyde at the top and the CH_2OH at the bottom. Both groups should point away from you. When examined from this perspective, the hydrogen is to the left and the hydroxyl to the right with both pointing toward you.

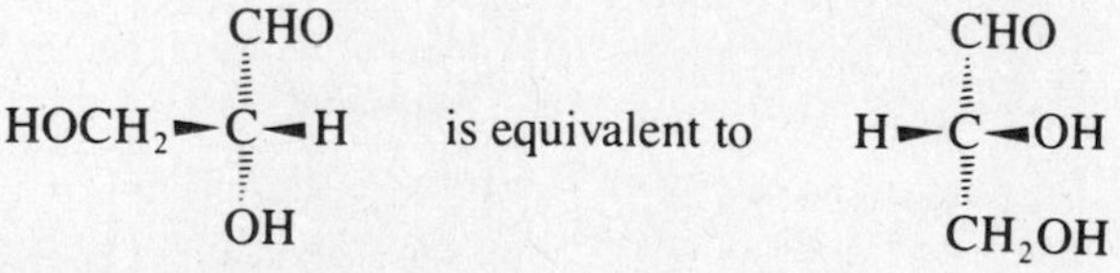

CHO | $HOCH_2$—C—H | OH is equivalent to CHO | H—C—OH | CH_2OH

The molecule is D-glyceraldehyde.

27.2 In order to match the compound with the Fischer projection formulas of the aldotetroses, first redraw it in an eclipsed conformation.

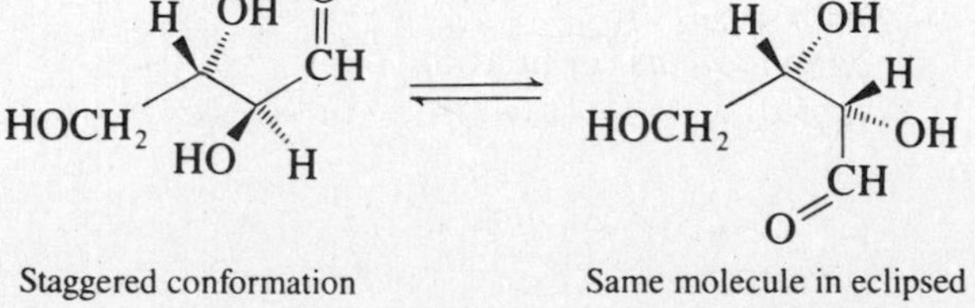

Staggered conformation ⇌ Same molecule in eclipsed conformation

The eclipsed conformation shown, when oriented so that the aldehyde carbon is at the top, vertical bonds back, and horizontal bonds pointing outward from their chiral centers, is readily transformed into the Fischer projection of L-erythrose.

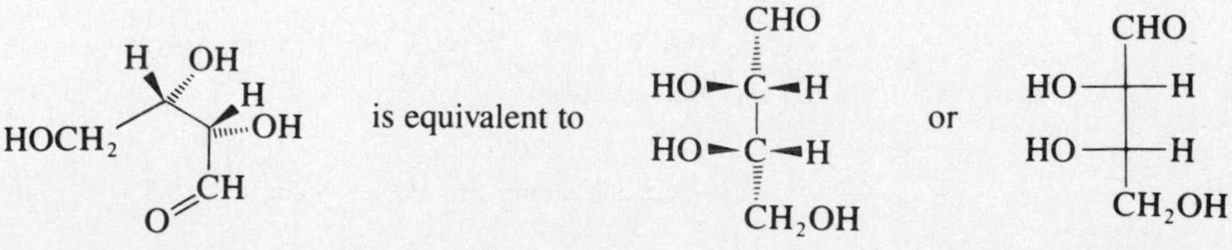

is equivalent to CHO | HO—C—H | HO—C—H | CH_2OH or CHO | HO—+—H | HO—+—H | CH_2OH

L-Erythrose

27.3 L-Arabinose is the mirror image of D-arabinose, the structure of which is given in text Figure 27.3. The configuration at *each* chiral center of D-arabinose must be reversed to transform it into L-arabinose.

D-(−)-Arabinose		
	CHO	
HO		H
H		OH
H		OH
	CH_2OH	

L-(+)-Arabinose		
	CHO	
H		OH
HO		H
HO		H
	CH_2OH	

27.4 The configuration at C-5 is opposite to that of D-(+)-glyceraldehyde. Therefore, this particular carbohydrate belongs to the L series. Comparing it with the Fischer projection formulas of the eight D-aldohexoses reveals it to be the mirror image of D-(+)-talose; it is L-(−)-talose.

27.5 (*b*) The Fischer projection formula of D-arabinose may be found in text Figure 27.3. The Fischer projection and the "coiled" form corresponding to it are:

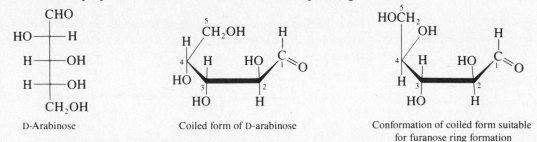

D-Arabinose		
	CHO	
HO		H
H		OH
H		OH
	CH_2OH	

Coiled form of D-arabinose

Conformation of coiled form suitable for furanose ring formation

Cyclic hemiacetal formation between the carbonyl group and the C-4 hydroxyl yields the α- and β-furanose forms of D-arabinose.

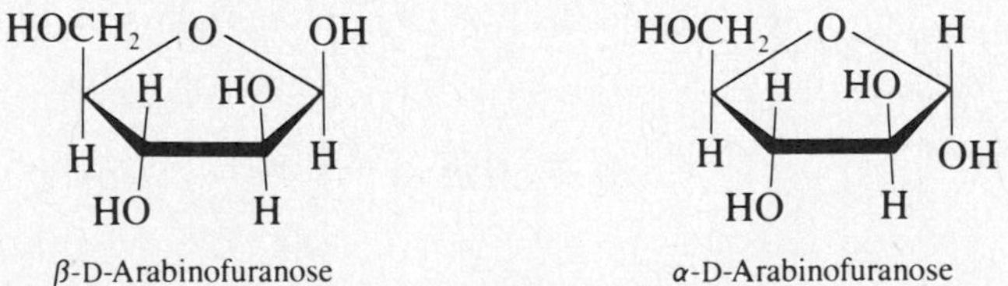

β-D-Arabinofuranose

α-D-Arabinofuranose

(*c*) The mirror image of D-arabinose [from (*b*)] is L-arabinose.

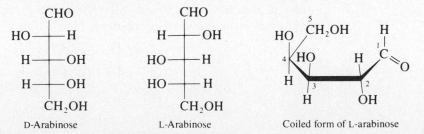

D-Arabinose		
	CHO	
HO		H
H		OH
H		OH
	CH_2OH	

L-Arabinose		
	CHO	
H		OH
HO		H
HO		H
	CH_2OH	

Coiled form of L-arabinose

The C-4 atom of the coiled form of L-arabinose must be rotated 120° in a clockwise sense so as to bring its hydroxyl group into the proper orientation for furanose ring formation.

rotate about C-3−C-4 bond

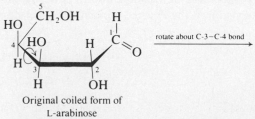

Original coiled form of L-arabinose

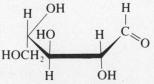

Conformation of coiled form suitable for furanose ring formation

Cyclization gives the α- and β-furanose forms of L-arabinose.

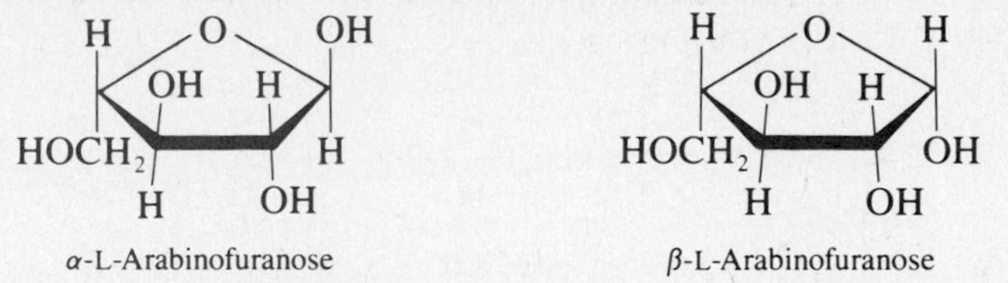

α-L-Arabinofuranose

β-L-Arabinofuranose

In the L series the anomeric hydroxyl is up in the α isomer and down in the β isomer.

(*d*) The Fischer projection formula for D-threose is given in Figure 27.3. Reorientation of that projection into a form that illustrates its potential for cyclization is shown.

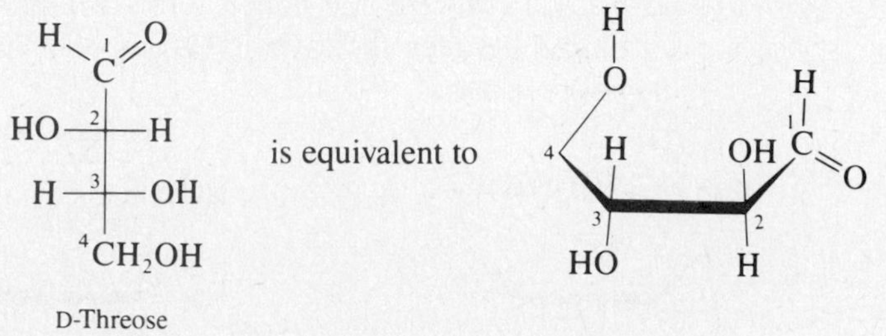

is equivalent to

D-Threose

Cyclization yields the two stereoisomeric furanose forms.

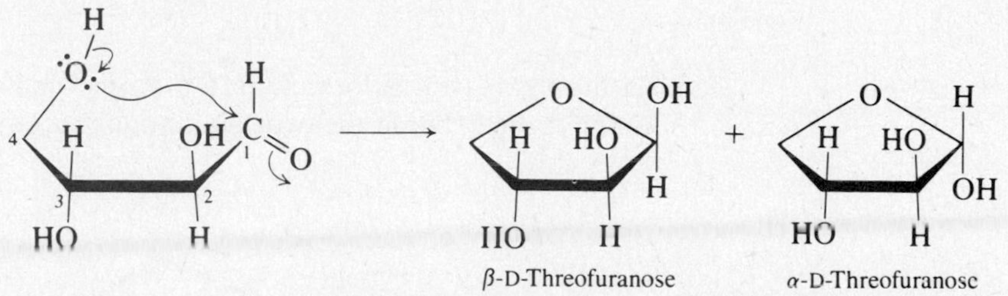

β-D-Threofuranose

α-D-Threofuranose

27.6 (*b*) The Fischer projection and Haworth formula for D-mannose are:

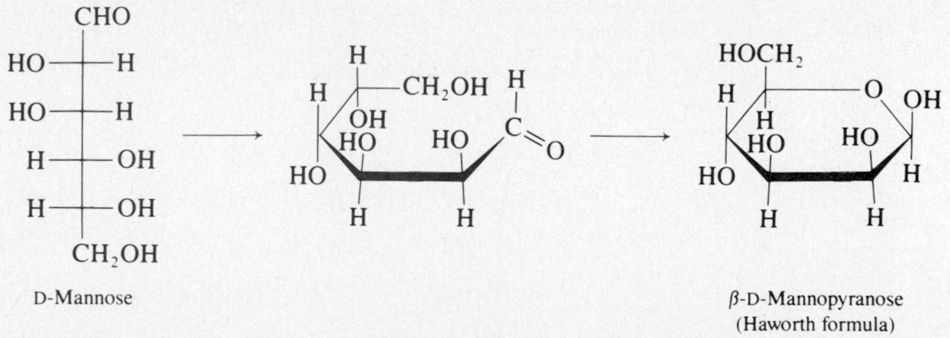

D-Mannose

β-D-Mannopyranose (Haworth formula)

The Haworth formula is more realistically drawn as the following chair conformation.

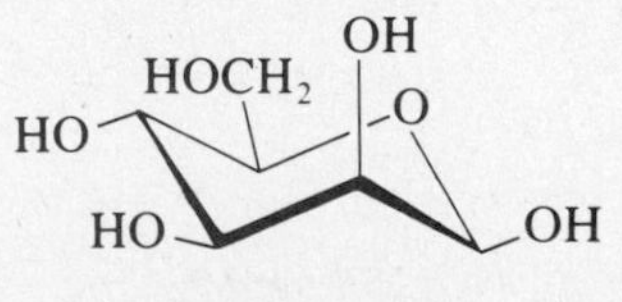

β-D-Mannopyranose

Mannose differs from glucose in configuration at C-2. All hydroxyl groups are equatorial in β-D-glucopyranose; the hydroxyl at C-2 is axial in β-D-mannopyranose.

(*c*) The conformational depiction of β-L-mannopyranose begins in the same way as that of β-D-mannopyranose. L-Mannose is the mirror image of D-mannose.

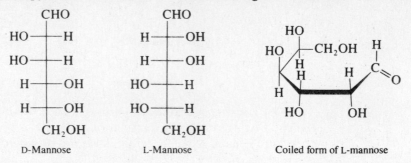

D-Mannose L-Mannose Coiled form of L-mannose

In order to rewrite the coiled form of L-mannose in a way that permits hemiacetal formation between the carbonyl group and the C-5 hydroxyl, C-5 is rotated 120° in the clockwise sense.

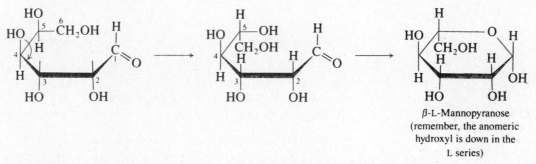

β-L-Mannopyranose (remember, the anomeric hydroxyl is down in the L series)

Translating the Haworth formula into a proper conformational depiction requires that a choice be made between the two chair conformations shown:

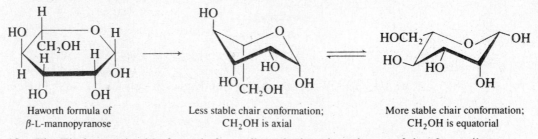

Haworth formula of β-L-mannopyranose

Less stable chair conformation; CH_2OH is axial

More stable chair conformation; CH_2OH is equatorial

(*d*) The Fischer projection formula for L-ribose is the mirror image of that for D-ribose.

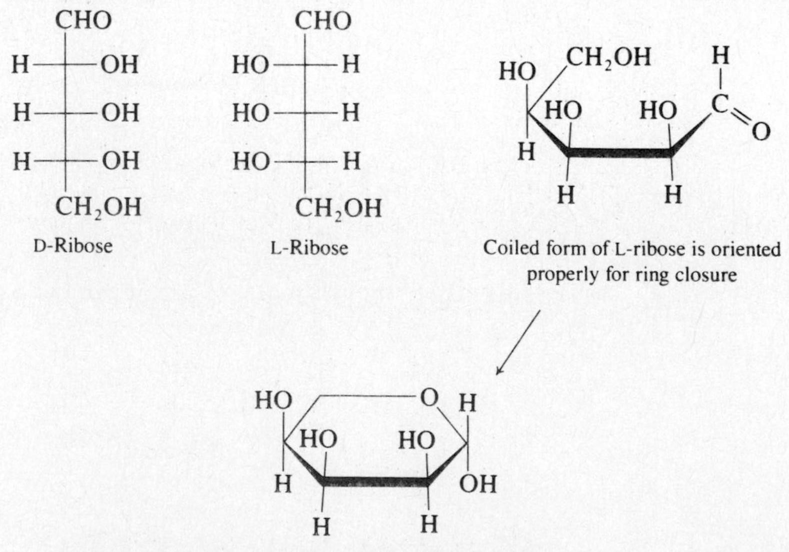

D-Ribose L-Ribose

Coiled form of L-ribose is oriented properly for ring closure

Haworth formula of β-L-ribopyranose

Of the two chair conformations of β-L-ribose, the one with the greater number of equatorial substituents is more stable.

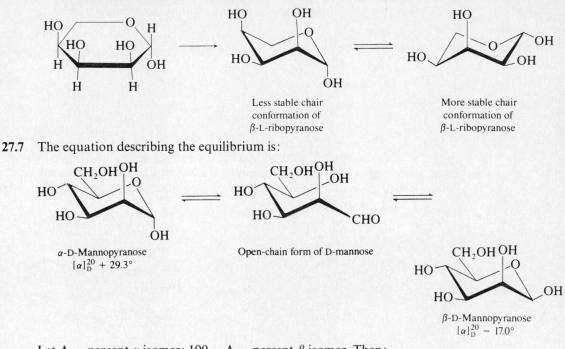

Less stable chair conformation of β-L-ribopyranose

More stable chair conformation of β-L-ribopyranose

27.7 The equation describing the equilibrium is:

α-D-Mannopyranose $[\alpha]_D^{20}$ + 29.3°

Open-chain form of D-mannose

β-D-Mannopyranose $[\alpha]_D^{20}$ − 17.0°

Let A = percent α isomer; 100 − A = percent β isomer. Then:

$$A\,(+29.3^\circ) + (100 - A)(-17.0^\circ) = 100\,(14.2^\circ)$$

$$46.3\ A = 3120$$

$$\text{Percent } \alpha \text{ isomer} = 67\%$$

$$\text{Percent } \beta \text{ isomer} = (100 - A) = 33\%$$

27.8 Review carbohydrate terminology by referring to Table 27.1. A *tetrulose* is a four-carbon ketose. Setting down a Fischer projection formula for a four-carbon ketose reveals that only one chiral center is present, and thus there are only two tetruloses. They are mirror images of each other and are known as D- and L-erythrulose.

```
      CH2OH             CH2OH
      |                 |
      C=O               C=O
      |                 |
  H --+-- OH       HO --+-- H
      |                 |
      CH2OH             CH2OH
  D-Erythrulose     L-Erythrulose
```

27.9 (*b*) Since L-fucose is 6-deoxy-L-galactose, first write the Fischer projection formula of D-galactose, then transform it to its mirror image, L-galactose. Transform the C-6 CH_2OH group to CH_3 to produce 6-deoxy-L-galactose.

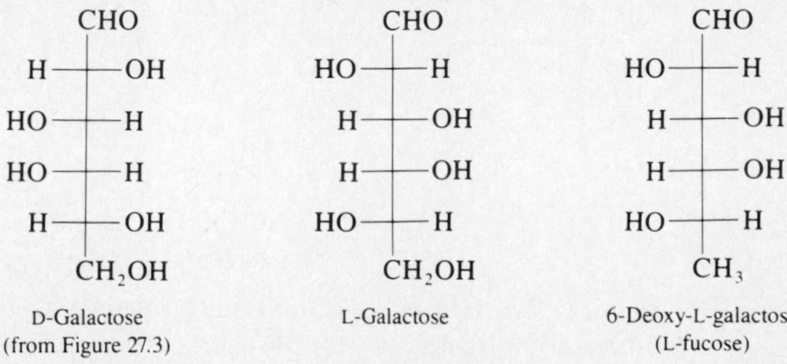

D-Galactose (from Figure 27.3) L-Galactose 6-Deoxy-L-galactose (L-fucose)

27.10 Acid-catalyzed addition of methanol to the glycal proceeds by regioselective protonation of the double bond in the direction that leads to the more stable carbocation. Here again, the more stable carbocation is the one stabilized by the ring oxygen.

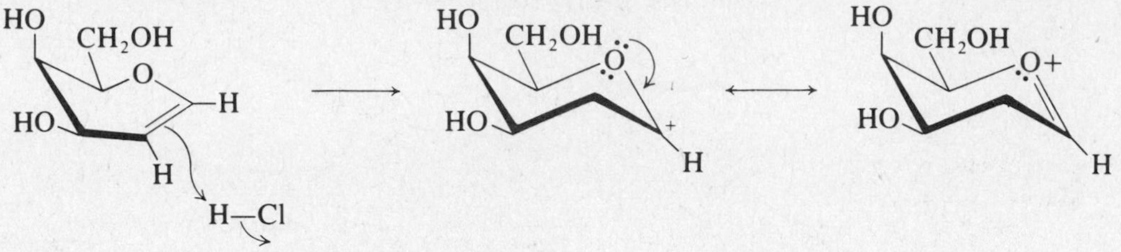

Capture of the carbocation by methanol yields the α and β methyl glycosides.

27.11 The hemiacetal function opens to give an intermediate containing a free aldehyde function. Cyclization of this intermediate can produce either the α or the β configuration at this center.

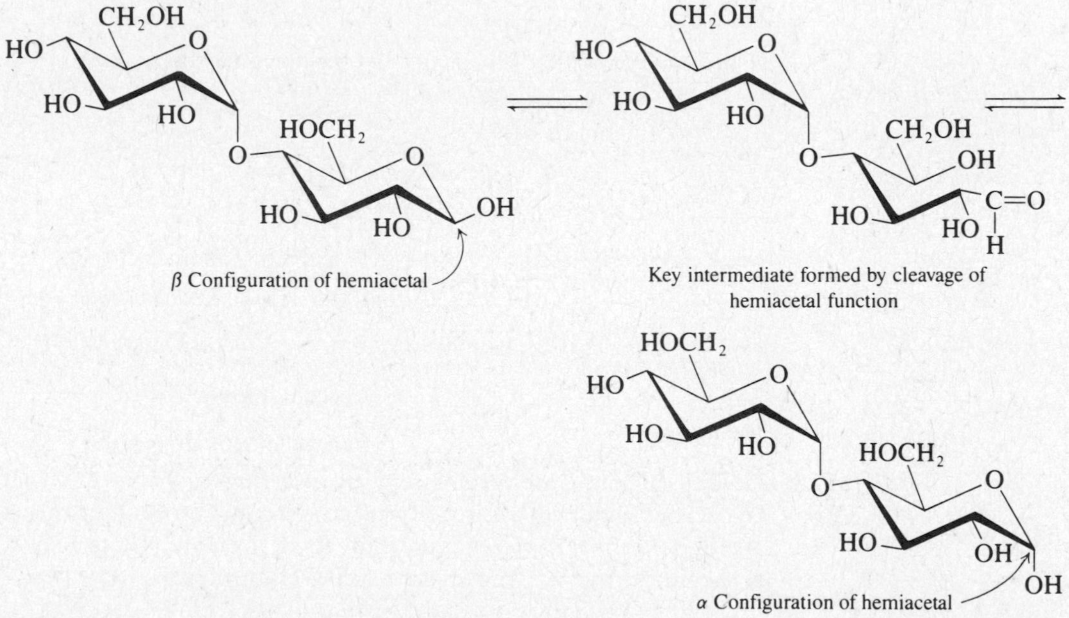

β Configuration of hemiacetal

Key intermediate formed by cleavage of hemiacetal function

α Configuration of hemiacetal

Only the configuration of the free hemiacetal function is affected in this process. The α configuration of the glycosidic linkage remains unchanged.

27.12 Cellulose is to cellobiose as amylose is to *maltose.* The polysaccharide cellulose and the disaccharide cellobiose contain β(1,4)-linked D-glucose units. Amylose contains α(1,4)-linked D-glucose units, as does the disaccharide *maltose* (Section 27.14). As cellulose can be thought of as a collection of cellobiose units, amylose can be thought of as a collection of maltose units.

27.13 Write the chemical equation so that you can clearly relate the product to the starting material.

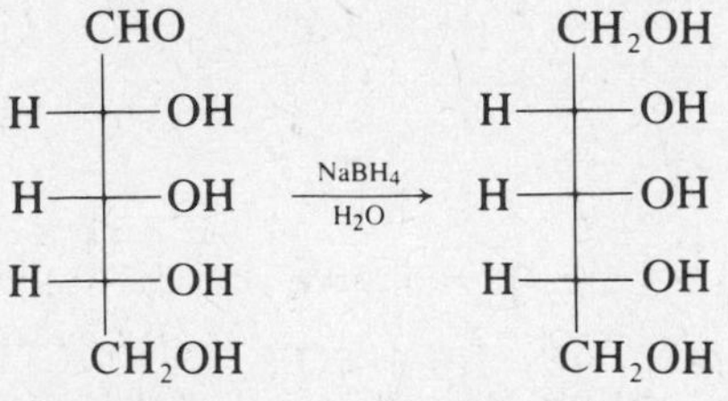

D-Ribose $\xrightarrow[H_2O]{NaBH_4}$ Ribitol

Ribitol is a meso form; it is not optically active. A plane of symmetry passing through C-3 bisects the molecule.

27.14 (*b*) Arabinose is a reducing sugar; it will give a positive test with Benedict's reagent because its open-chain form has a free aldehyde group capable of being oxidized by cupric ion.

(*c*) Benedict's reagent reacts with α-hydroxy ketones by way of an isomerization process involving an enediol intermediate.

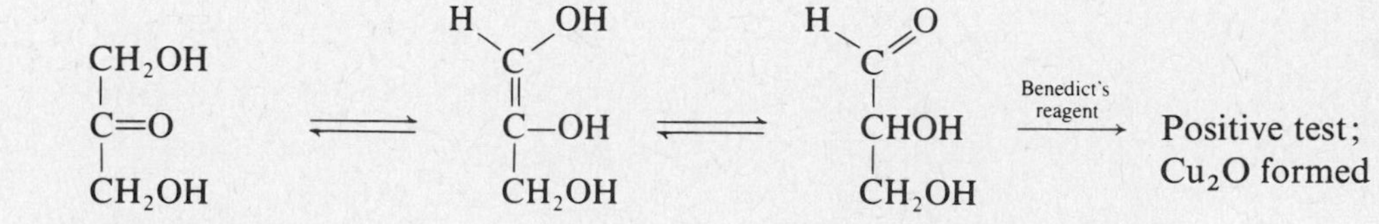

$$\underset{\text{1,3-Dihydroxyacetone}}{\begin{array}{c}CH_2OH\\ |\\ C{=}O\\ |\\ CH_2OH\end{array}} \rightleftharpoons \begin{array}{c}H\quad OH\\ \diagdown\ \diagup\\ C\\ \|\\ C{-}OH\\ |\\ CH_2OH\end{array} \rightleftharpoons \underset{\text{Glyceraldehyde}}{\begin{array}{c}H\quad O\\ \diagdown\ \diagup\!\!\!\!/\\ C\\ |\\ CHOH\\ |\\ CH_2OH\end{array}} \xrightarrow{\text{Benedict's reagent}} \text{Positive test; } Cu_2O \text{ formed}$$

1,3-Dihydroxyacetone gives a positive test with Benedict's reagent.

(*d*) D-Fructose is an α-hydroxy ketone and will give a positive test with Benedict's reagent.

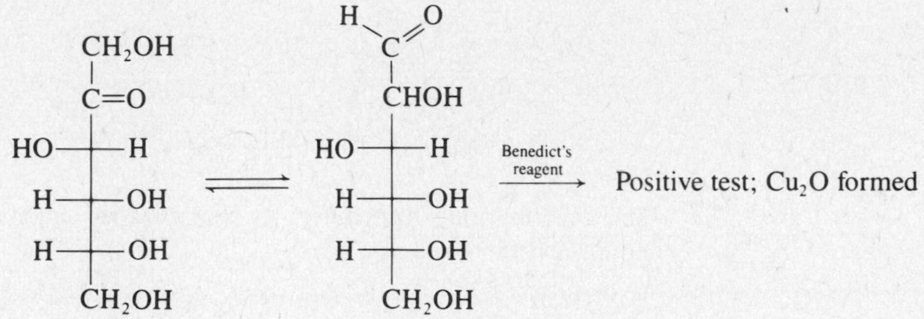

$$\underset{\text{D-Fructose}}{\begin{array}{c}CH_2OH\\ |\\ C{=}O\\ HO{-}C{-}H\\ H{-}C{-}OH\\ H{-}C{-}OH\\ |\\ CH_2OH\end{array}} \rightleftharpoons \begin{array}{c}H\quad O\\ \diagdown\ \diagup\!\!\!\!/\\ C\\ |\\ CHOH\\ HO{-}C{-}H\\ H{-}C{-}OH\\ H{-}C{-}OH\\ |\\ CH_2OH\end{array} \xrightarrow{\text{Benedict's reagent}} \text{Positive test; } Cu_2O \text{ formed}$$

(*e*) Lactose is a disaccharide and will give a positive test with Benedict's reagent by way of an open-chain isomer of one of the rings. Lactose is a reducing sugar.

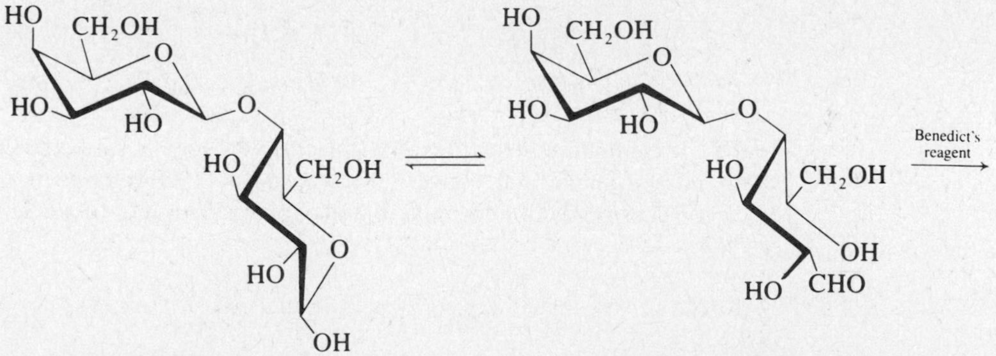

Lactose
(structure presented in Section 27.14)

⇌

Open-chain form

Benedict's reagent →

Positive test; Cu_2O formed

(*f*) Amylose is a polysaccharide. Its glycoside linkages are inert to Benedict's reagent, but the terminal glucose residues at the ends of the chain and its branches are hemiacetals in equilibrium with open-chain structures. A positive test is expected.

27.15 Since the groups at both ends of the carbohydrate chain are oxidized to carboxylic acid functions, two combinations of one CH_2OH with one CHO group are possible.

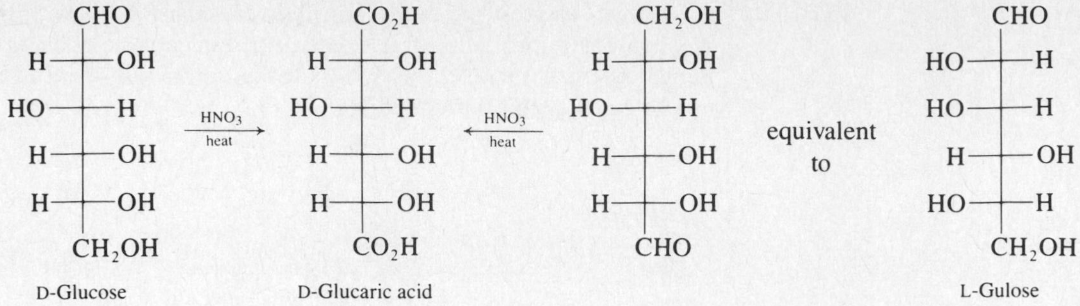

D-Glucose: CHO; H—OH; HO—H; H—OH; H—OH; CH_2OH

$\xrightarrow[\text{heat}]{HNO_3}$

D-Glucaric acid: CO_2H; H—OH; HO—H; H—OH; H—OH; CO_2H

$\xleftarrow[\text{heat}]{HNO_3}$

CH_2OH; H—OH; HO—H; H—OH; H—OH; CHO

equivalent to

L-Gulose: CHO; HO—H; HO—H; H—OH; HO—H; CH_2OH

L-Gulose yields the same aldaric acid on oxidation as does D-glucose.

27.16 (*b*) The phenylosazone derived from L-erythrose is the same as the one from L-threose. The two sugars differ in configuration only at C-2.

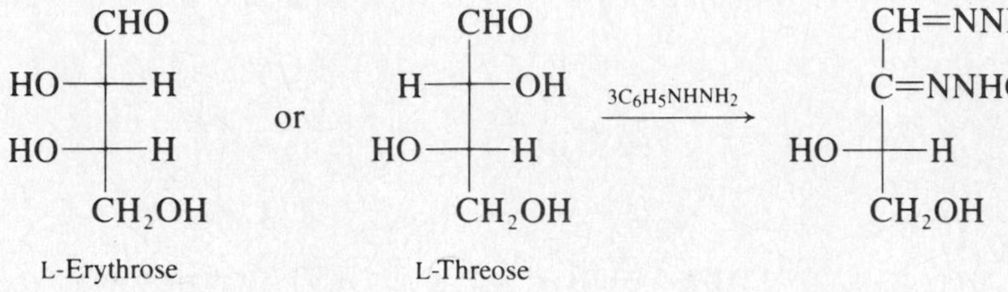

L-Erythrose: CHO; HO—H; HO—H; CH_2OH

or

L-Threose: CHO; H—OH; HO—H; CH_2OH

$\xrightarrow{3C_6H_5NHNH_2}$

$CH{=}NNHC_6H_5$; $C{=}NNHC_6H_5$; HO—H; CH_2OH

(*c*) The carbohydrate that differs in configuration from D-allose only at C-2 is D-altrose. Both yield the same osazone.

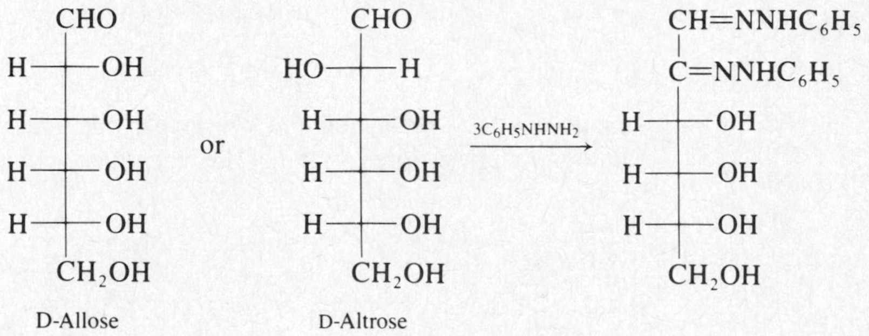

D-Allose: CHO; H—OH; H—OH; H—OH; H—OH; CH_2OH

or

D-Altrose: CHO; HO—H; H—OH; H—OH; H—OH; CH_2OH

$\xrightarrow{3C_6H_5NHNH_2}$

$CH{=}NNHC_6H_5$; $C{=}NNHC_6H_5$; H—OH; H—OH; H—OH; CH_2OH

(*d*) First find the structure of D-galactose in Figure 27.3, then write the structure of its mirror image and convert the terminal CH_2OH group to a CH_3 group to give the Fischer projection formula of 6-deoxy-L-galactose (L-fucose).

D-Galactose: CHO; H—OH; HO—H; HO—H; H—OH; CH_2OH

L-Galactose: CHO; HO—H; H—OH; H—OH; HO—H; CH_2OH

6-Deoxy-L-galactose (L-fucose): CHO; HO—H; H—OH; H—OH; HO—H; CH_3

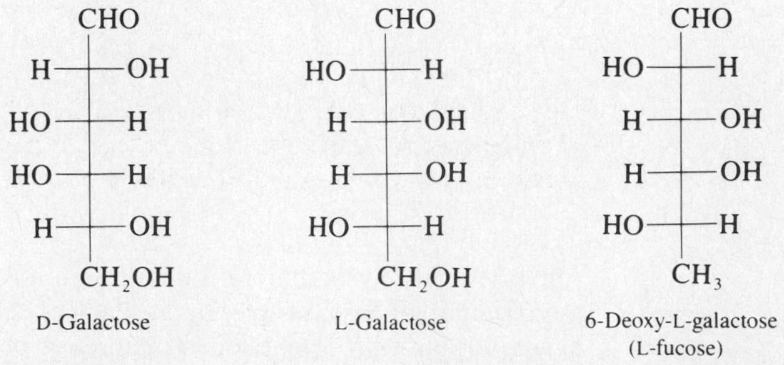

The compound that has the opposite configuration at C-2 to L-fucose and thus will give the same osazone is 6-deoxy-L-talose.

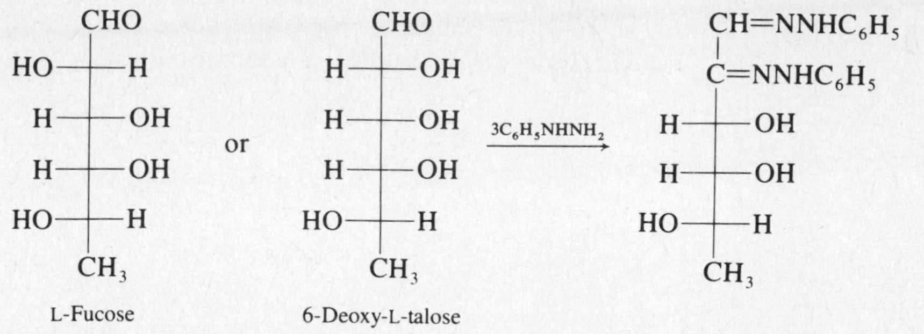

CHO / HO—H / H—OH / H—OH / HO—H / CH_3 — L-Fucose — or — CHO / H—OH / H—OH / H—OH / HO—H / CH_3 — 6-Deoxy-L-talose $\xrightarrow{3C_6H_5NHNH_2}$ $CH{=}NNHC_6H_5$ / $C{=}NNHC_6H_5$ / H—OH / H—OH / HO—H / CH_3

L-Fucose   6-Deoxy-L-talose

27.17 In analogy with the *D*-fructose ⇌ D-glucose interconversion, dihydroxyacetone phosphate and D-glyceraldehyde 3-phosphate can equilibrate by way of an enediol intermediate.

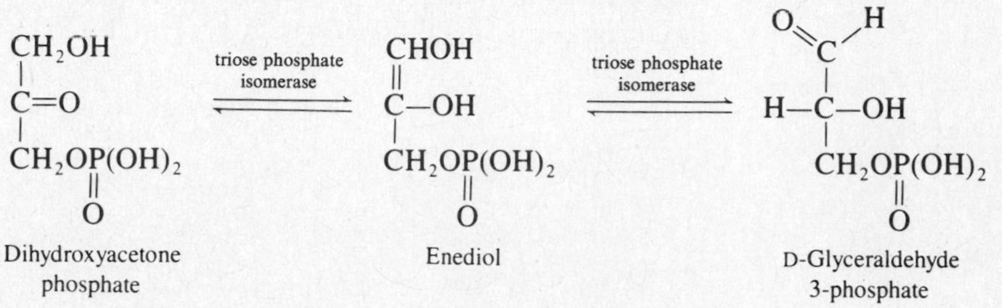

CH_2OH / C=O / $CH_2OP(OH)_2$ (P=O) $\underset{\text{isomerase}}{\overset{\text{triose phosphate}}{\rightleftharpoons}}$ CHOH / ‖ C—OH / $CH_2OP(OH)_2$ (P=O) $\underset{\text{isomerase}}{\overset{\text{triose phosphate}}{\rightleftharpoons}}$ O=C—H / H—C—OH / $CH_2OP(OH)_2$ (P=O)

Dihydroxyacetone phosphate   Enediol   D-Glyceraldehyde 3-phosphate

27.18 (*b*) The points of cleavage of D-ribose on treatment with periodic acid are as indicated.

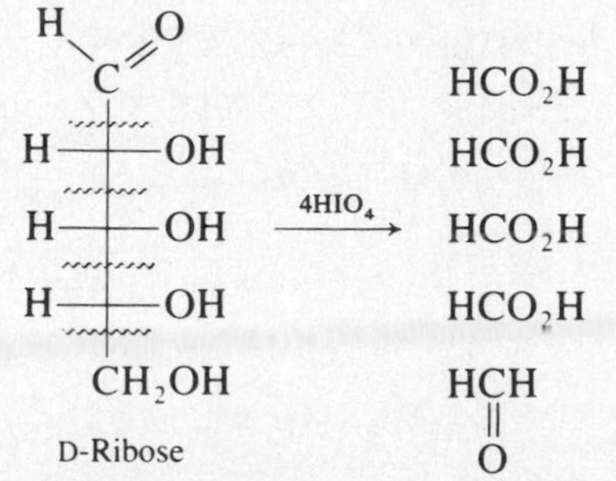

H—C=O / H—OH / H—OH / H—OH / CH_2OH (D-Ribose) $\xrightarrow{4HIO_4}$ HCO_2H / HCO_2H / HCO_2H / HCO_2H / HCH=O

D-Ribose

Four equivalents of periodic acid are required. Four equivalents of formic acid and one of formaldehyde are produced.

(*c*) Write the structure of methyl *β*-D-glucopyranoside so as to identify the adjacent alcohol functions.

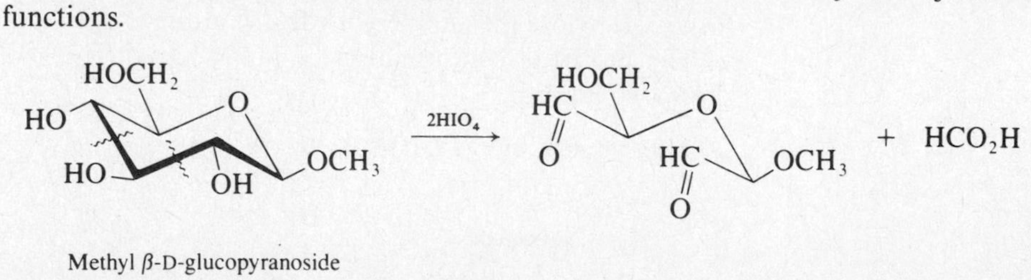

$HOCH_2$, HO, HO, OH, O, OCH_3 (Methyl β-D-glucopyranoside) $\xrightarrow{2HIO_4}$ $HOCH_2$, HC=O, O, HC=O, OCH_3 + HCO_2H

Methyl *β*-D-glucopyranoside

Two equivalents of periodic acid are required. One mole of formic acid is produced.

(*d*) There are two independent vicinal diol functions in this glycoside. Two moles of periodic acid are required per mole of substrate.

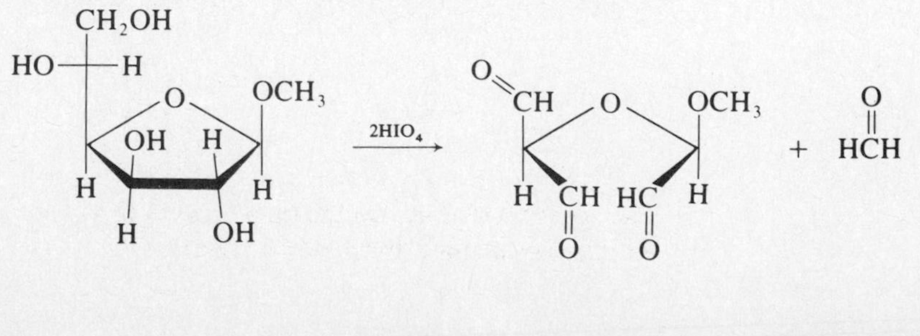

CH_2OH / HO—H / furanoside ring (O, OCH_3, OH, H, H, H, H, OH) $\xrightarrow{2HIO_4}$ O=CH, O, OCH_3, H, CH=O, HC=O, H + HCH=O

27.19 (*a*) The structure shown in Figure 27.3 is D-(+)-xylose; therefore (−)-xylose must be its mirror image and has the L-configuration at C-4.

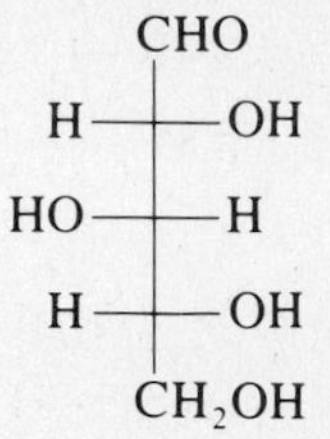

$$\begin{array}{rcl} & CHO & \\ H & - & OH \\ HO & - & H \\ H & - & OH \\ & CH_2OH & \end{array} \qquad\qquad \begin{array}{rcl} & CHO & \\ HO & - & H \\ H & - & OH \\ HO & - & H \\ & CH_2OH & \end{array}$$

D-(+)-Xylose (from Figure 27.3) L-(−)-Xylose

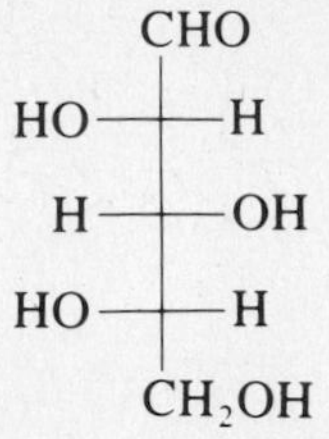

(*b*) Alditols are the reduction products of carbohydrates; D-xylitol is derived from D-xylose by conversion of the terminal $-CHO$ to $-CH_2OH$.

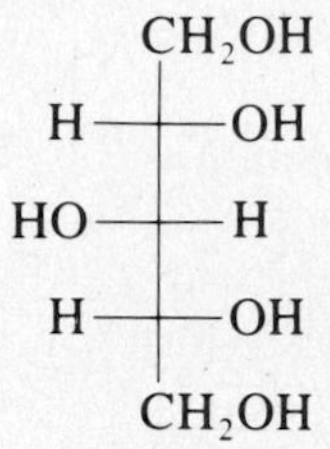

$$\begin{array}{rcl} & CH_2OH & \\ H & - & OH \\ HO & - & H \\ H & - & OH \\ & CH_2OH & \end{array}$$

D-Xylitol

(*c*) Redraw the Fischer projection of D-xylose in its coiled form.

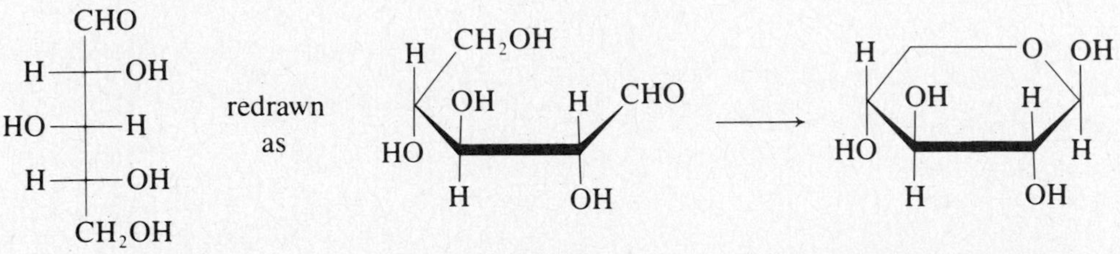

$$\begin{array}{rcl} & CHO & \\ H & - & OH \\ HO & - & H \\ H & - & OH \\ & CH_2OH & \end{array}$$

D-Xylose

redrawn as

Coiled form of D-xylose

Haworth formula of β-D-xylopyranose

The pyranose form arises by closure to a six-membered cyclic hemiacetal, with the C-5 hydroxyl group undergoing nucleophilic addition to the carbonyl. In the β-pyranose form of D-xylose the anomeric hydroxyl group is up.

The preferred conformation of β-D-xylopyranose is a chair with all the hydroxyl groups equatorial.

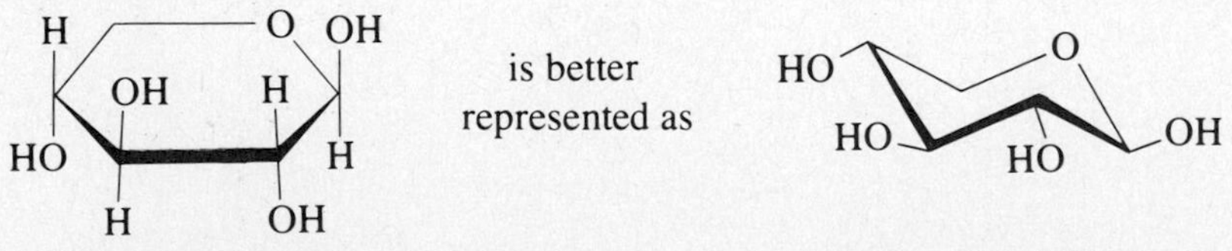

Haworth formula of β-D-xylopyranose

is better represented as

Chair conformation of β-D-xylopyranose

(*d*) L-Xylose is the mirror image of D-xylose.

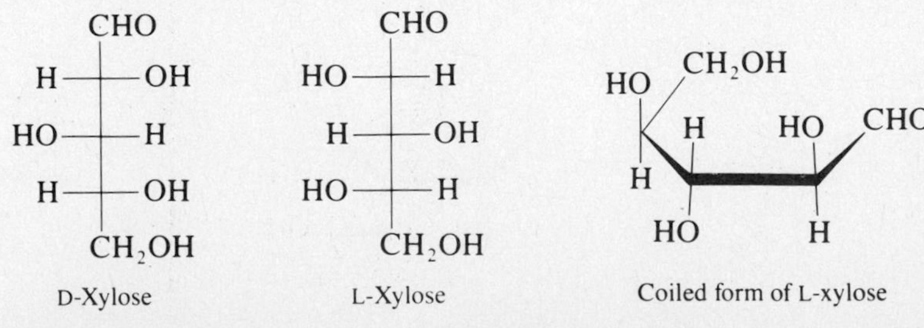

D-Xylose L-Xylose Coiled form of L-xylose

In order to construct the furanose form of L-xylose, the hydroxyl at C-4 needs to be brought into the proper orientation to form a five-membered ring.

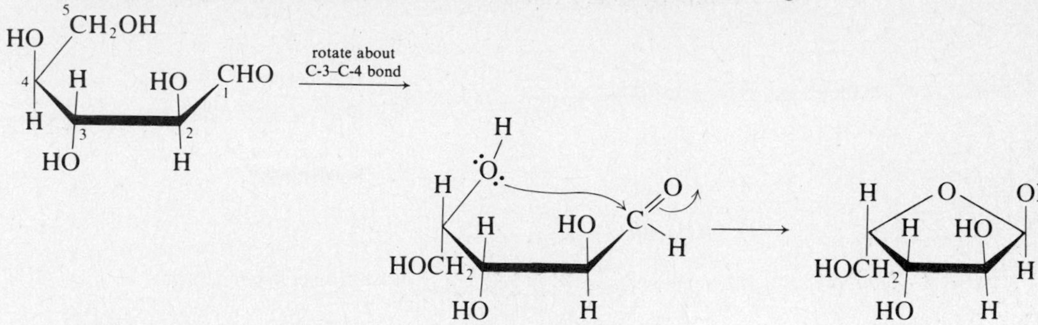

HO $\overset{5}{C}H_2OH$ / H HO CHO / H HO H (C-1 to C-4 labelled 1, 2, 3, 4)

$\xrightarrow{\text{rotate about C-3–C-4 bond}}$

H–Ö: (attacking C=O) / H H HO C(=O)H / $HOCH_2$ / HO H $\longrightarrow$ H O OH / H HO / $HOCH_2$ H / HO H

The anomeric hydroxyl group is up in the L series.

(*e*) Methyl α-L-xylofuranoside is the methyl glycoside corresponding to the structure just drawn.

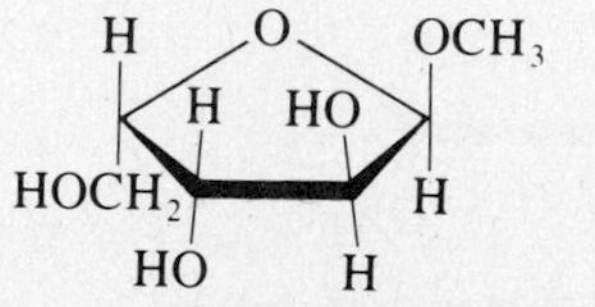

H O OCH_3 / H HO / $HOCH_2$ H / HO H

(*f*) Carbohydrates react with many of the compounds that undergo nucleophilic addition to carbonyl groups. Phenylhydrazine reacts with D-xylose to yield the corresponding phenylhydrazone.

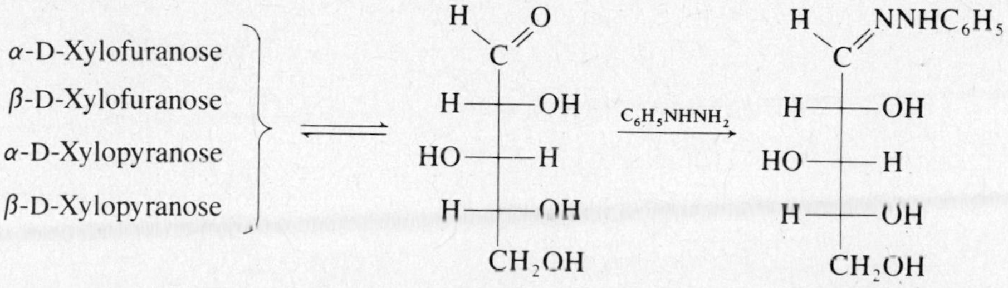

α-D-Xylofuranose
β-D-Xylofuranose
α-D-Xylopyranose
β-D-Xylopyranose

$\rightleftharpoons$

HC=O; H—OH; HO—H; H—OH; CH_2OH

D-Xylose

$\xrightarrow{C_6H_5NHNH_2}$

$HC=NNHC_6H_5$; H—OH; HO—H; H—OH; CH_2OH

D-Xylose phenylhydrazone

(*g*) With excess phenylhydrazine, D-xylose is converted to its osazone.

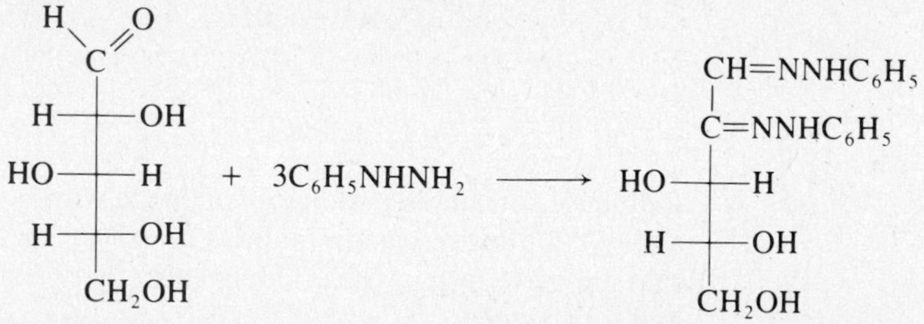

HC=O; H—OH; HO—H; H—OH; CH_2OH + $3C_6H_5NHNH_2$ $\longrightarrow$ $CH=NNHC_6H_5$; $C=NNHC_6H_5$; HO—H; H—OH; CH_2OH

D-Xylose D-Xylose phenylosazone

(*h*) Aldonic acids are derived from aldoses by oxidation of the terminal aldehyde to a carboxylic acid.

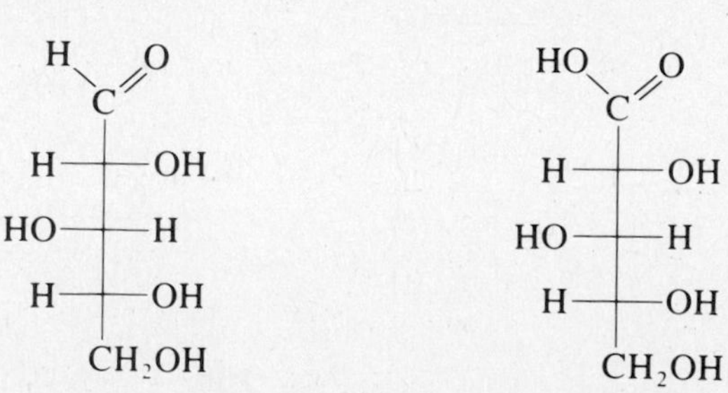

HC=O; H—OH; HO—H; H—OH; CH_2OH

D-Xylose

HOC=O; H—OH; HO—H; H—OH; CH_2OH

D-Xylonic acid

(*i*) Aldonic acids tend to exist as lactones. A δ-lactone has a six-membered ring.

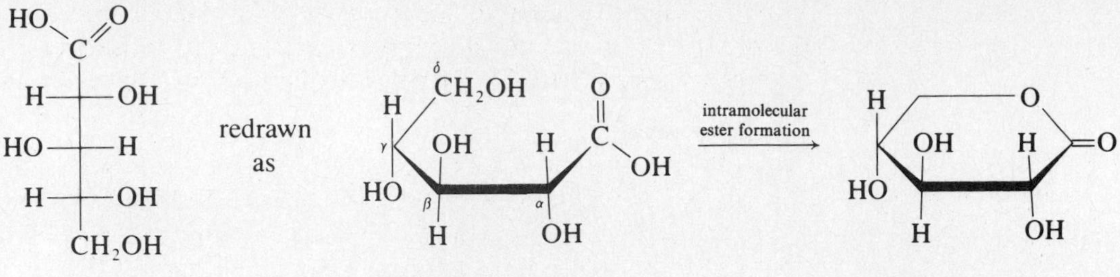

HO—C=O
H—OH
HO—H
H—OH
CH_2OH

D-Xylonic acid

redrawn as

Coiled form of D-xylonic acid

intramolecular ester formation →

δ-Lactone of D-xylonic acid

(*j*) A γ-lactone has a five-membered ring.

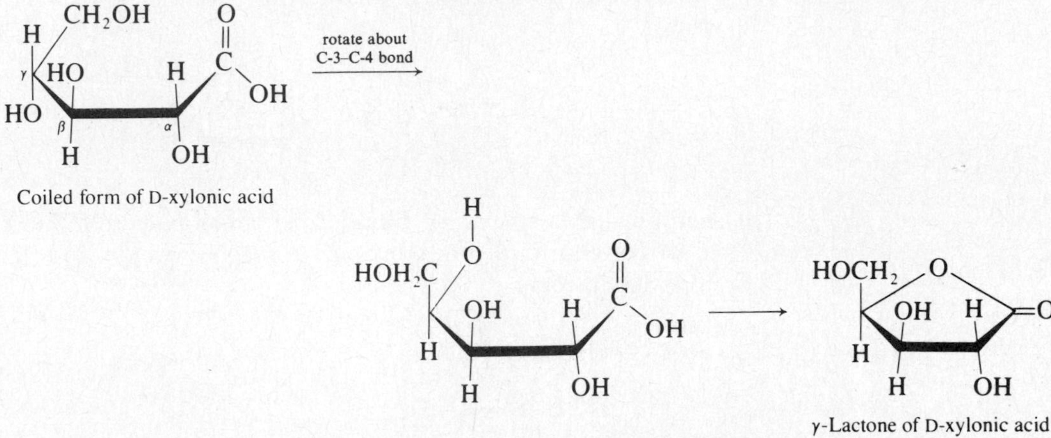

γ-Lactone of D-xylonic acid

(*k*) Aldaric acids have carboxylic acid groups at both ends of the chain.

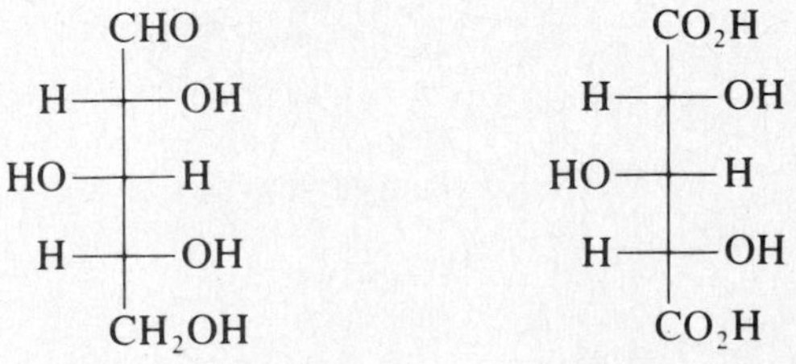

D-Xylose	D-Xylaric acid
CHO	CO_2H
H—OH	H—OH
HO—H	HO—H
H—OH	H—OH
CH_2OH	CO_2H

27.20 (*a*) Reduction of aldoses with sodium borohydride yields polyhydroxylic alcohols called alditols. Optically inactive alditols are those that have a plane of symmetry, i.e., those that are meso forms. The D-aldohexoses that yield optically inactive alditols are D-allose and D-galactose.

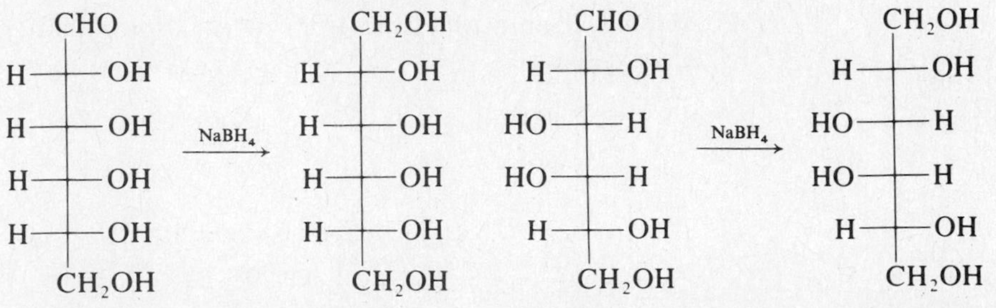

D-Allose	$\xrightarrow{NaBH_4}$	Allitol (meso compound)	D-Galactose	$\xrightarrow{NaBH_4}$	Galactitol (meso compound)
CHO		CH_2OH	CHO		CH_2OH
H—OH		H—OH	H—OH		H—OH
H—OH		H—OH	HO—H		HO—H
H—OH		H—OH	HO—H		HO—H
H—OH		H—OH	H—OH		H—OH
CH_2OH		CH_2OH	CH_2OH		CH_2OH

(*b*) All the aldonic acids and their lactones obtained on oxidation of the aldohexoses are optically active. The presence of a carboxyl group at one end of the carbon chain and a CH_2OH at the other precludes the existence of meso forms.

(*c*) Nitric acid oxidation of aldoses converts them to aldaric acids. The same D-aldoses found to yield optically inactive alditols in (*a*) yield optically inactive aldaric acids.

CHO; H—OH; H—OH; H—OH; H—OH; CH_2OH $\xrightarrow{HNO_3}$ CO_2H; H—OH; H—OH; H—OH; H—OH; CO_2H

D-Allose → Allaric acid (meso compound)

CHO; H—OH; HO—H; HO—H; H—OH; CH_2OH $\xrightarrow{HNO_3}$ CO_2H; H—OH; HO—H; HO—H; H—OH; CO_2H

D-Galactose → Galactaric acid (meso compound)

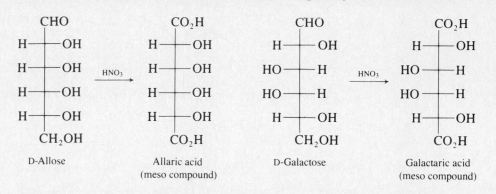

(*d*) Aldoses that differ in configuration only at C-2 enolize to the same enediol.

CHO; H—OH; H—OH; H—OH; H—OH; CH_2OH ⇌ CHOH; ‖ C—OH; H—OH; H—OH; H—OH; CH_2OH ⇌ CHO; HO—H; H—OH; H—OH; H—OH; CH_2OH

D-Allose ⇌ Enediol ⇌ D-Altrose

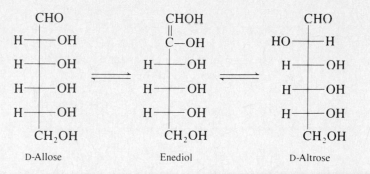

The chiral center at C-2 in the D-aldose becomes sp^2 hybridized in the enediol.

The other pairs of D-aldohexoses that form the same enediols are:

D-glucose and D-mannose
D-gulose and D-idose
D-galactose and D-talose

(*e*) When aldohexoses are converted to osazones, C-2 is converted to $\rangle C{=}NNHC_6H_5$ irrespective of its configuration. Thus, aldohexoses that differ in configuration only at C-2 yield the same osazone. The answers to this problem are the same as those of (*d*).

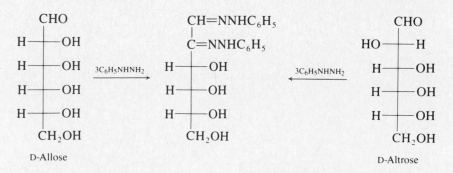

CHO; H—OH; H—OH; H—OH; H—OH; CH_2OH $\xrightarrow{3C_6H_5NHNH_2}$ $CH{=}NNHC_6H_5$; $C{=}NNHC_6H_5$; H—OH; H—OH; H—OH; CH_2OH $\xleftarrow{3C_6H_5NHNH_2}$ CHO; HO—H; H—OH; H—OH; H—OH; CH_2OH

D-Allose → osazone ← D-Altrose

The other pairs of D-aldohexoses that form the same osazones are:

D-glucose and D-mannose
D-gulose and D-idose
D-galactose and D-talose

27.21 (*a*) To unravel a pyranose form, locate the anomeric carbon and mentally convert the hemiacetal linkage to a carbonyl compound and a hydroxyl function.

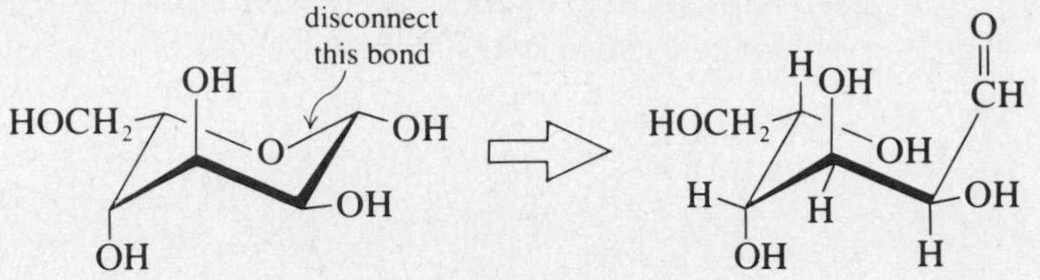

disconnect this bond

Convert the open-chain form to a proper Fischer projection.

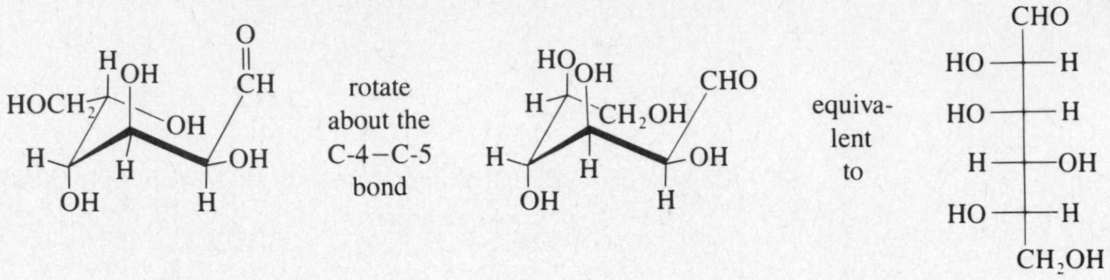

rotate about the C-4—C-5 bond

equivalent to

CHO
HO—H
HO—H
H—OH
HO—H
CH_2OH

(*b*) Proceed in the same manner as above and unravel the furanose sugar by disconnecting the hemiacetal function.

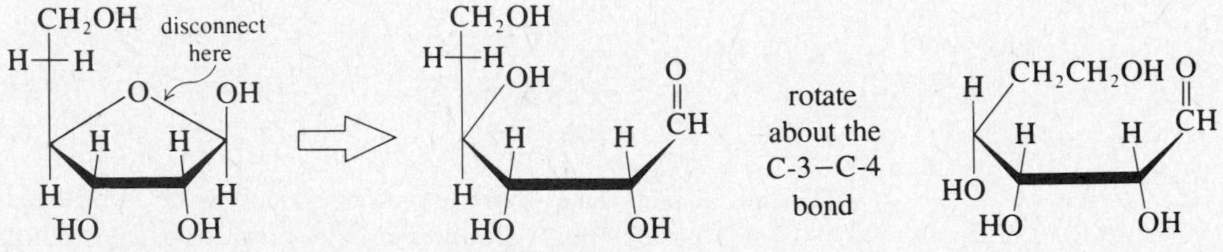

disconnect here

rotate about the C-3—C-4 bond

The Fischer projection is:

CHO
H—OH
H—OH
H—OH
H—H
CH_2OH

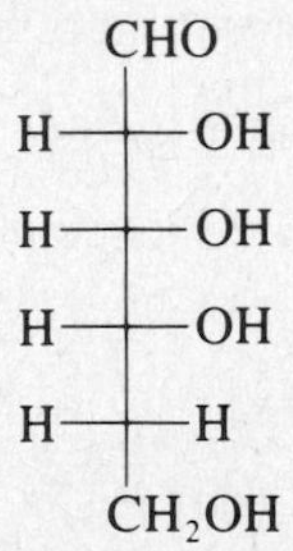

(*c*) By disconnecting and unraveling as before, the Fischer projection is revealed.

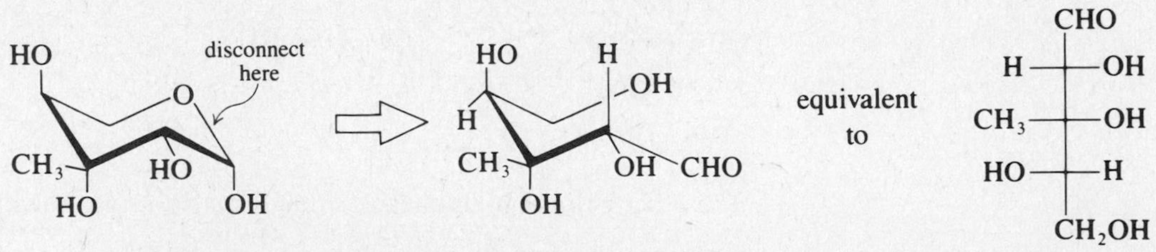

disconnect here

equivalent to

CHO
H—OH
CH_3—OH
HO—H
CH_2OH

(*d*) Remember in disconnecting cyclic hemiacetals that the anomeric carbon is the one that bears two oxygen substituents.

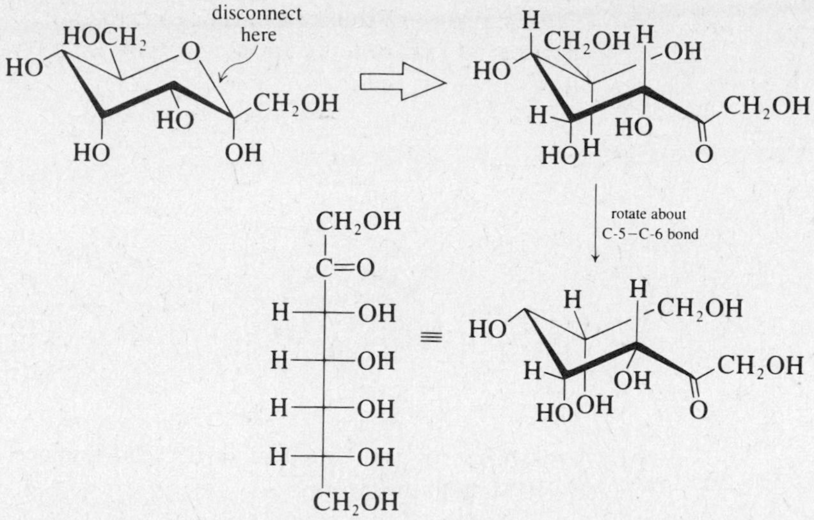

disconnect here

rotate about C-5–C-6 bond

27.22 (*a*) The L sugars have the hydroxyl group to the left at the highest-numbered chiral center in their Fischer projection. The L sugars are the ones shown in Problem 27.21(*a*) and (*c*).

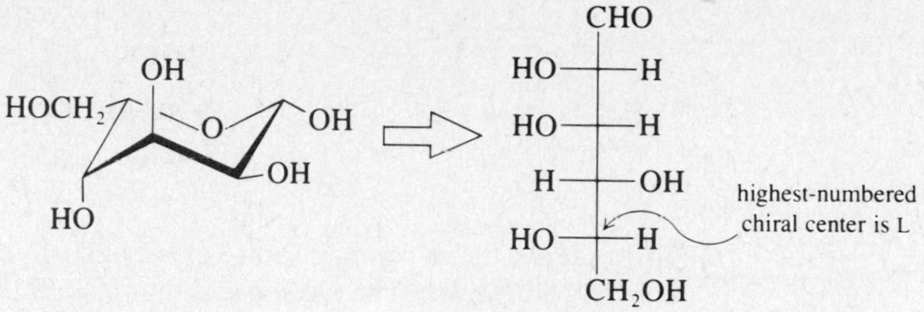

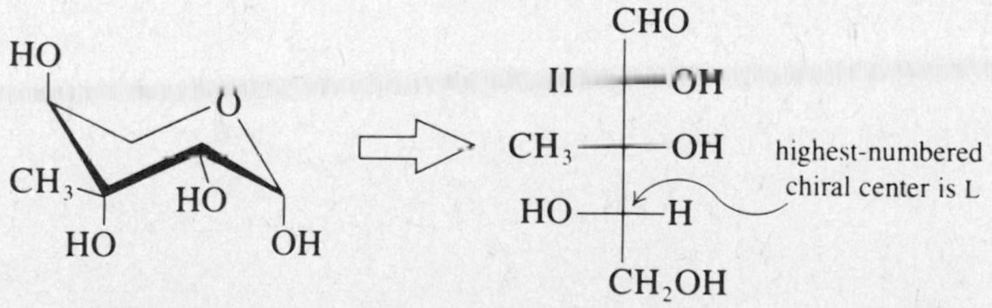

highest-numbered chiral center is L

highest-numbered chiral center is L

(*b*) Deoxy sugars are those that lack an oxygen substituent on one of the carbons in the main chain. The carbohydrate in Problem 27.21(*b*) is a deoxy sugar.

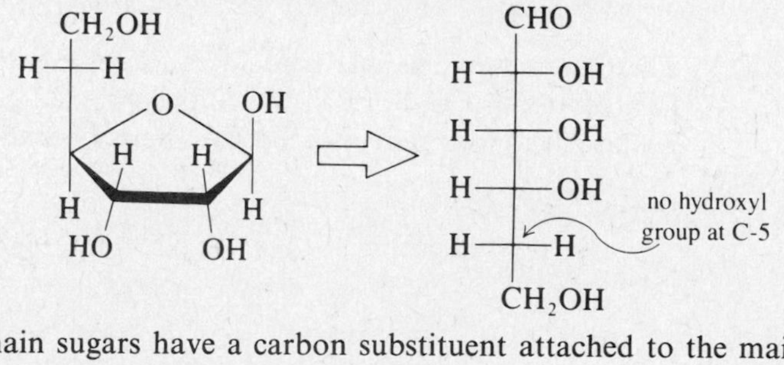

no hydroxyl group at C-5

(*c*) Branched-chain sugars have a carbon substituent attached to the main chain; the carbohydrate in Problem 27.21(*c*) fits this description.

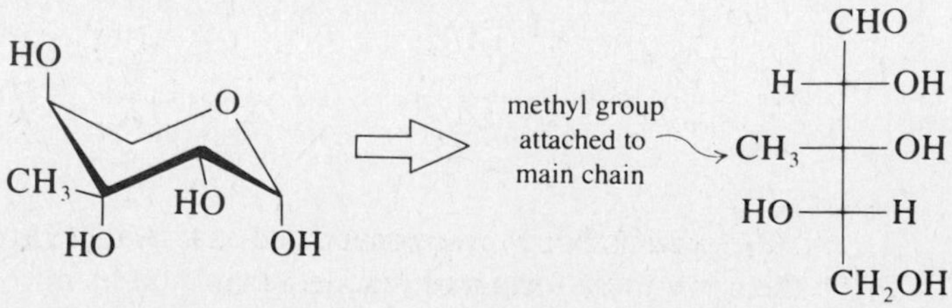

methyl group attached to main chain

(*d*) Only the sugar in Problem 27.21(*d*) is a ketose.

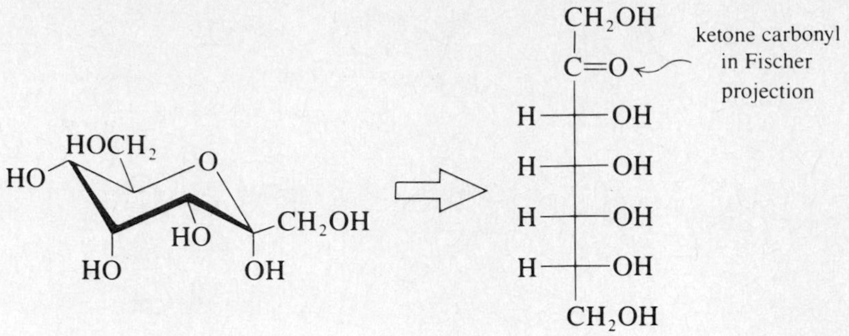

HOCH$_2$ HO O HO CH$_2$OH HO OH ⇨

	CH$_2$OH	
	C=O	← ketone carbonyl in Fischer projection
H	OH	
H	OH	
H	OH	
H	OH	
	CH$_2$OH	

(*e*) A furanose ring is a five-membered cyclic hemiacetal. Only the compound in Problem 27.21(*b*) is a furanose form.

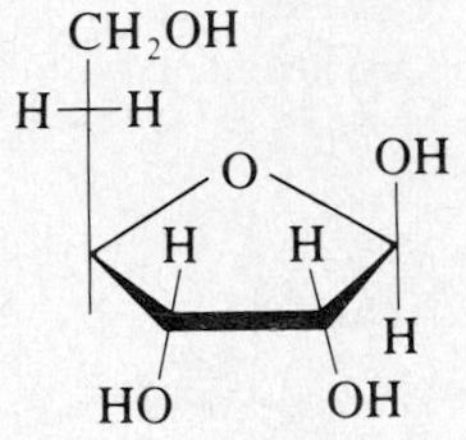

CH$_2$OH H—H O OH H H H HO OH

(*f*) In D sugars the α configuration corresponds to the hydroxyl group at the anomeric carbon being down. The α-D sugar is that in Problem 27.21(*d*).

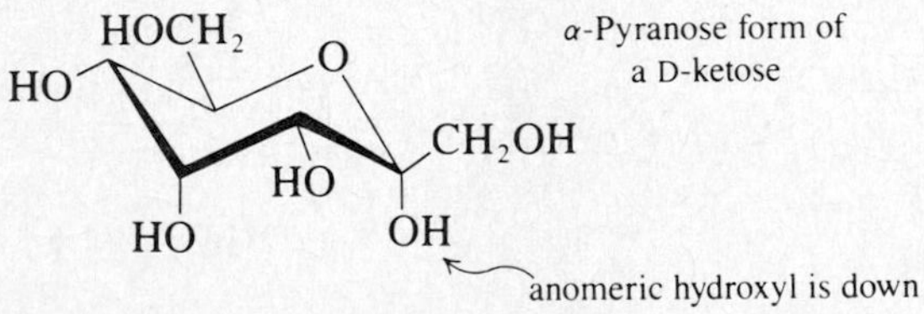

HOCH$_2$ HO O CH$_2$OH HO HO OH

α-Pyranose form of a D-ketose

anomeric hydroxyl is down

In the α-L series the anomeric hydroxyl is up. Neither of the L-sugars namely, those of Problem 27.21(*a*) and (*c*), is α; both are β.

27.23 There are seven possible pentuloses, i.e., five-carbon ketoses. The ketone carbonyl can be located at either C-2 or C-3. When the carbonyl group is at C-2, there are two chiral centers, giving rise to four stereoisomers (two pairs of enantiomers).

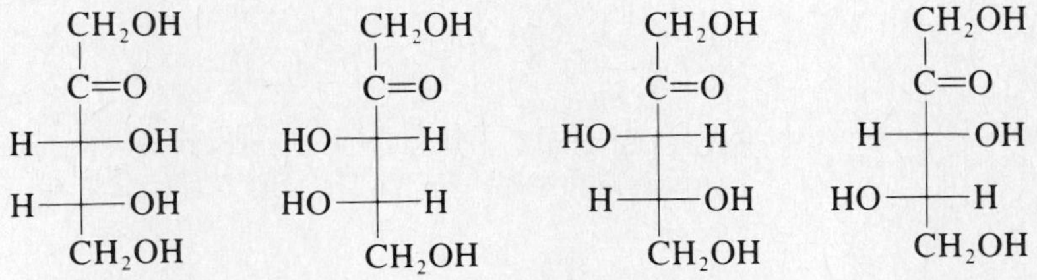

D-Erythropentulose (D-ribulose)	L-Erythropentulose (L-ribulose)	D-Threopentulose (D-xylulose)	L-Threopentulose (L-xylulose)
CH$_2$OH	CH$_2$OH	CH$_2$OH	CH$_2$OH
C=O	C=O	C=O	C=O
H—OH	HO—H	HO—H	H—OH
H—OH	HO—H	H—OH	HO—H
CH$_2$OH	CH$_2$OH	CH$_2$OH	CH$_2$OH

When the carbonyl group is located at C-3, there are only three stereoisomers because one of them is a meso form and is superposable on its mirror image.

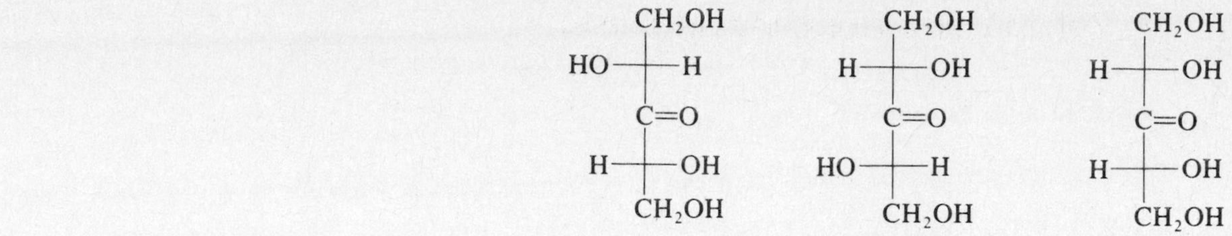

```
    CH2OH           CH2OH           CH2OH
HO──┼──H         H──┼──OH        H──┼──OH
    C=O             C=O             C=O
 H──┼──OH       HO──┼──H         H──┼──OH
    CH2OH           CH2OH           CH2OH
```

27.24 (*a*) Carbon-2 is the only chiral center in D-apiose.

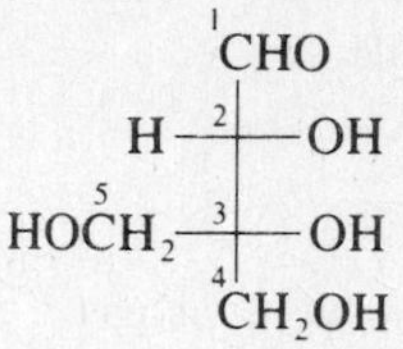

```
            ¹CHO
          H─²┼─OH
   ⁵HOCH2─³┼─OH          D-Apiose
            ⁴CH2OH
```

Carbon-3 is not a chiral center; it bears two identical CH_2OH substituents.

(*b*) When D-apiose is converted to an osazone, the chiral center at C-2 is transformed to $\rangle C{=}NNHC_6H_5$. The osazone of D-apiose is not optically active.

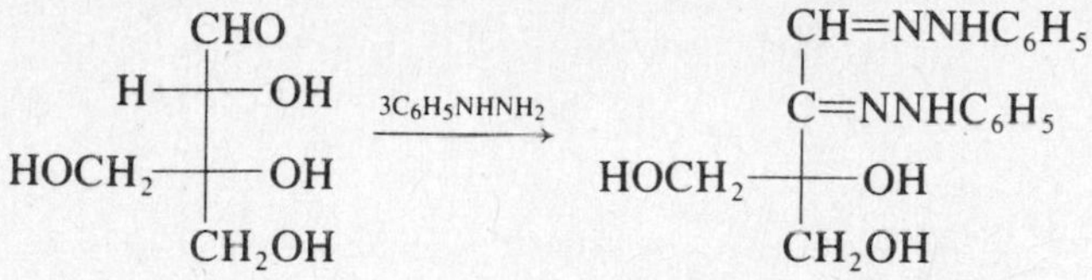

```
          CHO                                CH=NNHC6H5
       H──┼──OH      3C6H5NHNH2              C=NNHC6H5
   HOCH2──┼──OH     ──────────→       HOCH2──┼──OH
          CH2OH                              CH2OH
```

D-Apiose (optically active) → Phenylosazone of apiose (achiral; cannot be optically active)

(*c*) The alditol obtained on reduction of D-apiose retains the chiral center. It is chiral and optically active.

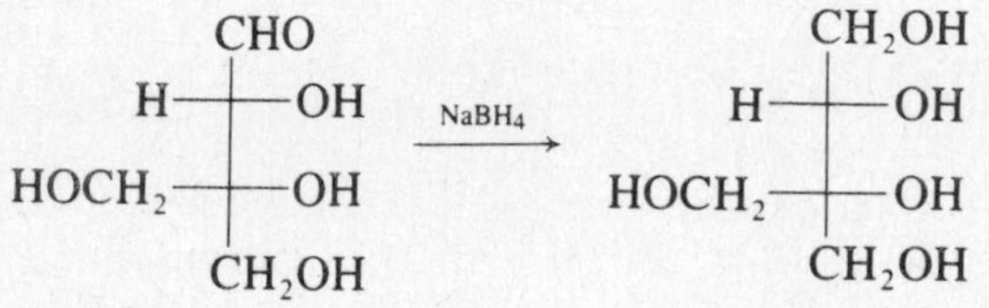

```
          CHO                        CH2OH
       H──┼──OH      NaBH4        H──┼──OH
   HOCH2──┼──OH     ──────→   HOCH2──┼──OH
          CH2OH                      CH2OH
```

D-Apiose (optically active) → D-Apitol (optically active)

(*d*), (*e*) Cyclic hemiacetal formation in D-apiose involves addition of a CH_2OH hydroxyl group to the aldehyde carbonyl.

```
H  OH         H
 5            |              O      OH              O      H
4 CH2OH H    C=O   ──→   4 5CH2OH  1     +     4 5CH2OH   1
                                H                    H
H 3       2                3   2  H             3     2  OH
HO        OH               HO     OH            HO      OH
```

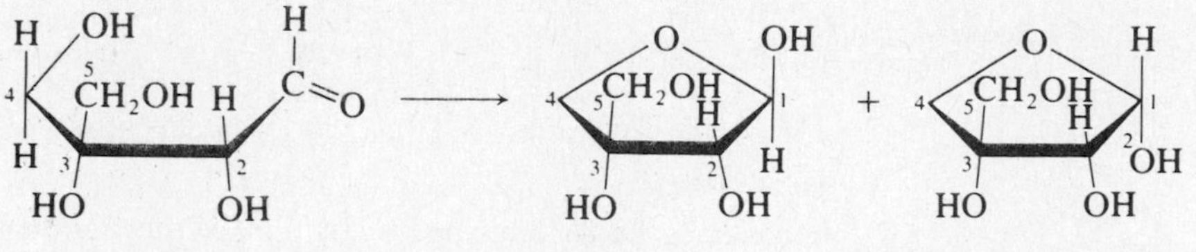

There are three chiral centers in the furanose form, are namely, the anomeric carbon C-1 and the original chiral center C-2, as well as a new chiral center at C-3.

In addition to the two furanose forms shown above, two more are possible. Instead of the CH_2OH group shown above reacting to form the cyclic hemiacetal, the other CH_2OH group may add to the aldehyde carbonyl.

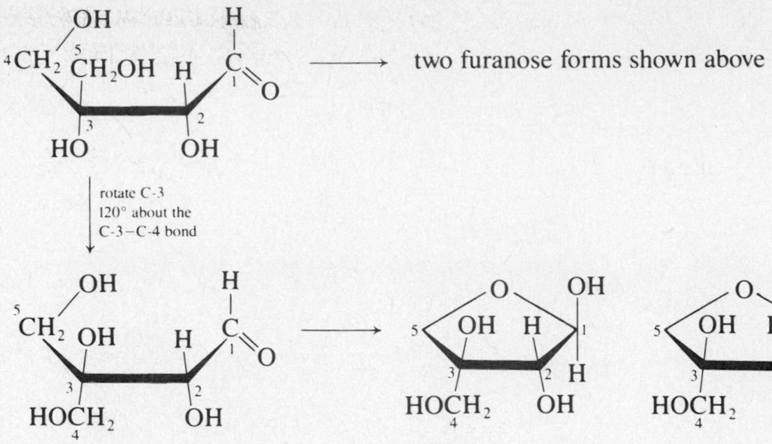

27.25 The most reasonable conclusion is that all four are methyl glycosides. Two are the methyl glycosides of the α- and β-pyranose forms of mannose and two are the methyl glycosides of the α- and β-furanose forms.

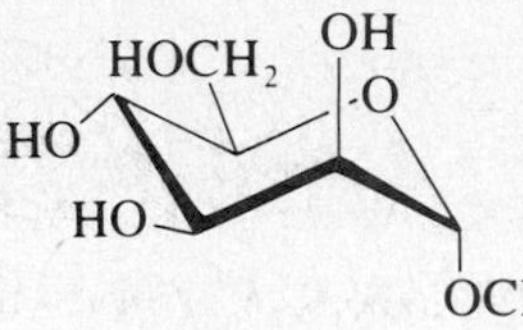

Methyl α-D-mannopyranoside

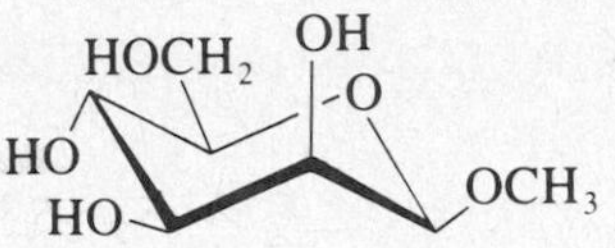

Methyl β-D-mannopyranoside

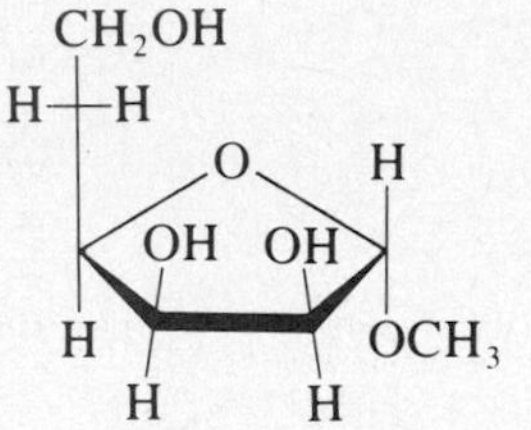

Methyl α-D-mannofuranoside

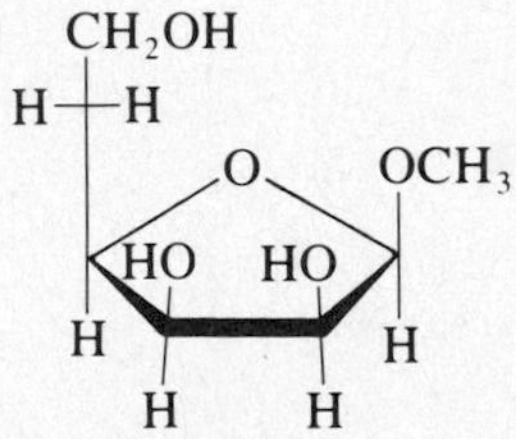

Methyl β-D-mannofuranoside

In the case of the methyl glycosides of mannose, comparable amounts of pyranosides and furanosides are formed. The major products are the α isomers.

27.26 (*a*) Disaccharides, by definition, involve an acetal linkage at the anomeric position; thus all the disaccharides must involve C-1. The bond to C-1 can be α or β. The available oxygen atoms in the second D-glucopyranosyl unit are located at C-1, C-2, C-3, C-4, and C-6. Thus, there are 11 possible disaccharides, including maltose and cellobiose, composed of D-glucopyranosyl units.

α,α(1,1)	α,β(1,1)	β,β(1,1)
α(1,2)		β(1,2)
α(1,3)		β(1,3)
α(1,4) (maltose)		β(1,4) (cellobiose)
α(1,6)		β(1,6)

(*b*) In order to be a reducing sugar, one of the anomeric positions must be a free hemiacetal. Therefore, all except α,α(1,1), α,β(1,1), and β,β(1,1) are reducing sugars.

27.27 Since gentiobiose undergoes mutarotation, it must have a free hemiacetal group. Formation of two molecules of D-glucose indicates that it is a disaccharide and because that hydrolysis is catalyzed by emulsin, the glycosidic linkage is β. The methylation data, summarized in the equation below, require that the glucose units be present in pyranose forms and be joined by a β(1,6) glycoside bond.

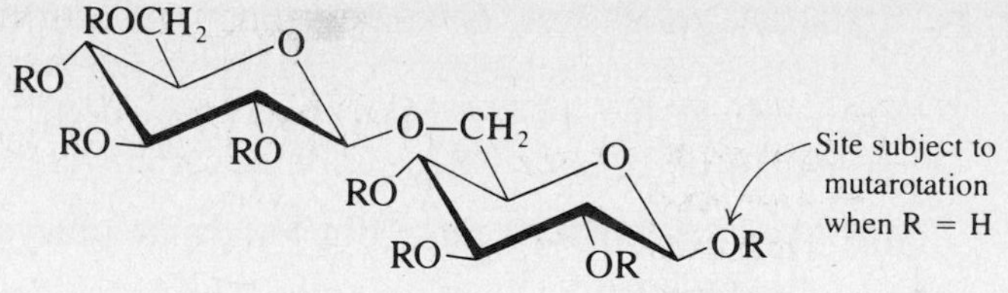

ROCH$_2$ … O … RO … RO … RO … O—CH$_2$ … O … RO … RO … OR … OR

Site subject to mutarotation when R = H

R = H: gentiobiose

↓

R = CH_3: gentiobiose octamethyl ether

↓ H_3O^+

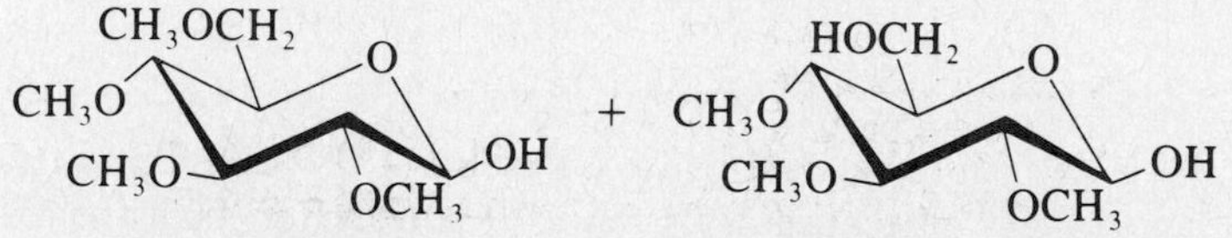

2,3,4,6-Tetra-O-methyl-D-glucose + 2,3,4-Tri-O-methyl-D-glucose

27.28 Like other glycosides, cyanogenetic glycosides are cleaved to a carbohydrate and an alcohol on hydrolysis.

(*a*) In the case of linamarin the alcohol is recognizable as the cyanohydrin of acetone. Once formed, this cyanohydrin dissociates to hydrogen cyanide and acetone.

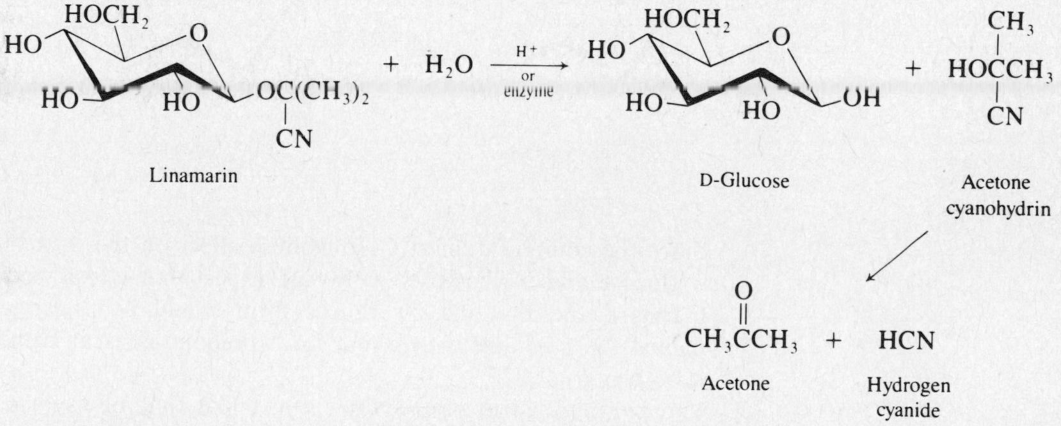

Linamarin + H_2O $\xrightarrow[\text{enzyme}]{H^+ \text{ or}}$ D-Glucose + Acetone cyanohydrin ($HOC(CH_3)_2CN$)

Acetone cyanohydrin → CH_3CCH_3 (Acetone, C=O) + HCN (Hydrogen cyanide)

(*b*) Laetrile undergoes an analogous hydrolytic cleavage to yield the cyanohydrin of benzaldehyde.

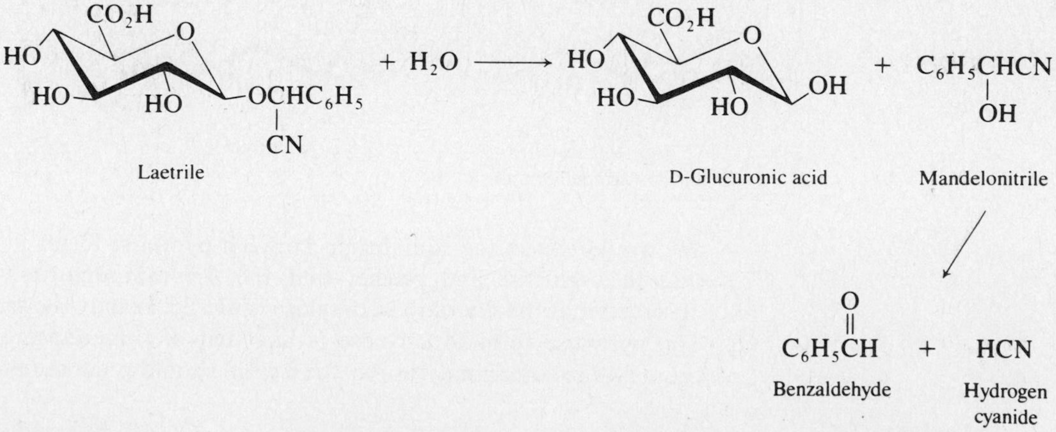

Laetrile + H_2O ⟶ D-Glucuronic acid + C_6H_5CHCN (OH) (Mandelonitrile)

Mandelonitrile → C_6H_5CH (Benzaldehyde, C=O) + HCN (Hydrogen cyanide)

27.29 Comparing D-glucose, D-mannose, and D-galactose, it can be said that the configuration of C-2 has a substantial effect on the relative energies of the α- and β-pyranose forms, but that the configuration of C-4 has virtually no effect. With this observation in mind, write the structures of the pyranose forms of the carbohydrates given in each part.

(*a*) The β-pyranose form of D-gulose is the same as that of D-galactose except for the configuration at C-3.

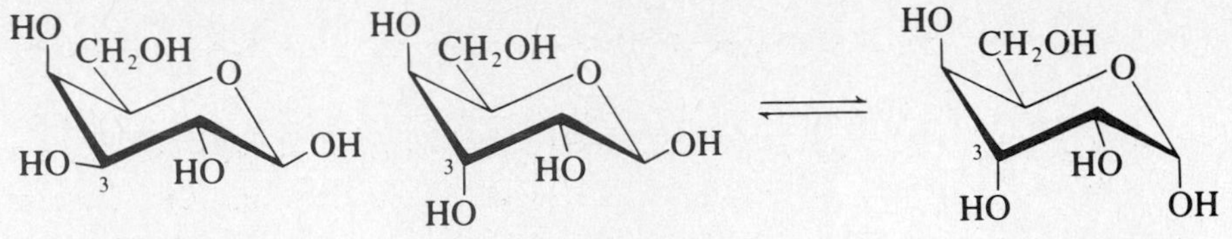

β-D-Galactopyranose (64% at equilibrium) β-D-Gulopyranose ⇌ α-D-Gulopyranose (1,3-diaxial repulsion between hydroxyl groups)

The axial hydroxyl group at C-3 destabilizes the α-pyranose form more than the β form because of its repulsive interaction with the axially disposed anomeric hydroxyl group. There should be an even higher β/α ratio in D-gulopyranose than in D-galactopyranose. This is so; the observed β/α ratio is 88 : 12.

(*b*) The β-pyranose form of D-talose is the same as that of D-mannose except for the configuration at C-4.

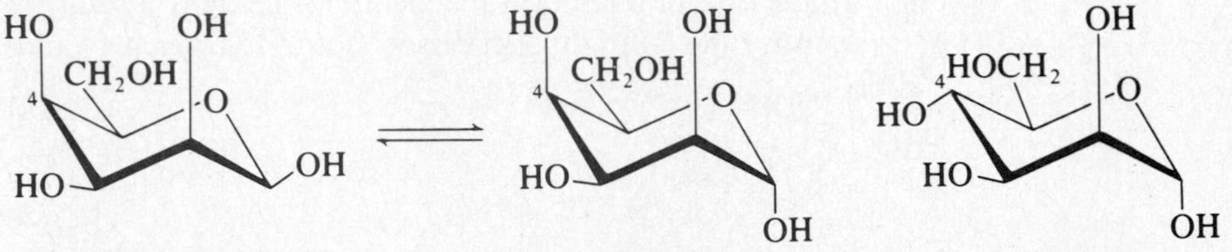

β-D-Talopyranose ⇌ α-D-Talopyranose α-D-Mannopyranose (68% at equilibrium)

Since the configuration at C-4 has little effect on the α- to β-pyranose ratio (compare D-glucose and D-galactose), we would expect that talose would behave very much like mannose and that the α-pyranose form would be preferred at equilibrium. This is indeed the case; the α-pyranose form predominates at equilibrium, the observed α/β ratio being 78 : 22.

(*c*) The pyranose form of D-xylose is just like that of D-glucose except that it lacks a CH_2OH group.

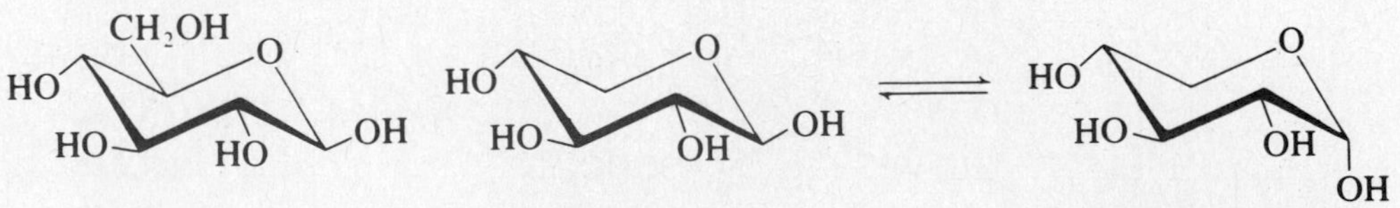

β-D-Glucopyranose (64% at equilibrium) β-D-Xylopyranose ⇌ α-D-Xylopyranose

We would expect the equilibrium between pyranose forms in D-xylose to be much like that in D-glucose and predict that the β-pyranose form would predominate. It is observed that the β/α ratio in D-xylose is 64 : 36, exactly the same as in D-glucose.

(*d*) The pyranose form of D-lyxose is like that of D-mannose except it lacks a CH_2OH group. As in D-mannopyranose, the α form should predominate over the β.

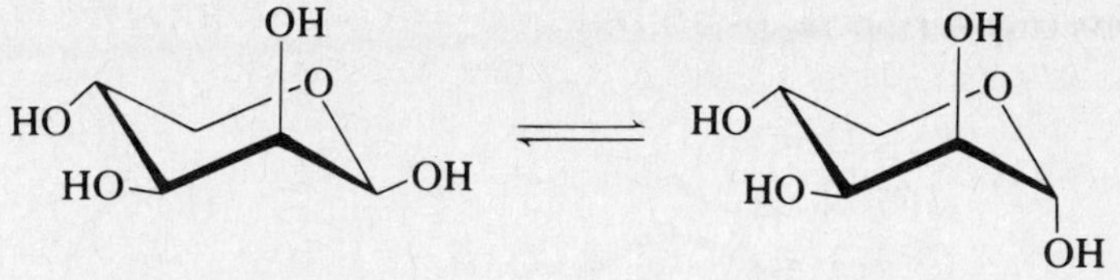

β-D-Lyxopyranose ⇌ α-D-Lyxopyranose

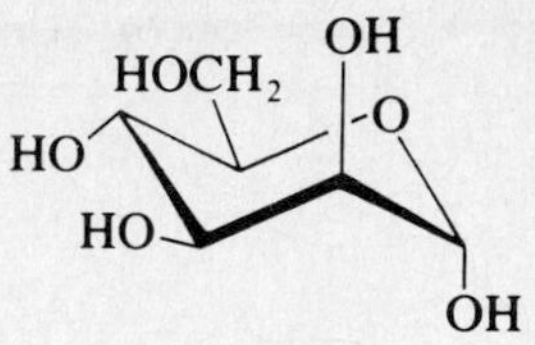

α-D-Mannopyranose (68% at equilibrium)

The observed α/β distribution ratio in D-lyxopyranose is 73 : 27.

27.30 (*a*) The rate-determining step in glycoside hydrolysis is carbocation formation at the anomeric position. The carbocation formed from methyl α-D-fructofuranoside (A) is tertiary and therefore more stable than the one from methyl α-D-glucofuranoside (B), which is secondary. The more stable a carbocation is, the more rapidly it will be formed.

Faster

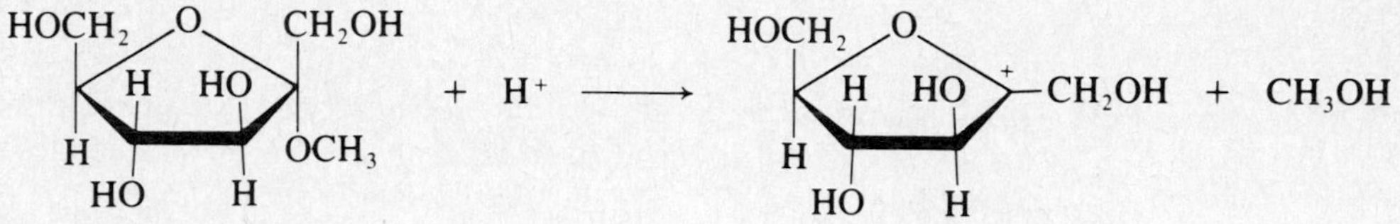

A + H^+ ⟶ Tertiary carbocation + CH_3OH

Slower

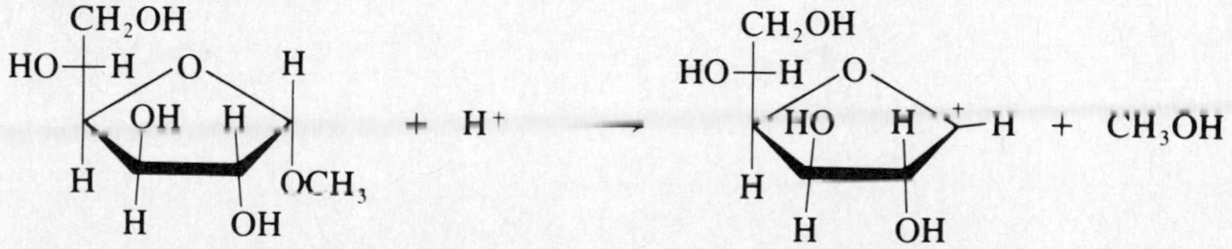

B + H^+ ⟶ Secondary carbocation + CH_3OH

(*b*) The carbocation formed from methyl β-D-glucopyranoside (D) is less stable than the one from its 2-deoxy analog (C) and is formed more slowly. It is less stable because it is destabilized by the electron-withdrawing inductive effect of the hydroxyl group at C-2.

Faster

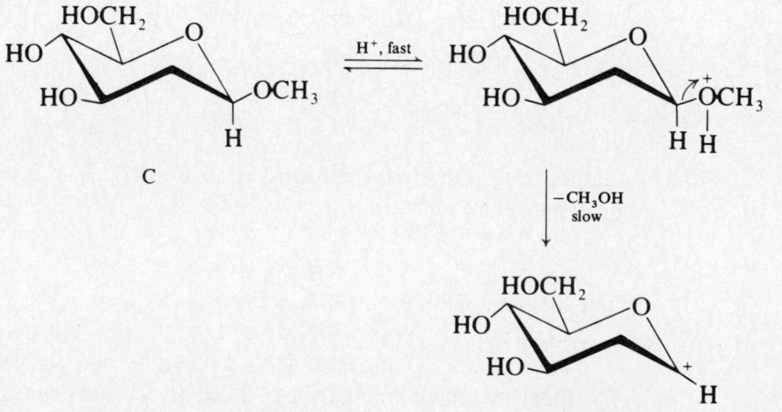

C ⇌ (H^+, fast) ⟶ ($-CH_3OH$, slow)

More stable

Slower

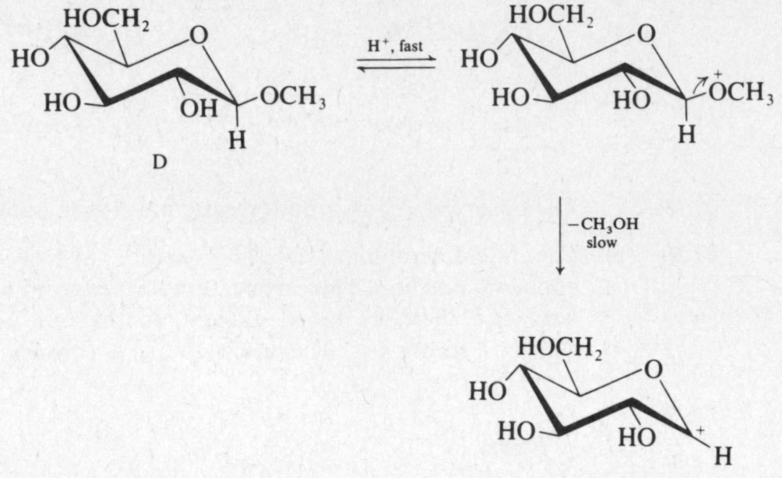

HOCH$_2$ HO HO OH O OCH$_3$ H D $\xrightleftharpoons{H^+, \text{ fast}}$ HOCH$_2$ HO HO HO O $\overset{+}{O}CH_3$ H

$-CH_3OH$ slow

HOCH$_2$ HO HO HO O + H

Less stable

27.31 D-Altrosan is a glycoside. The anomeric carbon—the one with two oxygen substituents—has an alkoxy group attached to it. Hydrolysis of D-altrosan follows the general mechanism for acetal hydrolysis.

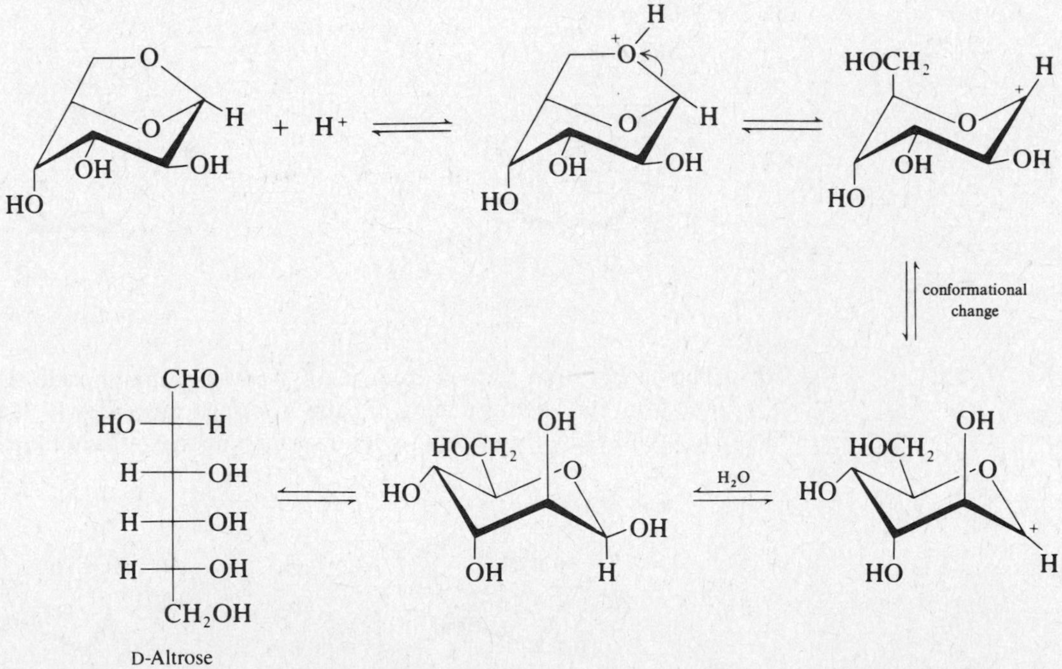

O O H OH OH HO + H^+ $\rightleftharpoons$ H $\overset{+}{O}$ O H OH OH HO $\rightleftharpoons$ HOCH$_2$ H + O OH OH HO

conformational change

CHO
HO—H
H—OH
H—OH
H—OH
CH$_2$OH

D-Altrose

$\rightleftharpoons$ OH HOCH$_2$ O HO OH OH H $\xleftarrow{H_2O}$ OH HOCH$_2$ O HO HO + H

27.32 Galactose has hydroxyl groups at carbons 2, 3, 4, 5, 6. Therefore, nine trimethyl ethers are possible. They are:

2, 3, 4	2, 4, 5	3, 4, 5	4, 5, 6
2, 3, 5	2, 4, 6	3, 4, 6	
2, 3, 6		3, 5, 6	

To find out which one of these is identical with the degradation product of E, carry compound E through the required transformations.

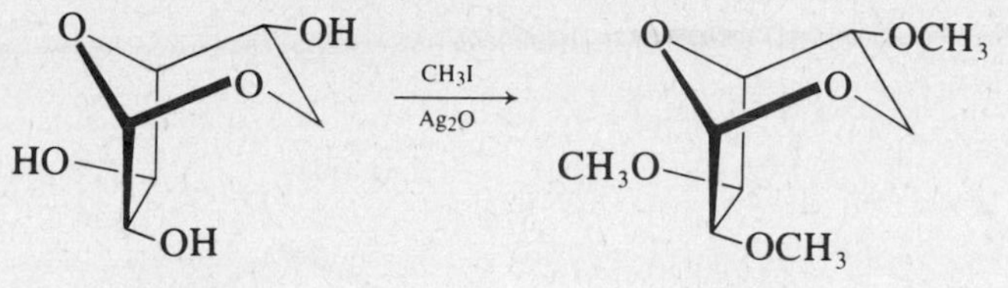

HO, OH, OH, O, O — Compound E $\xrightarrow[Ag_2O]{CH_3I}$ Tri-O-methyl ether of compound E

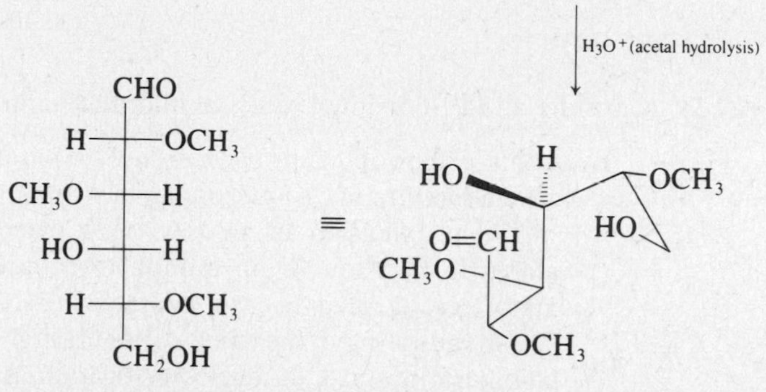

$\downarrow$ H_3O^+ (acetal hydrolysis)

CHO; H—OCH$_3$; CH$_3$O—H; HO—H; H—OCH$_3$; CH$_2$OH

2,3,5-Tri-O-methyl-D-galactose

27.33 The fact that phlorizin is hydrolyzed to D-glucose and compound F by emulsin indicates that it is a β-glucoside in which D-glucose is attached to one of the phenolic hydroxyls of G.

$$C_{21}H_{24}O_{10} + H_2O \xrightarrow{\text{emulsin}}$$

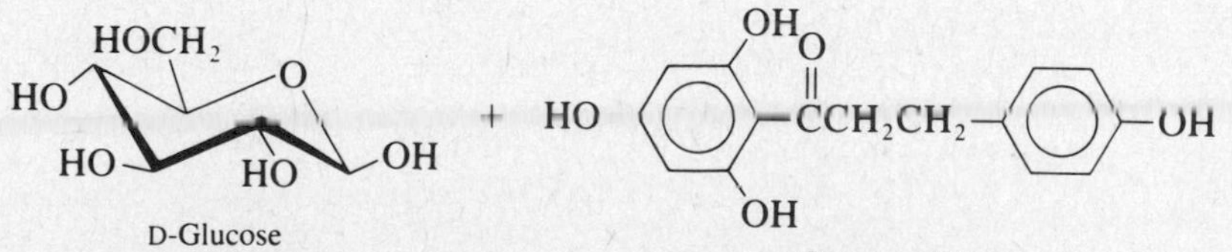

D-Glucose ($C_6H_{12}O_6$) + Compound F ($C_{15}H_{14}O_5$)

The methylation experiment reveals exactly to which hydroxyl glucose is attached. Excess methyl iodide reacts with all the available phenolic hydroxyl groups, but the glycosidic oxygen is not affected. Thus when the methylated phlorizin undergoes acid-catalyzed hydrolysis of its glycosidic bond, the oxygen in that bond is exposed as a phenolic hydroxyl group.

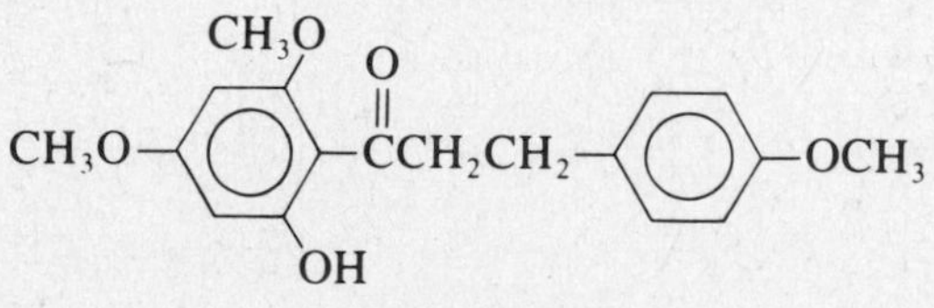

Compound G

This compound must arise by hydrolysis of

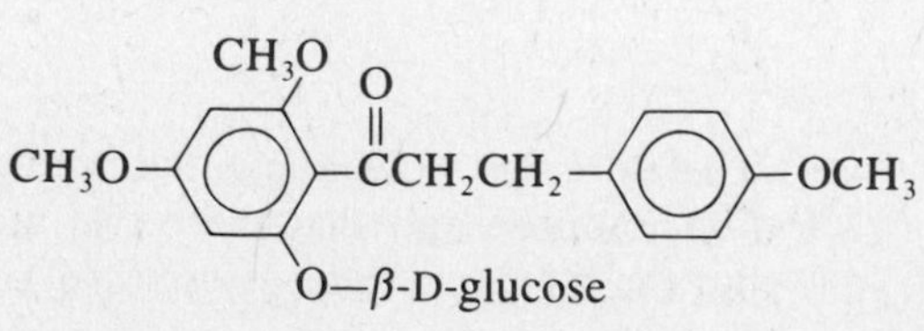

O—β-D-glucose

The structure of phlorizin is therefore:

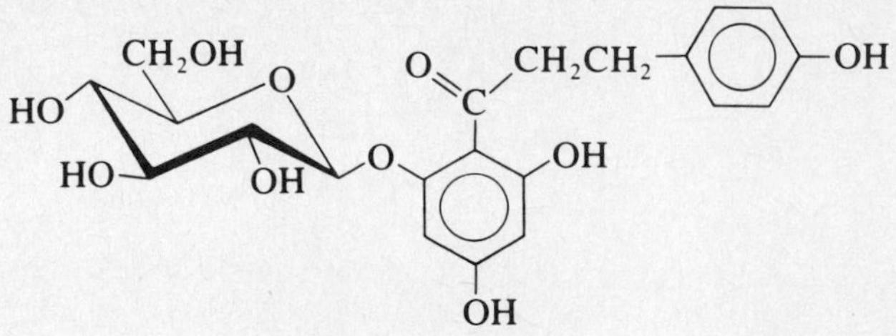

27.34 Consider all the individual pieces of information in the order in which they are presented.

1. Chain extension of the aldopentose (−)-arabinose by way of the derived cyanohydrin gave a mixture of (+)-glucose and (+)-mannose.

   Chain extension of aldoses takes place at the aldehyde end of the chain. The aldehyde function of an aldopentose becomes C-2 of an aldohexose, which normally results in two carbohydrates diastereomeric at C-2. Thus, (+)-glucose and (+)-mannose have the same configuration at C-3, C-4, and C-5; they have opposite configurations at C-2. The configuration at C-2, C-3, and C-4 of (−)-arabinose is the same as that at C-3, C-4, and C-5 of (+)-glucose and (+)-mannose.

2. Oxidation of (−)-arabinose with warm nitric acid gave an optically active aldaric acid.

   Since the hydroxyl group at C-4 of (−)-arabinose is at the right in a Fischer projection formula (evidence of part **1**, above), the hydroxyl at C-2 must be to the left in order for the aldaric acid to be optically active.

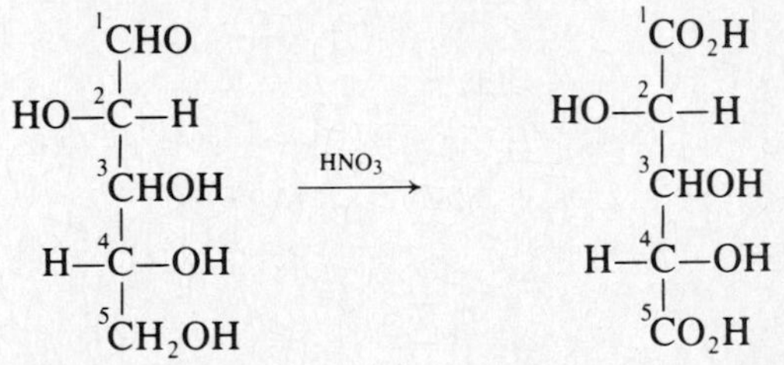

```
    ¹CHO                          ¹CO₂H
     |                             |
HO—²C—H                       HO—²C—H
     |           HNO₃              |
    ³CHOH      ——————→           ³CHOH
     |                             |
 H—⁴C—OH                       H—⁴C—OH
     |                             |
    ⁵CH₂OH                        ⁵CO₂H
```

Partial stereostructure of (−)-arabinose

Aldaric acid from (−)-arabinose; optically active irrespective of configuration at C-3

If the C-2 hydroxyl group had been to the right an optically inactive meso aldaric acid would have been produced.

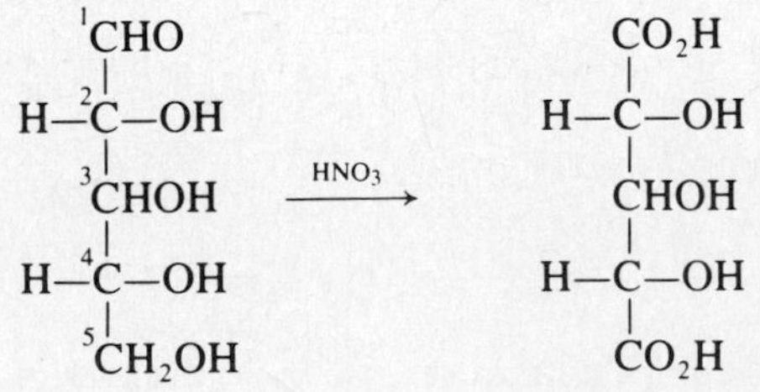

```
    ¹CHO                          CO₂H
     |                             |
 H—²C—OH                       H—C—OH
     |           HNO₃              |
    ³CHOH      ——————→           CHOH
     |                             |
 H—⁴C—OH                       H—C—OH
     |                             |
    ⁵CH₂OH                        CO₂H
```

Achiral meso form; cannot be optically active

Therefore we now know the configurations of C-3 and C-5 of (+)-glucose and (+)-mannose and that these two aldohexoses have opposite configurations at C-2, but the same (yet to be determined) configuration at C-4.

$$
\begin{array}{ccc}
{}^{1}\mathrm{CHO} & \qquad & {}^{1}\mathrm{CHO} \\
\mathrm{H{-}{}^{2}C{-}OH} & & \mathrm{HO{-}{}^{2}C{-}H} \\
\mathrm{HO{-}{}^{3}C{-}H} & & \mathrm{HO{-}{}^{3}C{-}H} \\
{}^{4}\mathrm{CHOH} & & {}^{4}\mathrm{CHOH} \\
\mathrm{H{-}{}^{5}C{-}OH} & & \mathrm{H{-}{}^{5}C{-}OH} \\
{}^{6}\mathrm{CH_2OH} & & {}^{6}\mathrm{CH_2OH}
\end{array}
$$

[One of these is (+)-glucose, the other is (+)-mannose]

3. Both (+)-glucose and (+)-mannose are oxidized to optically active aldaric acids with nitric acid.

Since both (+)-glucose and (+)-mannose yield optically active aldaric acids and both have the same configuration at C-4, the hydroxyl group must lie at the right in the Fischer projection at this carbon.

$$
\begin{array}{ccc}
{}^{1}\mathrm{CHO} & \qquad & {}^{1}\mathrm{CHO} \\
\mathrm{H{-}{}^{2}C{-}OH} & & \mathrm{HO{-}{}^{2}C{-}H} \\
\mathrm{HO{-}{}^{3}C{-}H} & & \mathrm{HO{-}{}^{3}C{-}H} \\
\mathrm{H{-}{}^{4}C{-}OH} & & \mathrm{H{-}{}^{4}C{-}OH} \\
\mathrm{H{-}{}^{5}C{-}OH} & & \mathrm{H{-}{}^{5}C{-}OH} \\
{}^{6}\mathrm{CH_2OH} & & {}^{6}\mathrm{CH_2OH}
\end{array}
$$

[One of these is (+)-glucose, the other is (+)-mannose]

The structures of the corresponding aldaric acids are:

$$
\begin{array}{ccc}
\mathrm{CO_2H} & \qquad & \mathrm{CO_2H} \\
\mathrm{H{-}C{-}OH} & & \mathrm{HO{-}C{-}H} \\
\mathrm{HO{-}C{-}H} & & \mathrm{HO{-}C{-}H} \\
\mathrm{H{-}C{-}OH} & & \mathrm{H{-}C{-}OH} \\
\mathrm{H{-}C{-}OH} & & \mathrm{H{-}C{-}OH} \\
\mathrm{CO_2H} & & \mathrm{CO_2H}
\end{array}
$$

Both are optically active. Had the C-4 hydroxyl group been to the left, one of the aldaric acids would have been a meso form.

$$
\begin{array}{ccc}
\mathrm{CO_2H} & \qquad & \mathrm{CO_2H} \\
\mathrm{H{-}C{-}OH} & & \mathrm{HO{-}C{-}H} \\
\mathrm{HO{-}C{-}H} & & \mathrm{HO{-}C{-}H} \\
\mathrm{HO{-}C{-}H} & & \mathrm{HO{-}C{-}H} \\
\mathrm{H{-}C{-}OH} & & \mathrm{H{-}C{-}OH} \\
\mathrm{CO_2H} & & \mathrm{CO_2H}
\end{array}
$$

(This aldaric acid is optically inactive)

4. There is another sugar, (+)-gulose, that gives the same aldaric acid on oxidation as does (+)-glucose.

This is the last piece in the puzzle, the one that permits one of the Fischer projections shown above to be assigned to (+)-glucose and the other to (+)-mannose. Consider first the structure:

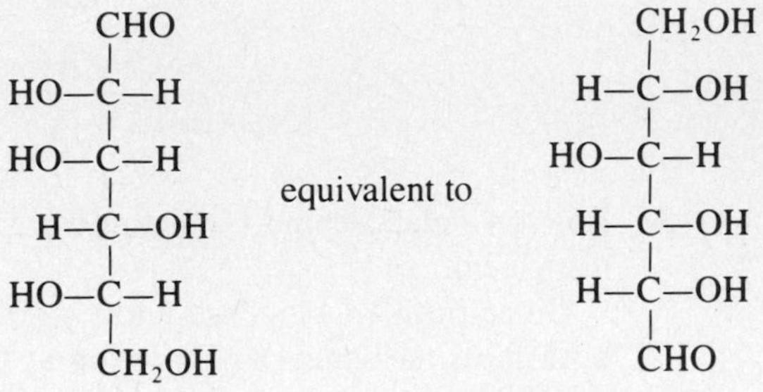

$$\begin{array}{c} \text{CHO} \\ | \\ \text{HO—C—H} \\ | \\ \text{HO—C—H} \\ | \\ \text{H—C—OH} \\ | \\ \text{HO—C—H} \\ | \\ \text{CH}_2\text{OH} \end{array} \quad \text{equivalent to} \quad \begin{array}{c} \text{CH}_2\text{OH} \\ | \\ \text{H—C—OH} \\ | \\ \text{HO—C—H} \\ | \\ \text{H—C—OH} \\ | \\ \text{H—C—OH} \\ | \\ \text{CHO} \end{array}$$

Oxidation gives the aldaric acid

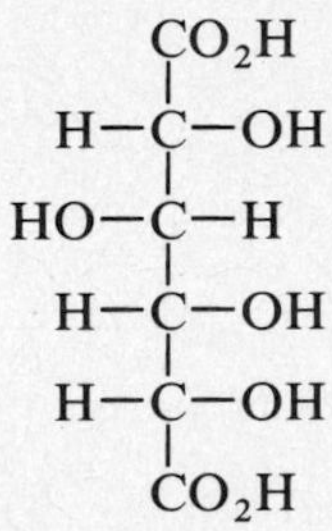

$$\begin{array}{c} \text{CO}_2\text{H} \\ | \\ \text{H—C—OH} \\ | \\ \text{HO—C—H} \\ | \\ \text{H—C—OH} \\ | \\ \text{H—C—OH} \\ | \\ \text{CO}_2\text{H} \end{array}$$

This is the same aldaric acid as that provided by one of the structures given as either (+)-glucose or (+)-mannose. Therefore that Fischer projection corresponds to (+)-glucose.

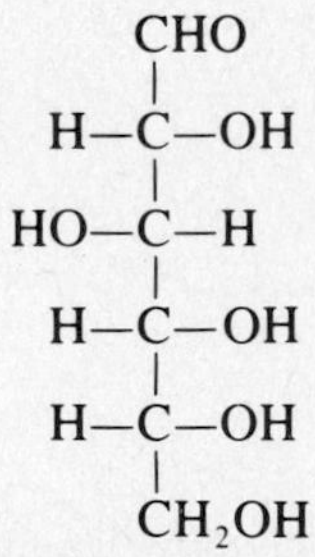

$$\begin{array}{c} \text{CHO} \\ | \\ \text{H—C—OH} \\ | \\ \text{HO—C—H} \\ | \\ \text{H—C—OH} \\ | \\ \text{H—C—OH} \\ | \\ \text{CH}_2\text{OH} \end{array}$$

This must be (+)-glucose

The structure of (+)-mannose is therefore:

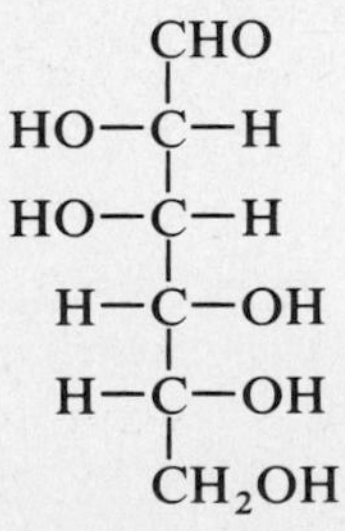

$$\begin{array}{c} \text{CHO} \\ | \\ \text{HO—C—H} \\ | \\ \text{HO—C—H} \\ | \\ \text{H—C—OH} \\ | \\ \text{H—C—OH} \\ | \\ \text{CH}_2\text{OH} \end{array}$$

A sugar that yields the same aldaric acid is:

```
       CH₂OH
        |
   HO—C—H
        |
   HO—C—H
        |
    H—C—OH
        |
    H—C—OH
        |
       CHO
```

This is, in fact, not a different sugar but simply (+)-mannose rotated through an angle of 180°.

SELF-TEST

PART A

A-1. Draw the structures indicated for each of the following:

(*a*) The enantiomer of D-erythrose:

```
      CHO
   H——+——OH
   H——+——OH
     CH₂OH
```

(*b*) An epimer of D-erythrose

(*c*) The α-furanose form of D-erythrose (use a Haworth formula)

(*d*) The anomer of (*c*)

A-2. The structure of D-mannose is shown below.

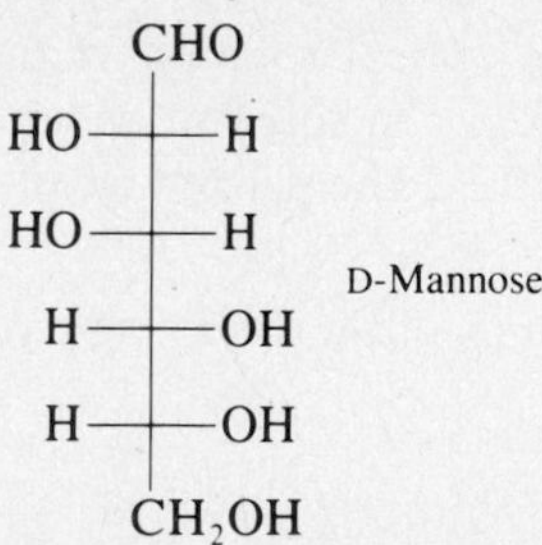

Using Fischer projections, draw the product of the reaction of D-mannose with:

(*a*) $NaBH_4$ in H_2O

(*b*) Benedict's reagent

(*c*) Excess phenylhydrazine, $C_6H_5NHNH_2$

(*d*) Excess periodic acid

PART B

B-1. The maximum number of stereoisomeric ketohexoses is

(*a*) 2 (*b*) 4 (*c*) 8 (*d*) 16

B-2. Choose the response which provides the best match between the terms given and the structures shown.

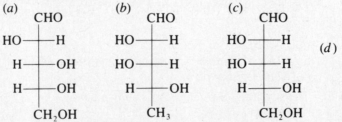

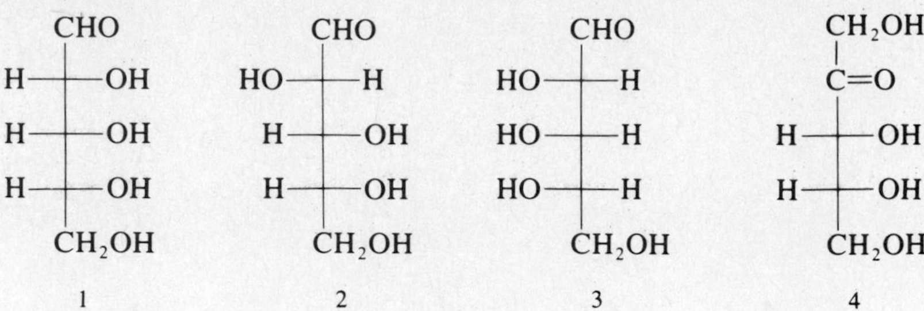

	Epimers	*Enantiomers*
(*a*)	1, 3, and 4	1 and 3
(*b*)	1 and 2	1 and 3
(*c*)	1, 2, and 3	1 and 3
(*d*)	1 and 4	1 and 2

B-3. A D carbohydrate is

(*a*) Always dextrorotatory
(*b*) Always levorotatory
(*c*) Always the anomer of the corresponding L carbohydrate
(*d*) None of the above

B-4. Two of the three compounds shown yield the same product on reaction with warm HNO_3. The *exception* is

(*a*)
CHO
HO—H
H—OH
H—OH
CH_2OH

(*b*)
CHO
HO—H
HO—H
H—OH
CH_3

(*c*)
CHO
HO—H
HO—H
H—OH
CH_2OH

(*d*) None of these (all yield the same product)

B-5. The optical rotation of the α form of a pyranose is +150.7°; that of the *β* form is +52.8°. In solution an equilibrium mixture of the anomers has an optical rotation of +80.2°. The percentage of the α form at equilibrium is

(*a*) 28% (*b*) 32% (*c*) 68% (*d*) 72%

B-6. Which of the following represents the anomer of the compound shown below?

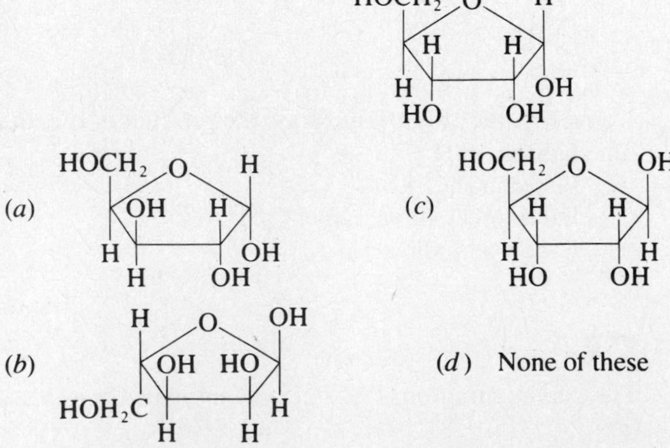

B-7. Which of the following aldoses yield an optically inactive substance on reaction with sodium borohydride?

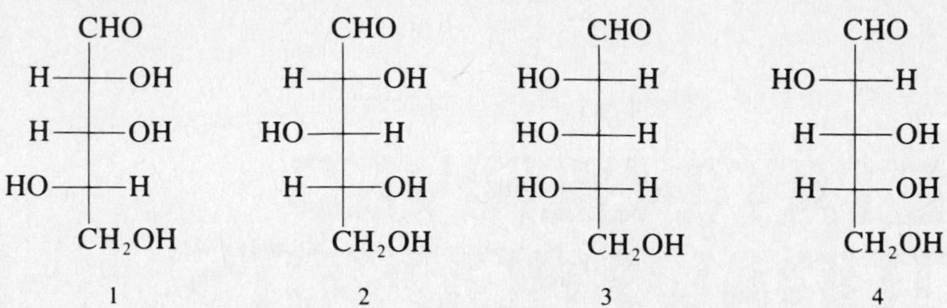

(*a*) 3 only (*b*) 1 and 4
(*c*) 2 and 3 (*d*) All of them (1, 2, 3, and 4)

B-8. Which set of terms correctly identifies the carbohydrate shown?

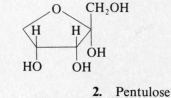

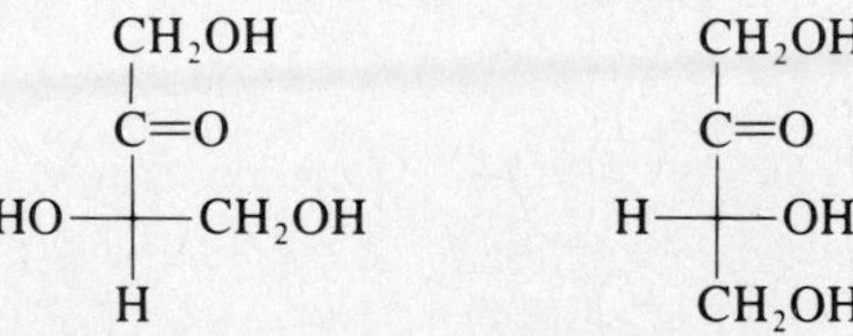

1. Pentose **2.** Pentulose
3. Hexulose **4.** Hexose
5. Aldose **6.** Ketose
7. Pyranose **8.** Furanose

(*a*) 2, 6, 8 (*b*) 2, 6,7
(*c*) 1, 5, 8 (*d*) A set of terms other than these

B-9. The following pair of Fischer projections

$$\begin{array}{c}CH_2OH\\ |\\ C{=}O\\ |\\ HO-\!\!\!+\!\!\!-CH_2OH\\ |\\ H\end{array}\qquad\qquad \begin{array}{c}CH_2OH\\ |\\ C{=}O\\ |\\ H-\!\!\!+\!\!\!-OH\\ |\\ CH_2OH\end{array}$$

represents structures that are:
(*a*) Epimers of each other (*b*) Enantiomers of each other
(*c*) Diastereomers of each other (*d*) Identical with each other

B-10. Which of the following will yield the same osazone on treatment with an excess of phenylhydrazine?

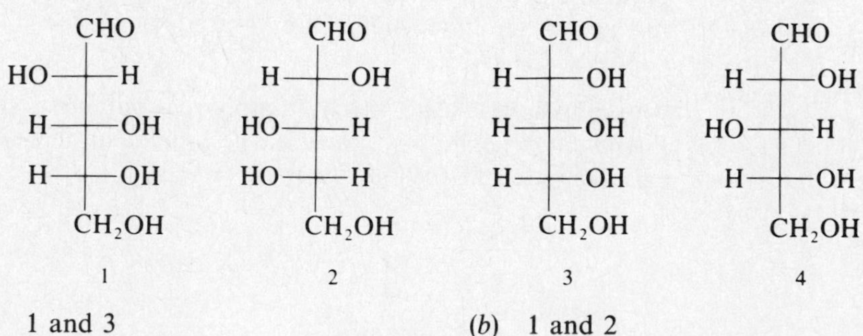

(*a*) 1 and 3 (*b*) 1 and 2
(*c*) 3 and 4 (*d*) 2 and 3

CHAPTER

28

ACETATE-DERIVED NATURAL PRODUCTS

IMPORTANT TERMS AND CONCEPTS

Fats and Fatty Acids (Secs. 28.1 to 28.3) Fats are a type of lipid. An operational definition of a *lipid* is a naturally occurring substance that is soluble in nonpolar solvents. Fats are biosynthesized from *acetate*. An important aspect of the biosynthesis of fats is the role of *acetyl coenzyme A*, a thioester which acts as an acyl transfer agent.

Fats and oils are naturally occurring triesters of glycerol, also known as triacylglycerols. *Tristearin* is a common component of animal fats.

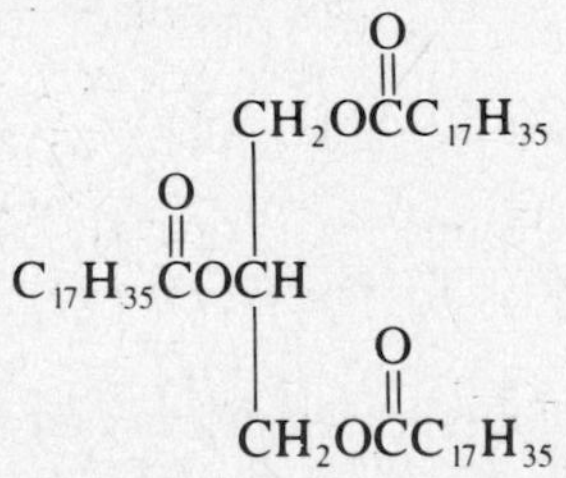

$$\begin{array}{l} \qquad\qquad\quad O \\ \qquad\qquad\quad \| \\ \qquad\quad CH_2OCC_{17}H_{35} \\ \quad\;\; O \quad | \\ \quad\;\; \| \quad\;\, | \\ C_{17}H_{35}COCH \\ \qquad\qquad | \quad\;\; O \\ \qquad\qquad | \quad\;\; \| \\ \qquad\quad CH_2OCC_{17}H_{35} \end{array}$$

Tristearin

The acyl groups have unbranched carbon chains and contain an even number of carbon atoms. Those having 14 to 20 carbon atoms are the most common. Unsaturation, when present, almost always occurs in the form of cis (or *Z*) alkene linkages. In the free acid form the acyl-group side chains are referred to as *fatty acids*; for example, stearic acid is $CH_3(CH_2)_{16}CO_2H$.

Phospholipids (Sec. 28.4) Phosphorus-containing lipids are known as *phospholipids. Phosphatidylcholine*, also known as *lecithin*, is an important phospholipid, which is a principal component of cell membranes.

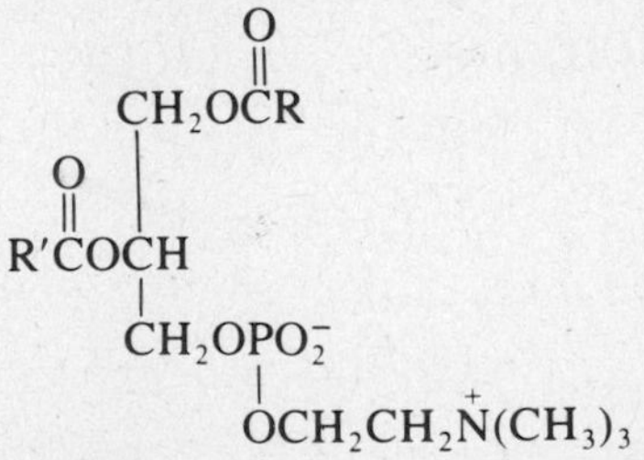

$$\begin{array}{l} \qquad\qquad\quad O \\ \qquad\qquad\quad \| \\ \qquad\quad CH_2OCR \\ \quad O \quad\;\; | \\ \quad \| \quad\;\; | \\ R'COCH \\ \qquad\quad | \\ \qquad\quad CH_2OPO_2^- \\ \qquad\qquad\quad | \\ \qquad\qquad\quad OCH_2CH_2\overset{+}{N}(CH_3)_3 \end{array}$$

Phosphatidylcholine (lecithin)

Cell membranes are thought to be made up of a *lipid bilayer*, as shown in text Figure 28.4.

Terpenes and Terpene Biosynthesis (Secs. 28.7 to 28.10) Terpenes are those naturally occurring substances derived biosynthetically from the *isoprene unit.*

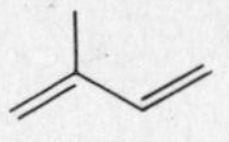

Isoprene

Terpenes are therefore often referred to as *isoprenoid* compounds. They are classified according to the number of carbon atoms they contain, as listed in text Table 28.2.

The biosynthesis of terpenes begins with acetic acid. The *isoprene unit* from which the terpenes are constructed is contained in the key intermediate *isopentenyl pyrophosphate.*

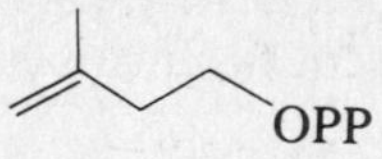

OPP

Isopentenyl pyrophosphate

Steroids (Secs. 28.11 to 28.15) Steroids are naturally occurring molecules characterized by the tetracyclic ring skeleton shown below.

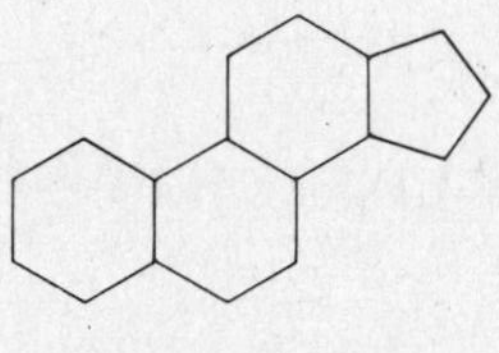

CH_3 CH_3 HO

Steroid ring skeleton

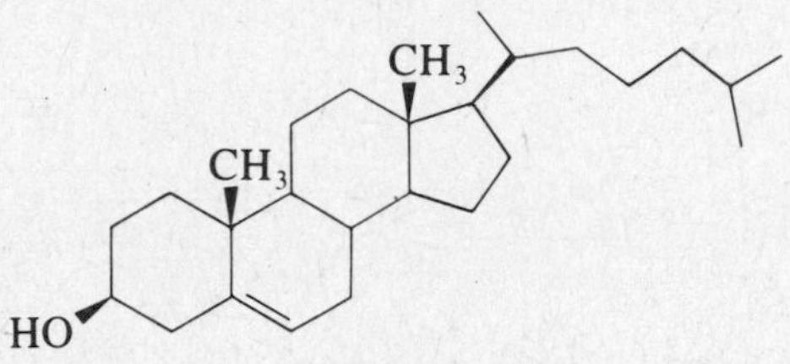

Cholesterol

An important steroid is *cholesterol.* The tetracyclic ring system is biosynthetically derived from the triterpene *squalene,* as shown in text Figure 28.11. A key intermediate, found in significant quantities in lanolin, is *lanosterol.*

Vitamin D and the *bile acids* are two examples of important natural compounds derived biologically from cholesterol. Also, the male sex hormone *testosterone* and the female sex hormones *estrogen* and *progesterone* are steroids formed biosynthetically from cholesterol.

SOLUTIONS TO TEXT PROBLEMS

28.1 The triacylglycerol shown in Figure 28.2*a*, with an oleyl group at C-2 of the glycerol unit and two stearoyl groups at C-1 and C-3, yields stearic and oleic acids in a 2 : 1 molar ratio on hydrolysis. A constitutionally isomeric structure in which the oleyl group is attached to C-1 of glycerol would yield the same hydrolysis products.

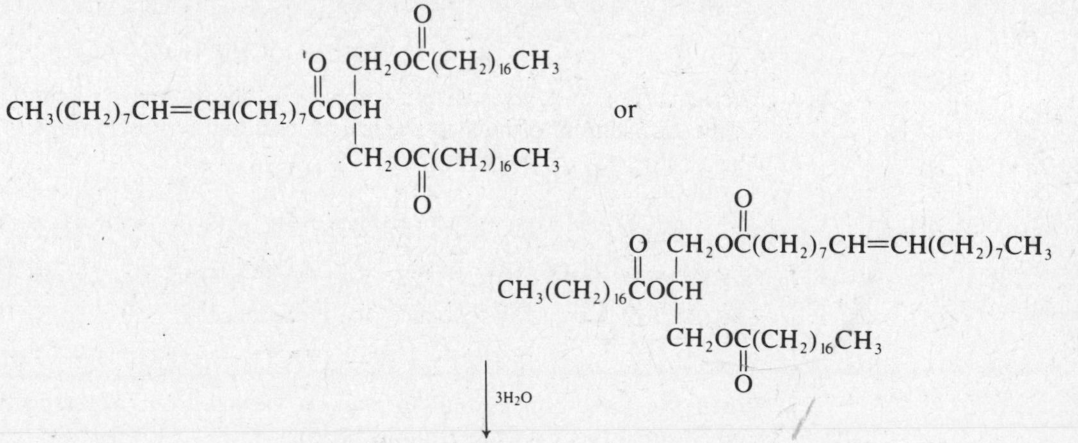

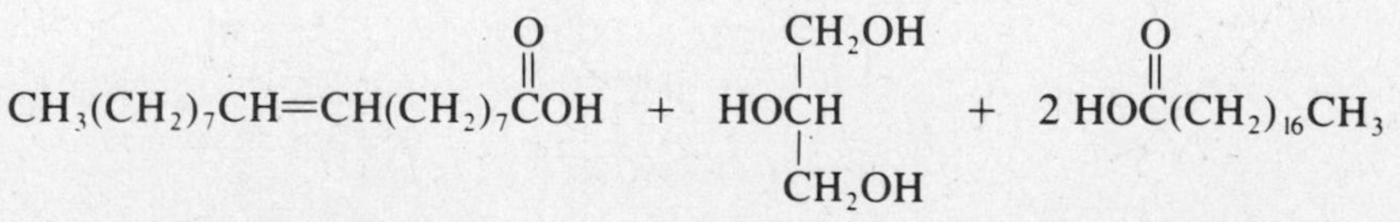

$$CH_3(CH_2)_7CH{=}CH(CH_2)_7\overset{O}{\overset{\|}{C}}OCH(CH_2O\overset{O}{\overset{\|}{C}}(CH_2)_{16}CH_3)_2$$

or

$$CH_3(CH_2)_{16}\overset{O}{\overset{\|}{C}}OCH\begin{cases} CH_2O\overset{O}{\overset{\|}{C}}(CH_2)_7CH{=}CH(CH_2)_7CH_3 \\ CH_2O\overset{O}{\overset{\|}{C}}(CH_2)_{16}CH_3 \end{cases}$$

$$\downarrow 3H_2O$$

$$CH_3(CH_2)_7CH{=}CH(CH_2)_7\overset{O}{\overset{\|}{C}}OH + HOCH(CH_2OH)_2 + 2\ HO\overset{O}{\overset{\|}{C}}(CH_2)_{16}CH_3$$

Oleic acid — Glycerol — Stearic acid

28.2 The reaction is an acyl transfer from the sulfur of acetyl coenzyme A to the sulfur of acyl carrier protein.

$$\underset{\text{Acetyl coenzyme A}}{CH_3\overset{O}{\overset{\|}{C}}SCoA} + \underset{\text{Acyl carrier protein}}{H\ddot{S}—ACP} \longrightarrow \underset{\text{Tetrahedral intermediate}}{CH_3\overset{OH}{\overset{|}{\underset{S—ACP}{\underset{|}{C}}}}—SCoA} \longrightarrow \underset{\text{S-Acetyl acyl carrier protein}}{CH_3\overset{O}{\overset{\|}{C}}S—ACP} + \underset{\text{Coenzyme A}}{HSCoA}$$

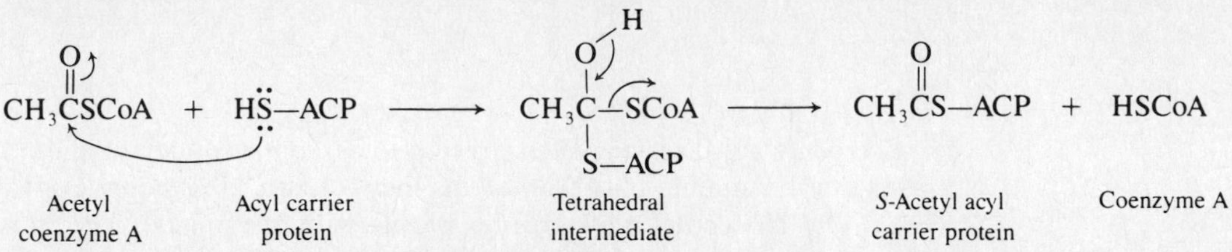

28.3 Conversion of acyl carrier protein–bound tetradecanoate to hexadecanoate proceeds through the series of intermediates shown.

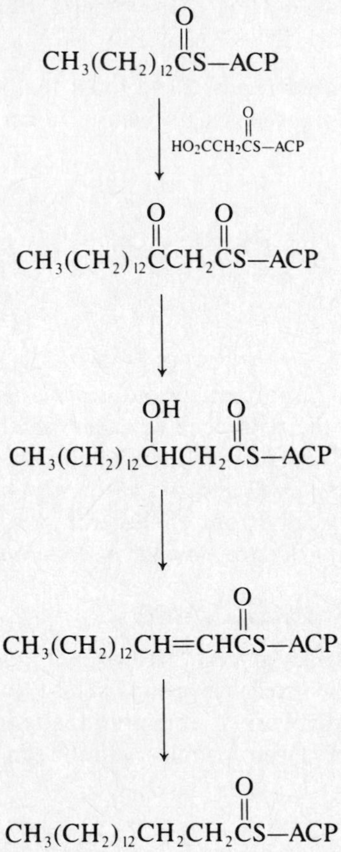

$$CH_3(CH_2)_{12}\overset{O}{\overset{\|}{C}}S—ACP$$

$$\downarrow \; HO_2CCH_2\overset{O}{\overset{\|}{C}}S—ACP$$

$$CH_3(CH_2)_{12}\overset{O}{\overset{\|}{C}}CH_2\overset{O}{\overset{\|}{C}}S—ACP$$

$$\downarrow$$

$$CH_3(CH_2)_{12}\overset{OH}{\overset{|}{C}}HCH_2\overset{O}{\overset{\|}{C}}S—ACP$$

$$\downarrow$$

$$CH_3(CH_2)_{12}CH=CH\overset{O}{\overset{\|}{C}}S—ACP$$

$$\downarrow$$

$$CH_3(CH_2)_{12}CH_2CH_2\overset{O}{\overset{\|}{C}}S—ACP$$

28.4 The structure of L-glycerol 3-phosphate is shown in a Fischer projection. Translate the Fischer projection formula to a three-dimensional representation:

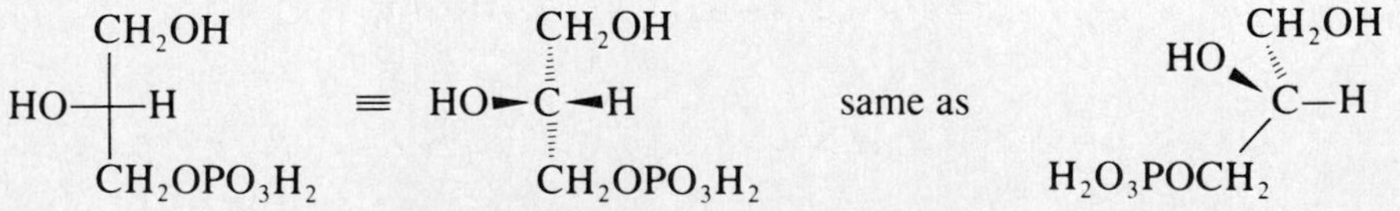

$$HO\overset{CH_2OH}{\underset{CH_2OPO_3H_2}{+}}H \equiv HO\blacktriangleright\overset{CH_2OH}{\underset{CH_2OPO_3H_2}{C}}\blacktriangleleft H \quad \text{same as} \quad \begin{matrix} HO\blacktriangle \; CH_2OH \\ C—H \\ H_2O_3POCH_2 \end{matrix}$$

The order of decreasing sequence rule precedence is:

$$HO— > H_2O_3POCH_2— > HOCH_2— > H—$$

When the three-dimensional formula is viewed from a perspective in which the lowest-ranked substituent is away from us, we see:

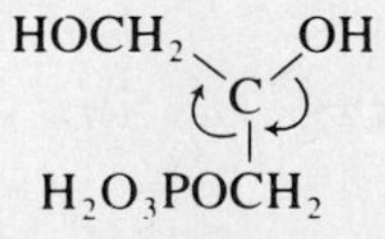

$$\begin{matrix} HOCH_2 \quad OH \\ C \\ H_2O_3POCH_2 \end{matrix}$$

Order of decreasing priority is clockwise, therefore *R*

The absolute configuration is *R*.

The conversion of L-glycerol 3-phosphate to a phosphatidic acid does not affect any of the bonds to the chiral center nor does it alter the sequence rule ranking of the substituents.

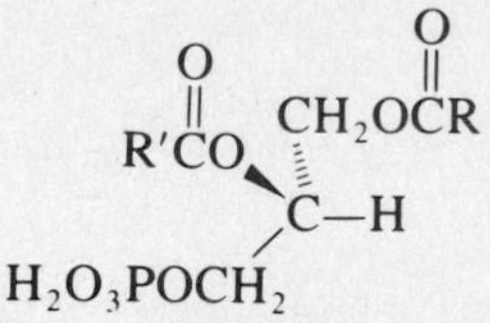

$$R'\overset{O}{\overset{\|}{C}}O- > H_2O_3POCH_2- > R\overset{O}{\overset{\|}{C}}OCH_2- > H$$

The absolute configuration is *R*.

28.5 Cetyl palmitate (hexadecyl hexadecanoate) is an ester in which both the acyl group and the alkyl group contain 16 carbon atoms.

$$CH_3(CH_2)_{14}\overset{O}{\overset{\|}{C}}O(CH_2)_{15}CH_3$$

Hexadecyl hexadecanoate

28.6 The biosynthetic scheme shown in Figure 28.6 illustrates the conversion of arachidonic acid to PGE_2.

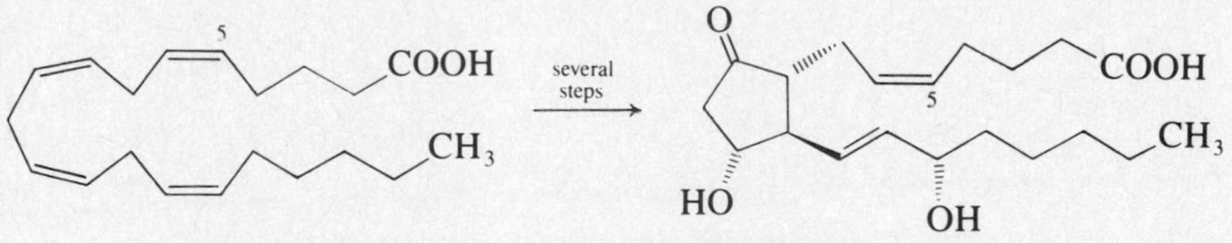

Arachidonic acid —several steps→ PGE_2

The structure of PGE_1 is identical to that of PGE_2 except that PGE_1 lacks the double bond at C-5. Therefore the fatty acid that leads to PGE_1 is analogous to arachidonic acid except for the absence of the C-5 double bond.

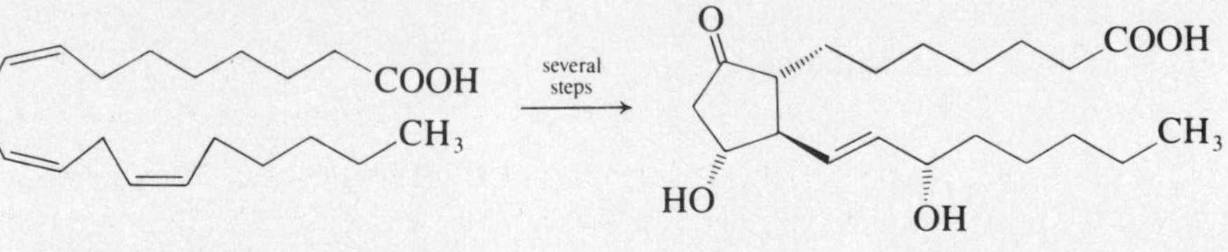

(8*Z*,11*Z*,14*Z*)-8,11,14-Icosatrienoic acid —several steps→ PGE_1

28.7 Isoprene units are 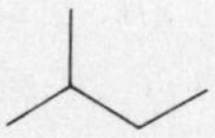 fragments in the carbon skeleton. Functional groups and multiple bonds are ignored when structures are examined for the presence of isoprene units.

α-Phellandrene (two equally correct answers)

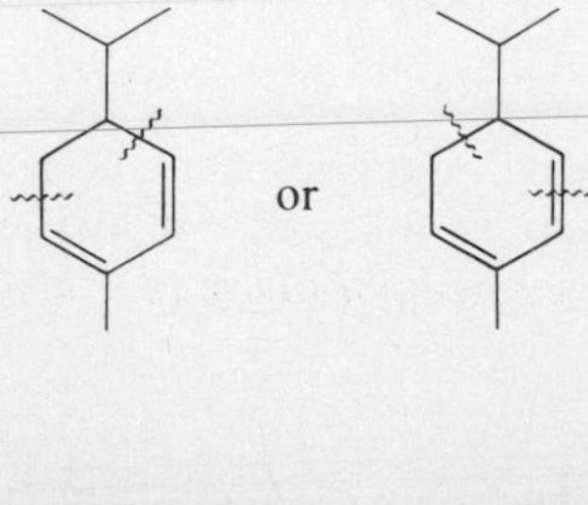

or

Menthol (same carbon skeleton as α-phellandrene but different functionality)

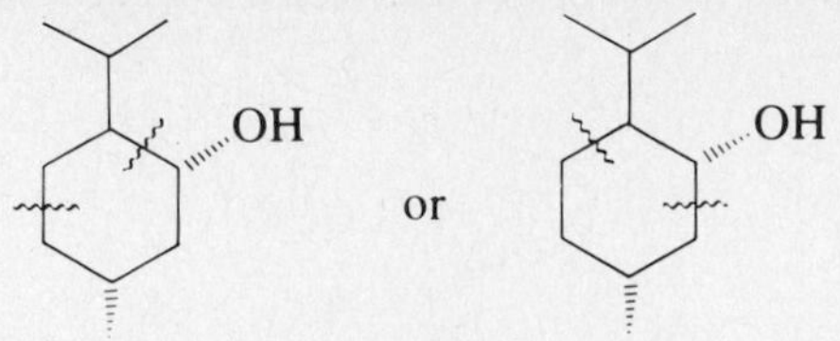

OH or OH

Citral

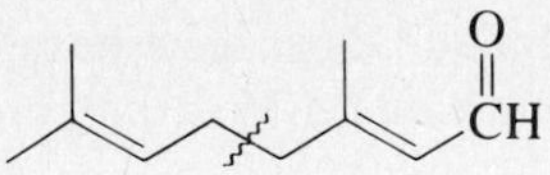

O ‖ CH

Farnesol

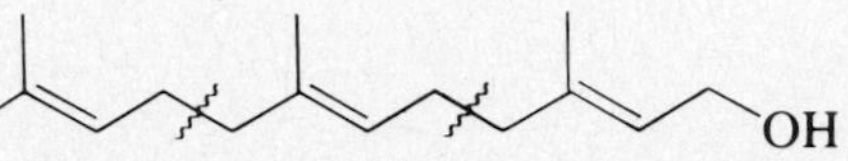

OH

Abscicic acid

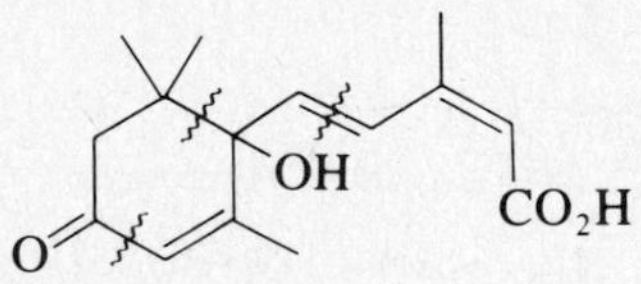

O OH CO_2H

Cembrene (two equally correct answers)

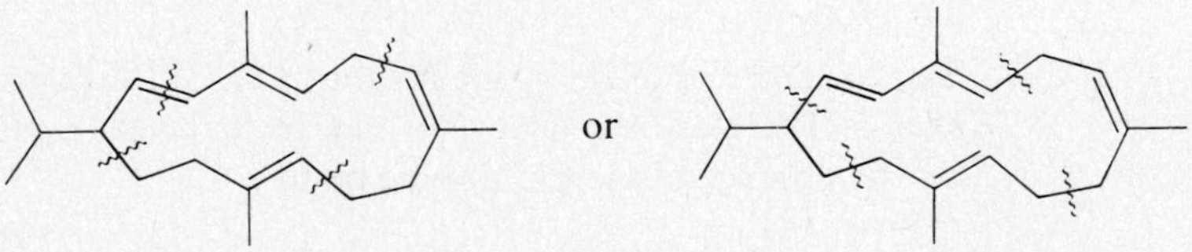

or

Vitamin A

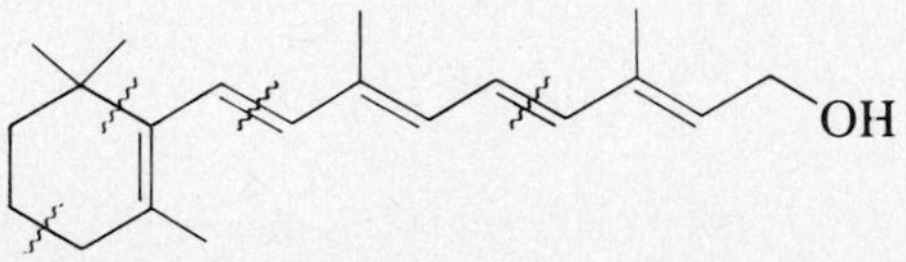

OH

28.8 *β*-Carotene is a tetraterpene, as it has 40 carbon atoms. The tail-to-tail linkage is at the midpoint of the molecule and connects two 20-carbon fragments.

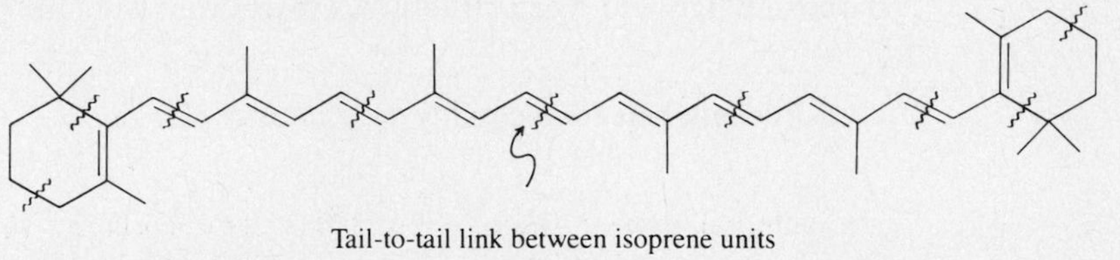

Tail-to-tail link between isoprene units

28.9 Isopentenyl pyrophosphate acts as an alkylating agent toward farnesyl pyrophosphate. Alkylation is followed by loss of a proton from the carbocation intermediate, giving geranylgeranyl pyrophosphate. Hydrolysis of the pyrophosphate yields geranylgeraniol.

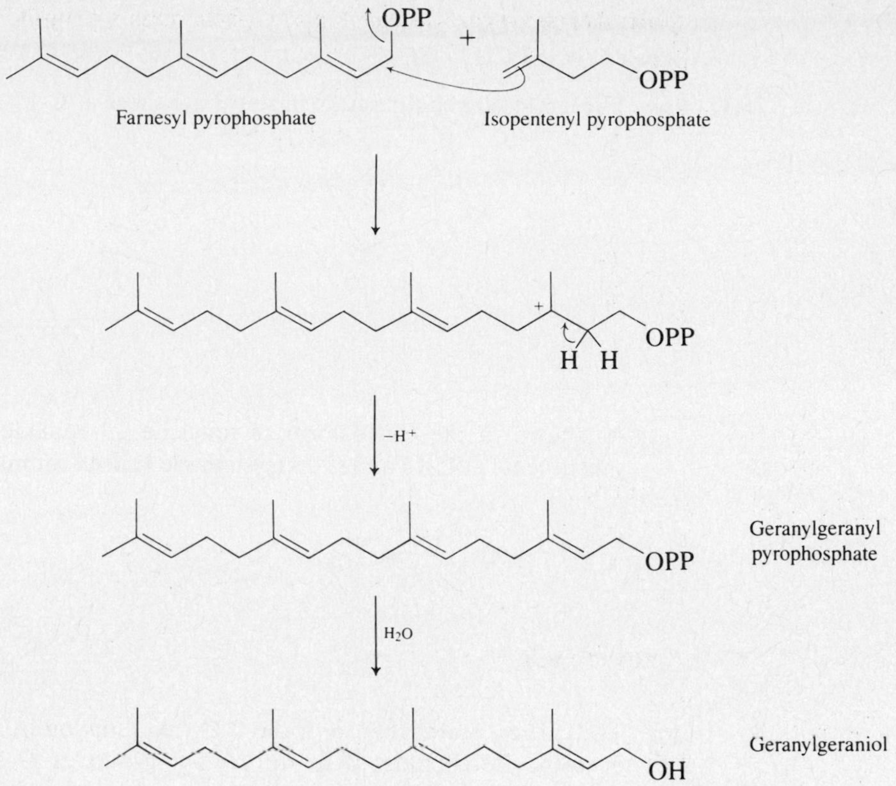

OPP +

OPP

Farnesyl pyrophosphate Isopentenyl pyrophosphate

OPP

H H

$-H^+$

OPP

Geranylgeranyl pyrophosphate

H_2O

OH

Geranylgeraniol

28.10 Borneol, the structure of which is given in Figure 28.8, is a secondary alcohol. Oxidation of borneol converts it to the ketone camphor.

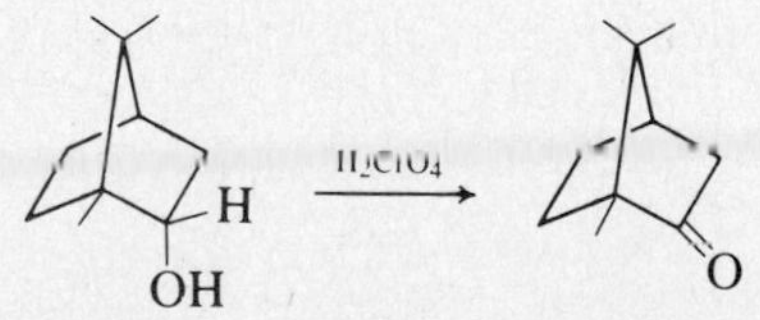

H

OH

H_2CrO_4

O

Borneol Camphor

Reduction of camphor with sodium borohydride gives a mixture of stereoisomeric alcohols, of which one is borneol and the other isoborneol.

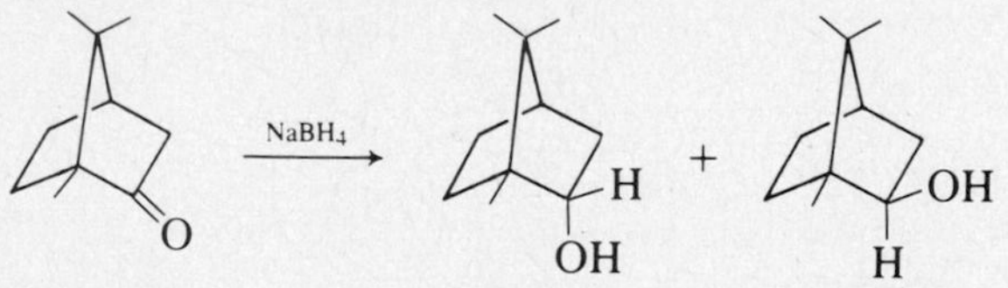

O

$NaBH_4$

H

OH

+

OH

H

Camphor Borneol Isoborneol

28.11 Figure 28.9 in the text describes the distribution of ^{14}C (denoted by *) in citronellal biosynthesized from acetate enriched with ^{14}C in its methyl group.

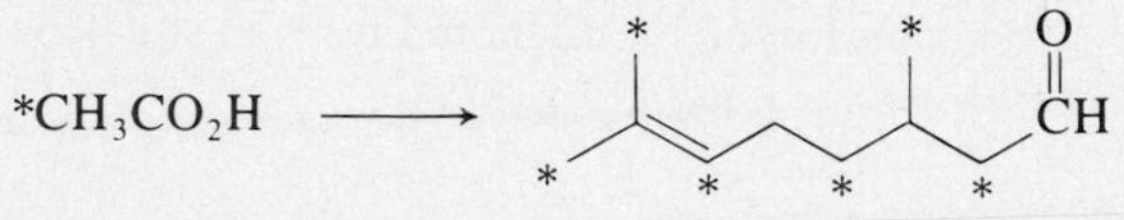

$$^*CH_3CO_2H \longrightarrow$$

O ‖ CH

If, instead, acetate enriched with ^{14}C at its carbonyl carbon were used, exactly the opposite distribution of the ^{14}C label would be observed.

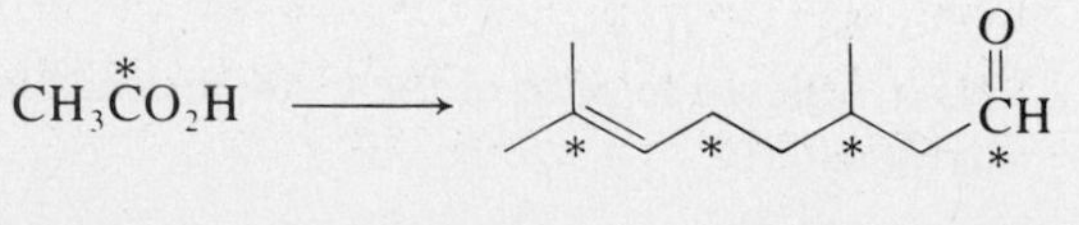

$$CH_3\overset{*}{C}O_2H \longrightarrow$$

O ‖ CH

When $^{14}CH_3CO_2H$ is used, C-2, C-4, C-6, C-8, and both methyl groups of citronellal are labeled. When $CH_3{}^{14}CO_2H$ is used, C-1, C-3, C-5, and C-7 are labeled.

28.12 (*b*) The hydrogens that migrate in step 3 are those at C-13 and C-17 (steroid numbering).

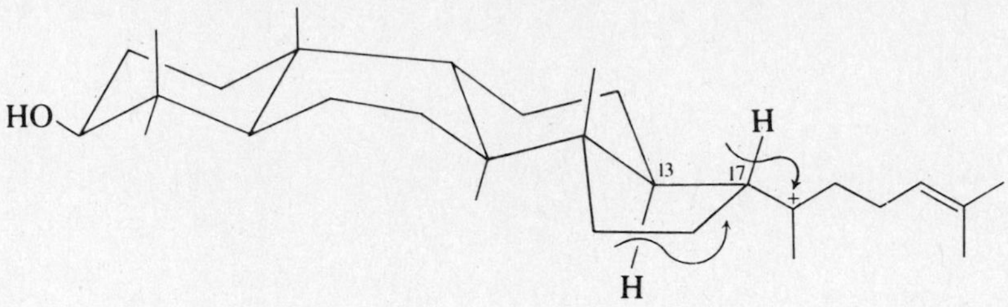

As shown in the coiled form of squalene 2,3-epoxide, these correspond to hydrogen substituents at C-14 and C-18 (systematic IUPAC numbering).

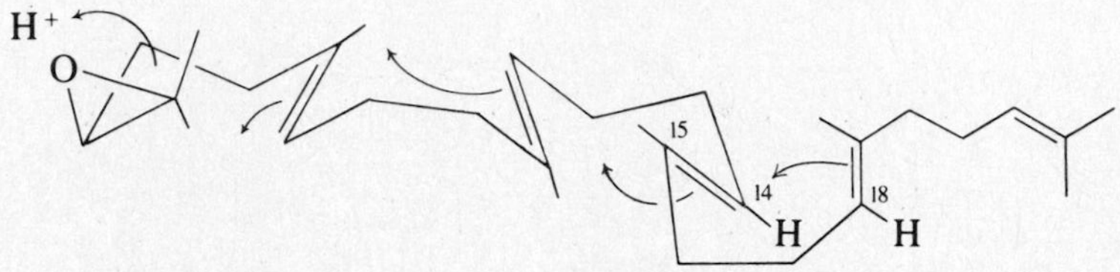

(*c*) The carbon atoms that form the C,D ring junction in cholesterol are C-14 and C-15 of squalene 2,3-epoxide. It is the methyl group at C-15 of squalene 2,3-epoxide that becomes the methyl group at this junction in cholesterol.

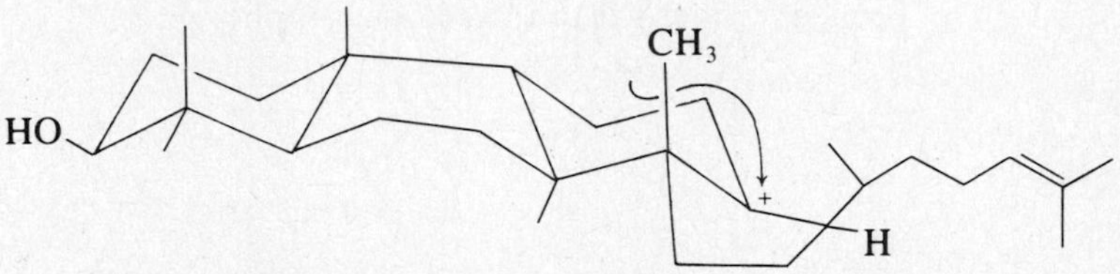

(*d*) The methyl groups that are lost are the methyl substituents at C-2 and C-10 plus the methyl group that is C-1 of squalene 2,3-epoxide.

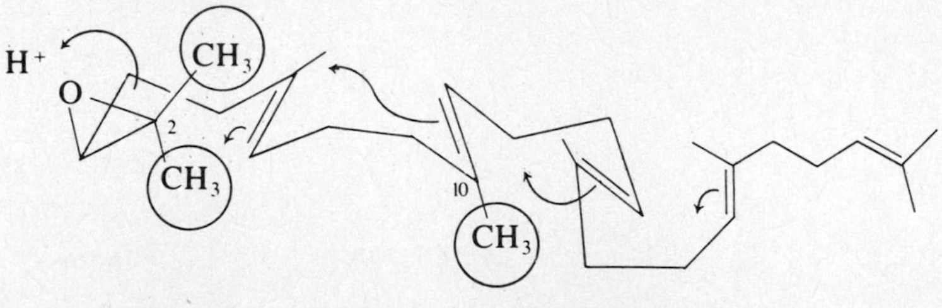

28.13 Tracking the ^{14}C label of $^{14}CH_3CO_2H$ through the complete biosynthesis of cholesterol requires a systematic approach. First, by analogy with Problem 28.11, we can determine the distribution of ^{14}C (denoted by *) in squalene 2,3-epoxide.

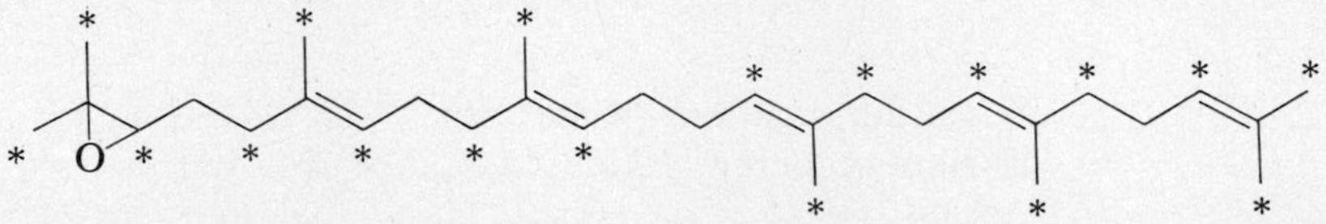

Next, follow the path of the C-14 enriched carbons in the cyclization of squalene 2,3-epoxide to lanosterol.

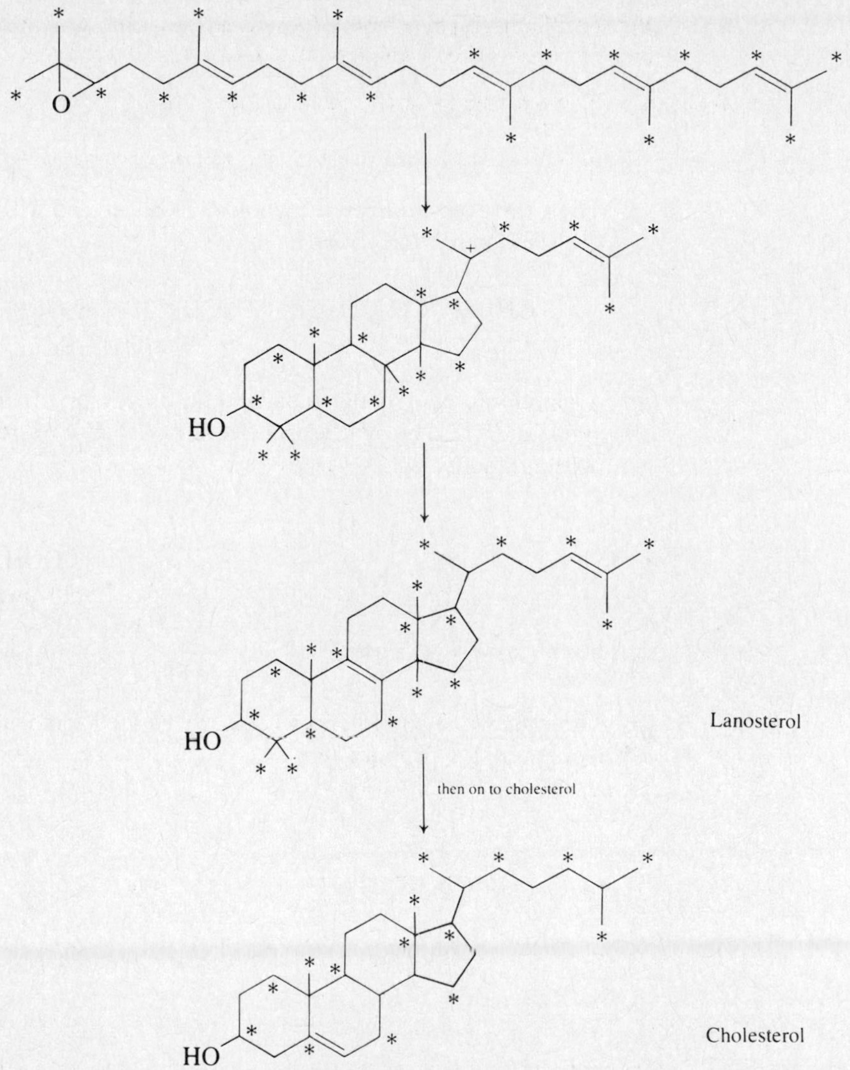

Lanosterol

then on to cholesterol

Cholesterol

28.14 By analogy to the reaction in which 7-dehydrocholesterol is converted to vitamin D_3, the structure of vitamin D_2 can be deduced from that of ergosterol.

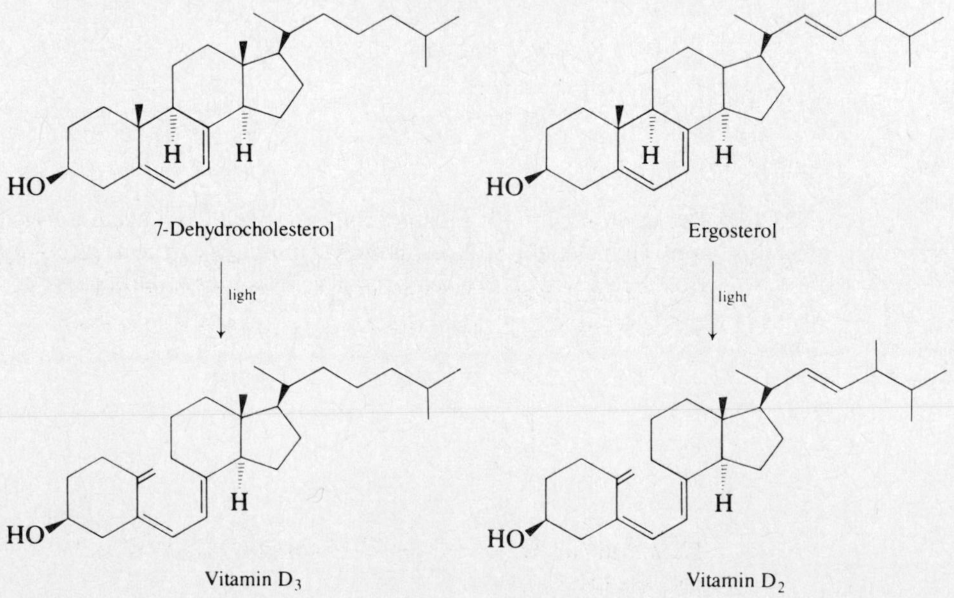

HO H H 7-Dehydrocholesterol

HO H H Ergosterol

light

light

HO H Vitamin D_3

HO H Vitamin D_2

28.15 (*a*) Fatty acid biosynthesis proceeds by the joining of acetate units.

$$\underset{\text{Acetyl coenzyme A}}{CH_3\overset{\overset{O}{\|}}{C}SCoA} \qquad \underset{\text{Acetoacetyl coenzyme A}}{CH_3\overset{\overset{O}{\|}}{C}CH_2\overset{\overset{O}{\|}}{C}SCoA} \qquad \underset{\text{Palmitoyl coenzyme A}}{CH_3(CH_2)_{14}\overset{\overset{O}{\|}}{C}SCoA}$$

Thus, the even-numbered carbons will be labeled with ^{14}C when palmitic acid is biosynthesized from $^{14}CH_3CO_2H$.

$$\overset{*}{C}H_3CH_2\overset{*}{C}H_2CH_2\overset{*}{C}H_2CH_2\overset{*}{C}H_2CH_2\overset{*}{C}H_2CH_2\overset{*}{C}H_2CH_2\overset{*}{C}H_2CH_2\overset{*}{C}H_2CO_2H$$

Palmitic acid

(*b*) Arachidonic acid is the biosynthetic precursor of PGE_2. The distribution of the ^{14}C label in PGE_2 biosynthesized from $^{14}CH_3CO_2H$ reflects the fatty acid origin of the prostaglandins.

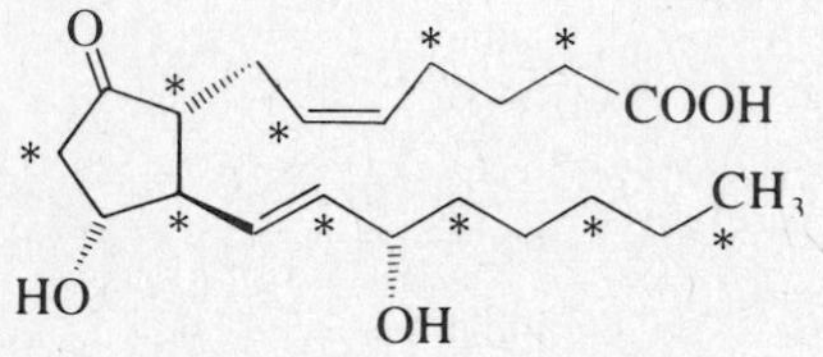

PGE_2

(*c*) Limonene is a monoterpene, biosynthesized from acetate by way of mevalonate and isopentenyl pyrophosphate.

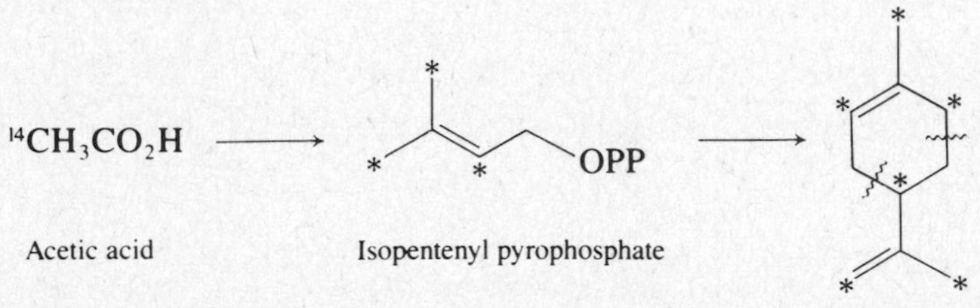

Acetic acid — Isopentenyl pyrophosphate — Limonene

(*d*) The distribution of the ^{14}C label in β-carotene becomes evident once its isoprene units are identified.

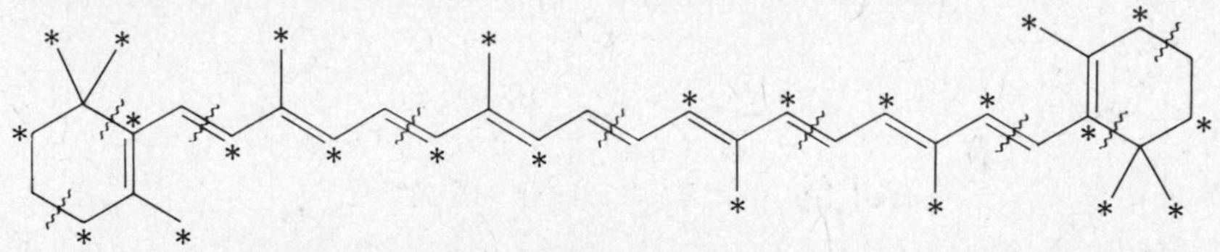

β-Carotene

28.16 The carbon chain of prostacyclin is derived from acetate by way of a C_{20} fatty acid. Trace a continuous chain of 20 carbons beginning with the carboxyl group. Even-numbered carbons are labeled with ^{14}C when prostacyclin is biosynthesized from $^{14}CH_3CO_2H$.

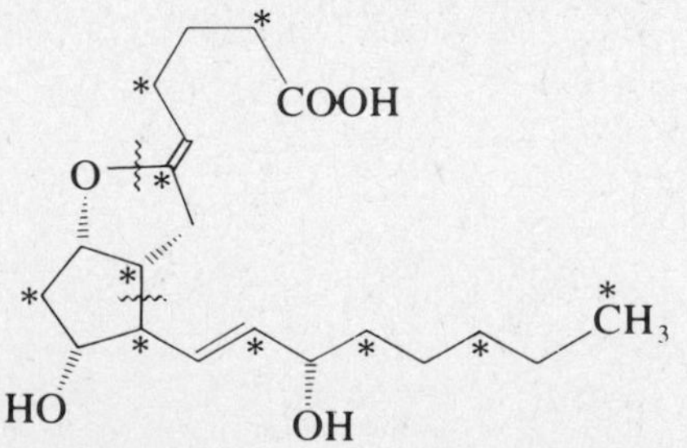

Prostacyclin

28.17 The isoprene units in the designated compounds are shown by disconnections in the structural formulas.

(*a*) Ascaridole:

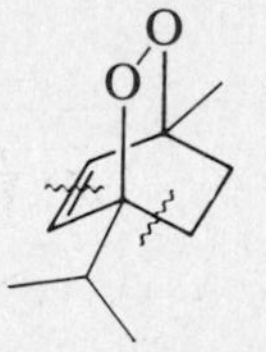

or

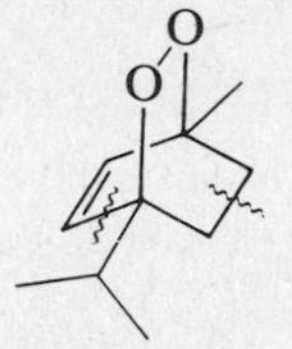

(*b*) Dendrolasin:

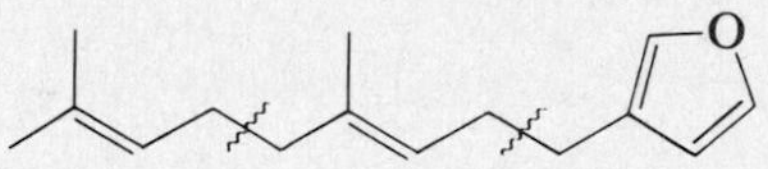

(*c*) γ-Bisabolene

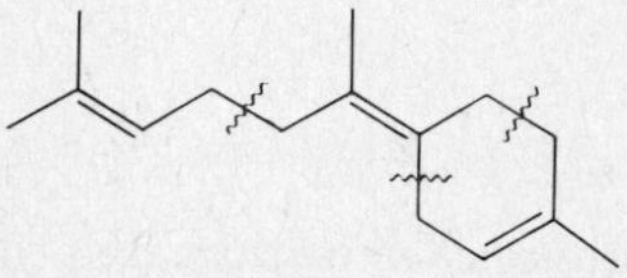

or

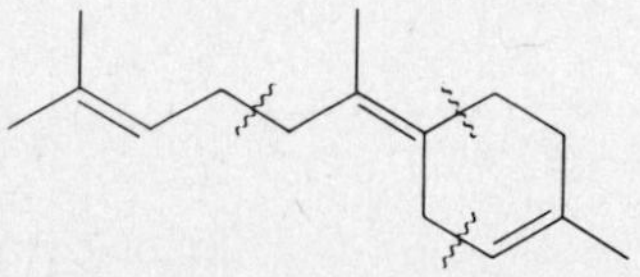

(*d*) α-Santonin

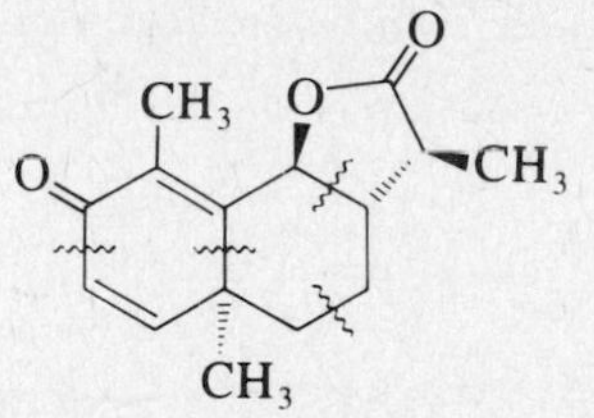

CH_3 O O CH_3 O CH_3

(*e*) Tetrahymanol

tail-to-tail linkage of isoprene units

OH CH_3 CH_3 H_3C CH_3 CH_3 CH_3 CH_3 CH_3

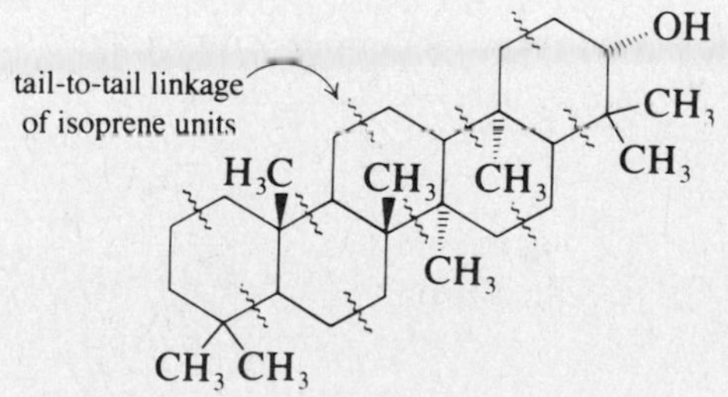

28.18 Of the four isoprene units of cubitene, three of them are joined in the usual head-to-tail fashion, but the fourth one is joined in an irregular way.

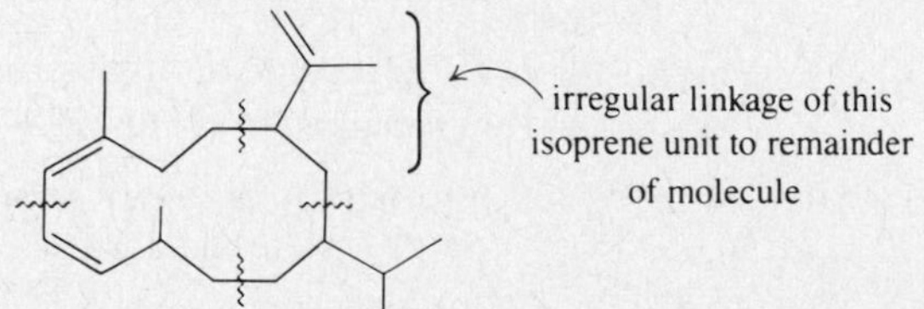

irregular linkage of this isoprene unit to remainder of molecule

28.19 (*a*) Cinerin I is an ester; the acyl portion of which is composed of two isoprene units, as shown below.

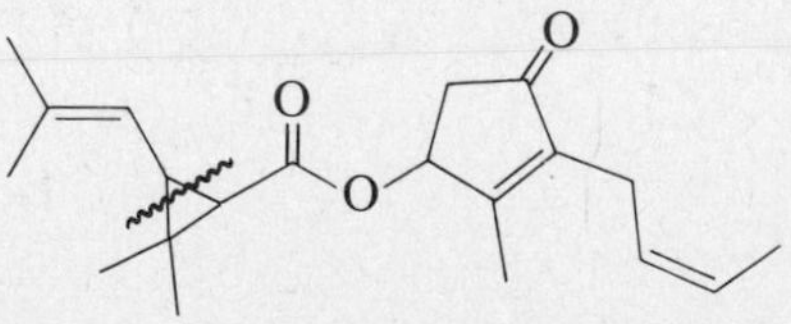

Cinerin I

(*b*) Hydrolysis of cinerin I involves cleavage of the ester unit.

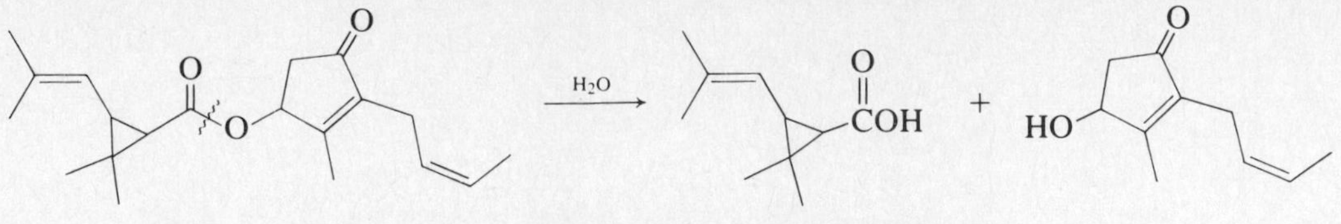

Cinerin I $\xrightarrow{H_2O}$ (+)-Chrysanthemic acid + [cyclopentenone alcohol]

Chrysanthemic acid has the constitution shown in the equation. Its stereochemistry is revealed by subsequent experiments.

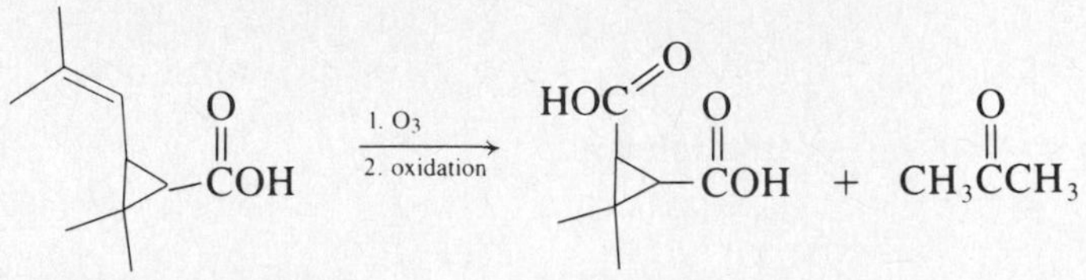

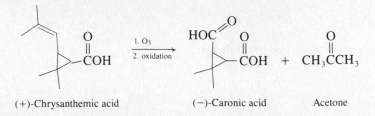

(+)-Chrysanthemic acid $\xrightarrow[\text{2. oxidation}]{\text{1. } O_3}$ (−)-Caronic acid + $CH_3\overset{O}{\overset{\|}{C}}CH_3$ (Acetone)

Since caronic acid is optically active, its carboxyl groups must be trans to one another. (The cis stereoisomer is an optically inactive meso form.) Therefore the structure of (+)-chrysanthemic acid must be either that shown below or its mirror image.

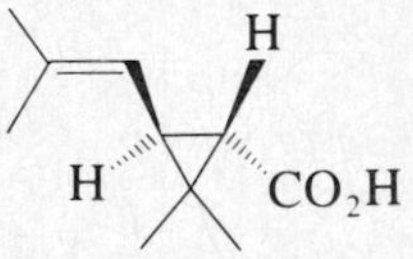

The carboxyl group and the 2-methyl-1-propenyl side chain must be trans to each other.

28.20 (*a*) Hydrolysis of phrenosine cleaves the glycosidic bond. The carbohydrate liberated by this hydrolysis is D-galactose.

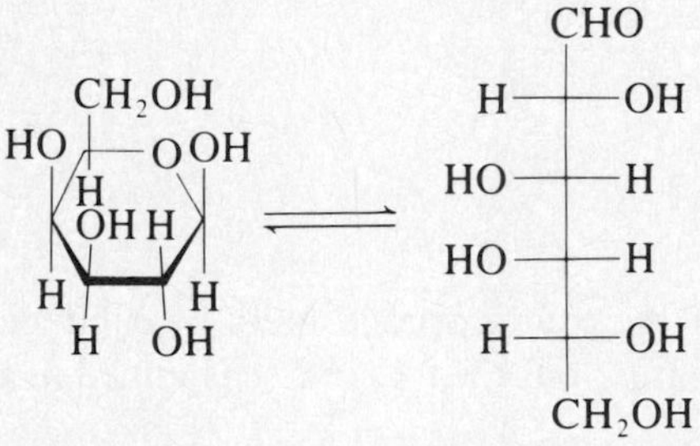

Phrenosine is a β-glycoside of D-galactose.

(*b*) The species that remains on cleavage of the galactose unit has the structure:

$$CH_3(CH_2)_{12}CH{=}CHCHOH\text{—}CH(CH_2OH)\text{—}NH\overset{O}{\overset{\|}{C}}CH(OH)(CH_2)_{21}CH_3$$

The two substances, sphingosine and cerebronic acid, that are formed along with D-galactose arise by hydrolysis of the amide bond.

$$CH_3(CH_2)_{12}CH{=}CHCHOH\text{—}CH(NH_2)\text{—}CH_2OH + HO\overset{O}{\overset{\|}{C}}CH(OH)(CH_2)_{21}CH_3$$

Sphingosine　　Cerebronic acid

28.21 (*a*) Catalytic hydrogenation over Lindlar palladium converts alkynes to cis alkenes.

$$CH_3(CH_2)_7C{\equiv}C(CH_2)_7COOH + H_2 \xrightarrow{\text{Lindlar Pd}} \overset{CH_3(CH_2)_7}{\underset{H}{}}\!\!\!\;C{=}C\;\overset{(CH_2)_7COOH}{\underset{H}{}}$$

9-Octadecynoic acid (stearolic acid) → (*Z*)-9-Octadecenoic acid (74%) (oleic acid)

(*b*) Carbon-carbon triple bonds are converted to trans alkenes by reduction with lithium and ammonia.

$$CH_3(CH_2)_7C{\equiv}C(CH_2)_7COOH \xrightarrow[\text{2. } H^+]{\text{1. Li, } NH_3} \overset{CH_3(CH_2)_7}{\underset{H}{}}\!\!\!\;C{=}C\;\overset{H}{\underset{(CH_2)_7COOH}{}}$$

9-Octadecynoic acid (stearolic acid) → (*E*)-9-Octadecenoic acid (97%) (elaidic acid)

(*c*) The carbon-carbon double bond is hydrogenated readily over a platinum catalyst. Hydrogenolysis of the ester function does not occur.

$$(Z)\text{-}CH_3(CH_2)_7CH{=}CH(CH_2)_7\overset{O}{\overset{\|}{C}}OCH_2CH_3 \xrightarrow{H_2,\ Pt} CH_3(CH_2)_{16}\overset{O}{\overset{\|}{C}}OCH_2CH_3$$

Ethyl (*Z*)-9-octadecenoate (ethyl oleate) → Ethyl octadecanoate (91%) (ethyl stearate)

(*d*) Lithium aluminum hydride reduces the ester function but leaves the carbon-carbon double bond intact.

$$(Z)\ CH_3(CH_2)_5\underset{OH}{\underset{|}{C}}HCH_2CH{=}CH(CH_2)_7\overset{O}{\overset{\|}{C}}OCH_3$$

Methyl (*Z*)-12-hydroxy-9-octadecenoate (methyl ricinoleate)

$$\downarrow \text{1. } LiAlH_4;\ \text{2. } H_2O$$

$$(Z)\text{-}CH_3(CH_2)_5\underset{OH}{\underset{|}{C}}HCH_2CH{=}CH(CH_2)_7CH_2OH + CH_3OH$$

(*Z*)-9-Octadecen-1,12-diol (52%) Methanol

(*e*) Epoxidation of the double bond occurs when an alkene is treated with a peroxy acid. The reaction is stereospecific; substituents that are cis to each other in the alkene remain cis in the epoxide.

$$(Z)\text{-}CH_3(CH_2)_7CH{=}CH(CH_2)_7COOH + C_6H_5CO_2OH$$

Oleic acid Peroxybenzoic acid

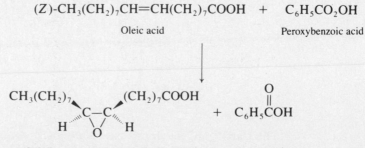

$$\downarrow$$

$$\overset{CH_3(CH_2)_7}{\underset{H}{}}\!\!\!\;C\overset{}{\underset{O}{-}}C\;\overset{(CH_2)_7COOH}{\underset{H}{}} + C_6H_5\overset{O}{\overset{\|}{C}}OH$$

cis-9,10-Epoxyoctadecanoic acid (62–67%) Benzoic acid

(*f*) Acid-catalyzed hydrolysis of the epoxide yields a diol; its stereochemistry corresponds to net anti hydroxylation of the double bond of the original alkene.

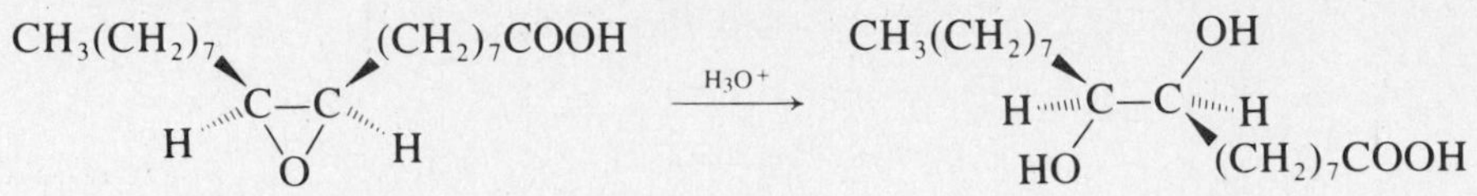

$CH_3(CH_2)_7$, $(CH_2)_7COOH$, H, H, O $\xrightarrow{H_3O^+}$ $CH_3(CH_2)_7$, OH, H, H, HO, $(CH_2)_7COOH$

cis-9,10-Epoxyoctadecanoic acid — 9,10-Dihydroxyoctadecanoic acid

The product is chiral but is formed as a racemic mixture containing equal amounts of the 9*R*,10*R* and 9*S*,10*S* stereoisomers.

(*g*) Hydroxylation of carbon-carbon double bonds with osmium tetraoxide proceeds with syn addition of hydroxyl groups.

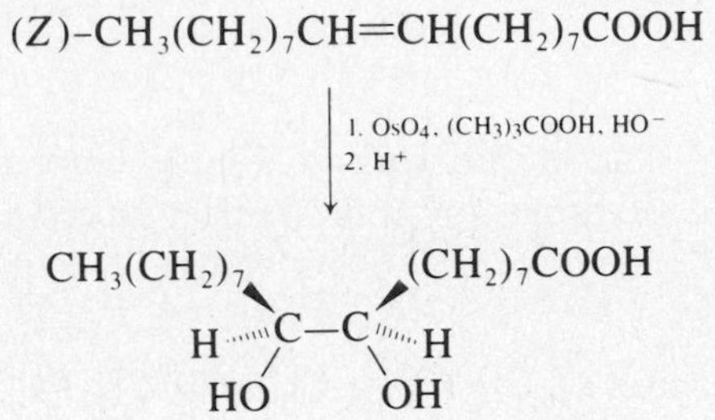

$$(Z)\text{-}CH_3(CH_2)_7CH{=}CH(CH_2)_7COOH$$

1. OsO_4, $(CH_3)_3COOH$, HO^-
2. H^+

$CH_3(CH_2)_7$, $(CH_2)_7COOH$, H, H, HO, OH

9,10-Dihydroxyoctadecanoic acid (70%)

The product is chiral but is formed as a racemic mixture containing equal amounts of the 9*R*,10*S* and 9*S*,10*R* stereoisomers.

(*h*) Hydrogenolysis of esters occurs at high temperatures and pressures in the presence of copper chromite as catalyst.

$$CH_3(CH_2)_{14}\overset{O}{\overset{\|}{C}}OCH_2(CH_2)_{14}CH_3 \xrightarrow[\text{copper chromite}]{H_2\text{ (high pressure), }250^\circ C} 2\,CH_3(CH_2)_{14}CH_2OH$$

Hexadecyl hexadecanoate — 1-Hexadecanol (97%)

Both the acyl group and the alkyl group of the ester are converted to 1-hexadecanol in this reduction.

(*i*) Hydroboration-oxidation effects the syn hydration of carbon-carbon double bonds with a regioselectivity contrary to Markovnikov's rule. The reagent attacks the less hindered face of the double bond of α-pinene.

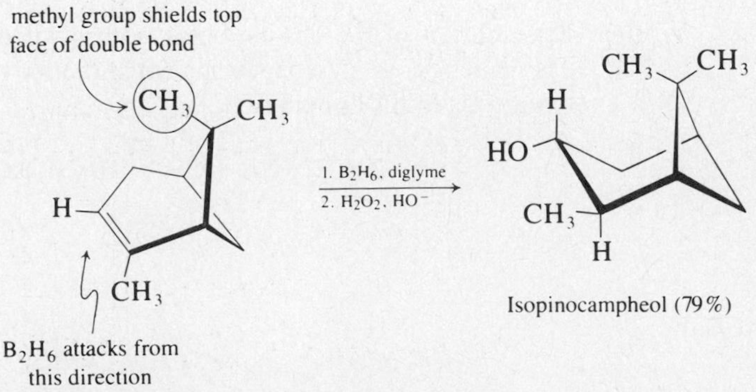

methyl group shields top face of double bond

CH_3, CH_3, H, CH_3

B_2H_6 attacks from this direction

1. B_2H_6, diglyme
2. H_2O_2, HO^-

CH_3, CH_3, H, HO, CH_3, H

Isopinocampheol (79%)

(*j*) The starting alkene in this case is β-pinene. As in the preceding exercise with α-pinene, diborane adds to the bottom face of the double bond.

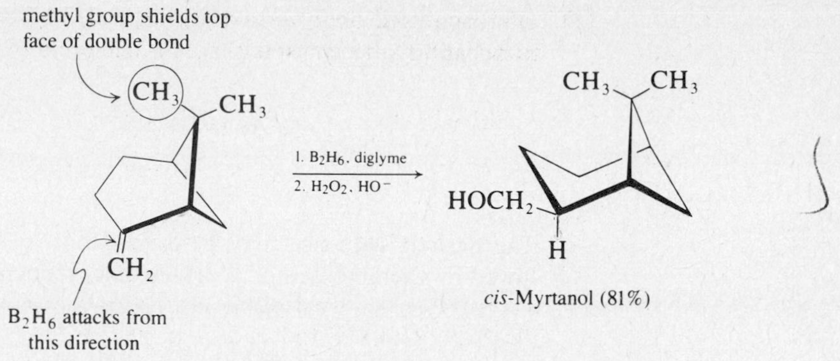

methyl group shields top face of double bond

CH_3 CH_3

CH_2

B_2H_6 attacks from this direction

1. B_2H_6, diglyme
2. H_2O_2, HO^-

CH_3 CH_3

$HOCH_2$ H

cis-Myrtanol (81%)

(*k*) The starting material is an acetal. It undergoes hydrolysis in dilute aqueous acid to give a ketone.

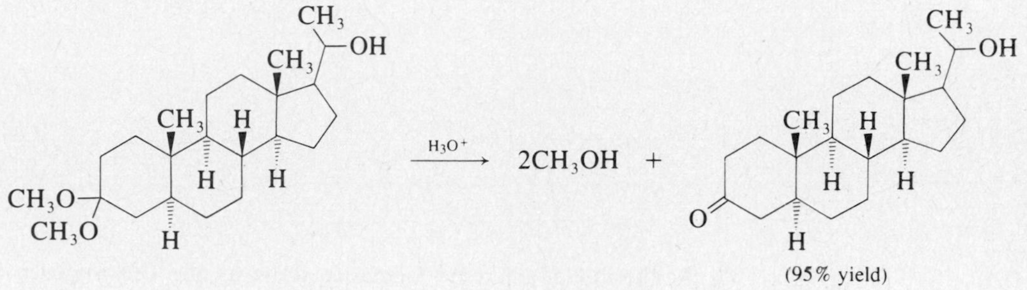

CH_3O, CH_3O (acetal steroid with CH_3, OH) $\xrightarrow{H_3O^+}$ $2CH_3OH$ + (ketone) (95% yield)

(*l*) The conditions described are those of the Reformatsky reaction.

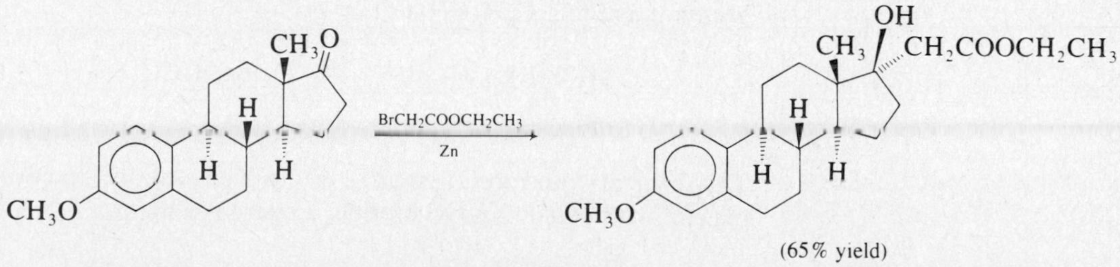

CH_3O (steroid ketone) $\xrightarrow[Zn]{BrCH_2COOCH_2CH_3}$ CH_3O (steroid with OH, $CH_2COOCH_2CH_3$) (65% yield)

The zinc enolate adds to the less hindered face of the carbonyl group, that is, trans to the methyl substituent on the adjacent carbon.

28.22 (*a*) There are no direct methods for the reduction of a carboxylic acid to an alkane. A number of indirect methods that may be used, however, involve first converting the carboxylic acid to an alkyl bromide via the corresponding alcohol.

$$\underset{\text{Octadecanoic acid}}{CH_3(CH_2)_{16}CO_2H} \xrightarrow[2.\ H_2O]{1.\ LiAlH_4} \underset{\text{1-Octadecanol}}{CH_3(CH_2)_{16}CH_2OH} \xrightarrow[\text{or } PBr_3]{HBr,\ heat} \underset{\text{1-Bromooctadecane}}{CH_3(CH_2)_{16}CH_2Br}$$

Once the alkyl bromide is in hand, it may be converted to an alkane by reduction with zinc in acid medium or by conversion to a Grignard reagent followed by addition of water.

$$\underset{\text{1-Bromooctadecane}}{CH_3(CH_2)_{16}CH_2Br} \xrightarrow[\text{acetic acid}]{Zn} \underset{\text{Octadecane}}{CH_3(CH_2)_{16}CH_3}$$

$$\underset{\text{1-Bromooctadecane}}{CH_3(CH_2)_{16}CH_2Br} \xrightarrow[\text{diethyl ether}]{Mg} CH_3(CH_2)_{16}CH_2MgBr \xrightarrow{H_2O} \underset{\text{Octadecane}}{CH_3(CH_2)_{16}CH_3}$$

Other routes are also possible (for example, E2 elimination from 1-bromooctadecane followed by hydrogenation of the resulting alkene), but require more steps.

(*b*) Retrosynthetic analysis reveals that the 18-carbon chain of the starting material must be attached to a benzene ring.

$$C_6H_5{-}(CH_2)_{17}CH_3 \Longrightarrow C_6H_6 + (CH_2)_{17}CH_3$$

1-Phenyloctadecane

The desired sequence may be carried out by a Friedel-Crafts acylation reaction, followed by Clemmensen or Wolff-Kishner reduction of the ketone.

$$CH_3(CH_2)_{16}CO_2H \xrightarrow{SOCl_2} CH_3(CH_2)_{16}\overset{O}{\overset{\|}{C}}Cl \xrightarrow[AlCl_3]{benzene} C_6H_5{-}\overset{O}{\overset{\|}{C}}(CH_2)_{16}CH_3$$

Octadecanoic acid — Octadecanoyl chloride — 1-Phenyl-1-octadecanone

$$\xrightarrow{Zn(Hg),\ HCl} C_6H_5{-}CH_2(CH_2)_{16}CH_3$$

1-Phenyloctadecane

Reduction of 1-phenyl-1-octadecanone by the Clemmensen method has been reported to proceed in 77 percent yield.

(*c*) In the synthesis of 1-phenylicosane, the 18-carbon chain of the starting material has been bonded to a $C_6H_5CH_2CH_2-$ group.

$$C_6H_5(CH_2)_{19}CH_3 \Longrightarrow C_6H_5CH_2CH_2- \ + \ -(CH_2)_{17}CH_3$$

1-Phenylicosane

This critical transformation may be carried out by reaction of an organocadmium reagent with octadecanoyl chloride, prepared as in (*b*).

$$(C_6H_5CH_2CH_2)_2Cd + Cl\overset{O}{\overset{\|}{C}}(CH_2)_{16}CH_3 \longrightarrow C_6H_5CH_2CH_2\overset{O}{\overset{\|}{C}}(CH_2)_{16}CH_3$$

Bis(2-phenylethyl)cadmium — Octadecanoyl chloride — 1-Phenyl-3-icosanone

The final step is reduction of the resulting ketone.

$$C_6H_5CH_2CH_2\overset{O}{\overset{\|}{C}}(CH_2)_{16}CH_3 \xrightarrow[\text{or } H_2NNH_2,\ HO^-]{Zn(Hg),\ HCl} C_6H_5CH_2CH_2CH_2(CH_2)_{16}CH_3$$

1-Phenyl-3-icosanone — 1-Phenylicosane

This sequence has been reported in the chemical literature. Both the coupling reaction of octadecanoyl chloride with the organocadmium reagent and Wolff-Kishner reduction of the resulting ketone were accomplished in 65 percent overall yield.

(*d*) First examine the structure of the target molecule 3-ethylicosane.

$$CH_3(CH_2)_{16}\underset{\underset{CH_2CH_3}{|}}{C}HCH_2CH_3$$

Retrosynthetic analysis reveals that two ethyl groups have been attached to a C_{18} unit.

$$CH_3(CH_2)_{16}\underset{\underset{CH_2CH_3}{|}}{C}H{-}CH_2CH_3 \Longrightarrow CH_3(CH_2)_{16}\underset{|}{C}H- \ + \ 2\ CH_3CH_2-$$

The necessary carbon-carbon bonds can be assembled by the reaction of an ester with two moles of a Grignard reagent.

$$\underset{\text{Ethyl octadecanoate (from octadecanoic acid and ethanol)}}{CH_3(CH_2)_{16}\overset{\overset{O}{\|}}{C}OCH_2CH_3} + \underset{\text{Ethylmagnesium bromide}}{2CH_3CH_2MgBr} \xrightarrow[2.\ H_3O^+]{1.\ \text{diethyl ether}} \underset{\text{3-Ethyl-3-icosanol}}{CH_3(CH_2)_{16}\overset{\overset{OH}{|}}{\underset{\underset{CH_2CH_3}{|}}{C}}CH_2CH_3}$$

With the correct carbon skeleton in place, all that is needed is to achieve the correct oxidation state. This can be accomplished by dehydration and reduction.

$$\underset{\text{3-Ethyl-3-icosanol}}{CH_3(CH_2)_{16}\overset{\overset{OH}{|}}{\underset{\underset{CH_2CH_3}{|}}{C}}CH_2CH_3} \xrightarrow[\text{heat}]{H_2SO_4} \underset{\text{3-Ethyl-2-icosene}}{CH_3(CH_2)_{16}\underset{\underset{CH_2CH_3}{|}}{C}{=}CHCH_3} + \underset{\text{3-Ethyl-3-icosene}}{CH_3(CH_2)_{15}CH{=}\underset{\underset{CH_2CH_3}{|}}{C}CH_2CH_3}$$

$$\xrightarrow{H_2,\ Pt} \underset{\text{3-Ethylicosane}}{CH_3(CH_2)_{16}\underset{\underset{CH_2CH_3}{|}}{C}HCH_2CH_3}$$

(*e*) Icosanoic acid contains two more carbon atoms than octadecanoic acid.

$$\underset{\text{Icosanoic acid}}{CH_3(CH_2)_{18}CO_2H} \Longrightarrow CH_3(CH_2)_{16}CH_2Br + \ddot{C}H_2CO_2H$$

A reasonable approach utilizes a malonic ester synthesis as a key step.

$$\underset{\text{1-Bromooctadecane [prepared as in 28.22(a)]}}{CH_3(CH_2)_{16}CH_2Br} + \underset{\text{Diethyl malonate}}{CH_2(CO_2CH_2CH_3)_2} \xrightarrow{NaOCH_2CH_3} \underset{\text{Diethyl 2-octadecylmalonate}}{CH_3(CH_2)_{16}CH_2CH(CO_2CH_2CH_3)_2}$$

$$\xrightarrow{1.\ HO^-;\ 2.\ H^+;\ 3.\ \text{heat}} \underset{\text{Icosanoic acid}}{CH_3(CH_2)_{16}CH_2CH_2CO_2H}$$

(*f*) The carbon chain must be shortened by one carbon atom in this problem. A Hofmann bromoamide rearrangement is indicated.

$$\underset{\text{Octadecanoic acid}}{CH_3(CH_2)_{16}CO_2H} \xrightarrow[2.\ NH_3]{1.\ SOCl_2} \underset{\text{Octadecanamide}}{CH_3(CH_2)_{16}\overset{\overset{O}{\|}}{C}NH_2} \xrightarrow{Br_2,\ HO^-} \underset{\text{1-Heptadecanamine}}{CH_3(CH_2)_{16}NH_2}$$

(*g*) Lithium aluminum hydride reduction of octadecanamide gives the corresponding amine.

$$\underset{\substack{\text{Octadecanamide}\\ \text{[from 28.22(}f\text{)]}}}{CH_3(CH_2)_{16}\overset{O}{\overset{\|}{C}}NH_2} \xrightarrow[2.\ H_2O]{1.\ LiAlH_4} \underset{\text{1-Octadecanamine}}{CH_3(CH_2)_{16}CH_2NH_2}$$

(*h*) Chain extension can be achieved via cyanide displacement of bromine from octadecyl bromide. Reduction of the cyano group completes the synthesis.

$$\underset{\substack{\text{1-Bromooctadecane}\\ \text{[from 28.22(}a\text{)]}}}{CH_3(CH_2)_{16}CH_2Br} \xrightarrow{KCN} \underset{\text{Nonadecanenitrile}}{CH_3(CH_2)_{16}CH_2C{\equiv}N}$$

$$\xrightarrow[2.\ H_2O]{1.\ LiAlH_4} \underset{\text{1-Nonadecanamine}}{CH_3(CH_2)_{16}CH_2CH_2NH_2}$$

28.23 First acylate the free hydroxyl group with an acyl chloride.

$$\text{(4-}CH_2OH\text{-2,2-dimethyl-1,3-dioxolane)} + R\overset{O}{\overset{\|}{C}}Cl \longrightarrow \text{(4-}CH_2O\overset{O}{\overset{\|}{C}}R\text{-2,2-dimethyl-1,3-dioxolane)}$$

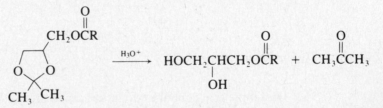

Treatment with aqueous acid brings about hydrolysis of the acetal function.

$$\text{(4-}CH_2O\overset{O}{\overset{\|}{C}}R\text{-2,2-dimethyl-1,3-dioxolane)} \xrightarrow{H_3O^+} HOCH_2\underset{OH}{\underset{|}{C}}HCH_2O\overset{O}{\overset{\|}{C}}R + CH_3\overset{O}{\overset{\|}{C}}CH_3$$

The two hydroxyl groups of the resulting diol are then esterified with two moles of the second acyl chloride.

$$HOCH_2\underset{OH}{\underset{|}{C}}HCH_2O\overset{O}{\overset{\|}{C}}R + 2\,R'\overset{O}{\overset{\|}{C}}Cl \longrightarrow R'\overset{O}{\overset{\|}{C}}OCH_2\underset{R'\underset{\|}{\underset{O}{C}}O}{\underset{|}{C}}HCH_2O\overset{O}{\overset{\|}{C}}R$$

28.24 The overall transformation

$$CH_2{=}C(CH_3)CH_2CH_2OH \quad \text{to} \quad CH_2{=}C(CH_3)CH_2CH_2SCH_3$$

simply requires conversion of the alcohol function to some suitable leaving group, followed by substitution by an appropriate nucleophile.

$$\underset{\text{3-Methyl-3-buten-1-ol}}{CH_2{=}C(CH_3)CH_2CH_2OH} \xrightarrow{PBr_3} \underset{\text{4-Bromo-2-methyl-1-butene}}{CH_2{=}C(CH_3)CH_2CH_2Br} \xrightarrow{NaSCH_3} \underset{\text{3-Methyl-3-butenyl methyl thioether}}{CH_2{=}C(CH_3)CH_2CH_2SCH_3}$$

As actually reported in the literature, the alcohol was converted to its corresponding *p*-toluenesulfonate ester and this substance then used as the substrate in the nucleophilic substitution step to produce the desired thioether in 76 percent yield.

28.25 The first transformation is an intramolecular aldol condensation. This reaction was carried out under conditions of base catalysis.

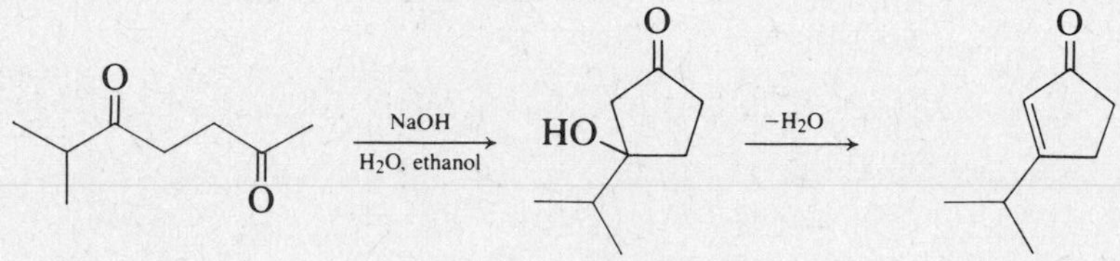

NaOH, H_2O, ethanol; $-H_2O$

6-Methyl-2,5-heptanedione — (not isolated) — 3-Isopropyl-2-cyclopentenone (71%)

The next step is reduction of a ketone to a secondary alcohol. Lithium aluminum hydride is suitable; it reduces carbonyl groups but leaves the double bond intact.

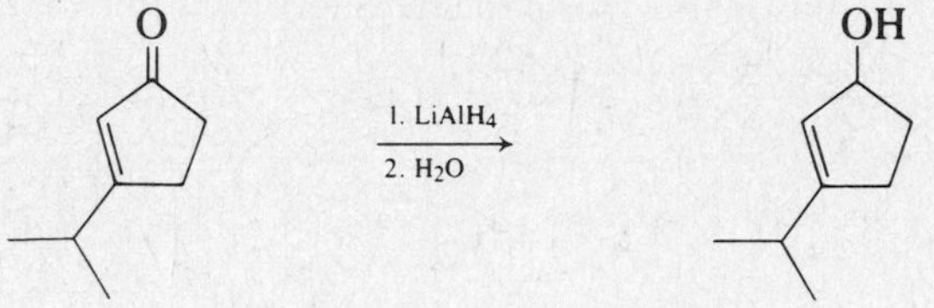

1. $LiAlH_4$ 2. H_2O

3-Isopropyl-2-cyclopentenone — 3-Isopropyl-2-cyclopenten-1-ol (97%)

Conversion of an alkene to a cyclopropane can be accomplished by using the Simmons-Smith reagent (iodomethylzinc iodide).

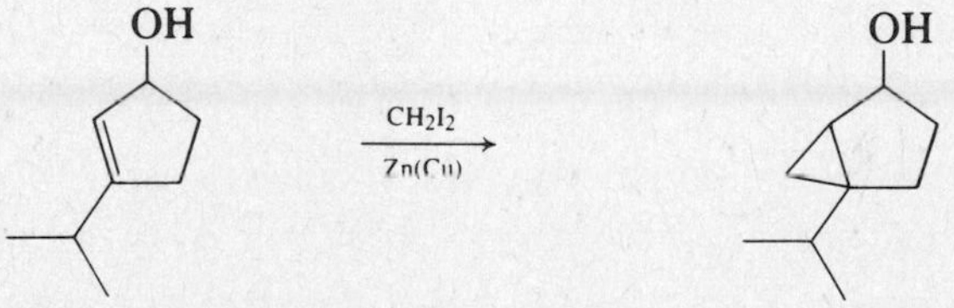

CH_2I_2, Zn(Cu)

3-Isopropyl-2-cyclopenten-1-ol — 5-Isopropylbicyclo[3.1.0]hexan-2-ol (66%)

Oxidation of the secondary alcohol to the ketone can be accomplished with any of a number of oxidizing agents. The chemists who reported this synthesis used chromic acid.

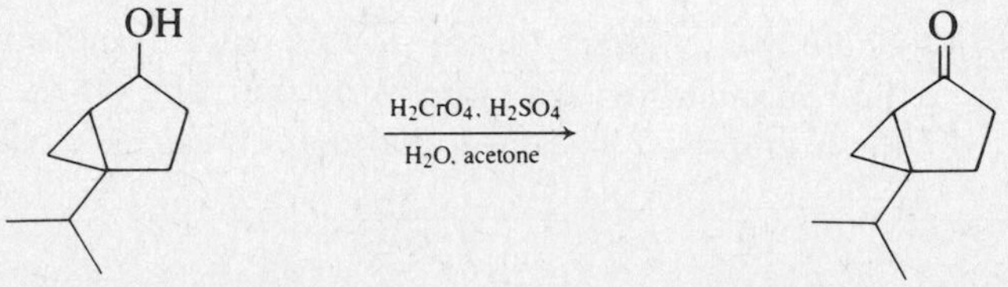

H_2CrO_4, H_2SO_4, H_2O, acetone

5-Isopropylbicyclo[3.1.0]hexan-2-ol — 5-Isopropylbicyclo[3.1.0]hexan-2-one (89%)

A Wittig reaction converts the ketone to sabinene.

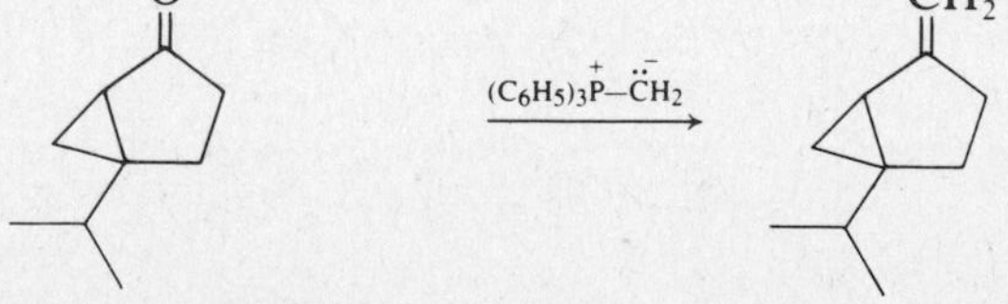

$(C_6H_5)_3\overset{+}{P}-\overset{..-}{C}H_2$

5-Isopropylbicyclo[3.1.0]hexan-2-one — Sabinene (70%)

28.26 The first step is a 1,4 addition of hydrogen bromide to the diene system of isoprene.

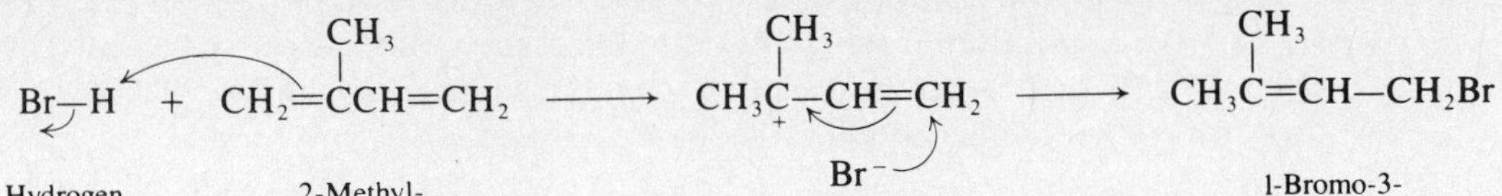

$$Br{-}H + CH_2{=}\underset{}{\overset{CH_3}{\overset{|}{C}}}CH{=}CH_2 \longrightarrow CH_3\overset{CH_3}{\overset{|}{\underset{+}{C}}}{-}CH{=}CH_2 \longrightarrow CH_3\overset{CH_3}{\overset{|}{C}}{=}CH{-}CH_2Br$$

Hydrogen bromide; 2-Methyl-1,3-butadiene; Br^-; 1-Bromo-3-methyl-2-butene

This is followed by Markovnikov addition of hydrogen bromide to the remaining double bond.

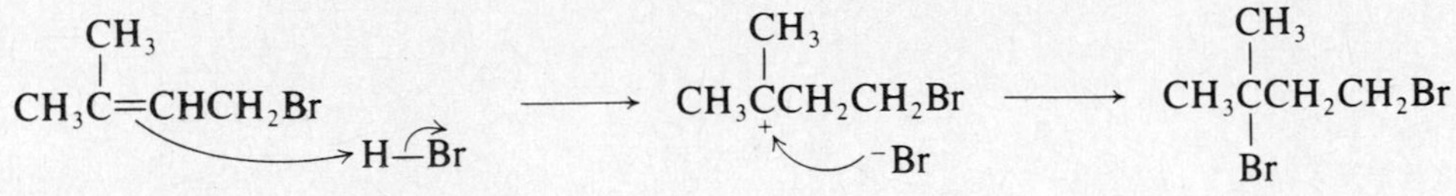

$$CH_3\overset{CH_3}{\overset{|}{C}}{=}CHCH_2Br + H{-}Br \longrightarrow CH_3\overset{CH_3}{\overset{|}{\underset{+}{C}}}CH_2CH_2Br \;\; {}^-Br \longrightarrow CH_3\overset{CH_3}{\overset{|}{\underset{Br}{\underset{|}{C}}}}CH_2CH_2Br$$

1-Bromo-3-methyl-2-butene; Hydrogen bromide; 1,3-Dibromo-3-methylbutane

28.27 A reasonable mechanism is protonation of the isolated carbon-carbon double bond, followed by cyclization.

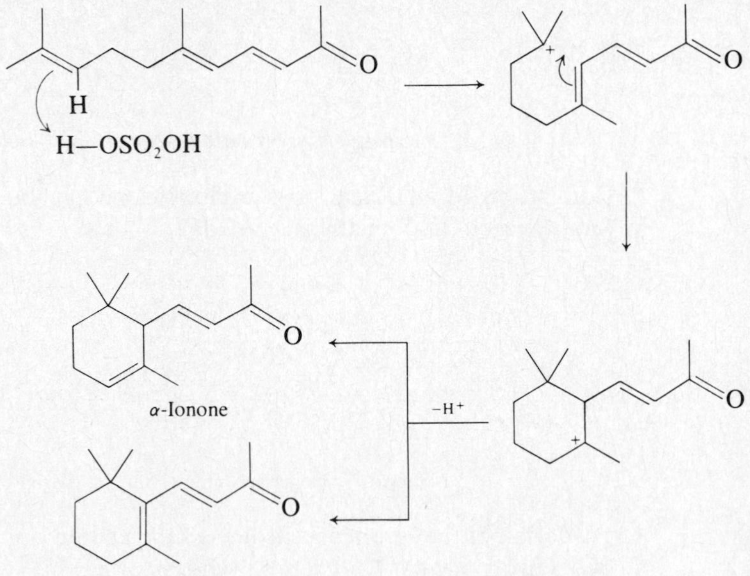

H—OSO$_2$OH

$-H^+$

α-Ionone

β-Ionone

28.28 There is a tendency for the double bond to move into conjugation with the carbonyl group. Two mechanisms are more likely than any others under conditions of acid catalysis. One of these simply involves protonation of the double bond followed by loss of a proton from C-4.

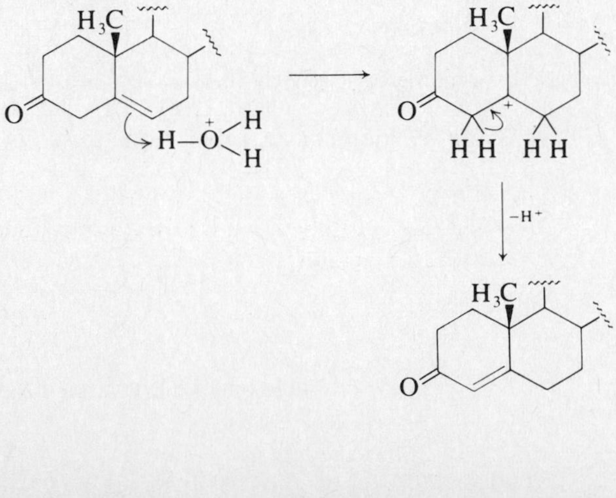

H_3C; $H{-}\overset{+}{O}H_2$; H_3C; H H H H; $-H^+$; H_3C

The other mechanism proceeds by enolization followed by proton-induced double bond migration.

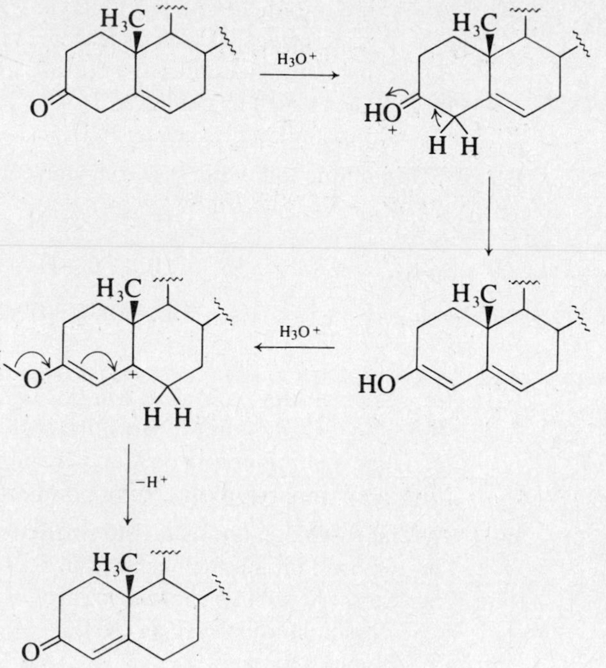

H_3C H_3O^+ H_3C HO H H

H_3C H_3O^+ H_3C H O HO H H

$-H^+$

H_3C O

SELF-TEST

PART A

A-1. Write a balanced chemical equation for the saponification of tristearin.

A-2. Both waxes and fats are lipids that contain the ester functional group. In what way do the structures of these lipids differ?

A-3. Classify each of the following isoprenoid compounds as a monoterpene, diterpene, etc. Indicate with dashed lines the isoprene units that make up each structure.

(*a*) α-Pinene:

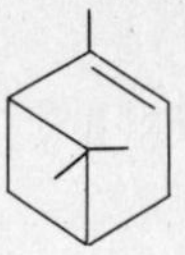

(*b*) Caryophyllene:

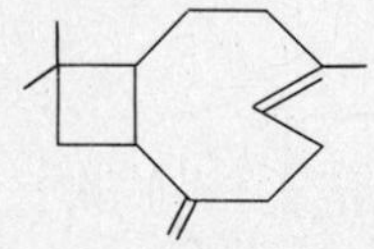

(*c*) Abietic acid: CO_2H

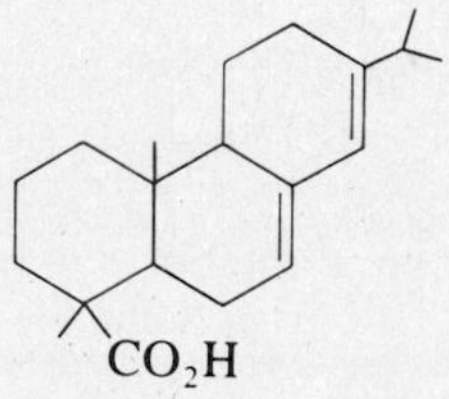

A-4. Propose a series of synthetic steps to carry out the preparation of oleic acid [(*Z*)-9-octadecenoic acid] from compound A. You may use any necessary organic or inorganic reagents.

$HC{\equiv}C(CH_2)_7CH$ (O, O)

A

PART B

B-1. A major component of a lipid bilayer is:
(*a*) A triacylglycerol such as tristearin
(*b*) Phosphatidylcholine, also known as lecithin
(*c*) A sterol such as cholesterol
(*d*) A prostaglandin such as PGE_1

B-2. Compare the following two triacylglycerols:

A	B
$CH_2O_2CC_{17}H_{35}$	$CH_2O_2CC_{17}H_{35}$
$CHO_2CC_{17}H_{35}$	$CHO_2CC_{17}H_{31}$
$CH_2O_2CC_{17}H_{35}$	$CH_2O_2CC_{17}H_{31}$

(*a*) The melting point of A will be higher
(*b*) The melting point of B will be higher
(*c*) The melting points of A and B will be the same
(*d*) No comparison of melting points can be made

B-3. An endoperoxide unit is an intermediate in the biosynthesis of:
(*a*) Phospholipids such as lecithin
(*b*) Steroids such as cholesterol
(*c*) Prostaglandins such as PGE_1
(*d*) None of these

B-4. Lanosterol, a biosynthetic precursor of cholesterol, exists naturally as a single enantiomer. How many *possible* stereoisomers having the lanosterol skeleton are there?

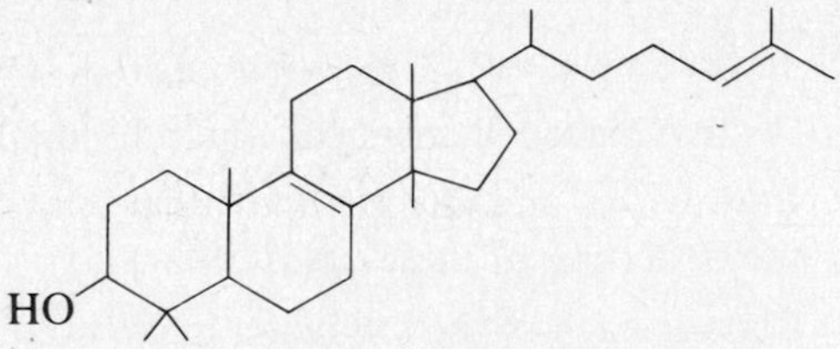

HO

Lanosterol

(*a*) 7 (*b*) 64
(*c*) 128 (*d*) 256

B-5. The compound whose carbon skeleton is shown, known as selinene, is found in celery.

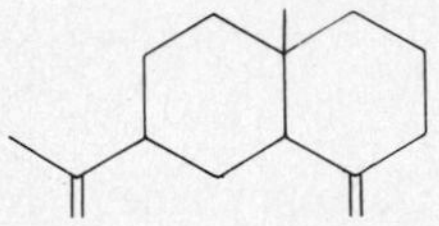

This substance is an example of
(*a*) Monoterpene (*b*) Diterpene
(*c*) Sesquiterpene (*d*) Triterpene

CHAPTER

29

AMINO ACIDS, PEPTIDES, AND PROTEINS. NUCLEIC ACIDS

IMPORTANT TERMS AND CONCEPTS

Amino Acids (Secs. 29.1 to 29.3) All the amino acids from which proteins are constructed are α-amino acids, and except for proline they contain a primary amino function. The 20 amino acids normally present in proteins are listed in text Table 29.1.

$$\underset{\overset{+}{N}H_3}{RCHCO_2^-}$$

Typical α-amino acid

Proline (CO_2^- on a pyrrolidine ring, $\overset{+}{N}$ bearing H H)

Proline

Except for glycine the amino acids in proteins are chiral and have the L configuration. With the exception of L-cysteine, the chiral amino acids in proteins have the *S* configuration as specified by the Cahn-Ingold-Prelog method.

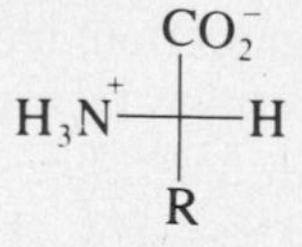

$$\begin{array}{c} CO_2^- \\ H_3\overset{+}{N}\text{—}\!\!\!+\!\!\!\text{—}H \\ R \end{array}$$

L Amino acid

The physical properties of a typical amino acid suggest that it is a salt. These properties are due to the fact that the amino acid exits as an inner salt, or *zwitterion*. Amino acids are *amphoteric*, that is, they contain both an acidic and a basic functional group. In a strongly acidic medium the predominant species is positively charged; in a strongly basic medium the predominant species is negatively charged.

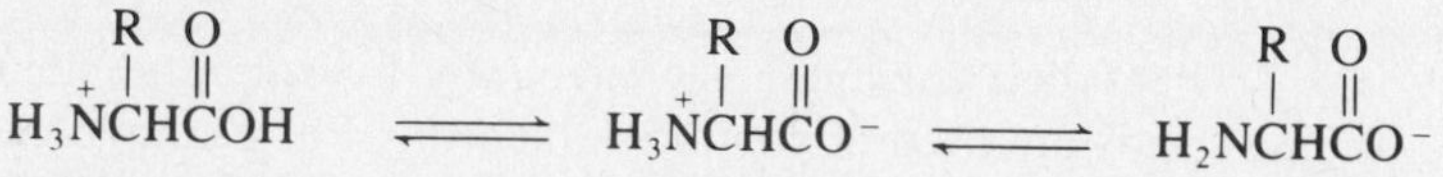

$$H_3\overset{+}{N}\overset{R}{\overset{|}{C}}H\overset{O}{\overset{\|}{C}}OH \rightleftharpoons H_3\overset{+}{N}\overset{R}{\overset{|}{C}}H\overset{O}{\overset{\|}{C}}O^- \rightleftharpoons H_2N\overset{R}{\overset{|}{C}}H\overset{O}{\overset{\|}{C}}O^-$$

Cationic species (present in strong acid) — Zwitterion (present at isoelectric point) — Anionic species (present in strong base)

The pH at which the zwitterion is the predominant species in solution is the *isoelectric point* of the amino acid. Each amino acid has a characteristic isoelectric point.

Peptides (Sec. 29.7) Peptides are constructed of amino acids linked together by amide bonds. The bond between the amino group of one amino acid and the carboxyl of another is known as a *peptide bond.* An example of a dipeptide consisting of two amino acids is glycylalanine. Using the abbreviations found in Table 29.1, the peptide is referred to as Gly-Ala.

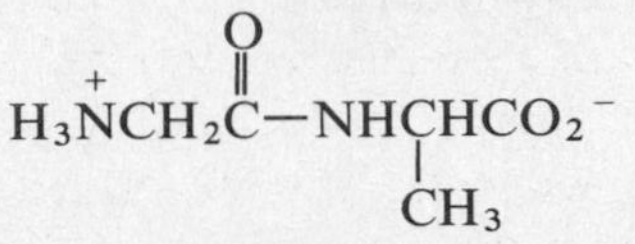

$$H_3\overset{+}{N}CH_2\overset{O}{\overset{\|}{C}}-NH\underset{CH_3}{\underset{|}{C}}HCO_2^-$$

Gly-Ala

Glycine forms the *N terminus* of the peptide; alanine is the *C terminus.* The *sequence* of amino acids in a peptide is important; Gly-Ala is not the same dipeptide as Ala-Gly.

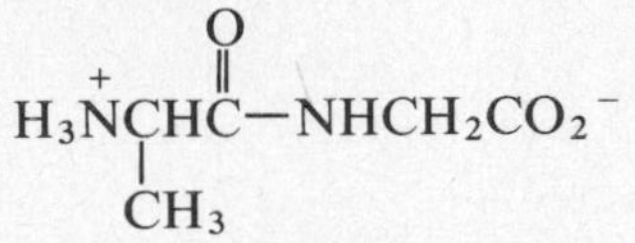

$$H_3\overset{+}{N}\underset{CH_3}{\underset{|}{C}}H\overset{O}{\overset{\|}{C}}-NHCH_2CO_2^-$$

Ala-Gly

Peptide Structure (Secs. 29.8 to 29.12, 29.18, 29.19) The structure of a peptide is described at several levels. The *primary structure* of a peptide is its constitution, that is, the amino acid sequence plus any disulfide links.

The stages of determining the sequence of amino acids in a peptide include end group analysis at both the N and the C terminus and selective hydrolysis.

End group analysis: N terminus

Reaction of the peptide with either Sanger's reagent or dansyl chloride, followed by hydrolysis of the peptide bonds, produces a derivative of the N-terminal amino acid.

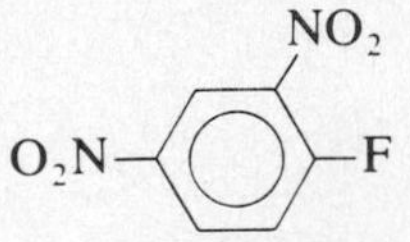

NO_2; O_2N–(benzene ring)–F

Sanger's reagent

$(CH_3)_2N$–(naphthalene)–SO_2Cl

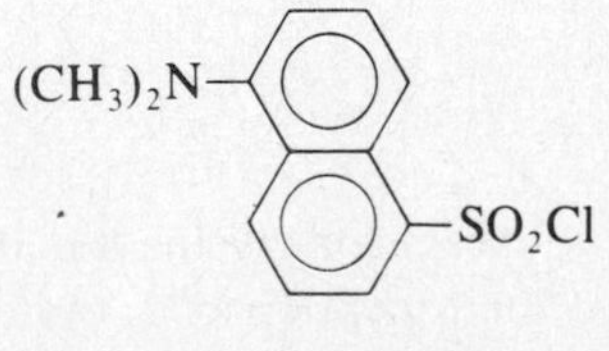

Dansyl chloride

The Edman degradation, shown in text Figure 29.9, allows a sequential analysis of the N terminus of a peptide to be undertaken. After each step, the *phenylhydantoin* derivative is isolated and characterized. The peptide chain minus the original N-terminal amino acid remains intact. This "new" peptide is then analyzed and the cycle repeated.

End group analysis: C terminus

Enzymes known as *carboxypeptidases* catalyze the hydrolysis of the C-terminal amino acid, allowing its isolation and analysis.

Selective hydrolysis

Peptidase enzymes that selectively hydrolyze amide bonds between specific pairs of amino acids are used in peptide sequencing. For example, *pepsin* catalyzes the hydrolysis of peptide bonds to the carbonyl groups of methionine and leucine.

The *secondary structure* of a peptide refers to the conformational relationship of nearest-neighbor amino acids with respect to each other. Hydrogen bonding interactions between the N—H group of one amino acid unit and the C=O group of another play an important role in determining the secondary structure of peptides and proteins.

The *tertiary structure* of a peptide or protein refers to the folding of the chain of amino acid units. The chain folding affects both the physical properties and the biological function of a protein.

Nucleic Acids (Secs. 29.23 to 29.28) The biological macromolecules that are involved in the transfer of genetic information and the control of protein biosynthesis are known as *nucleic acids.*

The major kinds of nucleic acids, DNA and RNA, contain heterocycles related to *pyrimidine* and *purine.*

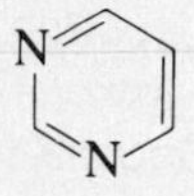

Pyrimidine 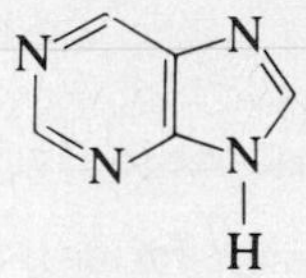 Purine

The pyrimidines that occur in RNA are uracil and cytosine; DNA contains thymine and cytosine. Both DNA and RNA contain the purine derivatives adenine and guanine.

A *nucleoside* consists of a purine or pyrimidine *base* attached to a carbohydrate. An example is adenosine.

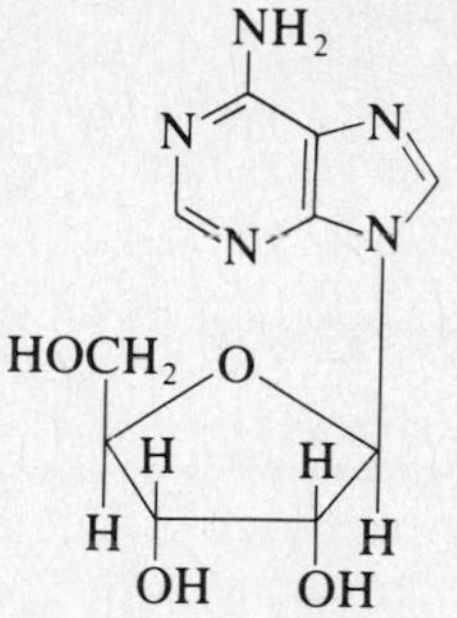

Adenosine

Nucleotides are phosphoric acid esters of nucleosides.

Nucleic acids such as DNA and RNA are *polynucleotides* in which a phosphate ester links one nucleotide unit with another. Text Figure 29.22 illustrates a generalized view of a polynucleotide chain. DNA exists as a *double helix* in which hydrogen bonds exist between complementary base pairs. Replication of the DNA strands is responsible for the transfer of genetic information.

RNA participates in the biosynthesis of proteins by controlling the sequence of amino acids that are brought together, using triplets of nucleotides known as *codons.*

IMPORTANT REACTIONS

Amino Acid Synthesis (Sec. 29.4)

Ammonia method:

$$\underset{\displaystyle\text{Br}}{\text{RCHCO}_2\text{H}} + 2\text{NH}_3 \xrightarrow{\text{H}_2\text{O}} \underset{\displaystyle{}^+\text{NH}_3}{\text{RCHCO}_2^-} + \text{NH}_4\text{Br}$$

Strecker synthesis:

$$\text{RCH}{=}\text{O} \xrightarrow[\text{NaCN}]{\text{NH}_4\text{Cl}} \underset{\displaystyle\text{NH}_2}{\text{RCHCN}} \xrightarrow[\text{2. HO}^-]{\text{1. H}_3\text{O}^+,\ \text{heat}} \underset{\displaystyle{}^+\text{NH}_3}{\text{RCHCO}_2^-}$$

Diethyl acetamidomalonate method:

$$CH_3\overset{O}{\overset{\|}{C}}NHCH(CO_2CH_2CH_3)_2 \xrightarrow[\text{2. RBr}]{\text{1. NaOCH}_2\text{CH}_3\text{, CH}_3\text{CH}_2\text{OH}} CH_3\overset{O}{\overset{\|}{C}}NH\underset{R}{\underset{|}{C}}(CO_2CH_2CH_3)_2$$

$$CH_3\overset{O}{\overset{\|}{C}}NH\underset{R}{\underset{|}{C}}(CO_2CH_2CH_3)_2 \xrightarrow{H_3O^+} H_3\overset{+}{N}\underset{R}{\underset{|}{C}}(CO_2H)_2 \xrightarrow[-CO_2]{\text{heat}} R\underset{^+NH_3}{\underset{|}{C}}HCO_2^-$$

Amino Acid Reactions (Sec. 29.5)

Amide formation:

$$H_3\overset{+}{N}\underset{R}{\underset{|}{C}}HCO_2^- + CH_3\overset{O}{\overset{\|}{C}}O\overset{O}{\overset{\|}{C}}CH_3 \longrightarrow CH_3\overset{O}{\overset{\|}{C}}NH\underset{R}{\underset{|}{C}}HCO_2H + CH_3CO_2H$$

Esterification:

$$R\underset{^+NH_3}{\underset{|}{C}}HCO_2^- + CH_3CH_2OH \xrightarrow{H^+} R\underset{^+NH_3}{\underset{|}{C}}HCO_2CH_2CH_3$$

Peptide Synthesis (Secs. 29.13 to 29.16)

General strategy:

$$XNH\underset{R}{\underset{|}{C}}H\overset{O}{\overset{\|}{C}}OH + H_2N\underset{R'}{\underset{|}{C}}H\overset{O}{\overset{\|}{C}}{-}Y \xrightarrow[\text{2. deprotect}]{\text{1. couple}} H_3\overset{+}{N}\underset{R}{\underset{|}{C}}H\overset{O}{\overset{\|}{C}}{-}NH\underset{R'}{\underset{|}{C}}HCO_2^-$$

Amino group protection:

Benzyloxycarbonyl group:

$$C_6H_5CH_2O\overset{O}{\overset{\|}{C}}Cl + H_3\overset{+}{N}\underset{R}{\underset{|}{C}}HCO_2^- \xrightarrow[\text{2. H}^+]{\text{1. NaOH, H}_2\text{O}} C_6H_5CH_2O\overset{O}{\overset{\|}{C}}NH\underset{R}{\underset{|}{C}}HCO_2H$$

$$(Z{-}NH\underset{R}{\underset{|}{C}}HCO_2H)$$

tert-*Butoxycarbonyl (Boc) group:*

$$(CH_3)_3CO\overset{O}{\overset{\|}{C}}Cl + H_3\overset{+}{N}\underset{R}{\underset{|}{C}}HCO_2^- \xrightarrow[\text{2. H}^+]{\text{1. NaOH, H}_2\text{O}} (CH_3)_3CO\overset{O}{\overset{\|}{C}}NH\underset{R}{\underset{|}{C}}HCO_2H$$

$$(Boc{-}NH\underset{R}{\underset{|}{C}}HCO_2H)$$

Deprotection:

$$Z{-}NHR \xrightarrow{H_2,\ Pd} H_2NR$$

$$Boc\text{-}NHR \xrightarrow{CF_3CO_2H} H_3\overset{+}{N}R$$

Carboxyl group protection:

Esterification:

$$\underset{\substack{|\\ R}}{H_3\overset{+}{N}CHCO_2^-} + C_6H_5CH_2OH \xrightarrow{H^+} \underset{\substack{|\\ R}}{H_3\overset{+}{N}CHCO_2CH_2C_6H_5}$$

Deprotection:

$$R\overset{O}{\overset{\|}{C}}OCH_2C_6H_5 \xrightarrow{H_2,\ Pd} RCO_2H$$

Peptide bond formation:

DCCI method:

$$\underset{\substack{|\\ R}}{ZNHCHCO_2H} + \underset{\substack{|\\ R'}}{H_2NCHCO_2CH_2C_6H_5} \xrightarrow[CHCl_3]{DCCI} ZNH\underset{\substack{|\\ R}}{CH}\overset{O}{\overset{\|}{C}}-NH\underset{\substack{|\\ R'}}{CH}CO_2CH_2C_6H_5$$

where DCCI is $C_6H_{11}-N{=}C{=}N-C_6H_{11}$ (cyclohexyl groups)

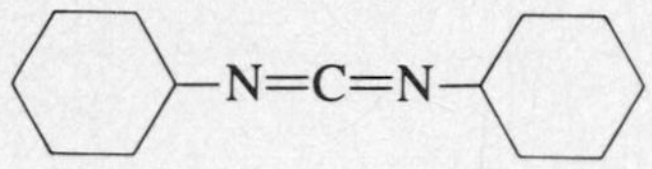

Active ester method:

$$ZNH\underset{\substack{|\\ R}}{CH}\overset{O}{\overset{\|}{C}}O-C_6H_4-NO_2 + \underset{\substack{|\\ R'}}{H_2NCHCO_2CH_2CH_3} \longrightarrow$$

$$ZNH\underset{\substack{|\\ R}}{CH}\overset{O}{\overset{\|}{C}}-NH\underset{\substack{|\\ R'}}{CH}CO_2CH_2CH_3 + HO-C_6H_4-NO_2$$

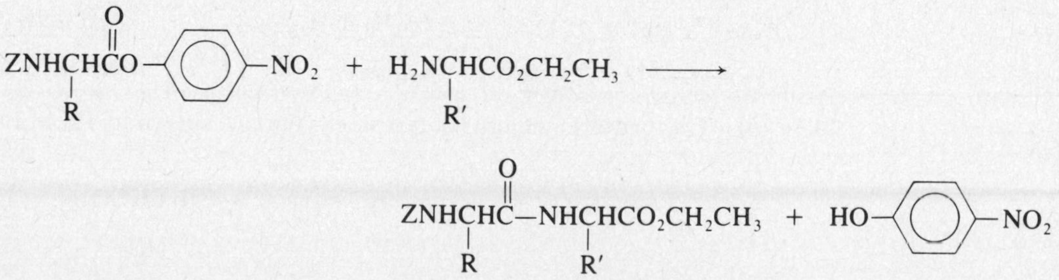

SOLUTIONS TO TEXT PROBLEMS

29.1 (*b*) L-Cysteine is the only amino acid in Table 29.1 that has the *R* configuration at its chiral center.

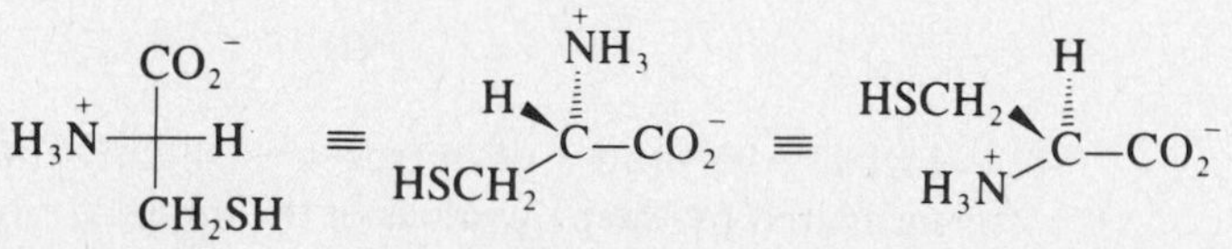

Fischer projection of L-Cysteine: CO_2^- at top, $H_3\overset{+}{N}$ on left, H on right, CH_2SH at bottom $\equiv$ wedge structures with C bearing $\overset{+}{N}H_3$, H, $HSCH_2$, CO_2^- $\equiv$ C bearing H, $HSCH_2$, $H_3\overset{+}{N}$, CO_2^-

L-Cysteine

The order of decreasing sequence rule precedence is:

$$H_3\overset{+}{N}- > HSCH_2- > -CO_2^- > H-$$

When the molecule is oriented so that the lowest-ranked substituent (H) is held away from us, the order of decreasing precedence traces a clockwise path.

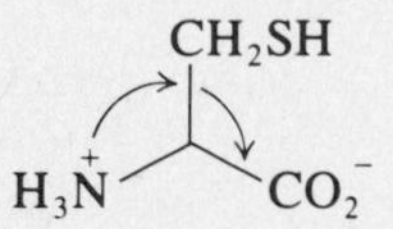

CH_2SH (top), $H_3\overset{+}{N}$ (left), CO_2^- (right) — Clockwise; therefore *R*

The reason why L-cysteine has the *R* configuration while all the other L-amino acids have the *S* configuration lies in the fact that the $-CH_2SH$ substituent is the only side chain that outranks $-CO_2^-$ according to the sequence rule. Remember, rank order is determined by atomic number at the first point of difference and $-C-S$ outranks $-C-O$. In all the other amino acids $-CO_2^-$ outranks the substituent at the chiral center. The reversal in the Cahn-Ingold-Prelog descriptor comes not from any change in the spatial arrangement of substituents at the chiral center but rather from a reversal in the relative ranks of the carboxylate group and the side chain.

(*c*) The order of decreasing sequence rule precedence in L-methionine is:

$$H_3\overset{+}{N}- > -CO_2^- > -CH_2CH_2SCH_3 > H-$$

Sulfur is one atom further removed from the chiral center, so C—O outranks C—C—S.

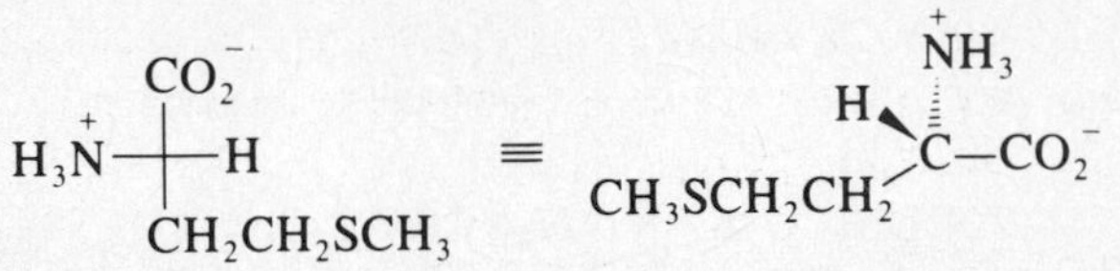

$$H_3\overset{+}{N}-C(CO_2^-)(H)(CH_2CH_2SCH_3) \equiv CH_3SCH_2CH_2-C(H)(\overset{+}{N}H_3)-CO_2^-$$

The absolute configuration is *S*.

29.2 The amino acids in Table 29.1 that have more than one chiral center are isoleucine and threonine. The chiral centers are marked with an asterisk in the structural formulas shown.

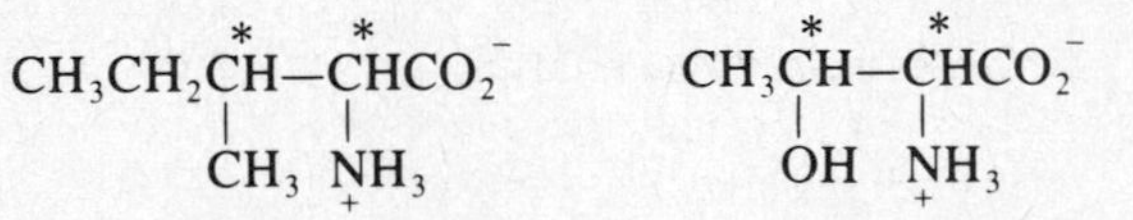

$$CH_3CH_2\overset{*}{C}H(CH_3)-\overset{*}{C}H(\overset{+}{N}H_3)CO_2^- \qquad CH_3\overset{*}{C}H(OH)-\overset{*}{C}H(\overset{+}{N}H_3)CO_2^-$$

Isoleucine Threonine

29.3 (*b*) The zwitterionic form of tyrosine is the one shown in Table 29.1.

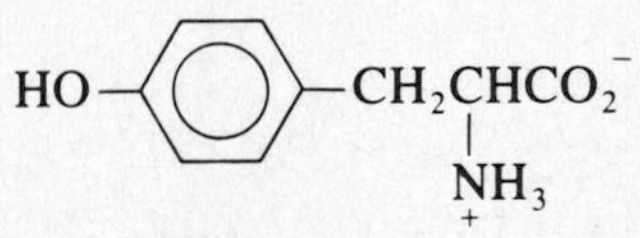

$$HO-C_6H_4-CH_2CH(\overset{+}{N}H_3)CO_2^-$$

(*c*) As base is added to the zwitterion, a proton is removed from either of two positions, the ammonium group or the phenolic hydroxyl. The acidities of the two sites are so close that it is not possible to predict with certainty which one is deprotonated preferentially. Thus there are two plausible structures for the monoanion.

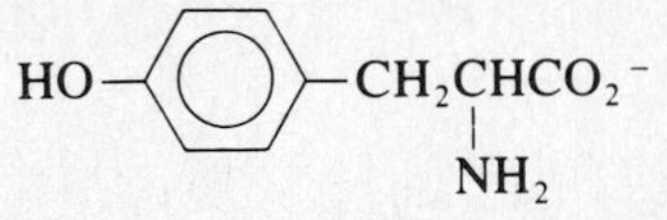

$$HO-C_6H_4-CH_2CH(NH_2)CO_2^- \quad \text{and} \quad {}^-O-C_6H_4-CH_2CH(\overset{+}{N}H_3)CO_2^-$$

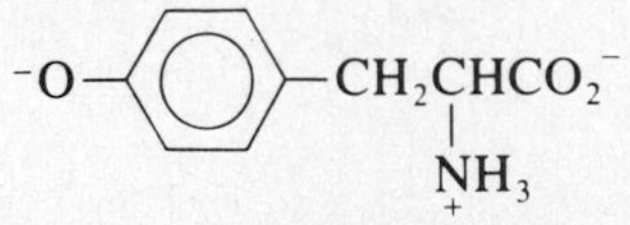

In fact, the proton on nitrogen is slightly more acidic than the phenolic hydroxyl, as measured by the pK_a values of the model compounds shown below.

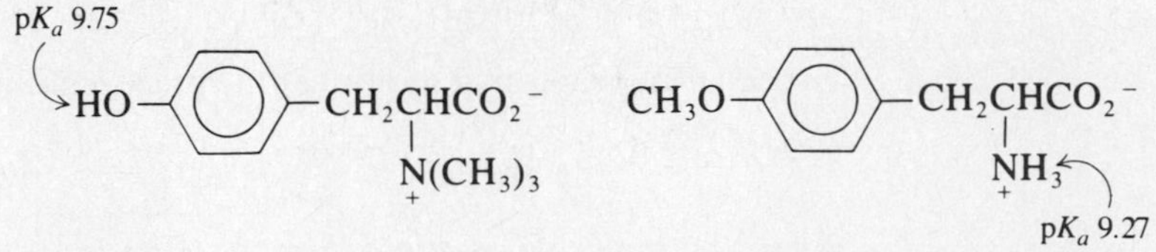

$$\underset{\text{p}K_a\ 9.75\ (\text{OH})}{HO-C_6H_4-CH_2CH(\overset{+}{N}(CH_3)_3)CO_2^-} \qquad \underset{\text{p}K_a\ 9.27\ (\text{NH}_3^+)}{CH_3O-C_6H_4-CH_2CH(\overset{+}{N}H_3)CO_2^-}$$

(*d*) On further treatment with base, both the monoanions in (*c*) yield the same dianion.

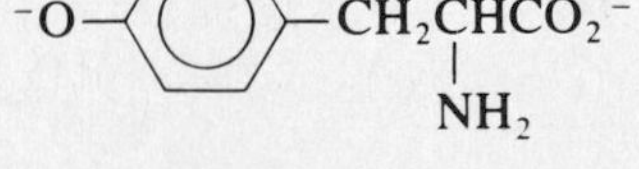

$${}^-O-C_6H_4-CH_2CH(NH_2)CO_2^-$$

29.4 At pH 1 the carboxylate oxygen and both nitrogens of lysine are protonated.

$$\overset{+}{H_3N}CH_2CH_2CH_2CH_2\underset{\underset{+}{NH_3}}{CH}CO_2H \quad \text{(principal form at pH 1)}$$

As the pH is raised, the carboxyl proton is removed first.

$$\overset{+}{H_3N}CH_2CH_2CH_2CH_2\underset{\underset{+}{NH_3}}{CH}CO_2H + HO^- \longrightarrow \overset{+}{H_3N}CH_2CH_2CH_2CH_2\underset{\underset{+}{NH_3}}{CH}CO_2^- + H_2O$$

The pK_a value for the first ionization of lysine is 2.18 (from Table 29.3), so this process is virtually complete when the pH is greater than this value. The second pK_a value for lysine is 8.95. This is a fairly typical value for the second pK_a of amino acids and likely corresponds to proton removal from the nitrogen on the α carbon. The species that results is the predominant one at pH 9.

$$\overset{+}{H_3N}CH_2CH_2CH_2CH_2\underset{\underset{+}{NH_3}}{CH}CO_2^- + HO^- \longrightarrow \overset{+}{H_3N}CH_2CH_2CH_2CH_2\underset{NH_2}{CH}CO_2^- + H_2O$$

(Principal form at pH 9)

The pK_a value for the third ionization of lysine is 10.53. This value is fairly high compared with those of most of the amino acids in Tables 29.1 to 29.3 and suggests that this proton is removed from the nitrogen of the side chain. The species that results is the major species present at pH values greater than 10.53.

$$\overset{+}{H_3N}CH_2CH_2CH_2CH_2\underset{NH_2}{CH}CO_2^- + HO^- \longrightarrow H_2NCH_2CH_2CH_2CH_2\underset{NH_2}{CH}CO_2^-$$

(Principal form at pH 13)

29.5 In order to convert 3-methylbutanoic acid to valine, a leaving group must be introduced at the α carbon prior to displacement by ammonia. This is best accomplished by bromination under the conditions of the Hell-Volhard-Zelinsky reaction.

$$(CH_3)_2CHCH_2CO_2H \xrightarrow[\text{or } Br_2,\ PCl_3]{Br_2,\ P} (CH_3)_2CH\underset{Br}{CH}CO_2H \xrightarrow{NH_3} (CH_3)_2CH\underset{\underset{+}{NH_3}}{CH}CO_2^-$$

3-Methylbutanoic acid — 2-Bromo-3-methylbutanoic acid — Valine

Valine has been prepared by this method. The Hell-Volhard-Zelinsky reaction was carried out in 88 percent yield, but reaction of the α-bromo acid with ammonia was not very efficient, valine being isolated in only 48 percent yield in this step.

29.6 In the Strecker synthesis an aldehyde is treated with ammonia and a source of cyanide ion. The resulting amino nitrile is hydrolyzed to an amino acid.

$$(CH_3)_2CH\overset{O}{\overset{\|}{C}}H \xrightarrow[HCN]{NH_3} (CH_3)_2CH\underset{NH_2}{CH}C{\equiv}N \xrightarrow[2.\ HO^-]{1.\ H_3O^+} (CH_3)_2CH\underset{\underset{+}{NH_3}}{CH}CO_2^-$$

2-Methylpropanal — 2-Amino-3-methylbutanenitrile — Valine

As actually carried out, the aldehyde was converted to the amino nitrile by treatment with an aqueous solution containing ammonium chloride and potassium cyanide. Hydrolysis was achieved in aqueous hydrochloric acid and gave valine as its hydrochloride salt in 65 percent overall yield.

29.7 The alkyl halide with which the anion of diethyl acetamidomalonate is treated is isopropyl bromide.

$$\underset{\text{Diethyl acetamidomalonate}}{CH_3\overset{O}{\overset{\|}{C}}NHCH(CO_2CH_2CH_3)_2} + \underset{\text{Isopropyl bromide}}{(CH_3)_2CHBr} \xrightarrow[CH_3CH_2OH]{NaOCH_2CH_3} \underset{\text{Diethyl acetamidoisopropylmalonate}}{CH_3\overset{O}{\overset{\|}{C}}NH\underset{CH(CH_3)_2}{\underset{|}{C}}(CO_2CH_2CH_3)_2}$$

This is the difficult step in the synthesis; it requires a nucleophilic substitution reaction of the S_N2 type involving a secondary alkyl halide. Competition of elimination with substitution results in only a 37 percent observed yield of alkylated diethyl acetamidomalonate.

Hydrolysis and decarboxylation of the alkylated derivative are straightforward and proceed in 85 percent yield to give valine.

$$\underset{\text{Diethyl acetamidoisopropylmalonate}}{CH_3\overset{O}{\overset{\|}{C}}NH\underset{CH(CH_3)_2}{\underset{|}{C}}(CO_2CH_2CH_3)_2} \xrightarrow[heat]{HBr,\ H_2O} \underset{\text{2-Aminoisopropylmalonic acid}}{H_3\overset{+}{N}\underset{CH(CH_3)_2}{\underset{|}{C}}(CO_2H)_2} \xrightarrow[-CO_2]{heat} \underset{\text{Valine}}{H_3\overset{+}{N}\underset{CH(CH_3)_2}{\underset{|}{C}}HCO_2^-}$$

29.8 Ninhydrin is the hydrate of a triketone and is in equilibrium with it.

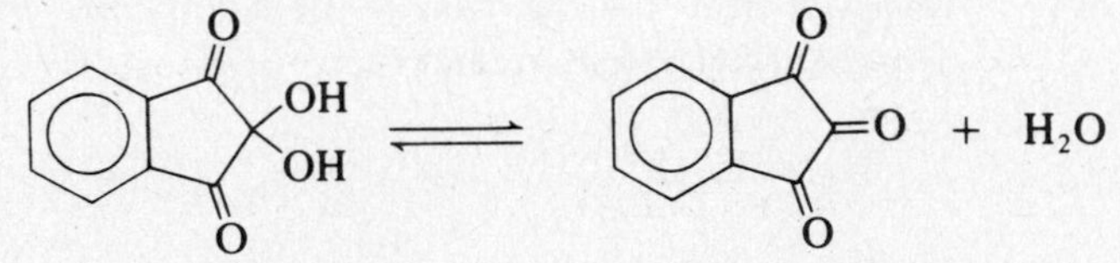

Hydrated form of ninhydrin ⇌ Triketo form of ninhydrin + H_2O

An amino acid reacts with this triketone to form an imine.

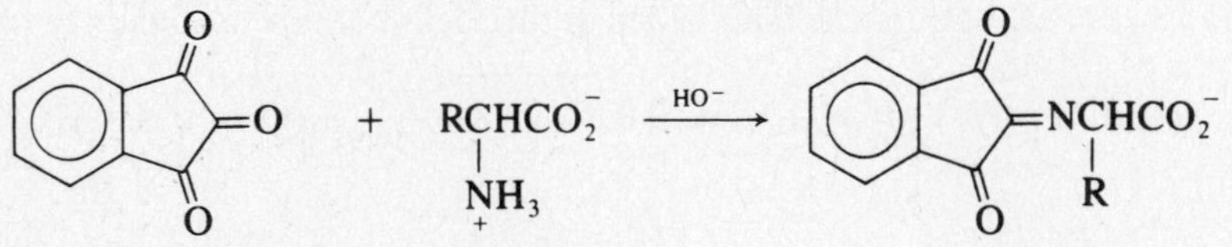

$$\underset{\text{Triketo form of ninhydrin}}{} + \underset{\alpha\text{-Amino acid}}{RCHCO_2^- \atop {|\atop \overset{+}{N}H_3}} \xrightarrow{HO^-} \underset{\text{Imine}}{=NCHCO_2^- \atop {|\atop R}}$$

This imine then undergoes decarboxylation.

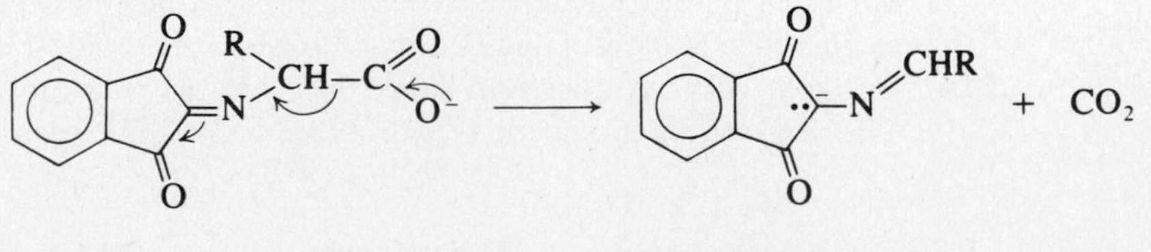

$$=N-CH(R)-C(=O)O^- \longrightarrow \ddot{\ }-N=CHR + CO_2$$

The anion that results from the decarboxylation step is then protonated. The product is shown as its diketo form but probably exists as an enol.

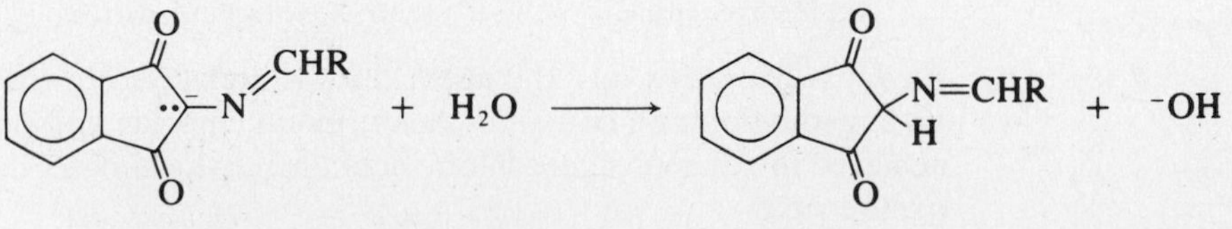

$$\ddot{\ }-N=CHR + H_2O \longrightarrow \underset{H}{\overset{}{C}}-N=CHR + {}^-OH$$

Hydrolysis of the imine function gives an aldehyde and a compound having a free amino group.

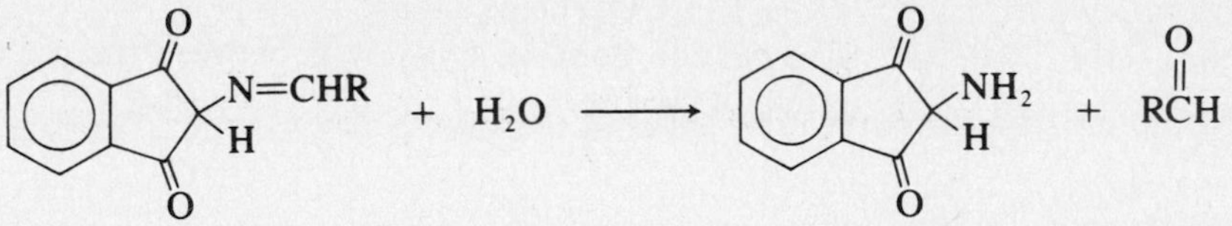

This amine then reacts with a second molecule of the triketo form of ninhydrin to give an imine.

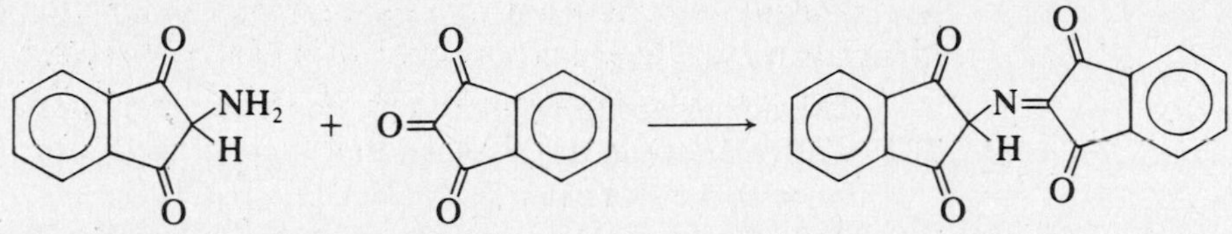

Proton abstraction from the neutral imine gives its conjugate base, which is a violet dye.

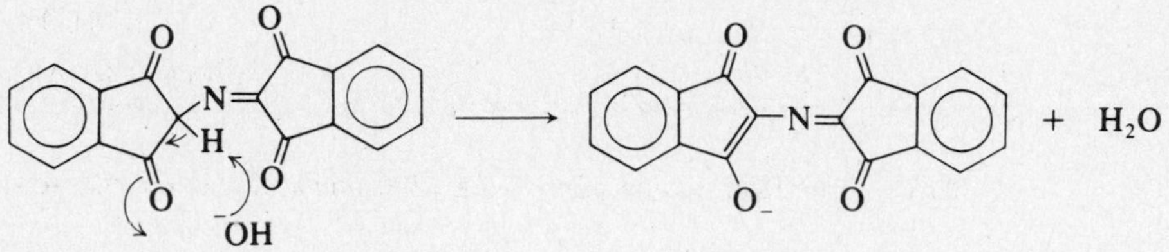

29.9 The carbon that bears the amino group of 4-aminobutanoic acid corresponds to the α carbon of an α-amino acid.

$$\underset{\text{4-Aminobutanoic acid}}{\underset{\overset{|}{^{+}NH_3}}{CH_2}CH_2CH_2CO_2^-} \quad \text{arises by decarboxylation of} \quad \underset{\text{Glutamic acid}}{^{-}O_2C\underset{\overset{|}{^{+}NH_3}}{C}HCH_2CH_2CO_2^-}$$

29.10 (*b*) Alanine is the N-terminal amino acid in Ala-Phe. Its carboxyl group is joined to the nitrogen of phenylalanine by a peptide bond.

$$H_3\overset{+}{N}\underset{\underset{\text{Alanine}}{CH_3}}{CH}\overset{O}{\overset{\|}{C}}\text{—}NH\underset{\underset{\text{Phenylalanine}}{CH_2C_6H_5}}{CH}CO_2^- \quad \text{Ala-Phe}$$

(*c*) The positions of the amino acids are reversed in Phe-Ala. Phenylalanine is the N terminus and alanine is the C terminus.

$$H_3\overset{+}{N}\underset{\underset{\text{Phenylalanine}}{C_6H_5CH_2}}{CH}\overset{O}{\overset{\|}{C}}\text{—}NH\underset{\underset{\text{Alanine}}{CH_3}}{CH}CO_2^- \quad \text{Phe-Ala}$$

(*d*) The carboxyl group of glycine is joined by a peptide bond to the amino group of glutamic acid.

$$\underset{\text{Glycine}}{H_3\overset{+}{N}CH_2\overset{O}{\overset{\|}{C}}}\text{—}NH\underset{\underset{\text{Glutamic acid}}{CH_2CH_2CO_2^-}}{CH}CO_2^- \quad \text{Gly-Glu}$$

The dipeptide is written in its anionic form because the carboxyl group of the side chain is ionized at pH 7. Alternatively, it could have been written as a neutral zwitterion with a $CH_2CH_2CO_2H$ side chain.

(*e*) The peptide bond in Lys-Gly is between the carboxyl group of lysine and the amino group of glycine.

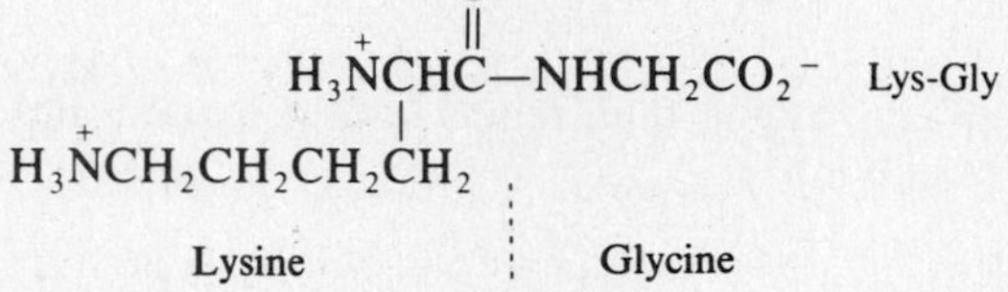

$$\overset{+}{H_3N}CH(CH_2CH_2CH_2CH_2\overset{+}{N}H_3)\overset{O}{\overset{\|}{C}}{-}NHCH_2CO_2^- \quad \text{Lys-Gly}$$

Lysine | Glycine

The amino group of the lysine side chain is protonated at pH 7 so the dipeptide is written above in its cationic form. It could have also been written as a neutral zwitterion with the side chain $H_2NCH_2CH_2CH_2CH_2$.

(*f*) Both amino acids are alanine in D-Ala-D-Ala. The fact that they have the D configuration has no effect on the constitution of the dipeptide.

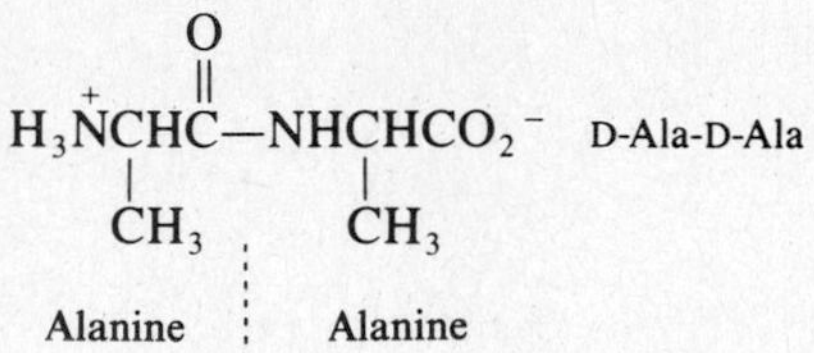

$$\overset{+}{H_3N}CH(CH_3)\overset{O}{\overset{\|}{C}}{-}NHCH(CH_3)CO_2^- \quad \text{D-Ala-D-Ala}$$

Alanine | Alanine

29.11 (*b*) When amino acid residues in a dipeptide are indicated without a prefix, it is assumed that the configuration at the α carbon atom is L. Therefore, the stereochemistry of Ala-Phe may be indicated for the zigzag conformation as shown.

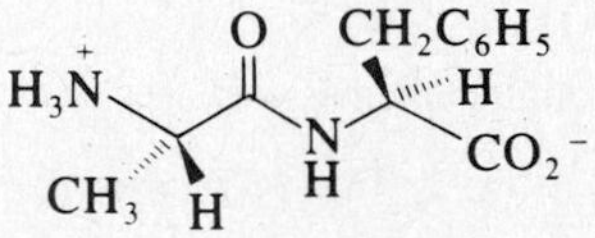

[Zigzag structure: $\overset{+}{H_3N}$, CH_3, H, O, N–H, $CH_2C_6H_5$, H, CO_2^-]

The L configuration corresponds to *S* for each of the chiral centers in Ala-Phe.

(*c*) Similarly, Phe-Ala has its substituent at the N-terminal amino acid directed away from us while the C-terminal side chain is pointing toward us, and the L configuration corresponds to *S* for each chiral center.

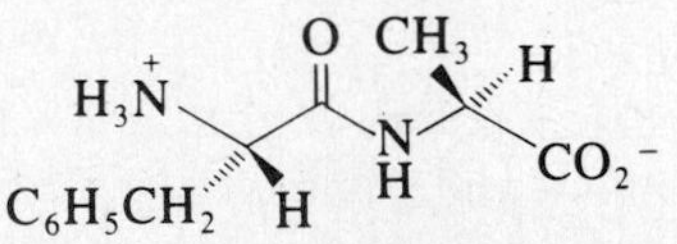

[Zigzag structure: $\overset{+}{H_3N}$, $C_6H_5CH_2$, H, O, N–H, CH_3, H, CO_2^-]

(*d*) There is only one chiral center in Gly-Glu. It has the L (or *S*) configuration.

[Zigzag structure: $\overset{+}{H_3N}$, O, N–H, $CH_2CH_2CO_2^-$, H, CO_2^-]

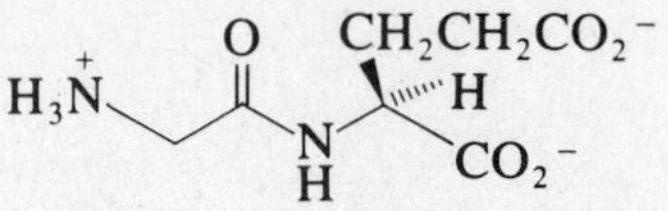

(*e*) In order for the N-terminal amino acid in Lys-Gly to have the L (or *S*) configuration, its side chain must be directed away from us in the conformation indicated.

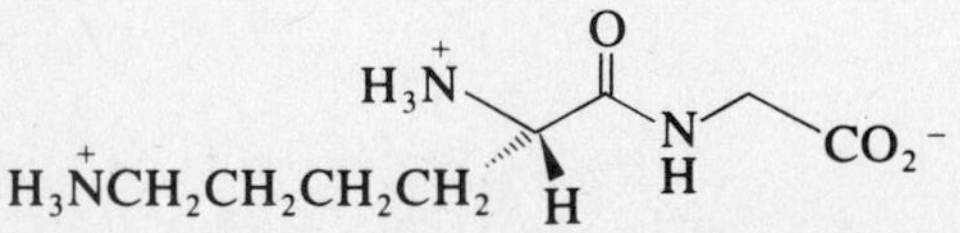

[Zigzag structure: $\overset{+}{H_3N}$, $H_3\overset{+}{N}CH_2CH_2CH_2CH_2$, H, O, N–H, CO_2^-]

(*f*) The configuration at both α carbon atoms in D-Ala-D-Ala is exactly the reverse of the configuration of the chiral centers in (*a*) through (*e*). Both chiral centers have the D (or *R*) configuration.

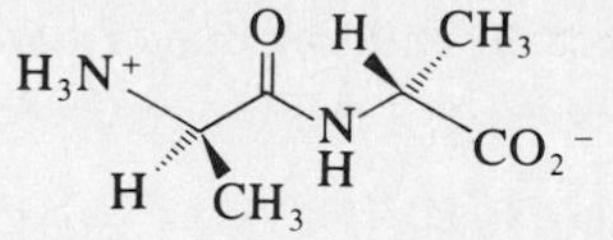

H_3N^+ ... O H CH_3 ... N H ... CO_2^- ... H CH_3

29.12 Figure 29.5 in the text gives the structure of leucine enkephalin. Methionine enkephalin differs from it only with respect to the C-terminal amino acid. The amino acid sequences of the two pentapeptides are:

Tyr-Gly-Gly-Phe-Leu	Tyr-Gly-Gly-Phe-Met
Leucine enkephalin	Methionine enkephalin

29.13 Unless a peptide is hydrolyzed completely so that all the amide bonds are cleaved, a mixture of smaller fragments is obtained corresponding to all possible cleavage modes.

Val-Phe-Gly-Ala

↓ hydrolysis

tripeptides: Val-Phe-Gly Phe-Gly-Ala
dipeptides: Val-Phe Phe-Gly Gly-Ala
amino acids: Val Phe Gly Ala

29.14 The Edman degradation removes the N-terminal amino acid, which is identified as a phenylthiohydantoin derivative. The first Edman degradation of Val-Phe-Gly-Ala gives the phenylthiohydantoin derived from valine; the second gives the phenylthiohydantoin derived from phenylalanine.

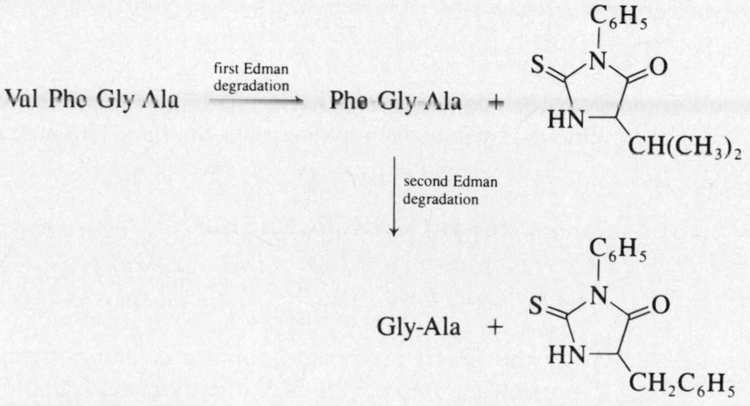

Val-Phe-Gly-Ala $\xrightarrow{\text{first Edman degradation}}$ Phe-Gly-Ala + (phenylthiohydantoin: C_6H_5–N, S=, =O, HN, $CH(CH_3)_2$)

↓ second Edman degradation

Gly-Ala + (phenylthiohydantoin: C_6H_5–N, S=, =O, HN, $CH_2C_6H_5$)

29.15 The dipeptide Asp-Phe has the structure shown. Aspartame is its C-terminal methyl ester.

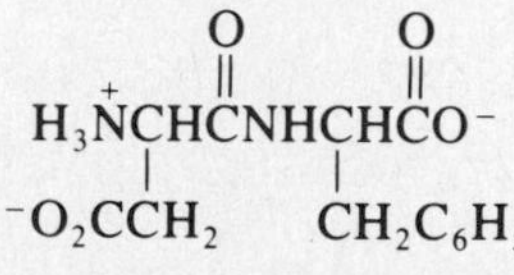

$$\underset{\underset{-O_2CCH_2}{|}}{H_3\overset{+}{N}CH}\overset{O}{\overset{\|}{C}}NH\underset{\underset{CH_2C_6H_5}{|}}{CH}\overset{O}{\overset{\|}{C}}O^- \qquad \underset{\underset{-O_2CCH_2}{|}}{H_3\overset{+}{N}CH}\overset{O}{\overset{\|}{C}}NH\underset{\underset{CH_2C_6H_5}{|}}{CH}\overset{O}{\overset{\|}{C}}OCH_3$$

Asp-Phe 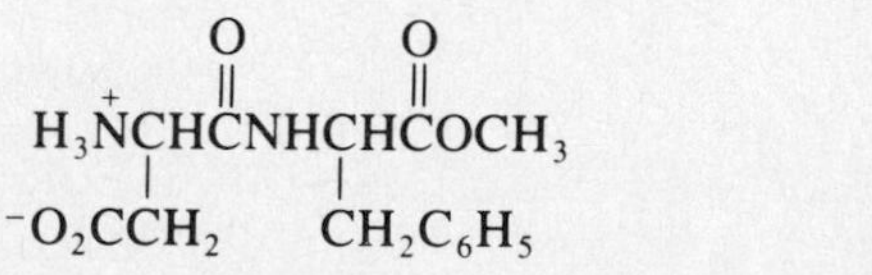 Aspartame

29.16 Lysine has two amino groups. Both amino functions are converted to amides on reaction with benzyloxycarbonyl chloride.

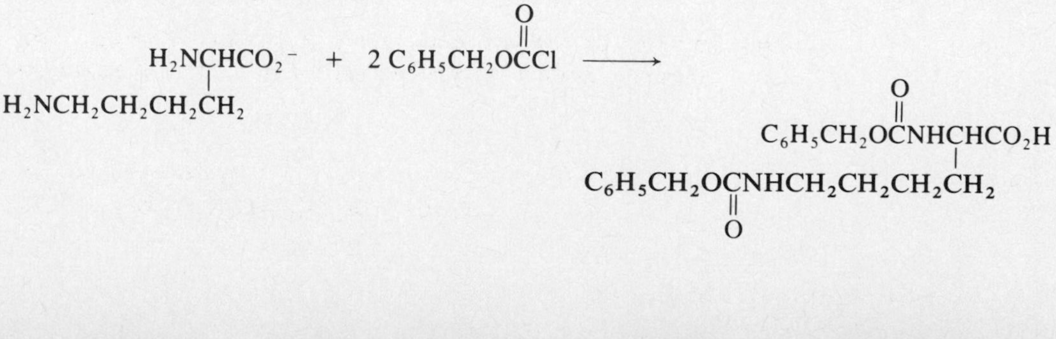

$$\underset{\underset{H_2NCH_2CH_2CH_2CH_2}{|}}{H_2NCHCO_2^-} + 2\,C_6H_5CH_2O\overset{O}{\overset{\|}{C}}Cl \longrightarrow \underset{\underset{C_6H_5CH_2O\underset{\underset{O}{\|}}{C}NHCH_2CH_2CH_2CH_2}{|}}{C_6H_5CH_2O\overset{O}{\overset{\|}{C}}NHCHCO_2H}$$

29.17 The peptide bond of Ala-Leu connects the carboxyl group of alanine and the amino group of leucine. Therefore we need to protect the amino group of alanine and the carboxyl group of leucine.

Protect the amino group of alanine as its benzyloxycarbonyl derivative:

$$\underset{\text{Alanine}}{H_3\overset{+}{N}\underset{\underset{CH_3}{|}}{C}HCO_2^-} + \underset{\text{Benzyloxycarbonyl chloride}}{C_6H_5CH_2O\overset{O}{\overset{\|}{C}}Cl} \longrightarrow \underset{\text{Z-Protected alanine}}{C_6H_5CH_2O\overset{O}{\overset{\|}{C}}NH\underset{\underset{CH_3}{|}}{C}HCO_2H}$$

Protect the carboxyl group of leucine as its benzyl ester:

$$\underset{\text{Leucine}}{H_3\overset{+}{N}\underset{\underset{(CH_3)_2CHCH_2}{|}}{C}HCO_2^-} + \underset{\text{Benzyl alcohol}}{C_6H_5CH_2OH} \xrightarrow[\text{2. HO}^-]{\text{1. H}^+\text{, heat}} \underset{\text{Leucine benzyl ester}}{H_2N\underset{\underset{(CH_3)_2CHCH_2}{|}}{C}HCO_2CH_2C_6H_5}$$

Coupling of the two amino acids is achieved by DCCI-promoted amide bond formation between the free amino group of leucine benzyl ester and the free carboxyl group of *Z*-protected alanine.

$$\underset{\text{Z-Protected alanine}}{C_6H_5CH_2O\overset{O}{\overset{\|}{C}}NH\underset{\underset{CH_3}{|}}{C}HCO_2H} + \underset{\text{Leucine benzyl ester}}{H_2N\underset{\underset{(CH_3)_2CHCH_2}{|}}{C}H\overset{O}{\overset{\|}{C}}OCH_2C_6H_5} \xrightarrow{\text{DCCI}}$$

$$\underset{\text{Protected dipeptide}}{C_6H_5CH_2O\overset{O}{\overset{\|}{C}}NH\underset{\underset{CH_3}{|}}{C}H\overset{O}{\overset{\|}{C}}NH\underset{\underset{CH_2CH(CH_3)_2}{|}}{C}H\overset{O}{\overset{\|}{C}}OCH_2C_6H_5}$$

Both the benzyloxycarbonyl protecting group and the benzyl ester protecting group may be removed by hydrogenolysis over palladium. This step completes the synthesis of Ala-Leu.

$$\underset{\text{Protected dipeptide}}{C_6H_5CH_2O\overset{O}{\overset{\|}{C}}NH\underset{\underset{CH_3}{|}}{C}H\overset{O}{\overset{\|}{C}}NH\underset{\underset{CH_2CH(CH_3)_2}{|}}{C}H\overset{O}{\overset{\|}{C}}OCH_2C_6H_5} \xrightarrow{H_2,\ Pd} \underset{\text{Ala-Leu}}{H_3\overset{+}{N}\underset{\underset{CH_3}{|}}{C}H\overset{O}{\overset{\|}{C}}NH\underset{\underset{CH_2CH(CH_3)_2}{|}}{C}HCO_2^-}$$

29.18 As in the DCCI-promoted coupling of amino acids, the first step is the addition of the *Z*-protected amino acid to *N,N'*-dicyclohexylcarbodiimide (DCCI) to give an *O*-acylisourea.

$$\underset{\text{Z-Protected amino acid}}{ZNH\underset{\underset{R}{|}}{C}H\overset{O}{\overset{\|}{C}}OH} + \underset{\text{DCCI}}{C_6H_{11}N{=}C{=}NC_6H_{11}} \longrightarrow \underset{\text{O-Acylisourea}}{ZNH\underset{\underset{R}{|}}{C}H\overset{O}{\overset{\|}{C}}O{-}C\begin{smallmatrix}{\diagup\!\!\diagup} NC_6H_{11}\\ \diagdown NHC_6H_{11}\end{smallmatrix}}$$

This *O*-acylisourea is attacked by *p*-nitrophenol to give the *p*-nitrophenyl ester of the *Z*-protected amino acid.

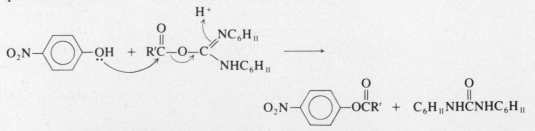

$$O_2N{-}C_6H_4{-}\ddot{O}H + R'\overset{O}{\overset{\|}{C}}{-}O{-}C\begin{smallmatrix}{\diagup\!\!\diagup} NC_6H_{11}\ (H^+)\\ \diagdown NHC_6H_{11}\end{smallmatrix} \longrightarrow O_2N{-}C_6H_4{-}O\overset{O}{\overset{\|}{C}}R' + C_6H_{11}NH\overset{O}{\overset{\|}{C}}NHC_6H_{11}$$

29.19 In order to add a leucine residue to the N terminus of the ethyl ester of Z-Phe-Gly, the benzyloxycarbonyl protecting group must first be removed. This can be accomplished by hydrogenolysis.

$$\underset{\text{Z-Protected ethyl ester of Phe-Gly}}{C_6H_5CH_2O\overset{O}{\overset{\|}{C}}NH\underset{C_6H_5CH_2}{\underset{|}{C}}H\overset{O}{\overset{\|}{C}}NHCH_2\overset{O}{\overset{\|}{C}}OCH_2CH_3} \xrightarrow{H_2,\ Pd} \underset{\text{Phe-Gly ethyl ester}}{H_2N\underset{C_6H_5CH_2}{\underset{|}{C}}H\overset{O}{\overset{\|}{C}}NHCH_2\overset{O}{\overset{\|}{C}}OCH_2CH_3}$$

The reaction shown has been carried out in 100 percent yield. Alternatively, the benzyloxycarbonyl protecting group may be removed by treatment with hydrogen bromide in acetic acid. This latter route has also been reported in the chemical literature and gives the hydrobromide salt of Phe-Gly ethyl ester in 82 percent yield.

Once the protecting group has been removed, the ethyl ester of Phe-Gly is allowed to react with the *p*-nitrophenyl ester of *Z*-protected leucine to form the protected tripeptide. Hydrogenolysis of the *Z*-protected tripeptide gives Leu-Phe-Gly as its ethyl ester.

$$\underset{\text{p-Nitrophenyl ester of Z-protected leucine}}{C_6H_5CH_2O\overset{O}{\overset{\|}{C}}NH\underset{(CH_3)_2CHCH_2}{\underset{|}{C}}H\overset{O}{\overset{\|}{C}}{-}O{-}C_6H_4{-}NO_2} + \underset{\text{Phe-Gly ethyl ester}}{H_2N\underset{C_6H_5CH_2}{\underset{|}{C}}H\overset{O}{\overset{\|}{C}}NHCH_2\overset{O}{\overset{\|}{C}}OCH_2CH_3}$$

$$\downarrow$$

$$\underset{\text{Z-Protected Leu-Phe-Gly ethyl ester}}{C_6H_5CH_2O\overset{O}{\overset{\|}{C}}NH\underset{(CH_3)_2CHCH_2}{\underset{|}{C}}H\overset{O}{\overset{\|}{C}}NH\underset{CH_2C_6H_5}{\underset{|}{C}}H\overset{O}{\overset{\|}{C}}NHCH_2\overset{O}{\overset{\|}{C}}OCH_2CH_3}$$

$$\downarrow H_2,\ Pd$$

$$\underset{\text{Leu-Phe-Gly ethyl ester}}{H_2N\underset{(CH_3)_2CHCH_2}{\underset{|}{C}}H\overset{O}{\overset{\|}{C}}NH\underset{CH_2C_6H_5}{\underset{|}{C}}H\overset{O}{\overset{\|}{C}}NHCH_2\overset{O}{\overset{\|}{C}}OCH_2CH_3}$$

29.20 Amino acid residues are added by beginning at the C terminus in the Merrifield solid-phase approach to peptide synthesis. Thus the synthesis of Phe-Gly requires glycine to be anchored to the solid support. Begin by protecting glycine as its Boc derivative.

$$\underset{\text{tert-Butoxycarbonyl chloride}}{(CH_3)_3CO\overset{O}{\overset{\|}{C}}Cl} + \underset{\text{Glycine}}{H_3\overset{+}{N}CH_2CO_2^-} \longrightarrow \underset{\text{Boc-Protected glycine}}{(CH_3)_3CO\overset{O}{\overset{\|}{C}}NHCH_2CO_2H}$$

The protected glycine is attached via its carboxylate anion to the solid support.

$$\underset{\text{Boc-Protected glycine}}{(CH_3)_3CO\overset{O}{\overset{\|}{C}}NHCH_2CO_2H} \xrightarrow[\text{2. ClCH}_2\text{-resin}]{\text{1. HO}^-} (CH_3)_3CO\overset{O}{\overset{\|}{C}}NHCH_2\overset{O}{\overset{\|}{C}}OCH_2\text{-resin}$$

The amino group of glycine is then exposed by removal of the protecting group. Typical conditions for this step involve treatment with hydrogen chloride in acetic acid.

$$\underset{\text{Boc-Protected, resin-bound glycine}}{(CH_3)_3CO\overset{O}{\overset{\|}{C}}NHCH_2\overset{O}{\overset{\|}{C}}OCH_2\text{-resin}} \xrightarrow[\text{acetic acid}]{HCl} \underset{\text{Resin-bound glycine}}{H_2NCH_2\overset{O}{\overset{\|}{C}}OCH_2\text{-resin}}$$

In order to attach phenylalanine to resin-bound glycine, we must first protect the amino group of phenylalanine. A Boc protecting group is appropriate.

$$\underset{\textit{tert}\text{-Butoxycarbonyl chloride}}{(CH_3)_3CO\overset{O}{\overset{\|}{C}}Cl} + \underset{\text{Phenylalanine}}{H_3\overset{+}{N}\underset{CH_2C_6H_5}{\underset{|}{C}}HCO_2^-} \longrightarrow \underset{\text{Boc-Protected phenylalanine}}{(CH_3)_3CO\overset{O}{\overset{\|}{C}}NH\underset{CH_2C_6H_5}{\underset{|}{C}}HCO_2H}$$

Peptide bond formation occurs when the resin-bound glycine and Boc-protected phenylalanine are combined in the presence of DCCI.

$$\underset{\text{Boc-Protected phenylalanine}}{(CH_3)_3CO\overset{O}{\overset{\|}{C}}NH\underset{CH_2C_6H_5}{\underset{|}{C}}HCO_2H} + \underset{\text{Resin-bound glycine}}{H_2NCH_2\overset{O}{\overset{\|}{C}}OCH_2\text{—resin}} \xrightarrow{DCCI}$$

$$\underset{\text{Boc-Protected, resin-bound Phe-Gly}}{(CH_3)_3CO\overset{O}{\overset{\|}{C}}NH\underset{CH_2C_6H_5}{\underset{|}{C}}H\overset{O}{\overset{\|}{C}}NHCH_2\overset{O}{\overset{\|}{C}}OCH_2\text{—resin}}$$

Remove the Boc group with HCl and then treat with HBr in trifluoroacetic acid to cleave Phe-Gly from the solid support.

$$\underset{\text{Boc-protected, resin-bound Phe-Gly}}{(CH_3)_3CO\overset{O}{\overset{\|}{C}}NH\underset{CH_2C_6H_5}{\underset{|}{C}}H\overset{O}{\overset{\|}{C}}NHCH_2\overset{O}{\overset{\|}{C}}OCH_2\text{—resin}} \xrightarrow[\text{2. HBr, trifluoroacetic acid}]{\text{1. HCl, acetic acid}} \underset{\text{Phe-Gly}}{H_3\overset{+}{N}\underset{CH_2C_6H_5}{\underset{|}{C}}H\overset{O}{\overset{\|}{C}}NHCH_2CO_2^-}$$

29.21 The numbering of the ring in uracil and its derivatives parallels that in pyrimidine.

Pyrimidine Uracil 5-Fluorouracil

29.22 (*b*) Cytidine is present in RNA and so is a nucleoside of D-ribose. The base is cytosine.

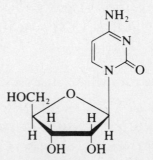

(*c*) Guanosine is present in RNA and so is a guanine nucleoside of D-ribose.

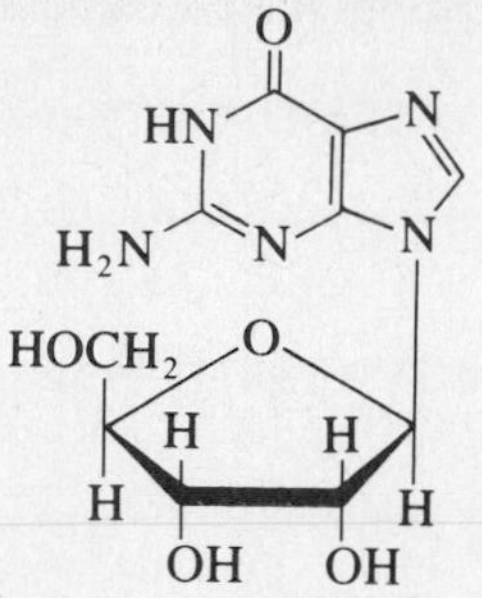

HN, N, H_2N, N, N, O, $HOCH_2$, O, H, H, H, H, OH, OH

29.23 Table 29.4 in the text lists the messenger RNA codons for the various amino acids. The codons for valine and for glutamic acid are:

Valine:	GUU	GUA	GUC	GUG
Glutamic acid:		GAA		GAG

As can be seen, the codons for glutamic acid (GAA and GAG) are very similar to two of the codons (GUA and GUG) for valine. Replacement of adenine in the glutamic acid codons by uracil causes valine to be incorporated into hemoglobin instead of glutamic acid and is responsible for the sickle cell trait.

29.24 The protonated form of imidazole represented by structure A is stabilized by delocalization of the lone pair of one of the nitrogens. The positive charge is shared by both nitrogens.

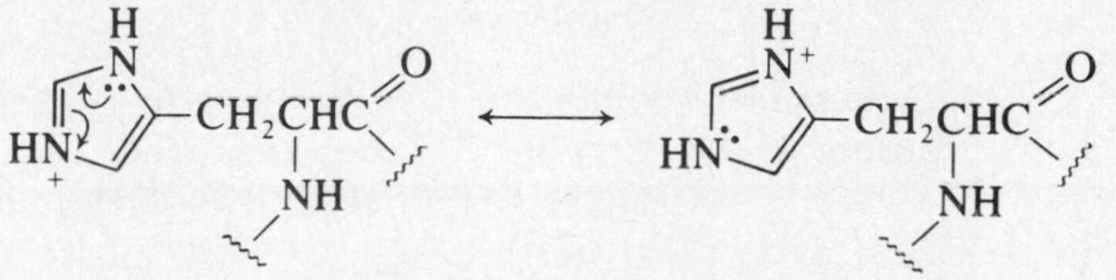

A

The positive charge in structure B is localized on a single nitrogen. Resonance stabilization of the type shown in A is not possible.

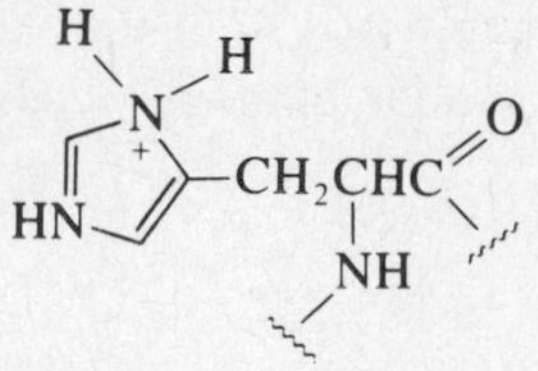

B

29.25 A synthesis of β-alanine in which conjugate addition to acrylonitrile plays a key role is outlined below.

$$CH_2{=}CHC{\equiv}N \xrightarrow{NH_3} H_2NCH_2CH_2C{\equiv}N \xrightarrow[\text{heat}]{H_2O,\ HO^-} H_3\overset{+}{N}CH_2CH_2CO_2^-$$

Acrylonitrile — 3-Aminopropanenitrile — β-Alanine

Addition of ammonia to acrylonitrile has been carried out in modest yield (31 to 33 percent). Hydrolysis of the nitrile group can be accomplished in the presence of either acids or bases. Hydrolysis in the presence of $Ba(OH)_2$ has been reported in the literature to give β-alanine in 85 to 90 percent yield.

29.26 The first step involves alkylation of diethyl malonate by 2-bromobutane.

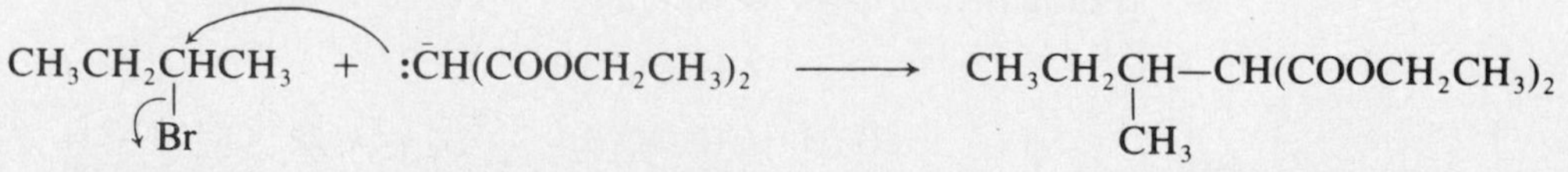

$$CH_3CH_2\underset{Br}{CH}CH_3 + :\bar{C}H(COOCH_2CH_3)_2 \longrightarrow CH_3CH_2\underset{CH_3}{CH}{-}CH(COOCH_2CH_3)_2$$

2-Bromobutane — Anion of diethyl malonate — Compound C

In the second step of the synthesis, compound C is subjected to ester saponification. Following acidification, the corresponding diacid (compound D) is isolated.

$$\underset{\text{Compound C}}{CH_3CH_2\underset{\displaystyle CH_3}{\underset{|}{C}}HCH(COOCH_2CH_3)_2} \xrightarrow[\text{2. HCl}]{\text{1. KOH}} \underset{\text{Compound D } (C_7H_{12}O_4)}{CH_3CH_2\underset{\displaystyle CH_3}{\underset{|}{C}}HCH(COOH)_2}$$

Compound D is readily brominated at its α carbon atom by way of the corresponding enol form.

$$\underset{\text{Compound D}}{(HOOC)_2CH-CH(CH_3)CH_2CH_3} \rightleftharpoons \underset{\text{Enol form}}{HOOC-C(=C(OH)OH)-CH(CH_3)CH_2CH_3} \xrightarrow{Br_2} \underset{\text{Compound E}(C_7H_{11}BrO_4)}{(HOOC)_2CBr-CH(CH_3)CH_2CH_3}$$

When compound E is heated, it undergoes decarboxylation to give an α-bromo carboxylic acid.

$$\underset{\text{Compound E}}{CH_3CH_2\underset{\displaystyle CH_3}{\underset{|}{C}}H-\underset{\displaystyle Br}{\underset{|}{C}}(COOH)_2} \xrightarrow{\text{heat}} \underset{\text{Compound F}}{CH_3CH_2\underset{\displaystyle CH_3}{\underset{|}{C}}H-\underset{\displaystyle Br}{\underset{|}{C}}HCOOH} + \underset{\text{Carbon dioxide}}{CO_2}$$

Treatment of compound F with ammonia converts it to isoleucine by nucleophilic substitution.

$$\underset{\text{Compound F}}{CH_3CH_2\underset{\displaystyle CH_3}{\underset{|}{C}}H-\underset{\displaystyle Br}{\underset{|}{C}}HCO_2H} + NH_3 \longrightarrow \underset{\text{DL-Isoleucine}}{CH_3CH_2\underset{\displaystyle CH_3}{\underset{|}{C}}H-\underset{\displaystyle \overset{+}{N}H_3}{\underset{|}{C}}HCO_2^-}$$

(*b*) The procedure just described can be adapted to the synthesis of other amino acids. The group attached to the α carbon atom is derived from the alkyl halide used to alkylate diethyl malonate. Benzyl bromide (or chloride or iodide) would be appropriate for the preparation of DL-phenylalanine.

$$\underset{\text{Benzyl bromide}}{C_6H_5CH_2Br} \longrightarrow \underset{\text{DL-Phenylalanine}}{C_6H_5CH_2\underset{\displaystyle \overset{+}{N}H_3}{\underset{|}{C}}HCO_2^-}$$

29.27 Acid hydrolysis of compound G converts all its ester functions to free carboxyl groups and cleaves both amide bonds.

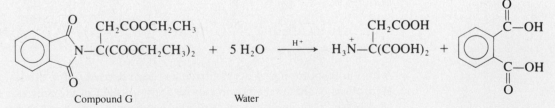

$$\underset{\text{Compound G}}{C_6H_4(CO)_2N-\underset{\displaystyle}{\overset{\displaystyle CH_2COOCH_2CH_3}{\overset{|}{C}}}(COOCH_2CH_3)_2} + \underset{\text{Water}}{5\,H_2O} \xrightarrow{H^+} H_3\overset{+}{N}-\overset{\displaystyle CH_2COOH}{\overset{|}{C}}(COOH)_2 + C_6H_4(COOH)_2$$

The hydrolysis product is a substituted derivative of malonic acid and undergoes decarboxylation on being heated. The product of this decarboxylation is aspartic acid (in its protonated form under conditions of acid hydrolysis).

$$H_3\overset{+}{N}-\overset{\displaystyle CH_2COOH}{\overset{|}{C}}(COOH)_2 \xrightarrow{\text{heat}} \underset{\text{Aspartic acid}}{H_3\overset{+}{N}-\overset{\displaystyle CH_2COOH}{\overset{|}{C}}HCOOH} + \underset{\text{Carbon dioxide}}{CO_2}$$

Aspartic acid is chiral, but is formed as a racemic mixture, so the product of this reaction is not optically active. The starting material (compound G) is achiral and cannot give an optically active product when it reacts with optically inactive reagents.

29.28 The amino acids leucine, phenylalanine, and serine each have one chiral center.

$$\underset{\underset{+}{\text{NH}_3}}{\text{RCHCO}_2^-}\qquad \begin{array}{ll}\text{Leucine:} & \text{R}- = (\text{CH}_3)_2\text{CHCH}_2-\\ \text{Phenylalanine:} & \text{R}- = \text{C}_6\text{H}_5\text{CH}_2-\\ \text{Serine:} & \text{R}- = \text{HOCH}_2-\end{array}$$

When prepared by the Strecker synthesis, each of these amino acids is obtained as a racemic mixture containing 50 percent of the D enantiomer and 50 percent of the L enantiomer.

$$\text{R}\overset{\text{O}}{\overset{\|}{\text{C}}}\text{H} + \text{NH}_3 + \text{HCN} \longrightarrow \underset{\text{NH}_2}{\text{RCHC}{\equiv}\text{N}} \xrightarrow{\text{H}_2\text{O}} \underset{+\text{NH}_3}{\text{RCHCO}_2^-}$$

Thus, preparation of the tripeptide Leu-Phe-Ser will yield a mixture of eight stereoisomers.

D-Leu-D-Phe-D-Ser	L-Leu-L-Phe-L-Ser
D-Leu-D-Phe-L-Ser	L-Leu-L-Phe-D-Ser
D-Leu-L-Phe-D-Ser	L-Leu-D-Phe-L-Ser
D-Leu-L-Phe-L-Ser	L-Leu-D-Phe-D-Ser

29.29 Bradykinin is a nonapeptide but contains only five different amino acids. Three of the amino acid residues are proline, two are arginine, and two are phenylalanine. There will be five peaks on the strip chart after amino acid analysis of bradykinin.

$$\text{Arg-Pro-Pro-Gly-Phe-Ser-Pro-Phe-Arg} \longrightarrow 2\,\text{Arg} + 3\,\text{Pro} + \text{Gly} + 2\,\text{Phe} + \text{Ser}$$

29.30 Asparagine and glutamine each contain an amide function in their side chain. Under the conditions of peptide bond hydrolysis that characterize amino acid analysis, the side-chain amide is also hydrolyzed, giving ammonia.

$$\underset{\text{Asparagine}}{\text{H}_2\text{N}\overset{\text{O}}{\overset{\|}{\text{C}}}\text{CH}_2\underset{\overset{+}{\text{NH}_3}}{\text{CHCO}_2^-}} + \underset{\text{Water}}{\text{H}_2\text{O}} \longrightarrow \underset{\text{Ammonia}}{\text{NH}_3} + \underset{\text{Aspartic acid}}{\text{HO}\overset{\text{O}}{\overset{\|}{\text{C}}}\text{CH}_2\underset{\overset{+}{\text{NH}_3}}{\text{CHCO}_2^-}}$$

$$\underset{\text{Glutamine}}{\text{H}_2\text{N}\overset{\text{O}}{\overset{\|}{\text{C}}}\text{CH}_2\text{CH}_2\underset{\overset{+}{\text{NH}_3}}{\text{CHCO}_2^-}} + \underset{\text{Water}}{\text{H}_2\text{O}} \longrightarrow \underset{\text{Ammonia}}{\text{NH}_3} + \underset{\text{Glutamic acid}}{\text{HO}\overset{\text{O}}{\overset{\|}{\text{C}}}\text{CH}_2\text{CH}_2\underset{\overset{+}{\text{NH}_3}}{\text{CHCO}_2^-}}$$

29.31 (*a*) 1-Fluoro-2,4-dinitrobenzene reacts with the amino group of the N-terminal amino acid in a nucleophilic aromatic substitution reaction of the addition-elimination type.

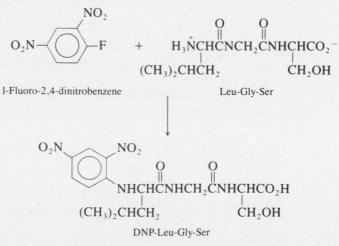

$$\underset{\text{1-Fluoro-2,4-dinitrobenzene}}{\text{O}_2\text{N}-\text{C}_6\text{H}_3(\text{NO}_2)-\text{F}} + \underset{\text{Leu-Gly-Ser}}{\overset{+}{\text{H}_3\text{N}}\underset{(\text{CH}_3)_2\text{CHCH}_2}{\text{CH}}\overset{\text{O}}{\overset{\|}{\text{C}}}\text{NCH}_2\overset{\text{O}}{\overset{\|}{\text{C}}}\text{NH}\underset{\text{CH}_2\text{OH}}{\text{CH}}\text{CO}_2^-}$$

$$\downarrow$$

$$\underset{\text{DNP-Leu-Gly-Ser}}{(\text{O}_2\text{N})(\text{NO}_2)\text{C}_6\text{H}_3-\text{NH}\underset{(\text{CH}_3)_2\text{CHCH}_2}{\text{CH}}\overset{\text{O}}{\overset{\|}{\text{C}}}\text{NHCH}_2\overset{\text{O}}{\overset{\|}{\text{C}}}\text{NH}\underset{\text{CH}_2\text{OH}}{\text{CH}}\text{CO}_2\text{H}}$$

(*b*) Hydrolysis of the product in (*a*) cleaves the peptide bonds. Leucine is isolated as its 2,4-dinitrophenyl (DNP) derivative, but glycine and serine are isolated as the free amino acids.

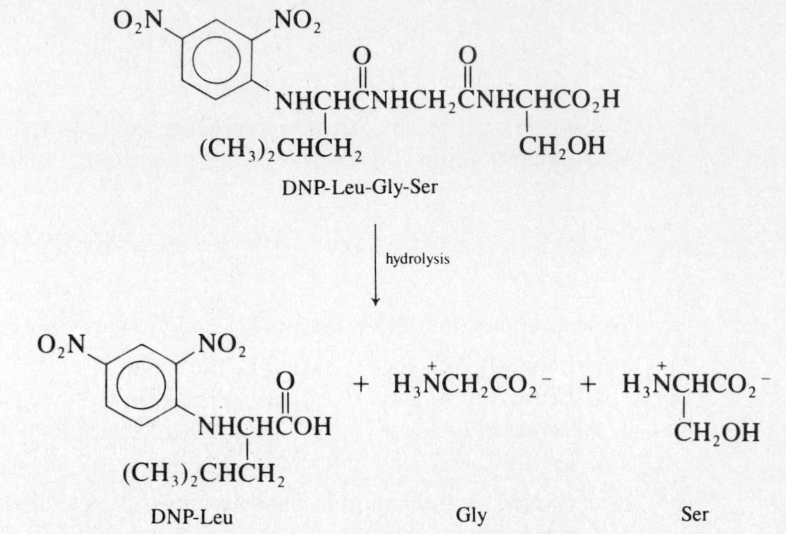

DNP-Leu-Gly-Ser

hydrolysis

DNP-Leu + $H_3\overset{+}{N}CH_2CO_2^-$ + $H_3\overset{+}{N}CHCO_2^-$ (CH_2OH)

DNP-Leu Gly Ser

(*c*) Dansyl chloride is a reagent used for labeling the N-terminal amino acid of a peptide. It is a sulfonyl chloride and reacts with the free amino group to give a sulfonamide.

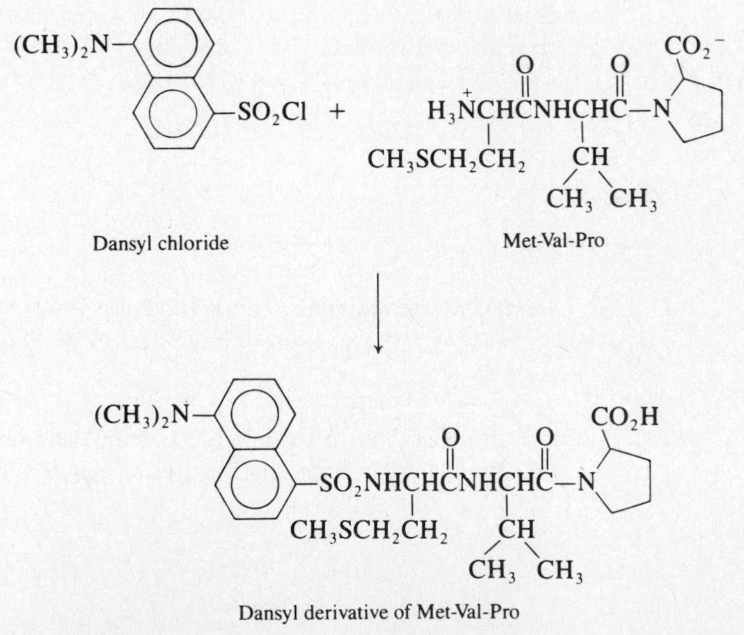

Dansyl chloride Met-Val-Pro

Dansyl derivative of Met-Val-Pro

Hydrolysis of the dansyl derivative of Met-Val-Pro gives the dansyl derivative of methionine along with the free amino acids valine and proline.

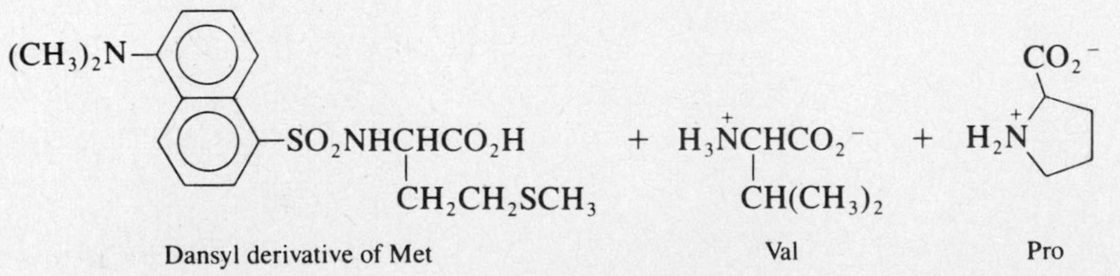

Dansyl derivative of Met Val Pro

(*d*) Phenyl isothiocyanate is a reagent used to identify the N-terminal amino acid of a peptide by the Edman degradation. The N-terminal amino acid is cleaved as a phenylthiohydantoin (PTH) derivative, the remainder of the peptide remaining intact.

$$\overset{+}{H_3N}CH(CH(CH_3)CH_2CH_3)\overset{O}{\overset{\|}{C}}NHCH(CH_2CH_2CO_2H)\overset{O}{\overset{\|}{C}}NHCH(CH_2C_6H_5)CO_2^-$$

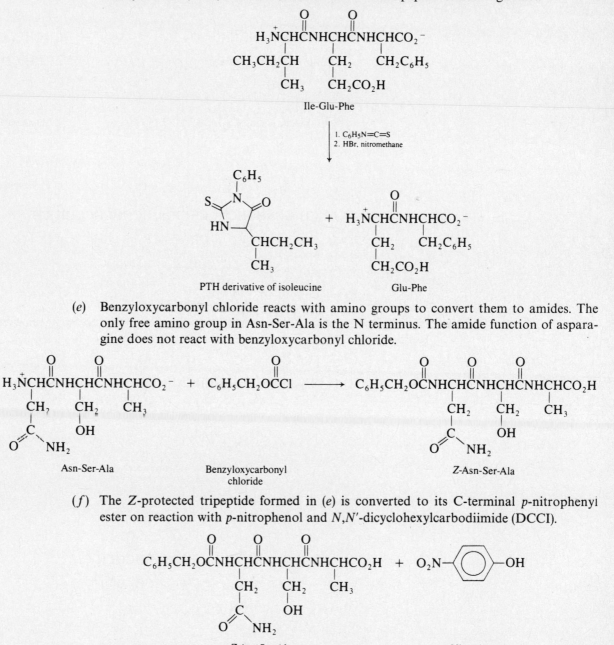

Ile-Glu-Phe

↓ 1. $C_6H_5N{=}C{=}S$ 2. HBr, nitromethane

PTH derivative of isoleucine (ring: C_6H_5–N, S=C, C=O, HN, CH–$CHCH_2CH_3$ with CH_3) + $\overset{+}{H_3N}CH(CH_2CH_2CO_2H)\overset{O}{\overset{\|}{C}}NHCH(CH_2C_6H_5)CO_2^-$

PTH derivative of isoleucine   Glu-Phe

(*e*) Benzyloxycarbonyl chloride reacts with amino groups to convert them to amides. The only free amino group in Asn-Ser-Ala is the N terminus. The amide function of asparagine does not react with benzyloxycarbonyl chloride.

$$\overset{+}{H_3N}CH(CH_2CONH_2)\overset{O}{\overset{\|}{C}}NHCH(CH_2OH)\overset{O}{\overset{\|}{C}}NHCH(CH_3)CO_2^- + C_6H_5CH_2O\overset{O}{\overset{\|}{C}}Cl \longrightarrow C_6H_5CH_2O\overset{O}{\overset{\|}{C}}NHCH(CH_2CONH_2)\overset{O}{\overset{\|}{C}}NHCH(CH_2OH)\overset{O}{\overset{\|}{C}}NHCH(CH_3)CO_2H$$

Asn-Ser-Ala   Benzyloxycarbonyl chloride   *Z*-Asn-Ser-Ala

(*f*) The *Z*-protected tripeptide formed in (*e*) is converted to its C-terminal *p*-nitrophenyl ester on reaction with *p*-nitrophenol and *N,N'*-dicyclohexylcarbodiimide (DCCI).

$$C_6H_5CH_2O\overset{O}{\overset{\|}{C}}NHCH(CH_2CONH_2)\overset{O}{\overset{\|}{C}}NHCH(CH_2OH)\overset{O}{\overset{\|}{C}}NHCH(CH_3)CO_2H + O_2N{-}C_6H_4{-}OH$$

Z-Asn-Ser-Ala   *p*-Nitrophenol

↓ DCCI

$$C_6H_5CH_2O\overset{O}{\overset{\|}{C}}NHCH(CH_2CONH_2)\overset{O}{\overset{\|}{C}}NHCH(CH_2OH)\overset{O}{\overset{\|}{C}}NHCH(CH_3)\overset{O}{\overset{\|}{C}}{-}O{-}C_6H_4{-}NO_2$$

Z-Asn-Ser-Ala *p*-nitrophenyl ester

(*g*) The *p*-nitrophenyl ester prepared in (*f*) is an "active ester." The *p*-nitrophenyl group is a good leaving group and can be displaced by the amino nitrogen of valine ethyl ester to form a new peptide bond.

$$C_6H_5CH_2O\overset{O}{\overset{\|}{C}}NH\underset{\substack{|\\CH_2\\|\\C(=O)NH_2}}{CH}\overset{O}{\overset{\|}{C}}NH\underset{\substack{|\\CH_2\\|\\OH}}{CH}\overset{O}{\overset{\|}{C}}NH\underset{\substack{|\\CH_3}}{CH}\overset{O}{\overset{\|}{C}}{-}O{-}C_6H_4{-}NO_2 + H_2N\underset{\substack{|\\CH(CH_3)_2}}{CH}\overset{O}{\overset{\|}{C}}OCH_2CH_3$$

Z-Asn-Ser-Ala *p*-nitrophenyl ester   Valine ethyl ester

$$\downarrow$$

$$C_6H_5CH_2O\overset{O}{\overset{\|}{C}}NH\underset{\substack{|\\CH_2\\|\\C(=O)NH_2}}{CH}\overset{O}{\overset{\|}{C}}NH\underset{\substack{|\\CH_2\\|\\OH}}{CH}\overset{O}{\overset{\|}{C}}NH\underset{\substack{|\\CH_3}}{CH}\overset{O}{\overset{\|}{C}}NH\underset{\substack{|\\CH(CH_3)_2}}{CH}\overset{O}{\overset{\|}{C}}OCH_2CH_3$$

Z-Asn-Ser-Ala-Val ethyl ester

(*h*) Hydrogenolysis of the *Z*-protected tetrapeptide ester formed in (*g*) removes the *Z* protecting group.

$$C_6H_5CH_2O\overset{O}{\overset{\|}{C}}NH\underset{\substack{|\\CH_2\\|\\C(=O)NH_2}}{CH}\overset{O}{\overset{\|}{C}}NH\underset{\substack{|\\CH_2\\|\\OH}}{CH}\overset{O}{\overset{\|}{C}}NH\underset{\substack{|\\CH_3}}{CH}\overset{O}{\overset{\|}{C}}NH\underset{\substack{|\\CH(CH_3)_2}}{CH}\overset{O}{\overset{\|}{C}}OCH_2CH_3$$

Z-Asn-Ser-Ala-Val ethyl ester

$$\downarrow H_2, Pd$$

$$H_2N\underset{\substack{|\\CH_2\\|\\C(=O)NH_2}}{CH}\overset{O}{\overset{\|}{C}}NH\underset{\substack{|\\CH_2\\|\\OH}}{CH}\overset{O}{\overset{\|}{C}}NH\underset{\substack{|\\CH_3}}{CH}\overset{O}{\overset{\|}{C}}NH\underset{\substack{|\\CH(CH_3)_2}}{CH}\overset{O}{\overset{\|}{C}}OCH_2CH_3$$

Asn-Ser-Ala-Val ethyl ester

29.32 Consider, for example, the reaction of hydrazine with a very simple dipeptide such as Gly-Ala. Hydrazine cleaves the peptide by nucleophilic attack on the carbonyl group of glycine.

$$H_3\overset{+}{N}CH_2\overset{O}{\overset{\|}{C}}{-}NH\underset{\substack{|\\CH_3}}{CH}CO_2^- + H_2\ddot{N}NH_2 \longrightarrow H_2NCH_2\overset{O}{\overset{\|}{C}}{-}NHNH_2 + H_3\overset{+}{N}\underset{\substack{|\\CH_3}}{CH}CO_2^-$$

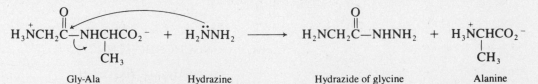

Gly-Ala   Hydrazine   Hydrazide of glycine   Alanine

It is the C-terminal residue that is cleaved as the free amino acid and identified in the hydrazinolysis of peptides.

29.33 Somatostatin is a tetradecapeptide and so is composed of 14 amino acids. The fact that Edman degradation gave the PTH derivative of alanine identifies this as the N-terminal amino acid. A major piece of information is the amino acid sequence of a hexapeptide obtained by partial hydrolysis.

Ala-Gly-Cys-Lys-Asn-Phe

Using this as a starting point and searching for overlaps with the other hydrolysis products gives the entire sequence.

```
Ala-Gly-Cys-Lys-Asn-Phe
            Asn-Phe-Phe-Trp-Lys
                Phe-Trp
                        Lys-Thr-Phe
                            Thr-Phe-Thr-Ser-Cys
                                    Thr-Ser-Cys
Ala-Gly-Cys-Lys-Asn-Phe-Phe-Trp-Lys-Thr-Phe-Thr-Ser-Cys
 1   2   3   4   5   6   7   8   9  10  11  12  13  14
```

The disulfide bridge in somatastatin is between cysteine 3 and cysteine 14. Thus the primary structure is:

```
            Lys-Asn-Phe-Phe-Trp-Lys
           /                      |
Ala-Gly-Cys                       |
           \                      |
            S-S-Cys-Ser-Thr-Phe-Thr
```

29.34 It is the C-terminal amino acid that is anchored to the solid support in the preparation of peptides by the Merrifield method. Refer to the structure of oxytocin in Figure 29.6 of the text and note that oxytocin, in fact, has no free carboxyl groups; all the acyl groups of oxytocin appear as amide functions. Thus, the carboxyl terminus of oxytocin has been modified by conversion to an amide. There are three amide functions of the type $\overset{O}{\overset{\|}{C}}NH_2$, two of which belong to side chains of asparagine and glutamine, respectively. The third amide belongs to the C-terminal amino acid, glycine, $-NHCH_2\overset{O}{\overset{\|}{C}}OH$, which in oxytocin has been modified so that it appears as $-NHCH_2\overset{O}{\overset{\|}{C}}NH_2$. Therefore, attach glycine to the solid support in the first step of the Merrifield synthesis. The carboxyl group can be modified to the required amide after all the amino acid residues have been added and the completed peptide is removed from the solid support.

29.35 Purine and its numbering system are as shown:

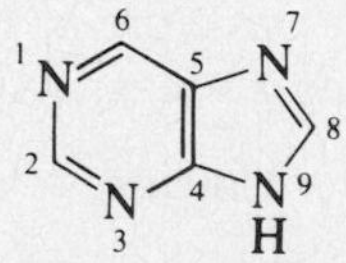

In nebularine, D-ribose in its furanose form is attached to position 9 of purine. The stereochemistry at the anomeric position is β.

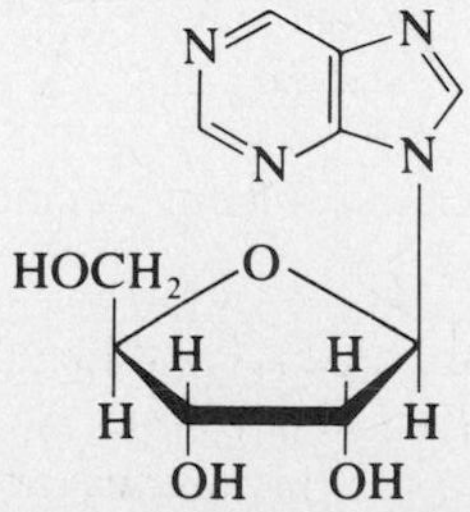

9-β-D-Ribofuranosylpurine (nebularine)

29.36 The problem states that vidarabine is the arabinose analog of adenosine. Arabinose and ribose differ only in their configuration at C-2.

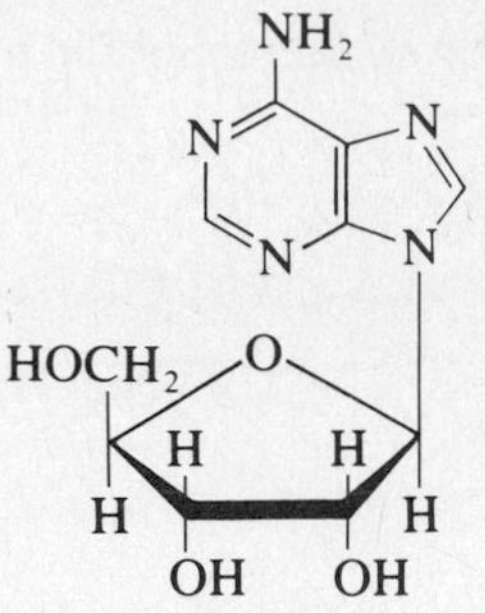

Adenosine 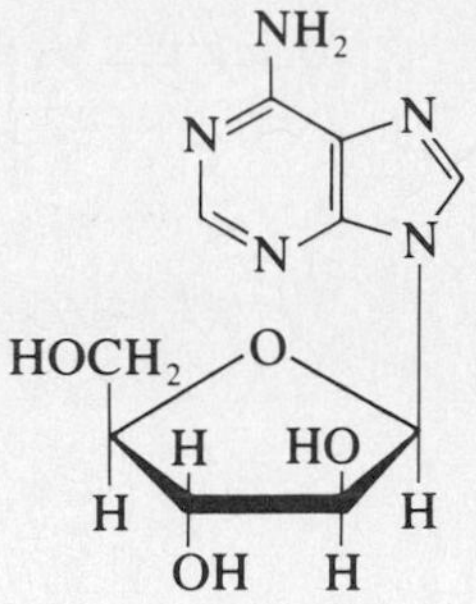   Vidarabine

29.37 All the bases in the synthetic messenger RNA prepared by Nirenberg were U; therefore, the codon is UUU. By referring to the codons in Table 29.4, we see that UUU codes for phenylalanine. A polypeptide in which all the amino acid residues were phenylalanine was isolated in Nirenberg's experiment.

SELF-TEST

PART A

A-1. Give the structure of the reactant, reagent, or product omitted from each of the following.

(*a*) $? \xrightarrow[\text{2. } H_3O^+\text{, heat} \quad \text{3. neutralize}]{\text{1. } NH_4Cl\text{, NaCN}} C_6H_5CH_2\underset{}{\overset{\overset{+}{N}H_3}{\overset{|}{C}}}HCO_2^-$

(*b*) $C_6H_5CH_2O\overset{O}{\overset{\|}{C}}Cl + \text{valine} \xrightarrow[\text{2. } H^+]{\text{1. } HO^-\text{, } H_2O} ?$

(*c*) $\text{Boc-Phe} + H_2NCH_2CO_2CH_2CH_3 \xrightarrow{?} \text{Boc-NH}\underset{\underset{CH_2C_6H_5}{|}}{C}H\overset{O}{\overset{\|}{C}}NHCH_2CO_2CH_2CH_3$

A-2. Give the structure of the derivative that would be obtained by treatment of Phe-Ala with Sanger's reagent followed by hydrolysis.

A-3. Outline a sequence of steps that would allow the following synthetic conversions to be carried out:

(*a*) $(CH_3)_2CHCH_2\overset{\overset{+}{N}H_3}{\overset{|}{C}}HCO_2^-$ (leucine) from $CH_3\overset{O}{\overset{\|}{C}}NHCH(CO_2CH_2CH_3)_2$

(*b*) Leu-Val from leucine and $(CH_3)_2CH\overset{\overset{+}{N}H_3}{\overset{|}{C}}HCO_2^-$ (valine)

A-4. The carboxypeptidase-catalyzed hydrolysis of a pentapeptide yielded phenylalanine (Phe). One cycle of an Edman degradation gave a derivative of leucine (Leu). Partial hydrolysis yielded the fragments Leu-Val-Gly and Gly-Ala among others. Deduce the structure of the peptide.

PART B

B-1. Which statement correctly describes the difference in the otherwise similar chemical constitutions of DNA and RNA?
(*a*) DNA contains uracil; RNA contains thymine.
(*b*) DNA contains guanine but not adenine; RNA contains both.
(*c*) DNA contains thymine; RNA contains uracil.
(*d*) None of these applies—the chemical constitution is the same.

B-2. Assume a particular amino acid has an isoelectric point of 6.0. In a solution of pH 1.0, which of the following species will predominate?

(*a*) $H_3\overset{+}{N}\overset{R}{\overset{|}{C}}HCO_2H$ (*b*) $H_2N\overset{R}{\overset{|}{C}}HCO_2H$
(*c*) $H_3\overset{+}{N}\overset{R}{\overset{|}{C}}HCO_2^-$ (*d*) $H_2N\overset{R}{\overset{|}{C}}HCO_2^-$

B-3. Choose the response which provides the best match of terms

	Purine	*Pyrimidine*
(*a*)	Adenine	guanine
(*b*)	Thymine	cytosine
(*c*)	Cytosine	adenine
(*d*)	Guanine	cytosinc

B-4. Which of the following reagents would be combined in the synthesis of Phe-Ala? [In phenylalanine (Phe), R in the generalized amino acid formula $H_2N\overset{R}{\overset{|}{C}}HCO_2H$ is $CH_2C_6H_5$ and in alanine (Ala) it is CH_3.]

1. $ZNH\overset{CH_3}{\overset{|}{C}}HCO_2H$ 2. $H_2N\overset{CH_3}{\overset{|}{C}}HCO_2CH_2C_6H_5$
3. $ZNH\overset{CH_2C_6H_5}{\overset{|}{C}}HCO_2H$ 4. $H_2N\overset{CH_2C_6H_5}{\overset{|}{C}}HCO_2CH_2C_6H_5$

(*a*) 1 and 2 (*b*) 1 and 4
(*c*) 2 and 3 (*d*) 3 and 4

B-5. Which amino acid is achiral?

(*a*) $\underset{\underset{+}{NH_3}}{\underset{|}{C}}H_2COO^-$ (*b*) $CH_3\underset{CH_3}{\underset{|}{C}}HCH_2\underset{\underset{+}{NH_3}}{\underset{|}{C}}HCOO^-$
(*c*) $CH_3\underset{HO}{\underset{|}{C}}H\underset{\underset{+}{NH_3}}{\underset{|}{C}}HCOO^-$ (*d*) $CH_3\underset{{}^{+}NH_3}{\underset{|}{C}}HCOO^-$

B-6. A nucleoside is a
(*a*) Phosphate ester of a nucleotide
(*b*) Unit having a sugar bonded to a purine or pyrimidine base
(*c*) Chain whose backbone consists of sugar units connected by phosphate groups
(*d*) Phosphate salt of a purine or pyrimidine base

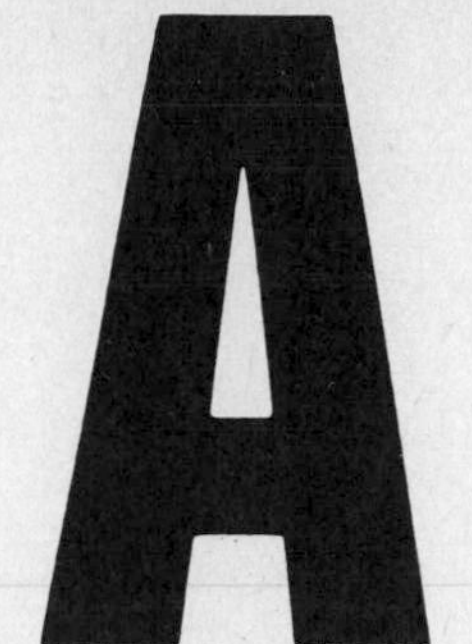

ANSWERS TO THE SELF-TESTS

CHAPTER 1

A-1. S^{2-}; $1s^2 2s^2 2p^6 3s^2 3p^6$

A-2. $C_{10}H_{14}N_2$

A-3. $[:\ddot{N}{=}C{=}\ddot{S}:]^{-1}$

Formal charge: −1 0 0

A-4.

$$\begin{array}{c} \quad H \\ \quad | \\ H{-}C{-}\ddot{N}{-}H \\ \quad | \quad\ \ | \\ \quad H \quad H \end{array}$$

A-5. $C_{12}H_{22}$

A-6.

$$\begin{array}{c} :O: \\ \| \\ S \\ \diagup\ \diagdown \\ :\ddot{O}.\ \ .\ddot{O}: \end{array} \longleftrightarrow \begin{array}{c} :\ddot{O}: \\ | \\ S \\ \diagup\ \diagdown\!\!\diagdown \\ :\ddot{O}.\ \ \ddot{O}\cdot \end{array} \longleftrightarrow \begin{array}{c} :\ddot{O}: \\ | \\ S \\ \diagup\!\!\diagup\ \diagdown \\ \cdot\ddot{O}\ \ .\ddot{O}: \end{array}$$

A-7. $C_{21}H_{28}O_5$

A-8. $C_{14}H_{24}O$

A-9. $[:\ddot{\underset{..}{O}}{-}C{\equiv}N:]^{-1}$

Formal charge: −1 0 0

A-10. Pyramidal; $:\!{-}N(Cl)_3$ (N bonded to three Cl)

B-1. (*b*)	**B-2.** (*c*)	**B-3.** (*d*)	**B-4.** (*a*)
B-5. (*a*)	**B-6.** (*b*)	**B-7.** (*c*)	**B-8.** (*b*)
B-9. (*d*)	**B-10.** (*d*)	**B-11.** (*b*)	

CHAPTER 2

A-1. $CH_3CH_2CH_2CH_2{-}$ — *n*-Butyl

$CH_3CH_2\underset{|}{C}HCH_3$ — *sec*-Butyl

$$\begin{array}{c} CH_3 \\ | \\ CH_3CHCH_2{-} \end{array}$$

Isobutyl

$$\begin{array}{c} CH_3 \\ | \\ CH_3{-}C{-} \\ | \\ CH_3 \end{array}$$

tert-Butyl

A-2. (*a*)

$$\begin{array}{c} CH_3CHCH_3 \\ | \\ CH_3CHCHCHCH_3 \\ |\quad\;\; | \\ CH_3\;\; CH_3 \end{array}$$

(*b*) Six methyl groups, three isopropyl groups

A-3. 3,4-Dimethylheptane

A-4. Four primary C, three secondary C, two tertiary C

A-5.

$$\begin{array}{c} CH_3 \\ | \\ CH_3CHCHCH_2CH_3 \equiv C_7H_{16} \\ | \\ CH_3 \end{array}$$

$$C_7H_{16} + 11\,O_2 \longrightarrow 7\,CO_2 + 8\,H_2O$$

A-6.

Cyclopentane

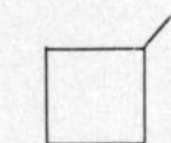

Methylcyclobutane

1,1-Dimethylcyclopropane

1,2-Dimethylcyclopropane

Ethylcyclopropane

A-7.

$$\begin{array}{c} CH_2CH_2CH_3 \\ | \\ CH_3CH{-}CCH_2CH_3 \\ |\qquad | \\ CH_3\;\; CH_3 \end{array}$$

No; 3-ethyl-2,3-dimethylhexane

A-8. $CH_3CH_2CH_2CH_2CH_2CH_2CH_2CH_3$

B-1. (*a*) **B-2.** (*d*) **B-3.** (*d*) **B-4.** (*d*)

B-5. (*b*) **B-6.** (*a*) **B-7.** (*c*) **B-8.** (*c*)

B-9. (*a*)

CHAPTER 3

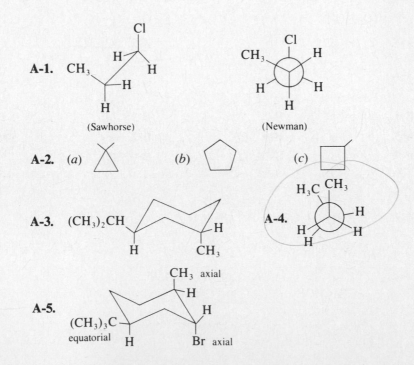

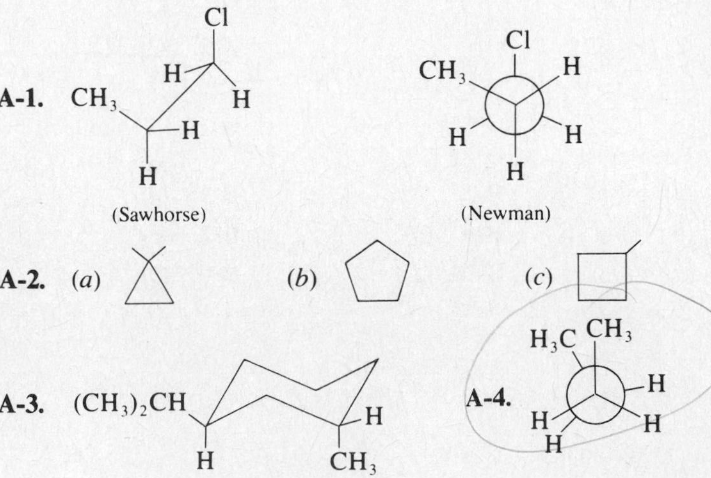

A-5.

CH$_3$ axial, H, H, (CH$_3$)$_3$C equatorial, H, Br axial

A-6. A is more strained; B is more stable.

A-7. 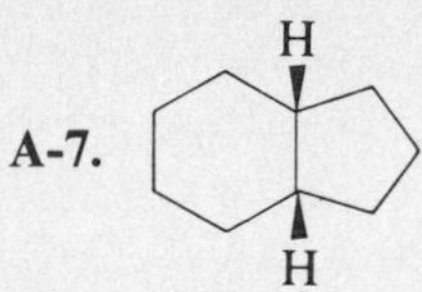

B-1. (*d*)	**B-2.** (*b*)	**B-3.** (*c*)	**B-4.** (*a*)
B-5. (*c*)	**B-6.** (*d*)	**B-7.** (*a*)	**B-8.** (*c*)

CHAPTER 4

A-1. (*a*) *trans*-1-Bromo-3-methylcyclopentane
(*b*) 2-Ethyl-4-methyl-1-hexanol

A-2. (*a*) $ICH_2C(CH_3)(Cl)CH_2CH_2CH_2CH_2CH_3$ (*b*) OH, $CH(CH_3)_2$

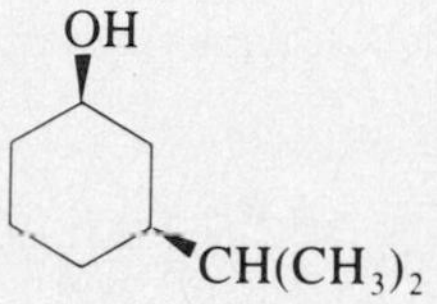

A-3. Conjugate acid $CH_3\overset{+}{O}H_2$; conjugate base CH_3O^-

A-4. $pK_a = 4.74$

A-5. (*a*) $(CH_3)_2CHCH_2Cl$ (*b*) $CH_3CH_2C(OH)(CH_3)_2$ (*c*) Na°

A-6. $CH_3CH_2O^- + NH_3 \rightleftharpoons CH_3CH_2OH + NH_2^-$ $(K < 1)$
Conjugate base; Conjugate acid; Stronger acid; Stronger base

A-7. (*a*) Three
(*b*), (*c*) $CH_3\dot{C}(CH_3)CH_2CH(CH_3)CH_3$ (most stable); $CH_3CH(CH_3)\dot{C}HCH(CH_3)CH_3$; $\dot{C}H_2CH(CH_3)CH_2CH(CH_3)CH_3$ (least stable)

A-8. $CH_3C(CH_3)_2CH_3 + Cl_2 \longrightarrow CH_3C(CH_3)_2CH_2Cl + HCl$

A-9. $Br^\cdot$ + [methylcyclohexane] $\longrightarrow$ HBr + [1-methylcyclohexyl radical]

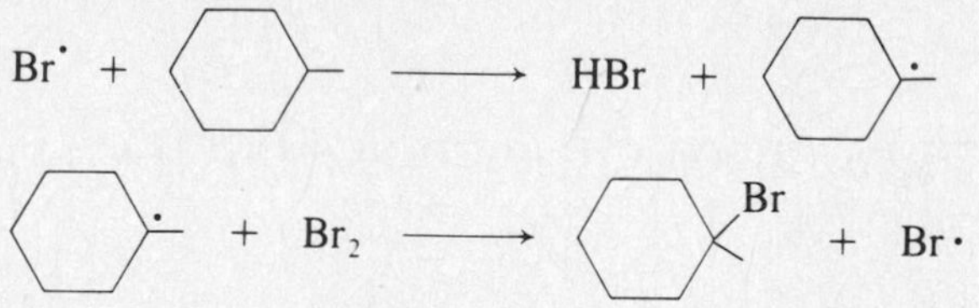

[1-methylcyclohexyl radical] + Br_2 $\longrightarrow$ [1-bromo-1-methylcyclohexane] + $Br\cdot$

B-1. (*c*)	**B-2.** (*b*)	**B-3.** (*d*)	**B-4.** (*c*)
B-5. (*c*)	**B-6.** (*d*)	**B-7.** (*a*)	**B-8.** (*c*)
B-9. (*d*)	**B-10.** (*a*)	**B-11.** (*c*)	**B-12.** (*c*)

CHAPTER 5

A-1. (*a*) 4,4-Dimethyl-2-pentene (*b*) (*E*)-3,5-Dimethyl-4-octene

A-2. (*a*) 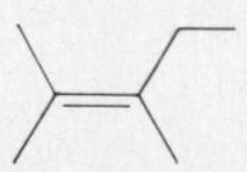 2,3-Dimethyl-2-pentene

(*b*) 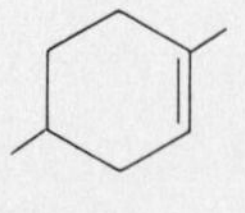 1,4-Dimethylcyclohexene

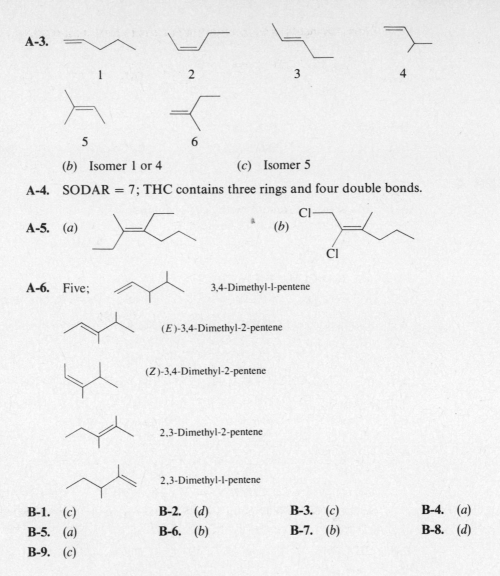

A-3.  1  2 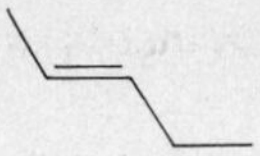 3 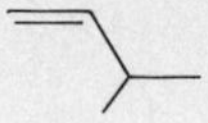 4

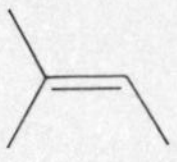

5 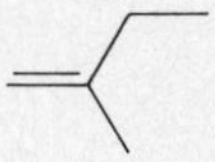 6

(*b*) Isomer 1 or 4 (*c*) Isomer 5

A-4. SODAR = 7; THC contains three rings and four double bonds.

A-5. (*a*) 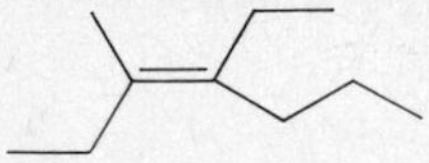 (*b*) Cl Cl

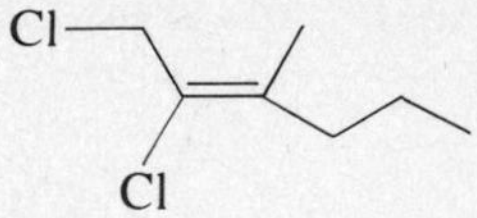

A-6. Five; 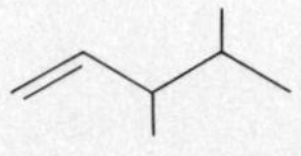 3,4-Dimethyl-1-pentene

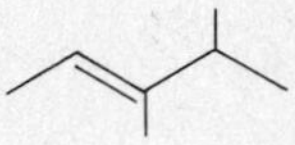

(*E*)-3,4-Dimethyl-2-pentene

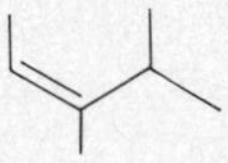

(*Z*)-3,4-Dimethyl-2-pentene

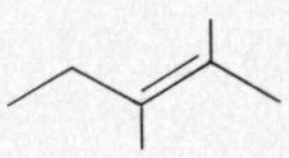

2,3-Dimethyl-2-pentene

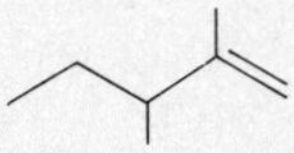

2,3-Dimethyl-1-pentene

B-1. (*c*) **B-2.** (*d*) **B-3.** (*c*) **B-4.** (*a*)
B-5. (*a*) **B-6.** (*b*) **B-7.** (*b*) **B-8.** (*d*)
B-9. (*c*)

CHAPTER 6

A-1. (*a*) $CH_2{=}C(CH_3)CH_2CH_2CH_3 + (CH_3)_2C{=}CHCH_2CH_3$ (major)

(*b*) CH_2 CH_3 + (major)

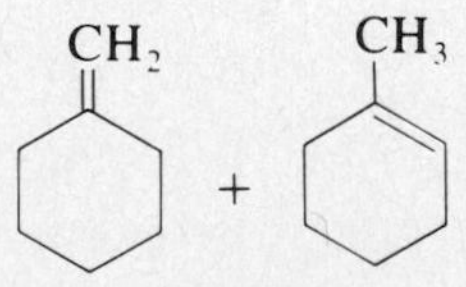

(*c*) $(CH_3)_2CHCH(X)CH(CH_3)_2$ (X = Cl, Br, I)

(*d*) $CH_2{=}C(CH_3)C(CH_3)_3$

A-2. $CH_3CH_2C(CH_3)_2{-}CH_2Br$

A-3. Rearrangement (hydride migration) occurs to form a more stable carbocation:

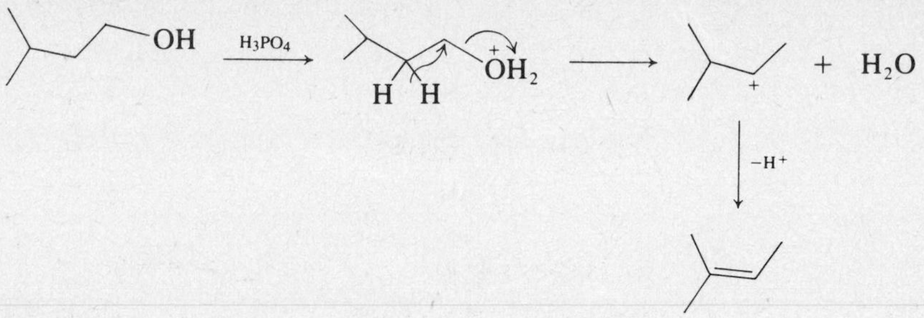

OH $\xrightarrow{H_3PO_4}$ H H $\overset{+}{O}H_2$ ⟶ + H_2O

$-H^+$

A-4. B: H CH_3 Br ⟶ CH_3 + Br^-

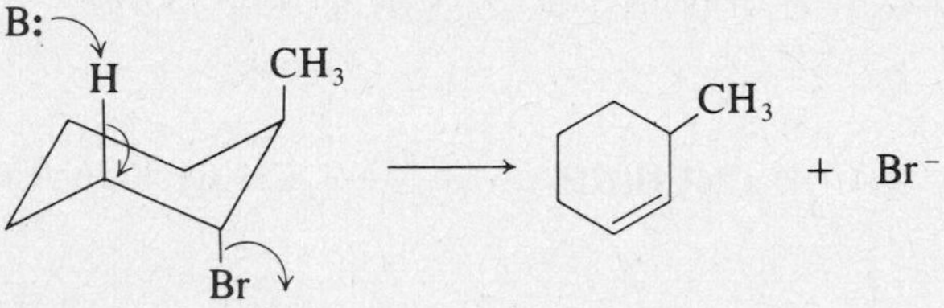

A-5. Cis isomer:

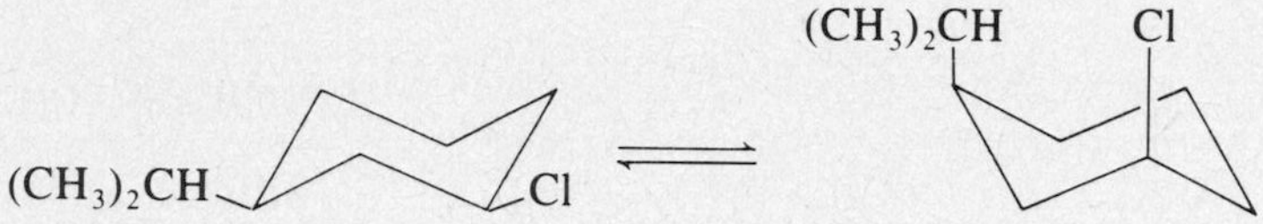

$(CH_3)_2CH$ Cl ⇌ $(CH_3)_2CH$ Cl

Trans isomer:

$(CH_3)_2CH$ Cl ⇌ $CH(CH_3)_2$ Cl

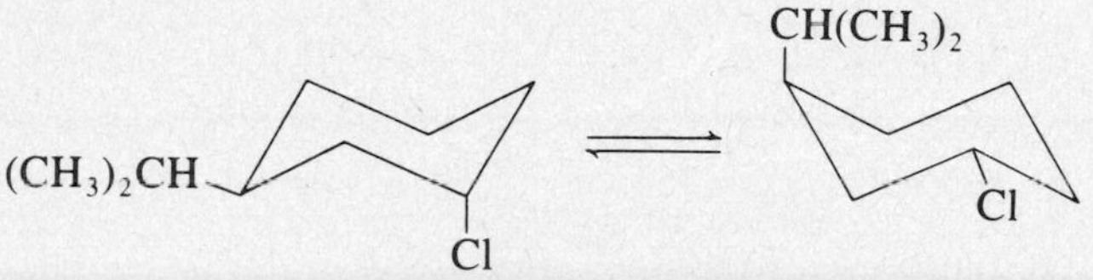

The trans isomer will react faster because its major conformational isomer (with the isopropyl group equatorial) has an axial Cl able to undergo E2 elimination.

A-6. $-CH_3$ $\xrightarrow[\text{light}]{Br_2}$ Br CH_3 $\xrightarrow[CH_3CH_2OH]{KOCH_2CH_3}$ $-CH_3$

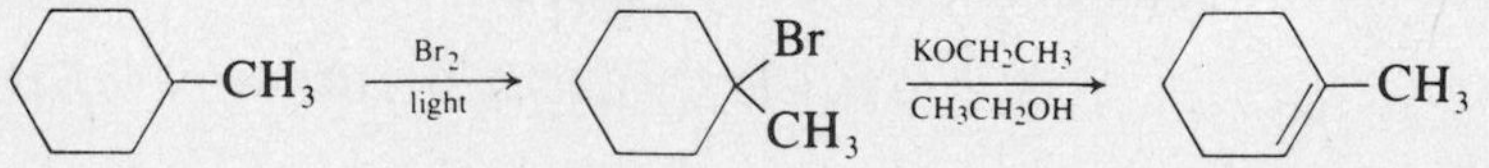

A-7. 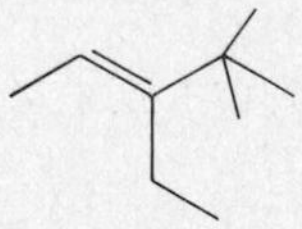 3-Ethyl-4,4-dimethyl-2-pentene

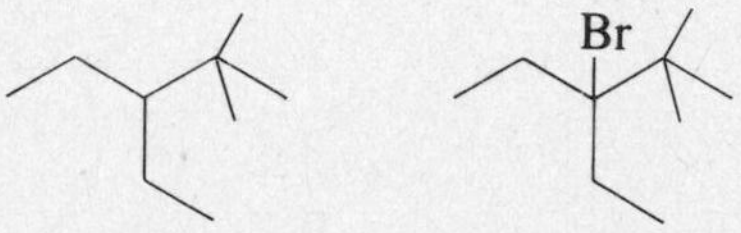

A B Br

B-1. (*a*) **B-2.** (*b*) **B-3.** (*a*) **B-4.** (*d*)
B-5. (*b*)

CHAPTER 7

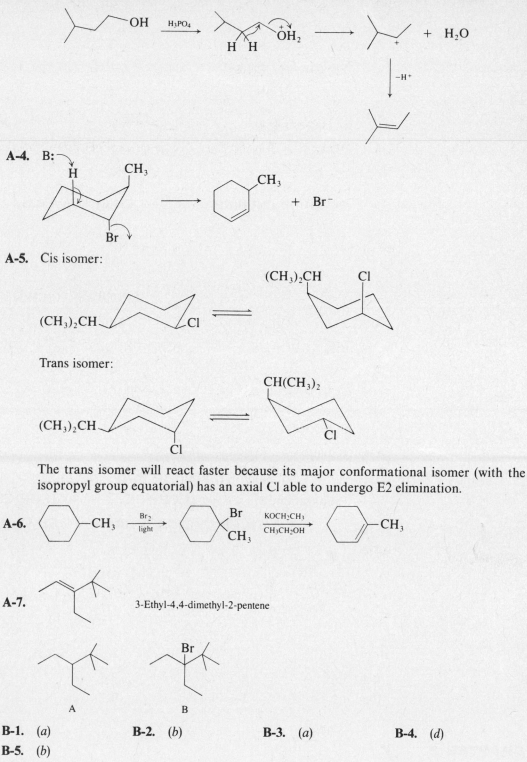

A-1. (*a*) $(CH_3)_2\overset{OH}{C}CH_2CH_3$ (*b*) HBr, peroxides

(*c*)

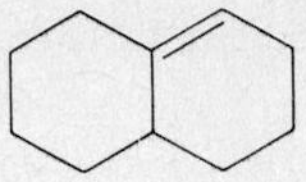

(*d*) 1. $Hg(OAc)_2$, THF—H_2O
2. $NaBH_4$, HO^-

(*e*) $CH_3CH_2C(CH_3)(OH)CH_2Br$

A-2. (*a*) 1-methylcyclohexanol $\xrightarrow[\text{(conc)}]{H_2SO_4}$ 1-methylcyclohexene $\xrightarrow[\text{2. } H_2O_2,\ HO^-]{\text{1. } B_2H_6}$ trans-2-methylcyclohexanol

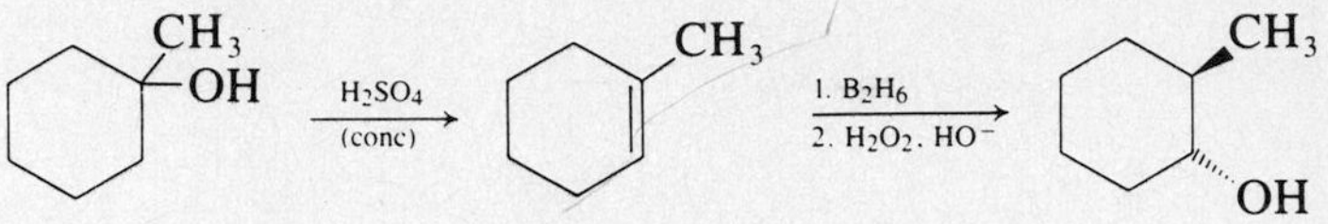

(*b*) $CH_3CH_2CHClCH(CH_3)_2 \xrightarrow[CH_3OH]{NaOCH_3} CH_3CH_2CH{=}C(CH_3)_2$

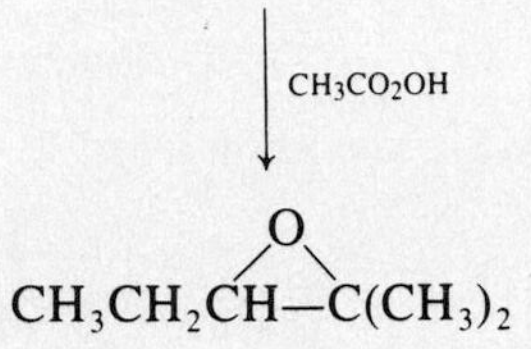

$\downarrow CH_3CO_2OH$

$CH_3CH_2\overset{O}{CH{-}C}(CH_3)_2$ (epoxide)

A-3. $(CH_3)_2CHC(CH_3){=}CH_2 \xrightarrow{HBr} (CH_3)_2CHCBr(CH_3)CH_3 \xrightarrow[CH_3OH]{NaOCH_3} (CH_3)_2C{=}C(CH_3)_2$

A B C

$C \xrightarrow[\text{2. } H_2O,\ Zn]{\text{1. } O_3} (CH_3)_2C{=}O$ (2 mol)

A-4. Initiation: $ROOR \xrightarrow[\text{heat}]{\text{light or}} 2\,RO\cdot$

$RO\cdot + HBr \longrightarrow ROH + Br\cdot$

Propagation: $Br\cdot + CH_3CH_2CH{=}CH_2 \longrightarrow CH_3CH_2\dot{C}HCH_2Br$

$CH_3CH_2\dot{C}HCH_2Br + HBr \longrightarrow CH_3CH_2CH_2CH_2Br + Br\cdot$

A-5. (*E*)-2-Butene $\xrightarrow{Cl_2}$ [chloronium ion, attacked by Cl^-] $\longrightarrow$ [2,3-dichlorobutane, shown in two equivalent projections, $\equiv$]

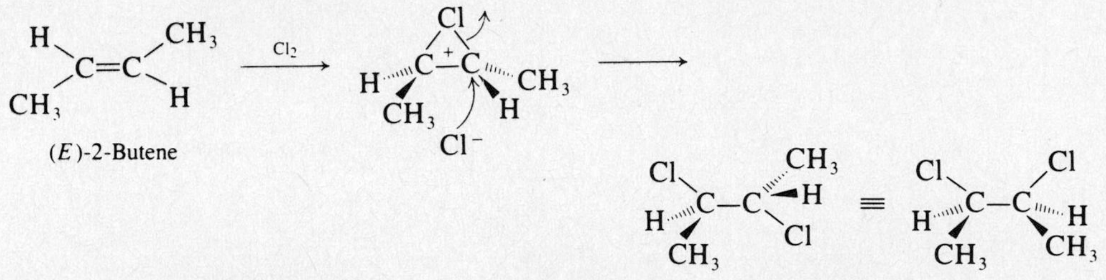

B-1. (*a*) **B-2.** (*c*) **B-3.** (*d*) **B-4.** (*b*)
B-5. (*d*) **B-6.** (*b*) **B-7.** (*d*) **B-8.** (*b*)

CHAPTER 8

A-1. (*a*) 1 and 2, both achiral; identical
(*b*) 3 and 4, both chiral; enantiomers
(*c*) 5 chiral, 6 achiral (meso); diastereomers
(*d*) 7 and 8, both chiral; diastereomers

A-2. 3: (*R*)-2-Chlorobutane; 4: (*S*)-2-Chlorobutane

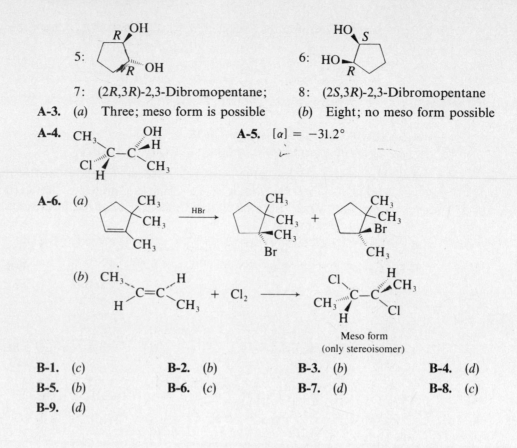

5: 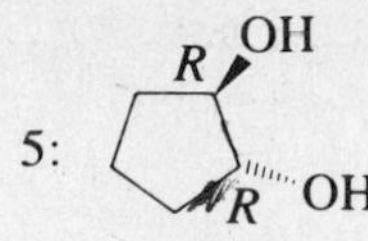

6: HO, S; HO, R (cyclopentane-1,2-diol)

7: (2*R*,3*R*)-2,3-Dibromopentane; 8: (2*S*,3*R*)-2,3-Dibromopentane

A-3. (*a*) Three; meso form is possible (*b*) Eight; no meso form possible

A-4. CH_3(Cl)(H)C–C(OH)(H)CH_3

A-5. $[\alpha] = -31.2°$

A-6. (*a*) 1,5,5-Trimethylcyclopentene $\xrightarrow{HBr}$ 1-bromo-1,2,2-trimethylcyclopentane (two stereoisomers) +

(*b*) $CH_3CH{=}CHCH_3$ (trans) + Cl_2 ⟶ $CH_3CHCl–CHClCH_3$

Meso form
(only stereoisomer)

B-1. (*c*) **B-2.** (*b*) **B-3.** (*b*) **B-4.** (*d*)
B-5. (*b*) **B-6.** (*c*) **B-7.** (*d*) **B-8.** (*c*)
B-9. (*d*)

CHAPTER 9

A-1. (*a*) $CH_3CH_2CH_2CH_2OCH_2CH_3$ (*b*) 1-X-2-methylcyclopentane (X = OTs, Br, I)

(*c*) $CH_3CH(CH_3)CH_2CH_2Cl + NaN_3 \longrightarrow CH_3CH(CH_3)CH_2CH_2N_3$

(*d*) Cyclohexene

A-2. $(CH_3)_2CHS^-Na^+ + CH_3CH_2CH_2Br$

A-3. $CH_3CH_2CH(CH_3)–OH \xrightarrow[\text{pyridine}]{CH_3C_6H_4SO_2Cl} CH_3CH_2CH(CH_3)–OTs \xrightarrow[\text{acetone}]{NaI} I–CH(CH_3)CH_2CH_3$

A-4. Step 1. Ionization to form a secondary carbocation:

$$CH_3C(CH_3)_2–CHClCH_3 \xrightarrow{H_2O} CH_3C(CH_3)_2–\overset{+}{C}HCH_3 + Cl^-$$

Step 2. Rearrangement by methyl migration to form a more stable tertiary carbocation:

$$CH_3C(CH_3)_2–\overset{+}{C}HCH_3 \longrightarrow CH_3\overset{+}{C}(CH_3)–CH(CH_3)CH_3$$

Step 3. Capture of the carbocation by water, followed by deprotonation:

$$CH_3\overset{+}{C}(CH_3)_2{-}CHCH_3 \xrightarrow{H_2O} CH_3C(\overset{+}{O}H_2)(CH_3){-}CH(CH_3)CH_3 \xrightarrow{-H^+} (CH_3)_2C(OH){-}CH(CH_3)_2$$

A-5. $(CH_3)_3CBr \xrightarrow{CH_3OH} (CH_3)_3COCH_3$
S_N1, unimolecular substitution; rate = $k[(CH_3)_3CBr]$

A-6. Sodium iodide is soluble in acetone whereas the by-product of the reaction, sodium bromide, is not. According to Le Chatelier's principle, the reaction will shift in the direction which will replace the component removed from solution, that is, toward product in this case.

B-1. (*b*) **B-2.** (*b*) **B-3.** (*d*) **B-4.** (*c*)
B-5. (*a*) **B-6.** (*a*) **B-7.** (*c*) **B-8.** (*d*)

CHAPTER 10

A-1. (*a*) 4,5-Dimethyl-2-hexyne (*b*) 4-Ethyl-3-propyl-1-heptyne

A-2. (*a*) $CH_3CH_2CH_2C(Cl){=}CH_2$ (*b*) H_2O, H_2SO_4, $HgSO_4$

(*c*) $(CH_3)(H)C{=}C(CH_3)(H)$ (*d*) $(CH_3)_2CHC{\equiv}CH$

A-3. Reaction II is effective; the desired product is formed by an S_N2 reaction:

$$CH_3CH_2CH(CH_3)C{\equiv}C^-Na^+ + CH_3I \longrightarrow CH_3CH_2CH(CH_3)C{\equiv}CCH_3 + NaI$$

Reaction I is not effective owing to E2 elimination from the secondary bromide:

$$CH_3CH_2CH(Br)CH_3 + CH_3C{\equiv}C^-Na^+ \longrightarrow CH_3CH{=}CHCH_3 + CH_3C{\equiv}CH + NaBr$$

A-4. (*a*) $CH_2{=}CHCH_2CH_3 + Br_2 \longrightarrow BrCH_2CH(Br)CH_2CH_3 \xrightarrow[NH_3]{NaNH_2}$

$$HC{\equiv}CCH_2CH_3 \xrightarrow{NaNH_2} Na^{+\,-}C{\equiv}CCH_2CH_3 \xrightarrow{CH_3CH_2Br} CH_3CH_2C{\equiv}CCH_2CH_3$$

(*b*) $HC{\equiv}CH \xrightarrow{NaNH_2} HC{\equiv}C^-Na^+ \xrightarrow{(CH_3)_2CHCH_2Br} HC{\equiv}CCH_2CH(CH_3)_2$

$$HC{\equiv}CCH_2CH(CH_3)_2 \xrightarrow[HgSO_4]{H_2O,\ H_2SO_4} CH_3\overset{O}{\overset{\|}{C}}CH_2CH(CH_3)_2$$

A-5. $(CH_3)(H)C{=}C(H)(CH_2CH_2CH_2CH_3)$ (*E*)-2-Heptene

B-1. (*c*) **B-2.** (*a*) **B-3.** (*a*) **B-4.** (*d*)
B-5. (*b*) **B-6.** (*b*)

CHAPTER 11

A-1. $CH_2{=}CHCH_2CH{=}CH_2$ $CH_2{=}CHCH{=}CHCH_3$ (conjugated)

$CH_2{=}CHC(CH_3){=}CH_2$ (Conjugated)

$CH_2{=}C{=}CHCH_2CH_3$, $CH_2{=}C{=}C(CH_3)_2$, $CH_3CH{=}C{=}CHCH_3$ } Allenes

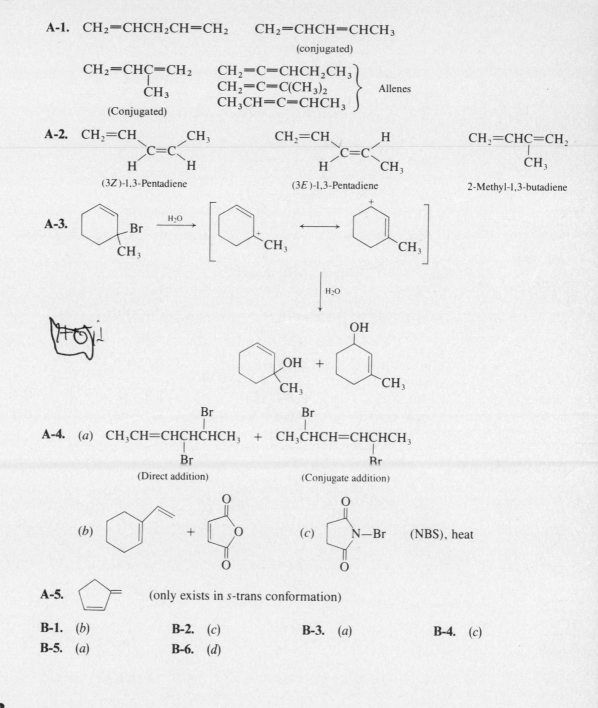

A-2. 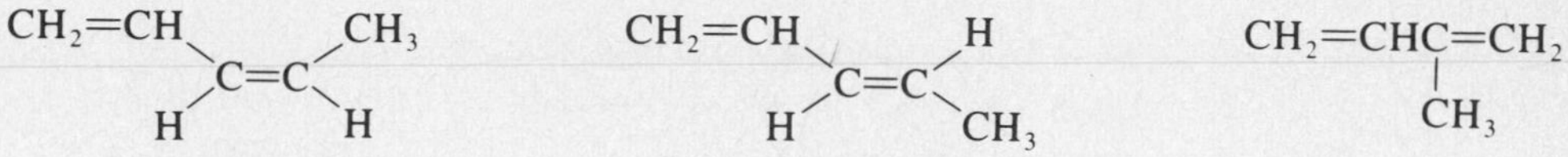 (3*Z*)-1,3-Pentadiene; (3*E*)-1,3-Pentadiene; $CH_2{=}CHC(CH_3){=}CH_2$ 2-Methyl-1,3-butadiene

A-3. 1-Bromo-1-methyl-2-cyclohexene $\xrightarrow{H_2O}$ [allylic carbocation $\longleftrightarrow$ resonance form] $\xrightarrow{H_2O}$ 1-methyl-2-cyclohexen-1-ol + 3-methyl-2-cyclohexen-1-ol

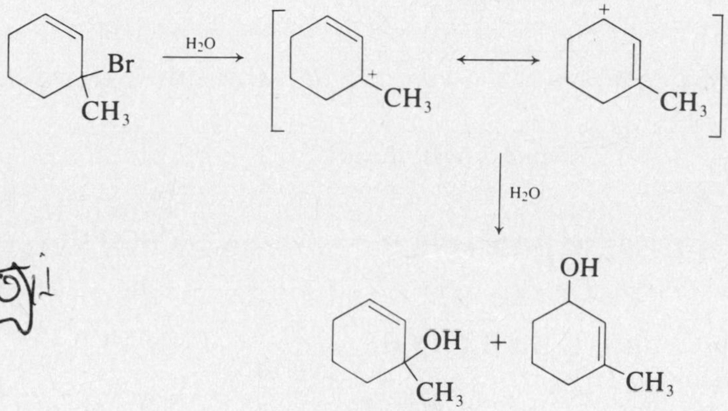

A-4. (*a*) $CH_3CH{=}CHCH(Br)CH(Br)CH_3$ (Direct addition) + $CH_3CH(Br)CH{=}CHCH(Br)CH_3$ (Conjugate addition)

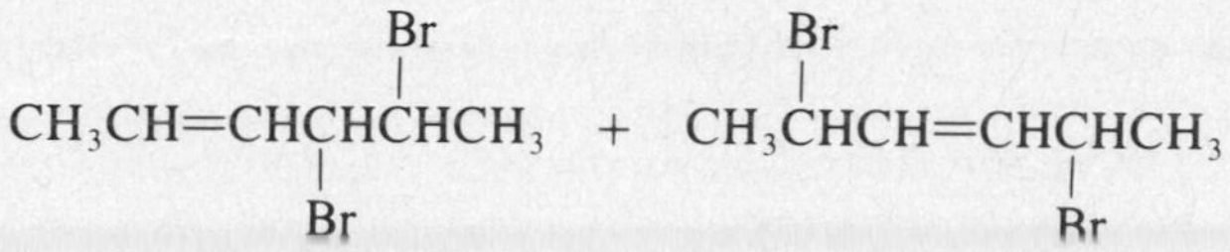

(*b*) 1-vinylcyclohexene + maleic anhydride (*c*) N-bromosuccinimide (NBS), heat

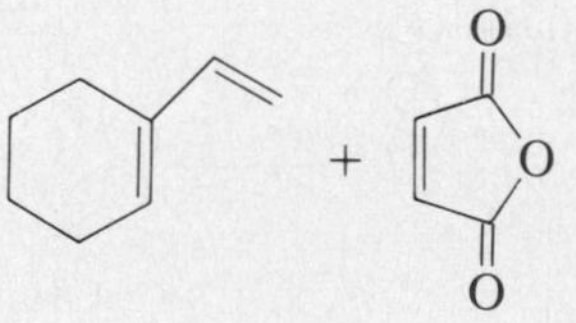

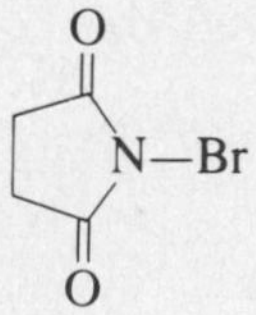

A-5. 3-methylenecyclopentene (only exists in *s*-trans conformation)

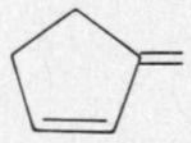

B-1. (*b*) **B-2.** (*c*) **B-3.** (*a*) **B-4.** (*c*)
B-5. (*a*) **B-6.** (*d*)

CHAPTER 12

A-1. (*a*) *m*-Bromotoluene (*b*) 2-Chloro-3-phenylbutane
(*c*) *o*-Chloroacetophenone (*d*) 2,4-Dinitrophenol

A-2. (*a*) m-chlorobenzoic acid (CO_2H, Cl) (*b*) p-nitroanisole (OCH_3, NO_2) (*c*) 2,4-dimethylaniline (NH_2, CH_3, CH_3)

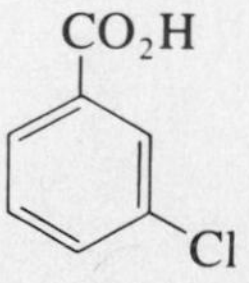

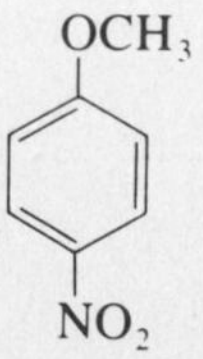

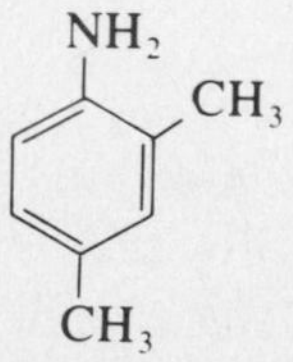

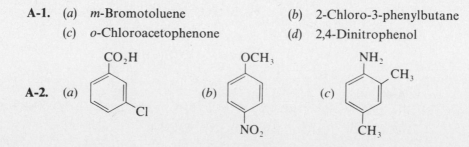

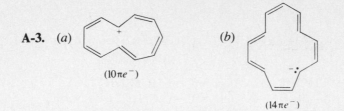

A-3. (*a*) 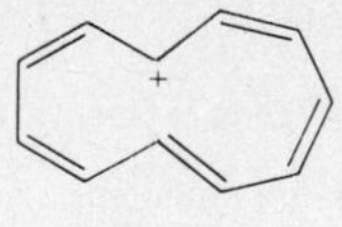 ($10\pi e^-$) (*b*) 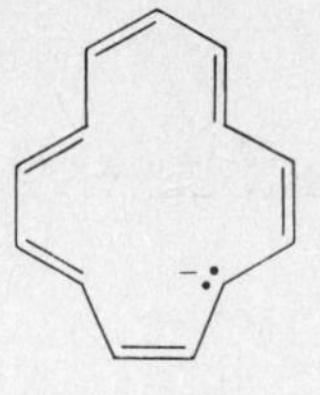 ($14\pi e^-$)

A-4. Eight π electrons. No, the substance is not aromatic.

A-5.

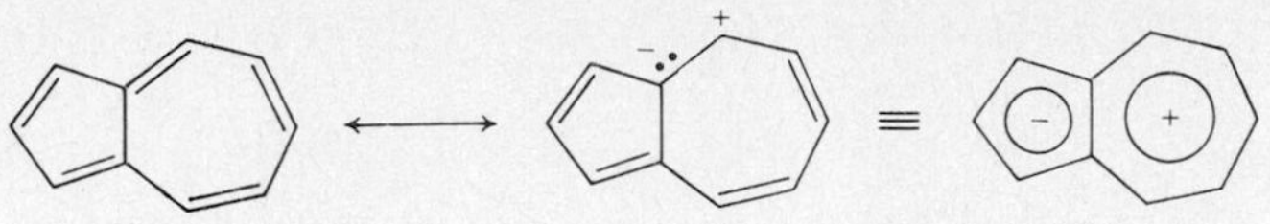

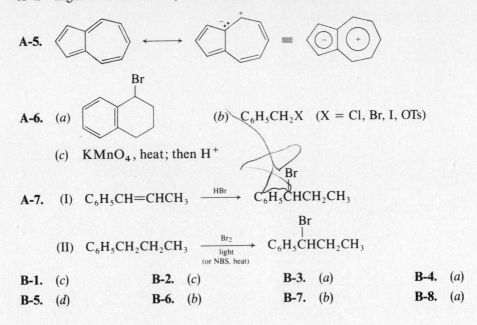

A-6. (*a*) Br (*b*) $C_6H_5CH_2X$ (X = Cl, Br, I, OTs)

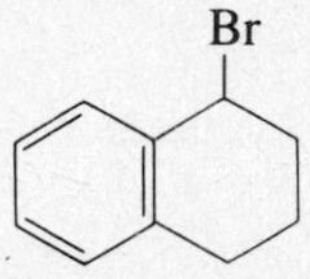

(*c*) $KMnO_4$, heat; then H^+

A-7. (I) $C_6H_5CH{=}CHCH_3 \xrightarrow{HBr} C_6H_5CHBrCH_2CH_3$

(II) $C_6H_5CH_2CH_2CH_3 \xrightarrow[\text{light (or NBS, heat)}]{Br_2} C_6H_5CHBrCH_2CH_3$

B-1. (*c*) **B-2.** (*c*) **B-3.** (*a*) **B-4.** (*a*)
B-5. (*d*) **B-6.** (*b*) **B-7.** (*b*) **B-8.** (*a*)

CHAPTER 13

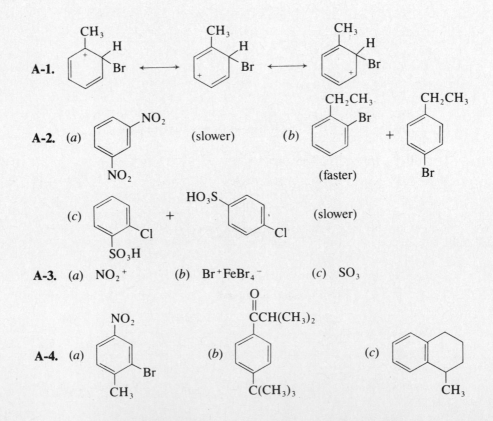

A-1. CH_3, H, Br ⟷ CH_3, H, Br ⟷ CH_3, H, Br

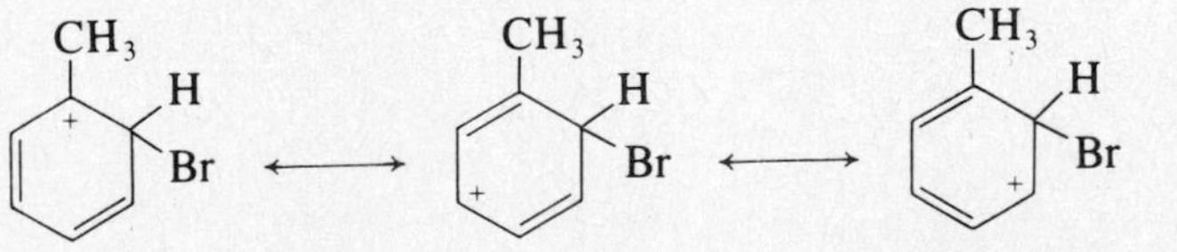

A-2. (*a*) NO_2, NO_2 (slower) (*b*) CH_2CH_3, Br + CH_2CH_3, Br (faster)

(*c*) Cl, SO_3H + HO_3S, Cl (slower)

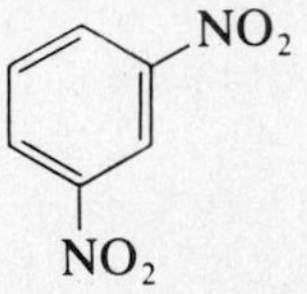

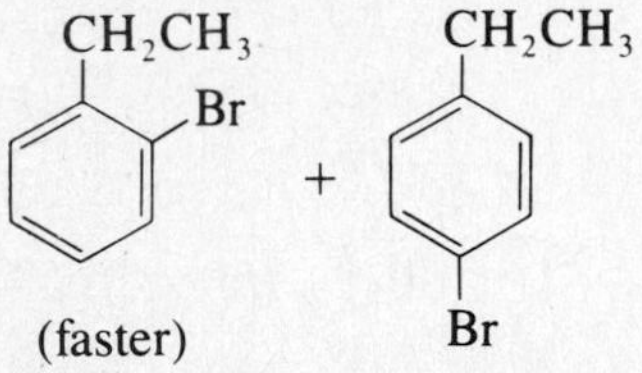

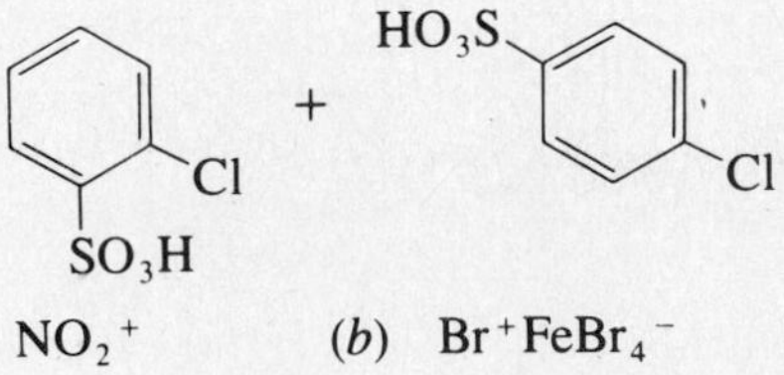

A-3. (*a*) NO_2^+ (*b*) $Br^+FeBr_4^-$ (*c*) SO_3

A-4. (*a*) NO_2, Br, CH_3 (*b*) O, $\overset{\|}{C}CH(CH_3)_2$, $C(CH_3)_3$ (*c*) CH_3

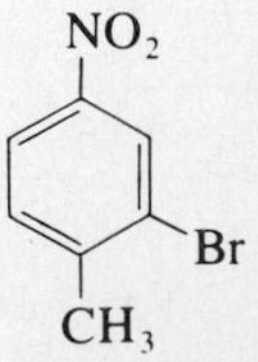

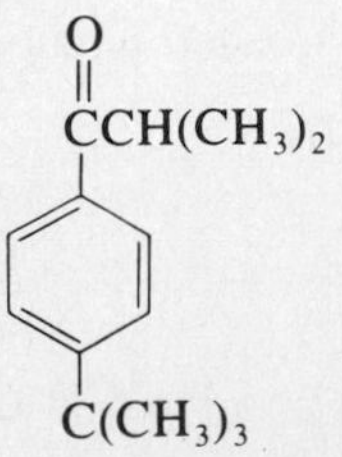

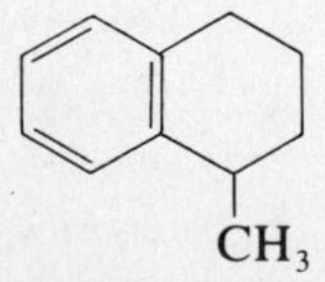

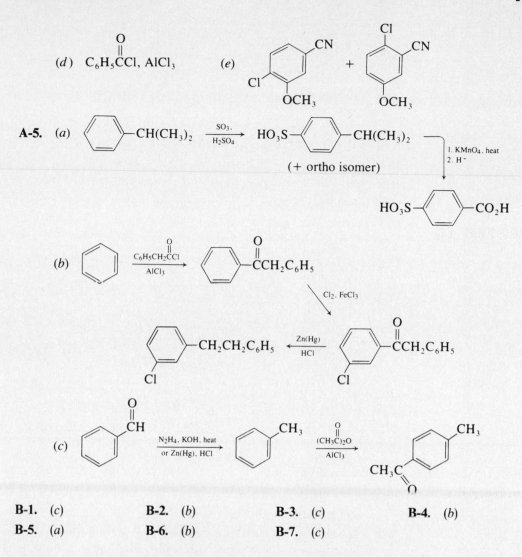

(*d*) $C_6H_5\overset{O}{\overset{\|}{C}}Cl$, $AlCl_3$ (*e*) 4-chloro-3-methoxybenzonitrile + 2-chloro-5-methoxybenzonitrile (Cl, CN, OCH_3 substituted rings)

A-5. (*a*) $C_6H_5CH(CH_3)_2 \xrightarrow[H_2SO_4]{SO_3,} HO_3S\text{–}C_6H_4\text{–}CH(CH_3)_2$ (+ ortho isomer) $\xrightarrow[2.\ H^+]{1.\ KMnO_4,\ heat} HO_3S\text{–}C_6H_4\text{–}CO_2H$

(*b*) $C_6H_6 \xrightarrow[AlCl_3]{C_6H_5CH_2\overset{O}{\overset{\|}{C}}Cl} C_6H_5\overset{O}{\overset{\|}{C}}CH_2C_6H_5 \xrightarrow{Cl_2,\ FeCl_3}$ (*m*-Cl)$C_6H_4\overset{O}{\overset{\|}{C}}CH_2C_6H_5 \xrightarrow[HCl]{Zn(Hg)}$ (*m*-Cl)$C_6H_4CH_2CH_2C_6H_5$

(*c*) $C_6H_5\overset{O}{\overset{\|}{C}}H \xrightarrow[\text{or } Zn(Hg),\ HCl]{N_2H_4,\ KOH,\ heat} C_6H_5CH_3 \xrightarrow[AlCl_3]{(CH_3\overset{O}{\overset{\|}{C}})_2O} p\text{-}CH_3\overset{O}{\overset{\|}{C}}\text{–}C_6H_4\text{–}CH_3$

B-1. (*c*) **B-2.** (*b*) **B-3.** (*c*) **B-4.** (*b*)
B-5. (*a*) **B-6.** (*b*) **B-7.** (*c*)

CHAPTER 14

A-1. 1: 6.10 ppm 2: 957 Hz
3: 60 MHz 4: 0.00 ppm

A-2. (*a*) Two signals $\underset{a}{BrCH_2}\underset{b}{CH_2}\underset{a}{CH_2Br}$
a: triplet b: pentet

(*b*) Two signals $\underset{a}{CH_3}\underset{b}{CH_2}\overset{Cl}{\underset{Cl}{C}}\underset{b}{CH_2}\underset{a}{CH_3}$
a: triplet b: quartet

(*c*) Three signals, all singlets

A-3. $CH_3\overset{O}{\overset{\|}{C}}OC(CH_3)_3$ A $CH_3O\overset{O}{\overset{\|}{C}}C(CH_3)_3$ B

A-4. (*a*) $C_6H_5CH_2\overset{O}{\overset{\|}{C}}CH_2CH_3$ (*b*) $(CH_3)_2\overset{HO}{C}\text{—}\overset{OH}{C}(CH_3)_2$

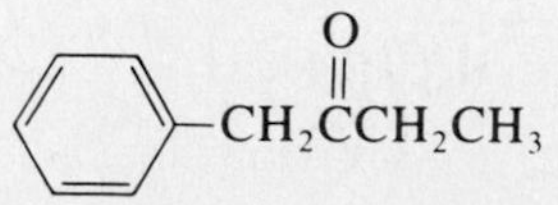

(*c*) $CH(\overset{O}{\overset{\|}{C}}OCH_2CH_3)_3$ (*d*) $(CH_3)_2\overset{OH}{\overset{|}{C}}-C\equiv N$

A-5. (a) Seven signals 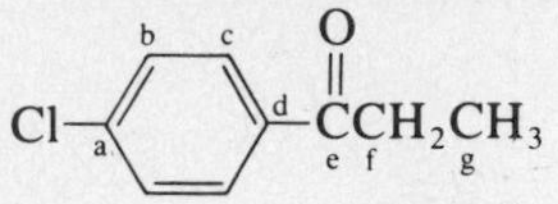

(*b*) a: singlet b: doublet c: doublet d: singlet
e: singlet f: triplet g: quartet

B-1. (*b*) **B-2.** (*d*) **B-3.** (*b*) **B-4.** (*b*)
B-5. (*a*) **B-6.** (*c*) **B-7.** (*a*) **B-8.** (*a*)

CHAPTER 15

A-1. (*a*) 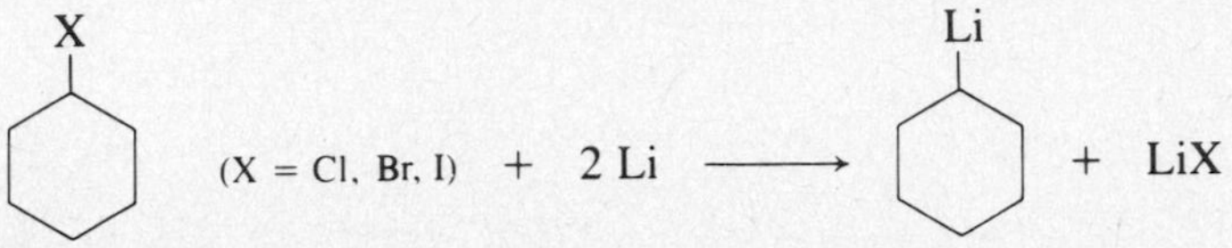

(*b*) $(CH_3)_3CBr + Mg \longrightarrow (CH_3)_3CMgBr$

(*c*) $2\ C_6H_5CH_2Li + \underset{(X\ =\ Cl,\ Br,\ I)}{CuX} \longrightarrow (C_6H_5CH_2)_2CuLi + LiX$

A-2. (*a*) $(C_6H_5)_2\overset{OH}{\overset{|}{C}}CH_3$ (*b*) $(CH_3)_2CHCH_2D$

(*c*) 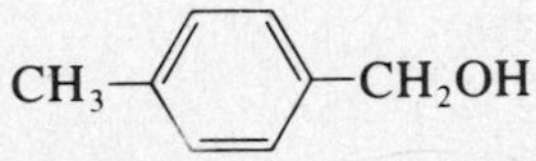 (*d*) 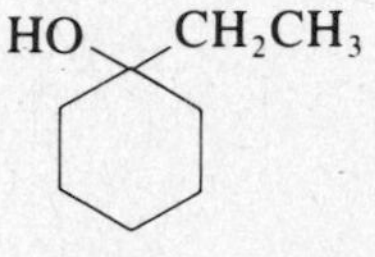

A-3. (*a*) $(CH_3CH_2CH_2)_2CuLi$ (*b*) $(CH_3)_2CHMgX$ (X = Cl, Br, I)
(*c*) CH_2I_2, Zn(Cu)

A-4. Solvents A, B, and E are suitable; they are all ethers. Solvents C and F have acidic hydrogens and will react with a Grignard reagent. Solvent D is an ester which will react with a Grignard reagent.

A-5. $CH_3(CH_2)_3OH \xrightarrow{PBr_3} CH_3(CH_2)_3Br \xrightarrow{2\,Li} CH_3(CH_2)_3Li + LiBr$

$2\ CH_3(CH_2)_3Li + CuBr \longrightarrow (C_4H_9)_2CuLi$

$(C_4H_9)_2CuLi \xrightarrow{CH_3(CH_2)_3Br} CH_3(CH_2)_6CH_3$

A-6. I. $(CH_3)_2CH\overset{O}{\overset{\|}{C}}CH_3 + CH_3MgBr$

II. $CH_3\overset{O}{\overset{\|}{C}}CH_3 + (CH_3)_2CHMgBr$

III. $(CH_3)_2CHCO_2CH_3 + 2\ CH_3MgBr$

$\xrightarrow[2.\ H_3O^+]{1.\ ether} (CH_3)_2CH\overset{OH}{\overset{|}{C}}(CH_3)_2$

A-7. $C_6H_5CH_2CH_3 \xrightarrow[\text{peroxides, heat}]{NBS} C_6H_5\overset{Br}{\overset{|}{C}}HCH_3 \xrightarrow[ether]{Mg} C_6H_5\overset{CH_3}{\overset{|}{C}}HMgBr$

$C_6H_5\overset{CH_3}{\overset{|}{C}}HMgBr \xrightarrow[2.\ H_3O^+]{1.\ CH_3\overset{O}{\overset{\|}{C}}H} C_6H_5\overset{CH_3}{\overset{|}{C}}H\underset{OH}{\underset{|}{C}}HCH_3$

B-1. (*c*) **B-2.** (*a*) **B-3.** (*b*) **B-4.** (*a*)
B-5. (*b*) **B-6.** (*c*)

CHAPTER 16

A-1. (*a*) O (cyclohexanone) (*b*) $C_6H_5CO_2CH_2CH_3$

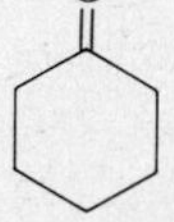

(*c*) 1. B_2H_6; 2. H_2O_2, HO^-
(*d*) $KMnO_4$, H_2O (cold); or OsO_4, $(CH_3)_3COOH$, $(CH_3)_3COH$, HO^-

A-2. (*a*) $C_6H_5CH_2\overset{O}{\overset{\|}{C}}H$ (*b*) $CH_3\overset{O}{\overset{\|}{C}}Cl$, pyridine; or $(CH_3\overset{O}{\overset{\|}{C}})_2O$; or CH_3CO_2H, H^+

(*c*) $(C_6H_5CH_2CH_2)_2O$ (*d*) $K_2Cr_2O_7$, H^+, H_2O

A-3. (I) $(CH_3)_2CHBr + Mg \longrightarrow (CH_3)_2CHMgBr \xrightarrow[\text{2. } H_3O^+]{\text{1. ethylene oxide}} (CH_3)_2CHCH_2CH_2OH$

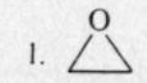

(II) $(CH_3)_2CHCH_2Br + Mg \longrightarrow (CH_3)_2CHCH_2MgBr$

$(CH_3)_2CHCH_2MgBr \xrightarrow[\text{2. } H_3O^+]{\text{1. } H_2C=O} (CH_3)_2CHCH_2CH_2OH$

A-4. (*a*) O, H, O (*b*) HO, OH

(*c*) CH_3, H, C=C, H, CH_3

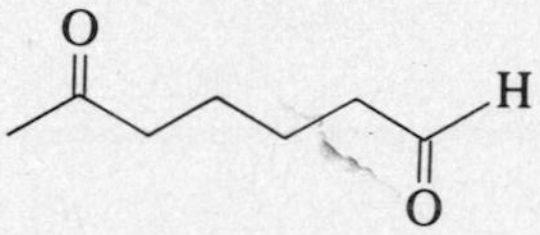

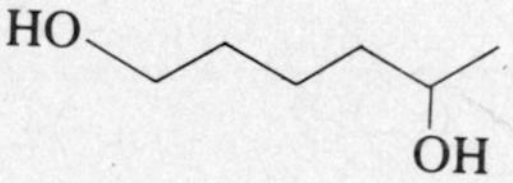

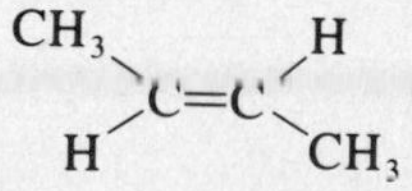

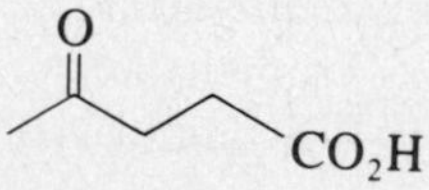

A-5.

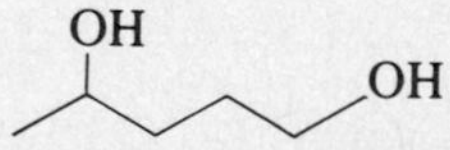

OH, OH — A

O, CO_2H — B

O, CO_2CH_3 — C

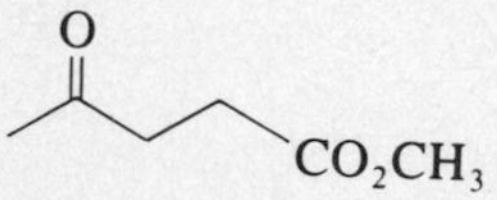

A-6. $C_6H_5CH_3 \xrightarrow[\text{peroxides, heat}]{\text{NBS}} C_6H_5CH_2Br \xrightarrow{\text{Mg}} C_6H_5CH_2MgBr \xrightarrow[\text{2. } H_3O^+]{\text{1. ethylene oxide}} C_6H_5CH_2CH_2CH_2OH$

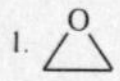

$C_6H_5CH_2CH_2CH_2OH \xrightarrow[H^+, H_2O]{K_2Cr_2O_7} C_6H_5CH_2CH_2CO_2H$

$C_6H_5CH_2CH_2CO_2H \xrightarrow{CH_3CH_2OH, H^+} C_6H_5CH_2CH_2CO_2CH_2CH_3$ (D)

B-1. (*b*) **B-2.** (*d*) **B-3.** (*c*) **B-4.** (*c*)
B-5. (*b*) **B-6.** (*b*) **B-7.** (*a*)

CHAPTER 17

A-1. $CH_3OCH_2CH_2CH_3$ Methyl propyl ether
$CH_3OCH(CH_3)_2$ Isopropyl methyl ether
$CH_3CH_2OCH_2CH_3$ Diethyl ether

A-2. (*a*) 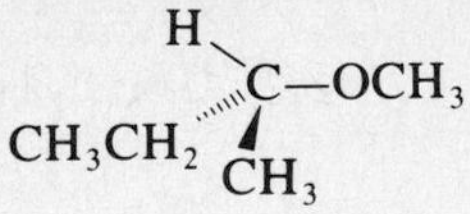 (*b*) 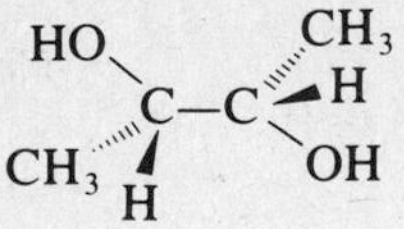

(*c*) $C_6H_5\overset{I}{C}HCH_2OH$ (*d*) 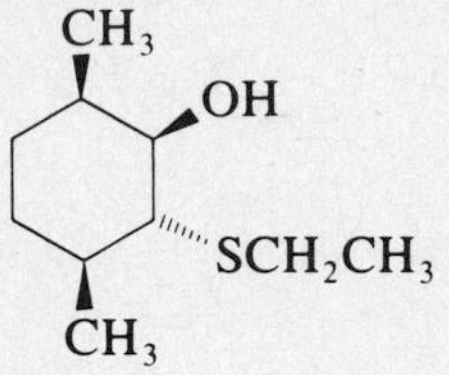

A-3. (*a*) 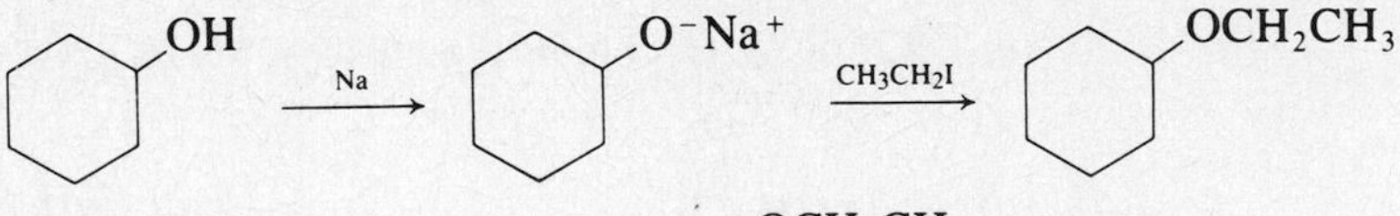
OH
Na
O^-Na^+
CH_3CH_2I
OCH_2CH_3

(*b*) 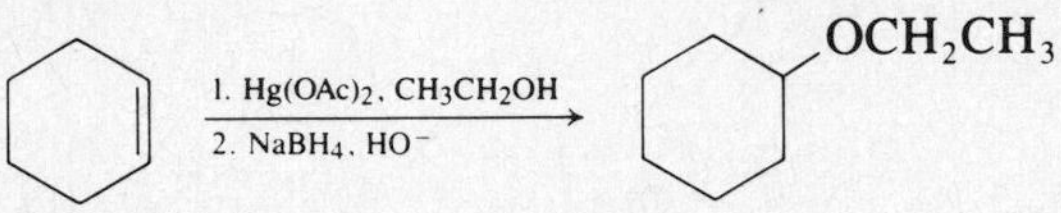
1. $Hg(OAc)_2$, CH_3CH_2OH
2. $NaBH_4$, HO^-
OCH_2CH_3

A-4. 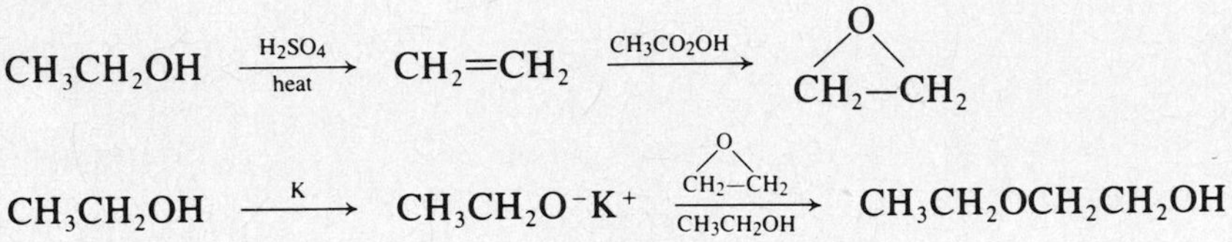
CH_3CH_2OH
H_2SO_4
heat
$CH_2{=}CH_2$
CH_3CO_2OH
O
$CH_2{-}CH_2$
CH_3CH_2OH
K
$CH_3CH_2O^-K^+$
CH_3CH_2OH
$CH_3CH_2OCH_2CH_2OH$

A-5. A: 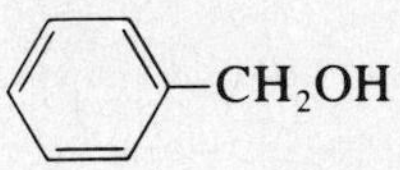 B: 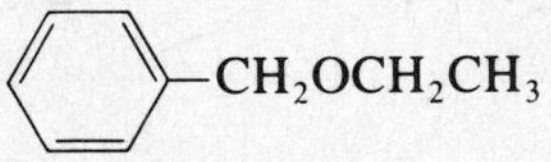

B-1. (*a*) **B-2.** (*a*) **B-3.** (*b*) **B-4.** (*d*)
B-5. (*d*)

CHAPTER 18

A-1. (*a*) 3,4-Dimethylhexanal
(*b*) 2,2,5-Trimethylhexan-3-one
(*c*) *trans*-4-Bromo-2-methylcyclohexanone

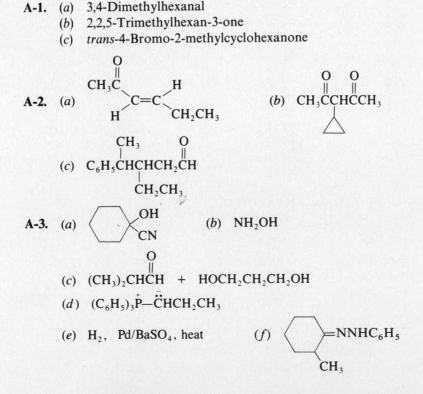

A-2. (*a*) 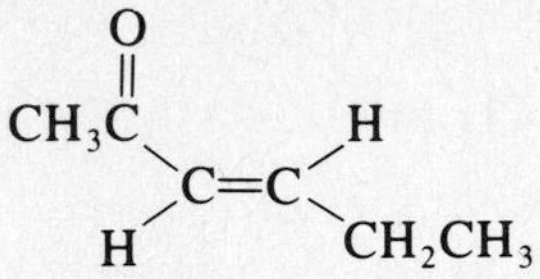 (*b*) 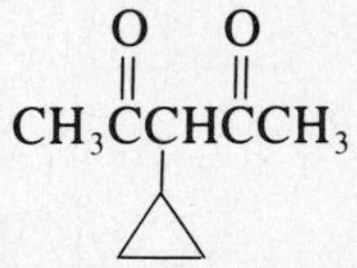

(*c*) 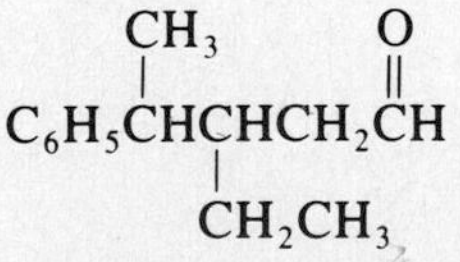

A-3. (*a*) 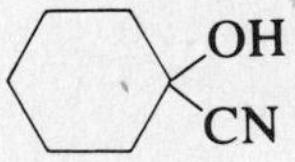 (*b*) NH_2OH

(*c*) $(CH_3)_2CH\overset{O}{\overset{\|}{C}}H$ + $HOCH_2CH_2CH_2OH$

(*d*) $(C_6H_5)_3\overset{+}{P}{-}\overset{-}{\ddot{C}}HCH_2CH_3$

(*e*) H_2, $Pd/BaSO_4$, heat (*f*) 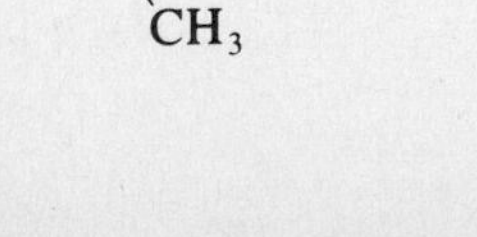

A-4. $(CH_3)_2C{=}C(CH_3)CH_2CH_3$ (A) $\quad CH_3CH_2\overset{O}{\overset{\|}{C}}CH_2CH_3$ (B)

A-5. (*a*) $CH_3CH_2I + (C_6H_5)_3P \longrightarrow (C_6H_5)_3\overset{+}{P}{-}CH_2CH_3\ I^-$ $\xrightarrow{C_4H_9Li}$ $(C_6H_5)_3\overset{+}{P}{-}\ddot{\overset{-}{C}}HCH_3$

$(CH_3)_2C{=}O + (C_6H_5)_3\overset{+}{P}{-}\overset{-}{C}HCH_3 \longrightarrow$

$(CH_3)_2C{=}CHCH_3 \xrightarrow{CH_3CO_2OH} (CH_3)_2\overset{O}{C{-}}CHCH_3$ (epoxide)

(*b*) 2-methyl-4-hydroxycyclohexanone $\xrightarrow[H^+]{HOCH_2CH_2OH}$ ethylene ketal (OH) $\xrightarrow[CH_2Cl_2]{(C_5H_5N)_2CrO_3}$ ketal ketone $\xrightarrow[2.\ H_3O^+]{1.\ CH_3MgI}$ 4-hydroxy-2,4-dimethylcyclohexanone (HO, CH_3)

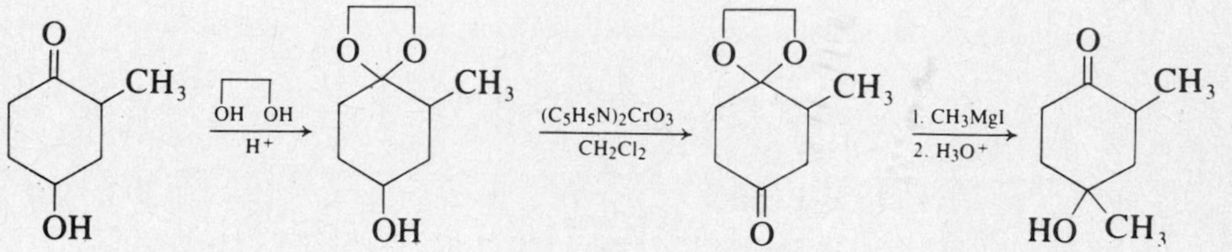

B-1. (*a*) **B-2.** (*b*) **B-3.** (*b*) **B-4.** (*b*)
B-5. (*a*)

CHAPTER 19

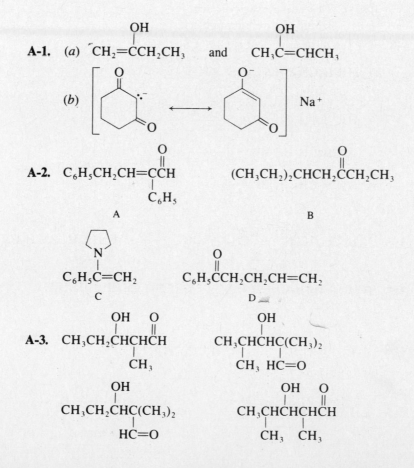

A-1. (*a*) $CH_2{=}\overset{OH}{C}CH_2CH_3$ and $CH_3\overset{OH}{C}{=}CHCH_3$

(*b*) [cyclohexane-1,3-dione carbanion $\longleftrightarrow$ enolate (O^-)] Na^+

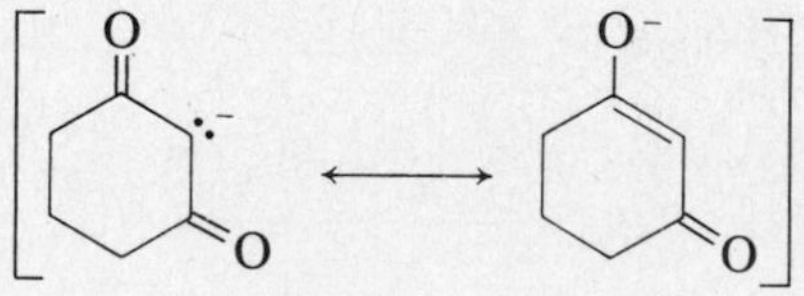

A-2. $C_6H_5CH_2CH{=}\overset{O}{\overset{\|}{C}}CH$ with C_6H_5 on C (A) $\quad (CH_3CH_2)_2CHCH_2\overset{O}{\overset{\|}{C}}CH_2CH_3$ (B)

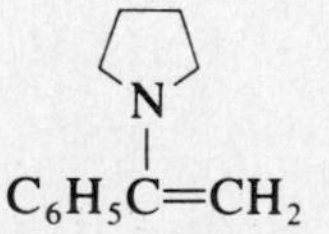

$C_6H_5C(N\text{-pyrrolidinyl}){=}CH_2$ (C) $\quad C_6H_5\overset{O}{\overset{\|}{C}}CH_2CH_2CH{=}CH_2$ (D)

A-3. $CH_3CH_2\overset{OH}{C}H\underset{CH_3}{C}H\overset{O}{\overset{\|}{C}}H$ $\quad CH_3\overset{OH}{C}H\underset{CH_3}{C}H\underset{HC=O}{C}(CH_3)_2$

$CH_3CH_2\overset{OH}{C}H\underset{HC=O}{C}(CH_3)_2$ $\quad CH_3\overset{OH}{C}H\underset{CH_3}{C}H\underset{CH_3}{C}H\overset{O}{\overset{\|}{C}}H$

A-4. $CH_3CH_2OH \xrightarrow[CH_2Cl_2]{(C_5H_5N)_2CrO_3} CH_3\overset{O}{\overset{\|}{C}}H$

$2\ CH_3\overset{O}{\overset{\|}{C}}H \xrightarrow{NaOH} CH_3\overset{OH}{\overset{|}{C}}HCH_2\overset{O}{\overset{\|}{C}}H \xrightarrow[\text{or 1. LiAlH}_4\text{ 2. H}_2\text{O}]{NaBH_4,\ CH_3OH} CH_3\overset{OH}{\overset{|}{C}}HCH_2CH_2OH$

A-5. 1,3-cyclopentanedione + $CH_3CH_2I \xrightarrow{K_2CO_3}$ 2-ethyl-1,3-cyclopentanedione (2-CH_2CH_3) $\xrightarrow{CH_2{=}CH\overset{O}{\overset{\|}{C}}CH_3,\ KOH}$ 2-ethyl-2-($CH_2CH_2\overset{O}{\overset{\|}{C}}CH_3$)-1,3-cyclopentanedione

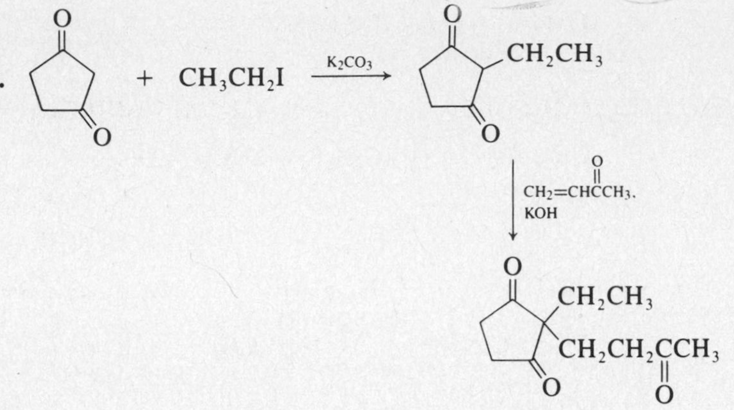

B-1. (*a*) **B-2.** (*b*) **B-3.** (*b*) **B-4.** (*a*)
B-5. (*b*) **B-6.** (*c*)

CHAPTER 20

A-1. (*a*) 4-Methyl-5-phenylhexanoic acid
(*b*) Cyclohexanecarboxylic acid
(*c*) 3-Bromo-2-ethylbutanoic acid

A-2. 4-Phenylbutanoic acid is $C_6H_5CH_2CH_2CH_2CO_2H$.

$C_6H_5CH_2CH_2CH(CO_2H)_2 \xrightarrow{heat}$

$C_6H_5CH_2CH_2CH_2\overset{O}{\overset{\|}{C}}CH_3 \xrightarrow[2.\ H^+]{1.\ X_2,\ HO^-\ \ (X = Cl, Br, I)}$

$C_6H_5CH_2CH_2CH_2Br \xrightarrow[2.\ H^+,\ H_2O,\ heat]{1.\ CN^-}$

$C_6H_5CH_2CH_2CH_2Br \xrightarrow[2.\ CO_2\ \ 3.\ H_3O^+]{1.\ Mg}$

A-3. $C_6H_5CH_2CO_2H + CH_3CH_2OH \xrightarrow{H^+(cat)} C_6H_5CH_2\overset{O}{\overset{\|}{C}}OCH_2CH_3 + H_2O$

A-4. $(CH_3)_2CHCH_2\overset{Br}{\overset{|}{C}}HCO_2H$ (A) $(CH_3)_2CHCH_2\overset{CN}{\overset{|}{C}}HCO_2H$ (B)

$C_6H_5\overset{CH_3}{\overset{|}{C}}HCO_2H$ (C) $C_6H_5\overset{O}{\overset{\|}{C}}\underset{CO_2H}{\underset{|}{C}}(CH_3)_2$ (D)

A-5. $CH_3CH_2\underset{Br}{\underset{|}{C}}HCO_2H$

B-1. (*b*) **B-2.** (*a*) **B-3.** (*c*) **B-4.** (*c*)
B-5. (*d*) **B-6.** (*c*)

CHAPTER 21

A-1. (*a*) Propyl butanoate (*b*) *N*-Methylbenzamide
(*c*) 4-Methylpentanoyl chloride

A-2. (*a*) $C_6H_5\overset{O}{\overset{\|}{C}}O\overset{O}{\overset{\|}{C}}C_6H_5$ (*b*) $CH_3\overset{O}{\overset{\|}{C}}NH\underset{CH_3}{\underset{|}{C}}HCH_2CH_3$ (*c*) $C_6H_5\overset{O}{\overset{\|}{C}}OC_6H_5$

A-3. (*a*) $SOCl_2$ (*b*) Br_2, NaOH, H_2O

A-4. (*a*) CH_3CO_2H + cyclohexanol (cyclohexyl–OH) (*b*) $C_6H_5\overset{O}{\overset{\|}{C}}O$–cyclopentyl

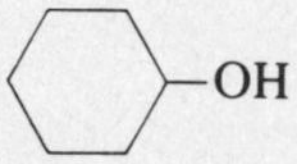

(*c*) benzene ring with ortho $\overset{O}{\overset{\|}{C}}OCH_2CH_3$ and $\underset{O}{\underset{\|}{C}}OH$

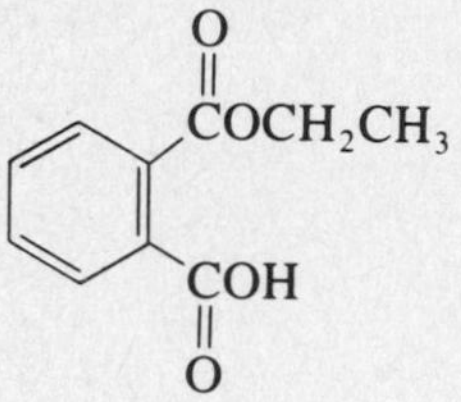

(*d*) $CH_3CH_2\overset{O}{\overset{\|}{C}}N(CH_3)_2$ + CH_3CH_2OH

(*e*) CH_3–C_6H_4–CO_2H (para) + $CH_3NH_3^+$ HSO_4^-

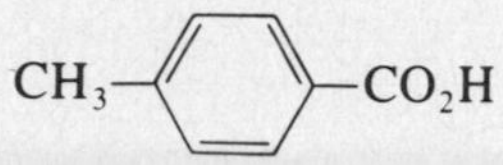

A-5. C_6H_5–C(OH) in a ring –O–CH_2–CH_2–N(CH_3)– from $C_6H_5\overset{O}{\overset{\|}{C}}$–O–$CH_2$–$CH_2$–$\ddot{N}H$–$CH_3$ (nitrogen lone pair attacks the carbonyl carbon)

A-6. trans-2-methylcyclohexanecarboxylic acid (CO_2H, CH_3) $\xrightarrow{SOCl_2}$ A (acyl chloride, $\overset{O}{\overset{\|}{C}}Cl$, CH_3) $\xrightarrow{NH_3}$ B (amide, $\overset{O}{\overset{\|}{C}}NH_2$, CH_3)

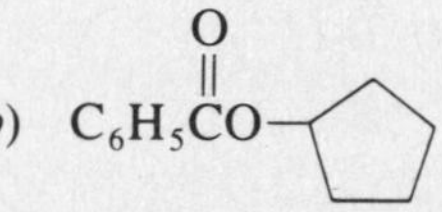

B $\xrightarrow[heat]{P_4O_{10}}$ D (nitrile, $C\equiv N$, CH_3)

B $\xrightarrow[H_2O]{Br_2,\ NaOH}$ C (trans-2-methylcyclohexylamine, NH_2, CH_3)

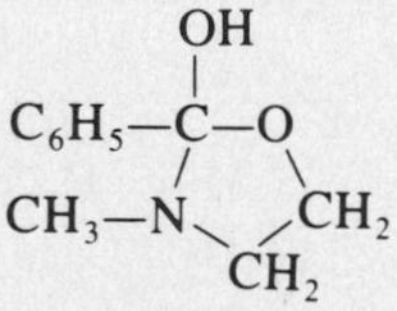

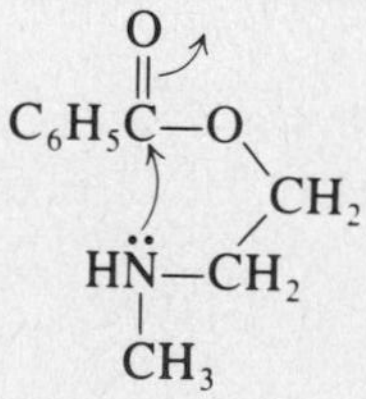

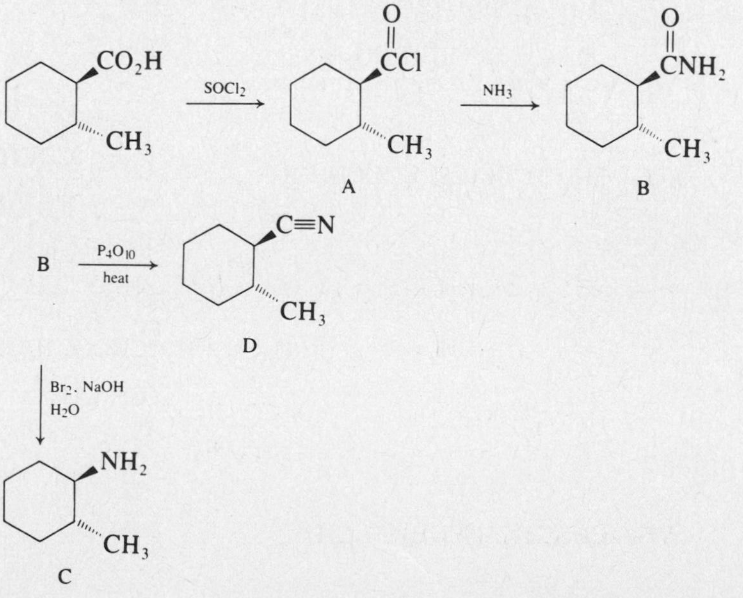

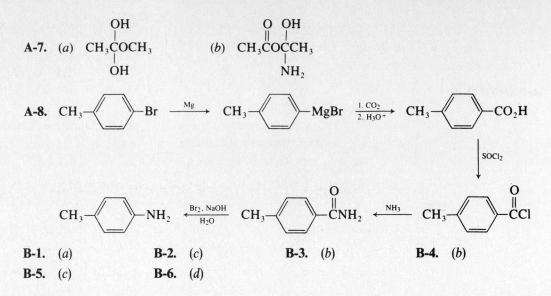

A-7. (*a*) $CH_3C(OH)_2CH_3$ (*b*) $CH_3C(=O)OC(OH)(NH_2)CH_3$

A-8. $CH_3\text{-}C_6H_4\text{-}Br \xrightarrow{Mg} CH_3\text{-}C_6H_4\text{-}MgBr \xrightarrow[2.\ H_3O^+]{1.\ CO_2} CH_3\text{-}C_6H_4\text{-}CO_2H \xrightarrow{SOCl_2} CH_3\text{-}C_6H_4\text{-}CCl(=O) \xrightarrow{NH_3} CH_3\text{-}C_6H_4\text{-}CNH_2(=O) \xrightarrow[H_2O]{Br_2,\ NaOH} CH_3\text{-}C_6H_4\text{-}NH_2$

B-1. (*a*)   **B-2.** (*c*)   **B-3.** (*b*)   **B-4.** (*b*)

B-5. (*c*)   **B-6.** (*d*)

CHAPTER 22

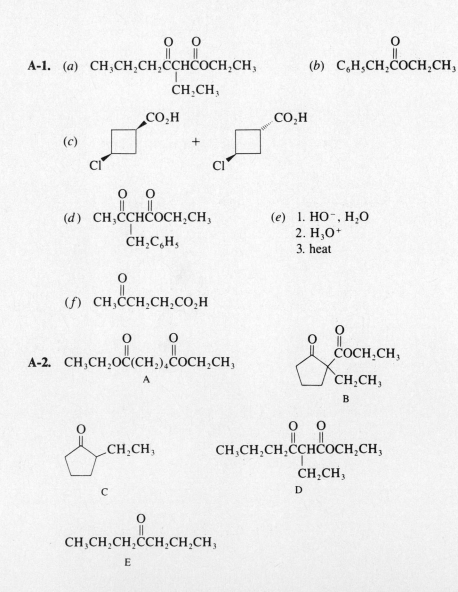

A-1. (*a*) $CH_3CH_2CH_2C(=O)CH(CH_2CH_3)C(=O)OCH_2CH_3$ (*b*) $C_6H_5CH_2C(=O)OCH_2CH_3$

(*c*) cis-3-chlorocyclobutanecarboxylic acid (CO_2H, Cl wedged) + trans-3-chlorocyclobutanecarboxylic acid (CO_2H hashed, Cl wedged)

(*d*) $CH_3C(=O)CH(CH_2C_6H_5)C(=O)OCH_2CH_3$ (*e*) 1. HO^-, H_2O
2. H_3O^+
3. heat

(*f*) $CH_3C(=O)CH_2CH_2CO_2H$

A-2. $CH_3CH_2OC(=O)(CH_2)_4C(=O)OCH_2CH_3$ (A)

2-oxocyclopentane with $C(=O)OCH_2CH_3$ and CH_2CH_3 on C-2 (B)

2-ethylcyclopentanone, CH_2CH_3 (C)

$CH_3CH_2CH_2C(=O)CH(CH_2CH_3)C(=O)OCH_2CH_3$ (D)

$CH_3CH_2CH_2C(=O)CH_2CH_2CH_3$ (E)

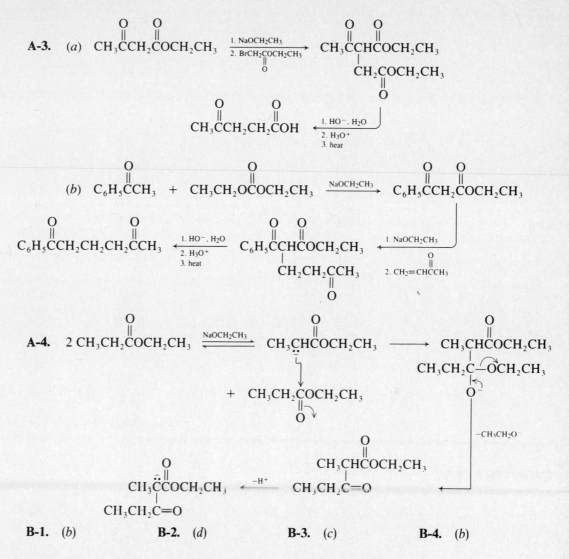

A-3. (*a*) $CH_3\overset{O}{\overset{\|}{C}}CH_2\overset{O}{\overset{\|}{C}}OCH_2CH_3 \xrightarrow[\text{2. } BrCH_2\overset{O}{\overset{\|}{C}}OCH_2CH_3]{\text{1. } NaOCH_2CH_3} CH_3\overset{O}{\overset{\|}{C}}\underset{CH_2\underset{O}{\underset{\|}{C}}OCH_2CH_3}{\underset{|}{C}H}\overset{O}{\overset{\|}{C}}OCH_2CH_3$

$\xrightarrow[\text{2. } H_3O^+ \text{ 3. heat}]{\text{1. } HO^-, H_2O} CH_3\overset{O}{\overset{\|}{C}}CH_2CH_2\overset{O}{\overset{\|}{C}}OH$

(*b*) $C_6H_5\overset{O}{\overset{\|}{C}}CH_3 + CH_3CH_2O\overset{O}{\overset{\|}{C}}OCH_2CH_3 \xrightarrow{NaOCH_2CH_3} C_6H_5\overset{O}{\overset{\|}{C}}CH_2\overset{O}{\overset{\|}{C}}OCH_2CH_3$

$\xrightarrow[\text{2. } CH_2{=}CH\overset{O}{\overset{\|}{C}}CH_3]{\text{1. } NaOCH_2CH_3} C_6H_5\overset{O}{\overset{\|}{C}}\underset{CH_2CH_2\underset{O}{\underset{\|}{C}}CH_3}{\underset{|}{C}H}\overset{O}{\overset{\|}{C}}OCH_2CH_3 \xrightarrow[\text{2. } H_3O^+ \text{ 3. heat}]{\text{1. } HO^-, H_2O} C_6H_5\overset{O}{\overset{\|}{C}}CH_2CH_2CH_2\overset{O}{\overset{\|}{C}}CH_3$

A-4. $2\ CH_3CH_2\overset{O}{\overset{\|}{C}}OCH_2CH_3 \underset{}{\overset{NaOCH_2CH_3}{\rightleftharpoons}} CH_3\ddot{C}H^-\overset{O}{\overset{\|}{C}}OCH_2CH_3 + CH_3CH_2\overset{}{\underset{O}{\underset{\|}{C}}}OCH_2CH_3 \longrightarrow CH_3CH\overset{O}{\overset{\|}{C}}OCH_2CH_3$ / $CH_3CH_2\underset{O^-}{\underset{|}{C}}{-}OCH_2CH_3$

$\xrightarrow{-CH_3CH_2O^-} CH_3\underset{CH_3CH_2C{=}O}{\underset{|}{C}H}\overset{O}{\overset{\|}{C}}OCH_2CH_3 \xrightarrow{-H^+} CH_3\underset{CH_3CH_2C{=}O}{\underset{|}{\ddot{C}^-}}\overset{O}{\overset{\|}{C}}OCH_2CH_3$

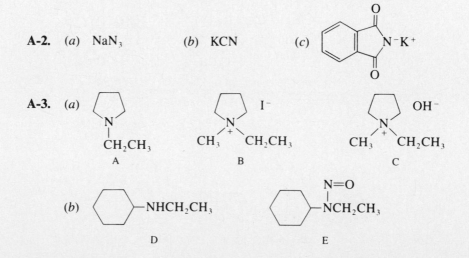

B-1. (*b*) **B-2.** (*d*) **B-3.** (*c*) **B-4.** (*b*)

CHAPTER 23

A-1. (*a*) 1,1-Dimethylpropylamine or 2-methyl-2-butanamine; primary
(*b*) *N*-Methylcyclopentylamine or *N*-methylcyclopentanamine; secondary

A-2. (*a*) NaN_3 (*b*) KCN (*c*) potassium phthalimide (N^-K^+ with two C=O groups on the benzo-fused five-membered ring)

A-3. (*a*) A: N-ethylpyrrolidine (N–CH_2CH_3); B: N-ethyl-N-methylpyrrolidinium iodide (N^+ with CH_3 and CH_2CH_3, I^-); C: N-ethyl-N-methylpyrrolidinium hydroxide (N^+ with CH_3 and CH_2CH_3, OH^-)

(*b*) D: cyclohexyl–$NHCH_2CH_3$; E: cyclohexyl–$\underset{}{N}(N{=}O)CH_2CH_3$

(*c*) F: H_3C, $^{+}N(CH_3)_2$, O^- G: H_3C

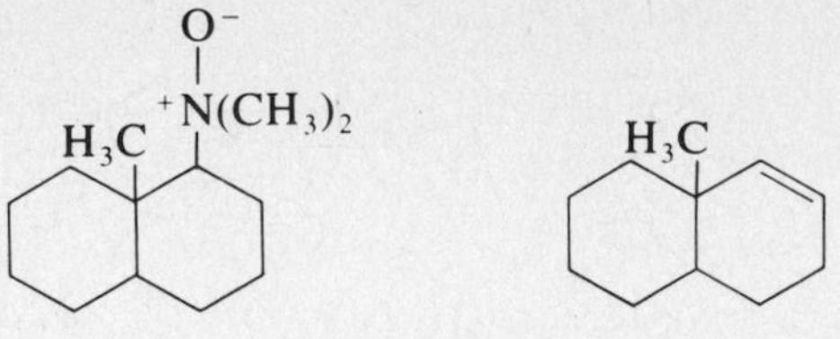

A-4. I undergoes anti elimination:

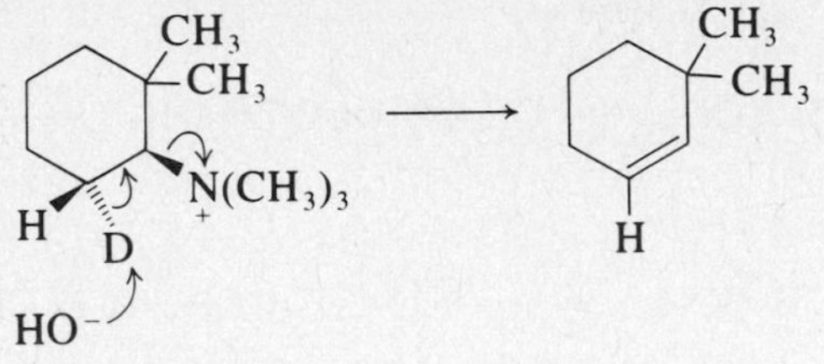

CH_3, CH_3, $\overset{+}{N}(CH_3)_3$, H, D, HO^- ⟶ CH_3, CH_3, H

II undergoes syn elimination:

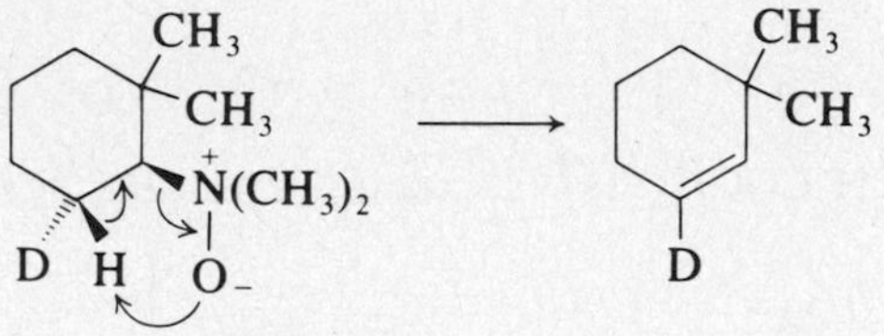

CH_3, CH_3, $\overset{+}{N}(CH_3)_2$, D H O^- ⟶ CH_3, CH_3, D

B-1. (*c*) **B-2.** (*a*) **B-3.** (*d*)

B-4. (*b*) **B-5.** (*d*) **B-6.** (*c*)

CHAPTER 24

A-1. (*a*) $CH_3-C_6H_4-N_2^{+}Cl^-$ (*b*) $CH_3-C_6H_4-Br$

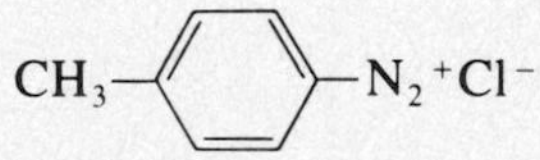

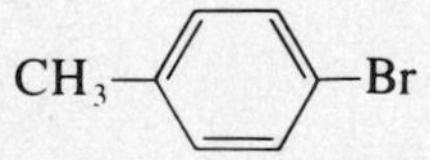

(*c*) H_3PO_2 (*d*) $(CH_3\overset{O}{\overset{\|}{C}})_2O$ or $CH_3\overset{O}{\overset{\|}{C}}Cl$

(*e*) $NH\overset{O}{\overset{\|}{C}}CH_3$, O_2N, CH_2CH_3 + $NH\overset{O}{\overset{\|}{C}}CH_3$, CH_2CH_3, NO_2

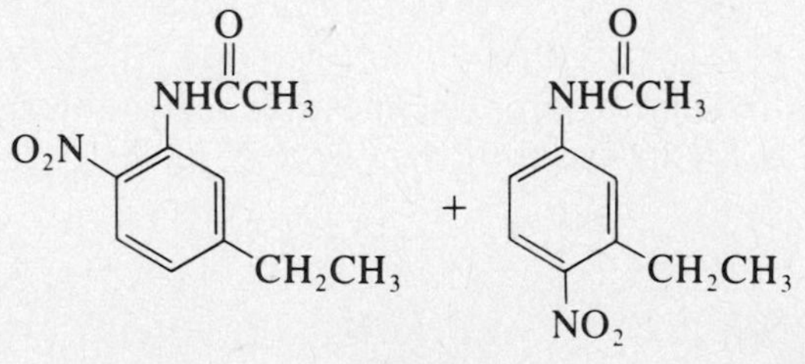

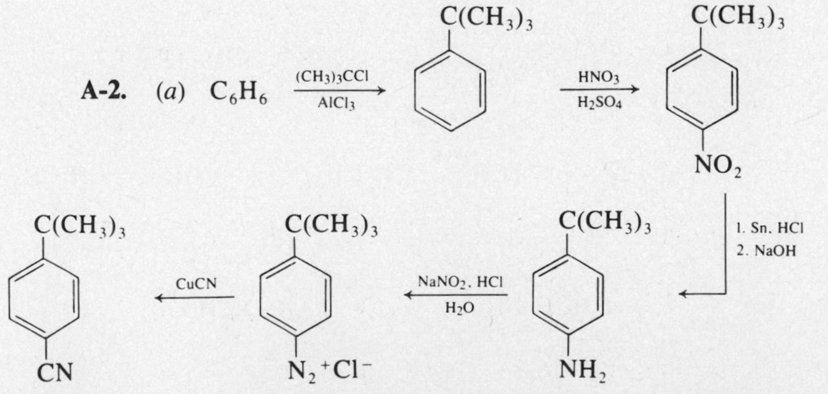

A-2. (*a*) C_6H_6 $\xrightarrow[AlCl_3]{(CH_3)_3CCl}$ $C_6H_5C(CH_3)_3$ $\xrightarrow[H_2SO_4]{HNO_3}$ $C(CH_3)_3$, NO_2 $\xrightarrow{1.\ Sn,\ HCl\quad 2.\ NaOH}$ $C(CH_3)_3$, NH_2 $\xrightarrow[H_2O]{NaNO_2,\ HCl}$ $C(CH_3)_3$, $N_2^{+}Cl^-$ $\xrightarrow{CuCN}$ $C(CH_3)_3$, CN (*b*) $C_6H_6 \xrightarrow[H_2SO_4]{HNO_3}$ nitrobenzene $\xrightarrow{Cl_2,\ FeCl_3}$ m-chloronitrobenzene $\xrightarrow[2.\ NaOH]{1.\ Sn,\ HCl}$ m-chloroaniline

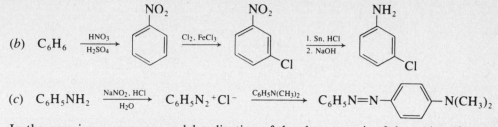

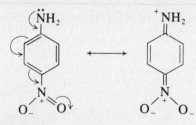

NO₂ / NO₂, Cl / NH₂, Cl

(*c*) $C_6H_5NH_2 \xrightarrow[H_2O]{NaNO_2,\ HCl} C_6H_5N_2{}^+Cl^- \xrightarrow{C_6H_5N(CH_3)_2} C_6H_5N{=}N{-}C_6H_4{-}N(CH_3)_2$

A-3. In the para isomer resonance delocalization of the electron pair of the amine nitrogen involves the nitro group.

$\ddot{N}H_2$ … N^+, O^-, O ⟷ $^+NH_2$ … N^+, O^-, O^-

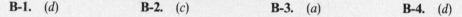

A-4. Strongest base: E, an alkylamine
Weakest base: F, a lactam (cyclic amide)

B-1. (*d*) **B-2.** (*c*) **B-3.** (*a*) **B-4.** (*d*)

CHAPTER 25

A-1. (*a*) m-$CF_3C_6H_4NH_2$ (CF₃, NH₂) (*b*) CH_3O, Cl, NO_2 (2-chloro-1-methoxy-4-nitrobenzene)

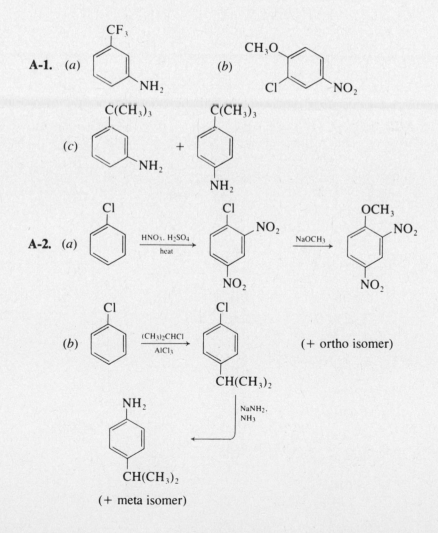

(*c*) m-$(CH_3)_3CC_6H_4NH_2$ + p-$(CH_3)_3CC_6H_4NH_2$ (C(CH₃)₃, NH₂)

A-2. (*a*) chlorobenzene $\xrightarrow[heat]{HNO_3,\ H_2SO_4}$ 1-chloro-2,4-dinitrobenzene (Cl, NO₂, NO₂) $\xrightarrow{NaOCH_3}$ 2,4-dinitroanisole (OCH₃, NO₂, NO₂)

(*b*) chlorobenzene $\xrightarrow[AlCl_3]{(CH_3)_2CHCl}$ p-$ClC_6H_4CH(CH_3)_2$ (Cl, CH(CH₃)₂) (+ ortho isomer)

$\xrightarrow{NaNH_2,\ NH_3}$ p-$H_2NC_6H_4CH(CH_3)_2$ (NH₂, CH(CH₃)₂)
(+ meta isomer)

A-3. [reaction scheme: ortho-nitrohalobenzene + Y^- → resonance-stabilized anion intermediate → substitution product]

NO_2-substituted ring with X, Y → NO_2-substituted ring with Y + X^-

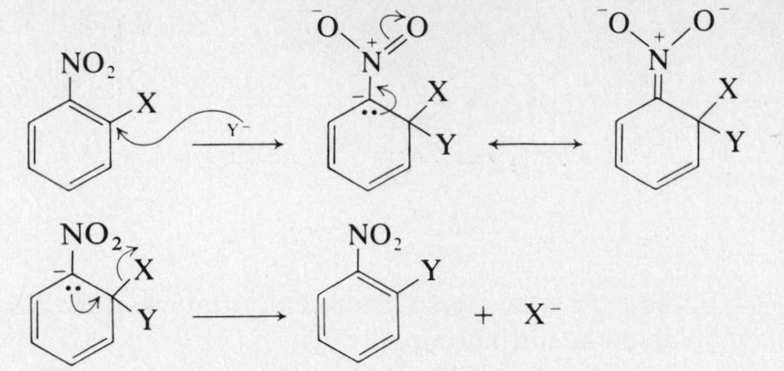

The mechanism for para substitution is similar.

B-1. (*a*) **B-2.** (*a*) **B-3.** (*c*) **B-4.** (*d*)
B-5. (*b*)

CHAPTER 26

A-1. *p*-Hydroxybenzaldehyde is the stronger acid. The phenoxide anion is stabilized by conjugation with the aldehyde carbonyl.

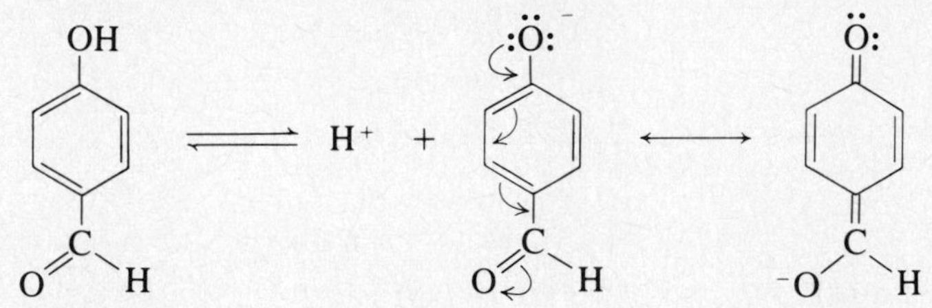

[structures: p-hydroxybenzaldehyde ⇌ H^+ + phenoxide anion ⟷ quinoid resonance form with ^-O on the former carbonyl carbon]

A-2. 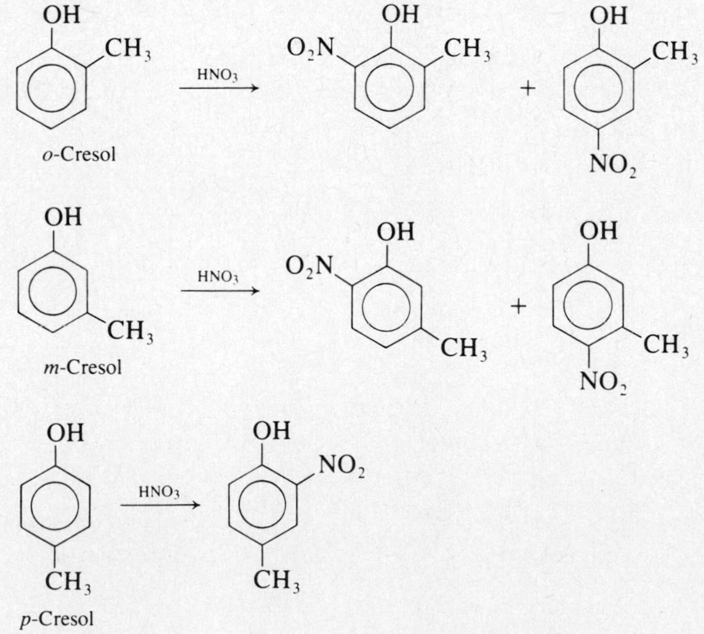 *o*-Cresol $\xrightarrow{HNO_3}$ 2-methyl-6-nitrophenol + 2-methyl-4-nitrophenol

m-Cresol $\xrightarrow{HNO_3}$ 5-methyl-2-nitrophenol + 3-methyl-4-nitrophenol

p-Cresol $\xrightarrow{HNO_3}$ 4-methyl-2-nitrophenol

A-3. *p*-cresol + $CH_3CH_2\overset{O}{\overset{\|}{C}}Cl$ $\xrightarrow{AlCl_3}$ 2-hydroxy-5-methylphenyl ketone ($\overset{O}{\overset{\|}{C}}CH_2CH_3$ ortho to OH, CH_3 para to OH)

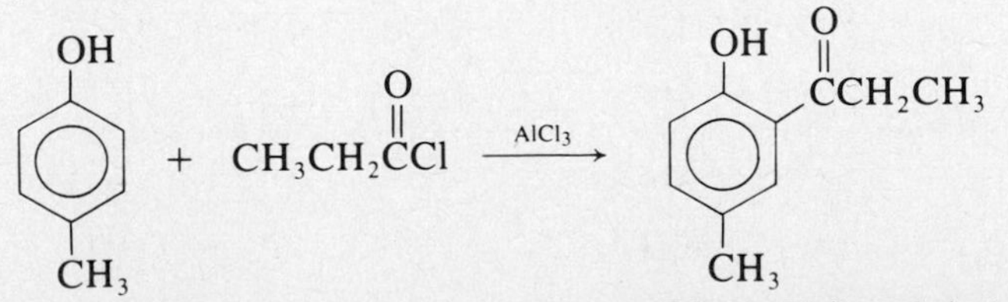

(Friedel–Crafts acylation)

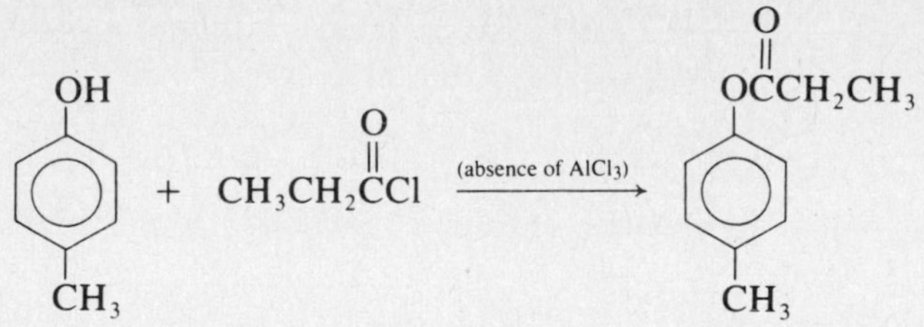

$p\text{-}CH_3C_6H_4OH + CH_3CH_2\overset{O}{\overset{\|}{C}}Cl \xrightarrow{\text{(absence of AlCl}_3)} p\text{-}CH_3C_6H_4O\overset{O}{\overset{\|}{C}}CH_2CH_3$

(esterification)

A-4. (*a*) C_6H_5—OH + $BrCH(CH_3)_2$

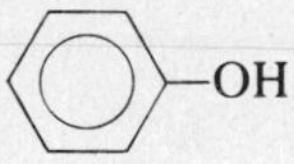

(*b*) $m\text{-}CH_3C_6H_4O^-Na^+$ + $BrCH_2CH(CH_3)_2$

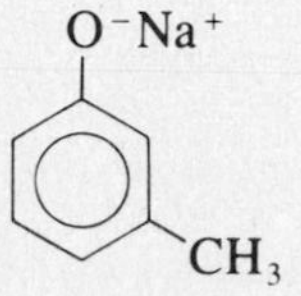

(*c*) CO_2, 125°C, 100 atm

B-1. (*d*) **B-2.** (*c*) **B-3.** (*b*) **B-4.** (*a*)

CHAPTER 27

A-1. (*a*) Fischer projection: CHO; HO—C—H; HO—C—H; CH_2OH — L-Erythrose

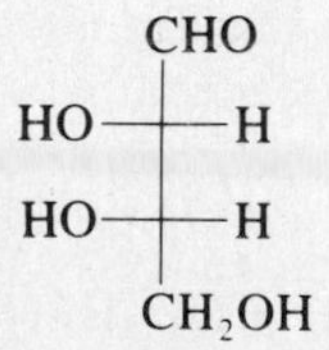

(*b*) CHO; H—C—OH; HO—C—H; CH_2OH (L-Threose) or CHO; HO—C—H; H—C—OH; CH_2OH (D-Threose)

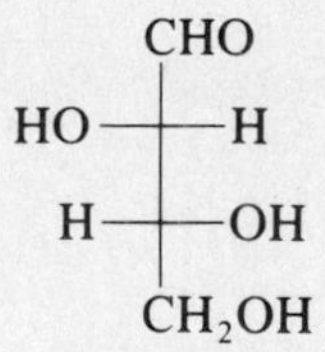

L-Threose D-Threose

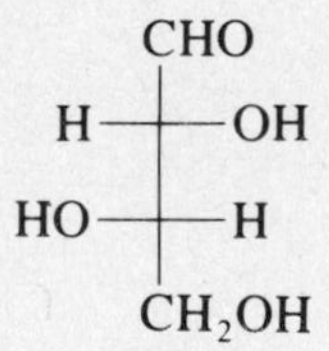

(*c*) Furanose ring (O; H, H; H, OH at C-1; OH, OH) — α-D-Erythrofuranose

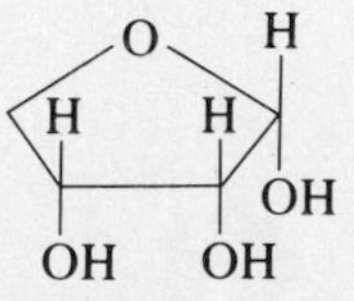

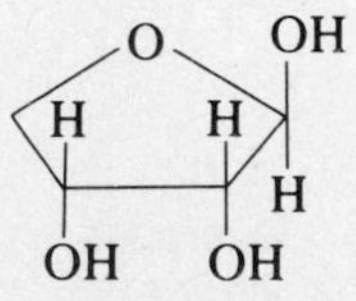

(*d*) Furanose ring (O; H, H; OH, H at C-1; OH, OH) — β-D-Erythrofuranose

A-2. (*a*) CH_2OH; HO—C—H; HO—C—H; H—C—OH; H—C—OH; CH_2OH

(*b*) CO_2^-; HO—C—H; HO—C—H; H—C—OH; H—C—OH; CH_2OH

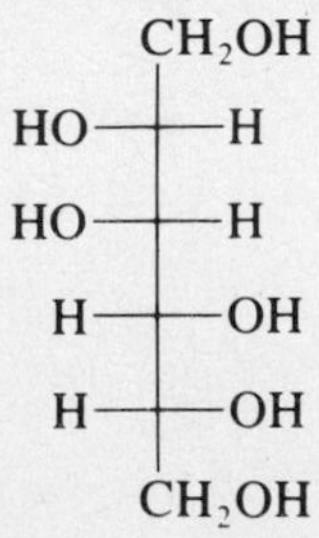

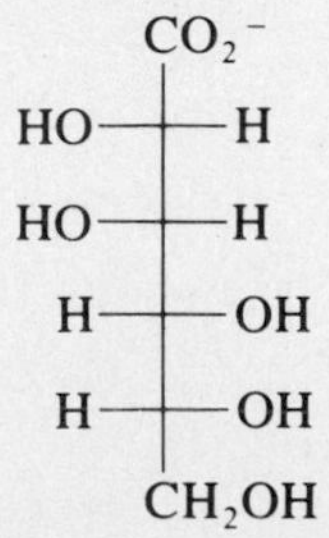

(*c*)

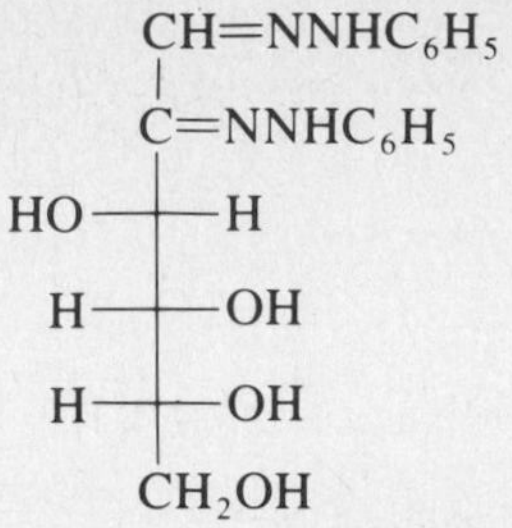

$$\begin{array}{c} CH{=}NNHC_6H_5 \\ | \\ C{=}NNHC_6H_5 \\ HO-C-H \\ H-C-OH \\ H-C-OH \\ | \\ CH_2OH \end{array}$$

(*d*) $4\ HCO_2H + H_2C{=}O$

B-1. (*c*) **B-2.** (*b*) **B-3.** (*d*) **B-4.** (*b*)
B-5. (*a*) **B-6.** (*c*) **B-7.** (*c*) **B-8.** (*a*)
B-9. (*d*) **B-10.** (*a*)

CHAPTER 28

A-1.

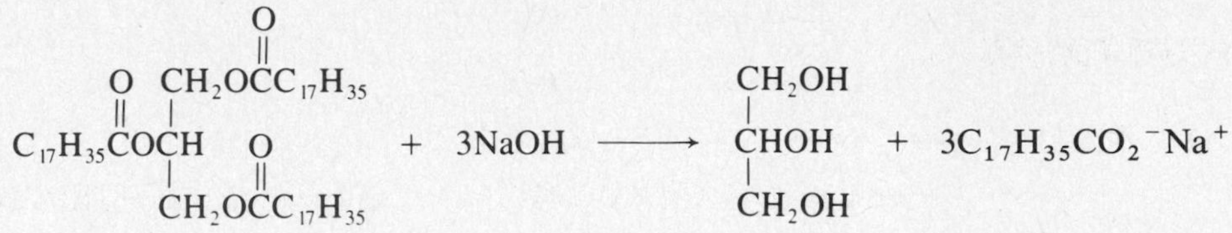

$$\begin{array}{c} CH_2OCC_{17}H_{35} \\ | \\ C_{17}H_{35}COCH \\ | \\ CH_2OCC_{17}H_{35} \end{array} + 3NaOH \longrightarrow \begin{array}{c} CH_2OH \\ | \\ CHOH \\ | \\ CH_2OH \end{array} + 3C_{17}H_{35}CO_2^-Na^+$$

(each C in COC bearing =O)

Tristearin

A-2. Fats are triesters of glycerol. A typical example is tristearin, shown above. A wax is usually a mixture of esters in which the alkyl and acyl group each contain 12 or more carbons. An example is hexadecyl hexadecanoate (cetyl palmitate):

$$C_{15}H_{31}\overset{O}{\overset{\|}{C}}OC_{16}H_{33}$$

A-3. (*a*) Monoterpene;

(*b*) Sesquiterpene;

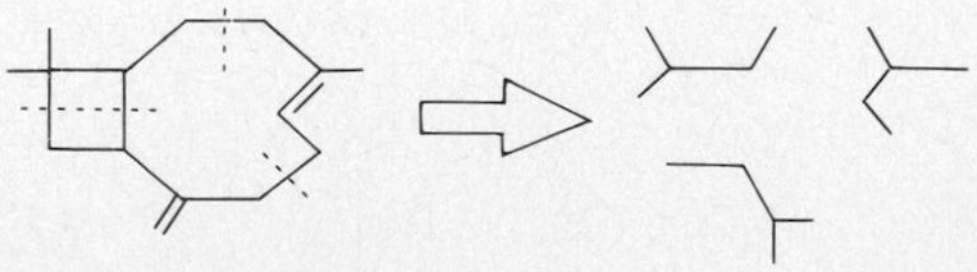

(*c*) Diterpene;

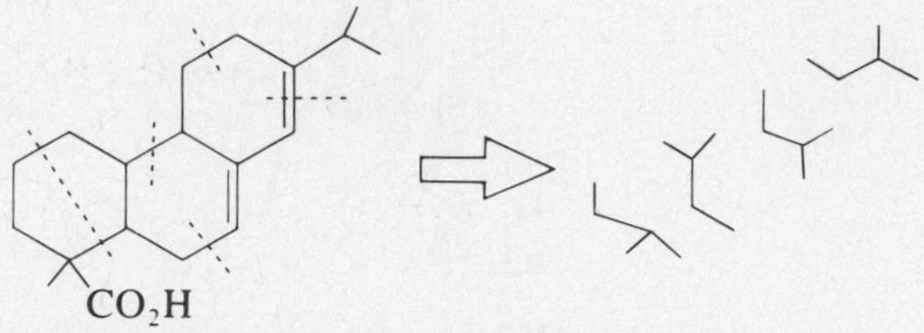

CO_2H

A-4.

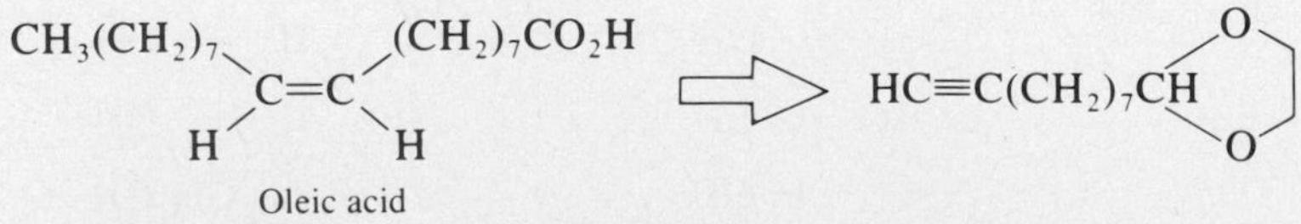

$$\underset{H}{\overset{CH_3(CH_2)_7}{}}\!C{=}C\underset{H}{\overset{(CH_2)_7CO_2H}{}}$$

Oleic acid

$\Longrightarrow$ $HC{\equiv}C(CH_2)_7CH$ (1,3-dioxolane ring: O–CH_2CH_2–O)

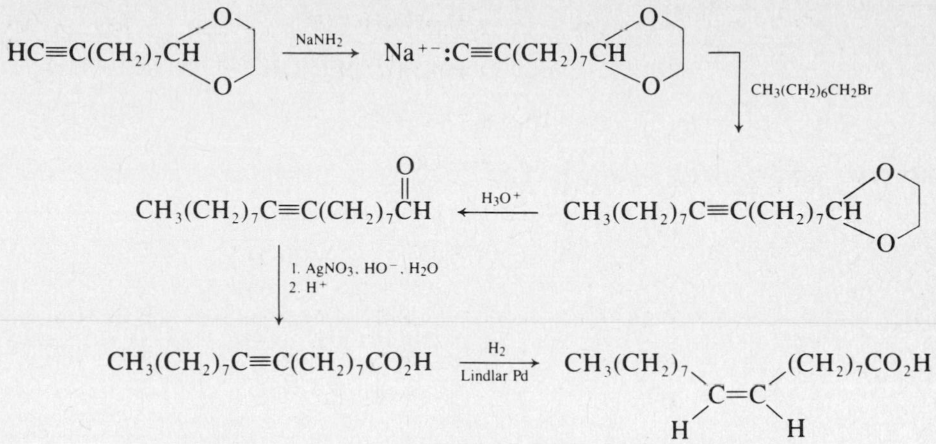

$$HC\equiv C(CH_2)_7CH\langle O_2(CH_2)_2\rangle \xrightarrow{NaNH_2} Na^{+-}{:}C\equiv C(CH_2)_7CH\langle O_2(CH_2)_2\rangle \xrightarrow{CH_3(CH_2)_6CH_2Br} CH_3(CH_2)_7C\equiv C(CH_2)_7CH\langle O_2(CH_2)_2\rangle$$

$$CH_3(CH_2)_7C\equiv C(CH_2)_7CH\langle O_2(CH_2)_2\rangle \xrightarrow{H_3O^+} CH_3(CH_2)_7C\equiv C(CH_2)_7\overset{O}{\overset{\|}{C}}H$$

$$\xrightarrow{1.\ AgNO_3,\ HO^-,\ H_2O;\ 2.\ H^+} CH_3(CH_2)_7C\equiv C(CH_2)_7CO_2H \xrightarrow[\text{Lindlar Pd}]{H_2} \underset{H\qquad\quad H}{CH_3(CH_2)_7CH{=}CH(CH_2)_7CO_2H}\ (cis)$$

B-1. (*b*) **B-2.** (*a*) **B-3.** (*c*) **B-4.** (*c*)
B-5. (*c*)

CHAPTER 29

A-1. (*a*) $C_6H_5CH_2\overset{O}{\overset{\|}{C}}H$ (*b*) $C_6H_5CH_2O\overset{O}{\overset{\|}{C}}NHCHCO_2H$ with $CH(CH_3)_2$ on the CH

(*c*) DCCI, $CHCl_3$

A-2. O_2N–$C_6H_3(NO_2)$–$NHCHCO_2H$ with $CH_2C_6H_5$ on the CH (2,4-dinitrophenyl)

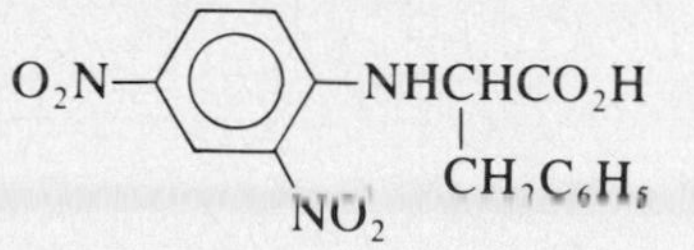

A-3. (*a*) $CH_3\overset{O}{\overset{\|}{C}}NHCH(CO_2CH_2CH_3)_2 \xrightarrow[\text{ethanol}]{NaOCH_2CH_3} CH_3\overset{O}{\overset{\|}{C}}NH\ddot{\bar{C}}(CO_2CH_2CH_3)_2 \xrightarrow{(CH_3)_2CHCH_2Br} CH_3\overset{O}{\overset{\|}{C}}NHC(CO_2CH_2CH_3)_2$ with $CH_2CH(CH_3)_2$ on C

$\xrightarrow{1.\ H_3O^+;\ 2.\ \text{heat}} H_3\overset{+}{N}CHCO_2^-$ with $CH_2CH(CH_3)_2$ on the CH

(*b*) Leu-Val = $H_3\overset{+}{N}CH\overset{O}{\overset{\|}{C}}{-}NHCHCO_2^-$ with $CH_2CH(CH_3)_2$ on the first CH and $CH(CH_3)_2$ on the second CH

N-Protect leucine: $C_6H_5CH_2O\overset{O}{\overset{\|}{C}}Cl + H_3\overset{+}{N}CHCO_2^-$ (with $CH_2CH(CH_3)_2$) $\xrightarrow{1.\ NaOH,\ H_2O;\ 2.\ H^+}$ $C_6H_5CH_2O\overset{O}{\overset{\|}{C}}NHCHCO_2H$ with $CH_2CH(CH_3)_2$ on the CH

(Z-Leu)

C-Protect valine: $C_6H_5CH_2OH + H_3\overset{+}{N}CHCO_2^-$ (with $CH(CH_3)_2$ side chain) $\xrightarrow{H^+}$ $H_3\overset{+}{N}CHCOCH_2C_6H_5$ (C=O; side chain $CH(CH_3)_2$)

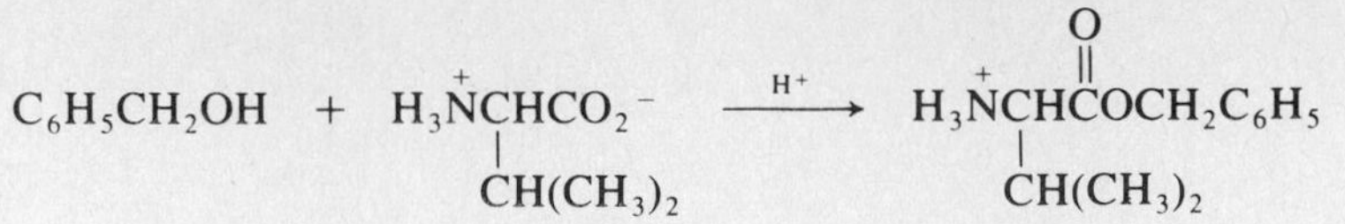

Couple: Z-Leu + $H_2NCHCOCH_2C_6H_5$ (C=O; side chain $CH(CH_3)_2$) $\xrightarrow{DCCI}$ $C_6H_5CH_2OCNHCHCNHCHCOCH_2C_6H_5$ (three C=O; side chains $CH_2CH(CH_3)_2$ and $CH(CH_3)_2$) $\xrightarrow[\text{(deprotect)}]{H_2, Pd}$ $H_3\overset{+}{N}CHCNHCHCO_2^-$ (C=O; side chains $CH_2CH(CH_3)_2$ and $CH(CH_3)_2$)

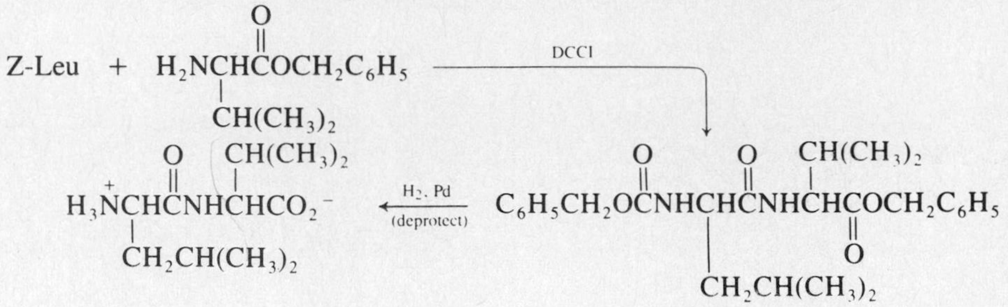

A-4. Leu-Val-Gly-Ala-Phe

B-1. (*c*) **B-2.** (*a*) **B-3.** (*d*) **B-4.** (*c*)

B-5. (*a*) **B-6.** (*b*)

APPENDIX

B

TABLES

Table B-1 Bond dissociation energies of some representative compounds

Bond	*Bond dissociation energy, kcal/mol*	*Bond*	*Bond dissociation energy, kcal/mol*
	Diatomic molecules		
H—H	104	H—F	136
F—F	38	H—Cl	103
Cl—Cl	58	H—Br	87.5
Br—Br	46	H—I	71
I—I	36		
	Alkanes		
CH_3—H	104	CH_3—CH_3	88
CH_3CH_2—H	98	CH_3CH_2—CH_3	85
$CH_3CH_2CH_2$—H	98	$(CH_3)_2CH$—CH_3	84
$(CH_3)_2CH$—H	94.5	$(CH_3)_3C$—CH_3	80
$(CH_3)_3C$—H	91		
	Alkyl halides		
CH_3—F	108	$(CH_3)_2CH$—F	105
CH_3—Cl	83.5	$(CH_3)_2CH$—Cl	81
CH_3—Br	70	$(CH_3)_2CH$—Br	68
CH_3—I	56	$(CH_3)_3C$—Cl	79
CH_3CH_2—Cl	81	$(CH_3)_3C$—Br	63
$CH_3CH_2CH_2$—Cl	82		
	Water and Alcohols		
HO—H	119	CH_3CH_2—OH	91
CH_3O—H	102	$(CH_3)_2CH$—OH	92
CH_3—OH	91	$(CH_3)_3C$—OH	91

Note: Bond dissociation energies refer to bonds indicated in structural formula for each substance.

Table B-2 Acid Dissociation Constants

Acid	*Formula*	*Conjugate base*	*Dissociation constant*	pK_a
Hydrogen fluoride	H—F	F^-	3.5×10^{-4}	3.5
Acetic acid	CH_3CO_2—H	$CH_3CO_2^-$	1.8×10^{-5}	4.7
Hydrogen cyanide	H—CN	CN^-	7.2×10^{-10}	9.1
Phenol	C_6H_5O—H	$C_6H_5O^-$	1.3×10^{-10}	9.8
Water	HO—H	HO^-	1.8×10^{-16}	15.7
Ethanol	CH_3CH_2O—H	$CH_3CH_2O^-$	10^{-16}	16
Alkyne (terminal; R = alkyl)	RC≡C—H	$RC≡C^-$	10^{-26}	26
Ammonia	NH_2—H	NH_2^-	10^{-36}	36
Alkene C—H	RCH=CH—H	$RCH=CH^-$	10^{-45}	45
Alkane C—H	RCH_2CH_2—H	$RCH_2CH_2^-$	10^{-62}	62

Note: Acid strength decreases from top to bottom of the table; conjugate base strength increases from top to bottom.

Table B-3 Chemical Shifts of Representative Types of Protons

Type of proton	*Chemical shift (δ), ppm**	*Type of proton*	*Chemical shift (δ), ppm**
H—C(—)(—)—R	0.9–1.8	H—C(—)(—)—NR	2.2–2.9
H—C(—)(—)—C=C	1.6–2.6	H—C(—)(—)—Cl	3.1–4.1
H—C(—)(—)—C(=O)—	2.1–2.5	H—C(—)(—)—Br	2.7–4.1
H—C≡C—	2.5	H—C(—)(—)—O	3.3–3.7
H—C(—)(—)—Ar	2.3–2.8	H—NR	1–3†
H—C(—)=C<	4.5–6.5	H—OR	0.5–5†
H—Ar	6.5–8.5	H—OAr	6–8†
H—C(=O)—	9–10	H—OC(=O)—	10–13†

* These are approximate values relative to tetramethylsilane; other groups within the molecule can cause a proton signal to appear outside of the range cited.

† The chemical shifts of protons bonded to nitrogen and oxygen are temperature- and concentration-dependent.

Table B-4 Chemical Shifts of Representative Carbons

Type of carbon	*Chemical shift (δ), ppm**	*Type of carbon*	*Chemical shift (δ), ppm**
RCH_3	0–35	$>C=C<$	100–150
R_2CH_2	15–40		
R_3CH	25–50	benzene ring	110–175
RCH_2NH_2	35–50		
RCH_2OH	50–65	$>C=O$	190–220
$-C\equiv C-$	65–90		

* Approximate values relative to tetramethylsilane.

Table B-5 Infrared Absorption Frequencies of Some Common Structural Units

Structural unit	*Frequency, cm^{-1}*	*Structural unit*	*Frequency, cm^{-1}*
Stretching vibrations			
Single bonds		*Double bonds*	
$-O-H$ (alcohols)	3200–3600	$>C=C<$	1620–1680
$-O-H$ (carboxylic acids)	2500–3600	$>C=O$	
$>N-H$	3350–3500	Aldehydes and ketones	1710–1750
sp C—H	3310–3320	Carboxylic acids	1700–1725
sp^2 C—H	3000–3100	Acid anhydrides	1800–1850 and 1740–1790
sp^3 C—H	2850–2950	Acyl halides	1770–1815
		Esters	1730–1750
sp^2 C—O	1200	Amides	1680–1700
sp^3 C—O	1025–1200	*Triple bonds*	
		$-C\equiv C-$	2100–2200
		$-C\equiv N$	2240–2280
Bending vibrations of diagnostic value			
Alkenes		*Substituted derivatives of benzene*	
Cis-disubstituted	665–730	Monosubstituted	730–770 and 690–710
Trans-disubstituted	960–980	Ortho disubstituted	735–770
Trisubstituted	790–840	Meta-disubstituted	750–810 and 680–730
		Para-disubstituted	790–840

APPENDIX C

ADDITIONAL READINGS

The list of articles that follows is intended to be selective rather than comprehensive. It is limited to short accounts in sources available in practically all college and university libraries. These include the *Journal of Chemical Education* (*JCE*), *Scientific American* (*SAM*), *Chemical and Engineering News* (*CEN*), and *Organic Reactions* (*OR*). More extensive treatments of many of the topics described in the text, along with additional references to the chemical literature, may be found in:

F. A. Carey and R. J. Sundberg, *Advanced Organic Chemistry. Part A: Structure and Mechanisms*, 2d ed., Plenum Press, New York, 1984.

F. A. Carey and R. J. Sundberg, *Advanced Organic Chemistry. Part B: Reactions and Synthesis*, 2d ed., Plenum Press, New York, 1983.

J. March, *Advanced Organic Chemistry. Reactions, Mechanisms, and Structure*, 3d ed., Wiley-Interscience, New York, 1985.

Acids and Bases

D. Kolb, "Acids and Bases," *JCE*, vol. **55**, pp. 459–464, 1978.

G. V. Calder and T. J. Barton, "Actual Effects Controlling the Acidity of Carboxylic Acids," *JCE*, vol. **48**, pp. 338–340, 1971.

Alcohols

R. L. Pruett, "Hydroformylation: An Old Yet New Industrial Route to Alcohols," *JCE*, vol. **63**, pp. 196–198, 1986.

A. W. Ingersoll, "The Resolution of Alcohols," *OR*, vol. **2**, pp. 376–414, 1944.

Aldehydes and Ketones

H. A. Wittcoff, "Acetaldehyde: A Chemical Whose Fortunes Have Changed," *JCE*, vol. **60**, pp. 1044–1047, 1983.

T. Mukaiyama, "The Directed Aldol Reaction," *OR*, vol. **28**, pp. 203–332, 1982.

H. Hart and M. Sasaoka, "Simple Enols: How Rare Are They?," *JCE*, vol. **57**, pp. 685–688, 1980.

E. Vedejs, "Clemmensen Reduction of Ketones in Anhydrous Organic Solvents," *OR*, vol. **22**, pp. 401–422, 1975.

B. P. Mundy, "The Synthesis of Fused Cycloalkenones via Annelation Methods," *JCE*, vol. **50**, 110–113, 1973.

H. Salzman, "Arthur Lapworth: The Genesis of Reaction Mechanism," *JCE*, vol. **49**, pp. 750–752, 1972.

M. J. Jorgenson, "Preparation of Ketones from the Reaction of Organolithium Reagents with Carboxylic Acids," *OR*, vol. **18**, pp. 1–98, 1970.

A. T. Nielsen and W. J. Houlihan, "The Aldol Condensation," *OR*, vol. **16**, pp. 1–438, 1968.

G. Jones, "The Knoevenagel Condensation," *OR*, vol. **15**, pp. 204–600, 1967.

A. Maercker, "The Wittig Reaction," *OR*, vol. **14**, pp. 270–490, 1965.

E. D. Bergmann, D. Ginsburg, and R. Pappo, "The Michael Reaction," *OR*, vol. **10**, pp. 179–556, 1959.

D. A. Shirley, "The Synthesis of Ketones from Acid Halides and Organometallic Compounds of Magnesium, Zinc, and Cadmium," *OR*, vol. **8**, pp. 28–58, 1954.

E. Mosettig, "The Synthesis of Aldehydes from Carboxylic Acids," *OR*, vol. **8**, pp. 218–257, 1954.

E. Mosettig and R. Mozingo, "The Rosenmund Reduction of Acid Chlorides to Aldehydes," *OR*, vol. **4**, pp. 362–377, 1948.

D. Todd, "The Wolff-Kishner Reduction," *OR*, vol. **4**, pp. 378–422, 1948.

W. S. Johnson, "The Formation of Cyclic Ketones by Intramolecular Acylation," *OR*, vol. **2**, pp. 114–177, 1944.

E. L. Martin, "The Clemmensen Reduction," *OR*, vol. **1**, pp. 155–209, 1942.

Alkanes

H. R. Henze and C. M. Blair, "The Number of Isomeric Hydrocarbons of the Methane Series," *J. Am. Chem. Soc.* vol. **53**, pp. 3077–3085, 1931.

Alkenes

C. C. Wamser and L. T. Scott, "The NBS Reaction: A Simple Explanation for the Predominance of Allylic Substitution Over Olefin Addition by Bromine at Low Concentrations," *JCE*, vol. **62**, pp. 650–652, 1985.

J. M. Tedder, "Who is Anti-Markovnikov?" *JCE*, vol. **61**, pp. 237–238, 1984.

W. C. Fernelius, H. Wittcoff, and R. E. Varnerin, "Ethylene: The Organic Chemical Industry's Most Important Building Block," *JCE*, vol. **56**, pp. 385–387, 1979.

M. A. Wilson, "Classification of the Electrophilic Addition Reactions of Olefins and Acetylenes," *JCE*, vol. **52**, pp. 495–498, 1975.

W. R. Dolbier, Jr., "Electrophilic Additions to Alkenes," *JCE*, vol. **46**, pp. 342–344, 1969.

N. Isenberg and M. Grdinic, "A Modern Look at Markovnikov's Rule and the Peroxide Effect," *JCE*, vol. **46**, pp. 601–605, 1969.

E. A. Walters, "Models for the Double Bond," *JCE*, vol. **43**, pp. 134–137, 1966.

J. G. Traynham, "The Bromonium Ion," *JCE*, vol. **40**, pp. 392–395, 1963.

G. Zweifel and H. C. Brown, "Hydration of Olefins, Dienes, and Acetylenes via Hydroboration," *OR*, vol. **13**, pp. 1–54, 1963.

D. Swern, "Epoxidation and Hydroxylation of Ethylenic Compounds with Organic Peracids," *OR*, vol. **7**, pp. 378–434, 1953.

Alkylation Reactions

B. Mundy, "Alkylations in Organic Chemistry," *JCE*, vol. **49**, pp. 91–96, 1972.

C. C. Price, "The Alkylation of Aromatic Compounds by the Friedel-Crafts Method," *OR*, vol. **3**, pp. 1–82, 1946.

Alkynes

T. L. Jacobs, "The Synthesis of Acetylenes," *OR*, vol. **5**, pp. 1–78, 1949.

Amines

J. S. Wishnok, "Formation of Nitrosamines in Food and in the Digestive System," *JCE*, vol. **54**, pp. 440–441, 1977.

A. C. Cope and E. R. Trumbull, "Olefins from Amines: The Hofmann Elimination Reaction and Amine Oxide Pyrolysis," *OR*, vol. **11**, pp. 317–494, 1960.

A. Roe, "Preparation of Aromatic Fluorine Compounds from Diazonium Fluoroborates: The Schiemann Reaction," *OR*, vol. **5**, pp. 193–228, 1949.

W. S. Emerson, "The Preparation of Amines by Reductive Alkylation," *OR*, vol. **4**, pp. 174–255, 1948.

E. S. Wallis and J. F. Lane, "The Hofmann Reaction," *OR*, vol. 3, pp. 267–306, 1946.

N. Kornblum, "Replacement of the Aromatic Primary Amino Group by Hydrogen," *OR*, vol. **2**, pp. 262–340, 1944.

Antibiotics

R. K. Robins, "Synthetic Antiviral Agents," *CEN*, pp. 28–40, January 27, 1986.

E. P. Abraham, "The β-Lactam Antibiotics," *SAM*, pp. 76–86, June 1981.

J. Webb, "Ionophores," *JCE*, vol. **56**, pp. 502–503, 1979.

Aromatic Compounds

R. C. Belioli, "The Misuse of the Circle Notation to Represent Aromatic Rings," *JCE*, vol. **60**, pp. 190–191, 1983.

J. G. Traynham, "Aromatic Substitution Reactions. When You've Said Ortho, Meta, and Para, You Haven't Said It All," *JCE*, vol. **60**, pp. 937–941, 1983.

J. F. Bunnett, "The Remarkable Reactivity of Aryl Halides with Nucleophiles," *JCE*, vol. **51**, 312–315, 1974.

E. Berliner, "The Friedel and Crafts Reaction with Aliphatic Dibasic Acid Anhydrides," *OR*, vol. **5**, pp. 229–289, 1949.

C. M. Suter and A. W. Weston, "Direct Sulfonation of Aromatic Hydrocarbons and Their Halogen Derivatives," *OR*, vol. **3**, pp. 141–197, 1946.

Bioorganic Chemistry

G. M. Bodner, "Metabolism Part I: Glycolysis or the Embden-Meyerhoff Pathway," *JCE*, vol. **63**, pp. 566–570, 1986.

G. M. Bodner, "Metabolism Part II: Tricarboxylic Acid (TCA), Citric Acid, or Krebs Cycle," *JCE*, vol. **63**, pp. 673–677, 1986.

G. M. Bodner, "Metabolism Part III: Lipids," *JCE*, vol. **63**, pp. 772–775, 1986.

M. S. Bretscher, "The Molecules of the Cell Membrane," *SAM*, pp. 100–108, October 1985.

S. H. Snyder, "The Molecular Basis of Communication Between Cells," *SAM*, pp. 132–141, October 1985.

M. J. Berridge, "The Molecular Basis of Communication Within the Cell," *SAM*, pp. 142–152, October 1985.

K. Folkers, "Perspectives from Research on Vitamins and Hormones," *JCE*, vol. **61**, pp. 747–756, 1984.

J. R. Holum, "The Chemical Composition of the Cell," *JCE*, vol. **61**, pp. 877–881, 1984.

S. Krishnamurthy, "The Intriguing Biological Role of Vitamin E," *JCE*, vol. **60**, pp. 465–467, 1983.

W. F. Wood, "Chemical Ecology. Chemical Communication in Nature," *JCE*, vol. **60**, pp. 531–539, 1983.

L. N. Ferguson, "Bio-Organic Mechanisms II: Chemoreception," *JCE*, vol. **58**, pp. 456–461, 1981.

L. N. Ferguson, "Bioactivity in Organic Chemistry Courses," *JCE*, vol. **57**, pp. 46–48, 1980.

D. A. Labianca, "On the Nature of Cyanide Poisoning," *JCE*, vol. **56**, pp. 385–387, 1979.

Cancer

D. A. Labianca, "The Chimney Sweepers' Cancer: An Interdisciplinary Approach to Carcinogenesis," *JCE*, vol. **59**, pp. 843–846, 1982.

L. S. Alexander and H. M. Goff, "Chemicals, Cancer, and Cytochrome P-450," *JCE*, vol. **59**, pp. 179–182, 1982.

P. Rademacher and H.-G. Hilde, "Chemical Carcinogens," *JCE*, vol. **53**, pp. 757–761, 1976.

L. N. Ferguson, "Cancer: How Can Chemists Help?", *JCE*, vol. **52**, pp. 688–694, 1975.

Carbanions and Enolates

S. Arseniyadis, K. S. Kyler, and D. S. Watt, "Addition and Substitution Reactions of Nitrile-Stabilized Carbanions," *OR*, vol. **31**, pp. 1–364, 1984.

M. W. Rathke, "The Reformatsky Reaction," *OR*, vol. **22**, pp. 423–460, 1975.

T. M. Harris and C. M. Harris, "The γ-Alkylation and γ-Arylation of Dianions of β-Dicarbonyl Compounds," *OR*, vol. **17**, pp. 155–212, 1969.

J. P. Schaefer and J. J. Bloomfield, "The Dieckmann Condensation," *OR*, vol. **15**, pp. 1–203, 1967.

A. C. Cope, H. L. Holmes, and H. O. House, "The Alkylation of Esters and Nitriles," *OR*, vol. **9**, pp. 107–331, 1957.

C. R. Hauser, F. W. Swamer, and J. T. Adams, "The Acylation of Ketones to Form β-Diketones or β-Keto Aldehydes," *OR*, vol. **8**, pp. 59–196, 1954.

R. L. Shriner, "The Reformatsky Reaction," *OR*, vol. **1**, pp. 1–37, 1942.

C. R. Hauser and B. E. Hudson, Jr., "The Acetoacetic Ester Condensation and Certain Related Reactions," *OR*, vol. **1**, pp. 266–302, 1942.

Carbenes

W. E. Parham and E. E. Schweizer, "Halocyclopropanes from Halocarbenes," *OR*, vol. **13**, pp. 55–90, 1963.

M. Jones, Jr., "Carbenes," *SAM*, pp. 101–113, February 1976.

Carbohydrates

N. Sharon, "Carbohydrates," *SAM*, pp. 90–116, November 1980.

C. S. Hudson, "Emil Fischer's Discovery of the Configuration of Glucose," *JCE*, vol. **18**, pp. 353–357, 1941.

Color

J. Alkema and S. L. Seager, "The Chemical Pigments of Plants," *JCE*, vol. **59**, pp. 183–186, 1982.

M. Sequin-Frey, "The Chemistry of Plant and Animal Dyes," *JCE*, vol. **58**, pp. 301–305, 1981.

Computers

H. W. Orf, "Computer-Assisted Instruction in Organic Synthesis," *JCE*, vol. **52**, pp. 464–467, 1975.

Cyclic Compounds

H. E. Simmons, T. L. Cairns, S. A. Vladuchick, and C. M. Hoiness, "Cyclopropanes from Unsaturated Compounds, Methylene Iodide, and Zinc-Copper Couple," *OR*, vol. **20**, pp. 1–132, 1973.

L. N. Ferguson, "Ring Strain and Reactivity of Alicycles," *JCE*, vol. **47**, pp. 46–53, 1970.

Diels-Alder Reaction

E. Ciganek, "The Intramolecular Diels-Alder Reaction," *OR*, vol. **32**, pp. 1–374, 1984.

L. W. Butz and A. W. Rytina, "The Diels-Alder Reaction: Quinones and Other Cyclenones," *OR*, vol. **5**, pp. 136–192, 1949.

M. C. Kloetzel, "The Diels-Alder Reaction with Maleic Anhydride," *OR*, vol. **4**, pp. 1–59, 1948.

H. L. Holmes, "The Diels-Alder Reaction: Ethylenic and Acetylenic Dienophiles," *OR*, vol. **4**, pp. 60–173, 1948.

Free Radicals

F. R. Mayo, "The Evolution of Free Radical Chemistry at the University of Chicago," *JCE*, vol. **63**, pp. 97–99, 1986.

C. Walling, "The Development of Free Radical Chemistry," *JCE*, vol. **63**, pp. 99–102, 1986.

F. W. Stacey and J. F. Harris, Jr., "Formation of Carbon-Hetero Atom Bonds by Free Radical Chain Additions to Carbon-Carbon Multiple Bonds," *OR*, vol. **13**, pp. 150–376, 1963.

Hydrogenation Reactions

A. J. Birch and D. H. Williamson, "Homogeneous Hydrogenation Catalysts in Organic Synthesis," *OR*, vol. **24**, pp. 1–186, 1976.

H. Adkins, "Catalytic Hydrogenation of Esters to Alcohols," *OR*, vol. **8**, pp. 1–27, 1954.

Insect Chemistry

W. R. Stine, "Chemical Communication by Insects," *JCE*, vol. **63**, pp. 603–605, 1986.

G. D. Prestwich, "The Chemical Defenses of Termites," *SAM*, pp. 78–87, August 1983.

W. F. Wood, "The Sex Pheromones of the Gypsy Moth and the American Cockroach," *JCE*, vol. **59**, pp. 35–36, 1982.

Lactones

H. Zaugg, "β-Lactones," *OR*, vol. **8**, pp. 305–363, 1954.

Mass Spectrometry

M. M. Campbell and O. Runquist, "Fragmentation Mechanisms in Mass Spectrometry," *JCE*, vol. **49**, pp. 104–108, 1972.

Molecular Formulas

V. Pellegrin, "Molecular Formulas of Organic Compounds. The Nitrogen Rule and Degree of Unsaturation," *JCE*, vol. **60**, pp. 626–633, 1983.

Molecular Models

Q. R. Petersen, "Some Reflections on the Use and Abuse of Molecular Models," *JCE*, vol. **47**, pp. 24–29, 1970.

Nuclear Magnetic Resonance

D. W. Brown, "A Short Set of ^{13}C-NMR Correlation Tables," *JCE*, vol. **62**, pp. 209–212, 1985.

R. S. Macomber, "A Primer on Fourier Transform NMR," *JCE*, vol. **62**, pp. 213–214, 1985.

R. G. Shulman, "NMR Spectroscopy of Living Cells," *SAM*, pp. 86–93, January 1983.

I. L. Pykett, "NMR Imaging in Medicine," *SAM*, pp. 78–89, May 1982.

H. C. Dorn, D. G. I. Kingston, and B. R. Simpers, "Interpretation of a ^{13}C Magnetic Resonance Spectrum," *JCE*, vol. **53**, pp. 584–585, 1976.

Nucleic Acids

W. M. Scovell, "Supercoiled DNA," *JCE*, vol. **63**, pp. 562–563, 1986.

G. Felsenfeld, "DNA," *SAM*, pp. 58–67, October 1985.

J. E. Darnell, Jr., "RNA," *SAM*, pp. 68–78, October 1985.

A. T. Wood, "Perspectives in Biochemistry. Methods for DNA Sequencing," *JCE*, vol. **61**, pp. 886–889, 1984.

R. E. Dickerson, "The DNA Helix and How It is Read," *SAM*, pp. 94–111, December 1983.

Nucleophilic Substitution Reactions

D. J. Raber and J. M. Harris, "Nucleophilic Substitution Reactions at Secondary Carbon Atoms," *JCE*, vol. **49**, pp. 60–64, 1972.

Organometallic Compounds

G. W. Parshall and R. E. Putscher, "Organometallic Chemistry and Catalysis in Industry," *JCE*, vol. **63**, pp. 189–191, 1986.

G. B. Kaufmann, "The Discovery of Ferrocene, the First Sandwich Compound," *JCE*, vol. **60**, pp. 185–186, 1983.

G. H. Posner, "Substitution Reactions Using Organocopper Reagents," *OR*, vol. **22**, pp. 253–400, 1975.

G. H. Posner, "Conjugate Addition Reactions of Organocopper Reagents," *OR*, vol. **19**, pp. 1–113, 1972.

Peptides and Proteins

R. F. Doolittle, "Proteins," *SAM*, pp. 88–99, October 1985.

N. M. Senozan and R. L. Hunt, "Hemoglobin: Its Occurrence, Structure, and Adaptation," *JCE*, vol. **59**, pp. 173–178, 1982.

F. E. Bloom, "Neuropeptides," *SAM*, pp. 148–168, October 1981.

J. F. Henahan, "R. Bruce Merrifield, Designer of Protein-Making Machine," *CEN*, pp. 22–26, August 2, 1971.

N. F. Albertson, "Synthesis of Peptides with Mixed Anhydrides," *OR*, vol. **12**, pp. 157–355, 1962.

Periodic Acid

E. L. Jackson, "Periodic Acid Oxidation," *OR*, vol. **2**, pp. 341–375, 1944.

Pharmaceutical Industry

N. M. Senozan and S. Nasser-Moaddeli, "Organic Nitrates and Nitrites as Heart Drugs," *JCE*, vol. **61**, pp. 674–677, 1984.

Y. Aharaonowitz and G. Cohen, "The Microbiological Production of Pharmaceuticals," *SAM*, pp. 140–152, September 1981.

Phenols

J. Cason, "Synthesis of Benzoquinones by Oxidation," *OR*, vol. **4**, pp. 305–361, 1948.

A. H. Blatt, "The Fries Reaction," *OR*, vol. **1**, pp. 342–369, 1942.

Phase-Transfer Catalysis

J. M. McIntosh, "Phase-Transfer Catalysis Using Quaternary Onium Salts," *JCE*, vol. **55**, pp. 235–238, 1978.

G. W. Gokel and W. P. Weber, "Phase Transfer Catalysis. Part I: General Principles," *JCE*, vol. **55**, pp. 350–354, 1978.

W. P. Weber and G. W. Gokel, "Phase Transfer Catalysis. Part II: Synthetic Applications," *JCE*, vol. **55**, pp. 429–433, 1978.

Polymers

B. L. Goodall, "The History and Current State of the Art of Propylene Polymerization Catalysts," *JCE*, vol. **63**, 191–195, 1986.

J. K. Smith and D. A. Hounshell, "Wallace H. Carothers and Fundamental Research at DuPont," *Science*, vol. **229**, p. 436, 1985.

C. Cummings, "Neoprene and Nylon Stockings: The Legacy of Wallace Hume Carothers," *JCE*, vol. **61**, pp. 241–242, 1984.

J. J. Eisch, "Karl Ziegler. Master Advocate for the Unity of Pure and Applied Research," *JCE*, vol. **60**, pp. 1009–1014, 1983.

F. W. Harris, "Introduction to Polymer Chemistry," *JCE*, vol. **58**, pp. 837–843, 1981.

J. E. McGrath, "Chain Reaction Polymerization," *JCE*, vol. **58**, pp. 844–861, 1981.

Prostaglandins

N. A. Nelson, R. C. Kelly, and R. A. Johnson, "Prostaglandins and the Arachidonic Acid Cascade," *CEN*, pp. 30–44, August 16, 1982.

Reducing Agents

H. C. Brown, "Hydride Reductions: A 40-Year Revolution in Organic Chemistry," *CEN*, pp. 24–29, March 5, 1979.

W. G. Brown, "Reductions by Lithium Aluminum Hydride," *OR*, vol. **6**, pp. 469–510, 1951.

Saccharin

D. S. Tarbell and A. T. Tarbell, "The Discovery of Saccharin," *JCE*, vol. **55**, pp. 161–162, 1978.

Stereochemistry

J. H. Brewster, "Stereochemistry and the Origins of Life," *JCE*, vol. **63**, pp. 667–670, 1986.

D. H. Rouvray, "Predicting Chemistry from Topology," *SAM*, pp. 40–47, September 1986.

J. E. Huheey, "A Novel Method for Assigning *R*, *S* Labels to Enantiomers," *JCE*, Vol. **62**, pp. 598–600, 1986.

M. P. Aalund and J. A. Pincock, "A Simple Hand Method for Cahn-Ingold-Prelog Assignment of *R* and *S* Configuration to Chiral Carbons," *JCE*, vol. **63**, pp. 600–601, 1986.

K. N. Carter, "L-Cysteine: The (*R*)-Amino Acid from Protein," *JCE*, vol. **63**, pp. 603–605, 1986.

H. Maehr, "A Proposed New Convention for Graphic Presentation of Molecular Geometry and Topography," *JCE*, vol. **62**, pp. 114–120, 1985.

D. L. Mattern, "Fingertip Assignment of Absolute Configuration," *JCE*, vol. **62**, p. 191, 1985.

P. S. Beauchamp, "'Absolutely' Simple Stereochemistry," *JCE*, vol. **61**, pp. 666–667, 1984.

J. J. McCullough, "Diastereomers, Geometrical Isomers, and Rotation About Bonds," *JCE*, vol. **59**, p. 37, 1982.

E. L. Eliel, "Stereochemical Non-Equivalence of Ligands and Faces (Heterotopicity)," *JCE*, vol. **57**, pp. 52–55, 1980.

J. K. O'Loane, "Optical Activity in Small Molecules, Nonenantiomorphous Crystals, and Nematic Liquid Crystals," *Chem. Rev.*, vol. **80**, pp. 41–61, 1980.

J. W. Cornforth, "Asymmetry and Enzyme Action," *Science*, vol. **193**, pp. 121–125, 1976.

H. A. M. Snelders, "The Reception of J. H. van't Hoff's Theory of the Asymmetric Carbon Atom," *JCE*, vol. **51**, pp. 2–7, 1974.

W. E. Elias, "The Natural Origin of Optically Active Compounds," *JCE*, vol. **49**, pp. 448–454, 1972.

J. E. Abernethy, "The Concept of Dissymmetric Worlds," *JCE*, vol. **49**, pp. 455–461, 1972.

Structure and Bonding

A. D. Baker and M. D. Baker, "A Geometric Method for Determining the Hückel Molecular Orbital Energy Levels of Open-Chain Fully Conjugated Molecules," *JCE*, vol. **61**, p. 770, 1984.

L. Pauling, "G. N. Lewis and the Chemical Bond," *JCE*, vol. **61**, pp. 201–203, 1984.

L. Salem, "A Faithful Couple: The Electron Pair," *JCE*, vol. **55**, pp. 344–348, 1978.

R. J. Gillespie, "The Electron-Pair Repulsion Model for Molecular Geometry," *JCE*, vol. **47**, pp. 18–23, 1970.

E. N. Hiebert, "The Experimental Basis of Kekule's Valence Theory," *JCE*, vol. **36**, pp. 320–327, 1959.

H. M. Leicester, "Contributions of Butlerov to the Development of Structural Theory," *JCE*, vol. **36**, pp. 328–329, 1959.

Substituent Effects

L. M. Stock, "The Origin of the Inductive Effect," *JCE*, vol. **49**, pp. 400–404, 1972.

Sulfur Compounds

E. Block, "The Chemistry of Garlic and Onions," *SAM*, pp. 114–119, March 1985.

K. K. Andersen and D. T. Bernstein, "1-Butanethiol and the Striped Skunk," *JCE*, vol. **55**, pp. 159–160, 1978.

Vision

P. S. Zurer, "The Chemistry of Vision," *CEN*, pp. 24–35, November 28, 1983.